MOLECULAR BIOLOGY OF
THE CELL
SECOND EDITION

MOLECULAR BIOLOGY OF
THE CELL
SECOND EDITION

Bruce Alberts · Dennis Bray
Julian Lewis · Martin Raff · Keith Roberts
James D. Watson

Garland Publishing, Inc.
New York & London

TEXT EDITOR: Miranda Robertson

GARLAND STAFF

Managing Editor: Ruth Adams
Project Editor: Alison Walker
Production Coordinator: Perry Bessas
Designer: Janet Koenig
Copy Editors: Lynne Lackenbach and Shirley Cobert
Editorial Assistant: Māra Abens
Art Coordinator: Charlotte Staub
Indexer: Maija Hinkle

Bruce Alberts received his Ph.D. from Harvard University and is currently Chairman of the Department of Biophysics and Biochemistry at the University of California Medical School in San Francisco. *Dennis Bray* received his Ph.D. from the Massachusetts Institute of Technology and is currently a Senior Scientist in the Medical Research Council Cell Biophysics Unit at King's College London. *Julian Lewis* received his D.Phil. from Oxford University and is currently a Senior Scientist in the Imperial Cancer Research Fund Developmental Biology Unit, Dept. of Zoology, Oxford University. *Martin Raff* received his M.D. degree from McGill University and is currently a Professor in the Biology Department at University College London. *Keith Roberts* received his Ph.D. from Cambridge University and is currently Head of the Department of Cell Biology at the John Innes Institute, Norwich. *James D. Watson* received his Ph.D. from Indiana University and is currently Director of the Cold Spring Harbor Laboratory. He is the author of *Molecular Biology of the Gene* and, with Francis Crick and Maurice Wilkins, won the Nobel Prize in Medicine and Physiology in 1962.

Library of Congress Cataloging-in-Publication Data

Molecular biology of the cell / Bruce Alberts ... [et al.].—2nd ed.
   p. cm.
  Includes bibliographies and index.
  ISBN 0-8240-3695-6.—ISBN 0-8240-3696-4 (pbk.)
  1. Cytology. 2. Molecular biology. I. Alberts, Bruce.
  [DNLM: 1. Cells. 2. Molecular Biology. QH 581.2 M718]
QH581.2.M64 1989
574.87—dc19
DNLM/DLC
for Library of Congress   88-38275
                       CIP

Published by Garland Publishing, Inc.
136 Madison Avenue, New York, NY 10016

Printed in the United States of America

15 14 13 12 11 10 9 8 7 6 5 4 3

Preface to the Second Edition

More than 50 years ago, E.B. Wilson wrote that "the key to every biological problem must finally be sought in the cell." Yet, until very recently, cell biology was usually taught to biology majors as a specialized upper-level course based largely on electron microscopy. And in most medical schools, many cell-biological topics, such as the mechanisms of endocytosis, chemotaxis, cell movement, and cell adhesion, were hardly taught at all—being regarded as too cellular for biochemistry and too molecular for histology. With the recent dramatic advances in our understanding of cells, however, cell biology is beginning to take its rightful place at the center of biological and medical teaching. In an increasing number of universities, it is a required full-year course for all undergraduate biology and biochemistry majors and is becoming the organizing theme for much of the first-year medical curriculum. The first edition of *Molecular Biology of the Cell* was written in anticipation of these much needed curriculum reforms and in the hope of catalyzing them. We will be gratified if the second edition helps to accelerate and spread these reforms.

In revising the book, we have found few cases where recent discoveries have proved old conclusions wrong. But in the six years since the first edition appeared, a fantastically rich harvest of new information about cells has uncovered new connections and in many places forced a radical change of emphasis. The present revision, therefore, goes deep: every chapter has undergone substantial changes, many have been almost entirely rewritten, and two new chapters—on the control of gene expression and on cancer—have been added.

Some commentators on the first edition, especially teachers, wanted more detailed discussion of the experimental evidence that supports the concepts discussed. We did not wish to disrupt the the conceptual flow or to enlarge an already very large book; but we agree that it is crucial to give students a sense of how advances are made. To this end, John Wilson and Tim Hunt have written a Problems Book to accompany the central portion of the main text (Chapters 5–14). As explained in the Note to the Reader on page xxxv, each section of *The Problems Book* covers a section of *Molecular Biology of the Cell* and takes as its principal focus a set of experiments drawn from the relevant research literature. These are the basis for a series of questions—some easy and some hard—that are meant to involve the reader actively in the reasoning that underlies the process of discovery.

The second edition, like the first, has been a long time in the making. As before, each chapter has been passed back and forth between the author who wrote the first draft and the other authors for criticism and extensive revision, so that every part of the book represents a joint composition; Tim Hunt and John Wilson often helped in this process. In addition, outside experts were invited to make suggestions for revision and, in a few cases, to contribute material for the text, which the authors reworked to fit with the rest of the book. We are especially indebted to James Rothman (Princeton University) for his contribution to Chapter 8 and to Jeremy Hyams (University College London), Tim Mitchison (University of California, San Francisco), and Paul Nurse (University of Oxford) for their contributions to Chapter 13. All sections of the revised text were read by outside experts, whose comments and suggestions were invaluable; a list of acknowledgments is on page xxxiii.

Miranda Robertson again played a major part in the creation of a readable book, by insisting that every page be lucid and coherent and rewriting many pages that were not. We are also indebted to the staff at Garland Publishing, and in particular to Ruth Adams, Alison Walker, and Gavin Borden, for their kindness, good humour, efficiency, and unfailingly generous support during the four years that it has taken to prepare this edition. Special thanks go to Carol Winter, for her painstaking care in typing the entire book and preparing the disks for the printer.

Finally, to our wives, families, colleagues, and students we again offer our gratitude and apologies for several years of impositions and neglect; without their help and tolerance this book could not have been written.

Preface to the First Edition

There is a paradox in the growth of scientific knowledge. As information accumulates in ever more intimidating quantities, disconnected facts and impenetrable mysteries give way to rational explanations, and simplicity emerges from chaos. Gradually the essential principles of a subject come into focus. This is true of cell biology today. New techniques of analysis at the molecular level are revealing an astonishing elegance and economy in the living cell and a gratifying unity in the principles by which cells function. This book is concerned with those principles. It is not an encyclopedia but a guide to understanding. Admittedly, there are still large areas of ignorance in cell biology and many facts that cannot yet be explained. But these unsolved problems provide much of the excitement, and we have tried to point them out in a way that will stimulate readers to join in the enterprise of discovery. Thus, rather than simply present disjointed facts in areas that are poorly understood, we have often ventured hypotheses for the reader to consider and, we hope, to criticize.

Molecular Biology of the Cell is chiefly concerned with eucaryotic cells, as opposed to bacteria, and its title reflects the prime importance of the insights that have come from the molecular approach. Part I and Part II of the book analyze cells from this perspective and cover the traditional material of cell biology courses. But molecular biology by itself is not enough. The eucaryotic cells that form multicellular animals and plants are social organisms to an extreme degree: they live by cooperation and specialization. To understand how they function, one must study the ways of cells in multicellular communities, as well as the internal workings of cells in isolation. These are two very different levels of investigation, but each depends on the other for focus and direction. We have therefore devoted Part III of the book to the behavior of cells in multicellular animals and plants. Thus developmental biology, histology, immunobiology, and neurobiology are discussed at much greater length than in other cell biology textbooks. While this material may be omitted from a basic cell biology course, serving as optional or supplementary reading, it represents an essential part of our knowledge about cells and should be especially useful to those who decide to continue with biological or medical studies. The broad coverage expresses our conviction that cell biology should be at the center of a modern biological education.

This book is principally for students taking a first course in cell biology, be they undergraduates, graduate students, or medical students. Although we assume that most readers have had at least an introductory biology course, we have attempted to write the book so that even a stranger to biology could follow it by starting at the beginning. On the other hand, we hope that it will also be useful to working scientists in search of a guide to help them pick their way through a vast field of knowledge. For this reason, we have provided a much more thorough list of references than the average undergraduate is likely to require, at the same time making an effort to select mainly those that should be available in most libraries.

This is a large book, and it has been a long time in gestation—three times longer than an elephant, five times longer than a whale. Many people have had a hand in it. Each chapter has been passed back and forth between the author who wrote the first draft and the other authors for criticism and revision, so that each chapter represents a joint composition. In addition, a small number of outside experts contributed written material, which the authors reworked to fit with the rest of the book, and all the chapters were read by experts, whose comments and corrections were invaluable. A full list of acknowledgments to these contributors and readers for their help with specific chapters is appended. Paul R. Burton (University of Kansas), Douglas Chandler (Arizona State University), Ursula Goodenough (Washington University), Robert E. Pollack (Columbia University), Robert E. Savage (Swarthmore College), and Charles F. Yocum (University of Michigan) read through all or some of the manuscript and made many helpful suggestions.

The manuscript was also read by undergraduate students, who helped to identify passages that were obscure or difficult.

Most of the advice obtained from students and outside experts was collated and digested by Miranda Robertson. By insisting that every page be lucid and coherent, and by rewriting many of those that were not, she has played a major part in the creation of a textbook that undergraduates will read with ease. Lydia Malim drew many of the figures for Chapters 15 and 16, and a large number of scientists very generously provided us with photographs: their names are given in the figure credits. To our families, colleagues, and students we offer thanks for forbearance and apologies for several years of imposition and neglect. Finally, we owe a special debt of gratitude to our editors and publisher. Tony Adams played a large part in improving the clarity of the exposition, and Ruth Adams, with a degree of good-humored efficiency that put the authors to shame, organized the entire production of the book. Gavin Borden undertook to publish it, and his generosity and hospitality throughout have made the enterprise of writing a pleasure as well as an education for us.

Contents in Brief

Introduction to the Cell

PART I

The Molecular Organization of Cells

PART II

From Cells to Multicellular Organisms

PART III

List of Topics

Introduction to the Cell

How Cells Are Studied

CHAPTER **4**

The Molecular Organization of Cells

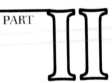

CHAPTER 5

Basic Genetic Mechanisms

The Plasma Membrane

Energy Conversion: Mitochondria and Chloroplasts

CHAPTER 7

CHAPTER (8)

Intracellular Sorting and the Maintenance of Cellular Compartments

CHAPTER 9

The Cell Nucleus

Control of Gene Expression

CHAPTER **10**

Cell Signaling   CHAPTER 12

Cell Growth and Division

CHAPTER **13**

From Cells to Multicellular Organisms

PART III

CHAPTER 15

Germ Cells and Fertilization

Cellular Mechanisms of Development

CHAPTER 17

Differentiated Cells and the Maintenance of Tissues

Maintenance of the Differentiated State — 952

The Immune System

CHAPTER **18**

The Nervous System

CHAPTER **20**

Special Features of Plant Cells

The Importance of the Cell Wall   1137

Transport Between Cells   1151

Interactions Between Plants and Other Organisms   1157

The Distinctive Internal Organization of Plant Cells   1161

Special Features

Acknowledgments

In writing this book we have benefited greatly from the advice of many biologists. In addition to those who advised us by telephone and those who helped with the first edition, we would like to thank the following for their written advice in preparing this edition:

Chapter 1   Hans Bode (University of California, Irvine), Tom Cavalier-Smith (Kings College, London), Elaine Robson (University of Reading, U.K.).

Chapter 2   Efraim Racker (Cornell University), Harry van der Westen (Wageningen, The Netherlands), Richard Wolfenden (University of North Carolina).

Chapter 3   Charles Cantor (Columbia University), Russell Doolittle (University of California, San Diego), Steven Harrison (Harvard University), Richard Wolfenden (University of North Carolina).

Chapter 4   Charles Cantor (Columbia University), Steven Harrison (Harvard University), Richard Henderson (MRC Laboratory of Molecular Biology, Cambridge, U.K.), George Ratcliffe (Oxford University), Peter Shaw (John Innes Institute, Norwich, U.K.).

Chapter 5   Tim Hunt (Cambridge University), Marilyn Kozak (University of Pittsburgh), Thomas Lindahl (Imperial Cancer Research Fund, South Mimms, U.K.), Harold Varmus (University of California, San Francisco).

Chapter 6   Mark Bretscher (MRC Laboratory of Molecular Biology, Cambridge, U.K.), Lewis Cantley (Harvard University), Stuart Cull-Candy (University College London), Anthony Gardner-Medwin (University College London), Walter Gratzer (Kings College London), Ari Helenius (Yale University), Richard Henderson (MRC Laboratory of Molecular Biology, Cambridge, U.K.), Regis Kelly (University of California, San Francisco), Mark Marsh (Institute of Cancer Research, London), Samuel Silverstein (Columbia University), Wilfred Stein (Hebrew University), Peter Walter (University of California, San Francisco), Judy White (University of California, San Francisco).

Chapter 7   Martin Brand (Cambridge University), Leslie Grivell (University of Amsterdam), Richard McCarty (Cornell University), David Nicholls (University of Dundee), Gottfried Schatz (University of Basel), Alison Smith (John Innes Institute, Norwich, U.K.).

Chapter 8   Regis Kelly (University of California, San Francisco), Stuart Kornfeld (Washington University, St. Louis), Paul Lazarow (Rockefeller University), Vishu Lingappa (University of California, San Francisco), George Palade (Yale University), James Rothman (Princeton University), Gottfried Schatz (University of Basel), Kai Simons (EMBO Laboratory, Heidelberg), Alex Varshavsky (Massachusetts Institute of Technology), Peter Walter (University of California, San Francisco), William Wickner (University of California, Los Angeles).

Chapter 9   Pierre Chambon (University of Strasbourg), Sarah Elgin (Washington University, St. Louis), Gary Felsenfeld (National Institutes of Health, Bethesda), Christine Guthrie (University of California, San Francisco), Harold Weintraub (Hutchinson Cancer Center, Seattle), Keith Yamamoto (University of California, San Francisco).

Chapter 10   Pierre Chambon (University of Strasbourg), Enrico Coen (John Innes Institute, Norwich, U.K.), Gary Felsenfeld (National Institutes of Health, Bethesda), Ira Herskowitz (University of California, San Francisco), Tim Hunt (Cambridge University), Robert Roeder (Rockefeller University), Harold Weintraub (Hutchinson Cancer Center, Seattle), John Wilson (Baylor University), Keith Yamamoto (University of California, San Francisco).

Chapter 11   Roger Cooke (University of California, San Francisco), Marc Kirschner (University of California, San Francisco), Michael Klymkowsky (University of Colorado, Boulder), Mark Mooseker (Yale University), Tom Pollard (Johns Hopkins University), Joel Rosenbaum (Yale University), Lewis Tilney (University of Pennsylvania), Klaus Weber (Max Planck Institute for Biophysical Chemistry, Göttingen).

Chapter 12   Henry Bourne (University of California, San Francisco), Philip Cohen (University of Dundee), Graham Hardie (University of Dundee), Daniel Koshland (University of California, Berkeley), Alex Levitzki (Hebrew University), Robert Mishell (University of Birmingham, U.K.), Anne Mudge (University College London), Michael Schramm (Hebrew University), Zvi Sellinger (Hebrew University), Tom Vanaman (University of Kentucky).

Chapter 13   Robert Brooks (King's College London), Leland Hartwell (University of Washington, Seattle), John Heath (Oxford University), Tim Hunt (Cambridge University), Jeremy Hyams (University College London), Marc Kirschner (University of California, San Francisco), Tim Mitchison (University of California, San Francisco), Andrew Murray (University of California, San Francisco), Paul Nurse (Oxford University).

Chapter 14   Michael Bennett (Albert Einstein College of Medicine), Benny Geiger (Weizmann Institute), Daniel Goodenough (Harvard Medical School), Barry Gumbiner (University of California, San Francisco), Richard Hynes (Massachusetts Institute of Technology), Tom Jessell (Columbia University), Louis Reichardt (University of California, San Francisco), Erkki Ruoslahti (La Jolla Cancer Research Foundation), Masatoshi Takeichi (Kyoto University), Robert Trelstad (UMDNJ—Robert Wood Johnson Medical School), Anne Warner (University College London).

Chapter 15   Adelaide Carpenter (University of California, San Diego), James Crow (University of Wisconsin, Madison), David Epel (Stanford University), Tim Hunt (Cambridge University), James Maller (University of Colorado Medical School), John Maynard Smith (University of Sussex), Anne McLaren (University College London), Montrose Moses (Duke University), Lewis Tilney (University of Pennsylvania), Victor Vacquier (University of California, San Diego).

Chapter 16   Marianne Bronner-Fraser (University of California, Irvine), Robert Horvitz (Massachusetts Institute of Technology), James Hudspeth (University of California, San Francisco), Philip Ingham (Imperial Cancer Research Fund, Oxford), Ray Keller (University of California, Berkeley), Cynthia Kenyon (University of California, San Francisco), Judith Kimble (University of Wisconsin, Madison), Tom Kornberg (University of California, San Francisco), Mark Krasnow (Stanford University), Peter Lawrence (MRC Laboratory of Molecular Biology, Cambridge, U.K.), Gail Martin (University of California, San Francisco), Patrick O'Farrell (University of California, San Francisco), Jonathan Slack (Imperial Cancer Research Fund, Oxford).

Chapter 17   Michael Banda (University of California, San Francisco), Alan Boyde (University College London), Michael Dexter (Paterson Institute for Cancer Research, Manchester, U.K.), Charles Emerson (University of Virginia), Howard Green (Harvard University), David Housman (Massachusetts Institute of Technology), Norman Iscove (Ontario Cancer Institute, Toronto), Anne Mudge (University College London), Michael Solursh (University of Iowa), Jim Till (Ontario Cancer Institute, Toronto), Fiona Watt (Imperial Cancer Research Fund, London).

Chapter 18   Fred Alt (Columbia University), Peter Lachmann (MRC Center, Cambridge, U.K.), Avrion Mitchison (University College London), William Paul (National Institutes of Health, Bethesda), Ronald Schwartz (National Institutes of Health, Bethesda), David Standring (University of California, San Francisco).

Chapter 19   Tim Bliss (National Institute for Medical Research, London), Michael Brown (Oxford University), Steven Burden (Massachusetts Institute of Technology), Gerald Fischbach (Washington University, St. Louis), James Hudspeth (University of California, San Francisco), Trevor Lamb (Cambridge University), Dale Purves (Washington University, St. Louis), James Schwartz (Columbia University), Charles Stevens (Yale University), Michael Stryker (University of California, San Francisco).

Chapter 20   Brian Gunning (Australian National University, Canberra), Andy Johnston (John Innes Institute, Norwich, U.K.), Clive Lloyd (John Innes Institute, Norwich, U.K.), Freiderick Meins (Freiderich Miescher Institut, Basel), Scott Stachel (University of California, Berkeley) Andrew Staehelin (University of Colorado, Boulder), Anthony Trewavas (Edinburgh University), Virginia Walbot (Stanford University), Patricia Zambryski (University of California, Berkeley).

Chapter 21   Michael Bishop (University of California, San Francisco), John Cairns (Harvard School of Public Health), Ruth Ellman (Institute of Cancer Research, Sutton, U.K.), Hartmut Land (Imperial Cancer Research Fund, London), Bruce Ponder (Institute of Cancer Research, Sutton, U.K.).

The cover photograph shows a set of human chromosomes at metaphase. Chromatin is shown red, fluorescently stained with propidium iodide, while the centromeres are revealed with an anti-kinetochore antiserum. The photograph was supplied by Bill Brinkley and is taken from Merry, D.E., Pathak, S., Hsu, T.C. and Brinkley, B.R. *Am. J. Hum. Genet.* 37:425–430, 1985.

A Note to the Reader

Although the chapters of this book can be read independently of one another, they are arranged in a logical sequence of three parts. The first three chapters of **Part I** cover elementary principles and basic biochemistry. They can serve either as an introduction for those who have not studied biochemistry or as a refresher course for those who have. Chapter 4, which concludes Part I, deals with the principles of the main experimental methods for investigating cells. It is not necessary to read this chapter in order to understand the later chapters, but a reader will find it a useful reference.

Part II represents the central core of cell biology and is concerned mainly with those properties that are common to most eucaryotic cells, beginning with the fundamental molecular mechanisms of heredity and concluding with cell adhesion and the extracellular matrix.

Part III follows the behavior of cells in the construction of multicellular organisms, starting with the formation of eggs and sperm and ending with the disruption of multicellular organization that occurs in cancer.

Chapter 4 includes several tables giving the dates of crucial developments along with the names of the scientists involved. Elsewhere in the book the policy has been to avoid naming individual scientists. The authors of major discoveries, however, can usually be identified by consulting the **lists of references** at the end of each chapter. These references frequently include the original papers in which important discoveries were first reported. **Superscript numbers** that accompany the text headings refer to the numbered citations in the reference lists, providing a convenient means of following up specific topics.

Throughout the book, **boldface type** has been used to highlight key terms at the point in a chapter where the main discussion of them occurs. This may or may not coincide with the first appearance of the term in the text. *Italics* are used to set off important terms with a lesser degree of emphasis.

The Problems Book is designed as a companion volume that will help the reader appreciate the elegance, the ingenuity, and the surprises of research. It provides problems to accompany the central portion of this book (Chapters 5-14). Each chapter of problems is divided into sections that correspond to the sections of the main textbook, the principal focus of each section being a set of research-oriented problems derived from the scientific literature.

Most of the research problems illustrate points in the main text and are flagged with a symbol in the margin next to the relevant concept heading. Thus 5-4 in the margin of this text refers to Problem 4 in Chapter 5 of *The Problems Book*. In addition, each section of *The Problems Book* begins with a set of short fill-in-the-blank and true-false questions intended to help the reader review the vocabulary and main concepts of the relevant topic. *The Problems Book* should be useful for homework assignments and as a basis for class discussion. It could even provide ideas for exam questions.

MOLECULAR BIOLOGY OF
THE CELL
SECOND EDITION

A sense of scale. These scanning
electron micrographs, taken at
progressively higher magnifications,
show bacterial cells on the point of an
ordinary domestic pin. (Courtesy of
Tony Brain and the Science Photo
Library.)

Introduction to the Cell

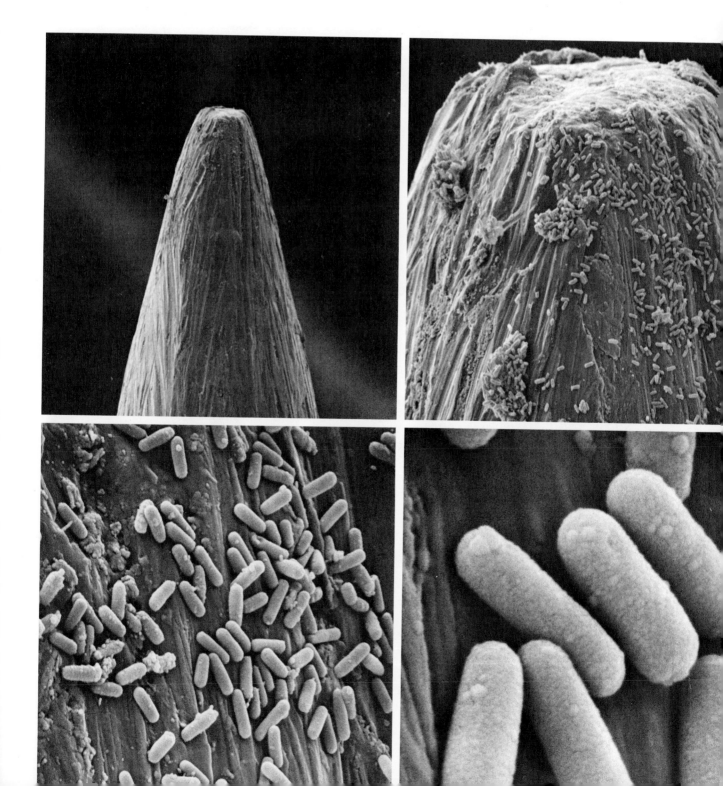

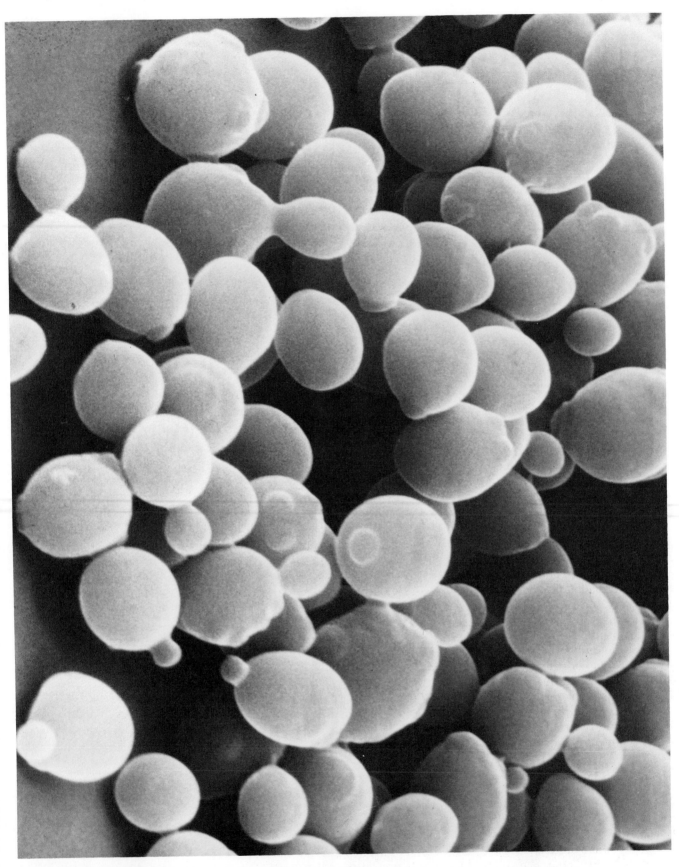

Scanning electron micrograph of growing yeast cells. These
unicellular eucaryotes bud off small daughter cells as they multiply.
(Courtesy of Ira Herskowitz and Eric Schabatach.)

The Evolution of the Cell

1

All living creatures are made of cells—small membrane-bounded compartments filled with a concentrated aqueous solution of chemicals. The simplest forms of life are solitary cells that propagate by dividing in two. Higher organisms, such as ourselves, are like cellular cities in which groups of cells perform specialized functions and are linked by intricate systems of communication. Cells occupy a halfway point in the scale of biological complexity. We study them to learn, on the one hand, how they are made from molecules and, on the other, how they cooperate to make an organism as complex as a human being.

All organisms, and all of the cells that constitute them, are believed to have descended from a common ancestor cell by *evolution*. Evolution involves two essential processes: (1) the occurrence of random *variation* in the genetic information passed from an individual to its descendants and (2) *selection* in favor of genetic information that helps its possessors to survive and propagate. Evolution is the central principle of biology, helping us to make sense of the bewildering variety in the living world.

This chapter, like the book as a whole, is concerned with the progression from molecules to multicellular organisms. It discusses the evolution of the cell, first as a living unit constructed from smaller parts and then as a building block for larger structures. Through evolution, we introduce the cell components and activities that are to be treated in detail, in broadly similar sequence, in the chapters that follow. Beginning with the origins of the first cell on earth, we consider how the properties of certain types of large molecules allow hereditary information to be transmitted and expressed and permit evolution to occur. Enclosed in a membrane, these molecules provide the essentials of a self-replicating cell. Following this, we describe the major transition that occurred in the course of evolution, from small bacteriumlike cells to much larger and more complex cells such as are found in present-day plants and animals. Lastly, we suggest ways in which single free-living cells might have given rise to large multicellular organisms, becoming specialized and cooperating in the formation of such intricate organs as the brain.

Clearly, there are dangers in introducing the cell through its evolution: the large gaps in our knowledge can be filled only by speculations that are liable to be wrong in many details. We cannot go back in time to witness the unique molecular events that took place billions of years ago. But those ancient events have left many traces for us to analyze. Ancestral plants, animals, and even bacteria are preserved as fossils. Even more important, every modern organism provides evidence of the character of living organisms in the past. Present-day biological molecules, in particular, are a rich source of information about the course of evolution, revealing fundamental similarities between the most disparate of living organisms and allowing us to map out the differences between them on an objective universal scale. These molecular similarities and differences present us with a problem like that which confronts the literary scholar who seeks to establish the original text of an ancient author by comparing a mass of variant manuscripts that have been corrupted through repeated copying and editing. The task is hard, and the evidence is incomplete, but it is possible at least to make intelligent guesses about the major stages in the evolution of living cells.

From Molecules to the First Cell[1]

Simple Biological Molecules Can Form Under Prebiotic Conditions

The conditions that existed on the earth in its first billion years are still a matter of dispute. Was the surface initially molten? Did the atmosphere contain ammonia, or methane? Everyone seems to agree, however, that the earth was a violent place with volcanic eruptions, lightning, and torrential rains. There was little if any free oxygen and no layer of ozone to absorb the ultraviolet radiation from the sun.

Simple organic molecules (that is, molecules containing carbon) are likely to have been produced under such conditions. The best evidence for this comes from laboratory experiments. If mixtures of gases such as CO_2, CH_4, NH_3, and H_2 are heated with water and energized by electrical discharge or by ultraviolet radiation, they react to form small organic molecules—usually a rather small selection, each made in large amounts (Figure 1–1). Among these products are a number of compounds, such as hydrogen cyanide (HCN) and formaldehyde (HCHO), that readily undergo further reactions in aqueous solution (Figure 1–2). Most important, the four major classes of small organic molecules found in cells—*amino acids, nucleotides, sugars,* and *fatty acids*—are generated.

Although such experiments cannot reproduce the early conditions on the earth exactly, they make it plain that the formation of organic molecules is surprisingly easy. And the developing earth had immense advantages over any human experimenter; it was very large and could produce a wide spectrum of conditions. But above all, it had much more time—hundreds of millions of years. In such circumstances it seems very likely that, at some time and place, many of the simple organic molecules found in present-day cells accumulated in high concentrations.

Polynucleotides Are Capable of Directing Their Own Synthesis

Simple organic molecules such as amino acids and nucleotides can associate to form large *polymers.* One amino acid can join with another by forming a peptide bond, and two nucleotides can join together by a phosphodiester bond. The repetition of these reactions leads to linear polymers known as **polypeptides** and **polynucleotides,** respectively. In present-day living organisms, polypeptides—known as *proteins*—and polynucleotides—in the form of both *ribonucleic acids (RNA)* and *deoxyribonucleic acids (DNA)*—are commonly viewed as the most important constituents. A restricted set of 20 amino acids constitute the universal building blocks of the proteins, while RNA and DNA molecules are each constructed from just four types of nucleotides. One can only speculate as to why these particular sets of monomers were selected for biosynthesis in preference to others that are chemically similar.

The earliest polymers may have formed in any of several ways—for example, by the heating of dry organic compounds or by the catalytic activity of high concentrations of inorganic polyphosphates. Under laboratory conditions the products of similar reactions are polymers of variable length and random sequence in which the particular amino acid or nucleotide added at any point depends mainly on chance (Figure 1–3). Once a polymer has formed, however, it can influence the formation of other polymers. Polynucleotides, in particular, have the ability to

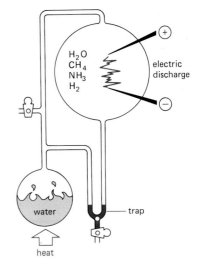

Figure 1–1 A typical experiment simulating conditions on the primitive earth. Water is heated in a closed apparatus containing CH_4, NH_3, and H_2, and an electrical discharge is passed through the vaporized mixture. Organic compounds accumulate in the U-tube trap.

HCHO	formaldehyde
HCOOH	formic acid
HCN	hydrogen cyanide
CH_3COOH	acetic acid
NH_2CH_2COOH	glycine
$CH_3CHCOOH$ $\|$ OH	lactic acid
$NH_2CHCOOH$ $\|$ CH_3	alanine
NH—CH_2COOH $\|$ CH_3	sarcosine
NH_2—C—NH_2 $\|\|$ O	urea
$NH_2CHCOOH$ $\|$ CH_2 $\|$ COOH	aspartic acid

Figure 1–2 A few of the compounds that might form in the experiment described in Figure 1–1. Compounds shown in color are important components of present-day living cells.

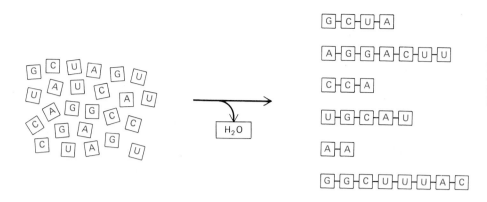

Figure 1–3 Nucleotides of four kinds (here represented by the single letters A, U, G, and C) can undergo spontaneous polymerization with the loss of water. The product is a mixture of polynucleotides that are random in length and sequence.

specify the sequence of nucleotides in new polynucleotides by acting as *templates* for the polymerization reactions. A polymer composed of one nucleotide (for example, polyuridylic acid, or poly U) can serve as a template for the synthesis of a second polymer composed of another type of nucleotide (polyadenylic acid, or poly A). Such templating depends on the fact that one polymer preferentially binds the other. By lining up the subunits required to make poly A along its surface, poly U promotes the formation of poly A (Figure 1–4).

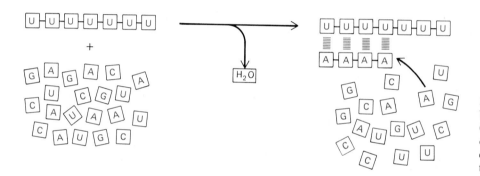

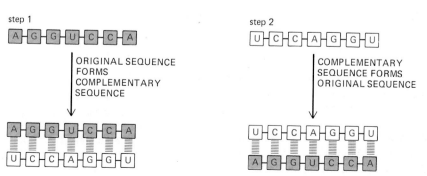

Figure 1–4 Preferential binding occurs between pairs of nucleotides (C with G and A with U) by relatively weak chemical bonds (*above*). This pairing enables one polynucleotide to act as a template for the synthesis of another (*left*).

Specific pairing between complementary nucleotides probably played a crucial part in the origin of life. Consider, for example, a polynucleotide such as RNA, made of a string of four nucleotides, containing the bases uracil (U), adenine (A), cytosine (C), and guanine (G). Because of complementary pairing between the bases A and U and between the bases G and C, when RNA is added to a mixture of activated nucleotides under conditions that favor polymerization, new RNA molecules are produced in which nucleotides are joined in a sequence that is complementary to the original. That is, the new molecules are rather like a mold of the original, with each A in the original corresponding to a U in the copy, and so on. The sequence of nucleotides in the original RNA strand contains information that is, in essence, preserved in the newly formed complementary strands: a second round of copying, with the complementary strand as a template, restores the original sequence (Figure 1–5).

Figure 1–5 Replication of a polynucleotide sequence (here an RNA molecule). In step 1, the original RNA molecule acts as a template to form an RNA molecule of complementary sequence. In step 2, this complementary RNA molecule itself acts as a template, forming RNA molecules of the original sequence. Since each templating molecule can produce many copies of the complementary strand, these reactions can result in the "multiplication" of the original sequence.

Such *complementary templating* mechanisms are elegantly simple, and they lie at the heart of information transfer processes in biological systems. Genetic information contained in every cell is encoded in the sequences of nucleotides in its polynucleotide molecules, and this information is passed on (inherited) from generation to generation by means of complementary base-pairing interactions.

Templating mechanisms, however, require catalysts to promote them: without specific catalysis, templated polymerization is slow and inefficient and other, competing reactions hinder the formation of accurate replicas. Today, the catalytic functions that replicate polynucleotides are provided by highly specialized proteins called *enzymes*, which would not have been present in the "prebiotic soup." On the primitive earth, metal ions and minerals such as clays would have provided some catalytic help. But most important, RNA itself can act as a catalyst: RNA molecules have both the templating properties essential for replication and the potential to fold up to form complex surfaces that catalyze specific reactions. As we shall now discuss, this special versatility of RNA molecules is likely to have provided the basis for the evolution of the first living systems.

Self-replicating Molecules Undergo Natural Selection[2]

Errors inevitably occur in any copying process, and imperfect copies of the original will be propagated. With repeated replication, therefore, the sequence of nucleotides in an original polynucleotide molecule will undergo substantial changes, generating in this way a variety of different molecules. In the case of RNA, these molecules are likely to have different functional properties. RNA molecules are not just strings of symbols that carry information in an abstract way. They also have chemical personalities that affect their behavior. In particular, the specific sequence of nucleotides governs how the molecule folds up in solution. Just as the nucleotides in a polynucleotide can pair with free complementary nucleotides in their environment to form a new polymer, so they can pair with complementary nucleotide residues within the polymer itself. A sequence GGGG in one part of a polynucleotide chain can form a relatively strong association with a CCCC sequence in another region of the same molecule. Such associations produce various three-dimensional folds, and the molecule as a whole takes on a unique shape that depends entirely on the sequence of its nucleotides (Figure 1–6).

The three-dimensional folded structure of a polynucleotide affects its stability, its actions on other molecules, and its ability to replicate, so not all polynucleotide shapes will be equally successful in a replicating mixture. In laboratory studies, replicating systems of RNA molecules have been shown to undergo a form of natural selection in which different favorable sequences eventually predominate, depending on the exact conditions.

An RNA molecule therefore has two special characteristics: it carries information encoded in its nucleotide sequence that it can pass on by the process of replication, and it has a unique folded structure that determines how it will interact with other molecules and respond to the ambient conditions. These two features—one informational, the other functional—are the two properties essential for evolution. The nucleotide sequence of an RNA molecule is analogous to the *genotype*—the hereditary information—of an organism. The folded three-dimensional structure is analogous to the *phenotype*—the expression of the genetic information on which natural selection operates.

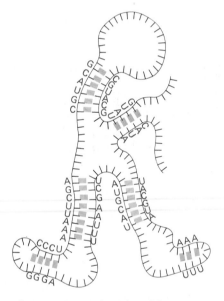

Figure 1–6 Nucleotide pairing between different regions of the same polynucleotide (RNA) chain causes the molecule to adopt a distinctive shape.

Specialized RNA Molecules Can Catalyze Biochemical Reactions[2,3]

Natural selection depends on the environment, and for a replicating RNA molecule a critical component of the environment is the set of other RNA molecules in the mixture. Besides acting as templates for their own replication, these can catalyze the breakage and formation of covalent bonds, including bonds between nucleotides. For example, some specialized RNA molecules can catalyze a change in other RNA molecules, cutting the nucleotide sequence at a particular point; and other types of RNA molecules spontaneously cut out a portion of their own nucleotide sequence and rejoin the cut ends (a process known as self-splicing). Each

RNA-catalyzed reaction depends on a specific arrangement of atoms that forms on the surface of the catalytic RNA molecule, causing particular chemical groups on one or more of its nucleotides to become highly reactive.

Certain catalytic activities would have had a cardinal importance in the primordial soup. Consider in particular an RNA molecule that catalyzes the process of templated polymerization, taking any given RNA molecule as template. This catalytic molecule, by acting on copies of itself, can replicate with heightened speed and efficiency (Figure 1–7A). At the same time, it can promote the replication of any other types of RNA molecules in its neighborhood. Some of these may have catalytic actions that help or hinder the survival or replication of RNA in other ways. If beneficial effects are reciprocated, the different types of RNA molecules, specialized for different activities, may evolve into a cooperative system that replicates with unusually great efficiency (Figure 1–7B).

Information Flows From Polynucleotides to Polypeptides[4]

It is suggested, therefore, that between 3.5 and 4 billion years ago, somewhere on earth, self-replicating systems of RNA molecules began the process of evolution. Systems with different sets of nucleotide sequences competed for the available precursor materials to construct copies of themselves, just as organisms now compete; success depended on the accuracy and the speed with which the copies were made and on the stability of those copies.

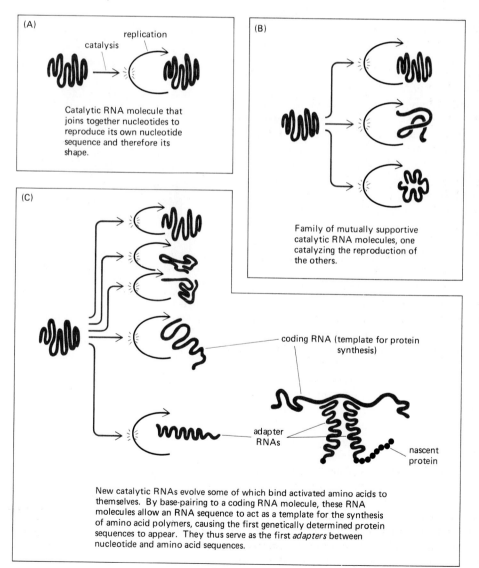

(A)

replication

catalysis

Catalytic RNA molecule that joins together nucleotides to reproduce its own nucleotide sequence and therefore its shape.

(B)

Family of mutually supportive catalytic RNA molecules, one catalyzing the reproduction of the others.

(C)

coding RNA (template for protein synthesis)

adapter RNAs

nascent protein

New catalytic RNAs evolve some of which bind activated amino acids to themselves. By base-pairing to a coding RNA molecule, these RNA molecules allow an RNA sequence to act as a template for the synthesis of amino acid polymers, causing the first genetically determined protein sequences to appear. They thus serve as the first *adapters* between nucleotide and amino acid sequences.

Figure 1–7 Diagram of three successive steps in the evolution of a self-replicating system of RNA molecules capable of directing protein synthesis.

However, while the structure of polynucleotides is well suited for information storage and replication, the catalytic capabilities of RNA molecules are apparently too limited to provide all the functions of a modern cell. Greater versatility is attained by polypeptides, which are composed of many different amino acids with chemically diverse side chains. As we shall discuss in Chapter 3, polypeptides are able to adopt diverse three-dimensional forms that bristle with reactive sites, and they are therefore ideally suited to a wide range of structural and chemical tasks. Even random polymers of amino acids produced by prebiotic synthetic mechanisms are likely to have had catalytic properties, some of which could have enhanced the replication of RNA molecules. Any polynucleotide that helped guide the synthesis of a useful polypeptide in its environment would have had a great advantage in the evolutionary struggle for survival.

But how could polynucleotides exert such control? How could the information encoded in their sequences specify the sequences of polymers of a different type? Clearly, the polynucleotides must act as catalysts to join selected amino acids together. In present-day organisms, a collaborative system of RNA molecules plays a central part in directing the synthesis of polypeptides—that is, **protein synthesis**—but the process is aided by other proteins synthesized previously. The biochemical machinery for protein synthesis is remarkably elaborate. One RNA molecule carries the genetic information for a particular polypeptide in the form of a code, while other RNA molecules act as adapters, each binding a specific amino acid. These two types of RNA molecules form complementary base pairs with one another to enable sequences of nucleotides in the coding RNA molecule to direct the incorporation of specific amino acids held on the adapter RNAs into a growing polypeptide chain. Precursors to these two types of RNA molecules presumably directed the first protein synthesis without the aid of proteins (Figure 1–7C).

Today, these events in the assembly of new proteins take place on the surface of *ribosomes*—complex particles composed of several large RNA molecules of yet another class, together with more than 50 different types of protein. In Chapter 5 we shall see that the ribosomal RNA in these particles plays a central catalytic role in the process of protein synthesis and forms more than 60% of the ribosome's mass. At least in evolutionary terms, it is the fundamental component of the ribosome.

It seems likely, then, that RNA guided the primordial synthesis of proteins, perhaps in a clumsy and primitive fashion. In this way, RNA was able to create tools—in the form of proteins—for more efficient biosynthesis, and some of these could have been put to use in the replication of RNA and in the process of tool production itself.

The synthesis of specific proteins under the guidance of RNA required the evolution of a code by which the polynucleotide sequence specifies the amino acid sequence that makes up the protein. This code—the *genetic code*—is spelled out in a "dictionary" of three-letter words: different triplets of nucleotides encode specific amino acids. The code seems to have been selected arbitrarily, and yet it is virtually the same in all living organisms. This strongly suggests that all present-day cells have descended from a single line of primitive cells that evolved the mechanism of protein synthesis.

Once the evolution of nucleic acids had advanced to the point of specifying powerful catalytic proteins, or *enzymes*, to aid in their own manufacture, the proliferation of the replicating system would have been greatly speeded up. The potentially explosive nature of such an autocatalytic process can be seen today in the life cycle of some bacterial viruses: after they have entered a bacterium, such viruses direct the synthesis of proteins that catalyze selectively their own replication, so that within a short time they take over the entire cell (Figure 1–8).

Membranes Defined the First Cell

One of the crucial events leading to the formation of the first cell must have been the development of an outer membrane. For example, the proteins synthesized under the control of a certain species of RNA would not facilitate reproduction of that species of RNA unless they remained in the neighborhood of the RNA; more-

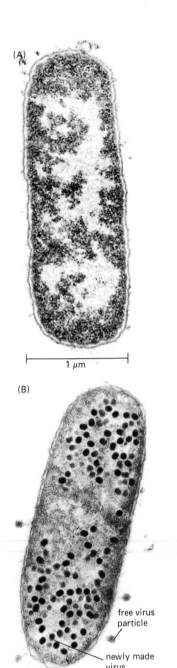

(A)

1 μm

(B)

free virus particle

newly made virus

Figure 1–8 Micrograph of a bacterial cell (*Escherichia coli*) (A) in its normal healthy state and (B) just over an hour after being infected with a virus (bacteriophage T4). The original free virus particles, some of which can be seen still adhering to the outside of the cell, injected their DNA into the cell; this then directed the synthesis of specific viral proteins, some of which degraded the host cell DNA while others catalyzed the replication of the virus. At the stage shown, the newly made viral DNA has been packaged into protein coats, seen as dense particles in the cytoplasm. The cell is about to burst open, releasing several hundred new virus particles to the surroundings. (Courtesy of E. Kellenberger.)

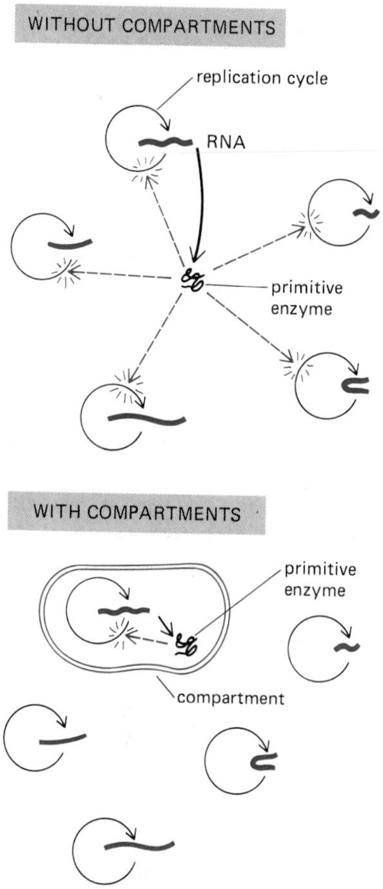

Figure 1–9 Schematic drawing showing the evolutionary significance of cell-like compartments. In a mixed population of self-replicating RNA molecules capable of protein synthesis (as illustrated in Figure 1–7), any improved form of RNA that is able to produce a more useful protein must share this protein with all of its competitors. However, if the RNA is enclosed within a compartment, such as a lipid membrane, then any protein it makes is retained for its own use; the RNA can therefore be selected on the basis of its making a better protein.

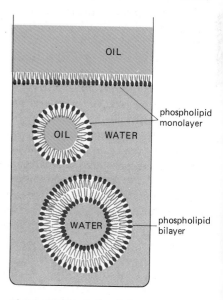

over, as long as these proteins were free to diffuse among the population of replicating RNA molecules, they could benefit equally any competing species of RNA that might be present. If a variant RNA arose that made a superior type of enzyme, the new enzyme could not contribute *selectively* to the survival of the variant RNA in its competition with its fellows. Selection of RNA molecules according to the quality of the proteins they generated could not begin until some form of compartment evolved to contain the proteins made by an RNA molecule and thereby make them available only to the RNA that had generated them (Figure 1–9).

The need for containment is easily fulfilled by another class of molecules, which has the simple physicochemical property of being *amphipathic*, that is, consisting of one part that is hydrophobic (water insoluble) and another part that is hydrophilic (water soluble). When such molecules are placed in water, they aggregate with their hydrophobic portions as much in contact with one another as possible and with their hydrophilic portions in contact with the water. Amphipathic molecules of appropriate shape spontaneously aggregate to form *bilayers*, creating small closed vesicles whose aqueous contents are isolated from the external medium (Figure 1–10). The phenomenon can be demonstrated in a test tube by simply mixing phospholipids and water together: under appropriate conditions, small vesicles will form. All present-day cells are surrounded by a **plasma membrane** consisting of amphipathic molecules—mainly phospholipids—in this configuration; in cell membranes, the lipid bilayer also contains amphipathic proteins. In the electron microscope such membranes appear as sheets about 5 nm thick, with a distinctive three-layered appearance due to the tail-to-tail packing of the phospholipid molecules.

It is not clear at what point in the evolution of biological catalysts the first cells were formed. They may have originated when phospholipid molecules in the prebiotic soup spontaneously assembled into membranous structures enclosing a self-replicating mixture of catalytic RNA molecules. It is more commonly assumed, however, that protein synthesis had to evolve before there were cells. In either case, once they were sealed within a closed membrane, RNA molecules could begin to evolve, not merely on the basis of their own structure, but also according to their effect on the other molecules in the same compartment: the nucleotide sequences of the RNA molecules could now be expressed in the character of the cell as a whole.

All Present-Day Cells Use DNA as Their Hereditary Material[4,5]

The picture we have presented is, of course, speculative: there are no fossil records that trace the origins of the first cell. Nevertheless, there is persuasive evidence from present-day organisms and from experiments that the broad features of this evolutionary story are correct. The prebiotic synthesis of small molecules, the self-replication of catalytic RNA molecules, the translation of RNA sequences into amino acid sequences, and the assembly of lipid molecules to form membrane-bounded compartments—all presumably occurred to generate primitive cells 3.5 or 4 billion years ago.

It is useful to compare these early cells with the simplest present-day cells, the **mycoplasmas.** Mycoplasmas are small bacteriumlike organisms that normally lead a parasitic existence in close association with animal or plant cells (Figure 1–11). Some have a diameter of about 0.3 μm and contain only enough nucleic acid to direct the synthesis of about 750 different proteins. Some of these proteins

Figure 1–10 Cell membranes are formed of phospholipids. Because these molecules have hydrophilic heads and lipophilic tails, they will align themselves at an oil-water interface with their heads in the water and their tails in the oil. In water, they will associate to form closed bilayer vesicles in which the lipophilic tails are in contact with one another and the hydrophilic heads are exposed to the water.

are enzymes, some are structural; some lie in the cell's interior, others are embedded in its membrane. Together they synthesize essential small molecules that are not available in the environment, redistribute the energy needed to drive biosynthetic reactions, and maintain appropriate conditions inside the cell.

The first cells on the earth presumably contained many fewer components than a mycoplasma and divided much more slowly. There was, however, a more fundamental difference between these primitive cells and a mycoplasma, or indeed any other present-day cell: the hereditary information in all cells alive today is stored in DNA rather than in the RNA that is thought to have stored the hereditary information in primitive cells. Both types of polynucleotides are found in present-day cells, but they function in a collaborative manner, each having evolved to perform specialized tasks. Small chemical differences fit the two kinds of molecules for distinct functions. As the permanent repository of genetic information, DNA is more stable than RNA. This is partly because DNA lacks a sugar hydroxyl group that makes RNA more susceptible to hydrolysis. But it is also because DNA, unlike RNA, exists principally in a double-stranded form, composed of a pair of complementary polynucleotide molecules. Its double-stranded structure not only makes DNA relatively easy to replicate (as will be explained in Chapter 3), it also permits a repair mechanism to operate that uses the intact strand as a template for the correction or repair of the associated damaged strand. DNA guides the synthesis of specific RNA molecules, again by the principle of complementary base-pairing, though now this pairing is between slightly different types of nucleotides. The resulting single-stranded RNA molecules then perform two primeval functions: they direct protein synthesis both as coding RNA molecules (*messenger* RNAs) and as RNA catalysts (*ribosomal* and other nonmessenger RNAs).

The suggestion, in short, is that RNA came first in evolution, having both genetic and catalytic properties; after efficient protein synthesis evolved, DNA took over the primary genetic function and proteins became the major catalysts, while RNA remained primarily as the intermediary connecting the two (Figure 1–12). DNA would have become necessary only when cells became so complex that they required an amount of genetic information greater than that which could be stably maintained in RNA molecules.

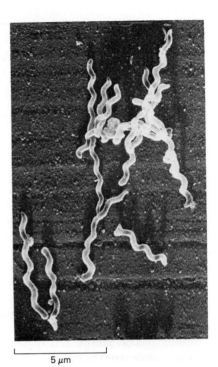

Figure 1–11 *Spiroplasma citrii*, a mycoplasma that grows in plant cells. (Courtesy of J. Burgess.)

Summary

Living cells probably arose on earth by the spontaneous aggregation of molecules about 3.5 billion years ago. From our knowledge of present-day organisms and the molecules they contain, it seems likely that the development of the autocatalytic mechanisms fundamental to living systems began with the evolution of families of RNA molecules that could catalyze their own replication. With time, one of these families of cooperating RNA catalysts developed the ability to direct synthesis of polypeptides. Early cells are likely to have relied heavily on both RNA and protein catalysis and to have contained only RNA as their hereditary material. Finally, as the accumulation of additional protein catalysts allowed more efficient and complex cells to evolve, the DNA double helix replaced RNA as a more stable molecule for storing the increased amounts of genetic information required by such cells.

From Procaryotes to Eucaryotes[6]

It is thought that all organisms living now on earth derive from a single primordial cell born more than three billion years ago. This cell, outreproducing its competitors, took the lead in the process of cell division and evolution that eventually covered the earth with green, changed the composition of its atmosphere, and made it the home of intelligent life. The family resemblances among all organisms seem too strong to be explained in any other way. One important landmark along this evolutionary road occurred about 1.5 billion years ago, when there was a transition from small cells with relatively simple internal structures—the so-called **procaryotic** cells, which include the various types of bacteria—to a flourishing of larger and radically more complex *eucaryotic* cells such as are found in higher animals and plants.

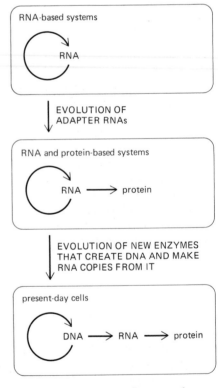

Figure 1–12 Suggested stages of evolution from simple self-replicating systems of RNA molecules to present-day cells, in which DNA is the repository of genetic information and RNA acts largely as a go-between to direct protein synthesis.

...mple but Biochemically Diverse[7]

...nd in most natural environments. They
...only several micrometers in linear di-
...s a tough protective coat, called a *cell*
...encloses a single cytoplasmic compart-
...small molecules. In the electron micro-
...of varying texture without any obvious
...–8A).

...ickly by simply dividing in two by *binary*
...f the fittest" generally means survival of
...ptimal conditions, a single procaryotic
...eby give rise to 5 billion cells (approxi-
...lation on earth) in less than 11 hours.
...ulations of bacteria to adapt rapidly to
changes in their environment. Under laboratory conditions, for example, a population of bacteria maintained in a large vat will evolve within a few weeks by
spontaneous mutation and natural selection to utilize new types of sugar molecules as carbon sources.

In nature, bacteria live in an enormous variety of ecological niches, and they
show a corresponding richness in their underlying biochemical composition. Two
distantly related groups can be recognized: the *eubacteria*, which are the commonly encountered forms that inhabit soil, water, and larger living organisms; and
the *archaebacteria*, which are found in such incommodious environments as bogs,
ocean depths, salt brines, and hot acid springs (Figure 1–14).

There exist species of bacteria that can utilize virtually any type of organic
molecule as food, including sugars, amino acids, fats, hydrocarbons, polypeptides,
and polysaccharides. Some are even able to obtain their carbon atoms from CO_2
and their nitrogen atoms from N_2. Despite their relative simplicity, bacteria have
survived for longer than any other organisms and still are the most abundant type
of cell on earth.

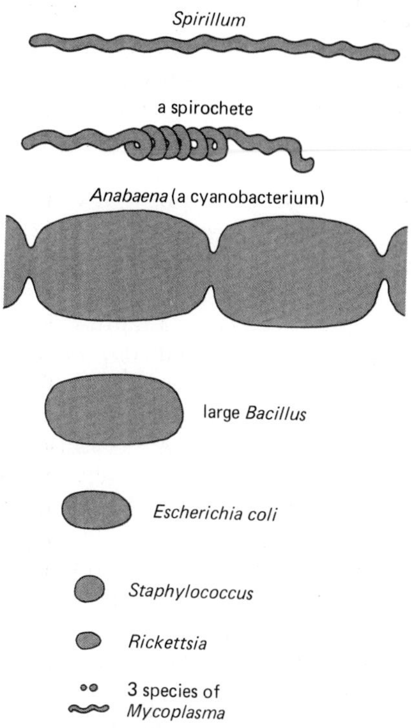

Metabolic Reactions Evolve[7,8]

A bacterium growing in a salt solution containing a single type of carbon source,
such as glucose, must carry out a large number of chemical reactions. Not only
must it derive from the glucose the chemical energy needed for many vital processes,
it must also use the carbon atoms of glucose to synthesize every type of organic
molecule that the cell requires. These reactions are catalyzed by hundreds of
enzymes working in reaction "chains" so that the product of one reaction is the
substrate for the next; such enzymatic chains, called *metabolic pathways*, will be
discussed in the following chapter.

Originally, when life began on earth, there was probably little need for such
metabolic reactions. Cells could survive and grow on the molecules in their surroundings—a legacy from the prebiotic soup. As these natural resources were
exhausted, organisms that had developed enzymes to manufacture organic mole-

Figure 1–13 Some procaryotic cells drawn to scale.

Spirillum

a spirochete

Anabaena (a cyanobacterium)

large *Bacillus*

Escherichia coli

Staphylococcus

Rickettsia

3 species of *Mycoplasma*

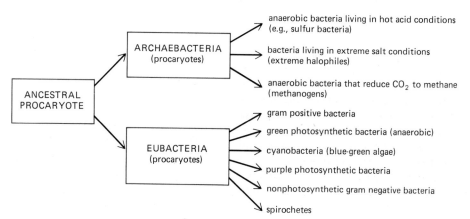

anaerobic bacteria living in hot acid conditions (e.g., sulfur bacteria)

bacteria living in extreme salt conditions (extreme halophiles)

anaerobic bacteria that reduce CO_2 to methane (methanogens)

gram positive bacteria

green photosynthetic bacteria (anaerobic)

cyanobacteria (blue-green algae)

purple photosynthetic bacteria

nonphotosynthetic gram negative bacteria

spirochetes

ARCHAEBACTERIA (procaryotes)

EUBACTERIA (procaryotes)

ANCESTRAL PROCARYOTE

Figure 1–14 Family relationships
between present-day bacteria; arrows
indicate probable paths of evolution.
The origin of eucaryotic cells is
discussed later in the text.

11 From Procaryotes to Eucaryotes

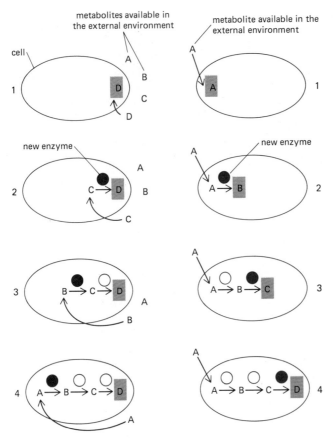

Figure 1–15 Schematic drawing showing two possible ways in which metabolic pathways might have evolved. The cell on the left is provided with a supply of related substances (A, B, C, and D) produced by prebiotic synthesis. One of these, substance D, is metabolically useful. As the cell exhausts the available supply of D, a selective advantage is obtained by the evolution of a new enzyme that is able to produce D from the closely related substance C. Fundamentally important metabolic pathways may have evolved by a series of similar steps. On the right, a metabolically useful compound A is available in abundance. An enzyme appears in the course of evolution that, by chance, has the ability to convert substance A to substance B. Other changes then occur within the cell that enable it to make use of the new substance. The appearance of further enzymes can build up a long chain of reactions.

cules had a strong selective advantage. In this way, the complement of enzymes possessed by cells is thought to have gradually increased, generating the metabolic pathways of present organisms. Two plausible ways in which a metabolic pathway could arise in evolution are illustrated in Figure 1–15.

If metabolic pathways evolved by the sequential addition of new enzymatic reactions to existing ones, the most ancient reactions should, like the oldest rings in a tree trunk, be closest to the center of the "metabolic tree," where the most fundamental of the basic molecular building blocks are synthesized. This position in metabolism is firmly occupied by the chemical processes that involve sugar phosphates, among which the most central of all is probably the sequence of reactions known as **glycolysis,** by which glucose can be degraded in the absence of oxygen (that is, *anaerobically*). The oldest metabolic pathways would have had to be anaerobic, because there was no free oxygen in the atmosphere of the primitive earth. Glycolysis occurs in virtually every living cell and drives the formation of the compound *adenosine triphosphate,* or *ATP,* which is used by all cells as a versatile source of chemical energy.

Linked to these core reactions of sugar phosphates are hundreds of other chemical processes. Some of these are responsible for the synthesis of small molecules, many of which in turn are utilized in further reactions to make the large polymers specific to the organism. Other reactions are used to degrade complex molecules, taken in as food, into simpler chemical units. One of the most striking features of these metabolic reactions is that they take place similarly in all kinds of organisms. Certainly differences exist: many specialized products of metabolism are restricted to certain genera or species. Even the common amino acid lysine is made in different ways in bacteria, in yeast, and in green plants and is not made at all in higher animals. But in broad terms the majority of reactions and most of the enzymes that catalyze them are found in all living things, from bacteria to people; for this reason they are believed to have been present in the primitive ancestral cells that gave rise to all these organisms.

The enzymes that catalyze the fundamental metabolic reactions, while continuing to serve the same essential functions, have undergone progressive modifications as organisms have evolved into divergent forms. For this reason, the

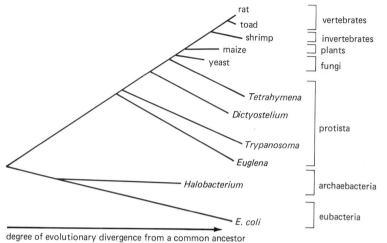

Figure 1–16 Evolutionary relationships of organisms deduced from the nucleotide sequences of their small-subunit ribosomal RNA genes. These genes contain highly conserved sequences, which change so slowly that they can be used to measure phylogenetic relationships spanning the entire range of living organisms. The data suggest that the plant, animal, and fungal lineages diverged from a common ancestor relatively late in the history of eucaryotic cells. (Adapted from M.L. Sogin, H.J. Elwood, and J.H. Gunderson, *Proc. Natl. Acad. Sci. USA* 83:1383–1387, 1986.)

amino acid sequence of the same type of enzyme in different living species provides a valuable indication of the evolutionary relationship between these species. The evidence obtained closely parallels that from other sources, such as the fossil record. An even richer source of information is locked in the living cell in the sequences of nucleotides in DNA, and modern methods of analysis allow these *DNA sequences* to be determined in large numbers and compared between species. Comparisons of highly conserved sequences, which have a central function and therefore change only slowly during evolution, can reveal relationships between organisms that diverged long ago (Figure 1–16), while very rapidly evolving sequences can be used to determine how more closely related species evolved (Figure 1–17). It is expected that continued application of these methods will enable the course of evolution to be followed with unprecedented accuracy.

Figure 1–17 Is a human being more closely related to the gorilla or to the orangutan? The question can be answered by comparing their DNA sequences, which gives the pedigree shown in (A). Mitochondrial DNA is usually chosen for such comparisons between closely related organisms, because this DNA undergoes evolutionary change about 5 to 10 times more rapidly than the bulk of the DNA in the cell (the nuclear DNA). (B) The first 75 nucleotides of the same gene (a mitochondrial gene coding for a subunit of the enzyme NAD dehydrogenase) are shown for each species. A colored letter marks each point where the gorilla or orangutan differs from humans. The boxes beneath the nucleotide sequences symbolize the amino acids in the corresponding proteins. The apes' amino acids are identified in color where they differ from those of the human. Analyzing additional DNA sequences in this way, one finds that the gorilla sequence differs from that of humans by 10% of its nucleotides, whereas orangutans differ both from humans and from gorillas by 17%. Assuming that these differences are due to mutations occurring at random and with a similar frequency in the lineage of each species, one can deduce the pedigree shown in (A). (Data from W.M. Brown, E.M. Prager, A. Wang, and A.C. Wilson, *J. Mol. Biol.* 18:225–239, 1982.)

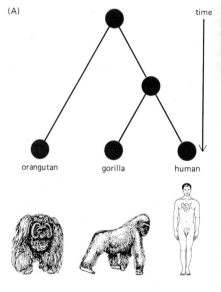

(B)

HUMAN

ATA ACC ATG CAC ACT ACT ATA ACC ACC CTA ACC CTG ACT TCC CTA ATT CCC CCC ATC CTT ACC ACC CTC GTT AAC

GORILLA

ATA ACT ATG TAC GCT ACC ATA ACC ACC TTA GCC CTA ACT TTC TTA ATT CCC CCT ATC CTT ACC ACC TTC ATC AAT

ORANGUTAN

ACA GCC ATG TTT ACC ACC ATA ACT GCC CTC ACC TTA ACT TCC CTA ATC CCC CCC ATT ACC GCT ACC CTC ATT AAC

Cyanobacteria Can Fix CO_2 and N_2[9]

If the earliest metabolic steps evolved to fill the gaps in the supply of organic molecules from earlier prebiotic synthesis, what happened when such compounds were exhausted? A strong selective advantage would then belong to those organisms able to utilize carbon and nitrogen atoms (in the form of CO_2 and N_2) from the atmosphere. But while they are abundantly available, CO_2 and N_2 are also very stable. It therefore requires a large amount of energy as well as a number of complicated chemical reactions to convert them to a usable form—that is, into organic molecules such as simple sugars.

In the case of CO_2, the major mechanism that evolved to achieve this transformation was **photosynthesis,** in which radiant energy captured from the sun drives the conversion of CO_2 into organic compounds. The interaction of sunlight with a pigment molecule, *chlorophyll,* excites an electron to a more highly energized state. As the electron drops back to a lower energy level, the energy it gives up drives chemical reactions that are facilitated and directed by protein molecules.

One of the first sunlight-driven reactions was probably the phosphorylation of nucleotides to form the high-energy compound ATP. Another would have been the generation of "reducing power." The carbon and nitrogen atoms in atmospheric CO_2 and N_2 are in an oxidized and inert state. One way to make them more reactive, so that they participate in biosynthetic reactions, is to reduce them—that is, to give them a larger number of electrons. Electrons are removed from a poor electron donor and transferred to a strong electron donor by chlorophyll in a reaction that requires light; in the process of reduction, the strong electron donor is then used to reduce CO_2 or N_2. Comparison of the mechanisms of photosynthesis in various present-day bacteria suggests that one of the first sources of electrons was H_2S, from which the primary waste product would have been elemental sulfur. Later the more difficult but ultimately more rewarding process of obtaining electrons from H_2O was accomplished, and O_2 was released in large amounts as a waste product.

Cyanobacteria (also known as blue-green algae) are today a major route by which both carbon and nitrogen are converted into organic molecules and thus enter the *biosphere.* They include the most self-sufficient organisms that now exist. Able to "fix" both CO_2 and N_2 into organic molecules, they are, to a first approximation, able to live on water, air, and sunlight alone; the mechanisms by which they do this have probably remained essentially constant for several billion years. Together with other bacteria that have some of these capabilities, they created the conditions in which more complex types of organisms could evolve: once one set of organisms had succeeded in synthesizing the whole gamut of organic cell components from inorganic raw materials, other organisms could subsist by feeding on the primary synthesizers and on their products.

Bacteria Can Carry Out the Aerobic Oxidation of Food Molecules[9]

Many people today are justly concerned about the environmental consequences of human activities. But in the past other organisms have caused revolutionary changes in the earth's environment (although very much more slowly). Nowhere is this more apparent than in the composition of the earth's atmosphere, which through oxygen-releasing photosynthesis was transformed from a mixture containing practically no molecular oxygen to one in which oxygen constitutes 21% of the total.

Since oxygen is an extremely reactive chemical that can interact with most cytoplasmic constituents, it must have been toxic to many early organisms, just as it is to many present-day anaerobic bacteria. However, this reactivity also provides a source of chemical energy, and, not surprisingly, this has been exploited by organisms during the course of evolution. By using oxygen, organisms are able to oxidize more completely the molecules they ingest. For example, in the absence of oxygen, glucose can be broken down only to lactic acid or ethanol, the end products of anaerobic glycolysis. But in the presence of oxygen, glucose can be completely degraded to CO_2 and H_2O. In this way, much more energy can be derived from each gram of glucose. The energy released in **respiration**—the aerobic oxidation of food molecules—is used to drive the synthesis of ATP in much the same way that photosynthetic organisms produce ATP from the energy of sunlight. In both processes there is a series of electron-transfer reactions that generates an H^+ gradient between the outside and inside of a membrane-bounded compartment; the H^+ gradient then serves to drive the synthesis of the ATP. Today, respiration is used by the great majority of organisms, including most procaryotes.

Eucaryotic Cells Contain Several Distinctive Organelles[10]

As molecular oxygen accumulated in the atmosphere, what happened to the remaining anaerobic organisms with which life had begun? In a world that was rich in oxygen, which they could not use, they were at a severe disadvantage. Some, no doubt, became extinct. Others either developed a capacity for respiration or found niches in which oxygen was largely absent, where they could continue an anaerobic way of life. It seems, however, that a third class discovered a strategy for survival more cunning and vastly richer in implications for the future: they are believed to have formed an intimate association with an aerobic type of cell, living with it in *symbiosis*. This is the most plausible explanation for the metabolic organization of present-day cells of the **eucaryotic** type (Panel 1–1, pp. 16–17), with which this book will be chiefly concerned.

Eucaryotic cells, by definition and in contrast to procaryotic cells, have a *nucleus* ("caryon" in Greek), which contains most of the cell's DNA, enclosed by a double layer of membrane (Figure 1–18). The DNA is thereby kept in a compartment separate from the rest of the contents of the cell, the **cytoplasm,** where most of the cell's metabolic reactions occur. In the cytoplasm, moreover, many distinctive *organelles* can be recognized. Prominent among these are two types of small bodies, the *chloroplasts* and *mitochondria* (Figures 1–19 and 1–20). Each of these is enclosed in its own double layer of membrane, which is chemically different from the membrane surrounding the nucleus. Mitochondria are an almost universal feature of eucaryotic cells, whereas chloroplasts are found only in those eucaryotic cells that are capable of photosynthesis—that is, in plants but not in animals or fungi. Both organelles are thought to have a symbiotic origin.

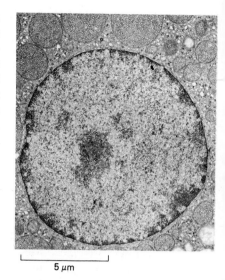

Figure 1–18 The nucleus contains most of the DNA of the eucaryotic cell. It is seen here in a thin section of a mammalian cell examined in the electron microscope. (Courtesy of Daniel S. Friend.)

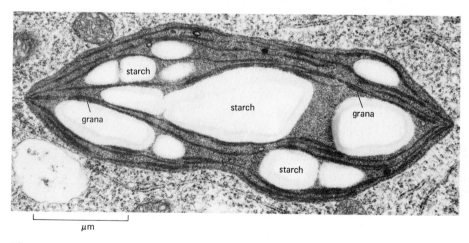

Figure 1–19 Electron micrograph of a chloroplast in a moss cell showing its extensive system of internal membranes. The flattened sacs of membrane contain chlorophyll and are arranged in stacks, or *grana*. This chloroplast also contains large accumulations of starch. (Courtesy of J. Burgess.)

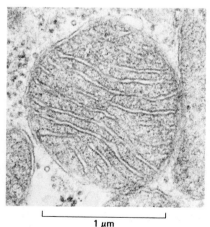

Figure 1–20 Mitochondria carry out the oxidative degradation of nutrient molecules in all eucaryotic cells. As seen in this electron micrograph, they possess a smooth outer membrane and a highly convoluted inner membrane. (Courtesy of Daniel S. Friend.)

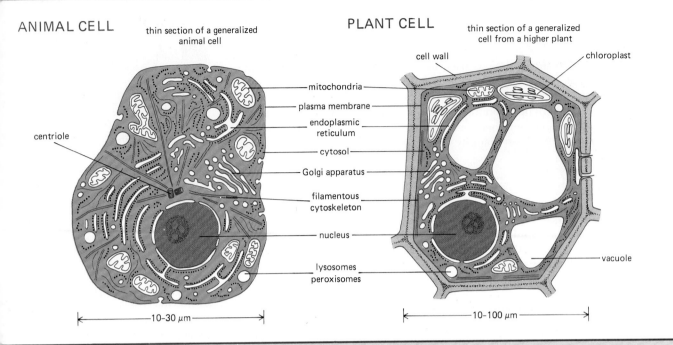

ANIMAL CELL

thin section of a generalized animal cell

centriole

PLANT CELL

thin section of a generalized cell from a higher plant

cell wall

chloroplast

mitochondria
plasma membrane
endoplasmic reticulum
cytosol
Golgi apparatus
filamentous cytoskeleton
nucleus
lysosomes
peroxisomes

vacuole

10-30 µm

10-100 µm

THE MEMBRANE SYSTEM OF THE CELL

PLASMA MEMBRANE

The outer boundary of the cell is the plasma membrane, a continuous sheet of phospholipid molecules about 4-5 nm thick in which various proteins are embedded.

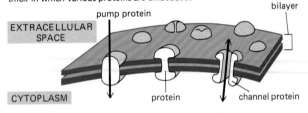

lipid bilayer

pump protein

EXTRACELLULAR SPACE

CYTOPLASM

protein

channel protein

Some of these proteins serve as pumps and channels for transporting specific molecules into and out of the cell.

GOLGI APPARATUS

A system of stacked, membrane-bounded, flattened sacs involved in modifying, sorting, and packaging macromolecules for secretion or for delivery to other organelles.

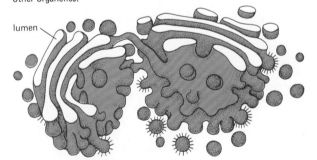

lumen

Around the Golgi apparatus are numerous small membrane-bounded vesicles (50 nm and larger). These are thought to carry material between the Golgi apparatus and different compartments of the cell.

ENDOPLASMIC RETICULUM

Flattened sheets, sacs, and tubes of membrane extend throughout the cytoplasm of eucaryotic cells, enclosing a large intracellular space. The ER membrane is in structural continuity with the outer membrane of the nuclear envelope and it specializes in the synthesis and transport of lipids and membrane proteins.

The rough endoplasmic reticulum (rough ER) generally occurs as flattened sheets and is studded on its outer face with ribosomes engaged in protein synthesis.

ribosomes

nucleus

lumen

The smooth endoplasmic reticulum (smooth ER) is generally more tubular and lacks attached ribosomes. A major function is in lipid metabolism.

lumen

LYSOSOMES

membrane-bounded vesicles that contain hydrolytic enzymes involved in intracellular digestions

0.2-0.5 µm

PEROXISOMES

membrane-bounded vesicles containing oxidative enzymes that generate and destroy hydrogen peroxide

0.2-0.5 µm

NUCLEUS

The nucleus is the most conspicuous organelle in the cell. It is separated from the cytoplasm by an envelope consisting of two membranes. All of the chromosomal DNA is held in the nucleus, packaged into chromatin fibers by its association with an equal mass of histone proteins. The nuclear contents communicate with the cytosol by means of openings in the nuclear envelope called nuclear pores.

3-10 μm

nuclear pores

nucleolus: a factory in the nucleus where the cell's ribosomes are assembled

chromatin

nuclear envelope
— inner membrane
— outer membrane

CYTOSKELETON

In the cytosol, arrays of protein filaments form networks that give the cell its shape and provide a basis for its movements. In animal cells the cytoskeleton is often organized from an area near the nucleus that contains the cell's pair of centrioles. Three main kinds of cytoskeletal filaments are:

1. microtubules

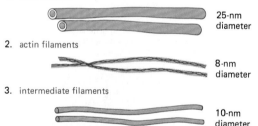

25-nm diameter

2. actin filaments

8-nm diameter

3. intermediate filaments

10-nm diameter

MITOCHONDRIA

About the size of bacteria, mitochondria are the power plants of all eucaryotic cells, harnessing energy obtained by combining oxygen with food molecules to make ATP.

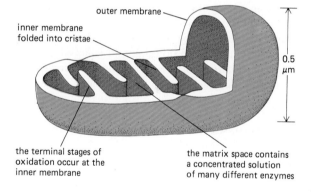

outer membrane

inner membrane folded into cristae

0.5 μm

the terminal stages of oxidation occur at the inner membrane

the matrix space contains a concentrated solution of many different enzymes

SPECIAL PLANT CELL ORGANELLES

chloroplasts—These chlorophyll-containing plastids are double-membrane-bounded organelles found in all higher plants. An elaborate membrane system in the interior of the chloroplast contains the photosynthetic apparatus.

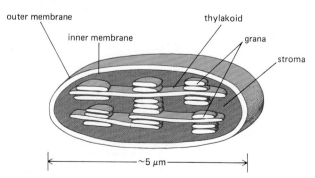

outer membrane

inner membrane

thylakoid

grana

stroma

~5 μm

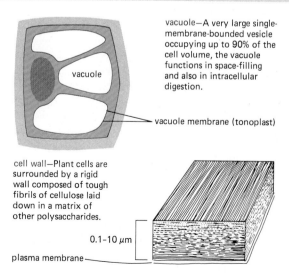

vacuole—A very large single-membrane-bounded vesicle occupying up to 90% of the cell volume, the vacuole functions in space-filling and also in intracellular digestion.

vacuole

vacuole membrane (tonoplast)

cell wall—Plant cells are surrounded by a rigid wall composed of tough fibrils of cellulose laid down in a matrix of other polysaccharides.

0.1-10 μm

plasma membrane

Eucaryotic Cells Depend on Mitochondria for Their Oxidative Metabolism[11]

Mitochondria show many similarities to free-living procaryotic organisms: for example, they often resemble bacteria in size and shape, they contain DNA, they make protein, and they reproduce by dividing in two. By breaking up eucaryotic cells and separating their component parts, it is possible to show that mitochondria are responsible for respiration and that this process occurs nowhere else in the eucaryotic cell. Without mitochondria, the cells of animals and fungi would be anaerobic organisms, depending on the relatively inefficient and antique process of glycolysis for their energy. Many present-day bacteria respire like mitochondria, and it seems probable that eucaryotic cells are descendants of primitive anaerobic organisms that survived, in a world that had become rich in oxygen, by engulfing aerobic bacteria—keeping them in symbiosis for the sake of their capacity to consume atmospheric oxygen and produce energy. Certain present-day microorganisms offer strong evidence of the feasibility of such an evolutionary sequence. There are several hundred species of single-celled eucaryotes that resemble the hypothetical ancestral eucaryote in that they live in oxygen-poor conditions (in the guts of animals, for example) and lack mitochondria altogether. Recently, comparative nucleotide sequence analyses have suggested that one group of these organisms, the *microsporidia*, may have diverged very early from the line leading to other eucaryotic cells. There is another eucaryote, the amoeba *Pelomyxa palustris*, that, while lacking mitochondria, nevertheless carries out oxidative metabolism by harboring aerobic bacteria in its cytoplasm in a permanent symbiotic relationship. Microsporidia and *Pelomyxa* therefore resemble two proposed stages in the evolution of eucaryotes such as ourselves.

Acquisition of mitochondria must have had many repercussions. The plasma membrane, for example, is heavily committed to energy metabolism in procaryotic cells but not in eucaryotic cells, where this crucial function has been relegated to the mitochondria. It seems likely that the separation of functions left the eucaryotic plasma membrane free to evolve important new features. In particular, because eucaryotic cells need not maintain a large H^+ gradient across their plasma membrane, as required for ATP production in procaryotes, it became possible to use controlled changes in the ion permeability of the plasma membrane for cell signaling purposes. Thus, at about the time that eucaryotes arose, a variety of ion channels appeared in the plasma membrane. Today these channels mediate the elaborate electrical signaling processes in higher organisms—notably in the nervous system—and they control much of the behavior of single-celled free-living eucaryotes such as protozoa (see below).

Chloroplasts Are the Descendants of an Engulfed Procaryotic Cell

Chloroplasts carry out photosynthesis in much the same way as procaryotic cyanobacteria, absorbing sunlight in the chlorophyll attached to their membranes. Some bear a close structural resemblance to the cyanobacteria, being similar in size and in the way that their chlorophyll-bearing membranes are stacked in layers (Figure 1–20). Moreover, chloroplasts reproduce by dividing, and they contain DNA that is nearly indistinguishable in nucleotide sequence from portions of a bacterial chromosome. All this strongly suggests that chloroplasts share a common ancestry with cyanobacteria and evolved from procaryotes that made their home inside eucaryotic cells. These procaryotes performed photosynthesis for their hosts in return for the sheltered and nourishing environment that their hosts provided for them. Symbiosis of photosynthetic cells with other cell types is, in fact, a common phenomenon, and some present-day eucaryotic cells contain authentic cyanobacteria (Figure 1–21).

Figure 1–22 shows the evolutionary origins of the eucaryotes according to the symbiotic theory. It must be stressed, however, that mitochondria and chloroplasts show important differences from, as well as similarities to, present-day aerobic bacteria and cyanobacteria. Their quantity of DNA is very small, for example, and most of the molecules from which they are constructed are synthesized elsewhere in the eucaryotic cell and imported into the organelle. Although they seem to

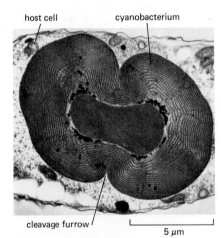

host cell   cyanobacterium

cleavage furrow

5 μm

Figure 1–21 A close relative of present-day cyanobacteria that lives in a permanent symbiotic relationship inside another cell (the two organisms are known jointly as *Cyanophora paradoxa*). The "cyanobacterium" is in the process of dividing. (Courtesy of Jeremy D. Pickett-Heaps.)

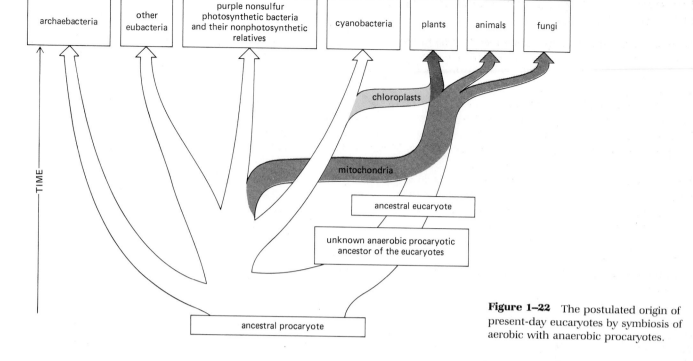

Figure 1-22 The postulated origin of present-day eucaryotes by symbiosis of aerobic with anaerobic procaryotes.

have originated as symbiotic bacteria, they have undergone large evolutionary changes and have become greatly dependent on their host cells.

The major existing eucaryotes have in common both mitochondria and a whole constellation of other features that distinguish them from procaryotes (Table 1-1). These function together to give eucaryotic cells a wealth of different capabilities, and one can only speculate as to which of them evolved first. But the acquisition of mitochondria by an anaerobic eucaryotic cell must have been a crucial step in the success of the eucaryotes, providing them with the means to tap an abundant source of energy to drive all their complex activities.

Table 1-1 Comparison of Procaryotic and Eucaryotic Organisms

	Procaryotes	**Eucaryotes**
Organisms	bacteria and cyanobacteria	protists, fungi, plants, and animals
Cell size	generally 1 to 10 μm in linear dimension	generally 10 to 100 μm in linear dimension
Metabolism	anaerobic or aerobic	aerobic
Organelles	few or none	nucleus, mitochondria, chloroplasts, endoplasmic reticulum, etc.
DNA	circular DNA in cytoplasm	very long linear DNA molecules containing many noncoding regions; bounded by nuclear envelope
RNA and protein	RNA and protein synthesized in same compartment	RNA synthesized and processed in nucleus; proteins synthesized in cytoplasm
Cytoplasm	no cytoskeleton: cytoplasmic streaming, endocytosis, and exocytosis all absent	cytoskeleton composed of protein filaments; cytoplasmic streaming; endocytosis and exocytosis
Cell division	chromosomes pulled apart by attachments to plasma membrane	chromosomes pulled apart by cytoskeletal spindle apparatus
Cellular organization	mainly unicellular	mainly multicellular, with differentiation of many cell types

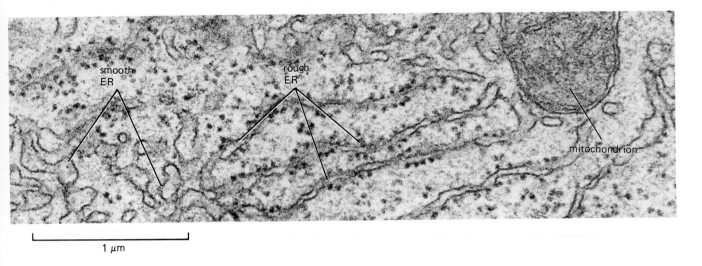

1 μm

Eucaryotic Cells Contain a Rich Array of Internal Membranes

Eucaryotic cells are usually much larger in volume than procaryotic cells, commonly by a factor of a thousand or more, and they carry a proportionately larger quantity of most cellular materials; for example, a human cell contains about one thousand times as much DNA as a typical bacterium. This large size creates problems. Since all the raw materials for the biosynthetic reactions occurring in the interior of a cell must ultimately enter and leave by passing through the plasma membrane covering its surface, and since the membrane is also the site of many important reactions, an increase in cell volume requires an increase in cell surface. But it is a fact of geometry that simply scaling up a structure increases the volume as the cube of the linear dimension while the surface area increases only as the square. Therefore, if the large eucaryotic cell is to keep as high a ratio of surface to volume as the procaryotic cell, it must supplement its surface area by means of convolutions, infoldings, and other elaborations of its membrane.

This probably explains in part the complex profusion of **internal membranes** that is a basic feature of all eucaryotic cells. Membranes surround the nucleus, the mitochondria, and (in plant cells) the chloroplasts. They form a labyrinthine compartment called the **endoplasmic reticulum** (Figure 1–23), where lipids and proteins of cell membranes, as well as materials destined for export from the cell, are synthesized. They also form stacks of flattened sacs constituting the **Golgi apparatus** (Figure 1–24), which is involved in the modification and transport of the molecules made in the endoplasmic reticulum. Membranes surround **lysosomes,** which contain stores of enzymes required for intracellular digestion and so prevent them from attacking the proteins and nucleic acids elsewhere in the cell. In the same way, membranes surround **peroxisomes,** where dangerously reactive hydrogen peroxide is generated and degraded during the oxidation of various molecules by O_2. Membranes also form small vesicles and, in plants, a large liquid-filled *vacuole.* All these membrane-bounded structures correspond to distinct internal compartments within the cytoplasm. In a typical animal cell, these compartments (or organelles) occupy nearly half the total cell volume. The remaining compartment of the cytoplasm, which includes everything other than the membrane-bounded organelles, is usually referred to as the **cytosol.**

All of the aforementioned membranous structures lie in the interior of the cell. How, then, can they help to solve the problem we posed at the outset and provide the cell with a surface area that is adequate to its large volume? The answer is that there is a continual exchange between the internal membrane-bounded compartments and the outside of the cell, achieved by *endocytosis* and *exocytosis,* processes unique to eucaryotic cells. In endocytosis, portions of the external surface membrane invaginate and pinch off to form membrane-bounded cytoplasmic vesicles that contain both substances present in the external medium and molecules previously adsorbed on the cell surface. Very large particles or even entire foreign cells can be taken up by *phagocytosis*—a special form of endocytosis.

Figure 1–23 Electron micrograph of a thin section of a mammalian cell showing both smooth and rough regions of the endoplasmic reticulum (ER). The smooth regions are involved in lipid metabolism; the rough regions, studded with ribosomes, are sites of synthesis of proteins that are destined to leave the cytosol and enter certain other compartments of the cell. (Courtesy of George Palade.)

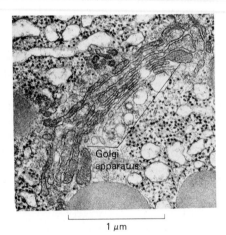

1 μm

Figure 1–24 Electron micrograph of a thin section of a mammalian cell showing the Golgi apparatus, which is composed of flattened sacs of membrane arranged in multiple layers (see also Panel 1–1, pp. 16–17). The Golgi apparatus is involved in the synthesis and packaging of molecules destined to be secreted from the cell, as well as in the routing of newly synthesized proteins to the correct cellular compartment. (Courtesy of Daniel S. Friend.)

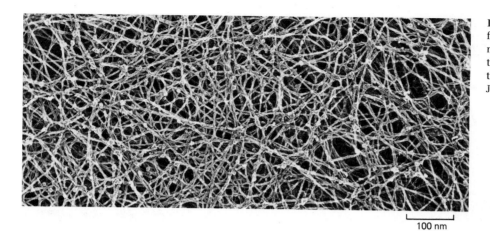

Figure 1–25 A network of actin filaments underlying the plasma membrane of an animal cell is seen in this electron micrograph prepared by the deep-etch technique. (Courtesy of John Heuser.)

100 nm

Exocytosis is the reverse process, whereby membrane-bounded vesicles inside the cell fuse with the plasma membrane and release their contents into the external medium. In this way, membranes surrounding compartments deep inside the cell serve to increase the effective surface area of the cell for exchanges of matter with the external world.

As we shall see in later chapters, the various membranes and membrane-bounded compartments in eucaryotic cells have become highly specialized, some for secretion, some for absorption, some for specific biosynthetic processes, and so on.

Eucaryotic Cells Have a Cytoskeleton

The larger a cell is, and the more elaborate and specialized its internal structures, the greater is its need to keep these structures in their proper places and to control their movements. All eucaryotic cells have an internal skeleton, the **cytoskeleton,** that gives the cell its shape, its capacity to move, and its ability to arrange its organelles and transport them from one part of the cell to another. The cytoskeleton is composed of a network of protein filaments, two of the most important of which are *actin filaments* and *microtubules* (Figure 1–25). These two must date from a very early epoch in evolution since they are found almost unchanged in all eucaryotes. Both are involved in the generation of cellular movements; actin filaments, for example, participate in the contraction of muscle, whereas microtubules are the main structural and force-generating elements in *cilia* and *flagella*—the long projections on some cell surfaces that beat like whips and serve as instruments of propulsion.

Actin filaments and microtubules are also essential for the internal movements that occur in the cytoplasm of all eucaryotic cells. Thus microtubules in the form of a *mitotic spindle* are a vital part of the usual machinery for partitioning DNA equally between the two daughter cells when a eucaryotic cell divides. Without microtubules, therefore, the eucaryotic cell could not reproduce. In this and other examples, movement by free diffusion would be either too slow or too haphazard to be useful. In fact, the organelles in a eucaryotic cell often appear to be attached, directly or indirectly, to the cytoskeleton and, when they move, to be propelled along cytoskeletal tracks.

Protozoa Include the Most Complex Cells Known[12]

The complexity that can be achieved by a single eucaryotic cell is nowhere better illustrated than in *protists* (Figure 1–26). These are free-living, single-celled eucaryotes that are evolutionarily diverse (see Figure 1–16) and exhibit a bewildering variety of different forms and behaviors: they can be photosynthetic or carnivorous, motile or sedentary. Their anatomy is often complex and includes such structures as sensory bristles, photoreceptors, flagella, leglike appendages, mouth parts, stinging darts, and musclelike contractile bundles. Although they are single cells, they can

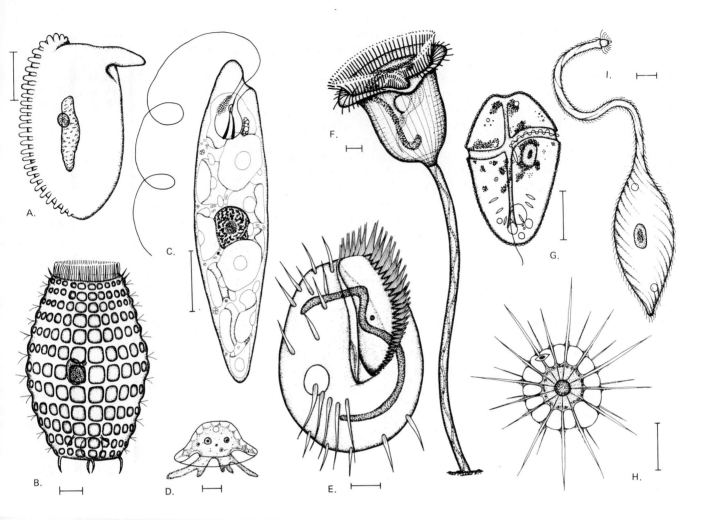

be as intricate and versatile as many multicellular organisms. This is particularly true of the group of protists known as **protozoa**—or "first animals."

Didinium is a carnivorous protozoan. It has a globular body, about 150 μm in diameter, encircled by two fringes of cilia; its front end is flattened except for a single protrusion rather like a snout (Figure 1–27). *Didinium* swims around in the water at high speed by means of the synchronous beating of its cilia. When it encounters a suitable prey, usually another type of protozoan, such as a *Paramecium*, it releases numerous small paralyzing darts from its snout region. Then the *Didinium* attaches to and devours the *Paramecium*, inverting like a hollow ball to engulf the other cell, which is as large as itself. Most of this complex behavior—swimming, and paralyzing and capturing its prey—is generated by the cytoskeletal structures lying just beneath the plasma membrane. Included in this *cell cortex*, for example, are the parallel bundles of microtubules that form the core of each cilium and enable it to beat.

Predatory behavior of this sort and the set of features on which it depends—large size, the capacity for phagocytosis, and the ability to move in pursuit of prey—are peculiar to eucaryotes. Indeed, it is probable that these features came very early in eucaryotic evolution, making possible the subsequent capture of bacteria for domestication as mitochondria and chloroplasts.

Genes Can Be Switched On and Off

The protozoa, for all their marvels, do not represent the peak of eucaryotic evolution. Greater things were achieved, not by concentrating every sort of complexity in a single cell, but by dividing the labor among different types of cells. *Multicellular organisms* evolved in which cells closely related by ancestry became differentiated from one another, some developing one feature to a high degree, others another, so forming the specialized parts of one great cooperative enterprise.

Figure 1–26 An assortment of protists, illustrating some of the enormous variety to be found among this class of single-celled organisms. These drawings are done to different scales, but in each case the bar denotes 10 μm. The organisms in (A), (B), (E), (F), and (I) are ciliates; (C) is an euglenoid; (D) is an amoeba; (G) is a dinoflagellate; (H) is a heliozoan. (From M.A. Sleigh, The Biology of Protozoa. London: Edward Arnold, 1973.)

The various specialized cell types in a single higher plant or animal often appear radically different (Panel 1–2, pp. 24–25). This seems paradoxical, since all of the cells in a multicellular organism are closely related, having recently descended from the same precursor cell—the fertilized egg. Common lineage implies similar genes; how then do the differences arise? In a few cases, cell specialization involves the loss of genetic material. An extreme example is the mammalian red blood cell, which loses its entire nucleus in the course of differentiation. But the overwhelming majority of cells in most plant and animal species retain all of the genetic information contained in the fertilized egg. Specialization depends on changes in *gene expression*, not on the loss or acquisition of genes.

Even bacteria do not make all of their types of protein all of the time but are able to adjust the level of synthesis according to external conditions. Proteins required specifically for the metabolism of lactose, for example, are made by many bacteria only when this sugar is available for use; and when conditions are unfavorable for cell proliferation, some bacteria arrest most of their normal metabolic processes and form *spores*, which have tough, impermeable outer walls and a cytoplasm of altered composition.

Eucaryotic cells have evolved far more sophisticated mechanisms for controlling gene expression, and these affect entire systems of interacting gene products. Groups of genes are activated or repressed in response to both external and internal signals. Membrane composition, cytoskeleton, secretory products, even metabolism—all these and other features must change in a coordinated manner when cells become differentiated. Compare, for example, a skeletal muscle cell specialized for contraction with an *osteoblast*, which secretes the hard matrix of bone in the same animal (Panel 1–2, pp. 24–25). Such radical transformations of cell character reflect stable changes in gene expression. The controls that bring about these changes have evolved in eucaryotes to a degree unmatched in procaryotes.

Eucaryotic Cells Have Vastly More DNA Than They Need for the Specification of Proteins

Eucaryotic cells contain a very large quantity of DNA. In human cells, as we have said, there is about a thousand times more DNA than in typical bacteria, while the cells of some amphibians have a DNA content more than 10 times that of humans (Figure 1–28). Yet it seems that only a small fraction of this DNA—perhaps 1% in human cells—carries the specifications for proteins that are actually made by the organism. Why then is the remaining 99% of the DNA there? One hypothesis is that much of it acts merely to increase the physical bulk of the nucleus. Another is that it is in large part parasitic—a collection of DNA sequences that have over the ages accumulated in the cell, exploiting the cell's machinery for their own reproduction and bringing no immediate benefit in return. Indeed, the DNA of many species has been shown to contain sequences called *transposable elements*, which have the ability to "jump" occasionally from one location to another in the DNA and even to insert additional copies of themselves at new sites. Transposable elements could thus proliferate like a slow infection, becoming an ever larger proportion of the genetic material.

But evolution is opportunistic. Whatever the origins of the DNA that does not code for protein, it is certain that it now has some important functions. Part of this DNA is structural, enabling portions of the genetic material to become condensed or "packaged" in specific ways, as described in the next section, and some of the DNA is regulatory and helps to switch on and off the genes that direct the synthesis of proteins, thus playing a crucial role in the sophisticated control of gene expression in eucaryotic cells.

In Eucaryotic Cells the Genetic Material Is Packaged in Complex Ways

The length of DNA in eucaryotic cells is so great that the risk of entanglement and breakage becomes severe. Probably for this reason, proteins unique to eucaryotes, the *histones*, have evolved to bind to the DNA and wrap it up into compact and

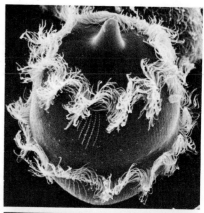
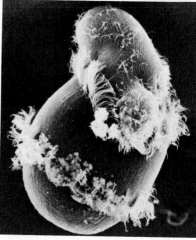

100 μm

Figure 1–27 Scanning electron micrographs showing one protozoan eating another. Protozoans are single-cell animals that show an amazing diversity of form and behavior. *Didinium* (*above*), a ciliated protozoan, has two circumferential rings of motile cilia and a snoutlike protuberance at its leading end, with which it captures its prey. Below, *Didinium* is shown engulfing another protozoan, *Paramecium*. (Courtesy of D. Barlow.)

CELL TYPES

There are over 200 different types of cell in the human body. These are assembled into a variety of different types of tissue such as

    epithelia

    connective tissue

    muscle

    nervous tissue

Most tissues contain a mixture of cell types.

EPITHELIA

Epithelial cells form coherent cell sheets called epithelia, which line the inner and outer surfaces of the body. There are many specialized types of epithelia.

Absorptive cells have numerous hairlike microvilli projecting from their free surface to increase the area for absorption.

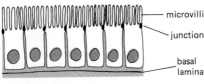

Adjacent epithelial cells are bound together by junctions that give the sheet mechanical strength and also make it impermeable to small molecules. The sheet rests on a basal lamina.

Ciliated cells have cilia on their free surface that beat in synchrony to move substances (such as mucus) over the epithelial sheet.

Secretory cells — most epithelial layers contain some specialized cells that secrete substances onto the surface of the cell sheet.

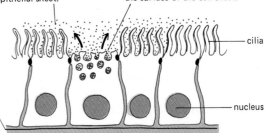

CONNECTIVE TISSUE

The spaces between organs and tissues in the body are filled with connective tissue made principally of a network of tough protein fibers embedded in a polysaccharide gel. This extracellular matrix is secreted mainly by fibroblasts.

fibroblasts in loose connective tissue

Two main types of extracellular protein fiber are collagen and elastin.

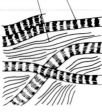

Bone is made by cells called osteoblasts. These secrete an extracellular matrix in which crystals of calcium phosphate are later deposited.

Calcium salts are deposited in the extracellular matrix.

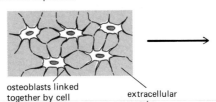

osteoblasts linked together by cell processes

extracellular matrix

Adipose cells are among the largest cells in the body. These cells are responsible for the production and storage of fat. The nucleus and cytoplasm are squeezed to the cell periphery by a large lipid droplet.

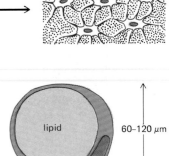

lipid

60–120 μm

NERVOUS TISSUE

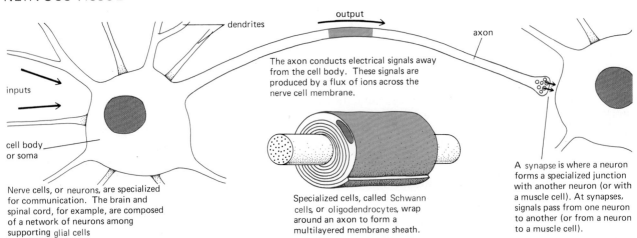

dendrites

output

axon

The axon conducts electrical signals away from the cell body. These signals are produced by a flux of ions across the nerve cell membrane.

inputs

cell body or soma

Nerve cells, or neurons, are specialized for communication. The brain and spinal cord, for example, are composed of a network of neurons among supporting glial cells

Specialized cells, called Schwann cells, or oligodendrocytes, wrap around an axon to form a multilayered membrane sheath.

A synapse is where a neuron forms a specialized junction with another neuron (or with a muscle cell). At synapses, signals pass from one neuron to another (or from a neuron to a muscle cell).

Secretory epithelial cells are often collected together to form a gland that specializes in the secretion of a particular substance. As illustrated, exocrine glands secrete their products (such as tears, mucus, and gastric juices) into ducts. Endocrine glands secrete hormones into the blood.

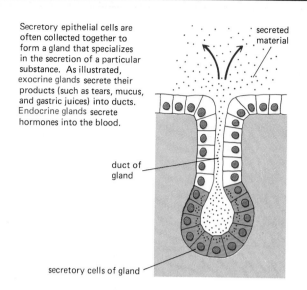

secreted material

duct of gland

secretory cells of gland

BLOOD

Erythrocytes (or red blood cells) are very small cells, usually with no nucleus or internal membranes, and are stuffed full of the oxygen-binding protein hemoglobin.

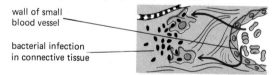

1 cm³ of blood contains 5,000,000,000 erythrocytes

Their normal shape is a biconcave disc.

White blood cells (leucocytes) protect against infections. Blood contains about one leucocyte for every 1000 red blood cells. Although leucocytes travel in the circulation, they can pass through the walls of blood vessels to do their work in the surrounding tissues.
There are several different kinds, including:
lymphocytes —responsible for immune responses such as the production of antibody and the rejection of tissue grafts.
macrophages and neutrophils —these cells move to sites of infection where they ingest bacteria and debris.

wall of small blood vessel

bacterial infection in connective tissue

GERM CELLS

Both sperm and egg are haploid i.e., they carry only one set of chromosomes. A sperm from the male fuses with an egg from the female, which then forms a new diploid organism by successive cell divisions.

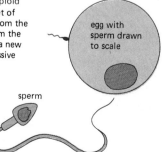

egg with sperm drawn to scale

sperm

MUSCLE

Muscle cells produce mechanical force by their contraction. In vertebrates there are three main types:

skeletal muscle — this moves joints by its strong and rapid contraction. Each muscle is a bundle of muscle fibers, each of which is an enormous multinucleated cell.

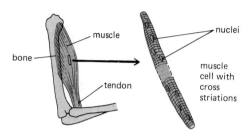

muscle

nuclei

bone

muscle cell with cross striations

tendon

smooth muscle — present in digestive tract, bladder, arteries, and veins, it is composed of thin elongated cells (not striated) each with a single nucleus.

cardiac muscle — intermediate in character between skeletal and smooth muscle, it produces the heart beat. Adjacent cells are linked by electrically conducting junctions that cause the cells to contract in synchrony.

SENSORY CELLS

Among the most strikingly specialized cells in the vertebrate body are those that detect external stimuli. Hair cells of the inner ear are primary detectors of sound. Modified epithelial cells, they carry special microvilli (stereocilia) on their surface. The movement of these in response to sound vibrations causes an electrical signal to pass to the brain.

stereocilia are very rigid because they are packed with actin filaments

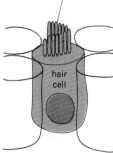

hair cell

Rod cells in the retina of the eye are specialized to respond to light. The photosensitive region contains many membranous discs in whose membranes the light-sensitive pigment rhodopsin is embedded. Light evokes an electrical signal, which is transmitted to nerve cells in the eye, which relay the signal to the brain.

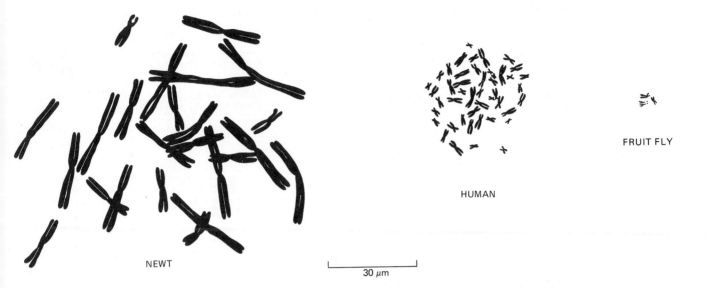

NEWT

HUMAN

FRUIT FLY

30 μm

manageable **chromosomes** (Figure 1–29). Tight packaging of the DNA in chromosomes is an essential part of the preparation for cell division in eucaryotes (Figure 1–30). All eucaryotes (with one minor exception) have histones bound to their DNA, and the importance of these proteins is reflected in their remarkable conservation in evolution: several of the histones of a pea plant are almost exactly the same, amino acid for amino acid, as those of a cow.

Many other proteins besides histones are bound to the DNA in eucaryotic cells. By altering the opportunities of the DNA to interact with other molecules, some of these DNA-binding proteins alter the patterns of gene expression from one type of specialized cell to another. For example, since genes contained in a tightly packed mass of DNA are not expressed, gene expression can be controlled through changes in the packaging of the DNA.

The membranes enclosing the nucleus in eucaryotic cells protect the structure of the DNA and its associated delicate control machinery, sheltering them from entanglement with the moving cytoskeleton and from many of the chemical changes that take place in the cytoplasm. They also allow the segregation of two crucial steps in gene expression: (1) the copying of DNA sequences into RNA sequences (*DNA transcription*) and (2) the use of these RNA sequences, in turn, to direct the synthesis of specific proteins (*RNA translation*). In procaryotic cells, there is no compartmentalization—the translation of RNA sequences into protein begins as soon as they are transcribed, even before their synthesis is completed. In eucaryotes, however (except in mitochondria and chloroplasts, which in this respect as in others are closer to bacteria), the two steps in the path from gene to protein are kept strictly separate: transcription occurs in the nucleus, translation in the cytoplasm. The RNA has to leave the nucleus before it can be used to guide protein synthesis. While in the nucleus, it undergoes elaborate changes in which some parts of the RNA molecule are discarded and other parts are modified (*RNA processing*).

Because of these complexities, the genetic material of a eucaryotic cell offers many more opportunities for control than are present in bacteria.

Summary

Present-day living cells are classified as procaryotic (bacteria and their close relatives) or eucaryotic. Although they have a relatively simple structure, procaryotic cells are biochemically versatile and diverse: for example, all of the major metabolic pathways can be found in bacteria, including the three principal energy-yielding processes of glycolysis, respiration, and photosynthesis. Eucaryotic cells are larger and more complex than procaryotic cells and contain more DNA, together with components that allow this DNA to be handled in elaborate ways. The DNA of the

Figure 1–28 A comparison of the chromosomes of a newt, a man, and a fruit fly, all drawn to the same scale. As illustrated by the total size of their chromosomes, man has more DNA per cell than the fruit fly, but less than the newt. Some plants, such as the lily, have more DNA than any animal. (Adapted from H.C. Macgregor, *Philos. Trans. R. Soc. Lond. (Biol.)* 283:309–318, 1978.)

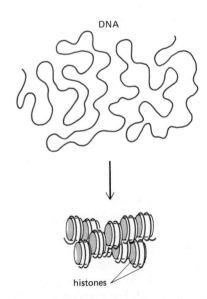

DNA

histones

Figure 1–29 Schematic illustration of how the positively charged proteins called histones mediate the folding of DNA in chromosomes.

eucaryotic cell is enclosed in a membrane-bounded nucleus, while the cytoplasm contains many other membrane-bounded organelles, including mitochondria, which carry out the oxidation of food molecules, and, in plant cells, chloroplasts, which carry out photosynthesis. Various lines of evidence suggest that mitochondria and chloroplasts are the descendants of earlier procaryotic cells that established themselves as internal symbionts of a larger anaerobic cell. Eucaryotic cells are also unique in containing a cytoskeleton of protein filaments that helps organize the cytoplasm and provides the machinery for movement.

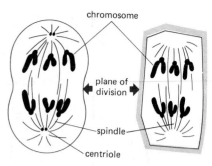

Figure 1–30 Schematic drawing of eucaryotic cells in mitosis. An animal cell is shown on the left and a plant cell on the right. The nuclear envelope has broken down, and the DNA, having replicated, has condensed into two complete sets of chromosomes. One set is distributed to each of the two newly forming cells by a mitotic spindle composed largely of microtubules.

From Single Cells to Multicellular Organisms[13]

Single-cell organisms, such as bacteria and protozoa, have been so successful in adapting to a variety of different environments that they comprise more than half of the total biomass on earth. Unlike higher animals, many of these unicellular organisms can synthesize all of the substances they need from a few simple nutrients, and some of them divide more than once every hour. What, then, was the selective advantage that led to the evolution of **multicellular organisms**?

The short answer is that multicellular organisms can exploit resources that no single cell could utilize so well. Multicellularity, for example, enables a tree to become physically large; to have roots in the ground, where one set of cells can take up water and nutrients; and to have leaves in the air, where another set of cells can efficiently capture radiant energy from the sun. Specialized cells in the trunk of the tree form channels for transporting water and nutrients between the roots and the leaves. Yet another set of specialized cells forms a layer of bark to prevent water loss and to provide a protected internal environment. The tree as a whole does not compete directly with unicellular organisms for its ecological niche; it has found a radically different way to survive and propagate.

As different animals and plants appeared, they changed the environment in which further evolution occurred. Survival in a jungle calls for different talents than survival in the open sea. Innovations in movement, sensory detection, communication, social organization—all enabled eucaryotic organisms to compete, propagate, and survive in ever more complex ways.

Single Cells Can Associate to Form Colonies

It seems likely that an early step in the evolution of multicellular organisms was the association of unicellular organisms to form colonies. The simplest way of achieving this is for daughter cells to remain associated after each cell division. Even some procaryotic cells show such social behavior in a primitive form. Myxobacteria, for example, live in the soil and feed on insoluble organic molecules that they break down by secreting degradative enzymes. They stay together in loose colonies in which the digestive enzymes secreted by individual cells are pooled, thus increasing the efficiency of feeding (the "wolf-pack" effect). These cells indeed represent a peak of sophistication among procaryotes, for when food supplies are exhausted, the cells aggregate tightly together and form a multicellular *fruiting body* (Figure 1–31), within which the bacteria differentiate into spores that can survive even in extremely hostile conditions. When conditions are more favorable, the spores in a fruiting body germinate to produce a new swarm of bacteria.

Green algae (not to be confused with the procaryotic "blue-green algae" or cyanobacteria) are eucaryotes that exist as unicellular, colonial, or multicellular forms (Figure 1–32). Different species of green algae can be arranged in order of complexity, illustrating the kind of progression that probably occurred in the evolution of higher plants and animals. Unicellular green algae, such as *Chlamydomonas*, resemble flagellated protozoa except that they possess chloroplasts, which enable them to carry out photosynthesis. In closely related genera, groups of flagellated cells live in colonies held together by a matrix of extracellular molecules secreted by the cells themselves. The simplest species (those of the genus *Gonium*) have the form of a concave disc made of 4, 8, 16, or 32 cells. Their flagella beat independently, but since they are all oriented in the same direction, they are able

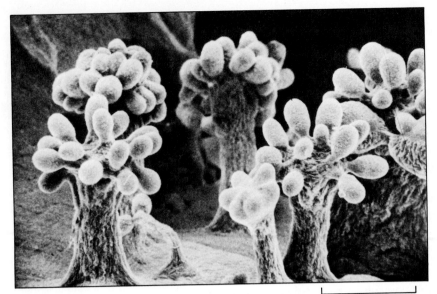

Figure 1–31 Scanning electron micrograph of the fruiting bodies formed by a myxobacterium (*Chondromyces crocatus*). Each fruiting body, packed with spores, is created by the aggregation and differentiation of about a million myxobacteria. (From P.L. Grilione and J. Pangborn, *J. Bacteriol.* 124:1558–1565, 1975.)

0.1 mm

to propel the colony through the water. Each cell is equivalent to every other, and each can divide to give rise to an entirely new colony. Larger colonies are found in other genera, the most spectacular being *Volvox*, some of whose species have as many as 50,000 or more cells linked together to form a hollow sphere. In *Volvox*, the individual cells forming a colony are connected by fine cytoplasmic bridges so that the beating of their flagella is coordinated to propel the entire colony along like a rolling ball (Figure 1–32). Within the *Volvox* colony there is some division of labor among cells, with a small number of cells being specialized for reproduction and serving as precursors of new colonies. The other cells are so dependent on each other that they cannot live in isolation, and the organism dies if the colony is disrupted.

Chlamydomonas

Gonium

The Cells of a Higher Organism Become Specialized and Cooperate

In some ways, *Volvox* is more like a multicellular organism than a simple colony. All of its flagella beat in synchrony as it spins through the water, and the colony is structurally and functionally polarized and can swim toward a distant source of light. The reproductive cells are usually confined to one end of the colony, where they divide to form new miniature colonies, which are initially sheltered inside the parent sphere. Thus, in a primitive way, *Volvox* displays the two essential features of all multicellular organisms: its cells become *specialized*, and they *cooperate*. By specialization and cooperation, the cells combine to form a coordinated single organism with more capabilities than any of its component parts.

Pandorina

Organized patterns of cell differentiation occur even in some procaryotes. For example, many kinds of cyanobacteria remain together after cell division, forming filamentous chains that can be as much as a meter in length. At regular intervals along the filament, individual cells take on a distinctive character and become able to incorporate atmospheric nitrogen into organic molecules. These few specialized cells perform nitrogen fixation for their neighbors and share the products with them. But eucaryotic cells appear to be very much better at this sort of organized division of labor; they, and not procaryotes, are the living units from which all the more complex multicellular organisms are constructed.

Multicellular Organization Depends on Cohesion Between Cells

To form a multicellular organism, the cells must be somehow bound together, and eucaryotes have evolved a number of different ways to satisfy this need. In *Volvox*, as noted above, the cells do not separate entirely at cell division but remain connected by cytoplasmic bridges. In higher plants, the cells not only remain

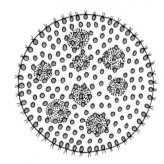

Volvox

Figure 1–32 Four closely related genera of green algae, showing a progression from unicellular to colonial and multicellular organization.

connected by cytoplasmic bridges (called *plasmodesmata*), they also are imprisoned in a rigid honeycomb of chambers walled with cellulose that the cells themselves have secreted (*cell walls*).

The cells of most animals do not have rigid walls, and cytoplasmic bridges are unusual. Instead, the cells are bound together by a relatively loose meshwork of large extracellular organic molecules (called the *extracellular matrix*) and by adhesions between their plasma membranes. In *sponges*, for example, which are commonly considered the most primitive of present-day animals, the body wall typically consists of a coherent sheet of cells comprising just five different specialized types; these form a system of channels and pores for the passage of water, from which food particles are filtered and ingested by the cells. Sponges grow indefinitely through cell proliferation, and their size and structure are not precisely fixed. They have no nervous system to coordinate the activities of their parts, and they have been described as "loose republics of cells"—as contrasted with the more strictly disciplined cell communities that constitute higher animals. Nevertheless, a sponge is far from being a totally chaotic structure. If the sponge is forced through a fine sieve so that its individual cells are mechanically separated from one another, the cells will often spontaneously reassemble into an intact sponge, aggregating initially into a large mass and then eventually rearranging themselves into a coherent multicellular sheet. Such a sheet of cells is called an **epithelium.**

Epithelial Sheets of Cells Enclose a Sheltered Internal Environment

Of all the ways in which animal cells are woven together into multicellular tissues, the epithelial arrangement is perhaps the most fundamentally important. The epithelial sheet has much the same significance for the evolution of complex multicellular organisms that the cell membrane has for the evolution of complex single cells.

The importance of epithelial sheets is well illustrated in another lowly group of animals, the *coelenterates*. These stand a rung higher in the scale of evolution than do the sponges, for they have a simple nervous system. The group includes sea anemones, jellyfish, and corals, as well as the small freshwater organism *Hydra*. Coelenterates are constructed from two layers of epithelium, the outer layer being the *ectoderm*, the inner being the *endoderm*. The endodermal layer surrounds a cavity, the *coelenteron*, in which food is digested (Figure 1–33). Among the en-

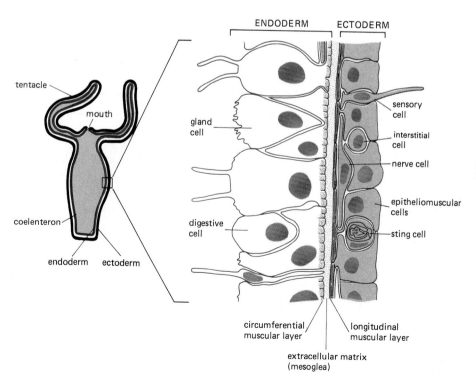

Figure 1–33 The body plan of *Hydra*. The outer layer of cells (ectoderm) has protective, predatory, and sensory functions, while cells of the inner layer (endoderm) function principally in digestion. Both epithelial sheets also have a contractile or muscular function, enabling the animal to move. The movements are coordinated by nerve cells that occupy a deep, protected position within each epithelium, forming an interconnected network.

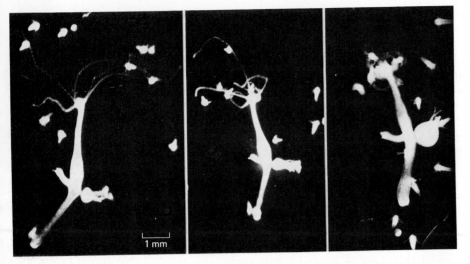

Figure 1–34 *Hydra* can perform a range of fairly complex activities. A single organism is photographed catching small water fleas in its tentacles; in the last panel it is stuffing these prey into its coelenteron for digestion. (Courtesy of Amata Hornbruch.)

dodermal cells are some that secrete digestive enzymes into the coelenteron, while other cells absorb and further digest the nutrient molecules that these enzymes release. By forming a tightly coherent epithelial sheet that prevents all these molecules from being lost to the exterior, the endodermal cells create for themselves an environment in the coelenteron that is suited to their own digestive tasks. Meanwhile, the ectodermal cells, facing the exterior, remain specialized for encounters with the outside world. In the ectoderm, for example, are cells that contain a poison capsule with a coiled dart that can be unleashed to kill the small animals on which *Hydra* feeds. The majority of other ectodermal and endodermal cells have musclelike properties, enabling *Hydra* to move, as a predator must.

Within the double layer of ectoderm and endoderm is another compartment, separate both from the coelenteron and from the outside world. Here the nerve cells lie, occupying narrow enclosed spaces between the epithelial cells, below the external surface where the specialized *cell junctions* between the epithelial cells form an impermeable barrier. The animal can change its shape and move by contractions of the musclelike cells in the epithelia, and it is the nerve cells that convey electrical signals to control and coordinate these contractions (Figures 1–33, 1–34, and 1–35). As we shall see later, the concentrations of simple inorganic ions in the medium surrounding a nerve cell are crucial for its function. Most nerve cells—our own included—are designed to operate when bathed in a solution with an ionic composition roughly similar to that of sea water. This presumably reflects the conditions under which the first nerve cells evolved. Most coelenterates still live in the sea, but not all. *Hydra*, in particular, lives in fresh water. It has evidently been able to colonize this new habitat only because its nerve cells are contained in a space that is sealed and isolated from the exterior within sheets of epithelial cells that maintain the internal environment necessary for nerve cell function.

Cell-Cell Communication Controls the Spatial Pattern of Multicellular Organisms[14]

The cells of *Hydra* are not only bound together mechanically and connected by junctions that seal off the interior from the exterior environment, they also communicate with one another along the length of the body. If one end of a *Hydra* is cut off, the remaining cells react to the absence of the amputated part by adjusting their characters and rearranging themselves so as to regenerate a complete animal. Evidently signals pass from one part of the organism to the other, governing the development of its body pattern—with tentacles and a mouth at one end and a foot at the other. Moreover, these signals are independent of the nervous system. If a developing *Hydra* is treated with a drug that prevents nerve cells from forming, the animal is unable to move about, catch prey, or feed itself. Its digestive system still functions normally, however, so that it can be kept alive by anyone with the

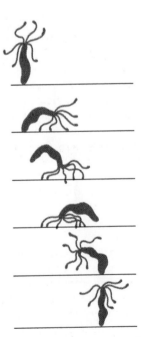

Figure 1–35 *Hydra* can swim, glide on its base, or, as shown here, travel by somersaulting.

patience to stuff its normal prey into its mouth. In such force-fed animals the body pattern is maintained, and lost parts are regenerated just as well as in an animal that has an intact nervous system.

The vastly more complex higher animals have evolved from humble ancestors resembling coelenterates, and these higher animals owe their complexity to more sophisticated exploitation of the same basic principles of cell cooperation that underlie the construction of *Hydra*. Epithelial sheets of cells line all external and internal surfaces in the body, creating sheltered compartments and controlled internal environments in which specialized functions are performed by differentiated cells. Specialized cells interact and communicate with one another, setting up signals to govern the character of each cell according to its place in the structure as a whole. To show how it is possible to generate multicellular organisms of such size, complexity, and precision as a human being, however, it is necessary to consider more closely the sequence of events in **development.**

Cell Memory Permits the Development of Complex Patterns

The cells of almost every multicellular organism are generated by repeated division from a single precursor cell; they constitute a *clone.* As proliferation continues and the clone grows, some of the cells, as we have seen, become differentiated from others, adopting a different structure, a different chemistry, and a different function, usually in response to cues from their neighbors. It is remarkable that eucaryotic cells and their progeny will usually persist in their differently specialized states even after the influences that originally directed their differentiation have disappeared—in other words, these cells have a *memory.* Consequently, their final character is not determined simply by their final environment, but rather by the entire sequence of influences to which they have been exposed in the course of development. Thus as the body grows and matures, progressively finer and finer details of the adult body pattern become specified, creating an organism of gradually increasing complexity whose ultimate form is the expression of a long developmental history that is remembered by the cells.

Basic Developmental Programs Tend to Be Conserved in Evolution[15]

The final structure of an animal reflects its evolutionary history, which, like development, presents a chronicle of progress from the simple to the complex. What then is the connection between the two perspectives, of evolution on the one hand and development on the other?

During evolution, many of the developmental devices that evolved in the simplest multicellular organisms have been conserved as basic principles for the construction of their more complex descendants. We have already mentioned, for example, the organization of cells into epithelia. It is notable also that some specialized cell types, such as nerve cells, are found throughout nearly the whole of the animal kingdom, from *Hydra* to humans. Furthermore, the early developmental stages of animals whose adult forms appear radically different are often surprisingly similar; it takes an expert eye to distinguish, for example, a young chick embryo from a young human embryo (Figure 1–36).

Such observations are not difficult to understand. Consider the process by which a new anatomical feature—say, an elongated beak—appears in the course of evolution. A random mutation occurs that changes the amino acid sequence of a protein or the timing of its synthesis and hence its biological activity. This alteration may, by chance, affect the cells responsible for the formation of the beak in such a way that they make one that is longer. But the mutation must also be compatible with the development of the rest of the organism; only then will it be propagated by natural selection. There would be little selective advantage in forming a longer beak if, in the process, the tongue was lost or the ears failed to develop. A catastrophe of this type is far more likely if the mutation affects events occurring early in development than if it affects those near the end. The early cells of an embryo are like cards at the bottom of a house of cards—a great deal depends on

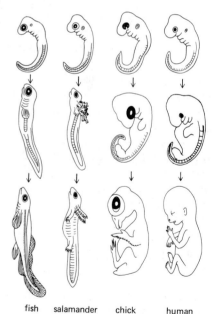

fish   salamander   chick   human

Figure 1–36 Comparison of the embryonic development of a fish, an amphibian, a bird, and a mammal. The early stages (*above*) are very similar; the later stages (*below*) are more divergent. The earliest stages are drawn roughly to scale; the later stages are not. (After E. Haeckel, Anthropogenie, oder Entwickelungsgeschichte des Menschen. Leipzig: Engelmann, 1874.)

them, and even small changes in their properties are likely to result in disaster. Fundamental steps have been "frozen" into developmental processes, just as the genetic code or protein synthesis mechanisms have become frozen into the basic biochemical organization of the cell. In contrast, cells produced near the end of development have more freedom to change. It is presumably for this reason that the embryos of different species so often resemble each other in their early stages and, as they develop, seem sometimes to replay the steps of evolution.

Eucaryotic Organisms Possess a Complex Machinery for Reproduction

Within the multicellular organism there must be some cells that serve as precursors for a new generation. In higher plants and animals these cells have a highly specialized character and are called *germ cells*. The propagation of the species depends on them, and there is a powerful selection pressure to adjust the structure of the organism as a whole to provide the germ cells with the best chance of survival. Other cells may die, but as long as germ cells survive, new organisms of the same sort will be produced. In this sense the most fundamental distinction to be drawn in a multicellular organism is the distinction between germ cells and the rest, that is, between germ cells and *somatic* cells.

Not all multicellular organisms reproduce by means of distinctive differentiated germ cells. Many simple animals, including sponges and coelenterates, can reproduce by budding off portions of their bodies, and many plants do likewise. Germ cells are, however, a necessity for **sexual reproduction.** This process is so familiar to us that we take it for granted, but it is by no means the obvious way to reproduce: it is far more complicated than asexual reproduction and requires a large diversion of resources. Two individuals of the same species but different sex produce germ cells of usually very different character—*eggs* from one, *sperm* from the other. An egg cell fuses with a sperm cell to form a *zygote*—the single precursor cell for the development of a new organism, whose genes represent a partly random reassortment of the genes of the two parents. While they may also reproduce in other ways, almost all eucaryotic species, unicellular as well as multicellular, are capable of reproducing sexually. Eucaryotic cells have evolved a complex machinery for sex; our lives revolve around it. Strong selective pressures must have operated to favor the evolution of sexual reproduction in preference to simpler strategies based on ordinary cell division. Although it is surprisingly difficult to say with certainty what those selection pressures were, it is at least plain that sexual reproduction brings new possibilities for manipulating and recombining the genes of a species. It may thus have played a crucial part in permitting the evolution of novel genes in novel combinations, and so in engendering the endless variety of forms and functions seen in plants and animals today.

The Cells of the Vertebrate Body Exhibit More Than 200 Different Modes of Specialization

The wealth of diverse specializations to be found among the cells of a higher animal is incomparably greater than any procaryote can show. In a vertebrate, more than 200 distinct **cell types** are plainly distinguishable, and many of these types of cells certainly include, under a single name, a large number of more subtly different varieties. Panel 1–2 (pp. 24–25) shows a small selection. In this profusion of specialized behaviors one can see displayed, in a single organism, the astonishing versatility of the eucaryotic cell. Much of our current knowledge of the general properties of eucaryotic cells has depended on the study of such specialized types of cells because they demonstrate to exceptionally good advantage particular features on which all cells depend in some measure. Each feature and each organelle of the prototype that we have outlined in Panel 1–1 (pp. 16–17) is developed to an unusual degree or revealed with special clarity in one cell type or another. To take one arbitrary example, consider the *neuromuscular junction*, where just three types of cells are involved: a muscle cell, a nerve cell, and a Schwann cell. Each has a very different role (Figure 1–37):

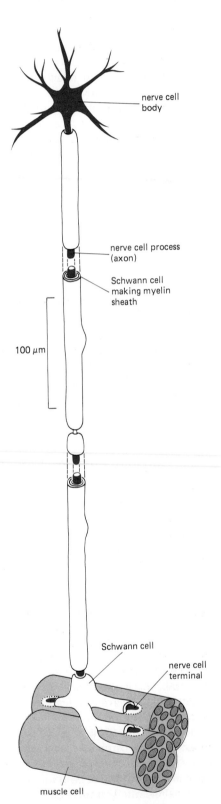

Figure 1–37 Schematic diagram showing a nerve cell, with its associated Schwann cells, contacting a muscle cell at a neuromuscular junction.

1. The muscle cell has made contraction its specialty. Its cytoplasm is packed with organized arrays of protein filaments, including vast numbers of actin filaments. There are also many mitochondria interspersed among the protein filaments, supplying ATP as fuel for the contractile apparatus.

2. The nerve cell stimulates the muscle to contract, conveying an excitatory signal to the muscle from the brain or spinal cord. The nerve cell is therefore extraordinarily elongated: its main body, containing the nucleus, may lie a meter or more from the junction with the muscle. The cytoskeleton is consequently well developed so as to maintain the unusual shape of the cell and to transport materials efficiently from one end of the cell to the other. The most crucial specialization of the nerve cell, however, is its plasma membrane, which contains proteins that act as ion *pumps* and ion *channels*, causing a movement of ions that is equivalent to a flow of electricity. Whereas all cells contain such pumps and channels in their plasma membranes, the nerve cell has exploited them in such a way that a pulse of electricity can propagate in a fraction of a second from one end of the cell to the other, conveying a signal for action.

3. Lastly, Schwann cells are specialists in the mass production of plasma membrane, which they wrap around the elongated portion of the nerve cell, laying down layer upon layer of membrane like a roll of tape, to form a *myelin sheath* that serves as insulation.

Cells of the Immune System Are Specialized for the Task of Chemical Recognition

Among all the cell systems that have evolved in higher animals, two stand out in different ways as pinnacles of complexity and sophistication: the *immune system* of the vertebrate is one, the *nervous system* is the other. Each far surpasses the performance of any artificial device—the vertebrate immune system in its capacity for chemical discrimination, the nervous system in its capacities for perception and control. Each system comprises a large number of different cell types and depends on complex interactions among them.

The protected and well-nourished environment in the interior of a multicellular animal is as inviting to foreign organisms as it is congenial to the animal's own cells. Hence there is a need for such animals to defend themselves against invading organisms—particularly viruses and bacteria. The primary task of the **immune system** is to destroy any such foreign microorganisms that may gain entry to the body.

As mentioned earlier, many eucaryotic cells are capable of phagocytosis: they can engulf and digest particles of matter from their surroundings. Among the differentiated cells in higher animals, there are professional phagocytic cells, such as *macrophages*, that specialize in this activity and can swallow up and destroy bacteria and other foreign cells (Figure 1–38). But there is a difficulty: it is good if the phagocytic cell attacks the foreign invader, but it would be disastrous if it were to attack also its own relatives and colleagues. The immune system therefore faces the problem of discriminating between the animal's own cells and those that are foreign—that is, of distinguishing between self and nonself.

The vertebrates have consequently evolved a specialized class of discriminatory cells, the *lymphocytes*. These are not themselves phagocytic; instead, they collaborate to provide the phagocytic cells with cues that tell them whether to attack or let live. In particular, certain of the lymphocytes (the B lymphocytes) manufacture specific protein molecules, or *antibodies*, that bind selectively to particular arrangements of atoms on the surfaces of invading organisms or on the toxic molecules they produce. To brand a new type of invader as foreign, new types of antibody must be produced; and since the variety of possible invaders is vast and essentially unpredictable, the B lymphocytes must be capable of making an almost endless variety of antibodies. On the other hand, the system must not produce antibodies that bind to the animal's own cells and molecules.

The vast diversity of antibodies is generated by unique genetic mechanisms that create millions of genetically different lymphocytes, each able to proliferate

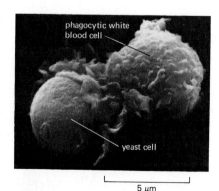

Figure 1–38 A scanning electron micrograph of a neutrophil—a type of professional phagocytic white blood cell—phagocytosing a yeast cell. (From J. Boyles and D.F. Bainton, *Cell* 24:905–914, 1981. © Cell Press.)

to form a clone whose members all produce the same distinctive antibody. Of these many potential clones, the ones that make antibodies that react with self molecules are destroyed or suppressed (by mechanisms still poorly understood), whereas those that make antibodies against foreign molecules are selected to survive and multiply. Thus the genesis of an individual animal's immune system, like the process of evolution, depends on a strategy of random variation followed by selection.

Nerve Cells Allow Rapid Adaptation to a Changing World[17,18]

The immune system of vertebrates places them in a class apart. Lower animals apparently do not have lymphocytes to help defend against invading microorganisms. A **nervous system,** on the other hand, is found in almost all multicellular animals and fulfills a still more fundamental need—the need for a quick adaptive response to external events.

Evolution acts over many generations to optimize the structure of an organism according to the environment in which it lives. In most ecological niches, though, there occur changes that are far too rapid for evolutionary adaptation to keep pace. The most successful organism, therefore, will be one that is capable of another sort of adaptation, requiring no genetic mutation yet producing optimal behavior when circumstances change. If the sequence of environmental changes is perfectly predictable, like the alternation of night and day or of summer and winter, the organism can be genetically programmed to change autonomously according to the appropriate timetable. Thus the photosynthetic protist *Gonyaulax* (belonging to the group of cells known as *dinoflagellates*) (Figure 1–39) shows a precise daily rhythm in its photosynthetic activities that continues even if the cell is maintained for weeks on end in conditions of constant lighting. Such biological clocks exist in many other organisms, but their mechanism remains a mystery.

Most environmental changes, however, are not so predictable. Bacteria in the gut, for example, will experience irregular fluctuations in the nature and quantity of food that is available to them, and any bacterium that can adjust its metabolism to these changes will have an advantage over one that cannot. These organisms consequently have evolved the ability to sense the concentrations of nutrients in their environment and to react by adjusting the rates at which they synthesize their metabolic enzymes. Special intracellular control molecules (such as *cyclic AMP*) serve to couple the environmental stimulus to the appropriate response.

In a multicellular organism, the signal that couples a sensation to a response must generally pass between cells. Thus metabolic adjustments are often mediated by *hormones* that are released by one set of cells and travel through the tissues via the bloodstream to produce a response in other sets of cells. But hormones take time to travel a long distance, and in doing so they diffuse widely. If a chemical signal is to be delivered fast, it must be released close to its target; and in that way it can also have a precisely localized action. But if the chemical signal is to be released close to its target, how can it be used to couple a sensation to a response in a remote part of the body? The highly elongated nerve cell provides the answer. At one end, it is itself sensitive to a chemical or physical stimulus; at

Figure 1–39 (A) Measurements of the free-running glow rhythm of a population of the photosynthetic protist *Gonyaulax polyedra* fitted to a theoretical curve. The mean period of the free-running rhythm is 22 hours 58 minutes, and individual cells in the population are accurate timekeepers to within ±18 minutes per day. (B) A scanning electron micrograph of *Gonyaulax*. (A, after D. Njus, V.D. Gooch, and J.W. Hastings, *Cell Biophys.* 3:223–231, 1981; B, courtesy of John Dodge.)

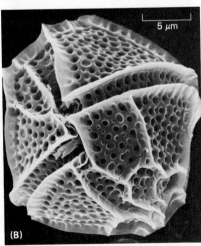

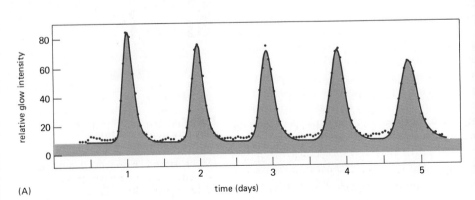

(A)

(B)

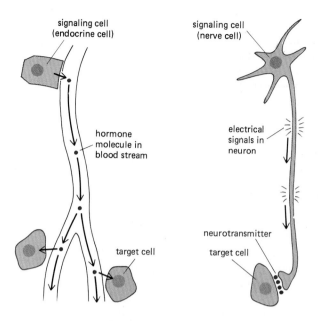

signaling cell
(endocrine cell)

signaling cell
(nerve cell)

hormone
molecule in
blood stream

electrical
signals in
neuron

target cell

neurotransmitter

target cell

Figure 1–40 Hormonal and neuronal signaling compared. Endocrine cells, when stimulated, release hormones into the bloodstream, and the circulating hormones evoke a response in any cell that is sensitive to them, no matter where it is located. A nerve cell, when stimulated, generates an action potential along its axon, rapidly triggering the release of neurotransmitter from the axon terminals; the neurotransmitter acts only on cells in the immediate neighborhood of the ax   minals.

the other end, it can in turn release a chemical signal, known as a *neurotransmitter* that acts on other cells (Figure 1–40). Stimulation at one end triggers an electrical excitation, which is propagated rapidly to the other end and, upon arriving there, triggers release of the neurotransmitter. This rapid signaling device enables multicellular animals to make rapid responses to the changing world around them. It also enables them to coordinate precisely the activities of widely separated parts of the body.

Nerve Cell Connections Determine Patterns of Behavior[18]

A single nerve cell of a human being is not very different from a single nerve cell of a worm. The superiority of the human nervous system lies in the enormous number of its nerve cells and, above all, in the way they are connected together to transmit, combine, and interpret sensory inputs and to coordinate complex patterns of activity. To understand the cellular basis for the evolution of the nervous system, one must therefore look to the mechanisms by which embryonic nerve cells develop their fantastically intricate shapes and form their precisely ordered patterns of connections (Figure 1–41). The visual system of a fly will serve to convey some notion of the astonishing complexity and orderliness of the networks that developing nerve cells weave (Figure 1–42). This entire structure is built to genetic specifications and will develop even in the absence of light. Yet when the events of neural development are analyzed, they are found to be based ultimately, in all species, on the same fundamental behaviors that can be observed in almost any eucaryotic cell: similar mechanisms are employed for cell movement, cell-to-cell adhesion, chemical signaling, and so forth. Evolution of a complex nervous system has depended above all on the evolution of complex controls to combine and orchestrate these basic cell activities.

The patterns of connections between nerve cells constrain the patterns of behavior of the animal. Without education, without need of experience, the male fly mates with the female, the spider spins its web, the hungry baby screws up her face and cries. All these activities are prescribed by the DNA of the species, acting through its control over the behavior of the individual cells as they build the nervous system in the embryo and play their part in its functioning in the adult.

But not all behavior is genetically determined. The experiences of an animal are important as well. Sensory deprivation during the development of a mammal can alter the microscopic structure of the brain, and mature animals of almost every species, from coelenterates to human beings, are to some degree capable of learning. Learning is by definition the outcome of experience, and therefore of

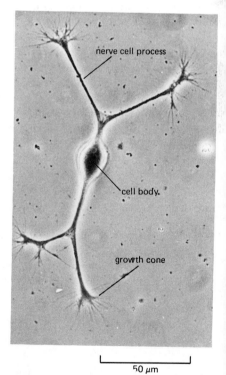

nerve cell process

cell body

growth cone

50 μm

Figure 1–41 Light micrograph of a developing nerve cell that has been isolated from a chick embryo and put into a tissue-culture dish containing a nutrient solution. The cell is beginning to grow elongated processes, each of which extends by means of a motile structure called a growth cone. (Courtesy of Zoltan Gabor.)

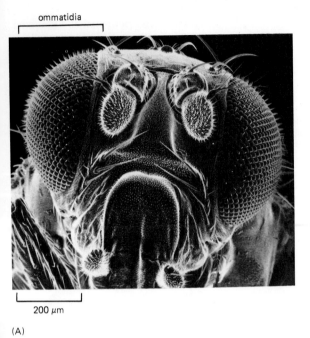

ommatidia

200 μm

(A)

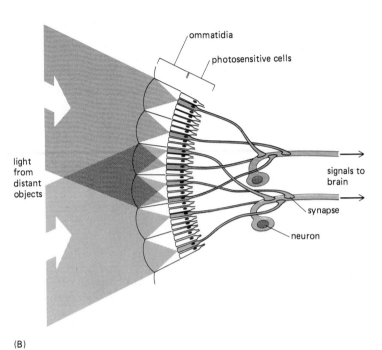

ommatidia

photosensitive cells

light
from
distant
objects

signals to
brain

synapse

neuron

(B)

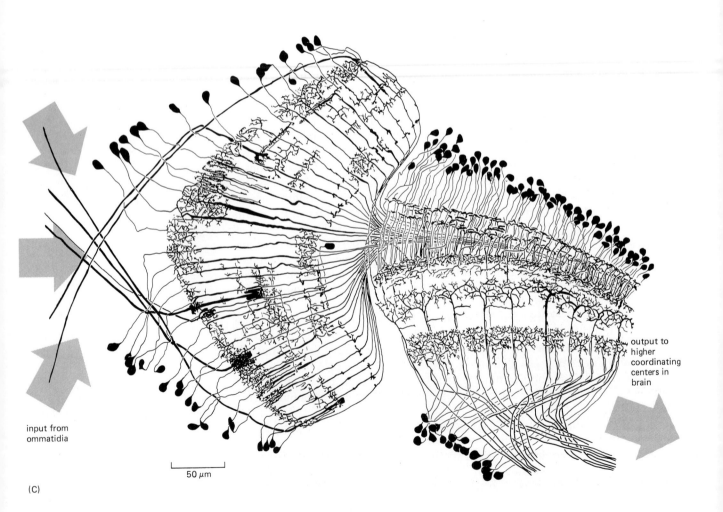

output to
higher
coordinating
centers in
brain

input from
ommatidia

50 μm

(C)

◄Figure 1–42 (A) The head of a fruit fly (*Drosophila*) seen in a scanning electron microscope. Situated on either side of the head are two huge compound eyes consisting of large numbers of units known as ommatidia. Each ommatidium has a separate lens that focuses the light onto a group of photosensitive receptor cells at its base. (B) Schematic diagram of the neuronal connections in the outermost layer of the fly eye as seen in a vertical section. Light enters each of the ommatidia of the compound eye and is focused onto one of eight photosensitive receptor cells at its base (only five of which are shown here). Because of the curvature of the compound eye, light from a distant point source is focused onto a different photosensitive receptor cell in different ommatidia. The interweaving of the short axons of the photoreceptor cells, however, connects the photoreceptor cells that are "looking" at the same point to the same bundle of nerve axons that passes into the insect brain. More than a thousand such axon bundles are present in each fly eye, and each bundle is precisely wired in the course of development to the correct set of photoreceptor cells. (C) Network of nerve cells in the portion of the fly's brain that receives and processes the visual input relayed from the ommatidia by the nerve cells illustrated in (B) (A, courtesy of Rudi Turner and Anthony Mahowald; C, redrawn from N. Strausfeld, Atlas of an Insect Brain. New York: Springer, 1976.)

electrical activity in nerve cells, and it must involve the production of lasting changes in neural connections. To decipher these mechanisms in detail is one of the central tasks of current neurobiological research.

Many of the brain connections that allow us to read and write and speak our native tongue are the outcome of education, and they represent an inheritance of a nongenetic kind. Learning and communication enable the human species to adapt over many generations in a way that for lower organisms is possible only through genetic evolution. Yet even these sophisticated capacities, on which all our culture and society depend, can be seen to rest on the minutiae of cell behavior—on the rules by which nerve cells make lasting adjustments of their interconnections as a consequence of electrical activity.

Of course, we can no more understand the multicellular organism by studying only single cells than we can understand the single cell by studying only isolated biological molecules. Yet if we do not understand the cell, we can never completely understand the organism. And if we do not understand the constituent molecules, we cannot properly understand the cell. Molecules, therefore, must be the starting point for our discussion of the living cell in the next chapter.

Summary

The evolution of large multicellular organisms depended on the ability of eucaryotic cells to express their hereditary information in many different ways and to function cooperatively in a single collective. One of the earliest developments was probably the formation of epithelial cell sheets, which separate the internal space of the animal from the exterior. In addition to epithelial cells, primitive differentiated cell types would have included nerve cells, muscle cells, and connective tissue cells, all of which can be found in very simple present-day animals.

In the evolution of higher animals (Figure 1–43), the same fundamental developmental strategies produced an increasing number of specialized cell types and more sophisticated methods of coordination among them. Two systems of cells in higher animals represent, in different ways, pinnacles of complexity in multicellular organization: one is the vertebrate immune system, the cells of which have the potential to produce millions of different protein antibodies; the other is the nervous system. In lower animals, the pattern of connections between nerve cells is for the most part rigidly specified genetically, and behavioral patterns evolve by genetic

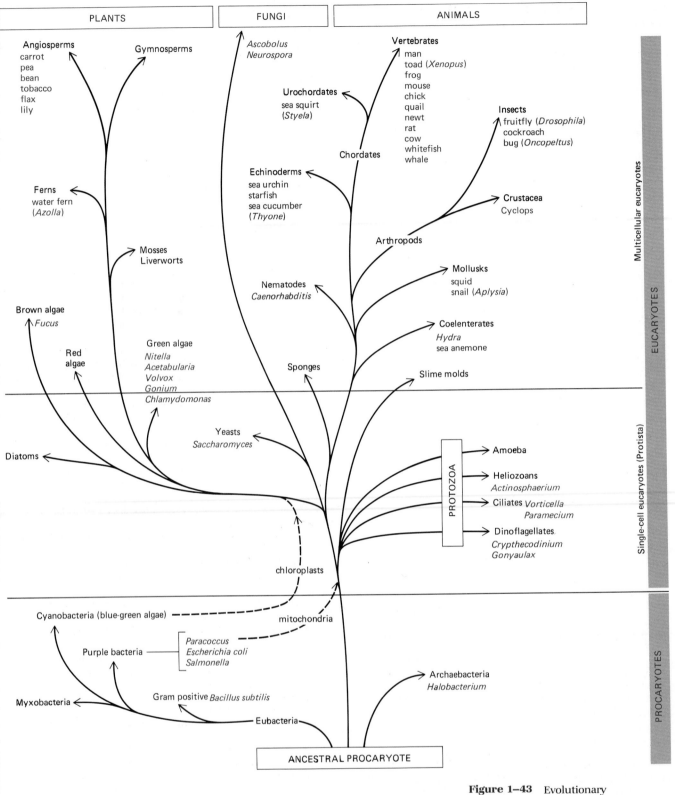

Figure 1–43 Evolutionary relationships among some of the organisms mentioned in this book. The branches of the tree show paths of common descent but do not indicate by their length the passage of time. (Note, similarly, that the vertical axis of the diagram shows major categories of organisms and not time.)

mutation. In higher animals, the performance and structure of the nervous system are increasingly subject to modification (learning), which is a consequence of the capacity of nerve cells to alter their connections in response to electrical activity caused by environmental influences.

References

General

Bendall, D.S., ed. Evolution from Molecules to Men. Cambridge, U.K.: Cambridge University Press, 1983.

Curtis, H. Biology, 4th ed. New York: Worth, 1983.

Darnell, J.E.; Lodish, H.F.; Baltimore, D. Molecular Cell Biology, Chapter 25. New York: W.H. Freeman, 1986.

Darwin, C. On the Origin of Species. London: Murray, 1859. Reprinted, New York: Penguin, 1984.

Evolution. Sci. Am. 239(3), 1978. (An entire issue devoted to the topic.)

Keeton, W.T.; Gould, J.W.; Gould, C.G. Biological Science, 4th ed. New York: Norton, 1986.

Maynard Smith, J. The Theory of Evolution, 3rd ed. New York: Penguin, 1975.

Watson, J.D.; Hopkins, N.H.; Roberts, J.W.; Steitz, J.A.; Weiner, A.M. Molecular Biology of the Gene, 4th ed., Chapter 28. Menlo Park, CA: Benjamin-Cummings, 1987.

Wilson, E.B. The Cell in Development and Heredity, 3rd ed. New York: Macmillan, 1925. (Reprinted, New York: Garland, 1987.)

Cited

1. Ferris, J.P.; Usher, D.A. Origins of life. In Biochemistry, 2nd ed. (G. Zubay, ed.), pp. 1120–1151. New York: Macmillan, 1988.
   Miller, S.M.; Orgel, L.E. The Origins of Life on the Earth. Englewood Cliffs, NJ: Prentice-Hall, 1974.
   Schopf, J.W.; Hayes, J.M.; Walter, M.R. Evolution of earth's earliest ecosystems: recent progress and unsolved problems. In Earth's Earliest Biosphere: Its Origin and Evolution (J.W. Schopf, ed.), pp. 361–384. Princeton, NJ: Princeton University Press, 1983.

2. Eigen, M.; Gardiner, W.; Schuster, P.; Winkler-Oswatitsch, R. The origin of genetic information. Sci. Am. 244(4):88–118, 1981.

3. Cech, T.R. RNA as an enzyme. Sci. Am. 255(5):64–75, 1986.

4. Alberts, B.M. The function of the hereditary materials: biological catalyses reflect the cell's evolutionary history. Am. Zool. 26:781–796, 1986.
   Darnell, J.E.; Doolittle, W.F. Speculations on the early course of evolution. Proc. Natl. Acad. Sci. USA 83:1271–1275, 1986.
   Orgel, L.E. RNA catalysis and the origin of life. J. Theor. Biol. 123:127–149, 1986.

5. Rogers, M; et al. Construction of the mycoplasma evolutionary tree from 5S rRNA sequence data. Proc. Natl. Acad. Sci. USA 82:1160–1164, 1985.

6. Vidal, G. The oldest eukaryotic cells. Sci. Am. 250(2):48–57, 1984.

7. Clarke, P.H. Enzymes in bacterial populations. In Biochemical Evolution (H. Gutfreund, ed.), pp. 116–149. Cambridge, U.K.: Cambridge University Press, 1981.

Doolittle, W.F. Archaebacteria coming of age. Trends Genet. 1:268–269, 1985.
Woese, C.R. Bacterial evolution. Microbiol. Rev. 51:221–271, 1987.

8. Wilson, A. The molecular basis of evolution. Sci. Am. 253(4):164–173, 1985.

9. Dickerson, R.E. Cytochrome c and the evolution of energy metabolism. Sci. Am. 242(3):136–153, 1980.

10. Cavalier-Smith, T. The origin of eukaryotic and archaebacterial cells. Ann. N.Y. Acad. Sci. 503:17–54, 1987.
    Margulis, L. Symbiosis in Cell Evolution. New York: W.H. Freeman, 1981.

11. Vossbrinck, C.R.; Maddox, J.V.; Friedman, S.; Debrunner-Vossbrinck, B.A.; Woese, C.R. Ribosomal RNA sequence suggests microsporidia are extremely ancient eukaryotes. Nature 326:411–414, 1987.
    Yang, D.; Oyaizu, Y.; Oyaizu, H.; Olsen, G.J.; Woese, C.R. Mitochondrial origins. Proc. Natl. Acad. Sci. USA 82:4443–4447, 1985.

12. Sleigh, M.A. The Biology of Protozoa. London: Edward Arnold, 1973.
    Sogin, M.L.; Elwood, H.J.; Gunderson, J.H. Evolutionary diversity of eukaryotic small-subunit rRNA genes. Proc. Natl. Acad. Sci. USA 83:1383–1387, 1986.

13. Buchsbaum, R. Animals Without Backbones, 2nd ed. Chicago: University of Chicago Press, 1976.
    Field, K.G.; et al. Molecular phylogeny of the animal kingdom. Science 239:748–753, 1988.
    Margulis, L.; Schwartz, K.V. Five Kingdoms: An Illustrated Guide to the Phyla of Life on Earth, 2nd ed. New York: W.H. Freeman, 1987.
    Shapiro, J.A. Bacteria as multicellular organisms. Sci. Am. 258(6):82–89, 1988.
    Valentine, J.W. The evolution of multicellular plants and animals. Sci. Am. 239(3):140–158, 1978.

14. Bode, P.M.; Bode, H.R. Patterning in Hydra. In Pattern Formation (G.M. Malacinski, S.V. Bryant, eds.), pp. 213–244. New York: Macmillan, 1984.

15. Raff, R.A.; Kaufman, T.C. Embryos, Genes, and Evolution. New York: Macmillan, 1983.

16. Maynard Smith, J. The Evolution of Sex. Cambridge, U.K.: Cambridge University Press, 1978.
    Michod, R.E.; Levin, B.R., eds. The Evolution of Sex: An Examination of Current Ideas. Sunderland, MA: Sinauer, 1988.

17. Winfree, A.T. The Timing of Biological Clocks. New York: W.H. Freeman, 1987.

18. Bullock, T.H.; Orkand, R.; Grinnell, A. Introduction to Nervous Systems. San Francisco: Freeman, 1977.

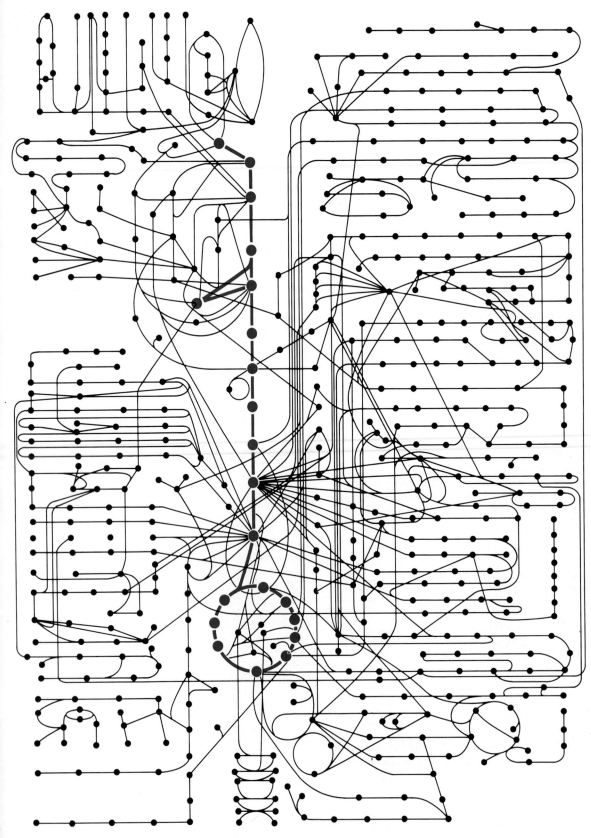

A maze illustrating the chemical reactions that interconvert
small molecules in cells.

Small Molecules, Energy, and Biosynthesis

2

"I must tell you that I can prepare urea without requiring a kidney or an animal, either man or dog." This sentence, written 150 years ago by the young German chemist Wöhler, signaled an end to the belief in a special *vital force* that exists in living organisms and gives rise to their distinctive properties and products. But what was a revelation in Wöhler's time is common knowledge today—living creatures are made of chemicals. There is no room in the contemporary view of life for vitalism—or for anything else outside the laws of chemistry and physics. This is not to say that no mysteries remain in biology: there are many areas of ignorance, as will become apparent in later chapters. But we should begin by emphasizing the enormous amount that is known.

We now have detailed information about the essential molecules of the cell—not just a small number of molecules, but thousands of them. In many cases we know their precise chemical structures and exactly how they are made and broken down. We know in general terms how chemical energy drives the biosynthetic reactions of the cell, how thermodynamic principles operate in cells to create molecular order, and how the myriad intracellular chemical changes occurring continuously within them are controlled and coordinated.

In this and the next chapter we briefly survey the chemistry of the living cell. Here we deal with the processes involving small molecules: those mechanisms by which the cell synthesizes its fundamental chemical ingredients and by which it obtains its energy. Chapter 3 describes the giant molecules of the cell, which are polymers of small molecules and whose properties are responsible for the specificity of biological processes and the transfer of biological information.

The Chemical Components of a Cell

Cell Chemistry Is Based on Carbon Compounds[1]

A living cell is composed of a restricted set of elements, four of which (C, H, N, and O) make up nearly 99% of its weight. This composition differs markedly from that of the earth's crust and is evidence of a distinctive type of chemistry (Figure 2–1). What is this special chemistry, and how did it evolve?

The most abundant substance of the living cell is water. It accounts for about 70% of a cell's weight, and most intracellular reactions occur in an aqueous environment. Life on this planet began in the ocean, and the conditions in that

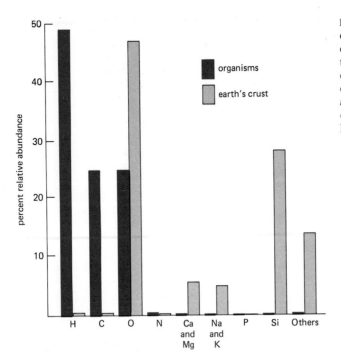

Figure 2–1 The relative abundance of chemical elements found in the earth's crust (the nonliving world) compared to that in the soft tissues of living organisms. The relative abundance is expressed as a percentage of the total *number* of atoms present. Thus, for example, nearly 50% of the atoms in living organisms are hydrogen atoms.

primeval environment put a permanent stamp on the chemistry of living things. All organisms have been designed around the special properties of water, such as its polar character, its ability to form hydrogen bonds, and its high surface tension. Some important properties of water are summarized in Panel 2–1 (pp. 46–47).

If we disregard water, nearly all of the molecules in a cell are carbon compounds, which are the subject matter of **organic chemistry.** Carbon is outstanding among all the elements on earth for its ability to form large molecules; silicon is a poor second. The carbon atom, because of its small size and four outer-shell electrons, can form four strong covalent bonds with other atoms. Most important, it can join to other carbon atoms to form chains and rings and thereby generate large and complex molecules with no obvious upper limit to their size. The other abundant atoms in the cell (H, N, and O) are also small and able to make very strong covalent bonds (Panel 2–2, pp. 48–49).

In principle, the simple rules of covalent bonding between carbon and other elements permit an infinitely large number of compounds. Although the number of different carbon compounds in a cell is very large, it is only a tiny subset of what is theoretically possible. In some cases we can point to good reasons why this compound or that performs a given biological function; more often it seems that the actual "choice" was one among many reasonable alternatives, and therefore something of an accident (Figure 2–2). Once established, certain chemical themes and patterns of reaction were preserved, with variations, during the course of evolution. Apparently the development of new classes of compounds was only rarely necessary or useful.

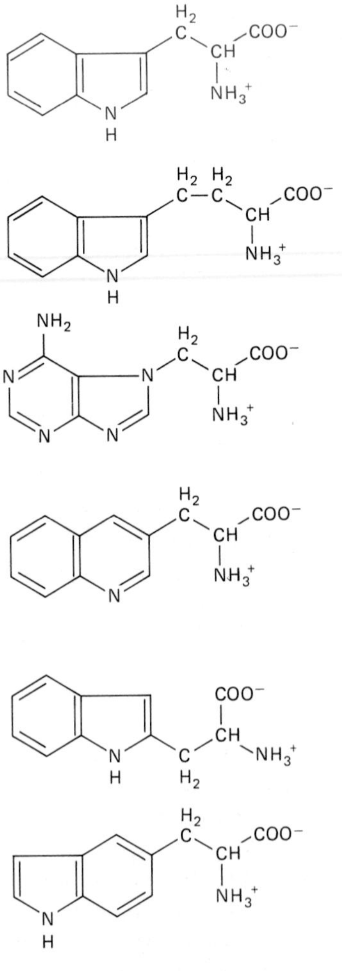

Cells Use Four Basic Types of Small Molecules[2]

Certain simple combinations of atoms—such as the methyl (—CH_3), hydroxyl (—OH), carboxyl (—COOH), and amino (—NH_2) groups—recur repeatedly in biological molecules. Each such group has distinct chemical and physical properties that influence the behavior of whatever molecule the group occurs in. The main types of chemical groups and some of their salient properties are summarized in Panel 2–2 (pp. 48–49).

The **small organic molecules** of the cell are carbon compounds with molecular weights in the range 100 to 1000 and containing up to 30 or so carbon atoms. They are usually found free in solution in the cytoplasm, where some of them form a pool of intermediates from which large polymers, called **macro-**

Figure 2–2 Living organisms synthesize only a small number of the organic molecules that they could in principle make. Of the six amino acids shown, only the top one (tryptophan) is made by cells.

Table 2–1   The Approximate Chemical Composition of a Bacterial Cell

	Percent of Total Cell Weight	Number of Types of Each Molecule
Water	70	1
Inorganic ions	1	20
Sugars and precursors	1	250
Amino acids and precursors	0.4	100
Nucleotides and precursors	0.4	100
Fatty acids and precursors	1	50
Other small molecules	0.2	~300
Macromolecules (proteins, nucleic acids, and polysaccharides)	26	~3000

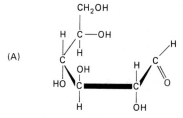

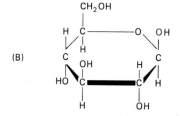

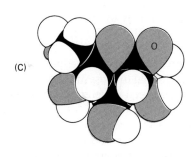

molecules, are made. They are also essential intermediates in the chemical reactions that transform energy derived from food into usable forms (see below).

The small molecules amount to about one-tenth of the total organic matter in a cell, and (at a rough estimate) only on the order of a thousand different kinds are present (Table 2–1). All biological molecules are synthesized from and broken down to the same simple compounds, synthesis and breakdown occurring through sequences of chemical changes that are limited in scope and follow definite rules. As a consequence, the compounds in a cell are chemically related and can be classified into a small number of distinct families. Since the macromolecules in a cell, which form the subject of Chapter 3, are assembled from the same small molecules, they belong to the same families.

Broadly speaking, cells contain just four major families of small organic molecules: the simple **sugars,** the **fatty acids,** the **amino acids,** and the **nucleotides.** Each of these families contains many different members with common chemical features. Although some cellular compounds do not fit into these categories, the four families, together with the macromolecules made from them, account for a surprisingly large fraction of the cell mass (Table 2–1).

Sugars Are Food Molecules of the Cell[3]

The simplest sugars—the **monosaccharides**—are compounds with the general formula $(CH_2O)_n$, where n is an integer from 3 through 7. *Glucose*, for example, has the formula $C_6H_{12}O_6$ (Figure 2–3). As shown in Figure 2–3, sugars can exist in either a ring or an open-chain form. In their open-chain form, sugars contain a number of hydroxyl groups and either one aldehyde ($_H{>}C{=}O$) or one ketone (${>}C{=}O$) group. The aldehyde or ketone group plays a special role. First, it can react with a hydroxyl group in the same molecule to convert the molecule into a ring; in the ring form, the carbon of the original aldehyde or ketone group can be recognized as the only one that is bonded to two oxygens. Second, once the ring is formed, this carbon can become further linked to one of the carbons bearing a hydroxyl group on *another* sugar molecule, creating a *disaccharide* (Panel 2–3, pp. 50–51). The addition of more monosaccharides in the same way results in **oligosaccharides** of increasing length (trisaccharides, tetrasaccharides, and so on) up to very large **polysaccharide** molecules with thousands of monosaccharide units (residues). Because each monosaccharide has several free hydroxyl groups that can form a link to another monosaccharide (or to some other compound), the number of possible polysaccharide structures is extremely large. Even a simple disaccharide consisting of two glucose residues can exist in 11 different varieties (Figure 2–4), while three different hexoses $(C_6H_{12}O_6)$ can join together to make several thousand different trisaccharides. For this reason it is very difficult to determine the structure of any particular polysaccharide; with present methods it takes longer to determine the arrangement of half a dozen linked sugars (for example, those in a glycoprotein) than to determine the nucleotide sequence of a DNA molecule containing many thousands of nucleotides.

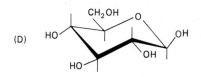

Figure 2–3   The structure of the monosaccharide glucose, a common hexose sugar. (A) The open-chain form of this sugar, which is in equilibrium with the more stable cyclic or ring form shown below it (B). (C) A space-filling model of this cyclic form (β-D-glucose). The chair form (D) is an alternative representation of the cyclic form that is frequently used instead of the cyclic form (B) because it more accurately reflects the structure. In all four representations, the colored O denotes the oxygen atom of the aldehyde group. For an outline of sugar structures and chemistry, see Panel 2–3 (pp. 50–51).

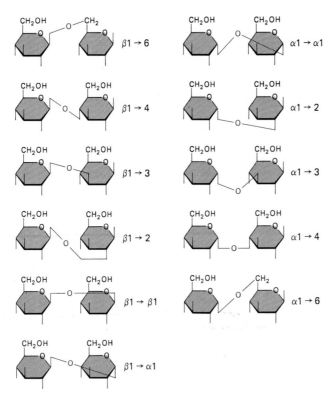

Figure 2–4 Eleven disaccharides consisting of two D-glucose units. Although these differ only in the type of linkage between the two glucose units, they are chemically distinct. Since the oligosaccharides associated with proteins and lipids may have six or more different kinds of sugar joined in both linear and branched arrangements through linkages such as those illustrated here, the number of possible distinct types of oligosaccharides is extremely large.

Glucose is the principal food compound of many cells. A series of oxidative reactions (see p. 61) leads from this hexose to various smaller sugar derivatives and eventually to CO_2 and H_2O. The net result can be written

$$C_6H_{12}O_6 + 6O_2 \rightarrow 6CO_2 + 6H_2O + \text{energy}$$

In the course of glucose breakdown, energy and "reducing power," both of which are essential in biosynthetic reactions, are salvaged and stored, mainly in the form of two crucial molecules, called **ATP** and **NADH,** respectively (see p. 66).

Simple polysaccharides composed only of glucose residues—principally *glycogen* in animal cells and *starch* in plant cells—are used to store energy for future use. But sugars do not function exclusively in the production and storage of energy. Important extracellular structural materials (such as cellulose) are composed of simple polysaccharides, and smaller but more complex chains of sugar molecules are often covalently linked to proteins in *glycoproteins* and to lipids in *glycolipids.*

Fatty Acids Are Components of Cell Membranes[4]

A fatty acid molecule, such as *palmitic acid* (Figure 2–5), has two distinct regions: a long hydrocarbon chain, which is hydrophobic (water insoluble) and not very reactive chemically, and a carboxylic acid group, which is ionized in solution (COO^-), extremely hydrophilic (water soluble), and readily forms esters and amides. In fact, almost all of the fatty acid molecules in a cell are covalently linked to other molecules by their carboxylic acid group. The many different fatty acids found in cells differ in such chemical features as the length of their hydrocarbon chains and the number and position of the carbon-carbon double bonds they contain (Panel 2–4, pp. 52–53).

Fatty acids are a valuable source of food since they can be broken down to produce more than twice as much usable energy, weight for weight, as glucose. They are stored in the cytoplasm of many cells in the form of droplets of *triglyceride* molecules, which consist of three fatty acid chains, each joined to a glycerol molecule (Panel 2–4, pp. 52–53); these molecules are the animal fats familiar from

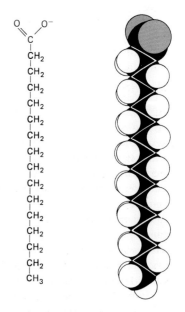

Figure 2–5 Palmitic acid. The carboxylic acid group *(color)* is shown in its ionized form. A space-filling model is presented on the right.

everyday experience. When required, the fatty acid chains can be released from triglycerides and broken down into two-carbon units. These two-carbon units, present as the acetyl group in a water-soluble molecule called *acetyl CoA,* are then further degraded in various energy-yielding reactions, which will be described below.

But the most important function of fatty acids is in the construction of cell membranes. These thin, impermeable sheets that enclose all cells and surround their internal organelles are composed largely of **phospholipids,** which are small molecules that resemble triglycerides in that they are constructed mostly from fatty acids and glycerol. However, in phospholipids the glycerol is joined to two rather than three fatty acid chains. The remaining site on the glycerol is coupled to a phosphate group, which is in turn attached to another small hydrophilic compound such as *ethanolamine, choline,* or *serine.*

Each phospholipid molecule, therefore, has a hydrophobic tail—composed of the two fatty acid chains—and a hydrophilic polar head group, where the phosphate is located. Thus phospholipid molecules are, in effect, detergents, and this is evident in their properties. A small amount of phospholipid will spread over the surface of water to form a *monolayer* of phospholipid molecules; in this thin film, the tail regions pack together very closely facing the air and the head groups are in contact with the water (Panel 2–4, pp. 52–53). Two such films can combine tail to tail to make a phospholipid sandwich, or **lipid bilayer,** which is the structural basis of all cell membranes.

Amino Acids Are the Subunits of Proteins

The common amino acids are chemically varied, but they all contain a carboxylic acid group and an amino group, both linked to a single carbon atom (Figure 2–6). They serve as subunits in the synthesis of **proteins,** which are long linear polymers of amino acids joined head to tail by a *peptide bond* between the carboxylic acid group of one amino acid and the amino group of the next (Figure 2–7). There are 20 common amino acids in proteins, each with a different *side chain* attached to the α-carbon atom (Panel 2–5, pp. 54–55). The same 20 amino acids occur over and over again in all proteins, including those made by bacteria, plants, and animals. Although the choice of precisely these 20 amino acids is probably an example of an evolutionary accident, the chemical versatility they provide is vitally important. For example, 5 of the 20 amino acids have side chains that can carry a charge (Figure 2–8), whereas the others are uncharged but reactive in specific ways (Panel 2–5, pp. 54–55). As we shall see, the properties of the amino acid side chains, in aggregate, determine the properties of the proteins they constitute and underlie all of the diverse and sophisticated functions of proteins.

Nucleotides Are the Subunits of DNA and RNA[5]

In nucleotides, one of several different nitrogen-containing ring compounds (often referred to as *bases* because they can combine with H^+ in acidic solutions) is linked to a five-carbon sugar (either *ribose* or *deoxyribose*) that carries a phosphate group. There is a strong family resemblance between the nitrogen-containing rings found in nucleotides. *Cytosine* (C), *thymine* (T), and *uracil* (U) are called **pyrimidine** compounds because they are all simple derivatives of a six-membered pyrimidine ring; *guanine* (G) and *adenine* (A) are **purine** compounds, with a second five-membered ring fused to the six-membered ring. Each nucleotide is named by reference to the unique base that it contains (Panel 2–6, pp. 56–57).

Nucleotides can act as carriers of chemical energy. The triphosphate ester of adenine, **ATP** (Figure 2–9), above all others, participates in the transfer of energy in hundreds of individual cellular reactions. Its terminal phosphate is added using energy from the oxidation of foodstuffs, and this phosphate can be readily split off by hydrolysis to release energy that drives energetically unfavorable biosynthetic reactions elsewhere in the cell. Other nucleotide derivatives serve as carriers for the transfer of particular chemical groups, such as hydrogen atoms or sugar residues, from one molecule to another. And a cyclic phosphate-containing ad-

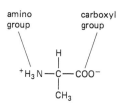

Figure 2–6 The amino acid alanine as it exists at pH 7 in its ionized form. When incorporated into a polypeptide chain, the charges on the amino and carboxyl groups of the free amino acid disappear. A space-filling model is shown below the structural formula.

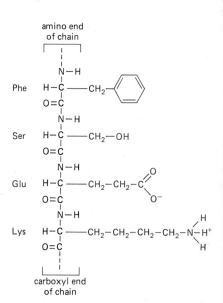

Figure 2–7 A small part of a protein molecule. The four amino acids shown are linked together by a type of covalent bond called a peptide bond. A protein is therefore also sometimes referred to as a polypeptide. The amino acid *side chains* are shown here in color.

HYDROPHILIC AND HYDROPHOBIC MOLECULES

Because of the polar nature of water molecules, they will cluster around ions and other polar molecules.

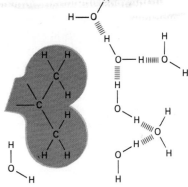

Nonpolar molecules interrupt the H-bonded structure of water without forming favorable interactions with water molecules. They are therefore hydrophobic and quite insoluble in water.

Molecules that can thereby be accommodated in water's hydrogen-bonded structures are hydrophilic and relatively water-soluble.

WATER

Although a water molecule has an overall neutral charge (having the same number of electrons and protons), the electrons are asymmetrically distributed, which makes the molecule polar.

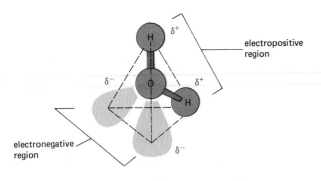

electropositive region

electronegative region

The oxygen nucleus draws electrons away from the hydrogen nuclei, leaving these nuclei with a small net positive charge. The excess of electron density on the oxygen atom creates weakly negative regions at the other two corners of an imaginary tetrahedron.

WATER STRUCTURE

Molecules of water join together transiently in a hydrogen-bonded lattice. Even at 37°C, 15% of the water molecules are joined to four others in a short-lived assembly known as a "flickering cluster."

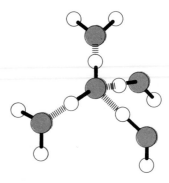

The cohesive nature of water is responsible for many of its unusual properties, such as high surface tension, specific heat, and heat of vaporization.

HYDROGEN BONDS

Because they are polarized, two adjacent H_2O molecules can form a linkage known as a hydrogen bond. Hydrogen bonds have only about 1/20 the strength of a covalent bond.

Hydrogen bonds are strongest when the three atoms lie in a straight line.

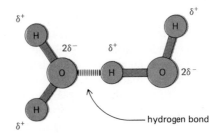

hydrogen bond

HYDROPHOBIC REPULSION CAN HOLD MOLECULES TOGETHER

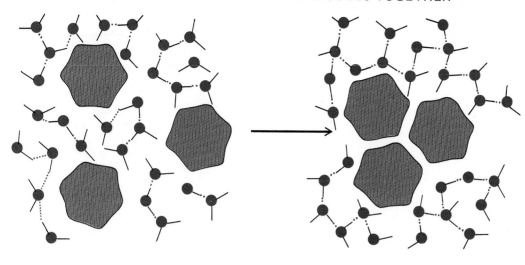

Small oil droplets coalesce into large oil drops in water since they thereby cause less disruption to the large hydrogen-bonded network of water molecules. For the same reason, hydrophobic molecules in aqueous solution will tend to be pushed together into larger aggregates by the water.

ACIDS AND BASES

An acid is a molecule that releases an H^+ ion (proton) in solution.

e.g.,

$$CH_3-C(=O)OH \rightleftharpoons CH_3-C(=O)O^- + H^+$$

    acid   base   proton

A base is a molecule that accepts an H^+ ion (proton) in solution.

e.g.,

$$CH_3-NH_2 + H^+ \rightleftharpoons CH_3-NH_3^+$$

  base   proton   acid

Water itself has a slight tendency to ionize and therefore can act as both a weak acid and as a weak base. When it acts as an acid, it releases a proton to form a hydroxyl ion. When it acts as a base, it accepts a proton to form a hydronium ion. Most protons in aqueous solutions exist as hydronium ions.

hydroxyl ion   hydronium ion

pH

The acidity of a solution is defined by the concentration of H^+ ions it possesses. For convenience we use the pH scale where

$$pH = -\log_{10}[H^+]$$

For pure water

$$[H^+] = 10^{-7} \text{ moles/liter}$$

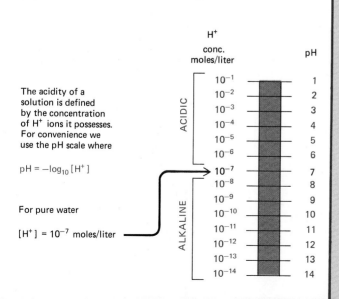

	H^+ conc. moles/liter	pH
ACIDIC	10^{-1}	1
	10^{-2}	2
	10^{-3}	3
	10^{-4}	4
	10^{-5}	5
	10^{-6}	6
	10^{-7}	7
ALKALINE	10^{-8}	8
	10^{-9}	9
	10^{-10}	10
	10^{-11}	11
	10^{-12}	12
	10^{-13}	13
	10^{-14}	14

OSMOSIS

If two aqueous solutions are separated by a membrane that allows only water molecules to pass, water will move into the solution containing the greatest concentration of solute molecules by a process known as osmosis.

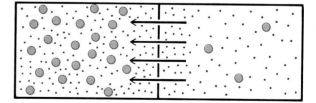

This movement of water from a hypotonic to a hypertonic solution can cause an increase in hydrostatic pressure in the hypertonic compartment. Two solutions that have identical solute concentrations and are therefore osmotically balanced are said to be isotonic.

CARBON SKELETONS

The unique role of carbon in the cell comes from its ability to form strong covalent bonds with other carbon atoms. Thus carbon atoms can join to form chains.

or branched trees

or rings

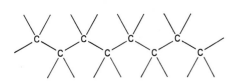

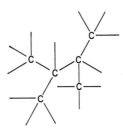

[also written as]

[also written as]

[also written as]

COVALENT BONDS

Atoms in biological molecules are usually joined by covalent bonds (formed by sharing pairs of electrons). Each atom can form a fixed number of such bonds in a definite spatial arrangement.

Double bonds exist and have a different spatial arrangement.

HYDROCARBONS

Carbon and hydrogen together make stable compounds called hydrocarbons. These are nonpolar, do not form hydrogen bonds, and are generally insoluble in water.

$$H-\overset{\overset{\displaystyle H}{|}}{\underset{\underset{\displaystyle H}{|}}{C}}-H \qquad H-\overset{\overset{\displaystyle H}{|}}{\underset{\underset{\displaystyle H}{|}}{C}}-$$

methane   methyl group

Part of a fatty acid chain:

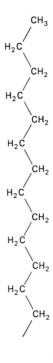

RESONANCE AND AROMATICITY

The carbon chain can include double bonds. If these are on alternate carbon atoms, the bonding electrons move within the molecule, stabilizing the structure by a phenomenon called resonance.

When resonance occurs throughout a ring compound, an aromatic ring is generated.

the truth is somewhere between these two structures

[often written as]

C–O COMPOUNDS

Many biological compounds contain a carbon bonded to an oxygen. For example:

alcohol

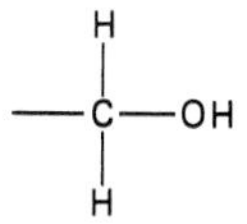

The –OH is called a hydroxyl group.

aldehyde

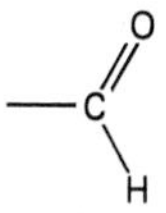

ketone

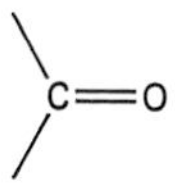

The C=O is called a carbonyl group.

carboxylic acid

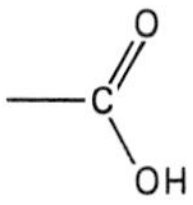

The –COOH is called a carboxyl group. In water this loses a H^+ ion to become –COO$^-$.

esters

Esters are formed by combining an acid and an alcohol:

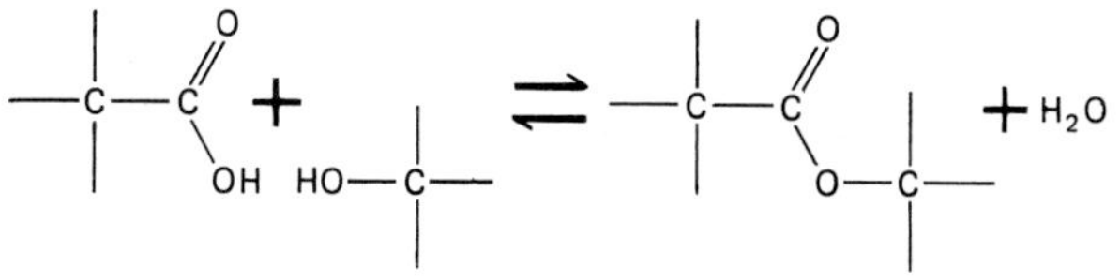

acid + alcohol ⇌ ester + H_2O

C–N COMPOUNDS

Amines and amides are two important examples of compounds containing a carbon linked to a nitrogen.

Amines in water combine with a H^+ ion to become positively charged.

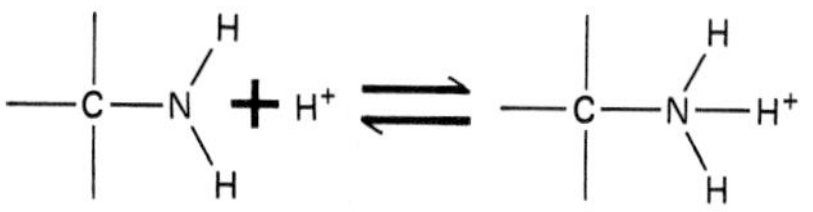

They are therefore basic.

Amides are formed by combining an acid and an amine. They are more stable than esters. Unlike amines, they are uncharged in water. An example is the peptide bond.

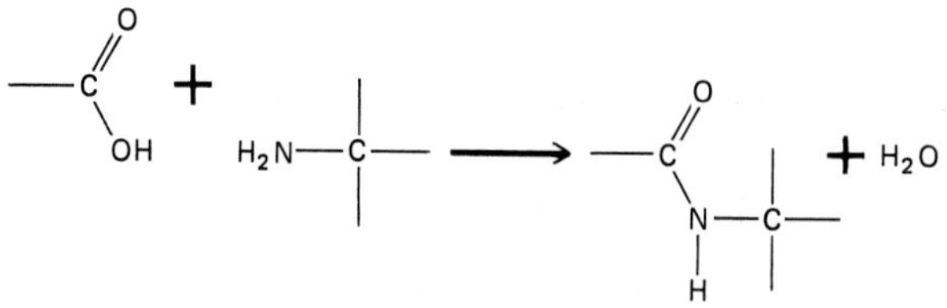

Nitrogen also occurs in several ring compounds, including important constituents of nucleic acids: purines and pyrimidines.

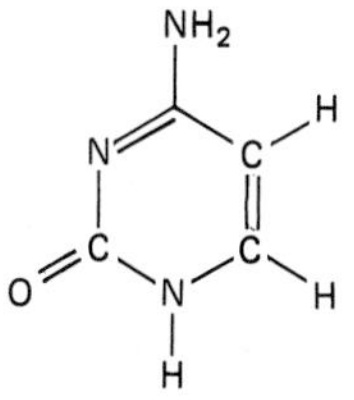

cytosine (a pyrimidine)

PHOSPHATES

Inorganic phosphate is a stable ion formed from phosphoric acid, H_3PO_4. It is often written as P_i.

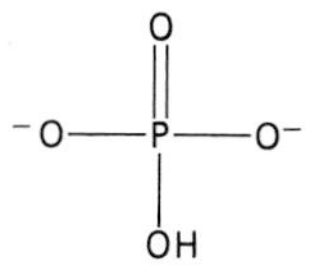

Phosphate esters can form between a phosphate and a free hydroxyl group.

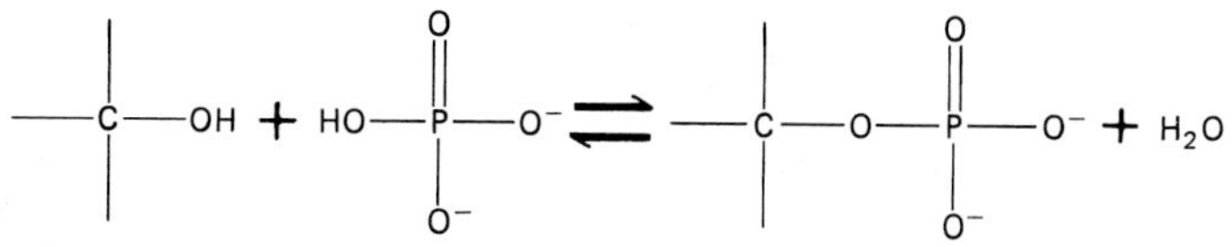

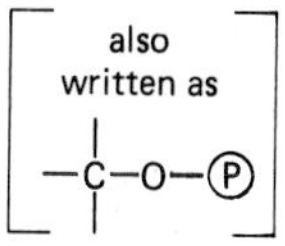

also written as

The combination of a phosphate and a carboxyl group, or two or more phosphate groups, gives an acid anhydride.

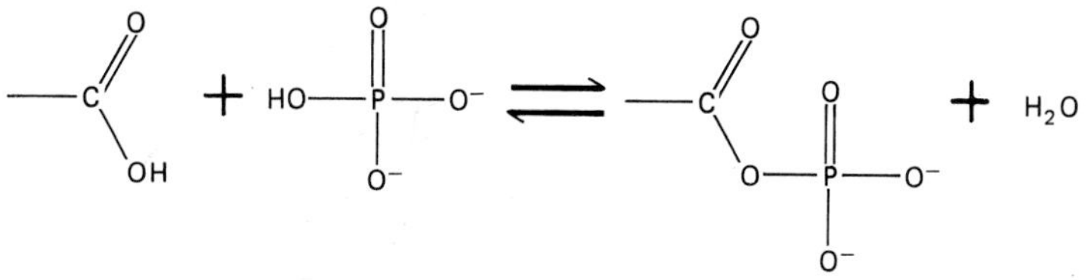

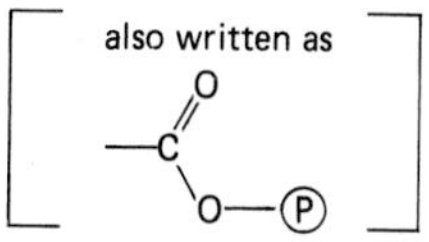

also written as

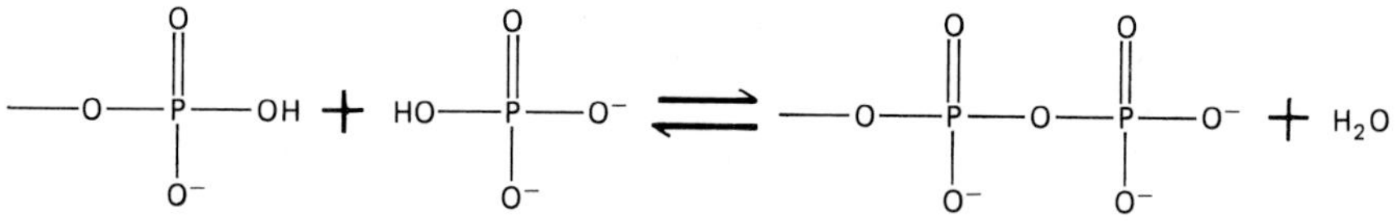

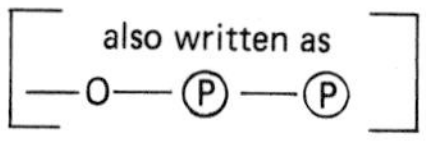

also written as

HEXOSES $n = 6$

Two common hexoses are
glucose   fructose

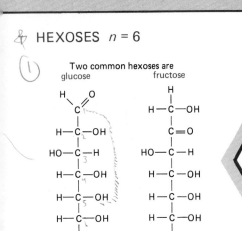

MONOSACCHARIDES

Monosaccharides are
aldehydes   or   ketones

$$\left(-C\!\!\begin{array}{c}O\\\diagdown H\end{array}\right) \quad \left(\begin{array}{c}\diagup\\\diagdown\end{array}C=O\right)$$

that also have two or more hydroxyl groups. Their general formula is $(CH_2O)_n$. The simplest are trioses ($n = 3$) such as

glyceraldehyde (an aldose)

$$CH_2OH$$
$$C=O$$
$$CH_2OH$$

dihydroxyacetone (a ketose)

PENTOSES $n = 5$

A common pentose is

ribose

RING FORMATION

The aldehyde or ketone group of a sugar can react with a hydroxyl group.

For the larger sugars ($n > 4$) this happens within the same molecule to form a 5- or 6-membered ring.

NUMBERING

The carbon atoms of a sugar are numbered from the end closest to the aldehyde or ketone.

D-glucose (open-chain form)

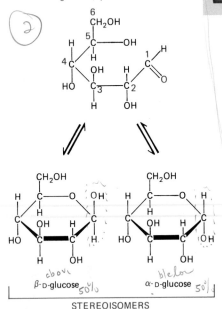

β-D-glucose 50%

α-D-glucose 50%

STEREOISOMERS

D-ribose (open-chain form)

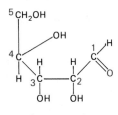

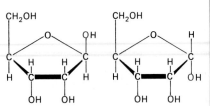

β-D-ribose   α-D-ribose

STEREOISOMERS

STEREOISOMERS

Monosaccharides have many isomers that differ only in the orientation of their hydroxyl groups — e.g., glucose, galactose, and mannose are isomers of each other.

glucose

mannose

galactose

D AND L FORMS

Two isomers that are mirror images of each other have the same chemistry and therefore are given the same name and distinguished by the prefix D or L.

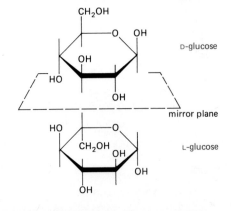

D-glucose

mirror plane

L-glucose

α- AND β-LINKS

The hydroxyl group on the carbon that carries the aldehyde or ketone can rapidly change from one position to another. These two positions are called α- and β-.

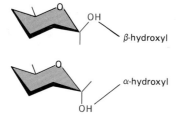

β-hydroxyl

α-hydroxyl

As soon as one sugar is linked to another, the α- or β-form is frozen.

SUGAR DERIVATIVES

The hydroxyl groups of a simple monosaccharide can be replaced by other groups. For example

D-glucuronic acid

D-glucosamine

N-acetyl-D-glucosamine

DISACCHARIDES

The carbon that carries the aldehyde or the ketone can react with any hydroxyl group on a second sugar molecule to form a glycosidic bond. Three common disaccharides are maltose (glucose α1,4 glucose), lactose (galactose β1,4 glucose), and sucrose (glucose α1,2 fructose).

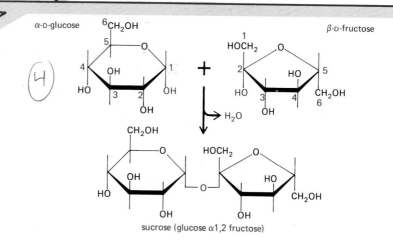

α-D-glucose   β-D-fructose

+ → H_2O

sucrose (glucose α1,2 fructose)

OLIGOSACCHARIDES AND POLYSACCHARIDES

Large linear and branched molecules can be made from simple repeating units. Short chains are called oligosaccharides, while long chains are called polysaccharides. Glycogen, for example, is a polysaccharide made entirely of glucose units joined together.

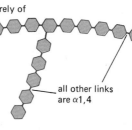

α1,6 links occur at branch points

glycogen

all other links are α1,4

COMPLEX OLIGOSACCHARIDES

In many cases a sugar sequence is nonrepetitive. Very many different molecules are possible. Such complex oligosaccharides are usually linked to proteins or to lipids

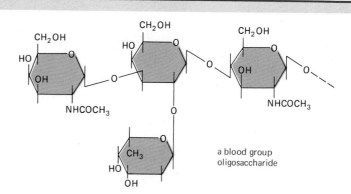

a blood group oligosaccharide

COMMON FATTY ACIDS

These are carboxylic acids with long hydrocarbon tails.

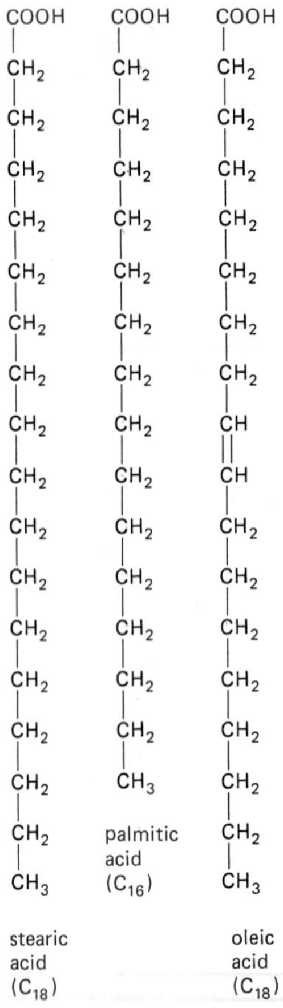

Hundreds of different kinds of fatty acids exist. Some have one or more double bonds and are said to be unsaturated.

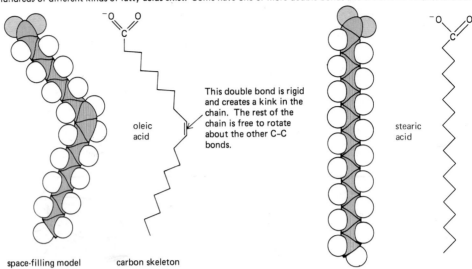

stearic acid

COOH	COOH	COOH
CH₂	CH₂	CH₂

```
COOH    COOH    COOH
 |       |       |
CH₂     CH₂     CH₂
 |       |       |
CH₂     CH₂     CH₂
 |       |       |
CH₂     CH₂     CH₂
 |       |       |
CH₂     CH₂     CH₂
 |       |       |
CH₂     CH₂     CH₂
 |       |       |
CH₂     CH₂     CH₂
 |       |       |
CH₂     CH₂     CH₂
 |       |       |
CH₂     CH₂     CH
 |       |       ‖
CH₂     CH₂     CH
 |       |       |
CH₂     CH₂     CH₂
 |       |       |
CH₂     CH₂     CH₂
 |       |       |
CH₂     CH₂     CH₂
 |       |       |
CH₂     CH₂     CH₂
 |       |       |
CH₂     CH₃     CH₂
 |    palmitic    |
CH₂    acid      CH₂
 |    (C₁₆)      |
CH₃             CH₃
```

stearic acid (C₁₈)   palmitic acid (C₁₆)   oleic acid (C₁₈)

oleic acid

This double bond is rigid and creates a kink in the chain. The rest of the chain is free to rotate about the other C–C bonds.

space-filling model   carbon skeleton

TRIGLYCERIDES

Fatty acids are stored as an energy reserve (fat) through an ester linkage to glycerol to form triglycerides.

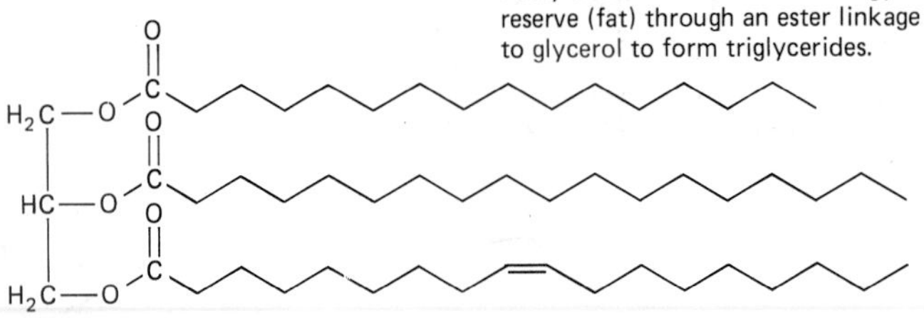

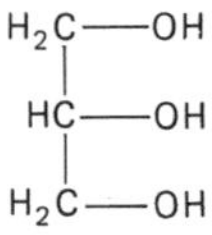

glycerol

CARBOXYL GROUP

If free, the carboxyl group of a fatty acid will be ionized.

But more usually it is linked to other groups to form either esters

or amides

PHOSPHOLIPIDS

Phospholipids are the major constituent of cell membranes.

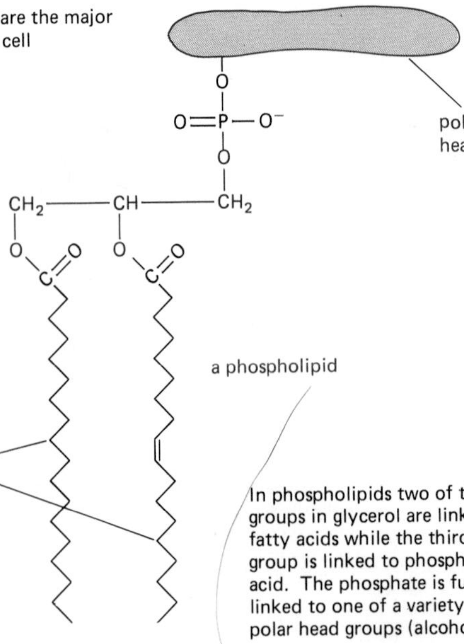

polar head group

a phospholipid

hydrophobic fatty acid "tails"

In phospholipids two of the —OH groups in glycerol are linked to fatty acids while the third —OH group is linked to phosphoric acid. The phosphate is further linked to one of a variety of small polar head groups (alcohols).

---

52   **Panel 2–4**   An outline of some of the types of fatty acids commonly encountered in cells and the structures that they form.

LIPID AGGREGATES

Fatty acids have a hydrophilic head and a hydrophobic tail.

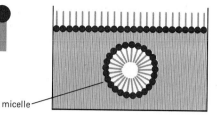

In water they can form a surface film or form small micelles.

micelle

Their derivatives can form larger aggregates held together by hydrophobic forces:

Triglycerides form large spherical fat droplets in the cell cytoplasm.

200 nm or more

Phospholipids and glycolipids form self-sealing lipid bilayers that are the basis for all cellular membranes.

5 nm

OTHER LIPIDS

Lipids are defined as the water-insoluble molecules in cells that are soluble in organic solvents. Two other common types of lipid are steroids and polyisoprenoids. Both are made from isoprene units.

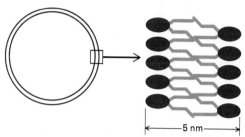

isoprene

STEROIDS

Steroids have a common multiple-ring structure.

HO

cholesterol — found in many membranes

OH

O

testosterone — male steroid hormone

GLYCOLIPIDS

Like phospholipids, these compounds are composed of a hydrophobic region, containing two long hydrocarbon tails, and a polar region, which now contains one or more sugar residues and no phosphate.

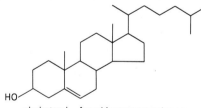

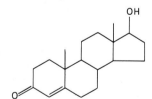

galactose

sugar residue

hydrophobic region

a simple glycolipid

POLYISOPRENOIDS

long chain polymers of isoprene

dolichol phosphate — used to carry activated sugars in the membrane-associated synthesis of glycoproteins and some polysaccharides.

THE AMINO ACID

The general formula of an amino acid is

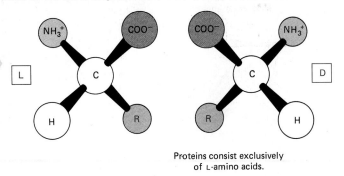

amino group — H_2N — C — COOH
H — α-carbon atom
carboxyl group
R — side-chain group

R is commonly one of 20 different side chains. At pH 7 both the amino and carboxyl groups are ionized.

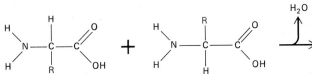

$\oplus H_3N$ — C — $COO^\ominus$
H
R

OPTICAL ISOMERS

The α-carbon atom is asymmetric, which allows for two mirror image (or stereo-) isomers, D and L.

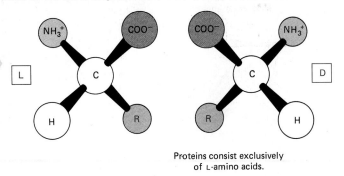

L

NH_3^+ COO^-
C
H   R

D

COO^- NH_3^+
C
R   H

Proteins consist exclusively of L-amino acids.

PEPTIDE BONDS

Amino acids are commonly joined together by an amide linkage, called a peptide bond.

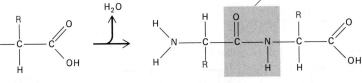

peptide bond

Proteins are long polymers of amino acids linked by peptide bonds, and they are always written with the N-terminus toward the left. The sequence of this tripeptide is His Cys Val.

amino or N-terminus

carboxyl or C-terminus

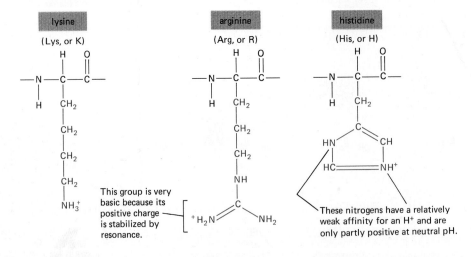

FAMILIES OF AMINO ACIDS

The common amino acids are grouped according to whether their side chains are

  acidic
  basic
  uncharged polar
  nonpolar

These 20 amino acids are given both three-letter and one-letter abbreviations.

Thus: alanine = Ala = A

BASIC SIDE CHAINS

lysine
(Lys, or K)

arginine
(Arg, or R)

histidine
(His, or H)

This group is very basic because its positive charge is stabilized by resonance.

These nitrogens have a relatively weak affinity for an H^+ and are only partly positive at neutral pH.

ACIDIC SIDE CHAINS

aspartic acid

(Asp, or D)

glutamic acid

(Glu, or E)

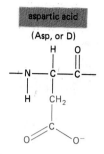

Amino acids with uncharged polar side chains are relatively hydrophilic and are usually on the outside of proteins, while the side chains on nonpolar amino acids tend to cluster together on the inside. Amino acids with basic and acidic side chains are very polar and they are nearly always found on the outside of protein molecules.

✳ UNCHARGED POLAR SIDE CHAINS ✳

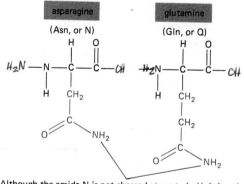

asparagine

(Asn, or N)

glutamine

(Gln, or Q)

Although the amide N is not charged at neutral pH, it is polar.

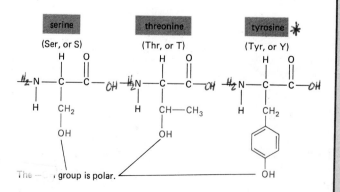

serine

(Ser, or S)

threonine

(Thr, or T)

tyrosine ✳

(Tyr, or Y)

The —OH group is polar.

✳ NONPOLAR SIDE CHAINS ✳

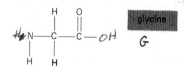

glycine

G

alanine

(Ala, or A)

valine

(Val, or V)

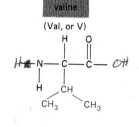

15.0347

leucine

(Leu, or L)

isoleucine

(Ileu, or I)

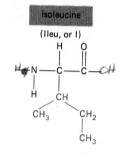

proline

(Pro, or P)

phenylalanine

(Phe, or F)

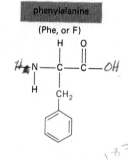

(actually an imino acid)

methionine

(Met, or M)

tryptophan

(Trp, or W)

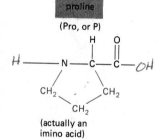

187.2203

56.0435

N 14.0067
9C 12.011
9H 1.0079

131.1768

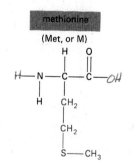

cysteine

(Cys, or C)

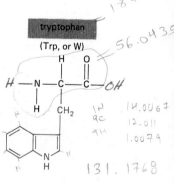

Paired cysteines allow disulfide bonds to form in proteins.

$$—CH_2—S—S—CH_2—$$

55

The bases are N-containing ring compounds, either purines or pyrimidines.

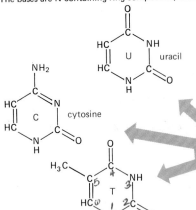

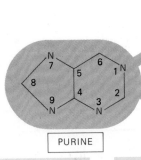

PYRIMIDINE

PURINE

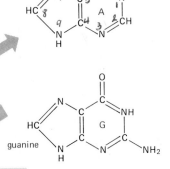

adenine

guanine

PHOSPHATES

The phosphates are normally joined to the C5 hydroxyl of the ribose or deoxyribose sugar. Mono-, di-, and triphosphates are common.

as in AMP

as in ADP

as in ATP

The phosphate makes a nucleotide negatively charged.

NUCLEOTIDES

A nucleotide consists of a nitrogen-containing base, a 5-carbon sugar, and one or more phosphate groups.

BASE

PHOSPHATE

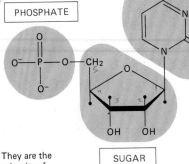

SUGAR

They are the subunits of the nucleic acids.

BASE–SUGAR LINKAGE

N-glycosidic bond

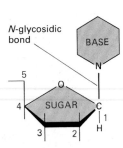

BASE

SUGAR

The base is linked to the same carbon (C1) used in sugar-sugar bonds.

SUGARS

PENTOSE

a 5-carbon sugar

two kinds are used

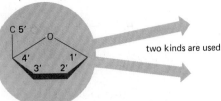

β-D-RIBOSE
used in ribonucleic acid

β-D-2-DEOXYRIBOSE
used in deoxyribonucleic acid

Each numbered carbon on the sugar of a nucleotide is followed by a prime mark; therefore, one speaks of the "5-prime carbon," etc.

NOMENCLATURE

The names can be confusing, but the abbreviations are clear.

$$BASE \; + \; SUGAR \; = \; NUCLEOSIDE$$
$$BASE \; + \; SUGAR \; + \; PHOSPHATE \; = \; NUCLEOTIDE$$

BASE	NUCLEOSIDE	ABBR.
adenine	adenosine	A
guanine	guanosine	G
cytosine	cytidine	C
uracil	uridine	U
thymine	thymidine	T

Nucleotides are abbreviated by three capital letters as follows:

AMP = adenosine monophosphate
dAMP = deoxyadenosine monophosphate
UDP = uridine diphosphate
ATP = adenosine triphosphate
etc.

NUCLEIC ACIDS

Nucleotides are joined together by a phosphodiester linkage between 5' and 3' carbon atoms to form nucleic acids. The linear sequence of nucleotides in a nucleic acid chain is commonly abbreviated by a one-letter code, A–G–C–T–T–A–C–A, with the 5' end of the chain written at the left.

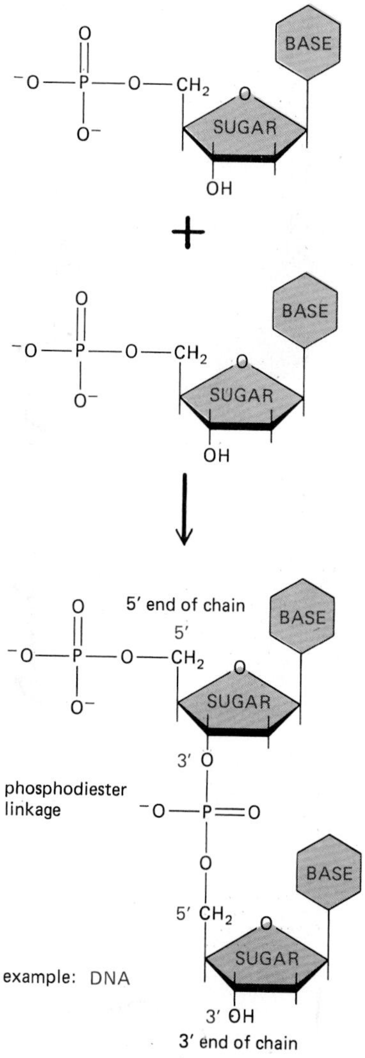

5' end of chain

phosphodiester linkage

3' end of chain

example: DNA

✳ NUCLEOTIDES HAVE MANY OTHER FUNCTIONS ✳

1. They carry chemical energy in their easily hydrolyzed acid-anhydride bonds.

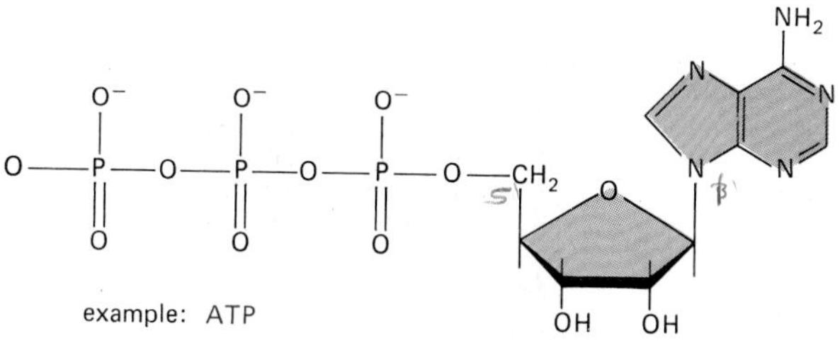

example: ATP

2. They combine with other groups to form coenzymes.

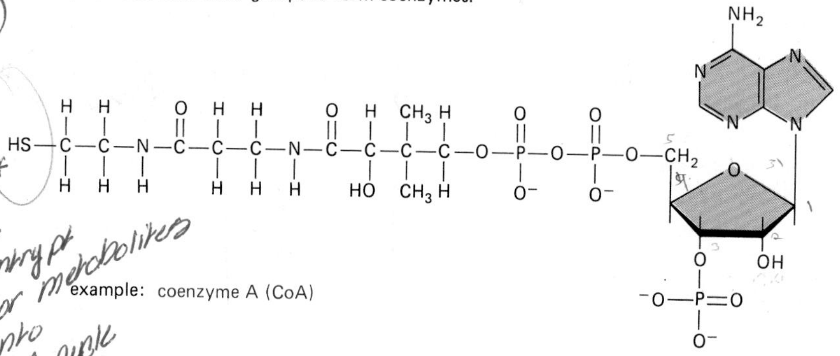

entry pt for metabolites into TCA cycle

example: coenzyme A (CoA)

3. They are used as specific signaling molecules in the cell.

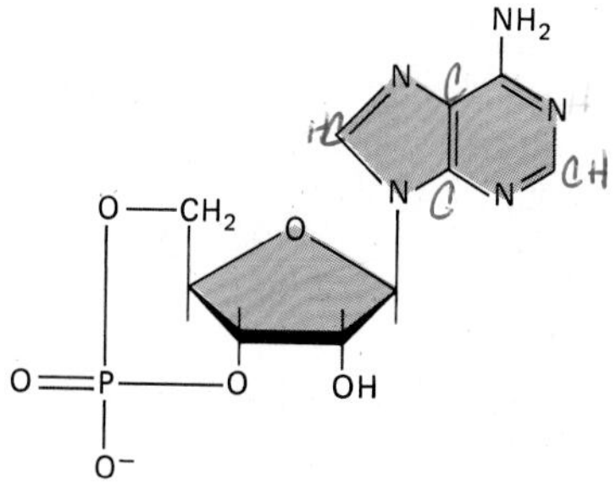

example: cyclic AMP

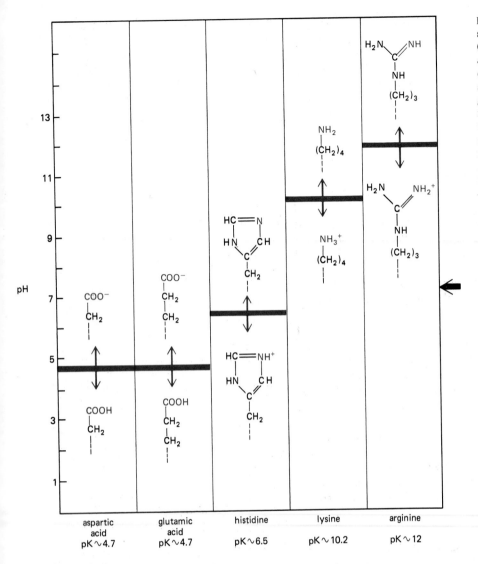

The chart shows amino acid side chains with their charged/uncharged forms at different pH values, with a pH axis on the left (values 1, 3, 5, 7, 9, 11, 13).

aspartic acid pK ∿4.7

glutamic acid pK ∿4.7

histidine pK ∿6.5

lysine pK ∿10.2

arginine pK ∿12

Figure 2–8 The charge on amino acid side chains depends on the pH. Carboxylic acids readily lose H$^+$ in aqueous solution to form a negatively charged ion that is denoted by the suffix "-ate," as in aspart*ate* or glutam*ate*. A comparable situation exists for amines, which in aqueous solution take up H$^+$ to form a positively charged ion (which does not have a special name). These reactions are rapidly reversible, and the amounts of the two forms, charged and uncharged, depend on the pH of the solution. At a high pH, carboxylic acids tend to be charged and amines uncharged. At a low pH, the opposite is true—the carboxylic acids are uncharged and amines are charged. The pH at which exactly *half* of the carboxylic acid or amine residues are charged is known as the pK of that amino acid.

In the cell the pH is close to 7, and almost all carboxylic acids and amines are in their fully charged form.

enine derivative, *cyclic AMP*, serves as a universal signaling molecule within cells and controls the rate of many different intracellular reactions.

As we have already mentioned in Chapter 1, the special significance of nucleotides is in the storage of biological information. Nucleotides serve as building blocks for the construction of **nucleic acids,** long polymers in which nucleotide subunits are covalently linked by the formation of a phosphate ester between the 3'-hydroxyl group on the sugar residue of one nucleotide and the 5'-phosphate group on the next nucleotide (Figure 2–10). There are two main types of nucleic acids, differing in the type of sugar in their polymeric backbone. Those based on the sugar *ribose* are known as **ribonucleic acids,** or **RNA,** and contain the bases A, U, G, and C. Those based on *deoxyribose* (in which the hydroxyl at the 2' position of ribose is replaced by a hydrogen) are known as **deoxyribonucleic acids,** or **DNA,** and contain the bases A, T, G, and C. The sequence of bases in a DNA or RNA polymer represents the genetic information of the living cell. The ability of the bases from different nucleic acid molecules to recognize each other by noncovalent interactions (called **base-pairing**)—G with C, and A with either T or U—underlies all of heredity and evolution, as will be explained in the following chapter.

Summary

Living organisms are autonomous, self-propagating chemical systems. They are made from a distinctive and restricted set of small carbon-based molecules that are essentially the same for every living species. The main categories are sugars,

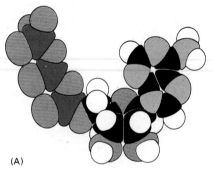

(A)

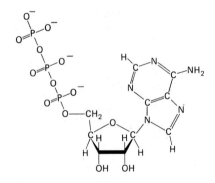

triphosphate   ribose   adenine

adenosine

(B)

Figure 2–9 Chemical structure of adenosine triphosphate (ATP). Both a space-filling model (A) and the structural formula (B) are shown.

fatty acids, amino acids, and nucleotides. Sugars are a primary source of chemical energy for cells and can be incorporated into polysaccharides for energy storage. Fatty acids are also important for energy storage, but their most significant function is in the formation of cell membranes. Polymers consisting of amino acids constitute the remarkable diverse and versatile macromolecules known as proteins. Nucleotides play a central part in energy transfer and are involved in intracellular signaling, but their unique role is as the subunits of the informational macromolecules, RNA and DNA.

Biological Order and Energy[6]

Cells must obey the laws of physics and chemistry. The rules of mechanics and of the conversion of one form of energy to another apply just as much to a cell as to a steam engine. It must be admitted, however, that there are puzzling features of a cell that, at first sight, seem to place it in a special category. It is common experience that things left to themselves eventually become disordered: buildings crumble, dead organisms decay, and so on. This general tendency is expressed in the *second law of thermodynamics*, which states that in any isolated system the degree of disorder can only increase.

The puzzle is that living organisms maintain, at every level, a very high degree of order; and as they feed, develop, and grow, they appear to create this order out of raw materials that lack it. Order is strikingly apparent in large structures such as a butterfly wing or an octopus eye, in subcellular structures such as a mitochondrion or a cilium, and in the shape and arrangement of molecules from which these structures are built. The constituent atoms have been captured, ultimately, from a relatively disorganized state in the environment and locked together into a precise structure. Even a nongrowing cell requires constant ordering processes for survival, since all of its organized structures are subject to spontaneous accidents and must be repaired continually. How is this possible thermodynamically? We shall see that the answer lies in the fact that the cell is constantly releasing heat to its environment, and therefore the cell is not an isolated system in the thermodynamic sense.

Biological Order Is Made Possible by the Release of Heat Energy from Cells[7]

Some important thermodynamic principles can be understood most easily if one considers the idealized example of a cell plus its immediate environment contained in a sealed box, sitting in a sea of matter representing the rest of the universe (Figure 2–11). As the cell grows and maintains itself, it creates order inside the box. As we have just seen, however, the second law of thermodynamics states that the amount of order in the entire system (that is, in the box plus the sea) must always decrease. Therefore, the increase in order inside the box must be more than compensated for by an even greater increase in *disorder* in the rest of the universe. Although molecules in our example cannot be exchanged between the box and the sea, heat can be, and there is a quantitative relationship between heat and order. Heat is energy in its most disordered form—the random commotion of molecules. If the cell releases heat to the sea, it increases the intensity of molecular motions in the sea—thereby increasing their randomness, or disorder.

The quantitative relationship between heat and order, first recognized in the late nineteenth century, enables us to calculate in principle exactly how much heat a cell must release in order to compensate for a given amount of ordering within it (such as the assembly of proteins from amino acids), with the net process increasing the total disorder of the universe. This relationship can be derived by considering the changes of molecular motions that result when a given amount of heat energy is transferred from a hot body to a cold one. The quantitative details of the relationship need not concern us here, but it is important to note that the cell will achieve nothing by producing heat unless the heat-generating reactions are linked with the processes that generate molecular order in the cell. Such linked

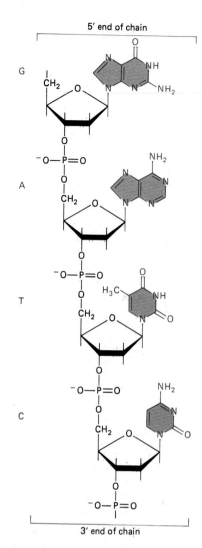

Figure 2–10 A short length of deoxyribonucleic acid, or DNA. DNA and its close relative RNA are the nucleic acids of the cell. This region of the DNA chain has the sequence G-A-T-C.

reactions are said to be *coupled*, as will be explained subsequently. It is the tight coupling of heat production to an increase in order that distinguishes the metabolism of the cell from the wasteful burning of fuel in a fire.

Figure 2–11 illustrates in a general way how coupled reactions release heat energy, which disorders the environment, thereby compensating for the increase in order they create in the cell. It is the resulting increase in the disorder of the universe that drives the coupled reactions in the direction of cell ordering.

Energy cannot be created or destroyed in chemical reactions. Therefore, in our example, the heat loss from the box that drives the production of biological order inside it requires an input of energy if it is to continue. This energy must be in a form other than heat. For plants, the energy is initially derived from the electromagnetic radiation of the sun; for animals, it is derived from the energy stored in the covalent bonds of the organic molecules they eat. However, since these organic nutrients are themselves produced by photosynthetic organisms, such as green plants, the sun is in fact the ultimate energy source for both types of organisms.

Photosynthetic Organisms Use Sunlight to Synthesize Organic Compounds[8]

Solar energy enters the living world (the *biosphere*) by means of the **photosynthesis** carried out by photosynthetic organisms—either plants or bacteria. In photosynthesis, electromagnetic energy is converted into chemical bond energy. At the same time, however, part of the energy of sunlight is converted into heat energy, and the release of this heat to the environment increases the disorder of the universe and thereby drives the photosynthetic process.

The reactions of photosynthesis are described in detail in Chapter 7. In broad terms, they occur in two distinct stages. In the first (the *light reactions*), the visible radiation excites an electron in a pigment molecule, which, in returning to a lower energy state, provides the energy needed for the synthesis of ATP and NADPH. In the second (the *dark reactions*), the ATP and NADPH are used to drive a series of "carbon-fixation" reactions in which CO_2 from the air is used to form sugar molecules (Figure 2–12).

The *net* result of photosynthesis, so far as the green plant is concerned, can be summarized by the equation

$$\text{energy} + CO_2 + H_2O \rightarrow \text{sugar} + O_2$$

which is the reverse of the oxidative decomposition of a sugar. However, this simple equation hides the complex nature of the dark reactions, which involve many linked reaction steps. Furthermore, although the initial fixation of CO_2 results in sugars, subsequent metabolic reactions soon convert these into the other small and large molecules essential to the plant cell.

Chemical Energy Passes from Plants to Animals

Animals and other nonphotosynthetic organisms cannot capture energy from sunlight directly and so have to survive on "second-hand" energy obtained by eating plants or on "third-hand" energy obtained by eating other animals. The organic molecules made by plant cells provide both building blocks and fuel to the organisms that feed on them. All types of plant molecules can serve this purpose—sugars, proteins, polysaccharides, lipids, and many others.

The transactions between plants and animals are not all one-way. Plants, animals, and microorganisms have existed together on this planet for so long that many of them have become an essential part of the others' environment. The oxygen released by photosynthesis is consumed in the combustion of organic molecules by nearly all organisms, and some of the CO_2 molecules that are "fixed" today into larger organic molecules by photosynthesis in a green leaf were yesterday released into the atmosphere by the respiration of an animal. Thus, carbon utilization is a cyclic process that involves the biosphere as a whole and crosses

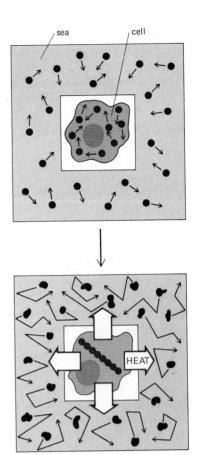

Figure 2–11 For a simple thermodynamic analysis of a living cell, it is useful to consider it and its immediate environment enclosed in a sealed box that allows heat but not molecules to be exchanged with the rest of the universe, which we have designated here as the "sea." In the upper diagram, the molecules of both the cell and the rest of the universe are depicted in a relatively disordered state. In the lower diagram, heat has been released from the cell by a reaction that orders the molecules it contains. The increase in random motion, including bond distortions, of the molecules in the rest of the universe creates a disorder that more than compensates for the increased order in the cell, as required by the laws of thermodynamics for spontaneous processes. In this way the release of heat by a cell to its surroundings allows it to become more highly ordered internally at the same time that the universe as a whole becomes more disordered.

boundaries between individual organisms (Figure 2–13). Similarly, atoms of nitrogen, phosphorus, and sulfur can, in principle, be traced from one biological molecule to another in a series of similar cycles.

Cells Obtain Energy by the Oxidation of Biological Molecules[9]

The carbon and hydrogen atoms in a cell are not in their most stable form. Because the earth's atmosphere contains a great deal of oxygen, the most energetically stable form of carbon is as CO_2 and that of hydrogen is as H_2O. A cell is therefore able to obtain energy from sugar or protein molecules by allowing their carbon and hydrogen atoms to combine with oxygen to produce CO_2 and H_2O, respectively. However, the cell does not oxidize molecules in one step, as occurs in a fire. It takes them through a large number of reactions that only rarely involve the direct addition of oxygen. Before we can consider these reactions and the driving force behind them, we need to discuss what is meant by the process of oxidation.

Oxidation, in the sense used above, does not mean only the addition of oxygen atoms; rather, it applies more generally to any reaction in which electrons are transferred from one atom to another. **Oxidation** in this sense refers to the removal of electrons, and **reduction**—the converse of oxidation—means the addition of electrons. Thus, Fe^{2+} is oxidized if it loses an electron to become Fe^{3+}, and a chlorine atom is reduced if it gains an electron to become Cl^-. The same terms are used when there is only a partial shift of electrons between atoms linked by a covalent bond. For example, when a carbon atom becomes covalently bonded to an electronegative atom such as oxygen, chlorine, or sulfur, it gives up more than its equal share of electrons—acquiring a partial positive charge—and so is said to be oxidized. Conversely, a carbon atom in a C-H linkage has more than its share of electrons, and so it is said to be reduced (Figure 2–14).

The combustion of food materials in a cell converts the C and H atoms in organic molecules (where they are both in a relatively electron-rich, or reduced, state) to CO_2 and H_2O, where they have given up electrons and are therefore highly oxidized. The shift of electrons from carbon and hydrogen to oxygen allows all of these atoms to achieve a more stable state and hence is energetically favorable.

The Breakdown of Organic Molecules Takes Place in Sequences of Enzyme-catalyzed Reactions[10]

Although the most energetically favorable form of carbon is as CO_2 and that of hydrogen is as H_2O, a living organism does not disappear in a puff of smoke for the same reason that the book in your hands does not burst into flame: the

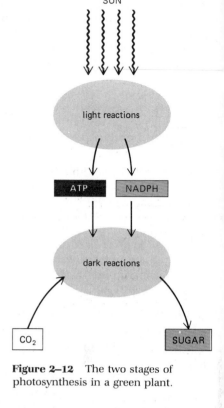

Figure 2–12 The two stages of photosynthesis in a green plant.

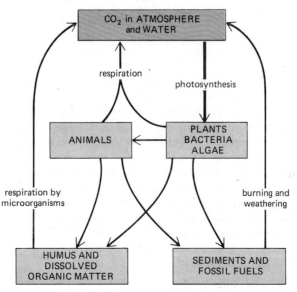

Figure 2–13 The carbon cycle. Individual carbon atoms are incorporated into organic molecules of the living world by the photosynthetic activity of plants, bacteria, and marine algae. They pass to animals, microorganisms, and organic material in soil and oceans in cyclic paths. CO_2 is restored to the atmosphere when organic molecules are oxidized by cells or burned by humans as fossil fuels.

molecules of both exist in metastable energy troughs (Figure 2–15) and require *activation energy* before they can pass to more stable configurations. In the case of the book, the activation energy can be provided by a lighted match. For a living cell, the combustion is achieved in a less sudden and destructive fashion. Highly specific protein catalysts, or **enzymes,** combine with biological molecules in such a way that they reduce the activation energy of particular reactions that the bound molecules can undergo. By selectively lowering the activation energy of one pathway or another, enzymes determine which of several alternative reaction paths is followed (Figure 2–16). In this way, each of the many different molecules in a cell is directed through specific reaction pathways, thereby determining the cell's chemistry.

The success of living organisms is attributable to the cell's ability to make enzymes of many different types, each with precisely specified properties. Each enzyme has a unique shape containing an *active site.* This active site binds a particular set of other molecules (called *substrates)* in such a way as to speed up a particular one of the many chemical reactions that the substrates can undergo, often by a factor of as much as 10^{14}. Like all other catalysts, enzyme molecules themselves are not changed after participating in a reaction and therefore can function over and over again.

Part of the Energy Released in Oxidation Reactions Is Coupled to the Formation of ATP[11]

Cells derive useful energy from the "burning" of glucose only because they burn it in a very complex and controlled way. By means of enzyme-directed reaction paths, the synthetic, or *anabolic,* chemical reactions that create biological order are closely coupled to the degradative, or *catabolic,* reactions that provide the energy. The crucial difference between a **coupled reaction** and an uncoupled reaction is illustrated by the mechanical analogy in Figure 2–17, where an energetically favorable chemical reaction is represented by rocks falling from a cliff. The kinetic energy of falling rocks would normally be entirely wasted in the form of heat generated when they hit the ground (section A). But, by careful design, part of the kinetic energy could be used to drive a paddle wheel that lifts a bucket of water (section B). Because the rocks can reach the ground only by moving the paddle wheel, we say that the spontaneous reaction of rock falling has been directly coupled to the nonspontaneous reaction of lifting the bucket of water. Note that because part of the energy is now used to do work in section B, the rocks hit the ground with less velocity than in section A, and therefore correspondingly less energy is wasted as heat.

In cells, enzymes play the role of paddle wheels in our analogy and couple the spontaneous burning of foodstuffs to reactions that generate the nucleoside triphosphate ATP. Just as the energy stored in the elevated bucket of water in Figure 2–17 can be dispensed in small doses to drive a wide variety of different hydraulic machines (section C), ATP serves as a convenient and versatile store, or currency, of energy to drive many different chemical reactions that the cell needs.

The Hydrolysis of ATP Generates Order in Cells[12]

How does ATP act as a carrier of chemical energy? Under the conditions existing in the cytoplasm, the breakdown of ATP by hydrolysis to release inorganic phosphate (P_i) occurs very readily and releases a great deal of usable energy (see p. 74). A chemical group that is linked by such a reactive bond is readily transferred to another molecule; for this reason, the terminal phosphate in ATP can be considered to exist in an *activated* state. The bond broken in this hydrolysis reaction is sometimes described as a *high-energy bond.* However, the important fact is that in aqueous solution the hydrolysis creates two molecules of much lower energy (ADP and P_i); there is nothing special about the covalent bond itself.

Other chemical reactions can be driven by the energy released by ATP hydrolysis through enzymes that couple the reactions to ATP hydrolysis. Among the many hundreds of these reactions are those involved in the synthesis of biological

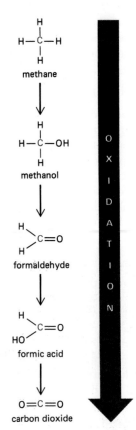

Figure 2–14 The carbon atom of methane may be converted to that of carbon dioxide by the successive removal of its hydrogen atoms. With each step, electrons are shifted away from the carbon as it passes to a more energetically stable state. Thus, in the process shown, the carbon atom becomes progressively more oxidized.

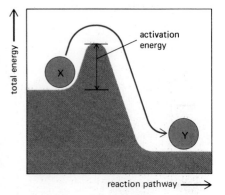

Figure 2–15 The principle of activation energy. Compound X can achieve a lower, more favorable energy state by being converted to compound Y. However, this transition will not take place unless X can acquire enough activation energy from its surroundings to undergo the reaction.

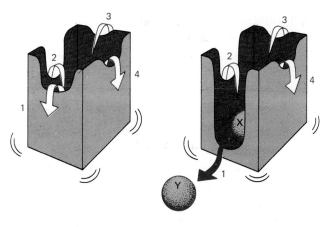

uncatalyzed

enzyme catalysis of pathway 1

Figure 2–16 A "jiggling-box" model illustrating how enzymes direct molecules along desired reaction pathways. In this model the colored ball represents a potential enzyme substrate (compound X) that is bouncing up and down due to the constant bombardment of colliding water molecules. The four walls of the box represent the activation energy barriers for four different chemical reactions that are energetically favorable. In the left-hand box, none of these reactions occurs because the energy available from collisions is insufficient to surmount any of the energy barriers. In the right-hand box, enzyme catalysis lowers the activation energy for reaction number 1 only. It thereby allows this reaction to proceed with available energies, causing compound Y to form from compound X.

molecules, in the active transport of molecules across cell membranes, and in the generation of force and movement. These three types of processes play a vital part in establishing biological order. The macromolecules formed in biosynthetic reactions carry information, catalyze specific reactions, and are assembled into highly ordered structures within cells and in the extracellular space. Membrane-bound pumps maintain the special internal composition of cells and permit signals to pass within and between cells. Finally, the production of force and movement enables the cytoplasmic contents of cells to become organized and the cells themselves to move about and assemble into organized tissues.

Summary

Living cells are highly ordered and must create order within themselves in order to survive and grow. This is thermodynamically possible only because of a continual input of energy, part of which is released from the cells to their environment as heat. The energy comes ultimately from the electromagnetic radiation of the sun, which drives the formation of organic molecules in photosynthetic organisms such as green plants. Animals obtain their energy by taking up these organic molecules and oxidizing them in a series of enzyme-catalyzed reactions that are coupled to the formation of ATP. ATP is a common currency of energy in all cells, and its hydrolysis is coupled to other reactions to drive a variety of energetically unfavorable processes that create order.

Figure 2–17 A mechanical model illustrating the principle of coupled chemical reactions. The spontaneous reaction shown in (A) might serve as an analogy for the direct oxidation of glucose to CO_2 and H_2O, which produces heat only. In (B) the same reaction is coupled to a second reaction; the second reaction might serve as an analogy for the synthesis of ATP. The more versatile form of energy produced in (B) can be used to drive other cellular processes, as in (C). ATP is the most versatile form of energy in cells.

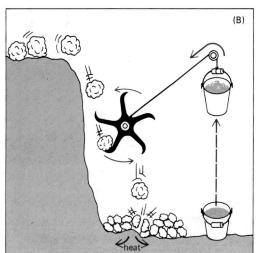

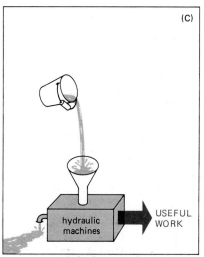

(A)

kinetic energy transformed into heat energy only

(B)

part of the kinetic energy is used to lift a bucket of water, and a correspondingly smaller amount is released as heat

(C)

the potential kinetic energy stored in the elevated bucket of water can be used to drive a wide variety of different hydraulic machines

Food and the Derivation of Cellular Energy[6,13]

Food Molecules Are Broken Down in Three Stages to Give ATP

The proteins, lipids, and polysaccharides that make up the major part of the food we eat must be broken down into smaller molecules before our cells can use them. The enzymatic breakdown, or catabolism, of these molecules may be regarded as proceeding in three stages (Figure 2–18). We shall give a short outline of these stages before discussing the last two of them in more detail.

In stage 1, called *digestion*, large polymeric molecules are broken down into their monomeric subunits—proteins into amino acids, polysaccharides into sugars, and fats into fatty acids and glycerol. These processes occur mainly outside cells through the action of secreted enzymes. In stage 2, the resultant small molecules enter cells and are further degraded in the cytoplasm. Most of the carbon and hydrogen atoms of sugars are converted into *pyruvate*, which then enters mitochondria, where it is converted to the acetyl groups of the chemically reactive compound *acetyl coenzyme A* (*acetyl CoA*, Figure 2–19). Major amounts of acetyl CoA are also produced by the oxidation of fatty acids. In stage 3, the acetyl group

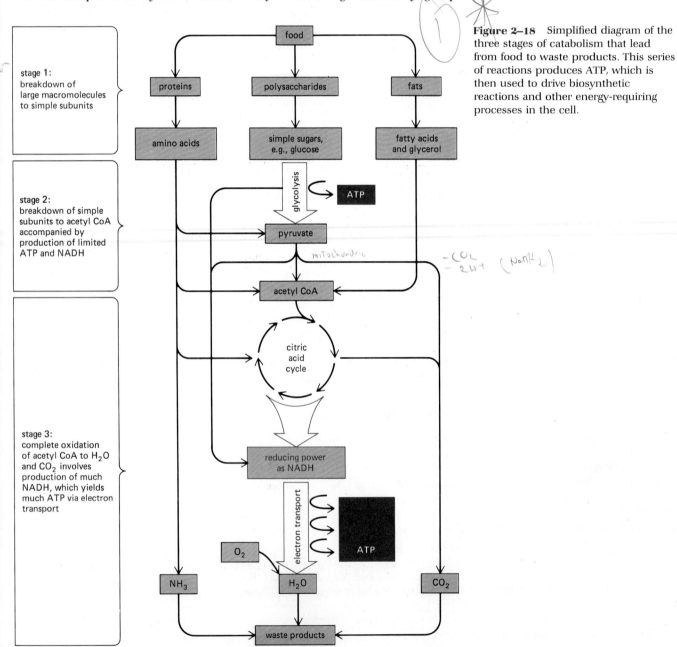

Figure 2–18 Simplified diagram of the three stages of catabolism that lead from food to waste products. This series of reactions produces ATP, which is then used to drive biosynthetic reactions and other energy-requiring processes in the cell.

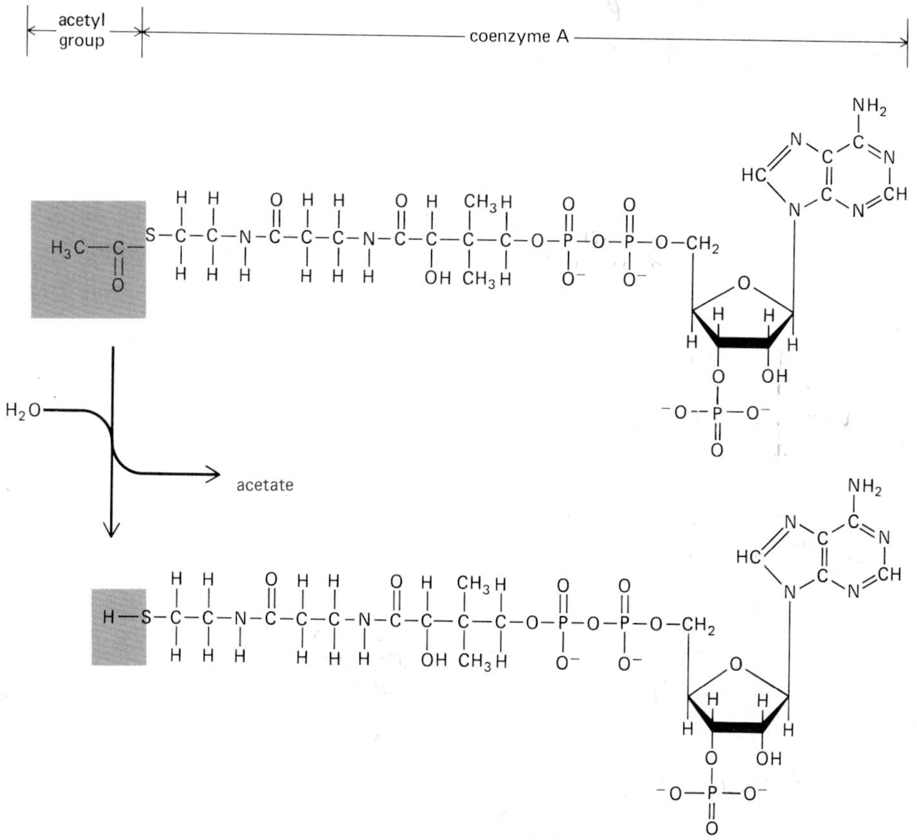

acetyl
group

coenzyme A

NH_2

H_3C-C- S-C-C-N-C-C-C-N-C-C-C-C-O-P-O-P-O-CH_2

H_2O

acetate

H-S-C-C-N-C-C-C-N-C-C-C-C-O-P-O-P-O-CH_2

NH_2

Figure 2–19 The structure of the crucial metabolic intermediate acetyl coenzyme A (acetyl CoA). Acetyl groups produced in stage 2 of catabolism (see Figure 2–18) are covalently linked to coenzyme A (CoA).

of acetyl CoA is completely degraded to CO_2 and H_2O. It is in this final stage that most of the ATP is generated. Through a series of coupled chemical reactions, more than half of the energy theoretically derivable from the combustion of carbohydrates and fats to H_2O and CO_2 is channeled into driving the energetically unfavorable reaction $P_i + ADP \rightarrow ATP$. Because the rest of the combustion energy is released by the cell as heat, this generation of ATP creates net disorder in the universe, in conformity with the second law of thermodynamics.

Through the production of ATP, the energy originally derived from the combustion of carbohydrates and fats is redistributed as a conveniently packaged form of chemical energy. Roughly 10^9 molecules of ATP are in solution throughout the intracellular space in a typical cell, where their energetically favorable hydrolysis back to ADP and phosphate provides the driving energy for a large number of different coupled reactions that would otherwise be energetically unfavorable.

Glycolysis Can Produce ATP Even in the Absence of Oxygen

The most important part of stage 2 of catabolism is a sequence of reactions known as **glycolysis**—the lysis (splitting) of glucose. In glycolysis, a glucose molecule with six carbon atoms is converted into two molecules of pyruvate, each with three carbon atoms. This conversion involves a sequence of nine enzymatic steps that create phosphate-containing intermediates (Figure 2–20). Logically, the sequence can be divided into three parts: (1) in steps 1 to 4, glucose is converted to two molecules of the three-carbon aldehyde *glyceraldehyde 3-phosphate*—a conversion that requires an investment of energy in the form of ATP hydrolysis to provide the two phosphates; (2) in steps 5 and 6, the aldehyde group of each glyceraldehyde 3-phosphate molecule is oxidized to a carboxylic acid, and the energy from this reaction is coupled to the synthesis of ATP from A·P and inorganic phosphate; and (3) in steps 7, 8, and 9, the same two phosphate molecules that were added to sugars in the first reaction sequence are transferred back to ADP to form ATP, thereby repaying the original investment of two ATP molecules hydrolyzed in the first reaction sequence.

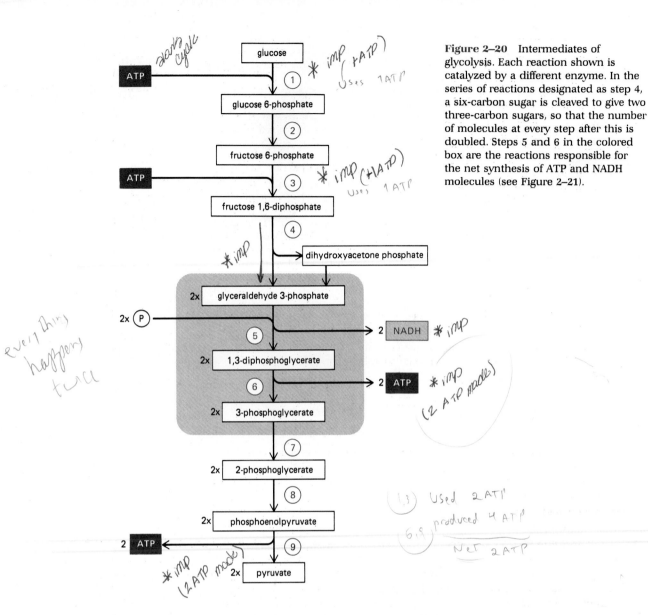

Figure 2–20 Intermediates of glycolysis. Each reaction shown is catalyzed by a different enzyme. In the series of reactions designated as step 4, a six-carbon sugar is cleaved to give two three-carbon sugars, so that the number of molecules at every step after this is doubled. Steps 5 and 6 in the colored box are the reactions responsible for the net synthesis of ATP and NADH molecules (see Figure 2–21).

At the end of glycolysis, the ATP balance sheet shows a net profit of the two molecules of ATP (per glucose molecule) that were produced in steps 5 and 6. As the only reactions in the sequence in which a high-energy phosphate linkage is created from inorganic phosphate, these two steps lie at the heart of glycolysis. They also provide an excellent illustration of the way in which reactions in the cell can be coupled together by enzymes to harvest the energy released by oxidations (Figure 2–21). The overall result is that an aldehyde group on a sugar is oxidized to a carboxylic acid and an inorganic phosphate group is transferred to a high-energy linkage on ATP; in addition, a molecule of NAD$^+$ is reduced to NADH (Figure 2–22). This elegant set of coupled reactions was probably among the earliest metabolic steps to appear in the evolving cell. Besides being central to glycolysis, these reactions are driven in reverse during photosynthesis by the large quantities of NADPH and ATP produced by the light-activated reactions; they thereby play a pivotal role in the photosynthetic carbon-fixation process by producing glyceraldehyde 3-phosphate (p. 370).

For most animal cells, glycolysis is only a prelude to stage 3 of catabolism, since the pyruvic acid that is formed quickly enters the mitochondria to be completely oxidized to CO_2 and H_2O. However, in the case of anaerobic organisms (those that do not utilize molecular oxygen) and for tissues, such as skeletal muscle, that can function under anaerobic conditions, glycolysis can become a major source of the cell's ATP. Here, instead of being degraded in mitochondria, the pyruvate molecules stay in the cytosol and, depending on the organism, can be

Figure 2–21 Steps 5 and 6 of glycolysis: the oxidation of an aldehyde to a carboxylic acid is coupled to the formation of ATP and NADH (see also Figure 2–20). As shown, step 5 begins when the enzyme *glyceraldehyde 3-phosphate dehydrogenase* forms a covalent bond to the carbon carrying the aldehyde group on glyceraldehyde 3-phosphate. Next, hydrogen (as a hydride ion— a proton plus two electrons) is removed from the enzyme-linked aldehyde group in glyceraldehyde 3-phosphate and transferred to the important hydrogen carrier NAD^+ (see Figure 2–22). This oxidation step creates a sugar carbonyl group attached to the enzyme in a high-energy linkage. This linkage is then broken by a phosphate ion from solution, creating a high-energy sugar-phosphate bond instead. In these last two reactions, the enzyme has *coupled* the energetically favorable process of oxidizing an aldehyde to the energetically unfavorable formation of a high-energy phosphate bond. Finally, in step 6 of glycolysis, the newly created reactive phosphate group is transferred to ADP to form ATP, leaving a free carboxylic acid group on the oxidized sugar.

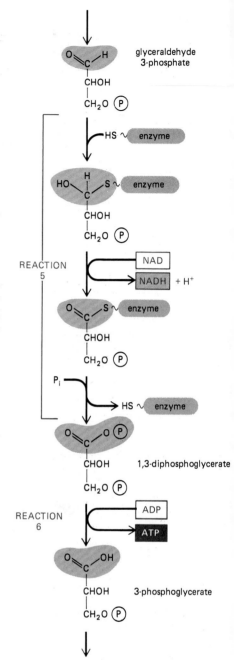

converted into ethanol plus CO_2 (as in yeast) or into lactate (as in muscle), which is then excreted. In such anaerobic energy-yielding reactions, called **fermentations,** the further reaction of pyruvate is required in order to use up the reducing power produced in reaction 5 of glycolysis, thereby regenerating the NAD^+ required for glycolysis to continue (see p. 381).

Oxidative Catabolism Yields a Much Greater Amount of Usable Energy[14]

The anaerobic generation of ATP from glucose through the reactions of glycolysis is relatively inefficient. The end products of anaerobic glycolysis still contain a great deal of chemical energy that can be released by further oxidation. The evolution of *oxidative catabolism* (cellular respiration) became possible only after molecular oxygen had accumulated in the earth's atmosphere as a result of photosynthesis by the cyanobacteria. Earlier, anaerobic catabolic processes had presumably dominated life on earth. The addition of an oxygen-requiring stage to the catabolic process (stage 3 in Figure 2–18) provided cells with a much more powerful and efficient method for extracting energy from food molecules. This third stage begins with the *citric acid cycle* (also called the tricarboxylic acid cycle, or the Krebs cycle) and ends with *oxidative phosphorylation*, both of which occur in aerobic bacteria and the mitochondria of eucaryotic cells.

Metabolism Is Dominated by the Citric Acid Cycle[15]

The primary function of the **citric acid cycle** is to oxidize acetyl groups that enter the cycle in the form of acetyl CoA molecules. The reactions form a cycle because the acetyl group is not oxidized directly, but only after it has been covalently added to a larger molecule, *oxaloacetate,* which is regenerated at the end of one turn of the cycle. As illustrated in Figure 2–23, the cycle begins with the reaction between acetyl CoA and oxaloacetate to form the tricarboxylic acid molecule called *citric acid* (or *citrate*). A series of reactions then occurs in which two of the six carbons of citrate are oxidized to CO_2, forming another molecule of oxaloacetate to repeat the cycle. (Because the two carbons that are newly added in each cycle enter a different part of the citrate molecule from the part oxidized to CO_2, it is only after several cycles that their turn comes to be oxidized.) The CO_2 produced in these reactions then diffuses from the mitochondrion (or from the bacterium) and leaves the cell.

The energy made available when the C—H and C—C bonds in citrate are oxidized is captured in several different ways in the course of the citric acid cycle. At one step in the cycle (succinyl CoA to succinate), a high-energy phosphate linkage is created by a mechanism resembling that described for glycolysis above. (Although this reaction produces GTP rather than ATP, all nucleoside triphosphates are equivalent energetically because of exchange reactions such as ADP +

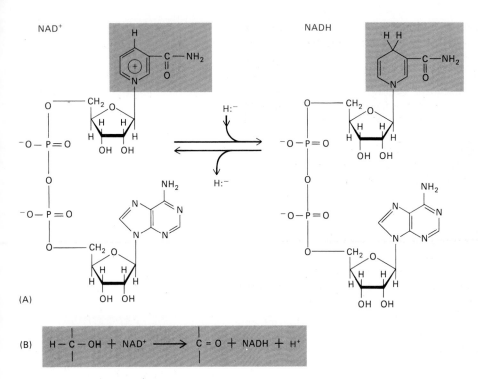

(A)

(B)

$$H - C - OH + NAD^+ \longrightarrow C = O + NADH + H^+$$

Figure 2–22 NADH and NAD$^+$ are the most important carriers of hydrogen in catabolic reactions. Their structures are shown in (A). NAD is an abbreviation for *nicotinamide adenine dinucleotide*, reflecting the fact that the right half of the molecule, as drawn, is adenosine monophosphate (AMP). The part of the NAD$^+$ molecule known as the nicotinamide ring (in the colored box) is able to accept a hydrogen atom together with an additional electron (a hydride ion, H$^-$), forming NADH. In this reduced form, the nicotinamide ring has a reduced stability because it is no longer stabilized by resonance. As a result, the added hydride ion is easily transferred to other molecules.

(B) An example of a reaction involving NAD$^+$ and NADH. In the biological oxidation of a substrate molecule such as an alcohol, two hydrogen atoms are lost from the substrate. One of these is added as a hydride ion to NAD$^+$, producing NADH, while the other is released into solution as a proton (H$^+$). (See also Figure 7–18, p. 351.)

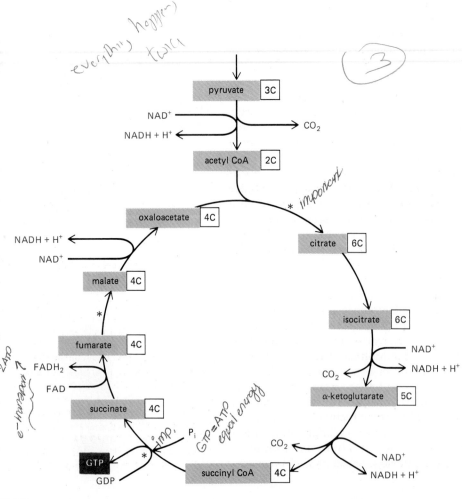

Figure 2–23 The citric acid cycle. In mitochondria and in aerobic bacteria, the acetyl groups produced from pyruvate are further oxidized. The carbon atoms of the acetyl groups are converted to CO_2, while the hydrogen atoms are transferred to the carrier molecules NAD$^+$ and FAD. Additional oxygen and hydrogen atoms enter the cycle in the form of water at the steps marked with an asterisk (*). For details, see Figure 7–14, p. 348.

GTP $\rightleftharpoons$ ATP + GDP.) All of the remaining energy of oxidation that is captured is channeled into the conversion of hydrogen- (or hydride ion-) carrier molecules to their reduced forms; for each turn of the cycle, three molecules of NAD^+ are converted to NADH and one *flavin adenine nucleotide* (FAD) is converted to $FADH_2$. The energy carried by the activated hydrogen on these carrier molecules will subsequently be harnessed through the reactions of *oxidative phosphorylation* (to be considered in more detail below), which require molecular oxygen from the atmosphere.

The additional oxygen atoms required to make CO_2 from the acetyl groups entering the citric acid cycle are supplied not by molecular oxygen but by water. Three molecules of water are split in each cycle, and their oxygen atoms are used to make CO_2. Some of their hydrogen atoms enter substrate molecules and are raised to a higher energy and removed (together with the hydrogen atoms of the acetyl groups) to carrier molecules such as NADH.

In the eucaryotic cell, the mitochondrion is the center toward which all catabolic processes lead, whether they begin with sugars, fats, or proteins. For, in addition to pyruvate, fatty acids and some amino acids also pass from the cytosol into mitochondria, where they are converted into acetyl CoA or one of the other intermediates of the citric acid cycle. The mitochondrion also functions as the starting point for some biosynthetic reactions by producing vital carbon-containing intermediates, such as *oxaloacetate* and *α-ketoglutarate*. These substances are transferred back from the mitochondrion to the cytosol, where they serve as precursors for the synthesis of essential molecules, such as amino acids.

In Oxidative Phosphorylation, the Transfer of Electrons to Oxygen Drives ATP Formation[9,16]

Oxidative phosphorylation is the last step in catabolism and the point at which the major portion of metabolic energy is released. In this process, molecules of NADH and $FADH_2$ transfer the electrons that they have gained from the oxidation of food molecules to molecular oxygen, O_2. The reaction, which is formally equivalent to the burning of hydrogen in air to form water, releases a great deal of chemical energy. Part of this energy is used to make the major portion of the cell's ATP; the rest is liberated as heat.

Although the overall chemistry of NADH and $FADH_2$ oxidation involves a transfer of hydrogen to oxygen, complete hydrogen atoms are not transferred directly. It is the *electrons* from the hydrogen atoms that are important. This is because a hydrogen atom can be readily dissociated into its constituent electron and proton (H^+). The electron can then be transferred separately to a molecule that accepts only electrons, while the proton remains in aqueous solution. By the same reasoning, if an electron alone is donated to a molecule with a strong affinity for hydrogen, then a hydrogen atom will be reconstituted automatically by the capture of a proton from solution. In the course of oxidative phosphorylation, electrons from NADH and $FADH_2$ pass down a chain of carrier molecules, but the presence or absence of intact hydrogen atoms depends on the nature of the carrier.

In a eucaryotic cell, this series of electron transfers along the **electron-transport chain** takes place on the inner membrane of the mitochondrion, in which all of the carrier molecules are embedded. At each step of the transfer, the electrons fall to a lower energy state, until at the end they are transferred to oxygen molecules. Oxygen molecules have the highest affinity of all of the carriers for electrons, and electrons bound to oxygen are thus in their lowest energy state. The energy released as these electrons fall to lower energy states is harnessed, in a way that is not fully understood, to pump protons from the inner mitochondrial compartment to the outside (Figure 2–24). An *electrochemical proton gradient* is thereby generated across the inner mitochondrial membrane. This gradient, in turn, drives a flux of protons back through a special enzyme complex in the same membrane, causing the enzyme (*ATP synthetase*) to add a phosphate group to ADP, and thereby generating ATP inside the mitochondrion. Finally, the newly

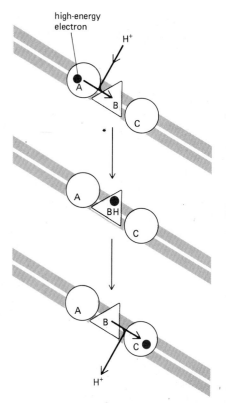

Figure 2–24 The generation of a H^+ gradient across a membrane by electron-transport reactions. A high-energy electron (derived, for example, from the oxidation of a metabolite) is passed sequentially by carriers A, B, and C to a lower energy state. In this diagram, carrier B is arranged in the membrane in such a way that it takes up H^+ from one side and releases it to the other as the electron passes. The resulting H^+ gradient represents a form of stored energy that is harnessed by other membrane proteins in the mitochondrion to drive the formation of ATP (see also Figure 7–35, p. 363).

made ATP is transferred from the mitochondrion to the rest of the cell, where it drives a variety of metabolic reactions.

The nature of the electron-transport chain and the mechanism of ATP synthesis will be described in detail in Chapter 7.

Amino Acids and Nucleotides Are Part of the Nitrogen Cycle

In our discussion so far, we have concentrated mainly on carbohydrate metabolism. We have not yet considered the metabolism of nitrogen or sulfur. These two elements are constituents of proteins and nucleic acids, which are the two most important classes of macromolecules in the cell and make up approximately two-thirds of its dry weight. Atoms of nitrogen and sulfur pass from compound to compound and between organisms and their environment in a series of reversible cycles.

Although molecular nitrogen is abundant in the earth's atmosphere, it is chemically unreactive. Only a few living species are able to incorporate it into organic molecules, a process called **nitrogen fixation.** Nitrogen fixation occurs in certain microorganisms and by some geophysical processes, such as lightning discharge. It is essential to the biosphere as a whole, for without it life would not exist on this planet. Only a small fraction of the nitrogenous compounds in today's organisms, however, represents fresh products of nitrogen fixation. Most organic nitrogen has been in circulation for some time, passing from one living organism to another. Thus nitrogen-fixing reactions can be said to perform a "topping-up" function for the total nitrogen supply.

Vertebrates receive virtually all of their nitrogen in their dietary intake of proteins and nucleic acids. In the body, these macromolecules are broken down to component amino acids and nucleotides, which are then repolymerized into new proteins and nucleic acids or utilized to make other molecules. About half of the 20 amino acids found in proteins are *essential amino acids* (Figure 2–25) for vertebrates: they cannot be synthesized from other ingredients of the diet. The others can be so synthesized, using a variety of raw materials, including intermediates of the citric acid cycle. The essential amino acids are made in other organisms, usually by long and energetically expensive pathways that have been lost in the course of vertebrate evolution.

The nucleotides needed to make RNA and DNA can be synthesized using specialized biosynthetic pathways: there are no "essential nucleotides" that must be provided in the diet. All of the nitrogens in the purine and pyrimidine bases (as well as some of the carbons) are derived from the plentiful amino acids glutamine, aspartic acid, and glycine, whereas the ribose and deoxyribose sugars are derived from glucose.

Amino acids that are not utilized in biosynthesis can be oxidized to generate metabolic energy. Most of their carbon and hydrogen atoms eventually form CO_2 or H_2O, whereas their nitrogen atoms are shuttled through various forms and eventually appear as urea, which is excreted. Each amino acid is processed differently, and a whole constellation of enzymatic reactions exists for their catabolism.

THE ESSENTIAL AMINO ACIDS
THREONINE
METHIONINE
LYSINE
VALINE
LEUCINE
ISOLEUCINE
HISTIDINE
PHENYLALANINE
TRYPTOPHAN

Figure 2–25 The nine essential amino acids, which cannot be synthesized by human cells and so must be supplied in the diet.

Summary

Animal cells can be considered to derive energy from food in three stages. In stage 1, proteins, polysaccharides, and fats are broken down by extracellular reactions to small molecules. In stage 2, these small molecules are degraded within cells to produce acetyl CoA and a limited amount of ATP and NADH. These are the only reactions that can yield energy in the absence of oxygen. In stage 3, the acetyl CoA molecules are degraded in mitochondria to give CO_2 and hydrogen atoms that are linked to carrier molecules such as NADH. Electrons from the hydrogen atoms are passed through a complex chain of membrane-bound carriers, finally being passed to molecular oxygen to form water. Driven by the energy released in these electron-transfer steps, hydrogen ions (H^+) are transported out of the mitochondria. The resulting electrochemical proton gradient across the inner mitochondrial membrane is harnessed to drive the synthesis of most of the cell's ATP.

Biosynthesis and the Creation of Order[17]

Thousands of different chemical reactions are occurring in a cell at any instant of time. The reactions are all linked together in chains and networks in which the product of one reaction becomes the substrate of the next. Most of the chemical reactions in cells can be roughly classified as being concerned either with catabolism or with biosynthesis. Having discussed the catabolic reactions, we now turn to the reactions of biosynthesis. These begin with the intermediate products of glycolysis and the citric acid cycle (and closely related compounds) and generate the larger and more complex molecules of the cell.

The Free-Energy Change for a Reaction Determines Whether It Can Occur[18]

Although enzymes speed up energetically favorable reactions, they cannot force energetically unfavorable reactions to occur. In terms of a water analogy, enzymes by themselves cannot make water run uphill. But cells must do just that in order to grow and divide; they must build highly ordered and energy-rich molecules from small and simple ones. We have seen that, in a general way, this is done through enzymes that couple energetically favorable reactions, which consume energy derived ultimately from the sun and produce heat, to energetically unfavorable reactions, which produce biological order. Let us examine in greater detail how such coupling is achieved.

First, we must consider more carefully the term "energetically favorable," which we have so far used loosely, without giving it a definition. As explained earlier, a chemical reaction can proceed spontaneously only if it results in a net increase in the disorder of the universe. Disorder increases when energy is dissipated as heat; and the criterion for an increase of disorder can conveniently be expressed in terms of a quantity called the **free energy, G.** This is defined in such a way that changes in its value, denoted by **ΔG,** measure the amount of disorder created in the universe when a reaction takes place, as explained in Panel 2–7 (pp. 72–73). "Energetically favorable" reactions, by definition, are those that release a large quantity of free energy, or in other words have a large *negative* ΔG and create much disorder. Such reactions have a strong tendency to occur spontaneously, although their rate will depend on other factors, such as the availability of specific enzymes (see below). Conversely, reactions with a *positive* ΔG, such as those in which two amino acids are joined together to form a peptide bond, by themselves create order in the universe and therefore cannot occur spontaneously. Energetically unfavorable reactions of this kind can take place only if they are coupled to a second reaction with a negative ΔG so large that the ΔG of the entire process is negative.

The course of most reactions can be predicted quantitatively. A large body of thermodynamic data has been collected that makes it possible to calculate the change in free energy for most of the important metabolic reactions of the cell. The overall free-energy change for a pathway is then simply the sum of the energy changes in each of its component steps. Consider two reactions

$$X \rightarrow Y \quad \text{and} \quad C \rightarrow D$$

where the ΔG values are $+1$ and -13 kcal/mole, respectively. (Recall that a mole is 6×10^{23} molecules of a substance.) If these two reactions can be coupled together, the ΔG for the coupled reaction will be -12 kcal/mole. Thus, the unfavorable reaction $X \rightarrow Y$, which will not occur spontaneously, can be driven by the favorable reaction $C \rightarrow D$, provided that a mechanism exists by which the two reactions can be coupled together.

Biosynthetic Reactions Are Often Directly Coupled to ATP Hydrolysis

Consider a typical biosynthetic reaction in which two monomers, A and B, are to be joined in a *dehydration* (also called *condensation*) reaction, in which water is released:

THE IMPORTANCE OF FREE ENERGY FOR CELLS

Life is possible because of the complex network of interacting chemical reactions occurring in every cell. In viewing the metabolic pathways that comprise this network, one might suspect that the cell has had the ability to evolve an enzyme to carry out any reaction that it needs. But this is not so. Although enzymes are powerful catalysts, they can speed up only those reactions that are thermodynamically possible; other reactions proceed in cells only because they are *coupled* to very favorable reactions that drive them. The question of whether a reaction can occur spontaneously, or instead needs to be coupled to another reaction, is central to cell biology. The answer is obtained by reference to a quantity called the *free energy*: the total change in free energy during a set of reactions determines whether or not the entire reaction sequence can occur. In this panel, we will explain some of the fundamental ideas—derived from a special branch of chemistry and physics called *thermodynamics*—that are required for understanding what free energy is and why it is so important to cells.

ENERGY RELEASED BY CHANGES IN CHEMICAL BONDING IS CONVERTED INTO HEAT

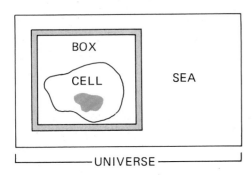

An *enclosed system* is defined as a collection of molecules that does not exchange matter with the rest of the universe (for example, the "cell in a box" described previously in Figure 2–11 and shown above). Any such system will contain molecules with a total energy E. This energy will be distributed in a variety of ways: some as the translational energy of the molecules, some as their vibrational and rotational energies, but most as the bonding energies between the individual atoms that make up the molecules. Suppose that a reaction occurs in the system. The first law of thermodynamics places a constraint on what types of reactions can occur in the system: it states that "in any process, the total energy of the universe remains constant." For example, suppose that reaction A → B occurs in the enclosed system and releases a great deal of chemical bond energy. This energy will initially increase the intensity of molecular motions (translational, vibrational, and rotational) in the system, which is equivalent to raising its temperature. However, these increased motions will soon be transferred out of the system by a series of molecular collisions that heat up first the walls of the box and then the outside world (represented by the sea in our example). In the end, the system returns to its initial temperature, by which time all the chemical bond energy released in the box has been converted into heat energy and transferred out of the box to the surroundings. According to the first law, the change in the energy in the box (ΔE_{box}, which we shall denote as ΔE) must be equal and opposite to the amount of heat energy transferred, which we shall designate as h: that is, $\Delta E = -h$. Thus, the energy in the box (E) decreases when heat leaves the system.

E also can change during a reaction due to work being done on the outside world. For example, suppose that there is a small increase in the volume (ΔV) of the box during a reaction. Since the walls of the box must push against the constant pressure (P) in the surroundings in order to expand, this does work on the outside world and requires energy. The energy used is $P(\Delta V)$, which according to the first law must decrease the energy in the box (E) by the same amount. In most reactions, chemical bond energy is converted into both work and heat. *Enthalpy* (H) is a composite function that includes both of these ($H = E + PV$). To be rigorous, it is the change in enthalpy (ΔH) in an enclosed system and not the change in energy that is equal to the heat transferred to the outside world during a reaction. Reactions in which H decreases release heat to the surroundings and are said to be "exothermic," while reactions in which H increases absorb heat from the surroundings and are said to be "endothermic." Thus, $-h = \Delta H$. However, the volume change is negligible in most biological reactions, so to a good approximation:

$$-h = \Delta H \cong \Delta E$$

THE SECOND LAW OF THERMODYNAMICS

Consider a container in which 1000 coins are all lying heads up. If the container is shaken vigorously, subjecting the coins to the types of random motions that all molecules experience due to their frequent collisions with other molecules, one will end up with about half of the coins oriented heads down. The reason for this reorientation is that there is only a single way in which the original state of the coins can be reinstated (every coin must lie heads up), whereas there are many different ways (about 10^{298}) to achieve a state in which there is an equal mixture of heads and tails; in fact, there are more ways to achieve a 50:50 state than to achieve any other state. Each state has a probability of occurrence that is proportional to the number of ways it can be realized. The second law of thermodynamics states that "systems will change spontaneously from states of lower probability to states of higher probability." Since states of lower probability are said to be more "ordered" than states of higher probability, the second law can be restated: "the universe constantly changes so as to become more disordered."

THE ENTROPY, S

The second law (but not the first law) allows one to predict the _direction_ of a particular reaction. But to make it useful for this purpose, one needs a convenient measure of the probability, or equivalently, the degree of disorder of a state. The entropy (S) is such a measure. It is a logarithmic function of the probability such that the _change in entropy_ (ΔS) that occurs when the reaction A → B converts one mole of A into one mole of B is

$$\Delta S = R \ln p_B/p_A$$

where p_A and p_B are the probabilities of the two states A and B, R is the gas constant (2 cal deg^{-1} mole^{-1}), and ΔS is measured in entropy units (eu). In our initial example of 1000 coins, the relative probability of all heads (state A) versus half heads and half tails (state B) is equal to the ratio of the number of different ways that the two results can be obtained. One can calculate that $p_A = 1$ and $p_B = 1000!(500! \times 500!) = 10^{298}$. Therefore, the entropy change for the reorientation of the coins when their container is vigorously shaken and an equal mixture of heads and tails is obtained is $R \ln (10^{298})$, or about 1370 eu per mole of such containers (6×10^{23} containers). We see that, because ΔS defined above is positive for the transition from state A to state B ($p_B/p_A > 1$), reactions with a large _increase_ in S (that is, for which $\Delta S > 0$) are favored and will occur spontaneously.

As discussed in the text, heat energy causes the random commotion of molecules. Because the transfer of heat from an enclosed system to its surroundings increases the number of different arrangements that the molecules in the outside world can have, it increases their entropy. It can be shown that the release of a fixed quantity of heat energy has a greater disordering effect at low temperature than at high temperature and that the value of ΔS for the surroundings, as defined above (ΔS_{sea}), is precisely equal to the amount of heat transferred to the surroundings from the system (h) divided by the absolute temperature (T):

$$\Delta S_{sea} = h/T$$

absolute temp

heat transferred to surroundings

THE GIBBS FREE ENERGY, G

When dealing with an enclosed biological system, one would like to have a simple way of predicting whether a given reaction will or will not occur spontaneously in the system. We have seen that the crucial question is whether the entropy change for the universe is positive or negative when that reaction occurs. In our idealized system, the cell in a box, there are two separate components to the entropy change of the universe—the entropy change for the system enclosed in the box and the entropy change for the surrounding "sea"—and both must be added together before any prediction can be made. For example, it is possible for a reaction to absorb heat and thereby decrease the entropy of the sea ($\Delta S_{sea} < 0$) and at the same time to cause such a large degree of disordering inside the box ($\Delta S_{box} > 0$) that the total $\Delta S_{universe} = \Delta S_{sea} + \Delta S_{box}$ is greater than 0. In this case, the reaction will occur spontaneously, even though the sea gives up heat to the box during the reaction. An example of such a reaction is the dissolving of sodium chloride in a beaker containing water (the "box"), which is a spontaneous process even though the temperature of the water drops as the salt goes into solution.

Chemists have found it useful to define a number of new "composite functions" that describe _combinations_ of physical properties of a system. The properties that can be combined include the temperature (T), pressure (P), volume (V), energy (E), and entropy (S). The enthalpy (H) is one such composite function. But by far the most useful composite function for biologists is the _Gibbs free energy, G_. It serves as an accounting device that allows one to deduce the entropy change of the universe resulting from a chemical reaction in the box, while avoiding any separate consideration of the entropy change in the sea. For a box of volume V at pressure P, the definition of G is

$$G = H - TS$$

where H is the enthalpy described above ($E + PV$), T is the absolute temperature, and S is the entropy. Each of these quantities applies to the inside of the box only. The change in free energy during a reaction in the box (the G of the products minus the G of the starting materials) is denoted as ΔG and, as we shall now demonstrate, is a direct measure of the amount of disorder that is created in the universe when the reaction occurs.

At constant temperature, the change in free energy (ΔG) during a reaction equals $\Delta H - T\Delta S$. Remembering that $\Delta H = -h$, the heat absorbed from the sea, we have

$$-\Delta G = -\Delta H + T\Delta S$$
$$-\Delta G = h + T\Delta S, \text{ and } -\Delta G/T = h/T + \Delta S$$

But h/T is equal to the entropy change of the sea (ΔS_{sea}), and the ΔS in the above equation is ΔS_{box}. Therefore

$$-\Delta G/T = \Delta S_{sea} + \Delta S_{box} = \Delta S_{universe}$$

We conclude that a reaction will proceed in the direction that causes the change in free energy (ΔG) to be _less than zero_, because in this case there will be a positive entropy change in the universe when the reaction occurs. Thus, the **free energy change is a direct measure of the entropy change of the universe.**

For a complex set of coupled reactions involving many different molecules, the total free-energy change can be computed simply by adding up the free energies of all the different molecular species after the reaction and comparing this value to the sum of free energies before the reaction; for common substances, the required free-energy values can be found from published tables. In this way one can predict the direction of a reaction and thereby readily disprove any proposed mechanism. Thus, for example, from the observed values for the magnitude of the electrochemical proton gradient across the inner mitochondrial membrane and the ΔG for ATP hydrolysis inside the mitochondrion, one can be certain that the enzyme ATP synthetase requires the passage of more than one proton for each molecule of ATP that it synthesizes (see p. 358). The value of ΔG for a reaction is a direct measure of the degree of displacement of the reaction from equilibrium. The large negative value for ATP hydrolysis in a cell merely reflects the fact that cells keep the ATP hydrolysis reaction as much as ten orders of magnitude away from equilibrium (see p. 354). If a reaction reaches equilibrium, $\Delta G = 0$, and the reaction then proceeds at precisely equal rates in the forward and backward direction. For ATP hydrolysis, equilibrium is reached when the vast majority of the ATP has been hydrolyzed, as occurs in a dead cell.

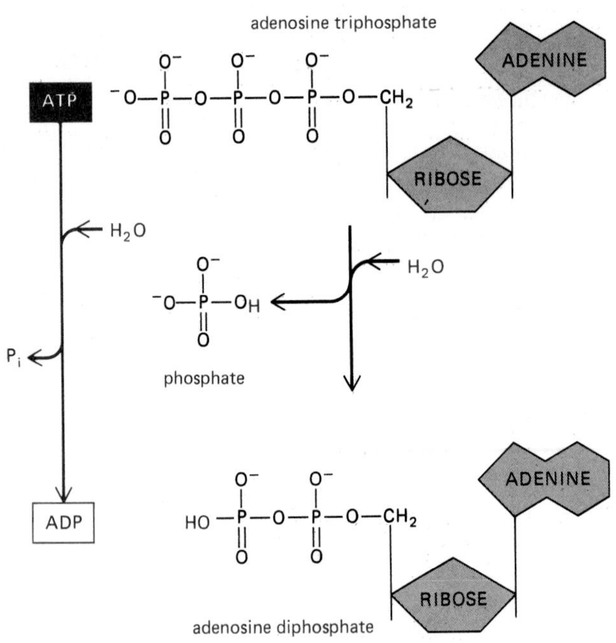

adenosine triphosphate

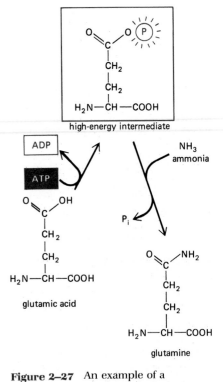

Figure 2–26 The hydrolysis of the terminal phosphate of ATP yields between 11 and 13 kcal/mole of usable energy, depending on the intracellular conditions. The large negative ΔG of this reaction arises from a number of factors. Release of the terminal phosphate group removes an unfavorable repulsion between adjacent negative charges. In addition, the inorganic phosphate ion released is stabilized by resonance and by favorable hydrogen bond formation with water.

phosphate

adenosine diphosphate

$$\text{A-H} + \text{B-OH} \rightarrow \text{A-B} + \text{H}_2\text{O}$$

Almost invariably the reverse reaction (called *hydrolysis*), in which water breaks the covalently linked compound A-B, will be the energetically favorable one. This is the case, for example, in the hydrolysis of proteins, nucleic acids, and polysaccharides into their subunits.

The general strategy that allows the cell to make A-B from A-H and B-OH involves a sequence of steps through which the energetically unfavorable synthesis of the desired compound is coupled to an even more energetically favorable reaction (see Figure 2–17). ATP hydrolysis has a large negative ΔG, as explained in Figure 2–26, and it is the usual source of the free energy used to drive the biosynthetic reactions in a cell. In the coupled pathway from A-H and B-OH to A-B, energy from ATP hydrolysis is first used to convert B-OH to a higher-energy intermediate compound, which then reacts directly with A-H to give A-B. The simplest mechanism involves the transfer of a phosphate from ATP to B-OH to make B-OPO$_3$ (or B-O-Ⓟ), in which case the reaction pathway contains only two steps:

1. B-OH + ATP → B-O-Ⓟ + ADP
2. A-H + B-O-Ⓟ → A-B + P$_i$

Since the intermediate B-O-Ⓟ is formed only transiently, the overall reactions that occur are

$$\text{A-H} + \text{B-OH} \rightarrow \text{A-B} \quad \text{and} \quad \text{ATP} \rightarrow \text{ADP} + \text{P}_i$$

The first reaction, which by itself is energetically unfavorable, is forced to occur by being directly coupled to the second energetically favorable reaction (ATP hydrolysis). An example of a coupled biosynthetic reaction of this kind, the synthesis of the amino acid glutamine, is shown in Figure 2–27.

The ΔG for the hydrolysis of ATP to ADP and inorganic phosphate (P$_i$) depends on the concentrations of all of the reactants, and under the usual conditions in a cell it is between -11 and -13 kcal/mole (see p. 354). In principle, this hydrolysis reaction can be used to drive an unfavorable reaction with a ΔG of, perhaps, $+10$ kcal/mole, provided that a suitable reaction path is available. For some biosynthetic reactions, however, even -13 kcal/mole may not be enough. In these cases, the path of ATP hydrolysis can be altered so that it initially produces AMP and pyrophosphate (PP$_i$) (Figure 2–28). Then pyrophosphate itself is hydrolyzed in a subsequent step. The whole process makes available a total free-energy change of about -26 kcal/mole.

high-energy intermediate

ADP

ATP

NH$_3$
ammonia

P$_i$

glutamic acid

glutamine

Figure 2–27 An example of a biosynthetic reaction of the dehydration type driven by ATP hydrolysis. Glutamic acid is first converted to a high-energy phosphorylated intermediate (corresponding to the compound B-O-Ⓟ described in the text), which then reacts with ammonia to form glutamine. In this example, both steps occur on the surface of the same enzyme, *glutamine synthetase*. Note that, for clarity, these molecules are shown in their uncharged forms.

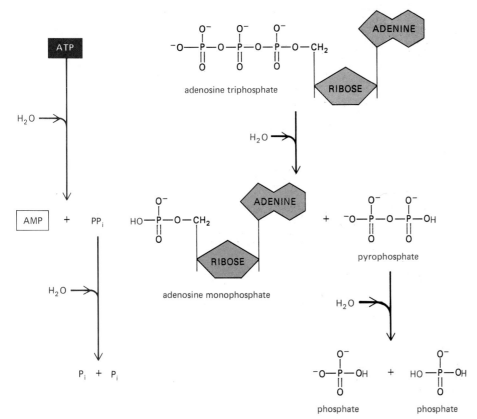

Figure 2–28 An alternative route for the hydrolysis of ATP, in which pyrophosphate is first formed and then hydrolyzed. This route releases about twice as much free energy as the reaction in Figure 2–26. The H atoms derived from water are shown attached to the phosphate groups following hydrolysis. However, at the pH of the cytoplasm, most of these dissociate to form free hydrogen ions, H^+.

How is the energy of pyrophosphate hydrolysis coupled to a biosynthetic reaction? One way can be illustrated by considering again the synthesis of compound A-B from A-H and B-OH. By an appropriate enzyme, B-OH can be converted to the higher-energy intermediate B-O-(P)-(P) by its reaction with ATP. The complete reaction now contains three steps:

1. B-OH + ATP → B-O-(P)-(P) + AMP
2. A-H + B-O-(P)-(P) → A-B + PP$_i$
3. PP$_i$ + H$_2$O → 2P$_i$

And the overall reactions are

$$A\text{-}H + B\text{-}OH \rightarrow A\text{-}B \quad \text{and} \quad ATP + H_2O \rightarrow AMP + 2P_i$$

Since an enzyme always speeds up equally the forward and backward directions of the reaction it catalyzes, the compound A-B can be destroyed by recombining with pyrophosphate (a reversal of step 2). But the energetically favorable reaction of pyrophosphate hydrolysis (step 3) greatly stabilizes compound A-B by keeping the concentration of pyrophosphate very low, essentially preventing the reversal of step 2. In this way, the energy of pyrophosphate hydrolysis is used to drive this reaction in the forward direction. An example of an important biosynthetic reaction of a related kind, polynucleotide synthesis, is given in Figure 2–29.

Coenzymes Are Involved in the Transfer of Specific Chemical Groups

Because the terminal phosphate linkage in ATP is easily cleaved, with release of free energy, ATP acts as an efficient phosphate donor in a large number of different phosphorylation reactions. A wide variety of other chemically labile linkages also work in this way. For example, specific carrier molecules contain acetyl or methyl groups linked by reactive bonds so that they can be transferred efficiently to other molecules (Table 2–2). The same carrier molecule will often participate in many different biosynthetic reactions in which its group is needed.

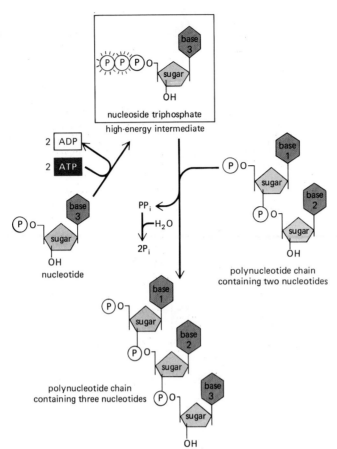

Figure 2–29 Synthesis of a polynucleotide, RNA or DNA, is a multistep process driven by ATP hydrolysis. In the first step a nucleoside monophosphate such as CMP is activated by the sequential transfer of the terminal phosphate groups from two ATP molecules. The high-energy intermediate formed—a nucleoside triphosphate such as CTP—exists free in solution until it reacts with the growing end of an RNA or DNA chain with release of pyrophosphate. Hydrolysis of the latter to inorganic phosphate is highly favorable and helps drive the overall reaction in the direction of polynucleotide synthesis.

Acetyl coenzyme A (acetyl CoA), which is produced in the breakdown of glucose, is an example of such a carrier molecule. It carries an acetyl group linked to CoA through a reactive thioester bond (see Figure 2–19). This acetyl group is readily transferred to another molecule, such as a growing fatty acid. Another important example is *biotin*, which transfers a carboxyl group in many biosynthetic reactions (Figure 2–30). Molecules such as acetyl CoA, biotin, and ATP are known as **coenzymes** because they are bound tightly to the surfaces of various enzymes and are essential for the activity of the enzymes. Many coenzymes cannot be synthesized by animals and must be obtained from plants or microorganisms in the diet. *Vitamins*—essential nutritional factors that animals need in trace amounts—are often the precursors of required coenzymes.

Table 2–2 Some Coenzymes Involved in Group-Transfer Reactions

Coenzyme	Group Transferred
ATP	phosphate
NADH, NADPH	hydrogen & electron (hydride ion)
Coenzyme A	acetyl
Biotin	carboxyl
S-Adenosylmethionine	methyl

Coenzymes are small molecules that are associated with some enzymes and are essential for their activity. Each one listed is a carrier molecule for a small chemical group, and it participates in various reactions in which that group is transferred to another molecule. Some coenzymes are covalently linked to their enzyme; others are less tightly bound. A covalently linked coenzyme is sometimes referred to as a *prosthetic group* of the enzyme.

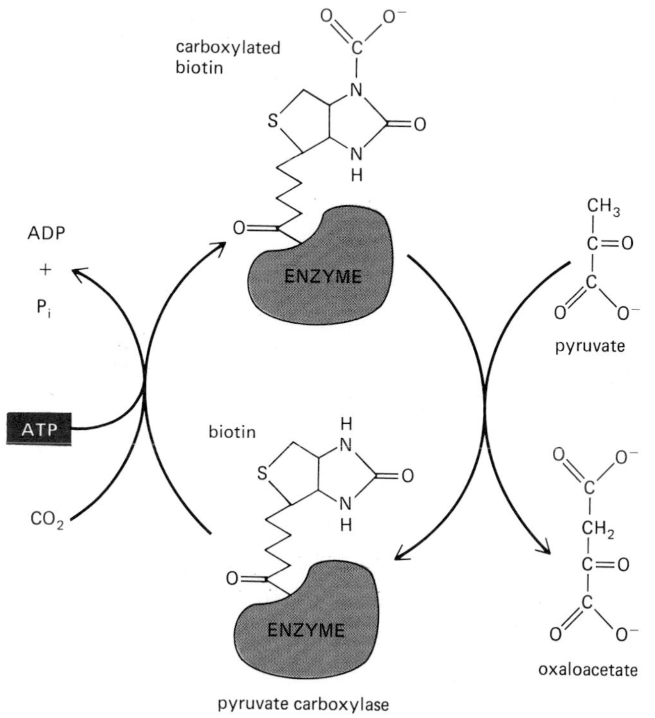

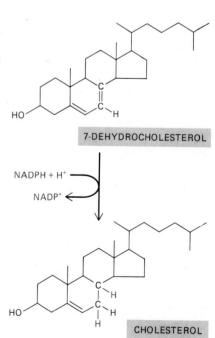

Figure 2–30 at top of page showing the biotin-mediated carboxyl transfer cycle.

carboxylated biotin

ADP
+
P$_i$

ATP

CO$_2$

ENZYME

biotin

ENZYME

pyruvate carboxylase

CH$_3$
C=O
pyruvate

oxaloacetate

Figure 2–30 Transfer of a carboxyl group by the coenzyme biotin. Biotin acts as a carrier molecule for the carboxyl group (—COO$^-$). In the sequence of reactions shown, biotin is covalently bound to the enzyme pyruvate carboxylase. An activated carboxyl group derived from a bicarbonate ion (HCO$_3^-$) is coupled to biotin in a reaction that requires an input of energy from the hydrolysis of an ATP molecule. Subsequently, this carboxyl group is transferred to the methyl group of pyruvate to form oxaloacetate.

Biosynthesis Requires Reducing Power

We have seen that oxidation and reduction reactions occur continuously in cells. The chemical energy in food molecules is released by oxidation reactions, while, in order to make biological molecules, the cell needs—among other things—to carry out a series of reduction reactions that require an input of chemical energy. By using the principle of coupled reactions described previously, cells directly channel chemical energy derived from catabolism into the synthesis of NADH (for example, see Figure 2–21). The high-energy bond between hydrogen and the nicotinamide ring in NADH then provides the energy for otherwise unfavorable enzyme reactions that transfer hydrogen (as a hydride ion) to another molecule. NADH, and the closely related NADPH to which it can be readily converted, are therefore said to carry "reducing power."

To see how this hydrogen transfer works in practice, consider just one biosynthetic step: the last reaction in a pathway for the synthesis of the lipid molecule *cholesterol*. In this reaction two hydrogen atoms are added to the polycyclic steroid ring in order to reduce a carbon-carbon double bond (Figure 2–31). As in most biosynthetic reactions, the constituents of the two hydrogen atoms required in this reaction are supplied as a hydride ion from NADPH and a proton (H$^+$) from the solution (H$^-$ + H$^+$ = 2H). As in NADH, the hydride ion to be transferred from NADPH is part of a nicotinamide ring and is easily lost because the ring can achieve a more stable aromatic state without it (see Figure 2–22). Therefore, NADH and NADPH both hold a hydride ion in a high-energy linkage from which it can be transferred to another molecule when a suitable enzyme is available to catalyze the transfer.

The difference between NADH and NADPH is trivial in chemical terms: NADPH has an extra phosphate group on a part of the molecule that is far from the active region (Figure 2–32). This phosphate group is of no importance to the hydride ion transfer reaction as such, but it serves as a handle for binding NADPH as a coenzyme to appropriate enzymes. As a general rule, NADH operates with enzymes catalyzing catabolic reactions, whereas NADPH operates with enzymes that catalyze biosynthetic reactions. This means that catabolic and biosynthetic pathways can be regulated separately by alterations in the levels of NADH and NADPH, respectively.

7-DEHYDROCHOLESTEROL

NADPH + H$^+$

NADP$^+$

CHOLESTEROL

Figure 2–31 The final stage in one of the biosynthetic routes leading to cholesterol. The reduction of the C=C bond is achieved by the transfer of a hydride ion from the carrier molecule NADPH, plus a proton (H$^+$) from the solution.

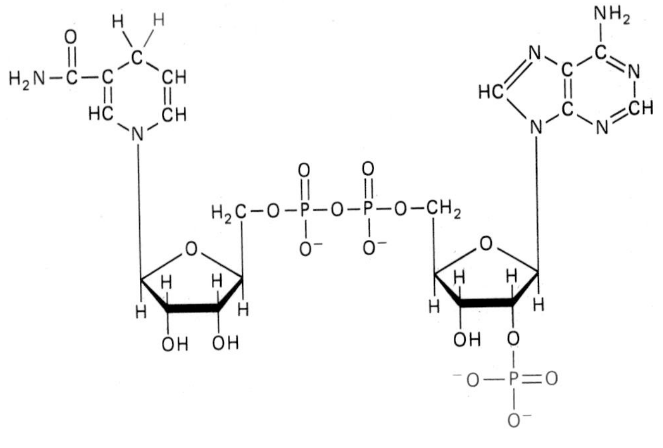

Figure 2–32 The structure of NADPH, which differs from NADH (Figure 2–22) only in the presence of an extra phosphate group that allows it to be selectively recognized by certain enzymes (usually those involved in biosynthesis).

Biological Polymers Are Synthesized by Repetition of Elementary Dehydration Reactions

The principal macromolecules synthesized by cells are polynucleotides (DNA and RNA), polysaccharides, and proteins. They are enormously diverse in structure and include the most complex molecules known. Despite this, they are synthesized from relatively few kinds of small molecules (referred to as either *monomers* or *subunits*) by a restricted repertoire of chemical reactions.

An oversimplified outline of the mechanism of addition of monomers to proteins, polynucleotides, and polysaccharides is shown in Figure 2–33. Although the synthetic reactions for each polymer involve a different kind of covalent bond and different enzymes and cofactors, there are strong underlying similarities. As indicated by the colored shading, the addition of subunits in each case occurs by a dehydration reaction, involving the removal of a molecule of water from the two reactants.

As in the general case discussed on page 74, the formation of these polymers requires the input of chemical energy, which is achieved by the standard strategy of coupling the biosynthetic reaction to the energetically favorable hydrolysis of a nucleoside triphosphate. For all three types of macromolecules, at least one nucleoside triphosphate is cleaved to produce pyrophosphate, which is subsequently hydrolyzed so that the driving force for the reaction is large. The mechanism used for polynucleotide synthesis was illustrated previously in Figure 2–29.

The activated intermediates in the polymerization reactions can be oriented in one of two ways, giving rise to either the head polymerization or the tail polymerization of monomers. In *head polymerization*, the reactive bond is carried on the end of the growing polymer and must therefore be regenerated each time a monomer is added. In this case, each monomer brings with it the reactive bond that will be used to react with the *next* monomer in the series (Figure 2–34). In *tail polymerization*, the reactive bond carried by each monomer is used instead for its own addition. Both types of polymerization are used for the synthesis of biological macromolecules. For example, the synthesis of polynucleotides and some simple polysaccharides occurs by tail polymerization, whereas the synthesis of proteins occurs by head polymerization.

Summary

The hydrolysis of ATP is commonly coupled to energetically unfavorable reactions, such as the biosynthesis of macromolecules, by the formation of reactive phosphorylated intermediates. Other reactive carrier molecules, called coenzymes, transfer other chemical groups in the course of biosynthesis: for example, NADPH transfers hydrogen as a proton plus two electrons (a hydride ion), whereas acetyl CoA transfers acetyl groups. Polymeric molecules such as proteins and nucleic acids are assembled from small activated precursor molecules by repetitive dehydration reactions.

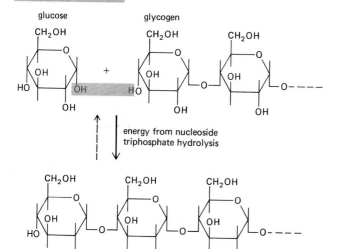

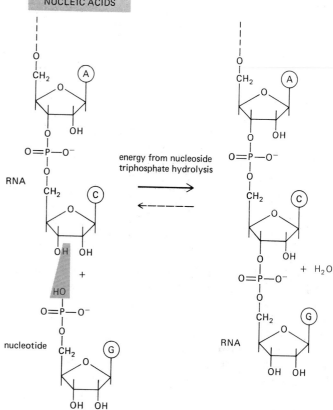

Figure 2–33 Outline of the polymerization reactions by which three kinds of biological polymer are synthesized, illustrating that synthesis in every case involves the loss of water (dehydration). Not shown is the consumption of high-energy nucleoside triphosphates that is required to activate each monomer prior to its addition. In contrast, the reverse reaction—the breakdown of all three types of polymer—occurs by the simple addition of water (hydrolysis).

PROTEINS

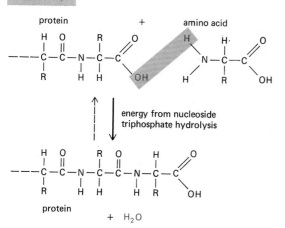

Figure 2–34 Head growth compared to the tail growth of polymers.

HEAD POLYMERIZATION (e.g., PROTEINS, FATTY ACIDS)

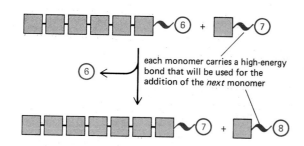

TAIL POLYMERIZATION (e.g., DNA, RNA, POLYSACCHARIDES)

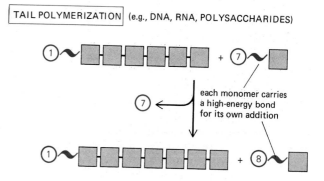

The Coordination of Catabolism and Biosynthesis [19]

Metabolism Is Organized and Regulated

Some idea of how intricate a cell is when viewed as a chemical machine can be obtained from Figure 2–35, which is a chart showing only some of the enzymatic pathways in a cell. All of these reactions occur in a cell that is less than 0.1 mm in diameter, and each requires an enzyme that is itself the product of a whole series of information-transfer and protein-synthesis reactions. For a typical small molecule—the amino acid *serine*, for example—there are half a dozen or more enzymes that can modify it chemically in different ways: it can be linked to AMP (adenylated) in preparation for protein synthesis, or degraded to glycine, or converted to pyruvate in preparation for oxidation; it can be acetylated by acetyl CoA or transferred to a fatty acid to make phosphatidyl serine. All of these different pathways compete for the same serine molecule, and similar competitions for thousands of other small molecules go on at the same time. One might think that the whole system would need to be so finely balanced that any minor upset, such as a temporary change in dietary intake, would be disastrous.

In fact, the cell is amazingly stable. Whenever it is perturbed, the cell reacts so as to restore its initial state. It can adapt and continue to function during starvation or disease. Mutations of many kinds can eliminate particular reaction pathways, and yet—provided that certain minimum requirements are met—the cell survives. It does so because an elaborate network of control mechanisms regulates and coordinates the rates of its reactions. Some of the higher levels of control will be considered in later chapters. Here we are concerned only with the simplest mechanisms that regulate the flow of small molecules through the various metabolic pathways.

Metabolic Pathways Are Regulated by Changes in Enzyme Activity [20]

The concentrations of the various small molecules in a cell are buffered against major changes by a process known as **feedback regulation**, which fine-tunes the flux of metabolites through a particular pathway by temporarily increasing or decreasing the activity of crucial enzymes. For example, the first enzyme of a series of reactions is usually inhibited by a *negative feedback* effect of the final product of that pathway: if large quantities of the final product accumulate, further entry of precursors into the reaction pathway is automatically inhibited (Figure 2–36). Where pathways branch or intersect, as they often do, there are usually multiple points of control by different final products. The complexity of such feedback control processes is illustrated in Figure 2–37, which shows the pattern of enzyme regulation observed in a set of related amino acid pathways.

Feedback regulation can work almost instantaneously and is reversible; in addition, a given end product may activate enzymes leading along other pathways, as well as inhibit enzymes that cause its own synthesis. The molecular basis for this type of control in cells is well understood, but since an explanation requires some knowledge of protein structure, it will be deferred until Chapter 3 (p. 128).

Catabolic Reactions Can Be Reversed by an Input of Energy [21]

By regulating a few enzymes at key points in a metabolic network, large-scale changes that affect the metabolism of the entire cell can be achieved. For example, a special pattern of feedback regulation enables a cell to switch from glucose degradation to glucose biosynthesis (denoted *gluconeogenesis*). The need for gluconeogenesis is especially acute in periods of violent exercise, when the glucose needed for muscle contraction is generated from lactic acid by liver cells, and also in periods of starvation, when glucose must be formed from the glycerol portion of fats and from amino acids for survival.

The normal breakdown of glucose to pyruvate during glycolysis is catalyzed by a number of enzymes acting in series. The reactions catalyzed by most of these enzymes are readily reversible, but three reaction steps (numbers 1, 3, and 9 in

Figure 2–35 Some of the chemical ▶ reactions occurring in a cell. (A) About 500 common metabolic reactions are shown diagrammatically, with each chemical species represented by a filled circle. The centrally placed reactions of the glycolytic pathway and the citric acid cycle are shown in solid color. A typical mammalian cell synthesizes more than 10,000 different proteins, a major proportion of which are enzymes. In the arbitrarily selected segment of this metabolic maze (color shaded), cholesterol is synthesized from acetyl CoA. To the right and below the maze, this segment is shown in detail in an enlargement (B).

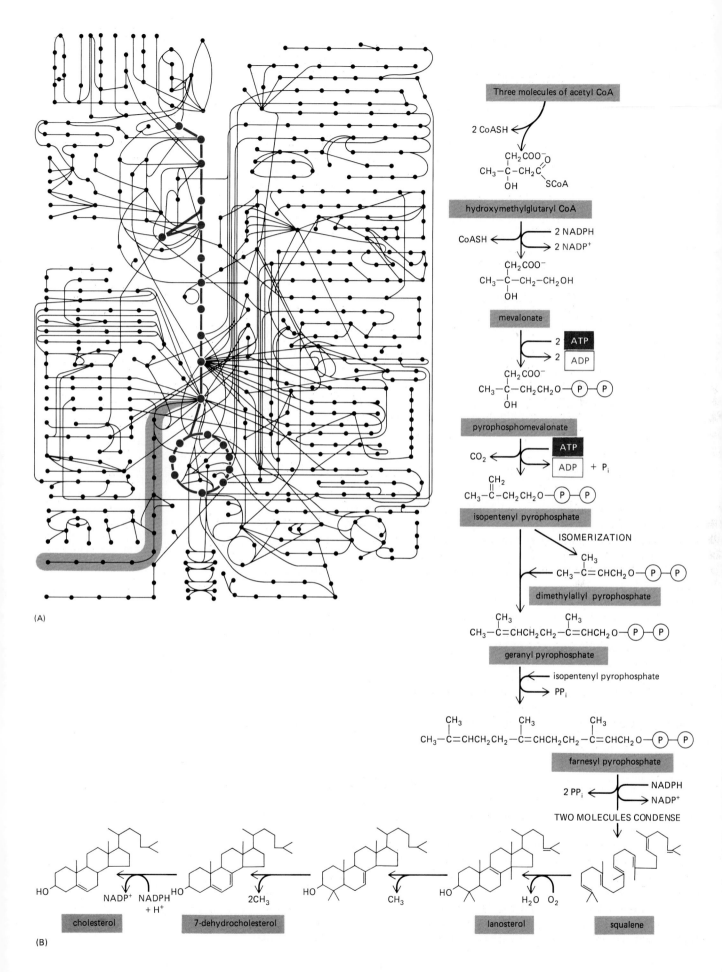

(A)

Three molecules of acetyl CoA

2 CoASH

hydroxymethylglutaryl CoA

CoASH   2 NADPH
          2 NADP⁺

mevalonate

2 ATP
2 ADP

pyrophosphomevalonate

CO₂   ATP
       ADP + P_i

isopentenyl pyrophosphate

ISOMERIZATION

dimethylallyl pyrophosphate

geranyl pyrophosphate

isopentenyl pyrophosphate
PP_i

farnesyl pyrophosphate

2 PP_i   NADPH
          NADP⁺

TWO MOLECULES CONDENSE

cholesterol   7-dehydrocholesterol   lanosterol   squalene

NADP⁺ NADPH
       + H⁺   2CH₃   CH₃   H₂O O₂

(B)

81

the sequence of Figure 2–20) are effectively irreversible. In fact, it is the large negative free-energy change that occurs in these reactions that normally drives the breakdown of glucose. For the reactions to proceed in the opposite direction and make glucose from pyruvate, each of these three reactions must be bypassed. This is achieved by substituting three alternative, enzyme-catalyzed bypass reactions that are driven in the uphill direction by an input of chemical energy (Figure 2–38). Thus, whereas two ATP molecules are generated as each molecule of glucose is degraded to two molecules of pyruvate, the reverse reaction during gluconeogenesis requires the hydrolysis of four ATP and two GTP molecules. This is equivalent, in total, to the hydrolysis of six molecules of ATP for every molecule of glucose synthesized.

The bypass reactions in Figure 2–38 must be controlled so that glucose is broken down rapidly when energy is needed but is synthesized when the cell is nutritionally replete. If both forward and reverse reactions were allowed to proceed without restraint, they would shuttle large quantities of metabolites backward and forward in futile cycles that would consume large amounts of ATP and generate heat for no purpose.

The elegance of the control mechanisms involved can be illustrated by a single example. Step 3 of glycolysis is one of the reactions that must be bypassed during glucose formation. Normally this step involves the addition of a second phosphate group to fructose 6-phosphate from ATP and is catalyzed by the enzyme *phosphofructokinase*. This enzyme is activated by AMP, ADP, and inorganic phosphate, whereas it is inhibited by ATP, citrate, and fatty acids. Therefore, the enzyme is activated by the accumulation of the products of ATP hydrolysis when energy supplies are low, and it is inactivated when energy (in the form of ATP) or food

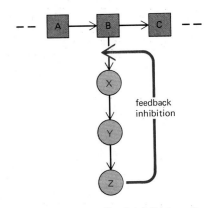

Figure 2–36 Feedback inhibition of a single biosynthetic pathway. The end product Z inhibits the first enzyme that is unique to its synthesis and thereby controls its own level in the cell.

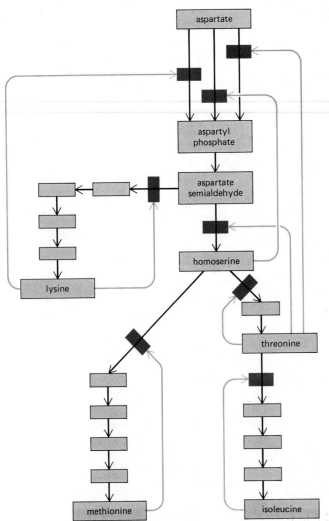

Figure 2–37 Feedback inhibition in the synthesis of the amino acids lysine, methionine, threonine, and isoleucine in bacteria. The colored arrows indicate positions at which products "feed back" to inhibit enzymes. Note that three different enzymes (called *isozymes*) catalyze the initial reaction, each inhibited by a different product.

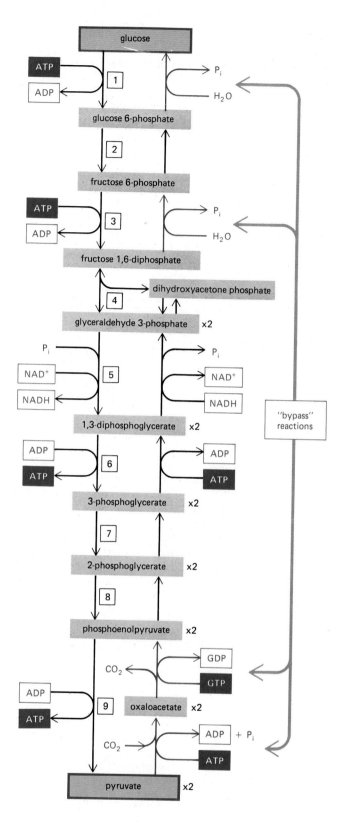

Figure 2–38 Comparison of the reactions that produce glucose during gluconeogenesis with those that degrade glucose during glycolysis. The degradative (glycolytic) reactions are energetically favorable (the free-energy change is less than zero), whereas the synthetic reactions require an input of energy. To synthesize glucose, different "bypass enzymes" are needed that bypass reactions 1, 3, and 9 of glycolysis. The overall flux of reactants between glucose and pyruvate is determined by feedback control mechanisms that operate to control the metabolism of the molecules that participate in these three glycolytic steps.

supplies such as fatty acids or citrate (derived from amino acids) are abundant. *Fructose diphosphatase* is the enzyme that catalyzes the reverse (bypass) reaction (the hydrolysis of fructose 1,6-diphosphate to fructose 6-phosphate, leading to the formation of glucose); this enzyme is regulated in the opposite way by the same feedback control molecules, so that it is stimulated when the phosphofructokinase is inhibited.

Note that phosphofructokinase is activated by ADP, which is a product of the reaction it catalyzes (ATP + fructose 6-phosphate → ADP + fructose 1,6-diphosphate), and is inhibited by ATP, which is one of its substrates. As a result, this enzyme can turn itself on, being subject to a complex form of positive feedback control. Under certain circumstances such feedback control gives rise to striking oscillations in the activity of the enzyme, causing corresponding oscillations in the concentrations of various glycolytic intermediates (Figure 2–39). Although the physiological significance of these particular oscillations is not known, they illustrate how a biological oscillator can be produced by a few enzymes. In principle, such oscillations could provide an internal clock, enabling a cell to "measure time" and, for example, to perform certain functions at fixed intervals.

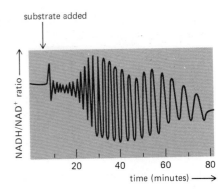

Figure 2–39   The abrupt addition of glucose to an extract containing the enzymes and cofactors required for glycolysis can produce large cyclic fluctuations in the levels of intermediates such as NADH. These metabolic oscillations arise, in part, from the positive feedback control of the glycolytic enzyme phosphofructokinase.

Enzymes Can Be Switched On and Off by Covalent Modification[22]

The types of feedback control just described permit the rates of reaction sequences to be continuously and automatically regulated in response to second-by-second fluctuations in metabolism. Cells have different devices for regulating enzymes when longer-lasting changes in activity, occurring over minutes or hours, are required. These involve reversible covalent modification of enzymes, which is often, but not always, accomplished by the addition of a phosphate group to a specific serine, threonine, or tyrosine residue in the enzyme. The phosphate comes from ATP, and its transfer is catalyzed by a family of enzymes known as *protein kinases*.

We shall describe in the following chapter how phosphorylation can alter the shape of an enzyme in such a way as to increase or inhibit its activity. The subsequent removal of the phosphate group, which reverses the effect of the phosphorylation, is achieved by a second type of enzyme, called a *phosphoprotein phosphatase*. Covalent modification of enzymes adds another dimension to metabolic control because it allows specific reaction pathways to be regulated by signals (such as hormones) that are unrelated to the metabolic intermediates themselves.

Reactions Are Compartmentalized Both Within Cells and Within Organisms[23]

Not all of a cell's metabolic reactions occur within the same subcellular compartment. Because different enzymes are found in different parts of the cell, the flow of chemical components is channeled physically as well as chemically.

The simplest form of such spatial segregation occurs when two enzymes that catalyze sequential reactions form an enzyme complex, and the product of the first enzyme does not have to diffuse through the cytoplasm to encounter the second enzyme. The second reaction begins as soon as the first is over. Some large enzyme aggregates carry out whole series of reactions without losing contact with the substrate. For example, the conversion of pyruvate to acetyl CoA proceeds in three chemical steps, all of which take place on the same large enzyme complex (Figure 2–40), and in fatty acid synthesis an even longer sequence of reactions is catalyzed by a single enzyme assembly. Not surprisingly, some of the largest en-

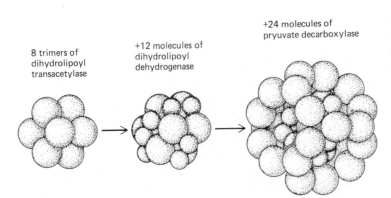

8 trimers of dihydrolipoyl transacetylase

+12 molecules of dihydrolipoyl dehydrogenase

+24 molecules of pryuvate decarboxylase

Figure 2–40   The structure of pyruvate dehydrogenase—an example of a large multienzyme complex in which reaction intermediates are passed directly from one enzyme to another. This enzyme complex catalyzes the conversion of pyruvate to acetyl CoA.

zyme complexes are concerned with the synthesis of macromolecules such as proteins and DNA.

The next level of spatial segregation in cells involves the confinement of functionally related enzymes within the same membrane or within the aqueous compartment of an organelle that is bounded by a membrane. The oxidative metabolism of glucose is a good example (Figure 2–41). After glycolysis, pyruvate is actively taken up from the cytosol into the inner compartment of the mitochondrion, which contains all of the enzymes and metabolites involved in the citric acid cycle. Moreover, the inner mitochondrial membrane itself contains all of the enzymes that catalyze the subsequent reactions of oxidative phosphorylation, including those involved in the transfer of electrons from NADH to O_2 and in the synthesis of ATP. The entire mitochondrion can therefore be regarded as a small ATP-producing factory. In the same way, other cellular organelles, such as the nucleus, the Golgi apparatus, and the lysosomes, can be viewed as specialized compartments where functionally related enzymes are confined to perform a specific task. In a sense, the living cell is like a city, with many specialized services concentrated in different areas that are extensively interconnected by various paths of communication.

Spatial organization in a multicellular organism extends beyond the individual cell. The different tissues of the body have different sets of enzymes and make distinct contributions to the chemistry of the organism as a whole. In addition to differences in specialized products such as hormones or antibodies, there are significant differences in the "common" metabolic pathways among various types of cells in the same organism. Although virtually all cells contain the enzymes of glycolysis, the citric acid cycle, lipid synthesis and breakdown, and amino acid metabolism, the levels of these processes in different tissues are differently regulated. Nerve cells, which are probably the most fastidious cells in the body, maintain almost no reserves of glycogen or fatty acids and rely almost entirely on a supply of glucose from the bloodstream. Liver cells supply glucose to actively contracting muscle cells and recycle the lactic acid produced by muscle cells back into glucose (Figure 2–42). All types of cells have their distinctive metabolic traits and cooperate extensively in the normal state, as well as in response to stress and starvation.

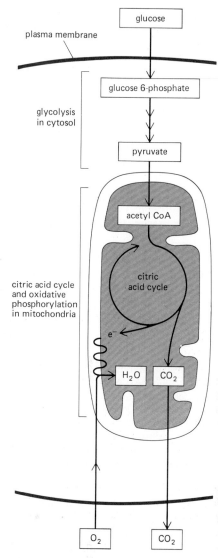

Figure 2–41 Segregation of the various steps in the breakdown of glucose in the eucaryotic cell. Glycolysis occurs in the cytosol, whereas the reactions of the citric acid cycle and oxidative phosphorylation take place only in mitochondria.

Summary

The many thousands of different chemical reactions carried out simultaneously by a cell are closely coordinated. A variety of control mechanisms regulate the activities of key enzymes in response to the changing conditions in the cell. One very common form of regulation is a rapidly reversible feedback inhibition exerted on the first enzyme of a pathway by the final product of that pathway. A longer-lasting form of regulation involves the chemical modification of one enzyme by another, often by phosphorylation. Combinations of regulatory mechanisms can produce major and long-lasting changes in the metabolism of the cell. Not all cellular reactions occur within the same intracellular compartment, and spatial segregation by internal membranes permits organelles to specialize in their biochemical tasks.

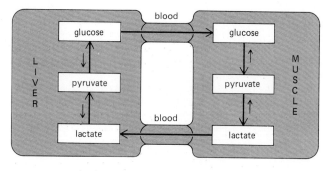

Figure 2–42 Schematic view of the metabolic cooperation between liver and muscle cells. The principal fuel of actively contracting muscle cells is glucose, much of which is supplied by liver cells. Lactic acid, the end product of anaerobic glucose breakdown by glycolysis in muscle, is converted back to glucose in the liver by the process of gluconeogenesis.

References

General

Herriott, J.; Jacobson, G.; Marmur, J.; Parsom, W. Papers in Biochemistry. Reading, MA: Addison-Wesley, 1984.

Lehninger, A.L. Principles of Biochemistry. New York: Worth, 1982.

Stryer, L. Biochemistry, 3rd ed. New York: W.H. Freeman, 1988.

Wood, W.B.; Wilson, J.H.; Benbow, R.M.; Hood, L.E. Biochemistry: A Problems Approach, 2nd ed. Menlo Park, CA: Benjamin-Cummings, 1981.

Cited

1. Henderson, L.J. The Fitness of the Environment. Boston: Beacon, 1927; reprinted 1958. (A classical and readable analysis.)

2. Ingraham, J.L.; Maaløe, O.; Neidhardt, F.C. Growth of the Bacterial Cell. Sunderland, MA: Sinauer, 1983.

3. Roehrig, K.L. Carbohydrate Biochemistry and Metabolism. Westport, CT: AUI Publishing, 1984.
   Sharon, N. Carbohydrates. *Sci. Am.* 243(5):90–116, 1980.

4. Robertson, R.N. The Lively Membranes. Cambridge, U.K.: Cambridge University Press, 1983.

5. Saenger, W. Principles of Nucleic Acid Structure. New York: Springer, 1984.

6. Hess, B.; Markus, M. Order and chaos in biochemistry. *Trends Biochem. Sci.* 12:45–48, 1987.
   Lehninger, A.L. Bioenergetics: The Molecular Basis of Biological Energy Transformations, 2nd ed. Menlo Park, CA: Benjamin-Cummings, 1971.
   Schrödinger, E. What is Life? Mind and Matter. Cambridge, U.K.: Cambridge University Press, 1969.

7. Dickerson, R.E. Molecular Thermodynamics. Menlo Park, Ca.: Benjamin, 1969.
   Klotz, I.M. Energy Changes in Biochemical Reactions. New York: Academic Press, 1967.

8. Raven, P.H.; Evert, R.F.; Eihhorn, S.E. Biology of Plants, 4th ed. New York: Worth, 1981.

9. Racker, E. From Pasteur to Mitchell: a hundred years of bioenergetics. *Fed. Proc.* 39:210–215, 1980.

10. Fothergill-Gilmore, L.A. The evolution of the glycolytic pathway. *Trends Biochem. Sci.* 11:47–51, 1986.
    Shulman, R.G. NMR Spectroscopy of living cells. *Sci. Am.* 248(1):86–93, 1983.

11. Lipmann, F. Wanderings of a Biochemist. New York: Wiley, 1971.

12. Schlenk, F. The ancestry, birth and adolescence of ATP. *Trends Biochem. Sci.* 12:367–368, 1987.

13. McGilvery, R.W. Biochemistry: A Functional Approach, 3rd ed. Philadelphia: Saunders, 1983.
    Racker, E. A New Look at Mechanisms in Bioenergetics. New York: Academic Press, 1976.

14. Kornberg, H.L. Tricarboxylic acid cycles. *Bioessays* 7:236–238, 1987.

15. Krebs, H.A. The history of the tricarboxylic acid cycle. *Perspect. Biol. Med.* 14:154–170, 1970.
    Krebs, H.A.; Martin, A. Reminiscences and Reflections. Oxford, U.K.: Clarendon Press; New York: Oxford University Press, 1981.

16. Pullmann, M.E.; et al. Partial resolution of the enzymes catalyzing oxidative phosphorylation. *J. Biol. Chem.* 235:3322–3329, 1960.

17. Prigogine I.; Stengers, I. Order out of Chaos: Man's New Dialogue with Nature. New York: Bantam Books, 1984.

18. Eisenberg, D.; Crothers, D. Physical Chemistry with Applications to the Life Sciences. Menlo Park, CA: Benjamin-Cummings, 1979.
    Reich, J.G.; Selkov, E.E. Energy Metabolism of the Cell-A Theoretical Treatise. New York: Academic Press, 1981.

19. Martin, B.R. Metabolic Regulation: A Molecular Approach. Oxford, U.K.; Blackwell Scientific, 1987.
    Newsholme, E.A.; Start, C. Regulation in Metabolism. New York: Wiley, 1973.

20. Pardee, A.B. Molecular basis of biological regulation: origins of feedback inhibition and allostery. *Bioessays* 2:37–40, 1985.

21. Hess, B. Oscillating reactions. *Trends Biochem. Sci.* 2:193–195, 1977.

22. Cohen, P. Control of Enzyme Activity, 2nd ed. London: Chapman and Hall, 1983.
    Koshland, D.E., Jr. Switches, thresholds and ultrasensitivity. *Trends Biochem. Sci.* 12:225–229, 1987.

23. Banks, P.; Bartley, W.; Birt, M. The Biochemistry of the Tissues, 2nd ed. New York: Wiley, 1976.
    Sies, H. Metabolic Compartmentation. New York: Academic Press, 1982.

Macromolecules: Structure, Shape, and Information

3

In moving from the small molecules of the cell to the giant macromolecules, we undergo a transition of more than size alone. Even though proteins, nucleic acids, and polysaccharides are made from amino acids, nucleotides, and sugars, respectively, they can have unique and truly astounding properties that bear little resemblance to those of their simple chemical precursors. Biological macromolecules are composed of many thousands—sometimes millions—of atoms linked together in precisely defined spatial arrangements. Each of these macromolecules carries specific information. Incorporated in its structure is a series of biological messages that can be "read" in its interactions with other molecules, enabling it to perform a precise function required by the cell.

In this chapter we shall examine the structures of macromolecules, emphasizing proteins and nucleic acids, and explain how they have adapted in the course of evolution to perform specific functions. We shall consider the principles by which these molecules catalyze chemical transformations, build complex multimolecular structures, generate movement, and—most fundamental of all—store and transmit hereditary information.

Molecular Recognition Processes[1]

Macromolecules typically have molecular weights between about 10,000 and 1 million and are intermediate in size between the organic molecules of the cell discussed in Chapter 2 and the large macromolecular assemblies and organelles that will be discussed in subsequent chapters (Figure 3–1). One small molecule, water, constitutes 70% of the total mass of a cell; nearly all of the remaining cell mass is due to macromolecules (Table 3–1).

As described in Chapter 2, a macromolecule is assembled from low-molecular-weight subunits that are added one after the other to form a long, chainlike polymer (see Figure 2–33). Usually only one family of subunits is used to construct each chain: amino acids are linked to other amino acids to form proteins, nucleotides are linked to other nucleotides to form nucleic acids, and sugars are linked to other sugars to form polysaccharides. Because the precise sequence of subunits is crucial to the function of a macromolecule, its biosynthesis requires mechanisms to ensure that the correct subunit goes into the polymer at each position in the chain.

sugars, amino acids,
and nucleotides ~0.5–1 nm

globular proteins ~2–10 nm

ribosome ~30 nm

Figure 3–1 The size of protein molecules compared to some other cell components. The ribosome is an important macromolecular assembly composed of about 60 protein and RNA molecules.

Table 3-1 Approximate Chemical Compositions of a Typical Bacterium and a Typical Mammalian Cell

Component	Percent of Total Cell Weight	
	E. Coli Bacterium	Mammalian Cell
H_2O	70	70
Inorganic ions (Na^+, K^+, Mg^{2+}, Ca^{2+}, Cl^-, etc.)	1	1
Miscellaneous small metabolites	3	3
Proteins	15	18
RNA	6	1.1
DNA	1	0.25
Phospholipids	2	3
Other lipids	—	2
Polysaccharides	2	2
Total cell volume:	2×10^{-12} cm^3	4×10^{-9} cm^3
Relative cell volume:	1	2000

Proteins, polysaccharides, DNA, and RNA are macromolecules. Lipids are not generally classed as macromolecules even though they share some of their features; for example, most are synthesized as linear polymers of a smaller molecule (the acetyl group on acetyl CoA) and self-assemble into larger structures (membranes). Note that water and protein comprise most of the mass of both mammalian and bacterial cells.

The Specific Interactions of a Macromolecule Depend on Weak Noncovalent Bonds[2]

A macromolecular chain is held together by *covalent* bonds, which are strong enough to preserve the sequence of subunits for long periods of time. Although the sequence of subunits determines the information content of a macromolecule, utilizing that information depends largely on much weaker, *noncovalent* bonds. These weak bonds form between different parts of the same macromolecule and between different macromolecules. They therefore play a major part in determining both the three-dimensional structure of macromolecular chains and how these structures interact with one another.

The noncovalent bonds encountered in biological molecules are usually classified into three types: **ionic bonds, hydrogen bonds,** and **van der Waals attractions.** Another important weak force is created by the three-dimensional structure of water, which tends to force hydrophobic groups together in order to minimize their disruptive effect on the hydrogen-bonded network of water molecules (see Panel 2–1, pp. 46–47). This expulsion from the aqueous solution generates what is sometimes thought of as a fourth kind of weak noncovalent bond. These four types of weak bonds are the subject of Panel 3–1, pp. 90–91.

In an aqueous environment each noncovalent bond is 30 to 300 times weaker than the typical covalent bonds that hold biological molecules together (Table 3–2) and only slightly stronger than the average energy of thermal collisions at 37°C. A single noncovalent bond—unlike a single covalent bond—is therefore too weak to withstand the thermal motions that tend to pull molecules apart, and large numbers of noncovalent bonds are needed to hold two molecular surfaces together. Large numbers of noncovalent bonds can form between two surfaces only when large numbers of atoms on the surfaces are precisely matched to each other (Figure 3–2), which accounts for the specificity of biological recognition, such as occurs between an enzyme and its substrates.

The weak noncovalent forces determine how different regions of the same macromolecule fit together, in addition to determining how that macromolecule will interact with other molecules. However, as explained at the top of Panel 3–1, atoms behave almost as if they were hard spheres with a definite radius (their

Table 3–2   Covalent and Noncovalent Chemical Bonds

Bond Type	Length (nm)	Strength (kcal/mole)	
		In Vacuum	In Water
Covalent	0.15	90	90
Ionic	0.25	80	3
Hydrogen	0.30	4	1
Van der Waals attraction (per atom)	0.35	0.1	0.1

The strength of a bond can be measured by the energy required to break it, here given in kilocalories per mole (kcal/mole). (*One kilocalorie* is the quantity of energy needed to raise the temperature of 1000 g of water by 1°C. An alternative unit in wide use is the kilojoule, kJ, equal to 0.24 kcal.) Individual bonds vary a great deal in strength, depending on the atoms involved and their precise environment, so that the above values are only a rough guide. Note that the aqueous environment in a cell will greatly weaken both the ionic and the hydrogen bonds between nonwater molecules (Panel 3–1, pp. 90–91). The bond length is the center-to-center distance between the two interacting atoms; the length given here for a hydrogen bond is that between its two nonhydrogen atoms.

"van der Waals radius"). The requirement that no two atoms overlap severely limits the number of three-dimensional arrangements of atoms (or **conformations**) that are possible for each polypeptide chain. Nevertheless, a long flexible chain such as a protein can still fold in an enormous number of ways, each conformation having a different set of weak intrachain interactions. In practice, however, most proteins in a cell fold stably in only one way: during the course of evolution, the sequence of amino acid subunits has been selected so that one conformation is able to form many more favorable intrachain interactions than any other.

A Helix Is a Common Structural Motif in Biological Structures Made from Repeated Subunits[3]

Biological structures are often formed by linking subunits that are very similar to each other—such as amino acids or nucleotides—into a long, repetitive chain (see p. 78). If all the subunits are identical, neighboring subunits in the chain will often fit together in only one way, adjusting their relative positions so as to minimize the free energy of the contact between them. Each subunit will be positioned in exactly the same way in relation to its neighboring subunits, so that subunit 3 will fit onto subunit 2 in the same way that subunit 2 fits onto subunit 1, and so on. Because it is very rare for subunits to join up in a straight line, this arrangement will generally result in a **helix**—a regular structure that resembles a spiral stair-

Figure 3–2   How weak bonds mediate recognition between macromolecules.

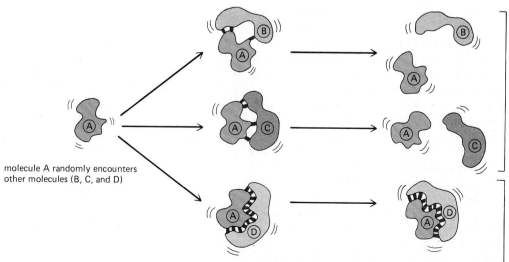

molecule A randomly encounters other molecules (B, C, and D)

the surfaces of molecules A and B, and A and C, are a poor match and are capable of forming only a few weak bonds; thermal motion rapidly breaks them apart

the surfaces of molecules A and D match well and therefore can form enough weak bonds to withstand thermal motion and stay bound to each other

VAN DER WAALS FORCES

At very short distances, any two atoms show a weak bonding inter-action due to their fluctuating electrical charges. This force is known as a van der Waals attraction. However, two atoms will very strongly repel each other if they are brought too close together. This van der Waals repulsion plays a major part in limiting the possible conformations of a molecule.

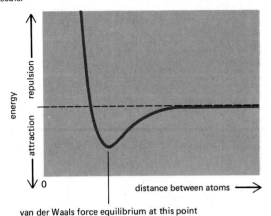

van der Waals force equilibrium at this point

Each type of atom has a radius, known as its van der Waals radius at which van der Waals forces are in equilibrium.

H	C	N	O
1.2 Å	2.0 Å	1.5 Å	1.4 Å
(0.12 nm)	(0.2 nm)	(0.15 nm)	(0.14 nm)

Two atoms will be attracted to each other by van der Waals forces until the distance between them equals the sum of their van der Waals radii. Although they are individually very weak, these van der Waals attractions can become important when two macromolecular surfaces fit very closely together.

WEAK CHEMICAL BONDS

Organic molecules can interact with other molecules through short-range noncovalent forces.

weak bond

Weak chemical bonds have less than 1/20 the strength of a strong covalent bond. They are strong enough to provide tight binding only when many of them are formed simultaneously.

HYDROGEN BONDS

A hydrogen atom is shared between two other atoms (both electronegative, such as O and N) to give a hydrogen bond.

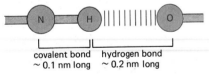

covalent bond hydrogen bond
~ 0.1 nm long ~ 0.2 nm long

Hydrogen bonds are strongest when the three atoms are in a straight line:

Examples in macromolecules:

Amino acids in polypeptide chains hydrogen-bonded together

Two bases, G and C, hydrogen-bonded in DNA or RNA

HYDROGEN BONDS IN WATER

Any molecules that can form hydrogen bonds to each other can alternatively form hydrogen bonds to water molecules. Because of this competition with water molecules, the hydrogen bonds formed between two molecules dissolved in water are relatively weak.

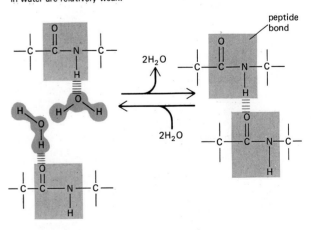

peptide bond

HYDROPHOBIC FORCES

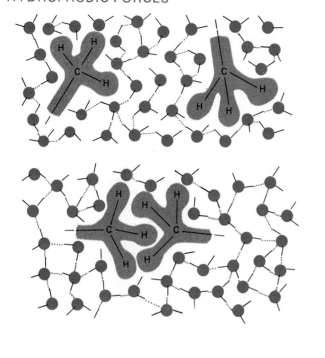

Water forces hydrophobic groups together in order to minimize their disruptive effects on the hydrogen-bonded water network. Hydrophobic groups held together in this way are sometimes said to be held together by "hydrophobic bonds," even though the attraction is actually caused by a repulsion from the water.

IONIC BONDS IN AQUEOUS SOLUTIONS

Charged groups are shielded by their interactions with water molecules. Ionic bonds are therefore quite weak in aqueous solution.

Ionic bonds are further weakened by the presence of salts, whose atoms form the counterions that cluster around ions of opposite charge.

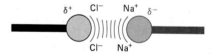

Measurement of the extent of destabilization of an interaction by salt provides a quantitative estimate of the total number of ionic bonds involved.

Despite being weakened by water and salt, ionic bonds are very important in biological systems; an enzyme that binds a positively charged substrate will often have a negatively charged amino acid side chain at the appropriate place.

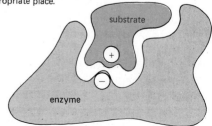

IONIC BONDS

Ionic interactions occur between either fully charged groups (ionic bond) or between partially charged groups.

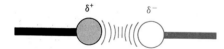

The force of attraction between the two charges δ^+ and δ^- is

$$\text{force} = \frac{\delta^+ \delta^-}{r^2 D} \quad \text{(Coulomb's law)}$$

where D = dielectric constant
(1 for vacuum; 80 for water)

r = distance of separation

In the absence of water, ionic forces are very strong. They are responsible for the strength of minerals such as marble and agate.

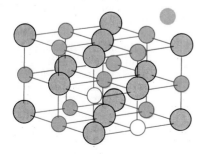

NaCl
crystal

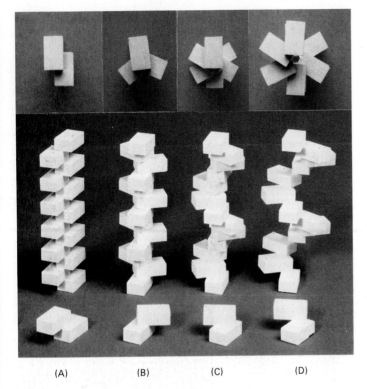

(A)   (B)   (C)   (D)

Figure 3–3 A helix will form when a series of subunits bind to each other in a regular way. In the foreground the interaction between two subunits is shown; behind it are the helices that result. These helices have (A) two, (B) three, and (C) and (D) six subunits per turn. At the top, the arrangement of subunits has been photographed from directly above the helix. Note that the helix in (D) has a wider path than that in (C).

left-handed   right-handed
helix   helix

Figure 3–4 Comparison of a left-handed and a right-handed helix. As a reference, it is useful to remember that standard screws, which insert when turned clockwise, are right-handed. Note that a helix preserves the same handedness when it is turned upside down.

case, as illustrated in Figure 3–3. Depending on the twist of the staircase, a helix is said to be either right-handed or left-handed (Figure 3–4). Handedness is not affected by turning the helix upside down, but it is reversed if the helix is reflected in a mirror.

Helices occur commonly in biological structures, whether the subunits are small molecules that are covalently linked together (as in DNA) or large protein molecules that are linked by noncovalent forces (as in actin filaments). This is not surprising. A helix is an unexceptional structure, generated simply by stacking many similar subunits next to each other, each in the same strictly repeated relationship to the one before.

Diffusion Is the First Step to Molecular Recognition[4]

Before two molecules can bind to each other, they must come into close contact. This is achieved by the thermal motions that cause molecules to wander, or *diffuse*, from their starting positions. As the molecules in a liquid rapidly collide and bounce off one another, an individual molecule moves first one way and then another, its path constituting a "random walk" (Figure 3–5). The average distance that each type of molecule travels from its starting point is proportional to the square root of the time involved: that is, if it takes a particular molecule 1 second on average to go 1 μm, it will go 2 μm in 4 seconds, 10 μm in 100 seconds, and so on. Diffusion is therefore an efficient way for molecules to move limited distances but an inefficient way for molecules to move long distances.

Experiments performed by injecting fluorescent dyes and other labeled molecules into cells show that the diffusion of small molecules through the cytoplasm is nearly as rapid as it is in water. A molecule the size of ATP, for example, requires only about 0.2 second to diffuse an average distance of 10 μm—the diameter of a

Figure 3–5 A random walk. Molecules in solution move in a random fashion due to the continual buffeting they receive in collisions with other molecules. This movement allows small molecules to diffuse from one part of the cell to another in a surprisingly short time: such molecules will generally diffuse across a typical animal cell in less than a second.

small animal cell. Large macromolecules, however, move much more slowly. Not only is their diffusion rate intrinsically slower, but their movement is retarded by frequent collisions with many other macromolecules that are held in place by molecular associations in the cytoplasm (Figure 3–6).

Thermal Motions Bring Molecules Together and Then Pull Them Apart[5]

Encounters between two macromolecules or between a macromolecule and a small molecule occur randomly through simple diffusion. An encounter may lead immediately to the formation of a complex, in which case the rate of complex formation is said to be *diffusion-limited*. Alternatively, the rate of complex formation may be slower, requiring some adjustment of the structure of one or both molecules before the interacting surfaces can fit together, so that most often the two colliding molecules will bounce off each other without sticking. In either case, once the two interacting surfaces have come sufficiently close together, they form multiple weak bonds with each other that persist until random thermal motion causes the molecules to dissociate again.

In general, the stronger the binding of the molecules in the complex, the slower their rate of dissociation. At one extreme the total energy of the bonds formed is negligible compared with that of thermal motion, and the two molecules dissociate as rapidly as they came together. At the other extreme the total bond energy is so high that dissociation rarely occurs. The precise strength of the bonding between two molecules is a useful index of the specificity of the recognition process.

To illustrate how the binding strength is measured, let us consider a reaction in which molecule A binds to molecule B. The reaction will proceed until it reaches an *equilibrium point*, at which the rates of formation and dissociation are equal. The concentrations of A, B, and the complex AB at this point can be used to determine an **equilibrium constant (K)** for the reaction, as explained in Figure 3–7. This constant is sometimes termed the **affinity constant** and is commonly employed as a measure of the strength of binding between two molecules: the *stronger* the binding, the *larger* is the value of the affinity constant.

The equilibrium constant of a reaction in which two molecules bind to each other is related directly to the standard free-energy change for the binding ($\Delta G°$) by the equation described in Table 3–3. The table also lists the $\Delta G°$ values corresponding to a range of K values. Affinity constants for simple binding interactions in biological systems often range between 10^3 and 10^{12} liters/mole; this corresponds to binding energies in the range 4–17 kcal/mole, which could arise from 4 to 17 average hydrogen bonds.

The strongest interactions occur whenever a biological function requires that two macromolecules remain tightly associated for a long period of time—for example, when a gene regulatory protein binds to DNA to turn off a gene (see p. 558). Weaker interactions occur when the function demands a rapid change in the structure of a complex—for example, when two interacting proteins change partners during the movements of a protein machine (see p. 131).

Atoms and Molecules Are in Constant Motion[6]

The chemical reactions in a cell occur at amazingly fast rates. A typical enzyme molecule, for example, will catalyze on the order of 1000 reactions per second, and rates of more than 10^6 reactions per second are achieved by some enzymes, such as catalase. Since each reaction requires a separate encounter between an enzyme and a substrate molecule, such rates are possible only because the molecules are moving so rapidly. Molecular motions can be classified broadly into three kinds: (1) the movement of a molecule from one place to another (translational motion), (2) the rapid back-and-forth movement of covalently linked atoms with respect to one another (vibrations), and (3) rotations. All of these motions are important in bringing the surfaces of interacting molecules together.

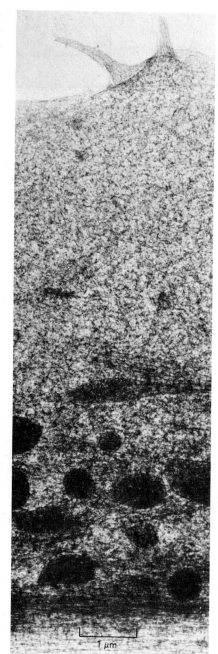

Figure 3–6 An electron micrograph of a region of the cytoplasm of an animal cell illustrating the high concentration of proteins it contains. Macromolecules diffuse relatively slowly in the cytoplasm because they interact with many other macromolecules; small molecules diffuse nearly as rapidly as they do in water. The cell shown in this micrograph was prepared by a special rapid freezing method that optimally preserves cytoplasmic structures. (Reproduced from P.C. Bridgman and T.S. Reese, *J. Cell Biol.* 99:1655–1668, 1984. By copyright permission of the Rockefeller University Press.)

$$(1) \quad \text{A} \text{B} \xrightarrow{\text{dissociation}} \text{A} + \text{B}$$

$$\text{dissociation rate} = \begin{pmatrix} \text{dissociation} \\ \text{rate constant} \end{pmatrix} \times \begin{pmatrix} \text{concentration} \\ \text{of AB} \end{pmatrix}$$

$$\text{dissociation rate} = k_{off} [AB]$$

$$(2) \quad \text{A} + \text{B} \xrightarrow{\text{association}} \text{A} \text{B}$$

$$\text{association rate} = \begin{pmatrix} \text{association} \\ \text{rate constant} \end{pmatrix} \times \begin{pmatrix} \text{concentration} \\ \text{of A} \end{pmatrix} \times \begin{pmatrix} \text{concentration} \\ \text{of B} \end{pmatrix}$$

$$\text{association rate} = k_{on} [A] [B]$$

---

(3) AT EQUILIBRIUM:

$$\text{association rate} = \text{dissociation rate}$$

$$k_{on} [A] [B] = k_{off} [AB]$$

$$\frac{[AB]}{[A] [B]} = \frac{k_{on}}{k_{off}} = K = \text{equilibrium constant}$$

EXAMPLE:

The concentration of a molecule present in only one copy in a typical mammalian cell (volume of 2000 μm^3) is about 10^{-12} M.
If such a cell contains 10^4 copies of protein molecule A and 10^6 copies of protein molecule B,

$$[A] = 10^{-8} \text{ M} \quad \text{and} \quad [B] = 10^{-6} \text{ M}$$

Suppose that protein A binds to protein B with $K = 10^7$ M^{-1}. The ratio of bound to unbound A will be [AB]/[A] and since

$$\frac{[AB]}{[A]} = K[B] = (10^7 \text{ M}^{-1})(10^{-6} \text{ M}) = 10$$

we expect one molecule of A to be free for every 10 molecules of A that are bound to B inside the cell. Repeating the calculation for $K = 10^4$ M^{-1} shows that only about one molecule of A in a hundred would be bound to B.

The rates of molecular motions can be measured by a variety of spectroscopic techniques, which indicate that a large globular protein is constantly tumbling, rotating about its axis about a million times per second. The rates of diffusional encounters due to translational movements are proportional to the concentration of the diffusing molecule. If ATP is present at its typical intracellular concentration of about 1 mM, for instance, each site on a protein molecule will be bombarded by about 10^6 random collisions with ATP molecules per second; for an ATP concentration tenfold lower, the number of collisions would drop to 10^5 per second, and so on.

Once two molecules have collided and are in the correct relative orientation, a chemical reaction can occur between them extremely rapidly. When one appreciates how quickly molecules move and react, the observed rates of enzymatic catalysis do not seem so amazing.

Molecular Recognition Processes Can Never Be Perfect[7]

All molecules possess energy—the kinetic energy of their translational movements, vibrations, and rotations, and the potential energy stored in their electron distributions. Through molecular collisions, this energy is randomly distributed to all of the atoms present, so that most atoms will have energy levels close to the average, with only a small proportion possessing very high energy. Although the favored conformations or states for a molecule will be those of lowest free energy (see p. 71), states of higher energy occur through unusually violent collisions. Given the temperature, it is possible to calculate the probability that an atom or a molecule will be in a particular energy state (see Table 3–3). The probability of a high energy state becomes smaller relative to a low energy state as the difference in free energy between the two increases. However, it reaches zero only when this energy difference becomes infinite.

Because of the random factor in molecular interactions, minor "side reactions" are bound to occur occasionally. As a consequence, a cell continually makes errors. Even reactions that are very energetically unfavorable will take place occasionally. Two atoms joined to each other by a covalent bond, for example, will eventually be subjected to an especially energetic collision and fall apart. Similarly, the specificity of an enzyme for its substrate cannot be absolute because the recognition of one molecule as distinct from another can never be perfect. Mistakes could be avoided completely only if the cell could evolve mechanisms with infinite energy differences between alternatives. Since this is not possible, cells are forced to tolerate a certain level of failure and have instead evolved a variety of repair reactions to correct those errors that are the most damaging (for example, see p. 217 and p. 223).

Figure 3–7 The principle of equilibrium. The equilibrium between molecules A and B and the complex AB is maintained by a balance between the two opposing reactions shown in (1) and (2). As shown in (3), the ratio of the rate constants for the association and the dissociation reactions is equal to the *equilibrium constant (K)* for the reaction. Molecules A and B must collide in order to react, and the synthesis rate in reaction (2) is therefore proportional to the product of their individual concentrations. As a result, the product [A] × [B] appears in the final expression for K, where [ ] indicates concentration.

As traditionally defined, the concentrations of products appear in the numerator and the concentrations of reactants appear in the denominator of the equation for an equilibrium constant. Thus the equilibrium constant in (3) is that for the *association* reaction A + B → AB, while the reciprocal of this constant would be the equilibrium constant for the *dissociation* reaction AB → A + B. When dealing with simple binding interactions, however, it is less confusing to speak of the *affinity constant* or *association constant* (in units of liters per mole); the larger the value of the association constant (K_a), the stronger is the binding between A and B. The reciprocal of K_a is the *dissociation constant* (in units of moles per liter); the smaller the value of the dissociation constant (K_d), the stronger is the binding between A and B.

On the other hand, errors are essential to life as we know it. If it were not for occasional mistakes in the maintenance of DNA sequences, evolution could not occur (see p. 97, below).

Summary

The sequence of subunits in a macromolecule contains information that determines the three-dimensional contours of its surface. These contours in turn govern the recognition between one molecule and another, or between different parts of the same molecule, by means of weak noncovalent bonds. Molecules are constantly in rapid motion, and they recognize each other by a process in which they first meet by random diffusion and then bind with a strength that can be expressed in terms of an equilibrium constant. Since the only way to make recognition infallible is to make the energy of binding infinitely large, living cells constantly make errors; those that are intolerable are corrected by specific repair processes.

Nucleic Acids[8]

Genes Are Made of DNA[9]

It has been obvious for as long as humans have sown crops or raised animals that each seed or fertilized egg must contain a hidden plan, or design, for the development of the organism. In modern times the science of genetics grew up around the premise of invisible information-containing elements, called **genes,** that are distributed to each daughter cell when a cell divides. Before dividing, therefore, a cell has to make a copy of its genes in order to give a complete set to each daughter cell. The genes in the sperm and egg cells carry the hereditary information from one generation to the next.

The inheritance of biological characteristics must involve patterns of atoms that follow the laws of physics and chemistry: in other words, genes must be formed from molecules. At first the nature of these molecules was hard to imagine. What kind of molecule could be stored in a cell and direct the activities of a developing organism and also be capable of accurate and almost unlimited replication?

By the end of the nineteenth century, biologists had recognized that the carriers of inherited information were the *chromosomes* that become visible in the nucleus as a cell begins to divide. But the evidence that the deoxyribonucleic acid (DNA) in these chromosomes is the substance of which genes are made came only much later, from studies on bacteria. In 1944 it was shown that adding purified DNA from one strain of bacteria to a second, slightly different bacterial strain conferred heritable properties characteristic of the first strain upon the second. Because it had been commonly believed that only proteins have enough conformational complexity to carry genetic information, this discovery came as a surprise, and it was not generally accepted until the early 1950s. Today the idea that DNA carries genetic information in its long chain of nucleotides is so fundamental to biological thought that it is sometimes difficult to realize the enormous intellectual gap that it filled.

DNA Molecules Consist of Two Long Chains Held Together by Complementary Base Pairs[10]

The difficulty that geneticists had in accepting DNA as the substance of genes is understandable, considering the simplicity of its chemistry. A DNA chain is a long, unbranched polymer composed of only four types of subunits. These are the deoxyribonucleotides containing the bases adenine (A), cytosine (C), guanine (G), and thymine (T). The nucleotides are linked together by covalent phosphodiester bonds that join the 5' carbon of one deoxyribose group to the 3' carbon of the next (Panel 2–6, pp. 56–57). The four kinds of bases are attached to this repetitive sugar-phosphate chain almost like four kinds of beads strung on a necklace.

Table 3–3 The Relationship Between Free-Energy Differences and Equilibrium Constants

Equilibrium Constant $\dfrac{[AB]}{[A][B]} = K$ (liters/mole)	Free Energy of AB Minus Free Energy of A + B (kcal/mole)
10^5	−7.1
10^4	−5.7
10^3	−4.3
10^2	−2.8
10	−1.4
1	0
10^{-1}	1.4
10^{-2}	2.8
10^{-3}	4.3
10^{-4}	5.7
10^{-5}	7.1

If the reaction A + B $\rightleftharpoons$ AB is allowed to come to equilibrium, the relative amounts of A, B, and AB will depend on the free-energy difference, $\Delta G°$, between them. The above values are given for 37°C (310° K) and are calculated from the equation

$$\Delta G° = -RT \ln \frac{[AB]}{[A][B]}$$

or

$$\frac{[AB]}{[A][B]} = e^{-\Delta G°/RT} = e^{-\Delta G°(1.623)}$$

Here $\Delta G°$ is in kilocalories per mole and represents the free-energy difference under standard conditions (where all components are present at a concentration of 1.0 mole/liter); T is the temperature in kelvins (K).

Similar principles apply to the even simpler case of a reaction A $\rightleftharpoons$ A*, where a molecule is interconvertible between two states A and A* differing in free energy by an amount $\Delta G°$. The quantities of molecules in the two states at equilibrium will be in the ratio

$$\frac{[A^*]}{[A]} = e^{-\Delta G°/RT}$$

which is the same as the ratio of probabilities for a single molecule to be found in one or the other of the two states.

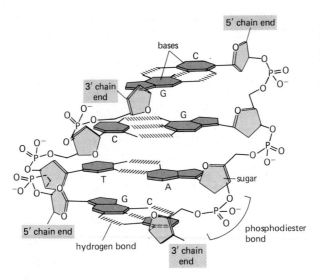

Figure 3–8 A short section of a DNA double helix. Four complementary base pairs are shown. The bases are shown in color, while the deoxyribose sugars are gray. Note that the two DNA strands run in opposite directions and that each base pair is held together by either two or three hydrogen bonds (see also Panel 3–2, pp. 98–99).

How can a long chain of nucleotides encode the instructions for an organism or even a cell? And how can these messages be copied from one generation of cells to the next? The answers lie in the structure of the DNA molecule.

Early in the 1950s, x-ray diffraction analyses of specimens of DNA pulled into fibers suggested that the DNA molecule is a helical polymer composed of two strands. The helical structure of DNA was not surprising since, as we have seen, a helix will often form if each of the neighboring subunits in a polymer is regularly oriented. But the finding that DNA is two-stranded was of crucial significance. It provided the clue that led, in 1953, to the construction of a model that fitted the observed x-ray diffraction pattern and thereby solved the puzzle of DNA structure and function (Figure 3–8 and Panel 3–2, pp. 98–99).

An essential feature of the model was that all of the bases of the DNA molecule are on the *inside* of the double helix, with the sugar phosphates on the outside. This demands that the bases on one strand be extremely close to those on the other, and the fit proposed required specific *base-pairing* between a large purine base (A or G, each of which has a double ring) on one chain and a smaller pyrimidine base (T or C, each of which has a single ring) on the other chain.

Both evidence from earlier biochemical experiments and conclusions derived from model building suggested that **complementary base pairs** (also called *Watson-Crick base pairs*) form between A and T and between G and C. Biochemical analyses of DNA preparations from different species had shown that, although the nucleotide composition of DNA varies a great deal (for example, from 13% A residues to 36% A residues in the DNA of different types of bacteria), there is a general rule that, quantitatively, $[G] = [C]$ and $[A] = [T]$. Model building revealed that the numbers of effective hydrogen bonds that could be formed between G and C or between A and T were greater than for any other combinations. The double-helical model for DNA thus neatly explained the quantitative biochemistry.

The Structure of DNA Provides an Explanation for Heredity[11]

A gene carries biological information in a form that must be precisely copied and transmitted from each cell to all of its progeny. The implications of the discovery of the DNA double helix were profound because the structure immediately suggested how information transfer could be accomplished. Since each strand contains a nucleotide sequence that is exactly complementary to the nucleotide sequence of its partner strand, both strands actually carry the same genetic information. If we designate the two strands A and A′, strand A can serve as a mold or *template* for making a new strand A′, while strand A′ can serve in the same way to make a new strand A. Thus, genetic information can be copied by a process in which strand A separates from strand A′ and each separated strand then serves as a template for the production of a new complementary partner strand.

As a direct consequence of the base-pairing mechanism, it becomes evident that DNA carries information by means of the linear sequence of its nucleotides. Each nucleotide—A, C, T, or G—can be considered a letter in a four-letter alphabet that is used to write out biological messages in a linear "ticker-tape" form. Organisms differ because their respective DNA molecules carry different nucleotide sequences and therefore different biological messages.

Since the number of possible sequences in a DNA chain n nucleotides long is 4^n, the biological variety that could in principle be generated using even a modest length of DNA is enormous. A typical animal cell contains a meter of DNA (3×10^9 nucleotides). Written in a linear alphabet of four letters, an unusually small human gene would occupy a quarter of a page of text (Figure 3–9), while the genetic information carried in a human cell would fill a book of more than 500,000 pages.

Although the principle underlying gene replication is both elegant and simple, the actual machinery by which this copying is carried out in the cell is complicated and involves a complex of proteins that forms a "replication machine" (see p. 232). The fundamental reaction is that shown in Figure 3–10, in which the enzyme *DNA polymerase* catalyzes the addition of a deoxyribonucleotide to the 3' end of a DNA chain. Each nucleotide added to the chain is a *deoxyribonucleoside triphosphate*; the release of pyrophosphate from this activated nucleotide and its subsequent hydrolysis provide the energy for the **DNA replication** reaction and make it effectively irreversible (see pp. 227–238).

Replication of the DNA helix begins with the local separation of its two complementary DNA strands. Each strand then acts as a template for the formation of a new DNA molecule by the sequential addition of deoxyribonucleoside triphosphates. The nucleotide to be added at each step is selected by a process that requires it to form a complementary base pair with the next nucleotide in the parental template strand, thereby generating a new DNA strand that is complementary in sequence to the template strand (Figure 3–10). Eventually the genetic information is duplicated in its entirety so that two complete DNA double helices are formed, each identical in nucleotide sequence to the parental DNA helix that served as the template. Since each daughter DNA molecule ends up with one of the original strands plus one newly synthesized strand, the mechanism of DNA replication is said to be *semiconservative* (Figure 3–11).

Errors in DNA Replication Cause Mutations[12]

One of the most impressive features of DNA replication is its accuracy. Several proofreading mechanisms are used to eliminate incorrectly positioned nucleotides; as a result, the sequence of nucleotides in a DNA molecule is copied with fewer than one mistake in 10^9 nucleotides added. Very rarely, however, the replication machinery skips or adds a few nucleotides, or puts a T where it should have put a C, or an A instead of a G. Any change of this kind in the DNA sequence constitutes a genetic mistake, called a **mutation,** which will be copied in all future cell generations, since "wrong" DNA sequences are copied as faithfully as "correct" ones. The consequence of such an error can be great, since even a single nucleotide change can have important effects on the cell, depending on where the mutation has occurred.

Geneticists demonstrated conclusively in the early 1940s that genes specify the structure of individual proteins. Thus a mutation in a gene, caused by an alteration in its DNA sequence, may lead to the inactivation of a crucial protein

Figure 3–9 The DNA sequence of the human gene for β-globin (one of the two subunits of the hemoglobin molecule that carries oxygen in the blood of all adults). Only one of the two DNA strands is shown (the "coding strand"), since the other strand has a precisely complementary sequence. The sequence should be read from left to right in successive lines down the page, as if it were normal English text.

```
CCCTGTGGAGCCACACCCTAGGGTTGGCCA
ATCTACTCCCAGGAGCAGGGAGGGCAGGAG
CCAGGGCTGGGCATAAAAGTCAGGGCAGAG
CCATCTATTGCTTACATTTGCTTCTGACAC
AACTGTGTTCACTAGCAACTCAAACAGACA
CCATGGTGCACCTGACTCCTGAGGAGAAGT
CTGCCGTTACTGCCCTGTGGGGCAAGGTGA
ACGTGGATGAAGTTGGTGGTGAGGCCCTGG
GCAGGTTGGTATCAAGGTTACAAGACAGGT
TTAAGGAGACCAATAGAAACTGGGCATGTG
GAGACAGAGAAGACTCTTGGGTTTCTGATA
GGCACTGACTCTCTCTGCCTATTGGTCTAT
TTTCCCACCCTTAGGCTGCTGGTGGTCTAC
CCTTGGACCCAGAGGTTCTTTGAGTCCTTT
GGGGATCTGTCCACTCCTGATGCTGTTATG
GGCAACCCTAAGGTGAAGGCTCATGGCAAG
AAAGTGCTCGGTGCCTTTAGTGATGGCCTG
GCTCACCTGGACAACCTCAAGGGCACCTTT
GCCACACTGAGTGAGCTGCACTGTGACAAG
CTGCACGTGGATCCTGAGAACTTCAGGGTG
AGTCTATGGGACCCTTGATGTTTTCTTTCC
CCTTCTTTTCTATGGTTAAGTTCATGTCAT
AGGAAGGGGAGAAGTAACAGGGTACAGTTT
AGAATGGGAAACAGACGAATGATTGCATCA
GTGTGGAAGTCTCAGGATCGTTTTAGTTTC
TTTTATTTGCTGTTCATAACAATTGTTTTC
TTTTGTTTAATTCTTGCTTTCTTTTTTTTT
CTTCTCCGCAATTTTTACTATTATACTTAA
TGCCTTAACATTGTGTATAACAAAAGGAAA
TATCTCTGAGATACATTAAGTAACTTAAAA
AAAAACTTTACACAGTCTGCCTAGTACATT
ACTATTTGGAATATATGTGTGCTTATTTGC
ATATTCATAATCTCCCTACTTTATTTTCTT
TTATTTTTAATTGATACATAATCATTATAC
ATATTTATGGGTTAAAGTGTAATGTTTTAA
TATGTGTACACATATTGACCAAATCAGGGT
AATTTTGCATTTGTAATTTTAAAAAATGCT
TTCTTCTTTTAATATACTTTTTTGTTTATC
TTATTTCTAATACTTTCCCTAATCTCTTTC
TTTCAGGGCAATAATGATACAATGTATCAT
GCCTCTTTGCACCATTCTAAAGAATAACAG
TGATAATTTCTGGGTTAAGGCAATAGCAAT
ATTTCTGCATATAAAATATTTCTGCATATAA
ATTGTAACTGATGTAAGAGGTTTCATATTG
CTAATAGCAGCTACAATCCAGCTACCATTC
TGCTTTTATTTTATGGTTGGGATAAGGCTG
GATTATTCTGAGTCCAAGCTAGGCCCTTTT
GCTAATCATGTTCATACCTCTTATCTTCCT
CCCACAGCTCCTGGGCAACGTGCTGGTCTG
TGTGCTGGCCCATCACTTTGGCAAAGAATT
CACCCCACCAGTGCAGGCTGCCTATCAGAA
AGTGGTGGCTGGTGTGGCTAATGCCCTGGC
CCACAAGTATCACTAAGCTCGCTTTCTTGC
TGTCCAATTTCTATTAAAGGTTCCTTTGTT
CCCTAAGTCCAACTACTAAACTGGGGGATA
TTATGAAGGGCCTTGAGCATCTGGATTCTG
CCTAATAAAAAACATTTATTTTCATTGCAA
TGATGTATTTAAATTATTTCTGAATATTTT
ACTAAAAAGGGAATGTGGGAGGTCAGTGCA
TTTAAAACATAAAGAAATGATGAGCTGTTC
AAACCTTGGGAAAATACACTATATCTTAAA
CTCCATGAAAGAAGGTGAGGCTGCAACCAG
CTAATGCACATTGGCAACAGCCCCTGATGC
CTATGCCTTATTCATCCCTCAGAAAAGGAT
TCTTGTAGAGGCTTGATTTGCAGGTTAAAG
TTTTGCTATGCTGTATTTTACATTACTTAT
TGTTTTAGCTGTCCTCATGAATGTCTTTTC
```

DNA and RNA The structure of RNA is shown in this half of the panel, while the structure of DNA is shown in the other half. Both DNA and RNA are linear polymers of nucleotides (see Panel 2-6, p. 56–57). RNA differs from DNA in three ways: (1) The sugar phosphate backbone contains ribose rather than deoxyribose; (2) it contains the base uracil (U) instead of thymine (T); and (3) it exists as a single strand rather than a double-stranded helix.

SUGAR-PHOSPHATE BACKBONE OF RNA

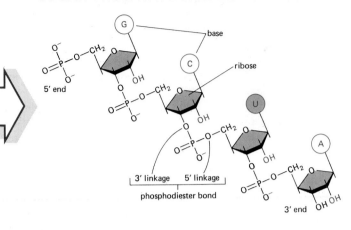

FOUR BASES OF RNA

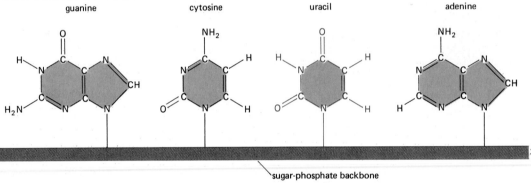

guanine   cytosine   uracil   adenine

sugar-phosphate backbone

RNA SINGLE STRAND

RNA is single-stranded, but it contains local regions of short complementary base-pairing that can form from a random matching process. Regions of base-pairing can be seen in the electron micrograph as branches off the stretched-out chain.

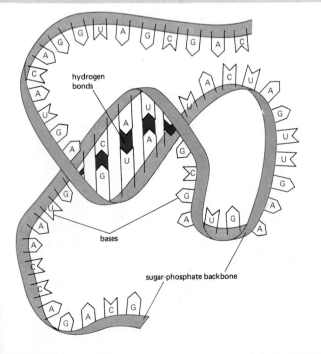

hydrogen bonds

bases

sugar-phosphate backbone

ELECTRON MICROGRAPH OF RNA

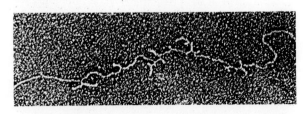

SUGAR-PHOSPHATE BACKBONE OF DNA

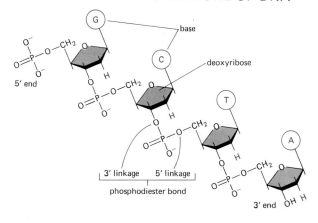

ELECTRON MICROGRAPH OF DNA

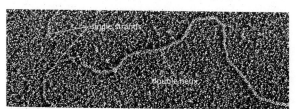

FOUR BASES AS BASE PAIRS OF DNA

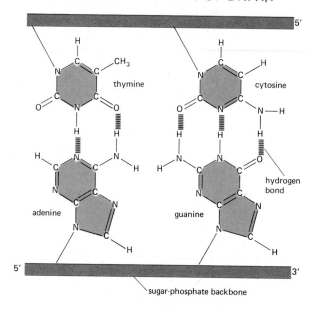

Specific hydrogen bonding between G and C and between A and T (A and U in RNA) generates complementary base-pairing.

DNA DOUBLE HELIX

In a DNA molecule, two antiparallel strands that are complementary in their nucleotide sequence are paired in a right-handed double helix with about 10 nucleotide pairs per helical turn. A schematic representation (top) and a space-filling model (bottom) are illustrated here.

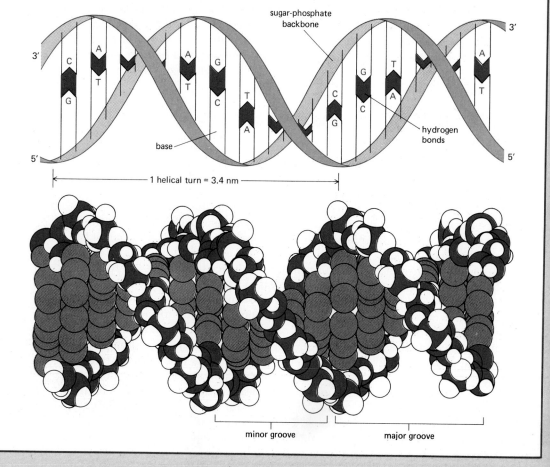

99

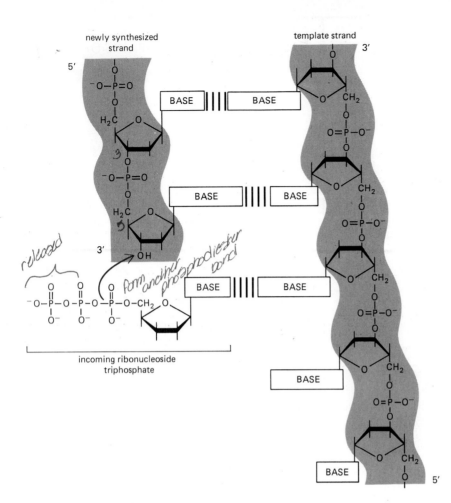

newly synthesized strand

5'

template strand

3'

BASE

BASE

BASE

BASE

BASE

BASE

released

form another phosphodiester bond

3'

OH

incoming ribonucleoside triphosphate

BASE

BASE

BASE

5'

Figure 3–10 Addition of a deoxyribonucleotide to the 3' end of a polynucleotide chain is the fundamental reaction by which DNA is synthesized. As shown, base-pairing between this incoming deoxyribonucleotide and an existing strand of DNA (the *template* strand) guides the formation of a new strand of DNA with a complementary nucleotide sequence.

and result in cell death, in which case the mutation is lost. On the other hand, a mutation may occur in a nonessential region and be without effect, a so-called *silent mutation*. Very rarely, a mutation creates a gene with an improved or novel useful function. In this case, organisms carrying the mutation will have an advantage, and the mutated gene may eventually replace the original gene in the population through natural selection.

The Nucleotide Sequence of a Gene Determines the Amino Acid Sequence of a Protein[13]

DNA is relatively inert chemically. The information it contains is expressed indirectly via other molecules: DNA directs the synthesis of specific RNA and protein molecules, which in turn determine the cell's chemical and physical properties.

At about the time that biophysicists were analyzing the three-dimensional structure of DNA by x-ray diffraction, biochemists were intensively studying the chemical structure of proteins. It was already known that proteins are chains of amino acids joined together by sequential peptide linkages; but it was only in the early 1950s, when the small protein *insulin* was sequenced (Figure 3–12), that it was discovered that each type of protein consists of a unique sequence of amino acids. Just as solving the structure of DNA was seminal in understanding the molecular basis of genetics and heredity, so sequencing insulin provided a key to understanding the structure and function of proteins. If insulin had a definite, genetically determined sequence, then presumably so did every other protein. It seemed reasonable to suppose, moreover, that the properties of a protein would depend on the precise order in which its constituent amino acids are arranged.

Both DNA and protein are composed of a linear sequence of subunits, and eventually the analysis of the proteins made by mutant genes demonstrated that the two sequences are *co-linear*—that is, the nucleotides in DNA are arranged in an order corresponding to the order of the amino acids in the protein they specify.

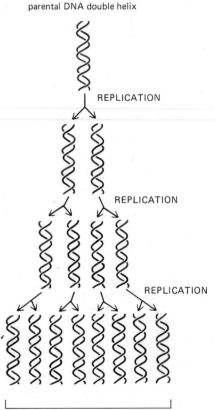

parental DNA double helix

REPLICATION

REPLICATION

REPLICATION

daughter DNA helices

Figure 3–11 The semiconservative replication of DNA. In each round of replication, each of the two strands of DNA is used as a template for the formation of a complementary DNA strand. The original strands therefore remain intact through many cell generations.

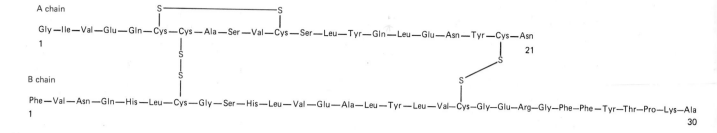

A chain

Gly—Ile—Val—Glu—Gln—Cys—Cys—Ala—Ser—Val—Cys—Ser—Leu—Tyr—Gln—Leu—Glu—Asn—Tyr—Cys—Asn
1   21

B chain

Phe—Val—Asn—Gln—His—Leu—Cys—Gly—Ser—His—Leu—Val—Glu—Ala—Leu—Tyr—Leu—Val—Cys—Gly—Glu—Arg—Gly—Phe—Phe—Tyr—Thr—Pro—Lys—Ala
1   30

It became evident that the DNA sequence contains a coded specification of the protein sequence. The central question in molecular biology then became how a cell translates a nucleotide sequence in DNA into an amino acid sequence in a protein.

Portions of DNA Sequence Are Copied into RNA to Make Protein[14]

The synthesis of proteins involves copying specific regions of DNA (the *genes*) into polynucleotides of a chemically and functionally different type known as ribonucleic acid, or RNA. RNA, like DNA, is composed of a linear sequence of nucleotides, but it has two small chemical differences: (1) the sugar-phosphate backbone of RNA contains ribose instead of a deoxyribose sugar, and (2) the base thymine (T) is replaced by uracil (U), a very closely related base that also pairs with A (see Panel 3–2, pp. 98–99).

RNA retains all of the information of the DNA sequence from which it was copied, as well as the base-pairing properties of DNA. Molecules of RNA are synthesized by a process known as **DNA transcription**, which is similar to DNA replication in that one of the two strands of DNA acts as a template on which the base-pairing abilities of incoming nucleotides are tested. When a good match is achieved with the DNA template, a ribonucleotide is incorporated as a covalently bonded unit. In this way the growing RNA chain is elongated one nucleotide at a time.

DNA transcription differs from DNA replication in a number of ways. For one, the RNA product does not remain as a strand annealed to DNA. Just behind the region where the ribonucleotides are being added, the original DNA helix re-forms and releases the RNA chain. Thus RNA molecules are single-stranded. Moreover, RNA molecules are relatively short compared to DNA molecules since they are copied from a limited region of the DNA—enough to make one or a few proteins (Figure 3–13). RNA transcripts that direct the synthesis of protein molecules are called **messenger RNA** (or **mRNA**) molecules, while other RNA transcripts serve

Figure 3–12 The amino acid sequence of bovine insulin. Insulin is a very small protein that consists of two polypeptide chains, each with a unique, genetically determined sequence of amino acids. The three-letter symbols used to specify amino acids are those listed in Panel 2–5, pages 54–55; the —S—S— bonds indicated are disulfide bonds between cysteine residues. The protein is made initially as a single long polypeptide chain (encoded by a single gene) that is subsequently cleaved to give the two chains (see Figure 3–48).

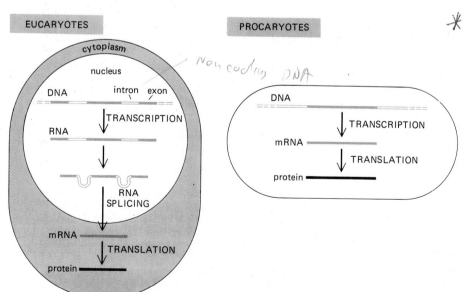

Figure 3–13 The transfer of information from DNA to protein proceeds by means of an RNA intermediate called messenger RNA (mRNA). In procaryotic cells the process is simpler than in eucaryotic cells. In eucaryotes the coding regions of the DNA (in the *exons*, shown in color) are separated by noncoding regions (called *introns*). As indicated, these introns must be removed by an enzymatically catalyzed RNA splicing reaction to form the mRNA.

as *transfer RNAs* (*tRNAs*, see p. 103) or form the RNA components of ribosomes (rRNA, see p. 104) or smaller ribonucleoprotein particles.

The amount of RNA made from a particular region of DNA is controlled by *gene regulatory proteins* that bind to specific sites on DNA close to the coding sequences of a gene (see p. 557). In any cell at any given time, some genes are used to make RNA in very large quantities, while other genes are not transcribed at all. For an active gene, thousands of RNA transcripts can be made from the same DNA segment in each cell generation. Because each mRNA molecule can be translated into many thousands of copies of a polypeptide chain, the information contained in a small region of DNA can direct the synthesis of millions of copies of a specific protein. The protein *fibroin*, for example, is the major component of silk. In each silk gland cell, a single fibroin gene makes 10^4 copies of mRNA, each of which directs the synthesis of 10^5 molecules of fibroin—producing a total of 10^9 molecules of fibroin in just 4 days.

Eucaryotic RNA Molecules Are Spliced to Remove Intron Sequences[15]

In bacterial cells most proteins are encoded by a single uninterrupted stretch of DNA sequence that is copied without alteration to produce an mRNA molecule. In 1977 molecular biologists were astonished by the discovery that most eucaryotic genes have their coding sequences (called *exons*) interrupted by noncoding sequences (called *introns*). To produce a protein, the entire length of the gene, including both its introns and its exons, is first transcribed into a very large RNA molecule—the *primary transcript*. Before this RNA molecule leaves the nucleus, a complex of RNA processing enzymes removes all of the intron sequences, thereby producing a much shorter RNA molecule. After this *RNA processing* step, called **RNA splicing,** has been completed, the RNA molecule moves to the cytoplasm as an mRNA molecule that directs the synthesis of a particular protein (see Figure 3–13).

This seemingly wasteful mode of information transfer in eucaryotes is presumed to have evolved because it makes protein synthesis much more versatile. For example, the primary RNA transcripts of some genes can be spliced in various ways to produce different mRNAs, depending on the cell type or stage of development. This allows different proteins to be produced from the same gene. Moreover, because the presence of numerous introns facilitates genetic recombination events between exons, this type of gene arrangement is likely to have been profoundly important in the early evolutionary history of genes—speeding up the process whereby organisms evolve new proteins from parts of preexisting ones instead of evolving totally new sequences (see p. 602).

Sequences of Nucleotides in mRNA Are "Read" in Sets of Three and Translated into Amino Acids[16]

The rules by which the nucleotide sequence of a gene is translated into the amino acid sequence of a protein, the so-called **genetic code,** were deciphered in the early 1960s. The sequence of nucleotides in the mRNA molecule that acts as an intermediate was found to be read in serial order in groups of three. Each triplet of nucleotides, called a **codon,** specifies one amino acid. In principle, each RNA sequence can be translated in any one of three different *reading frames* depending on where the decoding process begins (Figure 3–14). In almost every case, only one of these reading frames will produce a functional protein. Since there are no punctuation signals except at the beginning and end of the RNA message, the reading frame is set at the initiation of the translation process and is maintained thereafter.

Since RNA is a linear polymer of four different nucleotides, there are $4^3 = 64$ possible codon triplets (remember that it is the *sequence* of nucleotides in the triplet that is important). However, only 20 different amino acids are commonly found in proteins, so that most amino acids are specified by several codons; that is, the genetic code is *degenerate*. The code (shown in Figure 3–15) has been highly conserved during evolution: with a few minor exceptions, it is the same in organisms as diverse as bacteria, plants, and humans.

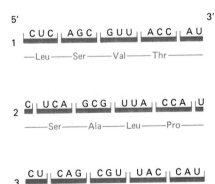

Figure 3–14 The three possible reading frames in protein synthesis. In the process of translating a nucleotide sequence into an amino acid sequence, the sequence of nucleotides in an mRNA molecule is read from the 5' to the 3' end in sequential sets of three nucleotides. In principle, therefore, the same RNA sequence can specify three completely different amino acid sequences, depending on the "reading frame."

handwritten note: *look at tRNA*

1st position (5' end)	2nd position U	C	A	G	3rd position (3' end)
U	Phe	Ser	Tyr	Cys	U
	Phe	Ser	Tyr	Cys	C
	Leu	Ser	STOP	STOP	A
	Leu	Ser	STOP	Trp	G
C	Leu	Pro	His	Arg	U
	Leu	Pro	His	Arg	C
	Leu	Pro	Gln	Arg	A
	Leu	Pro	Gln	Arg	G
A	Ile	Thr	Asn	Ser	U
	Ile	Thr	Asn	Ser	C
	Ile	Thr	Lys	Arg	A
	Met	Thr	Lys	Arg	G
G	Val	Ala	Asp	Gly	U
	Val	Ala	Asp	Gly	C
	Val	Ala	Glu	Gly	A
	Val	Ala	Glu	Gly	G

Figure 3–15 The genetic code. Sets of three nucleotides (codons) in an mRNA molecule are translated into amino acids in the course of protein synthesis according to the rules shown. For example, the codons GUG and GAG are translated into valine and glutamic acid, respectively. Note that those codons with U or C as the second nucleotide tend to specify the more hydrophobic amino acids (compare with Panel 2–5, pp. 54–55).

tRNA Molecules Match Amino Acids to Groups of Nucleotides[17]

The codons in an mRNA molecule do not directly recognize the amino acids they specify in the way that an enzyme recognizes a substrate. The **translation** of mRNA into protein depends on "adaptor" molecules that recognize both an amino acid and a group of three nucleotides. These adaptors consist of a set of small RNA molecules known as **transfer RNAs** (or **tRNAs**), each about 80 nucleotides in length.

A tRNA molecule has a folded three-dimensional conformation that is held together in part by noncovalent base-pairing interactions like those that hold together the two strands of the DNA helix. In the single-stranded tRNA molecule, however, the complementary base pairs form between nucleotide residues in the *same* chain, which causes the tRNA molecule to fold up in a unique way that is important for its function as an adaptor. Four short segments of the molecule contain a double-helical structure, producing a molecule that looks like a "cloverleaf" in two dimensions. This cloverleaf is in turn further compacted into a highly folded, L-shaped conformation that is held together by more complex hydrogen-bonding interactions (Figure 3–16). Two sets of unpaired nucleotide resi-

Figure 3–16 Phenylalanine tRNA of yeast. (A) The molecule is drawn with a "cloverleaf" shape to show the complementary base-pairing (gray) that occurs in the internal helical regions of the molecule. (B) The actual shape of the molecule, based on x-ray diffraction analysis, is shown schematically. Complementary base-pairs are indicated as gray bars. In addition, the nucleotides involved in unusual base-pair interactions that hold different parts of the molecule together are colored, and the pairs are numbered and connected by colored dotted lines that correspond to the colored lines in (A). (C) One of the unusual base-pair interactions. Here one base forms hydrogen-bond interactions with two others; several such "base triples" help fold up this tRNA molecule.

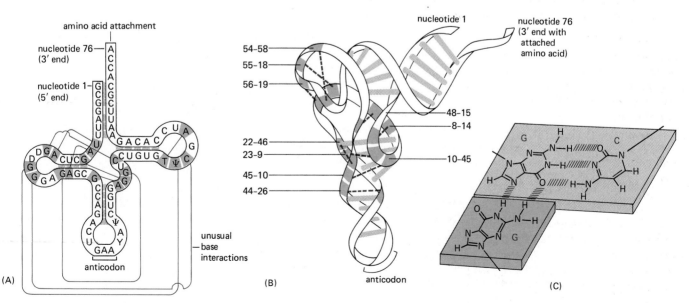

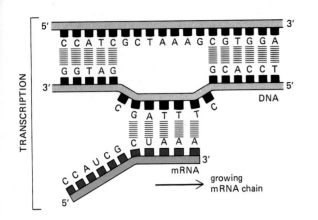

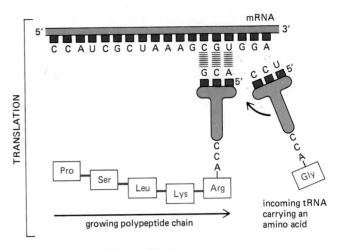

dues at either end of the "L" are especially important for the function of the tRNA molecule in protein synthesis: one forms the *anticodon* that base-pairs to a complementary triplet in an mRNA molecule (the codon), while the *CCA sequence* at the 3' end of the molecule is attached covalently to a specific amino acid (Figure 3–16A).

The RNA Message Is Read from One End to the Other by a Ribosome[18]

The codon recognition process by which genetic information is transferred from mRNA via tRNA to protein depends on the same type of base-pair interactions that mediate the transfer of genetic information from DNA to DNA and from DNA to RNA (Figure 3–17). But the mechanics of ordering the tRNA molecules on the mRNA are complicated and require a **ribosome,** a complex of more than 50 different proteins associated with several structural RNA molecules (rRNAs). Each ribosome is a large protein-synthesizing machine on which tRNA molecules position themselves so as to read the genetic message encoded in an mRNA molecule. The ribosome first finds a specific start site on the mRNA that sets the reading frame and determines the amino-terminal end of the protein. Then, as the ribosome moves along the mRNA molecule, it translates the nucleotide sequence into an amino acid sequence one codon at a time, using tRNA molecules to add amino acids to the growing end of the polypeptide chain (Figure 3–18). When a ribosome reaches the end of the message, both it and the freshly made carboxyl end of the protein are released from the 3' end of the mRNA molecule into the cytoplasm.

Ribosomes operate with remarkable efficiency: in 1 second a single bacterial ribosome adds about 20 amino acids to a growing polypeptide chain. Ribosome structure and the mechanism of protein synthesis are discussed in Chapter 5.

Figure 3–17 Information flow in protein synthesis. The nucleotides in an mRNA molecule are joined together to form a complementary copy of a segment of one strand of DNA. They are then matched three at a time to complementary sets of three nucleotides in the anticodon regions of tRNA molecules. At the other end of each type of tRNA molecule, a specific amino acid is held in a high-energy linkage, and when matching occurs, this amino acid is added to the end of the growing polypeptide chain. Thus, translation of the mRNA nucleotide sequence into an amino acid sequence depends on complementary base-pairing between codons in the mRNA and corresponding tRNA anticodons. The molecular basis of information transfer in translation is therefore very similar to that in DNA replication and transcription. Note that the mRNA is both synthesized and translated starting from its 5' end.

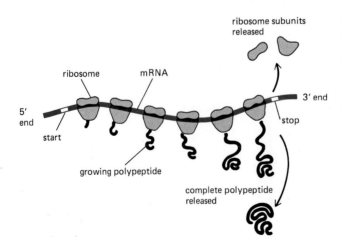

Figure 3–18 Synthesis of a protein by ribosomes attached to an mRNA molecule. Ribosomes become attached to a start signal near the 5' end of the mRNA molecule and then move toward the 3' end, synthesizing protein as they go. A single mRNA will usually have a number of ribosomes traveling along it at the same time, each making a separate but identical polypeptide chain; the entire structure is known as a *polyribosome.*

Some RNA Molecules Function as Catalysts[19]

RNA molecules have commonly been viewed as strings of nucleotides with a relatively uninteresting chemistry. In 1981 this view was shattered by the discovery of a catalytic RNA molecule with the type of sophisticated chemical reactivity that biochemists had previously associated only with proteins. The ribosomal RNA molecules of the ciliated protozoon *Tetrahymena* are initially synthesized as a large precursor from which one of the rRNAs had been shown to be produced by an RNA splicing reaction. The surprise came with the discovery that this splicing can occur *in vitro* in the absence of protein. It was subsequently shown that the intron sequence itself has an enzymelike catalytic activity that carries out the two-step reaction illustrated in Figure 3–19. The 400-nucleotide-long intron sequence was then synthesized in a test tube and shown to fold up to form a complex surface that can function like an enzyme in reactions with other RNA molecules. For example, it can bind two specific substrates tightly—a guanine nucleotide and an RNA chain—and catalyze their covalent attachment so as to sever the RNA chain at a specific site (Figure 3–20).

In this model reaction, which mimics the first step in Figure 3–19, the same intron sequence acts repeatedly to cut many RNA chains. Although RNA splicing is most commonly achieved by means that are not autocatalytic (see p. 533), self-splicing RNAs with intron sequences related to that in *Tetrahymena* have been discovered in other types of cells, including fungi and bacteria. This suggests that these RNA sequences may have arisen before the eucaryotic and procaryotic lineages diverged about 1.5 billion years ago.

Several other families of catalytic RNAs have recently been discovered. Most tRNAs, for example, are initially synthesized as a larger precursor RNA, and an RNA molecule has been shown to play the major catalytic role in an RNA-protein complex that recognizes these precursors and cleaves them at specific sites. A catalytic RNA sequence also plays an important part in the life cycle of many plant

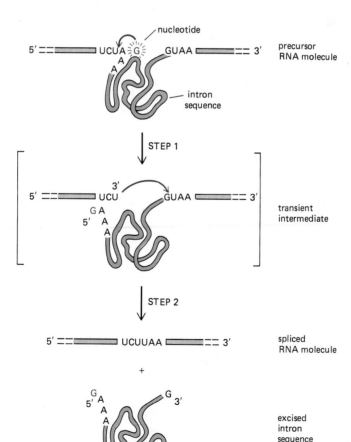

Figure 3–19 Diagram of the self-splicing reaction in which an intron sequence catalyzes its own excision from a *Tetrahymena* ribosomal RNA molecule. As shown, the reaction is initiated when a G nucleotide is added to the intron sequence, cleaving the RNA chain in the process; the newly created 3′ end of the RNA chain then attacks the other side of the intron to complete the reaction.

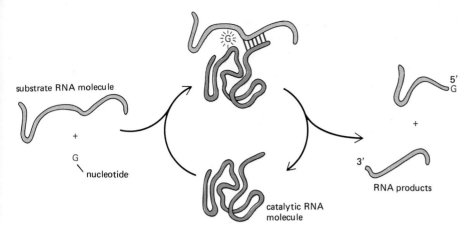

Figure 3–20 An enzymelike reaction catalyzed by the purified *Tetrahymena* intron sequence. In this reaction, which corresponds to the first step in Figure 3–19, both a specific substrate RNA molecule and a G nucleotide become tightly bound to the surface of the catalytic RNA molecule. The nucleotide is then covalently attached to the substrate RNA molecule, cleaving it at a specific site. The release of the resulting two RNA chains frees the intron sequence for further cycles of reaction.

viroids, and a similar sequence appears in a frog RNA molecule, although its role in the frog is unknown. Most remarkably, ribosomes are now suspected to function largely by RNA-based catalysis, with the ribosomal proteins playing a supporting role to the ribosomal RNAs, which make up more than half the mass of the ribosome.

How is it possible for an RNA molecule to act like an enzyme? The example of tRNA indicates that RNA molecules can fold up in highly specific ways. A proposed two-dimensional structure for the core of the self-splicing *Tetrahymena* intron sequence is shown in Figure 3–21. Interactions between different parts of this RNA molecule (analogous to the unusual hydrogen bonds in tRNA molecules—see Figure 3–16) are responsible for folding it further to create a complex three-dimensional surface with catalytic activity. An unusual juxtaposition of atoms can strain covalent bonds and thereby make selected atoms in the folded RNA chain unusually reactive.

As explained in Chapter 1, the discovery of catalytic RNA molecules has profoundly changed our views of how the first living cells arose (see p. 7).

Summary

Genetic information is carried in the linear sequence of nucleotides in DNA. Each molecule of DNA is a double helix formed from two complementary strands of nucleotides held together by hydrogen bonds between G-C and A-T base pairs. Duplication of the genetic information occurs by the polymerization of a new complementary strand onto each of the old strands of the double helix during DNA replication.

The expression of the genetic information stored in DNA involves the translation of a linear sequence of nucleotides into a co-linear sequence of amino acids in proteins. A limited segment of DNA is first copied into a complementary strand of RNA. This primary RNA transcript is spliced to remove intron sequences, pro-

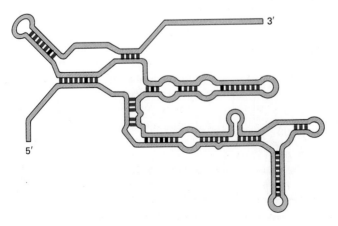

Figure 3–21 A two-dimensional view of the catalytic core of the intron RNA sequence illustrated in Figures 3–19 and 3–20. Normal complementary base pairs are shown in color, while weaker base-pair interactions, such as G-U pairs, are shown in black. This molecule is about 240 nucleotides long; it is normally tightly folded in three dimensions, but its precise conformation is unknown. Self-splicing RNAs with related structures have been discovered in the mitochondria of fungi and in a bacterial virus (bacteriophage T4).

ducing an mRNA molecule. Finally, the mRNA is translated into protein in a complex set of reactions that occur on a ribosome. The amino acids used for protein synthesis are first attached to a family of tRNA molecules, each of which recognizes, by complementary base-pairing interactions, particular sets of three nucleotides in the mRNA. The sequence of nucleotides in the mRNA is then read from one end to the other in sets of three, according to a universal genetic code.

Other RNA molecules in cells function as enzymelike catalysts. These RNA molecules fold up to create a surface containing nucleotides that have become unusually reactive.

Protein Structure[20]

To a large extent cells are made of protein, which constitutes more than half of their dry weight (see Table 3–1). Proteins determine the shape and structure of the cell and also serve as the main instruments of molecular recognition and catalysis. Although DNA stores the information required to make a cell, it has little direct influence on cellular processes. The gene for hemoglobin, for example, cannot carry oxygen: that is a property of the protein specified by the genes. In computer terminology, the DNA and the mRNA represent the "software"—instructions that a cell receives from its parent. Proteins and catalytic RNA molecules constitute the "hardware"—the machinery that executes the program stored in the memory.

DNA and RNA are chains of nucleotides that are chemically very similar to one another. In contrast, proteins are made from an assortment of 20 very different amino acids, each with a distinct chemical personality (see Panel 2–5, pp. 54–55). This variety allows for enormous versatility in the chemical properties of different proteins, and it presumably explains why evolution has selected proteins rather than RNA molecules to catalyze most cellular reactions.

The Shape of a Protein Molecule Is Determined by Its Amino Acid Sequence[21]

Many of the bonds in a long polypeptide chain allow free rotation of the atoms they join, giving the protein backbone great flexibility. In principle, then, any protein molecule could adopt an almost unlimited number of shapes (*conformations*). Most polypeptide chains, however, fold into only one particular conformation determined by their amino acid sequence. This is because the side chains of the amino acids associate with one another and with water to form various weak noncovalent bonds (see Panel 3–1, pp. 90–91). Provided that the appropriate side chains are present at crucial positions in the chain, large forces are developed that make one particular conformation especially stable.

Most proteins fold spontaneously into their correct shape. By treatment with certain solvents, a protein can be unfolded, or *denatured*, to give a flexible polypeptide chain that has lost its native conformation. When the denaturing solvent is removed, the protein will usually refold spontaneously into its original conformation, indicating that all the information necessary to specify the shape of a protein is contained in the amino acid sequence itself.

One of the most important factors governing the folding of a protein is the distribution of its polar and nonpolar side chains. The many hydrophobic side chains in a protein tend to be pushed together in the interior of the molecule, which enables them to avoid contact with the aqueous environment (just as oil droplets coalesce after being mechanically dispersed in water). By contrast, the polar side chains tend to arrange themselves near the outside of the protein molecule, where they can interact with water and with other polar molecules (Figure 3–22). Since the peptide bonds are themselves polar, they tend to interact both with one another and with polar side chains to form hydrogen bonds (Figure 3–23); nearly all polar residues buried within the protein are paired in this way. Hydrogen bonds thus play a major part in holding together different regions of

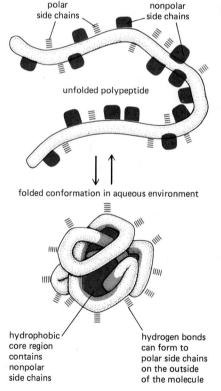

Figure 3–22 How a protein folds into a globular conformation. The polar amino acid side chains tend to gather on the outside of the protein, where they can interact with water; the nonpolar amino acid side chains are buried on the inside to form a hydrophobic core that is "hidden" from water.

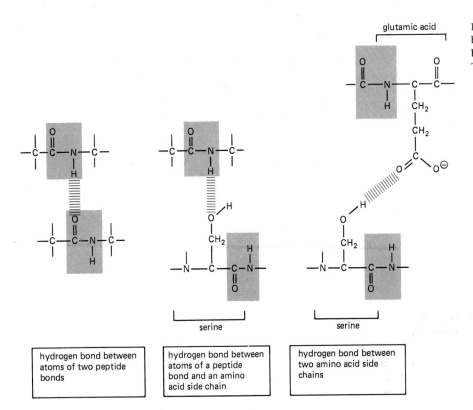

Figure 3–23 Some of the hydrogen bonds (shown in color) that can form between the amino acids in a protein. The peptide bonds are shaded in gray.

hydrogen bond between atoms of two peptide bonds

hydrogen bond between atoms of a peptide bond and an amino acid side chain

hydrogen bond between two amino acid side chains

polypeptide chain in a folded protein molecule, and they are crucially important for many of the binding interactions that occur on protein surfaces.

Secreted or cell-surface proteins often form additional *covalent* intrachain bonds. For example, the formation of **disulfide bonds** (also called S—S bonds) between the two —SH groups of neighboring cysteine residues in a folded polypeptide chain (Figure 3–24) often serves to stabilize the three-dimensional structure of extracellular proteins. These bonds are not required for the specific folding of proteins, since folding occurs normally in the presence of reducing agents that prevent S—S bond formation. In fact, S—S bonds are rarely, if ever, formed in protein molecules in the cytosol because the high cytosolic concentration of —SH reducing agents breaks such bonds (see p. 445).

The net result of all the individual amino acid interactions is that most protein molecules fold up spontaneously into precisely defined conformations, usually compact and globular but sometimes long and fibrous. The inner core is composed of clustered hydrophobic side chains—packed into a tight, nearly crystalline arrangement—while a very complex and irregular exterior surface is formed by the more polar side chains. The positioning and chemistry of the different atoms on this intricate surface make each protein unique and enable it to bind specifically to other macromolecular surfaces and to certain small molecules (see below). From both a chemical and a structural standpoint, proteins are the most sophisticated molecules known.

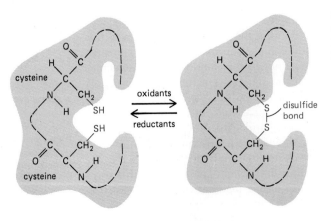

Figure 3–24 The formation of a covalent disulfide bond between the side chains of neighboring cysteine residues in a protein.

Common Folding Patterns Recur in Different Protein Chains[22]

Although all the information required for the folding of a protein chain is contained in its amino acid sequence, we have not yet learned how to "read" this information so as to predict the detailed three-dimensional structure of a protein whose sequence is known. Consequently, the folded conformation can be determined only by an elaborate *x-ray diffraction analysis* performed on crystals of the protein. So far, more than 100 types of proteins have been completely analyzed by this technique. Each of these proteins has a specific conformation so intricate and irregular that it would require a chapter to describe any one of them in full three-dimensional detail.

When the three-dimensional structures of different protein molecules are compared, it becomes clear that, although the overall conformation of each protein is unique, several folding patterns recur repeatedly in parts of these macromolecules. Two patterns are particularly common because they result from regular hydrogen-bonding interactions between the peptide bonds themselves rather than between the side chains of particular amino acids. Both patterns were correctly predicted in 1951 from model-building studies based on the different x-ray diffraction patterns of silk and hair. The two regular folding patterns discovered are now known as the *β sheet*, which occurs in the protein fibroin, found in silk; and the *α helix*, which occurs in the protein α-keratin, found in skin and its appendages, such as hair, nails, and feathers.

The core of most (but not all) globular proteins contains extensive regions of **β sheet.** In the example illustrated in Figure 3–25, which shows part of an antibody molecule, an *antiparallel β sheet* is formed when an extended polypeptide chain folds back and forth upon itself, with each section of the chain running in the direction opposite to that of its immediate neighbors. This gives a very rigid structure held together by hydrogen bonds that connect the peptide bonds in neighboring chains. The antiparallel β sheet and the closely related *parallel β sheet* (which is formed by regions of polypeptide chain that run in the same direction) frequently serve as the framework around which globular proteins are constructed.

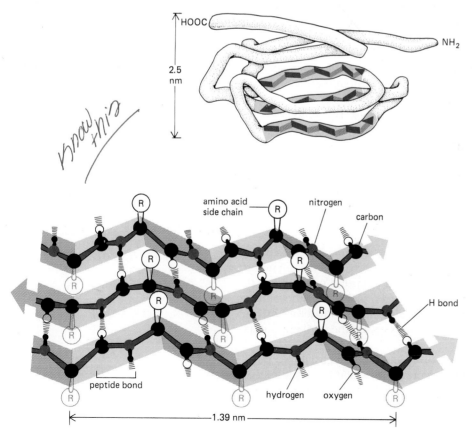

Figure 3–25 A β sheet is a common structure formed by parts of the polypeptide chain in globular proteins. At the top, a domain of 115 amino acids from an immunoglobulin molecule is shown; it consists of a sandwichlike structure of two β sheets, one of which is drawn in color. At the bottom, a perfect antiparallel β sheet is shown in detail. Note that every peptide bond is hydrogen-bonded to a neighboring peptide bond. The actual sheet structures in globular proteins are usually less regular than the β sheet shown here, and many sheets are slightly twisted (see Figure 3–27).

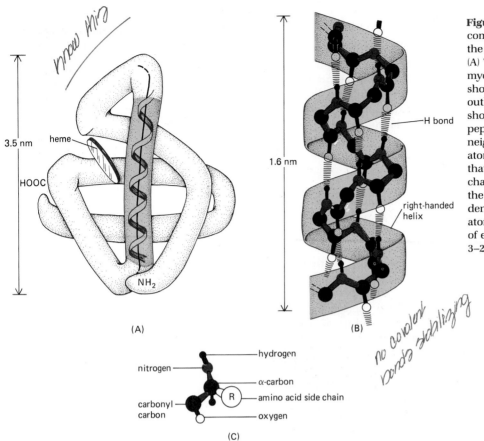

(A)

3.5 nm

heme

HOOC

NH₂

1.6 nm

H bond

right-handed helix

(B)

nitrogen

carbonyl carbon

hydrogen

α-carbon

R amino acid side chain

oxygen

(C)

Figure 3–26 An α helix is another common structure formed by parts of the polypeptide chain in proteins. (A) The oxygen-carrying molecule myoglobin (153 amino acids long) is shown, with one region of α helix outlined in color. (B) A perfect α helix is shown in detail. As in the β sheet, every peptide bond is hydrogen-bonded to a neighboring peptide bond. (C) The atoms in an amino acid residue. Note that for clarity in (B) both the side chains [which protrude radially along the outside of the helix and are denoted by R in (C)] and the hydrogen atom are omitted on the α-carbon atom of each amino acid (see also Figure 3–27).

An **α helix** is generated when a single polypeptide chain turns regularly about itself to make a rigid cylinder in which each peptide bond is regularly hydrogen-bonded to other peptide bonds nearby in the chain. Many globular proteins contain short regions of such α helices (Figure 3–26) and those portions of a trans-membrane protein that cross the lipid bilayer are nearly always α helices because of the constraints imposed by the hydrophobic lipid environment (see p. 285). In aqueous environments an isolated α helix is usually not stable on its own. However, two identical α helices that have a repeating arrangement of nonpolar side chains will twist around each other gradually to form a particularly stable structure known as a *coiled-coil* (see p. 617). Long rodlike coiled-coils are found in many fibrous proteins, such as the intracellular α-keratin fibers that reinforce skin and its appendages. Space-filling representations of an α helix and a β sheet from actual proteins are shown with and without their side chains in Figure 3–27.

Proteins Are Amazingly Versatile Molecules[23]

Because of the variety of their amino acid side chains, proteins are remarkably versatile with respect to the types of structures they can form. Contrast, for example, two abundant proteins secreted by cells in connective tissue—collagen and elastin—both present in the extracellular matrix. In **collagen** molecules three separate polypeptide chains, each rich in the amino acid proline and containing the amino acid glycine at every third residue, are wound around one another to generate a regular triple helix (see p. 809). These collagen molecules are, in turn, packed together into fibrils in which adjacent molecules are tied together by covalent cross-links between neighboring lysine residues, giving the fibril enormous tensile strength (Figure 3–28).

Elastin is at the opposite extreme. Its relatively loose and unstructured polypeptide chains are covalently cross-linked to generate a rubberlike elastic meshwork that enables tissues such as arteries and lungs to deform and stretch without damage (see p. 815). As illustrated in Figure 3–29, the elasticity is due to the ability

(A)

(B)

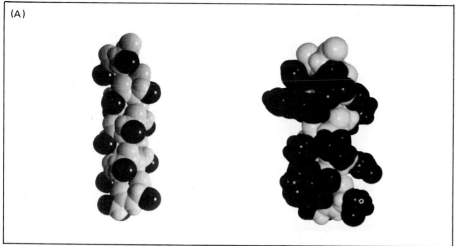
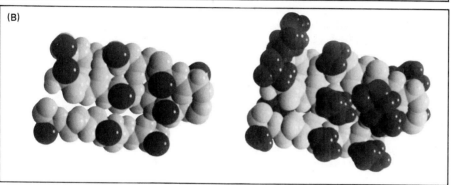

Figure 3–27 Space-filling models of an α helix and a β sheet with (*right*) and without (*left*) their amino acid side chains. (A) An α helix (part of the structure of myoglobin). (B) A region of β sheet (part of the structure of an immunoglobulin domain). In the photographs on the left, each side chain is represented by a single darkly shaded atom (the R groups in Figures 3–25 and 3–26); the entire side chain is shown on the right. (Courtesy of Richard J. Feldmann.)

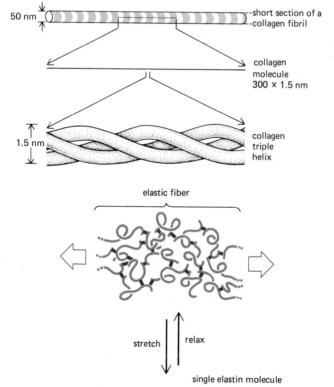

50 nm — short section of a collagen fibril

collagen molecule 300 × 1.5 nm

1.5 nm — collagen triple helix

Figure 3–28 Collagen is a triple helix formed by three extended protein chains that wrap around each other. Many rodlike collagen molecules are cross-linked together in the extracellular space to form inextensible collagen fibrils (*top*) that have the tensile strength of steel.

elastic fiber

stretch relax

single elastin molecule

cross-link

Figure 3–29 Elastin polypeptide chains are cross-linked together to form elastic fibers. Each elastin molecule uncoils into a more extended conformation when the fiber is stretched. The striking contrast between the physical properties of elastin and collagen is due entirely to their very different amino acid sequences.

of individual protein molecules to uncoil reversibly whenever a stretching force is applied.

It is remarkable that the same basic chemical structure—a chain of amino acids—can form so many different structures: a rubberlike elastic meshwork (elastin), an inextensible cable with the tensile strength of steel (collagen), or any of the wide variety of catalytic surfaces on the globular proteins that function as enzymes. Figure 3–30 illustrates and compares the range of shapes that could, in theory, be adopted by a polypeptide chain 300 amino acids long. As we have already emphasized, the conformation actually adopted depends on the amino acid sequence.

Proteins Have Different Levels of Structural Organization[24]

In describing the structure of a protein, it is helpful to distinguish various levels of organization. The amino acid sequence is called the **primary structure** of the protein. Regular hydrogen-bond interactions within contiguous stretches of polypeptide chain give rise to α helices and β sheets, which comprise the protein's **secondary structure.** Certain combinations of α helices and β sheets pack together to form compactly folded globular units, each of which is called a protein **domain.** Domains are usually constructed from a section of polypeptide chain that contains between 50 and 350 amino acids, and they seem to be the modular units from which proteins are constructed (see below). While small proteins may contain only a single domain, larger proteins contain a number of domains, which are usually connected by relatively open lengths of polypeptide chain. Finally, individual polypeptides often serve as subunits for the formation of larger molecules, sometimes called *protein assemblies* or *protein complexes*, in which the subunits are bound to one another by a large number of weak noncovalent interactions (see p. 88); in extracellular proteins these interactions are often stabilized by disulfide bonds.

The three-dimensional structure of a protein can be illustrated in various ways. Consider the unusually small protein basic pancreatic trypsin inhibitor (BPTI), which contains 58 amino acid residues folded into one domain. BPTI can be shown as a stereo pair displaying all of its nonhydrogen atoms (Figure 3–31A) or as an accurate space-filling model, where most of the details are obscured (Figure 3–31B). Alternatively, it can be shown more schematically, with all of the side chains and actual atoms omitted so that it is easier to follow the course of the main polypeptide chain (Figures 3–31C, D, and E). Such schematic drawings are essential for displaying the structure of proteins, which are usually much larger than BPTI, making it possible to trace the irregular course of the polypeptide chain within each domain (Figure 3–32).

Figure 3–33 shows how the structure of a large protein can be resolved into several levels of organization, each level constructed from the one below it in a hierarchical fashion. These levels of increased organizational complexity may correspond to the steps by which a newly synthesized protein folds into its final native structure inside the cell.

Relatively Few of the Many Possible Polypeptide Chains Would Be Useful

Since each of the 20 amino acids is chemically distinct and each can, in principle, occur at any position in a protein chain, there are $20 \times 20 \times 20 \times 20 = 160,000$ different possible polypeptide chains 4 amino acids long, or 20^n different possible polypeptide chains n amino acids long. For a typical protein length of about 300 amino acids, more than 10^{390} different proteins can be made.

We know, however, that only a very small fraction of these possible proteins would adopt a stable three-dimensional conformation. The vast majority would have many different conformations of roughly equal energy, each with different chemical properties. Proteins with such variable properties would not be useful and would therefore be eliminated by natural selection in the course of evolution.

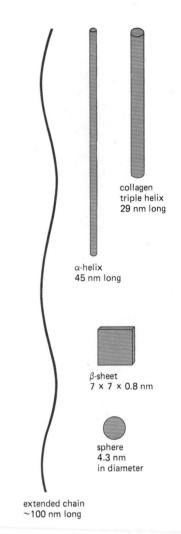

collagen
triple helix
29 nm long

α-helix
45 nm long

β-sheet
7 × 7 × 0.8 nm

sphere
4.3 nm
in diameter

extended chain
~100 nm long

Figure 3–30 Some possible sizes and shapes of a protein molecule 300 amino acid residues long. The structure formed is determined by the amino acid sequence. (Adapted from D.E. Metzler, Biochemistry. New York: Academic Press, 1977.)

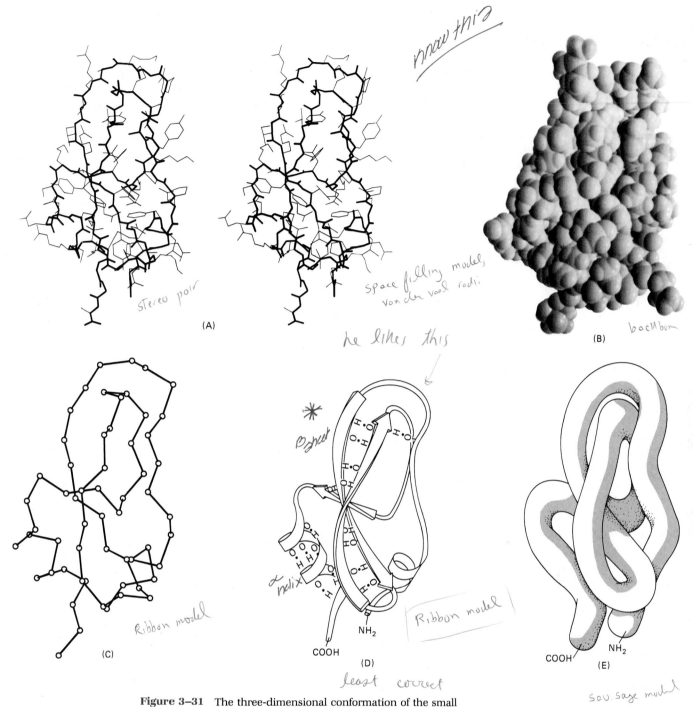

Handwritten annotations on figure:
draw this
stereo pair
space filling model, von der vaal radii
he likes this
backbone
β sheet
α helix
Ribbon model
Ribbon model
least correct
Sausage model

Figure 3–31 The three-dimensional conformation of the small protein, basic pancreatic trypsin inhibitor (BPTI) as seen in five commonly used representations. (A) A stereo pair illustrating the positions of all nonhydrogen atoms. The main chain is shown with heavy lines and the side chains with thin lines. (B) Space-filling model showing the van der Waals radii of all atoms (see Panel 3–1, pp. 90–91). (C) Backbone wire model composed of lines that connect each α-carbon along the polypeptide backbone. (D) "Ribbon model," which represents all regions of regular hydrogen-bonded interactions as either helices (α helices) or sets of arrows (β sheets) pointing toward the carboxyl-terminal end of the chain; in this example the hydrogen bonds are shown. (E) "Sausage model," which shows the course of the polypeptide chain but omits all detail. Note that the core of all globular proteins is densely packed with atoms. Thus the impression of an open structure produced by models (C), (D), and (E) is misleading. (B and C, courtesy of Richard J. Feldmann; A and D, courtesy of Jane Richardson.)

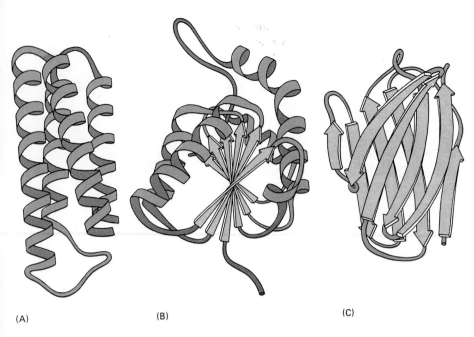

Figure 3–32 Ribbon models of the three-dimensional structure of several differently organized protein domains. (A) Cytochrome b$_{562}$, a single-domain protein composed almost entirely of α helices. (B) The NAD-binding domain of lactic dehydrogenase composed of a mixture of α helices and β sheets. (C) The variable domain of an immunoglobin light chain composed of a sandwich of two β sheets. In these examples the α helices and connecting strands are colored, while strands organized as β sheets are denoted by gray arrows. Note that the polypeptide chain generally traverses back and forth across the entire domain, making sharp turns only at the protein surface. (Drawings courtesy of Jane Richardson.)

(A)   (B)   (C)

Figure 3–33 The three-dimensional structure of a protein can be described in terms of different levels of folding, each of which is constructed from the preceding one in hierarchical fashion. These levels are illustrated here using the catabolite activator protein (CAP), a bacterial gene regulatory protein with two domains. When the large domain binds cyclic AMP, it causes a conformational change in the protein that enables the small domain to bind to a specific DNA sequence (see p. 560). The amino acid sequence is termed the *primary structure* and the first folding level the *secondary structure*. As indicated under the brackets at the bottom of this figure, the combination of the second and third folding levels shown here is commonly termed the *tertiary structure*, and the fourth level (the assembly of subunits) the *quaternary structure* of a protein. (Modified from a drawing by Jane Richardson.)

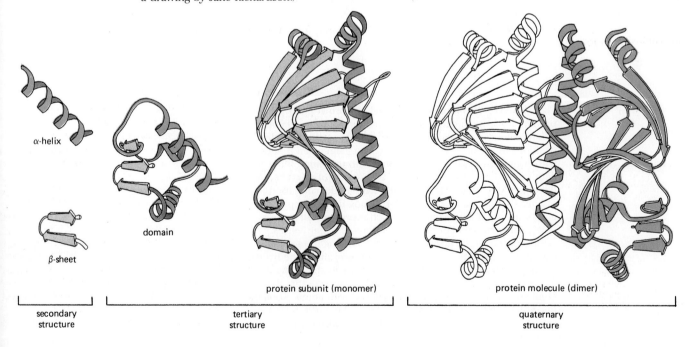

α-helix

domain

β-sheet

protein subunit (monomer)

protein molecule (dimer)

secondary
structure

tertiary
structure

quaternary
structure

Present-day proteins have an amazingly sophisticated structure and chemistry because of their unique folding properties. Not only is the amino acid sequence such that a single conformation is extremely stable, but this conformation has the precise chemical properties that enable the protein to perform a specific catalytic or structural function in the cell. Proteins are so precisely built that the change of even a few atoms in one amino acid can sometimes disrupt the structure and cause a catastrophic change in function.

New Proteins Usually Evolve by Minor Alterations of Old Ones[25]

Cells have genetic mechanisms that allow genes to be duplicated, modified, and recombined in the course of evolution (see p. 599). Consequently, once a protein with useful surface properties has evolved, its basic structure can be incorporated in many other proteins. Proteins of different but related function in present-day organisms often have similar amino acid sequences. Such families of proteins are believed to have evolved from a single ancestral gene that duplicated in the course of evolution to give rise to other genes in which mutations gradually accumulated to produce related proteins with new functions.

Consider the family of protein-cleaving (proteolytic) enzymes, the **serine proteases,** which includes the digestive enzymes chymotrypsin, trypsin, and elastase and some of the proteases in the blood-clotting and complement (see p. 1031) enzymatic cascades. When two of these enzymes are compared, about 40% of the positions in their amino acid sequences are found to be occupied by the same amino acid (Figure 3–34). The similarity of their three-dimensional conformations as determined by x-ray crystallography is even more striking: most of the detailed twists and turns in their polypeptide chains, which are several hundred amino acids long, are identical (Figure 3–35).

The various serine proteases nonetheless have quite distinct functions. Some of the amino acid changes that make these enzymes different were presumably selected in the course of evolution because they resulted in changes in substrate specificity and regulatory properties, giving them the different functional properties they have today. Other amino acid changes are likely to be "neutral," having neither a beneficial nor a damaging effect on the basic structure and function of the enzyme. Since mutation is a random process, there must also have been many deleterious changes that altered the three-dimensional structure of these enzymes sufficiently to inactivate them. Such inactive proteins would have been lost whenever the individual organisms making them were at enough of a disadvantage to be eliminated by natural selection. It is not surprising, then, that cells contain whole sets of structurally related polypeptide chains that have a common ancestry but different functions.

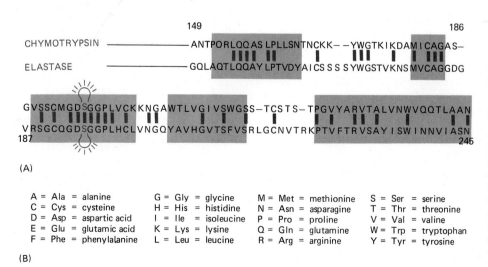

(A)

(B)

A = Ala = alanine	G = Gly = glycine	M = Met = methionine	S = Ser = serine
C = Cys = cysteine	H = His = histidine	N = Asn = asparagine	T = Thr = threonine
D = Asp = aspartic acid	I = Ile = isoleucine	P = Pro = proline	V = Val = valine
E = Glu = glutamic acid	K = Lys = lysine	Q = Gln = glutamine	W = Trp = tryptophan
F = Phe = phenylalanine	L = Leu = leucine	R = Arg = arginine	Y = Tyr = tyrosine

Figure 3–34 (A) Comparison of the amino acid sequences of two members of the serine protease family of enzymes. The carboxyl-terminal portions of the two proteins are shown (amino acids 149 to 245). Identical amino acids are connected by colored bars, and the serine residue in the active site at position 195 is highlighted. In the boxed sections of the polypeptide chains, each amino acid occupies a closely equivalent position in the three-dimensional structures of the two enzymes (see Figure 3–35). (B) The standard one-letter and three-letter codes for amino acids. (Modified from J. Greer, *Proc. Natl. Acad. Sci. USA* 77:3393–3397, 1980.)

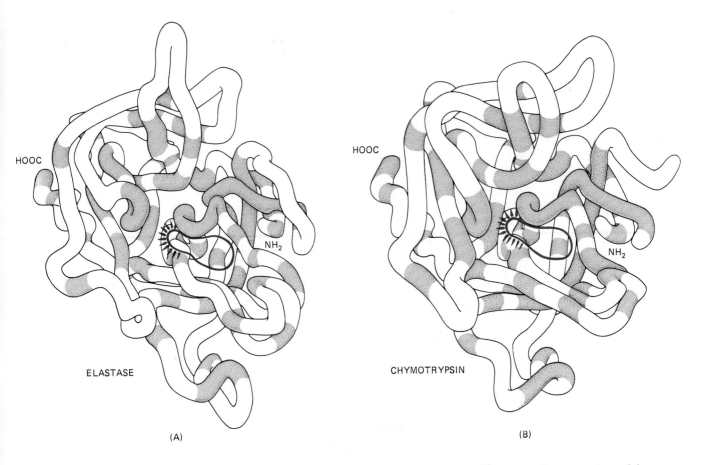

HOOC

NH₂

ELASTASE

(A)

HOOC

NH₂

CHYMOTRYPSIN

(B)

New Proteins Can Evolve by Recombining Preexisting Polypeptide Domains[26]

Once a number of stable protein surfaces have been made in a cell, new surfaces with different binding properties can be generated by joining two or more individual proteins together by noncovalent interactions between them. This combining of globular proteins to make larger, functional protein assemblies is common; although a typical polypeptide chain has a molecular weight of 40,000 to 50,000 (about 300 to 400 amino acids), and relatively few polypeptide chains are more than three times this size, many protein complexes have molecular weights of a million or more.

A related but distinct way of making a new protein from existing chains is to join the corresponding DNA sequences to make a gene that encodes a single large polypeptide chain (see p. 602). Proteins in which different parts of the polypeptide chain fold independently into separate globular domains are believed to have evolved in this way. Many proteins have such "multidomain" structures, and, as might be expected from the evolutionary considerations discussed above, an important binding site for another molecule frequently lies at the site where the separate domains are juxtaposed (Figure 3–36). The structure of one multidomain protein of this type is shown in Figure 3–37.

Figure 3–35 Comparison of the conformations of the two evolutionarily related proteases shown in Figure 3–34: elastase (A) and chymotrypsin (B). Although only those amino acid residues in the polypeptide chain shaded in color are the same in the two proteins, their conformations are very similar everywhere. The active site, which is circled, contains an activated serine residue (see Figure 3–47). Chymotrypsin contains more than two chain termini because it is formed by the proteolytic cleavage of chymotrypsinogen, an inactive precursor.

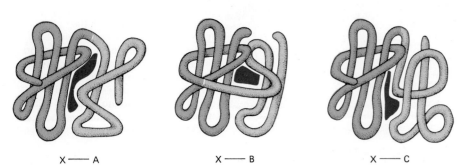

X ——— A   X ——— B   X ——— C

Figure 3–36 The general principle by which the juxtaposition of separate protein surfaces in the course of evolution has given rise to proteins that contain new binding sites for other molecules (*ligands*—see p. 122). As indicated here, the ligand-binding sites often lie at the interface between two protein domains.

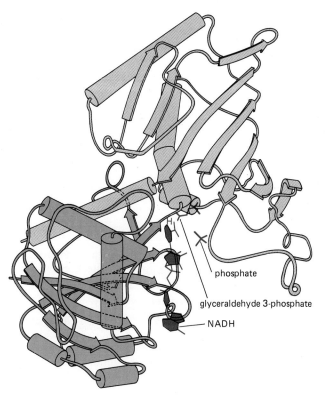

phosphate

glyceraldehyde 3-phosphate

NADH

Figure 3–37 The structure of the glycolytic enzyme glyceraldehyde 3-phosphate dehydrogenase. The protein is composed of two domains, each shown differently shaded, with regions of α helix represented by cylinders and regions of β sheet indicated by arrows. The details of the reaction catalyzed by the enzyme are shown in Figure 2–21. Note that the three bound substrates lie at an interface between the two domains. (Courtesy of Alan J. Wonacott.)

Another way of reutilizing an amino acid sequence is especially widespread among long fibrous proteins such as collagen (see Figure 3–28). In these cases a structure is formed from multiple internal repeats of an ancestral amino acid sequence (see p. 602). Putting together amino acid sequences by joining preexisting coding DNA sequences is clearly a much more efficient strategy for a cell than the alternative of deriving new protein sequences from scratch by random DNA mutation.

Structural Homologies Can Help Assign Functions to Newly Discovered Proteins[27]

The development of techniques for rapidly sequencing DNA molecules has made it possible to determine the amino acid sequences of many proteins from the nucleotide sequences of their genes (see p. 186). An ever-enlarging "protein data base" is therefore available that is routinely scanned by computers to search for possible sequence homologies between a newly sequenced protein and previously studied ones. Although sequences have so far been determined for only a few percent of the proteins in eucaryotic organisms, it is common to find that a newly sequenced protein is homologous to some other, known protein over part of its length, indicating that most proteins may have descended from relatively few ancestral types. As expected, the sequences of many large proteins often show signs of having evolved by the joining of preexisting domains in new combinations—a process called "domain shuffling" (Figure 3–38).

The discovery of domain homologies can also be useful in another way. It is much more difficult to determine the three-dimensional structure of a protein than to determine its amino acid sequence. But the conformation of a newly sequenced protein domain can be guessed if it is homologous to a domain of a protein whose conformation has already been determined by x-ray diffraction analysis. By assuming that the twists and turns of the polypeptide chain will be conserved in the two proteins despite the presence of discrepancies in amino acid sequence, one can often sketch the structure of the new protein with reasonable accuracy.

These protein comparisons are also important because related structures often imply related functions. Many years of experimentation can be saved by discovering an amino acid sequence homology with a protein of known function. For

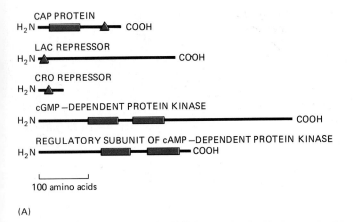

CAP PROTEIN

H_2N —▢— ▲ — COOH

LAC REPRESSOR

H_2N —▲——————— COOH

CRO REPRESSOR

H_2N —▲—

cGMP–DEPENDENT PROTEIN KINASE

H_2N ——▢—▢——————— COOH

REGULATORY SUBUNIT OF cAMP–DEPENDENT PROTEIN KINASE

H_2N ————▢—▢— COOH

⊢——⊣
100 amino acids

(A)

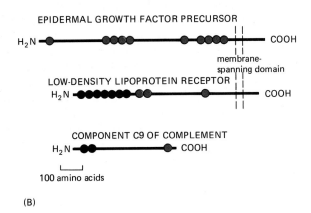

EPIDERMAL GROWTH FACTOR PRECURSOR

H_2N —●———●●●———●—●●●●—∣∣— COOH

membrane-
spanning domain

LOW-DENSITY LIPOPROTEIN RECEPTOR

H_2N —●●●●●●●—●●——————●————∣∣— COOH

COMPONENT C9 OF COMPLEMENT

H_2N —●●——————●— COOH

⊢——⊣
100 amino acids

(B)

example, such sequence homologies first indicated that certain cell cycle-regulatory genes in yeast cells (see p. 743) and certain genes that cause mammalian cells to become cancerous (see p. 1207) are protein kinases. In the same way, many of the proteins that control pattern formation in the fruit fly *Drosophila* were recognized to be gene regulatory proteins, while another protein involved in pattern formation was identified as a serine protease (see p. 937).

Many new protein sequences are being added to the data base each year, each one increasing the chance of finding useful homologies. Protein sequence comparisons will therefore become an increasingly important tool in cell biology.

Protein Subunits Can Self-assemble into Large Structures[28]

The same principles that enable several protein domains to associate to form binding sites operate to generate much larger structures in the cell. Supramolecular structures such as enzyme complexes, ribosomes, protein filaments, viruses, and membranes are not made as single, giant, covalently linked molecules; instead they are formed by the noncovalent assembly of preformed macromolecular subunits.

There are several advantages to the use of smaller subunits to build larger structures: (1) building a large structure from one or a few repeating smaller subunits reduces the amount of genetic information required; (2) both assembly and disassembly can be readily controlled, since the subunits associate through multiple bonds of relatively low energy; and (3) errors in the synthesis of the structure can be more easily avoided, since correction mechanisms can operate during the course of assembly to exclude malformed subunits.

A Single Type of Protein Subunit Can Interact with Itself to Form Geometrically Regular Assemblies[29]

If a protein has a binding site that is complementary to a region of its own surface, it will assemble spontaneously to form a larger structure. In the simplest case, a binding site recognizes itself and forms a symmetrical *dimer*. Many enzymes and other proteins form dimers of this kind, which frequently act as subunits in the formation of larger assemblies (Figures 3–39 and 3–40).

If the binding site of a protein is complementary to a region of its surface that does not include the binding site itself, a chain of subunits will be formed. For some orientations of the two binding sites, the chain will soon run into itself and

Figure 3–38 Evidence that an extensive shuffling of blocks of protein sequence has occurred during the evolution of proteins. Those portions of a protein denoted by the same shape and color are evolutionarily related but not identical. (A) The bacterial catabolite gene activator protein (CAP) contains one domain (*colored triangle*) that binds a specific DNA sequence and a second domain that binds cyclic AMP (see Figure 3–33). The DNA-binding domain is related to the DNA-binding domains of many other gene regulatory proteins, including the lac repressor and cro repressor proteins. In addition, two copies of the cyclic AMP-binding domain are found in eucaryotic protein kinases regulated by the binding of cyclic nucleotides. (B) Two domains of about 40 amino acids each are repeated in the three large vertebrate proteins shown. The low-density lipoprotein (LDL) receptor, for example, is a transmembrane protein of 839 amino acid residues and is responsible for the uptake of cholesterol into cells (see p. 328). It contains a number of domains found in other proteins, including seven copies of a cysteine-rich domain (*black ball*) involved in LDL binding and three copies of a domain of similar size (*colored ball*) whose function is unknown.

Figure 3–39 The formation of a dimer from a single type of protein subunit. A protein with a binding site that recognizes itself will often form symmetrical dimers. These often pair further with other subunits to form tetramers and larger assemblies.

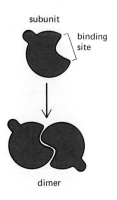

subunit

binding
site

dimer

terminate, forming a closed ring of subunits (Figure 3–41). More commonly, an indefinitely extended polymer of subunits will result, and, provided that each subunit is bound to its neighbor in an identical way, the subunits in the polymer will be arranged in a helix (see Figure 3–3). An *actin filament*, for example, is a helical structure formed from a single globular protein subunit called *actin*; actin filaments are major components in the cytosol of most eucaryotic cells (see p. 629). Where mechanical strength is especially important, supramolecular assemblies are usually made from fibrous rather than globular subunits, since fibrous subunits permit extensive regions of protein-protein contact when wound around one another as a multistranded helix (Figure 3–42A).

Protein subunits may also assemble into flat sheets in which the subunits are arranged in hexagonal arrays. Specialized membrane transport proteins are sometimes arranged in this way in lipid bilayers (see p. 293). With a slight change in the geometry of the individual subunits, a hexagonal sheet can be converted into a tube (Figure 3–42B); such cylindrical tubes are involved in the formation of the protein coats of some elongated viruses (see Figure 3–45).

The formation of closed structures, such as rings, tubes, or spheres, provides additional stability because it increases the number of bonds that can form between the protein subunits. Moreover, because such a structure is formed by mutually dependent, cooperative interactions between subunits, it can be driven to assemble or disassemble by a relatively small change that affects the subunits individually. These principles are dramatically illustrated in the protein *capsids* of many simple viruses, which take the form of a hollow sphere. These coats are often made of hundreds of identical protein subunits that enclose and protect the viral nucleic acid (Figure 3–43). The protein in such capsids must have a particularly adaptable structure, since it must make many kinds of contacts and also change its arrangement to let the nucleic acid out to initiate viral multiplication once the virus has entered a cell.

Some Self-assembling Structures Include Protein Subunits and Nucleic Acids[30]

Some large cellular structures, such as viruses and ribosomes, are constructed from a mixture of protein subunits and RNA or DNA molecules. The information for the assembly of many of these complex aggregates must be contained in the macromolecular subunits themselves, and under appropriate conditions the isolated subunits can spontaneously assemble in a test tube into the final structure. The first large macromolecular aggregate shown to be capable of self-assembly from its component parts was *tobacco mosaic virus (TMV)*. This virus is a long rod in which a cylinder of protein is arranged around a helical RNA core (Figures 3–44 and 3–45). If the dissociated RNA and protein subunits are mixed together in solution, they recombine to form fully active virus particles. The assembly process

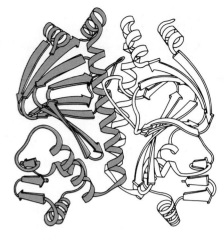

Figure 3–40 Ribbon model of a dimer formed from two identical protein subunits (monomers). The protein shown is the bacterial catabolite gene activator protein (CAP) illustrated previously in Figures 3–33 and 3–38A. (Courtesy of Jane Richardson.)

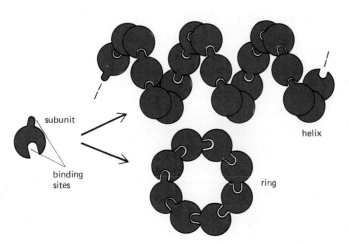

Figure 3–41 Rings or helices can form if a single type of protein subunit interacts with itself repeatedly in the manner shown. The formation of a helix was illustrated in Figure 3–3; a ring forms instead of a helix if the subunits run into one another, stopping further growth of the chain.

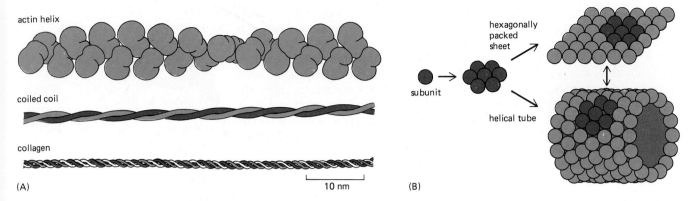

Figure 3–42 Some structures formed by the self-assembly of protein subunits. (A) Three common types of helical protein assemblies. In an *actin filament* there are about two globular protein subunits per turn, whereas many other cytoskeletal proteins contain rodlike regions in which two α helices are paired in a *coiled-coil* conformation. In a *collagen helix*, three extended protein chains intertwine over an extensive distance to give a very strong rodlike structure. (B) Hexagonally packed globular protein subunits can form either a flat sheet or a tube.

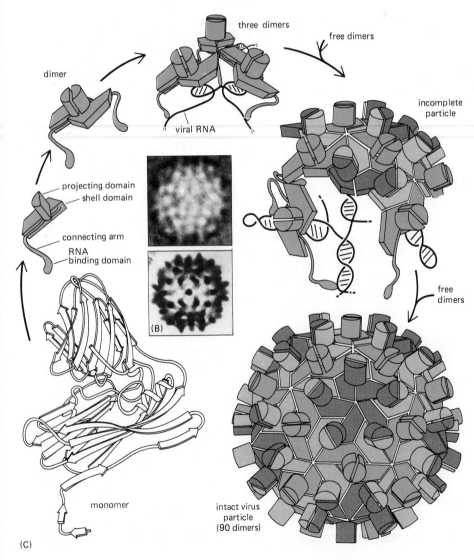

Figure 3–43 The structure of a spherical virus. In many viruses, identical protein subunits pack together to create a spherical shell (a *capsid*) that encloses the viral genome, composed of either RNA or DNA (see p. 250). For geometric reasons, no more than 60 identical subunits can pack together in a precisely symmetrical way. If slight irregularities are allowed, however, more subunits can be used to produce a larger capsid. Tomato bushy stunt virus (TBSV), for example, is a spherical virus about 33 nm in diameter that, in an electron micrograph (A) or a photograph of a three-dimensional reconstruction (B), can be seen to contain more than 60 subunits. Its postulated pathway of assembly and three-dimensional structure as determined by x-ray diffraction are shown in (C). The virus particle is formed from 180 identical copies of a 386-amino acid capsid protein plus an RNA genome of 4500 nucleotides. To form such a large capsid, the protein must be able to fit into three somewhat different environments, each of which is differently colored in the particle shown here. (Drawings courtesy of Steve Harrison; electron micrographs courtesy of John Finch.)

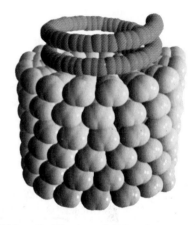

Figure 3–44 Electron micrograph of a tobacco mosaic virus (TMV), which consists of a single long RNA molecule enclosed in a cylindrical protein coat composed of a tight helical array of identical protein subunits. Fully infective virus particles can self-assemble in a test tube from purified RNA and protein molecules. (Courtesy of Robley Williams.)

is unexpectedly complex and includes the formation of double rings of protein, which serve as intermediates that add to the growing virus coat.

Another complex macromolecular aggregate that can reassemble from its component parts is the bacterial ribosome. These ribosomes are composed of about 55 different protein molecules and 3 different rRNA molecules (see p. 210). If the individual components are incubated under appropriate conditions in a test tube, they spontaneously re-form the original structure. Most important, such reconstituted ribosomes are able to carry out protein synthesis. As might be expected, the reassembly of ribosomes follows a specific pathway: certain proteins first bind to the RNA, and this complex is then recognized by other proteins, and so on until the structure is complete.

It is still not clear how some of the more elaborate self-assembly processes are regulated. For example, many structures in the cell appear to have a precisely defined length that is many times greater than that of their component macromolecules. How such length determination is achieved is in most cases a mystery. Three possible mechanisms are illustrated in Figure 3–46. In the simplest case a long core protein or other macromolecule provides a scaffold that determines the extent of the final assembly. This is the mechanism that determines the length of the TMV particle, where the RNA chain provides the core. Similarly, a core protein has been shown to determine the length of the long tails of some bacterial viruses (Figure 3–47).

Not All Biological Structures Form by Self-assembly[31]

Some cellular structures held together by noncovalent bonds are not capable of self-assembly. A mitochondrion, a cilium, or a myofibril, for example, cannot form spontaneously from a solution of their component macromolecules because part of the information for their assembly is provided by special enzymes and other cellular proteins that perform the function of jigs or templates but do not appear in the final assembled structure. Even small structures may lack some of the ingredients necessary for their own assembly. In the formation of some bacterial viruses, for example, the head structure, which is composed of a single protein

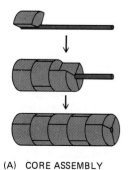

(A) CORE ASSEMBLY

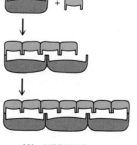

(B) ACCUMULATED STRAIN   (C) VERNIER

Figure 3–45 A model showing part of the structure of tobacco mosaic virus. A single-stranded RNA molecule of 6000 nucleotides is packaged in a helical coat constructed from 2130 copies of a coat protein 158 amino acid residues long.

Figure 3–46 Three possible ways in which a large protein assembly can be made to a fixed length: (A) coassembly along an elongated core protein or other macromolecule that acts as a measuring device; (B) termination of assembly because of strain that accumulates in the polymeric structure as additional subunits are added, so that beyond a certain length the energy required to fit another subunit onto the chain becomes excessively large; (C) a vernier type of assembly, in which two sets of rodlike molecules differing in length form a staggered complex that grows until their ends exactly match.

subunit, is assembled on a temporary scaffold composed of a second protein. The second protein is absent from the final virus particle, and so the head structure cannot spontaneously reassemble once it is taken apart. Other examples are known in which proteolytic cleavage is an essential and irreversible step in the assembly process. This is the case for the coats of some bacterial viruses and even for some simple protein assemblies, including the structural protein collagen and the hormone insulin (Figure 3–48). From these relatively simple examples, it seems very likely that the assembly of a structure as complex as a mitochondrion or a cilium will involve both temporal and spatial ordering imparted by other cellular components, as well as irreversible processing steps catalyzed by degradative enzymes.

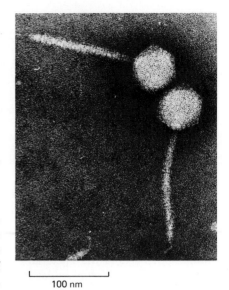

Summary

The three-dimensional conformation of a protein molecule is determined by its amino acid sequence. The folded structure is stabilized by noncovalent interactions between different parts of the polypeptide chain. The amino acids with hydrophobic side chains tend to cluster in the interior of the molecule, and local hydrogen-bond interactions between neighboring peptide bonds give rise to α helices and β sheets. Globular regions known as domains are the modular units from which many proteins are constructed; small proteins typically contain only a single domain, while large proteins contain several domains linked together by short lengths of polypeptide chain. As proteins evolved, domains were modified and combined with other domains to construct new proteins.

Proteins are brought together into larger structures by the same noncovalent forces that determine protein folding. Proteins with binding sites for their own surface can assemble into dimers, closed rings, spherical shells, or helical polymers. Although mixtures of proteins and nucleic acids can assemble spontaneously into complex structures in the test tube, many assembly processes involve irreversible steps, so that not all structures in the cell are capable of spontaneous reassembly if they are dissociated into their component parts.

Figure 3–47 Electron micrograph of bacteriophage lambda. The tip of the virus tail attaches to a specific protein on the surface of a bacterial cell, following which the tightly packaged DNA in the head is injected through the tail into the cell. The tail has a precise length, which is determined by the mechanism shown in Figure 3–46A.

Protein Function[32]

The chemical properties of a protein molecule depend almost entirely on its exposed surface residues, which are able to form weak noncovalent bonds with other molecules (see p. 88). When a protein molecule binds to another molecule, the second molecule is commonly referred to as a **ligand.** Because an effective interaction between a protein molecule and a ligand requires that many weak bonds be formed simultaneously between them, the only ligands that can bind tightly to a protein are those that fit precisely onto its surface.

The region of a protein that associates with a ligand, known as its **binding site,** usually consists of a cavity formed by a specific arrangement of amino acids on the protein surface. These amino acids often belong to widely separated regions of the polypeptide chain (Figure 3–49), and they represent only a minor fraction of the total amino acids present. The rest of the protein molecule is necessary to maintain the polypeptide chain in the correct position and to provide additional binding sites for regulatory purposes; the interior of the protein is often important only insofar as it gives the surface of the molecule the appropriate shape and rigidity.

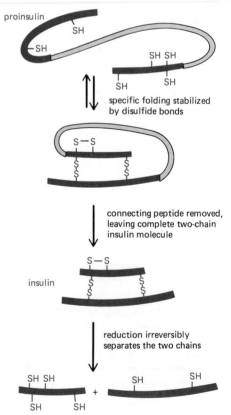

Figure 3–48 The polypeptide hormone insulin cannot spontaneously re-form if its disulfide bonds are disrupted. It is synthesized as a larger protein (*proinsulin*) that is cleaved by a proteolytic enzyme after the protein chain has folded into a specific shape. Excision of part of the proinsulin polypeptide chain causes an irretrievable loss of the information needed for the protein to fold spontaneously into its normal conformation.

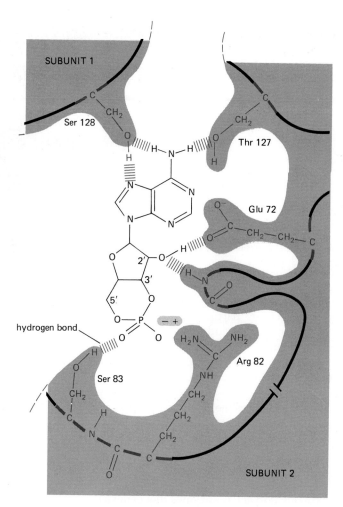

Figure 3–49 Hydrogen bonding between the catabolite gene activator protein (CAP) and its ligand, cyclic AMP, as determined by x-ray crystallographic analysis of the complex. As indicated, the two identical subunits of the dimer cooperate to form this binding site (see also Figure 3–40). (Courtesy of Tom Steitz.)

A Protein's Conformation Determines Its Chemistry[19,33]

Neighboring surface residues on a protein often interact in a way that alters the chemical reactivity of selected amino acid side chains. These interactions are of several types.

First, neighboring parts of the polypeptide chain may interact in a way that restricts the access of water molecules to other parts of the protein surface. Because water molecules tend to form hydrogen bonds, they compete with ligands for selected side chains on the protein surface. The tightness of hydrogen bonds (and ionic interactions) between proteins and their ligands is therefore greatly increased if water molecules are excluded. At first sight it is hard to imagine a mechanism that would exclude a molecule as small as water from a protein surface without affecting the access of the ligand itself. However, because of their strong tendency for hydrogen bonding, water molecules exist in a large hydrogen-bonded network (see Panel 2–1, pp. 46–47), and it is often energetically unfavorable for individual molecules to break away from this network to reach into a crevice on the protein surface.

Second, the clustering of neighboring polar amino acid side chains can alter their reactivity. For example, if a number of negatively charged side chains are forced together against their mutual repulsion by the way the protein folds, the affinity of the site for a positively charged ion is greatly increased. Selected amino acid side chains can also interact with one another through hydrogen bonds, which can activate normally unreactive side groups (such as the —CH$_2$OH on the serine shown in Figure 3–50) so that they are able to enter into reactions that make or break selected covalent bonds.

The surface of each protein molecule, therefore, has a unique chemical reactivity that depends not only on which amino acid side chains are exposed but

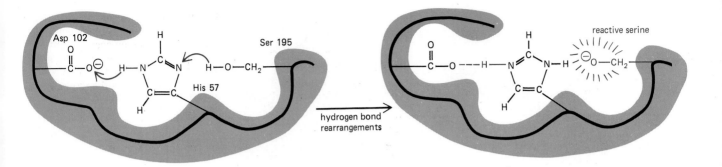

hydrogen bond
rearrangements

also on their exact orientation relative to one another. For this reason, even two slightly different conformations of the same protein molecule may differ greatly in their chemistry.

Where side-chain reactivities are insufficient, proteins often enlist the help of selected nonpolypeptide molecules that the proteins bind to their surface. These ligands serve as **coenzymes** in enzyme-catalyzed reactions, and they may be so tightly bound to the protein that they are effectively part of the protein itself. Examples are the iron-containing *hemes* in hemoglobin and cytochromes, *thiamine pyrophosphate* in enzymes involved in aldehyde-group transfers, and *biotin* in enzymes involved in carboxyl-group transfers (see p. 76). Each coenzyme is selected for the unique chemical reactivity it acquires when bound to a protein surface. Coenzymes are often very complex organic molecules, and their exact chemistry when protein-bound is not always understood. Besides its reactive center, a coenzyme has other residues that bind it to its host protein (Figure 3–51). A space-filling model of an enzyme bound to a coenzyme is shown in Figure 3–52A.

Substrate Binding Is the First Step in Enzyme Catalysis[34]

One of the most important functions of proteins is to act as enzymes that catalyze specific chemical reactions. The ligand in this case is the substrate molecule, and the binding of the substrate to the enzyme is an essential prelude to the chemical reaction (Figure 3–52B). Extremely high rates of chemical reactions are achieved by enzymes—far higher than for any synthetic catalysts. This efficiency is attributable to several factors. The enzyme serves, first, to increase the local concentration of substrate molecules at the catalytic site and to hold the appropriate atoms in the correct orientation for the reaction that is to follow. More important, however, some of the binding energy contributes directly to the catalysis. Substrate molecules pass through a series of intermediate forms of altered geometry and electron distribution before forming the ultimate products of the reaction, and the free energies of these intermediate forms—especially of those in the most unstable *transition states*—are the major determinants of the rate of reaction. Enzymes have a much greater affinity for these unstable substrate transition states than for the stable forms, and this binding interaction lowers the energies of the transition states, thereby greatly accelerating one particular reaction.

Some enzymes interact covalently with one of their substrates so that the substrate is linked to an amino acid or to a coenzyme molecule. Generally, one substrate enters the binding site, becomes covalently bound, and then reacts with a second substrate on the enzyme surface (Figure 3–53). At the end of each reaction cycle, the free enzyme is regenerated.

Because of the way enzymes work, there is a limit to the amount of substrate that a single enzyme can process in a given time. If the concentration of substrate is increased, the rate at which product is formed also increases up to a maximum value (Figure 3–54). At that point the enzyme molecule is saturated with substrate and the rate of reaction (denoted V_{max}) depends only on how rapidly the substrate molecule can be processed. This rate divided by the enzyme concentration is called the **turnover number,** which for many enzymes is about 1000 substrate molecules per second but can be as much as 10^6 or more per second in extreme cases.

Figure 3–50 An unusually reactive amino acid at the active site of an enzyme. The example shown is the "charge relay system" found in chymotrypsin, elastase, and other serine proteases (see Figure 3–35). The aspartic acid side chain induces the histidine to remove the proton from serine 195; this activates the serine to form a covalent bond with the enzyme substrate, hydrolyzing a peptide bond as illustrated in Figure 3–53.

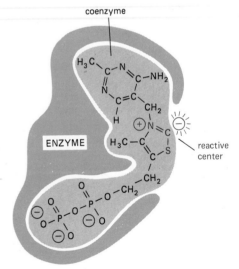

Figure 3–51 Coenzymes, such as thiamine pyrophosphate (TPP), shown here in gray, are small molecules that bind to an enzyme's surface and enable it to catalyze specific reactions. The reactivity of TPP depends on its "acidic" carbon atom, which readily exchanges its hydrogen atom for a carbon atom of a substrate molecule. Other regions of the TPP molecule probably act as "handles" by which the enzyme holds the coenzyme in the correct position.

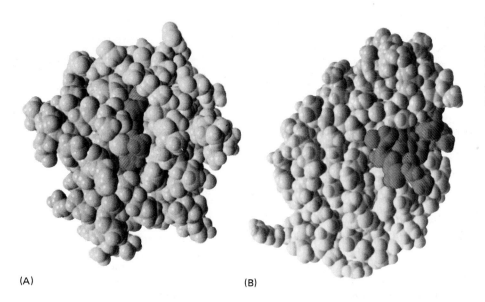

Figure 3–52 Computer-generated space-filling models of (A) cytochrome c with its bound heme coenzyme and (B) egg-white lysozyme with a bound oligosaccharide substrate. In both cases the bound ligand is shown in color. (Courtesy of Richard J. Feldmann.)

(A)

(B)

The other kinetic parameter frequently used to characterize an enzyme is its K_M, which is the substrate concentration that allows the reaction to proceed at one-half its maximum rate (see Figure 3–54). A *low* K_M value means that the enzyme reaches its maximum catalytic rate at a *low concentration* of substrate and generally indicates that the enzyme binds its substrate very tightly.

Enzymes Accelerate Chemical Reactions but Cannot Make Them Energetically More Favorable

No matter how sophisticated an enzyme is, it cannot make the chemical reaction it catalyzes either more or less energetically favorable. It cannot alter the free energy difference between the initial substrates and the final products of the reaction. Like the simple binding interactions already discussed (see p. 93), any given chemical reaction has an *equilibrium point*, at which the backward and forward reaction fluxes are equal, so that no net change occurs (see Figure 3–7).

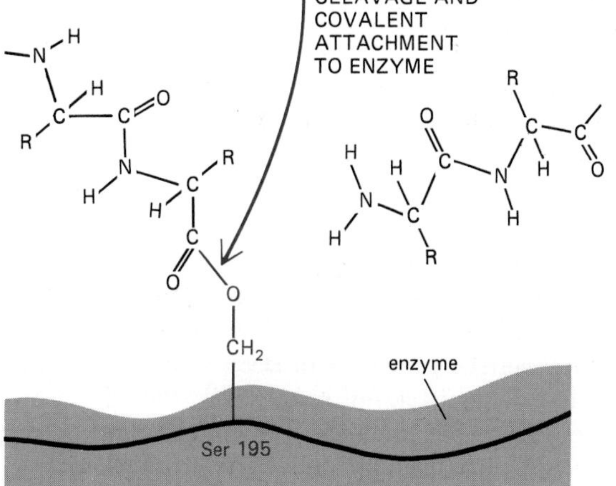

polypeptide substrate

CLEAVAGE AND COVALENT ATTACHMENT TO ENZYME

enzyme

Ser 195

Figure 3–53 Some enzymes form transient covalent bonds with their substrates. In the example shown, a carbonyl group in a polypeptide chain forms a covalent bond with an activated serine residue of a protease, which breaks the polypeptide chain. When the unbound portion of the polypeptide chain has diffused away, a second step occurs (not shown) in which a water molecule hydrolyzes the newly formed covalent bond, thereby releasing the portion of the polypeptide bound to the enzyme surface, freeing the serine at position 195 for another cycle of reaction (see also Figure 3–50).

If an enzyme speeds up the rate of the forward reaction, $A + B \rightarrow AB$, by a factor of 10^8, it must speed up the rate of the backward reaction, $AB \rightarrow A + B$, by a factor of 10^8 as well. The *ratio* of the forward to the backward rates of reaction depends only on the concentrations of A, B, and AB. The equilibrium point remains precisely the same whether or not the reaction is catalyzed by an enzyme.

Many Enzymes Make Reactions Proceed Preferentially in One Direction by Coupling Them to ATP Hydrolysis[35]

The living cell is a chemical system that is far from equilibrium. The product of each enzyme usually serves as a substrate for another enzyme in the metabolic pathway and is rapidly consumed. More important, by means of enzyme-catalyzed pathways, previously described in Chapter 2, many reactions are driven in one direction by being *coupled* to the energetically favorable hydrolysis of ATP to ADP and inorganic phosphate (see p. 74). To make this strategy effective, the ATP pool is itself maintained at a level far from its equilibrium point, with a high ratio of ATP to its hydrolysis products (see Chapter 7, p. 354). This ATP pool thereby serves as a "storage battery" that keeps energy and atoms continually passing through the cell, directed along pathways determined by the enzymes present. For a living system, approaching chemical equilibrium means decay and death.

Multienzyme Complexes Help to Increase the Rate of Cell Metabolism[36]

The efficiency of enzymes in accelerating chemical reactions is crucial to the maintenance of life. Cells, in effect, must race against the unavoidable processes of decay, which run downhill toward chemical equilibrium. If the rates of desirable reactions were not greater than the rates of competing side reactions, a cell would soon die. Some idea of the rate at which cellular metabolism proceeds can be obtained by measuring the rate of ATP utilization. A typical mammalian cell turns over (that is, completely degrades and replaces) its entire ATP pool once every one or two minutes. For each cell this turnover represents the utilization of roughly 10^7 molecules of ATP per second (or, for the human body, about a gram of ATP every minute).

The rates of cellular reactions are rapid because of the effectiveness of enzyme catalysis. Many important enzymes have become so efficient that there is no possibility of further useful improvement: the factor limiting the reaction rate is no longer the intrinsic speed of action of the enzyme but the frequency with which the enzyme collides with its substrate. Such a reaction is said to be *diffusion-limited* (see pp. 92–94).

If a reaction is diffusion-limited, its rate will depend on the concentration of both the enzyme and its substrate. For a sequence of reactions to occur very rapidly, each metabolic intermediate and enzyme involved must therefore be present in high concentration. Given the enormous number of different reactions carried out by a cell, there are limits to the concentrations of substrates that can be achieved. In fact, most metabolites are present in micromolar (10^{-6} M) concentrations, and most enzyme concentrations are much less. How is it possible, therefore, to maintain very fast metabolic rates?

The answer lies in the spatial organization of cell components. Reaction rates can be increased without raising substrate concentrations by bringing the various enzymes involved in a reaction sequence together to form a large protein assembly known as a **multienzyme complex**. In this way the product of enzyme A is passed directly to enzyme B, and so on to the final product, and diffusion rates need not be limiting even when the concentration of substrate in the cell as a whole is very low. Such enzyme complexes are very common (the structure of one, pyruvate dehydrogenase, was shown in Figure 2–40), and they are involved in nearly all aspects of metabolism, including the central genetic processes of DNA, RNA, and protein synthesis. In fact, it may be that few enzymes in eucaryotic cells diffuse freely in solution; instead, most may have evolved binding sites that concentrate them with other proteins of related function in particular regions of the cell,

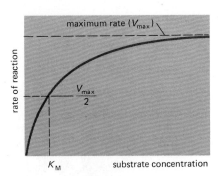

Figure 3–54 The rate of an enzyme reaction (V) increases as the substrate concentration increases until a maximum value (V_{max}) is reached. At this point all substrate-binding sites on the enzyme molecules are fully occupied, and the rate of reaction is limited by the rate of the catalytic process on the enzyme surface. For most enzymes the concentration of substrate at which the reaction rate is half-maximal (K_M) is a measure of how tightly the substrate is bound, with a large value of K_M corresponding to weak binding, and vice versa.

thereby increasing the rate and efficiency of the reactions they catalyze (see p. 668).

Cells have another way of increasing the rate of metabolic reactions, which depends on intracellular membranes.

Figure 3–55 A large increase in the concentration of interacting molecules can be achieved by confining them to the same membrane-bounded compartment in a eucaryotic cell.

Intracellular Membranes Increase the Rates of Diffusion-limited Reactions[37]

The extensive intracellular membranes of eucaryotic cells act in at least two ways to increase the rates of reactions that would otherwise be limited by the speed of diffusion. First, membranes can segregate certain substrates and the enzymes that act on them into the same membrane-bounded compartment, such as the endoplasmic reticulum or the cell nucleus. If, for example, the compartment occupies a total of 10% of the volume of the cell, the concentration of reactants in the compartment can be 10 times greater than in a similar cell with no compartmentalization (Figure 3–55).

A second way membranes can increase the rate of a reaction is by restricting the diffusion of reactants to the two dimensions of the membrane itself. Enzymes and their substrates confined to two dimensions will collide with one another much more frequently than if they were diffusing in three dimensions, even though the rate of diffusion of molecules in a membrane is about 100-fold slower than in aqueous solution (Figure 3–56). This process seems certain to operate in the case of the enzymes and substrates involved in the synthesis of lipid molecules, where the substrates dissolve directly in the lipid bilayer; it probably also operates to accelerate many other reactions that utilize membrane-bound enzymes.

A similar "diffusion to capture" mechanism has been shown to increase the rate at which some gene regulatory proteins find the specific DNA sequences they bind to on chromosomes. These proteins have a weak affinity for any region of DNA. Each time they encounter a chromosome, they "bind and slide," thereby scanning long lengths of DNA for the presence of their specific binding sites.

Protein Molecules Can Reversibly Change Their Shape[38]

Natural selection has generally favored the evolution of polypeptides that adopt specific stable conformations. However, many protein molecules—perhaps most—have two or more slightly different conformations available to them, and by shifting reversibly from one to another, they can alter their function. Such an **allosteric protein** may, for example, be able to form within itself several alternative sets of hydrogen bonds of roughly equal energy, each alternative set requiring different spatial relationships between two domains of the polypeptide chain. The alternative stabilized conformations will generally be separated from one another by unstable intermediate states, so that the molecule will "flip-flop" between the stabilized conformations.

Each distinct conformation of an allosteric protein has a somewhat different surface and thus a different ability to interact with other molecules. Often only

Figure 3–56 Reaction rates can increase when membranes convert diffusion in three dimensions to diffusion in two dimensions. The results of a series of theoretical calculations are shown here. (A) An average molecule diffusing in three dimensions will find a single "target" inside a sphere 10 μm in diameter within 30 minutes. (B) The diffusion time is greatly reduced if the target is fixed in a membrane; here it takes about 1 second for an average molecule to hit a large internal membrane, followed by a mean time of 2 minutes of diffusion in the membrane to find the target. (C) With an internal membrane that has a tenfold reduced surface area, an average molecule will require 10 seconds to find the membrane but will now hit the target by diffusion in the membrane about 10 times more quickly than in (B). Thus the efficiency of collision is nearly 100-fold greater in (C) than in (A).

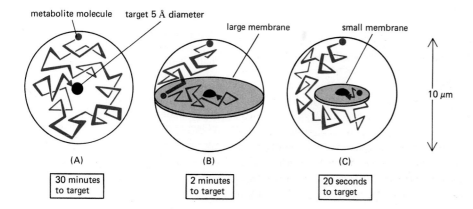

metabolite molecule   target 5 Å diameter   large membrane   small membrane

(A) 30 minutes to target   (B) 2 minutes to target   (C) 20 seconds to target   10 μm

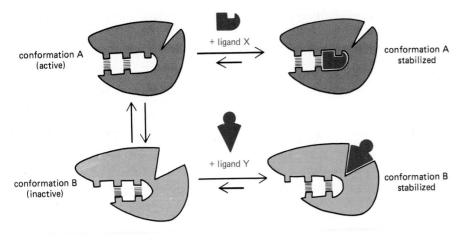

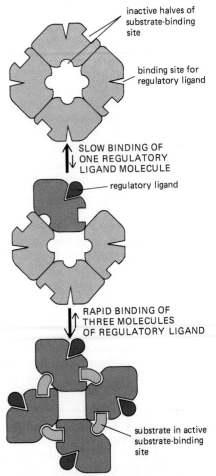

Figure 3–57 Each conformation of an allosteric protein will be stabilized by the binding of a ligand that it binds preferentially. The tight binding of a ligand to only one conformation of an allosteric protein will shift the protein into the conformation that best binds the ligand. Therefore, a high concentration of ligand X will activate the protein shown, whereas a high concentration of ligand Y will inactivate it.

one of two conformations has a high affinity for a particular ligand; in this case the presence or absence of the ligand will determine the conformation that the protein adopts (Figure 3–57). When there are two distinct ligands, each specific to a particular surface of the same protein, the concentration of one molecule will change the affinity of the protein for the other. Allosteric changes of this type are fundamental to the regulation of most biological processes.

Allosteric Proteins Help Regulate Metabolism[39]

Allosteric proteins are essential to the **feedback regulation** that controls the flux through a metabolic pathway (see p. 80). Enzymes that act early in a pathway, for example, are almost always allosteric proteins that can exist in two different conformations. One is an active conformation that binds substrate at its **active site** and catalyzes its conversion to the next substance in the pathway. The other is an inactive conformation that tightly binds the final product of the same pathway at a different place on the protein surface (the **regulatory site**). As the final product accumulates, it binds and converts the enzyme to its inactive conformation (*negative feedback*), which is stabilized because the product can bind to the enzyme only in this form. In other cases an enzyme involved in a metabolic pathway is *activated* by an allosteric transition that occurs when it binds a ligand that accumulates when the cell is deficient in a product of the pathway. Here the ligand binds preferentially to the active form of the enzyme (*positive feedback*), and this binding is required to flip the enzyme from an inactive to an active conformation (see Figure 3–57). As a result of both negative and positive feedback regulation, a cell makes a given product only when that product is needed, and a relatively constant concentration of each metabolite is thereby maintained.

Allosteric Proteins Are Vital for Cell Signaling[40]

As we have noted, allosteric proteins such as those involved in feedback regulation have at least two binding sites, one for the enzyme substrate and one or more for regulatory ligands. These sites occupy different regions of the protein surface, and the ligands they recognize can be totally different. Because the binding of a ligand to one site can affect another site by changing the protein's conformation, any metabolic process can be regulated by any other in the cell, regardless of its chemical nature. For example, the production and breakdown of glycogen in muscle cells is linked to the concentration of Ca^{2+} by means of allosteric enzymes whose activities are changed by changes in the concentration of Ca^{2+} in the cytosol (see p. 712).

Allosteric proteins mediate especially sensitive responses to signals when, as is often the case, they behave cooperatively as identical subunits in a symmetrical assembly. In these proteins the change in the conformation of one subunit caused by ligand binding can help the neighboring subunits to bind the same ligand (Figure 3–58), with the result that a relatively small change in ligand concentration

Figure 3–58 Schematic diagram illustrating how the conformation of one subunit influences that of its neighbors in a symmetrical protein composed of identical allosteric subunits. The binding of a single regulatory ligand molecule to one subunit of the enzyme changes the conformation of this subunit, as in Figure 3–57. Because this conformational change makes it easier for neighboring subunits to change their shape to the tight-binding conformation, the binding of the first molecule of ligand increases the affinity with which the other subunits bind the same ligand. The enzyme will therefore be activated by a relatively small increase in the concentration of the regulatory ligand (see Figure 3–59).

in the surrounding medium switches the whole assembly from an active to an inactive conformation, or vice versa. If the ligand binds preferentially to the active conformation of each enzyme subunit, this behavior will produce a sharp increase in enzyme activity as the ligand concentration is raised (Figure 3–59). The structure of one well-studied allosteric enzyme, aspartate transcarbamoylase, is shown in Figure 3–60.

Proteins Can Be Pushed or Pulled into Different Shapes[40,41]

All of the directed movements in cells depend on forces generated by proteins. But how can a protein molecule be made to move in a controlled fashion? Before we answer this question, we must discuss some of the ways in which cells regulate the conformation of allosteric proteins. Consider an allosteric protein that can adopt two alternative conformations—a low-energy form C (inactive) and a high-energy form C* (active)—that differ in energy by about 4.3 kcal/mole (the energy available from forming about four hydrogen bonds on a protein surface). Given this energy difference, conformation C will be favored by about 1000 to 1 (see Table 3–3, p. 95), and the protein will almost always be in its inactive conformation. However, there are two ways the protein can be forced to adopt the active conformation.

The molecule can, in a sense, be "pulled" into the active C* conformation by binding to a low-molecular-weight ligand. Provided this ligand binds only to C*, the energy of this conformation will be selectively reduced without affecting that of C. Since the ligand binds relatively weakly to the protein (most of its binding energy having been used up in pulling the shape of the protein to fit the ligand), it can readily dissociate, and so this conformational change in the protein is perfectly reversible.

Alternatively, an input of chemical energy can be used to convert conformation C to the active form C* in a less readily reversible manner. A common mechanism involves the transfer of a phosphate group from ATP to a serine, threonine, or tyrosine residue in the protein, forming a covalent linkage. Suppose that this phosphorylation reaction, driven by the highly favorable hydrolysis of ATP to ADP, creates a charge repulsion unfavorable to conformation C. If this repulsion is reduced in the active C* form, the change from C to C* will be greatly encouraged by the phosphorylation (Figure 3–61). Controlled protein phosphorylation is commonly observed to activate or inhibit the function of specific proteins in eucaryotic cells (see p. 417); in fact, about one-tenth of all the proteins made in a mammalian cell contain covalently bound phosphate.

It is sometimes observed that ATP can be made *in vitro* by adding ADP to such a phosphorylated protein, demonstrating directly that a great deal of the energy released by ATP hydrolysis has been stored in straining the protein's conformation during the initial phosphorylation event. But how do such energy-driven changes in protein conformation produce movement and, thereby, do useful work in a cell?

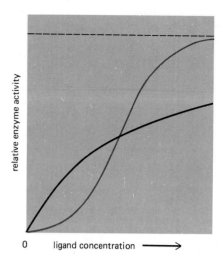

Figure 3–59 The activity of the multisubunit allosteric enzyme illustrated in Figure 3–58 would be expected to increase steeply with ligand concentration (as shown in color) due to the "cooperative" binding of the ligand molecules. In contrast, the activation of an allosteric enzyme containing only one subunit would be expected to display the simple kinetics shown in black. The dotted line indicates the maximum level of activity attained at very high concentrations of the ligand, which is assumed to be the same in both cases.

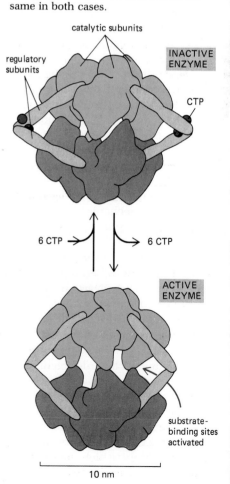

Figure 3–60 The enzyme *aspartate transcarbamoylase* is turned off in response to the binding of cytosine triphosphate (CTP). The enzyme complex contains six catalytic subunits and six regulatory subunits, and the structures of its inactive and active forms have been determined by x-ray diffraction analyses. Each of the regulatory subunits can bind one molecule of CTP, one of the final products in the pathway that begins when this enzyme catalyzes the formation of carbamoyl aspartate from carbamoyl phosphate and aspartic acid. By means of this negative feedback regulation, the enzyme is prevented from producing more CTP than the cell needs. (Based on K.L. Krause, K.W. Volz, and W.N. Lipscomb, *Proc. Natl. Acad. Sci. USA* 82:1643–1647, 1985.)

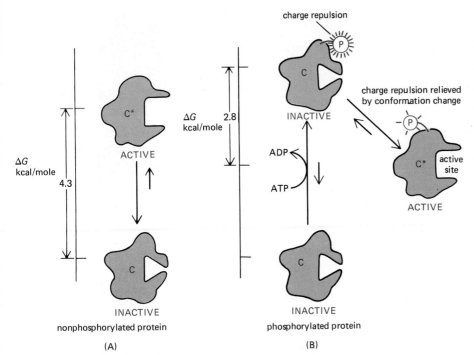

Figure 3–61 Phosphorylation by ATP can activate an allosteric protein. In this example the inactive conformation of the nonphosphorylated protein (A) is favored 1000 to 1 because of a free energy difference of 4.3 kcal/mole (see Table 3–3). When phosphorylated, the active conformation of the protein (B) is favored 100 to 1 (2.8 kcal/mole) because the phosphorylation produces an unfavorable charge repulsion, part of which is relieved by a shift to the active conformation C*. In this way phosphorylation "pushes" the enzyme into the active conformation. Alternatively, phosphorylation could create a charge attraction that brings together two separated parts of an allosteric protein.

Energy-driven Changes in Protein Conformations Can Do Useful Work[42]

Suppose that a protein is required to "walk" along a narrow thread, such as an actin filament or a DNA molecule. Figure 3–62 shows schematically how an allosteric protein might do this by adopting different conformations. With nothing to drive these conformational changes in an orderly way, the shape changes will be perfectly reversible and the protein will wander randomly back and forth along the thread.

Since directional movement of a protein does net work, the laws of thermodynamics demand that such movement depletes free energy from some other source (otherwise the protein could be used to make a perpetual motion machine). Therefore, no matter what modifications we make to the model in Figure 3–62, such as adding ligands that favor certain conformations, the protein molecule shown cannot go anywhere without an input of energy.

What is needed is some means of making the series of protein conformation changes unidirectional. For example, the entire cycle would proceed in one direction if any one of the steps could be made irreversible. One way to do this is through the mechanism just discussed for driving allosteric changes in a protein molecule by a phosphorylation-dephosphorylation cycle. Allosteric changes in proteins can also be driven by ATP hydrolysis without involving a phosphorylation of the protein. In the modified walking scheme shown in Figure 3–63, for example, ATP binding "pulls" the protein from conformation 1 to conformation 2; the bound ATP is then hydrolyzed to produce bound ADP and inorganic phosphate (P_i), causing a change from conformation 2 to conformation 3; and finally, the release of the bound ADP and P_i drives the protein back to conformation 1 again.

Because the transitions $1 \rightarrow 2 \rightarrow 3 \rightarrow 1$ are driven by the energy of ATP hydrolysis, and because the intracellular pools of ATP and its hydrolysis products are maintained far from equilibrium (see p. 354), this series of conformational changes will be effectively irreversible under physiological conditions (that is, the probability that ADP will recombine with P_i to form ATP by the route $1 \rightarrow 3 \rightarrow 2 \rightarrow 1$ is extremely low). Therefore, the entire cycle will go in only one direction, causing the protein molecule to move continuously to the right in this schematic example. Examples of proteins that generate directional movement in this way include *myosin* (which moves along actin filaments), *dynein* (which moves along microtubules), and *DNA helicase* (which moves along DNA).

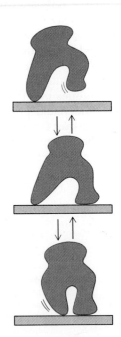

Figure 3–62 An allosteric "walking" protein. Although its three different conformations allow it to wander randomly back and forth while bound to a filament, the protein cannot move uniformly in a single direction.

Many proteins use similar mechanisms to create coherent movement. They all have the ability to go through cyclic changes in shape that are coupled to the hydrolysis of ATP; some are transiently phosphorylated in the process, whereas others are not.

ATP-driven Membrane-bound Allosteric Proteins Can Act as Pumps[43]

Besides generating mechanical force, allosteric proteins can use the energy of ATP hydrolysis to do other forms of work, such as pumping specific ions into or out of the cell. An important example is the **Na$^+$-K$^+$ ATPase** found in the plasma membrane of all animal cells, which pumps 3 Na$^+$ out of the cell and 2 K$^+$ in during each cycle of conformational change driven by ATP-mediated phosphorylation (see p. 305). This ATP-driven pump consumes more than 30% of the total energy requirement of most cells. By continuously pumping Na$^+$ out and K$^+$ in, it creates a cell interior in which the Na$^+$ concentration is low and the K$^+$ concentration is high with respect to the cell exterior, thereby generating two ion gradients across the plasma membrane, each the reverse of the other. The energy stored in these and other ion gradients is, in turn, used to drive conformational changes in a variety of other membrane-bound allosteric proteins, enabling them to do useful work for the cell.

Membrane-bound Allosteric Proteins Can Harness the Energy Stored in Ion Gradients to Do Useful Work[43,44]

Although critically important, ATP and the other nucleoside triphosphates are not the only sources of readily available energy that proteins can use to do useful work. Ion gradients across various cell membranes can store and release energy in a fashion analogous to differences of water pressure on either side of a dam. The large Na$^+$ gradient across the plasma membrane generated by the Na$^+$-K$^+$ ATPase, for example, is used to drive many other plasma membrane-bound protein pumps that transport glucose or specific amino acids into the cell (see p. 309).

Membrane-bound allosteric pumps driven by the hydrolysis of ATP can also work in reverse and employ the energy in the ion gradient to synthesize ATP. In fact, we shall see in Chapter 7 that the energy available in the H$^+$ (proton) gradient across the inner mitochondrial membrane is used in just such a fashion to synthesize most of the ATP required by animal cells.

Protein Machines Play Central Roles in Many Biological Processes[45]

Complex cellular processes such as DNA replication or protein synthesis are mediated by multienzyme complexes that function as sophisticated "protein machines." For example, the numerous protein parts of the DNA replication apparatus move relative to one another without disassembling, allowing the entire complex to travel rapidly in a zipperlike fashion along the DNA (see p. 232).

In such protein machines, the hydrolysis of bound nucleoside triphosphate molecules drives ordered conformational changes in the individual proteins, enabling groups of these proteins to move coordinately. In this way the appropriate enzymes are moved directly into the positions where they are needed to carry out each reaction in a series, instead of waiting for the random collision of each separate component that would otherwise be required. A simple mechanical analogy that captures the sense of such "high-technology" solutions to the cell's needs is illustrated in Figure 3–64. It seems likely that most of the major processes occurring in cells are mediated by such highly sophisticated, multicomponent protein machines.

Summary

The biological function of a protein depends on the detailed chemical properties of its surface. Binding sites are formed as surface cavities in which precisely positioned amino acid side chains are brought together by protein folding. By binding

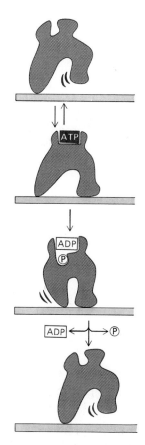

Figure 3–63 An allosteric "walking" protein in which an orderly transition among three conformations is driven by the hydrolysis of a bound ATP molecule. Because one of these transitions is coupled to the hydrolysis of ATP, the cycle is essentially irreversible. By repeated cycles, the protein moves consistently to the right along the filament.

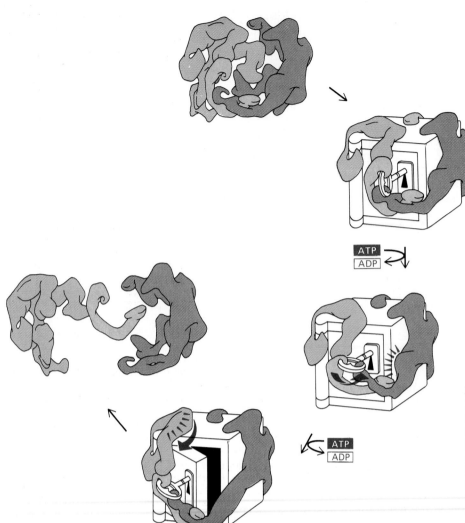

Figure 3–64 A "protein machine." Protein assemblies often contain one or more subunits that can move in an orderly way, driven by an energetically favorable change that occurs in a bound substrate molecule (see Figure 3–63). Protein movements of this type are especially useful to the cell if they occur in a large protein assembly in which, as illustrated here, the activities of several subunits can be coordinated.

unstable reaction intermediates especially tightly, enzymes catalyze chemical changes in bound substrate molecules, often employing small, tightly bound coenzyme molecules to extend their chemical versatility. The rates of enzyme reactions are often limited by diffusion and are increased when the enzyme and its substrate are confined to the same small compartment of the cell.

Allosteric proteins reversibly change their shape when ligands bind to their surface. The changes produced by one ligand often affect the binding of a second ligand, thereby providing a mechanism for regulating various cell processes. Such changes in protein shape can be driven in a unidirectional manner by the expenditure of chemical energy. For example, by coupling allosteric shape changes to ATP hydrolysis, proteins can do useful work—such as generating a mechanical force or pumping ions across a membrane. Highly efficient "protein machines" can be formed by incorporating into multienzyme complexes, proteins that move in coherent ways; protein assemblies of this type appear to carry out many central biological reactions.

References

General

Judson, H.F. The Eighth Day of Creation: Makers of the Revolution in Biology. New York: Simon & Schuster, 1979.

Lehninger, A.L. Principles of Biochemistry. New York: Worth, 1982.

Schulz, G.E.; Schirmer, R.H. Principles of Protein Structure. New York: Springer, 1979.

Stryer, L. Biochemistry, 3rd ed. New York: W.H. Freeman, 1988.

Watson, J.D.; et al. Molecular Biology of the Gene, 4th ed. Menlo Park, CA: Benjamin-Cummings, 1987.

Cited

1. Burley, S.K.; Petsko, G.A. Weakly polar interactions in proteins. *Adv. Prot. Chem.* 39:125–189, 1988.

   Cantor, C.R.; Schimmel, P.R. Biophysical Chemistry, Part I and Part III. New York: W.H. Freeman, 1980.

   Eisenberg, D.; Crothers, D. Physical Chemistry with Applications to the Life Sciences. Menlo Park, CA: Benjamin-Cummings, 1979.

   Pauling, L. The Nature of the Chemical Bond, 3rd ed. Ithaca, NY: Cornell University Press, 1960.

2. Fersht, A.R. The hydrogen bond in molecular recognition. *Trends Biochem. Sci.* 12:301–304, 1987.

3. Cohen, C.; Parry, D.A.D. α helical coiled coils. *Trends Biochem. Sci.* 11:245–248, 1986.

   Dickerson, R.E. The DNA helix and how it is read. *Sci. Am.* 249(6):94–111, 1983.

4. Berg, H.C. Random Walks in Biology. Princeton, NJ: Princeton University Press, 1983.

   Einstein, A. Investigations on the Theory of Brownian Movement. New York: Dover, 1956.

5. Lavenda, B.H. Brownian motion. *Sci. Am.* 252(2):70–85, 1985.

6. Karplus, M.; McCammon, J.A. The dynamics of proteins. *Sci. Am.* 254(4):42–51, 1986.

   McCammon, J.A.; Harvey, S.C. Dynamics of Proteins and Nucleic Acids. Cambridge, U.K.: Cambridge University Press, 1987.

7. Kirkwood, T.B.; Rosenberger, R.F.; Galas, D.J., eds. Accuracy in Molecular Processes: Its Control and Relevance to Living Systems. London: Chapman and Hall, 1986.

   Kurland, C.G.; Ehrenberg, M. Growth-optimized accuracy of gene expression. *Annu. Rev. Biophys. Biophys. Chem.* 16:291–317, 1987.

8. Rosenfield, I.; Ziff, E.; Van Loon, B. DNA for Beginners. London: Writers and Readers Publishing Cooperative. New York: Distributed in the USA by W.W. Norton, 1983.

   Saenger, W. Principles of Nucleic Acid Structure. Berlin: Springer, 1984.

9. Olby, R. The Path to the Double Helix. Seattle: University of Washington Press, 1974.

   Stent, G.S. Molecular Genetics: An Introductory Narrative. San Francisco: Freeman, 1971.

10. Watson, J.D.; Crick, F.H.C. Molecular structure of nucleic acids. A structure for deoxyribose nucleic acid. *Nature* 171:737–738, 1953.

11. Felsenfeld, G. DNA. *Sci. Am.* 253(4):58–66, 1985.

    Meselson, M.; Stahl, F.W. The replication of DNA in *E. coli. Proc. Natl. Acad. Sci. USA* 44:671–682, 1958.

    Watson, J.D.; Crick, F.H.C. Genetic implications of the structure of deoxyribonucleic acid. *Nature* 171:964–967, 1953.

12. Drake, J.W. The Molecular Basis of Mutation. San Francisco: Holden-Day, 1970.

    Drake, J.W.; Glickman, B.W.; Ripley, L.S. Updating the theory of mutation. *Am. Scientist* 71:621–630, 1983.

    Wilson, A.C. Molecular basis of evolution. *Sci. Am.* 253(4):164–173, 1985.

13. Sanger, F. Sequences, sequences, and sequences. *Annu. Rev. Biochem.* 57:1–28, 1988.

    Thompson, E.O.P. The insulin molecule. *Sci. Am.* 192(5):36–41, 1955.

    Yanofsky, C. Gene structure and protein structure. *Sci. Am.* 216(5):80–94, 1967.

14. Brenner, S.; Jacob, F.; Meselson, M. An unstable intermediate carrying information from genes to ribosomes for protein synthesis. *Nature* 190:576–581, 1961.

    Darnell, J.E., Jr. RNA. *Sci. Am.* 253(4):68–78, 1985.

15. Chambon, P. Split genes. *Sci. Am.* 244(5):60–71, 1981.

    Steitz, J.A. Snurps. *Sci. Am.* 258(6):58–63, 1988.

16. Witkowski, J.A. The discovery of "split" genes: a scientific revolution. *Trends Biochem. Sci.* 13:110–113, 1988.

17. Crick, F.H.C. The genetic code: III. *Sci. Am.* 215(4):55–62, 1966. The Genetic Code. *Cold Spring Harbor Symp. Quant. Biol.* 31, 1965.

18. Rich, A.; Kim, S.H. The three-dimensional structure of transfer RNA. *Sci. Am.* 238(1):52–62, 1978.

19. Lake, J.A. The ribosome. *Sci. Am.* 245(2):84–97, 1981.

    Watson, J.D. Involvement of RNA in the synthesis of proteins. *Science* 140:17–26, 1963.

    Zamecnik, P. The machinery of protein synthesis. *Trends Biochem. Sci.* 9:464–466, 1984.

20. Altman, S.; Baer, M.; Guerrier-Takada, C.; Viogue, A. Enzymatic cleavage of RNA by RNA. *Trends Biochem. Sci.* 11:515–518, 1986.

    Cech, T. RNA as an enzyme. *Sci. Am.* 255(5):64–75, 1986.

21. Cantor, C.R.; Schimmel, P.R. Biophysical Chemistry. Part I: The Conformation of Biological Macromolecules, Chapters 2 and 5. New York: W.H. Freeman, 1980.

    Creighton, T.E. Proteins: Structure and Molecular Properties. New York: W.H. Freeman, 1984.

    Dickerson, R.E.; Geis, I. The Structure and Action of Proteins. New York: Harper & Row, 1969.

    Schulz, G.E.; Schirmer, R.H. Principles of Protein Structure, Chapters 1 through 5. New York: Springer, 1979.

22. Anfinsen, C.B. Principles that govern the folding of protein chains. *Science* 181:223–230, 1973.

    Baldwin, R.L. Seeding protein folding *Trends Biochem. Sci.* 11:6–9, 1986.

    Creighton, T.E. Disulphide bonds and protein stability. *Bioessays* 8:57–63, 1988.

    Rupley, J.A.; Gratton, E.; Careri, G. Water and globular proteins. *Trends Biochem. Sci.* 8:18–22, 1983.

    Tanford, C. The Hydrophobic Effect: Formation of Micelles and Biological Membranes, 2nd ed. New York: Wiley, 1980.

23. Doolittle, R.F. Proteins. *Sci. Am.* 253(4):88–99, 1985.

    Milner-White, E.J.; Poet, R. Loops, bulges, turns and hairpins in proteins. *Trends Biochem Sci.* 12:189–192, 1987.

    Pauling, L.; Corey, R.B. Configurations of polypeptide chains with favored orientations around single bonds: two new pleated sheets. *Proc. Natl. Acad. Sci. USA* 37:729–740, 1951.

    Pauling, L.; Corey, R.B.; Branson, H.R. The structure of proteins: two hydrogen-bonded helical configurations of the polypeptide chain. *Proc. Natl. Acad. Sci. USA* 27:205–211, 1951.

    Richardson, J.S. The anatomy and taxonomy of protein structure. *Adv. Protein Chem.* 34:167–339, 1981.

23. Scott, J.E. Molecules for strength and shape. *Trends Biochem. Sci.* 12:318–321, 1987.

24. Hardie, D.G.; Coggins, J.R. Multidomain Proteins—Structure and Evolution. Amsterdam: Elsevier, 1986.

25. Doolittle, R.F. The genealogy of some recently evolved vertebrate proteins. *Trends Biochem. Sci.* 10:233–237, 1985.

    Doolittle, R.F. Protein evolution. In The Proteins, 3rd ed. (H. Neurath, R.L. Hill, eds.), Vol. 4, pp. 1–118. New York: Academic Press, 1979.

    Neurath, H. Proteolytic enzymes, past and present. *Fed. Proc.* 44:2907–2913, 1985.

26. Biesecker, G.; et al. Sequence and structure of D-glyceraldehyde 3-phosphate dehydrogenase from *Bacillus stearothermophilus. Nature* 266:328–333, 1977.

    Blake, C. Exons and the evolution of proteins. *Trends Biochem. Sci.* 8:11–13, 1983.

    Gilbert, W. Genes-in-pieces revisited. *Science* 228:823–824, 1985.

    McCarthy, A.D.; Hardie, D.G. Fatty acid synthase: an example of protein evolution by gene fusion. *Trends Biochem. Sci.* 9:60–63, 1984.

    Rossmann, M.G.; Argos, P. Protein folding. *Annu. Rev. Biochem.* 50:497–532, 1981.

27. Gehring, W.J. On the homeobox and its significance. *Bioessays* 5:3–4, 1986.

Hunter, T. The proteins of oncogenes. *Sci. Am.* 251(2):70–79, 1984.

Südhof, T.C.; et al. Cassette of eight exons shared by genes for LDL receptor and EGF precursor. *Science* 228:893–895, 1985.

Weber, I.T.; Takio, K.; Titani, K.; Steitz, T.A. The cAMP-binding domains of the regulatory subunit of cAMP-dependent protein kinase and the catabolite gene activator protein are homologous. *Proc. Natl. Acad. Sci. USA* 79:7679–7683, 1982.

28. Bajaj, M.; Blundell, T. Evolution and the tertiary structure of proteins. *Annu. Rev. Biophys. Bioeng.* 13:453–492, 1984.

Klug, A. From macromolecules to biological assemblies. *Biosci. Rep.* 3:395–430, 1983.

Metzler, D.E. Biochemistry. New York: Academic Press, 1977. (Chapter 4 describes how macromolecules pack together into large assemblies.)

29. Caspar, D.L.D.; Klug, A. Physical principles in the construction of regular viruses. *Cold Spring Harbor Symp. Quant. Biol.* 27:1–24, 1962.

Harrison, S.C. Virus structure: high resolution perspectives. *Adv. Virus Res.* 28:175–240, 1983.

30. Fraenkel-Conrat, H.; Williams, R.C. Reconstitution of active tobacco mosaic virus from its inactive protein and nucleic acid components. *Proc. Natl. Acad. Sci. USA* 41:690–698, 1955.

Harrison, S.C. Multiple modes of subunit association in the structures of simple spherical viruses. *Trends Biochem. Sci.* 9:345–351, 1984.

Hendrix, R.W. Tail length determination in double-stranded DNA bacteriophages. *Curr. Top. Microbiol. Immunol.* 136:21–29, 1988.

Hogle, J.M.; Chow, M.; Filman, D.J. The structure of polio virus. *Sci. Am.* 256(3):42–49, 1987.

Namba, K.; Caspar, D.L.D.; Stubbs, G.J. Computer graphics representation of levels of organization in tobacco mosaic virus structure. *Science* 227:773–776, 1985.

Nomura, M. Assembly of bacterial ribosomes. *Science* 179:864–873, 1973.

31. Mathews, C.K.; Kutter, E.M.; Mosig, G.; Berget, P.B. Bacteriophage T4, Chapters 1 and 4. Washington, DC: American Society of Microbiologists, 1983.

Steiner, D.F.; Kemmler, W.; Tager, H.S.; Peterson, J.D. Proteolytic processing in the biosynthesis of insulin and other proteins. *Fed. Proc.* 33:2105–2115, 1974.

32. Fersht, A. Enzyme Structure and Mechanism, 2nd ed. New York: W.H. Freeman, 1985.

33. Chothia, C. Principles that determine the structure of proteins. *Annu. Rev. Biochem.* 53:537–572, 1984.

34. Fersht, A.R.; Leatherbarrow, R.J.; Wells, T.N.C. Binding energy and catalysis. *Trends Biochem. Sci.* 11:321–325, 1986.

Lerner, R.A.; Tramontano, A. Catalytic antibodies. *Sci. Am.* 258(3):58–70, 1988.

Wolfenden, R. Analog approaches to the structure of the transition state in enzyme reactions. *Accounts Chem. Res.* 5:10–18, 1972.

35. Wood, W.B.; Wilson, J.H.; Benbow, R.M.; Hood, L.E. Biochemistry, A Problems Approach, 2nd ed. Menlo Park, CA: Benjamin-Cummings, 1981. (Chapters 9 and 15 and associated problems.)

36. Barnes, S.J.; Weitzman, P.D.J. Organization of citric acid cycle enzymes into a multienzyme cluster. *FEBS Lett.* 201:267–270, 1986.

Reed, L.J.; Cox, D.J. Multienzyme complexes. In The Enzymes, 3rd ed., (P.D. Boyer, ed.), Vol. 1, pp. 213–240. New York: Academic Press, 1970.

37. Adam, G.; Delbruck, M. Reduction of dimensionality in biological diffusion processes. In Structural Chemistry and Molecular Biology (A. Rich, N. Davidson, eds.), pp. 198–215. San Francisco: Freeman, 1968.

Berg, O.G.; von Hippel, P.H. Diffusion-controlled macromolecular interactions. *Annu. Rev. Biophys. Biophys. Biochem.* 14:131–160, 1985.

38. Cantor, C.R.; Schimmel, P.R. Biophysical Chemistry. Part III: The Behavior of Biological Macromolecules, Chapters 15 and 17. New York: W.H. Freeman, 1980.

Dickerson, R.E.; Geis, I. Hemoglobin: Structure, Function, Evolution and Pathology. Menlo Park, CA: Benjamin-Cummings, 1983.

Edelstein, S.J. Introductory Biochemistry. San Francisco: Holden-Day, 1973. (Chapter 10 on protein aggregates and allosteric interactions.)

Monod, J.; Changeux, J.-P.; Jacob, F. Allosteric proteins and cellular control systems. *J. Mol. Biol.* 6:306–329, 1963.

39. Koshland, D.E., Jr. Control of enzyme activity and metabolic pathways. *Trends Biochem. Sci.* 9:155–159, 1984.

Newsholme, E.A.; Start, C. Regulation in Metabolism. New York: Wiley, 1973.

40. Koch, K–W.; Stryer, L. Highly cooperative feedback control of retinal rod guanylate cyclase by calcium ions. *Nature* 334:64–65, 1988.

Nishizuka, Y. Protein kinases in signal transduction. *Trends Biochem. Sci.* 9:163–166, 1984.

41. Kantrowitz, E.R.; Lipscomb, W.N. *Escherichia coli* aspartate transcarbamoylase: the relation between structure and function. *Science* 241:669–674, 1988.

Schachman, H.K. Can a simple model account for the allosteric transition of aspartate transcarbamoylase? *J. Biol. Chem.* 263:18583–18586, 1988.

Sprang, S.; Goldsmith, E.; Fletterick, R. Structure of the nucleotide activation switch in glycogen phosphorylase a. *Science* 237:1012–1019, 1987.

42. Hill, T.L. Biochemical cycles and free energy transduction. *Trends Biochem. Sci.* 2:204–207, 1977.

Hill, T.L. A proposed common allosteric mechanism for active transport, muscle contraction, and ribosomal translocation. *Proc. Natl. Acad. Sci. USA* 64:267–274, 1969.

Johnson, K.A. Pathway of the microtubule-dynein ATPase and the structure of dynein: a comparison with actomyosin. *Annu. Rev. Biophys. Biophys. Chem.* 14:161–188, 1985.

43. Hokin, L.E. The molecular machine for driving the coupled transports of Na^+ and K^+ is an $(Na^+ + K^+)$-activated ATPase. *Trends Biochem. Sci.* 1:233–237, 1976.

Kyte, J. Molecular considerations relevant to the mechanism of active transport. *Nature* 292:201–204, 1981.

Tanford, C. Mechanism of free energy coupling in active transport. *Annu. Rev. Biochem.* 53:379–409, 1983.

44. Nicholls, D.G. Bioenergetics: An Introduction to the Chemiosmotic Theory. New York: Academic Press, 1982.

45. Alberts, B.M. Protein machines mediate the basic genetic processes. *Trends Genet.* 1:26–30, 1985.

How Cells Are Studied

4

Cells are small and complex. It is hard to see their structure, hard to discover their molecular composition, and harder still to find out how their various components function. What we can learn about cells depends on the tools at our disposal, and major advances have frequently sprung from the introduction of new techniques. To understand contemporary cell biology, therefore, it is necessary to know something of its methods.

In this chapter we shall review briefly some of the principal methods used to study cells. We shall start with techniques for examining the cell as a whole and then proceed to techniques for analyzing its constituent macromolecules. Microscopy will be our starting point, for cell biology began with the light microscope, and this is still an essential tool, along with more recent imaging devices based on beams of electrons and other forms of radiation. From passive observation we shall move to active intervention: we shall consider how cells of different types can be separated from tissues and grown outside the body, and how cells can be disrupted and their organelles and constituent macromolecules isolated in pure form. Last, we shall discuss how recombinant DNA technology has revolutionized the study of cells by making it possible to isolate, sequence, and manipulate genes and thereby to determine how they work inside the cell.

Microscopy[1]

A typical animal cell is 10 to 20 μm in diameter, or about five times smaller than the smallest visible particle. It was, therefore, not until good light microscopes became available, in the early part of the nineteenth century, that all plant and animal tissues were discovered to be aggregates of individual cells. This discovery, proposed as the **cell doctrine** by Schleiden and Schwann in 1838, marks the formal birth of cell biology.

Animal cells are not only tiny, they are colorless and translucent; consequently, the discovery of their main internal features depended on the development, in the latter part of the nineteenth century, of a variety of stains that provided sufficient contrast to render them visible. Likewise, introduction of the far more powerful electron microscope in the early 1940s required the development of new techniques for preserving and staining cells before the full complexities of

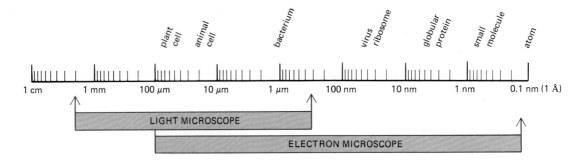

their internal fine structure could begin to emerge. To this day, microscopy depends as much on techniques for preparing the specimen as on the performance of the microscope itself. In the sections that follow, we shall therefore discuss both instruments and specimen preparation, beginning with the light microscope.

Figure 4–1 shows the fineness of detail that can be resolved with modern light microscopes in comparison with electron microscopes. Some of the landmarks in the development of light microscopy are outlined in Table 4–1.

Figure 4–1 Sizes of cells and their components drawn on a logarithmic scale, indicating the range of readily resolvable objects in the light and electron microscopes. The following units of length are commonly employed in microscopy:

μm (micrometer) $= 10^{-6}$ m
nm (nanometer) $= 10^{-9}$ m
Å (Ångström unit) $= 10^{-10}$ m

Table 4–1 Some Important Discoveries in the History of Light Microscopy

1611	**Kepler** suggested a way of making a compound microscope.
1655	**Hooke** used a compound microscope to describe small pores in sections of cork that he called "cells."
1674	**Leeuwenhoek** reported his discovery of protozoa. He saw bacteria for the first time 9 years later.
1833	**Brown** published his microscopic observations of orchids, clearly describing the cell nucleus.
1838	**Schleiden and Schwann** proposed the cell theory, stating that the nucleated cell is the unit of structure and function in plants and animals.
1857	**Kolliker** described mitochondria in muscle cells.
1876	**Abbé** analyzed the effects of diffraction on image formation in the microscope and showed how to optimize microscope design.
1879	**Flemming** described with great clarity chromosome behavior during mitosis in animal cells.
1881	**Retzius** described many animal tissues with a detail that has not been surpassed by any other light microscopist. In the next two decades, he, **Cajal**, and other histologists developed staining methods and laid the foundations of microscopic anatomy.
1882	**Koch** used aniline dyes to stain microorganisms and identified the bacteria that cause tuberculosis and cholera. In the following two decades, other bacteriologists, such as **Klebs** and **Pasteur**, identified the causative agents of many other diseases by examining stained preparations under the microscope.
1886	**Zeiss** made a series of lenses, to the design of **Abbé**, that enabled microscopists to resolve structures at the theoretical limits of visible light.
1898	**Golgi** first saw and described the Golgi apparatus by staining cells with silver nitrate.
1924	**Lacassagne** and collaborators developed the first autoradiographic method to localize radioactive polonium in biological specimens.
1930	**Lebedeff** designed and built the first interference microscope. In 1932, **Zernicke** invented the phase-contrast microscope. These two developments allowed unstained living cells to be seen in detail for the first time.
1941	**Coons** used antibodies coupled to fluorescent dyes to detect cellular antigens.
1952	**Nomarski** devised and patented the system of differential interference contrast for the light microscope that still bears his name.
1981	**Allen** and **Inoué** perfected video-enhanced contrast light microscopy.

Figure labels: plant cell, animal cell, bacterium, virus, ribosome, globular protein, small molecule, atom. Scale: 1 cm, 1 mm, 100 μm, 10 μm, 1 μm, 100 nm, 10 nm, 1 nm, 0.1 nm (1 Å). LIGHT MICROSCOPE, ELECTRON MICROSCOPE.

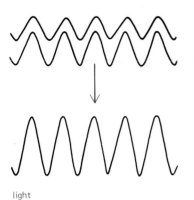

light

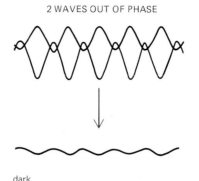

dark

Figure 4–2 Interference between light waves. When two light waves combine *in phase*, the amplitude of the resultant wave is larger and the brightness is increased. Two light waves that are *out of phase* partially cancel each other and produce a wave whose amplitude, and therefore brightness, is decreased.

The Light Microscope Can Resolve Details 0.2 μm Apart[2]

In general, radiation of a given wavelength cannot be used to probe structural details much smaller than its own wavelength: this represents a fundamental limitation of all microscopes. The ultimate limit to the resolution of a light microscope is, therefore, set by the wavelength of visible light, which ranges from about 0.4 μm (for violet) to 0.7 μm (for deep red). In practical terms, bacteria and mitochondria, which are about 500 nm (0.5 μm) wide, are generally the smallest objects that can be seen clearly in the light microscope; details smaller than this are obscured by effects resulting from the wave nature of light. To understand why this occurs, we must follow what happens to a beam of light waves as it passes through the lenses of a microscope.

Because of its wave nature, light does not follow exactly the idealized straight ray paths predicted by geometrical optics. Instead, light waves travel through an optical system by a variety of slightly different routes, so that they interfere with one another and cause *optical diffraction* effects. If two trains of waves reaching the same point by different paths are precisely in phase, with crest matching crest and trough matching trough, they will reinforce each other so as to increase brightness. On the other hand, if the trains of waves are out of phase, they will interfere with each other in such a way as to partially or entirely cancel each other (Figure 4–2). The interaction of light with an object will change the phase relationships of the light waves in a way that produces complex interference effects. At high magnification, for example, the shadow of a straight edge that is evenly illuminated with light of uniform wavelength appears as a set of parallel lines, whereas that of a circular spot appears as a set of concentric rings (Figure 4–3). For the same reason, a single point seen through a microscope appears as a blurred disc, and two point objects close together give overlapping images and may merge into one. No amount of refinement of the lenses can overcome this limitation imposed by the wavelike nature of light.

The limiting separation at which two objects can still be seen as distinct—the so-called **limit of resolution**—depends on both the wavelength of the light and the numerical aperture of the lens system used (Figure 4–4). Under the best conditions, with violet light (wavelength, $\lambda = 0.4$ μm) and a numerical aperture of 1.4, a limit of resolution of just under 0.2 μm can theoretically be obtained in the light microscope. This resolution was achieved by microscope makers at the end of the nineteenth century and is only rarely matched in practice in contemporary, factory-produced microscopes. Although it is possible to *enlarge* an image as much as one wants—for example, by projecting it onto a screen—it is never possible to resolve two objects in the light microscope that are separated by less than about 0.2 μm: such objects will appear as one.

The wave nature of light is not always a barrier to the study of cells: on the contrary, we shall see later how interference and diffraction can be exploited to study unstained cells in the living state. But first we shall discuss how permanent preparations of cells are made for viewing in the microscope and how chemical stains are used to enhance the visibility of the cell structures in such preparations.

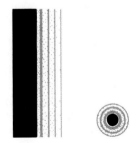

Figure 4–3 The interference effects observed at high magnification when light passes the edges of a solid object placed between the light source and the observer.

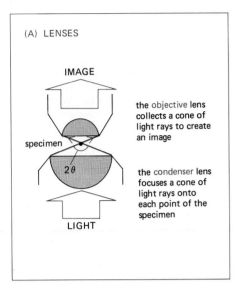

(A) LENSES

IMAGE

the objective lens collects a cone of light rays to create an image

specimen

2θ

the condenser lens focuses a cone of light rays onto each point of the specimen

LIGHT

(B) RESOLUTION: the resolving power of the microscope depends on the width of the cone of illumination and therefore on both the condenser and the objective lens. It is calculated using the formula

$$\text{resolution} = \frac{0.61\,\lambda}{n\,\sin\theta}$$

where:

θ = half the angular width of the cone of rays collected by the objective lens from a typical point in the specimen (since the maximum width is 180°, $\sin\theta$ has a maximum value of 1)

n = the refractive index of the medium (usually air or oil) separating the specimen from the objective and condenser lenses

λ = the wavelength of light used (for white light, a figure of 0.53 μm is commonly assumed)

Figure 4–4 The path of light rays passing through a transparent specimen in a microscope, illustrating the concept of numerical aperture and its relation to the limit of resolution.

(C) NUMERICAL APERTURE: $n\sin\theta$ in the equation above is called the numerical aperture of the lens (NA) and is a function of its light-collecting ability. For dry lenses this cannot be more than 1, but for oil-immersion lenses it can be as high as 1.4. The higher the numerical aperture, the greater the resolution and the brighter the image (brightness is important in fluorescence microscopy). However, this advantage is obtained at the expense of very short working distances and a very small depth of field.

Tissues Are Usually Fixed and Sectioned for Microscopy[2]

To make a permanent preparation that can be stained and viewed at leisure in the microscope, one must first treat cells with a fixative so as to immobilize, kill, and preserve them. In chemical terms, **fixation** makes cells permeable to staining reagents and cross-links their macromolecules so that they are stabilized and locked in position. Some of the earliest fixation procedures involved immersion in acids or in organic solvents such as alcohol. Current procedures usually include treatment with reactive aldehydes, particularly formaldehyde and glutaraldehyde, which form covalent bonds with the free amino groups of proteins and thereby cross-link adjacent molecules.

Most tissue samples are too thick for their individual cells to be examined directly at high resolution. After fixation, therefore, the tissues are usually cut into very thin slices (**sections**) with a *microtome*, a machine with a sharp metal blade that operates rather like a meat slicer (Figure 4–5). The sections (typically 1 to 10 μm thick) are then laid flat on the surface of a glass microscope slide.

Tissues are generally soft and fragile, even after fixation, and need to be **embedded** in a supporting medium before sectioning. The usual embedding media are waxes or resins. In liquid form these media will both permeate and surround the fixed tissue; they then can be hardened (by cooling or by polymerization) to a solid block, which is readily sectioned by the microtome.

There is a serious danger that any treatment used for fixation and embedding may alter the structure of the cell or its constituent molecules in undesirable ways. Rapid freezing provides an alternative method of preparation that to some extent avoids this problem by eliminating the need for fixation and embedding. The frozen tissue can be cut directly with a cryostat—a special microtome that is maintained in a cold chamber. Although **frozen sections** produced in this way avoid some artifacts, they suffer from others: the native structures of individual molecules such as proteins are well preserved, but the structure of the cell is often disrupted by ice crystals.

Once sections have been cut, by whatever method, the next step is usually to stain them.

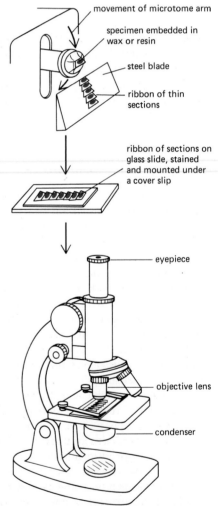

movement of microtome arm

specimen embedded in wax or resin

steel blade

ribbon of thin sections

ribbon of sections on glass slide, stained and mounted under a cover slip

eyepiece

objective lens

condenser

EXAMINATION WITH LIGHT MICROSCOPE

Figure 4–5 How an embedded tissue is sectioned with a microtome in preparation for examination in the light microscope.

Different Components of the Cell Can Be Selectively Stained[3]

There is little in the contents of most cells (which are 70% water by weight) to impede the passage of light rays. Thus most cells in their natural state, even if fixed and sectioned, are almost invisible in an ordinary light microscope. One way to make them visible is to stain them with dyes.

In the early nineteenth century, the demand for dyes to stain textiles led to a fertile period for organic chemistry. Some of the dyes were found to stain biological tissues and, unexpectedly, often showed a preference for particular parts of the cell—the nucleus or mitochondria, for example—making these internal structures clearly visible. Today a rich variety of organic dyes is available, with such colorful names as *Malachite green, Sudan black,* and *Coomassie blue,* each of which has some specific affinity for particular subcellular components. The dye *hematoxylin,* for example, has an affinity for negatively charged molecules and therefore reveals the distribution of DNA, RNA, and acidic proteins in a cell. The chemical basis for the specificity of many dyes, however, is not known.

An increased understanding of cell chemistry has led to more rational and selective staining procedures and, in particular, to methods that reveal specific proteins or other macromolecules in cells. A problem of sensitivity arises here. Since relatively few copies of most macromolecules are present in any given cell, one or two molecules of stain bound to each macromolecule will often be invisible. One way to solve this problem is to increase the number of stain molecules associated with a single macromolecule. Thus some enzymes can be located in cells through their catalytic activity: when supplied with appropriate substrate molecules, each enzyme molecule generates many molecules of a localized visible reaction product. An alternative approach to the problem of sensitivity is to utilize fluorescence. Here, instead of attempting to detect dye molecules by the small decrease they cause in the amount of light transmitted through a specimen, one detects special dyes against a dark background by means of the light that they—and they alone—emit when suitably excited, as we shall now explain.

Specific Molecules Can Be Located in Cells by Fluorescence Microscopy[4]

Fluorescent molecules absorb light at one wavelength and emit it at another, longer wavelength. If such a compound is illuminated at its absorbing wavelength and then viewed through a filter that allows only light of the emitted wavelength to pass, it is seen to glow against a dark background. The intensity and color of the light is a distinctive property of the fluorescent molecule used.

The fluorescent dyes used for staining cells are detected with the help of a **fluorescence microscope.** This microscope is similar to an ordinary light microscope except that the illuminating light, from a very powerful source, is passed through two sets of filters—one to intercept the light before it reaches the specimen and one to filter the light obtained from the specimen. The first filter is selected so that it passes only the wavelengths that excite the particular fluorescent dye, while the second filter blocks out this light and passes only those wavelengths emitted when the dye fluoresces (Figure 4–6).

Fluorescence microscopy is most often used to detect specific proteins or other molecules that have been made fluorescent by coupling them covalently to a fluorescent dye. For example, fluorescent dyes can be coupled to antibody molecules, which then serve as highly specific and versatile staining reagents by binding selectively to other macromolecules on the surface of a living cell or inside a fixed cell (see p. 177). Two fluorescent dyes that are commonly used for this purpose are *fluorescein,* which emits an intense yellow-green fluorescence when excited with blue light, and *rhodamine,* which emits a deep red fluorescence when excited with green-yellow light (Figure 4–7). By coupling one molecule to fluorescein and another to rhodamine, the distributions of different molecules can be followed in the same cell; the two molecules are detected separately in the microscope by switching back and forth between two sets of filters, each specific for one dye (Figure 4–8).

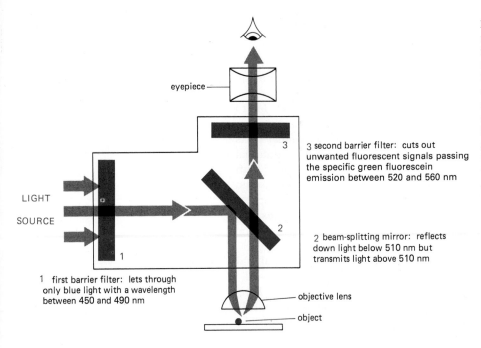

Figure 4–6 The optical system of a modern fluorescence microscope. A filter set consists of two barrier filters and a dichroic (beam-splitting) mirror. In this example the filter set for detection of fluorescein fluorescence is shown. High-numerical-aperture objective lenses are especially important in this type of microscopy since, for a given magnification, the brightness of the fluorescent image is proportional to the fourth power of the numerical aperture (see also Figure 4–4).

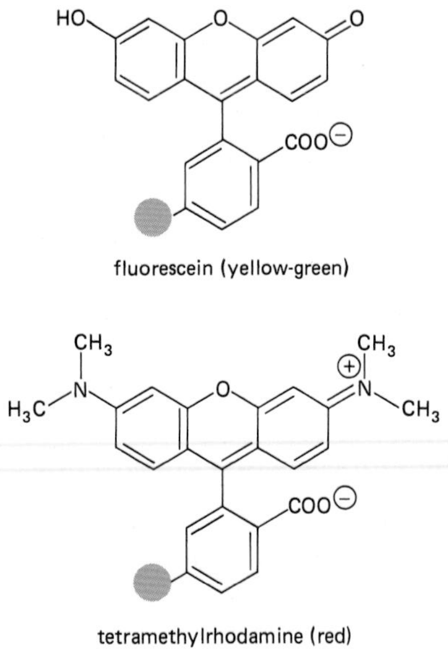

eyepiece

LIGHT

SOURCE

3 second barrier filter: cuts out unwanted fluorescent signals passing the specific green fluorescein emission between 520 and 560 nm

2 beam-splitting mirror: reflects down light below 510 nm but transmits light above 510 nm

1 first barrier filter: lets through only blue light with a wavelength between 450 and 490 nm

objective lens

object

Figure 4–7 The structures of fluorescein and tetramethylrhodamine, two dyes that are commonly used for fluorescence microscopy. Fluorescein emits yellow-green light when activated by light of the appropriate wavelength, whereas the rhodamine dye emits red light. The portion of each molecule shown in color denotes the position of a chemically reactive group; at this position a covalent bond is commonly formed between the dye and a protein (or other molecule). Commercially available versions of these dyes with different types of reactive groups allow the dye to be targeted either to an —SH group or to an —NH_2 group on a protein.

fluorescein (yellow-green)

tetramethylrhodamine (red)

Figure 4–8 Fluorescence micrographs of a portion of the surface of an early *Drosophila* embryo in which the microtubules have been labeled with an antibody coupled to fluorescein (*left panel*) and the actin filaments have been labeled with an antibody coupled to rhodamine (*middle panel*). In addition, the chromosomes have been labeled with a third dye that fluoresces only when it binds to DNA (*right panel*). At this stage, all the nuclei of the embryo share a common cytoplasm, and they are in the metaphase stage of mitosis. The three micrographs were taken of the same region of a fixed embryo using three different filter sets in the fluorescence microscope (see also Figure 4–6). (Courtesy of Tim Karr.)

Tubulin   Actin   DNA

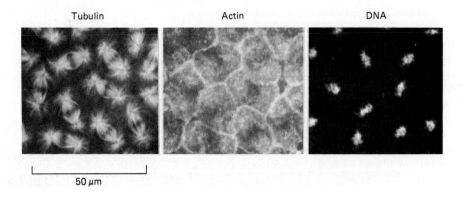

50 µm

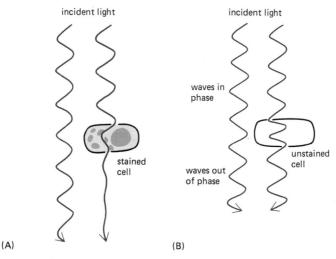

(A)   (B)

Figure 4–9 Two ways of increasing contrast in light microscopy. The stained portions of the cell in (A) reduce the amplitude of light waves of particular wavelengths passing through them. A colored image of the cell is thereby obtained that is visible in the ordinary way. Light passing through the unstained, living cell (B) undergoes scarcely any change in amplitude, and the structural details cannot be seen even if the image is highly magnified. However, the phase of the light is altered by its passage through the cell, and small phase differences can be made visible by exploiting interference effects using a phase-contrast or a differential-interference-contrast microscope.

Important new methods, to be discussed later, enable fluorescence microscopy to be used to monitor changes in the concentration and location of specific molecules inside *living* cells (see p. 142 and p. 156).

Living Cells Are Seen Clearly in a Phase-Contrast or a Differential-Interference-Contrast Microscope[2]

The possibility that some components of the cell may be lost or distorted during specimen preparation has always worried microscopists. The only certain way to avoid the problem is to examine cells while they are alive, without fixing or freezing. For this purpose, microscopes with special optical systems are useful.

When light passes through a living cell, the phase of the light wave is changed according to the cell's refractive index: light passing through a relatively thick or dense part of the cell, such as the nucleus, is retarded and its phase consequently shifted relative to light that has passed through an adjacent thinner region of the cytoplasm. Both the **phase-contrast microscope** and the **differential-interference-contrast microscope** exploit the interference effects produced when these two sets of waves recombine, thereby creating an image of the cell's structure (Figure 4–9). Both types of light microscopy are widely used to visualize living cells.

A simpler way to see some of the features of a living cell is to observe the light that is scattered by its various components. In the **dark-field microscope,** the illuminating rays of light are directed from the side, so that only scattered light enters the microscope lenses. Consequently, the cell appears as an illuminated object against a black background. Images of the same cell obtained by four different kinds of light microscopy are shown in Figure 4–10.

Figure 4–10 A fibroblast in tissue culture visualized with four types of light microscopy. The image in (A) was obtained by the simple transmission of light through the cell, a technique known as *bright-field microscopy.* The other images were obtained by techniques discussed in the text: (B) phase-contrast microscopy, (C) Nomarski differential-interference-contrast microscopy, and (D) dark-field microscopy. All four types of image can be obtained with most modern microscopes simply by interchanging optical components.

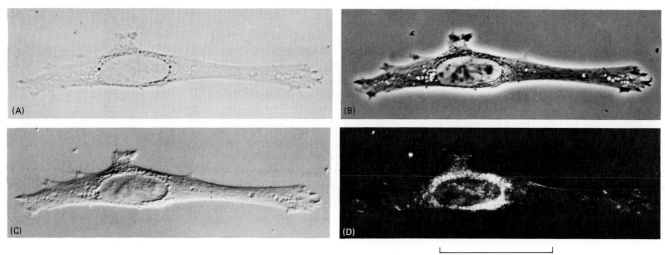

50 μm

One of the great advantages of phase-contrast, interference, and dark-field microscopy is that they make it possible to watch the movements involved in such processes as mitosis and cell migration. Since many cellular motions are too slow to be seen in real time, it is often helpful to take time-lapse motion pictures (*microcinematography*) or video recordings. Here, successive frames separated by a short time delay are recorded, so that when the resulting film or video tape is projected or played at normal speed, events appear greatly speeded up.

Images Can Be Enhanced and Analyzed by Electronic Techniques[5]

In recent years, video cameras and the associated technology of **image processing** have had a major impact on light microscopy. They have not only enabled certain practical limitations of microscopes—due to imperfections in the optical system—to be overcome, but they have also circumvented two fundamental limitations of the human eye. These limitations are that (1) the eye cannot see in extremely dim light and (2) it cannot perceive small differences in light intensity against a bright background. The first of these problems can be overcome by attaching highly light-sensitive video cameras (of the kind used in night surveillance) to a microscope. It is then possible to observe cells for long periods at very low light levels, thereby avoiding the damaging effects of prolonged bright light (and heat). Such *image-intensification systems* are especially important for viewing fluorescent molecules in living cells.

Because images produced by video cameras are in electronic form, they can be readily digitized, fed to a computer and processed in various ways to extract latent information. Such image processing makes it possible to compensate for various optical faults in microscopes so as to attain the theoretical limit of resolution. Moreover, by using video systems linked to image processors, contrast can be greatly enhanced so that the eye's limitations in detecting small differences are overcome. Although this processing also enhances the effects of random background irregularities in the optical system, this "noise" can be removed by electronically subtracting an image of a blank area of the field. Small transparent objects then become visible that were previously impossible to distinguish from the background.

The high contrast attainable by computer-assisted interference microscopy makes it possible to see even very small objects such as single microtubules (Figure 4–11), which have a diameter less than one-tenth the wavelength of light (0.025 μm). Individual microtubules can also be seen in a fluorescence microscope if they are fluorescently labeled (see Figure 4–56). In both cases, however, the unavoidable diffraction effects badly blur the image, so that the microtubules appear at least 0.2 μm wide, making it impossible to distinguish a single microtubule from a bundle of several microtubules. Such a distinction requires electron microscopes, which have pushed the limit of resolution well below the wavelength of visible light.

The Electron Microscope Resolves the Fine Structure of the Cell[6]

The relationship between the limit of resolution and the wavelength of the illuminating light (see Figure 4–4) holds true for any form of radiation, whether it is a beam of light or a beam of electrons; but with electrons the limit of resolution can be made very small. The wavelength of an electron decreases as its velocity increases. In an electron microscope with an accelerating voltage of 100,000 V, the wavelength of an electron is 0.004 nm, and in theory the resolution of such a microscope should be about 0.002 nm. However, because the aberrations of an electron lens are considerably harder to correct than those of a glass lens, the practical resolving power of most modern electron microscopes is, at best, 0.1 nm (1 Å) (Figure 4–12). Furthermore, problems of specimen preparation, contrast, and radiation damage effectively limit the normal resolution for biological objects to 2 nm (20 Å)—about 100 times better than the resolution of the light microscope. Some of the landmarks in the development of electron microscopy are outlined in Table 4–2.

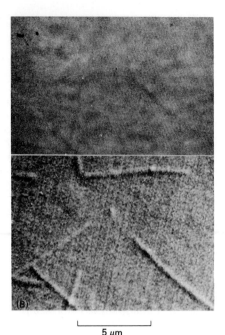

5 μm

Figure 4–11   Light-microscope images of unstained microtubules that have been visualized by differential-interference-contrast microscopy followed by electronic image processing. (A) shows the original unprocessed image; (B) shows the final result of an electronic process that greatly enhances contrast and reduces "noise." Although the microtubules are only 0.025 μm in diameter, they appear as much wider filaments because of diffraction effects. (Courtesy of Bruce Schnapp.)

Figure 4–12   Electron micrograph of a thin layer of gold showing the individual atoms as bright spots. The distance between adjacent gold atoms is about 0.2 nm (2 Å). (Courtesy of Graham Hills.)

Table 4–2 Major Events in the Development of the Electron Microscope and Its Applications to Cell Biology

1897	**J. J. Thomson** announced the existence of negatively charged particles, later termed *electrons*.
1924	**de Broglie** proposed that a moving electron has wavelike properties.
1926	**Busch** proved that is was possible to focus a beam of electrons with a cylindrical magnetic lens, laying the foundations of electron optics.
1931	**Ruska** and colleagues built the first transmission electron microscope.
1935	**Knoll** demonstrated the feasibility of the scanning electron microscope; three years later a prototype instrument was built by **Von Ardenne.**
1939	**Siemens** produced the first commercial transmission electron microscope.
1944	**Williams** and **Wyckoff** introduced the metal shadowing technique.
1945	**Porter, Claude,** and **Fullam** used the electron microscope to examine cells in tissue culture after fixing and staining them with OsO_4.
1948	**Pease** and **Baker** reliably prepared thin sections (0.1 to 0.2 μm thick) of biological material.
1952	**Palade, Porter,** and **Sjöstrand** developed methods of fixation and thin sectioning that enabled many intracellular structures to be seen for the first time. In one of the first applications of these techniques, **H. E. Huxley** showed that skeletal muscle contains overlapping arrays or protein filaments, supporting the "sliding filament" hypothesis of muscle contraction.
1953	**Porter and Blum** developed the first widely accepted ultramicrotome, incorporating many features introduced by **Claude** and **Sjöstrand** previously.
1956	**Glauert** and associates showed that the epoxy resin Araldite was a highly effective embedding agent for electron microscopy. **Luft** introduced another embedding resin, Epon, five years later.
1957	**Robertson** described the trilaminar structure of the cell membrane, seen for the first time in the electron microscope.
1957	Freeze-fracture techniques, initially developed by **Steere**, were perfected by **Moor and Mühlethaler**. Later (1966), **Branton** demonstrated that freeze-fracture allows the interior of the membrane to be visualized.
1959	**Singer** used antibodies coupled to ferritin to detect cellular molecules in the electron microscope.
1959	**Brenner and Horne** developed the negative staining technique, invented four years previously by **Hall,** into a generally useful technique for visualizing viruses, bacteria, and protein filaments.
1963	**Sabatini, Bensch, and Barrnett** introduced glutaraldehyde (usually followed by OsO_4) as a fixative for electron microscopy.
1965	**Cambridge Instruments** produced the first commercial scanning electron microscope.
1968	**de Rosier and Klug** described techniques for the reconstruction of three-dimensional structures from electron micrographs.
1975	**Henderson and Unwin** determined the first structure of a membrane protein by computer-based reconstruction from electron micrographs of unstained samples.
1979	**Heuser, Reese,** and colleagues developed a high-resolution, deep-etching technique based on very rapid freezing.

In overall design, the **transmission electron microscope (TEM)** is similar to a light microscope, although it is much larger and upside down (Figure 4–13). The source of illumination is a filament or cathode that emits electrons at the top of a cylindrical column about two meters high. Since electrons are scattered by collisions with air molecules, air must first be pumped out of the column to create a vacuum. The electrons are then accelerated from the filament by a nearby anode and allowed to pass through a tiny hole to form an electron beam that travels down the column. Magnetic coils placed at intervals along the column focus the electron beam, just as glass lenses focus the light in a light microscope. The specimen is put into the vacuum, through an airlock, into the path of the electron beam. Some of the electrons passing through the specimen are scattered according

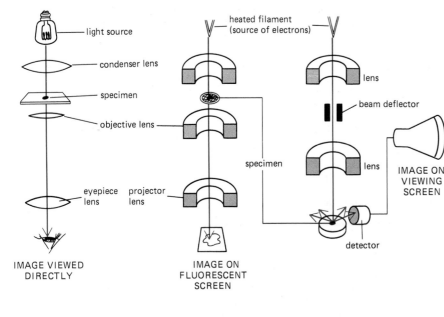

LIGHT
MICROSCOPE

TRANSMISSION
ELECTRON MICROSCOPE

SCANNING
ELECTRON MICROSCOPE

to the local density of the material; the remainder are focused to form an image—in a manner analogous to the way an image is formed in a light microscope—either on a photographic plate or on a phosphorescent screen. Because the scattered electrons are lost from the beam, the dense regions of the specimen show up in the image as areas of reduced electron flux.

Biological Specimens Require Special Preparation for the Electron Microscope[7]

In the early days of its application to biological materials, the electron microscope revealed many previously unimagined structures in cells. But before these discoveries could be made, electron microscopists had to develop new procedures for embedding, cutting, and staining tissues.

Since any specimen will be exposed to a very high vacuum in the electron microscope, there is no possibility of viewing it in the living, wet state. Tissues are usually preserved by fixation—first with *glutaraldehyde*, which covalently cross-links protein molecules to their neighbors, and then with *osmium tetroxide*, which binds to and stabilizes lipid bilayers as well as proteins (Figure 4–14). Since electrons have very limited penetrating power, the fixed tissues normally have to be cut into extremely thin sections (50 to 100 nm thick—about 1/200 of the thickness of a single cell) before they are viewed. This is achieved by dehydrating the specimen and permeating it with a monomeric resin that polymerizes to form a solid block of plastic; the block is then cut with a fine glass or diamond knife on a special microtome. These *thin sections*, free of water and other volatile solvents, are placed on a small circular metal grid for viewing in the microscope (Figure 4–15).

Contrast in the electron microscope depends on the atomic number of the atoms in the specimen: the higher the atomic number, the more electrons are scattered and the greater is the contrast. Biological molecules are composed of atoms of very low atomic number (mainly carbon, oxygen, nitrogen, and hydrogen). To make them visible, they are usually impregnated (before or after sectioning) with the salts of heavy metals such as osmium, uranium, and lead. Different cellular constituents are revealed with various degrees of contrast according to their degree of impregnation, or "staining," with these salts. For example, lipids tend to stain darkly with osmium, revealing the location of cell membranes (Figure 4–16).

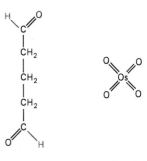

glutaraldehyde   osmium tetroxide

Figure 4–14 Glutaraldehyde and osmium tetroxide are common fixatives used for electron microscopy. The two reactive aldehyde groups of glutaraldehyde enable it to cross-link various types of molecules, forming covalent bonds between them. Osmium tetroxide is reduced by many organic compounds with which it forms cross-linked complexes. It is especially useful for fixing cell membranes since it reacts with the C=C double bonds present in many fatty acids.

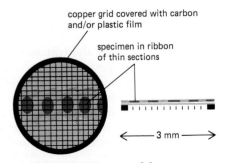

copper grid covered with carbon and/or plastic film

specimen in ribbon of thin sections

3 mm

Figure 4–15 Diagram of the copper grid used to support the thin sections of a specimen in the transmission electron microscope.

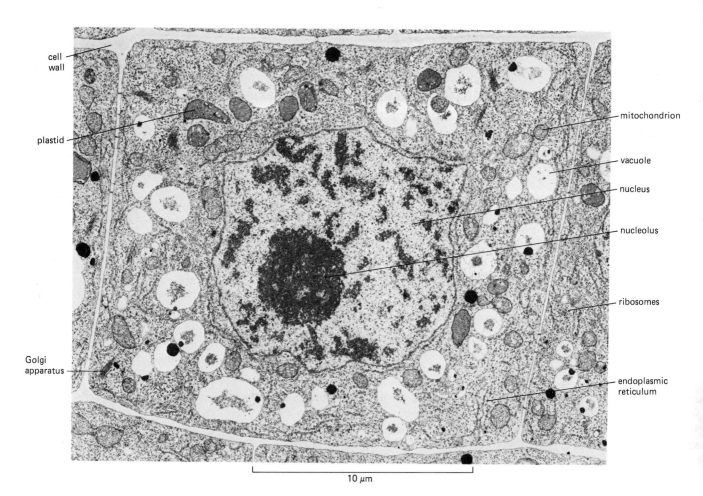

cell
wall

plastid

Golgi
apparatus

mitochondrion

vacuole

nucleus

nucleolus

ribosomes

endoplasmic
reticulum

10 μm

In some cases specific macromolecules can be located in thin sections by techniques adapted from light microscopy. Certain enzymes in cells can be detected by incubating the specimen with a substrate whose reaction leads to the local deposition of an electron-dense precipitate (Figure 4–17). Alternatively, as discussed on page 177, antibodies can be coupled to an indicator enzyme (usually peroxidase) or to an electron-dense marker (usually tiny spheres of metallic gold, which are referred to as *colloidal gold* particles) and then used to locate the macromolecules that the antibodies recognize.

Figure 4–16 Thin section of a root tip cell from a grass, stained with osmium and other heavy metal ions. The cell wall, nucleus, vacuoles, mitochondria, endoplasmic reticulum, Golgi apparatus, and ribosomes are easily seen. (Courtesy of Brian Gunning.)

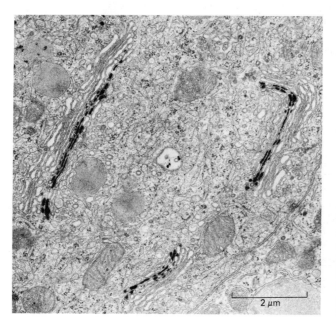

2 μm

Figure 4–17 Electron micrograph of a cell showing the location of a particular enzyme (nucleotide diphosphatase) in the Golgi apparatus. A thin section of the cell was incubated with a substrate that formed an electron-dense precipitate upon reaction with the enzyme. (Courtesy of Daniel S. Friend.)

Three-dimensional Images of Surfaces Can Be Obtained by Scanning Electron Microscopy[8]

Thin sections are effectively two-dimensional slices of tissue and fail to convey the three-dimensional arrangement of cellular components. Although the third dimension can be reconstructed from hundreds of serial sections (Figure 4–18), this is a lengthy and tedious process.

Fortunately, there are more direct means to obtain a three-dimensional image. One is to examine a specimen in a **scanning electron microscope (SEM),** which is usually a smaller and simpler device than a transmission electron microscope. Whereas the transmission electron microscope uses the electrons that have passed through the specimen to form an image, the scanning electron microscope uses electrons that are scattered or emitted from the specimen's surface. The specimen to be examined is fixed, dried, and coated with a thin layer of heavy metal. The specimen is then scanned with a very narrow beam of electrons. The quantity of electrons scattered or emitted as this primary beam bombards each successive point of the metallic surface is measured and used to control the intensity of a second beam, which moves in synchrony with the primary beam and forms an image on a television screen. In this way, a highly enlarged image of the surface as a whole is built up.

The SEM technique provides a tremendous depth of focus; moreover, since the amount of electron scattering depends on the angle of the surface relative to the beam, the image has highlights and shadows that give it a three-dimensional appearance (Figure 4–19). Only surface features can be examined, however, and in most forms of SEM the resolution attainable is not very high (about 10 nm, with an effective magnification of up to 20,000 times). As a result, the technique is usually used to study whole cells and tissues rather than subcellular organelles.

To a limited degree, a three-dimensional image of the interior of a cell can be obtained from conventionally stained thin sections by tilting the specimen in the electron beam of a transmission electron microscope and photographing it from two different angles. When the resulting pair of micrographs is examined through

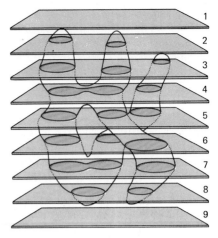

Figure 4–18 Single thin sections sometimes give misleading impressions. In this example, most sections through a cell containing a branched mitochondrion will appear to contain two or three separate mitochondria. Sections 4 and 7, moreover, might be interpreted as showing a mitochondrion in the process of dividing. However, the true three-dimensional shape can be reconstructed from serial sections.

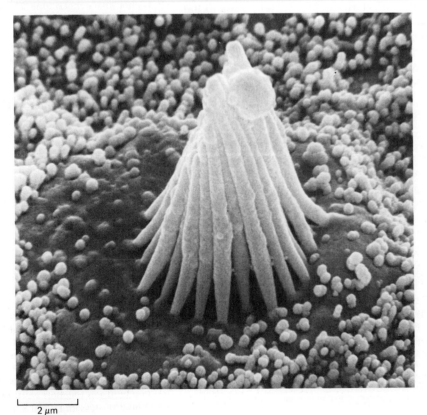

2 μm

Figure 4–19 Scanning electron micrograph of the organ pipe-like arrangement of stereocilia projecting from the surface of hair cells in the inner ear. (Courtesy of R. Jacobs and A.J. Hudspeth.)

stereo glasses, a three-dimensional image is simulated. The depth of specimen that can be examined in this way depends on the penetrating power of the electrons and hence on their energy. For this reason, **high-voltage electron microscopes** have been built that accelerate the electron beam through 1,000,000 V rather than 100,000 V. These giant machines allow one to examine sections that are several micrometers thick by transmission electron microscopy.

Metal Shadowing Allows Surface Features to Be Examined at High Resolution by Transmission Electron Microscopy[9]

The transmission electron microscope can be used to study the surface of a specimen at very high magnification, allowing individual macromolecules to be seen. As for scanning electron microscopy, a thin film of a heavy metal such as platinum is evaporated onto the dried specimen. The metal is sprayed on from an oblique angle, so as to deposit a coating that is thicker in some places than others—a process known as **shadowing** because a shadow effect is created that gives the image a three-dimensional appearance.

Some specimens coated in this way are thin enough or small enough for the electron beam to penetrate them directly; this is the case for individual molecules, viruses, and cell walls (Figure 4–20). But for thicker specimens the organic material of the cell must be dissolved away after shadowing so that only the thin metal *replica* of the surface of the specimen is left. The replica is reinforced with a film of carbon so that it can be placed on a grid and examined in the transmission electron microscope in the ordinary way (Figure 4–21).

Freeze-Fracture and Freeze-Etch Electron Microscopy Provide Unique Views of the Cell Interior[10]

Two methods that use metal replicas have been particularly useful in cell biology. One of these, **freeze-fracture** electron microscopy, provides a way of visualizing the interior of cell membranes. Cells are frozen at the temperature of liquid nitrogen ($-196°C$) in the presence of a *cryoprotectant* (antifreeze) to prevent distortion from ice crystal formation, and then the frozen block is cracked with a knife blade. The fracture plane often passes through the hydrophobic middle of lipid bilayers, thereby exposing the interior of cell membranes. The resulting fracture faces are shadowed with platinum, the organic material is dissolved away, and the replicas are floated off and viewed in the electron microscope (as in Figure 4–21). Such replicas are studded with small bumps, called *intramembrane particles*, which represent large membrane proteins. The technique provides a convenient and dramatic way to visualize the distribution of such proteins in the plane of a membrane (Figure 4–22).

Another important and related replica method is **freeze-etch** electron microscopy, which can be used to examine either the exterior or interior of cells. In this technique the cells are again frozen at a very low temperature, and the frozen block is cracked with a knife blade. But now the ice level is lowered around the cells (and to a lesser extent within the cells) by the sublimation of water in a vacuum as the temperature is raised (a process called *freeze-drying*) (Figure 4–23). The parts of the cell exposed by this *etching* process are then shadowed as before to make a platinum replica.

Cryoprotectants cannot be used in freeze-etching since they are nonvolatile and remain in the specimen as the water sublimes. Therefore, to obtain high-resolution images by this technique, one must prevent the formation of large ice crystals by freezing the specimen extremely rapidly (at a cooling rate greater than $20°C$ per millisecond). One way of achieving such *rapid freezing* is by using a special device to slam the sample against a copper block cooled to $-269°C$ with liquid helium. Particularly impressive results are achieved if rapidly frozen cells are *deep-etched* by prolonged freeze-drying. This technique exposes structures in the interior of the cell, revealing their three-dimensional organization with exceptional clarity (Figure 4–24).

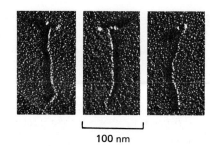

100 nm

Figure 4–20 Electron micrographs of individual myosin protein molecules that have been shadowed with platinum. Myosin is a major component of the contractile apparatus of muscle; as shown here, it is composed of two globular head regions linked to a common rodlike tail. (Courtesy of Arthur Elliot.)

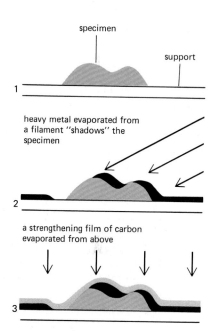

specimen

support

1

heavy metal evaporated from a filament "shadows" the specimen

2

a strengthening film of carbon evaporated from above

3

the replica is floated onto the surface of a powerful solvent to dissolve away the specimen

4

the replica is washed and picked up on a copper grid for examination

5

Figure 4–21 Preparation of a metal-shadowed replica of the surface of a specimen. Note that the thickness of the metal reflects the surface contours of the original specimen.

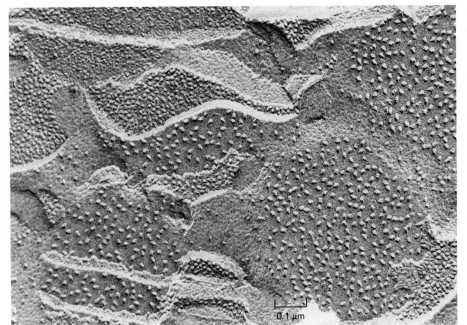

Figure 4–22 Freeze-fracture electron micrograph of the thylakoid membranes from the chloroplast of a plant cell. The thylakoid membranes, which carry out photosynthesis, are stacked up in multiple layers (see p. 367). The plane of the fracture has moved from layer to layer, passing through the middle of each lipid bilayer and exposing transmembrane proteins that have sufficient bulk in the interior of the bilayer to cast a shadow and show up as intramembrane particles in this platinum replica. The largest particles seen in the membrane are the complete photosystem II—a complex of multiple proteins. (Courtesy of L.A. Staehelin.)

Because a metal shadowed replica rather than the sample itself is viewed under vacuum in the microscope, both freeze-fracture and freeze-etch microscopy can be used to study frozen unfixed cells, thereby avoiding the risk of artifacts caused by fixation.

Negative Staining and Cryoelectron Microscopy Allow Macromolecules to Be Viewed at High Resolution[11]

Although isolated macromolecules, such as DNA or large proteins, can be visualized readily in the electron microscope if they are shadowed with a heavy metal to provide contrast (see Figure 4–20), finer detail can be seen by using **negative staining.** Here the molecules, supported on a thin film of carbon (which is nearly transparent to electrons), are washed with a concentrated solution of a heavy metal salt such as uranyl acetate. After the sample has dried, a very thin film of metal salt covers the carbon film everywhere except where it has been excluded by the presence of an adsorbed macromolecule. Because the macromolecule allows electrons to pass much more readily than does the surrounding heavy metal

Figure 4–23 Freeze-etch electron microscopy begins when the frozen specimen is fractured with a knife (A). The ice level is then lowered by the sublimation of water in a vacuum, exposing structures in the cell that were near the fracture plane (B). Following these steps, a replica of the still-frozen surface is prepared (as described in Figure 4–21) and examined in the transmission electron microscope.

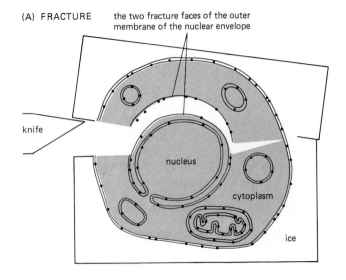

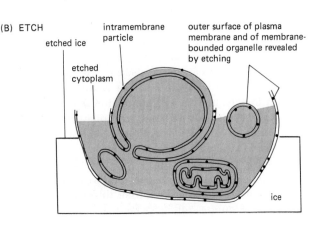

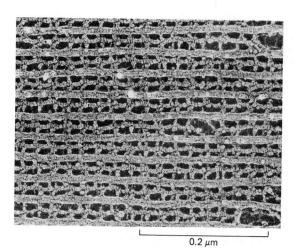

Figure 4–24 Regular array of protein filaments in an insect muscle. To obtain this image, the muscle cells were rapidly frozen in liquid helium, fractured through the cytoplasm, and subjected to deep etching. A metal replica was then prepared and examined at high magnification. (Courtesy of Roger Cooke and John Heuser.)

0.2 μm

stain, a reversed or negative image of the molecule is created. Negative staining is especially useful for viewing large macromolecular aggregates such as viruses or ribosomes and for seeing the subunit structure of protein filaments (Figure 4–25).

Both negative staining and shadowing are capable of providing high-contrast surface views of small macromolecular assemblies; but both are limited in resolution by the size of the metal particles in the shadow or stain, which only roughly outline the surface of a molecule or macromolecular assembly. Recent methods provide an alternative that has allowed even the interior features of three-dimensional structures such as viruses to be visualized directly at high resolution. In this technique, called **cryoelectron microscopy,** a very thin (~100 nm) layer of rapidly frozen hydrated sample is prepared on a microscope grid. A special sample holder is required to keep this hydrated specimen at −160°C in the vacuum of the microscope, where it can be viewed directly without fixation, staining, or drying. The homogeneity of the vitrified water layer and the use of underfocus phase contrast allow surprisingly clear images to be obtained of these unstained specimens (Figure 4–26).

Regardless of the method used, a single protein molecule gives only a weak and ill-defined image in the electron microscope. Efforts to get better information by prolonging the time of inspection or by increasing the intensity of the illuminating beam are self-defeating because they damage and disrupt the object under examination. To discover the details of molecular structure, therefore, it is necessary to combine the information obtained from many molecules in such a way as to average out the random errors in the individual images. This is possible for viruses or protein filaments, in which the individual subunits are present in regular repeating arrays (see p. 118); it is also possible for any substance that can be made to form a crystalline array in two dimensions in which large numbers of molecules are held in identical orientation and in regularly spaced positions. Given an electron micrograph of either type of array, one can use image-processing techniques to compute the average image of an individual molecule, revealing details obscured by the random "noise" in the original picture.

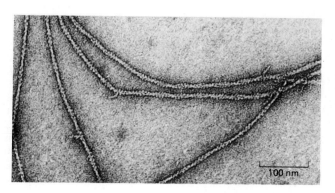

Figure 4–25 Electron micrograph of negatively stained actin filaments. Each filament is about 8 nm in diameter and is seen, on close inspection, to be composed of a helical chain of globular actin molecules. (Courtesy of Roger Craig.)

100 nm

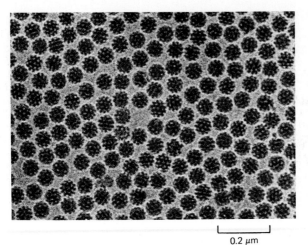

Figure 4–26 Semliki forest virus in a thin layer of unstained vitrified water viewed by cryoelectron microscopy at −160°C. A series of such micrographs are taken at different known degrees of underfocus to make use of phase contrast. A large number of these images can then be combined by image processing methods to produce a three-dimensional image at high resolution. (Courtesy of Jacques Dubochet; see also S.D. Fuller, *Cell* 48:923–934, 1987.)

Image reconstructions of this type have allowed the interior structure of an enveloped virus to be obtained to a resolution of 3.5 nm and have given the shape of an individual protein molecule to a resolution somewhat less than 0.5 nm (5 Å). But even in its most sophisticated forms, electron microscopy falls short of providing a full description of molecular structure, because the atoms in a molecule are separated by distances of only 0.1 or 0.2 nm. To resolve molecular structure in atomic detail, another technique is needed, using x-rays rather than electrons.

The Structure of an Object in a Crystalline Array Can Be Deduced from the Diffraction Pattern It Creates[12]

Like light, x-rays are a form of electromagnetic radiation, but because of their much shorter wavelength, they allow resolution of much finer detail. Unlike visible light or beams of electrons, however, x-rays cannot be focused to form an image of the usual sort after passing through a specimen. Instead, the structure of the specimen is deduced by the technique of **x-ray diffraction.**

Consider a single object (such as a single molecule) placed in a beam of radiation of any sort whose wavelength is small compared to the dimensions of the object. The object will scatter some of the radiation. The scattered radiation can be thought of as consisting of a family of overlapping waves, each emanating from a different part of the object. As the waves overlap, they undergo interference, producing a distribution of radiation known as a **diffraction pattern.** The diffraction pattern could be recorded on a photographic plate placed at some distance from the object and described in terms of the amounts of scattered radiation sent out by the object in different directions (Figure 4–27). The structure of the object determines the form of the diffraction pattern. Conversely, given a full description of the diffraction pattern, it is possible, in theory, to calculate the

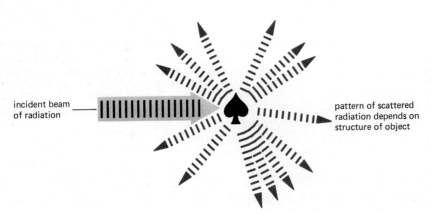

incident beam of radiation

pattern of scattered radiation depends on structure of object

Figure 4–27 The scattering of radiation by a single object whose dimensions are comparable to the wavelength of the radiation. Radiation falling on the object is scattered with different intensities in different directions. The intensity of the scattered beam in a given direction depends on the way in which radiation scattered from one part of the object interferes with that scattered from another part. In the diagram, the resultant intensity of scattering in the various possible directions is indicated by the number of colored arrows radiating from the object.

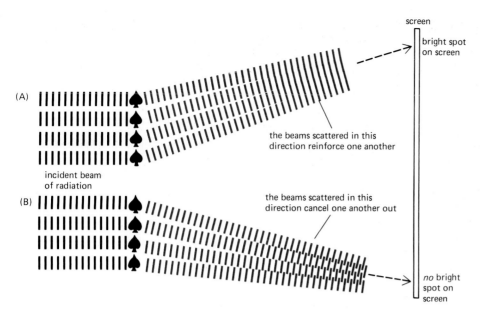

(A)

bright spot
on screen

the beams scattered in this
direction reinforce one another

incident beam
of radiation

(B)

the beams scattered in this
direction cancel one another out

no bright
spot on
screen

Figure 4–28 The scattering of radiation by a crystal. When many identical objects are arranged in a crystalline array, the radiation scattered by each object interferes with that scattered by the others. Only in certain directions (depending on the spacing of the objects in the array) do the individual scattered beams reinforce, producing bright spots in the diffraction pattern of the array. The intensity of each bright spot varies according to the intensity with which an individual object in the array would scatter radiation in that direction if the object were examined in isolation (see Figure 4–27).

structure of the object that produced it. In practice, the diffraction pattern due to a single molecule would be far too faint and erratic for such a purpose.

Suppose now that many identical objects are arranged in a crystalline array and again illuminated with a beam of radiation (Figure 4–28). Now the total amount of scattered radiation is much greater. However, the radiation scattered by each object will interfere with that scattered by the others. Only in certain directions, depending on the spacing of the objects in the array, will the individual scattered beams reinforce, producing a bright spot in the diffraction pattern. The complete diffraction pattern of the crystalline array will thus consist of many such discrete bright spots of differing intensities (Figure 4–29).

The relative intensities of the various spots in the diffraction pattern will depend on the scattering properties of the individual objects in the array. In fact, the intensity of a given spot is proportional to the average intensity of the radiation that would be scattered in the same direction from a representative single object standing alone. Thus, while the *positions* of the spots in the diffraction pattern depend on the arrangement of objects in the array, the *intensities* of the spots give information as to the internal structure of a representative single object. This information, moreover, is precise and plentiful because it is obtained by combining contributions from a very large number of equivalent sources. In fact, from a full description of the diffraction pattern of the crystalline array, it is possible to calculate the structure of the individual objects from which it is built.

X-ray Diffraction Reveals the Three-dimensional Arrangement of the Atoms in a Molecule[13]

If diffraction patterns are to be used to analyze molecular structure, the diffracted radiation must have a wavelength shorter than the distance between the atoms in a molecule. Since x-rays can have wavelengths of around 0.1 nm (which is the diameter of a hydrogen atom), they are ideally suited for investigating the arrangement of individual atoms in molecules—something that cannot be done with even the highest-resolution electron microscope. A further virtue of x-rays is that they have greater penetrating power than electrons; thus much thicker specimens can be used. Finally, because a vacuum is not required, thick hydrated specimens can be examined, and the distortions induced by the preparative procedures necessary for most forms of electron microscopy are thereby avoided.

For high-resolution studies, large, highly ordered crystals are required (Figure 4–30). The x-rays that penetrate the crystal are scattered chiefly by the electrons in its constituent atoms. Large atoms with many electrons scatter x-rays more than do small atoms, so atoms of C, N, O, and P are much more readily detected

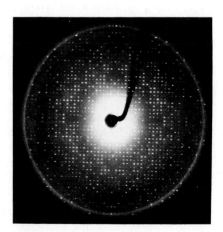

Figure 4–29 Part of the x-ray diffraction pattern obtained from a protein crystal. This particular crystal was used to help determine the atomic structure of the proteolytic enzyme trypsin. (Courtesy of Robert Stroud.)

than atoms of H; for the same reason, very heavy atoms create an intense beacon of scattered x-rays. Deducing the three-dimensional structure of a large molecule from the diffraction pattern of its crystal is a complex task and was not achieved for a protein molecule until 1960 (Table 4–3). It frequently requires measuring the positions and intensities of hundreds of thousands of spots, as well as the phases of the waves at each spot (which requires comparing the diffraction pattern of crystals with and without heavy atoms bound to specific sites on the molecule).

In recent years, x-ray diffraction analysis has become increasingly automated. The diffracted x-rays are measured electronically by sophisticated detectors that greatly speed the process of data collection, and powerful computers perform the huge numbers of calculations required. The slowest step now is obtaining suitable crystals of the macromolecule of interest; this requires large amounts of the pure macromolecule and often involves years of trial-and-error searching for the proper crystallization conditions.

Despite the difficulties, x-ray diffraction is a widely used technique because it is the only way to determine the detailed arrangement of atoms in most molecules. With reasonably good crystals, the structure of a protein can be calculated to a resolution of 0.3 nm, revealing the main course of the polypeptide chain but few other details. With very high-quality crystals (and much more work), a resolution of 0.15 nm is obtainable, revealing the position of almost all of the nonhydrogen atoms in the protein. The structures of more than a hundred proteins and of several small RNA and DNA molecules have now been determined in this way.

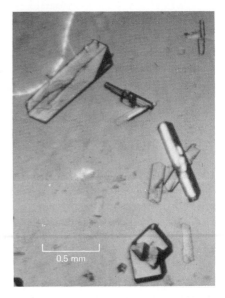

Figure 4–30 Crystals of the enzyme glycogen phosphorylase viewed in a light microscope. (Courtesy of Robert Fletterick.)

Table 4–3 Landmarks in the Development of X-ray Crystallography and Its Application to Biological Molecules

1864	**Hoppe-Seyler** crystallized, and named, the protein hemoglobin.
1895	**Röntgen** observed that a new form of penetrating radiation, which he named x-rays, was produced when cathode rays (electrons) hit a metal target.
1912	**Von Laue** obtained the first x-ray diffraction patterns by passing x-rays through a crystal of zinc sulfide.
	W. L. Bragg proposed a simple relationship between an x-ray diffraction pattern and the arrangement of atoms in a crystal that produces the pattern.
1926	**Summer** obtained crystals of the enzyme urease from extracts of jack beans and demonstrated that proteins possess catalytic activity.
1931	**Pauling** published his first essays on "The Nature of the Chemical Bond," detailing the rules of covalent bonding.
1934	**Bernal and Crowfoot** presented the first detailed x-ray diffraction patterns of a protein obtained from crystals of the enzyme pepsin.
1935	**Patterson** developed an analytical method for determining interatomic spacings from x-ray data.
1941	**Astbury** obtained the first x-ray diffraction pattern of DNA.
1951	**Pauling and Corey** proposed the structure of a helical conformation of a chain of L-amino acids—the α helix—and the structure of the β sheet, both of which were later found in many proteins.
1953	**Watson and Crick** proposed the double-helix model of DNA, based on x-ray diffraction patterns obtained by **Franklin and Wilkins.**
1954	**Perutz** and colleagues developed heavy-atom methods to solve the phase problem in protein crystallography.
1960	**Kendrew** described the first detailed structure of a protein (sperm whale myoglobin) to a resolution of 0.2 nm, and **Perutz** proposed a lower-resolution structure of the larger protein hemoglobin.
1966	**Phillips** described the structure of lysozyme, the first enzyme to be analyzed in detail.
1976	**Kim and Rich** and **Klug** and colleagues described the detailed three-dimensional structure of tRNA determined by x-ray diffraction.
1977–1978	**Holmes** and **Klug** determined the structure of tobacco mosaic virus (TMV), and **Harrison** and **Rossman** determined the structure of two small spherical viruses.

Summary

Many light microscopic techniques are available for observing cells. Stained fixed cells can be studied with conventional optics, while labeled antibodies can be used to locate specific molecules in cells with the fluorescence microscope. Cells in their natural living state can be seen with phase-contrast, interference, or dark-field optics. Such studies of living cells by light microscopy are facilitated by electronic image-processing techniques, which greatly enhance sensitivity and increase resolution.

Determining the detailed structure of the membranes and organelles in cells requires the higher resolution attainable in the transmission electron microscope. The shapes of isolated macromolecules that have been shadowed with a heavy metal or outlined by negative staining can also be visualized readily by electron microscopy. But the precise position of each atom in a molecule can be determined only if the molecule will assemble into large crystals, in which case the complete three-dimensional structure of the molecule can be deduced by x-ray diffraction analysis.

Probing Chemical Conditions in the Interior of Living Cells

The classical methods of microscopy give good views of cell architecture but not much information about cell chemistry. We have seen that antibodies can be used to locate specific macromolecules in cells, but it is also important to be able to investigate the distribution and concentration of the small molecules. The concentrations of fundamental metabolites such as ATP or glucose and of inorganic ions must be regulated precisely and rapidly in order to maintain life, and they can be very different in different regions of cells and tissues. Moreover, small molecules such as cyclic AMP, Ca^{2+}, and H^+ function as intracellular "messengers," and it is important to be able to monitor changes in their concentrations in response to extracellular signals. In this section we discuss some methods adapted from conventional chemistry that allow the chemical conditions inside a cell to be determined while the cell is alive.

Nuclear Magnetic Resonance (NMR) Can Be Used to Assay the Chemistry of Populations of Living Cells[14]

The nuclei of many atoms have a magnetic moment: that is, they have an intrinsic magnetization, like bar magnets. Because their magnetic behavior is influenced by surrounding atoms, these nuclei can be made to reveal the chemical nature of their environment by a method called **nuclear magnetic resonance (NMR)** spectroscopy, which is harmless to living cells. When atomic nuclei with a magnetic moment are placed in a strong magnetic field, they will adopt one of a limited number of permitted orientations. Each orientation will have an energy that depends on the strength of the field and on the chemical environment. If a set of atoms in identical chemical environments are irradiated with radio waves, the energy of these waves will be strongly absorbed when the waves have a precisely defined frequency that corresponds to the energy difference between two such nuclear orientations. This is the so-called *resonance frequency*. A sample of tissue, containing atoms in a variety of different molecules and environments, will absorb energy at a variety of different resonance frequencies. The graph of absorbance against resonance frequency for a given sample constitutes its *NMR spectrum*. This spectrum reflects the structure and relative quantities of each of the different molecules present that contain magnetic nuclei.

In chemistry laboratories, NMR is a routine analytical technique for determining the molecular structure of small molecules in solution. Improvements in instrumentation have made it possible to apply NMR methods to problems of biological interest. For example, the NMR signal from protons (hydrogen nuclei) is widely used to investigate proteins, nucleic acids, and other macromolecules in

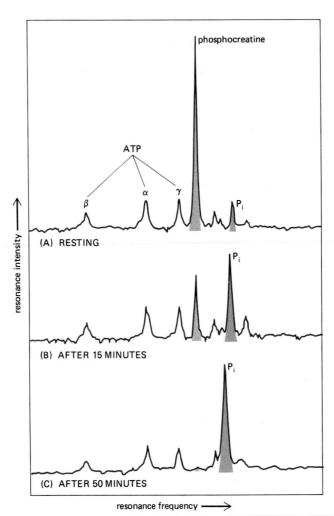

Figure 4–31 A ^{31}P NMR spectrum recorded from a frog muscle (A) in its resting condition, (B) after 15 minutes of stimulated activity under anaerobic conditions, and (C) after 50 minutes of such activity. The five labeled peaks in the spectrum represent signals from ^{31}P nuclei in different atomic environments: one peak corresponds to phosphocreatine, while α, β, and γ are the three phosphate groups in ATP, and P_i is inorganic phosphate. Phosphocreatine is at a high concentration in the resting muscle (A); it serves as an intermediate store of free energy in skeletal muscle because its phosphate is readily transferred to ADP to replenish the ATP hydrolyzed in muscle contraction (see p. 618). Thus in the fatigued muscle the store of phosphocreatine has been depleted while the concentration of inorganic phosphate (derived from ATP) has correspondingly risen. The P_i peak also shifts slightly to the left, reflecting a drop in the intracellular pH due to the accumulation of lactic acid, a by-product of anaerobic metabolism. From the position of the P_i peak compared with that of a known standard, it can be deduced that the pH has changed from 7.5 in (A) to 6.4 in (C). (Adapted by permission from M.J. Dawson, D.G. Gadian, and D.R. Wilkie, *Nature* 274:861–866, 1978. Copyright © 1978. Macmillan Magazines Limited.)

solution: the interactions between the parts of a macromolecule affect the NMR spectrum, which therefore contains detailed information about molecular structure and molecular movements. NMR, unlike x-ray diffraction, does not require a crystallized sample. However, because the molecules in solution must be tumbling very rapidly if they are to give an informative NMR spectrum, there is an upper limit (of about 20,000 daltons) to the size of the macromolecules whose conformations can be analyzed effectively.

Only a limited number of atoms have isotopes that will give a suitable NMR signal. The abundant isotopes ^{1}H, ^{23}Na, ^{31}P, and ^{39}K and the rare isotopes ^{13}C and ^{15}N are all useful for probing the molecules that are in solution inside living cells. Because of the central metabolic role of phosphorus compounds, ^{31}P NMR is used frequently. This isotope is naturally present in the phosphorus compounds in cells, and so its signals can be used to monitor changes in the intracellular concentrations of molecules such as ATP and inorganic phosphate during muscle contraction. NMR signals from ^{31}P can also be used to measure intracellular pH accurately, since the resonance frequency of inorganic phosphate depends on its state of ionization and hence on the pH of the solution (Figure 4–31).

The rare isotopes ^{13}C and ^{15}N are not normally found in cells in appreciable amounts. However, they can be introduced into specific molecules of biological interest, whose chemical transformations can then be followed by NMR. If ^{13}C glucose is fed to cells, for example, the rates of many of the reactions that glucose undergoes can be measured by determining the NMR spectrum of a sample of living cells as a function of time. By using other ^{13}C-labeled and ^{15}N-labeled compounds, one can, in principle, monitor the flow of carbon and nitrogen atoms along any metabolic pathway.

The main limitation of NMR is that it is a relatively insensitive technique. For example, for a substance to be detected by ^{31}P NMR in 1 g of living tissue with current techniques, it must be present at a concentration greater than about 0.2 mM. Many metabolites occur at concentrations below this limit. Moreover, since it generally takes several minutes to obtain an NMR spectrum, rapid changes in cell chemistry may be difficult to follow. On the other hand, NMR has the great virtue that it is completely noninvasive and harmless to living cells. It is this that makes it promising as an analytical tool for cell biology.

Ion Concentrations Can Be Measured with Intracellular Electrodes[15]

Methods that are more sensitive than NMR must be used to investigate cells individually. One approach is based on a technique developed by electrophysiologists to study voltages and current flows across the plasma membrane. For this purpose, intracellular *microelectrodes* are made from pieces of fine glass tubing that are pulled to a tip diameter of a fraction of a micrometer and filled with a conducting solution (usually a simple salt, such as KCl, in water). The tip of the microelectrode can be poked into the cytoplasm through the plasma membrane, which seals around the shaft, adhering tightly to the glass so that the cell is left relatively undisturbed.

Microelectrodes are useful for studying the cell interior in two ways: they can be converted into probes for measuring intracellular concentrations of common inorganic ions, such as H^+, Na^+, K^+, Cl^-, Ca^{2+}, and Mg^{2+}; and they can be used as micropipettes to inject molecules into cells. The principle that allows measurement of ion concentrations with a microelectrode is the same as that of the familiar pH meter. The tendency of ions to diffuse down their concentration gradient can be balanced by an opposing electric field: the greater the concentration gradient, the greater is the electric field required. The magnitude of the electric field required to maintain the concentration gradient as a stable equilibrium condition therefore gives a measure of the ion concentration gradient. To find the concentration of a specific ion, one must take a material that is permeable only to that ion, form the material into a sheet or barrier, and place it between the test solution and a solution that contains a known concentration of the ion. The electrical potential difference across the selectively permeable barrier when no current flows will then be a direct measure of the ratio of concentrations of the specific ion on the two sides. (The theory is discussed on pages 314–316 in connection with ion transport across the plasma membrane.) In practice, the tip of a microelectrode is filled with an appropriate organic compound to make a barrier that is selectively permeable to the chosen ion. The microelectrode is then inserted, together with a reference microelectrode, into the interior of a cell, as shown in Figure 4–32.

Figure 4–32 An ion-selective microelectrode can be used to measure the intracellular concentration of a specific ion. (A) shows the experimental arrangement. (B) illustrates the construction of a microelectrode selectively permeable to K^+. In general, the tip of the ion-sensitive intracellular microelectrode is either constructed from special glass or filled with a special organic compound to make it permeable only to a chosen ion. The rest of the tube is filled with an aqueous solution of the ion at a known concentration, with a metal conductor connected to one terminal of a voltmeter dipping into it. The other terminal of the voltmeter is connected in a similar way to a glass reference microelectrode that has an open tip and contains a simple conducting solution. The two microelectrodes are both poked through the plasma membrane of the cell to be studied. The voltage measured by the voltmeter, which equals the potential difference across the selectively permeable barrier, reveals the concentration of the ion in the cell (see text). Large cells are generally quite easy to impale with microelectrodes; cells smaller than about 10 μm in diameter are difficult to examine in this way.

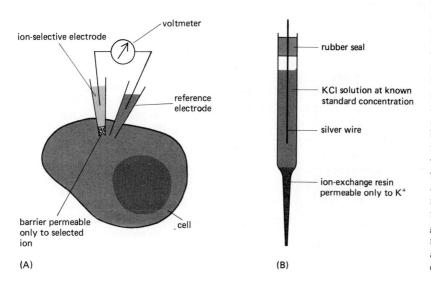

(A)   (B)

More recently, the microelectrode technique has been adapted to the study of ion transport through specialized channel proteins (called ion channels) contained in a small patch of plasma membrane. In this case, a glass microelectrode with a somewhat larger tip is pressed gently against the plasma membrane instead of being poked through it (Figure 4–33). It is then possible to study the electrical behavior of the small patch of membrane covering the tip of the electrode, either attached or free from the cell (Figure 4–34). This technique, known as **patch recording,** has revolutionized the study of ion channels. It is the only technique in cell biology that allows one to study the function of a single protein molecule in real time, as we shall discuss in Chapter 6.

Rapidly Changing Intracellular Ion Concentrations Can Be Measured with Light-emitting Indicators[16]

Ion-sensitive electrodes reveal the ion concentration only at one point in a cell, and for an ion that is present at very low concentration, such as Ca^{2+}, their responses are slow and somewhat erratic. Transient changes in the intracellular concentration of Ca^{2+} are very important in the response of cells to extracellular signals, and such changes can be analyzed with the use of intracellular indicators whose light emission reflects the local concentration of the ion. Some of these indicators are luminescent (emitting light spontaneously), while others are fluorescent (emitting light on exposure to light). Thus *aequorin* is a luminescent protein isolated from certain marine jellyfish; it emits light in the presence of Ca^{2+} and responds to changes in the Ca^{2+} concentration in the range 0.5 to 10 μM. If microinjected into an egg, for example, aequorin emits a flash of light in response to a sudden localized release of free Ca^{2+} into the cytoplasm that occurs when the egg is fertilized (Figure 4–35).

Fluorescent Ca^{2+} indicators have recently been synthesized that bind Ca^{2+} tightly and are excited at slightly longer wavelengths when free of Ca^{2+} than when

Figure 4–33 Micropipettes used for patch recording. A rod cell from the eye of a salamander is shown held by a suction pipette while a fine-tipped glass pipette, pressed against the cell so that the glass is sealed tightly to the plasma membrane, serves as a microelectrode. (From T.D. Lamb, H.R. Matthews, and V. Torre, *J. Physiol.* 37:315–349, 1986.)

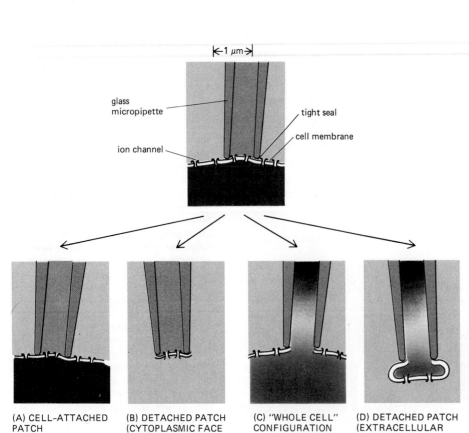

(A) CELL-ATTACHED PATCH

(B) DETACHED PATCH (CYTOPLASMIC FACE EXPOSED)

(C) "WHOLE CELL" CONFIGURATION

(D) DETACHED PATCH (EXTRACELLULAR FACE EXPOSED)

Figure 4–34 The four standard configurations used for patch recording. The mouth of the glass recording pipette is first pressed against the cell membrane so that a tight seal forms (*top*). Recordings of the current entering the pipette through the patch of membrane can then be made with the patch still attached to the cell (A) or pulled free from it, exposing the cytoplasmic surface of the plasma membrane (B). Alternatively, the patch can be ruptured by gentle suction so that the interior of the electrode communicates directly with the interior of the cell (C); in this latter "whole-cell" configuration, one can record the electrical behavior of the cell in the same way as with an intracellular electrode, with the added option that the internal chemistry of the cell can be altered by allowing substances to diffuse out of the relatively wide recording pipette into the cytoplasm. Configuration (D) is reached via configuration (C) by pulling the pipette away from the cell, thereby causing a fragment of the adjacent plasma membrane to fold back over the tip to form a seal. In (D) the exterior surface rather than the cytoplasmic surface of the membrane is exposed [compare with (B)].

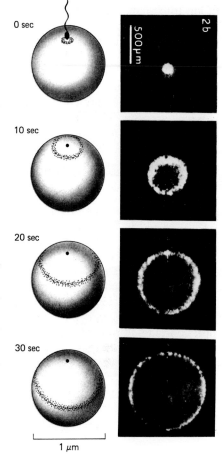

Figure 4–35 The luminescent protein aequorin emits light in the presence of free Ca^{2+}. Here an egg of the medaka fish has been injected with aequorin, which has diffused throughout the cytosol, and the egg has then been fertilized with a sperm and examined with the help of an image intensifier. The four photographs shown were taken at intervals of 10 seconds, looking down on the site of sperm entry, and reveal a wave of release of free Ca^{2+} into the cytosol from internal stores just beneath the plasma membrane. This wave sweeps across the egg starting from the site of sperm entry, as indicated in the diagrams on the left. (Photographs reproduced from J.C. Gilkey, L.F. Jaffe, E.B. Ridgway, and G.T. Reynolds, *J. Cell Biol.* 76:448–466, 1978. By copyright permission of the Rockefeller University Press.)

in their Ca^{2+}-bound form. By measuring the ratio of fluorescence intensity at two excitation wavelengths, the concentration ratio of the Ca^{2+}-bound indicator to the Ca^{2+}-free indicator can be determined; this provides an accurate measurement of the free Ca^{2+} concentration. Two popular indicators of this type, called *quin-2* and *fura-2*, are useful for second-by-second monitoring of changes in intracellular Ca^{2+} concentrations in the different parts of a cell viewed in a fluorescence microscope. Similar fluorescent indicators are available for measuring intracellular pH. Some of these indicators can enter cells by diffusion and so need not be microinjected; this makes it possible to monitor large numbers of individual cells simultaneously in a fluorescence microscope. By constructing new types of indicators and using them in conjunction with modern image-processing methods, it should be possible to develop similarly rapid and precise methods for analyzing changes in the concentrations of many different small molecules in cells.

There Are Several Ways of Introducing Membrane-impermeant Molecules into Cells[17]

It is often useful to be able to introduce membrane-impermeant molecules into a cell, whether they be light-emitting indicators (such as aequorin), normal cell proteins tagged with a fluorescent label, or molecules that influence cell behavior. One approach is to microinject the molecules into the cell through a glass micro-pipette. In an especially useful technique, a purified protein is coupled to a fluorescent dye and then microinjected into a cell; this allows the fate of the particular protein to be followed in a fluorescence microscope as the cell grows and divides (Figure 4–36).

Microinjection, although widely used, demands that each cell be injected individually; therefore it is possible to study at most a few hundred cells at a time. Other approaches allow large populations of cells to be permeabilized simultaneously. For example, one can partially disrupt the structure of the cell plasma

Figure 4–36 Fluorescence micrographs of a portion of an early *Drosophila* embryo that has been injected with rhodamine-labeled tubulin, the protein subunit of microtubules. The microtubules become labeled throughout the entire embryo because all of the nuclei share a common cytoplasm at this early stage of development. (A) The microtubules in the living embryo emanate from two bright spots on either side of each interphase nucleus; at the center of each spot is a centrosome. (B) The same embryo viewed a few minutes later, when all of the nuclei have synchronously entered mitosis. The microtubules still emanate from the two centrosomes, but they have now reorganized to form the mitotic spindle. (Courtesy of Douglas Kellogg.)

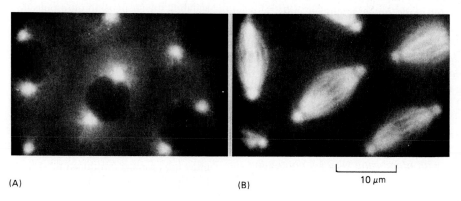

(A)

(B)

10 μm

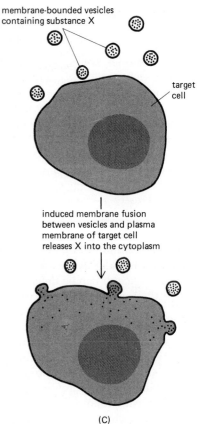

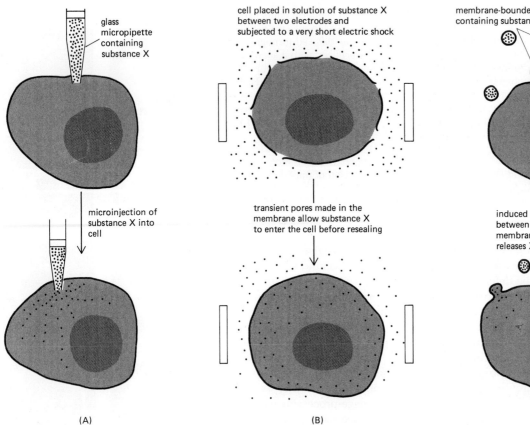

glass micropipette containing substance X	cell placed in solution of substance X between two electrodes and subjected to a very short electric shock	membrane-bounded vesicles containing substance X

target cell

microinjection of substance X into cell

transient pores made in the membrane allow substance X to enter the cell before resealing

induced membrane fusion between vesicles and plasma membrane of target cell releases X into the cytoplasm

(A)　　　　　　　　　　(B)　　　　　　　　　　(C)

Figure 4–37 Three methods used to introduce a membrane-impermeant substance, such as a protein, into a cell. In (A) the substance is injected through a micropipette, either by applying pressure or, if the substance is electrically charged, by applying a voltage that causes the substance to pass into the cell as an ionic current (a technique called *iontophoresis*). In (B) the cell membrane is made transiently permeable to the substance by disrupting the membrane structure with a brief but intense electric shock (for example, 2000 volts per centimeter for 200 microseconds). In (C) membrane fusion is used. Membrane-bounded vesicles can be loaded with a desired substance by mixing a concentrated solution of the substance with a suspension of phospholipids and agitating the suspension ultrasonically to create *liposomes* (see p. 278). Alternatively, red blood cells can be converted into loaded membrane-bounded vesicles by bursting them to remove their natural contents and then allowing their plasma membranes to reseal while immersed in a solution of the desired compound. Either the liposomes or these "red cell ghosts" can be induced to fuse with the target cells by the presence of certain viral fusion-inducing proteins (produced by viruses to help them get into cells).

membrane so as to make it more permeable, using a powerful electric shock or a chemical such as a low concentration of detergent. The electrical technique has the advantage of creating large pores in the plasma membrane without damaging intracellular membranes. The pores remain open for minutes or hours, depending on the cell type and size of the electric shock; even macromolecules can enter (and leave) the cytosol rapidly through the pores. With a limited treatment, a large fraction of the cells repair their plasma membrane and survive. A third method for introducing large molecules into cells is to cause membranous vesicles that contain these molecules to fuse with the cell's plasma membrane. These three methods are used widely in cell biology and are illustrated in Figure 4–37.

Summary

Measurements of the concentration and distribution of inorganic ions and other small molecules in cells generally must be made on intact living cells. Nuclear magnetic resonance provides a completely noninvasive technique for measuring the relative concentrations of many small molecules, but it requires a large sample. Fluorescent indicator dyes can be used to measure the concentration of specific ions in individual cells, and even in different parts of a cell. Glass microelectrodes, besides being indispensable for studying the electrical potentials and ionic currents across the plasma membrane, provide an alternative way of measuring the concentrations of specific intracellular ions; they can also be used as micropipettes to inject membrane-impermeant molecules into living cells. Alternatively, these molecules can be delivered to the cytosol by making the plasma membrane transiently permeable or by allowing the cell to fuse with membrane vesicles containing the molecules.

Cell Separation and Culture[18]

Although the structure of organelles and large molecules in a cell can be seen with microscopes, and specific molecules can be located by staining them, a molecular understanding of a cell requires detailed biochemical analysis. Unfortunately, biochemical procedures require cells in bulk and begin by disrupting them. If the sample is a piece of tissue, fragments of all its cells will be mixed together, creating confusion if the cells are of several different types, which is almost always the case. In order to preserve as much information as possible about each individual type of cell, cell biologists have developed ways of dissociating cells from tissues and separating the various cell types. The resulting, relatively homogenous population of cells then can be analyzed—either directly or after their number has been greatly increased by allowing the cells to proliferate in culture.

Cells Can Be Isolated from a Tissue and Separated into Different Types[19]

The first step in isolating cells of a uniform type from a tissue that contains a mixture of cell types is to reduce the tissue to a suspension of single cells. This is done by disrupting the extracellular matrix and intercellular junctions that hold the cells together. The best yields of viable dissociated cells are usually obtained from fetal or neonatal tissues, typically by treating them with proteolytic enzymes (such as trypsin and collagenase) and with agents that bind, or *chelate*, Ca^{2+} (such as ethylenediaminetetraacetic acid, or EDTA), on which cell-to-cell adhesion depends. The tissue can then be dissociated into single cells by gentle mechanical disruption.

Several approaches are used to separate the different cell types from a mixed cell suspension. One involves exploiting the differences in the physical properties of cells. For example, large cells can be separated from small cells and dense cells from light cells by centrifugation; the techniques will be described when we discuss the separation of organelles and macromolecules, for which they were originally developed. Another approach is based on the tendency of some cell types to adhere strongly to glass or plastic; these can be separated from cells that adhere less strongly.

An important refinement of this last technique depends on the specific binding properties of antibodies. Antibodies that bind specifically to the surface of only one cell type in a tissue can be coupled to various matrices—such as collagen, polysaccharide beads, or plastic—to form an "affinity surface" to which only cells recognized by the antibodies will adhere. The bound cells are then recovered by gentle shaking or, in the case of a digestible matrix (such as collagen), by degrading the matrix with enzymes (such as collagenase).

The most sophisticated cell-separation technique involves labeling specific cells with antibodies coupled to a fluorescent dye and then separating the labeled cells from the unlabeled ones in an electronic **fluorescence-activated cell sorter.** Here, individual cells traveling in single file in a fine stream are passed through a laser beam and the fluorescence of each cell is measured. Slightly farther downstream, tiny droplets, most containing either one or no cells, are formed by a vibrating nozzle. The droplets containing a single cell are automatically given a positive or a negative charge at the moment of formation, depending on whether the cell they contain is fluorescent; they are then deflected by a strong electric field into an appropriate container. Occasional clumps of cells, detected by their increased light scattering, are left uncharged and are discarded into a waste container (Figure 4–38). Such machines can select 1 cell in 1000 and sort about 5000 cells each second.

When a uniform population of cells has been obtained by any of the above methods, it can be used directly for biochemical analysis. Alternatively, it provides a suitable starting material for cell culture, allowing the complex behavior of cells to be studied under the strictly defined conditions of a culture dish.

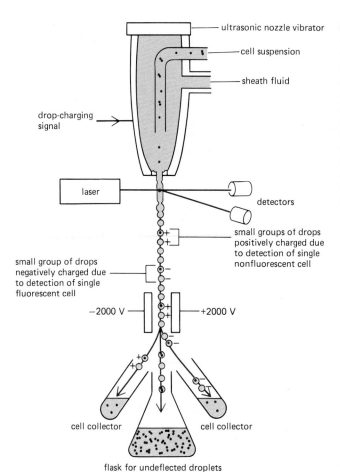

ultrasonic nozzle vibrator

cell suspension

sheath fluid

drop-charging signal

laser

detectors

small groups of drops positively charged due to detection of single nonfluorescent cell

small group of drops negatively charged due to detection of single fluorescent cell

−2000 V   +2000 V

cell collector   cell collector

flask for undeflected droplets

Figure 4–38 A fluorescence-activated cell sorter. When a cell passes through the laser beam, it is monitored for fluorescence. Droplets containing single cells are given a negative or positive charge, depending on whether the cell is fluorescent or not. Droplets are then deflected by an electric field into collection tubes according to their charge. Note that the cell concentration must be adjusted so that most droplets contain no cell; most of these flow to a waste container together with any cell clumps.

Cells Can Be Grown in a Culture Dish[20]

Given the appropriate conditions, most kinds of plant and animal cells will live, multiply, and even express differentiated properties in a tissue-culture dish. The cells can be watched under the microscope or analyzed biochemically, and the effects of adding or removing specific molecules such as hormones or growth factors can be explored. In addition, in a mixed culture the interactions between one cell type and another can be studied. Experiments on cultured cells are sometimes said to be carried out *in vitro* (literally, "in glass"); by contrast, experiments on intact organisms are said to be carried out *in vivo*. These two terms are used in a very different sense by most biochemists and cell biologists—for whom *in vitro* refers to biochemical reactions occurring outside living cells, while *in vivo* refers to any reaction taking place inside a living cell.

Tissue culture began in 1907 with an experiment designed to settle a controversy in neurobiology. The hypothesis under examination was known as the *neuronal doctrine*, which states that each nerve fiber is the outgrowth of a single nerve cell and not the product of the fusion of many cells. To test this contention, small pieces of spinal cord were placed on clotted tissue fluid in a warm, moist chamber and observed at regular intervals under the microscope. After a day or so, individual nerve cells could be seen extending long, thin processes into the clot. Thus the neuronal doctrine was validated, and the foundations for the cell-culture revolution were laid.

The original experiments in 1907 involved the culture of small tissue fragments, or **explants.** Today, cultures are more commonly made from suspensions of cells dissociated from tissues as described above. Unlike bacteria, most tissue cells are not adapted to living in suspension and require a solid surface on which to grow and divide. This mechanical support was originally provided by a clot of tissue fluid but is now usually the surface of a plastic tissue-culture dish (Figure

4–39). Cells vary in their requirements, however, and some will not grow or differentiate unless the culture dish is coated with specific extracellular matrix components, such as collagen.

Cultures prepared directly from the tissues of an organism, either with or without an initial cell fractionation step, are called **primary cultures.** In most cases, cells in primary cultures can be removed from the culture dish and used to form a large number of **secondary cultures;** they may be repeatedly subcultured in this way for weeks or months. Such cells often display many of the differentiated properties appropriate to their origin: fibroblasts continue to secrete collagen; cells derived from embryonic skeletal muscle fuse to form giant muscle fibers that spontaneously contract in the culture dish; nerve cells extend axons that are electrically excitable and make synapses with other nerve cells; and epithelial cells form extensive sheets with many of the properties of an intact epithelium. Since these phenomena occur in culture, they are accessible to study in ways that are not possible in intact tissues.

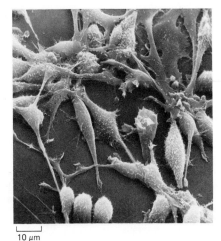

10 μm

Figure 4–39 Scanning electron micrograph of rat fibroblasts growing on a plastic surface in tissue culture. (Courtesy of Guenter Albrecht-Buehler.)

Chemically Defined Media Permit Identification of Specific Growth Factors[21]

Until the early 1970s, tissue culture was something of a blend of science and witchcraft. Although tissue fluid clots were replaced by dishes of liquid media containing well-defined mixtures of salts, amino acids, and vitamins, most media also included a proportion of some poorly defined biological material, such as horse serum or fetal calf serum or a crude extract made from chick embryos. Such media are still used today for most routine tissue culture (Table 4–4), but they are inappropriate for determining the specific requirements for the growth and differentiation of a particular type of cell.

This difficulty led to the development of various chemically defined media that support the growth of different types of cells. In these media, each component is a known molecule. In addition to small molecules, these media usually contain one or more of the various protein **growth factors** that most cells require in order to survive and proliferate in culture: for example, certain nerve cells need trace amounts of *nerve growth factor (NGF)* to differentiate and survive in culture as

Table 4–4 Composition of a Typical Medium Suitable for the Cultivation of Mammalian Cells

Amino Acids	Vitamins	Salts	Miscellaneous
Arginine	Biotin	NaCl	Glucose
Cystine	Choline	KCl	Penicillin
Glutamine	Folate	NaH_2PO_4	Streptomycin
Histidine	Nicotinamide	$NaHCO_3$	Phenol red
Isoleucine	Pantothenate	$CaCl_2$	Whole serum
Leucine	Pyridoxal	$MgCl_2$	
Lysine	Thiamine		
Methionine	Riboflavin		
Phenylalanine			
Threonine			
Tryptophan			
Tyrosine			
Valine			

Glucose is used at a concentration of 5 to 10 mM. The amino acids are all in the L form and, with one or two exceptions, are used at concentrations of 0.1 or 0.2 mM; vitamins are used at a 100-fold lower concentration, that is, about 1 μM. Serum, which is usually from horse or calf, is added to make up 10% of the total volume. Penicillin and streptomycin are antibiotics added to suppress the growth of bacteria. Phenol red is a pH indicator dye whose color is monitored to assure a pH of about 7.4.

Cultures are usually grown in a plastic or glass container with a suitably prepared surface that allows the attachment of cells. The containers are kept in a incubator at 37°C in an atmosphere of 5% CO_2, 95% air.

well as in the intact animal. Many other factors of this kind that play a vital role in the development and/or proliferation of specific cell types have also been discovered, and the search for new ones has been made very much easier by the availability of chemically defined media.

Eucaryotic Cell Lines Are a Widely Used Source of Homogeneous Cells[18]

Most vertebrate cells die after a finite number of divisions in culture: for example, human skin cells typically last for several months in culture, dividing only 50 to 100 times before they die. It has been suggested that this limited life-span is related to the limited life-span of the animal from which the cells are derived. Occasionally, however, variant cells arise in culture that are effectively immortal. Such cells can be propagated indefinitely as a **cell line** (Table 4–5). They usually grow best when attached to solid surfaces and typically cease growing when they have formed a continuous, or *confluent*, layer over the surface of the culture dish.

Cell lines prepared from cancer cells differ from those prepared from normal cells in several ways; for example, cancer cell lines often grow without attaching to a surface, and they proliferate to a very much higher density in a culture dish. Similar properties can be experimentally induced in normal cells by *transforming* them with a tumor-inducing virus or a chemical. The resulting **transformed cell lines** are capable of causing tumors if injected into an animal. Both transformed and untransformed cell lines are extremely useful in cell research as sources of very large numbers of cells of a uniform type, especially since they can be stored at −70°C for an indefinite period and are still viable when thawed. However, it is important to recognize that the cells in both types of cell lines nearly always differ in important ways from their normal ancestors in the tissues from which they were derived.

The genetic uniformity of a cell line can be improved by **cell cloning**, that is, by isolating a single cell and allowing it to proliferate to form a large colony. A *clone* is any such collection of cells that are all descendants of a single ancestor cell. One of the most important uses of cell cloning has been the isolation of mutant cell lines with defects in specific genes. Studying cells that are defective in a specific protein often reveals a good deal about the function of that protein in normal cells.

Cells Can Be Fused Together to Form Hybrid Cells[22]

It is possible to fuse one cell with another to form a combined cell with two separate nuclei, called a **heterocaryon.** The fusion is usually carried out by treating a suspension of cells with certain inactivated viruses or with polyethylene glycol, either of which alters the plasma membranes of cells in a way that induces them to fuse with each other. Heterocaryons provide a way of mixing the components of two separate cells in order to study their interactions. For example, the inert nucleus of a chicken red blood cell is reactivated to make RNA, and eventually to replicate its DNA, when it is exposed to the cytoplasm of a growing tissue-culture cell by fusion. And the first direct evidence that membrane proteins are able to move in the plane of the plasma membrane came from an experiment in which mouse cells and human cells were fused: although the mouse and human cell-surface proteins were initially confined to their own halves of the heterocaryon plasma membrane, they quickly diffused and mixed over the entire surface of the cell.

Eventually a heterocaryon will proceed to mitosis and produce a **hybrid cell** in which the two separate nuclear envelopes have been disassembled, allowing all the chromosomes to be brought together in a single large nucleus (Figure 4–40). Although such hybrid cells can be cloned to produce hybrid cell lines, the initial hybrid cells tend to be unstable and lose chromosomes. For unknown reasons, mouse-human hybrid cells predominantly lose human chromosomes. These chromosomes are lost at random, giving rise to a variety of different mouse-human hybrid cell lines, each of which contains only one or a few human chromosomes.

Table 4–5 Some Commonly Used Cell Lines

Cell Line	Cell Type and Origin
3T3	fibroblast (mouse)
BHK 21	fibroblast (Syrian hamster)
HeLa	epithelial cell (human)
PtK 1	epithelial cell (rat kangaroo)
L 6	myoblast (rat)
PC 12	chromaffin cell (rat)
SP 2	plasma cell (mouse)

Many of these cell lines were derived from tumors. All of them are capable of indefinite replication in culture and express at least some of the differentiated properties of their cell of origin. BHK 21 cells, HeLa cells, and SP 2 cells are capable of growth in suspension; the other cell lines require a solid culture substratum in order to multiply.

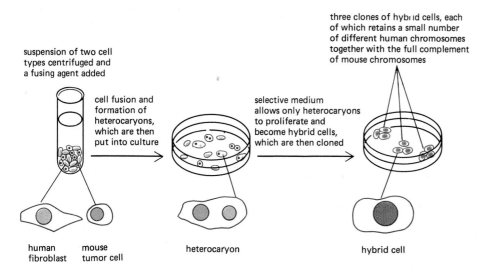

suspension of two cell types centrifuged and a fusing agent added

cell fusion and formation of heterocaryons, which are then put into culture

selective medium allows only heterocaryons to proliferate and become hybrid cells, which are then cloned

three clones of hybrid cells, each of which retains a small number of different human chromosomes together with the full complement of mouse chromosomes

human fibroblast   mouse tumor cell

heterocaryon

hybrid cell

Figure 4–40 How human cells and mouse cells are fused to produce heterocaryons (each with two or more nuclei) that lead eventually to hybrid cells (each with one fused nucleus). These particular hybrid cells are useful for mapping human genes on specific human chromosomes because most of the human chromosomes are quickly lost in a random manner, leaving clones that retain only one or a few. The hybrid cells produced by fusing other types of cells often retain most of their chromosomes.

This phenomenon has been put to good use in mapping the locations of genes in the human genome. For example, only hybrid cells containing human chromosome 11 synthesize human insulin, indicating that the gene encoding insulin is located on chromosome 11.

Some important steps in the development of tissue culture are outlined in Table 4–6.

Summary

Cells isolated from fetal or neonatal tissues serve as the usual starting material for the purification of individual cell types, which can be analyzed biochemically or used to establish cell cultures. Many animal and plant cells survive and proliferate in a culture dish if they are provided with a suitable nutrient medium; different cell types require different nutrients, including one or more specific protein growth factors. Although most animal cells die after a finite number of divisions, rare "immortal" variant cells arise spontaneously in culture and can be maintained indefinitely as cell lines. Clones derived from a single ancestor cell make it possible to isolate uniform populations of mutant cells with defects in a single protein. Two different types of cell can be fused to produce heterocaryons (cells with two nuclei), which eventually form hybrid cells (with one fused nucleus). Such cells are useful for studying the interactions between the components of two different cells and provide a convenient method for assigning genes to specific chromosomes.

Fractionation of Cell Contents[23]

Although biochemical analysis requires disruption of the delicate anatomy of the cell, gentle separation techniques have been devised that preserve the function of the various cell components. Just as a tissue can be separated into its living constituent cell types, so the cell can be separated into its functioning organelles and macromolecules. In this section we shall focus on the methods that allow organelles and proteins to be purified. Related methods for tracing and labeling molecules with radioisotopes and antibodies, as well as the powerful techniques available for analyzing DNA and gene function, are discussed in later sections.

Organelles and Macromolecules Can Be Separated by Ultracentrifugation[24]

Cells can be disrupted in various ways: they can be subjected to osmotic shock or ultrasonic vibration, or forced through a small orifice, or ground up. These procedures break many of the membranes of the cell (including the plasma membrane and membranes of the endoplasmic reticulum) into fragments that immediately reseal to form small closed vesicles. If carefully applied, however, the dis-

Table 4–6 Some Landmarks in the Development of Tissue and Cell Culture

1885	**Roux** showed that embryonic chick cells could be maintained alive in a saline solution outside the animal body.
1907	**Harrison** cultivated amphibian spinal cord in a lymph clot, thereby demonstrating that axons are produced as extensions of single nerve cells.
1910	**Rous** induced a tumor by using a filtered extract of chicken tumor cells, later shown to contain an RNA virus (Rous sarcoma virus).
1913	**Carrel** showed that cells could grow for long periods in culture provided they were fed regularly under aseptic conditions.
1948	**Earle** and colleagues isolated single cells of the L cell line and showed that they formed clones of cells in tissue culture.
1952	**Gey** and colleagues established a continuous line of cells derived from human cervical carcinoma, which later became the well-known HeLa cell line.
1954	**Levi-Montalcini** and associates showed that nerve growth factor (NGF) stimulated the growth of axons in tissue culture.
1955	**Eagle** made the first systematic investigation of the essential nutritional requirements of cells in culture and found that animal cells could propagate in a defined mixture of small molecules supplemented with a small proportion of serum proteins.
1956	**Puck** and associates selected mutants with altered growth requirements from cultures of HeLa cells.
1958	**Temin** and **Rubin** developed a quantitative assay for the infection of chick cells in culture by purified Rous sarcoma virus. In the following decade, the characteristics of this and other types of viral transformation were established by **Stoker, Dulbecco, Green,** and other virologists.
1961	**Hayflick and Moorhead** showed that human fibroblasts die after a finite number of divisions in culture.
1964	**Littlefield** introduced HAT medium for the selective growth of somatic cell hybrids. Together with the technique of cell fusion, this made somatic-cell genetics accessible.
	Kato and Takeuchi obtained a complete carrot plant from a single carrot root cell in tissue culture.
1965	**Ham** introduced a defined, serum-free medium able to support the clonal growth of certain mammalian cells.
	Harris and Watkins produced the first heterocaryons of mammalian cells by the virus-induced fusion of human and mouse cells.
1968	**Augusti-Tocco and Sato** adapted a mouse nerve cell tumor (neuroblastoma) to tissue culture and isolated clones that were electrically excitable and that extended nerve processes. A number of other differentiated cell lines were isolated at about this time, including skeletal-muscle and liver cell lines.
1975	**Köhler and Milstein** produced the first monoclonal antibody-secreting hybridoma cell lines.
1976	**Sato** and associates published the first of a series of papers showing that different cell lines require different mixtures of hormones and growth factors to grow in serum-free medium.

ruption procedures leave organelles such as nuclei, mitochondria, the Golgi apparatus, lysosomes, and peroxisomes largely intact. The suspension of cells is thereby reduced to a thick soup (homogenate) containing a variety of membrane-bounded particles, each with a distinctive size, charge, and density. Provided that the homogenization medium has been carefully chosen (by trial and error for each organelle), the various particles—including the vesicles derived from the endoplasmic reticulum, called microsomes—retain most of their original biochemical properties.

The various components of the homogenate must then be separated. This became possible only after the commercial development in the early 1940s of an instrument known as the **preparative ultracentrifuge,** in which preparations of broken cells are rotated at high speeds (Figure 4–41). This treatment separates cell components on the basis of size: in general, the largest units experience the largest centrifugal force and move the most rapidly. At relatively low speed, large components such as nuclei and unbroken cells sediment to form a pellet at the bottom of the centrifuge tube; at slightly higher speed, a pellet of mitochrondria is de-

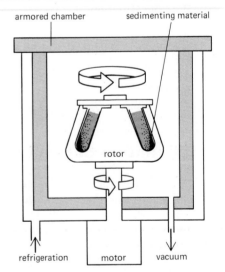

Figure 4–41 The preparative ultracentrifuge. The sample is contained in tubes that are inserted into a ring of cylindrical holes in a metal rotor. Rapid rotation of the rotor generates enormous centrifugal forces, which cause particles in the sample to sediment. The vacuum reduces friction, preventing heating of the rotor and allowing the refrigeration system to maintain the sample at 4°C.

posited; and at even higher speeds and with longer periods of centrifugation, first the small closed vesicles and then the ribosomes can be collected (Figure 4–42). All of these fractions are impure, but many of the contaminants can be removed by resuspending the pellet and repeating the centrifugation procedure several times.

The centrifugation method is the first step in most fractionations, but it separates only components that are very different in size. A finer degree of separation can be achieved by layering the extract as a narrow band on top of a dilute salt solution that fills a centrifuge tube. When centrifuged, the various components in the mixture move as a series of distinct bands through the salt solution, each at a different rate (Figure 4–43). For the procedure to work effectively, the bands must be protected from convective mixing, which would normally occur whenever a denser solution (for example, one containing organelles) finds itself on top of a lighter one (the salt solution). This is achieved by filling the centrifuge tube with a shallow gradient of sucrose prepared by a special mixing device; the resulting *density gradient*, with the dense end at the bottom of the tube, keeps each region of the salt solution denser than any solution above it and thereby prevents convective mixing from distorting the separation.

When sedimented through such dilute sucrose gradients, different cell components separate into distinct bands that can be collected individually (see Figure 4–43). The rate at which each component sediments depends primarily on its size and shape and is normally described in terms of its *sedimentation coefficient* or *s value* (see Table 4–7). Present-day ultracentrifuges rotate at speeds up to 80,000 rpm and produce forces up to 500,000 times gravity. With these enormous forces, even small macromolecules, such as tRNA molecules and simple enzymes, separate from one another on the basis of their size. Measurements of the sedimentation coefficients of macromolecular assemblies are routinely used to help determine the total mass of the assembly and, therefore, the number of subunits of each type that it contains.

The ultracentrifuge is also used to separate cellular components on the basis of their *buoyant density*, independent of their size and shape. In this case, the sample is usually sedimented through a steep density gradient that contains a very high concentration of sucrose or cesium chloride. Each cellular component begins to move down the gradient as in Figure 4–43, but it eventually reaches a position where the density of the solution is equal to its own density. At this point the component floats and can move no farther. A series of distinct bands is thereby produced in the centrifuge tube, with the bands closest to the bottom of the tube containing the components of highest buoyant density. The method is so sensitive that it is capable of separating macromolecules that have incorporated heavy isotopes, such as ^{13}C or ^{15}N, from the same macromolecules that have not. In fact, the cesium chloride method was developed in 1957 to separate the labeled and unlabeled DNA produced after exposure of a growing population of bacteria to nucleotide precursors containing ^{15}N; this classic experiment provided direct evidence for the semiconservative replication of DNA (see p. 97).

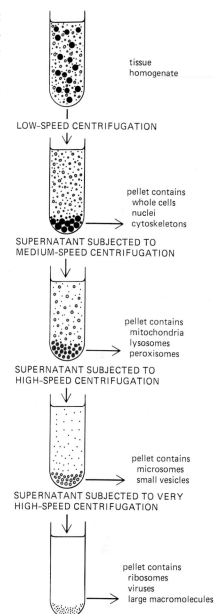

Figure 4–42 Repeated centrifugation at progressively higher speeds will fractionate extracts of cells into their components. In general, the smaller the subcellular component, the greater is the centrifugal force required to sediment it. Typical values for the various centrifugation steps referred to in the figure are

low speed:	1,000 times gravity for 10 minutes
medium speed:	20,000 times gravity for 20 minutes
high speed:	80,000 times gravity for 1 hour
very high speed:	150,000 times gravity for 3 hours

Table 4–7 Some Typical Sedimentation Coefficients

Particle or Molecule	Sedimentation Coefficient
Lysosome	9400S
Tobacco mosaic virus	198S
Ribosome	80S
Ribosomal RNA molecule	28S
tRNA molecule	4S
Hemoglobin molecule	4.5S

Sedimentation coefficients (s), in units of seconds, are given by $(dx/dt)/\omega^2x$, where x is the distance from the center of rotation in centimeters, dx/dt is the speed of sedimentation in centimeters per second, and ω is the angular rotation of the centrifuge rotor in radians per second. Because such coefficients are extremely small numbers, they are normally expressed in Svedberg units (S), where $1S = 1 \times 10^{-13}$ second.

The Molecular Details of Complex Cellular Processes Can Be Deciphered in Cell-free Systems[25]

Studies of organelles and other large subcellular components isolated in the ultracentrifuge have played an important part in determining the *functions* of different components of the cell. For example, experiments on mitochondria and chloroplasts purified by centrifugation demonstrated the central function of these organelles in energy interconversions. Similarly, resealed vesicles formed from fragments of rough and smooth endoplasmic reticulum (microsomes) have been separated from each other and analyzed as functional models of these compartments of the intact cell.

The study of isolated organelles is only the beginning. In order to establish the detailed molecular mechanisms involved in *any* biological process, the components must first be isolated so that the process can be studied free from all of the complex side reactions that occur in a cell. Fractionated cell extracts that maintain a biological function (called **cell-free systems**) are therefore widely used in cell biology. An early triumph of this approach was the elucidation of the mechanisms of protein synthesis. The starting point was a crude cell extract that could translate RNA molecules into protein. Fractionation of this extract, step by step, produced in turn the ribosomes, tRNA, and various enzymes that together constitute the protein synthetic machinery. Once individual pure components were available, they could be separately added or withheld and their exact role in the process defined. The same "*in vitro* translation system" was later used to decipher the genetic code by using synthetic polyribonucleotides of known sequence as the "messenger RNA" (mRNA) to be translated. Today, *in vitro* translation systems are used to determine how newly made proteins are sorted into various intracellular compartments (p. 439) as well as to identify the proteins encoded by purified preparations of mRNA—an important step in gene cloning procedures (see p. 265). Some landmarks in the development of methods for the preparation of fractionated cell extracts are outlined in Table 4–8.

Much of what we know about the molecular biology of the cell has been discovered by studying cell-free systems. As a few of many examples, they have been used to analyze the molecular details of DNA replication, DNA transcription, RNA splicing, muscle contraction, and particle transport along microtubules. The analysis of a cell-free system ultimately involves the complete separation of all of its individual macromolecular components and in particular all of its proteins. We shall now consider how this is achieved.

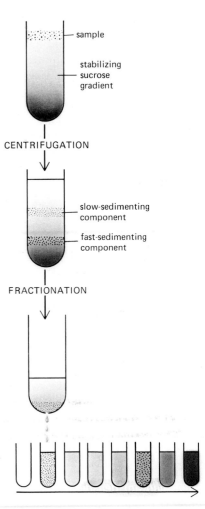

Figure 4–43 Subcellular components sediment at different speeds according to their size when layered over a dilute sucrose-containing solution. In order to stabilize the sedimenting bands against convective mixing caused by small differences in temperature or solute concentration, the tube contains a continuous gradient of sucrose that increases in concentration toward the bottom of the tube (typically from 5% to 20% sucrose). Following centrifugation, the different components may be collected individually, most simply by puncturing the plastic centrifuge tube and collecting drops from the bottom, as illustrated here.

Table 4–8 Some Major Events in the Development of the Ultracentrifuge and the Preparation of Cell-free Extracts

1897	**Buchner** showed that cell-free extracts of yeast can ferment sugars to form carbon dioxide and ethanol, laying the foundations of enzymology.
1926	**Svedberg** developed the first analytical ultracentrifuge and used it to estimate the molecular weight of hemoglobin as 68,000.
1935	**Pickels and Beams** introduced several new features of centrifuge design that led to its use as a preparative instrument.
1938	**Behrens** employed differential centrifugation to separate nuclei and cytoplasm from liver cells, a technique further developed for the fractionation of cell organelles by **Claude, Brachet, Hogeboom,** and others in the 1940s and early 1950s.
1949	**Szent-Györgyi** showed that isolated myofibrils from skeletal muscle cells contract upon the addition of ATP. In 1955, a similar cell-free system was developed for ciliary beating by **Hofmann-Berling.**
1951	**Brakke** used density-gradient centrifugation in sucrose solutions to purify a plant virus.
1954	**de Duve** isolated lysomes and, later, peroxisomes by centrifugation.
1954	**Zamecnik** and colleagues developed the first cell-free system to carry out protein synthesis. This was followed by a decade of intense research activity, during which the genetic code was elucidated.
1957	**Meselson, Stahl,** and **Vinograd** developed density-gradient centrifugation in cesium chloride solutions for separating nucleic acids.

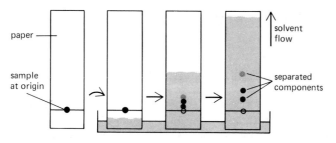

Figure 4–44 The separation of small molecules by paper chromatography. After the sample has been applied to one end of the paper (the "origin") and dried, a solution containing a mixture of two solvents is allowed to flow slowly through the paper by capillary action. Different components in the sample move at different rates in the paper according to their relative solubility in the solvent that is preferentially adsorbed by the paper. The development of this technique revolutionized biochemical analyses in the 1940s.

Proteins Can Be Separated by Chromatography[26]

One of the most generally useful methods for protein fractionation is **chromatography,** a technique originally developed to separate small molecules such as sugars and amino acids. A common type of chromatography, still widely used to separate small molecules, is *partition chromatography.* Typically, a drop of the sample is applied as a spot to a sheet of absorbent material, which may be paper (*paper chromatography*) or a sheet of plastic or glass covered with a thin layer of inert absorbent material, such as cellulose or silica gel (*thin-layer chromatography*). A mixture of solvents, such as water and an alcohol, is allowed to permeate the sheet from one edge; as the liquid moves across the sheet, it separates the molecules in the sample according to their relative solubilities in the two solvents. The solvents are selected so that one of them is held more strongly than the other by the absorbent material and forms a stationary solvent layer on the surface of the sheet. In each region of the sheet, molecules equilibrate between the stationary and moving solvents: those that are most soluble in the strongly adsorbed solvent are relatively retarded because they spend more time in the stationary layer, while those that are most soluble in the other solvent move more quickly. After a number of hours, the sheet is dried and stained to pinpoint the location of the various molecules (Figure 4–44).

Proteins are most often fractionated by *column chromatography,* in which a mixture of proteins in solution is passed through a column containing a porous solid matrix. The different proteins are retarded to different extents by their interaction with the matrix, and they can be separately collected, in their native functional state, as they flow out of the bottom of the column (Figure 4–45). According to the choice of matrix, proteins can be separated according to their

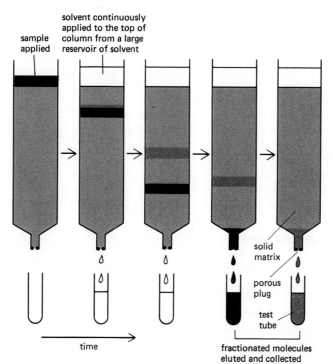

Figure 4–45 The separation of molecules by column chromatography. The sample is applied to the top of a cylindrical glass or plastic column filled with a permeable solid matrix, such as cellulose, immersed in solvent. Then a large amount of solvent is pumped slowly through the column and is collected in separate tubes as it emerges from the bottom. Various components of the sample travel at different rates through the column and are thereby fractionated into different tubes.

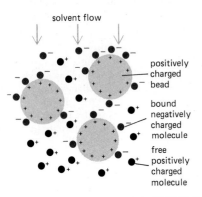

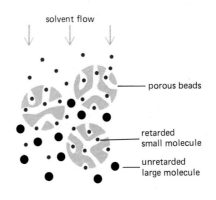

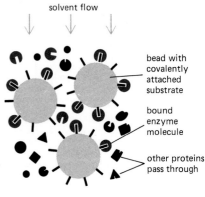

solvent flow

positively charged bead

bound negatively charged molecule

free positively charged molecule

(A) ION-EXCHANGE CHROMATOGRAPHY

solvent flow

porous beads

retarded small molecule

unretarded large molecule

(B) GEL-FILTRATION CHROMATOGRAPHY

solvent flow

bead with covalently attached substrate

bound enzyme molecule

other proteins pass through

(C) AFFINITY CHROMATOGRAPHY

charge (ion-exchange chromatography), their hydrophobicity (hydrophobic chromatography), their size (gel-filtration chromatography), or their ability to bind to particular chemical groups (affinity chromatography).

Many types of matrices are commercially available for these purposes (Figure 4–46). *Ion-exchange columns* are packed with small beads that carry either a positive or negative charge, so that proteins are fractionated according to the arrangement of charges on their surface. *Hydrophobic columns* are packed with beads from which hydrophobic side chains protrude, so that proteins with exposed hydrophobic regions are retarded. *Gel-filtration columns*, which separate proteins according to their size, are packed with tiny porous beads: molecules that are small enough to enter the pores linger inside successive beads as they pass down the column, while larger molecules remain in the solution flowing between the beads and therefore move more rapidly, emerging from the column first. Besides providing a means of separating molecules, gel-filtration chromatography is a convenient way to determine their size.

These types of column chromatography do not produce very highly purified fractions if one starts with a complex mixture of proteins: a single passage through the column generally increases the proportion of a given protein in the mixture by no more than twentyfold. Since most individual proteins represent less than 1/1000 of the total cellular protein, it is usually necessary to use several different types of column in succession to attain sufficient purity (Figure 4–47). A more efficient procedure, known as **affinity chromatography,** takes advantage of the biologically important binding interactions that occur on protein surfaces. For example, if an enzyme substrate is covalently coupled to an inert matrix such as a polysaccharide bead, the enzyme will often be specifically retained by the matrix and can then be eluted (washed out) in nearly pure form. Likewise, short DNA oligonucleotides of a specifically designed sequence (p. 191) can be immobilized in this way and used to purify DNA-binding proteins that normally recognize this sequence of nucleotides on chromosomes (see p. 489). Alternatively, specific antibodies can be coupled to a matrix in order to purify protein molecules recognized by the antibodies. Because of the great specificity of all such *affinity columns*, 1000- to 10,000-fold purifications can sometimes be achieved in a single pass.

The resolution of conventional column chromatography is limited by inhomogeneities in the matrices (such as cellulose), which cause an uneven flow of solvent through the column. Newer chromatography resins (usually silica-based) have been developed in the form of tiny spheres (3 to 10 μm in diameter) that can be packed with a special apparatus to form a uniform column bed. A high degree of resolution is attainable on such **high-performance liquid chromatography (HPLC)** columns.

Because they contain such tightly packed particles, HPLC columns have negligible flow rates unless high pressures are applied. For this reason, these columns are typically packed in steel cylinders and require an elaborate system of pumps and valves to force the solvent through them at sufficient pressure to produce the desired rapid flow rates of about one column volume per minute. In conventional column chromatography, flow rates must be kept slow (often about one column volume per hour) to give the solutes being fractionated time to equilibrate with

Figure 4–46 Three types of matrices used for chromatography. In ion-exchange chromatography (A), the insoluble matrix carries ionic charges that retard molecules of opposite charge. Matrices commonly used for separating proteins are diethylaminoethylcellulose (DEAE-cellulose), which is positively charged, and carboxymethylcellulose (CM-cellulose) and phosphocellulose, which are negatively charged. The strength of the association between the dissolved molecules and the ion-exchange matrix depends on both the ionic strength and the pH of the eluting solution, which may therefore be varied in a systematic fashion (as in Figure 4–47) to achieve an effective separation. In gel-filtration chromatography (B), the matrix is inert but porous. Molecules that are small enough to penetrate into the matrix are thereby delayed and travel more slowly through the column. Beads of cross-linked polysaccharide (dextran or agarose) are available commercially in a wide range of pore sizes, making them suitable for the fractionation of molecules of various molecular weights, from less than 500 to more than 5×10^6. Affinity chromatography (C) utilizes an insoluble matrix that is covalently linked to a specific ligand, such as an antibody molecule or an enzyme substrate, that will bind a specific protein. Enzyme molecules that bind to immobilized substrates on such columns can be eluted with a concentrated solution of the free form of the substrate molecule, while molecules that bind to immobilized antibodies can be eluted by dissociating the antibody-antigen complex with concentrated salt solutions or solutions of high or low pH. High degrees of purification are often achieved in a single pass through an affinity column.

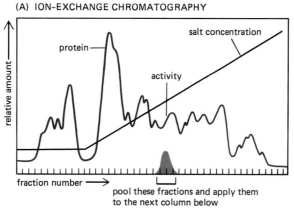

(A) ION-EXCHANGE CHROMATOGRAPHY

pool these fractions and apply them
to the next column below

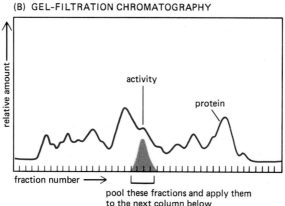

(B) GEL-FILTRATION CHROMATOGRAPHY

pool these fractions and apply them
to the next column below

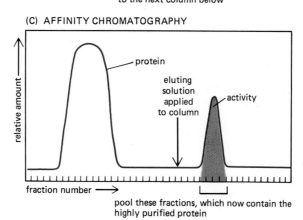

(C) AFFINITY CHROMATOGRAPHY

pool these fractions, which now contain the
highly purified protein

Figure 4–47 Typical results obtained when three different chromatographic steps are used in succession to purify a protein. In this example a whole-cell extract was first fractionated by allowing it to percolate through an ion-exchange resin packed into a column (A). The column was washed and the bound proteins were then eluted by passing a solution containing a gradually increasing concentration of salt onto the top of the column. Proteins with the lowest affinity for the ion-exchange resin passed directly through the column and were collected in the earliest fractions eluted from the bottom of the column. The remaining proteins were eluted in sequence according to their affinity for the resin—those proteins binding most tightly to the resin requiring the highest concentration of salt to remove them. The protein of interest eluted in a narrow peak and was detected by its enzymatic activity. The fractions with activity were pooled and then applied to a second, gel-filtration column (B). The elution position of the still-impure protein was again determined by its enzymatic activity and the active fractions pooled and purified to homogeneity on an affinity column (C) that contained an immobilized substrate of the enzyme.

the interior of the large matrix particles. In HPLC the solutes equilibrate very rapidly with the interior of the tiny spheres, so solutes with different affinities for the matrix are efficiently separated from each other even at fast flow rates. This allows most fractionations to be carried out in minutes, whereas hours are required to obtain a poorer separation by conventional chromatography. HPLC has therefore become the method of choice for separating many proteins and small molecules.

The Size and Subunit Composition of a Protein Can Be Determined by SDS Polyacrylamide-Gel Electrophoresis[27]

Proteins usually have a net positive or negative charge that reflects the mixture of charged amino acids they contain. If an electric field is applied to a solution containing a protein molecule, the protein will migrate at a rate that depends on its net charge and on its size and shape. This technique, known as **electrophoresis,** was originally used to separate mixtures of proteins either in free aqueous solution or in solutions held in a solid porous matrix such as starch.

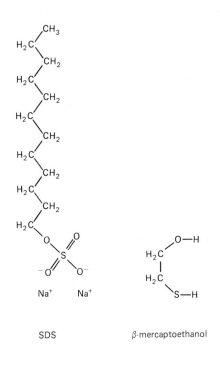

Figure 4–48 The detergent sodium dodecyl sulfate (SDS), here in its ionized form, and the reducing agent β-mercaptoethanol are two chemicals used in solubilizing proteins for SDS polyacrylamide-gel electrophoresis.

SDS

β-mercaptoethanol

In the mid-1960s, a modified version of this method—known as **SDS polyacrylamide-gel electrophoresis** (or **SDS-PAGE**)—was developed that has revolutionized the way proteins are routinely analyzed. It uses a highly cross-linked gel of polyacrylamide as the inert matrix through which the proteins migrate. The gel is usually prepared immediately before use by polymerization from monomers; the pore size of the gel can be adjusted so that it is small enough to retard the migration of the protein molecules of interest. The proteins themselves are not in a simple aqueous solution but in one that includes a powerful negatively charged detergent, **sodium dodecyl sulfate,** or **SDS** (Figure 4–48). Because this detergent binds to hydrophobic regions of the protein molecules, causing them to unfold into extended polypeptide chains, the individual protein molecules are released from their associations with other proteins or lipid molecules and rendered freely soluble in the detergent solution. In addition, a reducing agent such as *mercaptoethanol* (Figure 4–48) is usually added to break any S—S linkages in the proteins so that all of the constituent polypeptides in multisubunit molecules can be analyzed separately.

What happens when a mixture of SDS-solubilized proteins is electrophoresed through a slab of polyacrylamide gel? Each protein molecule binds large numbers of the negatively charged detergent molecules, which overwhelm the protein's intrinsic charge and cause it to migrate toward the positive electrode when a voltage is applied. Proteins of the same size tend to behave identically because (1) their native structure is completely unfolded by the SDS, so that their shapes are the same, and (2) they bind the same amount of SDS and therefore have the same amount of negative charge. Larger proteins, with more charge, will be subjected to larger electrical forces and also to a larger drag. In free solution the two effects would cancel out, but in the meshes of the polyacrylamide gel, which acts as a molecular sieve, large proteins are retarded much more severely than small ones. As a result, a complex mixture of proteins is fractionated into a series of discrete protein bands arranged in order of molecular weight. The major proteins are readily detected by staining the gel with a dye such as Coomassie blue, and even minor proteins are seen in gels treated with a silver stain (where as little as 10 ng of protein can be detected in a band). A specific protein can be identified on such gels by exposing all the proteins to a specific antibody that has been coupled to a radioactive isotope, to an enzyme, or to a fluorescent dye; this is often done after all the separated proteins present in the gel have been transferred (by "blotting") onto a sheet of nitrocellulose paper, as will be described later for nucleic acids (see p. 190). This protein-detection method is called *Western blotting*.

SDS polyacrylamide-gel electrophoresis is a more powerful procedure than any previous method of protein analysis principally because it can be used to separate all types of proteins, including those that are insoluble in water. Membrane proteins, protein components of the cytoskeleton, and proteins that are part of large macromolecular aggregates can all be resolved. Since the method separates polypeptides according to size, it also provides information about the molecular weight and the subunit composition of any protein complex (Figure 4–49). A photograph of a gel that has been used to analyze each of the successive stages in the purification of a protein is shown in Figure 4–50.

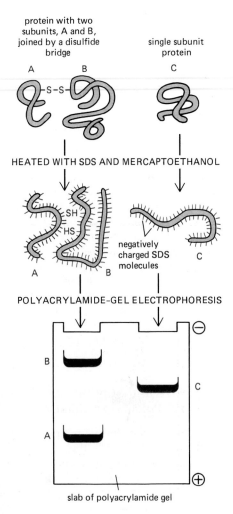

protein with two subunits, A and B, joined by a disulfide bridge

single subunit protein

HEATED WITH SDS AND MERCAPTOETHANOL

negatively charged SDS molecules

POLYACRYLAMIDE-GEL ELECTROPHORESIS

slab of polyacrylamide gel

Figure 4–49 SDS polyacrylamide-gel electrophoresis. Individual polypeptide chains form a complex with negatively charged molecules of sodium dodecyl sulfate (SDS) and therefore migrate as a negatively charged SDS-protein complex through a porous gel of polyacrylamide. Since the speed of migration under these conditions is greater the smaller the polypeptide, this technique can be used to determine the approximate molecular weight of a polypeptide chain as well as the subunit composition of a protein.

Figure 4–50 Analysis of protein samples by SDS polyacrylamide-gel electrophoresis. The photograph shows a gel that has been used to detect the proteins present at successive stages in the purification of an enzyme. The leftmost lane (lane 1) contains the complex mixture of proteins in the starting cell extract, and each succeeding lane analyzes the proteins obtained after a chromatographic fractionation of the protein sample analyzed in the previous lane (see Figure 4–47). The same total amount of protein (10 μg) was loaded onto the gel at the top of each lane. Individual proteins normally appear as sharp dye-stained bands; however, a band broadens when it contains too much protein. (Courtesy of Tim Formosa.)

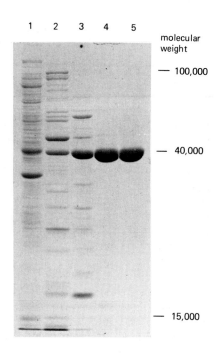

More Than 1000 Proteins Can Be Resolved on a Single Gel by Two-dimensional Polyacrylamide-Gel Electrophoresis[28]

Since closely spaced protein bands or peaks tend to overlap, one-dimensional separation methods, such as SDS polyacrylamide-gel electrophoresis or chromatography, can resolve only a relatively small number of proteins (generally fewer than 50). In contrast, **two-dimensional gel electrophoresis,** which combines two different separation procedures, can be used to resolve more than 1000 different proteins in the form of a two-dimensional protein map.

In the first step, the proteins are separated on the basis of their charge. The sample is dissolved in a small volume of a solution containing a nonionic (uncharged) detergent, together with mercaptoethanol and the denaturing reagent urea. This solution solubilizes, denatures, and dissociates all the polypeptide chains without changing their intrinsic charge. The polypeptide chains are then separated by a procedure called **isoelectric focusing,** which depends on the fact that the net charge on a protein molecule varies with the pH of the surrounding solution. For any protein there is a characteristic pH, called its *isoelectric point*, at which the protein has no net charge and therefore will not migrate in an electric field. In isoelectric focusing, proteins are electrophoresed in a narrow tube of polyacrylamide gel in which a gradient of pH is established by a mixture of special buffers. Each protein moves to a position in the gradient that corresponds to its isoelectric point and stays there (Figure 4–51). This is the first dimension of two-dimensional gel electrophoresis.

In the second step, the narrow gel containing the separated proteins is again subjected to electrophoresis but in a direction orthogonal to that used in the first step. This time SDS is added, and the proteins are separated according to their size as in one-dimensional SDS-PAGE: the original narrow gel is soaked in SDS and then placed on one edge of an SDS polyacrylamide-gel slab, through which each polypeptide chain migrates to form a discrete spot. This is the second dimension of two-dimensional polyacrylamide-gel electrophoresis. The only proteins left unresolved will be those that have both an identical size and an identical isoelectric point, a relatively rare situation. Even trace amounts of each polypep-

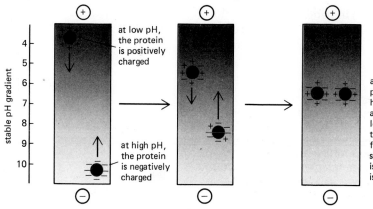

Figure 4–51 Separation of protein molecules by isoelectric focusing. At low pH (high H$^+$ concentration), the carboxylic acid groups of proteins tend to be uncharged (—COOH) and their nitrogen-containing basic groups fully charged (for example, —NH$_3$$^+$), giving most proteins a net positive charge. At high pH, the carboxylic acid groups are negatively charged (—COO$^-$) and the basic groups tend to be uncharged (for example, —NH$_2$), giving most proteins a net negative charge (see Figure 2–8, p. 58). At its *isoelectric pH*, a protein has no net charge since the positive and negative charges balance. Thus when a tube containing a fixed pH gradient is subjected to a strong electric field, each protein species present will migrate until it forms a sharp band at its isoelectric pH, as shown.

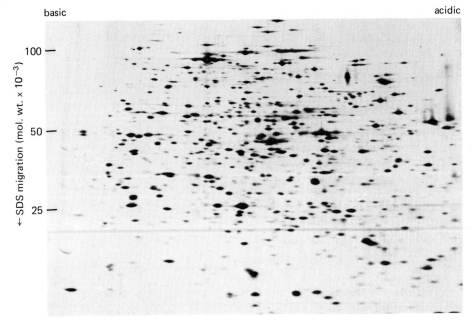

basic   acidic

100 —

50 —

25 —

← SDS migration (mol. wt. × 10⁻³)

Figure 4–52 Fractionation of all the proteins in an *E. coli* bacterial cell by two-dimensional polyacrylamide-gel electrophoresis. Each spot corresponds to a different polypeptide chain. The proteins were first separated according to their isoelectric points by isoelectric focusing from left to right. They were then further fractionated according to their polypeptide molecular weights by electrophoresis from top to bottom in the presence of SDS. Note that different proteins are present in very different amounts. (Courtesy of Patrick O'Farrell.)

tide chain can be detected on the gel by various staining procedures—or by autoradiography if the protein sample was initially labeled with a radioisotope (see p. 176). Up to 2000 individual polypeptide chains have been resolved on a single two-dimensional gel—which is almost the number of different proteins in a bacterium (Figure 4–52). The resolving power is so great that two proteins that differ in only a single charged amino acid can be readily distinguished.

Some landmarks in the development of chromatography and electrophoresis are outlined in Table 4–9.

Selective Cleavage of a Protein Generates a Distinctive Set of Peptide Fragments[29]

Although the molecular weight and isoelectric point are distinctive features of a protein, unambiguous identification ultimately depends on determining the amino acid sequence. The first stage of this process, which involves cleaving the protein into smaller fragments, can itself provide useful information about a protein. Proteolytic enzymes and chemical reagents are available that will cleave proteins between specific amino acid residues (Table 4–10). The enzyme *trypsin*, for instance, cuts on the carboxyl side of lysine or arginine residues, whereas the chemical *cyanogen bromide* cuts peptide bonds next to methionine residues. Since these enzymes and chemicals cleave at relatively few sites in a protein, they tend to produce rather large peptides. If such a mixture of peptides is separated by chromatographic or electrophoretic procedures, the resulting pattern, or **peptide map,** is diagnostic of the protein from which the peptides were generated and is sometimes referred to as the protein's "fingerprint" (Figure 4–53).

Protein fingerprinting was developed in 1956 as a way of comparing normal hemoglobin with the mutant form of the protein found in patients suffering from *sickle-cell anemia*. A single peptide difference was found and eventually traced to a single amino acid change, providing the first demonstration that a mutation can change a single amino acid in a protein.

Short Amino Acid Sequences Can Be Analyzed by Automated Machines[30]

Once a protein has been cleaved into smaller peptides, the next logical step in the analysis is to determine the amino acid sequence of each isolated peptide fragment. This is accomplished by a repeated series of chemical reactions originally devised in 1967. First the peptide is exposed to a chemical that forms a

Table 4–9 Landmarks in the Development of Chromatography and Electrophoresis and Their Application to Biological Molecules

1833	**Faraday** described the fundamental laws concerning the passage of electricity through ionic solutions.
1850	**Runge** separated inorganic chemicals by their differential adsorption to paper, a forerunner of later chromatographic separations.
1906	**Tswett** invented column chromatography, passing petroleum extracts of plant leaves through columns of powdered chalk.
1933	**Tiselius** introduced electrophoresis for separating proteins in solution.
1942	**Martin and Synge** developed partition chromatography, leading to paper chromatography two years later.
1946	**Stein and Moore** determined for the first time the amino acid composition of a protein, initially using column chromatography on starch and later developing chromatography on ion-exchange resins.
1955	**Smithies** used gels made of starch to separate serum proteins by electrophoresis.
	Sanger completed the analysis of the amino acid sequence of bovine insulin, the first protein to be sequenced.
1956	**Ingram** produced the first protein fingerprints, showing that the difference between sickle-cell hemoglobin and normal hemoglobin is due to a change in a single amino acid.
1959	**Raymond** introduced polyacrylamide gels, which are superior to starch gels for separating proteins by electrophoresis; improved buffer systems allowing high-resolution separations were developed in the next few years by **Ornstein** and **Davis**.
1966	**Maizel** introduced the use of sodium dodecyl sulfate (SDS) for improving polyacrylamide-gel electrophoresis of proteins.
1975	**O'Farrell** devised a two-dimensional gel system for analyzing protein mixtures in which SDS polyacrylamide-gel electrophoresis is combined with separation according to isoelectric point.
1984	**Schwartz and Cantor** developed pulsed field gel electrophoresis for the separation of very large DNA molecules.

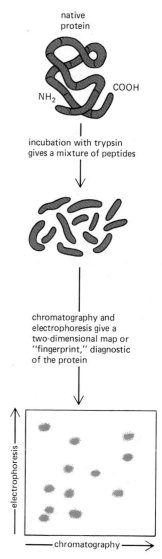

covalent bond only with the free amino group at the amino terminus of the peptide. This chemical is then further activated by exposure to a weak acid so that it specifically cleaves the peptide bond that attaches the amino-terminal amino acid to the peptide chain; the released amino acid is then identified by chromatographic methods. The remaining peptide, which is shorter by one amino acid, is then submitted to the same sequence of reactions, and so on, until every amino acid in the peptide has been determined.

The reiterative nature of these reactions lends itself to automation, and machines called **amino acid sequenators** are commercially available for automatic determination of the amino acid sequence of peptide fragments. The final step is to arrange the sequences of the various peptide fragments in the order in which

Table 4–10 Some Reagents Commonly Used to Cleave Peptide Bonds in Proteins

	Amino Acid 1	Amino Acid 2
Enzyme		
Trypsin	Lys or Arg	any
Chymotrypsin	Phe, Trp, or Tyr	any
V8 protease	Glu	any
Chemical		
Cyanogen bromide	Met	any
2-Nitro-5-thiocyanobenzoate	any	Cys

The specificity for the amino acids on either side of the cleaved bond is indicated. The carboxyl group of amino acid 1 is released by the cleavage; this amino acid is to the left of the peptide bond as normally written (see Panel 2–5, pp. 54–55).

Figure 4–53 Production of a peptide map, or fingerprint, of a protein. In this case the protein was digested with trypsin to generate a mixture of many small polypeptide fragments, which was then fractionated in two dimensions by electrophoresis and partition chromatography. The pattern of spots obtained is diagnostic of the protein analyzed.

they occur in the intact polypeptide chain. This was traditionally achieved by comparing the sequences of different sets of overlapping peptide fragments obtained by cleaving the same protein with different proteolytic enzymes.

Improvements in protein sequencing technology have greatly increased its speed and sensitivity, allowing analysis of minute samples; the sequence of several dozen amino acids at the amino-terminal end of a peptide can be obtained overnight from a few micrograms of protein—the amount available from a single band on an SDS polyacrylamide gel. This has been important for characterizing many minor cell proteins, such as the receptors for steroid and polypeptide hormones. Knowing the sequence of as few as 20 amino acids of a protein is frequently enough to allow a DNA probe to be designed that allows cloning of its gene (see p. 262). Once the gene has been isolated, the rest of the protein's amino acid sequence can be deduced from the DNA sequence by reference to the genetic code. This is a major advantage because, even with automation, the direct determination of the entire amino acid sequence of a protein is a major undertaking. A protein of 100 residues can often be sequenced in a month of hard work, but the difficulty increases steeply with the length of the polypeptide chain, and the chemical peculiarities of individual peptide fragments prevent the process from being routine. Since DNA sequencing can be done so quickly and simply (see below), the sequences of most proteins are now determined largely from the nucleotide sequences of their genes.

Summary

Populations of cells can be analyzed biochemically by disrupting them and fractionating their contents by ultracentrifugation. Further fractionations allow functional cell-free systems to be developed; such systems are required to determine the molecular details of complex cellular processes. For example, protein synthesis, DNA replication, RNA splicing, and various types of intracellular transport are all currently being studied in this way.

The major proteins in soluble cell extracts can be purified by column chromatography; depending on the type of column matrix, biologically active proteins can be separated according to their molecular weight, hydrophobicity, charge characteristics, or affinity for other molecules. In a typical purification the sample is passed through several different columns in turn—the enriched fractions obtained from one column being applied to the next. Once a protein has been purified to homogeneity, its biological activities can be examined in detail. In addition, a small part of the protein's amino acid sequence can be determined and its gene can be cloned; the remaining amino acid sequence is then obtained from the nucleotide sequence of the gene.

The molecular weight and subunit composition of even very small amounts of a protein can be determined by SDS polyacrylamide-gel electrophoresis. In two-dimensional gel electrophoresis, proteins are resolved as separate spots by isoelectric focusing in one dimension followed by SDS polyacrylamide-gel electrophoresis in a second dimension. These electrophoretic separations can be applied even to proteins that are normally insoluble in water.

Tracing Cellular Molecules with Radioactive Isotopes and Antibodies

Almost any property of a molecule—physical, chemical, or biological—can in principle be used as a means of detecting it. In cell biological studies, molecules are often monitored either by their optical properties—whether in their native state or after staining them with a dye—or by their biochemical activity. In this section we consider two other detection methods that have been particularly useful: those involving *radioisotopes* and those utilizing *antibodies*. Each of these methods is capable of detecting specific molecules in a complex mixture with great sensitivity: under optimal conditions they can detect fewer than 1000 molecules in a sample.

Radioactive Atoms Can Be Detected with Great Sensitivity[31]

Most naturally occurring elements are a mixture of slightly different *isotopes*. These differ from each other in the mass of their atomic nuclei, but because they have the same number of electrons, they have the same chemical properties. The nuclei of radioactive isotopes, or **radioisotopes,** are unstable and undergo random disintegration to produce different atoms. In the course of these disintegrations, energetic subatomic particles such as electrons or radiations such as γ-rays are given off.

Although naturally occurring radioisotopes are rare (because of their instability), radioactive atoms can be produced in large amounts in nuclear reactors in which stable atoms are bombarded with high-energy particles. As a result, many biologically important elements are readily available in radioisotopically labeled form (Table 4–11). The radiation they emit is detected in various ways. Electrons (β particles) can be detected in a *Geiger counter* by the ionization they produce in a gas. They can also be measured in a *scintillation counter* by the small flashes of light they induce in a scintillation fluid. These methods make it possible to measure the quantity of a particular radioisotope present in a biological specimen. It is also possible to localize the isotope by using *autoradiography* to detect its effect on the grains of silver halide in a photographic emulsion (see p. 176). All these methods of detection are capable of extreme sensitivity; in favorable circumstances, nearly every disintegration—and therefore every radioactive atom that decays—can be detected.

Radioisotopes Are Used to Trace Molecules in Cells and Organisms[32]

One of the earliest uses of radioactivity in biology was to trace the chemical pathway of carbon during photosynthesis. Unicellular green algae were maintained in an atmosphere containing radioactively labeled CO_2 ($^{14}CO_2$), and at various times after they had been exposed to sunlight their soluble contents were separated by paper chromatography. Small molecules containing ^{14}C atoms derived from CO_2 were detected by a sheet of photographic film placed over the dried paper chromatogram. In this way most of the principal components in the photosynthetic pathway from CO_2 to sugar were identified.

Radioactive molecules can be used to follow the course of almost any process in cells. In a typical experiment a precursor in radioactive form is added to cells so that the radioactive molecules mix with the preexisting unlabeled ones; both are treated identically by the cell, since they differ only in the weight of their atomic nuclei. Changes in the location or chemical form of the radioactive molecules can be followed as a function of time. The resolution of such experiments is often sharpened by using a **pulse-chase** labeling protocol, in which the radioactive material (the *pulse*) is added for only a very brief period and then washed away and replaced by nonradioactive molecules (the *chase*). Samples are taken at regular intervals, and the chemical form or location of the radioactivity is identified for each sample (Figure 4–54).

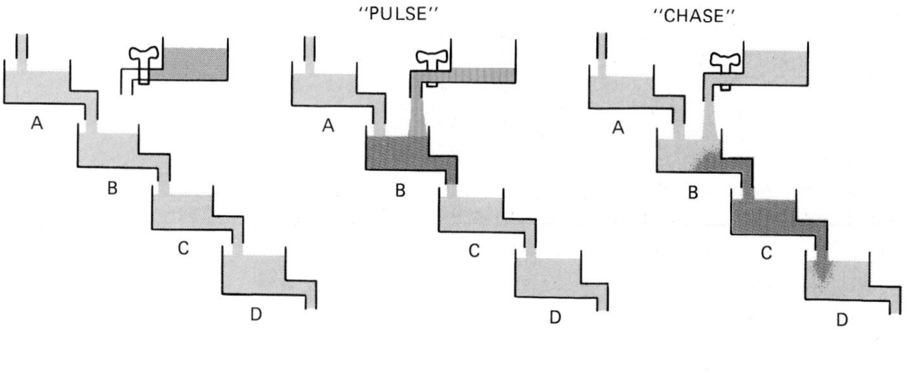

Table 4–11 Some Radioisotopes in Common Use in Biological Research

Isotope	Half-Life
^{32}P	14 days
^{131}I	8.1 days
^{35}S	87 days
^{14}C	5570 years
^{45}Ca	164 days
^{3}H	12.3 years

The isotopes are arranged in decreasing order of the energy of the β radiation (electrons) they emit. ^{131}I also emits γ radiation. The *half-life* is the time required for 50% of the atoms of an isotope to disintegrate.

Figure 4–54 The logic of a typical pulse-chase experiment using radioisotopes. The chambers labeled A, B, C, and D represent either different compartments in the cell (detected by autoradiography or by cell-fractionation experiments) or different chemical compounds (detected by chromatography or other chemical methods).

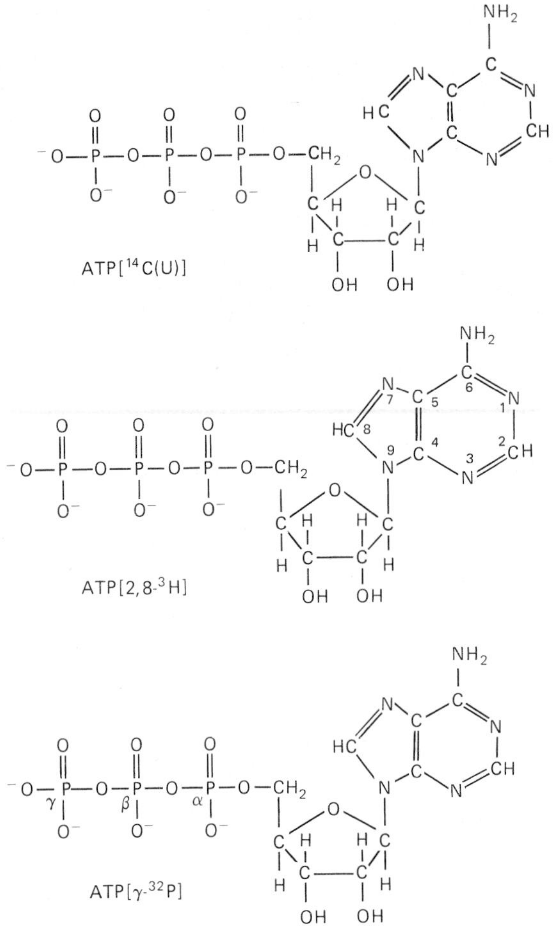

Figure 4–55 Three commercially available radioactive forms of ATP, with the radioactive atoms shown in color. The nomenclature used to identify the position and type of the radioactive atoms is also shown.

Radioisotopic labeling is uniquely valuable as a way of distinguishing between molecules that are chemically identical but have different histories—for example, those that differ in their time of synthesis. The use of radioactive tracers, in fact, has shown that almost all the molecules in a living cell are continually being degraded and replaced, even when the cell is not growing and is apparently in a steady state. Such "turnover" processes, which sometimes take place very slowly, would be almost impossible to detect without radioisotopes.

Today nearly all common small molecules are available in radioactive form from commercial sources, and virtually any biological molecule, no matter how complicated, can be radioactively labeled. Compounds are often made with radioactive atoms incorporated at particular positions in their structure, enabling the separate fates of different parts of the same molecule to be followed during biological reactions (Figure 4–55).

One of the important uses of radioactivity in cell biology is in the localization by **autoradiography** of radioactive compounds in sections of whole cells or tissues. In this procedure living cells are briefly exposed to a "pulse" of a specific radioactive compound and incubated for a variable period before being fixed and processed for light or electron microscopy. Each preparation is then overlaid with a thin film of photographic emulsion. After remaining in the dark for a number of days—during which time the radioisotope decays—the emulsion is developed. The position of the radioactivity in each cell can be determined by the position of the developed silver grains. For example, incubation of cells with a radioactive DNA precursor (*^{3}H-thymidine*) shows that DNA is made in the nucleus and remains there. By contrast, labeling cells with a radioactive RNA precursor (*^{3}H-uridine*) reveals that RNA is initially made in the cell nucleus and then moves rapidly into the cytoplasm.

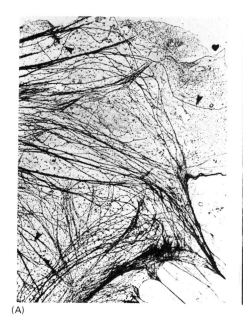

(A)

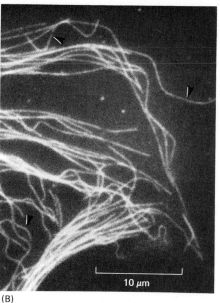

(B)

10 μm

Figure 4–56 (A) An electron micrograph of the periphery of a cultured epithelial cell showing the distribution of microtubules and other filaments. (B) The same area stained with a fluorescent antibody to tubulin, the protein subunit of microtubules, using the technique of indirect immunocytochemistry (see Figure 4–58). Arrows indicate individual microtubules that are readily recognizable in the two figures. (From M. Osborn, R. Webster, and K. Weber, *J. Cell Biol.* 77:R27–R34, 1978. Reproduced by copyright permission of the Rockefeller University Press.)

Antibodies Can Be Used to Detect and Isolate Specific Molecules[33]

Antibodies are proteins produced by vertebrates as a defense against infection (see Chapter 18). They are unique among proteins because they are made in millions of different forms, each with a different binding site that specifically recognizes the molecule (called an *antigen*) that induced its production. The precise antigen specificity of antibodies makes them powerful tools for the cell biologist. Labeled with fluorescent dyes, they are invaluable for locating specific molecules in cells by fluorescence microscopy (Figure 4–56); labeled with electron-dense particles such as colloidal gold spheres, they are used to locate particular molecules at high resolution in the electron microscope (Figure 4–57). As biochemical tools they are used to detect and quantify molecules in cell extracts and to identify specific proteins after they have been fractionated by electrophoresis in polyacrylamide gels. On a preparative scale, antibodies can be coupled to an inert matrix to produce an affinity column that is then used either to purify a specific molecule from a crude cell extract or, if the molecule is on the cell surface, to pick out specific types of living cells from a heterogeneous population.

The sensitivity of antibodies as probes for detecting specific molecules in cells and tissues is frequently enhanced by a signal-amplification method. For example, although a marker molecule such as a fluorescent dye can be linked directly to an antibody used for specific recognition (the *primary antibody*), a stronger signal is achieved by using an unlabeled primary antibody and then detecting it with a group of labeled *secondary antibodies* that bind to it (Figure 4–58A).

An alternative amplification system exploits the exceptionally high binding affinity of *biotin* (a small water-soluble vitamin) for *streptavidin* (a bacterial protein). If the primary antibody has been covalently coupled to biotin, streptavidin can be directly labeled with a marker and used in place of a secondary antibody. The streptavidin can also be used to link a single biotin-coupled antibody molecule to a whole network of biotin-coupled marker molecules (Figure 4–58B). Similar networks can be formed by extending the procedure in Figure 4–58A with a third layer of antibodies.

The most sensitive amplification methods use an enzyme as the marker molecule. For example, alkaline phosphatase produces inorganic phosphate, and coupling the enzyme to a secondary antibody permits a sensitive chemical test for phosphate to be employed to reveal the presence of an antibody-antigen complex. Since each enzyme molecule acts catalytically to generate many thousands of molecules of product, such an *enzyme-linked immunoassay (ELIZA)* allows even tiny amounts of antigen to be detected. These assays are frequently used in medicine as a sensitive test for various types of infections.

secretory vesicle

lysosome

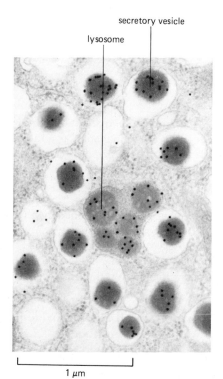

1 μm

Figure 4–57 Immunocytochemical localization of specific protein molecules in electron micrographs by labeling with antibodies coupled to colloidal gold particles. A thin section of an insulin-secreting cell is shown in which insulin molecules have been labeled with anti-insulin antibodies bound to tiny gold spheres (each seen as a black dot). Most of the insulin is stored in the dense cores of secretory vesicles; in addition, some cores are being degraded in lysosomes. (From L. Orci, *Diabetologia* 28:528–546, 1985.)

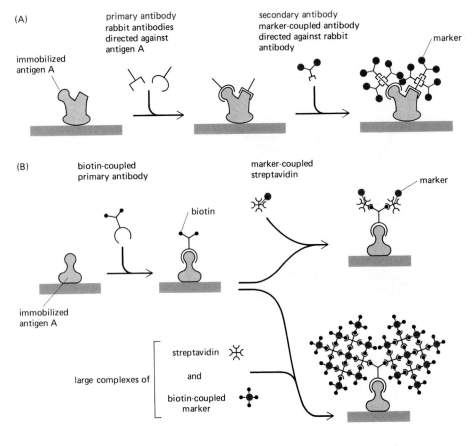

(A)

immobilized antigen A

primary antibody
rabbit antibodies
directed against
antigen A

secondary antibody
marker-coupled antibody
directed against rabbit
antibody

marker

(B)

biotin-coupled
primary antibody

biotin

marker-coupled
streptavidin

marker

immobilized
antigen A

large complexes of

streptavidin

and

biotin-coupled
marker

Figure 4–58 How antibodies are used to detect a particular molecule with great sensitivity. (A) The method illustrated here (known as *indirect immunocytochemistry*) has an enhanced sensitivity because the primary antibody (defined as the antibody molecule that binds to a recognized molecule of antigen) is itself recognized by many molecules of a second type of antibody. This secondary antibody is covalently coupled to a marker molecule that makes it readily detectable. Commonly used marker molecules include fluorescein or rhodamine dyes (for fluorescence microscopy), the enzyme horseradish peroxidase (for either bright-field light microscopy or electron microscopy), the iron-containing protein ferritin or colloidal gold spheres (for electron microscopy), and the enzyme alkaline phosphatase (for biochemical detection). (B) Modifications of the method in (A) that replace the secondary antibody by exploiting the high-affinity interaction between streptavidin and biotin. Because each streptavidin molecule can bind four biotin molecules, it can cross-link many biotinylated marker molecules into a large three-dimensional network. At the bottom an especially sensitive "sandwich technique" is shown that uses such networks to produce very heavy labeling of each primary antibody molecule.

Antibodies are made most simply by injecting a sample of the antigen several times into an animal such as a rabbit or a goat and then collecting the antibody-rich serum. This *antiserum* contains a heterogeneous mixture of antibodies, each produced by a different antibody-secreting cell (a B lymphocyte). The different antibodies recognize various parts of the antigen molecule as well as impurities in the antigen preparation. The specificity of an antiserum for a particular antigen sometimes can be sharpened by removing the unwanted antibody molecules that bind to other molecules; for example, an antiserum produced against protein X can be passed through an affinity column of antigens Y and Z to remove any contaminating anti-Y and anti-Z antibodies. Even so, the heterogeneity of such antisera has limited their usefulness.

Hybridoma Cell Lines Provide a Permanent Source of Monoclonal Antibodies[34]

In 1976 the problem of antiserum heterogeneity was overcome by the development of a technique that revolutionized the use of antibodies as tools in cell biology. The technique involves propagating a clone of cells from a single antibody-secreting B lymphocyte so that a homogeneous preparation of antibodies can be obtained in large quantities. But B lymphocytes normally have a limited life-span in culture. To overcome this limitation, individual antibody-producing B lymphocytes from an immunized mouse are fused with cells derived from an "immortal" B lymphocyte tumor. From the resulting heterogeneous mixture of hybrid cells, those hybrids that have both the ability to make a particular antibody and the ability to multiply indefinitely in tissue culture are selected. These **hybridomas** are propagated as individual clones, each of which provides a permanent and stable source of a single type of **monoclonal antibody** (Figure 4–59).

Since they are the product of a single B lymphocyte clone, monoclonal antibodies are made as a population of identical antibody molecules, each with an identical antigen-binding site. This site will recognize, for example, a particular

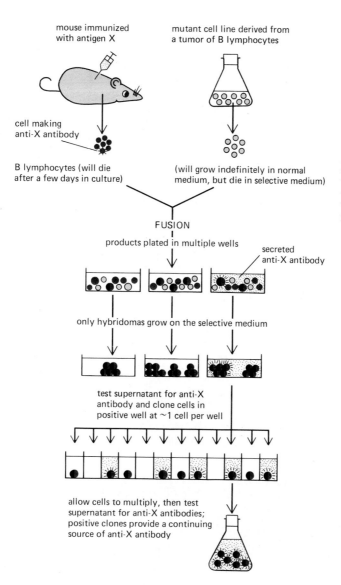

mouse immunized
with antigen X

mutant cell line derived from
a tumor of B lymphocytes

cell making
anti-X antibody

B lymphocytes (will die
after a few days in culture)

(will grow indefinitely in normal
medium, but die in selective medium)

FUSION
products plated in multiple wells

secreted
anti-X antibody

only hybridomas grow on the selective medium

test supernatant for anti-X
antibody and clone cells in
positive well at ~1 cell per well

allow cells to multiply, then test
supernatant for anti-X antibodies;
positive clones provide a continuing
source of anti-X antibody

Figure 4–59 Preparation of hybrid cells, or hybridomas, that secrete homogeneous monoclonal antibodies against a particular antigen (X). The selective growth medium used contains an inhibitor (aminopterin) that blocks the normal biosynthetic pathways by which nucleotides are made. The cells must therefore use a bypass pathway to synthesize their nucleic acids, and this pathway is defective in the mutant cell line to which the normal B lymphocytes are fused. Because neither cell type used for the initial fusion can grow on its own, only the cell hybrids survive.

conformation of a defined sequence of five or six amino acid side chains on a protein. Their uniform specificity alone makes monoclonal antibodies much more useful for most purposes than conventional antisera, which usually contain a mixture of antibodies that recognize a variety of different antigenic sites on even a small macromolecule.

But the most important advantage of the hybridoma technique is that monoclonal antibodies can be made against molecules that constitute only a minor component of a complex mixture. In an ordinary antiserum made against such a mixture, the number of antibody molecules that recognize the minor component would be too small to be useful. But if the B lymphocytes that produce all these antibodies are made into hybridomas and a desired hybridoma clone is selected from the large mixture of clones, it can be propagated indefinitely to produce the desired antibody in large quantities. In principle, therefore, a monoclonal antibody can be made against any protein in a biological sample; once the antibody is made, it can be used as a specific probe—both to track down and localize the protein that induced its formation and to purify the protein in order to study its structure and function. Since fewer than 5% of the estimated 10,000 proteins in a typical mammalian cell have thus far been isolated, many monoclonal antibodies made against impure protein mixtures in fractionated cell extracts identify new proteins. Using monoclonal antibodies and gene-cloning technology (see below), it is no longer difficult to identify and characterize novel proteins and genes; the problem is to determine their function.

Antibodies and Other Macromolecules Can Be Injected into Living Cells[35]

Antibody molecules can be used to determine the function of the molecules to which they bind. Antibodies to nerve growth factor, for example, when injected into a newborn mouse, prevent the development of certain classes of nerve cells that depend on the growth factor for their survival. Similarly, antibodies that react with molecules on the surface of a particular type of cell can be used to kill the cells; by specifically eliminating one cell type from a mixed population, one can establish the importance of that cell type for different biological functions.

Proteins inside living cells cannot be reached by antibodies added externally because the plasma membrane is impermeable to large molecules. Because the plasma membrane is self-sealing, however, it is possible to introduce antibodies and other macromolecules into the cytoplasm of eucaryotic cells in other ways—most commonly by injecting them through a fine glass needle (see p. 157). For example, anti-myosin antibodies injected into a sea urchin egg prevent the egg cell from dividing in two, even though nuclear division occurs normally. This observation demonstrates that myosin plays a crucial part in the contractile process that divides the cytoplasm during mitosis but suggests that it is not required for nuclear division. The high specificity of monoclonal antibodies and the ease of producing them in concentrated form make them particularly suitable for this type of application.

Summary

Any molecule in the cell can be "labeled" by the incorporation of one or more radioactive atoms. The unstable nuclei of these atoms disintegrate, emitting radiation that allows the molecule to be detected and its movements and metabolism traced in the cell. Applications of radioisotopes in cell biology include the analysis of metabolic pathways by pulse-chase methods and the determination of the location of individual molecules in a cell by autoradiography.

Antibodies are also versatile and sensitive tools for detecting and localizing specific biological molecules. Vertebrates make millions of different antibody molecules, each with a binding site that recognizes a specific region of a macromolecule. The hybridoma technique allows monoclonal antibodies of a single specificity to be obtained in virtually unlimited amounts. In principle, monoclonal antibodies can be made against any cell macromolecule; these can be used to locate and purify the molecule and, in some cases, to analyze its function.

Recombinant DNA Technology[36]

The central challenge in modern cell biology is to understand the workings of the cell in molecular detail. We have discussed a number of powerful techniques for purifying, analyzing, and tracing the proteins of a cell. In this final section we shall discuss how one examines the structure and function of the cell's DNA. Classically, the only way to investigate the information content of the DNA was through genetics, which allows gene functions to be deduced from the phenotypes of mutant organisms and their progeny. This approach continues to provide unique insights, but it has been supplemented in recent years by a set of techniques known collectively as "recombinant DNA technology." These techniques greatly facilitate genetic studies by allowing both direct control and detailed chemical analysis of the genetic material. In addition, by making even minor cell proteins available in large quantities, the same methods have revolutionized biochemical studies of protein structure and function.

Recombinant DNA Technology Has Revolutionized Cell Biology[37]

Until the early 1970s, DNA was the most difficult cellular molecule for the biochemist to analyze. Enormously long and chemically monotonous, the nucleotide sequence of DNA could be approached only by indirect means—such as through protein or RNA sequencing or by genetic analysis. Today the situation has changed entirely. From being the most difficult macromolecule of the cell to analyze, DNA has become the easiest. It is now possible to excise specific regions of DNA, to obtain them in virtually unlimited quantities, and to determine the sequence of their nucleotides at a rate of hundreds of nucleotides a day. By variations of the same techniques, an isolated gene can be altered (engineered) at will and transferred back into cells in culture or (with rather more difficulty) into the germ line of animals, where the modified gene becomes incorporated as a permanent functional part of the genome.

These technical breakthroughs have had a dramatic impact on cell biology by allowing the study of cells and their macromolecules in previously unimagined ways. For example, they have provided the means to determine the functions of many newly discovered proteins and their individual domains and to unravel the complex mechanisms by which eucaryotic gene expression is regulated. They have also made available large amounts of the rare proteins that regulate cell proliferation and development. In commercial laboratories, similar techniques offer great promise for the large-scale economical production of protein hormones and vaccines, previously available only with great labor and cost.

Recombinant DNA technology comprises a mixture of techniques, some new and some borrowed from other fields such as microbial genetics (Table 4–12). The most important of these techniques are (1) the specific cleavage of DNA by *restriction nucleases,* which greatly facilitates the isolation and manipulation of individual genes; (2) rapid *sequencing* of all the nucleotides in a purified DNA fragment, which makes it possible to determine the precise boundaries of a gene and the amino acid sequence it encodes; (3) *nucleic acid hybridization,* which makes it possible to find specific sequences of DNA or RNA with great accuracy and sensitivity on the basis of their ability to bind a complementary nucleic acid sequence; (4) *DNA cloning,* whereby a specific DNA fragment is integrated into a self-replicating genetic element (plasmid or virus) that inhabits bacteria so that a single DNA molecule can be reproduced to generate many billions of identical copies; and (5) *genetic engineering,* by which DNA sequences are altered to make modified versions of genes, which are reinserted back into cells or organisms.

In the sections that follow, we shall explain how recombinant DNA technology has generated new experimental approaches that have revolutionized cell biology. However, a proper appreciation of DNA cloning and genetic engineering requires an understanding of the natural mechanisms by which cells replicate and decode

Table 4–12   Some Major Steps in the Development of Recombinant DNA Technology

1869	**Miescher** isolated DNA for the first time.
1944	**Avery** provided evidence that DNA, rather than protein, carries the genetic information during bacterial transformation.
1953	**Watson and Crick** proposed the double-helix model for DNA structure based on x-ray results of **Franklin and Wilkins.**
1957	**Kornberg** discovered DNA polymerase, the enzyme now used to produce highly radioactive DNA probes.
1961	**Marmur and Doty** discovered DNA renaturation, establishing the specificity and feasibility of nucleic acid hybridization reactions.
1962	**Arber** provided the first evidence for the existence of DNA restriction nucleases, leading to their later purification and use in DNA sequence characterization by **Nathans and H. Smith.**
1966	**Nirenberg, Ochoa,** and **Khorana** elucidated the genetic code.
1967	**Gellert** discovered DNA ligase, the enzyme used to join DNA fragments together.
1972–1973	DNA cloning techniques were developed by the laboratories of **Boyer, Cohen, Berg,** and their colleagues at Stanford University and the University of California at San Francisco.
1975–1977	**Sanger and Barrell** and **Maxam and Gilbert** developed rapid DNA-sequencing methods.
1981–1982	**Palmiter and Brinster** produced transgenic mice; **Spradling and Rubin** produced transgenic fruit flies.

DNA. We shall therefore postpone our main discussion of the details of gene cloning and engineering until Chapter 5, where they will be explained after an introduction to the basic genetic mechanisms.

Restriction Nucleases Hydrolyze DNA Molecules at Specific Nucleotide Sequences[38]

Many bacteria make enzymes called **restriction nucleases,** which protect the bacteria by degrading the DNA molecules carried into the cell by viruses. Each enzyme recognizes a specific sequence of four to eight nucleotides in DNA. These sequences, where they occur in the genome of the bacterium itself, are "camouflaged" by methylation at an A or a C residue; where the sequences occur in foreign DNA, they are generally not methylated and so are cleaved by the restriction nuclease (Figure 4–60). Many restriction nucleases have been purified from different species of bacteria; more than 100, most of which recognize different nucleotide sequences, are now available commercially.

A particular restriction nuclease will cut any double-helical DNA molecule extracted from a cell into a series of specific DNA fragments known as **restriction fragments.** By comparing the sizes of the restriction fragments produced from a particular genetic region after treatment with a combination of different restriction nucleases, a **restriction map** of that region can be constructed showing the location of each cutting (restriction) site in relation to its neighboring restriction sites (Figure 4–61). Because each restriction nuclease recognizes a different short DNA sequence, a restriction map reflects the arrangement of specific nucleotide sequences in the region. This means that different regions of DNA can be compared (by comparing their restriction maps) without having to determine their nucleotide sequences. For example, by comparing the restriction maps illustrated in Figure 4–62, we know that the chromosomal regions that code for hemoglobin chains in humans, orangutans, and chimpanzees have remained largely unchanged during the 5 to 10 million years since these species first diverged. Restriction maps are also important in DNA cloning and genetic engineering, enabling one to locate a gene of interest on a particular restriction fragment, as will be explained later.

Figure 4–60 The DNA nucleotide sequences recognized by three widely used restriction nucleases. As in these examples, such sequences are often six base pairs long and "palindromic"—that is, the nucleotide sequences of the two strands are the same in the short region of helix that is recognized. The two strands of DNA are cut at or near the recognition sequence, often with a staggered cleavage that creates a cohesive end—as for Eco RI and Hind III. Restriction nucleases are obtained from various species of bacteria: Hpa I is from *Hemophilus parainfluenzae;* Eco RI is from *Escherichia coli;* and Hind III is from *Hemophilus influenzae.*

Any DNA Sequence Can Be Produced in Large Amounts by DNA Cloning[39]

Many restriction nucleases produce staggered cuts, which leave short single-stranded tails at the two ends of each fragment. These are known as *cohesive ends,* since each tail can form complementary base pairs with the tail at any other end produced by the same enzyme (Figure 4–63). The cohesive ends generated by restriction enzymes allow any two DNA fragments to be easily joined together, as long as the fragments were generated with the same restriction nuclease (or with another nuclease that produces the same cohesive ends). In this way a fragment of DNA from any source can be inserted into the purified DNA genome of a self-

Figure 4–61 A simple example illustrating how the cutting sites for different restriction nucleases (known as *restriction sites*) are positioned relative to each other on double helical DNA molecules to create a *restriction map.* (kb, kilobases; an abbreviation designating either 1000 nucleotides or 1000 nucleotide pairs.)

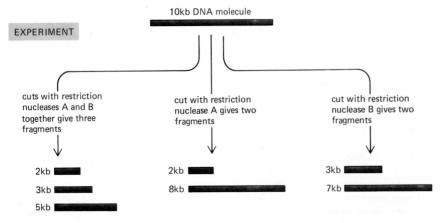

CONCLUSION: enzyme A cuts near one end of the molecule. Enzyme B must cut either near the same end or near the other end. The size of the fragments produced by both enzymes acting together rules out the first alternative and leads to the unambiguous order of restriction nuclease cutting sites shown below.

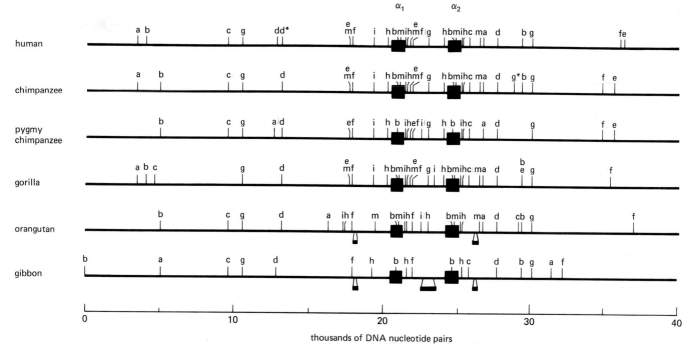

human

chimpanzee

pygmy chimpanzee

gorilla

orangutan

gibbon

thousands of DNA nucleotide pairs

replicating genetic element, which is generally either a plasmid or a bacterial virus (see p. 259). A clone of bacteria containing the resultant plasmid or virus can then serve as a factory for production of unlimited amounts of the DNA fragment—a process called **DNA cloning.** The initial DNA fragment may be derived directly from genomic DNA; or it may be made from *cDNA*, which is DNA that has been made by copying messenger RNA. The procedures will be described in detail in Chapter 5 and will be only briefly outlined here.

The first step in obtaining **genomic DNA clones** generally involves purifying DNA from the entire cellular genome and cleaving it with a restriction nuclease. This produces an enormous number of different DNA fragments: for example, between 10^5 and 10^7 fragments are generated from a mammalian genome. The cloning process will therefore produce millions of cell colonies (*clones*), most of which will carry a *different* DNA fragment. The most difficult part of genomic cloning is finding the one clone in a million that contains the DNA fragment of interest (see p. 262).

In order to obtain **cDNA clones,** one begins by purifying mRNA (or a subfraction of the mRNA) from cells. This mRNA is then used as a template for *reverse transcriptase,* an enzyme produced by certain viruses that synthesize DNA by copying an RNA sequence (the reverse of the usual transcription process, in which RNA is copied from a DNA sequence). The enzyme produces a complementary DNA copy (hence "cDNA") of each mRNA molecule present. These single-stranded cDNA molecules are then converted into double-stranded DNA molecules (see p. 260), which are then cloned using methods similar to those used to clone genomic DNA fragments.

As explained in Chapter 5 (p. 260), there are important differences between genomic and cDNA clones. In particular, because of the extensive RNA splicing that occurs in higher eucaryotes (p. 531), only cDNA clones are likely to contain an uninterrupted form of the nucleotide sequences that code for proteins.

Figure 4–62 Restriction maps of human and various primate DNAs in a cluster of genes coding for hemoglobin. The two squares in each map indicate the positions of the DNA corresponding to the α-globin genes. Each letter stands for a site cut by a different restriction nuclease. As in Figure 4–61, the location of each cut was determined by comparing the sizes of the DNA fragments generated by treating the DNAs with the various restriction nucleases, individually and in combinations. (Courtesy of Elizabeth Zimmer and Alan Wilson.)

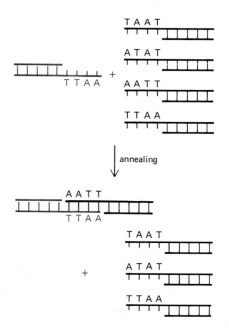

Figure 4–63 Many kinds of restriction nucleases produce DNA fragments with cohesive ends. DNA fragments with the same cohesive ends can readily join by complementary base-pairing between their cohesive ends as illustrated. The two DNA fragments that join in this example were both produced by the Eco RI restriction nuclease (see Figure 4–60).

Gel Electrophoresis Rapidly Separates DNA Molecules of Different Sizes[40]

In the early 1970s it was found that the length and purity of DNA molecules could be accurately determined by the same types of gel electrophoresis methods that had proved so useful in the analysis of protein chains (p. 170). The procedure is actually simpler than for proteins; because each nucleotide in a nucleic acid molecule already carries a single negative charge, there is no need for the negatively charged detergent SDS that is required to make protein molecules move in a uniform manner toward the positive electrode. For DNA fragments less than 500 nucleotides long, specially designed polyacrylamide gels allow molecules that differ in length by as little as a single nucleotide to be separated from each other (Figure 4–64A). However, the pores in polyacrylamide gels are too small to permit larger DNA molecules to pass; to separate these by size, the much more porous gels formed by dilute solutions of agarose (a polysaccharide isolated from seaweed) are used (Figure 4–64B). Both these DNA separation methods are widely used for analytical and preparative purposes.

A recent variation of agarose gel electrophoresis, called *pulsed-field gel electrophoresis*, makes it possible to separate even huge DNA molecules. Ordinary gel electrophoresis fails to separate these molecules because the steady electric field stretches them out into snakelike configurations that travel end-first through the gel at a rate that is independent of their length. But frequent alterations in the direction of the electric field force the molecules to reorient in order to move—a process that takes more time for larger molecules. Because entire bacterial or yeast chromosomes appear as discrete bands on such pulsed-field gels (Figure 4–64C), chromosome rearrangements can be detected directly. In addition, genes can be mapped to specific yeast chromosomes by using the hybridization of cloned DNA molecules from a gene to detect complementary DNA sequences in the gel (see p. 190).

The DNA bands on agarose or polyacrylamide gels will of course be invisible unless the DNA is labeled or stained in some way. One sensitive method of staining DNA is to soak the gel after electrophoresis in the dye *ethidium bromide*, which fluoresces under ultraviolet light when it is bound to DNA (Figure 4–64B and C).

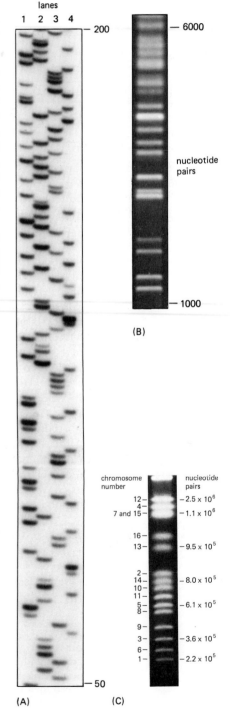

Figure 4–64 Gel electrophoresis is a powerful technique for separating DNA molecules according to their size. In the three examples shown, electrophoresis is from top to bottom, so that the largest DNA molecules are near the top of the gel. In (A) a polyacrylamide gel with small pores is used to fractionate single strands of DNA. In the size range 10 to 500 nucleotides, DNA molecules that differ in size by only a single nucleotide can be separated from each other. In this example, lanes 1 through 4 represent the products of four separate DNA sequencing reactions in which the chain-terminating dideoxyribonucleotides G, A, T, and C have been included, respectively (see Figure 4–68); since the DNA molecules used in these reactions are radiolabeled, their positions can be determined by autoradiography, as shown. In (B) an agarose gel with medium-sized pores is used to separate double-stranded DNA molecules. This method is most useful in the size range 300 to 10,000 nucleotide pairs. These DNA molecules are restriction fragments produced from the genome of a bacterial virus, and they have been detected by their fluorescence when stained with the dye ethidium bromide. In (C) the technique of pulsed-field agarose gel electrophoresis has been used to separate 16 different yeast (*Saccharomyces cerevisiae*) chromosomes that range in size from 220,000 to 2,500,000 DNA nucleotide pairs. DNA molecules as large as 10^7 nucleotide pairs can be separated on these gels. (A, courtesy of Leander Lauffer and Peter Walter; B, courtesy of Ken Kreuzer; C, from D. Vollrath and R.W. Davis, *Nucleic Acids Res.* 15:7876, 1987.)

An even more sensitive detection method involves incorporating a radioisotope into the DNA molecules before electrophoresis; ^{32}P is usually used since it can be incorporated into DNA phosphates and emits a very energetic β particle that is easy to detect by autoradiography (Figure 4–64A).

Purified DNA Molecules Can Be Labeled with Radioisotopes *in Vitro*[41]

Two procedures are widely used to radiolabel isolated DNA molecules. The first uses an *E. coli* enzyme, *DNA polymerase I*, to insert a large number of radioactive nucleotides (usually labeled with ^{32}P) into each DNA molecule (Figure 4–65A), thereby producing very radioactive "DNA probes" for nucleic acid hybridization reactions (see below). The second procedure uses the bacteriophage enzyme *polynucleotide kinase* to transfer a single ^{32}P-labeled phosphate from ATP to the 5' end of each DNA chain (Figure 4–65B). Because only one ^{32}P atom is incorporated by the kinase into each DNA strand, the DNA molecules are usually not radioactive enough to be used as DNA probes; but because they are labeled only at one end, they are invaluable for DNA sequencing and DNA footprinting, as we shall now discuss.

Isolated DNA Fragments Can Be Rapidly Sequenced[42]

Methods were developed in the late 1970s that allow the nucleotide sequence of any purified DNA fragment to be determined simply and quickly. As a result, the complete DNA sequences of many hundreds of mammalian genes have been obtained, including those coding for insulin, hemoglobin, interferon, and cytochrome c. The volume of DNA sequence information is already so large (many millions of nucleotides) that computers must be used to store and analyze it. Several continuous stretches of DNA sequence have been determined that contain

Figure 4–65 Two enzymatic procedures are used routinely for making DNA molecules radioactive. (A) DNA polymerase I labels all the nucleotides in a DNA molecule and can thereby produce highly radioactive DNA probes. (B) Polynucleotide kinase labels only the 5' ends of DNA strands; therefore, when labeling is followed by restriction nuclease cleavage, as shown, DNA molecules containing a single 5' end-labeled strand can be readily obtained.

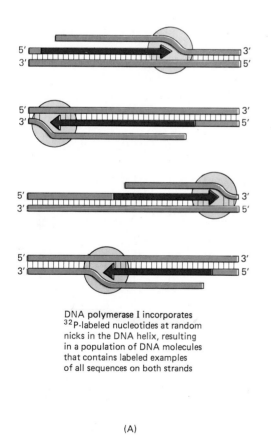

DNA polymerase I incorporates ^{32}P-labeled nucleotides at random nicks in the DNA helix, resulting in a population of DNA molecules that contains labeled examples of all sequences on both strands

(A)

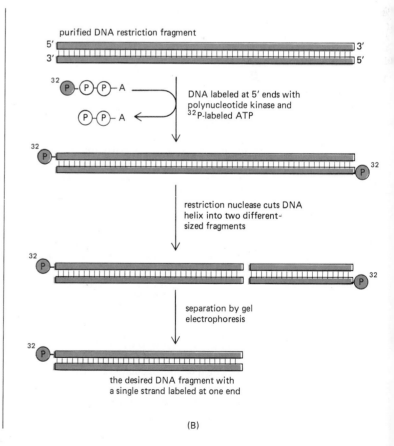

(B)

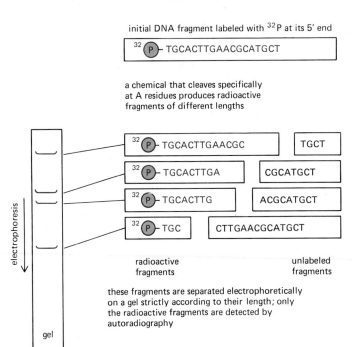

initial DNA fragment labeled with ^{32}P at its 5′ end

^{32} (P)– TGCACTTGAACGCATGCT

a chemical that cleaves specifically at A residues produces radioactive fragments of different lengths

radioactive fragments	unlabeled fragments
^{32} (P)– TGCACTTGAACGC	TGCT
^{32} (P)– TGCACTTGA	CGCATGCT
^{32} (P)– TGCACTTG	ACGCATGCT
^{32} (P)– TGC	CTTGAACGCATGCT

these fragments are separated electrophoretically on a gel strictly according to their length; only the radioactive fragments are detected by autoradiography

electrophoresis

gel

Figure 4–66 Generation of a family of DNA fragments by random cleavage of a 5′ end-labeled DNA chain at a particular type of nucleotide (in this case an A residue). The DNA chain to be analyzed is obtained by denaturing a double-stranded DNA molecule produced by the method outlined in Figure 4–65B. The chain is then cleaved by a mild chemical treatment that eliminates one nucleotide from the chain while leaving intact most of the nucleotides of the type eliminated. Because only the left-hand fragments possess a 5′-terminal, ^{32}P-phosphate group, only they are detected by autoradiography of the gel. This procedure makes possible the chemical method for DNA sequencing described in Figure 4–67.

more than 10⁵ nucleotide pairs; these include the entire genome of the Epstein-Barr virus (which infects humans and causes infectious mononucleosis) and the entire chloroplast genome of a plant (see p. 391). Two different **DNA sequencing** methods are now in widespread use; the principle underlying the *chemical method* is illustrated in Figures 4–66 and 4–67, and the *enzymatic method* is explained in Figure 4–68.

These DNA sequencing methods are so rapid and reliable that, as stated previously, the easiest and most accurate way to determine the amino acid sequence of a protein is to determine the nucleotide sequence of its gene: a cDNA clone is made from the appropriate mRNA, its nucleotide sequence is determined, and the genetic code is then used as a dictionary to convert this sequence back to an amino acid sequence. Although in principle there are six different reading frames in which a DNA sequence can be translated into protein (three on each strand), the correct one is generally recognizable as the only one lacking frequent *stop codons* (see p. 209). Usually a limited amount of amino acid sequence is determined from the purified protein to confirm the sequence determined from the DNA (p. 172).

A modification of the DNA sequencing technique illustrated in Figures 4–66 and 4–67 can be used to determine the nucleotide sequences recognized by DNA-binding proteins. Some of these proteins play a central part in determining which

Figure 4–67 The chemical method for sequencing DNA. The type of procedure described in Figure 4–66 is carried out simultaneously on four separate samples of the same 5′ end-labeled DNA molecule using chemicals that cleave DNA preferentially at T for the first sample, C for the second, G for the third, and A for the fourth. The resulting fragments are separated in parallel lanes of a gel like that shown in Figure 4–64A, giving a pattern of radioactive DNA bands from which the DNA sequence is read. The nucleotide closest to the 5′ end of the sequence is determined by looking across the gel at level 1 (at the bottom of the gel) and seeing in which lane a band appears (T). The same procedure is repeated for level 2, then level 3, and so on, to obtain the sequence. The method has been idealized here; the actual chemical treatments are less specific than shown.

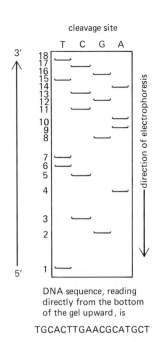

cleavage site

DNA sequence, reading directly from the bottom of the gel upward, is

TGCACTTGAACGCATGCT

(A)

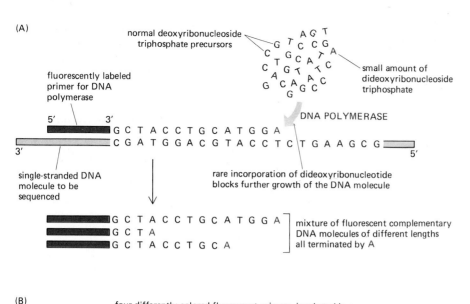

normal deoxyribonucleoside triphosphate precursors

small amount of dideoxyribonucleoside triphosphate

fluorescently labeled primer for DNA polymerase

DNA POLYMERASE

5' → 3'

G C T A C C T G C A T G G A

C G A T G G A C G T A C C T C │ T G A A G C G

3' ← 5'

single-stranded DNA molecule to be sequenced

rare incorporation of dideoxyribonucleotide blocks further growth of the DNA molecule

G C T A C C T G C A T G G A

G C T A

G C T A C C T G C A

mixture of fluorescent complementary DNA molecules of different lengths all terminated by A

(B)

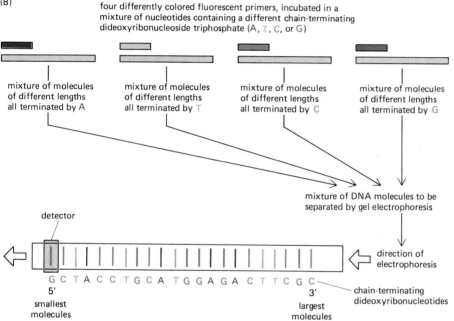

four differently colored fluorescent primers, incubated in a mixture of nucleotides containing a different chain-terminating dideoxyribonucleoside triphosphate (A, T, C, or G)

mixture of molecules of different lengths all terminated by A

mixture of molecules of different lengths all terminated by T

mixture of molecules of different lengths all terminated by C

mixture of molecules of different lengths all terminated by G

mixture of DNA molecules to be separated by gel electrophoresis

detector

direction of electrophoresis

G C T A C C T G C A T G G A G A C T T C G C

5'
smallest molecules

3'
largest molecules

chain-terminating dideoxyribonucleotides

The molecules terminated by each dideoxyribonucleotide are revealed as colored bands by their corresponding fluorescent primers. The sequence of the complementary DNA strand can therefore be read off directly as the bands in succession pass the detector.

Figure 4–68 A nucleic acid sequencing procedure based on the enzymatic incorporation of chain-terminating nucleotides. The key to this method is the use of dideoxyribonucleoside triphosphates in which the deoxyribose 3'-OH group present in normal nucleotides is missing; when such a modified nucleotide is incorporated into a DNA chain, it blocks the addition of the next nucleotide (see p. 224). (A) The primed *in vitro* synthesis of DNA molecules in the presence of a minor proportion of one of these nucleotides generates the type of ladder of DNA fragments shown previously in Figure 4–66. If radioactive DNA is used to produce such ladders, and four different synthesis reactions— each with a different chain-terminating nucleotide—are analyzed by electrophoresis in four parallel lanes of a gel, a DNA sequence can be derived as in Figure 4–67 (see also Figure 4–64A). Illustrated here, however, is a more recent automated version of the procedure, in which the four sets of differently labeled fragments are analyzed by their fluorescence as they move along a single gel lane (B).

genes are active in a particular cell by binding to regulatory DNA sequences, which are usually located outside the coding regions in a gene. To understand how such proteins function, it is important to identify the specific sequences to which they bind. The standard method used for this purpose is called **DNA footprinting.** A pure DNA fragment is first labeled at one end with ^{32}P and then is cleaved with a nuclease or a chemical that makes random cuts in the phosphodiester bonds of the DNA double helix. The resultant subfragments from the labeled strand are separated on a gel and detected by autoradiography so that the pattern of DNA bands from DNA cut in the presence of a DNA-binding protein can be compared with that from DNA cut in its absence. When the protein is present, it covers the nucleotides at its binding site and protects their phosphodiester bonds from cleavage. As a result, the labeled fragments containing the binding site will be missing, leaving a gap in the gel pattern called a "footprint" (Figure 4–69A). The footprint of a protein that activates the transcription of a eucaryotic gene is shown in Figure 4–69B.

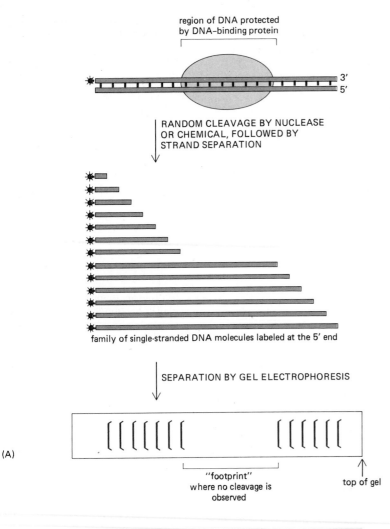

region of DNA protected
by DNA–binding protein

RANDOM CLEAVAGE BY NUCLEASE
OR CHEMICAL, FOLLOWED BY
STRAND SEPARATION

family of single-stranded DNA molecules labeled at the 5′ end

SEPARATION BY GEL ELECTROPHORESIS

"footprint"
where no cleavage is
observed

top of gel

(A)

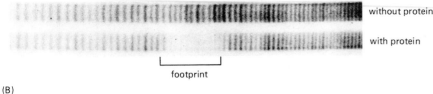

without protein

with protein

footprint

(B)

Figure 4–69 The DNA footprinting technique. (A) A protein binds tightly to a specific DNA sequence that is eight nucleotides long, thereby protecting these eight nucleotides from the cleaving agent. If the same reaction were carried out without the DNA-binding protein, a complete ladder of bands would be seen on the gel (not shown). (B) An actual footprint used to determine the binding site for a human protein that stimulates the transcription of specific eucaryotic genes. These results locate the binding site about 60 nucleotides upstream from the start site for RNA synthesis. The cleaving agent was a small, iron-containing organic molecule that normally cuts at every phosphodiester bond with nearly equal frequency. (B, courtesy of Michele Sawadogo and Robert Roeder.)

Nucleic Acid Hybridization Reactions Provide a Sensitive Way of Detecting Specific Nucleotide Sequences[43]

When an aqueous solution of DNA is heated at 100°C or exposed to a very high pH (pH ≥ 13), the complementary base pairs that normally hold the two strands of the double helix together are disrupted and the double helix rapidly dissociates into two single strands. This process, called *DNA denaturation*, was for many years thought to be irreversible. However, in 1961 it was discovered that complementary single strands of DNA will readily re-form double helices (a process called **DNA renaturation** or **hybridization**) if they are kept for a prolonged period at 65°C. Similar hybridization reactions will occur between any two single-stranded nucleic acid chains (DNA:DNA, RNA:RNA, or RNA:DNA), provided that they have a complementary nucleotide sequence.

The rate of double-helix formation during hybridization reactions is limited by the rate at which two complementary nucleic acid chains happen to collide, which depends on their concentration in the solution. Hybridization rates can therefore be used to determine the concentration of any desired DNA or RNA sequence in a mixture of other sequences. This assay requires a pure single-

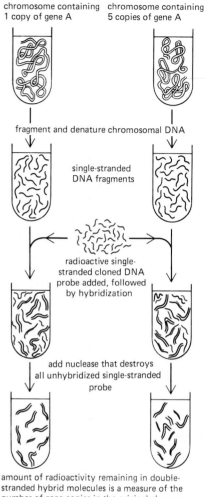

Figure 4–70 Measurement of the number of copies of a specific gene in a sample of DNA by means of DNA hybridization. The radioactive single-stranded DNA fragment used in such experiments is commonly referred to as a *DNA probe*; the chromosomal DNA is not radioactively labeled here.

chromosome containing 1 copy of gene A chromosome containing 5 copies of gene A

fragment and denature chromosomal DNA

single-stranded DNA fragments

radioactive single-stranded cloned DNA probe added, followed by hybridization

add nuclease that destroys all unhybridized single-stranded probe

amount of radioactivity remaining in double-stranded hybrid molecules is a measure of the number of gene copies in the original chromosome

stranded DNA fragment that is complementary in sequence to the nucleic acid (DNA or RNA) one wishes to detect; the DNA can be obtained by cloning, or if the sequence is short, it can be synthesized by chemical means. In either case the DNA fragment is heavily radiolabeled with ^{32}P (see Figure 4–65) so that its incorporation into double-stranded molecules can be followed during the course of a hybridization reaction. A single-stranded DNA molecule used as an indicator in this way is known as a **DNA probe;** it can be anywhere from 15 to thousands of nucleotides long.

Hybridization reactions using DNA probes are so sensitive and selective that complementary sequences present at a concentration as low as one molecule per cell can be detected (Figure 4–70). It is thus possible to determine how many copies of a particular DNA sequence (contained in the probe) are present in a cell's genome. The same technique can be used to search for related but non-identical genes; for example, once an interesting gene has been cloned from a mouse or a chicken, part of its sequence can be used as a probe to find the corresponding gene in a human.

Alternatively, DNA probes can be used in hybridization reactions with RNA rather than DNA to find out whether a cell is expressing a given gene. In this case a DNA probe that contains part of the gene's sequence is hybridized with RNA purified from the cell in question to see whether the RNA includes molecules matching the probe DNA and, if so, in what quantities. In somewhat more elaborate procedures the DNA probe is treated with specific nucleases after the hybridization is complete to determine the exact regions of the DNA probe that have paired with cellular RNA molecules. One can thereby determine the start and stop sites for RNA transcription (Figure 4–71); in the same way, one can identify the precise boundaries of the regions that are cut out of the RNA transcripts by *RNA splicing* (the intron sequences—see p. 533).

Large numbers of genes are switched on and off in elaborate patterns as an embryo develops. The hybridization of DNA probes to cellular RNAs allows one to determine whether a particular gene is off or on; moreover, when the expression of a gene changes, one can determine whether the change is due to controls that act on the transcription of DNA, the splicing of the gene's RNA, or the translation of its mature mRNA molecules into protein. Hybridization methods are in such wide use in cell biology today that it is difficult to imagine what it would be like to study gene structure and expression without them.

Northern and Southern Blotting Facilitate Hybridization with Electrophoretically Separated Nucleic Acid Molecules[44]

DNA probes are often used in conjunction with gel electrophoresis to detect the nucleic acid molecules with sequences that are complementary to all or part of the probe. The electrophoresis fractionates the many different RNA or DNA mole-

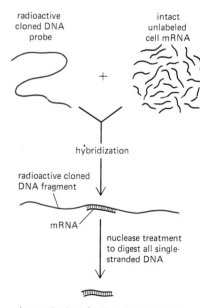

radioactive cloned DNA probe   intact unlabeled cell mRNA

hybridization

radioactive cloned DNA fragment

mRNA

nuclease treatment to digest all single-stranded DNA

characterization of radioactive nucleotides left in DNA probe reveals exact region translated into mRNA

Figure 4–71 The use of nucleic acid hybridization to determine the region of a cloned DNA fragment that is transcribed into mRNA. The method shown requires a nuclease that cuts the DNA chain only where it is not base-paired to a complementary RNA chain. Both the beginning and the end of an RNA molecule can be exactly mapped in this way; in addition, the positions of introns (intervening sequences) in eucaryotic genes are mapped by similar procedures.

cules in a crude mixture according to their size before the hybridization reaction is carried out; if molecules of only one or a few sizes become labeled with the probe, one can be certain that the hybridization was indeed specific. Moreover, the size information obtained can be invaluable in itself. An example will illustrate this point.

Suppose that one wishes to determine the nature of the defect in mutant mice that produces abnormally low amounts of *albumin*, a protein that liver cells normally secrete into the blood in large amounts. First one collects identical samples of liver tissue from defective and normal mice (the latter serving as controls) and disrupts the cells in a strong detergent to inactivate cellular nucleases that might otherwise degrade the nucleic acids. Next one separates the RNA and DNA from all of the other cell components: the proteins present are completely denatured and removed by repeated extractions with phenol—a potent organic solvent that is partly miscible with water; the nucleic acids, which remain in the aqueous phase, are then precipitated with alcohol to separate them from the small molecules of the cell. Then one separates the DNA from the RNA by their different solubilities in alcohols and degrades any contaminating nucleic acid of the unwanted type by treatment with highly specific enzymes—either RNase or DNase.

To analyze albumin-encoding RNAs with a DNA probe for such RNA, a technique called **Northern blotting** is used. First, the intact RNA molecules from defective and control liver cells are fractionated into a series of bands by gel electrophoresis. Then, to make the RNA molecules accessible to DNA probes, a replica of the gel is made by transferring ("blotting") the fractionated RNA molecules onto a sheet of nitrocellulose or nylon paper. The RNA molecules that hybridize to the radioactive DNA probe (because they contain part of the normal albumin gene sequence) are then located by incubating the paper with a solution containing the probe and detecting the hybridized probe by autoradiography (Figure 4–72). Because small nucleic acid molecules move more rapidly through the gel than large ones, the size of each RNA band that binds the probe can be determined by reference to the rates of migration of RNA molecules of known size

Figure 4–72 Northern and Southern blotting analyses. After the indicated mixture of RNA or DNA molecules is fractionated by electrophoresis through an agarose gel, the many different RNA or DNA molecules present are transferred to nitrocellulose or nylon paper by blotting. The paper sheet is then exposed to a radioactive DNA probe for a prolonged period under hybridization conditions. The sheet is washed thoroughly afterward, so that only those immobilized RNA or DNA molecules that hybridize to the probe become radioactively labeled and show up as bands on autoradiographs of the paper sheet.

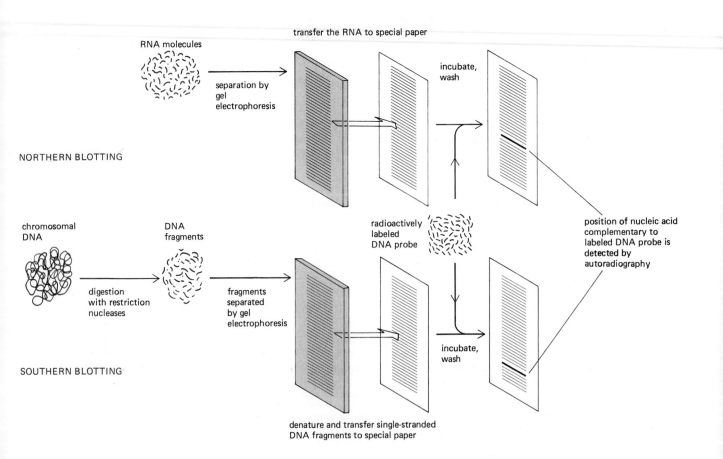

transfer the RNA to special paper

RNA molecules

separation by gel electrophoresis

incubate, wash

NORTHERN BLOTTING

chromosomal DNA

DNA fragments

radioactively labeled DNA probe

position of nucleic acid complementary to labeled DNA probe is detected by autoradiography

digestion with restriction nucleases

fragments separated by gel electrophoresis

incubate, wash

SOUTHERN BLOTTING

denature and transfer single-stranded DNA fragments to special paper

(*RNA standards*). In this way one might discover that liver cells from defective mice make albumin RNA in normal amounts and of normal size; alternatively, normal albumin RNA might be detected in greatly reduced amounts. Another possibility is that the mutant albumin RNA might be abnormally short and therefore move unusually quickly through the gel, in which case the gel blot could be retested with more selective DNA probes to reveal what part of the normal RNA is missing.

To characterize the structure of the albumin gene in the defective mice, an analogous method, called **Southern blotting,** which analyzes DNA rather than RNA, is used. Isolated DNA is first cut into readily separable fragments with restriction nucleases. The fragments are then separated according to their size by gel electrophoresis, and those complementary to the albumin DNA probe are identified by blotting and hybridization, as just described for RNA (see Figure 4–72). By repeating this procedure using different restriction nucleases, a detailed *restriction map* could be constructed for the genome in the region of the albumin gene (see p. 182). From this map one could determine if the albumin gene has been rearranged in the defective animals—for example, by the deletion or the insertion of a short DNA sequence.

Synthetic DNA Molecules Facilitate the Prenatal Diagnosis of Genetic Diseases[45]

At the same time that microbiologists were developing DNA cloning techniques, organic chemists were improving the methods for synthesizing short DNA chains. Today such *DNA oligonucleotides* are routinely produced by machines that can automatically synthesize any sequence up to 80 nucleotides long overnight. This ability to produce DNA molecules of a desired sequence makes it possible to redesign genes at will, an important aspect of genetic engineering, as will be explained in Chapter 5 (p. 266).

Another important use for DNA oligonucleotides is in the prenatal diagnosis of genetic diseases. More than 500 human genetic diseases are attributable to single-gene defects. In most of these the mutation is recessive: that is, it is harmful only when an individual inherits a defective copy of the gene from both parents. One goal of modern medicine is to identify those fetuses that carry two bad copies of the affected gene long before birth so that the mother, if she wishes, can have the pregnancy terminated. For example, in sickle-cell anemia the exact nucleotide change in the mutant gene is known (the sequence GAG is changed to GTG in the DNA strand that codes for the β chain of hemoglobin). For prenatal diagnosis, two DNA oligonucleotides are synthesized—one corresponding to the normal gene sequence in the region of the mutation and the other corresponding to the mutated sequence. By keeping these sequences short (about 20 nucleotides) and selecting a hybridization temperature where only the perfectly matched helix is stable, they can be used as radioactive probes to distinguish between the two forms of the gene. The diagnosis involves isolating DNA from fetal cells collected by amniocentesis and then using the oligonucleotide probes for Southern blotting (see Figure 4–72). A defective fetus can be readily recognized because its DNA will hybridize *only* with the oligonucleotide that is complementary to the mutant DNA sequence. For many genetic abnormalities, the exact nucleotide sequence change is not known. For an increasing number of these, prenatal diagnosis is still possible by using Southern blotting to assay for specific variations in the human genome (called *restriction fragment length polymorphisms,* or *RFLPs*) that are known to be closely linked to the defective gene.

Hybridization at Reduced Stringency Allows Distantly Related Genes to Be Identified[46]

New genes arise during evolution by the duplication and divergence of old genes and by the reutilization of portions of old genes in new combinations (p. 599). For this reason, most genes have a family of close relatives elsewhere in the genome, and some of them are likely to have a related function. Laborious methods are usually required to isolate a DNA clone corresponding to the first member of such

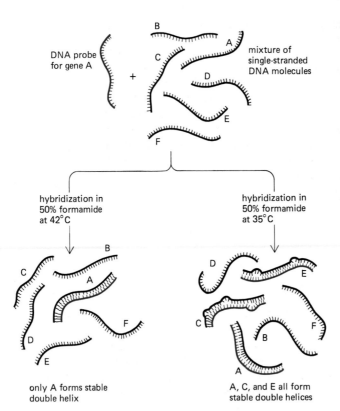

Figure 4–73 Comparison of stringent and reduced-stringency hybridization conditions. In the reaction on the left (stringent conditions), the solution is kept only a few degrees below the temperature at which a perfect DNA helix denatures (its *melting temperature*), so that the imperfect helices that can form under the conditions of reduced stringency on the right are unstable. Only the hybridization conditions on the right can be used to find genes that are nonidentical but related to gene A.

DNA probe for gene A

+

mixture of single-stranded DNA molecules

hybridization in 50% formamide at 42°C

hybridization in 50% formamide at 35°C

only A forms stable double helix

A, C, and E all form stable double helices

a *gene family* (see p. 262). However, additional genes that produce proteins with related functions can often be isolated relatively easily by using sequences from the first gene as DNA probes. Because the new genes are unlikely to have *identical* sequences, hybridizations with the DNA probe are usually carried out under conditions of "reduced stringency"—defined as conditions that allow even an imperfect match with the probe sequence to form a stable double helix (Figure 4–73).

Although using reduced stringency for hybridization carries the risk of obtaining a false signal from a chance region of short sequence homology in an unrelated DNA sequence, such hybridization represents one of the most powerful uses of recombinant DNA technology. For example, this approach has led to the isolation of a whole family of DNA-binding proteins that function as master regulators of gene expression during embryonic development in *Drosophila* (p. 937). It has also made it possible to isolate members of this same gene family from a variety of other organisms, including humans.

In Situ Hybridization Techniques Locate Specific Nucleic Acid Sequences in Chromosomes and Cells[47]

Nucleic acids, no less than other macromolecules, occupy precise positions in cells and tissues, and a great deal of potential information is lost when these molecules are extracted by homogenization. For this reason, techniques have been developed in which nucleic acid probes are used in much the same way as labeled antibodies to locate specific nucleic acid sequences *in situ*, a procedure called *in situ* **hybridization.** This can now be done both for DNA in chromosomes and for RNA in cells. Highly radioactive nucleic acid probes can be hybridized to chromosomes that have been exposed briefly to a very high pH to disrupt their DNA base pairs. The chromosomal regions that bind the radioactive probe during the hybridization step are visualized by autoradiography. The spatial resolution of this technique can be improved by labeling the DNA probes chemically instead of radioactively. Most commonly, the probes are synthesized with nucleotides that contain a biotin side chain (see p. 177), and the hybridized probes are detected by staining with a network of streptavidin and some type of marker molecule (Figure 4–74).

In situ hybridization methods have also been developed that reveal the distribution of specific RNA molecules in cells within tissues. In this case the tissues are not exposed to a high pH, so the chromosomal DNA remains double-stranded and cannot bind the probe. Instead the tissue is gently fixed so that its RNA is retained in an exposed form that will hybridize when the tissue is incubated with a complementary DNA probe. In this way striking patterns of differential gene expression have been observed in developing *Drosophila* embryos (Figure 4–75); these patterns have provided new insights into the mechanisms that distinguish between cells in different positions during development (p. 927).

Recombinant DNA Techniques Allow Even the Minor Proteins of a Cell to Be Studied[48]

Until very recently, the only proteins in a cell that could be studied easily were the relatively abundant ones. Starting with several hundred grams of cells, a major protein—one that constitutes 1% or more of the total cellular protein—can be purified by sequential chromatography steps to yield perhaps 0.1 g (100 mg) of pure protein. This quantity is sufficient for conventional amino acid sequencing, for detailed analysis of biological activity (if known), and for the production of antibodies, which can then be used to localize the protein in the cell. Moreover, if suitable crystals can be grown (usually a difficult task), the three-dimensional structure of the protein can be determined by x-ray diffraction techniques. In this way the structure and function of many abundant proteins—including hemoglobin, trypsin, immunoglobulin, and lysozyme—have been analyzed.

The vast majority of the thousands of different proteins in a eucaryotic cell, however, including many with crucially important functions, are present in only very small amounts. For most of them it is extremely difficult, if not impossible, to obtain more than a few micrograms of pure material. One of the most important contributions of DNA cloning and genetic engineering to cell biology is that they have made it possible to manufacture any of the cell's proteins, including the minor ones, in large amounts. This is done by first cloning the gene for the protein of interest and then inserting it into a special plasmid called an *expression vector*. The vector is engineered in such a way that when it is introduced into an appropriate type of bacterium, yeast, or mammalian cell, the inserted gene directs the synthesis of very large amounts of the protein of interest (see p. 265). In principle, therefore, any protein can now be made available for the kinds of detailed structural and functional studies that were previously possible only for a rare few.

Mutant Organisms Best Reveal the Function of a Gene[49]

Suppose that one has cloned a gene that codes for a newly discovered protein. How can one discover what the protein does in the cell? This has become a common problem in cell biology and one that is surprisingly difficult to solve, since neither the three-dimensional structure of the protein nor the complete nucleotide sequence of its gene is usually sufficient to deduce the protein's function. Moreover, many proteins—such as those that have a structural role in the

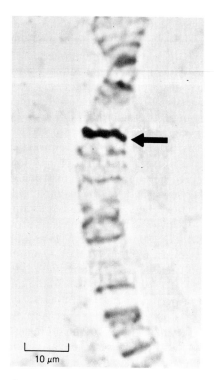

Figure 4–74 Localization of a *Drosophila* gene by *in situ* hybridization of a biotin-labeled cloned DNA probe with *Drosophila* polytene chromosomes. The DNA in this unusually large chromosome (see p. 508) has been partially denatured so that the probe can hybridize to it. After hybridization and washing, the location of the bound probe is detected by treating the chromosome with a network of horseradish peroxidase enzymes conjugated to streptavidin, as in Figure 4–58B. The peroxidase is made visible by the insoluble dark product deposited (*arrow*) when the treated chromosomes are incubated with hydrogen peroxide and a reactive dye molecule. (Courtesy of Todd Laverty and Gerald Rubin.)

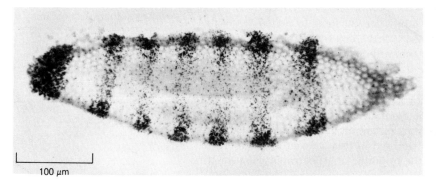

Figure 4–75 Autoradiograph of a section of a very young *Drosophila* embryo that has been subjected to *in situ* hybridization using a radioactive DNA probe complementary to a gene involved in segment development. The probe has been hybridized to RNA in the embryo, and the pattern of exposed silver grains reveals that the RNA made by the gene (called *ftz*) is localized in alternating stripes that are three or four cells wide across the embryo. At this stage of development (cellular blastoderm), the embryo contains about 6000 cells. (From E. Hafen, A. Kuriowa, and W.J. Gehring, *Cell* 37:833–841, 1984.)

cell or normally form part of a large multienzyme complex—will have no obvious activity when isolated from the other components of the functional unit.

One approach, mentioned earlier (see p. 180), is to inactivate the particular protein with a specific antibody and observe how cells are affected. Although this provides a powerful way to test protein function, for an intracellular protein the effect is transitory because the microinjected antibody is eventually diluted out during cell proliferation or destroyed by intracellular degradation.

Genetic approaches provide a more powerful solution to this problem. Mutants that lack a particular protein or, more usefully, synthesize a temperature-sensitive version of the protein (one that is inactivated by a small increase or decrease in temperature) may quickly reveal the function of the normal molecule. This approach has been immensely useful, for example, in demonstrating the function of enzymes involved in the principal metabolic pathways of bacteria and in discovering many of the gene products responsible for the orderly development of the *Drosophila* embryo. Traditionally, it has been most generally applicable to organisms that reproduce very rapidly—such as bacteria, yeasts, nematode worms, and fruit flies. By treating these organisms with agents that alter their DNA (*mutagens*), very large numbers of mutants can be isolated quickly and then screened for a particular defect of interest. For example, by screening populations of mutagen-treated bacteria for cells that stop making DNA when they are shifted from 30°C to 42°C, many temperature-sensitive mutants were isolated in the genes that encode the bacterial proteins required for DNA replication. These mutants were later used to identify and characterize the corresponding DNA replication proteins (see p. 238).

Humans do not reproduce rapidly, nor are they ever intentionally treated with mutagens. Moreover, any human with a serious defect in an essential process, such as DNA replication, would die long before birth. However, many mutations that are compatible with life—for example, tissue-specific defects in lysosomes or in cell-surface receptors—have arisen spontaneously in the human population. Analysis of the phenotypes of the affected individuals, together with studies of their cultured cells, have provided many unique insights into important cell functions (for example, see p. 329 and p. 464). Although such mutants are extremely rare, they are very efficiently discovered because of a unique human property: the mutant individuals call attention to themselves by seeking special medical care.

Cells and Organisms Containing Altered Genes Can Be Made to Order[50]

Although it is often not difficult to obtain mutants that are deficient in a particular process, such as DNA replication or eye development, it can take many years to trace the defect to a particular altered protein. Recently, recombinant DNA technology has made possible a different type of genetic approach in which one starts with a protein and creates a mutant cell or organism in which that protein (or its expression) is abnormal. Because the new approach reverses the traditional gene-to-protein direction of genetic analysis, it is commonly referred to as *reverse genetics*.

Reverse genetics begins with a protein with interesting properties that has been isolated from a cell. By methods that will be described in Chapter 5 (p. 258), the gene encoding the protein is cloned and its nucleotide sequence is determined; this sequence is then altered by biochemical means to create a mutant gene that codes for an altered version of the protein. The mutant gene is transferred into a cell, where it can integrate into a chromosome by genetic recombination (p. 239) to become a permanent part of the cell's genome. If the gene is expressed, the cell and all of its descendants will now synthesize an altered protein. If the original cell used for the gene transfer is a fertilized egg, whole multicellular organisms can be obtained that contain the mutant gene, and some of these **transgenic organisms** will pass the gene on to their progeny as a permanent part of their germ line (Figure 4–76). Such *genetic transformations* are now routinely performed with organisms as complex as fruit flies and mammals. Technically, even humans could now be transformed in this way, although such experiments are not performed for fear of the unpredictable aberrations that might occur in such individuals.

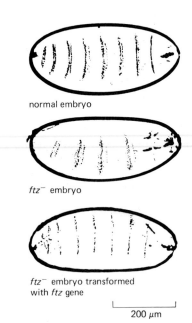

normal embryo

ftz⁻ embryo

ftz⁻ embryo transformed with *ftz* gene

|— 200 μm —|

Figure 4–76 Comparison of a normal *Drosophila* larva and two mutant larvae that contain defective *ftz* genes. One of the defective (*ftz⁻*) larvae has been transformed by the injection of a DNA clone containing the normal *ftz* gene sequence into the egg of one of its ancestors. This added DNA sequence has become permanently integrated into one of the fly's chromosomes and is therefore faithfully inherited and expressed. The *ftz* gene is required for normal development, and the addition of this gene to the genome is seen to restore the larval segments that are missing in the *ftz⁻* organism. For the technique used to make transgenic animals, see Figure 5–88, p. 268. (Courtesy of Walter Gehring.)

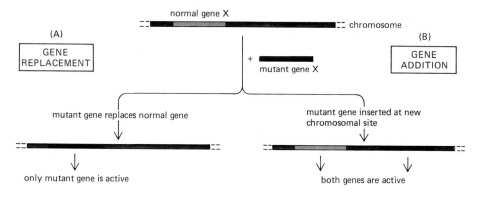

(A) **GENE REPLACEMENT**

normal gene X
chromosome

+ mutant gene X

(B) **GENE ADDITION**

mutant gene replaces normal gene

mutant gene inserted at new chromosomal site

only mutant gene is active

both genes are active

Figure 4–77 A gene whose nucleotide sequence has been altered can be inserted back into the chromosomes of an organism. In bacteria and yeast, it is possible to select for mutants that have undergone a gene *replacement* (A), in which a genetic recombination event inserts the mutant gene in place of the normal gene; in these cases, only the mutant gene remains in the cell. In higher eucaryotes, gene *addition* (B) generally occurs instead of gene replacement. The transformed cell or organism now contains the mutated gene in addition to the normal gene. Gene replacements are expected to be rare in organisms with large amounts of DNA, since the mutant gene must pair with the normal gene and replace it by genetic recombination despite a huge excess of other DNA sequences.

Engineered Genes Encoding Antisense RNA Can Create Specific Dominant Mutations[51]

In bacterial and yeast cells, which are generally haploid, an artificially introduced mutant gene will recombine with the normal gene often enough to make it possible to select cells in which the mutant gene has replaced the single copy of the normal gene (Figure 4–77A). In this way cells can be made to order that produce a specific protein only in its mutant form. The function of the normal protein can usually be determined from the phenotype of such cells. In higher eucaryotes, such as mammals or fruit flies, methods are not yet available for the easy replacement of a normal gene by the insertion of a cloned mutant gene. Genetic transformation in these organisms generally leads instead to insertion of the cloned gene at a random place in the genome, so that the cell (or the organism) ends up with the mutated gene in addition to its normal gene copies (Figure 4–77B).

It would be enormously useful to be able to create specific *dominant* mutations in higher eucaryotic cells by artificially inserting a mutant gene that would eliminate the activity of its normal counterparts in the cell. An ingenious and promising approach exploits the specificity of hybridization reactions between two complementary nucleic acid chains to accomplish this goal. Normally only one of the two DNA strands in a given portion of double helix is transcribed into RNA, and it is always the same strand for a given gene. If a cloned gene is engineered so that only the opposite DNA strand is transcribed, the result will be **antisense RNA** with a sequence complementary to the normal RNA transcripts. Antisense RNA, when synthesized in large enough amounts, will often hybridize with the "sense" RNA made by the normal genes and thereby inhibit the synthesis of the corresponding protein (Figure 4–78).

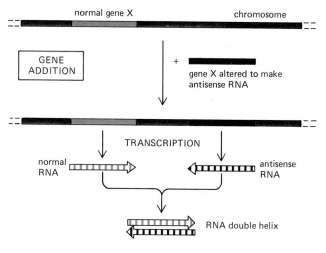

normal gene X
chromosome

GENE ADDITION

+ gene X altered to make antisense RNA

TRANSCRIPTION

normal RNA

antisense RNA

RNA double helix

formation of RNA/RNA helix prevents the synthesis of a protein product from normal gene X

Figure 4–78 The antisense RNA strategy for generating dominant mutations. As illustrated, mutant genes that have been engineered to produce RNA that is complementary in sequence to the RNA made by the normal gene can cause double-stranded RNA to form inside cells. If a large excess of the antisense RNA is produced, it can hybridize with—and thereby inactivate—most of the normal RNA produced by gene X. In the future it may become possible to inactivate any gene in this way. At present the technique seems to work for some genes but not others.

If a protein is required for the survival of the cell (or the organism), the dominant mutants just described will die, making it impossible to test the function of the protein. To avoid this problem, one can engineer a gene so that it produces antisense RNA only on command—for example, in response to an increase in temperature or to the presence of a specific signaling molecule. Cells or organisms containing such an *inducible* antisense gene can be deprived of the specific protein at a particular time, and the effect can then be followed. Although some technical problems remain to be worked out, in the future this technique for producing dominant mutations to inactivate specific genes is likely to be widely used to determine the functions of proteins in higher organisms.

Summary

Recombinant DNA technology has revolutionized the study of the cell. Any region of a cell's DNA can now be excised with restriction nucleases and inserted into a self-replicating genetic element (a plasmid or a virus) to produce a "genomic DNA clone." Alternatively, a DNA copy of any mRNA molecule can be used to produce a "cDNA clone." Unlimited amounts of a highly purified DNA molecule can thereby be obtained and its nucleotide sequence determined at a rate of hundreds of nucleotides per day, revealing the amino acid sequence of the protein it encodes. Genetic engineering techniques can then be used to create a mutant gene and insert it into a cell's chromosome so that it becomes a permanent part of the genome. If the cell used for this gene transfer is a fertilized egg, transgenic organisms can be produced that express the mutant gene and will pass it on to their progeny. Especially important for cell biology is the power that such technology gives the experimenter to alter cells in highly specific ways—allowing one to discern the effect on the cell of a change in a single protein that has been intentionally mutated by "reverse genetic" techniques.

The consequences of recombinant DNA technology are far-reaching in other ways as well. Bacteria, yeasts, or mammalian cells can be engineered to synthesize any desired protein in large quantities, making it possible to analyze the structure and function of the protein in detail, or to use the protein as a drug or a vaccine for medical purposes. In addition, highly specific DNA probes are readily obtained that—by means of nucleic acid hybridization reactions—allow determination of the spatial pattern of gene expression in a tissue, the localization of genes on chromosomes, and the discovery of genes with related functions in a gene family.

References

General

Cantor, C.R.; Schimmel, P.R. Biophysical Chemistry, 3 vols. New York: W.H. Freeman, 1980. (A comprehensive account of the physical principles underlying many biochemical and biophysical techniques.)

Freifelder, D. Physical Biochemistry, 2nd ed. New York: W.H. Freeman, 1982.

Van Holde, K.E. Physical Biochemistry, 2nd ed. Englewood, NJ: Prentice-Hall, 1985.

Cited

1. Bradbury, S. An Introduction to the Optical Microscope. Oxford, U.K.: Oxford University Press, 1984.
   Fawcett, D.W. A Textbook of Histology, 11th ed. Philadelphia: Saunders, 1986.
2. Spencer, M. Fundamentals of Light Microscopy. Cambridge, U.K.: Cambridge University Press, 1982.
3. Boon, M.E.; Drijver, J.S. Routine Cytological Staining Methods. London: Macmillan, 1986.
4. Ploem, J.S.; Tanke, H.J. Introduction to Fluorescence Microscopy. Royal Microscopical Society Microscopy Handbook No. 10. Oxford, U.K.: Oxford Scientific Publications, 1987.

Willingham, M.C.; Pastan, I. An Atlas of Immunofluorescence in Cultured Cells. Orlando, FL: Academic, 1985.

5. Allen, R.D. New observations on cell architecture and dynamics by video-enhanced contrast optical microscopy. *Annu. Rev. Biophys. Biophys. Chem.* 14:265–290, 1985.
   Inoué, S. Video Microscopy. New York: Plenum Press, 1986.
6. Pease, D.C.; Porter, K.R. Electron microscopy and ultramicrotomy. *J. Cell Biol.* 91:287s–292s, 1981.
7. Wischnitzer, S. Introduction to Electron Microscopy. 3rd ed. Elmsford, NY: Pergammon, 1981.
8. Everhart, T.E.; Hayes, T.L. The scanning electron microscope. *Sci. Am.* 226(1):54–69, 1972.
   Hayat, M.A. Introduction to Biological Scanning Electron Microscopy. Baltimore: University Park Press, 1978.
   Kessel, R.G.; Kardon, R.H. Tissues and Organs. New York: W.H. Freeman, 1979. (An atlas of vertebrate tissues seen by scanning electron microscopy.)
9. Sommerville, J.; Scheer, U., eds. Electron Microscopy in Molecular Biology: A Practical Approach. Washington, D.C.: IRL Press, 1987.
10. Heuser, J. Quick-freeze, deep-etch preparation of samples for 3-D electron microscopy. *Trends Biochem. Sci.* 6:64–68, 1981.

Pinto da Silva, P.; Branton, D. Membrane splitting in freeze-etching. *J. Cell Biol.* 45:598–605, 1970.

11. Chiu, W. Electron microscopy of frozen, hydrated specimens. *Annu. Rev. Biophys. Biophys. Chem.* 15:237–257, 1986.

Unwin, P.N.T.; Henderson, R. Molecular structure determination by electron microscopy of unstained crystalline specimens. *J. Mol. Biol.* 94:425–440, 1975.

12. Glusker, J.P.; Trueblood, K.N. Crystal Structure Analysis: A Primer. Oxford, U.K.: Oxford University Press, 1985.

13. Alzari, P.M.; Lascombe, M.B.; Poljak, R.J. Three-dimensional structure of antibodies. *Annu. Rev. Immunol.* 6:555–580, 1988.

Kendrew, J.C. The three-dimensional structure of a protein molecule. *Sci. Am.* 205(6):96-111, 1961.

Perutz, M.F. The hemoglobin molecule. *Sci. Am.* 211(5):64–76, 1964.

14. Cooke, R.M.; Cambell, I.D. Protein structure determination by NMR. *Bioessays* 8:52–56, 1988.

Shulman, R.G. NMR spectroscopy of living cells. *Sci. Am.* 248(1):86–93, 1983.

Wüthrich, K.; Wagner, G. Internal dynamics of proteins. *Trends Biochem. Sci.* 9:152–154, 1984.

15. Ammann, D. Ion Selective Microelectrodes: Principles, Design and Application. Berlin: Springer-Verlag, 1986.

Auerbach, A.; Sachs, F. Patch clamp studies of single ionic channels. *Annu. Rev. Biophys. Bioeng.* 13:269–302, 1984.

16. Grynkiewicz, G.; Poenie, M.; Tsien, R.Y. A new generation of Ca^{2+} indicators with greatly improved fluorescence properties. *J. Biol. Chem.* 260:3440–3450, 1985.

Tsien, R.Y.; Poenie, M. Fluorescence ratio imaging: a new window into intracellular ionic signalling. *Trends Biochem. Sci.* 11:450, 1986.

17. Celis, J.E.; Graessmann, A.; Logter, A., eds. Microinjection and Organelle Transplantation Techniques. London: Academic Press, 1986.

Gomperts, B.D.; Fernandez, J.M. Techniques for membrane permeabilization. *Trends Biochem. Sci.* 10:414–417, 1985.

Ostro, Marc J. Liposomes. *Sci. Am.* 256(1):102–111, 1987.

Ureta, T.; Radojkovic, J. Microinjected frog oocytes. *Bioessays* 2:221–225, 1985.

18. Freshney, R.I. Culture of Animal Cells: A Manual of Basic Technique. New York: Liss, 1987.

19. Herzenberg, L.A.; Sweet, R.G.; Herzenberg, L.A. Fluorescence-activated cell sorting. *Sci. Am.* 234(3):108–116, 1976.

Kamarck, M.E. Fluorescence-activated cell sorting of hybrid and transfected cells. *Methods Enzymol.* 151:150–165, 1987.

Nolan, G.P; Fiering, S.; Nicolas, J.F.; Herzenberg, L.A. Fluorescence- activated cell analysis and sorting of viable mammalian cells based on beta-D-galactosidase activity after transduction of *E. coli lac Z. Proc. Nat. Acad. Sci. USA.* 85:2603–2607, 1988.

20. Harrison, R.G. The outgrowth of the nerve fiber as a mode of protoplasmic movement. *J. Exp. Zool.* 9:787–848, 1910. (Possibly the first use of tissue culture.)

21. Ham, R.G. Clonal growth of mammalian cells in a chemically defined, synthetic medium. *PNAS* 53:288–293, 1965.

Loo, D.T.; Fuquay, J.I.; Rawson, C.L.; Barnes, D.W. Extended culture of mouse embryo cells without senescence: inhibition by serum. *Science* 236:200–202, 1987.

Sirabasku, D.A.; Pardee, A.B.; Sato, G.H., eds. Growth of Cells in Hormonally Defined Media. Cold Spring Harbor, NY: Cold Spring Harbor Laboratory, 1982.

22. Ruddle, F.H.; Creagan, R.P. Parasexual approaches to the genetics of man. *Annu. Rev. Genet.* 9:407–486, 1975.

23. Colowick, S.P.; Kaplan, N.O., eds. Methods in Enzymology, Vols. 1–.... San Diego, CA: Academic Press, 1955–1988. (A multivolume series containing general and specific articles on many procedures.)

Cooper, T.G. The Tools of Biochemistry. New York: Wiley, 1977.

24. Claude, A. The coming of age of the cell. *Science* 189:433–435, 1975.

deDuve, C.; Beaufay, H. A short history of tissue fractionation. *J. Cell Biol.* 91:293s–299s, 1981.

Meselson, M.; Stahl, F.W. The replication of DNA in *Escherichia coli. Proc. Natl. Acad. Sci. USA* 47:671–682, 1958. (Density gradient centrifugation was used to show the semiconservative replication of DNA.)

Palade, G. Intracellular aspects of the process of protein synthesis *Science* 189:347–358, 1975.

Sheeler, P. Centrifugation in Biology and Medical Science. New York: Wiley, 1981.

25. Morré, D.J; Howell, K.E.; Cook, G.M.W.; Evans, W.H., eds. Cell Free Analysis of Membrane Traffic. New York: Liss, 1986.

Nirenberg, N.W.; Matthaei, J.H. The dependence of cell free protein synthesis in *E. coli* on naturally occurring or synethetic polyribonucleotides. *Proc. Natl. Acad. Sci. USA* 47:1588–1602, 1961.

Racker, E. A New Look at Mechanisms in Bioenergetics. New York: Academic Press, 1976. (Cell-free systems in the working out of energy metabolism.)

Zamecnik, P.C. An historical account of protein synthesis with current overtones—a personalized view. *Cold Spring Harbor Symp. Quant. Biol.* 34:1–16, 1969.

26. Dean, P.D.G.; Johnson, W.S.; Middle, F.A. Affinity Chromatography: A Practical Approach. Arlington, VA: IRL Press, 1985.

Gilbert, M.T. High Performance Liquid Chromatography. Littleton, MA.: John Wright-PSG, 1987.

27. Andrews, A.T. Electrophoresis, 2nd ed. Oxford, U.K.: Clarendon Press, 1986.

Laemmli, U.K. Cleavage of the structural proteins during the assembly of the head of bacteriophage T4. *Nature* 227:680–685, 1970.

28. Celis, J.E.; Bravo, R., eds. Two-Dimensional Gel Electrophoresis of Proteins. New York: Academic Press, 1983.

O'Farrell, P.H. High-resolution two-dimensional electrophoresis of proteins. *J. Biol. Chem.* 250:4007–4021, 1975.

29. Cleveland, D.W.; Fisher, S.G.; Kirschner, M.W.; Laemmli, U.K. Peptide mapping by limited proteolysis in sodium dodecyl sulfate and analysis by gel electrophoresis. *J. Biol. Chem.* 252:1102–1106, 1977.

Ingram, V.M. A specific chemical difference between the globins of normal human and sickle-cell anemia hemoglobin. *Nature* 178:792–794, 1956.

30. Edman, P.; Begg, G. A protein sequenator. *Eur. J. Biochem.* 1:80–91, 1967. (First automated determination of sequencer.)

Hewick, R.M.; Hunkapiller, M.W.; Hood, L.E.; Dreyer, W.J. A gas-liquid-solid phase peptide and protein sequenator. *J. Biol. Chem.* 256:7990–7997, 1981.

Sanger, F. The arrangement of amino acids in proteins. *Adv. Protein Chem.* 7:1–67, 1952.

Walsh, K.A.; Ericsson, L.H.; Parmelee, D.C.; Titani, K. Advances in protein sequencing. *Annu. Rev. Biochem.* 50:261–284, 1981.

31. Chase, G.D.; Rabinowitz, J.L. Principles of Radio Isotope Methodology, 2nd ed. Minneapolis: Burgess, 1962.

Dyson, N.A. An Introduction to Nuclear Physics with Applications in Medicine and Biology. Chichester, U.K.: Horwood, 1981.

32. Calvin, M. The path of carbon in photosynthesis. *Science* 135:879–889, 1962. (One of the earliest uses of radio isotopes in biology.)

Rogers, A.W. Techniques of Autoradiography, 3rd ed. New York: Elsevier/ North Holland, 1979.

33. Anderton, B.H.; Thorpe, R.C. New methods of analyzing for antigens and glycoproteins in complex mixtures. *Immunol. Today* 2:122–127, 1980.

Coons, A.H. Histochemistry with labeled antibody. *Int. Rev. Cytol.* 5:1–23, 1956.

34. Milstein, C. Monoclonal antibodies. *Sci. Am.* 243(4):66–74, 1980.

    Yelton, D.E.; Scharff, M.D. Monoclonal antibodies: a powerful new tool in biology and medicine. *Annu. Rev. Biochem.* 50:657–680, 1981.

35. Mabuchi, I.; Okuno, M. The effect of myosin antibody on the division of starfish blastomeres. *J. Cell Biol.* 74:251–263, 1977.

    Mulcahy, L.S.; Smith, M.R.; Stacey, D.W. Requirement for ras proto-oncogene function during serum stimulated growth of NIH 3T3 cells. *Nature* 313:241–243, 1985.

36. Drlica, K. Understanding DNA and Gene Cloning. New York: Wiley, 1984.

    Sambrook, J.; Fritsch, E.F.; Maniatis, T. Molecular Cloning: A Laboratory Manual, 2nd ed. Cold Spring Harbor, NY: Cold Spring Harbor Laboratory, 1989.

    Watson, J.D.; Tooze, J. The DNA Story: A Documentary History of Gene Cloning. New York: W.H. Freeman, 1981.

    Watson, J.D.; Tooze, J.; Kurtz, D.T. Recombinant DNA: A Short Course. New York: W.H. Freeman, 1983.

37. Garoff, H. Using recombinant DNA techniques to study protein targeting in the eucaryotic cell. *Annu. Rev. Cell Biol.* 1:403–445, 1985.

    Jackson, Ian J. The real reverse genetics: targeted mutagenesis in the mouse. *Trends Genet.* 3:119, 1987.

    Kelly, J.H.; Darlington, G.J. Hybrid genes: molecular approaches to tissue-specific gene regulation. *Annu. Rev. Genet.* 19:273–296, 1985.

38. Nathans, D.; Smith, H.O. Restriction endonucleases in the analysis and restructuring of DNA molecules. *Annu. Rev. Biochem.* 44:273–293, 1975.

    Smith, H.O. Nucleotide sequence specificity of restriction endonucleases. *Science* 205:455–462, 1979.

39. Cohen, S.N. The manipulation of genes. *Sci. Am.* 233(1):24–33, 1975.

    Maniatis, T.; et al. The isolation of structural genes from libraries of eucaryotic DNA. *Cell* 15:687–701, 1978.

    Novick, R.P. Plasmids. *Sci. Am.* 243(6):102–127, 1980.

40. Cantor, C.R.; Smith, C.L.; Mathew, M.K. Pulsed-field gel electrophoresis of very large DNA molecules. *Annu. Rev. Biophys. Biophys. Chem.* 17:287–304, 1988.

    Maxam, A.M.; Gilbert, W. A new method of sequencing DNA. *Proc. Natl. Acad. Sci. USA* 74:560–564, 1977.

    Southern, E.M. Gel electrophoresis of restriction fragments. *Methods Enzymol.* 68:152–176, 1979.

41. Rigby, P.W.; Dieckmann, M.; Rhodes, C.; Berg, P. Labeling deoxyribonucleic acid to high specific acitivity *in vitro* by nick translation with DNA polymerase I. *J. Mol. Biol.* 113:237–251, 1977.

42. Galas, D.J.; Schmitz, A. DNAse footprinting: a simple method for the detection of protein-DNA binding specificity. *Nucleic Acids Res.* 5:3157–3170, 1978.

    Gilbert, W. DNA sequencing and gene structure. *Science* 214:1305–1312, 1981.

    Prober, J.M.; et al. A system for rapid DNA sequencing with fluorescent chain terminating dideoxynucleotides. *Science* 238:336–341, 1987.

    Sanger, F. Determinating of nucleotide sequences in DNA. *Science* 214:1205–1210, 1981.

    Tullius, T.D. Chemical "snapshots" of DNA: using the hydroxyl radical to study the structure of DNA and DNA protein complexes. *Trends Biochem. Sci.* 12:297–300, 1987.

43. Berk, A.J.; Sharp, P.A. Sizing and mapping of early adenovirus mRNAS by gel electrophoresis of S1 endonuclease digested hybrids. *Cell* 12:721–732, 1977.

Hood, L.E.; Wilson, J.H.; Wood, W.B. Molecular Biology of Eucaryotic Cells: A Problems Approach, pp. 56–61, 192–210. Menlo Park, CA: Benjamin-Cummings, 1975.

Wetmer, J.G. Hybridization and renaturation kinetics of nucleic acids. *Annu. Rev. Biophys. Bioeng.* 5:337–361, 1976.

44. Alwine, J.C.; Kemp, D.J.; Stark, G.R. A method for detection of specific RNAs in agarose gels by transfer to diazobenzyloxymethyl-paper and hybridization with DNA probes. *Proc. Natl. Acad. Sci. USA* 74:5350–5354, 1977.

    Southern, E.M. Detection of specific sequences among DNA fragments separated by gel electrophoresis. *J. Mol. Biol.* 98:503–517, 1975.

45. Itakura, K.; Rossi, J.J., Wallace, R.B. Synthesis and use of synthetic oligonucleotides. *Annu. Rev. Biochem.* 53:323–356, 1984.

    Ruddle, F.H. A new era in mammalian gene mapping: somatic cell genetics and recombinant DNA methodologies *Nature* 294:115–119, 1981.

    White, R.; Lalovel, J.M. Chromosome mapping with DNA markers. *Sci. Am.* 258(2):40–48, 1988.

46. McGinnis, W.; Garber, R.L.; Wirz, J.; Kuroiwa, A.; Ghering, W.J. A homologous protein-coding sequence in *Drosophila* homeotic genes and its conservation in other metazoans. *Cell* 37:403–408, 1984.

47. Gerhard, D.S.; Kawasaki, E.S.; Bancroft, F.C.; Szabo, P. Localization of a unique gene by direct hybridization. *Proc. Natl. Acad. Sci. USA* 78:3755–3759, 1981.

    Huten, E.; Kuroiwa, A.; Gehring, W.J. Spatial distribution of transcripts from the segmentation gene *fushi tarazu* during *Drosophila* embryonic development. *Cell* 37:833–841, 1984.

    Pardue, M.L.; Gall, J.G. Molecular hybridization of radioactive DNA to the DNA of cytological preparations. *Proc. Natl. Acad. Sci. USA* 64:600–604, 1969.

48. Abelson, J.; Butz, E., eds. Recombinant DNA. *Science* 209:1317–1438, 1980.

    Gilbert, W.; Villa-Komaroff, L. Useful proteins from recombinant bacteria. *Sci. Am.* 242(4):74–94, 1980.

49. Lederberg, J.; Lederberg, E.M. Replica plating and indirect selection of bacterial mutants. *J. Bacteriol.* 63:399–406, 1952.

    Nüsslein-Volhard, C.; Wieschaus, E. Mutations affecting segment number and polarity in *Drosophila*. *Nature* 287:795–801, 1980.

50. Palmiter, R.D.; Brinster, R.L. Germ line transformation of mice. *Annu. Rev. Genet.* 20:465–499, 1986.

    Pellicer, A.; et al. Altering genotype and phenotype by DNA-mediated gene transfer. *Science* 209:1414–1422, 1980.

    Rubin, G.M.; Spradling, A.C. Genetic transformation of *Drosophila* with transposable element vectors. *Science* 218:348–353, 1982.

    Shortle, D.; Nathans, D. Local mutagenesis: a method for generating viral mutants with base substitutions in preselected regions of the viral genome. *Proc. Natl. Acad. Sci. USA* 75:2170–2174, 1978.

    Struhl, K. The new yeast genetics. *Nature* 305:391–397, 1983.

51. Melton, D.A.; Rebagliati, M.R. Antisense RNA injections in fertilized eggs as a test for the function of localized mRNAs. *J. Embryol. Exp. Morphol.* 97:211–221, 1986.

    Rosenberg, V.B.; Preiss, A.; Seifert, E.; Jäckle, H.; Knipple, D.C. Production of phenocopies by Krüppel anti-sense RNA injection into *Drosophila* embryos. *Nature* 313:703–706, 1985.

    Weintraub, H.; Izant, J.G.; Harland, R.M. Anti-sense RNA as a molecular tool for genetic analysis. *Trends Genet.* 1:22–25, 1985.

The Molecular Organization of Cells

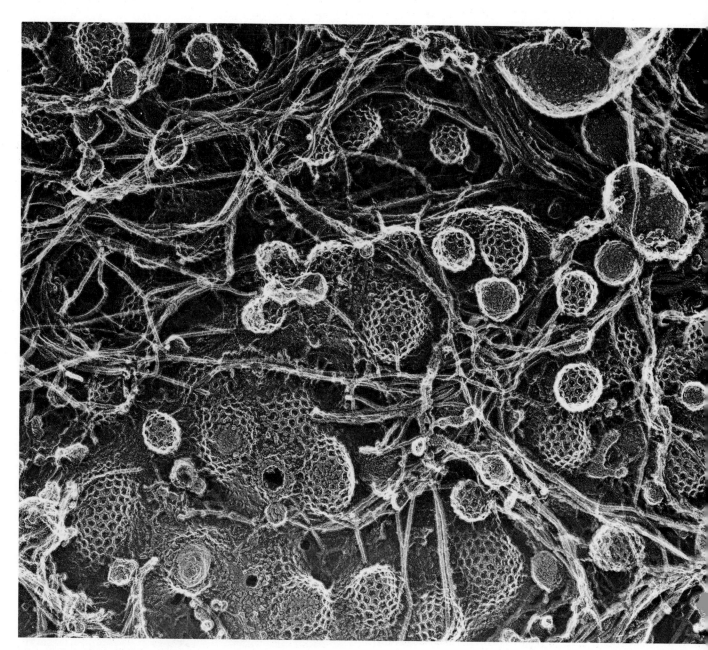

Electron micrograph of coated pits and vesicles on the inner
face of the plasma membrane of a mouse liver cell. Numerous
keratin filaments are also seen. (From N. Hirokawa and J. Heuser,
Cell 30:395–406, 1982.)

Electron micrograph of DNA spilling out of a single disrupted human chromosome. Only about half of the DNA in one chromosome is shown. (Courtesy of James Paulson and Ulrich Laemmli.)

Basic Genetic Mechanisms

The ability of cells to maintain a high degree of order in a chaotic universe stems from the genetic information that is expressed, maintained, replicated, and occasionally improved by the basic genetic processes—*RNA and protein synthesis, DNA repair, DNA replication,* and *genetic recombination.* These processes, which produce and maintain the proteins and nucleic acids of a cell, are all one-dimensional: the information in a linear sequence of nucleotides is used to specify either another linear chain of nucleotides (a DNA or RNA molecule) or a linear chain of amino acids (a protein molecule). Genetic events are therefore conceptually simple compared with most other cellular processes, which are largely the result of information expressed in the complex three-dimensional surfaces of protein molecules. Perhaps that is why we understand more about genetic mechanisms than about most other cellular processes.

In this chapter we examine the molecular machinery that repairs, replicates, and alters the DNA of the cell. We shall see that the machinery depends on enzymes that cut, copy, and recombine nucleotide sequences. We explain how these and other enzymes can be parasitized by viruses, plasmids, and transposable genetic elements, which not only direct their own replication but can alter the cell genome by genetic recombination events. Finally, we discuss how the basic genetic mechanisms can be exploited in the laboratory to isolate genes and their products and to make them to order in virtually unlimited quantities.

First, however, we reconsider a central topic mentioned briefly in Chapter 3: the mechanisms of RNA and protein synthesis.

RNA and Protein Synthesis

Proteins constitute more than half the total dry mass of a cell, and their synthesis is central to cell maintenance, growth, and development. Protein synthesis depends on the collaboration of several classes of RNA molecules and requires a series of preparatory steps. First, a molecule of *messenger RNA (mRNA)* must be copied from the DNA that encodes the protein to be synthesized. Meanwhile, in the cytoplasm, each of the 20 amino acids from which the protein is to be built must be attached to its specific *transfer RNA (tRNA)* molecule, and the subunits of the ribosome on which the new protein is to be made must be preloaded with auxiliary protein factors. Protein synthesis begins when all of these components come together in the cytoplasm to form a functional ribosome. As a single mole-cule of mRNA moves stepwise through each ribosome, the sequence of nucleotides in the mRNA molecule is translated into a corresponding sequence of amino acids to produce a distinctive protein chain. We shall begin by considering how the many different RNA molecules in a cell are made.

RNA Polymerase Copies DNA into RNA: the Process of DNA Transcription[1]

RNA is synthesized on a DNA template in a process known as **DNA transcription.** Transcription generates the mRNAs that carry the information for protein synthe-sis, as well as the transfer, ribosomal, and other RNA molecules that have structural and catalytic functions. These RNA molecules are synthesized by **RNA polymerase** enzymes, which make an RNA copy of a DNA sequence. In eucaryotes, different RNA polymerase molecules synthesize different types of RNA, but most of what we know about RNA polymerase comes from studies on bacteria, where a single species of the enzyme mediates all RNA synthesis.

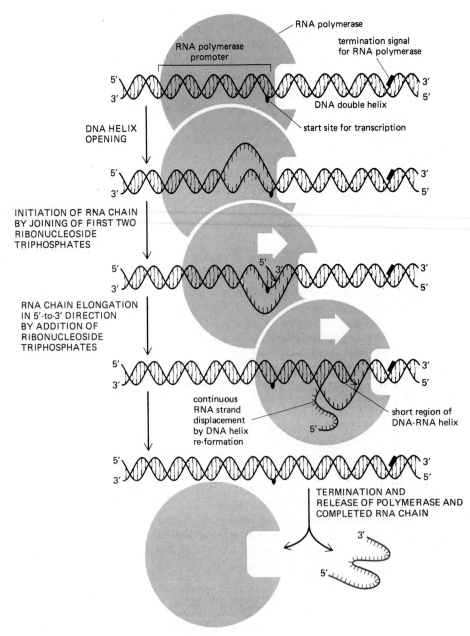

Figure 5–1 The synthesis of an RNA molecule by RNA polymerase. The enzyme begins its synthesis at a special start signal on the DNA called a promoter and completes its synthesis at a termination (stop) signal, whereupon both the polymerase and its completed RNA chain are released. During RNA chain elongation, polymerization rates average about 30 nucleotides per second at 37°C. Therefore, an RNA chain of 5000 nucleotides takes about 3 minutes to complete.

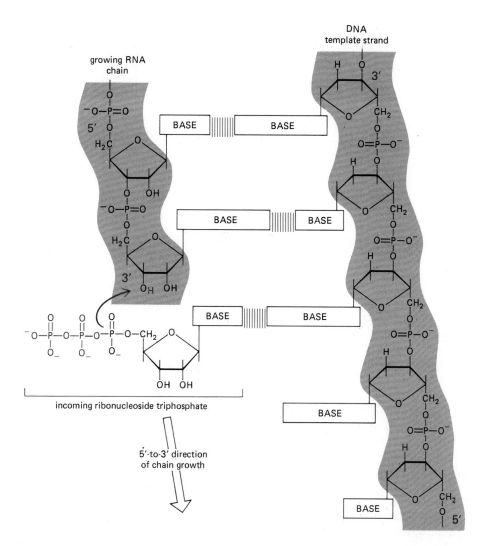

growing RNA
chain

DNA
template strand

BASE

BASE

BASE

BASE

BASE

BASE

BASE

incoming ribonucleoside triphosphate

5'-to-3' direction
of chain growth

Figure 5–2 The chain elongation reaction catalyzed by an RNA polymerase enzyme. In each step an incoming ribonucleoside triphosphate is selected for its ability to base-pair with the exposed DNA template strand; a ribonucleoside monophosphate is then added to the growing 3'-OH end of the RNA chain (*colored arrow*), and pyrophosphate is released (*colored atoms*). The new RNA chain therefore grows by one nucleotide at a time in the 5'-to-3' direction, and it is complementary in sequence to the DNA template strand. The reaction is driven both by the favorable free-energy change that accompanies the release of pyrophosphate and by the subsequent hydrolysis of the pyrophosphate to inorganic phosphate (see p. 74).

The bacterial RNA polymerase is a large multisubunit enzyme associated with several additional protein subunits that enter and leave the polymerase-DNA complex at different stages of transcription (see p. 524). Free RNA polymerase molecules collide randomly with the chromosome, sticking only weakly to most DNA. However, the polymerase binds very tightly when it collides with a specific DNA sequence, called the **promoter,** that contains the *start site* for RNA synthesis and signals where RNA synthesis should begin. The reactions that ensue are outlined in Figure 5–1. After binding to the promoter, the RNA polymerase opens up a local region of the double helix to expose the nucleotides on a short stretch of DNA on each strand. One of the two exposed DNA strands acts as a template for complementary base-pairing with incoming ribonucleoside triphosphate monomers, two of which are joined together by the polymerase to begin an RNA chain. The RNA polymerase molecule then moves stepwise along the DNA, unwinding the DNA helix just ahead to expose a new region of template for complementary base-pairing. In this way the growing RNA chain is extended by one nucleotide at a time in the 5'-to-3' direction (Figure 5–2). The chain elongation process continues until the enzyme encounters a second special sequence in the DNA, the **termination signal,** at which point the polymerase releases both the DNA template and the newly made RNA chain (see also Figure 5–6B).

As the enzyme moves, an RNA-DNA double helix is formed at the enzyme's active site. This helix is a very short one because the DNA-DNA helix immediately rewinds at the rear of the polymerase, displacing the RNA just made (Figure 5–3). As a result, each completed RNA chain is released from the DNA template as a free, single-stranded RNA molecule, typically between 70 and 10,000 nucleotides long.

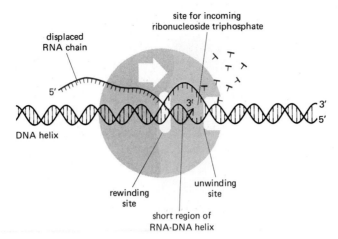

Figure 5–3 A moving RNA polymerase molecule is continuously unwinding the DNA helix ahead of the polymerization site while rewinding the two DNA strands behind this site to displace the newly formed RNA chain. A short region of RNA-DNA helix (about 17 nucleotide pairs for the bacterial enzyme) is therefore formed only transiently, and the final RNA product is released as a single-stranded copy of one of the two DNA strands.

The Promoter Sequence Defines Which DNA Strand Is to Be Transcribed[2]

In principle, any region of the DNA double helix could be copied into two different mRNA molecules—one from each of the two DNA strands. In reality, only one DNA strand is used as a template in each region, and the RNA made is therefore equivalent in nucleotide sequence to the opposite, nontemplate DNA strand. Which of the two strands is copied is determined by the promoter, which is an oriented DNA sequence that points the RNA polymerase in one direction or the other. Because RNA chains are synthesized only in the 5′-to-3′ direction, the promoter determines which DNA strand is copied (Figure 5–4). The DNA strand that is copied into RNA is often different for neighboring genes, as illustrated for a small region of a chromosome in Figure 5–5.

DNA footprinting studies (p. 187) indicate that the *E. coli* RNA polymerase covers a large region of DNA when it binds to a promoter; the polymerase "footprint" is about 60 base pairs long, extending from about 40 nucleotides "upstream" to 20 nucleotides "downstream" from the start site for transcription. Comparison of the DNA sequences of many strong promoters reveals that the polymerase primarily recognizes two highly conserved DNA sequences, each six nucleotides long, which are located upstream from the start site and separated from each other by about 17 nucleotides of unrecognized DNA (Figure 5–6A). Conserved sequences found in all examples of a particular type of regulatory region in DNA, such as promoters, are called **consensus sequences.** For example, when large numbers of *E. coli* promoters are compared, the two consensus hexanucleotide sequences are $T_{82} G_{78} A_{65} C_{54} A_{95}$ and $T_{80} A_{95} T_{45} A_{60} A_{50} T_{96}$ (where the number is the percentage of cases where the nucleotide is present at that position in the sequence; 100 means that the nucleotide is always present, whereas 25 means that it is present in one of four promoters—the frequency expected by chance). In general, strong promoters have sequences that match the two promoter consensus sequences closely, whereas weak promoters (those associated with genes that produce relatively small amounts of mRNA) match these sequences less well.

Eucaryotic cells have three different RNA polymerases. One of these makes all of the RNAs that code for proteins (that is, the mRNAs); the other two make RNA molecules with structural and catalytic roles (such as ribosomal RNAs and transfer

an RNA polymerase that moves from right to left makes RNA by using the top strand as a template

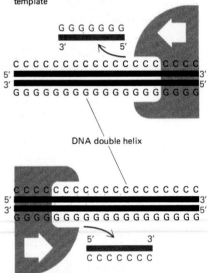

Figure 5–4 Because the DNA strand serving as template must be traversed from its 3′ end to its 5′ end (see Figure 5–2), the direction of RNA polymerase movement determines which of the two DNA strands will serve as a template for the synthesis of RNA, as shown in the diagram. The direction of polymerase movement is, in turn, determined by the orientation of the promoter sequence at which the RNA polymerase starts.

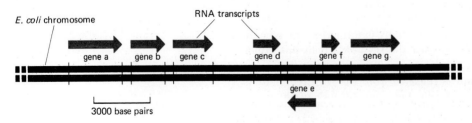

Figure 5–5 A short portion of a typical bacterial chromosome, illustrating the pattern of DNA transcription involved in the expression of several neighboring genes.

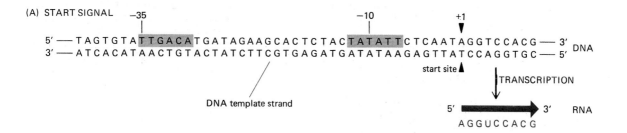

(A) START SIGNAL

(B) STOP SIGNAL

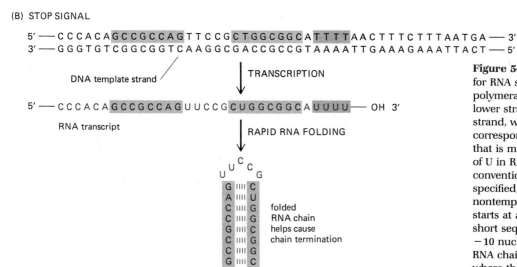

Figure 5–6 The start and stop signals for RNA synthesis by bacterial RNA polymerase of *E. coli*. Note that the lower strand of DNA is the template strand, whereas the upper strand corresponds in sequence to the RNA that is made (except for the substitution of U in RNA for T in DNA—see p. 98). By convention, when a DNA sequence is specified, it is presented for the nontemplate strand. (A) The polymerase starts at a promoter sequence. Two short sequences (*color*), about −35 and −10 nucleotides from the start of the RNA chain, are thought to determine where the polymerase binds; together with the start site, these sequences define a promoter. Major modifications in either of these two consensus sequences eliminate promoter activity, while modifications elsewhere do not. (B) The polymerase stops when it synthesizes a run of U residues from a complementary run of A residues on the template strand, provided that it has just synthesized a self-complementary RNA nucleotide sequence (shown in gray). This combination of DNA sequences therefore constitutes a termination signal. The precise sequence of nucleotides in the self-complementary region can vary; it is the hairpin helix that rapidly forms in this region of the newly synthesized RNA chain that is crucial for stopping transcription.

RNAs). All three are large multisubunit enzymes that resemble the bacterial enzyme, but the promoters each enzyme recognizes are more complex and not as well characterized (see p. 526). It is unclear why both bacterial and eucaryotic RNA polymerases are such complicated molecules, with multiple subunits and a total mass of more than 500,000 daltons, when some bacterial viruses encode single-chain RNA polymerases of one-fifth this mass that catalyze RNA synthesis at least as well as the host cell enzyme. Presumably the multiple subunit composition of the cellular RNA polymerases is important for regulatory aspects of cellular RNA synthesis that have not yet been well defined.

This outline of DNA transcription omits many details; usually other complex steps must occur before an mRNA molecule is produced. For example, *gene regulatory proteins* help to determine which regions of DNA are transcribed by the RNA polymerase and thereby play a major part in determining which proteins are made by a cell. Moreover, although mRNA molecules are produced directly by DNA transcription in procaryotes, in higher eucaryotic cells most RNA transcripts are altered extensively—by a process called *RNA splicing*—before they leave the cell nucleus and enter the cytoplasm as mRNA molecules (see p. 531). These aspects of mRNA production will be discussed in Chapters 9 and 10, when we consider the cell nucleus and the control of gene expression. For now, let us assume that functional mRNA molecules have been produced and proceed to examine how they direct protein synthesis.

Transfer RNA Molecules Act as Adaptors That Translate Nucleotide Sequences into Protein Sequences[3]

All cells contain a set of **transfer RNAs (tRNAs),** each of which is a small RNA molecule (most have a length between 70 and 90 nucleotides). The tRNAs, by binding at one end to a specific codon in the mRNA and at their other end to the

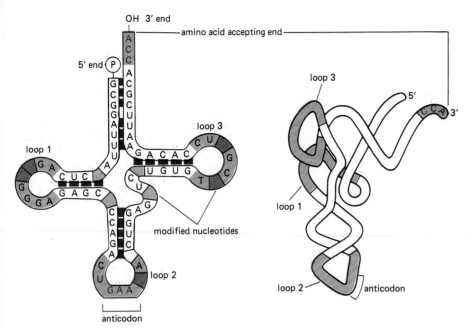

Figure 5–7 The structure of a typical tRNA molecule. The base-paired regions in the molecule are shown schematically at the left in a "cloverleaf" structure, and an outline of the overall three-dimensional conformation determined by x-ray diffraction is shown at the right. Note that the molecule is L-shaped, with one end designed to accept the amino acid while the other contains the three nucleotides of the anticodon. The amino acid is attached to the A residue of a CCA sequence at the 3' end of the molecule (see Figure 5–11).

amino acid specified by that codon, enable amino acids to line up according to the sequence of nucleotides in the mRNA. Each tRNA is designed to carry only one of the 20 amino acids used for protein synthesis: a tRNA that carries glycine is designated tRNA^{Gly}, and so on. Each of the 20 amino acids has at least one type of tRNA assigned to it, and most have several. Before an amino acid is incorporated into a protein chain, it is attached by its carboxyl end to the 3' end of an appropriate tRNA molecule. This attachment serves two purposes. First, and most important, it covalently links the amino acid to a tRNA containing the correct **anticodon**—the sequence of three nucleotides that is complementary to the three-nucleotide *codon* that specifies that amino acid on an mRNA molecule. Codon-anticodon pairings enable each amino acid to be inserted into a growing protein chain according to the dictates of the sequence of nucleotides in the mRNA, thereby allowing the genetic code to be used to translate nucleotide sequences into protein sequences. This is the essential "adaptor" function of the tRNA molecule: with one end attached to an amino acid and the other paired to a codon, the tRNA converts sequences of nucleotides into sequences of amino acids.

The second function of the amino acid attachment is to activate the amino acid by generating a high-energy linkage at its carboxyl end so that it can react with the amino group of the next amino acid in the sequence to form a *peptide bond*. The activation process is necessary for protein synthesis because nonactivated amino acids cannot be added directly to a growing polypeptide chain. (By contrast, the reverse process, in which a peptide bond is hydrolyzed by the addition of water, can occur spontaneously.)

The function of a tRNA molecule depends on its precisely folded three-dimensional structure. A few tRNAs have been crystallized and their complete structures determined by x-ray diffraction analyses. Both complementary base-pairings and unusual base interactions are required to fold a tRNA molecule (see Figure 3–16, p. 103). The nucleotide sequences of tRNA molecules from many different organisms reveal that tRNAs can form the loops and base-paired stems of a "cloverleaf" structure, and all are thought to fold further to adopt the L-shaped conformation detected in crystallographic analyses (Figure 5–7). In the native structure, the amino acid is attached to one end of the "L," while the anticodon is located at the other (Figure 5–8).

The nucleotides in a completed nucleic acid chain (like the amino acids in proteins, p. 416) can be covalently modified to modulate the biological activity of the nucleic acid molecule. Such posttranscriptional modifications are especially common in tRNA molecules, which contain many different modified nucleotides

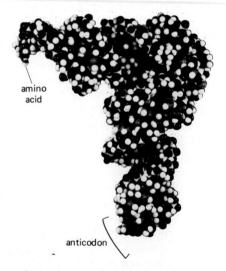

Figure 5–8 A space-filling model of a tRNA molecule with its bound amino acid. There are many different tRNA molecules, including at least one for each different amino acid. Although they differ in nucleotide sequence, they are all folded in a similar way. The particular tRNA molecule shown binds phenylalanine and is therefore denoted tRNA^{Phe}. (Courtesy of Sung-Hou Kim.)

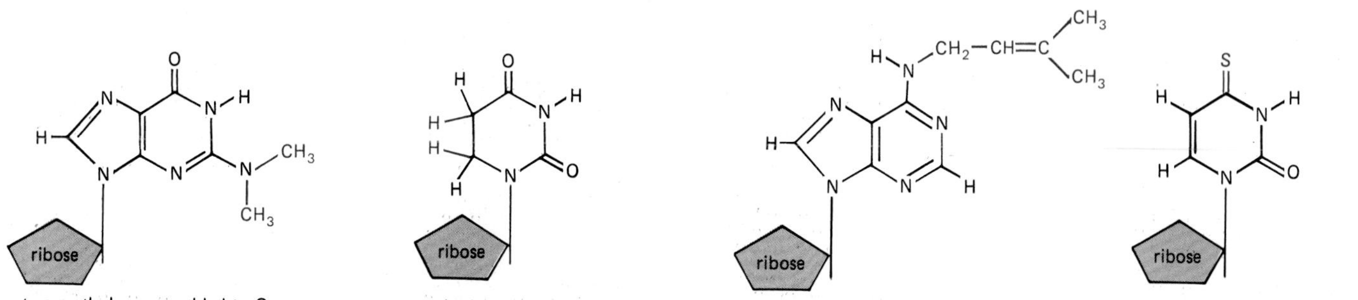

two methyl groups added to G
(N, N-dimethyl G)

two hydrogens added to U
(dihydro U)

isopentenyl group added to A
[N^6-(Δ^2-isopentenyl) A]

sulfur replaces oxygen in U
(4-thiouridine)

(Figure 5–9). Some of the modified nucleotides affect the conformation and base-pairing of the anticodon and thereby facilitate the recognition of the appropriate mRNA codon by the tRNA molecule.

Specific Enzymes Couple Each Amino Acid to Its Appropriate tRNA Molecule[4]

Only the tRNA molecule, and not its attached amino acid, determines where the amino acid is added during protein synthesis. This was established by an ingenious experiment in which an amino acid (cysteine) was chemically converted into a different amino acid (alanine) after it was already attached to its specific tRNA. When such "hybrid" tRNA molecules were used for protein synthesis in a cell-free system, the wrong amino acid was inserted at every point in the protein chain where that tRNA was used. Thus the success of decoding is crucially dependent on the accuracy of the mechanism that normally links each activated amino acid specifically to its corresponding tRNA molecules.

How does a tRNA molecule become covalently linked to the one amino acid out of 20 that is its appropriate partner? The mechanism depends on enzymes called **aminoacyl-tRNA synthetases** that couple each amino acid to its appropriate set of tRNA molecules. There is a different synthetase enzyme for every amino acid (20 synthetases in all): one attaches glycine to all tRNA^{Gly} molecules, another attaches alanine to all tRNA^{Ala} molecules, and so on. The coupling reaction that creates an **aminoacyl-tRNA** molecule is catalyzed in two steps, as illustrated in Figure 5–10. The structure of the amino acid-RNA linkage is shown in Figure 5–11.

Figure 5–9 A few of the unusual nucleotides found in tRNA molecules. These nucleotides are produced by covalent modification of a normal nucleotide after it has been incorporated into a polynucleotide chain. In most tRNA molecules about 10% of the nucleotides are modified (see Figure 5–7).

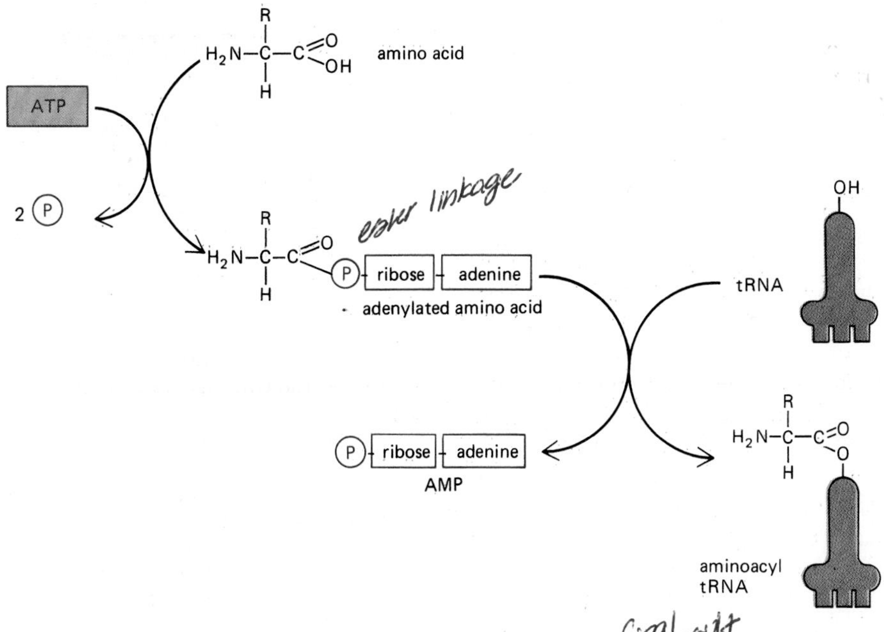

Figure 5–10 The two-step process in which an amino acid is activated for protein synthesis by an aminoacyl-tRNA synthetase enzyme. As indicated, the energy of ATP hydrolysis is used to attach each amino acid to its tRNA molecule in a high-energy linkage. The amino acid is first activated through the linkage of its carboxyl group directly to an AMP moiety, forming an *adenylated amino acid*; the linkage of the AMP, normally an unfavorable reaction, is driven by the hydrolysis of the ATP molecule that donates the AMP. Without leaving the synthetase enzyme, the AMP-linked carboxyl group on the amino acid is then transferred to a hydroxyl group on the sugar at the 3′ end of the tRNA molecule. This transfer joins the amino acid by an activated ester linkage to the tRNA and forms the final aminoacyl-tRNA molecule. The synthetase enzyme is not shown in these diagrams.

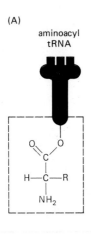

(A)

aminoacyl
tRNA

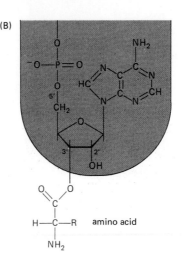

(B)

amino acid

Figure 5–11 The structure of the aminoacyl-tRNA linkage. The carboxyl end of the amino acid forms an ester bond to ribose. Because the hydrolysis of this ester bond is associated with a large favorable change in free energy, an amino acid held in this way is said to be *activated*. (A) Schematic of the structure. (B) Actual structure corresponding to boxed region in (A). As in Figure 5–10, the "R-group" on the amino acid indicates one of the 20 side chains (see pp. 54–55).

Although the tRNA molecules serve as the final adaptors in converting nucleotide sequences into amino acid sequences, the aminoacyl-tRNA synthetase enzymes are adaptors of equal importance to the decoding process. Thus the genetic code is translated by two sets of adaptors that act sequentially, each matching one molecular surface to another with great specificity; it is their combined action that associates each sequence of three nucleotides in the mRNA molecule—each codon—with its particular amino acid (Figure 5–12).

5-6 Amino Acids Are Added to the Carboxyl-Terminal End of a Growing Polypeptide Chain

The fundamental reaction of protein synthesis is the formation of a peptide bond between the carboxyl group at the end of a growing polypeptide chain and a free amino group on an amino acid. Consequently, a protein is synthesized stepwise from its amino-terminal end to its carboxyl-terminal end. Throughout the entire process, the growing carboxyl end of the polypeptide chain remains activated by its covalent attachment to a tRNA molecule (a *peptidyl-tRNA* molecule). This high-energy covalent linkage is disrupted in each cycle but is immediately replaced by the identical linkage on the most recently added amino acid (Figure 5–13). In this way, each amino acid added carries with it the activation energy for the addition of the *next* amino acid rather than the energy for its own addition—an example of the "head-growth" type of polymerization described in Chapter 2 (Figure 2–34, p. 79).

Figure 5–12 How the genetic code is translated by means of two linked "adaptors": the aminoacyl-tRNA synthetase enzyme, which couples a particular amino acid to its corresponding tRNA, and the tRNA molecule, which then binds to the appropriate nucleotide sequence on the mRNA.

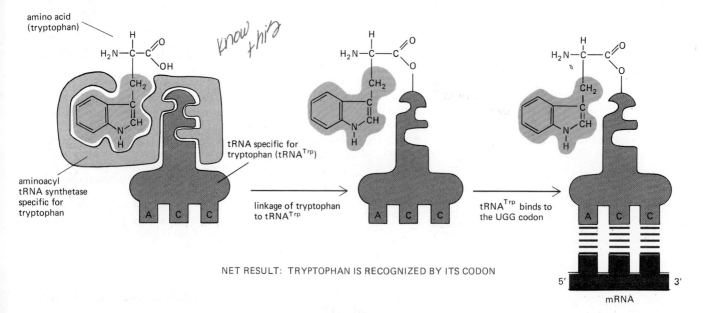

amino acid
(tryptophan)

aminoacyl tRNA synthetase specific for tryptophan

tRNA specific for tryptophan (tRNA^{Trp})

linkage of tryptophan to tRNA^{Trp}

tRNA^{Trp} binds to the UGG codon

NET RESULT: TRYPTOPHAN IS RECOGNIZED BY ITS CODON

mRNA

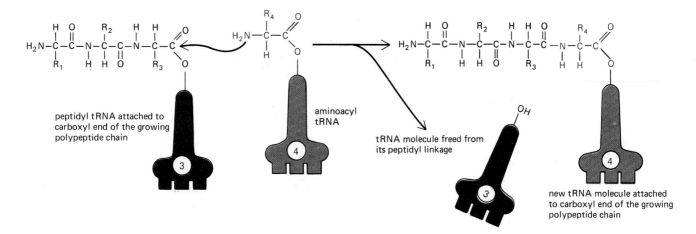

Figure 5–13 A polypeptide chain grows by the stepwise addition of amino acids to its carboxyl-terminal end. The formation of each peptide bond is energetically favorable because the growing carboxyl terminus has been activated by the covalent attachment of a tRNA molecule. The peptidyl-tRNA linkage that activates the growing end is regenerated in each cycle.

5-3 The Genetic Code Is Degenerate[5]

In the course of protein synthesis, the translation machinery moves in the 5'-to-3' direction along an mRNA molecule and the mRNA sequence is read three nucleotides at a time. As we have seen, each amino acid is specified by the triplet of nucleotides (**codon**) in the mRNA molecule that pairs with a sequence of three complementary nucleotides at the anticodon tip of a particular tRNA. Because only one of the many types of tRNA molecules in a cell can base-pair with each codon, the codon determines the specific amino acid residue to be added to the growing polypeptide chain end (Figure 5–14).

Since RNA is constructed from four types of nucleotides, there are 64 possible sequences composed of three nucleotides (4 × 4 × 4), and most of them occur somewhere in most mRNA molecules. Three of these 64 sequences do not code for amino acids but instead specify the termination of a polypeptide chain; they are known as *stop codons*. That leaves 61 codons to specify only 20 different amino acids. For this reason, most of the amino acids are represented by more than one codon (Figure 5–15), and the genetic code is said to be *degenerate*. Two amino acids, methionine and tryptophan, have only one codon each, and they are the least abundant amino acids in proteins.

The degeneracy of the genetic code implies that either there is more than one tRNA for each amino acid or a single tRNA molecule can base-pair with more than one codon. In fact, both situations occur. For some amino acids there is more than one tRNA molecule, and some tRNA molecules are constructed so that they require accurate base-pairing only at the first two positions of the codon and can tolerate a mismatch (or wobble) at the third. This *wobble base-pairing* explains why so many of the alternative codons for an amino acid differ only in their third nucleotide (see Figure 5–15). The standard wobble pairings make it possible to fit the 20 amino acids to 61 codons with as few as 31 different tRNA molecules; in animal mitochondria a more extreme "wobble" allows protein synthesis with only 22 tRNAs (see p. 392).

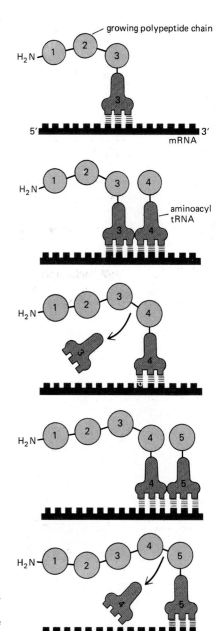

Figure 5–14 Each amino acid added to the growing end of a polypeptide chain is selected by complementary base-pairing between the anticodon on its attached tRNA molecule and the next codon on the mRNA chain.

GCA GCC GCG GCU	AGA AGG CGA CGC CGG CGU	GAC GAU	AAC AAU	UGC UGU	GAA GAG	CAA CAG	GGA GGC GGG GGU	CAC CAU	AUA AUC AUU	UUA UUG CUA CUC CUG CUU	AAA AAG	AUG	UUC UUU	CCA CCC CCG CCU	AGC AGU UCA UCC UCG UCU	ACA ACC ACG ACU	UGG	UAC UAU	GUA GUC GUG GUU	UAA UAG UGA
Ala	Arg	Asp	Asn	Cys	Glu	Gln	Gly	His	Ile	Leu	Lys	Met	Phe	Pro	Ser	Thr	Trp	Tyr	Val	stop
A	R	D	N	C	E	Q	G	H	I	L	K	M	F	P	S	T	W	Y	V	

The Events in Protein Synthesis Are Catalyzed on the Ribosome[6]

The protein synthesis reactions just described require a complex catalytic machinery to guide them. For example, the growing end of the polypeptide chain must be kept in register with the mRNA molecule to ensure that each successive codon in the mRNA engages precisely with the anticodon of a tRNA molecule and does not slip by one nucleotide, thereby changing the reading frame (see p. 102). This and the other events in protein synthesis are catalyzed by **ribosomes,** which are large complexes of RNA and protein molecules. Eucaryotic and procaryotic ribosomes are very similar in design and function. Each is composed of one large and one small subunit, which fit together to form a complex with a mass of several million daltons (Figure 5–16). The small subunit binds the mRNA and tRNAs, while the large subunit catalyzes peptide bond formation.

More than half of the weight of a ribosome is RNA, and there is increasing evidence that the **ribosomal RNA (rRNA)** molecules play a central part in its catalytic activities. Although the rRNA molecule in the small ribosomal subunit varies in size depending on the organism, its complicated folded structure is highly conserved (Figure 5–17); there are also close homologies between the rRNAs of the large ribosomal subunits in different organisms. Ribosomes contain a large number of proteins (Figure 5–18), but many of these have been relatively poorly conserved in sequence during evolution, and a surprising number seem not to be essential for ribosome function. Therefore, as discussed below (p. 219), it has been suggested that the ribosomal proteins mainly enhance the function of the rRNAs and that the RNA molecules rather than the protein molecules catalyze many of the reactions on the ribosome.

A Ribosome Moves Stepwise Along the mRNA Chain[6,7]

A ribosome contains three binding sites for RNA molecules: one for mRNA and two for tRNAs. One site, called the **peptidyl-tRNA binding site,** or **P-site,** holds the tRNA molecule that is linked to the growing end of the polypeptide chain. Another site, called the **aminoacyl-tRNA binding site,** or **A-site,** holds the

Figure 5–15 The genetic code. The standard one-letter abbreviation for each amino acid is presented below its three-letter abbreviation. Codons are written with the 5'-terminal nucleotide on the left. Note that most amino acids are represented by more than one codon and that variation is common at the third nucleotide (see also Figure 3–15, p. 103).

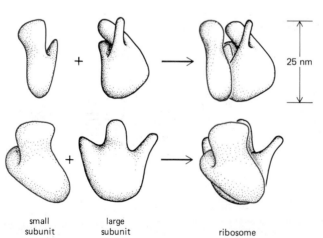

small subunit   large subunit   ribosome

25 nm

Figure 5–16 A three-dimensional model of the bacterial ribosome as viewed from two different angles. The positions of many ribosomal proteins in this structure have been determined by using an electron microscope to visualize the positions where specific antibodies bind, as well as by measuring the neutron scattering from ribosomes containing one or more deuterated proteins. (After J. A. Lake, *Ann. Rev. Biochem.* 54:507–530, 1985.)

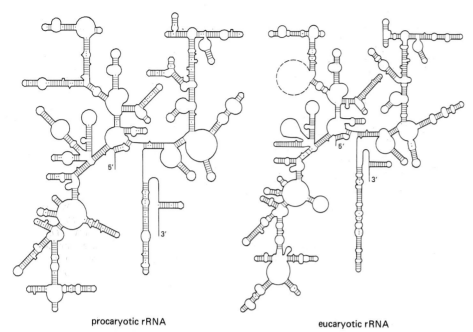

procaryotic rRNA

eucaryotic rRNA

Figure 5–17 The complex array of loops and base-paired stems in the folded structure of *E. coli* 16S rRNA (*left*) and yeast (*S. cerevisiae*) 18S rRNA (*right*). The major structural features appear to be shared by all known 16S-like rRNAs, including the rRNA of archaebacteria. Dots indicate postulated weak interactions between bases, such as G-U base pairs. (After R.R. Gutell, B. Weiser, C.R. Woese, and H.F. Noller, *Prog. Nucleic Acid Res. Mol. Biol.* 32:155–216, 1985.)

Figure 5–18 A comparison of the structures of procaryotic and eucaryotic ribosomes. Ribosomal components are commonly designated by their "S values," which indicate their rate of sedimentation in an ultracentrifuge (see p. 165). Note that, despite the differences in the number and size of their rRNA and protein components, both ribosomes have nearly the same structure and function in very similar ways. For example, although the 18S and 28S rRNAs of the eucaryotic ribosome contain many extra nucleotides not present in their bacterial counterparts, these nucleotides are present as multiple insertions that leave the basic structure of each rRNA largely unchanged (see Figure 5–17).

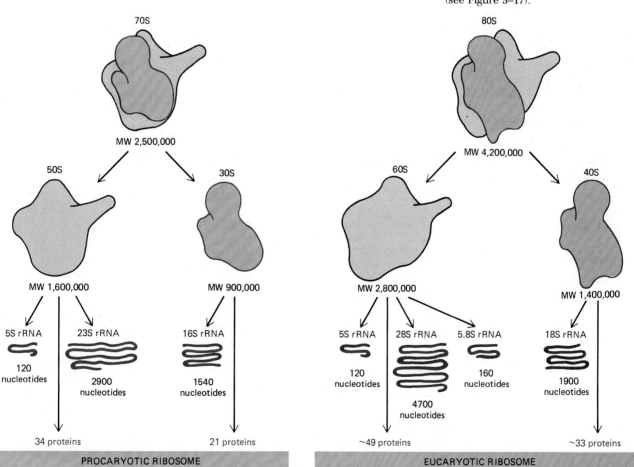

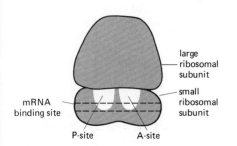

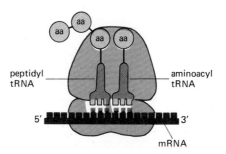

Figure 5–19 The three major RNA binding sites on a ribosome. An empty ribosome is shown at the left and a loaded ribosome at the right. The representation of a ribosome used here and in the next three figures is highly schematic; for a more accurate view, see Figures 5–16 and 5–23.

incoming tRNA molecule charged with an amino acid. A tRNA molecule is held tightly at either site only if its anticodon forms base pairs with a complementary codon on the mRNA molecule that is bound to the ribosome. The A- and P-sites are so close together that the two tRNA molecules are forced to form base pairs with adjacent codons in the mRNA molecule (Figure 5–19).

The process of polypeptide chain elongation on a ribosome can be considered as a cycle with three discrete steps (Figure 5–20). In step 1, an aminoacyl-tRNA molecule becomes bound to a vacant ribosomal A-site (adjacent to an occupied P-site) by forming base pairs with the three mRNA nucleotides exposed at the A-site. In step 2, the carboxyl end of the polypeptide chain is uncoupled from the tRNA molecule in the P-site and joined by a peptide bond to the amino acid linked to the tRNA molecule in the A-site. This reaction is catalyzed by **peptidyl transferase.** Its enzymelike activity requires the integrity of the ribosome and is thought to be mediated by a specific region of the major rRNA molecule in the large subunit. In step 3, the new peptidyl-tRNA in the A-site is translocated to the P-site as the ribosome moves exactly three nucleotides along the mRNA molecule. This step requires energy and is driven by a series of conformational changes induced in one of the ribosomal components by the hydrolysis of a GTP molecule (see p. 130).

As part of the translocation process of step 3, the free tRNA molecule that was generated in the P-site during step 2 is released from the ribosome to reenter the cytoplasmic tRNA pool. Therefore, upon completion of step 3, the unoccupied A-site is free to accept a new tRNA molecule linked to the next amino acid, which starts the cycle again. In a bacterium each cycle requires about one-twentieth of a second under optimal conditions, so that the complete synthesis of an average-sized protein of 400 amino acids is accomplished in about 20 seconds. Ribosomes move along an mRNA molecule in the 5' to 3' direction, which is also the direction of RNA synthesis (see Figure 5–2).

In most cells, protein synthesis consumes more energy than any other biosynthetic process. At least four high-energy phosphate bonds are split to make each new peptide bond: Two of these are required to charge each tRNA molecule with an amino acid (see Figure 5–10), and two more drive steps in the cycle of reactions occurring on the ribosome during synthesis itself—one for the aminoacyl-tRNA binding in step 1 (see p. 217) and one for the ribosome translocation in step 3.

5-5 A Protein Chain Is Released from the Ribosome When One of Three
5-8 Different Stop Codons Is Reached[6,8]

Three of the 64 possible codons (UAA, UAG, and UGA) in an mRNA molecule are **stop codons,** which terminate the translation process. Cytoplasmic proteins called

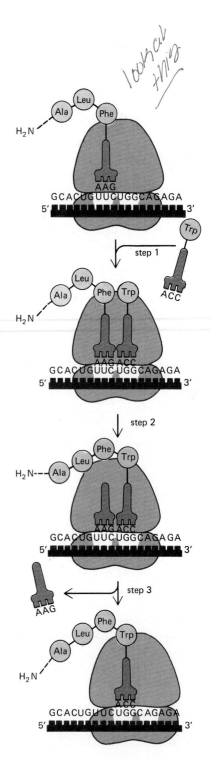

Figure 5–20 The elongation phase of protein synthesis on a ribosome. The three-step cycle shown is repeated over and over during the synthesis of a protein chain. An aminoacyl-tRNA molecule binds to the A-site on the ribosome in step 1; a new peptide bond is formed in step 2; and the ribosome moves a distance of three nucleotides along the mRNA chain in step 3, ejecting an old tRNA molecule and "resetting" the ribosome so that the next aminoacyl-tRNA molecule can bind.

release factors bind directly to any stop codon that reaches the A-site on the ribosome. This binding alters the activity of the peptidyl transferase, causing it to catalyze the addition of a water molecule instead of an amino acid to the peptidyl-tRNA. This reaction frees the carboxyl end of the growing polypeptide chain from its attachment to a tRNA molecule. Since only this attachment normally holds the growing polypeptide to the ribosome, the completed protein chain is immediately released into the cytoplasm (Figure 5–21). The ribosome then releases the mRNA and dissociates into its two separate subunits, which can assemble on another mRNA molecule to begin a new round of protein synthesis by the process to be described next.

5-4 The Initiation Process Sets the Reading Frame for Protein Synthesis[6,9]

In principle, an RNA sequence can be decoded in one of three different **reading frames,** each of which will specify a completely different polypeptide chain (see Figure 3–14, p. 102). Which of the three frames is actually read is determined by the way that a ribosome assembles on an mRNA molecule. During the *initiation* phase of protein synthesis, the two subunits of the ribosome are brought together at the exact spot on the mRNA where the polypeptide chain is to begin.

The initiation process is complicated, involving a number of steps catalyzed by proteins called **initiation factors (IF),** many of which are themselves composed of several polypeptide chains. Because the process is so complex, many of the details of initiation are still uncertain. It is clear, however, that each ribosome is assembled onto an mRNA chain in two steps; only after the small ribosomal subunit loaded with initiation factors finds the start codon does the large subunit bind.

Before a ribosome can begin a new protein chain, it must bind an aminoacyl tRNA molecule in its P-site, where normally only peptidyl tRNA molecules are bound (Figure 5–22). A special tRNA molecule is required for this purpose. This **initiator tRNA** provides the amino acid that starts a protein chain, and it always carries methionine (aminoformyl methionine in bacteria). In eucaryotes the initiator tRNA molecule is loaded onto the small ribosomal subunit before the mRNA is bound. An important initiation factor called *eucaryotic initiation factor 2 (eIF-2)* binds tightly to the initiator tRNA, and it is required to position the tRNA on the small subunit. In some cells the overall rate of protein synthesis is controlled by this factor (see below).

As described in the next section, the small ribosomal subunit helps its bound initiator tRNA molecule to find a special AUG codon (the *start codon*) on an mRNA molecule. Once this has occurred, the several initiation factors that were previously associated with the small ribosomal subunit are discharged to make way for the binding of a large ribosomal subunit to the small one. Because the initiator tRNA molecule is bound to the P-site of the ribosome, the synthesis of a protein chain can begin directly with the binding of a second aminoacyl-tRNA molecule to the A-site of the ribosome (Figure 5–22). Thus a complete functional ribosome is assembled, with the mRNA molecule threaded through it (Figure 5–23). Further steps in the elongation phase of protein synthesis then proceed as described previously (see step 2 of Figure 5–20). Because an initiator tRNA molecule has begun each polypeptide chain, all newly made proteins have a methionine (or the aminoformyl derivative of methionine in bacteria) as their amino-terminal residue. The methionine is often removed shortly after its incorporation by a specific aminopeptidase; this trimming process is important because the amino acid left at the amino terminus can determine the protein's lifetime in the cell by its effects on a ubiquitin-dependent degradation pathway (see p. 419).

Evidently the correct initiation site on the mRNA molecule has to be selected by the small subunit acting in concert with initiation factors but in the absence of the large subunit, which is presumably why all ribosomes are formed from two separate subunits. We shall now consider how this selection occurs.

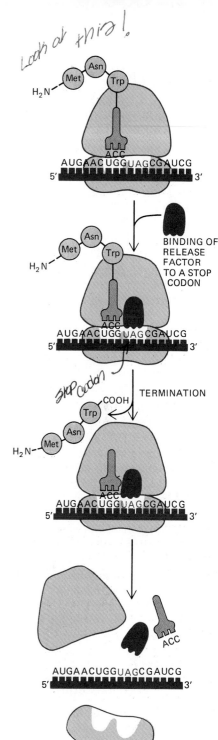

Figure 5–21 The final phase of protein synthesis. The binding of release factor to a stop codon terminates translation; the completed polypeptide is released, and the ribosome dissociates into its two separate subunits.

In Eucaryotes Only One Species of Polypeptide Chain Is Usually Synthesized from Each mRNA Molecule[10]

A messenger RNA molecule will typically contain many different AUG sequences, each of which can code for methionine. In eucaryotes, however, only one of these AUG sequences will normally be recognized by the initiator tRNA and thereby serve as a **start codon.** How is this start codon distinguished by the ribosome?

The mechanism for selecting a start codon is different in eucaryotes and procaryotes. The eucaryotic mRNAs (except those that are synthesized in mitochondria and chloroplasts) are extensively modified in the nucleus immediately after their transcription (see p. 531). Two general modifications are the addition of a unique "cap" structure, composed of a 7-methylguanosine residue linked to a triphosphate at the 5' end (Figure 5–24) and the addition of a run of about 200 adenylic residues ("poly A") at the 3' end. What part, if any, the poly A plays in the translation process is uncertain. But the 5' cap structure is essential for efficient protein synthesis. Experiments carried out with extracts of eucaryotic cells have shown that the small ribosomal subunit first binds at the 5' end of an mRNA chain, aided by recognition of the 5' cap (Figure 5–22). This subunit then moves along the mRNA chain, carrying its bound initiator tRNA in a search of an AUG start codon. The requirements for a start codon are apparently not very stringent, since it seems that only a few additional nucleotides besides the AUG itself are important. In most RNAs, only the first suitable AUG will be used; once a start codon near the 5' end has been selected, none of the many other AUG codons farther down the chain can serve as initiation sites. As a result, only a single species of polypeptide chain is usually synthesized from an mRNA molecule (for exceptions see p. 594).

In all these respects, procaryotic mRNAs are quite different from eucaryotic mRNAs (Figure 5–25). Bacterial mRNAs have no 5' cap structure. Instead they contain a specific initiation-site sequence, up to six nucleotides long, which can occur at multiple points in the same mRNA molecule. These sequences are located four to seven nucleotides upstream from an AUG, and they form base pairs with a specific region of the rRNA in a ribosome to signal the initiation of protein synthesis at this nearby start codon. Bacterial ribosomes, unlike eucaryotic ribosomes, bind directly to the middle of an mRNA molecule to recognize a start codon and initiate a polypeptide chain. As a result, bacterial messenger RNAs are commonly *polycistronic,* meaning that they encode multiple proteins separately translated from the same mRNA molecule. Eucaryotic mRNAs, in contrast, are typically *monocistronic,* only one species of polypeptide chain being translated per messenger molecule (see Figure 5–25).

The Binding of Many Ribosomes to an Individual mRNA Molecule Generates Polyribosomes[11]

The complete synthesis of a protein takes 20 to 60 seconds on average. But even during this very short period, multiple initiations usually take place on each mRNA molecule being translated. A new ribosome hops onto the 5' end of the mRNA molecule almost as soon as the preceding ribosome has translated enough of the amino acid sequence to get out of the way. Such mRNA molecules are thus present in **polyribosomes,** or **polysomes,** formed by several ribosomes spaced as close as 80 nucleotides apart along a single messenger molecule (Figures 5–26 and 5–27). In procaryotes (but *not* eucaryotes), RNA is accessible to ribosomes as soon as it is made. Thus, ribosomes will begin their synthesis at the 5' end of a new mRNA molecule and then follow behind the RNA polymerase as it completes an mRNA chain.

Polyribosomes are a common feature of cells. They can be isolated and separated from single ribosomes in the cytosol by ultracentrifugation after cell lysis (Figure 5–28). The mRNA purified from these polyribosomes can be used to determine if the protein encoded by a particular DNA sequence is being actively synthesized in the cells used to prepare the polyribosomes. These mRNA molecules can also serve as the starting material for the preparation of specialized cDNA libraries (see p. 261).

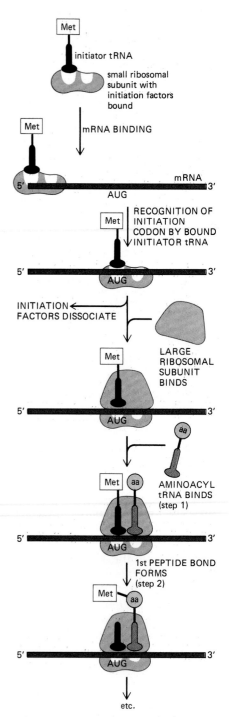

Figure 5–22 The initiation phase of protein synthesis. Events are illustrated as they occur in eucaryotes, but a very similar process occurs in bacteria. Step 1 and step 2 refer to steps in the elongation reaction shown in Figure 5–20.

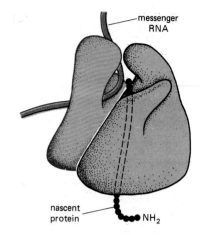

10-25 The Overall Rate of Protein Synthesis in Eucaryotes Is Controlled by Initiation Factors[12]

As we shall discuss in Chapter 13, the cells in a multicellular organism multiply only when they are in an environment where they are stimulated by specific growth factors. Although the mechanisms by which growth factors act are incompletely understood, one of their major effects must be to increase the overall rate of protein synthesis (p. 746). What determines this rate? Direct studies in tissues are difficult; but when eucaryotic cells in culture are starved of nutrients, there is a marked reduction of the rate of polypeptide-chain *initiation*, which can be shown to result from inactivation of the protein synthesis initiation factor eIF-2. In at least one type of cell (immature red blood cells), the activity of eIF-2 is reduced in a controlled way by the phosphorylation of one of its three protein subunits. This finding suggests that the rate of eucaryotic protein synthesis may be controlled by specific protein kinases that, when activated, reduce the frequency of initiation events. The action of growth factors may be mediated in part by intracellular regulatory molecules that inhibit or counteract the effect of such protein kinases.

The initiation factors required for protein synthesis are much more numerous and complex in eucaryotes than in procaryotes, even though they perform the same basic functions. Many of the extra components could be regulatory proteins that respond to different growth factors and help coordinate cell growth and proliferation in multicellular organisms. Such controls are not needed in bacteria, which generally grow as fast as the nutrients in their environment allow.

Figure 5–23 A three-dimensional model of a functioning bacterial ribosome. The small (*colored*) subunit and the large (*black*) subunit form a complex through which the messenger RNA is threaded. Although the exact paths of the mRNA and the nascent protein chain are unknown, the addition of amino acids occurs in the general region shown. (Modified from J.A. Lake, *Annu. Rev. Biochem.* 54:507–530, 1985.)

5-9 The Fidelity of Protein Synthesis Is Improved by Two Separate Proofreading Processes[13]

The error rate in protein synthesis can be estimated by monitoring the frequency of incorporation of an amino acid into a protein that normally lacks that amino acid. Error rates of about one amino acid misincorporated for every 10^4 amino acids polymerized are observed, which means that only about one in every 25 protein molecules of average size (400 amino acids) should contain an error. The fidelity of the decoding process depends on the accuracy of the two adaptor mechanisms previously discussed: the linking of each amino acid to its corresponding tRNA molecule and the base-pairing of the codons in mRNA to the anticodons in tRNA (see Figure 5–12). Not surprisingly, cells have evolved "proofreading" mechanisms to reduce the number of errors in both these crucial steps of protein synthesis.

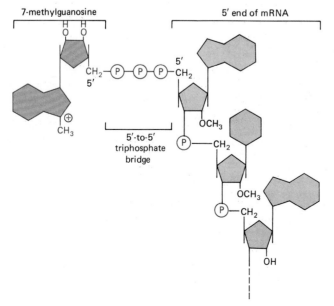

Figure 5–24 The structure of the cap at the 5′ end of eucaryotic mRNA molecules. Note the unusual 5′-to-5′ linkage to the positively charged 7-methylguanosine and the methylation of the 2′ hydroxyl group on the first ribose sugar in the RNA. (The second sugar is not always methylated.)

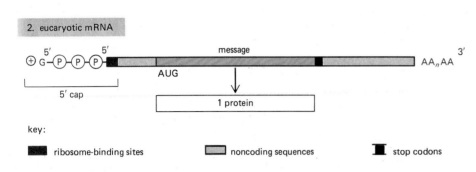

1. procaryotic mRNA

protein α   protein β   protein γ

message   message   message

5'
(P)(P)(P)   3'
AUG   AUG   AUG

2. eucaryotic mRNA

5'   5'
⊕ G-(P)-(P)-(P)   message   3'
                                                    AA_nAA
        AUG
5' cap
                            1 protein

key:

■ ribosome-binding sites   ▨ noncoding sequences   ▮ stop codons

Figure 5–25 A comparison of the structures of procaryotic and eucaryotic messenger RNA molecules. Although both mRNAs are synthesized with a triphosphate group at the 5' end, the eucaryotic RNA molecule immediately acquires a 5' cap. In eucaryotes the small ribosomal subunit specifically recognizes the 5' end of the mRNA, aided by this 5' cap. The synthesis therefore begins at a start codon near the 5' end (see Figure 5–22). In procaryotes, by contrast, the 5' end has no special significance, and there can be multiple ribosome-binding sites in the interior of an mRNA chain, each resulting in the synthesis of a different protein.

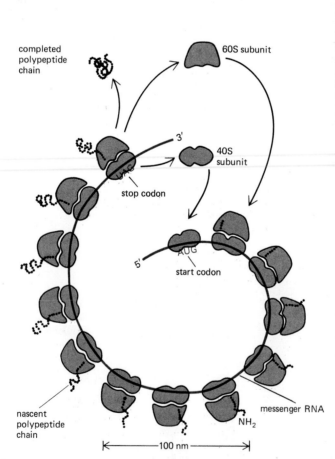

completed polypeptide chain

60S subunit

3'

40S subunit

stop codon

5'

AUG
start codon

nascent polypeptide chain

messenger RNA

NH₂

|←——— 100 nm ———→|

Figure 5–26 Schematic drawing of a polyribosome showing that a series of ribosomes simultaneously translates the same mRNA molecule. For each polypeptide chain being synthesized from an mRNA molecule in a eucaryotic cell, protein synthesis begins with the binding of a small ribosomal subunit to the single appropriate site on the mRNA molecule and proceeds from the 5' end to the 3' end of the mRNA chain. When a polypeptide chain is completed, the two ribosomal subunits dissociate from the mRNA.

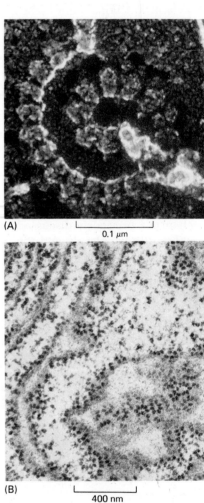

(A)
|—— 0.1 μm ——|

(B)
|—— 400 nm ——|

Figure 5–27 Electron micrographs of typical polyribosomes in action during protein synthesis in a eucaryotic cell. (A) Deep etch; (B) thin section. The cell cytoplasm is generally crowded with such polyribosomes, some free in the cytosol and some membrane-bound. (A, courtesy of John Heuser; B, courtesy of George Palade.)

Two fundamentally different mechanisms are used to reduce errors in the two steps, each representative of strategies used in other processes in the cell. Both involve expenditure of free energy, since, as discussed in Chapter 2 (p. 59), a price must be paid for any increase in order in the cell. A relatively simple mechanism is used to improve the accuracy of amino acid attachment to tRNA. Many aminoacyl tRNA synthetases have two separate active sites, one that carries out the loading reaction shown earlier (Figure 5–10) and one that recognizes an incorrect amino acid attached to its tRNA molecule and removes it by hydrolysis. The correction process is costly because, to be effective, it must remove an appreciable fraction of correctly attached amino acids as well. The same type of costly two-step proofreading process is used in DNA replication (see p. 229).

A more subtle "kinetic proofreading" mechanism is used to improve the fidelity of codon-anticodon pairing. Thus far we have given a simplified account of this pairing. In fact, once tRNA molecules have acquired an amino acid, they form a complex with an abundant protein called an *elongation factor (EF)*, which binds tightly to both the amino acid end of tRNAs and to a molecule of GTP. It is this complex, and not free tRNA, that pairs with the appropriate codon in an mRNA molecule. The bound elongation factor allows correct codon-anticodon pairing to occur but prevents the amino acid from being incorporated into the growing polypeptide chain. However, the initial codon recognition triggers the elongation factor to hydrolyze its bound GTP (to GDP and inorganic phosphate), whereupon the factor can dissociate from the ribosome without its tRNA, making it possible for protein synthesis to proceed. As illustrated in Figure 5–29, the elongation factor thereby introduces a short delay between codon-anticodon base-pairing and polypeptide chain elongation, which provides an opportunity for the bound tRNA molecule to exit from the ribosome. An incorrect tRNA molecule forms a smaller number of codon-anticodon hydrogen bonds than a correct one; it therefore binds more weakly to the ribosome and is more likely to dissociate during this period. Because the delay introduced by the elongation factor causes most incorrectly bound tRNA molecules to leave the ribosome without being used for protein synthesis, this factor increases the ratio of correct to incorrect amino acids incorporated into protein.

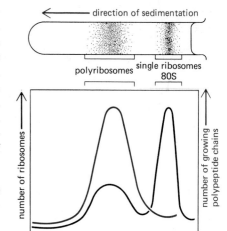

Figure 5–28 How polyribosomes are separated from single ribosomes (and their subunits) by sedimentation in a centrifuge. This method is based on the fact that large molecular aggregates move faster than small ones in a strong gravitational field. Generally the sedimentation is done through a gradient of sucrose to stabilize the solution against convective mixing (see p. 164).

Figure 5–29 A more detailed view of step 1 of the elongation phase of protein synthesis illustrates how the correct tRNA is selected on the ribosome. In the initial binding event, an aminoacyl-tRNA molecule bound to the elongation factor pairs transiently with the codon at the A-site. This pairing triggers GTP hydrolysis by the elongation factor, enabling the factor to dissociate from the aminoacyl-tRNA molecule, which can now fit correctly in the A-site and participate in chain elongation (see Figure 5–20). In general, only those tRNAs with the correct anticodon can remain paired to the mRNA long enough to be added to the growing polypeptide chain. The elongation factor, which is an abundant protein, is designated as EF-Tu in procaryotes and as EF-1 in eucaryotes.

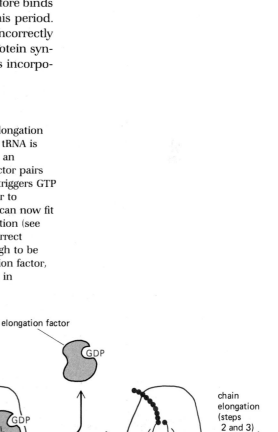

Table 5–1 Inhibitors of Protein or RNA Synthesis

Inhibitor	Specific Effect
*Acting Only on Procaryotes**	
Tetracycline	blocks binding of aminoacyl-tRNA to A-site of ribosome
Streptomycin	prevents the transition from initiation complex to chain-elongating ribosome and causes miscoding
Chloramphenicol	blocks the peptidyl transferase reaction on ribosomes (step 2 in Figure 5–20)
Erythromycin	blocks the translocation reaction on ribosomes (step 3 in Figure 5–20)
Rifamycin	blocks initiation of RNA chains by binding to RNA polymerase (prevents RNA synthesis)
Acting on Procaryotes and Eucaryotes	
Puromycin	causes the premature release of nascent polypeptide chains by its addition to growing chain end
Actinomycin D	binds to DNA and blocks the movement of RNA polymerase (prevents RNA synthesis)
Acting Only on Eucaryotes	
Cycloheximide	blocks the translocation reaction on ribosomes (step 3 in Figure 5–20)
Anisomycin	blocks the peptidyl transferase reaction on ribosomes (step 2 in Figure 5–20)
α-Amanitin	blocks mRNA synthesis by binding preferentially to RNA polymerase II

*The ribosomes of eucaryotic mitochondria (and chloroplasts) often resemble those of procaryotes in their sensitivity to inhibitors.

5-10 Many Inhibitors of Procaryotic Protein Synthesis Are Useful as Antibiotics[14]

Many of the most effective antibiotics used in modern medicine act by inhibiting bacterial protein synthesis. A number of these drugs exploit the structural and functional differences between procaryotic and eucaryotic ribosomes so as to affect procaryotic ribosomes preferentially. This selectivity enables some of these compounds to be used at relatively high concentrations in the human body without undue toxicity. Because different antibiotics bind to different regions of bacterial ribosomes, they often inhibit different steps in the synthetic process. Some of the more common compounds in this group, along with their specific effects, are listed in Table 5–1. This table also lists several other commonly used inhibitors of protein synthesis, some of which act on eucaryotic cells; the latter, of course, cannot be used as antibiotics.

Because they block specific steps in the processes that lead from DNA to protein, many of the compounds in Table 5–1 are useful for deciphering cellular mechanisms. Among the most commonly used drugs in such experimental studies are *chloramphenicol, cycloheximide*, and *puromycin*, all of which specifically inhibit protein synthesis. In a eucaryotic cell, for example, chloramphenicol inhibits protein synthesis only on the ribosomes in the mitochondria (and in chloroplasts of plants), presumably reflecting their procaryotic origins (see p. 398). Cycloheximide, on the other hand, affects only cytosolic ribosomes. The difference in the sensitivity of protein synthesis to these two drugs provides a powerful way to determine in which cell compartment a particular protein is translated. Puromycin is especially interesting because it is a structural analogue of a tRNA molecule linked to an amino acid; the ribosome mistakes it for an authentic amino acid and covalently incorporates it at the carboxyl terminus of the growing polypeptide chain, thereby causing the premature termination and release of this polypeptide. As might be expected, puromycin inhibits all types of protein synthesis.

How Did Protein Synthesis Evolve?[15]

The molecular processes underlying protein synthesis seem inexplicably complex. Although we can describe many of them, they do not make conceptual sense in the way that DNA transcription, DNA repair, and DNA replication do. As we have seen, protein synthesis in present-day organisms centers on a very large ribonucleoprotein machine, the ribosome, which consists of proteins arranged around a core of rRNA molecules. Why should rRNA molecules exist at all, and how did they come to play such a dominant part in the structure and function of the ribosome? The answer would undoubtedly help us to understand protein synthesis.

Before the discovery of mRNA in the early 1960s, it was suspected that the large amount of RNA in ribosomes served a "messenger" function, carrying genetic information from DNA to proteins. Now we know, however, that all of the ribosomes in a cell contain an identical set of rRNA molecules that have no such informational role. In bacterial ribosomes, small regions of the rRNA have been shown to have catalytic functions in protein synthesis; for example, the rRNA of the small procaryotic subunit forms a short base-paired helix with the initiation site sequence on mRNA molecules, which serves to position the neighboring AUG start codon at the P-site. Similar base-pair interactions may be formed between tRNA molecules and rRNAs, although these have not yet been demonstrated conclusively.

Protein synthesis also relies heavily on a large number of different proteins that are bound to the rRNAs in a ribosome. The complexity of a process with so many different interacting components has made many biologists despair of ever understanding the pathway by which protein synthesis evolved. However, the recent discovery of RNA molecules that act as enzymes (p. 105) has provided a new way of viewing the pathway. As discussed in Chapter 1, early biological reactions probably used RNA molecules rather than protein molecules as catalysts (see pp. 4–9). In the earliest cells, tRNA molecules on their own may have formed catalytic surfaces that allowed them to bind and activate specific amino acids without requiring aminoacyl-tRNA synthetase enzymes. Likewise, rRNA molecules may have served by themselves as the entire "ribosome," folding up in complex ways to generate an intricate set of surfaces that both guided tRNA pairings with mRNA codons and catalyzed the polymerization of the tRNA-linked amino acids (see Figure 1–7, p. 7). Over the course of evolution, individual proteins could have been added to this machinery, each one making the process a little more accurate and efficient. The large quantity of RNA in present-day ribosomes may therefore be a remnant of a very early stage in evolution, before proteins had come to dominate biological catalysis.

Summary

Before the synthesis of a particular protein can begin, the corresponding mRNA molecule must be produced by DNA transcription. Then a small ribosomal subunit binds to the mRNA molecule at a start codon that is recognized by a unique initiator tRNA molecule. A large ribosomal subunit binds to complete the ribosome and initiate the elongation phase of protein synthesis. During this phase, aminoacyl tRNAs, each bearing a specific amino acid, sequentially bind to the appropriate codon in mRNA by forming complementary base-pairs with the tRNA anticodon. Each amino acid is added to the carboxyl-terminal end of the growing polypeptide by means of a cycle of three sequential steps: aminoacyl-tRNA binding, followed by peptide bond formation, followed by ribosome translocation. The ribosome progresses from codon to codon in the 5'-to-3' direction along the mRNA molecule until one of three stop codons is reached. A release factor then binds to the stop codon, terminating translation and releasing the completed polypeptide from the ribosome.

Eucaryotic and procaryotic ribosomes are highly homologous, despite substantial differences in the number and size of their rRNA and protein components.

The predominant role of rRNA in ribosome structure and function may reflect the ancient origin of protein synthesis, which is thought to have evolved in an environment dominated by RNA-mediated catalysis.

DNA Repair Mechanisms[16]

The long-term survival of a species may be enhanced by changes in its genetic inheritance, but its survival in the short term demands accurate maintenance of the genetic record. Maintaining the genetic material requires not only an extremely accurate mechanism for copying DNA sequences before a cell divides but also a mechanism for repairing the many accidental lesions that occur continually in its DNA. Most such spontaneous changes in the DNA are transient because they are immediately erased by a correction process called *DNA repair*. Only rarely does one of the cell's DNA maintenance processes fail and allow a permanent change in the DNA. Such a change is called a **mutation,** and it can destroy an organism if the change occurs in a vital position in the DNA sequence.

Before examining the mechanisms of DNA repair, we shall discuss briefly the maintenance of DNA sequences from one generation to the next.

DNA Sequences Are Maintained with Very High Fidelity[17]

The rate at which stable changes occur in DNA sequences (the *mutation rate*) can be estimated only indirectly. One way is to compare the amino acid sequence of the same protein in several different species. The fraction of the amino acids that are different can then be compared with the estimated number of years since each pair of species diverged from a common ancestor, as determined from the fossil record. In this way one can calculate the average number of years that elapse before an inherited change becomes fixed in one of the protein's amino acids. Because each such change will commonly reflect the alteration of a single nucleotide in the DNA sequence of the gene coding for that protein, this value can be used to estimate the average number of years required to produce a single, stable mutation in the gene.

Such calculations will always substantially underestimate the actual mutation rate because most mutations will spoil the function of the protein and vanish from the population through natural selection. But there is one family of proteins whose sequence does not seem to matter, and so the genes that encode them can accumulate mutations without being selected against. These proteins are the **fibrinopeptides**—20-residue-long fragments that are discarded from the protein *fibrinogen* when it is activated to form *fibrin* during blood clotting. Since the function of fibrinopeptides apparently does not depend on their amino acid sequence, they can tolerate almost any amino acid change. Sequence analysis of the fibrinopeptides indicates that an average-sized protein 400 amino acids long would be randomly altered by an amino acid change roughly once every 200,000 years. More recently, DNA-sequencing technology (see p. 185) has made it possible to compare corresponding nucleotide sequences in regions of the genome that do not code for protein. Comparisons of such sequences in several different mammalian species produce estimates of the mutation rate during evolution that are in excellent agreement with those obtained from the fibrinopeptide studies.

The Observed Mutation Rates in Proliferating Cells Are Consistent with Evolutionary Estimates[18]

The mutation rate can be estimated more directly by observing the rate at which spontaneous genetic changes arise in a large population of cells followed over a relatively short period of time. This can be done either by estimating the frequency with which new mutants arise in very large animal populations (for example, fruit fly or mouse colonies) or by screening for changes in specific proteins in cells growing in tissue culture. Although they are only approximate, the numbers obtained in both cases are consistent with an error frequency of one base-pair change in roughly 10^9 base pairs for each cell generation. Consequently, a single gene that encodes an average-sized protein (containing about 10^3 coding base pairs) would suffer a mutation once in about 10^6 cell generations. This number is at least roughly consistent with the evolutionary estimate described above, in which one mutation appears in an average gene in the germ line every 200,000 years.

5-16 Most Mutations in Proteins Are Deleterious and Are Eliminated by Natural Selection[17]

When the number of amino acid differences in a particular protein is plotted for several pairs of organisms against the time since the organisms diverged, the result is a reasonably straight line. That is, the longer the period since divergence, the larger the number of differences. For convenience, the slope of this line can be expressed in terms of the "unit evolutionary time" for that protein, which is the average time required for one amino acid change to appear in a sequence of 100 amino acid residues. When various proteins are compared, each shows a different but characteristic rate of evolution (Figure 5–30). Since all DNA base pairs are thought to be subject to roughly the same rate of random mutation, these different rates must reflect differences in the probability that an organism with a random mutation in the given protein will survive and propagate. Evidently changes in amino acid sequence are much more harmful for some proteins than for others. From Table 5–2 we can estimate that about 6 of every 7 random amino acid changes are harmful in hemoglobin, about 29 of every 30 amino acid changes are harmful in cytochrome c, and virtually all amino acid changes are harmful in histone H4. It can be assumed that individuals who carry such harmful mutations are eliminated from the population by natural selection.

Low Mutation Rates Are Necessary for Life as We Know It[19]

Since most mutations are deleterious, no species can afford to allow them to accumulate at a high rate in its germ cells. We shall discuss later (p. 485) why the observed mutation frequency, low though it is, nevertheless is thought to limit the number of essential proteins that any organism can encode in its germ line to about 60,000. By an extension of the same arguments, a mutation frequency tenfold higher would limit an organism to about 6000 essential proteins. In this case evolution would probably have stopped at an organism no more complex than the fruit fly.

While germ cells must be protected against high rates of mutation in order to maintain the species, the other cells of a multicellular organism (its *somatic* cells) must be protected from genetic change to safeguard the individual. Nucleotide changes in somatic cells can facilitate a process of natural selection that leads to the survival of the fittest cells at the expense of the rest of the organism. In the extreme case the uncontrolled cell proliferation known as cancer results, which causes about 20% of deaths in the Western Hemisphere. There is good evidence that these deaths are due largely to the accumulation of changes in the DNA sequences of somatic cells (see Chapter 21, p. 1190). A tenfold increase in the mutation frequency would presumably cause a disastrous increase in the incidence of cancer by accelerating the rate at which somatic cell variants arise. Thus,

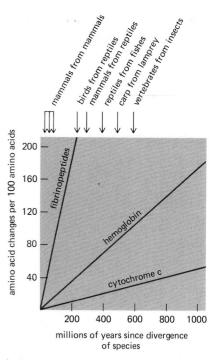

Figure 5–30 A comparison of the rates of amino acid change found in hemoglobin and cytochrome c with the rate found in the fibrinopeptides. Hemoglobin and cytochrome c have changed much more slowly during evolution than the fibrinopeptides. In determining rates of change per year (as in Table 5–2), it is important to realize that two organisms that diverged from a common ancestor 100 million years ago are separated by 200 million years of evolutionary time.

Table 5–2 Observed Rates of Change of the Amino Acid Sequences in Various Proteins over Evolutionary Time

Protein	Unit Evolutionary Time (in millions of years)
Fibrinopeptide	0.7
Hemoglobin	5
Cytochrome c	21
Histone H4	500

The "unit evolutionary time" is defined as the average time required for one *acceptable* amino acid change to appear in the indicated protein for every 100 amino acids that it contains.

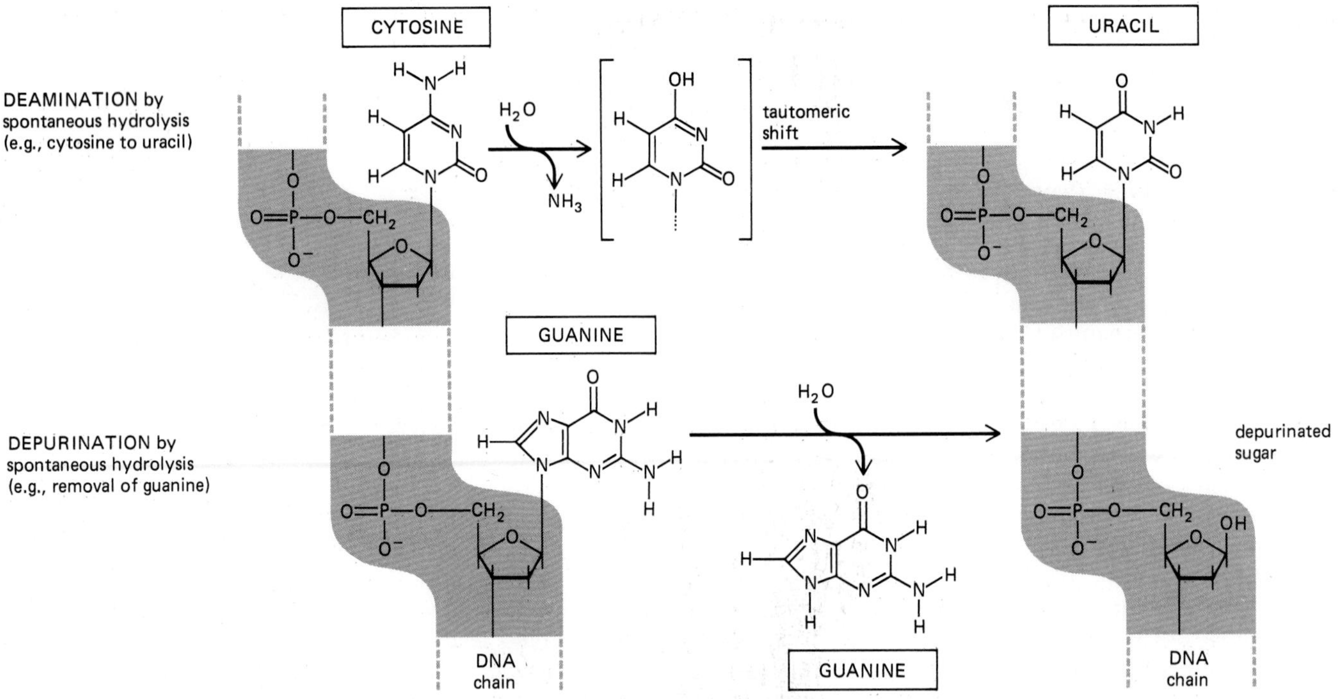

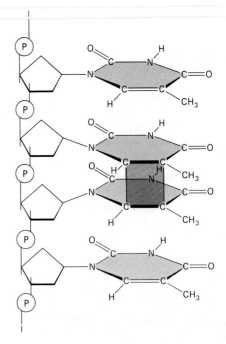

DEAMINATION by spontaneous hydrolysis (e.g., cytosine to uracil)

CYTOSINE → (tautomeric shift) → URACIL

DEPURINATION by spontaneous hydrolysis (e.g., removal of guanine)

GUANINE → depurinated sugar + GUANINE

DNA chain

Figure 5–31 Two frequent spontaneous chemical reactions known to create serious DNA damage in cells are deamination and depurination. One specific example is shown of each type of reaction.

both for the perpetuation of a species with 60,000 proteins (germ-cell stability) and for the prevention of cancer resulting from mutations in somatic cells (somatic-cell stability), eucaryotes depend on the remarkably high fidelity with which DNA sequences are maintained.

Low Mutation Rates Mean That Related Organisms Must Be Made from Essentially the Same Proteins[17]

Humans, as a genus distinct from the great apes, have existed for only a few million years. Each gene has therefore had the chance to accumulate relatively few nucleotide changes since our inception, and most of these have been eliminated by natural selection. A comparison of humans and monkeys, for example, shows that their cytochrome c molecules differ in only 1% and their hemoglobins in about 4% of their amino acid residues. Clearly, a great deal of our genetic heritage must have been formed long before *Homo sapiens* appeared, during the evolution of mammals (starting about 3×10^8 years ago) and even earlier. Because the proteins of mammals as different as whales and humans are very similar, the evolutionary changes that have produced such striking morphological differences must involve relatively few changes in the molecules from which we are made. Instead, it is thought that differences in the temporal and spatial pattern of gene expression during embryonic development are largely responsible for these morphological differences. In Chapter 10 we shall discuss how changes in gene expression of this type might arise (p. 606).

5-13 If Left Uncorrected, Spontaneous DNA Damage Would
5-17 Change DNA Sequences Rapidly[20]

The physicist Erwin Schroedinger pointed out in 1945 that, whatever its chemical nature (at that time unknown), a gene must be extremely small and composed of few atoms. Otherwise the very large number of genes thought to be necessary to generate an organism would not fit in the cell nucleus. On the other hand, because it was so small, a gene would be expected to undergo significant changes as a result of spontaneous reactions induced by random thermal collisions with solvent molecules. This poses a serious dilemma, since genetic data imply that genes are composed of a remarkably stable substance in which spontaneous changes (mutations) occur rarely.

Figure 5–32 The structure of the thymine dimer, a type of damage introduced into DNA in cells exposed to ultraviolet irradiation (as in sunlight). A similar dimer will form between any two neighboring pyrimidine bases (C or T residues) in DNA.

This dilemma is real. DNA does undergo major changes as a result of thermal fluctuations. We now know, for example, that about 5000 purine bases (adenine and guanine) are lost per day from the DNA of each human cell because of the thermal disruption of their *N*-glycosyl linkages to deoxyribose (*depurination*). Similarly, spontaneous *deamination* of cytosine to uracil in DNA is estimated to occur at a rate of 100 bases changed per genome per day (Figure 5–31). DNA bases are also subject to change by reactive metabolites that can alter their base-pairing abilities and by ultraviolet light from the sun, which promotes a covalent linkage of two adjacent pyrimidine bases in DNA (forming, for example, the *thymine dimers* shown in Figure 5–32). These are only a few of many changes that can occur in our DNA. Most of them would be expected to lead either to deletion of one or more base pairs in the daughter DNA chain after DNA replication or to a base-pair substitution (for example, each $C \rightarrow U$ deamination would eventually change a C-G base pair to a T-A base pair, since U closely resembles T and forms a complementary base pair with A; see p. 56). As we have seen, a high rate of such random changes would have disastrous consequences for living organisms.

5-14 The Stability of Genes Depends on DNA Repair[21]

Despite the thousands of random changes that heat energy causes in the DNA of a human cell every day, only a few stable changes (mutations) accumulate in the DNA sequence of an average cell in a year. We now know that fewer than one in a thousand accidental base changes in DNA causes a mutation; the rest are eliminated with remarkable efficiency by the process of **DNA repair.** There are a variety of repair mechanisms, but nearly all of them depend on the existence of two copies of the genetic information, one in each strand of the DNA double helix. Consequently, if the sequence in one strand is accidentally changed, information is not lost irretrievably because a second copy remains in the sequence of nucleotides in the other strand. The basic pathway for DNA repair is illustrated schematically in Figure 5–33; as indicated, it involves three steps:

1. The altered portion of a damaged DNA strand is recognized and removed by special enzymes called *DNA repair nucleases*, which hydrolyze the phosphodiester bonds that join the damaged nucleotides to the rest of the DNA molecule, leaving a gap in the DNA helix in this region.
2. Another enzyme, **DNA polymerase,** binds to the 3′-OH end of the cut DNA strand and fills in the gap one nucleotide at a time by copying the information stored in the "good" (*template*) strand.
3. The break or "nick" in the damaged strand left when the DNA polymerase has filled in the gap is sealed by a third type of enzyme, **DNA ligase,** which completes the restoration process.

Both DNA polymerase and DNA ligase play important general roles in DNA metabolism; for example, both function in DNA replication as well as in DNA repair. The reactions these two enzymes catalyze are illustrated in Figures 5–34 and 5–35, respectively.

5-18 Different Types of DNA Damage Are Recognized by
5-19 Different Enzymes[22]

The details of the excision step in DNA repair depend on the type of damage. For example, depurination, which is by far the most frequent lesion that occurs in DNA, leaves a deoxyribose sugar with a missing base (see Figure 5–31). This sugar is rapidly recognized by the enzyme **AP endonuclease,** which cuts the DNA phosphodiester backbone at the altered site. After enzymatic excision of a sugar phosphate residue, an undamaged DNA sequence is restored by the mechanism previously described (see Figure 5–33).

A related repair pathway involves a battery of different enzymes called **DNA glycosylases,** each of which recognizes a single type of altered base in DNA and catalyzes its hydrolytic removal. There are at least six types of these enzymes, including those that remove deaminated C's, deaminated A's, different types of alkylated bases, bases with opened rings, and bases in which a carbon-carbon

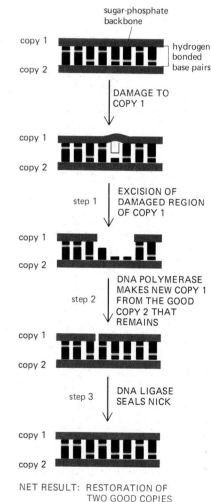

Figure 5–33 The three steps in DNA repair. In step 1 the damage is excised; in steps 2 and 3 the original DNA sequence is restored. DNA polymerase fills in the gap created by the excision events (step 2), and DNA ligase seals the nick left in the repaired strand (step 3). Nick sealing consists of the re-formation of a broken phosphodiester bond (see Figure 5–35).

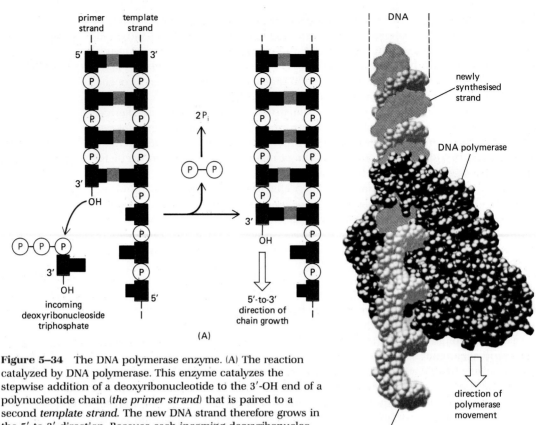

Figure 5–34 The DNA polymerase enzyme. (A) The reaction catalyzed by DNA polymerase. This enzyme catalyzes the stepwise addition of a deoxyribonucleotide to the 3'-OH end of a polynucleotide chain (*the primer strand*) that is paired to a second *template strand*. The new DNA strand therefore grows in the 5'-to-3' direction. Because each incoming deoxyribonucleoside triphosphate must pair with the template strand in order to be recognized by the polymerase, this strand determines which of the four possible deoxyribonucleotides (A, C, G, or T) will be added. As in the case of RNA polymerase, the reaction is driven by a large favorable free-energy change (see Figure 5–2). (B) The structure of an *E. coli* DNA polymerase molecule as determined by x-ray crystallography. The polymerase is shown as it would appear during DNA synthesis. (Courtesy of Tom Steitz.)

Figure 5–35 The DNA ligase enzyme seals a broken phosphodiester bond. As shown, DNA ligase uses a molecule of ATP to activate the 5' end at the nick (step 1) before forming the new bond (step 2). In this way the energetically unfavorable nick-sealing reaction is driven by being coupled to the energetically favorable process of ATP hydrolysis (see p. 74). In *Bloom's syndrome,* an inherited human disease, individuals are partially defective in DNA ligase and consequently are deficient in DNA repair; they also have a dramatically increased incidence of cancer.

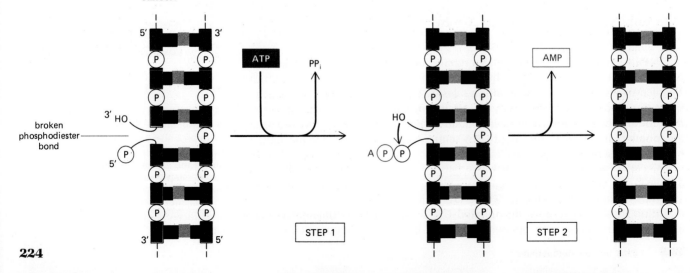

double bond has been accidentally converted to a carbon-carbon single bond. As an example of the general mechanism that operates in all cases, the removal of a deaminated C is shown in Figure 5–36. First the enzyme uracil DNA glycosylase removes the altered base (uracil), producing a deoxyribose sugar with a missing base. Because this sugar is the same substrate recognized by the AP endonuclease, the subsequent steps in the repair process proceed in the same way as for depurinated sites. The net result is that the U that was created by accidental deamination is restored to a C. The importance of removing accidentally deaminated DNA bases has been directly demonstrated. In mutant bacteria that lack the enzyme uracil DNA glycosylase, the normally low spontaneous rate of change of a C-G to a T-A base pair is increased about twentyfold.

Cells have a separate **"bulky lesion" repair** pathway capable of removing almost any type of DNA damage that creates a large change in the DNA double helix. Such bulky lesions include those created by the covalent reaction of DNA bases with large hydrocarbons (such as the carcinogen benzpyrene), as well as various pyrimidine dimers (T-T, T-C, and C-C) caused by sunlight (see Figure 5–32). In these cases a large multienzyme complex scans the DNA for a distortion in the double helix rather than for a specific base change. Once a bulky lesion is found, the phosphodiester backbone of the abnormal strand is cleaved on both sides of the distortion and the entire lesion is excised. The small gap produced in the DNA helix is then repaired in the usual manner.

The importance of these repair processes is indicated by the large investment that cells make in DNA repair enzymes. A comprehensive genetic analysis of a yeast suggests that these cells contain more than 50 different genes that code for DNA repair functions. DNA repair pathways are likely to be at least as complex in humans. Individuals with the genetic disease *xeroderma pigmentosum*, for example, are defective in a bulky lesion repair process that can be shown by genetic analysis to require at least seven different gene products. Such individuals develop severe skin lesions, including skin cancer, because of the accumulation of pyrimidine dimers in cells that are exposed to sunlight.

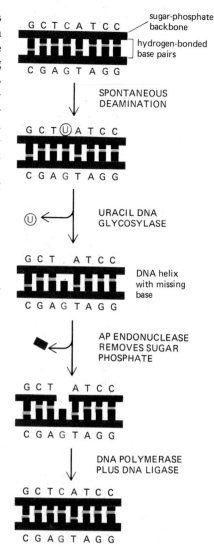

Figure 5–36 The DNA repair pathway involving the enzyme uracil DNA glycosylase, which leads to the restoration of an accidentally deaminated cytosine in DNA. After the action of a DNA glycosylase, the sugar phosphate with a missing base is cut out by the AP endonuclease, the same enzyme that initiates the repair of depurinated sites. Subsequent steps proceed as in Figure 5–33. The AP endonuclease derives its name from the fact that it recognizes any site in the DNA helix that contains a deoxyribose sugar with a missing base; such sites can arise either by the loss of a purine (*apurinic* sites) or by the loss of a pyrimidine (*apyrimidinic* sites).

5-15 Cells Can Produce DNA Repair Enzymes in Response to DNA Damage[23]

Cells have evolved a number of mechanisms to help them survive in a hazardous world. Often an extreme environmental insult activates a battery of genes whose products protect the cell from its effects. One such mechanism shared by all cells is the *heat-shock response*, which is evoked by the exposure of cells to unusually high temperatures; the induced "heat shock proteins" include some that are thought to help stabilize and repair partially denatured cell proteins (see p. 420).

Many cells also have mechanisms that enable them to synthesize DNA repair enzymes as an emergency response to severe DNA damage. The best-studied example is the **SOS response** in *E. coli*. In this bacterium, any block to DNA replication caused by DNA damage produces a signal (thought to be an excess of single-stranded DNA) that induces an increase in the transcription of more than 15 different genes, many of which code for proteins that function in DNA repair. The signal first activates the *E. coli* recA protein (see p. 242), which then destroys a negatively acting gene regulatory protein (a *repressor*) that normally suppresses the transcription of the entire set of SOS response genes. Studies of mutant bacteria deficient in different parts of the response indicate that the newly synthesized proteins have two effects. First, as would be expected, the induction of new DNA repair enzymes increases cell survival: when the mutants deficient in this part of the SOS response are treated with a DNA-damaging agent such as ultraviolet radiation, an unusually high proportion of them die. Second, by greatly increasing the number of errors made in copying DNA sequences, several of the induced proteins transiently increase the mutation rate and therefore the genetic variability in the bacterial population. This part of the response has little effect on short-term survival; it is presumably advantageous because it increases the chance that a mutant cell with increased fitness will arise.

The DNA repair system activated in the SOS response is not the only inducible

DNA repair system known. Bacteria have another that is activated specifically by the presence of methylated nucleotides in DNA, and there is at least one inducible DNA repair system in yeast cells. Some higher eucaryotic cells have been reported to adapt to DNA damage in similar ways.

The Structure and Chemistry of the DNA Double Helix Make It Easy to Repair

The DNA double helix seems to be optimally designed for repair. For example, RNA must have evolved before DNA (see p. 10), and one might wonder why the U in RNA has been replaced by T in DNA. It can be argued that the mechanism we have described for removing deaminated C's (Figure 5–36) would not work if U rather than T (which is 5-methyl U) were the fourth DNA nucleotide. As we have seen, spontaneous C deamination yields a U, and a repair enzyme designed to recognize and excise such accidents would be confused by the normal U's in a U-containing DNA molecule.

Similar considerations apply to the choice of G rather than hypoxanthine, which is the simplest purine base capable of pairing specifically with C. Hypoxanthine, however, is the direct deamination product of A (Figure 5–37). The addition of a second amino group to hypoxanthine produces G, which cannot be formed from A by spontaneous deamination. Every possible deamination event in DNA yields an unnatural base, which can therefore be directly recognized and removed by a specific DNA glycosylase (Figure 5–37).

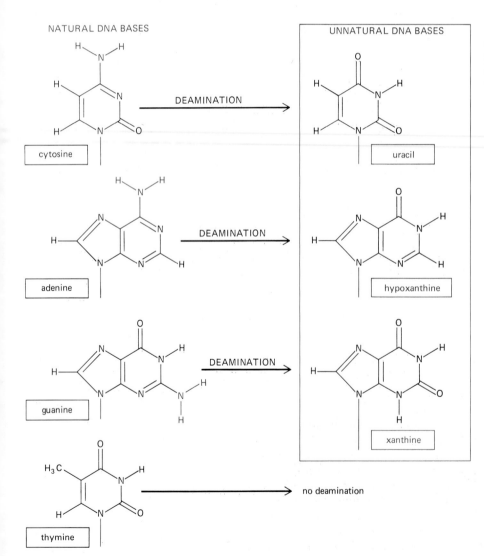

Figure 5–37 Illustration showing that all of the spontaneous deamination products of the DNA bases are recognizable as unnatural when they occur in DNA.

Whereas the chemistry of the bases ensures that deamination will be detected, accurate repair—and the fundamental answer to Schroedinger's dilemma—depends on the existence of separate copies of the genetic information in the two strands of the double helix. Only in the very unlikely event that both copies are damaged simultaneously at the same base pair is the cell left without one good copy to serve as a template for DNA repair. Even in this case mechanisms have evolved that are sometimes able to repair the damage; they involve genetic recombination events that are not yet well understood and require that a second DNA helix of the same sequence be present in the cell.

Genetic information is stored in single-stranded DNA or RNA molecules only in some very small viruses with genomes of a few thousand nucleotides (see p. 250). The types of repair processes that we have described cannot operate on such nucleic acids, and the chance of a nucleotide change occurring in these viruses is very high. It seems that only organisms with tiny genomes can afford to encode their genetic information in a structure other than a double helix.

Summary

The fidelity with which DNA sequences are maintained in higher eucaryotes can be estimated from the rates at which changes have occurred in nonessential protein and DNA sequences over evolutionary time. This fidelity is so high that a mammalian germ-line cell, with a genome of 3×10^9 base pairs, is subjected on average to only about 10 to 20 base-pair changes per year. But unavoidable chemical processes damage thousands of DNA nucleotides in a typical mammalian cell every day. Genetic information can be stored stably in DNA sequences only because a large variety of DNA repair enzymes continuously scan the DNA and replace the damaged nucleotides.

The process of DNA repair depends on the presence of a separate copy of the genetic information in each strand of the DNA double helix. An accidental lesion on one strand can therefore be cut out by a repair enzyme and a good strand resynthesized from the information in the undamaged strand.

DNA Replication Mechanisms[24]

Besides maintaining the integrity of DNA sequences by DNA repair, all organisms must duplicate their DNA accurately before every cell division. *DNA replication* involves polymerization rates of about 500 nucleotides per second in bacteria and about 50 nucleotides per second in mammals. Clearly, replication enzymes must be both accurate and fast. Speed and accuracy are achieved by means of a multi-enzyme complex that guides the process and constitutes an elaborate "replication machine."

Base-pairing Underlies DNA Replication as well as DNA Repair[25]

DNA templating is defined as a process in which the nucleotide sequence of DNA (or selected portions of DNA) is copied by complementary base-pairing (A with T or U, and G with C) into a complementary nucleic acid sequence (either DNA or RNA). The process entails the recognition of each nucleotide in DNA by an un-polymerized complementary nucleotide and requires that the two strands of the DNA helix be separated, at least transiently, so that the hydrogen-bond donor and acceptor groups on each base become exposed for base-pairing. The appropriate incoming single nucleotides are thereby aligned for their enzyme-catalyzed polymerization into a new nucleic acid chain. In 1957 the first such nucleotide polymerizing enzyme, **DNA polymerase,** was discovered. The substrates for this enzyme were found at that time to be deoxyribonucleoside triphosphates, which are polymerized on a single-stranded DNA template. The stepwise mechanism of this reaction is the one previously illustrated in Figure 5–34 in connection with DNA repair. This discovery later led to the isolation of *RNA polymerase*, which was correctly inferred to use ribonucleoside triphosphates as its substrates.

During DNA replication each old DNA strand serves as a template for the formation of an entire new strand. The enormously long DNA sequence is therefore said to be replicated "semiconservatively" by the DNA polymerase. Each of the two daughters of a dividing cell inherits a new DNA double helix containing one old and one new strand (see Figure 3–11, p. 100).

5-23 The DNA Replication Fork Is Asymmetrical[26]

Autoradiographic analyses carried out in the early 1960s on whole replicating chromosomes labeled with a short pulse of the radioactive DNA precursor ³H-thymidine revealed a localized region of replication that moves along the parental DNA helix. Because of its Y-shaped structure, this active region is called a **DNA replication fork.** At a replication fork the DNA of both new daughter helices is synthesized by a multienzyme complex that contains the DNA polymerase.

Initially the simplest mechanism of DNA replication appeared to be continuous growth of both new strands, nucleotide by nucleotide, at the replication fork as it moves from one end of a DNA molecule to the other. But because of the antiparallel orientation of the two DNA strands in the DNA double helix (see p. 99), this mechanism would require one daughter strand to grow in the 5'-to-3' direction and the other in the 3'-to-5' direction. Such a replication fork would require two different DNA polymerase enzymes. One would polymerize in the observed 5'-to-3' direction, where each incoming deoxyribonucleoside triphosphate carries the triphosphate activation needed for its own addition. The other would move in the 3'-to-5' direction and work by so-called head growth, in which the end of the growing DNA chain carries the triphosphate activation required for the addition of each subsequent nucleotide. Although "head growth" polymerization occurs elsewhere in biochemistry (see Figure 2–34, p. 79), it does not occur in DNA synthesis; no 3'-to-5' DNA polymerase has ever been found (Figure 5–38).

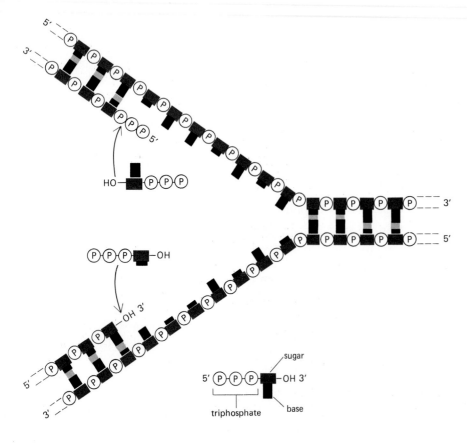

Figure 5–38 Although it might appear to be the simplest mechanism for DNA replication, the mechanism illustrated here is not the one cells use. Note that in this scheme both daughter DNA strands would grow continuously, which would require chain growth in both the 5' to 3' direction (*bottom*) and the 3' to 5' direction (*top*). No enzyme-catalyzing 3'-to-5' nucleotide polymerization has ever been found.

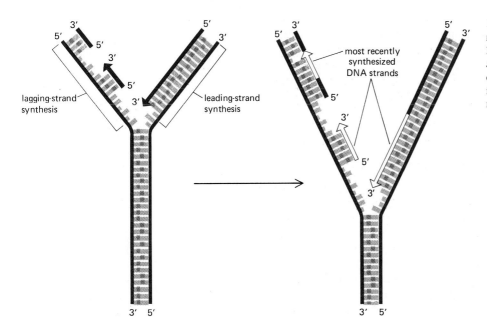

Figure 5–39 The structure of a DNA replication fork. Because both daughter DNA strands are synthesized in the 5'-to-3' direction, the DNA synthesized on the lagging strand must be made initially as a series of short DNA molecules, called *Okazaki fragments.*

How, then, is 3'-to-5' DNA synthesis achieved? The answer was first suggested in the late 1960s by experiments in which highly radioactive ^{3}H-thymidine was added to dividing bacteria for a few seconds so that only the most recently replicated DNA, just behind the replication fork, became radiolabeled. This selective labeling method revealed the transient existence of pieces of DNA that were 1000 to 2000 nucleotides long, now commonly known as *Okazaki fragments*, at the bacterial growing fork. (Such replication intermediates were later found in eucaryotes, where they are only 100 to 200 nucleotides long.) The Okazaki fragments were shown to be synthesized only in the 5'-to-3' chain direction and to be joined together after their synthesis to create long DNA chains by the same *DNA ligase* enzyme that seals nicks during DNA repair (see Figure 5–35).

A replication fork has the asymmetric structure shown in Figure 5–39. The DNA daughter strand that is synthesized continuously is known as the **leading strand,** and its synthesis slightly precedes the synthesis of the daughter strand that is synthesized discontinuously, which is known as the **lagging strand.** The synthesis of the lagging strand is delayed because it must wait for the leading strand to expose the template strand on which each Okazaki fragment is synthesized. The synthesis of the lagging strand by a discontinuous, "backstitching" mechanism means that only the 5'-to-3' type of DNA polymerase is needed for DNA replication.

5-24 The High Fidelity of DNA Replication Requires a Proofreading Mechanism[27]

The fidelity of copying during DNA replication is such that only about one error is made in every 10^9 base-pair replications, as required to maintain the mammalian genome of 3×10^9 DNA base pairs (p. 485). This fidelity is much higher than expected, given that the standard complementary base pairs are not the only ones possible. Rare tautomeric forms of the four DNA bases occur transiently in normal DNA in ratios of one part to 10^4 or 10^5. These forms mispair; for example, the rare tautomeric form of C pairs with A instead of G and thus causes a mutation (Figure 5–40). The high fidelity of DNA replication depends on "proofreading" mechanisms that remove errors brought about in this way.

One important proofreading process depends on special properties of the DNA polymerase enzyme. Unlike RNA polymerases, DNA polymerases do not begin a new polynucleotide chain by linking two nucleoside triphosphates together. They absolutely require the 3'-OH end of a base-paired *primer strand* on which

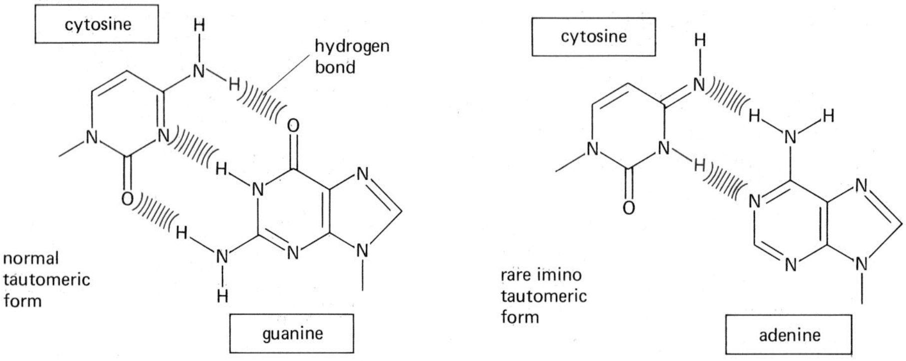

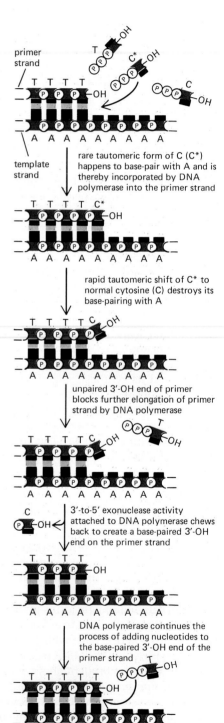

Figure 5–40 An example of an incorrect base pair that can occur transiently during DNA replication. When cytosine is in its unfavored tautomeric form, it can form effective hydrogen bonds with adenine.

cytosine

hydrogen bond

normal tautomeric form

guanine

cytosine

rare imino tautomeric form

adenine

primer strand

template strand

rare tautomeric form of C (C*) happens to base-pair with A and is thereby incorporated by DNA polymerase into the primer strand

rapid tautomeric shift of C* to normal cytosine (C) destroys its base-pairing with A

unpaired 3'-OH end of primer blocks further elongation of primer strand by DNA polymerase

3'-to-5' exonuclease activity attached to DNA polymerase chews back to create a base-paired 3'-OH end on the primer strand

DNA polymerase continues the process of adding nucleotides to the base-paired 3'-OH end of the primer strand

to add further nucleotides (see Figure 5–34). Moreover, DNA molecules with a mismatched (not base-paired) nucleotide at the 3'-OH end of the primer strand are not effective as templates. Bacterial DNA polymerase molecules are able to deal with such mismatched DNAs by means of a separate catalytic subunit or domain that clips off any unpaired residues at the primer terminus. Clipping by this *3'-to-5' exonuclease* activity continues until enough nucleotides have been removed from the 3' end to regenerate a base-paired terminus that can prime DNA synthesis. In this way DNA polymerase functions as a "self-correcting" enzyme that removes its own polymerization errors as it moves along the DNA. Figure 5–41 illustrates how this proofreading process would be expected to remove the rare tautomeric C-A base pair just described.

The requirement for a perfectly base-paired terminus is essential to the self-correcting properties of the DNA polymerase. For such an enzyme to start synthesis in the complete absence of a primer without losing any of its discrimination between base-paired and unpaired growing 3'-OH termini is apparently not possible. By contrast, the RNA polymerase enzymes involved in gene transcription (p. 202) need not be self-correcting: errors in DNA transcription are not passed on to the next generation, and an occasional defective molecule has no significance. RNA polymerases are able to start new polynucleotide chains without a primer, and an error frequency of about 1 in 10^4 is found, both in RNA synthesis and in the separate process of translating mRNA sequences into protein sequences.

5-22 DNA Replication in the 5'-to-3' Direction Allows Efficient Error Correction

The need for accuracy probably explains why DNA replication occurs only in the 5'-to-3' direction. If there were a DNA polymerase that added deoxyribonucleoside triphosphates in such a way as to cause chains to grow in the 3'-to-5' chain direction, the growing 5'-chain end rather than the incoming mononucleotide would carry the activating triphosphate. In this case, the mistakes in polymerization could not be simply hydrolyzed away, since the bare 5'-chain end thus created would immediately terminate DNA synthesis. It is, therefore, much easier to correct a mismatched base that has just been added to the 3' end than one that has just been added to the 5' end of a DNA chain. Although the type of mechanism for DNA replication shown in Figure 5–39 seems at first sight much more complex than the incorrect mechanism depicted in Figure 5–38, it is much more accurate because it involves DNA synthesis only in the 5'-to-3' direction.

Figure 5–41 Illustration of the proofreading process that removes errors during DNA synthesis in bacteria. A proofreading mechanism of this type is thought to operate in eucaryotic cells also.

5-25 A Special Nucleotide Polymerizing Enzyme Synthesizes Short Primer Molecules on the Lagging Strand[28]

For the leading strand, a special primer is needed only at the start of replication; once a replication fork is established, the DNA polymerase is continuously presented with a base-paired chain end on which to synthesize its new strand. But the DNA polymerase on the lagging side of the fork requires only about 4 seconds to complete each short DNA fragment, after which it must start synthesizing a completely new fragment at a site farther along the template strand (see Figure 5–39). A special mechanism is needed to produce the base-paired primer strand required by this DNA polymerase molecule. The mechanism involves an enzyme called *DNA primase*, which uses ribonucleoside triphosphates to synthesize short **RNA primers** (Figure 5–42). These primers are about 10 nucleotides long in eucaryotes, and they are made at intervals on the lagging strand, where they are elongated by the DNA polymerase to begin each Okazaki fragment. The synthesis of each Okazaki fragment ends when this DNA polymerase runs into the RNA primer attached to the 5′ end of the previous fragment. To produce a continuous DNA chain from the many DNA fragments made on the lagging strand, a special DNA repair system acts quickly to erase the old RNA primer and replace it with DNA. DNA ligase then joins the 3′ end of the new DNA fragment to the 5′ end of the previous one to complete the process (Figure 5–43).

Why might an erasable RNA primer be preferred to a DNA primer that need not be erased? The argument that a self-correcting polymerase cannot start chains *de novo* also implies its converse: an enzyme that starts chains *de novo* cannot be efficient at self-correction. Thus any enzyme that primes the synthesis of Okazaki fragments will of necessity make a relatively inaccurate copy (at least 1 error in 10^5). Even if the amount of this copy retained in the final product constitutes as little as 5% of the total genome (for example, 10 nucleotides per 200-nucleotide DNA fragment), the resulting increase in overall mutation rate would be enormous. It therefore seems likely that the evolution of RNA rather than DNA for priming entailed a powerful advantage, since the ribonucleotides in the primer automatically mark these sequences as "bad copy" to be removed.

5-26 Special Proteins Help Open Up the DNA Double Helix in
5-33 Front of the Replication Fork[29]

The DNA double helix must be opened up ahead of the replication fork so that the incoming deoxyribonucleoside triphosphates can form base-pairs with the template strand. However, the DNA double helix is very stable under normal conditions; the base pairs are locked in place so strongly that temperatures approaching that of boiling water (90°C) are required to separate the two strands in a test tube. For this reason, most DNA polymerases can copy DNA only when the template strand has already been separated from its complementary strand. Additional proteins are needed to help open the double helix and thus provide the appropriate exposed DNA template for the DNA polymerase to copy. Two types of replication proteins contribute to this process.

DNA helicases were first isolated as proteins that hydrolyze ATP when they are bound to single strands of DNA. As described in Chapter 3, the hydrolysis of ATP can change the shape of a protein molecule in a cyclic manner that allows the protein to perform mechanical work (p. 130). DNA helicases utilize this principle to move rapidly along a DNA single strand; where they encounter a region of double helix, they continue to move along their strand, thereby unwinding the helix (Figure 5–44). The unwinding of the template DNA helix at a replication fork could in principle be catalyzed by two DNA helicases acting in concert, one running along the leading strand and one along the lagging strand. These two helicases would need to move in opposite directions along a DNA single strand and therefore would have to be different enzymes. Both types of DNA helicase do in fact exist, but studies in bacteria show that the DNA helicase on the lagging strand plays the predominant role, for reasons that will become clear shortly.

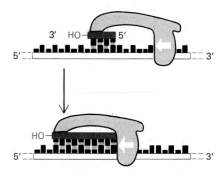

Figure 5–42 A schematic view of the reaction catalyzed by DNA primase, the enzyme that synthesizes the short RNA primers made on the lagging strand. Unlike DNA polymerase, this enzyme can start a new polynucleotide chain by joining two nucleoside triphosphates together. The primase stops after a short polynucleotide has been synthesized and makes the 3′ end of this primer available for the DNA polymerase.

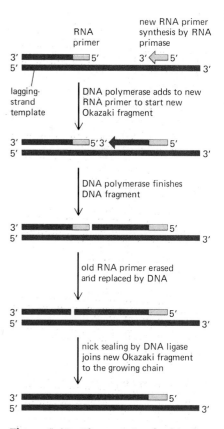

Figure 5–43 The steps involved in the synthesis of each DNA fragment on the lagging strand. In eucaryotes the RNA primers are made at intervals spaced by about 200 nucleotides on the lagging strand, and each RNA primer is 10 nucleotides long.

Helix-destabilizing proteins—also called *single-strand DNA-binding (SSB) proteins*—bind to exposed DNA strands without covering the bases, which therefore remain available for templating processes. These proteins are unable to open a long DNA helix directly, but they aid other helix-opening processes by stabilizing the unwound, single-stranded conformation; for example, they can help the DNA helicase expose strands at the replication fork. On the lagging strand template, their cooperative binding coats the regions of single-stranded DNA, thereby preventing formation of the short hairpin helices that would otherwise impede synthesis by the DNA polymerase (Figure 5–45).

The Proteins at a Replication Fork Cooperate to Form a Replication Machine[30]

Although we have discussed DNA replication as though it were carried out by a mixture of replication proteins that act independently, in reality most of the proteins are held together in a large multienzyme complex that moves rapidly along the DNA. This complex can be likened to a tiny sewing machine composed of protein parts and powered by nucleoside triphosphate hydrolyses. Although the replication complex has been well characterized so far only in the bacterium *E. coli* and several of its viruses, a very similar complex is thought to operate in eucaryotes (see p. 517).

The functions of the subunits of the "replication machine" are summarized in the two-dimensional diagram of the complete replication fork shown in Figure 5–46. Two identical DNA polymerase molecules work at the fork, one on the leading strand and one on the lagging strand. The DNA helix is opened by the DNA polymerase molecule on the leading strand acting in concert with a DNA helicase molecule running along the lagging strand; this process is aided by cooperatively bound molecules of helix-destabilizing protein. While the DNA polymerase molecule on the leading strand can proceed in a continuous fashion, the DNA polymerase molecule on the lagging strand must restart at intervals, using a short RNA primer made by a DNA primase molecule.

The efficiency of replication is greatly increased by the close association of all these protein components. The primase molecule is linked directly to the DNA helicase to form a unit on the lagging strand called a **primosome,** which moves with the fork, synthesizing RNA primers as it goes. Similarly, the DNA polymerase molecule that synthesizes DNA on the lagging strand moves in concert with the rest of the proteins, synthesizing a succession of new Okazaki fragments; to accommodate this arrangement, its DNA template strand is thought to be folded back in the manner shown in Figure 5–47. The replication proteins are thus linked together into a single large unit (total mass $> 10^6$ daltons) that moves rapidly along

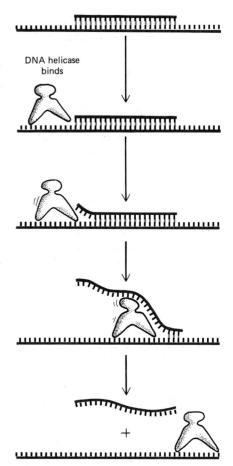

Figure 5–44 The assay used to test for DNA helicase enzymes. A short DNA fragment is annealed to a long DNA single strand to form a region of DNA double helix. This helix is melted as the helicase runs along the DNA single strand, releasing the short DNA fragment in a reaction that requires the presence of both the helicase protein and ATP. The movement of the helicase is powered by its ATP hydrolysis (see Figure 3–63, p. 131).

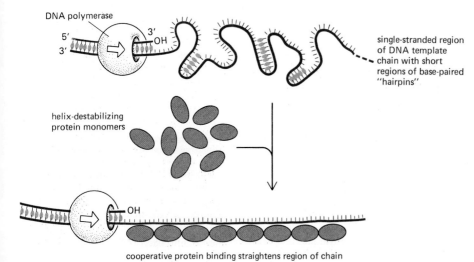

cooperative protein binding straightens region of chain

Figure 5–45 The effect of helix-destabilizing proteins on the structure of single-stranded DNA. Because each protein molecule prefers to bind next to a previously bound molecule (*cooperative binding,* see p. 492), long rows of this protein will form on a DNA single strand. This cooperative binding straightens out the DNA template and facilitates the DNA polymerization process. The "hairpin helices" shown in the bare single-stranded DNA result from a chance matching of short regions of complementary nucleotide sequence; they are similar to the short helices that form in all RNA molecules (see p. 98).

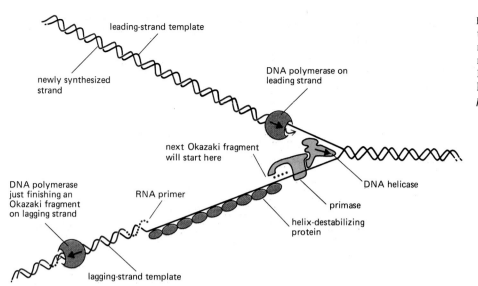

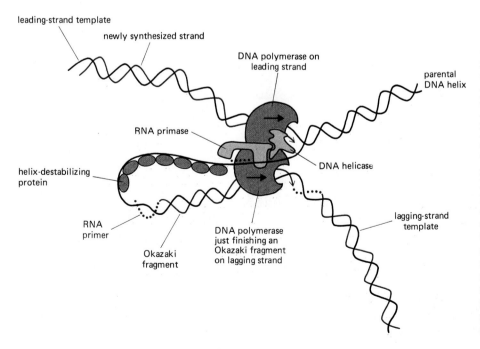

Figure 5–46 Summary of the major types of proteins that act at a DNA replication fork, showing their locations relative to the DNA. The complex of the DNA primase and DNA helicase on the lagging strand is known as the *primosome*.

the DNA, enabling DNA to be synthesized on both sides of the fork in a coordinated and efficient manner.

This DNA replication machine leaves behind a series of unsealed Okazaki fragments on the lagging strand, which still contain the RNA that primed their synthesis at their 5′ ends. This RNA must be removed and the fragments joined up by the DNA repair enzymes that operate behind the replication fork (see Figure 5–43).

A Mismatch Proofreading System Removes Replication Errors in Bacterial Cells[31]

Bacteria such as *E. coli* are capable of dividing once every 30 minutes, so it is relatively easy to screen large populations to find rare mutants that are altered in a specific process. One interesting class of mutants contains so-called mutator genes, which greatly increase the rate of spontaneous mutation. Not surprisingly, one mutator gene encodes a defective form of the 3′-to-5′ proofreading exonuclease that is a subunit of the DNA polymerase enzyme (see p. 230). When this

Figure 5–47 A current view of how the replication proteins are arranged at a replication fork when the fork is moving. The two-dimensional structure of Figure 5–46 has been altered by folding the DNA on the lagging strand to bring the lagging-strand DNA polymerase molecule into a complex with the leading-strand DNA polymerase molecule. This folding process also brings the 3′ end of each completed Okazaki fragment close to the start site for the next Okazaki fragment (compare with Figure 5–46). Because the lagging-strand DNA polymerase molecule is held to the rest of the replication proteins, it can be continually reused at the same fork; thus it is about to let go of its completed DNA fragment and move to the new RNA primer nearby, as required to start synthesizing the next fragment. Note that one daughter DNA helix extends toward the bottom right and the other toward the top left in this diagram.

protein is defective, the DNA polymerase no longer proofreads effectively and many replication errors that would otherwise have been removed accumulate in the DNA.

The study of *E. coli* mutants containing mutator genes has uncovered another proofreading system that normally removes replication errors missed by the proofreading exonuclease. This **mismatch proofreading** system (also called a *mismatch repair* system) differs from the DNA repair systems discussed previously in that it does not depend on the presence of abnormal nucleotides in the DNA that can be recognized and excised. Instead, it detects the distortion on the outside of the helix that results from the misfit between normal but noncomplementary base pairs. If the proofreading system simply recognized a mismatch in newly replicated DNA and randomly excised one of the two mismatched nucleotides, it would make the mistake of "correcting" the original template strand to match the error exactly half the time and not change the error rate on average. To be effective at proofreading, the proofreading system must be able to distinguish and remove the mismatched nucleotide on the new strand (the replication error) specifically.

The recognition system used by the mismatch proofreading system in *E. coli* depends on the methylation of selected A residues in the DNA. Methyl groups are added to all A residues in the sequence GATC, but not until some time after the A has been incorporated into a newly synthesized DNA chain. Because only the new strands just behind a replication fork will contain GATC sequences that have not yet been methylated, these new DNA strands can be distinguished from old ones. Mismatch proofreading is carried out by a large multienzyme complex that scans each of the two strands of the double helix. The excision function in this complex specifically requires a mismatched nucleotide but it remains dormant unless an unmethylated GATC has also been encountered on the same strand. Therefore, only the mismatches on the new strand will be excised and replication errors are selectively removed (Figure 5–48).

Neither of the two types of DNA proofreading processes found in bacteria has been demonstrated conclusively in higher eucaryotic cells. Nevertheless, the overall fidelity of DNA replication is about the same in mammals and in *E. coli*, and it therefore seems likely that both types of proofreading occur. Because there are no methylated A residues in mammalian DNA, its mismatch repair system must rely on some other marker to recognize the newly synthesized DNA strand.

Figure 5–48 Diagram of an experiment carried out with the mismatch proofreading system that corrects DNA replication errors in bacteria. A special protein complex removes a base-pair mismatch on the newly synthesized DNA strand behind a replication fork; this repair complex specifically recognizes the new strand by scanning for unmethylated GATC sequences. In the experiment illustrated, three different DNA molecules were prepared, each containing the same mismatched base pair—but with both strands (A), neither strand (B), or only one strand (C) methylated. When these DNA molecules were treated with a cell extract containing the mismatch proofreading complex, the results shown were obtained. The DNA molecule in (C) mimics the situation just behind the replication fork, where the bottom strand represents the new strand that is not yet methylated.

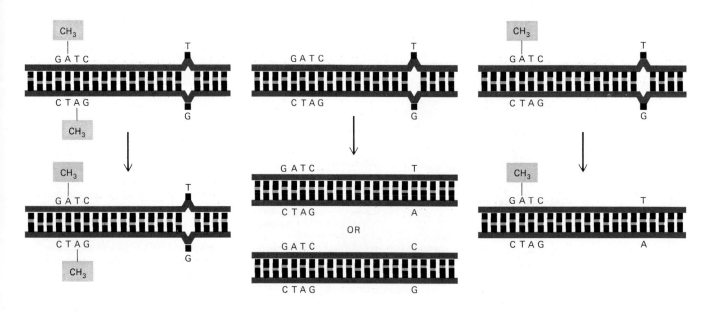

(A) NO REPAIR OF MISMATCH

(B) EITHER NUCLEOTIDE REPLACED WITH EQUAL PROBABILITY

(C) NUCLEOTIDE ON UNMETHYLATED STRAND REPLACED ONLY

Figure 5–49 An outline of the processes involved in the initiation of replication forks at replication origins. (See also Figure 5–50.)

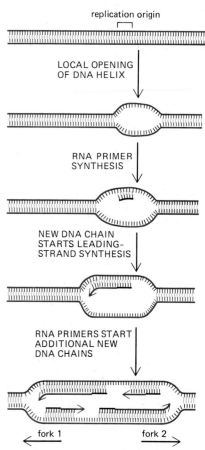

5-27 Replication Forks Initiate at Replication Origins[32]

In both bacteria and mammals, replication forks originate at a structure called a *replication bubble*, a local region where the two strands of the parental DNA helix have separated from each other to serve as templates for DNA synthesis (Figure 5–49). For bacteria, as well as for several viruses that grow in eucaryotic cells, replication bubbles have been shown to form at special DNA sequences called **replication origins,** which can be as long as 300 nucleotides. Analogous replication origins are thought to exist in eucaryotic chromosomes, but so far they have not been as well characterized (see p. 516).

For several well-defined replication origins, it has been possible to reproduce the fork initiation reaction *in vitro*. The *in vitro* studies reveal that fork initiation in bacteria and bacterial viruses starts in the manner indicated in Figure 5–50. Multiple copies of an *initiator protein* bind to specific sites at the replication origin to form a large protein complex. This complex then binds the DNA helicase and positions it onto an exposed DNA single strand in an adjacent region of helix. The DNA primase also binds, forming the primosome, which moves away from the origin and makes an RNA primer that starts the first DNA chain. This quickly leads to assembly of the remaining proteins to create two replication protein complexes moving away from the origin in opposite directions (see Figure 5–49); these continue to synthesize DNA until all of the DNA template downstream of each fork has been replicated.

The available information concerning replication fork initiation in eucaryotic chromosomes will be discussed in Chapter 9, when we describe the cell nucleus.

two complete replication forks are formed, one moving leftward with leading (top) and lagging (bottom) strands, and one moving rightward with leading (bottom) and lagging (top) strands

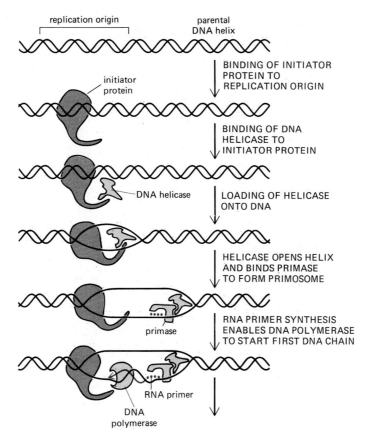

Figure 5–50 A simplified outline of the initial steps leading to the formation of replication forks at the *E. coli* and bacteriophage lambda replication origins. Discovery of the indicated mechanism required *in vitro* studies utilizing a mixture of highly purified proteins. Subsequent steps result in the initiation of three more DNA chains (Figure 5–49) by a pathway that is not yet clear. For *E. coli* DNA replication, the initiator protein is the dnaA protein, and the primosome is composed of the dnaB (DNA helicase) and dnaG (RNA primase) proteins.

DNA Topoisomerases Prevent DNA Tangling During Replication[33]

When we draw the DNA helix (incorrectly) as a flat "ladder," we are ignoring the "winding problem" that arises during DNA replication. Every 10 base pairs replicated at the fork correspond to one complete turn about the axis of the parental double helix. Therefore, for a replication fork to move, the entire chromosome ahead of the fork would normally have to rotate rapidly (Figure 5–51), which would require large amounts of energy for long chromosomes. An alternative strategy is used during DNA replication: a "swivel" is formed in the DNA helix by proteins known as **DNA topoisomerases.**

A DNA topoisomerase can be viewed as a "reversible nuclease" that adds itself covalently to a DNA phosphate, thereby breaking a phosphodiester bond in the DNA chain. Because the covalent linkage that joins a topoisomerase to a DNA phosphate retains the energy of the broken phosphodiester bond, the breakage reaction is reversible; resealing is rapid and does not require additional energy input. The rejoining mechanism is different in this respect from that of the enzyme DNA ligase, discussed previously (see Figure 5–35).

One type of topoisomerase (*topoisomerase I*) causes a *single-strand break* (or *nick*), which allows the two sections of DNA helix on either side of the nick to rotate freely relative to each other, using the phosphodiester bond in the strand opposite the nick as a swivel point (Figure 5–52). Any tension in the DNA helix will drive this rotation in the direction that relieves the tension. As a result, DNA replication can occur with the rotation of only a short length of helix—the part just ahead of the fork. The analogous problem that arises during DNA transcription is solved in a similar way.

A second type of DNA topoisomerase (*topoisomerase II*) forms a covalent linkage to both strands of the DNA helix at the same time, making a transient *double-strand break* in the helix. These enzymes are activated by sites on chromosomes where two double helices cross over one another. When the topoisomerase binds to such a crossing site, it (1) breaks one double helix reversibly to create a DNA "gate," (2) causes the second nearby double helix to pass through this break, and (3) reseals the break and dissociates from the DNA. In this way type II DNA topoisomerases can efficiently separate two interlocked DNA circles (Figure 5–53). The same reaction prevents the severe DNA tangling problems that would otherwise arise during DNA replication. For example, when temperature-sensitive mutant yeast cells that produce a version of topoisomerase II that is inactive at 37°C are warmed to this temperature, their chromosomes remain intertwined at mitosis and are unable to separate. The usefulness of topoisomerase II for untangling chromosomes can readily be appreciated by anyone who has struggled to remove a tangle from a fishing line without the aid of scissors.

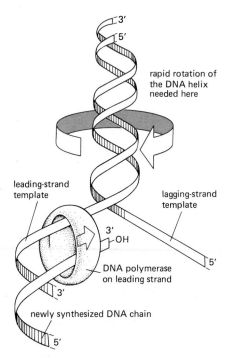

Figure 5–51 The "winding problem" that arises during DNA replication. For a bacterial replication fork moving at 500 nucleotides per second, the parental DNA helix ahead of the fork must rotate at 50 revolutions per second.

5-28 DNA Replication Is Basically Similar in Eucaryotes and Procaryotes[34]

Most of what we know about DNA replication comes from studies of purified bacterial and bacteriophage multienzyme systems capable of DNA replication *in vitro*. The development of these systems in the 1970s was greatly facilitated by the use of mutants in a variety of replication genes that could be exploited to identify and purify the corresponding replication proteins (Figure 5–54).

Much less is known about the detailed enzymology of DNA replication in eucaryotes, largely because it is difficult to obtain replication-deficient mutants. Nevertheless, the basic mechanisms of DNA replication, including the geometry of the replication fork and the use of an RNA primer, seem to be identical for procaryotes and eucaryotes. The major difference is that eucaryotic DNA is replicated not as bare DNA but as *chromatin*, in which the DNA is complexed with tightly bound proteins called *histones*. As described in Chapter 9, histones form disclike structures around which the eucaryotic DNA is wound, creating a repeating structural unit called a *nucleosome*. Nucleosomes are spaced at intervals of about 200 base pairs along the DNA, which may be why new Okazaki fragments are synthesized on the lagging strand at intervals of 100 to 200 nucleotides in eucaryotes, instead of at intervals of 1000 to 2000 nucleotides as in bacteria.

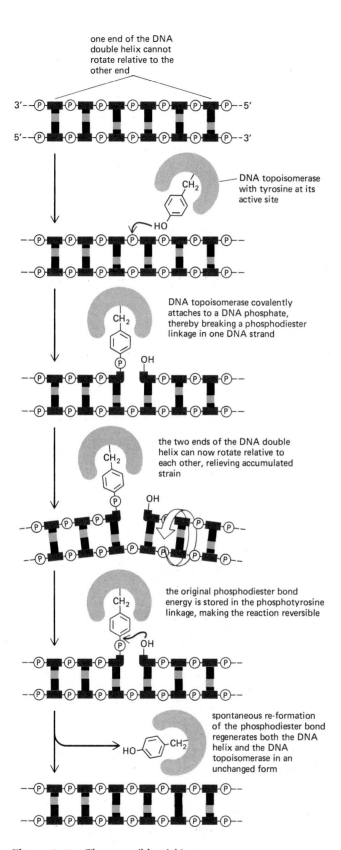

one end of the DNA
double helix cannot
rotate relative to the
other end

3'--(P)(P)(P)(P)(P)(P)(P)--5'

5'--(P)(P)(P)(P)(P)(P)(P)--3'

DNA topoisomerase
with tyrosine at its
active site

DNA topoisomerase covalently
attaches to a DNA phosphate,
thereby breaking a phosphodiester
linkage in one DNA strand

the two ends of the DNA double
helix can now rotate relative to
each other, relieving accumulated
strain

the original phosphodiester bond
energy is stored in the phosphotyrosine
linkage, making the reaction reversible

spontaneous re-formation
of the phosphodiester bond
regenerates both the DNA
helix and the DNA
topoisomerase in an
unchanged form

Figure 5–52 The reversible nicking
reaction catalyzed by a eucaryotic DNA
topoisomerase I enzyme. As indicated,
these enzymes form a transient covalent
bond with DNA.

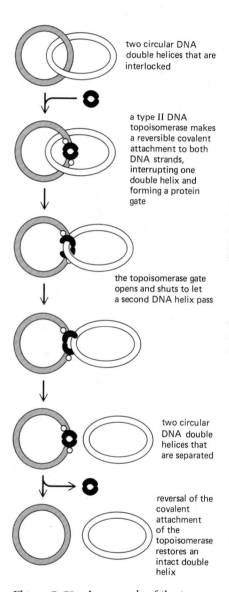

two circular DNA
double helices that are
interlocked

a type II DNA
topoisomerase makes
a reversible covalent
attachment to both
DNA strands,
interrupting one
double helix and
forming a protein
gate

the topoisomerase gate
opens and shuts to let
a second DNA helix pass

two circular
DNA double
helices that
are separated

reversal of the
covalent
attachment
of the
topoisomerase
restores an
intact double
helix

Figure 5–53 An example of the type
of DNA-helix-passing reaction catalyzed
by a type II DNA topoisomerase. Unlike
type I topoisomerases, these enzymes
require ATP hydrolysis for their
function, and some of them can
introduce superhelical tension
into DNA (see p. 581). Type II
topoisomerases are found in both
procaryotes and eucaryotes and
probably function in many reactions
involving DNA.

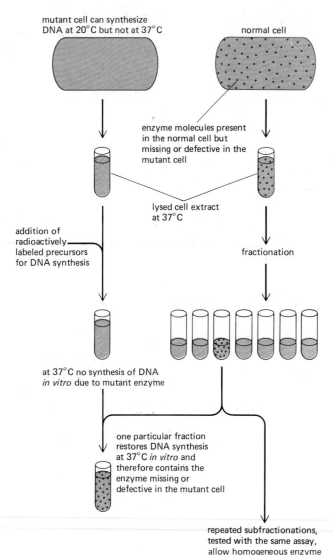

mutant cell can synthesize
DNA at 20°C but not at 37°C

normal cell

enzyme molecules present
in the normal cell but
missing or defective in the
mutant cell

lysed cell extract
at 37°C

addition of
radioactively
labeled precursors
for DNA synthesis

fractionation

at 37°C no synthesis of DNA
in vitro due to mutant enzyme

one particular fraction
restores DNA synthesis
at 37°C *in vitro* and
therefore contains the
enzyme missing or
defective in the mutant cell

repeated subfractionations,
tested with the same assay,
allow homogeneous enzyme
to be obtained

Figure 5–54 The availability of bacterial and bacteriophage mutants defective in DNA replication made it possible to detect and purify proteins of unknown function essential for procaryotic DNA replication. The temperature-sensitive mutants used here are "conditional mutants"; usually the mutant protein functions normally at low temperature but not at high temperature. Nonconditional DNA replication-defective mutants could not synthesize DNA at any temperature and, therefore, could not survive. Related "*in vitro* complementation assays" have allowed the proteins that catalyze many other biological processes to be purified.

Nucleosomes may also act as barriers that interfere with the movement of DNA polymerase molecules, which could explain why eucaryotic replication forks move only one-tenth as fast as bacterial replication forks.

Summary

A self-correcting DNA polymerase catalyzes nucleotide polymerization in a 5'-to-3' direction, copying a DNA template with remarkable fidelity. Since the two strands of a DNA helix are antiparallel, this 5'-to-3' DNA synthesis can take place continuously on only one of the strands at a replication fork (the leading strand). On the lagging strand, short DNA fragments are made by a "backstitching" process. Because the self-correcting DNA polymerase cannot start a new chain, these lagging-strand DNA fragments are primed by short RNA primer molecules that are subsequently erased and replaced with DNA.

DNA replication requires the cooperation of many proteins, including (1) DNA polymerase and DNA primase to catalyze nucleoside triphosphate polymerizations, (2) DNA helicases and helix-destabilizing proteins to help open up the DNA helix to be copied, (3) DNA ligase and an enzyme that degrades RNA primers to seal together the discontinuously synthesized lagging-strand DNA fragments, (4) DNA topoisomerases to help relieve helical winding and tangling problems, and (5) initiator proteins that bind to specific DNA sequences at a replication origin and catalyze the formation of a replication fork at that site. At a replication origin, a complex of

DNA helicase and DNA primase (a primosome) is loaded onto the DNA template; other proteins are then added to form the multienzyme "replication machine" that catalyzes DNA synthesis.

Genetic Recombination Mechanisms[35]

In the two preceding sections we discussed the mechanisms by which DNA sequences in cells are maintained from generation to generation with very little change. Although such genetic stability is crucial for survival in the short term, in the longer term the survival of organisms may depend on genetic variation, through which they can adapt to a changing environment. Thus an important property of the DNA in cells is its ability to undergo rearrangements that can vary the particular combination of genes present in any individual genome, as well as the timing and the level of expression of these genes. These DNA rearrangements are caused by **genetic recombination.** Two broad classes of genetic recombination are commonly recognized: general recombination and site-specific recombination.

In *general recombination*, genetic exchange takes place between homologous DNA sequences, usually located on two copies of the same chromosome. One of the most important examples is the exchange of sections of homologous chromosomes (homologues) in the course of *meiosis*. This "crossing-over" occurs between tightly apposed chromosomes early in the development of eggs and sperm (p. 848), and it allows different versions (*alleles*) of the same gene to be tested in new combinations with other genes, increasing the chance that at least some members of a mating population will survive in a changing environment (p. 847). Although meiosis occurs only in eucaryotes, the advantage of this type of gene mixing is so great that the mating and reassortment of genes by general recombination is also widespread in bacteria.

DNA homology is not required in *site-specific recombination*. Instead, exchange occurs at short, specific nucleotide sequences (on either one or both of the two participating DNA molecules) that are recognized by a site-specific recombination enzyme. Site-specific recombination therefore alters the relative positions of the nucleotide sequences in genomes. In some cases these changes are scheduled and organized, as when an integrated bacterial virus is induced to leave a chromosome of a bacterium under stress (p. 253); in others they are haphazard, as when the DNA sequence of a transposable element is inserted at a randomly selected site in a chromosome (p. 605).

As for DNA replication, most of what we know about the biochemistry of genetic recombination has come from studies of simple organisms, especially of *E. coli* and its viruses.

5-31 General Recombination Is Guided by Base-pairing Interactions Between Complementary Strands of Two Homologous DNA Molecules[36]

General recombination involves DNA strand-exchange intermediates that require some effort to understand. Although the exact pathway followed is likely to be different in different organisms, detailed genetic analyses of viruses, mating bacteria, and fungi suggest that the major outcome of general recombination is always the same: (1) Two homologous DNA molecules "cross over"; that is, their double helices break and the two broken ends join to their opposite partners to re-form two intact double helices, each composed of parts of the two initial DNA molecules (Figure 5–55). (2) The site of exchange (that is, where a colored double helix is joined to a black double helix in Figure 5–55) can occur anywhere in the homologous nucleotide sequences of the two participating DNA molecules. (3) At the site of exchange, a strand of one DNA molecule becomes base-paired to a strand of the second DNA molecule to create a *staggered joint* (usually called a *heteroduplex joint*) between the two different double helices (Figure 5–56). The heteroduplex region can be thousands of base pairs long; we shall explain later how it forms. (4) No nucleotide sequences are altered at the site of exchange; the

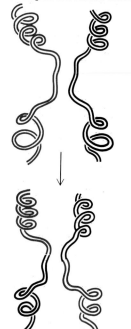

two homologous DNA double helices

DNA molecules that have crossed over

Figure 5–55 In general recombination the breakage and reunion of two homologous DNA double helices creates two "crossed over" DNA molecules.

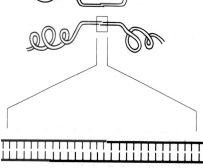

DNA molecules that have crossed over

heteroduplex joint where strands from two different DNA helices have base-paired

Figure 5–56 A heteroduplex joint unites two DNA molecules where they have crossed over. Such a joint is often thousands of nucleotides long.

breakage and reunion events occur so precisely that not a single nucleotide is lost, gained, or changed.

The general recombination pathway is designed to ensure that two regions of DNA double helix undergo an exchange reaction only if they have extensive sequence homology. The formation of a heteroduplex joint requires that such homology be present because it involves a long region of complementary base-pairing between a strand from one of the two original double helices and a strand from the other. But how does this heteroduplex joint arise, and how do the two homologous regions of DNA at the site of crossing-over recognize each other? As we shall see, recognition takes place by means of a direct base-pairing interaction. The formation of base pairs between complementary strands from the two DNA molecules then guides the general recombination process, allowing it to occur only between long regions of matching DNA sequence. Despite this matching requirement, general recombination often rearranges DNA sequences; the heteroduplex joint can contain a small number of mismatched base pairs, and, more important, the two DNAs that cross over are usually not exactly the same on either side of the joint.

5-32 General Recombination Can Be Initiated at a Nick in One Strand of a DNA Double Helix[36]

Each of the two strands in a DNA molecule is helically wound around the other. As a result, extensive base-pair interactions can occur between two homologous DNA double helices only if a nick is first made in a strand of one of them, freeing that strand for the unwinding and rewinding events required to form a heteroduplex with another DNA molecule. For the same reason, any *exchange* of strands between two DNA double helices requires at least two nicks, one in a strand of each interacting double helix. Finally, to produce the heteroduplex joint illustrated in Figure 5–56, each of the four strands present must be cut to allow each to be joined to a different partner. In general recombination these nicking and resealing events are coordinated so that they occur only when two DNA helices share an extensive region of matching DNA sequence.

There is evidence from a number of sources that a single nick in only one strand of a DNA molecule is sufficient to initiate general recombination. For example, chemical agents or types of irradiation that introduce single strand nicks will trigger a genetic recombination event. Moreover, one of the special proteins required for recombination in *E. coli*—the *recBCD protein*—has been shown to make single strand nicks in DNA molecules. The recBCD protein is also a DNA-dependent ATPase that acts like a DNA helicase, traveling along a DNA helix transiently exposing its strands. By combining its nuclease and helicase activities, the recBCD protein will create a single-stranded "whisker" on the DNA double helix (Figure 5–57). Figure 5–58 shows how such a whisker could initiate a base-pairing interaction between two complementary stretches of DNA double helix.

5-33 DNA Hybridization Reactions Provide a Simple Model for the Base-pairing Step in General Recombination[29,37]

In its simplest form, the base-pairing interaction central to general recombination can be mimicked in a test tube by allowing a DNA double helix to re-form from its separated single strands. This process, called **DNA renaturation** or **hybridization,** occurs when a rare random collision juxtaposes complementary nucleotide sequences on two matching DNA single strands, allowing the formation of a short stretch of double helix between them. This relatively slow *helix nucleation* step is followed by a very rapid "zippering" step as the region of double helix is extended to maximize the number of base-pairing interactions (Figure 5–59).

Formation of a new double helix in this way requires that the annealing strands be in an open, unfolded conformation. For this reason, *in vitro* hybridization reactions are carried out at high temperature or in the presence of an organic solvent such as formamide; these conditions "melt out" the short hairpin helices formed

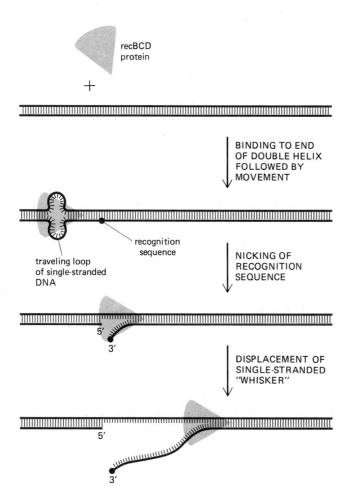

Figure 5–57 A reaction catalyzed by the recBCD protein, an enzyme required for general genetic recombination in *E. coli*. The protein enters the DNA from one end of a double helix and then uses energy derived from the hydrolysis of bound ATP molecules to propel itself in one direction along the DNA at a rate of about 300 nucleotides per second. A special recognition site (a DNA sequence of eight nucleotides scattered throughout the *E. coli* chromosome) is cut in the traveling loop of DNA created by the recBCD protein, and thereafter a single-stranded whisker is thrown out of the helix, as shown. This whisker could initiate genetic recombination by pairing with a homologous helix (see Figure 5–58).

where base-pairing interactions occur within a single strand that folds back on itself. Bacterial cells could not survive such harsh conditions and instead use a helix-destabilizing protein, the *SSB protein*, to open their helices. This protein is essential for DNA replication as well as for general recombination in *E. coli* (see p. 232); it binds tightly and cooperatively to the sugar-phosphate backbone of all single-stranded regions of DNA, holding them in an extended conformation with their bases exposed (see Figure 5–45). In this extended conformation, a DNA single strand can base-pair with either a nucleoside triphosphate molecule (in DNA replication) or a complementary section of another DNA single strand (in genetic recombination). When hybridization reactions are carried out *in vitro* under conditions that mimic those inside a cell, the SSB protein speeds up the rate of DNA helix nucleation and thereby the overall rate of strand annealing by a factor of more than 1000.

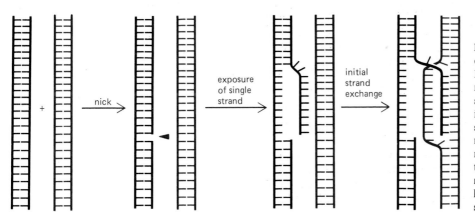

Figure 5–58 The initial strand exchange between two homologous DNA double helices undergoing general recombination. A nick in a single DNA strand frees the strand, which then invades the second helix to form a short pairing region. Only two DNA molecules that are complementary in nucleotide sequence can base-pair in this way and thereby initiate a general recombination event. Enzymes are known that can catalyze all of the steps shown here (see Figures 5–57 and 5–60).

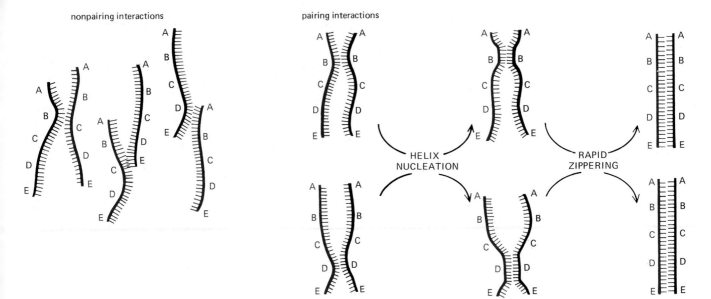

nonpairing interactions   pairing interactions

HELIX NUCLEATION

RAPID ZIPPERING

5-34 The recA Protein Enables DNA Single Strands to Pair with a Homologous Region of DNA Double Helix in *E. coli*[38]

General genetic recombination is more complex than the simple hybridization reactions just described. In the course of general recombination, one DNA double helix must be invaded by a single DNA strand derived from another (see Figure 5–58). In *E. coli* this requires the **recA protein,** produced by the *recA* gene that was identified in 1965 as being essential for chromosome pairing. Long sought by biochemists, this elusive gene product was finally purified to homogeneity in 1976 and shown to be a protein of 38,000 daltons. Like a helix-destabilizing protein, it binds tightly and in large cooperative clusters to single-stranded DNA, but it also has several distinctive properties. For example, it has more than one DNA-binding site, and it can therefore hold a single strand and a double helix together. These sites allow the recA protein to catalyze a multistep reaction (called **synapsis**) between a DNA double helix and a homologous region of single-stranded DNA, as shown in Figure 5–60. The crucial step in synapsis occurs when a region of homology is identified by an initial base-pairing between complementary nucleotide sequences (step 2 in Figure 5–60). This interaction begins the pairing process shown previously in Figure 5–58 and so initiates the exchange of strands between two recombining DNA double helices. Studies *in vitro* suggest that the *E. coli* SSB protein cooperates with the recA protein to facilitate these pairing reactions, and genetic recombination is greatly reduced in bacteria that are defective in either protein.

Once synapsis has occurred, the short heteroduplex region where the strands from two different DNA molecules have begun to pair is enlarged through *protein directed branch migration*, which is also catalyzed by the recA protein. **Branch migration** can take place at any point where two single DNA strands with the same sequence are attempting to pair with the same complementary strand; an unpaired region of one of the single strands will displace a paired region of the other single strand, moving the branch point without changing the total number of DNA base pairs. Spontaneous branch migration proceeds equally in both directions, and so it is unlikely to complete recombination efficiently (Figure 5–61A). Because the recA protein catalyzes unidirectional branch migration, it readily produces a region of heteroduplex that is thousands of base pairs long (Figure 5–61B).

The catalysis of branch migration depends on a further property of the recA protein. In addition to having a complex DNA-binding site, the recA protein (like the recBCD protein) is a DNA-dependent ATPase, with an additional site for binding and hydrolyzing ATP. The protein associates much more tightly with DNA

Figure 5–59 During DNA hybridization reactions *in vitro*, DNA double helices re-form from their separated strands. Helix re-formation depends on the random collision of two complementary strands (see p. 188). Most such collisions are not productive, as shown at the left, but a few result in a short region where complementary base pairs have formed (helix nucleation). A rapid zippering then completes each helix. A DNA strand can use this trial-and-error process to find its complementary partner in the midst of millions of nonmatching DNA strands. Trial-and-error recognition of a complementary partner DNA sequence appears to initiate all general recombination events.

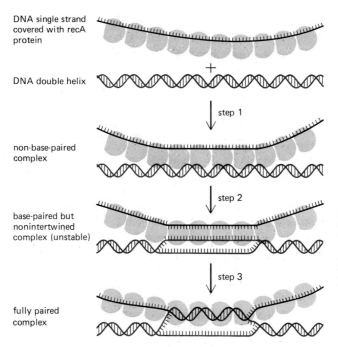

DNA single strand covered with recA protein

DNA double helix

step 1

non-base-paired complex

step 2

base-paired but nonintertwined complex (unstable)

step 3

fully paired complex

Figure 5–60 *In vitro* experiments show that several types of complexes are formed between a DNA single strand covered with recA protein and a DNA double helix. First a non-base-paired complex is formed (step 1), which is converted to a base-paired but nonintertwined complex as soon as a region of homologous sequence is found (step 2). This complex is unstable because it involves an unusual form of DNA in which the two complementary strands are either not helically wound or are wound in alternating stretches of right-handed (normal) and left-handed helix. Step 3 stabilizes the strand exchange. It requires a transient nick (not shown) in one of the two strands forming the helix, followed by the helical winding of one strand about the other.

when it has ATP bound than when it has ADP bound. Moreover, new recA molecules with ATP bound are preferentially added at one end of the recA protein filament, and the ATP is then hydrolyzed to ADP. The recA protein filaments that form on DNA may therefore share many of the dynamic assembly properties displayed by the cytoskeletal filaments formed from actin or tubulin (p. 639); for example, an ability of the protein to "treadmill" unidirectionally along a DNA strand could drive the branch migration reaction shown in Figure 5–61B.

5-35 General Genetic Recombination Usually Involves a Cross-Strand Exchange[39]

Exchanging a single strand between two double helices is presumed to be the slow and difficult step in a general recombination event (see Figure 5–58, above). After this initial exchange, extending the region of pairing and establishing further strand exchanges between the two closely apposed helices is thought to proceed rapidly. During these events, a limited amount of nucleotide excision and local

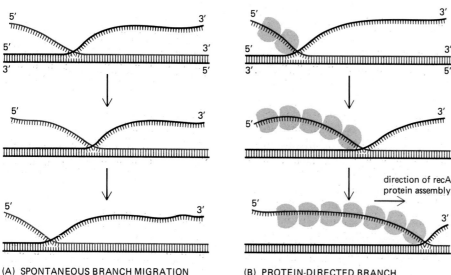

(A) SPONTANEOUS BRANCH MIGRATION

(B) PROTEIN-DIRECTED BRANCH MIGRATION

direction of recA protein assembly

Figure 5–61 Two types of DNA branch migration observed in experiments *in vitro*. Spontaneous branch migration is a back-and-forth, random-walk type of process, and it therefore makes little progress over long distances. In contrast, recA protein-directed branch migration proceeds at a uniform rate in one direction; it may be driven by the polarized assembly of the recA protein filament on a DNA single strand, which occurs in the direction indicated.

DNA resynthesis often occurs, resembling some of the events in DNA repair. Because of the large number of possibilities, different organisms are likely to follow different pathways at this stage. In most cases, however, an important intermediate structure, the **cross-strand exchange,** will be formed by the two participating DNA helices. One of the simplest ways in which this structure can form is shown in Figure 5–62.

In the cross-strand exchange (also called a *Holliday junction*), the two homologous DNA helices that initially paired are held together by mutual exchange of two of the four strands present, one originating from each of the helices. No disruption of base-pairing is necessary to maintain this structure, which has two important properties: (1) The point of exchange between the two homologous DNA double helices (where the two strands cross in Figure 5–62) can migrate rapidly back and forth along the helices by a double branch migration. (2) The cross-strand exchange contains pairs of strands: two crossing and two noncrossing. However, the structure can *isomerize* by undergoing the series of rotational movements shown in Figure 5–63. This isomerization alters the positions of the two pairs of strands so that the two original noncrossing strands become crossing strands, and vice versa.

In order to regenerate two separate DNA helices and thus terminate the pairing process, the two crossing strands must be cut. If the crossing strands are cut *before* isomerization, the two original DNA helices separate from each other nearly unaltered, with only a very short piece of single-stranded DNA exchanged (Figure 5–63 *top*). If the crossing strands are cut *after* isomerization, however, one section of each original DNA helix is linked by a heteroduplex joint to a section of the other DNA helix; in other words, the two DNA helices have crossed over (Figure 5–63 *bottom*).

The isomerization of the cross-strand exchange is thought to be required before two chromosomes can cross over. Figure 5–64 illustrates this process more realistically, as it might occur between two sister chromatids in mitotic cells or between two homologous nonsister chromatids in meiotic cells (see p. 847). The isomerization should occur spontaneously at some rate, but it may also be enzymatically driven or otherwise regulated by cells. Some kind of control probably operates during meiosis, when the two DNA double helices that pair are constrained in an elaborate structure called the *synaptonemal complex* (see p. 848).

Gene Conversion Results from Combining General Recombination and Limited DNA Synthesis[40]

It is a fundamental law of genetics that each parent makes an equal genetic contribution to the offspring, one complete set of genes being inherited from the father and one from the mother. Thus, when a diploid cell undergoes meiosis to produce four haploid cells (see p. 845), exactly half of the genes in these cells should be maternal (genes that the diploid cell inherited from its mother) and the other half paternal (genes that the diploid cell inherited from its father). In a complex animal such as a human, it is not possible to check this prediction directly. But in other organisms, such as fungi, where it is possible to recover and analyze all four of the daughter cells produced from a single cell by meiosis, one finds many cases in which the standard genetic rules have apparently been violated. For example, occasionally meiosis yields three copies of the maternal version (allele) of a gene and only one copy of the paternal allele, revealing that one of the two copies of the paternal allele has been changed to a copy of the maternal allele. This phenomenon is known as **gene conversion.** It often occurs in association with general genetic recombination events, and it is thought to be important in the evolution of certain genes (see p. 600). Gene conversion is believed to be a straightforward consequence of the mechanisms of general recombination and DNA repair.

During meiosis, heteroduplex joints are formed at the sites of crossing-over between homologous maternal and paternal chromosomes. If the maternal and paternal DNA sequences are slightly different, the heteroduplex joint may include some mismatched base pairs. The resulting fault in the double helix may then be corrected by the DNA repair machinery (see p. 223), which can either erase

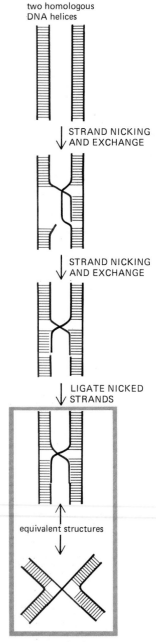

two homologous
DNA helices

STRAND NICKING
AND EXCHANGE

STRAND NICKING
AND EXCHANGE

LIGATE NICKED
STRANDS

equivalent structures

two DNA molecules
joined by a cross-strand
exchange

Figure 5–62 The formation of a cross-strand exchange. There are many possible pathways that can lead from the structure formed in Figure 5–58 (a single-strand exchange) to a cross-strand exchange, but only one is shown. The top-most representation of the cross-strand exchange may most accurately represent its structure; however, the bottom-most representation provides the clearest starting point for explaining the isomerization reaction, as illustrated in Figure 5–63.

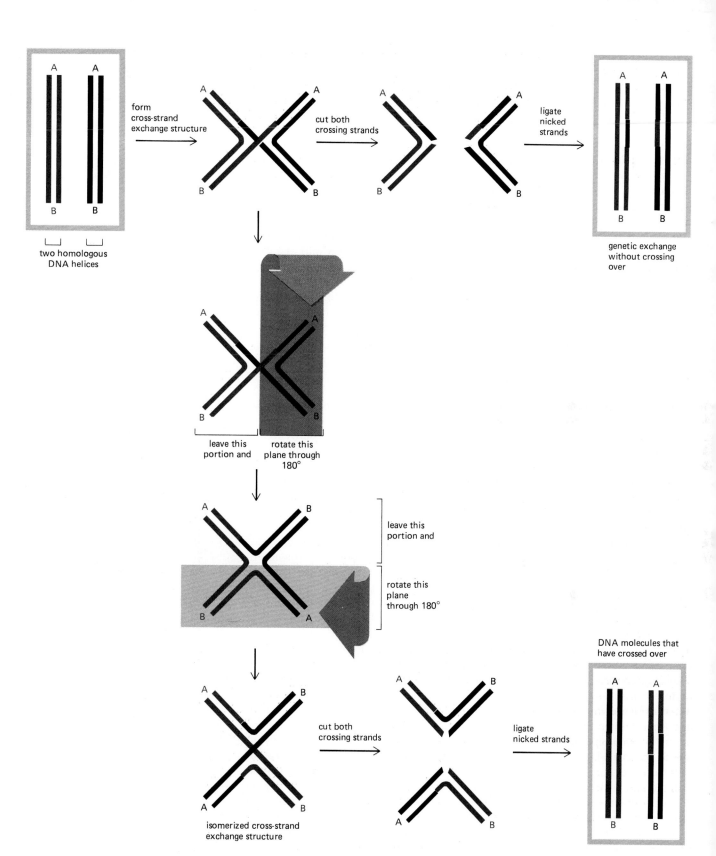

two homologous
DNA helices

form
cross-strand
exchange structure

cut both
crossing strands

ligate
nicked
strands

genetic exchange
without crossing
over

leave this
portion and

rotate this
plane through
180°

leave this
portion and

rotate this
plane
through 180°

isomerized cross-strand
exchange structure

cut both
crossing strands

ligate
nicked strands

DNA molecules that
have crossed over

Figure 5–63 The isomerization of a cross-strand exchange. Without isomerization, the cutting of the two crossing strands terminates the exchange without crossing over (*top*). With isomerization, the cutting of the two crossing strands creates two DNA molecules that have crossed over (*bottom*). Isomerization is, therefore, thought to be required for the breakage and reunion of two homologous DNA double helices that result from general genetic recombination.

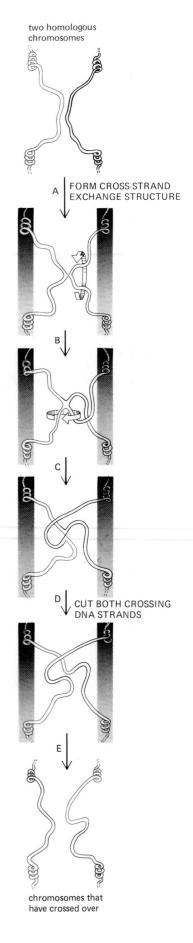

Figure 5–64 A general genetic recombination event between two homologous chromosomes that leads to crossing-over. Isomerization of the cross-strand exchange occurs as described in Figure 5–63.

two homologous chromosomes

A | FORM CROSS-STRAND EXCHANGE STRUCTURE

B |

C |

D | CUT BOTH CROSSING DNA STRANDS

E |

chromosomes that have crossed over

nucleotides on the paternal strand and replace them with nucleotides that match the maternal strand, or vice versa. The consequence of this mismatch repair will be a gene conversion. Gene conversion can also take place by a number of other mechanisms, but they all require some type of general recombination event that brings two copies of a closely related DNA sequence together. Because an extra copy of one of the two DNA sequences is generated, a limited amount of DNA synthesis must also be involved. Genetic studies show that usually only small sections of DNA undergo gene conversion, and in many cases only part of a gene is changed.

Gene conversion can also occur in mitotic cells, but it does so more rarely. As in meiotic cells, some gene conversions in mitotic cells probably result from a mismatch repair process operating on heteroduplex DNA. Another likely mechanism in both types of cells is illustrated in Figure 5–65.

5-36 Site-Specific Recombination Enzymes Move Special DNA Sequences in and out of Genomes[41]

Site-specific genetic recombination, unlike general recombination, is guided by a recombination enzyme that recognizes specific nucleotide sequences present on one or both of the recombining DNA molecules. Base-pairing between the recombining DNA molecules need not be involved; and even when it is, the heteroduplex joint that is formed is only a few base pairs long. This type of recombination enables various types of mobile DNA sequences to move about within and between chromosomes.

Site-specific recombination was first discovered as the means by which a bacterial virus, bacteriophage *lambda*, moves its genome in and out of the *E. coli* chromosome. In its integrated state the virus is hidden in the bacterial chromosome and replicated as part of the host's DNA (see p. 253). When the virus enters a cell, a virus-encoded enzyme called *lambda integrase* is synthesized. This enzyme catalyzes a recombination process that begins when multiple copies of the integrase protein bind tightly to a specific DNA sequence on the circular bacteriophage chromosome. The resulting DNA-protein complex can now bind to a different specific DNA sequence on the bacterial chromosome, bringing the bacterial and bacteriophage chromosomes close together (Figure 5–66). The integrase then catalyzes the required DNA cutting and resealing reactions, using a short region of sequence homology to form a tiny heteroduplex joint at the point of union (Figure 5–67A). The integrase has DNA topoisomerase activity (p. 236), but the various recombination steps occur so quickly that it is very difficult to detect the intermediate DNA forms.

The same type of site-specific recombination mechanism can also be carried out in *reverse* by the lambda bacteriophage, enabling it to exit from its integration site in the *E. coli* chromosome in order to multiply rapidly within the bacterial cell. This excision reaction is catalyzed by a complex of the integrase enzyme with a second bacteriophage protein, which is produced by the virus only when its host cell is stressed (see Figure 9–20, p. 495).

Many other enzymes that catalyze site-specific recombination resemble lambda integrase in requiring a short region of identical DNA sequence on the two regions of DNA helix to be joined. Because of this requirement, each enzyme in this class is relatively fastidious with respect to the DNA sequences that it recombines.

A second class of site-specific recombination enzyme is less fastidious. Like the lambda integrase, each of these enzymes recognizes a specific DNA sequence in the particular mobile genetic element whose recombination it catalyzes. Unlike the lambda enzyme, however, these enzymes do not require a specific "target" DNA sequence and they do not form a heteroduplex joint. Instead they introduce

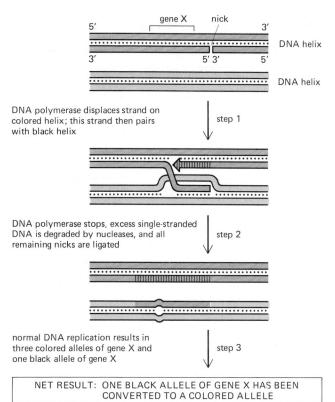

gene X · nick

5′   3′   DNA helix

3′   5′ 3′   5′

DNA helix

DNA polymerase displaces strand on
colored helix; this strand then pairs
with black helix   step 1

DNA polymerase stops, excess single-stranded
DNA is degraded by nucleases, and all
remaining nicks are ligated   step 2

normal DNA replication results in
three colored alleles of gene X and
one black allele of gene X   step 3

NET RESULT: ONE BLACK ALLELE OF GENE X HAS BEEN
CONVERTED TO A COLORED ALLELE

Figure 5–65 A proposed general recombination pathway that causes gene conversion. In step 1, DNA polymerase begins the synthesis of an extra copy of a strand in the colored helix, displacing the original copy as a single strand. This single strand then pairs with the homologous region of the black helix in the manner shown in Figure 5–60. In step 2, the short region of unpaired black strand produced in step 1 is degraded, completing the transfer of nucleotide sequences. The result is normally seen in the next cell cycle, after DNA replication has separated the two nonmatching strands (step 3).

a staggered cut into the target DNA molecule; this cut creates protruding DNA tails in the target, which are then joined covalently to the specific DNA sequence on the mobile genetic element (Figure 5–67B). In this way the entire element is inserted into the site where the target DNA molecule was cut. The two short single-stranded gaps left in the recombinant DNA molecule, one at each end of the mobile element, are filled in by DNA polymerase to complete the recombination process. As illustrated in Figure 5–67B, this creates a short duplication of the adjacent target DNA sequence; such flanking duplications are the hallmark of this type of site-specific recombination event.

An enzyme of the nonfastidious type has been purified in active form from bacteriophage Mu. Like a DNA topoisomerase, it carries out all of its cutting and rejoining reactions without requiring an energy source (such as ATP). Although this enzyme belongs to a bacterial virus, similar enzymes are presumed to be present in organisms as diverse as bacteria, fruit flies, and humans—all these cells contain similar mobile genetic elements, as we shall discuss next.

Summary

Genetic recombination mechanisms allow large sections of DNA double helix to move from one chromosome to another. The reaction pathways that have evolved cause minimal disruption of the two DNA helices as they are broken and rejoined so that two intact chromosomes are restored. There are two broad classes of recombination events. In general recombination the initial reactions rely on extensive base-pairing interactions between strands of the two DNA double helices that will recombine. As a result, general recombination occurs only between two homologous DNA molecules, and although it moves sections of DNA back and forth between chromosomes, it does not normally change the arrangement of the genes in a chromosome. In site-specific recombination, on the other hand, the pairing reactions depend on a protein-mediated recognition of the two DNA sequences that will recombine, and extensive sequence homology is not required. As a result, this class of recombination event alters the relative positions of nucleotide sequences in chromosomes.

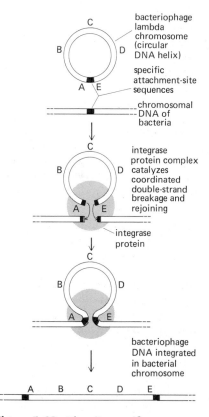

bacteriophage
lambda
chromosome
(circular
DNA helix)

specific
attachment-site
sequences

chromosomal
DNA of
bacteria

integrase
protein complex
catalyzes
coordinated
double-strand
breakage and
rejoining

integrase
protein

bacteriophage
DNA integrated
in bacterial
chromosome

A   B   C   D   E

Figure 5–66 The site-specific recombination event that inserts bacteriophage lambda DNA into the *E. coli* host chromosome. The specific sites recognized by the lambda integrase enzyme (*gray*) are the DNA sequences shown as colored squares (see also Figure 5–74 and Figure 9–19, p. 495).

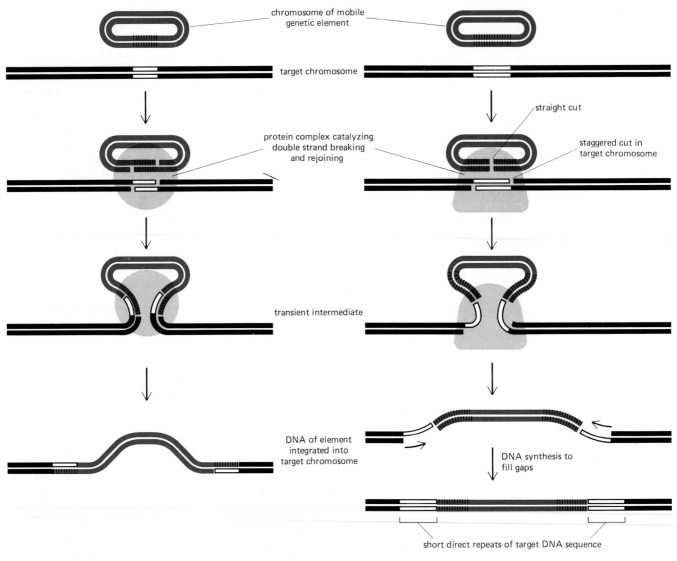

chromosome of mobile genetic element

target chromosome

straight cut

protein complex catalyzing double strand breaking and rejoining

staggered cut in target chromosome

transient intermediate

DNA of element integrated into target chromosome

DNA synthesis to fill gaps

short direct repeats of target DNA sequence

(A) SOME HOMOLOGY REQUIRED

(B) NO HOMOLOGY REQUIRED

Viruses, Plasmids, and Transposable Genetic Elements[42]

In our description of the basic genetic mechanisms, we have so far focused on their selective advantage for the cell. We saw that the short-term survival of the cell depends absolutely on the maintenance of genetic information by DNA repair, while the multiplication of the cell requires rapid and accurate DNA replication. On a longer time scale the appearance of genetic variants, on which evolution of the species depends, is greatly facilitated by the reassortment of genes and the occasional rearrangement of DNA sequences caused by genetic recombination. We shall now examine a group of genetic elements that seem to act as parasites, subverting the genetic mechanisms of the cell for their own advantage.

Under certain circumstances, DNA sequences can replicate independently of the rest of the genome. Such sequences have widely different degrees of independence from their host cells. Of these, virus chromosomes are the most independent because they have a protein coat that allows them to move freely from cell to cell. To varying degrees the viruses are closely related to more host cell-dependent DNA sequences known as plasmids and transposable elements, which lack a coat and are therefore confined to replicate with a single cell and its progeny. More primitive still are some DNA sequences that are suspected of being mobile because they are repeated many times in a cell's chromosome—but that

Figure 5–67 Comparison of the mechanisms used by two different classes of site-specific recombination enzymes. In each case a specific enzyme (*gray*) binds to the cross-hatched DNA sequence on the mobile genetic element and holds this sequence close to a site on the target chromosome. In (A) the enzyme makes a staggered cut on either side of a very short homologous DNA sequence on both chromosomes (12 nucleotides in the case of lambda integrase) and then switches the partner strands so as to form a short heteroduplex joint. In (B) the enzyme makes a staggered cut in the target chromosome and joins the protruding ends directly to the evenly cut ends of the mobile element DNA. This type of enzyme leaves short repeats of target DNA sequence (3 to 12 nucleotides in length, depending on the enzyme) on either side of the integrated DNA segment.

move or multiply so rarely that it is not clear if they should be considered as separate genetic elements at all.

All these quasi-independent genetic elements must heavily exploit the metabolism of the host cell in order to multiply, and they have thereby served as important tools for investigating the normal cell machinery. We shall begin our discussion with the viruses, which are the best understood of the mobile genetic elements. Then we shall describe the properties of plasmids and transposable elements, some of which bear a remarkable resemblance to viruses and may in fact have been their ancestors.

5-39 Viruses Are Mobile Genetic Elements[43]

Viruses were first described as disease-causing agents that can multiply only in cells and that by virtue of their tiny size pass through ultrafine filters that hold back even the smallest bacteria. Before the advent of the electron microscope, their nature was obscure, although it was suspected that they might be naked genes that had somehow acquired the ability to move from one cell to another. The use of ultracentrifuges in the 1930s made it possible to separate viruses from host cell components, and by the early 1940s the generalization emerged that all viruses contain nucleic acids. The idea that viruses and genes carry out similar functions was confirmed by studies on the bacterial viruses (*bacteriophages*). In 1952 it was shown for the bacteriophage T4 that only the phage DNA, and not the phage protein, enters the bacterial host cell and initiates the replication events that lead to the production of several hundred progeny viruses in every infected cell.

These observations led to the notion of viruses as genetic elements enclosed by a protective coat that enables them to move from one cell to another. Virus multiplication per se is often lethal to the cells in which it occurs; in many cases the infected cell breaks open (*lyses*) and thereby allows the progeny viruses access to nearby cells. Many of the clinical manifestations of viral infection reflect this cytolytic effect of the virus. For example, both the cold sores formed by herpes simplex virus and the lesions caused by smallpox reflect the killing of the epithelial cells in a local area of the skin.

The type of nucleic acid in a virus, the structure of its coat, its mode of entry into the host cell, and its mechanism of replication once inside all vary considerably from one type of virus to another. Electron micrographs illustrating some structural differences among viruses are presented in Figure 5–68.

The Outer Coat of a Virus May Be a Protein Capsid or a Membrane Envelope[44]

Initially it was thought that the outer coat of a virus might be constructed from a single type of protein molecule. Viral infections were believed to start with the dissociation of the viral chromosome (its nucleic acid) from its protein coat inside the host cell, followed by replication of the chromosome to form many identical

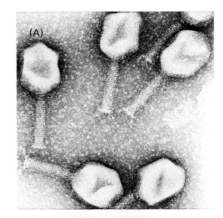

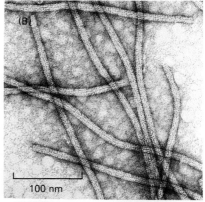

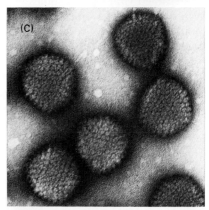

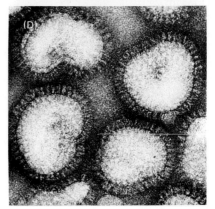

Figure 5–68 Electron micrographs (all at the same scale) of negatively stained virus particles. (A) *Bacteriophage T4*, a large DNA-containing virus that infects *E. coli*. The DNA is stored in the bacteriophage head and injected into the bacterium through the cylindrical tail. (Courtesy of James Paulson.) (B) *Potato virus X*, a filamentous plant virus that contains an RNA genome. (Courtesy of Graham Hills.) (C) *Adenovirus*, a DNA-containing virus that can infect human cells. The protein capsid forms the outer surface of this virus. (Courtesy of Mei Lie Wong.) (D) *Influenza virus*, a large RNA-containing animal virus whose protein capsid is further enclosed in a lipid bilayer-based envelope containing protruding spikes of viral glycoprotein. (Courtesy of R.C. Williams and H.W. Fisher.)

copies. After the synthesis of new copies of the virus-specific coat protein from virally encoded messenger RNA molecules, formation of the progeny virus particles would occur by the spontaneous assembly of these coat protein molecules around the progeny viral chromosomes (Figure 5–69).

It is now known that these ideas vastly oversimplify the diversity of virus life cycles. For example, the protein shell that surrounds the nucleic acid of most viruses (the **capsid**) contains more than one type of polypeptide chain, often arranged in several layers. In many viruses, moreover, the protein capsid is further enclosed by a lipid bilayer membrane that contains proteins. Many of these *enveloped viruses* acquire their envelope in the process of budding from the plasma membrane (Figure 5–70). This budding process allows the virus particles to leave the cell without disrupting the plasma membrane and, therefore, without killing the cell. While their lipid components are identical to those found in the plasma membrane of the host cell, the proteins in the lipid bilayer are virus-specific. The assembly of viral envelopes at the plasma membrane is discussed in Chapter 8 (p. 469), and the assembly of a viral capsid is illustrated in Figure 3–43 (p. 120).

5-45 Viral Genomes Come in a Variety of Forms and Can Be Either RNA or DNA[45]

When the DNA double helix was discovered, it seemed logical that genetic information should be stored only in this form since, as we have seen, it has obvious advantages for DNA stability and repair. If one polynucleotide chain is accidentally damaged, its complementary chain permits the damage to be readily corrected. This concern with repair, however, need not bother small viral chromosomes that contain only several thousand nucleotides—for the chance of accidental damage is very small compared with the risk to a cell genome containing millions of nucleotides.

The genetic information of a virus can, therefore, be carried in a variety of unusual forms, including RNA instead of DNA. A viral chromosome may be a single-stranded RNA chain (tobacco mosaic virus), a double-stranded RNA helix (reovirus), a circular single-stranded DNA chain (M13 and φX174 bacteriophages), or a linear single-stranded DNA chain (parvoviruses). Moreover, although the first well-studied viral chromosomes were simple linear DNA double helices, circular DNA double helices and more complex linear DNA double helices are also common. For example, several viruses have protein molecules covalently attached to the 5′ ends of their DNA strands, and the DNA double helices from the very large poxviruses have their opposite strands at each end covalently joined through phosphodiester linkages (Figure 5–71). Each type of genome requires unique enzymatic tricks for its replication and thus must encode not only the viral coat protein but also one or more of the enzymes needed to replicate the viral nucleic acid.

5-40 A Viral Chromosome Codes for Enzymes Involved in the Replication of Its Nucleic Acid[46]

The amount of information that a virus brings into a cell to ensure its own selective replication varies greatly. For example, the DNA of the relatively large bacteriophage T4 codes for at least 30 different enzymes that ensure the rapid replication of the T4 chromosome in preference to the DNA of its *E. coli* host cell (Figure 5–72). Some of these proteins mediate continuous rounds of T4 DNA replication, with the unusual feature that 5-hydroxymethylcytosine is incorporated in place of cytosine in the DNA. The unusual base composition of the T4 DNA makes it distinguishable from host DNA and selectively protects it from nucleases also encoded in the T4 genome that thus degrade only the *E. coli* DNA. Still other proteins alter host cell RNA polymerase molecules so that they transcribe different sets of bacteriophage genes at different stages of infection, appropriately to the needs of the phage.

Smaller DNA viruses, such as the monkey virus SV40 and the tiny bacteriophage φX174, carry much less genetic information. They rely more on host cell

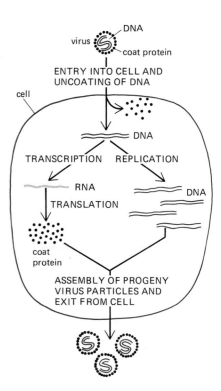

Figure 5–69 The simplest of all viral life cycles. The hypothetical virus shown consists of a small double-stranded DNA molecule that codes for only a single viral capsid protein. No known virus is this simple.

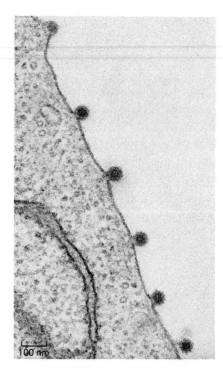

Figure 5–70 Electron micrograph of a thin section of an animal cell from which several copies of an enveloped virus (Semliki forest virus) are budding. This virus contains a single-stranded RNA genome. (Courtesy of M. Olsen and G. Griffiths.)

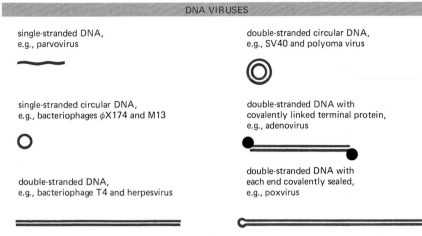

single-stranded RNA, e.g., tobacco mosaic virus, bacteriophage R17, and poliovirus

double-stranded RNA, e.g., reovirus

single-stranded DNA, e.g., parvovirus

single-stranded circular DNA, e.g., bacteriophages φX174 and M13

double-stranded DNA, e.g., bacteriophage T4 and herpesvirus

double-stranded circular DNA, e.g., SV40 and polyoma virus

double-stranded DNA with covalently linked terminal protein, e.g., adenovirus

double-stranded DNA with each end covalently sealed, e.g., poxvirus

enzymes to carry out their protein and DNA synthesis, parasitizing host cell DNA replication proteins, including the DNA polymerase enzyme. Even the smallest DNA viruses, however, code for proteins that selectively initiate the synthesis of their own DNA, recognizing a particular nucleotide sequence in the virus that serves as a *replication origin*. This is essential because a virus must override the cellular control signals that would otherwise cause the viral DNA to replicate in pace with the host cell DNA, doubling only once in each cell cycle. We do not yet understand how eucaryotic cells regulate their own DNA synthesis (see p. 732), and the mechanisms used by viruses to escape from this regulation—which are much more accessible to study—should provide insights into the host regulatory mechanisms.

RNA viruses have very specialized requirements for replication since, to reproduce their genomes, they must copy RNA molecules, which means polymerizing nucleoside triphosphates on an RNA template. Cells normally do not have enzymes to carry out this reaction, so even the smallest RNA viruses must encode their own RNA-dependent nucleic acid polymerase enzymes in order to replicate.

We shall now look in more detail at the replication mechanisms of the various types of viruses.

5-42   Both RNA Viruses and DNA Viruses Replicate Through the Formation of Complementary Strands[47]

The replication of the genomes of RNA viruses, like DNA replication, occurs through the formation of complementary strands. For most RNA viruses this process is catalyzed by specific RNA-dependent RNA polymerase enzymes (*replicases*). These enzymes are encoded by the viral RNA chromosome and are often incorporated into the progeny virus particles, so that upon entry of the virus into a cell, they immediately begin replicating the viral RNA. Replicases are always packaged into the capsid of the so-called *negative-strand RNA viruses*, such as influenza or vesicular stomatitis virus. Negative-strand viruses are so called because the infecting strand does not code for protein; only its complementary strand carries the coding sequences. Thus the infecting strand remains impotent without a preformed replicase. In contrast, the viral RNA of *positive-strand RNA viruses*, such as poliovirus, can serve as mRNA and therefore the naked genome itself is infectious.

Figure 5–71 Schematic drawings (not to scale) of several types of viral genomes. The smallest viruses contain only a few genes and can have an RNA or a DNA genome; the largest viruses contain hundreds of genes and have a double-stranded DNA genome. The peculiar ends on some of these DNA molecules (as well as the circular forms) overcome the difficulty of replicating the last few nucleotides at the end of a DNA chain (see p. 519).

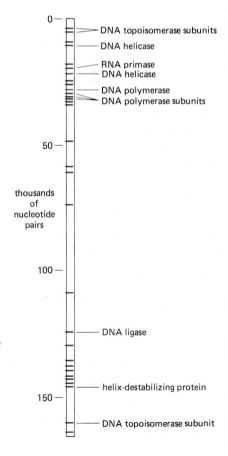

Figure 5–72 The T4 bacteriophage chromosome, showing the positions of the more than 30 genes involved in T4 DNA replication. The genome of bacteriophage T4 consists of more than 160,000 nucleotide pairs and encodes more than 200 different proteins, including those involved in DNA replication (some of which are labeled). The remaining proteins include many that are involved in the bacteriophage head and tail assemblies (see Figure 5–68A).

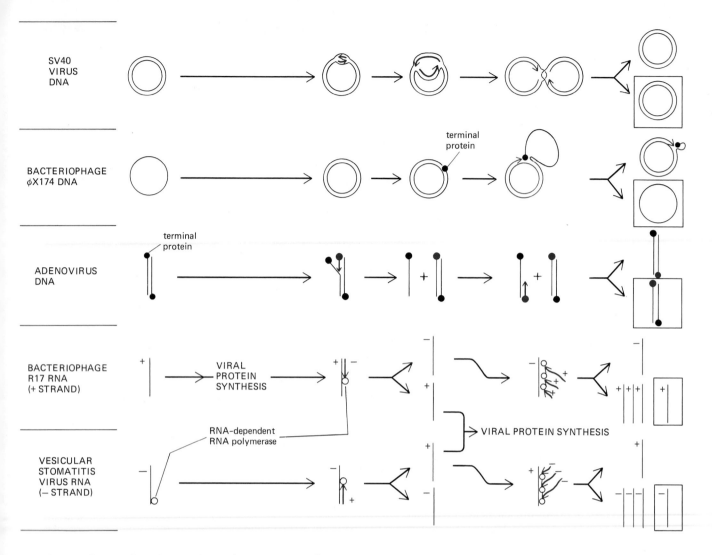

The synthesis of viral RNA always begins at the 3′ end of the RNA template (starting at the 5′ end of the new viral RNA molecule) and progresses until the 5′ end of the template is reached. There are no error-correcting mechanisms for viral RNA synthesis, and error rates are similar to those in DNA transcription (about one error in 10^4 nucleotides synthesized). This is not a serious deficiency as long as the RNA chromosome is relatively short; for this reason the genomes of all RNA viruses are small relative to those of the large DNA viruses.

All DNA viruses begin their replication at a replication origin, which binds special initiator proteins that attract the replication enzymes of the host cell (see p. 235). However, there are many different replication pathways. The complexity of these diverse replication schemes reflects, in part, the problem of replicating the ends of a simple linear DNA molecule, given a DNA polymerase enzyme that cannot begin synthesis without a primer (p. 229). DNA viruses have solved this problem in a variety of ways: some have circular DNA genomes and thus no ends; others have linear DNA genomes that repeat their terminal sequences or end in loops; while still others have special terminal proteins that serve to prime the DNA polymerase directly.

Some of the ways in which viruses are known to replicate their genomes are outlined in Figure 5–73.

Figure 5–73 Some of the diverse strategies used by different viruses to replicate their genomes. Where indicated, *terminal proteins* are covalently attached to the ends of DNA chains; these proteins play an important role in the replication process. A major difference between the life cycles of *positive-strand* and *negative-strand* RNA viruses is that the latter must synthesize a positive RNA strand before making viral proteins. A negative-strand virus, therefore, must carry within its capsid one or more molecules of the viral RNA-dependent RNA polymerase (replicase). The final RNA or DNA product, which is identical to the infecting viral genome at the left, is boxed.

Viral Chromosomes Can Integrate into Host Chromosomes[48]

The end result of the entry of a viral chromosome into a cell is not always its immediate multiplication to produce large numbers of progeny. Many viruses enter a *latent* state, in which their genomes are present but inactive in the cell and no progeny are produced. Viral latency was discovered when it was found that exposure to ultraviolet light induced many apparently uninfected bacteria to produce progeny bacteriophages. Subsequent experiments showed that these *lysogenic bacteria* carry in their chromosomes a dormant but complete viral chromosome. Such integrated viral chromosomes are called **proviruses.**

Bacteriophages that can integrate their DNA into bacterial chromosomes are known as *lysogenic bacteriophages*. The best-known example is the bacteriophage lambda, whose integrase protein we have already discussed. When lambda infects a suitable *E. coli* host cell, it normally multiplies to produce several hundred progeny particles that are released when the bacterial cell lyses; this is called a *lytic infection*. More rarely, the free ends of the linear infecting DNA molecules join to form a DNA circle that becomes integrated into the circular host *E. coli* chromosome by a site-specific recombination event (see p. 246). The resulting lysogenic bacterium, carrying the proviral lambda chromosome, multiples normally until it is subjected to an environmental insult, such as exposure to ultraviolet light or ionizing radiation. The resulting cell debilitation induces the integrated provirus to leave the host chromosome and begin a normal cycle of viral replication. In this way the integrated provirus need not perish with its damaged host cell but has a chance to escape to a nearby normal *E. coli* cell (Figure 5–74).

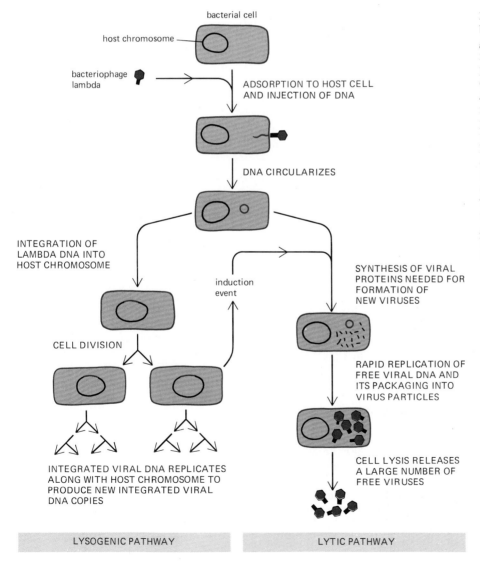

bacterial cell

host chromosome

bacteriophage lambda

ADSORPTION TO HOST CELL AND INJECTION OF DNA

DNA CIRCULARIZES

INTEGRATION OF LAMBDA DNA INTO HOST CHROMOSOME

induction event

SYNTHESIS OF VIRAL PROTEINS NEEDED FOR FORMATION OF NEW VIRUSES

CELL DIVISION

RAPID REPLICATION OF FREE VIRAL DNA AND ITS PACKAGING INTO VIRUS PARTICLES

INTEGRATED VIRAL DNA REPLICATES ALONG WITH HOST CHROMOSOME TO PRODUCE NEW INTEGRATED VIRAL DNA COPIES

CELL LYSIS RELEASES A LARGE NUMBER OF FREE VIRUSES

LYSOGENIC PATHWAY

LYTIC PATHWAY

Figure 5–74 The life cycle of bacteriophage lambda. The lambda genome contains about 50,000 nucleotide pairs and encodes about 50 proteins. Its double-stranded DNA can exist in both linear and circular forms. As shown, the bacteriophage can multiply by either a lytic or a lysogenic pathway in the *E. coli* bacterium. When the bacteriophage is growing in the lysogenic state, damage to the cell causes the integrated viral DNA (provirus) to exit from the host chromosome and shift to lytic growth. The entrance and exit of the DNA from the chromosome are site-specific genetic recombination events catalyzed by the lambda *integrase* protein (see p. 495).

The Continuous Synthesis of Viral Proteins Can Make Cells Cancerous[49]

Animal cells, like bacteria, can offer viruses an alternative to lytic growth. *Permissive* cells permit DNA viruses to multiply lytically and kill the cell. *Nonpermissive* cells may allow the DNA virus to enter but not to replicate lytically; in a small percentage of such cells, the viral chromosome either becomes integrated into the host cell genome, where it is replicated along with the host chromosomes, or forms a plasmid—a circular DNA molecule—that replicates in a controlled fashion without killing the cell. This sometimes results in a genetic change in the nonpermissive cells, causing them to proliferate in an ill-controlled way and thus transforming them into their cancerous equivalents. In this case the DNA virus is called a *DNA tumor virus* and the process is called virus-mediated *neoplastic transformation*. The most extensively studied DNA tumor viruses are two papovaviruses, SV40 and polyoma. Their transforming ability has been traced to several viral proteins that cooperate to drive quiescent cells from G_0 to S phase (see p. 517). In permissive cells the shift to S phase provides the virus with all of the host cell replication enzymes required for viral DNA synthesis. The synthesis of these viral proteins by a provirus in a nonpermissive cell overrides some of the normal growth control mechanisms in the cell and in all of its progeny.

RNA Tumor Viruses Are Retroviruses[50]

For one group of RNA viruses, the so-called *RNA tumor viruses*, the infection of a permissive cell often leads simultaneously to a nonlethal release of progeny virus from the cell surface by budding and a permanent genetic change in the infected cell that makes it cancerous. How RNA virus infection could lead to a permanent genetic alteration was unclear until the discovery of the enzyme *reverse transcriptase*, which transcribes the infecting RNA chains of these viruses into complementary DNA molecules that integrate into the host cell genome. RNA tumor viruses—which include the first well-known tumor virus, the Rous sarcoma virus—are members of a large class of viruses known as **retroviruses.** These viruses are so named because they reverse the normal process in which DNA is transcribed into RNA as part of their normal life cycle. The virus that causes acquired immune deficiency syndrome (AIDS) is also a retrovirus.

The life cycle of a retrovirus is outlined in Figure 5–75. The enzyme **reverse transcriptase** is an unusual DNA polymerase that uses either RNA or DNA as a template; it is encoded by the retrovirus RNA and is packaged inside the viral capsid. When the single-stranded RNA of the retrovirus enters a cell, the reverse transcriptase first makes a DNA copy of the RNA strand to form a DNA-RNA hybrid helix, which is then used by the same enzyme to make a double helix with two DNA strands. This DNA copy of the RNA genome then circularizes and integrates into a host cell chromosome. The integration is aided by a virus-encoded site-specific recombination enzyme that recognizes a particular viral DNA sequence and helps to catalyze the insertion of the viral DNA into virtually any site on a host cell chromosome (see Figure 5–67B). The next step in the infectious process is transcription of the integrated viral DNA by host cell RNA polymerase, producing large numbers of viral RNA molecules identical to the original infecting genome. Finally, these RNA molecules are translated to produce the capsid, envelope, and reverse transcriptase proteins that are assembled with the RNA into new enveloped virus particles, which bud from the plasma membrane (see Figure 5–75).

Both RNA and DNA tumor viruses transform cells because the permanent presence of the viral DNA in the cell causes the synthesis of new proteins that alter the control of host cell proliferation. The genes that code for such proteins are called *oncogenes*. Unlike DNA tumor viruses, whose oncogenes typically encode normal viral proteins essential for viral multiplication, the oncogenes carried by RNA tumor viruses are modified versions of normal host cell genes that are not required for viral replication. Since only a limited amount of RNA can be packed into the capsid of a retrovirus, the acquired oncogene sequences often replace an essential part of the retroviral genome, making the virus defective. We shall discuss

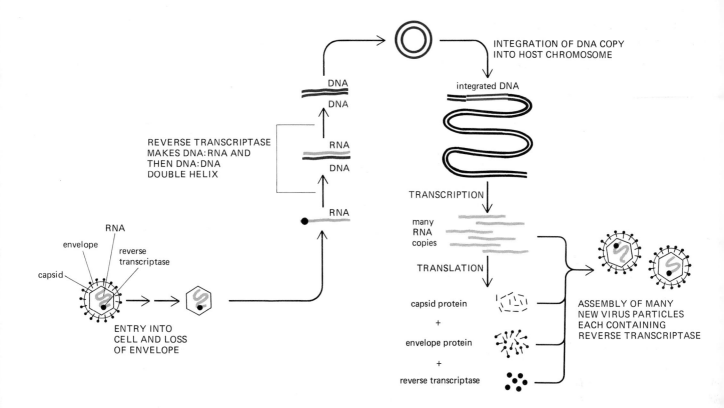

elsewhere how viral oncogenes have provided important clues to the causes and nature of cancer, as well as to the normal mechanisms that control cell growth and division in multicellular animals (see p. 754 and p. 1205).

Some Transposable Elements Are Close Relatives of Retroviruses[51]

Because many viruses can move in and out of their host chromosomes, any large genome is likely to contain a number of different proviruses. These genomes are also likely to house a variety of mobile DNA sequences that do not form viral particles and cannot leave the cell. Such **transposable elements** range in length from a few hundred to tens of thousands of base pairs, and they are usually present in multiple copies per cell. One can consider these elements as tiny parasites hidden in chromosomes. Each transposable element is occasionally activated to move to another DNA site in the same cell by a process called *transposition*, catalyzed by its own site-specific recombination enzyme. These enzymes, referred to as *transposases*, are often encoded in the DNA of the element itself. Since most transposable elements move only very rarely (once in 10^5 cell generations for many elements in bacteria), it is often difficult to discriminate them from nonmobile parts of the chromosome. It is not known what suddenly triggers their movement.

Transposition can occur by a variety of mechanisms. One large family of transposable elements uses a mechanism that is indistinguishable from part of a retrovirus life cycle. These elements, called *retrotransposons*, are present in organisms as diverse as yeast, *Drosophila*, and mammals. One of the best-understood retrotransposons is the so-called Ty1 element of yeast. The first step in its transposition is the transcription of the entire transposable element, producing an RNA copy of the element that is more than 5000 nucleotides long. This transcript encodes a reverse transcriptase enzyme that makes a circular double-stranded DNA copy of the RNA molecule via a RNA-DNA hybrid intermediate, precisely mimicking the early stages of infection by a retrovirus (see Figure 5–75). The analogy continues as the DNA circle integrates into a randomly selected site on the chromosome. As for a retrovirus, integration proceeds by the type of site-specific mechanism shown in Figure 5–67B, presumably utilizing a transposase that is also encoded by the

Figure 5–75 The life cycle of a retrovirus. The retrovirus genome consists of an RNA molecule of about 8500 nucleotides; two such molecules are packaged into each viral particle. The enzyme *reverse transcriptase* is a DNA polymerase that first makes a DNA copy of the viral RNA molecule and then a second DNA strand, generating a double-stranded DNA copy of the RNA genome. The integration of this DNA double helix into the host chromosome, catalyzed by a viral protein, is required for the synthesis of new viral RNA molecules by the host cell RNA polymerase.

long RNA transcript. Although the resemblance to a retrovirus is striking, without a functional protein coat the Ty1 element is constrained to move within a single cell and its progeny.

5-44 Other Transposable Elements Transfer Themselves Directly from One Site in the Genome to Another[52]

Unlike retrotransposons, many transposable elements seem never to exist free of the host chromosome; the transposases that catalyze their movement act on the DNA of the element while it is still integrated in the host genome. The transposase is thought to bind to a short sequence that is repeated in reverse orientation at each end of the element, thereby holding these two ends close together while catalyzing the subsequent recombination event (Figure 5–76). For some transposable elements the transposition mechanism involves only the breaking and rejoining of DNA, with the two ends of the element being inserted into a staggered nick made elsewhere on a chromosome (see Figure 5–67B). Such a transposable element moves directly from one chromosomal site to another without DNA replication. However, the DNA sequence is often altered when the break in the vacated chromosome reseals, causing a mutation at the old chromosomal site.

Other transposable elements replicate when they move. In the best-studied example, site-specific recombination triggers a localized synthesis of DNA; one copy of the replicated transposable element is inserted at a randomly selected new chromosomal site, while the other copy remains at the old one (Figure 5–77). The mechanism is closely related to the nonreplicative mechanism just described; indeed, some transposable elements can move by either pathway.

In addition to moving themselves, all types of transposable elements occasionally move or rearrange neighboring DNA sequences of the host genome. For example, they frequently cause deletions of adjacent nucleotide sequences or carry them to another site. The presence of transposable elements makes the arrangement of the DNA sequences in chromosomes much less stable than previously thought, and it is likely that they have been responsible for many important evolutionary changes in genomes (see pp. 605–609).

Are the transposable elements also of evolutionary importance as the most ancient ancestors of viruses? Although the precursors of retroviruses were almost certainly retrotransposons, all present-day transposable elements rely heavily on

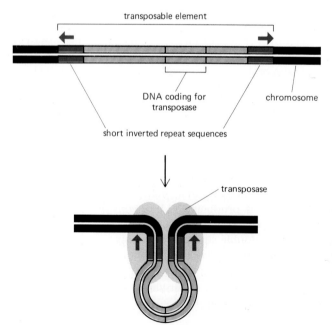

Figure 5–76   The structure of a transposable element that moves directly from one chromosomal site to another. Transposable elements of this type can be recognized by the "inverted repeat DNA sequences" at their ends. Experiments show that the repetition of these sequences, which can be as short as 20 nucleotides, is all that is necessary for the DNA between them to be transposed by the particular transposase enzyme associated with the element. The protein complex that forms in the first stage of movement is shown. Subsequent events involve cleavage of the DNA at the ends of the inverted repeat sequences and occur by modifications of the recombination scheme shown previously in Figure 5–67B (see also Figure 5–77).

DNA metabolism. But very early cells are thought to have had RNA rather than DNA genomes (see p. 10), so we must look to RNA metabolism for the ultimate origin of viruses.

Most Viruses Probably Evolved from Plasmids[53]

Even the largest viruses depend heavily on their host cells for biosynthesis; for example, no known virus makes its own ribosomes or generates the ATP it requires. Clearly, therefore, cells must have evolved before viruses. The precursors of the first viruses were probably small nucleic acid fragments that developed the ability to multiply independent of the chromosomes of their host cells. Such independently replicating elements, called **plasmids,** can replicate indefinitely outside the host chromosome. Plasmids occur in both DNA and RNA forms (see p. 259), and, like viruses, they contain a special nucleotide sequence that serves as an origin of replication. Unlike viruses, however, they cannot make a protein coat and therefore cannot move from cell to cell readily. Many also cannot integrate into chromosomes.

The first RNA plasmids may have resembled the *viroids* found in some plant cells. These small RNA circles, only 300 to 400 nucleotides long, are replicated despite the fact that they do not code for any protein (see Fig. 10–61, p. 598). Having no protein coat, viroids exist as naked RNA molecules and pass from plant to plant only when the surfaces of both donor and recipient cells are damaged so that there is no membrane barrier for the viroid to pass. Under the pressure of natural selection, such independently replicating elements could be expected to acquire nucleotide sequences from the host cell that would facilitate their own multiplication, including sequences that code for proteins. Some present-day plasmids are indeed quite complex, encoding proteins and RNA molecules that regulate their replication, as well as proteins that control their partitioning into daughter cells. The largest known plasmids are double-stranded DNA circles more than 100,000 base pairs long.

The first virus probably appeared when an RNA plasmid acquired a gene coding for a capsid protein. But a capsid can enclose only a limited amount of nucleic acid; therefore a virus is limited in the number of genes it can contain. Forced to make optimal use of their limited genomes, some small viruses (like φX174) evolved *overlapping genes,* in which part of the nucleotide sequence encoding one protein is used (in the same or a different reading frame) to encode a second protein. Other viruses evolved larger capsids and consequently could accommodate more genes.

With their unique ability to transfer nucleic acid sequences across species barriers, viruses have almost certainly played an important part in the evolution of the organisms they infect. Many recombine frequently with their host cell genome and with one another and in this way can pick up small pieces of host chromosome at random and carry them to different cells or organisms. Moreover, integrated copies of viral DNA (proviruses) have become a normal part of the genome of most organisms. Examples of such proviruses include the lambda family of bacteriophages and the so-called endogenous retroviruses found in numerous copies in vertebrate genomes. The integrated viral DNA can become altered so that it cannot produce a complete virus but can still encode proteins, some of which may be useful to the host cell. Therefore, viruses, like sexual reproduction, can speed up evolution by promoting the mixing of gene pools.

The process in which DNA sequences are transferred between different host cell genomes by means of a virus is called *DNA transduction,* and several viruses that transduce DNA with particularly high frequencies are commonly used by researchers to move genes from one cell to another. Viruses and their close relatives, plasmids and transposable elements, have also been important to cell biology in many other ways. Because of their relative simplicity, for example, studies of their reproduction have progressed unusually rapidly and have illuminated many of the basic genetic mechanisms in cells. In addition, both viruses and plasmids have been crucial elements in the development of the recombinant DNA technologies that we shall consider next.

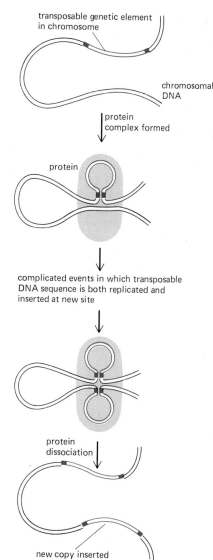

Figure 5–77 Schematic diagram illustrating the movement of one type of transposable element within a chromosome. The element shown replicates during transposition, its movement occurring without it being excised from its original site. The two inverted repeat DNA sequences that commonly flank the two ends of transposable elements are shown as colored squares. At the start of transposition, the transposase cuts one of the two DNA strands at each end of the element, and the element then serves as a template for DNA synthesis that begins by the the addition of nucleotides to the 3' ends of chromosomal DNA sequences. Further details are known, but the process is too complex to be illustrated here.

uses are infectious particles that consist of a DNA or RNA molecule (the viral genome) packaged in a protein capsid, which in the enveloped viruses is surrounded by a lipid bilayer-based membrane. Both the structure of the viral genome and its mode of replication vary widely among viruses. A virus can multiply only inside a host cell, whose genetic mechanisms it subverts for its own reproduction. A common outcome of a viral infection is the lysis of the infected cell and release of infectious viral particles. In some cases, however, the viral chromosome instead integrates into a host cell chromosome, where it is replicated as a provirus along with the host genome. Many viruses are thought to have evolved from plasmids, which are self-replicating DNA or RNA molecules that lack the ability to wrap themselves in a protein coat.

Transposable elements are DNA sequences that differ from viruses in being able to multiply only in their host cell and its progeny; like plasmids, they cannot leave the cell. Unlike plasmids, they normally replicate only as an integral part of a chromosome. However, some transposable elements are closely related to retroviruses and can move from place to place in the genome by the reverse transcription of an RNA intermediate. Other transposable elements can move without ever detaching from the chromosomes. Although both viruses and transposable elements can be viewed as parasites, many of the DNA sequence rearrangements they cause are important for the evolution of cells and organisms.

DNA Cloning and Genetic Engineering[54]

The discoveries we have discussed in this chapter resulted from curiosity about the cell and the fundamental mechanisms of heredity. In recent years, however, this basic knowledge has been put to practical use. The techniques of *DNA cloning* and *genetic engineering* enable specific genes to be isolated in quantity, redesigned, and then inserted back into cells and organisms. They form part of the panel of methods previously introduced in Chapter 4, known collectively as *recombinant DNA technology.* Recombinant DNA technology has revolutionized the study of living cells. It has also provided medicine and industry with an efficient means of producing specific proteins in large amounts that previously were available only in extremely small quantities, if at all.

Restriction Nucleases Facilitate the Cloning of Genes[55]

[-48]
[-49]
[-50] In the 1960s the goal of isolating a single gene from a large chromosome seemed impossibly remote. Unlike a protein, a gene does not exist as a discrete entity in cells, but rather as a small region of a much larger DNA molecule. Although the DNA molecules in a cell can be randomly broken into small pieces by mechanical force, a fragment containing an individual gene in a mammalian genome would still be only one part in a million of the total DNA fragments and indistinguishable in size from other genes. How could such a gene be purified? Since all DNA molecules consist of an approximately equal mixture of the same four nucleotides, they cannot be readily separated, as proteins can, on the basis of their different charges and binding properties (see p. 167). Moreover, even if a purification scheme could be devised, vast amounts of DNA would be needed to yield enough of any particular gene to be useful for further experiments.

The solution to all of these problems began to appear with the discovery of **restriction nucleases.** These enzymes, which can be purified from bacteria, cut the DNA double helix at specific sequences of four to eight nucleotides, producing DNA fragments of strictly defined sizes that are known as **restriction fragments** (see p. 182). Different species of bacteria make restriction nucleases with different sequence specificities, and it is relatively simple to find a restriction nuclease that will create a small DNA fragment that includes a particular gene. The size of the restriction fragment can then be used as a basis for partially purifying the gene from a mixture.

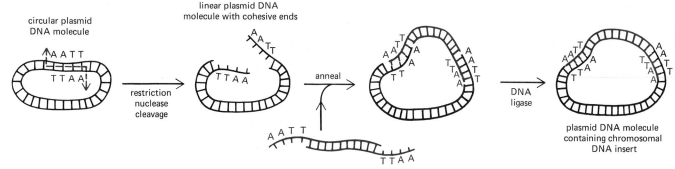

circular plasmid DNA molecule

linear plasmid DNA molecule with cohesive ends

restriction nuclease cleavage

anneal

DNA ligase

plasmid DNA molecule containing chromosomal DNA insert

one of many DNA fragments produced by cutting chromosomal DNA with the same restriction nuclease

Figure 5–78 The cohesive ends produced by many kinds of restriction nucleases (see Figure 4–63) allow two DNA fragments to be joined by complementary base-pair interactions. DNA fragments joined in this way can be covalently linked in a highly efficient reaction catalyzed by the enzyme DNA ligase. In this example, a recombinant plasmid DNA molecule containing a chromosomal DNA insert is formed.

Another property of restriction nucleases that is helpful for gene cloning is that many of them produce staggered cuts that leave short, single-stranded tails on both ends of the DNA fragment. These are known as *cohesive ends* since they can form complementary base pairs with any other end produced by the same enzyme (see p. 182). The cohesive ends generated by a restriction nuclease make it possible to join two double-helical DNA fragments from different genomes by complementary base-pairing (Figure 5–78). For example, a DNA fragment containing a human gene can be joined in a test tube to the chromosome of a bacterial virus and the new *recombinant DNA molecule* then introduced into a bacterial cell. Starting with only one such recombinant DNA molecule that infects a single cell, the normal replication mechanism of the virus can produce more than 10^{12} identical virus DNA molecules in less than a day, thereby amplifying the amount of the attached human DNA fragment by the same factor. A virus used in this way is known as a *cloning vector*.

5-51 A DNA Library Can Be Made Using Either Viral or Plasmid Vectors[56]

In order to clone a gene, one begins by constructing a *DNA library* in either a virus or a plasmid vector. The principles underlying the methods used for cloning genes are the same for either cloning vector, although the details may be different. For simplicity, in this chapter we shall ignore these differences and illustrate the methods discussed in reference to plasmid vectors.

The **plasmid vectors** used for gene cloning are small circular molecules of double-stranded DNA derived from larger plasmids that occur naturally in bacteria, yeast, and mammalian cells (see p. 257). They generally account for only a minor fraction of the total host cell DNA, but they can easily be separated on the basis of their small size from chromosomal DNA molecules, which are large and form a pellet upon centrifugation. For use as cloning vectors, the purified plasmid DNA circles are first cut with a restriction nuclease to create linear DNA molecules. The cellular DNA to be used in constructing the library is cut with the same restriction nuclease, and the resulting restriction fragments (including those containing the gene to be cloned) are then added to the cut plasmids and annealed to form recombinant DNA circles. These recombinant molecules containing foreign DNA inserts are then covalently sealed with the DNA ligase enzyme described previously (see p. 223) to form intact DNA circles (see Figure 5–78).

In the next step in preparing the library, the recombinant DNA circles are introduced into cells (usually bacterial or yeast cells) that have been made transiently permeable to DNA; such cells are said to be *transfected* with the plasmids. As these cells grow and divide, the recombinant plasmids also replicate to produce an enormous number of copies of DNA circles containing the foreign DNA (Figure 5–79). Many bacterial plasmids carry genes for antibiotic resistance, a property that can be exploited to select those cells that have been successfully transfected; if the bacteria are grown in the presence of the antibiotic, only cells containing plasmids will survive. These surviving bacteria are said to contain a DNA library. However, only a few of these bacteria will harbor the particular recombinant plas-

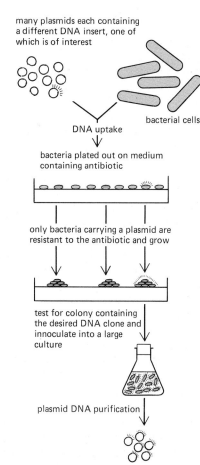

many plasmids each containing a different DNA insert, one of which is of interest

bacterial cells

DNA uptake

bacteria plated out on medium containing antibiotic

only bacteria carrying a plasmid are resistant to the antibiotic and grow

test for colony containing the desired DNA clone and innoculate into a large culture

plasmid DNA purification

Figure 5–79 Purification and amplification of a specific DNA sequence by DNA cloning in a bacterium.

mids that contain the gene to be isolated. One needs to be able to identify these cells in order to recover the DNA of interest in pure form and in useful quantities. Before discussing how this is achieved, we need to describe a second way of generating a DNA library that is commonly used in gene cloning.

Two Types of DNA Libraries Serve Different Purposes[57]

Cleaving the entire genome of a cell with a specific restriction nuclease as just described is sometimes called the "shotgun" approach to gene cloning. It produces a very large number of DNA fragments—on the order of a million for a mammalian genome—which will generate millions of different colonies of transfected cells. Each of these colonies will be composed of a *clone* derived from a single ancestor cell and therefore harbor a recombinant plasmid with the same inserted genomic DNA sequence. Such a plasmid is said to contain a **genomic DNA clone,** and the entire collection of plasmids is said to comprise a **genomic DNA library.** But because the genomic DNA is cut into fragments at random, only some fragments will contain genes; many will contain only a portion of a gene, while most of the genomic DNA clones obtained from the DNA of a higher eucaryotic cell will contain only noncoding DNA, which comprises most of the DNA in such genomes (see p. 485).

An alternative strategy begins the cloning process by selecting only those DNA sequences that are transcribed into RNA and thus are presumed to correspond to genes. This is done by extracting the mRNA (or a purified subfraction of the mRNA) from cells and then making a **complementary DNA (cDNA)** copy of each mRNA molecule present; this reaction is catalyzed by the *reverse transcriptase* enzyme of retroviruses, which synthesizes a DNA chain on an RNA template (see p. 254). The single-stranded DNA molecules synthesized by the reverse transcriptase are converted into double-stranded DNA molecules by DNA polymerase, and these molecules are inserted into plasmids and cloned (Figure 5–80). Each clone obtained in this way is called a **cDNA clone,** and the entire collection of clones derived from one mRNA preparation constitutes a **cDNA library.**

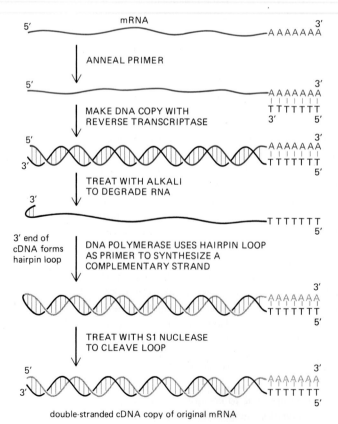

double-stranded cDNA copy of original mRNA

Figure 5–80 The synthesis of cDNA. A DNA copy (cDNA) of an mRNA molecule is produced by the enzyme reverse transcriptase (see p. 254), thereby forming a DNA/RNA hybrid helix. Treating the DNA/RNA hybrid with alkali selectively degrades the RNA strand into nucleotides. The remaining single-stranded cDNA is then copied into double-stranded cDNA by the enzyme DNA polymerase. As indicated, both reverse transcriptase and DNA polymerase require a primer to begin their synthesis. For reverse transcriptase a small oligonucleotide is used; in this example oligo(dT) has been annealed with the long poly A tract at the 3' end of most mRNAs (see p. 528).

Note that the double-stranded cDNA molecule produced here lacks cohesive ends; such "blunt-ended" DNA molecules can be cloned by one of several procedures analogous to that shown in Figure 5–78 but less efficient. For example, synthetic oligonucleotides containing restriction-enzyme cutting sites can be ligated onto the DNA ends, or single-stranded DNA "tails" can be added enzymatically to facilitate the insertion of the cDNA molecule into a cloning vector.

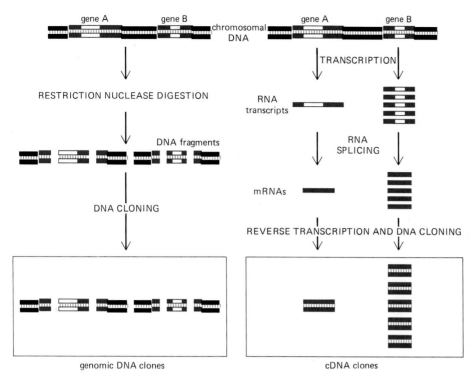

Figure 5–81 Diagrammatic representation of the differences between cDNA clones and genomic DNA clones. In this example gene A is infrequently transcribed while gene B is frequently transcribed. Both types of RNA transcripts are processed by RNA splicing to remove a single intron sequence during the formation of the mRNA. Most genes contain many introns (see Table 9–1, p. 486).

There are important differences between genomic DNA clones and cDNA clones, as illustrated in Figure 5–81. Genomic clones represent a random sample of all of the DNA sequences in an organism and with very rare exceptions will be the same regardless of the cell type used to prepare them. By contrast, cDNA clones contain only those regions of the genome that have been transcribed into mRNA, and—since the cells of different tissues produce distinct sets of mRNA molecules (see p. 553)—a different cDNA library will be obtained for each type of cell used to prepare the library.

The use of a cDNA library for gene cloning has several advantages. First, many proteins are produced in very large quantities by specialized cells, in which case the mRNA encoding the protein is produced in such excess that a cDNA library prepared from the cells will be highly enriched for the cDNA molecules encoding the protein (Figure 5–81). The abundance of the cDNA of interest greatly reduces the problem of identifying the desired clone in the library. Hemoglobin, for example, is made in large amounts by developing erythrocytes, and, for this reason, the globin genes were among the first to be cloned.

A second advantage of cDNA clones is that they contain the uninterrupted coding sequence of a gene. Eucaryotic genes, as we have seen (p. 102), usually consist of coding sequences of DNA separated by noncoding sequences, so that the production of mRNA from these genes entails the removal of the noncoding sequences (intron sequences) from the initial RNA transcript and the splicing together of the remaining coding sequences. Neither bacterial nor yeast cells will make these modifications to the RNA produced from a gene of a higher eucaryotic cell. Thus, if the aim of the cloning is either to deduce the amino acid sequence of the protein from the DNA or to produce the protein in bulk by expressing the cloned gene in a bacterial or yeast cell, it is preferable to start with cDNA.

Genomic and cDNA libraries are inexhaustable resources that are widely shared among investigators, and an increasing number of such libraries are available from commercial sources.

10-5 cDNA Libraries Can Be Prepared from Selected Populations of mRNA Molecules[58]

When cDNAs are prepared from cells that express the gene of interest at extremely high levels, the majority of cDNA clones may contain the gene sequence, which can therefore be selected with minimal effort. For less abundantly transcribed genes, various methods can be used to enrich for particular mRNAs before making the cDNA library. For example, if an antibody against the protein is available, it can be used to precipitate selectively those isolated polyribosomes (see p. 214) that contain the appropriate growing polypeptide chains. Since these ribosomes will also contain the mRNA coding for the protein, the precipitate may be enriched in the desired mRNA by as much as 1000-fold.

Substrative hybridization provides a powerful alternative way of enriching for particular nucleotide sequences prior to cDNA cloning. This selection procedure can be used, for example, if two closely related cell types are available from the same species of organism but only one type produces the protein or proteins of interest. An early use was to identify cell-surface receptor proteins present on T lymphocytes but not on B lymphocytes (see p. 1037). It can also be used where a cell that expresses the protein has a mutant counterpart that does not. The first step is to synthesize cDNA molecules using the mRNA from the cell type that makes the protein of interest. These cDNAs are then hybridized with a large excess of mRNA molecules from the second cell type. Those rare cDNA sequences that fail to find a complementary mRNA partner, and therefore are likely to represent mRNA sequences present only in the first cell type, are then purified by a simple biochemical procedure that separates single-stranded from double-stranded nucleic acids (Figure 5–82). Besides providing a powerful way to clone genes whose products are known to be restricted to a specific differentiated cell type, cDNA libraries prepared after subtractive hybridization are useful for defining the differences in gene expression between any two related cells.

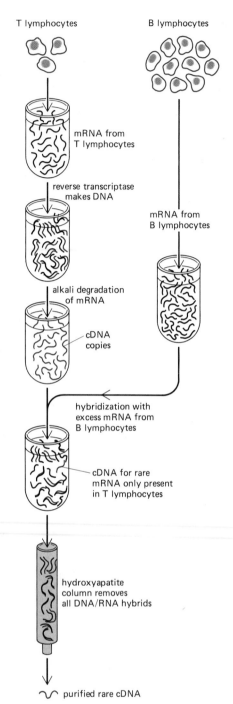

Figure 5–82 Subtractive hybridization is used in this example to purify rare cDNA clones corresponding to mRNA molecules present in T lymphocytes but not in B lymphocytes. Because the two cell types are very closely related, most of the mRNAs will be common to both cell types, and this technique is thus a powerful way to enrich for those specialized molecules that distinguish the two cells.

5-52 Hybridization with a Radioactive DNA Probe Can Be Used to Identify the Clones of Interest in a DNA Library[59]

The most difficult part of gene cloning is often the identification of the rare colonies in the library that contain the DNA fragment of interest. This is especially true in the case of a genomic library, where one has to identify one bacterial cell in a million to select a specific mammalian gene. The technique most frequently used is a form of *in situ* hybridization that takes advantage of the exquisite specificity of the base-pairing interactions between two complementary nucleic acid molecules (see p. 192). Culture dishes containing the growing bacterial colonies are blotted with a piece of filter paper, to which some members of each bacterial colony adhere. The adhering colonies, known as *replicas*, are treated with alkali and then incubated with a radioactive DNA probe containing part of the sequence of the gene being sought (Figure 5–83). If necessary, millions of bacterial clones can be screened in this way to find the one clone that hybridizes with the probe.

The way the specific DNA probes are made will depend on the information available about the gene to be cloned. In many cases the protein of interest has been identified by biochemical studies and purified in small amounts. Even a few micrograms of pure protein are enough to determine the sequence of its first 30 or so amino acids. From this amino acid sequence, the corresponding nucleotide sequence can be deduced using the genetic code. Two sets of DNA oligonucleotides, each about 20 nucleotides long, are then synthesized by chemical methods and radioactively labeled. The two sets of oligonucleotides are chosen to match different parts of the predicted nucleotide sequence of the gene (Figure 5–84). Colonies of cells that hybridize with both sets of DNA probes are strong candidates for containing the desired gene and are saved for further characterization (see below).

The Selection of Overlapping DNA Clones Allows One to "Walk" Along the Chromosome to a Nearby Gene of Interest[60]

Many of the most interesting genes—for example, those that control development—are known only from genetic analysis of mutants in such organisms as the fruit fly *Drosophila* and the nematode *C. elegans*. The protein products of these genes are unknown and may be present in very small quantities in a few cells or

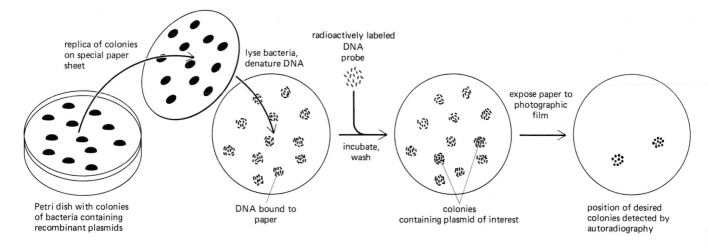

produced only at one stage of development. However, a study of the genetic linkage between different mutations can be used to generate *chromosome maps*, which give the relative locations of the genes. Once one mapped gene has been cloned, the clones in a genomic DNA library that correspond to neighboring genes can be identified using a technique known as **chromosome walking.** For this method, two different genomic DNA libraries can be used, each prepared by fragmenting the same DNA with a different restriction nuclease. Figure 5–85 illustrates how a clone in one library can be used as a DNA probe to find an overlapping clone in the other library. This new clone is then used to prepare a DNA probe, which is used to find another overlapping clone in the first library, and so on. In this way one can walk along a chromosome one clone at a time, in steps of 30,000 base pairs or more. But how does one know when the gene of interest (identified originally by a deleterious mutation) has been reached? The standard way is continually to compare the size of the DNA restriction fragments produced from mutant and normal chromosomes by Southern blot analysis (see p. 191), using each new clone as a probe as the walk proceeds. Some of the mutants will have

Figure 5–83 An efficient technique commonly used to detect a bacterial colony carrying a particular DNA clone (see also Figure 5–79). Each bacterial cell carrying a recombinant plasmid develops into a colony of identical cells, visible as a white spot on the nutrient agar. A replica of the culture is then made by pressing a piece of absorbent paper against the surface. This replica is treated with alkali (to break open the adherent cells and denature the plasmid DNA) and then hybridized to a highly radioactive DNA probe. Those bacterial colonies that have bound the probe are identified by autoradiography.

Figure 5–84 Selecting regions of a known amino acid sequence to make synthetic oligonucleotide probes. Only one nucleotide sequence will actually code for the protein. But the degeneracy of the genetic code means that several different nucleotide sequences will give the same amino acid sequence, and it is impossible to tell in advance which is the correct one. Because it is desirable to have as large a fraction of the correct nucleotide sequence as possible in the mixture of oligonucleotides to be used as a probe, those regions with the fewest possibilities are chosen, as illustrated. After the oligonucleotide mixture is synthesized by chemical means, the 5′ end of each oligonucleotide is radioactively labeled (see Figure 4–65B).

known portion of amino acid sequence

H₂N--- Gly Val Arg Met Asp Trp Asn Tyr Glu Pro Leu Ser Thr Trp Glu Met Asn Gln Trp Phe Val Arg Ala --COOH

possible codons

5′	GGA	GUA	AGA	AUG	GAC	UGG	AAC	UAC	GAA	CCA	UUA	AGC	ACA	UGG	GAA	AUG	AAC	CAA	UGG	UUC	GUA	AGA	GCA	3′
	GGC	GUC	AGG		GAU		AAU	UAU	GAG	CCC	UUG	AGU	ACC		GAG		AAU	CAG		UUU	GUC	AGG	GCC	
	GGG	GUG	CGA							CCG	CUA	UCA	ACG								GUG	CGA	GCG	
	GGU	GUU	CGC							CCU	CUC	UCC	ACU								GUU	CGC	GCU	
			CGG								CUG	UCG										CGG		
			CGU								CUU	UCU										CGU		

regions of coding sequence with least ambiguity

synthetic oligonucleotides used as probes

AUGGA ᶜ/ᵤ UGGAA ᶜ/ᵤ UA ᶜ/ᵤ GA ᴬ/ᵍ CC

[16 possibilities]

UGGGA ᴬ/ᵍ AUGAA ᶜ/ᵤ CA ᴬ/ᵍ UGGUU

[8 possibilities]

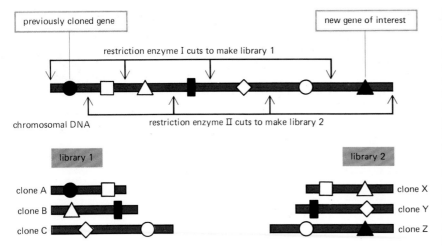

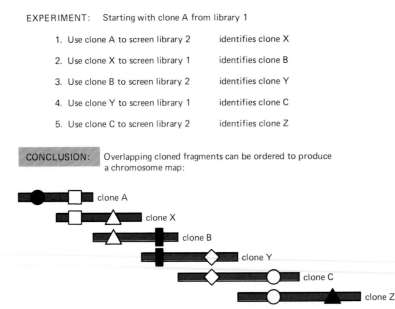

EXPERIMENT: Starting with clone A from library 1

1. Use clone A to screen library 2   identifies clone X
2. Use clone X to screen library 1   identifies clone B
3. Use clone B to screen library 2   identifies clone Y
4. Use clone Y to screen library 1   identifies clone C
5. Use clone C to screen library 2   identifies clone Z

CONCLUSION: Overlapping cloned fragments can be ordered to produce a chromosome map:

Figure 5–85 The use of overlapping DNA clones to find a new gene by "chromosome walking." To speed up the walk, genomic libraries containing very large cloned DNA molecules are optimal. To probe for the next clone in the walk, one uses a short ^{32}P-labeled DNA fragment from one end of the previously identified clone: If a "right-handed" end is used, for example, the walk will go in the "rightward" direction, as shown in this example. By using a small end fragment as a probe, one also reduces the probability that the probe will contain a repeated DNA sequence that would hybridize with many clones from different parts of the genome and thereby interrupt the walk.

been caused by small deletions or insertions of DNA sequences in the relevant gene, and these can be readily identified. When the deleterious mutations in a typical human gene are analyzed, for example, about 1 in 10 turns out to be a deletion that is easily detected by Southern blotting, and the precise molecular defects responsible for human genetic diseases are increasingly being identified in this way.

By using related methods, it has been possible to order (map) a nearly complete set of large genomic clones along the chromosomes of the nematode *C. elegans*. Such large clones, each about 30,000 base pairs in length, are prepared in bacteriophage lambda vectors called *cosmids*, which are specially designed to accept only large DNA inserts. It takes a few thousand cosmid clones to cover the entire genome of an organism such as *C. elegans* or *Drosophila*. To map the entire human genome in this way would require ordering more than 100,000 such cosmid clones, which is time-consuming but technically feasible. Moreover, human DNA fragments that are 10 times larger than cosmid clones (300,000 base pairs) can be cloned as artificial chromosomes in yeast cells; in principle, the human genome could be mapped as 10,000 clones of this type (see Fig. 9–5, p. 485).

In the near future, ordered sets of genomic clones will no doubt be available from centralized DNA libraries for use by all research workers. A complete library will be available for each commonly studied organism, with each DNA insert cataloged according to its chromosome of origin and numbered sequentially with respect to the positions of all other DNA fragments derived from the same chromosome. One will then begin a "chromosome walk" simply by obtaining from the library all the clones covering the region of the genome that contains the mutant gene of interest. These clones would then be used to make DNA probes to locate the altered gene precisely. Many of the mutant genes that cause genetic diseases in humans will eventually be isolated in this way.

In Vitro Translation Facilitates Identification of the Correct DNA Clone[61]

Despite the power of *in situ* hybridization methods to find specific cDNA or genomic clones from a DNA library, they usually pick out many "false positive" clones. Further ingenuity is required to discriminate between these and authentic positive clones. The task is easiest when the desired clone encodes a protein that has already been characterized by other means. In this case, each candidate cloned DNA fragment is used to purify its complementary mRNA molecules from a mixture of cellular mRNAs by a process called **hybrid selection,** in which an excess of the DNA fragment is separated into single strands and immobilized on a filter that is then used to select complementary mRNA molecules by RNA-DNA hybridization (Figure 5–86). The mRNA purified this way is then allowed to direct protein synthesis in a cell-free system using radioactive amino acids. Finally, the radioactive protein produced is characterized and compared with the expected protein product of the desired clone. A match in such a test is normally a prerequisite for concluding that a cloned DNA fragment encodes the given protein.

Expression Vectors Allow cDNA Clones to Be Used to Overproduce Proteins[62]

Very often there is no biochemical information about the protein encoded by a cloned DNA fragment. This is usually the case, for example, when the clone has been identified by subtractive hybridization or by chromosome walking to a mutant gene. In these cases, moreover, the mRNA for the protein in question is often present in such low abundance and in such a limited group of cells that hybrid selection of the complementary mRNA is not feasible. Other methods must then be used to characterize the protein product of the cloned gene. One method is to synthesize a short protein fragment (an *oligopeptide*) corresponding to the deduced amino acid sequence of the protein product of the sequenced cDNA molecule, and then to raise antibodies against the oligopeptide. In many cases the antibodies will recognize the same amino acid sequence when it occurs as part of the natural protein molecule, and these then provide a means to detect, locate, and purify the protein encoded by the original cDNA. In conjunction with cloning by subtractive hybridization, this immunological approach is a powerful way to identify cell type-specific proteins and investigate the development, properties, and functions of each type of cell in a multicellular organism.

The most direct way to characterize the protein encoded by a cloned cDNA, however, is to allow the cDNA itself to direct protein synthesis in a host cell. Plasmids or viruses used for this purpose are called **expression vectors,** and they are constructed so that the cDNA clone is connected directly to a DNA sequence that acts as a strong promoter for DNA transcription. A variety of expression vectors are available, each engineered to function in the type of cell in which the protein is to be made. By means of such *genetic engineering*, bacteria, yeast, or mammalian cells can be induced to make vast quantities of useful proteins—such as human growth hormone, interferon, and viral antigens for vaccines. Bacterial cells with plasmids or viral vectors that have been engineered in this way are especially adept at protein production, and it is common for an engineered gene to produce more than 10% of the total cell protein. Because the production of such a large quantity of a single protein frequently kills the cell, special "inducible" promoters have been designed to allow transcription of the gene to be delayed until a few hours before the cells are collected for protein isolation. Some plasmid expression vectors, for example, contain a bacteriophage lambda promoter that is kept silent by a heat-sensitive gene-repressor protein; the promoter can be turned on at the desired time by increasing the temperature of the bacteria to 42°C, and large amounts of the protein of interest will suddenly be made.

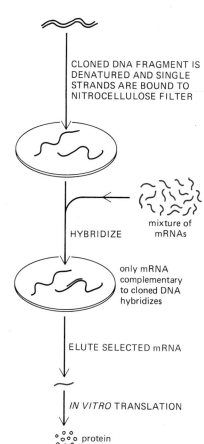

CLONED DNA FRAGMENT IS DENATURED AND SINGLE STRANDS ARE BOUND TO NITROCELLULOSE FILTER

HYBRIDIZE

mixture of mRNAs

only mRNA complementary to cloned DNA hybridizes

ELUTE SELECTED mRNA

IN VITRO TRANSLATION

protein

Figure 5–86 The technique of hybrid selection. The purified mRNA molecules are eluted from the filter by subjecting it to conditions that separate the two strands of an RNA-DNA helix.

If a cDNA library is prepared in an expression vector, each clone will in general produce a different protein, and it is then possible to identify the clones of interest by their protein product instead of by their nucleic acid sequence. This is usually done by probing the clones with radioactively labeled antibodies. Alternatively, one can sometimes test directly for the biological activity of the gene product. This is an especially powerful way to search for eucaryotic genes coding for secreted growth factors. For example, cDNA clones from cells that produce a growth factor can be prepared in an expression vector that replicates in mammalian cells. A mixture of transfected cells containing many different such clones is grown in a small dish, where any cell expressing the gene will secrete the growth factor into the culture medium. Samples of the medium are then tested for the growth factor by adding the medium to cultures of other cells known to respond to that particular factor. The test is so sensitive that it will usually be positive even if only one cell in a thousand in the original culture contains a gene encoding the growth factor. Repeated rounds of testing allow the search to be narrowed down to the single clone in the mixture that is producing the growth factor. In this way previously unrecognized growth factors have been discovered and isolated in a few months; isolating growth factors using standard biochemical techniques can require tedious purifications of more than a millionfold that take years.

5-53 ## Genes Can Be Redesigned to Produce Proteins of Any Desired Sequence[63]

Both the coding sequence of a gene and its regulatory regions can be redesigned to change the functional properties of the protein product, the amount of protein made, or the particular cell type in which the protein is produced.

The coding sequence of a gene can be extensively altered—for example, by fusing part of it to the coding sequence of a different gene to produce a novel hybrid gene that encodes a *fusion protein*. Such proteins are frequently used to test the function of different domains of a protein molecule. For example, most nuclear proteins contain specific short sequences of amino acids that are recognized as signals for their direct import into the cell nucleus. By artificially attaching different parts of nuclear proteins to a cytoplasmic protein using gene fusion techniques, the "signal peptides" responsible for nuclear import have been identified (see p. 424).

Special techniques are required to alter a gene in more subtle ways, so that the protein it encodes differs from the original by one or a few amino acids. The first step is the chemical synthesis of a short DNA molecule containing the altered portion of the gene's nucleotide sequence. This synthetic DNA oligonucleotide is hybridized with single-stranded plasmid DNA that contains the DNA sequence to be altered, using conditions that allow imperfectly matched DNA strands to pair (Figure 5–87). The synthetic oligonucleotide will now serve as a primer for DNA synthesis by DNA polymerase, thereby generating a DNA double helix that incorporates the altered sequence into the gene. The modified gene is then inserted into an expression vector so that the redesigned protein can be produced in large enough quantities for detailed studies. By changing selected amino acids in a protein in this way, one can analyze which parts of the polypeptide chain are important in such fundamental processes as protein folding, protein-ligand interactions, and enzymatic catalysis.

To determine which parts of a eucaryotic gene are responsible for regulating its expression, suspected regulatory DNA sequences of the gene (see p. 564) can be joined to the coding sequence of a gene that encodes a readily detected marker enzyme not normally present in eucaryotic cells. One of the most commonly used marker enzymes is the bacterial protein *chloramphenicol acetyltransferase (CAT)*. The resulting recombinant DNA molecule is then inserted into a eucaryotic cell to test whether the suspected regulatory sequences promote expression of the CAT gene. This is most conveniently done by transfecting cells in culture and testing for CAT enzyme activity after an overnight incubation. Although many of the transfected cells will transiently express the foreign gene, very few will retain the gene permanently. To obtain *permanent transformants* it is necessary to select

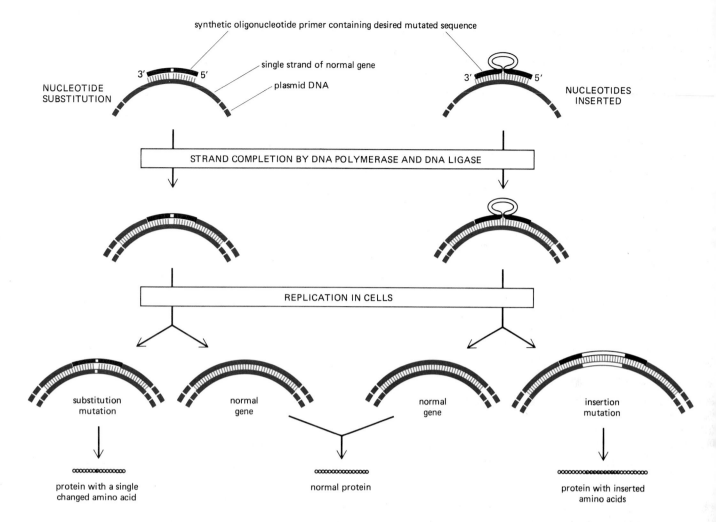

the rare cell clones that have stably incorporated the recombinant DNA molecules into their chromosomes. By transfecting different cell types with such recombinant DNA molecules, it has been possible to identify regulatory DNA sequences that enable a gene to be expressed in one specific cell type but not in others (see p. 565).

Figure 5–87 The use of synthetic oligonucleotides to modify the protein-coding regions of genes. Only two of the many types of changes that can be engineered in this way are shown. For example, with an appropriate oligonucleotide, more than one amino acid substitution can be made at a time, or one or more amino acids can be deleted. As indicated, because only one of the two DNA strands in the original recombinant plasmid is altered by this procedure, only half of the transfected cells will end up with a plasmid that contains the desired mutant gene. Note that most of the plasmid sequence is not illustrated here.

Engineered Genes Can Be Permanently Inserted into the Germ Line of Mice or Fruit Flies to Produce Transgenic Animals[64]

The ultimate test of the function of an altered gene is to reinsert it into an organism and see what effect it has. Ideally one would like to be able to replace the normal gene with the altered one so that the function of the mutant protein can be analyzed in the absence of the normal protein. The only eucaryotic organism in which this can be done routinely is yeast. DNA fragments transferred into growing yeast cells are efficiently integrated as single copies into the homologous chromosomal sites by general recombination events (see p. 239), so that a redesigned gene can be inserted in place of the endogenous copy of the gene. This provides a powerful way to study the function of yeast genes (see p. 195).

When DNA is transferred into mammalian cells, by contrast, there is presently no way to control how and where it becomes integrated into a chromosome. Only about one in a thousand integrative events leads to gene replacement. Instead linear DNA fragments are rapidly ligated end to end by enzymes in the cell to form long tandem arrays, which usually become integrated into a chromosome at an apparently randomly selected site. Fertilized mammalian eggs behave like other mammalian cells in this respect. A mouse egg injected with 200 copies of a linear DNA molecule will often develop into a mouse containing a tandem array of copies of the injected gene integrated at a single random site in one of its chromosomes

(Figure 5–88). If the modified chromosome is also present in the germ line cells (eggs or sperm), the mouse will pass the newly adopted genes on to its progeny: animals that have been permanently altered in this way are called **transgenic.** Because the normal gene generally remains present, most transgenic mammals cannot be used with the elegance of yeast cells to test the effects of altered genes, but they have already provided important insights into how mammalian genes are regulated (see p. 580) and how certain altered genes (called oncogenes) cause cancer (see p. 1210).

Similar experiments can be performed with the fruit fly *Drosophila* using a technique whereby single copies of any gene can be inserted at random into the fly genome. The trick in this case is first to insert the DNA fragment between the two terminal sequences of a particular *Drosophila* transposable element, called the *P element*. The terminal sequences enable the P element to integrate into

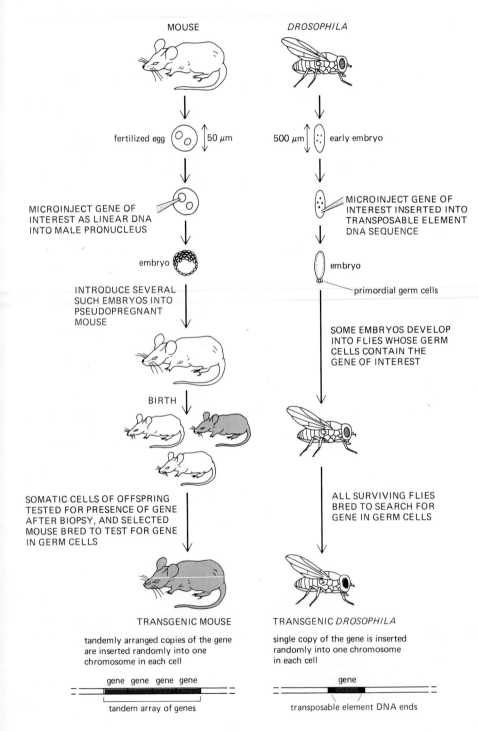

Figure 5–88 Comparison of the standard procedures used to make transgenic mice and *Drosophila*. In these examples, the gene injected into the mouse egg causes a change in coat color, whereas the gene injected into the fly embryo causes a change in eye color. In both organisms, some of the transgenic animals are found to have DNA insertions at more than one chromosomal site.

MOUSE

DROSOPHILA

fertilized egg   50 μm

500 μm   early embryo

MICROINJECT GENE OF INTEREST AS LINEAR DNA INTO MALE PRONUCLEUS

MICROINJECT GENE OF INTEREST INSERTED INTO TRANSPOSABLE ELEMENT DNA SEQUENCE

embryo

embryo

primordial germ cells

INTRODUCE SEVERAL SUCH EMBRYOS INTO PSEUDOPREGNANT MOUSE

SOME EMBRYOS DEVELOP INTO FLIES WHOSE GERM CELLS CONTAIN THE GENE OF INTEREST

BIRTH

SOMATIC CELLS OF OFFSPRING TESTED FOR PRESENCE OF GENE AFTER BIOPSY, AND SELECTED MOUSE BRED TO TEST FOR GENE IN GERM CELLS

ALL SURVIVING FLIES BRED TO SEARCH FOR GENE IN GERM CELLS

TRANSGENIC MOUSE

TRANSGENIC *DROSOPHILA*

tandemly arranged copies of the gene are inserted randomly into one chromosome in each cell

single copy of the gene is inserted randomly into one chromosome in each cell

gene gene gene gene

gene

tandem array of genes

transposable element DNA ends

Drosophila chromosomes if the P element transposase enzyme is also present (see p. 256). To make transgenic fruit flies, therefore, the appropriately modified DNA fragment is injected into a very young fruit fly embryo along with a separate plasmid containing the gene encoding the transposase. When this is done, the injected gene often enters the germ line in a single copy as the result of a transposition event (Figure 5–88).

Direct gene replacements have been achieved recently in mice by a laborious, indirect pathway. A DNA fragment containing a desired mutant gene is first transfected into a special line of pluripotent embryo-derived stem cells that grow in cell culture. After a period of cell proliferation, the rare colonies of cells in which general recombination mechanisms have caused a gene replacement to occur are identified by Southern blotting, and individual cells are then implanted into an early mouse embryo. In this new environment the embryo-derived stem cells often proliferate to produce major portions of a normal mouse (see p. 897). Mice that thereby acquire the gene replacement in their germ line are bred to produce both a male and a female animal, each heterozygous for the gene replacement (that is, they have one normal and one mutant copy of the gene). When these two mice are bred, one-fourth of their progeny will be homozygous for the altered gene. Studies of these homozygotes allow the function of an altered gene to be examined in the absence of the corresponding normal gene.

The ability to prepare transgenic mice and *Drosophila* is a powerful new tool for dissecting the function of genes in an intact organism.

Selected DNA Segments Can Be Cloned in a Test Tube by a Polymerase Chain Reaction[65]

The availability of purified DNA polymerases and chemically synthesized DNA oligonucleotides (see p. 191) has made it possible to clone specific DNA sequences rapidly without the need of a living cell. The technique, called the **polymerase chain reaction (PCR)**, allows the DNA from a selected region of a genome to be amplified by more than a millionfold, provided that at least part of its nucleotide sequence is already known. Portions of the sequence that surround the region to be amplified are used to design two synthetic DNA oligonucleotides, one complementary to each strand of the DNA double helix. These oligonucleotides serve as primers for *in vitro* DNA synthesis, which is catalyzed by a DNA polymerase, and they determine the ends of the final DNA fragment that is obtained (Figure 5–89A).

The principle of the PCR technique is illustrated in Figure 5–89B. Each cycle of the reaction requires a brief heat treatment to separate the two strands of the DNA double helix (step 1). A subsequent cooling of the DNA in the presence of a large excess of the two DNA oligonucleotides allows their specific hybridization to complementary DNA sequences (step 2). The annealed mixture is then incubated with DNA polymerase and the four deoxyribonucleoside triphosphates so that the regions of DNA downstream from the primers are selectively synthesized (step 3). For effective DNA amplification, 20 to 30 cycles of reaction are required. Each cycle doubles the amount of DNA synthesized in the previous cycle. A single cycle requires only about 5 minutes, and an automated procedure permits "cell-free molecular cloning" of a DNA fragment in a few hours, compared to the several days required for standard cloning procedures.

Figure 5–89 The polymerase chain reaction for amplifying specific nucleotide sequences *in vitro*. (A) DNA isolated from cells is heated to separate its complementary strands. These strands are then annealed with an excess of two DNA oligonucleotides (each 15 to 20 nucleotides long) that have been chemically synthesized to match sequences separated by X nucleotides (where X is generally between 50 and 2000). The two oligonucleotides serve as specific primers for *in vitro* DNA synthesis catalyzed by DNA polymerase, which copies the DNA between the sequences corresponding to the two oligonucleotides. (B) After multiple cycles of reaction, a large amount of a single DNA fragment, X nucleotides long, is obtained, provided that the original DNA sample contains the DNA sequence that was anticipated when the two oligonucleotides were designed.

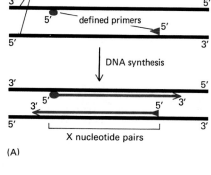

(A)

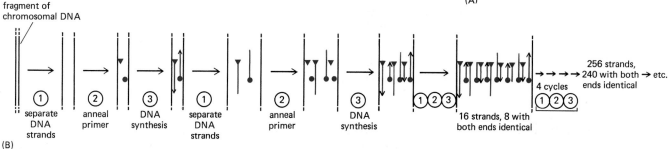

(B)

269   DNA Cloning and Genetic Engineering

The PCR method is extremely sensitive: it can detect a single DNA molecule in a sample. Trace amounts of RNA can be analyzed in a similar way by converting them to DNA sequences with reverse transcriptase (see p. 260). The PCR cloning technique is rapidly replacing Southern blotting for prenatal diagnosis of genetic diseases and for the detection of low levels of viral infection. It also has great promise for forensic medicine, as it allows the unambiguous identification of the human source of a single cell.

10-31
Recombinant DNA Methods Greatly Facilitate Genetic Approaches to the Mapping and Analysis of Large Genomes[66]

Recently developed methods make it possible to prepare detailed maps of very large genomes. There are two categories of maps: (1) *Physical maps* are based on the DNA molecules that comprise each chromosome. They include restriction maps (see p. 182) and ordered libraries of genomic DNA clones (see p. 264). (2) *Genetic linkage maps* are based on the frequency of coinheritance of two or more features that serve as *genetic markers*, each different in the mother and father and attributable to a specific chromosomal site. Traditionally these markers were genes whose expression is detected by means of their effects (as with genes that cause genetic diseases such as muscular dystrophy). More recently, recombinant DNA methods have made it possible to use as genetic markers short DNA sequences that contain a restriction nuclease cutting site and differ between individuals; such a sequence is particularly useful for genetic mapping because it creates a **restriction fragment length polymorphism,** or **RFLP**, which can be readily detected by Southern blotting (see p. 191) with a nearby DNA probe (Figure 5–90).

If two genetic markers are on different chromosomes, their inheritance will be *unlinked*—that is, they will have only a 50:50 chance of being inherited together. The same is true for markers at opposite ends of a single chromosome because of the high probability that they will be separated during the extensive crossing over that occurs during meiosis in the development of eggs and sperm (see p. 847). The closer together two markers are on the same chromosome, the larger is the chance that they will not be separated by crossover events and will therefore be coinherited. By screening large family groups for the coinheritance of a gene of interest (such as one associated with a disease) and a large number of individual RFLPs, a few RFLP markers can be identified that surround the gene. In this way DNA sequences in the vicinity of the gene can be located and eventually the DNA corresponding to the gene itself can be found (Figure 5–91). Many genes that cause

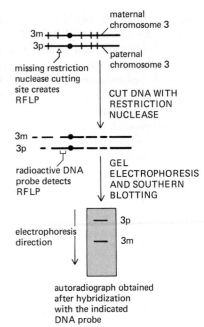

Figure 5–90 Detection of a restriction fragment length polymorphism (RFLP) by Southern blotting. For simplicity the chromosomes are shown to contain only a few restriction-enzyme cutting sites, whereas in reality they contain many thousands of such sites. If the PCR technique (see p. 269) is used to amplify the appropriate DNA region before the restriction nuclease treatment, the same test can be applied without radioisotopes and the blotting step can be omitted.

Figure 5–91 Genetic linkage analysis detects the coinheritance of a gene that causes a specific human phenotype (here a genetic disease) with a nearby RFLP marker. Analyses of this type are allowing many human genes to be cloned and their gene products to be analyzed.

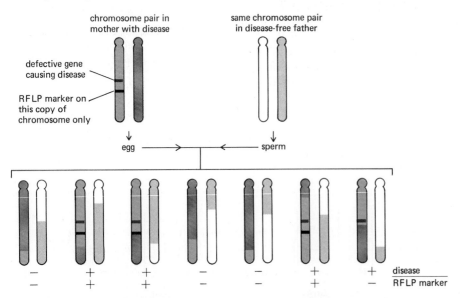

CONCLUSION: Gene causing disease is co-inherited with RFLP marker from diseased mother in 75% of the progeny. If result still holds after examining other families, the gene causing disease is mapped to this chromosome close to RFLP marker.

human diseases are being located in this way. Once the gene is isolated, the protein it encodes can be analyzed in detail (see p. 193).

To facilitate these and other studies, an intensive effort is underway to prepare a high-resolution RFLP map of the human genome, with thousands of RFLP markers spaced an average of 10^6 nucleotide pairs apart (on average, two markers separated by this distance will be coinherited by 99 of every 100 progeny). With such a map it will become possible to use genetic linkage studies to locate genes that have been identified only by their effect on humans to one or a few large DNA clones in an ordered genomic DNA library. Isolating the gene could then be accomplished in most cases in a matter of months.

Given the rapid advances in recombinant DNA technology, it should soon be possible to sequence very long stretches of DNA at reasonable cost. This will make it possible to sequence the entire genomes of bacteria, yeasts, nematodes, and *Drosophila*, as well as major portions of the genomes of various higher plants and vertebrates, including humans.

Summary

In DNA cloning the DNA fragment containing a gene of interest is typically identified by using a radiolabeled DNA probe or, following expression of the gene in host cells, by using an antibody to detect the protein the gene encodes. The cells carrying this DNA fragment are then allowed to proliferate, producing large amounts of the gene or gene product. In genetic engineering, the nucleotide sequence of such a cloned DNA fragment is altered, joined to other DNA sequences, and then inserted back into cells. Together, DNA cloning and genetic engineering provide powerful research tools for the cell biologist. In principle, a gene encoding a protein of any desired amino acid sequence can be constructed and attached to a promoter DNA sequence that causes the timing and pattern of the gene's expression to be controlled in a desired way. The new gene can then be inserted either into cultured cells or into the germ line of a mouse or a fruit fly. In transgenic animals one can examine the effect of the expression of the gene on many different cells and tissues.

References

General

Judson, H.F. The Eighth Day of Creation: Makers of the Revolution in Biology. New York: Simon & Schuster, 1979. (History derived from personal interviews.)

Lewin, B. Genes, 3rd ed. New York: Wiley, 1987.

Stent, G.S. Molecular Genetics: An Introductory Narrative. San Francisco: Freeman, 1971. (Clear descriptions of the classical experiments.)

Watson, J.D.; Hopkins, N.H.; Roberts, J.W.; Steitz, J.A.; Weiner, A.M. Molecular Biology of the Gene, 4th ed. Menlo Park, CA: Benjamin-Cummings, 1987.

Cited

1. Chamberlin, M. Bacterial DNA-dependent RNA polymerases. In The Enzymes, 3rd ed., Vol. 15B (P. Boyer, ed.), pp. 61–108. New York: Academic Press, 1982.
   McClure, W. Mechanism and control of transcription initiation in prokaryotes. *Annu. Rev. Biochem.* 54:171–204, 1985.
   Yager, T.D.; von Hippel, P.H. Transcript elongation and termination in *E. coli*. In *Escherichia coli* and *Salmonella typhimurium*: Cellular and Molecular Biology (F.C. Niedhardt, ed.), pp. 1241–1275. Washington, DC: American Society for Microbiology, 1987.
2. Hawley, D.K.; McClure, W.R. Compilation and Analysis of *Escherichia coli* promoter DNA sequences. *Nucleic Acids Res.* 11:2237–2255, 1983.

Siebenlist, U.; Simpson, R.; Gilbert, W. *E. coli* RNA polymerase interacts homologously with two different promoters. *Cell* 20:269–281, 1980.
3. Rich, A.; Kim, S.H. The three-dimensional structure of transfer RNA. *Sci. Am.* 238(1):52–62, 1978.
   Schimmel, P.R.; Söll, D.; Abelson, J.N., eds. Transfer RNA: Structure, Properties and Recognition, and Biological Aspects (2 volumes). Cold Spring Harbor, NY: Cold Spring Harbor Laboratory, 1980.
4. Schimmel, P. Aminoacyl tRNA synthetases: general scheme of structure-function relationships in the polypeptides and recognition of transfer RNAs. *Annu. Rev. Biochem.* 56:125–158, 1987.
5. Crick, F.H.C. The genetic code. III. *Sci. Am.* 215(4):55–62, 1966.
   The Genetic Code. *Cold Spring Harbor Symp. Quant. Biol.*, Vol. 31, 1966. (The original experiments that defined the code.)
6. Moore, P.B. The ribosome returns. *Nature* 331:223–227, 1988.
   Noller, H.F. Structure of ribosomal RNA. *Annu. Rev. Biochem.* 53:119–162, 1984.
   Spirin, A.S. Ribosome Structure and Protein Synthesis. Menlo Park, CA: Benjamin-Cummings, 1986.
7. Clark, B. The elongation step of protein biosynthesis. *Trends Biochem. Sci.* 5:207–210, 1980.
   Watson, J.D. The involvement of RNA in the synthesis of proteins. *Science* 140:17–26, 1963. (A description of how ribosome function was initially deciphered.)
8. Caskey, C.T. Peptide chain termination. *Trends Biochem. Sci.* 5:234–237, 1980.

Craigen, W.J.; Caskey, C.T. The function, structure and regulation of *E. coli* peptide chain release factors. *Biochemie* 69:1031–1041, 1987.

9. Gold, L. Posttranscriptional regulatory mechanisms in *Escherichia coli*. *Annu. Rev. Biochem.* 57:199–233, 1988.

Hunt, T. The initiation of protein synthesis. *Trends Biochem. Sci.* 5:178–181, 1980.

Pain, V.M. Initiation of protein synthesis in mammalian cells. *Biochem. J.* 235:625–637, 1986.

10. Kozak, M. Comparison of initiation of protein synthesis in procaryotes, eucaryotes, and organelles. *Microbiol. Rev.* 47:1–45, 1983.

Kozak, M. An analysis of 5′ noncoding sequences from 699 vertebrate messenger RNAs. *Nucleic Acids Res.* 15:8125–8147, 1987.

Steitz, J.A. Genetic signals and nucleotide sequences in messenger RNA. In Biological Regulation and Development (R. Goldberger, ed.), Vol. 1, pp. 349–399. New York: Plenum, 1979.

11. Rich, A. Polyribosomes. *Sci. Am.* 209(6):44–53, 1963.

12. Hunt, T. Phosphorylation and the control of protein synthesis. *Philos. Trans. R. Soc. Lond. (Biol.)* 302:127–134, 1983.

Jagus, R.; Anderson, W.F.; Safer, B. The regulation of initiation of mammalian protein synthesis. *Prog. Nucleic Acid Res. Mol. Biol.* 25:127–185, 1981.

13. Kirkwood, T.B.; Rosenberger, R.F.; Galas, D.J., eds. Accuracy in Molecular Processes: Its Control and Relevance to Living Systems. London: Chapman & Hall, 1986. (Chapters 4, 5, 6, and 11.)

Thompson, R.C. EFTu provides an internal kinetic standard for translational accuracy. *Trends Biochem. Sci.* 13:91–93, 1988.

14. Jiménez, A. Inhibitors of translation. *Trends Biochem. Sci.* 1:28–30, 1976.

15. Alberts, B.M. The function of the hereditary materials: biological catalyses reflect the cell's evolutionary history. *Am. Zool.* 26:781–796, 1986.

Orgel, L.E. RNA catalysis and the origin of life. *J. Theor. Biol.* 123:127–149, 1983.

Weiner, A.M.; Maizels, N. tRNA-like structures tag the 3′ ends of genomic RNA molecules for replication: implications for the origin of protein synthesis. *Proc. Natl. Acad. Sci. USA* 84:7383–7387, 1987.

16. Friedberg, E.C. DNA Repair. San Francisco: Freeman, 1985.

Lindahl, T. DNA repair enzymes. *Annu. Rev. Biochem.* 51:61–88, 1982.

17. Dickerson, R.; Geis, I. The Structure and Action of Proteins, pp. 59–66. New York: Harper & Row, 1969.

Wilson, A.C.; Carlson, S.S.; White, T.J. Biochemical evolution. *Annu. Rev. Biochem.* 46:573–639, 1977.

Wilson, A.C.; Ochman, H.; Prager, E.M. Molecular time scale for evolution. *Trends Genet.* 3:241–247, 1987.

18. Drake, J.W. Comparative rates of spontaneous mutation. *Nature* 221:1132, 1969.

19. Cairns, J. Cancer: Science and Society. San Francisco: Freeman, 1978. (Chapter 7 discusses the role of mutation in the development of cancer.)

Ohta, T.; Kimura, M. Functional organization of genetic material as a product of molecular evolution. *Nature* 233:118–119, 1971. (The mutation rate limits the maximum number of genes in an organism.)

20. Schrödinger, E. What is Life? Cambridge, U.K.: Cambridge University Press, 1945.

21. Kornberg, A. DNA Replication. San Francisco: Freeman, 1980. (Chapter 4 describes DNA polymerase, Chapter 8 describes DNA ligase, and Chapter 16 covers the general enzymology of DNA repair.)

22. Cleaver, J.E.; Karentz, D. DNA repair in man: regulation by a multigene family and association with human disease. *Bioessays* 6:122–127, 1987.

Coulondre, C.; Miller J.H.; Farabaugh, P.J.; Gilbert, W. Molecular basis of base substitution hotspots in *E. coli*. *Nature* 274:775–780, 1978.

Lindahl, T. New class of enzymes acting on damaged DNA. *Nature* 259:64–66, 1976.

Sancar, A.Z.; Sancar, G.B. DNA repair enzymes. *Annu. Rev. Biochem.* 57:29–67, 1988.

23. Lindahl, T.; Sedgwick, B.; Sekiguchi, M.; Nakabeppu, Y. Regulation and expression of the adaptive response to alkylating agents. *Annu. Rev. Biochem.* 57:133–157, 1988.

Ossanna, N.; Peterson, K.R.; Mount, D.W. Genetics of DNA repair in bacteria. *Trends Genet.* 2:55–58, 1986.

Walker, G.C. Inducible DNA repair systems. *Annu. Rev. Biochem.* 54:425–457, 1985.

24. Alberts, B.M. Prokaryotic DNA replication mechanisms. *Philos. Trans. R. Soc. Lond. (Biol.)* 317:395–420, 1987.

Kornberg, A. DNA Replication. San Francisco: Freeman, 1980.

McMacken, R.; Silver, L.; Georgopoulos, C. DNA replication. In *Escherichia coli* and *Salmonella typhimurium*: Cellular and Molecular Biology (F.C. Neidhardt, ed.), pp. 564–612. Washington, DC: American Society for Microbiology, 1987.

25. Joyce, C.M.; Steitz, T.A. DNA polymerase I: from crystal structure to function via genetics. *Trends Biochem. Sci.* 12:288–292, 1987.

Kornberg, A. Biologic synthesis of deoxyribonucleic acid. *Science* 131:1503–1508, 1960. (Nobel prize address.)

Meselson, M.; Stahl, F.W. The replication of DNA in *E. coli*. *Proc. Natl. Acad. Sci. USA* 44:671–682, 1958.

26. Cairns, J. The bacterial chromosome and its manner of replication as seen by autoradiography. *J. Mol. Biol.* 6:208–213, 1963.

Inman, R.B.; Schnös, M. Structure of branch points in replicating DNA: presence of single-stranded connections in lambda DNA branch parts. *J. Mol. Biol.* 56:319–325, 1971. (Electron microscopy demonstrates that the replication fork is asymmetric.)

Ogawa, T.; Okazaki, T. Discontinuous DNA replication. *Annu. Rev. Biochem.* 49:421–457, 1980.

27. Brutlag, D.; Kornberg, A. Enzymatic synthesis of deoxyribonucleic acid: a proofreading function for the 3′ to 5′ exonuclease activity in DNA polymerases. *J. Biol. Chem.* 247:241–248, 1972.

Fersht, A.R. Enzymatic editing mechanisms in protein synthesis and DNA replication. *Trends Biochem. Sci.* 5:262–265, 1980.

Topal, M.; Fresco, J.R. Complementary base pairing and the origin of substitution mutations. *Nature* 263:285–289, 1976.

28. Rowen, L.; Kornberg, A. Primase, the *dnaG* protein of *Escherichia coli*: an enzyme which starts DNA chains. *J. Biol. Chem.* 253:758–764, 1978.

29. Abdel-Monem, M.; Hoffmann-Berling, H. DNA unwinding enzymes. *Trends Biochem. Sci.* 5:128–130, 1980.

Alberts, B.M.; Frey, L. T4 bacteriophage gene 32: a structural protein in the replication and recombination of DNA. *Nature* 227:1313–1318, 1970.

Lohman, T.M.; Bujalowski, W.; Overman, L.B. *E. coli* single strand binding protein: a new look at helix-destabilizing proteins. *Trends Biochem. Sci.* 13:250–254, 1988.

30. Alberts, B.M. Protein machines mediate the basic genetic processes. *Trends Genet.* 1:26–30, 1985.

Kornberg, A. DNA replication. *J. Biol. Chem.* 263:1–4, 1988.

LeBowitz, J.; McMacken, R. The *E. coli* dnaB replication protein is a DNA helicase. *J. Biol. Chem.* 261:4738–4748, 1986.

31. Modrich, P. DNA mismatch correction. *Annu. Rev. Biochem.* 56:435–466, 1987.

Radman, M.; Wagner, R. The high fidelity of DNA duplication. *Sci. Am.* 259(2):40–47, 1988.

32. Bramhill, D.; Kornberg, A. Duplex opening by dnaA protein at novel sequences in initiation of replication at the origin of the *E. coli* chromosome. *Cell* 52:743–755, 1988.

Dodson, M.; et al. Specialized nucleoprotein structures at the origin of replication of bacteriophage lambda: localized unwinding of duplex DNA by a six-protein reaction. *Proc. Natl. Acad. Sci. USA* 83:7638–7642, 1986.

Zyskind, J.W.; Smith, D.W. The bacterial origin of replication, ori C. *Cell* 46:489–490, 1986.

33. DiNardo, S. Voelkel, K.; Sternglanz, R. DNA topoisomerase II mutant of *Saccharomyces cerevisiae:* topoisomerase II is required for segregation of daughter molecules at the termination of DNA replication. *Proc. Natl. Acad. Sci. USA* 81:2616–2620, 1984.

Wang, J.C. DNA topoisomerases. *Sci. Am.* 247(1):94–109, 1982.

Wang, J.C. DNA topoisomerases. *Annu. Rev. Biochem.* 54:665–698, 1985.

34. Campbell, J.L. Eucaryotic DNA replication. *Annu. Rev. Biochem.* 55:733–771, 1986.

35. Low, K.B., ed. The Recombination of Genetic Material. San Diego: Academic Press, 1988.

Stahl, F.W. Genetic recombination. *Sci. Am.* 256(2):90–101, 1987.

Whitehouse, H.L.K. Genetic Recombination: Understanding the Mechanisms. New York: Wiley, 1982.

36. Dressler, D.; Potter, H. Molecular mechanisms of genetic recombination. *Annu. Rev. Biochem.* 51:727–762, 1982.

Smith, G.R. Mechanism and control of homologous recombination in *Escherichia coli. Annu. Rev. Genet.* 21:179–201, 1987.

37. Wetmur, J.G.; Davidson, N. Kinetics of renaturation of DNA. *J. Mol. Biol.* 31:349–370, 1968.

38. Cox, M.M.; Lehman, I.R. Enzymes of general recombination. *Annu. Rev. Biochem.* 56:229–262, 1987.

Kowalczykowski, S.C. Mechanistic aspects of the DNA strand exchange activity of *E. coli* recA protein. *Trends Biochem. Sci.* 12:141–145, 1987.

Radding, C.M. Recombination activities of *E. coli* recA protein. *Cell* 25:3–4, 1981.

39. Holliday, R. A mechanism for gene conversion in fungi. *Genet. Res.* 5:282–304, 1964.

Meselson, M.S.; Radding, C.M. A general model for genetic recombination. *Proc. Natl. Acad. Sci. USA* 72:358–361, 1975.

Sigal, N.; Alberts, B. Genetic recombination: the nature of a crossed-strand exchange between two homologous DNA molecules. *J. Mol. Biol.* 71:789–793, 1972.

40. Fincham, J.R.S.; Day, P.R.; Radford, A. Fungal Genetics, 4th ed. Berkeley: University of California Press, 1979.

Kourilsky, P. Molecular mechanisms of gene conversion in higher cells. *Trends Genet.* 2:60–62, 1986.

41. Campbell, A. Some general questions about movable elements and their implications. *Cold Spring Harbor Symp. Quant. Biol.* 45:1–9, 1980.

Craig, N. The mechanism of conservative site-specific recombination. *Annu. Rev. Genet.* 22:77–105, 1988.

Grindley, N.D.; Reed, R.R. Transpositional recombination in prokaryotes. *Annu. Rev. Biochem.* 54:863–896, 1985.

Mizuuchi, K.; Craigie, R. Mechanism of bacteriophage mu transposition. *Annu. Rev. Genet.* 20:385–429, 1986.

42. Joklik, W.K., ed. Virology, 2nd ed. Norwalk, CT: Appleton & Lange, 1985.

Shapiro, J.A., ed. Mobile Genetic Elements. New York: Academic Press, 1983.

Watson, J.D.; Hopkins, N.H.; Roberts, J.W.; Steitz, J.A.; Weiner, A.M. Molecular Biology of the Gene, 4th ed. Menlo Park, CA: Benjamin-Cummings, 1987. (Chapter 24 provides a modern overview of eucaryotic viruses.)

43. Hershey, A.D.; Chase, M. Independent functions of viral protein and nucleic acid in growth of bacteriophage. *J. Gen. Physiol.* 36:39–56, 1952.

Luria, S.E.; Darnell, J.E.; Baltimore, D.; Campbell, A. General Virology, 3rd ed. New York: Wiley, 1978. (The introductory chapter provides a historical overview of the development of virology.)

44. Simons, K.; Garoff, H.; Helenius, A. How an animal virus gets in and out of its host cell. *Sci. Am.* 246(2):58–66, 1982.

45. Gierer, A.; Schramm, G. Infectivity of ribonucleic acid from tobacco mosaic virus. *Nature* 177:702–703, 1956.

46. Cohen, S.S. Virus-Induced Enzymes. New York: Columbia University Press, 1968.

47. Kornberg, A. DNA Replication. San Francisco: Freeman, 1980. (Chapters 14 and 15 provide up-to-date, beautifully illustrated descriptions of the varieties of ways phage and animal viral DNA's replicate.)

48. Campbell, A.M. Bacteriophage lambda as a model system. *Bioessays* 5:277–280, 1986.

Daniels, D.; Sanger, F.; Coulson, A.R. Features of bacteriophage lambda: analysis of the complete nucleotide sequence. *Cold Spring Harbor Symp. Quant. Biol.* 47:1009–1024, 1982.

Lwoff, A. Lysogeny. *Bacteriol. Rev.* 17:269–337, 1952.

49. Watson, J.D.; Hopkins, N.H.; Roberts, J.W.; Steitz, J.A.; Weiner, A.M. Molecular Biology of the Gene, 4th ed. Menlo Park, CA: Benjamin-Cummings, 1987. (Chapter 26 describes the viruses that can cause tumors.)

50. Baltimore, D. Viral RNA-dependent DNA polymerase. *Nature* 226:1209–1211, 1970.

Gallo, R.C. The AIDS virus. *Sci. Am.* 256(1):46–56, 1987.

Temin, H.M.; Mizutani S. RNA-dependent DNA polymerase in virions of Rous sarcoma virus. *Nature* 226:1211–1213, 1970.

Varmus, H. Retroviruses. *Science* 240:1427–1435, 1988.

51. Boeke, J.D.; Garfinkel, D.J.; Styles, C.A.; Fink, G.R. Ty elements transpose through an RNA intermediate. *Cell* 40:491–500, 1985.

Fink, G.R.; Boeke, J.D.; Garfinkel, D.J. The mechanisms and consequences of retrotransposition. *Trends Genet.* 2:118–123, 1986.

52. Cohen, S.N.; Shapiro, J.A. Transposable genetic elements. *Sci. Am.* 242(2):40–49, 1980.

Mizuuchi, K. Mechanism of transposition of bacteriophage Mu: polarity of the strand transfer reaction at the initiation of transposition. *Cell* 39:395–404, 1984.

53. Novick, R.P. Plasmids. *Sci. Am.* 243(6):102–127, 1980.

Reisner, D.; Gross, H.J. Viroids. *Annu. Rev. Biochem.* 54:531–564, 1985.

Rossmann, M.G. The evolution of RNA viruses. *Bioessays* 7:99–103, 1987.

54. Drlica, K. Understanding DNA and Gene Cloning. New York: Wiley, 1984.

Sambrook, J.; Fritsch, E.F.; Maniatis, T. Molecular Cloning: A Laboratory Manual, 2nd ed. Cold Spring Harbor, NY: Cold Spring Harbor Laboratory, 1989.

Watson, J.D.; Tooze, J.; Kurtz, D.T. Recombinant DNA: A Short Course. New York: W.H. Freeman, 1983.

55. Smith, H.O. Nucleotide sequence specificity of restriction endonucleases. *Science* 205:455–462, 1979.

56. Cohen, S.; Chang, A.; Boyer, H.; Helling, R. Construction of biologically functional bacterial plasmids *in vitro. Proc. Natl. Acad. Sci. USA* 70:3240–3244, 1973.

57. Maniatis, T.; et al. The isolation of structural genes from libraries of eucaryotic DNA. *Cell* 15:687–701, 1978.

Maniatis, T.; Kee, S.G.; Efstratiadis, A.; Kafatos, F.C. Amplification and characterization of a β-globin gene synthesized *in vitro. Cell* 8:163–182, 1976.

58. Hedrick, S.M.; Cohen, D.I.; Nielsen, E.A.; Davis, M.M. Isolation of cDNA clones encoding T cell-specific membrane-associated proteins. *Nature* 308:149–153, 1984.

59. Grunstein, M.; Hogness, D.S. Colony hybridization: a method for the isolation of cloned DNAs that contain a specific gene. *Proc. Natl. Acad. Sci. USA* 72:3961–3965, 1975.

Suggs, S.V.; Wallace, R.B.; Hirose, T.; Kawashima, E.H.; Itakura, K. Use of synthetic oligonucleotides as hybridization probes: isolation of cloned cDNA sequences for human β₂-microglobulin. *Proc. Natl. Acad. Sci. USA* 78:6613–6617, 1981.

60. Coulson, A.; Sulston, J.; Brenner, S.; Karn, J. Toward a physical map of the genome of the nematode *Caenorhabditis elegans. Proc. Natl. Acad. Sci. USA* 83:7821–7825, 1986.

Proustka, A.; Lehrach, H. Jumping libraries and linking libraries: the next generation of molecular tools in mammalian genetics. *Trends Genet.* 2:174–179, 1986.

61. Pelham, H.R.B.; Jackson, R.J. An efficient mRNA-dependent translation system from reticulocyte lysates. *Eur. J. Biochem.* 67:247–256, 1976.

62. Geysen, H.M.; et al. Chemistry of antibody binding to a protein. *Science* 235:1184–1190, 1987.

Gilbert, W.; Villa-Komaroff, L. Useful proteins from recombinant bacteria. *Sci. Am.* 242(4):74–94, 1980.

Yokata, T.; et al. Use of a cDNA expression vector for isolation of mouse interleukin 2 cDNA clones: expression of T-cell growth-factor activity after transfection of monkey cells. *Proc. Natl. Acad. Sci. USA* 82:68–72, 1985.

Young, R.A.; Davis, R.W. Efficient isolation of genes by using antibody probes. *Proc. Natl. Acad. Sci. USA* 80:1194–1198, 1980.

63. Smith, M. *In vitro* mutagenesis. *Annu. Rev. Genet.* 19:423–462, 1985.

64. Hinnen, A.; Hicks, J.B.; Fink, G.R. Transformation of yeast. *Proc. Natl. Acad. Sci. USA* 75:1929–1933, 1978.

Palmiter, R.D.; Brinster, R.L. Germ line transformation of mice. *Annu. Rev. Genet.* 20:465–499, 1986.

Spradling, A.C.; Rubin, G.M. Transposition of cloned P elements into *Drosophila* germ line chromosomes. *Science* 218:341–347, 1982.

Thomas, K.R.; Capecchi, M.R. Site-directed mutagenesis by gene targeting in mouse embryo-derived stem cells. *Cell* 51:503–512, 1987.

65. Marx, J.L. Multiplying genes by leaps and bounds. *Science* 240:1408–1410, 1988.

Saiki, R.K.; et al. Primer-directed enzymatic amplification of DNA with a thermostable DNA polymerase. *Science* 239:487–491, 1988.

66. Botstein, D.; White, R.L.; Skolnick, M.; Davis, R.W. Construction of a genetic linkage map in man using restriction fragment length polymorphisms. *Am. J. Hum. Genet.* 32:314–331, 1980.

Mapping and Sequencing the Human Genome. Washington, DC: National Academy Press, 1988.

White, R.; LaLouel, J.-M. Chromosome mapping with DNA markers. *Sci. Am.* 258(2):40–49, 1988.

The Plasma Membrane

A **plasma membrane** encloses every cell, defining the cell's extent and maintaining the essential differences between its contents and the environment. This membrane is a highly selective filter and a device for active transport; it controls the entry of nutrients and the exit of waste products, and it generates differences in ion concentrations between the interior and exterior of the cell. The plasma membrane also acts as a sensor of external signals, allowing the cell to change in response to environmental cues.

All biological membranes, including the plasma membrane and the internal membranes of eucaryotic cells, have a common general structure: they are assemblies of lipid and protein molecules held together mainly by noncovalent interactions. Cell membranes are dynamic, fluid structures, and most of their lipid and protein molecules are able to move about in the plane of the membrane. As shown in Figure 6–1, the lipid molecules are arranged as a continuous double layer about 5 nm thick. This *lipid bilayer* provides the basic structure of the membrane and serves as a relatively impermeable barrier to the passage of most water-soluble molecules. The protein molecules, usually "dissolved" in the lipid bilayer, mediate most of the other functions of the membrane. Some membrane proteins, for example, serve to transport specific molecules into or out of the cell, whereas others are enzymes that catalyze membrane-associated reactions. Other types of mem-

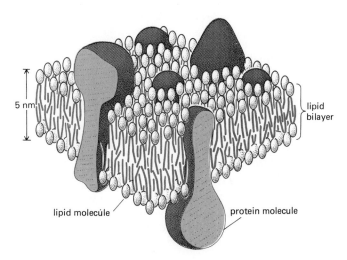

Figure 6–1 Three-dimensional view of a small section of a cell membrane.

5 nm

lipid bilayer

lipid molecule

protein molecule

brane proteins serve as structural links that connect the plasma membrane to the cytoskeleton and/or to either the extracellular matrix or an adjacent cell. Still others serve as receptors for receiving and transducing chemical signals in the cell's environment, and so on. As would be expected, membranes are asymmetrical structures: the lipid and protein compositions of the outside and inside faces differ from one another in ways that reflect the different functions performed at the two surfaces of the membrane.

Although the specific lipid and protein components vary greatly from one type of membrane to another, most of the basic structural and functional concepts discussed in this chapter are applicable to the various internal membranes in cells as well as to plasma membranes. We shall consider first the structure and organization of the main constituents of all biological membranes—the lipids, proteins, and carbohydrates. We shall then discuss the mechanisms by which cells transport small molecules across their plasma membranes as well as the very different mechanisms they use to import and export macromolecules and larger particles. Some additional functions of the plasma membrane will be considered in later chapters: for example, its role in cell adhesion is discussed in Chapter 14, and its role in cell signaling is discussed in Chapter 12.

The Lipid Bilayer[1]

The first indication that the lipid molecules in biological membranes are organized in a bilayer came from an elegantly simple experiment performed in 1925. Lipids from red blood cell membranes were extracted with acetone and floated on the surface of a water trough. The area they occupied was then decreased by means of a movable barrier until a continuous monomolecular film (a monolayer) was formed. This monolayer occupied a final area about twice the surface area of the original red blood cells. Because the only membrane in a red blood cell is the plasma membrane, the experimenters concluded that the lipid molecules in this membrane must be arranged largely as a continuous *bilayer*—a conclusion that had a profound influence on cell biology.

The **lipid bilayer** has since been firmly established as the basis for cell-membrane structure by more sophisticated methods. For example, x-ray diffraction analyses have demonstrated the existence of lipid bilayers in the highly ordered stacks of cell membrane that form the insulating myelin sheath surrounding nerve cell processes (see p. 1073). Evidence that all biological membranes contain a lipid bilayer is provided by freeze-fracture electron microscopy: all cell membranes can be split down the middle, between the two lipid monolayers, when they are frozen (see p. 291). The bilayer structure is attributable to the special properties of the lipid molecules, which cause them to assemble spontaneously into bilayers even in simple artificial conditions.

6-4 Membrane Lipids Are Amphipathic Molecules That Spontaneously Form Bilayers[2]

Lipid molecules are insoluble in water but dissolve readily in organic solvents. They constitute about 50% of the mass of most animal cell plasma membranes, nearly all of the remainder being protein. There are approximately 5×10^6 lipid molecules in a 1 μm × 1 μm area of lipid bilayer, or about 10^9 lipid molecules in the plasma membrane of a small animal cell. The three major types of lipids in cell membranes are *phospholipids* (the most abundant), *cholesterol*, and *glycolipids*. All three are *amphipathic* (or amphiphilic)—that is, they have a hydrophilic ("water-loving," or polar) end and a hydrophobic ("water-hating," or nonpolar) end. For example, the typical **phospholipid** molecule illustrated in Figure 6–2 has a *polar head group* and two *hydrophobic hydrocarbon tails*. The tails can differ in length (they normally contain between 14 and 24 carbon atoms), and one tail usually has one or more *cis*-double bonds (that is, it is *unsaturated*), while the other tail does not (that is, it is *saturated*). As indicated in Figure 6–2, each double

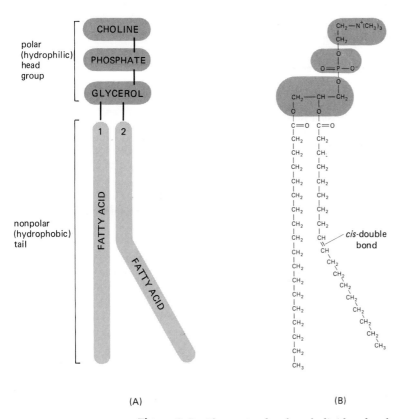

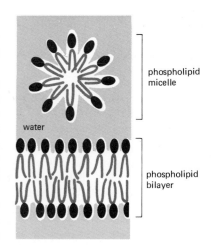

(D)

(A)   (B)   (C)

Figure 6–2 The parts of a phospholipid molecule, phosphatidylcholine, represented schematically (A), in formula (B), as a space-filling model (C), and as a symbol (D). In order to distinguish the unsaturated fatty acid chain from the saturated one, in this drawing (and those that follow) the unsaturated fatty acid chain is drawn with a distinct bend in it rather than with a small kink. Actually, only the double bond is rigid in the unsaturated fatty acid. Because the single carbon-carbon bonds in the rest of the chain are free to rotate, both the saturated and unsaturated fatty acid chains tend to pack in parallel arrays in each phospholipid monolayer.

bond creates a small kink in the tail. Differences in tail length and saturation are important because they influence the ability of phospholipid molecules to pack against one another and thereby affect the fluidity of the membrane (see below).

It is the amphipathic nature of the lipid molecules that causes them to form bilayers spontaneously in aqueous solution. When amphipathic molecules are surrounded on all sides by an aqueous environment, they tend to aggregate so that their hydrophobic tails are buried in the interior and their hydrophilic heads are exposed to water. They can do this in either of two ways: they can form spherical *micelles*, with the tails inward, or they can form bimolecular sheets, or *bilayers*, with the hydrophobic tails sandwiched between the hydrophilic head groups. These two possibilities are illustrated in Figure 6–3.

Most phospholipids and glycolipids form bilayers spontaneously in aqueous environments. Moreover, these lipid bilayers tend to close on themselves to form sealed compartments, thereby eliminating free edges where the hydrophobic tails would be in contact with water. For the same reason, compartments formed by lipid bilayers tend to reseal themselves when they are torn. Besides its self-sealing properties, a lipid bilayer has other characteristics that make it an ideal structure for cell membranes. One of the most important of these is its fluidity, which, as we shall see, is crucial to many of its functions.

Figure 6–3 A phospholipid micelle and a phospholipid bilayer seen in cross-section. Phospholipid molecules form such structures spontaneously in water.

6-6 The Lipid Bilayer Is a Two-dimensional Fluid[3]

Surprisingly, it was only in the early 1970s that researchers first recognized that individual lipid molecules are able to diffuse freely within lipid bilayers. The initial demonstration came from studies of synthetic lipid bilayers. Two types of synthetic bilayers have been very useful in experimental studies: (1) bilayers made in the form of spherical vesicles, called **liposomes,** which can vary in size from about 25 nm to 1 μm in diameter depending on how they are produced (Figure 6–4); and (2) planar bilayers, called **black membranes,** formed across a hole in a partition between two aqueous compartments (Figure 6–5).

A variety of techniques have been used to measure the motion of individual lipid molecules and of their different parts. For example, one can construct a lipid molecule whose polar head group carries a "spin label," such as a nitroxyl group ($>$N—O); this contains an unpaired electron whose spin creates a paramagnetic signal that can be detected by electron spin resonance (ESR) spectroscopy. (The principles of this technique are similar to those of nuclear magnetic resonance; see p. 153.) The motion and orientation of a spin-labeled lipid in a bilayer can be deduced from the ESR spectrum. Such studies show that lipid molecules in synthetic bilayers very rarely migrate from the monolayer on one side to that on the other; this process, called "flip-flop," occurs less often than once a month for any individual lipid molecule (Figure 6–6). On the other hand, lipid molecules readily exchange places with their neighbors *within* a monolayer ($\sim 10^7$ times a second). This gives rise to a rapid lateral diffusion, with a diffusion coefficient (D) of about 10^{-8} cm^2/sec, which means that an average lipid molecule diffuses the length of a large bacterial cell (~ 2 μm) in about 1 second. These studies have also shown that individual lipid molecules rotate very rapidly about their long axes and that their hydrocarbon chains are flexible, the greatest degree of flexion occurring near the center of the bilayer and the smallest adjacent to the polar head group (see Figure 6–6).

Similar studies have been carried out with labeled lipid molecules in isolated biological membranes and in relatively simple whole cells such as mycoplasma, bacteria, and nonnucleated red blood cells. The results are generally the same as for synthetic bilayers, and they demonstrate that the lipid component of a biological membrane is a two-dimensional liquid in which the constituent molecules are free to move laterally. As in synthetic bilayers, individual lipid molecules are normally confined to their own monolayer. There is an exception, however: in membranes where lipids are actively synthesized, such as the endoplasmic reticulum, there must be a rapid flip-flop of specific lipids across the bilayer, and membrane-bound enzymes called *phospholipid translocators* (see p. 449) are present to catalyze it.

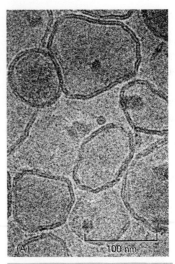

(A)   100 nm

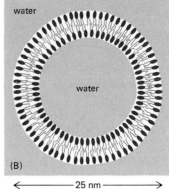

water

water

(B)

←———— 25 nm ————→

Figure 6–4   (A) An electron micrograph of unfixed, unstained phospholipid vesicles (liposomes) in water. Note that the bilayer structure of the vesicles is readily apparent. (B) A drawing of a small spherical liposome seen in cross-section. Liposomes are commonly used as model membranes in experimental studies. (A, courtesy of Jean Lepault.)

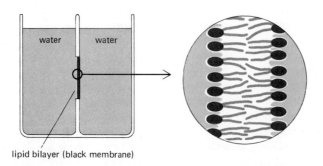

water   water

lipid bilayer (black membrane)

Figure 6–5   A cross-sectional view of a synthetic lipid bilayer, called a black membrane, formed across a small hole in a partition separating two aqueous compartments. Black membranes are used to measure the permeability properties of artificial membranes.

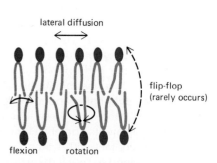

lateral diffusion

flip-flop
(rarely occurs)

flexion   rotation

Figure 6–6   The types of movement possible for phospholipid molecules in a lipid bilayer.

The Fluidity of a Lipid Bilayer Depends on Its Composition[4]

A synthetic lipid bilayer made from a single type of phospholipid changes from a liquid state to a rigid crystalline (or gel) state at a characteristic freezing point. This change of state is called a *phase transition*, and the temperature at which it occurs is lower (that is, the membrane becomes more difficult to freeze) if the hydrocarbon chains are short or have double bonds. A shorter chain length reduces the tendency of the hydrocarbon tails to interact with one another, and *cis*-double bonds produce kinks in the hydrocarbon chains that make them more difficult to pack together (Figure 6–7).

In synthetic bilayers containing a mixture of phospholipids with varying degrees of saturation (and therefore different phase-transition points), *phase separations* can occur: when their freezing points are reached, individual phospholipid molecules of the same type aggregate spontaneously within the bilayer to form frozen patches. In biological membranes, saturated and unsaturated fatty acid chains are usually bonded together in the same lipid molecule (that is, one chain is unsaturated while the other is not), so that phase separations of this kind are unlikely to occur.

Another determinant of membrane fluidity is **cholesterol.** Eucaryotic plasma membranes contain large amounts of cholesterol, up to one molecule for every phospholipid molecule. Cholesterol molecules orient themselves in the bilayer with their hydroxyl groups close to the polar head groups of the phospholipid molecules; their rigid, platelike steroid rings interact with—and partly immobilize—those regions of the hydrocarbon chains that are closest to the polar head groups, leaving the rest of the chain flexible (Figure 6–8). Although cholesterol tends in this way to make lipid bilayers less fluid, at the high concentrations found in most eucaryotic plasma membranes it also prevents the hydrocarbon chains from coming together and crystallizing. In this way cholesterol inhibits possible phase transitions.

In addition to affecting fluidity, cholesterol decreases the permeability of lipid bilayers to small water-soluble molecules and is thought to enhance both the flexibility and the mechanical stability of the bilayer. Changes in membrane shape that would require the two sides of the lipid bilayer to compress or expand to very different degrees are facilitated because cholesterol, unlike phospholipids, can readily redistribute (flip-flop) between the two monolayers in response to such forces. The flip-flop of cholesterol molecules occurs with a low energy barrier, and is therefore rapid, because its small polar head group (a hydroxyl group) can pass relatively easily through the center of the bilayer.

The importance of cholesterol in maintaining the mechanical stability of higher eucaryotic cell membranes is suggested by mutant animal cell lines that are unable to synthesize cholesterol. Such cells rapidly lyse (break open and release their

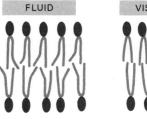
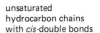

unsaturated
hydrocarbon chains
with *cis*-double bonds

saturated straight
hydrocarbon chains

Figure 6–7 Double bonds in unsaturated hydrocarbon chains increase the fluidity of a phospholipid bilayer by making it more difficult to pack the chains together.

Figure 6–8 Cholesterol represented by a formula (A) and a schematic drawing (B), and depicted interacting with two phospholipid molecules in a monolayer (C).

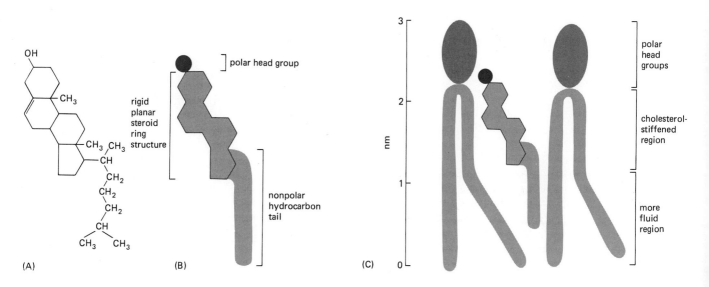

contents) unless cholesterol is added to the culture fluid. The added cholesterol is incorporated into the plasma membrane, thereby enabling the cells to survive.

The precise fluidity of the plasma membrane must be biologically important, since bacteria, yeast, and other poikilothermic organisms, whose temperatures fluctuate with that of their environment, change the fatty acid composition of their plasma membrane so as to maintain a relatively constant fluidity. As the temperature falls, for example, fatty acids with more *cis*-double bonds are synthesized, so that the decrease in bilayer fluidity that would otherwise result from the drop in temperature is avoided. Certain membrane transport processes and enzyme activities can be shown to cease when the bilayer viscosity increases beyond a threshold level.

The Lipid Bilayer Serves as a Solvent for Membrane Proteins[5]

The lipid compositions of several biological membranes are compared in Table 6–1. Note that bacterial plasma membranes are often composed of one main type of phospholipid and contain no cholesterol; in the absence of cholesterol, mechanical stability is provided by the overlying cell wall (see Figure 6–54, p. 312). The plasma membranes of most eucaryotic cells, however, contain not only large amounts of cholesterol but also a variety of phospholipids. The plasma membrane of many mammalian cells, for example, contains four major phospholipids—*phosphatidylcholine, sphingomyelin, phosphatidylserine,* and *phosphatidylethanolamine*. The structures of these molecules are shown in Figure 6–9. Note that only phosphatidylserine carries a net negative charge; the other three are electrically neutral at physiological pH, carrying one positive and one negative charge. Together these four phospholipids constitute more than half the mass of lipid in most membranes (Table 6–1). Other phospholipids, such as the *inositol phospholipids* (see p. 702), are functionally important but present in relatively small quantities. The crucial role of the inositol phospholipids in cell signaling is discussed in Chapter 12.

One may wonder why the eucaryotic plasma membrane contains such a variety of phospholipids, with head groups that differ in size, shape, and charge. In seeking an answer it may be helpful to think of the membrane lipids as constituting a two-dimensional solvent for the proteins in the membrane, just as water constitutes a three-dimensional solvent for proteins in an aqueous solution. It may be that some membrane proteins can function only in the presence of specific phospholipid head groups, just as many enzymes in aqueous solution require a

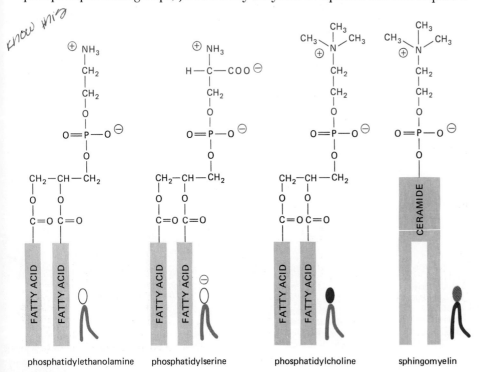

phosphatidylethanolamine   phosphatidylserine   phosphatidylcholine   sphingomyelin

Figure 6–9 Formulas and symbols of four major phospholipids in mammalian plasma membranes. Note that different head groups are represented by different symbols in this figure and the next. All of the lipid molecules shown are derived from glycerol except for sphingomyelin, which is derived from serine. The structure and formation of ceramide are shown in Figure 6–11.

Table 6–1 Approximate Lipid Compositions of Different Cell Membranes

Lipid	Percentage of Total Lipid by Weight					
	Liver Plasma Membrane	Erythrocyte Plasma Membrane	Myelin	Mitochondrion (inner and outer membranes)	Endoplasmic Reticulum	E. coli
Cholesterol	17	23	22	3	6	0
Phosphatidyl-ethanolamine	7	18	15	35	17	70
Phosphatidylserine	4	7	9	2	5	trace
Phosphatidyl-choline	24	17	10	39	40	0
Sphingomyelin	19	18	8	0	5	0
Glycolipids	7	3	28	trace	trace	0
Others	22	13	8	21	27	30

particular ion for activity. Consistent with this view is the finding that some membrane proteins, when inserted into a synthetic lipid bilayer, will function optimally only if the bilayer contains certain specific phospholipids.

6-5 The Lipid Bilayer Is Asymmetrical[6]
6-6

The lipid compositions of the two halves of the lipid bilayer in those membranes that have been analyzed are strikingly different. In the human red blood cell membrane, for example, almost all of the lipid molecules that have choline—$(CH_3)_3N^+CH_2CH_2OH$—in their head group (that is, phosphatidylcholine and sphingomyelin) are in the outer half of the lipid bilayer, while almost all of the phospholipid molecules that contain a terminal primary amino group (phosphatidylethanolamine and phosphatidylserine) are in the inner half (Figure 6–10). The fatty acid tails of phosphatidylcholine and sphingomyelin are more saturated than those of phosphatidylethanolamine and phosphatidylserine; therefore, the asymmetry in the distribution of the head groups is accompanied by an asymmetry in the distribution of hydrocarbon tails that may make the inner monolayer somewhat more fluid than the outer monolayer. In addition, because the negatively charged phosphatidylserine is located in the inner monolayer, there is a significant difference in charge between the two halves of the bilayer.

Most of the membranes in a eucaryotic cell, including the plasma membrane, are synthesized in the endoplasmic reticulum (ER), and it is here that the phospholipid asymmetry is generated by the phospholipid translocators in the ER that move specific phospholipid molecules from one monolayer to the other (see p. 449). Although the function of the lipid asymmetry is largely unknown, there are membrane-bound enzymes that make use of it: when *protein kinase C* is activated, for example, it is bound to the cytoplasmic face of the plasma membrane, where phosphatidylserine is concentrated, and it requires this negatively

EXTRACELLULAR SPACE

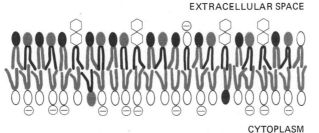

CYTOPLASM

Figure 6–10 The asymmetrical distribution of phospholipids and glycolipids in the lipid bilayer of human red blood cells. The symbols used for the phospholipids are those introduced in Figure 6–9. Glycolipids are drawn with hexagonal polar head groups. Cholesterol (not shown) is thought to be distributed about equally in both monolayers.

charged phospholipid in order to act (see p. 704). The importance of having specific inositol phospholipids concentrated in the cytoplasmic leaflet of the plasma membrane lipid bilayer, where they can be used to generate intracellular mediators in response to extracellular signals, is discussed in Chapter 12 (see p. 702).

Glycolipids Are Found on the Surface of All Plasma Membranes, but Their Function Is Unknown[7]

The lipid molecules that show the most striking and consistent asymmetry in distribution in the plasma membranes of animal cells are the oligosaccharide-containing lipid molecules called **glycolipids.** These intriguing molecules are found only in the outer half of the bilayer, and their sugar groups are exposed at the cell surface (see Figure 6–10), suggesting some role in interactions of the cell with its surroundings. The asymmetric distribution of glycolipids in the bilayer derives from the addition of sugar groups to the lipid molecules in the lumen of the Golgi apparatus, which is topologically equivalent to the exterior of the cell (see p. 408).

Glycolipids probably occur in all animal cell plasma membranes, and they generally constitute about 5% of the lipid molecules in the outer monolayer. They differ remarkably from one animal species to another and even among tissues in the same species. In bacteria and plants almost all glycolipids are derived from glycerol-based lipids, as is the common phospholipid phosphatidylcholine. In animal cells, however, they are almost always produced from *ceramide*, as is the phospholipid sphingomyelin (see Figure 6–9). These *glycosphingolipids* have a general structure that is similar to that of the glycerol-based lipids, having a polar head group and two hydrophobic fatty acid chains. However, one of the fatty acid chains is initially coupled to serine to form the amino alcohol *sphingosine*, to which the second fatty acid chain is then linked to form ceramide (Figure 6–11).

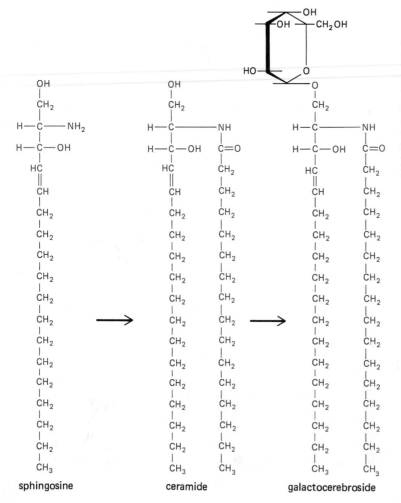

sphingosine   ceramide   galactocerebroside

Figure 6–11 Final steps in the synthesis of the simple glycosphingolipid galactocerebroside. Sphingosine is formed by condensing the amino acid serine with one fatty acid; a second fatty acid is then added to form ceramide, as shown. Ceramide is made in the endoplasmic reticulum, and the carbohydrate is added in the Golgi apparatus (see p. 449). Ceramide is also used to form the major phospholipid sphingomyelin (see Figure 6–9).

Glycolipids are distinguished from one another by their polar head group, which consists of one or more sugar residues. Among the most widely distributed glycolipids in the plasma membranes of both eucaryotic and procaryotic cells are the **neutral glycolipids,** whose polar head groups consist of anywhere from 1 to 15 or more neutral (uncharged) sugars, depending on the organism and cell type. One example is *galactocerebroside,* one of the simplest glycolipids, which has only galactose as its polar head group (see Figure 6–11). It is the main glycolipid in *myelin,* which consists of many concentric layers of plasma membrane wound around a nerve cell process (an axon) by a specialized myelinating cell (see p. 1073). The myelinating cells can be distinguished by the large amount of galactocerebroside in their plasma membrane, where it constitutes almost 40% of the outer monolayer. Galactocerebroside is largely absent from the membranes of most other cells, and it is thought to play an important part in the specific interaction between the myelinating cell and the axon.

The most complex of the glycolipids, the **gangliosides,** contain one or more sialic acid residues (also known as *N*-acetylneuraminic acid, or NANA), which gives them a net negative charge (Figure 6–12). Gangliosides are most abundant in the plasma membrane of nerve cells, where they constitute 5–10% of the total lipid mass, although they are found in much smaller quantities in most cell types. So far more than 40 different gangliosides have been identified. Some common examples are shown in Figure 6–13, where the nomenclature used to describe them is also introduced.

There are only hints as to what the functions of glycolipids might be. For example, the ganglioside G_{M1} (Figure 6–13) acts as a cell-surface receptor for the bacterial toxin that causes the debilitating diarrhea of cholera; cholera toxin binds to and enters only those cells with G_{M1} on their surface, including intestinal epithelial cells. The entry of cholera toxin into the cell leads to a prolonged increase in the concentration of intracellular cyclic AMP, which in turn causes a large efflux of Na^+ and water into the intestine (see p. 697). Although binding bacterial toxins cannot be the *normal* function of gangliosides, such observations suggest that these glycolipids may also serve as receptors for normal signaling between cells.

Summary

Biological membranes consist of a continuous double layer of lipid molecules in which various membrane proteins are embedded. This lipid bilayer is fluid, with individual lipid molecules able to diffuse rapidly within their own monolayer. However, most types of lipid molecules very rarely flip-flop spontaneously from one

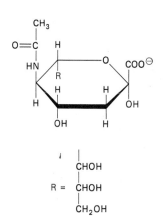

Figure 6–12 The structure of sialic acid (*N*-acetylneuraminic acid, or NANA). In cells this acid exists in its ionized form (—COO⁻), as shown.

Figure 6–13 Some representative gangliosides with their standard designations. In G_{M1}, G_{M2}, G_{M3}, G_{D1}, and G_{T1}, the letters M, D, and T refer to the number of sialic acid residues (mono, di, and tri, respectively), while the number that follows the letter is determined by subtracting the number of uncharged sugar residues from 5. NANA = *N*-acetylneuraminic (sialic) acid; Gal = galactose; Glc = glucose; GalNAc = *N*-acetylgalactosamine. Gal, Glc, and GalNAc are all uncharged, whereas NANA carries a negative charge (see Figure 6–12).

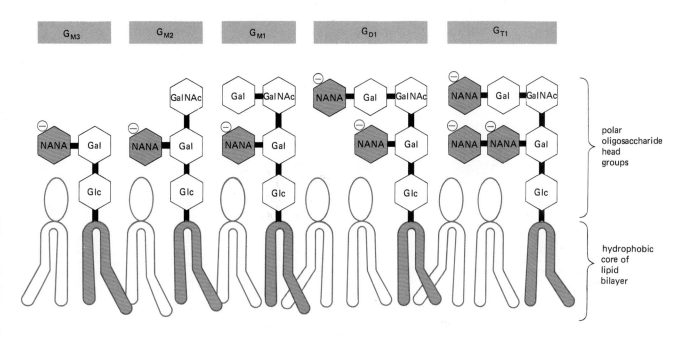

monolayer to the other. Membrane lipid molecules are amphipathic and assemble spontaneously into bilayers when placed in water; the bilayers form sealed compartments that reseal if torn. There are three major classes of lipid molecules in the plasma membrane bilayer—phospholipids, cholesterol, and glycolipids—and the lipid compositions of the inner and outer monolayers are different. Different mixtures of lipids are found in the plasma membranes of cells of different types, as well as in the various internal membranes of a single eucaryotic cell. For the most part the functional significance of the different lipid compositions of different membranes is unknown.

Membrane Proteins

Although the basic structure of biological membranes is provided by the lipid bilayer, most of the specific functions are carried out by proteins. Accordingly, the amounts and types of proteins in a membrane are highly variable: in the myelin membrane, which serves mainly to insulate nerve cell axons, less than 25% of the membrane mass is protein, whereas in the membranes involved in energy transduction (such as the internal membranes of mitochondria and chloroplasts), approximately 75% is protein; the usual plasma membrane is somewhere in between, with about 50% of the mass being protein. Because lipid molecules are small in comparison to protein molecules, there are always many more lipid molecules than protein molecules in membranes—about 50 lipid molecules for each protein molecule in a membrane that is 50% protein by mass.

6-10 The Polypeptide Chain of Many Membrane Proteins
6-11 Crosses the Lipid Bilayer One or More Times[8]

Many membrane proteins extend across the lipid bilayer (examples 1 and 2 in Figure 6–14). Like their lipid neighbors, these so-called **transmembrane proteins** are amphipathic: they have hydrophobic regions that pass through the membrane and interact with the hydrophobic tails of the lipid molecules in the interior of the bilayer and hydrophilic regions that are exposed to water on both sides of the membrane. The hydrophobicity of some of these membrane proteins is increased by the covalent attachment of a fatty acid chain that is inserted in the cytoplasmic leaflet of the bilayer (see example 1 in Figure 6–14). Some intracellular membrane proteins are associated with the bilayer only by means of such a fatty acid chain (see example 3 in Figure 6–14 and p. 417), while some cell-surface proteins are attached to the bilayer only by a covalent linkage (via a specific oligosaccharide) to phosphatidylinositol, a minor phospholipid, in the outer lipid monolayer of the plasma membrane (see example 4 in Figure 6–14 and p. 448).

Figure 6–14 Five ways in which membrane proteins can be associated with the lipid bilayer. Transmembrane proteins extend across the bilayer as a single α helix (1) or as multiple α helices (2); some of these "single-pass" and "multipass" proteins have a covalently attached fatty acid chain inserted in the cytoplasmic monolayer (1). Other membrane proteins are attached to the bilayer solely by a covalently attached lipid—either a fatty acid chain in the cytoplasmic monolayer (3) or, less often, via an oligosaccharide, to a minor phospholipid, phosphatidylinositol, in the noncytoplasmic monolayer (4). Finally, many proteins are attached to the membrane only by noncovalent interactions with other membrane proteins (5). The details of how membrane proteins become associated with the lipid bilayer in these ways are discussed in Chapter 8.

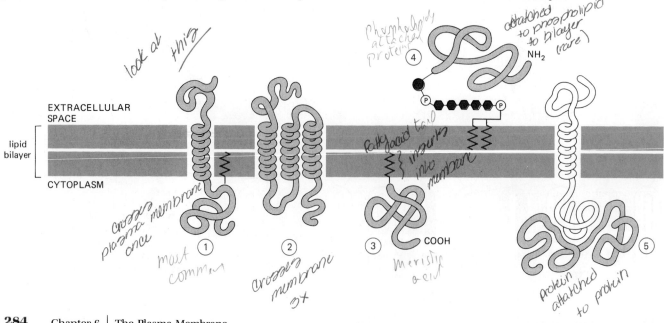

EXTRACELLULAR SPACE

lipid bilayer

CYTOPLASM

COOH

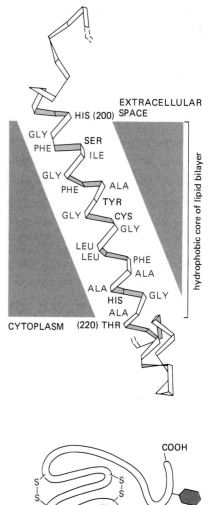

Figure 6–15 A segment of a transmembrane polypeptide chain crossing the lipid bilayer as an α helix, as determined by x-ray diffraction analyses of crystals of a membrane protein. Only the α-carbon backbone of the polypeptide chain is shown, and the hydrophobic amino acids are indicated in color. The protruding nonpolar side chains of the amino acid residues (not shown) interact with the hydrophobic fatty acid chains in the interior of the lipid bilayer, while the polar peptide bonds shield themselves from the hydrophobic environment of the bilayer by forming hydrogen bonds with one another (not shown). The polypeptide segment shown is part of the bacterial photosynthetic reaction center illustrated in Figure 6–32, p. 294. (Based on data from J. Deisenhofer *et al.*, *Nature* 318:618–624, 1985, and H. Michel *et al.*, *EMBO J.* 5:1149–1158, 1986.)

Other proteins associated with membranes do not extend into the hydrophobic interior of the lipid bilayer at all but are bound to one or other face of the membrane by noncovalent interactions with other membrane proteins (see example 5 in Figure 6–14). Many of these can be released from the membrane by relatively gentle extraction procedures, such as exposure to solutions of very high or low ionic strength or extreme pH, which interfere with protein-protein interactions but leave the lipid bilayer intact; these proteins are referred to operationally as **peripheral membrane proteins.** By contrast, transmembrane proteins, proteins linked to phosphatidylinositol, and some proteins held in the bilayer by a fatty acid chain, as well as some other tightly bound proteins, can be released only by disrupting the bilayer with detergents or organic solvents and are called **integral membrane proteins.**

In transmembrane proteins the parts of the polypeptide chain that are buried in the hydrophobic environment of the lipid bilayer are composed largely of amino acid residues with nonpolar side chains. But the peptide bonds themselves are polar, and because water is absent, all peptide bonds in the bilayer are driven to form hydrogen bonds with one another (see Figure 3–26, p. 110). The hydrogen bonding between peptide bonds is maximized if the polypeptide chain forms a regular α helix as it crosses the bilayer, and this is how most polypeptide chains traverse the membrane (Figure 6–15). In those cases where multiple strands of the polypeptide chain cross the bilayer, peptide bonds could, in principle, satisfy their hydrogen-bonding requirements if the strands were arranged as a β sheet (see p. 109). However, in such "multipass" membrane proteins the polypeptide chain usually crosses the bilayer as a series of α helices rather than as a β sheet (see example 2 in Figure 6–14). The strong drive to maximize hydrogen bonding in the absence of water also means that a polypeptide chain that enters the bilayer is likely to pass entirely through it before changing direction, since chain bending requires a loss of regular hydrogen-bonding interactions. Probably for this reason, there is still no established example of a membrane protein in which the polypeptide chain extends only partway across the lipid bilayer.

A transmembrane protein always has a unique orientation in the membrane. This reflects both the asymmetrical manner in which it is synthesized and inserted into the lipid bilayer in the endoplasmic reticulum (see p. 441) and the different functions of its cytoplasmic and extracellular domains. The great majority of transmembrane proteins are glycosylated. As in the case of glycolipids, the oligosaccharide chains are always present on the extracellular side of the membrane, since the sugar residues are added in the lumen of the endoplasmic reticulum and Golgi apparatus (see p. 454). A further asymmetry involves protein sulfhydryl (SH) groups, which remain reduced as cysteines in cytoplasmic domains but often participate in the formation of inter- or intrachain disulfide (S—S) bonds in extracellular domains (Figure 6–16).

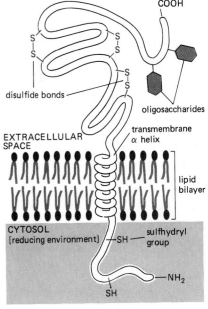

Figure 6–16 A typical "single-pass" transmembrane glycoprotein. Note that the polypeptide chain traverses the lipid bilayer as a right-handed α helix and that the oligosaccharide chains and disulfide bonds are all on the outside of the cell. Disulfide bonds do not form between the sulfhydryl groups on the cytoplasmic domain of the protein because of the reducing environment in the cytosol (see p. 445).

Membrane Proteins Can Be Solubilized and Purified in Detergents[9]

In general, transmembrane proteins (and some other tightly bound membrane proteins) can be solubilized only by agents that disrupt hydrophobic associations and destroy the bilayer. The most useful among these for the membrane biochemist are **detergents,** which are small amphipathic molecules that tend to form micelles in water (Figure 6–17). When mixed with membranes, the hydrophobic ends of detergents bind to the hydrophobic regions of the membrane proteins, thereby displacing the lipid molecules. Since the other end of the detergent molecule is polar, this binding tends to bring the membrane proteins into solution as detergent-protein complexes (although some tightly bound lipid molecules also remain) (Figure 6–18). The polar ends of detergents can be either charged (ionic), as in the case of *sodium dodecyl sulfate (SDS)*, or uncharged (nonionic), as in the case of the *Triton* detergents. The structures of these two common detergents are illustrated in Figure 6–19.

With strong ionic detergents such as SDS, even the most hydrophobic membrane proteins can be solubilized. This allows them to be analyzed by *SDS polyacrylamide-gel electrophoresis* (see p. 169), a procedure that has revolutionized the study of membrane proteins. Such strong detergents unfold (denature) proteins by binding to their internal "hydrophobic cores" (see Figure 3–22, p. 107), thereby rendering them inactive and unusable for functional studies. Nonetheless, proteins can be readily purified in their SDS-denatured form, and in some cases the purified protein can be renatured, with recovery of functional activity, by removing the detergent.

Less hydrophobic membrane proteins can be solubilized by a low concentration of a mild detergent, enabling the solubilized protein to be purified in an active, if not entirely normal, form. If the detergent is then removed in the absence of phospholipid, the membrane protein molecules usually aggregate and precipitate

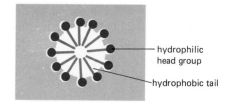

Figure 6–17 A detergent micelle in water, shown in cross-section. Because they have both polar and nonpolar ends, detergent molecules are amphipathic.

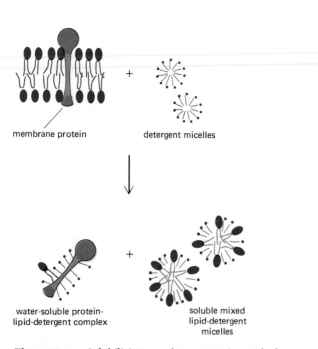

Figure 6–18 Solubilizing membrane proteins with detergent. The detergent disrupts the lipid bilayer and brings the proteins into solution as protein-lipid-detergent complexes. The phospholipids in the membrane are also solubilized by the detergent.

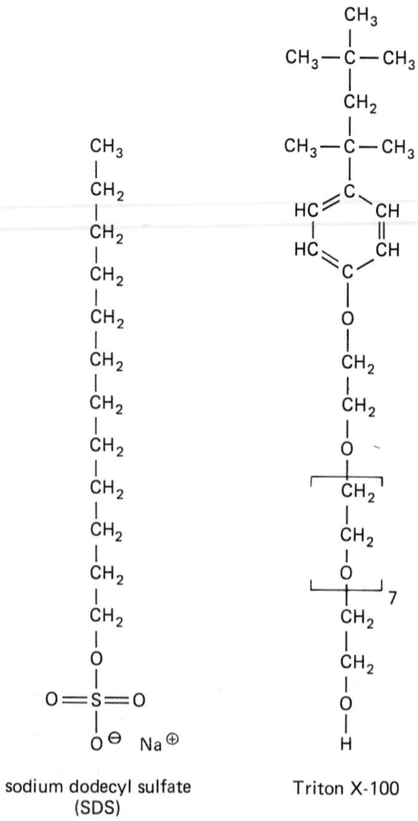

Figure 6–19 The structure of two commonly used detergents: sodium dodecyl sulfate (SDS), an anionic detergent, and Triton X-100, a nonionic detergent. Note that the bracketed portion of Triton X-100 is repeated seven times.

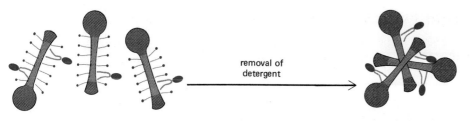

Figure 6–20 When the detergent is removed from detergent-solubilized membrane proteins, the relatively naked protein molecules (with only a few attached lipid molecules) tend to bury their hydrophobic regions by clustering together, forming large aggregates that precipitate from solution.

water-soluble protein-lipid-detergent complexes

aggregated proteins

out of solution (Figure 6–20). If, however, the purified protein is mixed with phospholipids before the detergent is removed, the active protein will usually insert into the lipid bilayer formed by the phospholipids (Figure 6–21). In this way functionally active membrane protein systems can be reconstituted from purified components, providing a powerful means to analyze their activities. For example, if a purified protein can be shown to pump ions across a synthetic lipid bilayer in the absence of other proteins, then that protein is unequivocally an ion pump; moreover, by controlling the access of the protein to ATP and ions, its detailed mechanism of action can be dissected (see p. 304).

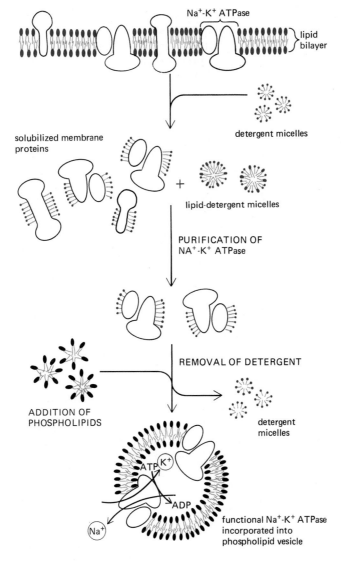

Na$^+$-K$^+$ ATPase

lipid bilayer

solubilized membrane proteins

detergent micelles

+

lipid-detergent micelles

PURIFICATION OF NA$^+$-K$^+$ ATPase

ADDITION OF PHOSPHOLIPIDS

REMOVAL OF DETERGENT

detergent micelles

ATP K$^+$

ADP

Na$^+$

functional Na$^+$-K$^+$ ATPase incorporated into phospholipid vesicle

Figure 6–21 Solubilizing, purifying, and reconstituting functional Na$^+$-K$^+$ ATPase molecules into phospholipid vesicles. The Na$^+$-K$^+$ ATPase is an ion pump that is present in the plasma membrane of most animal cells; it uses the energy of ATP hydrolysis to pump Na$^+$ out of the cell and K$^+$ in. The reconstitution is carried out in the presence of high Na$^+$ and ATP so that these molecules end up in sufficiently high concentration inside the vesicles to allow the ATPase to function as a pump. The detergent is generally removed either by prolonged dialysis or by any of various forms of chromatography.

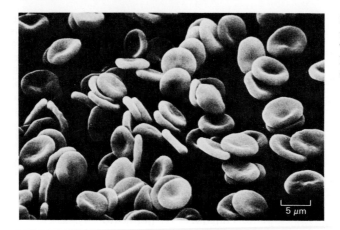

Figure 6–22 A scanning electron micrograph of human red blood cells. The cells have a biconcave shape and lack nuclei. (Courtesy of Bernadette Chailley.)

6-12 The Cytoplasmic Side of Membrane Proteins Can Be Studied in Red Blood Cell Ghosts[10]

More is known about the plasma membrane of the human red blood cell (Figure 6–22) than about any other eucaryotic membrane. There are a number of reasons for this: (1) Red blood cells are available in large numbers (from blood banks, for example) relatively uncontaminated by other cell types. (2) Since these cells have no nucleus or internal organelles, the plasma membrane is their only membrane, and it can be isolated without contamination by internal membranes (thus avoiding a serious problem encountered in plasma membrane preparations from other cell types, in which the plasma membrane typically constitutes less than 5% of the total membrane [see Table 8–2, p. 408]). (3) It is easy to prepare empty red blood cell membranes, or "ghosts," by exposing the cells to a hypotonic salt solution. Because the solution has a lower salt concentration than the cell interior, water flows into the red cells (see p. 308), causing them to swell and burst (lyse) and release their hemoglobin (the major nonmembrane protein). (4) Membrane ghosts can be studied while they are still leaky (in which case any reagent can interact with molecules on both faces of the membrane), or they can be allowed to reseal so that water-soluble reagents cannot reach the internal face. Moreover, since sealed *inside-out* vesicles can also be prepared from red blood cell ghosts (Figure 6–23), the external side and internal (cytoplasmic) side of the membrane can be studied separately. The use of sealed and unsealed red cell ghosts first made it possible to demonstrate that some membrane proteins extend all the way through the lipid bilayer (see below) and that the lipid compositions of the two halves of the bilayer are different. Like most of the basic principles initially demonstrated in red blood cell membranes, these findings have since been extended to the membranes of nucleated cells.

The "sidedness" of a membrane protein can be determined in several ways. One is to use a covalent labeling reagent (for example, one carrying a radioactive or fluorescent marker) that is water soluble and therefore cannot penetrate the lipid bilayer; it attaches covalently to specific groups only on the exposed side of

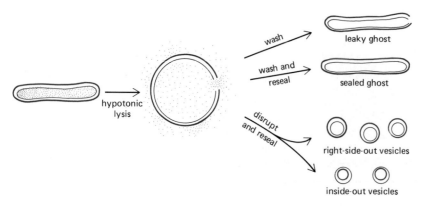

Figure 6–23 The preparation of sealed and unsealed red blood cell ghosts and of right-side-out and inside-out vesicles. As indicated, there is evidence that the red cells rupture in only one place, giving rise to ghosts with a single hole in them. The smaller vesicles are produced by mechanically disrupting the ghosts; the orientation of the membrane in these vesicles can be either right-side-out or inside-out, depending on the ionic conditions used during the disruption procedure.

the membrane. The membranes are then solubilized, the proteins separated by SDS polyacrylamide-gel electrophoresis, and the labeled proteins detected either by their radioactivity (by autoradiography of the gel) or by their fluorescence (by exposing the gel to ultraviolet light). By using such *vectorial labeling* it is possible to determine how a particular protein, detected as a band on a gel, is oriented in the membrane: if it is labeled from both the external side (when intact cells or sealed ghosts are labeled) and the internal (cytoplasmic) side (when sealed inside-out vesicles are labeled), then it must be a transmembrane protein. An alternative approach is to expose either the external or internal surface to membrane-impermeant proteolytic enzymes: if a protein is partially digested from both surfaces, it must be a transmembrane protein. In addition, labeled antibodies can be used to determine if a specific part of a transmembrane protein is exposed on one side of the membrane or the other.

When the plasma membrane proteins of the human red blood cell are studied by SDS polyacrylamide-gel electrophoresis, approximately 15 major protein bands are detected, varying in molecular weight from 15,000 to 250,000. Three of these proteins—*spectrin, glycophorin,* and *band 3*—account for more than 60% (by weight) of the total membrane protein (Figure 6–24). Each of these proteins is arranged in the membrane in a different manner. We shall, therefore, use them as examples of three major ways that proteins are associated with membranes.

6-14 Spectrin Is a Cytoskeletal Protein Noncovalently Associated with the Cytoplasmic Side of the Red Blood Cell Membrane[11]

Most of the protein molecules associated with the human red blood cell membrane are peripheral membrane proteins associated with the cytoplasmic side of the lipid bilayer. The most abundant of these proteins is **spectrin,** a long, thin, flexible rod about 100 nm in length that constitutes about 25% of the membrane-associated protein mass (about 2.5×10^5 copies per cell). It is the principal component of the protein meshwork (the *cytoskeleton*) that underlies the red blood cell membrane, maintaining the structural integrity and biconcave shape of this membrane (see Figure 6–22); if the cytoskeleton is extracted from red blood cell ghosts in low-ionic-strength solutions, the membrane fragments into small vesicles.

Spectrin is composed of two very large polypeptide chains, α spectrin (~240,000 daltons) and β spectrin (~220,000 daltons). Each chain is thought to be made up of many α-helical segments interwound in groups of three and connected by nonhelical regions (Figure 6–25). The spectrin heterodimers self-associate head-to-head to form 200-nm-long tetramers. The tail ends of five or six tetramers are linked together by binding to short actin filaments and to another protein (*band 4.1*) in a "junctional complex," forming a deformable, netlike meshwork that underlies the entire cytoplasmic surface of the membrane (Figure 6–26). It is this spectrin-based cytoskeleton that enables the red cell to withstand the stress on its membrane as it is forced through narrow capillaries. Anemic mice and humans with a genetic abnormality of spectrin have red cells that are spherical (instead of concave) and abnormally fragile; the severity of the anemia increases with the degree of the spectrin deficiency.

The protein mainly responsible for attaching the spectrin cytoskeleton to the red cell plasma membrane was identified by binding radiolabeled spectrin to red cell membranes from which spectrin and various other peripheral proteins had been removed. These experiments showed that the rebinding of spectrin depends on a large intracellular attachment protein called **ankyrin,** which binds both to β spectrin and to the cytoplasmic domain of the transmembrane protein band 3 (see Figure 6–26). By connecting band 3 protein to spectrin, ankyrin links the spectrin network to the membrane; it also greatly reduces the rate of diffusion of the band 3 molecules in the lipid bilayer. The spectrin-based cytoskeleton may also be attached to the membrane by a second mechanism: the cytoskeletal protein band 4.1—which binds to spectrin and actin—has been shown to bind to the cytoplasmic domain of glycophorin, the other major transmembrane protein in red blood cells.

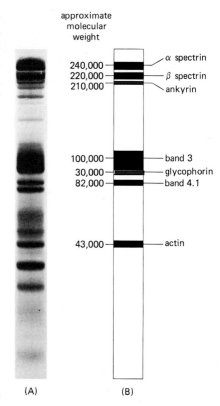

approximate molecular weight

240,000 — α spectrin
220,000 — β spectrin
210,000 — ankyrin

100,000 — band 3
30,000 — glycophorin
82,000 — band 4.1

43,000 — actin

(A)   (B)

Figure 6–24 SDS polyacrylamide-gel electrophoresis pattern of the proteins in the human red blood cell membrane stained with Coomassie blue (A). The positions of some of the major proteins in the gel are indicated in the drawing in (B); glycophorin is shown in color to distinguish it from band 3. Other bands in the gel are omitted from the drawing. The large amount of carbohydrate in glycophorin molecules slows their migration so that they run almost as slowly as the much larger band 3 molecules. (A, courtesy of Ted Steck.)

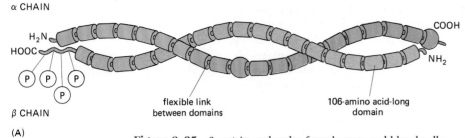

α CHAIN

H₂N
HOOC

P   P   P
    P

β CHAIN

(A)

COOH

NH₂

flexible link
between domains

106-amino acid-long
domain

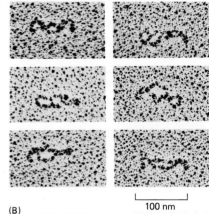

(B)

100 nm

Figure 6–25 Spectrin molecules from human red blood cells shown schematically in (A) and in electron micrographs in (B). Each spectrin heterodimer consists of two antiparallel, loosely intertwined, flexible polypeptide chains; these are attached noncovalently to each other at multiple points, including both ends. The phosphorylated "head" end, where two dimers associate to form a tetramer, is on the left. Both the α and β chains are composed largely of repeating domains 106 amino acids long. It has been proposed that each domain is organized into groups of three α helices (not shown) that are linked by nonhelical regions. In (B) the spectrin molecules have been shadowed with platinum. (A, adapted from D.W. Speicher and V.T. Marchesi, *Nature* 311:177–180, 1984; B, courtesy of D.M. Shotton, with permission from D.M. Shotton, B.E. Burke, and D. Branton, *J. Mol. Biol.* 131:303–329, 1979, © Academic Press Inc. [London] Ltd.)

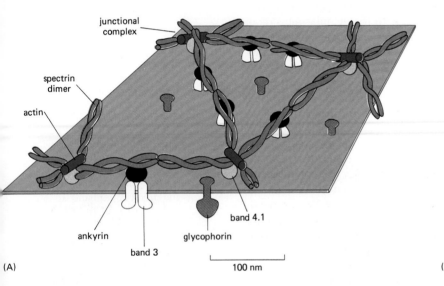

junctional
complex

spectrin
dimer

actin

ankyrin

band 3

glycophorin

band 4.1

100 nm

(A)

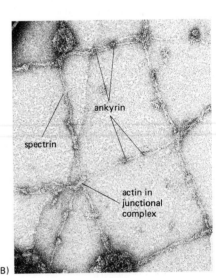

ankyrin

spectrin

actin in
junctional
complex

(B)

Figure 6–26 Schematic drawing (A) and electron micrograph (B) of the spectrin-based cytoskeleton on the cytoplasmic side of the human red blood cell membrane. The arrangement shown in (A) has been deduced mainly from studies on the interactions of purified proteins *in vitro*. Spectrin dimers associate head-to-head to form tetramers that are linked together into a netlike meshwork by junctional complexes composed of short actin filaments (containing about 15 actin monomers) and band 4.1 (and two or three other proteins that are not shown). This cytoskeleton is linked to the membrane by the indirect binding of spectrin tetramers to band 3 proteins via ankyrin molecules and may also be linked by the binding of band 4.1 proteins to glycophorin (not shown). The electron micrograph in (B) shows the cytoskeleton on the cytoplasmic side of a red blood cell membrane after fixation and negative staining. The spectrin meshwork has been purposely stretched out to allow the details of its structure to be seen; in the normal cell the meshwork shown would occupy only about one-tenth of this area. (Courtesy of T. Byers and D. Branton, *Proc. Natl. Acad. Sci. USA* 82:6153–6157, 1985.)

An analogous but more elaborate and complicated cytoskeletal network exists beneath the plasma membrane of nucleated cells. This network, which constitutes the cortical region (or *cortex*) of the cytoplasm, is rich in actin filaments that are thought to be attached to the plasma membrane in numerous ways (see p. 632). Proteins that are structurally homologous to spectrin, ankyrin, and band 4.1 are present in the cortex, but their organization and functions are as yet poorly understood (see p. 641).

Glycophorin Extends Through the Red Blood Cell Lipid Bilayer as a Single α Helix[12]

Glycophorin is one of the two major proteins exposed on the outer surface of the human red blood cell and was the first membrane protein for which the complete amino acid sequence was determined. It is a small transmembrane glycoprotein (131 amino acid residues) with most of its mass on the external surface of the membrane, where its hydrophilic amino-terminal end is located. This part of the protein carries all of the carbohydrate (about 100 sugar residues on 16 separate oligosaccharide side chains), which accounts for 60% of the molecule's mass. In fact, the great majority of the total red blood cell surface carbohydrate (including more than 90% of the sialic acid and, therefore, most of the negative charge of the surface) is carried by glycophorin molecules. The hydrophilic carboxyl-terminal tail of glycophorin is exposed to the cytosol, while a hydrophobic α-helical segment 23 amino acids long spans the lipid bilayer.

Despite there being more than 6×10^5 glycophorin molecules per cell, their function remains unknown. Indeed, individuals whose red cells lack a major subset of these molecules appear to be perfectly healthy. Although glycophorin itself is found only in red blood cells, its structure is representative of a common class of transmembrane glycoproteins that traverse the lipid bilayer as a single α helix—so-called *single-pass membrane proteins* (see example 1 in Figure 6–14 and Figure 6–16). A variety of cell-surface receptors (see p. 706), for example, belongs to this class.

6-13 Band 3 of the Human Red Blood Cell Membrane Is an Anion Transport Protein[13]

Unlike glycophorin, the **band 3 protein** is known to play an important part in cell function. It derives its name from its position relative to the other membrane proteins after electrophoresis in SDS polyacrylamide gels (see Figure 6–24). Like glycophorin, band 3 is a transmembrane protein, but it is a *multipass membrane protein*, traversing the membrane in a highly folded conformation: the polypeptide chain (about 930 amino acids long) extends across the bilayer at least 10 times. Each red blood cell contains about 10^6 band 3 polypeptide chains, which are thought to form dimers and possibly tetramers in the membrane.

The main function of red blood cells is to carry O_2 from the lungs to the tissues and CO_2 from the tissues to the lungs. Band 3 protein is instrumental in this exchange. Red cells dispose of the CO_2 they accumulate in the tissues by ejecting HCO_3^- in exchange for Cl^- as they move through the lungs. This exchange takes place through an anion transport protein and can be blocked by specific inhibitors that bind to the transport protein. By modifying the inhibitors so that they radioactively label the protein to which they bind, the anion transport protein has been identified as band 3. More recently, anion transport has been reconstituted *in vitro* using purified band 3 protein incorporated into phospholipid vesicles. A closely related anion transporter is found in many nucleated cells, where it helps to control intracellular pH (see p. 309).

Band 3 proteins can be seen as distinct *intramembrane particles* by the technique of **freeze-fracture electron microscopy,** in which cells are frozen in liquid nitrogen and the resulting block of ice is fractured with a knife (see p. 147). The fracture plane tends to pass through the hydrophobic middle of membrane lipid bilayers, separating them into their two monolayers. The exposed *fracture faces* are then shadowed with platinum, and the resulting platinum replica is examined with an electron microscope. As illustrated in Figure 6–27, two different fracture

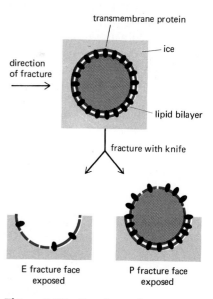

Figure 6–27 How freeze-fracture electron microscopy provides images of the hydrophobic interior of the cytoplasmic (or protoplasmic) half of the bilayer (called the P face) and the external half of the bilayer (called the E face). After the fracturing process shown here, the exposed fracture faces are shadowed with platinum and carbon, the organic material is digested away, and the resulting platinum replica is examined in the electron microscope (see also Figure 4–23, p. 148).

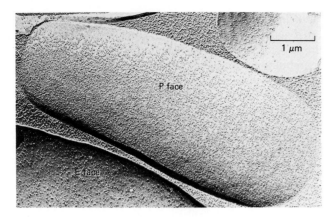

Figure 6–28 Freeze-fracture electron micrograph of human red blood cells. Note that the density of intramembrane particles on the protoplasmic (P) face is higher than on the external (E) face. (Courtesy of L. Engstrom and D. Branton.)

faces are exposed and replicated in this technique—the face representing the hydrophobic interior of the cytoplasmic (or protoplasmic) half of the bilayer (called the **P face**) and the face representing the hydrophobic interior of the external half of the bilayer (called the **E face**). When examined in this way, human red blood cell membranes are studded with intramembrane particles that are relatively homogeneous in size (7.5 nm in diameter), randomly distributed, and more concentrated on the P face than on the E face (Figure 6–28). These are thought to be principally band 3 molecules: when synthetic lipid bilayers are reconstituted with purified band 3 protein molecules, typical 7.5-nm intramembrane particles are observed when the bilayers are fractured. Figure 6–29 illustrates why band 3 molecules are seen in freeze-fracture electron microscopy of red blood cell membranes but glycophorin molecules probably are not.

In a general way, it is not difficult to imagine how a transmembrane protein such as band 3, with much of its mass in the lipid bilayer, could mediate the passive transport of polar molecules across the nonpolar bilayer. The band 3 protein (or its dimer or tetramer) could provide a transmembrane hydrophilic pathway along which the Cl^- and HCO_3^- ions are transported without having to make contact with the hydrophobic interior of the lipid bilayer (see Figure 6–43A, p. 302). It is difficult to see how a molecule such as glycophorin, which spans the bilayer as a single α helix, could mediate this type of transport on its own.

An understanding of how membrane transport proteins work requires precise information about their three-dimensional structure in the bilayer. The first plasma membrane transport protein for which such detail became known was *bacteriorhodopsin*, a protein that serves as a light-activated proton (H^+) pump in the

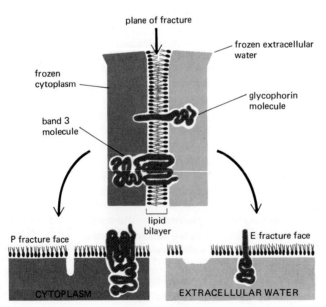

Figure 6–29 Probable fates of band 3 and glycophorin molecules in the human red blood cell membrane during freeze-fracture. When the lipid bilayer is split, either the inside or outside half of each transmembrane protein is pulled out of the frozen monolayer with which it is associated; the protein tends to remain with the monolayer that contains the main bulk of the protein. For this reason band 3 molecules usually remain with the inner (P) fracture face; since they have sufficient mass above the fracture plane, they are readily seen as intramembrane particles. Glycophorin molecules usually remain with the outer (E) fracture face, but it is thought that their cytoplasmic tails have insufficient mass to be seen.

plasma membrane of certain bacteria. The structure of bacteriorhodopsin is similar to that of various other membrane proteins (see p. 705), and it merits a brief digression here.

Figure 6–30 Schematic drawing of the bacterium *Halobacterium halobium* showing the patches of purple membrane that contain bacteriorhodopsin molecules. These bacteria, which live in saltwater pools where they are exposed to a large amount of sunlight, have evolved a variety of light-activated proteins, including bacteriorhodopsin—a light-activated proton pump in their plasma membrane.

Bacteriorhodopsin Is a Proton Pump That Traverses the Bilayer as Seven α Helices[14]

The "purple membrane" of the bacterium *Halobacterium halobium* is a specialized patch in the plasma membrane (Figure 6–30) that contains a single species of protein molecule, **bacteriorhodopsin.** Each bacteriorhodopsin molecule contains a single light-absorbing prosthetic group, or chromophore (called *retinal*), which is related to vitamin A and is identical to the chromophore found in rhodopsin of the vertebrate retinal rod cell (see p. 1106). Retinal is covalently linked to a lysine side chain of the protein; when it is activated by a single photon of light, the excited chromophore causes a conformational change in the protein that results in the transfer of one or two H^+ from the inside to the outside of the cell. This transfer establishes a H^+ and voltage gradient across the plasma membrane, which in turn drives the production of ATP by a second protein in the cell's plasma membrane.

Because the numerous bacteriorhodopsin molecules in the cell membrane are arranged in a planar crystalline lattice (like a two-dimensional crystal), it has been possible to determine the three-dimensional structure and orientation of bacteriorhodopsin in the membrane to a resolution of 0.7 nm by a combination of low-intensity electron microscopy and low-angle electron diffraction analysis. The latter procedure is analogous to the study of three-dimensional crystals of soluble proteins by x-ray diffraction analysis, although less structural detail can be obtained. As illustrated in Figure 6–31, these studies have shown that each bacteriorhodopsin molecule is folded into seven closely packed α helices (each containing about 25 amino acids), which pass roughly at right angles through the lipid bilayer. It seems likely that the protons are passed by the chromophore along a relay system set up by the side chains of the α helices; however, the molecular details are unknown.

Bacteriorhodopsin is a member of a family of membrane proteins with similar structures but different functions. For example, the light receptor protein *rhodopsin* in rod cells of the vertebrate retina and a number of cell-surface receptor proteins that bind specific hormones are also folded into seven transmembrane α helices (see p. 705). These proteins function as signal transducers rather than as transporters; each responds to an extracellular signal by activating another plasma membrane protein, which generates a chemical signal in the cytosol (see p. 685).

To understand the function of bacteriorhodopsin in molecular detail, it will be necessary to locate each of its atoms precisely, which will require x-ray diffraction studies of crystals of the protein. Because of their amphipathic nature, membrane proteins are extremely difficult to crystallize, and it was only in 1985 that the first one—a bacterial photosynthetic reaction center—was successfully studied by x-ray diffraction techniques. The results of these studies are of general importance to membrane biology because they show for the first time how multiple polypeptides can associate in a membrane to form a complex protein machine.

Four Different Polypeptide Chains in a Membrane-bound Complex Form a Bacterial Photosynthetic Reaction Center[15]

In Chapter 3 we discussed how different polypeptides associate to form large multienzyme assemblies that can carry out complex reactions with great efficiency because the subunits cooperate (see p. 126). Similar protein complexes are found in membranes, and the first of these to be understood in detail is a bacterial **photosynthetic reaction center.** This protein complex is located in the plasma membrane of the purple photosynthetic bacterium *Rhodopseudomonas viridis*, where it uses captured light energy to create a high-energy electron that it trans-

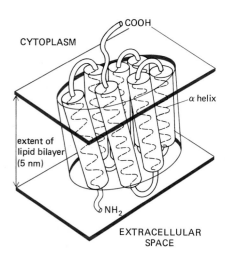

Figure 6–31 The structure of one bacteriorhodopsin molecule and its relationship to the lipid bilayer. The polypeptide chain crosses the bilayer as seven α helices. (Based on data from R. Henderson and P.N.T. Unwin, *Nature* 257:28–32, 1975.)

fers across the membrane in less than a nanosecond. The electron is then passed to other electron carriers in the membrane, which use some of the energy released by the electron-transport process to synthesize ATP in the cytosol. The reaction center is a tetramer composed of four different polypeptides: L, M, H, and a cytochrome. It has been solubilized in detergent, crystallized as a protein-detergent complex, and analyzed by x-ray diffraction to obtain its complete three-dimensional structure.

The protein complex contains four molecules of chlorophyll and eight molecules of other coenzymes that carry electrons. The determination of the precise location of each of these coenzymes was a major advance in our understanding of photosynthesis, as discussed in Chapter 7 (see p. 375). Equally important, and more relevant to our present discussion, was the determination of the precise organization of the four protein subunits in the transmembrane protein complex. The L and M subunits are homologous, each containing five α helices that traverse the lipid bilayer of the plasma membrane (Figure 6–32). These two subunits form a heterodimer that is the core of the reaction center, its 10 α helices surrounding the electron carriers. The H subunit has a single transmembrane α helix, while the rest of the polypeptide chain is folded into a globular domain that protrudes on the cytoplasmic face of the membrane where it binds to the L-M heterodimer. The cytochrome is a peripheral membrane protein that is bound to the L-M heterodimer on the noncytoplasmic side of the membrane (see Figure 6–32).

The two "extra" subunits greatly increase the efficiency of the photosynthetic reaction catalyzed by the L-M heterodimer: the cytochrome keeps the heterodimer supplied with electrons, while the H subunit is believed to couple the reaction center to an array of light-harvesting proteins in the cell interior. The L-M heterodimer has been highly conserved during evolution, and a closely related pair of proteins is thought to form the core of one of the photosynthetic reaction centers in green plants (see p. 377).

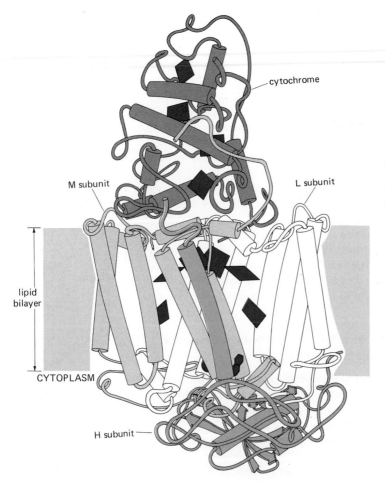

Figure 6–32 The structure of the photosynthetic reaction center of the bacterium *Rhodopseudomonas viridis* as determined by x-ray diffraction analysis of crystals of this transmembrane protein complex. The protein complex consists of four subunits, L, M, H, and a cytochrome: the L and M subunits form the core of the reaction center, and each contains five α helices that span the lipid bilayer. The locations of the electron carrier coenzymes are shown in solid color. (Adapted from a drawing by J. Richardson based on data from J. Deisenhofer, O. Epp, K. Miki, R. Huber, and H. Michel, *Nature* 318:618–624, 1985.)

6-15 Many Membrane Proteins Diffuse in the Plane of the Membrane[16]

Like membrane lipids, membrane proteins do not tumble (*flip-flop*) across the bilayer, and they rotate about an axis perpendicular to the plane of the bilayer (*rotational diffusion*). In addition, many membrane proteins are able to move laterally within the membrane (*lateral diffusion*). The first direct evidence that some plasma membrane proteins are mobile in the plane of the membrane was provided in 1970 by an experiment using hybrid cells (*heterocaryons*, see p. 162) that were artificially produced by fusing mouse cells with human cells. Two differently labeled antibodies were used to distinguish selected mouse and human plasma membrane proteins. Although at first the mouse and human proteins were confined to their own halves of the newly formed heterocaryon, the two sets of proteins diffused and mixed over the entire cell surface within half an hour or so (Figure 6–33). Further evidence for membrane protein mobility was soon provided by the discovery of a process called *patching* (see p. 333): when antibodies bind to specific proteins on the surface of cells, the proteins tend to become cross-linked into large clusters, indicating that the proteins are able to move laterally in the lipid bilayer.

The lateral diffusion rates of membrane proteins can be quantitated using the technique of *fluorescence recovery after photobleaching (FRAP)*. This method was first used to study the diffusion of individual rhodopsin molecules in the disc membranes of vertebrate rod cells. As we have seen, rhodopsin has a structure similar to bacteriorhodopsin, and it contains the same chromophore (retinal). The

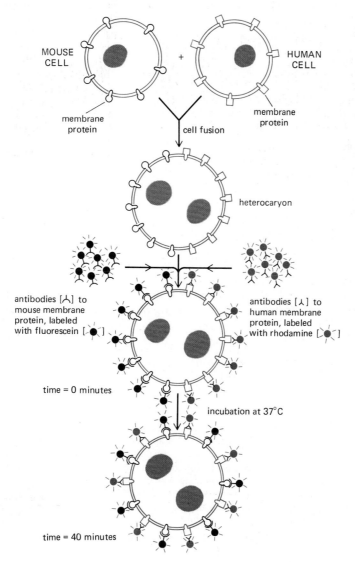

Figure 6–33 The experiment demonstrating the mixing of plasma membrane proteins on mouse-human hybrid cells. The mouse and human proteins are initially confined to their own halves of the newly formed heterocaryon plasma membrane, but they intermix with time. The two antibodies used to visualize the proteins can be distinguished in a fluorescence microscope because fluorescein is green whereas rhodamine is red. (Based on observations of L.D. Frye and M. Edidin, *J. Cell Sci.* 7:319–335, 1970, by permission of The Company of Biologists.)

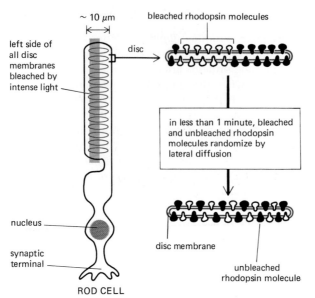

Figure 6–34 Measuring the rate of lateral diffusion of rhodopsin molecules in the disc membranes of a retinal photoreceptor (rod) cell by the FRAP technique. After the retinal chromophores in the rhodopsin molecules are bleached on one side of the cell, the rate at which the bleached and unbleached rhodopsin molecules mix by diffusion is measured. (Based on observations of M. Poo and R.A. Cone, *Nature* 247:438–441, 1974.)

diffusion of the rhodopsin molecules can be measured by bleaching the chromophore in the rhodopsin molecules on one side of a photoreceptor cell with an intense, highly focused beam of light and measuring the time taken for the bleached and unbleached rhodopsin molecules to mix by diffusion (Figure 6–34). The rate of diffusion, measured as a *diffusion coefficient (D)*, is about 5×10^{-9} cm²/sec; this is only half the diffusion coefficient of a phospholipid molecule in a membrane (see p. 278) and is the highest diffusion coefficient of any membrane protein known.

The same technique can be used to study membrane proteins that do not contain chromophores by first attaching fluorescent ligands to them. Most commonly these are fluorescent monovalent antibodies (fragments of antibodies that have only one antigen-binding site and therefore cannot cross-link neighboring molecules—see p. 1013). The tightly bound ligands are then bleached in a small area by a laser beam, and the time taken for adjacent membrane proteins carrying unbleached fluorescent antibody molecules to diffuse into the bleached area is measured (Figure 6–35). The diffusion rates of various plasma membrane glycoproteins measured in this way have usually been found to be at least 5 to 50 times slower than those of rhodopsin molecules. These relatively slow diffusion rates are not an intrinsic property of the individual glycoprotein molecules themselves, as the same glycoproteins diffuse much more rapidly when inserted into synthetic lipid bilayers. The reasons for the slow diffusion rates measured by the FRAP technique for glycoproteins in plasma membranes are uncertain. One possibility is that the bulky oligosaccharide chains on the extracellular domains of such molecules interact with those on other glycoproteins in the membrane to slow diffusion: in at least some cases the removal of the carbohydrate greatly increases the diffusion rate of the protein.

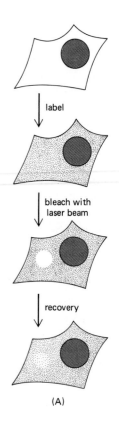

label

bleach with laser beam

recovery

(A)

Figure 6–35 Measuring the rate of lateral diffusion of a plasma membrane glycoprotein by the FRAP technique. (A) A specific glycoprotein is labeled on the cell surface with a fluorescent monovalent antibody that binds only to that protein. After the antibodies are bleached in a small area using a laser beam, the fluorescence intensity recovers as the bleached molecules diffuse away and unbleached molecules diffuse into the irradiated area. (B) A graph showing the rate of recovery. The greater the diffusion coefficient of the membrane glycoprotein, the faster is the recovery.

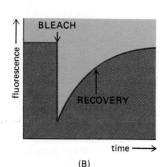

(B)

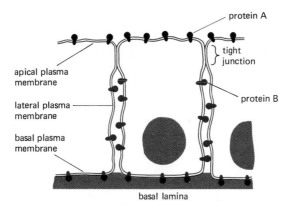

apical plasma
membrane

lateral plasma
membrane

basal plasma
membrane

protein A

tight
junction

protein B

basal lamina

Figure 6–36 Diagram of an epithelial cell showing how a plasma membrane protein is restricted to a particular domain of the membrane. Protein A (in the apical membrane) and protein B (in the basal and lateral membranes) can diffuse laterally in their own domains but are prevented from entering the other domain, probably by the specialized cell junction called a *tight junction*. Lipid molecules in the outer (noncytoplasmic) monolayer of the plasma membrane are also unable to diffuse between the two domains, whereas those in the inner (cytoplasmic) monolayer are able to do so.

Cells Can Confine Proteins and Lipids to Specific Domains Within a Membrane[17]

The recognition that biological membranes are two-dimensional fluids was a major advance in understanding membrane structure and function, but it has become clear that the picture of a membrane as a lipid sea in which all proteins float freely is greatly oversimplified. Many cells have ways of confining membrane proteins to specific domains in a continuous lipid bilayer. For example, in epithelial cells, such as those that line the gut or the tubules of the kidney, certain plasma membrane enzymes and transport proteins are confined to the apical surface of the cells, whereas others are confined to the basal and lateral surfaces (Figure 6–36). As we shall discuss later, this asymmetric distribution of membrane proteins is often essential for the function of the epithelium (see p. 310). The lipid compositions of these two membrane domains are also different, demonstrating that epithelial cells can prevent the diffusion of lipid as well as protein molecules between the domains, although experiments with labeled lipids suggest that only lipid molecules in the outer monolayer of the membrane are confined in this way. The separation of both protein and lipid molecules is thought to be maintained, at least in part, by the barriers set up by a specific type of intercellular junction (called a *tight junction*, see p. 793). As discussed below, the membrane proteins that form intercellular junctions do not diffuse laterally in the interacting membranes (see Figure 6–38D).

A cell can also create membrane domains without using intercellular junctions. The mammalian spermatozoon, for instance, is a single cell that consists of two structurally and functionally distinct parts—a head and a tail (see p. 864)—covered by a continuous plasma membrane. When a sperm cell is examined by immunofluorescence microscopy using a variety of antibodies that react with cell-surface antigens, the plasma membrane is found to consist of at least three distinct domains (Figure 6–37). In some cases at least, the antigens are able to diffuse within the confines of their own domain; it is not known how they are confined.

In the two examples just considered, the diffusion of protein and lipid molecules is confined to specialized domains within a continuous plasma membrane.

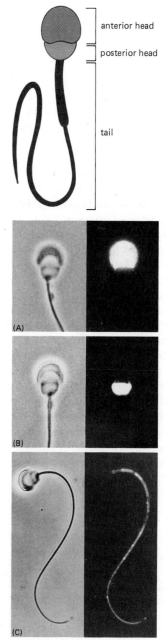

anterior head

posterior head

tail

(A)

(B)

(C)

Figure 6–37 Three domains in the plasma membrane of guinea pig sperm defined with monoclonal antibodies (see p. 178). A guinea pig sperm is shown schematically in the upper drawing, while each of the three pairs of micrographs shown in (A), (B), and (C) shows cell-surface immunofluorescence staining with a different monoclonal antibody next to a phase-contrast micrograph of the same cell. The antibody shown in (A) labels only the anterior head, that in (B) only the posterior head, whereas that in (C) labels only the tail. (A and B, from D.G. Myles, P. Primakoff, and A.R. Bellvé, *Cell* 23:434–439, 1981, by copyright permission of Cell Press. C, from P. Primakoff and D.G. Myles, *Dev. Biol.* 98:417–428, 1983.)

Cells also have more drastic ways of immobilizing certain membrane proteins. One is exemplified by the purple membrane of *Halobacterium*, where the bacteriorhodopsin molecules assemble into large two-dimensional crystals in which the individual protein molecules are relatively fixed in relationship to one another; large aggregates of this kind diffuse very slowly. A more common way of restricting the lateral mobility of specific membrane proteins is to tether them to macromolecular assemblies inside or outside the cell. We have seen that some red blood cell membrane proteins are anchored to the cytoskeleton inside; in other cell types, plasma membrane proteins can be anchored to the cytoskeleton or to the extracellular matrix or to both. The four known ways of immobilizing specific membrane proteins are summarized in Figure 6–38.

Summary

Whereas the lipid bilayer determines the basic structure of biological membranes, proteins are responsible for most membrane functions, serving as specific receptors, enzymes, transport proteins, and so on. Most membrane proteins extend across the lipid bilayer: in some of these transmembrane proteins the polypeptide chain crosses the bilayer as a single α helix (single-pass proteins); in others the polypeptide chain crosses the bilayer multiple times as a series of α helices (multipass proteins). Other membrane-associated proteins do not span the bilayer but instead are attached to one or the other side of the membrane. Many of these are bound by noncovalent interactions with transmembrane proteins, but others are covalently attached to lipid molecules. Like the lipid molecules in the bilayer, many membrane proteins are able to diffuse in the plane of the membrane. On the other hand, cells have ways of immobilizing specific membrane proteins and of confining both membrane protein and lipid molecules to particular domains in a continuous lipid bilayer.

Membrane Carbohydrate

All eucaryotic cells have carbohydrate on their surface, both as oligosaccharide and polysaccharide chains covalently bound to membrane proteins and as oligosaccharide chains covalently bound to lipids (glycolipids). The total carbohydrate in plasma membranes constitutes between 2% and 10% of the membrane's total weight. Most plasma membrane protein molecules exposed at the cell surface carry sugar residues, whereas somewhat fewer than 1 in 10 lipid molecules in the outer lipid monolayer of most plasma membranes contain carbohydrate (see p. 282). Although the fiftyfold excess of lipid to protein molecules in membranes means that there are more lipid than protein molecules carrying carbohydrate in a typical membrane, there is more total carbohydrate attached to proteins: a single glycoprotein, such as glycophorin, can have many oligosaccharide side chains, whereas each glycolipid molecule has only one. Moreover, many plasma membranes contain integral *proteoglycan* molecules. Proteoglycans, which consist of long polysaccharide chains linked to a protein core, are found mainly outside the cell as part of the extracellular matrix (see p. 806); in the case of some integral membrane proteoglycans, however, the core is thought to extend across the lipid bilayer.

6-18 ## The Carbohydrate in Biological Membranes Is Confined Mainly to the Noncytosolic Surface[18]

As we have seen, biological membranes are strikingly asymmetrical: the lipids of the outer and inner lipid monolayers are different, as are the exposed polypeptides on the two surfaces. The distribution of carbohydrate is also asymmetrical, since the carbohydrate chains of the major glycolipids, glycoproteins, and proteoglycans of both internal and plasma membranes are located exclusively on the noncytosolic surface: in plasma membranes the sugar residues are exposed on the outside

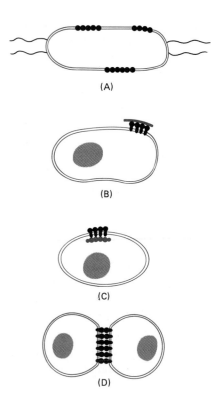

Figure 6–38 Four ways in which the lateral mobility of specific plasma membrane proteins can be restricted. The proteins can self-assemble into large aggregates (such as bacteriorhodopsin in the purple membrane of *Halobacterium*) (A), they can be tethered by interactions with assemblies of macromolecules outside (B) or inside (C) the cell, or they can interact with proteins on the surface of another cell (D).

Table 6–2 Commonly Used, Commercially Available Plant Lectins and the Specific Sugar Residues They Recognize

Lectin	Sugar Specificity
Concanavalin A (from jack beans)	α-D-glucose and α-D-mannose
Soybean lectin	α-galactose and N-acetyl-D-galactosamine
Wheat germ lectin	N-acetylglucosamine
Lotus seed lectin	fucose

of the cell, whereas in internal membranes they face inward toward the lumen of the membrane-bounded compartment (see p. 454).

There are two distinct ways in which oligosaccharides are attached to membrane glycoproteins: they may be *N-linked* to an asparagine residue in the polypeptide chain or *O-linked* to a serine or threonine residue (see p. 446). The *N*-linked oligosaccharides usually contain about 12 sugars and are constructed around a common core of mannose residues (see p. 452), whereas the *O*-linked oligosaccharides tend to be shorter (about 4 sugars long).

One convenient way to demonstrate the presence of cell-surface sugars is to use carbohydrate-binding proteins called **lectins.** These are proteins with binding sites that recognize a specific sequence of sugar residues. They were originally isolated from plants, where they are found in large quantities in many seeds; some of them are highly toxic and serve to deter animals from eating the seeds. More recently, lectins have been demonstrated in many other organisms, including mammals; some of them occur on cell surfaces and are thought to be involved in cell-cell recognition (for example, see Figure 15–42, p. 870). Since lectins bind to cell-surface glycoproteins, proteoglycans, and glycolipids, they are widely used as biochemical tools in cell biology to localize and isolate sugar-containing plasma membrane molecules. Some commonly used plant lectins and their sugar specificities are listed in Table 6–2.

The term *cell coat* or *glycocalyx* is often used to describe the carbohydrate-rich peripheral zone on the outside surface of most eucaryotic cells. This zone can be visualized by a variety of stains, such as ruthenium red (Figure 6–39), as well as by labeled lectins. Although the carbohydrate is attached mainly to intrinsic plasma membrane molecules, the glycocalyx can also contain both glycoproteins and proteoglycans that have been secreted and then adsorbed on the cell surface (Figure 6–40). Some of these adsorbed macromolecules are components of the extracellular matrix, so that where the plasma membrane ends and the extracellular matrix begins is largely a matter of semantics.

Although the high concentration of cell-surface carbohydrate must have important influences on many functions of the plasma membrane, the nature of these influences is not yet understood. The complexity of some of the oligosaccharides, taken together with their exposed position on the cell surface, suggests

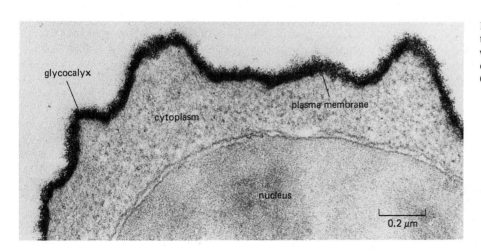

Figure 6–39 Electron micrograph of the surface of a lymphocyte stained with ruthenium red to show the cell coat (glycocalyx). (Courtesy of A.M. Glauert and G.M.W. Cook.)

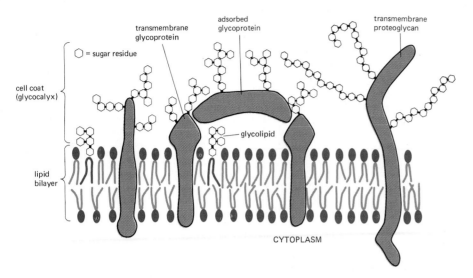

Figure 6–40 Diagram of the cell coat (glycocalyx), which is made up of the oligosaccharide side chains of glycolipids and integral membrane glycoproteins, and polysaccharide chains on integral proteoglycans. In addition, adsorbed glycoproteins and adsorbed proteoglycans (not shown) contribute to the glycocalyx in some cells. Note that all of the carbohydrate is on the outside of the membrane. Some integral glycoproteins and proteoglycans are covalently attached to phosphatidylinositol in the outer monolayers of the plasma membrane via a specific oligosaccharide (not shown, but see Figure 6–14–4, p. 284).

that they may play an important part in cell-cell and cell-matrix recognition processes. Although there is strong indirect evidence for this in a few cases (see pp. 869 and 875), for the most part this function of cell-surface carbohydrate has been difficult to demonstrate unambiguously.

Summary

In the plasma membrane of all eucaryotic cells, most of the proteins exposed on the cell surface and some of the lipid molecules in the outer lipid monolayer have oligosaccharide chains covalently attached to them. Some plasma membranes also contain integral proteoglycan molecules in which several polysaccharide chains are covalently linked to a transmembrane or lipid-linked core protein. Although the function of cell-surface carbohydrate remains uncertain, it is likely that at least some of it plays a part in cell-cell and cell-matrix recognition processes.

Membrane Transport of Small Molecules[19]

Because of its hydrophobic interior, the lipid bilayer is a highly impermeable barrier to most polar molecules and therefore prevents most of the water-soluble contents of the cell from escaping. For this very reason, however, cells have evolved special ways of transferring water-soluble molecules across their membranes. Cells must ingest essential nutrients and excrete metabolic waste products. They must also regulate intracellular ion concentrations, which means transporting specific ions into or out of the cell. Transport of small water-soluble molecules across the lipid bilayer is achieved by specialized transmembrane proteins, each of which is responsible for the transfer of a specific molecule or a group of closely related molecules. Cells have also evolved the means to transport macromolecules such as proteins and even large particles across their plasma membrane, but the mechanisms involved are very different from those used for transferring small molecules and will be discussed in a later section (see p. 323).

In this section we shall see that a combination of selective permeability and active transport across the plasma membrane creates large differences in the ionic composition of the cytosol compared with the extracellular fluid (Table 6–3). This enables cell membranes to store potential energy in the form of ion gradients. These transmembrane ion gradients are used to drive various transport processes, to convey electrical signals, and (in mitochondria, chloroplasts, and bacteria) to make ATP. Before discussing membrane transport proteins and the ion gradients that some of them generate, it is important to know something of the permeability properties of protein-free, synthetic lipid bilayers.

Table 6–3 Comparison of Ion Concentrations Inside and Outside a Typical Mammalian Cell

Component	Intracellular Concentration (mM)	Extracellular Concentration (mM)
Cations		
Na$^+$	5–15	145
K$^+$	140	5
Mg^{2+}	0.5	1–2
Ca^{2+}	10^{-4}	1–2
H$^+$	8×10^{-5} ($10^{-7.1}$ M or pH 7.1)	4×10^{-5} ($10^{-7.4}$ M or pH 7.4)
Anions*		
Cl$^-$	5–15	110

*Because the cell must contain equal + and − charges (that is, be electrically neutral), the large deficit in intracellular anions reflects the fact that most cellular constituents are negatively charged (HCO$_3^-$, PO$_4^{3-}$, proteins, nucleic acids, metabolites carrying phosphate and carboxyl groups, etc.). The concentrations of Ca^{2+} and Mg^{2+} given are for the free ions. There is a total of about 20 mM Mg^{2+} and 1−2 mM Ca^{2+} in cells, but this is mostly bound to proteins and other substances and, in the case of Ca^{2+}, stored within various organelles.

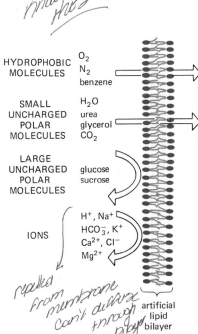

Figure 6–41 The relative permeability of a synthetic lipid bilayer to different classes of molecules. The smaller the molecule, and, more important, the fewer hydrogen bonds it makes with water, the more rapidly the molecule diffuses across the bilayer.

Protein-free Lipid Bilayers Are Impermeable to Ions but Freely Permeable to Water[20]

Given enough time, virtually any molecule will diffuse across a protein-free, synthetic lipid bilayer down its concentration gradient. The rate at which a molecule diffuses across such a lipid bilayer, however, varies enormously, depending largely on the size of the molecule and its relative solubility in oil. In general, the smaller the molecule and the more soluble it is in oil (that is, the more hydrophobic or nonpolar it is), the more rapidly it will diffuse across a bilayer. *Small nonpolar* molecules, such as O$_2$ (32 daltons), readily dissolve in lipid bilayers and therefore rapidly diffuse across them. *Uncharged polar* molecules also diffuse rapidly across a bilayer if they are small enough. For example, CO$_2$ (44 daltons), ethanol (46 daltons), and urea (60 daltons) cross rapidly; glycerol (92 daltons) less rapidly; and glucose (180 daltons) hardly at all (Figure 6–41). Importantly, water (18 daltons) diffuses very rapidly across lipid bilayers even though water molecules are relatively insoluble in oil. This is because water molecules have a very small volume and are uncharged.

In contrast, lipid bilayers are highly impermeable to all *charged* molecules (ions), no matter how small: the charge and high degree of hydration of such molecules prevents them from entering the hydrocarbon phase of the bilayer. Consequently, synthetic bilayers are 10^9 times more permeable to water than to even such small ions as Na$^+$ or K$^+$ (Figure 6–42).

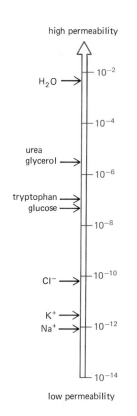

Figure 6–42 Permeability coefficients (cm/sec) for the passage of various molecules through synthetic lipid bilayers. The rate of flow of a solute across the bilayer is directly proportional to the difference in its concentration on the two sides of the membrane. Multiplying this concentration difference (in mol/cm^3) by the permeability coefficient (cm/sec) gives the flow of solute in moles per second per square centimeter of membrane. For example, a concentration difference of tryptophan of 10^{-4} mol/cm^3 ($10^{-4}/10^{-3}$ L = 0.1 M) would cause a flow of 10^{-4} mol/cm$^3 \times 10^{-7}$ cm/sec = 10^{-11} mol/sec through 1 cm^2 of membrane, or 6×10^4 molecules/sec through 1 μm^2 of membrane.

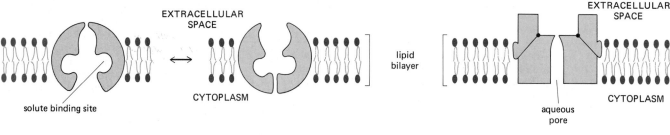

EXTRACELLULAR SPACE

CYTOPLASM

solute binding site

(A) CARRIER PROTEIN

lipid bilayer

EXTRACELLULAR SPACE

CYTOPLASM

aqueous pore

(B) CHANNEL PROTEIN

6-19 Membrane Transport Proteins Can Act as Carriers or Channels[19]

Like synthetic lipid bilayers, cell membranes allow water and nonpolar molecules to permeate by simple diffusion. Cell membranes, however, are also permeable to various polar molecules such as ions, sugars, amino acids, nucleotides, and many cell metabolites that pass across synthetic lipid bilayers only very slowly. Special membrane proteins are responsible for transferring such solutes across cell membranes. These proteins, referred to as **membrane transport proteins,** occur in many forms and in all types of biological membranes. Each protein is designed to transport a particular class of molecule (such as ions, sugars, or amino acids) and often only certain molecular species of the class. The specificity of transport proteins was first indicated by studies in which single gene mutations were found to abolish the ability of bacteria to transport specific sugars across their plasma membrane. Similar mutations have now been discovered in humans suffering from a variety of inherited diseases that affect the transport of a specific solute in the kidney or intestine or both. For example, individuals with the inherited disease *cystinuria* are unable to transport certain amino acids (including cystine, the disulfide-linked dimer of cysteine) from either the urine or the intestine into the blood; the resulting accumulation of cystine in the urine leads to the formation of cystine "stones" in the kidneys.

All membrane transport proteins that have been studied in sufficient detail to establish their orientation in the membrane have been found to be multipass transmembrane proteins—that is, their polypeptide chain traverses the lipid bilayer multiple times. By forming a continuous protein pathway across the membrane, these proteins enable the specific solutes they transport to pass across the membrane without coming into direct contact with the hydrophobic interior of the lipid bilayer.

There are two major classes of membrane transport proteins: carrier proteins and channel proteins. **Carrier proteins** (also called *carriers* or *transporters*) bind the specific solute to be transported and undergo a conformational change in order to transfer the solute across the membrane. **Channel proteins,** on the other hand, form water-filled pores that extend across the lipid bilayer; when these pores are open, they allow specific solutes (usually inorganic ions of appropriate size and charge) to pass through them and thereby cross the membrane (Figure 6–43).

Figure 6–43 A simplified schematic view of the two classes of membrane transport proteins. A *carrier protein* alternates between two conformations, so that the solute binding site is sequentially accessible on one side of the bilayer and then on the other. In contrast, a *channel protein* forms a water-filled pore across the bilayer through which specific ions can diffuse.

Active Transport Is Mediated by Carrier Proteins Coupled to an Energy Source[21]

All channel proteins and many carrier proteins allow solutes to cross the membrane only passively ("downhill")—a process called **passive transport** (or **facilitated diffusion**). If the transported molecule is uncharged, only the difference in its concentration on the two sides of the membrane (its *concentration gradient*) determines the direction of passive transport. If the solute carries a net charge, however, both its concentration gradient and the electrical potential difference across the membrane (the *membrane potential*) influence its transport. The concentration gradient and the electrical gradient together constitute the **electrochemical gradient** for each solute (Panel 6–2, p. 315). All plasma membranes have an electrical potential difference (voltage gradient) across them, with the inside negative compared to the outside. This potential favors the entry of positively charged ions into cells but opposes the entry of negatively charged ions.

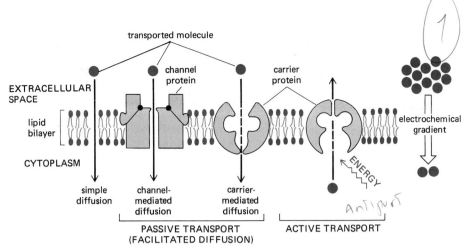

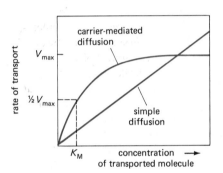

Figure 6-44 Schematic diagram of *passive transport* down an electrochemical gradient and *active transport* against an electrochemical gradient. Whereas simple diffusion and passive transport by membrane transport proteins (facilitated diffusion) occur spontaneously, active transport requires an input of metabolic energy. Only nonpolar molecules and small, uncharged polar molecules can cross the lipid bilayer directly by simple diffusion; the transfer of other polar molecules occurs at significant rates only through specific carrier proteins or channel proteins.

Cells also require transport proteins that will actively pump certain solutes across the membrane against their electrochemical gradient ("uphill"); this process, known as **active transport,** is always mediated by carrier proteins. In active transport the pumping activity of the carrier protein is directional because it is tightly coupled to a source of metabolic energy, such as ATP hydrolysis or an ion gradient, as will be discussed below. Thus transport by carrier proteins can be either active or passive, whereas transport by channel proteins is always passive (Figure 6–44).

6-20 Carrier Proteins Behave like Membrane-bound Enzymes[19]

The process by which a carrier protein specifically binds and transfers a solute molecule across the lipid bilayer resembles an enzyme-substrate reaction, and the carriers involved behave like specialized membrane-bound enzymes. Each type of carrier protein has a specific binding site for its solute (substrate). When the carrier is saturated (that is, when all these binding sites are occupied), the rate of transport is maximal. This rate, referred to as V_{max}, is characteristic of the specific carrier. In addition, each carrier protein has a characteristic binding constant for its solute, K_M, equal to the concentration of solute when the transport rate is half its maximum value (Figure 6–45). As with enzymes, the solute binding can be blocked specifically by competitive inhibitors (which compete for the same binding site and may or may not be transported by the carrier) or by noncompetitive inhibitors (which bind elsewhere and specifically alter the structure of the carrier). The analogy with an enzyme-substrate reaction is limited, however, since the transported solute is usually not covalently modified by the carrier protein.

Some carrier proteins simply transport a single solute from one side of the membrane to the other; they are called **uniports.** Others function as **coupled transporters,** in which the transfer of one solute depends on the simultaneous or sequential transfer of a second solute, either in the same direction (**symport**) or in the opposite direction (**antiport**) (Figure 6–46). Most animal cells, for ex-

Figure 6-45 Kinetics of simple diffusion compared to carrier-mediated diffusion. Whereas the rate of the former is always proportional to the solute concentration, the rate of the latter reaches a maximum (V_{max}) when the carrier protein is saturated. The solute concentration when transport is at half its maximal value approximates the binding constant (K_M) of the carrier for the solute and is analogous to the K_M of an enzyme for its substrate (see p. 94).

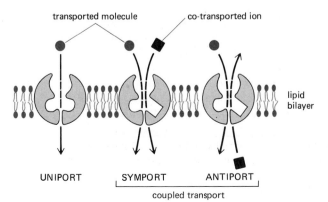

Figure 6-46 Schematic diagram of carrier proteins functioning as uniports, symports, and antiports.

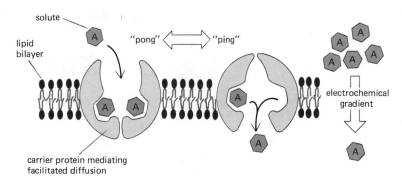

solute

lipid bilayer

"pong" ⟷ "ping"

carrier protein mediating facilitated diffusion

electrochemical gradient

Figure 6–47 A hypothetical model showing how a conformational change in a carrier protein could mediate the facilitated diffusion of a solute A. The carrier protein shown can exist in two conformational states: in state "pong" the binding sites for solute A are exposed on the outside of the bilayer; in state "ping" the same sites are exposed on the other side of the bilayer. The transition between the two states is proposed to occur randomly and to be completely reversible. Therefore, if the concentration of A is higher on the outside of the bilayer, more A will bind to the carrier protein in the pong conformation than in the ping conformation, and there will be a net transport of A down its electrochemical gradient.

ample, must take up glucose from the extracellular fluid, where the concentration of the sugar is relatively high, by *passive* transport through glucose carriers that operate as uniports. By contrast, intestinal and kidney cells must take up glucose from the lumen of the intestine and kidney tubules, respectively, where the concentration of the sugar is low. These cells actively transport glucose by symport with Na^+, whose extracellular concentration is very high (see p. 304). As previously discussed, the anion carrier (band 3 protein) of the human red blood cell operates as an antiport to exchange Cl^- for HCO_3^- (see p. 291).

Although the molecular details are unknown, carrier proteins are thought to transfer the solute across the bilayer by undergoing a reversible conformational change that alternately exposes the solute-binding site first on one side of the membrane and then on the other. A schematic model of how such a carrier protein might operate is shown in Figure 6–47. Because carriers are now known to be multipass transmembrane proteins, it is highly unlikely that they ever tumble in the membrane or shuttle back and forth across the lipid bilayer as was once believed.

As we shall discuss below, it requires only a relatively minor modification of the model shown in Figure 6–47 to link the carrier protein to a source of energy, such as ATP hydrolysis (see Figure 6–49 below), or to an ion gradient (see Figure 6–51 below) to pump a solute uphill against its electrochemical gradient. An important example of a carrier protein that uses the energy of ATP hydrolysis to pump ions is the Na^+-K^+ *pump*, which plays a crucial part in generating and maintaining the Na^+ and K^+ gradients across the plasma membranes of animal cells.

6-23 The Plasma Membrane Na^+-K^+ Pump Is an ATPase[22]

The concentration of K^+ is typically 10 to 20 times higher inside cells than outside, whereas the reverse is true of Na^+ (see Table 6–3, p. 301). These concentration differences are maintained by a **Na^+-K^+ pump** that is found in the plasma membrane of virtually all animal cells. The pump operates as an antiport, actively pumping Na^+ out of the cell against its steep electrochemical gradient and pumping K^+ in. As explained below, the Na^+ gradient due to the pump regulates cell volume through its osmotic effects and is also exploited to drive transport of sugars and amino acids into the cell. Almost one-third of the energy requirement of a typical animal cell is consumed in fueling this pump; in electrically active nerve cells, which are repeatedly gaining small amounts of Na^+ and losing small amounts of K^+ during the propagation of action potentials (see below), this figure approaches two-thirds of the cell's energy requirement.

A major advance in understanding the Na^+-K^+ pump came with the discovery in 1957 that an enzyme that hydrolyzes ATP to ADP and phosphate requires Na^+ and K^+ for optimal activity. An important clue linking this **Na^+-K^+ ATPase** with the Na^+-K^+ pump was the observation that a known inhibitor of the pump, *ouabain*, also inhibits the ATPase. But the crucial evidence that ATP hydrolysis provides the energy for driving the pump came from studies of resealed red blood

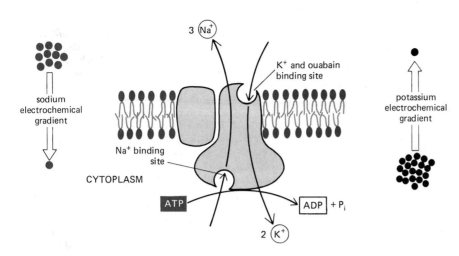

Figure 6-48 The Na$^+$-K$^+$ ATPase actively pumps Na$^+$ out and K$^+$ into a cell against their electrochemical gradients. For every molecule of ATP hydrolyzed inside the cell, three Na$^+$ are pumped out and two K$^+$ are pumped in. The specific pump inhibitor ouabain and K$^+$ compete for the same site on the external side of the ATPase.

cell ghosts (see p. 288), in which the concentrations of ions, ATP, and drugs on either side of the membrane could be varied and the effects on ion transport and ATP hydrolysis observed. It was found that (1) the transport of Na$^+$ and K$^+$ is tightly coupled to ATP hydrolysis, so that one cannot occur without the other; (2) ion transport and ATP hydrolysis can occur only when Na$^+$ and ATP are present inside the ghosts and K$^+$ is present on the outside; (3) ouabain is inhibitory only when present outside the ghosts, where it competes for the K$^+$ binding site; and (4) for every molecule of ATP hydrolyzed (100 ATP molecules can be hydrolyzed by each ATPase molecule each second), three Na$^+$ are pumped out and two K$^+$ are pumped in (Figure 6-48).

Although these experiments provided compelling evidence that ATP supplies the energy for pumping Na$^+$ and K$^+$ ions across the plasma membrane, they did not explain how ATP hydrolysis is coupled to ion transport. A partial explanation was provided by the finding that the terminal phosphate group of the ATP is transferred to an aspartic acid residue of the ATPase in the presence of Na$^+$. This phosphate group is subsequently hydrolyzed in the presence of K$^+$, and it is this last step that is inhibited by ouabain. The Na$^+$-dependent phosphorylation is coupled to a change in the conformation of the ATPase, which results in the transport of Na$^+$ out of the cell, whereas the K$^+$-dependent dephosphorylation, which occurs subsequently, results in the transport of K$^+$ into the cell during the return of the ATPase to its original conformation (Figure 6-49).

The Na$^+$-K$^+$ pump in red blood cell ghosts can be driven in reverse to produce ATP. When the Na$^+$ and K$^+$ gradients are experimentally increased to such an extent that the energy stored in their electrochemical gradients is greater than the chemical energy of ATP hydrolysis, these ions move down their electrochemical gradients and ATP is synthesized from ADP and phosphate by the Na$^+$-K$^+$ ATPase. Thus the phosphorylated form of the ATPase (step 2 in Figure 6-49) can relax either by donating its phosphate to ADP (step 2 to step 1) or by changing its conformation (step 2 to step 3). Whether the overall change in free energy is used to synthesize ATP (see p. 129) or to pump Na$^+$ out of the ghost depends on the relative concentrations of ATP, ADP and phosphate, and on the electrochemical gradients for Na$^+$ and K$^+$.

The Na$^+$-K$^+$ ATPase has been purified and found to consist of a large, multipass, transmembrane catalytic subunit (about 1000 amino acid residues long) and an associated smaller glycoprotein. The former has binding sites for Na$^+$ and ATP on its cytoplasmic surface and binding sites for K$^+$ and ouabain on its external surface, and is reversibly phosphorylated and dephosphorylated. The function of the glycoprotein is unknown. A functional Na$^+$-K$^+$ pump can be reconstituted from the purified complex: the ATPase is solubilized in detergent, purified, and mixed with appropriate phospholipids; when the detergent is removed, membrane vesicles are formed that pump Na$^+$ and K$^+$ in opposite directions in the presence of ATP (see Figure 6-21, p. 287).

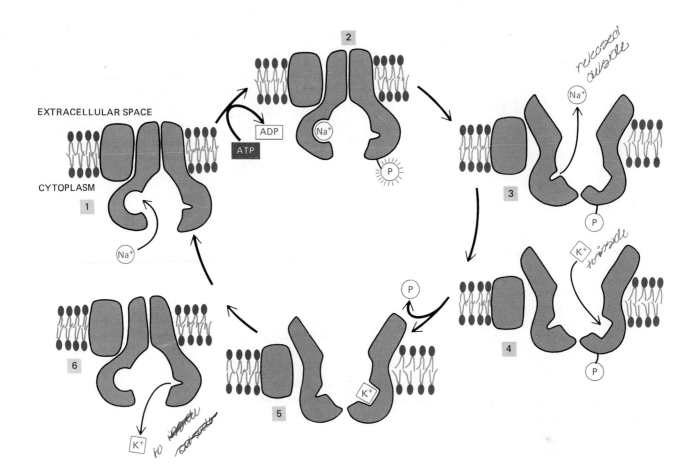

Figure 6-49 handwritten annotations: "released outside" (near Na+ top right), "K+ to inside" (middle right), "K+ to inside" (bottom left)

The Na$^+$-K$^+$ ATPase Is Required to Maintain Osmotic Balance and Stabilize Cell Volume[23]

Since the Na$^+$-K$^+$ ATPase drives three positively charged ions out of the cell for every two it pumps in, it is "electrogenic"; that is, it drives a net current across the membrane, tending to create an electrical potential, with the inside negative relative to the outside. This effect of the pump, however, seldom contributes more than 10% to the membrane potential. The remaining 90%, as we shall see, depends on the pump only indirectly; its immediate cause lies in an electrical force related to the unequal concentrations of K$^+$ on either side of the membrane. These K$^+$ concentrations are governed in turn by the need for K$^+$ inside the cell to balance the large negative charge carried by the cell's *fixed anions*—the many negatively charged organic molecules that are confined inside the cell, unable to cross the plasma membrane.

The Na$^+$-K$^+$ ATPase has a more direct role in regulating cell volume: it controls the solute concentration inside the cell, thereby regulating the osmotic forces that can make a cell swell or shrink (Figure 6-50). As explained in Panel 6-1 (see p. 308), the solutes inside a cell—including its fixed anions and the accompanying cations required for charge balance—create a large osmotic gradient that tends to "pull" water in. For animal cells this effect is counteracted by an opposite osmotic gradient due to a high concentration of inorganic ions—chiefly Na$^+$ and Cl$^-$—in the extracellular fluid. The Na$^+$-K$^+$ ATPase maintains osmotic balance by pumping out the Na$^+$ that leaks in down its steep electrochemical gradient; the Cl$^-$ is kept out by the membrane potential, as will be explained below (see p. 314).

The importance of the Na$^+$-K$^+$ ATPase in controlling cell volume is indicated by the observation that many animal cells swell, and may burst, if they are treated with ouabain, which inhibits the Na$^+$-K$^+$ ATPase. There are, of course, other ways for a cell to cope with its osmotic problems. Plant cells and many bacteria are prevented from bursting by the semirigid cell wall that surrounds their plasma membrane; in amoebae the excess water that flows in osmotically is collected in

Figure 6-49 A schematic model of the Na$^+$-K$^+$ ATPase. The binding of Na$^+$ (1) and the subsequent phosphorylation by ATP (2) of the cytoplasmic face of the ATPase induce the protein to undergo a conformational change that transfers the Na$^+$ across the membrane and releases it on the outside (3). Then the binding of K$^+$ on the external surface (4) and the subsequent dephosphorylation (5) return the protein to its original conformation, which transfers the K$^+$ across the membrane and releases it into the cytosol (6). These changes in conformation are analogous to the ping ⇌ pong transitions shown in Figure 6-47 except that here the Na$^+$-dependent phosphorylation and the K$^+$-dependent dephosphorylation of the protein cause the conformational transitions to occur in an orderly manner, enabling the protein to do useful work. Although for simplicity only one Na$^+$ and one K$^+$ binding site are shown, in the real pump there are thought to be three Na$^+$ and two K$^+$ binding sites.

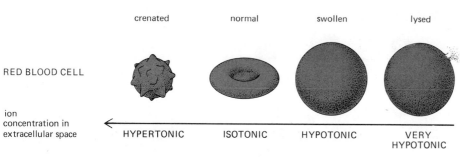

crenated · normal · swollen · lysed

RED BLOOD CELL

ion concentration in extracellular space ←

HYPERTONIC · ISOTONIC · HYPOTONIC · VERY HYPOTONIC

Figure 6–50 Response of a human red blood cell to changes in osmolarity (also called *tonicity*) of the extracellular fluid. Because the plasma membrane is freely permeable to water, water will move into or out of cells down its concentration gradient, a process called *osmosis*. If cells are placed in a *hypotonic solution* (i.e., a solution having a low solute concentration and therefore a high water concentration), there will be a net movement of water into the cells, causing them to swell and burst (lyse). Conversely, if cells are placed in a *hypertonic solution*, they will shrink (see also Panel 2–1, p. 47).

contractile vacuoles, which periodically discharge their contents to the exterior (see Panel 6–1, p. 308). But for most cells in multicellular animals, the Na^+-K^+ ATPase is crucial.

Some Ca^{2+} Pumps Are Also Membrane-bound ATPases[24]

Eucaryotic cells maintain very low concentrations of free Ca^{2+} in their cytosol ($\sim 10^{-7}$ M) in the face of very much higher extracellular Ca^{2+} concentrations ($\sim 10^{-3}$ M). Even a small influx of Ca^{2+} increases significantly the concentration of free Ca^{2+} in the cytosol, and the flow of Ca^{2+} down its steep concentration gradient in response to extracellular signals is one means of transmitting these signals rapidly across the plasma membrane (see p. 701). The Ca^{2+} gradient is in part maintained by Ca^{2+} pumps in the plasma membrane that actively transport Ca^{2+} out of the cell. One of these is known to be an ATPase, while the other is an antiporter that is driven by the Na^+ electrochemical gradient (see below).

The best-understood Ca^{2+} pump is a membrane-bound ATPase in the *sarcoplasmic reticulum* of muscle cells. The sarcoplasmic reticulum forms a network of tubular sacs in the cytoplasm of muscle cells and serves as an intracellular store of Ca^{2+}. (When an action potential depolarizes the muscle cell membrane, Ca^{2+} is released from the sarcoplasmic reticulum into the cytosol, stimulating the muscle to contract—see p. 624.) The Ca^{2+} pump is responsible for pumping Ca^{2+} from the cytosol into the sarcoplasmic reticulum. Like the Na^+-K^+ pump, it is an ATPase that is phosphorylated and dephosphorylated during its pumping cycle, and it pumps two Ca^{2+} ions into the sarcoplasmic reticulum for every ATP molecule hydrolyzed. Since the Ca^{2+} ATPase accounts for about 90% of the membrane protein of the sarcoplasmic reticulum, it has been relatively easy to purify. It is a single, large, multipass transmembrane polypeptide chain (containing about 1000 amino acid residues); and when it is incorporated into phospholipid vesicles (as described above for the Na^+-K^+ ATPase), it pumps Ca^{2+} as it hydrolyzes ATP. DNA cloning and sequencing experiments indicate that the Ca^{2+} ATPase is homologous in amino acid sequence to the large catalytic subunit of the Na^+-K^+ ATPase, revealing that these two ion pumps are evolutionarily related.

In nonmuscle cells an organelle equivalent to the sarcoplasmic reticulum likewise contains a Ca^{2+} ATPase that enables it to take up Ca^{2+} from the cytosol; the Ca^{2+} sequestered in this way is released back into the cytosol in response to specific extracellular signals, as discussed in Chapter 12 (see p. 701).

Membrane-bound Enzymes That Synthesize ATP Are Transport ATPases Working in Reverse[25]

The plasma membrane of bacteria, the inner membrane of mitochondria, and the thylakoid membrane of chloroplasts all contain an enzyme that is analogous to the two transport ATPases discussed above, but it normally works in reverse. Instead of ATP hydrolysis driving ion transport (as is usually the case for the Na^+-K^+ ATPase and Ca^{2+} ATPase), H^+ gradients across these membranes drive the synthesis of ATP from ADP and phosphate. The H^+ gradients are generated during the electron-transport steps of oxidative phosphorylation (in aerobic bacteria and mitochondria) or photosynthesis (in chloroplasts) or by the light-activated H^+

To transport Ca^{2+} out

1. ATPase

2. antiporter that is driven by Na^+ e-chemical gradient

SOURCES OF INTRACELLULAR OSMOLARITY

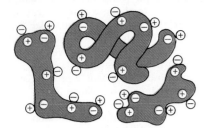

Macromolecules themselves contribute very little to the osmolarity of the cell interior since, despite their large size, each one counts only as a single molecule and there are relatively few of them compared to the number of small molecules in the cell. However, most biological macromolecules are highly charged, and they attract many inorganic ions of opposite charge. Because of their large numbers, these counterions make a major contribution to intracellular osmolarity.

As the result of active transport and metabolic processes, the cell contains a high concentration of small organic molecules, such as sugars, amino acids, and nucleotides, to which its plasma membrane is impermeable. Because most of these metabolites are charged, they also attract counterions. Both the small metabolites and their counterions make a further major contribution to intracellular osmolarity.

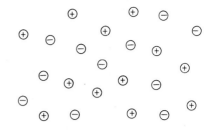

The osmolarity of the extracellular fluid is usually due mainly to small inorganic ions. These leak slowly across the plasma membrane into the cell. If they were not pumped out, and if there were no other molecules inside the cell that interacted with them so as to influence their distribution, they would eventually come to equilibrium with equal concentrations inside and outside the cell. However, the presence of charged macromolecules and metabolites in the cell that attract these ions gives rise to the Donnan effect: it causes the total concentration of inorganic ions (and therefore their contribution to the osmolarity) to be greater inside than outside the cell at equilibrium.

THE PROBLEM

Because of the above factors, a cell that does nothing to control its osmolarity will have a higher total concentration of solutes inside than outside. As a result, water will be higher in concentration outside the cell than inside. This difference in water concentration across the plasma membrane will cause water to move continuously into the cell by osmosis, causing it to rupture.

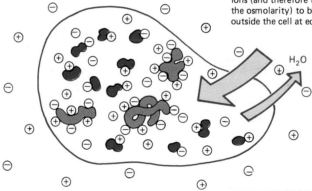

THE SOLUTION

Animal cells and bacteria control their intracellular osmolarity by actively pumping out inorganic ions, such as Na^+, so that their cytoplasm contains a lower total concentration of inorganic ions than the extracellular fluid, thereby compensating for their excess of organic solutes.

Plant cells are prevented from swelling by their rigid walls and so can tolerate an osmotic difference across their plasma membranes: an internal turgor pressure is built up, which at equilibrium forces out as much water as enters.

Many protozoa avoid becoming swollen with water, despite an osmotic difference across the plasma membrane, by periodically extruding water from special contractile vacuoles.

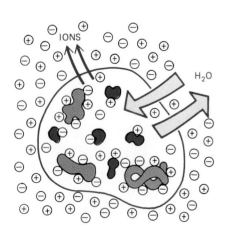

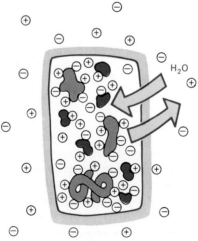

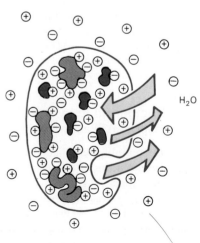

pump (bacteriorhodopsin) in *Halobacterium*. The enzyme that normally synthesizes ATP, called *ATP synthetase*, can, like the transport ATPases, work in either direction, depending on the conditions: it can hydrolyze ATP and pump H^+ across the membrane, or it can synthesize ATP when H^+ flows through the enzyme in the reverse direction. ATP synthetase is responsible for producing nearly all of the ATP in most cells, and it is discussed in detail in Chapter 7 (see p. 356).

Active Transport Can Be Driven by Ion Gradients[26]

Many active transport systems are driven by the energy stored in ion gradients rather than by ATP hydrolysis. All of these function as coupled transporters—some as symports, others as antiports. In animal cells, Na^+ is the usual co-transported ion whose electrochemical gradient provides the driving force for the active transport of a second molecule. The Na^+ that enters the cell with the solute is pumped out by the Na^+-K^+ ATPase, which, by maintaining the Na^+ gradient, indirectly drives the transport. Intestinal and kidney epithelial cells, for instance, contain a variety of symport systems that are driven by the Na^+ gradient across the plasma membrane, each system specific for importing a small group of related sugars or amino acids into the cell. In these systems the solute and Na^+ bind to different sites on a carrier protein; because the Na^+ tends to move into the cell down its electrochemical gradient, the sugar or amino acid is, in a sense, "dragged" into the cell with it. The greater the Na^+ gradient, the greater the rate of solute entry; conversely, solute transport stops if the Na^+ concentration in the extracellular fluid is markedly reduced. A hypothetical (and highly simplified) model of how such a symport system could work is shown in Figure 6–51.

In bacteria and plants, most active transport systems driven by ion gradients depend on H^+ rather than Na^+ gradients. The active transport of many sugars and amino acids into bacterial cells, for example, is driven by the H^+ gradient across the plasma membrane. The *lactose carrier* (permease) is the most extensively studied example. It is a single transmembrane polypeptide (composed of about 400 amino acid residues) that is thought to traverse the lipid bilayer at least nine times. It functions as a H^+ symporter: one proton is co-transferred for every lactose molecule transported into the cell.

Ion gradients also can be used to drive antiport systems. Two important examples are the antiports that function together to regulate intracellular pH in many animal cells.

Antiports in the Plasma Membrane Regulate Intracellular pH[27]

Almost all vertebrate cells have a Na^+-driven antiport, called a **Na^+-H^+ exchange carrier,** in their plasma membrane that plays a crucial part in maintaining intracellular pH (pH_i, usually around 7.1 or 7.2). This carrier couples the efflux of H^+ to the influx of Na^+, and it thereby removes excess H^+ ions produced as a result of acid-forming reactions in the cell. The Na^+-H^+ exchanger is regulated by pH_i: when pH_i rises above 7.7 in chick muscle cells, for example, the exchanger is inactive; as pH_i falls, the activity of the exchanger increases, reaching half-maximal activity at around pH 7.4. This regulation is mediated by the binding of H^+ to a regulatory site on the cytoplasmic surface of the exchanger. The importance of the Na^+-H^+ exchanger in pH_i control is demonstrated by the fate of mutant fibroblasts that lack the exchanger: they rapidly die when exposed to a large acid load that has little effect on the survival of normal fibroblasts.

A **Cl^--HCO_3^- exchanger,** similar to the band 3 protein in the membrane of red blood cells (see p. 291), is also thought to play an important part in pH_i regulation in many nucleated cells. Like the Na^+-H^+ exchanger, the Cl^--HCO_3^- exchanger is regulated by pH_i, but in the opposite direction. Its activity increases as pH_i rises, increasing the rate at which HCO_3^- is ejected from the cell in exchange for Cl^-, thereby decreasing pH_i whenever the cytosol becomes too alkaline.

There is evidence that the Na^+-H^+ exchanger may be involved in transducing extracellular signals into intracellular ones, as well as in pH_i regulation. For example, most protein growth factors activate this antiport system in the course of

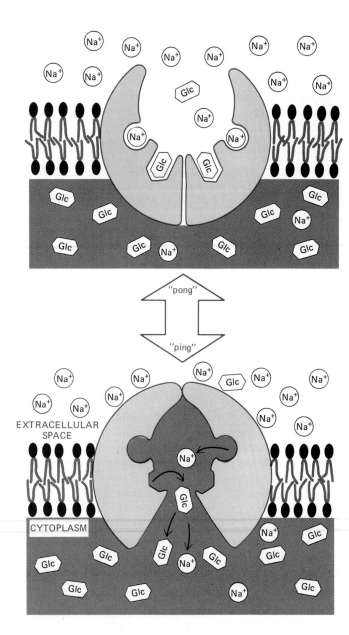

Figure 6–51 How a glucose pump could, in principle, be driven by a Na$^+$ gradient. The pump oscillates randomly between two states, "pong" and "ping," as in Figure 6–47. Although Na$^+$ binds equally well to the protein in either state, the binding of Na$^+$ induces an allosteric transition in the protein that greatly increases its affinity for glucose. Since the Na$^+$ concentration is higher in the extracellular space than in the cytosol, glucose is much more likely to bind to the pump in the "pong" state; therefore, both Na$^+$ and glucose enter the cell (via a pong → ping transition) much more often than they leave it (via a ping → pong transition). As a result, the system carries both glucose and Na$^+$ into the cell. By generating and maintaining the Na$^+$ gradient, the Na$^+$-K$^+$ ATPase indirectly provides the energy for such a symport system. For this reason, ion-driven carriers are said to mediate *secondary active transport* while transport ATPases are said to mediate *primary active transport*.

stimulating cell proliferation, increasing the pH$_i$ from 7.1 or 7.2 to about 7.3. In some cases at least, they do so indirectly through the activation of a specific protein kinase (protein kinase C—see p. 704), which is thought to phosphorylate the exchanger, thereby increasing the affinity of its regulatory binding site for H$^+$ so that it remains active at a higher pH. Mutant cells that are deficient in the Na$^+$-H$^+$ exchanger, as well as cells that are treated with the drug *amiloride*, which inhibits the exchanger, fail to respond to these growth factors. These findings suggest that the activation of the exchanger and the resulting increase in pH$_i$ play an important part in initiating cell proliferation. Similarly, following the fertilization of sea urchin eggs, an increase in pH$_i$ caused by activation of the Na$^+$-H$^+$ exchanger activates both protein and DNA synthesis in the egg (see p. 874); it is not known which intracellular proteins respond to the increase in pH$_i$ to cause these activations.

An Asymmetrical Distribution of Carrier Proteins in Epithelial Cells Underlies the Transcellular Transport of Solutes[28]

In some epithelial cells, such as those involved in absorbing nutrients from the gut, carrier proteins are distributed asymmetrically in the plasma membrane and thereby contribute to the **transcellular transport** of absorbed solutes. As shown

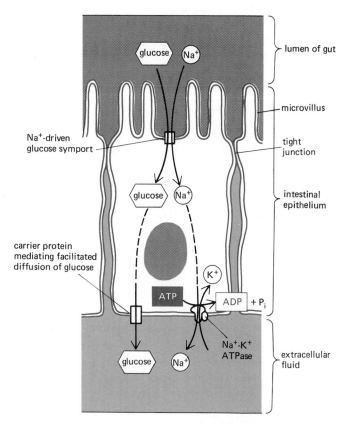

glucose   Na+

lumen of gut

microvillus

Na+-driven
glucose symport

tight
junction

glucose   Na+

intestinal
epithelium

carrier protein
mediating facilitated
diffusion of glucose

K+

ATP

ADP + P_i

glucose   Na+

Na+-K+
ATPase

extracellular
fluid

Figure 6–52 The asymmetrical distribution of transport proteins in the plasma membrane of an intestinal epithelial cell results in the transcellular transport of glucose from the gut lumen to the extracellular fluid (from where it passes into the blood). Glucose is pumped into the cell through the apical domain of the membrane by a Na^+-powered glucose symport, and glucose passes out of the cell (down its concentration gradient) by facilitated diffusion mediated by a different glucose carrier protein in the basal and lateral domain. The Na^+ gradient driving the glucose symport is maintained by the Na^+-K^+ ATPase in the basolateral plasma membrane domain, which keeps the internal concentration of Na^+ low.

Adjacent cells are connected by impermeable junctions (called *tight junctions*—see p. 793), which have a dual function in the transport process illustrated. These junctions prevent solutes from crossing the epithelium between cells, allowing a concentration gradient of glucose to be maintained across the cell sheet. The tight junctions are thought to serve also as diffusion barriers within the plasma membrane, confining the various carrier proteins to their respective membrane domains (see Figure 6–36, p. 297).

in Figure 6–52, Na^+-linked symports, located in the apical (absorptive) domain of the plasma membrane, actively transport nutrients into the cell, building up substantial concentration gradients, while Na^+-independent transport proteins in the basal and lateral (basolateral) domain allow nutrients to leave the cell passively down these concentration gradients. The Na^+-K^+ ATPase that maintains the Na^+ gradient across the plasma membrane of these cells is located in the basolateral domain. Related mechanisms are thought to be used by kidney and intestinal epithelial cells to pump water from one extracellular space to another.

In many of these epithelial cells, the plasma membrane area is greatly increased by the formation of thousands of **microvilli,** which extend as thin, fingerlike projections from the apical surface (see Figure 6–52). Such microvilli can increase the total absorptive area of a cell by as much as 25-fold, thereby greatly increasing its transport capabilities. Because the apical surface of gut epithelial cells is also a site where immobilized hydrolytic enzymes involved in the final stages of food digestion are located, both digestion and absorption are greatly enhanced by the increase in surface area afforded by the microvilli in this epithelium.

Active Transport in Bacteria Can Occur by "Group Translocation"[29]

Thus far we have seen that active transport can be driven by light (as in bacteriorhodopsin), by ATP hydrolysis, or by ion gradients. A fourth strategy, which operates in many bacteria, is to "trap" a molecule that has entered the cell passively by modifying it in such a way that it cannot escape by the same route. In some bacteria, for example, sugars are phosphorylated after their transfer across the plasma membrane. Because they are ionized and cannot leak out, the resulting sugar phosphates accumulate in the cell. Moreover, sugars entering the cell are phosphorylated immediately, so that the concentration of unphosphorylated sugars inside the cell is kept very low and sugar from outside can continue to enter down its concentration gradient. Because a phosphate group is transferred to the solute after its transport, this type of active transport is called **group translocation.** In the most extensively studied example, the phosphorylation mechanism is complex

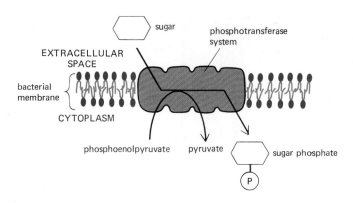

EXTRACELLULAR SPACE

bacterial membrane

CYTOPLASM

sugar

phosphotransferase system

phosphoenolpyruvate   pyruvate   sugar phosphate

P

Figure 6–53 Active transport of sugars into bacteria by group translocation. A special "phosphotransferase system" of proteins in the bacterial membrane phosphorylates the sugar after the sugar's transport through the membrane. Phosphoenolpyruvate rather than ATP is the phosphate donor.

and highly regulated, involving at least four separate membrane proteins and phosphoenolpyruvate (rather than ATP) as the high-energy-phosphate donor (Figure 6–53).

Bacteria with Double Membranes Have Transport Systems That Depend on Water-soluble Substrate-binding Proteins[30]

As mentioned previously, the plasma membranes of all bacteria contain carrier proteins that use the H$^+$ gradient across the membrane to pump a variety of nutrients into the cell. But many bacteria, including E. coli, also have a surrounding outer membrane, through which solutes of up to 600 daltons can diffuse relatively freely through a variety of channel-forming proteins (known collectively as porins) (Figure 6–54). In these bacteria some sugars, amino acids, and small peptides are transported across the inner (plasma) membrane via a two-component transport system that utilizes water-soluble proteins located in the periplasmic space between the two membranes. These **periplasmic substrate-binding proteins** bind the specific molecule to be transported and, as a consequence, undergo a conformational change that enables them to bind to the second component in the transport system, which is a transmembrane carrier protein located in the inner membrane (Figure 6–55). It is thought that the substrate-binding proteins pass the bound solute to the carrier, which then uses the energy of ATP hydrolysis to transfer the solute across the inner membrane into the cell. The same periplasmic substrate-binding proteins serve as receptors in chemotaxis, an adaptive response that enables bacteria to swim toward an increasing concentration of a specific nutrient (see p. 720).

We now turn from carrier proteins to channel proteins.

Channel Proteins Form Aqueous Pores in the Plasma Membrane[31]

Unlike carrier proteins, **channel proteins** form water-filled pores across membranes. But whereas the channel-forming proteins of the outer membranes of bacteria (and of mitochondria and chloroplasts) have large, relatively unselective pores, channel proteins in the plasma membranes of animal and plant cells have

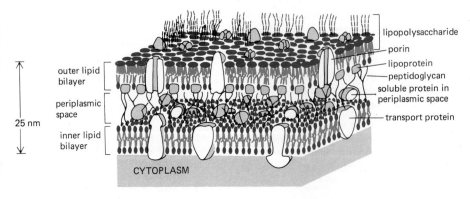

outer lipid bilayer

periplasmic space

25 nm

inner lipid bilayer

CYTOPLASM

lipopolysaccharide
porin
lipoprotein
peptidoglycan
soluble protein in periplasmic space
transport protein

Figure 6–54 Schematic view of a small section of the double membrane of an E. coli bacterium. The inner membrane is the cell's plasma membrane. Between the inner and outer lipid bilayer membranes there is a highly porous, rigid peptidoglycan composed of protein and polysaccharide that constitutes the bacterial cell wall; it is attached to lipoprotein molecules in the outer membrane and fills the periplasmic space. This space also contains a variety of soluble protein molecules. The dashed black threads at the top represent the polysaccharide chains of the special lipopolysaccharide molecules that form the external monolayer of the outer membrane; for clarity, only a few of these chains are shown. Bacteria with double membranes are called gram negative because they do not retain the dark blue dye used in the gram staining procedure. Bacteria with single membranes (but thicker cell walls), such as staphylococci and streptococci, retain the blue dye and therefore are called gram positive; their single membrane is analogous to the inner (plasma) membrane of gram-negative bacteria.

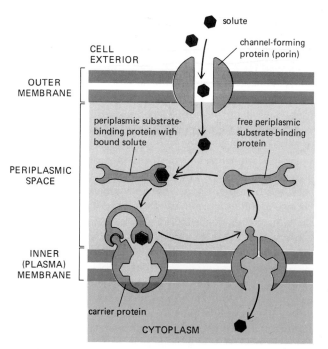

solute

CELL
EXTERIOR

channel-forming
protein (porin)

OUTER
MEMBRANE

periplasmic substrate-
binding protein with
bound solute

free periplasmic
substrate-binding
protein

PERIPLASMIC
SPACE

INNER
(PLASMA)
MEMBRANE

carrier protein

CYTOPLASM

Figure 6–55 The transport system that depends on *periplasmic substrate-binding proteins* in bacteria with double membranes. The solute diffuses through channel-forming proteins (called *porins*) in the outer membrane and binds to a periplasmic substrate-binding protein. As a result, the substrate-binding protein undergoes a conformational change that enables it to bind to a carrier protein in the plasma membrane, which then picks up the solute and actively transfers it across the bilayer in a reaction driven by ATP hydrolysis. The peptidoglycan is omitted for simplicity; its porous structure allows the substrate-binding proteins and water-soluble solutes to move through it by simple diffusion.

small, highly selective pores. Almost all of the latter proteins are concerned specifically with ion transport and so are referred to as **ion channels.** More than 10^6 ions can pass through such a channel each second, which is a rate more than 100 times greater than the transport mediated by any known carrier protein. On the other hand, ion channels cannot be coupled to an energy source, so the transport they mediate is always passive ("downhill"), allowing specific ions, mainly Na^+, K^+, Ca^{2+}, or Cl^-, to diffuse down their electrochemical gradients across the lipid bilayer.

The channel proteins in plasma membranes show *ion selectivity,* permitting some ions to pass but not others. This suggests that their pores must be narrow enough in places to force permeating ions into intimate contact with the walls of the channel so that only ions of appropriate size and charge can pass. It is thought that the permeating ions have to shed most or all of their associated water molecules in order to get through the narrowest part of the channel. This both limits their maximum rate of passage and acts as a selective filter, letting only certain ions pass through. Thus, as ion concentrations are increased, the flux of ions through a channel increases proportionally but then levels off (saturates) at a certain maximum rate.

Another way in which ion channels differ from simple aqueous pores is that they are not continuously open. Instead they have "gates," which open briefly and then close again, as shown schematically in Figure 6–56. In most cases the gates open in response to a specific perturbation of the membrane. The main types of perturbations that are known to cause ion channels to open are a change in the voltage across the membrane (*voltage-gated channels*), mechanical stimulation

Figure 6–56 Schematic drawing of a gated ion channel in its closed and open conformations. A transmembrane protein, seen in cross-section, forms an aqueous pore across the lipid bilayer only when the gate is open. Hydrophilic amino acid side chains are thought to line the wall of the pore; hydrophobic side chains interact with the lipid bilayer. The pore narrows to atomic dimensions in one region (the "ion-selective filter"), where the ion selectivity of the channel is determined. A transient opening of the gate is caused by a specific perturbation of the membrane, which is different for different channels, as discussed in the text. The location of the gate and ion-selective filter, shown here on the external side of the membrane, is unknown for most channels.

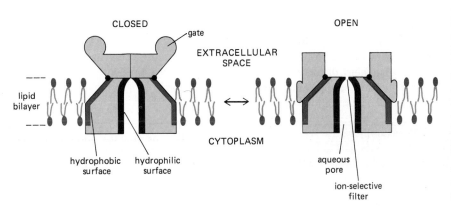

CLOSED

gate

OPEN

EXTRACELLULAR
SPACE

lipid
bilayer

CYTOPLASM

hydrophobic
surface

hydrophilic
surface

aqueous
pore

ion-selective
filter

(*mechanically gated channels*—see p. 1103), or the binding of a signaling molecule (*ligand-gated channels*). The signaling ligand can be either an extracellular mediator, called a *neurotransmitter* (*transmitter-gated channels*), or an intracellular mediator, such as an ion (*ion-gated channels*—see p. 1189), a nucleotide (*nucleotide-gated channels*—see p. 713), or a GTP-binding regulatory protein (*G-protein-gated channels*—see p. 706).

Approximately 50 types of ion channels have been described thus far, and new ones are still being discovered. They are responsible for the electrical excitability of nerve and muscle cells and mediate most forms of electrical signaling in the nervous system. A single nerve cell typically contains more than five kinds of ion channels. But these channels are not restricted to electrically excitable cells. They are present in all animal cells and are found in plant cells and microorganisms: they propagate the leaf-closing response of the mimosa plant, for example, and allow the single-celled paramecium to reverse direction after a collision.

Perhaps the most common ion channels are those that are permeable mainly to K^+ and are found in the plasma membrane of almost all animal cells. Because they seem not to require a specific membrane perturbation in order to open, they are sometimes called K^+ *leak channels*. Although poorly characterized and heterogeneous, they are the reason that most plasma membranes are much more permeable to K^+ than to other ions, and they play a critical part in maintaining the *membrane potential*—the voltage difference that is present across all plasma membranes.

The Membrane Potential Depends Largely on K^+ Leak Channels and the K^+ Gradient Across the Membrane

A **membrane potential** arises when there is a difference in the electric charge on the two sides of a membrane, due to a slight excess of positive ions over negative on one side and a slight deficit on the other. Such charge differences can result both from passive ion diffusion and from active electrogenic pumping. We shall see in Chapter 7 (p. 351) that most of the membrane potential of the mitochondrion is generated by electrogenic proton pumps in the mitochondrial inner membrane. Electrogenic pumps also generate most of the plasma membrane potential in plants and fungi. In typical animal cells, however, ion diffusion makes the largest contribution to the potential across the plasma membrane.

As explained earlier, the Na^+-K^+ ATPase helps to maintain osmotic balance across the animal cell membrane by keeping the intracellular concentration of Na^+ low. Because there is little Na^+ inside the cell, other cations have to be plentiful there to balance the charge carried by the cell's fixed anions—the negatively charged organic molecules that are confined inside the cell. This balancing role is performed largely by K^+, which is actively pumped into the cell by the Na^+-K^+ ATPase but can also move freely in or out through the **K^+ leak channels.** Thanks to the K^+ leak channels, K^+ comes very nearly to an equilibrium in which the electrical force due to negative charges attracting K^+ into the cell balances the tendency of K^+ to leak out down its concentration gradient. The membrane potential is the manifestation of this electrical force, and its size can be calculated from the steepness of the K^+ concentration gradient. The following argument may help to make this clear.

Suppose that initially there is no voltage gradient across the plasma membrane (the membrane potential is zero) but the concentration of K^+ is high inside the cell and low outside. K^+ will tend to leave the cell through the K^+ leak channels, driven by its concentration gradient. As K^+ moves out of the cell, it will leave behind negative charge, thereby creating an electrical field, or membrane potential, that will tend to oppose the further efflux of K^+. The net efflux of K^+ will halt if the membrane potential reaches a value where this electrical driving force on K^+ exactly balances the effect of its concentration gradient—that is, when the electrochemical gradient for K^+ is zero. Cl^- ions also equilibrate across the membrane, but because their charge is negative, the membrane potential keeps most of these diffusible ions out of the cell. This equilibrium condition, in which there is no net

THE NERNST EQUATION AND ION FLOW

The flow of any ion through a membrane channel protein is driven by the **electrochemical gradient** for that ion. This gradient represents the combination of two influences: the voltage gradient and the concentration gradient of the ion across the membrane. When these two influences just balance each other, the electrochemical gradient for the ion is zero and there is no *net* flow of the ion through the channel. The voltage gradient (membrane potential) at which this equilibrium is reached is called the **equilibrium potential** for the ion. It can be calculated from an equation that will be derived below, called the **Nernest equation**.

The Nernst equation is

$$V = \frac{RT}{zF} \ln \frac{C_o}{C_i}$$

where

V = the equlibrium potential in volts (internal potential minus the external potential)

C_o and C_i = outside and inside concentrations of the ion, respectively

R = the gas constant (2 cal mol^{-1} °K^{-1})

T = the absolute temperature (°K)

F = Faraday's constant (2.3 × 10^4 cal V^{-1} mol^{-1})

z = the valence (charge) of the ion

The Nernst equation is derived as follows:

A molecule in solution (a solute) always moves from a region of high concentration to a region of low concentration simply due to the pressure of numbers. Consequently, movement down a concentration gradient is accompanied by a favorable free-energy change ($\Delta G < 0$), whereas movement up a concentration gradient is accompanied by an unfavorable free-energy change ($\Delta G > 0$). (Free energy is introduced and discussed in Panel 2–7, pp. 72–73.) The free-energy change per mole of solute moved across the plasma membrane (ΔG_{conc}) is equal to $-RT \ln C_o/C_i$. If the solute is an ion, moving it into a cell across a membrane whose inside is at a voltage V relative to the outside will cause an additional free-energy change (per mole of solute moved) of $\Delta G_{volt} = zFV$. At the point where the concentration and voltage gradients just balance, $\Delta G_{conc} + \Delta G_{volt} = 0$ and the ion distribution is at equilibrium across the membrane. Thus,

$$zFV - RT \ln \frac{C_o}{C_i} = 0$$

and therefore,

$$V = \frac{RT}{zF} \ln \frac{C_o}{C_i} = 2.3 \frac{RT}{zF} \log_{10} \frac{C_o}{C_i}$$

For a univalent ion,

$$2.3 \frac{RT}{F} = 58 \text{ mV at 20 °C} \quad \text{and} \quad 61.5 \text{ mV at 37°C}$$

Therefore, when the membrane potential has a value of 61.5 $\log_{10}([K^+]_o/[K^+]_i)$ millivolts (-89 mV when $[K^+]_o = 5$ mM and $[K^+]_i = 140$ mM), which is the K$^+$ equilibrium potential (V_K), there is no net flow of K$^+$ across the membrane. Similarly, when the membrane potential has a value of 61.5 $\log_{10}([Na^+]_o/[Na^+]_i)$, the Na$^+$ equilibrium potential (V_{Na}), there is no net flow of Na$^+$. For a typical cell, V_K is between -70 mV and -100 mV and V_{Na} is between $+50$ mV and $+65$ mV.

For any particular membrane potential V_M, the net force tending to drive a particular type of ion out of the cell is proportional to the difference between V_M and the equilibrium potential for the ion: hence, for K$^+$ it is $V_M - V_K$ and for Na$^+$ it is $V_M - V_{Na}$. The actual current carried by each type of ion depends not only on this driving force but also on the ease with which that ion passes through its membrane channels, which is a function of the *conductance* of the channels. If the conductances of the sets of channels for K$^+$ and Na$^+$ are, respectively, g_K and g_{Na}, from Ohm's law the K$^+$ and Na$^+$ currents are, respectively, $g_K(V_M - V_K)$ and $g_{Na}(V_M - V_{Na})$. (**Conductance** is the reciprocal of resistance, and the unit of measurement is the reciprocal of ohm, or siemens, S. Most individual ion channels have conductances in the range of 1 to 150 × 10^{-12} S, or 1 to 150 pS.) At the resting membrane potential of most cells, the K$^+$ leak channels are the main type of channel open and, therefore, g_K dominates the total conductance of the membrane. During an action potential in nerve and muscle cells, large numbers of voltage-gated Na$^+$ channels briefly open, and g_{Na} comes to dominate the total membrane conductance for a moment.

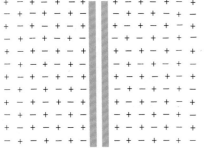

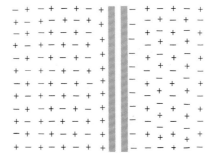

exact balance of charges on each side of the membrane; membrane potential = 0

a few of the positive ions (color) cross the membrane from right to left, leaving their negative counterions behind (color); this sets up a nonzero membrane potential

Figure 6–57 A small flow of ions carries sufficient charge to cause a large change in the membrane potential. The ions that give rise to the membrane potential lie in a surface layer close to the membrane, held there by their electrical attraction to their oppositely charged counterparts (counterions) on the other side of the membrane. For a typical cell, 1 microcoulomb of charge (6×10^{12} univalent ions) per square centimeter of membrane, transferred from one side of the membrane to the other, will change the membrane potential by roughly 1 V. This means, for example, that in a spherical cell of diameter 10 μm, the number of K^+ ions that have to flow out to alter the membrane potential by 100 mV is only about 1/100,000 of the total number of K^+ ions in the cytosol.

flow of electric current across the membrane, defines the **resting membrane potential** for the cell. A simple but very important formula, the **Nernst equation,** expresses the equilibrium condition quantitatively and, as explained in Panel 6–2, makes it possible to calculate the resting membrane potential if the ratio of internal and external ion concentrations is known.

The number of ions that must move across the membrane to set up the membrane potential is minute. Thus one can think of the membrane potential as arising from movements of charge that leave ion *concentrations* practically unaffected and result in only a very slight discrepancy in the number of positive and negative ions on the two sides of the membrane (Figure 6-57). Moreover, these movements of charge are generally rapid, taking only a few milliseconds or less.

It is illuminating to consider what happens if the Na^+-K^+ ATPase is suddenly inactivated. First, there is an immediate slight drop in the membrane potential. This is because the pump is electrogenic and, when active, makes a small direct contribution to the membrane potential by driving a current out of the cell (see p. 306). Switching off the pump, however, does not abolish the major component of the resting potential, which is generated by the K^+ equilibrium mechanism outlined above. This persists as long as the K^+ concentration inside the cell stays high—typically for many minutes. The plasma membrane, however, is somewhat permeable to all small ions (including Na^+). Therefore, the ion gradients set up by pumping will eventually run down without the Na^+-K^+ ATPase, and the membrane potential established by diffusion through K^+ channels will fall as well. At the same time, the osmotic balance is upset (see p. 306); but if the cell does not burst, it eventually comes to a new resting state where Na^+, K^+, and Cl^- are all at equilibrium across the membrane. The membrane potential in this state is much less than it was in the normal cell with an active Na^+-K^+ pump.

The potential difference across the plasma membrane of a cell at rest varies between −20 mV and −200 mV, depending on the organism and cell type. Although the K^+ gradient always has a major influence on this potential, the gradients of other ions (and the disequilibrating effects of ion pumps) also have a significant effect: the more permeable the membrane for a given ion, the more strongly the membrane potential tends to be driven toward the equilibrium value for that ion. Consequently, almost any change of a membrane's permeability to ions causes a change in the membrane potential. This is the key principle relating the electrical excitability of cells to the activities of ion channels.

6-25 Voltage-gated Ion Channels Are Responsible for the Electrical Excitability of Nerve and Muscle Cells[33]

The plasma membranes of electrically excitable cells (mainly nerve and muscle cells) contain **voltage-gated ion channels,** which are responsible for generating **action potentials**—rapid, transient, self-propagating electrical excitations of the membrane. An action potential is triggered by a *depolarization* of the membrane—that is, by a shift in the membrane potential to a less negative value. A stimulus

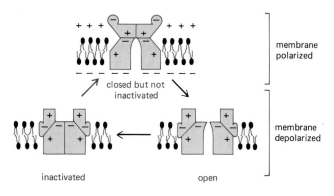

closed but not inactivated

membrane polarized

inactivated   open

membrane depolarized

Figure 6–58 The voltage-gated Na^+ channel can adopt at least three conformations (states). Internal forces, represented here by attractions between charges on different parts of the channel, stabilize each state against small disturbances, but a sufficiently violent collision with other molecules can cause the channel to flip from one of these states to another. The state of lowest energy depends on the membrane potential because the different conformations have different charge distributions. When the membrane is at rest (highly polarized), the *closed but not inactivated* conformation has the lowest free energy and is therefore most stable; when the membrane is depolarized, the energy of the *open* conformation is lower and so the channel opens. But the free energy of the *inactive* conformation is lower still, and so, after a randomly variable period spent in the open state, the channel becomes inactivated. Thus the open conformation corresponds to a metastable state that can exist only transiently. The black arrows indicate the sequence that follows a sudden depolarization, while the red arrow indicates the return to the original conformation as the lowest energy state after the membrane is repolarized.

that causes a momentary partial depolarization promptly causes **voltage-gated Na^+ channels** to open, allowing a small amount of Na^+ to enter the cell. The influx of positive charge depolarizes the membrane further, thereby opening more Na^+ channels, which admit more Na^+ ions, causing still further depolarization. This process continues in a self-amplifying fashion until the potential in the local region of membrane has shifted from its resting value of about -70 mV all the way to the Na^+ equilibrium potential of about $+50$ mV (see Panel 6–2, p. 311). At this point, when the net electrochemical driving force for the flow of Na^+ is zero, the cell would come to a new resting state with all of its Na^+ channels permanently open if the open conformation of the channel were stable.

The cell is saved from such a permanent electrical spasm because the Na^+ channels have an automatic inactivating mechanism, and they rapidly reclose even though the membrane is still depolarized. In this *inactivated* state, the channel cannot open again until a few milliseconds after the membrane potential returns to its initial negative value. A schematic illustration of these three distinct states of the voltage-gated Na^+ channel—closed but not inactivated, open, and inactivated—is shown in Figure 6–58. How they contribute to the rise and fall of the action potential is shown in Figure 6–59.

The description just given of an action potential concerns only a small patch of plasma membrane. However, the self-amplifying depolarization of the patch is sufficient to depolarize neighboring regions of membrane, which then go through the same cycle. In this way the action potential spreads from the initial site of depolarization to involve the entire plasma membrane. A more detailed discussion

Figure 6–59 The triggering of an action potential by a brief pulse of current (shown in the upper graph), which partially depolarizes the membrane (middle graph). In the plot of membrane potential, the solid curve shows the course of an action potential due to the opening and subsequent inactivation of voltage-gated Na^+ channels. The return of the membrane potential to its initial value of -70 mV is automatic once the Na^+ channels close because of the continuous efflux of K^+ through K^+ channels (see p. 314). The membrane cannot fire a second action potential until the Na^+ channels, whose state is shown at the bottom, have returned to the closed but not inactivated conformation (see Figure 6–58); until then the membrane is refractory to stimulation. The dashed curve shows how the membrane potential would have simply relaxed back to the resting value after the initial depolarizing stimulus if there had been no voltage-gated ion channels in the membrane.

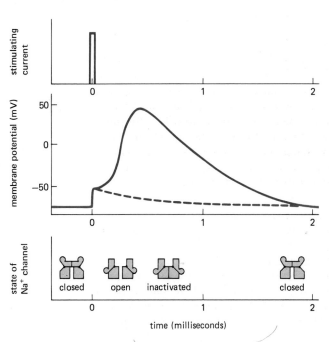

stimulating current

membrane potential (mV)

50

0

−50

state of Na^+ channel

closed   open   inactivated   closed

time (milliseconds)

refractory

of the properties and functions of action potentials is given in Chapter 19 (see pp. 1065–1074).

Neurons and muscle cell membranes contain many thousands of voltage-gated Na^+ channels, and the current crossing the membrane is the sum of the currents flowing through all of these. This aggregate current can be recorded with a microelectrode, as described in Chapter 19 (see p. 1071). However, it is also possible to record current flowing through individual channels by means of *patch-clamp recording*, which provides a much more detailed picture of the properties of these channels.

6-27 Patch-Clamp Recording Indicates That Individual Na^+ Channels Open in an All-or-None Fashion[34]

Patch-clamp recording has revolutionized the study of ion channels. With this technique, transport through a single molecule of channel protein can be studied in a small patch of membrane covering the mouth of a micropipette (Figure 6–60). This makes it possible to record from ion channels in all sorts of cell types, including those that are not electrically excitable. Many of these cells, such as yeasts, are too small to be investigated by the traditional electrophysiologist's method of impalement with an intracellular electrode.

Patch-clamp recording indicates that individual Na^+ channels open in an all-or-nothing fashion: when open, the channel always has the same conductance, although the times of its opening and closing are random. Therefore, the aggregate current crossing the membrane of an entire cell through a large population of Na^+ channels does not indicate the *degree* to which a typical individual channel is open, but rather the average *probability* that it will open (Figure 6–61).

The phenomenon of voltage gating can be understood in terms of simple physical principles. The interior of the resting nerve or muscle cell is at an electric potential about 50–100 mV more negative than the external medium. This potential difference may seem small, but since it exists across a plasma membrane only about 5 nm thick, the resulting voltage gradient is about 100,000 V/cm. Proteins in the membrane are thus subjected to a very large electrical field. Membrane proteins, like all others, have a number of charged groups on their surface and polarized bonds between their various atoms. The electrical field therefore exerts forces on the molecular structure. For many membrane proteins the effects of changes in the membrane electrical field are probably insignificant. But voltage-gated ion channels have evolved a delicately poised sensitivity to the field: they can adopt a number of alternative conformations, whose stabilities depend on the strength of the field. Each conformation is stable against small disturbances, but it can "flip" to another conformation if given a sufficient jolt by the random thermal movements of the surroundings (see Figure 6–58).

The function of the voltage-gated Na^+ channel is blocked specifically by two paralytic poisons, *tetrodotoxin (TTX)*, made by puffer fish, and *saxitoxin*, made by certain marine dinoflagellates. Because of their high affinity and specificity, these toxins have been invaluable for pharmacological studies, for counting the number of Na^+ channels in a membrane, and for purifying them. In a skeletal muscle plasma membrane, for example, it can be shown that there are only a few hundred Na^+ channels per square micrometer, which is about one for every 10,000 phospholipid molecules. Despite this low density, the membranes are electrically excitable because each channel has a very large conductance, allowing more than 8000 ions to pass per millisecond.

The DNA sequence that encodes a voltage-gated Na^+ channel (in eels) was determined in 1984. It was found to code for a single very large polypeptide chain (about 1800 amino acid residues) containing four homologous transmembrane domains (each containing six putative membrane-spanning α helices), which are thought to cluster together to form the walls of an aqueous pore. More recently, the DNA sequence that codes for a voltage-gated Ca^{2+} channel was determined and shown to encode a large polypeptide with a structure very similar to that of the Na^+ channel, suggesting that voltage-gated ion channels belong to a family of evolutionarily and structurally related proteins. In each of these channels, one of the putative transmembrane segments contains regularly spaced, positively charged

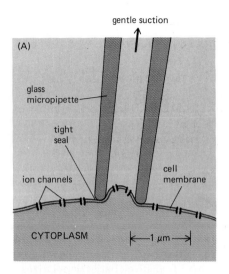

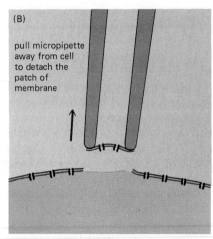

Figure 6–60 Patch-clamp recording. Because of the extremely tight seal between the micropipette and the membrane, current can enter or leave the micropipette only by passing through the channels in the patch of membrane covering its tip. Recordings of the current through these channels can be made with the patch still attached to the rest of the cell, as in (A), or detached, as in (B). The advantage of the detached patch is that it is easy to alter the composition of the solution on either side of the membrane to test the effect of various solutes on channel behavior. The detached patch can also be in the opposite orientation, with the cytoplasmic surface of the membrane facing the lumen of the pipette. (See also Figures 4–33 and 4–34, p. 156).

Figure 6–61 Recordings of current through a single voltage-gated Na$^+$ channel in a tiny patch of plasma membrane detached from an embryonic rat muscle cell (see Figure 6–60). The membrane was depolarized by an abrupt shift of potential, as indicated in (A). The three current records shown in (B) are from three experiments performed on the same patch of membrane. Each major current step in (B) represents the opening and closing of a single channel. Comparison of the three records shows that, whereas the times of channel opening and closing vary greatly, the rate at which current flows through an open channel is practically constant. The minor fluctuations in the current record arise largely from electrical noise in the recording apparatus. The sum of the currents measured in 144 repetitions of the same experiment is shown in (C). This aggregate current is equivalent to the usual Na$^+$ current that would be observed flowing through a relatively large region of membrane containing 144 channels. Comparison of (B) and (C) reveals that the time course of the aggregate current reflects the probability that any individual channel will be in the open state; this probability decreases with time as the channels in the depolarized membrane adopt their inactivated conformation. The kinetics of channel opening and inactivation are much slower for this embryonic muscle cell than for a typical nerve cell. (Based on data from J. Patlak and R. Horn, *J. Gen. Physiol.* 79:333–351, 1982, by copyright permission of the Rockefeller University Press.)

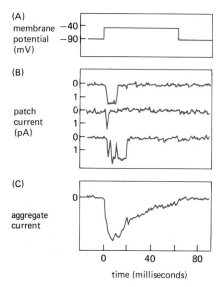

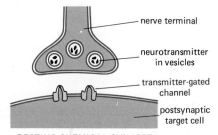

amino acid residues, which together are thought to function as a voltage sensor, ensuring that the channel opens when the membrane is sufficiently depolarized (see Figure 6–58).

But more is known about the structure of another class of ion channels, which open in response to the binding of a specific neurotransmitter rather than to a change in membrane potential. These *transmitter-gated ion channels* also belong to a family of closely related proteins; unlike the voltage-gated Na$^+$ and Ca^{2+} channels, however, each of which is constructed from a single large polypeptide, all of the transmitter-gated ion channels studied so far are constructed from multiple homologous subunits.

The Acetylcholine Receptor Is a Transmitter-gated Cation Channel[35]

Transmitter-gated ion channels are specialized for converting extracellular chemical signals into electrical signals and are located at specialized junctions (called *chemical synapses*) between nerve cells and their target cells. The channels are concentrated in the target cell plasma membrane in the region of the synapse. They open transiently in response to the binding of a *neurotransmitter* released by the nerve terminal, thereby producing a permeability change in the postsynaptic membrane of the target cell (Figure 6–62). Unlike the voltage-gated channels responsible for action potentials, transmitter-gated channels are relatively insensitive to the membrane potential and, therefore, cannot by themselves produce a self-amplifying excitation. Instead, they produce permeability changes, and hence changes of membrane potential, that are graded according to how much neurotransmitter is released at the synapse and how long it persists there. An action potential can be triggered only if voltage-gated channels are also present in the same target cell membrane.

In addition to having a characteristic ion selectivity, each transmitter-gated channel has a highly selective binding site for its particular neurotransmitter. The best-studied example of such a transmitter-gated channel is the **acetylcholine receptor** of skeletal muscle cells. This channel is opened transiently by *acetylcholine*, the neurotransmitter released from the nerve terminal at a *neuromuscular junction* (see p. 1075). The acetylcholine receptor has a special place in the history of ion channels. It was the first ion channel to be purified, the first to have its complete amino acid sequence determined, the first to be functionally reconstituted into synthetic lipid bilayers, and the first for which the electrical signal of a single open channel was recorded. Its gene was also the first channel protein gene

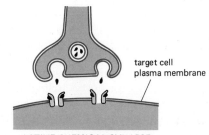

Figure 6–62 A chemical synapse. When an action potential reaches the nerve terminal, it stimulates the terminal to release its neurotransmitter; the neurotransmitter is contained in secretory vesicles and is released to the cell exterior when the vesicles fuse with the plasma membrane of the nerve terminal. The released neurotransmitter binds to and opens the transmitter-gated ion channels concentrated in the plasma membrane of the target cell at the synapse. The resulting ion flows alter the membrane potential of the target cell, thereby transmitting a signal from the excited nerve.

RESTING CHEMICAL SYNAPSE

ACTIVE CHEMICAL SYNAPSE

to be isolated, cloned, and sequenced. There are at least two reasons for the rapid progress in purifying and characterizing this receptor. First, there is an unusually rich source of the receptor in the electric organs of electric fish and rays; these organs are modified muscles designed to deliver a large electric shock to prey. Second, neurotoxins (such as α-bungarotoxin) in the venom of certain snakes bind with high affinity ($Ka = 10^9$ liters/mole) and specificity to the receptor and therefore can be used to purify it by affinity chromatography (see p. 168). Fluorescent or radiolabeled α-bungarotoxin can also be used to localize and count acetylcholine receptors; in this way it has been shown that the receptors are densely packed in the muscle cell plasma membrane at a neuromuscular junction (about 20,000 such receptors/μm^2), with relatively few receptors elsewhere in the same membrane.

The acetylcholine receptor is a glycoprotein composed of five transmembrane polypeptides, two of one kind and three others, encoded by four separate genes. The four genes are strikingly homologous, implying that they evolved from a single ancestral gene. The two identical polypeptides in the pentamer each have binding sites for acetylcholine. When two acetylcholine molecules bind to the pentameric complex, they induce a conformational change that opens the channel. The channel remains open for about 1 millisecond and then closes; it is thought that, like the voltage-gated Na^+ channel, the open form of the channel is short-lived and quickly flips to a closed state of lower free energy (Figure 6–63). Subsequently, the acetylcholine molecules dissociate from the receptor and are hydrolyzed by a specific enzyme (acetylcholinesterase) located in the neuromuscular junction. Once freed of its bound neurotransmitter, the acetylcholine receptor reverts to its initial resting state.

The general shape of the acetylcholine receptor and the likely arrangement of its subunits have been determined by a combination of electron microscopy and low-angle x-ray diffraction. Although it is uncertain how the subunits combine to form a hydrophilic transmembrane channel, several models have been proposed, based largely on the amino acid sequences of the subunits. One of these models is shown in Figure 6–64. Clusters of negatively charged amino acid residues at the mouth of the channel are thought to explain why negative ions are excluded, while any positive ion of diameter less than 0.65 nm can pass through. The normal traffic consists chiefly of Na^+ and K^+, together with some Ca^{2+}. Since there is little selectivity among cations, their relative contributions to the current through the channel depend chiefly on their concentrations and on the electrochemical driving forces. When the muscle cell membrane is at its resting potential, the net driving force for K^+ is near zero, since the voltage gradient nearly balances the K^+ concentration gradient across the membrane (see Panel 6–2, p. 315). For Na^+, on the other hand, the voltage gradient and the concentration gradient both act in the same direction to drive the ion into the cell. (The same is true for Ca^{2+}, but the extracellular concentration of Ca^{2+} is so much lower than that of Na^+ that Ca^{2+} makes only a small contribution to the total inward current.) Therefore, the opening of the acetylcholine receptor channels leads to a large net influx of Na^+ (peak rate of about 30,000 ions per channel each millisecond). This influx causes a membrane depolarization that signals the muscle to contract, as described below.

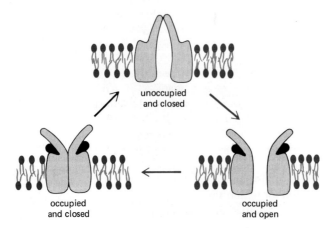

unoccupied
and closed

occupied
and closed

occupied
and open

Figure 6–63 Three conformations of the acetylcholine receptor. The binding of two acetylcholine molecules opens this transmitter-gated ion channel. But even with acetylcholine bound, the receptor is thought to stay in the open conformation only briefly, before the channel recloses. The acetylcholine then dissociates from the receptor, which enables the receptor to return to its original conformation.

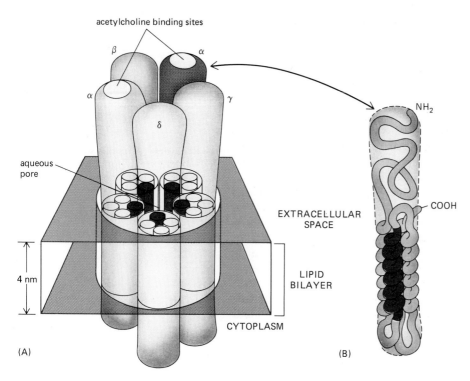

acetylcholine binding sites

EXTRACELLULAR
SPACE

LIPID
BILAYER

4 nm

CYTOPLASM

(A)

aqueous
pore

NH₂

COOH

(B)

The DNA sequences that encode the subunits of several other transmitter-gated ion channels have been determined. The deduced amino acid sequences are homologous to one another and to those of the subunits of the acetylcholine receptor, suggesting that these ion channels are evolutionarily related.

Neuromuscular Transmission Involves the Sequential Activation of at Least Four Different Sets of Gated Ion Channels[36]

The importance of gated ion channels to electrically excitable cells can be illustrated by following the process whereby a nerve impulse stimulates a muscle cell to contract. This apparently simple response requires the sequential activation of at least four different sets of gated ion channels—all within less than a second (Figure 6–65).

1. The process is initiated when a nerve impulse reaches the nerve terminal and depolarizes its plasma membrane. The depolarization transiently opens voltage-gated Ca^{2+} channels in this membrane. Since the Ca^{2+} concentration outside cells is more than 1000 times greater than the free Ca^{2+} concentration inside, Ca^{2+} flows into the nerve terminal. The increase in Ca^{2+} concentration

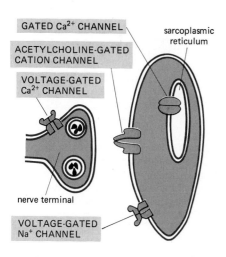

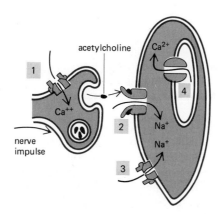

GATED Ca^{2+} CHANNEL

ACETYLCHOLINE-GATED CATION CHANNEL

VOLTAGE-GATED Ca^{2+} CHANNEL

sarcoplasmic reticulum

VOLTAGE-GATED Na^+ CHANNEL

nerve terminal

RESTING NEUROMUSCULAR JUNCTION

acetylcholine

nerve impulse

ACTIVATED NEUROMUSCULAR JUNCTION CAUSING MUSCLE CONTRACTION

Figure 6–65 Schematic diagram of a neuromuscular junction illustrating some of the gated ion channels involved in the stimulation of muscle contraction by a nerve impulse. The various channels are numbered in the sequence in which they open, as described in the text. The gating mechanism for the Ca^{2+} channel in the sarcoplasmic reticulum is unknown.

in the cytosol of the nerve terminal triggers the localized release of acetylcholine into the synaptic cleft.

2. The released acetylcholine binds to acetylcholine receptors in the postsynaptic muscle cell plasma membrane, transiently opening the cation channels associated with them. The resulting influx of Na^+ causes a localized membrane depolarization.

3. The depolarization of the muscle cell plasma membrane opens voltage-gated Na^+ channels in this membrane, allowing more Na^+ to enter, which further depolarizes the membrane. This in turn opens more voltage-gated Na^+ channels and results in a self-propagating depolarization (an action potential) that spreads to involve the entire plasma membrane.

4. The generalized depolarization of the muscle cell plasma membrane causes ion channels in the sarcoplasmic reticulum to open transiently and release Ca^{2+} into the cytosol. It is the sudden increase in the cytosolic Ca^{2+} concentration that causes the myofibrils in the muscle cell to contract (see p. 621). It is not known how the voltage change in the muscle plasma membrane signals the sarcoplasmic reticulum. One possibility is that the action potential spreads to the sarcoplasmic reticulum membrane, causing voltage-dependent Ca^{2+} channels in the membrane to open. Another possibility is that the depolarization of the muscle cell plasma membrane activates the inositol phospholipid cell-signaling pathway discussed in Chapter 12 (see p. 702).

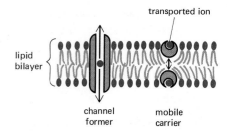

Figure 6–66 A mobile ion carrier and a channel-forming ionophore. In both cases net ion flow occurs only down an electrochemical gradient.

Ionophores Increase the Ion Permeability of Membranes[37]

Ionophores are small hydrophobic molecules that dissolve in lipid bilayers and increase their ion permeability. Most are synthesized by microorganisms (presumably as biological weapons to weaken their competitors), and some have been used as antibiotics. They have been widely employed by cell biologists to increase membrane permeability to specific ions in studies on synthetic bilayers, cells, and cell organelles. There are two classes of ionophores—**mobile ion carriers** and **channel formers** (Figure 6–66). Both types operate by shielding the charge of the transported ion so that it can penetrate the hydrophobic interior of the lipid bilayer. Since ionophores are not coupled to energy sources, they permit net movement of ions only down their electrochemical gradients.

Valinomycin is an example of a mobile ion carrier. It is a ring-shaped polymer that increases the permeability of a membrane to K^+. The ring has a hydrophobic exterior (made up of valine side chains), which contacts the hydrocarbon core of the lipid bilayer, and a polar interior where a single K^+ can fit precisely (Figure 6–67). Valinomycin transports K^+ down its electrochemical gradient by picking up K^+ on one side of the membrane, diffusing across the bilayer, and releasing K^+ on the other side.

The ionophore *A23187* is another example of a mobile ion carrier, but it transports divalent cations such as Ca^{2+} and Mg^{2+}. This ionophore normally acts as an ion-exchange shuttle, carrying two H^+ out of the cell for every divalent cation it carries in. When cells are exposed to A23187, Ca^{2+} enters the cytosol down a steep electrochemical gradient. Accordingly, this ionophore is widely used in cell biology to increase the concentration of free Ca^{2+} in the cytosol, thereby mimicking certain cell-signaling mechanisms (see p. 704).

If the temperature of the membrane is lowered to below its freezing point, mobile carriers can no longer diffuse across the lipid bilayer and ion transport stops. This temperature dependence serves to identify an ionophore as a mobile carrier, since, by contrast, channel-forming ionophores continue to transport ions when the bilayer is frozen.

Gramicidin A is an example of a channel-forming ionophore. It is a linear peptide of 15 amino acids, all with hydrophobic side chains. Two such molecules are thought to come together in the bilayer to form a transmembrane channel that selectively allows monovalent cations (H^+ most readily, K^+ somewhat less readily, and Na^+ less readily still) to flow down their electrochemical gradients. These dimers are unstable and are constantly forming and dissociating, so that the average open time for a channel is about 1 second. With a large electrochemical gradient, Gramicidin A can transport about 20,000 cations per open channel each

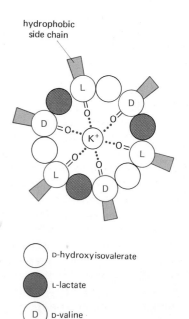

Figure 6–67 A valinomycin molecule with a K^+ ion bound by six oxygen atoms at the center of the ring structure.

millisecond, which is a thousand times more than can be transported by a single mobile carrier molecule in the same time. Gramicidin is an antibiotic made by certain bacteria to kill other microorganisms; it kills them by collapsing the H^+, Na^+, and K^+ gradients that are essential for cell survival.

Summary

Lipid bilayers are highly impermeable to most polar molecules. In order to transport small water-soluble molecules into or out of cells, plasma membranes contain various transport proteins, each of which is responsible for transferring a particular solute across the membrane. There are two classes of membrane transport proteins—carriers and channels; both form continuous protein pathways across the lipid bilayer.

Carrier proteins bind specific solutes and transfer them across the bilayer by undergoing a conformational change that exposes the solute-binding site sequentially on one side of the membrane and then on the other. Some carrier proteins transport solutes only "downhill," whereas others are driven through a series of conformational changes by ATP hydrolysis or ion binding and are thereby able to act as pumps to transport the bound solute "uphill" against its electrochemical gradient.

Channel proteins form aqueous pores across the bilayer and allow inorganic ions of appropriate size and charge to cross the membrane down their electrochemical gradients at rates that are at least 100 times greater than transfer by any known carrier. These ion channels are "gated" and usually open transiently in response to a specific perturbation in the membrane, such as the binding of a neurotransmitter (transmitter-gated channels) or a change in membrane potential (voltage-gated channels).

Membrane Transport of Macromolecules and Particles: Exocytosis and Endocytosis

The transport proteins that mediate the passage of small polar molecules across the plasma membrane cannot transport macromolecules such as proteins, polynucleotides, or polysaccharides. Yet most cells are able to ingest and secrete macromolecules, and some specialized cells even manage to ingest large particles. The mechanisms by which cells do this are very different from those that mediate small solute and ion transport, and they involve the sequential formation and fusion of membrane-bounded vesicles. For example, in order to secrete insulin across their plasma membranes, insulin-producing cells package insulin molecules in specialized *secretory vesicles*; in response to extracellular signals, these vesicles fuse with the plasma membrane and open to the extracellular space, thereby releasing the insulin to the exterior. This fusion process is called **exocytosis.** Cells ingest macromolecules and particles by a similar mechanism but with the sequence reversed. The substance to be ingested is progressively enclosed by a small portion of the plasma membrane, which first invaginates and then pinches off to form an intracellular vesicle containing the ingested material. This process is called **endocytosis.** Exocytosis and endocytosis are compared in Figure 6–68. Both involve the fusion of initially separate regions of lipid bilayer and occur in at least two steps: first the two bilayers come into close apposition (*bilayer adherence*), and then they fuse (*bilayer joining*). Both steps are thought to be mediated by specialized proteins, as will be discussed later (see p. 336).

An important feature of both exocytosis and endocytosis is that the secreted or ingested macromolecules are sequestered in vesicles and do not mix with most of the other macromolecules or organelles in the cytoplasm. The vesicles are designed to fuse only with specific membranes, causing a directed transfer of macromolecules between the outside and inside of the cell. A very similar process operates in the transfer of newly synthesized macromolecules from the endoplasmic reticulum to the Golgi apparatus and then to other compartments of the cell (see Chapter 8). Although it is clear that the rapid, large-scale formation and

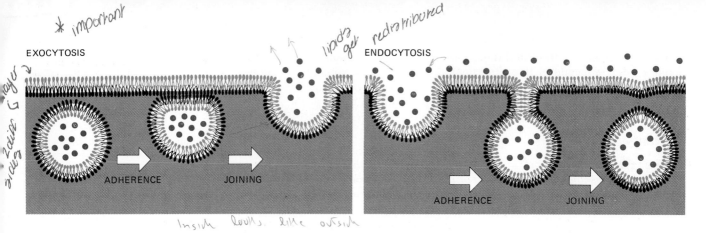

EXOCYTOSIS

ADHERENCE   JOINING

ENDOCYTOSIS

ADHERENCE   JOINING

fusion of vesicles is a fundamental feature of all eucaryotic cells, much remains to be learned about the molecular mechanisms involved in driving and guiding this traffic along specific paths.

There Are Two Pathways of Exocytosis, One Constitutive and One Regulated[38]

In all eucaryotic cells, transport vesicles are continually carrying new plasma membrane components from the Golgi apparatus to the plasma membrane by the process of exocytosis. In addition, cells secrete various types of molecules by exocytosis; some of these molecules may adhere to the cell surface and become part of the cell coat, others are incorporated into the extracellular matrix, whereas still others diffuse into the interstitial fluid and/or blood to nourish or signal other cells.

As described in Chapter 8, proteins to be secreted are synthesized on ribosomes attached to the rough endoplasmic reticulum (rough ER). They pass into the lumen of the ER and are transported to the Golgi apparatus by ER-derived *transport vesicles.* In the Golgi apparatus the proteins are modified, concentrated, sorted, and finally packaged into vesicles that pinch off into the cytosol and subsequently fuse with the plasma membrane. By contrast, small molecules to be secreted, such as histamine (see below), are actively transported from the cytosol into preformed vesicles, where they are often complexed to specific macromolecules (proteoglycans in the case of histamine) so that they can be stored at high concentration without generating an excessive osmotic gradient.

Some proteins are continuously secreted by the cells that make them, being packaged into transport vesicles in the Golgi apparatus and then carried promptly to the plasma membrane. Such molecules are said to follow the *constitutive pathway of secretion.* In other cells, selected proteins and/or small molecules are stored in special **secretory vesicles,** which fuse with the plasma membrane only when the cell is triggered by an extracellular signal; they are said to follow the *regulated pathway of secretion* (Figure 6–69). The constitutive pathway operates in all cells,

Figure 6–68 Bilayer adherence and bilayer joining in exocytosis and endocytosis. The extracellular space is at the top, separated from the cytoplasm below by the plasma membrane. Note that because of the bilayer adherence step, endocytosis and exocytosis are not simply the reverse of each other: in exocytosis two cytoplasmic-side monolayers of the plasma membrane initially adhere, whereas in endocytosis two noncytoplasmic-side monolayers initially adhere. In both cases the asymmetric character of the membranes is conserved and the cytoplasmic-side monolayer remains in contact with the cytosol.

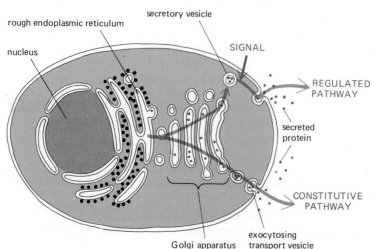

rough endoplasmic reticulum

nucleus

secretory vesicle

SIGNAL

REGULATED PATHWAY

secreted protein

CONSTITUTIVE PATHWAY

Golgi apparatus

exocytosing transport vesicle

Figure 6–69 Two pathways that can be followed by secreted proteins. Some secreted proteins are packaged in transport vesicles and secreted continually (the *constitutive pathway*), whereas others are stored in special secretory vesicles and released only when the cell is triggered by an extracellular signal (the *regulated pathway*). The constitutive pathway operates in all eucaryotic cells, but the regulated pathway operates only in cells that are specialized for secretion (secretory cells).

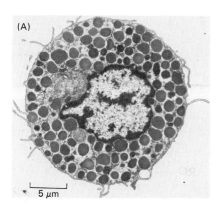

(A)

5 µm

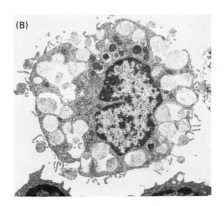

(B)

Figure 6–70 Electron micrographs of exocytosis in rat mast cells. The cell in (A) has not been stimulated. The cell in (B) has been activated to secrete its stored histamine by a soluble extracellular ligand. Histamine-containing secretory vesicles are dark; those that have released their histamine are light. The material remaining in the spent vesicles consists of a network of proteoglycans to which the stored histamine is normally bound.

Once a secretory vesicle has fused with the plasma membrane, the secretory vesicle membrane often serves as a target to which a limited number of other secretory vesicles fuse. Consequently, many secretory vesicles in mast cells open to the extracellular space indirectly through fusion with other open vesicles. Thus the cell in (B) contains several large cavities lined by the fused membranes of many spent secretory vesicles, which are now in continuity with the plasma membrane. This continuity is not always apparent in one plane of section through the cell. (From D. Lawson, C. Fewtrell, B. Gomperts, and M. Raff, *J. Exp. Med.* 142:391–402, 1975. By copyright permission of the Rockefeller University Press.)

but the regulated pathway is found mainly in cells that are specialized for secreting their products rapidly on demand—products such as hormones, neurotransmitters, or digestive enzymes. In these specialized *secretory cells* the signal to secrete is often a chemical messenger, such as a hormone, that binds to receptors on the cell surface. The resulting activation of the receptors generates one or more intracellular signals, often including a transient increase in the concentration of free Ca^{2+} in the cytosol (see p. 701). By an unknown mechanism, these signals initiate the process of exocytosis, causing the secretory vesicles to fuse with the plasma membrane, thereby releasing their contents to the extracellular space.

During exocytosis the vesicle membrane becomes incorporated into the plasma membrane (see Figure 6–68). In the case of the regulated pathway at least, the distinctive protein and lipid components of the secretory vesicle membrane are later specifically retrieved by endocytosis and reincorporated into new secretory vesicles. The amount of secretory vesicle membrane that is temporarily added to the plasma membrane can be enormous: in a pancreatic acinar cell discharging digestive enzymes, about 900 μm^2 of vesicle membrane is inserted into the apical plasma membrane (whose area is only 30 μm^2) when the cell is stimulated to secrete.

Regulated Exocytosis Is a Localized Response of the Plasma Membrane and Its Underlying Cytoplasm[39]

Mast cells secrete *histamine* (see Table 12–1, p. 685) when triggered by specific ligands that bind to receptors on their surface. It is the histamine secreted by mast cells that is responsible for many of the unpleasant symptoms, such as itching and sneezing, that accompany allergic reactions. When mast cells are incubated in fluid containing a soluble stimulant, exocytosis occurs all over the cell surface (Figure 6–70). However, if the stimulating ligand is artificially attached to a solid bead so that it can interact only with a localized region of the mast cell surface, exocytosis is restricted to the region where the cell contacts the bead (Figure 6–71). Clearly, the mast cell does not respond as a whole when it is triggered; the activation of receptors, the resulting intracellular signals, and the subsequent exocytosis must all be localized in the particular region of the cell that has been excited. This demonstrates an important property of plasma membranes: individual segments can function independently of the rest of the membrane. As we shall see, this property is as important in endocytosis as it is in exocytosis.

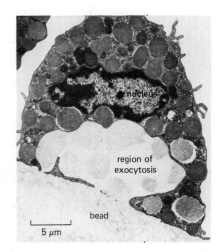

nucleus

region of exocytosis

bead

5 µm

Figure 6–71 Electron micrograph of a mast cell that has been activated to secrete histamine by a stimulant coupled to a large solid bead. Exocytosis has occurred only in the region of the cell that is in contact with the bead. (From D. Lawson, C. Fewtrell, and M. Raff, *J. Cell Biol.* 79:394–400, 1978. By copyright permission of the Rockefeller University Press.)

There Are Two Types of Endocytosis: Pinocytosis and Phagocytosis[40]

Two main types of endocytosis are distinguished on the basis of the size of the endocytic vesicles formed: **pinocytosis** ("cellular drinking"), which involves the ingestion of fluid and solutes via small vesicles (<150 nm in diameter); and **phagocytosis** ("cellular eating"), which involves the ingestion of large particles, such as microorganisms or cell debris, via large vesicles (called *phagosomes* or *vacuoles*, generally >250 nm in diameter). As we shall see, the two types of endocytosis are mediated by different mechanisms. Although most eucaryotic cells are continually ingesting fluid and solutes by pinocytosis, large particles are ingested mainly by specialized phagocytic cells. For this reason the terms pinocytosis and endocytosis tend to be applied interchangeably to nonphagocytic cells.

Many of the particles and molecules ingested by cells by either phagocytosis or pinocytosis end up in *lysosomes*. Large particles are taken up into phagosomes, which are thought to fuse with lysosomes to form *phagolysosomes*. Fluid and macromolecules ingested by pinocytosis are initially transferred to intermediate membrane-bounded organelles called *endosomes*, from where they are eventually carried to lysosomes unless they are specifically retrieved. Because lysosomes contain a wide variety of degradative (hydrolytic) enzymes (see p. 459), most of the material from phagosomes and endosomes that ends up in lysosomes is rapidly broken down; the small breakdown products, such as amino acids, sugars, and nucleotides, are transported across the lysosomal membrane into the cytosol, where they can be used by the cell. Many of the membrane constituents of the original endocytic vesicles are retrieved from phagolysosomes and endosomes and recycled to the plasma membrane by exocytosis.

Pinocytic Vesicles Form from Coated Pits in the Plasma Membrane[41]

Virtually all eucaryotic cells continually ingest bits of their plasma membrane in the form of small endocytic (pinocytic) vesicles that are later returned to the cell surface. This endocytic cycle begins at specialized regions of the plasma membrane called **coated pits.** In conventional electron micrographs these regions appear in cross-section as depressions of the plasma membrane coated with bristlelike structures on their cytoplasmic surface, and in various cultured cells they occupy about 2% of the total area of the plasma membrane. The lifetime of coated pits is short: within a minute or so of being formed, they invaginate into the cell and pinch off to form **coated vesicles** (Figure 6–72). It has been estimated that about 2500 coated vesicles leave the plasma membrane of a cultured fibroblast every minute. Coated vesicles are even more transient than coated pits: within seconds of being formed, they shed their coat and are now able to fuse with endosomes. Unless retrieved from endosomes, the contents of these endocytic vesicles ultimately end up in lysosomes.

Figure 6–72 Electron micrographs illustrating the probable sequence of events in the formation of a coated vesicle from a coated pit. The coated pits and vesicles shown are involved in taking up lipoprotein particles into a very large hen oocyte to form yolk. They are larger than those seen in normal-sized cells. (Courtesy of M.M. Perry and A.B. Gilbert, from *J. Cell Sci.* 39:257–272, 1979, by permission of The Company of Biologists.)

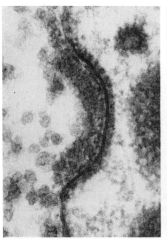 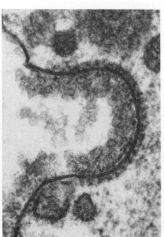 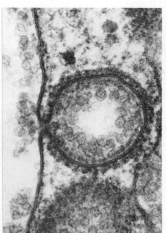 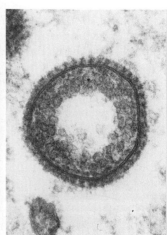

0.1 μm

6-30 Coated Pits Contain Clathrin[42]

In electron micrographs of cells studied by the rapid-freeze–deep-etch technique, the coat on coated pits and vesicles appears as a network of polygons (Figure 6–73). But what is the coat made of, and what is its function? When coated vesicles formed from coated pits are isolated, their membranes are found to contain several major proteins. The best characterized is **clathrin,** a protein complex that has been highly conserved in evolution. It is composed of three large polypeptide chains and three smaller polypeptide chains that together form a three-legged structure (a *triskelion*). A number of triskelions assemble into a basketlike network of hexagons and pentagons on the cytoplasmic surface of membranes, thereby forming a characteristic polyhedral coat on the surface of coated pits and vesicles (Figure 6–74). Under appropriate conditions, isolated triskelions will spontaneously reassemble into typical polyhedral baskets (even in the absence of membrane vesicles). Other proteins are more intimately associated with the coated vesicle membrane and are required to bind the clathrin coat to the vesicle and to capture various plasma membrane receptors (see below).

The invagination of a coated pit is believed to be driven by forces generated by the assembly of clathrin and other coat proteins on the cytoplasmic surface of the plasma membrane. Once a coated vesicle forms, the clathrin and associated proteins dissociate from the vesicle membrane and return to the plasma membrane to form a new coated pit. How coated pit formation is initiated, how a coated pit pinches off to form a coated vesicle, and how the coat is subsequently shed are still unknown. A member of the hsp70 family of stress-response proteins acts *in vitro* as an *uncoating ATPase* that removes the coat from clathrin-coated vesicles (see p. 463). However, some additional control mechanism must operate to prevent the clathrin coat from being removed from a pit before it has had time to form a vesicle, especially since the coat on the pit persists much longer than the coat on the vesicle.

There Are at Least Two Types of Coated Vesicles[43]

In most cells, clathrin-coated pits and vesicles provide the major pathway for pinocytic uptake of both extracellular fluid and membrane-bound ligands (see p. 333). Pinocytic pathways involving other types of vesicles also exist, but very

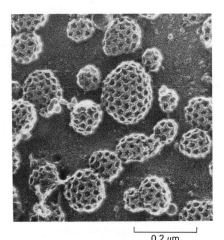

Figure 6–73 Electron micrograph of numerous coated pits and vesicles on the inner surface of the plasma membrane of cultured fibroblasts. The cells were rapidly frozen in liquid helium, fractured, and deep-etched to expose the cytoplasmic surface of the plasma membrane. (Reproduced from J. Heuser, *J. Cell Biol.* 84:560–583, 1980. By copyright permission of the Rockefeller University Press.)

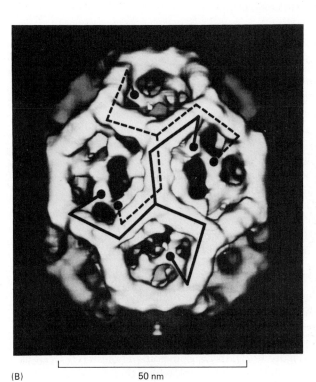

(A)   50 nm   (B)   50 nm

Figure 6–74 The structure of a clathrin coat. (A) Electron micrographs of clathrin triskelions shadowed with platinum. Although this feature cannot be seen in these micrographs, each triskelion is composed of 3 clathrin heavy chains and 3 clathrin light chains. (B) A three-dimensional reconstruction of a clathrin coat composed of 36 triskelions organized in a network of 12 pentagons and 8 hexagons. The contours of two triskelions are sketched on the photograph. Note that each leg of a triskelion runs along two neighboring polygonal edges and then turns inward so that its amino-terminal domain (black ball) forms an inner shell. The overlapping arrangement of the flexible triskelion arms provides both mechanical strength and flexibility. Other sizes and shapes of clathrin coats are similarly constructed from 12 pentagons plus a variable number of hexagons. (A, from E. Ungewickell and D. Branton, *Nature* 289:420–422, 1981, © 1981 Macmillan Journals Ltd.; B, from G.P.A. Vigers, R.A. Crowther, and B.M.F. Pearse, *EMBO J.* 5:2079–2085, 1986.)

little is known about their relative importance. Some endothelial cells lining small blood vessels are thought to transport substances out of the bloodstream and into the surrounding extracellular fluid by means of endocytic vesicles that are not clathrin-coated; these vesicles shuttle from one surface of the cell to the other in a process called *transcytosis*. In most other cell types where transcytosis occurs, however, it is mediated by clathrin-coated pits and vesicles (see p. 332).

Not all coated vesicles in cells are derived from the plasma membrane. As will be discussed in Chapter 8, it is thought that many are formed continually from the endoplasmic reticulum and Golgi apparatus and mediate an extensive vesicular traffic between these and other organelles. There are at least two types of coated vesicles: (1) *clathrin-coated* vesicles, which are involved both in endocytosis and in vesicular transport from the *trans* Golgi network of the Golgi apparatus to endolysosomes (see below) and secretory vesicles (see p. 465); and (2) *non-clathrin-coated* vesicles, which are involved in vesicular transport from the endoplasmic reticulum to the Golgi apparatus, from one Golgi cisterna to another, and from the Golgi apparatus to the plasma membrane (see p. 472); the molecules that form the coat on these vesicles are still uncharacterized. Clathrin-coated pits and vesicles may be more sophisticated than non-clathrin-coated ones, since they are able to select specific macromolecules for transport, whereas non-clathrin-coated vesicles may not (see p. 472).

6-35 Receptor-mediated Endocytosis Serves as a Concentrating Device for Internalizing Specific Extracellular Macromolecules

In most animal cells, clathrin-coated pits and vesicles provide an efficient pathway for taking up specific macromolecules from the extracellular fluid, a process called **receptor-mediated endocytosis.** The macromolecules bind to complementary cell-surface receptors, accumulate in coated pits, and enter the cell as receptor-macromolecule complexes in endocytic vesicles. Since extracellular fluid is trapped in coated pits as they invaginate to form coated vesicles, substances dissolved in the extracellular fluid are also internalized, but at a very much lower rate—a process called **fluid-phase endocytosis.** Receptor-mediated endocytosis provides a selective concentrating mechanism that increases the efficiency of internalization of particular ligands more than 1000-fold, so that even minor components of the extracellular fluid can be specifically taken up in large amounts without taking in a correspondingly large volume of extracellular fluid.

6-33 6-34 Cells Import Cholesterol by Taking up Low-Density Lipoproteins (LDL) by Receptor-mediated Endocytosis[44]

An important process known to occur through receptor-mediated endocytosis in many animal cells is the uptake of cholesterol. Cholesterol uptake provides these cells with most of the cholesterol they require to make new membrane. If the uptake is blocked, cholesterol accumulates in the blood and can contribute to the formation of atherosclerotic plaques in blood vessel walls. Most cholesterol is transported in the blood bound to protein in the form of complexes known as **low-density lipoproteins,** or **LDL.** Each of these large spherical particles (22 nm in diameter) contains a core of about 1500 cholesterol molecules esterified to long-chain fatty acids; the cholesteryl ester core is surrounded by a lipid monolayer containing a single large protein molecule that organizes the particle (Figure 6–75).

When the cell needs cholesterol for membrane synthesis, it makes receptor proteins for LDL and inserts them into its plasma membrane. Once in the plasma membrane, the LDL receptors diffuse until they associate with coated pits that are in the process of forming (Figure 6–76A). Since coated pits constantly pinch off to form coated vesicles, any LDL particles bound to LDL receptors in the coated pits are rapidly internalized. After shedding their clathrin coats, the endocytic vesicles deliver their contents to endosomes. In endosomes the LDL particles and their receptors part company: the receptors are returned to the plasma membrane, while the LDL ends up in lysosomes (Figure 6–77). In lysosomes the cholesteryl esters in the LDL particles are hydrolyzed to free cholesterol, which thereby be-

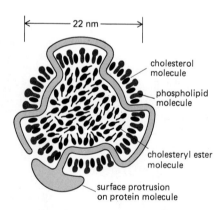

Figure 6–75 A low-density lipoprotein (LDL) particle seen in cross-section. Each spherical particle has a mass of 3×10^6 daltons. It contains about 1500 molecules of cholesteryl ester in its core surrounded by a lipid monolayer composed of about 800 phospholipid and 500 unesterified cholesterol molecules. A single molecule of a 500,000-dalton protein organizes the particle and is responsible for the specific binding of LDL to cell-surface receptor proteins.

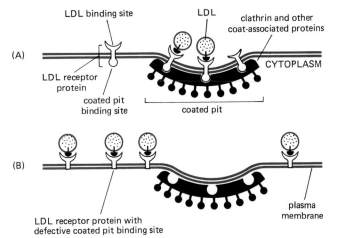

Figure 6–76 LDL receptor proteins binding to a coated pit in the plasma membrane of a normal cell (A). The human LDL receptor is a single-pass transmembrane glycoprotein composed of about 840 amino acid residues, only 50 of which are on the cytoplasmic side of the membrane. (B) A mutant cell in which the LDL receptor proteins are abnormal and lack the site in the cytoplasmic domain that enables them to bind to coated pits. Such cells bind LDL but cannot ingest it. In most human populations, one in 500 individuals inherits one defective LDL receptor gene and, as a result, often dies prematurely from heart attacks.

comes available to the cell for new membrane synthesis. If too much free cholesterol accumulates in a cell, it shuts off both the cell's own cholesterol synthesis and the synthesis of LDL receptor proteins, so that less cholesterol is made and less is taken up by the cell.

This regulated pathway for the uptake of cholesterol is disrupted in certain individuals who inherit defective genes for making LDL receptor proteins and whose cells, consequently, are deficient in the capacity to take up LDL from the blood. The resulting high levels of blood cholesterol predispose these individuals to premature atherosclerosis, and most die at an early age of heart attacks resulting from coronary artery disease. The abnormality can be traced to the LDL receptor, which can be lacking or defective—either in the extracellular binding site for LDL or in the intracellular binding site that attaches the receptor to coated pits (see Figure 6–76B). In the latter case, normal numbers of LDL-binding receptor proteins are present, but they are not localized in the coated regions of the plasma membrane; although LDL binds to the surface of these mutant cells, it is not internalized, directly demonstrating the importance of coated pits in the receptor-mediated endocytosis of cholesterol.

More than 25 different receptors have been observed to participate in receptor-mediated endocytosis of different types of molecules, and they all apparently utilize the same coated pit pathway; many of these receptors enter coated pits

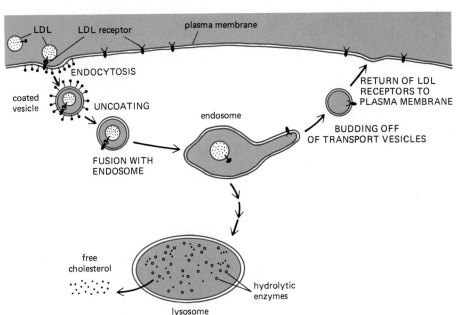

Figure 6–77 Receptor-mediated endocytosis of LDL. Note that the LDL dissociates from its receptors in the acidic environment of the endosome. Although the mechanism of transport is uncertain (see p. 331), the LDL ends up in lysosomes, where it is degraded to release free cholesterol. The LDL receptor proteins are returned to the plasma membrane via transport vesicles that bud off from the tubular region of the endosome, as shown. For simplicity, only one LDL receptor is shown entering the cell and returning to the plasma membrane. Whether it is occupied or not, an LDL receptor typically makes one round trip into and out of the cell every 10 minutes, making a total of several hundred trips in its 20-hour life-span.

irrespective of whether they have bound their specific ligands. Not all plasma membrane proteins accumulate in coated pits, however, indicating that the pits function as molecular filters, collecting certain plasma membrane proteins (receptors) and excluding others. Electron microscopic studies of cultured cells exposed to different ligands (labeled to make them visible in the electron microscope) have demonstrated that many kinds of receptors cluster in the same coated pit. The plasma membrane of one coated pit can probably accommodate about 1000 receptors of assorted varieties, and all the receptor-ligand complexes that utilize the clathrin-coated endocytic pathway apparently are delivered to the same endosomal compartment. The subsequent fates of the endocytosed molecules, however, vary depending on the receptor.

The Contents of Endosomes End up in Lysosomes Unless They Are Specifically Retrieved[45]

The **endosomal compartment** can be recognized as a distinct entity in the electron microscope after it has been marked by exposing cells briefly to a labeled ligand. The internalized ligand reveals the endosomal compartment as a complex set of heterogeneous membrane-bounded tubes and vesicles extending from the periphery of the cell to the perinuclear region, where it is often close to but separate from the Golgi apparatus (Figure 6–78). Two sets of endosomes can be readily distinguished in such labeling experiments: endocytosed macromolecules appear in *peripheral endosomes* just beneath the plasma membrane within a minute or so and in *perinuclear (internal) endosomes* after 5–15 minutes. The interior of the endosomal compartment is acidic (pH 5–6) due to the presence of ATP-driven H^+ pumps in the endosomal membrane that pump H^+ into the lumen from the cytosol, and, in general, internal endosomes are more acidic than peripheral endosomes. As we shall see, the acidic environment plays a crucial part in the function of these organelles. A similar or identical *vacuolar H^+ ATPase* is thought to acidify all endocytic and exocytic organelles, including phagosomes, lysosomes, selected compartments of the Golgi apparatus, and many transport and secretory vesicles.

Most of the contents of perinuclear endosomes end up in lysosomes. However, many molecules are specifically recycled from the peripheral (and probably peri-

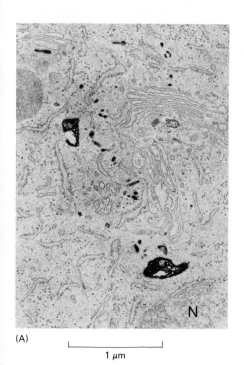

(A)

1 μm

(B)   1 μm

Figure 6–78 Electron micrograph (A) and drawing of a serial reconstruction (B) of perinuclear endosomes in baby hamster kidney (BHK) cells in culture. The cells were incubated in a solution containing the enzyme peroxidase for 15 minutes at 37°C, which was long enough for the peroxidase to be taken up by fluid-phase endocytosis and delivered to endosomes (and endolysosomes—see text) but not long enough for it to be delivered to lysosomes. After the cells were fixed and exposed to a peroxidase substrate (diaminobenzidine), the product of the enzymatic reaction was made electron dense by fixation with osmium tetroxide. The reconstruction in (B) was drawn from 18 serial thin sections. The nucleus is indicated by N in (A). (From M. Marsh, G. Griffiths, G. Dean, I. Mellman, and A. Helenius, *Proc. Natl. Acad. Sci. USA* 83:2899–2903, 1986.)

nuclear) endosomes to the plasma membrane via specific transport vesicles that bud off from these endosomes. As a result, only those endocytosed molecules that are not specifically retrieved from endosomes are degraded.

The pathway from endosome to lysosome is now known to be more complicated than was originally believed. The hydrolytic enzymes destined for lysosomes are initially delivered from the Golgi apparatus (via transport vesicles) to a special prelysosomal compartment located close to the perinuclear endosomes but distinct from them. Although this membrane-bounded organelle can be viewed as the final stage in endosome maturation, it is where the process of hydrolytic degradation, which will be completed in lysosomes, begins. We shall therefore refer to this organelle as an **endolysosome** (Figure 6–79).

It is uncertain how endocytosed molecules move from one endosomal compartment to another so as to end up in lysosomes. One extreme view is that peripheral endosomes slowly move inward to become perinuclear endosomes, which, as a result of the fusion of transport vesicles from the Golgi apparatus, continuous membrane retrieval, and increasing acidification, are converted into endolysosomes and then lysosomes. At the other extreme, each endosomal and lysosomal compartment may be a permanent structure, like a cisterna in a Golgi stack (see p. 457), with transport between them occurring via transport vesicles, similar to transport between adjacent Golgi cisternae (see p. 473). We shall return to this question of how lysosomes are formed in Chapter 8 (see p. 460).

Ligand-Receptor Complexes Are Sorted in the Acidic Environment of Endosomes[46]

The endosomal compartment acts as the main sorting station in the endocytic pathway, just as the Golgi apparatus (especially the *trans* Golgi network) serves this function in the biosynthetic-secretory pathway (see p. 457). The acidic environment of the endosome plays a crucial part in the sorting process by altering many receptor-ligand complexes in a way that affects their fate. When the pH drops, these receptors change their conformation and, as a result, dissociate from their ligand. Ligands that dissociate from their receptors in the endosome usually are doomed to destruction in lysosomes along with the other non-membrane-bound contents of the endosome; other ligands remain bound to, and thereby share the fate of, their receptors. The routes taken by specific receptors differ according to the function of the receptor. Receptors can follow one of at least three pathways from the endosomal compartment: (1) they can return to the same plasma membrane domain from which they came, which is the most common pathway; (2) they can travel to lysosomes; or (3) they can return to a different domain of the plasma membrane, thereby mediating the process of *transcytosis* (Figure 6–80).

The LDL receptor described on page 328 follows the first pathway. The receptor dissociates from its ligand LDL in the endosome and is recycled to the plasma membrane for reuse, while the discharged LDL is carried to lysosomes (see Figure 6–77). A similar but more complex recycling occurs following the endocytosis of **transferrin,** a protein that carries iron in the blood. Cell-surface transferrin receptors deliver transferrin with its bound iron to peripheral endosomes by receptor-mediated endocytosis. The low pH in the endosome induces transferrin to release its bound iron, but the iron-free transferrin itself (called apotransferrin) remains bound to its receptor and is recycled back to the plasma membrane as a receptor-apotransferrin complex. When it has returned to the neutral pH of the extracellular fluid, the apotransferrin dissociates from the receptor and is thereby freed to pick up more iron and begin the cycle again. Thus the transferrin protein shuttles back and forth between the extracellular fluid and the endosomal compartment, avoiding lysosomes and delivering the iron that cells need to grow.

The second pathway that endocytosed receptors can follow from endosomes is taken by the receptor that binds *epidermal growth factor (EGF)*, a small protein that stimulates epidermal and various other cells to divide. These receptors are unusual because they accumulate in coated pits only after binding EGF, suggesting

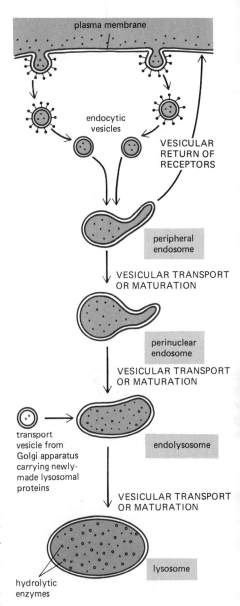

Figure 6–79 Molecules that are endocytosed via coated pits are delivered to peripheral endosomes and then move sequentially to perinuclear endosomes, endolysosomes, and lysosomes unless they are specifically retrieved from endosomes or endolysosomes. Hydrolytic enzymes destined for lysosomes are initially delivered from the Golgi apparatus to endolysosomes. As indicated, it is uncertain how the endocytosed molecules that are not retrieved move from one compartment to another so as to end up in lysosomes, where they are degraded.

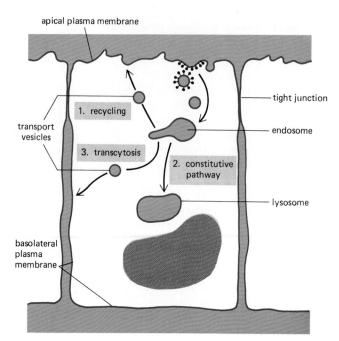

apical plasma membrane

1. recycling

transport
vesicles

3. transcytosis

2. constitutive
pathway

basolateral
plasma
membrane

tight junction

endosome

lysosome

Figure 6–80 Three pathways from the endosomal compartment in an epithelial cell. Most molecules that are not specifically retrieved from endosomes (or endolysosomes—not shown) follow the constitutive pathway from the endosomal compartment to lysosomes. Retrieved molecules are returned either to the same plasma membrane domain from which they came (recycling) or to a different domain of the plasma membrane (transcytosis).

that a ligand-induced conformational change is required for them to bind to the pits. Moreover, EGF dissociates from its receptor at a lower pH than is required for the dissociation of many other receptor-ligand complexes. Perhaps for this reason, many of the ingested EGF receptors end up in lysosomes, where they are degraded along with the ingested EGF. EGF binding therefore leads to a decrease in the concentration of EGF receptors on the cell surface—a process called *receptor down-regulation*. As a result, the concentration of signaling ligand in the extracellular fluid regulates the number of its complementary receptor molecules on the target cell surface (see p. 718)—a mechanism quite different from the regulation of LDL receptors, whose numbers depend on the intracellular concentration of cholesterol (see p. 329).

Macromolecules Can Be Transferred Across Epithelial Cell Sheets by Transcytosis[47]

Some receptors on the surface of polarized epithelial cells transfer specific macromolecules from one extracellular space to another by a process called **transcytosis.** These receptors follow the third pathway from endosomes. A newborn rat, for example, obtains antibodies from its mother's milk (which help protect it against infection) by transporting them across the epithelium of its gut. The lumen of the gut is somewhat acidic, and at this low pH the antibodies in the milk bind to specific receptors on the apical (absorptive) surface of the gut epithelial cells and are internalized via coated pits. The receptor-antibody complexes remain intact in endosomes and are retrieved in transport vesicles that fuse with the basolateral domain of the plasma membrane. Upon exposure to the neutral pH of the extracellular fluid, the antibodies dissociate from their receptors and eventually enter the newborn's bloodstream. In the mother the secretion of these antibodies into the milk also occurs by transcytosis, but in the reverse direction, from blood to milk (see Figure 18–20, p. 1016).

The variety of pathways that different receptors follow from endosomes implies that, in addition to a binding site for its ligand and a binding site for coated pits, many receptors also contain sorting signals that guide them into the appropriate type of transport vesicle leaving the endosome and thereby to the appropriate target membrane in the cell. The nature of these signals is unknown. Since more is known about the similar process by which the *trans* Golgi network sorts different proteins into different transport vesicles, we shall defer further discussion of intracellular sorting until Chapter 8 (see p. 460).

6-31 Coated Pits and Vesicles Provide the Major Pathway
6-32 for Fluid-Phase Endocytosis in Many Cells[48]

The rates at which cells internalize their plasma membrane by endocytosis can be quantified by introducing a labeled tracer into the extracellular fluid for a short time and then measuring its rate of uptake into endocytic vesicles. Two kinds of tracers can be used: those that remain dissolved in the extracellular fluid and are taken up by *fluid-phase endocytosis* (see Figure 6–78), and those that bind to cell-surface receptors and are taken up by *receptor-mediated endocytosis*. The estimated rates of plasma membrane internalization using either type of tracer are usually similar. It is therefore believed that the coated pit–coated vesicle pathway, in addition to being responsible for receptor-mediated endocytosis, is also the major pathway for fluid-phase endocytosis in many cells.

The rate of plasma membrane internalization varies from cell type to cell type, but it is usually surprisingly large. A macrophage, for example, ingests 25% of its own volume of fluid each hour. This means that it must ingest 3% of its plasma membrane each minute, or 100% in about half an hour. Fibroblasts endocytose at a somewhat lower rate, whereas some amoebae ingest their plasma membrane even more rapidly. Since a cell's surface area and volume remain unchanged during this process, it is clear that as much membrane is being added to the cell surface by exocytosis as is being removed by endocytosis.

6-8 The Endocytic Cycle May Play a Role in Cell Locomotion and in the Phenomenon of "Capping"[49]

Coated pits are distributed more or less randomly on most cells, and so the internalization of plasma membrane (specific receptor proteins plus lipid) occurs over the whole cell surface. In nonpolarized cells the internalized membrane probably is returned randomly over the surface of the cell. In polarized cells, however, such as fibroblasts crawling along a substratum, the centrosome and its associated Golgi apparatus are oriented toward the front of the cell, and the internalized membrane is returned preferentially to the cell's leading edge. This spatial asymmetry in the endocytic-exocytic cycle may help the motile cell to extend its leading edge forward in the process of locomotion (see p. 672). Moreover, because the sites of endocytosis and exocytosis are not coincident in such cells, there will be a net flow of lipid and receptors along the plasma membrane backward from the leading edge (Figure 6–81). This *membrane flow* may explain why objects such as carbon particles, when attached to the upper surface of a cultured fibroblast, are observed to migrate from the front toward the back of the cell as the cell crawls forward.

When multivalent ligands such as antibodies or lectins bind to specific plasma membrane proteins exposed on the surface of cells, they tend to cross-link them into large clusters—a process called **patching.** Once such clusters have formed on the surface of a cell capable of locomotion, such as a white blood cell, they rapidly collect over one pole of the cell to form a "cap" (Figure 6–82). This **capping** process takes several minutes, requires ATP, and, in white blood cells, results in a cap forming at the tail end of the cell. Although cell locomotion is not required for capping, only cells capable of crawling along a surface are able to form caps, suggesting that the mechanism may be related to the mechanisms used in cell locomotion. One hypothesis is that the membrane flow resulting from the endocytic cycle described above sweeps large cross-linked patches of protein to the back of the cell, while proteins that are not cross-linked diffuse rapidly enough to maintain a nearly random distribution despite the flow. An alternative hypothesis

Figure 6–81 Membrane flow as postulated to occur in the plasma membrane of motile cells as a result of the asymmetry in the endocytic cycle in these cells. A fibroblast is shown in cross section moving from left to right. Endocytosis of lipid bilayer containing receptors occurs by means of coated pits, which are distributed randomly over the cell surface. The endocytosed membrane is retrieved from the endosomal compartment (not shown) in the form of vesicles and reinserted by exocytosis preferentially at the cell's leading edge. The pattern of randomly distributed endocytosis and directed exocytosis during vesicle recycling would cause a net backward flow of the endocytosed membrane components along the plasma membrane, as indicated by the large colored arrows. (Adapted from M.S. Bretscher, *Science* 224:681–686, 1984. Copyright 1984 by the AAAS.)

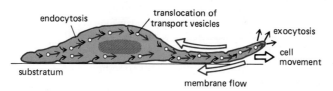

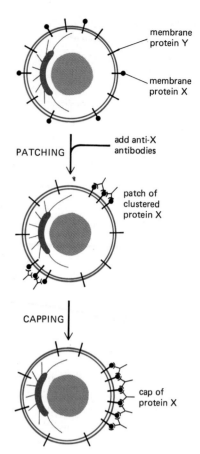

Figure 6–82 Antibody-induced patching and capping of a cell-surface protein on a white blood cell. Antibodies cross-link the protein molecules to which they bind, causing them to cluster into large patches, which are actively swept to the tail end of the cell to form a "cap." Note that the head end of the cell is determined by the position of the centrosome, and cross-linked proteins are capped to the opposite end, even when the cell is in suspension and not migrating.

membrane protein Y

membrane protein X

PATCHING

add anti-X antibodies

patch of clustered protein X

CAPPING

cap of protein X

is that the clustered membrane proteins interact directly or indirectly with a contractile system of intracellular actin filaments (located in the cortex, just under the plasma membrane) that pulls the patches to the tail of the cell. There are still no definitive experiments that exclude either of these hypotheses.

Specialized Phagocytic Cells Ingest Particles That Bind to Specific Receptors on Their Surface[50]

Phagocytosis is a special form of endocytosis in which large particles such as microorganisms and cell debris are ingested via large endocytic vesicles called *phagosomes*. In protozoa, phagocytosis is a form of feeding: large particles taken up into phagosomes end up in lysosomes, and the products of the subsequent digestive processes pass into the cytosol to be utilized as food. Most cells in multicellular organisms are not able to ingest large particles efficiently. In the gut, large particles of food are broken down extracellularly before import into cells; and in the body at large, phagocytosis is a task left to "professional" phagocytes. In mammals there are two classes of white blood cells that act as phagocytes: **macrophages** (which are widely distributed in tissues as well as in blood) and **neutrophils.** These two types of cells develop from a common precursor cell (see p. 981), and they defend us against infection by ingesting invading microorganisms. Macrophages also play an important part in scavenging senescent and damaged cells and cellular debris. In quantitative terms the latter function is far more important: macrophages phagocytose more than 10^{11} senescent red blood cells in each of us every day.

The endocytic vesicles that form from coated pits are relatively small (~150 nm in diameter). **Phagosomes**, in contrast, have diameters that are determined by the size of the particle being ingested, and they can be almost as large as the phagocytic cell itself (Figure 6–83). The phagosomes fuse with lysosomes to form large *phagolysosomes*, where the ingested material is degraded; indigestible substances remain in phagolysosomes, forming *residual bodies.* As for endosomes, some of the internalized plasma membrane components are retrieved from the phagosome by transport vesicles and returned to the plasma membrane. In certain macrophages, peptides derived from degraded endocytosed proteins are returned to the cell surface in a complex with recycled major histocompatibility complex glycoproteins (see p. 1046). Because these macrophages are scrutinized by T lymphocytes of the immune system, they act as antigen-presenting cells: when the peptides come from a foreign invader, they activate the T lymphocytes to mount an immune response against the foreign protein (see Chapter 18, p. 1046).

In order to be phagocytosed, particles must first bind to the surface of the phagocyte. Not all particles that bind are ingested, however. Phagocytes have a variety of specialized surface receptors that are functionally linked to the phagocytic machinery of the cell. Unlike pinocytosis, which is a constitutive process that occurs continuously, phagocytosis is a triggered process in which activated receptors transmit signals to the cell interior to initiate the response. The best-characterized triggers are antibodies (see p. 1015), which protect us against infectious microorganisms by binding to the surface of microorganisms to form a coat in which the tail region of each antibody molecule (called the Fc region—see p. 1013) is exposed on the exterior. This antibody coat is then recognized by specific *Fc receptors* on the surface of macrophages and neutrophils. The binding of antibody-coated particles to these receptors induces the phagocytic cell to extend

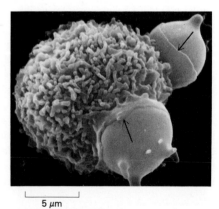

5 µm

Figure 6–83 Scanning electron micrograph of a mouse macrophage phagocytosing two chemically altered red blood cells. The arrows point to edges of thin processes (pseudopods) of the macrophage that are extending as collars to engulf the red cells. (Courtesy of Jean Paul Revel.)

pseudopods that engulf the particle and fuse at their tips to form a phagosome (Figure 6–84).

Two other classes of receptors are known to promote phagocytosis: those that recognize complement components (see p. 1034) and those that recognize oligosaccharides on the surface of certain microorganisms. In addition, macrophages can recognize and ingest senescent or damaged cells, but the receptors involved are unknown.

Phagocytosis Is a Localized Response That Proceeds by a "Membrane-zippering" Mechanism[51]

It is possible to treat red blood cells in such a way that they will bind to the surface of macrophages without being phagocytosed. If macrophages that have bound such red blood cells are then allowed to phagocytose antibody-coated bacteria, only the bacteria are ingested; even red blood cells bound immediately next to a site of active phagocytosis are not ingested. This indicates that phagocytosis, like triggered exocytosis in mast cells (see p. 325), is a localized response of a region of the plasma membrane and its underlying cytoplasmic structures.

If a macrophage is exposed to target cells that have been evenly coated with antibody, it ingests the coated cells. If the antibody molecules have been confined to one pole of the target cell by capping (see p. 333), however, the plasma membrane of the macrophage spreads over the surface of the target cell only in the region of the cap and the cell is not ingested (Figure 6–85). It is clear, therefore, that the initial interaction of antibody-coated target cells with the Fc receptors on the macrophage surface is insufficient to trigger ingestion. Instead, it initiates a spreading process of membrane apposition that requires the continuous contact of receptors with antibody if the target cell is to be fully enclosed by a phagosome. This suggests that phagocytosis proceeds by a continuous *membrane-zippering* mechanism.

Neither the mechanism by which the binding of antibody to the Fc receptor initiates the movements of ingestion nor the nature of the motile force that extends the pseudopods is known. However, actin and various actin-binding proteins accumulate in the pseudopods, and cytochalasin (a drug that prevents actin polymerization) inhibits phagocytosis, suggesting an actin-dependent mechanism of pseudopod extension. Clathrin is sometimes present on the cytoplasmic surface of phagosomes formed in macrophages that are ingesting antibody-coated particles, raising the possibility that clathrin plays a role in phagocytosis as well as in pinocytosis. The mechanisms that mediate these two forms of endocytosis clearly differ, however, because cytochalasin does not inhibit pinocytosis.

How do the pseudopods enclosing a particle fuse to form a phagosome? This question brings us back to a fundamental problem that we touched on briefly at the outset of this account of exocytosis and endocytosis.

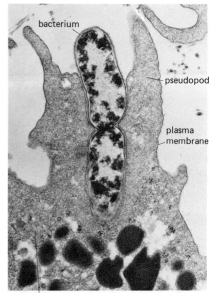

bacterium

pseudopod

plasma membrane

phagocytic white blood cell

1 μm

Figure 6–84 Electron micrograph of a neutrophil phagocytosing a bacterium, which is in the process of dividing. (Courtesy of Dorothy F. Bainton.)

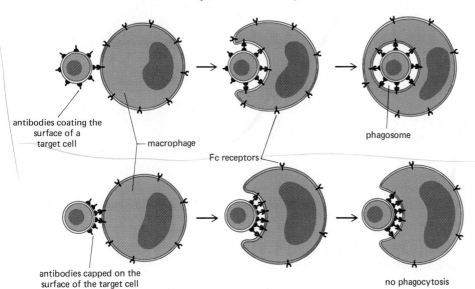

antibodies coating the surface of a target cell

macrophage

Fc receptors

phagosome

antibodies capped on the surface of the target cell

no phagocytosis

Figure 6–85 Schematic drawing of an experiment suggesting that phagocytosis proceeds by a "membrane-zippering" mechanism. (Based on observations by F.M. Griffin, Jr., J.A. Griffin, and S.C. Silverstein, *J. Exp. Med.* 144:788–809, 1976.)

CELL FUSION   CELL DIVISION

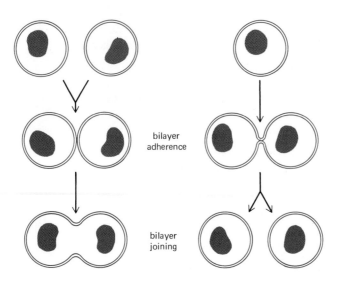

bilayer
adherence

bilayer
joining

Figure 6–86 Membrane fusion
(consisting of bilayer adherence and
bilayer joining) in the processes of cell
division and cell fusion. Natural cell
fusions occur between sperm and egg
during fertilization and between
myoblasts in the formation of
multinucleated skeletal muscle cells.

Membrane Fusion During Exocytosis and Endocytosis Is Probably Catalyzed by Fusogenic Proteins[52]

Bilayer adherence and *bilayer joining*, which together constitute membrane fusion, are fundamental cell-membrane phenomena that occur not only in exocytosis and endocytosis, but also during cell division and cell fusion (Figure 6–86); they also go on continually within cells as transport vesicles bud from and fuse with various organelles. In none of these cases is the mechanism of membrane fusion understood, but some clues can be gleaned from the way enveloped viruses fuse with the cells they infect.

Cellular membranes do not fuse spontaneously. In order to fuse, water must be displaced to allow the interacting lipid bilayers to approach within 1.5 nm of each other, a process that is energetically highly unfavorable. For this reason it seems likely that all membrane fusions in cells are catalyzed by specialized *fusogenic proteins*. So far such cellular proteins have not been directly identified, but viral fusogenic proteins are known to play a crucial part in permitting the entry of enveloped viruses (which have a lipid-bilayer-based membrane coat) into the cells they infect (see p. 469). Viruses such as the influenza virus, for example, enter the cell by receptor-mediated endocytosis and are delivered to endosomes. The low pH in endosomes activates a fusogenic protein (fusogen) in the viral envelope that catalyzes the fusion of the viral and endosomal membranes, thereby allowing the viral nucleic acid to escape into the cytosol (Figure 6–87).

The genes encoding several viral fusogenic proteins have been cloned and used to transfect eucaryotic cells in culture. These transfected cells express the viral proteins on their surface and, when briefly exposed to a low pH, fuse to form giant multinucleated cells. In the best-studied case, that of the influenza virus, the three-dimensional structure of the fusogenic protein has been determined by x-ray diffraction techniques (see p. 446). It has been shown that the low pH induces a large conformational change in the fusogenic protein, exposing a previously buried hydrophobic region on the surface of the protein that can interact with the lipid bilayer of a target membrane. A cluster of such hydrophobic regions on

Figure 6–87 Schematic drawing showing how fusogenic
proteins on the surface of many enveloped viruses are thought
to catalyze the fusion of viral and endosomal membranes. These
viruses enter the target cell by receptor-mediated endocytosis
and are delivered to endosomes; the low pH in the endosome
activates the fusogenic protein to catalyze the membrane fusion
that allows the virus capsid to escape into the cytosol, where the
virus nucleic acid can replicate.

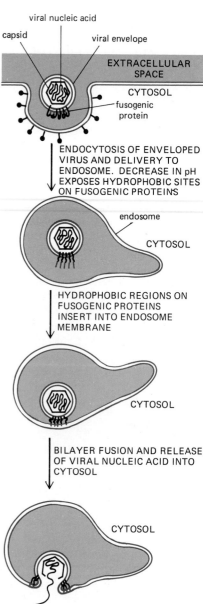

viral nucleic acid

capsid   viral envelope

EXTRACELLULAR
SPACE

CYTOSOL

fusogenic
protein

ENDOCYTOSIS OF ENVELOPED
VIRUS AND DELIVERY TO
ENDOSOME. DECREASE IN pH
EXPOSES HYDROPHOBIC SITES
ON FUSOGENIC PROTEINS

endosome

CYTOSOL

HYDROPHOBIC REGIONS ON
FUSOGENIC PROTEINS
INSERT INTO ENDOSOME
MEMBRANE

CYTOSOL

BILAYER FUSION AND RELEASE
OF VIRAL NUCLEIC ACID INTO
CYTOSOL

CYTOSOL

closely spaced fusogenic protein molecules is thought to bring the two lipid bilayers into close apposition and to destabilize them so that the bilayers fuse (see Figure 6–87).

Summary

Most cells secrete and ingest macromolecules by exocytosis and endocytosis, respectively. In exocytosis the contents of either transport vesicles or secretory vesicles are released to the extracellular space when the vesicles fuse with the plasma membrane. In endocytosis the sequence is reversed: localized regions of the plasma membrane invaginate and pinch off to form endocytic vesicles; many of the endocytosed particles and molecules end up in lysosomes, where they are degraded. Exocytosis and endocytosis occur both constitutively and as triggered responses to extracellular signals.

Most cells continually ingest bits of their plasma membrane and then return them to the cell surface in a large-scale endocytic–exocytic cycle that is largely mediated by clathrin-coated pits and vesicles. Many cell-surface receptors that bind specific extracellular macromolecules become localized in clathrin-coated pits and consequently are internalized in coated vesicles—a process called receptor-mediated endocytosis. The coated endocytic vesicles rapidly shed their clathrin coats and fuse with endosomes, where the receptors and ligands are sorted. Most ligands dissociate from their receptors in the acidic environment of the endosome and eventually end up in lysosomes, while most receptors are recycled via transport vesicles back to the cell surface for reuse. Other receptor-ligand complexes follow other pathways from the endosomal compartment. In some cases both the receptor and the ligand end up being degraded in lysosomes. In other cases both are transferred through the cell, and the ligand is released by exocytosis at a different surface of the cell from where it originated—a process called transcytosis.

References

General

Bretscher, M.S. The molecules of the cell membrane. *Sci. Am.* 253(4):100–109, 1985.

Datta, D.B. A Comprehensive Introduction to Membrane Biochemistry. Madison: Floral, 1987.

Finean, J.B.R.; Coleman, R.; Michell, R.H. Membranes and Their Cellular Functions, 3rd ed. Oxford: Blackwell, 1984.

Singer, S.J.; Nicolson, G.L. The fluid mosaic model of the structure of cell membranes. *Science* 175:720–731, 1972.

Yeagle, P. The Membranes of Cells. Orlando: Academic, 1987.

Cited

1. Branton, D. Fracture faces of frozen membranes. *Proc. Natl. Acad. Sci. USA* 55:1048–1056, 1966.

   Ceve, G.; Marsh, D. Phospholipid Bilayers: Physical Principles and Models. New York: Wiley, 1987.

   Gorter, E.; Grendel, F. On bimolecular layers of lipoids on the chromocytes of the blood. *J. Exp. Med.* 41:439–443, 1925.

2. Hawthorne, J.N.; Ansell, G.B. Phospholipids. New Comprehensive Biochemistry, Vol. 4. Amsterdam: Elsevier, 1982.

   Storch, J.; Kleinfeld, A.M. The lipid structure of biological membranes. *Trends Biochem. Sci.* 10:418–421, 1982.

3. Bangham, A.D. Models of cell membranes. In Cell Membranes: Biochemistry, Cell Biology and Pathology (G. Weissmann, R. Claiborne, eds.), pp. 24–34. New York: Hospital Practice, 1975.

   Edidin, M. Rotational and lateral diffusion of membrane proteins and lipids: phenomena and function. *Curr. Top. Memb. Transp.* 29:91–127, 1987.

Kornberg, R.D.; McConnell, H.M. Lateral diffusion of phospholipids in a vesicle membrane. *Proc. Natl. Acad. Sci. USA* 68:2564–2568, 1971.

4. Chapman, D. Lipid dynamics in cell membranes. In Cell Membranes: Biochemistry, Cell Biology and Pathology (G. Weissmann, R. Claiborne, eds.), pp. 13–22. New York: Hospital Practice, 1975.

   Chapman, D.; Benga, G. Biomembrane fluidity—studies of model and natural membranes. In Biological Membranes (D. Chapman, ed.), Vol. 5, pp. 1–56. London: Academic, 1984.

   Kates, M.; Manson, L.A., eds. Membrane Fluidity. Biomembranes, Vol. 12. New York: Plenum, 1984.

   Kimelberg, H.K. The influence of membrane fluidity on the activity of membrane-bound enzymes. In Dynamic Aspects of Cell Surface Organization. Cell Surface Reviews (G. Poste, G.L. Nicolson, eds.), Vol. 3, pp. 205–293. Amsterdam: Elsevier, 1977.

5. Carruthers, A.; Melchior, D.L. How bilayer lipids affect membrane protein activity. *Trends Biochem. Sci.* 11:331–335, 1986.

   de Kruiff, B.; et al. Lipid polymorphism and membrane function. In The Enzymes of Biological Membranes, Vol 1, Membrane Structure and Dynamics, 2nd ed. (A.N. Martonosi, ed.), pp. 131–204. New York: Plenum, 1985.

   Kleinfeld, A. Current views of membrane structure. Curr. Top. Memb. Transp. 29:1–27, 1987.

6. Bretscher, M. Membrane structure: some general principles. *Science* 181:622–629, 1973.

   Rothman, J.; Lenard, J. Membrane asymmetry. *Science* 195:743–753, 1977.

7. Hakomori, S. Glycosphingolipids. *Sci. Am.* 254(5):44–53, 1986.
   Weigandt, H. The gangliosides. *Adv. Neurochem.* 4:149–223, 1982.

8. Eisenberg, D. Three-dimensional structure of membrane and surface proteins. *Annu. Rev. Biochem.* 53:595–623, 1984.
   Engelman, D.M.; Steitz, T.A.; Goldman, A. Identifying nonpolar transbilayer helices in amino acid sequences of membrane proteins. *Annu. Rev. Biophys. Biophys. Chem.* 15:321–353, 1986.
   Henderson, R. The structure of bacteriorhodopsin and its relevance to other membrane proteins. In Membrane Transduction Mechanisms (R.A. Cone, J.E. Dowling, eds.), pp. 3–15. New York: Raven, 1979.
   Sefton, B.M.; Buss, J.E. The covalent modification of eukaryotic proteins with lipid. *J. Cell Biol.* 104:1449–1453, 1987.
   Unwin, N.; Henderson, R. The structure of proteins in biological membranes. *Sci. Am.* 250(2):78–94, 1984.

9. Helenius, A.; Simons, K. Solubilization of membranes by detergents. *Biochim. Biophys. Acta* 415:29–79, 1975.
   Montal, M. Functional reconstitution of membrane proteins in planar lipid bilayer membranes. In Techniques of the Analysis of Membrane Proteins (C.I. Ragan, R.J. Cherry, eds.), pp. 97–128. London: Chapman & Hall, 1986.
   Racker, E. Reconstitution of Transporters, Receptors and Pathological States. Orlando: Academic, 1985.

10. Bretscher, M. Membrane structure: some general principles. *Science* 181:622–629, 1973.
    Steck, T.L. The organization of proteins in the human red blood cell membrane. *J. Cell Biol.* 62:1–19, 1974.

11. Bennet, V. The membrane skeleton of human erythrocytes and its implications for more complex cells. *Annu. Rev. Biochem.* 54:273–304, 1985.
    Branton, D.; Cohen, C.M.; Tyler, J. Interaction of cytoskeletal proteins on the human erythrocyte membrane. *Cell* 24:24–32, 1981.
    Byers, T.J.; Branton, D. Visualization of the protein associations in the erythrocyte membrane skeleton. *Proc. Natl. Acad. Sci. USA* 82:6153–6157, 1985.
    Marchesi, V.T. Stabilizing infrastructure of cell membranes. *Annu. Rev. Cell Biol.* 1:531–561, 1985.
    Shen, B.W.; Josephs, R.; Steck, T.L. Ultrastructure of the intact skeleton of the human erythrocyte membrane. *J. Cell Biol.* 102:997–1006, 1986.

12. Marchesi, V.T.; Furthmayr, H.; Tomita, M. The red cell membrane. *Annu. Rev. Biochem.* 45:667–698, 1976.

13. Jay, D.; Cantley, L. Structural aspects of the red cell anion exchange protein. *Annu. Rev. Biochem.* 55:511–538, 1986.
    Kopito, R.R.; Lodish, H.F. Primary structure and transmembrane orientation of the murine anion exchange protein. *Nature* 316:234–238, 1985.

14. Henderson, R.; Unwin, P.N.T. Three-dimensional model of purple membrane obtained by electron microscopy. *Nature* 257:28–32, 1975.
    Stoeckenius, W.; Bogomolni, R.A. Bacteriorhodopsin and related pigments of Halobacteria. *Annu. Rev. Biochem.* 51:587–616, 1982.

15. Deisenhofer, J.; Epp, O.; Miki, K.; Huber, R.; Michel, H. The structure of the protein subunits in the photosynthetic reaction centre of *Rhodopseudomonas viridis* at 3 Å resolution. *Nature* 318:618–624, 1985.

16. de Petris, S.; Raff, M.C. Normal distribution, patching and capping of lymphocyte surface immunoglobulin studied by electron microscopy. *Nature New Biol.* 241:257–259, 1973.
    Edidin, M. Rotational and lateral diffusion of membrane proteins and lipids: phenomena and function. *Curr. Top. Memb. Transp.* 29:91–127, 1987.

Frye, L.D.; Edidin, M. The rapid intermixing of cell surface antigens after formation of mouse-human heterokaryons. *J. Cell Sci.* 7:319–335, 1970.
McCloskey, M.; Poo, M-M. Protein diffusion in cell membranes: some biological implications. *Int. Rev. Cytol.* 87:19–81, 1984.
Poo, M.; Cone, R.A. Lateral diffusion of rhodopsin in the photoreceptor membrane. *Nature* 247:438–441, 1974.
Wier, M.; Edidin, M. Constraint of the translational diffusion of a membrane glycoprotein by its external domains. *Science* 242:412–414, 1988.

17. Gumbiner, B.; Louvard, D. Localized barriers in the plasma membrane: a common way to form domains. *Trends Biochem. Sci.* 10:435–438, 1985.
    Myles, D.G.; Primakoff, P. Sperm surface domains. In Hybridoma Technology in the Biosciences and Medicine (T.A. Springer, ed.), pp. 239–250. New York: Plenum, 1985.
    Simons, K.; Fuller, S.D. Cell Surface Polarity in Epithelia. *Annu. Rev. Cell Biol.* 1:243–288, 1985.

18. Hirano, H.; Parkhouse, B.; Nicolson, G.L.; Lennox, E.S.; Singer, S.J. Distribution of saccharide residues on membrane fragments from a myeloma-cell homogenate: its implications for membrane biogenesis. *Proc. Natl. Acad. Sci. USA* 69:2945–2949, 1972.
    Kornfeld, R.; Kornfeld, S. Structure of glycoproteins and their oligosaccharide units. In The Biochemistry of Glycoproteins and Proteoglycans (W.J. Lennarz, ed.), pp. 1–84. New York: Plenum, 1980.
    Lis, H.; Sharon, N. Lectins as molecules and tools. *Annu. Rev. Biochem.* 55:35–67, 1986.
    Olden, K.; Bernard, B.A., Humphries, M.J., Yeo, T-K.; Yeo, K.T.; White, S.L.; Newton, S.A.; Bauer, H.C.; Parent, J.B. Function of glycoprotein glycans. *Trends Biochem. Sci.* 10:78–82, 1985.

19. Hille, B. Ionic Channels of Excitable Membranes. Sunderland, MA.: Sinauer, 1984.
    Martonosi, A.N., ed. The Enzymes of Biological Membranes, Vol. 3, Membrane Transport, 2nd ed. New York: Plenum, 1985.
    West, I.C. The Biochemistry of Membrane Transport. London: Chapman & Hall, 1983.
    Stein, W.D. Transport and Diffusion Across Cell Membranes Orlando: Academic, 1986.

20. Andersen, O.S. Permeability properties of unmodified lipid bilayer membranes. In Membrane Transport in Biology, Vol. 1 (G. Giebisch, D.C. Tosteson, H.H. Ussing, eds.), pp. 369–446. New York: Springer-Verlag, 1978.
    Finkelstein, A. Water movement through membrane channels. *Curr. Top. Memb. Trans.* 21:295–308, 1984.
    Walter, A.; Gutknecht, J. Permeability of small nonelectrolytes through lipid bilayer membranes. *J. Membr. Biol.* 90:207–217, 1986.

21. Hobbs, A.S.; Albers, R.W. The structure of proteins involved in active membrane transport. *Annu. Rev. Biophys. Bioeng.* 9:259–291, 1980.
    Kyte, J. Molecular considerations relevant to the mechanism of active transport. *Nature* 292:201–204, 1981.
    Stein, W.D. Ed. Ion pumps: Structure, function and regulation. New York: Alan Liss, 1988.
    Tanford, C. Mechanism of free energy coupling in active transport. *Annu. Rev. Biochem.* 52:379–409, 1983.
    Wilson, D.B. Cellular transport mechanisms. *Annu. Rev. Biochem.* 47:933–965, 1978.

22. Cantley, L.C. Structure and mechanisms of the (Na,K) ATPase. *Curr. Topics Bioenerget.* 11:201–237, 1981.
    Glynn, I.M.; Ellory, C., eds. The Sodium Pump. Cambridge, U.K.: The Company of Biologists, 1985.

Shull, G.E.; Schwartz, A.; Lingrel, J.B. Amino acid sequence of the catalytic subunit of the (Na$^+$-K$^+$) ATPase deduced from a complementary DNA. *Nature* 316:691–695, 1985.

23. Sweadner, K.J.; Goldin, S.M. Active transport of sodium and potassium ions: mechanism, function and regulation. *N. Engl. J. Med.* 302:777–783, 1980.

Glynn, I.M. The Na$^+$-K$^+$ transporting adenosine triphosphatase. In The Enzymes of Biological Membranes, 2nd ed. (A. Martonosi, ed.) Vol. 3, pp. 34–114. New York: Plenum, 1985.

24. Hasselbach, W.; Oetliker, H. Energetics and electrogenicity of the sarcoplasmic reticulum calcium pump. *Annu. Rev. Physiol.* 45:325–339, 1983.

MacLennan, D.H.; Brandl, C.J.; Korczak, B.; Green, N.M. Amino acid sequence of a Ca^{2+}, Mg^{2+}-dependent ATPase from rabbit muscle sarcoplasmic reticulum, deduced from its complementary DNA sequence. *Nature* 316:696–700, 1985.

Schatzman, H.J. The red cell calcium pump. *Annu. Rev. Physiol.* 45:303–312, 1983.

25. Hinkle, P.C.; McCarty, R.E. How cells make ATP. *Sci. Am.* 238(3):104–123, 1978.

Nicholls, D.G. An Introduction to the Chemiosmotic Theory, 2nd ed. New York: Academic, 1987.

26. Scott, D.M. Sodium cotransport systems: cellular, molecular and regulatory aspects. *Bioessays* 7:71–78, 1987.

Wright, J.K.; Seckler, R.; Overath, P. Molecular aspects of sugar:ion transport. *Annu. Rev. Biochem.* 55:225–248, 1986.

27. Grinstein, S.; Rothstein, A. Mechanisms of regulation of the Na$^+$/H$^+$ exchanger. *J. Membrane Biol.* 90:1–12, 1986.

Olsnes, S.; Tønnessen, T.I.; Sandvig, K. pH-regulated anion antiport in nucleated mammalian cells. *J. Cell Biol.* 102:967–971, 1986.

Pouysségur, J.; Franchi, A; Kohno, M.; L'Allemain, G.; Paris, S. Na$^+$-H$^+$ exchange and growth control in fibroblasts: a genetic approach. *Curr. Top. Memb. Transp.* 26:201–220, 1986.

Rozengunt, E.; Mendoza, S. Early stimulation of Na$^+$-H$^+$ antiport, Na$^+$-K$^+$ pump activity, and Ca^{2+} fluxes in fibroblast mitogenesis. *Curr. Top. Memb. Transp.* 27:163–191, 1986.

28. Almers, W.; Stirling, C. Distribution of transport proteins over animal cell membranes. *J. Membrane Biol.* 77:169–186, 1984.

Semenza, G. Anchoring and biosynthesis of stalked brush border membrane proteins: Glycosidases and peptidases of Enterocytes and renal tubuli. *Annu. Rev. Cell Biol.* 2:255–313, 1986.

29. Postma, P.W.; Lengeler, J.W. Phosphoenolpyruvate: carbohydrate phosphotransferase system of bacteria. *Microbiol. Rev.* 49:232–269, 1985.

30. Ames, G.F.L. Bacterial periplasmic transport systems: structure, mechanism and evolution. *Annu. Rev. Biochem.* 55:397–425, 1986.

31. Hille, B. Ionic Channels of Excitable Membranes, Sunderland, MA: Sinauer, 1984.

32. Baker, P.F.; Hodgkin, A.L.; Shaw, T.L. The effects of changes in internal ionic concentration on the electrical properties of perfused giant axons. *J. Physiol.* 164:355–374, 1962.

Hodgkin, A.L.; Keynes, R.D. Active transport of cations in giant axons from *Sepia* and *Loligo*. *J. Physiol.* 128:26–60, 1955.

Kuffler, S.W.; Nicholls, J.G.; Martin, A.R. From Neuron to Brain, pp. 111–125. Sunderland, MA: Sinauer, 1984.

33. Hodgkin, A.L.; Huxley, A.F. A quantitative description of membrane current and its application to conduction and excitation in nerve. *J. Physiol.* 117:500–544, 1952.

Kuffler, S.W.; Nicholls, J.G.; Martin, A.R. From Neuron to Brain, pp. 125–152. Sunderland, MA: Sinauer, 1984.

34. Catterall, W.A. Molecular properties of voltage-sensitive sodium channels. *Annu. Rev. Biochem.* 55:953–985, 1986.

Noda, M.; et al. Primary structure of *Electrophorus electricus* sodium channel deduced from cDNA sequence. *Nature* 312:121–127, 1984.

Sigworth, F.J.; Neher, E. Single Na$^+$ channel currents observed in cultured rat muscle cells. *Nature* 287:447–449, 1980.

Stevens, C.F. Biophysical studies of ion channels. *Science* 225:1346–1350, 1984.

Tanabe, T.; et al. Primary structure of the receptor for calcium channel blockers from skeletal muscle. *Nature* 328:313–318, 1987.

35. Grenningloh, G.; et al. The strychnine-binding subunit of the glycine receptor shows homology with nicotinic acetylcholine receptors. *Nature* 328:215–220, 1987.

Guy, H.R.; Hucho, F. The ion channel of the nicotinic acetylcholine receptor. *Trends Neurosci.* 10:318–321, 1987.

Hille, B. Ionic Channels of Excitable Membranes, pp. 117–138 and 335–360. Sunderland, MA: Sinauer, 1984.

Noda, M.; et al. Structural homology of *Torpedo californica* acetylcholine receptor subunits. *Nature* 302:528–532, 1983.

Schofield, P.R.; et al. Sequence and functional expression of the GABA$_A$ receptor shows a ligand-gated receptor super family. *Nature* 328:221–227, 1987.

36. Stevens, C.F. The neuron. *Sci. Am.* 241(3):54–65, 1979.

37. Gomperts, B.D. The Plasma Membrane: Models for Its Structure and Function, pp. 109–212. New York: Academic Press, 1976.

Pressman, B.C. Biological applications of ionophores, *Annu. Rev. Biochem.* 45:501–530, 1976.

38. Burgess, T.L.; Kelly, R.B. Constitutive and regulated secretion of proteins. *Annu. Rev. Cell Biol.* 3:243–294, 1987.

39. Lawson, D.; Fewtrell, C.; Raff, M. Localized mast cell degranulation induced by concanavalin A-sepharose beads: implications for the Ca^{2+} hypothesis of stimulus-secretion coupling. *J. Cell Biol.* 79:394–400, 1978.

40. Steinman, R.M.; Mellman, I.S.; Muller, W.A.; Cohn, Z. Endocytosis and recycling of plasma membrane. *J. Cell Biol.* 96:1–27, 1983.

41. Goldstein, J.L.; Anderson, R.G.W.; Brown, M.S. Coated pits, coated vesicles, and receptor-mediated endocytosis. *Nature* 279:679–685, 1979.

Roth, T.F.; Porter, K.R. Yolk protein uptake in the oocyte of the mosquito, *Aedes aegypti.* L *J. Cell Biol.* 20:313–332, 1964.

42. Pearse, B.M.F.; Bretscher, M.S. Membrane recycling by coated vesicles. *Annu. Rev. Biochem.* 50:85–101, 1981.

Pearse, B.M.F.; Crowther, R.A. Structure and assembly of coated vesicles. *Ann. Rev. Biophys. Biophys. Chem.* 16:49–68, 1987.

Schmid, S.L.; Rothman, J.E. Enzymatic dissociation of clathrin in a two-stage process. *J. Biol. Chem.* 260:10044–10049, 1985.

43. Orci, L.; Glick, B.S.; Rothman, J.E. A new type of coated vesicular carrier that appears not to contain clathrin: its possible role in protein transport within the Golgi stack. *Cell* 46:171–184, 1986.

Simionescu, N.; Simionescu, M.; Palade, G.E. Permeability of muscle capillaries to small heme-peptides: evidence for the existence of patent transendothelial channels. *J. Cell Biol.* 64:586–607, 1975.

44. Brown, M.S.; Goldstein, J.L. How LDL receptors influence cholesterol and atherosclerosis. *Sci. Am.* 251(5):58–66, 1984.

Brown, M.S.; Goldstein, J.L. A receptor-mediated pathway for cholesterol homeostasis. *Science* 232:34–48, 1986.

45. Brown, W.J.; Goodhouse, J.; Farquhar, M.G. Mannose-6-phosphate receptors for lysosomal enzymes cycle between the Golgi complex and endosomes. *J. Cell Biol.* 103:1235–1247, 1986.

Griffiths, G.; Hoflack, B.; Simons, K.; Mellman, I.; Kornfeld, S. The mannose-6-phosphate receptor and the biogenesis of lysosomes. *Cell* 52:329–341, 1988.

Helenius, A.; Mellman, I.; Wall, D.; Hubbard, A. Endosomes. *Trends Biochem. Sci.* 8:245–250, 1983.

Mellman, I.; Howe, C.; Helenius, A. The control of membrane traffic on the endocytic pathway. *Curr. Top. Memb. Transp.* 29:255–288, 1987.

46. Carpenter, G. Receptors for epidermal growth factor and other polypeptide mitogens. *Annu. Rev. Biochem.* 56:881–914, 1987.

Dautry-Varsat, A.; Ciechanover, A.; Lodish, H.F. pH and the recycling of transferrin during receptor-mediated endocytosis. *Proc. Natl. Acad. Sci. USA* 80:2258–2262, 1983.

Goldstein, J.L.; Brown, M.S.; Anderson, R.G.W.; Russell, D.W.; Schneider, W.J. Receptor-mediated endocytosis. *Annu. Rev. Cell Biol.* 1:1–39, 1985.

Mellman, I.; Fuchs, R.; Helenius, A. Acidification of the endocytic and exocytic pathways. *Annu. Rev. Biochem.* 55:663–700, 1986.

47. Mostov, K.E.; Simister, N.E. Transcytosis. *Cell* 43:389–390, 1985.

Rodewald, R.; Abrahamson, D.R. Receptor-mediated transport of IgG across the intestinal epithelium of the neonatal rat. In Membrane Recycling (Ciba Foundation Symposium 92), pp. 209–232. London: Pitman, 1982.

48. Helenius, A.; Marsh, M. Endocytosis of enveloped animal viruses. In Membrane Recycling (Ciba Foundation Symposium 92), pp. 59–76. London: Pitman, 1982.

Thilo, L. Quantification of endocytosis-derived membrane traffic. *Biochim. Biophys. Acta* 822:243–266, 1985.

49. Abercrombie, M.; Heaysman, J.E.M.; Pegrum, S.M. The locomotion of fibroblasts in culture. III. Movements of particles on the dorsal surface of the leading lamella. *Exp. Cell Res.* 62:389–398, 1970.

Bretscher, M.S. Endocytosis: relation to capping and cell locomotion. *Science* 224:681–686, 1984.

Taylor, R.B.; Duffus, W.P.H.; Raff, M.C.; de Petris, S. Redistribution and pinocytosis of lymphocyte surface immunoglobulin molecules induced by anti-immunoglobulin antibody. *Nature New Biol.* 233:225–229, 1971.

50. Wright, S.D.; Silverstein, S.C. Overview: the function of receptors in phagocytosis. In Handbook of Experimental Immunology, 4th ed. (D.M. Weir, L. Herzenberg, eds.), pp. 41-1–41-14. Oxford, U.K.: Blackwell, 1983.

51. Aggeler, J.; Werb, Z. Initial events during phagocytosis by macrophages viewed from outside and inside the cell: Membrane-particle interactions and clathrin. *J. Cell Biol.* 94:613–623, 1982.

Griffin, F.M., Jr.; Griffin, J.A.; Silverstein, S.C. Studies on the mechanism of phagocytosis. II. The interaction of macrophages with anti-immunoglobulin IgG-coated bone marrow-derived lymphocytes. *J. Exp. Med.* 144:788–809, 1976.

Griffin, F.M., Jr.; Silverstein, S.C. Segmental response of the macrophage plasma membrane to a phagocytic stimulus. *J. Exp. Med* 139:323–336, 1974.

52. Blumenthal R. Membrane Fusion. *Curr. Top. Memb. Transp.* 29:203–254, 1987

Gething, M.J.; Doms, R.; York, D.; White, J. Studies on the mechanism of membrane fusion: site-specific mutagenesis of the hemagglutinin of influenza virus. *J. Cell Biol.* 102:11–23, 1986.

White, J.; Kielian, M.; Helenius, A. Membrane fusion proteins of enveloped animal viruses. *Quart. Rev. Biophys.* 16:151–195, 1983.

Wiley, D.C.; Skehel, J.J. The structure and function of the hemagglutinin membrane glycoprotein of influenza virus. *Annu. Rev. Biochem* 56:365–394, 1987.

Energy Conversion: Mitochondria and Chloroplasts

7

Mitochondria, which are present in all eucaryotic cells, and **plastids** (most notably **chloroplasts**), which occur only in plants, convert energy to forms that can be used to drive cellular reactions. Their specialized function is reflected in the morphological feature that most strikingly distinguishes mitochondria and chloroplasts—the large amount of internal membrane they contain. This membrane plays two crucial roles in the function of these *energy-converting organelles.* First, it provides the framework for electron-transport processes used to convert the energy of oxidation reactions (see p. 61) into more useful forms—particularly ATP. Second, it encloses a large internal compartment containing enzymes that catalyze other cellular reactions.

Without mitochondria, an animal cell would be dependent on anaerobic glycolysis to make all of its ATP. But when glucose is converted to pyruvate by glycolysis (p. 65), only a small fraction of the total free energy potentially available from the oxidation of the sugar is released. In mitochondria the metabolism of sugars (and of fatty acids) is completed: the pyruvate (and fatty acids) are oxidized by molecular oxygen (O_2) to CO_2 and H_2O. The energy released by this oxidation is harnessed so efficiently that about 36 molecules of ATP are produced for each molecule of glucose oxidized. By contrast, only 2 molecules of ATP are produced by glycolysis alone. Chloroplasts are also very efficient ATP-producing machines, although their ultimate source of energy is sunlight rather than sugars or fatty acids. Despite this fundamental difference, mitochondria and chloroplasts are organized similarly, and they synthesize ATP in the same way. This striking conclusion emerged from painstaking studies carried out over the past 25 years that established the central pathway by which energy is harnessed.

The common pathway by which mitochondria, chloroplasts, and even bacteria harness energy for biological purposes operates by a process known as *chemiosmotic coupling.* It begins when strong electron donors pass their "high-energy" electrons along a series of electron carriers embedded in an ion-impermeable membrane. In the course of traveling along this **electron-transport chain,** the electrons, which have been either excited by sunlight or derived from the oxidation of electron-rich foodstuffs, fall to successively lower energy levels. Part of the energy released is harnessed to pump protons from one side of the membrane to the other, and this generates an *electrochemical proton gradient* across the membrane. The energy stored in this gradient is then used to drive reactions catalyzed by enzymes embedded in the membrane (Figure 7–1). In mitochondria and chloroplasts, most of the energy is used to convert ADP and P_i to ATP, although some

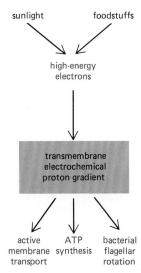

Figure 7–1 Chemiosmotic coupling is used by all cells to harness energy. Energy from sunlight or the oxidation of foodstuffs is first used to create an electrochemical proton gradient across a membrane. This gradient serves as a versatile energy store and is used to drive a variety of reactions in mitochondria, chloroplasts, and bacteria.

is used to transport specific metabolites into and out of the organelles. In bacteria, by contrast, the electrochemical gradient itself is as important a store of directly usable energy as is the ATP it generates; it not only drives many transport processes, it also drives the rapid rotation of the bacterial flagellum, which allows the bacterium to swim (p. 720).

It is generally believed that the energy-converting organelles of eucaryotes evolved from procaryotes that were engulfed by primitive eucaryotic cells and developed a symbiotic relationship with them about 1.5×10^9 years ago. This would explain why mitochondria and chloroplasts contain their own DNA, which codes for some of their proteins. Since then, however, these organelles have lost much of their own genomes and have become heavily dependent on proteins that are encoded by genes in the nucleus, synthesized in the cytosol, and then imported into the organelle. Conversely, the host cells have become dependent on these organelles for much of the ATP they need to carry out biosyntheses, ion and solute pumping, and movement—as well as for the catalysis of selected biosynthetic reactions that occur inside each organelle.

Figure 7–2 Rapid changes of shape are observed when a mitochondrion is visualized in a living cell.

The Mitochondrion[1]

Mitochondria occupy a substantial fraction of the cytoplasm of virtually all eucaryotic cells. Although they are large enough to be seen in the light microscope and were first identified in the nineteenth century, real progress in understanding their function depended on procedures developed in 1948 for isolating intact mitochondria. For technical reasons, most biochemical studies have been carried out with mitochondria purified from liver; each liver cell contains 1000 to 2000 mitochondria, which in total occupy roughly a fifth of the cell volume.

Mitochondria are usually depicted as stiff, elongated cylinders with a diameter of 0.5 to 1 μm, resembling bacteria. Time-lapse microcinematography of living cells, however, shows that mitochondria are remarkably mobile and plastic organelles, constantly changing their shape (Figure 7–2) and even fusing with one another and then separating again. As they move about in the cytoplasm, they often appear to be associated with microtubules (Figure 7–3), which may determine the unique orientation and distribution of mitochondria in different types of cells. Thus the mitochondria in some cells form long moving filaments or chains, while in others they remain fixed in one position where they provide ATP directly to a site of unusually high ATP consumption—packed between adjacent myofibrils in a cardiac muscle cell, for example, or wrapped tightly around the flagellum in a sperm (Figure 7–4).

Figure 7–3 (A) Light micrograph of chains of elongated mitochondria in a living mammalian cell in culture. The cell was stained with a vital fluorescent dye (rhodamine 123) that specifically labels mitochondria. (B) Immunofluorescence micrograph of the same cell stained (after fixation) with fluorescent antibodies that bind to microtubules. Note that the mitochondria tend to be aligned along microtubules. The bar indicates 10 μm. (Courtesy of Lan Bo Chen.)

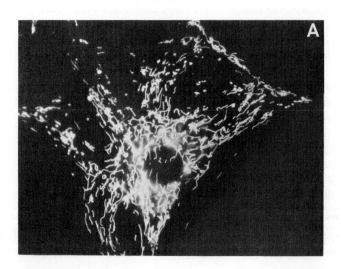
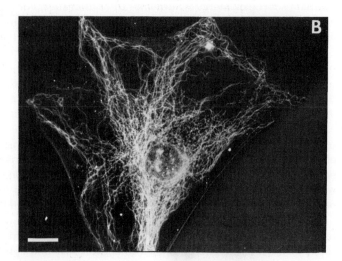

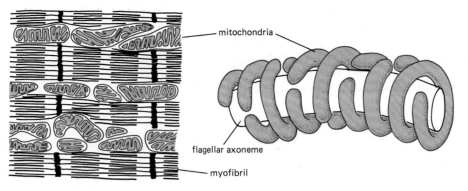

mitochondria

flagellar axoneme

myofibril

CARDIAC MUSCLE   SPERM TAIL

Figure 7–4 Localization of mitochondria near sites of high ATP utilization in cardiac muscle and a sperm tail. During the development of the flagellum of the sperm tail, microtubules wind helically around the axoneme, where they are thought to help localize the mitochondria in the tail; these microtubules then disappear.

The Mitochondrion Contains an Outer Membrane and an Inner Membrane That Define Two Internal Compartments[2]

Each mitochondrion is bounded by two highly specialized membranes that play a crucial part in its activities. Together they create two separate mitochondrial compartments: the internal **matrix space** and a much narrower **intermembrane space.** If purified mitochondria are gently disrupted and then fractionated into separate components (Figure 7–5), the biochemical composition of each of the two membranes and of the the spaces enclosed by them can be determined. As described in Figure 7–6, each contains a unique collection of proteins.

The **outer membrane** contains many copies of a transport protein called *porin*, which forms large aqueous channels through the lipid bilayer. This membrane thus resembles a sieve that is permeable to all molecules of 10,000 daltons or less, including small proteins. Such molecules can enter the intermembrane space, but most of them cannot pass the impermeable inner membrane. Thus, while the intermembrane space is chemically equivalent to the cytosol with respect to the small molecules it contains, the matrix space contains a highly selected set of small molecules.

As we shall explain in detail later, the major working part of the mitochondrion is the matrix space and the **inner membrane** that encloses it. The inner membrane is highly specialized. It contains a high proportion of the "double" phospholipid *cardiolipin* (p. 398), which is thought to help make the membrane especially impermeable to ions. It also contains a variety of transport proteins that make it selectively permeable to those small molecules that are metabolized or required by the many mitochondrial enzymes concentrated in the matrix space. The matrix enzymes include those that metabolize pyruvate and fatty acids to produce acetyl CoA and those that oxidize acetyl CoA in the citric acid cycle. The principal end products of this oxidation are CO_2, which is released from the cell, and NADH, which is the main source of electrons for transport along the **respiratory chain**—the name given to the electron-transport chain in mitochondria. The enzymes of the respiratory chain are embedded in the inner mitochondrial membrane, and they are essential to the process of *oxidative phosphorylation*, which generates most of the animal cell's ATP.

Figure 7–5 Techniques for fractionating purified mitochondria into separate components have made it possible to study the different proteins in each mitochondrial compartment. The method shown, which allows the processing of large numbers of mitochondria at the same time, takes advantage of the fact that in media of low ionic strength, water flows into mitochondria and greatly expands the matrix space. While the cristae of the inner membrane allow it to unfold to accommodate the expansion, the outer membrane—which has no folds to begin with—breaks, releasing a structure composed of only the inner membrane and the matrix.

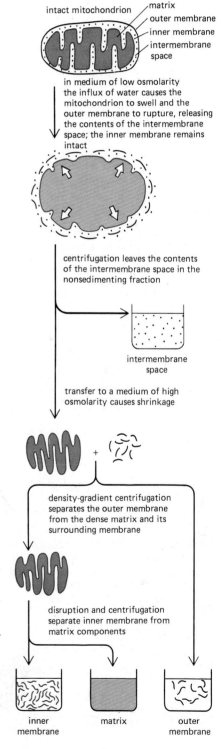

intact mitochondrion — matrix
outer membrane
inner membrane
intermembrane space

in medium of low osmolarity the influx of water causes the mitochondrion to swell and the outer membrane to rupture, releasing the contents of the intermembrane space; the inner membrane remains intact

centrifugation leaves the contents of the intermembrane space in the nonsedimenting fraction

intermembrane space

transfer to a medium of high osmolarity causes shrinkage

density-gradient centrifugation separates the outer membrane from the dense matrix and its surrounding membrane

disruption and centrifugation separate inner membrane from matrix components

inner membrane   matrix   outer membrane

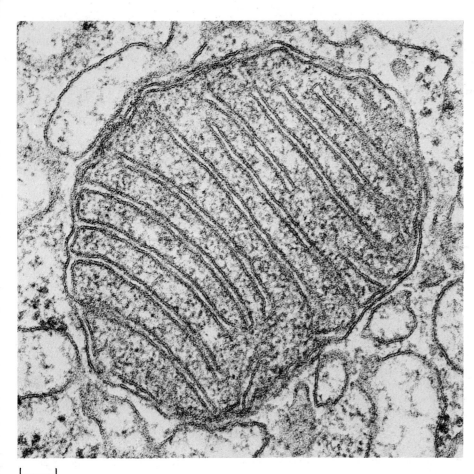

Figure 7–6 The general organization
of a mitochondrion. In the liver
an estimated 67% of the total
mitochondrial protein is located in the
matrix, 21% is located in the inner
membrane, 6% in the outer membrane,
and 6% in the intermembrane space.
As indicated below, each of these four
regions contains a special set of
proteins that mediate distinct
functions. (Micrograph courtesy of
Daniel S. Friend.)

100 nm

Matrix. The matrix contains a highly concentrated mixture of
hundreds of enzymes, including those required for the oxidation
of pyruvate and fatty acids and for the citric acid cycle. The
matrix also contains several identical copies of the mitochondrial
DNA genome, special mitochondrial ribosomes, tRNAs, and
various enzymes required for expression of the mitochondrial
genes.

Inner Membrane. The inner membrane is folded into
numerous cristae, which greatly increase its total surface area. It
contains proteins with three types of functions: (1) those that
carry out the oxidation reactions of the respiratory chain, (2) an
enzyme complex called *ATP synthetase* that makes ATP in the
matrix, and (3) specific transport proteins that regulate the
passage of metabolites into and out of the matrix. Since an
electrochemical gradient that drives the ATP synthetase is
established across this membrane by the respiratory chain, it is
important that the membrane be impermeable to most small
ions.

Outer membrane. Because it contains a large channel-forming
protein (called porin), the outer membrane is permeable to all
molecules of 10,000 daltons or less. Other proteins in this
membrane include enzymes involved in mitochondrial lipid
synthesis and enzymes that convert lipid substrates into forms
that are subsequently metabolized in the matrix.

Intermembrane Space. This space contains several enzymes
that use the ATP passing out of the matrix to phosphorylate
other nucleotides.

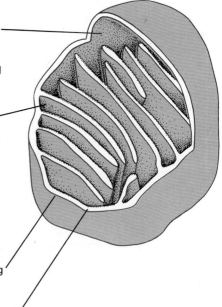

The Inner Membrane Is Folded into Cristae[3]

The inner membrane is usually highly convoluted, forming a series of infoldings, known as **cristae,** in the matrix space. These convolutions greatly increase the area of the inner membrane, so that in liver cell mitochondria, for example, it constitutes about a third of the total membrane of the cell (see Table 8–2, p. 408). The number of cristae is three times greater in the mitochondrion of a cardiac muscle cell than in the mitochondrion of a liver cell, presumably because of the greater demand for ATP in heart cells. Mitochondrial cristae also show striking morphological differences in different cell types, although the significance of these differences is unknown (Figure 7–7).

In addition to morphological distinctions, there are substantial differences in the mitochondrial enzymes of different cell types. In this chapter, however, we shall largely ignore the differences and focus instead on the enzymes and properties that are common to all mitochondria.

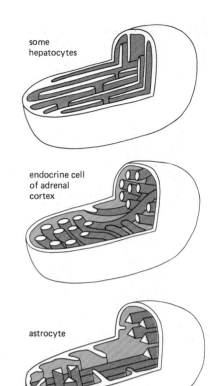

some
hepatocytes

endocrine cell
of adrenal
cortex

astrocyte

Figure 7–7 Some morphologically distinct mitochondrial cristae found in various cells in the rat. The effects of these differences on mitochondrial function are not known.

7-3 Mitochondrial Oxidation Begins When Large Amounts of Acetyl CoA Are Produced in the Matrix Space from Fatty Acids and Pyruvate[4]

Oxidative metabolism in mitochondria is fueled primarily by fatty acids and by the pyruvate produced in the cytosol during glycolysis. These compounds are selectively transported from the cytosol into the mitochondrial matrix, where they are broken down into the two-carbon acetyl group on acetyl coenzyme A (acetyl CoA, Figure 7–8); the acetyl group is then fed into the citric acid cycle for further degradation, and the process ends with the passage of acetyl-derived high-energy electrons along the respiratory chain.

To ensure a continuous supply of fuel for oxidative metabolism, animal cells store fatty acids in the form of fats and glucose in the form of glycogen, which can be broken down to produce pyruvate. Quantitatively, fat is far more important than glycogen, in part because its oxidation releases more than six times as much energy as the oxidation of an equal mass of glycogen in its hydrated form. An average adult human stores enough glycogen for only about a day of normal activities, but enough fat to last for nearly a month. If our main fuel reservoir were carried as glycogen instead of fat, body weight would be increased by an average of about 60 pounds.

Most of our fat is stored in adipose tissue, from which it is released into the bloodstream for other cells to utilize as needed. The need arises after a period of not eating; even a normal overnight fast results in the mobilization of fat, so that

Figure 7–8 The central intermediate produced during the breakdown of foodstuffs in the mitochondrion is acetyl coenzyme A (acetyl CoA). A space-filling model is shown above a common abbreviation (see also Figure 2–19, p. 65). The atom labeled S is a sulfur atom, which forms a thioester linkage to acetate. Because this is a high-energy linkage, the acetate group can be readily transferred to other molecules, such as oxaloacetate (see Figure 7–14).

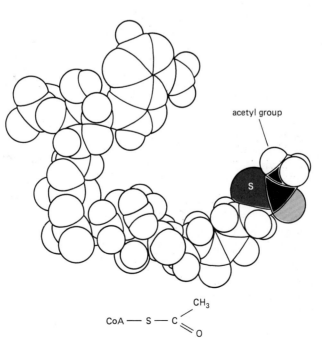

acetyl group

$$CoA - S - C {\overset{CH_3}{\underset{O}{\diagup}}}$$

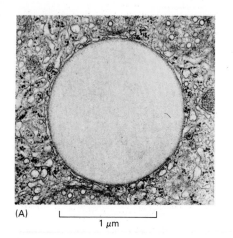

(A)

1 μm

(B)

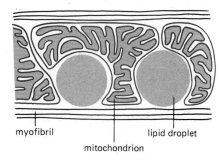

Figure 7–9 (A) Electron micrograph of a lipid droplet in the cytoplasm that contains triacylglycerols, the main form of stored fat. (B) The structure of triacylglycerol, with its glycerol portion in color. (A, courtesy of Daniel S. Friend.)

in the morning most of the acetyl CoA entering the citric acid cycle is derived from fatty acids rather than from glucose. After a meal, however, most of the acetyl CoA entering the citric acid cycle comes from glucose derived from foodstuffs, and any excess glucose is used to replenish depleted glycogen stores or to synthesize fats. (Note that while sugars are readily converted to fats in animal cells, fatty acids cannot be converted to sugars.)

A fat molecule is composed of three molecules of fatty acid held in ester linkage to glycerol. Such **triacylglycerols** (*triglycerides*) have no charge and are virtually insoluble in water, coalescing into droplets in the cytosol (Figure 7–9). A single very large fat droplet accounts for most of the volume of *adipocytes* (fat cells), the large cells specialized for fat storage in adipose tissue. Much smaller fat droplets are common in cells, such as cardiac muscle cells, that rely on the breakdown of fatty acids for their energy supply; these droplets are often closely associated with mitochondria (Figure 7–10). In all cells, enzymes in the outer and inner mitochondrial membranes mediate the movement of fatty acids derived from fat molecules into the mitochondrial matrix. In the matrix, each fatty acid molecule (as *fatty acyl CoA*) is broken down completely by a cycle of reactions that trims two carbons at a time from its carboxyl end, generating one molecule of acetyl CoA in each turn of the cycle (Figure 7–11). The acetyl CoA produced is fed into the citric acid cycle to be oxidized further.

Figure 7–10 Fat droplets in a cardiac muscle cell are surrounded by mitochondria that oxidize the fatty acids derived from triacylglycerols.

myofibril · lipid droplet · mitochondrion

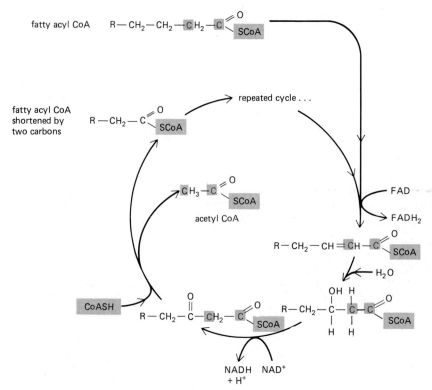

fatty acyl CoA

fatty acyl CoA shortened by two carbons

repeated cycle . . .

acetyl CoA

FAD

FADH₂

H₂O

CoASH

NADH + H⁺ · NAD⁺

Figure 7–11 The fatty acid oxidation cycle, which is catalyzed by a series of four enzymes in the mitochondrial matrix. Each turn of the cycle shortens the fatty acid chain by two carbons (shown in color), as indicated, and generates one molecule of acetyl CoA and one molecule each of NADH and FADH₂. The NADH is freely soluble in the matrix. In contrast, the FADH₂ remains tightly bound to the enzyme *fatty acyl-CoA dehydrogenase;* its two electrons will be rapidly transferred to the ubiquinone pool in the mitochondrial inner membrane (see p. 360), regenerating FAD. The four-step fatty acid oxidation pathway shown is identical in its chemistry to the one used to break many other carbon-carbon bonds (see Figure 7–14, for example).

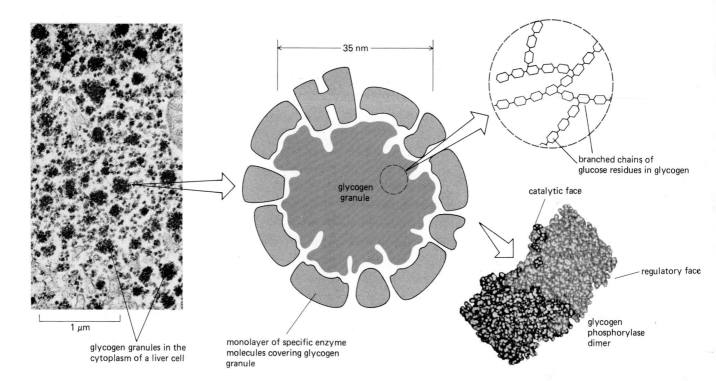

Glycogen is a large, branched polymer of glucose that is contained in granules in the cell cytoplasm (Figure 7–12); its synthesis and degradation are highly regulated according to need (see p. 709). When the need arises, glycogen is broken down to release glucose 1-phosphate. The reactions of glycolysis convert the six-carbon glucose molecule (and related sugars) to two three-carbon pyruvate molecules (see p. 66), which still retain most of the energy that can be derived from the complete oxidation of sugars. This energy is harvested only after the pyruvate is transported from the cytosol into the mitochondrial matrix, where it encounters a multienzyme complex larger than a ribosome, the *pyruvate dehydrogenase complex*. This complex—containing multiple copies of three enzymes, five coenzymes, and two regulatory proteins—rapidly converts pyruvate to acetyl CoA, releasing CO_2 as a by-product (Figure 7–13). This acetyl CoA joins the acetyl CoA produced from fatty acids to fuel the citric acid cycle.

Figure 7–12 Electron micrograph and schematic drawing of a glycogen granule, the major storage form of carbohydrate in vertebrate cells. Glycogen is a polymer of glucose, and each glycogen granule is a single, highly branched molecule. The synthesis and degradation of glycogen are catalyzed by enzymes bound to the granule surface, including the synthetic enzyme *glycogen synthase* and the degradative enzyme *glycogen phosphorylase*. (Courtesy of Robert Fletterick and Daniel S. Friend.)

7-4 The Citric Acid Cycle Oxidizes the Acetyl Group on Acetyl CoA to Generate NADH and FADH₂ for the Respiratory Chain[5]

In the nineteenth century, biologists noticed that in the absence of air (anaerobic conditions), cells produce lactic acid (or ethanol), while in its presence (aerobic conditions), they use O_2 to produce CO_2 and H_2O. Efforts to define the pathways of aerobic metabolism eventually focused on the oxidation of pyruvate and led in 1937 to the discovery of the **citric acid cycle,** also known as the *tricarboxylic acid cycle* or the *Krebs cycle*. The citric acid cycle accounts for about two-thirds of the total oxidation of carbon compounds in most cells, and its major end products are CO_2 and NADH. The CO_2 is released as a waste product, while the NADH

Figure 7–13 The reactions carried out by the pyruvate dehydrogenase complex that convert pyruvate to acetyl CoA in the mitochondrial matrix; NADH is also produced in this reaction. A, B, and C are the three enzymes *pyruvate dehydrogenase, dihydrolipoyl transacetylase,* and *dihydrolipoyl dehydrogenase,* whose activities are coupled as shown. The structure of the complex is shown in Figure 2–40, page 84; the complex also contains a protein kinase and a protein phosphatase that regulate the activity of pyruvate dehydrogenase, turning it off whenever ATP levels are high.

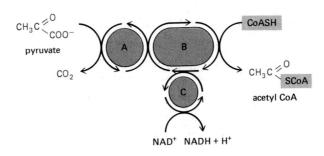

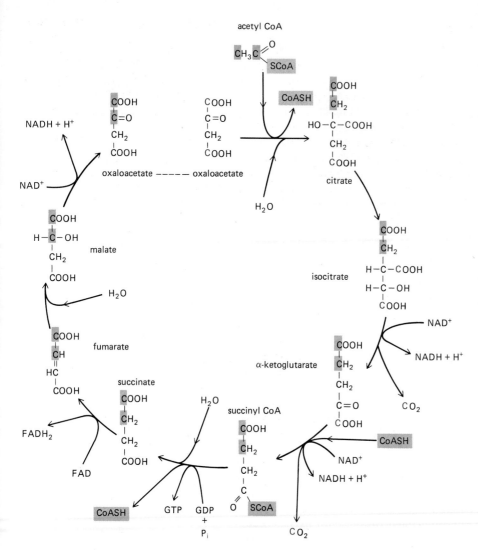

Figure 7–14 The citric acid cycle. The intermediates are shown as their free acids, although the carboxyl groups are actually ionized. Each of the indicated steps is catalyzed by a different enzyme located in the mitochondrial matrix. The two carbons from acetyl CoA that enter this turn of the cycle (shadowed in color) will be converted to CO_2 in subsequent turns of the cycle. A colored C denotes the two carbon atoms that are converted to CO_2 in this cycle. Three molecules of NADH are formed. The GTP molecule produced can be converted to ATP by the exchange reaction GTP + ADP → GDP + ATP. The molecule of $FADH_2$ formed remains protein-bound as part of the *succinate dehydrogenase complex* in the mitochondrial inner membrane; this complex feeds the electrons acquired by $FADH_2$ directly to ubiquinone.

molecules pass their high-energy electrons to the respiratory chain; at the end of the chain, these electrons are used to reduce O_2 to H_2O.

The citric acid cycle begins when the acetyl CoA formed from fatty acids or pyruvate reacts with the four-carbon compound *oxaloacetate* to produce the six-carbon *citric acid* for which the cycle is named. Then, as a result of seven sequential enzyme-mediated reactions, two carbon atoms are removed as CO_2 and oxaloacetate is regenerated. Each such turn of the cycle produces two CO_2 molecules from two carbon atoms that entered in *previous* cycles (Figure 7–14). But the net result, insofar as the acetyl group on acetyl CoA is concerned, is

$$CH_3COOH \text{ (as acetyl CoA)} + 2H_2O + 3NAD^+ + \text{protein-bound FAD} \rightarrow 2CO_2$$
$$+ 3H^+ + 3NADH + \text{protein-bound } FADH_2$$

This reaction also produces one molecule of ATP (via GTP) through a *substrate-level phosphorylation* reaction of the kind that occurs in glycolysis (see p. 66).

The most important contribution of the citric acid cycle to metabolism is the extraction of high-energy electrons during the oxidation of the two acetyl carbon atoms to CO_2 (see p. 69). These electrons, which are transiently held by NADH and $FADH_2$, are quickly passed to the respiratory chain in the mitochondrial inner membrane. $FADH_2$, which is part of the succinate dehydrogenase complex in the inner membrane, passes its electrons directly to the respiratory chain. The NADH, in contrast, forms a soluble pool of reducing equivalents in the mitochondrial matrix and passes on its electrons after a random collision with a membrane-bound dehydrogenase enzyme. We shall now consider how the energy stored in these electrons is used to synthesize ATP.

A Chemiosmotic Process Converts Oxidation Energy into ATP on the Mitochondrial Membrane[6]

Although the citric acid cycle constitutes part of aerobic metabolism, none of the reactions leading to the production of NADH and FADH$_2$ makes direct use of molecular oxygen; only the final catabolic reactions that take place on the mitochondrial inner membrane do so. Nearly all of the energy available from burning carbohydrates, fats, and other foodstuffs in the earlier stages of oxidation is initially saved in the form of high-energy electrons carried by NADH and FADH$_2$. Then these electrons, which have been removed from substrates by NAD$^+$ and FAD, are combined with molecular oxygen by means of the respiratory chain. Because the large amount of energy released is harnessed by the enzymes in the inner membrane to drive the conversion of ADP + P$_i$ to ATP, the term **oxidative phosphorylation** is used to describe these last reactions (Figure 7–15).

As previously mentioned, the generation of ATP by oxidative phosphorylation via the respiratory chain depends on a chemiosmotic process. When it was first proposed in 1961, this mechanism explained a long-standing puzzle in cell biology. Nonetheless, the idea was so novel that it was some years before enough supporting evidence accumulated to make it generally accepted. It was originally believed that the energy for ATP synthesis via the respiratory chain was supplied by the same process that operates during substrate-level phosphorylations: that is, the energy of oxidation was thought to generate a high-energy bond between a phosphate group and some intermediate compound, and the conversion of ADP to ATP was thought to be driven by the energy released when this bond was broken (p. 66). Despite intensive efforts, however, the expected intermediates could not be detected.

According to the *chemiosmotic hypothesis*, the high-energy chemical intermediates are replaced by a link between chemical processes ("chemi") and transport processes ("osmotic"—from the Greek *osmos*, push)—hence **chemiosmotic coupling** (Table 7–1). As the high-energy electrons from the hydrogens on NADH and FADH$_2$ are transported down the respiratory chain in the mitochondrial inner membrane, the energy released as they pass from one carrier molecule to the next is used to pump protons (H$^+$) across the inner membrane from the mitochondrial matrix into the intermembrane space. This creates an *electrochemical proton gradient* across the mitochondrial inner membrane, and the backflow of H$^+$ down this gradient is in turn used to drive a membrane-bound enzyme, *ATP synthetase*, which catalyzes the conversion of ADP + P$_i$ to ATP, completing the process of oxidative phosphorylation (Figure 7–16).

In the remainder of this section we shall briefly outline the type of reactions that make oxidative phosphorylation possible, saving the details for later (see pp. 356–365).

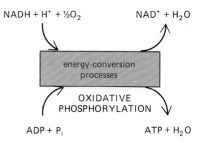

Figure 7–15 The major net energy conversion catalyzed by the mitochondrion. In this process of *oxidative phosphorylation*, the mitochondrial inner membrane serves as an energy-conversion machine, changing part of the energy of NADH (and FADH$_2$) oxidation into phosphate-bond energy in ATP.

Table 7–1 Chemiosmotic Coupling

The chemiosmotic hypothesis, as proposed in the early 1960s, consisted of four independent postulates. In terms of mitochondrial function they were as follows:

1. The mitochondrial respiratory chain in the inner membrane is proton-translocating; it pumps H$^+$ out of the matrix space when electrons are transported along the chain.
2. The mitochondrial ATP synthetase complex also translocates protons across the inner membrane: being reversible, it can use the energy of ATP hydrolysis to pump H$^+$ across the membrane, but if a large enough electrochemical proton gradient is present, protons flow in the reverse direction through the complex and drive ATP synthesis.
3. The mitochondrial inner membrane is impermeable to H$^+$, OH$^-$, and generally to anions and cations.
4. The mitochondrial inner membrane is equipped with a set of carrier proteins that mediate the entry and exit of essential metabolites and selected inorganic ions.

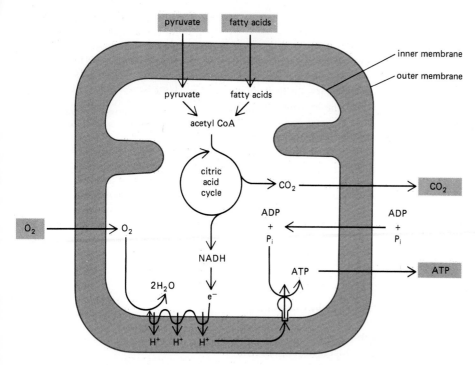

Figure 7–16 A summary of the flow of major reactants into and out of the mitochondrion. Pyruvate and fatty acids enter the mitochondrion and are metabolized by the citric acid cycle, which produces NADH. In the process of oxidative phosphorylation, high-energy electrons from NADH are then passed to oxygen by means of the respiratory chain in the inner membrane, producing ATP by a chemiosmotic mechanism.

NADH generated by glycolysis in the cytosol also passes electrons to the respiratory chain (not shown). Since NADH cannot pass across the mitochondrial inner membrane, the electron transfer from cytosolic NADH must be accomplished indirectly by means of one of several "shuttle" systems that transport another reduced compound into the mitochondrion; after being oxidized, this compound is returned to the cytosol, where it is reduced by NADH again.

Electrons Are Transferred from NADH to Oxygen Through Three Large Respiratory Enzyme Complexes[7]

Although the mechanism by which energy is harvested by the respiratory chain differs from that in other catabolic reactions, the principle is the same. The reaction $H_2 + \frac{1}{2}O_2 \rightarrow H_2O$ is made to occur in many small steps, so that most of the energy released can be converted into a storage form instead of being lost to the environment as heat. As in the formation of ATP and NADH in glycolysis or the citric acid cycle, this involves employing an indirect pathway for the reaction. The respiratory chain is unique in that the hydrogen atoms are first separated into protons and electrons. The electrons pass through a series of electron carriers in the mitochondrial inner membrane. When the electrons reach the end of this electron-transport chain, the protons are returned to neutralize the negative charges created by the final addition of the electrons to the oxygen molecule (Figure 7–17).

We shall outline the oxidation process starting from NADH, the major collector of reactive electrons derived from the oxidation of food molecules. Each hydrogen atom (which we shall denote as H·) consists of one electron (e^-) and one proton (H^+). The mechanism by which electrons are acquired by NADH was discussed earlier (p. 68) and is shown in greater detail in Figure 7–18. As this example makes clear, each molecule of NADH carries a *hydride ion* (a hydrogen atom plus an extra electron, $H:^-$) rather than a single hydrogen atom. However, because protons are freely available in aqueous solutions, carrying the hydride ion on NADH is equivalent to carrying two hydrogen atoms, or a hydrogen molecule ($H:^- + H^+ \rightarrow H_2$).

Electron transport begins when the hydride ion is removed from NADH to regenerate NAD^+ and is converted into a proton and two electrons ($H:^- \rightarrow H^+ + 2e^-$). The two electrons are passed to the first of the more than 15 different electron carriers in the respiratory chain. The electrons start with very high energy and gradually lose it as they pass along the chain. For the most part, the electrons pass from one metal atom to another, each metal atom being tightly bound to a protein molecule, which alters the electron affinity of the metal atom. The various types of electron carriers in the respiratory chain will be discussed in detail later (p. 359). Most important, the many proteins involved are grouped into three large *respiratory enzyme complexes*, each containing transmembrane proteins that hold the complex firmly in the mitochondrial inner membrane (see p. 360). Each complex in the chain has a greater affinity for electrons than its predecessor, and electrons pass sequentially from one complex to another until they are finally transferred to oxygen, which has the greatest affinity of all for electrons.

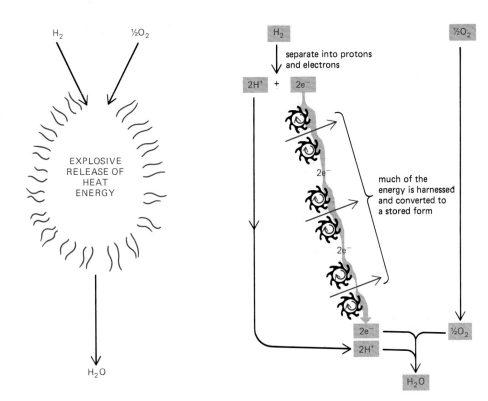

Figure 7–17 Illustration showing how most of the energy that would be released as heat if hydrogen were burned (*left panel*) is instead harnessed and stored in a form useful to the cell by means of the electron-transport chain in the mitochondrial inner membrane (*right panel*). The rest of the oxidation energy is released as heat by the mitochondrion. In reality the protons and electrons shown are removed from hydrogen atoms that are covalently linked to NADH or $FADH_2$ molecules (see Figure 7–18).

Energy Released by the Passage of Electrons Along the Respiratory Chain Is Stored as an Electrochemical Proton Gradient Across the Inner Membrane[8]

The close association of the electron carriers with protein molecules makes oxidative phosphorylation possible. The proteins guide the electrons along the respiratory chain so that the electrons move sequentially from one enzyme complex to another—with no short circuits. Most important, the transfer of electrons is coupled to allosteric changes in selected protein molecules, so that the energetically favorable flow of electrons pumps protons (H^+) across the inner membrane, from the matrix space to the intermembrane space and from there to the outside of the mitochondrion. This movement of protons has two major consequences.

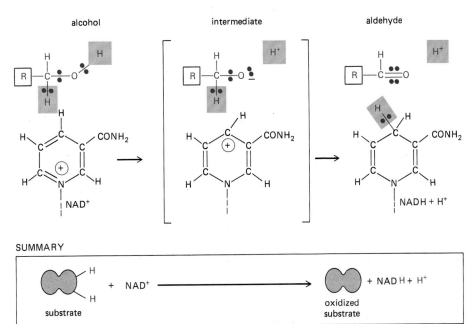

Figure 7–18 The biological oxidation of an alcohol to an aldehyde is thought to proceed in the manner shown. The components of two complete hydrogen atoms are lost from the alcohol: a hydride ion is transferred to NAD^+ and a proton escapes to the aqueous solution. Only the nicotinamide ring portion of the NAD^+ and NADH molecules is shown here (see Figure 2–22). The steps illustrated occur on a protein surface, being catalyzed by specific chemical groups on the enzyme alcohol dehydrogenase (not shown). (Modified with permission from P.F. Cook, N.J. Oppenheimer, and W.W. Cleland, *Biochemistry* 20:1817–1825, 1981. Copyright 1981 American Chemical Society.)

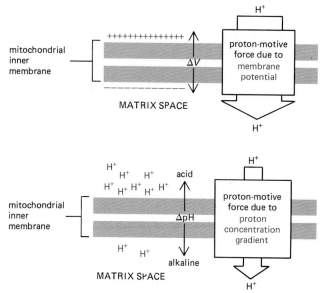

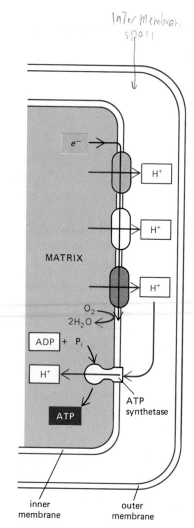

Figure 7–19   The two components of the electrochemical proton gradient. The total proton-motive force across the mitochondrial inner membrane consists of a large force due to the membrane potential (traditionally designated $\Delta\psi$ by experts, but designated ΔV in this text) and a smaller force due to the proton concentration gradient (ΔpH). Both forces act to drive protons into the matrix space.

(1) It generates a pH gradient across the inner mitochondrial membrane, with the pH higher in the matrix than in the cytosol, where the pH is generally close to 7. (Since small molecules equilibrate freely across the outer membrane of the mitochondrion, the pH in the intermembrane space is the same as in the cytosol.) (2) It generates a voltage gradient (membrane potential) across the inner mitochondrial membrane, with the inside negative and the outside positive (as a result of the net outflow of positive ions).

The pH gradient (ΔpH) acts to drive H^+ back into the matrix and OH^- out of the matrix and thus reinforces the effect of the membrane potential (ΔV), which acts to attract any positive ion into the matrix and to push any negative ion out. Together, these two forces are said to constitute an **electrochemical proton gradient** (Figure 7–19).

The electrochemical proton gradient exerts a **proton-motive force,** which can be measured in units of millivolts (mV). Since each ΔpH of 1 pH unit has an effect equivalent to a membrane potential of about 60 mV, the total proton-motive force equals $\Delta V - 60(\Delta pH)$. In a typical cell the proton-motive force across the inner membrane of a respiring mitochondrion is about 220 mV and is made up of a membrane potential of about 160 mV and a pH gradient of about -1 pH unit.

7-6   The Energy Stored in the Electrochemical Proton Gradient Is Used to Produce ATP and to Transport Metabolites and Inorganic Ions into the Matrix Space[9]

The mitochondrial inner membrane contains an unusually high proportion of protein, being approximately 70% protein and 30% phospholipid by weight. Many of the proteins belong to the electron-transport chain, which establishes the electrochemical proton gradient across the membrane. Another major component is the enzyme *ATP synthetase*, which catalyzes the synthesis of ATP. This is a large protein complex through which protons flow down their electrochemical gradient into the matrix. Like a turbine, ATP synthetase converts one form of energy to another, synthesizing ATP from ADP and P_i in the mitochondrial matrix in a reaction that is coupled to the inward flow of protons (Figure 7–20).

ATP synthesis is not the only process that is driven by the electrochemical proton gradient. The enzymes in the mitochondrial matrix, where the citric acid cycle and other metabolic reactions take place, must be supplied with high concentrations of substrates, and ATP synthetase must be supplied with ADP and phosphate. Thus many charged substrates must be transported across the inner membrane. This is achieved by various membrane carrier proteins (see p. 303), many of which actively transport specific molecules against their electrochemical

Figure 7–20   The general mechanism of oxidative phosphorylation. As a high-energy electron is passed along the electron-transport chain, some of the energy released is used to drive three respiratory enzyme complexes that pump protons out of the matrix space. These protons create an electrochemical proton gradient across the inner membrane that drives protons back through the ATP synthetase, a transmembrane protein complex that uses the energy of the proton flow to synthesize ATP from ADP and P_i in the matrix.

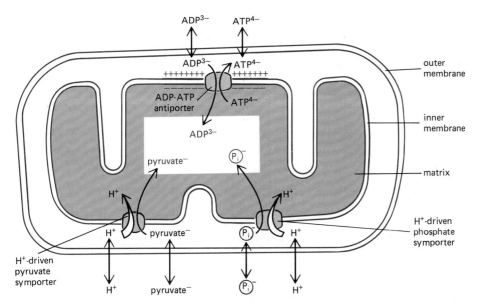

Figure 7–21 Some of the active transport processes driven by the electrochemical proton gradient across the mitochondrial inner membrane. The charge on each of the transported molecules is indicated for reference to the membrane potential, which is negative inside, as shown. The outer membrane is freely permeable to all of these compounds. See Chapter 6 for a discussion of symport and antiport transport mechanisms.

gradients, a process that requires an input of energy. For most metabolites the energy comes from *co-transporting* another molecule down its electrochemical gradient (see p. 309). The transport of ADP into the matrix space, for example, is mediated by an ADP-ATP antiport system: for each ADP molecule that moves in, an ATP molecule moves out down its electrochemical gradient. The transport of phosphate into the matrix space is mediated by a symport system that couples the inward movement of phosphate to the inward flow of protons down their electrochemical gradient so that the phosphate is dragged with them. Pyruvate is transported into the matrix in the same way (Figure 7–21). The electrochemical proton gradient is also used to import Ca^{2+}, which is thought to be important in regulating the activity of selected mitochondrial enzymes; the import of Ca^{2+} into mitochondria may also be important for removing Ca^{2+} from the cytosol when cytosolic Ca^{2+} levels become dangerously high (see p. 701).

The more energy from the electrochemical proton gradient is used to transport molecules and ions into the mitochondrion, the less there is to drive the ATP synthetase. If isolated mitochondria are incubated in a high concentration of Ca^{2+}, for example, they cease ATP production completely; all the energy in their electrochemical proton gradient is diverted to pumping Ca^{2+} into the matrix. Similarly, in certain specialized cells the electrochemical proton gradient is short-circuited so that their mitochondria produce heat instead of ATP (see p. 365). The use of the energy stored in the electrochemical proton gradient can be regulated by cells so that it is directed toward those activities most needed at the time.

7-8 7-38 The Rapid Conversion of ADP to ATP in Mitochondria Maintains a High Ratio of ATP to ADP in Cells[10]

Because of the antiporter in the inner membrane that pumps ADP into the matrix space in exchange for ATP (see Figure 7–21), ADP molecules produced by ATP hydrolysis in the cytosol rapidly enter mitochondria for recharging, while the ATP molecules formed in the mitochondrial matrix by oxidative phosphorylation are rapidly pumped into the cytosol, where they are needed. A typical ATP molecule in the human body shuttles in and out of a mitochondrion for recharging (as ADP) thousands of times a day, keeping the concentration of ATP in a cell about 10 times higher than that of ADP.

As discussed in Chapter 2, biosynthetic enzymes in cells guide their substrates along specific reaction paths, often driving energetically unfavorable reactions by coupling them to the energetically favorable hydrolysis of ATP (see Figure 2–27, p. 74). The highly charged ATP pool is thereby used to drive cellular processes in much the same way that a battery can be used to drive electric engines: if the activity of the mitochondria is halted, ATP levels fall and the cell's battery runs down, so that, eventually, energetically unfavorable reactions can no longer be driven by ATP hydrolysis.

It might seem that this state would not be reached until the concentration of ATP is zero, but in fact it is reached much sooner than that, at a concentration of ATP that depends on the concentrations of ADP and P_i. To explain why, we must consider some elementary principles of thermodynamics.

7-5 The Difference Between $\Delta G°$ and ΔG: A Large Negative Value of ΔG is Required for ATP Hydrolysis to Be Useful to the Cell[11]

The second law of thermodynamics states that chemical reactions proceed spontaneously in the direction that corresponds to an increase in the disorder of the *universe*. In Chapter 2 we noted that reactions that release energy to their surroundings as heat (such as the hydrolysis of ATP) tend to increase the disorder of the universe by increasing random molecular motions. Reactions will also affect the degree of disorder by changing the concentrations of the reactant and product molecules. The net change of disorder in the universe due to a reaction is proportional to the **change in free energy, ΔG,** that is associated with the reaction: reactions that bring about a large *decrease* in free energy (so that ΔG is very negative) create the most disorder in the universe and proceed most readily (see Panel 2–7, pp. 72–73).

When ATP is hydrolyzed to ADP and P_i under the conditions that normally exist in a cell, the free-energy change is roughly -11 to -13 kcal/mole. This extremely favorable ΔG requires that the concentration of ATP in the cell be kept high compared to the concentration of ADP and P_i. At "standard conditions," where ATP, ADP, and P_i are all present at the same concentration of 1 mole/liter, the ΔG for ATP hydrolysis—called the **standard free-energy change,** or $\Delta G°$, of the reaction—is only -7.3 kcal/mole. At still lower concentrations of ATP relative to ADP and P_i, ΔG will become equal to zero. At this point the rate at which ADP and P_i will join to form ATP will be equal to the rate at which ATP hydrolyzes to form ADP and P_i. In other words, when $\Delta G = 0$, the reaction is at *equilibrium* (Figure 7–22).

At a fixed temperature, the value of $\Delta G°$ is a constant that depends only on the *nature* of the reactants. The value of ΔG, in contrast, is a variable that depends on the *concentrations* of the reactants, and its value indicates how far a reaction is from equilibrium. Therefore it is ΔG, not $\Delta G°$, that determines if a reaction can be used to drive other reactions. Because the efficient conversion of ADP to ATP in mitochondria maintains such a high concentration of ATP relative to ADP and P_i, the ATP-hydrolysis reaction in cells is kept very far from equilibrium and ΔG is correspondingly very negative. Without this disequilibrium, ATP hydrolysis could not be used to direct the reactions of the cell, and many biosynthetic reactions would run backward rather than forward.

7-6 Cellular Respiration Is Remarkably Efficient
7-7

By means of oxidative phosphorylation, each pair of electrons in NADH provides energy for the formation of about 3 molecules of ATP. The pair of electrons in $FADH_2$, being at a lower energy, generates only about 2 ATP molecules. In all, about 12 molecules of ATP can be formed from each molecule of acetyl CoA that enters the citric acid cycle, which means that 24 ATP molecules are produced from 1 molecule of glucose and 96 ATP molecules from 1 molecule of palmitate, a 16-carbon fatty acid. If one includes the energy-yielding reactions that occur before acetyl CoA is formed, the complete oxidation of 1 molecule of glucose gives a net yield of about 36 ATPs, while about 129 ATPs are obtained from the complete oxidation of 1 molecule of palmitate. These numbers are approximate maximal values, since the actual amount of ATP made in the mitochondrion depends on what fraction of the electrochemical gradient energy is used for purposes other than ATP synthesis.

1.

$$\text{ATP} \xrightarrow{\text{hydrolysis}} \text{ADP} + \text{P}_i$$

$$\text{hydrolysis rate} = \left(\begin{array}{c}\text{hydrolysis}\\\text{rate constant}\end{array}\right) \times \left(\begin{array}{c}\text{concentration}\\\text{of ATP}\end{array}\right)$$

2.

$$\text{ADP} + \text{P}_i \xrightarrow{\text{synthesis}} \text{ATP}$$

$$\text{synthesis rate} = \left(\begin{array}{c}\text{synthesis}\\\text{rate constant}\end{array}\right) \times \left(\begin{array}{c}\text{conc. of}\\\text{phosphate}\end{array}\right) \times \left(\begin{array}{c}\text{conc. of}\\\text{ADP}\end{array}\right)$$

3. AT EQUILIBRIUM: synthesis rate = hydrolysis rate

$$\left(\begin{array}{c}\text{synthesis}\\\text{rate constant}\end{array}\right) \times \left(\begin{array}{c}\text{conc. of}\\\text{phosphate}\end{array}\right) \times \left(\begin{array}{c}\text{conc. of}\\\text{ADP}\end{array}\right) = \left(\begin{array}{c}\text{hydrolysis}\\\text{rate constant}\end{array}\right) \times \left(\begin{array}{c}\text{conc. of}\\\text{ATP}\end{array}\right)$$

$$\frac{\left(\begin{array}{c}\text{conc. of}\\\text{ADP}\end{array}\right) \times \left(\begin{array}{c}\text{conc. of}\\\text{phosphate}\end{array}\right)}{(\text{conc. of ATP})} = \left(\dfrac{\begin{array}{c}\text{hydrolysis}\\\text{rate constant}\end{array}}{\begin{array}{c}\text{synthesis}\\\text{rate constant}\end{array}}\right) = \text{equilibrium constant } K$$

$$\frac{[\text{ADP}]\,[\text{P}_i]}{[\text{ATP}]} = K$$

4.

For the reaction

$$\text{ATP} \longrightarrow \text{ADP} + \text{P}_i$$

the following equation applies:

$$\Delta G = \Delta G° + RT \ln \frac{[\text{ADP}]\,[\text{P}_i]}{[\text{ATP}]}$$

Where ΔG and $\Delta G°$ are in kilocalories per mole, R is the gas constant (2×10^{-3} kcal/mole °K), T is the absolute temperature (°K), and all the concentrations are in moles per liter.
When the concentrations of all reactants are at 1 M, $\Delta G = \Delta G°$ (since $RT \ln 1 = 0$). $\Delta G°$ is thus a constant defined as the standard free-energy change for the reaction.

At equilibrium the reaction has no net effect on the disorder of the universe, so $\Delta G = 0$. Therefore, at equilibrium,

$$-RT \ln \frac{[\text{ADP}]\,[\text{P}_i]}{[\text{ATP}]} = \Delta G°$$

But the concentrations of reactants at equilibrium must satisfy the equilibrium equation:

$$\frac{[\text{ADP}]\,[\text{P}_i]}{[\text{ATP}]} = K$$

Therefore, at equilibrium,

$$\Delta G° = -RT \ln K$$

We thus see that whereas $\Delta G°$ indicates the equilibrium point for a reaction, ΔG reveals *how far* the reaction is from equilibrium

When the free-energy changes for burning fats and carbohydrates directly to CO_2 and H_2O are compared to the total amount of energy generated and stored in the phosphate bonds of ATP during the corresponding biological oxidations, it is seen that the efficiency with which oxidation energy is converted into ATP bond energy is often greater than 50%. This is considerably better than the efficiency of most nonbiological energy-conversion devices. If cells worked with the efficiency of an electric motor or a gasoline engine (10–20%), an organism would have to eat voraciously in order to maintain itself. Moreover, since wasted energy is liberated as heat, large organisms would need much more efficient mechanisms for giving up heat to the environment.

Students sometimes wonder why the chemical interconversions in cells follow such complex pathways. The oxidation of sugars to CO_2 plus H_2O could certainly be accomplished more directly, eliminating the citric acid cycle and many of the steps in the respiratory chain. Although this would have made respiration easier to learn, it would have been a disaster for the cell. Oxidation produces huge amounts of free energy, which can be utilized efficiently only in small bits. The complex oxidative pathways involve many intermediates, each differing only slightly from its predecessor. As a result, the energy released is parceled out into small packets that can be efficiently converted to high-energy bonds in useful molecules such as ATP and NADH by means of coupled reactions (see Figure 2–17, p. 63).

Figure 7–22 The basic relationship between free-energy changes and equilibrium, as illustrated by the ATP hydrolysis reaction. The equilibrium constant shown here, K, is in units of moles per liter. (See Panel 2–7, pp. 72–73, for a discussion of free energy, and Figure 3–7, p. 94, for a definition of the equilibrium constant.)

Summary

The mitochondrion carries out most cellular oxidations and produces the bulk of the animal cell's ATP. The mitochondrial matrix space contains a large variety of enzymes, including those that oxidize pyruvate and fatty acids to acetyl CoA and those that oxidize this acetyl CoA to CO_2 through the citric acid cycle. Large amounts of NADH (and $FADH_2$) are produced by these oxidation reactions. The energy available from combining oxygen with the reactive electrons carried by NADH and $FADH_2$

is harnessed by an electron-transport chain in the mitochondrial inner membrane called the respiratory chain. The respiratory chain pumps protons out of the matrix to create a transmembrane electrochemical proton gradient, which includes contributions from both a membrane potential and a pH difference. The transmembrane gradient is in turn used both to synthesize ATP and to drive the active transport of selected metabolites across the mitochondrial inner membrane. The combination of these reactions is responsible for an efficient ATP-ADP exchange between the mitochondrion and the cytosol that keeps the cell's ATP pool highly charged.

The Respiratory Chain and ATP Synthetase[12]

Having considered in general terms how mitochondria function, let us now look in more detail at the respiratory chain—the electron-transport chain that is so crucial to all oxidative metabolism. Most of the elements of the chain are intrinsic components of the inner mitochondrial membrane, and they provide some of the clearest examples of the complex interactions that can occur among the individual proteins in a biological membrane.

Functional Inside-out Particles Can Be Isolated from Mitochondria[13]

The respiratory chain is relatively inaccessible to experimental manipulation in intact mitochondria. By disrupting mitochondria with ultrasound, however, it is possible to isolate functional *submitochondrial particles*, which consist of broken cristae that have resealed into small closed vesicles about 100 nm in diameter (Figure 7–23). When negatively stained submitochondrial particles are examined in an electron microscope, their outside surfaces are seen to be studded with tiny spheres attached to the membrane by stalks (Figure 7–24). In intact mitochondria these lollipoplike structures are located on the *inner* (matrix) side of the inner membrane. Thus the submitochondrial particles are inside-out vesicles of inner membrane, with what was previously their matrix-facing surface exposed to the surrounding medium. As a result, they can readily be provided with the membrane-impermeable metabolites that would normally be present in the matrix space. When NADH, ADP, and inorganic phosphate are added, such particles transport electrons from NADH to O_2 and couple this oxidation to ATP synthesis. This cell-free system provides an assay that makes it possible to purify the many proteins responsible for oxidative phosphorylation in a functional form.

7-11 ATP Synthetase Can Be Purified and Added Back to Membranes[14]

The first experiments to show that the various membrane proteins that catalyze oxidative phosphorylation can be separated without destroying their activity were performed in 1960. The tiny protein spheres studding the surface of submitochondrial particles were stripped from the particles and purified in soluble form. The stripped particles could still oxidize NADH in the presence of oxygen, but they could no longer synthesize ATP. On the other hand, the purified spheres on their own acted as ATPases, hydrolyzing ATP to ADP and P_i. When purified spheres (referred to as $F_1ATPases$) were added back to stripped submitochondrial particles, the reconstituted particles once again made ATP from ADP and P_i.

Subsequent work showed that the $F_1ATPase$ is part of a larger transmembrane complex (about 500,000 daltons) containing at least nine different polypeptide chains, which is now known as **ATP synthetase** (also called $F_0F_1ATPase$). ATP synthetase comprises about 15% of the total inner membrane protein, and very similar enzyme complexes are present in both chloroplast and bacterial membranes. This protein complex contains a transmembrane proton carrier, and it synthesizes ATP when protons pass through it down their electrochemical gradient.

One of the most convincing demonstrations of the function of ATP synthetase came from an experiment performed in 1974. By that time, methods had been developed for transferring detergent-solubilized integral membrane proteins into

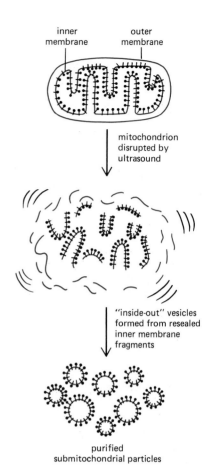

Figure 7–23 Outline of the procedure used to prepare submitochondrial particles from purified mitochondria. The particles are pieces of broken-off cristae that form closed vesicles.

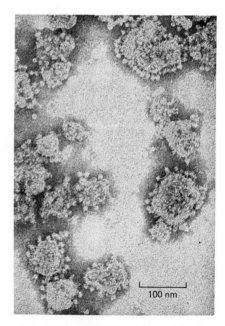

Figure 7–24 Electron micrograph of submitochondrial particles. (Courtesy of Efraim Racker.)

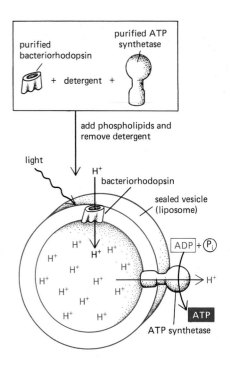

Figure 7–25 Outline of an important experiment demonstrating that the ATP synthetase can be driven by a simple proton flow. By combining a light-driven bacterial proton pump (bacteriorhodopsin), an ATP synthetase purified from ox heart mitochondria, and phospholipids, vesicles were produced that synthesized ATP in response to light.

lipid vesicles (liposomes) formed from purified phospholipids (see p. 278). It thus became possible to form a hybrid membrane that contained both a purified mitochondrial ATP synthetase and bacteriorhodopsin—a bacterial light-driven proton pump (see p. 293). When these vesicles were exposed to light, the protons pumped into the vesicle lumen by the bacteriorhodopsin flowed back out through the ATP synthetase, causing ATP to be made in the medium outside (Figure 7–25). Because a direct interaction between a bacterial proton pump and a mammalian ATP synthetase seems highly unlikely, this experiment strongly suggests that in mitochondria the proton translocation driven by electron transport and the ATP synthesis are separate events.

ATP Synthetase Can Function in Reverse to Hydrolyze ATP and Pump H$^+$ [15]

ATP synthetase is a reversible enzyme complex: it can either use the energy of ATP hydrolysis to pump protons across the inner mitochondrial membrane, or it can harness the flow of protons down an electrochemical proton gradient to make ATP (Figure 7–26). It thereby acts as a *reversible coupling device*, interconverting electrochemical-proton-gradient and chemical-bond energies. Its direction of action depends on the balance between the steepness of the electrochemical proton gradient and the local ΔG for ATP hydrolysis.

Figure 7–26 ATP synthetase is a reversible coupling device that interconverts the energies of the electrochemical proton gradient and chemical bonds. Composed of at least nine different polypeptide chains, it is also known as the F_0F_1ATPase. Five of its polypeptide chains make up the spherical head of the complex, which is called the F_1ATPase. The ATP synthetase can either synthesize ATP by harnessing the proton-motive force (*top*) or pump protons against their electrochemical gradient by hydrolyzing ATP (*bottom*). As explained in the text, the direction of operation at any given instant depends on the net free-energy change for the coupled processes of proton translocation across the membrane and the synthesis of ATP from ADP and P_i.

We have previously shown how the free-energy change (ΔG) for ATP hydrolysis depends on the concentrations of the three reactants ATP, ADP, and P_i (Figure 7–22); the ΔG for ATP synthesis is the negative of this value. The ΔG for proton translocation across the membrane is the sum of (1) the ΔG for moving a mole of any ion through a difference in membrane potential ΔV and (2) the ΔG for moving a mole of a molecule between any two compartments in which its concentration differs. The equation for the proton-motive force on p. 352 combines the same two components, replacing the concentration difference by an equivalent increment in the membrane potential to produce an "electrochemical potential" for the proton. Thus the ΔG for proton translocation and the proton-motive force measure the same potential, one in kilocalories and the other in millivolts. The conversion factor between them is the faraday. Thus,

$$\Delta G_{H+} = -0.023 \text{ (proton-motive force)}$$

where ΔG_{H+} is in kilocalories per mole (kcal/mole) and the proton-motive force is in millivolts (mV). For an electrochemical proton gradient of 220 mV, $\Delta G_{H+} = -5.06$ kcal/mole.

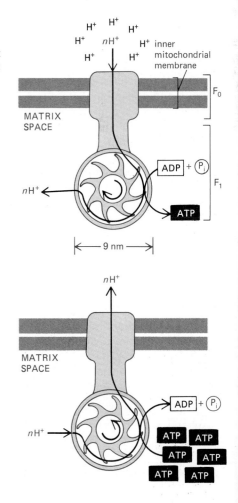

ATP synthetase is so-called because it is normally driven by the large electrochemical proton gradient maintained by the respiratory chain (see Figure 7–20) to make most of the cell's ATP. The exact number of protons needed to make each ATP molecule is not known with certainty. To facilitate the calculations to be described below, however, we shall assume that one molecule of ATP is made by the ATP synthetase for every three protons driven through it.

Whether the ATP synthetase works in its ATP-synthesizing or its ATP-hydrolyzing direction at any instant depends on the exact balance between the favorable free-energy change for moving the three protons across the membrane into the matrix space (ΔG_{3H+}, which is less than zero) and the unfavorable free-energy change for ATP *synthesis* in the matrix ($\Delta G_{ATP\ synthesis}$, which is greater than zero). As previously discussed, the value of $\Delta G_{ATP\ synthesis}$ depends on the exact concentrations of the three reactants ATP, ADP, and P_i in the mitochondrial matrix space (see Figure 7–22). The value of ΔG_{3H+}, on the other hand, is proportional to the value of the proton-motive force across the inner mitochondrial membrane. The following example will help to explain how the balance between these two free-energy changes affects the ATP synthetase.

As explained in the legend to Figure 7–26, a single proton moving into the matrix down an electrochemical gradient of 220 mV liberates 5.06 kcal/mole of free energy, while the movement of three protons liberates three times this much free energy ($\Delta G_{3H+} = -15.2$ kcal/mole). Thus, if the proton-motive force remains constant at 220 mV, the ATP synthetase will synthesize ATP until a ratio of ATP to ADP and P_i is reached where $\Delta G_{ATP\ synthesis}$ is just equal to $+15.2$ kcal/mole (here $\Delta G_{ATP\ synthesis} + \Delta G_{3H+} = 0$). At this point there will be no further net ATP synthesis or hydrolysis by the ATP synthetase.

Suppose that a great deal of ATP is suddenly hydrolyzed by energy-requiring reactions in the cytosol—causing the ATP:ADP ratio in the matrix to fall. Now the value of $\Delta G_{ATP\ synthesis}$ will decrease (see Figure 7–22), and ATP synthetase will begin to synthesize ATP again to restore the original ATP:ADP ratio. Alternatively, if the proton-motive force drops suddenly and is then maintained at a constant 200 mV, ΔG_{3H+} will change to -13.8 kcal/mole. As a result, ATP synthetase will start hydrolyzing some of the ATP in the matrix—until a new balance of ATP to ADP and P_i is reached (where $\Delta G_{ATP\ synthesis} = +13.8$ kcal/mole)—and so on.

In many bacteria, ATP synthetase is routinely reversed in a transition between aerobic and anaerobic metabolism, as we shall see later. The reversibility of the ATP synthetase is a property shared by other membrane proteins that couple ion movement to ATP synthesis or hydrolysis. For example, both the Na^+-K^+ pump and the Ca^{2+} pump, described in Chapter 6, hydrolyze ATP and use the energy released to pump specific ions across a membrane (see p. 305). If either of these pumps is exposed to an abnormally steep gradient of the ions it transports, it will act in reverse—synthesizing ATP from ADP and P_i instead of hydrolyzing it. Thus, like ATP synthetase, such pumps are able to convert the electrochemical energy stored in a transmembrane ion gradient directly into phosphate bond energy in ATP.

The Respiratory Chain Pumps H$^+$ Across the Inner Mitochondrial Membrane[16]

The ATP synthetase does not normally transport H^+ out of the matrix space across the inner mitochondrial membrane. Instead, the respiratory chain embedded in this membrane normally does, thereby generating the electrochemical proton gradient that drives ATP synthesis. The ability of the respiratory chain to translocate H^+ outward from the matrix space can be demonstrated experimentally under special conditions. For example, a suspension of isolated mitochondria can be provided with a suitable substrate for oxidation, and the H^+ flow through ATP synthetase can be blocked. In the absence of air, the injection of a small amount of oxygen into such a preparation causes a brief burst of respiration, which lasts for 1 to 2 seconds before all the oxygen is consumed. During this respiratory burst, a sudden acidification of the medium resulting from the extrusion of H^+ from the matrix space can be measured with a sensitive pH electrode.

A similar experiment can be carried out with a suspension of submitochondrial particles. In this case the medium becomes more basic when oxygen is injected, since protons are pumped *into* each vesicle because of its inside-out orientation.

7-12 Spectroscopic Methods Have Been Used to Identify Many Electron Carriers[17]

Many of the electron carriers in the respiratory chain absorb visible light and change color when they are oxidized or reduced. In general, each has an absorption spectrum and reactivity that is distinct enough to allow its behavior to be traced spectroscopically even in crude mixtures. It was therefore possible to purify these components long before their exact functions were known. Thus the *cytochromes* were discovered in 1925 as compounds that undergo rapid oxidation and reduction in living organisms as disparate as bacteria, yeasts, and insects. By observing cells and tissues with a spectroscope, three types of cytochromes were identified by their distinctive absorption spectra and designated cytochromes a, b, and c. This nomenclature has survived even though cells are now known to contain several cytochromes of each type, and the classification into types is not functionally important.

The **cytochromes** constitute a family of colored proteins that are related by the presence of a bound *heme group*, whose iron atom changes from the ferric (Fe III) to the ferrous (Fe II) state whenever it accepts an electron. The heme group consists of a *porphyrin* ring with a tightly bound iron atom held by four nitrogen atoms at the corners of a square (Figure 7–27). A related porphyrin ring is responsible for the red color of blood and the green color of leaves, being bound to iron in hemoglobin (see p. 601) and to magnesium in chlorophyll (see p. 373). The best understood of the many proteins in the respiratory chain is *cytochrome c*, whose three-dimensional structure has been determined by x-ray crystallography (Figure 7–28).

Iron-sulfur proteins are a second major family of electron carriers. In these proteins either two or four iron atoms are bound to an equal number of sulfur atoms and to cysteine side chains, forming an **iron-sulfur center** on the protein (Figure 7–29). There are more iron-sulfur centers than cytochromes in the respi-

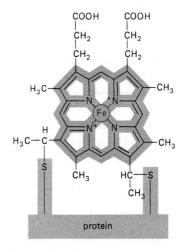

Figure 7–27 The structure of the heme group attached covalently to cytochrome c. Four of the six coordination positions on the iron are occupied by the porphyrin ring. The fifth and sixth coordination positions on the iron are perpendicular to the plane of the ring. In nearly all cytochromes these two positions are occupied by amino acid side chains, so that no other ligand can be bound. The exception is cytochrome a_3; as for hemoglobin and myoglobin, the sixth coordination position of the iron in this cytochrome oxidase component is free, so that it can bind oxygen.

There are five different cytochromes in the respiratory chain. Because the hemes in different cytochromes have slightly different structures and are held by their respective proteins in different ways, each of the cytochromes has a different affinity for an electron.

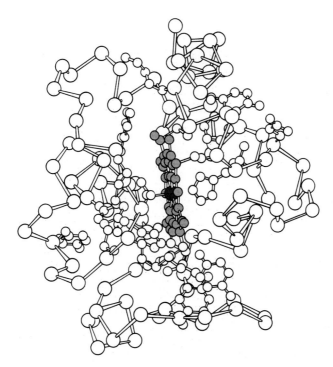

Figure 7–28 The three-dimensional structure of cytochrome c, an electron carrier in the electron-transport chain. This small protein contains just over 100 amino acids and is held loosely on the membrane by ionic interactions (see Figure 7–35). The iron atom (*dark color*) on the bound heme (*light color*) can carry a single electron (see also Figure 3–52A).

ratory chain, but their spectroscopic detection requires electron spin resonance (ESR) spectroscopy, and they are less well characterized.

The simplest of the electron carriers is a small hydrophobic molecule dissolved in the lipid bilayer known as *ubiquinone*, or *coenzyme Q*. A **quinone (Q)** can pick up or donate either one or two electrons, and it temporarily picks up a proton from the medium along with each electron that it carries (Figure 7–30).

In addition to six different hemes linked to cytochromes, more than six iron-sulfur centers, and ubiquinone, there are two copper atoms and a flavin serving as electron carriers tightly bound to respiratory-chain proteins in the pathway from NADH to oxygen. The pathway involves about 40 different proteins in all. The order of the individual electron carriers in the chain has been determined by sophisticated spectroscopic measurements (Figure 7–31), and many of the proteins were initially isolated and characterized as individual polypeptides. Real understanding of the respiratory chain, however, depended on the later realization that the proteins are organized into three large enzyme complexes.

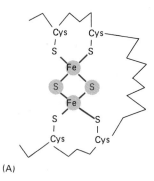

(A)

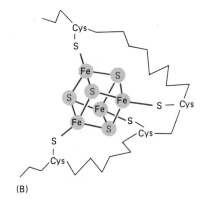

(B)

Figure 7–29 The structures of two types of iron-sulfur centers. (A) A center of the 2Fe2S type. (B) A center of the 4Fe4S type. Although they contain multiple iron atoms, each iron-sulfur center can carry only one electron at a time. There are more than six different iron-sulfur centers in the respiratory chain.

7-13 The Respiratory Chain Contains Three Large Enzyme Complexes Embedded in the Inner Membrane[18]

Membrane proteins are difficult to purify as intact complexes because they are insoluble in most aqueous solutions, and the agents required to solubilize them— such as detergents or urea—can destroy normal protein-protein interactions (see p. 286). In the early 1960s, however, it was found that relatively mild ionic detergents, such as deoxycholate (Figure 7–32), will solubilize selected components of the mitochondrial inner membrane in their native form. This permitted the identification and purification of the three major membrane-bound **respiratory enzyme complexes** in the pathway from NADH to oxygen (Figure 7–33).

1. The **NADH dehydrogenase complex** is the largest of the respiratory enzyme complexes, with a mass of about 800,000 daltons and more than 22 polypeptide chains. It accepts electrons from NADH and passes them through a flavin and at least five iron-sulfur centers to ubiquinone, a small lipid-soluble molecule (see Figure 7–30) that transfers its electrons to a second respiratory enzyme complex, the *b-c₁ complex*.

2. The **b-c₁ complex** contains at least 8 different polypeptide chains and is thought to function as a dimer of about 500,000 daltons. Each monomer contains three hemes bound to cytochromes and an iron-sulfur protein. The complex accepts electrons from ubiquinone and passes them on to cytochrome c, a small peripheral membrane protein (see Figure 7–28) that carries its electron to the *cytochrome oxidase complex*.

3. The **cytochrome oxidase complex** (cytochrome aa₃) is the best characterized of the three complexes. It is composed of at least 8 different polypeptide chains and is isolated as a dimer of about 300,000 daltons; each monomer contains two cytochromes and two copper atoms. The complex accepts electrons from cytochrome c and passes them to oxygen.

Figure 7–30 Quinones are important carriers in electron-transport chains. A quinone picks up one proton from the aqueous environment for every electron it accepts, and it can carry either one or two electrons. When it donates its electrons to the next carrier in the chain, these protons are released. In mammalian mitochondria the quinone is ubiquinone (coenzyme Q), shown here; the long hydrophobic tail, which confines ubiquinone to the membrane, commonly consists of 10 five-carbon isoprene units. The corresponding electron carrier in plants is plastoquinone, which is almost identical. For simplicity, both ubiquinone and plastoquinone will normally be referred to as *quinone* and abbreviated as Q.

ubiquinone

ubisemiquinone
(free radical)

ubiquinol
(dihydroubiquinone)

hydrophobic hydrocarbon tail

$e^- + H^+$

Figure 7–31 The general methods used to determine the path of electrons along an electron-transport chain. The extent of oxidation of electron carriers a, b, c, and d is continuously monitored by following their distinct spectra, which differ in their oxidized and reduced states. (A) Under normal conditions, all carriers are in a partially oxidized state. Addition of a specific inhibitor causes the downstream carriers to become more oxidized and the upstream carriers to become more reduced. (B) In the absence of oxygen, all carriers are in their fully reduced state. The sudden addition of oxygen converts each carrier to its partially oxidized form with a delay that is greatest for the most upstream carriers.

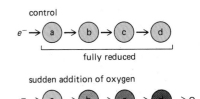

(A) NORMAL CONDITIONS

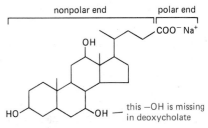

(B) ANAEROBIC CONDITIONS

The cytochromes, iron-sulfur centers, and copper atoms can carry only one electron at a time. Yet each NADH donates two electrons, and each O_2 molecule must receive four electrons to produce water. There are several electron-collecting and electron-dispersing points along the electron-transport chain where these changes in electron number are accommodated. Thus, for example, the cytochrome oxidase complex accepts four electrons, one at a time, from cytochrome c molecules and ultimately transfers them to a single bound O_2 molecule, thereby forming two water molecules. At the intermediate stages in this process, two electrons enter the heme of cytochrome a and a protein-linked copper atom, Cu_a, before being passed to the oxygen binding site, which in turn involves another copper atom and the heme of cytochrome a_3. Exactly how each bound oxygen molecule reacts with four electrons and four protons to generate two molecules of water is not known.

The cytochrome oxidase reaction is estimated to account for 90% of the total oxygen uptake in most cells. The toxicity of the poisons cyanide and azide is due to their ability to bind tightly to this complex and thereby block all electron transport.

Electron Transfers Are Mediated by Random Collisions Between Diffusing Donors and Acceptors in the Mitochondrial Inner Membrane[19]

The two components that carry electrons between the three major enzyme complexes of the respiratory chain—*ubiquinone* and *cytochrome c*—diffuse rapidly in the plane of the membrane. Collisions between these mobile carriers and the enzyme complexes can account for the observed rates of electron transfer (each complex donates and receives an electron about once every 5 to 20 milliseconds). Thus there is no need to postulate a structurally ordered chain of electron-transfer proteins in the lipid bilayer; indeed, the three enzyme complexes appear to exist as independent entities in the plane of the inner membrane, and the ordered transfer of electrons is due entirely to the specificity of the functional interactions among the components of the chain.

This view is supported by the observation that the various components of the respiratory chain are present in quite different amounts. For each molecule of

Figure 7–32 The structure of the relatively mild anionic detergents cholate and deoxycholate. While both are strong enough to solubilize membrane proteins, they can often be used without damaging enzyme activities.

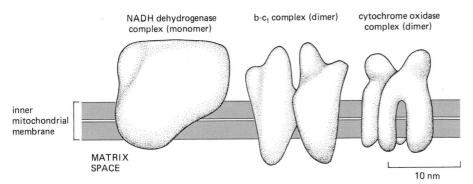

Figure 7–33 The relative sizes and shapes of the three respiratory enzyme complexes. These low-resolution three-dimensional structures were obtained from images of two-dimensional crystals (crystalline sheets) viewed in the electron microscope at various tilt angles.

NADH dehydrogenase complex in heart mitochondria, for example, there are estimated to be 3 molecules of b-c$_1$ complex, 7 molecules of cytochrome oxidase complex, 9 molecules of cytochrome c, and 50 molecules of ubiquinone; very different ratios are found in some other cells.

7-14 A Large Drop in Redox Potential Across Each of the Three Respiratory Enzyme Complexes Provides the Energy for Proton Pumping[12,20]

Pairs of compounds such as H$_2$O and ½O$_2$, or NADH and NAD$^+$, are called **conjugate redox pairs,** since one compound is converted to the other by adding one or more electrons plus one or more protons—the protons being readily available in any aqueous solution. Thus, for example,

$$\text{½O}_2 + 2e^- + 2\text{H}^+ \rightarrow \text{H}_2\text{O}$$

It is well known that a 50:50 mixture of the members of a *conjugate acid-base pair* acts as a buffer, maintaining a defined "H$^+$ pressure," or pH, that is a measure of the dissociation constant of the acid. In exactly the same way, a 50:50 mixture of the members of a conjugate redox pair maintains a defined "electron pressure," or **redox** (oxidation-reduction) **potential**, *E,* that is a measure of the electron carrier's affinity for electrons.

By placing electrodes in contact with solutions that contain the appropriate conjugate redox pairs, one can measure the redox potential of each of the various electron carriers that participate in biological oxidation-reduction reactions. Those pairs of compounds that have the most negative redox potentials have the weakest affinity for electrons and therefore contain carriers with the strongest tendency to donate electrons and the least tendency to accept them. Thus a 50:50 mixture of NADH and NAD$^+$ has a redox potential of -320 mV, indicating that NADH has a strong tendency to donate electrons; a 50:50 mixture of H$_2$O and ½O$_2$ has a redox potential of $+820$ mV, indicating that O$_2$ has a strong tendency to accept electrons.

Redox potentials can be determined for all the electron carriers in the respiratory chain that can be distinguished by their spectra. For example, small molecules can be added that short-circuit the respiratory chain by readily donating and accepting electrons. Spectral measurements are then used to determine the ratio of each oxidized to reduced carrier as stepwise changes are made in the redox potential of the solution causing the short circuit. As expected, the redox potentials increase as one passes along the chain of electron carriers. Most cytochromes have higher redox potentials than iron-sulfur centers. Thus, the cytochromes generally serve as electron carriers near the O$_2$ end, whereas the iron-sulfur proteins serve as carriers near the NADH end of the respiratory chain.

An outline of the redox potentials measured along the respiratory chain is shown in Figure 7–34. The potentials drop in three large steps, one across each major enzyme complex. The change in redox potential between any two electron carriers is directly proportional to the free energy released by an electron transfer between them (see Figure 7–34). Each complex acts as an energy-conversion device to harness this free-energy change, pumping H$^+$ across the inner membrane to create an electrochemical proton gradient as electrons pass through. This conversion can be demonstrated by incorporating each purified complex separately into liposomes (see Figure 7–25). When any one of the three isolated complexes is presented with an appropriate electron donor and acceptor so that electrons can pass through it, protons are translocated across the liposome membrane.

The energy-conversion mechanism underlying oxidative phosphorylation requires that each protein complex be inserted across the inner mitochondrial membrane in a fixed orientation, so that all protons are pumped in the same direction out of the matrix space (Figure 7–35). Such a *vectorial organization* of membrane proteins has been demonstrated by using membrane-impermeable probes to label each complex from only one side of the membrane or the other (see p. 288). A unique orientation in the bilayer is a feature of all membrane proteins and is essential to their function.

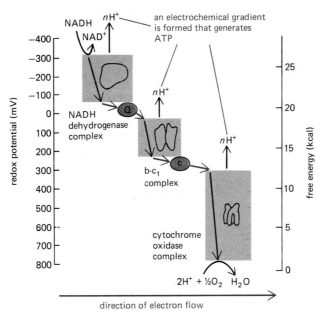

Figure 7–34 The redox potential (denoted E_0' or E_h) increases as electrons flow down the respiratory chain to oxygen. The standard free-energy change, $\Delta G°$, for the transfer of each of the two electrons donated by an NADH molecule can be obtained from the right-hand ordinate ($\Delta G = -n(0.023)\Delta E_0'$, where n is the number of electrons transferred across a redox potential change of $\Delta E_0'$ mV). Electrons flow through an enzyme complex by passing in sequence to the four or more electron carriers in each complex. As indicated, part of the favorable free-energy change is harnessed by each enzyme complex to pump protons across the mitochondrial inner membrane. The number of protons pumped per electron (n) is not known with certainty.

The two electrons transported from $FADH_2$, generated by fatty acid oxidation (see Figure 7–11) and by the citric acid cycle (see Figure 7–14), produce less useful energy than the two electrons transported from NADH. Because the redox potential of $FADH_2$ is in the range of 0 mV, its electrons are passed to ubiquinone by a process that does not conserve energy (not shown). Electron transport from $FADH_2$ to oxygen therefore causes protons to be pumped at only two sites rather than three.

The Mechanisms of Respiratory Proton Pumping Are Not Well Understood[21]

In oxidative phosphorylation, a maximum of about three ATP molecules can be synthesized for each NADH molecule oxidized (that is, for every *two* electrons that pass through all three of the enzyme complexes in the respiratory chain). With the assumption made previously that three protons flow back through the ATP synthetase for every ATP molecule it synthesizes (see p. 358), about 1.5 protons would be pumped by an average enzyme complex for each electron it transports (that is, some enzyme complexes pump one proton per electron, while others pump two).

The molecular mechanism by which electron transport is coupled to H^+ pumping is likely to be different for different respiratory enzyme complexes. Allosteric changes in protein conformations driven by electron transport can, in principle, pump protons, just as protons are pumped when ATP is hydrolyzed by the ATP synthetase running in reverse (see p. 357). In addition, as mentioned previously, a quinone picks up a proton from the aqueous medium along with each electron it carries and liberates it when it releases the electron (see Figure 7–30). Since ubiquinone is freely mobile in the lipid bilayer, it can accept electrons near the

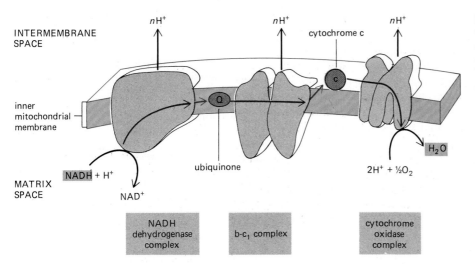

Figure 7–35 The flow of electrons through the three major respiratory enzyme complexes during the transfer of two electrons from NADH to oxygen. Ubiquinone and cytochrome c serve as carriers between the complexes.

inside surface of the membrane and donate them to the b-c$_1$ complex near the outside surface, thereby transferring one H$^+$ across the bilayer for every electron transported. More elaborate models can account for the observed stoichiometry of two protons pumped per electron in the b-c$_1$ complex by recycling ubiquinone through the complex in an ordered way.

In contrast, there is no obvious electron carrier that also carries a proton for the cytochrome oxidase complex, and it seems that electron transport induces an orderly allosteric change in protein conformation that causes part of the protein complex itself to pump protons.

7-15 H$^+$ Ionophores Dissipate the H$^+$ Gradient and Thereby Uncouple Electron Transport from ATP Synthesis[22]

Since the 1940s, several lipophilic weak acids have been known to act as *uncoupling agents*, uncoupling electron transport from ATP synthesis. The addition of these low-molecular-weight organic compounds to cells stops ATP synthesis by mitochondria without blocking their uptake of oxygen. In the presence of an uncoupling agent, electron transport continues at a rapid rate, but no H$^+$ gradient is generated. The explanation for this effect is both simple and elegant: uncoupling agents act as H$^+$ carriers (H$^+$ ionophores) and provide a pathway in addition to the ATP synthetase for the flow of H$^+$ across the inner mitochondrial membrane. As a result of this "short-circuiting," the proton-motive force is completely dissipated, and ATP can no longer be synthesized. The mechanism involved is illustrated for the widely used uncoupler 2,4-dinitrophenol in Figure 7–36.

7-16 Respiratory Control Normally Restrains Electron Flow Through the Chain[23]

When an uncoupler such as dinitrophenol is added to cells, mitochondria increase their oxygen uptake substantially because of an increased rate of electron transport. This increase reflects the existence of **respiratory control.** The control is thought to act via a direct inhibitory influence of the electrochemical proton gradient on the rate of electron transport. When the gradient is collapsed by an uncoupler, electron transport is free to run unchecked at the maximal rate. As the gradient increases, electron transport becomes more difficult and the process slows. Moreover, if an artificially large electrochemical gradient is experimentally created across the inner membrane, normal electron transport stops completely and a *reverse electron flow* can be detected in some sections of the respiratory chain. This observation suggests that respiratory control reflects a simple balance between the free-energy change for electron-transport-linked proton pumping and the free-energy change for electron transport: that is, the magnitude of the electrochemical proton gradient affects both the rate and the direction of electron transport, just as it affects the directionality of the ATP synthetase (see p. 358).

Respiratory control is just one part of an elaborate interlocking system of feedback controls that coordinates the rates of glycolysis, fatty acid breakdown, the citric acid cycle, and electron transport. The rates of all of these processes are adjusted to the ATP:ADP ratio, increasing whenever increased utilization of ATP causes the ratio to fall. The ATP synthetase in the inner mitochondrial membrane, for example, works faster as the concentrations of its substrates ADP and P$_i$ increase. As it speeds up, the enzyme lets more protons flow into the matrix and thereby dissipates the electrochemical proton gradient more rapidly. The falling gradient, in turn, enhances the rate of electron transport.

Similar controls, including feedback inhibition of several key enzymes by ATP (for example, see Figure 7–13), act to adjust the rates of NADH production to the rate of NADH utilization by the respiratory chain, and so on. As a result of these many control mechanisms, the body oxidizes fats and sugars 5 to 10 times more rapidly during a period of strenuous exercise than during a period of rest.

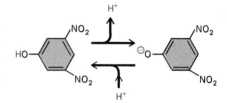

high pH causes protons to dissociate from DNP molecules

low pH causes protons to bind to DNP molecules

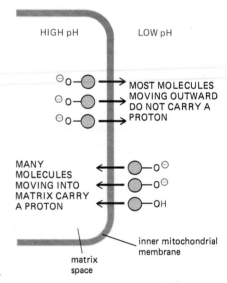

Figure 7–36 Proton conduction through the inner mitochondrial membrane by the uncoupling agent 2,4-dinitrophenol (DNP). The uncharged (protonated) form of DNP can diffuse freely across the lipid bilayer, and the movement of the charged species across the membrane is thought to be facilitated by an anion-transport protein (not shown). As illustrated, the electrochemical proton gradient will cause the dinitrophenol molecules to carry more protons in than out until the proton-motive force has completely disappeared.

Natural Uncouplers Convert the Mitochondria in Brown Fat into Heat-generating Machines[24]

In some specialized fat cells, mitochondrial respiration is normally uncoupled from ATP synthesis. In these cells, known as brown fat cells, most of the energy of oxidation is dissipated as heat rather than being converted into ATP. The inner membranes of the large mitochondria in these cells contain a special transport protein that allows protons to move down their electrochemical gradient without activating ATP synthetase. As a result, the cells oxidize their fat stores at a rapid rate and produce more heat than ATP. Tissues containing brown fat thereby serve as "heating pads" that revive hibernating animals and protect sensitive areas of newborn human babies from the cold.

7-17 All Bacteria Use Chemiosmotic Mechanisms to Harness Energy[25]

Bacteria use enormously diverse energy sources. Some, like animal cells, synthesize ATP from sugars that they oxidize to CO_2 and H_2O by glycolysis and the citric acid cycle; they have a respiratory chain in their plasma membrane similar to that in the mitochondrial inner membrane. Others are strict anaerobes, deriving their energy from glycolysis alone (fermentation) or from an electron-transport chain that employs a molecule other than oxygen as the final electron acceptor. As examples, the alternative electron acceptor can be a nitrogen compound (nitrate or nitrite), a sulfur compound (sulfate or sulfite), or a carbon compound (fumarate or carbonate). The electrons are transferred to these acceptors by a series of electron carriers in the plasma membrane that are comparable to those present in mitochondrial respiratory chains.

Despite this diversity, the plasma membrane of the vast majority of bacteria contains an ATP synthetase that is very similar to that in mitochondria (and chloroplasts). In anaerobic bacteria that lack an electron-transport chain, this ATP synthetase works in reverse, using the ATP produced by glycolysis to establish a proton-motive force across the bacterial plasma membrane. In aerobic bacteria, an electron-transport chain establishes the proton-motive force that drives the ATP synthetase to make ATP.

Most bacteria, including the strict anaerobes, maintain a proton-motive force across their plasma membrane. The electrochemical proton gradient is harnessed to drive a flagellar motor that enables bacteria to swim (see p. 720), and Na^+ is pumped out of bacteria by a proton-driven Na^+-H^+ antiport system that takes the place of the Na^+-K^+ ATPase of eucaryotic cells. Most inward transport across the plasma membrane in animal cells is driven by a Na^+ gradient established by the Na^+-K^+ ATPase (see p. 309). In contrast, the active transport of nutrients into bacteria tends to be mediated by H^+-driven symport systems in which the desired metabolite is dragged into the cell along with one or more protons by means of a specific carrier protein. Most amino acids and many sugars are transported in this way (Figure 7-37).

Among the diverse types of bacteria are some that have been able to adapt to extremely hostile environments. Some, for example, live in very alkaline environments and must maintain their cytoplasm at a more acidic pH than the cell exterior in order to protect their alkali-labile molecules, such as RNA. For these cells, any attempt to generate an electrochemical proton gradient would be opposed by a large proton concentration gradient in the wrong direction (H^+ higher inside than outside). Presumably for this reason, an alkali-tolerant marine bacterium, *Vibrio alginolyticus*, substitutes Na^+ for H^+ in all of its chemiosmotic mechanisms. Its repiratory chain pumps Na^+ out of the cell, its transport systems are coupled to inward Na^+ symport, its flagellar motor is driven by an inward flux of Na^+, and a Na^+-driven ATP synthetase seems to be responsible for much of its ATP synthesis. The existence of such bacteria demonstrates that the principle of chemiosmosis is more fundamental than the electrochemical proton gradient on which it is normally based.

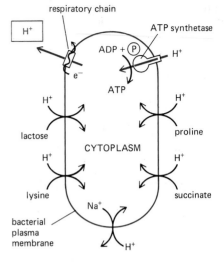

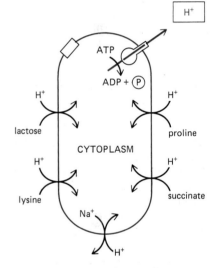

Figure 7-37 A proton-motive force generated across the plasma membrane of bacterial cells pumps nutrients into the cell and expels sodium. In (A) the electrochemical proton gradient is being generated in an aerobic bacterium by a respiratory chain and is then used by ATP synthetase to make ATP. It is also used to transport some nutrients into the bacterium. In (B) the same bacterium growing under anaerobic conditions can derive its ATP from glycolysis. Part of this ATP is hydrolyzed by ATP synthetase to establish the transmembrane proton-motive force that drives transport processes. (As described in the text, other bacteria have electron-transport chains that pump protons out of the cell under anaerobic conditions, using a molecule other than oxygen as the final acceptor for electrons.)

Summary

The respiratory chain in the inner mitochondrial membrane contains three major enzyme complexes that are involved in transferring electrons from NADH to O_2. Each of these can be purified, inserted into synthetic lipid vesicles, and then shown to pump protons when electrons are transported through it. In the native membrane the mobile electron carriers ubiquinone and cytochrome c complete the electron-transport chain by shuttling between the enzyme complexes. The path of electron flow is NADH $\rightarrow$ NADH dehydrogenase complex $\rightarrow$ ubiquinone $\rightarrow$ b-c_1 complex $\rightarrow$ cytochrome c $\rightarrow$ cytochrome oxidase complex $\rightarrow$ molecular oxygen (O_2).

The respiratory enzyme complexes couple the energetically favorable transport of electrons to the pumping of protons out of the matrix. The resulting electrochemical proton gradient is harnessed to make ATP by another transmembrane protein complex, ATP synthetase, through which the protons flow back into the matrix. The ATP synthetase is a reversible coupling device that normally converts a backflow of protons into ATP phosphate-bond energy, but it can also hydrolyze ATP to pump protons in the opposite direction if the electrochemical proton gradient is reduced. Its universal presence in mitochondria, chloroplasts, and bacteria testifies to the central importance of chemiosmotic mechanisms in cells.

Chloroplasts and Photosynthesis[26]

All animals and most microorganisms rely on the continual uptake of large amounts of organic compounds from their environment. These compounds provide both the carbon skeletons for biosynthesis and the metabolic energy that drives all cellular processes. It is believed that the first organisms on the primitive earth had access to an abundance of organic compounds produced by geochemical processes (see p. 4) but that most of these original compounds were used up billions of years ago. Since that time, virtually all of the organic materials required by living cells have been produced by *photosynthetic organisms*, including many types of photosynthetic bacteria. The most advanced photosynthetic bacteria are the cyanobacteria, which have minimal nutrient requirements. They use electrons from water and the energy of sunlight to convert atmospheric CO_2 into organic compounds. Moreover, in the course of splitting water [in the reaction $nH_2O + nCO_2 \xrightarrow{\text{light}} (CH_2O)_n + nO_2$], they liberate into the atmosphere the oxygen required for oxidative phosphorylation. As we shall explain later, it is thought that the evolution of cyanobacteria from more primitive photosynthetic bacteria first made possible the development of aerobic life forms.

In plants, which developed later, photosynthesis is carried out in a specialized intracellular organelle—the chloroplast. Chloroplasts carry out photosynthesis during the daylight hours. At night, when the production of energy-rich metabolites by photosynthesis ceases, the plant cell relies on its mitochondria (which closely resemble their counterparts in animal cells) to generate ATP.

Biochemical evidence suggests that chloroplasts are descendants of oxygen-producing photosynthetic bacteria that were endocytosed and lived in symbiosis with primitive eucaryotic cells. Mitochondria are also generally believed to be descended from endocytosed bacteria. The many differences between chloroplasts and mitochondria are thought to reflect their different bacterial ancestors as well as subsequent evolutionary divergence. Nevertheless, the fundamental mechanisms involved in light-driven ATP synthesis in chloroplasts and in respiration-driven ATP synthesis in mitochondria are very similar.

Chloroplasts Resemble Mitochondria but Have an Extra Compartment[27]

Chloroplasts carry out their energy interconversions by chemiosmotic mechanisms in much the same way that mitochondria do, and they are organized on the same principles (Figures 7–38 and 7–39). They have a highly permeable outer

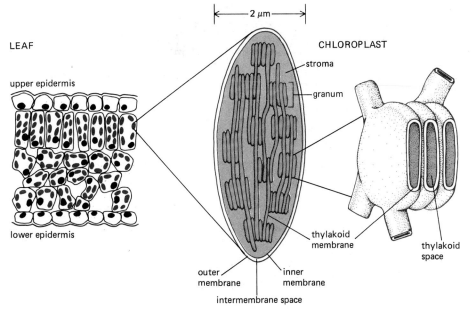

LEAF

upper epidermis

lower epidermis

2 μm

CHLOROPLAST

stroma

granum

thylakoid membrane

outer membrane

inner membrane

intermembrane space

thylakoid space

Figure 7–38 The chloroplast contains three distinct membranes (the outer membrane, the inner membrane, and the thylakoid membrane) that define three separate internal compartments— the intermembrane space, the stroma, and the thylakoid space. The thylakoid membrane contains all the energy-generating systems of the chloroplast. In electron micrographs this membrane appears to be broken up into separate units that enclose individual flattened vesicles (see Figure 7–39), but these are probably joined into a single, highly folded membrane in each chloroplast. As indicated, the individual thylakoids are interconnected, and they tend to stack to form aggregates called grana.

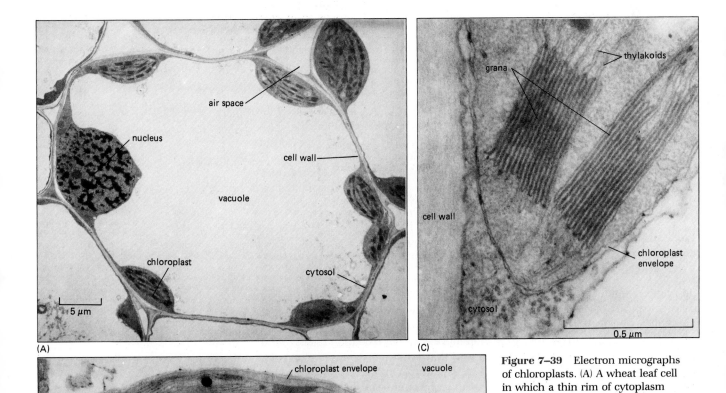

(A) air space / nucleus / cell wall / vacuole / chloroplast / cytosol / 5 μm

(C) grana / thylakoids / cell wall / chloroplast envelope / cytosol / 0.5 μm

(B) chloroplast envelope / vacuole / thylakoids / lipid / starch / grana / cell wall / 1 μm

Figure 7–39 Electron micrographs of chloroplasts. (A) A wheat leaf cell in which a thin rim of cytoplasm containing chloroplasts surrounds a large vacuole. (B) A thin section of a single chloroplast, showing the starch granules and lipid droplets that have accumulated in the stroma as a result of the biosyntheses occurring there. (C) A high-magnification view of a granum, showing its stacked thylakoid membrane. (Courtesy of K. Plaskitt.)

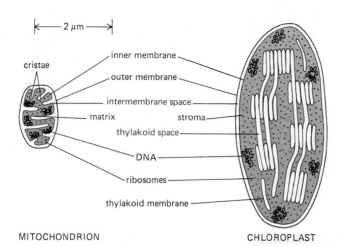

Figure 7-40 Comparison of a mitochondrion and a chloroplast. The chloroplast is generally much larger and contains a thylakoid membrane and thylakoid space. The mitochondrial inner membrane is folded into cristae.

membrane; a much less permeable inner membrane, in which special carrier proteins are embedded; and a narrow intermembrane space between. The inner membrane surrounds a large space called the **stroma,** which is analogous to the mitochondrial matrix and contains various enzymes, ribosomes, RNA, and DNA.

There is, however, an important difference between the organization of mitochondria and that of chloroplasts. The inner membrane of the chloroplast is not folded into cristae and does not contain an electron-transport chain. Instead, the photosynthetic light-absorbing system, the electron-transport chain, and an ATP synthetase are all contained in a third distinct membrane that forms a set of flattened disclike sacs, the **thylakoids** (see Figure 7-38). The lumen of each thylakoid is thought to be connected with the lumen of other thylakoids, thereby defining a third internal compartment called the *thylakoid space*, which is separated from the stroma by the *thylakoid membrane*.

The structural similarities and differences between mitochondria and chloroplasts are illustrated in Figure 7-40. In a general way, one might view the chloroplast as a greatly enlarged mitochondrion in which the cristae are converted into a series of interconnected submitochondrial particles in the matrix space. The knobbed end of the chloroplast ATP synthetase, where ATP is made, protrudes from the thylakoid membrane into the stroma, just as it protrudes into the matrix from the membrane of each mitochondrial crista (see Figure 7-51).

7-21 ## Two Unique Reactions in Chloroplasts: The Light-driven Production of ATP and NADPH and the Conversion of CO_2 to Carbohydrate[26]

The many reactions that occur during photosynthesis can be grouped into two broad categories. (1) In the **photosynthetic electron-transfer reactions** (sometimes called the "light reactions"), energy derived from sunlight energizes an electron in *chlorophyll*, enabling the electron to move along an oxidation chain in the thylakoid membrane in much the same way that an electron moves along the respiratory chain in mitochondria. This electron-transport process pumps protons across the thylakoid membrane, and the resulting proton-motive force drives the synthesis of ATP in the stroma. At the same time, the electron-transfer reactions generate high-energy electrons that convert $NADP^+$ to NADPH; in this process, water is oxidized to provide the electrons donated to NADPH, and O_2 is liberated. All of these reactions are confined to the chloroplast. (2) In the **carbon-fixation reactions** (sometimes called the "dark reactions"), the ATP and NADPH produced by the photosynthetic electron-transfer reactions serve as the source of energy and reducing power, respectively, to drive the conversion of CO_2 to carbohydrate. These reactions, which begin in the chloroplast stroma and continue in the cytosol, produce sucrose in the leaves of the plant, from where it is exported to other tissues as a source of both organic molecules and energy for growth.

Thus the formation of oxygen (which directly requires light energy) and the conversion of carbon dioxide to carbohydrate (which requires light energy only indirectly) are separate processes (Figure 7–41). As we shall see, however, elaborate feedback mechanisms interconnect the two in order to balance biosynthesis. Changes in the cell's ATP and NADPH requirements, for example, regulate the production of these molecules in the thylakoid membrane, and several of the chloroplast enzymes required for carbon fixation are inactivated in the dark and reactivated by light-stimulated electron-transport processes.

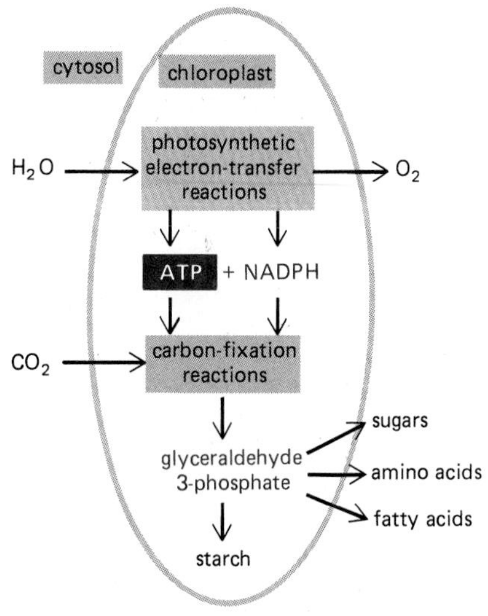

Carbon Fixation Is Catalyzed by Ribulose Bisphosphate Carboxylase[28]

We have seen earlier in this chapter how cells produce ATP by using the large amount of free energy released when carbohydrates are oxidized to CO_2 and H_2O. Clearly, therefore, the reverse reaction, in which CO_2 and H_2O combine to make carbohydrate, must be a very unfavorable one. Consequently, this synthesis must be coupled to other, very favorable reactions to drive it.

The central reaction, in which an atom of inorganic carbon is converted to organic carbon, is illustrated in Figure 7–42: CO_2 from the atmosphere combines with the five-carbon compound ribulose 1,5-bisphosphate plus water to give two molecules of the three-carbon compound 3-phosphoglycerate. This "carbon-fixing" reaction, which was discovered in 1948, is catalyzed in the chloroplast stroma by a large enzyme called *ribulose bisphosphate carboxylase* (~500,000 daltons). Since each molecule of the enzyme works sluggishly (processing about 3 molecules of substrate per second compared to 1000 molecules per second for a typical enzyme), many copies are needed. Ribulose bisphosphate carboxylase often represents more than 50% of the total chloroplast protein and is widely claimed to be the most abundant protein on earth.

Three Molecules of ATP and Two Molecules of NADPH Are Consumed for Each CO_2 Molecule That Is Fixed in the Carbon-Fixation Cycle[29]

The actual reaction in which CO_2 is fixed is energetically favorable, but it requires a continuous supply of the energy-rich compound *ribulose 1,5-bisphosphate*, to which each molecule of CO_2 is added (see Figure 7–42). The elaborate pathway by which this compound is regenerated was worked out in one of the most successful early applications of radioisotopes. As outlined in Figure 7–43, three molecules of CO_2 are fixed by ribulose bisphosphate carboxylase to produce six molecules of 3-phosphoglycerate (containing $6 \times 3 = 18$ carbon atoms in all: 3 from the CO_2 and 15 from ribulose 1,5-bisphosphate). The 18 carbon atoms then undergo a cycle of reactions that regenerate the three molecules of ribulose 1,5-bisphosphate used in the initial carbon-fixation step (containing $3 \times 5 = 15$ carbon atoms). This leaves one molecule of *glyceraldehyde 3-phosphate* (3 carbon atoms) as the net gain. In this **carbon-fixation cycle** (or Calvin-Benson cycle), three

Figure 7–41 Photosynthesis in a chloroplast can be broadly separated into photosynthetic electron-transfer reactions and carbon-fixation reactions. Water is oxidized and oxygen is released in the first set of reactions, while carbon dioxide is assimilated (fixed) to produce organic molecules in the second set of reactions.

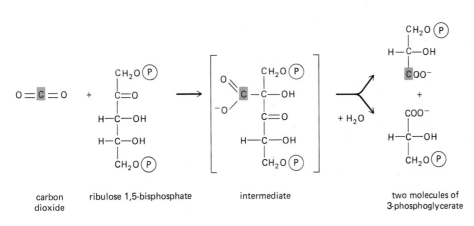

carbon dioxide   ribulose 1,5-bisphosphate   intermediate   two molecules of 3-phosphoglycerate

Figure 7–42 The initial reaction in the process by which carbon dioxide is converted into organic carbon. This reaction is catalyzed in the chloroplast stroma by the abundant enzyme *ribulose bisphosphate carboxylase*, and it forms 3-phosphoglycerate, which is also an important intermediate in glycolysis (see Figure 2–20, p. 66). The two colored carbon atoms are used to produce *phosphoglycolate* when the enzyme adds oxygen instead of CO_2 (see p. 371, below).

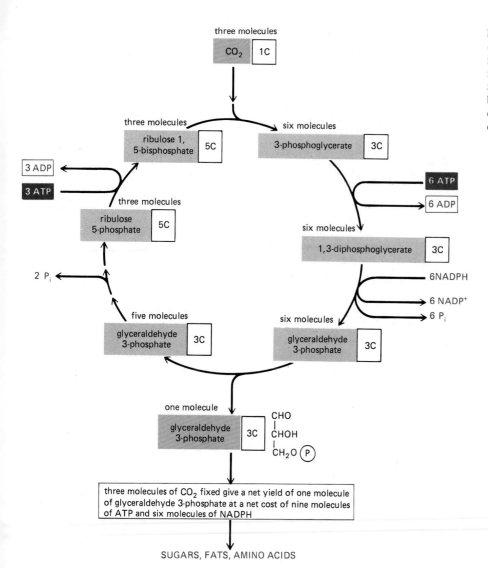

Figure 7–43 The carbon-fixation cycle, which forms organic molecules from CO_2 and H_2O. There are many intermediates between glyceraldehyde 3-phosphate and ribulose 5-phosphate, but they have been omitted here for clarity. The entry of water into the cycle is also not shown.

molecules of ATP and two molecules of NADPH are consumed for each CO_2 molecule converted into carbohydrate. The net equation is

$$3CO_2 + 9ATP + 6NADPH + \text{water} \rightarrow \text{glyceraldehyde 3-phosphate}$$
$$+ 8P_i + 9ADP + 6NADP^+$$

Thus both *phosphate-bond energy* (as ATP) and *reducing power* (as NADPH) are required for the formation of organic molecules from CO_2 and H_2O. We shall return to this important point later.

The glyceraldehyde 3-phosphate produced in chloroplasts by the carbon-fixation cycle is a three-carbon sugar that serves as a central intermediate in glycolysis (p. 65). Much of it is exported to the cytoplasm, where it can be rapidly converted into fructose 6-phosphate and glucose 1-phosphate by reversal of several reactions in glycolysis (see p. 83). Glucose 1-phosphate is then converted to the sugar nucleotide UDP-glucose, and this combines with fructose 6-phosphate to form sucrose phosphate, the immediate precursor of the disaccharide **sucrose.** Sucrose is the major form in which sugar is transported between plant cells, acting in plant cells as glucose acts in animal cells: just as glucose is transported in the blood of animals, sucrose is exported from the leaves via vascular bundles (see Figure 7–45, below), providing the carbohydrate required by the rest of the plant.

Most of the glyceraldehyde 3-phosphate that remains in the chloroplast is converted to *starch* in the stroma. Like glycogen in animal cells, **starch** is a large polymer of glucose that serves as a carbohydrate reserve. The production of starch is regulated so that it is produced and stored as large grains in the chloroplast stroma (see Figure 7–39B) during periods of excess photosynthetic capacity. This occurs through reactions in the stroma that are the reverse of those in glycolysis: they convert glyceraldehyde 3-phosphate to glucose 1-phosphate, which is then used to produce the sugar nucleotide ADP-glucose, the immediate precursor of starch. At night the starch is broken down to help support the metabolic needs of the plant.

7-20 Carbon Fixation in Some Tropical Plants Is Compartmentalized
7-22 to Facilitate Growth at Low CO_2 Concentrations[30]

Although ribulose bisphosphate carboxylase preferentially adds CO_2 to ribulose 1,5-bisphosphate, if the concentration of CO_2 is low, it will add O_2 instead. This is an apparently wasteful pathway, which produces one molecule of 3-phospho-glycerate and one molecule of the two-carbon compound phosphoglycolate rather than two molecules of 3-phosphoglycerate (see Figure 7–42). The phosphoglyco-late is converted to glycolate and shuttled into peroxisomes, which begin the process of converting two molecules of glycolate into one molecule of 3-phos-phoglycerate (three carbons) plus one molecule of CO_2. Because the entire process uses up O_2 and liberates CO_2, it is termed *photorespiration*. In many plants about one-third of the CO_2 fixed is lost again as CO_2 because of photorespiration. It is not known whether photorespiration has some function in the plant or is simply a means of returning to the carbon-fixation pathway some of the carbon diverted into phosphoglycolate by the undesirable reactivity of oxygen with ribulose 1,5-bisphosphate.

Photorespiration can be a serious liability for plants in hot, dry conditions, where they close their stomata—the gas exchange pores in their leaves—to avoid excessive water loss. This causes the CO_2 levels in the leaf to fall precipitously and favors photorespiration. However, a special adaptation occurs in the leaves of many plants, such as corn and sugar cane, that live in hot, dry environments. In these plants the carbon-fixation cycle shown in Figure 7–43 occurs only in the chloroplasts of specialized *bundle-sheath cells*, which contain all of the plant's ribulose bisphosphate carboxylase. These cells are protected from the air and are surrounded by a specialized layer of mesophyll cells that "pump" CO_2 into the bundle-sheath cells, supplying the ribulose bisphosphate carboxylase with a high concentration of CO_2, which greatly reduces photorespiration.

The CO_2 pump is a reaction cycle that begins with a CO_2-fixation step cata-lyzed in the cytosol of the mesophyll cells by an enzyme that binds carbon dioxide (as bicarbonate) with high affinity. The product is a four-carbon compound that diffuses into the bundle-sheath cells, where it is broken down to give one molecule of CO_2 and one molecule of a three-carbon compound. The latter is then returned to the mesophyll cells where, in a reaction requiring ATP hydrolysis, it is converted to an activated form that can pick up another CO_2 molecule to start the CO_2 pumping cycle again (Figure 7–44).

When a pulse of radioactive $^{14}CO_2$ is given to a plant that pumps CO_2, it is assimilated in the CO_2-fixation step in the mesophyll cells so that the first organic compound that is labeled contains four carbons. In all other plants the three-carbon compound 3-phosphoglycerate is labeled first (see Figure 7–43). Conse-quently, CO_2-pumping plants are called C_4 *plants* and all other plants are called C_3 *plants* (Figure 7–45).

As for any vectorial transport process, pumping CO_2 into the bundle-sheath cells in C_4 plants costs energy. However, in hot, dry environments this cost is often much less than the energy lost by photorespiration in C_3 plants, and so C_4 plants are at an advantage. Moreover, because C_4 plants can carry out photosynthesis at a lower concentration of CO_2 inside the leaf, they need to open their stomata less and can therefore fix about twice as much net carbon as C_3 plants per unit of water lost.

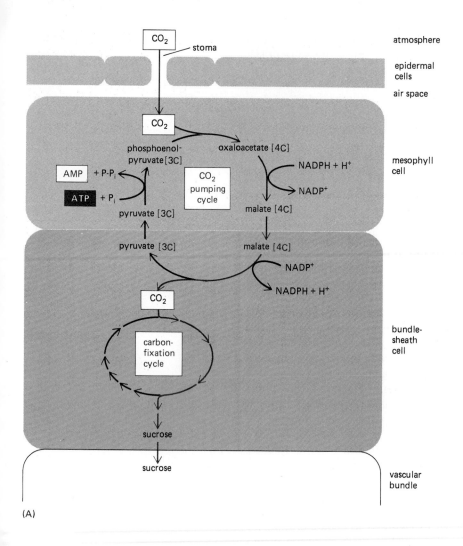

(A)

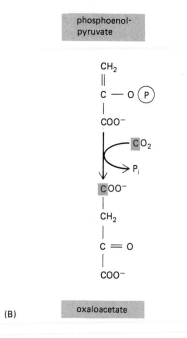

(B)

Figure 7–44 The CO_2 pumping cycle in a plant such as corn. (A) The net production of carbohydrate is confined to bundle-sheath cells, which contain all of the ribulose bisphosphate carboxylase. The mesophyll cells initiate the CO_2 pumping cycle, which involves the four-carbon and three-carbon compounds indicated. Variations of this pumping cycle are used in other CO_2-pumping plants. (B) The reaction of phosphoenolpyruvate with CO_2 in mesophyll cells.

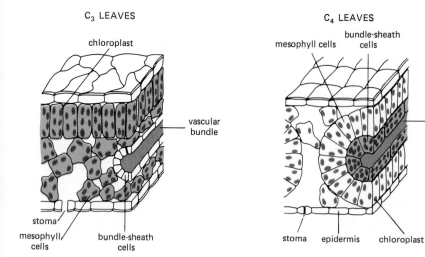

Figure 7–45 A comparison of the anatomy of the leaf in a C_3 plant and a C_4 plant. The colored cells contain chloroplasts that carry out the normal carbon-fixation cycle. In C_4 plants, the mesophyll cells are specialized for CO_2 pumping rather than for carbon fixation, and they create a high $CO_2:O_2$ ratio in the bundle-sheath cells, which are the only cells in these plants where the carbon-fixation cycle occurs (see Figure 7–44). The vascular bundles carry the sucrose made in the leaf to other tissues.

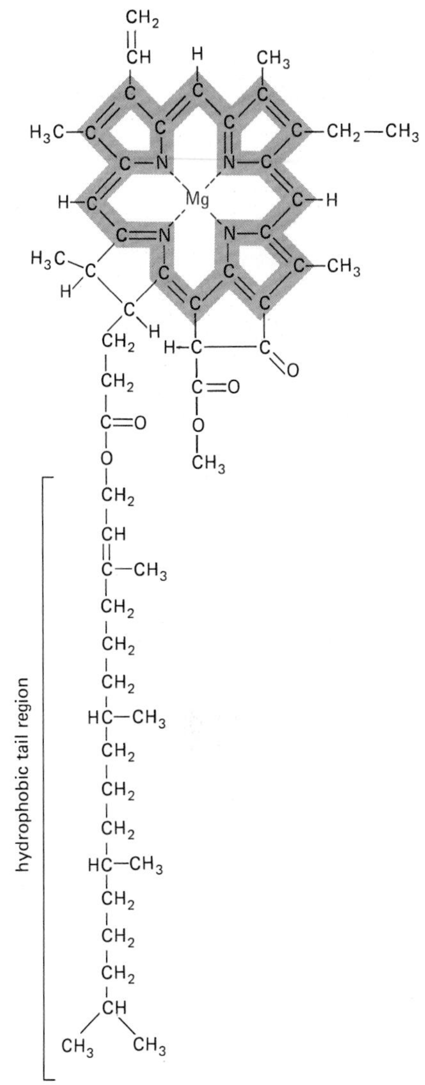

Figure 7–46 The structure of chlorophyll. A magnesium atom is held in a porphyrin ring, which is related to the porphyrin ring that binds iron in heme (compare with Figure 7–27). Electrons are delocalized over the bonds shown in color.

7-23 Photosynthesis Depends on the Photochemistry of Chlorophyll Molecules[31]

Having discussed the carbon-fixation reactions, we now return to the question of how the photosynthetic electron-transfer reactions in the chloroplast generate the ATP and the NADH needed to drive the production of carbohydrates from CO_2 and H_2O (see Figure 7–41). The required energy is derived from sunlight absorbed by **chlorophyll** molecules (Figure 7–46). The process of energy conversion begins when a chlorophyll molecule is excited by a quantum of light (a photon) and an electron is moved from one molecular orbital to another of higher energy. Such an excited molecule is unstable and will tend to return to its original, unexcited state in one of three ways: (1) by converting the extra energy into heat (molecular motions) or to some combination of heat and light of a longer wavelength (fluorescence), as when light energy is absorbed by an isolated chlorophyll molecule in solution; (2) by transferring the energy—but not the electron—directly to a neighboring chlorophyll molecule by a process called *resonance energy transfer*; or (3) by transferring the high-energy electron to another nearby molecule (an *electron acceptor*) and then returning to its original state by taking up a low-energy electron from some other molecule (an *electron donor*, Figure 7–47). The last two mechanisms play an essential part in photosynthesis.

A Photosystem Contains a Reaction Center Plus an Antenna Complex[32]

Multiprotein complexes called **photosystems** catalyze the conversion of the light energy captured in excited chlorophyll molecules to useful forms. A photosystem consists of two closely linked components: a *photochemical reaction center* and an *antenna complex* (Figure 7–48).

The **antenna complex** is important for light harvesting. In chloroplasts it consists of a cluster of several hundred chlorophyll molecules linked together by proteins that hold them tightly on the thylakoid membrane. Depending on the

Figure 7–47 Three possible ways for an excited chlorophyll molecule (a molecule containing a high-energy electron) to return to its original, unexcited state. The light energy absorbed by an isolated chlorophyll molecule is completely released as light and heat by process 1. In photosynthesis, by contrast, chlorophylls undergo process 2 in the antenna complex and process 3 in the reaction center, as described in the text.

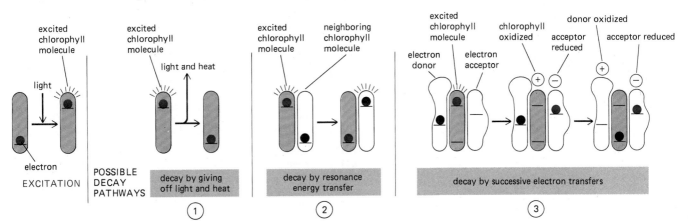

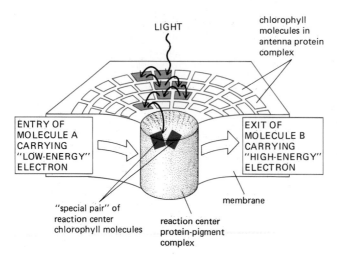

LIGHT

chlorophyll molecules in antenna protein complex

ENTRY OF MOLECULE A CARRYING "LOW-ENERGY" ELECTRON

EXIT OF MOLECULE B CARRYING "HIGH-ENERGY" ELECTRON

membrane

"special pair" of reaction center chlorophyll molecules

reaction center protein-pigment complex

Figure 7–48 A photosystem consists of a reaction center and an antenna. The reaction center is a transmembrane protein complex that holds a "special pair" of chlorophyll molecules in a fixed spatial relation to other electron carriers (see Figure 7–49). It catalyzes process 3 in Figure 7–47. If viewed as an enzyme, the substrates of a reaction center would be a weak election donor (molecule A) and a weak electron acceptor (molecule B), and its products would be a strong electron acceptor (oxidized molecule A) and a strong electron donor (reduced molecule B). The antenna complex contains most of the chlorophylls in the thylakoid membrane, and it serves as a funnel for transferring the energy of an excited electron to the reaction center. Many of these energy transfers are between identical chlorophyll molecules, through which the excitation wanders randomly (process 2 in Figure 7–47). Nevertheless, the average time needed for the excitation quantum to encounter and "fall into" the reaction center is only 10^{-10} to 10^{-9} seconds, so very few of the quanta absorbed are lost by the random decay pathway shown as process 1 in Figure 7–47.

plant, varying amounts of accessory pigments called *carotenoids*, which can help collect light of other wavelengths, are also located in each complex. When a chlorophyll molecule in the antenna complex is excited, the energy is rapidly transferred from one molecule to another by resonance energy transfer until it reaches a special pair of chlorophyll molecules in the photochemical reaction center. Each antenna complex thereby acts as a "funnel," collecting light energy and directing it to a specific site where it can be used effectively (see Figure 7–48).

The **photochemical reaction center** is a transmembrane protein-pigment complex that lies at the heart of photosynthesis. It is thought to have evolved more than 3 billion years ago in primitive photosynthetic bacteria. The special pair of chlorophyll molecules in the reaction center acts as an irreversible trap for excitation quanta because its excited electron is immediately passed to a chain of electron acceptors that are precisely positioned as neighbors in the same protein complex (Figure 7–49). By moving the high-energy electron rapidly away from the chlorophylls, the reaction center transfers it to an environment where it is much more stable. The electron is thereby suitably positioned for subsequent photochemical reactions, which require more time to complete. As we shall see, the net result of these slower reactions is the conversion of a "low-energy" electron on a weak electron donor (such as water) to a "high-energy" electron on a strong electron donor (such as a *quinone*).

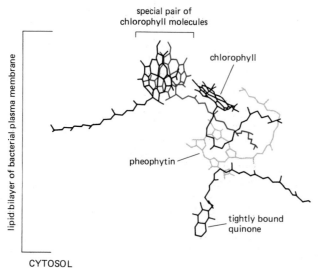

special pair of chlorophyll molecules

chlorophyll

pheophytin

tightly bound quinone

lipid bilayer of bacterial plasma membrane

CYTOSOL

Figure 7–49 The arrangement of the electron carriers in a bacterial photochemical reaction center as determined by x-ray crytallography. The pigment molecules shown are held in the interior of a transmembrane protein and are surrounded by the lipid bilayer, as indicated. An electron in the special pair is excited by resonance from an antenna complex chlorophyll (process 2 in Figure 7–47), and the excited electron is then transferred stepwise from the special pair to the quinone (see Figure 7–50).

In a Reaction Center, Light Energy Captured by Chlorophyll Creates a Strong Electron Donor from a Weak One[33]

The electron transfers involved in the photochemical reactions just outlined have been analyzed extensively by rapid spectroscopic methods, especially in the photosystem of purple bacteria, which is simpler than the evolutionarily related photosystem in chloroplasts. The bacterial reaction center is a large protein-pigment complex that can be solubilized with detergent and purified in active form. In 1985 its complete three-dimensional structure was determined by x-ray crystallography (see Figure 6–32, p. 294, and Figure 7–49). This structure, combined with kinetic data, provides the best picture we have of the initial electron-transfer reactions that underlie photosynthesis.

The sequence of transfers that take place in the reaction center of purple bacteria is shown diagrammatically in Figure 7–50. An electron excited by the absorption of light is passed rapidly from the special pair of reaction-center chlorophylls through the other pigments in Figure 7–49 to a tightly bound quinone electron acceptor, designated as Q_A. This electron transfer, which occurs in less than 10^{-9} seconds and is virtually irreversible, leaves a positively charged "hole" with a very high affinity for electrons in the chlorophyll, which is refilled by the capture of an electron from a nearby cytochrome (normally a weak electron donor). The high-energy electron held by Q_A is subsequently passed to a second quinone, Q_B, from which it leaves the reaction center and passes to a mobile quinone molecule (Q) in the photosynthetic membrane. This reduced quinone is a strong electron donor whose reducing power can be harnessed to pump protons.

The essential principle illustrated here is that a photosystem enables light to cause a net electron transfer from a weak electron donor—that is, a molecule with a strong affinity for electrons (in this case a cytochrome)—to a molecule such as a quinone, which is a strong electron donor in its reduced form. In this way the excitation energy that would otherwise be released as fluorescence and/or heat is used instead to raise the energy of an electron and create a strong electron donor where none had been before. In the chloroplasts of higher plants, as we shall see, water, rather than cytochrome, serves as the initial electron donor, which is why oxygen is released by photosynthesis in plants. But before returning to consider the processes by which the more complex photosystems of chloroplasts ultimately drive the production of ATP and NADPH, we shall consider how these end products are generated in a less sophisticated way, although by essentially similar machinery, in purple bacteria.

Bacterial Photosynthesis Establishes an Electrochemical Proton Gradient Across the Plasma Membrane That Drives the Production of Both ATP and NADPH[34]

The energy stored in the electrons carried by reduced quinones in the plasma membranes of purple photosynthetic bacteria is used in two different ways to generate ATP and NADPH, respectively. ATP is produced by a proton-pumping mechanism very similar to the one we have already encountered in mitochondria (see p. 352): protons are pumped across the bacterial plasma membrane when the quinone transfers its high-energy electrons through a *b-c complex* embedded in this membrane. The b-c complex then transfers electrons to a soluble cytochrome

Figure 7–50 The electron transfers that occur in the photochemical reaction center of a purple bacterium. A similar set of reactions is believed to occur in the evolutionarily related photosystem II in plants. At the top right is a schematic diagram showing the molecules that carry electrons, which are those in Figure 7–49, plus an exchangeable quinone (Q_B) and a freely mobile quinone dissolved in the lipid bilayer (Q). Electron carriers 1 through 5 are bound in a specific position on a 596-amino-acid transmembrane protein formed from two separate subunits (see Figure 6–32, p. 294). Following excitation by a photon of light, a high-energy electron passes from pigment molecule to pigment molecule, creating a charge separation as shown in the sequence in steps B through D below, where the pigment molecule carrying high-energy electrons is indicated in color. Once released into the bilayer, the quinone with two electrons picks up two protons and loses its charge (see Figure 7–30).

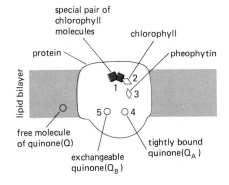

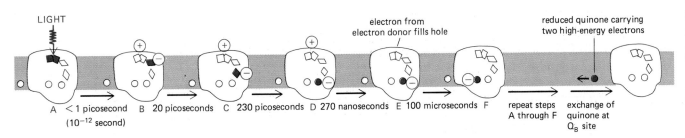

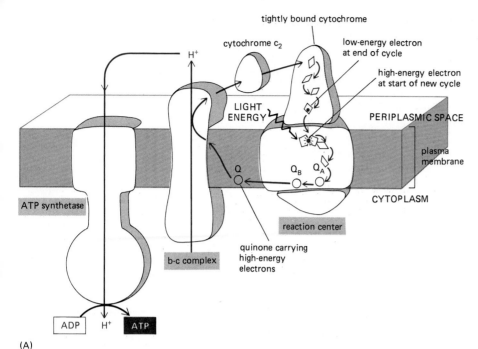

tightly bound cytochrome

cytochrome c₂

low-energy electron
at end of cycle

high-energy electron
at start of new cycle

H⁺

LIGHT
ENERGY

PERIPLASMIC SPACE

plasma
membrane

Q

Q_B   Q_A

ATP synthetase

CYTOPLASM

reaction center

b-c complex

quinone carrying
high-energy
electrons

ADP   H⁺   ATP

(A)

Figure 7–51 Two photosynthetic electron-transfer reactions in purple bacteria. These reactions occur in the inner (plasma) membrane of the bacterium, and cytochrome c₂ exists in soluble form in the periplasmic space beneath the outer membrane (see Figure 6–54, p. 312). (A) Cyclic electron flow produces an electrochemical proton gradient across the plasma membrane. This gradient is used here to drive the production of ATP by an ATP synthetase in the bacterial plasma membrane. (B) Reverse electron flow through NADH dehydrogenase, which is also driven by the large electrochemical proton gradient produced in (A), produces NADH. The reversibility of the reactions linked to the electrochemical gradient is discussed on p. 364.

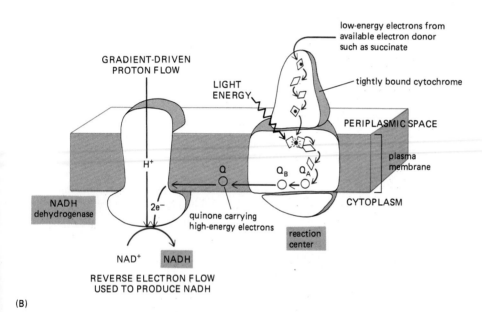

low-energy electrons from
available electron donor
such as succinate

GRADIENT-DRIVEN
PROTON FLOW

LIGHT
ENERGY

tightly bound cytochrome

PERIPLASMIC SPACE

plasma
membrane

H⁺

Q

Q_B   Q_A

NADH
dehydrogenase

2e⁻

CYTOPLASM

quinone carrying
high-energy electrons

reaction
center

NAD⁺   NADH

REVERSE ELECTRON FLOW
USED TO PRODUCE NADH

(B)

that returns the electrons (now at low energy) to the reaction center complex via a tightly bound cytochrome, completing a cyclic process (Figure 7–51A). The electrochemical proton gradient created by this cyclic electron transport is then used by an ATP synthetase in the plasma membrane to produce ATP, just as in mitochondria (Figure 7–51A).

NADPH is generated by a second electron-transport reaction, in which high-energy electrons from quinones, instead of cycling through the b-c complex, are transferred to NAD; the NADH produced is then converted to NADPH by a transhydrogenase. Because the high-energy electrons carried by quinones are held at a lower energy level than are the electrons in NADH (recall that, in mitochondria, electrons are transferred to quinone from NADH and not vice versa—see Figure 7–34), the production of NADH from NAD requires a source of energy. In purple photosynthetic bacteria the energy is provided by the electrochemical proton gradient across the plasma membrane, which drives H^+ back into the cell through an NADH-dehydrogenase complex, enabling this complex to catalyze the energetically unfavorable **reverse electron flow** from quinone to NAD (Figure 7–51B).

In summary, the purple photosynthetic bacterium uses its reaction center to produce a large pool of reduced quinone molecules in its plasma membrane. Some of this quinone is used to generate a large electrochemical proton gradient across the bacterial plasma membrane. This gradient is then used in two ways: (1) to drive an ATP synthetase to make ATP and (2) to drive a reverse electron flow from the remaining reduced quinone to NAD, thereby generating the reducing power needed for the synthesis of the bacterium's organic molecules.

7-24 7-25 7-26 In Plants and Cyanobacteria, Noncyclic Photophosphorylation Produces Both NADPH and ATP[31,35]

Photosynthesis in plants and cyanobacteria is more complex. It produces both ATP and NADPH directly by a two-step process called **noncyclic photophosphorylation.** Because two photosystems in series are used to energize an electron, the electron can be transferred all the way from water to NADPH. As the high-energy electrons pass through the coupled photosystems to generate NADPH, some of their energy is siphoned off for ATP synthesis.

In the first of the two photosystems—called *photosystem II* for historical reasons—the oxygens of two water molecules bind to a cluster of manganese atoms in a poorly understood water-splitting enzyme, and electrons are removed one at a time to fill the holes created by light in reaction-center chlorophyll molecules. As soon as four electrons have been removed (requiring four quanta of light), O_2 is released by the enzyme; photosystem II thus catalyzes the reaction $2H_2O \rightarrow 4H^+ + 4e^- + O_2$.

The core of the reaction center in photosystem II is homologous to the bacterial reaction center just described, and it likewise produces strong electron donors in the form of reduced quinone molecules in the membrane. The quinones pass their electrons to a b_6-*f complex*, which closely resembles the b-c complex of bacteria and the b-c$_1$ complex in the respiratory chain of mitochondria. As in mitochondria, the complex pumps protons into the thylakoid space across the thylakoid membrane (in chloroplasts) or out of the cytosol across invaginations of the plasma membrane (in cyanobacteria), and the resulting electrochemical gradient drives the synthesis of ATP by an ATP synthetase (Figures 7–52 and 7–53). The final electron acceptor in this electron-transport chain is the second photosystem in the scheme (*photosystem I*), which accepts the electron into the hole left by light excitation of its reaction-center chlorophyll molecule. Whereas the electrons energized by photosystem II are at too low an energy to be passed to $NADP^+$, each electron that leaves photosystem I has been boosted to a very high energy level by the two quanta of light that have sequentially activated it. Consequently, these electrons can be passed to the iron-sulfur center in ferredoxin to drive the reduction of $NADP^+$ to NADPH (Figure 7–53), which also involves the uptake of a proton from the medium.

The zigzag scheme for photosynthesis shown in Figure 7–53 is known as the **Z scheme.** By means of its two electron-energizing steps, one catalyzed by each photosystem, an electron is passed from water, which normally holds onto its electrons very tightly (redox potential $= +820$ mV), to NADPH, which normally holds onto its electrons rather loosely (redox potential $= -320$ mV). There is not enough energy in a single quantum of visible light to energize an electron all the way from the bottom of photosystem II to the top of photosystem I, which is probably the energy change required to pass an electron efficiently from water to $NADP^+$. The use of two separate photosystems in series also means that there is enough energy left over to enable the electron-transport chain that links the two photosystems to pump H^+ across the thylakoid membrane (or the plasma membrane of cyanobacteria), which allows ATP synthetase to harness some of the light-derived electron energy for ATP production.

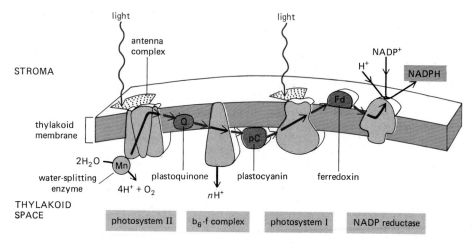

Figure 7–52 Electron flow during photosynthesis in the thylakoid membrane. The mobile electron carriers in the chain are plastoquinone (which closely resembles the ubiquinone of mitochondria), plastocyanin (a small copper-containing protein), and ferredoxin (a small protein containing an iron-sulfur center). The $b_6\text{-}f$ *complex* closely resembles the b-c_1 complex of mitochondria and the b-c complex of bacteria (see Figure 7–63): all three complexes accept electrons from quinones and pump protons. Note that the H^+ released by water oxidation and the H^+ taken up during NADPH formation also contribute to the generation of the electrochemical proton gradient that drives ATP synthesis.

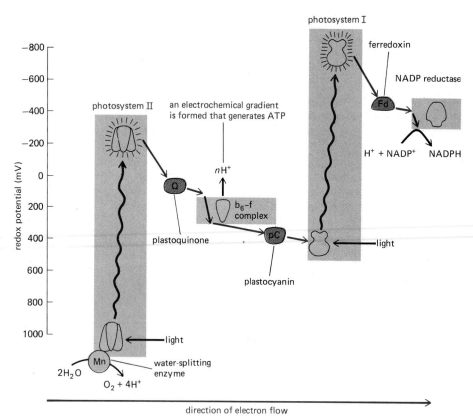

Figure 7–53 Redox potential changes for the passage of electrons during the photosynthetic production of NADPH and ATP in plants and cyanobacteria. Photosystem II closely resembles the reaction center in purple bacteria (see Figure 7–50), to which it is evolutionarily related. Photosystem I differs and is thought to be evolutionarily related to the photosystems of a different class of bacteria, the green bacteria; it passes electrons from its excited chlorophyll through a series of tightly bound iron-sulfur centers. The net electron flow through the two photosystems joined in series is from water to $NADP^+$, and it produces NADPH. In addition, ATP is synthesized by an ATP synthetase (not shown) that harnesses the electrochemical proton gradient produced by the electron-transport chain linking photosystem II and photosystem I. This *Z scheme* for ATP production is called *noncyclic photophosphorylation* to distinguish it from the cyclic scheme shown in Figure 7–54 (see also Figure 7–52).

Chloroplasts Can Make ATP by Cyclic Photophosphorylation Without Making NADPH[31,36]

In the noncyclic photophosphorylation scheme just discussed, high-energy electrons leaving photosystem II are harnessed to generate ATP, while those leaving photosystem I drive the production of NADPH. This produces slightly more than one molecule of ATP for every pair of electrons that passes from H_2O to $NADP^+$ to generate a molecule of NADPH. But considerably more ATP than NADPH is needed for carbon fixation (see Figure 7–43). To produce the extra ATP, chloroplasts can switch photosystem I into a cyclic mode in which its energy is directed into the synthesis of ATP instead of NADPH. This process, called **cyclic photophosphorylation,** involves an electron flow much like that used by photosynthetic bacteria to make ATP (see Figure 7–51A). Here the high-energy electrons from photosystem I are transferred back to the b_6-f complex rather than being passed

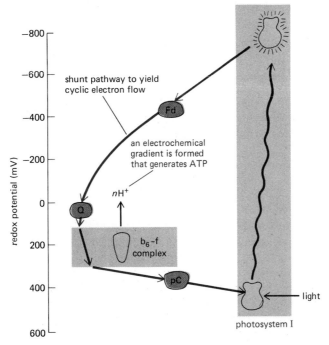

Figure 7–54 The path of electron flow in cyclic photophosphorylation. This pathway allows ATP to be made without producing either NADPH or O_2. Whether noncyclic or cyclic electron flow occurs depends on whether ferredoxin (Fd) donates its reactive electron to $NADP^+$, as in Figure 7–53, or to components leading back to the b_6-f complex. Whenever NADPH accumulates, $NADP^+$ levels will be low, tending to favor the cyclic scheme; other, less direct controls also ensure that appropriate proportions of ATP and NADPH are produced by photosynthesis.

on to $NADP^+$, so that protons are pumped across the thylakoid membrane, and the resulting electrochemical gradient drives the synthesis of ATP (Figure 7–54).

To review briefly, noncyclic photophosphorylation involves the photoreduction of $NADP^+$ by water, is mediated by the combined action of photosystem I and photosystem II, and produces NADPH, ATP, and O_2. Cyclic photophosphorylation, in contrast, involves only photosystem I and produces ATP without the formation of either NADPH or O_2. Thus the relative activities of cyclic and noncyclic electron flows determine how much light energy is converted into reducing power (NADPH) and how much into high-energy phosphate bonds (ATP). The balance is regulated according to the need for NADPH. Whether the electron flow is noncyclic or cyclic depends on whether ferredoxin donates its reactive electron to $NADP^+$ or to components that lead back to the b_6-f complex (compare Figures 7–53 and 7–54). At low $NADP^+$ concentrations, caused by an accumulation of NADPH, the cyclic scheme that produces only ATP is favored.

The effect of NADPH levels on cyclic photophosphorylation is part of an extensive regulatory network that balances the activities of photosystem I and photosystem II. An excess activity of photosystem II, for example, will increase the ratio of reduced to oxidized quinones in the thylakoid membrane, while an excess activity of photosystem I will have the opposite effect (see Figure 7–53). Whenever the ratio of reduced to oxidized quinones increases above a certain threshold, however, a protein kinase is activated that phosphorylates the major light-harvesting pigment protein in the antenna complex. This tends to dissociate the antenna complex from photosystem II and may even cause it to move in the thylakoid membrane from the stacked regions (the grana), where photosystem II is concentrated, to the unstacked regions, where photosystem I is concentrated. An increased proportion of light energy is thereby transferred to photosystem I until the quinone pool returns to normal.

7-27 The Geometry of Proton Translocation Is Similar in Mitochondria and Chloroplasts[37]

The presence of the thylakoid space separates a chloroplast into three rather than two internal compartments, making it seem quite different from a mitochondrion. However, the geometry of H^+ translocation in the two organelles is similar. As illustrated in Figure 7–55, in chloroplasts protons are pumped out of the stroma (pH 8) into the thylakoid space (pH about 5), creating a gradient of 3 to 3.5 pH units. This represents a proton-motive force of about 200 mV across the thylakoid membrane (nearly all of which is contributed by the pH gradient rather than by a membrane potential), which drives ATP synthesis by the ATP synthetase embedded in this membrane.

Like the stroma, the mitochondrial matrix has a pH of about 8, but this is created by pumping protons out of the mitochondrion into the cytosol (pH about 7), rather than into an interior space in the organelle. Thus the pH gradient is relatively small, and most of the proton-motive force across the mitochondrial inner membrane, which is about the same as that across the chloroplast thylakoid membrane, is caused by the resulting membrane potential (see p. 352). For both mitochondria and chloroplasts, however, the catalytic site of the ATP synthetase is at a pH of about 8 and is located in a large organelle compartment (matrix or stroma) packed full of soluble enzymes. Consequently, it is here that all of the organelle's ATP is made (Figure 7–55).

Although there are many similarities between mitochondria and chloroplasts, the structure of chloroplasts makes their electron- and proton-transport processes easier to study: by breaking both the inner and outer membranes of a chloroplast, isolated thylakoid discs can be obtained intact. These thylakoids resemble submitochondrial particles in that they have a membrane whose electron-transport chain has its $NADP^+$-, ADP-, and phosphate-utilization sites all freely accessible to the outside. But isolated thylakoids retain their undisturbed native structure and are much more active than isolated submitochondrial particles. For this reason, several of the experiments that first demonstrated the central role of chemiosmotic mechanisms were carried out with chloroplasts rather than with mitochondria.

Like the Mitochondrial Inner Membrane, the Chloroplast Inner Membrane Contains Carrier Proteins That Facilitate Metabolite Exchange with the Cytosol[38]

Although the photosynthetic electron- and proton-transfer reactions of photosynthesis are most readily studied in chloroplast preparations in which the inner and outer membranes have been broken or removed, such chloroplasts fail to carry out photosynthetic CO_2 fixation because of the absence of important substances that are normally present in the stroma. Chloroplasts can also be isolated in a way that leaves their inner membrane intact. In such chloroplasts the inner membrane can be shown to have a selective permeability, reflecting the presence of specific carrier proteins. Most notably, much of the glyceraldehyde 3-phosphate produced by CO_2 fixation in the chloroplast stroma is transported out of the chloroplast by an efficient antiport system that exchanges three-carbon sugar-phosphates for inorganic phosphate.

Glyceraldehyde 3-phosphate normally provides the cytosol with an abundant source of carbohydrate, which is used by the cell as the starting point for many other biosyntheses—including the production of sucrose for export. But this is not all it provides. Once the glyceraldehyde 3-phosphate reaches the cytosol, it is readily converted (by part of the glycolytic pathway) to 3-phosphoglycerate, generating one molecule of ATP and one of NADH. (The same two-step reaction working in reverse forms glyceraldehyde 3-phosphate in the carbon-fixation cycle—see Figure 7–43.) As a result, the export of glyceraldehyde 3-phosphate from the chloroplast provides not only the main source of fixed carbon to the rest of the cell, but also the reducing power and ATP needed for metabolism outside the chloroplast.

Chloroplasts Carry out Other Biosyntheses[39]

The chloroplast carries out many biosyntheses in addition to photosynthesis. All of the cell's fatty acids and a number of amino acids, for example, are made by enzymes in the chloroplast stroma. Similarly, the reducing power of light-activated electrons drives the reduction of nitrite (NO_2^-) to ammonia (NH_3) in the chloroplast; this ammonia provides the plant with nitrogen for the synthesis of amino acids and nucleotides. The metabolic importance of the chloroplast for plants and algae therefore extends far beyond its role in photosynthesis.

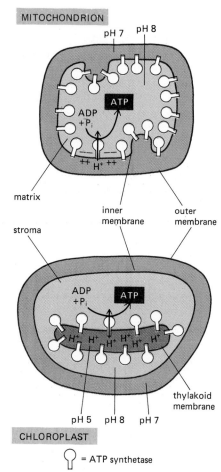

Figure 7–55 Comparison of proton flows and ATP synthetase orientations in mitochondria and chloroplasts. Those compartments with a similar pH have been colored similarly. The proton-motive force across the thylakoid membrane consists almost entirely of the pH gradient; the high permeability of this membrane to Mg^{2+} and Cl^- ions allows the flow of these ions to dissipate most of the membrane potential. Presumably, mitochondria could not tolerate having their matrix at pH 10, as would be required to generate their proton-motive force without a membrane potential.

Summary

Chloroplasts and photosynthetic bacteria obtain high-energy electrons by means of photosystems that capture the electrons excited when sunlight is absorbed by chlorophyll molecules. Photosystems are composed of an antenna complex attached to a photochemical reaction center, which is a precisely ordered complex of proteins and pigments in which the photochemistry of photosynthesis occurs. By far the best-understood photochemical reaction center is that of the purple photosynthetic bacteria, for which the complete three-dimensional structure is known. In these bacteria a single photosystem produces an electrochemical proton gradient that is used to drive both ATP and NADPH synthesis. There are two photosystems in chloroplasts and cyanobacteria. Depending on the cell's needs, two types of electron flow occur in different ratios: (1) a noncyclic flow, mediated by the two photosystems linked in series, transfers electrons from water to $NADP^+$ to produce NADPH, with the concomitant production of ATP; and (2) a cyclic flow, mediated by a single photosystem through which electrons circulate in a closed loop, produces only ATP. In chloroplasts, all electron-transport processes occur in the thylakoid membrane: to make ATP, protons are pumped into the thylakoid space, and a backflow of protons through an ATP synthetase then produces the ATP in the stroma.

The ATP and NADPH made by photosynthesis drive many biosynthetic reactions in the chloroplast stroma, including the all-important carbon-fixation cycle, which creates carbohydrate from CO_2. This carbohydrate is exported to the cell cytosol where—as glyceraldehyde 3-phosphate—it provides organic carbon, ATP, and reducing power to the rest of the cell.

The Evolution of Electron-Transport Chains[40]

Much of the structure, function, and evolution of cells and organisms can be related to their need for energy. We have seen that the fundamental mechanisms for harnessing energy from such disparate sources as light and the oxidation of glucose are the same. Apparently, an effective method for synthesizing ATP arose early in evolution and has since been conserved with only small variations. How did the crucial individual components—ATP synthetase, redox-driven proton pumps, and photosystems—first arise? Hypotheses about events occurring on an evolutionary time scale are difficult to test. But clues abound, both in the many different primitive electron-transport chains that survive in some present-day bacteria and in geological evidence concerning the environment of the earth billions of years ago.

The Earliest Cells Probably Produced ATP by Fermentation[41]

As explained in Chapter 1, the first living cells are thought to have arisen roughly 3.5×10^9 years ago, when the earth was about 10^9 years old. Because the environment lacked oxygen but was rich in geochemically produced organic molecules, the earliest metabolic pathways for producing ATP presumably resembled present-day forms of fermentation.

In the process of **fermentation**, ATP is made by a substrate-level phosphorylation event (see p. 66) that harnesses the energy released by a reaction pathway in which a hydrogen-rich organic molecule, such as glucose, is partly oxidized. Without oxygen to serve as a hydrogen acceptor, the hydrogens lost from the oxidized molecules must be transferred (via NADH or NADPH) to a different organic molecule (or to a different part of the same molecule), which thereby becomes more reduced. At the end of the fermentation process, one (or more) of the organic molecules produced is excreted into the medium as a metabolic waste product; others, such as pyruvate, are retained by the cell for biosynthesis.

The excreted end products are different in different organisms, but they tend to be organic acids (carbon compounds that carry a COOH group). The most important of such products in bacterial cells include lactic acid (which also ac-

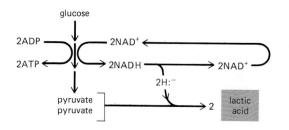

(A) FERMENTATION LEADING TO EXCRETION OF LACTIC ACID

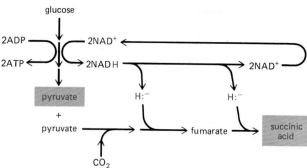

(B) FERMENTATION LEADING TO EXCRETION OF SUCCINIC ACID

Figure 7–56 Two types of fermentation processes, with the end products highlighted by colored boxes. (A) The two molecules of NAD$^+$ used for each molecule of glucose that undergoes glycolysis are regenerated by the transfer of hydride ions from NADH to pyruvate to produce two molecules of lactic acid (the reverse of the type of reaction shown previously in Figure 7–18). The lactic acid is excreted. (B) The two molecules of NAD$^+$ used for each molecule of glucose that undergoes glycolysis are regenerated by successive transfers of hydride ions from two NADH molecules to compounds derived from pyruvate that produce succinic acid. For each molecule of succinic acid excreted, a molecule of pyruvate (in colored box) is saved for biosyntheses inside the cell. In both (A) and (B) an organic acid must be excreted to re-form NAD$^+$ and thereby enable glycolysis to continue in the absence of oxygen.

cumulates in anaerobic mammalian glycolysis; see p. 67) and formic, acetic, propionic, butyric, and succinic acids. Two fermentation pathways of present-day bacteria are illustrated in Figure 7–56.

7-28 The Evolution of Energy-conserving Electron-Transport Chains Enabled Anaerobic Bacteria to Use Nonfermentable Organic Compounds as a Source of Energy[42]

The early fermentation processes would have provided not only the ATP but also the reducing power (as NADH or NADPH) required for essential biosyntheses, and many of the major metabolic pathways probably evolved while fermentation was the only mode of energy production. With time, however, the metabolic activities of these procaryotic organisms must have changed the local environment, so that they were forced to evolve new biochemical pathways. The accumulation of waste products of fermentation might have resulted in the following series of changes:

Stage 1. Because of the continuous excretion of organic acids, the pH of the environment was lowered, and proteins that function as transmembrane proton pumps evolved to pump H$^+$ out of the cell to prevent death from intracellular acidification. One of these pumps may have used the energy available from ATP hydrolysis and could have been the ancestor of the present-day ATP synthetase.

Stage 2. At the same time that nonfermentable organic acids were accumulating and favoring the evolution of an ATP-consuming proton pump, the supply of geochemically generated fermentable nutrients, which provided the energy both for the pumps and for all other cellular processes, was dwindling. The resulting selective pressures strongly favored bacteria that could excrete H$^+$ without hydrolyzing ATP, allowing the ATP to be conserved for other cellular activities. Such pressures might have led to the first membrane-bound proteins that could use electron transport between molecules of different redox potential as the energy source for transporting H$^+$ across the plasma membrane. Some of these proteins must have found their electron donors and electron acceptors among the nonfermentable organic acids that had accumulated. Many such electron-transport proteins can, in fact, be found in present-day bacteria: for example, some present-

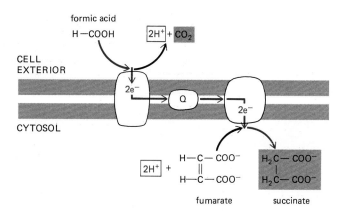

Figure 7–57 The oxidation of formic acid by fumarate is mediated by an energy-conserving electron-transport chain in the plasma membrane of some present-day bacteria that grow anaerobically, including *E. coli.* As indicated, the products are succinate and CO_2. Note that protons are used up inside the cell and generated outside the cell, which is equivalent to pumping protons to the cell exterior. Thus this membrane-bound electron-transport system can generate an electrochemical proton gradient across the plasma membrane. The redox potential of the formic-acid–CO_2 pair is -420 mV, while that of the fumarate-succinate pair is $+30$ mV.

day bacteria that grow on formic acid pump protons by using the relatively small amount of redox energy derived from the transfer of electrons from formic acid to fumarate (Figure 7–57). Other bacteria have developed similar electron-transport components devoted solely to the oxidation and reduction of inorganic substrates (for example, see Figure 7–59, below).

Stage 3. Eventually some bacteria evolved H^+-pumping electron-transport systems that were efficient enough to harness more redox energy than they needed just to maintain their internal pH. A large electrochemical proton gradient generated by excessive H^+ pumping allowed protons to leak back into the cell through the ATP-driven proton pumps, thereby running them in reverse so that they functioned as ATP synthetases to make ATP. Because such bacteria required much less of the increasingly scarce supply of fermentable nutrients, they proliferated at the expense of their neighbors.

These three hypothetical stages in the evolution of oxidative phosphorylation mechanisms are summarized in Figure 7–58.

Photosynthetic Bacteria, by Providing an Inexhaustible Source of Reducing Power, Overcame a Major Obstacle in the Evolution of Cells[43]

The evolutionary steps just outlined would have solved the problem of maintaining both a neutral intracellular pH and an abundant store of energy, but they would have left unsolved another equally serious problem. The depletion of fermentable organic nutrients meant that some alternative source of carbon had to be found to make the sugars that served as the precursors of so many other cellular molecules. The carbon dioxide in the atmosphere provided an abundant potential carbon source. But to convert carbon dioxide into an organic molecule such as a carbohydrate requires that the fixed carbon dioxide be reduced by a strong electron donor, such as NADH or NADPH, which can provide the high-energy electrons needed to generate each (CH_2O) unit from CO_2 (see Figure 7–43). Early in cellular evolution, such strong reducing agents would have been plentiful as products of fermentation. But as the supply of fermentable nutrients dwindled and a membrane-bound ATP synthetase began to produce most of the ATP, the plentiful supply of NADH and other reducing agents would also have disappeared. It thus became imperative for cells to evolve a new way of generating a source of strong reducing power.

After most of the fermentable molecules in the environment had been used up, the main electron donors still available were the organic acids produced by the anaerobic metabolism of carbohydrates, inorganic molecules such as hydrogen sulfide (H_2S) generated geochemically, and water. But the reducing power of all of these molecules is far too weak to be useful for carbon dioxide fixation. An early supply of strong electron donors was probably generated by using the electrochemical proton gradient across the plasma membrane to drive a reverse electron flow, requiring the evolution of membrane-bound enzyme complexes resembling

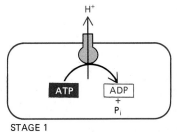

STAGE 1

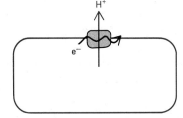

STAGE 2

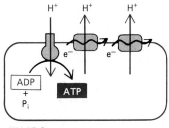

STAGE 3

Figure 7–58 One possible sequence for the evolution of oxidative phosphorylation mechanisms (see text).

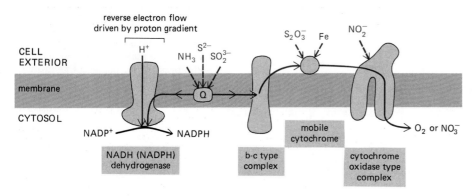

CELL
EXTERIOR

membrane

CYTOSOL

reverse electron flow
driven by proton gradient

H⁺

NH₃ S²⁻ SO₃²⁻

S₂O₃⁻ Fe   NO₂⁻

Q

NADP⁺ → NADPH

NADH (NADPH)
dehydrogenase

b-c type
complex

mobile
cytochrome

cytochrome
oxidase type
complex

O₂ or NO₃⁻

Figure 7–59 Some of the electron-transport pathways in present-day bacteria that generate all the ATP and reducing power they require for growth from the oxidation of inorganic molecules—such as iron, ammonia, nitrite, and sulfur compounds. As indicated, some species can grow anaerobically by substituting nitrate for oxygen as the terminal electron acceptor. Most use the carbon-fixation cycle and synthesize their organic molecules entirely from carbon dioxide. The forward electron flows cause protons to be pumped out of the cell, and the resulting proton gradient drives the production of ATP by an ATP synthetase (not shown). In addition, the NADPH required for carbon fixation is produced by reverse electron flow, as indicated (see also Figure 7–51B).

a NADH dehydrogenase (Figure 7–59). But the major evolutionary breakthrough in energy metabolism was the development of photochemical reaction centers that could produce molecules such as NADH directly. It is thought that this first occurred more than 3×10^9 years ago in the ancestors of the green sulfur bacteria. Present-day green sulfur bacteria use light energy to transfer hydrogen atoms (as an electron plus a proton) from hydrogen sulfide to NADPH, thereby creating the strong reducing power required for carbon fixation (Figure 7–60). Because the electrons removed from H_2S are at a much more negative redox potential than those of water (-230 mV compared to $+820$ mV for water), one quantum of light absorbed by the single photosystem in these bacteria is sufficient to achieve a high enough redox potential to generate NADPH via a relatively simple photosynthetic electron-transport chain.

The More Complex Photosynthetic Electron-Transport Chains of Cyanobacteria Probably Produced the First Atmospheric Oxygen[44]

The next step, which is thought to have occurred with the development of the cyanobacteria about 3×10^9 years ago, was the evolution of organisms capable of using water as the hydrogen source for carbon dioxide reduction. This entailed the addition of a second photosystem, acting in series with the first, to bridge the enormous gap in redox potential between H_2O and NADPH. Present-day structural homologies between photosystems suggest that this change involved the cooperation of a photosystem derived from green bacteria (photosystem I) with a photosystem derived from purple bacteria (photosystem II). The biological consequences of this evolutionary step were far-reaching. For the first time there were organisms that made only very minimal chemical demands on their environment and could therefore spread and evolve in ways denied the earlier photosynthetic bacteria, which needed H_2S or organic acids as a source of electrons. Conse-

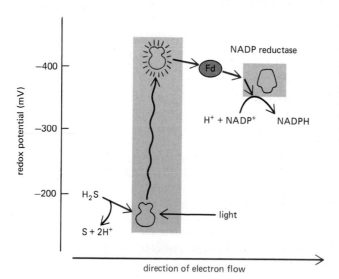

redox potential (mV)

−400

−300

−200

NADP reductase

Fd

H⁺ + NADP⁺   NADPH

H₂S

S + 2H⁺

light

direction of electron flow

Figure 7–60 The general flow of electrons in a relatively primitive form of noncyclic photosynthesis observed in present-day green sulfur bacteria. The photosystem in green bacteria resembles photosystem I in plants and cyanobacteria in using a series of iron-sulfur centers as primary electron acceptors that eventually donate their high-energy electrons to ferredoxin (Fd).

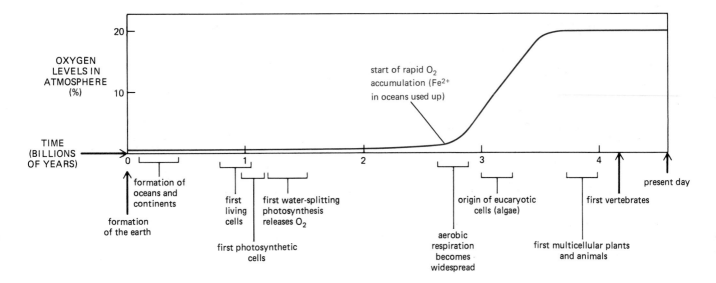

Figure 7–61 The relationship between
changes in atmospheric oxygen levels
and some of the major stages that are
believed to have occurred during the
evolution of living organisms on earth.
As indicated, geological evidence
suggests that there was more than a
billion-year delay between the rise of
cyanobacteria (thought to be the first
organisms to release oxygen) and the
time that high oxygen levels began to
accumulate in the atmosphere. This
delay was due largely to the rich supply
of dissolved ferrous ion in the oceans,
which reacted with the released oxygen
to form enormous iron oxide deposits.

quently, large amounts of biologically synthesized, reduced organic materials accumulated. Moreover, oxygen entered the atmosphere for the first time.

Oxygen is highly toxic because the oxidation reactions it catalyzes can inactivate enzymes. Many present-day anaerobic bacteria, for example, are rapidly killed when exposed to air. Thus organisms on the primitive earth would have had to evolve protective measures against the rising O_2 levels in the environment. Late evolutionary arrivals, such as ourselves, have numerous detoxifying mechanisms that protect our enzymes from oxygen's ill effects.

The increase in atmospheric oxygen was very slow at first. The early seas contained large amounts of ferrous iron (Fe[II]), and nearly all the oxygen produced by early photosynthetic bacteria was utilized in converting Fe(II) to Fe(III). This conversion caused the precipitation of huge amounts of ferric oxides, and the extensive banded iron formations beginning about 2.7×10^9 years ago help to date the rise of the cyanobacteria. By about 2×10^9 years ago, the supply of ferrous iron was exhausted and the deposition of further iron precipitates ceased. The geological evidence suggests that oxygen levels in the atmosphere then began to rise, reaching current levels between 0.5 and 1.5×10^9 years ago (Figure 7–61).

The availability of oxygen made possible the development of bacteria that relied on aerobic metabolism to make their ATP. These organisms could harness the large amount of energy released by breaking down carbohydrates and other reduced organic molecules all the way to CO_2 and H_2O. Components of preexisting electron-transport complexes were modified to produce a cytochrome oxidase, so that the electrons obtained from organic or inorganic substrates could be transported to O_2 as the terminal electron acceptor. Many present-day purple photosynthetic bacteria can switch between photosynthesis and respiration, depending on the availability of light and oxygen, by surprisingly minor reorganizations of their electron-transport chains.

As organic materials accumulated on the earth as a result of photosynthesis, some photosynthetic bacteria (including the precursors of *E. coli*) lost their ability to survive on light energy alone and came to rely entirely on respiration. It is believed that mitochondria first arose some 1.5×10^9 years ago, when such a respiration-dependent bacterium became an endosymbiont in a primitive eucaryotic cell (see p. 398). Later, descendants of this early aerobic eucaryotic cell endocytosed a photosynthetic bacterium that became the precursor of chloroplasts; however, present-day chloroplasts are sufficiently unique in different types of algae to suggest that chloroplasts evolved separately in these different lineages. Figure 7–62 outlines some of the suspected evolutionary pathways just discussed.

Evolution is always conservative, taking parts of the old and building upon them to create something new. Thus parts of the electron-transport chains that were derived to service anaerobic bacteria more than 3 billion years ago probably survive, in altered form, in the mitochondria and chloroplasts of today's higher

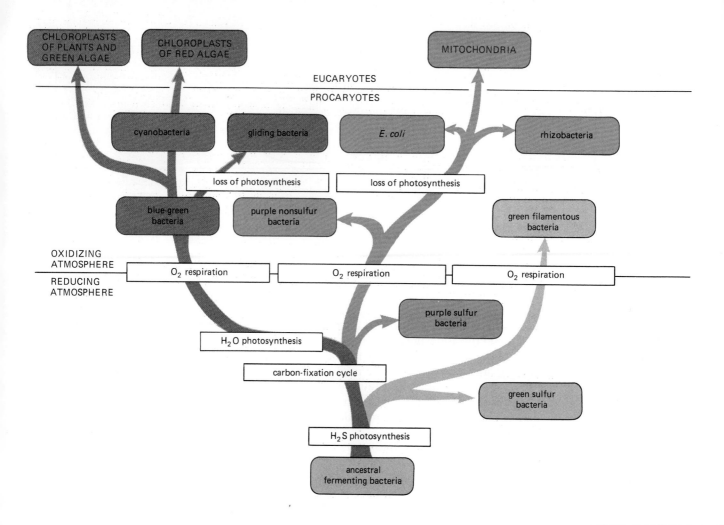

CHLOROPLASTS OF PLANTS AND GREEN ALGAE

CHLOROPLASTS OF RED ALGAE

MITOCHONDRIA

EUCARYOTES

PROCARYOTES

cyanobacteria

gliding bacteria

E. coli

rhizobacteria

loss of photosynthesis

loss of photosynthesis

blue-green bacteria

purple nonsulfur bacteria

green filamentous bacteria

OXIDIZING ATMOSPHERE

REDUCING ATMOSPHERE

O_2 respiration

O_2 respiration

O_2 respiration

H_2O photosynthesis

purple sulfur bacteria

carbon-fixation cycle

green sulfur bacteria

H_2S photosynthesis

ancestral fermenting bacteria

eucaryotes. As one example, there is a striking homology in structure and function between the enzyme complexes that pump protons in the central segment of the mitochondrial respiratory chain (the b-c$_1$ complex) and the corresponding segments of the electron-transport chains of both bacteria and chloroplasts (Figure 7–63).

Summary

Early cells are believed to have been bacteriumlike organisms living in an environment rich in highly reduced organic molecules that had been formed by geochemical processes over the course of hundreds of millions of years. They probably derived most of their ATP by converting the reduced organic molecules to a variety of organic acids, which were then released as waste products. By acidifying the environment, these fermentations may have led to the evolution of the first membrane-bound H$^+$ pumps, which could maintain a neutral pH in the cell interior. The properties of present-day bacteria suggest that an electron-transport-driven H$^+$ pump and an ATP-driven H$^+$ pump first arose in this anaerobic environment. Reversal of the ATP-driven pump would have allowed it to function as an ATP synthetase. Therefore as more effective electron-transport chains developed, the energy released by redox reactions between inorganic molecules, accumulated nonfermentable compounds, or both, could be harnessed for generating ATP.

Because preformed organic molecules were replenished by geochemical processes only very slowly, the proliferation of bacteria that used them as the source of both carbon and reducing power could not go on forever. The depletion of fermentable organic nutrients presumably led to the evolution of bacteria that could use carbon dioxide to make carbohydrates. By combining parts of the electron-transport chains that had developed earlier, light energy was harvested by a single photosystem in photosynthetic bacteria to generate the NADPH required for carbon

Figure 7–62 A phylogenetic tree of the probable evolution of mitochondria and chloroplasts and their bacterial ancestors. Oxygen respiration is thought to have begun developing about 2×10^9 years ago; as indicated, it seems to have evolved independently in the green, purple, and blue-green (cyanobacterial) lines of photosynthetic bacteria. It is thought that an aerobic purple bacterium that had lost its ability to photosynthesize gave rise to the mitochondrion, while several different blue-green bacteria gave rise to chloroplasts. Detailed nucleotide sequence analyses suggest that mitochondria arose from bacteria that resembled the rhizobacteria, agrobacteria, and rickettsias—three closely related species known to form intimate associations with present-day eucaryotic cells (see p. 1157 and p. 1159).

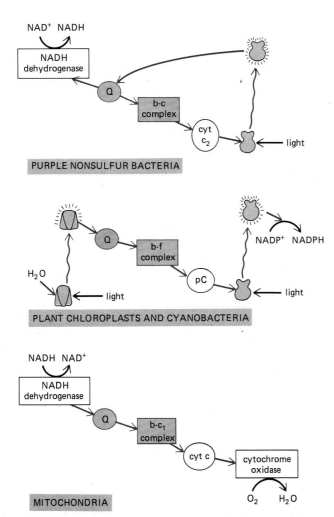

Figure 7–63 A summary of the three electron-transport chains discussed extensively in this chapter. Bacteria, chloroplasts, and mitochondria all contain a membrane-bound enzyme complex that closely resembles the b-c$_1$ complex of mitochondria. These complexes all accept electrons from a quinone carrier (designated as Q) and pump protons across their respective membranes. Moreover, in reconstituted *in vitro* systems the different complexes can substitute for one another, and the amino acid sequences of their protein components indicate that they are evolutionarily related.

fixation. The subsequent appearance of the more complex photosynthetic electron-transport chains of the cyanobacteria allowed water to be used as the electron donor for NADPH formation, rather than the much less abundant electron donors required by other photosynthetic bacteria. Life could then proliferate over large areas of the earth, so that reduced organic molecules accumulated again. About 2 billion years ago, the oxygen released by photosynthesis in cyanobacteria began to accumulate in the atmosphere. Once both organic molecules and oxygen were abundant, electron-transport chains became adapted for the transport of electrons from NADH to oxygen, and efficient aerobic metabolism developed in many bacteria. Exactly the same aerobic mechanisms operate in the mitochondria of eucaryotes, and there is considerable evidence that both mitochondria and chloroplasts evolved from aerobic bacteria that were endocytosed by primitive eucaryotic cells.

The Genomes of Mitochondria and Chloroplasts[45]

Cells must generate new cytoplasmic organelles to keep pace with cell growth and division. They must also replenish those organelles that are degraded as part of the continual process of organelle turnover in nonproliferating cells. Organelle biosynthesis requires the ordered synthesis of the requisite proteins and lipids and the delivery of each component to the correct organelle subcompartment. In Chapter 8 we discuss the transfer of selected proteins and lipids made elsewhere into mitochondria and chloroplasts. Here we describe the contributions these organelles make to their own biogenesis.

The biosynthesis of mitochondria and chloroplasts involves the contribution of two separate genetic systems. While most of their proteins are encoded by

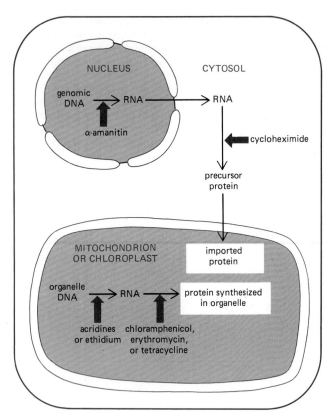

Figure 7–64 An overview of the biosynthesis of mitochondrial and chloroplast proteins. Each thick arrow indicates the site of action of an inhibitor that is specific for either organelle or cytosolic protein synthesis.

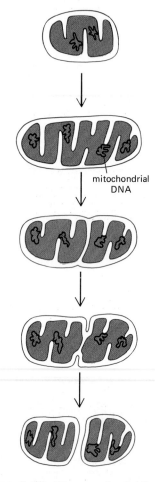

Figure 7–65 Diagram of a dividing mitochondrion. The pathway shown has been postulated from static views of dividing mitochondria like that in Figure 7–66.

nuclear DNA and imported into the organelle from the cytosol after they are synthesized on cytosolic ribosomes, some are encoded by organelle DNA and synthesized on ribosomes within the organelle. The protein traffic between the cytosol and these organelles seems to be unidirectional, inasmuch as no protein is known to be exported from mitochondria or chloroplasts to the cytosol.

The contributions from the two genetic systems to the construction of mitochondria and chloroplasts are closely coordinated (see p. 396). The coupling is not absolute, however, and isolated organelles in a test tube continue to make organelle DNA, RNA, and proteins for brief periods, thereby providing one means of determining which proteins are encoded in organelle DNA and which in nuclear DNA. Another approach uses specific inhibitors on intact cells. The drug *cycloheximide*, for example, inhibits cytosolic protein synthesis but does not inhibit organelle protein synthesis. Conversely, various antibiotics—such as chloramphenicol, tetracycline, and erythromycin—inhibit protein synthesis in mitochondria and chloroplasts but have little effect on cytosolic protein synthesis (Figure 7–64). These inhibitors are widely used in studies of the functions of these organelles.

7-33 Organelle Growth and Division Maintain the Number of Mitochondria and Chloroplasts in a Cell[46]

Mitochondria and chloroplasts are never made de novo. They always arise by the growth and division of existing mitochondria and chloroplasts. Observations of living cells indicate that mitochondria not only divide but also fuse with one another. On average, however, each organelle must double in mass and then divide in half once in each cell generation. Electron microscopic studies suggest that organelle division begins by an inward furrowing of the inner membrane, as occurs in cell division in many bacteria (Figures 7–65 and 7–66), implying that it is a controlled process rather than a chance pinching in two.

In most cells, individual energy-converting organelles divide throughout interphase, out of phase with either the division of the cell or with each other. Similarly, the replication of organelle DNA is not limited to the S phase, when nuclear DNA replicates, but occurs throughout the cell cycle. Individual organelle

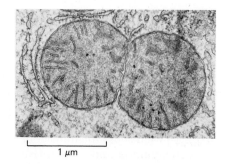

Figure 7–66 Electron micrograph of a dividing mitochondrion in a liver cell. (Courtesy of Daniel S. Friend.)

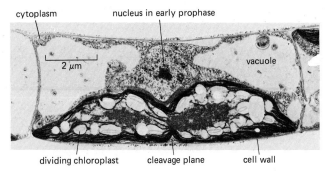

cytoplasm   nucleus in early prophase

2 μm

vacuole

dividing chloroplast   cleavage plane   cell wall

Figure 7–67 Chloroplast division occurs at a fixed point early in mitosis in the simple filamentous alga *Klebsormidium*. This cell has only one chloroplast, and its plane of division coincides with the plane of the subsequent division of the cell. (First published in J.D. Pickett-Heaps, *Cytobios* 6:167–183, 1972.)

DNA molecules seem to be selected for replication at random, so that in a given cell cycle some may replicate more than once and others not at all. Nonetheless, under constant conditions the process is regulated to ensure that the total number of organelle DNA molecules doubles in every cell cycle, so that each cell type maintains a constant amount of organelle DNA.

The number of organelles per cell can be regulated according to need; a large increase in mitochondria (as much as five- to tenfold), for example, is observed if a resting skeletal muscle is repeatedly stimulated to contract for a prolonged period. Moreover, in special circumstances, organelle division is precisely controlled by the cell: thus in some algae that contain only one or a few chloroplasts, the organelle divides just prior to cytokinesis in a plane that is identical to the future plane of cell division (Figure 7–67). The regulatory mechanisms required to explain these observations are not understood at the molecular level.

The Genomes of Chloroplasts and Mitochondria Are Usually Circular DNA Molecules[47]

Organelle DNA molecules are relatively small and simple, and, except for the mitochondrial genomes of some algae and protozoans, they are circular. The size of the chloroplast genome is similar in all organisms examined, but the mitochondrial genome is very much larger in plants than in animals (Table 7–2). Many organelle DNA molecules are about the same size as viral DNAs. In mammals, for example, the mitochondrial genome is a DNA circle of about 16,500 base pairs (less than 10^{-5} times the size of the nuclear genome). It is nearly the same size

Table 7–2   The Size of Organelle Genomes*

Type of DNA	Size (thousands of nucleotide pairs)
Chloroplast DNA	
Higher Plants	120–200
Chlamydomonas (green alga)	180
Mitochondrial DNA	
Animals (including flatworms, insects, and mammals)	16–19
Higher plants	150–2500
Fungi	
Schizosaccharomyces pombe (fission yeast)	17
Aspergillus nidulans	32
Neurospora crassa	60
Saccharomyces cerevisiae (budding yeast)	78
Chlamydomonas (green alga)	16 (linear molecule)
Protozoa	
Trypanosoma brucei	22
Paramecium	40 (linear molecule)

*These genomes are circular DNA molecules unless indicated otherwise.

in animals as diverse as *Drosophila* and sea urchins (Figure 7–68). Plants, however, contain a circular mitochondrial genome that is 10 to 150 times larger, depending on the plant. The largest of these are about half the size of typical bacterial genomes, which are also circular DNA molecules.

All mitochondria and chloroplasts contain multiple copies of the organelle DNA molecule (Table 7–3). These DNA molecules are usually distributed in several clusters in the matrix of the mitochondrion and in the stroma of the chloroplast, where they are thought to be attached to the inner membrane. Although it is not known how the DNA is packaged, the genome structure is likely to resemble that in bacteria rather than eucaryotic chromatin. As in bacteria, for example, there are no histones.

In mammalian cells, mitochondrial DNA makes up less than 1% of the total cellular DNA. However, in other cells—such as the leaves of higher plants or the very large egg cells of amphibia—a much larger fraction of the cellular DNA may be present in the energy-converting organelles (see Table 7–3), and a larger fraction of RNA and protein synthesis takes place there.

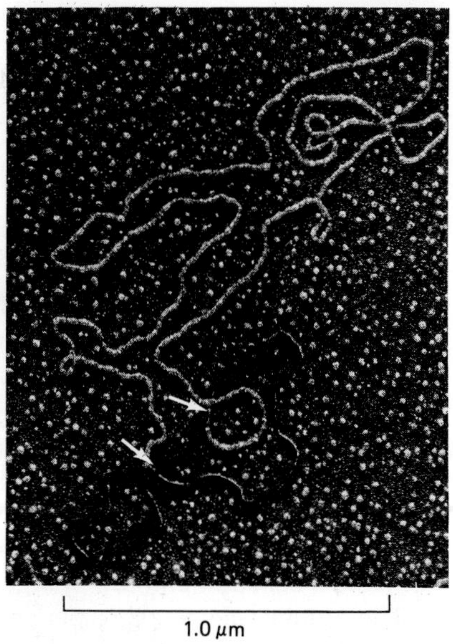

Mitochondria and Chloroplasts Contain Complete Genetic Systems[48]

Despite the small number of proteins encoded in their genomes, energy-converting organelles carry out their own DNA replication, DNA transcription, and protein synthesis. These processes take place in the matrix in mitochondria and in the stroma in chloroplasts. The proteins that mediate these genetic processes are unique to the organelle, but most of them are encoded in the nuclear genome rather than in the organelle DNA (see p. 400). This is all the more surprising because the protein-synthesis machinery of the organelles resembles that of bacteria rather than that of eucaryotes. The resemblance is particularly close in the case of chloroplasts:

1. Chloroplast ribosomes are very similar to *E. coli* ribosomes, both in their sensitivity to various antibiotics (such as chloramphenicol, streptomycin, erythromycin, and tetracycline) and in their structure. Not only are the nucleotide sequences of the ribosomal RNAs of chloroplasts and *E. coli* strikingly similar, but chloroplast ribosomes are able to use bacterial tRNAs in protein synthesis. In all these respects, chloroplast ribosomes differ from those found in the cytosol of the same plant cell.
2. Protein synthesis in chloroplasts starts with *N*-formylmethionine, as in bacteria, and not with methionine, as in the cytosol of eucaryotic cells.
3. Unlike nuclear DNA, chloroplast DNA can be transcribed by the RNA polymerase enzyme from *E. coli* to produce chloroplast mRNAs, which are efficiently translated by an *E. coli* protein-synthesizing system.

1.0 μm

Figure 7–68 Electron micrograph of an animal mitochondrial DNA molecule caught during the process of DNA replication. The circular DNA genome has replicated only between the two points marked by arrows (*colored strands*). (Courtesy of David Clayton.)

Table 7–3 Relative Amounts of Organelle DNA in Some Cells and Tissues

Organism	Tissue or Cell Type	DNA Molecules per Organelle	Organelles per Cell	Organelle DNA as Percent of Total Cellular DNA
Mitochondrial DNA				
Rat	liver	5–10	1000	1
Mouse	L-cell line	5–10	100	<1
Yeast*	vegetative	2–50	1–50	15
Frog	egg	5–10	10^7	99
Chloroplast DNA				
Chlamydomonas	vegetative	80	1	7
Maize	leaves	20–40	20–40	15

*The large variation in the number and size of mitochondria per cell in yeast is due to mitochondrial fusion and fragmentation.

Pass your cell biology exams with flying colors with the help of

THE PROBLEMS BOOK
for
MOLECULAR BIOLOGY OF THE CELL
SECOND EDITION

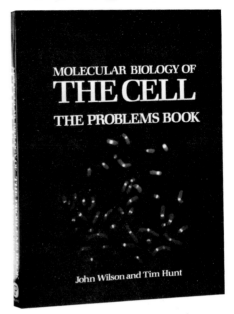

Keyed to the textbook's main chapters (5-14), THE PROBLEMS BOOK...

☐ is a great study aid when you are facing exams

☐ helps you see the elegance, ingenuity, and surprises of the research process

☐ shows you the kinds of questions that you are likely to encounter on tests

Detailed answers are provided for about half the problems to demonstrate how to analyze experimental observations and draw conclusions from them. The other problems are not supplied with answers, but may be used for homework assignments, class discussions, or even as exam questions.

A new study aid for the Second Edition...

THE PROBLEMS BOOK

By John Wilson and Tim Hunt

384 pages   312 illustrations
$8\frac{1}{2}$ x 11   softcover

Comments by students:

"I found these problems sets very effective preparation for the final exam in the course."
—Mark Krieger, Harvard University

"Well worded and doable."
—Diana Scouras, University of Colorado

"I think the problems overall are excellent. They focus on experimental aspects of cell biology, which is the best way to learn."
—Nancy Mayer, Harvard University

"I found doing the problem sets very enjoyable. The problem sets definitely enhanced my understanding of the material. I liked the fact that many of the problems were analyses of actual experiments. I also thought that the true-false questions were very good and thought-provoking."
—Emily Chan, Harvard University

"My opinion is that these problems are an indispensable learning aid."
—Kenneth J. Rhodes, Boston University School of Medicine

Although mitochondrial genetic systems are much less similar to those of present-day bacteria than are the genetic systems of chloroplasts, their ribosomes are also sensitive to antibacterial antibiotics, and protein synthesis in mitochondria also starts with *N*-formylmethionine.

The Chloroplast Genome of Higher Plants Contains About 120 Genes[49]

The best-studied chloroplast genomes are those of plants and green algae, whose chloroplasts are very similar. They are circular DNA molecules, and the complete nucleotide sequence has been determined for the chloroplasts of tobacco and liverwort. The results indicate that these two distantly related higher plants contain nearly identical chloroplast genes. In addition to four ribosomal RNAs, these genomes encode about 20 chloroplast ribosomal proteins, selected subunits of the chloroplast RNA polymerase, several proteins that are part of photosystems I and II, subunits of the ATP synthetase, portions of enzyme complexes in the electron-transport chain, one of the two subunits of ribulose bisphosphate carboxylase, and 30 tRNAs (Figure 7–69). In addition, the DNA sequences present seem to encode at least 40 proteins whose functions are unknown. Paradoxically, all of the known proteins encoded in the chloroplast are part of larger protein complexes that also contain one or more subunits encoded in the nucleus. Possible reasons will be discussed later (see p. 400).

The similarities between the genomes of chloroplasts and bacteria are striking. The basic regulatory sequences, such as transcription promoters and terminators, are virtually identical in the two cases. Protein sequences encoded in chloroplasts are clearly recognizable as bacterial, and several clusters of genes with related functions (for example, those encoding ribosomal proteins) are organized in the same way in the genomes of chloroplasts, *E. coli*, and cyanobacteria.

Detailed comparisons of homologous nucleotide sequences will be required to trace the evolutionary pathway from bacteria to chloroplasts, but several conclusions can already be drawn: (1) Chloroplasts in higher plants arose from photosynthetic bacteria. (2) The chloroplast genome has been stably maintained for at least several hundred million years, the estimated time of divergence of liverwort and tobacco. (3) Many of the genes of the original bacterium can be identified in the nuclear genome, where they have been transferred and stably maintained. In higher plants, for example, two-thirds of the 60 or so chloroplast ribosomal proteins are encoded in the cell nucleus, although the genes have a clear bacterial ancestry, and the chloroplast ribosomes retain their original bacterial properties.

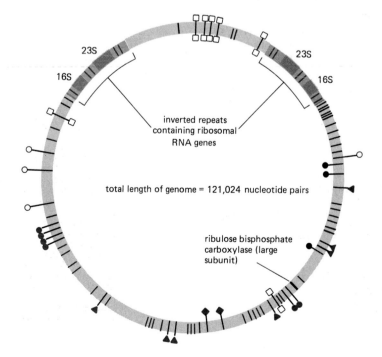

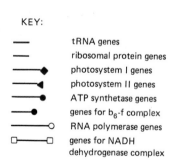

KEY:

— tRNA genes
— ribosomal protein genes
—◆ photosystem I genes
—◀ photosystem II genes
—● ATP synthetase genes
—● genes for b₆-f complex
—○ RNA polymerase genes
□—□ genes for NADH dehydrogenase complex

Figure 7–69 The organization of the liverwort chloroplast genome, whose complete nucleotide sequence has been determined. The organization of the chloroplast genome is very similar in all higher plants. However, the size of these circular DNA molecules varies from species to species depending on how much of the DNA surrounding the genes encoding the chloroplast's 16S and 23S ribosomal RNAs is present in two copies.

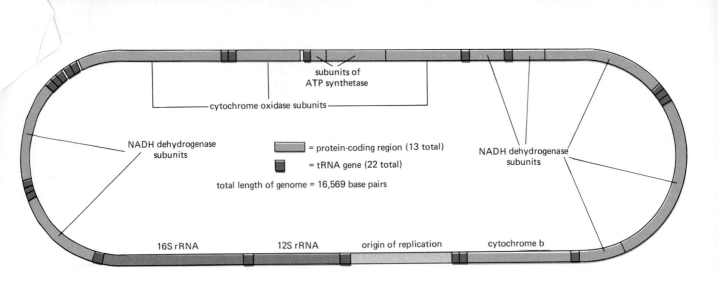

Figure 7–70 The organization of the human mitochondrial genome, based on determination of its complete nucleotide sequence. The genome contains 2 rRNA genes, 22 tRNA genes, and 13 protein-coding sequences. The DNAs of the bovine and mouse mitochondrial genomes have also been completely sequenced and have the same genes and gene organization.

7-34
Mitochondrial Genomes Have Several Surprising Features[50]

The chloroplast genome was not the first organelle genome to be completely sequenced. The relatively small size of the human mitochondrial genome made it a particularly attractive target for molecular geneticists equipped with newly devised DNA-sequencing techniques (see p. 185), and in 1981 the complete sequence of its 16,569 nucleotides was published. By comparing this sequence with known mitochondrial tRNA sequences and with the partial amino acid sequences available for proteins encoded by the mitochondrial DNA, it has been possible to locate all of the human mitochondrial genes on the circular DNA molecule (Figure 7–70).

Compared to nuclear, chloroplast, and bacterial genomes, the human mitochondrial genome has several surprising features: (1) Unlike other genomes, nearly every nucleotide appears to be part of a coding sequence, either for a protein or for one of the rRNAs or tRNAs. Since these coding sequences run directly into each other, there is very little room left for regulatory DNA sequences. (2) Whereas at least 31 tRNAs specify amino acids in the cytosol, and 30 in chloroplasts, only 22 tRNAs are required for mitochondrial protein synthesis. The normal codon-anticodon pairing rules are relaxed in mitochondria, so that many tRNA molecules recognize any one of the four nucleotides in the third (wobble) position (see p. 209). Such "2 out of 3" pairing allows one tRNA to pair with any one of four codons and permits protein synthesis with fewer tRNA molecules. (3) Perhaps most surprising, comparison of mitochondrial gene sequences and the amino acid sequences of the corresponding proteins indicates that the genetic code is altered, so that 4 of the 64 codons have "meanings" different from those they have in other genomes (Table 7–4).

The observation that the genetic code is nearly the same in all organisms provides strong evidence that all cells have evolved from a common ancestor. How,

Table 7–4 Some Differences Between the "Universal" Code and Mitochondrial Genetic Codes*

Codon	"Universal" Code	Mitochondrial Codes			
		Mammals	*Drosophila*	Yeasts	Plants
UGA	STOP	*Trp*	*Trp*	*Trp*	STOP
AUA	Ile	*Met*	*Met*	*Met*	Ile
CUA	Leu	Leu	Leu	*Thr*	Leu
AGA ⎤ AGG ⎦	Arg	*STOP*	*Ser*	Arg	Arg

*Italics and color shading indicate that the code differs from the "universal" code.

then, does one explain the few differences in the genetic code in mitochondria? A hint comes from the recent finding that the mitochondrial genetic code is different in different organisms. Thus UGA, which is a stop codon elsewhere, is read as tryptophan in mitochondria of mammals, fungi, and protozoans but as *stop* in plant mitochondria. Similarly, the codon AGG normally codes for arginine, but it codes for *stop* in the mitochondria of mammals and for serine in *Drosophila* (see Table 7–4). Such variation suggests that a random drift can occur in the genetic code in mitochondria. Presumably the unusually small number of proteins encoded by the mitochondrial genome makes an occasional change in the meaning of a rare codon tolerable, whereas such a change in a large genome would alter the function of many proteins and thereby destroy the cell.

Animal Mitochondria Contain the Simplest Genetic Systems Known[51]

Comparisons of DNA sequences in different organisms reveal that the rate of nucleotide substitution during evolution has been 10 times greater in mitochondrial genomes than in nuclear genomes (see p. 220), which is presumably due to a reduced fidelity of mitochondrial DNA replication, DNA repair, or both. Because only about 16,500 DNA nucleotides need to be replicated and expressed as RNAs and proteins in animal cell mitochondria, the error rate per nucleotide copied by DNA replication, maintained by DNA repair, transcribed by RNA polymerase, or translated into protein by mitochondrial ribosomes can be relatively high without adversely affecting the organelle. This is thought to explain why the mechanisms that carry out these processes are relatively simple compared to those used for the same purpose elsewhere in cells. For example, although this has not yet been tested adequately, one would expect that the presence of only 22 tRNAs and the unusually small size of the rRNAs (less than two-thirds the size of the *E. coli* rRNAs) would reduce the fidelity of protein synthesis in mitochondria.

The relatively high rate of evolution of mitochondrial genes makes mitochondrial DNA sequence comparisons especially useful for estimating the dates of relatively recent evolutionary events, such as the steps in primate development (see p. 13).

7-35 Why Are Plant Mitochondrial Genomes So Large?[52]

Mitochondrial genomes are much larger in plant than in animal cells and vary remarkably in their DNA content, ranging from about 150,000 to about 2.5×10^6 nucleotide pairs. Yet these genomes seem to encode only a few more proteins than do animal mitochondrial genomes. The paradox is compounded by the observation that in one family of plants, the cucurbits, mitochondrial genomes vary in size by as much as sevenfold. The green alga *Chlamydomonas* has a linear mitochondrial genome of only 16,000 nucleotide pairs, the same size as in animals.

Although very little sequence information is available for higher plant mitochondrial DNA molecules, almost all of the 78,000 nucleotide pairs in the large mitochondrial genome of the yeast *Saccharomyces cerevisiae* have been sequenced, and only about one-third of them code for protein. This finding raises the possibility that much of the extra DNA in yeast mitochondria, and possibly in plant mitochondria as well, is "junk DNA" of little consequence to the organism.

Some Organelle Genes Contain Introns[53]

The processing of precursor RNAs plays an important role in the two mitochondrial systems studied in most detail—human and yeast. In human cells both strands of the mitochondrial DNA are transcribed at the same rate from a single promoter region on each strand, producing two different giant RNA molecules, each containing a full-length copy of one DNA strand. Transcription is, therefore, completely symmetric. The transcripts made on one strand—called the *heavy strand (H strand)* because of its density in CsCl—are extensively processed by nuclease cleavage to yield the two rRNAs, most of the tRNAs, and about 10 poly-A-containing RNAs. In contrast, the *light strand (L strand)* transcript is processed to produce only eight tRNAs and one small poly-A-containing RNA; the remaining 90% of this transcript apparently contains no useful information (being complementary to coding sequences synthesized on the other strand) and is degraded. The poly-A-containing RNAs are the mitochondrial mRNAs: although they lack a cap structure at their 5' end, they carry a poly-A tail at their 3' end that is added post-transcriptionally by a mitochondrial poly-A polymerase.

Unlike human mitochondrial genes, some plant and fungal (including yeast) mitochondrial genes contain *introns*, which must be removed by RNA splicing (see p. 102). Introns have also been found in about 20 plant chloroplast genes. Many of the introns in organelle genes consist of related nucleotide sequences that are capable of splicing themselves out of the RNA transcripts by RNA-mediated catalysis (see p. 538), although these self-splicing reactions are generally aided by proteins. The presence of introns in organelle genes is surprising in view of the endosymbiont theory of the origin of the energy-converting organelles, since similar introns have not been found in the genes of the bacteria whose ancestors are thought to have given rise to mitochondria and plant chloroplasts.

In yeasts the same mitochondrial gene may have an intron in one strain but not in another. Such "optional introns" seem to be able to move in and out of genomes like transposable elements. On the other hand, introns in other yeast mitochondrial genes have been found in a corresponding position in the mitochondria of *Aspergillus* and *Neurospora*, implying that they were inherited from a common ancestor of these three fungi. It seems likely that the intron sequences themselves are of ancient origin and that, while they have been lost from many bacteria, they have been preferentially retained in those organelle genomes where RNA splicing is regulated to help control gene expression (see p. 397 and p. 603).

7-36 Mitochondrial Genes Can Be Distinguished from Nuclear Genes by Their Non-Mendelian (Cytoplasmic) Inheritance[54]

Most experiments on the mechanisms of mitochondrial biogenesis are performed with *Saccharomyces carlsbergensis* (brewer's yeast) and *Saccharomyces cerevisiae* (baker's yeast). There are several reasons for this. First, when grown on glucose, these yeasts have a unique ability to live by glycolysis alone and can therefore survive without functional mitochondria, which are required for oxidative phosphorylation. This makes it possible to grow cells with mutations in mitochondrial or nuclear DNA that drastically interfere with mitochondrial biogenesis; such mutations are lethal in nearly all other organisms. Second, yeasts are simple unicellular eucaryotes that are easy to grow and characterize biochemically. Finally, these yeast cells normally reproduce asexually by budding (asymmetrical mitosis), but they can also reproduce sexually. During sexual reproduction, two haploid cells mate and fuse to form a diploid zygote, which can either grow mitotically or divide by meiosis to produce new haploid cells. The ability to control the alternation between asexual and sexual reproduction in the laboratory (see p. 739) greatly facilitates genetic analyses. Because mutations in mitochondrial genes are not inherited according to the Mendelian rules that govern the inheritance of nuclear genes, genetic studies reveal which of the genes involved in mitochondrial function are located in the nucleus and which in the mitochondria.

An example of **non-Mendelian (cytoplasmic) inheritance** of mitochondrial genes in a haploid yeast cell is illustrated in Figure 7–71. The mutant gene makes mitochondrial protein synthesis resistant to chloramphenicol, so yeast cells that contain the mutant gene can be detected by their ability to grow in the presence of chloramphenicol on a substrate, such as glycerol, that cannot be used for glycolysis. With glycolysis blocked, ATP must be provided by functional mitochondria, and therefore only cells that carry chloramphenicol-resistant mitochondria will grow. When a chloramphenicol-resistant haploid cell mates with a chloramphenicol-sensitive wild-type haploid cell, the resulting diploid zygote will contain a mixture of mutant and wild-type mitochondria. But when the zygote undergoes mitosis to produce a diploid daughter by budding, only a limited number of mitochondria enter the bud. With continuing mitotic division, an occasional bud will receive all mutant or all wild-type mitochondria. Thereafter, all of the progeny

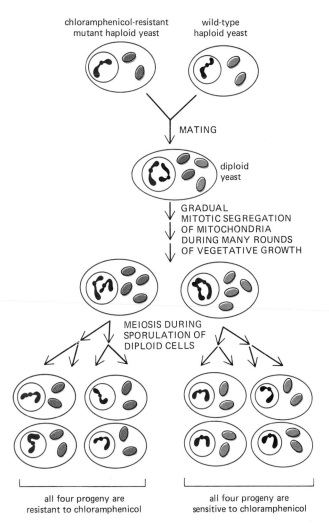

chloramphenicol-resistant
mutant haploid yeast

wild-type
haploid yeast

MATING

diploid
yeast

GRADUAL
MITOTIC SEGREGATION
OF MITOCHONDRIA
DURING MANY ROUNDS
OF VEGETATIVE GROWTH

MEIOSIS DURING
SPORULATION OF
DIPLOID CELLS

all four progeny are
resistant to chloramphenicol

all four progeny are
sensitive to chloramphenicol

Figure 7–71 The difference in the pattern of inheritance between mitochondrial and nuclear genes of yeast. For each nuclear gene, two of the four cells that result from meiosis inherit the gene from one of the original haploid parent cells and the remaining two cells inherit the gene from the other (*Mendelian inheritance*). In contrast, because of the gradual mitotic segregation of mitochondria during vegetative growth (see text), it is possible for all four of the cells that result from meiosis to inherit their mitochondrial genes from only one of the two original haploid cells (*non-Mendelian* or *cytoplasmic inheritance*). In this example the mitochondrial gene is one that can mutate to make protein synthesis in the mitochondrion resistant to chloramphenicol, a protein synthesis inhibitor that acts specifically on energy-converting organelles and bacteria (see p. 218).

from that bud will have mitochondria that are genetically identical. Eventually this random process, called *mitotic segregation*, produces diploid yeast progeny with only a single type of mitochondrial DNA. When such diploid cells undergo meiosis to form four haploid daughter cells, each of the four daughters receives the same mitochondrial genes. This type of inheritance is called *non-Mendelian* or *cytoplasmic* to contrast it with the Mendelian inheritance of nuclear genes (see Figure 7–71). When it occurs, it demonstrates that the gene in question is located outside the nuclear chromosomes and, therefore, probably in the yeast mitochondria.

Organelle Genes Are Maternally Inherited in Many Organisms[55]

The consequences of cytoplasmic inheritance are more profound for some organisms, including ourselves, than they are for yeasts. In yeasts, when two haploid cells mate, they are equal in size and contribute equal amounts of mitochondrial DNA to the zygote (see Figure 7–71). Mitochondrial inheritance in yeasts is therefore *biparental:* both parents contribute equally to the mitochondrial gene pool of the progeny (although, as we have just seen, after several generations of vegetative growth the *individual* progeny often contain mitochondria from only one parent). In higher animals, by contrast, the egg cell always contributes much more cytoplasm to the zygote than does the sperm, and in some animals the sperm may contribute no cytoplasm at all. One would expect mitochondrial inheritance in higher animals, therefore, to be *uniparental* (or more precisely, *maternal*). Such *maternal inheritance* has been demonstrated in laboratory animals using two strains that differ in their mitochondrial DNA. When animals carrying type A mitochondrial DNA are crossed with animals carrying type B, the progeny contain only the maternal type of mitochondrial DNA. Similarly, by following the distribution of variant mitochondrial DNA sequences in large families, human mitochondrial DNA has been shown to be maternally inherited.

In about two-thirds of higher plants, the chloroplasts from the male parent (contained in pollen grains) do not enter the zygote, so that chloroplast as well as mitochondrial DNA is maternally inherited. In other plants the pollen chloroplasts enter the zygote, making chloroplast inheritance biparental. In such plants, defective chloroplasts are a cause of *variegation:* a mixture of normal and defective chloroplasts in a zygote may sort out by mitotic segregation during plant growth and development, thereby producing alternating green and white patches in leaves; the green patches contain normal chloroplasts, while the white patches contain defective chloroplasts.

Petite Mutants in Yeasts Demonstrate the Overwhelming Importance of the Cell Nucleus in Mitochondrial Biogenesis[56]

Genetic studies of yeasts have played a crucial part in the analysis of mitochondrial biogenesis. A striking example is provided by studies of yeast mutants that contain large deletions in their mitochondrial DNA so that all mitochondrial protein synthesis is abolished. Not surprisingly, these mutants cannot make respiring mitochondria. A rare but important subclass of such mutants lacks mitochondrial DNA altogether. Because they form unusually small colonies when grown in media with low glucose, all mutants with such defective mitochondria are called *cytoplasmic petite mutants.*

Although petite mutants cannot synthesize proteins in their mitochondria and therefore cannot make mitochondria that produce ATP, they nevertheless contain mitochondria that have a normal outer membrane and an inner membrane with poorly developed cristae (Figure 7–72). These mitochondria contain virtually all of the mitochondrial proteins that are specified by nuclear genes and imported from the cytosol, including DNA and RNA polymerases, all of the citric acid cycle enzymes, and most inner membrane proteins. Such mutants dramatically demonstrate the overwhelming importance of the nucleus in mitochondrial biogenesis. They also show that an organelle that divides by fission can replicate indefinitely in the cytoplasm of proliferating eucaryotic cells even in the complete absence of its own genome. Many biologists believe that peroxisomes normally replicate in this way (see p. 433).

For chloroplasts the nearest equivalent to yeast mitochondrial petite mutants are mutants of unicellular algae such as *Euglena.* Cells in which no chloroplast protein synthesis occurs still contain chloroplasts and are perfectly viable if oxidizable substrates are provided. However, if the development of mature chloroplasts is blocked in plants, either by raising the plants in the dark (see p. 1162) or because chloroplast DNA is defective or absent, the plants die as soon as their food stores run out.

7-37 Nuclear Gene Products Regulate the Synthesis of Mitochondria and Chloroplasts[57]

Nuclear and organelle genetic systems must communicate in order to coordinate their contributions to the formation of the energy-converting organelles. Overall control clearly resides in the nucleus, inasmuch as mitochondria and chloroplasts are made in normal amounts, although not with normal functions, in mutants in which organelle protein synthesis is blocked. In some of these functionally defective organelles, DNA synthesis and some RNA synthesis can also continue normally, showing that the proteins required for these processes are all encoded by nuclear genes.

The nucleus must regulate the number of mitochondria and chloroplasts in a cell according to need; it must also control the amount of protein made on organelle ribosomes so that a proper balance is maintained between nuclear and organelle contributions. Although these regulatory aspects are crucial to our understanding of eucaryotic cells, we know relatively little about them.

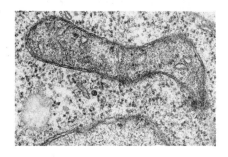

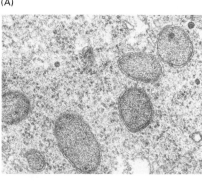

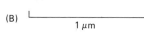
(A)

(B) 1 μm

Figure 7–72 Electron micrographs of yeast cells showing the structure of normal mitochondria (A) and mitochondria in a petite mutant in which all of the mitochondrion-encoded gene products are missing (B). In the latter case the organelle is constructed entirely from nucleus-encoded proteins. (Courtesy of Barbara Stevens.)

The nuclear regulation of mitochondrial protein synthesis has been analyzed extensively in yeast mutants. In *Saccharomyces cerevisiae*, large numbers of nuclear (as well as mitochondrial—see p. 394) gene mutants have been isolated that produce nonrespiring mitochondria. Each of these *nuclear petite mutants* has a defect in one nucleus-encoded protein that is required for mitochondrial function. The effect of each of these nuclear mutations on mitochondrial gene expression can be determined by labeling yeast cells with radioactive amino acids in the presence of cycloheximide so that only organelle-encoded proteins are made (see p. 388). Most of the mutations are found to have no effect on the synthesis of proteins in the mitochondrion, as expected for a mutant nuclear gene encoding a mitochondrial protein with a direct function in respiration, such as an ATP synthetase subunit or one of the citric acid cycle enzymes. Other mutations block the synthesis of all proteins in the mitochondrion, as expected for a mutant nuclear gene encoding a mitochondrial ribosomal protein or one of the subunits of the mitochondrial RNA polymerase.

It is a third class of yeast nuclear petite mutants, in which one or a few products of mitochondrial genes are absent or altered, that is most relevant to the regulatory process. More than 50 of these nuclear genes have been discovered, and several of them that are required for the expression of a single mitochondrial gene have been cloned and characterized. Some of these genes encode proteins that appear to act directly on a specific mitochondrial mRNA molecule to increase its stability or the efficiency with which it is used for protein synthesis in the mitochondrion. Others help catalyze mitochondrial RNA splicing and are therefore required for the expression of those yeast mitochondrial genes that contain introns. Both types of nuclear genes have been postulated to help regulate the function of mitochondrion-encoded proteins according to the cell's metabolic needs, although how the regulatory network operates is unknown.

Despite the dominance of the nucleus, there is evidence that the interactions between the nuclear and organelle genetic systems occur in both directions. When mitochondrial protein synthesis is blocked in intact cells, for example, some imported enzymes involved in mitochondrial DNA, RNA, and protein synthesis are overproduced, as though the cell were trying to overcome the block. The nature of the signal from the mitochondria to the nucleus remains to be determined.

The Energy-converting Organelles Contain Tissue-specific Proteins[58]

Mitochondrial function is also regulated by the cell in more conventional ways. The *urea cycle*, for example, is the central metabolic pathway in mammals for disposing of cellular breakdown products that contain nitrogen. These products are excreted in the urine as urea. Nuclear-encoded enzymes in the mitochondrial matrix carry out several steps in the cycle. Urea synthesis occurs in only a few tissues, such as the liver, and these enzymes are synthesized and imported into mitochondria only in these tissues. In addition, the respiratory enzyme complexes in the mitochondrial inner membrane of mammals contain several tissue-specific, nuclear-encoded subunits, which are thought to act as regulators of electron transport. Thus some humans with a genetic muscle disease have a defective subunit of cytochrome oxidase; since the subunit is specific to skeletal muscle cells, their heart muscle cells function normally, allowing the individuals to survive. As would be expected, tissue-specific differences are also found among the nuclear-encoded proteins in chloroplasts.

We shall now consider the general question of how specific cytosolic proteins are imported into mitochondria and chloroplasts, a subject that will be discussed in more detail in Chapter 8.

Proteins Are Imported into Mitochondria and Chloroplasts by an Energy-requiring Process[59]

The finding that most chloroplasts and mitochondrial proteins are imported from the cell cytosol (see p. 426) raises two related questions: how does the cell direct proteins to the appropriate organelle, and how do they enter the organelle?

The general answer was first provided by studies on the import of the small subunit (S) of the abundant enzyme *ribulose bisphosphate carboxylase* into the chloroplast stroma. When mRNA isolated from the cytoplasm of either the unicellular alga *Chlamydomonas* or pea leaves is translated into protein *in vitro*, one of the many proteins produced is the precursor of the S protein, called pro-S, which is larger than mature S by about 50 amino acids. When the completed pro-S protein is incubated with intact chloroplasts, it is taken up into the organelle and converted into mature S by an endopeptidase in the chloroplast. Mature S then associates with the large subunit of ribulose 1,5-bisphosphate carboxylase, which is made on chloroplast ribosomes, to form the active enzyme in the chloroplast stroma. As expected for a process of this type, the translocation of the pro-S protein into chloroplasts requires energy, and, as discussed in Chapter 8, the energy is provided by ATP hydrolysis (see p. 430).

The import of proteins into mitochondria is generally similar. If purified yeast mitochondria are incubated with cell extracts containing newly synthesized yeast proteins in a radioactive form, the nuclear-encoded mitochondrial proteins are specifically incorporated into the mitochondria in a manner that faithfully mimics their selective uptake within the cell. The outer membrane proteins, inner membrane proteins, matrix proteins, and proteins of the intermembrane space each find their way to their own special compartments within the mitochondrion (see Figure 8–30, p. 429).

The transport of proteins through the mitochondrial and chloroplast membranes seems to occur at special *contact sites* (also called contact zones) where the inner and outer membranes are joined (Figure 7–73), and it involves precursor proteins that contain a special *signal peptide*. The transported proteins must be unfolded in order to move into the organelle at these sites (see p. 428).

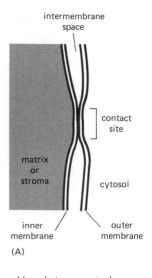

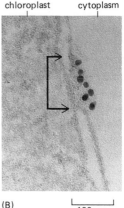

Figure 7–73 Contact sites. (A) Drawing of a small region of a mitochondrion or chloroplast containing a membrane contact site. These sites appear to be involved in selective protein import into these organelles. Contact sites (also called contact zones) have recently been isolated, and their special protein components are being identified. The proteins imported through these sites are those encoded by the cell nucleus and synthesized in the cytosol. (B) Electron micrograph of a small portion of a pea chloroplast, in which a contact site (*arrow*) has been labeled with a gold-conjugated antibody that is thought to localize an integral membrane protein involved in protein import. (B, from D. Pain, Y.S. Kanwar, and G. Blobel, *Nature* 331:232–237, 1988.)

Mitochondria Import Most of Their Lipids; Chloroplasts Make Most of Theirs[60]

The biosynthesis of new mitochondria and chloroplasts requires lipids in addition to nucleic acids and proteins. Chloroplasts tend to make the lipids they require. In spinach leaves, for example, all cellular fatty acid synthesis takes place in the chloroplast, although desaturation of the fatty acids occurs elsewhere. The major glycolipids of the chloroplast are also synthesized locally.

Mitochondria, on the other hand, import most of their lipids. In animal cells the phospholipids phosphatidylcholine and phosphatidylserine are synthesized in the endoplasmic reticulum and then transferred to the outer membrane of mitochondria. Although proof is lacking, the transfer reactions are thought to be mediated by phospholipid exchange proteins (see p. 450); the imported lipids then move into the inner membrane, presumably at contact sites. In addition to decarboxylating imported phosphatidylserine to phosphatidylethanolamine, the main reaction of lipid biosynthesis catalyzed by the mitochondria themselves is the conversion of imported lipids to cardiolipin (diphosphatidylglycerol). Cardiolipin is a "double" phospholipid that contains four fatty-acid tails; it is found mainly in the mitochondrial inner membrane, where it constitutes about 20% of the total lipid.

7-38 Both Mitochondria and Chloroplasts Probably Evolved from Endosymbiotic Bacteria[61]

As discussed in Chapter 1, the procaryotic character of the organelle genetic systems, especially striking in chloroplasts, suggests that mitochondria and chloroplasts evolved from bacteria that were at one time endocytosed. According to the **endosymbiont hypothesis,** eucaryotic cells started out as anaerobic organisms without mitochondria or chloroplasts and then established a stable endosymbiotic relation with a bacterium, whose oxidative phosphorylation system they subverted for their own use (Figure 7–74). The endocytic event that led to the development of mitochondria is presumed to have occurred when oxygen entered the atmosphere in substantial amounts, about 1.5×10^9 years ago, before animals and

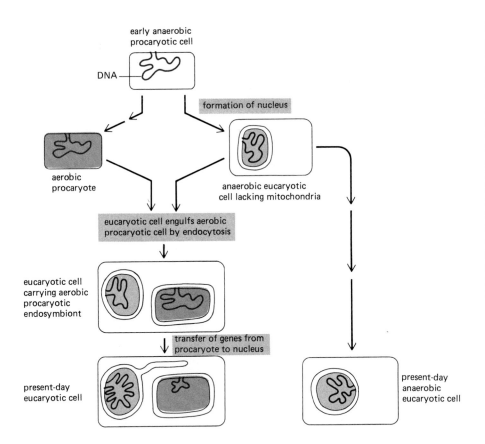

early anaerobic
procaryotic cell

DNA

formation of nucleus

aerobic
procaryote

anaerobic eucaryotic
cell lacking mitochondria

eucaryotic cell engulfs aerobic
procaryotic cell by endocytosis

eucaryotic cell
carrying aerobic
procaryotic
endosymbiont

transfer of genes from
procaryote to nucleus

present-day
eucaryotic cell

present-day
anaerobic
eucaryotic cell

Figure 7–74 A suggested evolutionary pathway for the origin of mitochondria (shown in color). Although a single origin for all mitochondria is sometimes postulated, separate endosymbiotic events may have led to the mitochondria found in such distantly related eucaryotes as trypanosomes and euglenoids (see Figure 1–16, p. 13). *Microsporidia* are present-day anaerobic single-celled eucaryotes (protozoa) without mitochondria that live in the gut of many animals. Because they have an rRNA sequence that suggests a great deal of evolutionary distance from all other known eucaryotes, it has been postulated that their ancestors were also anaerobic and resembled the eucaryote that first engulfed the precursors of mitochondria (see p. 18).

plants separated (see Figure 7–61). Plant and algal chloroplasts seem to have been derived later from an endocytic event involving an oxygen-evolving photosynthetic bacterium. In order to explain the different pigments and properties of the chloroplasts found in present-day higher plants and green algae, red algae, and brown algae (see Figure 7–62), it is usually assumed that at least three separate events of this kind occurred.

Since most of the genes encoding present-day mitochondrial and chloroplast proteins are in the cell nucleus, it seems that an extensive transfer of genes from organelle to nuclear DNA has occurred during eucaryote evolution. This would explain why some of the nuclear genes encoding mitochondrial proteins resemble bacterial genes: the amino acid sequence of the amino terminus of the chicken mitochondrial enzyme *superoxide dismutase*, for example, resembles the corresponding segment of the bacterial enzyme much more than it resembles the superoxide dismutase found in the cytosol of the same eucaryotic cells. Further evidence that such DNA transfers have occurred during evolution comes from the discovery of some noncoding DNA sequences in nuclear DNA that seem to be of recent mitochondrial origin; they have apparently integrated into the nuclear genome as "junk DNA."

What type of bacterium gave rise to the mitochondrion? Complete amino acid sequence and three-dimensional x-ray crystallographic analyses of the c-type cytochromes isolated from many types of bacteria provided an early clue by showing that these proteins are all closely related to one another and to the cytochrome c of animal and plant mitochondrial respiratory chains. These findings, together with more recent nucleotide-sequence analyses, provide the main evidence for the evolutionary tree shown previously in Figure 7–62. It appears that mitochondria are descendants of a particular type of purple photosynthetic bacterium, which had previously lost its ability to carry out photosynthesis and was left with only a respiratory chain. However, as for chloroplasts, it is not clear that all mitochondria have originated from a single endosymbiotic event. While the mitochondria from protozoans have distinctly procaryotic features, for example, some of them are sufficiently different from plant and animal mitochondria to suggest a separate origin.

Why Do Mitochondria and Chloroplasts Have Their Own Genetic Systems?[62]

Why do mitochondria and chloroplasts require their own separate genetic systems, when other organelles—such as peroxisomes and lysosomes—do not? The question is not trivial, because maintaining a separate genetic system is costly: more than 90 proteins—including many ribosomal proteins, aminoacyl-tRNA synthetases, DNA and RNA polymerases, and RNA processing and modifying enzymes—must be encoded by nuclear genes specifically for this purpose (Figure 7–75). The amino acid sequences of most of these proteins in mitochondria and chloroplasts differ from those of their counterparts in the nucleus and cytosol, and there is reason to think that these organelles have relatively few proteins in common with the rest of the cell. This means that the nucleus must provide at least 90 genes just to maintain each organelle genetic system. The reason for such a costly arrangement is not clear, and the hope that the nucleotide sequences of mitochondrial and chloroplast genomes would provide the answer has proved unfounded. We cannot think of compelling reasons why the proteins made in mitochondria and chloroplasts should be made there rather than in the cytosol.

At one time it was suggested that some proteins have to be made in the organelle because they are too hydrophobic to get to their site in the membrane from the cytosol. More recent studies, however, make this explanation implausible. In many cases even highly hydrophobic subunits are synthesized in the cytosol. Moreover, although the individual protein subunits in the various mitochondrial

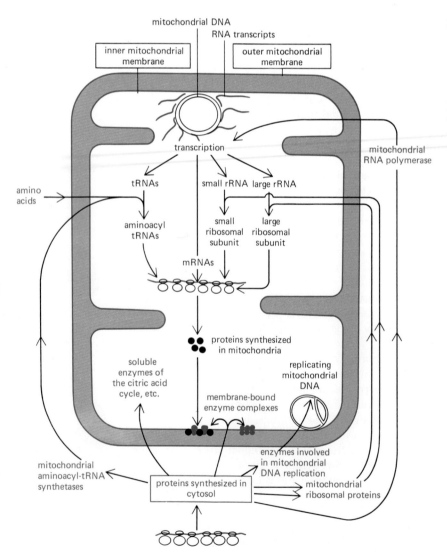

Figure 7–75 The proteins synthesized in the cytosol and then imported into the mitochondrion play a major part in creating the genetic system of the mitochondrion in addition to contributing most of the organelle protein. The mitochondrion itself contributes only mRNAs, rRNAs, and tRNAs to its genetic system.

enzyme complexes are highly conserved in evolution, their site of synthesis is not. The diversity in the location of the genes coding for the subunits of functionally equivalent proteins in different organisms is difficult to explain by any hypothesis that postulates a specific evolutionary advantage of present-day mitochondrial or chloroplast genetic systems.

Perhaps the organelle genetic systems are an evolutionary dead end. In terms of the endosymbiont hypothesis, this would mean that the process whereby the endosymbionts transferred most of their genes to the nucleus stopped before it was complete, perhaps becoming frozen—in the case of mitochondria—by recent alterations in the mitochondrial genetic code. Such alterations would probably make the remaining mitochondrial genes nonfunctional if they were transferred to the nucleus.

Summary

Mitochondria and chloroplasts grow and divide in two in a coordinated process that requires the contribution of two separate genetic systems—that of the organelle and that of the cell nucleus. Most of the proteins in these organelles are encoded by nuclear DNA, synthesized in the cytosol, and then imported individually into the organelle. However, some organelle proteins—along with their RNAs—are encoded by the organelle DNA and are synthesized in the organelle itself. The human mitochondrial genome contains about 16,500 nucleotides and encodes 2 ribosomal RNAs, 22 transfer RNAs, and 13 different polypeptide chains. Chloroplast genomes are about 10 times larger and contain about 120 genes. But partially functional organelles will form in normal numbers even in mutants that lack a functional organelle genome, demonstrating the overwhelming importance of the nucleus for the biogenesis of both organelles.

The ribosomes of chloroplasts closely resemble bacterial ribosomes, while mitochondrial ribosomes show both similarities and differences that make their origin more difficult to trace. Protein similarities, however, suggest that both organelles originated when a primitive eucaryotic cell entered into a stable endosymbiotic relationship with a bacterium: a purple bacterium is thought to have given rise to the mitochondrion, and (later) a relative of a cyanobacterium is thought to have given rise to the plant chloroplast. Although many of the genes of these ancient bacteria still function to make organelle proteins, most of them have become integrated into the nuclear genome, where they encode bacterial-like enzymes that are synthesized on cytosolic ribosomes and then imported into the organelle.

References

General

Becker, W.M. The World of the Cell, pp. 117–284, Menlo Park, CA: Benjamin-Cummings, 1986.

Ernster, L., ed. Bioenergetics. New York: Elsevier, 1984.

Harold, F.M. The Vital Force: A Study of Bioenergetics. New York: W.H. Freeman, 1986.

Lehninger, A.L. Principles of Biochemistry, Chapters 16, 17, 23. New York: Worth, 1982.

Nicholls, D.G. Bioenergetics: An Introduction to the Chemiosmotic Theory. New York: Academic Press, 1982.

Stryer, L. Biochemistry, 3rd ed., Chapters 16, 17, and 22. New York: W.H. Freeman, 1988.

Cited

1. Ernster, L.; Schatz, G. Mitochondria: a historical review. *J. Cell Biol.* 91:227s–255s, 1981.

    Fawcett, D.W. The Cell, 2nd ed., pp. 410–485. Philadelphia: Saunders, 1981.

Harold, F.M. The Vital Force: A Study of Bioenergetics, Chapter 7. New York: W.H. Freeman, 1986.

Tzagoloff, A. Mitochondria. New York: Plenum, 1982.

2. DePierre, J.W.; Ernster, L. Enzyme topology of intracellular membranes. *Annu. Rev. Biochem.* 46:201–262, 1977.

    Srere, P.A. The structure of the mitochondrial inner membrane-matrix compartment. *Trends Biochem. Sci.* 7:375–378, 1982.

3. Krstic, R.V. Ultrastructure of the Mammalian Cell, pp. 28–57. New York: Springer-Verlag, 1979.

    Pollak, J.K.; Sutton, R. The differentiation of animal mitochondria during development. *Trends Biochem. Sci.* 5:23–27, 1980.

4. Geddes, R. Glycogen: a metabolic viewpoint. *Biosci. Rep.* 6:415–428, 1986.

    McGilvery, R.W. Biochemistry: A Functional Approach, 3rd ed. Philadelphia: Saunders, 1983.

    Newsholme, E.A.; Start, C. Regulation in Metabolism. New York: Wiley, 1973.

5. Baldwin, J.E.; Krebs, H. The evolution of metabolic cycles. *Nature* 291:381–382, 1981.

Krebs, H.A. The history of the tricarboxylic acid cycle. *Perspect. Biol. Med.* 14:154–170, 1970.

Reed, I.J.; Damuni, Z.; Merryfield, M.L. Regulation of mammalian pyruvate and branched-chain α-keto acid dehydrogenase complexes by phosphorylation-dephosphorylation. *Curr. Top. Cell Regul.* 27:41–49, 1985.

Willamson, J.R.; H. Cooper, R.H. Regulation of the citric acid cycle in mammalian systems. *FEBS Lett.*, Suppl. 117:K73–K85, 1980.

6. Mitchell, P. Coupling of phosphorylation to electron and hydrogen transfer by a chemi-osmotic type of mechanism. *Nature* 191:144–148, 1961.

Racker, E. From Pasteur to Mitchell: a hundred years of bioenergetics. *Fed. Proc.* 39:210–215, 1980.

7. Hatefi, Y. The mitochondrial electron transport and oxidative phosphorylation system. *Annu. Rev. Biochem.* 54:1015–1070, 1985.

8. Nicholls, D.G. Bioenergetics: An Introduction to the Chemiosmotic Theory, Chapter 3. New York: Academic Press, 1982.

Wood, W.B.; Wilson, J.H.; Benbow, R.M.; Hood, L.E. Biochemistry: A Problems Approach, 2nd ed. Menlo Park, CA: Benjamin-Cummings, 1981. (See problems, Chapters 9, 12, and 14).

9. Al-Awqati, Q. Proton-translocating ATPases. *Annu. Rev. Cell Biol.* 2:179–199, 1986.

Hinkle, P.C.; McCarty, R.E. How cells make ATP. *Sci. Am.* 238(3):104–123, 1978.

10. Durand, R.; Briand, Y.; Touraille, S.; Alziari, S. Molecular approaches to phosphate transport in mitochondria. *Trends Biochem. Sci.* 6:211–214, 1981.

Klingenberg, M. The ADP, ATP shuttle of the mitochondrion. *Trends Biochem. Sci.* 4:249–252, 1979.

LaNoue, K.F.; Schoolwerth, A.C. Metabolite transport in mitochondria. *Annu. Rev. Biochem.* 48:871–922, 1979.

11. Eisenberg, D.; Crothers, D. Physical Chemistry with Applications to the Life Sciences, Chapters 4 and 5. Menlo Park, CA: Benjamin-Cummings, 1979.

12. Hatefi, Y. The mitochondrial electron transport and oxidative phosphorylation system. *Annu. Rev. Biochem.* 54:1015–1069, 1985.

Wikstrom, M.; Saraste, M. The mitochondrial respiratory chain. In Bioenergetics (L. Ernster, ed.), pp. 49–94. New York: Elsevier, 1984.

13. Racker, E. A New Look at Mechanisms in Bioenergetics. New York: Academic Press, 1976. (A personal account of the concepts and history.)

14. Amzel, L.M.; McKinney, M.; Narayanan, P.; Pedersen, P.L. Structure of the mitochondrial F$_1$ ATPase at 9-Å resolution. *Proc. Natl. Acad. Sci. USA* 79:5852–5856, 1982.

Futai, M.; Kanazawa, H. Structure and function of proton-translocating adenosine triphosphatase (F$_o$F$_1$): biochemical and molecular biological approaches. *Microbiol. Rev.* 47:285–312, 1983.

Racker, E.; Stoeckenius, W. Reconstitution of purple membrane vesicles catalyzing light-driven proton uptake and adenosine triphosphate formation. *J. Biol. Chem.* 249:662–663, 1974.

Schneider, E.; Altendorf, K. The proton-translocating portion (F$_o$) of the *E. coli* ATP synthase. *Trends Biochem. Sci.* 9:51–53, 1984.

15. Hammes, G.G. Mechanism of ATP synthesis and coupled proton transport: studies with purified chloroplast coupling factor. *Trends Biochem. Sci.* 8:131–134, 1983.

Ogawa, S.; Lee, T.M. The relation between the internal phosphorylation potential and the proton motive force in mitochondria during ATP synthesis and hydrolysis. *J. Biol. Chem.* 259:10004–10011, 1984.

Pederson, P.L.; Carafoli, E. Ion motive ATPases. II. Energy coupling and work output. *Trends Biochem. Sci* 12:186–189, 1987.

Senior, A.E. ATP synthesis by oxidative phosphorylation. *Physiol. Rev.* 68:177–231, 1988.

16. Fillingame, R.H. The proton-translocating pumps of oxidative phosphorylation. *Annu. Rev. Biochem.* 49:1079–1113, 1980.

17. Chance, B.; Williams, G.R. A method for the localization of sites for oxidative phosphorylation, *Nature* 176:250–254, 1955.

Dickerson, R.E. The structure and history of an ancient protein. *Sci. Am.* 226(4):58–72, 1972. (The conformation and evolution of cytochrome *c*.)

Keilin, D. The History of Cell Respiration and Cytochromes. Cambridge, U.K.: Cambridge University Press, 1966.

Spiro, T.G., ed. Iron-Sulfur Proteins. New York: Wiley-Interscience, 1982.

18. Capaldi, R.A.; Darley-Usmar, V.; Fuller, S.; Millet, F. Structural and functional features of the interaction of cytochrome *c* with complex III and cytochrome *c* oxidase. *FEBS Lett.* 138:1–7, 1982.

Casey, R.P. Membrane reconstitution of the energy-conserving enzymes of oxidative phosphorylation. *Biochim. Biophys. Acta* 768:319–347, 1984.

Leonard, K.; Haiker, H.; Weiss, H. Three-dimensional structure of NADH: ubiquinone reductase (complex I) from *Neurospora* mitochondria determined by electron microscopy of membrane crystals. *J. Mol. Biol.* 194:277–286, 1987.

Weiss, H.; Linke, P.; Haiker, H.; Leonard, K. Structure and function of the mitochondrial ubiquinol: cytochrome *c* reductase and NADH: ubiquinone reductase. *Biochem. Soc. Trans.* 15:100–102, 1987.

19. Hackenbrock, C.R. Lateral diffusion and electron transfer in the mitochondrial inner membrane. *Trends Biochem. Sci.* 6:151–154, 1981.

20. Dutton, P.L; Wilson, D.F. Redox potentiometry in mitochondrial and photosynthetic bioenergetics. *Biochim. Biophys. Acta* 346:165–212, 1974.

Hamamoto, T.; Carrasco, N.; Matsushita, K.; Kabak, H.R.; Montal, M. Direct measurement of the electrogenic activity of O-type cytochrome oxidase from *E. coli* reconstituted into planar lipid bilayers. *Proc. Natl. Acad. Sci. USA* 82:2570–2573, 1985.

Lehninger, A.L. Bioenergetics: The Molecular Basis of Biological Energy Transformations, 2nd ed. Menlo Park, CA: Benjamin-Cummings, 1971.

21. Prince, R.C. The proton pump of cytochrome oxidose. *Trends Biochem. Sci.* 13:159–160, 1988.

Slater, E.C. The Q Cycle, an ubiquitous mechanism of electron transfer. *Trends Biochem. Sci.* 8:239–242, 1983.

22. Hanstein, W.G. Uncoupling of oxidative phosphorylation. *Trends Biochem. Sci.* 1:65–67, 1976.

23. Brand, M.D.; Murphy M.P. Control of electron flux through the respiratory chain in michondria and cells. *Biol. Rev. Cambridge Philsophic Soc.* 62:141–193, 1987.

Erecinska, A.; Wilson, D.F. Regulation of cellular energy metabolism. *J. Membr. Biol.* 70:1–14, 1982.

Racker, E. A New Look at Mechanisms in Bioenergetics. New York: Academic Press, 1976.

24. Klingenberg, M. Principles of carrier catalysis elucidated by comparing two similar membrane translocators from mitochondria, the ADP/ATP carrier and the uncoupling protein. *Ann. N.Y. Acad. Sci.* 456:279–288, 1985.

Nicholls, D.G.; Rial, E. Brown fat mitochondria. *Trends Biochem. Sci.* 9:489–491, 1984.

25. Gottschalk, G. Bacterial Metabolism, 2nd ed. New York: Springer-Verlag, 1986.

MacNab, R.M. The bacterial flagellar motor. *Trends Biochem. Sci.* 9:185–189, 1984.

Neidhardt, F.C.; et al, eds. *Escherichia coli* and *Salmonella typhimurium*: Cellular and Molecular Biology. Washington, DC: American Society for Microbiology, 1987.

Skulachev, V.P. Sodium bioenergetics. *Trends Biochem. Sci.* 9:483–485, 1984.

Thauer, R.; Jungermann, K.; Decker, K. Energy conservation in chemotrophic anaerobic bacteria. *Bacteriol. Rev.* 41:100–180, 1977.

26. Bogorad, L. Chloroplasts. *J. Cell Biol.* 91:256s–270s, 1981. (A historical review.)

Clayton, R.K. Photosynthesis: Physical Mechanisms and Chemical Patterns. Cambridge, U.K.: Cambridge University Press, 1980 (Excellent general treatment.)

Haliwell, B. Chloroplast Metabolism—The Structure and Function of Chloroplasts in Green Leaf Cells. Oxford, U.K.: Clarendon, 1981.

Hoober, J.K. Chloroplasts. New York: Plenum, 1984.

27. Cramer, W.A.; Widger, W.R.; Herrmann, R.G.; Trebst, A. Topography and function of thylakoid membrane proteins. *Trends Biochem. Sci.* 10:125–129, 1985.

Miller, K.R. The photosynthetic membrane. *Sci. Am.* 241(4)102–113, 1979.

28. Akazawa, T.; Takabe, T.; Kobayashi, H. Molecular evolution of ribulose-1,5-bisphosphate carboxylase/oxygenase (RuBisCO). *Trends Biochem. Sci.* 9:380–383, 1984.

Barber, J. Structure of key enzyme refined. *Nature* 325:663–664, 1987.

Lorimer, G.H. The carboxylation and oxygenation of ribulose-1, 5-bisphosphate: the primary events in photosynthesis and photorespiration. *Annu. Rev. Plant Physiol.* 32:349–383, 1981.

29. Bassham, J.A. The path of carbon in photosynthesis. *Sci. Am.* 206(6):88–100, 1962.

Preiss, J. Starch, sucrose biosynthesis and the partition of carbon in plants are regulated by orthophosphate and triose-phosphates. *Trends Biochem. Sci.* 9:24–27, 1984.

30. Bjorkman, O.; Berry J. High-efficiency photosynthesis. *Sci. Am.* 229(4):80–93, 1973. (C⁴ plants.)

Chollet, R. The biochemistry of photorespiration. *Trends Biochem. Sci.* 2:155–159, 1977.

Edwards, G.; Walker, D. C³, C⁴ Mechanisms, and Cellular and Environmental Regulation of Photosynthesis. Berkeley: University of California Press, 1983.

Heber, U.; Krause, G.H. What is the physiological role of photorespiration? *Trends Biochem. Sci.* 5:32–34, 1980.

31. Clayton, R.K. Photosynthesis: Physical Mechanisms and Chemical Patterns. Cambridge, U.K.: Cambridge University Press, 1980.

Parson, W.W. Photosynthesis and other reactions involving light. In Biochemistry (G. Zubay, ed.), 2nd ed., pp. 564–597. New York: Macmillan, 1988.

32. Barber, J. Photosynthethic reaction centres: a common link. *Trends Biochem. Sci.* 12:321–326, 1987.

Govindjee; Govindjee, R. The absorption of light in photosynthesis. *Sci. Am.* 231(6):68–82, 1974.

Li, J. Light-harvesting chlorophyll *a/b*-protein: three-dimensional structure of a reconstituted membrane lattice in negative stain. *Proc. Natl. Acad. Sci. USA* 82:386–390, 1985.

Zuber, H. Structure of light-harvesting antenna complexes of photosynthetic bacteria, cyanobacteria, and red algae. *Trends Biochem. Sci.* 11:414–419, 1986.

33. Deisenhofer, J. Epp, O.; Miki, K.; Huber, R.; Michel, H. Structure of the protein subunits in the photosynthetic reaction centre of *Rhodopseudomonas virdis* at 3Å resolution *Nature* 318:618–624, 1985.

Deisenhofer, J.; Michel, H.; Huber, R. The structural basis of photosynthetic light reactions in bacteria. *Trends Biochem. Sci.* 10:243–248, 1985.

Knaff, D.B. Reaction centers of photsynthetic bacteria. *Trends Biochem. Sci.* 13:157–158, 1988.

Michel, H.; Epp, O.; Deisenhofer, J. Pigment-protein interactions in the photosynthetic reaction center from *Rhodopseudomonas viridis*. *EMBO J.* 5:2445–2451, 1986. (Three-dimensional structure by x-ray diffraction.)

34. Govindjee, ed. Photosynthesis: Energy Conversion by Plants and Bacteria. New York: Academic Press, 1982. (Two volumes of review articles at an advanced level.)

Nugent, J.H.A. Photosynthetic electron transport in plants and bacteria. *Trends Biochem Sci.* 9:354–357, 1984.

Scolnik, P.A.; Marrs, B.L. Genetic research with photosynthetic bacteria. *Annu. Rev. Microbiol.* 41:703–726, 1987.

35. Blankenship, R.E.; Prince, R.C. Excited-state redox potentials and the Z scheme of photosynthesis. *Trends Biochem. Sci.* 10:382–383, 1985.

Prince, R.C. Manganese at the active site of the chloroplast oxygen-evolving complex. *Trends Biochem. Sci.* 11:491–492, 1986.

36. Anderson, J.M. Photoregulation of the composition, function, and structure of thylakoid membranes. *Annu. Rev. Plant Physiol.* 37:93–136, 1986.

Carrillo, N., Vallejos, R.H. The light-dependent modulation of photosynthetic electron transport. *Trends Biochem. Sci.* 8:52–56, 1983.

Miller, K.R.; Lyon, M.K. Do we really know why chloroplast membranes stack? *Trends Biochem. Sci.* 10:219–222, 1985.

37. Hinkle, P.C.; McCarty, R.E. How cells make ATP. *Sci. Am.* 238(3):104–123, 1978.

Jagendorf, A.T. Acid-base transitions and phosphorylation by chloroplasts. *Fed. Proc.* 26:1361–1369, 1967.

38. Flügge, U.I.; Heldt, H.W. The phosphate-triose phosphate-phosphoglycerate translocator of the chloroplast. *Trends Biochem. Sci.* 9:530–533, 1984.

Heber, U.; Heldt, H.W. The chloroplast envelope: structure, function and role in leaf metabolism. *Annu. Rev. Plant Physiol.* 32:139–168, 1981.

39. Raven, J.A. Division of labor between chloroplast and cytoplasm. In The Intact Chloroplast (J. Barber, ed.), pp. 403–443. Amsterdam: Elsevier, 1976.

40. Wilson, T.H.; Lin. E.C.C. Evolution of membrane bioenergetics. *J. Supramol. Struct.* 13:421–446, 1980.

Woese, C.R. Bacterial Evolution *Microbiol. Rev.* 51:221–271, 1987.

41. Gest, H. The evolution of biological energy-transducing systems. *FEMS Microbiol. Lett.* 7:73–77, 1980.

Gottschalk, G. Bacterial Metabolism, 2nd ed. New York: Springer-Verlag, 1986. (Chapter 8 covers fermentations.)

Miller, S.M.; Orgel, L.E. The Origins of Life on the Earth. Englewood Cliffs, NJ: Prentice-Hall, 1974.

42. Danson, M.J. Archaebacteria: the comparative enzymology of their central metabolic pathways. *Adv. Microb. Physiol.* 29:165–231, 1988.

Knowles, C.J., ed. Diversity of Bacterial Respiratory Systems, Vol. 1. Boca Raton, FL: CRC Press, 1980.

43. Clayton, R.K.; Sistrom, W.R., eds. The Photosynthetic Bacteria. New York: Plenum, 1978.

Deamer, D.W., ed. Light Transducing Membranes: Structure, Function and Evolution. New York: Academic Press, 1978.

Gromet-Elhanan, Z. Electrochemical gradients and energy coupling in photosynthetic bacteria. *Trends Biochem. Sci.* 2:274–277, 1977.

Olson, J.M.; Pierson, B.K. Evolution of reaction centers in photosynthetic prokaryotes. *Int. Rev. Cytol.* 108:209–248, 1987.

44. Dickerson, R.E. Cytochrome *c* and the evolution of energy metabolism. *Sci. Am.* 242(3):136–153, 1980.

Gabellini, N. Organization and structure of the genes for the cytochrome *b/c* complex in purple photosynthetic bacteria. A phylogenetic study describing the homology of the b/c1 subunits between prokaryotes, miotochondria, and chloroplasts. *J. Bioenerg. Biomembr.* 20:59–83, 1988.

Schopf, J.W.; Hayes, J.M.; Walter, M.R. Evolution of earth's earliest ecosystems: recent progress and unsolved problems. In Earth's Earliest Biosphere: Its Origin and Evolution (J.W. Schopf, ed.), pp. 361–384. Princeton, NJ: Princeton University Press, 1983.

45. Attardi, G.; Schatz, G. Biogenesis of mitochondria. *Annu. Rev. Cell Biol.* 4:289–333, 1988.

Ellis, R.J., ed. Chloroplast biogenesis. Cambridge, U.K.: Cambridge University Press, 1984.

46. Clayton, D.A. Replication of animal mitochondrial DNA. *Cell* 28:693–705, 1982.

Posakony, J.W.; England, J.M.; Attardi, G. Mitochondrial growth and division during the cell cycle in HeLa cells. *J. Cell Biol.* 74:468–491, 1977.

47. Attardi, G.; Borst, P.; Slonimski, P.P. Mitochondrial Genes. Cold Spring Harbor, NY: Cold Spring Harbor Laboratory, 1982.

Borst, P.; Grivell, L.A.; Groot, G.S.P. Organelle DNA. *Trends Biochem. Sci.* 9:128–130, 1984.

Palmer, J.D. Comparative organization of chloroplast genomes. *Annu. Rev. Genet.* 19:325–354, 1985.

48. Grivell, L.A. Mitochondrial DNA. *Sci. Am.* 248(3):60–73, 1983.

Hoober, J.K. Chloroplasts, New York: Plenum, 1984.

49. Ohyama, K.; et al. Chloroplast gene organization deduced from complete sequence of liverwort. *Marchantia polymorpha* chloroplast DNA. *Nature* 322:572–574, 1986.

Rochaix, J.D. Molecular genetics of chloroplasts and mitochondria in the unicellular green alga. *Chlamydomonas*. *FEMS Microbiol. Rev.* 46:13–34, 1987.

Shinozaki, K.; et al. The complete nucleotide sequence of the tobacco chloroplast genome: its gene organization and expression. *EMBO J.* 5:2034–2049, 1986.

Umesono, K.; Ozeki, H. Chloroplast gene organization in plants. *Trends Genet.* 3:281–287, 1987.

50. Anderson, S.; et al. Sequence and organization of the human mitochondrial genome. *Nature* 290:457–465, 1981.

Bibb, M.J.; Van Etten, R.A.; Wright, C.T.; Walberg, M.W.; Clayton, D.A. Sequence and gene organization of mouse mitochondrial DNA. *Cell* 26:167–180, 1981.

Breitenberger, C.A.; RajBhandary, U.L. Some highlights of mitochondrial research based on analysis of *Neurospora crassa* mitchondrial DNA. *Trends Biochem. Sci.* 10:478–482, 1985.

Fox, T.D. Natural variation in the genetic code. *Annu. Rev. Genet.* 21:67–91, 1987.

51. Attardi, G. Animal mitochondrial DNA: an extreme example of genetic economy. *Int. Rev. Cytol.* 93:93–145, 1985.

Wilson, A. The molecular basis of evolution. *Sci. Am.* 253(4):164–173, 1985.

52. Levings, C.S. The plant mitochondrial genome and its mutants. *Cell* 32:659–661, 1983.

Mulligan, R.M.; Walbot, V. Gene expression and recombination in plant mitochondrial genomes. *Trends Genet.* 2:263–266, 1986.

Newton, K.J. Plant mitochondrial genomes: organization, expression and variation. *Annu. Rev. Plant Physiol. Plant Mol. Biol.* 39:503–532, 1988.

53. Clayton, D.A. Transcription of the mammalian mitochondrial genome. *Annu. Rev. Biochem.* 53:573–594, 1984.

Gruissem, W.; Barken, A.; Deng, S.; Stern, D. Transcriptional and post-transcriptional control of plastid mRNA in higher plants. *Trends Genet.* 4:258–262, 1988.

Mullet, J.E. Chloroplast development and gene expression. *Annu. Rev. Plant Physiol. Plant Mol. Biol.* 39:475–502, 1988.

Tabak, H.F.; Grivell, L.A. RNA catalysis in the excision of yeast mitochondrial introns. *Trends Genet.* 2:51–55, 1986.

54. Birky, C.W., Jr. Transmission genetics of mitochondria and chloroplasts. *Annu. Rev. Genet.* 12:471–512, 1978.

55. Giles, R.E.; Blanc, H.; Cann, H.M.; Wallace, D.C. Maternal inheritance of human mitochondrial DNA. *Proc. Natl. Acad. Sci. USA* 77:6715–6719, 1980.

56. Bernardi, G. The petite mutation in yeast. *Trends Biochem. Sci.* 4:197–201, 1979.

Locker, J.; Lewin, A.; Rabinowitz, M. The structure and organization of mitochondrial DNA from petite yeast. *Plasmid* 2:155–181, 1979.

Montisano, D.F.; James, T.W. Mitochondrial morphology in yeast with and without mitochondrial DNA. *J. Ultrastruct. Res.* 67:288–296, 1979.

57. Attardi, G.; Schatz, G. Biogenesis of mitochondria. *Annu. Rev. Cell Biol.* 4:289–333, 1988.

Fox, T.D. Nuclear gene products required for translation of specific mitochondrially coded mRNAs in yeast. *Trends Genet.* 2:97–99, 1986.

Tzagoloff, A.; Myers, A.M. Genetics of mitochondrial biogenesis. *Annu. Rev. Biochem.* 55:249–285, 1986.

58. Capaldi, R.A. Mitochondrial myopathies and respiratory chain proteins. *Trends Biochem. Sci.* 13:144–148, 1988.

DiMauro, S.; et al. Mitochondrial myopathies. *J. Inherited Meta. Dis.* 10(Suppl. 1):113–28, 1987.

59. Chua, N.-H.; Schmidt, G.W. Post-translational transport into intact chloroplasts of a precursor to the small subunit of ribulose-1,5-bisphosphate carboxylase. *Proc. Natl. Acad. Sci. USA* 75:6110–6114, 1978.

Douglas, M.G.; McCammon, M.T.; Vassarotti A. Targeting proteins into mitochondria. *Microbiol. Rev.* 50:166–178, 1986.

Keegstra, K.; Bauerle, C. Targeting of proteins into chloroplasts. *Bioessays* 9:15–19, 1988.

Schmidt, G.W.; Mishkind, M.L. The transport of proteins into chloroplasts. *Annu. Rev. Biochem.* 55:879–912, 1986.

60. Bishop, W.R.; Bell, R.M. Assembly of phospholipids into cellular membranes: biosynthesis, transmembrane movement, and intracellular translocation. *Annu. Rev. Cell Biol.* 4:579–611, 1988.

61. Cavalier-Smith, T. The origin of eukaryotic and archaebacterial cells. *Ann. N.Y. Acad. Sci.* 503:17–54, 1987.

Butow, B.A.; Doeherty, R.; Parikh, V.S. A path from mitochondria to the yeast nucleus. *Philos. Trans. R. Soc. Lond. (Biol.)* 319:127–133, 1988.

Gellissen, G.; Michaelis, G. Gene transfer: Mitochondria to nucleus. *Ann. N.Y. Acad. Sci.* 503:391–401, 1987.

Margulis, I. Symbiosis in Cell Evolution. New York: W.H. Freeman, 1981.

Schwartz, R.M.; Dayhoff, M.O. Origins of prokaryotes, eukaryotes, mitochondria, and chloroplasts. *Science* 199:395–403, 1978.

Whatley, J.M.; John, P.; Whatley, F.R. From extracellular to intracellular: the establishment of mitochondria and chloroplasts. *Proc. R. Soc. Lond. (Biol.)* 204:165–187, 1979.

62. von Heijne, G. Why mitochondria need a genome. *FEBS Lett.* 198:1–4, 1986

Intracellular Sorting and the Maintenance of Cellular Compartments

8

Unlike a bacterium, which generally consists of a single compartment surrounded by a plasma membrane, a eucaryotic cell is elaborately subdivided into functionally distinct, membrane-bounded compartments. Each compartment, or *organelle*, contains its own distinct set of enzymes and other specialized molecules, and complex distribution systems convey specific products from one compartment to another. To understand the eucaryotic cell, it is essential to know what occurs in each of these compartments, how molecules move between them, and how the compartments themselves are created and maintained.

Proteins play a central part in the compartmentalization of a eucaryotic cell. They catalyze the reactions that occur in each organelle and selectively transport small molecules into and out of its interior, or *lumen*. They also serve as organelle-specific surface markers that direct new deliveries of proteins and lipids to the appropriate organelle. A mammalian cell contains about 10 billion (10^{10}) protein molecules of perhaps 10,000 kinds, and the synthesis of almost all of these begins in the cytosol, the common space that surrounds the organelles. Each newly synthesized protein is then delivered specifically to the cellular compartment that requires it. By tracing the protein traffic from one compartment to another, one can begin to make sense of the otherwise bewildering maze of intracellular membranes. We shall therefore make the intracellular traffic of proteins the central theme of this chapter. Although almost every compartment of the cell will be discussed, most attention will be devoted to the endoplasmic reticulum (ER) and Golgi apparatus, which together play a crucial role in modifying, sorting, and dispatching many of the newly synthesized proteins.

The Compartmentalization of Higher Cells

In this introductory section we give a brief overview of the compartments of the cell and of the relationships between them. In doing so, we consider how the traffic of proteins between compartments can be traced experimentally and review the general strategies by which protein molecules pass from one compartment to another.

405

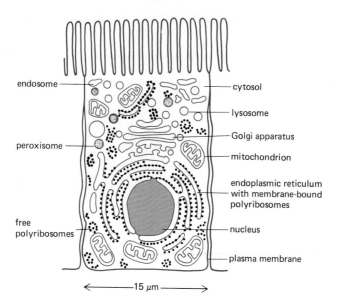

Figure 8–1 A drawing of an animal cell emphasizing the major intracellular compartments. The cytosol, endoplasmic reticulum, Golgi apparatus, nucleus, mitochondrion, endosome, lysosome, and peroxisome are distinct compartments isolated from the rest of the cell by at least one selectively permeable membrane.

In the figure the following labels appear: endosome, peroxisome, free polyribosomes, cytosol, lysosome, Golgi apparatus, mitochondrion, endoplasmic reticulum with membrane-bound polyribosomes, nucleus, plasma membrane, 15 μm

All Eucaryotic Cells Have a Basic Set of Membrane-bounded Organelles[1]

Many vital biochemical processes take place in or on membrane surfaces. Oxidative phosphorylation and photosynthesis, for example, both require a semipermeable membrane in order to couple the transport of H^+ to the synthesis of ATP. Moreover, membranes themselves provide the framework for the synthesis of new membrane components. The internal membranes of eucaryotic cells not only provide extra membrane area but also allow functional specialization of different membranes, which, as we shall see, is crucial in separating the many different processes that occur in the cell.

The major intracellular compartments common to eucaryotic cells are illustrated in Figure 8–1. The *nucleus* contains the main genome and is the principal site of DNA and RNA synthesis. The surrounding *cytoplasm* consists of the *cytosol* and the cytoplasmic organelles suspended in it. The cytosol constitutes a little more than half the total volume of the cell, and it is the site not only of protein synthesis but also of most of the cell's intermediary metabolism—that is, the many reactions by which some small molecules are degraded and others are synthesized to provide the building blocks of macromolecules (see p. 64). About half the total area of membrane in a cell is utilized to enclose the labyrinthine spaces of the *endoplasmic reticulum (ER)*. The ER has many ribosomes bound to its cytosolic surface; these are engaged in the synthesis of integral membrane proteins and soluble proteins destined for secretion or for various other organelles. The ER also produces the lipid for the rest of the cell. The *Golgi apparatus* consists of organized stacks of disclike compartments called Golgi *cisternae;* it receives lipids and proteins from the ER and dispatches these molecules to a variety of intracellular destinations, usually covalently modifying them en route. *Mitochondria* and (in plants) *chloroplasts* generate most of the ATP utilized to drive biosynthetic reactions that require an input of free energy. *Lysosomes* contain digestive enzymes that degrade defunct intracellular organelles, as well as macromolecules and particles taken up from outside the cell by endocytosis. On their way to lysosomes, endocytosed macromolecules and particles must first pass through a series of compartments called *endosomes*. Finally, *peroxisomes* (also known as *microbodies*) are small vesicular compartments that contain enzymes utilized in a variety of oxidative reactions.

In addition to the major membrane-bounded organelles just described, cells contain many small vesicles that act as carriers between organelles, as well as other vesicles that communicate with the plasma membrane in the secretory and the endocytic pathways.

In general, each membrane-bounded organelle has a set of properties that are the same in all cell types, as well as properties that vary from cell type to cell type

Table 8–1 The Relative Volumes Occupied by the Major Intracellular Compartments in a Typical Liver Cell (Hepatocyte)

Intracellular Compartment	Percent of Total Cell Volume	Approximate Number per Cell*
Cytosol	54	1
Mitochondria	22	1700
Rough ER cisternae	9	1
Smooth ER cisternae plus Golgi cisternae	6	
Nucleus	6	1
Peroxisomes	1	400
Lysosomes	1	300
Endosomes	1	200

*All the cisternae of the rough and smooth endoplasmic reticulum are thought to be joined to form a single large compartment. The Golgi apparatus, in contrast, is organized into a number of discrete sets of stacked cisternae in each cell, and the extent of interconnection between these sets has not been clearly established.

and contribute to the specialized functions of differentiated cells. Together these organelles occupy nearly half the volume of the cell (Table 8–1).

As might be expected, a large amount of intracellular membrane is required to make all of these organelles. In the two mammalian cells analyzed in Table 8–2, for example, the endoplasmic reticulum has a total membrane surface that is, respectively, 25 times and 12 times that of the plasma membrane. In terms of its area and mass, the plasma membrane is only a minor membrane in most eucaryotic cells (Figure 8–2).

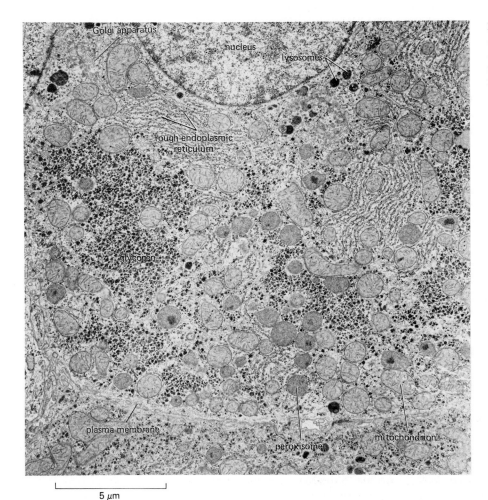

Figure 8–2 Electron micrograph of part of a liver cell seen in cross-section. Examples of most of the major intracellular compartments are indicated. (Courtesy of Daniel S. Friend.)

5 μm

Table 8–2 Relative Amounts of Membrane Types in Two Eucaryotic Cells

Membrane Type	Percent of Total Cell Membrane	
	*Liver Hepatocyte**	*Pancreatic Exocrine Cell**
Plasma membrane	2	5
Rough ER membrane	35	60
Smooth ER membrane	16	<1
Golgi apparatus membrane	7	10
Mitochondria		
Outer membrane	7	4
Inner membrane	32	17
Nucleus		
Inner membrane	0.2	0.7
Secretory vesicle membrane	not determined	3
Lysosome membrane	0.4	not determined
Peroxisome membrane	0.4	not determined
Endosome membrane	0.4	not determined

*These two cells are of very different sizes, since the average hepatocyte has a volume of about 5000 μm^3 compared with about 1000 μm^3 for the pancreatic exocrine cell. Total cell membrane areas are estimated at about 110,000 μm^2 and 13,000 μm^2, respectively.

The Topological Relationships of Membrane-bounded Organelles Can Be Interpreted in Terms of Their Evolutionary Origins[2]

To understand the relationships between the compartments of the cell, it is helpful to consider how they may have evolved. The precursors of the first eucaryotic cells are thought to have been organisms resembling bacteria, which generally have a plasma membrane but no internal membranes. The plasma membrane in such cells therefore provides all membrane-dependent functions, including the pumping of ions, ATP synthesis, and lipid synthesis. Present-day eucaryotes are, however, 10 to 30 times larger in linear dimension than a typical bacterium such as *E. coli*, and the profusion of internal membranes can be seen in part as an adaptation to this increase in size: the larger cell has a smaller ratio of surface to volume, and its area of plasma membrane is presumably too small to sustain the many vital functions for which membranes are required.

The evolution of internal membranes has evidently gone hand in hand with specialization of membrane function. In some present-day bacteria there are specialized patches of the plasma membrane in which a selected set of membrane proteins come together to carry out a group of related functions (Figure 8–3A). These specialized membrane patches, such as the "purple membrane" containing bacteriorhodopsin in *Halobacterium* (see p. 293) and the chromatophores in photosynthetic bacteria (see p. 383), represent primitive organelles. In some photosynthetic bacteria, the patches have become elaborated into extensive invaginations of the plasma membrane (Figure 8–3B), while in others the invaginations seem to have pinched off completely, forming sealed membrane-bounded vesicles specialized for photosynthesis. The interior surface of these vesicles is topologically equivalent to the exterior surface of the cell (Figure 8–3C).

A eucaryotic organelle that originated by the type of pathway illustrated in Figure 8–3 might also be expected to have an interior that is topologically equivalent to the exterior of the cell. We shall see that this is the case for the ER, the Golgi apparatus, the endosome, and the lysosome—as well as for the many vesicular intermediates in the secretory and endocytic pathways (see pp. 433–475).

As discussed in Chapter 7, mitochondria and chloroplasts differ from the other membrane-bounded organelles in that they contain their own genomes. The nature of these genomes and the close resemblance of the proteins in these organelles to those in some present-day bacteria strongly suggest that mitochondria and chloroplasts evolved from bacteria that were engulfed by other cells with which they initially lived in symbiosis (see p. 398). According to the hypothetical

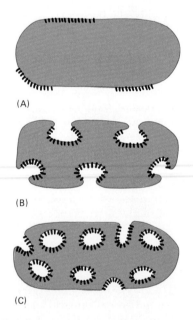

(A)

(B)

(C)

Figure 8–3 Some specialized membranes in bacteria. (A) Membrane patches on the cell surface consisting of clusters of specialized membrane proteins. (B) Invaginated patches of plasma membrane that increase the amount of membrane available for a specialized function such as photosynthesis. (C) Internalization of the specialized invaginated membrane to form vesicles whose interior surface is topologically equivalent to the exterior surface of the cell. Membrane-bounded vesicles of this type are present in some types of photosynthetic bacteria; their topological relationship to the cell surface is similar to that of the ER, Golgi apparatus, endosomes, and lysosomes in eucaryotic cells.

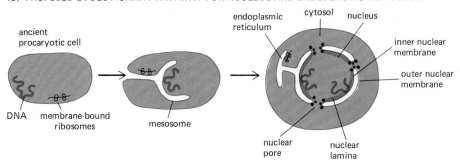

(A) PROPOSED EVOLUTIONARY PATHWAY FOR MITOCHONDRIA

procaryotic cell

ancient
eucaryotic cell

(B) PROPOSED EVOLUTIONARY PATHWAY FOR NUCLEUS AND ENDOPLASMIC RETICULUM

ancient
procaryotic cell

endoplasmic   cytosol   nucleus
reticulum

inner nuclear
membrane

outer nuclear
membrane

DNA   membrane-bound
ribosomes

mesosome

nuclear   nuclear
pore   lamina

Figure 8–4 Hypotheses for the evolutionary origins of mitochondria, chloroplasts, ER, and the cell nucleus that explain the topological relationships of these intracellular compartments in eucaryotic cells. (A) Mitochondria (and chloroplasts) are thought to have originated when a bacterium was engulfed by a eucaryotic cell. This may explain why the lumen of these organelles remains isolated from the extensive vesicular traffic that interconnects the lumens of many other intracellular compartments. (B) A possible pathway for the evolution of the ER and cell nucleus. In some bacteria the DNA is attached to an invagination of the plasma membrane, called a *mesosome.* Such an invagination in a very ancient procaryotic cell could have spread to form an envelope around the DNA while still allowing access of the DNA to the cell cytosol (as is required for DNA to direct protein synthesis). This envelope is presumed to have eventually pinched off completely from the plasma membrane, producing a nuclear compartment surrounded by a double membrane. As illustrated, the nuclear envelope is organized by a fibrous shell called the *nuclear lamina* and is penetrated by communicating channels called *nuclear pores.* Because of these pores, the lumen of the nucleus is topologically equivalent to the cytosol. The lumen of the ER is continuous with the space between the inner and outer nuclear membranes and topologically equivalent to the extracellular space.

scheme shown in Figure 8–4A, the inner membrane of mitochondria and chloroplasts corresponds to the original plasma membrane of the bacterium, while the lumen of these organelles evolved from the bacterial cytoplasm. Accordingly, these two organelles remain isolated from the extensive vesicular traffic that connects the interior of most other organelles to one another and to the outside of the cell.

The origin of the cell nucleus, with its distinctive double-membrane organization, is more mysterious. One possibility, based on the observation that the single bacterial chromosome is attached at special sites to the inside of the procaryotic plasma membrane, is that the double-layered nuclear envelope may have originated as a deep invagination of the plasma membrane, as shown in Figure 8–4B. This scheme would explain why the interior of the nucleus—its "lumen"—is topologically equivalent to the cytosol. In fact, during mitosis in higher eucaryotic cells, the nuclear envelope breaks down and the nuclear contents are completely mixed with the cytosol, a situation that never occurs for the contents of any other membrane-bounded organelle. Thus during mitosis the cell temporarily reverts to a procaryotelike state, in which the chromosomes do not have a separate compartment.

This evolutionary scheme groups the main intracellular compartments in eucaryotic cells into five sets: (1) the nucleus and the cytosol, which communicate through the *nuclear pores* (see p. 422) and are thus topologically continuous (although functionally distinct); (2) the mitochondria; (3) the chloroplasts (in plants only); (4) the peroxisomes (whose evolutionary origins will be discussed on p. 431); and (5) the other membrane-bounded organelles—including the ER, Golgi apparatus, endosomes, and lysosomes—whose interiors all communicate with one another and with the outside of the cell via *transport vesicles* that bud off from one organelle and fuse with another (see p. 471). All the spaces enclosed in this last set of organelles are topologically equivalent to one another and to the cell exterior, and they can be regarded as parts of a single functionally connected complex.

The Intracellular Traffic of Proteins Through the ER and Golgi Can Be Traced by Autoradiography[3]

Methods for tracing the traffic of proteins through the cell were developed in the 1960s and were first applied in studies that revealed the pathway from the ER to the cell exterior in protein-secreting cells of the pancreas.

Specialized secretory cells, such as the acinar cells of the pancreas, store large amounts of their products awaiting secretion in **secretory vesicles,** which rapidly release their contents to the cell exterior by *exocytosis* when the cell is stimulated by an extracellular signal. This process is known as *regulated secretion,* to distin-

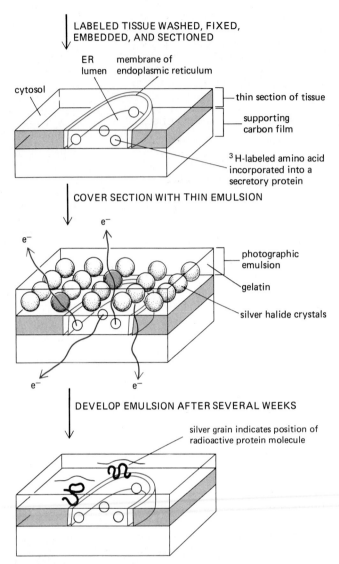

LABELED TISSUE WASHED, FIXED, EMBEDDED, AND SECTIONED

ER membrane of
lumen endoplasmic reticulum

cytosol

thin section of tissue

supporting carbon film

³H-labeled amino acid incorporated into a secretory protein

COVER SECTION WITH THIN EMULSION

e⁻

e⁻

photographic emulsion

gelatin

silver halide crystals

e⁻

e⁻

DEVELOP EMULSION AFTER SEVERAL WEEKS

silver grain indicates position of radioactive protein molecule

Figure 8–5 The principles of electron microscopic autoradiography, used here to trace secretory proteins by locating the ³H-labeled amino acids they have incorporated. It is important that all free radioactive amino acids are washed away prior to embedding, so that the only radioactivity remaining in the tissue is that incorporated into proteins. The tissue is then covered with a thin layer of photographic emulsion. As indicated, electrons emitted by ³H decay activate silver grains in a range of positions slightly displaced from the point of emission. (The developed silver grains appear as dark squiggles in the electron microscope.) Therefore the localization of a radioactive molecule is imprecise in comparison with the resolution of the electron microscope.

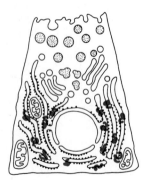

3 minutes:
silver grains over the ER

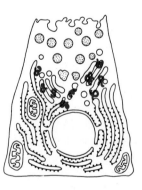

20 minutes:
silver grains over the Golgi apparatus

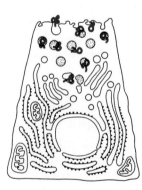

90 minutes:
silver grains over secretory vesicles

Figure 8–6 Results obtained when electron microscopic autoradiography (see Figure 8–5) is used to examine a pancreatic acinar cell that has been briefly pulse-labeled with ³H-amino acids and then incubated in unlabeled media ("chased") for various lengths of time. With time, the activated silver grains (*colored*) are found ever closer to the cell exterior, indicating the path followed by newly synthesized secretory protein molecules. This cell is unusual in that about 85% of the protein it synthesizes is secreted; the proteins are stored in secretory vesicles until the cell is stimulated to secrete. A stimulated cell is shown.

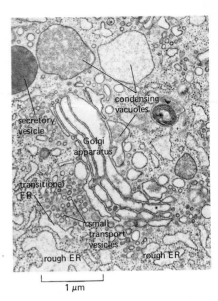

Figure 8–7 Electron micrograph of the Golgi apparatus in a pancreatic acinar cell, showing secretory vesicles in various stages of maturation. Immature secretory vesicles (called *condensing vacuoles* in these cells) form by budding from the Golgi apparatus. The secretory proteins become progressively concentrated in these vesicles (hence the name "condensing" vacuole) to form mature, dense-cored secretory vesicles (*top left*). These secretory vesicles are called *zymogen granules* in pancreatic acinar cells. (Courtesy of George Palade.)

guish it from *constitutive secretion*, which is another form of exocytosis that occurs continuously in the absence of a stimulatory signal (see p. 465). The acinar cell of the pancreas secretes a variety of digestive enzymes (such as amylase, lipase, deoxyribonuclease, and ribonuclease) as well as enzyme precursors called *zymogens* (such as trypsinogen and chymotrypsinogen), which are activated by specific proteolytic cleavages after they are secreted into ducts that lead to the small intestine.

Because the bulk of the protein synthesized in the pancreatic acinar cell is destined for secretion, the problem of tracing the flow of traffic is simplified, and the path followed by secreted proteins from their site of synthesis to their discharge from the cell can be revealed by using autoradiography in conjunction with electron microscopy. The autoradiographic technique developed for this purpose is described in Figure 8–5. When cells are pulse-labeled with ^{3}H-amino acids, followed by a "chase" of varying length in nonradioactive medium, the newly synthesized radioactive proteins are seen first in the ER and then in the Golgi apparatus (Figure 8–6). The ^{3}H-proteins are next found in large, immature secretory vesicles close to the Golgi stack (Figure 8–7), where they become progressively concentrated to form mature secretory vesicles, which are easily identified in electron micrographs by the high density of their contents (Figures 8–7 and 8–8).

The secretory vesicles are stored in the apical region of the acinar cell (the side facing the lumen of the duct system) between the Golgi apparatus and the plasma membrane. They release their contents to the outside by exocytosis when the vesicles fuse with the plasma membrane (see p. 000). The vesicles fuse only with the apical portions of the plasma membrane, thereby avoiding fruitless and dangerous discharges into the spaces between cells or into other organelles in the cell. Furthermore, exocytosis occurs only in response to an appropriate extracellular chemical signal, which is released by nerves or by intestinal cells when pancreatic enzymes are needed for digestion.

Many secreted proteins have been studied in a variety of cell types, and all of them follow a similar pathway: ribosome → ER → Golgi → cell exterior. Moreover, those proteins destined to reside in the plasma membrane, in lysosomes, or in various cisternae of the Golgi apparatus are likewise initially imported into the ER—passing from there via the Golgi apparatus to their final locations.

More complicated techniques are needed to reveal the finer details of the pathways that begin in the ER. While some of the imported proteins will remain in the ER to catalyze ER functions, most are packaged into *transport vesicles* (typically 50–100 nm in diameter) that pinch off from specialized regions of the ER called the *transitional elements* (Figure 8–9) and fuse specifically with nearby cisternal elements of the Golgi apparatus. After fusion the constituents of the vesicle membrane become part of the Golgi membrane, while the soluble proteins in the vesicle lumen are delivered to the lumen of a Golgi cisterna (Figure 8–10). In this way soluble proteins are selectively carried from one membrane-bounded compartment to another without actually passing across a membrane. By similar cycles of vesicle budding and fusion, proteins are thought to be transported from one Golgi cisterna to another and then from the Golgi apparatus to the various final destinations appropriate to their different functions, passing through a whole series of spaces that are topologically equivalent to one another and to the cell exterior (Figure 8–11).

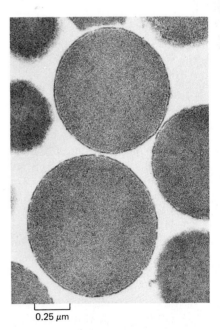

Figure 8–8 Electron micrograph of a purified preparation of large secretory vesicles. Such vesicles are found only in cells specialized for secretion. (Courtesy of Daniel S. Friend.)

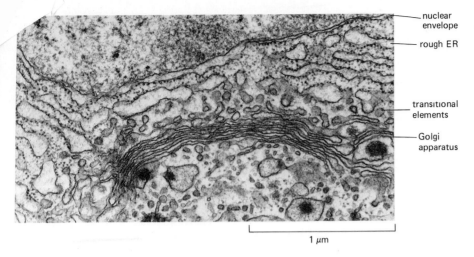

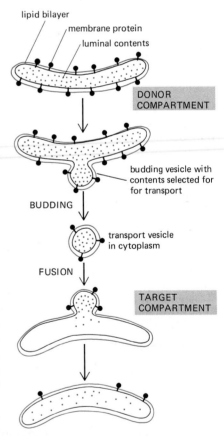

Figure 8–9 Transport vesicles bud from specialized regions of the ER that are nearly free of ribosomes, termed *transitional elements*. The vesicles fuse with a nearby Golgi cisterna, thereby transferring newly made proteins and lipids from the ER to the Golgi apparatus. (Courtesy of Brij J. Gupta.)

Proteins face choices at every step in this pathway—to remain in the cytosol or enter the ER, to remain in the ER or move on to the Golgi apparatus, to become a passenger in a transport vesicle heading for a lysosome or embark instead in one targeted to the cell surface, and so on. These choices depend on the amino acid sequence of the protein molecule, which contains *sorting signals* that determine the path that the protein follows through the cell.

8-3 The Traffic of Proteins Has Two Divergent Branches—One Cytosolic, the Other via the ER[4]

The major pathways of protein traffic are outlined in Figure 8–12. All proteins originate on ribosomes in the cytosol (although a few are synthesized on the ribosomes of mitochondria and chloroplasts), but the traffic from this point onward diverges into two main branches. The proteins in one branch are released into the cytosol after their synthesis has been completed. Some of these proteins have sorting signals that cause them subsequently to be exported from the cytosol into mitochondria, chloroplasts (in plants), the nucleus, or peroxisomes; others—the majority—have no specific sorting signal and consequently remain in the cytosol as permanent residents.

The other main branch of the traffic consists of proteins destined to be secreted from the cell—plus those destined to reside in the ER, Golgi apparatus, plasma membrane, or lysosomes. All of these proteins are transferred into the ER as they are being synthesized because of a sorting signal that is generally located at their amino terminus. The ribosomes that make these proteins become attached to the membrane of the ER shortly after the synthesis of the polypeptide chain begins. As each region of the polypeptide is made, it is threaded through the lipid bilayer of this membrane. Some proteins synthesized in this way will be released into solution in the ER lumen when their synthesis is complete, while others will remain partially embedded in the ER membrane as transmembrane proteins (see p. 444).

It generally takes no more than a minute or two for a protein that has been released into the cytosol to be imported into the appropriate organelle. But whereas the proteins directed to the nucleus, mitochondria or chloroplasts, and peroxisomes have finished their journey once they have arrived at their organelle, most of the proteins that are sorted into the ER have only just begun theirs. Once a protein has reached the ER, it is subjected to further transport steps, which are thought to be mediated by transport vesicles that bud off from one membrane and fuse with another. Since there are generally several such steps before the final destination is reached, a protein synthesized on the ER membrane may take up to an hour to reach its final destination.

Figure 8–10 The "sidedness" of membranes is preserved during vesicle-mediated transport. Note that the original orientation of both proteins and lipids in the donor-compartment membrane is preserved in the target-compartment membrane and that soluble materials are transferred from lumen to lumen.

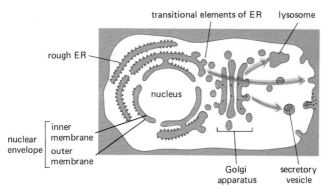

Figure 8–11 Sketch of a cell in which topologically equivalent spaces are shown in color. In principle, cycles of vesicle budding and fusion permit any lumen to communicate with any other and with the cell exterior. Some organelles, however, most notably mitochondria and (in plant cells) chloroplasts, do not communicate with other organelles by means of transport vesicles. They are therefore isolated from the traffic between organelles shown here.

Proteins Can Move Between Compartments in Two Fundamentally Different Ways[5]

To understand the general principles by which sorting signals operate, it is important to distinguish two fundamentally different ways in which proteins move from one compartment to another. First, they may be directly transported across a membrane, thereby passing from a space topologically equivalent to the cytosol into a space topologically equivalent to the exterior of the cell, or vice versa. This requires a special *protein translocator* in the membrane, and the protein molecule must generally be unfolded in order to snake through. The initial movement of selected proteins from the cytosol into the ER lumen is an example. Second, proteins can be ferried between compartments by means of *transport vesicles*. These vesicles capture a cargo of molecules from the lumen of one compartment as they pinch off from its membrane and then discharge that cargo into another compartment as they fuse with it, as in the transfer of soluble proteins from the ER to the Golgi apparatus (see Figure 8–10). In this type of *vesicular transport*, the transported proteins do not cross any membrane; they are therefore transferred only between compartments that are topologically equivalent to one another.

Both types of transfer process are selectively controlled according to the sorting signals possessed by the proteins. For a protein to be directly transferred across a membrane, its sorting signal must be recognized by the translocator in the membrane to be crossed. A protein will enter a transport vesicle if its sorting signal binds to a special receptor protein in the vesicle membrane (see p. 462); other transport vesicles, however, seem to be nonselective and take up any protein that lacks a sorting signal that directs the protein elsewhere (see p. 467). In either case the newly formed transport vesicle carries only appropriate proteins.

We shall see that some of the sorting signals on proteins are known, whereas most of their complementary membrane receptors are not. In addition, we know almost nothing about transport vesicles, of which there are thought to be numerous kinds, corresponding to the many different transfers that must be effected. To

Figure 8–12 A simplified "road map" of biosynthetic protein traffic. The signals that direct a given protein's movement through the system, and thereby determine its eventual location in the cell, are contained in its amino acid sequence. The journey begins with the synthesis of a protein on a ribosome and terminates when the final destination is reached. At each intermediate station (boxes), a decision is made as to whether to retain the protein or to transport it further. In principle, a signal could be required either for retention in or for leaving each of the compartments shown, with the alternative fate being the "default pathway" (one that requires no signal). In this diagram the routes that are most likely to require special signals are drawn in color; the routes that are most likely to be taken by default are drawn in black. Because the biosynthesis of endosomes is poorly understood, no route to endosomes is shown.

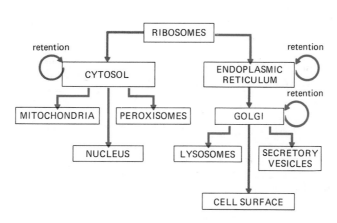

perform its function, each vesicle must take up only the appropriate proteins and must fuse only with the appropriate target membrane: a vesicle carrying cargo from the ER to the Golgi apparatus, for example, must exclude proteins that are to remain in the ER and fuse only with the Golgi apparatus and not with other organelles. We shall consider how transport vesicles might achieve this selectivity later in the chapter.

Signal Peptides and Signal Patches Specify a Protein's Fate[6]

There are thought to be two types of sorting signals on proteins that direct them step by step through the branching pathways outlined in Figure 8–12. For some steps the sorting signal resides in a continuous stretch of amino acid sequence, typically 15–60 residues long. This **signal peptide** is often (but not always) re-moved from the finished protein once the sorting decision has been executed. The signal for other sorting steps is thought to consist of a particular three-dimensional arrangement of atoms on the protein's surface that forms when the protein folds up. The amino acid residues that comprise this **signal patch** may be quite distant from one another in the linear amino acid sequence, and they generally remain in the finished protein (Figure 8–13). *Signal peptides* are used to direct proteins from the cytosol into the ER, mitochondria, chloroplasts, and nu-cleus; they are also used to retain certain proteins in the ER. *Signal patches* are thought to be used for some other sorting steps, including the recognition of certain lysosomal proteins by a special sorting enzyme in the Golgi apparatus (see p. 463).

Different types of signal peptides are used to specify different destinations in the cell (Table 8–3). Proteins destined for initial transfer to the ER usually have amino-terminal signal peptides with a central part of the sequence composed of 5–10 hydrophobic amino acid residues. Most of these proteins will pass from the ER to the Golgi apparatus; those with a specific sequence of four amino acids at their carboxyl terminus, however, are retained as permanent ER residents. Proteins destined for mitochondria have signal peptides in which positively charged amino acid residues alternate with hydrophobic ones. Many proteins destined for the nucleus carry signal peptides formed from a cluster of positively charged amino acid residues. Finally, some cytosolic proteins have signal peptides that cause a fatty acid to be covalently attached to them, which directs the proteins to mem-branes without insertion into the ER (see p. 417).

The importance of each of these signal peptides for protein targeting has been shown by experiments in which the peptide is transferred from one protein to another by genetic engineering techniques: placing the amino-terminal ER signal peptide at the beginning of a cytosolic protein, for example, redirects the protein to the ER. The signal peptides attached to all proteins having the same destination are functionally interchangeable, even though their amino acid sequences may vary greatly. Physical properties, such as hydrophobicity, often appear to be more important in the signal-recognition process than the exact amino acid sequence.

Figure 8–13 Two ways that a transport signal can be built into a protein. (A) The signal is in a single discrete stretch of amino acid sequence, called a *signal peptide*, that is exposed in the folded protein. Signal peptides often occur at the end of the polypeptide chain (as shown), but they can also be located elsewhere. They are normally detected experimentally by their effect on the intracellular sorting of other proteins when they are attached to them by recombinant DNA methods (see p. 266). (B) A *signal patch* can be formed by the juxtaposition of amino acids from regions that are physically separated before the protein folds (as shown); alternatively, separate "patches" on the surface of the folded protein that are spaced a fixed distance apart could form the signal. In either case the transport signal depends on the three-dimensional conformation of the protein. For this reason it is very difficult to locate this type of signal precisely.

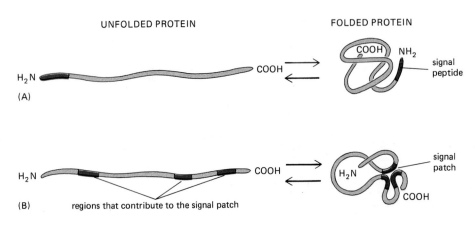

UNFOLDED PROTEIN   FOLDED PROTEIN

(A)

(B) regions that contribute to the signal patch

Table 8–3 Typical Signal Peptide Sequences

Function of Signal Peptide	Example of Signal Peptide
Import into ER	H₃N-Met-Met-Ser-Phe-Val-Ser- $\boxed{\text{Leu-Leu-Leu-Val}}$ $\oplus$ $\boxed{\text{Gly-Ile-Leu-Phe-Trp-Ala}}$ -Thr-Glu-Ala-Glu- $\ominus$ $\ominus$ Gln-Leu-Thr-Lys-Cys-Glu-Val-Phe-Gln- $\oplus$ $\ominus$
Retain in lumen of ER	-Lys-Asp-Glu-Leu-COO⁻ $\oplus$ $\ominus$ $\ominus$ $\ominus$
Import into mitochondria	H₃N-Met-Leu-Ser-Leu-Arg-Gln-Ser-Ile-Arg-Phe- $\oplus$ $\oplus$ $\oplus$ Phe-Lys-Pro-Ala-Thr-Arg-Thr-Leu-Cys-Ser- $\oplus$ $\oplus$ Ser-Arg-Tyr-Leu-Leu- $\oplus$
Import into nucleus	-Pro-Pro-Lys-Lys-Arg-Lys-Val- $\oplus$ $\oplus$ $\oplus$ $\oplus$ $\oplus$
Attach to membranes via the covalent linkage of a myristic acid to the amino terminus	H₃N-Gly-Ser-Ser-Lys-Ser-Lys-Pro-Lys- $\oplus$ $\oplus$ $\oplus$ $\oplus$

Charged residues are indicated by a $\oplus$ or a $\ominus$. An extended block of hydrophobic residues is enclosed in a box. H₃N- indicates the amino terminus of a protein; —COO⁻ indicates the carboxyl terminus.

Signal patches are far more difficult to analyze than signal peptides; consequently, much less is known about their structure. Because they result from a complex three-dimensional protein-folding pattern, they cannot simply be transferred from one protein to another. Moreover, experimental alteration of a signal patch will often disturb the conformation of the protein as a whole.

Cells Cannot Construct Their Membrane-bounded Organelles de Novo: They Require Information in the Organelle Itself[7]

When a cell reproduces and divides, it has to duplicate its membrane-bounded organelles. In general, cells do this by enlarging these organelles by incorporating new molecules into them; the enlarged organelles then divide and are distributed to the two daughter cells. It is unlikely that cells could make all of these organelles de novo. If the ER were completely removed from a cell, for example, how could the cell reconstruct it? The membrane proteins that define the ER and carry out many of its key functions are themselves products of the ER: without an existing ER—at the very least, without a membrane that contains the translocators required to import proteins into the ER (and that lacks the translocators required to import proteins into other organelles)—a new ER could not be made.

Thus it seems that the information required to construct a membrane-bounded organelle does not reside exclusively in the DNA that specifies the organelle proteins. "Epigenetic" information in the form of at least one distinct protein in the organelle membrane is also required, and this information is passed from parent cell to progeny cell in the form of the organelle itself. Presumably, such information is essential for the propagation of the cell's compartmental organization, just as the information in DNA is essential for the propagation of nucleotide and amino acid sequences.

Summary

Eucaryotic cells contain intracellular membranes that enclose nearly half their total volume in separate intracellular compartments. The main types of membrane-bounded organelles in all eucaryotic cells are the endoplasmic reticulum, Golgi apparatus, nucleus, mitochondria, lysosomes, endosomes, and peroxisomes; plant

cells also contain chloroplasts. Each organelle contains distinct proteins that mediate its unique functions.

Each newly synthesized organelle protein finds its way from the ribosome to
the organelle by following a specific pathway, guided by a signal in its amino acid
sequence that functions either as a signal peptide or as a signal patch. Protein
sorting begins with a primary segregation event in which the protein either remains
in the cytosol or is transferred to another compartment (such as the nucleus, a
mitochondrion, or the endoplasmic reticulum). Proteins that enter the ER undergo
secondary sorting processes as they are transported to the Golgi apparatus and
from the Golgi apparatus to lysosomes, to secretory vesicles, or to the plasma
membrane. Some resident proteins are retained specifically in the ER and in the
various cisternae of the Golgi apparatus. Those destined for other compartments
are thought to enter transport vesicles, which bud from one compartment and fuse
with another.

The Cytosolic Compartment

The **cytosol**—that is, the part of the cytoplasm that occupies the space between
the membrane-bounded organelles—generally accounts for about half of the total
cell volume (see Table 8–1). It teems with thousands of enzymes involved in intermediary metabolism and is packed with ribosomes making proteins. About half
of the proteins made on these ribosomes are destined to remain in the cytosol as
permanent residents. In this section we discuss the fate of these cytosolic proteins
as well as some of the devices that are used to control their lifetime and to direct
them to specific locations within the cytosol.

The Cytosol Is Organized by Protein Filaments[8]

As we shall discuss in Chapter 11, the cytosol contains a variety of protein filaments arranged in a fibrous *cytoskeleton*. The cytoskeleton imparts shape to the
cell, mediates coherent cytoplasmic movements, and provides a general framework that could help organize enzymatic reactions. Moreover, inasmuch as about
20% of its weight is protein, it is more appropriate to think of the cytosol as a
highly organized gelatinous mass rather than as a simple solution of enzymes.
Studies of diffusion rates, however, have shown that small molecules and some
small proteins diffuse nearly as quickly in the cytosol as they do in pure water.
This suggests that, from the point of view of intermediary metabolism (in which
the substrates and products are all small molecules), we can treat the cytosol as
a simple solution.

Large particles such as transport vesicles and organelles, on the other hand,
diffuse very slowly, partly because they collide frequently with components of the
cytoskeleton. To move at useful rates, therefore, such particles usually are actively
transported by protein "motors" that hydrolyze ATP and use the energy released
to propel the particles along microtubules or actin filaments (see p. 660 and
p. 632). In principle, one could imagine a highly organized cytosol (see p. 668), in
which specific filaments act as "highways" to direct each type of transport vesicle
to its appropriate target membrane for fusion. Most cell biologists, however, believe
that the cytoskeleton generally plays a less specific role and that most of the
specificity of vesicular traffic resides in receptor systems located on the cytosolic
surface of the vesicles themselves (see p. 463).

Many Proteins Are Covalently Modified in the Cytosol[9]

More than 100 different *post-translational modifications* of amino acid side chains
have been described in proteins. The functions of most of these modifications are
not known; some are chemical accidents and have no function at all, but many
must be important for the operation of the cell, since they are tightly controlled
by specific enzymes. We shall see later that some of these modifications occur in

the ER and the Golgi apparatus. Glycosylating enzymes in these organelles, for example, add a complex series of sugar residues to proteins to produce glycoproteins (see p. 446). The only type of protein glycosylation known to occur in the cytosol of mammalian cells is the much simpler attachment of a single *N*-acetylglucosamine to an occasional protein (Figure 8–14). Many other covalent modifications, however, occur primarily in the cytosol. Some are permanent and required for activity, such as the covalent attachment of coenzymes (for example, biotin, lipoic acid, or pyridoxal phosphate) to some enzymes (see p. 75). Other covalent modifications that occur in the cytosol, such as phosphorylation, are reversible and serve to regulate the activity of many proteins (Figure 8–15).

From the perspective of protein targeting, one type of covalent modification that takes place in the cytosol is especially important. The attachment of a fatty acid to a protein can direct the protein to a specific membrane surface that is exposed to the cytosol.

Some Cytosolic Proteins Are Attached to the Cytoplasmic Face of Membranes by a Fatty Acid Chain[10]

Cells have a special mechanism for directing selected water-soluble proteins from the cytosol to membranes. The protein is covalently attached to a fatty acid chain, which inserts into the cytoplasmic leaflet of the lipid bilayer, anchoring the protein to the membrane. Attaching a protein to a membrane by a fatty acid can have important functional consequences. The *src* oncogene of Rous sarcoma virus, for example, encodes a tyrosine-specific protein kinase that is normally bound to membranes by a covalently attached myristic acid chain (an unsaturated fatty acid with 14 carbons). In this configuration the kinase can transform a cell into a cancer cell. If the attachment of this fatty acid is prevented by converting the aminoterminal glycine of the protein to an alanine, the src protein is still perfectly active as a protein kinase, but it remains in the cytosol and does not transform the cell. Evidently the kinase needs to be attached to a membrane in order to find its substrates efficiently (see p. 757). Similar experiments suggest that a different oncogene product, the ras protein (see p. 705), must be bound to membranes by a covalently attached palmitic acid chain (an unsaturated fatty acid with 16 carbons) in order to transform cells.

What determines whether a given cytosolic protein receives a fatty acid chain and whether it is myristic acid or palmitic acid? The enzymes that catalyze these modifications recognize distinct signal peptides in the protein: a myristic acid

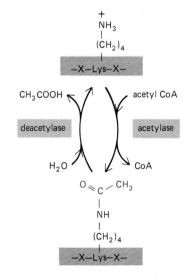

Figure 8–14 The attachment of *N*-acetylglucosamine to serine (or threonine) residues of a protein is the only form of protein glycosylation known to occur in the cytosol of mammalian cells. Many gene regulatory proteins and several nuclear pore proteins are modified in this way; the function of this modification is not known. Much more complex glycosylations occur in the ER and Golgi apparatus (see Figure 8–52).

Figure 8–15 Three types of reversible covalent modifications that occur in proteins to regulate their activity. Each modification shown changes the charge on an amino acid side chain. The most common modification is the phosphorylation of an OH group on a serine, threonine, or tyrosine side chain of a protein; it has been estimated that about 10% of the cytosolic proteins in animal cells are modified in this way (see p. 129).

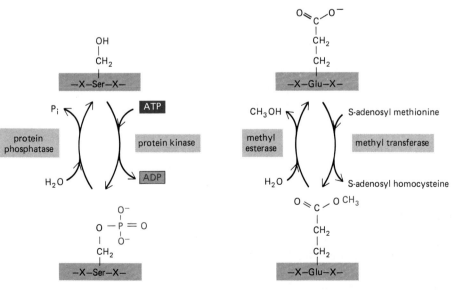

(A) REVERSIBLE PHOSPHORYLATION

(B) REVERSIBLE METHYLATION

(C) REVERSIBLE ACETYLATION

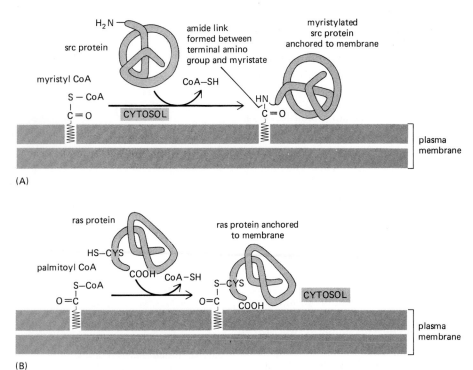

(A)

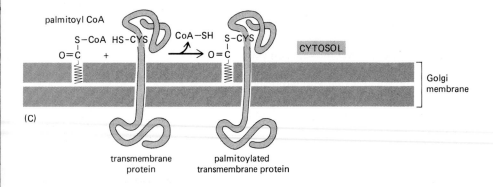

(B)

(C)

transmembrane
protein

palmitoylated
transmembrane protein

Figure 8–16 The covalent attachment of a fatty acid to a protein can localize a water-soluble protein to a membrane. (A) An amide linkage between an amino-terminal glycine and myristic acid anchors the src protein to the cytoplasmic side of a membrane after its synthesis in the cytosol. (B) A thiolester linkage between palmitic acid and a cysteine near the carboxyl terminus anchors the ras protein to the plasma membrane after its synthesis in the cytosol. (C) Palmitic and other fatty acids are frequently attached by a thiolester bond to a selected cysteine residue located in the cytoplasmic domains of a transmembrane protein. This occurs during the transport of the protein from the ER to the Golgi apparatus. The acylated cysteine is usually preceded by three hydrophobic amino acids (X) in the sequence NH_2-...-X-X-X-Cys-...-COOH. The protein in (C), unlike those in (A) and (B), will remain membrane-bound even without the fatty acid attachment.

chain (Figure 8–16A) is added to an amino-terminal glycine residue that is located in a particular context (see Table 8–3), while a palmitic acid chain is added to a cysteine side chain four residues from the carboxyl terminus in another signal peptide (Figure 8–16B). In addition, a reaction in the cytosol, which is catalyzed by a different enzyme, adds a palmitic acid chain to the exposed cytosolic tails of many transmembrane proteins as they pass from the ER to the Golgi apparatus on their way to the plasma membrane and elsewhere (Figure 8–16C).

8-8 Some Cytosolic Proteins Are Programmed for Rapid Destruction[11]

In addition to carrying signals that determine their location, proteins in cells carry signals that determine their lifetime. Cellular proteins are subject to continuous turnover: some molecules of each protein are randomly selected for degradation and are replaced by new copies. Most of the resident proteins of the cytosol are relatively long-lived, functioning for several days before they are degraded. Others, however, are degraded much more rapidly, sometimes within a few minutes of their synthesis. Among the short-lived proteins are enzymes that catalyze a rate-determining step in a metabolic pathway; the rates of synthesis of these enzymes are usually regulated according to environmental conditions to promote efficient use of the metabolic pathway. Other short-lived proteins are the products of cellular oncogenes such as *fos* and *myc*, which are thought to play important parts in the control of cell growth and division (see p. 757). Because these types of proteins are continuously and rapidly degraded, their concentrations can be quickly changed by changing their rates of synthesis (see p. 714). In most cases such regulation also requires an unusually rapid turnover of the mRNAs that encode these proteins (see p. 595).

Most misfolded, denatured, or otherwise abnormal proteins are also rapidly degraded in the cytosol. They are generally destroyed within minutes, while intact copies of the same proteins are spared. Abnormal proteins arise by accidents of protein synthesis in which incorrect amino acids are incorporated, and they are also produced by chemical damage such as the oxidation of certain amino acid side chains. Various mutant forms of normal proteins can also be recognized as abnormal. There is increasing evidence that both abnormal proteins and the proteins that are genetically programmed for rapid turnover are ultimately destroyed by the same proteolytic machinery in the cytosol.

8-6 A Ubiquitin-dependent Proteolytic Pathway Is Responsible for Selective Protein Turnover in Eucaryotes[12]

The cytosolic proteins that are programmed for rapid destruction carry signals that are recognized by the proteolytic machinery responsible for their degradation. One of these signals is remarkably simple, consisting only of the first amino acid in the polypeptide chain. When present as the amino-terminal residue, the amino acids Met, Ser, Thr, Ala, Val, Cys, Gly, and Pro are stabilizing, while the remaining 12 amino acids attract a proteolytic attack. These *destabilizing* amino acids are virtually never found at the amino terminus of stable cytosolic proteins. They are often present, however, at the amino terminus of proteins that are transported into other compartments, such as the ER; since the cytosolic degradation machinery is not present in the lumen of the ER or the Golgi apparatus, such proteins are normally long-lived in their respective compartments. A destabilizing amino-terminal amino acid on these noncytosolic proteins may provide the cell with a simple way of disposing of copies that have been misrouted: those molecules that fail to be rapidly translocated out of the cytosol are rapidly degraded. A similar single-residue code is apparently used in bacteria to signal the rapid destruction of specific proteins.

The proteolytic machinery responsible for selective protein degradation is complex and guarantees the complete destruction of a protein once the machinery is engaged. In eucaryotes a **ubiquitin-dependent pathway** is used. In this pathway numerous copies of a small protein called *ubiquitin* (Figure 8–17) are covalently linked to the target protein to be degraded. The ubiquitination of target proteins is catalyzed by a multienzyme complex that is thought to bind to the amino terminus of proteins that have a destabilizing amino-terminal amino acid. The enzyme complex attaches a ubiquitin molecule to a nearby exposed lysine residue in the polypeptide chain and then adds a series of additional ubiquitin molecules to this first ubiquitin, producing a branched multiubiquitin chain (Figure 8–18). Subsequently a large ATP-dependent protease rapidly degrades these proteins. Because only proteins that contain branched multiubiquitin chains appear to be substrates for the protease, proteins that contain a single ubiquitin molecule attached to a lysine, such as certain histones are spared.

8-7 A Protein's Stability Can Be Determined by Enzymes That Alter Its Amino Terminus[13]

Since the amino-terminal amino acid of a cytosolic protein determines whether the protein will be degraded by the ATP-dependent protease, it is important to know how this crucial amino acid residue is acquired. Proteins that are genetically programmed for rapid destruction seem to acquire a destabilizing amino acid at their amino terminus immediately after their synthesis. As discussed in Chapter 5 (see p. 213), all proteins are initially synthesized with methionine as their amino-terminal amino acid (formyl methionine in bacteria). This methionine, which is a

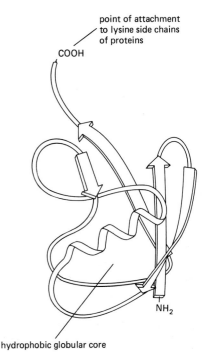

point of attachment to lysine side chains of proteins

COOH

NH₂

hydrophobic globular core

Figure 8–17 The three-dimensional structure of ubiquitin, a heat-stable protein of 76 amino acid residues. The attachment of a single ubiquitin molecule to a protein is a reversible modification with a regulatory role (see also Figure 8–15). The addition of a branched ubiquitin chain to a protein, however, condemns that protein to immediate and complete degradation (see Figure 8–18). (Based on S. Vijay-Kumar, C.E. Bugg, K.D. Wilkinson and W.J. Cook, *Proc. Natl. Acad. Sci. USA* 82:3582–3585, 1985.)

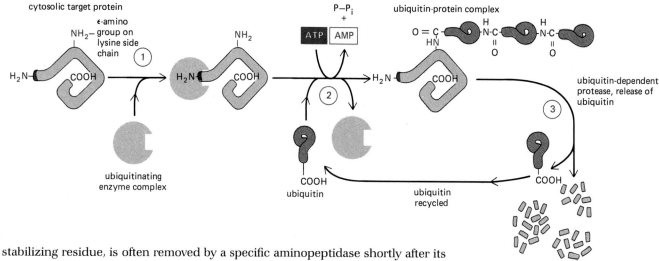

cytosolic target protein

ε-amino group on lysine side chain

ubiquitinating enzyme complex

ubiquitin-protein complex

ubiquitin-dependent protease, release of ubiquitin

ubiquitin recycled

ubiquitin

amino acids

Figure 8–18 Ubiquitin-dependent protein degradation. A target protein (containing a "destabilizing" amino-terminal amino acid that signals its degradation) is recognized by the ubiquitinating enzyme complex (step 1). Then, in a repeated series of biochemical reactions (step 2), ubiquitin molecules are joined together to produce a branched multiubiquitin chain attached to the ε-amino group of a nearby lysine side chain on the target protein. This reaction requires flexibility in the target protein and thus may be accelerated if the protein is misfolded. Next (step 3), a large protease that cleaves only proteins marked with the branched chain of ubiquitins cuts the target protein into a series of small fragments.

stabilizing residue, is often removed by a specific aminopeptidase shortly after its incorporation into the protein. In addition, enzymes known as *aminoacyl-tRNA-protein transferases* add a single destabilizing amino acid to the amino terminus of selected proteins. The net result of these trimming and end-addition reactions will be to leave either a stabilizing or a destabilizing amino acid at the amino terminus of the protein. The rules governing these reactions are poorly understood.

Denatured or misfolded proteins and those containing oxidized or otherwise abnormal amino acids are preferential substrates for ubiquitination and degradation, even when their amino-terminal amino acid is stabilizing. In the degradation of a misfolded or denatured protein, the first step might involve the recognition of a group of hydrophobic amino acids that are normally buried in the core of a properly folded protein but are transiently exposed in the incorrectly folded protein (see p. 107). This recognition step might then trigger protein cleavage or modification reactions that create a new amino-terminal residue that is destabilizing; the ubiquitin-dependent proteolytic machinery would then be activated and the abnormal protein destroyed. A major problem in any scheme for detecting denatured or misfolded proteins is the necessity of distinguishing between those *completed* proteins that have the wrong conformation and the many growing polypeptide chains on ribosomes that will likewise be "abnormally" folded. The problem can be demonstrated experimentally: if the protein-synthesis inhibitor puromycin (see p. 218) is added to cells, the prematurely terminated polypeptides that are formed are rapidly degraded by a ubiquitin-dependent pathway. The metabolic stability of both long-lived and otherwise short-lived proteins during their synthesis on ribosomes may be due to their temporary protection by the translation machinery.

The amino-terminal residue of a protein is often found to be resistant to hydrolysis by the reiterative chemical reactions used in amino acid sequenators (see p. 172). These proteins usually have been modified by the acetylation of their amino terminus, which is then said to be "blocked." Some of the proteins modified in this way are thought to be especially resistant to intracellular proteolysis and therefore unusually long-lived in the cell; they include many of the proteins of the cytoskeleton and the histones that package the DNA in the cell nucleus. How proteins are selected for this modification is unknown.

Stress-Response Proteins Help Prevent the Accumulation of Protein Aggregates in a Cell[14]

Mammalian cells in culture are normally grown at 37°C. If they are briefly "heat shocked" by raising the temperature (usually to about 43°C), they begin to synthesize a special set of abundant proteins. Most of these **heat-shock** or **stress-response proteins** are also synthesized in response to other harmful treatments, and they seem to help the cell survive severe insults. Similar proteins are induced in *Drosophila*, yeast, and even bacterial cells. DNA sequencing studies suggest that there are three main families of stress-response proteins: the 25,000 dalton proteins, the 70,000 dalton proteins, and the 90,000 dalton proteins. In each family, closely related abundant proteins have been discovered in normal cells.

The stress-response proteins are thought to help solubilize and refold denatured or misfolded proteins. They also have other functions (see p. 428 and p. 440). Members of the 90,000 dalton family, for example, have been shown to be bound to the inactive forms of steroid-hormone-receptor proteins and tyrosine-specific protein kinases, suggesting that they might regulate the function of these proteins. The best-studied stress-response proteins are those in the 70,000 dalton family. These **hsp70 proteins** bind to selected proteins, including abnormal protein complexes and aggregates, from which they can be released by the addition of ATP. It has been suggested that they help solubilize and refold aggregated and misfolded proteins by cycles of ATP binding and hydrolysis. Although these proteins are needed in all cells, a severe stress such as a heat shock greatly increases the amount of damaged protein in a cell and consequently creates a need for more stress-response proteins, which are provided by activating the transcription of selected stress-response genes. The yeast *S. cerevisiae*, for example, contains eight hsp70-related genes, some of which are transcribed under all conditions, whereas the others are expressed only when the cells are exposed to an elevated temperature or otherwise stressed.

Summary

The cytosol, which generally represents about half the volume of a eucaryotic cell, comprises all the intracellular space outside the membrane-bounded organelles. Most intermediary metabolism and protein synthesis occur in the cytosol. Newly synthesized proteins remain in the cytosol if they lack a signal for transport to a cell organelle. Certain of these proteins are quickly destroyed after their synthesis: a single "destabilizing" amino acid at their amino terminus facilitates the attachment of multiple copies of ubiquitin to specific lysine residues in the target proteins, which in turn activates a ubiquitin- and ATP-dependent protease to degrade the protein. The rapid degradation of selected proteins allows their concentrations to be changed quickly in response to regulatory signals. Abnormal copies of most cytosolic proteins are also degraded by the same ubiquitin-dependent pathway.

Many proteins are covalently modified in the cytosol. Some of these modifications are permanent; others, such as phosphorylation, are reversible and play an important role in regulating the activity of the proteins. Fatty acids are covalently attached to selected proteins; this modification can cause an otherwise soluble protein to attach to the cytoplasmic face of a cell membrane.

The Transport of Proteins and RNA Molecules into and out of the Nucleus[15]

The **nuclear envelope** encloses the DNA and defines the nuclear compartment. It is formed from two concentric membranes. The spherical **inner nuclear membrane** contains specific proteins that act as binding sites for the *nuclear lamina* that supports it (see p. 665), and it contacts the chromosomes and nuclear RNAs. This membrane is surrounded by the **outer nuclear membrane,** which closely resembles the membrane of the endoplasmic reticulum, which is continuous with it (Figure 8–19). In fact, the outer nuclear membrane can be regarded as a specialized region of the ER membrane. Like the membrane of the rough ER (see p. 434), its outer surface is generally studded with ribosomes engaged in protein synthesis. The proteins made on these ribosomes are transported into the space between the inner and outer nuclear membranes (the *perinuclear space*), which is in turn continuous with the ER lumen (see Figure 8–19).

The nucleus contains many proteins that help mediate its unique functions. These proteins, which include histones, DNA and RNA polymerases, gene regulatory proteins, and RNA-processing proteins, are imported from the cytosol, where

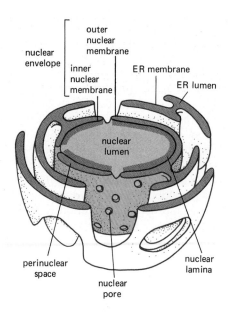

Figure 8–19 A three-dimensional sketch of the double-membrane envelope that surrounds the nucleus. The nuclear envelope is penetrated by nuclear pores and is continuous with the endoplasmic reticulum.

they are made. They must pass through both the outer and inner nuclear membranes to reach the inside of the nucleus (the *nuclear lumen*). This transport process is selective: many proteins made in the cytosol are excluded from the nucleus.

Nuclear Pores Perforate the Nuclear Double Membrane[16]

The nuclear envelope in all eucaryotes, from yeasts to humans, is perforated by **nuclear pores.** Each pore is embedded in a large disclike structure known as the **nuclear pore complex,** which has an estimated molecular weight of 50 to 100 million. Each complex is formed from a set of large protein granules arranged in an octagonal pattern (Figures 8–20A and 8–21). The pore complex perforates the double membrane, bringing the lipid bilayers of the inner and outer membranes together around the margins of each pore (Figure 8–20B). Despite this continuity, which would seem to provide a pathway for the diffusion of membrane compo-

Figure 8–20 The arrangement of the nuclear pore complexes in the nuclear envelope. (A) A top view and a central vertical section. The "central granule" shown in color is seen in some pores but not others; these granules may be part of the pore, or they may be large complexes caught in transit through it. (B) Three-dimensional sketch of a small region of the nuclear envelope.

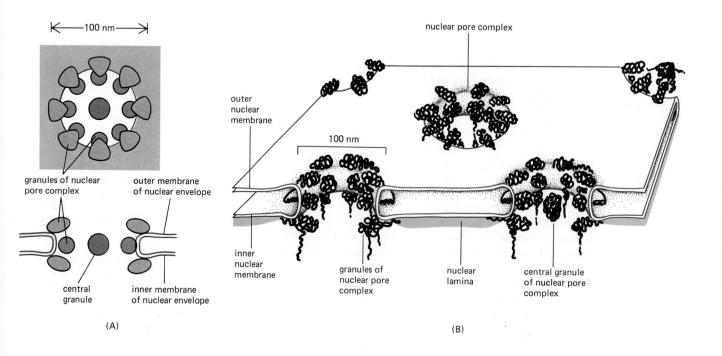

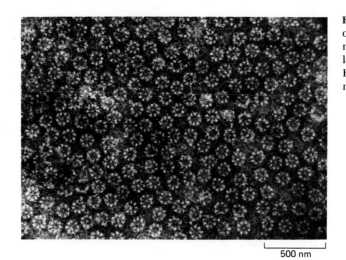

Figure 8–21 Electron micrograph of a negatively stained preparation of nuclear pore complexes. A ring of eight large granules surrounds each pore. Each granule is about the size of a ribosome. (Courtesy of A.C. Fabergé.)

500 nm

nents between the inner and outer membranes, the two membranes remain chemically distinct.

The "hole" in the center of each complex (the nuclear pore) is an aqueous channel through which water-soluble molecules shuttle between the nucleus and the cytoplasm. This hole often appears to be plugged by a large central granule believed to consist of newly made ribosomes or other particles caught in transit (see Figure 8–20B). The effective size of the pore in a resting state has been determined by injecting into the cytosol labeled molecules that are not nuclear components and measuring their rate of diffusion into the nucleus. Small molecules (5000 daltons or less) diffuse in so fast that the nuclear envelope can be considered to be freely permeable to them. A protein of 17,000 daltons equilibrates between cytoplasm and nucleus within 2 minutes; a protein of 44,000 daltons takes 30 minutes to equilibrate, while a globular protein larger than about 60,000 daltons seems hardly able to enter the nucleus at all. A quantitative analysis of such data suggests that the nuclear pore complex contains a water-filled cylindrical channel about 9 nm in diameter and 15 nm long (Figure 8–22), dimensions that are compatible with the size of the irregular channel seen in some electron micrographs.

The nuclear envelope seems designed to shield the contents of the nuclear compartment (the *nucleoplasm*) from many of the particles, filaments, and large molecules that function in the cytoplasm. Mature cytoplasmic ribosomes, for example, are too large to pass through the putative 9-nm channels, thus ensuring that all protein synthesis is confined to the cytoplasm. But how does the nucleus import the large molecules that it needs, such as DNA and RNA polymerases, which have subunit molecular weights of 100,000 to 200,000? Recent evidence indicates that these and many other nuclear proteins interact with receptor proteins located on the pore margin that actively transport the proteins into the nucleus while enlarging the pore channel.

Proteins Are Actively Transported into the Nucleus Through Nuclear Pores[17]

When proteins are extracted from the nucleus and microinjected back into the cytoplasm, even the very large ones efficiently reaccumulate in the nucleus. One of the best-studied examples is an abundant nuclear protein called *nucleoplasmin*, which can be proteolytically cleaved into a head and a tail piece. Tail pieces are taken up by nuclei in microinjection experiments; head pieces are not (Figure 8–23). When tails are linked to 20-nm-diameter colloidal gold spheres, which are much larger than the inner diameter of the resting nuclear pores, the gold spheres accumulate in the nucleus and are seen in the nuclear pores during transit (Figures 8–23 and 8–24). Evidently nuclear pores can open up to accommodate an

9 nm

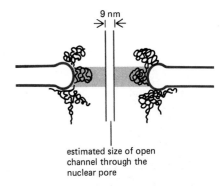

estimated size of open channel through the nuclear pore

Figure 8–22 Highly schematic cross-section of a nuclear pore complex. The idealized cylinder that has been superimposed on the pore represents the relative size of the open channel estimated from transport measurements. In some electron micrographs, poorly visible strands appear to fill most of the interior of the pore. These may be responsible for confining the effective opening to the 9-nm channel shown.

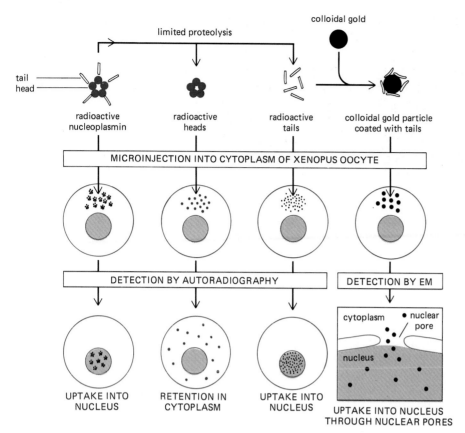

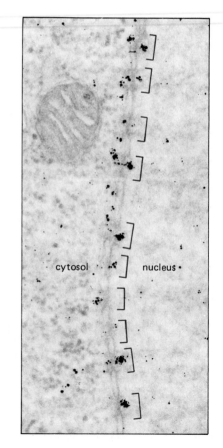

Figure 8–23 Experiments showing the uptake of selected proteins into nuclei via nuclear pores. *Nucleoplasmin* is a large pentameric nuclear protein with distinct head and tail domains. The heads can be cleaved from the tails by limited proteolysis. When injected into the cytoplasm of a frog oocyte, intact nucleoplasmin molecules rapidly accumulate in the nucleus even though they are too large to diffuse passively through the small channel in the center of a nuclear pore complex. The signal for this nuclear import apparently resides in the tail domains, since injected tails are taken up by the nucleus but heads are not. The role of nuclear pores in this signal-directed import is demonstrated by electron microscopy using nucleoplasmin tails coupled to spheres of colloidal gold, which are easily visualized because of their high electron density. The attached nucleoplasmin tails direct the entry of the gold particles via nuclear pores (see Figure 8–24).

object as large and as alien as a gold particle. The pore appears to function like a close-fitting diaphragm that opens to just the right extent when activated by a signal on an appropriate large protein. How this occurs at the molecular level is a mystery.

Proteins like nucleoplasmin are actively transported through the pores, probably while still in their folded forms. Experiments in which active nuclear transport is reconstituted *in vitro* suggest that the energy required is derived from ATP hydrolysis.

8-11 Only Proteins That Contain Nuclear Import Signals
8-12 Are Actively Transported into the Nucleus[18]
8-13

The selectivity of nuclear transport resides in *nuclear import signals*, which are present only in nuclear proteins. The experiments described above demonstrate that a signal of this type resides in the tail piece of nucleoplasmin, but nuclear import signals have been more precisely defined in some other nuclear proteins using genetic engineering techniques. The signal, which can be located anywhere in the protein, consists of a short peptide (typically from four to eight amino acid residues) that is rich in the positively charged amino acids lysine and arginine

Figure 8–24 Electron micrograph showing colloidal gold spheres coated with nucleoplasmin (see Figure 8–23) entering the nucleus by means of nuclear pores, whose positions are indicated by the colored brackets. The same result is obtained when the gold spheres are coated with the tail regions of nucleoplasmin molecules. These gold particles are larger in diameter than a resting pore, implying that a pore has been induced to open to permit their passage. (From C. Feldherr, E. Kallenbach, and N. Schultz, *J. Cell Biol.* 99:2216–2222, 1984.)

(A) LOCALIZATION OF T-ANTIGEN CONTAINING WILD-TYPE NUCLEAR IMPORT SIGNAL

Pro—Pro—Lys—Lys—Lys—Arg—Lys—Val—

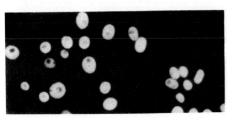

(B) LOCALIZATION OF T-ANTIGEN CONTAINING A MUTATED NUCLEAR IMPORT SIGNAL

Pro—Pro—Lys—Thr—Lys—Arg—Lys—Val—

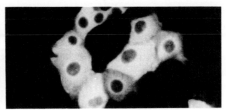

Figure 8–25 The localization of SV40 virus T-antigen containing or lacking a signal peptide that specifies nuclear localization. The wild-type T-antigen protein contains the lysine-rich sequence indicated and is imported to its site of action in the nucleus, as indicated by immunofluorescence staining with antibody against the T-antigen (A). T-antigens with an altered signal peptide (such as the replacement of a lysine with a threonine) remain in the cytoplasm (B). (From D. Kalderon, B. Roberts, W. Richardson, and A. Smith, *Cell* 39:499–509, 1984.)

and usually contains proline. It was first identified in the SV40 virus-encoded protein called *T-antigen*, a large protein of 90,000 daltons that is needed for viral DNA replication in the nucleus. The T-antigen normally accumulates in the nucleus shortly after being synthesized in the cytosol. However, a mutation in a single amino acid prevents nuclear transport and causes the mutant protein to remain in the cytoplasm (Figure 8–25). On the assumption that this mutation is in a nuclear import signal sequence, short lengths of the DNA encoding this region of the normal T-antigen were fused to a gene coding for a cytoplasmic protein. The shortest sequence that caused the resulting "fusion" protein to be imported into the nucleus was determined, and it indicated that the T-antigen nuclear import signal was a stretch of eight contiguous amino acids located in an internal region of the polypeptide chain (see Table 8–3, p. 415). Further experiments showed that the signal sequence can also function if it is synthesized as a short peptide and chemically linked to randomly selected lysine side chains on a protein that would otherwise remain in the cytoplasm. Thus the location of the nuclear import signal on a protein does not seem to be important.

The mechanism of nuclear protein import is fundamentally different from the import mechanisms that we will describe for the other membrane-bounded organelles in that it occurs through a controlled aqueous pore rather than through one or more membranes. Moreover, when a nucleus disassembles in mitosis, its contents mix with those of the cytosol and the nuclear proteins escape. When reassembly occurs, groups of chromosomes are first enveloped within their own double membrane, which is so closely applied that it excludes soluble proteins, among them many former residents of the nucleus (see p. 777). The enveloped chromosomes then fuse to form a single nucleus that must reimport the proteins it requires from the cytosol. Possibly because a nuclear protein molecule needs to be repeatedly imported, its nuclear import signal peptide is not cleaved off after transport into the nucleus. In contrast, once a protein molecule has been imported by any of the other membrane-bounded organelles, it is passed on from generation to generation within that compartment and need never be translocated again; the signal peptide on these molecules is often removed following protein translocation.

Nuclear Pores Transport Selected RNAs out of the Nucleus[19]

The nuclear envelope of a typical mammalian cell contains 3000 to 4000 pores (about 11 pores/μm^2 of membrane area). If the cell is synthesizing DNA, it needs to import about 10^6 histone molecules from the cytoplasm every 3 minutes in order to package newly made DNA into chromatin, which means that, on average, each pore needs to transport about 100 histone molecules per minute. If the cell is growing rapidly, each pore also needs to transport about three newly assembled ribosomes per minute to the cytoplasm, since ribosomes are produced in the nucleus but function in the cytosol (see p. 541). And that is only a very small part of the total traffic that passes through the nuclear pores.

The export of new ribosomal subunits is particularly problematic. Since these particles (about 15 nm in diameter) are much too large to pass through the 9-nm channels, it seems likely that they are specifically exported through the nuclear pores by an active transport system. Similarly, messenger RNA molecules, complexed with special proteins to form ribonucleoprotein particles, are thought to be actively exported from the nucleus. If 20-nm-diameter gold spheres, similar to those used for the nucleoplasmin experiments (see Figure 8–24), are coated with small RNA molecules (tRNA or 5S RNA) and then injected into the nucleus of a frog oocyte, they are rapidly transported through the pores into the cytoplasm. If they are injected into the cytoplasm of the oocyte, on the other hand, they remain there. It seems that, in addition to receptors that recognize nuclear protein import signals, the pore contains one or more receptors that recognize RNA molecules (or the proteins bound to them) destined for the cytosol; when these receptors are occupied, the pore catalyzes active transport outward instead of inward. Although several nuclear pore proteins have recently been isolated, including an abundant 190,000-dalton membrane protein, it is still not known how the pores function.

Summary

The nucleus is enclosed by an envelope consisting of a concentric pair of membranes. The outer nuclear membrane is continuous with the ER membrane, and the space between it and the inner nuclear membrane is continuous with the ER lumen. RNA molecules and ribosomes are made in the nucleus and are exported to the cytosol, while all of the proteins that function in the nucleus are synthesized in the cytosol and are imported. The exchange of materials between nucleus and cytosol occurs through nuclear pores that provide a direct passageway through the inner and outer nuclear membranes.

Proteins containing nuclear import signals are actively transported inward through the pores, being recognized by their short, positively charged signal peptide; because this signal peptide is not removed, nuclear proteins can be imported repeatedly, as is required each time the nucleus reassembles following mitosis. RNA molecules and probably ribosomal subunits are actively transported outward through the pores.

The Transport of Proteins into Mitochondria and Chloroplasts[20]

As discussed in detail in Chapter 7, mitochondria and chloroplasts are double-membrane-bounded organelles that specialize in the synthesis of ATP—by electron transport and oxidative phosphorylation in mitochondria and by photosynthetic phosphorylation in chloroplasts. Although both contain their own DNA and machinery for protein synthesis, most of their proteins are encoded by the cell nucleus and imported from the cytosol. Moreover, each imported protein must reach the particular subcompartment in which it functions. For mitochondria there are four subcompartments: the **matrix space,** the **inner membrane,** the **intermembrane space,** and the **outer membrane** that faces the cytosol (Figure 8–26A). For chloroplasts there are, in addition, the *thylakoid membrane* and the *thylakoid space* (Figure 8–26B). Each of these subcompartments contains a distinct set of proteins. The growth of mitochondria and chloroplasts by the import of cytoplasmic proteins is therefore a major feat, involving selective translocation across one, two, or (in chloroplasts) even three membranes in succession.

The relatively few proteins encoded in the genomes of these organelles are located mostly in the inner membrane in mitochondria and in the thylakoid membrane in chloroplasts. The organelle-encoded polypeptides generally form subunits of protein complexes whose other subunits are encoded by nuclear genes and are imported from the cytosol. The formation of such hybrid protein complexes requires a balanced synthesis of the two types of subunits; how protein synthesis is coordinated on different types of ribosomes located two membranes apart is still largely a mystery (see p. 396).

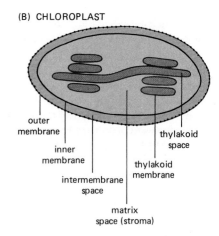

(A) MITOCHONDRION

outer membrane
inner membrane
matrix space
intermembrane space

(B) CHLOROPLAST

outer membrane
inner membrane
intermembrane space
thylakoid space
thylakoid membrane
matrix space (stroma)

Figure 8–26 The major subcompartments of mitochondria and chloroplasts. The topology of the chloroplast can be derived from that of the mitochondrion in a simple way: pinching off the invaginations of the inner mitochondrial membrane to create vesicles would generate a compartment that is topologically equivalent to the thylakoid vesicle in chloroplasts.

8-16 Mitochondrial Signal Peptides Are Amphipathic Amino Acid Sequences[21]

The study of mitochondrial biogenesis has been facilitated by the use of yeasts, into which hybrid genes encoding fusion proteins (produced by recombinant DNA techniques) can be introduced efficiently. Much more is known about import mechanisms in mitochondria than in chloroplasts, but it seems that the mechanisms are virtually identical in the two organelles—although chloroplasts have an extra internal membrane-bounded compartment, the thylakoid.

Proteins imported into the mitochondrial matrix are usually taken up from the cytosol within a minute or two of their release from free polyribosomes. These proteins almost always have a signal peptide (20–80 residues long) at their amino terminus. After being imported, the signal peptide is rapidly removed by a specific protease (a *signal peptidase*) in the mitochondrial matrix and then probably is degraded to amino acids in the matrix. The signal peptide can be remarkably simple. Molecular genetic experiments in which the signal peptide is progressively reduced in length have shown that, for one mitochondrial protein, only 12 amino acids at the amino terminus are needed to signal mitochondrial import. These 12 residues can be attached to any cytoplasmic protein and will direct the protein into the mitochondrial matrix. Physical studies of full-length signal peptides suggest that they can form amphipathic α-helical structures (Figure 8–27) in which positively charged residues all line up on one side of the helix while uncharged hydrophobic residues line up toward the opposite side.

The mitochondrial outer membrane is believed to contain receptor proteins that bind the mitochondrial signal peptides, aiding the translocation process, but these putative import receptors have not yet been well characterized.

Translocation into the Mitochondrial Matrix Depends on Both the Electrochemical Gradient Across the Inner Membrane and ATP Hydrolysis[22]

Almost everything we know about the molecular mechanism of protein import into mitochondria has been learned from analysis of cell-free, reconstituted transport systems. Mitochondria are first purified by differential centrifugation of homogenized cells and then incubated with radiolabeled proteins destined for mitochondria (**mitochondrial precursor proteins**). Purified precursor proteins are generally taken up into such mitochondria rapidly and efficiently.

All forms of vectorial movement and transport require energy. In most cases the energy is supplied in the form of ATP. In the case of mitochondrial import, however, an electrochemical gradient across the mitochondrial inner membrane is also required. This gradient is formed as protons are pumped from the matrix to the intermembrane space during electron transport (see p. 351). The mitochondrial outer membrane is freely permeable to ions, so no gradient is maintained across it. The energy in the electrochemical gradient across the inner membrane is tapped like a battery to drive most of the cell's ATP synthesis, but it is also used to help drive the import of proteins bearing the positively charged mitochondrial import signal peptide: when ionophores that collapse the mitochondrial membrane potential are added (see p. 364), import is blocked. How does the electrochemical gradient help drive protein translocation? The answer is still uncertain.

Mitochondrial Proteins Are Imported into the Matrix at Contact Sites That Join the Two Membranes[23]

Does a protein reach the mitochondrial matrix by crossing the two membranes one at a time, or does it pass through both at once? To find out, a cell-free import system can be cooled to ice temperature, trapping the proteins at an intermediate step in the act of translocation. The proteins that accumulate at this step have their amino termini in the matrix space (as indicated by the finding that their amino-terminal signal peptides have already been removed by the matrix pro-

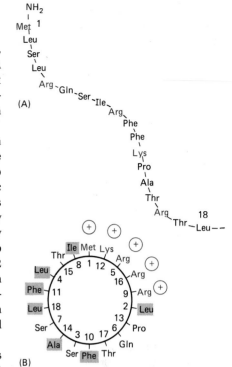

Figure 8–27 A signal peptide for mitochondrial protein import. Cytochrome oxidase is a large multiprotein complex located in the mitochondrial inner membrane, where it functions as the terminal enzyme in the electron-transport chain (see p. 360). (A) The first 12 amino acids of the precursor to subunit IV of this enzyme suffice as a signal peptide for the import of this subunit into the mitochondrion. (B) When the full-length signal peptide is folded as an α helix with 3.6 residues per turn and viewed from the top, the positively charged residues (*color*) are seen to be clustered on one face of the helix while the nonpolar residues (*boxed*) are clustered on the opposite face. Mitochondrial signal peptide sequences almost always have the potential to form such an amphipathic helix. A helix of this type is thought to be important in aiding protein translocation across the mitochondrial membranes.

tease); yet the bulk of the protein can still be attacked from outside the mito-chondria by externally added proteolytic enzymes. This result demonstrates that the precursor proteins pass through both mitochondrial membranes at once to enter the matrix. Electron microscopists have noted numerous **contact sites** at which the inner and outer mitochondrial membranes appear to be joined, suggesting that these are the sites where translocation into the matrix occurs. Recently these translocation contact sites have been identified biochemically by their association with partially translocated precursor proteins and have been partially purified.

When cooled mitochondria containing partly translocated intermediates are warmed up, import is rapidly completed (Figure 8–28), even if the membrane potential across the inner membrane is collapsed. The membrane potential is apparently required only for the initial penetration, which occurs even at low temperature. The remainder of the transport process, however, requires the pres-ence of ATP outside the mitochondrion. These findings imply that the import process normally occurs in two stages: (1) an electrically driven penetration of the signal peptide and adjoining sequences across both mitochondrial membranes and (2) movement of the remainder of the chain into the matrix, which requires both ATP hydrolysis and physiological temperatures (Figure 8–29).

Proteins Unfold as They Are Imported into the Mitochondrial Matrix[24]

Precursor proteins presumably unfold before crossing the two mitochondrial membranes at a contact site. It is difficult to envisage how a folded, water-soluble protein could straddle two (or even one) lipid bilayers while retaining its native three-dimensional conformation. It is also hard to imagine a protein pore opening to accommodate globular proteins that vary greatly in size and shape without letting protons pass through it and thereby collapsing the electrochemical gra-dient across the inner membrane. On the other hand, all proteins have a similar conformation when unfolded, so they could be translocated by a common mech-anism. But since the folded state of a protein is of lower free energy than the unfolded state (which is why polypeptides spontaneously fold up), unfolding a protein requires energy. ATP hydrolysis is believed to provide this energy.

To test whether precursor proteins unfold as they cross the mitochondrial membrane, a hybrid gene was engineered to encode a fusion protein in which a mitochondrial signal peptide was linked to the amino terminus of dihydrofolate reductase (DHFR), which is a cytosolic enzyme. The fusion protein had nearly full enzyme activity, implying that the DHFR portion was in its native three-dimen-sional conformation. When presented to isolated mitochondria, the fusion protein was imported into the matrix. If the fusion protein was first treated with the drug methotrexate, however, which binds so tightly to the active site of the enzyme that it makes the DHFR much more difficult to unfold, import was severely inhibited.

Genetic studies in yeasts suggest that certain members of the hsp70 family of stress-response proteins (see p. 420) are essential for an ATP-dependent protein-

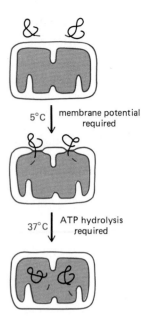

Figure 8–28 Proteins transiently span both the inner and outer mitochondrial membranes during their translocation into the matrix. When isolated mitochondria are incubated with a precursor protein at 5°C, the precursor is only partially translocated. The amino-terminal signal peptide is cleaved off in the matrix; most of the polypeptide chain remains outside the mitochondria (accessible to proteolytic enzymes). Upon warming to 37°C, the translocation is completed. The initial penetration of the mitochondrial membranes at 5°C is driven by the voltage difference across the inner membrane. The subsequent translocation can occur without this membrane potential, but it requires ATP on the cytoplasmic side of the inner membrane. The hydrolysis of ATP is thought to be used to unfold the polypeptide chain to allow the protein to pass through the membrane.

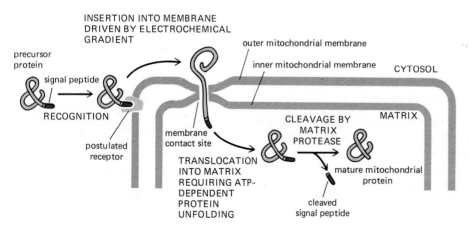

Figure 8–29 Protein import by mitochondria. The amino-terminal signal peptide of the precursor protein is recognized by a receptor postulated to reside in the outer membrane. The protein is translocated across both mitochondrial membranes at special contact sites, driven first by the electrochemical gradient across the inner membrane. The signal peptide is cleaved off by a specific protease in the matrix to form the mature protein.

unfolding reaction on the cytosolic side of both mitochondrial and ER membranes during protein import (see p. 440): when the genes encoding these proteins are inactivated, both mitochondrial precursor proteins and proteins destined for the ER fail to be transported across their respective membranes and accumulate in the cytosol instead.

8-17 Protein Transport into the Mitochondrial Intermembrane Space Requires Two Signals[25]

The transport of several precursor proteins to the mitochondrial intermembrane space begins with their initial transfer into the matrix, as illustrated in Figure 8–29. A very hydrophobic amino acid sequence, however, is strategically placed after the amino-terminal signal peptide that initiates import; once the amino-terminal signal is cleaved by the matrix protease, the hydrophobic sequence functions as a signal peptide to reinsert the protein into the inner membrane. This transfer from the matrix presumably occurs by a mechanism similar to that used for protein import into the ER membrane (see p. 439), and it is also the mechanism used to insert proteins that are encoded in the mitochondrion into the inner membrane (Figure 8–30A).

After proteins destined for the intermembrane space have been inserted into the inner membrane, they are cleaved by a protease in the intermembrane space to release the mature polypeptide chain as a soluble protein (Figure 8–30B). Many of these proteins ultimately become attached as peripheral membrane proteins to the outer surface of the inner membrane, where they form subunits of protein complexes that also contain transmembrane proteins.

The transport of proteins from the cytosol to the mitochondrial inner membrane also requires a hydrophobic signal peptide. It might occur by the mechanism outlined in Figure 8–30A, but this has not been demonstrated directly. Such a two-step pathway is difficult to distinguish experimentally from an alternative pathway in which the transport at a contact site is aborted when the hydrophobic signal is reached, leaving the protein imbedded in the bilayer of the inner membrane.

Transfer from the Cytosol to the Mitochondrial Outer Membrane also Requires Protein Unfolding[26]

The mitochondrial outer membrane has an unusual structure, which is reminiscent of the outer membrane of gram-negative bacteria (see p. 312) in that its lipid bilayer contains large amounts of a pore-forming protein called *porin*. For this reason the outer membrane is freely permeable to inorganic ions and metabolites and to protein molecules smaller than about 10,000 daltons. But the outer membrane does present a permeability barrier to larger proteins, and it therefore helps to keep proteins in the intermembrane space from leaking back into the cytosol.

The insertion of nuclear-encoded proteins such as porin into the outer membrane occurs by an ATP-dependent mechanism but does not involve the use of a cleaved signal peptide. Membrane potential is not required. Very little is known about how this transfer takes place. The amino terminus of at least one outer membrane protein seems to contain a normal matrix translocation signal followed by an anchoring sequence that somehow aborts translocation at the outer membrane.

Figure 8–30 Import of proteins from the cytosol to the mitochondrial intermembrane space or inner membrane requires multiple signals. A pathway that requires two signal peptides and two translocation events is known to be used to move some proteins from the cytosol to the intermembrane space. The protein is first imported into the matrix space as in Figure 8–29. Cleavage of the signal peptide used for the initial translocation, however, unmasks an adjacent hydrophobic signal peptide at the new amino terminus. This signal causes the protein to be inserted into the inner membrane by the same pathway that is used to insert proteins encoded by the mitochondrial genome into this membrane (A). This mechanism is presumably similar to the one used by the bacterial ancestor of the mitochondrion to insert proteins into its plasma membrane, and it is thought to resemble the ER insertion mechanism (see p. 439). The transport of proteins into the intermembrane space requires a third step, in which a protease with its active site in the intermembrane space cleaves the bulk of the protein from its transmembrane signal peptide in the inner membrane (B). The pathway in (A) may also be used to transfer proteins from the cytosol to the inner membrane, which likewise requires a hydrophobic peptide.

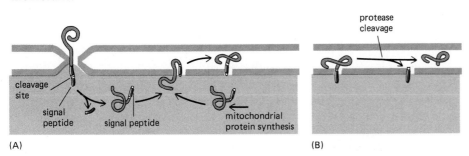

protease
cleavage

cleavage
site

signal
peptide   signal peptide

mitochondrial
protein synthesis

(A)   (B)

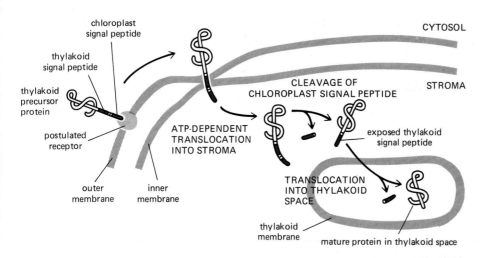

Figure 8–31 Translocation into the thylakoid space of chloroplasts requires two signal peptides and two translocation events. The precursor polypeptide contains an amino-terminal chloroplast signal peptide followed immediately by a thylakoid signal peptide. The chloroplast signal peptide initiates translocation into the stroma through a membrane contact site (see Figure 7–73, p. 398) by a mechanism similar to that used for translocation into the mitochondrial matrix (Figure 8–29). The signal peptide is then cleaved off, unmasking the thylakoid signal peptide, which initiates translocation across the thylakoid membrane by a mechanism that presumably resembles the one used to move proteins into the ER (see p. 439).

8-18 Two Signal Peptides Are Required to Direct Proteins to the Thylakoid Membrane in Chloroplasts[27]

Protein transport into chloroplasts resembles transport into mitochondria in many respects: both occur post-translationally, both require energy, and both utilize hydrophilic amino-terminal signal peptides that are removed after use. There is at least one important difference, however: mitochondria exploit the electrochemical gradient across their inner membrane to help drive the transport, whereas chloroplasts, which have an electrochemical gradient across their thylakoid but not their inner membrane (see Chapter 7), appear to employ only ATP hydrolysis to power import across their double-membrane outer envelope.

Although the signal peptides for import into chloroplasts resemble those described previously for import into mitochondria, mitochondria and chloroplasts are both present in the same plant cells, and proteins must choose appropriately between them. In plants, for example, a bacterial enzyme is targeted specifically to mitochondria if it is joined (by recombinant DNA methods) to an amino-terminal sequence of a mitochondrial protein; the same enzyme joined to an amino-terminal sequence of a chloroplast protein ends up in chloroplasts.

Chloroplasts have an extra membrane-bounded compartment, the **thylakoid.** Many chloroplast proteins, including protein subunits of the photosynthetic system and of the ATP synthetase, are imported into the thylakoid membrane from the cytosol. Like some mitochondrial precursors, these proteins are transported to their final destination in two steps. First they pass across the double-membrane envelope into the matrix space of the chloroplast (termed the **stroma**) and then are translocated into the thylakoid membrane (or across this membrane into the thylakoid space). The precursors of these proteins have a hydrophobic thylakoid signal peptide following the amino-terminal chloroplast signal peptide (see p. 398). After the amino-terminal signal peptide has been used to import the protein into the stroma, it is removed by a stromal protease (analogous to the matrix protease in mitochondria). This cleavage unmasks the thylakoid signal peptide, which then initiates transport across the thylakoid membrane (Figure 8–31). As for mitochondria, the second step is the pathway used to insert chloroplast-encoded proteins into the thylakoid membrane; the protein translocator required presumably originated in the chloroplast's bacterial ancestor.

Summary

Mitochondria and chloroplasts import most of their proteins from the cytosol using similar mechanisms. The transport processes involved have been most extensively studied in mitochondria, especially in yeasts. A protein is translocated into the mitochondrial matrix space by passing through sites of adhesion between the outer and inner membranes called contact sites. Translocation is driven by both ATP hydrolysis and the electrochemical gradient across the inner membrane, and the transported protein is unfolded as it crosses the mitochondrial membranes. Only proteins that contain a specific signal peptide are translocated into mitochondria or chloroplasts. The signal peptide is usually located at the amino terminus and is cleaved off after import. Transport to the inner membrane can occur as a second step if a hydrophobic signal peptide is also present in the imported protein; this second signal peptide is unmasked when the first signal peptide is cleaved. In the case of chloroplasts, import from the stroma into the thylakoid likewise requires a second signal peptide.

Peroxisomes[28]

Peroxisomes (also called *microbodies*) differ from mitochondria and chloroplasts in many ways. Most notably, these organelles are surrounded by only a single membrane, and they do not contain DNA or ribosomes. In spite of these differences, peroxisomes are thought to be formed by a similar process of selective protein (and lipid) import from the cytosol. Because peroxisomes have no genome, however, all of their proteins must be supplied from the cytosol. Peroxisomes thus resemble the ER in being a self-replicating membrane-bounded organelle that exists without a genome of its own.

Because we do not discuss peroxisomes elsewhere, we shall pause to consider some of the functions of this diverse family of organelles before discussing their biosynthesis. Although they are present in all eucaryotic cells, peroxisomes differ widely in function in different types of cells.

By the early 1960s a combination of biochemistry and electron microscopy had identified a unique organelle with a diameter of about 0.5 μm as a concentrated source of at least three oxidative enzymes in liver cells: D-*amino acid oxidase*, *urate oxidase*, and *catalase*. In mammals, peroxisomes of this size are confined to a few cell types. They sometimes stand out in electron micrographs because of the presence of a "crystalloid" core composed of urate oxidase (Figure 8–32). With the later development of a histochemical stain for catalase, an enzyme that constitutes up to 40% of the total peroxisomal protein, it was discovered that all cells contain peroxisomes. In most cells the peroxisomes are smaller (0.15–0.25 μm in diameter) than in liver cells.

Like the mitochondrion, the peroxisome is a major site of oxygen utilization. One hypothesis is that the peroxisome is a vestige of an ancient organelle that carried out all of the oxygen metabolism in the primitive ancestors of eucaryotic cells. When the oxygen produced by photosynthetic bacteria first began to accumulate in the atmosphere (see p. 385), it would have been highly toxic to most cells. Peroxisomes may have served to lower the concentration of oxygen in such cells while also exploiting its chemical reactivity to carry out useful oxidative reactions. According to this view, the later development of mitochondria rendered the peroxisome largely obsolete because many of the same reactions—which had formerly been carried out in peroxisomes without producing energy—were now coupled to ATP formation by means of oxidative phosphorylation. The oxidative reactions carried out by peroxisomes in present-day cells would therefore be those that have remained useful despite the presence of mitochondria.

Peroxisomes Use Molecular Oxygen and Hydrogen Peroxide to Carry Out Oxidative Reactions[29]

Peroxisomes are so called because they usually contain one or more enzymes that use molecular oxygen to remove hydrogen atoms from specific organic substrates (designated here as R) in an oxidative reaction that produces hydrogen peroxide (H_2O_2):

$$RH_2 + O_2 \rightarrow R + H_2O_2$$

Catalase utilizes the H_2O_2 generated by other enzymes in the organelle to oxidize a variety of other substrates—including phenols, formic acid, formaldehyde, and alcohol—by the "peroxidative" reaction: $H_2O_2 + R'H_2 \rightarrow R' + 2H_2O$. This type of

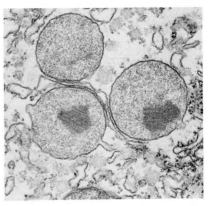

Figure 8–32 Electron micrograph of three peroxisomes in a rat liver cell. The paracrystalline electron-dense inclusions are the enzyme urate oxidase. (Courtesy of Daniel S. Friend.)

200 nm

oxidative reaction is particularly important in liver and kidney cells, whose peroxisomes detoxify various toxic molecules that enter the bloodstream. Almost half of the ethanol we drink is oxidized to acetaldehyde in this way. In addition, when excess H_2O_2 accumulates in the cell, catalase converts H_2O_2 to H_2O ($2H_2O_2 \rightarrow 2H_2O + O_2$).

Peroxisomes are unusually diverse organelles and even in the different cells of a single organism may contain very different sets of enzymes. They also can adapt remarkably to changing conditions. Yeast cells grown on sugar, for example, have small peroxisomes. But when some yeasts are grown on methanol, they develop large peroxisomes that oxidize methanol; when grown on fatty acids, they develop large peroxisomes that break down fatty acids to acetyl CoA.

Peroxisomes have especially important roles in plants. Two very different types have been studied extensively. One type is present in leaves (Figure 8–33A), where it catalyzes the oxidation of a side product of the reaction that fixes CO_2 in carbohydrate; this process is called *photorespiration* because it uses up O_2 and liberates CO_2 (see p. 371). The other type of peroxisome is present in germinating seeds (Figure 8–33B), where it plays an essential role in converting the fatty acids stored in seed lipids into the sugars needed for the growth of the young plant. Because this conversion of fats to sugars is accomplished by a series of reactions known as the *glyoxylate cycle*, these peroxisomes are also called *glyoxysomes*. In the glyoxylate cycle, two molecules of acetyl CoA produced by fatty acid breakdown in the peroxisome are used to make succinic acid, which leaves the peroxisome and is converted into glucose. The glyoxylate cycle does not occur in animal cells, and animals are thus unable to convert the fatty acids in fats into carbohydrates.

Figure 8–33 Electron micrographs of two types of peroxisomes found in plant cells. (A) A leaf peroxisome with a paracrystalline core in a tobacco leaf mesophyll cell. Its close association with chloroplasts is thought to facilitate the exchange of materials between these organelles during photorespiration (see p. 371). (B) Peroxisomes in a fat-storing cotyledon cell of a tomato seed, 4 days after germination. Here the peroxisomes (*glyoxysomes*) are associated with the lipid bodies where fat is stored, reflecting their central role in fat mobilization and gluconeogenesis during seed germination. (A, courtesy of P.J. Gruber and E.H. Newcomb; B, courtesy of S.E. Frederick and E.H. Newcomb.)

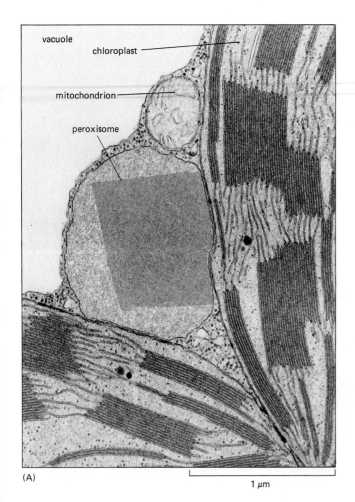

(A)

1 μm

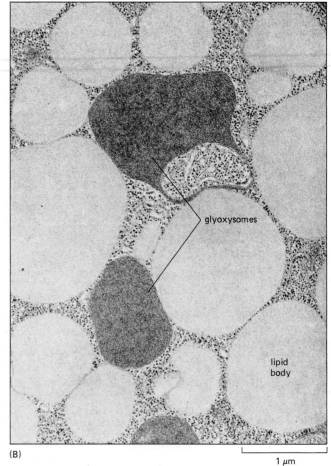

(B)

1 μm

All the Components of Peroxisomes Are Imported from the Cytosol[30]

Peroxisomes can be thought of as a collection of organelles with a constant membrane and a variable content. As mentioned previously, they contain no DNA or ribosomes; all of their proteins are encoded by nuclear genes and are synthesized in the cytosol. The proteins in both the lumen and the membrane of the peroxisome are taken up post-translationally from the cytosol. *Catalase* has been the most thoroughly studied of the peroxisomal proteins. It is a tetrameric heme-containing protein that is made in the cytosol as heme-free monomers; the monomers are imported into the lumen of peroxisomes, where they assemble into tetramers in the presence of heme. Although catalase does not contain a signal sequence that is cleaved after use, it must contain a signal that directs it to the peroxisome. Recent evidence suggests that at least part of the signal is a specific sequence of three amino acids located near the carboxyl terminus of many peroxisomal proteins.

Peroxisomes presumably have at least one unique protein exposed on their cytosolic surface to act as a receptor that recognizes a signal on the imported proteins (see p. 415). At one time it was thought that the membrane "shell" of the peroxisome forms by budding from the ER, while the "content" is imported from the cytosol. However, there is now evidence suggesting that new peroxisomes always arise from preexisting ones, being formed by organelle growth and fission, as described previously for mitochondria and chloroplasts (see p. 388). It is thought that all of the membrane proteins of the peroxisome, including the postulated receptor protein(s), are imported directly from the cytosol (Figure 8–34). Presumably, the lipids required to make new peroxisome membrane are also imported from the cytosol, possibly being carried by phospholipid exchange proteins from sites of synthesis in the ER membrane (see p. 450).

Figure 8–34   A model for how peroxisomes are assembled. The peroxisome membrane contains specific import receptor proteins. All peroxisomal proteins, including new copies of the import receptor, are synthesized by cytosolic ribosomes and then imported from the cytosol. Thus peroxisomes form only from preexisting peroxisomes by a process of growth and fission; like mitochondria and chloroplasts, they continually import new components from the cytosol.

Summary

Peroxisomes are specialized for carrying out oxidative reactions using molecular oxygen. They generate hydrogen peroxide, which they also use for oxidative purposes—destroying the excess by means of the catalase they contain. Like mitochondria and chloroplasts, peroxisomes are believed to be self-replicating organelles. However, they contain no DNA or ribosomes. They are thought to contain a unique membrane receptor that allows them to import all of their proteins (including the receptor itself) by selective transfer from the cytosol.

The Endoplasmic Reticulum[31]

All eucaryotic cells have an **endoplasmic reticulum (ER).** Its highly convoluted single membrane typically constitutes more than half of the total membrane of the cell (see Table 8–2). Despite being organized into a netlike meshwork that extends throughout the cytoplasm (Figure 8–35), the ER membrane is thought to

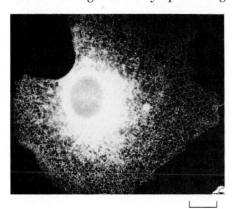

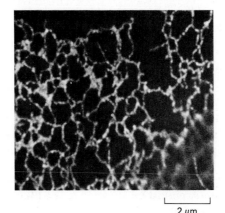

Figure 8–35   Fluorescence micrograph of a cultured mammalian cell stained with an antibody that binds to a protein retained in the ER. The ER extends like a net over the whole cell cytoplasm, so that all regions of the cytosol are close to some portion of the ER membrane. (Courtesy of Hugh Pelham.)

10 μm

2 μm

433   The Endoplasmic Reticulum

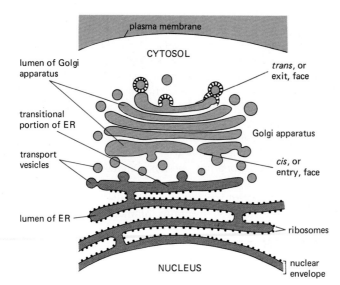

Figure 8–36 The relationship of the ER lumen to other intracellular compartments with which the ER communicates. The ER lumen is separated from both the cytosol and the nucleus by only a single membrane, whereas it is separated from the lumen of each of the stacked cisternae of the Golgi apparatus by two membranes. For most purposes the ER and the Golgi apparatus can be thought of as forming a single functional unit whose parts communicate via transport vesicles.

form a continuous sheet enclosing a single internal space. This highly convoluted space is called the **ER lumen** or the *ER cisternal space*, and it often occupies more than 10% of the total cell volume (see Table 8–1). The ER membrane separates the ER lumen from the cytosol, and it mediates the rapid transfer of selected molecules between these two compartments. By contrast, the lumen of the ER and the lumen of each cisterna of the Golgi apparatus are separated from each other by two membranes plus the intervening cytosol, and the extensive macro-molecular traffic between these two organelles is mediated by transport vesicles (Figure 8–36).

The ER plays a central part in cell biosyntheses. The transmembrane proteins and lipids of the ER, Golgi, lysosome, and plasma membranes all begin their synthesis in association with the ER membrane. The ER membrane also makes a major contribution to the mitochondrial and peroxisomal membranes by producing most of the lipids for these organelles (see p. 449). In addition, all of the newly synthesized proteins that will end up in the *lumen* of the ER, Golgi apparatus, or lysosome—as well as those to be secreted to the cell exterior—are initially delivered to the ER lumen. Since the ER is the starting point for the synthesis of all secreted proteins, it is also the site where the formation of the extracellular matrix begins.

Membrane-bound Ribosomes Define the Rough ER[32]

The ER removes selected proteins from the cytosol as they are being synthesized. These proteins are of two types: (1) transmembrane proteins, which are only partly translocated across the ER membrane and become embedded in it, and (2) water-soluble proteins, which are fully translocated across the ER membrane and are released into the lumen of the ER. The transmembrane proteins are destined to reside in the plasma membrane or the membrane of another organelle, whereas the water-soluble proteins are destined either for the lumen of an organelle or for secretion. All of these proteins are translocated across the ER membrane by the same mechanism after recognition of the same kind of signal peptide.

In mammalian cells the import of proteins into the ER begins before the polypeptide chain is completely synthesized—that is, it occurs *co-translationally*. This distinguishes the process from import into mitochondria, chloroplasts, nuclei, and peroxisomes, which is post-translational and requires different signal peptides. Since a protein is usually translocated into the ER as the polypeptide chain is being made, the ribosome responsible must be attached to the ER membrane. These *membrane-bound ribosomes* coat the surface of the ER, creating regions termed **rough endoplasmic reticulum** (Figure 8–37).

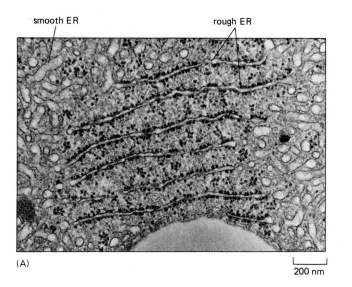

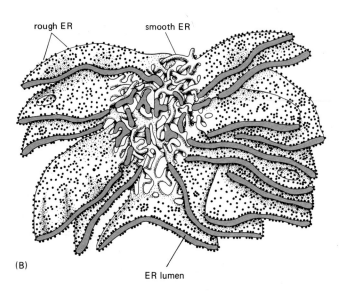

smooth ER   rough ER

rough ER   smooth ER

(A)   (B)

200 nm

ER lumen

There are therefore two spatially separate populations of ribosomes in the cytoplasm. **Membrane-bound ribosomes,** attached to the cytosolic side of the ER membrane, are engaged in the synthesis of proteins that are being concurrently translocated into the ER. **Free ribosomes,** unattached to any membrane, make all other proteins encoded by the nucleus. Membrane-bound and free ribosomes are structurally and functionally identical. They differ only in the proteins they are making at any given time. When a ribosome happens to be making a protein with an ER signal peptide, the signal directs the ribosome to the ER membrane. Since many ribosomes can bind to a single mRNA molecule, a **polyribosome** is usually formed, which becomes attached to the membrane by multiple growing polypeptide chains. The individual ribosomes associated with such an mRNA molecule can return to the cytosol as they finish translation near the 3′ end of the mRNA molecule. The mRNA itself, however, tends to remain attached to the ER membrane by a changing population of membrane-bound ribosomes (Figure 8–38). In contrast, if an mRNA molecule encodes a protein that lacks an ER signal peptide, the polyribosome that forms remains free in the cytosol and its protein product is discharged there. Therefore, only those mRNA molecules that encode proteins with an ER signal peptide bind to rough ER membranes; those mRNA molecules that encode all other proteins remain free in the cytosol. The individual ribosomes are thought to move randomly between these two segregated populations of mRNA molecules.

Figure 8–37 (A) Electron micrograph showing the very different morphologies of the rough and smooth ER. The Leydig cell shown produces steroid hormones in the testis and has an unusually extensive smooth ER. Part of a large spherical lipid droplet is also seen. (B) Three-dimensional reconstruction of a region of the smooth and rough ER in a liver cell. The rough ER receives its name from the many ribosomes on its cytosolic surface; it forms oriented stacks of flattened cisternae, each having a luminal space 20 to 30 nm wide. The smooth ER membrane is connected to these cisternae and forms a fine network of tubules 30 to 60 nm in diameter. It is thought that the ER membrane is continuous and encloses a single lumen. (A, courtesy of Daniel S. Friend; B, after R.V. Krstić, Ultrastructure of the Mammalian Cell. New York: Springer-Verlag, 1979.)

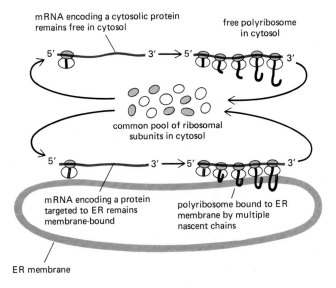

mRNA encoding a cytosolic protein remains free in cytosol

free polyribosome in cytosol

5′   3′   5′   3′

common pool of ribosomal subunits in cytosol

5′   3′   5′   3′

mRNA encoding a protein targeted to ER remains membrane-bound

polyribosome bound to ER membrane by multiple nascent chains

ER membrane

Figure 8–38 A common pool of ribosomes is used to synthesize both the proteins that stay in the cytosol and those that are transported into the ER. It is the ER signal peptide on a newly formed polypeptide chain that directs the engaged ribosome to the ER membrane. The mRNA molecule may remain permanently bound to the ER as part of a polyribosome, while the ribosomes that move along it are recycled; at the end of each round of protein synthesis, the ribosomal subunits are released and rejoin the common pool in the cytosol.

Smooth ER Is Abundant in Some Specialized Cells[33]

Regions of ER that lack bound ribosomes are called *smooth ER*. The great majority of cells contain little, if any, true smooth ER; instead, a small region of the ER is partly smooth and partly rough. This region is called the **transitional elements** because it is from here that transport vesicles carrying newly synthesized proteins and lipids bud off for transport to the Golgi apparatus (see Figure 8–9). In certain specialized cells, however, the smooth ER is abundant and has other functions. In particular, the **smooth endoplasmic reticulum** may predominate in cells that specialize in lipid metabolism. Cells that synthesize steroid hormones from cholesterol, for example, have an expanded smooth ER compartment to accommodate the enzymes needed both to make cholesterol and to modify it to form the hormones (see Figure 8–37A).

The main cell type in the liver, the *hepatocyte*, provides another example. It is the principal site of production of lipoprotein particles for export. The enzymes that synthesize the lipid components of lipoproteins are located in the membrane of the smooth ER, which also contains enzymes that catalyze a series of reactions to detoxify both lipid-soluble drugs and harmful compounds produced by metabolism. The most extensively studied of the *detoxification reactions* are catalyzed by the *cytochrome P450* family of enzymes. These proteins use high-energy electrons, transferred from NADPH by a reductase enzyme in the ER, to add hydroxyl groups to any one of a variety of potentially harmful water-insoluble hydrocarbons dissolved in the bilayer. Other enzymes in the ER membrane then add negatively charged water-soluble molecules (such as sulfate or glucuronic acid) to these hydroxyl groups. After a series of such reactions, a water-insoluble drug or metabolite that would otherwise accumulate in cell membranes is rendered sufficiently water-soluble to leave the cell and be excreted in the urine. Because the rough ER alone cannot house enough of these and other necessary enzymes, a major portion of the membrane in a hepatocyte normally consists of smooth ER (see Table 8–2).

When large quantities of certain compounds, such as the drug phenobarbitol, enter the circulation, detoxification enzymes are synthesized in the liver in unusually large amounts, and the smooth ER doubles in surface area within a few days. Once the drug disappears, the excess smooth ER membranes appear to be removed specifically by a lysosome-dependent process (involving intermediate structures called *autophagosomes*—see p. 460), and the smooth ER returns to normal within 5 days. How these dramatic changes are regulated is not known.

Muscle cells have a specialized and "smooth-ER-like" organelle called the *sarcoplasmic reticulum*, which sequesters Ca^{2+} from the cytosol. The Ca^{2+}-ATPase that pumps in Ca^{2+} (see p. 307) is the major membrane protein in the sarcoplasmic reticulum. The release of Ca^{2+} from this organelle and its subsequent reuptake from the cytosol mediates the rapid contraction and relaxation of the myofibrils during each round of muscle contraction (see p. 621). A similar Ca^{2+}-sequestering organelle of smaller volume is present in most eucaryotic cells, liberating Ca^{2+} into the cytosol in response to extracellular signals. Although originally thought to be the ER compartment itself, it now appears to be a separate membrane-bounded compartment of unknown origin (see p. 701).

We shall now discuss the two major roles of the ER: the synthesis and modification of proteins, and the synthesis of lipids.

Rough and Smooth Regions of ER Can Be Separated by Centrifugation[34]

In order to study the functions and biochemistry of the endoplasmic reticulum, it is necessary to separate the ER membranes from other components of the cell. Intuitively this seems a hopeless task, since the ER is interleaved extensively with other components of the cytoplasm (see Figure 8–35). Fortunately, when tissues or cells are disrupted by homogenization, the ER is fragmented into many small (~100 nm in diameter) closed vesicles called **microsomes,** which are relatively easy to purify.

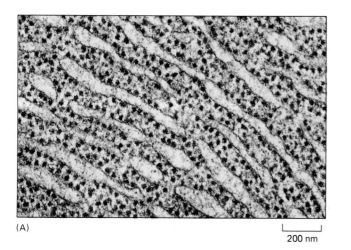

(A)

200 nm

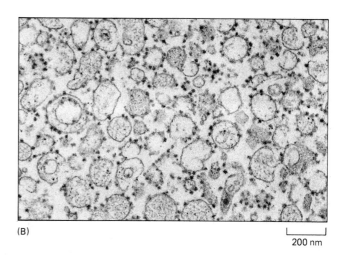

(B)

200 nm

Microsomes derived from rough ER are studded with ribosomes and are called *rough microsomes*. The ribosomes are always found on the *outside* surface, the interior being biochemically equivalent to the luminal space of the ER (Figure 8–39). Many vesicles of a size similar to that of rough microsomes, but lacking attached ribosomes, are also found in these homogenates. Such *smooth microsomes* are derived in part from smooth portions of the ER and in part from vesiculated fragments of plasma membrane, Golgi apparatus, endosomes, and mitochondria (the ratio depending on the tissue). Thus, while rough microsomes can be equated with rough portions of ER, the origins of smooth microsomes cannot be so easily assigned. An outstanding exception is the liver. Because of the exceedingly large quantities of smooth ER in the hepatocyte, most of the smooth microsomes in liver homogenates are derived from smooth ER.

Ribosomes, because they contain large amounts of RNA, make rough microsomes more dense than smooth microsomes. As a result, the rough and smooth microsomes can be separated from each other by sedimenting the mixture to equilibrium in a density gradient prepared from high concentrations of sucrose (Figure 8–40). When the separated rough and smooth microsomes of liver are compared with respect to such properties as enzyme activity or polypeptide composition, they are remarkably similar (but not identical): apparently most of the components of the ER membrane can diffuse freely between rough and smooth regions of the ER membrane, as would be expected for a continuous fluid membrane.

Because they can be readily purified in functional form, rough microsomes are especially useful for studying the many processes carried out by the ER. The

Figure 8–39 When cells are disrupted by homogenization, the cisternae of rough ER (A) break up into small closed vesicles called *rough microsomes* (B). Similarly, the smooth ER breaks up into small vesicles that lack ribosomes and are called *smooth microsomes*. (A, electron micrograph courtesy of Daniel S. Friend; B, electron micrograph courtesy of George Palade.)

Figure 8–40 The isolation procedure used to purify rough and smooth microsomes from the ER.

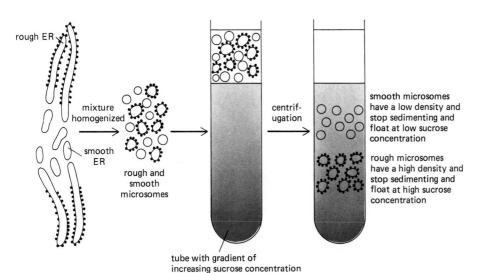

rough ER

mixture homogenized

smooth ER

rough and smooth microsomes

tube with gradient of increasing sucrose concentration

centrifugation

smooth microsomes have a low density and stop sedimenting and float at low sucrose concentration

rough microsomes have a high density and stop sedimenting and float at high sucrose concentration

vesicles are sealed, enclosing a lumen equivalent to that of the rough ER, while their cytoplasmic surface is easily accessible to components that can be added *in vitro*. To the biochemist, rough microsomes represent small authentic versions of the rough endoplasmic reticulum, still capable of protein synthesis, protein glycosylation, and lipid synthesis.

Rough Regions of ER Contain Proteins Responsible for Binding Ribosomes[35]

Because the ER membrane, like all membranes, is a two-dimensional fluid, most proteins and lipids will equilibrate freely between rough and smooth regions in the absence of special restraints. When isolated from liver, however, rough microsomes contain more than 20 proteins that are not present in smooth microsomes, showing that some restraining mechanism must in fact exist. Some of the nonequilibrating proteins in the rough ER membrane help bind ribosomes to it, while others presumably produce its flattened shape (see Figure 8–37). It is not clear whether these membrane proteins are retained by forming large two-dimensional aggregates in the lipid bilayer or whether they are held in place by interactions with a network of structural proteins on one or the other face of the ER membrane (see p. 297).

The ribosomes of the rough ER are held on the membrane in part by their growing polypeptide chains, which are threaded across the ER membrane as they are synthesized (see below). However, when the synthesis of polypeptide chains is terminated with a drug (such as puromycin) that releases the nascent chains from the ribosomes, the ribosomes still retain some affinity for the membrane of rough microsomes. This affinity is artificially increased at low salt concentrations, and when purified ribosomes are mixed under these conditions with rough microsomal membranes from which ribosomes have been removed, the "stripped" membrane regains the same number of ribosomes it had when originally isolated. The binding site on the ribosome is located on the large ribosomal subunit, but it is not yet clear to which of the many proteins in the rough ER membrane the ribosome binds. We shall see, however, that an additional, more specific attachment is required to bind ribosomes to the ER membrane under physiological conditions, where a nascent protein that contains a signal peptide is required.

Signal Peptides Were First Discovered in Proteins Imported into the ER[36]

Signal peptides (and the signal peptide strategy of protein import) were first discovered in the early 1970s in secreted proteins that are translocated across the ER membrane prior to their transport to the Golgi apparatus and eventual discharge from the cell. The observations that led to this discovery came from an experiment in which the mRNA encoding the secreted protein was translated by free ribosomes *in vitro*. When microsomes were omitted from this cell-free system, the protein synthesized was slightly larger than the normal secreted protein, the extra length being due to the presence of an amino-terminal *leader peptide*. In the presence of microsomes derived from the rough endoplasmic reticulum, however, a protein of the correct size was produced. These results were explained by the **signal hypothesis,** which postulates that the leader serves as a signal peptide that directs the secreted protein to the ER membrane and is then cleaved off by a special protease in the ER membrane before the polypeptide chain is completed (Figure 8–41).

According to the signal hypothesis, the secreted protein should be extruded into the lumen of the microsome during its *in vitro* synthesis. This can be demonstrated with protease treatment: a newly synthesized protein made in the absence of microsomes is degraded by the addition of a protease, whereas the same protein made in the presence of microsomes remains intact because of the protection afforded by the microsomal membrane. When proteins without ER signal peptides are similarly translated *in vitro*, they are not imported into microsomes and therefore remain susceptible to protease treatment.

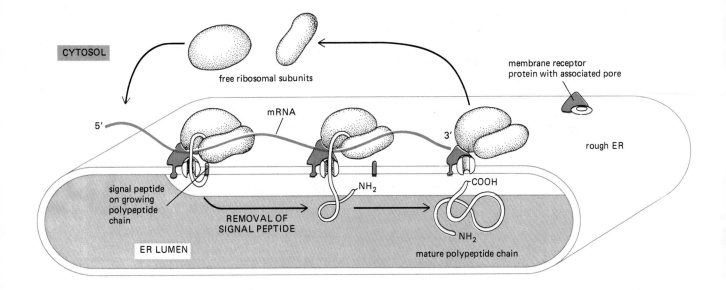

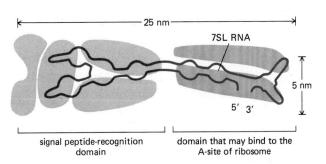

The signal hypothesis has been thoroughly tested by genetic and biochemical experiments and found to apply to both plant and animal cells, as well as to protein translocations across procaryotic plasma membranes. Moreover, amino-terminal leader peptides have been found not only in secreted proteins but also in precursors of plasma membrane and lysosomal proteins, which are also imported by the ER. As mentioned earlier, the signaling function of these leader peptides has been demonstrated directly by using recombinant DNA techniques to attach signal sequences to proteins that do not normally have them; the resulting fusion proteins are directed to the ER.

Cell-free systems in which protein import occurs have provided powerful assay procedures for identifying, purifying, and studying the various components of the molecular machinery responsible for the ER import process.

Figure 8–41 A simplified view of protein translocation across the ER membrane, as proposed in the original "signal hypothesis." When the signal peptide emerges from the ribosome, it directs the ribosome to a receptor protein on the ER membrane. As it is synthesized, the polypeptide is postulated to be translocated across the ER membrane through a protein pore associated with the receptor. The signal peptide is clipped off during translation, and the mature protein is released into the lumen of the ER immediately after being synthesized.

8-25 A Signal-Recognition Particle Directs ER Signal Peptides to a Specific Receptor in the ER Membrane[37]

The signal peptide is guided to the ER membrane by at least two components: a **signal-recognition particle (SRP),** which cycles between the ER membrane and the cytosol and binds to the signal peptide, and an *SRP receptor,* also known as a *docking protein,* in the ER membrane. The SRP was discovered when it was found that washing microsomes with salt eliminated their ability to import secreted proteins. Import could be restored by adding back the supernatant containing the salt extract. The "translocation factor" in the salt extract was then purified and found to be a complex particle consisting of six different polypeptide chains bound to a single molecule of 7SL RNA (Figure 8–42).

The signal-recognition particle binds to the signal peptide as soon as the peptide emerges from the ribosome. This causes a pause in protein synthesis and sometimes stops it completely. The translational pause presumably gives the ribo-

Figure 8–42 A highly schematic drawing of a signal-recognition particle (SRP). It is an elongated complex containing six polypeptide chains (*shaded areas*) and one molecule of 7SL RNA. One end of the particle binds to a ribosome; the other binds to an ER signal peptide on a nascent polypeptide chain. It has been suggested that a portion of the 7SL RNA may fold into a tRNA-like structure that competes at the A-site of the ribosome with incoming aminoacyl tRNAs, thereby causing a pause in translation. (Based on V. Siegel and P. Walter, *Nature* 320:82–84, 1986.)

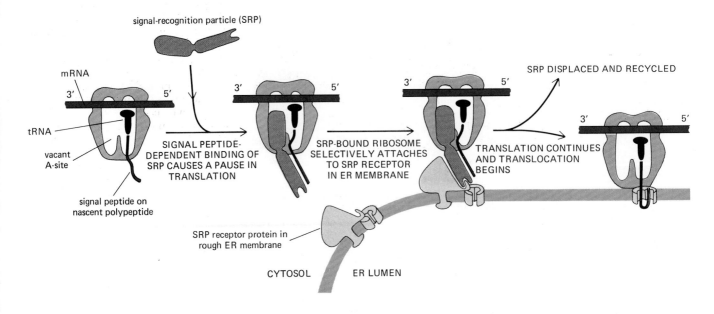

signal-recognition particle (SRP)

mRNA

3' 5'

tRNA

vacant A-site

signal peptide on nascent polypeptide

SIGNAL PEPTIDE-DEPENDENT BINDING OF SRP CAUSES A PAUSE IN TRANSLATION

3' 5'

SRP-BOUND RIBOSOME SELECTIVELY ATTACHES TO SRP RECEPTOR IN ER MEMBRANE

3' 5'

SRP receptor protein in rough ER membrane

TRANSLATION CONTINUES AND TRANSLOCATION BEGINS

SRP DISPLACED AND RECYCLED

3' 5'

CYTOSOL   ER LUMEN

some a chance to engage the ER membrane before the polypeptide is completed, thereby avoiding the inappropriate release of the protein into the cytosol.

The SRP is an elongated molecule consisting of two groups of proteins bound to an RNA scaffold (see Figure 8–42). In one possible model the SRP straddles the ribosome, binding both to the signal peptide as it emerges from the large ribosomal subunit and to the ribosomal site for aminoacyl tRNAs (see p. 212), thus halting translation by preventing the next aminoacyl tRNA from entering the ribosome (Figure 8–43).

Translational arrest is lifted when the ribosome-carrying SRP binds to the **SRP receptor,** which is exposed on the cytosolic surface of the rough ER membrane. The SRP receptor, like the SRP, was initially identified as a component needed to reconstitute *in vitro* protein translocation into the ER. It is a two-chain integral membrane protein that interacts with SRP-bound ribosomes in such a way that the SRP is displaced and translation resumes. Simultaneously the ribosome becomes bound to the ER membrane, and its growing polypeptide chain is transferred to a poorly understood translocation apparatus in the membrane that includes a second signal-peptide-receptor protein distinct from the SRP (see Figure 8–43). This mechanism ensures that a ribosome that begins synthesizing a protein with an ER signal peptide will bind to the ER membrane and begin protein translocation across it.

Figure 8–43 The signal-recognition particle and the SRP receptor protein are thought to act in concert to direct a ribosome synthesizing a protein with an ER signal peptide to the ER. The SRP binds to the exposed signal peptide and to the ribosome, probably covering the A-site. Because entry of the next aminoacyl-tRNA is blocked, translation pauses. The SRP receptor in the ER membrane binds the SRP-ribosome complex, and, in a complex reaction that is poorly understood, SRP is displaced and translation continues, with the ribosome now located on the ER membrane. The mechanism that initially inserts the polypeptide chain into the membrane involves a separate transmembrane protein that binds to the signal peptide (a signal-peptide receptor) as well as other protein components involved in translocation that have not been well characterized.

8-26 Translocation Across the ER Does Not Always Require Ongoing Polypeptide Chain Elongation[38]

As we have seen, translocation of proteins into mitochondria, chloroplasts, and peroxisomes occurs *post-translationally,* after the protein is completed and released into the cytosol, whereas translocation across the ER membrane usually occurs during translation (*co-translationally*). This explains why ribosomes are bound to the ER membrane but not to the cytoplasmic surface of the other organelles. For many years it was thought that the ribosomes of the rough ER might use the energy released during protein synthesis to "eject" their growing polypeptide chains through the ER membrane. Recent studies of ER translocation *in vitro,* however, have shown that selected protein precursors can be imported into the ER after their synthesis has been completed. The import requires ATP hydrolysis but not ongoing protein synthesis (Figure 8–44). As for mitochondrial protein import (see p. 428), the ATP hydrolysis is thought to be required to unfold the protein as it passes through the membrane, and both genetic and biochemical experiments in yeasts indicate that a subclass of hsp70 stress-response proteins is required (see p. 420).

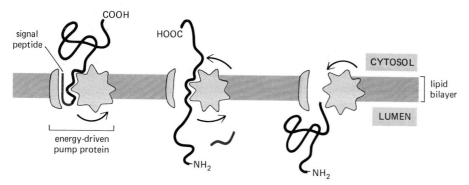

Figure 8–44 One view of the translocation of a protein across a membrane. After a receptor recognizes some special feature of an amino-terminal peptide, an energy-driven protein pump is activated that forces the entire protein through the membrane; in the process the polypeptide chain is transiently unfolded. An alternative possibility is that the unfolding of the protein is catalyzed by an ATP-dependent process on the cytosolic side of the membrane and that only the free energy of refolding on the luminal side drives the protein through the membrane.

Most ER precursor proteins in mammalian cells, however, cannot be imported into the ER once their synthesis has progressed beyond a certain point. It seems that these proteins fold up in a way that either masks the signal peptide or makes it impossible for the ER translocation machinery to unfold the protein. Co-translational import may have allowed these proteins to evolve without the folding constraints that presumably exist for proteins imported into other organelles.

8-27 Combinations of Start- and Stop-Transfer Peptides Can Determine the Many Different Topologies Observed for Transmembrane Proteins[39]

Although most amino-terminal signal peptides are removed by a specific **signal peptidase** bound to the ER membrane, the signal peptide is not in itself a sufficient cue for the peptidase to act: removal of the signal peptide requires an adjacent cleavage site that is not necessary for translocation; indeed, some proteins have signal peptides well within the polypeptide chain, where they are never cleaved.

Uncleaved signal peptides are thought to be essential to achieve the various modes of membrane insertion found in transmembrane proteins (see p. 284). All modes of insertion can be considered as variants of the sequence of events by which a soluble protein is transferred into the lumen of the ER. According to the current view of this process, the hydrophobic ER signal peptide of a soluble protein, in addition to its other functions, serves as a **start-transfer signal** that remains anchored to the membrane throughout translocation, while the rest of the protein is threaded continuously through the membrane in the form of a large loop (Figure 8–45A). Once the carboxyl terminus of the protein has passed through the membrane, only the signal peptide keeps the protein membrane-bound. Therefore, if this peptide is cleaved off, the protein is released into the ER lumen.

Membrane proteins have more complex requirements because some parts of the polypeptide chain are translocated, whereas others are not. In the simplest case the protein is translocated by exactly the same means as the soluble proteins just described—except that the signal peptide lacks an adjacent cleavage site and so is not removed by the signal peptidase. As a result, the translocated protein remains as a single-pass transmembrane protein anchored to the ER membrane at its amino terminus, where the signal peptide forms a membrane-spanning segment of 20–30 hydrophobic amino acid residues in the form of an α helix (see Figure 8–48A).

A more elaborate mechanism is needed in the case of single-pass transmembrane proteins that have their amino terminus, rather than their carboxyl terminus, on the luminal side of the ER. For these proteins, too, an amino-terminal signal peptide initiates translocation; but now an additional hydrophobic segment in the protein stops the process before the entire polypeptide chain is translocated. In these proteins it is this **stop-transfer peptide** that anchors the protein in the membrane, and the signal (start-transfer) peptide is cleaved off (Figure 8–45B).

Many "multipass" transmembrane proteins are known, in which the polypeptide chain passes repeatedly back and forth across the lipid bilayer (see

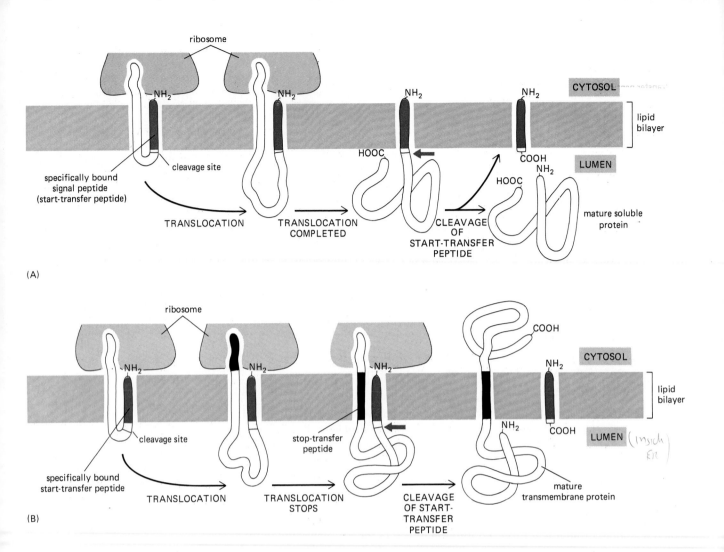

(A)

(B)

Figure 8–45 The topology of protein translocation across the ER membrane is illustrated for two simple cases. The translocational intermediate is thought to contain a loop of polypeptide chain, in which the signal peptide (also called the start-transfer peptide) forms half of the stem of the loop and the region of the polypeptide being translocated across the membrane at any instant forms the other half. When there is one start-transfer peptide and no stop-transfer peptide, the entire polypeptide is translocated across the membrane, and cleavage of the start-transfer peptide releases the mature protein into the lumen of the ER as a soluble protein (A). When one start-transfer peptide and one stop-transfer peptide are present, translocation stops when the stop peptide enters the stem of the loop while protein synthesis continues on the cytosolic side of the membrane; after cleavage of the start-transfer peptide, the mature protein is left spanning the lipid bilayer of the ER with one domain protruding on each side (B).

p. 293). It is thought that an internal signal peptide serves as a start-transfer signal in these proteins and initiates translocation, with each translocation event proceeding until the next stop-transfer peptide is reached. Thus the fundamental unit translocated is a loop of polypeptide between two hydrophobic segments (one start-transfer peptide and one stop-transfer peptide), with both of these peptides serving as α-helical membrane-spanning domains in the mature protein. A possible mechanism that might insert one such loop into the membrane is illustrated in Figure 8–46. In complex multipass transmembrane proteins, in which many hydrophobic α helices span the bilayer, a second start-transfer peptide would reinitiate translocation farther down the chain until the next stop-transfer peptide halted it once again, and so on for subsequent start-transfer and stop-transfer peptides (see Figure 8–48D).

8-28 ## The General Conformation of a Transmembrane Protein Can Often Be Predicted from the Distribution of Its Hydrophobic Amino Acids[40]

Although stop-transfer peptides are generally more hydrophobic than start-transfer peptides, they can sometimes act as start-transfer peptides if their location in a protein is changed. Thus the distinction between hydrophobic start-transfer and stop-transfer peptides results in part from the order in which they occur in the nascent polypeptide chain. It seems that the translocation machinery in the ER membrane begins scanning an unfolded polypeptide chain for hydrophobic segments at its amino terminus and proceeds toward the carboxyl terminus—in the same direction as the protein is synthesized. The SRP recognizes the first appropriate segment and thereby sets the "reading frame." The next appropriate hy-

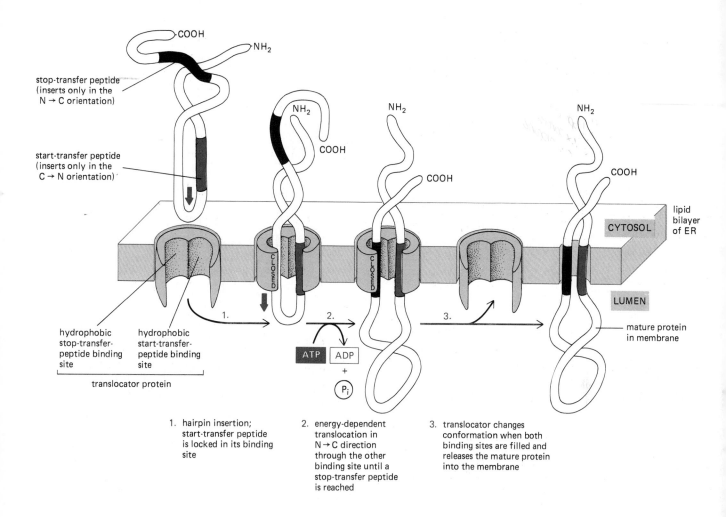

stop-transfer peptide
(inserts only in the
N → C orientation)

start-transfer peptide
(inserts only in the
C → N orientation)

COOH

NH₂

NH₂

COOH

NH₂

COOH

NH₂

COOH

CYTOSOL

lipid
bilayer
of ER

LUMEN

mature protein
in membrane

hydrophobic
stop-transfer-
peptide binding
site

hydrophobic
start-transfer-
peptide binding
site

translocator protein

ATP   ADP
+
Pᵢ

1.   2.   3.

1. hairpin insertion;
    start-transfer peptide
    is locked in its binding
    site

2. energy-dependent
    translocation in
    N → C direction
    through the other
    binding site until a
    stop-transfer peptide
    is reached

3. translocator changes
    conformation when both
    binding sites are filled and
    releases the mature protein
    into the membrane

drophobic segment is recognized as a stop-transfer peptide, causing the region of the polypeptide chain in between to be threaded through the membrane (see Figure 8–46). It has been postulated that a similar process continues until all of the hydrophobic regions in the protein are inserted into the membrane.

This mechanism for membrane insertion means that one can often predict the topography of a membrane protein from its amino acid sequence. One begins by scanning for contiguous regions of about 20–30 amino acid residues with a high degree of hydrophobicity. These segments are long enough to span a membrane as an α helix, and they can often be identified by means of a *hydropathy plot* (Figure 8–47). The topology of the polypeptide chain can then be predicted on the assumption that the first segment, scanning from the amino terminus, has a start-transfer function and that stop-transfer peptides thereafter alternate with further start-transfer peptides. Four examples are illustrated in Figure 8–48.

Because membrane proteins are always inserted from the cytosolic side of the ER in this programmed manner, all copies of the same polypeptide chain will have the same overall orientation in the bilayer. This generates an asymmetrical ER membrane in which the protein domains exposed on one side are different from those domains exposed on the other. This asymmetry is maintained during the many membrane budding and fusion events that transport the proteins made in the ER to other cell membranes (see Figure 8–10), and it therefore determines the orientation of the proteins in these membranes as well.

When proteins are dissociated from a membrane and reconstituted into artificial lipid vesicles (see p. 287), a random mixture of right-side-out and inside-out protein orientations usually results. Thus the protein asymmetry observed in cell membranes is thought to result solely from the process by which proteins are inserted into the ER membrane from the cytosol.

Figure 8–46 A hypothetical model for the insertion of an internal loop of polypeptide chain into the lipid bilayer of the ER. A protein translocator is postulated to have two conformations, "closed" and "open." On binding the start-transfer peptide of a protein to be translocated, the translocator enters the closed state and becomes active in translocation. But it flips back to an inactive, open conformation and discharges its protein as soon as a stop-transfer peptide enters its other binding site.

(A) GLYCOPHORIN

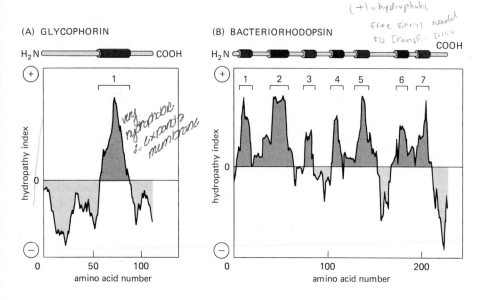

(B) BACTERIORHODOPSIN

(+) = hydrophobic
Free Energy needed
to Transfer To H₂O

1 2 3 4 5 6 7

very
hydrophobic
so expands
membrane

(A) glycophorin (see p. 291)
(B) bacteriorhodopsin (see p. 293)

Figure 8–47 Localization of potential hydrophobic (membrane-spanning) segments in a polypeptide chain through the use of hydropathy plots. The free energy needed to transfer successive segments of a polypeptide chain from a nonpolar solvent to water is calculated from the amino acid composition of each segment using data on model compounds. This calculation is made for segments of a fixed size (usually around 10 amino acid residues), beginning with each successive amino acid in the chain. The "hydropathy index" of the segment is plotted on the Y axis as a function of its location in the chain. A positive value indicates that free energy is required for transfer to water (that is, the segment is hydrophobic), and the value assigned is an index of the amount of energy needed. Peaks in the hydropathy index appear at the positions of hydrophobic segments in the amino acid sequence. Two examples are shown:
(A) glycophorin (see p. 291) has a single membrane-spanning hydrophobic domain and one corresponding peak in the hydropathy plot;
(B) bacteriorhodopsin (see p. 293) has seven membrane-spanning helices and seven corresponding peaks in the hydropathy plot. (Adapted from D. Eisenberg, *Annu. Rev. Biochem.* 53:595–624, 1984.)

Figure 8–48 The topology of a membrane protein is dictated by alternating stop-transfer and signal (start-transfer) peptides. In all the examples shown, signal peptides are not cleaved. The hypothetical protein translocator is assumed to function in the manner illustrated previously in Figure 8–46. (A) When an amino-terminal signal peptide is not cleaved and no stop-transfer peptide is present, a membrane protein with a single carboxyl-terminal domain that protrudes on the luminal side of the ER membrane is generated. (B) When the signal peptide is internal, a membrane protein with an amino-terminal cytoplasmic domain and a carboxyl-terminal luminal domain is produced. (C) When a stop-transfer peptide follows an internal signal peptide, three separate domains will protrude from a membrane protein. (D) Membrane proteins that span the bilayer many times can be generated by a simple extension of the same principles, employing alternating signal peptides and stop-transfer peptides.

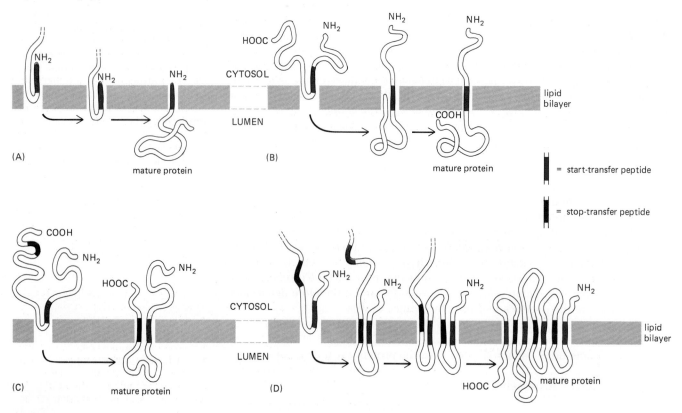

Translocated Polypeptide Chains Refold in the Lumen of the ER[41]

The folding of polypeptide chains in the lumen of the ER might pose a special problem not encountered by proteins that fold up in the cytosol. The ER lumen is filled primarily with proteins in transit that are in the process of folding, whereas the cytosol is filled mainly with resident proteins that are already folded. Folded proteins have an internal hydrophobic core when they are folded up (see p. 107), but before folding, the hydrophobic residues that comprise this core are exposed to water. When unfolded polypeptide chains are present at even low concentrations, they tend to aggregate with one another and with other proteins in order to bury these hydrophobic groups and, as a result, precipitate out of solution. In a complex mixture of many unfolded proteins, as in the lumen of the ER, an amorphous and heterogeneous precipitate might be expected to form. The amount of time spent in the ER before exiting to the Golgi apparatus varies greatly for different proteins; one possibility is that this variability largely reflects the amount of time it takes a protein to dissociate from this precipitate (and thereby become soluble) and fold properly.

The ER lumen contains a high concentration of a **binding protein (BiP)** that seems to recognize incorrectly folded proteins, possibly by binding to their exposed hydrophobic surfaces. BiP contains a signal peptide of four amino acids at its carboxyl terminus that is responsible for retaining the protein in the ER (see Table 8–3, p. 415). It seems likely that, by binding to unfolded chains, BiP helps to keep improperly folded proteins in the ER (and thus out of the Golgi). It might also help to catalyze protein refolding. BiP binds ATP and is structurally related to the hsp70 family of *stress-response proteins*, which function in protein import (see p. 428 and p. 440) and are thought to help solubilize proteins that have been damaged by heat shock or other insults (see p. 420).

Protein Disulfide Isomerase Speeds Up the Formation of Correct Disulfide Bonds in the ER Lumen[42]

The cytosol contains a mixture of thiol-containing reducing agents that prevent the formation of S—S linkages (disulfide bonds) by maintaining the cysteine residues of cytosolic proteins in reduced (—SH) form (Figure 8–49). The ER lumen does not contain these reducing agents, and S—S linkages can form there. In a sea of folding polypeptide chains in a nonreducing environment, many improper S—S bonds are likely to form. An enzyme in the lumen of the ER helps to correct these errors. **Protein disulfide isomerase** is an abundant protein loosely attached to the luminal side of the ER membrane; it contains the same ER retention signal as BiP. It acts repetitively to cleave S—S bonds, allowing proteins to search rapidly through many different arrangements until the one with the lowest overall free energy is found (Figure 8–50). At this point the newly synthesized protein is folded correctly. Although the same final state would eventually be reached without the enzyme, the isomerase greatly accelerates the process.

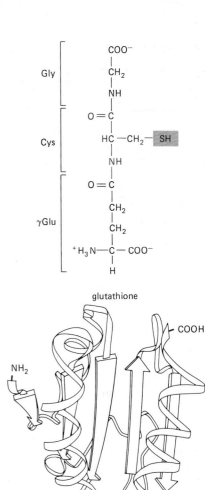

glutathione

thioredoxin

Figure 8–49 Because high concentrations of reducing agents such as the tripeptide *glutathione* and the small protein *thioredoxin* are present in the cytosol and not in the lumen of compartments such as the ER, disulfide bonds can form in the ER lumen but not in the cytosol. Both glutathione and thioredoxin are kept highly reduced in the cytosol by enzymes that transfer electrons from NADPH to convert any disulfide bonds that form in them to cysteines. As indicated, the three-dimensional structure of *E. coli* thioredoxin has been determined by x-ray crystallography. It contains two closely spaced cysteines, shown here in their disulfide-linked form.

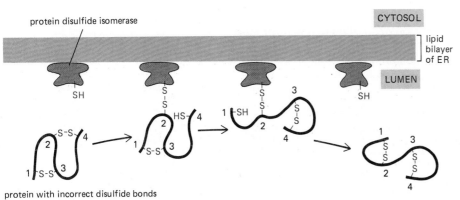

protein with incorrect disulfide bonds

protein with correct disulfide bonds

Figure 8–50 In the ER lumen the enzyme *protein disulfide isomerase* acts repetitively to cleave intrachain S—S bonds until their arrangement has achieved the lowest overall free energy, at which point the protein is folded correctly. In this way the enzyme facilitates the folding of the newly synthesized proteins that enter the ER.

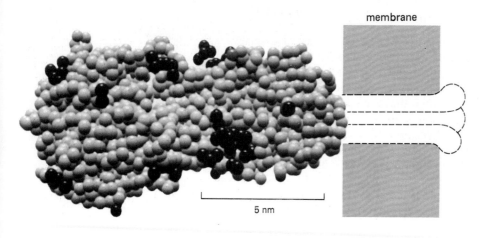

Figure 8–51 The three-dimensional structure of a membrane glycoprotein, the influenza virus hemagglutinin, showing the location of its covalently attached oligosaccharide chains (*dark atoms*). Glycoproteins are either inserted into cell membranes or secreted from cells, and they may contain from as little as 1% to as much as 85% carbohydrate by weight. Some carbohydrate-rich glycoproteins contain tens or even hundreds of attached oligosaccharide chains per molecule. The membrane-spanning regions of the three identical polypeptide subunits in this viral envelope protein would be at the bottom of the illustrated structure; they are not shown because this region was cleaved off in preparing the protein molecules for x-ray crystallographic analysis. (Photograph courtesy of Richard J. Feldmann.)

Most Proteins Synthesized in the Rough ER Are Glycosylated by the Addition of a Common *N*-linked Oligosaccharide[43]

The covalent addition of sugars to proteins is one of the major biosynthetic functions of the ER. Most of the proteins that are sequestered in the lumen of the ER before being transported to the Golgi apparatus, lysosomes, plasma membrane, or extracellular space are **glycoproteins** (Figure 8–51). In contrast, very few proteins in the cytosol are glycosylated, and those that are carry a different sugar modification (see p. 417).

An important advance in understanding the process of **protein glycosylation** was the discovery that a single species of oligosaccharide (composed of *N*-acetylglucosamine, mannose, and glucose and containing a total of 14 sugar residues) is transferred to proteins in the ER. Because it is always transferred to the NH$_2$ group on the side chain of an asparagine residue of the protein, this oligosaccharide is said to be *N-linked* or *asparagine-linked* (Figure 8–52). The transfer is catalyzed by a membrane-bound enzyme with its active site exposed on the luminal surface of the ER membrane, which explains why cytosolic proteins are not glycosylated in this way. The preformed precursor oligosaccharide is transferred *en bloc* to the target asparagine residue in a single enzymatic step almost as soon as that residue emerges in the ER lumen during protein translocation (Figure 8–53). Since most proteins are co-translationally imported into the ER, *N*-linked oligosaccharides are almost always added during protein synthesis, ensuring maximum access to the target asparagine residues, which are those in the sequences *Asn-X-Ser* or *Asn-X-Thr* (where X is any amino acid except proline). These two sequences thus function as signals for *N*-linked glycosylation. They occur much less frequently in glycoproteins than in nonglycosylated cytoplasmic proteins. Evidently there has been selective pressure against these sequences, presumably because glycosylation at too many sites would interfere with protein folding.

The precursor oligosaccharide is held in the ER membrane by a special lipid molecule, **dolichol.** The oligosaccharide is linked to the dolichol by a high-energy pyrophosphate bond, which provides the activation energy for the glycosylation reaction. The oligosaccharide is built up sugar by sugar on this membrane-bound lipid molecule prior to transfer to a protein. Sugars are first activated in the cytosol by the formation of *nucleotide-sugar intermediates,* which then donate their sugar (directly or indirectly) to the lipid in an orderly sequence. Partway through this process, the lipid-linked oligosaccharide is flipped from the cytosolic to the luminal side of the ER membrane (Figure 8–54). Dolichol is long and very hydrophobic:

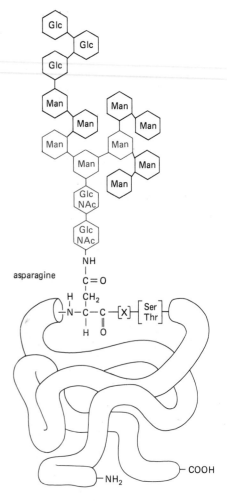

Figure 8–52 The structure of the asparagine-linked oligosaccharide that is added to most proteins on the luminal side of the ER membrane. The sugars shown in color form the "core region" of this oligosaccharide. For many glycoproteins, only the core sugars survive the extensive oligosaccharide trimming process in the Golgi apparatus (see Figure 8–63). Note that the asparagine is in the sequence Asp-X-Ser or Asp-X-Thr, where X can be any amino acid other than proline.

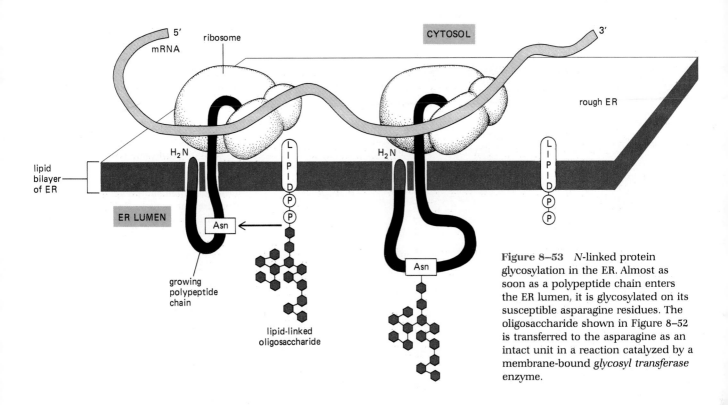

Figure 8–53 *N*-linked protein glycosylation in the ER. Almost as soon as a polypeptide chain enters the ER lumen, it is glycosylated on its susceptible asparagine residues. The oligosaccharide shown in Figure 8–52 is transferred to the asparagine as an intact unit in a reaction catalyzed by a membrane-bound *glycosyl transferase* enzyme.

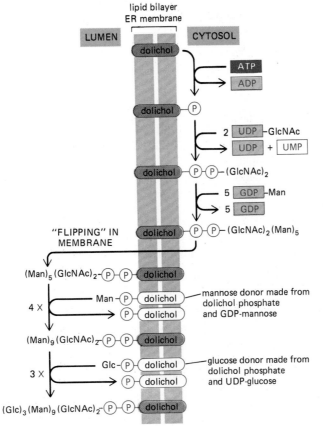

Figure 8–54 Synthesis of the lipid-linked oligosaccharide that is transferred to asparagine residues of nascent polypeptides on the luminal side of the ER membrane. The oligosaccharide is assembled sugar by sugar onto the carrier lipid dolichol (a polyisoprenoid—see Panel 2–4), to which the first sugar is linked by a pyrophosphate bridge. This high-energy bond activates the oligosaccharide for its transfer from the lipid to an asparagine side chain. The synthesis of the oligosaccharide starts on the cytosolic side of the ER membrane and continues on the luminal face after the Man_5-$GlcNAc_2$ lipid intermediate is flipped across the bilayer. All of the glycosyl transfer reactions on the luminal side of the ER involve transfers from dolichol-P-glucose and dolichol-P-mannose. These activated, lipid-linked monosaccharides are synthesized from dolichol phosphate and UDP-glucose or GDP-mannose (as appropriate) on the cytosolic side of the ER and are then thought to be flipped across the ER membrane. Abbreviations: GlcNAc, *N*-acetylglucosamine; Man, mannose; Glc, glucose.

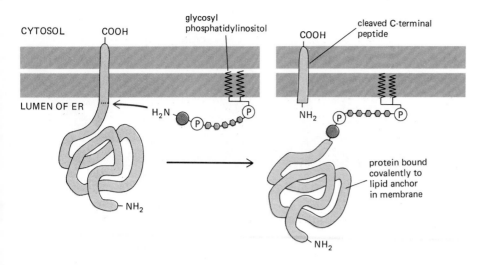

CYTOSOL COOH

glycosyl
phosphatidylinositol

LUMEN OF ER

H₂N

NH₂

cleaved C-terminal
peptide

COOH

NH₂

protein bound
covalently to
lipid anchor
in membrane

NH₂

Figure 8–55 The synthesis of proteins linked to membranes by phosphatidylinositol anchors. Immediately after the completion of protein synthesis, the precursor protein remains anchored to the ER membrane only by a hydrophobic carboxyl-terminal sequence of 15 to 20 amino acids, with the rest of the protein in the ER lumen. Within less than a minute, an enzyme in the ER cuts the protein free from its membrane-bound carboxyl terminus while simultaneously attaching the new carboxyl terminus to an amino group on a preassembled glycosyl-phosphatidylinositol intermediate. Because of this covalently linked lipid anchor, the protein remains membrane bound; all of its amino acids are exposed on the luminal side of the ER and will therefore protrude on the cell exterior if the protein is transported to the plasma membrane. The exact structure of the glycolipid head group is not known.

its 22 five-carbon units can span the thickness of a lipid bilayer more than three times, so that the attached oligosaccharide is firmly anchored in the membrane.

All of the diversity of the *N*-linked oligosaccharide structures on mature glycoproteins results from extensive modification of the original precursor structure. While still in the ER, three glucose residues and one mannose residue are quickly removed from the oligosaccharides of most glycoproteins (see Figure 8–63). This oligosaccharide "trimming" or "processing" continues in the Golgi apparatus and is discussed on page 452.

The *N*-linked oligosaccharides are by far the most common ones found in glycoproteins. Less frequently, oligosaccharides are linked to the hydroxyl group on the side chain of a serine, threonine, or hydroxylysine residue. These *O-linked oligosaccharides* are formed in the Golgi apparatus by pathways that are not yet fully understood (see p. 456).

Some Membrane Proteins Exchange a Carboxyl-Terminal Transmembrane Tail for a Covalently Attached Inositol Phospholipid Shortly After They Enter the ER[44]

As discussed previously, several cytosolic enzymes catalyze the covalent addition of a single fatty acid to selected proteins (see p. 417). It has recently been discovered that a related process is catalyzed by enzymes in the ER: the carboxyl terminus of some plasma membrane proteins is covalently attached to a sugar residue of a glycolipid. This linkage forms in the lumen of the ER by the mechanism illustrated in Figure 8–55, and it adds a glycosylated phosphatidylinositol molecule, which contains two fatty acids, to the protein. An increasing number of plasma membrane proteins have been shown to be modified in this way, including one form of the neural cell adhesion molecule N-CAM and the major coat protein of a trypanosome (see p. 830). Since these proteins are attached to the exterior of the plasma membrane only by this means, in principle they could be released from cells in soluble form in response to signals that activate a specific phospholipase in the plasma membrane (see p. 702); so far such release has not been demonstrated.

Most Membrane Lipid Bilayers Are Assembled in the ER[45]

The ER membrane produces nearly all of the lipids required for the elaboration of new cellular membranes, including both phospholipids and cholesterol. The major phospholipid made is *phosphatidylcholine* (also called lecithin), which can be formed in three steps from two fatty acids, glycerol phosphate, and choline (Figure 8–56). Each step is catalyzed by enzymes in the ER membrane that have their active sites facing the cytosol, where all of the required metabolites are found.

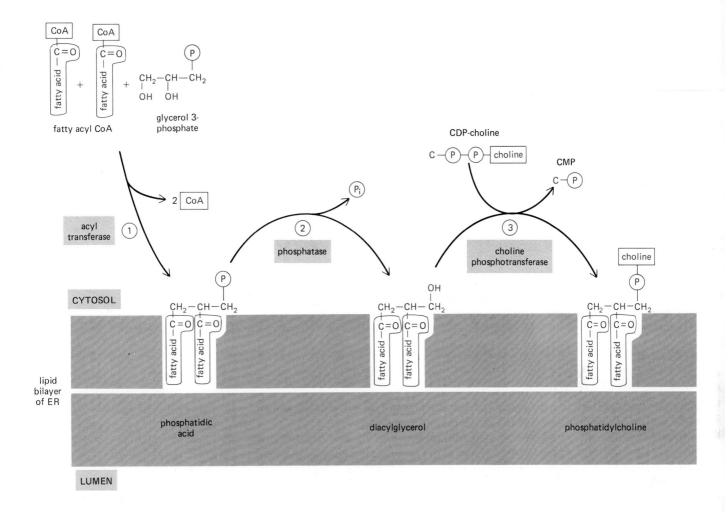

In the first step, acyl transferases successively add two fatty acids to glycerol phosphate to produce phosphatidic acid, a compound sufficiently water-insoluble to remain in the lipid bilayer after it has been synthesized. It is this step that enlarges the lipid bilayer. The later steps determine the head group of a newly formed lipid molecule, and therefore the chemical nature of the bilayer, but do not result in net membrane growth (see Figure 8–56). The major membrane phospholipids—phosphatidylcholine (PC), phosphatidylethanolamine (PE), phosphatidylserine (PS), and phosphatidylinositol (PI)—are all synthesized in this way.

The initial formation of phosphatidic acid and its subsequent modifications to form the different types of phospholipid molecules all take place in the cytosolic half of the ER lipid bilayer. This process might eventually turn the lipid bilayer into a monolayer if it were not for a mechanism that transfers some of the newly formed phospholipid molecules to the other half of the ER bilayer. In synthetic lipid bilayers, lipids do not "flip-flop" in this way (see p. 278). In the ER, however, phospholipids equilibrate across the membrane within minutes, which is almost 100,000 times faster than can be accounted for by spontaneous "flip-flop." This rapid transbilayer movement is thought to be mediated by *phospholipid translocators* that are head-group-specific. In particular, the ER membrane seems to contain a translocator (a "*flippase*") that transfers choline-containing phospholipids—but not ethanolamine-, serine-, or inositol-containing phospholipids—between cytoplasmic and luminal faces. This means that PC reaches the luminal face much more readily than PE, PS, or PI. The translocator maintains the bilayer and is responsible for the asymmetric distribution of the lipids in it (Figure 8–57).

The ER also produces cholesterol and ceramide. The ceramide is exported to the Golgi apparatus, where it serves as the precursor for the synthesis of two types of lipids: oligosaccharide chains are added to form *glycosphingolipids* (see p. 282),

Figure 8–56 The synthesis of phospholipids occurs on the cytosolic side of the ER membrane. Each enzyme in the pathway is an integral membrane protein of the ER with its active site facing the cytosol, where all the intermediates required for phospholipid assembly are located. In the pathway illustrated here, *phosphatidylcholine* is synthesized from fatty acyl-coenzyme A, glycerol 3-phosphate, and cytidine-diphosphocholine (CDP-choline).

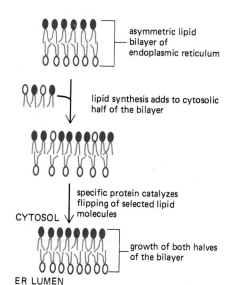

Figure 8–57 The growth of both halves of the ER lipid bilayer requires the protein-catalyzed "flipping" of phospholipid molecules from one half of the ER bilayer to the other. Since new lipid molecules are added only to the cytosolic half and lipid molecules do not flip spontaneously from one monolayer to the other, membrane-bound phospholipid translocator proteins ("flippases") are required to transfer selected lipid molecules to the luminal half so that the membrane grows as a bilayer. Because these proteins preferentially recognize and transfer only certain species of lipid, an asymmetric bilayer is generated in the ER. In particular, the luminal monolayer (which produces the outer half of the plasma membrane bilayer) becomes highly enriched for phosphatidylcholine.

and phosphocholine head groups are transferred from phosphatidylcholine to other ceramide molecules to form *sphingomyelin* (see p. 280). Thus both glycolipids and sphingomyelin are produced relatively late in the process of membrane synthesis. Because they are produced by enzymes exposed to the Golgi lumen, they are found exclusively in the noncytosolic half of the lipid bilayers that contain them.

8-29 Phospholipid Exchange Proteins May Transport Phospholipids from the ER to Mitochondria and Peroxisomes[46]

The plasma membrane and the membranes of the Golgi apparatus and lysosomes all form part of a membrane system that communicates with the ER by means of transport vesicles that supply both their proteins and their lipids. Mitochondria and peroxisomes do not belong to this system and require different mechanisms for the import of proteins and lipids for growth. We have already seen that most of the proteins in these organelles are imported post-translationally from the cytosol. Although mitochondria modify some of the lipids they import (see p. 398), their lipids have to be transferred from the ER, where they are synthesized, or obtained less directly from the ER by means of other cellular membranes.

Water-soluble carrier proteins—called *phospholipid exchange proteins* (or *phospholipid transfer proteins*)—have been shown in *in vitro* experiments to have the ability to transfer individual phospholipid molecules between membranes. Each exchange protein recognizes only specific types of phospholipids. Transfer between membrane bilayers is achieved when the protein "extracts" a molecule of phospholipid from a membrane and diffuses away with the lipid buried within its binding site. When it encounters another membrane, the exchange protein tends to discharge the bound phospholipid molecule into the new lipid bilayer (Figure 8–58). It has been proposed that phosphatidylserine is imported into mitochondria in this way and then decarboxylated to yield phosphatidylethanolamine, while phosphatidylcholine is imported intact.

Exchange proteins act to distribute phospholipids at random among all membranes present. In principle, such a random exchange process can result in a net transport of lipids from a lipid-rich to a lipid-poor membrane, allowing phosphatidylcholine and phosphatidylserine molecules to be transferred from the ER, where they are synthesized, to a mitochondrial or peroxisomal membrane. It may be that mitochondria and peroxisomes are the only "lipid-poor" organelles in the cytoplasm and that such an exchange process is sufficient, although other more specific mechanisms may exist for transporting phospholipids to these organelles.

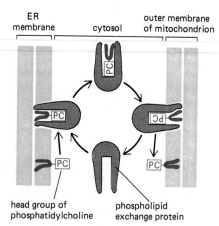

Figure 8–58 Soluble *phospholipid exchange proteins* can redistribute phospholipids between membrane-bounded compartments. Phospholipids are insoluble in water, so their passage between membranes requires a carrier protein. The exchange proteins carry a single molecule of phospholipid at a time and can pick up a lipid molecule from one membrane and release it at another. The transfer of phosphatidylcholine (PC) from ER to mitochondria can in principle occur spontaneously because the concentration of PC is high in the ER membrane (where it is made) and low in the mitochondrial outer membrane.

Summary

The ER serves as a factory for the production of the protein and lipid components of many organelles. Its extensive membrane contains numerous biosynthetic enzymes, including those responsible for almost all of the cell's lipid synthesis and for the addition of an N-linked oligosaccharide to many proteins on the luminal side of the ER. Newly synthesized proteins destined for secretion, as well as those destined for the ER itself and for the Golgi apparatus, the lysosomes, and the plasma membrane, must first be imported into the ER from the cytosol. Only proteins that carry a special hydrophobic signal peptide are imported into the ER. The signal peptide is recognized by a signal-recognition particle (SRP), which binds the nascent polypeptide chain and the ribosome and directs them to a receptor protein on the surface of the ER membrane. This binding to the membrane initiates an ATP-dependent translocation process that threads a loop of polypeptide chain across the ER membrane.

Soluble proteins destined for the ER lumen, for secretion, or for transfer to the lumen of other organelles pass completely into the ER lumen. Transmembrane proteins destined for the ER or for other cell membranes are translocated across the ER membrane but are not released into the lumen; instead, they remain anchored in the bilayer by one or more membrane-spanning α-helical regions in their polypeptide chain. These hydrophobic portions of the protein can act either as start-transfer or stop-transfer peptides during the translocation process. When a polypeptide contains multiple alternating start-transfer and stop-transfer peptides, it will pass back and forth across the bilayer many times.

The asymmetry of lipid synthesis, protein insertion, and glycosylation in the ER establishes the polarity of the membranes of all of the other organelles that the ER supplies with lipids and membrane proteins.

8-32 The Golgi Apparatus[47]

The **Golgi apparatus** (also called the *Golgi complex*) is usually located near the cell nucleus, and in animal cells it is often close to the centrosome, or cell center. It contains a collection of flattened membrane-bounded *cisternae* resembling a stack of plates. These **Golgi stacks** (called *dictyosomes* in plants) usually consist of four to six cisternae, each typically about 1 μm in diameter (Figure 8–59). The number of Golgi stacks per cell varies greatly depending on the cell type: some cells contain one large one, while others contain hundreds of very small ones.

Swarms of small vesicles (~50 nm in diameter) are associated with the Golgi stacks, clustered on the side abutting the ER and along the dilated rims of each cisterna (see Figure 8–59). These *Golgi vesicles* are thought to transport proteins and lipids to and from the Golgi and between the Golgi cisternae. Many vesicles

Figure 8–59 (A) Three-dimensional drawing of a Golgi apparatus, derived from electron micrographs of a secretory animal cell. The stacks of flattened Golgi cisternae have dilated edges from which small vesicles appear to be budding. The large secretory vesicles form by budding from the *trans* Golgi. (B) Electron micrograph of a Golgi apparatus in a plant cell seen in cross-section (the green alga *Chlamydomonas*). In plant cells the Golgi apparatus is generally more distinct and more clearly separated from other intracellular membranes than in animal cells. (A, after R.V. Krstić, Ultrastructure of the Mammalian Cell. New York: Springer-Verlag, 1979; B, courtesy of George Palade.)

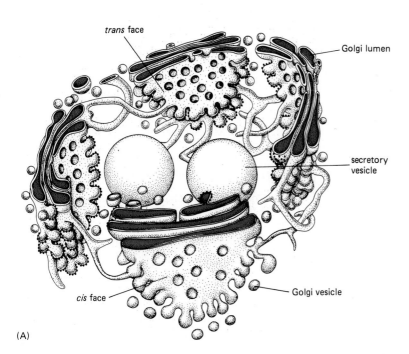

(A)

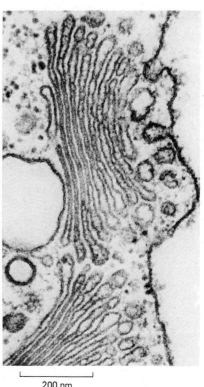

200 nm

(B)

Figure 8–60 A goblet cell of the small intestine. This cell is specialized for secreting mucus, which is a mixture of glycoproteins and proteoglycans synthesized in the ER and Golgi apparatus. The Golgi apparatus is highly polarized, which facilitates the discharge of mucus by exocytosis at the apical surface. (After R.V. Krstić, Illustrated Encyclopedia of Human Histology. New York: Springer-Verlag, 1984.)

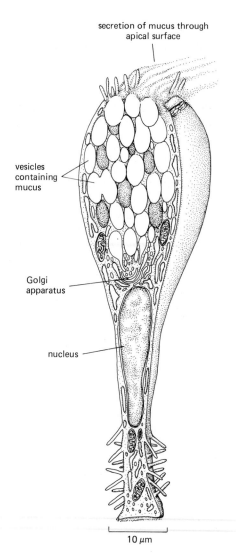

secretion of mucus through apical surface

vesicles containing mucus

Golgi apparatus

nucleus

10 μm

are *coated*, some by clathrin and others by other types of coat protein (see p. 472). Such **coated vesicles** are often seen budding from the Golgi cisternae.

The Golgi stack has two distinct faces: a **cis face** (or entry face) and a **trans face** (or exit face). The *cis* face is closely associated with the *transitional elements* of the ER (see p. 411); the *trans* face is distended into a tubular reticulum called the *trans Golgi network* (TGN). Proteins and lipids enter a Golgi stack in small vesicles from the ER on the *cis* side and exit for various destinations in vesicles that form on the *trans* side; these transported molecules undergo an ordered series of modifications as they pass from one Golgi cisterna to another.

The Golgi apparatus is prominent in cells that are specialized for secretion, such as the goblet cells of the intestinal epithelium, which secrete large amounts of mucus into the gut. In such cells, unusually large vesicles form from the *trans* side of the Golgi apparatus, which faces the plasma membrane domain where secretion occurs (Figure 8–60).

8-33 Oligosaccharide Chains Are Processed in the Golgi Apparatus[48]

As described previously, a single species of **N-linked oligosaccharide** is attached to many proteins in the ER, and this oligosaccharide is extensively trimmed while the protein is still in the ER (see p. 446). Further modifications occur in the Golgi apparatus.

Two broad classes of *N*-linked oligosaccharides, the *complex oligosaccharides* and the *high-mannose oligosaccharides*, are found in mature glycoproteins (Figure

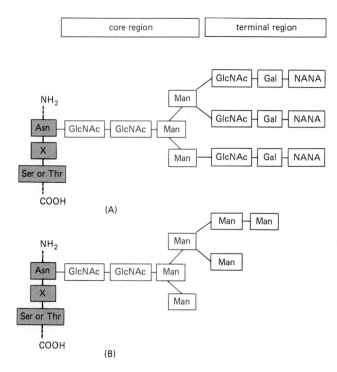

Figure 8–61 Examples of the two main classes of asparagine-linked (*N*-linked) oligosaccharides found in mature glycoproteins: a "complex" oligosaccharide (A) and a "high-mannose" oligosaccharide (B). Many variations occur on these themes. For example, while the complex oligosaccharide shown has three terminal branches, two and four branches are also common, depending on the glycoprotein and the cell in which it is made. "Hybrid" oligosaccharides with one Man branch and one GlcNAc-Gal branch are also found. The three colored amino acids constitute the sequence recognized by the enzyme that adds the initial oligosaccharide to the protein. Abbreviations: Asn, asparagine; Man, mannose; GlcNAc, *N*-acetylglucosamine; NANA, *N*-acetylneuraminic acid (sialic acid); Gal, galactose; X, any amino acid.

8–61). Sometimes both types are attached (in different places) to the same poly-peptide chain. **High-mannose oligosaccharides** have no new sugars added to them in the Golgi apparatus. They contain just two *N*-acetylglucosamines and many mannose residues, often approaching the number originally present in the lipid-linked oligosaccharide precursor added in the ER. **Complex oligosaccha-rides,** by contrast, can contain more than the original two *N*-acetylglucosamines as well as a variable number of galactose and sialic acid residues and, in some cases, fucose. Sialic acid is of special importance because it is the only sugar residue of glycoproteins that bears a net negative charge (see p. 283).

The complex oligosaccharides are generated by a combination of further trim-ming of the original oligosaccharide added in the ER and the addition of further sugars. Thus each complex oligosaccharide consists of a *core region*, derived from the original *N*-linked oligosaccharide and typically containing two *N*-acetylglu-cosamine and three mannose residues (shown in color in Figure 8–61), plus a *terminal region* consisting of a variable number of *N*-acetylglucosamine–galactose–sialic acid trisaccharide units linked to the core mannose residues. Fre-quently the terminal region is truncated, containing only *N*-acetylglucosamine and galactose, or even just *N*-acetylglucosamine. In addition, a fucose residue may or may not be added, usually to the core *N*-acetylglucosamine residue attached to the asparagine. All of the terminal-region sugars are added in the *trans* Golgi by a series of glycosyl transferases that act in a rigidly determined sequence (Figure 8–62). The substrates are specific sugars activated by linkage to a nucleotide. These are transported into the lumen of the Golgi from the cytosol by a set of membrane-bound carriers that also export the nucleotide by-products of the glycosylation. One of the glycosyl transferases (*galactosyl transferase*) is commonly used as a marker to identify membrane vesicles derived from the Golgi after they have been purified by differential centrifugation of tissue homogenates (see p. 163).

The processing that generates complex oligosaccharide chains follows the highly ordered pathway shown in Figure 8–63. A number of drugs and antibiotics that inhibit specific steps have helped in defining the details of this pathway (Table 8–4). Whether a given oligosaccharide remains high-mannose or is processed is determined largely by its configuration on the protein to which it is attached: if the oligosaccharide is sterically accessible to the processing enzymes in the Golgi, it is likely to be converted to a complex form; if it is inaccessible, it is likely to remain in a high-mannose form.

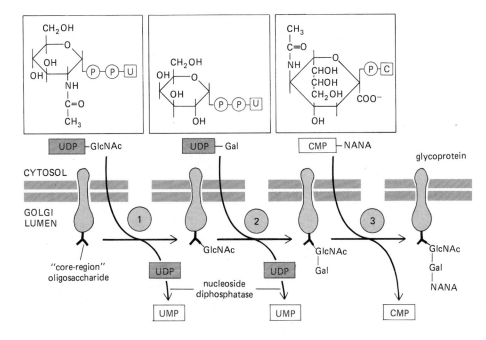

Figure 8–62 The stepwise addition of sugar residues that occurs to finish the synthesis of complex oligosaccharides in the cisternal compartments of the Golgi apparatus. Three types of glycosyl transferase enzymes act sequentially, using sugar substrates that have been activated by linkage to the indicated nucleotide. The membranes of the Golgi cisternae contain specific transmembrane carrier proteins (see p. 302) that allow each sugar nucleotide to enter in exchange for its nucleoside monophosphate product, thus permitting glycosylations to occur on the luminal face. The structures of the sugar nucleotides UDP-*N*-acetylglucosamine, UDP-galactose, and CMP-*N*-acetylneuraminic acid are shown at the top. For abbreviations used, see legend for Figure 8–61.

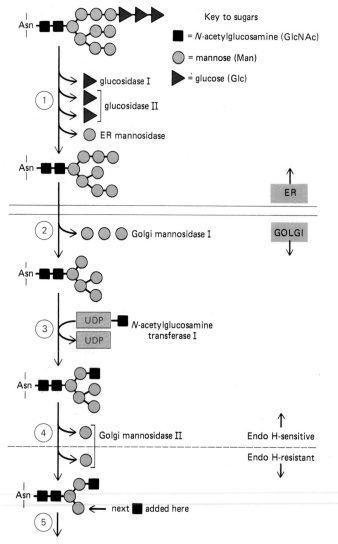

Figure 8–63 The oligosaccharide processing that takes place in the ER and the Golgi apparatus. The pathway is highly ordered so that each step shown is dependent on the previous reaction in the series. Processing begins in the ER with the removal of the glucose residues from the oligosaccharide initially transferred to the protein. All three glucose residues can be removed even before protein synthesis is completed. Then a mannosidase in the ER membrane removes a particular mannose residue. In the Golgi stack, mannosidase I removes three more mannose residues and N-acetylglucosamine transferase I adds a residue of GlcNAc, which enables mannosidase II to remove two additional mannose residues. This yields the final core of three mannose residues that is present in a "complex" type of oligosaccharide. At this stage the bond between the two GlcNAc residues in the core becomes resistant to attack by a highly specific endoglycosidase (*Endo H*). Since all later structures in the pathway are also Endo H-resistant, treatment with this enzyme is widely used to distinguish complex from high-mannose oligosaccharides. Finally, additional GlcNAc, galactose, and sialic acid residues are added. The extent of processing depends on the protein and on the location of the asparagine residue within the protein to which the oligosaccharide is attached. Some oligosaccharides will escape processing in the Golgi apparatus, whereas others will follow the pathway shown to varying extents. For abbreviations used, see legend for Figure 8–61.

The Carbohydrate in Cell Membranes Faces the Side of the Membrane That Is Topologically Equivalent to the Outside of the Cell

Because all oligosaccharide chains are added on the luminal side of the ER and Golgi apparatus (see p. 447), the distribution of carbohydrate on membrane proteins and lipids is asymmetrical. As with the asymmetry of the lipid bilayer itself, the asymmetric orientation of these glycosylated molecules is maintained during transport to the plasma membrane, secretory vesicles, or lysosomes. As a result, the oligosaccharides of all the glycoproteins and glycolipids in the corresponding intracellular membranes face the luminal side, while those in the plasma membrane face the outside of the cell (Figure 8–64).

Table 8–4 Drugs That Inhibit Steps in N-linked Glycosylation

Drug(s)	Step(s) Inhibited
Tunicamycin	dolichol-P $\rightarrow$ dol-P-P-GlcNAc
Castanospermine and N-Methyl deoxynojirimycin	glucose$_3$-Man$_9$-GlcNAc$_2$-Asn $\rightarrow$ glucose$_2$-Man$_9$-GlcNAc$_2$-Asn
Bromoconduritol	glucose$_2$-Man$_9$-GlcNAc$_2$-Asn $\rightarrow$ Man$_9$-GlcNAc$_2$-Asn
Deoxymannonojirimycin	Man$_8$-GlcNAc$_2$-Asn $\rightarrow$ Man$_5$-GlcNAc$_2$-Asn
Swainsonine	GlcNAc-Man$_5$-GlcNAc$_2$-Asn $\rightarrow$ GlcNAc-Man$_3$-GlcNAc$_2$-Asn

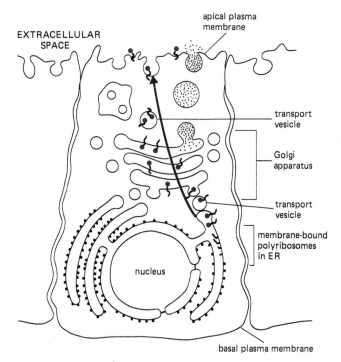

EXTRACELLULAR SPACE

apical plasma membrane

transport vesicle

Golgi apparatus

transport vesicle

membrane-bound polyribosomes in ER

nucleus

basal plasma membrane

Figure 8–64 The orientation of a transmembrane protein in the ER membrane is preserved when that protein is transported to other membranes. The colored ball on the end of each glycoprotein molecule represents the *N*-linked oligosaccharide that is added to proteins in the ER lumen. Note that these sugar residues are confined to the lumen of each of the internal organelles and become exposed to the extracellular space after a transport vesicle fuses with the plasma membrane.

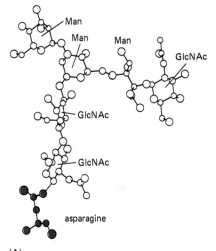

Man
Man
Man
GlcNAc
GlcNAc
GlcNAc
asparagine

(A)

What Is the Purpose of *N*-linked Glycosylation?[49]

There is an important difference between the construction of an oligosaccharide and the synthesis of other macromolecules such as DNA, RNA, and protein. Whereas nucleic acids and proteins are copied from a template in a repeated series of identical steps using the same enzyme(s), complex carbohydrates require a different enzyme at each step, each product being recognized as the exclusive substrate for the next enzyme in the series. Given the complicated pathways that have evolved to synthesize them, it seems likely that the oligosaccharides on glycolipids and glycoproteins have important functions, but for the most part these functions are not known.

N-linked glycosylation, for example, is prevalent in all eucaryotes, including yeasts, but is absent from eubacteria. Because one or more *N*-linked oligosaccharides are present on most proteins transported through the ER and Golgi apparatus—a process that is unique to eucaryotic cells—it was once thought that their function was to aid this transport process. Drugs that block steps in glycosylation (Table 8–4), however, do not generally interfere with transport (with the important exception of transport to lysosomes, which will be discussed below—see p. 459), and mutant cells in culture that are blocked in various glycosylation steps in the Golgi are nevertheless viable and transport proteins normally. Although some proteins do not fold correctly without their normal oligosaccharide and therefore precipitate in the ER and fail to be transported, most proteins retain their normal activities in the absence of glycosylation.

Because chains of sugars have limited flexibility, even a small *N*-linked oligosaccharide protrudes from the surface of a glycoprotein (Figure 8–65) and can

Figure 8–65 The three-dimensional structure of a small *N*-linked oligosaccharide, as determined by x-ray crystallographic analysis of a glycoprotein. This oligosaccharide contains only 6 sugar residues, whereas there are 14 sugar residues in the *N*-linked oligosaccharide that is initially transferred to proteins in the ER (see Figure 8–52). (A) Backbone model showing all atoms except hydrogens; (B) space-filling model, with the asparagine shown as dark atoms. (Courtesy of Richard Feldmann.)

(B)

thus limit the approach of other macromolecules to the surface of the glycoprotein. In this way, for example, the presence of oligosaccharide tends to make a glycoprotein relatively resistant to protease digestion. It may be that the oligosaccharides originally provided an ancestral eucaryotic cell with a protective coat that, unlike the rigid bacterial cell wall, allowed the cell freedom to change shape and move. They may have since become modified to serve other purposes as well.

Proteoglycans Are Assembled in the Golgi Apparatus[50]

It is not only the *N*-linked oligosaccharide chains on proteins that are altered as the proteins pass through the Golgi cisternae en route from the ER to their final destinations; many proteins are also modified in other ways. As mentioned earlier, for example, some proteins have sugars added to selected serine or threonine side chains. This **O-linked glycosylation,** like the extension of *N*-linked oligosaccharide chains, is catalyzed by a series of glycosyl transferase enzymes that use the sugar nucleotides in the Golgi lumen to add one sugar residue at a time to a protein. Usually *N*-acetylgalactosamine is added first, followed by a variable number of additional sugar residues, ranging from just a few to 10 or more.

The most heavily glycosylated proteins of all are some *proteoglycan core proteins*, which are modified in the Golgi apparatus to produce *proteoglycans*. As discussed in Chapter 14 (see p. 806), this involves the polymerization of one or more glycosaminoglycan chains (long unbranched polymers composed of repeating disaccharide units) to serines on the core protein, with xylose rather than *N*-acetylgalactosamine added first. Many proteoglycans are secreted as components of the extracellular matrix, while others remain anchored to the plasma membrane as integral membrane proteoglycans (see p. 808). In addition, the mucus that is secreted to form a protective coating over many epithelia consists of a concentrated mixture of proteoglycans and heavily glycosylated glycoproteins (see, for example, Figure 8–60).

The sugars incorporated into glycosaminoglycans are heavily sulfated immediately after these polymers are made in the Golgi apparatus, which helps to give proteoglycans their high negative charge. The sulfate is added from the activated sulfate donor 3'-phosphoadenosine–5'-phosphosulfate (PAPS), which is transported from the cytosol to the lumen of a late Golgi compartment. A more subtle protein modification carried out in the Golgi is the addition of sulfate from PAPS to the hydroxyl group of selected tyrosine residues in proteins. Sulfated tyrosines are frequently found in secreted proteins and occasionally in the extracellular domains of plasma membrane proteins.

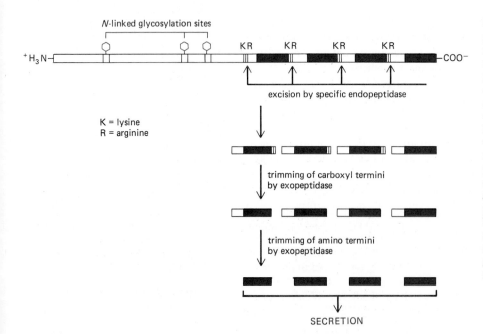

Figure 8–66 An example of a *polyprotein* that is cleaved to produce multiple copies of the same peptide signaling molecule. As indicated, the processing generally begins with cleavages at pairs of basic amino acids (here Lys-Arg pairs), which are catalyzed by a specific membrane-bound protease located in secretory vesicles or possibly in the *trans* Golgi network. Shown here is the processing pathway that produces the 13-amino acid α-factor in the yeast *Saccharomyces cerevisiae*, a secreted peptide that controls the mating behavior of this single-celled eucaryote. (After R. Fuller, A. Brake, and J. Thorner, in Microbiology 1986 [L. Lieve, ed.], pp. 273–278. Washington, D.C.: American Society for Microbiology, 1986.)

Proteins Are Often Proteolytically Processed During the Formation of Secretory Vesicles[51]

The most drastic of the many modifications that can occur to proteins before they are secreted are made last. Many polypeptide hormones and neuropeptides are synthesized as inactive protein precursors from which the active molecules are liberated by proteolysis. These cleavages are thought to begin in the *trans Golgi network*, and they continue in the secretory vesicles that form from this network to store these proteins. The initial cleavages are made by membrane-bound proteases that cut next to pairs of basic amino acid residues (Lys-Arg, Lys-Lys, Arg-Lys, or Arg-Arg pairs), and trimming reactions then produce the final secreted product (Figure 8–66). In the simplest case a polypeptide often will have a single amino-terminal *pro-piece* that is cleaved off shortly before secretion to yield the mature protein. These proteins are thus synthesized as *pre-pro-proteins*, the *pre-piece* consisting of the ER signal peptide that is cleaved off in the rough ER (see p. 438). Somewhat more complex are peptide signaling molecules that are made as *polyproteins* containing multiple copies of the same amino acid sequence (see Figure 8–66). Finally, a variety of peptide signaling molecules are synthesized as parts of a polyprotein that acts as a precursor for multiple end products, which are individually cleaved from the initial polypeptide chain. In these cases the same polyprotein can be processed in various ways to produce different peptides in different cell types, thereby increasing the diversity of molecules that can be used for chemical signaling between cells.

Why is this type of delayed proteolytic processing observed for so many polypeptides? Some of the peptides produced in this way, such as the enkephalins (five-amino acid neuropeptides), are undoubtedly too short in their mature forms to be synthesized efficiently by ribosomes, while even much longer peptides might be expected to lack the necessary signals for packaging into secretory vesicles (see p. 465). Moreover, delaying the production of an active protein until it reaches a secretory vesicle has the potential advantage of preventing it from acting inside the cell that synthesizes it.

The Golgi Cisternae Are Organized as a Sequential Series of Processing Compartments[52]

The processing pathways just outlined are highly organized in the Golgi stack. Each cisterna is a distinct compartment, with its own set of processing enzymes, so that the stack forms a multistage processing unit. Proteins are modified in successive stages as they move from compartment to compartment across the stack.

Proteins exported from the ER enter the first of the Golgi compartments (the **cis compartment**), then move to the next compartment (the **medial compartment,** consisting of the central cisternae of the stack), and finally move to the **trans compartment** (consisting of the last cisternae), where glycosylation is completed. From the *trans* compartment the proteins move to the **trans Golgi network (TGN);** in this tubular reticulum they are segregated into different transport vesicles and dispatched to their final destination—the plasma membrane, lysosomes, or secretory vesicles.

The functional differences between the *cis, medial,* and *trans* subdivisions of the Golgi stack were first discovered by localizing the enzymes involved in processing *N*-linked oligosaccharides in distinct regions of the stack, both by physical fractionation of the organelle and by immunoelectron microscopy. The removal of mannose residues and the addition of *N*-acetylglucosamine, for example, were shown to occur in the *medial* compartment, while the addition of galactose and sialic acid was found to occur in the *trans* compartment (Figures 8–67 and 8–68).

The proteins that enter the Golgi from the ER (except those destined to remain in one of the Golgi compartments) flow together across the stack from *cis* to *medial* to *trans* compartments, permitting their stepwise processing. Although the mechanism of protein and lipid transfer from one cisterna to another is uncertain, coated vesicles are thought to bud from the cisternal rims to carry this traffic from

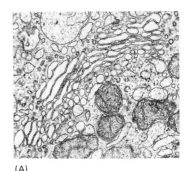

(A)

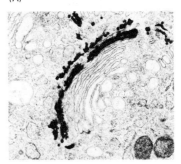

(B)

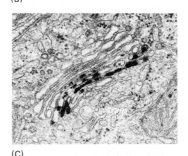

(C)

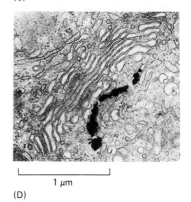

1 µm

(D)

Figure 8–67 Histochemical stains demonstrate that the Golgi apparatus is biochemically polarized. (A) Unstained. (B) Osmium is preferentially reduced by the cisternae of the *cis* compartment. (C) The enzyme nucleoside diphosphatase (see Figure 8–62) is found in the *trans* Golgi cisternae; this enzyme was formerly called "thiamine pyrophosphatase." (D) The enzyme acid phosphatase marks the *trans* Golgi network (formerly referred to as the "Golgi endoplasmic reticulum lysosomes," or "GERL"). (Courtesy of Daniel S. Friend.)

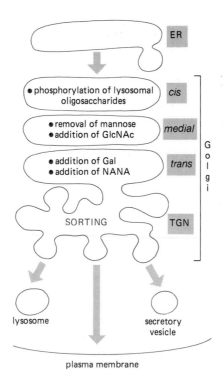

Figure 8–68 The compartmentalization of the Golgi apparatus. The covalent modification of proteins occurs sequentially during their passage through the clustered compartments of the Golgi stack; the *trans* Golgi network (TGN) is a tubular reticulum that acts primarily as a sorting station. The localization of each processing step shown was determined by a combination of techniques including subfractionation of the Golgi apparatus membranes and electron microscopy after staining with antibodies specific to some of the processing enzymes. The locations of many other processing reactions have not yet been determined.

cisterna to cisterna across the stack. Since experiments demonstrate that exported proteins move unidirectionally from *cis* to *medial* to *trans* compartments and never skip an intervening compartment, this scheme requires that the vesicles be programmed to fuse only with the next compartment in the sequence. Although only three functionally distinct cisternal compartments have so far been demonstrated, each of these sometimes consists of a block of two or more cisternae in sequence, and it is possible that there are finer subdivisions still to be discovered. Alternatively, it may be that there are only three fundamental compartments and that the several cisternae within the *cis*, *medial*, and *trans* divisions are simply multiple copies of the same functional unit.

Summary

The Golgi apparatus receives newly synthesized proteins and lipids from the ER and distributes them to the plasma membrane, lysosomes, and secretory vesicles. It is a polarized structure made up of one or more stacks of disc-shaped cisternae surrounded by a swarm of small vesicles. The cisternae are organized as a series of at least three distinct and sequential processing compartments, termed cis, medial, and trans Golgi. Proteins are transferred from the lumen and membrane of the ER to the cis face of the Golgi stack by means of transport vesicles. Proteins targeted for secretory vesicles, the plasma membrane, and lysosomes move across the stack in the cis-to-trans direction, passing from one cisterna to the next in series. Finally, they reach the trans Golgi network, from which each type of protein departs for its final destination in a particular type of vesicle.

The Golgi, unlike the ER, contains many sugar nucleotides; a variety of glycosyl transferase enzymes use these substrates to carry out glycosylation reactions on both lipid and protein molecules as they pass through the Golgi apparatus. N-linked oligosaccharides on proteins, for example, are often trimmed by removal of mannose residues, and additional sugars—including N-acetylglucosamine, galactose, and sialic acid residues—are added. In addition, the Golgi is the site of both O-linked glycosylation and the conversion of proteoglycan core proteins to proteoglycans. Sulfation of the sugars in proteoglycans and of selected tyrosines on proteins also occurs in a late Golgi compartment.

Transport of Proteins from the Golgi Apparatus to Lysosomes

All of the proteins that pass through the Golgi apparatus, except those that are retained there as permanent residents, are thought to be sorted in the *trans* Golgi network according to their final destination. The mechanism of sorting is particularly well understood for those proteins destined for the lumen of lysosomes, and in this section we shall consider this selective transport process. We begin with a brief account of lysosomal structure and function.

Lysosomes Are the Principal Sites of Intracellular Digestion[53]

In the course of studies of enzymes in liver homogenates in 1949, certain irregularities were noticed in the assay of acid phosphatase. The activity of the enzyme, for example, was higher in extracts prepared with distilled water than in extracts prepared with an osmotically balanced sucrose solution. It was also higher in aged than in fresh preparations, and in aged preparations it was no longer associated with sedimentable particles. Similar findings were soon reported for several other hydrolytic enzymes and led to the discovery of a new organelle, named the **lysosome,** a membranous bag of hydrolytic enzymes used for the controlled intracellular digestion of macromolecules. Damage to lysosomal membranes in cell extracts, induced by osmotic lysis or aging, releases the enzymes in a nonsedimentable form.

About 40 hydrolytic enzymes are now known to be contained in lysosomes. They include proteases, nucleases, glycosidases, lipases, phospholipases, phosphatases, and sulfatases. All are **acid hydrolases,** optimally active near the pH 5 maintained within lysosomes. The membrane of the lysosome normally keeps the enzymes out of the cytosol (whose pH is about ~7.2), but the acid dependence of the enzymes protects the contents of the cytosol against damage even if leakage should occur.

Like all other intracellular organelles, the lysosome not only contains a unique collection of enzymes but also has a unique surrounding membrane. The lysosomal membrane, for example, contains transport proteins that allow the final products of the digestion of macromolecules to escape so that they can be either excreted or reutilized by the cell. It also contains a H$^+$ pump that utilizes the energy of ATP hydrolysis to pump H$^+$ into the lysosome, thereby maintaining the lumen at its pH of about 5 (Figure 8–69). Most of the lysosomal membrane proteins are unusually highly glycosylated, which may help protect them from the lysosomal proteases in the lumen.

Lysosomes Are Heterogeneous Organelles[54]

Lysosomes were seen clearly in the electron microscope about a decade after they had first been described. They are extraordinarily diverse in shape and size but can be identified as a single family of organelles by histochemistry, using the precipitate formed by the reaction product of a hydrolase with its substrate to show which organelles contain the enzyme (Figure 8–70). By this criterion, lysosomes are found in all eucaryotic cells.

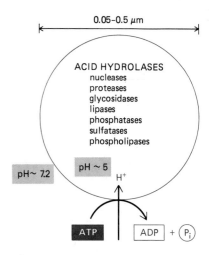

Figure 8–69 Lysosomes are defined as membrane-bounded vesicles involved in intracellular digestion that contain a variety of hydrolytic enzymes that are active under acidic conditions. The lumen is maintained at an acidic pH (around 5) by a H$^+$ pump in the membrane that uses the energy of ATP hydrolysis to pump H$^+$ into the vesicle.

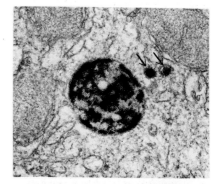

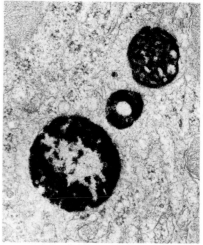

Figure 8–70 Electron micrographs of two sections of a cell stained to reveal the location of acid phosphatase, a marker enzyme for lysosomes. The larger membrane-bounded organelles, containing dense precipitates of lead phosphate, are lysosomes, whose diverse morphology reflects variations in the amount and nature of the material they are digesting. Two small vesicles thought to be carrying hydrolases from the Golgi apparatus are indicated by arrows in the top panel. The precipitates are produced when tissue fixed with glutaraldehyde (to fix the enzyme in place) is incubated with a phosphatase substrate in the presence of lead ions. (Courtesy of Daniel S. Friend.)

The heterogeneity of lysosomal morphology contrasts with the relatively uniform structures of most other cellular organelles. The diversity reflects the wide variety of digestive functions mediated by acid hydrolases, including the digestion of intra- and extracellular debris, the digestion of phagocytosed microorganisms, and even cell nutrition (since lysosomes are the principal site of cholesterol assimilation from endocytosed serum lipoprotein—see p. 328). For this reason, lysosomes are sometimes viewed as a heterogeneous collection of distinct organelles whose common feature is a high content of hydrolytic enzymes.

Lysosomes Can Obtain the Materials They Degrade by at Least Three Pathways[55]

The traffic of materials to lysosomes follows different paths depending on the source of the materials to be digested. The best-studied pathway is used to digest materials taken up by *endocytosis* in the pathway that leads from *coated pits* to *endosomes* to lysosomes. As discussed in Chapter 6 (see p. 323), materials endocytosed by this pathway pass sequentially from a *peripheral* to a *perinuclear* endosomal compartment. Those materials that have not been specifically retrieved from these endosomes for recycling to the plasma membrane then enter a third distinct "intermediate compartment," which receives newly synthesized lysosomal hydrolases and lysosomal membrane proteins from the Golgi apparatus. Because this *endolysosomal* compartment is mildly acidic, it is thought to be the site where the hydrolytic digestion of endocytosed materials begins. Conversion of the endolysosome into a mature lysosome requires that it lose its distinct endosomal membrane components and further decrease its internal pH. It is not known how the conversion occurs (see p. 331).

All cells have a second pathway for supplying materials to lysosomes for degradation, in which obsolete parts of the cell itself can be destroyed—a process called *autophagy*. In a liver cell, for example, an average mitochondrion has a lifetime of about 10 days, and electron microscopic images of normal cells reveal lysosomes containing (and presumably digesting) recognizable organelles such as mitochondria and secretory vesicles. The digestion process appears to involve the enclosure of an organelle by membranes derived from the ER, creating an *autophagosome*. The autophagosome is thought to fuse then with a lysosome (or endolysosome), initiating the digestion of its contents in an *autophagolysosome*. The process is highly regulated, and selected cell components can be targeted for destruction during cell remodeling: the smooth ER that has proliferated in a liver cell in response to a drug, for example, is selectively removed by autophagy when the drug is withdrawn (see p. 436).

The third pathway that provides materials to lysosomes occurs only in cells that are specialized for the *phagocytosis* of large particles and microorganisms. Cells such as macrophages and neutrophils can engulf such large objects to form a *phagosome* (see p. 334). The phagosome is thought to be converted to a *phagolysosome* in the manner discussed for the autophagosome. These three pathways are summarized in Figure 8–71; the three types of lysosomes that result may not be different from one another except for the materials they digest. All are normally referred to simply as *lysosomes*.

8-37 Lysosomal Enzymes Are Sorted in the Golgi Apparatus by a Membrane-bound Receptor Protein That Recognizes Mannose 6-Phosphate[56]

The biosynthesis of lysosomes requires the synthesis of specialized lysosomal hydrolases and membrane proteins. Both classes of proteins are synthesized in the ER and transported through the Golgi apparatus. Transport vesicles that deliver these proteins to the endolysosome—and then to lysosomes—bud from the *trans* Golgi network. These vesicles must incorporate lysosomal proteins while excluding the many other proteins being packaged into different transport vesicles for delivery elsewhere.

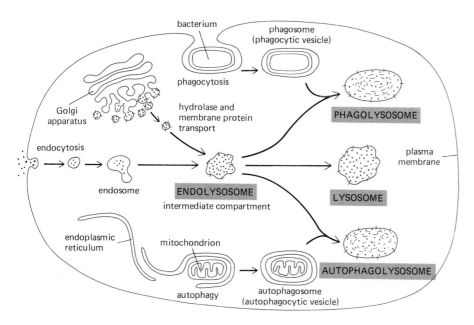

Figure 8–71 Three ways in which lysosomes are thought to form. Each pathway leads to the intracellular digestion of materials derived from a different source and produces a morphologically distinct lysosome. At the center of these pathways is an "intermediate compartment," designated here as an *endolysosome* (see p. 331).

Until recently it was traditional to make a distinction between *primary lysosomes* and *secondary lysosomes*, with the former term being used to describe newly formed lysosomal vesicles that have not yet encountered materials to be digested. Recent evidence, however, shows that lysosomal hydrolases and lysosomal membrane proteins are sorted by different receptors, raising the possibility that they leave the Golgi apparatus in separate transport vesicles and first meet in the endolysosome, which already contains endocytosed materials for digestion.

How are lysosomal proteins recognized and selected with the required accuracy? The same question can be asked for many other vesicle-mediated sorting processes in the cell. At the molecular level the answer is known in only one case—that of the lysosomal hydrolases. These carry a unique marker in the form of *mannose 6-phosphate (M6P)* groups, which are added exclusively to the N-linked oligosaccharides of these soluble lysosomal enzymes. This reaction occurs in the lumen of the *cis* Golgi. Complementary **M6P receptor proteins,** which cluster in the membrane and become concentrated in clathrin-coated vesicles budding from the *trans* Golgi network, have also been isolated and characterized. These receptors are transmembrane proteins that bind the lysosomal enzymes, thereby separating them from all of the other proteins present and concentrating them in coated transport vesicles. These vesicles rapidly lose their coats and fuse with an endolysosome, delivering their contents to this organelle.

In some cells a small fraction of the mannose 6-phosphate receptors are present in the plasma membrane, where they function in receptor-mediated endocytosis of lysosomal enzymes that have been released into the extracellular medium. These receptors transport the enzymes via coated pits into endosomes, from where the enzymes are eventually delivered to lysosomes. This *scavenger pathway* recaptures lysosomal hydrolases that have escaped the normal packaging process in the *trans* Golgi network and have therefore been transported to the cell surface and secreted.

The Mannose 6-Phosphate Receptor Shuttles Back and Forth Between Specific Membranes[57]

The mannose 6-phosphate receptor protein has been purified and characterized in *in vitro* studies. It binds its specific oligosaccharide at pH 7 and releases it at pH 6, which is the pH in the interior of the endolysosome. Thus the lysosomal enzymes dissociate from the M6P receptor proteins in the endolysosome and begin to digest the endocytosed material delivered from endosomes. Having released their enzymes, the receptors are retrieved and returned to the membrane of the *trans* Golgi network, probably by coated-vesicle-mediated transport (Figure 8–72). This process of *membrane recycling* from endolysosome back to the Golgi apparatus is similar to the recycling that occurs between endosomes and the plasma membrane during receptor-mediated endocytosis (see p. 331). Indeed, the receptor-mediated transport of lysosomal hydrolases from the Golgi apparatus to endolysosomes is analogous to the receptor-mediated endocytosis of extracellular molecules from the plasma membrane to endosomes. In both processes, receptors

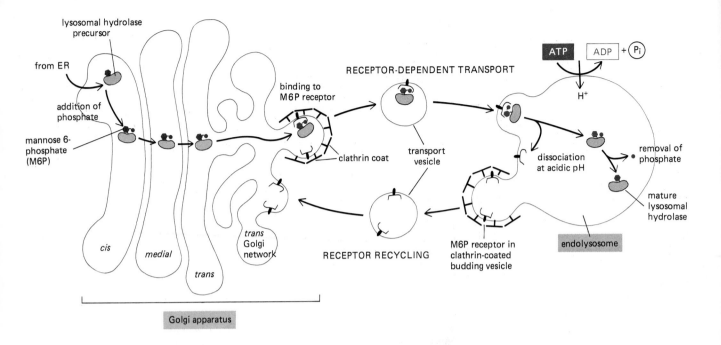

Figure 8–72 The transport of newly synthesized lysosomal hydrolases to lysosomes. The precursors of lysosomal hydrolases are tagged with mannose 6-phosphate groups in the *cis* Golgi and segregated from all other types of proteins in the *trans* Golgi network. The segregation occurs because clathrin-coated vesicles budding from the *trans* Golgi network concentrate mannose 6-phosphate-specific receptors, which bind the lysosomal hydrolases. These coated vesicles lose their coats and fuse with endolysosomes (see Figure 8–71). At the low pH of the endolysosome, the hydrolases dissociate from the receptors, which are recycled to the Golgi apparatus for further rounds of transport. The removal of phosphate from the mannose on the hydrolases further decreases the chance that the hydrolases will return to the Golgi apparatus with the receptor. Although there are two structurally distinct M6P receptor glycoproteins of very different size, they have a related amino acid sequence and appear to have similar functions.

cluster in clathrin-coated regions of membrane (called coated pits); these regions bud off to form clathrin-coated vesicles that transport ligands to a second compartment, which is acidic, from where the receptors are recycled back to their membrane of origin.

The recycling of the mannose 6-phosphate receptor has been followed using specific antibodies to locate the protein in the cell. The receptors are normally found in membranes of the Golgi apparatus and endolysosomes but not in mature lysosomes. When some cultured cells are treated with a weak base such as *ammonia* or *chloroquine*, which specifically accumulates in the interior of acidified organelles and raises their pH toward neutrality, the receptors disappear from the Golgi apparatus and accumulate in endolysosomes. The return of the receptors to the Golgi can be triggered in such cells either by removing the weak base or by adding high concentrations of mannose 6-phosphate to the culture medium. Both treatments cause the receptors to release their bound enzymes in the endolysosome, in one case by reacidifying the organelle and in the other by competitive binding of endocytosed mannose 6-phosphate to the receptors. These experiments suggest that transport back to the Golgi apparatus is facilitated by a conformational change in the receptor that occurs when it releases its bound hydrolase.

The shuttle system for the mannose 6-phosphate receptor shown in Figure 8–72 is specific—the vesicles carrying the receptor fuse with their specific target organelles but not, for example, with the ER membrane. The clathrin coat on the forming vesicles is thought to act as a "molecular filter," sequestering the receptor and its ligand into vesicles (see p. 328), but it cannot be responsible for the specificity of vesicle targeting because the coat is quickly removed after the vesicle is formed. *In vitro* experiments suggest that the removal of clathrin is catalyzed by an hsp70-like protein in a reaction that requires ATP hydrolysis. One or more of the proteins left exposed on the outer surface of the vesicle membrane then presumably serve as a specific "docking marker" that is recognized by a complementary "acceptor" on the target organelle membrane. A highly schematic view of such a process is illustrated in Figure 8–73.

The sorting of lysosomal hydrolases is the best model available for understanding the many transport-vesicle-mediated sorting events that occur in a eucaryotic cell. Although an oligosaccharide marker is not likely to be used elsewhere, "cargo" recognition by a membrane-bound receptor during vesicle budding, fusion of the vesicle with a specific target membrane, cargo release in the target compartment, and recycling of the empty receptor back to the original compartment are all likely to be common themes.

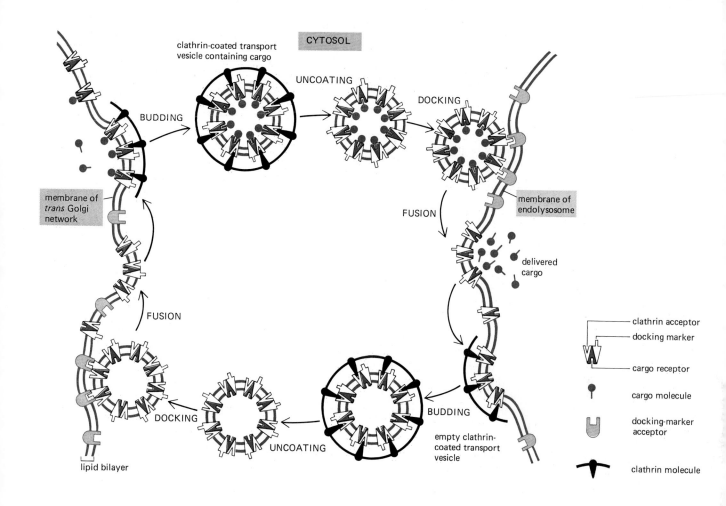

The Attachment of Multiple Mannose 6-Phosphate Groups to a Lysosomal Enzyme Amplifies the Sorting Signal[58]

The sorting system that segregates lysosomal hydrolases and dispatches them to endolysosomes works because M6P groups are added only to the appropriate glycoproteins in the Golgi apparatus. This requires specific recognition of the hydrolases by the Golgi enzyme responsible for adding M6P. Since all glycoproteins arrive in the *cis* Golgi with identical *N*-linked oligosaccharide chains, the signal for adding the M6P units to oligosaccharides must reside somewhere in the polypeptide chain of each hydrolase.

Two enzymes act sequentially to catalyze the addition of M6P groups to lysosomal hydrolases: *N*-acetylglucosamine phosphotransferase (*GlcNAc-P-transferase*) transfers the GlcNAc-P portion of the sugar nucleotide UDP-GlcNAc to a mannose unit on the oligosaccharide while the hydrolase is in the *cis* Golgi; a second enzyme—a phosphoglycosidase—then removes the terminal GlcNAc, exposing the phosphate to form the final M6P marker (Figure 8–74). The phosphotransferase specifically binds the hydrolase by means of a *recognition site* that is separate from the *catalytic site* for the reaction (Figure 8–75). The signal recognized by the recognition site is believed to be a conformation-dependent *signal patch* rather than a signal peptide (see p. 414), since recognition is virtually eliminated when the hydrolase is partially unfolded.

Once the phosphotransferase has recognized the signal on the hydrolase, it adds GlcNAc-P to one or two of the mannoses on each oligosaccharide chain. Since most lysosomal hydrolases have multiple oligosaccharides, they acquire many M6P residues; this greatly amplifies the signal. Thus, while a lysosomal hydrolase typically binds to the recognition site of the phosphotransferase with an affinity constant (K_a) of about 10^5 liters/mole, the multiply phosphorylated hydrolase binds to the M6P receptor with a K_a of about 10^9 liters/mole, a 10,000-fold amplification.

Figure 8–73 A possible mechanism for directing clathrin-coated transport vesicles back and forth between two specific membranes. In this hypothetical model the "cargo molecules" are lysosomal hydrolases and the "cargo receptor" is the mannose 6-phosphate receptor protein. The postulated "docking marker" and "docking-marker acceptor" molecules have not been well characterized, although the recent discovery of clathrin-binding proteins that also bind to the cytoplasmic tails of selected membrane proteins fits the description given here of the docking marker acceptor. (See Pearse, B.M., *EMBO J.* 7:3331–3336, 1988.)

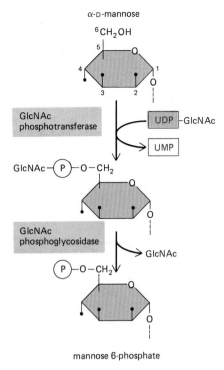

Figure 8–74 Synthesis of the mannose 6-phosphate marker on a lysosomal hydrolase occurs in two steps. First, GlcNAc phosphotransferase transfers P-GlcNAc residues to the 6 position of several mannose residues on the *N*-linked oligosaccharides of the lysosomal precursor glycoprotein. Second, a phosphoglycosidase cleaves off the GlcNAc residue, creating the mannose 6-phosphate marker. The first enzyme is specifically activated by a signal patch present on lysosomal hydrolases (see Figure 8–75), while the phosphoglycosidase is a nonspecific enzyme. This modification of selected mannose residues in the *cis* Golgi compartment protects these mannoses from removal by the mannosidases that will be encountered later in the *medial* Golgi compartment.

8-38 Defects in the GlcNAc Phosphotransferase Cause a Lysosomal Storage Disease in Humans[59]

Lysosomal storage diseases played a crucial part in the discovery of the lysosomal hydrolase-sorting mechanism. These diseases are caused by genetic defects that affect one or more of the lysosomal hydrolases and result in accumulation of their undigested substrates in lysosomes, with profound pathological consequences. They usually result from a mutation in a structural gene that codes for an individual lysosomal hydrolase. The most dramatic form of lysosomal storage disease, however, is a very rare disorder called inclusion cell disease (*I-cell disease*). In this disease almost all of the hydrolytic enzymes are missing from the lysosomes of fibroblasts, and their undigested substrates accumulate as large "inclusions" in the patients' cells. I-cell disease is due to a single gene defect, and like most genetic enzyme deficiencies, it is recessive—that is, only individuals with two bad copies of the gene have the disease.

In I-cell disease all the hydrolases missing from lysosomes are found in the blood, indicating that the structural genes encoding them are unaffected. The abnormality results from a missorting in the Golgi apparatus that causes the hydrolases to be secreted rather than transported to lysosomes. The missorting has been traced to a defective or missing GlcNAc-P-transferase. Because lysosomal enzymes are not phosphorylated in the *cis* Golgi, they are not segregated by M6P receptors into coated vesicles in the *trans* Golgi network. Instead they are carried to the cell surface and secreted. The oligosaccharides that would contain M6P in normal lysosomal enzymes are converted to the "complex" type, containing GlcNAc, Gal, and sialic acid. This indicates that the phosphorylation of mannose in the *cis* Golgi normally prevents the subsequent processing of the oligosaccharides on the hydrolases to complex forms in the *medial* and *trans* Golgi.

Experiments on I-cell disease in the late 1960s provided the first clue that all lysosomal enzymes have a common recognition marker. The marker was identified as mannose 6-phosphate in the late 1970s, when the same hydrolase was com-

Figure 8–75 The GlcNAc phosphotransferase enzyme that recognizes lysosomal hydrolases in the Golgi apparatus has separate catalytic and recognition sites. The catalytic site binds both high-mannose *N*-linked oligosaccharides and UDP-GlcNAc. The recognition site binds to a signal patch that is present only on the surface of lysosomal hydrolases and their precursors.

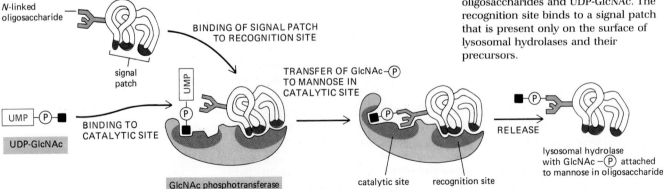

pared in normal and mutant individuals. This led quickly to the purification of mannose 6-phosphate receptors and of GlcNAc-P-transferase and to an understanding of the role of the Golgi apparatus in the lysosomal hydrolase-sorting pathway.

In I-cell disease the lysosomes in some cell types, such as hepatocytes, contain a normal complement of lysosomal enzymes. This implies that there is another pathway for directing hydrolases to lysosomes that is used by some cell types but not others. The nature of this M6P-independent pathway is unknown. Perhaps in this case the hydrolases are sorted by direct recognition of their signal patch. Similarly, the lysosomal membrane proteins are sorted from the *trans* Golgi network to endolysosomes by an M6P-independent pathway in all cells. It is unclear why cells should need more than one sorting pathway to construct a lysosome.

Summary

Lysosomes are specialized for intracellular digestion. They contain unique membrane proteins and a wide variety of hydrolytic enzymes that operate best at pH 5, the internal pH of lysosomes. The acidic pH in lysosomes is maintained by an ATP-driven proton pump in their membranes. Newly synthesized lysosomal proteins are transferred into the lumen of the ER, transported through the Golgi apparatus, and then transported from the trans *Golgi network to an intermediate compartment (an endolysosome) by means of transport vesicles.*

The lysosomal hydrolases contain N-linked oligosaccharides that are processed in a unique way in the cis *Golgi so that their mannose residues are phosphorylated. These mannose 6-phosphate (M6P) groups are recognized by an M6P receptor protein in the* trans *Golgi network that segregates the hydrolases and helps to package them into budding clathrin-coated vesicles, which quickly lose their coats. These transport vesicles containing the mannose 6-phosphate receptor act as shuttles that move the receptor back and forth between the* trans *Golgi network and endolysosomes. The low pH in the endolysosome dissociates the lysosomal hydrolases from this receptor, making the transport of the hydrolases unidirectional.*

Transport from the Golgi Apparatus to Secretory Vesicles and to the Cell Surface[60]

Transport vesicles designed for immediate fusion with the plasma membrane normally leave the Golgi apparatus in a steady stream. The transmembrane proteins and lipids in these vesicles provide the new proteins and lipids for the cell's plasma membrane, while the soluble proteins in the vesicles are secreted to the cell exterior. In this way, for example, cells produce the proteoglycans and proteins of the extracellular matrix (see p. 802).

Whereas all cells require this **constitutive secretory pathway,** specialized secretory cells have a second secretory pathway in which soluble proteins and other substances are stored in secretory vesicles for later release—the so-called *triggered,* or **regulated secretory pathway** (Figure 8–76).

In this section we shall consider the role of the Golgi apparatus in these two secretory pathways and compare the mechanisms involved. We shall also consider how viruses exploit the sorting apparatus of their host cells and can be used to reveal the diversity of intracellular transport pathways.

8-41 Secretory Vesicles Bud from the *Trans* Golgi Network[61]

In cells in which secretion occurs in response to an extracellular signal, secreted proteins are concentrated and stored in **secretory vesicles** (frequently called *secretory granules* because of their dense cores), from which they are released by exocytosis in response to the signal. Secretory vesicles form by budding from the *trans* Golgi network. Their formation is thought to involve clathrin and its associated coat proteins, since a portion of the surface of forming secretory vesicles is

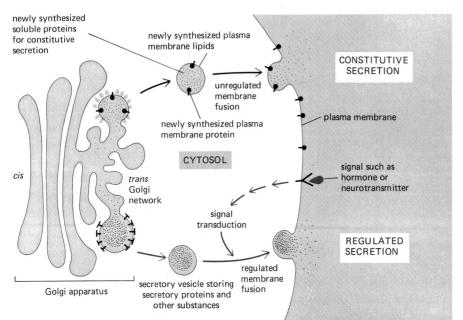

newly synthesized
soluble proteins
for constitutive
secretion

newly synthesized plasma
membrane lipids

unregulated
membrane
fusion

CONSTITUTIVE
SECRETION

newly synthesized plasma
membrane protein

plasma membrane

CYTOSOL

cis

trans
Golgi
network

signal such as
hormone or
neurotransmitter

signal
transduction

REGULATED
SECRETION

Golgi apparatus

secretory vesicle storing
secretory proteins and
other substances

regulated
membrane
fusion

Figure 8–76 The regulated and constitutive pathways of secretion diverge in the *trans* Golgi network. Many soluble proteins are continually secreted from the cell by the *constitutive secretory pathway*, which operates in all cells. This pathway also supplies the plasma membrane with newly synthesized lipids and transmembrane proteins. Specialized secretory cells also have a *regulated secretory pathway*, by which selected proteins in the *trans* Golgi network are diverted into secretory vesicles, where the proteins are concentrated and stored until an extracellular signal stimulates their secretion.

usually clathrin coated. The coat is lost by the time the vesicle is fully formed (Figure 8–77).

Like the lysosomal hydrolases discussed in the preceding section, proteins destined for secretory vesicles (often called *secretory proteins*) must be sorted and packaged into appropriate vesicles in the *trans* Golgi network. In this case the mechanism is believed to involve the selective aggregation of secretory proteins, which can be detected in the electron microscope as electron-dense material in the lumen of the *trans* Golgi network. The "sorting signal" that directs proteins into such an aggregate is not known, but it is thought to be a signal patch that is shared by many secretory proteins: when a gene encoding a secretory protein is transferred to a different type of secretory cell that normally does not make the protein, the foreign protein is appropriately packaged into secretory vesicles.

How the aggregates containing secretory proteins are segregated into forming secretory vesicles is also unclear. Secretory vesicles contain unique membrane proteins, some of which might serve as receptors (in the *trans* Golgi network) to bind the aggregated material that will be packaged. Secretory vesicles are larger than the transport vesicles that carry lysosomal hydrolases, and the aggregates they contain are much too large for each molecule of the secreted protein to be bound by a receptor in the vesicle membrane—as proposed for transport of the lysosomal enzymes (see Figure 8–73); the uptake of the aggregates into secretory vesicles may more closely resemble the uptake of particles by phagocytosis at the cell surface (see p. 335), which can also be mediated by clathrin-coated membranes.

After the immature secretory vesicles bud from the *trans* Golgi network, the clathrin coat is removed and their contents become greatly condensed. The condensation occurs suddenly and is believed to be caused by an acidification of the vesicle lumen induced by an ATP-driven H^+ pump in the vesicle membrane. The aggregation of secreted proteins (or other compounds, see p. 325) and their subsequent condensation in secretory vesicles can concentrate these substances by as much as 200-fold from the Golgi lumen, enabling the secretory cell to release large amounts of material on demand.

Secretory-Vesicle Membrane Components Are Recycled[62]

Many secretory cells, such as the pancreatic acinar cell, are polarized, and exocytosis occurs only at the apical surface, which typically faces the lumen of a duct system that collects the secretions. When a secretory vesicle fuses with the plasma membrane, its contents are discharged from the cell by exocytosis and its mem-

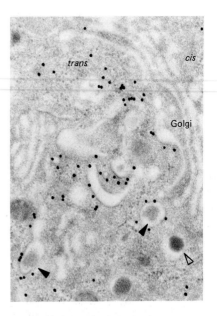

Figure 8–77 Electron micrograph of secretory vesicles forming from the *trans* Golgi network in an insulin-secreting cell of the pancreas. An antibody conjugated to gold spheres (*black dots*) has been used to locate clathrin molecules. The immature secretory vesicles (*black arrowheads*), which contain proinsulin molecules, are coated with clathrin. The clathrin coat is rapidly shed once the vesicle has formed and is absent from the mature secretory vesicles (*open arrowhead*). (Micrograph courtesy of Lelio Orci.)

brane becomes part of the plasma membrane (see p. 324). Although this should greatly increase the surface area of the plasma membrane, it does so only transiently because membrane components are removed from the surface (or *recycled*) by endocytosis almost as fast as they are added by exocytosis (Figure 8–78). There is evidence that this removal returns the proteins of the secretory vesicle membrane to the Golgi apparatus, where they can be used again. Such recycling maintains a steady-state distribution of membrane components among the various cellular compartments.

Proteins and Lipids Seem to Be Carried Automatically from the ER and Golgi Apparatus to the Cell Surface in Unpolarized Cells[63]

In a cell capable of regulated secretion, at least three types of proteins must be separated before they leave the *trans* Golgi network: those destined for endolysosomes, those destined for secretory vesicles, and those destined for immediate delivery to the cell surface. Proteins destined for endolysosomes are selected for packaging into departing vesicles on the basis of positive signals (M6P for lysosomal hydrolases), and a special signal must also be present on each of the proteins packaged into secretory vesicles. The proteins transferred to the cell surface could therefore in principle be transported by a nonselective "default pathway," provided that there is no need to target a particular plasma membrane protein to a selected region of the cell (Figure 8–79). Thus, in an unpolarized cell such as a white blood cell or most cells in culture, it has been proposed that any protein in the ER will automatically be carried through the Golgi apparatus to the cell surface by the constitutive secretory pathway unless it is either specifically retained as a resident of the ER or Golgi or selected for transport elsewhere. An attraction of this model lies in its simplicity and in the opportunity that such a default pathway offers for discarding damaged or misdirected proteins to the outside of the cell.

In an attempt to test for such an unselected "bulk-flow" pathway to the plasma membrane, cultured cells were incubated with a simple tripeptide (Asn-Tyr-Thr) containing the glycosylation code Asn-X-Thr. This small peptide was able to diffuse into cells and across intracellular membranes, and it became glycosylated on its asparagine residue in the lumen of the ER. The addition of the *N*-linked oligosaccharide prevented the tripeptide from diffusing back into the cytosol. Instead it was transported unidirectionally from the ER through the Golgi and to the cell surface in about 10 minutes, which is the rate of transport of the fastest normal plasma membrane proteins. This result is consistent with the hypothesis that there is an unselected flow of luminal fluid that automatically carries any soluble molecule in the lumen of the ER promptly to the cell surface unless it is specifically retained or directed elsewhere.

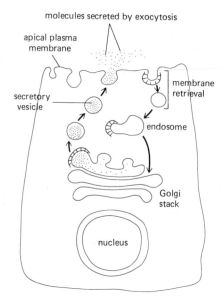

Figure 8–78 The membrane of a secretory vesicle is recycled after it fuses with the apical plasma membrane. The amount of membrane added during regulated secretion can be massive, but because membrane is retrieved in clathrin-coated vesicles and returned to the *trans* Golgi, probably via endosomes, the apical surface maintains an essentially constant area.

Figure 8–79 The best-understood pathways of protein sorting in the *trans* Golgi network. Proteins with the mannose 6-phosphate marker are diverted to lysosomes (via endolysosomes) in clathrin-coated vesicles (see Figure 8–72). Proteins with signals directing them to secretory vesicles are concentrated in large clathrin-coated vesicles that lose their coats to become secretory vesicles—a pathway that occurs only in specialized secretory cells. In unpolarized cells, proteins with no special features are thought to be delivered to the cell surface by default via the constitutive secretory pathway. In polarized cells, however, secreted and plasma membrane proteins are selectively directed to either the apical or the basolateral plasma membrane domain, so that at least one of these two pathways must be signal mediated.

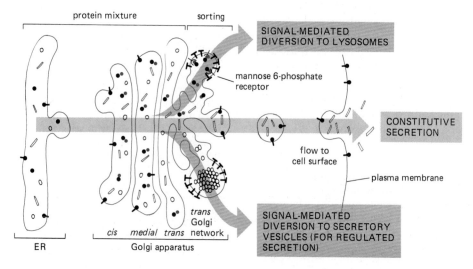

In principle, the same type of unselected flow could carry transmembrane proteins and lipids that lack sorting signals automatically from the ER to the cell surface. It could also transport proteins destined for secretory vesicles and lysosomes from the ER to the end of the Golgi apparatus; only in the *trans* Golgi network would specific signals be required to separate these proteins from those going to the plasma membrane. A special sorting mechanism would also be needed to retain all proteins intended for permanent residence in the ER or Golgi apparatus (see p. 413). The observation that several ER resident proteins (including BiP and protein disulfide isomerase) contain a signal peptide that is responsible for their retention in the ER (see p. 415) supports this general view of protein traffic through the cell.

Some constitutively secreted proteins take a long time to leave the ER and require many hours to be secreted. To reconcile this observation with the unselected-flow hypothesis, it has been suggested that these proteins take a long time to fold correctly in the ER and are therefore delayed there, either because they stick to the ER membrane or because they are bound by special proteins such as BiP (see p. 445); once folded correctly, these retained proteins would also be carried forward in the unselected flow.

Polarized Cells Must Direct Proteins from the Golgi Apparatus to the Appropriate Domain of the Plasma Membrane[64]

Most cells in tissues are *polarized* and have two (and sometimes more) distinct plasma membrane domains (see p. 310). A typical epithelial cell, for example, has two physically continuous but compositionally distinct plasma membrane domains (see Figure 6–36, p. 297): the *apical domain* faces the lumen and often has specialized features such as cilia or a brush border of microvilli; the *basolateral domain* covers the rest of the cell. The two domains are joined at their border by a ring of *tight junctions* (see p. 793), which prevent proteins (and lipids in the outer leaflet of the lipid bilayer) from diffusing between the two membrane domains. Therefore, although the two domains appear as one continuous membrane when viewed in the electron microscope, the tight junctions effectively separate the domains so that the apical membrane contains one set of proteins while the basolateral membrane contains another. The lipid composition of the two bilayers is also different; in particular, glycolipids are found only in the apical membrane domain.

Epithelial cells can also secrete one set of proteins at their apical surface and a different set from their basolateral surface. Thus polarized cells must have ways of directing both membrane-bound and secreted molecules specifically to each plasma membrane domain. It has been shown in polarized cells in culture that proteins destined for different domains pass together from the ER to the *trans* Golgi network, where they are separated and dispatched in secretory or transport vesicles to the appropriate plasma membrane domain. It is possible that both the basolateral and apical proteins have distinct sorting signals that direct them to the appropriate domain; alternatively, only one of these pathways may require a sorting signal, with the other operating by default. The nature of the signals is unknown.

Viruses Exploit the Sorting Mechanisms of Their Host Cells[65]

Many animal viruses have only a small amount of nucleic acid in their genome and contain no more than four or five genes. Most of these genes code for structural proteins of the mature viral particle (or virion), so these viruses must parasitize host-cell pathways for most of the steps in their replication (see p. 248). Because viral products are usually synthesized in large amounts during infection, and because during its life cycle the virus follows a sequential route through the compartments of the host cell, virus-infected cells have been very useful for tracing the pathways of intracellular transport and for studying how essential biosynthetic reactions are compartmentalized in eucaryotic cells.

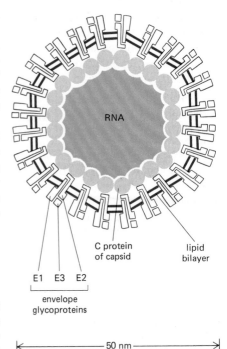

C protein
of capsid

lipid
bilayer

E1 E3 E2

envelope
glycoproteins

|← 50 nm →|

(A)

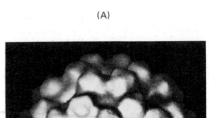

(B)

Figure 8–80 The structure of Semliki forest virus seen in a schematic drawing of a cross-section of the virus (A) and in a three-dimensional reconstruction of the surface of the virus derived from cryoelectron micrographs of unstained specimens (B). There are 180 copies of the capsid C protein (arranged as 60 trimers) and 240 copies of each of the three envelope proteins (arranged as 80 trimers) in each virus particle. The outer envelope of the virus consists of the envelope proteins embedded in a lipid bilayer membrane. The virus has a total mass of 46 million daltons. (B, from S.D. Fuller, *Cell*, 48:923–934, 1987.)

Enveloped animal viruses, in which the genome is enclosed in a lipid bilayer membrane (see p. 249), have exploited the compartmentalization of the cell to an especially fine degree. To follow the life cycle of an enveloped virus is to take a tour through the cell. A well-studied example is *Semliki forest virus*, which consists of an RNA genome surrounded by a **capsid** formed by a regularly arranged icosahedral (20-faced) shell of a protein (called C protein). The *nucleocapsid* (genome + capsid) is surrounded by a closely apposed lipid bilayer that contains only three proteins (called E1, E2, and E3). These **envelope proteins** are glycoproteins that span the lipid bilayer and interact with the C protein of the nucleocapsid, linking the membrane and nucleocapsid together (Figure 8–80A). The glycosylated portions of the envelope proteins are always on the outside of the lipid bilayer, and complexes of these proteins form "spikes" that can be seen in electron micrographs projecting outward from the surface of the virus (Figure 8–80B).

Infection is initiated when the virus binds to receptor proteins on the host-cell plasma membrane. The virus uses the cell's normal endocytic pathway to enter the cell by receptor-mediated endocytosis and is delivered to endosomes (see p. 329). But instead of being delivered to lysosomes, the virus escapes from the endosome by virtue of the special properties of one of its envelope proteins. At the acidic pH of the endosome, this protein causes the viral envelope to fuse with the endosome membrane, releasing the bare nucleocapsid into the cytosol (Figure 8–81). The nucleocapsid is "uncoated" in the cytosol, releasing the viral RNA, which is then translated by host-cell ribosomes to produce a virus-coded RNA polymerase. This in turn makes many copies of the RNA, some of which serve as mRNA molecules to direct the synthesis of the four structural proteins of the virus—the capsid C protein and the three envelope proteins E1, E2, and E3.

The capsid and envelope proteins follow separate pathways through the cytoplasm. The envelope proteins, like normal cellular membrane glycoproteins, are

Figure 8–81   The life cycle of Semliki forest virus. The virus parasitizes the host cell for most of its biosyntheses.

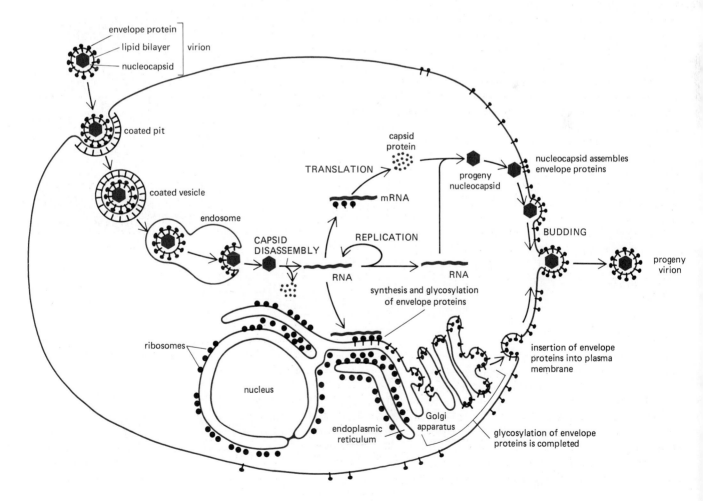

synthesized by ribosomes that are bound to the rough ER; the capsid protein, like a normal cytosolic protein, is synthesized by ribosomes that are not membrane bound. The newly synthesized capsid protein binds to the recently replicated viral RNA to form new nucleocapsids. The envelope proteins, in contrast, are inserted into the membrane of the ER, where they are glycosylated, transported to the Golgi apparatus (where their oligosaccharides are modified), and then delivered to the plasma membrane.

The viral nucleocapsids and envelope proteins finally meet at the plasma membrane (see Figure 8–81). As the result of a specific interaction with a cluster of envelope proteins, the nucleocapsid becomes wrapped in a portion of the plasma membrane, forming a bud that is highly enriched in the envelope proteins but contains host-cell lipids. Finally, the bud pinches off and a free virus is released on the outside of the cell. The clustering of envelope proteins in the lipid bilayer as they assemble around the nucleocapsid during viral budding has been suggested as a model for the segregation of specific membrane proteins into coated vesicles.

Viral Envelope Proteins Carry Signals That Direct Them to Specific Intracellular Membranes[66]

Different enveloped viruses bud from different host-cell membranes. Their envelope proteins, which are transmembrane proteins synthesized in the ER, must therefore carry signals directing them from the ER to the appropriate cell membrane. Some epithelial cell lines form polarized cell sheets when they are cultured on an appropriate surface such as a collagen-coated porous filter. When viruses infect such polarized cells, which maintain distinct apical and basolateral plasma membrane domains, some of them (such as influenza virus) bud exclusively from the apical plasma membrane, whereas others (such as Semliki forest virus and vesicular stomatitis virus) bud only from the basolateral plasma membrane (Figure 8–82). This polarity of budding reflects the presence on the envelope proteins of distinct apical or basolateral sorting signals, which direct the proteins to only one cell-surface domain; the proteins in turn cause the virus to assemble in that domain.

Other viruses have envelope proteins with different kinds of sorting signals. *Herpes virus*, for example, is a DNA virus that replicates in the nucleus, where its nucleocapsid assembles, and then acquires an envelope by budding through the inner nuclear membrane into the ER lumen; the envelope proteins therefore must

Figure 8–82 Electron micrographs showing that one type of enveloped virus buds from the apical plasma membrane, while another type buds from the basolateral plasma membrane, of the same epithelial cell line grown in culture. These cells grow with their basal surface attached to the culture dish. (Courtesy of E. Rodriguez-Boulan and D.D. Sabatini.)

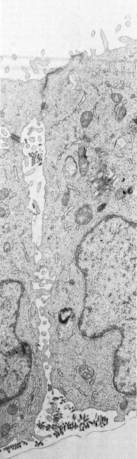

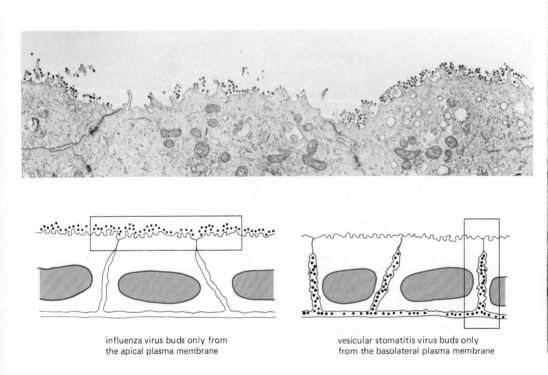

influenza virus buds only from the apical plasma membrane

vesicular stomatitis virus buds only from the basolateral plasma membrane

be specifically transported from the ER membrane to the inner nuclear membrane. *Flavivirus*, in contrast, buds directly into the ER lumen, and *bunyavirus* buds into the Golgi apparatus, indicating that their envelope proteins carry signals for retention in the ER and Golgi membranes, respectively. The herpes virus, flavivirus, and bunyavirus particles are soluble in the ER and Golgi lumen, and they move outward toward the cell surface exactly as if they were secreted proteins; in the *trans* Golgi network they are incorporated into transport vesicles and secreted from the cell by the constitutive secretory pathway.

Summary

Proteins can be secreted from cells by exocytosis in either a constitutive or a regulated fashion. In the regulated pathway, molecules are stored in secretory vesicles, which do not fuse with the plasma membrane to release their contents until an extracellular signal is received. A selective condensation of the proteins targeted to secretory vesicles accompanies their packaging into these vesicles in the trans Golgi network. The regulated pathway operates only in specialized secretory cells, but a constitutive secretory pathway operates in all cells, mediated by a continual process of vesicular transport from the trans Golgi network to the plasma membrane. For unpolarized cells there is evidence that proteins made in the ER are automatically delivered to the trans Golgi network and then to the plasma membrane by this constitutive pathway unless they are otherwise diverted or retained by specific sorting signals. In polarized cells, however, the transport pathways from the trans Golgi network to the plasma membrane must operate selectively to ensure that different sets of membrane proteins, secreted proteins, and lipids are delivered to the apical and basolateral domains.

Vesicular Transport and the Maintenance of Compartmental Identity

Intracellular sorting seems to require at least 10 distinct types of transport vesicles, each with a unique set of "molecular address labels" on its surface that allows it to deliver its contents only to specific cell membranes. Thus the transport vesicles leaving the ER must fuse only with the *cis* Golgi compartment, those leaving the *cis* Golgi compartment must fuse only with the *medial* Golgi compartment, and so on. In each step, vesicle budding, vesicle docking, and vesicle fusion are involved, each requiring highly specific recognition events. At present we know little about the molecular mechanisms involved (see Figure 8–73). In this final section we examine some speculative views of the mechanisms that maintain compartmental identity and describe some new experimental approaches to these problems.

The ER and Golgi Compartments Retain Selected Proteins as Permanent Residents[67]

In ferrying cargo from one compartment to another, transport vesicles necessarily transfer membrane as well as their luminal content. Yet in the face of this homogenizing influence, the vital differences of membrane composition are maintained between the different compartments: the SRP receptor protein (see p. 440) is found only in the ER membrane, whereas glycosyltransferases and oligosaccharide-processing enzymes are located only in the membranes of the correct Golgi cisterna, and so on. The membranes of the ER and each type of Golgi cisterna must therefore have special mechanisms for maintaining their unique compositions. One possible mechanism would be to have each step in the forward movement through the ER and Golgi compartments selected by a signal, in the way that plasma membrane proteins that enter the cell by receptor-mediated endocytosis are selected by coated pits. As discussed previously, however, biosynthetic transport through the ER and Golgi apparatus is thought to work in the opposite way, with forward movement being automatic and retention requiring signals. In this view, each permanent resident of the ER or a Golgi compartment must carry a sorting signal that is responsible for its selective retention. As described previously, a signal peptide that causes the selective retention of proteins in the ER has been identified (see p. 415), although the mechanism of retention is unknown. A strategy of automatic forward movement and selective retention is attractive in part because the number of proteins passing through the ER and Golgi en route to other destinations greatly exceeds the number retained. Moreover, this strategy allows proteins that have either lost their sorting signals or been misdirected in an earlier step to be secreted to the cell exterior, as observed (see p. 464). Finally, if forward transport required specific signals, a protein presumably would need a different signal for each forward transfer step it undergoes—from ER to Golgi and from each Golgi cisterna to the next. The structure of many proteins might then be overly constrained by the large number of sorting signals required on their surface.

8-45 There Are at Least Two Kinds of Coated Vesicles[68]

Clathrin is found on the cytoplasmic surface of both the plasma membrane and the *trans* Golgi network and seems to be associated with transport decisions that are signaled. From the plasma membrane, clathrin-coated vesicles carry receptor-mediated endocytic traffic to endosomes; from the *trans* Golgi network, they are known to carry receptor-mediated traffic to endolysosomes. In the former case the vesicles are known to transport a highly selected set of cell-surface receptors (see p. 328), and the same is presumably true for each of the different types of clathrin-coated vesicles that bud from the Golgi apparatus.

Clathrin-coated vesicles do not bud from the ER or from *cis* or *medial* cisternae of the Golgi apparatus. Instead, another type of coated vesicle buds from these locations (as well as from the *trans* Golgi network). The coats of these vesicles do not stain with anti-clathrin antibodies, and they appear different from clathrin coats in the electron microscope (Figure 8–83). The coat proteins of these vesicles have not yet been identified.

A possible model for protein transport in unpolarized cells is presented in Figure 8–84. In this model, proteins move by default from ER to Golgi, from Golgi cisterna to Golgi cisterna, and from the *trans* Golgi network to the cell surface in an unselected process that is postulated to be mediated by non-clathrin-coated vesicles. In the *trans* Golgi network the signal-mediated diversion to lysosomes is mediated by clathrin-coated vesicles, while the signal-mediated diversion to secretory vesicles is mediated by a membrane that is partly clathrin coated. How can one test such ideas?

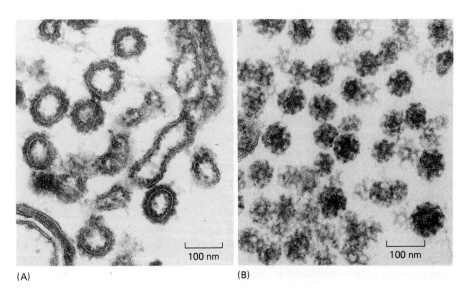

(A)   (B)

Figure 8–83 Comparison of clathrin- and non-clathrin-coated vesicles. (A) Electron micrograph of Golgi cisternae from a cell-free system in which non-clathrin-coated vesicles bud in the test tube. The clathrin-coated vesicles shown in (B) have a much more regular structure. (Electron micrographs courtesy of Lelio Orci, from L. Orci, B. Glick, and J. Rothman, *Cell* 46:171–184, 1986.)

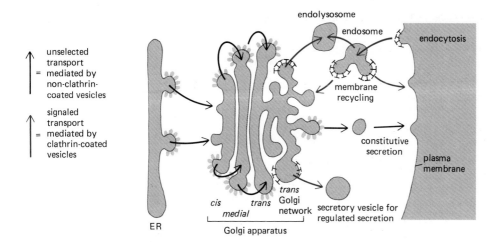

Figure 8–84 A hypothetical model for protein transport in nonpolarized cells. All unselected (constitutive) transport is postulated to be mediated by non-clathrin-coated vesicles. The various forms of signal-mediated transport are postulated to be carried out by clathrin-coated vesicles budding from the *trans* Golgi network, the endosomes, and the plasma membrane. In polarized cells an additional signaled pathway from the *trans* Golgi network is required.

Mutants in the Secretory Pathway Help to Identify the Transport Machinery[69]

Since vesicular transport is crucial to a eucaryotic cell, a mutant cell defective in a central component of this process—such as a docking marker or an acceptor protein (see Figure 8–73)—would probably die. If, however, the mutant protein is defective only at high temperatures, the mutant cell would survive at normal temperatures. More than 25 such *temperature-sensitive* mutations in genes involved in the secretory pathway have been identified in yeasts. When grown at high temperatures, some of these mutants fail to transport proteins from the ER to the Golgi apparatus, while others fail to transport proteins from one Golgi cisterna to another, or from the Golgi apparatus to the vacuole (the yeast lysosome) or to the plasma membrane.

The wild-type versions of these essential yeast genes can be readily cloned by transfecting wild-type DNA into the temperature-sensitive secretory mutants and then screening the transfected cells for survival at high temperatures. This approach is extremely powerful because one is led directly to the central proteins in the transport machinery without needing to know how secretion works. One gene required in the yeast secretory pathway that was isolated in this way is the *sec 4* gene, whose nucleotide sequence suggests that it encodes a GTP-binding protein of the *ras* family (see p. 705 and p. 757). Biochemical experiments in mammalian cells suggest that a similar GTP-binding protein functions in mammalian vesicular transport; it seems to regulate the uncoating of non-clathrin-coated vesicles prior to their fusion with membranes. These studies have depended on the ability to reconstitute vesicular transport in a cell-free system.

Cell-free Systems Provide Another Powerful Way to Analyze the Molecular Mechanisms of Vesicular Transport[70]

The key to understanding the molecular mechanisms that underlie the flow of traffic between membrane compartments is to elucidate the working parts of transport vesicles. How do transport vesicles bud off from membranes? What guides them to their targets? How do they fuse? In addition to the genetic approaches just described, a crucial step in trying to answer these questions is to reconstitute vesicular transport in a cell-free system. This was first achieved for the Golgi stack. When Golgi stacks are isolated from cells and incubated with cytosol and with ATP as a source of energy, non-clathrin-coated vesicles bud from their rims and apparently transport proteins between the cisternae (see Figure 8–83A). By following the progressive processing of the oligosaccharides on a glycoprotein as it moves from one Golgi compartment to the next, it was possible to reconstruct the process of vesicular transport *in vitro* (Figure 8–85).

The budding of the non-clathrin-coated vesicles in this cell-free system is found to require both ATP and a mixture of proteins from the cytosol, establishing

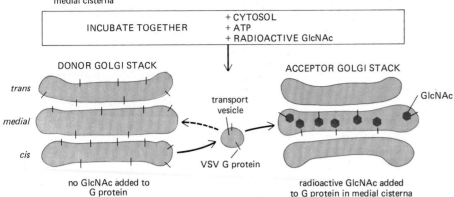

1. DONOR: Golgi stack isolated from VSV-infected mutant cell is unable to add GlcNAc to G protein in medial cisterna

+

2. ACCEPTOR: Golgi stack isolated from an uninfected normal cell can add GlcNAc to proteins in the medial cisterna

INCUBATE TOGETHER

+ CYTOSOL
+ ATP
+ RADIOACTIVE GlcNAc

DONOR GOLGI STACK

ACCEPTOR GOLGI STACK

trans

medial

cis

transport vesicle

VSV G protein

GlcNAc

no GlcNAc added to G protein

radioactive GlcNAc added to G protein in medial cisterna

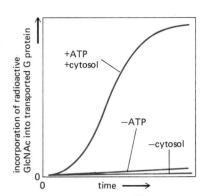

Figure 8–85   A cell-free system that reconstitutes the vesicular transport of a viral protein between the cisternal compartments of the Golgi apparatus. When isolated Golgi stacks are incubated with a cytosolic extract and ATP, non-clathrin-coated vesicles bud from the cisternae. Vesicular stomatitis virus (VSV) encodes an envelope glycoprotein called *G protein*. To measure the transport of this transmembrane protein between the cisternae of the Golgi apparatus, two populations of Golgi stacks are incubated together. The "donor" population is from VSV-infected cells of a mutant type lacking GlcNAc transferase I. As a result, the donor Golgi cannot incorporate GlcNAc residues into the *N*-linked oligosaccharides on the G protein (see Figure 8–63). The "acceptor" Golgi stacks are from uninfected wild-type cells and thus contain a good copy of GlcNAc transferase I but lack G protein. The addition of GlcNAc to G protein therefore requires the transport of G protein from the *cis* compartment of the donor Golgi stack to the *medial* compartment of the acceptor Golgi stack. This transport-coupled glycosylation is monitored by measuring the incorporation into G protein of ^{3}H-GlcNAc from UDP-^{3}H-GlcNAc added to the incubation medium. As shown on the right, transport occurs only when both ATP and cytosol are added, which are the same conditions needed for the non-clathrin-coated vesicles to bud. This type of scheme, first used to measure transport from *cis* to *medial* Golgi, has also allowed *in vitro* reconstitutions of transport from ER to *cis* Golgi, *medial* to *trans* Golgi, *trans* Golgi to plasma membrane, endosomes to lysosomes, and *trans* Golgi to endolysosomes.

that vesicle budding is an active, rather than a simple, self-assembly process. The necessary components have been highly conserved during evolution, since the cytosol from yeasts and from plants can be substituted for animal cell cytosol in promoting transport in the Golgi stacks isolated from animal cells. This suggests that it will be possible to combine the genetic approach in yeasts and mammalian cell-free systems to identify the molecules that constitute the transport machinery.

Summary

The ER and each of the compartments in the Golgi apparatus contain their own unique sets of proteins. It seems likely that these proteins are selectively retained in these organelles by a mechanism that depends on "retention signals" in the proteins, while much of the flow of materials from ER to the Golgi apparatus, through the Golgi stack, and from the trans Golgi network to the cell surface occurs by default. It has been postulated that transport along the unselected default pathway is mediated by non-clathrin-coated transport vesicles that are nonselective with regard to their content, while signal-directed transport (sorting) is mediated by clathrin-coated vesicles. To test these ideas, the molecular mechanisms involved in transport vesicle budding, targeting, and fusion will have to be deciphered. Genetic studies in yeast cells have identified more than 25 genes whose products are required for particular steps in the pathway. In addition, reconstituted cell-free mammalian systems have been developed in which selective vesicular budding and fusion occur. The combination of genetics and biochemistry should allow the many proteins that mediate these processes to be identified and obtained in pure form so that their structures and activities can be ascertained.

References

General

Burgess, T.L.; Kelly, R.B. Constitutive and regulated secretion of proteins. *Annu. Rev. Cell Biol.* 3:243–293, 1987.

Dingwall, C.; Laskey, R.A. Protein import into the cell nucleus. *Annu. Rev. Cell Biol.* 2:367–390, 1986.

Kornfeld, S. Trafficking of lysosomal enzymes. *FASEB J.* 1:462–468, 1987.

Pfeffer, S.R.; Rothman, J.E. Biosynthetic protein transport and sorting by the endoplasmic reticulum and Golgi. *Annu. Rev. Biochem.* 56:829–852, 1987.

Verner, K.; Schatz, G. Protein translocation across membranes. *Science* 241:1307–1313, 1988.

Cited

1. Bolender, R.P. Stereological analysis of the guinea pig pancreas. *J. Cell Biol.* 61:269–287, 1974.

   Palade, G.E.; Farquhar, M.G. Cell biology. In Pathophysiology: The Biological Principles of Disease (L.H. Smith, S.D. Thier, eds.), pp. 1–56. Philadelphia: Saunders, 1981.

   Weibel, E.R.; Staubli, W.; Gnagi, H.R.; Hess, F.A. Correlated morphometric and biochemical studies on the liver cell. *J. Cell Biol.* 42:68–91, 1969.

2. Blobel, G. Intracellular protein topogenesis. *Proc. Natl. Acad. Sci. USA* 77:1496–1500, 1980.

   Gray, M.W.; Doolittle, W.F. Has the endosymbiont hypothesis been proven? *Microbiol. Rev.* 46:1–42, 1982.

   Schwarz, R.M.; Dayhoff, M.O. Origins of prokaryotes, eukaryotes, mitochondria, and chloroplasts. *Science* 199:395–403, 1978.

3. Palade, G. Intracellular aspects of the process of protein synthesis. *Science* 189:347–358, 1975.

4. Kelly, R.B. Pathways of protein secretion in eukaryotes. *Science* 230:25–31, 1985.

   Sabatini, D.D.; Kreibich, G.; Morimoto, T.; Adesnik, M. Mechanisms for the incorporation of proteins in membranes and organelles. *J. Cell Biol.* 92:1–22, 1982.

5. Pfeffer, S.R.; Rothman, J.E. Biosynthetic protein transport and sorting by the endoplasmic reticulum and Golgi. *Annu. Rev. Biochem.* 56:829–852, 1987.

   Wickner, W.T.; Lodish, H.E. Multiple mechanisms of protein insertion into and across membranes. *Science,* 230:400–406, 1985.

6. Blobel, G. Intracellular protein topogenesis. *Proc. Natl. Acad. Sci. USA* 77:1496–1500, 1980.

   Garoff, H. Using recombinant DNA techniques to study protein targeting in the eucaryotic cell. *Annu. Rev. Cell Biol.* 1:403–445, 1985.

7. Warren, G. Membrane traffic and organelle division. *Trends Biochem. Sci.* 10:439–443, 1985.

8. Allen, R.D. The microtubule as an intracellular engine. *Sci. Am.* 238(2):42–49, 1987.

Fulton, A.B. How crowded is the cytoplasm? *Cell* 30:345–347, 1982.

Luby-Phelps, K.; Taylor, D.L.; Lanni, F. Probing the structure of cytoplasm. *J. Cell Biol.* 102:2015–2022, 1986.

Vale, R.D. Intracellular transport using microtubule-based motors. *Annu. Rev. Cell Biol.* 3:347–378, 1987.

9. Chock, P.B.; Rhee, S.G.; Stadtman, E.R. Interconvertible enzyme cascades in cellular regulation. *Annu. Rev. Biochem.* 49:813–843, 1980.

   Holt, G.D.; et al. Nuclear pore complex glycoproteins contain cytoplasmically disposed O-linked N-acetylglucosamine. *J. Cell Biol.* 104:1157–1164, 1987.

   Wold, F. *In vivo* chemical modification of proteins (post-translational modification). *Annu. Rev. Biochem.* 50:783–814, 1981.

10. Kamps, M.P.; Buss, J.E.; Sefton, B.M. Mutation of NH_2-terminal glycine of p60[src] prevents both myristoylation and morphological transformation. *Proc. Natl. Acad. Sci. USA* 82:4625–4628, 1985.

    Schultz, A.M.; Henderson, L.E.; Orozlan, S. Fatty acylation of proteins. *Annu. Rev. Cell Biol.* 4:611–648, 1988.

    Willumsen, B.M.; Norris, K.; Papageorge, A.G.; Hubbert, N.L.; Lowy, D.R. Harvey murine sarcoma virus p21 *ras* protein: biological and biochemical significance of the cysteine nearest the carboxy terminus. *EMBO J.* 3:2582–2585, 1984.

11. Dice, J.F. Molecular determinants of protein half-lives in eucaryotic cells. *FASEB J.* 1:349–357, 1987.

    Goldberg, A.L.; Goff, S.A. The selective degradation of abnormal proteins in bacteria. In Maximizing Gene Expression (W. Reznikoff, L. Gold, eds.), pp. 287–314. Stoneham, MA: Butterworth, 1986.

12. Bachmair, A.; Finley, D.; Varshavsky, A. *In vivo* half-life of a protein is a function of its amino-terminal residue. *Science* 234:179–186, 1986.

    Hershko, A.; Ciechanover, A. The ubiquitin pathway for the degradation of intracellular proteins. *Prog. Nucleic Acid Res. Mol. Biol.* 33:19–56, 1986.

    Rechsteiner, M. Ubiquitin-mediated pathways for intracellular proteolysis. *Annu. Rev. Cell Biol.* 3:1–30, 1987.

13. Augen, J.; Wold, F. How much sequence information is needed for the regulation of amino-terminal acetylation of eukaryotic proteins? *Trends Biochem. Sci.* 11:494–497, 1986.

    Ferber, S.; Ciechanover, A. Role of arginine-tRNA in protein degradation by the ubiquitin pathway. *Nature* 326:808–811, 1987.

    Varshavsky, A.; Bachmair, A.; Finley, D.; Gonda, D.; Wünning, I. The N-end rule of selective protein turnover: mechanistic aspects and functional implications. In Ubiquitin (M. Rechsteiner, ed.), pp. 284–324. New York: Plenum, 1988.

14. Craig, E.A. The heat-shock response. *CRC Crit. Rev. Biochem.* 18:239–280, 1985.

    Lindquist, S. The heat-shock response. *Annu. Rev. Biochem.* 55:1151–1191, 1986.

Pelham, H.R.B. Speculations on the functions of the major heat shock and glucose-regulated proteins. *Cell* 46:959–961, 1986.

15. Franke, W.W.; Scheer, U.; Krohne, G.; Jarasch, E.D. The nuclear envelope and the architecture of the nuclear periphery. *J. Cell Biol.* 91:39s–50s, 1981.

Newport, J.W.; Forbes, D.J. The nucleus: structure, function, and dynamics. *Annu. Rev. Biochem.* 56:535–565, 1987.

16. Bonner, W.M. Protein migration and accumulation in nuclei. In The Cell Nucleus (H. Busch, ed.), Vol. 6, Part C, pp. 97–148. New York: Academic, 1978.

Lang, I.; Scholz, M.; Peters, R. Molecular mobility and nucleocytoplasmic flux in hepatoma cells. *J. Cell Biol.* 102:1183–1190, 1986.

17. Dingwall, C.; Laskey, R.A. Protein import into the cell nucleus. *Annu. Rev. Cell Biol.* 2:367–390, 1986.

Feldherr, C.M.; Kallenbach, E.; Schultz, N. Movement of a karyophilic protein through the nuclear pores of oocytes, *J. Cell Biol.* 99:2216–2222, 1984.

Newmeyer, D.D.; Forbes, D.J. Nuclear import can be separated into distinct steps *in vitro*: nuclear pore binding and translocation. *Cell* 52:641–653, 1988.

18. Goldfarb, D.S.; Gariépy, J.; Schoolnik, G.; Kornberg, R.D. Synthetic peptides as nuclear localization signals. *Nature* 322:641–644, 1986.

Kalderon, D.; Roberts, B.L.; Richardson, W.D.; Smith, A.E. A short amino acid sequence able to specify nuclear location. *Cell* 39:499–509, 1984.

Lanford, R.E.; Butel, J.S. Construction and characterization of an SV40 mutant defective in nuclear transport of T antigen. *Cell* 37: 801–813, 1984.

19. Clawson, G.A.; Feldherr, C.M.; Smuckler, E.A. Nucleocytoplasmic RNA transport. *Mol. Cell. Biochem.* 67:87–100, 1985.

Dworetzky, S.I.; Feldherr, C.M. Translocation of RNA-coated gold particles through the nuclear pores of oocytes. *J. Cell Biol.* 106:575–584, 1988.

20. Attardi, G.; Schatz, G. Biogenesis of mitochondria. *Annu. Rev. Cell Biol.* 4:289–333, 1988.

Tzagoloff, A. Mitochondria. New York: Plenum, 1982.

21. Hawlitschek, G.; et al. Mitochondrial protein import: identification of processing petidase and of PEP, a processing enhancing protein. *Cell* 53:795–806, 1988.

Hurt, E.C.; van Loon, A.P.G.M. How proteins find mitochondria and intramitochondrial compartments. *Trends Biochem. Sci.* 11:204–207, 1986.

Pfanner, N.; Neupert, W. Biogenesis of mitochondrial energy transducing complexes. *Curr. Top. Bioenerg.* 15:177–219, 1987.

Roise, D.; Schatz, G. Mitochondrial presequences. *J. Biol. Chem.* 263:4509–4511, 1988.

22. Eilers, M.; Schatz, G. Protein unfolding and the energetics of protein translocation across biological membranes. *Cell* 52:481–483, 1988.

Pfanner, N.; Neupert, W. Transport of proteins into mitochondria: a potassium diffusion potential is able to drive the import of ADP/ATP carrier. *EMBO J.* 4:2819–2825, 1985.

Roise, D.; Horvath, S.J.; Tomich, J.M.; Richards, J.H.; Schatz, G. A chemically synthesized pre-sequence of an imported mitochondrial protein can form an amphiphilic helix and perturb natural and artificial phospholipid bilayers. *EMBO J.* 5:1327–1334, 1986.

23. Schleyer, M.; Neupert, W. Transport of proteins into mitochondria: translocational intermediates spanning contact sites between outer and inner membranes. *Cell* 43: 339–350, 1985.

Schwaiger, M.; Herzog, V.; Neupert, W. Characterization of tranlocation contact sites involved in the import of mitochondrial proteins. *J. Cell Biol.* 105:235–246, 1987.

24. Deshaies, R.J.; Koch, B.D.; Werner-Washburne, M.; Craig, E.A.; Schekman, R. A subfamily of stress proteins facilitates translocation of secretory and mitochondrial precursor polypeptides. *Nature* 332:800–805, 1988.

Eilers, M.; Schatz, G. Binding of a specific ligand inhibits import of a purified precursor protein into mitochondria. *Nature* 322:228–232, 1986.

Pfanner, N.; Tropschug, M.; Neupert, W. Mitochondrial protein import: nucleoside triphosphates are involved in conferring import competence to precursors. *Cell* 49:815–823, 1987.

25. Hartl, F.U.; Ostermann, J.; Guiard, B.; Neupert, W. Successive translocation into and out of the mitochondrial matrix: targeting of proteins to the intermembrane space by a bipartite signal peptide. *Cell* 51:1027–1037, 1987.

van Loon, A.P.G.M.; Brandli, A.W.; Schatz, G. The presequences of two imported mitochondrial proteins contain information for intracellular and intramitochondrial sorting. *Cell* 44:801–812, 1986.

26. Pfaller, R.; Neupert, W. High-affinity binding sites involved in the import of porin into mitochondria. *EMBO J.* 6:2635–2642, 1987.

Pfanner, N.; et al. Role of ATP in mitochondrial protein import. *J. Biol. Chem.* 263:4049–4051, 1988.

27. Boutry, M.; Nagy, F.; Poulsen, G.; Aoyagi, K.; Chua, N.H. Targeting of bacterial chloramphenicol acetyltransferase to mitochondria in transgenic plants. *Nature* 328:340–342, 1987.

Pain, D.; Kanwar, Y.S.; Blobel, G. Identification of a receptor for protein import into chloroplasts and its localization to envelope contact zones. *Nature* 331:232–237, 1988.

Schmidt, G.W.; Mishkind, M.L. The transport of proteins into chloroplasts. *Annu. Rev. Biochem.* 55:879–912, 1986.

Smeekens, S.; Bauerie, C.; Hageman, J.; Keegstra, K.; Weisbeek, P. The role of the transit peptide in the routing of precursors toward different chloroplast compartments. *Cell* 46:365–375, 1986.

28. de Duve, C. Microbodies in the living cell. *Sci. Am.* 248(5):74–84, 1983.

de Duve, C.; Baudhuin, P. Peroxisomes (microbodies and related particles). *Physiol. Rev.* 46:323–357, 1966.

Fahimi, H.D.; Sies, H., eds. Peroxisomes in Biology and Medicine. Heidelberg: Springer, 1987.

29. Tolbert, N.E.; Essner, E. Microbodies: peroxisomes and glyoxysomes. *J. Cell Biol.* 91:271s–283s, 1981.

Veenbuis, M.; Van Dijken, J.P.; Harder, W. The significance of peroxisomes in the metabolism of one-carbon compounds in yeasts. *Adv. Microb. Physiol.* 24:1–82, 1983.

30. Gould, S.J.; Keller, G.A.; Subramani, S. Identification of a peroxisomal targeting signal at the carboxy terminus of four peroxisomal proteins. *J. Cell Biol.* 107:897–905, 1988.

Imanaka, T.; Small, G.M.; Lazarow, P.B. Translocation of acyl-CoA oxidase into peroxisomes requires ATP hydrolysis but not a membrane potential. *J. Cell Biol.* 105:2915–2922, 1987.

Lazarow, P.B.; Fujiki, Y. Biogenesis of peroxisomes. *Annu. Rev. Cell Biol.* 1:489–530, 1985.

31. DePierre, J.W.; Dallner, G. Structural aspects of the membrane of the endoplasmic reticulum. *Biochim. Biophys. Acta* 415: 411–472, 1975.

Fawcett, D. The Cell, 2nd ed., pp. 303–352. Philadelphia: Saunders, 1981.

Lee, C.; BoChen, L. Dynamic behavior of endoplasmic reticulum in living cells. *Cell* 54:37–46, 1988.

32. Adelman, M.R.; Sabatini, D.D.; Blobel, G. Ribosome-membrane interaction: nondestructive disassembly of rat liver rough microsomes into ribosomal and membranous components. *J. Cell Biol.* 56:206–229, 1973.

Blobel, G.; Dobberstein, B. Transfer of proteins across membranes. *J. Cell Biol.* 67:852–862, 1975.

33. Jones, A.L.; Fawcett, D.W. Hypertrophy of the agranular endoplasmic reticulum in hamster liver induced by phenobarbital. *J. Histochem. Cytochem.* 14:215–232, 1966.

Mori, H.; Christensen, A.K. Morphometric analysis of Leydig cells in the normal rat testis. *J. Cell Biol.* 84:340–354, 1980.

34. Dallner, G. Isolation of rough and smooth microsomes—general. *Methods Enzymol.* 31:191–201, 1974.

de Duve, C. Tissue fractionation past and present. *J. Cell Biol.* 50:20d–55d, 1971.

35. Hortsch, M.; Avossa, D.; Meyer, D.I. Characterization of secretory protein translocation: ribosome-membrane interaction in endoplasmic reticulum. *J. Cell Biol.* 103:241–253, 1986.

Kreibich, G.; Ulrich, B.L.; Sabatini, D.D. Proteins of rough microsomal membranes related to ribosome binding. *J. Cell Biol.* 77:464–487, 1978.

36. Blobel, G.; Dobberstein, B. Transfer of proteins across membranes. *J. Cell Biol.* 67:835–851, 1975.

Garoff, H. Using recombinant DNA techniques to study protein targeting in the eucaryotic cell. *Annu. Rev. Cell Biol.* 1:403–445, 1985.

Milstein, C.; Brownlee, G.; Harrison, T.; Mathews, M.B. A possible precursor of immunoglobulin light chains. *Nature New Biol.* 239:117–120, 1972.

von Heijne, G. Signal sequences: the limits of variation. *J. Mol. Biol.* 184:99–105, 1985.

37. Meyer, D.I.; Krause, E.; Dobberstein, B. Secretory protein translocation across membranes—the role of the "docking protein." *Nature* 297:647–650, 1982.

Tajima, S.; Lauffer, L.; Rath, V.L.; Walter, P. The signal recognition particle receptor is a complex that contains two distinct polypeptide chains. *J. Cell Biol.* 103:1167–1178, 1986.

Walter, P.; Blobel, G. Signal recognition particle contains a 7S RNA essential for protein translocation across the endoplasmic reticulum. *Nature* 299:691–698, 1982.

Walter, P.; Lingappa, V.R. Mechanism of protein translocation across the endoplasmic reticulum membrane. *Annu. Rev. Cell Biol.* 2:499–516, 1986.

Wiedmann, M.; Kurzchalia, T.V.; Hartmann, E.; Rapoport, T.A. A signal sequence receptor in the endoplasmic reticulum membrane. *Nature* 328:830–833, 1987.

38. Chirico, W.J.; Waters, M.G.; Blobel, G. 70K heat shock related proteins stimulate protein translocation into microsomes. *Nature* 332:805–810, 1988.

Perara, E.; Rothman, R.E.; Lingappa, V.R. Uncoupling translocation from translation: implications for transport of proteins across membranes. *Science* 232:348–352, 1986.

Zimmermann, R.; Meyer, D.I. 1986: A year of new insights into how proteins cross membranes. *Trends Biochem. Sci.* 11:512–515, 1986.

39. Rapoport, T.A. Extensions of the signal hypothesis—sequential insertion model versus amphipathic tunnel hypothesis. *FEBS Lett.* 187:1–10, 1985.

Wickner, W.T.; Lodish, H.F. Multiple mechanisms of protein insertion into and across membranes. *Science* 230:400–406, 1985.

40. Engelman, D.M.; Steitz, T.A.; Goldman, A. Identifying nonpolar transbilayer helices in amino acid sequences of membrane proteins. *Annu. Rev. Biophys. Biophys. Chem.* 15:321–353, 1986.

Kaiser, C.A.; Preuss, D.; Grisafi, P.; Botstein, D. Many random sequences functionally replace the secretion signal sequence of yeast invertase. *Science* 235:312–317, 1987.

Kyte, J.; Doolittle, R.F. A simple method for displaying the hydropathic character of a protein. *J. Mol. Biol.* 157:105–132, 1982.

Zerial, M.; Huylebroeck, D.; Garoff, H. Foreign transmembrane peptides replacing the internal signal sequence of transferrin receptor allow its translocation and membrane binding. *Cell* 48:147–155, 1987.

41. Bole, D.G.; Hendershot, L.M.; Kearney, J.F. Posttranslational association of immunoglobulin heavy chain binding protein with nascent heavy chains in nonsecreting and secreting hydridomas. *J. Cell Biol.* 102:1558–1566, 1986.

Lodish, H.F. Transport of secretory and membrane glycoproteins from the rough endoplasmic reticulum to the Golgi. *J. Biol. Chem.* 263:2107–2110, 1988.

Munro, S.; Pelham, H.R.B. A C-terminal signal prevents secretion of luminal ER proteins. *Cell* 48:899–907, 1987.

42. Freedman, R. Native disulphide bond formation in protein biosynthesis: evidence for the role of protein disulphide isomerase. *Trends Biochem. Sci.* 9:438–441, 1984.

Holmgren, A. Thioredoxin. *Annu. Rev. Biochem.* 54:237–272, 1985.

43. Hirschberg, C.B.; Snider, M.D. Topography of glycosylation in the rough endoplasmic reticulum and Golgi apparatus. *Annu. Rev. Biochem.* 56:63–87, 1987.

Kornfeld, R.; Kornfeld, S. Assembly of asparagine-linked oligosaccharides. *Annu. Rev. Biochem.* 54:631–664, 1985.

Torres, C.; Hart, G. Topography and polypeptide distribution of terminal *N*-acetylglucosamine residues on the surface of intact lymphocytes. *J. Biol. Chem.* 259:3308–3317, 1984.

44. Cross, G.A.M. Eukaryotic protein modification and membrane attachment via phosphatidylinositol. *Cell* 48:179–181, 1987.

Ferguson, M.A.J.; Williams, A.F. Cell-surface anchoring of proteins via glycosyl-phosphatidylinositol structures. *Annu. Rev. Biochem.* 57:285–320, 1988.

Low, M.G.; Saltiel, A.R. Structural and functional roles of glycosyl-phosphatidylinositol in membranes. *Science* 239:268–275, 1988.

45. Bishop, W.R.; Bell, R.M. Assembly of phospholipids into cellular membranes: biosynthesis, transmembrane movement, and intracellular translocation. *Annu. Rev. Cell Biol.* 4:579–610, 1988.

Bishop, W.R.; Bell, R.M. Assembly of the endoplasmic reticulum phospholipid bilayer: the phosphatidylcholine transporter. *Cell* 42:51–60, 1985.

Dawidowicz, E.A. Dynamics of membrane lipid metabolism and turnover. *Annu. Rev. Biochem.* 56:43–61, 1987.

Pagano, R.E.; Sleight, R.G. Defining lipid transport pathways in animal cells. *Science* 229:1051–1057, 1985.

Rothman, J.E.; Lenard, J. Membrane asymmetry. *Science* 195:743–753, 1977.

46. Dawidowicz, E.A. Lipid exchange: transmembrane movement, spontaneous movement, and protein-mediated transfer of lipids and cholesterol. *Curr. Top. Memb. Transp.* 29:175–202, 1987.

Yaffe, M.P.; Kennedy, E.P. Intracellular phospholipid movement and the role of phospholipid transfer proteins in animal cells. *Biochemistry* 22:1497–1507, 1983.

47. Farquhar, M.G.; Palade, G.E. The Golgi apparatus (complex)—(1954–1981)—from artifact to center stage. *J. Cell Biol.* 91:77s–103s, 1981.

Pavelka, M. Functional morphology of the Golgi apparatus. *Adv. Anat. Embryol. Cell Biol.* 106:1–94, 1987.

Rothman, J.E. The compartmental organization of the Golgi apparatus. *Sci. Am.* 253(3):74–89, 1985.

48. Hubbard, S.C.; Ivatt, R.J. Synthesis and processing of asparagine-linked oligosaccharides. *Annu. Rev. Biochem.* 50:555–583, 1981.

Kornfeld, R.; Kornfeld, S. Assembly of asparagine-linked oligosaccharides. *Annu. Rev. Biochem.* 54:631–664, 1985.

Schachter, H.; Roseman, S. Mammalian glycosyltransferases: their role in the synthesis and function of complex carbohydrates and glycolipids. In The Biochemistry of Glycoproteins and Proteoglycans (W.J. Lennarz, ed.), Chapter 3. New York: Plenum, 1980.

49. Elbein, A.D. Inhibitors of the biosynthesis and processing of N-linked oligosaccharide chains. *Annu. Rev. Biochem.* 56:497–534, 1987.

Stanley, P. Glycosylation mutants and the functions of mammalian carbohydrates. *Trends Genet.* 3:77–81, 1987.

West, C.M. Current ideas on the significance of protein glycosylation. *Mol. Cell. Biochem.* 72:3–20, 1986.

50. Hassell, J.R.; Kimura, J.H.; Hascall, V.C. Proteoglycan core protein families. *Annu. Rev. Biochem.* 55:539–567, 1986.

Huttner, W.B. Tyrosine sulfation and the secretory pathway. *Annu. Rev. Physiol.* 50:363–376, 1988.

Ruoslahti, F. Structure and biology of proteoglycans. *Annu. Rev. Cell Biol.* 4:229–255, 1988.

Wagh, P.V.; Bahl, O.P. Sugar residues on proteins. *CRC Crit. Rev. Biochem.* 10:307–377, 1981.

51. Douglass, J.; Civelli, O.; Herbert, E. Polyprotein gene expression: generation of diversity of neuroendocrine peptides. *Annu. Rev. Biochem.* 53:665–715, 1984.

Orci, L.; et al. Conversion of proinsulin to insulin occurs coordinately with acidification of maturing secretory vesicles. *J. Cell Biol.* 103:2273–2281, 1986.

52. Dunphy, W.G.; Rothman, J.E. Compartmental organization of the Golgi stack. *Cell* 42:13–21, 1985.

53. Bainton, D. The discovery of lysosomes. *J. Cell Biol.* 91:66s–76s, 1981.

de Duve, C. Exploring cells with a centrifuge. *Science* 189:186–194, 1975.

54. Holzman, E. Lysosomes: A Survey. New York: Springer-Verlag, 1976.

55. Griffiths, G.; Hoflack, B.; Simons, K.; Mellman, I.; Kornfeld, S. The mannose 6-phosphate receptor and the biogenesis of lysosomes. *Cell* 52:329–341, 1988.

Helenius, A.; Mellman, I.; Wall, D.; Hubbard, A. Endosomes. *Trends Biochem. Sci.* 8:245–250, 1983.

Mayer, R.J.; Doherty, F. Intracellular protein catabolism: state of the art. *FEBS Lett.* 198:181–193, 1986.

Mellman, I.; Fuchs, R.; Helenius, A. Acidification of the endocytic and exocytic pathways. *Annu. Rev. Biochem.* 55:663–700, 1986,

Silverstein, S.C.; Steinman, R.M.; Cohn, Z.A. Endocytosis. *Annu. Rev. Biochem.* 46:669–722, 1977.

56. Dahms, N.M.; Lobel, P.; Breitmeyer, J.; Chirgwin, J.M.; Kornfeld, S. 46 kd mannose 6-phosphate receptor: cloning, expression, and homology to the 215 kd mannose 6-phosphate receptor. *Cell* 50:181–192, 1987.

Kornfeld, S. Trafficking of lysosomal enzymes. *FASEB J.* 1:462–468, 1987.

Pfeffer, S.R. Mannose 6-phosphate receptors and their role in targeting of proteins to lysosomes. *J. Membr. Biol.* 103:7–16, 1988.

von Figura, K.; Hasilik, A. Lysosomal enzymes and their receptors. *Annu. Rev. Biochem.* 55:167–193, 1986.

57. Brown, W.J.; Goodhouse, J.; Farquhar, M.G. Mannose 6-phosphate receptors for lysosomal enzymes cycle between the Golgi complex and endosomes. *J. Cell Biol.* 103:1235–1247, 1986.

Duncan, J.R.; Kornfeld, S. Intracellular movement of two mannose 6-phosphate receptors: return to the Golgi apparatus. *J. Cell Biol.* 106:617–628, 1988.

Geuze, H.J.; Slot, J.W.; Strous, G.J.A.M.; Hasilik, A.; von Figura, K. Possible pathways for lysosomal enzyme delivery. *J. Cell Biol.* 101:2253–2262, 1985.

Rothman, J.E.; Schmid, S.L. Enzymatic recycling of clathrin from coated vesicles. *Cell* 46:5–9, 1986.

58. Lang, L.; Reitman, M.; Tang, J.; Roberts, R.M.; Kornfeld, S. Lysosomal enzyme phosphorylation. *J. Biol. Chem.* 259:14663–14671, 1984.

Reitman, M.L.; Kornfeld, S. Lysosomal enzyme targeting. N-acetylglucosaminylphosphotransferase selectively phosphorylates native lysosomal enzymes. *J. Biol. Chem.* 256:11977–11980, 1981.

59. Kornfeld, S. Trafficking of lysosomal enzymes in normal and disease states. *J. Clin. Invest.* 77:1–6, 1986.

Neufeld, E.F.; Lim, T.W.; Shapiro, L.J. Inherited disorders of lysosomal metabolism. *Annu. Rev. Biochem.* 44:357–376, 1975.

60. Burgess, T.L.; Kelly, R.B. Constitutive and regulated secretion of proteins. *Annu. Rev. Cell Biol.* 3:243–293, 1987.

61. Griffiths, G.; Simons, K. The *trans*-Golgi network: sorting at the exit site of the Golgi complex. *Science* 234:438–443, 1986.

Orci, L.; et al. The trans-most cisternae of the Golgi complex: a compartment for sorting of secretory and plasma membrane proteins. *Cell* 51:1039–1051, 1987.

62. Herzog, V.; Farquhar, M.G. Luminal membrane retrieved after exocytosis reaches most Golgi cisternae in secretory cells. *Proc. Natl. Acad. Sci. USA* 74:5073–5077, 1977.

Snider, M.D.; Rogers, O.C. Membrane traffic in animal cells: cellular glycoproteins return to the site of Golgi mannosidase I. *J. Cell Biol.* 103:265–275, 1986.

63. Lodish, H.F. Transport of secretory and membrane glycoproteins from the rough endoplasmic reticulum to the Golgi. *J. Biol. Chem.* 263:2107–2110, 1988.

Rothman, J.E. Protein sorting by selective retention in the endoplasmic reticulum and Golgi stack. *Cell* 50:521–522, 1987.

Wieland, F.T.; Gleason, M.L.; Serafini, T.A.; Rothman, J.E. The rate of bulk flow from the endoplasmic reticulum to the cell surface. *Cell* 50:289–300, 1987.

64. Bartles, J.R.; Hubbard, A.L. Plasma membrane protein sorting in epithelial cells: do secretory pathways hold the key? *Trends Biochem. Sci.* 13:181–184, 1988.

Matlin, K.S. The sorting of proteins to the plasma membrane in epithelial cells. *J. Cell Biol.* 103:2565–2568, 1986.

Mostov, K.E.; Breitfeld, P.; Harris, J.M. An anchor-minus form of the polymeric immunoglobulin receptor is secreted predominantly apically in Madin-Darby canine kidney cells. *J. Cell Biol.* 105:2031–2036, 1987.

Simons, K.; Fuller, S.D. Cell surface polarity in epithelia. *Annu. Rev. Cell Biol.* 1:243–288, 1985.

65. Simons, K.; Garoff, H.; Helenius, A. How an animal virus gets into and out of its host cell. *Sci. Am.* 246(2):58–66, 1982.

Simons, K.; Warren, G. Semliki forest virus: a probe for membrane traffic in the animal cell. *Adv. Protein Chem.* 36:79–132, 1984.

66. Rodriguez-Boulan, E.J. Membrane biogenesis, enveloped RNA viruses, and epithelial polarity. In Modern Cell Biology. Vol. 1 (J.R. McIntosh, B.H. Satir, eds.), pp.119–170, 1983.

Rodriguez-Boulan, E.J.; Sabatini, D.D. Asymmetric budding of viruses in epithelial monolayers: a model system for the study of epithelial polarity. *Proc. Natl. Acad. Sci. USA* 75:5071–5075, 1978.

Roth, M.G.; Sirnivas, R.V.; Compans, R.W. Basolateral maturation of retroviruses in polarized epithelial cells. *J. Virol.* 45:1065–1073, 1983.

Strauss, E.G.; Strauss, J.H. Assembly of enveloped animal viruses. In Virus Structure and Assembly (S. Casjens, ed.), Chapter 6. Boston: Jones and Bartlett, 1985.

67. Rothman, J.E. Protein sorting by selective retention in the endoplasmic reticulum and Golgi stack. *Cell* 50:521–522, 1987.

68. Griffiths, G.; Pfeiffer, S.; Simons, K.; Matlin, K. Exit of newly synthesized membrane proteins from the *trans* cisterna of the Golgi complex to the plasma membrane. *J. Cell Biol.* 101:949–964, 1985.

Orci, L.; Glick, B.S.; Rothman, J.E. A new type of coated vesicular carrier that appears not to contain clathrin: its possible role in protein transport within the Golgi stack. *Cell* 46:171–184, 1986.

69. Bourne, H. Do GTPases direct membrane traffic in secretion? *Cell* 53:669–671, 1988.

Novick, P.; Field, C.; Schekman, R. Identification of 23 complementation groups required for post-translational events in the yeast secretory pathway. *Cell* 21:205–215, 1980.

Schekman, R. Protein localization and membrane traffic in yeast. *Annu. Rev. Cell Biol.* 1:115–143, 1985.

70. Balch, W.E.; Dunphy, W.G.; Braell, W.A.; Rothman, J.E. Reconstitution of the transport of protein between successive compartments of the Golgi measured by the coupled incorporation of *N*-acetylglucosamine. *Cell* 39:405–416, 1984.

Dunphy, W.G.; et al. Yeast and mammals utilize similar cytosolic components to drive protein transport through the Golgi complex. *Proc. Natl. Acad. Sci. USA* 83:1622–1626, 1986.

Fries, E.; Rothman, J.E. Transport of vesicular stomatitis virus glycoprotein in a cell-free extract. *Proc. Natl. Acad. Sci. USA* 77:3870–3874, 1980.

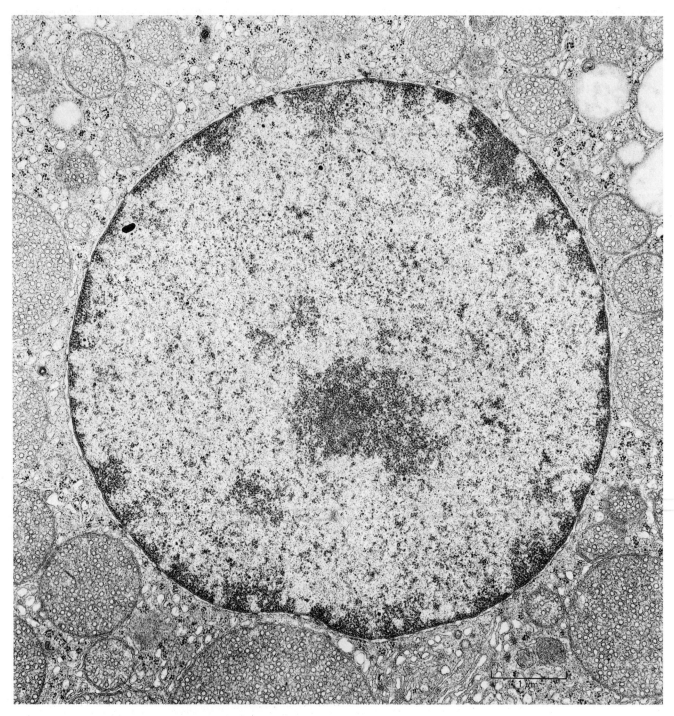

Thin-section electron micrograph of an adrenal cortex cell
nucleus showing some of its fine structure. Since most of the
chromatin is in its extended interphase form, the individual
chromosomes cannot be discerned. (Courtesy of Daniel
S. Friend.)

The Cell Nucleus

Nearly all of a eucaryotic cell's DNA is sequestered in the nucleus, which occupies about 10% of the total cell volume. The nucleus is delimited by a *nuclear envelope* formed by two concentric membranes punctured at intervals by nuclear pores, which allow selected molecules to be actively transported to and from the cytoplasm. The nuclear envelope is directly connected to the endoplasmic reticulum (see p. 421), and it is enmeshed in two networks of intermediate filaments: a thin shell of such filaments just inside the nucleus (the *nuclear lamina*) supports the inner nuclear membrane, while less regularly organized intermediate filaments surround the outer nuclear membrane (Figure 9–1).

Like modern procaryotes, the ancestors of eucaryotic cells almost certainly lacked a nucleus (see p. 408), and one can only speculate on why a separate nuclear compartment evolved. Two special features of eucaryotic cells suggest possible reasons for segregating the DNA from the activities in the cytoplasm. One of these features is the cytoskeletal system of cytoplasmic filaments—mainly microtubules and actin filaments—that mediates the movements of eucaryotic cells

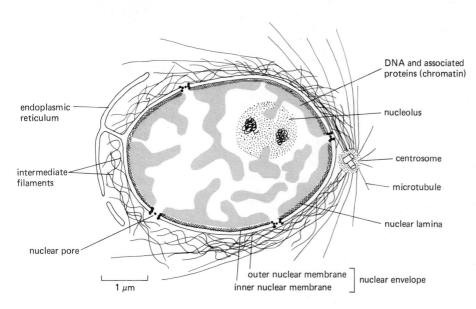

Figure 9–1 Cross-section of a typical cell nucleus. The nuclear envelope consists of two membranes, the outer one being continuous with the endoplasmic reticulum membrane (see also Figure 8–19, p. 422). The lipid bilayers of the inner and outer nuclear membranes are fused at the nuclear pores. Two networks of ropelike intermediate filaments (*colored lines*) provide mechanical support for the nuclear envelope; the filaments inside the nucleus form a sheetlike *nuclear lamina.*

(see Chapter 11). Bacteria, whose DNA is in direct contact with the cytoplasm, lack these filaments and move by means of external structures. One function of the nuclear envelope may therefore be to protect the long, fragile DNA molecules from the mechanical forces generated by the cytoplasmic filaments in eucaryotes.

A second distinguishing feature of eucaryotic cells is the extensive processing that RNA undergoes before it is translated into protein. Eucaryotic cells have evolved numerous membrane-bounded compartments that segregate their various chemical reactions so as to make them more efficient (see p. 406), and the nucleus is one such compartment. In procaryotic cells, RNA synthesis (*transcription*) and protein synthesis (*translation*) occur concurrently: ribosomes translate the 5′ end of an RNA molecule into protein while the RNA molecule is still being synthesized at its 3′ end. Consequently, there is relatively little opportunity to alter the RNA transcripts before they are translated into protein. In eucaryotes, by contrast, transcription (in the nucleus) is separated both temporally and spatially from translation (in the cytoplasm). The RNA transcripts are immediately packaged into ribonucleoprotein complexes that facilitate their extensive alteration by *RNA splicing*, in which certain portions of the nucleotide sequence are removed; this is an important intermediate step in the transfer of genetic information in eucaryotes (see p. 531). Only when splicing is complete are the packaging proteins removed and the RNA molecules transported out of the nucleus to the cytoplasm, where ribosomes begin translating the RNA into protein (Figure 9–2).

RNA splicing allows a single gene to make several different proteins (see p. 589) and also has other advantages for the cell (see p. 602). This might help explain why eucaryotic cells have a nucleus, where splicing can occur without interference from ribosomes (Figure 9–3). A hypothetical evolutionary pathway for the formation of the nucleus was presented earlier (see Figure 8–4, p. 409).

The nuclear envelope was discussed in detail in Chapter 8 in connection with the selective transport of macromolecules into and out of the nucleus (see p. 421). In this chapter we describe how proteins package eucaryotic DNA into chromosomes, how chromosomes are folded and organized in the nucleus, and how they are replicated during the S phase of each cell cycle. We then discuss RNA synthesis and RNA splicing, the most prominent events that take place continuously in the interphase nucleus. The mechanisms that control eucaryotic gene expression are discussed in Chapter 10.

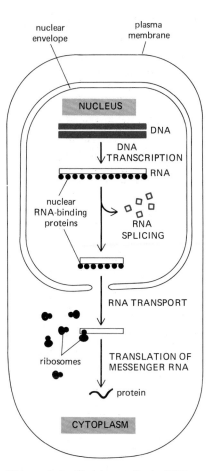

Figure 9–2 Protein synthesis (DNA → RNA → protein) in eucaryotes. The nuclear envelope keeps functional ribosomes out of the nucleus, so that RNA transcripts can be extensively processed (spliced) before they are transported out of the nucleus into the cytoplasm, where they are translated into protein by ribosomes. Thus RNA processing and transport steps are interposed between DNA transcription and RNA translation.

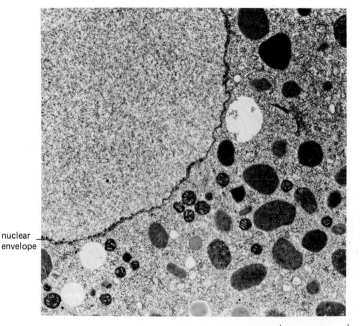

Figure 9–3 The nuclear envelope keeps the nuclear compartment free from cytoplasmic organelles. This electron micrograph shows a thin section of a sea urchin egg, which has a nucleus that stains unusually evenly and a cytoplasm densely packed with organelles. (Courtesy of David Begg and Tim Hunt.)

Chromosomal DNA and Chromosomal Proteins[1]

For the first 40 years of this century, biologists tended to dismiss the possibility that DNA could carry the genetic information in chromosomes, partly because nucleic acids were erroneously believed to contain only a simple repeating tetranucleotide sequence (such as *AGCTAGCTAGCT*...). We now know, however, that a DNA molecule is an enormously long, unbranched, linear polymer that can contain many millions of nucleotides arranged in an irregular but nonrandom sequence, and that the genetic information is contained in the linear order of the nucleotides. The genetic code, written in "words" of three nucleotides (*codons* that specify an amino acid, see p. 209), neatly solves the problem of storing a large amount of genetic information in a small amount of space: every million "letters" (nucleotides) take up a linear distance of only 3.4×10^5 nm (0.034 cm) and occupy a total volume of about 10^6 nm^3 (10^{-15} cm^3).

Each DNA molecule is packaged in a separate **chromosome,** and the total genetic information stored in the chromosomes of an organism is said to constitute its **genome.** The genome of the *E. coli* bacterium contains 4.7×10^6 nucleotide pairs of DNA, present in a single DNA molecule (one chromosome). The human genome, in contrast, contains about 3×10^9 nucleotide pairs, present in 24 chromosomes (22 different autosomes and 2 different sex chromosomes), and thus consists of 24 kinds of DNA molecules. In diploid organisms such as ourselves, there are two copies of each kind of chromosome, one inherited from the mother and one from the father (except for the sex chromosomes in males, where a Y chromosome is inherited from the father and an X from the mother). A typical human cell thus contains a total of 46 chromosomes and about 6×10^9 nucleotide pairs of DNA. Other mammals have genomes of similar size. This amount of DNA could in theory be packed into a cube 1.9 μm on each side. By comparison, 6×10^9 letters in this book would occupy more than a million pages, thus requiring more than 10^{17} times as much space.

In this section we consider the relationship between DNA molecules and chromosomes and discuss the various proteins that bind to a DNA molecule to form a functional eucaryotic chromosome. Some of these proteins control the expression of the genetic information stored in DNA molecules, by regulating the synthesis of RNA molecules from selected regions of the genome. Others, most notably the *histones*, fold up each long DNA molecule so that it is compact and orderly while leaving the required genetic information accessible.

9-3 Each Chromosome Is Formed from a Single Long DNA Molecule[2]

An individual human chromosome contains from 50×10^6 to 250×10^6 nucleotide pairs of DNA. DNA molecules of this size are 1.7 to 8.5 cm long when uncoiled, and even the slightest mechanical force will break them once the chromosomal proteins have been removed. Intact chromosomal DNA molecules, however, can be isolated readily from some simpler eucaryotic organisms, such as the yeast *Saccharomyces cerevisiae*, which contains much shorter chromosomes. The technique of pulsed gel electrophoresis (see Figure 4–64C, p. 184) indicates that each yeast chromosome is formed from a single linear DNA molecule. This finding agrees with more complex measurements of DNA size (based on rates of elastic recoil) made earlier on *Drosophila* chromosomes, whose DNA molecules are roughly the same size as human chromosomes. For these and other reasons, it is thought that all chromosomes contain only a single DNA molecule.

9-4 Each DNA Molecule That Forms a Chromosome Must Contain a
9-5 Centromere, Two Telomeres, and Replication Origins[3]

In order for a DNA molecule to form a functional chromosome, it must be able to replicate, segregate its two copies at mitosis, and maintain itself between cell generations. The ability to manipulate yeast chromosomes by recombinant DNA methods has made it possible to isolate and define the minimal DNA sequence elements required for these chromosomal functions. Two of the three elements

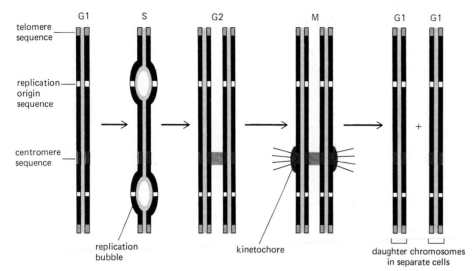

Figure 9–4 The functions of the three DNA sequence elements needed to produce a stable linear eucaryotic chromosome. *Telomere sequences* are required to counteract the shortening of the chromosome that would otherwise occur during each round of DNA replication (see p. 519). *Centromeres* are required to align the DNA molecule on the mitotic spindle during M phase (see p. 768). *Replication origins* are needed to form replication forks during S phase (see p. 515).

were identified by studying small circular DNA molecules that can replicate as *plasmids* (see p. 257) in cells of the yeast *Saccharomyces cerevisiae.* In order to replicate, such a DNA molecule requires a specific nucleotide sequence to act as a **DNA replication origin** (see p. 235); the many origins in each yeast chromosome can be identified by their ability to allow a test DNA molecule to replicate when free of the host chromosome (see p. 516). A second sequence element, called a **centromere,** attaches any DNA molecule that contains it to the mitotic spindle during M phase (see p. 768). Each yeast chromosome contains a single centromere. When this sequence is inserted into a plasmid, it guarantees that each daughter cell will receive one of the two copies of the newly replicated plasmid DNA molecule when the yeast cell divides.

The third required DNA sequence element is a **telomere,** which is needed at each end of a linear chromosome. If a circular plasmid that contains a replication origin and a centromere is broken at a single site to create two free ends in the double helix, it will still replicate and attach to the mitotic spindle, but it will eventually be lost from the progeny cells. This is because replication on the lagging strand of a replication fork requires the presence of some DNA ahead of the sequence to be copied that can serve as the template for an RNA primer (see Figure 5–43, p. 231). Since there can never be such a template for the last few nucleotides of a linear DNA molecule, each such DNA strand would become shorter with each replication cycle. Bacteria and many viruses solve this "end-replication problem" by having a circular DNA molecule as their chromosome. Eucaryotic cells have instead evolved a special telomeric DNA sequence at each chromosome end. This simple repeating sequence is periodically extended by a special enzyme (see p. 519), thus compensating for the loss of a few nucleotides of telomeric DNA in each cycle and permitting a linear chromosome to be completely replicated.

Figure 9–4 summarizes the functions of the three DNA sequence elements that are required to produce a stable linear chromosome in a yeast cell. The same types of elements are thought to be required to maintain a chromosome in human cells. But the human replication origin and centromere sequences have not yet been well defined, and the yeast versions of these sequences do not function in the cells of higher eucaryotes.

On the other hand, recombinant DNA methods allow the required yeast sequence elements to be added to human DNA molecules, which can then replicate in yeast cells as artificial chromosomes. In this way yeast cells can be used to prepare human genomic DNA libraries (see p. 260) in which each DNA clone (propagated as an artificial chromosome) contains as many as a million nucleotide pairs of human DNA sequence (Figure 9–5).

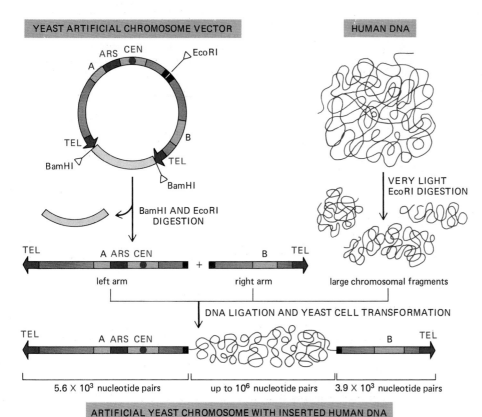

Figure 9–5 A yeast artificial chromosome vector (YAC vector) that allows the cloning of very large DNA molecules. TEL, CEN, and ARS are the telomere, centromere, and replication origin DNA sequences, respectively, for the yeast *Saccharomyces cerevisiae*. (ARS stands for "autonomously replicating sequence." A replication origin enables a plasmid to replicate free of the host cell's chromosomes.) BamH1 and EcoR1 are restriction nucleases that cut the DNA double helix at the specific sites indicated. The sequences denoted as "A" and "B" encode enzymes that serve as selectable markers to allow the easy isolation of transformed yeast cells carrying the artificial chromosome. (Adapted from D.T. Burke, G.F. Carle, and M.V. Olson, *Science* 236:806–812, 1987.)

Most Chromosomal DNA Does Not Code for Essential Proteins or RNAs[4]

There seems to be a large excess of DNA in the genomes of higher organisms. Long before it was possible to examine the nucleotide sequences of chromosomal DNA directly, it was evident that the relative amounts of DNA in the haploid genomes of different organisms has no systematic relationship to the complexity of the organism: human cells, for example, contain about 700 times more DNA than the bacterium *E. coli*, while some amphibian and plant cells contain 30 times more DNA than human cells (Figure 9–6). Moreover, the genomes of different species of amphibians can vary 100-fold in their DNA content.

Population biologists have tried to estimate how much of the DNA in higher organisms codes for essential proteins (or regulates genes that code for such proteins) on the basis of the following indirect argument. Each gene is inevitably subject to a small risk of mutation—an accidental process in which randomly selected nucleotides in the DNA are altered. The greater the number of genes, the greater the probability that a mutation will occur in at least one of them. Since most mutations will impair the function of the gene in which they occur (see p. 221), the mutation rate sets a limit to the number of essential genes that can be maintained by any organism. Using this argument and the observed mutation rate, it has been estimated that no more than a few percent of the mammalian genome can be involved in regulating or encoding essential proteins. We shall see later that other evidence supports this conclusion.

The most important implication of this estimate is that, although the mammalian genome contains enough DNA, in principle, to code for nearly 3 million average-sized proteins (3×10^9 nucleotides), the limited fidelity with which DNA sequences can be maintained means that no mammal (or any other organism) is likely to be constructed from more than perhaps 60,000 essential proteins (ignoring, for the moment, the important consequences of alternative RNA splicing—see p. 589). Thus, from a genetic point of view, humans are unlikely to be more than

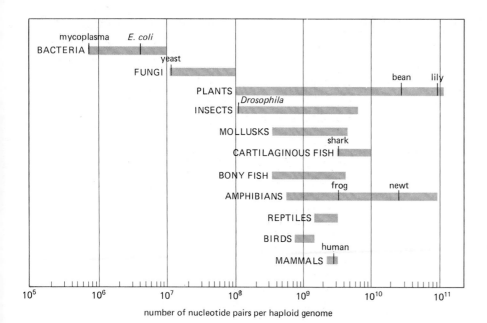

number of nucleotide pairs per haploid genome

Figure 9–6 The amount of DNA in a haploid genome varies over a 100,000-fold range from the smallest procaryotic cell, the mycoplasma, to the large cells of some plants and amphibia. Note that the genome size of humans (3 × 10⁹ nucleotide pairs) is much smaller than that of some other organisms.

about 10 times more complex than the fruit fly *Drosophila*, which is estimated to have about 5000 essential genes (see p. 510).

Whatever the nonessential DNA in higher eucaryotic chromosomes may do (see Chapter 10, p. 607), the data shown in Figure 9–6 make it clear that it is not a great handicap for a higher eucaryotic cell to carry a large amount of extra DNA. Indeed, even the essential coding regions are often interrupted by long stretches of noncoding DNA.

Each Gene Is a Complex Functional Unit for the Regulated Production of an RNA Molecule

The primary function of the genome is to produce RNA molecules. Selected portions of the DNA nucleotide sequence are copied into a corresponding RNA nucleotide sequence, which either (as mRNA) encodes a protein or forms a "structural" RNA, such as a tRNA or rRNA molecule. Each region of the DNA helix that produces a functional RNA molecule constitutes a **gene.**

Genes in a chromosome of a higher eucaryote can contain as many as 2 million DNA nucleotide pairs, and genes more than 100,000 nucleotide pairs in length are common (Table 9–1); yet only about 1000 nucleotide pairs are required to encode

Table 9–1 The Size of Some Human Genes in Thousands of Nucleotides

	Gene Size	mRNA Size	Number of Introns
β-Globin	1.5	0.6	2
Insulin	1.7	0.4	2
Protein kinase C	11	1.4	7
Albumin	25	2.1	14
Catalase	34	1.6	12
LDL receptor	45	5.5	17
Factor VIII	186	9	25
Thyroglobulin	300	8.7	36
Dystrophin*	more than 2000	17	more than 50

The size specified here for a gene includes both its transcribed portion and nearby regulatory DNA sequences. (Compiled from data supplied by Victor McKusick.)

*An altered form of this gene causes Duchenne muscular dystrophy.

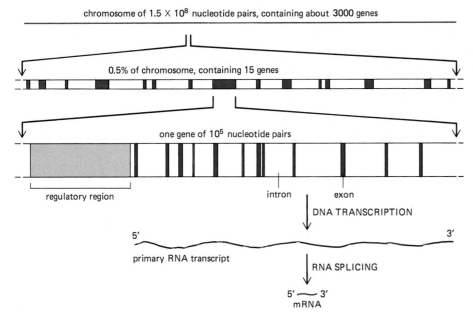

chromosome of 1.5 × 10⁸ nucleotide pairs, containing about 3000 genes

0.5% of chromosome, containing 15 genes

one gene of 10⁵ nucleotide pairs

regulatory region

intron   exon

DNA TRANSCRIPTION

5′   3′

primary RNA transcript

RNA SPLICING

5′ ⌒ 3′
mRNA

Figure 9–7 The organization of genes on a typical vertebrate chromosome. Proteins that bind to the DNA in regulatory regions determine whether a gene is transcribed; although often located on the 5′ side of a gene, as shown here, regulatory regions can also be located in introns, in exons, or on the 3′ side of a gene. The intron sequences are removed from the primary RNA transcripts that encode protein molecules to produce a messenger RNA (mRNA) molecule. The figure given here for the number of genes per chromosome is only a minimal estimate.

a protein of average size (one containing 300 to 400 amino acid residues). Most of the extra length consists of long stretches of noncoding DNA that interrupt the relatively short segments of coding DNA. The coding sequences are called **exons,** the intervening (noncoding) sequences are called **introns.** The RNA molecule (called a *primary RNA transcript*) synthesized from such a gene is altered to remove the intron sequences during its conversion to an mRNA molecule (see Figure 9–2) in the process of *RNA splicing* (see p. 531).

Large genes consist of a long string of alternating exons and introns, with most of the gene consisting of introns. In addition, each gene contains *regulatory DNA sequences*, which bind *gene regulatory proteins* that control transcription of the gene. Many regulatory sequences are located "upstream" (on the 5′ side) of the site where the RNA transcript begins, but they can also be located in introns, "downstream" (on the 3′ side) of the site where the RNA transcript ends, or even in exons. A typical vertebrate chromosome is illustrated schematically in Figure 9–7, along with one of its many genes.

9-6 Comparisons Between the DNAs of Related Organisms Distinguish Conserved and Nonconserved Regions of DNA Sequence[5]

Technical improvements in DNA sequencing are expected to allow the routine sequencing of stretches of chromosomal DNA that are millions of nucleotide pairs long, so that one can foresee the eventual determination of the sequence of all 3 × 10⁹ nucleotides of the human genome. If more than 90% of this sequence is unimportant, however, it will be crucial to have some way of identifying the small proportion of sequence that is important. One way to achieve this is by the simultaneous sequencing of the corresponding regions of a related genome, such as that of the mouse. Human beings and mice are thought to have diverged from a common mammalian ancestor about 80 × 10⁶ years ago, which is long enough for roughly two out of every three nucleotides to have been changed by random mutational events (see p. 220). Consequently, the only regions that will have remained closely similar (*conserved* regions) in the two genomes are those where mutations would impair function. (The organisms with these deleterious mutations would have been eliminated from the population by natural selection—see p. 221.) Thus, in general, *nonconserved* regions represent noncoding DNA—both between genes and in introns—whose DNA sequence is not critical for function. Conserved regions, in contrast, represent functionally important exons and regulatory regions. By revealing in this way the results of a very long natural "experiment," comparative DNA sequencing studies highlight the most interesting regions in genomes. Such studies also provide strong support for the conclusion that the sequence of nucleotides is not important for more than 90% of the DNA in vertebrates.

Chromosomes Contain a Variety of Proteins Bound to Specific DNA Sequences

The information stored in DNA is organized, replicated, and read by a variety of *DNA-binding proteins*. Some of these proteins bind relatively nonspecifically to the DNA along most of its length and help to package it without preventing the access of other DNA-binding proteins; we will discuss these packaging proteins later (see p. 496). Other proteins bind to specific short DNA sequences, which are thus often evolutionarily conserved between different genomes (see Figure 10–34, p. 576). These **sequence-specific DNA-binding proteins** have a variety of functions: some are thought to help fold the long, continuous DNA molecule into distinct domains; some help initiate DNA replication; and many control gene transcription. Each type of cell in a multicellular organism contains a different mixture of the latter *gene regulatory proteins*, which act in combinations to cause the expression of different genes (see p. 566).

The control of gene expression will be discussed in Chapter 10. Here we consider the ways in which proteins bind to DNA. All of the sequence-specific DNA-binding proteins that have been characterized thus far recognize their particular DNA sequence from outside the helix and bind to the DNA without disturbing base-pairing. This is possible because portions of each base pair are exposed in two separate "grooves," called the *major* (wide) and the *minor* (narrow) grooves (Figure 9–8A). In the major groove, each of the four possible base-pair arrangements (A-T, T-A, G-C, or C-G) can be uniquely recognized by the specific arrangement of the atoms that protrude, whereas less sequence information is exposed in the minor groove. As we shall see later, the amino acids at the binding site of each sequence-specific DNA-binding protein are arranged so as to maximize the protein's electrostatic and hydrogen bond interactions with a particular DNA sequence. As expected, most of the hydrogen bonding to base pairs seems to occur in the major groove. A typical interaction between one amino acid side chain and one base pair in this groove is illustrated in Figure 9–8B.

Figure 9–8 Recognition of DNA base pairs by sequence-specific DNA-binding proteins. (A) The DNA double helix (*B*-form) with its major and minor grooves. The edges of the base pairs protrude into these grooves, enabling DNA-binding proteins to recognize specific DNA sequences from outside the helix through hydrogen bond interactions. A sample interaction in the major groove between an amino acid and an A-T base pair is illustrated in (B), viewed along the helix axis; the pattern of hydrogen bond donor and acceptor groups is different in this groove for each of the four possible base-pair arrangements. Note that the *B*-form DNA helix is right-handed (see Figure 3–4, p. 92). Sequence-specific DNA-binding proteins may recognize a set of as few as 4 or as many as 50 base pairs.

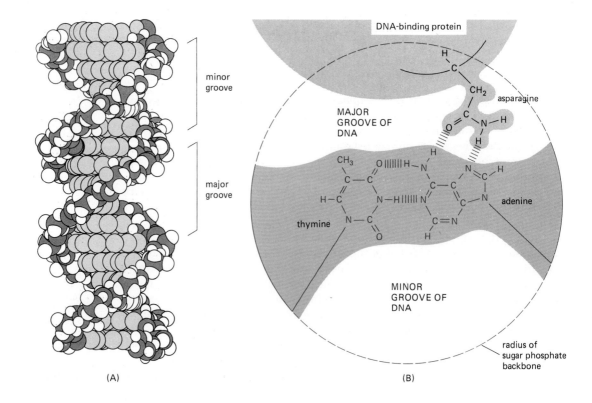

(A)   (B)

¹⁰⁻³ Gel Retardation Assays Detect Sequence-specific
¹⁰⁻¹² DNA-binding Proteins in Cell Extracts[6]

A DNA molecule is highly negatively charged and will therefore move rapidly toward a positive electrode when it is subjected to an electric field. When analyzed by polyacrylamide-gel electrophoresis, DNA molecules are separated according to their size because smaller molecules are able to penetrate the fine gel meshwork more easily than large ones (see p. 184). Protein molecules bound to a DNA molecule will cause it to move more slowly through the gel; in general, the larger the bound protein, the greater the retardation of the DNA molecule. This phenomenon provides the basis for a **gel retardation assay** (also called a *gel–mobility shift assay*), which allows even trace amounts of a sequence-specific DNA-binding protein to be readily detected. In this assay a short DNA fragment of specific length and sequence (produced either by DNA cloning [see p. 258] or by chemical synthesis [see p. 191]) is radioactively labeled and mixed with a cell extract; the mixture is then loaded onto a polyacrylamide gel and subjected to electrophoresis. If the DNA fragment corresponds to a chromosomal region where multiple sequence-specific proteins bind, autoradiography will reveal a series of DNA bands, each retarded to a different extent and representing a distinct DNA-protein complex. The proteins responsible for each band on the gel can then be separated from one another by subsequent fractionations of the cell extract (Figure 9–9).

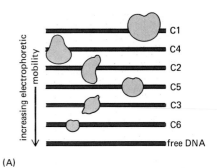

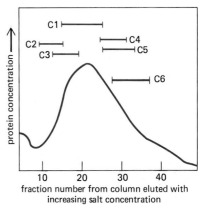

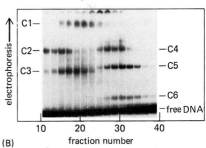

Sequence-specific DNA-binding Proteins Can Be Isolated and Characterized by Exploiting Their Specific DNA Affinity[7]

Sequence-specific DNA-binding proteins were first discovered in bacteria, where genetic analyses revealed the existence of gene regulatory proteins, such as the *E. coli* lactose repressor protein (see p. 568) and the bacteriophage lambda repressor and cro proteins (see p. 574). These proteins were eventually isolated by fractionating cell extracts on a series of chromatography columns (see p. 167), and their specific binding sites on the genes they control were then identified by a combination of genetics and DNA-footprinting methods (see p. 187). The first sequence-specific DNA-binding proteins discovered in eucaryotes—the T-antigen protein of the monkey virus SV40 (see p. 517), the transcription factor *TFIIIA* (see p. 490), and steroid hormone receptor proteins (see p. 691)—were purified and characterized by similar methods.

Much more powerful purification techniques are now available for the isolation of such sequence-specific DNA-binding proteins. These methods often begin with a gel retardation assay that identifies a site on a DNA fragment that binds an uncharacterized protein in a cell extract (see Figure 9–9, for example). A double-stranded oligonucleotide with a sequence corresponding to this binding site is then synthesized by chemical methods and used in one of two ways. In *DNA-affinity chromatography*, the oligonucleotide is linked to an insoluble porous matrix such as agarose; the matrix with the oligonucleotide attached is then used to construct a column that selectively binds proteins that recognize the particular DNA sequence (see p. 168). Purifications as great as 10,000-fold can be achieved by this means with relatively little effort.

Although most proteins that bind to a specific DNA sequence are present in a few thousand copies per higher eucaryotic cell (and generally represent only about one part in 50,000 of the total cell protein), enough pure protein can usually be isolated by affinity chromatography to obtain an amino-terminal amino acid sequence. This allows one to synthesize an oligonucleotide probe, which can be used to identify the corresponding cDNA clone (see p. 262). The clone in turn provides the complete amino acid sequence of the protein as well as the means to produce the protein in unlimited amounts.

In some cases the cDNA clone that encodes a sequence-specific DNA-binding protein has been obtained more directly by a second method, which is even more powerful than DNA-affinity chromatography. This method begins with a cDNA library cloned in an appropriately designed expression vector (see p. 265). An individual colony of bacteria (if the expression vector is a plasmid) or bacterio-

Figure 9–9 A gel retardation assay. The principle of the assay is shown schematically in (A). An extract of an antibody-producing cell line is mixed with a radioactive DNA fragment containing the sequence from −131 to +36 (relative to the RNA start site at +1) of a gene encoding the light chain of the antibody made by the cell line. The effect of the proteins in the extract on the mobility of the DNA fragment is analyzed by polyacrylamide-gel electrophoresis followed by autoradiography. The free DNA fragments run rapidly to the bottom of the gel, while those fragments bound to proteins are retarded; the finding of six retarded bands suggests that there are six different sequence-specific DNA-binding proteins (indicated as C1–C6) in the extract. In (B) the extract was fractionated by a standard chromatographic technique, and each fraction was mixed with the radioactive DNA fragment, applied to one lane of a polyacrylamide gel, and analyzed as in (A). (B, modified from C. Scheidereit, A. Heguy, and R.G. Roeder, *Cell* 51:783–793, 1987.)

phage plaque (if the expression vector is a virus) will produce large amounts of the protein that is encoded by the cDNA it contains. To find the rare colony that produces the protein of interest, the oligonucleotide with the desired binding site is radioactively labeled (see p. 185) and used to probe paper replicas that carry aliquots of thousands of individual colonies (see p. 262). Those few colonies that produce proteins that specifically bind the radiolabeled oligonucleotide are selectively grown and tested further to find the one that produces the desired protein.

Because these powerful methods have only recently been developed, only a small fraction of the many hundreds of sequence-specific DNA-binding proteins thought to be present in higher eucaryotic cells have been isolated so far (see Table 10–1, p. 566).

There Are Common Motifs in Many Sequence-specific DNA-binding Proteins[8]

The eucaryotic *transcription factor III A* (**TFIIIA**) is required to initiate the synthesis of a small ribosomal RNA—the 5S rRNA—and it binds as a monomer to a sequence of about 50 nucleotide pairs near the middle of the very small 5S rRNA gene. The amino acid sequence of this protein suggests that it is organized as a series of 9 repeating domains, each containing 30 amino acids folded into a single structural unit around a Zn atom that links 2 cysteines and 2 histidines. A number of other presumptive gene regulatory proteins contain a smaller number of domains that are at least superficially similar. Such domains are commonly referred to as **zinc fingers.** The proteins include several that are implicated in pattern formation in the early *Drosophila* embryo, more than five involved in yeast mating, the common mammalian transcription factor Sp1 (see p. 566), and the large family of steroid hormone receptor proteins (see Figure 10–25, p. 568).

No three-dimensional structures have yet been determined for this type of protein, so their exact mode of DNA binding is unknown. A plausible general model, supported by DNA-footprinting experiments, is illustrated in Figure 9–10.

Symmetric Dimeric DNA-binding Proteins Often Recognize a Symmetric DNA Sequence[8]

Determining the three-dimensional structure of a protein generally requires x-ray diffraction analysis of large crystals of the protein, which are often difficult to obtain (see p. 151). One of the first gene regulatory proteins to be analyzed in this

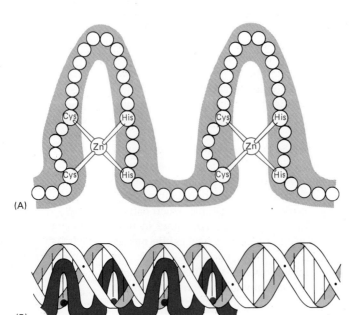

(A)

(B)

Figure 9–10 The "zinc finger" family of sequence-specific DNA-binding proteins. A highly schematic model for the general conformation of the DNA-binding domain is shown in (A), with each amino acid represented by a sphere; the polypeptide chain in each finger is actually thought to be folded into a complex globular conformation. (B) Schematic view of how four such fingers might bind to a specific DNA sequence. Each finger is postulated to recognize a specific sequence of about five nucleotide pairs. (Adapted from A. Klug and D. Rhodes, *Trends in Biochem. Sci.* 12:464–469, 1987.)

way was the bacteriophage lambda **cro protein**, a small protein (66 amino acid residues) that does not contain zinc fingers but binds to a cluster of specific DNA sequences, each 17 nucleotide pairs in length. One of these sequences is illustrated in Figure 9–11. Note that the portion of the sequence shown in color is symmetric—that is, the sequence is the same if the DNA helix is flipped through 180°. Many binding sites for sequence-specific DNA-binding proteins are similarly symmetric. The structure of the cro protein determined by x-ray diffraction shows why.

The cro protein is a symmetric homodimer that is thought to bind to DNA in the manner shown in Figure 9–12. Because the twofold symmetry axis of the protein is positioned on the twofold symmetry axis of the DNA sequence, each identical half of the dimer can make identical contacts to the DNA base pairs that it recognizes. Whenever a DNA-binding site is symmetric in its sequence, the DNA-binding protein that recognizes it is likely to bind as a dimer (or a larger symmetric assembly).

Figure 9–11 A specific DNA sequence recognized by the bacteriophage lambda cro protein. The colored nucleotides in this sequence are arranged symmetrically, allowing each half of the site to be recognized in the same way by half of the dimeric protein (see Figure 9–12).

The cro Protein Is a Member of a "Helix-Turn-Helix" Family of DNA-binding Proteins[9]

The site on the cro protein monomer that contacts DNA is formed by a stretch of 20 amino acids arranged as two α helices separated by a short turn. This **helix-turn-helix** motif has been found in several other bacterial sequence-specific DNA-binding proteins whose three-dimensional structures have been determined (Figure 9–13). Moreover, amino acid sequence homologies suggest that this motif is also present in many other gene regulatory proteins in bacteria, yeasts, and *Drosophila*.

All of the helix-turn-helix proteins shown in Figure 9–13 are symmetric homodimers. One of the α helices in each monomer, termed the *recognition helix*, lies in the major groove of the DNA helix—where its protruding amino acid side chains can hydrogen-bond to specific DNA base pairs. A crucial feature is the separation of the two identical recognition helices in the dimer by exactly one turn of the DNA helix (3.4 nm). This separation allows each recognition helix to pair in the same way with the symmetrically related base pairs in the binding site, as illustrated in Figure 9–14.

An allosteric effector molecule that binds to this type of protein can dramatically increase or decrease its affinity for DNA by changing the distance between the two recognition helices. In analogous ways, allosteric effectors can alter the fit of other types of gene regulatory proteins to the DNA. Such changes are important for turning genes on and off in response to environmental cues (see p. 568).

Figure 9–12 The structure of the bacteriophage lambda cro protein bound to DNA. The protein is a symmetric homodimer that binds to the symmetric DNA sequence shown in Figure 9–11. The mode of DNA binding has been inferred from model-building studies. (A) Wire model bound to a schematic DNA helix. (B) Space-filling model of the DNA-protein complex shown in (A). In this model each amino acid in the protein is represented by a sphere, and the colored balls represent the DNA backbone. (Courtesy of Brian W. Matthews, redrawn from W.F. Anderson, D.H. Ohlendorf, Y. Takeda, and B.W. Matthews, *Nature* 290:754–758, 1981. © 1981 Macmillan Journals Ltd.)

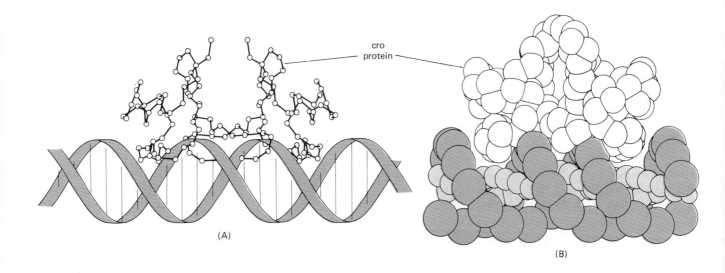

(A)

(B)

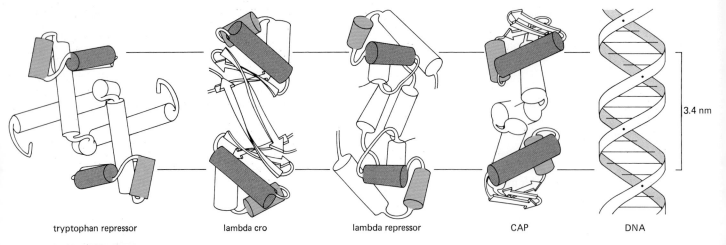

| tryptophan repressor | lambda cro | lambda repressor | CAP | DNA |

Protein Molecules Often Compete or Cooperate with One Another When They Bind to DNA[10]

Most genetic processes depend on interactions between protein molecules that bind simultaneously to nearby sites on the DNA. In the simplest case, two sequence-specific DNA-binding proteins whose DNA sites partially or totally overlap compete with each other for occupation of a DNA site (Figure 9–15A). Repressor proteins, for example, can inhibit gene transcription by blocking the binding of an activating protein to the DNA. But proteins can also help each other hold onto the DNA more tightly when they bind. Such **cooperative binding** can occur either between two different protein molecules (Figure 9–15B) or between two copies of a single type of molecule. In the latter case a protein will tend to bind in an all-or-none fashion to form clusters on the DNA, and its binding to DNA will increase sharply when its concentration is raised (Figure 9–15C). Examples of cooperative binding proteins of this type are the helix-destabilizing protein and recA protein, discussed in Chapter 5 (see p. 232 and p. 242), and the histone H1 protein, to be discussed below (see p. 499).

We shall encounter many examples of cooperative and competitive binding interactions involving two different sequence-specific DNA-binding proteins when we consider the regulation of gene transcription in Chapter 10.

DNA Helix Geometry Depends on the Nucleotide Sequence[11]

For 20 years after the discovery of the DNA double helix in 1953, DNA was thought to have the same monotonous structure, with exactly 36° of helical twist between its adjacent base pairs (10 nucleotide pairs per helical turn) and a uniform helix geometry. Subsequent experiments have shown that DNA is much more polymorphic than this, with a variation in form that depends on its sequence. This variation can have important consequences for its interactions with proteins.

There are several types of variants, some of which involve major changes in helix geometry. By far the most stable form of DNA is so-called **B-form DNA,** whose right-handed helical structure is illustrated throughout this book (Figure 9–16A). X-ray diffraction studies of small synthetic DNAs of unusual sequence suggest that local regions of special sequence preferentially form a different right-handed helix known as **A-form DNA.** This form of DNA contains highly tilted base pairs and generates a shorter, wider helix than B form DNA (Figure 9–16B). It is likely to be important in any circumstance in which a DNA strand is paired with RNA, as in the primer portion of Okazaki fragments, for example (see p. 231), since the extra sugar hydroxyl group on a ribose sugar prevents an RNA-DNA helix from assuming the B form. RNA-RNA helices will also be arranged as an A-form helix for this reason. The A-form helix is thus clearly important biologically, being present in the hairpin helical regions of all single-stranded RNA chains (see p. 98). The biological importance of a third helical form of DNA is less certain. DNA sequences composed of alternating purines and pyrimidines, such as GCGCGCGC, readily

Figure 9–13 A family of dimeric helix-turn-helix DNA-binding proteins. These four gene regulatory proteins act in bacterial systems: the lambda repressor and cro proteins control bacteriophage lambda gene expression (see p. 574), and the catabolite activator protein (CAP) controls the expression of a set of *E. coli* genes that can be turned on only in the absence of glucose (see p. 560). In each case an x-ray diffraction analysis of protein structure reveals that the two copies of the recognition helix (*brown cylinder*) are separated by exactly one turn of the DNA helix (3.4 nm). See also Figure 9–14.

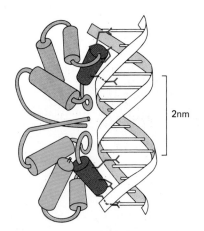

Figure 9–14 The pairing of amino acid side chains (shown in dark red) with DNA base pairs during the recognition of a specific DNA sequence by a bacteriophage gene regulatory protein. This dimeric helix-turn-helix protein is the bacteriophage 434 repressor, whose specific DNA-protein complex has been investigated by x-ray diffraction analysis. (Adapted from J.E. Anderson, M. Ptashne, and S.C. Harrison, *Nature* 326:846–852, 1987.)

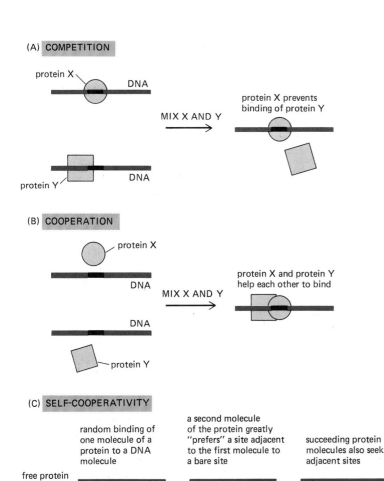

(A) COMPETITION

protein X

DNA

MIX X AND Y

protein X prevents
binding of protein Y

protein Y

DNA

(B) COOPERATION

protein X

DNA

MIX X AND Y

protein X and protein Y
help each other to bind

DNA

protein Y

(C) SELF-COOPERATIVITY

random binding of
one molecule of a
protein to a DNA
molecule

a second molecule
of the protein greatly
"prefers" a site adjacent
to the first molecule to
a bare site

succeeding protein
molecules also seek
adjacent sites

free protein
molecules

NET RESULT:
one region of the
DNA is covered
with the protein,
and the other lacks
the protein

Figure 9–15 Examples of competitive and cooperative protein-protein interactions during the binding of proteins to DNA. (A) Competition occurs when the binding of proteins X and Y to their specific DNA sites is mutually exclusive. (B) Cooperation occurs when the binding of one protein to DNA increases the affinity of a second protein for a nearby site. (C) Self-cooperativity causes a single protein to bind in a cluster of molecules, and it tends to produce "all-or-none" binding to a region of DNA.

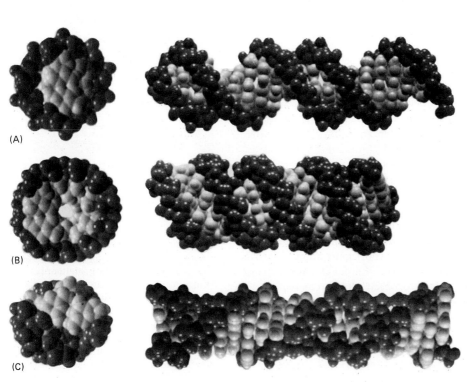

(A)

(B)

(C)

Figure 9–16 Three helical forms of DNA, each containing 22 nucleotide pairs. All are formed from two antiparallel DNA strands held together by complementary base-pairing, and each is shown in both side and top views. The sugar-phosphate backbone (dark gray) and the base pairs (light gray) are shaded differently. (A) *B-form DNA,* which is by far the most common form in cells. (B) *A-form DNA,* which becomes the predominant form for all DNA sequences when most of the water is removed by drying DNA fibers. (C) *Z-form DNA,* which can form for selected DNA sequences under special circumstances. Whereas the *B*-form and *A*-form helices are right-handed, the *Z*-form helix is left-handed (see Figure 3–4, p. 92). (Courtesy of Richard Feldmann.)

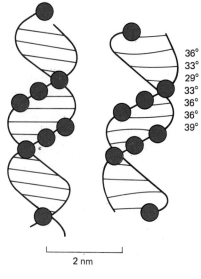

Figure 9–17 The conformation of DNA observed in the crystalline bacteriophage 434 repressor-DNA complex (*right*) compared with an idealized *B*-form helix (*left*). The DNA phosphates that are contacted by the protein are shown as colored circles. The helical twist angles between base pairs are indicated for the irregular helix; the distortions from a perfect helix are thought to be required for tight protein binding. Part of the binding energy depends on the "nonspecific" interactions between N-H groups of the protein and DNA phosphates; only an irregular *B*-form helix—with the minor groove in the middle of the binding site narrowed and the helix slightly bent—can make these favorable contacts. Because A-T base pairs are required in this central region to permit the observed bending and compression of the helix, their replacement by G-C base pairs greatly weakens protein binding. Thus, the central base pairs are "recognized" by the repressor even though the protein does not contact these base pairs directly (see also Figure 9–14). (Adapted from J.E. Anderson, M. Ptashne, and S.C. Harrison, *Nature* 326:846–852, 1987.)

form a left-handed double helix known as **Z-form DNA** (Figure 9–16C). Short regions of such drastically altered helical geometry are expected to be rare in chromosomes, but it is possible that they could be specifically recognized by proteins and thereby have important biological roles.

While both *A* DNA and *Z* DNA represent major distortions of the usual *B* form, less drastic alterations of the *B* helix are common and of undisputed biological importance. In *B*-form DNA, both the exact tilt of the bases and the helical twist angle between base pairs depend on which nucleotides are adjacent to each other in the sequence. Because of these variations, the atoms in the helix (including those in DNA phosphates) are usually displaced from their idealized positions. Even DNA-binding proteins that do not specifically recognize particular base pairs can sense these displacements and therefore bind preferentially to the helical conformations in particular regions, so that the exact conformation of *B* DNA makes an important contribution to protein-DNA interactions.

The importance of variations in *B*-form helix geometry for DNA recognition by sequence-specific DNA-binding proteins has been clearly demonstrated in the case of a bacteriophage repressor protein. The structure of the *B*-form DNA helix when the repressor protein is bound (as determined by x-ray diffraction analysis) is compared to an idealized *B*-form structure in Figure 9–17.

9-7 Some DNA Sequences Are Highly Bent[12]

The DNA helix has enough conformational freedom to be constantly undergoing springlike bending and rotatory motions; but because the helix is stiff, an average DNA molecule requires 200 nucleotide pairs or so to turn gradually through 90°. Some DNA sequences, however, are unusually flexible and adopt such a curved form much more easily.

Other DNA sequences have a strong inherent tendency to remain permanently bent. These include sequences containing repeats of particular dinucleotides and segments containing the sequence AAAAANNNNN (where N can be any nucleotide) repeated every 10 to 11 nucleotides. In these DNA molecules the sequence-induced bending of the helix axis in each helical turn reinforces that from the neighboring regions, causing the DNA to appear unusually curved when viewed in an electron microscope (Figure 9–18). This bending represents an extreme form of variation in local helix geometry; it is not yet clear whether the bend in each helical turn results from the cumulative effects of small tilts between particular base pairs, from a large bend at a junction between short regions of different helical type, or from both types of effects.

Even without special sequences, the DNA double helix can be distorted enough by protein binding to allow the formation of the highly condensed protein-DNA complexes that we consider next.

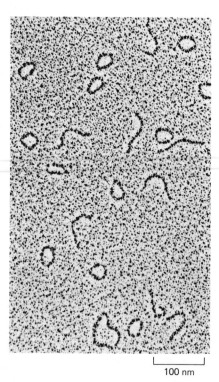

Figure 9–18 Electron micrograph of fragments of a highly bent segment of DNA helix. The DNA fragments, derived from the kinetoplast DNA minicircles of the trypanosome *Crithidia fasciculata*, are only about 200 nucleotide pairs long, but many of them have bent to form a complete circle. A normal DNA helix of this length would bend only enough to produce one-fourth of a circle (one smooth right-angle turn) on average. (From J. Griffith, M. Bleyman, C.A. Rauch, P.A. Kitchin, and P.T. Englund, *Cell* 46:717–724, 1986.)

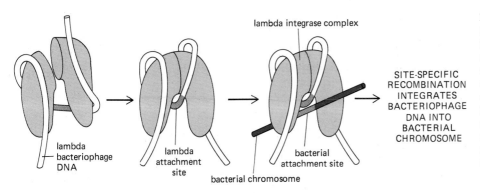

Figure 9–19 Diagram of the protein-DNA complex formed by lambda integrase, the enzyme that mediates entry of bacteriophage DNA into the *E. coli* host chromosome. This complex catalyzes a site-specific recombination event by breaking and rejoining the lambda and bacterial DNA helices at specific sites called attachment sites (see Figure 5–66, p. 247). (Adapted from E. Richet, P. Abcarian, and H.A. Nash, *Cell* 52:9–17, 1988.)

Proteins Can Wrap DNA into a Tight Coil[13]

Many proteins bend the DNA helix when they bind to it, and multiple DNA-protein contacts of this type can wrap the DNA helix into a tight coil around a protein complex to produce a nucleoprotein particle. In bacteria, such nucleoprotein particles are known to form when initiator proteins bind to a DNA replication origin (see p. 235) and when the lambda integrase protein binds to DNA to catalyze a site-specific recombination event (Figure 9–19). Both competitive and cooperative protein-protein interactions are expected in such complicated three-dimensional assemblies. These types of interactions are exploited to regulate the catalytic activity of the nucleoprotein particle, as illustrated for a protein complex that contains the lambda integrase in Figure 9–20. Since the expression of eucaryotic genes involves the binding of clusters of gene regulatory proteins to specific regulatory

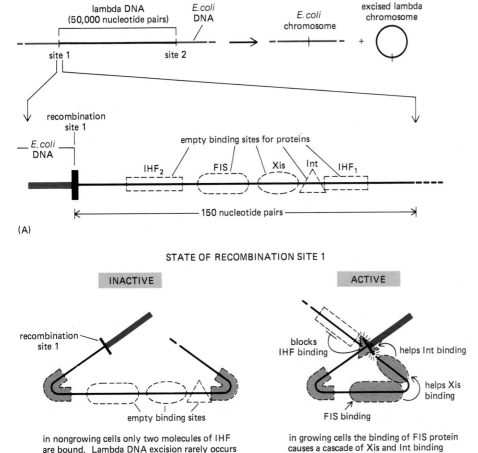

Figure 9–20 Excision of bacteriophage lambda DNA from the bacterial chromosome is controlled by cooperative and competitive interactions between sequence-specific DNA-binding proteins. The reaction is catalyzed by the lambda integrase and is the reverse of the site-specific recombination event in Figure 9–19. (A) Outline of the reaction and illustration of some of the protein-binding sites involved (not all sites are shown). Excision requires breaking and rejoining the DNA double helices at recombination sites 1 and 2 to form a circular lambda chromosome. Int is lambda integrase (see p. 246), Xis is lambda excisionase, and IHF and FIS are proteins produced by the bacterial host cell. (B) Activation of excision by FIS protein; the events shown are those expected at low concentrations of the Int and Xis proteins. As indicated, several proteins strongly bend the DNA when they bind. Although the Int protein catalyzes the site-specific recombination reaction, its action is controlled by the other proteins. (Courtesy of Arthur Landy.)

regions on the DNA, similar protein complexes may be involved in the control of DNA transcription in eucaryotes (see Figure 10–23, p. 567).

Any complex involved in the regulation of specific genes would be expected to be relatively rare. The structure of eucaryotic chromosomes is dominated by a different type of nucleoprotein particle, the *nucleosome*, which plays a major role in packing and organizing all of the DNA in the cell nucleus.

Histones Are the Principal Structural Proteins of Eucaryotic Chromosomes[14]

By far the best understood of the structural proteins in chromosomes are the *histones*, which are found only in eucaryotic cells. These proteins are so abundant that it is traditional to divide the DNA-binding proteins in eucaryotes into two general classes: the **histones** and the **nonhistone chromosomal proteins.** The complex of both classes of proteins with the nuclear DNA of eucaryotic cells is known as **chromatin.** Histones are present in such enormous quantities (about 60 million molecules of each type per cell, compared to 10,000 molecules per cell for a typical sequence-specific DNA-binding protein) that their total mass in chromatin is about equal to that of the DNA.

Histones are relatively small proteins with a very high proportion of positively charged amino acids (lysine and arginine); the positive charge helps the histones bind tightly to DNA, regardless of its nucleotide sequence. Histones probably only rarely dissociate from the DNA, and they are therefore likely to have an influence on any reaction that occurs on chromosomes.

The five types of histones fall into two main groups: (1) the *nucleosomal histones* and (2) the *H1 histones.* The **nucleosomal histones** are small proteins (102–135 amino acid residues) responsible for coiling the DNA into *nucleosomes,* as discussed in the next section. These four histones are designated **H2A, H2B, H3,** and **H4.** H3 and H4 form the inner core of the nucleosome, and they are among the most highly conserved of all known proteins: there are only two differences in the amino acid sequences of histone H4 in peas and cows, for example (Figure 9–21). This evolutionary conservation suggests that the functions of these two histones involve nearly all of their amino acids, so that a change in any position is deleterious to the cell (see p. 221).

The **H1 histones** are larger (containing about 220 amino acids) and have been less conserved during evolution than the nucleosomal histones. The yeast *Saccharomyces cerevisiae* appears to lack H1 entirely (see p. 583).

The Association of Histones with DNA Leads to the Formation of Nucleosomes, the Unit Particles of Chromatin[15]

If it were stretched out, the DNA double helix in each human chromosome would span the cell nucleus thousands of times. Histones play a crucial part in packing this very long DNA molecule in an orderly way into a nucleus only a few micrometers in diameter. Their role in DNA folding is also important for a second reason. As we shall see, not all the DNA is folded in exactly the same way, and the manner in which a region of the genome is packaged into chromatin in a particular cell seems to influence the activity of the genes it contains (see p. 577).

A major advance in our understanding of chromatin structure came in 1974 with the discovery of the fundamental packing unit known as the **nucleosome,** which gives chromatin a "beads-on-a-string" appearance in electron micrographs taken after treatments that unfold higher-order packing (Figure 9–22). The nucleosome "bead" can be removed from the long DNA "string" by digestion with enzymes that degrade DNA, such as a bacterial enzyme, micrococcal nuclease. (Enzymes that degrade both DNA and RNA are called *nucleases;* enzymes that degrade only DNA are *deoxyribonucleases,* or *DNases.*) After digestion for a short period with micrococcal nuclease, only the DNA between the nucleosome beads is degraded. The rest is protected from digestion and remains as double-stranded DNA fragments 146 nucleotide pairs long bound to a specific complex of 8 nucleosomal histones (the *histone octamer*). The nucleosome beads obtained in this way have been crystallized and analyzed by x-ray diffraction. Each is a disc-shaped

Figure 9–21 The amino acid sequence of histone H4, one of the four nucleosomal histones. The amino acids are designated by their single-letter abbreviations, with the positively charged amino acids colored for emphasis. As for the other nucleosomal histones, an elongated amino-terminal "tail" is reversibly modified in the cell by the acetylation of selected lysines, as indicated (see p. 417). The bovine sequence is shown; the sequence is the same in peas, except that one valine is changed to an isoleucine and one lysine is changed to an arginine.

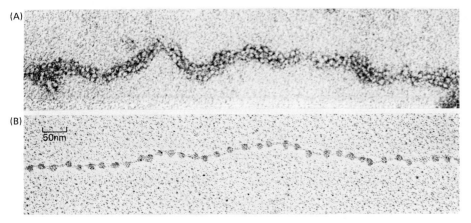

(A)

(B)

50nm

Figure 9–22 Electron micrographs of chromatin strands before and after treatments that stretch, or "decondense," the native structure to produce the "beads-on-a-string" form. (A) shows the native structure, known as the 30-nm fiber (see p. 498). In (B) the decondensed, beads-on-a-string form of chromatin is shown at the same magnification. For a schematic drawing of the relation between these two chromatin forms, see Figure 9–38. These electron micrographs were taken by modifications of the procedure outlined in Figure 9–71. (A, courtesy of Barbara Hamkalo; B, courtesy of Victoria Foe.)

particle with a diameter of about 11 nm containing 2 copies of each of the 4 nucleosomal histones, H2A, H2B, H3, and H4. This **histone octamer** forms a protein core around which the double-stranded DNA helix is wound twice (Figure 9–23).

In undigested chromatin the DNA extends as a continuous thread from nucleosome to nucleosome. Each nucleosome bead is separated from the next by a region of *linker DNA*, which can vary in length between 0 and 80 nucleotide pairs. On average, nucleosomes repeat at intervals of about 200 nucleotide pairs (see Figure 9–23). Thus a eucaryotic gene of 10,000 nucleotide pairs will be associated with 50 nucleosomes, and each human cell with 6×10^9 DNA nucleotide pairs contains 3×10^7 nucleosomes.

9-10 Some Nucleosomes Are Specifically Positioned on DNA[16]

In vitro experiments with isolated chromatin suggest that the histone octamers generally remain fixed in one position under physiological conditions, inasmuch as their tight binding to DNA prevents them from sliding back and forth along the DNA helix. But are these octamers randomly placed on the DNA, so that any particular DNA sequence will be wound tightly around the histone octamer in one cell while remaining more accessible in the linker regions between nucleosome beads in another?

The positions of nucleosomes in cells can be inferred by lightly treating cell nuclei with an enzyme or chemical that cuts DNA and then analyzing the sites that are protected from digestion in a manner analogous to DNA footprinting (see p. 187). While the majority of nucleosomes seem not to be precisely positioned, striking examples of precise nucleosome positioning are known. Up to 15 precisely positioned nucleosomes surround the centromeric DNA (CEN sequence) of the yeast *Saccharomyces cerevisiae*, for example (see p. 769); in addition, the single nucleosome bound to the tiny 5S rRNA gene has a unique location, and there is at least one uniquely placed nucleosome just upstream from the start site for β-globin RNA synthesis. Because the spacing between nucleosomes is more or

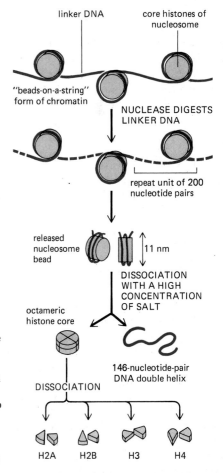

Figure 9–23 The nature of the nucleosome. The nucleosome consists of two full turns of DNA (83 nucleotide pairs per turn) wound around an octameric histone core, plus the adjacent "linker DNA." The part of the nucleosome referred to here as the "nucleosome bead" is released from chromatin by digestion of the linker DNA (plus part of one DNA turn) with micrococcal nuclease. In each nucleosome bead, 146 nucleotide pairs of DNA double helix (about 1.8 turns) remain wound around an octameric histone core. The histone octamer is composed of two each of histones H2A, H2B, H3, and H4. These histones contain 102–135 amino acids each, and the octamer has a total mass of about 100,000 daltons. In the "beads-on-a-string" form of chromatin, each nucleosome is connected to its neighbors by its exposed "string" of linker DNA.

less regular, the exact placement of these nucleosomes on the DNA can also affect the placement of their neighbors.

What determines this preferential positioning of nucleosomes? At least for some of them, such as those bound to 5S rRNA genes, a mixture of the four purified nucleosomal histones will re-form a nucleosome *in vitro* at the exact position where the nucleosome is located *in vivo*. Part of the explanation is that nucleosomes tend to bind so as to maximize A-T-rich minor grooves on the inside of the DNA coil. This preference stems from the difficulty of bending the DNA double helix into two tight turns around the outside of the histone octamer, which requires a substantial compression of the minor groove of the DNA helix (Figure 9–24). We have already seen in the case of a bacteriophage repressor protein (see Figure 9–17) that a cluster of two or three A-T base pairs in the minor groove makes this compression easier. Other, unknown features of the DNA sequence must also contribute to the preferential placement of nucleosomes.

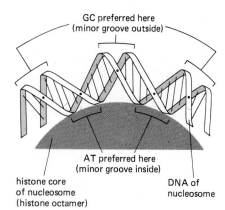

Figure 9–24 The bending of DNA in a nucleosome. The DNA helix makes two tight turns around the histone octamer, each with 83 nucleotide pairs per turn. This diagram is drawn approximately to scale to illustrate how the minor groove is compressed on the inside of the turn. As previously explained (see Figure 9–17), A-T base pairs are preferentially accommodated in a narrow minor groove.

Specific Sites on Chromosomes Lack Nucleosomes[17]

Some regions of DNA lack a nucleosome even though they are hundreds of nucleotide pairs long. These regions can be detected by treating cell nuclei with trace amounts of a deoxyribonuclease (DNase I) that at low concentrations will digest long stretches of nucleosome-free DNA but not the short stretches of linker DNA between nucleosomes. Chromatin treated in this way is preferentially cut at specific DNA sequences that seem to lack a nucleosome; these sites are generally spaced thousands of nucleotide pairs apart.

The first evidence that such **nuclease-hypersensitive sites** are biologically important came from experiments with the monkey virus SV40, whose circular DNA chromosome is organized by the histones produced by its host cells. The SV40 chromosome often contains a single nucleosome-free region about 300 nucleotide pairs long that is rapidly digested by DNase I. The exposed region is located very near the DNA sequences at which both viral DNA replication and RNA synthesis begin, and it contains several sequence-specific DNA-binding proteins that protect only very short regions of the DNA from nuclease digestion while apparently excluding nucleosomes. Similarly, many of the DNase-hypersensitive sites in cell chromatin are located in the regulatory regions of a gene (Figure 9–25), and more of these sites are present in cells where the gene is active than in other cells. Most such sites are believed to represent regions from which a nucleosome has been displaced by sequence-specific DNA-binding proteins involved in eucaryotic gene regulation (see Figure 9–27).

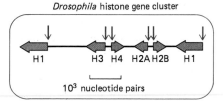

Figure 9–25 The location of nuclease-hypersensitive sites (*colored arrows*) in the regulatory regions of active genes. Although these specially exposed sites in chromatin are often present on the 5′ side of a gene, as illustrated here for the cluster of genes that encode histones (H1, H2A, H2B, H3, and H4) in *Drosophila*, they are also located elsewhere (see Figure 10–40A, p. 581).

Nucleosomes Are Usually Packed Together to Form Regular Higher-Order Structures[18]

In the living cell, chromatin probably rarely adopts the extended "beads-on-a-string" form. Instead, the nucleosomes are packed upon one another to generate regular arrays in which the DNA is more highly condensed. When nuclei are very gently lysed onto an electron microscope grid, most of the chromatin is seen to be in the form of a fiber with a diameter of about 30 nm (see Figure 9–22A). One

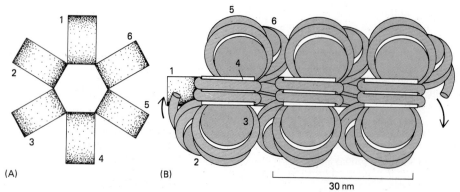

Figure 9–26 A model suggested to explain how the "beads-on-a-string" form of nucleosomes is packed to form the 30-nm fiber seen in electron micrographs (see Figure 9–22A). (A) top view, (B) side view. This type of packing requires one molecule of histone H1 per nucleosome (not shown). Although the position where the H1 attaches to the nucleosome has been defined (see Figure 9–28), the location of the H1 molecules in this fiber is unknown. (See also Figure 9–27.)

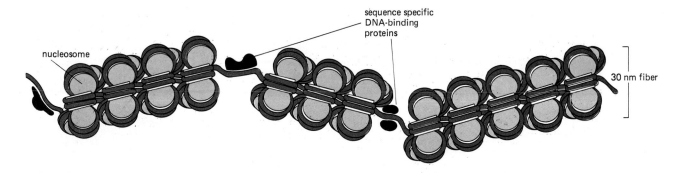

nucleosome

30 nm fiber

of several models proposed to explain how nucleosomes are packed in the **30-nm chromatin fiber** is illustrated in Figure 9–26. Such models represent an idealized structure, since both the range of linker lengths created by preferred nucleosome positioning and the presence of occasional nucleosome-free sequences will punctuate the 30-nm fiber with irregular features (Figure 9–27).

As a 30-nm fiber, the chromatin of a typical human chromosome would be about 0.1 cm long if it were extended and thus could span the nucleus more than 100 times. Electron microscopic studies of intact chromosomes, however, suggest that the 30-nm fiber inside cells is further folded to produce chromatin strands as thick as 100 nm; how the nucleosomes are arranged in these structures is not known.

Figure 9–27 A schematic section of chromatin illustrating the interruption of its regular nucleosomal structure by short regions where the chromosomal DNA is unusually vulnerable to digestion by DNase I. At each of these *nuclease-hypersensitive sites*, a nucleosome appears to have been displaced from the DNA by one or more sequence-specific DNA-binding proteins.

Histone H1 Proteins Help Pack Nucleosomes Together[19]

There are about six closely related varieties (subtypes) of *H1 histones* in a mammalian cell, differing somewhat in their amino acid sequence. These **histone H1** molecules appear to be responsible for packing nucleosomes into the 30-nm fiber. The H1 molecule has an evolutionarily conserved globular central region linked to extended amino-terminal and carboxyl-terminal "arms," whose amino acid sequence has evolved much more rapidly. Each H1 molecule binds through its globular portion to a unique site on a nucleosome and has arms that are thought to extend to contact other sites on the histone cores of adjacent nucleosomes, so that the nucleosomes are pulled together into a regular repeating array (Figure 9–28). Some models place the H1 molecules on the inside of the 30-nm chromatin fiber, while other models place them on the outside.

The histone octamer that forms each nucleosome core is a symmetrical structure, whereas the single histone H1 molecule that binds to each nucleosome is not. The binding of H1 molecules to chromatin is thus believed to create a local polarity that the chromatin would otherwise lack (Figure 9–29).

Experiments *in vitro* suggest that histone H1 tends to bind to DNA in clusters of eight or more histone H1 molecules, an example of the *cooperative binding* discussed previously (see p. 492). It has been proposed that chromatin is organized by cooperative interactions of this kind and that their disruption by gene regulatory proteins may therefore be responsible for the local decondensation of chromatin that occurs around active genes; as we shall discuss shortly, such regions of "active chromatin" seem to have an unusually low affinity for histone H1 (see p. 511). In this view, a local region of chromatin may resemble a tiny crystal that is capable of changing its conformation in an all-or-none fashion during gene activation processes (Figure 9–30).

Figure 9–28 The way histone H1 (220 amino acids) is thought to help pack adjacent nucleosomes together. The globular core of H1 binds to each nucleosome near the site where the DNA helix enters and leaves the histone octamer. When H1 is present, two full turns of the DNA (166 nucleotide pairs) are protected from micrococcal nuclease digestion (an extra 20 nucleotide pairs—see Figure 9–23). Neither the three-dimensional structure of histone H1 nor the precise sites of interaction of its extended amino-terminal and carboxyl-terminal arms with the nucleosome are known.

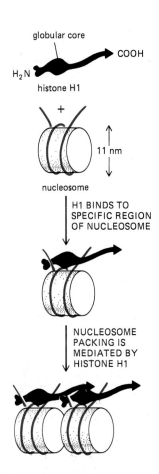

globular core

COOH

H₂N

histone H1

+

11 nm

nucleosome

H1 BINDS TO
SPECIFIC REGION
OF NUCLEOSOME

NUCLEOSOME
PACKING IS
MEDIATED BY
HISTONE H1

Figure 9–29 The polarity that is thought to be imparted to chromatin by the binding of histone H1. (A) Without histone H1, chromatin has no polarity inasmuch as each nucleosome is a symmetric structure. (B) With histone H1, chromatin acquires a polarity; the significance of this polarity is unknown.

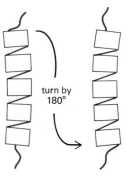

chromatin structure without H1 is the same approached from either end of DNA helix

(A)

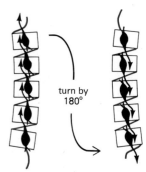

H1 makes chromatin different when approached from either end

(B)

RNA Synthesis Can Proceed Through Nucleosomes[20]

Biochemical tests indicate that most of the DNA remains bound up in nucleosomes during DNA transcription, and electron micrographs of spread chromatin generally show a regular pattern of nucleosome beads throughout both its transcribed and untranscribed regions (Figure 9–31). It seems that the histone octamer is so firmly bound to DNA that the association is normally permanent. But it is difficult to imagine RNA polymerase transcribing the DNA bound up in a nucleosome without some temporary change in the conformation of the nucleosome itself. As the RNA polymerase passes, the nucleosomal DNA presumably uncoils from the histone octamer without releasing it completely. Perhaps the whole nucleosome opens up transiently, splitting the histone octamer into two halves; alternatively, the octamer could remain intact and "rock" back and forth to let the polymerase pass (Figure 9–32). The difficulty of performing such maneuvers may help to explain why the amino acid sequences of the nucleosomal histones have been so highly conserved during evolution (see p. 221): if a nucleosome were simply a DNA-packing device, one would expect many sequence variations in its histones to be permissible.

As we have discussed, most nucleosomes in the cell are packaged into a 30-nm chromatin fiber, which is then folded further. It is inconceivable that chromatin in this form could be transcribed by RNA polymerase without a major disruption of the nucleosome packing (Figure 9–33). Several types of experiments suggest that some such disruption occurs (see, for example, Figure 9–50 and p. 580).

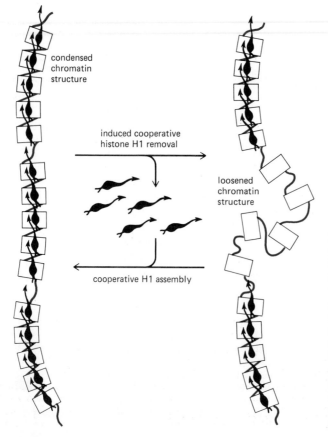

condensed chromatin structure

induced cooperative histone H1 removal

loosened chromatin structure

cooperative H1 assembly

Figure 9–30 A local region of chromatin may act as a structural unit because of the cooperative packing of its nucleosomes caused by histone H1. This highly schematic diagram depicts a unit of chromatin (as a 30-nm fiber or more compact structure) being decondensed suddenly upon receiving some external regulatory signal. (H1 histones are shown in groups of five for simplicity; in fact they are likely to bind in much larger clusters.) A local chromatin decondensation of this type may accompany gene activation (see Figure 9–50).

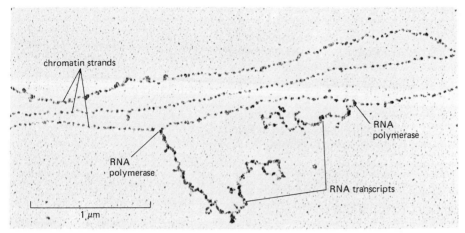

Figure 9–31 A region of chromatin in its "beads-on-a-string" form, showing three chromatin strands, on one of which two RNA polymerase molecules are engaged in DNA transcription. Most of the chromatin in a higher eucaryotic nucleus does not contain active genes and therefore lacks RNA transcripts. Note that nucleosomes are present on both transcribed and nontranscribed regions and that they are bound to the DNA immediately in front of and just behind the moving RNA polymerase molecules. (Courtesy of Victoria Foe.)

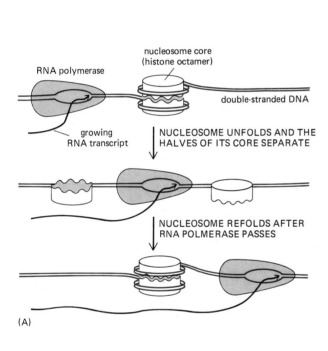

(A)

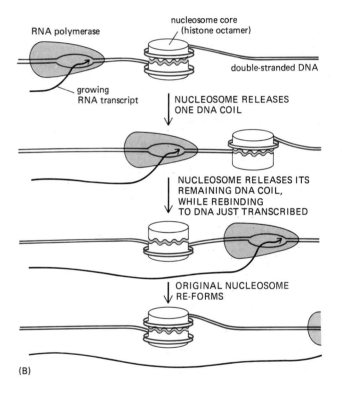

(B)

Figure 9–32 Two possible models to explain how RNA polymerase can transcribe chromatin without displacing nucleosomes. (A) Transcription past transiently unfolded "half-nucleosomes." (B) Transcription past an intact histone octamer.

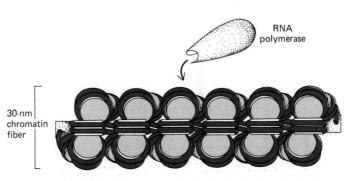

Figure 9–33 Schematic diagram of an RNA polymerase molecule approaching a 30-nm chromatin fiber, drawn roughly to scale. The number of nucleosomes shown would cover about 7000 nucleotide pairs of DNA—about the size of a modest-sized human gene (see Table 9–1, p. 486). Somehow the polymerase must gain access to all of this DNA without displacing the histone octamers from the chromatin. A major unfolding of the chromatin is clearly required to make this possible.

Summary

A gene is defined as a nucleotide sequence that acts as a functional unit for the production of an RNA molecule. A chromosome is formed from a single, enormously long DNA molecule that contains a series of genes. A chromosomal DNA molecule also contains three other types of functionally important nucleotide sequences: replication origins and telomeres allow the DNA molecule to be replicated, while a centromere is needed to attach the DNA molecule to the mitotic spindle. The human haploid genome contains 3×10^9 DNA nucleotide pairs, divided among 22 different autosomes and 2 sex chromosomes. Only a few percent of this DNA is thought to code for proteins.

The DNA in eucaryotes is tightly bound to an equal mass of histones, which serve to form a repeating array of DNA-protein particles called nucleosomes. Nucleosomes are usually packed together in regular arrays to form a 30-nm fiber (and more compact structures) with the aid of histone H1 molecules. In regions of DNA that contain genes, however, the histone octamer that forms each nucleosome must compete with a large variety of sequence-specific DNA-binding proteins for sites on the DNA. The occasional regions where a nucleosome has been displaced by the binding of one or more of these proteins can generally be detected as nuclease-hypersensitive sites in the chromatin and are usually diagnostic of regions active in DNA transcription.

There are hundreds of different sequence-specific DNA-binding proteins in a eucaryotic cell, each of which recognizes a short DNA sequence by hydrogen-bonding interactions with base pairs and by sensing helix geometry. These proteins cooperate and compete with one another during the formation of protein complexes at selected DNA sites.

The Complex Global Structure of Chromosomes

Having discussed the DNA and protein molecules from which chromosomes are made, we now review what is known about chromosome organization on a more global scale. We shall see that the DNA in a chromosome is not only packaged with histones into regularly repeating nucleosomes but is also elaborately folded and organized by other proteins into a series of subdomains of distinct character. This higher-order structure is a fascinating feature of the chromatin of eucaryotic cells, although its biological function is still a mystery.

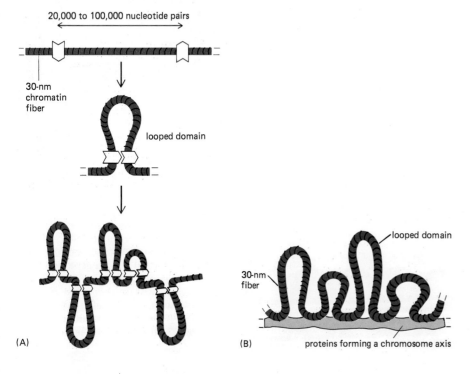

(A)

(B)

Figure 9–34 Schematic views of a section of a chromosome folded into a series of looped domains, each containing perhaps 20,000 to 100,000 nucleotide pairs of double-stranded DNA condensed in a 30-nm chromatin fiber. (A) A folding model requiring sequence-specific DNA-binding proteins at each end of a loop. (B) A folding model requiring a chromosome axis. The actual mode of folding is unknown, although cytological studies reveal that the axial region of isolated mitotic chromosomes (where the ends of the loops are located) is highly enriched for the abundant enzyme DNA topoisomerase II (see p. 236).

Chromosomes Appear to Be Folded into a Series of Looped Domains[21]

The nucleus is typically about 5 μm (5 × 10⁻⁴ cm) in diameter. As previously mentioned, the packaging of DNA into a 30-nm chromatin fiber leaves a human chromosome about 0.1 cm long, so there must be several higher orders of folding. The probable nature of one further level of folding was originally suggested by the appearance of specialized chromosomes—the *lampbrush chromosomes* in many oocytes and the *polytene chromosomes* of certain insect cells. We shall see that these two types of chromosomes seem to contain a series of **looped domains**—loops of chromatin that extend at an angle from the main chromosome axis. The *E. coli* bacterial chromosome—a circular DNA molecule about 0.1 cm in length and lacking histones—is also folded into a similar series of loops, suggesting that the presence of loops may be a general feature of chromosomes.

In principle, looped domains in chromatin could be established and maintained by DNA-binding proteins that clamp two regions of the 30-nm fiber together by recognizing specific DNA sequences that will form the neck of each loop (Figure 9–34A). Alternatively, they could be formed by the binding of the DNA at the base of the loop to a chromosome axis (Figure 9–34B). Organisms as different as flies and humans seem, from indirect estimates, to have looped domains of similar average size, with a typical loop containing about 20,000 to 100,000 nucleotide pairs—corresponding to an average length of about 0.5 μm of the 30-nm fiber. If organized predominantly into looped domains, a typical human chromosome might contain more than 2000 of them.

Mitotic Chromosomes Are Formed from Chromatin in Its Most Condensed State[22]

Most chromosomes are too extended, thin, and entangled for their loops to be readily detectable except during mitosis, when the chromosomes coil up to form much more condensed structures. This coiling, which reduces a 5-cm length of DNA to about 5 μm, is accompanied by the phosphorylation of all of the histone H1 in the cell at five of its serine residues. Because histone H1 helps to pack nucleosomes together (see Figure 9–28), its phosphorylation during mitosis seems likely to have a causal role in chromosome condensation (see p. 777).

Figure 9–35 depicts a typical chromosome at the metaphase stage of mitosis. The two daughter DNA molecules are separately folded to produce two sister chromosomes (called *sister chromatids*) held together at their *centromeres* (see p. 738). These chromosomes are normally covered with a variety of molecules, including large amounts of ribonucleoproteins (see p. 544). Once this covering is stripped away, each chromatid can be seen in electron micrographs to be organized into loops of chromatin emanating from a central axis (Figures 9–36 and 9–37). Several types of experiments demonstrate that the order of visible features on a mitotic chromosome at least roughly reflects the order of the genes along the DNA molecule. To illustrate how such an organized folding of the long DNA helix might be achieved, we have drawn each chromatid as a closely packed series of looped domains wound in a tight helix in Figure 9–38, which also presents a schematic view of all the different folding processes that would contribute to this structure.

The chromatin in mitotic chromosomes is transcriptionally inert: all RNA synthesis ceases as the chromosomes condense. Presumably the condensation prevents RNA polymerase from gaining access to the DNA, although other control factors might also be involved.

Each Mitotic Chromosome Contains a Characteristic Pattern of Very Large Domains[23]

The display of the 46 human chromosomes at mitosis is called the human **karyotype.** Cytological methods developed since 1970 permit unambiguous identification of each individual chromosome. Some of these methods involve staining mitotic chromosomes with dyes that fluoresce only when they bind to certain

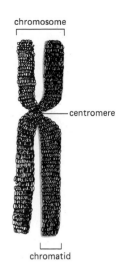

Figure 9–35 Drawing of a typical chromosome at the metaphase stage of mitosis. Each sister chromatid contains one of two identical daughter DNA molecules (one of which is colored here) generated earlier in the cell cycle by DNA replication.

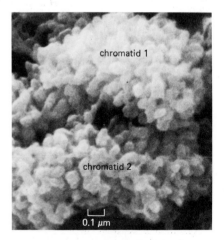

Figure 9–36 Scanning electron micrograph of a region near one end of a typical mitotic chromosome. Each knoblike projection is believed to represent the tip of a separate looped domain. Note that the two identical paired chromatids diagrammed in Figure 9–35 can be distinguished clearly. (From M.P. Marsden and U.K. Laemmli, *Cell* 17:849–858, 1979. © Cell Press.)

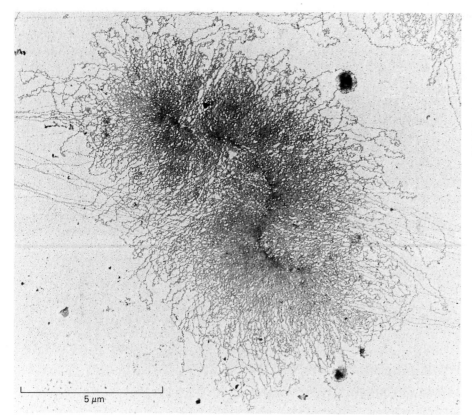

Figure 9–37 Electron micrograph of a single chromatid of a mitotic chromosome from an insect (*Oncopeltus*) treated to reveal loops of chromatin fibers that emanate from a central axis of the chromatid. Such micrographs have been used to support the idea that the chromatin in all chromosomes is folded into a series of looped domains. (Courtesy of Victoria Foe.)

5 μm

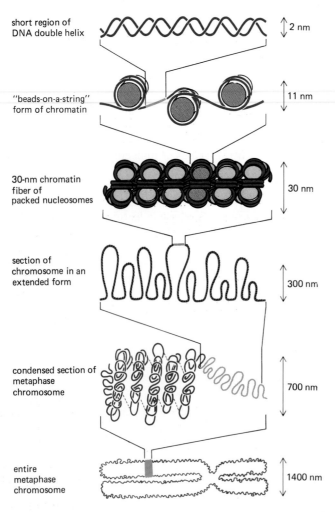

short region of
DNA double helix

2 nm

"beads-on-a-string"
form of chromatin

11 nm

30-nm chromatin
fiber of
packed nucleosomes

30 nm

section of
chromosome in an
extended form

300 nm

condensed section of
metaphase
chromosome

700 nm

entire
metaphase
chromosome

1400 nm

Figure 9–38 Schematic illustration of some of the many orders of chromatin packing postulated to give rise to the highly condensed metaphase chromosome.

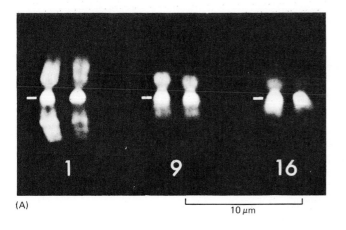

(A)

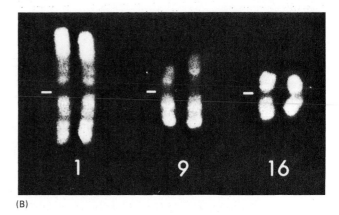

(B)

10 μm

types of DNA sequences. Although these dyes have very low specificity and appear mainly to distinguish DNA that is rich in A-T nucleotide pairs (**G bands**) from DNA that is rich in G-C nucleotide pairs (**R bands**), they produce a striking and reproducible pattern of bands along each mitotic chromosome (Figure 9–39). The pattern of bands on each type of chromosome is unique, allowing each chromosome to be identified and numbered, as illustrated in Figure 9–40.

By examining human chromosomes very early in mitosis, when they are less condensed than at metaphase, it has been possible to estimate that the total haploid genome contains about 2000 distinct A-T-rich bands, which progressively

Figure 9–39 Fluorescence micrographs of three pairs of human mitotic chromosomes stained with (A) the A-T-base-pair-specific dye Hoescht 33258, which stains the G bands, and (B) the G-C-base-pair-specific dye olivomycin, which stains the R bands. The bars indicate the position of the centromere. Note that these banding patterns are the reverse of each other, since the bands that are bright in (A) are dark in (B), and vice versa. The G bands also stain with Giemsa stain— hence their name—while the R bands were named by their "reverse" pattern to G bands. (From K.F. Jorgenson, J.H. van de Sande, and C.C. Lin, *Chromosoma* 68:287–302, 1978.)

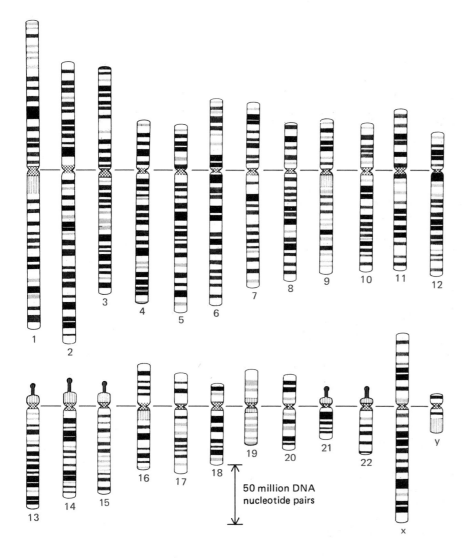

50 million DNA nucleotide pairs

Figure 9–40 A standard map of the banding pattern of each chromosome in the human karyotype, as determined at the prometaphase stage of mitosis. Chromosomes 1 through 22 are labeled in the approximate order of their size; a diploid cell contains two of each of these chromosomes plus two X chromosomes (female) or an X and a Y chromosome (male). The 850 bands shown here are G bands, which stain with reagents that appear to be specific for A-T-rich DNA sequences. The colored knobs on chromosomes 13, 14, 15, 21, and 22 indicate the positions of the genes that encode the large ribosomal RNAs; the colored lines mark the centromere on each chromosome. (Adapted from U. Franke, *Cytogenet. Cell Genet.* 31:24–32, 1981.)

coalesce as condensation proceeds during mitosis to produce fewer and thicker bands.

Chromosome bands are detected in mitotic chromosomes from species as diverse as humans and flies. Moreover, the exact pattern of bands in a chromosome has remained unchanged over long periods of evolutionary time. Every human chromosome, for example, has a clearly recognizable counterpart with a nearly identical banding pattern in the chromosomes of the chimpanzee, gorilla, and orangutan (although there has been a single chromosome fusion that gives humans 46 chromosomes instead of the apes' 48). This suggests that the long-range spatial organization of chromosomes may be important for cell function. But why such bands exist at all is a mystery. Even the thinnest of the bands diagrammed in Figure 9–40 should contain 30 or more looped domains, and the average nucleotide-pair composition would be expected to be random over such long stretches of DNA sequence (more than a million nucleotide pairs, or nearly the size of a bacterial genome). Both the A-T-rich bands and the G-C-rich bands are known to contain genes.

9-13 The DNA in Lampbrush Chromosomes Is Organized into a Series of Distinct Domains During Interphase[24]

Despite the higher-order packing of chromatin discussed on page 498, the interphase chromatin strands are generally too fine and too tangled to allow an entire interphase chromosome to be visualized clearly. The overall structure of chromosomes in interphase can, however, be discerned in a few special cell types. The meiotically paired chromosomes in growing oocytes (immature eggs), for example, are highly active in RNA synthesis, and they form unusually stiff and extended chromatin loops that are covered with newly transcribed RNA packed into dense RNA-protein complexes. Because of this coating on the DNA, these so-called **lampbrush chromosomes** are clearly visible even in the light microscope (Figure 9–41).

The organization of the lampbrush chromosome is illustrated schematically in Figure 9–42. Large decondensed chromatin loops extend outward from the chromosome axis. Nucleic acid hybridization experiments show that a given loop always contains the same DNA sequence and that it remains extended in the same manner as the oocyte grows. Thus these loops apparently correspond to fixed units of chromatin folding that have decondensed and become transcriptionally active. Since an average-sized loop can be estimated to contain roughly 100,000 nucleotide pairs of DNA, each may correspond to one of the looped domains of chromatin described previously (see p. 503). Many of the loops are transcribed

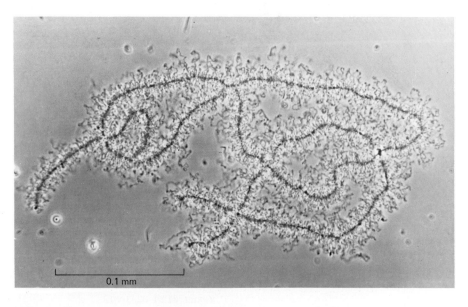

0.1 mm

Figure 9–41 Light micrograph of lampbrush chromosomes in an amphibian oocyte. Early in oocyte differentiation, each chromosome replicates to begin meiosis, and the homologous replicated chromosomes pair to form this highly extended structure containing a total of four chromatids. The lampbrush chromosome stage may persist for months or years as the oocyte builds up a supply of mRNA and other materials required for its ultimate development into a new individual (see p. 857). Note that each chromosome axis is about 400 μm long, whereas most mitotic chromosomes are less than 10 μm long. (Courtesy of Joseph G. Gall.)

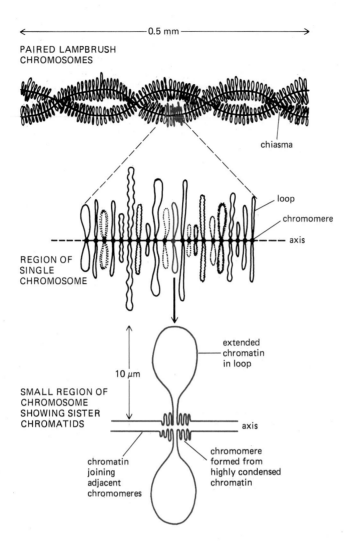

Figure 9–42 Lampbrush chromosome structure. The set of lampbrush chromosomes in many amphibians contains a total of about 10,000 chromatin loops, although most of the DNA remains highly condensed in the *chromomeres*. Each loop corresponds to a particular DNA sequence. Four copies of each loop are present in each cell, since the structure shown at the top consists of two paired homologous chromosomes and each chromosome is composed of two closely apposed sister chromatids. This four-stranded structure is characteristic of this stage of development of the oocyte (the diplotene stage of meiosis—see p. 852).

continuously from end to end; others, however, include an extended section of chromatin that is not transcribed at all. Most of the chromatin is not in loops and remains highly condensed in the *chromomeres* (see Figure 9–42); in general this chromatin is not transcribed. Short regions of chromatin that are neither highly condensed nor highly transcriptionally active connect the adjacent chromomeres along a well-defined chromosomal axis.

Lampbrush chromosomes are unusual in that the rate of transcription is higher and most of the RNA transcripts produced are longer than on most chromosomes. There is evidence, however, that the DNA molecule in other interphase chromosomes is similarly subdivided into distinct regions, each separated from its neighbors by a boundary that allows the chromatin in each region to be folded differently (for example, as loops, chromomeres, or the axial chromatin between chromomeres).

Orderly Domains of Interphase Chromatin Can Also Be Seen in Polytene Chromosomes[25]

Because of a different specialization, chromatin structure on the scale of single looped domains is also clearly visible in certain insect cells. Many of the cells of the larvae of flies grow to an enormous size through multiple cycles of DNA synthesis without cell division. The resulting giant cells contain as much as several thousand times the normal DNA complement. Cells with more than the normal DNA complement are said to be *polyploid* when, as is usually the case, they contain increased numbers of standard chromosomes. In several types of secretory cells of fly larvae, however, all the homologous chromosome copies remain side by side,

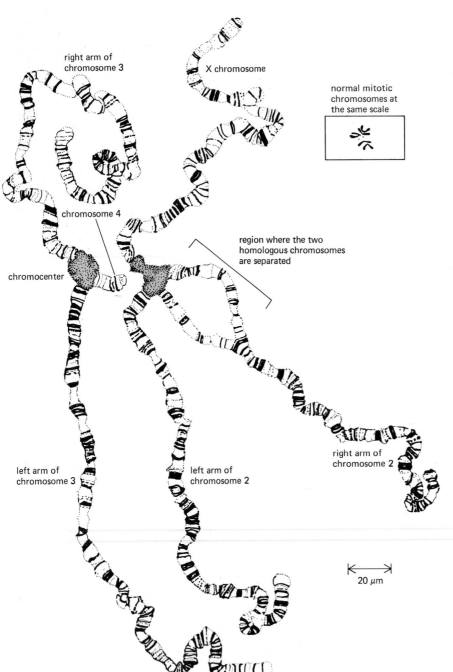

right arm of
chromosome 3

X chromosome

normal mitotic
chromosomes at
the same scale

chromosome 4

region where the two
homologous chromosomes
are separated

chromocenter

left arm of
chromosome 3

left arm of
chromosome 2

right arm of
chromosome 2

20 μm

Figure 9–43 A detailed sketch of the entire set of polytene chromosomes in one *Drosophila* salivary cell. These chromosomes have been spread out for viewing by squashing them against a microscope slide. There are four different chromosome pairs present. Each chromosome is tightly paired with its homolog (so that each pair appears as a single structure), which is not the case in most interphase nuclei. The four chromosome pairs are linked together by regions near their centromeres that have aggregated to create a single large "chromocenter" (*colored region*). The chromocenter has been split into two halves by the squashing procedure used. Note that the exact side-by-side alignment of many chromatin strands has greatly uncoiled each DNA molecule compared to its packing in a normal mitotic chromosome. (Modified from T.S. Painter, *J. Hered.* 25:465–476, 1934.)

10 μm

Figure 9–44 Light micrograph of a portion of a polytene chromosome from *Drosophila* salivary glands showing the distinct patterns recognizable in different chromosome bands. The bands are regions of increased chromatin concentration; they occur in interphase chromosomes and are a special property of the giant polytene chromosomes. They should not be confused with the much coarser bands diagrammed in Figure 9–40, which are revealed by special staining techniques on normal mitotic chromosomes. (Courtesy of Joseph G. Gall.)

creating a single giant *polytene chromosome*. The fact that some large insect cells can undergo a direct polytene-to-polyploid conversion demonstrates that these two chromosomal states are closely related and that the basic structure of a polytene chromosome must be similar to that of a normal chromosome.

Polytene chromosomes are easy to see in the light microscope because they are so large and because the precisely aligned side-to-side adherence of individual chromatin strands greatly elongates the chromosome axis and prevents tangling. Like lampbrush chromosomes, these are interphase chromosomes active in RNA synthesis. Polyteny has been most studied in the salivary gland cells of *Drosophila* larvae, in which the DNA in each of the four *Drosophila* chromosomes has been replicated through 10 cycles without separation of the daughter chromosomes, so that 1024 (= 2^{10}) identical strands of chromatin are lined up side by side.

When polytene chromosomes are viewed in the light microscope, distinct alternating dark *bands* and light bands (known as *interbands*) are visible (Figures 9–43 and 9–44). Each band and interband represents a set of 1024 identical DNA sequences arranged in register. About 85% of the DNA in polytene chromosomes is in bands, and 15% is in interbands. The chromatin in each band stains darkly because it is much more condensed than the chromatin in the interbands (Figure 9–45) and is thought to represent a looped domain that is highly folded, as illustrated schematically in Figure 9–46. Depending on their size, individual bands are estimated to contain 3000 to 300,000 nucleotide pairs per chromatin strand. Since the bands can be recognized by their different thicknesses and spacings, each one has been given a number to generate a polytene chromosome "map." There are approximately 5000 bands and 5000 interbands in the total *Drosophila* genome.

Individual Chromatin Domains in Polytene Chromosomes Can Unfold and Refold as a Unit[26]

Long before anything was known about chromatin structure, studies of polytene chromosomes suggested that a major change in DNA packing accompanies gene transcription, since individual chromosome bands often expand when the genes they contain become active and recondense when these genes become quiescent.

The regions on a polytene chromosome being transcribed at any instant can be identified by labeling the cells briefly with the radioactive RNA precursor ^{3}H-uridine and locating the growing RNA transcripts by autoradiography (Figure 9–47). This analysis reveals that the most active chromosomal regions are decondensed, forming distinctive **chromosome puffs.**

One of the main factors controlling the activity of genes in polytene chromosomes of *Drosophila* is the insect hormone *ecdysone*, the levels of which rise and fall periodically during larval development, inducing the transcription of various genes coding for proteins that the larva requires for each molt and for pupation. As the organism progresses from one developmental stage to another, new puffs arise and old puffs recede as transcription units are activated and deactivated, and different mRNAs and proteins are made (Figure 9–48). From inspection of each puff when it is relatively small and the banding pattern of the chromosome is still discernable, it seems that most puffs arise from the uncoiling of a single chromosome band (Figure 9–49). Electron microscopy of thin sections of such puffs (Figure 9–50) shows that the DNA in the chromatin is much less condensed than it would be in the 30-nm chromatin fiber. These observations suggest that a looped domain, which is thought to be folded to form a chromosome band (see Figure 9–46), can decondense as a unit during transcription.

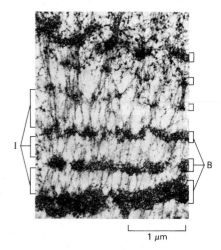

Figure 9–45 Electron micrograph of a small section of a *Drosophila* polytene chromosome, seen in thin section. Bands of very different thickness can be readily distinguished (B), separated by interbands (I), which contain less condensed chromatin. (Courtesy of Viekko Sorsa.)

region of single chromosome showing looped domains

MULTIPLE CYCLES OF DNA REPLICATION PRODUCE NUMEROUS ALIGNED SINGLE CHROMOSOMES

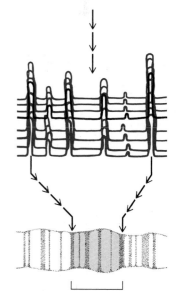

6 of the 5000 bands of the *Drosophila* polytene chromosomes (formed from the loops of 1024 single chromosomes packed side to side)

Figure 9–46 Diagram indicating how the bands seen on polytene chromosomes are thought to be generated by the side to-side packing of homologous looped domains. Within each band the loop of 30-nm chromatin fiber appears to be folded tightly to form a more condensed structure than is indicated here.

Both Bands and Interbands in Polytene Chromosomes Are Likely to Contain Genes[27]

The fixed pattern of bands and interbands in a *Drosophila* polytene chromosome suggested to early cytogeneticists that each band might correspond to a single gene. Consistent with this view, mutational analyses have allowed geneticists to estimate that *Drosophila* contains only about 5000 essential genes, a number that corresponds roughly to the number of chromosome bands. When an intensive effort was made to isolate as many mutants as possible in a small chromosomal region, for example, about 50 essential genes were genetically identified in a region that contains about 50 visible bands. Although it is not possible with these techniques to determine whether a particular gene lies in a band or an interband region, these observations suggested that an average band might contain the DNA coding sequences for only one essential protein.

More recent data, however, have made the simple "one-band, one-gene" hypothesis seem unlikely. For example, a continuous 315,000-nucleotide-pair region of the *Drosophila* genome has been cloned, and fragments have been used as DNA probes to catalog the mRNAs produced from the region. Three times as many separate mRNAs as bands were identified. While most of these mRNAs are likely to represent a gene that confers a selective advantage to flies in the wild, mutations in many of them may be missed in standard genetic tests because the loss of the gene has little or no effect on flies grown under laboratory conditions, where predators are absent and both food and a mate are provided. It has also been shown recently that mRNAs are produced from both interbands and bands.

Although the divisions in chromosomes can be seen most clearly in unusual interphase states (the lampbrush and polytene chromosomes), all higher eucaryotic chromosomes are probably subdivided in a similar way. The studies with *Drosophila* polytene chromosomes suggest that genes can be located in both the condensed chromatin in bands and in the less condensed chromatin in interbands and that each band can contain several genes. Why, then, is the single long strand of chromatin in each chromosome subdivided into so many distinct regions? Although the answer is not known, it is possible that this type of organization helps to (1) keep the DNA organized, (2) isolate genes from their neighbors and thereby prevent biological "crosstalk," or (3) regulate gene transcription (for example, constitutively expressed, "housekeeping" genes could be located in interbands, whereas cell-type-specific genes could be confined to bands).

Figure 9–47 Synthesis of RNA along a giant polytene chromosome from the salivary gland of the insect *Chironomus tentans*. In this autoradiograph of a chromosome labeled with ³H-uridine, sites of RNA synthesis are covered with dark silver grains in proportion to their activity. (From C. Pelling, *Chromosoma* 15:71–122, 1964.)

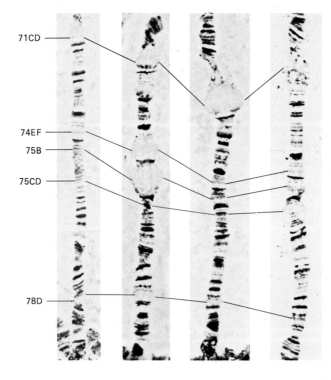

71CD

74EF

75B

75CD

78D

Figure 9–48 A temporal series of photographs illustrating how puffs arise and recede in the polytene chromosomes of *Drosophila melanogaster* during larval development. A region of the left arm of chromosome 3 is shown, which exhibits five very large puffs in salivary gland cells, each active for only a short developmental period. The series of changes shown occur over a period of 22 hours, appearing in a reproducible pattern as the organism develops. (Courtesy of Michael Ashburner.)

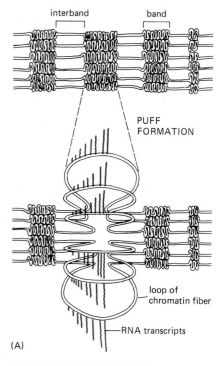

(A)

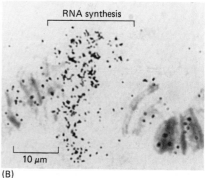

RNA synthesis

10 μm

(B)

Figure 9–49 (A) A highly schematic view of the process of puff formation in a polytene chromosome. (B) Autoradiogram of a single puff, in which the indicated portion is undergoing RNA synthesis and has therefore become labeled with ³H-uridine. (B, courtesy of Jose Bonner.)

9-14 Chromatin Is Less Condensed in Regions That Are Transcriptionally Active[28]

Treatment of isolated nuclei from vertebrate cells with DNase I (the same nuclease that at lower concentrations reveals the nuclease-hypersensitive sites lacking nucleosomes—see p. 498) provides evidence that the chromatin of active genes is usually selectively decondensed—extending the observations made on the puffs of insect polytene chromosomes. At an appropriate concentration of DNase I, about 10% of a vertebrate genome is preferentially degraded. Different DNA sequences are digested by the nuclease in different cells of the same organism, in a pattern corresponding to the different pattern of RNAs that the cells make: hybridization studies with specific DNA probes (see p. 188) show that the degraded sequences correspond mainly to the regions of the genome that are transcribed in the particular cell type tested. Remarkably, even genes that are transcribed only a few times in every cell generation are nuclease sensitive, indicating that a special state of the chromatin, rather than the direct process of DNA transcription itself, makes these regions unusually accessible to digestion (Figure 9–51). Chromatin in this accessible state is often termed **active chromatin,** and its nucleosomes are thought to be altered in a way that makes their packing less condensed. Even if the nuclease digestion is carried out on mitotic chromosomes, the active chromatin has been reported to be preferentially digested, suggesting that some difference between active and inactive chromatin still remains at mitosis, when the entire genome is unusually highly folded.

Usually no difference can be detected in the nucleosomes of active chromatin in the electron microscope. A striking exception, however, is found in certain insect embryos that display conspicuous tandem arrays of apparently unbeaded chromatin in the region of activated rRNA genes (Figure 9–52). Because these genes remain covered by protein, it has been suggested that they represent regions of the chromatin where each nucleosome has opened up into an extended conformation in order to facilitate transcription by unwrapping the two coils of DNA helix that are normally wound around its core.

9-15 Active Chromatin Is Biochemically Distinct[29]

An understanding of the differently condensed forms of chromatin observed in cells will require purification and characterization of the chromosomal proteins that are unique to each chromatin conformation. Some progress has been made toward this goal with respect to active chromatin. By separating chromatin into fractions that have different solubility or binding properties and monitoring them by hybridization with appropriate DNA probes, one can enrich for active chro-

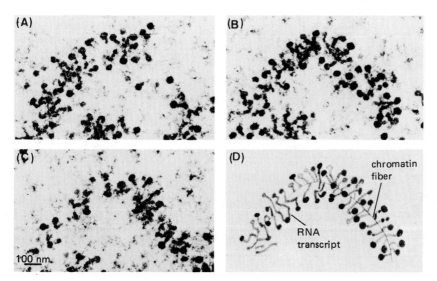

(A)

(B)

(C)

(D)

chromatin fiber

RNA transcript

100 nm

Figure 9–50 Electron micrographs of serial thin sections through a large chromosome puff, showing the conformation of a prominent, unusually long transcription unit. A three-dimensional reconstruction of part of the transcription unit is shown in (D), revealing RNA transcripts with knobbed ends attached to a single chromatin fiber. The knobs near the 5′ ends of most RNA transcripts are due to the packaging of the RNA molecule into ribonucleoprotein particles (see p. 532). (From K. Andersson, B. Björkroth, and B. Daneholt, *Exp. Cell Res.* 130:313–326, 1980.)

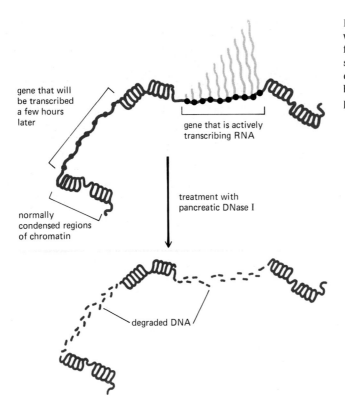

Figure 9–51 Digestion of chromatin with pancreatic DNase I. The enzyme first cuts at nuclease-hypersensitive sites (not shown) and then selectively degrades the entire DNA sequence of both actively transcribing genes and potentially active genes.

matin and then analyze the chromosomal proteins in the enriched fractions. In this way the following observations have been made. (1) Histone H1 seems to be less tightly bound to at least some active chromatin, and particular subtypes of this histone may be preferentially represented. (2) Although the four nucleosomal histones are present in normal amounts in active chromatin, they appear to be unusually highly acetylated on lysines near their amino terminus. The acetyl groups (whose addition removes a plus charge—see Figure 8–15, p. 417) are constantly being added to these histones by the enzyme *histone acetylase* and removed by *histone deacetylase*, and each acetyl group persists on average for only about 10 minutes. (3) The nucleosomes in active chromatin selectively bind two closely related small chromosomal proteins, HMG 14 and HMG 17. Consistent with their presence only in active chromatin, these "high-mobility-group (HMG) proteins" are present at about the level required to bind to 1 in every 10 nucleosomes in total chromatin; moreover, both of their amino acid sequences have been highly conserved during evolution—implying that they have important functions. (4) In at least some organisms, such as the ciliate *Tetrahymena*, active chromatin is highly enriched in a minor variant form of histone H2A. This histone variant, called hv1 in *Tetrahymena*, has also been identified in *Drosophila*, chickens, and humans.

Any or all of these changes might play an important part in uncoiling the chromatin of active genes, helping to make the DNA available as a template for

Figure 9–52 Electron micrograph of spread chromatin from an insect embryo (*Oncopeltus*), showing a visible change in the chromatin structure of two tandemly arranged rRNA genes. A change from a beaded to an unbeaded form of chromatin appears to precede the initiation of rRNA synthesis in these embryos. The unbeaded chromatin is thought to represent a region where the DNA has uncoiled from around the nucleosome beads in preparation for transcription. Note that the chromatin structure has been changed in the gene on the left even though it is not presently being transcribed. (From V. E. Foe, *Cold Spring Harbor Symp. Quant. Biol.* 42:723–740, 1978.)

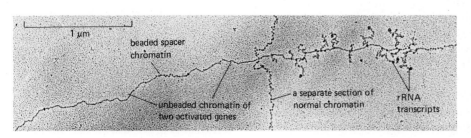

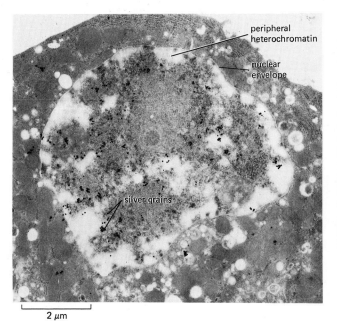

peripheral
heterochromatin

nuclear
envelope

silver grains

2 μm

Figure 9–53 Autoradiograph of a thin section of a cell nucleus from a cell that has been pulse-labeled with ³H-uridine in order to label sites of RNA synthesis (silver grains). The white areas are regions of heterochromatin, which tend to pack along the inside of the nuclear envelope. Normally, heterochromatin stains unusually darkly in electron micrographs (for example, see Figure 9–100); but because of the particular method used for sample preparation here, areas containing heterochromatin have been bleached. As can be seen, most of the RNA synthesis occurs in the euchromatin that borders these heterochromatic areas. (Courtesy of Stan Fakan.)

RNA synthesis, but more direct experiments are needed to test this idea. New techniques for purifying active chromatin, which exploit its selective binding to an affinity matrix containing an antibody that is specific for either an HMG protein or acetylated lysine (see p. 168), should facilitate future biochemical studies. In Chapter 10 we discuss mechanisms by which the formation of active chromatin may be controlled (see p. 582).

Heterochromatin Is Highly Condensed and Transcriptionally Inactive[30]

Evidence for a second unusual form of chromatin was obtained in the 1930s when light microscopic observations on interphase nuclei of higher eucaryotic cells distinguished a specially condensed form of chromatin called **heterochromatin** (the rest of the chromatin, which was less condensed, was called *euchromatin*). Heterochromatin was originally identified because it remains unusually compact during interphase, apparently maintaining the type of structure adopted by most chromatin only during mitosis. It was later found that, like mitotic chromatin, heterochromatin is transcriptionally inactive (Figure 9–53). In most cells about 90% of the chromatin is thought to be transcriptionally inactive. While its relative resistance to nuclease digestion shows that this chromatin has a more condensed conformation than the 10% of the chromatin in transcriptionally active regions, only a fraction of it (perhaps 10% to 20%) is packed in the highly condensed conformation known as heterochromatin. Heterochromatin is therefore thought to be a special class of transcriptionally inactive chromatin that can have specific functions (see p. 577). In mammals and many other higher eucaryotes, the DNA surrounding each centromere is composed of relatively simple, repeating nucleotide sequences; these "satellite DNAs" (see p. 604) constitute a major portion of the heterochromatin in these organisms.

In *Drosophila*, both the centromeric chromatin and occasional short regions along the chromosome arms are organized as heterochromatin. These heterochromatic regions fail to replicate well during the early cycles of DNA synthesis that give rise to the polytene chromosomes, and they are therefore underrepresented in these chromosomes. Their distinctive nature can be demonstrated at a molecular level by staining with an antibody that detects a chromosomal protein found only in heterochromatin (Figure 9–54). The recent cloning of the gene that encodes this protein represents the beginning of a biochemical analysis of heterochromatin, whose structure is otherwise unknown.

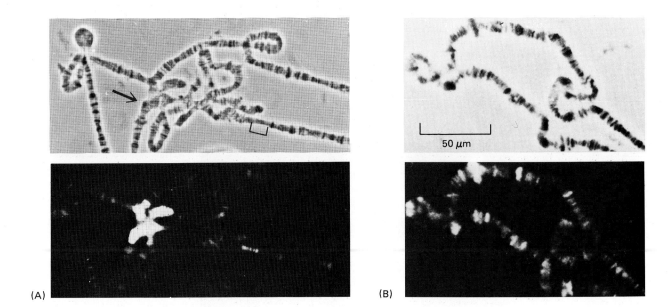

(A)   (B)

Figure 9–54 The same region of a *Drosophila* polytene chromosome viewed by phase-contrast (*top*) and fluorescence (*bottom*) microscopy after immunofluorescence labeling of either heterochromatin (*left*) or active chromatin (*right*). (A) Staining with a monoclonal antibody that detects a protein confined to heterochromatic regions; the chromocenter (*arrow*) and a few selected bands are stained. (B) Staining with a monoclonal antibody that detects a protein enriched in active chromatin; a subset of the bands are stained. (A, from T.C. James and S.C.R. Elgin, *Mol. Cell Biol.* 6:3862–3872, 1986; B, from G.C. Howard, S.M. Abmayr, L.A. Shinefeld, V.L. Sato, and S.C.R. Elgin, *J. Cell Biol.* 88:219–225, 1981. © 1981 Rockefeller University Press.)

Summary

All chromosomes adopt a highly condensed conformation during mitosis. When they are specially stained, these mitotic chromosomes have a banded structure that allows each individual chromosome to be recognized unambiguously; these bands contain millions of DNA nucleotide pairs, and they reflect a coarse heterogeneity of chromosome structure that is not understood.

Chromosomes generally become sufficiently decondensed during interphase that their structure is largely masked. Notable exceptions are the specialized lampbrush chromosomes of vertebrate oocytes and the polytene chromosomes of giant insect secretory cells. Studies of these two types of interphase chromosomes suggest that each long DNA molecule in a chromosome is divided into a large number of discrete domains that are folded differently. A major feature is the presence of many looped domains of chromatin, most containing 20,000 to 100,000 DNA nucleotide pairs. In both lampbrush and polytene chromosomes the regions that are actively synthesizing RNA are least condensed. Likewise, as judged by nuclease sensitivity, about 10% of the DNA in interphase vertebrate cells is in a relatively decondensed conformation that correlates with DNA transcription in these regions. Such "active chromatin" is biochemically distinct from the more condensed regions of chromatin.

Chromosome Replication

Before a cell can divide, it must produce a new copy of each of its chromosomes. Thus cell division is preceded by a transition from the normal interphase state in which most cells reside, the G_1 phase, to a special "DNA-synthesis phase" called the **S phase.** In a typical higher eucaryotic cell the S phase lasts for about 8 hours. By its end each chromosome has been copied to produce two complete chromosomes, which remain joined together at their centromeres until the M phase that soon follows (see Figure 9–35). Chromosome duplication requires both the replication of the long DNA molecule in each chromosome and the assembly of a new set of chromosomal proteins onto the DNA to form chromatin. In Chapter 5 we discussed the enzymology of DNA replication and described the structure of the *DNA replication fork,* where DNA synthesis occurs by a semiconservative process (see Figure 5–39, p. 229). In Chapter 13 we consider how entry into the S phase is regulated as part of a more general discussion of cell-cycle control (see p. 732). In this section we describe the timing and pattern of eucaryotic chromosome replication and its relation to chromosome structure.

Replication Origins Are Activated in Clusters
on Higher Eucaryotic Chromosomes[31]

We have previously described how **DNA replication forks** are initiated in bacteria at special DNA sequences called *replication origins* (see p. 235). Two forks begin at each origin and proceed in opposite directions; each moves away from the origin at a rate of about 500 nucleotides per second until all of the DNA in the circular bacterial chromosome is replicated. The bacterial genome is so small that these two replication forks can duplicate it in less than 40 minutes.

A method for determining the general pattern of eucaryotic chromosome replication was developed in the early 1960s. Human cells growing in culture are labeled for a short time with ^{3}H-thymidine so that the DNA synthesized during this period becomes highly radioactive. The cells are then gently lysed and the DNA is streaked onto the surface of a glass slide, which is coated with a photographic emulsion so that the pattern of labeled DNA can be determined by autoradiography. Since the time allotted for radioactive labeling is chosen to allow each replication fork to move several micrometers along the DNA, the replicated DNA can be detected as lines of silver grains in the light microscope, even though the DNA molecule itself is too thin to be visible. In this way both the rate and the direction of replication fork movement can be determined (Figure 9–55). From the rate at which tracks of replicated DNA increase in length with increasing labeling time, the replication forks are estimated to travel at a speed of about 50 nucleotides per second. This is one-tenth the rate at which bacterial replication forks move, possibly reflecting the increased difficulty of replicating DNA packaged in chromatin.

As discussed previously, an average human chromosome is thought to be composed of a single DNA molecule containing about 150 million nucleotide pairs. To replicate such a DNA molecule from end to end with a single replication fork moving at a rate of 50 nucleotides per second would require $0.02 \times 150 \times 10^6 = 3.0 \times 10^6$ seconds (about 800 hours). As expected, therefore, the autoradiographic experiments just described reveal that many forks are moving simultaneously on each eucaryotic chromosome. Moreover, many forks are often found close together in the same DNA region, while other regions of the same chromosome have none. Further experiments of this type have shown the following. (1) Replication origins tend to be activated in clusters (called **replication units**) of perhaps 20 to 80 origins. (2) New replication units seem to be activated throughout the S phase until all of the DNA is replicated. (3) Within a replication unit, individual origins are spaced at intervals of 30,000 to 300,000 nucleotide pairs from one another; thus, on average, there may be only about one replication origin per looped domain of chromatin. (4) As in bacteria, replication forks are formed in pairs and create a *replication bubble* as they move in opposite directions away from a common point of origin, stopping only when they collide head-on with a replication

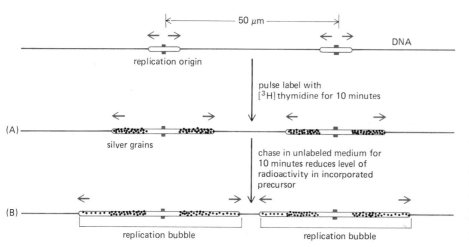

Figure 9–55 The experiments that demonstrated the pattern in which replication forks move during the S phase. The new DNA made in human cells in culture was briefly labeled with a pulse of highly radioactive thymidine (^{3}H-thymidine). In the experiment illustrated in (A), the cells were lysed and the DNA was stretched out on a glass slide that was subsequently covered with a photographic emulsion. After several months the emulsion was developed, revealing a line of silver grains over the radioactive DNA. The experiment in (B) was the same except that a further incubation in unlabeled medium allowed additional DNA, with a lower level of radioactivity, to be replicated. The pairs of dark tracks in (B) were found to have silver grains tapering off in opposite directions, demonstrating bidirectional fork movement from a central replication origin (see Figure 5–49, p. 235). The colored DNA in this figure is shown only to help with the interpretation of the autoradiograph; the unlabeled DNA is invisible in such experiments.

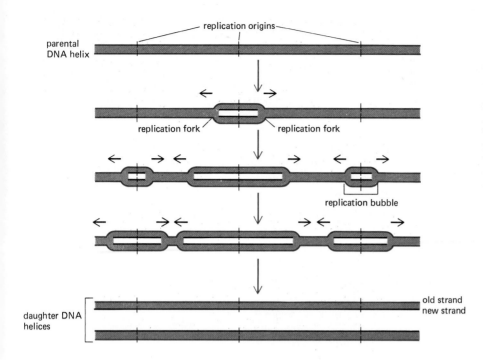

Figure 9–56 The pattern by which DNA is replicated in eucaryotic chromosomes as initially determined by the type of experiment described in Figure 9–53. The replication origins are spaced at intervals of 30,000 to 300,000 nucleotide pairs in most cells. A replication fork is thought to stop only when it encounters a replication fork moving in the opposite direction or when it reaches the end of the chromosome; in this way all the DNA is eventually replicated.

fork moving in the opposite direction (or reach a chromosome end). In this way many replication forks can operate independently on each chromosome and yet form two complete daughter DNA helices (Figure 9–56).

9-19 Specific DNA Sequences Serve as Replication Origins[32]
9-21

Replication origins have been precisely defined as particular DNA sequences in bacteria and several animal viruses. By analogy, one would expect the replication origins in eucaryotic chromosomes to be specific DNA sequences too. The search for replication origins in the chromosomes of eucaryotic cells has been most successful in the yeast *Saccharomyces cerevisiae*, where powerful selection methods have been devised to find them. These methods make use of yeast cells that are defective for an essential gene, so that they will survive in a selective medium only if they are provided with a plasmid carrying a functional copy of the missing gene. If a bacterial plasmid that carries the essential yeast gene is transfected into the defective yeast cells, it is unable to replicate free of the host chromosome because the bacterial replication origin in the plasmid DNA cannot initiate replication in a yeast cell. If random pieces of yeast DNA are inserted into the plasmid before transfection of the yeast cells, however, a small proportion of the plasmid DNA molecules will contain a yeast replication origin and thereby will be able to replicate. The yeast cells that carry such plasmids will be able to proliferate since they will have been provided with the essential gene (Figure 9–57). The DNA sequences identified by their presence in the plasmids isolated from these cells, called *autonomously replicating sequences (ARS)*, are present in the yeast genome at a frequency similar to that expected for yeast replication origins (about once every 40,000 nucleotide pairs).

An 11-nucleotide "core consensus sequence" [(A or T)TTTAT(A or G)TTT(A or T)] has been shown to be essential for yeast ARS function. All ARSs appear to be composed of multiple near matches to this sequence spaced throughout a region of about 100 DNA nucleotides. An artificially constructed tandem array of core sequences is fully active as an ARS in the yeast plasmid assay, suggesting that an initiator protein recognizes closely spaced copies of this sequence to begin DNA replication in a yeast cell. Specific DNA sequences that seem to function as replication origins in higher eucaryotic cells have been tentatively identified, but they have not yet been well characterized.

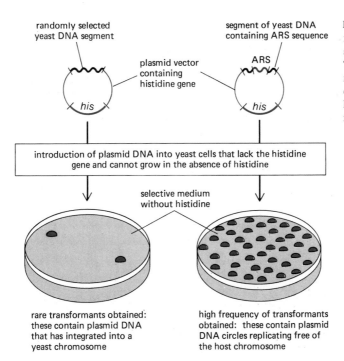

randomly selected
yeast DNA segment

segment of yeast DNA
containing ARS sequence

plasmid vector
containing
histidine gene

ARS

his

his

introduction of plasmid DNA into yeast cells that lack the histidine
gene and cannot grow in the absence of histidine

selective medium
without histidine

rare transformants obtained:
these contain plasmid DNA
that has integrated into a
yeast chromosome

high frequency of transformants
obtained: these contain plasmid
DNA circles replicating free of
the host chromosome

Figure 9–57 The strategy used to identify autonomously replicating sequences (ARS elements) in yeasts. These sequences act as origins of replication and enable plasmids that contain them to replicate freely in the host cell without having to be incorporated into its chromosomes.

9-20 A Mammalian Cell-free System Replicates the Chromosome of SV40 Virus[33]

In procaryotic replication systems a multienzyme complex that includes DNA polymerase, DNA primase, and DNA helicase moves the replication fork (see p. 232); this complex is assembled at a replication origin in a reaction involving an initiator protein that binds specifically to the origin DNA (see Figure 5–50, p. 235).

Many attempts have been made to reconstruct mammalian DNA replication in a test tube, so that all the enzymes required can be identified and their actions at a replication fork defined. The most successful *in vitro* system developed thus far replicates the small circular chromosome of the monkey virus, SV40. This virus parasitizes the host cell for all but one of the replication proteins the virus needs to replicate; the one exception is the *SV40 T-antigen*, a large multifunction protein (90,000 daltons) that allows the virus to avoid the normal DNA-re-replication block (see p. 522) and thereby to replicate faster than the host cell. Multiple copies of the **T-antigen** bind specifically to the SV40 replication origin and act both as an initiator protein and as a DNA helicase to open the DNA helix at that site. A complex of DNA polymerase (polymerase alpha) and DNA primase, both provided by the host cell, then binds to a single strand of DNA exposed at the origin to begin DNA synthesis. Although the replication fork formed is similar to a procaryotic replication fork (see p. 229), the SV40 experiments suggest that two distinct types of DNA polymerase are needed in eucaryotes: *DNA polymerase alpha* on the lagging strand and *DNA polymerase delta* on the leading strand. In procaryotes, by contrast, separate copies of a single type of DNA polymerase function on each side of the replication fork (see Figure 5–47, p. 233).

The protein that initiates replication on eucaryotic-cell chromosomes has not yet been discovered. It is therefore not known whether it resembles the T-antigen in functioning as a DNA helicase at the fork, or whether—as in procaryotes—a separate initiator protein and DNA helicase are involved.

New Histones Are Assembled into Chromatin as DNA Replicates[34]

A large amount of new histone, approximately equal in mass to the newly synthesized DNA, is required to make new chromatin in each cell cycle. For this reason most organisms possess multiple copies of the gene for each histone.

Vertebrate cells, for example, have about 20 repeated sets, each set containing all five histone genes.

Unlike most proteins, which are made continuously throughout interphase (see p. 731), the histones are synthesized mainly in the S phase, during which the level of histone mRNA increases about fiftyfold as a result of both increased transcription and decreased mRNA degradation. Because of their unusual 3' ends (see p. 596), the major histone mRNAs become unusually unstable and are degraded within minutes when DNA synthesis stops at the end of S phase (or when inhibitors are added to stop DNA synthesis prematurely). In contrast, the histone molecules themselves are remarkably stable and may survive for the entire life of a cell. The tight linkage between DNA synthesis and histone synthesis may be due, at least in part, to a feedback mechanism that monitors the level of free histone to ensure that the amount of histone made is appropriate for the amount of new DNA synthesized.

Once they are assembled in nucleosomes, histone molecules rarely, if ever, leave the DNA to which they are bound. As a replication fork advances, therefore, it must somehow pass through the parental nucleosomes, which seem to have been designed to allow both transcription and replication to proceed past them (see also p. 500). According to one hypothesis, each nucleosome transiently unfolds into two "half-nucleosomes" during DNA replication, thereby allowing the DNA polymerase to copy the uncoiled nucleosomal DNA (Figure 9–58).

The newly synthesized DNA behind a replication fork inherits the old histones, but it also needs to bind an equal amount of new histones to complete its packaging into chromatin (see Figure 9–58). These new histones are normally deposited in the first few minutes after the replication fork passes. Newly formed chromatin, however, requires as much as an hour to become fully mature; until then, for example, the new histones are more susceptible than usual to histone-modifying enzymes. The chemical differences between mature and immature chromatin are unknown, although it is thought that maturation involves both covalent modifications of histones and changes in the binding of other proteins to the chromatin.

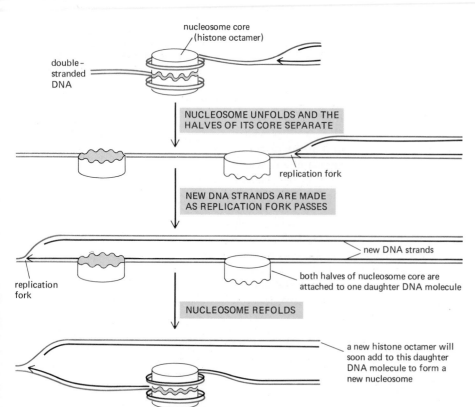

Figure 9–58 Speculative model showing how a nucleosome might open up to permit DNA replication and then reassemble as the replication fork passes, allowing the histones of the nucleosome core to remain permanently bound to the DNA. Although in this diagram the old nucleosome has been inherited intact by the DNA helix made on the leading strand, its actual pattern of inheritance is disputed. This is only one of many possible models.

nucleosome core
(histone octamer)

double-
stranded
DNA

NUCLEOSOME UNFOLDS AND THE
HALVES OF ITS CORE SEPARATE

replication fork

NEW DNA STRANDS ARE MADE
AS REPLICATION FORK PASSES

new DNA strands

both halves of nucleosome core are
attached to one daughter DNA molecule

replication
fork

NUCLEOSOME REFOLDS

a new histone octamer will
soon add to this daughter
DNA molecule to form a
new nucleosome

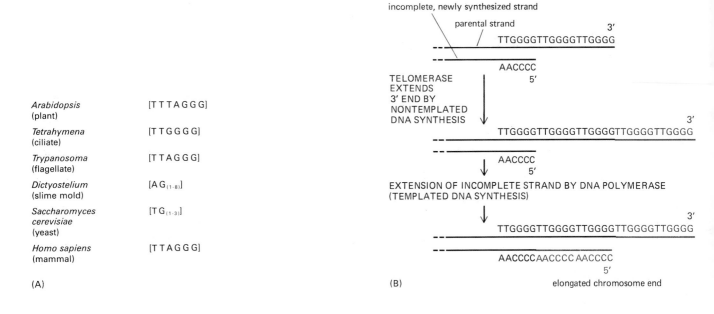

		incomplete, newly synthesized strand

parental strand

3′
TTGGGGTTGGGGTTGGGG

AACCCC
5′

TELOMERASE
EXTENDS
3′ END BY
NONTEMPLATED
DNA SYNTHESIS

3′
TTGGGGTTGGGGTTGGGGTTGGGGTTGGGG

AACCCC
5′

EXTENSION OF INCOMPLETE STRAND BY DNA POLYMERASE
(TEMPLATED DNA SYNTHESIS)

3′
TTGGGGTTGGGGTTGGGGTTGGGGTTGGGG

AACCCCAACCCC AACCCC
5′

elongated chromosome end

Arabidopsis (plant)	[T T T A G G G]
Tetrahymena (ciliate)	[T T G G G G]
Trypanosoma (flagellate)	[T T A G G G]
Dictyostelium (slime mold)	[A G$_{(1-8)}$]
Saccharomyces cerevisiae (yeast)	[T G$_{(1-3)}$]
Homo sapiens (mammal)	[T T A G G G]

(A)

(B)

9-5 Telomeres Consist of Short G-rich Repeats That Are Added to Chromosome Ends[35]

We have already described how the inability to replicate completely the end of a linear DNA molecule with DNA polymerase has led to the evolution of special DNA sequences, called **telomeres** (see p. 484), at the ends of eucaryotic chromosomes. These sequences are similar in organisms as diverse as protozoa, fungi, plants, and mammals. They consist of many tandem repeats of a short sequence that contains a block of neighboring G nucleotides (Figure 9–59A). The G-rich strand of the telomere always forms the 3′ end of the DNA molecule, and it seems to fold to form a special structure that protects the chromosome end. The mechanism of DNA replication at a telomere, deduced from studies of the ciliated protozoa *Tetrahymena*, is outlined in Figure 9–59B.

Different Regions on the Same Chromosome Replicate at Distinct Times During S Phase[36]

The replication of DNA in the region between one replication origin and the next should normally require only about an hour to complete, given the rate at which a replication fork moves and the largest distances measured between the replication origins in a replication unit. Yet the S phase usually lasts for about 8 hours in a mammalian cell. This implies that the replication origins are not all activated simultaneously and that the DNA in each *replication unit* (which contains a cluster of perhaps 20 to 80 replication origins) is replicating for only a small part of the total S-phase interval.

Are different replication units activated at random, or is there a definite order in which different regions of the genome are replicated? One way to answer this question is to use the thymidine analogue 5-bromodeoxyuridine (BrdU) to label synchronized cell populations for different short periods throughout the S phase. Later, in M phase, those regions of the mitotic chromosomes that have incorporated BrdU into their DNA can be recognized by their decreased staining with certain dyes or by means of anti-BrdU antibodies. Such experiments show that chromosomal regions are replicated in large units (as one would expect from the clusters of replication forks seen in DNA autoradiographs), and that different regions of each chromosome are replicated in a reproducible order during S phase (Figure 9–60).

Figure 9–59 (A) The repeating G-rich sequences that form the telomeres of diverse eucaryotic organisms. (B) Outline of the reactions involved in the formation of telomeres, as suggested by experiments in *Tetrahymena*. The incomplete, newly synthesized strand is that made on the lagging side of a replication fork (see Figure 5–39, p. 229). The *Tetrahymena telomerase* recognizes a structural feature unique to the G-rich sequence in binding to the telomere. Because it lacks a template and uses nucleoside triphosphates as substrates, the nucleotide at the 3′ end must inform the telomerase which nucleotide (G or T) to add at each step. The reactions that shorten and restore the telomere sequence are not precisely balanced, so each chromosome end contains a variable number of the repeats, which generally extend for hundreds of nucleotide pairs.

Highly Condensed Chromatin Replicates Late in S Phase[37]

In higher eucaryotic cells, some regions of the DNA are more condensed than others. We have seen, for example, that *heterochromatin* remains in a highly condensed conformation (similar to that at mitosis) during interphase, while *active chromatin* assumes an especially decondensed conformation, which is apparently required to allow RNA synthesis (see p. 580).

One important clue to the mechanism that determines the timing of DNA replication is the observation that the blocks of heterochromatin, including the regions near the centromere that remain condensed throughout interphase, are replicated very late in the S phase. Late replication could thus be related to the packing of the DNA in chromatin. This conclusion is supported by the timing of replication of the two X chromosomes in a female mammalian cell. While these two chromosomes contain essentially the same DNA sequences, one is active and the other is not (see p. 577). Nearly all the inactive X chromosome is condensed into heterochromatin and its DNA replicates late in the S phase, whereas its active homologue is less condensed and replicates throughout S phase. It therefore seems that those regions of the genome whose chromatin is least condensed during interphase, and therefore most accessible to the replication machinery, are replicated first.

Autoradiography shows that replication forks move at comparable rates throughout S phase, so that the extent of chromosome condensation does not appear to influence replication forks once they have formed. It seems, however, that the order in which replication origins are activated depends, at least in part, on the chromatin structure in which the origins reside.

Genes in Active Chromatin Replicate Early in S Phase[38]

The suggested relationship between chromatin structure and the time of DNA replication is supported by studies in which the replication times of specific genes are measured. In these studies a growing cell population is briefly labeled with a pulse of BrdU, and the cells are immediately separated by centrifugation according to size. Because cells grow as they proceed through the cell cycle, the larger cells will be "older" and their DNA will therefore have been labeled later in S phase. For each size class of cells, the BrdU-labeled DNA is isolated and analyzed by hybridization with a series of specific DNA probes for the genes it contains. (Because DNA containing BrdU is denser, it can be easily separated from normal DNA by sedimentation to equilibrium in a cesium chloride density gradient—see p. 165.)

This method gives the replication time for any gene for which a DNA probe is available. The results show that so-called "housekeeping" genes, which are those active in all cells (see p. 585), replicate very early in S phase in all cells tested. Genes that are active in only a few cell types, in contrast, generally replicate early in the cells in which they are active and later in other types of cells. When a continuous stretch of 300,000 nucleotide pairs of an immunoglobulin gene was studied in this way, for example, all regions of its chromatin completed their replication near the beginning of S phase in cells in which the gene was active, suggesting that there are several replication origins within the gene and that these are all activated at about the same time. When replication times were measured with the same DNA probes in cells where no immunoglobulin is made, a single replication fork appeared to enter from one end of this chromosomal region about an hour after the start of S phase and then move steadily across the DNA at the expected rate of about 3000 nucleotides per minute.

A simple model that can explain these results is shown in Figure 9–61. In this view, all replication origins located in active chromatin are utilized very early in S phase. Because the replication forks formed at these origins will eventually move into adjacent chromosomal regions that have a more condensed chromatin structure, any gene located less than a million nucleotide pairs away from a replication origin in active chromatin will replicate by mid S phase. To explain how inactive chromosomal regions located far away from a patch of active chromatin eventually

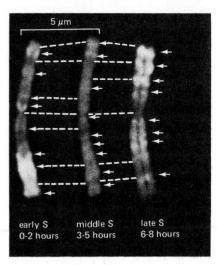

Figure 9–60 Light micrographs of stained mitotic chromosomes in which the replicating DNA has been differentially labeled during different defined intervals of the preceding S phase. In these experiments, cultured cells grown in the presence of the synthetic nucleoside 5-bromodeoxyuridine (BrdU) were briefly pulsed with thymidine during early, middle, or late S phase. Because the DNA made during the thymidine pulse is a double helix with thymidine on one strand and BrdU on the other, it stains more darkly than the remaining DNA (which has BrdU on both strands) and shows up as a bright band (*arrows*) on these negatives. Dashed lines connect corresponding positions on the three copies of the chromosome shown. (Courtesy of Elton Stubblefield.)

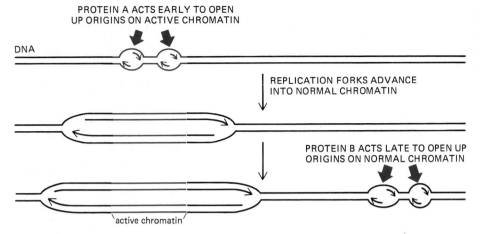

PROTEIN A ACTS EARLY TO OPEN
UP ORIGINS ON ACTIVE CHROMATIN

DNA

REPLICATION FORKS ADVANCE
INTO NORMAL CHROMATIN

PROTEIN B ACTS LATE TO OPEN UP
ORIGINS ON NORMAL CHROMATIN

active chromatin

Figure 9–61 One model to explain why active chromatin (*colored*) is replicated early in S phase, whereas normal ("inactive") chromatin is replicated late in S phase. Different initiator proteins are postulated to act on the replication origins in active chromatin and in normal chromatin. Alternatively, the two sets of origins might be utilized by the same molecular machinery but at different times simply because the condensed structure of the normal chromatin delays access of the requisite replication proteins.

replicate (as with the DNA in the inactive X chromosome in females), it is necessary to postulate that a second group of replication origins is activated in mid to late S phase that can initiate replication forks in any form of chromatin.

The Late-replicating Replication Units Coincide with the A-T-rich Bands on Metaphase Chromosomes[23]

Many of the replication units seem to correspond to distinct chromosome bands made visible by the various fixation and staining procedures used for karyotyping. As discussed previously (see p. 503), as many as 2000 dark-staining *A-T-rich bands (G bands)* can be detected early in mitosis in the haploid set of mammalian chromosomes, and these are separated by an equal number of light-staining *G-C-rich bands (R bands)*. It is intriguing that the A-T-rich DNA and the G-C-rich DNA differ in the time of their replication during S phase. Experiments like the one shown in Figure 9–60 suggest that most G-C-rich bands replicate during the first half of S phase, while most A-T-rich bands replicate during the second half of S phase. It has therefore been suggested that housekeeping genes are located mostly in G-C-rich bands, while many cell-type-specific genes—the vast majority of which will be inactive in most cells—are located in A-T-rich bands. As stated earlier, it is a complete mystery why the mammalian genome should be segregated into such large alternating blocks of chromatin—many nearly equal in size to an entire bacterial genome. It is also not known how the many replication origins present in each replication unit are activated all at once. Perhaps the chromatin in a late-replicating unit remains condensed even after the end of M phase and decondenses only in mid S phase, making all the replication origins in the unit simultaneously accessible. In this case the all-or-none coordinated replication of the DNA in a single replication unit could reflect the cooperative nature of the chromatin decondensation process (see p. 499).

Does the Controlled Timing of Utilization of Replication Origins Have a Function?[39]

The S phase is completed extremely rapidly in the cleaving eggs of many species, where large stores of chromatin components (such as histones) are present, as required for the rapid manufacture of new nuclei (see p. 881). As illustrated in Figure 9–62, a short S phase also requires the use of an exceptionally large number of replication origins spaced at intervals of only a few thousand nucleotide pairs (rather than the tens or hundreds of thousands of nucleotide pairs found between the replication origins later in development). Since any foreign DNA injected into a fertilized frog egg is replicated (including small circular fragments of bacterial DNA), if a specific DNA sequence is required to form a replication origin in this cell, it must be a very short one that is present in any DNA molecule.

Figure 9–62 Electron micrograph of spread chromatin from an early *Drosophila* embryo showing that replication bubbles (*arrows*) are extremely closely spaced. Only about 10 minutes elapse between some of the successive nuclear divisions in this embryo. Since the replication origins used here are so closely spaced (separated by a few thousand nucleotide pairs), it should require only about a minute to replicate all the DNA between them. (Courtesy of Victoria Foe.)

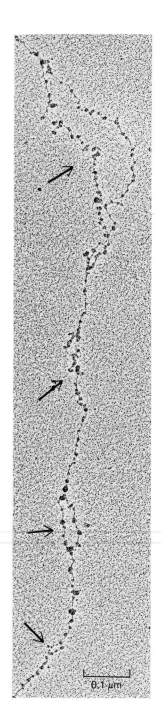

0.1 µm

Thus DNA replication can be viewed as a potentially very rapid process that in most cells is subject to a complex system of regulation that restrains the initiation of replication forks in a way that causes different portions of the genome to replicate at very different times. One possible advantage of such a prolonged S phase may be that it allows replication time to influence chromatin structure. It has been suggested, for example, that the chromatin that replicates early is assembled in part from a special store of chromosomal proteins produced during the G1 phase, which could help to maintain active chromatin and thereby facilitate the transcription of expressed genes.

9-21 9-22 Chromatin-bound Factors Ensure That Each Region of the DNA Is Replicated Only Once During Each S Phase, Providing a DNA Re-replication Block[40]

In a normal S phase the whole genome must be replicated exactly once and no more. As we have just seen, DNA replication in most eucaryotic cells is an asynchronous process that takes a relatively long time to complete. Because the replication origins are used at different times in different chromosomal regions, in the middle of S phase some parts of a chromosome will not yet have begun replication, while other parts will have replicated completely. An enormous "bookkeeping" problem therefore arises during the middle and late stages of the S phase. Those replication origins already used have been duplicated and, at least with respect to their DNA sequences, are presumably identical to other replication origins not yet used. But each replication origin must be used only once in each S phase. How is this accomplished?

Cell fusion experiments have provided an important clue. When an S-phase cell is fused with a G_1-phase cell, DNA synthesis is induced in the G_1-phase nucleus, suggesting that the transition from G_1 to S phase is mediated by a diffusible activator of DNA synthesis (see p. 732). In contrast, when the S phase cell is fused with a G_2-phase cell (that is, a cell that has just completed S phase—see p. 728), the G_2 nucleus is not stimulated to synthesize DNA. Since DNA synthesis continues undisturbed in the S-phase nucleus, the G_2 nucleus—having replicated all its DNA once—acts as though it is prevented from entering further rounds of replication by some nondiffusable inhibitor that is tightly bound to its DNA. Such an inhibitor, if applied locally during S phase in the wake of each replication fork, would neatly solve the bookkeeping problem: by modifying the chromatin of freshly replicated DNA, it would ensure that once-replicated DNA is not replicated again in the same S period (Figure 9–63A). As an equally plausible alternative, a mechanism based on tightly bound initiator proteins that are inactivated by the passage of a replication fork has been proposed (Figure 9–63B). Whatever its nature, the **DNA re-replication block** must be removed at or near the time of mitosis, since after cell division the DNA in the G_1 nuclei that emerge in the daughter cells is no longer protected.

The fragments of bacterial DNA that replicate when injected into a fertilized frog egg can be shown to be affected by the re-replication block. Therefore the mechanism responsible for the block cannot require a highly specific replication origin. The block does not affect the SV40 virus, presumably because its T-antigen supplies both initiator and DNA helicase functions to substitute for analogous host cell components that are as yet uncharacterized (see p. 517).

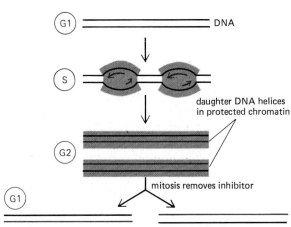

(A) INHIBITOR-ADDITION MODEL

G1 ━━━━━━━━━━ DNA

S

daughter DNA helices
in protected chromatin

G2

G1

mitosis removes inhibitor

DNA in G1 nucleus can replicate again

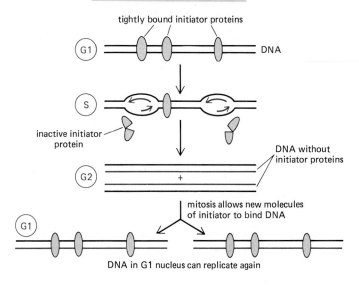

(B) ACTIVATOR-REMOVAL MODEL

tightly bound initiator proteins

G1 ━━━━━━━━━━ DNA

S

inactive initiator
protein

G2   +   DNA without
initiator proteins

mitosis allows new molecules
of initiator to bind DNA

G1

DNA in G1 nucleus can replicate again

Summary

In vitro *studies using the monkey virus SV40 as a model system suggest that, in eucaryotes as in procaryotes, DNA replication begins with the loading of a DNA helicase onto the DNA by an initiator protein bound to a replication origin. A replication bubble forms at such an origin as two replication forks move away from each other. During S phase in higher eucaryotes, neighboring replication origins appear to be activated in clusters known as replication units, with the origins spaced about one looped domain of chromatin apart. Since the replication fork moves at about 50 nucleotides per second, only about an hour should be required to complete the DNA synthesis in a replication unit. Throughout a typical 8-hour S phase, different replication units are activated in a sequence determined in part by their chromatin structure, the most condensed regions of chromatin being replicated last. The correspondence between replication units and the bands containing millions of nucleotide pairs seen on mitotic eucaryotic chromosomes suggests that replication units may correspond to structurally distinct domains in interphase chromatin.*

After the replication fork passes, chromatin structure is re-formed by the addition of new histones and other chromosomal proteins to the old histones inherited on the daughter DNA molecules. A DNA re-replication block of unknown nature acts locally to prevent a second round of replication from occurring until a chromosome has passed through mitosis; this block is needed to ensure that each region of the DNA is replicated only once in each S phase.

Figure 9–63 Two possible mechanisms that have been proposed to explain the "re-replication block," which protects replicated DNA from further replication in the same cell cycle. This block is crucial to replication bookkeeping, but its molecular nature is not known. Normally the block is removed at mitosis, but in a few specialized types of cells (the salivary gland cells of *Drosophila* larvae, for example), it is removed without mitosis, leading to the formation of giant *polytene* chromosomes (see p. 507). (A) A model based on the addition of an inhibitor to all newly replicated chromatin; (B) a model based on tightly bound initiator proteins that act only once and that can be added to the DNA only during mitosis.

RNA Synthesis and RNA Processing[41]

We have thus far considered how chromosomes are organized as very large DNA-protein complexes and how they are duplicated before a cell divides. But the main function of a chromosome is to act as a template for the synthesis of RNA molecules, since only in this way does the genetic information stored in chromosomes become directly useful to the cell. RNA synthesis is extensive: the total rate at which nucleotides are incorporated into RNA during interphase is about 20 times the rate at which nucleotides are incorporated into DNA during S phase.

RNA synthesis (**DNA transcription**) is a highly selective process. In most mammalian cells, for example, only about 1% of the DNA nucleotide sequence is copied into functional RNA sequences (mature messenger RNA or structural RNA). The selectivity occurs at two levels, which we discuss in turn in this section: (1) only part of the DNA sequence is transcribed to produce nuclear RNAs, and

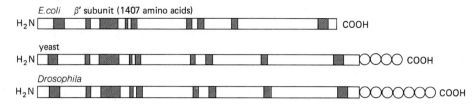

E.coli   β' subunit (1407 amino acids)

H₂N [████████████████████████████████████] COOH

yeast

H₂N [████████████████████████████████████]OOOO COOH

Drosophila

H₂N [████████████████████████████████████]OOOOOOOO COOH

Figure 9–64 Amino acid sequence similarities between RNA polymerase subunits suggest a common evolutionary origin for the bacterial and eucaryotic enzymes. The largest subunit of the *E. coli*, yeast, and *Drosophila* polymerases (compared here) is thought to bind to DNA. The shaded regions of the sequence are more than 70% identical between yeast and *Drosophila* and more than 40% identical between *Drosophila* and *E. coli*. A uniquely eucaryotic sequence of unknown function (Ser-Pro-Ser-Tyr-Ser-Pro-Thr) is repeated 26 times at the carboxyl terminus of the yeast subunit and more than 40 times in the *Drosophila* subunit (indicated here by circles). (After A.L. Greenleaf et al., in RNA Polymerase and the Regulation of Transcription [W.S. Reznikoff et al., eds.], pp. 459–464. New York: Elsevier, 1987.)

(2) only a minor proportion of the nucleotide sequences in nuclear RNAs survives the RNA processing steps that precede the export of RNA molecules to the cytoplasm. We begin by describing RNA polymerases, the enzymes that catalyze all DNA transcription.

RNA Polymerase Exchanges Subunits as It Begins Each RNA Chain[42]

A general outline of DNA transcription was given in Chapter 5. Transcription begins when an *RNA polymerase* molecule binds to a *promoter* DNA sequence. First, in a complicated initiation step, the two strands of the DNA are locally separated to form an *open complex*, in which the template strand is exposed. After this complex has formed, the polymerase begins to move along the DNA, extending its growing RNA chain in the 5'-to-3' direction by the stepwise addition of ribonucleoside triphosphates until it reaches a stop (termination) signal, at which point the newly synthesized RNA chain and the polymerase are released from the DNA. Each RNA molecule thus represents a single-strand copy of the nucleotide sequence of one DNA strand in a relatively short region of the genome (see Figure 5–1, p. 202).

RNA polymerases are generally formed from multiple polypeptide chains and have masses of 500,000 daltons or more. The enzymes in bacteria and eucaryotes are evolutionarily related (Figure 9–64). Since the bacterial enzyme has been far easier to study, its properties provide a basis for understanding its eucaryotic relatives. The *E. coli* enzyme contains five subunits, α, β, β', σ, and ω, there being two copies of α and one each of the others. The complete amino acid sequence of each subunit has been determined from the nucleotide sequence of its gene, and DNA-footprinting studies indicate that the enzyme covers 60 nucleotide pairs when it binds to DNA.

The sigma (σ) subunit of the *E. coli* polymerase has a specific role as an **initiation factor** for transcription: it enables the enzyme to find the consensus promoter sequences described on page 204. A subset of RNA polymerases that recognize different promoters contain variant forms of the sigma subunit (see p. 562). Once bound to a promoter, the enzyme undergoes a series of reactions to start an RNA chain (Figure 9–65). After about eight nucleotides of an RNA molecule

Figure 9–65 Highly schematic diagram of the steps in the initiation of RNA synthesis catalyzed by RNA polymerase. The steps indicated have been discovered in studies of the *E. coli* enzyme. A DNA molecule containing a promoter sequence for the *E. coli* polymerase is shown (see Figure 5–6, p. 205). The enzyme first forms a *closed complex* in which the two DNA strands remain fully base-paired. In the next step the enzyme catalyzes the opening of a little more than one turn of the DNA helix to form an *open complex*, in which a template is exposed for the initiation of an RNA chain. The polymerase containing the bound sigma subunit, however, behaves as though it is tethered to the promoter site: it seems unable to proceed with the elongation of the RNA chain and on its own frequently reverts to the closed complex with the release of a short RNA chain. As indicated, the conversion to an actively elongating polymerase requires the release of initiation factors (the sigma subunit in the case of the *E. coli* enzyme) and generally involves the binding of other proteins that serve as elongation factors (for example, the *E. coli* nusA protein). (Modified from D.C. Straney and D.M. Crothers, *J. Mol. Biol.* 193:267–278, 1987.)

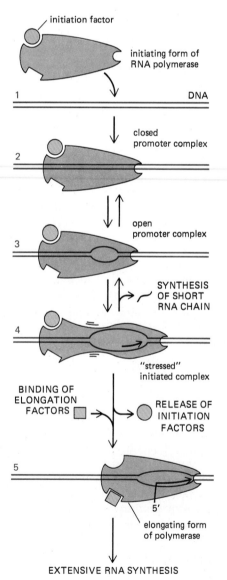

initiation factor

initiating form of RNA polymerase

1   DNA

closed promoter complex

2

open promoter complex

3

SYNTHESIS OF SHORT RNA CHAIN

4

"stressed" initiated complex

BINDING OF ELONGATION FACTORS ☐ →   RELEASE OF INITIATION FACTORS

5

5'

elongating form of polymerase

EXTENSIVE RNA SYNTHESIS

have been synthesized (step 4 in Figure 9–65), the sigma subunit dissociates and a number of **elongation factors**—important for chain elongation and termination—become associated with the enzyme instead. The elongation factors include several proteins that, although well characterized, have functions that are incompletely understood.

In all organisms the various active genes are transcribed at very different rates. At some promoters a new RNA chain begins once every 1 or 2 seconds, while at others more than an hour is required to start an RNA chain. As described in detail in Chapter 10, *gene regulatory proteins* that affect transcriptional initiation usually determine the level of transcription of each gene. In general, these proteins act by accelerating or retarding one or more of the steps illustrated in Figure 9–65 (see p. 563).

9-25 Three Different RNA Polymerases Make RNA in Eucaryotes[43]

Although the mechanism of DNA transcription is similar in eucaryotes and procaryotes such as *E. coli*, the machinery is considerably more complex in eucaryotes. In eucaryotes as diverse as yeasts and humans, for example, there are three types of RNA polymerase, each responsible for transcribing different sets of genes. These enzymes—denoted as RNA polymerases I, II, and III—are structurally similar to one another and have some common subunits, although other subunits are unique (Figure 9–66). Each is more complex than *E. coli* RNA polymerase and is thought to contain 10 or more polypeptide chains. The most important distinction, however, is that, whereas the purified bacterial enzyme binds directly to the promoter, the eucaryotic enzymes can bind to their promoters only in the presence of additional protein factors already on the DNA. Partly for this reason, it was not until 1979 that systems became available in which eucaryotic initiation mechanisms could be analyzed *in vitro*.

The three eucaryotic RNA polymerases were initially distinguished by their chemical differences during purification and by their sensitivity to α-amanitin, a poison isolated from mushrooms. RNA polymerase I is unaffected by α-amanitin; RNA polymerase II is very sensitive to this poison; and RNA polymerase III is moderately sensitive to it. The sensitivity of RNA synthesis to α-amanitin is still used to determine which polymerase transcribes a gene. Such studies indicate that only **RNA polymerase II** transcribes the genes whose RNAs will be translated into proteins. The other two polymerases synthesize only RNAs that have structural or catalytic roles, chiefly as part of the protein synthetic machinery: **polymerase I** makes the large ribosomal RNAs and **polymerase III** makes a variety of very small, stable RNAs—including the small 5S ribosomal RNA and the transfer RNAs. Most of the small RNAs that form snRNPs (see p. 532), however, are made by polymerase II.

Mammalian cells typically contain about 40,000 molecules of RNA polymerase II, about the same number of RNA polymerase I molecules, and about 20,000 molecules of RNA polymerase III—although studies with cultured cells indicate that the concentrations of the RNA polymerases are regulated individually according to the rate of cell growth.

9-26 Transcription Factors Form Stable Complexes
9-27 at Eucaryotic Promoters[44]

Eucaryotic RNA polymerases, as mentioned previously, do not recognize their promoters on purified DNA molecules. Instead, one or more sequence-specific DNA-binding proteins must be bound to the DNA to form a functional promoter. These are called **transcription factors** (**TF**) and are necessary for the initiation of RNA synthesis. They are distinct from procaryotic initiation factors (the sigma factors) in that they bind to DNA independently of the RNA polymerase. Polymerases I, II, and III recognize different promoters and for the most part require different transcription factors—designated by the prefixes **TFI, TFII,** and **TFIII,** respectively, followed by an alphabetical suffix assigned according to their order of discovery. TFIIIA, for example, was discussed previously as the prototype for a

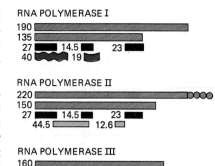

Figure 9–66 Some of the protein subunits that form the three eucaryotic RNA polymerases in yeasts. Only the best-characterized subunits are shown. Related subunits are colored the same, and their molecular mass is given in kilodaltons. The three subunits with identical mass (shown in black) are shared by all three enzymes.

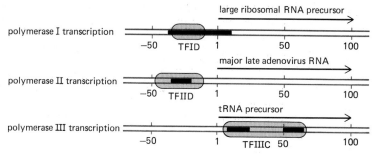

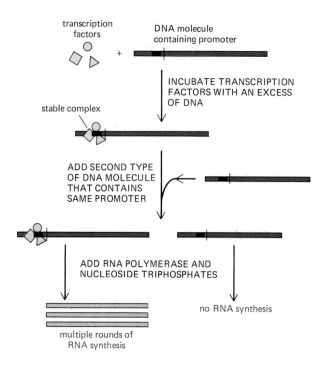

polymerase I transcription
large ribosomal RNA precursor
−50 TFID 1 50 100

polymerase II transcription
major late adenovirus RNA
−50 TFIID 1 50 100

polymerase III transcription
tRNA precursor
−50 1 TFIIIC 50 100

Figure 9–67 Some stable complexes formed by transcription factors on eucaryotic genes. The black boxes indicate the location of the specific DNA sequences thought to be essential for the function of each promoter. The approximate region of each gene covered by the bound factors (as determined by DNA-footprinting studies—see Figure 4–69, p. 188) is shown in color. Each of the transcription factors shown is a protein molecule that may contain multiple subunits, although the precise structures of these factors are not known. The numbers indicate positions on the DNA molecule (in nucleotide pairs) relative to the start site for transcription, which is set at +1.

"zinc finger" family of sequence-specific DNA-binding proteins (see p. 490); it was the first transcription factor to be characterized that acts on a polymerase-III-transcribed gene (the 5S rRNA gene).

Transcription factors *in vitro* appear to form relatively stable transcription complexes that selectively attract RNA polymerase molecules to their promoter. The different factors bind at different positions in relation to the transcription start site. Both polymerase I and polymerase II form a complex with a transcription factor that binds just upstream from the transcription start site. The major transcription factor for polymerase III genes, however, binds just *downstream* from the start site, so RNA polymerase III must transcribe over the protein without displacing it from the DNA (Figure 9–67). This factor (TFIIIC) is thought to fold the DNA around itself to form a large nucleoprotein particle.

Although many of the transcription factors have been difficult to purify to homogeneity, it has been possible to compare the transcriptional activity of DNA molecules that have been preincubated with transcription factors with that of DNA molecules that have not been preincubated. This has shown that a DNA molecule complexed with the appropriate factors is used repeatedly for RNA synthesis by RNA polymerase molecules, whereas a second DNA molecule that is otherwise identical is ignored (Figure 9–68). The observed pattern of transcription from DNA molecules injected into *Xenopus* eggs makes it very likely that the same types of complexes form *in vivo* as well.

A crucial transcription factor for many polymerase II promoters is TFIID. This is a large protein complex that is more commonly called the **TATA factor** because it can bind to a conserved A-T-rich sequence called the **TATA box**, centered about

transcription factors

DNA molecule containing promoter

+

INCUBATE TRANSCRIPTION FACTORS WITH AN EXCESS OF DNA

stable complex

ADD SECOND TYPE OF DNA MOLECULE THAT CONTAINS SAME PROMOTER

ADD RNA POLYMERASE AND NUCLEOSIDE TRIPHOSPHATES

no RNA synthesis

multiple rounds of RNA synthesis

Figure 9–68 An *in vitro* experiment that demonstrates the importance of the formation of a stable transcription complex at a eucaryotic promoter. This type of experiment has been performed with transcription factors and promoters that are specific for each of the three eucaryotic RNA polymerases. Examples of the cloned genes used for such studies are shown in Figure 9–67.

Figure 9-69 Minimum requirements for promoter recognition by a eucaryotic RNA polymerase II molecule. A TATA box-binding factor (TFIID) must form a stable transcription complex before the promoter can be recognized by the polymerase. The consensus sequence of the TATA box is $T_{82} A_{97} T_{93} A_{85} (A \text{ or } T)_{100} A_{83} (A \text{ or } T)_{83}$, where the subscript indicates the percent occurrence of the indicated nucleotide. See also Figure 10-27, page 569. (After data in J.L. Workman and R.G. Roeder, *Cell* 51:613-622, 1987.)

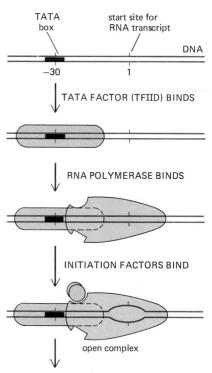

TATA factor remains bound at the TATA box during RNA synthesis, so that it facilitates the use of the promoter by many RNA polymerase molecules

25 nucleotides upstream from the start site for transcription. The activity of the TATA factor in stimulating polymerase II transcription is outlined in Figure 9-69.

Because RNA polymerase II makes all of the mRNA precursors and thus determines which proteins a cell will make, we shall focus most of our discussion on the synthesis and fate of the RNA transcripts synthesized by this enzyme.

9-28 RNA Polymerase II Transcribes Some DNA Sequences Much More Often Than Others[45]

Although experiments with purified polymerases and transcription factors *in vitro* are essential for establishing the mechanism of transcription, much can also be learned about the pattern of DNA transcription in a cell by using the electron microscope to examine genes in action, with their bound RNA polymerases caught in the act of transcription.

Ordinary thin-section electron micrographs of interphase nuclei show granular clumps of chromatin (see Figure 9-100), but they reveal very little about how genes are transcribed. A much more detailed picture emerges if the nucleus is ruptured and its contents spilled out onto an electron microscope grid (Figures 9-70 and 9-71). At the farthest point from the center of the lysed nucleus, the chromatin is diluted sufficiently to make individual chromatin strands visible in the expanded, beads-on-a-string form shown previously in Figure 9-22B.

RNA polymerase molecules actively engaged in transcription appear as globular particles with a single RNA molecule trailing behind. Particles representing active RNA polymerase II molecules are usually seen as single units, without nearby neighbors. This indicates that most genes are transcribed into mRNA precursors only infrequently, so that one polymerase finishes transcription before another one begins. Occasionally, however, many polymerase molecules (and their associated RNA transcripts) are seen clustered together. These clusters occur on the relatively few genes that are transcribed at high frequency (Figure 9-72). The length of the attached RNA molecules in such a cluster increases in the direction of transcription, producing a characteristic pattern. This pattern defines the RNA polymerase II start site and stop site for a specific **transcription unit** (Figure 9-73).

Biochemical studies have confirmed and extended the results obtained by electron microscopy, leading to three major conclusions:

1. Eucaryotic RNA polymerase molecules, like those in procaryotes, begin and end transcription at specific sites on the chromosome.
2. The average length of the finished RNA molecule produced by RNA polymerase II in a transcription unit is about 8000 nucleotides, and RNA molecules 10,000 to 20,000 nucleotides long are quite common. These lengths, which are much longer than the 1200 nucleotides of RNA needed to code for an average protein of 400 amino acid residues, reflect the complex structure of eucaryotic genes, and in particular the presence of long introns, as will be discussed later.
3. Although chain elongation rates of about 30 nucleotides per second are observed for all RNAs, different RNA polymerase II start sites function with very different efficiencies, so that some genes are transcribed at much higher rates than others. The pattern of transcription observed in electron micrographs agrees well with the results of biochemical studies showing that while many different messenger RNA molecules accumulate in a cell, most of them are present at relatively low frequency (Table 9-2).

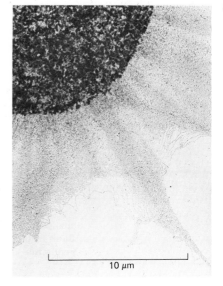

Figure 9-70 A typical cell nucleus visualized by electron microscopy using the procedure shown in Figure 9-71. An enormous tangle of chromatin can be seen spilling out of the lysed nucleus; only the chromatin at the outermost edge of this tangle will be sufficiently dilute for meaningful examination at higher power. (Courtesy of Victoria Foe.)

The Precursors of Messenger RNA Are Covalently Modified at Both Ends[46]

The RNA polymerase II transcripts in the nucleus are known as **heterogeneous nuclear RNA (hnRNA)** molecules because one of the first characteristics used to distinguish them from other RNAs in the nucleus was the heterogeneity of their sizes. Many of these transcripts are destined to leave the nucleus as **messenger RNA (mRNA)** molecules, and as they are being synthesized, they are covalently modified at both their 5′ end and their 3′ end in ways that clearly distinguish them from transcripts made by other RNA polymerases. These modifications will be useful later for their function as mRNA molecules in the cytoplasm.

The 5′ end of the RNA molecule (which is the end synthesized first during transcription) is first *capped* by the addition of a methylated G nucleotide. Capping occurs almost immediately, after about 30 nucleotides of RNA have been synthesized, and it involves condensation of the triphosphate group of a molecule of GTP with a diphosphate left at the 5′ end of the initial transcript (Figure 9–74). This **5′ cap** will later play an important part in the initiation of protein synthesis (see p. 214); it also seems to protect the growing RNA transcript from degradation.

The 3′ end of most polymerase II transcripts is defined not by the termination of transcription (which overshoots this point) but by a second modification, in which the growing transcript is cleaved at a specific site and a **poly-A tail** is added by a separate polymerase to the cut 3′ end. The signal for the cleavage is the appearance in the RNA chain of the sequence AAUAAA located 10 to 30 nucleotides upstream from the site of cleavage, plus a less well-defined downstream sequence. Immediately after cleavage, a *poly-A polymerase* enzyme adds 100 to 200 residues of adenylic acid (as *poly A*) to the 3′ end of the RNA chain to complete the **primary RNA transcript.** Meanwhile the polymerase fruitlessly continues transcribing for hundreds or thousands of nucleotides, until termination occurs at one of several later sites; the extra piece of RNA transcript thus generated presumably lacks a 5′ cap and is rapidly degraded (Figure 9–75).

The function of the poly-A tail is not known for certain, but it may play a role in the export of mature mRNA from the nucleus; there is also evidence that it helps to stabilize at least some mRNA molecules by retarding their degradation in the cytoplasm.

Even though polymerase II transcripts comprise more than half of the RNA synthesized by a cell, we shall see below that most of the RNA in these transcripts is unstable and therefore short-lived. Consequently, hnRNA in the cell nucleus and the cytoplasmic mRNA derived from it constitute only a minor fraction of the total RNA in a cell (Table 9–3). Despite their relative scarcity, these RNA molecules can be readily purified because of the long stretch of poly A at their 3′ ends. When the total cellular RNA is passed through a column containing poly dT linked to a

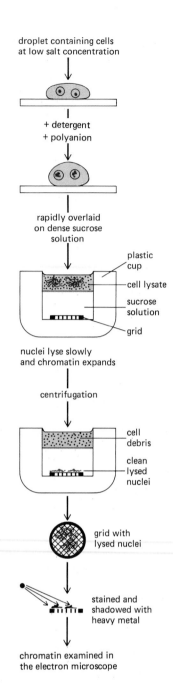

droplet containing cells at low salt concentration

+ detergent
+ polyanion

rapidly overlaid on dense sucrose solution

plastic cup
cell lysate
sucrose solution
grid

nuclei lyse slowly and chromatin expands

centrifugation

cell debris
clean lysed nuclei

grid with lysed nuclei

stained and shadowed with heavy metal

chromatin examined in the electron microscope

Figure 9–71   A method for examining the chromatin of a cell nucleus by electron microscopy after it has been gently spread out and freed from cellular debris.

Table 9–2   The Population of mRNA Molecules in a Typical Mammalian Cell

	Copies per Cell of Each mRNA Sequence		Number of Different mRNA Sequences in Each Class		Total Number of mRNA Molecules in Each Class
Abundant class	12,000	×	4	=	48,000
Intermediate class	300	×	500	=	150,000
Scarce class	15	×	11,000	=	165,000

This division of mRNAs into just three discrete classes is somewhat arbitrary, and in many cells a more continuous spread in abundances is seen. However, a total of 10,000 to 20,000 different mRNA species is normally observed in each cell, most species being present at a low level (5 to 15 molecules per cell). Most of the total cytoplasmic RNA is rRNA and only 3% to 5% is mRNA, a ratio consistent with the presence of about 10 ribosomes per mRNA molecule. This particular cell type contains a total of about 360,000 mRNA molecules in its cytoplasm.

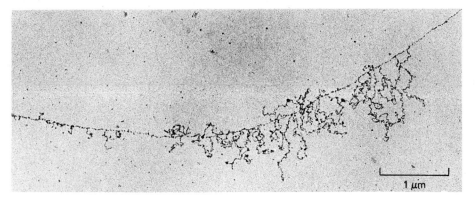

Figure 9–72 A region of chromatin containing a gene being transcribed at unusually high frequency, so that many RNA polymerase II molecules with their growing RNA transcripts are visible at the same time. The direction of transcription is from left to right (see Figure 9–73). (From V.E. Foe, L.E. Wilkinson, and C.D. Laird, *Cell* 9:131–146, 1976. © 1976 Cell Press.)

solid support, the complementary base-pairing between T and A residues selectively binds the molecules with poly-A tails to the column; the bound molecules can then be released for further analysis. This procedure is widely used to separate the hnRNA and mRNA molecules from the ribosomal and transfer RNA molecules that predominate in cells.

9-30 Capping and Poly-A Addition Require RNA Polymerase II[47]

Only RNA polymerase II transcripts have 5′ caps and 3′ poly-A tails. A gene normally transcribed by polymerase II can be separated from its promoter by recombinant DNA methods and fused to a promoter recognized by polymerase I or by polymerase III. If the modified gene is inserted back into a cell, it is transcribed by the polymerase that recognizes the promoter, but the RNA molecule produced is neither capped nor polyadenylated. It therefore seems that both the capping and the cleavage plus poly-A addition reactions are mediated by enzymes that bind selectively to polymerase II and function only when associated with it. These enzymes can thus be viewed as elongation factors for polymerase II. The requirement for capping and polyadenylation of mRNA precursors may explain why these RNAs are synthesized by a separate type of RNA polymerase molecule in eucaryotes.

Early hints that the cleavage plus poly-A addition reactions might require an elongation factor bound to the polymerase came from studies implying that each transcribing polymerase II molecule is able to generate only one polyadenylated 3′ end—that is, all AAUAAA signals encountered by the polymerase downstream from the initial cleavage and poly-A addition site are ignored. This suggests that the polymerase carries a factor that is lost upon completion of the cleavage and polyadenylation reaction (see Figure 9–75).

Most of the RNA Synthesized by RNA Polymerase II Is Rapidly Degraded in the Nucleus[48]

The first evidence that RNA polymerase II transcripts in the nucleus are unstable came from studies of cultured cells exposed to ^{3}H-uridine for a short period. This exposure introduced radioactivity into the hnRNA molecules, which could then be followed over a longer period of time. These and later experiments resulted in two remarkable discoveries:

1. The length of the newly made hnRNA molecules decreases rapidly, reaching the size of cytoplasmic mRNA molecules after only about 30 minutes. The primary RNA transcripts contain about 6000 nucleotides on average, while the mRNA molecules contain about 1500 nucleotides.
2. After about 30 minutes, radioactively labeled RNA molecules begin to leave the nucleus as mRNA molecules. Only about 5% of the mass of the labeled hnRNA, however, ever reaches the cell cytoplasm. The remainder is degraded into small fragments in the cell nucleus over a period of about an hour.

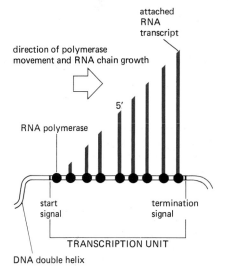

Figure 9–73 An idealized transcription unit, showing how the electron microscope appearance demonstrates the direction of transcription, as well as the start and stop sites of the unit.

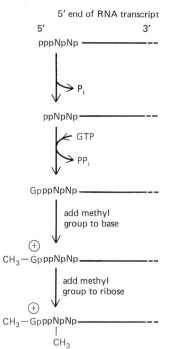

5′ end of RNA transcript

5′   3′

pppNpNp ————— ---

↓ ➔ P_i

ppNpNp ————— ---

↓ ↙ GTP
↓ ➔ PP_i

GpppNpNp ————— ---

↓ add methyl group to base

⊕
CH_3—GppNpNp ————— ---

↓ add methyl group to ribose

⊕
CH_3—GpppNpNp —————— --
|
CH_3

Figure 9–74 The reactions that cap the 5′ end of each RNA molecule synthesized by RNA polymerase II. The final cap contains a novel 5′-to-5′ linkage between the positively charged 7-methyl G residue and the original 5′ end of the RNA transcript (see Figure 5–24, p. 215). At least some of the enzymes required for this process are thought to be bound to polymerase II, since polymerase I and III transcripts are not capped and the indicated reaction occurs almost immediately following initiation of each RNA chain. The letter N is used here to represent any one of the four ribonucleotides, although the nucleotide that starts an RNA chain is usually a purine (an A or a G). (After A.J. Shatkin, *Bioessays* 7:275–277, 1987.)

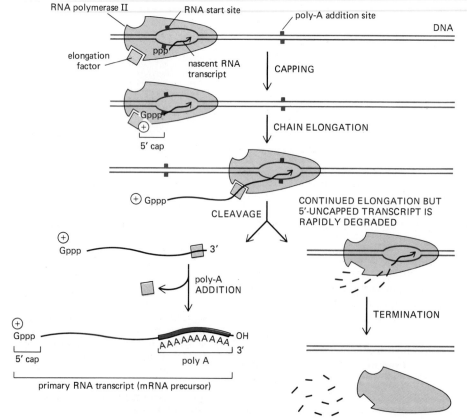

Figure 9–75 Synthesis of an hnRNA molecule (an mRNA precursor) by RNA polymerase II. This diagram starts with a polymerase that has just begun synthesizing an RNA chain (step 5 of Figure 9–65). Recognition of a poly-A addition signal in the growing RNA transcript causes the chain to be cleaved and then polyadenylated as shown. In yeasts the polymerase terminates its RNA synthesis almost immediately thereafter, but in higher eucaryotes it often continues transcription for thousands of nucleotides. It seems likely that the polymerase changes its properties once one RNA chain cleavage has occurred; thus it cannot cause poly-A addition to the downstream RNA transcript, and it seems to have a greater probability of responding to the sequences that cause chain termination and polymerase release. The simplest hypothesis, represented here, is that an elongation factor (or factors) is released from the polymerase after cleavage of the transcript.

Table 9–3 Selected Data on Amounts of RNA in a Typical Mammalian Cell

	Steady-State Amount (percent of total cell RNA)	**Percent of Total RNA Synthesis**
Nuclear rRNA precursors ↓	4	39
Cytoplasmic rRNA	71	—
Nuclear hnRNA ↓	7	58
Cytoplasmic mRNA	3	—
Small stable RNAs (mostly tRNAs)	15	3

The figures shown here were derived from the analysis of a mouse fibroblast cell line (L cells) in culture. Each cell contained 26 pg of RNA (5×10^{10} nucleotides of RNA), of which about 14% was located in the cell nucleus. (The cell nucleus thus contains about twice as much DNA as RNA.) An average of about 200×10^6 nucleotides is polymerized into RNA every minute during interphase. This is about 20 times the average rate at which DNA is synthesized during S phase. Note that although most of the RNA synthesized is hnRNA, most of this RNA is rapidly degraded in the nucleus. As a result, the mRNA produced from the hnRNA is only a minor fraction of the total RNA in the cell. (Modified from B. P. Brandhorst and E. H. McConkey, *J. Mol. Biol.* 85:451–563, 1974.)

When, in the early 1970s, both mRNAs and hnRNAs were discovered to have poly-A tails at their 3' ends, it was natural to assume that mRNAs are derived from hnRNAs by extensive degradation at 5' ends—in other words, that most of an hnRNA molecule consists of a very long "5' leader sequence" upstream of the coding sequence. But the hypothesis had to be abandoned when the 5' caps were discovered and shown to be largely preserved during the conversion of hnRNA molecules to mRNA. With hindsight, the logical conclusion would have been that the *middle* of an hnRNA molecule is removed, leaving both the 3' and the 5' ends intact. At the time, however, such a postulate seemed absurd. Moreover, researchers were at a loss to explain why a cell should discard most of the RNA that it synthesizes. The solution to this puzzle came only when it became possible to compare the nucleotide sequence of a particular mRNA molecule with the sequence of the genomic DNA that encodes it.

RNA Processing Removes Long Nucleotide Sequences from the Middle of RNA Molecules[48]

The discovery of interrupted genes in 1977 was entirely unexpected. Previous studies in bacteria had shown that their genes are composed of a continuous string of the nucleotides needed to encode the amino acids of a protein, and there seemed to be no obvious reason why a gene should be organized in any other way. The first indication that eucaryotic genes are not continuous like bacterial genes came when the new methods allowing an accurate comparison of mRNA and DNA sequences were applied to mRNAs produced by a human *adenovirus* (a large DNA virus). The region of the viral DNA producing these RNAs turned out to contain sequences that are not present in the mature RNAs. The possibility that this situation was unique to viruses was quickly eliminated by the finding of similar interruptions in the ovalbumin and β-globin genes of vertebrates. As discussed earlier, the sequences present in the DNA but omitted from the mRNA are called *intron* sequences, while those present in the mRNA are called *exon* sequences (Figures 9–76 and 9–77).

It then remained to determine how exon and intron sequences are sorted out by the cell: only then did the significance of the hnRNA become clear. We now know that the primary RNA transcript is a faithful copy of the gene, containing both exon and intron sequences, and that the latter sequences are cut out of the middle of the RNA transcript to produce an mRNA molecule that codes directly for a protein (see Figure 3–13, p. 101). Because the coding RNA sequences on either

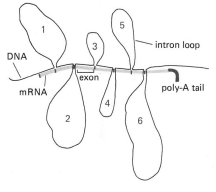

Figure 9–76 Early evidence for the existence of introns in eucaryotic genes was provided by the "R-loop technique," in which a base-paired complex between mRNA and DNA molecules is visualized in the electron microscope. An unusually abundant mRNA molecule, such as β-globin mRNA or ovalbumin mRNA, is readily purified from the specialized cells that produce it. When this single-stranded mRNA preparation is annealed in a suitable solvent to a cloned double-stranded DNA molecule containing the gene that encodes the mRNA, the RNA can displace a DNA strand wherever the two sequences match and form regions of RNA-DNA helix. Regions of DNA where no match to the mRNA sequence is possible are clearly visible as large loops of double-stranded DNA. Each of these loops (numbered 1 to 6) represents an intron in the gene sequence.

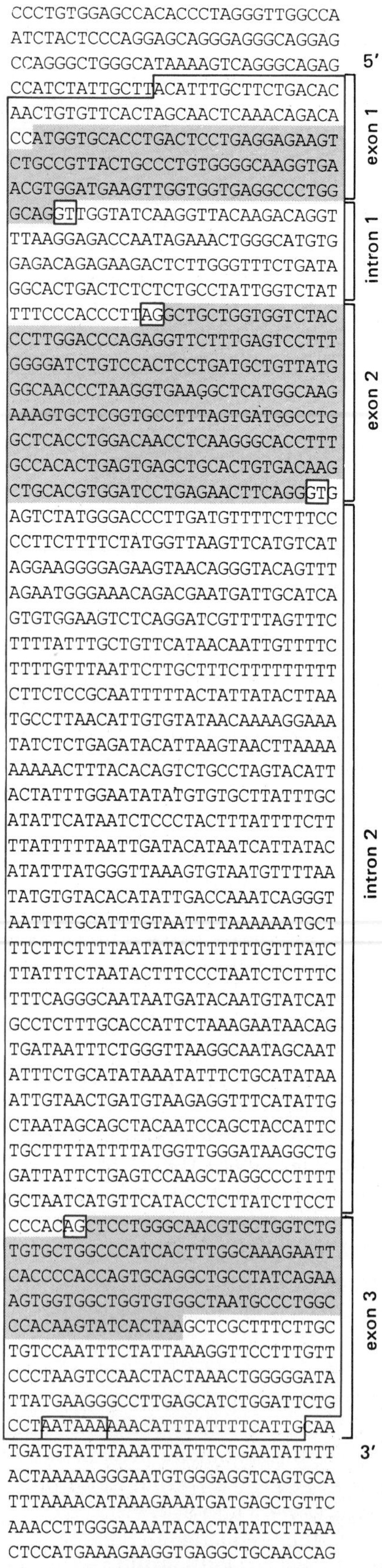

Figure 9–77 The transcribed portion of the human β-globin gene. The sequence of the DNA strand corresponding to the mRNA sequence is given, with the primary RNA transcript surrounded by a colored line and the nucleotides in the three coding regions (exons) shaded. Note that exon 1 includes a *5'-leader sequence* and that exon 3 includes a *3'-trailer sequence*; although these sequences are included in the mRNA, they do not code for amino acids. The highly conserved GT and AG nucleotides at the ends of each intron are boxed (see Figure 9–79), along with the cleavage and polyadenylation signal near the 3' end of the gene (AATAAA, see Figure 9–75).

side of an intron sequence are joined to each other after the intron sequence has been cut out, this reaction is known as **RNA splicing.** RNA splicing occurs in the cell nucleus, out of reach of the ribosomes, and RNA is exported to the cytoplasm only when processing is complete (see p. 538).

Because most mammalian genes contain much more intron than exon sequence (see Table 9–1, p. 486), RNA splicing can account for the conversion of the very long nuclear hnRNA molecules (up to more than 50,000 nucleotides) to the much shorter cytoplasmic mRNA molecules (usually 500 to 3000 nucleotides).

Before discussing the distribution of introns in eucaryotic genes and some of their consequences for cell function, it is necessary to explain how intron sequences are recognized and removed by the splicing machinery.

hnRNA Transcripts Are Immediately Coated with Proteins and snRNPs[49]

Newly made RNA in eucaryotes, unlike that in bacteria, appears to become immediately condensed into a series of closely spaced protein-containing particles. These particles consist of about 500 nucleotides of RNA wrapped around an abundant protein complex that serves to condense and package each growing RNA transcript in a manner reminiscent of the DNA-protein complexes of nucleosomes. The resulting **hnRNP particles** (*heterogeneous nuclear ribonucleoprotein particles*) can be purified after nuclei have been treated with ribonucleases at levels just sufficient to destroy the linker RNA between them. The particles sediment at about 30S and have a diameter about twice that of nucleosomes (20 nm). Their protein core is correspondingly larger and more complex, being composed of a set of at least 8 different proteins of mass 34,000 to 120,000 daltons. Except for histones, the proteins in this core are the most abundant proteins in the cell nucleus. Those characterized thus far contain one or more copies of a short sequence of amino acids that is shared by many RNA-binding proteins (Figure 9–78).

The hnRNP particles are generally distorted by the standard spreading techniques used to view transcribing genes in the electron microscope (see Figure 9–70). These micrographs, however, reveal especially stable particles of a less common type on many RNA transcripts, whose location strongly implicates a role for them in RNA splicing. The stable particles form very quickly at the junctions between intron and exon sequences, and, as the RNA transcript elongates, these particles coalesce in pairs to form a larger assembly that is thought to be the *spliceosome* that catalyzes RNA splicing (Figure 9–79).

Biochemical analysis has revealed that the cell nucleus contains many complexes of proteins with small RNAs (generally RNAs of 250 nucleotides or less), which have arbitrarily been designated U1, U2,...,U12 RNAs. These complexes, called **small nuclear ribonucleoproteins (snRNPs),** resemble ribosomes in that each contains a set of proteins complexed to a stable RNA molecule. They are much smaller than ribosomes, however—about 250,000 daltons compared to 4.5 million daltons for a ribosome—and have higher protein-to-RNA ratios. Some proteins are present in several types of snRNPs, whereas others are unique to one type. This was first demonstrated using serum from patients with the disease *systemic lupus erythematosus*, who make antibodies directed against one or more snRNP proteins: a single antibody was found that binds the U1, U2, U5, and U4/U6 snRNPs, for example, suggesting that they all contain a common protein.

```
CCCTGTGGAGCCACACCCTAGGGTTGGCCA
ATCTACTCCCAGGAGCAGGGAGGGCAGGAG
CCAGGGCTGGGCATAAAAGTCAGGGCAGAG        5'
CCATCTATTGCTTACATTTGCTTCTGACAC
AACTGTGTTCACTAGCAACTCAAACAGACA
CCATGGTGCACCTGACTCCTGAGGAGAAGT
CTGCCGTTACTGCCCTGTGGGGCAAGGTGA
ACGTGGATGAAGTTGGTGGTGAGGCCCTGG
GCAGGTTGGTATCAAGGTTACAAGACAGGT
TTAAGGAGACCAATAGAAACTGGGCATGTG
GAGACAGAGAAGACTCTTGGGTTTCTGATA
GGCACTGACTCTCTCTGCCTATTGGTCTAT
TTTCCCACCCTTAGGCTGCTGGTGGTCTAC
CCTTGGACCCAGAGGTTCTTTGAGTCCTTT
GGGGATCTGTCCACTCCTGATGCTGTTATG
GGCAACCCTAAGGTGAAGGCTCATGGCAAG
AAAGTGCTCGGTGCCTTTAGTGATGGCCTG
GCTCACCTGGACAACCTCAAGGGCACCTTT
GCCACACTGAGTGAGCTGCACTGTGACAAG
CTGCACGTGGATCCTGAGAACTTCAGGGTG
AGTCTATGGGACCCTTGATGTTTTCTTTCC
CCTTCTTTTCTATGGTTAAGTTCATGTCAT
AGGAAGGGGAGAAGTAACAGGGTACAGTTT
AGAATGGGAAACAGACGAATGATTGCATCA
GTGTGGAAGTCTCAGGATCGTTTTAGTTTC
TTTTATTTGCTGTTCATAACAATTGTTTTC
TTTTGTTTAATTCTTGCTTTCTTTTTTTT
CTTCTCCGCAATTTTTACTATTATACTTAA
TGCCTTAACATTGTGTATAACAAAAGGAAA
TATCTCTGAGATACATTAAGTAACTTAAAA
AAAAACTTTACACAGTCTGCCTAGTACATT
ACTATTTGGAATATATGTGTGCTTATTTGC
ATATTCATAATCTCCCTACTTTATTTTCTT
TTATTTTTAATTGATACATAATCATTATAC
ATATTTATGGGTTAAAGTGTAATGTTTTAA
TATGTGTACACATATTGACCAAATCAGGGT
AATTTTGCATTTGTAATTTTAAAAAATGCT
TTCTTCTTTTAATATACTTTTTTGTTTATC
TTATTTCTAATACTTTCCCTAATCTCTTTC
TTTCAGGGCAATAATGATACAATGTATCAT
GCCTCTTTGCACCATTCTAAAGAATAACAG
TGATAATTTCTGGGTTAAGGCAATAGCAAT
ATTTCTGCATATAAATATTTCTGCATATAA
ATTGTAACTGATGTAAGAGGTTTCATATTG
CTAATAGCAGCTACAATCCAGCTACCATTC
TGCTTTTATTTTATGGTTGGGATAAGGCTG
GATTATTCTGAGTCCAAGCTAGGCCCTTTT
GCTAATCATGTTCATACCTCTTATCTTCCT
CCCACAGCTCCTGGGCAACGTGCTGGTCTG
TGTGCTGGCCCATCACTTTGGCAAAGAATT
CACCCCACCAGTGCAGGCTGCCTATCAGAA
AGTGGTGGCTGGTGTGGCTAATGCCCTGGC
CCACAAGTATCACTAAGCTCGCTTTCTTGC
TGTCCAATTTCTATTAAAGGTTCCTTTGTT
CCCTAAGTCCAACTACTAAACTGGGGGATA
TTATGAAGGGCCTTGAGCATCTGGATTCTG
CCTAATAAAAAACATTTATTTTCATTGCAA
TGATGTATTTAAATTATTTCTGAATATTTT        3'
ACTAAAAAGGGAATGTGGGAGGTCAGTGCA
TTTAAAAACATAAAGAAATGATGAGCTGTTC
AAACCTTGGGAAAATACACTATATCTTAAA
CTCCATGAAAGAAGGTGAGGCTGCAACCAG
CTAATGCACATTGGCAACAGCCCCTGATGC
CTATGCCTTATTCATCCCTCAGAAAAGGAT
TCTTGTAGAGGCTTGATTTGCAGGTTAAAG
TTTTGCTATGCTGTATTTTACATTACTTAT
TGTTTTAGCTGTCCTCATGAATGTCTTTTC
```

exon 1

intron 1

exon 2

intron 2

exon 3

Figure 9–78 An amino acid sequence found in many eucaryotic RNA-binding proteins. This consensus—found in proteins from organisms as diverse as yeasts, *Drosophila*, and humans—is present in the proteins of hnRNP particles, in the protein bound to the poly-A tail of hnRNAs, in several snRNP proteins, and in the abundant nucleolar protein, nucleolin. When this sequence is found in a protein of unknown function, it suggests that the protein binds to RNA.

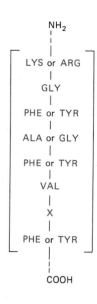

Individual snRNPs are believed to recognize specific nucleic acid sequences through RNA-RNA base-pair complementarity. Some mediate RNA splicing, some are involved in the cleavage reactions that generate the 3′ ends of some newly formed RNAs (see p. 596), while the function of others is unknown. The evidence for the role of snRNPs in RNA splicing comes from experiments on RNA processing *in vitro*.

Intron Sequences Are Removed as Lariat-shaped RNA Molecules[50]

Introns range in size from about 80 nucleotides to 10,000 nucleotides or more. They differ dramatically from exons in that their exact nucleotide sequences seem to be unimportant. Thus introns have accumulated mutations rapidly during evolution, and it is often possible to alter most of the nucleotide sequence of an intron without greatly affecting gene function. This has led to the suggestion that intron sequences have no function at all and are largely genetic "junk," a proposition we shall examine later (see p. 602). The only highly conserved sequences in introns are those required for intron removal. Thus there are consensus sequences at each end of an intron that are nearly the same in all known intron sequences, and these cannot be altered without affecting the splicing process that normally removes the intron sequence from the primary RNA transcript. These conserved boundary sequences at the **5′ splice site** (**donor site**) and the **3′ splice site** (**acceptor site**) are shown in Figure 9–80. The RNA breaking and rejoining reactions must be carried out precisely because an error of even one nucleotide would shift the reading frame in the resulting mRNA molecule and make nonsense of its message.

The pathway by which the intron sequences are removed from primary RNA transcripts has been elucidated by *in vitro* studies in which a pure RNA species containing a single intron is prepared by incubating an appropriately designed DNA fragment with a highly active, purified RNA polymerase (Figure 9–81). When these RNA molecules are added to a cell extract, they become spliced in a two-step enzymatic reaction that requires prolonged incubation with ATP, selected proteins in the extract, and the U1, U2, U5, and U4/U6 snRNPs; these components assemble into a large multicomponent ribonucleoprotein complex, or **spliceosome.** Characterization of the RNA species that appear as intermediates during the reaction, as well as the snRNPs required to produce them, led to the discovery

Figure 9–79 Electron micrograph of a chromatin spread showing large ribonucleoprotein particles assembling at the 5′ and 3′ splice sites to form a *spliceosome*. In the micrograph (A) a gene encoding a *Drosophila* chorion protein has been identified, so that the positions of the splice sites on the primary RNA transcript are known. (B) Most of the RNA transcripts have either one or two large RNP particles near their 5′ ends. When there are two particles on a transcript [open circles in (B)], they average 25 nm in diameter and occur at or very near the positions of the 5′ and 3′ splice sites for the single small intron sequence (228 nucleotides long) near the 5′ end of the transcripts. The more mature, longer transcripts on the two genes frequently display a single larger particle [colored circles in (B)] in the region of the intron, which probably results from the stable association of the two smaller particles and represents the assembled spliceosome. Since splicing occurs in some cases while the 3′ end of the RNA chain is still being transcribed, the poly A at the 3′ end of hnRNA molecules cannot be required for splicing. Most of the major hnRNP proteins have been removed from these transcripts by the spreading conditions used. (Adapted from Y.N. Osheim, O.L. Miller, and A.L. Beyer, *Cell* 43:143–151, 1985.)

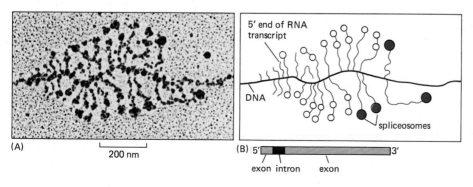

(A) |———| 200 nm

(B) 5′ ▭ 3′

exon intron exon

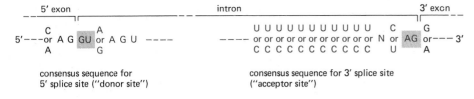

Figure 9–80 Consensus sequences for the 5' and 3' splice sites used in RNA splicing. The sequence given is that for the RNA chain; the nearly invariant GU and AG dinucleotides at either end of the intron are shaded in color (see also Figure 9–77).

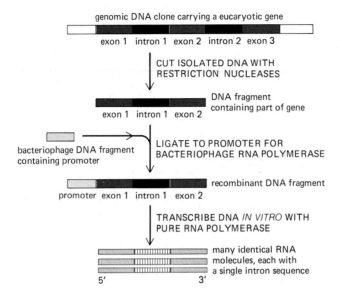

Figure 9–81 Outline of the procedure used to produce abundant amounts of pure RNA molecules for the analysis of RNA splicing *in vitro*. The method depends on the ability to produce large amounts of any desired DNA sequence by genetic engineering and DNA cloning (see p. 258) as well as on the availability of relatively simple RNA polymerases from bacteriophages T7 or SP6, which transcribe DNA with high efficiency *in vitro*. By coupling a eucaryotic DNA fragment to a bacteriophage promoter, a bacteriophage RNA polymerase can be used to generate *in vitro* large amounts of the RNA encoded by the eucaryotic DNA fragments. The 5' cap present on hnRNAs can be incorporated into such RNAs by using a chemically synthesized, capped nucleotide to initiate the transcription process (not shown).

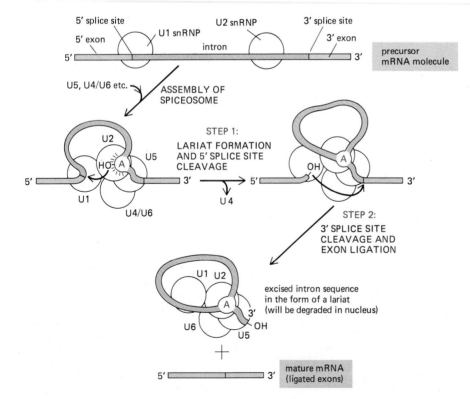

Figure 9–82 Catalysis of RNA splicing by a spliceosome formed from the assembly of U1, U2, U5, and U4/U6 snRNPs (shown as circles), plus other components (not shown). After assembly of the spliceosome, the reaction occurs in two steps: in step 1 a special A nucleotide in the intron sequence located close to the 3' splice site attacks the 5' splice site, which is cleaved; the cut 5' end of the intron sequence joins covalently to this A nucleotide, forming the branched nucleotide shown in Figure 9–83. In step 2 the 3'-OH end of the first exon, which was exposed in the first step, adds to the beginning of the second exon, cleaving the RNA molecule at the 3' splice site; the two exon sequences are thereby joined to each other and the intron sequence is released as a lariat. The complete spliceosome complex sediments at 60S, indicating that it is nearly as large as a ribosome. These splicing reactions occur in the nucleus and generate mRNA molecules from primary RNA transcripts (mRNA precursor molecules).

that the intron is excised in the form of a *lariat*, according to the splicing pathway shown in Figures 9–82 and 9–83.

Individual roles have been defined for several of the snRNPs. The U1 snRNP, for example, binds to the 5′ splice site, guided by a nucleotide sequence in the U1 RNA that is complementary to the nine-nucleotide splice-site consensus sequence (see Figure 9–80). Since RNA is capable of acting like an enzyme (see p. 105), either the RNA or the protein components of the spliceosome could be responsible for catalyzing the breakage and formation of covalent bonds required for RNA splicing.

9-32 Multiple Intron Sequences Are Usually Removed from Each RNA Transcript[51]

Because the spliceosome seems mainly to recognize a consensus sequence at each intron boundary, the *5′ splice site (donor site)* at the end of any one intron can in principle be joined to the 3′ *splice site (acceptor site)* of any other intron in the splicing process. Thus, when the 5′ and 3′ halves of two different introns are experimentally combined, the resulting hybrid intron sequence is often recognized by the RNA-splicing enzymes and removed.

In view of this result, it is surprising that vertebrate genes can contain as many as 50 introns (see Table 9–1, p. 486). If any two 5′ and 3′ splice sites were mispaired for splicing, some functional mRNA sequences would be lost, with disastrous consequences. Somehow such mistakes are avoided: the RNA processing machinery normally guarantees that each 5′ splice site pairs only with the 3′ splice site that is closest to it in the downstream (5′-to-3′) direction of the linear RNA sequence (Figure 9–84). How this sequential pairing of splice sites is accomplished is not known, although the assembly of the spliceosome while the RNA transcript is still growing (see Figure 9–79) is presumed to play a major part in ensuring an orderly pairing of the appropriate splice sites. There is also evidence that the exact three-dimensional conformations adopted by the intron and exon sequences in the RNA transcript are important. We shall see, however—both below and in Chapter 10—that splicing can be controlled, and in selected cases the simple pattern of 5′-to-3′ splicing does not hold.

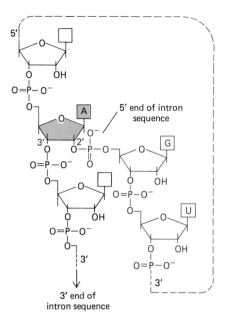

Figure 9–83 Structure of the branched RNA chain that forms during nuclear RNA splicing. The shaded A is the nucleotide highlighted in Figure 9–82, and the branch is formed in step 1 of the splicing reaction illustrated there. In this step the 5′ end of the intron sequence is cleaved and its phosphate group couples covalently to the 2′-OH ribose group of the A nucleotide, which is located about 30 nucleotides from the 3′ end of the intron sequence. The branched chain remains in the final excised intron sequence and is responsible for its lariat form (see Figure 9–82).

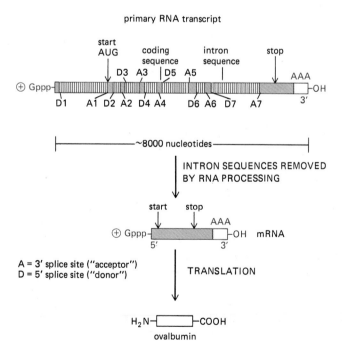

Figure 9–84 The primary RNA transcript for the chicken ovalbumin gene, showing the organized removal of seven introns required to obtain a functional mRNA molecule. The 5′ splice sites (donor sites) are denoted by D, and 3′ splice sites (acceptor sites) are denoted by A.

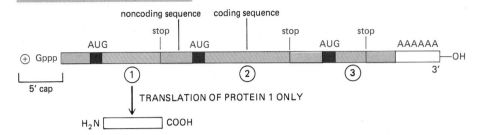

FULL-LENGTH mRNA MOLECULES

noncoding sequence   coding sequence

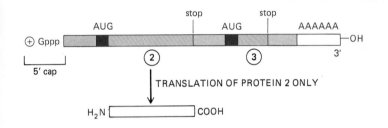

FRACTION OF mRNA MOLECULES THAT HAVE THE
CAP MOVED TO CODING SEQUENCE 2 BY RNA SPLICING

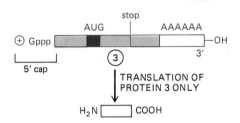

FRACTION OF mRNA MOLECULES THAT HAVE THE
CAP MOVED TO CODING SEQUENCE 3 BY RNA SPLICING

9-33 The Same RNA Transcript Can Be Processed in Different Ways to Produce mRNAs Coding for Several Proteins[52]
10-23

Although most intron sequences themselves appear to have no specific function, the existence of RNA splicing makes it possible to generate several different mRNAs, and thereby several different proteins, from the same RNA transcript, thus conferring extra genetic flexibility on the cell. We shall see in Chapter 10 that changes in the pattern of splicing of many RNA transcripts occur in the course of cell differentiation, so that the same DNA coding sequences are put to different uses as cells develop (see p. 589).

The versatility conferred by RNA splicing was first discovered in the adenovirus in which splicing itself was discovered. The adenovirus genome directs the synthesis of some very long RNA transcripts that contain the coding sequences for several proteins. This does not normally occur in a eucaryotic cell, where an individual mRNA molecule generally codes for only one protein (see p. 214); translation is initiated only near the 5' cap site (see p. 528), and it usually will stop as soon as the first stop codon is encountered. In adenovirus, however, this limitation is overcome by the RNA-splicing machinery, which can treat coding sequences as introns and remove them, so that the same 5' cap is spliced to each of the downstream coding sequences in a proportion of the mRNAs produced by the virus. This *alternative RNA splicing* allows the same 5' cap to serve as the initiation signal for the synthesis of different proteins (Figure 9–85). This device is widely used by viruses to enable only a few different RNA transcripts to code for a much larger number of proteins.

Figure 9–85 For some viruses the same primary RNA transcript is spliced in several ways to produce three (or more) different mRNA molecules, each coding for a different protein. In each case, only the coding sequence closest to the 5' cap is translated from the mRNA molecule.

9-34 The mRNA Changes in Thalassemia Reveal How RNA Splicing Can Allow New Proteins to Evolve[53]

Recombinant DNA techniques have made human mutants an increasingly important source of material for genetic studies of cellular mechanisms (see p. 181). In a group of human genetic diseases called the **thalassemia syndromes,** for example, patients have an abnormally low level of hemoglobin—the oxygen-carrying protein in red blood cells. The change in the DNA sequence has been determined for more than 50 such mutants, and a large proportion of these cause alterations in the pattern of RNA splicing. Thus single nucleotide changes have been detected that either inactivate a splice site or create a new splice site by changing a se-

(A) NORMAL ADULT β-GLOBIN PRIMARY
RNA TRANSCRIPT

normal mRNA is formed from three exons

(B) SINGLE NUCLEOTIDE CHANGES THAT CREATE
A NEW SPLICE SITE

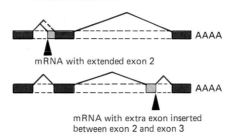

mRNA with extended exon 2

mRNA with extra exon inserted
between exon 2 and exon 3

(C) SINGLE NUCLEOTIDE CHANGES THAT
DESTROY A NORMAL SPLICE SITE WILL
ACTIVATE CRYPTIC SPLICE SITES

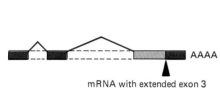

multiple mRNAs with both
shortened and extended exon 1

(D) SINGLE NUCLEOTIDE CHANGE THAT
DESTROYS NORMAL POLYADENYLATION SIGNAL

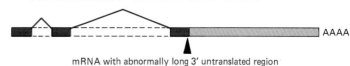

mRNA with abnormally long 3' untranslated region

mRNA with extended exon 3

quence in an intron or an exon into a consensus splice site. Surprisingly, analysis of the mRNAs produced in these mutant individuals reveals that the loss of a splice site does not prevent splicing but instead causes its normal partner site to seek out and become joined to a new "cryptic" site nearby; often a number of alternative splices are made in these mutants, causing the mutant gene to produce a set of altered proteins rather than just one (Figure 9–86). These results demonstrate that RNA splicing is a very flexible process in higher eucaryotic cells.

Because a simple mutation will often cause a gene to produce a variety of new proteins, the cell can test possible improvements to its genetic dowry in a very efficient manner. For this reason the flexibility of RNA splicing might have played a crucial part in the evolution of higher eucaryotes. Lower eucaryotes, such as yeasts, have more highly specified splicing rules, which should greatly constrain the evolution of new mRNAs through changes in patterns of RNA splicing; perhaps as a result, their rate of divergence in form and function seems to have been very much slower than that of the higher eucaryotes.

Spliceosome-catalyzed RNA Splicing Probably Evolved from Self-splicing Mechanisms[54]

When the lariat intermediate in nuclear RNA splicing was first discovered, it puzzled molecular biologists. Why was this bizarre pathway used, rather than the apparently simpler alternative of bringing the 5' and 3' splice sites together in an initial step, followed by their direct cleavage and rejoining? The answer seems to lie in the way the spliceosome evolved.

As explained in Chapter 1, it is thought that early cells used RNA molecules rather than proteins as their major catalysts and stored their genetic information in RNA rather than DNA sequences (see p. 7). RNA-catalyzed splicing reactions presumably played important roles in these early cells, and some self-splicing RNA introns remain today—for example, in the nuclear rRNA genes of *Tetrahymena* (see p. 105), in bacteriophage T4, and in some mitochondrial and chloroplast genes (see p. 393). In these cases, large parts of the intron sequence have been highly conserved because of their need to fold to create a catalytic surface in the RNA molecule. Two major classes of self-splicing introns can be readily distinguished: *group I introns* begin the splicing reaction by binding a G nucleotide to

Figure 9–86 Examples of the abnormal processing of the β-globin primary RNA transcript that have been observed in mutant humans with β thalassemia. The site of each mutation is denoted by the black arrowhead. The colored boxes represent the three normal exons illustrated previously in Figure 9–77, and the colored lines join the 5' and 3' splice sites utilized in splicing the primary RNA transcript produced by the gene. The open boxes depict new nucleotide sequences included in the final mRNA molecule as a result of a mutation. Note that when a mutation leaves a normal splice site without a partner, one or more abnormal "cryptic" splice sites nearby are used as the partner site, as in (C). (After S.H. Orkin, in The Molecular Basis of Blood Diseases [G. Stanatoyannopoulos et al., eds.], pp. 106–126. Philadelphia: Saunders, 1987.)

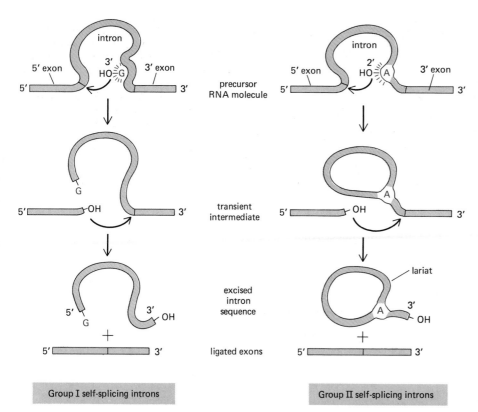

intron	intron
5' exon 3' exon	5' exon 3' exon
3'	2'
HO—G	HO—A

precursor
RNA molecule

transient
intermediate

excised
intron
sequence

lariat

ligated exons

Group I self-splicing introns   Group II self-splicing introns

Figure 9–87 The two known classes of self-splicing introns. The group I introns bind a free G nucleotide to a specific site to initiate splicing (see Figure 3–19, p. 105), while the group II introns use a specially reactive A nucleotide in the intron sequence itself for the same purpose. The two mechanisms have been drawn in a way that emphasizes their similarities. Both are normally aided by proteins that speed up the reaction, but the catalysis is nevertheless mediated by the RNA in the intron. The mechanism used by group II introns forms a lariat and resembles the pathway catalyzed by the spliceosome (compare to Figure 9–82.) (After T.R. Cech, *Cell* 44:207–210, 1986.)

the intron sequence; the G is thereby activated to form the attacking group that will break the first of the phosphodiester bonds cleaved during splicing (the bond at the 5' splice site); in *group II introns* a specially reactive A residue in the intron sequence has this role, and a lariat intermediate is generated. Otherwise the reaction pathways are the same, and both are presumed to represent vestiges of very ancient mechanisms (Figure 9–87).

In the evolution of nuclear RNA splicing, the reaction pathway used by the group II self-splicing introns seems to have been retained, but the catalytic role of the intron sequences has been replaced by separate spliceosome components. Thus the small RNAs U1 and U2, for example, may well be remnants of catalytic sequences that were originally present in introns. Shifting the catalysis from intron to spliceosome presumably lifted most of the constraints on the evolution of introns, allowing many new intron sequences to evolve.

The Transport of mRNAs to the Cytoplasm Is Delayed Until Splicing Is Complete[55]

As discussed in Chapter 8, finished mRNA molecules are thought to be recognized by receptor proteins in the nuclear pore complex that help move the mRNAs to the cytoplasm by an active transport process (see p. 425). The major proteins of the hnRNP particles and various processing molecules bound to the RNA in the nucleus, however, never seem to leave the nucleus, which suggests that they are stripped off the RNA as it passes through the nuclear pore (Figure 9–88). Studies of mutant yeasts suggest that, for RNAs that have splice sites, this transport process can occur only after the splicing reaction has been completed. In conditionally lethal yeast mutants that fail to splice their RNAs at high temperature because of a defect in their splicing machinery, all unspliced mRNA precursors remain in the nucleus, while those mRNAs that do not require splicing (which includes most of the mRNAs in this single-celled eucaryote) are transported normally to the cytoplasm. This observation is consistent with the idea that RNAs are retained by their bound spliceosome components, which seem to form numerous large aggregates throughout the interior of the nucleus of higher eucaryotes. These aggregates may serve as "splicing islands," although it is not known how they form or function (Figure 9–89). They could, however, be analogous to the *nucleolus*, a much larger

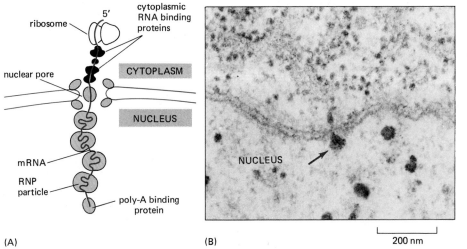

(A)   (B)

ribosome
5'
cytoplasmic
RNA binding
proteins
nuclear pore
CYTOPLASM
NUCLEUS
mRNA
RNP
particle
poly-A binding
protein

NUCLEUS

200 nm

Figure 9–88 The movement of mRNA molecules through nuclear pores. (A) Schematic illustration of the change thought to occur in the proteins bound to the RNA molecule as it moves out of the nucleus. (B) Electron micrograph of a large mRNA molecule produced in an insect salivary gland cell; this molecule has apparently been caught in the process of moving to the cytoplasm (*arrow*). (B, from B.J. Stevens and H. Swift, *J. Cell Biol.* 31:55–77, 1966.)

and more prominent structure in the nucleus, whose organization and function are better understood.

The nucleolus is the site where ribosomal RNA (rRNA) molecules are processed from a larger precursor RNA and assembled into ribosomes by the binding of ribosomal proteins. Before discussing nucleolar structure, however, we need to consider how the precursor rRNA molecules are synthesized from rRNA genes.

10-23 Ribosomal RNAs Are Made on Tandemly Arranged Sets of Identical Genes[56]

Many of the most abundant proteins of a differentiated cell, such as hemoglobin in the red blood cell and myoglobin in a muscle cell, are synthesized from genes that are present in only a single copy per haploid genome. These proteins are abundant because each of the many mRNA molecules transcribed from the gene can be translated into as many as 10 protein molecules per minute. This will normally produce more than 10,000 protein molecules per mRNA molecule in each cell generation. Such an amplification step is not available for the synthesis of the RNA components of ribosomes, however, since these RNA molecules are the final gene products. Yet a growing higher eucaryotic cell must synthesize 10 million copies of each type of ribosomal RNA molecule in each cell generation in order to construct its 10 million ribosomes. Adequate quantities of ribosomal RNAs can, in fact, be produced only because the cell contains multiple copies of the genes that code for ribosomal RNAs (**rRNA genes**).

Even *E. coli* needs seven copies of its rRNA genes to keep up with the the cell's need for ribosomes. Human cells contain about 200 rRNA gene copies per haploid genome, spread out in small clusters on five different chromosomes; while cells of the frog *Xenopus* contain about 600 rRNA gene copies per haploid genome in a single cluster on one chromosome. In eucaryotes the multiple copies of the highly conserved rRNA genes on a given chromosome are located in a tandemly arranged series in which each gene (8,000 to 13,000 nucleotide pairs long, depending on the organism) is separated from the next by a nontranscribed region known as *spacer DNA*, which can vary greatly in length and sequence. We shall see in Chapter 10 that such multiple copies of tandemly arranged genes tend to co-evolve (see p. 600).

Because of their repeating arrangement, and because they are transcribed at a very high rate, the tandem arrays of rRNA genes can easily be seen in spread chromatin preparations. The RNA polymerase molecules and their associated transcripts are so densely packed (typically about 100 per gene) that the transcripts fan out perpendicularly from the DNA to give each transcription unit a "Christmas tree" appearance (Figure 9–90). As noted earlier (see Figure 9–73), the tip of each of these "trees" represents the point on the DNA at which transcription begins and where the transcripts are thus shortest, while the other end of the rRNA transcription unit is sharply demarcated by the sudden disappearance of RNA polymerase molecules and their transcripts.

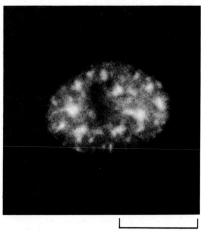

10 μm

Figure 9–89 Immunofluorescence staining of a human fibroblast nucleus with a monoclonal antibody that detects the snRNP particles involved in nuclear splicing of mRNA precursor molecules. The snRNP particles are present in large aggregates, which could function as "splicing islands." The antibody detects specific proteins that are present in several of the snRNPs that function in the spliceosome. (Courtesy of N. Ringertz.)

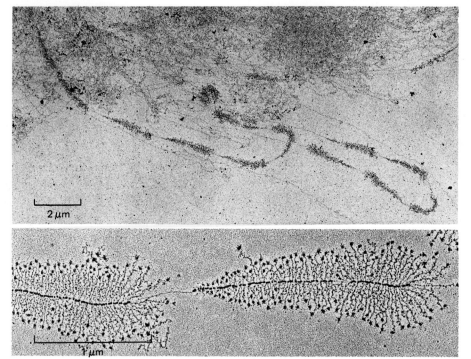

Figure 9–90 Transcription from tandemly arranged rRNA genes, as visualized in the electron microscope. The pattern of alternating transcribed gene and nontranscribed spacer is readily seen in the lower-magnification view in the upper panel. The large particles at the 5′ end of each rRNA transcript (*lower panel*) are believed to reflect the beginning of ribosome assembly; RNA polymerase molecules are also clearly visible. (Upper panel, from V.E. Foe, *Cold Spring Harbor Symp. Quant. Biol.* 42:723–740, 1978; lower panel, courtesy of Ulrich Scheer.)

The rRNA genes are transcribed by RNA polymerase I, and each gene produces the same primary RNA transcript. In humans this RNA transcript, known as *45S rRNA*, is about 13,000 nucleotides long. Before it leaves the nucleus in assembled ribosomal particles, the 45S rRNA is cleaved to give one copy each of the 28S rRNA (about 5000 nucleotides), the 18S rRNA (about 2000 nucleotides), and the 5.8S rRNA (about 160 nucleotides) of the final ribosome (see p. 221). The derivation of these three rRNAs from the same primary transcript ensures that they will be made in equal quantities. The remaining part of each primary transcript (about 6000 nucleotides) is degraded in the nucleus (Figure 9–91). Some of these extra RNA sequences are thought to play a transient part in ribosome assembly, which begins immediately as specific proteins bind to the growing 45S rRNA transcripts.

Another set of tandemly arranged genes with nontranscribed spacers codes for the 5S rRNA of the large ribosomal subunit (the only rRNA that is transcribed separately). The 5S rRNA genes are only about 120 nucleotide pairs in length, and like a number of other genes encoding small stable RNAs (most notably the tRNA genes), they are transcribed by RNA polymerase III. Humans have about 2000 5S rRNA genes tandemly arranged in a single cluster far from all the other rRNA genes. It is not known why this one type of rRNA is transcribed separately.

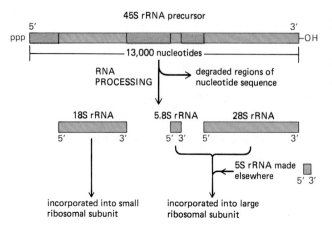

Figure 9–91 The pattern of processing of a 45S rRNA precursor molecule into three separate ribosomal RNAs. Nearly half of the nucleotide sequences in this precursor are degraded in the nucleus.

The Nucleolus Is a Ribosome-producing Machine[57]

The continuous transcription of multiple gene copies ensures an adequate supply of the rRNAs, which are immediately packaged with ribosomal proteins to form ribosomes. The packaging takes place in the nucleus, in a large, distinct structure called the **nucleolus.** The nucleolus contains large loops of DNA emanating from several chromosomes, each of which contains a cluster of rRNA genes. Each such gene cluster is known as a **nucleolar organizer** region. Here the rRNA genes are transcribed at a rapid rate by RNA polymerase I. The beginning of the rRNA packaging process can be seen in electron micrographs of these genes: the 5' tail of each transcript is encased by a protein-rich granule (see Figure 9–90). These granules, which do not appear on other types of RNA transcripts, presumably reflect the first of the protein-RNA interactions that take place in the nucleolus.

The biosynthetic functions of the nucleolus can be traced by briefly labeling newly made RNA with ^{3}H-uridine. After various intervals of further incubation in unlabeled media, a cell fractionation procedure can be used to break the rRNA genes free of their chromosomes, thereby allowing the radioactive nucleoli to be isolated in relatively pure form (Figure 9–92). Such experiments show that the intact 45S transcript is first packaged into a large complex containing many different proteins imported from the cytoplasm, where all proteins are synthesized. Most of the 70 different polypeptide chains that will make up the ribosome, as well as the 5S rRNAs, are incorporated at this stage. Other molecules are needed to guide the assembly process. Thus the nucleolus also contains other RNA-binding proteins and certain small ribonucleoprotein particles (including U3 snRNP) that are believed to help catalyze the construction of ribosomes. These components remain in the nucleolus when the ribosomal subunits are exported to the cytoplasm in finished form. An especially notable component is *nucleolin*, an abundant, well-characterized RNA-binding protein that seems to coat only ribosomal transcripts; this protein stains with silver in the characteristic manner of the nucleolus itself.

As the 45S rRNA molecule is processed, it gradually loses some of its RNA and protein and then splits to form separate precursors of the large and small ribosomal subunits (Figure 9–93). Within 30 minutes of radioactive pulse labeling, the first mature small ribosomal subunits, containing their 18S rRNA, emerge from the nucleolus and appear in the cytoplasm. Assembly of the mature large ribosomal subunit, with its 28S, 5.8S, and 5S rRNAs, takes about an hour to complete. The nucleolus therefore contains many more incomplete large ribosomal subunits than small ones.

The last steps in ribosome maturation occur only as these subunits are transferred to the cytoplasm. This delay prevents functional ribosomes from gaining access to the incompletely processed hnRNA molecules in the nucleus.

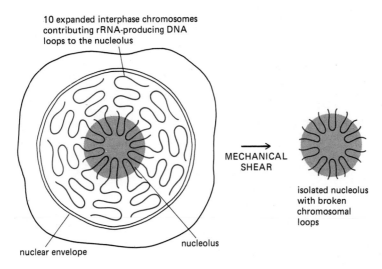

10 expanded interphase chromosomes contributing rRNA-producing DNA loops to the nucleolus

MECHANICAL SHEAR

isolated nucleolus with broken chromosomal loops

nuclear envelope

nucleolus

Figure 9–92 Highly schematic view of a human cell, showing the contributions to a single large nucleolus of loops of chromatin containing rRNA genes from 10 separate chromosomes. Purified nucleoli are very useful for biochemical studies of nucleolar function; to obtain such nucleoli, the loops of chromatin are mechanically sheared from their chromosomes, as shown.

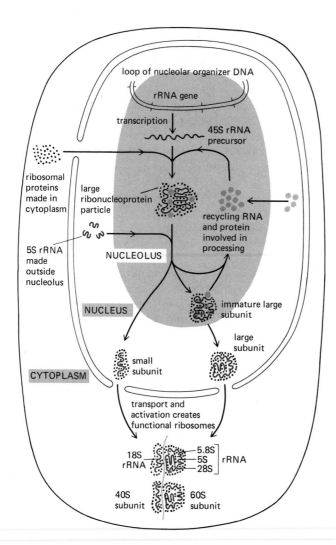

loop of nucleolar organizer DNA

rRNA gene

transcription

45S rRNA precursor

ribosomal proteins made in cytoplasm

large ribonucleoprotein particle

recycling RNA and protein involved in processing

5S rRNA made outside nucleolus

NUCLEOLUS

NUCLEUS

immature large subunit

large subunit

small subunit

CYTOPLASM

transport and activation creates functional ribosomes

18S rRNA

5.8S
5S
28S

rRNA

40S subunit

60S subunit

Figure 9–93 The function of the nucleolus in ribosome synthesis. The 45S rRNA transcript is packaged in a large ribonucleoprotein particle containing many ribosomal proteins imported from the cytoplasm. While this particle remains in the nucleolus, selected pieces are discarded as it is processed into immature large and small ribosomal subunits. These two subunits are thought to attain their final functional form only as they are individually transported through the nuclear pores into the cytoplasm.

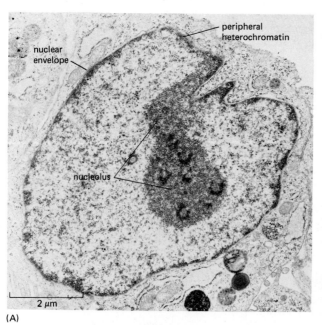

peripheral heterochromatin

nuclear envelope

nucleolus

2 μm

(A)

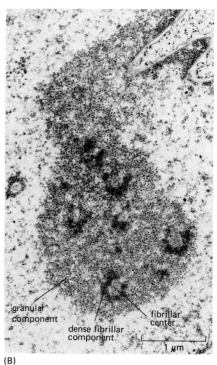

granular component

dense fibrillar component

fibrillar center

1 μm

(B)

Figure 9–94 Electron micrograph of a thin section of a nucleolus in a human fibroblast, showing its three distinct zones. (A) View of entire nucleus. (B) High-power view of the nucleolus. (Courtesy of E.G. Jordan and J. McGovern.)

The Nucleolus Is a Highly Organized Subcompartment of the Nucleus[57]

As seen in the light microscope, the large spheroidal nucleolus is the most obvious structure in the nucleus of a nonmitotic cell. Consequently, it was so closely scrutinized by early cytologists that an 1898 review could list some 700 references. By the 1940s cytologists had demonstrated that the nucleolus contains high concentrations of RNA and proteins; but its major function in ribosomal RNA synthesis and ribosome assembly was not discovered until the 1960s.

Some of the details of nucleolar organization can be seen in the electron microscope. Unlike the cytoplasmic organelles, the nucleolus is not bounded by a membrane; instead, it seems to be constructed by the specific binding of unfinished ribosome precursors to each other to form a large network. In a typical electron micrograph, three partially segregated regions can be distinguished (Figure 9–94): (1) a pale-staining *fibrillar center*, which contains DNA that is not being actively transcribed; (2) a *dense fibrillar component*, which contains RNA molecules in the process of transcription; and (3) a *granular component*, which contains maturing ribosomal precursor particles.

The size of the nucleolus reflects its activity and therefore varies greatly in different cells and can change in a single cell. It is very small in some dormant plant cells, for example, but can occupy up to 25% of the total nuclear volume in cells that are making unusually large amounts of protein. The differences in size are due largely to differences in the amount of the granular component, which is probably controlled at the level of ribosomal gene transcription: electron microscopy of spread chromatin shows that both the fraction of activated ribosomal genes and the rate at which each gene is transcribed can vary according to circumstances.

The Nucleolus Is Reassembled on Specific Chromosomes After Each Mitosis[58]

The appearance of the nucleolus changes dramatically during the cell cycle. As the cell approaches mitosis, the nucleolus first decreases in size and then disappears as the chromosomes condense and all RNA synthesis stops, so that generally there is no nucleolus in a metaphase cell. When ribosomal RNA synthesis restarts at the end of mitosis (in telophase), tiny nucleoli reappear at the chromosomal locations of the ribosomal RNA genes (Figure 9–95).

In humans the ribosomal RNA genes are located near the tips of each of 5 different chromosomes, as shown previously in Figure 9–40 (that is, on 10 of the 46 chromosomes in a diploid cell). Correspondingly, 10 small nucleoli form after mitosis in a human cell, although they are rarely seen as separate entities because they quickly grow and fuse to form the single large nucleolus typical of many interphase cells (Figure 9–96).

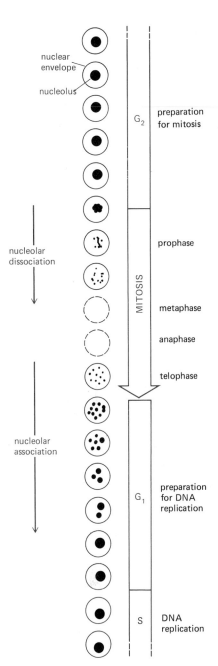

Figure 9–95 Changes in the appearance of the nucleolus in a human cell during the cell cycle. Only the cell nucleus is represented in this diagram.

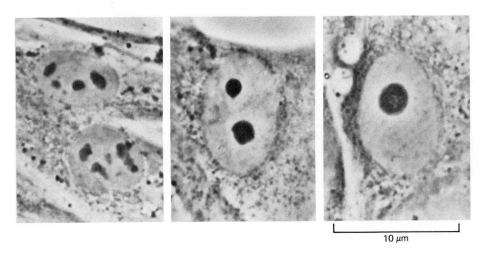

Figure 9–96 Light micrographs of human fibroblasts grown in culture, showing various stages of nucleolar fusion. (Courtesy of E.G. Jordan and J. McGovern.)

10 μm

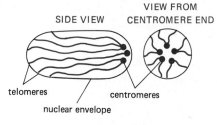

Figure 9–97 The polarized orientation of chromosomes in interphase cells of the early *Drosophila* embryo. (A) Diagrams of the *Rabl orientation*, with all centromeres facing one nuclear pole and all telomeres pointing toward the opposite pole. In the embryo each nucleus is elongated as shown. (B) Low-magnification light micrograph of a *Drosophila* embryo at the cellular blastoderm stage, in which the chromosomes in each interphase nucleus have been stained with a fluorescent dye. Note that the most brightly staining region (the chromocenter), which is known to contain the centromeric regions of each of the four chromosomes (see Figure 9–43), is oriented toward the outer surface of the embryo and thus faces the apical plasma membrane of every cell. (Courtesy of John Sedat.)

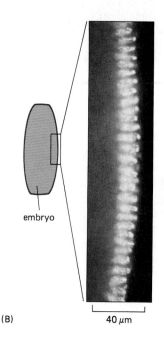

What happens to the RNA and protein components of the disassembled nucleolus during mitosis? It seems that at least some of them become distributed over the surface of all of the metaphase chromosomes and are carried as cargo to each of the two daughter cell nuclei. As the chromosomes decondense at telophase, these "old" nucleolar components help reestablish the newly emerging nucleoli.

Individual Chromosomes Occupy Discrete Territories in the Nucleus During Interphase[59]

As we have just seen, specific genes from separate interphase chromosomes are brought together at a single site in the nucleus when the nucleolus forms. This raises the question of whether other parts of chromosomes are also nonrandomly ordered in the nucleus. First raised by biologists in the late nineteenth century, this fundamental question has still not been answered satisfactorily.

A certain degree of chromosomal order results from the configuration that the chromosomes always have at the end of mitosis. Just before a cell divides, the condensed chromosomes are pulled to each spindle pole by microtubules attached to the centromeres; thus the centromeres lead the way and the distal arms of the chromosomes (terminating in telomeres) lag behind (see p. 773). The chromosomes in many nuclei tend to retain this so-called *Rabl orientation* throughout interphase, with their centromeres facing one pole of the nucleus and their telomeres pointing toward the opposite pole (Figure 9–97A). In some cases the nuclear poles are specifically oriented in the cell: in the early *Drosophila* embryo, for example, all the centromeres face apically (Figure 9–97B). Such fixed nuclear orientations might have important effects on cell polarity, but it is difficult to design experiments to test this possibility.

In most cells the various chromosomes are indistinguishable from one another during interphase, so that it is difficult to assess their arrangement in more detail than just described. The giant interphase chromosomes of the polytene

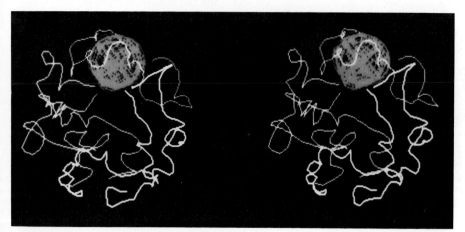

Figure 9–98 A stereo pair that displays the three-dimensional arrangement of the polytene chromosomes in a single nucleus of a *Drosophila* larval gland cell. The large ball is the nucleolus, and the course of each chromosome arm is represented by a line running along the chromosome axis. The telomeres tend to be on the surface of the nuclear envelope opposite the surface that is nearest the nucleolus, where all the centromeres are located. The chromosomes in such nuclei are never entangled, but their detailed foldings and neighbors are different in otherwise identical nuclei. (Courtesy of Mark Hochstrasser and John W. Sedat.)

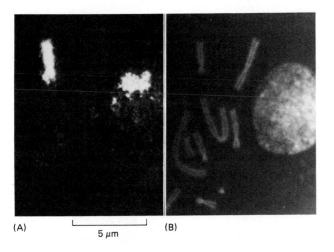
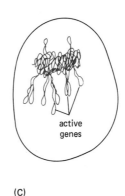

(A)

5 µm

(B)

(C)

Figure 9–99 Selective labeling of a single chromosome in a cultured mammalian cell nucleus during interphase. (A) The results of *in situ* hydridization (using a fluorescent probe) to outline the single human chromosome in a human-hamster hybrid cell line. The same preparation is shown with all of the DNA fluorescently labeled in (B). In both (A) and (B), an interphase nucleus freed from its cytoplasm is shown at the right with scattered mitotic chromosomes released from a second cell on the left. (C) Schematic drawing of the human chromosome in the interphase nucleus shown in (A). (A and B, courtesy of Joyce A. Kobori and David R. Cox.)

cells of *Drosophila* larvae, however, are an exception. Here the individual chromosome bands can be resolved clearly enough to determine the precise positions of specific genes in intact nuclei by optical-sectioning and reconstruction techniques. The results of such analyses suggest that the interphase chromosome set is not highly ordered: although the Rabl orientation tends to be maintained, two apparently identical cells often have different chromosomes as nearest neighbors.

These analyses of polytene chromosomes have also indicated that each chromosome occupies its own territory in the interphase nucleus—that is, the individual chromosomes are not extensively intertwined (Figure 9–98). Other experiments have shown that nonpolytene chromosomes also tend to occupy discrete domains in interphase nuclei. *In situ* hybridization experiments with an appropriate DNA probe, for example, can outline a single chromosome in hybrid mammalian cells grown in culture (Figure 9–99). Most of the DNA of such a chromosome is seen to occupy only a small portion of the interphase nucleus, suggesting that each individual chromosome remains compact and organized while allowing portions of its DNA to be active in RNA synthesis.

How Well Ordered Is the Nucleus?[60]

The interior of the nucleus is not a random jumble of its many RNA, DNA, and protein components. We have seen that the nucleolus is organized as an efficient ribosome-construction machine, and clusters of spliceosome components are apparently organized as discrete RNA-splicing islands (see Figure 9–89). Order is also seen in the electron microscope when one focuses on the regions around nuclear pores: the chromatin that lines the inner nuclear membrane (which is unusually condensed chromatin and therefore clearly visible in electron micrographs) is excluded from a considerable region beneath and around each nuclear pore, clearing a path between the cytoplasm and the nucleoplasm (Figure 9–100). In some

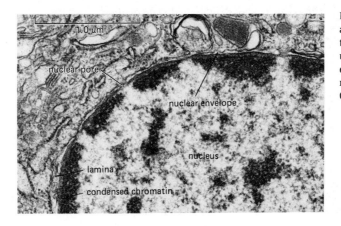

Figure 9–100 Electron micrograph of a mammalian cell nucleus, showing that the condensed chromatin underlying the nuclear envelope is excluded from regions around the nuclear pores. (Courtesy of Larry Gerace.)

special cases, moreover, the nuclear pores are found to be highly organized in the nuclear envelope (Figure 9–101). Such ordering presumably reflects a corresponding organization within the nuclear lamina to which the pores are attached.

Is there an intranuclear framework, analogous to the cytoskeleton, on which nuclear components are organized? Many cell biologists believe there is. The *nuclear matrix* or *scaffold* has been defined as the insoluble material left in the nucleus after a series of biochemical extraction steps. The proteins that constitute it can be shown to bind specific DNA sequences called *SARs* or *MARs* (for scaffold- or matrix-associated regions). Such DNA sequences have been postulated to form the base of chromosomal loops (see Figure 9–34). By means of such chromosomal attachment sites, the matrix might help organize chromosomes, localize genes (see, for example, p. 583), and regulate DNA transcription and replication within the nucleus. Because the structural components of the matrix have not yet been identified, however, it remains uncertain whether the matrix isolated by cell biologists represents a structure that is present in intact cells.

1 μm

Figure 9–101 Freeze-fracture electron micrograph of the elongated nuclear envelope of a fern spore, illustrating the ordered arrangement of the nuclear pore complexes in parallel rows. In other cells, either concentrated clusters of nuclear pores or unusual areas free of nuclear pores have been detected in the nuclear envelope, and these are specifically oriented with respect to other structures in the cell. (Courtesy of Don H. Northcote; from K. Roberts and D.H. Northcote, *Microsc. Acta* 71:102–120, 1971.)

Summary

RNA polymerase, the enzyme that catalyzes DNA transcription, is a complex molecule containing many polypeptide chains. In eucaryotic cells there are three RNA polymerases, designated polymerases I, II, and III; they are evolutionarily related to one another and to bacterial RNA polymerase, and they have some subunits in common. After initiating transcription, each enzyme is thought to release one or more subunits (called initiation factors) and to bind other subunits (called elongation factors) that are required for RNA chain elongation, termination, and modification. The elongation factors are presumably different for each type of polymerase, which would explain why the transcripts that each enzyme synthesizes are differently modified.

Most of the cell's mRNA is produced by a complex process beginning with the synthesis of heterogeneous nuclear RNA (hnRNA). The primary hnRNA transcript is made by RNA polymerase II. It is then capped by the addition of a special nucleotide to its 5' end and is cleaved and then polyadenylated at its 3' end. The modified RNA molecules are usually then subjected to one or more RNA splicing events, in which intron sequences are removed from the middle of the hnRNA by a reaction catalyzed by a large ribonucleoprotein complex known as a spliceosome. In this process most of the mass of the primary RNA transcript is removed and degraded in the nucleus. As a result, although the rate of production of hnRNA typically accounts for about half of a cell's RNA synthesis, the mRNA produced represents only about 3% of the steady-state quantity of RNA in a cell.

Unlike genes that code for proteins, which are transcribed by polymerase II, the genes that code for most structural RNAs are transcribed by polymerase I and III. These genes are usually repeated many times in the genome and are often clustered in tandem arrays. RNA polymerase III makes a variety of stable small RNAs, including the tRNAs and the small 5S rRNA of the ribosome. RNA polymerase I makes the large rRNA precursor molecule (45S rRNA) containing the major rRNAs. Except for those in mitochondria and chloroplasts, all the cell's ribosomes are assembled in the nucleolus—a distinct intranuclear organelle that is formed around the tandemly arranged rRNA genes, which are brought together from several chromosomes.

References

General

Lewin, B. Gene Expression, Vol. 2: Eucaryotic Chromosomes, 2nd ed. New York: Wiley, 1980.
Lewin, B. Genes, 3rd ed. New York: Wiley, 1987.

Newport, J.W.; Forbes, D.J. The nucleus: structure, function, and dynamics. *Annu. Rev. Biochem.* 56:535–566, 1987.
Watson, J.D.; Hopkins, N.H.; Roberts, J.W.; Steitz, J.A.; Weiner, A.M. Molecular Biology of the Gene, 4th ed. Menlo Park, CA: Benjamin-Cummings, 1987.

Cited

1. Adolph, K.W., ed. Chromosomes and Chromatin, Vols. 1–3. Boca Raton, FL: CRC Press, 1988.

   Felsenfeld, G. DNA. *Sci. Am.* 253(4):58–67, 1985.

   Hsu, T.C. Human and Mammalian Cytogenetics: A Historical Perspective. New York: Springer-Verlag, 1979.

2. Kavenoff, R.; Klotz, L.C.; Zimm, B.H. On the nature of chromosome-sized DNA molecules. *Cold Spring Harbor Symp. Quant. Biol.* 38:1–8, 1974.

3. Burke, D.T.; Carle, G.F.; Olson, M.V. Cloning of large segments of exogenous DNA into yeast by means of artificial chromosome vectors. *Science* 236:806–812, 1987.

   Murray, A.W. Chromosome structure and behavior. *Trends Biochem. Sci.* 10:112–115, 1985.

4. Gall, J.G. Chromosome structure and the C-value paradox. *J. Cell Biol.* 91:3s–14s, 1981.

   Ohta, T.; Kimura, M. Functional organization of genetic material as a product of molecular evolution. *Nature* 233:118–119, 1971.

5. Mapping and Sequencing the Human Genome. Washington, DC: National Academy Press, 1988.

   Wilson, A.C.; Ochman, H.; Prager, E.M. Molecular time scale for evolution. *Trends Genet.* 3:241–247, 1987.

6. Fried, M.; Crothers, D.M. Equilibria and kinetics of *lac* repressor-operator interactions by polyacrylamide gel electrophoresis. *Nucleic Acids Res.* 9:6505–6525, 1981.

7. Kadonaga, J.T.; Tjian, R. Affinity purification of sequence-specific DNA binding proteins. *Proc. Natl. Acad. Sci. USA* 83:5889–5893, 1986.

   Rosenfeld, P.J.; Kelly, T.J. Purification of nuclear factor I by DNA recognition site affinity chromatography. *J. Biol. Chem.* 261:1398–1408, 1986.

   Staudt, L.M.; et al. Cloning of a lymphoid-specific cDNA encoding a protein binding the regulatory octamer DNA motif. *Science* 241:577–580, 1988.

8. Pabo, C.T.; Sauer, R.T. Protein-DNA interactions. *Annu. Rev. Biochem.* 53:293–321, 1984.

   Schleif, R. DNA binding by proteins. *Science* 241:1182–1187, 1988.

9. Ptashne, M. A Genetic Switch: Gene Control and Phage Lambda. Palo Alto, CA: Blackwell, 1986.

   Takeda, Y.; Ohlendorf, D.H.; Anderson, W.F.; Matthews, B.W. DNA-binding proteins. *Science* 221:1020–1026, 1983.

10. Berg, O.G.; von Hippel, P.H. Selection of DNA binding sites by regulatory proteins. *Trends Biochem. Sci.* 13:207–211, 1988.

    von Hippel, P.H.; Bear, D.G.; Morgan, W.D.; McSwiggen, J.A. Protein-nucleic acid interaction in transcription. *Annu. Rev. Biochem.* 53:389–446, 1984.

11. Dickerson, R.E. The DNA helix and how it is read. *Sci. Am.* 249(6):94–111, 1983.

    Drew, H.R.; McCall, M.J.; Callandine, C.R. Recent studies of DNA in the crystal. *Annu. Rev. Cell Biol.* 4:1–20, 1988.

12. Koo, H.S.; Crothers, D.M. Calibration of DNA curvature and a unified description of sequence-directed bending. *Proc. Natl. Acad. Sci. USA* 85:1763–1767, 1988.

    Lilley, D. Bent molecules—how and why? *Nature* 320:487, 1986.

    Marini, J.C.; Levene, S.D.; Crothers, D.M.; Englund, P.T. Bent helical structure in kinetoplast DNA. *Proc. Natl. Acad. Sci. USA* 80:7678–7682, 1982.

13. Echols, H. Multiple DNA-protein interactions governing high-precision DNA transactions. *Science* 233:1050–1056, 1986.

    Thompson, J.F.; de Vargas, L.M.; Koch, C.; Kahmann, R.; Landy, A. Cellular factors couple recombination with growth phase: characterization of a new component in the lambda site-specific recombination pathway. *Cell* 50:901–908, 1987.

14. Isenberg, I. Histones. *Annu. Rev. Biochem.* 48:159–191, 1979.

    von Holt, C. Histones in perspective. *Bioessays* 3:120–124, 1985.

    Wells, D.E. Compilation analysis of histones and histone genes. *Nucleic Acids Res.* 14:r119–r149, 1986.

    Wu, R.S.; Panusz, H.T.; Hatch, C.L.; Bonner, W.M. Histones and their modifications. *CRC Crit. Rev. Biochem.* 20:201–263, 1986.

15. Chromatin. *Cold Spring Harbor Symp. Quant. Biol.*, Vol. 42, 1978.

    Kornberg, R.D.; Klug A. The nucleosome. *Sci. Am.* 244(2):52–64, 1981.

    McGhee, J.D.; Felsenfeld, G. Nucleosome structure. *Annu. Rev. Biochem.* 49:1115–1156, 1980.

    Richmond, T.J.; Finch, J.T.; Rushton, B.; Rhodes, D.; Klug, A. Structure of the nucleosome core particle at 7 Å resolution. *Nature* 311:532–537, 1984.

16. Simpson, R.T. Nucleosome positioning *in vivo* and *in vitro*. *Bioessays* 4:172–176, 1986.

    Travers, A.A. DNA bending and nucleosome positioning. *Trends Biochem. Sci.* 12:108–112, 1987.

17. Eissenberg, J.C.; Cartwright, I.L.; Thomas, G.H.; Elgin, S.C. Selected topics in chromatin structure. *Annu. Rev. Genet.* 19:485–536, 1985.

    Emerson, B.M.; Lewis, C.D.; Felsenfeld, G. Interaction of specific nuclear factors with the nuclease-hypersensitive region of the chicken adult β-globin gene: nature of the binding domain. *Cell* 41:21–30, 1985.

    Gross, D.S.; Garrard, W.T. Nuclease hypersensitive sites in chromatin. *Annu. Rev. Biochem.* 57:159–198, 1988.

18. Belmont, A.S.; Sedat, J.W.; Agard, D.A. A three-dimensional approach to mitotic chromosome structure: evidence for a complex hierarchical organization. *J. Cell Biol.* 105:77–92, 1987.

    Pederson, D.S.; Thoma, F.; Simpson, R. Core particle, fiber, and transcriptionally active chromatin structure. *Annu. Rev. Cell Biol.* 2:117–147, 1986.

19. Allan, J.; Hartman, P.G.; Crane-Robinson, C.; Aviles, F.X. The structure of histone H1 and its location in chromatin. *Nature* 288:675–679, 1980.

    Clark, D.J.; Thomas, J.O. Salt-dependent cooperative interaction of histone H1 with linear DNA. *J. Mol. Biol.* 187:569–580, 1986.

    Coles, L.S.; Robins, A.J.; Madley, L.K.; Wells, J.R. Characterization of the chicken histone H1 gene complement: generation of a complete set of vertebrate H1 protein sequences. *J. Biol. Chem.* 262:9656–9663, 1987.

    Thoma, F.; Koller, T.; Klug, A. Involvement of histone H1 in the organization of the nucleosome and of the salt-dependent superstructures of chromatin. *J. Cell Biol.* 83:403–427, 1979.

20. De Bernardin, W.; Koller, T.; Sogo, J.M. Structure of *in vivo* transcribing chromatin as studied in simian virus 40 minichromosomes. *J. Mol. Biol.* 191:469–482, 1986.

    Losa, R.; Brown, D.D. A bacteriophage RNA polymerase transcribes *in vitro* through a nucleosome core without displacing it. *Cell* 50:801–808, 1987.

21. Benyajati, C.; Worcel, A. Isolation, characterization, and structure of the folded interphase genome of *Drosophila melanogaster*. *Cell* 9:393–408, 1976.

    Gasser, S.M.; Laemmli, U.K. A glimpse at chromosomal order. *Trends Genet.* 3:16–22, 1987.

    Schmid, M.B. Structure and function of the bacterial chromosome. *Trends Biochem. Sci.* 13:131–135, 1988.

22. Marsden, M.; Laemmli, U.K. Metaphase chromosome structure: evidence for a radial loop model. *Cell* 17:849–858, 1979.

    Georgiev, G.P.; Nedospasov, S.A.; Bakayev, V.V. Supranucleosomal levels of chromatin organization. In The Cell Nucleus (H. Busch, ed.), Vol. 6, pp. 3–34. New York: Academic Press, 1978.

23. Holmquist, G. DNA sequences in G-bands and R-bands. In Chromosomes and Chromatin Structure (K.W. Adolph, ed.), Vol. 2, pp. 75–122. Boca Raton, FL: CRC Press, 1988.

Lewin, B. Gene Expression, Vol. 2: Eucaryotic Chromosomes, 2nd ed., pp. 428–440. New York: Wiley, 1980.

24. Bostock, C.J.; Sumner, A.T. The Eucaryotic Chromosome, pp. 347–374. Amsterdam: North-Holland, 1978.

Callan, H.G. Lampbrush chromosomes. *Proc. R. Soc. Lond. (Biol.)* 214:417–448, 1982.

Roth, M.B.; Gall, J.G. Monoclonal antibodies that recognize transcription unit proteins on newt lampbrush chromosomes. *J. Cell Biol.* 105:1047–1054, 1987.

25. Agard, D.A.; Sedat, J.W. Three-dimensional architecture of a polytene nucleus. *Nature* 302:676–681, 1983.

Beermann, W. Chromosomes and genes. In Developmental Studies on Giant Chromosomes (W. Beermann, ed.), pp. 1–33. New York: Springer-Verlag, 1972.

26. Ashburner, M.; Chihara, C.; Meltzer, P.; Richards, G. Temporal control of puffing activity in polytene chromosomes. *Cold Spring Harbor Symp. Quant. Biol.* 38:655–662, 1974.

Lamb, M.M.; Daneholt, B. Characterization of active transcription units in Balbiani rings of *Chironomus tentans*. *Cell* 17:835–848, 1979.

27. Bossy, B.; Hall, L.M.; Spierer, P. Genetic activity along 315 kb of the *Drosophila* chromosome. *EMBO J.* 3:2537–2541, 1984.

Hill, R.J.; Rudkin, G. Polytene chromosomes: the status of the band-interband question. *Bioessays* 7:35–40, 1987.

Judd, B.H.; Young, M.W. An examination of the one cistron: one chromomere concept. *Cold Spring Harbor Symp. Quant. Biol.* 38:573–579, 1974.

28. Garel, A.; Zolan, M.; Axel, R. Genes transcribed at diverse rates have a similar conformation in chromatin. *Proc. Natl. Acad. Sci. USA* 74:4867–4871, 1977.

Weintraub, H.; Groudine, M. Chromosomal subunits in active genes have an altered conformation. *Science* 193:848–856, 1976.

Yaniv, M.; Cereghini, S. Structure of transcriptionally active chromatin. *CRC Crit. Rev. Biochem.* 21:1–26, 1986.

29. Allis, C.D.; et al. hv1 is an evolutionarily conserved H2A variant that is preferentially associated with active genes. *J. Biol. Chem.* 261:1941–1948, 1986.

Dorbic, T.; Wittig, B. Chromatin from transcribed genes contains HMG17 only downstream from the starting point of transcription. *EMBO J.* 6:2393–2399, 1987.

Hebbes, T.R.; Thorne, A.W.; Crane-Robinson, C. A direct link between core histone acetylation and transcriptionally active chromatin. *EMBO J.* 7:1395–1402, 1988.

Rose, S.M.; Garrard, W.T. Differentiation-dependent chromatin alterations precede and accompany transcription of immunoglobulin light chain genes. *J. Biol. Chem.* 259:8534–8544, 1984.

30. Brown, S.W. Heterochromatin. *Science* 151:417–425, 1966.

James, T.C.; Elgin, S.C.R. Identification of a nonhistone chromosomal protein associated with heterochromatin in *Drosophila melanogaster* and its gene. *Mol. Cell. Biol.* 6:3862–3872, 1986.

Pimpinelli, S.; Bonaccorsi, S.; Gatti, M.; Sandler, L. The peculiar genetic organization of *Drosophila* heterochromatin. *Trends Genet.* 2:17–20, 1986.

31. Hand, R. Eucaryotic DNA: organization of the genome for replication. *Cell* 15:317–325, 1978.

Huberman, J.A.; Riggs, A.D. On the mechanism of DNA replication in mammalian chromosomes. *J. Mol. Biol.* 32:327–341, 1968.

32. Campbell, J. Eucaryotic DNA replication: yeast bares its ARSs. *Trends Biochem. Sci.* 13:212–217, 1988.

Palzkill, T.G.; Newlon, C.S. A yeast replication origin consists of multiple copies of a small conserved sequence. *Cell* 53:441–450, 1988.

Struhl, K.; Stinchcomb, D.T.; Sherer, S.; Davis, R.W. High-frequency transformation of yeast: autonomous replication of hybrid DNA molecules. *Proc. Natl. Acad. Sci. USA* 76:1035–1039, 1979.

33. Dodson, M.; Dean, F.B., Bullock, P.; Echols, H.; Hurwitz, J. Unwinding of duplex DNA from the SV40 origin of replication by T antigen. *Science* 238:964–967, 1987.

Li, J.J.; Kelly, T.J. Simian virus 40 DNA replication *in vitro*. *Proc. Natl. Acad. Sci. USA* 81:6973–6977, 1984.

So, A.G.; Downey, K.M. Mammalian DNA polymerases alpha and delta: current status in DNA replication. *Biochemistry* 27:4591–4595, 1988.

Stahl, H.; Dröge, P.; Knippers, R. DNA helicase activity of SV40 large T antigen. *EMBO J.* 5:1939–1944, 1986.

34. Russev, G.; Hancock, R. Assembly of new histones into nucleosomes and their distribution in replicating chromatin. *Proc. Natl. Acad. Sci. USA* 79:3143–3147, 1982.

Sariban, E.R.; Wu, R.S.; Erickson, L.C.; Bonner, W.M. Interrelationships of protein and DNA synthesis during replication in mammalian cells. *Mol. Cell. Biol.* 5:1279–1286, 1985.

Weintraub, H.; Worcel, A.; Alberts, B.M. A model for chromatin based upon two symmetrically paired half-nucleosomes. *Cell* 9:409–417, 1976.

Worcel, A.; Han, S.; Wong, M.L. Assembly of newly replicated chromatin. *Cell* 15:969–977, 1978.

35. Blackburn, E.H.; Szostak, J.W. The molecular structure of centromeres and telomeres. *Annu. Rev. Biochem.* 53:163–194, 1984.

Greider, C.W.; Blackburn, E.H. The telomere terminal transferase of *Tetrahymena* is a ribonucleoprotein enzyme with two kinds of primer specificity. *Cell* 51:887–898, 1987.

36. Stubblefield, E. Analysis of the replication pattern of Chinese hamster chromosomes using 5-bromodeoxyuridine suppression of 33258 Hoechst fluorescence. *Chromosoma* 53:209–221, 1975.

37. Lima-de-Faria, A.; Jaworska, H. Late DNA synthesis in heterochromatin. *Nature* 217:138–142, 1968.

38. Brown, E.H.; et al. Rate of replication of the murine immunoglobulin heavy-chain locus: evidence that the region is part of a single replicon. *Mol. Cell. Biol.* 7:450–457, 1987.

39. Callan, H.G. DNA Replication in the chromosomes of eukaryotes. *Cold Spring Harbor Symp. Quant. Biol.* 38:195–203, 1974.

Kriegstein, H.J.; Hogness, D.S. Mechanism of DNA replication in *Drosophila* chromosomes: structure of replication forks and evidence for bidirectionality. *Proc. Natl. Acad. Sci. USA* 71:135–139, 1974.

Mechali, M.; Kearsey, S. Lack of specific sequence requirement for DNA replication in *Xenopus* eggs compared with high sequence specificity in yeast. *Cell* 38:55–64, 1984.

40. Blow, J.J.; Laskey, R.A. A role for the nuclear envelope in controlling DNA replication within the cell cycle. *Nature* 332:546–548, 1988.

Harland, R. Initiation of DNA replication in eukaryotic chromosomes. *Trends Biochem. Sci.* 6:71–74, 1981.

Rao, P.N.; Johnson, R.T. Mammalian cell fusion: studies on the regulation of DNA synthesis and mitosis. *Nature* 225:159–164, 1970.

41. Watson, J.D.; Hopkins, N.H.; Roberts, J.W.; Steitz, J.A.; Weiner, A.M. Molecular Biology of the Gene, 4th ed. Menlo Park, CA: Benjamin-Cummings, 1987. (Chapters 13, 20, and 21.)

42. Chamberlin, M. Bacterial DNA-dependent RNA polymerases. In The Enzymes, 3rd ed. (P. Boyer, ed.), Vol. 15B, pp. 61–108. New York: Academic Press, 1982.

Greenblatt, J.; Li, J. Interaction of the sigma factor and the *nusA* gene protein of *E. coli* with RNA polymerase in the initiation-termination cycle of transcription. *Cell* 24:421–428, 1981.

McClure, W. Mechanism and control of transcription initiation in prokaryotes. *Annu. Rev. Biochem.* 54:171–204, 1985.

Yager, T.D.; von Hippel, P.H. Transcript elongtation and termination in *E. coli.* In *Escherichia coli* and *Salmonella typhimurium:* Cellular and Molecular Biology (F.C. Neidhardt, ed.), pp. 1241–1275. Washington, DC: American Society for Microbiology, 1987.

43. Chambon, P. Eucaryotic nuclear RNA polymerases. *Annu. Rev. Biochem.* 44:613–638, 1975.

Geiduschek, E.P.; Tocchini-Valentini, G.P. Transcription by RNA polymerase III. *Annu. Rev. Biochem.* 57:873–914, 1988.

Sentenac, A. Eucaryotic RNA polymerases. *CRC Crit. Rev. Biochem.* 18:31–91, 1985.

Sollner-Webb, B.; Tower, J. Transcription of cloned eucaryotic ribosomal RNA genes. *Annu. Rev. Biochem.* 55:801–830, 1986.

44. Brown, D.D. The role of stable complexes that repress and activate eucaryotic genes. *Cell* 37:359–365, 1984.

Workman, J.L.; Roeder, R.G. Binding of transcription factor TFIID to the major late promoter during *in vitro* nucleosome assembly potentiates subsequent initiation by RNA polymerase II. *Cell* 51:613–622, 1987.

45. Foe, V.E.; Wilkinson, L.E.; Laird, C.D. Comparative organization of active transcription units in *Oncopeltus fasciatus. Cell* 9:131–146, 1976.

Hastie, N.D.; Bishop, J.O. The expression of three abundance classes of mRNA in mouse tissues. *Cell* 9:761–774, 1976.

Lewin, B. Gene Expression, Vol. 2: Eucaryotic Chromosomes, 2nd ed., pp. 708–719. New York: Wiley, 1980.

Miller, O.L. The nucleolus, chromosomes, and visualization of genetic activity. *J. Cell Biol.* 91:15s–27s, 1981. (A review.)

46. Birnstiel, M.L.; Busslinger, M.; Strub, K. Transcription termination and 3′ processing: the end is in site! *Cell* 41:349–359, 1985.

Friedman, D.I.; Imperiale, M.J.; Adhya, S.L. RNA 3′ end formation in the control of gene expression. *Annu. Rev. Genet.* 21:453–488, 1987.

Nevins, J.R. The pathway of eukaryotic mRNA formation. *Annu. Rev. Biochem.* 52:441–466, 1983.

Takagaki, Y.; Ryner, L.C.; Manley, J.L. Separation and characterization of a poly(A) polymerase and a cleavage/specificity factor required for pre-mRNA polyadenylation. *Cell* 52:731–742, 1988.

47. Sisodia, S.S.; Sollner-Webb, B.; Cleveland, D.W. Specificity of RNA maturation pathways: RNAs transcribed by RNA polymerase III are not substrates for splicing or polyadenylation. *Mol. Cell Biol.* 7:3602–3612, 1987.

Smale, S.T.; Tjian, R. Transcription of herpes simplex virus tk sequences under the control of wild-type and mutant human RNA polymerase I promoters. *Mol. Cell. Biol.* 5:352–362, 1985.

48. Chambon, P. Split genes. *Sci. Am.* 244(5):60–71, 1981.

Crick, F. Split genes and RNA splicing. *Science* 204:264–271, 1979.

Darnell, J.E., Jr. Variety in the level of gene control in eucaryotic cells. *Nature* 297:365–371, 1982.

Perry, R.P. RNA processing comes of age. *J. Cell Biol.* 91:28s–38s, 1981. (Includes a historical review.)

49. Guthrie, C.; Patterson, B. Spliceosomal snRNAs. *Annu. Rev. Genet.* 22:387–419, 1988.

Dreyfuss, G.; Swanson, M.S.; Piñol-Roma, S. Heterogeneous nuclear ribonucleoprotein particles and the pathway of mRNA formation. *Trends Biochem. Sci.* 13:86–91, 1988.

Osheim, Y.N.; Miller, O.L.; Beyer, A.L. RNP particles at splice junction sequences on *Drosophila* chorion transcripts. *Cell* 43:143–151, 1985.

Samarina, O.P.; Krichevskaya, A.A.; Georgiev, G.P. Nuclear ribonucleoprotein particles containing messenger ribonucleic acid. *Nature* 210:1319–1322, 1966.

Steitz, J.A. "Snurps." *Sci. Am.* 258(6):56–63, 1988.

50. Edmonds, M. Branched RNA. *Bioessays* 6:212–216, 1987.

Maniatis, T.; Reed, R. The role of small nuclear ribonucleoprotein particles in pre-mRNA splicing. *Nature* 325:673–678, 1987.

Padgett, R.A.; Grabowski, P.J.; Konarska, M.M.; Seiler, S.; Sharp, P.A. Splicing of messenger RNA precursors. *Annu. Rev. Biochem.* 55:1119–1150, 1986.

51. Aebi, M.; Weissman, C. Precision and orderliness in splicing. *Trends Genet.* 3:102–107, 1987.

52. Andreadis, A.; Gallego, M.E.; Nadal-Ginard, B. Generation of protein isoform diversity by alternative splicing: mechanistic and biological implications. *Annu. Rev. Cell Biol.* 3:207–242, 1987.

53. Orkin, S.H.; Kazazian, H.H. The mutation and polymorphism of the human β-globin gene and its surrounding DNA. *Annu. Rev. Genet.* 18:131–171, 1984.

54. Cech, T.R. The generality of self-splicing RNA: relationship to nuclear mRNA splicing. *Cell* 44:207–210, 1986.

55. Ringertz, N.; et al. Computer analysis of the distribution of nuclear antigens: studies on the spatial and functional organization of the interphase nucleus. *J. Cell Sci.* Suppl. 4:11–28, 1986.

Warner, J. Applying genetics to the splicing problem. *Genes Dev.* 1:1–3, 1987.

56. Long, E.O.; Dawid, I.B. Repeated genes in eucaryotes. *Annu. Rev. Biochem.* 49:727–764, 1980.

Miller, O.L. The nucleolus, chromosomes, and visualization of genetic activity. *J. Cell Biol.* 91:15s–27s, 1981.

57. Hadjiolov, A.A. The Nucleolus and Ribosome Biogenesis. New York: Springer-Verlag, 1985.

Jordan, E.G.; Cullis, C.A., eds. The Nucleolus. Cambridge, U.K.: Cambridge University Press, 1982.

Sommerville, J. Nucleolar structure and ribosome biogenesis. *Trends Biochem. Sci.* 11:438–442, 1986.

58. Anastassova-Kristeva, M. The nucleolar cycle in man. *J. Cell Sci.* 25:103–110, 1977.

McClintock, B. The relation of a particular chromosomal element to the development of the nucleoli in Zea Mays. *Z. Zellforsch. Mikrosk. Anat.* 21:294–323, 1934.

59. Comings, D.E. Arrangement of chromatin in the nucleus. *Hum. Genet.* 53:131–143, 1980.

Cremer, T.; et al. Rabl's model of the interphase chromosome arrangement tested in Chinese hamster cells by premature chromosome condensation and laser-UV-microbeam experiments. *Hum. Genet.* 60:46–56, 1980.

Hochstrasser, M.; Sedat, J.W. Three-dimensional organization of *Drosophila melanogaster* interphase nuclei. II. Chromosome spatial organization and gene regulation. *J. Cell Biol.* 104:1471–1483, 1987.

Manuelidis, L. Individual interphase chromosome domains revealed by *in situ* hybridization. *Hum. Genet.* 71:288–293, 1985.

60. Gasser, S.M.; Laemmli, U.K. A glimpse at chromosome order. *Trends Genet.* 3:16–22, 1987.

Gerace, L.; Burke, B. Functional organization of the nuclear envelope. *Annu. Rev. Cell Biol.* 4:335–374, 1988.

Newport, J.W.; Forbes, D.J. The nucleus: structure, function, and dynamics. *Annu. Rev. Biochem.* 56:535–566, 1987.

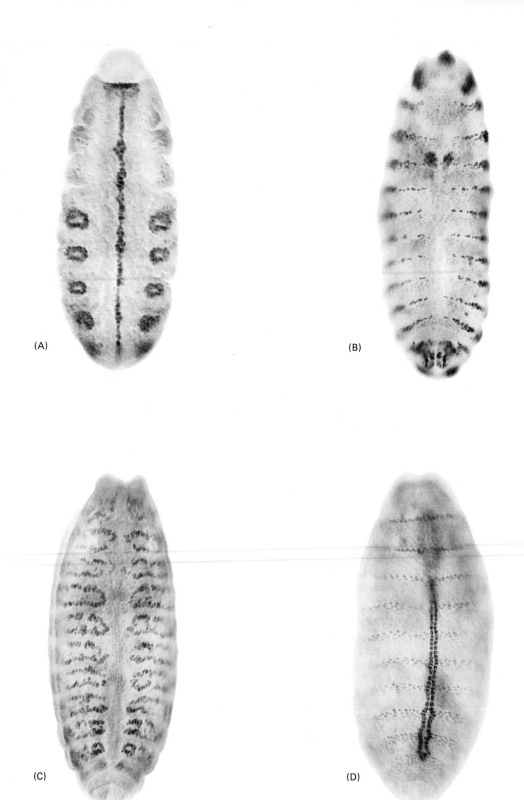

Position-dependent patterns of gene expression in four different transgenic *Drosophila* embryos. Each embryo has a single copy of a bacterial B-galactosidase gene inserted into its genome. Depending on the chromosomal insertion site, nearby *Drosophila* enhancers cause the gene to be expressed in a pattern that corresponds either to (A) trachea and mesectoderm, (B) sensory organs, (C) muscle, or (D) dorsal epidermal stripe plus a segmentally repeated pattern. (Courtesy of Yuh-Nung Jan and Larry Ackerman.)

Control of Gene Expression

10

An organism's DNA sequences encode all of the RNA and protein molecules that are available to construct its cells. Yet a complete description of the DNA sequence of a genome—be it the few million nucleotides of a bacterium or the 3 billion nucleotides of a human—would provide relatively little understanding of the organism itself. It has been said that the genome represents a complete "dictionary" for an organism, containing all of the "words" available for its construction. But we can no more reconstruct the logic of an organism from such a dictionary than we can reconstruct a play by Shakespeare from a dictionary of English words. In both cases the problem is to know how the elements in the dictionary are used; the number of possible combinations of elements is so vast that obtaining the dictionary itself is the relatively easy part and only a start toward solving the problem.

Of course, we are still very far from being able to "write" an organism from the sequence of its genome. This will require a much more complete understanding of all of cell biology, including knowledge of how the thousands of large and small molecules in a cell behave once they have been synthesized. In this chapter we discuss a more limited aspect of the problem of how the genome determines the form and behavior of an organism: we consider the rules by which a subset of the genes are selectively activated in each cell. We shall see that the mechanisms that control the *expression* of genes operate at a variety of levels, and in the central sections of this chapter we discuss the different levels in turn. At the end of the chapter we shall consider how genes and their elaborate regulatory networks might have evolved. We begin, however, with an overview of some basic principles of gene control in higher organisms.

Strategies of Gene Control

The different cell types in a higher organism often differ dramatically in both morphology and function. If we compare a mammalian neuron with a lymphocyte, for example (see Figure 13–29, p. 750), the differences are so extreme that it is difficult to imagine that the two cells contain the same genome. For this reason, and because cell differentiation is usually irreversible, biologists originally suspected that genes might be selectively lost when a cell differentiates. We now know, however, that cell differentiation generally depends on changes in gene expression rather than on gene loss.

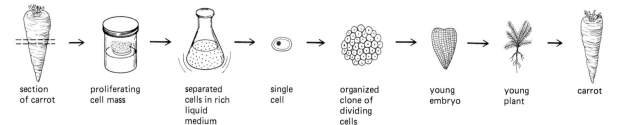

section of carrot → proliferating cell mass → separated cells in rich liquid medium → single cell → organized clone of dividing cells → young embryo → young plant → carrot

The Different Cell Types of a Multicellular Organism Contain the Same DNA[1]

The cell types in a multicellular organism become different from one another because they synthesize and accumulate different sets of RNA and protein molecules. They do this without changing their DNA in an irreversible way. The best evidence for the preservation of the genome during cell differentiation comes from a classic set of experiments in frogs. When the nucleus of a fully differentiated frog cell is injected into a frog egg whose nucleus has been removed, the injected "donor" nucleus is capable of programming the recipient egg to produce a normal tadpole (see p. 891). Because the tadpole contains a full range of differentiated cells that derived their DNA sequences from the nucleus of the original donor cell, it follows that the differentiated donor cell cannot have lost any important DNA sequences. A similar conclusion has been reached in experiments done with various plants. Here differentiated pieces of tissue were cultured in synthetic media and then dissociated into single cells. Often, one of these individual cells can regenerate an entire adult plant (Figure 10–1).

Further evidence that large blocks of DNA are not lost or rearranged during vertebrate development comes from comparing the detailed banding patterns detectable in condensed chromosomes at mitosis in different cell types (see Figure 9–40, p. 505). By this criterion the chromosome sets of all differentiated cells in the human body appear to be identical. Moreover, comparisons of the genomes of different cells based on recombinant DNA technology (see p. 180) have shown, as a general rule, that the changes in gene expression that underlie the development of multicellular organisms are not accompanied by changes in the DNA sequences of the corresponding genes (for an important exception, see p. 1024).

Figure 10–1 An experiment that can be carried out in many types of plants. Differentiated plant cells often retain the ability to "dedifferentiate," so that a single cell can then form a clone of progeny cells that can later give rise to an entire plant (see Chapter 20).

Different Cell Types Synthesize Different Sets of Proteins[2]

As a first step in trying to understand cell differentiation, one would like to know how many differences there are between one cell type and another. Although we still do not know the answer to this fundamental question, certain general statements can be made:

1. Many processes are common to all cells, and any two cells in a single organism therefore have many proteins in common. These include some abundant proteins that are easy to analyze, such as the major structural proteins of the cytoskeleton and of chromosomes, some of the proteins essential to the endoplasmic reticulum and Golgi membranes, ribosomal proteins, and so on. Many nonabundant proteins, such as various enzymes involved in the central reactions of metabolism (see p. 81), are also the same in all cell types.

2. Some proteins are abundant in the specialized cells in which they function and cannot be detected elsewhere, even by sensitive tests. Hemoglobin, for example, can be detected only in red blood cells.

3. If the 2000 or so most abundant proteins (those present in quantities of 50,000 or more copies per cell) are compared among different cell types of the same organism using two-dimensional polyacrylamide-gel electrophoresis (see p. 171), remarkably few differences are found. Whether the comparison is between two cell lines grown in culture (such as muscle and nerve cells) or between cells of two young rodent tissues (such as liver and lung), the great majority of the proteins detected are synthesized in both cell types and at rates that differ by less than a factor of five; only a few percent of the proteins are present in very different amounts in the two cell types.

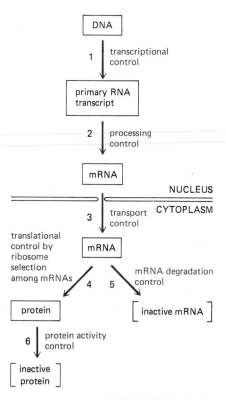

Figure 10–2 Six steps at which gene expression can be controlled in eucaryotes. Only control steps 1 through 5 are discussed in this chapter. After a protein has been synthesized, its activity can be controlled (step 6) by regulated protein degradation (irreversible inactivation, see p. 419), by reversible protein modifications such as phosphorylation (see p. 417), and by targeting the protein to selected locations in the cell (see p. 412).

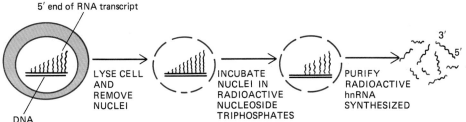

5' end of RNA transcript

DNA

LYSE CELL AND REMOVE NUCLEI

INCUBATE NUCLEI IN RADIOACTIVE NUCLEOSIDE TRIPHOSPHATES

PURIFY RADIOACTIVE hnRNA SYNTHESIZED

3'
5'

Figure 10–3   Obtaining radioactively labeled, newly synthesized hnRNA by "nuclear run on." The RNA polymerase molecules caught in the act of transcription in the cell continue elongating the same RNA molecules *in vitro*, thus providing a highly radioactive sample of the cell's transcribed sequences (*primary RNA transcripts*).

Studies of the number of different mRNA sequences in a cell suggest that a typical higher eucaryotic cell synthesizes 10,000 to 20,000 different proteins (Table 9–2, p. 528). Most of these are too rare to be detected by two-dimensional gel electrophoresis of cell extracts. If these minor cell proteins differ between cells to the same extent as the more abundant proteins, as is commonly assumed, only a small number of protein differences (perhaps several hundred) suffice to create very large differences in cell morphology and behavior.

10-3 Gene Expression Can Be Regulated at Each Step in the Pathway from DNA to RNA to Protein[3]

If the differences between different cell types depend on the particular genes that they express, at what level is the control of gene expression exercised? There are many steps in the pathway leading from DNA to protein, and there is evidence that all of them can be regulated (Figure 10–2). Thus a cell can control the proteins it makes by (1) controlling when and how often a given gene is transcribed (**transcriptional control**), (2) controlling how the primary RNA transcript is spliced or otherwise processed (**RNA processing control**), (3) selecting which completed mRNAs in the cell nucleus are exported to the cytoplasm (**RNA transport control**), (4) selecting which mRNAs in the cytoplasm are translated by ribosomes (**translational control**), (5) selectively destabilizing certain mRNA molecules in the cytoplasm (**mRNA degradation control**), or (6) selectively activating, inactivating, or compartmentalizing specific protein molecules after they have been made (**protein activity control**).

For most genes, transcriptional controls are paramount. Early evidence for the importance of transcriptional controls in vertebrate development came from a comparison of the mRNA molecules present in the liver with those present in the brain. A number of these RNAs were found to be liver-specific, and the cloned cDNA probes made from them could be used to show that brain cells lacked not only the corresponding mRNA, but also the primary RNA transcripts (hnRNA—see p. 528) from which the mRNA is derived. To demonstrate this, nuclei from liver and brain cells were isolated and incubated with highly radioactive RNA precursors (ribonucleoside triphosphates), so that the RNA transcripts being synthesized at the time became radioactively labeled. This process, which is known as *nuclear run on*, is illustrated in Figure 10–3. The labeled RNA molecules were then assayed as described in Figures 10–4 and 10–5. Within the limited sensitivity of the assay, no brain cell hnRNA sequences could be found that contained sequences from any one of 11 liver-specific mRNAs tested. The hnRNA from the liver cells served as a control that does contain sequences that hybridize to the liver cDNAs. Therefore, the absence of the liver-specific mRNAs (and thus the proteins they encode) from the brain cell is primarily due to a lack of transcription of the corresponding genes in the brain cell nuclei. In other words, for the proteins encoded by these 11 mRNAs, the differences between a brain cell and a liver cell are largely determined by transcriptional controls.

10-4 Gene Regulatory Proteins Can Either Activate or Repress Gene Transcription

Eucaryotic cells contain a large set of sequence-specific DNA-binding proteins whose main function is to turn genes on or off. Each of these **gene regulatory proteins** is present in relatively few copies per cell (about 1 molecule per 3000

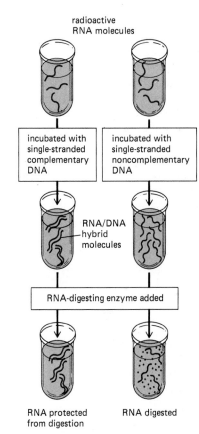

radioactive RNA molecules

incubated with single-stranded complementary DNA

incubated with single-stranded noncomplementary DNA

RNA/DNA hybrid molecules

RNA-digesting enzyme added

RNA protected from digestion

RNA digested

Figure 10–4   A general method to detect radioactive RNA molecules of specific sequence. If the RNA molecules can form RNA/DNA-hybrid double helices with the specific single-stranded DNA segment used as a probe, they will not be digested by the enzyme treatment and can be detected by their radioactivity. In the experiment described in the text, pure DNA segments corresponding to a variety of genes were obtained by cloning the cDNA molecules produced by reverse transcription of a mixture of liver mRNAs (see p. 260). Those DNAs corresponding to liver-specific mRNAs were then selected, and each was used as a probe in a separate experiment (see Figure 10–5).

nucleosomes, or 10^4 copies per mammalian cell) and recognizes a particular DNA sequence that is usually about 8 to 15 nucleotides long (see p. 566). The binding of these proteins to the DNA can either facilitate (**positive regulation**) or inhibit (**negative regulation**) transcription of an adjacent gene (Figure 10–6). We shall describe some of the mechanisms involved later (see p. 564). The different cell types in a multicellular organism have different mixtures of gene regulatory proteins, which causes each cell type to transcribe different sets of genes.

Combinations of a Few Gene Regulatory Proteins Could Specify a Large Number of Cell Types[4]

The essence of **combinatorial gene regulation** is illustrated in Figure 10–7, in which each numbered element represents a different gene regulatory protein. In this purely hypothetical scheme, one initial cell type gives rise to two types of cells, A and B, which differ only in the presence of gene regulatory protein ① in one but not in the other. The subsequent development of each of these cells leads to the additional production in some cells first of gene regulatory proteins ② and ③ and later of gene regulatory proteins ④ and ⑤. In the end, 8 cell types (G through cell N) have been created with 5 different gene regulatory proteins. With the addition of 2 more gene regulatory proteins to the scheme shown in Figure 10–7 (⑥ and ⑦), 16 cell types could be generated at the next step. By the time 10 more such steps occurred, slightly more than 10,000 cell types could have been specified, in principle, through the action of only 25 different gene regulatory proteins.

Combinatorial regulation of this sort is thus a very efficient way to generate biological complexity with relatively few regulatory elements. In Chapter 16 we shall see how it is used during the development of the *Drosophila* embryo to determine specific cell fates, as the interactions of a set of regulatory genes define a progressive subdivision of the early embryo into distinct regions (see p. 925).

Input from Several Different Regulatory Proteins Usually Determines the Activity of a Gene[5]

The scheme for combinatorial gene regulation shown in Figure 10–7, at first glance, seems to predict simple additive differences between the cells of successive generations. One might imagine, for example, that the addition of regulatory protein ② to cell C and to cell E would add to each of these cells the same set of additional proteins—namely, those encoded by genes activated by the regulatory protein ②. Such a view is incorrect for a simple reason. Combinatorial gene regulation is more complicated than that because different gene regulatory proteins interact with one another. Even in bacteria the interaction of two different regulatory proteins is often needed to turn on a single gene (see p. 560). In higher eucaryotes whole clusters of gene activator proteins generally act in concert to determine whether a gene is to be transcribed (see p. 567). By interacting with gene regulatory protein ①, for example, protein ② can turn on a different set of genes in cell E from those it turns on in cell C. This is presumably why a single steroid-hormone-receptor protein (an example of a gene regulatory protein) modulates the synthesis of different sets of proteins in different types of mammalian cells (see p. 692). The particular change in gene expression caused by the synthesis of a given gene

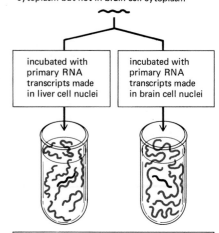

single-stranded DNA segment complementary to an mRNA molecule found in liver cell cytoplasm but not in brain cell cytoplasm

| incubated with primary RNA transcripts made in liver cell nuclei | incubated with primary RNA transcripts made in brain cell nuclei |

synthesis of complementary RNA molecules detected only in liver nuclei. CONCLUSION: the DNA sequence corresponding to this liver-specific mRNA molecule is not transcribed in brain cell nuclei

Figure 10–5 An experiment showing that gene expression in vertebrate cells is controlled largely at the level of gene transcription (see also Figure 10–4).

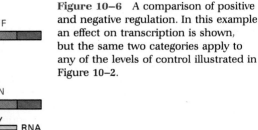

Figure 10–6 A comparison of positive and negative regulation. In this example an effect on transcription is shown, but the same two categories apply to any of the levels of control illustrated in Figure 10–2.

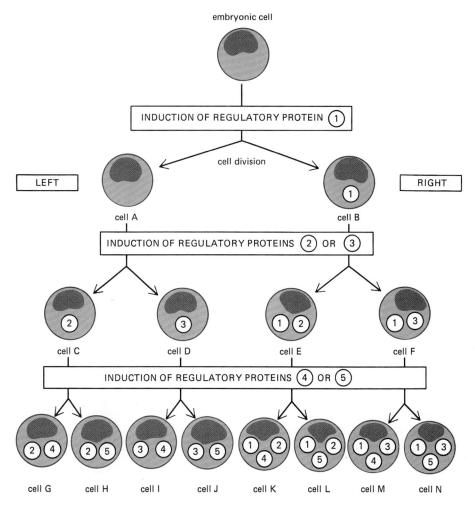

embryonic cell

INDUCTION OF REGULATORY PROTEIN ①

cell division

LEFT   RIGHT

cell A   cell B

INDUCTION OF REGULATORY PROTEINS ② OR ③

cell C   cell D   cell E   cell F

INDUCTION OF REGULATORY PROTEINS ④ OR ⑤

cell G   cell H   cell I   cell J   cell K   cell L   cell M   cell N

Figure 10–7 A highly schematic scheme for cell development illustrating how combinations of a few gene regulatory proteins can generate many cell types in embryos.

   In this simple scheme a "decision" to make one of a pair of different gene regulatory proteins (shown as numbered circles) is made after each cell division. Sensing its relative position in an embryonic field, the daughter cell toward the left side of the embryo (the left side of the page) is always induced to synthesize the even-numbered protein of each pair, while the daughter cell toward the right side of the embryo is induced to synthesize the odd-numbered protein. The production of each gene regulatory protein is assumed to be self-perpetuating (see Figures 10–33 and 10–35). Therefore, the cells in the enlarging clone contain an increasing number of regulatory proteins, each of which is assumed to control a whole battery of genes.

regulatory protein in a developing cell will depend, in general, on the cell's past history, since this history will determine which gene regulatory proteins are already present in the cell (Figure 10–8).

Master Gene Regulatory Proteins Activate Multiple Genes[6]

As we have just seen, a cell contains many gene regulatory proteins, each of which acts in combination with others to control numerous other genes. Thus a branching network of interactions is possible in which each gene regulatory protein controls genes that produce other gene regulators, and so on.

   Not all gene regulatory proteins are equal, however. The regulatory network contains **master gene regulatory proteins,** each of which has a decisive coordinating effect in controlling many other genes (Figure 10–9). We shall see in Chapter 16, for example, that a number of single gene mutations in *Drosophila* convert one part of the fly's body to another. Mutations that bring about such transformations are known as *homeotic mutations.* One such mutation in *Drosophila,* called *Antennapedia,* produces a single master gene regulatory protein aberrantly in the group of cells that would normally make an antenna and thereby causes these cells to switch to making a leg instead, so that the adult fly has a leg growing out of its head (see p. 920). Early evidence for a similar role for master gene regulatory proteins in vertebrates came from the observation that the absence of a single gene regulatory protein (the receptor protein for the steroid hormone testosterone) causes a human with a male (XY) genotype to develop as an almost perfect female (see p. 693). The most direct evidence for the use of master gene regulators in vertebrate development, however, comes from recent studies on the development of skeletal muscle cells.

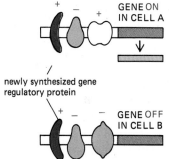

GENE ON
IN CELL A

newly synthesized gene
regulatory protein

GENE OFF
IN CELL B

Figure 10–8 The effect of a newly synthesized gene regulatory protein on a cell. The effect depends on the regulatory proteins already present and, therefore, on the cell's past history. In this diagram the same gene is illustrated in cells A and B. This gene is initially off in both cells; the production of the leftmost protein, however, turns on the gene in cell A but fails to do so in cell B. For simplicity, each gene regulatory protein individually is assumed to have either a positive or negative effect on transcription, and the effects combine to determine the transcription of the gene. In reality, the net effect need not simply be additive; in some cases, for example, two gene regulatory proteins interact with each other when they bind to the DNA to change their individual activities.

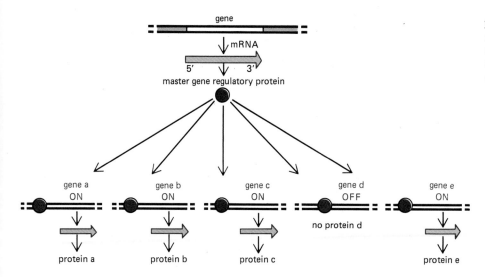

Figure 10–9 A schematic illustration of how the "decision" to produce a single master gene regulatory protein can affect the production of many different proteins in a cell.

10-5 A Single Master Gene Regulatory Protein Can Convert a Fibroblast to a Myoblast[7]

A mammalian skeletal muscle cell is typically extremely large and contains many nuclei. It is formed by the fusion of many muscle cell precursor cells called *myoblasts* (see p. 983). The mature muscle cell is distinguished from other cells by a large number of characteristic proteins, including specific types of actin, myosin, tropomyosin, and troponin (all part of the contractile apparatus), creatine phosphokinase (for the specialized metabolism of muscle cells), and acetylcholine receptors (to make the membrane sensitive to nerve stimulation). In proliferating myoblasts these muscle-specific proteins and their mRNAs are absent, or are present in very low concentrations. As myoblasts begin to fuse with one another, the production of these proteins increases in parallel with an increase in the concentration of their mRNAs, which indicates that the expression of the corresponding genes is controlled at the level of transcription. The rate of synthesis of many of the muscle-specific proteins increases by a factor of at least 500. Studies using two-dimensional polyacrylamide-gel electrophoresis indicate that the rates of production of many other proteins change at the same time: some are switched off, others rise to a peak and then fall, others shift from one steady level to another, and so on.

This entire series of events, including cell fusion, can be triggered in cultured skin fibroblasts—cells that never express these muscle-specific genes otherwise—by transfecting them with DNA that contains a gene that has been engineered to produce a gene regulatory protein called **myoD1,** which normally is expressed only in myoblasts and skeletal muscle cells (Figure 10–10). Apparently, myoD1 is a master gene regulatory protein that normally specifies "myoblast"; when expressed at a high enough concentration in a fibroblast, it subverts the normal gene controls of the fibroblast and converts it to a muscle cell.

The myoD1 protein is concentrated in the cell nucleus, and its amino acid sequence has been deduced from DNA-sequencing studies. Moreover this protein has been shown to bind to the regulatory regions of several muscle-specific genes. Discovery of this remarkable protein raises the possibility that a small number of master gene regulatory proteins also specify other cell types.

Summary

The cell types in multicellular organisms become different by expressing different genes from the same genome, although surprisingly few differences in protein content distinguish one cell type from another. The expression of most genes is controlled predominantly at the transcriptional level, although post-transcriptional controls are also important. Transcriptional controls depend on gene regulatory

20 µm

Figure 10–10 Immunofluorescence micrograph of skin fibroblasts from a chick embryo that have been converted to muscle cells by the experimentally induced expression of the *myoD1* gene product. These fibroblasts were grown in culture and transfected 3 days earlier with a recombinant DNA plasmid containing the *myoD* coding sequence linked to a viral promoter/enhancer that is active in chick fibroblasts. A few percent of the fibroblasts take up the DNA and produce the myoD protein; these cells have fused to form elongated myotubes, which are stained here with an antibody that detects a muscle-specific protein. The stained cells are intermixed with a confluent layer of fibroblasts, which are not visible in this micrograph. Control cultures transfected with another plasmid contain no muscle cells. (Courtesy of Stephen Tapscott, Andrew Lassar, Robert Davis, and Harold Weintraub.)

proteins that bind to specific DNA sequences. These proteins can help to turn a gene either on (positive control) or off (negative control). Genes in higher eucaryotes usually are regulated by the combinatorial effects of several such positive and negative gene regulatory proteins. Master gene regulatory proteins play a special part in this gene control network by regulating large sets of genes: the experimentally induced expression of the myoD1 protein, for example, can convert a fibroblast to a myoblast.

Controlling the Start of Transcription[8]

Only 40 years ago the idea that a gene could be specifically turned on or off was revolutionary. This concept, which was a major advance in our understanding of cells, originated from studies of *E. coli* growing in a mixture of glucose and lactose (a disaccharide). Given this choice of carbon source, the bacteria first used up all the glucose and only then began to metabolize the lactose. The switch to lactose utilization was accompanied by a pause in bacterial growth, during which the enzyme *β-galactosidase*, which hydrolyzes lactose to glucose and galactose, was synthesized. The isolation and characterization of mutant bacteria with specific defects in the regulation of this switch led to biochemical studies that, in 1966, resulted in the identification and isolation of the *lactose repressor protein*.

Biochemical and genetic studies of the lactose repressor, the bacteriophage lambda repressor (see p. 492), and several other bacterial gene regulatory proteins quickly led to a general model for transcriptional regulation in procaryotes. Sequence-specific DNA-binding proteins were presumed either to inhibit or to stimulate the initiation of RNA synthesis from a gene by binding next to the **promoter,** where RNA polymerase binds to start transcription (see p. 524). Changes in the binding of these gene regulatory proteins to the DNA were presumed to turn genes on and off.

For many years it remained unclear how well this bacterial model of gene control might apply to eucaryotic cells. The DNA of eucaryotes, unlike that of procaryotes, is packaged into nucleosomes by histone proteins, whose mass is equal to that of the DNA (see p. 496). The presence of nucleosomes suggested new possibilities for gene regulation; besides, some novel mechanism seemed to be needed to explain why eucaryotic gene regulatory proteins often bind thousands of nucleotide pairs away from the promoters they affect. Progress in elucidating the mechanisms of eucaryotic gene control was slow, largely because most eucaryotic gene regulatory proteins are present in small quantities (about one part in 50,000 of total cell protein). As late as 1983 only two unusually abundant (and perhaps atypical) examples—the SV40 virus *large T-antigen* and the 5S rRNA gene transcription factor *TFIIIA*—had been well characterized.

Soon thereafter the situation changed dramatically. New methods based on recombinant DNA technology made large numbers of eucaryotic gene regulatory proteins available for detailed biochemical and genetic studies. In addition, studies in bacteria provided clear examples of gene regulatory proteins affecting genes from a distance and revealed their mechanism of action. In this section, therefore, we emphasize the similarities rather than the differences between the mechanisms that regulate gene transcription in procaryotes and eucaryotes. Additional mechanisms that may be unique to eucaryotes are discussed later when we consider some of the mechanisms responsible for creating different cell types (see p. 577).

10-8 Bacterial Gene Repressor Proteins Bind Near Promoters and Thereby Inhibit Transcription of Specific Genes[9]

The chromosome of *E. coli* consists of a single circular DNA molecule of about 4.7 $\times$ 10^6 nucleotide pairs. This is enough DNA to code for about 4000 different proteins, though only a fraction of these proteins are made at any one time. *E. coli* regulates the expression of many of its genes according to the intracellular

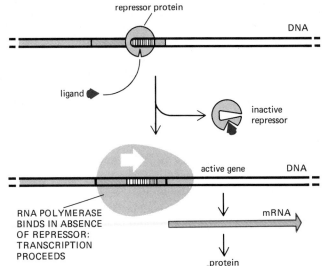

repressor protein

DNA

ligand

inactive repressor

active gene   DNA

RNA POLYMERASE
BINDS IN ABSENCE
OF REPRESSOR:
TRANSCRIPTION
PROCEEDS

mRNA

protein

Figure 10–11 Gene derepression in bacteria. A specific small molecule binds to a repressor protein, thereby altering the conformation of the protein, which dissociates from the DNA, allowing the adjacent gene(s) to be transcribed. In the example in the text, allolactose binds to the lactose repressor protein to allow the synthesis of a single RNA transcript that codes for three proteins involved in lactose metabolism (β-galactosidase, galactoside permease, and galactoside acetylase). Because the three genes that encode these proteins are adjacent and coordinately controlled as part of the same transcription unit, the gene cluster is called the lactose *operon*.

levels of specific metabolites, which often vary depending on the food sources in the cell's environment.

By 1959, genetic studies of *E. coli* lactose utilization had provided strong indirect evidence for the existence of a repressor protein that binds to a transcription unit called the *lac* operon to turn β-galactosidase production off when lactose is absent. The subsequent purification of the **lactose repressor protein** allowed the mechanism of this regulation to be deciphered. In experiments *in vitro* with purified components, the repressor was shown to inhibit transcription of the *lac* operon by binding to a specific DNA sequence of 21 nucleotide pairs (called the *operator*) that overlaps an adjacent RNA polymerase binding site at which RNA synthesis begins (the *promoter*). When the repressor binds to the operator sequence, it prevents RNA polymerase from starting RNA synthesis at the promoter, thereby blocking transcription of the adjacent region of DNA (see Figure 10–14).

The lactose repressor protein acts to adjust the production of β-galactosidase to the needs of the cell. In the presence of lactose a small sugar molecule called *allolactose* is formed in the cell. Allolactose binds to the repressor protein, and when it reaches a high enough concentration, it induces an allosteric conformational change (see p. 127) that causes the repressor to loosen its hold on the DNA so that transcription can proceed: the gene is then said to be *derepressed*. As a result, an *E. coli* cell is able to make the enzymes it needs for the breakdown of lactose only when lactose is present (Figure 10–11).

Figure 10–12 The binding of tryptophan to the tryptophan repressor protein changes the conformation of the repressor. The conformational change enables this gene regulatory protein to bind tightly to a specific DNA sequence, thereby blocking transcription of the genes encoding the enzymes required to produce tryptophan (the *trp* operon). The three-dimensional structure of this bacterial helix-turn-helix protein, as determined by x-ray diffraction with and without tryptophan bound, is illustrated. Tryptophan binding increases the distance between the two recognition helices (*colored cylinders*) in the dimer (see p. 492), allowing the formation of symmetrically arranged hydrogen-bond interactions, indicated schematically here as colored rays. (Adapted from R. Zhang et al., *Nature* 327:591–597, 1987.)

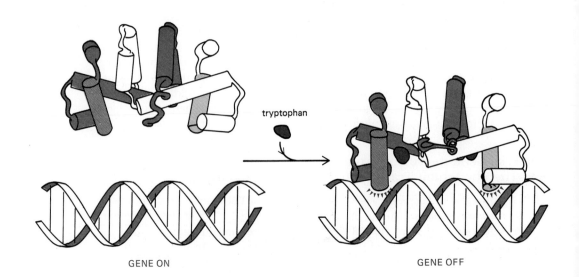

tryptophan

GENE ON

GENE OFF

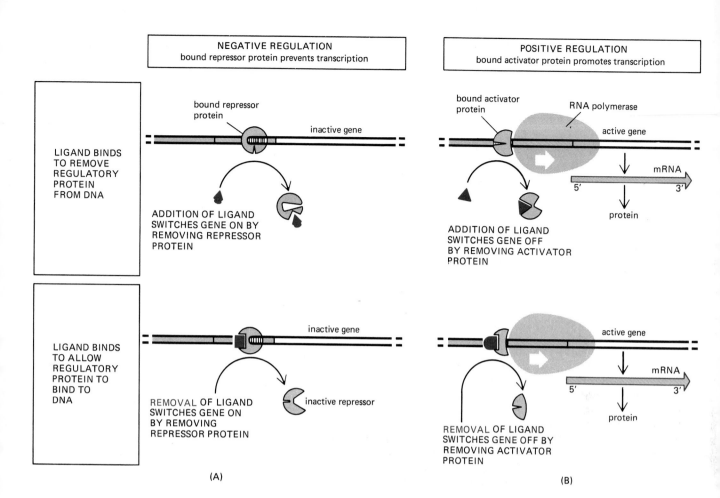

LIGAND BINDS TO REMOVE REGULATORY PROTEIN FROM DNA

bound repressor protein

inactive gene

ADDITION OF LIGAND SWITCHES GENE ON BY REMOVING REPRESSOR PROTEIN

bound activator protein

RNA polymerase

active gene

mRNA

5'

3'

protein

ADDITION OF LIGAND SWITCHES GENE OFF BY REMOVING ACTIVATOR PROTEIN

LIGAND BINDS TO ALLOW REGULATORY PROTEIN TO BIND TO DNA

inactive gene

REMOVAL OF LIGAND SWITCHES GENE ON BY REMOVING REPRESSOR PROTEIN

inactive repressor

active gene

mRNA

5'

3'

protein

REMOVAL OF LIGAND SWITCHES GENE OFF BY REMOVING ACTIVATOR PROTEIN

(A)

(B)

Many other examples of this type of specific gene repression in bacteria are now known. In each case the binding of a **gene repressor protein** to a specific DNA sequence turns a gene (or transcription unit) off. The binding is always regulated by specific signaling molecules, such as allolactose. Sometimes, as for the lactose repressor, the presence of the signaling molecule in the cell turns on a gene or transcription unit by decreasing the affinity of the repressor protein for its specific DNA sequence. But a signaling molecule can equally be used to turn *off* a gene or transcription unit by means of a repressor protein. An allosteric change caused by binding of the signaling molecule to a repressor, for example, can *increase*, instead of decrease, the affinity with which the repressor binds to its specific DNA sequence. This mechanism operates in the control of a set of five adjacent genes (the *trp* operon) encoding enzymes producing the amino acid tryptophan in *E. coli.* The synthesis of the single large mRNA molecule that encodes these five proteins is controlled by the **tryptophan repressor protein,** whose binding to DNA requires that tryptophan (the signaling molecule that turns off this operon) be bound to the repressor (Figure 10–12).

For the *lac* operon an increase in the concentration of a signaling molecule releases the repressor from the DNA and activates transcription, whereas for the *trp* operon it induces DNA binding and suppresses transcription. Because in both cases the binding of the regulatory protein suppresses transcription, this type of gene control is called **negative regulation** (Figure 10–13A).

Figure 10–13 Summary of the mechanisms by which specific gene regulatory proteins control gene transcription in procaryotes. (A) Negative regulation; (B) positive regulation. Note that the addition of an inducing ligand can turn on a gene either by removing a gene repressor protein from the DNA (*upper left panel*) or by causing a gene activator protein to bind (*lower right panel*). Likewise, the addition of an inhibitory ligand can turn off a gene either by removing a gene activator protein from the DNA (*upper right panel*) or by causing a gene repressor protein to bind (*lower left panel*).

10-10 Bacterial Gene Activator Proteins Contact the RNA Polymerase and Help It Start Transcription[10]

For the *negative regulation* discussed so far, a gene repressor protein binds near the promoter and interferes with the activity of the RNA polymerase. In **positive regulation,** by contrast, a **gene activator protein** facilitates the local action of

RNA polymerase. Positive regulation is important for some RNA transcription units in *E. coli* that have relatively weak promoters, which on their own do not readily bind polymerase. Such transcription units are activated by binding an activator protein to an adjacent specific DNA sequence, from which the protein can touch the RNA polymerase in a way that increases its likelihood of initiating transcription.

In other respects gene activator proteins closely resemble repressor proteins. In fact, some bacterial gene regulatory proteins function both as repressors and as activators: they bind at several sites in the genome and repress transcription at some sites while activating it at others. Like repressors, activators often bind to specific signaling ligands that either increase or decrease the affinity of the activator protein for DNA and thereby turn genes on or off, respectively. This type of gene control is called *positive regulation* because more transcription occurs in the presence of the gene regulatory protein than in its absence (Figure 10–13B).

A particularly well-studied example of a gene activator protein is the **catabolite activator protein (CAP)** of *E. coli*, whose structure was described in Chapter 3 (see p. 114). This protein enables the bacterium to use alternative carbon sources only if glucose, a preferred carbon source, is not available.

In the case of the *lac* operon, the lactose repressor protein and CAP cooperate to produce the pattern of gene expression observed. We have seen that, in the presence of lactose, allolactose releases the lactose repressor from the DNA. This is not sufficient to activate transcription of the *lac* operon, however, because the *lac* promoter is a very poor match to the *E. coli* promoter consensus sequence (see p. 204) and thus binds RNA polymerase only weakly. Transcription from this promoter requires that RNA polymerase binding be enhanced by the binding of CAP just upstream from the promoter, as illustrated in Figure 10–14. The same applies to the promoters for the genes encoding maltose, galactose, and several other sugar-metabolizing enzymes.

The DNA binding of CAP is regulated by glucose to ensure that alternative carbon sources are used only in the absence of glucose. Glucose starvation induces an increase in the intracellular levels of cyclic AMP, which acts as an intracellular signaling molecule in bacteria as well as in eucaryotic cells. Cyclic AMP binds to CAP, inducing a conformational change in the protein that enables it to bind to its specific DNA sequence and thereby activate transcription of adjacent genes. When glucose is plentiful, cyclic AMP levels drop; cyclic AMP therefore dissociates from CAP, which reverts to an inactive form that can no longer bind DNA, so that the cell switches to metabolize only glucose (see Figure 10–14).

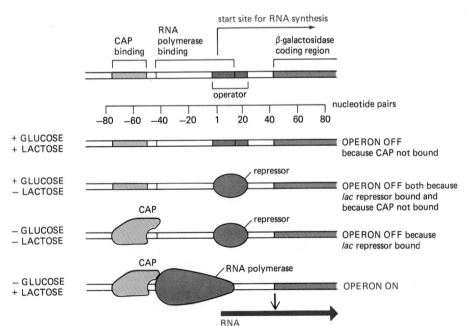

Figure 10–14 Glucose and lactose levels control the initiation of transcription of the *lac* operon through their effects on the lactose repressor protein and CAP. Lactose addition increases the concentration of allolactose, which removes the repressor protein from the DNA (see Figure 10–11). Glucose addition decreases the concentration of cyclic AMP; because cyclic AMP no longer binds to CAP, this gene activator protein dissociates from the DNA, turning off the operon. Binding sites and proteins are drawn approximately to scale; as indicated, CAP is thought to touch the polymerase to help it to begin RNA synthesis.

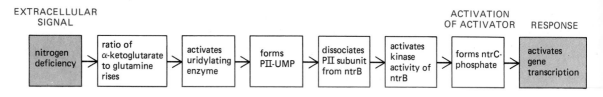

| nitrogen deficiency | ratio of α-ketoglutarate to glutamine rises | activates uridylating enzyme | forms PII-UMP | dissociates PII subunit from ntrB | activates kinase activity of ntrB | forms ntrC-phosphate | activates gene transcription |

10-9 Changes in Protein Phosphorylation Can Regulate Genes[11]

The first gene regulatory proteins identified were all bacterial. Like the lactose repressor, the tryptophan repressor, and CAP, these proteins are controlled by the reversible binding of specific small molecules. While small intracellular signaling molecules, such as cyclic AMP, also control the activity of gene regulatory proteins in eucaryotic cells, they usually do so indirectly, often by affecting protein phosphorylation and dephosphorylation (see p. 695). Although phosphorylation is used much less extensively for regulation in bacteria, there is one bacterial regulatory system dependent on protein phosphorylation that has been especially well studied. We shall use it to introduce several aspects of gene regulation that may help in understanding the more complicated regulatory systems in higher eucaryotes.

Evolutionarily related proteins control aspects of nitrogen metabolism, phosphate metabolism, membrane protein synthesis, chemotaxis and sporulation in various bacteria. We shall focus on nitrogen metabolism in *E. coli*, where—when nitrogen becomes limiting—the synthesis of a number of proteins is increased. These proteins include the enzyme *glutamine synthetase*, which is the most important enzyme of nitrogen assimilation, catalyzing the reaction glutamic acid + ammonia → glutamine.

Gene activation in nitrogen metabolism involves two central regulatory components: the *ntrC protein*, a gene activator protein that turns on genes only in its phosphorylated form (ntrC-phosphate), and an enzyme, the *ntrB protein*, which can either phosphorylate (through a kinase activity) or dephosphorylate (through a phosphatase activity) the ntrC protein. The degree of ntrC phosphorylation is determined by an enzymatic cascade that is triggered by a drop in nitrogen levels to add uridine monophosphate (UMP) residues to a regulatory subunit of ntrB. This modification increases the relative kinase activity of the ntrB enzyme.

A need for nitrogen is measured as an increase in the ratio of α-ketoglutarate to glutamine, which stimulates the phosphorylation of ntrC and thereby induces transcription of the glutamine synthetase gene (Figure 10–15). Similar cascades of protein modification reactions occur in response to changes in the environment of higher eucaryotic cells and lead to the phosphorylation or dephosphorylation of gene regulatory proteins, although the details are less well understood.

10-10 DNA Flexibility Allows Gene Regulatory Proteins That Are Bound to a Remote Site to Affect Gene Transcription[11,12]

A second aspect of the two-component bacterial regulatory system just described is even more similar to another important feature of eucaryotic gene control. The ntrC protein has two strong binding sites in DNA that are located 100 nucleotide pairs or more upstream from the glutamine synthetase gene promoter (Figure 10–16). When the phosphorylated form of ntrC (ntrC-phosphate) is bound to them,

Figure 10–15 Part of the regulatory cascade that allows nitrogen deficiency to activate the genes needed for nitrogen metabolism in bacteria. The protein PII serves as a regulatory subunit of the ntrB protein. This multistep regulatory pathway has the advantage that it can produce a large change in metabolism in response to a relatively small change in nitrogen availability. Side branches of the pathway also reversibly modify other enzymes to control their catalytic activities in response to nitrogen.

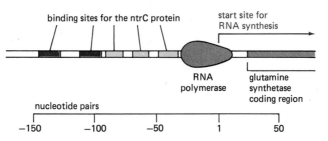

Figure 10–16 The regulatory region of the bacterial glutamine synthetase gene. Glutamine synthetase catalyzes the reaction glutamic acid + ammonia → glutamine. The two darkly colored sites bind the ntrC protein with particularly high affinity and are essential for transcriptional activation.

these two sites stimulate transcription efficiently. Moreover, similar stimulation occurs if the sites are moved (by recombinant DNA techniques) more than 1000 nucleotide pairs away from the promoter.

While such "action at a distance" is not unusual in procaryotes, it is especially prevalent in eucaryotes (see p. 564). In procaryotes there is good evidence that proteins bound to DNA at a distance act in much the same way as proteins bound at adjacent sites—for example, the ntrC protein enhances the ability of the RNA polymerase to bind to the promoter and form an open initiation complex (see Figure 9–65, p. 524). The intervening DNA loops out to allow the proteins bound far upstream to interact with polymerase (Figure 10–17). This interaction occurs readily because the DNA acts as a "tether," causing a protein bound even thousands of nucleotide pairs away to collide repeatedly with the promoter—in effect by increasing the protein's local concentration there (Figure 10–18).

Different Sigma Factors Allow the Bacterial RNA Polymerase to Recognize Different Promoters[11,13]

Another type of transcriptional control was mentioned in Chapter 9. When discussing the structure of the bacterial RNA polymerase, we pointed out that one of its five major polypeptide chains—the **sigma (σ) factor**—functions as an *initiation factor* that is ejected once the polymerase has successfully begun RNA synthesis (see p. 524). The consensus promoter sequence for transcription by the bacterial polymerase (see p. 204) is the one recognized by the major form of the polymerase, which contains the σ70 subunit. Minor forms of the polymerase, however, also exist; each of these contains a different sigma subunit, which recognizes a different promoter sequence. The σ54 subunit, for example, is the product of the *ntrA* gene and is specifically required (along with the ntrB and ntrC proteins) for transcription of the glutamine synthetase gene. Thus environmental changes can turn on a specific set of genes by causing the production of a particular sigma factor.

How can an RNA polymerase initiation factor, such as σ54, that affects promoter recognition be distinguished from a gene regulatory protein, such as ntrC? Sigma factors bind tightly to RNA polymerase but do not bind to specific DNA sequences on their own; gene regulatory proteins bind to specific DNA sequences but do not bind tightly to free RNA polymerase. Since only a single sigma subunit can bind to an RNA polymerase molecule at a time, another test is to measure transcription *in vitro* while varying the ratio of the known sigma factor, σ70, to the suspected new sigma factor; if σ70 inhibits transcription in proportion to this ratio (competitive inhibition), the factor is judged to be a sigma subunit.

The general types of mechanisms that are known to regulate the initiation of bacterial gene transcription are summarized in Figure 10–19. We shall return to bacteria again when we discuss how the competition between two gene regulatory proteins can create a molecular switch (see p. 574).

A Requirement for General Transcription Factors Adds an Extra Element to the Control of Eucaryotic Gene Transcription[14]

Transcription in eucaryotes is more complex than in procaryotes. While the three eucaryotic RNA polymerases (polymerases I, II, and III) are evolutionarily related to the bacterial enzyme, they contain more subunits (see p. 525). As yet, these enzyme complexes have not been completely defined. In addition, as discussed in Chapter 9, whereas the bacterial polymerase recognizes a DNA sequence, the corresponding eucaryotic enzymes normally recognize a protein-DNA complex formed by general *transcription factors* (see p. 526).

In the rest of this section we are primarily concerned with RNA polymerase II, which synthesizes all of the precursors of mRNAs. An important part of the protein-DNA complex recognized by this polymerase is the **TATA factor**, also known as *transcription factor IID (TFIID)*, a protein that binds to the consensus "TATA box" sequence, TATAAA. This sequence is generally located at position -25 to -30 relative to the RNA start site; for many genes that encode proteins it is critical both for promoter activity and for determining the exact point of RNA

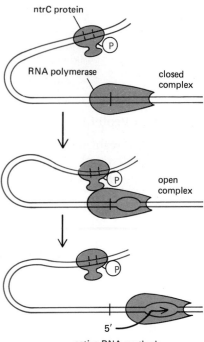

Figure 10–17 Model for the enhancer effect of ntrC protein on RNA synthesis at the *E. coli* glutamine synthetase gene. By binding to upstream DNA sequences, this gene regulatory protein increases the rate of RNA transcript initiation by the RNA polymerase. Although the protein binds to these sites even when not phosphorylated, only the phosphorylated form (ntrC-phosphate) can activate transcription. Its effects appear to be mediated through contacts with the polymerase that increase this enzyme's ability to unwind the DNA and form an open complex (see Figure 9–65, p. 524), as indicated (see also Figure 10–18).

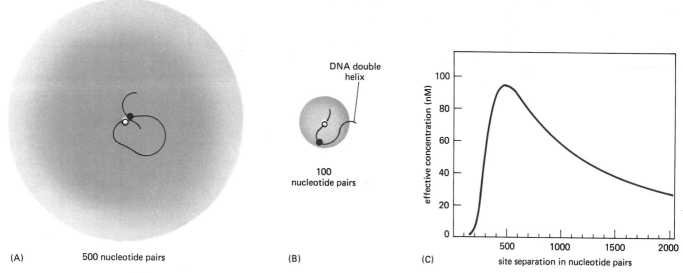

is positioned above with the following labels:

(A) 500 nucleotide pairs

DNA double helix

100 nucleotide pairs

(B)

(C)

Figure 10–18 Binding of two proteins to separate sites on the DNA double helix can greatly increase their probability of interacting. (A) The tethering of one protein to the other via an intervening DNA loop of 500 nucleotide pairs increases their frequency of collision. The intensity of coloring reflects the probability that the colored protein will be located at each position in space relative to the white protein. (B) The flexibility of DNA is such that an average sequence makes a smoothly graded 90° bend (a curved turn) about once every 200 nucleotide pairs. Thus, when two proteins are tethered by only 100 nucleotide pairs, their contact is relatively restricted. In such cases the protein interaction is facilitated when the two protein binding sites are separated by a multiple of about 10 nucleotide pairs, which places both proteins on the same side of the DNA helix (which has 10 nucleotides per turn) and thus on the inside of the DNA loop, where they can best reach each other. (C) Theoretical effective concentration of the colored protein at the site where the white protein is bound as a function of their separation. (C, courtesy of Gregory Bellomy, modified from M.C. Mossing and M.T. Record, *Science* 233:889–892, 1986.)

chain initiation. Like gene regulatory proteins, the TATA factor is a sequence-specific DNA-binding protein. Because it is required for the transcription of most (possibly all) genes transcribed by polymerase II, it is considered to be part of the general transcription machinery rather than a gene regulatory molecule. *In vitro* experiments suggest that once the TATA factor has bound to the DNA at a promoter, it tends to remain in a **stable transcription complex** that can mediate multiple rounds of transcription by RNA polymerase II molecules (see Figure 9–69, p. 527).

It seems likely that all the mechanisms bacteria use to control RNA polymerase activity at a promoter are used by eucaryotic cells as well (see Figure 10–19). But the formation of a stable transcription complex on the DNA by the TATA factor adds an extra dimension—and an extra complication—to considerations of eucaryotic gene regulation. Thus *in vitro* experiments suggest that the major function of some eucaryotic gene activator proteins is to help assemble the TATA factor onto the DNA at a promoter (see Figure 10-27).

Figure 10–19 Mechanisms by which proteins control the initiation of RNA transcription in bacterial cells. The proteins can achieve their effect by either stimulating or blocking any of the several steps required to begin productive RNA synthesis (see Figure 9–65, p. 524). Mechanism (D) was first discovered in studies of the arabinose operon in *E. coli*. It is a combination of mechanisms (A) and (B) and is the only one of the four mechanisms not discussed thus far in the text. We encounter a version of it later in eucaryotes (see Figure 10–27).

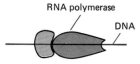

(A) direct contact with the polymerase from a specific DNA binding site at the promoter

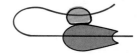

(B) contact with the polymerase from a distant DNA site

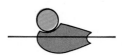

(C) initiation factor bound to polymerase

(D) contact from a distant DNA site with another gene regulatory protein molecule at the promoter

RNA polymerase

DNA

Most Higher Eucaryotic Genes Are Regulated by a Nearby Upstream Promoter Element and Distant Enhancer Elements[15]

As described in Chapters 4 and 5, the development of new methods based on recombinant DNA technology has had an explosive impact on our understanding of eucaryotic genes. The methods now available allow the nucleotide sequence of any segment of a gene to be redesigned, obtained in unlimited amounts, and joined to any other gene segment to create a new gene. The function of this modified gene can then be tested either by transfecting cultured cells with it or (with more difficulty) by producing a transgenic animal that contains the gene stably integrated into one of its chromosomes (see pp. 266–269).

This type of analysis, combined with *in vitro* studies in cell-free systems (see p. 166), has made it possible to identify the regulatory regions of eucaryotic genes even in the absence of any direct knowledge of the gene regulatory proteins that bind to them. Initial studies entailed cutting and joining together different portions of a gene's DNA sequence. They showed that DNA sequences near the RNA start site are required for gene transcription and led to the recognition of an **upstream promoter element,** which generally extends for about 100 nucleotide pairs and includes the TATA box. More surprising was the finding that sequences further upstream, called enhancers, are also required to give the same level of expression as the normal gene; like the ntrC-protein binding sites in bacteria, enhancers affect transcription from a distance.

The first enhancer studied was a small segment of the genome of an animal virus, simian virus 40 (SV40). The normal function of the SV40 enhancer is to increase the transcription of viral genes whose products are required early in infection; the enhancer works by providing binding sites for gene regulatory proteins that increase transcription from the SV40 "early" promoter. In 1981 it was shown that, if attached to a β-globin gene, this small segment of the SV40 genome could stimulate the level of transcription of the β-globin gene by more than 100-fold—even when placed more than 3000 nucleotide pairs away from the start site for globin RNA synthesis. Through a series of manipulations of the position and orientation of this region of the virus, an **enhancer element (enhancer)** was defined as a regulatory DNA sequence that

1. Activates transcription from a promoter linked to it, with synthesis beginning at the normal RNA start site.
2. Operates in both orientations (normal or flipped).
3. Functions even when moved more than 1000 nucleotide pairs from the promoter and from either an upstream or a downstream position.

We now know that most genes in higher eucaryotes are regulated by the combination of a nearby upstream promoter element plus one or more enhancers farther away from the RNA start site (see Figure 10–22A). Although enhancers are generally located within a few thousand nucleotide pairs from the promoter they regulate, some may be more than 20,000 nucleotide pairs away (see p. 580). Both enhancers and upstream promoter elements bind sequence-specific DNA-binding proteins that mediate their effects. As we discuss next, some of these gene regulatory proteins seem to be present in all tissues, whereas others are confined to particular cell types.

Most Enhancers and Upstream Promoter Elements Are Modular Sequences That Bind Multiple Proteins Engaged in Combinatorial Controls[15,16]

The activity of each enhancer and upstream promoter element can generally be pinpointed to a segment of DNA that is 100 to 200 nucleotide pairs long. An element of this size contains binding sites for multiple proteins, and it generally can be dissected by recombinant DNA methods into smaller, partially active DNA segments. The function of both kinds of regulatory elements depends on the binding of specific gene regulatory proteins that are often restricted to particular cell types. Only some (such as the SV40 virus enhancer) will function in almost any cell type; most function best in the specific cell types that express the gene

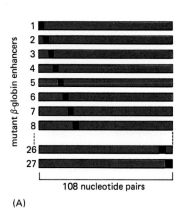

(A)

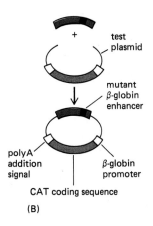

(B)

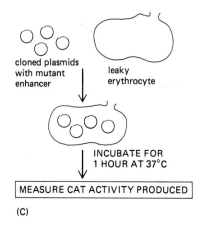

(C)

Figure 10–20 (A) Construction of a set of regulatory region mutants by oligonucleotide mutagenesis (see p. 266). A block of four nucleotide pairs is changed in each mutant; testing the 108 nucleotides in the β-globin enhancer thus requires a set of 27 mutants. (B) Inserting a mutant β-globin enhancer in a test plasmid. The oligonucleotide and the cloning site are joined together directly by DNA ligase (see p. 259). The enzyme produced by the recombinant gene is the bacterial enzyme *chloramphenicol acetyl transferase* (CAT), which is easily assayed. (C) Testing the mutant enhancers for their effects on RNA synthesis. RNA synthesis is measured indirectly by measuring the amount of protein that the recombinant gene produces (as CAT enzyme activity).

with which they are normally associated. An example is the enhancer for the chicken β-globin gene.

The chicken β-globin enhancer lies downstream from the β-globin transcription unit. It forms a nuclease-hypersensitive site in cells of the red blood cell (erythrocyte) lineage but not in other cell types; this suggests that gene regulatory proteins are bound to the enhancer in these blood cells, displacing a nucleosome (see p. 498). To identify these gene regulatory proteins, one must first determine the exact nucleotide sequences important for the function of the enhancer. For this purpose a series of mutated versions of the enhancer sequence can be joined to a reporter gene whose product is easily measured, so that the effect of each mutation on transcription can be tested. Each of these recombinant DNAs is then introduced into chicken erythrocytes, and the efficiency of expression of the reporter gene is measured (Figure 10–20). Those nucleotides required for enhancer function in this test are then identified as binding sites for specific proteins by means of gel-mobility shift (see p. 489) and DNA-footprinting studies (see p. 187) in cell extracts. In this way three proteins have been found to be required for the activity of the chick β-globin enhancer (Figure 10–21). Each of these gene regulatory proteins is present only in very small amounts in the cell. Because their exact binding sites have been identified, recently devised methods should allow them to be purified and the DNA sequences that encode them to be cloned, which would make the proteins available in unlimited amounts (see p. 489).

These and other studies of enhancers and upstream promoter elements have led to the following general conclusions:

1. Each of these regulatory elements has a modular design, consisting of a series of specific nucleotide sequences—each 8 to 15 nucleotides long—that bind a corresponding series of gene regulatory proteins. Some of these proteins are found only in a restricted set of cell types, whereas others are present in most or all cell types.

2. Some gene regulatory proteins activate transcription when they bind; others inhibit it. The net effect of a regulatory element depends on the combination

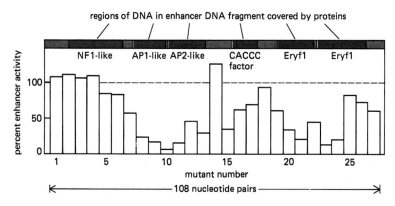

Figure 10–21 Results of the assay for mutant enhancer activity described in Figure 10–20C, correlated with data showing the location of binding sites for proteins in the normal enhancer. The colored bars in the top line indicate portions of the normal enhancer DNA sequence covered by the indicated proteins, as judged by mixing the normal enhancer sequence with extracts prepared from chick erythrocytes and performing gel-mobility shift (see p. 489) and footprinting (see p. 187) experiments. The activity data is plotted as "percent enhancer activity": 100 means that the indicated mutant enhancer stimulates RNA synthesis to the same level as the normal enhancer, whereas 0 means that RNA synthesis is no greater than that observed with no enhancer. The results indicate that the AP1-like, AP2-like, and Eryf1 proteins make the largest contributions to the enhancer's stimulation of transcription but that no single protein suffices to produce strong enhancer activity (see Figure 10–23). (From M. Reitman and G. Felsenfeld, *Proc. Natl. Acad. Sci. USA* 85:6267–6271, 1988.)

of proteins bound, and for a given element the effect can change as a cell develops. An enhancer element, for example, can either activate or inhibit transcription (see Figure 10–22B). In this sense the original name given to these elements is a misnomer.

3. Enhancer elements and upstream promoter elements seem to bind many of the same proteins, which may mean that both types of element affect transcription by a similar mechanism.

4. When a newly defined regulatory element from a vertebrate gene is analyzed, many of the proteins that bind to it turn out to be previously described regulators of other genes. This suggests that a relatively small number of gene regulatory proteins may control transcription in higher eucaryotes (Table 10–1).

The proteins bound to the upstream promoter element of a gene cooperate with those bound to enhancers in controlling transcription. In general, their net effect on gene activity will reflect the result of opposing activating and repressing influences (Figure 10–22). A change in the balance of positively and negatively acting gene regulatory proteins is thought to cause the rise and fall in transcription of the β-globin gene at various stages of chicken erythrocyte development, as outlined in Figure 10–23.

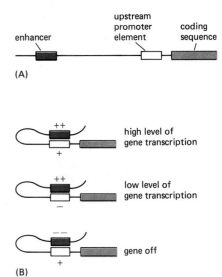

Figure 10–22 The gene regulatory proteins bound to the upstream promoter element act synergistically with those bound to the enhancer to determine the transcriptional activity of a gene. (A) Typical locations of the two types of regulatory elements relative to a coding sequence; (B) examples of the combined action of proteins bound to these two elements. Depending on the proteins bound, each individual element in this example has either a weak positive (+), a strong positive (+ +), a weak negative (−), or a strong negative (− −) effect on transcription, as indicated.

Table 10–1 Some Well-characterized Mammalian Gene Regulatory Proteins and Their Specific DNA-binding Sites

Protein Name[a]	Protein Mass (kilodaltons)[b]	DNA-binding Site[c]	Comments
jun (or AP1)	36	TGANTCA ACTNAGT	proto-oncogene product that can associate with the fos protein, the product of another proto-oncogene (see p. 1207). Activity induced by C-kinase activation; related proteins encoded by multiple genes
AP2	48	CCCCAGGC GGGGTCCG	activity induced by C-kinase activation
ATF (or CREB)	43	$(^T_A$ or $^G_C)(^T_A$ or $^A_T)$CGTCA $(^A_T$ or $^C_G)(^A_T$ or $^T_A)$GCAGT	activity induced by A-kinase activation
SP1	85	GGGCGG CCCGCC	binds to the promoters of many housekeeping genes
OTF1	90	ATTTGCAT TAAACGTA	ubiquitous in mammalian cells; one of several proteins that bind to the same octamer sequence
NF1 (or CTF)	52–66	GCCAAT CGGTTA	several proteins encoded by alternatively spliced RNAs from one gene
SRF	67	GATGCCCATA CTACGGGTAT	mediates transient transcriptional activation of *fos* proto-oncogene (and other genes) by growth factors; binds as a dimer to a symmetric DNA sequence (only a monomer site is shown)

[a]The names given are abbreviations for arbitrary designations, such as "nuclear factor 1" (NF1).
[b]The DNA that encodes each of these proteins has been cloned and the amino acid sequence thereby determined (see p. 185).
[c]In many cases only part of the recognized DNA sequence is known; N indicates any nucleotide.

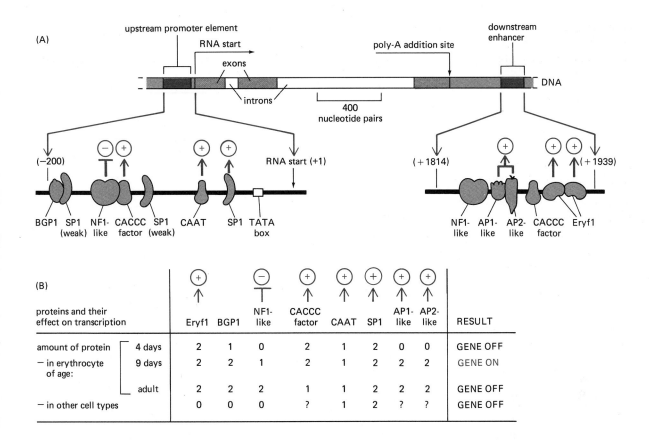

(A)

upstream promoter element

RNA start

exons

introns

poly-A addition site

downstream enhancer

400 nucleotide pairs

DNA

(−200)

RNA start (+1)

(+1814)

(+1939)

BGP1   SP1   NF1-   CACCC   SP1   CAAT   SP1   TATA
       (weak) like   factor (weak)   box

NF1-   AP1-   AP2-   CACCC   Eryf1
like   like   like   factor

(B)

proteins and their effect on transcription

		Eryf1	BGP1	NF1-like	CACCC factor	CAAT	SP1	AP1-like	AP2-like	RESULT
amount of protein — in erythrocyte of age:	4 days	2	1	0	2	1	2	0	0	GENE OFF
	9 days	2	2	1	2	1	2	2	2	GENE ON
	adult	2	2	2	1	1	2	2	2	GENE OFF
— in other cell types		0	0	0	?	1	2	?	?	GENE OFF

10-12 Most Gene Regulatory Proteins Contain Domains
10-13 with Distinct Functions[17]

Although it is still unclear exactly how most gene regulatory proteins work, recombinant DNA techniques have been used to show that their effects on transcription require more than just DNA binding. The *gal4 protein*, for example, has separate DNA-binding and transcription-activating domains. This yeast gene activator protein turns on the transcription of several genes that are required to convert galactose to glucose when yeast cells are grown in a medium containing galactose. The gal4 protein functions by binding to a specific 17-nucleotide-pair sequence that is the yeast equivalent of an enhancer—called an *upstream activating sequence (UAS)* when (as in this case) it promotes gene expression. The gal4 protein contains nearly 900 amino acid residues, but the amino-terminal 73 of these, which form a series of "zinc fingers" (see p. 490), are sufficient on their own to bind the specific 17-nucleotide-pair sequence. The DNA-binding domain alone, however, cannot activate transcription, and the rest of the protein likewise has no activity in the absence of a DNA-binding domain. A "domain swap" experiment demonstrates that the DNA-binding and transcription-activating domains of this protein have separate functions (Figure 10–24).

Related results are obtained with mammalian gene regulatory proteins. Well-studied examples are found in the large *steroid hormone receptor family* of gene regulators. These receptor proteins enable cells to respond to various lipid-soluble hormones by activating or repressing specific genes (see p. 690). They possess a central DNA-binding domain of about 100 amino acids; as in the case of gal4, this domain contains a series of zinc fingers and recognizes a specific DNA sequence. For some members of this family, a transcription-activating domain has been located in an amino-terminal region. In addition, these receptors all contain a specific hormone-binding domain in the carboxyl-terminal part of the protein (Figure 10–25). Domain swap experiments show that these domains can function interchangeably in hybrid receptors. Replacing the DNA-binding domain of a glucocorticoid-receptor protein with the DNA-binding domain of an estrogen receptor, for example, causes the modified glucocorticoid receptor to bind to the enhancers

Figure 10–23 The known gene regulatory proteins that control the expression of the chick β-globin gene during normal erythrocyte development. This summary incorporates the results of the experiment described in Figures 10–20 and 10–21, a similar experiment with mutant upstream promoter elements, and the results from other types of analyses. (A) Binding sites for gene regulatory proteins and their effects. Note that the enhancer for this gene is located downstream of the coding sequence. Those proteins not marked with + (activation) or − (repression) appear not to have a major effect on transcription, as judged by a lack of an effect on transcription when their binding site is mutated (see Figure 10–21). (B) Relative amounts of each gene regulatory protein present at different times of development. The numbers 0, 1, and 2 indicate none, intermediate, and high levels, respectively, on a nonlinear scale. The data available thus far do not necessarily explain why the gene turns on between 4 and 9 days of development, but they suggest that it may turn off due to a large increase in the NF1-like protein combined with a decrease in the CACCC factor concentration. (Courtesy of Gary Felsenfeld.)

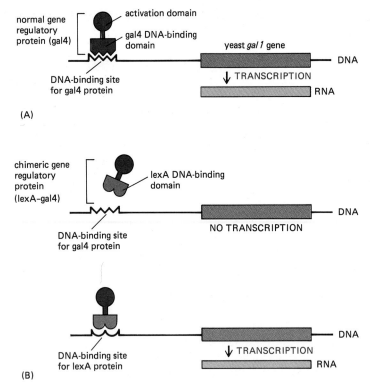

(A)

normal gene regulatory protein (gal4)

activation domain

gal4 DNA-binding domain

DNA-binding site for gal4 protein

yeast *gal 1* gene

DNA

↓ TRANSCRIPTION

RNA

(B)

chimeric gene regulatory protein (lexA–gal4)

lexA DNA-binding domain

DNA-binding site for gal4 protein

NO TRANSCRIPTION

DNA

DNA-binding site for lexA protein

↓ TRANSCRIPTION

DNA

RNA

Figure 10–24 Outline of a "domain swap" experiment that reveals the presence of independent DNA-binding and transcription-activating domains in the yeast gal4 gene activator protein. A functional activator can be reconstituted from the carboxyl-terminal portion of the gal4 protein if it is attached to the DNA-binding domain of a bacterial gene regulatory protein (the lexA protein) by gene fusion techniques. When the resulting bacterial-yeast hybrid protein is produced in yeast cells, it will activate transcription from yeast genes, provided that the specific DNA-binding site for the bacterial protein has been inserted next to them. (A) The normal activation of gene transcription caused by the gal4 protein. (B) The chimeric gene regulatory protein requires the lexA-protein DNA-binding site for its activity. Several other eucaryotic gene regulatory proteins have been shown to have separable domains by this type of experiment.

that normally activate estrogen-responsive genes; as a result, these genes are now transcribed in response to glucocorticoids instead of estrogens. Similarly, adding the hormone-binding domain of the glucocorticoid receptor to an unrelated gene regulatory protein causes the modified regulatory protein to require glucocorticoids for its activity. Each of the domains in these receptor proteins is encoded by one or more separate exons, and sequence comparisons between the various members of the steroid-hormone receptor family (Figure 10–25) imply that each protein has evolved as the result of chromosomal accidents that have combined domains from other regulatory proteins (see p. 602).

10-5 Eucaryotic Gene Regulatory Proteins Can Be Turned On and Off[18]

Like the other members of the steroid-hormone receptor family, the glucocorticoid and estrogen receptors resemble the bacterial CAP protein (see p. 114) in that they are activated by the binding of a small signaling molecule to a separate ligand-binding domain. Other eucaryotic gene regulatory proteins seem to be activated and inactivated by protein phosphorylation or by binding to other proteins that change their activity (see Table 10–1, p. 566). These and other known ways of controlling the activity of proteins that regulate gene expression in eucaryotic cells are summarized in Figure 10–26. Some of these mechanisms control the ability of the regulatory protein to bind DNA; others resemble the effect of phosphorylation on the bacterial ntrC protein in affecting only the function of a transcription-activating domain.

The Mechanism of Enhancer Action Is Uncertain[15,19]

The mechanisms that regulate eucaryotic genes have been highly conserved during evolution. The yeast gal4 protein, for example, will stimulate transcription from mammalian genes in human cells in culture, provided that these genes contain the *gal4* upstream activating sequence. Likewise, when the human estrogen receptor protein is produced in a yeast cell, yeast genes that have been linked to a human estrogen-response enhancer are turned on by the addition of estrogen.

This conservation of function is not reflected in a conservation of amino acid sequence. Amino acid sequences are known for many yeast and mammalian gene

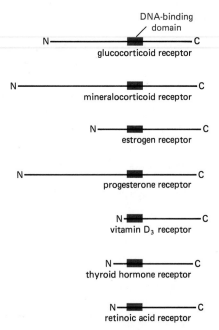

DNA-binding domain

N ———■——— C
glucocorticoid receptor

N ———■——— C
mineralocorticoid receptor

N ——■—— C
estrogen receptor

N ——■—— C
progesterone receptor

N ■— C
vitamin D₃ receptor

N —■— C
thyroid hormone receptor

N ■— C
retinoic acid receptor

Figure 10–25 Evolutionarily related members of the steroid-hormone receptor family of gene regulatory proteins. The short DNA-binding domain in each receptor is colored. Domain swap experiments suggest that many of the hormone-binding, transcription-activating, and DNA-binding domains in these receptors can function as interchangeable modules.

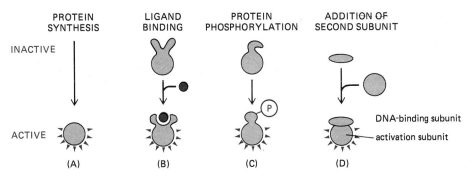

PROTEIN SYNTHESIS LIGAND BINDING PROTEIN PHOSPHORYLATION ADDITION OF SECOND SUBUNIT

INACTIVE

ACTIVE

(A)   (B)   (C)   (D)

DNA-binding subunit
activation subunit

Figure 10–26 Various ways of activating gene regulatory proteins in eucaryotic cells. Examples of all of these mechanisms are known. (A) The gene regulatory protein is synthesized only when needed and is rapidly degraded by proteolysis so that it does not accumulate. (B) Ligand binding. (C) Phosphorylation or other covalent modification. (D) Formation of a complex with a separate protein that serves as a transcription-activating domain.

activator proteins, but their transcription-activating domains reveal no obvious sequence similarities. Reasonable levels of gene activation are obtained in both yeast and mammalian cells, moreover, if the transcription-activating domain of a gene regulatory protein is replaced by a region that is merely rich in acidic amino acids (glutamic acid and aspartic acid). This suggests that the mechanism of activation may be relatively simple.

One likely mechanism of enhancer action is that suggested by some bacterial systems, where DNA looping allows gene regulatory proteins bound at a distance to help RNA polymerase start at the promoter (see p. 562). This is consistent with the observation that enhancer elements usually function best when they are close to a promoter, their activity falling off progressively with distance. Figure 10–27 illustrates two variants of the DNA-looping model for enhancer action. So little is known, however, that a number of alternative mechanisms have been proposed to explain how an enhancer might work: (1) An enhancer could exert its long-range effects by activating a DNA topoisomerase that uses the energy of ATP hydrolysis to introduce torsional stress into a large loop of DNA. (2) It could affect transcription by acting as an DNA entry site for mobile proteins that bind and then travel along the DNA. (3) It could bind proteins that cause its nearby gene to adhere to a selected region of the nucleus where transcription factors are sequestered.

We shall see later that some DNA sites have been shown to have important effects on transcription from distances that are 50,000 nucleotide pairs or more away from a promoter (see p. 580). Such sites could function by very different mechanisms from those of the enhancers thus far described. Some of the alternative mechanisms proposed for enhancer action might well be the basis for such long-range effects (see Figure 10–41).

Figure 10–27 Two models for enhancer action in eucaryotic cells based on the DNA looping observed in bacteria (see Figure 10–17). In this example the enhancer is located upstream from the coding sequence, but a loop can form just as well from an enhancer located at a downsteam site, as in the β-globin gene example in Figure 10–23.

Summary

In procaryotes, gene regulatory proteins usually bind to specific DNA sequences close to RNA polymerase start sites and thereby either repress or activate the transcription of adjacent genes. The flexibility in the DNA helix, however, also allows

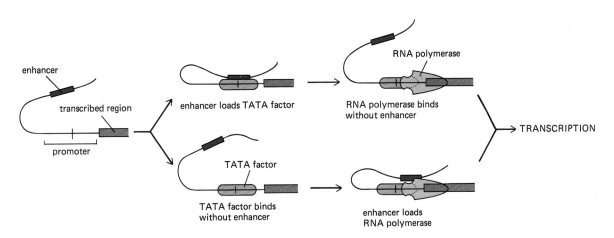

enhancer
transcribed region
promoter

enhancer loads TATA factor

RNA polymerase binds without enhancer

RNA polymerase

TRANSCRIPTION

TATA factor

TATA factor binds without enhancer

enhancer loads RNA polymerase

proteins bound to distant sites to affect the RNA polymerase at the promoter by forming a DNA loop. Action at a distance is extremely common in eucaryotic cells, where enhancer elements located thousands of nucleotide pairs away from the promoter often control gene expression.

Many protein-coding genes in eucaryotes, which are transcribed by RNA polymerase II, contain a "TATA box" sequence just upstream from the RNA polymerase start site. An important transcription factor (the TATA factor, TFIID) must form a stable transcription complex at this site before RNA synthesis can begin. Other sequences located close to the TATA box form an upstream promoter element that is required for transcription. Both enhancer and upstream promoter elements contain a series of short nucleotide sequences that bind a corresponding series of gene regulatory proteins, and the proteins bound to these two types of regulatory elements cooperate to turn genes on or off. Mutation of the gene regulatory proteins by recombinant DNA techniques shows that they often contain several domains, each with a distinct function.

The Molecular Genetic Mechanisms That Create Different Cell Types

In the initial section of this chapter we considered the gene regulation strategies that allow a multicellular organism to produce a large number of distinct cell types. The inherent economy of controlling genes by means of different combinations of gene regulatory proteins (see p. 554) presumably explains why multiple gene regulatory proteins are observed to bind to the enhancer and upstream promoter elements of most higher eucaryotic genes (see p. 567). The observation that cells exist as discrete types implies that the activity of the gene regulatory proteins in this combinatorial network is linked by a hierarchy of feedback controls, which seem to be dominated by the activity of a relatively small number of master gene regulatory proteins (see p. 555). We still do not understand, however, the logic of these complex gene regulatory networks. How, for example, does the transient inappropriate production of the *Drosophila* master gene regulatory protein Antennapedia change an antenna into a leg?

We shall not attempt to answer this question in this section. Instead, we shall explore the general problem of how gene control networks create *molecular switches*, which once set are often remembered through many subsequent cell generations. This phenomenon of *cell memory* is a fundamental prerequisite for the creation and maintenance of stably differentiated cell types.

We begin our discussion of molecular switches and cell memory by considering some of the well-understood cell-differentiation mechanisms that occur in bacterial and yeast cells, even though some of them are not generally used in the differentiation of higher eucaryotic cells.

0-16 Local DNA Sequences Are Rearranged in Many Bacteria to Turn Genes On and Off in a Switchlike Fashion[20]

We have seen that the differentiation of higher eucaryotic cells usually occurs without detectable changes in DNA sequences (see p. 552). But a stably inherited pattern of gene regulation is attained in many procaryotes by specific DNA rearrangements that activate or inactivate specific genes. Since all changes in a DNA sequence will be faithfully copied during subsequent DNA replications, an altered state of gene activity will be inherited by all the progeny of the cell in which the rearrangement occurred. Some of these rearrangements are reversible in principle, and over long enough periods they can produce an alternating pattern of gene activity.

A well-studied example of this type of cell-differentiation mechanism occurs in *Salmonella* bacteria, in which a specific 1000-nucleotide-pair piece of DNA occasionally becomes inverted by a reaction catalyzed by a site-specific recombination enzyme (Figure 10–28). The DNA at this site performs a "flip-flop" between

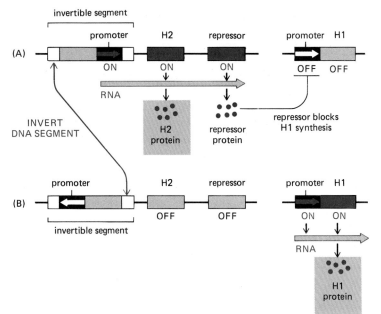

Figure 10–28 Alternating transcription of two genes in a *Salmonella* bacterium caused by a simple site-specific recombination event that inverts a small DNA segment. The recombination mechanism is described elsewhere (see p. 246) and is activated only rarely (about once in every 10^5 cell divisions). Therefore, the alternative production of the H2 flagellin (A) or the H1 flagellin (B) tends to be faithfully inherited in each clone of cells.

two states. The effect of the inversion on gene expression is due to a promoter within the 1000 nucleotide pairs that causes the bacteria to synthesize one type of protein (called flagellin) on its surface when the DNA is in one orientation and another type when it is in the other. Because flip-flops occur only rarely, whole clones of bacteria will grow up with one type of flagellin or the other. This phenomenon, known as **phase variation,** almost certainly evolved because it protects the bacterial population against the immune response of its vertebrate hosts. If the host makes antibodies against one type of flagellin, a few bacteria whose flagellin has been altered by gene inversion will still be able to survive and multiply.

Bacteria isolated from the wild very often exhibit phase variation for one or more phenotypic traits. These "instabilities" are usually eliminated from standard laboratory strains, and only some of the mechanisms underlying them have been studied. Not all involve DNA inversion. A bacterium that causes a common sexually transmitted human disease (*Neisseria gonorrhoeae*), for example, avoids immune attack by means of an inherited variation in its surface properties that is generated by gene conversion (see p. 244). This mechanism is dependent on the recA recombination protein, and it transfers DNA sequences to a master gene from silent "gene cassettes" (see p. 572); it has the advantage of creating more than 100 variants of the major bacterial surface protein.

10-17 A Master Regulatory Locus Determines Mating Types in Yeasts[21]
10-18

Yeasts are single-cell eucaryotes that exist in either a haploid or a diploid state. Diploid cells are formed by a process known as **mating,** in which two haploid cells fuse (see Figure 13–17, p. 739). In order for two haploid cells to mate, they must differ in *mating type* (sex). In the common baker's yeast, *Saccharomyces cerevisiae*, there are two mating types, α and **a.** These two types of cells are specialized for mating with each other: each produces a specific diffusible signaling molecule and a receptor protein that enable it both to signal to and to recognize and fuse with its opposite cell type. The resulting diploid cells, called **a**/α, have features that are distinct from those of either parent: they are unable to mate but can form spores (sporulate), giving rise to haploid cells by meiosis when they run out of food.

The genetic mechanisms that create these three yeast cell types provide a clear example of the action of master regulatory genes. The master gene regulatory proteins are produced by a single genetic locus, the **mating-type locus (*MAT*),** and they control cell type by acting in combinations to determine the transcription of many other genes. As described in Figure 10–29, the mechanisms are known,

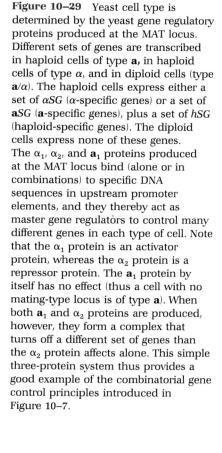

MATING-TYPE LOCUS PRESENT	PATTERN OF GENE EXPRESSION		PHENOTYPE
	MATING-TYPE LOCUS	OTHER GENES	
a	**a1** ON → NO EFFECT	aSG ON / αSG OFF / hSG ON	haploid cell mating-type a
α	α1 ON / α2 ON	aSG OFF / αSG ON / hSG ON	haploid cell mating-type α
a/α	α1 OFF / α2 ON / **a1** ON	aSG OFF / αSG OFF / hSG OFF	diploid cell neither a-specific nor α-specific genes expressed

Figure 10–29 Yeast cell type is determined by the yeast gene regulatory proteins produced at the MAT locus. Different sets of genes are transcribed in haploid cells of type **a**, in haploid cells of type α, and in diploid cells (type **a**/α). The haploid cells express either a set of αSG (α-specific genes) or a set of **a**SG (**a**-specific genes), plus a set of hSG (haploid-specific genes). The diploid cells express none of these genes. The α₁, α₂, and **a**₁ proteins produced at the MAT locus bind (alone or in combinations) to specific DNA sequences in upstream promoter elements, and they thereby act as master gene regulators to control many different genes in each type of cell. Note that the α₁ protein is an activator protein, whereas the α₂ protein is a repressor protein. The **a**₁ protein by itself has no effect (thus a cell with no mating-type locus is of type **a**). When both **a**₁ and α₂ proteins are produced, however, they form a complex that turns off a different set of genes than the α₂ protein affects alone. This simple three-protein system thus provides a good example of the combinatorial gene control principles introduced in Figure 10–7.

and they provide a good illustration of the combinatorial control principles introduced earlier (see p. 554).

The production of the three master regulatory proteins that control yeast mating type (the α₁, α₂, and **a**₁ proteins) is itself regulated, since haploid yeast cells regularly switch their mating type. The molecular mechanism responsible for this developmental switch is apparently peculiar to yeasts: the cell type depends on which of two DNA sequences, the α-type sequence (encoding the α₁ and α₂ proteins) or the **a**-type sequence (encoding the **a**₁ protein), occupies the MAT locus, and the change in mating type results from a change in the DNA at this locus. Each time an α-type cell changes to an **a**-type cell, the α-type gene in the MAT locus is excised and replaced by a newly synthesized **a**-type gene copied from a silent **a**-type gene elsewhere in the genome. Because the change involves the removal of one gene from an active "slot" and its replacement by another, this mechanism is called the *cassette mechanism*. The change is reversible because, although the original α-type gene at the MAT locus is discarded, a silent α-type gene remains in the genome. In alternation, new DNA copies made from the silent **a**-type and the silent α-type genes function as disposable cassettes that are inserted into the MAT locus (which acts as a "playing head") (Figure 10–30).

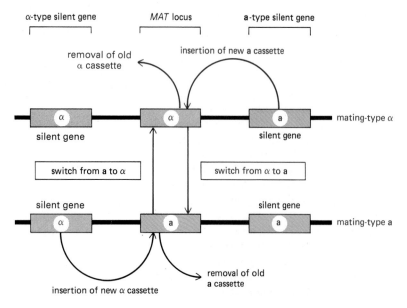

Figure 10–30 Cassette model of yeast mating-type switching. Cassette switching occurs by a gene-conversion process that is triggered when the HO nuclease makes a double-stranded cut at a specific DNA sequence in the MAT locus. The DNA near the cut is then excised and replaced by a copy of the silent cassette of opposite mating type.

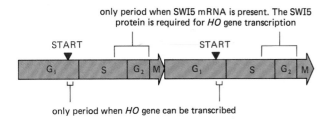

only period when SWI5 mRNA is present. The SWI5
protein is required for *HO* gene transcription

only period when *HO* gene can be transcribed

RESULT: A CELL CAN ONLY PRODUCE *HO* mRNA — AND THEREFORE THE *HO*
ENDONUCLEASE REQUIRED FOR MATING-TYPE SWITCHING — IF IT HAS
INHERITED THE SWI5 PROTEIN IN THE PREVIOUS CELL DIVISION

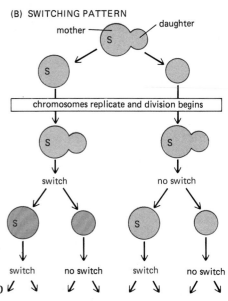

(B) SWITCHING PATTERN

The Ability to Switch Mating Type Is Asymmetrically Inherited[22,23]

Mating-type switching is initiated by a site-specific endonuclease (called the HO
endonuclease) that is the product of the *HO* gene. This enzyme makes a double-
stranded cut in the DNA at the *MAT* locus, causing the DNA there to be excised
and then replaced by means of a special type of DNA resynthesis that uses the
silent gene of opposite mating type as the template (see Figure 10–30). The transcription of the *HO* gene, which determines when and where switching occurs,
is strictly controlled. Genetic analysis has shown that at least six regulatory genes
(*SWI1* to *SWI6*) control *HO* gene transcription. Because yeast cells divide asymmetrically by budding, one of the two cells produced in each division (the "mother")
is large compared to the other (the "daughter"). Most new mother cells switch
their mating type during the subsequent cell cycle, but the newly formed daughter
cells (which arise from the bud) produce no *HO* gene product and are unable to
switch until after they have divided as a mother cell (Figure 10–31). This asymmetry
in switching has been attributed to an asymmetric inheritance of the SWI5 protein,
which binds upstream of the *HO* gene and is required for *HO* gene transcription.
The SWI5 protein (or its active form) is believed to be inherited only by the mother
cell. How this protein is excluded from the bud is unknown, but its inheritance
provides an attractive model for the asymmetric segregation of developmental
potential seen in higher eucaryotes (see p. 967).

A Specific Silencer Mechanism Seems to "Close" a Region of Yeast Chromatin[23,24]

If the coding sequences of the two silent mating-type genes lacked promoters,
their movement to the *MAT* locus could activate them by a simple mechanism, as
in phase variation. DNA-sequencing studies, however, show that the silent genes
contain all the regulatory sites they need for transcription but are kept silent by
an adjacent DNA sequence some distance downstream that somehow blocks their
expression. How these silencer sequences work is not known, but some of the
protein machinery has been identified because mutants defective in any one of
four *SIR (silent information regulator)* genes allow expression of the silent cassettes. The SIR proteins require the silencer DNA sequence for their action and
will repress all genes located up to several thousand nucleotide pairs away. The
same repression mechanism normally prevents the HO endonuclease from cutting
the DNA in the silent mating-type genes and thus ensures that DNA sequences
are transferred from a silent locus to the *MAT* locus and not in the reverse direction.

The finding that the SIR proteins inhibit both transcription and HO cutting
suggests that they may cause a change in yeast chromatin structure that somehow
"closes" the entire adjacent region of chromatin, making it inaccessible to a variety
of enzymes. Two other observations suggest that there is something unusual about
the silencing mechanism: the establishment of repression seems to require a round
of DNA replication, and a replication origin sequence (ARS sequence—see p. 516)

Figure 10–31 An explanation for the
pattern of mating-type switching
observed in haploid *S. cerevisiae* cells.
(A) Gene expression during the yeast
cell cycle. For the reason shown, the
mating type can switch only in a cell
that has inherited the SWI5 protein, a
sequence-specific DNA-binding
protein that is required for *HO* gene
transcription. Moreover, switching can
take place only in G₁ phase, which is
the only time in the cell cycle when the
HO endonuclease can be synthesized
(this protein rapidly disappears when
its synthesis ceases and is presumably
quickly proteolyzed). Since DNA
replication follows after each switch, a
dividing cell will always produce two
cells of the same mating type. (B)
Pattern of switching. Cell division
occurs by budding. In each division a
large "mother" cell and a smaller
"daughter" cell can be distinguished.
The newly formed daughter cells are
found to be unable to switch their
mating type (indicated here by different
colors) until they have passed through a
mitosis as a mother cell and entered
the next G₁ phase. As schematically
indicated, this asymmetry is postulated
to be caused by an asymmetric
inheritance of the SWI5 gene regulatory
protein (here designated S).

forms an essential part of the silencer site. Working out the details of this novel gene control mechanism may help to explain how chromatin structure affects gene activity in higher eucaryotic cells (see p. 577).

Two Bacteriophage Proteins That Repress Each Other's Synthesis Can Produce a Stable Molecular Switch[25]

We have discussed thus far several switchlike changes in cell type that are caused by DNA rearrangements. The observation that a whole vertebrate or plant can form from the genetic information present in a single somatic cell nucleus (see p. 552) eliminates *irreversible* changes in DNA sequence as the major mechanism in the differentiation of higher eucaryotic cells (although they are a crucial part of lymphocyte differentiation—see p. 1024). *Reversible* DNA sequence changes—resembling those just described for *Salmonella* and yeasts—could still be responsible for some of the inherited changes in gene expression observed in higher organisms, but there is currently no evidence that such mechanisms are used.

We shall see later that many other types of mechanisms are capable of producing an inherited pattern of gene regulation with discrete stable states. One especially well-understood example is the switchlike mechanism that determines whether the *lambda bacteriophage* will multiply in the *E. coli* cytoplasm and kill its host, or whether it will instead become integrated into the host cell DNA, to be replicated automatically each time the bacterium divides. The switch is mediated by proteins encoded by the bacteriophage genome, which contains a total of about 50 genes. These bacteriophage genes are transcribed very differently in the two states. Thus, for example, a virus destined to integrate must produce the lambda *integrase* protein, which is needed to insert the lambda DNA into the bacterial chromosome (see p. 246), but must repress production of the viral proteins responsible for virus multiplication, which would be lethal to the host cell. Once one transcriptional pattern or the other has been established, it is stably maintained. As a result, an integrated lambda virus can remain hidden in the *E. coli* genome for thousands of cell generations.

We cannot go into the details of this complex gene regulatory system here, but we shall outline a few of its general features. At the heart of the system are two gene regulatory proteins synthesized by the virus: the **lambda repressor protein (cI protein)** and the **cro protein,** each of which blocks the synthesis of the other when bound to the operator site (see p. 558) of the other's gene. The predominance of one or the other of these proteins in turn controls a set of other genes, giving rise to just two stable states. In state 1 (the *lysogenic state*) the lambda

Figure 10–32 A simplified version of the switchlike regulatory system that determines the mode of growth of bacteriophage lambda in the *E. coli* host cell. In stable state 1 (the lysogenic state) large amounts of lambda repressor protein are synthesized. This gene regulatory protein turns off the synthesis of several bacteriophage proteins, including the cro protein. As a result, the viral DNA becomes integrated in the *E. coli* chromosome and is duplicated automatically as the bacterium grows. In stable state 2 (the lytic state) large amounts of cro protein are synthesized. This gene regulatory protein turns off the synthesis of the lambda repressor protein. As a result, many bacteriophage proteins are made and the viral DNA replicates freely in the *E. coli* cell, eventually producing many new bacteriophage particles and killing the cell. Cooperative and competitive interactions (see p. 492) between the lambda repressor and cro proteins at a cluster of repeated DNA-binding sites helps to produce the observed all-or-none switch between these two states.

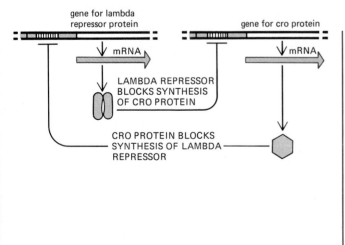

UNSTABLE INITIAL STATE

gene for lambda repressor protein

gene for cro protein

mRNA

mRNA

LAMBDA REPRESSOR BLOCKS SYNTHESIS OF CRO PROTEIN

CRO PROTEIN BLOCKS SYNTHESIS OF LAMBDA REPRESSOR

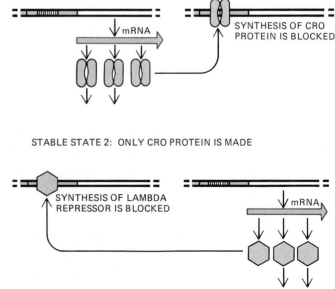

STABLE STATE 1: ONLY LAMBDA REPRESSOR IS MADE

mRNA

SYNTHESIS OF CRO PROTEIN IS BLOCKED

STABLE STATE 2: ONLY CRO PROTEIN IS MADE

SYNTHESIS OF LAMBDA REPRESSOR IS BLOCKED

mRNA

repressor dominates, and the lambda repressor, but not the cro protein, is synthesized. In state 2 (the *lytic state*) the cro protein dominates, and the cro protein, but not the lambda repressor, is synthesized (Figure 10–32). In state 1 most of the DNA of the stably integrated bacteriophage is not transcribed; in state 2 it is extensively transcribed, replicated, packaged into new bacteriophage, and released by host cell lysis.

The mode of growth adopted by an infecting lambda bacteriophage is not randomly determined. When the host bacteria are growing well, the virus tends to adopt state 1, allowing the DNA of the virus to multiply quickly along with the host chromosome. When an integrated virus finds itself in a sickly cell, it converts from state 1 to state 2 in order to multiply in the cell cytoplasm and make a quick exit (see p. 253). Other regulatory protein networks relay the information about the status of the host cell to affect the lambda repressor and cro protein switch.

An important lesson to be learned from bacteriophage lambda is that a sophisticated pattern of behavior can be achieved with only a few gene regulatory proteins that reciprocally affect each other's synthesis and activities. With the number of proteins available to eucaryotes, the possibilities for gene regulation are staggering.

Eucaryotic Gene Regulatory Proteins Can Also Generate Alternative Stable States

As discussed in Chapter 16, genetic analysis of *Drosophila* development has led to the identification and molecular characterization of more than 30 proteins that are required to establish the basic body plan of the embryo (see p. 920). Many of these are master gene regulatory proteins, which bind to enhancer elements and control the transcription of multiple genes. Several of the master gene regulatory proteins stimulate their own transcription, thereby forming a positive feedback loop that promotes their continued synthesis; at the same time they repress the transcription of similar master genes expressed in other portions of the early embryo (see p. 928). The dominant role of enhancers in the control of eucaryotic gene transcription allows multiple positive and negative inputs to affect a single gene from distant sites, and it has led to the evolution of large networks of interacting gene regulatory proteins (see Figure 10–73). A simple two-gene network analogous to the lambda switch, however, suffices to illustrate the general principles involved in this kind of cell memory (Figure 10–33).

Figure 10–33 How two master gene regulatory proteins can create a molecular switch in eucaryotes. Note that the same protein can have either an activating or a repressing effect on transcription, depending on the DNA sequence to which it is bound. Examples of this type of behavior are thought to occur among the DNA-binding proteins that control early developmental decisions in *Drosophila* (see p. 928).

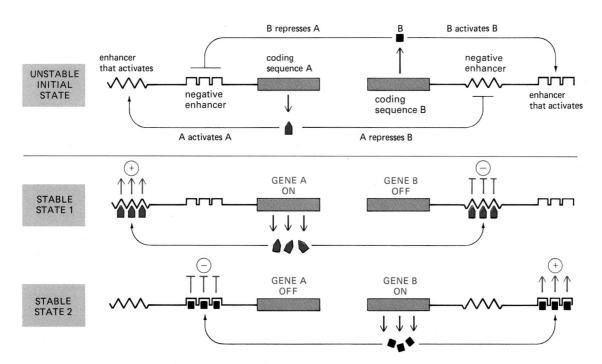

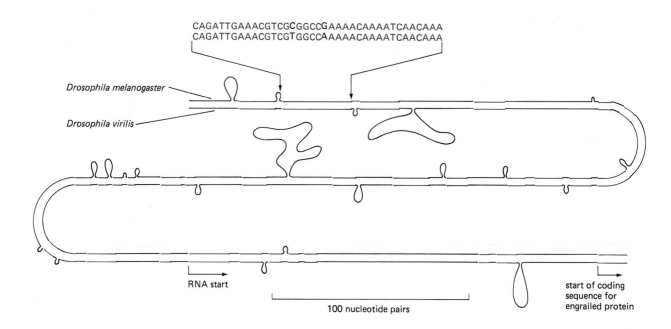

CAGATTGAAACGTCGCGGCCGAAAACAAAATCAACAAA
CAGATTGAAACGTCGTGGCCAAAAACAAAATCAACAAA

Drosophila melanogaster

Drosophila virilis

RNA start

100 nucleotide pairs

start of coding
sequence for
engrailed protein

A sense of the complexity of the regulatory networks in higher eucaryotes can be gained from the DNA sequence of the regulatory region of a master gene that controls the development of many of the major body parts in *Drosophila*. Comparison of the very long regulatory region of this gene in two distantly related *Drosophila* species reveals an elaborate pattern of more than 30 short evolutionarily conserved sequences, each of which is thought to be an important binding site for a gene regulatory protein (Figure 10–34; see p. 487).

Figure 10–34 A comparison of part of the regulatory region upstream from the *engrailed* gene in two species of *Drosophila*. A sequence comparison between *Drosophila melanogaster* and *Drosophila virilis* is shown, with regions of 90% identity colored. One example of an actual sequence match is illustrated in detail at the top. Loops indicate places where insertions or deletions of nucleotides have occurred since these two species evolved from a common ancestor about 60 million years ago. (Courtesy of Judith A. Kassis and Patrick H. O'Farrell.)

-19 Cooperatively Bound Clusters of Gene Regulatory Proteins Can, in Principle, Be Directly Inherited[26]

Other mechanisms can also cause an inherited pattern of gene regulation. One of the simplest schemes is based on a gene regulatory protein that binds cooperatively in multiple copies (or as part of a cooperatively formed large protein complex) to a specific site in chromatin. If some of the protein molecules remain bound to the DNA during DNA replication, part of the complex could be inherited by each of the two daughter DNA helices. The cooperativity of protein binding would then encourage each inherited part of the complex to bind the missing components, thereby restoring the complete functional complex. In this way the regulatory state of a gene could be directly inherited via its bound chromosomal proteins (Figure 10–35). Although in principle such directly inherited protein complexes can act either to turn off or to turn on a gene in a stable manner, there are as yet no proven examples of this type of gene regulatory mechanism. There are, however, some phenomena that suggest that it may have an important role.

Figure 10–35 A general scheme that permits the direct inheritance of states of gene expression during DNA replication. In this hypothetical model, portions of a cooperatively bound cluster of gene regulatory proteins are transferred directly from the parental DNA helix to both daughter helices. The inherited cluster then causes each of the daughter DNA helices to bind additional copies of the same gene regulatory proteins. Because the binding is cooperative (see Figure 9–15, p. 493), DNA synthesized from an identical parental DNA helix that lacks the bound gene regulatory proteins will remain free of them. If the bound gene regulatory proteins turn off gene transcription, then the inactive gene state will be directly inherited, as in the case of X-chromosome inactivation discussed in the text (see p. 577). If the bound gene regulatory proteins turn on gene transcription, the active gene state will be directly inherited as indicated here.

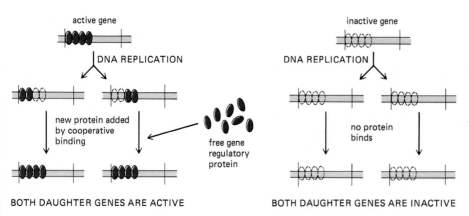

active gene

DNA REPLICATION

new protein added by cooperative binding

free gene regulatory protein

BOTH DAUGHTER GENES ARE ACTIVE

inactive gene

DNA REPLICATION

no protein binds

BOTH DAUGHTER GENES ARE INACTIVE

Heterochromatin Contains Specially Condensed Regions of DNA in Higher Eucaryotic Cells[27]

Unlike the situation in bacteria, the gene regulatory proteins that bind to specific DNA sequences in eucaryotic cells must interact with DNA molecules densely packed with nucleosomes. The necessity of transcribing DNA as chromatin clearly complicates the mechanisms controlling transcription, but we know very little about the molecular details. It is virtually certain, however, that changes in DNA packaging influence gene expression in eucaryotes. We have seen that a transcriptional silencer somehow "closes" a neighboring region of chromatin in a yeast cell, making it unavailable for either transcription or cutting by an endonuclease (p. 573). Long before this effect was discovered, however, studies of higher eucaryotic cells had produced much more dramatic examples of chromatin closing that could be correlated with visible features of chromatin structure.

Light microscope studies in the 1930s revealed that some regions of chromosomes fail to decondense during interphase, appearing to maintain the highly condensed conformation of a metaphase chromosome (see p. 503). These regions were called **heterochromatin** to distinguish them from the rest of the chromatin in the interphase nucleus (which was called *euchromatin*).

Some chromosomal regions are condensed into heterochromatin in all cells. In humans this *constitutive heterochromatin* is localized around the centromere of each mitotic chromosome, where it can be detected as darkly staining bands (Figure 10–36). In some other mammals, constitutive heterochromatin is also located in specific bands on chromosome arms. During interphase the regions of constitutive heterochromatin can aggregate to form *chromocenters* (see Figure 9–43, p. 508). In mammals the number and arrangement of such chromocenters varies with cell type and developmental stage. Most regions of constitutive heterochromatin contain relatively simple, serially repeated DNA sequences, called *satellite DNAs*. These highly repeated DNA sequences are not transcribed, and their function, and that of the condensed interphase chromatin structure that they form, is obscure (see p. 604).

Other regions of DNA are unusually condensed during interphase in some cell types of an organism but not in others. These regions also are thought not to be transcribed, but they do not consist of simple sequences. The total amount of this *facultative heterochromatin* is very different in different cell types: embryonic cells seem to have very little, whereas some highly specialized cells have a great deal. This suggests that, as cells develop, progressively more genes can be packaged in a condensed form in which they are no longer accessible to gene activator proteins. More than one mechanism may underlie the condensation of DNA in this type of heterochromatin. Much of what is known about facultative heterochromatin, however, comes from studies of the inactivation of one of the two X chromosomes in female mammalian cells.

An Inactive X Chromosome Is Inherited[28]

In mammals all female cells contain two X chromosomes, while male cells contain one X and one Y chromosome. Presumably because a double dose of X-chromosome products would be lethal, the female cells have evolved a mechanism for permanently inactivating one of the two X chromosomes in each cell. In mice this occurs between the third and the sixth day of development, when one or the other of the two X chromosomes in each cell is chosen at random and condensed into heterochromatin. This chromosome is seen in the light microscope during interphase as a distinct structure known as a *Barr body*, and is located near the nuclear membrane. It replicates late in S phase, and most of its DNA is not transcribed. Because the inactive X chromosome is faithfully inherited, every female is a mosaic composed of clonal groups of cells in which only the paternally inherited X chromosome (X_p) is active and a roughly equal number of groups of cells in which only the maternally inherited X chromosome (X_m) is active. In general, the cells expressing X_p and those expressing X_m are distributed in small clusters in the adult animal, reflecting the tendency of sister cells to remain close neighbors

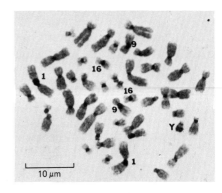

Figure 10–36 Human chromosomes at metaphase, stained by a special technique that darkens the constitutive heterochromatin. The numbers and letter mark specific chromosomes. (Courtesy of James German.)

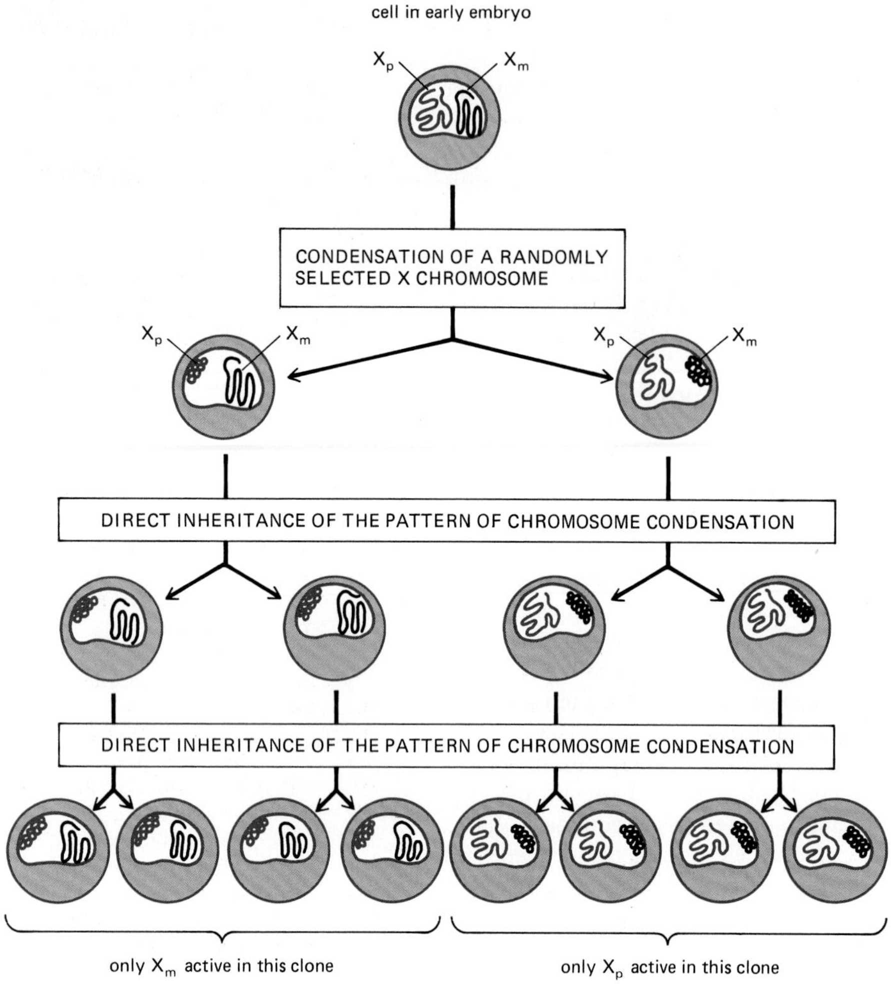

cell in early embryo

X_p X_m

CONDENSATION OF A RANDOMLY
SELECTED X CHROMOSOME

X_p X_m X_p X_m

DIRECT INHERITANCE OF THE PATTERN OF CHROMOSOME CONDENSATION

DIRECT INHERITANCE OF THE PATTERN OF CHROMOSOME CONDENSATION

only X_m active in this clone   only X_p active in this clone

Figure 10–37 The clonal inheritance of a condensed inactive X chromosome that occurs in female mammals.

during the later stages of embryonic development and growth after an early period when there is more mixing (Figure 10-37).

The condensation process that forms heterochromatin in an X chromosome has a tendency to spread continuously along a chromosome. This has been demonstrated by studies with mutant animals in which one of the X chromosomes has become joined to a portion of an *autosome* (a nonsex chromosome). In the mutant chromosomes, regions of the autosome adjacent to an inactivated X chromosome are often condensed into heterochromatin and the genes they contain thereby inactivated in a heritable way. This suggests that X-chromosome inactivation occurs by a cooperative process, which can be thought of as chromatin "crystallization" spreading from a nucleation site on the X chromosome. Once the condensed chromatin structure is established, a process resembling that in Figure 10–35 could cause it to be faithfully inherited during all subsequent replications of the DNA. The condensed X chromosome is reactivated in the formation of germ cells in the female; thus, no permanent change can have occurred in its DNA.

Drosophila Genes Can Also Be Turned Off by Heritable Features of Chromatin Structure[29]

A phenomenon analogous to X-chromosome inactivation occurs in *Drosophila*, and genetics can be used to study its mechanism. In flies with chromosomal rearrangements, breakage and rejoining events that place the middle of a region of heterochromatin next to a region of euchromatin tend to inactivate the nearby euchromatic genes. The situation is analogous to fusing a mammalian autosome to an inactive X chromosome as just described, and the inactivation events are

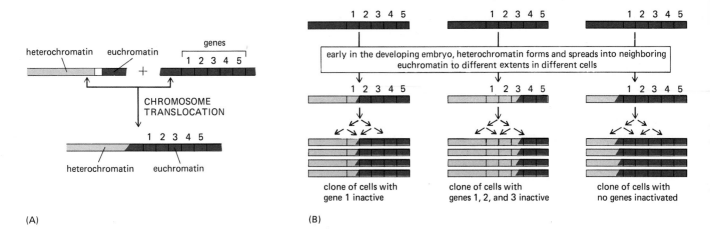

(A)

(B)

similarly patterned: the zone of inactivation spreads from the chromosome break-point to cover one or more genes in order. Moreover, while the extent of the "spreading effect" is different in different cells, the inactivated zone established in an embryonic cell is stably inherited by all of the cell's progeny (Figure 10–38).

Studies of this **position effect variegation** in *Drosophila* have revealed that the extent of spreading is reduced in flies bred to contain extra amounts of constitutive heterochromatin, which suggests that the supply of special proteins needed to produce heterochromatin can become limiting in flies with extra heterochromatin. Many single gene mutations have a similar effect. Since several of these genes are being cloned and sequenced, some of the proteins that mediate the formation of heterochromatin in *Drosophila* should soon be characterized.

Regardless of its molecular basis, the packing of selected regions of the genome into heterochromatin is a type of genetic regulatory mechanism that is not available to bacteria. The crucial feature of this uniquely eucaryotic form of gene regulation is the storing of the stable memory of gene states in an inherited chromatin structure rather than in a stable feedback loop of self-regulating gene regulatory proteins that can diffuse from place to place in the nucleus. Whether mechanisms of this type operate only to inactivate large regions of chromosomes or whether they can also operate at the level of one or a few genes is not known. The evidence to be described next, however, suggests that the expression of individual genes is often regulated by nearby control sequences and is not absolutely dependent on the more global chromosomal environment.

Figure 10–38 The phenomenon of position-effect variegation in *Drosophila*. Heterochromatin (*gray*) is normally prevented from spreading into adjacent regions of euchromatin (*color*) by special boundary sequences of unknown nature. In flies that inherit certain chromosomal translocations, however, this boundary is no longer present (A). During the early development of such flies, the heterochromatin now spreads into neighboring chromosomal DNA, proceeding for different distances in different cells (B). The spreading soon stops, but the established pattern of heterochromatin is inherited, so that large clones of progeny cells are produced that have the same neighboring genes condensed into heterochromatin and thereby inactivated (hence the "variegated" appearance of some of these flies). This phenomenon shares many features with X-chromosome inactivation in mammals.

A Specific Chromosomal Environment Is Often Required for the Optimal Expression of a Gene but Not for It to Be Turned On in the Correct Cells[30]

Transplanted genes can be transcribed in a cell-type-specific manner from many chromosomal locations. They therefore must carry with them the information needed for their selective expression in appropriate cell types. As a striking example, a *Drosophila* gene (the "salivary glue" gene *Sgs-3*) requires only 600 nucleotide pairs of upstream sequence for its correct transcription in salivary glands, and it seems to be transcribed at normal levels from almost any site (as long as it is not in heterochromatin) on a polytene chromosome. Likewise, *transgenic* mammals that carry a fragment of a cell-type-specific gene will usually express the gene only in the appropriate cells, provided that the fragment is long enough to contain most of its normal enhancer sequences.

Most genes transferred to transgenic animals (*transgenes*) are found to require much more flanking DNA than does *Sgs-3* for appropriate tissue-specific expression. Different levels of gene activity, moreover, are observed in different transgenic animals. Since these variations depend on the exact site where the transgene has inserted into the host animal's chromosomes, they are called *chromosomal position effects*. Some results obtained when the mouse alpha-fetoprotein gene was

Table 10–2 The Expression of a Gene Transferred to a Mouse Generally Shows Chromosome Position Effects, with Different Levels of Gene Activity in Independently-derived Transgenic Animals

	Percent of Total mRNA in				**Gene Copies per Cell**
	Yolk Sac	**Liver**	**Gut**	**Brain**	
Endogenous gene	**20**	**5**	**0.1**	**0**	**2**
Transgenic animal 1	3.4	1	0.1	0	4
Transgenic animal 2	4.8	30	1.3	0	4
Transgenic animal 3	4.4	13	4.7	0	4
Transgenic animal 4	0.4	0.4	0	0	12

In these experiments, carried out with the mouse alpha-fetoprotein gene, the DNA fragment injected into the fertilized mouse egg included 14,000 nucleotide pairs of upstream (5′-flanking) sequence, where three enhancers that affect the expression of this gene are located. Hybridization was used to compare the level of mRNA produced by the injected gene to that normally produced by the endogenous mouse alpha-fetoprotein gene in the indicated fetal tissues. (Data from R.E. Hammer et al., *Science* 235:53–58, 1987.)

inserted at random sites into the mouse genome are presented as an illustration in Table 10–2. While expression is restricted to the correct range of tissues in this example, the level of transcription is generally five- to tenfold less than it should be in tissues where the gene is normally highly expressed; conversely, it can be abnormally high in tissues that normally show lower levels of expression.

It has recently been discovered that the strong position effects observed for the human β-globin gene in the erythrocytes of transgenic mice are eliminated and full transcriptional activity recovered if a DNA fragment that is normally located 50,000 nucleotide pairs away from the promoter is attached to the gene. This fragment, which contains a cluster of six nuclease-hypersensitive sites (see p. 498), seems to affect an entire cluster of globin genes. We therefore shall refer to it as a **domain control region.** The general observation of position effects makes it likely that many vertebrate genes will require specific distant sequences for quantitatively correct expression. We next discuss what these sequences may do.

An Initial Local Chromatin Decondensation Step Might Be Required Before Eucaryotic Genes Can Be Activated[31]

People with a particular form of thalassemia, an inherited type of anemia (see p. 536), have a large deletion of DNA upstream from the β-globin gene. The deleted region (about 100,000 nucleotide pairs) contains several other β-like globin genes, as well as the domain control region identified in the transgenic mouse experiments (Figure 10–39A). Although the β-globin gene is intact, its level of transcription is greatly reduced. In nuclease digestion experiments this gene, unlike a normal β-globin gene, is digested at the same slow rate as bulk chromatin and is thus presumably not packaged into active chromatin (see p. 511). The normal

Figure 10–39 The cluster of β-like globin genes in humans. (A) The organization of the genes. The large chromosomal region shown spans 100,000 nucleotide pairs. It contains a cluster of sites that are hypersensitive to nuclease cleavage that create a "domain control locus." This control locus and all but one of these globin genes are deleted in γ-β-thalassemia 1. (B) Changes in the expression of the β-like globin genes at various stages of human development. Each of the globin chains encoded by these genes combines with an α-globin chain to form the hemoglobin in red cells. In the thalassemia patients the level of β-globin synthesis is greatly reduced (not shown). (A, after F. Grosveld, G.B. van Assendelft, D.R. Greaves, and G. Kollias, *Cell* 51:975–985, 1987.)

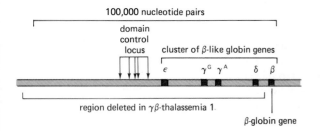

(A)

(B)

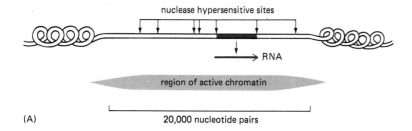

(A)

nuclease hypersensitive sites

region of active chromatin

20,000 nucleotide pairs

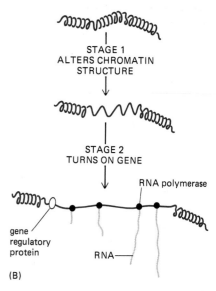

STAGE 1
ALTERS CHROMATIN
STRUCTURE

STAGE 2
TURNS ON GENE

RNA polymerase

gene
regulatory
protein

RNA

(B)

Figure 10–40 The chromatin changes in active genes. (A) The chromatin structure of the lysozyme gene in the oviduct of laying hens, where the gene is active. As indicated, a region of generally decondensed, active chromatin (detected as a region of general DNAse sensitivity) extends for 24,000 nucleotide pairs; it contains seven nuclease-hypersensitive sites, representing specific DNA sequences where a nucleosome is thought to have been displaced by other DNA-binding proteins (see p. 498). (B) The two stages postulated to be involved in eucaryotic gene activation. In stage 1 the structure of a local region of chromatin is modified to decondense it in preparation for transcription. In stage 2, gene regulatory proteins bind to specific sites on the altered chromatin to induce RNA synthesis. According to this view, transcription in procaryotes provides a model only for stage 2 of the eucaryotic gene control process. (A, data from A.E. Sipple et al., in Structure and Function of Eucaryotic Chromosomes [W. Hennig, ed.]. Berlin: Springer Verlag, 1987.)

homologue in the same erythrocyte lacks this deletion, and its entire cluster of β-like globin genes—including at least 90,000 nucleotide pairs of DNA—seems to be decondensed as active chromatin by the time the first of these genes (the ε-globin gene) begins to be transcribed (see Figure 10–39B).

Results such as these have been used to support a two-step model for the induction of transcription in higher eucaryotic genes. In stage 1 all chromatin in a region tens of thousands of nucleotide pairs long is converted into a relatively decondensed, "active" form (Figure 10–40). This step might be triggered by a special type of gene regulatory protein that causes a structural change in neighboring chromatin that is propagated from a domain control region throughout an entire looped domain (see p. 503) of chromatin. In stage 2, gene regulatory proteins that act at enhancer and upstream promoter elements regulate the transcription of specific genes within regions of exposed active chromatin. Because of these more local controls, initially the human ε-globin gene is expressed in the embryonic yolk sac, then the two γ-globin genes are expressed in the fetal liver, and, finally, the β-globin gene is turned on near the time of birth (see Figure 10–39B).

Superhelical Tension in DNA Allows Action at a Distance[32]

It is not known how long-range controls are mediated in chromosomes. There are several ways, however, of transmitting a signal from one site on a DNA molecule to another site on the same molecule many thousands of nucleotide pairs away. One involves topological changes in a closed loop of DNA double helix that can lead to the formation of DNA supercoils, a conformation that DNA adopts in response to *superhelical tension.* DNA supercoiling is most readily studied in small circular DNA molecules, such as the chromosomes of some viruses and plasmids (see, for example, Figure 5–71, p. 251). The same considerations apply, however, to any region of DNA bracketed by two ends that are unable to rotate freely—as, for example, in a loop of chromatin that is tightly clamped at its base (see Figure 9–34, p. 502).

A simple way of visualizing the topological constraints that cause DNA supercoiling is given in Figure 10–41: in a DNA double helix with two fixed ends, one DNA supercoil forms to compensate for each 10 nucleotide pairs that are opened (unwound) in the helix. The formation of the supercoil restores the normal helical twist to the base-paired regions that remain, which would otherwise need to increase from 10 to about 11 nucleotide pairs per turn in the example shown (Figure 10–41B); the DNA helix resists such a deformation in a springlike fashion, preferring to relieve this superhelical tension by bending into supercoiled loops.

In bacteria such as *E. coli* a special type II DNA topoisomerase, called **DNA gyrase,** uses the energy of ATP hydrolysis to pump supercoils continuously into the DNA, thereby maintaining the DNA under constant tension. These supercoils are *negative supercoils,* having the opposite handedness from the *positive supercoils,* shown in Figure 10–41, that form when a region of DNA helix opens. Therefore, rather than creating a positive supercoil, a negative supercoil is removed from bacterial DNA when a region of helix opens. Because superhelical tension in the DNA is reduced during this event, the opening of the DNA helix in *E. coli* is energetically favored compared to helix opening in DNA that is not supercoiled.

DNA gyrase is not known to be present in eucaryotic cells, and the eucaryotic type I and type II DNA topoisomerases remove superhelical tension rather than generate it (see p. 236). Consequently, most of the DNA in eucaryotic cells is not under tension. But helix unwinding is associated with the initiation step of DNA transcription (see Figure 9–65, p. 524). Moreover, a moving RNA polymerase molecule (as well as other proteins tracking along DNA) will tend to generate positive superhelical tension in the DNA in front of it and negative superhelical tension behind it (Figure 10–42). Through such topological effects, an event that occurs at a single DNA site can produce forces that are felt throughout an entire looped domain of chromatin. It is not yet known, however, whether an effect of this kind triggers further changes, as would be required for it to have any causative role in the control of gene expression in eucaryotes.

The Mechanisms That Form Active Chromatin Are Not Understood

The model for gene activation outlined in Figure 10–40 implies that some of the gene regulatory proteins in higher eucaryotes have functions that differ from those of their bacterial analogues. Instead of loading RNA polymerase (or its transcription factors) directly onto a nearby promoter sequence (see Figure 10–27), some sequence-specific DNA-binding proteins may function solely to decondense the chromatin in a local chromosomal domain or to remove a nucleosome from an adjacent enhancer or promoter in order to provide access for gene regulatory proteins of a more familiar kind. We are not certain, however, that this is so; it remains possible that the observed differences in the chromatin structure of active genes are an automatic consequence of the assembly of transcription factors and/or RNA polymerase onto a promoter sequence rather than being a required prerequisite for any of these events.

The identification of specific DNA sequences that act as a domain control region for the human β-globin gene should allow the proteins that bind to these sequences to be isolated and their genes cloned (see p. 489). At present we can only guess how they might act. It is possible that these regions simply serve as unusually potent enhancer elements of the conventional type (see p. 569). Three other possibilities are outlined in Figure 10–43.

The very different types of models proposed in Figure 10–43 indicate how far we are from understanding the transition from inactive to active chromatin. It is not known how many forms of chromatin there are or exactly what structural feature makes some regions more condensed than others. Simple views of chromatin function fail to explain why the amino acid sequences of the histones (especially H3 and H4) have been so highly conserved (see p. 221). Some unique chemical features of active chromatin have been discovered (see p. 511), however, and monoclonal antibodies that recognize either acetylated histones or the HMG 17 protein can now be used to separate the nucleosomes (with their associated DNA sequences) present in active chromatin from all other nucleosomes. If applied to the chromatin prepared from transgenic animals, such tools should in principle make it possible to distinguish those regulatory regions that create active chromatin from those that do not, which seems an essential step in deciphering exactly how higher eucaryotic genes are controlled.

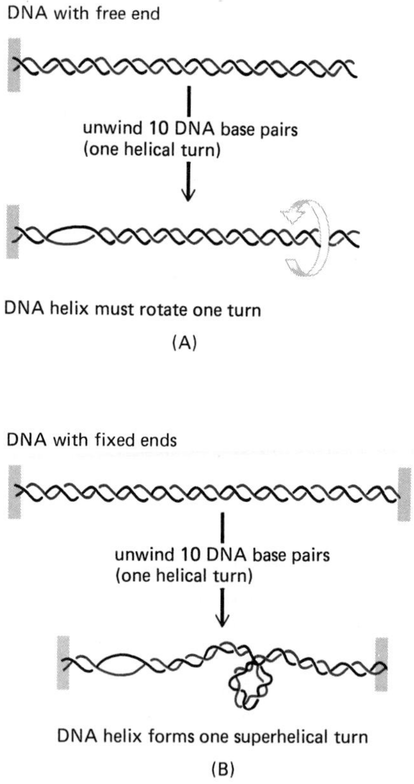

Figure 10–41 Superhelical tension in DNA causes DNA supercoiling. (A) For a DNA molecule with one free end (or a nick in one strand that serves as a swivel—see p. 236), the DNA double helix rotates by one turn for every 10 base pairs opened. (B) If rotation is prevented, superhelical tension is introduced into the DNA by helix opening. As a result, one DNA supercoil forms in the DNA double helix for every 10 base pairs opened. The supercoil formed here is a *positive supercoil* (see Figure 10–42).

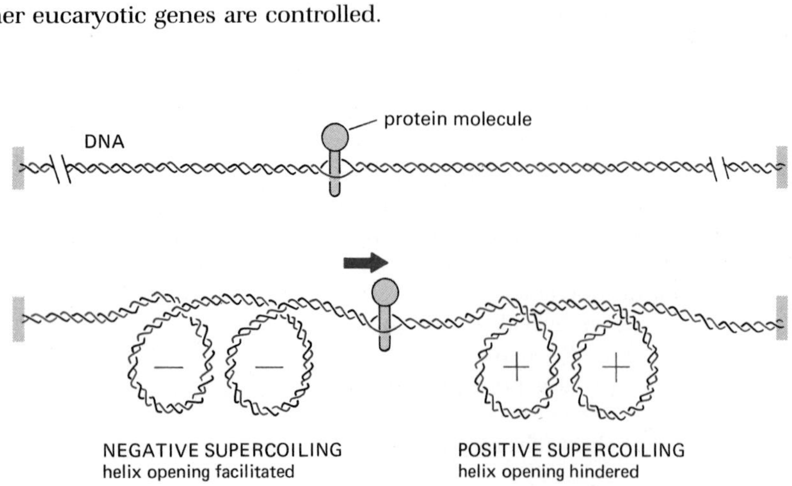

Figure 10–42 Supercoiling of a DNA segment by a protein tracking along the DNA double helix. The two ends of the DNA are fixed as in Figure 10–41B, and the protein molecule is assumed to be anchored relative to these ends or prevented by frictional forces from rotating freely as it moves. The movement thereby causes an excess of helical turns to accumulate in the DNA helix ahead of the protein and a deficit of helical turns to arise in the DNA behind the protein. Experimental evidence suggests that a moving RNA polymerase molecule causes supercoiling in this way; the positive superhelical tension ahead of it makes the DNA helix somewhat more difficult to open, but this tension should facilitate the unwrapping of the DNA from around nucleosomes, as required for the polymerase to transcribe chromatin (see Figure 9–32, p. 501).

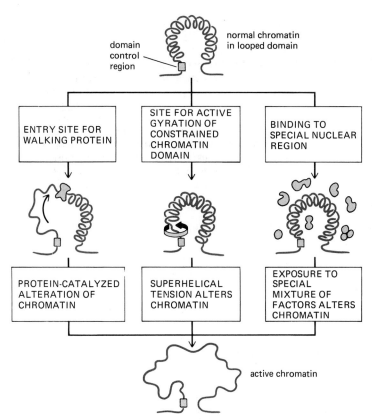

domain control region

normal chromatin in looped domain

| ENTRY SITE FOR WALKING PROTEIN | SITE FOR ACTIVE GYRATION OF CONSTRAINED CHROMATIN DOMAIN | BINDING TO SPECIAL NUCLEAR REGION |

| PROTEIN-CATALYZED ALTERATION OF CHROMATIN | SUPERHELICAL TENSION ALTERS CHROMATIN | EXPOSURE TO SPECIAL MIXTURE OF FACTORS ALTERS CHROMATIN |

active chromatin

Figure 10–43 Three models proposed to explain the long-range influence of a domain control region on gene activity. The loop of chromatin shown usually is postulated to represent an entire chromosomal looped domain, which can contain 100,000 or more nucleotide pairs of DNA. The actual mechanism that causes long-range effects is unknown, and other mechanisms are also conceivable. Evidence for a chromatin decondensation step that is independent of DNA transcription has come from observations of the polytene chromosome band that contains the *Drosophila Sgs-3* gene. Several mutants that fail to produce RNA from this region nevertheless form a chromosomal puff from the band as the larva develops (see Figure 9–48, p. 510).

New Layers of Gene Control Have Been Added During the Evolution of Multicellular Eucaryotes[33]

Yeast cells provide important models for understanding higher eucaryotic cells. In fact, the high degree of functional relatedness between yeast and human proteins has been one of the great surprises of recent years. Not all aspects of gene control can be studied in yeasts, however. Yeast cells do not seem to contain H1 histone, and nearly all of their chromatin appears to be in an active form. Likewise, control regions of DNA that influence genes from·great distances have not been observed. Nor do yeast cells make use of the alternative RNA splicing to be discussed shortly (see p. 589).

Invertebrates such as *Drosophila* have larger genomes than yeasts and contain H1 histone, heterochromatin, and at least some form of active chromatin (see Figure 9–54, p. 514). They also utilize alternative RNA splicing. But invertebrates lack at least one level of gene control that seems to be unique to vertebrates: they do not possess a general repression system based on DNA methylation.

10-20 ## The Pattern of DNA Methylation Is Inherited When Vertebrate Cells Divide[34]

The bases in the DNA helix can be covalently modified. We have discussed previously how the methylation of A in the sequence GATC makes possible a mismatch proofreading system in bacterial DNA replication (see p. 233), while the methylation of either an A or a C at a specific site protects a bacterium from the action of its own restriction nuclease (see p. 182). Vertebrate DNAs contain **5-methylcytosine (5-methyl C)**, which has the same relation to cytosine that thymine has to uracil and likewise has no effect on base-pairing (Figure 10–44A). The methylation is restricted to C bases in the sequence CG; since this sequence is base-paired to exactly the same sequence (in opposite orientation) on the other strand of the DNA helix, a simple mechanism permits a preexisting pattern of DNA methylation to be inherited directly by a templating process. An enzyme called a *maintenance methylase* acts only on those CG sequences that are base-paired with

cytosine

5-methylcytosine

methylation

(A)

5-azacytosine

no methylation

(B)

Figure 10–44 (A) Formation of 5-methylcytosine occurs by methylation of a cytosine base in the DNA double helix. In vertebrates this event is confined to selected cytosine (C) residues located in the sequence CG. (B) A synthetic nucleotide carrying the base 5-azacytosine (5-aza C) cannot be methylated. Furthermore, when small amounts of 5-aza C are incorporated into DNA, they inhibit the methylation of normal C residues.

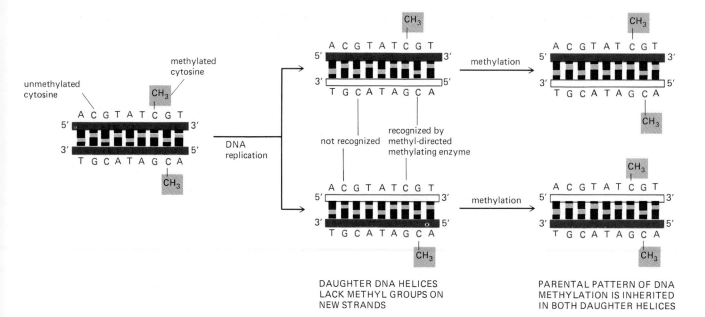

DAUGHTER DNA HELICES
LACK METHYL GROUPS ON
NEW STRANDS

PARENTAL PATTERN OF DNA
METHYLATION IS INHERITED
IN BOTH DAUGHTER HELICES

a CG sequence that is *already* methylated. As a result, the preexisting pattern of DNA methylation will be inherited directly following DNA replication (Figure 10–45).

Certain restriction nucleases cut DNA at sequences containing an unmethylated CG dinucleotide and require the absence of methylation for cutting. HpaII, for example, cuts the sequence CCGG but fails to cleave it if the central C is methylated. Thus the susceptibility of a DNA molecule to cleavage by HpaII can be used to detect whether specific DNA sites are methylated. The enzyme that normally protects the bacterium against its own HpaII enzyme is the HpaII-methylase. Treatment with this methylase thus can be used to introduce 5-methyl C bases into specific CG sequences (those in the sequence CCGG) on a cloned DNA molecule (DNA molecules cloned in *E. coli* lose all of their CG methylation). Experiments based on these techniques confirm that the maintenance methylase works as expected: each individual methylated CG is generally retained through many cell divisions in a cultured vertebrate cell, whereas unmethylated CG sequences remain unmethylated.

The automatic inheritance of 5-methyl C residues raises a "chicken and egg" problem: where is the methyl group first added in a vertebrate organism? Experiments show that methyl groups will be added to nearly every CG site in a fully unmethylated DNA molecule that is injected into a fertilized mouse egg (although an important exception will be described below). Thus the vast majority of a vertebrate genome starts out heavily methylated. Since the maintenance methylase normally cannot methylate fully unmethylated DNA, a novel *establishment methylase* activity must be present in the egg. Since the establishment methylase soon disappears, the DNA in the cells of developing tissues relies on the maintenance methylase for retention of its methylated nucleotides.

Figure 10–45 How DNA methylation patterns are faithfully inherited. In vertebrate DNAs a large fraction of the cytosine bases in the sequence CG are methylated (see Figure 10–44). Because of the existence of a methyl-directed methylating enzyme (the maintenance methylase), once a pattern of DNA methylation is established, each site of methylation is inherited in the progeny DNA, as shown. This means that changes in DNA methylation patterns will be perpetuated in a clonally inherited manner.

10-5 DNA Methylation Reinforces Developmental Decisions in Vertebrate Cells[35]

<small>0-20</small>

What is the effect of CG methylation? Tests with the HpaII enzyme indicate that, in general, the DNA of inactive genes is more heavily methylated than that of active genes. Moreover, an inactive gene that contains methylated DNA usually will lose many of its methyl groups after the gene has been activated. Evidence that the change in methylation affects gene expression comes from experiments in which a nucleoside containing the base analogue 5-azacytosine (5-aza C, see Figure 10–44B) is added for a brief period to cells in culture. The 5-aza C, which cannot be methylated, is incorporated into DNA, where it acts as an inhibitor of the maintenance methylase, thereby reducing the general level of DNA methylation. In cells treated in this way, selected genes that were previously inactive become active and, at the same time, acquire unmethylated C residues. Once activated, the active state of these genes usually is maintained for many cell generations in the absence of 5-aza C, implying that the initial methylation of the genes helped to maintain their inactivity.

When the effect of 5-aza C on cultured cells was discovered, it was proposed that DNA methylation might play a dominant part in generating different cell types. This would require that cell specialization occur by different mechanisms in vertebrates and invertebrates. Subsequent studies have suggested that DNA methylation plays a more subsidiary part in cell diversification. The important developmental decisions apparently are made by gene regulatory proteins that can turn genes on or off regardless of their methylation status. The female X chromosome, for example, is first condensed and inactivated and only later acquires an increased level of methylation on some of its genes. Conversely, several liver-specific genes are turned on during development while they are fully methylated; only later does their level of methylation decrease.

DNA transfection experiments have helped to reconcile these contrasting observations on the role of DNA methylation in gene expression. A tissue-specific gene coding for muscle actin, for example, can be prepared in both its fully methylated and fully unmethylated form. When these two versions of the gene are introduced into cultured muscle cells, both are transcribed at the same high rate. When they are introduced into fibroblasts, which normally do not transcribe the gene, the unmethylated gene is transcribed at a low level but still one much higher than either the exogenously added methylated gene or the endogenous gene of the fibroblast (which is also methylated). These experiments suggest that DNA methylation is used in vertebrates to reinforce developmental decisions made in other ways.

In a few cases it has been possible to test with high sensitivity whether a DNA sequence that is transcribed at high levels in one vertebrate cell type is transcribed at all in another cell type. Such experiments have demonstrated rates of gene transcription differing between two cell types by a factor of more than 10^6. Unexpressed vertebrate genes are much less "leaky" in terms of transcription than are unexpressed genes in bacteria, in which the largest known differences in transcription rates between expressed and unexpressed gene states are about 1000-fold. By further reducing the transcription of genes that are turned off in other ways, DNA methylation appears to account for at least part of this difference.

CG-rich Islands Reveal About 30,000 "Housekeeping Genes" in Mammals[36]

Because of the way DNA repair enzymes work, methylated C residues in the genome tend to be eliminated in the course of evolution. Accidental deamination of an unmethylated C gives rise to U, which is not normally present in DNA and thus is recognized easily by the DNA repair enzyme uracil DNA glycosylase, excised, and then replaced with a C (see p. 225). But accidental deamination of a 5-methyl C cannot be repaired in this way, for the deamination product is a T and so indistinguishable from the other, nonmutant T residues in the DNA. Thus those C residues in the genome that are methylated tend to mutate to T over evolutionary time.

Since the divergence of vertebrates from invertebrates about 400 million years ago, more than three out of every four CGs have been lost in this way, leaving vertebrates with a remarkable deficiency of this dinucleotide. The CG sequences that remain are very unevenly distributed in the genome; they are present at 10 to 20 times their average density in selected regions that are 1000 to 2000 nucleotide pairs long, called **CG islands.** These islands surround the promoters of so-called "**housekeeping genes**"—those genes that encode the many proteins that are essential for cell viability and are therefore expressed in most cells (Figure 10–46). These genes are to be contrasted with **tissue-specific genes,** which encode proteins needed only in selected types of cells.

The distribution of CG islands can be explained readily as a secondary effect of the introduction of CG methylation as a way of reducing the expression of

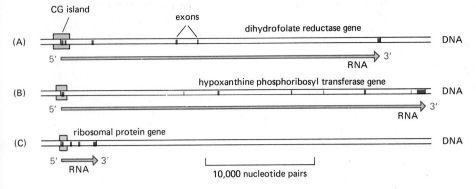

Figure 10–46 The CG islands in three mammalian housekeeping genes. Boxes show the extent of each island, which are seen to surround the promoter of each gene. Note also that, as for most genes in mammals, the exons (*color*) are short relative to the introns. (Adapted from A.P. Bird, *Trends Genet.* 3:342–347, 1987.)

inactive genes in vertebrates (Figure 10–47). In germ cells, all tissue-specific genes (except those specific to eggs and sperm) are inactive and methylated; over long periods of evolutionary time, their methylated CG sequences are lost through accidental deamination events that are not correctly repaired. The CG sequences in the regions surrounding the promoters of all active genes in germ cells, however, including all housekeeping genes, are kept demethylated, and so they can be readily repaired after spontaneous deamination events in the germ line. These genes are thought to be recognized by sequence-specific DNA-binding proteins present in the germ cells that remove any methylation near their promoters. Experiments with cloned genes show that only the CGs in CG islands remain unmethylated when fully unmethylated DNAs are injected into a fertilized mouse egg.

The mammalian genome (about 3×10^9 nucleotide pairs) contains an estimated 30,000 CG islands, each about a thousand or so nucleotide pairs in length. Most of the islands mark the 5' ends of a transcription unit and thus, presumably, a gene. Since it is possible to clone specifically the DNA surrounding the CG islands, it is relatively easy to identify and characterize housekeeping genes. Presumably, some tens of thousands of other genes are cell-type-specific and not expressed in germ cells. Because they have lost most of their CG sequences, these tissue-specific genes are more difficult to find in the genome.

Complex Patterns of Gene Regulation Are Required to Produce a Multicellular Organism[37]

As we shall discuss in Chapter 16, the cells in an embryo resemble tiny computers in that they constantly receive information about their present location and integrate it with remembered information from their past in order to act appropriately at each stage of development (see p. 914). Genetic studies in *Drosophila* indicate that a relatively small number (perhaps 100 or so) of interacting master regulatory genes play major roles in establishing and maintaining the basic body plan (see p. 923). In any multicellular organism the vast majority of genes, including both housekeeping genes and tissue-specific genes, are presumably regulated by complex control pathways that lead from such master regulatory genes. If mechanisms very different from those known in bacteria play a central role in eucaryotic

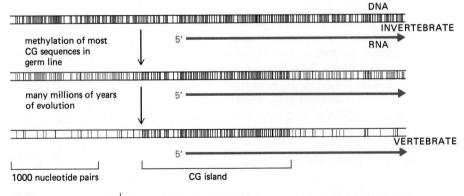

Figure 10–47 A mechanism to explain both the marked deficiency of CG sequences and the presence of CG islands in vertebrate genomes. A black line marks the location of an unmethylated CG dinucleotide in the DNA sequence, while a red line marks the location of a methylated CG dinucleotide. No examples of methylated invertebrate genes are known.

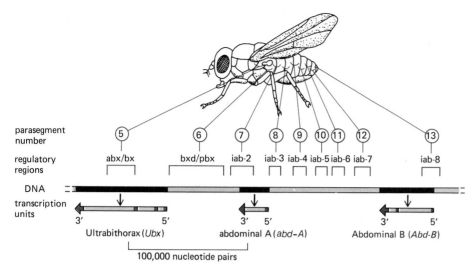

parasegment number
⑤ ⑥ ⑦ ⑧ ⑨ ⑩⑪ ⑫ ⑬

regulatory regions
abx/bx   bxd/pbx   iab-2   iab-3 iab-4 iab-5 iab-6 iab-7   iab-8

DNA

transcription units
3'   5'   3'   5'   3'   5'
Ultrabithorax (Ubx)   abdominal A (abd-A)   Abdominal B (Abd-B)

100,000 nucleotide pairs

Figure 10–48 Organization of the *Drosophila* bithorax complex. This important chromosomal region of 300,000 nucleotide pairs contains three genes—*Ubx*, *abd-A*, and *Abd-B*—encoding master gene regulatory proteins that control the development of the thoracic and abdominal regions of the fly. Homeotic mutations (see p. 930) have further defined nine groups of regulatory DNA sequences. Each is required for the development of the indicated parasegment as well as more posterior parasegments. These regulatory DNA sequences are believed to act as enhancers to control the expression of one of the nearby genes, and their order on the DNA corresponds to the order of the body segments that they affect (see Figure 10–49). (Adapted from M. Peifer, F. Karch, and W. Bender, *Genes Devel.* 1:891–898, 1987.)

gene regulation—for example, mechanisms that depend on a directly inherited chromatin structure (see p. 576)—we might expect to find these mechanisms controlling some of the master regulatory genes.

Although nothing is known about how master regulatory genes are controlled in vertebrates, we are beginning to learn how these genes are controlled in *Drosophila*. The homeotic genes that determine the different identities of the fly's body segments, for example, are present in two complex loci, known as the *Antennapedia complex* and the *bithorax complex* (see p. 929). The bithorax complex, which is responsible for the differentiation of two thoracic and eight abdominal segments, contains three transcription units—called *Ubx*, *abd-A*, and *Abd-B*. Each transcription unit appears to encode a family of gene regulatory proteins produced by alternative RNA splicing. Whereas the protein-coding regions of the complex are thought to contain fewer than 20,000 nucleotide pairs, the regulatory regions span about 300,000 nucleotide pairs. Remarkably, the regulatory regions seem to be formed from sets of enhancers that are arranged on the chromosome in the same order as the body segments that the enhancers affect (Figure 10–48). This finding, taken together with the properties of a mutant fly in which a single enhancer has been translocated from one part of the bithorax complex to another, has suggested a model for gene regulation within the bithorax complex based on changes in chromatin structure. In this model, successive chromatin domains are exposed along the bithorax complex in the cells of progressively more posterior segments, allowing the enhancers in these domains to be successively activated (Figure 10–49). It seems likely that the mechanisms controlling master regulatory genes in vertebrates will be at least as complex.

Figure 10–49 A model explaining the exact correspondence between the chromosomal position of each regulatory region in the bithorax complex and the location in the fly of the most anterior parasegment that is affected by a mutation in that region. (A) Control by effects on chromatin structure. The chromatin is postulated to become decondensed or otherwise activated progressively in each of the more posterior parasegments, so that only the abx/bx regulatory region is exposed in parasegment 5, while all of the regulatory regions are exposed in the most posterior parasegment affected by the complex (parasegment 13, see Figure 10–48). Only three of the parasegments are illustrated here. The *Ubx* gene can produce several different transcripts (symbolized here by squares and circles), whose selection is governed by the regulatory regions; the bithorax complex therefore produces a different mixture of gene regulatory proteins in each parasegment (B).

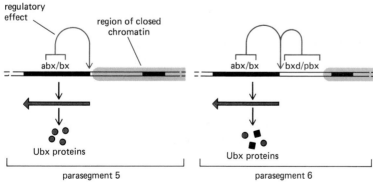

regulatory effect
region of closed chromatin
abx/bx
Ubx proteins
parasegment 5

abx/bx   bxd/pbx
Ubx proteins
parasegment 6

abx/bx   bxd/pbx iab-2
Ubx proteins   abd-A proteins
parasegment 7
(A)

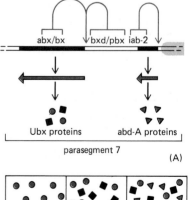

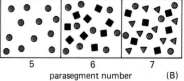

5   6   7
parasegment number   (B)

Summary

The many types of cells in animals and plants are largely created through mechanisms that cause different genes to be transcribed in different cells. Since many specialized animal cells can maintain their unique character when grown in culture, the gene regulatory mechanisms involved in creating them must be stable once established and heritable when the cell divides. Procaryotes and yeasts provide unusually accessible model systems in which to study gene regulatory mechanisms, some of which may be relevant to the creation of specialized cell types in higher eucaryotes. One such mechanism involves a competitive interaction between two (or more) master gene regulatory proteins, each of which stimulates its own synthesis while inhibiting the synthesis of the other.

Studies of the expression of engineered genes and gene fragments that have become integrated at randomly selected sites in the genomes of transgenic animals suggest that most higher eucaryotic genes are controlled by a mixture of diffusible gene regulatory proteins that is unique in each type of cell. In addition, gene expression in higher eucaryotes can be affected by transitions in chromatin structure to more or less condensed forms, and—in vertebrates—DNA methylation suppresses the transcription of inactive genes. For most genes, however, these additional controls seem to be either determined or overridden by the diffusible gene regulatory proteins. We do not yet know, however, how the master regulatory genes that determine cell type in vertebrates are themselves controlled.

Posttranscriptional Controls

Although controls on the initiation of gene transcription are the predominant forms of regulation for most genes, a variety of controls can act later in the pathway from RNA to protein. In fact, every step in gene expression that can be controlled in principle is likely to be regulated under some circumstances for some genes. Much of what is known about such **posttranscriptional regulation** has been learned only recently, and many of the mechanisms are incompletely understood.

In this section we describe how the expression of a gene can be regulated after RNA polymerase has bound to the gene's promoter and begun RNA synthesis. The discussion is organized according to the sequence in which regulatory mechanisms might act on an RNA molecule after its transcription has begun.

Transcription Attenuation Causes the Transcription of Some RNA Molecules to Be Terminated Prematurely[38]

Transcription attenuation has been studied extensively in bacteria, where it regulates many of the genes involved in amino acid synthesis. This regulatory process requires a particular nucleotide sequence organization near the beginning of the RNA chain that allows the growing RNA molecule to flip-flop between two alternative conformations. The most stable conformation has an RNA loop that serves as a termination signal for the bacterial RNA polymerase, so that RNA synthesis terminates prematurely (see Figure 5–6, p. 205). When a regulatory molecule binds to a specific site on the growing RNA molecule, however, the RNA chain now adopts a conformation that lacks the termination signal, producing a long, functional mRNA molecule (Figure 10–50). In bacteria this regulatory component is usually a ribosome that has stalled during translation of the nascent RNA chain. The specificity of the response arises because multiple codons are present at the stall site for the amino acid whose synthesis is catalyzed by the products of the downstream genes. For other genes a regulatory protein functions instead of a ribosome.

In eucaryotes there is evidence that transcription attenuation contributes to the regulation of a small proportion of genes. Since there are no functional ribosomes in the cell nucleus, a regulatory molecule that binds to a specific RNA sequence seems likely to be involved, but the mechanism of attenuation in eucaryotic cells is unknown.

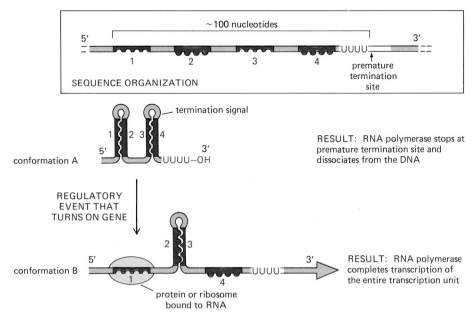

As the RNA transcript grows by the addition of nucleotides to its 3' end (see p. 203), it folds into one of the two conformations shown. The most stable form is conformation A, which contains two hairpin helices formed by complementary base-pairing. Since the hairpin helix formed by pairing sites 3 and 4 is followed by a series of U nucleotides, only it serves as a termination signal for the bacterial RNA polymerase, causing the polymerase to stop transcription prematurely (see Figure 5–6, p. 205). Conformation B is formed if a regulatory protein or a ribosome is bound to site 1 on the RNA transcript; this frees site 2 to pair with site 3, thereby disrupting the termination signal so that a long, functional RNA transcript that turns on the expression of the gene is produced.

10-23 RNA Splicing Can Be Regulated to Produce Different Forms of a Protein from the Same Gene[39]

RNA splicing was initially discovered in a virus that produces several differently spliced mRNAs from a single primary RNA transcript and thereby makes several distinct proteins (see p. 536). A large proportion of higher eucaryotic genes produce multiple proteins by means of such **alternative RNA splicing.** When different splicing possibilities exist at several positions of the transcript, a single gene can produce dozens of different proteins; usually, however, the splice alternatives are more limited, and only a few different proteins are synthesized from each transcription unit.

In some cases alternative RNA splicing occurs because there is an "intron ambiguity": the standard spliceosome mechanism for removing introns (see p. 533) is unable to distinguish cleanly between two or more alternative pairings of 5' and 3' splice sites, so that different choices are made haphazardly on different occasions. This *constitutive* form of alternative splicing is presumably responsible for the formation of various abnormal mRNAs from a mutant β-globin gene in some humans with *β thalassemia* (see Figure 9–86, p. 537). For other genes an ambiguity of this type occurs normally, so that several versions of the protein encoded by the gene are made in all tissues in which the gene is expressed.

In many cases alternative RNA splicing is *regulated* rather than constitutive. Splice-site selections are determined by the cell. Thus the same primary RNA transcript can produce a different protein (or set of proteins) in different cell types, according to the needs of the organism. Many proteins are produced in tissue-specific forms in this way, including components of the (1) extracellular matrix, such as fibronectin (see p. 818); (2) cytoskeleton, such as tropomyosin (see p. 621); (3) plasma membrane, such as N-CAM (see p. 830) and K$^+$ ion channels (see p. 1089); (4) nucleus (see Table 10–1, p. 704); and (5) intracellular cell-signaling pathways, such as C-kinase (see p. 704) and the tyrosine protein kinase encoded by the *src* proto-oncogene (Figure 10–51).

In general, the exon changes caused by alternative RNA splicing do not produce radically different proteins. Instead, they produce a set of proteins of comparable function, called **protein isoforms,** which are modified to suit the particular tissue. The changes may determine which other proteins the molecule interacts with, for example, while leaving the catalytic or structural domains unaltered. There are exceptions to this rule, however. For example, a major change of coding exons resulting from alternative RNA splicing leads to the production of two entirely different polypeptide hormones from the same transcription unit—calcitonin in the thyroid gland and calcitonin-gene-related peptide (CGRP) in neural tissue.

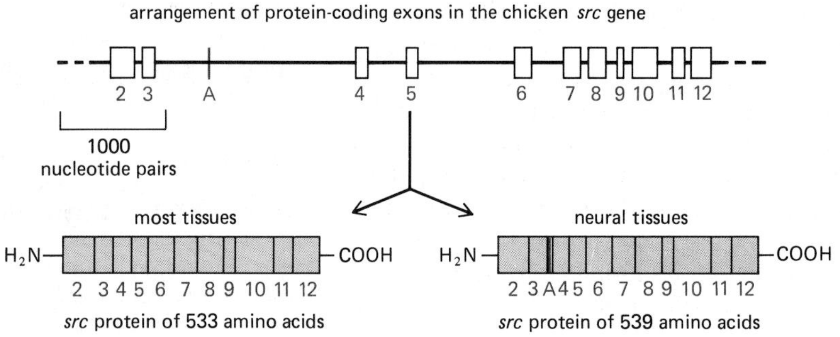

arrangement of protein-coding exons in the chicken *src* gene

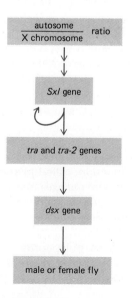

Figure 10–51   Regulated alternative RNA splicing produces two slightly different tyrosine protein kinases from the *src* gene (see p. 754); exon A is included only in neural tissues. Since this tissue-specific splicing difference has been conserved evolutionarily (being found in both birds and mammals), the resulting difference in the src protein is assumed to be important for the biological function of this regulatory protein. Only the protein-coding exons are shown (exon 1 forms the 5' leader on the mRNA). (After J.B. Levy et al., *Mol. Cell Biol.* 7:4142–4145, 1987.)

Alternative RNA Splicing Can Be Used to Turn Genes On and Off[40]

Some genes are constantly transcribed in all cells, but because the constitutive RNA-splicing mechanism produces an mRNA that codes for a nonfunctional protein, the gene is expressed only in selected cells in which a specialized splicing reaction occurs.

This type of gene regulation has been especially well characterized in *Drosophila*. The ability of the P element (see p. 268) to transpose only in germ cells, for example, is due to its failure to produce a functional transposase in somatic cells; this failure has in turn been traced to the presence of an intron in the transposase mRNA that seems to be removed only in germ cells. As another example, genetic analyses indicate that the sex of the fly is determined by a cascade of gene activations, each leading to the production of a protein that makes possible the correct splicing of the RNA that is synthesized by the next gene in the series (Figure 10–52). The DNA sequences encoding several of these sex-determining proteins have been cloned and sequenced, which should greatly facilitate studies of the mechanisms regulating the splice-site choice.

The Mechanisms Responsible for Splice-Site Choice in Regulated RNA Splicing Are Not Understood

Regulated changes in the choice of RNA splice sites are presumed to be mediated by the binding of tissue- and gene-specific proteins or RNA molecules to the growing RNA transcripts. Since the splice sites selected in both the constitutive

Figure 10–52   The cascade of changes in gene expression that determine the sex of a fly depends on alternative RNA splicing. In *Drosophila*, individuals with an autosome to X-chromosome ratio of one (normally, two sets of autosomes and two X chromosomes) develop as females, while those with a ratio of two (normally, two sets of autosomes and one X chromosome) develop as males. The ratio is assessed early in development and is remembered in each cell thereafter. The function of the genes shown is to transmit the information about this ratio to the many other genes that are involved in creating the sex-related phenotypes. These other genes function as two alternative sets: those that specify female features and those that specify male features. The *dsx* gene derives its name (*doublesex*) from the observation that, in mutants that do not express the gene, both the genes specifying female features and those specifying male features are expressed.

Events are shown as they occur in females, with arrows indicating the effects of each functional gene. In males the *Sxl*, *tra*, and *tra2* genes are transcribed but produce only nonfunctional mRNAs, and the *dsx* gene transcript is spliced to produce a protein that turns off the genes that specify female features. In females the *Sxl* transcript is spliced in a different way, so that it now produces a splicing-control protein that both maintains its own synthesis and turns on the two *tra* genes (*arrows*). The products of the *tra* genes in turn cooperate to change the pattern of RNA splicing for the *dsx* gene transcript; the resulting *dsx* mRNA produces an altered form of the dsx protein that turns off the genes that specify male features.

and regulated pathways of RNA splicing all seem to share the standard consensus sequences described previously (see Figure 9–80, p. 534), the binding of the specific component must change the conformation of the RNA transcript so as preferentially to mask or expose preexisting splice sites. The mechanism is likely to be complex, since no simple scheme, such as the masking of a splice site by the binding of a protein to it, seems able to account for the variety of splicing alternatives observed (Figure 10–53). Deciphering the molecular details of alternative RNA splicing will require that regulated splicing be reconstructed in a cell-free system so that all of the required components can be isolated and the effect of each of them on the spliceosome can be analyzed (see p. 533).

A Change in the Site of RNA Transcript Cleavage and Poly-A Addition Can Change the Carboxyl Terminus of a Protein[41]

In eucaryotes the 3' end of an mRNA molecule is not determined by the termination of RNA synthesis by the RNA polymerase; instead, it is determined by an RNA cleavage reaction that is catalyzed by additional factors while the transcript is elongating (see p. 528). The site of this cleavage can be controlled so as to change the carboxyl terminus of the resultant protein (which is encoded by the 3' end of the mRNA). In procaryotes, producing a longer RNA transcript can only add more amino acids onto a protein chain. In eucaryotes, however, RNA splicing can create mRNAs that cause the original carboxyl terminus of a protein to be removed entirely and to be replaced with a new one after a longer transcript is made.

A well-studied change of this type mediates the switch from the synthesis of membrane-bound to secreted antibody molecules during the development of B lymphocytes. Early in the life history of a B cell, the antibody it produces is anchored in the plasma membrane, where it serves as a receptor for antigen. Antigen stimulation causes these cells both to multiply and to start secreting their antibody (see p. 1028). The secreted form of the antibody is identical to the membrane-bound form except at the extreme carboxyl terminus, where the membrane-bound form has a long string of hydrophobic amino acids that traverses the lipid bilayer of the membrane, and the secreted form has a much shorter string of water-soluble amino acids. The switch from membrane-bound to secreted antibody therefore requires a different nucleotide sequence at the 3' end of the mRNA.

The membrane-bound form of the protein is generated by the transcription of all of the coding DNA sequences into RNA to make a long primary transcript. The nucleotides coding for the hydrophobic carboxyl terminus of the membrane-bound protein are located in the last exon of this long transcript (left side of Figure 10–54). The intron that precedes this exon contains the nucleotides coding for the water-soluble tail of the secreted molecule, and they are therefore removed by RNA splicing during production of the mRNA.

The secreted form of the molecule is generated from a shorter primary transcript that ends before the beginning of the last exon. Thus no acceptor splice site is present in this transcript to combine with the donor splice site just upstream from the nucleotides coding for the water-soluble tail; these nucleotides therefore remain in the final mRNA molecule (right side of Figure 10–54).

It is not known how the cleavage reaction is controlled to produce the kind of programmed switch in RNA transcript length just described.

The Definition of a Gene Has Had to Be Modified Since the Discovery of Alternative RNA Splicing[42]

The discovery that eucaryotic genes usually contain introns and that their coding sequences can be put together in more than one way has raised new questions about the definition of a gene. A gene was first clearly defined in molecular terms in the early 1940s from work on the biochemical genetics of the fungus *Neurospora*. Until then, a **gene** had been defined operationally as a region of the genome that segregates as a single unit during meiosis and gives rise to a definable phenotypic trait, such as a red or a white eye in *Drosophila* or a round or wrinkled seed in peas. After the work on *Neurospora* it became clear that most genes correspond to a region of the genome that directs the synthesis of a single enzyme.

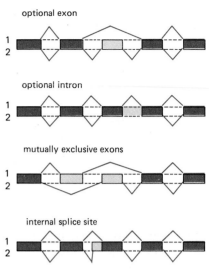

Figure 10–53 Four patterns of alternative RNA splicing that have been observed. In each case a single type of RNA transcript is spliced in two alternative ways to produce two distinct mRNAs (1 and 2). The darkly colored boxes mark RNA sequences that are retained in both mRNAs; the lightly colored boxes mark sequences that are included in only one of the mRNAs. Adjacent boxes are joined by colored lines denoting intron sequences. There seems to be no simple mechanism or rule that can explain all of these choices. (Adapted from A. Andreadis, M.E. Gallego, and B. Nadal-Ginard, *Annu. Rev. Cell Biol.* 3:207–242, 1987.)

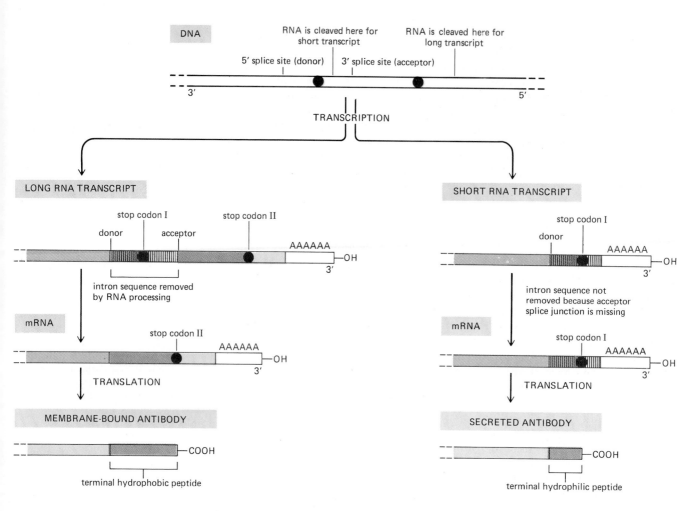

This led to the hypothesis that one gene encodes one polypeptide chain. The hypothesis proved to be extremely fruitful for subsequent research; and, as more was learned about the mechanism of gene expression in the 1960s, a gene became identified as that stretch of DNA that was transcribed into the RNA coding for a single polypeptide chain (or a single structural RNA such as a tRNA or rRNA molecule). The discovery of split genes in the late 1970s could be readily accommodated by the original definition of a gene, provided that a single polypeptide chain was specified by the RNA transcribed from any one DNA sequence. But it is now clear that many DNA sequences in higher eucaryotic cells produce two or more distinct proteins by means of alternative RNA splicing. How then is a gene to be defined?

In those relatively rare cases in which two very different eucaryotic proteins are produced from a single transcription unit, the two proteins are considered to be produced by distinct genes that overlap on the chromosome. It seems unnecessarily complex, however, to consider most of the modified proteins produced by alternative RNA splicing as being derived from overlapping genes. A more sensible alternative is to modify the original definition to include as a gene any DNA sequence that is transcribed as a single unit and encodes one set of closely related polypeptide chains (*protein isoforms*).

Figure 10–54 Regulation of the site of RNA cleavage and poly-A addition has an important effect on antibody synthesis. In unstimulated B cells (*left side*) a long RNA transcript is produced, and the intron sequence near its 3′ end is removed by RNA splicing to give rise to an mRNA molecule that codes for a membrane-bound antibody molecule. In contrast, after antigen stimulation (*right side*) the primary RNA transcript is cleaved upstream from the acceptor splice site of the last exon. As a result, some of the intron sequence that is removed from the long transcript remains as coding sequence in the short transcript. These are the sequences that encode the hydrophilic carboxyl-terminal portion of the secreted antibody molecule.

RNA Transport from the Nucleus Can Be Regulated[43]

An average primary RNA transcript seems to be at most 10 times longer than the mature mRNA molecule generated from it by RNA splicing. Yet it has been estimated that only about one-twentieth of the total mass of the hnRNA made ever leaves the cell nucleus (see p. 531). It seems, therefore, that a substantial fraction of the primary transcripts (perhaps half) may be completely degraded in the nucleus without ever generating an mRNA molecule for export. The discarded RNAs may be those whose sequence cannot be made into an mRNA molecule; on the other hand, some may represent potential mRNA molecules that are appropriately processed into mRNA only in other cell types.

RNA export through the nuclear pores is an active process (see p. 425). If this export depends on the specific recognition of the transported RNA molecule (or of a protein or RNA molecule bound to it) by a receptor protein in the nuclear pore complex, RNAs that lack this recognition signal would be selectively retained in the nucleus. Alternatively, RNA export may not require recognition signals; all RNAs might be automatically transported unless they are specifically retained. A third possibility is that a combination of selective export and selective retention operates. Since an RNA molecule seems to be selectively retained in the nucleus until all of the spliceosome components have dissociated from it (see p. 538), selective retention could be caused by a mechanism that prevents the completion of RNA splicing on particular RNA molecules. At present, however, these ideas remain largely speculative, and it seems unlikely that RNA export from the nucleus provides an important regulatory point for the expression of most eucaryotic genes.

Because a virus parasitizes normal intracellular pathways, studies of viral development often help to decipher these pathways (see p. 469). Adenovirus, for example, has a double-stranded DNA genome, which is replicated and transcribed in the host-cell nucleus. Late in infection the transport of host-cell RNAs from the nucleus is blocked, so that most of the RNAs that reach the cytoplasm are encoded by the adenovirus. Genetic analysis has shown that two adenovirus proteins that are produced early in infection are required for this change in the selectivity of transport of RNAs from the nucleus, providing a promising model system for analyzing how RNA transport is controlled.

10-24 ## Proteins That Bind to the 5′ Leader Region of mRNAs Mediate Negative Translational Control[44]

Not all mRNA molecules that reach the cytoplasm are translated into protein. The translation of some is blocked by specific *translation repressor proteins* that bind near their 5′ end, where translation would otherwise begin (Figure 10–55). This type of mechanism was first discovered in bacteria, where it enables excess ribosomal proteins to repress the translation of their own mRNAs.

In eucaryotic cells a particularly well-studied form of **negative translational control** allows the synthesis of the intracellular iron storage protein *ferritin* to be adjusted rapidly to the level of soluble iron atoms present. The ferritin mRNA in the cytoplasm can be shown to shift from an inactive ribonucleoprotein complex to a translationally active polyribosome complex after exposure of a cell to iron. Recombinant DNA experiments indicate that the iron regulation depends on a sequence of about 30 nucleotides in the 5′ leader of the ferritin mRNA molecule. This *iron-response element* folds into a stem-loop structure (see Figure 10–60B) that binds a regulatory iron-binding protein when that protein is not bound to iron. When the protein is bound to the iron-response element, the translation of any RNA sequence downstream is repressed (see Figure 10–55). The addition of iron dissociates the protein from the mRNA, increasing the rate of translation of the mRNA by as much as 100-fold.

The Translation-Enhancer Sequences in Some Viral mRNAs Suggest a Mechanism for Positive Translational Control[45]

In principle, **positive translational control** could be mediated by a special "translation-enhancer" region in an mRNA molecule that preferentially attracts ribosomes. Certain RNA viruses—the *picornaviruses*—can be shown to contain such a region, whose presence can cause translation to begin at internal AUG sites that would not otherwise be used to start protein synthesis in a eucaryotic cell (Figure 10–56).

Positive translational control has also been demonstrated in yeast cells. Genetic studies have identified specific proteins required to activate the translation of the mRNA produced from a yeast gene called *GCN4;* without them the mRNA remains untranslated. The *GCN4* mRNA resembles a class of poorly translated mRNAs from higher eucaryotes, whose translation is suspected to be similarly

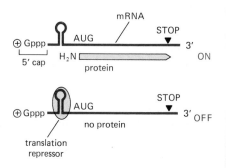

NEGATIVE TRANSLATIONAL CONTROL

Figure 10–55 Negative translational control mediated by a sequence-specific RNA-binding protein (a *translation repressor*). Binding of the protein to an mRNA molecule decreases the translation of the mRNA. Several cases of this type of translational control are known; the illustration is modeled on the mechanism that causes more ferritin protein to be synthesized when the free iron concentration rises in the cell (see also Figure 10–60).

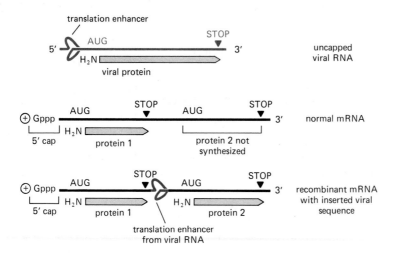

Figure 10–56 Diagram of an experiment that demonstrates the presence of an RNA sequence that serves as a translation enhancer in the genomes of certain RNA viruses. Picornaviruses (which includes poliovirus) are positive-strand RNA viruses—that is, their genomes can serve directly as mRNAs for the synthesis of viral proteins (see p. 251). Because of their mode of RNA synthesis, they lack the 5′ caps that are required for the initiation of protein synthesis on most mRNAs (see p. 215). By measuring the protein synthesis catalyzed by various recombinant RNAs, a translation-enhancer sequence several hundred nucleotides long can be identified in the viral RNA molecule. As illustrated, moving this sequence to an internal position in an mRNA chain will cause a ribosome to start translation at a neighboring internal AUG, so that the rules that normally cause only the first AUG in the mRNA chain to be used to start protein synthesis (see p. 214) are bypassed.

controlled. (This class contains about 5% of all mRNAs characterized thus far.) These RNAs have an unusually long 5′ leader sequence that contains a series of AUGs that interfere with the translation of the major coding sequence downstream by initiating the synthesis of short peptides; a stop codon occurring before the major coding sequence prevents readthrough. By analogy with the picornaviruses, the translation of the major coding sequence in these mRNAs might depend on the binding of translation-activator molecules to a translation-enhancer sequence next to the appropriate AUG, so that translation reinitiates (Figure 10–57). The mechanism responsible for these types of translational activations, however, is not known.

Many mRNAs Are Subject to Translational Control[46]

How widespread is translational control in higher eucaryotes? According to one estimate, the expression of about one gene in ten is strongly regulated in this way. Translational control enables a cell to change the concentration of a protein rapidly and reversibly without rapid turnover of its mRNA (see p. 714). The expression of some proto-oncogenes seems to be regulated at this level.

Translational controls are especially important in many fertilized eggs, which need to switch from making proteins required to maintain a quiescent oocyte to making proteins required for rapid cell division. These eggs have a large store of mRNA that is built up during the development of the oocyte. Many of these *maternal mRNAs* are not translated until the egg has been fertilized. Studies with clam eggs have shown that different sets of mRNAs are associated with ribosomes before and after fertilization. The translation of these mRNAs in a cell-free system produces proteins corresponding to the appropriate stage of clam egg development—but only if the RNA is left in its native, ribonucleoprotein form. If the RNA is stripped of its bound proteins and then translated, the differences between the proteins synthesized from the mRNAs taken from different stages of development disappear. Therefore, the controls that determine whether a particular mRNA is translated must depend on how the RNA is packaged with regulatory molecules.

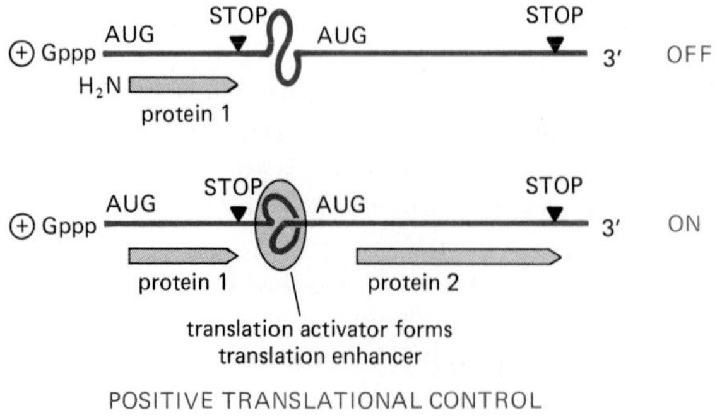

POSITIVE TRANSLATIONAL CONTROL

Figure 10–57 A model for positive translational control in which the binding of a protein (a *translation activator*) is needed for extensive translation of an mRNA. Although positive controls are known to cause the translation of specific mRNAs, the mechanisms involved are unclear. However, positive control is associated with the synthesis of short peptides whose translation is initiated upstream from the correct AUG. This suggests an analogy with the picornavirus mechanism in Figure 10–56, on which this illustration is modeled.

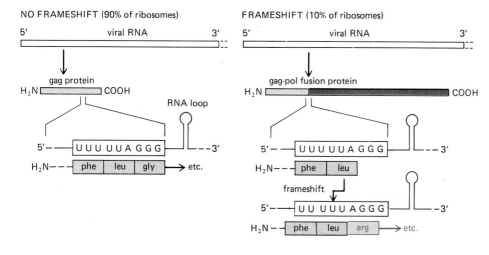

NO FRAMESHIFT (90% of ribosomes)

5' viral RNA 3'

gag protein
H₂N [   ] COOH
RNA loop

5'— [U U U U U A G G G] —-3'
H₂N —— [phe | leu | gly] → etc.

FRAMESHIFT (10% of ribosomes)

5' viral RNA 3'

gag-pol fusion protein
H₂N [   ] COOH

5'—— [U U U U U A G G G] —-3'
H₂N —- [phe | leu]

frameshift
5'—— [U U U U U A G G G] —-3'
H₂N — [phe | leu | arg] → etc.

Figure 10–58 Translational frameshifting is necessary to produce the reverse transcriptase of a retrovirus. The viral reverse transcriptase and integrase are produced by cleavage of the large gag-pol fusion protein, whereas the viral capsid proteins are produced by cleavage of the more abundant gag protein. Both proteins start identically, but the gag protein terminates at an in-frame stop codon; the indicated shift to the −1 frame upstream from this codon allows the synthesis of the fusion protein. The frameshift occurs because features in the local RNA structure (including the RNA loop shown) cause the tRNA^Leu attached to the carboxyl terminus of the growing polypeptide chain (see p. 208) occasionally to slip backward by one nucleotide on the ribosome, so that it pairs with a UUU codon instead of the UUA codon that had specified its incorporation. The sequence shown is from the human immunodeficiency virus (HIV-1), which causes AIDS. (Adapted from T. Jacks et al., *Nature* 331:280–283, 1988.)

5-8 ## Translational Frameshifting Allows Two Proteins to Be Made from One mRNA Molecule[47]

The translational controls thus far discussed affect the rate at which new protein chains are initiated on an mRNA molecule. Usually, the completion of the synthesis of a protein is automatic once this synthesis has begun. In special cases, however, a process called **translational frameshifting** can alter the final protein that is made.

Translational frameshifting is commonly used by retroviruses; it allows different amounts of two or more proteins to be synthesized from a single mRNA. These viruses commonly make a large *polyprotein* that is cleaved by a viral protease to produce a group of capsid proteins (*gag proteins*) and the viral reverse transcriptase and integrase (*pol proteins*). In many cases the *gag* and *pol* genes are in different reading frames, so that a translational frameshift is required to produce the much less abundant *pol* proteins. The frameshift occurs at a particular codon in the mRNA and requires specific sequences, some of which are upstream and some downstream from this site (Figure 10–58).

10-26 ## Gene Expression Can Be Controlled by a Change in mRNA Stability[48]
10-27

Most mRNAs in a bacterial cell are very unstable, having a half-life of about 3 minutes. Because bacterial mRNAs are both rapidly synthesized and rapidly degraded, a bacterium can adjust its gene expression quickly in response to environmental changes.

The mRNAs in eucaryotic cells are more stable. Some, such as that encoding β globin, have a half-life of more than 10 hours. Others have a half-life of only 30 minutes or less. The unstable mRNAs often code for regulatory proteins whose levels change rapidly in cells, such as growth factors and the products of the proto-oncogenes *fos* and *myc* (see p. 1207). The 3' untranslated region of many of these unstable mRNAs contains a long sequence rich in A and U nucleotides, which seems to be responsible for their instability (Figure 10–59).

The stability of an mRNA can be changed in response to extracellular signals. Thus, for example, steroid hormones affect a cell not only by increasing the transcription of specific genes (see p. 690), but also by increasing the stability of several of their mRNAs. Conversely, the addition of iron to cells decreases the stability of the mRNA that encodes the *transferrin receptor*, causing less of this iron-scavenging protein to be made. Interestingly, the destabilization of the transferrin receptor mRNA seems to be mediated by the same iron-sensitive RNA-binding protein that controls ferritin mRNA translation; in the case of the transferrin receptor, the protein binds to the opposite end of the mRNA (the 3' untranslated region) and causes an increase rather than a decrease in protein production (Figure 10–60).

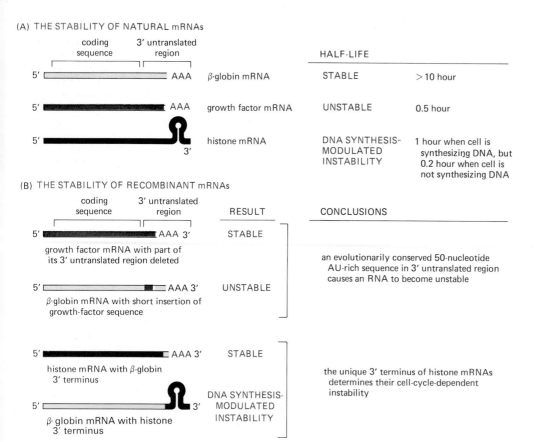

(A) THE STABILITY OF NATURAL mRNAs

	HALF-LIFE	
β-globin mRNA	STABLE	> 10 hour
growth factor mRNA	UNSTABLE	0.5 hour
histone mRNA	DNA SYNTHESIS-MODULATED INSTABILITY	1 hour when cell is synthesizing DNA, but 0.2 hour when cell is not synthesizing DNA

coding sequence / 3' untranslated region

(B) THE STABILITY OF RECOMBINANT mRNAs

coding sequence / 3' untranslated region

RESULT — CONCLUSIONS

growth factor mRNA with part of its 3' untranslated region deleted — STABLE

β-globin mRNA with short insertion of growth-factor sequence — UNSTABLE

an evolutionarily conserved 50-nucleotide AU-rich sequence in 3' untranslated region causes an RNA to become unstable

histone mRNA with β-globin 3' terminus — STABLE

β-globin mRNA with histone 3' terminus — DNA SYNTHESIS-MODULATED INSTABILITY

the unique 3' terminus of histone mRNAs determines their cell-cycle-dependent instability

Ongoing Protein Synthesis Is Required for Selective mRNA Degradation[49]

The control of mRNA stability in eucaryotic cells is best understood for the mRNAs that encode histones. These mRNAs have a half-life of about 1 hour during the S phase of the cell cycle (when new histones are needed) but are degraded within minutes when DNA synthesis stops. If DNA synthesis during S phase is inhibited with a drug, histone mRNAs immediately become unstable, perhaps because the accumulation of free histones in the absence of new DNA to bind them specifically increases the degradation rate of their mRNAs.

The regulation of histone mRNA stability depends on a short 3' stem and loop that replaces the poly-A tail at the 3' end of other mRNAs. A special cleavage reaction, which requires base-pairing to a small RNA in a ribonucleoprotein particle (U7 snRNA), creates this 3' end after the histone mRNA is synthesized by RNA polymerase II. If the 3' end is transferred to other mRNAs by recombinant DNA methods, they also become unstable when DNA synthesis stops (see Figure 10–59). Thus, as for other types of mRNAs, the degradation rate is strongly influenced by signals near the 3' end, where mRNA degradation is thought to begin.

If a stop codon is inserted in the middle of a coding sequence in a histone mRNA, the mRNA is no longer rapidly degraded. It has therefore been suggested that the nuclease responsible for degrading mRNAs is bound to the ribosome and that most of the histone mRNA has to be translated before the nuclease can begin digesting the 3' end of the mRNA. This hypothesis would explain why most unstable mRNAs are selectively stabilized when cells are treated with the protein synthesis inhibitor *cycloheximide*. It is not clear why the mRNA degradation system should be linked to the ribosome in this way.

Some mRNAs Are Localized to Specific Regions of the Cytoplasm[50]

The technique of *in situ* hybridization allows specific mRNA molecules to be located in cells (see p. 192). We shall see in Chapter 16 that several mRNAs that play a part in early pattern formation are concentrated in a particular region of the

Figure 10–59 Special signal sequences responsible for the instability (rapid degradation) of some mRNAs. (A) Three normal mRNAs with very different half-lives. The continuous rapid degradation of the mRNA molecules that encode various growth factors allows their concentration to be changed rapidly in response to extracellular signals (see p. 714). Histones are mainly needed to form the new chromatin produced during DNA synthesis; a large change in the stability of their mRNAs helps to confine histone synthesis to the S phase of the cell cycle. (B) Special sequences in the 3' untranslated region of unstable mRNAs are responsible for their unusually rapid degradation. The results shown were obtained from experiments in which the indicated altered RNA sequences were synthesized in cells from genes that had been modified by genetic engineering (see p. 266).

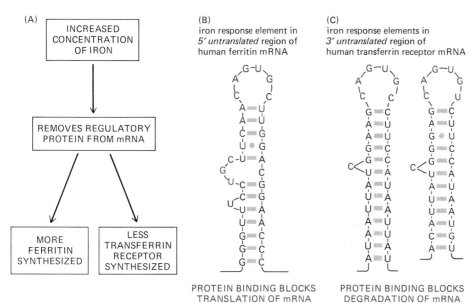

(A)

INCREASED
CONCENTRATION
OF IRON

↓

REMOVES REGULATORY
PROTEIN FROM mRNA

↓

MORE
FERRITIN
SYNTHESIZED

LESS
TRANSFERRIN
RECEPTOR
SYNTHESIZED

(B) iron response element in
5' untranslated region of
human ferritin mRNA

PROTEIN BINDING BLOCKS
TRANSLATION OF mRNA

(C) iron response elements in
3' untranslated region of
human transferrin receptor mRNA

PROTEIN BINDING BLOCKS
DEGRADATION OF mRNA

Figure 10–60 In response to an increase in iron concentration, a cell increases its synthesis of ferritin (in order to bind the extra iron) and decreases its synthesis of transferrin receptors (in order to import less iron.) Both responses are thought to be mediated by the same iron-responsive regulatory protein, which recognizes common features in the mRNAs encoding ferritin and transferrin receptor. (A) The regulatory protein dissociates from the mRNA when it binds iron. (B) The iron response element in the 5' untranslated region of ferritin mRNA. The colored nucleotides are identical in the ferritin mRNAs of mammals, birds, and amphibians. Inserting this sequence into the 5' untranslated region of other mRNAs causes their translation to be similarly regulated by iron. (C) Two of five similar iron response sequences in the 3' untranslated region of the mRNA that encodes the transferrin receptor. The multiple binding sites for the regulatory protein are required to observe the iron-regulated control of mRNA stability. The colored nucleotides, plus a helical stem whose sequence is less important, are believed to be crucial for protein recognition. Because the transferrin receptor and ferritin are regulated by different types of mechanisms, their levels respond oppositely to iron concentrations even though they are regulated by the same iron-responsive regulatory protein. (Adapted from M.W. Hentze et al., *Science* 238:1570–1573, 1987; J.L. Casey et al., *Science* 240:924–928, 1988.)

cytoplasm of the egg (see Figure 16–22, p. 894, and Figure 16–60, p. 925). Although it is not known what the mRNA is bound to, in several cases the localized distribution depends on a long 3' untranslated region of the mRNA.

While it is easiest to demonstrate the localization of mRNA molecules in large egg cells, there are examples of specific mRNA localizations in somatic cells as well. In the future, recombinant DNA methods should make it possible to alter the position of specific mRNA molecules in the cytoplasm in order to determine how their function is affected by this change.

10-28 RNA Editing Changes the Meaning of the RNA Message[51]

Biologists have been surprised continually by the discovery of novel molecular mechanisms used by cells. A recent example is the discovery that all mRNAs in trypanosomes possess a common 5' capped leader sequence that is transcribed separately and added to the 5' ends of hnRNA transcripts by RNA splicing of two initially unconnected molecules. Such *trans* splicing is also used to add a 5' leader to several mRNAs in nematodes and to combine the separate RNA transcripts that form the coding sequence of some chloroplast proteins. The cutting and pasting between transcripts could provide a shortcut to the evolution of new proteins, and the few cases where exons are known to be joined in this way may be evolutionary remnants of a much more extensive process that dominated ancient cells.

Another curious mechanism that extensively alters RNA transcripts that code for proteins has been found in the mitochondria of trypanosomes. In this **RNA-editing** process, one or more U nucleotides are either added or removed from selected regions of a transcript, causing modifications in the original reading frame that change the meaning of the message. It is not known how this editing is controlled so that a useful protein-coding sequence results.

RNA editing of a much more limited kind occurs in mammals, where it allows the *apolipoprotein-B* gene to produce two types of transcripts: in one of the transcripts a DNA-encoded C is changed to a U, creating a stop codon that causes a truncated version of this large protein to be made in a tissue-specific manner. Although this is the only example reported thus far, it seems very unlikely that RNA editing is confined to a single gene in mammals.

RNA-catalyzed Reactions in Cells Are Likely to Be of Extremely Ancient Origin[52]

All of the posttranscriptional control mechanisms discussed in this section depend on the specific recognition of a particular RNA molecule, so that the molecule is selected for special treatment, such as splicing or degradation. This recognition is achieved by a large number of sequence-specific RNA-binding molecules, most of which are still uncharacterized. The sites on the RNA to which they bind usually include a group of nucleotides whose bases are exposed in the single-stranded RNA chain (see Figure 10–60); this sequence-specific binding therefore differs from that which occurs on DNA, where the nucleotide sequence is generally recognized with the bases paired in a double helix (see p. 488). Moreover, whereas the known molecules that bind to specific DNA sequences are all proteins, the binding to specific RNA sequences can be mediated either by proteins or by other RNA molecules, which use complementary RNA-RNA base-pairing as part of their recognition mechanism. Thus, in attempting to dissect posttranscriptional mechanisms, we have largely entered an RNA world.

The reactions catalyzed by RNA molecules have been more difficult to study than the reactions catalyzed by proteins. Large RNA molecules are degraded readily during isolation by traces of ribonucleases, and they are difficult to purify to homogeneity in functional form. Recombinant DNA technology, however, now allows large amounts of pure RNAs of any sequence to be produced *in vitro* with purified RNA polymerases (see Figure 9–81, p. 534). This has made it possible to study the detailed chemistry of RNA-catalyzed self-splicing reactions (see p. 105) and to define the minimal sequences required for the RNA-mediated self-cleavage of a plant viroid (Figure 10–61). Another reaction catalyzed directly by RNA in both procaryotes and eucaryotes is the cleavage of tRNA precursors by a protein-RNA complex known as *RNAse P*. The RNA components of some of the snRNP particles in the spliceosome may likewise make and break covalent bonds (although this has not yet been demonstrated), and the active center of the *peptidyl transferase* enzyme complex that polymerizes amino acids on the ribosome is suspected to reside in the rRNA (see p. 212).

RNA molecules also have regulatory roles in cells. The *antisense RNA* strategy for experimentally manipulating cells so that they fail to express a particular gene (see p. 195) mimics a normal mechanism that is known to regulate the expression of a few selected genes in bacteria and may be used much more widely than is now realized. A particularly well-understood example of this kind of mechanism provides a feedback control on the initiation of DNA replication for a large family of bacterial DNA plasmids. The control system limits the copy number of the plasmid, thereby preventing the plasmid from killing its host cell (Figure 10–62).

Studies of RNA-catalyzed reactions are of special interest from an evolutionary perspective. As discussed in Chapter 1 (see p. 7), the first cells are thought to have contained no DNA and may have contained very few, if any, proteins. Many of the RNA-catalyzed reactions in present-day cells may represent molecular fossils—descendants of the complex network of RNA-mediated reactions that are presumed to have dominated cellular metabolism more than 3.5 billion years ago. By understanding these reactions, biologists may be able to trace the paths by which a living cell first evolved.

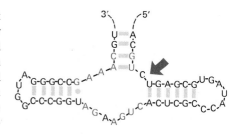

Figure 10–61 Structure of an active center of a plant viroid RNA. This short RNA molecule cleaves itself at the colored arrow. The colored nucleotides are identical in seven self-cleaving RNAs, six of which are found in plants (viroids and virusoids) and one of which is present in an animal (newts). This RNA-catalyzed reaction converts a tandemly repeated single-stranded RNA sequence that forms as a replication intermediate in viroidlike RNA molecules into a single-copy linear RNA molecule, which then circularizes. (After A.C. Forster et al., *Nature* 334:265–267, 1988).

Figure 10–62 A regulatory interaction between two RNA molecules maintains a constant plasmid copy number in the ColE1 family of bacterial DNA plasmids. RNA 1 (about 100 nucleotides long) is a regulatory RNA that inhibits the activity of RNA 2 (about 500 nucleotides long) in the initiation of plasmid DNA replication. RNA 1 is complementary in sequence to the 5′ end of RNA 2, and its concentration increases in proportion to the number of plasmid DNA molecules in a cell. In RNA 2, sequence 2 is complementary to both sequence 1 and sequence 3 (compare with Figure 10–50), and it is displaced from one to the other by the binding of RNA 1; RNA 1 thereby alters the conformation of sequence 4, inactivating RNA 2. (After H. Masukata and J. Tomizawa, *Cell* 44:125–136, 1986.)

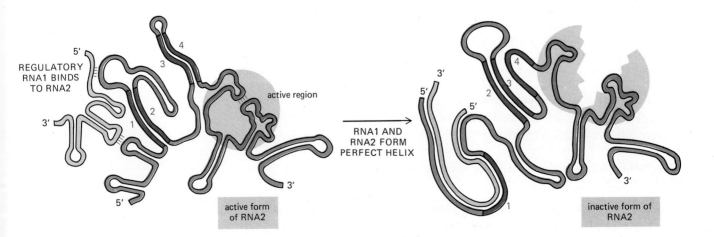

Summary

Many steps in the pathway from RNA to protein are regulated by cells to control gene expression. Most genes are thought to be regulated at multiple levels, although the control of the initiation of transcription usually predominates. Some genes, however, are transcribed at a constant level and turned on and off solely by processes that subsequently affect the RNA. These posttranscriptional regulatory processes include (1) attenuation of the transcript by its premature termination, (2) alternative splice-site selection, (3) control of 3'-end formation by cleavage and poly-A addition, (4) control of translational initiation, and (5) regulated mRNA degradation. Most of these control processes require the recognition of specific sequences or structures in the RNA molecule being regulated. This recognition can be accomplished by either a regulatory protein or a regulatory RNA molecule.

The Organization and Evolution of the Nuclear Genome[53]

Much of evolutionary history is recorded in the genomes of present-day organisms and can be deciphered from a careful analysis of their DNA sequences. Tens of millions of DNA nucleotides have been sequenced thus far, and we can now see in outline how the genes coding for certain proteins have evolved over hundreds of millions of years. Studies of the occasional changes that occur in present-day chromosomes provide additional clues to the mechanisms that have brought about evolutionary change in the past. In this section we present some of the general principles that have emerged from such molecular genetic studies, with emphasis on the organization and evolution of the nuclear genome in higher eucaryotes.

10-32 Genomes Are Fine-Tuned by Point Mutation and Radically
10-34 Remodeled or Enlarged by Genetic Recombination[54]

DNA nucleotide sequences must be accurately replicated and conserved. In Chapter 5 we discussed the elaborate DNA-replication and DNA-repair mechanisms that enable DNA sequences to be inherited with extraordinary fidelity: only about one nucleotide pair in a thousand is randomly changed every 200,000 years (see p. 220). Even so, in a population of 10,000 individuals, every possible nucleotide substitution will have been "tried out" on about fifty occasions in the course of a million years, which is a short span of time in relation to the evolution of species. If a variant sequence is advantageous, it will be rapidly propagated by natural selection. Consequently, it can be expected that in any given species the function of most genes will be optimized with respect to variation by point mutation.

While point mutation is an efficient mechanism for fine-tuning the genome, evolutionary progress in the long term must depend on more radical types of genetic change. Genetic recombination causes major rearrangements of the genome with surprising frequency: the genome can expand or contract by duplication or deletion, and its parts can be transposed from one region to another to create new combinations. Component parts of genes—their individual exons and regulatory elements—can be shuffled as separate modules to create proteins that serve entirely new roles. In addition, duplicated copies of genes tend to diverge by further mutation and become specialized and individually optimized for subtly different functions. By these means the genome as a whole can evolve to become increasingly complex and sophisticated. In a mammal, for example, multiple variant forms of almost every gene exist—different actin genes for the different types of contractile cells, different rhodopsin genes for the perception of lights of different colors, different collagen genes for the different types of connective tissues, and so on. The expression of each gene is regulated according to its own precise and specific rules. Moreover, DNA sequencing reveals that many genes share related modular segments but are otherwise very different. Thus certain portions of the rhodopsin genes share a common ancestry with portions of the genes for certain hormone and neurotransmitter receptors (see p. 706), and a common sequence motif can likewise be recognized in many other proteins (see p. 117).

Genetic recombination is fundamental to the creation of such families of genes and gene segments. Previously we have discussed the molecular mechanisms of both general recombination (see pp. 239–246) and site-specific recombination (see pp. 246–248). Here we consider some of their effects on the genome.

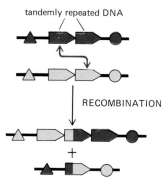

Tandemly Repeated DNA Sequences Tend to Remain the Same[55]

Gene duplications are usually attributed to rare accidents catalyzed by some of the enzymes that mediate normal recombination processes. Higher eucaryotes, however, contain an efficient enzymatic system that joins the two ends of a broken DNA molecule together. Therefore, duplications (as well as inversions, deletions, and translocations of DNA segments) can also arise as a consequence of the erratic rejoining of fragments of chromosomes that have somehow become broken in more than one place. When duplicated DNA sequences are joined head to tail, they are said to be *tandemly repeated*. Once a single tandem repeat appears, it can be extended readily into a long series of tandem repeats by unequal crossover events between two sister chromosomes, inasmuch as the large amount of matching sequence provides an ideal substrate for general recombination (Figure 10–63). DNA duplication followed by sequential unequal crossing-over underlies *DNA amplification*, a process that often contributes to the formation of cancer cells (see Figure 21–26, p. 1211).

Tandemly repeated genes both increase and decrease in number due to unequal crossing-over (see Figure 10–63). They therefore would be expected to be maintained by natural selection in large numbers only if the extra copies were beneficial to the organism. We have already discussed the hundreds of tandemly repeated genes that code for the vertebrate large ribosomal RNA precursor; these are needed to keep up with a growing cell's demand for new ribosomes (see p. 539). Similarly, vertebrates have clusters of tandemly repeated genes that encode other structural RNAs, including 5S rRNA and the U1 and U2 snRNAs, as well as clusters of repeated histone genes, which produce the large amounts of histones required during each S phase.

One might expect that in the course of evolution the sequences of the genes in a tandem array—and of the nontranscribed *spacer DNA* between them—would drift apart. With many copies of the same gene, there should be little selection against random mutations that alter one or a few of the copies, and most nucleotide changes in the long nontranscribed spacer regions would have no functional consequence. In fact, however, the sequences of the tandemly repeated genes and their spacer DNAs are generally almost identical. Two mechanisms are thought to account for this. First, recurring unequal crossing-over events will cause the continued expansion and contraction of tandem arrays, and computer simulations show that this will tend to keep the sequences the same (Figure 10–64A). Second,

Figure 10–63 A family of tandemly repeated genes frequently loses and gains gene copies due to unequal crossing-over between sister chromosomes containing the genes. This type of event is frequent because the long regions of homologous DNA sequence are good substrates for the general genetic recombination process (see p. 240).

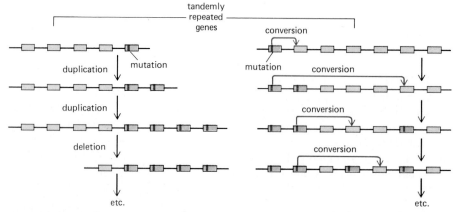

(A) HOMOGENIZATION BY UNEQUAL CROSSOVERS BETWEEN SISTER CHROMOSOMES

(B) HOMOGENIZATION BY GENE CONVERSION

Figure 10–64 Two types of events that help to keep all DNA sequences in a tandem array very similar to one another. (A) The continual expansion and contraction of the number of gene copies in a tandem array caused by unequal crossing-over (see Figure 10–63) tends to homogenize all of the gene sequences in the array. (B) In gene conversion, one gene copy acts as a template that passes all or part of its DNA sequence to another gene copy (see p. 244). In higher eucaryotes this process seems to be largely confined to genes that are next to each other on the chromosome; in lower eucaryotes such as fungi—where gene conversion is most readily studied— it is not confined to neighboring genes.

gene conversion (see p. 244) can act to homogenize related DNA sequences (Figure 10–64B). Although gene conversion does not require that the genes be tandemly repeated, in higher eucaryotes it seems to occur mainly between genes that are close to each other. The movement of one gene copy in a tandem array to a new chromosomal location will protect it from both of these homogenizing influences, thus allowing it to evolve independently and facilitating its acquisition of a modified function.

The Evolution of the Globin Gene Family Shows How Random DNA Duplications Contribute to the Evolution of Organisms[56]

In addition to generating a few sets of tandemly repeated genes, DNA duplications have played a more important general role in the evolution of new proteins. The globin gene family provides a good example because its evolutionary history has been particularly well worked out. The unmistakable homologies in amino acid sequence and structure among the present-day globin genes indicate that they must all derive from a common ancestral gene, even though some now occupy widely separated locations in the mammalian genome.

We can reconstruct some of the past events that produced the various types of oxygen-carrying hemoglobin molecules by considering the different forms of the protein in organisms at different levels on the phylogenetic scale. A molecule like hemoglobin became necessary to allow multicellular animals to grow to a large size, since large animals could no longer rely on the simple diffusion of oxygen through the body surface to oxygenate their tissues adequately; consequently, hemoglobinlike molecules are found in all vertebrates and in many invertebrates. The most primitive oxygen-carrying molecule in animals is a globin polypeptide chain of about 150 amino acids, which is found in many marine worms, insects, and primitive fish. The hemoglobin molecule in higher vertebrates, however, is composed of two kinds of globin chains. It appears that about 500 million years ago, during the evolution of higher fish, a series of gene mutations and duplications occurred. (The suspected role of diploidy in allowing the creation of new genes through mutation followed by duplication is discussed on p. 843.) These events initially established two slightly different globin genes, coding for the α- and β-globin chains in the genome of each individual. In modern higher vertebrates each hemoglobin molecule is a complex of two α chains and two β chains (Figure 10–65). This structure is much more efficient than a single-chain globin molecule. The four oxygen-binding sites in the $\alpha_2\beta_2$ molecule interact. This interaction causes a cooperative allosteric change (see p. 128) in the molecule as it binds and releases oxygen so that a much larger fraction of the bound oxygen is delivered to the tissues.

Still later, during the evolution of mammals, the β-chain gene apparently underwent mutation and duplication to give rise to a second β-like chain that is synthesized specifically in the fetus. The resulting hemoglobin molecule has a higher affinity for oxygen than adult hemoglobin and thus helps in the transfer of oxygen from the mother to the fetus. The gene for the new β-like chain subsequently mutated and duplicated again to produce two new genes, ε and γ, the ε chain being produced earlier in development (to form $\alpha_2\varepsilon_2$) than the fetal γ chain, which forms $\alpha_2\gamma_2$ (see Figure 10–39B). A duplication of the adult β-chain gene occurred still later, during primate evolution, to give rise to a δ-globin gene and thus to a minor form of hemoglobin ($\alpha_2\delta_2$) found only in adult primates (Figure 10–66). Each of these duplicated genes has been modified by point mutations that affect the properties of the final hemoglobin molecule, as well as by changes in regulatory regions that determine the timing and level of expression of the gene (see Figure 10–73).

Figure 10–66 An evolutionary scheme for the globin chains that carry oxygen in the blood of animals, emphasizing the β-like globin gene family (see also Figure 10–39). A relatively recent gene duplication of the γ-chain gene produced γ^G and γ^A, which are fetal β-like chains of identical function.

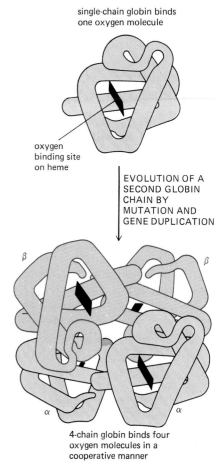

single-chain globin binds one oxygen molecule

oxygen binding site on heme

EVOLUTION OF A SECOND GLOBIN CHAIN BY MUTATION AND GENE DUPLICATION

4-chain globin binds four oxygen molecules in a cooperative manner

Figure 10–65 A comparison of the structure of 1-chain and 4-chain globins. The 4-chain globin shown is hemoglobin, which is a complex of two α- and two β-globin chains. The 1-chain globin in some primitive vertebrates forms a dimer that dissociates when it binds oxygen, representing an intermediate in the evolution of the 4-chain globin.

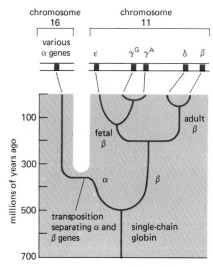

The end result of the gene duplication processes that have given rise to the diversity of globin chains is seen clearly in the genes that arose from the original β gene, which are arranged as a series of homologous DNA sequences located within 50,000 nucleotide pairs of one another (see Figure 10–39A). A similar cluster of α-globin genes is located on a separate human chromosome. Because the α- and β-globin gene clusters are on separate chromosomes in birds and mammals but are together in the frog *Xenopus*, it is believed that a translocation event separated the two genes about 300 million years ago (see Figure 10–66). Such translocations probably help stabilize duplicated genes with distinct functions by protecting them from the homogenizing processes that act on closely linked genes of similar DNA sequence (see Figure 10–64).

There are several duplicated globin DNA sequences in the α- and β-globin gene clusters that are not functional genes. They are examples of *pseudogenes*, which have a close homology to the functional genes but have been disabled by mutations that prevent their expression. The existence of such pseudogenes should not be surprising since not every DNA duplication would be expected to lead to a new functional gene and nonfunctional DNA sequences are not rapidly discarded, as indicated by the large excess of noncoding DNA in mammalian genomes (see p. 485).

A great deal of our evolutionary history will be discernible in our chromosomes once the DNA sequences of many gene families have been compared in animals of increasing complexity (see also Figure 4–62, p. 183).

Genes Encoding New Proteins Can Be Created by the Recombination of Exons[54]

The role of DNA duplication in evolution is not confined to the generation of large gene families. It can also be important in generating new single genes. The proteins encoded by genes so generated can be recognized by the presence of repeating, similar protein domains, which are covalently linked to one another in series. The immunoglobulins (Figure 10–67) and albumins, for example, as well as most fibrous proteins (such as spectrins and collagens) are encoded by genes that have evolved by repeated duplications of a primordial DNA sequence.

In genes that have evolved in this way, as well as in many other genes, each separate exon often encodes an individual protein folding unit, or domain (see p. 112). It is believed that the organization of DNA coding sequences as a series of such exons separated by long introns has greatly facilitated the evolution of new proteins. The duplications necessary to form a single gene coding for a protein with repeating domains, for example, can occur by breaking and rejoining the DNA anywhere in the long introns on either side of an exon encoding a useful protein domain; without introns there would be only a few sites in the original gene at which a recombinational exchange between sister DNA molecules could duplicate the domain. By enabling the duplication to occur at many potential recombination sites rather than at just a few, introns greatly increase the probability of a favorable duplication event.

For the same reason the presence of introns greatly increases the probability that a chance recombination event will join two initially separated DNA sequences that code for different protein domains (for example, see Figure 10–71). The presumed results of such recombinations are seen in many present-day proteins (see Figure 3–38, p. 118). Thus the large separation between the exons encoding individual domains in higher eucaryotes is thought to accelerate the process by which random genetic-recombination events generate useful new proteins. This could help to explain the successful evolution of these very complex organisms.

Most Proteins Probably Originated from Highly Split Genes Containing a Series of Small Exons[57]

The discovery of split genes in 1977 was unexpected. Previously all genes analyzed in detail were bacterial genes, which lack introns. Bacteria also lack nuclei and internal membranes and have smaller genomes than eucaryotic cells, and traditionally they were considered to resemble the simpler cells from which eucaryotic

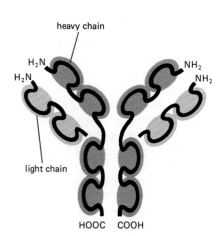

Figure 10–67 Schematic view of an antibody (immunoglobulin) molecule. This molecule is a complex of two identical heavy chains and two identical light chains (*colored*). Each heavy chain contains four similar, covalently linked domains; each light chain contains two such domains. Each domain is encoded by a separate exon, and all of the exons are thought to have evolved by the serial duplication of a single ancestral exon.

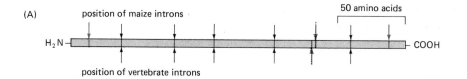

(A)

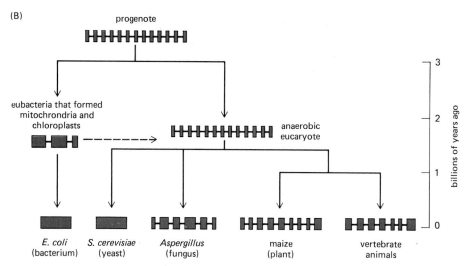

(B)

Figure 10–68 The ancient evolution of split genes. (A) A comparison of the exon structure of the *triosephosphate isomerase* gene in plants and animals. The intron positions that are identical in maize (corn) and vertebrates are marked with black arrows, while the intron positions that differ are marked with red arrows. Since plants and animals are thought to have diverged from a common ancestor about one billion years ago, the introns that they share must be of very ancient origin. (B) An outline of how a particular gene may have evolved. The exon sequences are shown in color and the intron sequences in black. The gene illustrated here codes for a protein that is required in all cells. Like *triosephosphate isomerase*, this protein must have evolved to its final three-dimensional structure before the eubacterial, archaebacterial, and eucaryotic lineages split off from a common ancestor cell—designated here as a "progenote." The dotted line marks the approximate time of the endosymbiotic events that gave rise to mitochondria and chloroplasts (see p. 398). (A, after W. Gilbert, M. Marchionni, and G. McKnight, *Cell* 46:151–154, 1987.)

cells must have been derived. Not surprisingly, most biologists initially assumed that introns were a bizarre and late evolutionary addition to the eucaryotic line. It now seems likely, however, that split genes are the ancient condition and that bacteria lost their introns only after most of their proteins had evolved.

The idea that introns are very old is consistent with current concepts of protein evolution by the trial-and-error recombination of separate exons that encode distinct protein domains. Moreover, evidence for the ancient origin of introns has been obtained by examination of the gene that encodes the ubiquitous enzyme *triosephosphate isomerase*. Triosephosphate isomerase has an essential role in the metabolism of all cells, catalyzing the interconversion of glyceraldehyde-3 phosphate and dihydroxyacetone phosphate—a central step in glycolysis and gluconeogenesis (see Figure 2–38, p. 83). By comparing the amino acid sequence of this enzyme in various organisms, it is possible to deduce that the enzyme evolved before the divergence of procaryotes and eucaryotes from a common ancestor; the human and bacterial amino acid sequences are 46% identical. The gene encoding the enzyme contains six introns in vertebrates (chickens and humans), and five of these are in precisely the same positions in maize. This implies that these five introns were present in the gene before plants and animals diverged in the eucaryotic lineage, an estimated 10^9 years ago (Figure 10–68).

In general, small unicellular organisms are under a strong selection pressure to reproduce by cell division at the maximum rate permitted by the levels of nutrients in the environment. For this, they must minimize the amount of unnecessary DNA that they have to synthesize in each cell division cycle. For larger organisms that live by predation, where size is an advantage, and for multicellular organisms in general, where rates of cell division are constrained by other requirements, there will not be such strong selection pressure to eliminate superfluous DNA from the genome. This argument may help to explain why bacteria should have lost their introns while eucaryotes retain them. It also tallies with another finding from the study of triosephosphate isomerase: whereas the multicellular fungus *Aspergillus* has five introns in its gene for this enzyme, its unicellular relative, the yeast *Saccharomyces*, has none.

What is the mechanism for loss of introns? While it is possible to lose introns by piecemeal random deletions of short segments of DNA, eucaryotic cells (and perhaps the ancestors of bacteria also) appear to have ways of precisely and selectively deleting entire introns from their genomes. Whereas most vertebrates contain only a single insulin gene with two introns, for example, rats contain a second, neighboring insulin gene with only one intron. The second gene apparently arose by gene duplication relatively recently and subsequently lost one of its introns. Because intron loss requires the exact rejoining of DNA coding sequences, it is presumed to arise from the rare incorporation into the genome of a DNA copy of the mRNA of the appropriate gene, from which the introns have been precisely removed. Such intronless copies presumably arise on occasion through the activity of reverse transcriptases (see p. 260), and it is thought that recombination enzymes allow them to become paired with the original sequence, which is then "corrected" to an intronless form by a gene-conversion type of event (see p. 244).

Reverse transcriptases are produced in cells by specific transposable elements (see Table 10–3, p. 605) as well as by all retroviruses, and the generation of DNA copies of segments of the genome by reverse transcription has also contributed in other ways to the evolution of the genomes of higher organisms (see p. 608).

A Major Fraction of the DNA of Higher Eucaryotes Consists of Repeated, Noncoding Nucleotide Sequences[58]

Eucaryotic genomes contain not only introns but also large numbers of copies of other seemingly nonessential DNA sequences that do not code for protein. The presence of such repeated DNA sequences in higher eucaryotes was first revealed by a hybridization technique that measures the number of gene copies (see p. 188). In this procedure the genome is broken mechanically into short fragments of DNA double helix about 1000 nucleotide pairs long, and the fragments are then denatured to produce DNA single strands. The speed with which the single-stranded fragments in the mixture reanneal under conditions in which the double-helical conformation is stable depends on how many complementary strands each fragment finds. For the most part, the reaction is very slow. The haploid genome of a mammalian cell, for example, is represented by about 6 million different 1000-nucleotide-long DNA fragments, and any fragment whose sequence is present in only one copy must randomly collide with 6 million noncomplementary strands for every complementary partner strand that it happens to find.

When the DNA from a human cell is analyzed in this way under conditions that require near perfect matching (high stringency conditions, see p. 191), about 70% of the DNA strands reanneal as slowly as one would expect for a large collection of unique (nonrepeated) DNA sequences, requiring days for complete annealing. But most of the remaining 30% of the DNA strands anneal much more quickly. These strands contain sequences that are repeated many times in the genome, and they thus collide with a complementary partner relatively rapidly. Most of these highly repeated DNA sequences do not encode proteins, and they are of two types: about one-third are the tandemly repeated *satellite DNAs*, to be discussed next; the rest are *interspersed repeated DNAs*. Most of the latter DNAs derive from a few transposable DNA sequences that have multiplied to especially high copy numbers in our genome (see p. 608).

Satellite DNA Sequences Have No Known Function[59]

The most rapidly annealing DNA strands in an experiment of the type just described usually consist of very long tandem repetitions of a short nucleotide sequence (Figure 10–69). The repeat unit in a sequence of this type may be composed of only one or two nucleotides, but most repeats are longer, and in mammals they are typically composed of variants of a short sequence organized into a repeat of a few hundred nucleotides. These tandem repeats of a simple sequence are called **satellite DNAs** because the first DNAs of this type to be discovered had an unusual ratio of nucleotides that made it possible to separate them from the bulk of the cell's DNA as a minor component (or "satellite"). Satellite DNA sequences generally

Figure 10–69 A simple satellite DNA sequence consisting of many serially arranged repetitions of a sequence seven nucleotide pairs long. This particular DNA sequence is found in *Drosophila*.

are not transcribed and are located most often in the heterochromatin associated with the centromeric regions of chromosomes (see p. 577). In some mammals a single type of satellite DNA sequence constitutes 10% or more of the DNA and may even occupy a whole chromosome arm, so that the cell contains millions of copies of the basic repeated sequence.

Satellite DNA sequences seem to have changed unusually rapidly and even to have shifted their positions on chromosomes in the course of evolution. When two homologous mitotic chromosomes of any human are compared, for example, some of the satellite DNA sequences usually are found arranged in a strikingly different manner on the two chromosomes. Moreover, in contrast to the high degree of conservation of DNA sequences elsewhere in the genome, generally there are marked differences in the satellite DNA sequences of two closely related species. No function has yet been found for satellite DNA sequences: tests designed to demonstrate a role in chromosome pairing or nuclear organization have failed thus far to reveal any evidence for such a role. It has therefore been suggested that they are an extreme form of "selfish DNA" sequences, whose properties ensure their own retention in the genome but which do nothing to help the survival of the cells containing them. Other sequences that are commonly viewed as selfish are the *transposable elements*, which we discuss next.

10-36 The Evolution of Genomes Has Been Accelerated by Transposable Elements of at Least Three Types[60]

Genomes generally contain many varieties of **transposable elements.** These elements were first discovered in maize, where several have been sequenced and characterized. Transposable elements have been studied most extensively in *Drosophila*, where more than 30 varieties are known, varying in length between 2000 and 10,000 nucleotide pairs; most are present in 5 to 10 copies per diploid cell.

At least three broad classes of transposable elements can be distinguished by the peculiarities of their sequence organization (Table 10–3). Some elements move from place to place within chromosomes directly as DNA, while many others move via an RNA intermediate, as described in Chapter 5 (see p. 255). In either case they can multiply and spread from one site in a genome to a multitude of other sites, sometimes behaving as disruptive parasites.

Transposable elements seem to make up at least 10% of higher eucaryotic genomes. Although most of these elements move only very rarely, so many elements are present that their movement has a major effect on the variability of a species. More than half of the spontaneous mutations examined in *Drosophila*,

Table 10–3 Three Major Families of Transposable Elements

Structure	Genes in Complete Element	Mode of Movement	Examples
① short inverted repeats at each end	encodes transposase	moves as DNA, either excising or following a replicative pathway	P element (*Drosophila*) Ac-Ds (maize) tn3 and IS1 (*E. coli*) Tam3 (*Antirrhinum*)
② directly repeated long terminal repeats (LTRs) at ends	encodes reverse transcriptase and resembles retrovirus	moves via an RNA intermediate produced by promoter in LTR	Copia Ty THE-1 bs1
③ Poly A at 3′ end of RNA transcript; 5′ end is often truncated	encodes reverse transcriptase	moves via an RNA intermediate that is presumably produced from a neighboring promotor	P element (*Drosophila*) L.1 (human) cin4 (maize)

These elements range in length from 2000 to about 12,000 nucleotide pairs; each family contains many members, only a few of which are listed here.

605 The Organization and Evolution of the Nuclear Genome

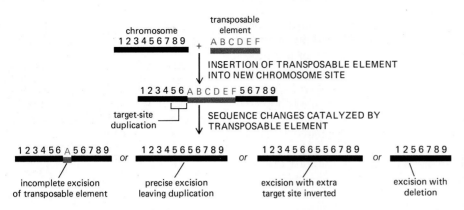

Figure 10–70 Some changes in chromosomal DNA sequences caused by transposable elements. The insertion of a transposable element always produces a short target-site duplication of the chromosomal sequence, which is generally 3 to 12 nucleotide pairs in length. The site-specific recombination enzymes associated with the element can also cause its subsequent excision. This excision often fails to restore the original chromosomal DNA sequence, as in the four examples shown.

for example, are due to the insertion of a transposable element in or near the mutant gene.

Mutations can occur either when an element inserts into a gene or when it exits to move elsewhere. All known transposable elements cause a short "target-site duplication" because of their mechanism of insertion (see Figure 5–67B, p. 248); when they exit, they generally leave behind part of this duplication—often with other local sequence changes as well (Figure 10–70). Thus, as transposable elements move in and out of chromosomes, they cause a variety of short additions and deletions of nucleotide sequences.

Transposable elements have also contributed to genome diversity in another way. When two transposable elements that are recognized by the same site-specific recombination enzyme (*transposase*) integrate into neighboring chromosomal sites, the DNA between them can become a substrate for transposition by the transposase. Because this provides a particularly effective pathway for the duplication and movement of exons, these elements may help to create new genes (Figure 10–71).

Transposable Elements Can Affect Gene Regulation[61]

The DNA sequence rearrangements caused by transposable elements often alter the pattern of expression of nearby genes, thereby affecting various aspects of animal or plant development, such as pigmentation (Figure 10–72) or morphogenesis (the shape of an eye or a flower, for example). While most of these changes in gene regulation would be expected to be detrimental to an organism, some of them will bring benefits.

Several aspects of the mutations caused by transposable elements are unusual and seem to distinguish them from the mutations caused by errors in DNA rep-

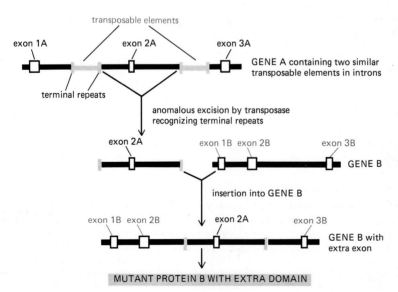

Figure 10–71 An example of the exon shuffling that can be caused by transposable elements. When two elements of the same type (*colored DNA*) happen to insert near each other in a chromosome, the transposition mechanism may occasionally use the ends of two different elements (instead of the two ends of the same element) and thereby move the chromosomal DNA between them to a new chromosomal site. Since introns are very large relative to exons (see Figure 9–7, p. 487), the illustrated insertion of a new exon into a preexisting intron is not an improbable outcome.

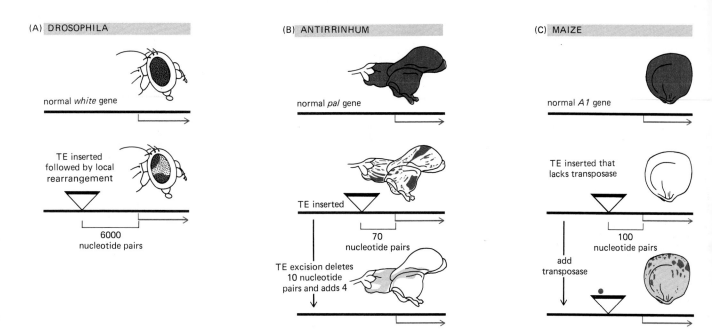

normal *white* gene

TE inserted
followed by local
rearrangement

6000
nucleotide pairs

normal *pal* gene

TE inserted

70
nucleotide pairs

TE excision deletes
10 nucleotide
pairs and adds 4

normal *A1* gene

TE inserted that
lacks transposase

100
nucleotide pairs

add
transposase

lication or DNA repair. One major difference is that the movement of a transposable element often will bring to the vicinity of a gene new sequences that act as binding sites for sequence-specific DNA-binding proteins, including a transposase and the proteins that regulate the transcription of the transposable element DNA. These sequences can thereby act as enhancers and affect the transcription of genes located thousands of nucleotide pairs away. An example of this type of effect on the expression of a pigment gene in maize is illustrated in Figure 10–72. Similar effects commonly contribute to the evolution of cancer cells, where oncogenes can be created by the transposition of such regulatory sequences into the neighborhood of a proto-oncogene (see p. 1208).

The organization of higher eucaryotic genomes, with long noncoding DNA sequences interspersed with comparatively short coding sequences, provides an accommodating "playground" for the integration and excision of mobile DNA sequences. Because gene transcription can be regulated from distances that are tens of thousands of nucleotide pairs away from a promoter (see p. 580), many of the resulting changes in the genome would be expected to affect gene expression; by contrast, relatively few would be expected to disrupt the short exons that contain the coding sequences.

Might the vast excess of noncoding DNA in higher eucaryotes have been favored by selection during evolution because of the regulatory flexibility that it has provided to organisms with a large variety of transposable elements? What is known about the regulatory systems that control higher eucaryotic genes is consistent with this possibility. Enhancers, like exons, seem to function as separate modules, and the activity of a gene depends on a summation of the influences received at its promoter from a set of enhancers (Figure 10–73). Transposable elements, by moving such enhancer modules around in a genome, may allow gene regulation to be optimized for the long-term survival of the organism.

10-36 Transposition Bursts Cause Cataclysmic Changes in Genomes and Increase Biological Diversity[62]

Another unique feature that distinguishes transposable elements as mutagens is their tendency to undergo long quiescent periods, during which they remain fixed in their chromosomal positions, followed by a period of intense movement. Their transposition, and therefore their mutagenic action, is activated from time to time in a few individuals in a population of organisms. Such cataclysmic changes in genomes, called **transposition bursts,** can involve near simultaneous transpositions of several types of transposable elements. Transposition bursts were first observed in developing maize plants that were subjected to repeated chromosome

Figure 10–72 Striking changes in gene regulation can be caused by transposable elements. Examples of heritable changes in the pigmentation pattern caused by transposable-element (TE) insertion into the regulatory regions of genes are shown for each of three organisms; similar types of events can cause morphological changes in the organism by affecting cell growth and differentiation. (A) An insertion into an upstream regulatory region of the *white* gene causes red eye pigmentation to appear only near the dorsal and the ventral edges of the fly eye. (B) An insertion into the upstream promoter of a gene needed for pigment production produces a snapdragon flower that lacks red pigment everywhere except in those patches of cells where the element has been excised by a transposition event. A subsequent excision of the element from the genome of the entire plant creates the spatially restricted pattern of pale pigmentation shown in the bottom panel. (C) An example of a regulated change in the pigmentation of the maize kernel that is caused by a transposable element. In this case the transposase serves as a gene regulatory protein that restores some pigmentation to all cells in an otherwise unpigmented kernel. In addition, the transposase catalyzes the occasional excision of the element, thereby producing isolated patches of more darkly colored cells. (A, after G.M. Rubin et al., *Cold Spring Harbor Symp. Quant. Biol.* 50:329–335, 1985; B, after E.S. Coen, R. Carpenter, and C. Martin, *Cell* 47:285–296, 1986; C, after Zs. Schwarz-Sommer et al., *EMBO J.* 6:287–294, 1987.)

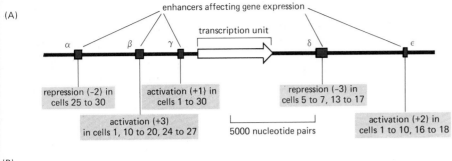

(A)

enhancers affecting gene expression

transcription unit

α   β   γ   δ   ε

repression (–2) in cells 25 to 30

activation (+1) in cells 1 to 30

repression (–3) in cells 5 to 7, 13 to 17

activation (+3) in cells 1, 10 to 20, 24 to 27

5000 nucleotide pairs

activation (+2) in cells 1 to 10, 16 to 18

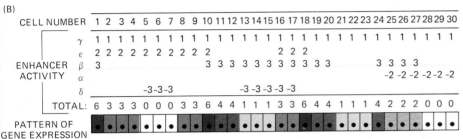

(B)

CELL NUMBER	1	2	3	4	5	6	7	8	9	10	11	12	13	14	15	16	17	18	19	20	21	22	23	24	25	26	27	28	29	30
γ	1	1	1	1	1	1	1	1	1	1	1	1	1	1	1	1	1	1	1	1	1	1	1	1	1	1	1	1	1	1
ε	2	2	2	2	2	2	2	2	2	2						2	2	2												
β	3									3	3	3	3	3	3	3	3	3	3	3				3	3	3	3			
α																									–2	–2	–2	–2	–2	–2
δ					–3	–3	–3						–3	–3	–3	–3	–3													
TOTAL:	6	3	3	3	0	0	0	3	3	6	4	4	1	1	1	3	3	6	4	4	1	1	1	4	2	2	2	0	0	0

ENHANCER ACTIVITY

PATTERN OF GENE EXPRESSION IN CELL SHEET

Figure 10–73 How the combined action of separate enhancer modules creates cell-specific patterns of gene expression. Because the mixture of gene regulatory proteins that binds to each enhancer varies from cell to cell, the effect of an enhancer is different in different cells. This example is modeled after results obtained in *Drosophila*, where the many enhancers that control a single gene can be analyzed for their separate effects in transgenic flies. For simplicity we have given each of the stimulatory (+ values) and inhibitory (– values) effects caused by each enhancer (α, β, γ, δ, or ε) a number between +3 and −3 and assumed that these numbers can be arithmetically summed to obtain a total enhancer activity, which determines the level of gene expression.

breakage. They also are observed in crosses between certain strains of flies—a phenomenon known as *hybrid dysgenesis*. When they occur in the germ line, they induce multiple changes in the genome of an individual progeny fly or plant.

By simultaneously changing several properties of an organism, transposition bursts increase the probability that two new traits that are useful together but of no selective value by themselves will appear in a single individual in a population. In several types of plants there is evidence that transposition bursts can be activated by a severe environmental stress, creating a variety of randomly modified progeny organisms, some of which may be better suited than the parent to survive in the new conditions. It seems that, at least in these plants, a mechanism has evolved to activate transposable elements to serve as mutagens that create an enhanced range of variant organisms when this variation is most needed. Thus transposable elements are not necessarily just disruptive parasites; rather, they may on occasion act as useful symbionts that aid the long-term survival of the species whose genomes they inhabit.

10-37 About 10% of the Human Genome Consists of Two Families of Transposable Elements That Appear to Have Multiplied Relatively Recently[63]

Primate DNA is unusual in at least one respect: it contains a remarkably large number of copies of two transposable DNA sequences that seem to have overrun our chromosomes. Both of these sequences move by an RNA-mediated process that requires a reverse transcriptase. One is the **L1 transposable element,** which resembles the F element in *Drosophila* and the cin4 element in maize and is thought to encode a reverse transcriptase (see Table 10–3, p. 605). Transposable elements have generally evolved with feedback control systems that severely limit their numbers in each cell (thereby saving the cell from potential disaster); the L1 element in humans, however, constitutes about 4% of the mass of the genome.

Even more unusual is the *Alu* **sequence,** which is very short (about 300 nucleotide pairs) and moves like a transposable element, creating target-site duplications when it inserts. It was derived, however, from an internally deleted host cell 7SL RNA gene, which encodes the RNA component of the signal-recognition particle (SRP) that functions in protein synthesis (see p. 439); it is therefore not clear whether the *Alu* sequence should be considered as a transposable element or as an unusually mobile pseudogene. It is present in about 500,000 copies in the haploid genome and constitutes about 5% of human DNA; thus it is present on average about once every 5000 nucleotide pairs. The *Alu* DNA is transcribed from the 7SL RNA promoter, a polymerase-III promoter that is internal to the transcript (see p. 526), so that it carries the information necessary for its own transcription wherever it moves. It needs to use a borrowed reverse transcriptase, however, to transpose.

Comparisons of the sequence and locations of the L1- and *Alu*-like sequences in different mammals suggest that these sequences have multiplied to high copy numbers relatively recently (Figure 10–74). It is hard to imagine that these highly abundant sequences scattered throughout our genome have not had major effects on the expression of many nearby genes. How many of our uniquely human qualities, for example, do we owe to these parasitic elements?

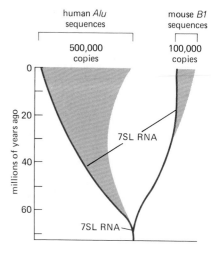

Figure 10–74 The proposed pattern of evolution of the abundant *Alu*-like sequences found in human and mouse genomes. Both of these transposable DNA sequences are thought to have evolved from the essential 7SL RNA gene. Based on the species distribution and sequence homology of these highly repeated elements, however, the major expansion in copy numbers seems to have occurred independently. (Adapted from P.L. Deininger and G.R. Daniels, *Trends in Genetics* 2:76–80, 1986.)

Summary

The functional DNA sequences in the genomes of higher eucaryotes appear to be constructed from small genetic modules of at least two kinds. Modules of coding sequence are combined in myriad ways to produce proteins, whereas modules of regulatory sequences are scattered throughout long stretches of noncoding sequences and regulate the expression of genes. Both the coding sequences (exons) and the regulatory sequences (enhancers) are typically less than a few hundred nucleotide pairs long. A variety of genetic-recombination processes occur in genomes, causing the random duplication and translocation of DNA sequences. Some of these changes create duplicates of entire genes, which can then evolve new functions. Others produce new proteins by shuffling exons or alter the expression of genes by exposing them to new combinations of enhancers. This type of sequence shuffling, which is of great importance for the evolution of organisms, is greatly facilitated by the split structure of higher eucaryotic genes and by the fact that these genes are subject to multiple activating and repressing influences from a combination of distant enhancers.

Many types of transposable elements are present in genomes. Collectively, they constitute more than 10% of the mass of both Drosophila and vertebrate genomes. Occasionally, "transposition bursts" occur in germ cells and cause many heritable changes in gene expression in the same individual. Transposable elements are thought to have had a special evolutionary role in the generation of organismal diversity.

References

General

Lewin, B. Genes, 3rd ed. New York: Wiley, 1987.

Schleif, R. Genetics and Molecular Biology. Reading, MA: Addison-Wesley, 1986.

Stent, G.S. Molecular Genetics: An Introductory Narrative. San Francisco: Freeman, 1971.

Watson, J.D.; Hopkins, N.H.; Roberts, J.W.; Steitz, J.A.; Weiner, A.M. Molecular Biology of the Gene, 4th ed. Menlo Park, CA: Benjamin-Cummings, 1987.

Cited

1. Gurdon, J.B. The developmental capacity of nuclei taken from intestinal epithelium cells of feeding tadpoles. *J. Embryol. Exp. Morphol.* 10:622–640, 1962.
   Steward, F.C.; Mapes, M.O.; Mears, K. Growth and organized development of cultured cells. *Am. J. Bot.* 45:705–713, 1958.
2. Garrels, J.I. Changes in protein synthesis during myogenesis in a clonal cell line. *Dev. Biol.* 73:134–152, 1979.
3. Darnell, J.E., Jr. Variety in the level of gene control in eucaryotic cells. *Nature* 297:365–371, 1982.
   Derman, E.; et al. Transcriptional control in the production of liver-specific mRNAs. *Cell* 23:731–739, 1981.
4. Gierer. A. Molecular models and combinatorial prinicples in cell differentiation and morphogenesis. *Cold Spring Harbor Symp. Quant. Biol.* 38:951–961, 1974.

Scott, M.P.; O'Farrell, P.H. Spatial programming of gene expression in early *Drosophila* embryogenesis. *Annu. Rev. Cell Biol.* 2:49–80, 1986.

5. Maniatis, T.; Goodbourn, S.; Fischer, J.A. Regulation of inducible and tissue-specific gene expression. *Science* 236:1237–1244, 1987.
   Yamamoto, K. Steroid receptor regulated transcription of specific genes and gene networks. *Annu. Rev. Genet.* 19:209–252, 1985.
6. Gehring, W.J.; Hiromi, Y. Homeotic genes and the homeobox. *Annu. Rev. Genet.* 20:147–173, 1986.
7. Blau, H.M.; et al. Plasticity of the differentiated state. *Science* 230:758–766, 1985.
   Davis, R.L.; Weintraub, H.; Lassar, A.B. Expression of a single transfected cDNA converts fibroblasts to myoblasts. *Cell* 51:987–1000, 1987.
8. Miller, J.H.; Reznikoff, W.S., eds. The Operon. Cold Spring Harbor, NY: Cold Spring Harbor Laboratory, 1978.
   Neidhardt, F.C.; et al., eds. *Escherichia coli* and *Salmonella typhimurium*: Cellular and Molecular Biology, Vol. 2, pp. 1439–1526. Washington, DC: American Society for Microbiology, 1987. (Paradigms of operon regulation in bacteria.)
   Ptashne, M. A Genetic Switch. Palo Alto, CA: Blackwell, 1986.
9. Gilbert, W.; Müller-Hill, B. The *lac* operator is DNA. *Proc. Natl. Acad. Sci. USA* 58:2415–2421, 1967.

Gottesman, S. Bacterial regulation: global regulatory networks. *Annu. Rev. Genet.* 18:415–441, 1984.

Jacob, F.; Monod, J. Genetic regulatory mechanisms in the synthesis of proteins. *J. Mol. Biol.* 3:318–356, 1961.

Reznikoff, W.S.; Siegele, D.A.; Cowing, D.W.; Gross, C.A. The regulation of transcription initiation in bacteria. *Annu. Rev. Genet.* 19:355–388, 1985.

10. de Combrugghe, B.; Busby, S.; Buc, H. Cyclic AMP receptor protein: role in transcription activation. *Science* 224:831–838, 1984.

Hochschild, A.; Irwin, N.; Ptashne, M. Repressor structure and the mechanism of positive control. *Cell* 32:319–325, 1983.

Raibaud, O.; Schwartz, M. Positive control of transcription initiation in bacteria. *Annu. Rev. Genet.* 18:173–206, 1984.

11. Keener, J.; Wong, P.; Popham, D.; Wallis, J.; Kustu, S. A sigma factor and auxiliary proteins required for nitrogen-regulated transcription in enteric bacteria. In RNA Polymerase and the Regulation of Transcription. (W.S. Reznikoff, et al, eds.), pp. 159–175. New York: Elsevier, 1987.

Ninfa, A.J.; Reitzer, L.J.; Magasanik, B. Initiation of transcription at the bacterial *glnAp2* promoter by purified *E. coli* components is facilitated by enhancers. *Cell* 50:1039–1046, 1987.

12. Dunn, T.M.; Hahn, S.; Ogden, S.; Schleif, R.F. An operator at −280 base pairs that is required for the repression of araBAD operon promoter. *Proc. Natl. Acad. Sci. USA* 81:5017–5020, 1984.

Griffith, J.; Hochschild, A.; Ptashne, M. DNA loops induced by cooperative binding of lambda repressor. *Nature* 322:750–752, 1986.

Mossing, M.C.; Record, M.T. Upstream operators enhance repression of the *lac* promoter. *Science* 233:889–892, 1986.

13. Helmann, J.D.; Chamberlin, M.J. Structure and function of bacterial sigma factors. *Annu. Rev. Biochem.* 57:839–872, 1988.

14. Davison, B.L.; Egly, J.M.; Mulvihill, E.R.; Chambon, P. Formation of stable preinitiation complexes betwen eucaryotic class B transcription factors and promoter sequences. *Nature* 301:680–686, 1983.

Sawadogo, M.; Roeder, R.G. Interaction of a gene-specific transcription factor with the adenovirus major late promoter upstream of the TATA box region. *Cell* 43:165–175, 1985.

Workman, J.L.; Roeder, R.G. Binding of transcription factor TFIID to the major late promoter during *in vitro* nucleosome assembly potentiates subsequent initiation by RNA polymerase II. *Cell* 51:613–622, 1987.

15. Atchison, M.L. Enhancers: mechanisms of action and cell specificity. *Annu. Rev. Cell Biol.* 4:127–153, 1988.

Maniatis, T.; Goodbourn, S.; Fischer, J. Regulation of inducible and tissue-specific gene expression. *Science* 236:1237–1245, 1987.

McKnight, S.L.; Kingsbury, R. Transcriptional control signals of a eukaryotic protein-coding gene. *Science* 217:316–324, 1982.

Serfling, E.; Jasin, M.; Schaffner, W. Enhancers and eucaryotic gene transcription. *Trends Genet.* 1:224–230, 1985.

16. Emerson, B.M.; Nickol, J.M.; Jackson, P.D.; Felsenfeld, G. Analysis of the tissue-specific enhancer at the 3' end of the chicken adult β-globin gene. *Proc. Natl. Acad. Sci. USA* 84:4786–4790, 1987.

Evans, T.; Reitman, M.; Felsenfeld, G. An erythrocyte-specific DNA-binding factor recognizes a regulatory sequence common to all chicken globin genes. *Proc. Natl. Acad. Sci. USA* 85:5976–5980, 1988.

Jones, N.C.; Rigby, P.W.J.; Ziff, E.B. *Trans*-acting protein factors and the regulation of eukaryotic transcription: lessons from studies on DNA tumor viruses. *Genes Dev.* 2:267–281, 1988.

Nomiyama, H.; Fromental, C.; Xiao, J.H.; Chambon, P. Cell-specific activity of the constituent elements of the Simian virus 40 enhancer. *Proc. Natl. Acad. Sci. USA* 84:7881–7885, 1987.

17. Brent. R.; Ptashne, M. A eukaryotic transcriptional activator bearing the DNA specificity of a prokaryotic repressor. *Cell* 43:729–736, 1985.

Evans, R.M. The steroid and thyroid hormone receptor superfamily. *Science* 240:889–895, 1988.

Godowski, P.J.; Picard, D.; Yamamoto, K. Signal transduction and transcriptional regulation by glucocorticoid receptor—lex A fusion proteins. *Science* 241:812–816, 1988.

Kumar, V.; et al. Functional domains of the human estrogen receptor. *Cell* 51:941–951, 1987.

18. Sen, R.; Baltimore, D. Inducibility of kappa immunoglobin enhancer-binding protein NF-kappa B by a posttranslational mechanism. *Cell* 47:921–928, 1986.

Yamamoto, K.K.; Gonzalez, G.A.; Biggs, W.H.; Montminy, M.R. Phosphorylation-induced binding and transcriptional efficacy of nuclear factor CREB. *Nature* 334:494–498, 1988.

Zimarino, V.; Wu, C. Induction of sequence-specific binding of *Drosophila* heat shock activator protein without protein synthesis *Nature* 327:727–730, 1987.

19. Metzger, D.; White, J.H.; Chambon, P. The human estrogen receptor functions in yeast. *Nature* 334:31–36, 1988.

Ptashne, M. Gene regulation by proteins acting nearby and at a distance. *Nature* 322:697–701, 1986.

Struhl, K. Promoters, activator proteins, and the mechanism of transcriptional initiation in yeast. *Cell* 49:295–297, 1987.

20. Borst, P.; Greaves, D.R. Programmed gene rearrangements altering gene expression. *Science* 235:658–667, 1987.

Meyer, T.F. Molecular basis of surface antigen variation in *Neisseria*. *Trends Genet.* 3:319–324, 1987.

Simon, M.; Zieg, J.; Silverman, M.; Mandel, G.; Doolittle, R. Phase variation: evolution of a controlling element. *Science* 209:1370–1374, 1980.

21. Cross, F.; Hartwell, L.H.; Jackson, C.; Konopka, J.B. Conjugation in *Saccharomyces cerevisiae*. *Annu. Rev. Cell Biol.* 4:429–457, 1988.

Herskowitz, I. Master regulatory loci in yeast and lambda. *Cold Spring Harbor Symp. Quant. Biol.* 50:565–574, 1985.

Kushner, P.J.; Blair, L.C.; Herskowitz, I. Control of yeast cell types by mobile genes: a test. *Proc. Natl. Acad. Sci. USA* 76:5264–5268, 1979.

22. Kostriken, R.; Strathern, J.N.; Klar, A.; Hicks, J.B.; Heffron, F. A site-specific endonuclease essential for mating-type switching in *Saccharomyces cerevisiae*. *Cell* 35:167–174, 1983.

23. Nasmyth, K.; Shore, D. Transcriptional regulation in the yeast life cycle. *Science* 237:1162–1170, 1987.

24. Brand, A.H.; Breeden, L.; Abraham, J.; Sternglanz, R.; Nasmyth, K. Characterization of a "silencer" in yeast: a DNA sequence with properties opposite to those of a transcriptional enhancer. *Cell* 41:41–48, 1985.

25. Friedman, D.I.; et al. Interactions of bacteriophage and host macromolecules in the growth of bacteriophage lambda. *Microbiol. Rev.* 48:299–325, 1984.

Ptashne, M.; et al. How the lambda repressor and cro work. *Cell* 19:1–11, 1980.

26. Brown, D.D. The role of stable complexes that repress and activate eucaryotic genes. *Cell* 37:359–365, 1984.

Weintraub, H. Assembly and propagation of repressed and derepressed chromosomal states. *Cell* 42:705–711, 1985.

27. Brown, S.W. Heterochromatin. *Science* 151:417–425, 1966.

Hsu, T. C.; Cooper, J.E.K., Mace, M.L., Brinkley, B.R. Arrangement of centromeres in mouse cells. *Chromosoma* 34:73–87, 1971.

28. Gartler, S.M.; Riggs, A.D. Mammalian X-chromosome inactivation. *Annu. Rev. Genet.* 17:155–190, 1983.

Lock, L.F.; Takagi, N.; Martin G.R. Methylation of the *Hprt* gene on the inactive *X* occurs after chromosome inactivation. *Cell* 48:39–46, 1987.

Lyon, M.F. X-chromosome inactivation and developmental patterns in mammals. *Biol. Rev.* 47:1–35, 1972.

29. Baker, W.K. Position-effect variegation. *Adv. Genet.* 14:133–169, 1968.

Spofford, J.B. Position-effect variegation in *Drosophila*. In The Genetics and Biology of *Drosophila* (M. Ashburner, E. Novitski, eds.), Vol. 1C, pp. 955–1018. New York: Academic Press, 1976.

30. Goldberg, D.A.; Posakony, J.W.; Maniatis, T. Correct developmental expression of a cloned alcohol dehydrogenase gene transduced into the *Drosophila* germ line. *Cell* 34:59–73, 1983.

Grosveld, F.; van Assendelft, G.B.; Greaves, D.R.; Kollias, G. Position-independent, high-level expression of the human β-globin gene in transgenic mice. *Cell* 51:975–985, 1987.

Meyerowitz, E.M.; Raghavan, K.V.; Mathers, P.H.; Roark, M. How *Drosophila* larvae make glue: control of *Sgs-3* gene expression. *Trends Genet.* 3:288–293, 1987.

Palmiter, R.D.; Brinster, R.L. Germ-line transformation of mice. *Annu. Rev. Genet.* 20:465–499, 1986.

31. Ephrussi, A.; Church, G.M.; Tonegawa, S.; Gilbert, W. B lineage-specific interactions of an immunolglobulin enhancer with celluar factors *in vivo*. *Science* 227:134–140, 1985.

Garel, A.; Zolan, M.; Axel, R. Genes transcribed at diverse rates have a similar conformation in chromatin. *Proc. Natl. Acad. Sci. USA* 74:4867–4871, 1977.

Karlsson, S.; Nienhuis, A.W. Developmental regulation of human globin genes. *Annu. Rev. Biochem.* 54:1071–1108, 1985.

Weintraub, H.; Groudine, M. Chromosomal subunits in active genes have an altered conformation. *Science* 193:848–856, 1976.

32. Brill, S.J.; Sternglanz, R. Transcription-dependent DNA supercoiling in yeast DNA topoisomerase mutants. *Cell* 54:403–411, 1988.

Wang, J.C. Superhelical DNA. *Trends Biochem. Sci.* 5:219–221, 1980.

Wang, J.C.; Giaever, G.N. Action at a distance along a DNA. *Science* 240:300–304, 1988.

33. Guarente, L. Regulatory proteins in yeast. *Annu. Rev. Genet.* 21:425–452, 1987.

34. Razin, A.; Cedar, H.; Riggs, A.D., eds. DNA Methylation: Biochemistry and Biological Significance. New York: Springer-Verlag, 1984.

35. Cedar, H. DNA methylation and gene activity. *Cell* 53:3–4, 1988.

Ivarie, R.D.; Schacter, B.S.; O'Farrell, P.H. The level of expression of the rat growth hormone gene in liver tumor cells is at least eight orders of magnitude less than that in anterior pituitary cells. *Mol. Cell. Biol.* 3:1460–1467, 1983.

Yisraeli, J.; et al. Muscle-specific activation of a methylated chimeric actin gene. *Cell* 46:409–416, 1986.

36. Bird, A.P. CpG islands as gene markers in the vertebrate nucleus. *Trends Genet.* 3:342–347, 1987.

37. Duncan, I. The bithorax complex. *Annu. Rev. Genet.* 21:285–319, 1987.

Peifer, M.; Karchi, F.; Bender, W. The bithorax complex: control of segmental identity. *Genes Dev.* 1:891–898, 1987.

38. Landrick, R.; Yanofsky, C. Transcription attenuation. In *Escherichia coli* and *Salmonella typhimurium*: Cellular and Molecular Biology (F.C. Neidhardt; et al, eds.), Vol. 2, pp. 1276–1301, Washington, DC: American Society for Microbiology, 1987.

Platt, T. Transcription termination and the regulation of gene expression. *Annu. Rev. Biochem.* 55:339–372, 1986.

Yanofsky, C. Operon-specific control by transcription attenuation. *Trends Genet.* 3:356–360, 1987.

39. Andreadis, A.; Gallego, M.E.; Nadal-Ginard, B. Generation of protein isoform diversity by alternative splicing: mechanistic and biological implications. *Annu. Rev. Cell Biol.* 3:207–242, 1987.

Leff, S.; Rosenfeld, M.; Evans, R. Complex transcriptional units: diversity in gene expression by alternative RNA processing. *Annu. Rev. Biochem.* 55:1091–1117, 1986.

Schwarz, T.L.; Tempel, B.L.; Papazian, D.M.; Jan, Y.N.; Jan, L.Y. Multiple potassium-channel components are produced by alternative splicing at the *Shaker* locus in *Drosophila*. *Nature* 331:137–142, 1988.

40. Baker, B.S.; Belote, J.M. Sex determination and dosage compensation in *Drosophila melanogaster*. *Annu. Rev. Genet.* 17:345–393, 1983.

Bingham, P.M.; Chou, T.; Mims, I.; Zachar, Z. On/off regulation of gene expression at the level of splicing. *Trends Genet.* 4:134–138, 1988.

Boggs, R.T.; Gregor, P.; Idriss, S.; Belote, J.M.; McKeown, M. Regulation of sexual differentiation in *D. .melanogaster* via alternative splicing of RNA from the transformer gene. *Cell* 50:739–747, 1987.

Laski, F.A.; Rio, D.C.; Rubin, G.M. Tissue specificity of *Drosophila* P element transposition is regulated at the level of RNA splicing. *Cell* 44:7–19, 1986.

41. Early, P.; et al. Two mRNAs can be produced from a single immunoglobulin μ gene by alternative RNA processing pathways. *Cell* 20:313–319, 1980

Peterson, M.L.; Perry, R.P. Regulated production of μ_m and μ_s mRNA requires linkage of the poly(A) addition sites and is dependent on the length of the μ_m-μ_s intron. *Proc. Natl. Acad. Sci. USA* 83:8883–8887, 1986.

42. Beadle, G. Genes and the chemistry of the organism. *Am. Sci.* 34:31–53, 1946.

43. Newport, J.W.; Forbes, D.J. The nucleus: structure, function, and dynamics. *Annu. Rev. Biochem.* 56:535–565, 1987.

Schneider, R.J.; Shenk, T. Impact of virus infection on host cell protein synthesis. *Annu. Rev. Biochem.* 56:317–332, 1987.

44. Aziz, N.; Munro, H.N. Iron regulates ferritin mRNA translation through a segment of its 5' untranslated region. *Proc. Natl. Acad. Sci. USA* 84:8478–8482, 1987.

Gold, L. Posttranscriptional regulatory mechanisms in *Escherichia coli*. *Annu. Rev. Biochem.* 57:199–234, 1988.

Nomura, M.; Gourse, R.; Baughman, G. Regulation of the synthesis of ribosomes and ribosomal components. *Annu. Rev. Biochem.* 53:75–117, 1984.

Walden, W.E.; et al. Translational repression in eukaryotes: partial purification and characterization of a repressor of ferritin in RNA translation. *Proc. Natl. Acad. Sci. USA* 85:9503–9507, 1988.

45. Hunt, T. False starts in translational control of gene expression. *Nature* 316:580–581, 1985.

Kozak, M. Bifunctional messenger RNAs in eucaryotes. *Cell* 47:481–483, 1986.

Pelletier, J.; Sonenberg, N. Internal initiation of translation of eukaryotic mRNA directed by a sequence derived from poliovirus RNA. *Nature* 334:320–325, 1988.

46. Ilan, J., ed. Translational Regulation of Gene Expression. New York: Plenum Press, 1987.

Rosenthal, E.T.; Hunt, T.; Ruderman, J.V. Selective translation of mRNA controls the pattern of protein synthesis during early development of the surf clam, *Spisula solidissima*. *Cell* 20:487–494, 1980.

Walden, W.E.; Thach, R.E. Translational control of gene expression in a normal fibroblast: characterization of a subclass of mRNAs with unusual kinetic properties. *Biochemistry* 25:2033–2041, 1986.

47. Craigen, W.J.; Caskey, C.T. Translational frameshifting: where will it stop? *Cell* 50:1–2, 1987.

48. Casey, J.L.; et al. Iron-responsive elements: regulatory RNA

sequences that control mRNA levels and translation. *Science* 240:924–928, 1988.

Raghow, R. Regulation of messenger RNA turnover in eukaryotes. *Trends Biochem. Sci.* 12:358–360, 1987.

Shaw, G.; Kamen, R. A conserved AU sequence from the 3' untranslated region of GM-CSF mRNA mediaters selective mRNA degradation. *Cell* 46:659–667, 1986.

49. Graves, R.A.; Pandey, N.B.; Chodchoy, N.; Marzluff, W.F. Translation is required for regulation of histone mRNA degradation. *Cell* 48:615–626, 1987.

Marzluff, W.F.; Pandey, N.B. Multiple regulatory steps control histone mRNA concentrations. *Trends Biochem. Sci.* 13:49–52, 1988.

Mowry, K.L.; Steitz, J.A. Identification of the human U7 snRNP as one of several factors involved in the 3' end maturation of histone premessenger RNAs. *Science* 238:1682–1687, 1987.

50. Driever, W.; Nüsslein-Volhard, C. A gradient of bicoid protein in *Drosophila* embryos. *Cell* 54:83–93, 1988.

Lawrence, J.B.; Singer, R.H. Intracellular localization of messenger RNAs for cytoskeletal proteins. *Cell* 45:407–415, 1986.

Weeks, D.L.; Melton, D.A. A maternal mRNA localized to the vegetal hemisphere in *Xenopus* eggs codes for a growth factor related to TGF-beta. *Cell* 51:861–867, 1987.

51. Borst, P. Discontinous transcription and antigenic variation in trypanosomes. *Annu. Rev. Biochem.* 55:701–732, 1986.

Eisen, H. RNA editing: who's on first? *Cell* 53:331–332, 1988.

Powell, L.M.; et al. A novel form of tissue-specific RNA processing produces apolipoprotein-B48 in intestine. *Cell* 50:831–840, 1987.

Sharp, P.A. *Trans* splicing: variation on a familiar theme. *Cell* 50:147–148, 1987.

52. McClain, W.H.; Guerrier-Takada, C.; Altman, S. Model substrates for an RNA enzyme. *Science* 238:527–530, 1987.

Pines, O.; Inouye, M. Antisense RNA regulation in prokaryotes. *Trends Genet.* 2:284–287, 1986.

Tomizawa, J. Control of ColE1 plasmid replication: binding of RNA I to RNA II and inhibition of primer formation. *Cell* 47:89–97, 1986.

Wu, H.N.; Uhlenbeck, O.C. Role of a bulged A residue in a specific RNA-protein interaction. *Biochemistry* 26:8221–8227, 1987.

53. Clarke, B.C.; Robertson, A.; Jeffreys, A.J., eds. The Evolution of DNA Sequences. London: The Royal Society, 1986.

Nei, M.; Koehn, R.K., eds. Evolution of Genes and Proteins. Sunderland, MA: Sinauer, 1983.

54. Doolittle, R.F. Proteins. *Sci. Am.* 253(4):88–99, 1985.

Holland, S.K.; Blake, C.C. Proteins, exons, and molecular evolution. *Biosystems* 20:181–206, 1987.

Maeda, N.; Smithies; O. The evolution of multigene families: human haptoglobin genes. *Annu. Rev. Genet.* 20:81–108, 1986.

55. Kourilsky, P. Molecular mechanisms for gene conversion in higher cells. *Trends Genet.* 2:60–63, 1986.

Roth, D.B.; Porter, T.N.; Wilson, J.H. Mechanisms of nonhomologous recombination in mammalian cells. *Mol. Cell. Biol.* 5:2599–2607, 1985.

Smith, G.P. Evolution of repeated DNA sequences by unequal crossovers. *Science* 191:528–535, 1976.

Stark, G.R.; Wahl, G.M. Gene amplification. *Annu. Rev. Biochem.* 53: 447–491, 1984.

56. Dickerson, R.E.; Geis, I. Hemoglobin: Structure, Function, Evolution, and Pathology. Menlo Park, CA: Benjamin-Cummings, 1983.

Efstratiadis, A.; et al. The structure and evolution of the human β-globin gene family. *Cell* 21:653–668, 1980.

Vollrath, D.; Nathans, J.; Davis, R.W. Tandem array of human visual pigment genes at Xq28. *Science* 240:1669–1672, 1988.

57. Doolittle, W.F. RNA mediated gene conversion? *Trends Genet.* 1:64–65, 1985.

Gilbert, W.; Marchionni, M.; McKnight, G. On the antiquity of introns. *Cell* 46:151–153, 1986.

Sharp, P. On the origin of RNA splicing and introns. *Cell* 42:397–400, 1985.

58. Britten, R.J.; Kohne, D.E. Repeated sequences in DNA. *Science* 161:529–540, 1968.

Jelinek, W.R.; Schmid, C.W. Repetitive sequences in eukaryotic DNA and their expression. *Annu. Rev. Biochem.* 51:813–844, 1982.

59. Craig-Holmes, A.P.; Shaw, M.W. Polymorphism of human constitutive heterochromatin. *Science* 174:702–704, 1971.

Hsu, T.C. Human and Mammalian Cytogenetics: A Historical Perspective. New York: Springer-Verlag, 1979.

John B.; Miklos, G.L.G. Functional aspects of satellite DNA and heterochromatin. *Int. Rev. Cytol.* 58:1–114, 1979.

Orgel L.E.; Crick, F.H.C. Selfish DNA: the ultimate parasite. *Nature* 284:604–607, 1980.

60. Berg, D.E.; Howe, M.M.; eds. Mobile DNA. Washington, DC: American Society for Microbiology, 1989.

Döring, H.-P.; Starlinger, P. Molecular genetics of transposable elements in plants. *Annu. Rev. Genet.* 20:175–200, 1986.

Finnegan, D.J. Transposable elements in eukaryotes. *Int. Rev. Cytol.* 93:281–326, 1985.

McClintock, B. Controlling elements and the gene. *Cold Spring Harbor Symp. Quant. Biol.* 21:197–216, 1956.

61. Coen, E.S.; Carpenter, R. Transposable elements in *Antirrhinum majus*: generators of genetic diversity. *Trends Genet.* 2:292–296, 1986.

Georgiev, G.P. Mobile genetic elements in animal cells and their biological significance. *Eur. J. Biochem.* 145:203–220, 1984.

O'Kane, C.J.; Gehring, W. Detection *in situ* of genomic regulatory elements in *Drosophila*. *Proc. Natl. Acad. Sci. USA* 84:9123–9127, 1987.

Hiromi, Y.; Gehring, W.J. Regulation and function of the Drosophila segmentation gene *fushi tarazu*. *Cell* 50:963–974, 1987.

62. Gerasimova, T.I.; Mizrokhi, L.J.; Georgiev, G.P. Transposition bursts in genetically unstable. *Drosophila*. *Nature* 309:714–716, 1984.

McClintock, B. The significance of responses of the genome to challenge. *Science* 226:792–801, 1984.

Walbot, V.; Cullis, C.A. Rapid genomic change in higher plants. *Annu. Rev. Plant Physiol.* 36:367–396, 1985.

63. Deininger, P.L.; Daniels, G.R. The recent evolution of mammalian repetitive DNA elements. *Trends Genet.* 2:76–80, 1986.

Ruffner, D.E.; Sprung, C.N.; Minghetti, P.P.; Gibbs, P.E.; Dugaiczyk, A. Invasion of the human albumin-α-fetoprotein gene family by Alu, Kpn, and two novel repetitive DNA elements. *Mol. Biol. Evol.* 4:1–9, 1987.

Weiner, A.M.; Deininger, P.L.; Efstratiadis, A. Nonviral retroposons: genes, pseudogenes, and transposable elements generated by the reverse flow of genetic information. *Annu. Rev. Biochem.* 55:631–661, 1986.

The Cytoskeleton

11

The ability of eucaryotic cells to adopt a variety of shapes and to carry out coordinated and directed movements depends on the **cytoskeleton,** a complex network of protein filaments that extends throughout the cytoplasm. The cytoskeleton might equally well be called the "cytomusculature," because it is directly responsible for such movements as the crawling of cells on a substratum, muscle contraction, and the many changes in shape of a developing vertebrate embryo; it also provides the machinery for actively moving organelles from one place to another in the cytoplasm. Since the cytoskeleton is apparently absent from bacteria, it may have been a crucial factor in the evolution of eucaryotic cells.

The diverse activities of the cytoskeleton depend on just three principal types of protein filaments: *actin filaments, microtubules,* and *intermediate filaments.* Each type of filament is formed from a different protein monomer and can be built into a variety of structures according to its associated proteins. Some of the associated proteins link filaments to one another or to other cell components, such as the plasma membrane. Others control where and when actin filaments and microtubules are assembled in the cell by regulating the rate and extent of their polymerization. Yet other associated proteins interact with filaments to produce movements, the two best-understood examples being muscle contraction, which depends on actin filaments, and the beating of cilia, which depends on microtubules.

We begin this chapter by considering structures built from actin filaments, proceeding from the specialized *myofibril* in a muscle cell to the ubiquitous actin-rich *cortex* beneath the plasma membrane of all animal cells. We then examine microtubules, proceeding from those in organized bundles that are responsible for ciliary beating to those that extend throughout the cytoplasm of cells and control organelle movements and define cell polarity. After discussing the diverse family of intermediate filaments, which give the cell tensile strength and form the nuclear lamina, we finally consider how the cytoskeleton functions as an integrated network to control and coordinate the movements and shapes of single cells and tissues.

Muscle Contraction[1]

Many of the protein molecules that constitute the actin-based structures common to all cells were first discovered in muscle, and muscle contraction is the most familiar and the best understood of all the kinds of movement of which animals are capable. In vertebrates, for example, running, walking, swimming, and flying all depend on the ability of *skeletal muscle* to contract rapidly on its scaffolding

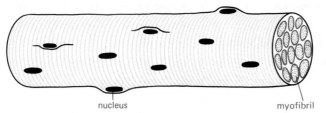

Figure 11–1   Schematic drawing of a short section of a skeletal muscle cell (also called a muscle fiber). In an adult human these huge multinucleated cells are typically 50 μm in diameter, and they can be up to 500,000 μm (500 mm) long.

nucleus   myofibril

of bone, while involuntary movements, such as heart pumping and gut peristalsis, depend on the contraction of *cardiac* and *smooth muscle*, respectively.

Muscle contraction is mediated by a sophisticated and powerful intracellular protein apparatus that is present in a more rudimentary form in almost all eucaryotic cells. During the evolution of muscle cells, parts of the cytoskeleton became greatly amplified and specialized to make the contractile machinery in muscle unusually stable and efficient. In *striated muscles* (including skeletal muscle, cardiac muscle, and comparable invertebrate tissues such as insect flight muscle) this machinery is so highly organized that its contraction can be visualized directly, revealing some important properties of its molecules.

Myofibrils Are the Contractile Elements of a Skeletal Muscle Cell

The long thin *muscle fibers* of skeletal muscle are huge single cells formed during development by the fusion of many separate cells (see p. 983). The nuclei of the contributing cells are retained in this large cell and lie just beneath the plasma membrane. But the bulk of the cytoplasm (about two-thirds of its dry mass) is made up of *myofibrils*—cylindrical elements 1 to 2 μm in diameter, which are often as long as the muscle cell itself (Figure 11–1). Isolated myofibrils have a series of prominent bands along their length that are responsible for the striated appearance of skeletal muscle cells. If ATP and Ca^{2+} are added to isolated myofibrils, the myofibrils instantly contract, indicating that they are the force generators in muscle cells. Each myofibril consists of a chain of tiny contractile units composed of miniature, precisely arranged assemblies of thick and thin filaments.

Myofibrils Are Composed of Repeating Assemblies of Thick and Thin Filaments

Each of the regular repeating units, or *sarcomeres*, that give the vertebrate myofibril its striated appearance is about 2.5 μm long. At high magnification a series of broad light and dark bands can be seen in each sarcomere; a dense line in the center of each light band separates one sarcomere from the next and is known as the Z line or Z *disc* (Figure 11–2).

Figure 11–2   (A) Low-magnification electron micrograph of a longitudinal section through a skeletal muscle cell of a rabbit, showing the regular pattern of cross-striations. The cell contains many myofibrils aligned in parallel (see Figure 11–1). (B) Detail of the skeletal muscle cell shown in (A), showing portions of two adjacent myofibrils and the definition of a sarcomere. (C) Schematic diagram of a single sarcomere, showing the origin of the dark and light bands seen in the electron micrographs. The dark bands are sometimes referred to as *A bands* because they appear anisotropic in polarized light (that is, their refractive index changes with the plane of polarization). The light bands are relatively *isotropic* in polarized light and are sometimes called *I bands*. (A and B, courtesy of Roger Craig.)

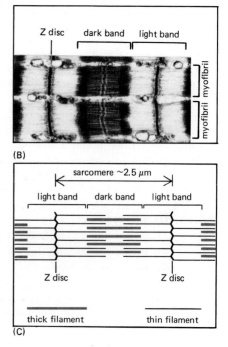

(B)

(C)

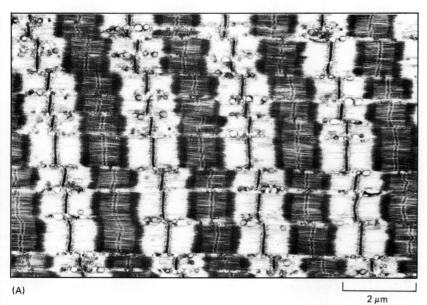

(A)

2 μm

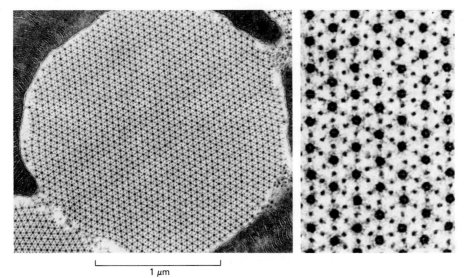

Figure 11–3 Electron micrographs of an insect flight muscle viewed in cross-section, showing how the thick and thin filaments are packed together with crystalline regularity. Unlike their vertebrate counterparts, these thick filaments have a hollow center, as seen in the enlargement on the right. The geometry of the hexagonal lattice is slightly different in vertebrate muscle. (From J. Auber, *J. de Microsc.* 8:197–232, 1969.)

1 μm

The molecular basis of the cross-striations, and a strong clue to their functional significance, was revealed in 1953 in one of the first applications of the electron microscope to biological material. Each sarcomere was found to contain two sets of parallel and partly overlapping filaments: *thick filaments*, extending from one end of the dark band to the other, and *thin filaments*, extending across each light band and partway into the two neighboring dark bands (Figure 11–2C). When the region of the dark band where thick and thin filaments overlap was viewed in cross-section, the thick filaments were seen to be arranged in a regular hexagonal lattice, with the thin filaments placed regularly between them (Figure 11–3).

11-3 Contraction Occurs as the Thick and Thin Filaments Slide Past Each Other[2]

If a source of monochromatic light is directed through a living muscle cell, a series of interference fringes is produced that provides a sensitive measure of sarcomere spacing. Such measurements reveal that each sarcomere shortens proportionately as the muscle contracts: if a myofibril containing a chain of 20,000 sarcomeres contracts from 5 cm to 4 cm (that is, by 20%), the length of each sarcomere decreases correspondingly from 2.5 to 2.0 μm.

When a sarcomere shortens, only the light band decreases in length; the dark band remains unchanged. This is easily explained if the contraction is caused by thick filaments sliding past the thin filaments with no change in the length of either type of filament (Figure 11–4). This *sliding filament model,* first proposed in 1954, was crucial to understanding the contractile mechanism. In particular, it

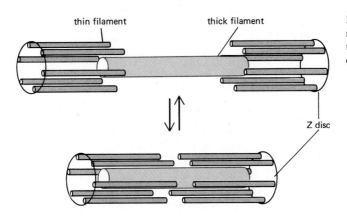

thin filament   thick filament

Z disc

Figure 11–4 The sliding filament model of muscle contraction, in which the thin and thick filaments slide past one another without shortening.

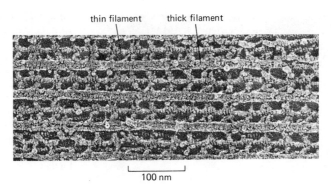

thin filament   thick filament

Figure 11–5 Electron micrograph of a longitudinal section through an insect flight muscle that has been subjected to rapid freezing, fracturing, and deep etching. Note the nearly crystalline array of thick myosin filaments and thin actin filaments. The cross-bridges spanning the two types of filaments are myosin heads. (Courtesy of John Heuser and Roger Cooke.)

100 nm

directed attention to the molecular interactions between adjacent thick and thin filaments that cause them to slide.

The sliding filament model is supported by several lines of evidence. Electron microscopic studies show that thick and thin filaments do not change in length as a muscle shortens, while x-ray diffraction analyses demonstrate that the internal packing of the subunit molecules in the filaments also remains unchanged. As the muscle shortens, mechanical tension increases in proportion to the amount of overlap between the thick and thin filaments, as would be expected if the tension is generated by sliding interactions between them.

The ultrastructural basis for the force-generating interaction is visible at very high magnification in electron micrographs. Thick filaments are seen to possess numerous tiny side arms, or *cross-bridges*, that extend about 13 nm to make contact with adjacent thin filaments (Figure 11–5). We shall see that when a muscle contracts, the thick and thin filaments are pulled past each other by the cross-bridges acting cyclically, like banks of tiny oars.

The interacting proteins of the thin and thick filaments have been identified as *actin* and *myosin*, respectively. Actin is the most abundant of all the cytoskeletal proteins, and it is often associated with myosin in structures that produce contraction. Although these proteins are present in virtually all eucaryotic cells, most of what we know about their properties was originally learned from biochemical experiments on the actin and myosin extracted from muscle.

Thin Filaments Are Composed Mainly of Actin[2]

All eucaryotic species (including single-celled organisms such as yeasts) contain **actin**. Actin genes have been very highly conserved in evolution, and actin molecules from widely divergent sources are functionally interchangeable in tests performed *in vitro*. The principal properties of actin molecules extracted from skeletal muscle, for example, are common to actin molecules from all other sources as well.

Actin is usually isolated by treating dry powdered muscle with very dilute salt solutions, which dissociates the actin filaments into their globular subunits. Each subunit is composed of a single polypeptide 375 amino acids long, sometimes known as globular actin or *G actin*, and is associated with one molecule of non-covalently bound ATP. The terminal phosphate of the ATP is hydrolyzed after the actin polymerizes to form actin filaments, also called filamentous actin or *F actin*. Polymerization can be induced simply by raising the salt concentration to a level closer to that found in cells: the actin solution, which is only slightly more viscous than water, then undergoes a rapid increase in viscosity as the actin molecules form filaments.

Although the bound ATP is hydrolyzed during the polymerization of actin, the polymerization process does not require energy: actin will polymerize even when the bound nucleotide is ADP or a nonhydrolyzable analogue of ATP. ATP hydrolysis, however, does have important effects on the dynamic behavior of actin filaments, as will be explained later when we discuss cellular activities that, unlike muscle contraction, depend on the controlled polymerization and depolymerization of actin.

Actin filaments appear in electron micrographs as uniform threads about 8 nm wide (Figure 11–6). They have been identified as the principal components

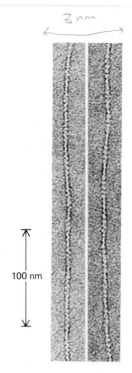

8 nm

100 nm

Figure 11–6 Electron micrographs of negatively stained actin filaments. (Courtesy of Roger Craig.)

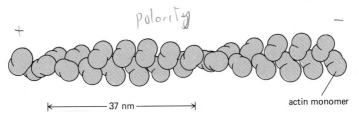

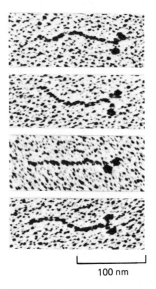

Figure 11–7   The arrangement of globular actin molecules in an actin filament. The molecules are packed into a tight helix with about two actin monomers per turn. Although this arrangement may give the appearance of two helical strands of actin molecules twisting around each other every 37 nm, this appearance is misleading inasmuch as the hypothetical single "actin strand" cannot exist on its own.

of thin filaments in skeletal muscle by their appearance in the electron microscope, their x-ray diffraction pattern, and their staining with anti-actin antibodies. The thin filaments of muscle are composed of more than just actin, however, as described below (see p. 622).

Actin filaments consist of a tight helix of uniformly oriented actin monomers (Figure 11–7). They are polar structures, with two structurally different ends. The polarity is essential to their function in cell motility and is most readily detectable by the oriented complex that each monomer in the filament forms with myosin. Before discussing this crucial interaction, however, we need to consider some of the properties of myosin molecules.

Thick Filaments Are Composed of Myosin[2]

Myosin is found in almost every cell type in the vertebrate body and is always present where actin filaments form contractile bundles in the cytoplasm. It has been much less highly conserved in evolution than actin and occurs in several forms. The filaments formed when skeletal muscle myosin polymerizes spontaneously *in vitro*, for example, are much larger than the assemblies formed *in vitro* by nonmuscle cell myosins.

Myosin can be specifically extracted from skeletal muscle by treatment with concentrated salt solutions, which causes the thick filaments to depolymerize into their constituent myosin molecules (Figure 11–8). Each molecule consists of six polypeptide chains: two identical *heavy chains* and two pairs of *light chains* (Figure 11–9).

The proteolytic enzyme papain cleaves the myosin molecule into a long α-helical section, called the *myosin rod* (or *myosin tail*), and two separate globular *myosin heads*, also called *subfragment-1*, or *S1 fragments* (Figure 11–10). These two parts of the myosin molecule mediate different functions: the tail is responsible for the spontaneous assembly of myosin molecules into thick filaments, whereas the heads are responsible for moving the molecules against adjacent actin filaments. We shall first discuss the structure and spontaneous assembly of the tails and then consider how the heads generate force.

Figure 11–8   Electron micrographs of myosin molecules shadowed with platinum. Note that each molecule is composed of two globular heads attached to a single fibrous tail. (Courtesy of David Shotton.)

11-4 ## Myosin Tails Assemble Spontaneously into a Bipolar Thick Filament[3]

Myosin tails, like many cytoskeletal proteins, are long rodlike molecules. The rigid structure of these proteins depends on a common structural motif in which two α helices with a characteristic spacing of hydrophobic residues coil around each other to form a **coiled coil** (Figure 11–11). In myosin, and many other cytoskeletal proteins, the two helices run in parallel (that is, in the same direction from amino to carboxyl terminal) to give rise to a filament that has a diameter of about 2 nm.

While the structure of individual myosin molecules depends on hydrophobic interactions between the two α-helical heavy chains (see Figure 11–11A), the structure of the thick filaments that myosin molecules form in muscle depends on

Figure 11–9   A myosin molecule is composed of two heavy chains (each about 2000 amino acid residues long) and four light chains. The light chains are of two types (one containing about 190 and the other about 170 amino acid residues), and one molecule of each type is present on each myosin head.

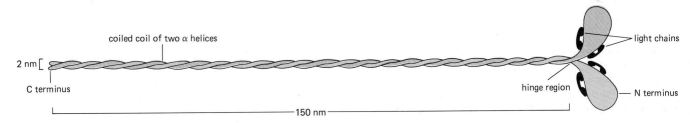

ionic interactions between the tails of the individual molecules. This is why solutions of high salt concentration, which disrupt ionic interactions but do not affect hydrophobic interactions, release individual myosin molecules from muscle. As the salt concentration is reduced to physiological ionic strength, the tails of the myosin molecules associate to form large filaments that may closely resemble muscle thick filaments. In muscle cells these interactions are stabilized by various accessory proteins, and the thick filaments that form are composed of hundreds of myosin tails packed together in a regular staggered array from which the myosin heads project in a repeating pattern (Figure 11–12). The structure is bipolar, with a bare central region where two oppositely oriented sets of myosin tails come together. The globular heads of the myosin molecules interact with actin, forming the crossbridges between the thick and thin filaments.

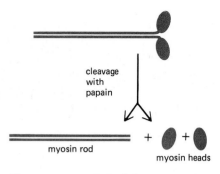

Figure 11–10 Limited digestion with the proteolytic enzyme papain cleaves the myosin molecule into a rod and two heads.

1-5 ATP Hydrolysis Drives Muscle Contraction[4]

Skeletal muscle converts chemical energy into mechanical work with very high efficiency—only 30% to 50% of the energy is wasted as heat. (An automobile engine, by contrast, typically wastes 80% to 90% of the energy available from gasoline.)

The energy for muscle contraction comes from ATP hydrolysis. Yet one detects no major difference in ATP levels between a resting muscle and one that is actively contracting, because a muscle cell has a very efficient backup system for regenerating ATP. The enzyme *phosphocreatine kinase* catalyses a reaction between an even more reactive phosphate compound, *phosphocreatine* (Figure 11–13), and ADP to form creatine and ATP. It is the intracellular level of phosphocreatine that drops after a short burst of muscle activity, even though the contractile machinery itself consumes ATP. The pool of phosphocreatine acts like a battery—storing ATP energy and recharging itself from the new ATP that is generated by cellular oxidations when the muscle is resting.

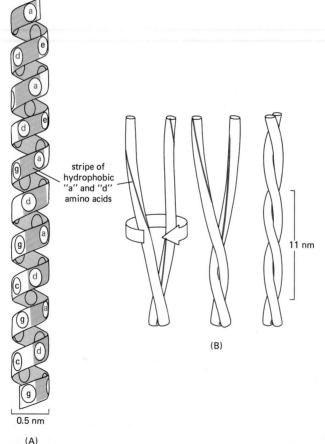

stripe of hydrophobic "a" and "d" amino acids

11 nm

0.5 nm

(A)

(B)

Figure 11–11 Topology of a coiled coil. In (A) a single α helix is represented as a cylinder with successive amino acid side chains labeled in a sevenfold sequence "abcdefg" (from bottom to top). Amino acids "a" and "d" in such a sequence lie close together on the cylinder surface, forming a "stripe" that winds slowly around the α helix (*shaded in color*). Proteins that form coiled coils typically have hydrophobic amino acids at positions "a" and "d." Consequently, as shown in (B), the two α helices can wrap around each other with the hydrophobic side chains of one α helix intercalated in the spaces between the hydrophobic side chains of the other, while the more hydrophilic amino acid side chains are left exposed to the aqueous environment.

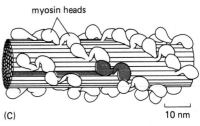

(C)

10 nm

(A)

500 nm

bare zone   myosin heads

(B)

Myosin Is an Actin-activated ATPase[5]

ATP hydrolysis during muscle contraction is a direct consequence of the inter-action between myosin and actin. Even on its own, myosin acts as an ATPase. But purified myosin by itself works relatively slowly. Each molecule takes about 30 seconds to complete a cycle in which an ATP molecule is hydrolyzed to com-pletion. The rate-limiting step is not the initial binding of ATP to the myosin nor the hydrolysis of its terminal phosphate group—both of which occur rapidly—but the release of the products of ATP hydrolysis (ADP and inorganic phosphate, P_i), which remain noncovalently bound to the myosin molecule and prevent fur-ther ATP binding and hydrolysis.

In the presence of actin filaments, the ATPase activity of myosin is greatly stimulated. Each myosin molecule now hydrolyzes 5 to 10 molecules of ATP every second, which is comparable to the rates measured in contracting muscle. The stimulation of myosin ATPase by actin filaments reflects a physical association between the two that is at the heart of muscle contraction. The binding of myosin to an actin filament causes a rapid release of ADP and P_i from the myosin molecule, which is thus freed to bind another molecule of ATP and start the cycle again.

Myosin Heads Bind to Actin Filaments[6]

It is the globular head of the myosin molecule that both binds to actin filaments and hydrolyzes ATP. Isolated myosin heads, which can be obtained separately after papain digestion (see Figure 11–10), retain both the ATPase activity and the actin-filament-binding properties of the intact myosin molecule and therefore can be used to analyze the interaction between actin and myosin.

Each actin molecule in an actin filament is capable of binding one myosin head to form a complex that reveals the structural polarity of the actin filament. With negative staining, such complexes can be seen in the electron microscope to have a regular and distinctive form: each myosin head forms a lateral projection, and the superimposed image of many such projections gives the appearance of arrowheads along the actin filament (Figure 11–14). Because the heads bind with the same orientation to each actin subunit, all of the actin molecules face in the same direction along the filament. The actin filament therefore has two structurally distinct ends, which are called the **minus end** (also called the *pointed end* because it corresponds to the point of the arrow after decoration) and the **plus end** (also called the *barbed end* because it corresponds to the barb of the arrow). The ter-minology "plus" and "minus" derives from the observation that the two ends of an actin filament elongate at different rates *in vitro* (see p. 637).

As shown in Figure 11–12, myosin heads face in opposite directions on either side of the bare central region of a thick filament. Since the heads must interact

Figure 11–12 The myosin thick filament. (A) Electron micrograph of a myosin thick filament isolated from scallop muscle. Note the central bare zone. (B) Schematic diagram, not drawn to scale. The myosin molecules aggregate together by means of their tail regions, with their heads projecting to the outside. The bare zone in the center of the filament consists entirely of myosin tails. (C) A small section of a thick filament as reconstructed from electron micrographs. An individual myosin molecule is highlighted in color. (A, courtesy of Roger Craig; C, based on R.A. Crowther, R. Padron, and R. Craig, *J. Mol. Biol.* 184:429–439, 1985.)

phosphocreatine

Figure 11–13 Phosphocreatine acts as a reserve source of high-energy phosphate groups in vertebrate muscle and other tissues. The high-energy phosphate group (*color*) is transferred to ADP by the enzyme creatine kinase to produce ATP when needed.

with thin filaments in the region of overlap, the thin filaments on either side of the sarcomere should be of opposite polarity. This has been demonstrated by using myosin heads to decorate the actin filaments attached to isolated Z discs: all the myosin arrowheads are found to point away from the Z disc. Therefore, the plus end of each actin filament is embedded in the Z disc, while the minus end points toward the thick filaments (Figure 11–15).

A Myosin Head "Walks" Toward the Plus End of an Actin Filament[7]

Muscle contraction is driven by the interaction between myosin heads and adjacent actin filaments. During this interaction, the myosin head hydrolyzes ATP. The ATP hydrolysis and subsequent dissociation of the tightly bound products (ADP and P_i) produce an ordered series of allosteric changes in the conformation of myosin. As a result, part of the energy released is coupled to the production of movement. For a discussion of the general principles involved in coupling ATP hydrolysis to the directed movement of protein molecules, see page 130.

Kinetic analyses of ATP hydrolysis during muscle contraction, together with electron microscopic and x-ray diffraction studies, suggest the sequence of events illustrated in Figure 11–16. A free myosin head binds ATP (state 1) and hydrolyzes it; the process is reversible because the energy of ATP hydrolysis is initially stored in a highly strained protein conformation with ADP and P_i bound (state 2). While alternating between these two states, a myosin head, as a result of random motions, can move to a neighboring actin subunit and bind to it weakly; this triggers the release of P_i, which causes the head to bind very tightly to the actin filament (state 3). Once bound in this way, the head undergoes a conformational change that generates a "power stroke" that pulls on the rest of the thick filament. At the end of the power stroke (state 4), ADP is released and a fresh molecule of ATP binds to the head, detaching it from the actin filament and returning the head to state 1. Hydrolysis of the bound ATP then prepares the myosin head for a second cycle.

Because each turn of the cycle illustrated in Figure 11–16 results in the hydrolysis and release of one ATP molecule, the series of conformational changes just described is driven by a large favorable change in free energy, making it unidirectional (see p. 130). Each individual myosin head, therefore, "walks" in a single direction along an adjacent actin filament, always moving toward the filament's plus end (see Figure 11–15). As it undergoes its cyclical change in conformation, the myosin head pulls against the actin filament, causing this filament to slide against the thick filament. Once an individual myosin head has detached from the actin filament, it is carried along by the action of other myosin heads in the same thick filament, so that a snapshot of an entire thick filament in a contracting muscle would show some of the myosin heads attached to actin filaments and others unattached. (A certain amount of springlike elasticity in the myosin molecule is essential to allow this to happen.) Each thick filament has about 500 myosin heads, and each head cycles about five times per second in the course of a rapid contraction—sliding the thick and thin filaments past each other at rates up to 15 μm/second.

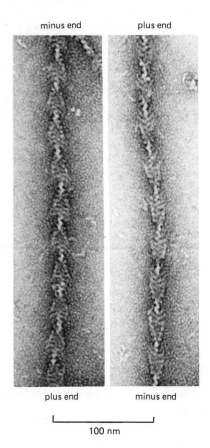

minus end   plus end

plus end   minus end

100 nm

Figure 11–14 Electron micrograph of actin filaments decorated with isolated myosin heads. The helical arrangement of the bound myosin heads, which are tilted in one direction, gives the appearance of arrowheads and indicates the polarity of the actin filament. The pointed end is called the *minus end*, the barbed end the *plus end* because of the different rates of assembly of actin monomers at the two ends (see Figure 11–40). (Courtesy of Roger Craig.)

Figure 11–15 The thick and thin filaments of a sarcomere overlap with the same relative polarity on either side of the midline.

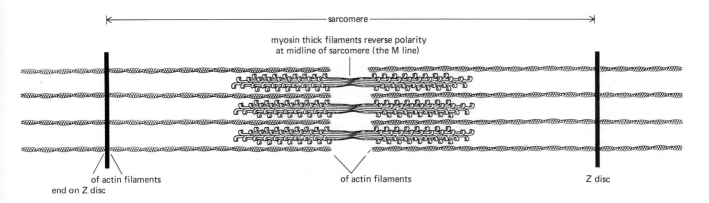

sarcomere

myosin thick filaments reverse polarity at midline of sarcomere (the M line)

of actin filaments end on Z disc

of actin filaments

Z disc

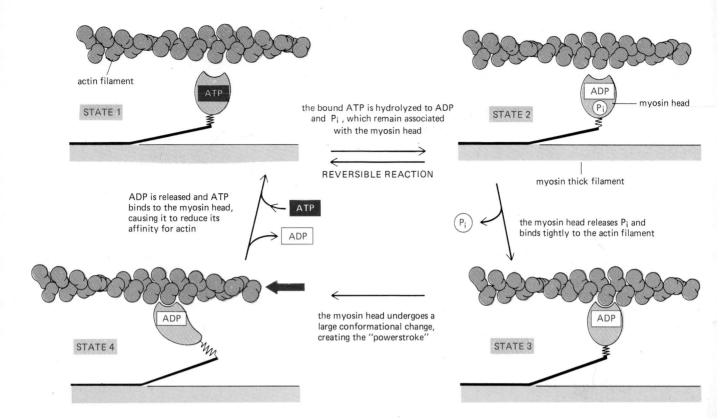

the bound ATP is hydrolyzed to ADP and P_i, which remain associated with the myosin head

REVERSIBLE REACTION

actin filament

STATE 1

ATP

ADP is released and ATP binds to the myosin head, causing it to reduce its affinity for actin

ATP

ADP

STATE 4

ADP

STATE 2

ADP
P_i

myosin head

myosin thick filament

P_i

the myosin head releases P_i and binds tightly to the actin filament

STATE 3

ADP

the myosin head undergoes a large conformational change, creating the "powerstroke"

Muscle Contraction Is Initiated by a Sudden Rise in Cytosolic Ca^{2+}[8]

The force-generating molecular interaction just described takes place only when a signal passes to the skeletal muscle from its motor nerve. The signal from the nerve triggers an action potential in the muscle cell plasma membrane, and this electrical excitation spreads rapidly into a series of membranous folds, the *transverse tubules*, or *T tubules*, that extend inward from the plasma membrane around each myofibril. The signal is then somehow relayed to the *sarcoplasmic reticulum*, an adjacent sheath of anastomosing flattened vesicles that surrounds each myofibril like a net stocking (Figure 11–17).

The gap between the T tubule and the sarcoplasmic reticulum is only 10–20 nm, but it is unclear how the signal passes between them. When the T tubules are electrically excited, large Ca^{2+} *release channels* in the sarcoplasmic reticulum membrane (see Figure 11–17) are somehow opened, allowing Ca^{2+} to escape into the cytosol from the sarcoplasmic reticulum, where Ca^{2+} is stored in large quantities. The resulting sudden rise in free Ca^{2+} concentration in the cytosol initiates the contraction of each myofibril. Because the signal from the muscle-cell plasma membrane is passed within milliseconds (via the T tubules and sarcoplasmic reticulum) to every sarcomere in the cell, all of the myofibrils in the cell contract at the same time. The increase in Ca^{2+} concentration in the cytosol is transient because the Ca^{2+} is rapidly pumped back into the sarcoplasmic reticulum by an abundant Ca^{2+}-ATPase in its membrane (see p. 307). Typically, the cytosolic Ca^{2+} concentration is restored to resting levels within 30 milliseconds, causing the myofibrils to relax.

Troponin and Tropomyosin Mediate the Ca^{2+} Regulation of Skeletal Muscle Contraction[9]

The Ca^{2+} dependence of vertebrate skeletal muscle contraction, and hence its dependence on motor commands transmitted via nerves, is due entirely to a set of specialized accessory proteins closely associated with actin filaments. If myosin is mixed with pure actin filaments in a test tube, myosin ATPase is activated whether or not Ca^{2+} is present; in a normal myofibril, on the other hand, where

Figure 11–16 Diagram showing how a myosin molecule is thought to use the energy of ATP hydrolysis to move from the minus to the plus end of an actin filament. In the transition from state 2 to state 3, an initial binding to actin causes the myosin head to release its bound phosphate and bind more tightly to the actin filament. The myosin head then undergoes a poorly understood change in shape accompanied by the release of ADP, which makes the myosin head pull against the actin filament (the power stroke). Each of the two heads on a myosin molecule is thought to cycle independently of the other.

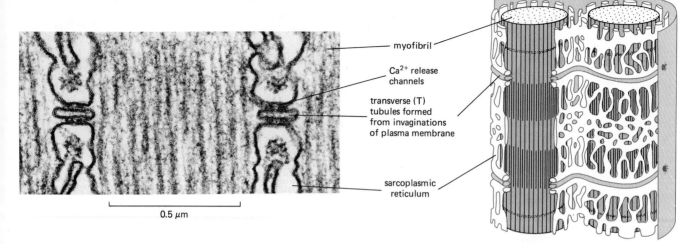

0.5 μm

Figure 11–17 The system of membranes involved in relaying the signal to contract from the muscle cell plasma membrane to all of the myofibrils in the cell. The electron micrograph illustrates two T tubules and shows the large Ca²⁺ release channels in the sarcoplasmic reticulum membrane that look like square-shaped "feet" and seem to connect to the adjacent T-tubule membrane. (Micrograph courtesy of Clara Franzini-Armstrong.)

the actin filaments are associated with accessory proteins, the activation of the myosin ATPase depends on Ca^{2+}.

One of these accessory proteins is a rigid, 41-nm-long rod-shaped molecule, called *tropomyosin* because of similarities to myosin in its x-ray diffraction pattern. Like the myosin tail, tropomyosin is a dimer of two identical α-helical chains (284 amino acids each), which wind around each other in a coiled coil (see p. 618). By binding along the length of an actin filament, tropomyosin stabilizes and stiffens the filament (Figure 11–18).

The other major accessory protein involved in Ca^{2+} regulation in vertebrate skeletal muscle is *troponin*, a complex of three polypeptides—troponins T, I, and C (named for their *T*ropomyosin-binding, *I*nhibitory, and *C*alcium-binding activities). The troponin complex has an elongated shape, with subunits C and I forming a globular head region and T forming a long tail. The tail of *troponin T* binds to tropomyosin and is thought to be responsible for positioning the complex on the thin filament (see Figure 11–18). *Troponin I* binds to actin, and when it is added to troponin T and tropomyosin, the complex inhibits the interaction of actin and myosin, even in the presence of Ca^{2+}.

The further addition of *troponin C* completes the troponin complex and makes its effects sensitive to Ca^{2+}. Troponin C binds up to four molecules of Ca^{2+}, and with Ca^{2+} bound, it relieves the inhibition of myosin binding to actin produced by the other two troponin components. Troponin C is closely related to *calmodulin*, which mediates Ca^{2+}-signaled responses in all cells, including the activation of smooth muscle myosin (see p. 626). Troponin C may therefore be regarded as a specialized form of calmodulin that has evolved permanent binding sites for troponin I and troponin T, thereby ensuring that the myofibril responds extremely rapidly to an increase in Ca^{2+} concentration.

There is only one molecule of the troponin complex for every seven actin monomers in an actin filament (see Figure 11–18). Structural studies suggest that in a resting muscle the binding of troponin I to actin moves the tropomyosin molecules to a position on the actin filaments that in an actively contracting muscle is occupied by the myosin heads and thus inhibits the interaction of actin

Figure 11–18 A muscle thin filament showing the position of tropomyosin and troponin along the actin filament. Each tropomyosin molecule has seven evenly spaced regions of homologous sequence, each of which is thought to bind to an actin monomer as shown. Note that the ends of adjacent tropomyosin molecules overlap slightly as they polymerize in a head-to-tail fashion along the actin filament. (Adapted from G.N. Phillips, J.P. Fillers, and C. Cohen, *J. Mol. Biol.* 192:111–131, 1986.)

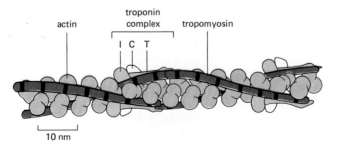

actin

troponin complex

tropomyosin

I   C   T

10 nm

and myosin. When the level of Ca^{2+} is raised, troponin C causes the troponin I to release its hold on actin, thereby allowing the tropomyosin molecules to shift their position slightly so that the myosin heads can bind to the actin filament (Figure 11–19).

Other Accessory Proteins Maintain the Architecture of the Myofibril and Provide It with Elasticity[10]

The remarkable speed and power of muscle contraction depend on the filaments of actin and myosin in each myofibril being held at the optimal distance from each other and in correct alignment. The precise organization of the myofibril is maintained by a number of structural proteins, more than a dozen of which have so far been identified (Table 11–1); the positions of most of these proteins in the sarcomere have been determined by immunocytochemical methods (see p. 177).

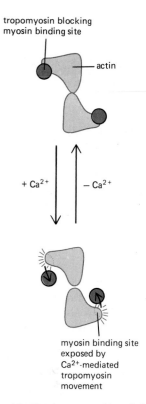

Figure 11–19 A cross-sectional view of a thin filament showing how, in the absence of Ca^{2+}, tropomyosin is thought to block the interaction of the myosin head with actin.

Table 11–1 Major Protein Components of Vertebrate Skeletal Myofibrils

Protein	Percent Total Protein	MW (kDa)	Subunits (kDa)	Function
Myosin	44	510	2 × 223 (heavy chains) 22 + 18 (light chains)	Major component of thick filaments. Interacts with actin filaments with hydrolysis of ATP to develop mechanical force
Actin	22	42	—	Major component of muscle thin filaments, against which muscle thick filaments slide during muscle contraction
Tropomyosin	5	64	2 × 32	Rodlike protein that binds along the length of actin filaments
Troponin	5	78	30 (Tn-T) 30 (Tn-I) 18 (Tn-C)	Complex of three muscle proteins positioned at regular intervals along actin filaments and involved in the Ca^{2+} regulation of muscle contraction
Titin	9	about 2500	—	Very large flexible protein that forms an elastic network linking thick filaments to Z discs
Nebulin	3	600	—	Elongated, inextensible protein attached to Z disc, oriented parallel to actin filaments
α-Actinin	1	190	2 × 95	Actin-bundling protein that links actin filaments together in the region of the Z disc
Myomesin	1	185	—	Myosin-binding protein present at the central "M line" of the muscle thick filament
C protein	1	140	—	Myosin-binding protein found in distinct stripes on either side of the thick-filament M line

The vertebrate striated myofibril also contains at least 20 other proteins not included in this table.

Actin filaments are anchored by their plus ends to the Z disc, where they are held in a square lattice arrangement by other proteins. One of the best characterized of these is *α-actinin*, an actin-binding protein that is found in most animal cells (see p. 642). In muscle cells it is located in the Z disc. Purified α-actinin is a bipolar rod-shaped molecule (Figure 11–20), which can bundle actin filaments together into parallel arrays. An equivalent function may be performed for myosin by a molecule called *myomesin*, which cross-links adjacent myosin filaments at the "M line" (midway along the bipolar thick filaments) to produce a hexagonal packing arrangement. The packing of myosin is also stabilized by other myosin-binding proteins, which can be detected in antibody-labeling studies as a series of 11 regularly spaced, faint stripes on either side of the M line.

Muscle cells also contain a system of highly insoluble protein filaments, which can be isolated after concentrated solutions of potassium iodide have been used to extract all actin and myosin from the sarcomere. One set of these filaments, composed of a large protein called *titin*, runs parallel to the thick and thin filaments in the sarcomere and connects the thick filaments to the Z discs. The titin filaments are very elastic and are thought to act like springs to help keep the thick filaments centered between the Z discs (Figure 11–21). Another set of insoluble filaments are the intermediate filaments (see p. 661) that run between the Z discs of adjacent myofibrils. They are thought to keep the sarcomeres in register and to connect the myofibrils to the muscle plasma membrane.

There Are Three Principal Types of Vertebrate Muscle[11]

Thus far we have described only one of the three major types of muscle present in vertebrates—*skeletal muscle*. The others are *heart* (or *cardiac*) *muscle*, which contracts about 3 billion times in the course of an average human life-span, and *smooth muscle*, which produces the slower and longer-lasting contractions characteristic of organs such as the intestines. All three types of muscle cells, together with another class of contractile cells known as *myoepithelial cells* (see p. 983), contract by an actin and myosin sliding filament mechanism.

Like skeletal muscle, **heart muscle** is striated, reflecting a very similar organization of actin filaments and myosin filaments. It is also triggered to contract by a similar mechanism: an action potential in the T tubules somehow triggers the sarcoplasmic reticulum to release Ca^{2+}, which activates contraction by means of a troponin-tropomyosin complex. Heart muscle cells, however, are not multinucleated, and they are joined end to end by special structures called *intercalated discs* (Figure 11–22). The intercalated discs serve at least three functions. (1) They attach one cell to the next by means of desmosomes (see p. 797). (2) They connect the thin filaments of the myofibrils of adjacent cells (performing a function analogous to that of the Z discs inside the cells). (3) They contain gap junctions (see p. 798), which allow an action potential to spread rapidly from one cell to the next, synchronizing the contractions of the heart muscle cells.

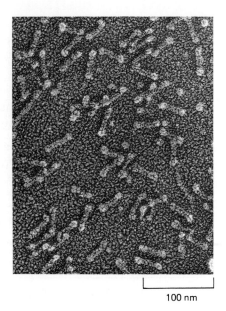

100 nm

Figure 11–20 Electron micrograph of purified α-actinin molecules. (Courtesy of John Heuser.)

Figure 11–21 The network of titin filaments that is postulated to link myosin thick filaments to Z discs in the sarcomeres of skeletal muscle. The elastic titin filaments are thought to be anchored along the surface of the thick filaments, with only the segment between the tip region of the thick filaments and the Z discs free to change length and contribute to the elastic restoring force of the sarcomere. Such an elastic network keeps thick filaments centered between the Z discs and allows muscles to be stretched beyond the region of thick and thin filament overlap without destroying the sarcomere. The cross-bridges between thick and thin filaments have been omitted for clarity. (After K. Wang and J. Wright, *J. Cell Biol.* 107:2199–2212, 1988.)

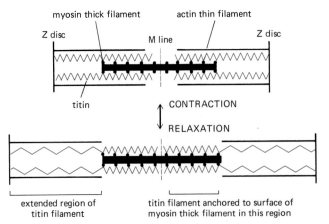

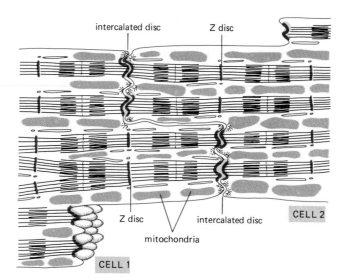

intercalated disc

Z disc

Z disc

intercalated disc

mitochondria

CELL 2

CELL 1

Figure 11–22 The structure of heart muscle. Heart muscle is composed of many discrete cells, each with its own nucleus. The cells are joined end to end by specialized junctions known as intercalated discs. Actin filaments from sarcomeres in adjacent cells insert into the dense material associated with the plasma membrane in the region of each intercalated disc as though they were Z discs. Thus the myofibrils continue across the muscle, ignoring cell boundaries.

The most "primitive" muscle, in the sense of being most like nonmuscle cells, has no striations and is therefore called **smooth muscle.** It forms the contractile portion of the stomach, intestine, and uterus, the walls of arteries, the ducts of secretory glands, and many other regions in which slow and sustained contractions are needed. It is composed of sheets of elongated spindle-shaped cells, each with a single nucleus (see p. 983). The cells contain both thick and thin filaments, but these are not arranged in the strictly ordered pattern found in skeletal and cardiac muscle and do not form distinct myofibrils. Instead, the filaments form a more loosely arranged contractile apparatus, which is roughly aligned with the long axis of the cell but is attached obliquely to the plasma membrane at disclike junctions connecting groups of cells together.

Although the contractile apparatus in smooth muscle does not contract as rapidly as the myofibrils in a striated muscle cell, it has the advantage of permitting a much greater degree of shortening and therefore can produce large movements even though it lacks the leverage provided by attachments to bones. The organization of the actin filaments and myosin that makes this possible is poorly understood; one model is presented in Figure 11–23.

The Activation of Myosin in Both Smooth Muscle and Nonmuscle Cells Depends on Myosin Light-Chain Phosphorylation[12]

The highly specialized contractile mechanisms that we have described in muscle cells evolved from the simpler force-generating mechanisms found in all eucaryotic cells. Not surprisingly, the myosin in nonmuscle cells most resembles the myosin in smooth muscle cells, the least specialized type of muscle. As in skeletal

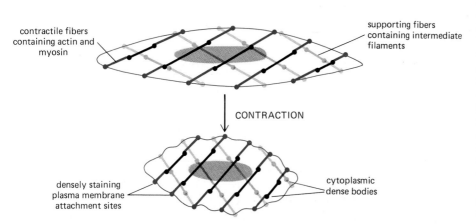

contractile fibers containing actin and myosin

supporting fibers containing intermediate filaments

CONTRACTION

densely staining plasma membrane attachment sites

cytoplasmic dense bodies

Figure 11–23 A model for the contractile apparatus in a smooth muscle cell. In this hypothetical view, bundles of contractile filaments containing actin and myosin are anchored at one end to dense bodies in the plasma membrane and at the other end, through cytoplasmic dense bodies, to noncontractile bundles of intermediate filaments. The contractile actin-myosin bundles are oriented obliquely to the long axis of the cell (which is generally much more elongated than shown), and their contraction greatly shortens the cell. Only a few of the many bundles are shown.

and cardiac muscle, contraction is triggered in both smooth muscle and non-muscle cells by a rise in cytosolic Ca^{2+}, but in neither cell type does the Ca^{2+} act through a troponin-tropomyosin complex. Instead, contraction is initiated mainly by phosphorylation of one of the two myosin light chains, which in turn controls the interaction of myosin with actin.

The two myosin light chains on each myosin head (see Figure 11–9) are different, and only one of them is phosphorylated during nonmuscle and smooth muscle contraction. When this light chain is phosphorylated, the myosin head can interact with an actin filament and thereby cause contraction; when it is dephosphorylated, the myosin head tends to dissociate from actin and becomes inactive. The phosphorylation is catalyzed by the enzyme *myosin light-chain kinase*, whose action requires the binding of a Ca^{2+}-calmodulin complex (see p. 711). As a result, contraction is controlled by the level of cytosolic Ca^{2+}, as in cardiac and skeletal muscle (Figure 11–24). The phosphorylation occurs relatively slowly, so that maximum contraction often requires nearly a second (compared to the few milliseconds required for a striated muscle cell), but rapid activation of contraction is not important in smooth muscle or nonmuscle cells.

The myosins in both smooth muscle and nonmuscle cells hydrolyze ATP about 10 times more slowly than skeletal muscle myosin, producing a slow cross-bridge cycle that allows these cells to contract only slowly. A smooth muscle cell, however, should not be regarded as a slow, poorly constructed version of a skeletal muscle cell that is adequate only because fewer demands are made on it. It is designed specifically for slow sustained contraction, being able to maintain tension for prolonged periods while hydrolyzing five- to tenfold less ATP than would be required by a skeletal muscle cell performing the same task. Moreover, smooth muscle contraction is controlled by a wide variety of stimuli, including autonomic nerve impulses and hormones such as epinephrine. Many of these controls act via the myosin light-chain kinase. In some cells, for example, epinephrine, by increasing cyclic AMP levels, induces the phosphorylation of this kinase, which greatly reduces the affinity of the kinase for the Ca^{2+}-calmodulin complex; in this way epinephrine inhibits myosin light-chain phosphorylation, causing the smooth muscle to relax.

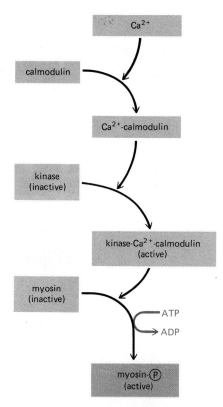

Figure 11–24 The contraction of smooth muscle is activated in the presence of Ca^{2+} by *myosin light-chain kinase*, which catalyzes the phosphorylation of a particular site on one of the two types of myosin light chains. Nonmuscle myosin molecules are regulated by the same mechanism (see Figure 11–25).

Light-Chain Phosphorylation Enables Nonmuscle Myosin Molecules to Assemble into Filaments[13]

Although myosin is present in nearly all eucaryotic cells, it is found in stable thick filaments only in skeletal and heart muscle. Myosin molecules in nonmuscle cells form smaller assemblies on demand; intracellular signals dictate both the size and the location of the contractile assemblies formed. An important factor influencing the state of myosin aggregation is its phosphorylation by myosin light-chain kinase, which affects not only the ATPase activity of the myosin as just described, but also its shape and ability to self-assemble.

If nonmuscle myosin is dephosphorylated by treatment with a phosphatase, it becomes freely soluble. Sedimentation analysis shows that the soluble single myosin molecules have folded into a compact configuration, and electron microscopy suggests that each myosin tail has folded back on itself to adhere to a "sticky patch" on the head. In this folded configuration the myosin molecules are not able to assemble into filaments efficiently. Myosin light-chain kinase phosphorylates the head so that it is no longer "sticky"; the tail is then free to extend and to associate lengthwise with the tails of other myosin molecules, generating a bipolar myosin filament (Figure 11–25 and Figure 11–26).

"Musclelike" Assemblies Can Form Transiently in Nonmuscle Cells[14]

Control over myosin assembly is very important in nonmuscle cells, where contractile bundles of actin filaments and myosin often form for a specific function and then disassemble. For example, a beltlike bundle of actin filaments and myosin known as the **contractile ring** appears beneath the plasma membrane during

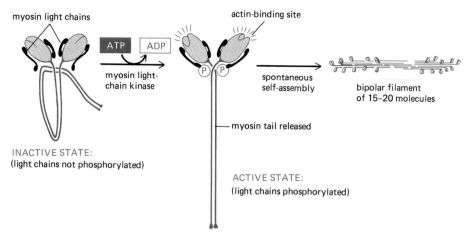

myosin light chains

actin-binding site

ATP → ADP

myosin light-chain kinase

myosin tail released

INACTIVE STATE:
(light chains not phosphorylated)

spontaneous self-assembly

bipolar filament of 15-20 molecules

ACTIVE STATE:
(light chains phosphorylated)

Figure 11–25 The assembly of nonmuscle myosin is controlled by phosphorylation of its light chains. The phosphorylation has two effects: it causes a change in the conformation of the myosin head, exposing its actin-binding site; and it releases the myosin tail from a "sticky patch" on the myosin head, thereby allowing the myosin molecules to assemble into short bipolar filaments. Smooth muscle myosin behaves in the same way.

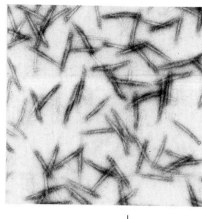

1 μm

Figure 11–26 Electron micrograph of negatively stained short filaments of nonmuscle myosin that have been induced to assemble by the phosphorylation of their light chains. (Courtesy of John Kendrick-Jones.)

cell division; forces generated by this ring constrict the middle of the cell, leading to the eventual separation of the two daughter cells (Figure 11–27; see also p. 780). Since the contractile ring is not a permanent structure, it must be assembled at the start of cell division. The process can be monitored by staining dividing cells with fluorescent anti-myosin antibodies. In sea urchin eggs that are about to divide, for example, myosin molecules are at first distributed evenly beneath the plasma membrane and then move to the equatorial region as the contractile ring forms. Once cell division is complete, the myosin molecules disperse. It is not known how this process is controlled.

Another example of temporary contractile bundles of actin filaments and myosin are **stress fibers,** which are prominent components of the cytoskeleton of fibroblast cells in culture (see Figure 11–27). They resemble tiny myofibrils in their structure and function (Figure 11–28). At one end they insert into the plasma membrane at special junctions called *focal contacts* (see p. 635), which are similar in ultrastructure and composition to the plasma membrane attachment sites for actin filaments in a smooth muscle cell (see Figure 11–23). The other end of a stress fiber can insert either into a meshwork of intermediate filaments that surrounds the cell nucleus (see Figure 9–1) or into a second focal contact. Stress fibers form in response to tension generated across a cell and are disassembled at mitosis when the cell rounds up and loses its attachments to the substratum. They also disappear rapidly if tension is released by suddenly detaching one end of the stress fiber from the focal contact by means of a laser beam. Stress fibers within fibroblasts in tissues are thought to contract to allow the cells to exert tension on the matrix of collagen surrounding them—an essential process in both wound healing and morphogenesis (see p. 813).

Not all contractile assemblies of actin filaments in nonmuscle cells are transitory. Those associated with intercellular anchoring junctions called *adhesion belts,* for example, are more lasting. Adhesion belts are found near the apical surface of epithelial cells (see p. 796). Among other functions, they are thought to play an important part in the folding of epithelial cell sheets during embryogenesis (see p. 674).

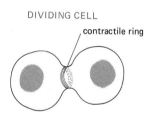

DIVIDING CELL

contractile ring

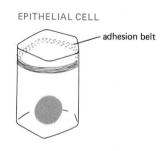

EPITHELIAL CELL

adhesion belt

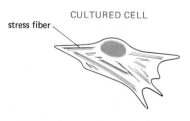

CULTURED CELL

stress fiber

Figure 11–27 Examples of contractile bundles of actin filaments that contain myosin in nonmuscle cells.

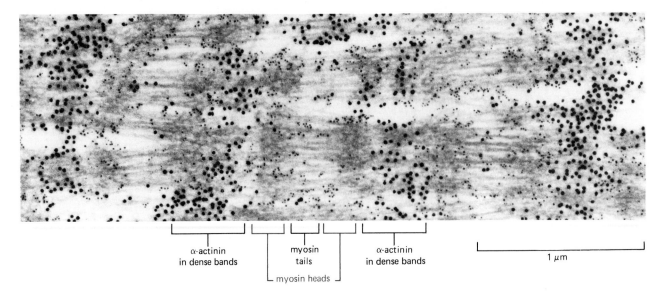

α-actinin in dense bands | myosin tails | α-actinin in dense bands

myosin heads

1 μm

Muscle Proteins Are Encoded by Multigene Families[15]

We have seen that a very similar type of contractile apparatus is present in three kinds of muscle cells as well as in most nonmuscle cells. The important differences between the types of contractions that occur in these cell types depend in part on the tissue-specific expression of the genes that encode the proteins of the contractile apparatus. A mammal, for example, has at least six actin genes, six myosin heavy-chain genes, three tropomyosin genes, and three troponin T genes. In some cases the different genes encode proteins that are known to have somewhat different functions; in others, functional differences have not yet been detected.

Of the six actin proteins expressed in mammals, one is restricted to skeletal muscle and another to heart muscle; two are restricted to smooth muscle (one to vascular and the other to nonvascular smooth muscle); and two, known as non-muscle or cytoplasmic actins, appear to be universal components of the cytoskeleton and are abundant in most nonmuscle cells. These closely related species of actin, also known as actin "isoforms," are very similar in sequence. Muscle and cytoplasmic actins, for example, differ in fewer than 7% of their amino acids. Apart from some differences in the amino-terminal end of the molecule that may have a subtle influence on actin polymerization, it is not certain that the amino acid differences have any functional consequence. If a heart-muscle actin gene is expressed in a cultured fibroblast, it makes a protein that integrates readily with the indigenous population of actin molecules without changing the shape or behavior of the cell. By contrast, as just discussed, the variation among the myosins affects the rate of contraction, the regulatory mechanisms that control contraction, and the extent of myosin self-assembly in the cell.

Muscle Proteins Are Further Diversified by Alternative RNA Splicing[16]

The variations of muscle type described so far give only a partial indication of the diversity that exists. An adult human skeletal muscle, for example, contains a mixture of three types of muscle cells: *white* muscle cells specialized for fast, anaerobic contraction, in which the ATP for contraction is generated mainly by glycolysis; *red* muscle cells specialized for slower, longer-lasting contraction, which use mainly aerobic metabolism and whose color is due to the high concentration of the oxygen-carrying protein myoglobin; and an *intermediate* type of muscle cell, which utilizes both aerobic and anaerobic metabolism (see p. 65). Within each category there are further subtypes that "fine-tune" the muscle to a particular metabolic and physiological function. The same muscle in the fetus is different again.

Each of these types of muscle has somewhat different proteins. A particularly rich source of protein variation is provided by tissue-specific regulation of RNA splicing, which can combine different sets of exons to make slightly different RNA

Figure 11–28 An electron micrograph of a cultured fibroblast, showing the musclelike arrangement of proteins in a stress fiber revealed by immunogold staining. The positions of two types of actin-binding protein molecules are shown: α-actinin (large gold particles) is seen in association with periodic dense bands in the stress fiber (α-actinin is associated with the Z disc in striated muscle), while myosin heads (small gold particles) are seen on either side of the α-actinin-containing bands. This pattern is reminiscent of a sarcomere (compare with Figure 11–2) and indicates that the myosin molecules are arranged in filaments. (Courtesy of M. de Brabander, J. de Mey, and G. Langanger.)

molecules from the same gene (see p. 589). Hormonal, neuronal, and other influences can alter the pattern of RNA splicing and thereby change the amino acid sequences of particular muscle proteins according to the tissue and stage of development. The single skeletal muscle gene that encodes troponin T, for example, can produce at least 10 distinct forms of the protein by alternative RNA splicing. This variability is likely to modify interactions of troponin T with troponin C and tropomyosin, and hence to provide subtle variations in the regulation of muscle contraction.

Summary

Muscle contraction is produced by the sliding of actin filaments against myosin filaments. The head regions of myosin molecules, which project from myosin filaments, engage in an ATP-driven cycle in which they attach to adjacent actin filaments, undergo a conformational change that pulls the myosin filament against the actin filament, and then detach. This cycle is facilitated by special accessory muscle proteins that hold the actin and myosin filaments in parallel overlapping arrays with the correct orientation and spacing for sliding to occur. Two other accessory proteins—troponin and tropomyosin—allow the contraction of skeletal and cardiac muscle to be regulated by Ca^{2+}.

Actin and myosin are also found in smooth muscle cells and in most nonmuscle cells, where they produce contraction in fundamentally the same way as in skeletal and cardiac muscle. The contractile units, however, are smaller and less highly ordered in such cells, and their activity and state of assembly is controlled by the Ca^{2+}-regulated phosphorylation of one of the myosin light chains.

The contractile apparatus in muscle and nonmuscle cells is fine-tuned to fit each cell type by the tissue-specific expression of different genes encoding muscle proteins and by tissue-specific regulation of RNA splicing, by which the same gene can produce slightly different forms of a protein.

Actin Filaments and the Cell Cortex[17]

Actin is the most abundant protein in many eucaryotic cells, often constituting 5% or more of the total cell protein. While actin is distributed throughout the cytoplasm, most animal cells possess an especially dense network of actin filaments and associated proteins just beneath the plasma membrane. This network constitutes the **cell cortex,** which gives mechanical strength to the surface of the cell and enables the cell to change its shape and to move. The precise form of the cortex varies from cell to cell and in different regions of the same cell. In some cells it consists of a thick three-dimensional network of cross-linked actin filaments that excludes large particles and organelles in the underlying cytoplasm (Figure 11–29); in other cells the cortex is a thinner, more two-dimensional net-

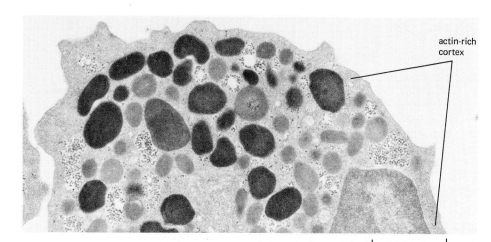

actin-rich cortex

Figure 11–29 The actin cortex in a thin-section electron micrograph of a white blood cell. Although various kinds of granules fill the cytoplasm, they are excluded from the layer just beneath the plasma membrane (the cortex), which contains a network of actin filaments and associated proteins that determines the surface movements of the cell. (Courtesy of Dorothy Bainton.)

1 μm

work. In selected areas of animal cells, small bundles of actin filaments project outward from the cortex to form the stiff core of cell-surface extensions; while in other areas, actin filaments pull inward on the membrane. Because the plasma membrane is so closely integrated with the cortical actin network, for some purposes these two entities are best considered as a single functional unit.

Approximately 50% of the actin molecules in most animal cells is unpolymerized, existing either as free monomers or as small complexes with other proteins. A dynamic equilibrium exists between this pool of unpolymerized actin molecules and actin filaments, which helps drive many of the surface movements of cells. In this section we shall discuss how actin-binding proteins control the assembly of actin filaments, link them together into bundles or networks, and determine their position, length, and other properties.

Abundant Actin-binding Proteins Cross-link Actin Filaments into Large Networks[18]

Actin filaments are often linked together into a stiff three-dimensional network by *cross-linking proteins*. The most abundant of these is **filamin**, a long flexible molecule composed of two identical polypeptide chains joined head to head, with a binding site for actin filaments at each tail end (Figure 11–30). Proteins of this type

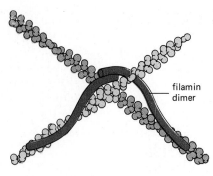

Figure 11–30 By forming a flexible link between two adjacent actin filaments, filamin produces a three-dimensional network of actin filaments with the physical properties of a gel. Each filamin dimer is about 160 nm long when fully extended.

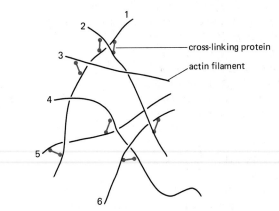

(A) actin-filament-based gel

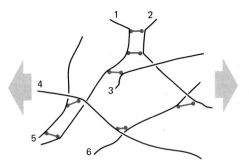

(B) gel resists sudden pull because of cross-linking proteins

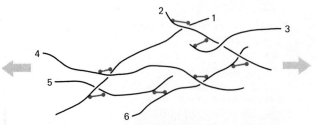

(C) gel slowly deforms with continuous pull, as cross-linking proteins rearrange their contacts with actin filaments

Figure 11–31 The mechanical properties of an actin filament gel formed by actin cross-linking proteins. A rapid deformation (B) is strongly resisted by the gel because cross-linking proteins do not have time to dissociate from their bound actin filaments. A slow deformation (C) encounters little resistance because the cross-linking protein molecules have time to dissociate and reassociate repeatedly, readjusting their positions. (After M. Sato, W.H. Schwartz, and T.D. Pollard, *Nature* 325:828–830, 1987.)

constitute almost 1% of the total protein of many cells (about 1 dimer for every 50 actin monomers).

Gels made *in vitro* from actin filaments and a cross-linking protein have the interesting mechanical property of preserving their form when subjected to a sudden force but being readily deformed by a slow steady pressure. Figure 11–31 suggests a molecular basis for this distinctive property, which is also displayed by the cortical cytoplasm. It is presumably because of such actin networks that the cell surface recoils elastically from small rapid insults, while being capable of large deformations if a slight force is steadily applied.

Gelsolin Fragments Actin Filaments When Activated by Ca^{2+} [19]

Extracts prepared from many types of animal cells form a gel in the presence of ATP when they are warmed to 37°C. Although this gelation depends on both actin filaments and a cross-linking protein such as filamin, the gels exhibit more complex behavior than simple mixtures of actin filaments and filamin. If the Ca^{2+} concentration is raised above 10^{-7} M, for example, the semisolid actin gel begins to liquefy—a process known as solation—and regions of the solating gel show vigorous local streaming when examined under a microscope. Clearly there must be components besides actin and filamin in the extracts to account for its Ca^{2+}-dependent solation and streaming. These components are likely to be involved in the *cytoplasmic streaming* observed in some large cells, where vigorous flowing movements are required to maintain an even distribution of metabolites and other cytoplasmic components. These movements seem to be associated with a sudden local change in the cytoplasm from a solid gel-like consistency to a more fluid state.

A number of proteins have been isolated from cell extracts that, when added to a gel of actin filaments and filamin, cause it to change to a more fluid state in the presence of Ca^{2+}. The best characterized of these is **gelsolin,** a compact protein of 90,000 daltons. When activated by the binding of Ca^{2+}, gelsolin severs an actin filament and forms a cap on the newly exposed plus end of the filament, thus breaking up the cross-linked network of actin filaments. Similar proteins are found in the cortex of many types of vertebrate cells. These *severing proteins* are activated by concentrations of Ca^{2+} (about 10^{-6} M) that occur only transiently in the cytosol, and they are believed to help mediate responses of the cell cortex to extracellular signals. When a phagocytic white blood cell contacts a microorganism, for example, the network of actin filaments in the white cell's cortex locally disassembles, enabling the surface of the cell to move and engulf the microorganism. We shall return later to the actin-based mechanism underlying such movements.

Myosin Can Mediate Cytoplasmic Movements[20]

While an artificial mixture of actin filaments, filamin, and gelsolin is capable of undergoing Ca^{2+}-dependent gel-to-sol transitions, it will not contract or show the streaming movements displayed by the cruder actin-rich gels obtained from cells. These activities seem to require *myosin*: if myosin is selectively removed from the crude actin-rich gels, contractions and streaming no longer occur, suggesting that an interaction between actin and myosin generates the force for cytoplasmic streaming.

How can actin and myosin produce coherent movements when the actin filaments in extracts are initially distributed in an apparently random three-dimensional network? As we have seen, an actin filament has a well-defined polarity, and myosin heads can bind and move along it only if they are oriented correctly with regard to its polarity. The small bipolar assemblies of nonmuscle myosin molecules (see Figure 11–26) may be able to generate some long-range order in the solution simply by pulling one set of actin filaments against another, even though the actin filaments and myosin assemblies are initially not well oriented (Figure 11–32).

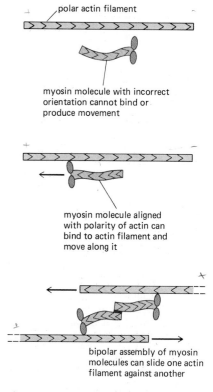

polar actin filament

myosin molecule with incorrect orientation cannot bind or produce movement

myosin molecule aligned with polarity of actin can bind to actin filament and move along it

bipolar assembly of myosin molecules can slide one actin filament against another

Figure 11–32 A bipolar assembly of nonmuscle myosin molecules (see Figure 11–26) produces a sliding of two actin filaments of opposite polarity, as in muscle. In this way myosin can cause contraction even in a randomly oriented network of actin filaments.

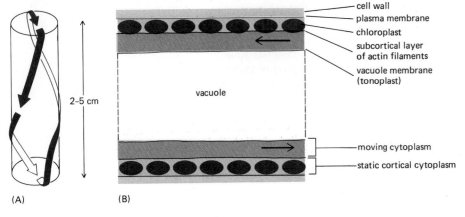

(A)   (B)

2–5 cm

cell wall
plasma membrane
chloroplast
subcortical layer
of actin filaments
vacuole membrane
(tonoplast)

vacuole

moving cytoplasm
static cortical cytoplasm

Figure 11–33 The pattern of streaming cytoplasm in the giant algal cell *Nitella*. (A) The path followed by the moving cytoplasm in the cylindrical cell. For clarity, the cell's diameter is exaggerated relative to its length. (B) A longitudinal section through part of such a cell, showing the arrangement of the static and moving layers of cytoplasm. The static cortical cytoplasm contains the chloroplasts, which are attached to underlying bundles of actin filaments. Internal to the actin filaments is the moving layer of cytoplasm containing the nuclei, mitochondria, and other organelles. The relative size of the vacuole is much larger than is shown here.

Cytoplasmic Streaming in Giant Algal Cells Depends on Actin and Myosin[21]

As explained in Chapter 20 (see p. 1170), plant cells, because of their central vacuole and stiff cell walls, can grow much larger than animal cells. Because diffusion is relatively ineffective over long distances, a very large cell requires extensive **cytoplasmic streaming** to ensure an effective mixing of its cytoplasm. In the giant multinucleated cells of the green algae *Chara* and *Nitella* (which are 2 to 5 cm in length), a protein molecule would typically take many weeks to diffuse from one end of the cell to the other. Not surprisingly, these cells provide the most dramatic examples of cytoplasmic streaming. A continuous ribbon of cytoplasm streams along a gentle helical path down one side of the cell and back across the other in an endless belt that surrounds a huge central vacuole (Figure 11–33A). The streaming cytoplasm moves internal membranes, mitochondria, nuclei, and cytosol around the cell at speeds of up to 75 μm/second.

The motility-generating system in these giant cells lies between the moving layer of cytoplasm and a monolayer of chloroplasts that is fixed in place just beneath the plasma membrane (Figure 11–33B). Electron microscopy reveals large bundles of actin filaments in this intervening region, all aligned with the same polarity. The polarity of the actin filaments is such that the directional movement of myosin molecules along them (from the minus to the plus end) could produce the observed cytoplasmic streaming. Moreover, if the giant *Nitella* cells are opened up to expose the network of actin bundles on the surface of the chloroplast layer and latex beads coated with myosin are then added to this preparation, the beads move in an ATP-dependent manner with characteristics similar to the movements of organelles in the intact cell. It seems very likely, therefore, that myosin mediates the cytoplasmic streaming.

Because continuous directional movement rather than contraction is involved, bipolar myosin filaments are not required for the type of cytoplasmic streaming found in *Nitella*. In addition to the filamentous nonmuscle myosin illustrated in Figure 11–26, most cells contain low-molecular-weight *minimyosins*. Minimyosins were first isolated from extracts of a large amoeba (*Acanthamoeba*). They consist of a single globular myosin head with only a short flexible tail, which, instead of interacting with a second tail to form a coiled coil, binds to membranes either directly or via a linker protein. Purified minimyosins will associate with membrane-bounded organelles *in vitro* and propel them along oriented bundles of actin filaments, and it seems likely that this is how many organelles are actively moved in cells.

The Cortex Is Organized by the Association of Actin Filaments with the Plasma Membrane

We have thus far ignored an issue that is central to understanding the structure and function of the actin cortex—the nature of the association of actin filaments with the plasma membrane. Specific proteins embedded in the membrane are thought to serve as organizing centers for the actin network. These or other plasma membrane proteins must transmit the forces generated by cortical actin filaments responsible for cell-surface movements. Little is known about these proteins and how they interact with actin. It is clear, however, that there are at least three functional types of actin attachments to the plasma membrane: those that primarily strengthen and give shape to the membrane, those that allow actin filaments to pull inward on the membrane, and those that allow actin filaments to move the membrane rapidly outward. We shall consider each in turn.

Membrane-attached Cytoskeletal Networks Provide Mechanical Support to the Plasma Membrane[22]

As discussed in Chapter 6 (see p. 289), the proteins **spectrin** and **ankyrin** were first discovered as prominent components of the membrane-associated cytoskeleton of mammalian red blood cells (erythrocytes). In these unusual cells the nucleus and internal membranes have been lost, and so the plasma membrane is the only membrane. It is supported by a two-dimensional network of spectrin tetramers, which are connected at their ends by very short actin filaments and linked to a transmembrane protein (band 3) by ankyrin bridges (see Figure 6–26, p. 289). Close relatives of both spectrin and ankyrin are found in the cortex of many vertebrate cells. It has been suggested that the unusually thin (about 20 nm) cytoskeletal network underlying the erythrocyte plasma membrane may develop from the thicker cortex in nucleated erythrocyte precursor cells by the progressive depolymerization of actin filaments (Figure 11–34). Thus the detailed arrangement of proteins in the erythrocyte cortex also provides a model for the actin-based cytoskeletal network that supports the plasma membrane in all other animal cells.

11-10 ## Microvilli Illustrate How Bundles of Cross-linked Actin Filaments Can Stabilize Local Extensions of the Plasma Membrane[23]

Microvilli are fingerlike extensions found on the surface of many animal cells. They are especially abundant on those epithelial cells that require a very large surface area to function efficiently. A single epithelial cell in the human small intestine, for example, has several thousand microvilli on its apical surface; each

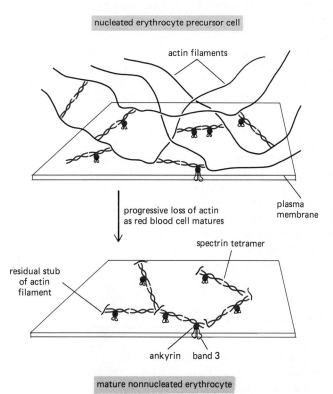

Figure 11–34 A hypothetical sequence by which the cytoskeleton underlying the membrane of a mammalian erythrocyte might develop from a more conventional actin-based cortex. The nucleated erythrocyte precursor cell is postulated to have a membrane-associated cortex consisting of actin filaments held together by spectrin tetramers and other cross-linking proteins. When the precursor cell matures into an erythrocyte, most of the actin is lost. As a result, the actin filaments depolymerize and most of the cross-linking proteins dissociate, leaving only a thin two-dimensional network of spectrin and short stubs of actin filaments bound to the plasma membrane.

Figure 11–35 A bundle of parallel actin filaments, held together by actin-bundling proteins, forms the core of a microvillus. Lateral arms connect the sides of the actin filament bundle to the overlying plasma membrane. The plus ends of the actin filaments are all at the tip of the microvillus, where they are embedded in an amorphous, densely staining substance of unknown composition.

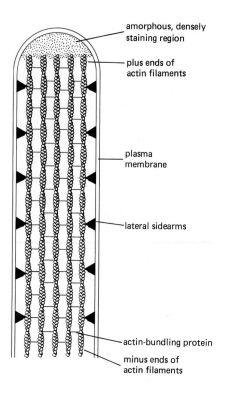

amorphous, densely staining region

plus ends of actin filaments

plasma membrane

lateral sidearms

actin-bundling protein

minus ends of actin filaments

is about 0.08 μm wide and 1 μm long, making the cell's absorptive surface area 20 times greater than it would be without them. The plasma membrane that covers these microvilli is highly specialized, containing a thick extracellular coat of polysaccharide and digestive enzymes.

The core of each intestinal microvillus contains a rigid bundle of 20 to 30 parallel actin filaments that extends from the tip of the microvillus down into the cell cortex. The actin filaments in the bundle are all oriented with their plus ends pointing away from the cell body and are held together at regular intervals by several *actin-bundling proteins*, including *fimbrin* and *fascin* (Figure 11–35). In contrast to filamin and other flexible actin-cross-linking proteins that tie actin filaments into loose networks (see p. 630), these actin-bundling proteins are relatively small, compact molecules with two distinct actin-binding sites on a single polypeptide chain. Consequently, they tie actin filaments into tight bundles, with adjacent actin filaments held rigidly about 10 nm apart in parallel arrays.

The lower portion of the actin filament bundle in the microvillus core is anchored in the specialized cortex at the apex of the intestinal epithelial cell. This cortex, known as the *terminal web*, contains a dense network of spectrin molecules that overlies a layer of intermediate filaments (Figure 11–36); it is thought to stiffen

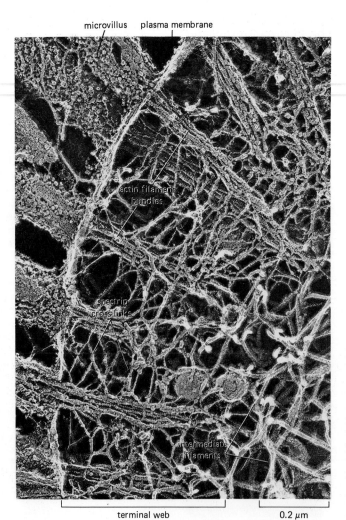

microvillus   plasma membrane

actin filament bundles

spectrin cross-links

intermediate filaments

terminal web   0.2 μm

Figure 11–36 Freeze-etch electron micrograph of an intestinal epithelial cell, showing the terminal web beneath the apical plasma membrane. Bundles of actin filaments forming the core of microvilli extend into the terminal web, where they are linked together primarily by spectrin. Beneath the terminal web is a layer of intermediate filaments. (From N. Hirokawa and J.E. Heuser, *J. Cell Biol.* 91:399–409, 1981. Reproduced by permission of the Rockefeller University Press.)

the microvilli above it, keeping their actin bundles projecting outward at a right angle to the apical cell surface.

How is the bundle of actin filaments at the core of a microvillus attached to the overlying plasma membrane? Electron microscopy reveals two distinct regions of membrane association: a helical "staircase" of lateral arms extends from the sides of the actin filament bundle to make contact with the plasma membrane, and an amorphous "cap" of densely staining material connects the plus ends of the actin filaments to the plasma membrane at the tip of the microvillus (see Figure 11–35).

If the plasma membrane is removed from such an epithelial cell with a non-ionic detergent, the lateral arms remain attached to the denuded cytoskeleton. They can, however, be removed by subsequent exposure to ATP; each arm is then found to consist of a minimyosin molecule tightly bound to the calcium-binding protein calmodulin. The minimyosin binds with its ATP-dependent head region to actin filaments in the microvillus core and by its truncated tail region to the plasma membrane. Why a motility-generating protein is used to form this linkage is not known; but since myosin molecules move toward the plus ends of actin filaments, those in the microvillus would be expected to carry membrane components toward the microvillus tip. Perhaps this movement assists in the continual release ("sloughing") of plasma membrane from the microvillus into the lumen of the intestine, where the digestive enzymes associated with the plasma membrane continue their action.

Less is known about the amorphous, densely staining material at the tip of a microvillus than about the lateral arms. As we shall see, there are reasons to suspect that proteins in this region control the length and diameter of the actin filament bundle and thereby determine the size of the microvillus (see p. 676). If so, the mode of actin filament insertion into the membrane at the microvillus tip is likely to be complex.

Focal Contacts Allow Actin Filaments to Pull Against the Substratum[24]

Bundles of actin filaments often bind to the plasma membrane in a way that allows them to pull on the extracellular matrix or on another cell. We have already mentioned such attachment sites for the ends of actin filament bundles in fibroblasts and smooth muscle cells (see p. 627 and p. 625). Attachments of this type are mediated by transmembrane linker glycoproteins in the plasma membrane. Those formed by cultured fibroblasts with the extracellular matrix are the best characterized. When fibroblasts grow on a culture dish, most of their cell surface is separated from the substratum by a gap of more than 50 nm. In certain regions called **focal contacts,** or *adhesion plaques,* however, this gap is reduced to 10 to 15 nm. These regions appear as dark areas in an interference reflection microscope, in which only the light reflected from the under surface of the cell is collected. They can be shown by staining with anti-actin antibodies to be the sites where the ends of stress fibers attach to the plasma membrane (Figure 11–37).

Figure 11–37 The relation between focal contacts and stress fibers in cultured fibroblasts. Focal contacts are best seen in living cells by reflection-interference microscopy (A). In this technique, light is reflected from the lower surface of a cell attached to a glass slide, and the focal contacts appear as dark patches. (B) Staining of the same cell (after fixation) with antibodies to actin shows that most of the cell's actin-filament bundles (or stress fibers) terminate at or close to a focal contact. (Courtesy of Grenham Ireland.)

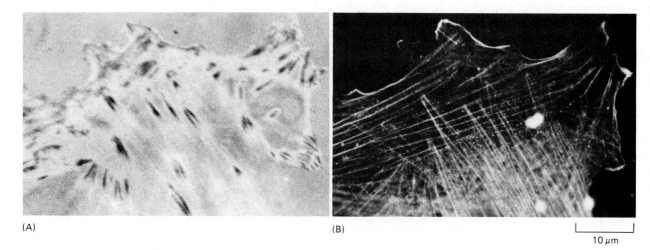

(A)   (B)

10 μm

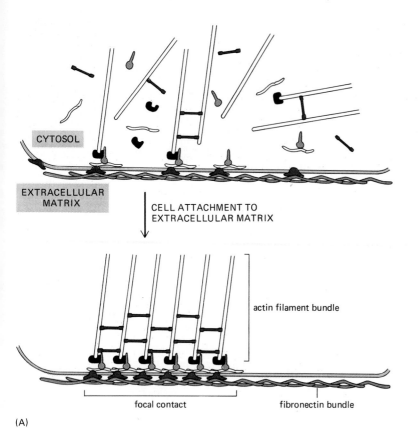

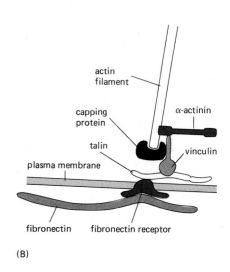

(B)

(A)

The cell cortex at the site of a focal contact is attached by transmembrane linker proteins to components of the extracellular matrix, such as *fibronectin* (see p. 816), that have become adsorbed to the dish. The prototype transmembrane linker protein of this type is the *fibronectin receptor*, a two-chain glycoprotein of the *integrin* family (see p. 821), whose external domain binds to fibronectin while the cytoplasmic domain is linked to actin filaments in stress fibers (Figure 11–38A). The linkage is indirect and is mediated by at least four *attachment proteins*, including *talin* and *vinculin*. Talin binds both to the cytoplasmic domain of the fibronectin receptor and to vinculin. Neither protein, however, binds directly to actin filaments. This function is presumably served by other attachment proteins, including an actin-filament-capping protein, which binds to the plus end of an actin filament, and α-actinin, which in muscle is attached near the plus ends of the actin filaments at the Z disc (see p. 624). A possible arrangement of the various proteins at a focal contact is shown in Figure 11–38B.

Focal contacts are closely related to the actin-filament attachment sites in the plasma membrane of a smooth muscle cell (see p. 625). They are more distantly related to *adhesion belts*, which help join epithelial cells into sheets by means of contractile bundles of actin filaments that interact across adjacent plasma membranes (see p. 796). These cell-cell junctions contain vinculin and α-actinin but not talin, suggesting a somewhat different mode of attachment of actin filaments to the plasma membrane than in focal contacts.

So far we have discussed examples of actin filaments attaching to the plasma membrane where the filaments have a structural role (as in a microvillus) or where they can pull on the membrane (as at a focal contact). In these cases, relatively stable arrangements of the actin filaments give rise to persistent structures. Many rapid changes of shape at the cell surface, however, such as those that occur during cell movement, depend on the transient and regulated polymerization of actin. Before considering the rapid outward movements of the plasma membrane that are driven by actin polymerization, we need to digress briefly to examine what has been learned about how actin molecules polymerize *in vitro*.

Figure 11–38 A model for how transmembrane linker glycoproteins in the plasma membrane connect intracellular actin filaments to the extracellular matrix at a focal contact. The formation of a focal contact occurs when the binding of matrix glycoproteins on the outside of the cell causes the transmembrane linker glycoproteins to cluster at the contact site, as illustrated schematically in (A). A possible arrangement of the intracellular attachment proteins that mediate the linkage between a transmembrane linker glycoprotein (such as a fibronectin receptor) and actin filaments is shown in (B).

11-11 An Actin Filament Grows Mainly at Its Plus End[25]

We have already noted that actin molecules in a test tube can assemble spontaneously into filaments in the presence of ATP. Filament formation is typically initiated by increasing the ionic strength of the actin solution, and it is most easily monitored by measuring the resulting large increase either in light scattering or in the fluorescence of a covalently bound dye.

Polymerization begins after a lag phase, which reflects a slow and difficult initial step in which three actin molecules must come together in a specific geometric configuration (Figure 11–39). This step is known as *nucleation*. Once nucleation has been achieved, addition of further actin molecules to the two ends of the filament proceeds rapidly. The assembly reaction is reversible, and eventually the concentration of actin monomers falls to a point where the monomers add to the filament at the same rate as they dissociate; this concentration is known as the *critical concentration*. If the actin solution is now suddenly diluted so that the monomer concentration falls below the critical concentration, actin filaments begin to depolymerize, and they continue to do so until the critical concentration of monomers is restored. As we shall see, both the nucleation and the polymerization of actin in cells are controlled by various actin-binding proteins, which thereby determine both the position and length of actin filaments.

An additional feature of actin polymerization explains the directionality of the filament growth observed when actin polymerization generates cell movements. We have seen that actin filaments have a structural polarity. An important consequence of this polarity is that the kinetics of polymerization are different at the two ends of the filament. The difference can be detected by decorating short lengths of actin filaments with myosin head fragments to mark the polarity of the filaments before exposing them to actin monomers under polymerizing conditions. If the growing actin filaments are then fixed after a short period and examined in the electron microscope, the *plus end* of the filament is found to have grown much more than the *minus end* (Figure 11–40). By carrying out a series of experiments of this kind, it is possible to measure the rates of growth at the two ends as a function of the concentration of free actin monomers. Under ionic conditions equivalent to those in the cell, the growth of purified actin filaments is 5 to 10 times faster at the plus end than at the minus end. It appears that the growth of actin filaments *in vivo* is almost always at the plus end; therefore, by anchoring the plus end of a filament in a specific orientation, the cell can determine the rate and direction of growth of the filament.

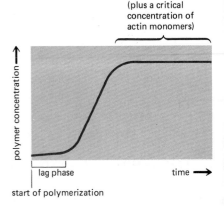

Figure 11–39 Time course of actin polymerization *in vitro*. Polymerization is normally initiated by increasing the ionic strength of the actin solution and usually proceeds after an initial lag phase.

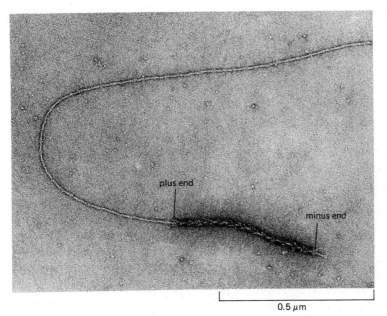

Figure 11–40 Asymmetry of actin filament growth. If a "seed filament" decorated with myosin heads is used to nucleate actin polymerization, assembly occurs much more rapidly at the plus end (also known as the "barbed end") of the seed filament than at its minus end (or "pointed end"). (Electron micrograph courtesy of M.S. Runge and T.D. Pollard.)

2 Actin Filaments Can Undergo a "Treadmilling" of Subunits[25]

In vitro experiments like that shown in Figure 11–40 indicate that the actin concentration at which filament growth ceases—the critical concentration—is different at the plus and minus ends of the filament. Consequently, when actin polymerizes in a test tube, a steady state is reached at which actin monomers dissociate predominantly at the minus end and associate predominantly at the plus end. At this steady state the rate of subunit addition at the plus end equals the rate of loss from the minus end, so that a constant concentration of the actin monomer is maintained. Even though the net length of the polymer does not change, individual actin molecules are continually transferred from one end of the filament to the other, a process called **treadmilling** (Figure 11–41).

The treadmilling of actin filaments requires energy, since otherwise it could be used to create a perpetual motion machine that does work without an input of energy, violating the laws of thermodynamics. The energy in this case is provided by ATP hydrolysis. Each actin monomer binds a molecule of ATP, which is hydrolyzed shortly after the monomer is added to a filament. As explained in Panel 11–1, this hydrolysis of ATP makes treadmilling possible. As we shall see shortly, treadmilling may be one mechanism by which actin filaments and components associated with them generate movements in cells (see p. 640).

Many Cells Extend Dynamic Actin-containing Microspikes and Lamellipodia from Their Surface[26]

Dynamic surface extensions containing actin filaments are a common feature of animal cells, especially when the cells are migrating or changing shape. Cells in culture, for example, often extend many thin, stiff protrusions called **microspikes,** which are about 0.1 μm wide and 5 to 10 μm long and contain a loose bundle of about 20 actin filaments oriented with their plus ends pointing outward. The growing tip (growth cone) of a developing nerve cell axon extends even longer microspikes, called *filopodia,* which can be up to 50 μm long (see p. 1115). These extensions are motile structures that can form and retract with great speed. It is thought that they act as feelers by which cells explore their environment: those microspikes that adhere strongly to their surroundings act to steer the cell toward the adhesive region; those that fail to adhere are carried backward over the upper surface of the cell and are retracted.

In addition to microspikes, crawling cells and growth cones periodically extend thin sheetlike processes known as **lamellipodia** from their advancing edge (the *leading edge*). Like microspikes, some of these adhere to the substratum, while others fail to adhere, curl back over the top of the cell, and are swept backward as a "ruffle" (Figure 11–42).

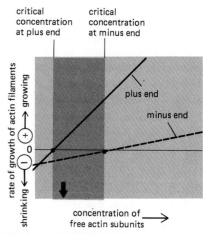

Figure 11–41 The rate of growth observed at the two ends of an actin filament at different concentrations of free actin molecules. This graph shows that the two ends have different growth rates and different critical concentrations. As a consequence, there is a range of free actin concentrations (*dark color*) in which the plus end of the actin filament is polymerizing while the minus end is depolymerizing. When free actin is in apparent equilibrium with actin filaments *in vitro*, no net growth occurs because the rate at which actin molecules are coming off the minus end is equal to the rate at which they are adding to the plus end. At this concentration of free actin subunits (intermediate between the critical concentrations of the two ends—*colored arrow*), actin filaments do not change their length, but individual actin molecules continually move from one end of the polymer to the other in a process called treadmilling (see Panel 11–1).

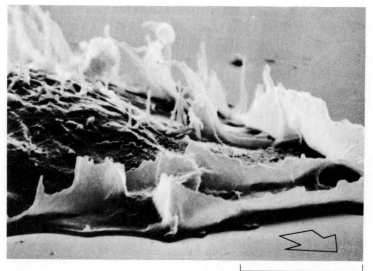

5 μm

Figure 11–42 Scanning electron micrograph of lamellipodia and microspikes at the leading edge of a human fibroblast migrating in culture. The arrow shows the direction of cell movement. As the cell moves forward, lamellipodia and microspikes sweep backward over its dorsal surface—a movement known as ruffling. (Courtesy of Julian Heath.)

DIFFERENT GROWTH RATES AT THE PLUS AND MINUS ENDS

The assembly (polymerization) and disassembly (depolymerization) of actin filaments occurs by the addition and removal of actin subunits from the filament ends. During assembly, one end of the filament grows faster than the other. The fast-growing end is called the plus end, while the slow-growing end is called the minus end. The difference in the rates of growth at the two ends is made possible by changes in the conformation of each subunit as it enters the polymer
($\square \rightarrow \square\rangle$).
free subunit   subunit in polymer

This conformational change affects subunit addition at one end preferentially.

SLOW ADDITION TO MINUS END

FAST ADDITION TO PLUS END

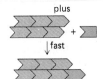

THE CRITICAL CONCENTRATION OF SUBUNITS

The number of subunits that add to the polymer per second will be proportional to the concentration of free subunit ($k_{on}[C]$), but the subunits will leave the polymer end at a constant rate (k_{off}) that does not depend on [C]. As the polymer grows, subunits are used up and [C] drops until it reaches a constant value, called the critical concentration (C_c), at which the rate of subunit addition equals the rate of subunit loss.

At this equilibrium point,

$$k_{on}[C_c] = k_{off}$$

so that

critical conc. $= C_c = \dfrac{k_{off}}{k_{on}} = \dfrac{1}{K}$	(where **K** is the equilibrium constant for subunit addition)

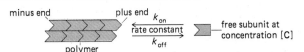

Even though k_{on} and k_{off} will have different values for the plus and minus ends of the polymer, their ratio, k_{off}/k_{on}—and hence C_c—must be the same at both ends. This is because exactly the same subunit interactions are broken when a subunit is lost at either end, and the final state of the subunit after dissociation is identical. Therefore the ΔG for subunit loss, which determines the equilibrium constant for its association with the end (see Table 3–3, p. 95), is identical at both ends: if the plus end grows four times faster than the minus end, it must also shrink four times faster.

Thus: for [C] > C_c both ends grow
          for [C] < C_c both ends shrink   (see Figure 11–41)

ATP HYDROLYSIS CHANGES THE EQUILIBRIUM CONSTANT AT EACH END

Each actin molecule carries a tightly bound ATP molecule that is hydrolyzed to a tightly bound ADP molecule soon after the conformational change occurs that accompanies subunit assembly into the polymer.

free subunit	subunit in polymer	subunit in polymer
actin (ATP)	$\rightarrow$ actin (ATP)	$\rightarrow$ actin (ADP) + P$_i$

These changes will be abbreviated as
$\square_T \rightarrow \square_T\rangle \rightarrow \square_D\rangle$

ATP hydrolysis reduces the binding affinity of the subunit for neighboring subunits and makes it more likely to dissociate from each end of the filament. It is usually the $\square_T\rangle$ form that adds to the filament and the $\square_D\rangle$ form that leaves (changing to $\square_D$ in solution), so that the reaction that takes place when the

subunit leaves the polymer is no longer the exact reverse of the reaction that occurs on assembly.

Considering events at the plus end only:

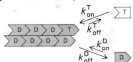

As before, the polymer will grow until [C] = C_c. For illustrative purposes, we can ignore k^D_{on} and k^T_{off} as insignificant, so that polymer growth ceases when

$$k^T_{on}[C_c] = k^D_{off} \qquad \text{or} \qquad C_c = \frac{k^D_{off}}{k^T_{on}} = \frac{1}{K}$$

This is a steady state and not a true equilibrium because the ATP that is hydrolyzed must be replenished by a nucleotide-exchange reaction on the free subunit ($\square_D \rightarrow \square_T$).

TREADMILLING

Since k^D_{off} and k^T_{on} refer to different reactions, the ratio k^D_{off}/k^T_{on} need not be the same at the two ends of the polymer. Thus the ATP hydrolysis is found to create a different critical concentration at each end, with

$$C_c \,(\textit{minus end}) > C_c \,(\textit{plus end})$$

Assuming that both ends of the polymer are exposed, polymerization will proceed until [C] reaches a value that is above C_c for the plus end but below C_c for the minus end (see Figure 11–41). At this steady state, subunits will assemble at the plus end and disassemble at the minus end at an identical rate, so that the polymer maintains a constant length even though there is a net flux of subunits through the polymer known as treadmilling:

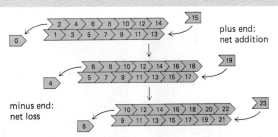

ATP hydrolysis is not required for filament assembly. Actin subunits that are bound to a nonhydrolyzable analogue of ATP will polymerize under appropriate conditions. The filaments formed in these cases, however, will not treadmill; treadmilling is made possible by the ATP hydrolysis that accompanies polymerization.

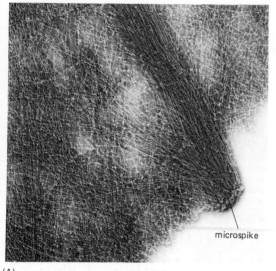

(A)

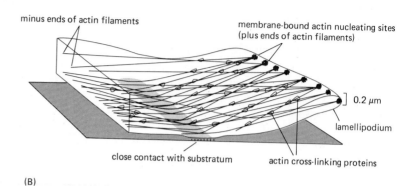

minus ends of actin filaments

membrane-bound actin nucleating sites
(plus ends of actin filaments)

0.2 μm

lamellipodium

close contact with substratum

actin cross-linking proteins

(B)

microspike

A lamellipodium can be viewed as a two-dimensional version of a microspike; indeed, short microspikes often delineate its edge. When carefully fixed and stained for examination in an electron microscope, the actin filaments in the lamellipodium of a moving cell appear to be more organized than they are in other regions of the cell cortex; the filaments project outward in an orderly array, with their plus ends inserted into the leading edge of the plasma membrane (Figure 11–43).

Studies on moving cells in culture indicate that actin molecules undergo treadmilling along actin filaments at a cell's leading edge at much faster rates than they do in actin filaments that form in a test tube. Movements of actin molecules in cells can be detected by fluorescence bleaching experiments. When fluorescently labeled actin monomers are injected into a moving cell, they are quickly incorporated into all of the actin filaments. And when a laser beam is used to destroy the fluorescent label in a small spot in the thin leading edge of the cell, the spot moves steadily backward from the margin of the cell at a speed of about 0.8 μm/minute, suggesting that there is a rapid treadmilling of actin monomers along the actin filaments in both the lamellipodium and the microspikes (Figure 11–44). The monomers are apparently added continuously to the plus ends at the plasma membrane and removed at the minus ends in the cell interior.

When a spot on an actin filament is similarly bleached elsewhere in the cell, it recovers its fluorescence within a few minutes without seeming to move. Apparently there is something special about the behavior or organization of the actin at the leading edge. It has been suggested that the continuous polymerization of the actin filaments just beneath the membrane can be harnessed to extend the leading edge and thereby help move the cell forward (see p. 671). A more dramatic case where actin polymerization is thought to help extend the leading edge of a cell occurs in some invertebrate sperm.

Figure 11–43 Actin filaments at the leading edge of a fibroblast in culture. (A) Wholemount electron micrograph of the leading edge of a cultured cell that has been extracted with nonionic detergent to remove the plasma membrane and most of the soluble proteins. Note the oriented network of actin filaments in the lamellipodium, in which a microspike is embedded. A schematic view of the actin filaments in the lamellipodium is shown in (B). (Electron micrograph from J.V. Small, *J. Cell Biol.* 91:695–705, 1981. Reproduced by permission of the Rockefeller University Press.)

Figure 11–44 Actin treadmilling in a cultured fibroblast. Fluorescent actin molecules were microinjected into the cell, where they became incorporated into actin filaments. A small spot on the actin filaments at the leading edge of the cell was bleached with a laser beam. The cell was then photographed at intervals using a fluorescence microscope equipped with an image intensifier (see p. 142). The rapid backward movement of the bleached spot suggests that actin molecules are continuously treadmilling along the actin filaments away from the leading edge. (Courtesy of Y.L. Wang.)

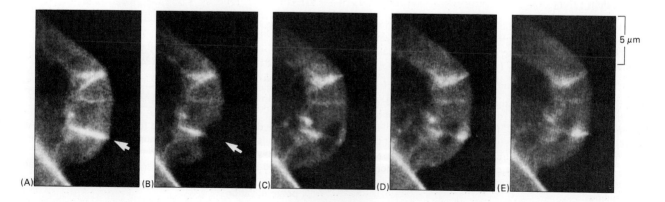

5 μm

(A)   (B)   (C)   (D)   (E)

11-13 An Explosive Polymerization of Actin Helps Drive the Formation of an Acrosomal Process in Some Invertebrate Sperm[27]

Although the protrusion of many cell-surface extensions are thought to depend on the polymerization of actin filaments, very little is known about how the process is initiated. An exception is the *acrosomal reaction* in some invertebrate sperm (see p. 868). The acrosomal region of a sea cucumber sperm head, for example, is packed full of unpolymerized actin bound to a small protein known as **profilin,** which is abundant in most animal cells. By forming a one-to-one complex with actin molecules, profilin blocks the nucleation step in the spontaneous polymerization of new actin filaments without having a major effect on polymerization at the plus end of a preformed actin filament. When the sperm makes contact with the outer coat of a sea cucumber egg, the intracellular pH rises, allowing actin molecules to dissociate from profilin. An explosive polymerization of actin ensues, which helps drive the rapid formation of a long, thin, membrane-covered *acrosomal process*, which punctures the egg coat like a harpoon, enabling the sperm and egg membranes to fuse (Figure 11–45). In addition to the polymerization of actin, an influx of water into the sperm head creates a hydrostatic pressure that is thought to help extend the acrosomal process.

The formation of the acrosomal process suggests that actin polymerization can rapidly push out the plasma membrane without disrupting membrane continuity. While it is unlikely that microspikes and lamellipodia are formed by changes in the available actin monomer pool in this way, they probably form similarly by local actin polymerization at the plasma membrane.

Actin Assembly Is Controlled from the Plasma Membrane[28]

Actin-containing extensions of the cell surface must be regulated so that they form only when and where they are needed. A macrophage, for example, extends pseudopods to surround a bacterium only in the region of cell contact (see p. 335), and the actin polymerization that helps drive the extension of the acrosomal process in an activated sea cucumber sperm occurs only when the head of the sperm encounters the jelly layer surrounding the egg.

The timing and location of such actin-mediated responses seem to be controlled by the regulation of nucleation sites for actin filament growth at the plasma membrane. This can be seen in the chemotactic response of neutrophils. In response to a local bacterial infection, these phagocytic white blood cells migrate in large numbers through the walls of capillaries and crawl through the surrounding tissues until they reach the site of infection. They are guided to their target by diffusible molecules released at the site of infection. In this process, known as *chemotaxis*, the cell detects a gradient of a chemical substance and responds to it by changing its direction of movement (see p. 720). Neutrophils have receptor proteins on their surface that enable them to detect very low concentrations ($\sim 10^{-10}$ M) of specific chemoattractants. Under optimal conditions they can detect a 1% difference in the concentration of a chemoattractant peptide on one side of the cell compared to the other and reorient their leading edge in order to migrate up the gradient.

A clue to the link between receptor binding and reorientation is obtained by studying the effects of exposure to a sudden increase in the concentration of a chemoattractant peptide: the neutrophil forms microspikes and lamellipodia over its entire surface, while the proportion of actin that is polymerized in the cell rises from 30% to close to 60% in 20 seconds. This sudden polymerization of actin appears to be due to the generation of nucleating sites in the plasma membrane: when cells are treated with the chemoattractant and then lysed, they are found to have an increased ability to nucleate the growth of actin filaments *in vitro;* moreover, each activated cell-surface receptor induces many actin filaments to grow.

We shall return to the part played by controlled actin polymerization in the directed migration of animal cells later, when we discuss cell locomotion as an integrated response of the whole cell (see p. 669).

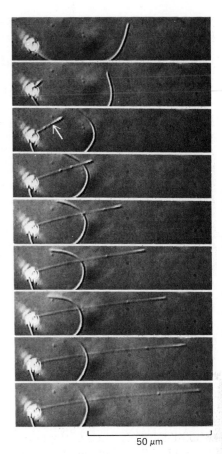

50 μm

Figure 11–45 Light micrographs of sequential stages in the elongation of the acrosomal process of a sea cucumber sperm. The photographs were taken at intervals of 0.75 second, beginning 2 seconds after the sperm was artificially activated to undergo the acrosomal reaction. The arc to the right of the sperm head is a portion of its tail that has curved around out of the field of the micrograph. (From L.G. Tilney and S. Inoué, *J. Cell Biol.* 93:820–827, 1982. Reproduced by permission of the Rockefeller University Press.)

Specific Drugs Change the State of Actin Polymerization and Thereby Affect Cell Behavior[29]

One would expect that many of the movements produced by the cell cortex, such as phagocytosis and cell locomotion, would depend on the dynamic equilibrium between actin molecules and actin filaments. But compared to the explosive changes that take place in activated sperm, changes in actin polymerization during the course of such movements are usually small and transient and therefore difficult to detect. The importance of actin polymerization and depolymerization in these cell movements is indicated by the effects of drugs that prevent changes in the the state of actin polymerization and thereby disrupt such movements. For example, the **cytochalasins** (Figure 11–46), a family of metabolites excreted by various molds, paralyze many kinds of vertebrate cell movement—including cell locomotion, phagocytosis, cytokinesis, the production of microspikes and lamellipodia, and the folding of epithelial sheets into tubes. These drugs do not inhibit chromosome separation on the mitotic spindle, which depends principally on microtubules; or muscle contraction, which depends on stable actin filaments that do not undergo assembly and disassembly. The principal action of the cytochalasins is to bind specifically to the fast-growing plus ends of actin filaments, preventing the addition of actin molecules there.

Phalloidin is a highly poisonous alkaloid produced by the toadstool *Amanita phalloides*. By contrast with the cytochalasins, it stabilizes actin filaments and inhibits their depolymerization. Also unlike the cytochalasins, it does not readily cross the plasma membrane and therefore must be injected into a cell in order

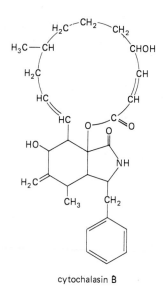

cytochalasin B

Figure 11–46 Chemical structure of cytochalasin B.

Figure 11–47 Some major classes of actin-binding proteins that are found in most vertebrate cells.

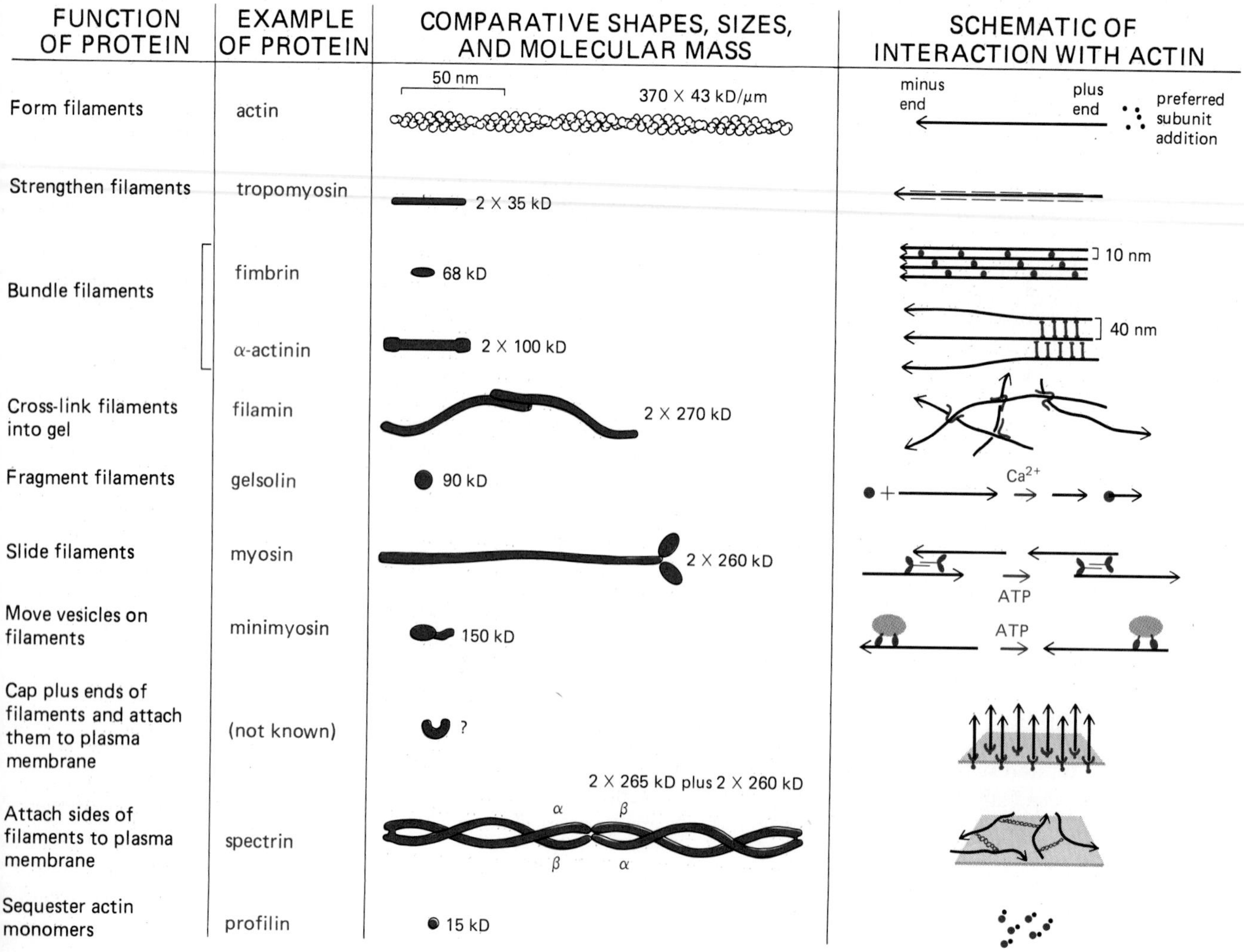

FUNCTION OF PROTEIN	EXAMPLE OF PROTEIN	COMPARATIVE SHAPES, SIZES, AND MOLECULAR MASS	SCHEMATIC OF INTERACTION WITH ACTIN
Form filaments	actin	50 nm / 370 × 43 kD/μm	minus end → plus end / preferred subunit addition
Strengthen filaments	tropomyosin	2 × 35 kD	
Bundle filaments	fimbrin	68 kD	10 nm
	α-actinin	2 × 100 kD	40 nm
Cross-link filaments into gel	filamin	2 × 270 kD	
Fragment filaments	gelsolin	90 kD	Ca²⁺
Slide filaments	myosin	2 × 260 kD	ATP
Move vesicles on filaments	minimyosin	150 kD	ATP
Cap plus ends of filaments and attach them to plasma membrane	(not known)	?	
Attach sides of filaments to plasma membrane	spectrin	2 × 265 kD plus 2 × 260 kD α β β α	
Sequester actin monomers	profilin	15 kD	

to act efficiently. When this is done, phalloidin blocks the migration of both amoeba and vertebrate cells in culture, suggesting that the dynamic assembly and disassembly of actin filaments is crucial for these movements. In addition, because phalloidin binds specifically to actin filaments, fluorescent derivatives of the drug are often used instead of anti-actin antibodies to stain actin filaments in cells.

The Behavior of the Cell Cortex Depends on a Balance of Cooperative and Competitive Interactions Among a Large Set of Actin-binding Proteins

Much more is known about actin-binding proteins than about the proteins that associate with microtubules or with intermediate filaments, the other two major classes of cytoskeletal filaments. A summary of the actin-binding proteins in vertebrate nonmuscle cells that we have discussed in this chapter is presented in Figure 11–47. The list is far from complete. Each of the functional categories indicated is thought to have several members, each with slightly different properties. Moreover, there are some actin-binding proteins with no obvious function, and there are undoubtedly other actin-binding proteins that have not yet been identified. In particular, cells must have mechanisms by which they can distinguish the two ends of an actin filament and control its orientation and location in the cytoplasm. These mechanisms presumably involve *"capping proteins"* that bind selectively to the plus end of an actin filament, hold it close to the plasma membrane, and regulate the addition of actin monomers. In contrast, since the minus ends of actin filaments are relatively inactive with respect to both polymerization and depolymerization, it is possible that these ends are often left free. Much remains to be learned about the nature of the membrane-bound organizing centers for actin: virtually nothing is known, for example, about the molecular nature of the densely staining material at the tip of a microvillus, which is presumably responsible for organizing the actin core and controlling the growth and regeneration of the microvillus (see p. 634).

Even after all the components of the actin filament network have been defined, there will remain the difficult task of deciphering the consequences of their many interactions. The proteins shown in Figure 11–47 do not bind simultaneously or randomly to actin filaments; they cooperate and compete to produce ordered patterns of interactions that have profound consequences for the cell (Figure 11–48). Moreover, the interactions among the components of the network are modulated by local changes in the concentration of ions in the cytosol and by physical forces that stretch or compress the network and hence move its constituent molecules in relation to one another. It is not suprising, therefore, that we are only beginning to understand this crucial part of the cytoskeleton.

Figure 11–48 Some examples of competitive and cooperative interactions between actin-binding proteins. Tropomyosin and filamin both bind strongly to actin filaments, but their binding is competitive. Because tropomyosin binds cooperatively to actin filaments, either tropomyosin or filamin will predominate over large regions of the actin filament network. Other actin-binding proteins, such as α-actinin or myosin, will be excluded from specific sites by a competitive interaction; thus, for example, α-actinin binds all along pure actin filaments *in vitro*, but it binds relatively weakly to actin filaments in cells, where it is largely confined to sites near the plus ends because of competition with other proteins. Alternatively, binding can be enhanced through cooperative interaction; thus tropomyosin appears to enhance the binding of myosin to actin filaments. Multiple interactions of these types between the actin-binding proteins in Figure 11–47 (and others) are thought to be responsible for the complex variety of actin networks found in all eucaryotic cells.

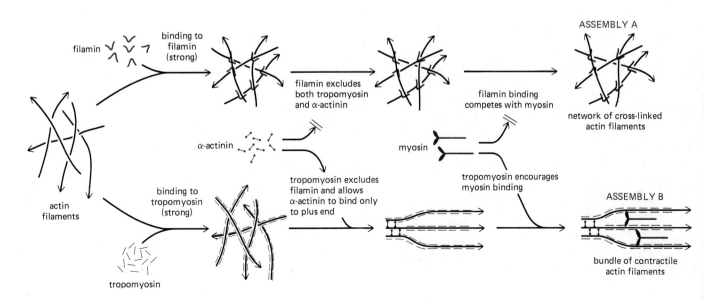

Summary

Eucaryotic cells possess a distinct cortical layer of actin filaments under their plasma membrane. Generally this layer is cross-linked into a uniform three-dimensional network with the physical properties of a gel. The cortical actin filaments also form local specialized structures, however. Bundles of actin filaments complexed with myosin, for example, attach to the plasma membrane to provide the cell with musclelike assemblies capable of contraction. At other sites, controlled polymerization at the plus ends of actin filaments can push out the plasma membrane, creating dynamic surface protrusions.

The varied forms and functions of the cortex depend on a versatile repertoire of actin-binding proteins that cross-link actin filaments into loose gels, tie them into stiff bundles, move along them with the generation of force, or attach them to the plasma membrane. Some of the latter proteins cap the plus ends of the filaments and thereby control the polymerization and breakdown of actin filaments in the cell. These proteins are thought to play a vital part in the complex movements of the cell surface, such as those involved in cell crawling and phagocytosis.

Ciliary Movement[30]

Next to muscle contraction, the best-understood cellular movement is ciliary beating. **Cilia** are tiny hairlike appendages about 0.25 μm in diameter that are constructed from *microtubules*, the second of the three major cytoskeletal filaments. Cilia extend from the surface of many kinds of cells and are found in most animal species and some lower plants. Their primary function is to move fluid over the surface of the cell or to propel single cells through a fluid. Protozoa, for example, use cilia both to collect food particles and for locomotion. On the epithelial cells lining the human respiratory tract, huge numbers of cilia (10^9/cm^2 or more) sweep layers of mucus, together with trapped particles of dust and dead cells, up toward the mouth, where they are swallowed and eliminated. Cilia also help to sweep ova along the oviduct, and a related structure, the flagellum, propels human sperm.

Just as the study of muscle contraction contributes to our understanding of actin- and myosin-based movements in nonmuscle cells, so our knowledge of how cilia beat provides an important insight into how arrays of microtubules might produce the other types of movement they are responsible for, such as intracellular transport and mitosis.

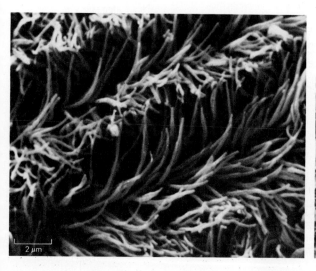

Figure 11–49 Scanning electron micrographs of a field of cilia in the gut of a marine worm. Although the cilia are uniformly distributed, the beat of adjacent cilia is coordinated, giving rise to a series of unidirectional waves. These waves can be seen most clearly at a lower magnification (*right*). (From J.S. Mellor and J.S. Hyams, *Micron* 9:91–94, 1978.)

Figure 11–50 The contrasting motions of beating cilia and flagella. (A) The beat of a cilium such as that on an epithelial cell from the human respiratory tract resembles the breast stroke in swimming. The power stroke (stages 1 and 2), in which fluid is driven over the surface of the cell, is followed by a distinct recovery stroke (stages 3, 4, and 5). Each cycle typically requires 0.1 to 0.2 second and generates a force perpendicular to the axis of the axoneme. For comparison, the wavelike movements of the flagellum of a sperm cell from a tunicate are shown in (B). The cell was photographed on moving film with stroboscopic illumination at 400 flashes per second. Note that waves of constant amplitude move continuously from the base to the tip of a flagellum. The cell is thereby pushed directly away from its axoneme, a distinctly different effect than that caused by a cilium. (B, courtesy of C.J. Brokaw.)

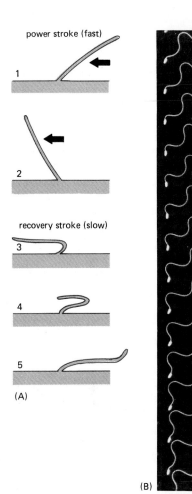

(A)

(B)

Cilia and Flagella Propagate Bending Movements

Fields of cilia bend in coordinated unidirectional waves (Figure 11–49). Each cilium moves with a whiplike motion: a forward active stroke, in which the cilium is fully extended and beating against the surrounding liquid, is followed by a recovery phase, in which the cilium returns to its original position with an unrolling movement that minimizes viscous drag (Figure 11–50A). The cycles of adjacent cilia are almost but not quite in synchrony, creating the wavelike patterns that can be seen in fields of beating cilia under the microscope.

The simple **flagella** of sperm and of many protozoa are much like cilia in their internal structure, but they are usually very much longer. Further, instead of making whiplike movements, they propagate quasi-sinusoidal waves (Figure 11–50B). Nevertheless, the molecular basis for their movement is the same as that in cilia. The flagella of bacteria, on the other hand, are completely different from the cilia and flagella of eucaryotic cells, as described on page 720.

The Core of a Cilium Contains a Bundle of Parallel Microtubules in a "9 + 2" Arrangement

The movement of a cilium or a flagellum is produced by the bending of the core, or **axoneme**—a complex structure composed entirely of microtubules and their associated proteins. Microtubules are normally long hollow tubes of protein with an outer diameter of 25 nm (see below). In the axoneme they are modified and arranged in a pattern whose curious and distinctive appearance was one of the most striking revelations of early electron microscopy: nine special doublet microtubules are arranged in a ring around a pair of single microtubules (Figure 11–51). This "9 + 2" array is characteristic of almost all forms of cilia and eucaryotic flagella, from those of protozoa to those found in humans. The microtubules extend continuously for the length of the axoneme, which is usually about 10 μm long but may be as long as 200 μm in some cells.

While each member of the pair of single microtubules (the *central pair*) is a complete microtubule, each of the outer doublets is composed of one complete and one partial microtubule—known as the *A and B tubules*, respectively—fused together so that they share a common tubule wall. In transverse sections, each complete microtubule appears to be formed from a ring of 13 subunits, while the incomplete B tubule of the outer doublet is formed from only 11.

Figure 11–51 Electron micrograph of the flagellum of a green algal cell (*Chlamydomonas*) shown in cross-section. The distinctive "9 + 2" arrangement of microtubules occurs in almost all cilia and flagella in eucaryotic cells. A schematic diagram of this structure seen in cross-section, illustrating the principal structural components, is shown in Figure 11–53. (Micrograph courtesy of Lewis Tilney.)

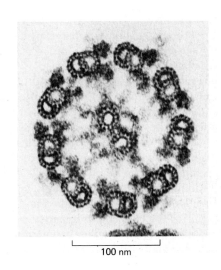

100 nm

Figure 11–52 (A) Electron micrograph of a microtubule seen in cross-section, with its ring of 13 distinct subunits, each of which corresponds to a separate tubulin molecule. (B) Electron micrograph of a negatively stained microtubule. (C and D) Schematic diagrams of a microtubule, showing how the tubulin molecules pack together to form the cylindrical wall. (C) shows the 13 molecules in cross-section. (D) shows a side view of a short section of a microtubule, with the tubulin molecules aligned into rows, or *protofilaments*. Each of the 13 protofilaments is composed of a series of tubulin molecules, each an α/β heterodimer. A microtubule is a polar structure, with a different end of the tubulin molecule (α or β) facing each end of the microtubule. (A, courtesy of Richard Linck; B, courtesy of Robley Williams; C, drawn from data supplied by Linda Amos.)

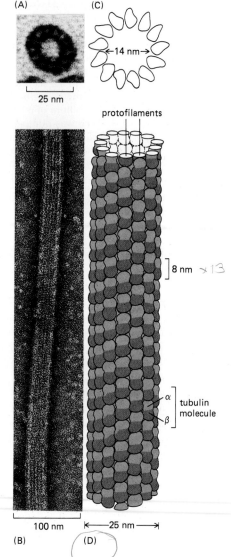

Microtubules Are Hollow Tubes Formed from Tubulin Dimers[31]

Microtubules are formed from molecules of *tubulin*, each of which is a heterodimer consisting of two tightly linked globular subunits. These subunits are related proteins (each composed of about 450 amino acid residues) called *α-tubulin* and *β-tubulin*. Although tubulin is present in virtually all eucaryotic cells, the most abundant source for biochemical studies is vertebrate brain. Most extraction procedures yield 10–20% of the total soluble protein in brain as tubulin, reflecting the high density of microtubules in the elongated axons and dendrites of nerve cells.

When tubulin molecules assemble into microtubules, they form linear "protofilaments," with the β-tubulin subunit of one tubulin molecule in contact with the α-tubulin subunit of the next. In a complete microtubule there are 13 protofilaments arranged side by side around a central core that appears to be empty in electron micrographs (Figure 11–52). Since all protofilaments are aligned in parallel with the same polarity, the microtubule—like an actin filament—is a polar structure with a plus (fast-growing) end and a minus (slow-growing) end (see Panel 11–2, p. 655); the plus end is located at the distal tip of cilia.

In common with actin and many other cytoskeletal proteins, tubulin is encoded in most organisms by a family of closely related genes. The green alga *Chlamydomonas* has two genes for α-tubulin and two for β-tubulin, whereas the fruitfly *Drosophila* has four genes for α-tubulin and four for β-tubulin. Tubulins from different species are closely related in sequence, although they are not as highly conserved as the actins. The β-tubulin from yeast, for example, is about 70% identical to chicken β-tubulin, while actins from these species are more than 90% identical.

The unusual evolutionary conservation of actin and tubulin may in part reflect the structural constraints imposed by the large number of different proteins to which they bind. Tubulin molecules, like actin, interact not only among themselves but also with many associated proteins. As we shall see, the latter modify the properties of microtubules and attach them to other structures in the cell. Presumably most random mutational changes perturb at least one function of the microtubule or actin filament and therefore are deleterious to the organism.

All known tubulins copolymerize into the same microtubules if they are mixed together *in vitro*. It is likely, however, that some of the variation in the tubulins in an organism has functional implications. The carboxyl terminus of both α- and β-tubulin from higher vertebrates, for example, which contains an unusually large number of acidic amino acids, shows distinct differences in different tissues. This region of the molecule is apparently involved in binding to *microtubule accessory proteins*, and the changes in amino acid sequence may alter function by affecting this binding.

A Long Thin Filament Runs Along the Wall of the Microtubule Doublet[32]

Most microtubules are thought to be formed exclusively from tubulin subunits. Additional proteins must be present, however, in order to form special types of microtubules, such as the doublet microtubules in a cilium. When the microtubules from a cilium or flagellum are dissociated in a dilute salt solution, fragments of the A tubule can be isolated as especially stable ribbons of two to four protofilaments. In addition to tubulin, these protofilaments contain a protein called *tektin*, arranged in long, 2- to 3-nm diameter filaments that seem to be related to intermediate filaments. The tektin filaments run longitudinally along the wall of the doublet microtubules and seem to help form the shared wall of the A and B tubules. It is thought that either they or some unknown filamentous molecules determine the positions on the microtubule of the periodic structures to be described next.

The Axoneme Contains Protein Links, Spokes, and Side Arms[33]

The microtubules of the axoneme are linked to structures that generate force and harness it to produce wavelike movements. Perhaps the most important of these structures are sets of arms that join neighboring doublets in the outer ring (Figure 11–53). Each arm is composed of a protein called *dynein*, and pairs of inner and outer arms are spaced all along each A tubule at regular 24-nm intervals. Dynein plays an essential part in the movement of cilia and flagella, as described below. At more widely spaced intervals, thin links of *nexin* extend between the adjacent doublets; these nexin links are thought to function like rubber bands to resist the sliding between adjacent doublet microtubules.

Projecting inward from each doublet are *radial spokes* extending to an *inner sheath* that surrounds the central pair of single microtubules (Figure 11–53). Looked at from the side of the axoneme, each of these structures—the dynein side arms, the nexin links, the radial spokes, and the arms of the inner sheath—can be seen as a series of regular projections, each with its own periodicity (Table 11–2).

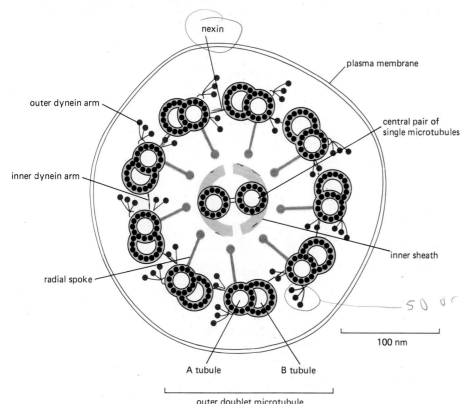

Figure 11–53 A cilium shown in cross-section. (An electron micrograph of this type of structure is shown in Figure 11–51.) The various projections from the microtubules occur at regular intervals along the cilium as described in Table 11–2.

Table 11–2 Major Protein Structures of the Ciliary Axoneme

Axoneme Component (periodicity along axoneme)	Function
Tubulin (8 nm)	principal component of microtubules
Dynein arms (24 nm)	project from microtubule doublets and interact with adjacent doublets to produce bending
Nexin links (86 nm)	hold adjacent microtubule doublets together
Radial spokes (29 nm)	extend from each of the nine outer doublets inward to the central pair
Sheath projections (14 nm)	project as a series of side arms from the central pair of microtubules; together with the radial spokes these regulate the form of the ciliary beat

18 The Axoneme Moves by a Sliding Microtubule Mechanism[34]

If a flagellum is severed from a cell by a laser beam, the isolated structure continues to propagate bending movements in a normal way, indicating that the motile machinery is contained in the axoneme itself and its movements do not depend on a motor at its base (unlike the bacterial flagellum—see p. 720). Indeed, even an isolated axoneme will propagate bending movements if it is placed in a salt solution containing ATP.

The bending force is produced by the sliding of microtubules. This has been shown by exposing isolated axonemes to proteolytic enzymes, which disrupt both the nexin links and the radial spokes while leaving the dynein arms and the microtubules themselves intact. If this partially digested structure is exposed to as little as 10 μM ATP, the axoneme elongates until it is up to nine times its original length, the component fibers in the axoneme telescoping out of the loosened structure (Figure 11–54). It seems that the adjacent outer doublets can actively slide against each other once they have been freed of their lateral cross-links (such as those made of nexin). In the intact structure this sliding movement is converted to bending, as shown diagramatically in Figure 11–55.

Dynein Is Responsible for the Sliding[35]

Since the adjacent outer doublets actively slide against each other, a force must be generated between them. The nexin links cannot be responsible for generating the force, since they are destroyed by protease treatment that leaves the sliding unimpaired. This leaves the *dynein arms*, which in most micrographs do not quite extend all the way from one doublet to the next; if the cilium is allowed to use

$\overline{20\ \mu m}$

Figure 11–54 Electron micrograph of an isolated axoneme (from a cilium of *Tetrahymena*) that has been briefly exposed to the proteolytic enzyme trypsin to loosen the protein ties that normally hold it together. Following treatment with ATP, the individual microtubule doublets slide against each other, causing the original structure to increase in length by as much as nine times. (From F.D. Warner and D.R. Mitchell, *J. Cell Biol.* 89:35–44, 1981. Reproduced by permission of the Rockefeller University Press.)

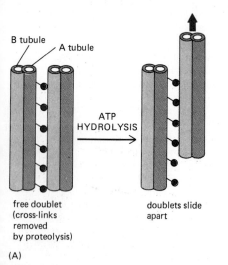

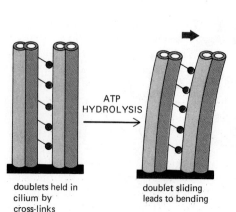

(A)

(B)

Figure 11–55 The sliding of outer microtubule doublets against each other (A) causes bending if the doublets are tied to each other at one end (B). Note that the base of the dynein molecule attaches only to the A tubule, leaving the heads free to contact the adjacent B tubule. Apparently the different structure of the B tubule prevents the base of a dynein molecule from binding to it. The resulting asymmetry in the arrangement of the dynein molecules is required to prevent a fruitless tug-of-war between neighboring microtubules, which presumably explains why each of the nine outer microtubules is an A-B doublet.

up all its ATP, however, the dynein arms make contact with the adjacent doublet (Figure 11–56).

Dynein is a large protein complex containing two or three globular heads (depending on the source) linked to a common root by thin flexible strands (Figure 11–57). Each globular head has an ATPase activity that is stimulated about sixfold by its association with microtubules. A dynein arm is formed from one dynein molecule. Kinetic studies suggest that the heads drive the sliding of microtubules in a cilium by a process that is fundamentally similar to the way myosin heads operate in muscle (see p. 620): a regular cycle of conformational changes in each head, driven by ATP binding and hydrolysis, moves the dynein heads along a microtubule in a single direction—from its plus end toward its minus end. This movement generates a force that drives the adjacent microtubule doublet toward the tip of the axoneme (see Figure 11–55).

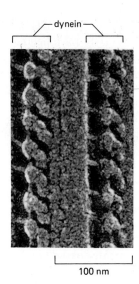

100 nm

Figure 11–56 Freeze-etch electron micrograph of a cilium, showing the dynein arms projecting at regular intervals from the doublet microtubules. (Courtesy of John Heuser.)

11-17 Microtubule Sliding Must Be Regulated to Generate Ciliary Bending[36]
11-19

If all the dynein became active at the same time—as the myosin heads do in contracting muscle—the axoneme would simply twist into a tight helix. Consequently, in order to produce a local bend that propagates from the base of a cilium to the tip in a coherent wave, there must be controls that coordinate the movements of the dynein molecules. These controls cannot depend on fluxes of Ca^{2+} or other ions, since the axoneme will beat without its plasma membrane. Instead, the activation of individual dynein arms depends on the mechanical movements of other parts of the axoneme, relayed through protein-protein interactions. The elastic resistance of the flagellum is also important, tending to restore the equilibrium configuration of the structure in the absence of activity.

11-20 The Axoneme Can Be Dissected Genetically[37]

When axonemes are purified and their component proteins analyzed, nearly 200 different polypeptide chains can be detected by two-dimensional polyacrylamide-gel electrophoresis (see p. 171). The analysis of the function of these proteins and their positions in the axoneme has been facilitated by examination of mutant organisms. A favorite object for such studies is the unicellular green alga *Chlamydomonas reinhardtii*, which has two flagella that propel it through the water (Figure 11–58). Many mutants with defects in motility have been isolated: in some the mechanism of flagellar assembly is defective and the flagella are absent or rudimentary; in others, flagella are present but either nonmotile or slow moving. Various structural abnormalities are apparent in electron micrographs of such mutant flagella. In one class of nonmotile mutants the only detectable change is the loss of the dynein arms. In a second class the mutants lack only the radial spokes, while in a third they lack both the central pair of microtubules and the inner sheath. In all three classes the isolated membrane-free axonemes fail to move in the presence of ATP.

The most informative class of flagellar mutants are those that move even though they are missing particular components of the axoneme. Slow-moving mutants can be isolated, for example, that lack either the outer or the inner dynein arms, demonstrating that either type of dynein molecule suffices to produce movement but that both normally contribute to flagellar motility. More surprising are second-site mutations that restore motility to nonmotile flagella that lack the central pair or radial spokes without restoring the defective component. The existence of such mutants suggests that the absence of a normal central pair or radial-spoke complex switches off the activity of the dynein arms, "freezing" the motor. A mutation in a second gene removes the control switch and allows the axoneme to beat. It is interesting that the motile double mutants with a defective core can move only in the flagellar mode, being incapable of the whiplike ciliary beat that is normal for *Chlamydomonas*. It thus seems that the ciliary form of movement requires a more complex machinery than does the wavelike movement of a more typical flagellum (see Figure 11–50).

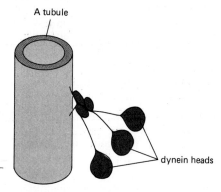

A tubule

dynein heads

Figure 11–57 Dynein is a large protein assembly (nearly 2 million daltons) composed of 9 to 12 polypeptide chains, the largest of which are about 450,000 daltons. The base of the molecule binds tightly to an A tubule in an ATP-independent manner, while the large globular heads have an ATP-dependent binding site for a second microtubule. When the heads hydrolyze their bound ATP, they move toward the minus end of this second microtubule, thereby producing a sliding force between the adjacent microtubule doublets in a cilium or flagellum (see Figure 11–55).

Defects in axonemes also occur in humans, where various hereditary forms of male sterility have been shown to be due to nonmotile sperm. Depending on the particular mutation, the sperm flagella lack dynein arms, the radial-spoke heads, or the inner sheath together with one or both of the central pair of microtubules. Exactly the same defects occur in the respiratory cilia of such individuals, who commonly have long histories of respiratory tract disease—with recurrent bronchitis and chronic sinusitis—because their nonmotile respiratory cilia are unable to clear mucus from their lungs and sinuses. Remarkably, about half of the individuals with this *immotile cilia syndrome* also have an extremely rare condition known as situs inversus, in which the normal asymmetry of the body—as seen in the position of the heart, intestine, liver, and appendix—is reversed (the whole complex of abnormalities is known as *Kartagener's syndrome*). It has therefore been suggested that the unidirectional beating of cilia during early human development could play a critical part in determining the normal left-right asymmetry of the body.

Having discussed how flagella and cilia move, we shall now consider how they are formed.

Centrioles Play Two Distinct Roles in the Cell[38]

If the twin flagella of *Chlamydomonas* are sheared from the cell, the centrioles remain and the flagella rapidly re-form. Nearly all of the necessary protein components can be found in a soluble form in the cytoplasm of the cell, and these are utilized to construct the new flagella. A few of the individual steps of the assembly can take place in cell-free extracts: tubulin molecules will polymerize into microtubules, as we shall see in greater detail in the following section, and dynein arms can be added back to an axoneme stripped of its own arms by extraction in salt solutions of high ionic strength. By themselves, however, axonemal proteins cannot generate the distinctive 9 + 2 pattern of the axoneme; a nucleating structure is needed to serve as a template on which futher growth can take place. In the cell the required template is provided by a *centriole*.

A **centriole** is a small cylindrical organelle about 0.2 μm wide and 0.4 μm long. Nine groups of three microtubules, fused into triplets, form the wall of the centriole, each triplet being tilted inward toward the central axis at an angle of about 45° to the circumference, like the blades of a turbine (Figure 11–59). Adjacent triplets are linked at intervals along their length, while faint protein spokes can often be seen in electron micrographs to radiate out to each triplet from a central core, forming a pattern like a cartwheel (see Figure 11–59A). Centrioles often exist in pairs, with the two centrioles at right angles to each other (Figure 11–60).

The centriole is a permanent feature of the ciliary axoneme, where it is traditionally called a *basal body*. Appendages known as *striated rootlets* are often attached to this centriole, linking it to other components of the cytoskeleton. During the formation or regeneration of the cilium, each doublet microtubule of the axoneme grows from two of the microtubules in the triplet microtubules of the centriole so that the ninefold symmetry of the centriolar microtubules is pre-

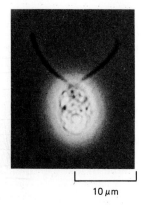

Figure 11–58 The unicellular green alga *Chlamydomonas reinhardtii*. This organism swims by means of its two flagella, which cooperate to produce a cycle of whiplike movements reminiscent of the breast stroke of human swimmers (unlike the flagella of sperm). (Courtesy of John Hopkins.)

Figure 11–59 (A) Electron micrograph of a cross-section through three centrioles in the cortex of a protozoan. Each centriole (also called a basal body—see text) forms the lower portion of a ciliary axoneme. (B) Schematic drawing of a centriole. It is composed of nine sets of *triplet* microtubules, each triplet containing one complete microtubule (the A tubule) fused to two incomplete microtubules (the B and C tubules). Other proteins form links that hold the cylindrical array of microtubules together (*color*). (Micrograph courtesy of D.T. Woodrum and R.W. Linck.)

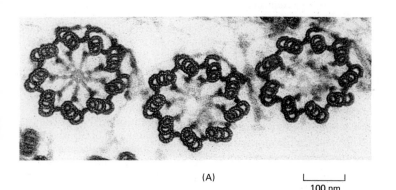

(A)

100 nm

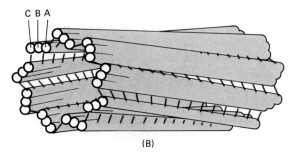

C B A

(B)

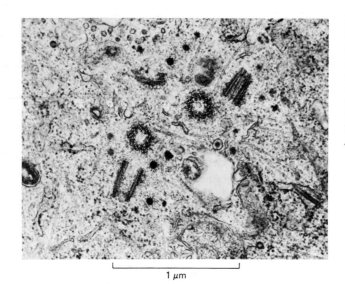

Figure 11–60 An electron micrograph showing a newly replicated pair of centrioles. One centriole of each pair has been cut in cross-section and the other in longitudinal section, indicating that the two members of each pair are aligned at right angles to each other. (From M. McGill, D.P. Highfield, T.M. Monahan, and B.R. Brinkley, *J. Ultrastruct. Res.* 57:43–53, 1976.)

1 μm

served in the ciliary axoneme. Autoradiographic evidence suggests that the addition of tubulin and other proteins of the axoneme takes place at the distal tip of the structure, at the plus end of the microtubules. How the central pair of single microtubules forms in the axoneme is not known; there is no central pair in centrioles.

The centrioles that form the basal bodies of cilia are performing a specialized function in the cell, just as the cilia themselves are specialized structures. Nearly all animal cells, however, have a pair of centrioles that acts as a focal point for the *centrosome*. The centrosome (also called the *cell center*—see p. 763) organizes the array of cytoplasmic microtubules during interphase and duplicates at mitosis to nucleate the two poles of the mitotic spindle, as we shall see in the next section. Sometimes centrioles can serve first one function and then another in turn: prior to each division in *Chlamydomonas*, for example, the two flagella resorb and the basal bodies leave their position to act as mitotic spindle poles.

Centrioles Usually Arise by the Duplication of Preexisting Centrioles[39]

The otherwise continuous increase in cell mass throughout the animal cell cycle is punctuated by two discrete duplication events: the replication of DNA and the doubling of the centrioles. In cultured fibroblasts, centriole doubling commences at around the time that DNA synthesis begins. First the two members of a pair separate; then a daughter centriole is formed perpendicular to each original centriole (see Figure 11–60). An immature centriole contains a ninefold symmetric array of *single* microtubules; each microtubule then presumably acts as a template for the assembly of the triplet microtubule of mature centrioles.

In a ciliated vertebrate cell, which may contain hundreds of cilia, the centrioles of the precursor cell give rise to the many basal bodies required to nucleate the cilia in the mature cell. During the differentiation of the ciliated epithelial cells that line the oviduct and the trachea, for example, the centriole pair migrates from its normal location near the nucleus to the apical region of the cell where the cilia will form. There, instead of forming a single daughter centriole in the typical manner, each centriole in the pair forms numerous electron-dense "satellites." Many basal bodies then arise from these satellites and migrate to the membrane to initiate the formation of cilia.

There are also cases where centrioles seem to arise de novo. For example, although unfertilized eggs of many animals lack functional centrioles and use the sperm centriole for the first mitotic division (see p. 874), under certain experimental conditions—such as extreme ionic imbalance or electrical stimulation—the unfertilized egg can produce a variable number of centrioles. Each of these centrioles nucleates the formation of a small aster, one of which can be used by the egg for division, so that a haploid organism develops by a process called parthenogenesis (see p. 761). Apparently, centriole precursors are stored in the cytoplasm of unfertilized eggs and can be activated to form a new centriole under special circumstances.

The unusual mode of duplication of centrioles and their continuity over many generations led to the earlier suggestion that centrioles might be fully autonomous, self-replicating organelles. Although it is now known that this is not the case and that under certain circumstances centrioles can arise de novo in the cytoplasm, it is still possible that some information necessary for centriole formation is usually carried in the centriole itself—just as the replication of mitochondria and chloroplasts depends on extrachromosomal genes carried in the organelle (see p. 387). In *Chlamydomonas* a set of genes that encode proteins involved in basal body structure and flagellar assembly is carried on a discrete genetic element that segregates independently of the major chromosomes. The nature and location of this genetic element, however, remains to be identified.

Summary

Cilia and eucaryotic flagella have a core containing a cylindrical bundle of nine outer doublet microtubules that cause the cilium or flagellum to bend by sliding against each other. Dynein side arms extend between adjacent microtubule doublets and hydrolyze ATP to generate a sliding force between the doublets. Accessory proteins bundle the ring of microtubule doublets together and limit the extent of their sliding. Other accessory proteins create a mechanically activated relay system that regulates the activity of the dynein so that the cyclic ciliary bends underlying ciliary beating are produced. The complex structure of the ciliary axoneme forms by the self-assembly of its component proteins and is nucleated by a centriole (basal body), which serves as a template for the distinct 9 + 2 pattern of microtubules in the axoneme.

Cytoplasmic Microtubules[40]

Although both actin and tubulin are major proteins in nearly all eucaryotic cells, tubulin is generally present in smaller amounts. Furthermore, because a microtubule has a greater diameter than an actin filament, a mass of about 10 times more tubulin than actin is required to make the same length of polymer (see Table 11–4, p. 654). Consequently the total length of the actin filaments in a cell is at least 30 times greater than the total length of the microtubules, reflecting a fundamental difference in the organization and function of these two cytoskeletal polymers. Whereas actin filaments form cross-linked networks and small bundles in the peripheral cytoplasm, microtubules usually exist as single filaments that radiate outward throughout the cytoplasm from a position close to the nucleus. By providing a system of fibers along which vesicles and other membrane-bounded organelles can travel, microtubules can impart polarity to the cell and can regulate cell shape, cell movement, and the plane of cell division.

Microtubules Are Highly Labile Structures That Are Sensitive to Specific Antimitotic Drugs[41]

Many of the microtubule arrays in cells are labile and depend on this lability for their function. One of the most striking examples is the mitotic spindle, which forms after the cytoplasmic microtubules disassemble at the onset of mitosis (see p. 768).

The microtubules of the mitotic spindle are in a state of unusually rapid assembly and disassembly, which explains the extreme sensitivity of the spindle to various drugs that bind to tubulin (see p. 763). *Colchicine,* an alkaloid extracted from the meadow saffron, has been used medicinally since ancient Egyptian times.

Each molecule of colchicine (Figure 11–61) binds tightly to one tubulin molecule and prevents its polymerization, so that the exposure of a dividing cell to colchicine causes the disappearance of the mitotic spindle and blocks the cell in mitosis within a few minutes. Compounds with this action are known as *antimitotic drugs* (Table 11–3). In many cases their action is reversible, so that removal of the drug allows the spindle to re-form and mitosis to proceed. Because the disruption of spindle microtubules preferentially kills many rapidly dividing cells, antimitotic drugs, such as vinblastine and vincristine, are widely used for cancer therapy.

The drug *taxol* has the opposite effect. It binds tightly to microtubules and stabilizes them; when added to cells, it causes much of the free tubulin to assemble into microtubules. Just as the stabilization of actin filaments by phalloidin halts cell migration, so the stabilization of microtubules by taxol arrests dividing cells in mitosis.

Like actin filament polymerization, microtubule assembly has complex kinetics that play an important part in many cellular processes. Most of what we know about the dynamic behavior of microtubules has come from studying the polymerization of purified tubulin molecules *in vitro*.

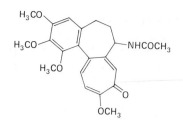

Figure 11–61 Chemical structure of colchicine.

The Two Ends of a Microtubule Are Different and Grow at Different Rates[42]

As outlined below (p. 655), microtubules have an inherent polarity because their tubulin subunits are arranged in a specific orientation in the polymer. As with actin filaments, this structural polarity makes the two ends of the polymer different in ways that are important for understanding how microtubules form in cells. If purified tubulin molecules are allowed to polymerize for a short time on fragments of a ciliary axoneme and the mixture is then examined in the electron microscope, one end can be seen to elongate at three times the rate of the other. The fast-growing end is thereby defined as the *plus end* and the other as the *minus end*.

It is possible to detect the polarity of microtubules in cross-section by adding free tubulin molecules to existing microtubules under special conditions: rather than adding to the ends of the microtubules, the tubulin monomers form curved protofilament sheets on the sides. In cross-section the sheets resemble hooks and, depending on the polarity of the microtubule, will appear to point either clockwise or counterclockwise (Figure 11–62). In this way, for example, it has been shown that all of the microtubules in nerve cell axons and in cilia have the same polarity: the ends of the microtubules farthest from the center of the cell are always the plus ends. The plus ends also extend away from the region that organizes microtubules—the centrosome or the spindle poles (Figure 11–63).

11-23 ## Nucleotide Hydrolysis Helps to Make Slowly Growing Microtubules Unstable[43]

The assembly of tubulin into microtubules resembles actin polymerization in several ways. It occurs spontaneously *in vitro* and is normally accompanied by the delayed hydrolysis of one molecule of bound nucleotide, which in the case of tubulin is GTP rather than ATP (Table 11–4). The energy provided by nucleotide hydrolysis is not required for polymerization: tubulin will assemble into a micro-

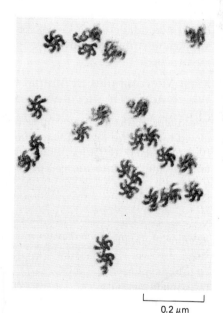

0.2 µm

Figure 11–62 Microtubule polarity as revealed by the hook-decoration method. All the microtubules in this group (seen in cross-section in this electron micrograph) have the same polarity. The hooks formed by the added tubulin curve clockwise, which indicates that the microtubules are being viewed as though looking along each filament from its plus end toward its minus end. Microtubule polarity can also be determined by decoration with dynein molecules (not shown). (Adapted from U. Euteneuer, *Cell Muscle Motil.* 5:1–82, 1984.)

Table 11–3 Some Drugs That Bind to Tubulin (Antimitotic Drugs)

Drug	Mode of Action
Colchicine, colcemid, nocadazole	inhibit the addition of tubulin molecules to microtubules, leading to microtubule depolymerization
Vinblastine, vincristine	induce the formation of paracrystalline aggregates of tubulin, leading to microtubule depolymerization
Taxol	stabilizes microtubules by binding tightly to polymer

tubule even when the only nucleotide present is the nonhydrolyzable GTP analogue GTPγS. The nucleotide hydrolysis, however, makes the critical concentration of the free subunit required for net polymerization lower at the plus end than at the minus end: that is, the plus end binds more tightly to free actin or tubulin molecules (see Panel 11–1, p. 639). For actin filaments this difference makes possible a continual treadmilling of subunits through the polymer (see p. 640). Although in principle treadmilling can also occur in microtubules, a different phenomenon, called *dynamic instability*, seems to predominate instead.

Dynamic instability can be demonstrated directly if microtubules assembling from purified tubulin on a glass slide are observed by light microscopy using video-enhanced dark-field optics. Under these conditions one sees that the ends of individual microtubules are either slowly growing or rapidly shrinking and that they alternate between these two states in a seemingly random fashion, with each state persisting for many seconds (Figure 11–64). A likely explanation for this dynamic instability is presented in Panel 11–2. Dynamic instability is thought to be a consequence of the delayed hydrolysis of GTP after tubulin assembly. When a microtubule grows rapidly, tubulin molecules add to a polymer end faster than the nucleotide they carry can be hydrolyzed. This results in the presence of a "GTP cap" on the end of the microtubule; and because tubulin molecules carrying GTP bind to one another with higher affinity than tubulin molecules carrying GDP, the GTP cap will encourage a growing microtubule to continue growing. Conversely, once a microtubule has lost its GTP cap—for example, if the rate of polymerization slows down—it will start to shrink and then tend to go on shrinking. Other factors besides the GTP cap, however, such as a persistent distortion of the tubulin lattice at the end, also appear to be important in creating shrinking microtubules (see Panel 11–2).

As we shall now see, dynamic instability allows a cell to control the position of its microtubules by means of special structures in the cytoplasm that bind and stabilize microtubule ends.

1-24 Most Microtubules in Animal Cells Grow from the Centrosome, Which Acts as a Microtubule-organizing Center[44]

The microtubules in the cytoplasm of an interphase cell in culture can be visualized by staining the cell with fluorescent anti-tubulin antibodies after the cells have been fixed. The microtubules are seen in greatest density around the nucleus and radiate out into the cell periphery in fine lacelike threads (Figure 11–65A).

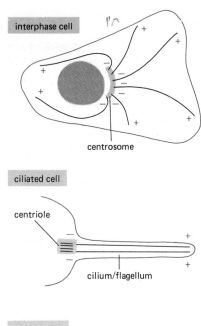
interphase cell

centrosome

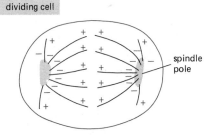
ciliated cell

centriole

cilium/flagellum

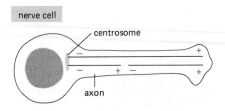
dividing cell

spindle pole

nerve cell

centrosome

axon

Figure 11–63 The minus ends of microtubules in cells are generally embedded in a microtubule-organizing center, while the plus ends are often near the plasma membrane.

Table 11–4 Comparison of Actin and Tubulin Polymers

	Actin	Tubulin
Polypeptide molecular weight	42,000	50,000 (α-tubulin) 50,000 (β-tubulin)
Unpolymerized form	globular, monomer	globular, dimer (one α + one β)
Bound nucleotide (in unpolymerized state)	ATP (one per monomer)	GTP (two per dimer)
Form of polymer	helical filament	hollow tube composed of 13 protofilaments
Diameter of polymer	8 nm	25 nm
Subunits per micrometer of polymer length	370 monomers	1,600 dimers
Molecular weight per micrometer of polymer length	$370 \times 42,000$ $= 15.5 \times 10^6$	$1,600 \times 100,000$ $= 160 \times 10^6$

THE OBSERVED BEHAVIOR OF MICROTUBULES

Microtubules are dynamic structures, and their assembly and disassembly determines where and when they form in the cell. Microtubules, like actin filaments, grow by the reversible addition of subunits, accompanied by nucleotide hydrolysis and conformational change (see Panel 11–1, p. 639). Microtubule polymerization, however, exhibits distinctive features that are not observed with actin filaments.

The minus ends of microtubules in cells are bound tightly to microtubule-organizing centers, which prevent (or control) the assembly and disassembly of subunits at this end (see p. 654). Therefore we shall consider only the plus ends. The plus ends of individual microtubules *in vitro* switch sponta-

neously from a *slowly growing* to a *rapidly shrinking* state, with each state persisting for many seconds. At any instant the total population of microtubules, therefore, consists of two types of polymer, which interconvert only slowly:

This behavior, known as dynamic instability, can be explained by taking into account the effect of delayed GTP hydrolysis on the polymerization.

A STRUCTURAL BASIS FOR DYNAMIC INSTABILITY

The changes that take place during the polymerization of tubulin molecules are essentially the same as those that occur in actin molecules. We shall therefore use the same conventions for the different forms of the subunits as in Panel 11–1—that is, $\boxed{T}$ represents the GTP-bound monomer, $\boxed{T}$ represents the subunits of the polymer after they have undergone their conformational change but before hydrolysis of the bound GTP, and $\boxed{D}$ represents the GDP-bound subunits of the polymer.

Since GTP hydrolysis occurs only *after* each subunit is polymerized, $\boxed{T}$ subunits will tend to be found only at the growing end of a microtubule; this "cap" of $\boxed{T}$ subunits is expected to be larger the faster the rate of new subunit addition:

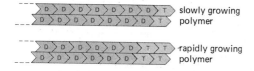

Since $\boxed{D}$ subunits dissociate more readily than $\boxed{T}$ subunits, once a $\boxed{T}$ cap is lost from the end of a microtubule, the end will tend to lose subunits more rapidly. (Microtubules depolymerize about 100 times faster from a $\boxed{D}$ than from a $\boxed{T}$ end.) Thus, once rapid depolymerization begins, the $\boxed{T}$ cap on a microtubule is difficult to regain. In addition, it seems that a further structural transition can occur in which the end of the microtubule frays and so is unable to add new tubulin subunits easily.

This type of shrinking microtubule often never regains its cap and shrinks until it depolymerizes completely.

THE ABSENCE OF A STEADY STATE FOR INDIVIDUAL MICROTUBULES

At steady state, the free subunit concentration, [C], is constant, so that

[fraction shrinking microtubules] × [depolymerization rate]
= [fraction growing microtubules] × [polymerization rate]

Since the depolymerization rate is much greater than the polymerization rate, many microtubules grow and only a few

shrink at any instant. There is no critical concentration of free tubulin at which an individual microtubule maintains a constant length; instead, a mixture of growing and shrinking microtubules is observed over a wide range of [C]. Although at high [C] the ratio of growing to shrinking microtubules is observed to increase, an individual microtubule, unlike an actin filament, never attains a stable "steady state."

THE IMPORTANCE OF CAPTURING THE POLYMER ENDS

Because of its dynamic instability, a newly formed microtubule will persist only if both of its ends are protected from depolymerizing. In cells the minus ends of microtubules are generally protected by the organizing centers from which these filaments grow, and it is thought that selected plus ends are captured by specific proteins that control the stability, and therefore the locations, of microtubules in cells. In the diagram, for example, a nonpolarized cell is depicted (A) with new microtubules growing and shrinking from a centrosome in all directions randomly. This array of microtubules then encounters structures in a selected region of the cell cortex that can cap the free plus ends of the microtubules (B). The selective stabilization of those microtubules that happen by chance to encounter these capping structures will lead to a rapid redistribution of the arrays and convert the cell to a polarized form (C and D).

In principle, actin filaments might also be expected to exhibit some dynamic instability. The ends of these filaments, however, are much less complex than the ends of microtubules, and actin filaments are thought to interconvert so rapidly between growing and shrinking phases that they behave instead in the manner explained in Panel 11–1.

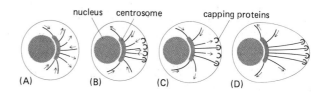

The origin of the microtubules is seen most clearly if they are depolymerized with colchicine and then allowed to regrow (Figure 11–65B). Regenerating microtubules first appear in a small, perinuclear, starlike structure called an *aster*; they then elongate toward the cell periphery until the original distribution is reestablished. If the microtubules in cultured cells are decorated with tubulin hooks to determine their polarity, they are all seen to have their plus ends facing away from the **microtubule-organizing center (MTOC)** where the aster originates.

There is one major microtubule-organizing center in almost all animal cells, called the *cell center* or **centrosome.** The centrosome is located to one side of the nucleus and contains a pair of centrioles arranged at right angles to each other in an L-shaped configuration (see Figure 11–60). Not all microtubule-organizing centers, however, contain centrioles. In mitotic cells of higher plants, for example, the microtubules terminate in poorly defined regions of electron density that are completely devoid of centrioles. Similarly, centrioles are not present in the meiotic spindle of mouse oocytes, although they appear later in the developing embryo.

Therefore, unlike the ciliary axoneme, which grows directly onto the end of the centriole (see p. 650), microtubules are not directly nucleated by the centriole itself: the essential component for nucleation is the cloud of amorphous material that surrounds the centriole (Figure 11–66). The exact composition of this material is unknown, but antibodies against one of its protein components can be used to stain the microtubule-organizing centers in both plant and animal cells, indicating that the protein has been highly conserved in evolution (Figure 11–67). If centrosomes (centrioles together with their pericentriolar material) are isolated from the cell and mixed with purifed tubulin, they will nucleate microtubule assembly rapidly *in vitro*. In these microtubules, as in microtubules *in vivo*, the minus ends are embedded in the pericentriolar material. Moreover, the number of microtubules that can be nucleated by any given isolated centrosome seems to be fixed and is similar to the number of microtubules attached to the centrosome in the cell from which it originated (about 250 microtubules in an interphase fibroblast). This suggests that centrosomes contain a fixed number of nucleating elements.

Different forms of microtubule-organizing centers are common in plants (see p. 1177) and in protozoa. The cytopharangeal basket in ciliates, for example, is a complex structure that acts as a rudimentary gullet. It is built from palisades of microtubules that radiate from the lower surface of a flat three-layered plate (Figure 11–68). As soon as these microtubules start to assemble, they are regularly spaced in a hexagonal array, suggesting that this microtubule-organizing center contains a regular array of nucleating elements.

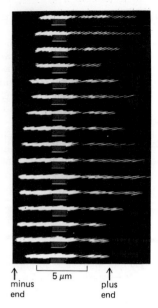

Figure 11–64 Fluctuations in length of a single microtubule as seen by video-enhanced dark-field microscopy. Images of the same microtubule were recorded at intervals of 1 to 2 minutes and displayed in sequential order on a monitor screen. The two ends go through cycles of elongation and shortening independently, with the plus end showing the greatest fluctuations. (From T. Horio and H. Hotani, *Nature* 321:605–607, 1986.)

11-25 The Dynamic Instability of Microtubules Provides an Organizing Principle for Cell Morphogenesis[45]

Cytoplasmic microtubules in animal cells tend to radiate out in all directions from the centrosome where their minus ends are anchored. Most animal cells, however, are polarized, and the assembly of tubulin molecules is controlled so that many microtubules extend toward specific regions of the cell. It is not known for certain how this is achieved, but it seems likely that the mechanisms depend on the dynamic instability of microtubules.

We have seen that individual microtubules *in vitro* tend to exist in one of two states: steady growth or rapid, "catastrophic" disassembly. Microtubules in a cell

Figure 11–65 Immunofluorescence micrographs showing the arrangement of microtubules in cultured cells as revealed by staining with anti-tubulin antibodies. A normal tissue-culture cell is shown in (A). The cells shown in (B) were treated with colcemid for 1 hour to depolymerize their microtubules and were then allowed to recover; microtubules appear first in starlike asters and then elongate toward the periphery of the cell. (A, courtesy of Eric Karsenti and Marc Kirschner; B, from M. Osborn and K. Weber, *Proc. Natl. Acad. Sci. USA* 73:867–871, 1976.)

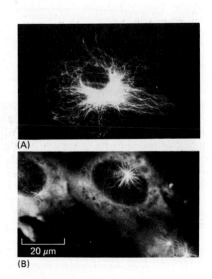

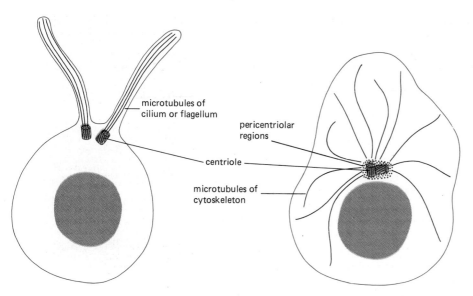

Figure 11–66 Centrioles can act as organizing centers for very different types of microtubule arrays. The two centrioles in the cell on the left serve as *basal bodies* for ciliary axonemes, whereas the two in the cell on the right form part of an organizing center (the *centrosome*) for an array of microtubules that radiates out into the cytoplasm. Note that in the centrosome microtubules grow not from the centriole itself but from the amorphous pericentriolar material.

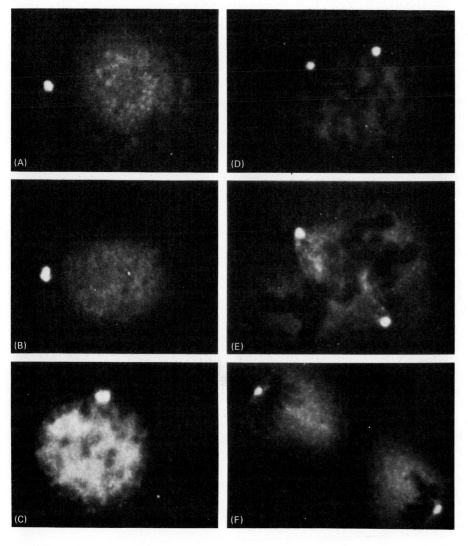

Figure 11–67 Centrosomes in interphase and mitotic animal cells stained by immunofluorescence with an antibody to a centrosomal protein. The micrographs illustrate the doubling and separation of centrosomes associated with cell division. (A), (B) Interphase; (C), (D) prophase; (E) metaphase; (F) telophase. For a description of the stages of mitosis, see page 766. From S.L. Brenner and B.R. Brinkley, *Cold Spring Harbor Symp. Quant. Biol.* 46:241–254, 1981. (Courtesy of Bill Brinkley.)

also seem to exist in these two states. The average lifetime of a microtubule in cultured fibroblasts in interphase, for example, is short—less than 10 minutes. Thus the array of microtubules radiating from the centrosome is continually changing, as new microtubules grow and replace others that have depolymerized.

The inherent instability of microtubules helps to explain how they can be induced to grow in specific directions in a cell—toward the leading edge of a crawling cell, for example, or toward a condensed chromosome in a dividing cell (see p. 770). A naked microtubule with both ends free in the cytoplasm rapidly disappears. But an organizing center continually produces new microtubules in random directions, with their minus ends anchored in the organizing center and thus protected from depolymerization. A microtubule that grows from such a center can be stabilized if its plus end is somehow capped so as to prevent its depolymerization. If capped by a structure in a particular region of the cell, it will establish a relatively stable link between that structure and the organizing center. Microtubules originating in the organizing center can thus be selectively stabilized by events elsewhere in the cell. Cell polarity is thought to be determined in this way by unknown structures or factors localized in specific regions of the cell cortex that "capture" the plus ends of microtubules. The plane of cell division may be determined in a similar way (see p. 779).

In many cells the initial stabilization of microtubules at their plus ends is consolidated to produce a more permanent polarization of the cell, as we shall now discuss.

Microtubules Undergo a Slow "Maturation" as a Result of Posttranslational Modifications of Their Tubulin Subunits[46]

The continual formation and loss of microtubules is characteristic of cells undergoing a major internal reorganization—such as cells that are dividing or crawling over a substratum. When cells have become part of an established tissue, the microtubules they contain become relatively permanent features, especially in those cells, such as nerve cells, that no longer divide after they differentiate. This microtubule "maturation" depends partly on the posttranslational modification of the tubulin molecules and partly on the interaction of microtubules with specific microtubule-associated proteins.

A number of enzymes modify selected amino acids in tubulin. One of these is *tubulin acetyl transferase*, which acetylates a specific lysine of the α-tubulin subunit. In *Chlamydomonas* this enzyme is located principally in the flagellar axoneme and appears to acetylate tubulin molecules after they assemble into the distal tip of the axoneme (see p. 645). A specific deacetylating enzyme is located in the cytoplasm and removes acetyl groups from unpolymerized tubulin. As a consequence of the localization of these two enzymes in *Chlamydomonas*, tubulin molecules in the axoneme microtubules are acetylated whereas those in cytoplasmic microtubules, most of which are turning over rapidly, are mainly non-acetylated. We shall discuss the possible consequence of tubulin acetylation below.

A second, more unusual form of modification is the removal of the carboxyl-terminal tyrosine residue from an α-tubulin molecule that has become incorporated into a microtubule. This modification is catalyzed by a detyrosinating enzyme in the cytoplasm of many vertebrate cells, and, as in the case of acetylation, there is an oppositely-acting enzyme that restores the tyrosine to unpolymerized tubulin. In cells with highly unstable microtubules, tubulin is generally not present in polymerized microtubules long enough to be detyrosinated, and so "tyrosine-tubulin" is the predominant form. By contrast, "detyrosinated tubulin" becomes enriched in any "older" microtubules that survive the normal rapid turnover of newly formed microtubules. Both detyrosination and acetylation thus mark the conversion of transiently stabilized microtubules into a much more permanent form (Figure 11–69). When cultured fibroblasts are treated with drugs that depolymerize microtubules (see p. 652), the small population of microtubules that is spared during the drug treatment can be selectively labeled with antibodies that recognize either acetylated or detyrosinated tubulin.

0.2 μm

Figure 11–68   A large and unusually highly ordered microtubule-organizing center found in the cytopharyngeal basket of the ciliate *Nassula*. Microtubules grow in a regular hexagonal array from the lower surface of a flat three-layered sheet, forming one element of the complex mouthpart of this cell. (From J.B. Tucker, *J. Cell Sci.* 6:385–429, 1970.)

Figure 11–69 Immunofluorescence micrographs showing that while most of the microtubules in the cytoplasm of a cultured cell are highly dynamic (A), others are relatively stable (B). The single cell shown was injected with tubulin that was covalently coupled to the small molecule biotin; after 1 hour the cell was labeled first with anti-biotin antibodies coupled to a fluorescent molecule (fluorescein) and then with anti-tubulin antibodies coupled to another fluorescent molecule (rhodamine). Microtubules that are rapidly depolymerizing and repolymerizing incorporate the biotinylated tubulin and are therefore labeled by the anti-biotin antibodies (A), which cover the microtubules and prevent anti-tubulin antibodies from binding. Stable microtubules, on the other hand, do not incorporate biotinylated tubulin and therefore are not stained by the anti-biotin antibodies, but they are stained by the anti-tubulin antibodies (B). The stable microtubules are also found to stain with antibodies that recognize detyrosinated or acetylated tubulin. (Courtesy of Eric Schulze and Marc Kirschner.)

(A)

(B)

Microtubules formed *in vitro* from either acetylated or detyrosinated tubulin are not detectably more stable than microtubules formed from unmodified tubulin. These modifications are therefore thought to act as signals for the binding of specific proteins that stabilize microtubules and modify their properties inside the cell.

The Properties of Cytoplasmic Microtubules Are Modified by Microtubule-associated Proteins (MAPs)[47]

Posttranslational modification of tubulin appears to mark certain microtubules as "mature" and to add to their stability. But the most far-reaching and versatile modifications of microtubules are believed to be those conferred by other proteins. These **microtubule-associated proteins,** or **MAPs,** serve both to stabilize microtubules against disassembly and to mediate their interaction with other cell components. As one might expect from the diverse functions of microtubules, there are many kinds of MAPs.

Two major classes of MAPs can be isolated from brain in association with microtubules: *HMW proteins* (high-molecular-weight proteins), which have molecular weights of 200,000 to 300,000 or more; and *tau proteins*, with molecular weights of 40,000 to 60,000. Both classes of proteins have two domains, one of which binds to microtubules; because this domain binds to several unpolymerized tubulin molecules simultaneously, MAPs speed up the nucleation step of tubulin polymerization *in vitro*. The other domain is thought to be involved in linking the microtubule to other cell components (Figure 11–70). Antibodies to HMW and tau proteins show that both proteins bind along the entire length of cytoplasmic microtubules.

Many other MAPs have been isolated as proteins that bind selectively to microtubules. The functions of most of these are unknown. Some presumably act as structural components to stabilize microtubules and provide permanent links to other cell components (including other parts of the cytoskeleton and selected organelle membranes). Others are responsible for moving organelles along microtubules.

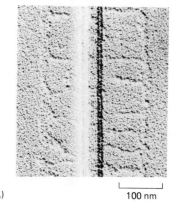

(A) |—————————| 100 nm

— microtubule

|——| 25 nm

(B)

Figure 11–70 The regularly spaced side arms formed on a microtubule by a large microtubule-associated protein (known as MAP-2) isolated from vertebrate brain. The electron micrograph in (A) shows a portion of a microtubule to which many molecules of MAP-2 are bound. Portions of the protein project away from the microtubule, as shown schematically in (B). (Electron micrograph courtesy of William Voter and Harold Erickson.)

11-32 ## The Transport of Organelles in the Cytoplasm Is Often Guided by Microtubules[48]

If a living vertebrate cell is observed in a phase-contrast or a differential-interference-contrast microscope (see p. 141), its cytoplasm is seen to be in continual motion. Over the course of minutes, mitochondria and smaller membrane-bounded organelles change their positions by periodic *saltatory movements*, which are too sustained and directional to be confused with the continual small Brownian movements caused by random thermal motions. Many of these intracellular movements take place in close association with microtubules. If a cell containing a moving organelle is rapidly fixed and sectioned for electron microscopy, the organelle membrane is often found to be connected by thin filamentous strands to a cytoplasmic microtubule. This suggests that microtubules may play an important part in such movements, although, as discussed previously (see p. 632), some movements of vesicles in the cytoplasm seem to take place along actin filaments rather than microtubules. The clearest demonstration of a transport role for microtubules comes from studies of *fast axonal transport* in nerve cells, where there is an extensive traffic of membrane-bounded vesicles in both directions along the axon between the cell body and the nerve terminus.

Kinesin and Cytoplasmic Dynein Hydrolyze ATP to Move Vesicles in Opposite Directions Along Axonal Microtubules[49]

The squid giant axon can be dissected from the animal and its cytoplasm extruded like toothpaste from a tube. If thin regions of the extruded axoplasm are squashed beneath a coverslip and examined by video-enhanced light microscopy (see p. 142), organelles are seen to move along fine filamentous tracks. A combination of immunofluorescence and electron microscopy shows that these "tracks" are single microtubules.

Extracted vesicles, or even artificial particles such as polystyrene beads, will associate with the microtubules in extruded axoplasm and move along them in a manner very similar to that seen in the living cell. Since the extruded axoplasm no longer has a plasma membrane, its content of ions and small molecules can be controlled experimentally and the effects on transport examined. In this way it was found that a nonhydrolyzable analogue of ATP, called AMPPNP, arrests transport and freezes the organelles in position, still bound to microtubules.

The AMPPNP effect provides a means of identifying the components responsible for the vesicle movement. By analyzing extracts of axoplasm for proteins that bind to microtubules in the presence of AMPPNP but are released by addition of ATP, a large protein complex called **kinesin** was isolated. Kinesin is an ATPase that uses the energy of ATP hydrolysis to move vesicles unidirectionally along a microtubule (see p. 130). The movements occur at rates of 0.5 to 2 μm/second, without the characteristic interruptions and hesitations seen in the intact axon, which are believed to arise from the repeated collisions of the vesicle with elements in the crowded cytoplasm.

The direction of movement produced by kinesin has been identified by allowing it to move polystyrene beads on microtubules polymerized on centrosomes *in vitro*. Whereas crude extracts of axoplasm drive transport in both directions, purified kinesin drives transport only outward, toward the plus ends of the microtubules (recall that ciliary dynein moves in the opposite direction—see p. 649). Since the microtubules in axons are known to be arrayed with their plus ends away from the cell body (see p. 654), the movement produced by kinesin would be expected to carry vesicles from the cell body to the axon terminus. It seems likely that a dyneinlike protein is responsible for transport in the other direction, from the terminus back toward the cell body. Recently a high-molecular-weight MAP with these properties has been purified from several sources, including mammalian brain; this protein is called **cytoplasmic dynein** (Figure 11–71).

Microtubules Determine the Location of the Endoplasmic Reticulum and the Golgi Apparatus in Cells[50]

Motile proteins that resemble kinesin and dynein are not confined to neurons. They seem to be present in all cells that contain microtubules, and they appear to mediate many of the important organizing effects of these filaments. Recent *in vitro* studies have shown that kinesin attaches to the membrane of the endoplasmic reticulum and can stretch this organelle into its typical lacelike network by pulling it along oriented microtubules. Similarly, studies on intact cells indicate

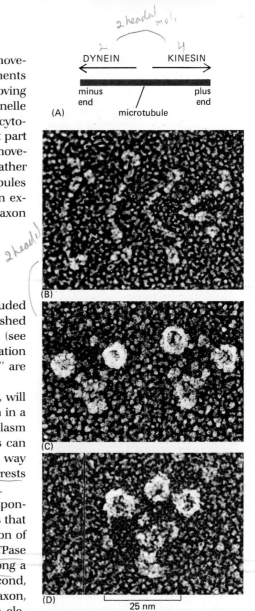

Figure 11–71 Kinesin and cytoplasmic dynein are large microtubule-associated proteins (MAPs) that use the energy of ATP hydrolysis to move in opposite directions along a microtubule (A). These proteins are complexes composed of two identical "heavy chains" plus several smaller "light chains." Each heavy chain forms a globular head region that attaches the protein to microtubules in an ATP-dependent fashion. Therefore, like myosin, kinesin [four molecules shown in (B)] and cytoplasmic dynein [two molecules shown in (C)] are two-headed molecules. In contrast, the ciliary dynein molecule shown in the electron micrograph in (D) has three heads (see Figure 11–57). (Free-etch electron micrographs prepared by John Heuser from proteins isolated by Trina Schroer, Jeff Gelles, and Michael Scheetz (B); Eric Steuer (C); and Ursula Goodenough (D).)

Figure 11–72 The alignment of the ER membrane with microtubules in a cultured cell. (A) The immunofluorescence localization of an ER protein outlines the ER cisterna as a lacelike network in the cell periphery; (B) immunofluorescence localization of the microtubules in the same cell. C, D, and E show the effect of microtubules on the Golgi apparatus. When a cultured cell is stained for both microtubules (C) and Golgi membranes (D), immunofluorescence reveals a congregation of the Golgi cisternae around the centrosome. However, when the microtubules are depolymerized by treatment of the cells with nocodazole, the Golgi membranes become dispersed throughout the cytoplasm (E), as they do in mitosis. (Courtesy of Mark Terasaki and Lan Bo Chen, C–E, courtesy of Viki Allan and Thomas Kreis.)

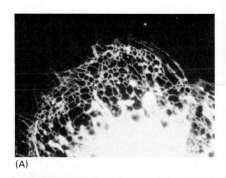
(A)

(B)

10 μm

(C)

(D)

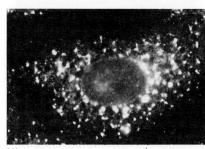
(E)

20 μm

that the ER is stretched outward from the centrosome along microtubules, as expected from a kinesin-mediated process. Finally, immunofluorescence shows that the thin strands of ER cisterna are often aligned along microtubules in the peripheral regions of a cell (Figure 11–72A, B).

Whereas the ER membrane is actively moved away from the centrosome on tracks of microtubules (while remaining attached to the nuclear envelope at its other end), the membranes of the Golgi cisternae seem to be moved in the opposite direction, as if attached to dyneinlike proteins, so that the Golgi apparatus becomes localized near the centrosome (Figure 11–72C, D). Both the ER and the Golgi apparatus become highly fragmented during mitosis (see p. 784), and when the cytoplasmic microtubules re-form in interphase, their orientation presumably guides the reassembly of these two organelles from smaller vesicles and membrane fragments (Figure 11–72E).

Summary

Microtubules form by the polymerization of tubulin molecules, which is quickly followed by the hydrolysis of a molecule of tightly bound GTP. Slowly growing microtubules are unstable and liable to catastrophic disassembly, but they can be stabilized by association with other structures that cap their two ends. Microtubule-organizing centers such as centrosomes continually nucleate the formation of new microtubules, which grow out in random directions. Any microtubule that happens to encounter a structure that caps its free plus end will be selectively stabilized, while other microtubules will depolymerize. It is thought that this process largely determines the polarity and position of the microtubule arrays in a cell.

The tubulin subunits in microtubules that have formed in the correct location are modified by acetylation and detyrosination. These alterations are thought to label the microtubule as "mature" and provide sites for the binding of specific microtubule-associated proteins (MAPs), which further stabilize the microtubule against disassembly and adapt its properties to specific functions in the cell. A special class of MAPs uses the energy of ATP hydrolysis to move unidirectionally along a microtubule, providing a mechanism for the spatial organization and directed movements of organelles in the cytoplasm.

Intermediate Filaments[51]

Intermediate filaments (IFs) are tough and durable protein fibers in the cytoplasm of most higher eucaryotic cells. Constructed like woven ropes, they are typically between 8 and 10 nm in diameter, which is "intermediate" between the thin and thick filaments in muscle cells, where they were first described; their diameter is also intermediate between actin filaments and microtubules. In most animal cells they form a "basket" around the nucleus and extend out in gently curving arrays to the cell periphery. They are particularly prominent where cells are subject to mechanical stress, such as in epithelia, where they are linked from cell to cell at desmosomal junctions (see p. 797), along the length of axons, and throughout the cytoplasm of smooth muscle cells. When cells are extracted with salt solutions of high or low ionic strength or with nonionic detergents, intermediate filaments remain behind while most of the rest of the cytoskeleton is lost. In fact, the term "cytoskeleton" was originally coined to describe these unusually stable and insoluble fibers.

Intermediate Filaments Are Formed from Four Types of Fibrous Polypeptides[51,52]

Unlike actin and tubulin, which are globular proteins, the subunits of intermediate filaments are fibrous proteins. These elongated IF protein molecules associate side by side in overlapping arrays to form long filaments with high tensile strength. Since only a part of each IF protein molecule is involved in the lateral interactions that form the filament, the remaining portion of the molecule can vary considerably without affecting the overall filament structure. Thus intermediate filaments, in contrast to actin filaments and microtubules, are composed of polypeptides of a surprisingly wide range of sizes (from about 40,000 to about 130,000 daltons)—the size varying with different cell types.

Intermediate filaments are classified on the basis of their amino acid sequence into four broad categories (Table 11–5). *Type I* IF proteins are found primarily in epithelial cells and include two subfamilies of **keratins,** *acidic keratins* and *neutral or basic keratins.* Keratin filaments are always heteropolymers formed from an equal number of subunits from each of these two keratin subfamilies. The keratins are by far the most complex class of IF proteins, with at least 19 distinct forms in human epithelia and 8 more in the keratins of hair and nails. Many morphologically and functionally distinct epithelia can be distinguished by the types of keratin molecules they make.

Type II IF proteins include (1) vimentin, (2) desmin, and (3) glial fibrillary acidic protein. **Vimentin** is widely distributed in cells of mesenchymal origin, including fibroblasts, blood vessel endothelial cells, and white blood cells, and is often expressed by cells in culture and transiently by developing cells. **Desmin** is found in both striated and smooth muscle cells, and **glial fibrillary acidic protein** forms *glial filaments* in some types of glial cells (astrocytes and some Schwann cells) in the nervous system. Each of these IF proteins will assemble spontaneously *in vitro*

Table 11–5   Principal Types of Intermediate Filament Proteins

Type of Intermediate Filament	Component Polypeptide (mass in daltons)	Cellular Location
Type I	acidic keratins (40,000 − 70,000)	epithelial cells and epidermal derivatives such as hair and nails
	neutral and basic keratins (40,000 − 70,000)	
Type II	vimentin (53,000)	many cells of mesenchymal origin, often expressed by cells in culture
	desmin (52,000)	muscle cells
	glial fibrillary acidic protein (45,000)	glial cells (astrocytes and some Schwann cells)
Type III	neurofilament proteins (about 130,000[*]; 100,000[*]; and 60,000)	neurons
Type IV	nuclear lamins A, B, and C (65,000 − 75,000)	nuclear lamina of all cells

[*]Because these proteins migrate abnormally slowly on SDS gels, they were initially thought to be larger.

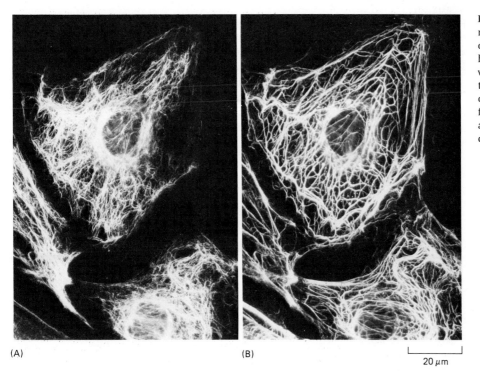

(A)   (B)   20 μm

Figure 11–73 Immunofluorescence micrographs of rat kangeroo epithelial cells (PtK2 cells) in interphase. The cells have been labeled with antibodies to vimentin (A) as well as with antibodies to keratin (B). Note that the cells contain separate arrays of vimentin filaments and keratin filaments, although the two arrays have a similar distribution. (Courtesy of Mary Osborn.)

to form homopolymers and will also co-assemble with the other Type II IF proteins to form heteropolymers. Indeed, co-polymers of vimentin and desmin, or of vimentin and glial fibrillary acidic protein, are found in some types of cells in tissues.

Type III IF proteins assemble into *neurofilaments*, a major cytoskeletal element in nerve axons and dendrites, and consequently are called **neurofilament proteins.** They consist of three distinct polypeptides in vertebrates, the so-called neurofilament triplet. The *type IV* IF proteins are the **nuclear lamins** (see p. 665), which have an amino acid sequence similar to the other IF proteins but differ from them in several ways. Most notably, the lamins form highly organized two-dimensional sheets of filaments, which rapidly disassemble and reassemble at specific stages of mitosis.

All eucaryotic cells make nuclear lamins and usually at least one type of cytoplasmic IF protein. Some cells make two types of cytoplasmic intermediate filaments, which form separate arrays in the cell. Some epithelial cells, for example, have distinct arrays of keratin and vimentin intermediate filaments (Figure 11–73).

11-28 Intermediate Filaments Are Formed from a Dimeric Subunit with a Central Rodlike Domain[53]

Figure 11–74 All IF proteins share a similar central region (about 310 amino acid residues) that forms an extended α helix with three short interruptions. The amino-terminal and carboxyl-terminal domains are non-α-helical and vary greatly in size and sequence in different IF proteins.

Despite the large differences in their size, all cytoplasmic IF proteins are encoded by members of the same multigene family. Their amino acid sequences indicate that each IF polypeptide chain contains a homologous central region of about 310 amino acid residues that forms an extended α helix with three short non-α-helical interruptions (Figure 11–74). Moreover, long stretches of this central region have

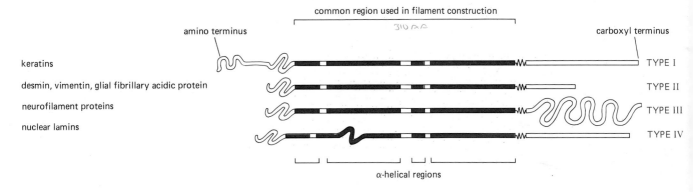

the distinctive sequence motif of a polypeptide that forms a coiled coil (see p. 618). As in tropomyosin or the tail of muscle myosin, the coiled coil is double-stranded, with two identical IF polypeptides contributing to form a dimer. The two polypeptide chains in an IF homodimer line up in parallel to form a central rodlike domain with two globular domains at each end. In assembling into an intermediate filament, the rodlike domains interact with one another to form the uniform core of the filament, while the globular domains, which vary in size in different intermediate filament proteins, project from the filament surface. One view of the assembly of the dimeric subunits into intermediate filaments is shown in Figure 11–75.

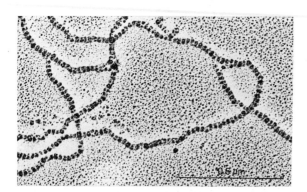

Figure 11–75 A current model of intermediate filament assembly. The monomer shown in (A) pairs with an identical monomer to form a dimer (B) in which the conserved α-helical central regions are aligned in parallel and are wound together into a coiled coil. Two dimers then line up side by side to form a 48-nm by 3-nm *protofilament* containing four polypeptide chains (C). These protofilaments then associate in a staggered manner to form successively larger structures (D and E). The final 10-nm diameter of the intermediate filament is thought to be composed of 8 protofilaments (32 polypeptide chains) joined end on end to neighbors by staggered overlaps to form the long ropelike filament (F). An electron micrograph of the final filament is shown above. It is not known whether intermediate filaments are polar structures (like actin and tubulin) or nonpolar (like the DNA double helix), or whether the two coiled coils in the four-chain protofilament are aligned in a parallel (polar) or an antiparallel (nonpolar) orientation. (Micrograph courtesy of N. Geisler and K. Weber.)

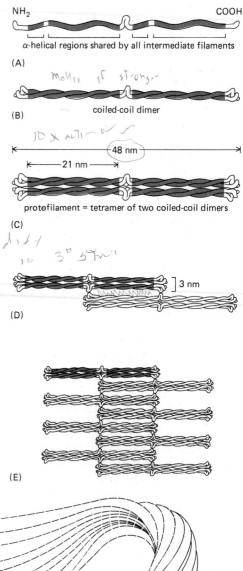

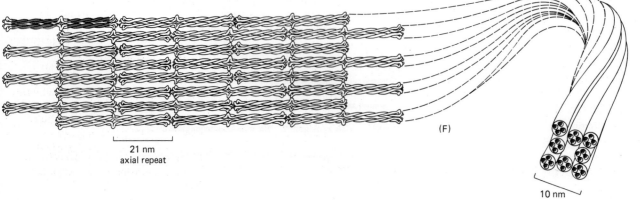

Intermediate Filaments Extend from the Nuclear Envelope to the Cell Periphery[54]

If a cell in culture is stained with an antibody to a cytoplasmic IF protein such as vimentin, a delicate network of threadlike filaments is usually seen surrounding the nucleus and extending through the cytoplasm (see Figure 11–73). The pattern is different from that of other elements of the cytoskeleton, although in places it appears to be co-linear with cytoplasmic microtubules. The organization of cytoplasmic intermediate filaments appears to depend on an interaction with microtubules, since the depolymerization of microtubules by drugs such as colchicine leads to the collapse of the entire intermediate filament network into a perinuclear cap. This suggests that many cytoplasmic intermediate filaments may be linked to the nuclear envelope and that they are normally pulled outward toward the cell periphery by attachments to microtubules.

The organization of cytoplasmic intermediate filaments may also depend on interactions with the plasma membrane. In avian red blood cells (which, unlike their mammalian counterparts, have a nucleus and intermediate filaments), vimentin is thought to bind indirectly to the plasma membrane by binding to ankyrin (see p. 289). In epithelial cells, keratin intermediate filaments are attached to the plasma membrane at desmosomes—specialized intracellular junctions that help hold neighboring cells together (see p. 797). Because the keratin filaments in each cell are connected via desmosomes to those of its neighbors, they form a continuous network throughout the entire epithelium.

11-29 Intermediate Filament Assembly May Be Controlled by Phosphorylation[55]

Isolated intermediate filaments are extremely stable under the ionic conditions in the cytoplasm. Moreover, there is not a large pool of unpolymerized intermediate filament proteins in the cell, as there is for actin and tubulin. Yet cells can clearly regulate the number, length, and position of the intermediate filaments they contain, indicating that they can control the assembly and disassembly of these filaments. An important factor in this control is the phosphorylation of specific residues on the IF proteins. Vimentin, for example, exists in a nonphosphorylated as well as a phosphorylated form. If isolated vimentin filaments are phosphorylated by protein kinases, the filaments disassemble into smaller units. The most convincing example of the importance of phosphorylation in the control of intermediate filament disassembly, however, is provided by the nuclear lamins, which disassemble each time the cell enters mitosis.

The Nuclear Lamina Is Constructed from a Special Class of Intermediate Filaments[56]

The **nuclear lamina** is a protein meshwork (typically 10 to 20 nm thick) that lines the inside surface of the inner nuclear membrane (see p. 481). It consists of a square lattice of intermediate filaments (Figure 11–76A), which in mammals are composed of three IF proteins, known as *lamins A, B, and C* (see Figure 11–74 and Table 11–3). The lamins form dimers that have a rodlike domain and two globular heads at one end (Figure 11–76B). Under appropriate conditions of pH and ionic strength, the dimers spontaneously associate into filaments that have a diameter and repeating structure similar to those of cytoplasmic intermediate filaments.

The nuclear lamins differ from the cytoplasmic IF proteins in several respects, however. The most obvious difference is that the filaments they form assemble into a square lattice (see Figure 11–76A), although this presumably requires their association with other proteins. In addition, the nuclear lamina is a very dynamic structure. In mammalian cells undergoing mitosis, the transient phosphorylation of several serine residues on the lamins causes the lamina to reversibly disassemble into tetramers of hyperphosphorylated lamin A and lamin C and membrane-associated lamin B. As the cell reenters interphase, the lamins are dephosphorylated and an intact nuclear envelope re-forms around the separated chromosomes (see p. 777).

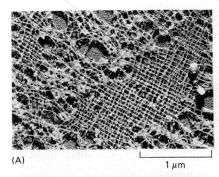

Figure 11–76 (A) Electron micrograph of a portion of the nuclear lamina in a *Xenopus* oocyte prepared by freeze-drying and metal shadowing. The lamina is formed by a highly organized square lattice of intermediate filaments composed of nuclear lamins. (B) Electron micrograph of metal-shadowed isolated lamin dimers (marked L). They have an overall form similar to muscle myosin (marked M), but they are much smaller molecules. The globular heads are formed from the two large carboxyl-terminal domains. (Courtesy of Ueli Aebi.)

Keratin Filaments Are Remarkably Diverse[52]

Among the most stable and long-lived types of intermediate filaments are those formed from keratins. They are also the most diverse. The simplest epithelia, such as those found in developing embryos and in some adult tissues such as the liver, contain just two types of keratin, one acidic and one neutral. Epithelia in other locations (for example, tongue, bladder, and sweat glands) contain six or more keratins—the particular "blend" depending on their anatomical location. Because of the variety and stability of keratin filaments, they provide a distinctive "fingerprint," which is useful for tracing the origins of tumors to particular types of epithelial cells.

The diversity of keratins is even more pronounced in the epidermis of the skin, which consists of a tough, stratified epithelium (see p. 968). Distinct sets of keratin proteins are expressed by the cells in the different layers of the epidermis. The keratin filaments eventually become covalently cross-linked to one another and to associated proteins, and as cells in the outermost layers of the epidermis die, the cross-linked keratins persist as a major part of the protective outer layer of the animal. Specialized epithelial cells at particular locations in the skin provide regional variation by generating surface appendages such as hairs, nails, and feathers. Thus the intermediate filaments help provide the animal with its primary barrier against heat and water loss, as well as supplying it with camouflage, armament, and ornamentation.

What Is the Function of Intermediate Filaments?

Animal cells can survive without cytoplasmic intermediate filaments. The glial cells that make myelin in the central nervous system do not have any, and those in cultured fibroblasts can be disrupted by an intracellular injection of antibodies against IF proteins without apparent effects on cell organization or behavior. It seems likely that the principal function of most intermediate filaments is to provide mechanical support to the cell and its nucleus. Intermediate filaments in epithelia form a transcellular network that seems designed to resist external forces. The neurofilaments in the nerve cell axon probably resist stresses caused by the motion of the animal, which would otherwise break these long, thin cylinders of cytoplasm. Desmin filaments provide mechanical support for the sarcomeres in muscle cells, and vimentin filaments surround (and probably support) the large fat droplets in fat cells.

But if the function of intermediate filaments is simply to resist tension, why are there so many types of subunit proteins? And what is the function of the variable parts of the molecule, which do not appear to be involved in formation of the filament itself? A detailed answer to these questions cannot be given at present, but it is clear from our examples that the type of support provided by intermediate filaments, and the way it is harnessed to other components, varies greatly among cell types. The desmin filaments that appear to "tie" the edges of the Z discs together in striated muscle cells are likely to have binding sites for specific proteins in the Z disc. Neurofilaments are subjected to lesser forces but may have to be linked together side by side to provide a continuous "rope" a meter or more in length; it is perhaps for this reason that neurofilaments have large projections along their length while other intermediate filaments do not (Figure 11–77).

The different requirements for binding to other proteins must be provided by the variable regions of the IF proteins. By modulating the properties of the intermediate filament, the variable regions determine not only its ability to self-associate but also how it interacts with other cellular components, such as microtubules or the plasma membrane. This strategy contrasts with that used by the two other major elements of the cytoskeleton—actin filaments and microtubules. As we have seen, these polymers are largely invariant in structure, and their properties are adapted to different functions by diverse sets of actin-binding proteins and microtubule-associated proteins. In a sense, therefore, the variable regions of

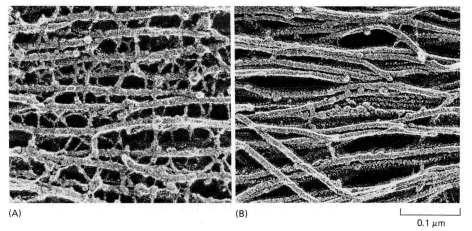

Figure 11–77 Electron micrographs of two types of intermediate filaments seen in neural tissues prepared by fast-freezing and deep-etching. (A) Neurofilaments in a nerve cell axon are extensively cross-linked through protein cross-bridges—an arrangement believed to provide great tensile strength in this long cell process. The cross-links are thought to be formed by the long nonhelical extensions at the carboxyl terminus of the largest neurofilament protein (see Figure 11–74). (B) The intermediate filaments (called glial filaments) in an astrocyte are subjected to less mechanical stress. They are smooth and have few cross-bridges. (Courtesy of N. Hirokawa.)

(A)

(B)

0.1 μm

IF proteins may serve a function similar to some of the accessory proteins of actin filaments and microtubules, with the primary difference that they are covalently linked to the subunit of the filament itself rather than being a separate protein.

Summary

Intermediate filaments are ropelike polymers of fibrous polypeptides that are thought to play a structural or tension-bearing role in the cell. A variety of tissue-specific forms are known that differ in the type of polypeptide they contain: these include the keratin filaments of epithelial cells, the neurofilaments of nerve cells, the glial filaments of astrocytes and Schwann cells, the desmin filaments of muscle cells, and the vimentin filaments of fibroblasts and many other cell types. The nuclear lamins, which form the fibrous lamina that underlies the nuclear envelope, are a separate family of intermediate filament proteins that are present in all eucaryotic cells.

The polypeptides of the different types of intermediate filaments differ in amino acid sequence and have very different molecular weights. But they all contain a homologous central domain that forms a rigid coiled-coil structure when the protein dimerizes. These dimeric subunits associate with one another in large overlapping arrays to form the intermediate filament. The rodlike domains of the subunits form the structural core of the intermediate filaments, while the globular domains at either end project from the surface and allow the individual filaments to vary in character. This variation allows the mechanical properties of the intermediate filaments and their associations with other components of the cell to match the requirements of the particular cell type.

Organization of the Cytoskeleton[57]

Up to this point we have discussed microtubules, actin filaments, and intermediate filaments as though they were independent cytoskeletal components in the cell. But the different parts of the cytoskeleton must, of course, be linked together and their functions coordinated in order to mediate changes in cell shape and produce various types of cell movement. When a fibroblast in culture rounds up to divide, for example, the entire cytoskeleton is reorganized: stress fibers and cytoplasmic microtubules are disassembled, while a mitotic spindle and then a contractile ring are formed, all as part of a controlled sequence of events.

In this section we shall discuss the interactions among the major filament systems of the cytoskeleton in relation to three of its functions. First we shall examine how the cytoskeleton helps to organize the contents of the cytoplasm, including components that are usually thought to be freely soluble. Next we shall consider how the coordinated behavior of the cytoskeleton enables an animal cell to crawl in a directed fashion over a solid surface. Lastly we shall discuss how the cytoskeleton generates the many morphological changes involved in the development of an embryo. In the course of this account, it will become apparent that we have only a rudimentary understanding of the molecular mechanisms involved in these basic processes.

The Cytoplasm Contains a Complex Three-dimensional Network of Protein Filaments[58]

We have seen that the cortex of many animal cells contains a meshwork of cross-linked actin filaments. Similar meshworks, formed by interactions among actin filaments, microtubules, and intermediate filaments, occur throughout the cytoplasm. They are most clearly seen when cells are extracted with a nonionic detergent to remove the phospholipids and soluble proteins. Metal replicas of cells treated in this way and then quickly frozen and deep-etched give a particularly striking view of the cytoplasm (Figure 11–78). The different types of protein filaments can be identified by their diameter and, in some cases, by the arrangement of their protein subunits. Adjacent filaments are often seen to be connected by thinner strands, and in some cases these have been identified as specific proteins by staining with antibodies. The connections include various kinds of microtubule-associated proteins and the long flexible side arms projecting from some types of intermediate filament subunits (see Figure 11–77A). The proteins that form the majority of the cross-links, however, are unknown.

The structure of the cytoplasm in cells that have not been treated with detergent is even more complex. Spaces between cytoskeletal filaments are filled with a granular "ground substance," which is believed to correspond to the very concentrated mixture of "soluble" proteins in the living cell. Membrane-bounded organelles of many kinds are also seen to be embedded in this dense matrix and linked to the cytoskeleton by thin strands of protein. Both the granular material and the organelles are found in greatest abundance in more central regions of the cell, where microtubules and intermediate filaments are concentrated and where the majority of cytoplasmic transport takes place, as seen by video-enhanced light microscopy. In the more peripheral regions there is a much denser network of actin filaments, which appears to exclude most of the membrane-bounded organelles and possibly some of the granular material (Figure 11–79). This dense network is anchored to the plasma membrane and corresponds to the actin-rich cortex discussed previously (see p. 629).

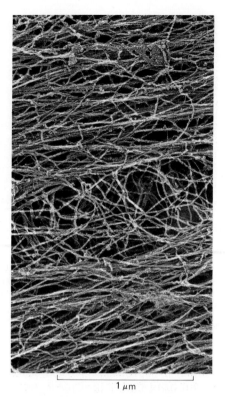

1 µm

Figure 11–78 Deep-etch electron micrograph of the cytoplasm of a fibroblast extracted with a nonionic detergent. Most of the straighter filaments arranged in loose bundles running from left to right are actin filaments; those that criss-cross (center of the micrograph) are predominantly intermediate filaments. (Courtesy of John Heuser and Marc Kirschner.)

How Well Organized Is the Cytoplasm?[59]

As they move rapidly from place to place in the cytoplasm, membrane-bounded organelles are propelled along protein tracks to which they are linked by cross-bridges. As previously discussed, kinesin and dyneinlike proteins can generate movement along microtubules (p. 660), while myosinlike proteins can generate movements along actin filaments (p. 632). Clusters of cytosolic ribosomes are also frequently observed in association with filaments; and when cells are extracted with nonionic detergents, much of the protein-synthesizing machinery remains behind with the cytoskeleton. Even soluble enzymes—including some of those involved in glycolysis—appear to be bound to specific sites on myofibrils in muscle cells and on stress fibers in fibroblasts, where they can be detected by fluorescent antibodies.

The degree of organization in the cytosol is currently a matter of debate. What we know of the function of the cytosol comes mainly from biochemical studies that begin with the homogenization of the cell so that its enzymes can be assayed and purified. The success of this approach has led many biochemists to view the cytosol as merely a solution of enzymes. But others have suggested that many, if not most, of the enzymes in the cytosol are physically clustered, pathway by pathway, and attached to the cytoskeleton to permit a more efficient channeling of intermediates within each pathway. Because these attachments would presumably be weak and easily disrupted, novel methods, such as the injection of fluorescently labeled proteins into living cells, will probably be needed to demonstrate them (see p. 157).

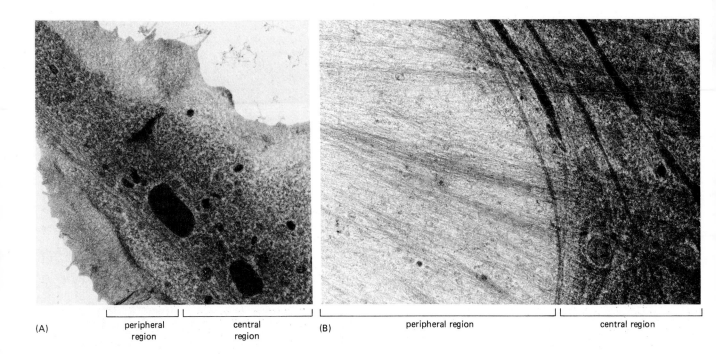

(A) peripheral region   central region   (B) peripheral region   central region

Stretching the Actin-rich Cortex Can Polarize a Cell for Directed Migration[60]

In addition to providing a framework for the attachment and movement of cytoplasmic components, the cytoskeleton is responsible for cell migration. Despite the many recent advances in understanding the structure and functions of the cytoskeleton, the mechanism by which an animal cell crawls along a surface is still poorly understood. Effective movement requires that the cell be polarized so that its plasma membrane is relatively quiescent everywhere except at its **leading edge,** where lamellipodia and microspikes periodically project outward as the cell crawls forward.

When a nonpolarized migratory cell first attaches to the surface of a culture dish, microspikes and lamellipodia extend in all directions, pulling the cell one way or another without resulting in net migration. When two opposing lamellipodia chance to become strongly engaged, however, the stretching of the cell cortex in between is seen to inhibit the further formation of microspikes and lamellipodia in the stretched region—probably because the actin filaments in this region lie parallel to the plasma membrane (Figure 11–80), rather than perpendicular to it, as required for the formation of lamellipodia or microspikes (see p. 638).

Figure 11–79 Transmission electron micrograph of a directly frozen freeze-substituted fibroblast reveals an organelle-free region in the periphery of the cell and an organelle-rich region toward its center (A). The boundary between the two regions is seen more clearly at higher magnification (B). (From P.C. Bridgman, B. Kachar, and T.S. Reese, *J. Cell Biol.* 102:1510–1521, 1986. Reproduced by permission of the Rockefeller University Press.)

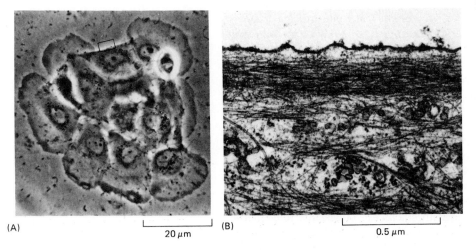

(A)   20 μm   (B)   0.5 μm

Figure 11–80 The arrangement of actin filaments in a region of stretched cell cortex. The pulling action of opposing lamellipodia in the group of eight epithelial cells shown in the light micrograph (A) generates a quiescent area in one cell where protrusive activity is suppressed (*boxed*). When this region is examined by electron microscopy (B), bundles of actin filaments are found that run parallel to the plasma membrane and therefore perpendicular to the orientation needed to generate lamellipodia or microspikes. (Courtesy of John Kolega, *J. Cell Biol.* 102:1400–1411, 1986. Reproduced by permission of the Rockefeller University Press.)

Figure 11–81 Generation of polarity in a migrating cell. Although illustrated here for a cell in culture, a similar process is thought to play a fundamental role in the morphogenesis of many tissues. Upon attaching to the surface of a culture dish, the cell shown extends lamellipodia in all directions. As these pull on the cell, they create intervening regions in which the cortex is stretched and further formation of lamellipodia is suppressed (as illustrated in Figure 11–80). The tug-of-war continues until one lamellipodium becomes dominant—either by chance, because it has encountered a particularly adherent region of the substratum, or because a slight chemotactic gradient favors protrusions toward one side of the cell (see p. 641). The cell then adopts a unipolar form and migrates in the direction of the successful lamellipodium. By means of a similar tug-of-war, a cell in a developing tissue will move toward those cells or elements of the extracellular matrix to which it is most adherent, often choosing between alternatives that differ only slightly in their degree of adhesiveness (see p. 832).

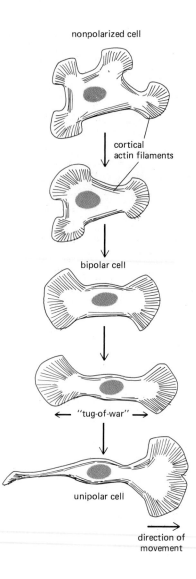

nonpolarized cell

cortical actin filaments

bipolar cell

"tug-of-war"

unipolar cell

direction of movement

Mechanical effects alone, therefore, can explain why protrusive activity is locally blocked when the cortex is stretched.

As the competition among lamellipodia continues, they engage in a tug-of-war: those that pull less strongly detach and become part of the stretched, quiescent cortex. Eventually only one active region of the plasma membrane will remain to become the leading edge of the now polarized cell, which will begin to move in one direction (Figure 11–81).

On a uniform planar surface, selection of successful lamellipodia and microspikes will be a matter of chance, depending on random variations in the arrangement of cytoskeletal components. On a nonuniform surface, however, such as a culture substratum coated with a gradient of an adhesive substance, lamellipodia and microspikes will be selected by features of the environment and will therefore act as "feelers" that determine the eventual direction of migration.

11-34 The Relative Importance of Actin Filaments and Microtubules in Cell Migration Varies in Different Cell Types[61]

When a cell migrates, microtubules and the actin-based cytoskeleton normally act in concert, so that their individual roles are difficult to discern. The extent to which microtubules contribute to the directed movement of a cell varies among cell types. If the microtubules in a fibroblast are depolymerized with colchicine, the cell ceases to move directionally and instead begins to extend lamellipodia in seemingly random directions. The directed migration of a neutrophil, in contrast, is hardly affected by colchicine. Moreover, if a neutrophil is heated briefly to 42°C, a major portion of its cortical cytoplasm detaches from the rest of the cell and "walks away." Even though these motile cell fragments lack microtubules or a cell nucleus, they are able to move about on a culture dish for a day or more before they disintegrate (Figure 11–82). Initially the fragments are as effective as the parent neutrophil in chemotaxis: they will move in a directed manner up a gradient of N-formylated peptides, for example (see p. 641). Migration stops if the actin filaments in neutrophils (or in fragments derived from neutrophils) are depolymerized by treatment with cytochalasin. In these cells it seems that directed cell migration requires actin filaments but not microtubules.

For the growth cones of nerve cells, however, migration seems to depend on both microtubules and actin filaments. Nerve cells contain an unusually highly ordered array of microtubules in their axons and dendrites (collectively called *neurites*). Each neurite is normally pulled out of the nerve cell body under tension generated by the growth cone at the growing tip of the neurite. Neurite outgrowth stops if nerve cells in culture are treated with colchicine, suggesting that microtubules are required for growth cone migration. The growth cone is equivalent to

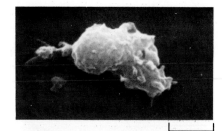

Figure 11–82 Scanning electron micrograph of a neutrophil that has been briefly heated to 42°C. As a result of the heat treatment, a large cell fragment has been produced that is capable of persistent movement and chemotaxis, even though it lacks a nucleus and microtubules. (From S.E. Malawista and A. De Boisfleury Chevance, *J. Cell Biol.* 95:960–973, 1982. Reproduced by permission of the Rockefeller University Press.)

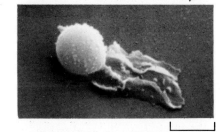

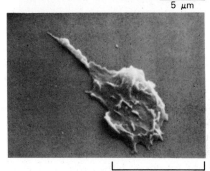

the leading edge of a neutrophil or fibroblast, and the formation of its lamellipodia and microspikes depends on actin filaments and ceases when cells are treated with cytochalasin. Nonetheless, provided that nerve cells are on an especially adhesive surface (such as a culture dish coated with the polycation polylysine), neurites will continue to grow in the presence of cytochalasin, although the direction of this growth is abnormally erratic. The lamellipodia and microspikes are apparently required to steer the growth cone and to allow it to migrate over poorly adhesive surfaces. The arrangements of microtubules and actin filaments in the growth cone and the leading edge of a fibroblast are compared in Figure 11–83.

Tension in the Actin-rich Cortex May Help Drive Animal Cell Locomotion[62]

What is the molecular mechanism that propels a migrating cell forward? The answer to this crucial question is unknown. One hypothesis, which emphasizes the role of the actin-rich cortex, is illustrated in Figure 11–84. There is good evidence that the cortex in animal cells is under tension. This cortical tension, which has been measured in very large cells such as sea urchin eggs, tends to cause cells in suspension to adopt a spherical shape so as to minimize their total surface area. Although the mechanism is unknown, cells appear to be able to relax the cortical tension at specific regions of their surface, such as at the leading edge. Thus lamellipodia extend periodically from the leading edge, presumably as the result of the activation of special capping proteins in the plasma membrane in

Figure 11–83 The arrangement of microtubules and actin filaments in two types of motile cells. Staining with a fluorescein-coupled antibody to tubulin (A) reveals microtubules as a fine network throughout the cytoplasm of a fibroblast and as a densely packed bundle in the center of a nerve axon. When the same cells are stained with rhodamine-phalloidin (B), actin filaments are seen in bundles that extend to the leading edge of the fibroblast and are concentrated in the microspikes of the growth cone of the axon. (Courtesy of Peter Hollenbeck.)

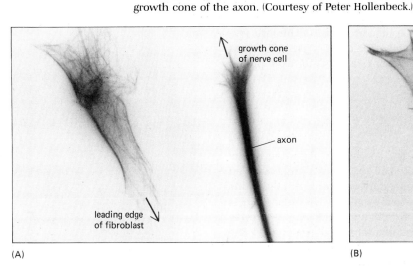

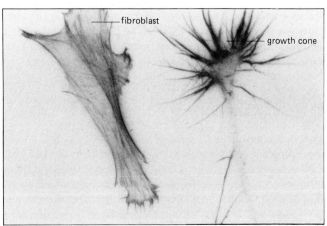

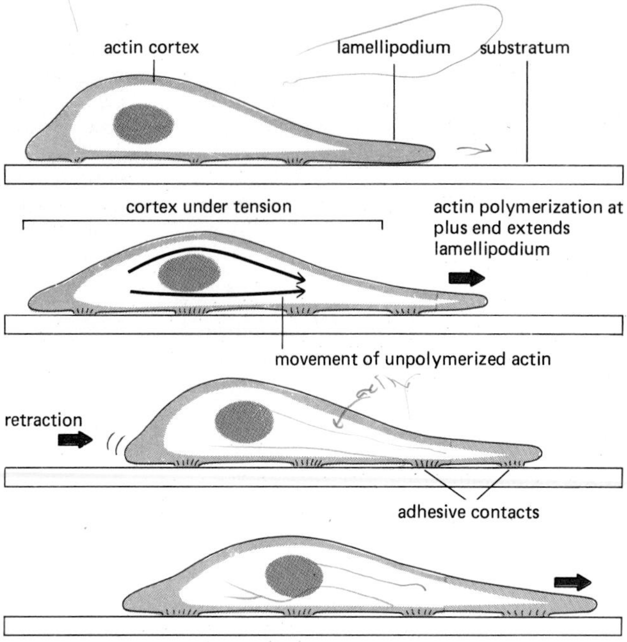

actin cortex | lamellipodium | substratum

cortex under tension | actin polymerization at plus end extends lamellipodium

movement of unpolymerized actin

retraction

adhesive contacts

Figure 11–84 One model of how forces generated in the actin-rich cortex might move a cell forward. The actin-dependent extension and firm attachment of a lamellipodium at the leading edge stretches the actin cortex. The cortical tension then draws the body of the cell forward to relax some of the tension. The same cycle can be repeated over and over again, moving the cell forward in a stepwise fashion.

this region that allow the addition of actin subunits to the plus ends of actin filaments (see p. 643). It is hypothesized that those lamellipodia that fail to attach firmly to the substratum are pulled backward by the cortical tension to produce a "ruffle" (see Figure 11–42). But where a lamellipodium adheres tightly, transmembrane connections will anchor its actin filaments to the substratum (see Figure 11–43): now the same cortical tension that pulls an unattached lamellipodium toward the rear of the cell will instead move the body of the cell toward the new point of attachment, pulling the whole cell forward (see Figure 11–84).

By analogy with muscle, the best-characterized actin-based motile system, one might expect that the contractile forces in the cortex would be generated by an interaction between actin and myosin filaments. Experiments with the cellular slime mold *Dictyostelium discoideum* (see p. 825), however, argue against this possibility. By isolating the gene that codes for filamentous myosin in this organism, modifying it to remove the long sequence that encodes the protein, and transferring it back into cells so that it replaces the normal myosin gene (see p. 194), mutant organisms that lacked this myosin were obtained. Not surprisingly, the mutant cells were unable to form a contractile ring and therefore developed into multinucleated giant cells, which divided only occasionally by tearing themselves in two. Remarkably, however, the cells could migrate and respond chemotactically to a source of cyclic AMP (see p. 826), although both processes were somewhat impaired. It seems that neither coherent cell migration nor the development of cortical tension is solely dependent on bipolar myosin filaments. Perhaps the cortical tension can also be provided by an elastic net of actin filaments (resembling a rubber sheet) or by other pulling forces generated, for example, by actin disassembly or by minimyosin.

The Endocytic Cycle May Help Extend the Leading Edge of Migrating Cells[63]

As we discussed in Chapter 6, all animal cells are continually ingesting bits of their plasma membrane and returning them to the cell surface in a process called the *endocytic cycle* (see p. 323). In polarized cells crawling along a substratum, there is evidence that membrane is internalized all over the cell surface but is preferentially returned to the leading edge. This asymmetry in the endocytic cycle is thought to help extend the leading edge of the migrating cell (see p. 333). The return of recycling membrane to the leading edge is likely to depend on the oriented microtubules and actin filaments in a polarized cell, both of which can

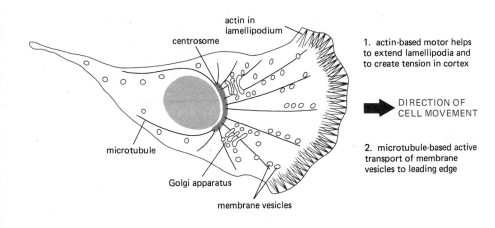

actin in lamellipodium

centrosome

microtubule

Golgi apparatus

membrane vesicles

1. actin-based motor helps to extend lamellipodia and to create tension in cortex

DIRECTION OF CELL MOVEMENT

2. microtubule-based active transport of membrane vesicles to leading edge

Figure 11–85 Two directional motors in a migrating cell—one based on actin filaments in the cortex and the other based on the recycling of membrane vesicles along microtubules (and perhaps actin filaments)—are thought to cooperate to drive movement in one direction.

bind accessory proteins that direct the active transport of membrane vesicles toward their plus ends (see p. 660 and p. 620). Thus migratory cells seem to have at least two types of directional motors to power cell locomotion—one based on actin filaments in the cell cortex, which extends lamellipodia and generates cortical tension, and another requiring oriented microtubules or actin filaments (or both) in the cell interior, which actively transports recycling membrane vesicles to the cell's leading edge (Figure 11–85).

Microtubules Can Organize Other Cytoskeletal Filaments

In addition to their role in cell movement, microtubules play a crucial part in controlling cell shape. All eucaryotic cells have a distinct geometry, both in terms of their external features and in terms of the position of their organelles. While all components of the cytoskeleton participate in this geometry, microtubules often seem to play a unique part in determining it. It is a common observation, for example, that microtubules are aligned with the long axis of elongated cells, and in many cases their presence is essential for the maintenance of the cell's elongated shape.

As mentioned previously, microtubules determine the location of the Golgi apparatus and the endoplasmic reticulum in each cell (see p. 660), and they influence the distribution of intermediate filaments—which collapse into a perinuclear cap if a cell is treated with colchicine (see p. 652). Microtubules can also determine the distribution of actin filaments in a cell. The actin-based contractile ring, whose action pinches the cell in half to complete the process of cell division, always forms in a plane perpendicular to the equator of the microtubule-based mitotic spindle. If the forming mitotic spindle is displaced mechanically, the site at which the contractile ring subsequently forms is correspondingly changed, indicating that the position of the contractile ring is determined by the position of the spindle (see p. 779).

The mechanisms by which microtubules influence the positioning of intermediate filaments and cortical actin filaments are unknown, although they presumably involve proteins that link together the different types of protein filaments.

Cytoskeletal Organization Can Be Transmitted Across Cell Membranes[64]

The cytoskeleton of one cell can influence that of its neighbors, and this form of intercellular communication is thought to play a major part in determining the morphology of tissues and organs. One of the simplest interactions between the cytoskeleton of one cell and that of another occurs when the leading edges of two migrating cells come into contact. In most cell types the encounter causes an immediate paralysis of the leading edge of each cell—a phenomenon known as **contact inhibition of movement.** Thus two fibroblasts that collide in culture

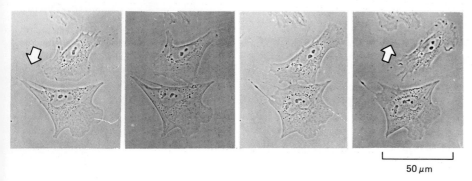

Figure 11–86 Contact inhibition of fibroblast movement. If two fibroblasts crawling along the surface of a culture dish collide, the lamellipodia in the region of contact are paralyzed. After a period of 10 to 15 minutes, the cells typically move away from each other in different directions, as shown here. (Courtesy of Graham Dunn.)

50 μm

cease generating microspikes and lamellipodia in the region of contact and begin to produce them elsewhere, so that eventually the two cells move away from each other along new trajectories (Figure 11–86). This response presumably involves rapid changes in the actin-based cortical cytoskeleton at the region of contact, but the molecular mechanism is unknown.

Contact inhibition of movement should not be confused with the *contact inhibition of cell division* seen in cultured cells that have proliferated until they cover the entire surface of a culture dish. As we shall see in Chapter 13, cessation of growth and proliferation under these circumstances does not depend solely on contact between the cells; it also depends on the shapes the cells are forced to adopt when crowded and on their increasingly limited supply of nutrients (see p. 748).

Contact inhibition of movement is an essential feature of wound healing in animals. Sheets of epithelial cells at the margin of a wound move rapidly out over the wounded area by extending lamellipodia; this movement ceases as soon as cells from different margins make contact across the gap created by the wound. Once a continuous cell sheet has been reestablished, junctions form between the newly adjacent cells, providing anchor points for protein filaments that connect the cytoskeletons of the cells in the sheet (see p. 795). Contact inhibition of movement may also contribute to the selective bundling, or "fasciculation," of axons in the developing nervous system. Growth cones from neurons of the central nervous system (CNS) have been found to cease their advance and retract when they make contact with axons from neurons of the peripheral nervous system, although they grow readily along axons from other CNS neurons.

Another mechanism by which the cytoskeleton of one cell can influence that of its neighbors depends on interactions between the cytoskeleton and the extracellular matrix that a cell secretes. As described in Chapter 14, a cell with an oriented cytoskeleton tends to secrete a similarly oriented extracellular matrix, and this, in turn, will influence the orientation of the cytoskeleton in other cells in contact with the matrix (see Figure 14–54, p. 823). Because of the cell-cell interactions mediated by cell junctions and the extracellular matrix, a cell's cytoskeleton is often organized according to the pattern of the entire tissue, rather than cell-autonomously.

Concerted Cytoskeletal Contractions Underlie the Folding of Epithelial Cell Sheets[65]

Communication between the cytoskeletons of adjacent cells underlies a fundamental process in animal morphogenesis—the folding of cell sheets. One possibility is that this communication is mediated by mechanical stimuli passing from cell to cell. Imagine, for example, a sheet of epithelial cells in which the apical bundles of actin filaments in *adhesion belts* (see p. 796) contract in response to a transient rise in tension in the epithelial sheet. If one cell in the sheet constricts its adhesion belt, tension will be generated in its neighboring cells, which could cause the adhesion belts in these cells to contract as well. A wave of contraction would spread across the cell sheet. Computer models based on this simple mechanism generate a remarkably rich variety of final forms, depending on the geometry

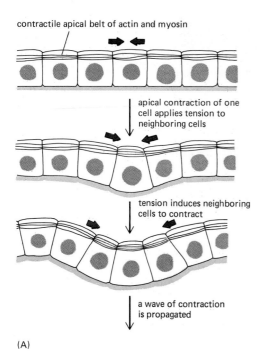

contractile apical belt of actin and myosin

apical contraction of one cell applies tension to neighboring cells

tension induces neighboring cells to contract

a wave of contraction is propagated

(A)

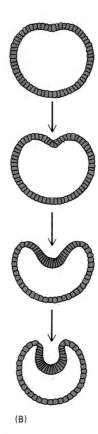

(B)

of the initial arrangement. Thus an epithelial cell sheet surrounding a spherical cavity will invaginate to form a two-layered structure in a process that resembles gastrulation in sea urchin embryos (Figure 11–87).

These models are entirely speculative, but there is no doubt that the cytoskeleton plays a fundamental part in shaping tissues. The complex form and arrangement of cells in the tissues and organs of the vertebrate body all depend on cytoskeletal proteins in highly individual arrangements. Often it is the microtubules that seem to guide the morphogenetic movements. In other cases, actin filaments play the starring role, as in the development of the hair cells in the vertebrate ear.

Development of Hair Cells in the Cochlea Depends on Precisely Controlled Actin Polymerization[66]

Hair cells are specialized epithelial cells found in the cochlea and vestibule of the inner ear. They are exquisitely sensitive to very small movements, whether caused by sound vibrations or by the displacement of fluid in the semicircular canals produced by changing the position of the head. Such movements are detected by hexagonally packed clusters of *stereocilia* on the surface of each hair cell; within each cluster the stereocilia decrease stepwise in length, row by row, giving the appearance of a pipe organ (Figure 11–88). Sounds are detected when tiny movements of stereocilia are converted into electrical signals in the hair cell, which transmits the signal to the brain (see p. 1102).

In addition to the precise arrangement of stereocilia on the surface of each hair cell, their size, number, and orientation vary in a systematically graded fashion from one end of the cochlea to the other. Thus each hair cell at the proximal end of the chicken cochlea has more than 250 stereocilia with an average length of 1.7 μm, while at the opposite end, each has 50 stereocilia with an average length of 5.2 μm. These and other gradations in structure reflect the functional organization of hair cells, which are tuned to respond to low frequencies at one end of the cochlea and to high frequencies at the other. The dimensions of the array of sterocilia on any hair cell are so amazingly precise and reproducible that they can be used to pinpoint the position of the cell in two axes on the cochlea.

Stereocilia are very large, specialized microvilli and are therefore unrelated to true cilia. They are formed as plasma membrane protrusions and contain a bundle

Figure 11–87 Computer model of gastrulation based on a wave of cytoskeletal contraction that passes from cell to cell. Each cell in the epithelial sheet contains an adhesion belt (a cell junction containing a contractile bundle of actin filaments) that is postulated to contract in response to stretching. Because the adhesion belt is situated at the apical end of the cell, its contraction changes the shape of the cell from a cylinder to a cone. In a sheet of contiguous cells, contraction at one point will pull on neighboring cells, inducing them to contract in succession (A). Operation of this principal over a spherical or cylindrical cell sheet will then generate an invagination such as that shown in (B). The precise form of the invagination will depend on the mechanical properties assumed in the model: that shown in (B) was designed to mimic gastrulation in a sea urchin embryo. (From G. Odell, G. Oster, P. Alberch, and B. Burnside, *Dev. Biol.* 85:446–462, 1981.)

Figure 11–88 Scanning electron micrograph of the surface of hair cells, showing the "pipe-organ" arrangement of stereocilia. (Courtesy of Lewis Tilney.)

of actin filaments in their core (Figure 11–89). The formation of the cochlea, therefore, provides a remarkable demonstration of the ability of cells to control the number, position, and length of their actin filaments in a tissue-wide pattern.

Scanning electron microscopy of chick embryos shows that stereocilia grow in three distinct stages (Figure 11–90). First, short stereocilia appear on each hair cell, emerging on the hair cell surface more or less simultaneously. These then begin to elongate—those destined to be the longest in the final array form first, the second-longest next, and so on. Once a row of stereocilia has started to grow, it continues to elongate at the same slow rate of about 0.5 μm/day, apparently adding actin monomers to the distal tip (the *plus end,* see p. 639) of each actin filament. After three or four days of this steady growth, each hair cell carries a truncated version of the final pipe-organ array on its surface.

For the next six days, actin monomers are thought to add only to the base of the stereocilia—thereby increasing the length of actin filaments in the "rootlet" that extends into the cortex of the hair cell. Growth at this stage evidently takes place on the minus ends of the actin filaments, showing that, if necessary, this nonpreferred end for growth can be utilized by the cell. At the same time, more actin filaments are added to the core bundle in stereocilia at selected locations on the cochlea, increasing the diameter of these stereocilia.

Finally, at the third stage of growth (from embryonic day 17 to well after hatching), the process of adding actin monomers to the plus ends of the actin filament bundles is resumed, and the stereocilia again increase in length. Now the hair cells at the distal end of the cochlea show more extensive growth than those at the proximal end, stopping when the mature gradient of stereocilium lengths has been produced.

The mechanisms that allow cells to assemble actin filaments with such precision are unknown. What specifies the initial number of stereocilia on the surface of each hair cell so accurately? And how does the cell control the final length of each actin filament bundle so as to create the precise gradation of stereocilium lengths observed on each cell? These are just two of the many intriguing questions that challenge scientists studying the cytoskeleton.

Summary

Actin filaments, microtubules, intermediate filaments, and their associated proteins assemble spontaneously into a complex network of protein filaments that organizes the contents of the cytoplasm. The cytoskeleton also plays the preeminent role in determining the shape and polarity of the cell and the nature of its movements. During animal cell locomotion, the assembly of actin filaments periodically pushes out lamellipodia and microspikes on one side of the cell (the leading edge), stretching the cell cortex in a way that polarizes the cell and helps it move forward. This polarity is reinforced by microtubules or actin filaments that direct the traffic of recycling plasma membrane components to the cell's leading edge.

The cytoskeleton of one cell can influence that of its neighbors either through intercellular junctions or by its effect on the extracellular matrix. In this way, changes in cell shape are coordinated during the development of tissues and organs.

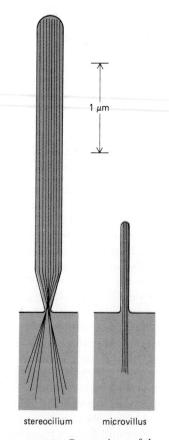

1 μm

stereocilium   microvillus

Figure 11–89 Comparison of the sizes of a typical stereocilium and a typical microvillus, both of which are supported by bundles of actin filaments in their core.

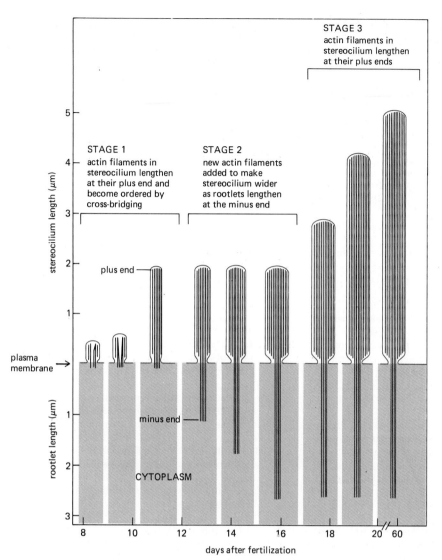

Figure 11–90   Three stages of growth of a single stereocilium in the chicken cochlea. All stereocilia go through the same three stages, but the time of initiation and duration of growth vary both among individual stereocilia on each hair cell and from one hair cell to another in the cochlea—thereby generating a remarkably precise gradation of staggered lengths. The embryos hatch at 21 days. (Adapted from L.G. Tilney and M.S. Tilney, *Hearing Research* 22:55–77, 1986.)

STAGE 3
actin filaments in stereocilium lengthen at their plus ends

STAGE 1
actin filaments in stereocilium lengthen at their plus end and become ordered by cross-bridging

STAGE 2
new actin filaments added to make stereocilium wider as rootlets lengthen at the minus end

plus end

plasma membrane

minus end

CYTOPLASM

stereocilium length (µm)

rootlet length (µm)

days after fertilization

References

General

Bershadsky, A.D.; Vasiliev, J.M. Cytoskeleton. New York: Plenum Press, 1988.

Lackie, J.M. Cell Movement and Cell Behaviour. London: Allen and Unwin, 1985.

Schliwa, M. The Cytoskeleton. New York: Springer-Verlag, 1986.

Sheterline, P. Mechanisms of Cell Motility. London: Academic Press, 1983.

Cited

1. Amos, L.A. Structure of muscle filaments studied by electron microscopy. *Annu. Rev. Biophys. Biophys. Chem.* 14:291–313, 1985.

   Huxley, A.F. Reflections on Muscle. Princeton, NJ: Princeton University Press, 1980.

   Squire, J.M. Muscle: Design, Diversity and Disease. Menlo Park, CA: Benjamin-Cummings, 1986.

2. Cooke, R. The mechanism of muscle contraction. *CRC Crit. Rev. Biochem.* 21:53–118, 1986.

   Huxley, H.E. The mechanism of muscular contraction. *Science* 164:1356–1366, 1969.

3. Cohen, C.; Parry, D.A.D. α-helical coiled coils—a widespread motif in proteins. *Trends Biochem. Sci.* 11:245–248, 1986.

   Davis, J.S. Assembly processes in vertebrate skeletal thick filament formation. *Annu. Rev. Biophys. Biophys. Chem.* 17:217–239, 1988.

4. Bessman, S.P.; Carpenter, C.L. The creatine-creatine phosphate energy shuttle. *Annu. Rev. Biochem.* 54:831–862, 1985.

5. Warrick, H.M.; Spudich, J.A. Myosin structure and function in cell motility. *Annu. Rev. Cell Biol.* 3:379–421, 1987.

6. Pollard, T.D.; Cooper, J.A. Actin and actin-binding proteins. A critical evaluation of mechanisms and functions. *Annu. Rev. Biochem.* 55:987–1035, 1986.

7. Irving, M. Muscle mechanics and probes of the crossbridge cycle. In Fibrous Protein Structure (J.M. Squire, P.J. Vibert, eds.). San Diego. CA: Academic Press, 1987.

   Pollard, T.D. The myosin crossbridge problem. *Cell* 48:909, 910, 1987.

8. Katz, B. Nerve, Muscle and Synapse. New York: McGraw-Hill, 1966.

   Lai, F.A.; Erickson, H.P.; Rousseau, E.; Liu, Q-Y.; Meissner, G. Purification and reconstitution of the calcium release channel from skeletal muscle. *Nature* 331:315–319, 1988.

Saito, A.; Inui, M.; Radermacher, M.; Frank, J.; Fleischer, S. Ultrastructure of the calcium release channel of sarcoplasmic reticulum. *J. Cell Biol.* 107:211–219, 1988.

9. Murray, J.M.; Weber, A. The cooperative action of muscle proteins. *Sci. Am.* 230(2):59–71, 1974.

Phillips, G.N.; Fillers, J.P.; Cohen, C. Tropomyosin crystal structure and muscle regulation. *J. Mol. Biol.* 192:111–131, 1986.

Zot, A.S.; Potter, J.D. Structural aspects of troponin-tropomyosin regulation of skeletal muscle contraction. *Annu. Rev. Biophys. Biophys. Chem.* 16:535–559, 1987.

10. Wang, K. Sarcomere-associated cytoskeletal lattices in striated muscle. Review and hypothesis. In Cell and Muscle Motility (J.W. Shay, ed.), Vol. 6, pp. 315–369. New York: Plenum Press, 1985.

11. Fawcett, D.W. A Textbook of Histology, 11th ed. Philadelphia: Saunders, 1986.

12. Korn, E.D.; Hammer, J.A. Myosins of nonmuscle cells. *Annu. Rev. Biophys. Biophys. Chem.* 17:23–45, 1988.

Sellers, J.R.; Adelstein, R.S. Regulation of contractile activity. In The Enzymes (P. Boyer, E.G. Krebs, ed.), Vol. 18, pp. 381–418. San Diego, CA: Academic Press, 1987.

13. Citi, S.; Kendrick-Jones, J. Regulation of non-muscle myosin structure and function. *Bioessays* 7:155–159, 1987.

14. Byers, H.R.; Fujiwara, K. Stress fibers in cells*in situ*: immunofluorescence visualization with anti-actin, anti-myosin and anti-alpha-actinin. *J. Cell Biol.* 93:804–811, 1982.

Langanger, B.; et al. The molecular organization of myosin in stress fibers of cultured cells. *J. Cell Biol.* 102:200–209, 1986.

Schroeder, T.E. Actin in dividing cells: contractile ring filaments bind heavy meromyosin. *Proc. Natl. Acad. Sci. USA* 70:1688–1692, 1973.

15. Buckingham, M.E. Actin and myosin multigene families: their expression during the formation of skeletal muscle. *Essays Biochem.* 20:77–109, 1985.

Emerson, C.P.; Bernstein, S.I. Molecular genetics of myosin. *Annu. Rev. Biochem.* 56:695–726, 1987.

Otey, C.A.; Kalnoski, M.H.; Bulinski, J.C. Identification and quantification of actin isoforms in vertebrate cells and tissues. *J. Cell. Biochem.* 34:113–124, 1987.

16. Breitbart, R.E.; Andreadis, A.; Nadal-Ginard, B. Alternative splicing: a ubiquitous mechanism for the generation of multiple protein isoforms from single genes. *Annu. Rev. Biochem.* 56:467–95, 1987.

17. Bray, D.; Heath, J.; Moss, D. The membrane-associated "cortex" of animal cells: its structure and mechanical properties. *J. Cell Sci.*, Suppl. 4:71–88, 1986.

Korn, E.D. Actin polymerization and its regulation by proteins from nonmuscle cells. *Physiol. Rev.* 62:672–737, 1982.

Pollard, T.D.; Cooper, J.A. Actin and actin-binding proteins. A critical evaluation of mechanisms and functions. *Annu. Rev. Biochem.* 55:987–1035, 1986.

Tilney, L.G. Interactions between actin filaments and membranes give spatial organization to cells. In Modern Cell Biology. Vol. 2: Spatial Organization of Eukaryotic Cells (J.R. McIntosh, B.H. Satir ed.) pp. 163–199, New York: Liss, 1983.

18. Sato, M.; Schwartz, W.H.; Pollard, T.D. Dependence of the mechanical properties of actin/alpha-actinin gels on deformation rate. *Nature* 325:828–830, 1987.

Stossel, T.P.; et al. Non-muscle actin binding proteins. *Annu. Rev. Cell Biol.* 1:353–402, 1985.

19. Matsudaira, P.; Janmey, P. Pieces in the actin-severing protein puzzle. *Cell* 54:139–140, 1988.

Yin, H.L. Gelsolin: calcium and polyphosphoinositide-regulated actin modulating protein. *Bioessays* 7:176–179, 1987.

20. Korn, E.D.; Hammer, J.A. Myosins of nonmuscle cells. *Annu. Rev. Biophys. Biophys. Chem.* 17:23–45, 1988.

Warrick, H.M.; Spudich, J.A. Myosin structure and function in cell motility. *Annu. Rev. Cell Biol.* 3:379–421, 1987.

21. Adams, R.J.; Pollard. T.D. Propulsion of organelles isolated from *Acanthamoeba* along actin filaments by myosin-I. *Nature* 322:754–756, 1986.

Sheetz, M.P., Spudich, J.A. Movement of myosin-coated fluorescent beads on actin cables *in vitro*. *Nature* 303:31–35, 1983.

22. Bennett, V. The membrane skeleton of human erythrocytes and its implications for more complex cells. *Annu. Rev. Biochem.* 54:273–304, 1985.

23. Conzelman, K.A.; Mooseker, M.S. The 110-kD protein-calmodulin complex of the intestinal microvillus in an actin-activated MgATPase. *J. Cell Biol.* 105:313–324, 1987.

Mooseker, M.S. Organization, chemistry, and assembly of the cytoskeletal apparatus of the intestinal brush border. *Annu. Rev. Cell Biol.* 1:209–241, 1985.

24. Burridge, K.; et al. Focal adhesions: transmembrane junctions between the extracellular matrix and the cytoskeleton. *Annu. Rev. Cell Biol.* 4:487–525, 1988.

Horwitz, A.; Duggan, K.; Buck, C.; Beckerle, M.C.; Burridge, K. Interaction of plasma membrane fibronectin receptor with talin—a transmembrane linkage. *Nature* 320:531–533, 1986.

25. Bonder, E.M.; Fishkind, D.J.; Mooseker, M.S. Direct measurement of critical concentrations and assembly rate constants at the two ends of an actin filament. *Cell* 34:491–501, 1983.

Korn, E.D.; Carlier, M-F.; Pantaloni, D. Actin polymerization and ATP hydrolysis. *Science* 238:638–644, 1987.

26. Abercrombie, M. The crawling movement of metazoan cells. *Proc. R. Soc. Lond. (Biol.)* 207:129–147, 1980.

Small, J.V.; Rinnerthaler, G.; Hinssen, H. Organization of actin meshworks in cultured cells: the leading edge. *Cold Spring Harbor Symp. Quant. Biol.* 46:599–611, 1982.

Wang, Y. Exhange of actin subunits at the leading edge of living fibroblasts: possible role of treadmilling. *J. Cell Biol.* 101:597–602, 1985.

27. Tilney, L.G.; Inoué, S. Acrosomal reaction of *Thyone* sperm. II. The kinetics and possible mechanism of acrosomal process elongation. *J. Cell Biol.* 93:820–827, 1982.

28. Carson, M.; Weber, A.; Zigmond, S.H. An actin-nucleating activity in polymorphonuclear leukoyctes is modulated by chemotactic peptides. *J. Cell Biol.* 103:2707–2714, 1986.

Devreotes, P.; Zigmond, S. Chemotaxis in eucaryotic cells. *Annu. Rev. Cell Biol.* 4:649–686, 1988.

Tilney, L.G.; Bonder, E.M.; DeRosier, D.J. Actin filaments elongate from their membrane-associated ends. *J. Cell Biol.* 90:485–494, 1981.

29. Cooper, J.A. Effects of cytochalasin and phalloidin on actin. *J. Cell Biol.* 105:1473–1478, 1987.

30. Dustin, P. Microtubules, 2nd ed., pp. 127–164. New York: Springer-Verlag, 1984.

Gibbons, I.R. Cilia and flagella of eukaryotes. *J. Cell Biol.* 91:107s–124s, 1981.

Roberts, K.; Hyams, J.S.; eds. Microtubules. New York: Academic Press, 1979.

Satir, P. How cilia move. *Sci. Am* 231(4):44–63, 1974.

31. Amos, L.A.; Baker, T.S. The three dimensional structure of tubulin protofilaments. *Nature* 279:607–612, 1979.

Mandelkow, E-M., Schultheiss, R., Rapp, R., Müller, M., Mandelkow, E. On the surface lattice of microtubules: helix starts, protofilament number, seam and handedness. *J. Cell Biol.* 102:1067–1073, 1986.

Raff, E.C. Genetics of microtubule systems. *J. Cell Biol.* 99:1–10, 1984.

Sullivan, K.F. Structure and utilization of tubulin isotypes. *Annu. Rev. Cell Biol.* 4:687–716, 1988.

32. Linck, R.W.; Amos, L.A.; Amos, W.B. Localization of tektin filaments in microtubules of sea urchin sperm flagella by immunoelectron microscopy. *J. Cell Biol.* 100:126–135, 1985.

33. Goodenough, U.W.; Heuser, J.E. Substructure of inner dynein arms, radial spokes, and the central pair/projection complex of cilia and flagella. *J. Cell Biol.* 100:2008–2018, 1985.

34. Summers, K.E.; Gibbons, I.R. ATP-induced sliding of tubules in trypsin-treated flagella of sea urchin sperm. *Proc. Natl. Acad. Sci, USA* 68:3092–3096, 1971.

    Warner, F.D.; Satir, P. The structural basis of ciliary bend formation. *J. Cell Biol.* 63:35–63, 1974.

35. Johnson, K.A. Pathway of the microtubule-dynein ATPase and structure of dynein: a comparison with actomyosin. *Annu. Rev. Biophys. Biophys. Chem.* 14:161–188, 1985.

36. Brokaw, C.J. Future directions for studies of mechanisms for generating flagellar bending waves. *J. Cell Sci.*, Suppl. 4:103–113, 1986.

    Brokaw, C.J.; Luck, D.J.L.; Huang, B. Analysis of the movement of *Chlamydomonas* flagella: the function of the radial-spoke system is revealed by comparison of wild-type and mutant flagella. *J. Cell Biol.* 92:722–732, 1982.

37. Afzelius, B.A. The immotile-cilia syndrome: a microtubule-associated defect. *CRC Crit. Rev. Biochem.* 19:63–87, 1985.

    Huang, B. *Chlamydomonas reinhardtii*: a model system for genetic analysis of flagellar structure and motility. *Int. Rev. Cytol.* 99:181–215, 1986.

    Luck, D.J.L. Genetic and biochemical dissection of the eucaryotic flagellum. *J. Cell Biol.* 98:789–794, 1984.

38. Lefebvre, P.A.; Rosenbaum, J.L. Regulation of the synthesis and assembly of ciliary and flagellar proteins during regeneration. *Annu. Rev. Cell Biol.* 2:517–546, 1986.

    Wheatley, D.N. The Centriole: A Central Enigma of Cell Biology. New York: Elsevier, 1982.

39. Karsenti, E.; Maro, B. Centrosomes and the spatial distribution of microtubules in animal cells. *Trends Biochem. Sci.* 11:460–463, 1986.

    Ramanis, Z.; Luck, D.J.L. Loci affecting flagellar assembly and function map to an unusual linkage group in *Chlamydomonas reinhardtii. Proc. Natl. Acad. Sci. USA* 83:423–436, 1986.

    Vorobjev, I.A.; Chentsov, Y.S. Centrioles in the cell cycle. 1. Epithelial cells. *J. Cell. Biol.* 93:938–949, 1982.

40. Dustin, P. Microtubules, 2nd ed. Berlin: Springer-Verlag, 1984.

41. De Brabander, M.; et al. Microtubule dynamics during the cell cycle: the effects of taxol and nocodazole on the microtubule system of Ptk2 cells at different stages of the mitotic cycle. *Int. Rev. Cytol.* 101:215–274, 1986.

    Inoué, S. Cell division and the mitotic spindle. *J. Cell Biol.* 91:131s–147s, 1981.

    Salmon, E.D.; McKeel, M.; Hays, T. Rapid rate of tubulin dissociation from microtubules in the mitotic spindle *in vivo* measured by blocking polymerization with colchicine. *J. Cell Biol.* 99:1066–1075, 1984.

42. Farrell, K.W.; Jordan, M.A.; Miller, H.P.; Wilson, L. Phase dynamics at microtubule ends: the coexistence of microtubule length changes and treadmilling. *J. Cell Biol.* 104:1035–1046, 1987.

    McIntosh, J.R.; Euteneuer, U. Tubulin hooks as probes for microtubule polarity: an analysis of the method and evaluation of data on microtubule polarity in the mitotic spindle. *J. Cell Biol.* 98:525–533, 1984.

43. Carlier, M-F. Role of nucleotide hydrolysis in the polymerization of actin and tubulin. *Cell Biophys.* 12:105–117, 1988.

    Horio, H. Hotani, H. Visualization of the dynamic instability of individual microtubules by dark field microscopy. *Nature* 321:605–607, 1986.

    Mitchison, T., Kirschner, M. Dynamic instability of microtubule growth. *Nature* 312:237–242, 1984.

44. Karsenti, E.; Maro, B. Centrosomes and the spatial distribution of microtubules in animal cells. *Trends Biochem. Sci.* 11:460–463, 1986.

    Mitchison, T., Kirschner, M. Microtubule assemby nucleated by isolated centrosomes. *Nature* 312:232–237, 1984.

45. Kirschner, M.; Mitchison, T. Beyond self-assembly: from microtubules to morphogenesis. *Cell* 45:329–342, 1986.

    Sammak, P.J. Borisy, G.G. Direct observation of microtubule dynamics in living cells. *Nature* 332:724–726, 1988.

46. Barra, H.S.; Arce, C.A.; Argarana, C.E. Posttranslational tyrosination detyrosination of tubulin. *Molec. Neurobiol.* 2:133–153, 1988.

    Gundersen, G.G., Khawja, S., Bulinski, J.C. Postpolymerization detyrosination of [α]-tubulin: a mechanism for subcelluar differentiation of microtubules. *J. Cell Biol.* 105:251–264, 1987.

    Maruta, H., Greer, K., Rosenbaum, J.L. The acetylation of α-tubulin and its relationship to the assembly and disassembly of microtubules. *J. Cell Biol.* 103:571–579, 1986.

    Schulze, E.; Asai, D.J.; Bulinski, J.C.; Kirschner, M. Post-translational modification and microtubule stability. *J. Cell Biol.* 105:2167–2177, 1987.

47. Olmsted, J.B. Microtubule-associated proteins. *Annu. Rev. Cell Biol.* 2:421–457, 1986.

    Vallee, R.B.; Bloom, G.S.; Theurkauf, W.E. Microtubule-associated proteins: subunits of the cytomatrix. *J. Cell Biol.* 99:38s–44s, 1984.

48. Allen, R.D. The microtubule as an intracellular engine. *Sci. Am.* 256(2):42–49, 1987.

    Allen, R.D.; et al. Gliding movement of and bidirectional transport along single native microtubules from squid axoplasm: evidence for an active role of microtubules in cytoplasmic transport. *J. Cell Biol.* 100:1736–1752, 1985.

49. Vale, R. Intracelluar transport using microtubule-based motors. *Annu. Rev. Cell Biol.* 3:347–378, 1987.

    Vale, R.D.; Reese, T.S.; Sheetz, M.P. Indentification of a novel force-generating protein, kinesin, involved in microtubule-based motility. *Cell* 42:39–50, 1985.

    Vallee, R.B.; Wall, J.S.; Paschal, B.M.; Shpetner, H.S. Microtubule-associated protein 1C from brain is a two-headed cytosolic dynein. *Nature* 332:561–563, 1988.

50. Allan, V.J.; Kreis, T.E. A microtubule-binding protein associated with membranes of the Golgi apparatus. *J. Cell Biol.* 103:2229–2239, 1986.

    Dabora, S.L.; Sheetz, M.P. The microtubule-dependent formation of a tubulovesicular network with characteristics of the ER from cultured cell extracts. *Cell* 54:27–35, 1988.

    Lee, C.; Chen, L.B. Dynamic behavior of endoplasmic reticulum in living cells. *Cell* 54:37–46, 1988.

    Lucocq, J.M.; Warren, G. Fragmentation and partitioning of the Golgi apparatus during mitosis in Hela cells. *EMBO J.* 6:3239–3246, 1987.

51. Geiger, B. Intermediate filaments: looking for a function. *Nature* 329:392–393, 1987.

    Steinert, P.M.; Roop, D.R. Molecular and cellular biology of intermediate filaments. *Annu. Rev. Biochem.* 57:593–626, 1988.

    Traub, P. Intermediate Filaments: A Review. New York: Springer-Verlag, 1985.

    Wang, E.; Fischman, D.; Liem, R.K.H.; Sun, T.-T., eds. Intermediate Filaments. *Ann. N.Y. Acad. Sci.* 455, 1985.

52. Osborn, M.; Weber, K. Tumor diagnosis by intermediate filament typing: a novel tool for surgical pathology. *Lab. Invest.* 48:372–394, 1983.

53. Ip. W.; Hartzer, M.K.; Pang, S.Y.-Y.; Robson, R.M. Assembly of vimentin *in vitro* and its implications concerning the structure of intermediate filaments. *J. Mol. Biol.* 183:365–375, 1985.

Quinlan, R.A.; et al. Characterization of dimer subunits of intermediate filament proteins. *J. Mol. Biol.* 192:337–349, 1986.

54. Geuens, G.; De Brabander, M.; Nuydens, R.; De Mey, J. The interaction between microtubules and intermediate filaments in cultured cells treated with taxol and nocodazole. *Cell Biol. Int. Rep.* 7:35–47, 1983.

    Goldman, R.; et al. Intermediate filaments: possible functions as cytoskeletal connecting links between the nucleus and the cell surface. *Ann. N.Y. Acad. Sci.* 455:1–17, 1985.

55. Geisler, N.; Weber, K. Phosphorylation of desmin *in vitro* inhibits formation of intermediate filaments: identification of three kinase A sites in the aminoterminal head domain. *EMBO J.* 7:15–20, 1988.

    Inagaki, M.; Nishi, Y.; Nishizawa, K.; Matsuyama, M.; Sato, C. Site-specific phosphorylation induces disassembly of vimentin filaments *in vitro. Nature* 328:649–652, 1987.

56. Aebi, U.; Cohn, J.; Buhle, L.; Gerace, L. The nuclear lamina is a meshwork of intermediate-type filaments. *Nature* 323:560–564, 1986.

    McKeon, F.D.; Kirschner, M.W.; Caput, D. Homologies in both primary and secondary structure between nuclear envelope and intermediate filament proteins. *Nature* 319:463–468, 1986.

57. Abercrombie, M. The crawling movment of metazoan cells. *Proc. R. Soc. Lond. (Biol.)* 207:129–147, 1980.

    Bridgman, P. Structure of cytoplasm as revealed by modern electron microscopy techniques. *Trends Neurosci.* 10:321–325, 1987.

    Singer, S.J.; Kupfer, A. The directed migration of eukaryotic cells. *Annu. Rev. Cell Bio.* 2:337–365, 1986.

    Trinkaus, J.P. Cells into Organs: The Forces that Shape the Embryo. 2nd ed. Englewood Cliffs, NJ: Prentice-Hall, 1984.

58. Bridgman, P.C.; Reese, T.S. The structure of cytoplasm in directly frozen cultured cells. 1. Filamentous meshworks and the cytoplasmic ground substance. *J. Cell Biol.* 99:1655–1668, 1984.

    Heuser, J.; Kirschner, M.W. Filament organization revealed in platinum replicas of freeze-dried cytoskeletons. *J. Cell Biol.* 86:212–234, 1980.

59. Fulton, A.B. How crowded is the cytoplasm? *Cell* 30:345–347, 1982.

    Luby-Phelps, K.; Taylor, D.L.; Lanni, F. Probing the structure of the cytoplasm. *J. Cell Biol.* 102:2015–2022, 1986.

60. Kolega, J. Effects of mechanical tension on protrusive activity and microfilament and intermediate filament organization in an epidermal epithelium moving in culture. *J. Cell Biol.* 102:1400–1411, 1986.

Trinkaus, J.P. Cells into Organs: The Forces That Shape the Embryo, 2nd. ed., pp. 157–244. Englewood Cliffs, NJ: Prentice-Hall, 1984.

61. Bray, D.; Hollenbeck, P.J. Growth cone motility and guidance. *Annu. Rev. Cell Biol.* 4:43–62, 1988.

    Euteneuer, U.; Schliwa, M. Persistent, directional motility of cells and cytoplasmic fragments in the absence of microtubles. *Nature* 310:58–61, 1984.

    Malawista, S.E.; De Boisfleury Chevance, A. The cytokinetplast: purified, stable, and functional motile machinery from human blood polymorphonuclear leukocytes. *J. Cell Biol.* 95:960–973, 1982.

    Marsh, L.; Letourneau, P.C. Growth of neurites without filopodial or lamellipodial activity in the presence of cytochalasin B. *J. Cell Biol.* 99:2041–2047, 1984.

    Vasiliev, J.M.; et al. Effect of colcemid on the locomotion of fibroblasts. *J. Embryol. Exp. Morphol.* 24:625–640, 1970.

62. Bray, D.; White, J.G. Cortical flow in animal cells. *Science* 239:883–888, 1988.

    De Lozanne, A.; Spudich, J.A. Disruption of the *Dictyostelium* myosin heavy chain gene by homologous recombination. *Science* 236:1086–1091, 1987.

    Knecht, D.A.; Loomis, W.F. Antisense RNA inactivation of myosin heavy chain gene expression in *Dictyostelium discoideum. Science* 236:1081–1086, 1987.

63. Bergmann, J.E.; Kupfer, A.; Singer, S.J. Membrane insertion at the leading edge of motile fibroblasts. *Proc. Natl. Acad. Sci. USA* 80:1367–1371, 1983.

    Bretscher, M.S. How animal cells move. *Sci. Am.* 257(6):72–90, 1987.

64. Lackie, J.M. Cell Movement and Cell Behaviour. pp. 253–275 London: Allen and Unwin, 1986.

    Trinkaus, J.P. Cells into Organs: The Forces That Shape the Embryo. 2nd ed., pp. 157–244. Englewood Cliffs, NJ: Prentice-Hall, 1984.

65. Odell, G.M.; Oster, G.; Alberch, P.; Burnside, B. The mechanical basis of morphogenesis. 1. Epithelial folding and invagination. *Dev. Biol.* 85:446–462, 1981.

66. Tilney LG; Tilney MS; Cotanche DA. Actin filaments, stereocilia, and hair cellls of the bird cochlea. V. How the staircase pattern of stereociliary lengths is generated. *J Cell Biol.* 106:355–365, 1988.

    Tilney, L.G.; De Rosier, D.J. Actin filaments, stereocilia, and hair cells of the bird cochlea. IV. How the actin filaments become organized in developing stereocilia and in the cuticular plate. *Dev. Biol.* 116:119–129, 1986.

Cell Signaling

12

Cells in a multicellular organism need to communicate with one another in order to regulate their development and organization into tissues, to control their growth and division, and to coordinate their functions. Animal cells communicate in three ways: (1) they secrete chemicals that signal to cells some distance away; (2) they display plasma-membrane-bound signaling molecules that influence other cells in direct physical contact; and (3) they form gap junctions that directly join the cytoplasms of the interacting cells, thereby allowing exchange of small molecules (Figure 12–1).

Communication that depends on cell-cell contact through gap junctions will be discussed in Chapter 14. In this chapter we shall be concerned primarily with communication at a distance that is mediated by secreted chemical signals. This

REMOTE SIGNALING BY SECRETED MOLECULES

Figure 12–1   Three ways in which cells communicate with one another.

CONTACT SIGNALING BY PLASMA-MEMBRANE-BOUND MOLECULES

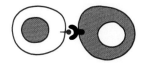

CONTACT SIGNALING VIA GAP JUNCTIONS

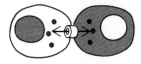

emphasis reflects the state of current knowledge. Secreted molecules are very much easier to study than those that are membrane-bound, and much is known about how they work. Contact-dependent signaling via membrane-bound molecules, although harder to demonstrate and less well studied, may nonetheless be important, especially during development and in immune responses (see p. 1037); its molecular basis is thought to be closely related to that of signaling at a distance. Specialized aspects of chemical signaling in the nervous system and the unique principles that apply to signaling in plants will be discussed separately, in Chapters 19 and 20, respectively.

Three Strategies of Chemical Signaling: Endocrine, Paracrine, and Synaptic

Chemical signaling mechanisms vary in the distances over which they operate: (1) In **endocrine signaling,** specialized endocrine cells secrete **hormones,** which travel through the bloodstream to influence target cells that are distributed widely throughout the body. (2) In **paracrine signaling,** cells secrete **local chemical mediators,** which are so rapidly taken up, destroyed, or immobilized that the mediators act only on cells in the immediate environment, perhaps within a millimeter or so. (3) In **synaptic signaling,** which is confined to the nervous system, cells secrete **neurotransmitters** at specialized junctions called *chemical synapses;* the neurotransmitter diffuses across the synaptic cleft, typically a distance of about 50 nm, and acts only on the adjacent postsynaptic target cell (Figure 12–2). In each case the target cell responds to a particular extracellular signal by means of specific proteins, called **receptors,** that bind the signaling molecule and initiate the response. Many of the same signaling molecules and receptors are used in endocrine, paracrine, and synaptic signaling. The crucial differences lie in the speed and selectivity with which the signals are delivered to their targets.

Endocrine Cells and Nerve Cells Are Specialized for Different Types of Chemical Signaling[1]

Endocrine cells and nerve cells work together to coordinate the diverse activities of the billions of cells in a higher animal. The endocrine cells are usually organized in discrete glands, and they secrete their hormone molecules into the extracellular (interstitial) fluid that surrounds all cells in tissues. From there the molecules diffuse into capillaries to enter the bloodstream, which carries them to tissues throughout the body. Within each tissue, hormone molecules escape from capillaries into the interstitial fluid, where they can bind to their target cells. Because endocrine signaling relies on diffusion and blood flow, it is relatively slow: it usually takes minutes for a hormone to reach its target cells after secretion. Moreover, the specificity of signaling in the endocrine system depends entirely on the chemistry of the signal and the receptors on the target cell: each type of endocrine cell secretes a different hormone into the blood, and each cell that has complementary receptors will respond in a way appropriate for the cell type (Figure 12–3A).

Figure 12–2 Three forms of signaling mediated by secreted molecules. Not all neurotransmitters act in the strictly synaptic mode shown; some act in a paracrine mode as local chemical mediators that influence multiple target cells in the area.

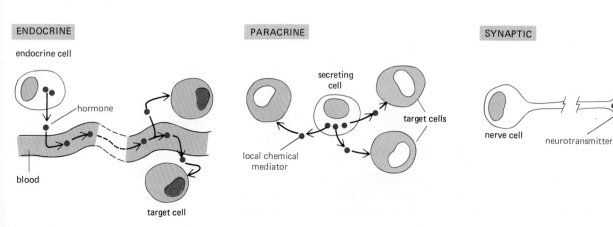

Figure 12–3 The contrast between endocrine (A) and synaptic signaling (B). Endocrine cells secrete many different hormones into the blood and signal specific target cells, which have receptors for binding specific hormones and thereby "pull" the appropriate hormones from the extracellular fluid. In synaptic signaling, by contrast, the specificity arises from the contacts between nerve processes and the specific target cells they signal: only a target cell that is in synaptic contact with a nerve cell is exposed to the neurotransmitter released from the nerve terminal. Different endocrine cells must use different hormones in order to communicate specifically with their target cells, but many nerve cells can use the same neurotransmitter and still communicate in a specific manner.

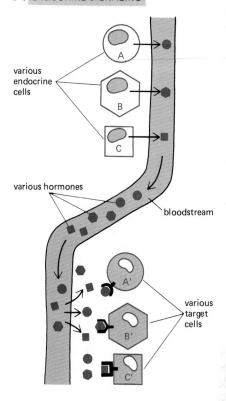

(A) ENDOCRINE SIGNALING

various endocrine cells

various hormones

bloodstream

various target cells

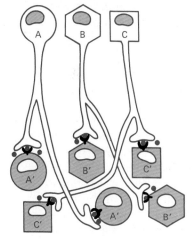

(B) SYNAPTIC SIGNALING

various neurons

various target cells

Nerve cells, by contrast, can achieve much greater speed and precision. They can transmit information over long distances by means of electrical impulses that carry signals along nerve processes at rates of up to 100 meters per second. Only when a neurotransmitter is released at the nerve terminal is the electrical impulse converted into a chemical signal. Chemical signals released by nerve cells can act in either the paracrine or the synaptic mode. In the paracrine mode the neurotransmitter functions as a local chemical mediator, diffusing outward to influence any target cells in the neighborhood that have receptors for the released molecule. Signaling in the synaptic mode is much more precise, and the effect of the neurotransmitter is confined to a single target cell even if adjacent cells have receptors for the same neurotransmitter (see Figure 12–3B); here the neurotransmitter has to diffuse less than 100 nm to the target cell, a process that takes less than a millisecond (see Figure 12–2).

Hormones are greatly diluted in the bloodstream and interstitial fluid and therefore must be able to act at very low concentrations (typically $<10^{-8}$ M), but neurotransmitters are diluted much less in their short journey to the target cell and can achieve high local concentrations. For example, the concentration of the neurotransmitter acetylcholine in the synaptic cleft of an active neuromuscular junction is about 5×10^{-4} M. Correspondingly, in synaptic signaling the neurotransmitter receptors have a relatively low affinity for their ligand, and, as a result, they do not respond significantly to the low concentrations of transmitter that reach them by diffusion from neighboring synapses. Moreover, after being secreted, the neurotransmitter is quickly removed from the synaptic cleft either by specific hydrolytic enzymes or by specific membrane transport proteins that pump the neurotransmitter back into the nerve terminal. Rapid removal ensures not only spatial precision of the signal but also temporal precision: a brief pulse of neurotransmitter release evokes a prompt and brief response, so that the timing of the signal can be faithfully relayed from cell to cell (see p. 1081).

Neuroendocrine Cells in the Hypothalamus Regulate the Endocrine System[1,2]

The endocrine system and nervous system in vertebrates are physically and functionally linked by a specific region of the brain called the **hypothalamus.** The hypothalamus communicates directly with the *pituitary gland* via a bridge called the *pituitary stalk.* The linking function of the hypothalamus is mediated by cells that have properties of both nerve cells and endocrine cells; for this reason they are called *neuroendocrine cells.* Most types of hypothalamic neuroendocrine cells respond to stimulation by other nerve cells in the brain by secreting a specific peptide hormone into the special blood vessels of the pituitary stalk; on reaching the pituitary gland, each hormone specifically stimulates or suppresses the secretion of a second hormone into the main bloodstream. (Other hypothalamic neuroendocrine cells send their axons along the pituitary stalk and discharge their secretions directly into the main bloodstream.) Because many of the pituitary hormones released under the control of the hypothalamus stimulate another endocrine gland to secrete a third hormone into the blood, the hypothalamus serves as the main regulator of the endocrine system in vertebrates. As an example, Figure 12–4 illustrates how the hypothalamus regulates the secretion of *thyroid hormone*.

Selected examples of local chemical mediators, neurotransmitters, and hormones are listed in Table 12–1, together with their sites of origin, structures, and principal actions. It can be seen that signaling molecules are as varied in structure as they are in function. They include small peptides, larger proteins and glycoproteins, amino acids and related compounds, steroids (molecules derived from cholesterol and closely related in structure), and fatty acid derivatives. Although each signaling molecule is listed in only one category in Table 12–1, many of them can act in more than one mode. Many peptide hormones, for example, also act as neurotransmitters (in the paracrine mode) in the vertebrate brain.

Different Cells Respond in Different Ways to the Same Chemical Signal

Most cells in mature animals are specialized to perform one primary function, and they contain a characteristic array of receptor proteins that allows them to respond to each of the different chemical signals that initiate or modulate that function. Many of these chemical signals act at very low concentration (typically $\leq 10^{-8}$ M), and their complementary receptors usually bind the signaling molecule with high affinity (affinity constant $K_a \geq 10^8$ liters/mole; see p. 98).

A single signaling molecule often has different effects in different target cells. Acetylcholine, for example, stimulates the contraction of skeletal muscle cells but decreases the rate and force of contraction in heart muscle cells. This is because the acetylcholine receptor proteins on skeletal muscle cells are different from those on heart muscle cells. But receptor differences are not always the explanation for the different effects. In many cases the same signaling molecule binds to identical receptor proteins and yet produces very different responses in different types of target cells (Figure 12–5). This indicates that the responses to a signaling molecule are programmed in two ways—through the receptors the target cells carry, and through the internal machinery to which the receptors are coupled.

Some Cellular Responses to Chemical Signals Are Rapid and Transient, While Others Are Slow and Long-lasting[2,3]

In coordinating the responses of cells to changes in an animal's environment, chemical signals generally induce rapid and transient responses. An increase in blood glucose levels, for instance, stimulates endocrine cells in the pancreas to secrete the protein hormone *insulin* into the blood. Within minutes the resulting increase in insulin concentration stimulates fat and muscle cells to take up more glucose, and, consequently, blood glucose levels fall. There are three parts to the response, none of which requires new protein synthesis: (1) In the pancreas the elevated glucose levels trigger the exocytic release of stored insulin. (2) In fat and muscle cells, extra membrane-bound glucose transport proteins are stored in intracellular vesicles and the elevated insulin levels cause these proteins to be added by exocytosis to the plasma membrane, where they increase the rate of glucose uptake. This causes blood glucose levels to drop, and, as a result, the rate of insulin secretion decreases. (3) Since the extra glucose carriers are rapidly removed from the cell surface by endocytosis and returned to the intracellular pool, when insulin levels decrease, the rate of glucose uptake by fat and muscle cells returns to its previous level. In this way insulin helps to maintain a relatively constant blood glucose concentration. Neurotransmitters elicit even more rapid responses than hormones do: skeletal muscle cells contract and relax again within milliseconds in response to acetylcholine released from nerve terminals at neuromuscular junctions.

Chemical signals also play an important part in animal development, often influencing when and how certain cells differentiate. Some of these effects are slow in onset and long-lasting. For example, the female sex hormone *estradiol*, a steroid, is secreted in large amounts by cells in the ovary around the time of puberty. It induces changes in a wide variety of cells in different parts of the body,

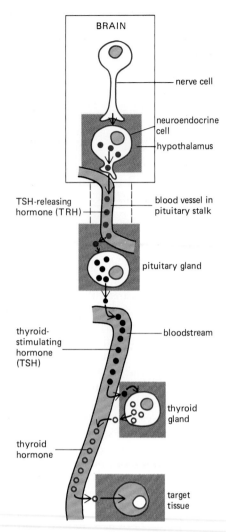

Figure 12–4 Thyroid hormone secretion is regulated indirectly by the nervous system. When stimulated by nerve cells in higher centers of the brain, specific neuroendocrine cells in the hypothalamus secrete TSH-releasing hormone (TRH) into blood vessels of the pituitary stalk, which transport the hormone to the pituitary gland. Here the TRH stimulates specific cells to release TSH (thyroid-stimulating hormone) into the main bloodstream, which transports the hormone to the thyroid gland. TSH in turn stimulates the cells in the thyroid gland to synthesize and secrete thyroid hormone, which is transported in the blood to most cells in the body. Thyroid hormone then stimulates a variety of metabolic processes in these cells. The secretion of both TRH and TSH is suppressed by increased concentrations of thyroid hormone in the blood (not shown). This *feedback inhibition* prevents the level of thyroid hormone in the blood from rising too high. Many hormones are regulated by a similar feedback mechanism.

Table 12–1   Some Examples of Extracellular Signaling Molecules

Local Chemical Mediators	Site of Origin	Structure	Major Effects
Proteins Nerve growth factor	skin; all tissues innervated by sympathetic nerves	2 identical chains of 118 amino acids	survival and growth of sensory and sympathetic neurons and some neurons in central nervous system
Small Peptides Eosinophil chemotactic factor	mast cells	4 amino acids	chemotactic signal for a particular class of white blood cells (eosinophils)
Amino Acid Derivatives Histamine	mast cells		causes blood vessels to dilate and become leaky
Fatty Acid Derivatives Prostaglandin E_2	many cell types		contraction of smooth muscle

Neurotransmitters*	Site of Origin	Structure	Major Effects
Amino Acids and Related Compounds Glycine	nerve terminals	$^+H_3N-CH_2-COO^-$	inhibitory transmitter in central nervous system
Norepinephrine (noradrenaline)	nerve terminals		excitatory and inhibitory transmitter in central and peripheral nervous system
γ-Aminobutyric acid (GABA)	nerve terminals	$^+H_3N-CH_2-CH_2-CH_2-COO^-$	inhibitory transmitter in central nervous system
Acetylcholine	nerve terminals		excitatory transmitter at neuromuscular junction; excitatory and inhibitory transmitter in central and peripheral nervous system
Small Peptides Enkephalin	nerve terminals	5 amino acids	morphinelike action (inhibits pain pathways in central nervous system)

*Norepinephrine and enkephalin act in paracrine rather than synaptic signaling; acetylcholine can act in either mode.
Excitatory neurotransmitters stimulate target cell activity, whereas inhibitory neurotransmitters suppress target cell activity.

(Continued)

Table 12–1 *Continued*

Hormones**	Site of Origin	Structure	Major Effects
Proteins			
Insulin	beta cells of pancreas	protein α-chain = 21 amino acids β-chain = 30 amino acids	utilization of carbohydrate (including uptake of glucose into cells); stimulation of protein synthesis; stimulation of lipid synthesis in fat cells
Somatotropin (growth hormone)	anterior pituitary	protein 191 amino acids	stimulation of liver to produce somatomedin-1, which in turn causes growth of muscle and bone; stimulation of fat, muscle, and cartilage cell differentiation
Somatomedin-1 (insulinlike growth factor-1)	mainly liver	protein 70 amino acids	growth of bone and muscle; influences metabolism of Ca^{2+}, phosphate, carbohydrate, and lipid
Adrenocorticotropic hormone (ACTH)	anterior pituitary	protein 39 amino acids	stimulation of adrenal cortex to produce cortisol; fatty acid release from fat cells
Parathormone	parathyroid	protein 84 amino acids	increase in bone resorption, thereby increasing blood Ca^{2+} and phosphate; increase in resorption of Ca^{2+} and Mg^{2+} and decrease in resorption of phosphate in kidney tubules
Follicle-stimulating hormone (FSH)	anterior pituitary	glycoprotein α-chain = 92 amino acids β-chain = 118 amino acids	stimulation of ovarian follicles to grow and secrete estradiol; stimulation of spermatogenesis in testis
Luteinizing hormone (LH)	anterior pituitary	glycoprotein α-chain = 92 amino acids β-chain = 115 amino acids	stimulation of oocyte maturation and ovulation and progesterone secretion from ovary; stimulation of testis to produce testosterone
Epidermal growth factor	unknown	protein 53 amino acids	stimulation of epidermal and other cells to divide
Thyroid-stimulating hormone (TSH)	anterior pituitary	glycoprotein α-chain = 92 amino acids β-chain = 112 amino acids	stimulation of thyroid to produce thyroid hormone; fatty acid release from fat cells
Small Peptides			
TSH-releasing hormone (TRH)	hypothalamus	3 amino acids	stimulation of anterior pituitary to secrete thyroid-stimulating hormone (TSH)

**Many of the nonsteroid hormones are also made by some nerve cells in the brain.

(Continued)

Table 12–1 *Continued*

Hormones	Site of Origin	Structure	Major Effects
LH-releasing hormone	hypothalamus	10 amino acids	stimulation of anterior pituitary to secrete luteinizing hormone (LH)
Vasopressin (antidiuretic hormone, ADH)	posterior pituitary	9 amino acids	elevation of blood pressure by constriction of small blood vessels; increase in water resorption in kidney tubules
Somatostatin	hypothalamus	14 amino acids	inhibition of somatotropin release from anterior pituitary
Amino Acid Derivatives Epinephrine (adrenaline)	adrenal medulla		increase in blood pressure and heart rate; increase in glycogenolysis in liver and muscle; fatty acid release from fat cells
Thyroid hormone (thyroxine)	thyroid		increase in metabolic activity in most cells
Steroids Cortisol	adrenal cortex		effects on metabolism of proteins, carbohydrates and lipids in most tissues; suppression of inflammatory reactions
Estradiol	ovary, placenta		development and maintenance of secondary female sex characteristics; maturation and cyclic function of accessory sex organs; development of duct system in mammary glands
Testosterone	testis		development and maintenance of secondary male sex characteristics; maturation and normal function of accessory sex organs
Progesterone	ovary (corpus luteum), placenta		preparation of uterus for pregnancy; maintenance of pregnancy; development of alveolar system in mammary glands

changes that eventually lead to the development of secondary female characteristics such as breast enlargement. Although this effect is slowly reversed if estradiol secretion stops, some of the responses to steroid sex hormones during very early mammalian development are irreversible. Similarly, a tenfold increase in thyroid hormone levels in the blood of a tadpole induces the dramatic and irreversible changes that result in its transformation into a frog (Figure 12–6).

Only Lipid-soluble Signaling Molecules Can Enter Cells Directly

Most hormones and local chemical mediators, and all known neurotransmitters, are water-soluble. There are some exceptions, however, that form a distinctive class of signaling molecules. Important examples are the water-insoluble steroid and thyroid hormones, which are made soluble for transport in the bloodstream by binding to specific carrier proteins. This difference in solubility gives rise to a fundamental difference in the mechanism by which the two classes of molecules influence target cells. Water-soluble molecules are too hydrophilic to pass directly through the lipid bilayer of a target-cell plasma membrane; instead they bind to specific receptor proteins on the cell surface. The steroid and thyroid hormones, on the other hand, are lipid-soluble, and once released from their carrier proteins, they can pass easily through the plasma membrane of the target cells; these hormones then bind to specific receptor proteins *inside* the cell (Figure 12–7).

Another important difference between these two classes of signaling molecules is the length of time that they persist in the bloodstream or tissue fluids. Most water-soluble hormones are removed and/or broken down within minutes of entering the blood, and local chemical mediators and neurotransmitters are removed from the extracellular space even faster—within seconds or milliseconds. Steroid hormones, by contrast, persist in the blood for hours, and thyroid hormone for days. Consequently, water-soluble signaling molecules usually mediate responses of short duration, whereas the water-insoluble molecules tend to mediate longer lasting responses. As usual, there are exceptions to these general rules: *prostaglandins*, for example, are local chemical mediators that are hydrophobic, yet they bind to cell-surface receptors and mediate rapid, short-lasting responses, as we shall see below.

Local Chemical Mediators Are Rapidly Destroyed, Retrieved, or Immobilized After They Are Secreted[4]

The molecules that mediate paracrine signaling act only on cells in the immediate vicinity of the cells that secrete them. These *local chemical mediators* are so rapidly taken up by cells or destroyed by extracellular enzymes or immobilized in the extracellular matrix that they generally do not enter the blood in significant amounts.

Some cells are specialized for paracrine signaling. For example, *histamine* (a derivative of the amino acid histidine, see Table 12–1) is secreted mainly by *mast cells*. These cells, which are found in connective tissues throughout the body, store histamine in large secretory vesicles and release it rapidly by exocytosis when stimulated by injury, local infection, or certain immunological reactions (see p. 1016). Histamine causes local blood vessels to dilate and become leaky, which facilitates the access of both serum proteins (such as antibodies and components of the complement system, see Chapter 18) and phagocytic white blood cells to the site of injury. Mast cells also release two tetrapeptides that attract a class of white blood cells called *eosinophils* from the blood to the site of tetrapeptide release; eosinophils contain a variety of enzymes that help inactivate histamine and other chemical mediators released by mast cells and thereby help to terminate the response.

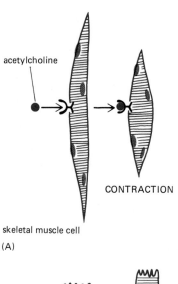

acetylcholine

CONTRACTION

skeletal muscle cell

(A)

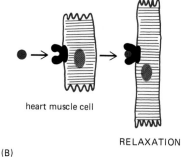

heart muscle cell

RELAXATION

(B)

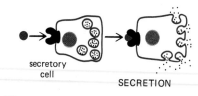

secretory cell

SECRETION

(C)

Figure 12–5 The same signaling molecule can induce different responses in different target cells. In some cases this is because the signaling molecule binds to different receptor proteins, as illustrated in (A) and (B). In other cases the signaling molecule binds to identical receptor proteins but these activate different response pathways in different cells, as illustrated in (B) and (C).

Figure 12–6 Various stages in the metamorphosis of a tadpole into a frog. All of the dramatic changes shown are signaled by thyroid hormone. If the presumptive thyroid gland is removed from a developing embryo, the animal fails to undergo metamorphosis and continues to grow as a tadpole. If thyroid hormone is injected into such a giant tadpole, the tadpole transforms into a frog.

Figure 12–7 Depending on their solubility, extracellular signaling molecules bind to either cell-surface receptors or intracellular receptors. Hydrophilic signaling molecules are unable to cross the plasma membrane directly and bind to receptors on the surface of the target cell. Many hydrophobic signaling molecules are able to diffuse across the plasma membrane and bind to receptors inside the target cell—either in the cytoplasm or in the nucleus (as shown). Because they are insoluble in aqueous solutions, hydrophobic hormones are transported in the bloodstream bound to specific carrier proteins from which they dissociate before entering the target cell.

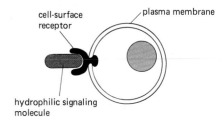

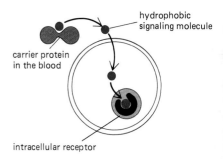

Some local chemical mediators are rapidly immobilized rather than eliminated after they have been secreted. Fibronectin, proteoglycans, and other macromolecules of the extracellular matrix are a case in point. These secreted macromolecules can be considered as special types of local mediators because they signal neighboring cells to alter their behavior (see p. 803). Unlike other local chemical mediators, they assemble into large insoluble networks in the extracellular space and thus become immobilized near the site where they are produced; thus their effects, although local, can be long-lasting. The extracellular matrix may also bind soluble signaling molecules, immobilizing them so that they act only in a particular location. *Fibroblast growth factor (FGF)*, for example, which is a small protein that stimulates a wide variety of cells to divide in culture, binds strongly to a matrix proteoglycan in a test tube and might be immobilized in this way in tissues.

Cells in All Mammalian Tissues Continuously Release Prostaglandins[5]

Many local chemical mediators are secreted by cells that are specialized for this purpose, but others are of more widespread origin. The **prostaglandins,** a family of 20-carbon fatty acid derivatives, are an important example, being made by cells in all mammalian tissues. These local mediators are continuously synthesized in membranes from precursors cleaved from membrane phospholipids by phospholipases (Figure 12–8). They are also continuously degraded by enzymes in extra-

Figure 12–8 Prostaglandins are continuously synthesized in membranes from 20-carbon fatty acid chains that contain at least three double bonds, as shown for the synthesis of PGE_2. The subscript refers to the two carbon-carbon double bonds outside the ring of PGE_2. Prostaglandins, together with the chemically related signaling molecules *thromboxanes*, *leukotrienes*, and *lipoxins*, are all made mainly from arachidonic acid and are collectively called *eicosanoids*. This metabolic pathway is an important target for therapeutic drugs, since eicosanoids play an important part in inflammation. Corticosteroid hormones such as cortisone, for example, are widely used clinically as drugs to treat noninfectious inflammatory diseases such as some forms of arthritis. One way they might act is by inducing white blood cells to synthesize and/or secrete local chemical mediators called *lipocortins* (or calpactins), proteins that somehow inhibit the activity of the phospholipase in the first step of the eicosanoid synthesis pathway shown. Nonsteroid anti-inflammatory drugs, such as aspirin, block the oxidation steps of prostaglandin synthesis. Both corticosteroids and aspirin are used in the treatment of arthritis.

membrane phospholipid

phospholipase

arachidonic acid (20 carbons), extended conformation

COOH

arachidonic acid, folded conformation

OXIDATION STEPS

prostaglandin (PGE_2)

cellular fluid. There are 16 or more different prostaglandins, falling into 9 classes (designated PGA, PGB, PGC, ... PGI), which produce a variety of biological effects by binding to specific cell-surface receptors.

Unlike most signaling molecules, prostaglandins are not stored but are continuously released to the cell exterior as soon as they are produced. However, when cells are activated by tissue damage or by some types of chemical signals, the rate of prostaglandin synthesis is increased; the resulting increase in the local level of prostaglandins influences both the cell that makes them (so-called *autocrine stimulation*) and its immediate neighbors. Such autocrine signaling in response to another chemical signal may provide a mechanism for amplifying and/ or prolonging a response to the initial stimulus, as well as for spreading the response through a local population of similar cells.

A wide variety of biological activities have been ascribed to prostaglandins. They cause contraction of smooth muscle, aggregation of platelets, and inflammation. Certain prostaglandins produced in large amounts in the uterus at the time of childbirth, for example, are thought to be important in stimulating the contraction of the uterine smooth muscle cells; these prostaglandins are now widely used as pharmacological agents to induce abortion. Conversely, drugs such as aspirin work by inhibiting prostaglandin biosynthesis at sites of inflammation.

Summary

Extracellular signaling molecules mediate three kinds of intercellular communication, distinguished by the distance over which they act: (1) in endocrine signaling, hormones are carried in the blood to target cells throughout the body; (2) in paracrine signaling, local chemical mediators are rapidly taken up, destroyed, or immobilized so that they act only on local cells; and (3) in synaptic signaling, neurotransmitters act only on the adjacent postsynaptic cell. Each cell type in the body contains a distinctive set of receptor proteins that enables it to bind and respond to a complementary set of signaling molecules in a preprogrammed and characteristic way.

Signaling molecules can also be classified according to their solubility in water. Small hydrophobic signaling molecules, such as steroid and thyroid hormones, pass through the plasma membrane of the target cell and activate receptor proteins inside the cell. In contrast, hydrophilic signaling molecules, including all neurotransmitters and the great majority of hormones and local chemical mediators, activate receptor proteins on the surface of the target cell.

Signaling Mediated by Intracellular Receptors: Mechanisms of Steroid Hormone Action

Many developmental and physiological processes in organisms ranging from fungi to humans are regulated by a small number of **steroid hormones,** all of which are synthesized from cholesterol. Being relatively small (molecular weight ≈ 300) hydrophobic molecules, they are thought to cross the plasma membrane by simple diffusion. Once inside the target cell, each type of steroid hormone binds tightly but reversibly to a complementary receptor protein. The binding of the hormone causes the receptor protein to undergo a conformational change (a process called *receptor activation*) that increases its affinity for DNA and enables it to bind to specific genes in the nucleus and regulate their transcription. Thyroid hormones work in the same way and bind to very similar receptor proteins.

Activated Steroid-Hormone Receptors Regulate Gene Transcription by Binding to Specific DNA Sequences

A typical target cell contains about 10,000 steroid receptors, each of which reversibly binds one molecule of a specific steroid hormone with high affinity (affinity constant $K_a = 10^8$–10^{10} liters/mole). Because the receptors constitute less than

0.01% of the total mass of protein in the cell, they have been extremely difficult to characterize and purify. Recently, however, the DNA sequences coding for several vertebrate steroid hormone receptors have been cloned and sequenced and shown to encode receptor proteins of very similar structure. The polypeptide chain (about 800 amino acid residues long) is folded into at least three distinct domains; the carboxyl-terminal domain, which binds hormone, and the middle domain, which binds to DNA, and an amino-terminal domain, which activates gene transcription (Figure 12–9).

Some types of steroid hormone receptor are located primarily in the cytosol when they are inactive (unliganded), whereas others are located primarily in the nucleus. In either case, hormone binding increases the affinity of the receptor protein for DNA, enabling it to bind tightly to specific nucleotide sequences in the genes that the hormone regulates. The binding of the receptor-hormone complexes to these specific sites activates (or occasionally suppresses) transcription of these genes.

Direct demonstration that activated steroid hormone receptors bind to specific genes proved to be difficult and was not achieved until 1983. The demonstration depended on the advent of recombinant DNA technology, which made it possible to clone genes that are regulated by steroid hormones, making large amounts of specific DNA sequences available. It was also necessary to purify the receptor proteins, a tedious and time-consuming process. Once the receptors were purified, the precise DNA sequences involved in receptor binding to a gene could be mapped *in vitro* by *DNA footprinting* experiments (see p. 187), which showed that receptor binding protects a number of specific nucleotide sequences from degradation when the DNA is lightly digested with nucleases or chemicals. When these short *recognition sequences* are deleted from a gene, the relevant steroid hormone no longer activates transcription of the gene. In addition, when a short DNA fragment that contains a recognition sequence is fused with a different ("reporter") gene and then put into cells containing the receptor protein, the appropriate steroid hormone activates transcription of the reporter gene. These experiments demonstrate that the DNA sequences recognized by activated steroid-hormone receptors *in vitro* mediate receptor action in cells. Several sets of recognition sequences are generally found in a steroid-hormone-responsive gene, in flanking regions upstream (and sometimes downstream) from the coding region, and elsewhere in the gene (Figure 12–10). Given the structural homologies among the various steroid-hormone receptor proteins, it is perhaps not surprising that the recognition sequences they bind to are closely related.

Only a small number of genes in any target cell are directly influenced by steroid hormones. For example, 30 minutes after cultured rat liver cells are exposed to the steroid hormone *cortisol*, only 6 of the 1000 proteins that can be distinguished by two-dimensional gel electrophoresis have increased in amount, and 1 has decreased. The effect is reversible and the rate of synthesis of these proteins returns to normal when the hormone is removed. Since it is thought that about 10% of a cell's proteins can be detected on these gels, the binding of cortisol to 10,000 molecules of its receptor in a liver cell probably directly alters the transcription of only about 50 genes—far fewer than expected from the measured number of binding sites on the DNA. Thus it appears that most activated receptors bind to DNA at sites where their presence has no effect.

How does the binding of receptor-hormone complexes to a gene activate its transcription? It has been shown that the DNA recognition sites associated with steroid-hormone-responsive genes function as receptor-dependent *transcriptional*

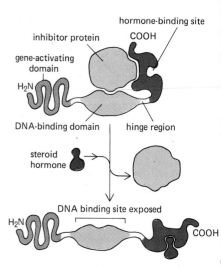

Figure 12–9 A model of a steroid-hormone receptor protein. In its inactive state the receptor is thought to be bound to an inhibitor protein that blocks the DNA-binding domain of the receptor. The binding of hormone to the receptor causes the inhibitor protein to dissociate, thereby activating the receptor. The model shown is based on the receptor for cortisol (glucocorticoids), but the receptors for estrogen, testosterone, progesterone, aldosterone, thyroid hormone, retinoic acid, and vitamin D have a similar structure (see Figure 10–25, p. 569); together these proteins form the *steroid receptor superfamily*. In the case of the cortisol and estrogen receptors, the inhibitor protein is thought to be the 90,000 dalton heat-shock protein hsp90 (see p. 420).

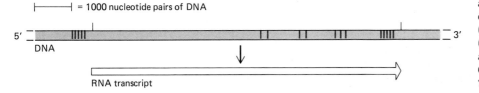

Figure 12–10 The location of the DNA sequences in the mammary tumor virus gene known to be recognized by the activated cortisol receptor protein. Each colored bar represents one binding site (eight nucleotides long) for the receptor. (Based on K.R. Yamamoto, in Transfer and Expression of Eukaryotic Genes [H. Ginsberg and H.J. Vogel, eds.], pp. 79–92. New York: Academic Press, 1984.)

enhancers, which can stimulate transcription even when they are located thousands of nucleotides away from the promoter where RNA synthesis begins. The mechanisms by which enhancers are thought to act are discussed in Chapter 10 (see p. 568).

₂Steroid Hormones Often Induce Both Primary and Secondary Responses[7]

In many cases the response to a steroid hormone takes place in two steps. The direct induction of transcription of a few specific genes is known as the *primary response*. The products of these genes may then in turn activate other genes and produce a delayed *secondary response*. Thus a simple hormonal trigger can cause a very complex change in the pattern of gene expression.

A striking example is seen in the fruit fly *Drosophila*. Within 5 to 10 minutes after injection of the steroid molting hormone *ecdysone*, six major new sites of RNA synthesis (seen as *puffs*) are induced on the giant polytene chromosomes of the salivary gland (see p. 509). After a delay, some of the proteins produced during this primary response induce an additional 100 or so sites of RNA synthesis, leading to the synthesis of a large group of proteins characteristic of the secondary response. The response is controlled by negative feedback through one or more of the first proteins synthesized, which shuts off further transcription of all of the primary-response genes (Figure 12–11). It is likely that similar mechanisms provide for both amplification and control in many of the responses of mammalian cells to hormones.

Steroid Hormones Regulate Different Genes in Different Target Cells[8]

The responses to steroid hormones, like hormonal responses in general, are determined as much by the nature of the target cell as by the nature of the hormone (see Figure 12–5). In principle, this observation has two possible explanations. Either different types of cells have different receptors for the same steroid hormone, or the receptors are the same but the genes they activate are different. There is strong evidence that the latter explanation is correct.

The evidence comes both from molecular genetic experiments, which have shown that the receptor proteins for estradiol, cortisol, and progesterone are each

Figure 12–11 Early primary response (A) and delayed secondary response (B) to ecdysone in *Drosophila* cells. Some of the primary-response proteins turn on secondary-response genes, whereas others turn off the primary-response genes. The actual number of primary- and secondary-response genes is greater than shown.

(A) EARLY PRIMARY RESPONSE TO ECDYSONE

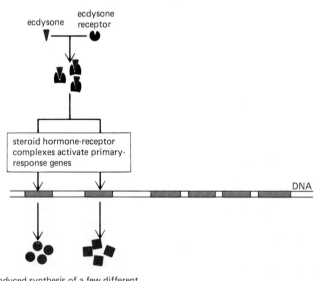

induced synthesis of a few different proteins in the primary response

(B) DELAYED SECONDARY RESPONSE TO ECDYSONE

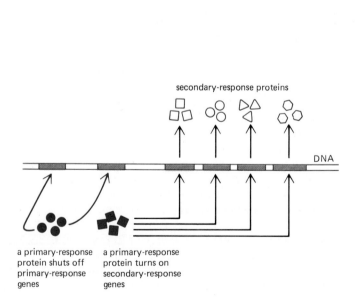

a primary-response protein shuts off primary-response genes

a primary-response protein turns on secondary-response genes

encoded by a single gene, and from studies of mammalian mutants that have defective receptors for the male steroid hormone *testosterone*. All mammals develop female features unless they are exposed to testosterone during embryonic development. The mutant individuals are genetically male, with normal testosterone-secreting testes, but their tissues fail to respond to the hormone. They therefore develop all of the secondary sexual characteristics of females, and their testes remain in the abdomen and fail to descend. This **testicular feminization syndrome** occurs in mice, rats, and cattle, as well as in humans. Although only the gene coding for testosterone receptors is abnormal, all of the various cell types in the body that are normally influenced in different ways by testosterone are affected (Figure 12–12).

Why is one set of genes activated by a steroid hormone receptor in one type of a cell and a different set activated by the same receptor in another type of cell? As described in Chapter 10, more than one type of gene regulatory protein generally must bind to a eucaryotic gene in order to activate its transcription (see p. 554). Therefore a steroid hormone receptor can activate a gene only if the right combination of other gene regulatory proteins is also present, and some of these are cell-type specific.

Thus each steroid hormone induces a characteristic set of responses because (1) only certain types of cells have receptors for it and (2) each of these cell types contains a different combination of other cell-type-specific gene regulatory proteins that collaborate with the activated steroid receptor to influence the transcription of specific sets of genes.

Summary

Steroid hormones are small hydrophobic molecules derived from cholesterol that are solubilized by binding reversibly to carrier proteins in the blood. Once released from their carrier proteins, they diffuse through the plasma membrane of the target cell and bind reversibly to specific steroid-hormone-receptor proteins in the cytoplasm or nucleus. The binding of hormone activates the receptor, enabling it to bind with high affinity to specific DNA sequences that act as transcriptional enhancers. This binding increases the level of transcription from certain nearby genes. The products of some of these activated genes may, in turn, activate other genes and produce a delayed secondary response, thereby extending the initial effect of the hormone. Each steroid hormone is recognized by a different member of a family of homologous receptor proteins. The same receptor protein regulates different genes in different target cells, presumably because other DNA-binding proteins required for specific gene transcription differ in these cells.

Mechanisms of Transduction by Cell-Surface Receptor Proteins[9]

All water-soluble signaling molecules (including neurotransmitters, protein hormones, and growth factors), as well as some lipid-soluble ones (see p. 668), bind to specific receptor proteins on the surface of the target cells they influence. These cell-surface receptor proteins bind the signaling molecule (the ligand) with high affinity and convert this extracellular event into one or more intracellular signals that alter the behavior of the target cell.

By using receptor-binding ligands labeled with radioactive atoms, fluorescent dyes, or electron-dense particles (such as colloidal gold), it has been shown that the distribution of cell-surface receptors for a specific ligand can be either diffuse or localized to specific regions of the plasma membrane and that their number can vary from 500 to more than 100,000 per cell, depending on the receptor. Like other membrane proteins, cell-surface receptors are difficult to purify or to study in isolation, especially since they usually constitute less than 0.1% of the total protein mass of the plasma membrane. Techniques for cloning the DNA sequences encoding cell-surface receptors make it possible to circumvent many of the diffi-

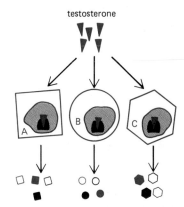

NORMAL MALES

testosterone

testosterone causes different target cells to produce different proteins, even though they all contain the same receptor protein

MALES WITH TESTICULAR FEMINIZATION SYNDROME

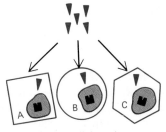

since all target cells have the same mutant receptor protein, they fail to respond to testosterone

Figure 12–12 Target cells that respond differently to testosterone contain the same receptor protein. Thus, in the *testicular feminization syndrome*, a single gene defect that leads to an abnormal testosterone receptor results in all target cells failing to respond to testosterone.

culties and have revolutionized our understanding of receptor structure and function.

Unlike intracellular receptors for steroid and thyroid hormones, cell-surface receptors do not regulate gene expression directly. Instead, they relay a signal across the plasma membrane, and the influence they exert on events in the cytosol or nucleus generally depends on the production of new intracellular signals. It might be imagined that cell-surface receptors could simply transfer the extracellular signaling molecule across the membrane into the cytosol to serve there as an intracellular signal, but this is not the case. Even though many protein signaling molecules, such as insulin, are ingested by receptor-mediated endocytosis (see p. 718), they do not escape from the endosomal or lysosomal compartments into the cytosol. The task of the extracellular ligand seems to be simply to force an appropriate conformational change in the cell-surface receptor protein. In fact, antibodies that bind to the receptor can often mimic the effects of the normal ligand—a phenomenon that underlies some disease states. The usual cause of hyperthyroid disease in humans, for example, is the abnormal production of antibodies that bind to thyroid-stimulating-hormone (TSH) receptors, activating the receptor and causing too much thyroid hormone to be produced.

In this section we consider how a conformational change induced in a cell-surface receptor protein by the binding of an extracellular ligand enables the receptor to act directly or indirectly as a transducer, converting the extracellular signal into a signal inside the cell.

There Are at Least Three Known Classes of Cell-Surface Receptor Proteins: Channel-linked, G-Protein-linked, and Catalytic[9]

Most cell-surface receptor proteins belong to one of three classes, which are defined by the transduction mechanism used. **Channel-linked receptors** are transmitter-gated ion channels (see p. 1074) involved mainly in rapid synaptic signaling between electrically excitable cells. This type of signaling is mediated by a small number of neurotransmitters that transiently open or close the ion channel to which they bind, briefly changing the ion permeability of the plasma membrane and thereby the excitability of the postsynaptic cell. DNA sequencing studies have shown that channel-linked receptors belong to a family of homologous, multipass

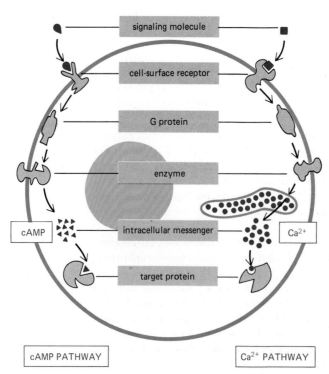

Figure 12–13 Two major pathways by which G-protein-linked cell-surface receptors generate intracellular messengers. In both cases the binding of an extracellular ligand alters the conformation of the cytoplasmic domain of the receptor so that it binds to a G protein, which in turn activates (or inactivates) a plasma membrane enzyme. In other cases the G protein binds to an ion channel rather than to an enzyme. In the cyclic AMP pathway, the enzyme produces cyclic AMP. In the Ca^{2+} pathway, the enzyme produces a soluble mediator that releases Ca^{2+} from an intracellular storage site. Both cyclic AMP and Ca^{2+} bind to other specific proteins in the cell, thereby altering their activity.

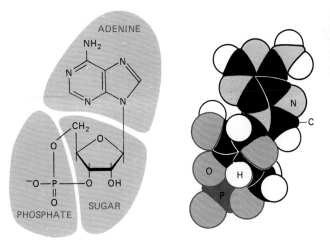

Figure 12–14   Cyclic AMP shown as a formula and as a space-filling model. (C, H, N, O, and P indicate carbon, hydrogen, nitrogen, oxygen, and phosphorus atoms, respectively.)

transmembrane proteins. These receptors are discussed in Chapter 19 (see p. 1074) and will not be considered further here.

Catalytic receptors, when activated by their ligand, operate directly as enzymes. Almost all of the known catalytic receptors are transmembrane proteins with a cytoplasmic domain that functions as a tyrosine-specific protein kinase.

G-protein-linked receptors indirectly activate or inactivate a separate plasma-membrane-bound enzyme or ion channel. The interaction between the receptor and the enzyme or ion channel is mediated by a third protein, called a *GTP-binding regulatory protein* (or *G protein*). The G-protein-linked receptors usually activate a chain of events that alters the concentration of one or more small intracellular signaling molecules, often referred to as **intracellular messengers** (or **intracellular mediators**). These intracellular messengers act in turn to alter the behavior of yet other target proteins in the cell. Two of the most important intracellular messengers are *cyclic AMP (cAMP)* and Ca^{2+}. Cyclic AMP and Ca^{2+} signals are generated by different pathways, both involving G proteins, and are used by almost all animal cells (Figure 12–13). We shall discuss these pathways before returning to consider catalytic receptors with tyrosine-specific protein kinase activity. We begin with the experiments that led to the discovery of cyclic AMP and paved the way to our present understanding of the coupling of its production to extracellular signals.

Cyclic AMP Is a Ubiquitous Intracellular Messenger in Animal Cells[10]

When muscle or liver cells are exposed to the hormone *epinephrine* (*adrenaline*), they are stimulated to break down their stores of glycogen. It was found that epinephrine causes activation of the enzyme *glycogen phosphorylase,* which catalyzes glycogen breakdown. It was then shown that treatment of the isolated membranes of liver cells with epinephrine (in the presence of ATP) induced the production of a small heat-stable mediator that could activate the phosphorylase present in a membrane-free extract of liver cells. The mediator was identified in 1959 as **cyclic AMP** (Figure 12–14), which has since been found to regulate intracellular reactions in all procaryotic and animal cells that have been studied.

The identification of cyclic AMP led to the study of the enzymes that make and degrade it. For cyclic AMP to function as an intracellular mediator, its intracellular concentration (normally $\leq 10^{-6}$ M) must be able to change rapidly up or down in response to extracellular signals: upon hormonal stimulation, cyclic AMP levels can change by fivefold in seconds. As explained on page 714, such responsiveness requires that rapid synthesis of the molecule be balanced by rapid breakdown or removal. Cyclic AMP is synthesized from ATP by the plasma-membrane-bound enzyme **adenylate cyclase,** and it is rapidly and continuously destroyed by one or more **cyclic AMP phosphodiesterases,** which hydrolyze cyclic AMP to adenosine 5′-monophosphate (5′-AMP) (Figure 12–15).

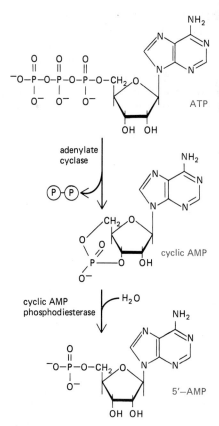

Figure 12–15   The synthesis and degradation of cyclic AMP. A pyrophosphatase makes the synthesis of cyclic AMP an irreversible reaction by hydrolyzing the released pyrophosphate (Ⓟ—Ⓟ).

Table 12–2 Some Hormone-induced Cellular Responses Mediated by Cyclic AMP

Target Tissue	Hormone	Major Response
Thyroid	thyroid-stimulating hormone (TSH)	thyroid hormone synthesis and secretion
Adrenal cortex	adrenocorticotropic hormone (ACTH)	cortisol secretion
Ovary	luteinizing hormone (LH)	progesterone secretion
Muscle, liver	epinephrine	glycogen breakdown
Bone	parathormone	bone resorption
Heart	epinephrine	increase in heart rate and force of contraction
Kidney	vasopressin	water resorption
Fat	epinephrine, ACTH, glucagon, TSH	triglyceride breakdown

6 Receptor and Adenylate Cyclase Molecules Are Separate Proteins That Functionally Interact in the Plasma Membrane

Many hormones and local chemical mediators work by controlling cyclic AMP levels, and they do so by activating (or in some cases by inhibiting) adenylate cyclase rather than by altering phosphodiesterase activity. Just as the same steroid hormone produces different effects in different target cells, so different target cells respond very differently to external signals that change intracellular cyclic AMP levels (Table 12–2). All ligands that activate adenylate cyclase in a given type of target cell, however, usually produce the same effect. For example, at least four hormones activate adenylate cyclase in fat cells, and all of them stimulate the breakdown of triglyceride (the storage form of fat) to fatty acids (see Table 12–2). The different receptors for these hormones seem to activate a common pool of adenylate cyclase molecules. That the receptors and the adenylate cyclase are separate molecules can be shown by receptor "transplantation" experiments. Epinephrine receptors isolated from detergent-solubilized plasma membranes have no adenylate cyclase activity, but when they are transplanted to the plasma membrane of cells that do not have their own epinephrine receptors, the transplanted receptors are able to interact functionally with adenylate cyclase molecules of the recipient cell, causing their activation in the presence of the hormone (Figure 12–16).

Receptors Activate Adenylate Cyclase Molecules via a Stimulatory G Protein (G$_s$)[12]

In bacteria, receptors and adenylate cyclase molecules interact directly, but in animal cells another protein intervenes to couple the activated receptor to the enzyme. The first hint of this complication came from the observation that hormonal activation of adenylate cyclase in disrupted animal cells requires GTP. Later, mutant cell lines were isolated in which binding of epinephrine failed to activate adenylate cyclase in spite of normal levels of epinephrine receptors and of adenylate cyclase. By mixing plasma membrane preparations from such "uncoupled" cells with detergent extracts of plasma membranes from normal cells, a hormone-sensitive adenylate cyclase system requiring GTP could be reconstituted. The detergent extracts proved to contain a **GTP-binding regulatory protein,** or **G protein,** that was missing from the "uncoupled" mutant cells. Because this G protein is involved in enzyme *activation,* it is called **stimulatory G protein (G$_s$).** Individuals who are genetically deficient in G$_s$ have decreased responses to many hormones and, consequently, fail to grow or mature sexually, are mentally retarded, and have many metabolic abnormalities. Recently an epinephrine-activated adenylate cyclase system has been reconstituted in synthetic phospholipid vesicles (see p. 286)

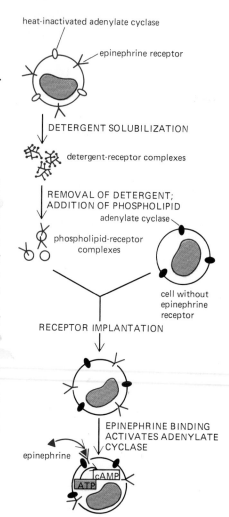

Figure 12–16 Functional epinephrine receptors can be extracted from cells (in which adenylate cyclase molecules have been inactivated by heat) and implanted into the plasma membrane of cells that lack such receptors. When activated by epinephrine, the transplanted receptors activate adenylate cyclase molecules in the recipient cell plasma membrane. The intermediary G protein is not shown. More recently, purified epinephrine receptors have been used in these experiments with the same results. There are at least three types of epinephrine receptors (also called adrenergic receptors)—α_1, α_2, and β; only β-adrenergic receptors activate adenylate cyclase.

from purified epinephrine receptor, G_s, and adenylate cyclase molecules, indicating that no other proteins are required for the activation process.

If the G_s protein is to relay a signal from the receptor to the cyclase, it must have some way of changing its structure when a signal has been received. This is the function of the GTP. When G_s is activated by the receptor-hormone complex, it simultaneously binds a molecule of GTP. Carrying the GTP, it now activates an adenylate cyclase molecule. The G_s keeps the cyclase active so long as the GTP is intact. Eventually, hydrolysis of the GTP to GDP by the G_s protein (which is a GTPase) terminates the activation of the cyclase.

12-17 ### The G_s Protein is a Heterotrimer That Is Thought to Disassemble When Activated[13]

The function of a G protein depends critically on its subunit structure. G_s is composed of three polypeptides: an *α chain* ($G_{s\alpha}$), which binds and hydrolyzes GTP and activates adenylate cyclase, and a tight complex of a *β chain* and a *γ chain* ($G_{\beta\gamma}$), which anchors G_s to the cytoplasmic face of the plasma membrane. A current model of how G_s couples receptor activation to adenylate cyclase activation is shown in Figure 12–17. In its inactive form, G_s exists as a trimer with GDP bound to $G_{s\alpha}$. When activated by binding to a receptor-hormone complex, the guanyl-nucleotide-binding site on $G_{s\alpha}$ is altered, allowing GTP to bind in place of GDP. The binding of GTP is thought to cause $G_{s\alpha}$ to dissociate from $G_{\beta\gamma}$, allowing $G_{s\alpha}$ to bind tightly to an adenylate cyclase molecule, which is thus activated to produce cyclic AMP. Within less than a minute, the $G_{s\alpha}$ hydrolyzes its bound GTP to GDP, causing $G_{s\alpha}$ to dissociate from the adenylate cyclase (which thereby becomes inactive) and reassociate with $G_{\beta\gamma}$ to reform an inactive G_s molecule.

The adenylate cyclase system in bacteria lacks a G_s intermediate. Why, then, have animal cells evolved such a complex multistep mechanism for signal transduction, with a G protein interposed between the receptor and the enzyme it is to activate? One reason may lie in a need for amplification of the signal (see p. 713), another in a need for additional levels of control.

G_s allows for two sorts of amplification. Most simply, a single activated receptor protein can, in principle, collide with and activate many molecules of G protein, thereby activating many molecules of adenylate cyclase. In some cases, however, the extracellular ligand may not remain bound to its receptor long enough for this amplification mechanism to operate: some ligands, for example, may dissociate from their receptors in less than a second. G_s itself, however, is thought to remain active for up to 10 or 15 seconds before hydrolyzing its bound GTP. In this way it can keep an adenylate cyclase molecule active long after such an extracellular ligand has dissociated. This amplification effect can be demonstrated in an exaggerated form if cells are broken open and exposed to an analogue of GTP in which the terminal phosphate cannot be hydrolyzed. Hormone treatment then has a greatly prolonged effect on cyclic AMP production.

In addition to amplification, G proteins provide an important step where the activation process can be regulated. In principle, the efficiency of coupling between receptors and enzyme can be altered by covalently modifying the G protein or by changing its concentration in the plasma membrane. This is illustrated most dramatically by the effects of the bacterial toxin that is responsible for the symptoms of cholera. **Cholera toxin** is an enzyme that catalyzes the transfer of ADP-ribose from intracellular NAD^+ to the α subunit of the G_s protein, altering it so that it can no longer hydrolyze its bound GTP. An adenylate cyclase molecule activated by such an altered G_s protein thus remains in the active state indefinitely. The resulting prolonged elevation in cyclic AMP levels within intestinal epithelial cells causes a large efflux of Na^+ and water into the gut, which is responsible for the severe diarrhea that is characteristic of cholera.

G_s is only one member of a large family of G proteins that couple receptors to a variety of enzymes and ion channels in eucaryotic cell membranes. As we shall now discuss, one of these G proteins inhibits rather than activates adenylate cyclase.

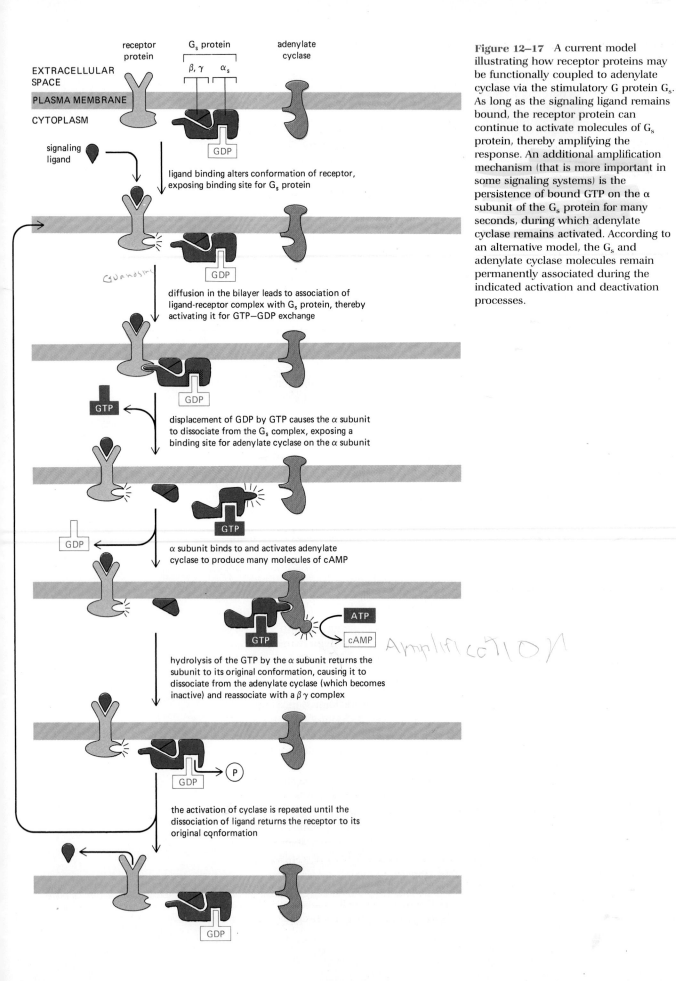

Figure 12–17 A current model illustrating how receptor proteins may be functionally coupled to adenylate cyclase via the stimulatory G protein G_s. As long as the signaling ligand remains bound, the receptor protein can continue to activate molecules of G_s protein, thereby amplifying the response. An additional amplification mechanism (that is more important in some signaling systems) is the persistence of bound GTP on the α subunit of the G_s protein for many seconds, during which adenylate cyclase remains activated. According to an alternative model, the G_s and adenylate cyclase molecules remain permanently associated during the indicated activation and deactivation processes.

Receptors Inactivate Adenylate Cyclase via an Inhibitory G Protein (G$_i$)[13]

The same signaling molecule can either increase or decrease the intracellular concentration of cyclic AMP depending on the type of receptor to which it binds. For example, there are several types of epinephrine (adrenergic) receptors: β-adrenergic receptors activate adenylate cyclase, whereas α$_2$-adrenergic receptors inhibit it. The difference is due to the G proteins that couple these receptors to the cyclase. The β receptors are functionally coupled to adenylate cyclase by G$_s$; the α$_2$ receptors are coupled to the same enzyme by an **inhibitory G protein (G$_i$)**, which contains the same G$_{\beta\gamma}$ complex as G$_s$ but a different α subunit (G$_{i\alpha}$). When activated, α$_2$-adrenergic receptors bind to G$_i$, causing G$_{i\alpha}$ to bind GTP in place of GDP. This is thought to cause G$_{i\alpha}$ to dissociate from the G$_{\beta\gamma}$, and both the released G$_{i\alpha}$ and G$_{\beta\gamma}$ are believed to contribute to the inhibition of adenylate cyclase: G$_{i\alpha}$ inhibits the cyclase directly, whereas G$_{\beta\gamma}$ acts indirectly by binding to free α$_s$ subunits, thereby preventing them from activating cyclase molecules.

Just as cholera toxin maintains high levels of cyclic AMP by ADP-ribosylating G$_{s\alpha}$ and inactivating its GTPase activity, so *pertussis toxin*, made by the bacterium that causes whooping cough, produces the same effect by ADP-ribosylating G$_{i\alpha}$. In this case, however, the G$_i$ complex is prevented from interacting with receptors and therefore fails to inhibit adenylate cyclase in response to receptor activation.

Although G proteins were first discovered because of their effects on adenylate cyclase, they can also act in other ways, as summarized in Table 12–3. In particular, by activating phospholipase C (see p. 702), other G proteins can couple receptor activation to changes in the concentration of Ca^{2+} in the cytosol, and Ca^{2+} is even more widely used as an intracellular messenger than cyclic AMP.

Ca^{2+} Is Stored in a Special Intracellular Calcium-sequestering Compartment[14]

The concentration of free Ca^{2+} in the cytosol of any cell is extremely low (~10^{-7} M), whereas its concentration in the extracellular fluid (>10^{-3} M) and in a specialized intracellular calcium-sequestering compartment is high. Thus there is a large gradient tending to drive Ca^{2+} into the cytosol across the plasma membrane and the membrane of the intracellular compartment. When a signal tran-

Table 12–3 Some GTP-binding Regulatory Proteins Involved in Cell Signaling

Type of G Protein	α Subunit*	Function	Modified by Bacterial Toxin
G$_s$	α$_s$	activates adenylate cyclase	cholera
G$_i$	α$_i$	inactivates adenylate cyclase	pertussis
G$_p$	?	activates phosphoinositide-specific phospholipase C	pertussis (only in some cells)
G$_o$	α$_o$	main G protein in brain; may regulate ion channels	pertussis
Transducin	T$_\alpha$	activates cyclic GMP phosphodiesterase in vertebrate rod cells (see p. 705)	pertussis and cholera
ras proteins	—**	involved in unknown way in growth-factor stimulation of cell proliferation (see pp. 705 and 757)	no

*Except for *ras* proteins (and G$_p$, whose structure is unknown), G proteins are heterotrimers composed of an α chain loosely bound to a βγ dimer. All of the known α subunits (40,000–50,000 daltons) are homologous, and several of the G proteins share the same (or very similar) β (35,000 daltons) and γ (8000 daltons) subunits.

**The *ras* proteins are single polypeptide chains (21,000 daltons) with relatively little homology to the subunits of the other G proteins; it is not known if they act as intermediaries in receptor coupling the way the other listed G proteins do.

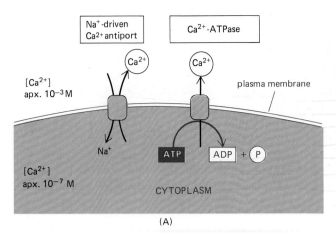

Na⁺-driven
Ca²⁺antiport

Ca²⁺-ATPase

Ca²⁺

Ca²⁺

plasma membrane

[Ca²⁺]
apx. 10^{-3} M

Na⁺

ATP

ADP + P

[Ca²⁺]
apx. 10^{-7} M

CYTOPLASM

(A)

Figure 12–18 The main ways in which cells maintain a very low concentration of free Ca^{2+} in the cytosol in the face of high concentrations of Ca^{2+} in the extracellular fluid. Ca^{2+} is actively pumped out of the cytosol to the cell exterior (A) and into a special internal compartment, the calcium-sequestering compartment (B). In addition, various molecules in the cell bind free Ca^{2+} tightly. Mitochondria can also pump Ca^{2+} out of the cytosol, but they do so efficiently only when Ca^{2+} levels are extremely high—usually as a result of cell damage.

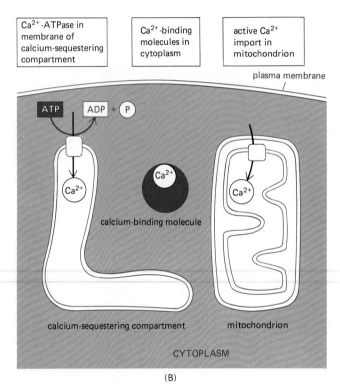

Ca²⁺-ATPase in
membrane of
calcium-sequestering
compartment

Ca²⁺-binding
molecules in
cytoplasm

active Ca²⁺
import in
mitochondrion

plasma membrane

ATP

ADP + P

Ca²⁺

Ca²⁺

Ca²⁺

calcium-binding molecule

calcium-sequestering compartment

mitochondrion

CYTOPLASM

(B)

siently opens Ca^{2+} channels in either of these membranes, Ca^{2+} rushes into the cytosol, dramatically increasing the local Ca^{2+} concentration and activating Ca^{2+}-sensitive response mechanisms in the cell.

For this signaling mechanism to work, the concentration of Ca^{2+} in the cytosol must be kept low, and this is achieved in several ways (Figure 12–18). All eucaryotic cells have a Ca^{2+}-ATPase in their plasma membrane that uses the energy of ATP hydrolysis to pump Ca^{2+} out of the cytosol. Muscle and nerve cells, which make extensive use of Ca^{2+} signaling, have an additional Ca^{2+} pump in their plasma membrane that couples the efflux of Ca^{2+} to the influx of Na⁺. This Na⁺-Ca^{2+} exchanger has a relatively low affinity for Ca^{2+} and therefore begins to operate efficiently only when cytosolic Ca^{2+} levels rise to about 10 times their normal level, as occurs after repeated muscle or nerve cell stimulation.

A Ca^{2+} pump in the membrane of the specialized intracellular compartment also plays an important part in keeping the cytosolic Ca^{2+} concentration low: this Ca^{2+}-ATPase enables the intracellular compartment to take up large amounts of Ca^{2+} from the cytosol against a steep concentration gradient, even when Ca^{2+} levels in the cytosol are low. This Ca^{2+} is stored in the lumen of the compartment, loosely bound to a Ca^{2+}-binding protein called *calsequestrin,* which has a low affinity ($K_a \simeq 10^3$ liters/mole) but high capacity (~50 calcium ions/molecule) for Ca^{2+}. When antibodies against calsequestrin and the Ca^{2+}-ATPase are used to

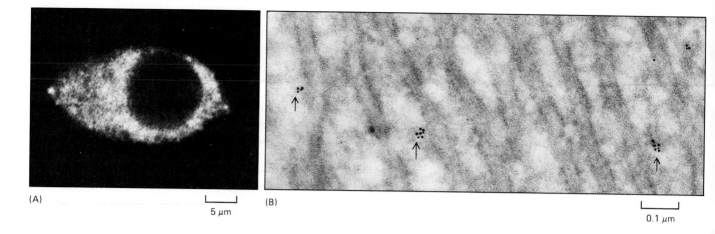

(A)

(B)

|——————|
5 μm

|——————|
0.1 μm

stain cells, the two antibodies label the same membrane-bounded compartment, which is distinct from the rough endoplasmic reticulum and much more limited in extent (Figure 12–19). This recently discovered compartment is homologous with the more extensive sarcoplasmic reticulum in muscle cells (see p. 621) and, like the sarcoplasmic reticulum, is specialized for calcium storage and release. We shall therefore refer to it as the **calcium-sequestering compartment.**

Normally the concentration of free Ca^{2+} in the cytosol varies from about 10^{-7} M when the cell is at rest to about 5×10^{-6} M when the cell is activated by an extracellular signal. But when a cell is damaged and cannot pump Ca^{2+} out of the cytosol efficiently, the Ca^{2+} concentration can rise to dangerously high levels ($>10^{-5}$ M). In these circumstances a low-affinity, high-capacity Ca^{2+} pump in the inner mitochondrial membrane comes into action and uses the electrochemical gradient generated across this membrane during the electron-transfer steps of oxidative phosphorylation to take up Ca^{2+} from the cytosol (see p. 353).

Ca^{2+} Functions as a Ubiquitous Intracellular Messenger[15]

The first direct evidence that Ca^{2+} functions as an intracellular mediator came from an experiment done in 1947 showing that the intracellular injection of a small amount of Ca^{2+} causes a skeletal muscle cell to contract. In recent years it has become clear that Ca^{2+} acts as an intracellular messenger in a wide variety of cellular responses, including secretion and cell proliferation. Two pathways of Ca^{2+} signaling have been defined (Figure 12–20), one used mainly by electrically

Figure 12–19 The intracellular calcium-sequestering compartment revealed by labeling with antibodies against the Ca^{2+}-binding protein calsequestrin. (A) Immunofluorescence micrograph of a rat neural cell in culture, showing that the compartment is distributed throughout the cytoplasm. (B) Immunogold electron micrograph of a frozen thin section of rat liver, showing that the calcium-sequestering compartment is not the rough endoplasmic reticulum (ER), although it might be part of the smooth ER; the gold-coupled antibodies are indicated by arrows. Quantitative analyses of such electron micrographs indicate that the compartment occupies less than 1% of the cell volume. (A, from P. Volpe et al. *Proc. Natl. Acad. Sci. USA*, 85:1091–1095, 1988. B, courtesy of J. Meldolesi.)

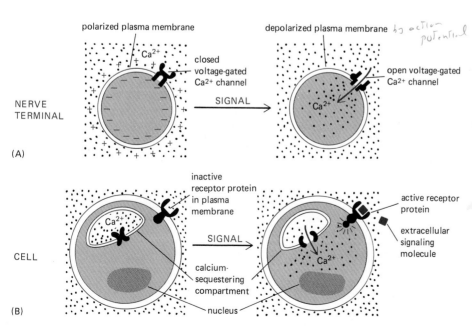

Figure 12–20 Two common pathways by which Ca^{2+} can enter the cytosol to act as an intracellular mediator of extracellular signals. In (A), Ca^{2+} enters a nerve terminal from the extracellular fluid through voltage-gated Ca^{2+} channels when the nerve terminal membrane is depolarized by an action potential. In (B), the binding of an extracellular signaling molecule to a cell-surface receptor stimulates the release of Ca^{2+} from the calcium-sequestering compartment inside the cell.

active cells and the other used by almost all eucaryotic cells. The first of these pathways has been particularly well studied in nerve cells, in which depolarization of the plasma membrane causes an influx of Ca^{2+} into the nerve terminal, initiating the secretion of neurotransmitter; the Ca^{2+} enters through voltage-gated Ca^{2+} channels that open when the plasma membrane of the nerve terminal is depolarized by an invading action potential (see p. 1077). In the second, ubiquitous pathway the binding of extracellular signaling molecules to cell-surface receptors causes the release of Ca^{2+} from the calcium-sequestering compartment; the events at the cell surface are coupled to the opening of Ca^{2+} channels in the internal membrane through yet another intracellular messenger molecule, *inositol trisphosphate*. Production of this messenger involves the rapid hydrolysis of a minor class of phospholipid molecules (inositol phospholipids) in the plasma membrane.

Inositol Trisphosphate (InsP₃) Couples Receptor Activation to Ca²⁺ Release from the Calcium-sequestering Compartment[16]

A role for inositol phospholipids (*phosphoinositides*) in signal transduction was first suggested in 1953, when it was found that some extracellular signaling molecules stimulate the incorporation of radioactive phosphate into **phosphatidyl-inositol (PI)**, a minor phospholipid in cell membranes. It was later shown that this incorporation results from the breakdown and subsequent resynthesis of inositol phospholipids triggered by a receptor protein that activates an enzyme called **phosphoinositide-specific phospholipase C.** The inositol phospholipids found to be most important in signal transduction were two phosphorylated derivatives of PI, *PI-phosphate (PIP)* and *PI-bisphosphate (PIP₂)*, which are thought to be located mainly in the inner leaflet of the plasma membrane (Figure 12–21). Although PIP₂ is less plentiful in animal cell membranes than PI, it is the hydrolysis of PIP₂ that matters most.

The chain of events linking extracellular signals to intracellular responses via PIP₂ breakdown begins with the binding of the signaling molecule to its receptor in the plasma membrane. More than 25 different cell-surface receptors have been shown to utilize the inositol-phospholipid transduction pathway (Table 12–4). Although the details of the activation process are not as well understood as they are in the cyclic AMP pathway, there is increasing evidence that the same type of multistep mechanism operates in the plasma membrane. An activated receptor is thought to activate a G protein (tentatively called **G_p**) that in turn activates the phospholipase C. In less than a second, this enzyme cleaves PIP₂ to generate two products: *inositol trisphosphate* and *diacylglycerol* (Figure 12–22). At this step the signaling pathway splits into two branches. Since both molecules play crucial parts in signaling the cell, we shall consider them in turn.

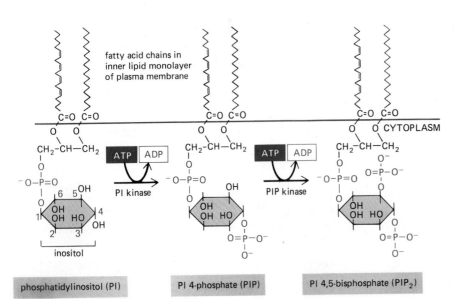

phosphatidylinositol (PI)   PI 4-phosphate (PIP)   PI 4,5-bisphosphate (PIP₂)

Figure 12–21   The inositol phospholipids (phosphoinositides) comprise less than 10% of the total phospholipids in mammalian cell membranes. The polyphosphoinositides (PIP and PIP₂) are produced by the phosphorylation of phosphatidylinositol (PI). Although all three inositol phospholipids may be broken down in the signaling response, it is the breakdown of PIP₂ that is most critical, even though it is the least abundant, constituting less than 10% of the total inositol lipids and less than 1% of the total phospholipids.

Table 12–4 Some Cellular Responses Mediated by the Inositol-Phospholipid Signaling Pathway

Target Tissue	Signaling Molecule	Major Response
Liver	vasopressin	glycogen breakdown
Pancreas	acetylcholine	amylase secretion
Smooth muscle	acetylcholine	contraction
Pancreatic β cells	acetylcholine	insulin secretion
Mast cells	antigen	histamine secretion
Blood platelets	thrombin	serotonin and PDGF secretion

Inositol trisphosphate (InsP$_3$) is a small water-soluble molecule that releases Ca^{2+} from the calcium-sequestering compartment (see Figure 12–22). When InsP$_3$ is added to permeabilized cells or to isolated intracellular vesicles, it causes the efflux of Ca^{2+} into the medium. Apparently the binding of InsP$_3$ to a receptor protein on the cytoplasmic surface of the intracellular compartment opens gated Ca^{2+} channels in its membrane. Two mechanisms operate to make the Ca^{2+} response transient: (1) the Ca^{2+} that enters the cytosol is rapidly pumped out, mainly out of the cell (to be replaced by an influx across the plasma membrane when the stimulus is removed); and (2) some of the InsP$_3$ is rapidly dephosphorylated (and thereby inactivated) by a specific phosphatase. Not all of the InsP$_3$ is dephosphorylated: some is instead phosphorylated to form inositol 1,3,4,5-tetrakisphosphate (InsP$_4$), which may mediate slower and more prolonged responses in the cell. In some cells the Ca^{2+} response occurs as a series of transient increases or "spikes," each lasting 10 seconds or more.

Diacylglycerol Produced by PIP$_2$ Hydrolysis Activates Protein Kinase C[17]

While the InsP$_3$ produced by hydrolysis of PIP$_2$ is increasing the concentration of Ca^{2+} in the cytosol, the other cleavage product of PIP$_2$, **diacylglycerol**, is exerting quite different effects. It has two potential signaling roles. It can be further cleaved

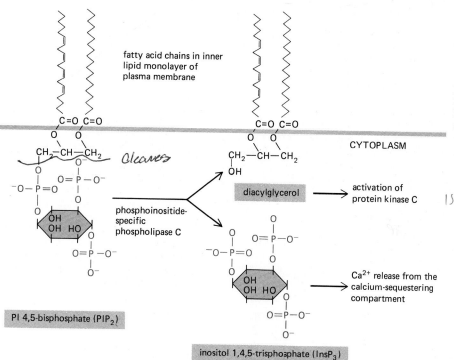

Figure 12–22 The hydrolysis of PIP$_2$ produces *inositol trisphosphate (InsP$_3$)*, which diffuses through the cytosol and releases Ca^{2+} from the calcium-sequestering compartment. The other product, *diacylglycerol*, also plays an important part in intracellular signaling by activating the enzyme, protein kinase C (see below).

to release arachidonic acid, which can be used in the synthesis of prostaglandins and related lipid signaling molecules (see Figure 12–8); or, more important, it can activate a specific protein kinase, which can then phosphorylate a number of proteins with different functions in the target cell.

The enzyme activated by diacylglycerol is called **protein kinase C (C-kinase)** because it is Ca^{2+}-dependent. The diacylglycerol produced by receptor activation, together with the phospholipid *phosphatidylserine* in the cytoplasmic half of the plasma membrane, binds to C-kinase, thereby increasing the affinity of the enzyme for Ca^{2+} so that the C-kinase becomes active at the usual low concentrations of Ca^{2+} in the cytosol. In many cells, however, it seems likely that C-kinase is normally activated by the cooperative effect of diacylglycerol and an increase in cytosolic Ca^{2+} brought about by InsP$_3$. The activation of C-kinase is transient because, within seconds, diacylglycerol is phosphorylated to form phosphatidate or further cleaved to arachidonic acid.

When activated by diacylglycerol and Ca^{2+}, C-kinase transfers the terminal phosphate group from ATP to specific serine or threonine residues on target proteins that vary depending on the cell. In many animal cells, for example, C-kinase is thought to phosphorylate and thereby activate the plasma membrane Na^+-H^+ exchanger that controls intracellular pH; the resulting increase in intracellular pH may help signal some cells to proliferate (see p. 309). The highest concentrations of C-kinase are found in the brain, where (among other things) it phosphorylates ion channels in nerve cells, thereby changing their properties and altering the excitability of the nerve cells (see p. 1092). In some cells the activation of C-kinase increases the transcription of specific genes. The promoters of at least some of these genes contain a common transcriptional enhancer sequence that is recognized by a gene regulatory protein whose activity is stimulated by C-kinase activation (see Table 10–1, p. 566); it is not known, however, whether the protein is directly phosphorylated (and thereby activated) by C-kinase or whether it is activated indirectly by a protein kinase cascade.

Each of the two branches of the inositol phospholipid signaling pathway can be mimicked by the addition of specific pharmacological agents to intact cells. The effects of InsP$_3$ can be mimicked by using a *calcium ionophore*, such as A23187 or ionomycin, which allows Ca^{2+} to move into the cytosol from the extracellular fluid (see p. 322). The effects of diacylglycerol can be mimicked by monoacyl derivatives of diacylglycerol or by *phorbol esters*, plant products that bind to C-kinase and activate it directly (Figure 12–23). Using these reagents it has been shown that the two branches of the pathway often collaborate in producing a full

Figure 12–23 The two branches of the inositol phospholipid pathway. It is thought that the activated receptor binds to a specific G protein (G$_p$), causing the putative α subunit to dissociate and activate the phospholipase C that cleaves PIP$_2$ to generate InsP$_3$ and diacylglycerol. The actual structure of G$_p$ is unknown. The diacylglycerol (together with bound Ca^{2+} and phosphatidylserine—not shown) activates C-kinase. Both phospholipase C and C-kinase are present in cells in both water-soluble and membrane-bound forms, and it is likely that one or both translocate from the cytosol to the inner face of the plasma membrane in the process of being activated. The effects of InsP$_3$ can be mimicked experimentally in intact cells by treatment with Ca^{2+} ionophores, while the effects of diacylglycerol can be mimicked by treatment with monoacyl derivatives of diacylglycerol or phorbol esters, which bind to C-kinase and activate it.

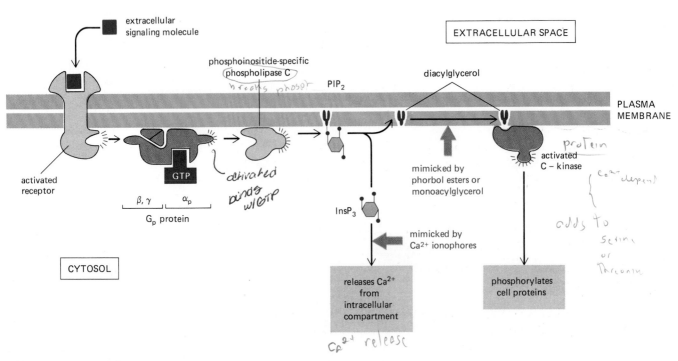

cellular response. For example, a number of cell types can be stimulated to proliferate in culture when treated with both a calcium ionophore and a C-kinase activator but not when treated with either reagent alone.

When cells in the body are stimulated to proliferate excessively, they are liable to attract attention. Thus the effects of phorbol esters on cell proliferation were first discovered because of their activity as "tumor promoters" in whole animals. Tumor promoters are substances that do not cause cancer on their own but can induce tumor growth if they are administered to an animal following a dose of carcinogen that would normally be ineffective by itself. The further discovery that the phorbol ester tumor promoters directly activate C-kinase thus helped to delineate one of the intracellular pathways by which cells respond to growth signals. In fact, basic research on the causes of cancer, and particularly on viruses that cause tumors in animals, has been responsible for a number of important insights into intracellular signaling pathways.

The *ras* Oncogenes Encode a Novel Class of G Proteins Involved in the Regulation of Cell Proliferation[18]

Several signaling pathways seem to be involved in the control of cell growth and division. Some of the genes encoding normal components of these pathways have become accidentally incorporated, often in modified form, into the genomes of a class of *retroviruses* known as RNA tumor viruses. These cell-derived viral genes in some cases confer on the viruses the ability to induce uncontrolled proliferation in infected cells and thus to produce tumors in infected animals: such genes are known as viral **oncogenes.** Study of the tumors led to discovery of the viruses and hence of the oncogenes, and this in turn led to discovery of the normal cellular genes, known as *proto-oncogenes*, from which the viral oncogenes are derived (see p. 754).

A major class of oncogenes and proto-oncogenes discovered in this way, called the *ras* genes (because they were first discovered in viruses that cause *rat* sarcomas), encode G proteins that are located on the inner face of the plasma membrane, where they bind and hydrolyze GTP. The **ras proteins** encoded by viral *ras* oncogenes differ from normal ras proteins (encoded by proto-oncogenes) by amino acid substitutions in one of two positions in the proteins, which is usually enough to impair their GTPase activity and thereby interfere with their normal shut-off mechanism (see p. 697). When antibodies against the products of *ras* proto-oncogenes are injected into cells in culture, the cells will no longer divide in response to growth factors. The ras proteins are therefore thought to be involved in coupling growth factor receptors to effector proteins in the cell. The nature of the effector proteins and of the coupling mechanism is uncertain, although there is increasing evidence that the effector proteins may regulate the inositol phospholipid pathway as at least part of their activity. It is very unlikely that the G_p protein(s) that couples receptors to phospholipase C is encoded by a *ras* proto-oncogene, because then G_p would be very different from all of the other G proteins discussed in this chapter, which are homologous heterotrimers; the ras proteins are single polypeptides (see Table 12–3, p. 699). It is much more likely that the *ras* proto-oncogene family, which has at least three members, represents a special class of G proteins with novel functions. Despite this uncertainty, it is clear that the pursuit of oncogenes has provided an important approach to the study of the normal pathways of cell signaling via G proteins and, as we shall see later, of other cell signaling mechanisms as well.

A Large Family of Homologous, Seven-Pass Transmembrane Glycoproteins Activate G Proteins[19]

The signaling systems that utilize G proteins are remarkably diverse. We have seen that some G proteins (G_s and G_i) couple receptors to adenylate cyclase, whereas others (G_p) couple receptors to phospholipase C. In the vertebrate eye a G protein called *transducin* couples the reception of a photon of light by a rhodopsin molecule to the activation of a phosphodiesterase enzyme that hydrolyzes *cyclic GMP* (see p. 713); the resulting fall in the concentration of this intracellular mediator causes an electrical change in the photoreceptor cell (see p. 1006). Whereas all of these G proteins can modify ion channels indirectly, some G proteins can interact with ion channels directly. The binding of acetylcholine to receptors on heart muscle cells, for example, activates a G protein (related to G_i) that directly activates a K^+ channel in the plasma membrane. (These receptors, which are sensitive to the fungal alkaloid muscarine, are called *muscarinic acetylcholine receptors* to distinguish them from the very different *nicotinic acetylcholine receptors*, which are channel-linked receptors on skeletal muscle cells—see p. 319.)

The G proteins involved in these disparate systems that have been well characterized are evolutionarily related, with similar subunit structure and amino acid sequence; the α subunits of transducin and G_i, for example, are about 65% identical in amino acid sequence. Perhaps not surprisingly, many, if not all, of the receptors that interact with these G proteins are themselves homologous, as has become clear from DNA sequencing experiments. The deduced amino acid sequences of an increasing number of these receptors reveal a common structure consisting of a single polypeptide chain that threads back and forth across the lipid bilayer seven times. This family of seven-pass transmembrane receptor proteins includes β-adrenergic receptors (Figure 12–24), muscarinic acetylcholine receptors, several neuropeptide receptors, and even rhodopsin. It seems likely that these glycoproteins are part of a very large family of evolutionarily related receptors. The structural motif probably arose early in evolution, as it is shared by bacteriorhodopsin, a bacterial light-activated proton pump (which, however, does not act via a G protein—see p. 293) and by receptor proteins used by yeasts during mating (see p. 571). Not all cell-surface receptors are multipass transmembrane proteins, however, and we turn now to another class: the tyrosine-specific protein kinase family of receptors.

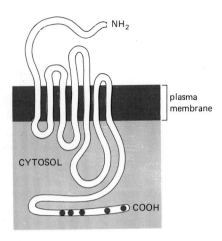

Figure 12–24 The β-adrenergic receptor as it is thought to be oriented in the plasma membrane. The colored regions of the cytoplasmic tail indicate the location of serine residues, which are potential sites for the phosphorylations that mediate receptor desensitization (see p. 718). Other G-protein-linked receptor proteins have been shown to have a similar seven-pass transmembrane structure. (Based on data from R.A.F. Dixon et al., *Nature* 321:75–79, 1986.)

Many Catalytic Receptors Are Single-Pass Transmembrane Glycoproteins with Tyrosine-specific Protein Kinase Activity[20]

Many cell-surface receptors seem to convert an extracellular signal into an intracellular one by regulating the activity of a G protein, but some seem to signal the cell more directly. These are catalytic receptor proteins, and the best-studied examples in animal cells are single-pass transmembrane **tyrosine-specific protein kinases** with their catalytic domain exposed on the cytoplasmic side of the plasma membrane. When activated by ligand binding, they transfer the terminal phosphate group from ATP to the hydroxyl group on a tyrosine residue of selected proteins in the target cell. Included in this family of receptors are those for *insulin* and for a number of growth factors, including *platelet-derived growth factor* (*PDGF*, see p. 746) and *epidermal growth factor* (*EGF*, which stimulates epidermal cells and a variety of other cell types to divide) (Figure 12–25). Most other protein kinases phosphorylate serine or (less often) threonine residues on proteins, so that less than 0.1% of phosphorylated proteins in cells contain phosphotyrosine. In all cases studied, receptor proteins with tyrosine kinase activity phosphorylate themselves when activated; in the case of the insulin receptor, this autophosphorylation enhances the activity of the kinase—an example of *positive feedback regulation*.

How does binding of a ligand to the extracellular domain of these receptors activate the catalytic domain on the other side of the plasma membrane? It is difficult to imagine how a conformational change could propagate across the lipid bilayer through the single transmembrane α helix. In the case of the EGF receptor, ligand binding induces a conformational change in the extracellular domain of the receptor protein that causes the receptor to assemble into dimers. It is possible that the resulting interaction between the two adjacent cytoplasmic domains in such a dimer activates the catalytic activity.

There is good evidence that the kinase activity of these receptors is important in the signaling process. For example, cells containing a mutant insulin receptor with a single amino acid change that selectively inactivates the kinase activity are unresponsive to insulin. However, it has been exceedingly difficult to identify the key substrates that become phosphorylated besides the receptor itself, so the exact

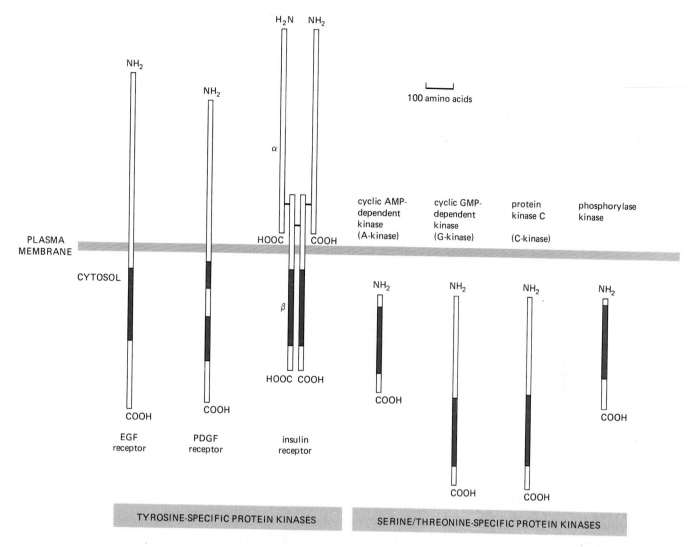

role of tyrosine phosphorylation in signal transduction remains uncertain. In the case of the PDGF receptor, however, one substrate seems to be the kinase that phosphorylates phosphatidylinositol (*PI-kinase*—see Figure 12–21). This may explain the paradoxical finding that PDGF causes a slow activation of the inositol phospholipid signaling pathway, even though it binds to a receptor that is not thought to be coupled to G_p.

Following ligand binding, many catalytic receptors are endocytosed via coated vesicles as receptor-ligand complexes (see p. 328). In some cases, this receptor-mediated endocytosis seems to depend on receptor autophosphorylation, and it can play an important role both in degrading the signaling molecule and in regulating the concentration of the receptor on the target-cell surface (see p. 718). It also translocates the tyrosine kinase domain to new locations in the cell, which could be important in the signaling process, but this has not been demonstrated.

Some Oncogenes Encode Abnormal Catalytic Receptors with Constitutive Kinase Activity[21]

The first tyrosine-specific protein kinase was discovered in 1979. It was not a cell-surface receptor but an intracellular product of a viral oncogene, a protein called pp60[v-src] (see p. 754). The EGF receptor was the first receptor protein to be shown to be a tyrosine-specific kinase (in 1982), and several years later the viral *erbB* oncogene was shown to encode a truncated version of an EGF receptor. This truncated protein lacks the extracellular EGF-binding domain and has an intracellular tyrosine kinase domain that is constitutively active: cells carrying this faulty receptor behave as though they are constantly being signaled to proliferate.

Figure 12–25   Some of the protein kinases discussed in this chapter, showing the size and location of their catalytic domain. In each case the catalytic domain (*colored*) is about 250 amino acid residues long and is similar in amino acid sequence, suggesting that they have all evolved from a common primordial kinase. The three tyrosine-specific kinases shown are transmembrane receptor proteins that, when activated by the binding of specific extracellular ligands, phosphorylate proteins (including themselves) on tyrosine residues inside the cell. Both chains of the insulin receptor are encoded by a single gene, which produces a precursor protein that is cleaved into the two disulfide-linked chains. The extracellular domain of the PDGF receptor is thought to be folded into five immunoglobulin (Ig)-like domains, suggesting that this protein belongs to the Ig superfamily (see p. 1053). The regulatory subunits normally associated with A-kinase (see Figure 12–27) and with phosphorylase kinase (see Figure 12–31) are not shown.

More recently, the *neu* oncogene, expressed by some chemically induced tumors of the nervous system in rats, was found to encode an abnormal receptor that is a tyrosine kinase, although the ligand (suspected to be a growth factor) for the normal receptor has not yet been identified. In this case the abnormal and normal receptors differ in only one amino acid residue, which is in the single transmembrane segment of the protein and suffices to make the tyrosine kinase constitutively active. These studies emphasize the importance of the tyrosine kinase receptors just discussed for the control of cell proliferation.

Many other links have been discovered between oncogenes and the normal signaling pathways involved in cell proliferation. The *sis* oncogene, for example, encodes a functionally active subunit of PDGF, while the *erbA* oncogene encodes a variant form of the thyroid hormone receptor. As we shall discuss at greater length in Chapter 13, studies of oncogenes provide a promising route toward the discovery and understanding of the full set of mechanisms by which the signals for cell proliferation exert their effects.

Nearly half of all oncogenes so far discovered encode protein kinases that phosphorylate target proteins on tyrosine, serine, or threonine residues. This is not surprising since phosphorylation plays a major part in the signaling pathways activated by all catalytic and G-protein-linked receptors, and a very large family of protein kinases has evolved to bring it about. More than 70 protein kinases have so far been defined, all of which are thought to have a common evolutionary origin since they have homologous catalytic domains (see Figure 12–25). In fact, it is possible to predict whether a protein is a kinase and whether it phosphorylates either serine and threonine or tyrosine residues simply by examining its amino acid sequence. We shall see in the next section that two of the principal intracellular mediators, cyclic AMP and Ca^{2+}, exert many of their effects by activating protein kinases of the serine/threonine-specific class.

Summary

There are three known families of cell-surface receptors, each of which transduces extracellular signals in a different way. Channel-linked receptors are neurotransmitter-gated ion channels that open or close briefly in response to neurotransmitter binding and thereby transiently alter the electrical excitability of the cell. Catalytic receptors are mainly tyrosine-specific protein kinases that directly phosphorylate specific target cell proteins on tyrosine residues. G-protein-linked receptors indirectly activate or inactivate plasma-membrane-bound enzymes or ion channels via GTP-binding regulatory proteins (G proteins) that shut themselves off by slowly hydrolyzing their bound GTP. Some G-protein-linked receptors activate or inactivate adenylate cyclase, thereby altering the intracellular concentration of the intracellular messenger cyclic AMP. Others activate a phosphoinositide-specific phospholipase C, which hydrolyzes phosphatidylinositol bisphosphate (PIP₂) to generate two intracellular messengers: (1) inositol trisphosphate (InsP₃) releases Ca^{2+} from an internal calcium-sequestering compartment and thereby increases the concentration of Ca^{2+} in the cytosol; and (2) diacylglycerol remains in the plasma membrane and activates protein kinase C, which in turn phosphorylates various cell proteins. The various responses mediated by the G-protein-linked receptors are rapidly turned off when the extracellular signaling ligand is removed. Thus G proteins self-inactivate, InsP₃ is rapidly dephosphorylated by a phosphatase, diacylglycerol is rapidly converted to phosphatidate or cleaved to arachidonic acid, cyclic AMP is hydrolyzed by a phosphodiesterase, and Ca^{2+} is rapidly pumped out of the cytosol.

The Mode of Action of Cyclic AMP and Calcium Ions[22]

Since cyclic AMP and Ca^{2+} are perhaps the most important intracellular mediators produced as a result of cell-surface receptor activation, their modes of action merit special consideration. Both of these molecules act as *allosteric effectors*, activating specific proteins by binding to them and changing their conformation.

Cyclic AMP Activates a Cyclic AMP-dependent Protein Kinase[23]

Cyclic AMP exerts its effects in animal cells mainly by activating an enzyme called **cyclic AMP-dependent protein kinase (A-kinase)**, which catalyzes the transfer of the terminal phosphate group from ATP to specific serine or threonine residues of selected proteins in the target cell. The residues phosphorylated by A-kinase are marked by the presence of two or more basic amino acids on their amino-terminal side. Covalent phosphorylation of the appropriate residues in turn regulates the activity of the protein.

Cyclic AMP-mediated protein phosphorylation was first demonstrated in studies of glycogen metabolism in skeletal muscle cells. Glycogen is the major storage form of glucose, and we have already mentioned that its degradation in muscle cells is regulated by epinephrine. In fact, epinephrine controls both the synthesis and the degradation of glycogen in skeletal muscle cells. For example, when an animal is frightened or otherwise stressed, the adrenal gland secretes epinephrine into the blood, "alerting" various tissues in the body. Among other effects, the circulating epinephrine induces muscle cells to break down glycogen to glucose 1-phosphate and at the same time to stop synthesizing glycogen. Glucose 1-phosphate is converted to glucose 6-phosphate, which is then oxidized by glycolysis to provide ATP for sustained muscle contraction. In this way epinephrine prepares the muscle cells for strenuous activity.

As discussed previously, epinephrine binds to *β-adrenergic receptors* on the muscle cell, which activate adenylate cyclase and increase the level of cyclic AMP in the cytosol. The cyclic AMP in turn activates A-kinase, which phosphorylates another enzyme, *phosphorylase kinase*, which was the first protein kinase to be purified (in 1959). It in turn phosphorylates the enzyme *glycogen phosphorylase*, thereby activating the phosphorylase to release glucose residues from the glycogen molecule (Figure 12–26). The activated A-kinase also increases the phosphorylation of *glycogen synthase*, the enzyme that performs the final step in glycogen synthesis from glucose, thereby inhibiting its activity and shutting off glycogen synthesis. By means of this cascade of interactions, an increase in cyclic AMP levels both stimulates glycogen breakdown and inhibits glycogen synthesis, thus maximizing the amount of glucose available to the cell.

In some animal cells an increase in cyclic AMP activates the transcription of specific genes. In neuroendocrine cells of the hypothalamus (see p. 683), for example, cyclic AMP turns on the gene that encodes the peptide hormone *soma-*

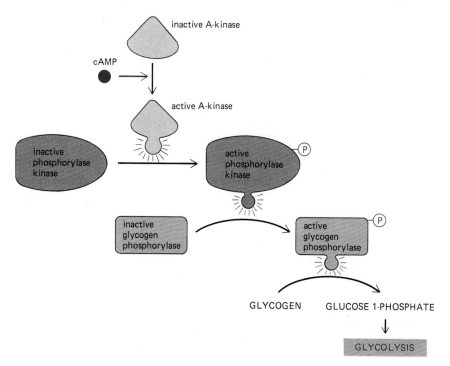

Figure 12–26 How an increase in cyclic AMP in skeletal muscle cells (induced by epinephrine binding to cell-surface β-adrenergic receptors) stimulates glycogen breakdown. The binding of cyclic AMP to A-kinase activates this enzyme to phosphorylate and thereby activate phosphorylase kinase, which in turn phosphorylates and activates glycogen phosphorylase, the enzyme that breaks down glycogen. The A-kinase also directly and indirectly increases the phosphorylation of glycogen synthase, which inhibits the enzyme, thereby shutting off glycogen synthesis (not shown).

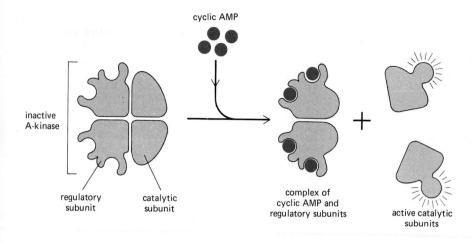

cyclic AMP

inactive
A-kinase

regulatory
subunit

catalytic
subunit

complex of
cyclic AMP and
regulatory subunits

active catalytic
subunits

Figure 12–27 The activation of cyclic AMP-dependent protein kinase (A-kinase). The binding of cyclic AMP to the regulatory subunits induces a conformational change, causing these subunits to dissociate from the complex and thereby activating the catalytic subunits. Each regulatory subunit has two cyclic AMP-binding sites, and the release of the catalytic subunits is a cooperative process requiring the binding of more than two cyclic AMP molecules to the tetramer. This greatly sharpens the response of the kinase to changes in cyclic AMP concentration, as discussed on page 716. Many cells have two types of A-kinase, with identical catalytic subunits but different regulatory subunits.

tostatin. The promoter region of the somatostatin gene contains a short DNA sequence (about 30 nucleotides long) that is also found in the promoter region of several other genes that are activated by cyclic AMP. This sequence is recognized by a specific gene regulatory protein that activates transcription from these genes when it is phosphorylated by A-kinase.

In the inactive state the A-kinase consists of a complex of two regulatory subunits that bind cyclic AMP and two catalytic subunits. The binding of cyclic AMP alters the conformation of the regulatory subunits, causing them to dissociate from the complex. The released catalytic subunits are thereby activated to phosphorylate substrate protein molecules (Figure 12–27).

A-kinase is found in all animal cells and is thought to account for almost all of the effects of cyclic AMP in these cells. While most of the substrates for the kinase have not yet been characterized, it is clear that many of them differ in different cell types, explaining why the effects of cyclic AMP vary depending on the target cell.

Cyclic AMP Inhibits an Intracellular Protein Phosphatase[24]

Since the effects of cyclic AMP are usually transient, it is clear that cells must dephosphorylate the proteins that have been phosphorylated by A-kinase. The dephosphorylation is catalyzed by two main *protein phosphatases*, one of which is itself regulated by cyclic AMP. The level of phosphorylation at any instant will depend on the balance between the kinase and phosphatase activities.

In skeletal muscle cells, the cyclic AMP-regulated protein phosphatase is most active in the absence of cyclic AMP, and it dephosphorylates each of the three key enzymes in the glycogen pathway that were mentioned earlier—phosphorylase kinase, glycogen phosphorylase, and glycogen synthase. These dephosphorylation reactions tend to counteract the protein phosphorylations stimulated by cyclic AMP. However, when A-kinase is activated by cyclic AMP, it also phosphorylates a specific *phosphatase inhibitor protein*, which is thereby activated. This activated inhibitor protein binds to the protein phosphatase and inactivates it (Figure 12–28). By both activating phosphorylase kinase and inhibiting the opposing action of the protein phosphatase, the A-kinase causes a rise in cyclic AMP levels to have a much larger and sharper effect on glycogen metabolism than could be obtained if the A-kinase acted on one of these enzymes alone.

Calmodulin Is a Ubiquitous Intracellular Receptor for Ca^{2+}[25]

Since the free Ca^{2+} concentration in the cytosol is usually about 10^{-7} M and generally does not rise above 5×10^{-6} M even when the cell is activated by an influx of Ca^{2+}, any structure in the cell that is to serve as a direct target for Ca^{2+}-dependent regulation must have an affinity constant (K_a) for Ca^{2+} of around 10^6 liters/mole. Moreover, since the concentration of free Mg^{2+} in the cytosol is relatively constant at about 10^{-3} M, these Ca^{2+}-binding sites must have a selectivity

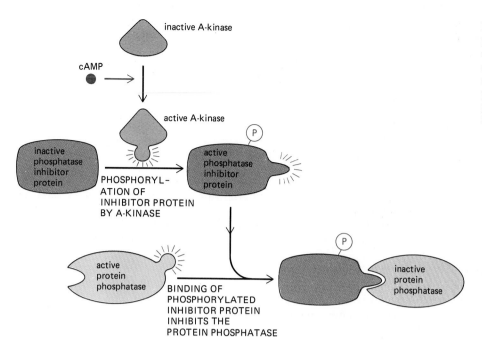

inactive A-kinase

cAMP

active A-kinase

P

inactive phosphatase inhibitor protein

PHOSPHORYL-ATION OF INHIBITOR PROTEIN BY A-KINASE

active phosphatase inhibitor protein

P

active protein phosphatase

BINDING OF PHOSPHORYLATED INHIBITOR PROTEIN INHIBITS THE PROTEIN PHOSPHATASE

inactive protein phosphatase

for Ca^{2+} over Mg^{2+} of at least 1000-fold. Several specific Ca^{2+}-binding proteins fulfill these criteria.

The first such protein to be discovered was *troponin C* in skeletal muscle cells; its role in muscle contraction has been discussed in Chapter 11 (see p. 621). A closely related Ca^{2+}-binding protein, known as calmodulin, is found in all animal and plant cells that have been examined. A typical animal cell contains more than 10^7 molecules of calmodulin, which can constitute as much as 1% of the total protein mass of the cell. Calmodulin functions as a multipurpose intracellular Ca^{2+} receptor, mediating most Ca^{2+}-regulated processes. It is a highly conserved, single polypeptide chain of about 150 amino acid residues, with four high-affinity Ca^{2+}-binding sites; and it undergoes a large conformational change when it binds Ca^{2+} (Figure 12–29).

The allosteric activation of calmodulin by Ca^{2+} is analogous to the allosteric activation of A-kinase by cyclic AMP, except that Ca^{2+}-calmodulin complexes have no enzyme activity themselves but act by binding to other proteins. In some cases calmodulin serves as a permanent regulatory subunit of an enzyme complex (as in the case of phosphorylase kinase—see below), but in most cases the binding of Ca^{2+} induces calmodulin to bind to various target proteins in the cell and thereby alter their activity (Figure 12–30).

Among the targets regulated by Ca^{2+}-calmodulin complexes are many enzymes and membrane transport proteins. Prominent among these are **Ca^{2+}/calmodulin-dependent protein kinases (Ca-kinases)**, which phosphorylate serine and threonine residues on proteins. The first Ca-kinases to be discovered, including *myosin light-chain kinase* and *phosphorylase kinase*, had narrow substrate specificities. More recently a broad-specificity Ca-kinase has been identified; called *Ca^{2+}/calmodulin-regulated "multifunctional" kinase* (or *Ca-kinase II*), it may mediate many of the actions of Ca^{2+} in mammalian cells. As in the case of cyclic AMP, the response of a target cell to an increase in free Ca^{2+} concentration in the cytosol depends on which Ca^{2+}-calmodulin regulated target proteins are present in the cell.

12-23 **The Cyclic AMP and Ca^{2+} Pathways Interact**[26]

The cyclic AMP and Ca^{2+} second messenger pathways interact in at least three ways. First, intracellular Ca^{2+} and cyclic AMP levels can influence each other. In some cells, for example, Ca^{2+}-calmodulin complexes bind to and regulate en-

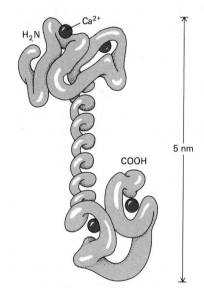

Ca^{2+}

H$_2$N

5 nm

COOH

Figure 12–29 The structure of calmodulin based on x-ray diffraction studies. The molecule has a "dumbbell" shape, with two globular ends connected by a long, exposed α helix. Each end has two Ca^{2+}-binding domains, each with a loop of 12 amino acid residues in which aspartic acid and glutamic acid side chains form ionic bonds with Ca^{2+}. The two Ca^{2+}-binding sites in the carboxyl-terminal part of the molecule have a tenfold higher affinity for Ca^{2+} than those in the amino-terminal part. (Based on data from Y.S. Babu et al., *Nature* 315:37–40, 1985.)

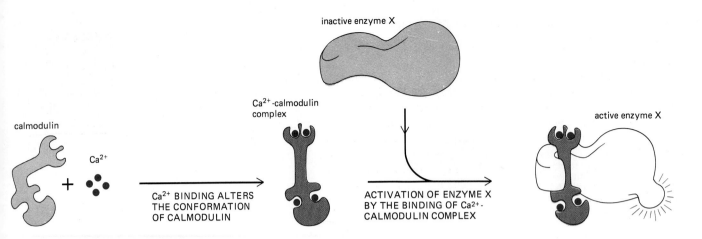

inactive enzyme X

calmodulin

Ca²⁺

Ca²⁺-calmodulin complex

active enzyme X

Ca²⁺ BINDING ALTERS THE CONFORMATION OF CALMODULIN

ACTIVATION OF ENZYME X BY THE BINDING OF Ca²⁺-CALMODULIN COMPLEX

zymes that break down and make cyclic AMP—cyclic AMP phosphodiesterase and adenylate cyclase, respectively. Conversely, A-kinase can phosphorylate some Ca^{2+} channels and pumps and alter their activity. Second, some Ca-kinases are phosphorylated by A-kinase. Third, A-kinase and Ca-kinases frequently phosphorylate different sites on the same proteins, which are thereby regulated by both cyclic AMP and Ca^{2+}.

As an example of how Ca^{2+} and cyclic AMP pathways can interact, consider the *phosphorylase kinase* of skeletal muscle, whose role in glycogen degradation we have already discussed. This kinase phosphorylates glycogen phosphorylase, which breaks down glycogen (see Figure 12–26). It is a multisubunit enzyme, but only one of its four subunits actually catalyzes the phosphorylation reaction: the other three subunits are regulatory and enable the enzyme complex to be activated both by cyclic AMP and by Ca^{2+}. The four subunits are designated α, β, γ, and δ, and each is present in four copies in the phosphorylase kinase complex. The γ subunit carries the catalytic activity; the δ subunit is calmodulin and is largely responsible for the Ca^{2+} dependence of the enzyme. The α and β subunits are the targets for cyclic AMP-mediated regulation, both being phosphorylated by the A-kinase (Figure 12–31).

The same Ca^{2+} signal that initiates muscle contraction ensures that there is adequate glucose to power the contraction. The large influx of Ca^{2+} into the cytosol from the sarcoplasmic reticulum that initiates myofibril contraction (see p. 621) also alters the conformation of the δ subunit, increasing the activity of phosphorylase kinase and thereby increasing the rate of glycogen breakdown several hundredfold within seconds. In addition, the Ca^{2+} influx activates two Ca-kinases that phosphorylate and inhibit glycogen synthase, thereby shutting off glycogen synthesis. The epinephrine-induced phosphorylations previously discussed adjust muscle cell metabolism in anticipation of an increased energy demand; for example, the phosphorylation of phosphorylase kinase by A-kinase allows the enzyme to be activated when fewer calcium ions are bound to calmodulin, thereby making the enzyme more sensitive to Ca^{2+}.

Hundreds of different phosphorylated proteins are revealed by two-dimensional gel electrophoresis in a eucaryotic cell. Only a small fraction of these are phosphorylated by known protein kinases, which suggests that most kinases remain to be discovered. Indeed, it has been estimated that a single mammalian cell may contain more than 100 distinct protein kinases, which regulate the myriad reaction pathways in the cell. There are also many other reversible covalent modifications that regulate the activity of proteins: methylation-demethylation, acetylation-deacetylation, uridylation-deuridylation, and adenylation-deadenylation, among others. It seems likely that most cellular proteins that catalyze the rate-limiting steps in biological processes are regulated in one or another of these ways. In view of the complexities of the feedback loops known to exist in the glycogen pathways alone, it is clear that the unraveling of the mechanisms and kinetics of these regulatory processes poses a formidable challenge.

Figure 12–30 How an increase in free Ca^{2+} in the cytosol indirectly activates an enzyme by altering the conformation of calmodulin molecules. As is the case for cyclic AMP binding to A-kinase, the binding of Ca^{2+} to calmodulin is cooperative, requiring more than one Ca^{2+} to activate the protein. This sharpens the response of calmodulin to changes in Ca^{2+} concentration (see p. 716).

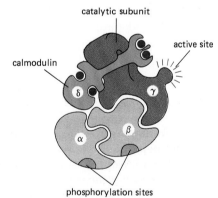

catalytic subunit

active site

calmodulin

δ

γ

α

β

phosphorylation sites

Figure 12–31 Highly schematized drawing of the four subunits of the enzyme *phosphorylase kinase* from mammalian muscle. The γ subunit has the catalytic activity of the active enzyme; the α and β subunits and the δ subunit (calmodulin) mediate the regulation of the enzyme by cyclic AMP and Ca^{2+}, respectively. The actual enzyme complex contains four copies of each subunit.

Cyclic GMP Is Also an Intracellular Messenger[27]

Cyclic AMP is not the only cyclic nucleotide to participate in intracellular signaling. Most animal cells also contain **cyclic guanosine monophosphate (cyclic GMP)** (Figure 12–32), although at a concentration one-tenth or less that of cyclic AMP. Cyclic GMP is known to activate a specific protein kinase (*G-kinase*) that phosphorylates target proteins in the cell (see Figure 12–25), but the role of cyclic GMP in signaling by cell-surface receptors is, with a few exceptions, unclear. Unlike adenylate cyclase, **guanylate cyclase,** which catalyzes the production of cyclic GMP from GTP, is usually a soluble enzyme that is not obviously coupled to cell-surface receptors, although cyclic GMP levels often increase when the inositol-phospholipid pathway is activated. However, recently plasma-membrane-bound forms of guanylate cyclase have been demonstrated in some cells that act as, or are coupled to, cell-surface receptors.

The signaling role of cyclic GMP is especially well understood in the response to light of rod cells in the vertebrate retina, where cyclic GMP acts directly on Na^+ channels in the plasma membrane of the rod cells. In the absence of a light signal, cyclic GMP is bound to the Na^+ channels, keeping them open. Light activates rhodopsin in the disc membrane of the rod cell, and activated rhodopsin then binds to and activates the G protein *transducin*. The α subunit of transducin (T_α) in turn activates a *cyclic GMP phosphodiesterase* that hydrolyzes cyclic GMP, so that cyclic GMP levels in the cytosol drop and the cyclic GMP bound to the Na^+ channels dissociates—allowing the channels to close. In this way the light signal is converted into an electrical signal, as explained in detail in Chapter 19 (see p. 1006). This direct effect of cyclic GMP on an ion channel is one of the few known examples where a cyclic nucleotide acts independently of a protein kinase in a vertebrate cell. But it is not the only example: olfactory receptor cells in the nose provide another. The binding of some odorants to specific olfactory receptor proteins on these cells activates adenylate cyclase via a G_s protein, increasing the concentration of cyclic AMP in the cytosol. The cyclic AMP then directly opens Na^+ channels in the plasma membrane, causing an excitatory depolarization.

Despite the differences in molecular details, all the signaling systems that are triggered by G-protein-dependent receptors share certain features and are governed by similar general principles. All of them, for example, depend on complex cascades or relay chains of intracellular messengers. What are the advantages of such seemingly complex systems that cause them to be used by so many types of cells for such a wide range of purposes?

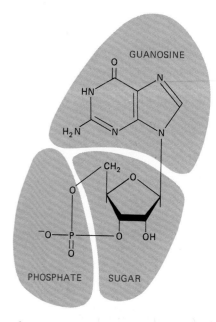

Figure 12–32 Cyclic GMP.

Extracellular Signals Are Greatly Amplified by the Use of Intracellular Messengers and Enzymatic Cascades[28]

By contrast with more direct signaling systems, such as those involving steroid hormones, catalytic cascades of intracellular mediators provide numerous opportunities for amplifying and regulating the responses to extracellular signals. As illustrated in Figure 12–33, for example, when a ligand activates adenylate cyclase indirectly by binding to a receptor, each receptor protein can activate many molecules of G_s protein, each of which can activate a cyclase molecule. Each cyclase molecule, in turn, catalyzes the conversion of a large number of ATP molecules to cyclic AMP molecules. The same type of amplification operates in the inositol-phospholipid pathway. As a result, a nanomolar (10^{-9} M) concentration of an extracellular signal often induces micromolar (10^{-6} M) concentrations of an intracellular second messenger such as cyclic AMP or Ca^{2+}. Since these molecules themselves function as allosteric effector molecules to activate specific enzymes, a single extracellular signaling molecule can cause many thousands of molecules to be altered within the target cell. Moreover, each regulatory protein in the relay chain of signals can be a separate target for metabolic control, as, for example, in the glycogen breakdown cascade in skeletal muscle cells (see p. 709).

Such metabolically explosive cascades require tight regulation. Therefore, it is not surprising that cells have efficient mechanisms for rapidly degrading cyclic

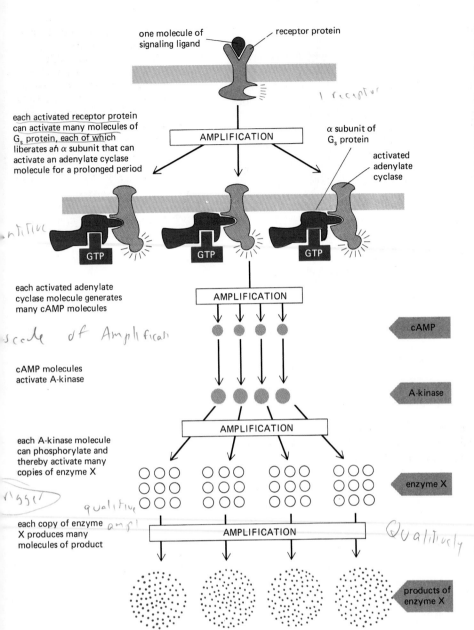

one molecule of signaling ligand

receptor protein

1 receptor

each activated receptor protein can activate many molecules of G_s protein, each of which liberates an α subunit that can activate an adenylate cyclase molecule for a prolonged period

AMPLIFICATION

α subunit of G_s protein

activated adenylate cyclase

ntitive

GTP   GTP   GTP

each activated adenylate cyclase molecule generates many cAMP molecules

AMPLIFICATION

scale of Amplificat

cAMP

cAMP molecules activate A-kinase

A-kinase

each A-kinase molecule can phosphorylate and thereby activate many copies of enzyme X

AMPLIFICATION

trigger

qualitive

enzyme X

each copy of enzyme X produces many molecules of product *ampl*

AMPLIFICATION

Qualituely Amp

products of enzyme X

Figure 12–33 Besides transducing an extracellular signal into an intracellular one, the coupling of cell-surface receptors to adenylate cyclase activation via a G protein greatly amplifies the initial signal. The same type of amplification occurs when receptors activate the inositol phospholipid pathway. In the first amplification step the signaling ligand may or may not remain bound to the receptor long enough for the complex to activate many G_s molecules (see p. 697).

AMP and for buffering and sequestering Ca^{2+}, as well as for inactivating the responding enzymes and transport proteins once they have been activated. This is clearly essential for turning a response off; it may not be as obvious that it is just as important in allowing a response to be rapidly turned on.

12-25 The Concentration of a Molecule Can Be Adjusted Quickly Only If the Lifetime of the Molecule Is Short[29]

If it is important for a cell to be able to alter very rapidly the concentration of a molecule it contains (such as cyclic AMP or Ca^{2+}), then the molecule must be constantly degraded or otherwise promptly removed—in other words, it must undergo rapid turnover. Consider, for example, two intracellular molecules X and Y, both of which are normally maintained at a concentration of 1000 molecules per cell. Molecule X has a slow turnover rate: it is synthesized and degraded at a rate of 10 molecules per second, so that each molecule has an average "lifetime" in the cell of 100 seconds. Molecule Y turns over 10 times as quickly: it is synthesized and degraded at a rate of 100 molecules per second, with each molecule having an average lifetime of 10 seconds. If the rates of synthesis of both X and Y

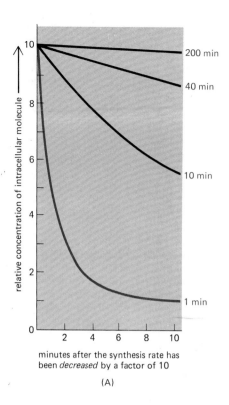

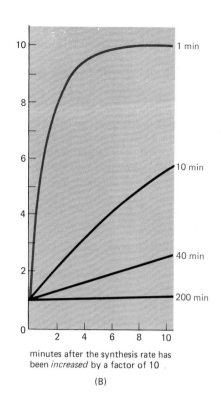

Figure 12–34 Predicted relative rates of change in the intracellular concentrations of molecules with differing turnover times when their rates of synthesis are either decreased (A) or increased (B) suddenly by a factor of 10. In both cases the concentrations of those molecules that are normally being rapidly degraded in the cell (*colored lines*) change quickly, whereas the concentrations of those that are normally being slowly degraded change proportionally more slowly. The numbers in color on the right-hand side are the half-lives assumed for each of the different molecules.

are increased tenfold without any change in the molecular lifetimes, at the end of 1 second the concentration of Y will have increased by nearly 900 molecules per cell (10 × 100 − 100) while the concentration of X will have increased by only 90 molecules per cell. In fact, after its synthesis rate has been either increased or decreased abruptly, the time required for a molecule to shift halfway from its old to its new equilibrium concentration is equal to its normal half-life—that is, it is equal to the time that would be required for its concentration to fall by half if all synthesis were stopped (Figure 12–34).

The same principles apply to proteins as well as to small molecules, and to molecules in the extracellular space as well as to those in cells. Many intracellular proteins that are rapidly degraded or covalently modified have half-lives of 2 hours or less, and some survive less than 10 minutes; in most cases these are proteins with key regulatory roles, whose concentrations are rapidly regulated in the cell by changes in their rates of synthesis.

Cells Can Respond Suddenly to a Gradually Increasing Concentration of an Extracellular Signal[30]

Some cellular responses to signaling ligands are smoothly graded in simple proportion to the concentration of the ligand. The primary responses to steroid hormones often follow this pattern, presumably because each hormone receptor protein binds a single molecule of hormone and each specific DNA recognition sequence in a steroid-hormone-responsive gene (see p. 691) acts independently. As the concentration of hormone increases, the concentration of hormone-receptor complexes increases proportionally, as does the number of complexes bound to specific recognition sequences in the responsive genes; therefore the cellular response is a gradual and linear one.

Other responses to signaling ligands, however, begin more abruptly as the concentration of ligand increases. Some may even occur in a nearly "all-or-none" manner, being undetectable below a threshold concentration of ligand and then reaching a maximum as soon as this concentration is exceeded. Some growth factors, for example, act as all-or-none signals that stimulate cells to begin replicating their DNA as a prelude to cell division. What might be the molecular basis for such steep or even switchlike responses to graded signals?

One possible mechanism for steepening the response exploits the principle of *cooperativity*, in which more than one intracellular effector molecule (or molecular complex of effector with receptor) must bind to some target macromolecule in order to induce a response. In some steroid-hormone-induced responses, for example, it appears that more than one hormone-receptor complex must bind simultaneously to specific recognition sequences in the DNA in order to activate a particular gene. As a result, gene activation begins more abruptly as the hormone concentration rises than it would if only one bound complex were sufficient for activation (Figure 12–35). A similar cooperative mechanism operates in the activations mediated by A-kinase and calmodulin. For example, two or more Ca^{2+} ions must bind before calmodulin adopts its activating conformation (see Figure 12–30); as a result, a fiftyfold increase in activation occurs when the free intracellular Ca^{2+} concentration increases only tenfold. Such cooperative responses become sharper as the number of cooperating molecules increases, and if the number is large enough, responses approaching the all-or-none type can be achieved (Figures 12–36 and 12–37).

Responses are also greatly sharpened when a ligand activates one enzyme and at the same time inhibits another that catalyzes the opposite reaction. We have already discussed one example of this common type of regulation in the stimulation of glycogen breakdown in skeletal muscle cells, where a rise in the intracellular cyclic AMP level both activates phosphorylase kinase and inhibits the opposing action of phosphoprotein phosphatase (see p. 709–710).

Another mechanism by which a cell could respond in an all-or-none way to a gradual increase in a signaling ligand involves positive feedback. By this mechanism, nerve and muscle cells generate all-or-none *action potentials* in response to neurotransmitters. The activation of acetylcholine receptors at a neuromuscular junction, for example, opens cation channels in the muscle-cell plasma membrane. The result is a net influx of Na^+ that locally depolarizes the membrane. This causes voltage-gated Na^+ channels to open in the same membrane region, producing a further influx of Na^+, which further depolarizes the membrane and thereby opens more Na^+ channels. If the initial depolarization exceeds a certain threshold, this positive feedback has an explosive "runaway" effect, producing an action potential that propagates to involve the entire muscle membrane.

The same type of accelerating positive feedback mechanism can operate through a receptor protein that is an enzyme rather than an ion channel. Suppose that a

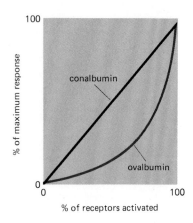

Figure 12–35 Observed response of chick oviduct cells to the steroid hormone estradiol. When activated, estradiol receptors turn on the transcription of several genes. Dose-response curves for two of these genes, one coding for the egg protein *conalbumin* and the other coding for the egg protein *ovalbumin*, are shown. The linear response curve for conalbumin indicates that each activated receptor molecule that binds to the conalbumin gene increases the activity of the gene by the same amount. In contrast, the lag followed by the steep increase in the response curve for ovalbumin suggests that more than one activated receptor (in this case two receptors) must bind simultaneously to the ovalbumin gene in order to initiate its transcription. (Adapted from E.R. Mulvihill and R.D. Palmiter, *J. Biol. Chem.* 252:2060 –2068, 1977.)

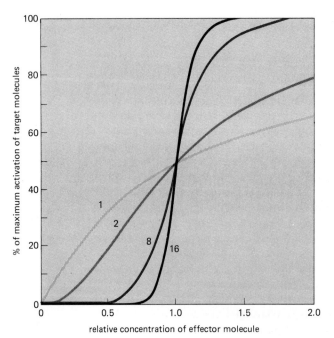

Figure 12–36 Sharp, "all-or-none" activation curves that result if many effector molecules must bind simultaneously to a target macromolecule in order to activate it. The curves shown are those expected if the activation requires the simultaneous binding of 1, 2, 8, and 16 effector molecules, respectively.

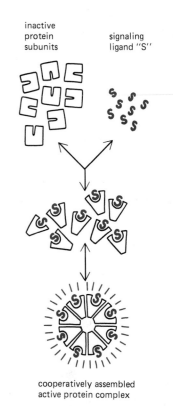

Figure 12–37 Highly schematized illustration of one type of signaling mechanism expected to show a steep thresholdlike response. Here the simultaneous binding of eight molecules of a signaling ligand is required to form an active protein complex. At low ligand concentrations the number of active complexes will increase roughly in proportion to the eighth power of the ligand concentration.

inactive protein subunits

signaling ligand "S"

cooperatively assembled active protein complex

particular signaling ligand activates an enzyme and that two or more molecules of the product of the enzymatic reaction bind back to the enzyme to activate it further (Figure 12–38). The consequence will be a very low rate of synthesis of the enzyme product in the absence of the ligand, increasing slowly with the concentration of ligand until, at some threshold level of ligand, enough of the product is being synthesized to activate the enzyme in a self-accelerating, runaway fashion; the concentration of the enzyme product then suddenly increases to a much higher level. In this way the cell can translate a gradual change in the concentration of a signaling ligand into a switchlike change in the level of a particular enzyme product.

The type of mechanism illustrated in Figure 12–38 is relatively simple in principle, but it has an important corollary that makes it unsuitable for some purposes and uniquely valuable for others. If such a system has been switched on by raising the concentration of signaling ligand above threshold, it will generally remain switched on even when the signal drops back below that threshold: instead of faithfully reflecting the current level of signal, the response system displays a memory. An example is provided by the Ca^{2+}/calmodulin-regulated multifunctional kinase mentioned previously (see p. 711), which is thought to function as a Ca^{2+}-triggered molecular switch. The kinase phosphorylates itself as well as other cell proteins when it is activated by Ca^{2+}-calmodulin complexes. The resulting autophosphorylation causes the enzyme to become independent of Ca^{2+}, thereby prolonging the duration of kinase activity beyond the duration of the initial activating Ca^{2+} signal; the Ca^{2+}-independent autophosphorylation will maintain the active state until levels of cellular phosphatases rise to shut the enzyme off.

Summary

The continuous rapid removal of both free Ca^{2+} and cyclic AMP from the cytosol makes possible both rapid increases and rapid decreases in the concentrations of these intracellular mediators when cells respond to extracellular signals. Rising cyclic AMP levels affect cells by stimulating cyclic AMP-dependent protein kinase (A-kinase) to phosphorylate specific target proteins. The effects are reversible because the phosphorylated proteins are rapidly dephosphorylated when cyclic AMP levels fall. In a similar way, rising free Ca^{2+} levels affect cells by altering the conformation of the Ca^{2+}-binding protein calmodulin; Ca^{2+}-calmodulin complexes activate specific target proteins, including Ca^{2+}/calmodulin-dependent protein kinases (Ca-kinases). Each type of cell has characteristic sets of target proteins that are regulated in these ways, enabling the cell to make its own distinctive response to a change in intracellular cyclic AMP or Ca^{2+} level. Through the signaling cascades mediated by cyclic AMP or Ca^{2+}, the responses to extracellular signals can be greatly amplified as well as regulated by a variety of inputs.

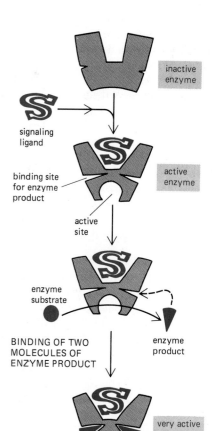

inactive enzyme

signaling ligand

binding site for enzyme product

active enzyme

active site

enzyme substrate

enzyme product

BINDING OF TWO MOLECULES OF ENZYME PRODUCT

very active enzyme

Figure 12–38 Schematic illustration of an "accelerating positive feedback" mechanism. The initial binding of the signaling ligand activates the enzyme to generate a product that binds back to the enzyme, further increasing the enzyme's activity.

Target Cell Adaptation

In responding to almost any type of stimulus, cells and organisms typically can detect the same percent change in a signal over a very wide range of stimulus intensities. At the cellular level this requires that, when target cells are exposed to a stimulus for a prolonged period, they lose the ability to respond to the stimulus with their original sensitivity. By this process of **adaptation** or **desensitization,** a cell reversibly adjusts its sensitivity to the level of the stimulus. In the case of chemical signaling, desensitization enables cells to respond to *changes* in the concentration of a signaling ligand (rather than to the absolute concentration of the ligand) over a very wide range of absolute concentrations.

Desensitization to chemical signals can occur in various ways. In some cases it results from a decrease in the number of specific cell-surface receptor proteins or from the inactivation of such receptors. In other cases it is due to changes in the proteins involved in transducing the signal following receptor activation.

Some Forms of Desensitization Result from Receptor-mediated Endocytosis[31]

After protein hormones and growth factors have bound to receptors on the surface of target cells, they are often ingested by receptor-mediated endocytosis (see p. 328). Since endocytic vesicles eventually deliver their contents to lysosomes, the ligand and its receptor will be degraded by hydrolytic enzymes unless they are specifically retrieved. This process not only represents a major pathway for the breakdown of some signaling ligands, it also plays an important part in regulating the concentration of certain receptors on the surface of target cells. Although receptor degradation and replacement take place continuously, in the absence of ligand a receptor typically has a half-life of a day or so. By inducing endocytosis, some ligands markedly increase the rate of receptor degradation. For example, when human fibroblasts are grown in culture in the absence of EGF, EGF receptors have a half-life of about 10 hours; when excess EGF is added, the receptors are degraded with a half-life of 1 hour. At high concentrations of such ligands, the number of cell-surface receptors decreases, with a concomitant decrease in the sensitivity of the target cell to the ligand. By this type of mechanism, known as **receptor down-regulation,** a cell can adjust its sensitivity to the concentration of a stimulating ligand.

Most types of endocytosed receptors, however, are not degraded in lysosomes. Instead, they discharge their ligand in endosomes and are retrieved and recycled back to the plasma membrane for reuse (see p. 331). Even in this case, however, when the concentration of ligand increases, a larger fraction of the recycling receptors will be inside the cell at any one time and inaccessible to extracellular ligand. This form of desensitization is called **receptor sequestration.**

Desensitization Often Involves Receptor Phosphorylation[32]

The desensitization of many cell-surface receptors depends on their becoming reversibly phosphorylated. As mentioned previously, the endocytosis (and therefore sequestration or down-regulation) of some catalytic receptors depends on their phosphorylating themselves on tyrosine residues (see p. 707). Another example is provided by frog red blood cells, which become desensitized when they are exposed to epinephrine for prolonged periods. Their β-adrenergic receptors gradually become modified so that after a few hours they are no longer able to activate adenylate cyclase; the inactivated receptors become sequestered, probably in endosomes. The modification has been shown to involve the phosphorylation of several specific serine residues on the receptor protein (see Figure 12–24), mediated by a specific protein kinase that can phosphorylate only the activated form of the receptor. There is some evidence that phosphorylation shuts off the function of the β receptors indirectly by enabling them to bind an inhibitory protein that blocks their ability to activate G_s proteins. Rhodopsin, which is structurally related to β-adrenergic receptors (see p. 706), has been shown to be inactivated by this type of mechanism after it has been activated by light: a specific rhodopsin kinase phosphorylates the activated rhodopsin molecules, enabling them to bind an inhibitory protein (called *arrestin*) that blocks their ability to activate the G protein transducin. This, however, is only one of the mechanisms that photoreceptors use to adapt to light (see p. 1107).

Some Forms of Desensitization Are Due to Alterations in G Proteins Rather Than in Receptor Proteins[33]

Although most known mechanisms of desensitization involve changes in receptor proteins, desensitization could, in principle, result from a change in any of the components in the signaling pathway. There are several cases in which target cell desensitization has been shown to involve a change in a G protein. Thus, when fibroblasts in culture are exposed to the prostaglandin PGE_1, which normally activates adenylate cyclase via a G_s protein, the cells soon become insensitive not only to PGE but also to other ligands binding to other receptors that act through the G_s-adenylate cyclase pathway. (This is called *heterologous desensitization*, to distinguish it from *homologous desensitization*, in which receptors are endocytosed or inactivated so that cells are desensitized only to ligands that bind to those receptors.) When the G_s proteins from the desensitized fibroblasts are transplanted into a mutant cell lacking its own G_s protein, they are found to be inefficient at activating adenylate cyclase compared to normal G_s proteins. The nature of the alteration in the G_s protein in PGE_1-desensitized fibroblasts is not yet known, but it would not be surprising if it involved phosphorylation.

Changes in G proteins may also account for some aspects of morphine addiction. In morphine addicts, target cells in the brain become desensitized to morphine (Figure 12–39), so the addicts require much higher doses of morphine than normal individuals to relieve pain or to feel euphoric. The desensitized cells, however, have normal levels of functional cell-surface morphine (opiate) receptors. The mechanism of desensitization has been studied using morphine-sensitive neural cell lines in culture. Morphine receptors on the surface of these cells activate G_i proteins (see p. 699) that inhibit adenylate cyclase and thereby cause a decrease in intracellular cyclic AMP levels. Cultured cells maintained in the presence of a constant concentration of morphine become desensitized, so that both adenylate cyclase activity and intracellular cyclic AMP levels return to normal even though morphine is still bound to cell-surface receptors. If morphine is now removed from the culture medium, there is a marked increase in adenylate cyclase activity, which causes cyclic AMP concentrations to rise to abnormally high levels. A similar increase in cyclic AMP concentration in target cells may be responsible for the extremely unpleasant withdrawal symptoms (anxiety, sweating, tremors, etc.) experienced by morphine addicts who go "cold-turkey." The mechanism of adaptation to morphine is uncertain, but it has been suggested that it involves alterations to G_i proteins.

The various mechanisms of target cell desensitization that we have discussed are summarized in Figure 12–40.

Figure 12–39 The structure of morphine, which comes from poppy seeds. Why do some of our cells have receptors for a drug like morphine? Pharmacologists long suspected that morphine may mimic some endogenous signaling molecule that regulates pain perception and mood. In 1975, two pentapeptides with morphinelike activity, called **enkephalins,** were isolated from pig brain, and soon thereafter larger polypeptides with similar activity, called **endorphins,** were isolated from the pituitary gland and other tissues. All of these so-called *endogenous opiates* contain a common sequence of four amino acids and bind to the same cell-surface receptors as morphine (and related narcotics). Unlike morphine, however, they are rapidly degraded after release and so do not accumulate in quantities large enough to induce the tolerance seen in morphine addicts.

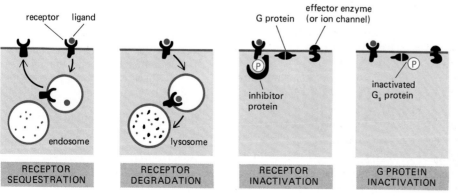

Figure 12–40 Four ways in which target cells can become desensitized after prolonged exposure to a signaling molecule. The receptor shown normally activates or inhibits an effector enzyme (or ion channel) via a G protein. Although the inactivation mechanisms illustrated here for both receptor and G protein involve phosphorylation, other types of modifications can also occur (as we shall see when we discuss bacterial chemotaxis, see p. 720). Moreover, receptor inactivation by phosphorylation does not always involve an inhibitor protein.

Adaptation Plays a Crucial Role in Bacterial Chemotaxis[34]

Many of the mechanisms involved in chemical signaling between cells in multicellular animals are likely to have evolved from mechanisms used by unicellular organisms to respond to chemical changes in their environment. In fact, as we have already mentioned, some of the same intracellular mediators are used by both types of organisms. Among the best-studied reactions of unicellular organisms to extracellular signals are chemotactic responses, in which cell movement is oriented toward or away from a source of some chemical in the environment. The chemotaxis of eucaryotic cells is discussed on page 826 for the cellular slime mold *Dictyostelium discoideum* and on page 670 for human neutrophils. We conclude the present chapter with an account of bacterial chemotaxis, which, largely through the power of genetic analysis, provides a particularly clear and elegant illustration of the crucial role of adaptation in the response to chemical signals.

Motile bacteria will swim toward higher concentrations of nutrients (*attractants*), such as sugars, amino acids, and small peptides, and away from higher concentrations of various noxious chemicals (*repellents*) (Figure 12–41). This relatively simple but highly adaptive chemotactic behavior has been most studied in *E. coli* and *Salmonella typhimurium*. We shall concentrate here chiefly on chemotaxis toward attractants; chemotaxis away from repellents depends on essentially the same mechanisms operating in reverse.

Bacteria swim by means of flagella that are much simpler than the flagella of eucaryotic cells (see p. 645) and consist of a helical tube containing a single type of protein subunit, called *flagellin*. Each flagellum is attached at its base, by a short flexible hook, to a small protein disc embedded in the bacterial membrane. Incredible though it may seem, this disc is part of a tiny "motor" that uses the energy stored in the transmembrane H⁺ gradient to rotate rapidly and turn the helical flagellum (Figure 12–42).

Because the flagella on the bacterial surface have an intrinsic "handedness," different directions of rotation have different effects on movement. Counterclockwise rotation allows all the flagella to draw together into a coherent bundle so that the bacterium swims uniformly in one direction. Clockwise rotation causes them to fly apart, so that the bacterium tumbles chaotically without moving forward (Figure 12–43). In the absence of any environmental stimulus, the direction of rotation of the disc reverses every few seconds, producing a characteristic pattern of movement in which smooth swimming in a straight line is interrupted by abrupt changes in direction caused by tumbling (Figure 12–44A).

The normal swimming behavior of bacteria is modified by chemotactic attractants or repellents, which bind to specific receptor proteins and affect the frequency of tumbling by increasing or decreasing the time that elapses between

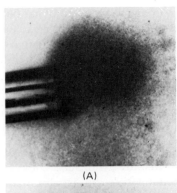

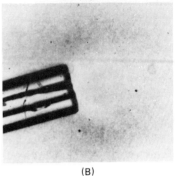

(A)

(B)

Figure 12–41 Photographs of *Salmonella typhimurium* bacteria being attracted to a small glass capillary tube containing the amino acid serine (A) and repelled from a capillary tube containing phenol (B). The photographs were taken 5 minutes after the capillary tubes had been introduced into the culture dishes containing the bacteria. This capillary tube assay is a simple method of demonstrating bacterial chemotaxis. (From B.A. Rubik and D.E. Koshland, *Proc. Natl. Acad. Sci. USA* 75:2820–2824, 1978.)

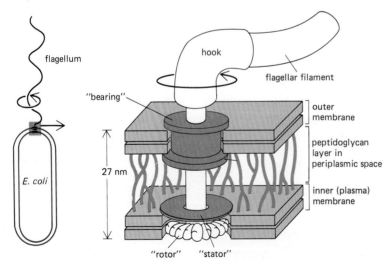

Figure 12–42 Schematic drawing of the flagellar rotatory "motor" of *E. coli*. The "rotor" is a protein disc integrated into the plasma membrane. Driven by the H⁺ gradient across this membrane, it rotates rapidly (~100 revolutions/second) in the lipid bilayer against another protein disc (the "stator"). A rod links the "rotor" to a hook and flagellum, thereby causing them to rotate. The protein "bearing" serves to seal the outer membrane as the rotating rod passes through it. In this illustration the parts of the assembly that remain stationary are shown in color, while the rotating parts are white. (Adapted from M.L. de Pamphilis and J. Adler, *J. Bacteriol.* 105:384–395 and 396–407, 1971.)

successive changes in direction of flagellar rotation. When bacteria are swimming in a favorable direction (toward a higher concentration of an attractant or away from a higher concentration of a repellent), they tumble less frequently than when they are swimming in an unfavorable direction (or when no gradient is present). Since the periods of smooth swimming are longer when a bacterium is traveling in a favorable direction, it will gradually progress in that direction—toward an attractant (Figure 12–44B) or away from a repellent.

In its natural environment, a bacterium detects a spatial gradient of attractants or repellents in the medium by swimming at a constant velocity and comparing the concentration of chemicals over *time*. (It does not monitor changes in concentration by using a *spatial* separation of receptors over its length; this would be extremely difficult given the very small size of a bacterium.) Changes over time can be produced artificially in the laboratory by the sudden addition or removal of a chemical to the culture medium. When an attractant is added in this way, tumbling is suppressed within a few tenths of a second, as expected. But after some time, even in the continuing presence of the attractant, tumbling frequency returns to normal. The bacteria remain in this adapted state as long as there is no increase or decrease in the concentration of the attractant; addition of more attractant will briefly suppress tumbling, whereas removal of the attractant will briefly enhance tumbling until the bacteria again adapt to the new level. Adaptation is a crucial part of the chemotactic response in that it enables bacteria to respond to *changes* in concentration rather than to steady-state levels of an attractant and therefore to prolong swimming when moving in a favorable direction.

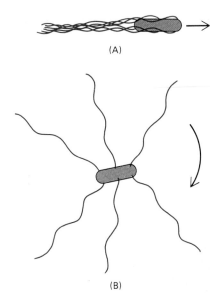

Figure 12–43 Positions of the flagella on *E. coli* during swimming. When the flagella rotate counterclockwise (A), they are drawn together into a single bundle, which acts as a propeller to produce smooth swimming. When the flagella rotate clockwise (B), they fly apart and produce tumbling.

12-31 Bacterial Chemotaxis Is Mediated by a Family of Four Homologous Transmembrane Receptors[35]

The unraveling of the molecular mechanisms responsible for bacterial chemotaxis has depended largely on the isolation and analysis of mutants with defective chemotactic behavior. In this way it has been shown that chemotaxis to a number of chemicals depends on a small family of closely related transmembrane receptor proteins that are responsible for transmitting chemotactic signals across the plasma membrane. These **chemotaxis receptors** are methylated during adaptation (see below) and are therefore often called *methyl-accepting chemotaxis proteins (MCPs)* (Figure 12–45).

There are four types of plasma membrane chemotaxis receptors, each concerned with the response to a small group of chemicals. Two mediate responses

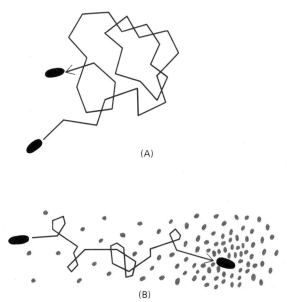

Figure 12–44 The tracks of a swimming bacterium. In the absence of a chemotactic signal (A), periods of smooth swimming (runs) are interrupted by brief tumbles that randomly change the direction of swimming. Thus runs and tumbles occur in alternating sequence, each run constituting a step in a three-dimensional random walk. In the presence of a chemotactic attractant (B), tumbling is partially suppressed whenever the bacterium happens to be swimming toward a higher concentration of the attractant, so that it gradually moves in the direction of the attractant—a biased random walk.

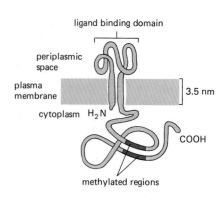

Figure 12–45 Structure of a chemotaxis receptor protein. The two regions of the polypeptide chain that become methylated consist of highly conserved 13-amino acid sequences. Although this drawing is based on the structure of the aspartate receptor, all four types of chemotaxis receptors have similar structures. (Based on A.F. Russo and D.E. Koshland, Jr., *Science* 220:1016–1020, 1983. Copyright 1983 by the AAAS.)

to serine and aspartate, respectively, by binding these amino acids directly and transducing the binding event into a signal in the cytosol. The other two receptors mediate responses to sugars and dipeptides, respectively, but are activated indirectly through *periplasmic substrate-binding proteins*, which also serve to mediate the transport of sugars and dipeptides across the plasma membrane (see p. 312). These proteins, which are dissolved in the periplasmic space (the space between the outer membrane and the plasma membrane), specifically bind the sugars and dipeptides and then form a complex with the appropriate chemotaxis receptors in the plasma membrane to activate them (Figure 12–46). Although the transport and chemotaxis systems for these sugars and dipeptides use common periplasmic substrate-binding proteins, the other parts of the machinery are different, as indicated by mutations that inactivate transport without affecting chemotaxis, and vice versa.

Receptor Methylation Is Responsible for Adaptation[35]

There is strong evidence that adaptation in bacterial chemotaxis results from the covalent methylation of the chemotaxis receptor proteins. When methylation is blocked by mutation, adaptation is markedly inhibited and exposure of the mutant bacteria to an attractant results in the suppression of tumbling for days instead of for a minute or so. Therefore, the activation of chemotaxis receptors by a chemoattractant has two separable consequences: (1) a rapid excitation occurs because the activated receptor generates an intracellular signal that causes the flagellar motor to continue to rotate counterclockwise, resulting in the suppression of tumbling and continuous smooth swimming; (2) a slower adaptation occurs because, while activated, the receptor is methylated by enzymes in the cytoplasm, reversing its activation over a period of a few minutes (Figure 12–47).

Figure 12–46 The steps in signal transduction during bacterial chemotaxis. Chemical attractants bind to type 1 or type 2 chemotaxis receptors in the plasma membrane or to periplasmic substrate-binding proteins that then bind to type 3 or type 4 chemotaxis receptors. This binding activates the chemotaxis receptors to produce an intracellular signal that causes the flagellar motor to continue to rotate counterclockwise, thereby suppressing tumbling and causing continuous smooth swimming. The attractants diffuse into the periplasmic space from outside the cell through large channels in the outer membrane (not shown).

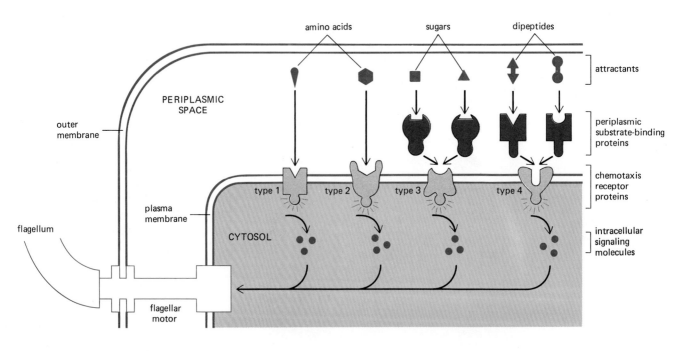

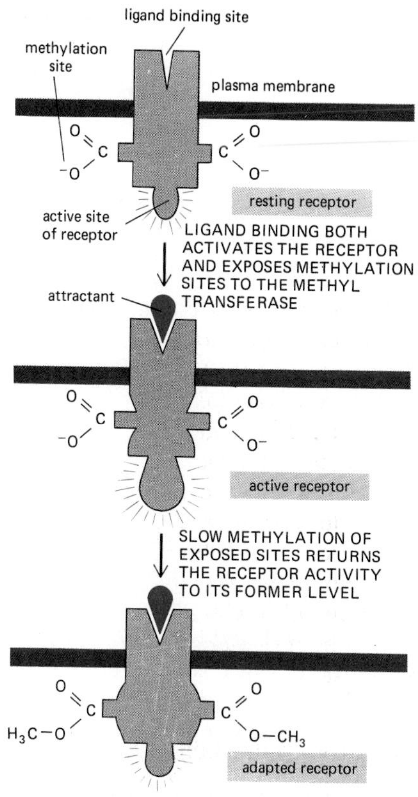

Figure 12-47 The sequential activation and adaptation (via methylation) of a chemotaxis receptor. Note that the state of the receptor, and therefore the tumbling frequency of the bacterium, is the same in the resting and adapted states. The receptor is shown with two methylation sites for simplicity; in fact, there are four methylation sites on each receptor. As the concentration of ligand increases, the fraction of time that the receptor is occupied by the ligand increases. A higher level of ligand will thereby initially cause a greater change in the conformation of the receptor than a low level, pushing the receptor more toward its fully altered state. However, a slower increase in methylation ensues, so that within minutes the conformational strain on the receptor is exactly reversed—with more methyl groups being present at higher attractant concentrations. The receptor has now adapted. Although the ligand is shown here binding directly to the receptor, in some cases it binds first to a periplasmic substrate-binding protein, which then binds to the receptor.

Receptor methylation is catalyzed by a soluble enzyme (*methyl transferase*) that transfers a methyl group to a free carboxyl group on a glutamic acid residue of the activated receptor protein (Figure 12–48). As many as four methyl groups can be transferred to a single receptor, the extent of methylation increasing at higher concentrations of attractant (where each receptor spends a larger proportion of its time with ligand bound). When the attractant is removed, the receptor is demethylated by a soluble demethylating enzyme (see Figure 12–48). Although the level of methylation changes during chemotactic responses, it remains constant once a bacterium is adapted because an exact balance is reached between the rates of methylation and demethylation.

A Cascade of Protein Phosphorylation Couples Receptor Activation to Changes in Flagellar Rotation[36]

The activation of chemotaxis receptors by attractants or repellents must lead to the generation of an intracellular signal that affects the direction of rotation of the flagellar motor. Genetic studies indicate that four cytoplasmic proteins—CheA, CheW, CheY, and CheZ—are involved in this intracellular signaling process. CheY and CheZ act at the effector end of the pathway to control the direction of flagellar rotation, apparently by binding to the flagellar motor. CheY signals the motor to rotate clockwise, resulting in tumbling; mutants that lack this protein swim continuously without tumbling. CheZ antagonizes the action of CheY, causing the motor to rotate counterclockwise, resulting in smooth swimming. CheA, and possibly CheW, are thought to relay the signal from the chemotaxis receptors to CheY and CheZ by a mechanism that involves protein phsophorylation and dephosphorylation.

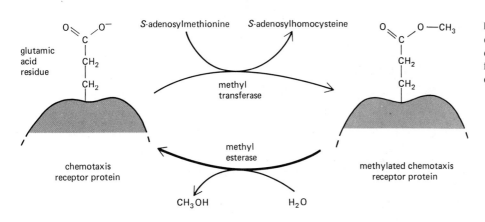

Figure 12–48 Methylation and demethylation reactions involving chemotaxis receptor proteins. Up to four methyl groups can be added to each receptor by ester linkages.

Figure 12–49 The phosphorylation relay system that is thought to enable the chemotaxis receptors to control the flagellar motor. The binding of a repellent activates the receptor, leading to the transient phosphorylation of CheA. CheA quickly transfers its covalently-bound, high-energy phosphate directly to CheY to generate CheY-phosphate, which binds to the flagellar motor and causes it to rotate clockwise, resulting in tumbling. The binding of an attractant has the opposite effect, leading to a decrease in the phosphorylation of CheA and CheY, counterclockwise flagellar rotation, and smooth swimming. CheZ accelerates the dephosphorylation of CheY-phosphate, thereby antagonizing the action of CheY. Each of these phosphorylated intermediates decays in about 10 seconds, enabling the bacterium to respond very quickly to changes in its environment (see Figure 12–34). It is not known how the chemotaxis receptors communicate with CheA or what role CheW plays in the process.

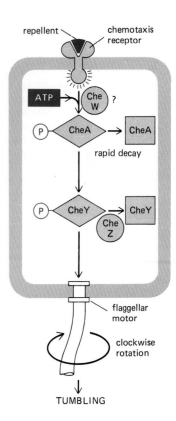

In vitro studies with purified proteins show that CheA is a protein kinase that phosphorylates itself in the presence of ATP and then quickly transfers the phosphate to CheY. The phosphorylated CheY is then dephosphorylated by a reaction that is accelerated by CheZ, which, as mentioned above, antagonizes CheY function *in vivo*. CheY is apparently activated (to cause tumbling) by CheA-mediated phosphorylation and inactivated (to cause smooth swimming) by CheZ-mediated dephosphorylation. It is thought that the phosphorylation of CheA and CheY decreases when chemotaxis receptors bind an attractant; the resulting inactivation of CheY leads to a decrease in tumbling and prolonged smooth swimming. By contrast, repellents are thought to activate CheA-mediated phosphorylation of CheY, thereby activating the protein to induce tumbling (Figure 12–49).

The same network of proteins also regulates the adaptation process. CheA phosphorylates the enzyme that demethylates the chemotaxis receptors (see Figure 12–48), increasing its activity and thereby providing feedback regulation to the chemotaxis receptors.

All of the genes and proteins involved in bacterial chemotaxis may now have been identified, and in most cases the proteins have been sequenced and are available in large quantities. It seems that we are rapidly approaching an almost complete molecular understanding of this highly adaptive behavior.

Summary

By adapting to high concentrations of a signaling ligand in a time-dependent, reversible manner, cells can adjust their sensitivity to the level of the stimulus and thereby respond to changes in a ligand's concentration instead of to its absolute level. Adaptation occurs in various ways: (1) ligand binding can induce the internalization of receptors, which are then transiently sequestered inside the cell or are degraded in lysosomes; (2) activated receptors can be reversibly inactivated by being phosphorylated or methylated; (3) nonreceptor proteins in the signal transduction pathway (such as G proteins) can be reversibly inactivated by mechanisms that are still uncertain. At a molecular level the best-understood example of adaptation occurs in bacterial chemotaxis, in which the reversible methylation of key signal-transducing proteins in the plasma membrane helps the cell to swim toward an optimal environment.

References

Cited

1. Smith, E.L.; et al. Principles of Biochemistry: Mammalian Biochemistry, 7th ed., pp. 355–619. New York: McGraw-Hill, 1983.

   Snyder, S.H. The molecular basis of communication between cells. *Sci. Am.* 253(4):132–140, 1985.

2. Norman, A.W.; Litwack, G. Hormones. San Diego, CA: Academic, 1987.

   Wilson, J.D.; Foster, D.W. Williams' Textbook of Endocrinology, 7th ed. Philadelphia: Saunders, 1985.

3. Simpson, I.A.; Cushman, S.W. Hormonal regulation of mammalian glucose transport. *Annu. Rev. Biochem.* 55:1059–1089, 1986.

4. Beer, D.J.; Matloff, S.M.; Rocklin, R.E. The influence of histamine in immune and inflammatory responses. *Adv. Immunol.* 35:209–268, 1984.

   Gospodarowicz, D.; Cheng, J.; Lui, G.M.; Baird, A.; Bohlen, P. Isolation of brain fibroblast growth factor by heparin sepharose affinity chromatography: identity with pituitary fibroblast growth factor. *Proc. Natl. Acad. Sci. USA* 81:6963–6967, 1984.

5. Smith, W.L.; Borgeat, P. The eiocosanoids: prostaglandins, thromboxanes, leukotrienes, and hydroxyeicosaenoic acids. In Biochemistry of Lipids and Membranes (D.E. Vance, J.E. Vance, eds.), pp. 325–360. Menlo Park, CA: Benjamin-Cummings, 1985.

6. Evans, R.M. The steroid and thyroid hormone receptor superfamily. *Science* 240:889–895, 1988.

   Gehring, U. Steroid hormone receptors: biochemistry, genetics, and molecular biology. *Trends Biochem. Sci.* 12:399–402, 1987.

   Ivarie, R.D.; O'Farrell, P.H. The glucocorticoid domain: steroid-mediated changes in the rate of synthesis of rat hepatoma proteins. *Cell* 13:41–55, 1978.

   Yamamoto, K.R. Steroid receptor regulated transcription of specific genes and gene networks. *Annu. Rev. Genet.* 19:209–252, 1985.

7. Ashburner, M.; Chihara, C.; Meltzer, P.; Richards, G. Temporal control of puffing activity in polytene chromosomes. *Cold Spring Harbor Symp. Quant. Biol.* 38:655–662, 1974.

8. Attardi, B.; Ohno, S. Physical properties of androgen receptors in brain cytosol from normal and testicular feminized (Tfm/y♂) mice. *Endocrinology* 103:760–770, 1978.

9. Berridge, M. The molecular basis of communication within the cell. *Sci. Am.* 253(4):142–152, 1985.

   Kahn, C.R. Membrane receptors for hormones and neurotransmitters. *J. Cell Biol.* 70:261–286, 1976.

   Levitski, A. Receptors: A Quantitative Approach. Menlo Park, CA: Benjamin-Cummings, 1984.

   Rees Smith, B.; Buckland, P.R. Structure-function relations of the thyrotropin receptor. In Receptors, Antibodies and Disease, Ciba Foundation Symposium 90 (D. Evered, J. Whelan, eds.), pp. 114–132. London: Pitman, 1982.

   Snyder, S.H. The molecular basis of communication between cells. *Sci. Am.* 253(4):132–140, 1985.

10. Pastan, I. Cyclic AMP. *Sci. Am.* 227(2):97–105, 1972.

    Sutherland, E.W. Studies on the mechanism of hormone action. *Science* 177:401–408, 1972.

11. Schramm, M.; Selinger, Z. Message transmission: receptor controlled adenylate cyclase system. *Science* 225:1350–1356, 1984.

12. Casperson, G.F.; Bourne, H.R. Biochemical and molecular genetic analysis of hormone-sensitive adenylate cyclase. *Annu. Rev. Pharmacol. Toxicol.* 27:371–384, 1987.

    Feder, D.; et al. Reconstitution of beta$_1$-adrenoceptor-dependent adenylate cyclase from purified components. *EMBO J.* 5:1509–1514, 1986.

    Rodbell, M. The role of hormone receptors and GTP-regulatory proteins in membrane transduction. *Nature* 284:17–22, 1980.

13. Gilman, A.G. G proteins and dual control of adenylate cyclase. *Cell* 36:577–579, 1984.

    Gilman, A.G. G proteins: transducers of receptor-generated signals. *Annu. Rev. Biochem.* 56:615–649, 1987.

    Lai, C.-Y. The chemistry and biology of cholera toxin. *CRC Crit. Rev. Biochem.* 9:171–206, 1980.

    Levitzki, A. From epinephrine to cyclic AMP. *Science* 241:800–806, 1988.

    Stryer, L; Bourne, H.R. G proteins: a family of signal transducers. *Annu. Rev. Cell Biol.* 2:391–419, 1986.

14. Carafoli, E. Intracellular calcium homeostasis. *Annu. Rev. Biochem.* 56:395–433, 1987.

    Carafoli, E.; Penninston, J.T. The calcium signal. *Sci. Am.* 253(5):70–78, 1985.

    Evered, D.; Whelan, J., eds. Calcium and the Cell, Ciba Foundation Symposium 122. Chichester, U.K.: Wiley, 1986.

    Volpe, P.; et al. "Calciosome," a cytoplasmic organelle: the inositol 1,4,5-trisphosphate-sensitive Ca^{2+} store of nonmuscle cells? *Proc. Natl. Acad. Sci. USA* 85:1091–1095, 1988.

15. Augustine, G.J.; Charlton, M.P.; Smith, S.J. Calcium action in synaptic transmitter release. *Annu. Rev. Neurosci.* 10:633–693, 1987.

    Heilbrunn, L.V.; Wiercenski, F.J. The action of various cations on muscle protoplasm. *J. Cell. Comp. Physiol.* 29:15–32, 1947.

16. Berridge, M.J. Inositol lipids and calcium signalling. *Pro. R. Soc. Lond. (Biol.)* 234:359–378, 1988.

    Cockcroft, S. Polyphosphoinositide phosphodiesterase: regulation by a novel guanine nucleotide binding protein, Gp. *Trends Biochem. Sci.* 12:75–78, 1987.

    Majerus, P.W.; et al. The metabolism of phosphoinositide-derived messenger molecules. *Science* 234:1519–1526, 1986.

    Michell, R.H.; Putney, J.W., eds. Inositol Lipids in Cellular Signaling. Current Communications in Molecular Biology. Cold Spring Harbor, NY: Cold Spring Harbor Laboratory, 1987.

    Sekar, M.C.; Hokin, L.E. The role of phosphoinositides in signal transduction. *J. Memb. Biol.* 89:193–210, 1986.

    Woods, N.M.; Cuthbertson, K.S.R.; Cobbold, P.H. Repetitive transient rises in cytoplasmic free calcium in hormone-stimulated hepatocytes. *Nature* 319:600–602, 1986.

17. Angel, P.; et al. Phorbol ester-inducible genes contain a common *cis* element recognized by a TPA-modulated *trans*-acting factor. *Cell* 49:729–739, 1987.

    Bell, R.M. Protein kinase C activation by diacylglycerol second messengers. *Cell* 45:631–632, 1986.

    Lee, W.; Mitchell, P.; Tijan, R. Purified transcription factor AP-1 interacts with TPA-inducible enhancer elements. *Cell* 49:741–752, 1987.

    Nishizuka, Y. Studies and perspectives of protein kinase C. *Science* 233:305–312, 1986.

    Parker, P.J.; et al. The complete primary structure of protein kinase C—the major phorbol ester receptor. *Science* 233:853–859, 1986.

18. Barbacid, M. *ras* genes. *Annu. Rev. Biochem.* 56:779–827, 1987.

19. Dohlman, H.G.; Caron, M.G.; Lefkowitz, R.J. A family of receptors coupled to guanine nucleotide regulatory proteins. *Biochemistry* 26: 2657–2664, 1987.

Dunlap, K.; Holz, G.G.; Rane, S.G. G proteins as regulators of ion channel function. *Trends Neurosci.* 10:241–244, 1987.

Kubo, T.; et al. Cloning, sequencing and expression of complementary DNA encoding the muscarinic acetylcholine receptor. *Nature* 323:411–416, 1986.

Masu, Y.; et al. cDNA cloning of bovine substance-K receptor through oocyte expression system. *Nature* 329:836–838, 1987.

Stryer, L. The molecules of visual excitation. *Sci. Am.* 257(1):42–50, 1987.

20. Carpenter, G. Receptors for epidermal growth factor and other polypeptide mitogens. *Annu. Rev. Biochem.* 56:881–914, 1987.

Kaplan, D.R.; et al. Common elements in growth factor stimulation and oncogenic transformation: 85 kd phosphoprotein and phosphatidylinositol kinase activity. *Cell* 50:1021–1029, 1987.

Rosen, O.M. After insulin binds. *Science* 237:1452–1458, 1987.

Schlessinger, J. Allosteric regulation of the epidermal growth factor receptor kinase. *J. Cell Biol.* 103:2067–2072, 1986.

Yarden, Y.; Ullrich, A. Growth factor receptor tyrosine kinases. *Annu. Rev. Biochem.* 57:443–478, 1988.

21. Deuel, T.F. Polypeptide growth factors: roles in normal and abnormal cell growth. *Annu. Rev. Cell Biol.* 3:443–492, 1987.

Hanks, S.K.; Quinn, A.M.; Hunter, T. The protein kinase family: conserved features and deduced phylogeny of the catalytic domains. *Science* 241:42–52, 1988.

Hunter, T. A thousand and one protein kinases. *Cell* 50:823–829, 1987.

Ullrich, A.; et al. Human insulin receptor and its relationship to the tyrosine kinase family of oncogenes. *Nature* 313:756–761, 1985.

22. Cohen, P. Control of Enzyme Activity, 2nd ed. London: Chapman & Hall, 1983.

Edelman, A.M.; Blumenthal, D.K.; Krebs, E.G. Protein serine/threonine kinases. *Annu. Rev. Biochem.* 56:567–613, 1987.

23. Cohen, P. Protein phosphorylation and the control of glycogen metabolism in skeletal muscle. *Philos. Trans. R. Soc. Lond. (Biol.)* 302:13–25, 1983.

Montminy, M.R.; Bilezikjian, L.M. Binding of a nuclear protein to the cyclic-AMP response element of the somatostatin gene. *Nature* 328:175–178, 1987.

Pilkis, S.J.; El-Maghrabi, M.R; Claus, T.H. Hormonal regulation of hepatic gluconeogenesis and glycolysis. *Annu. Rev. Biochem.* 57:755–784, 1988.

Smith, S.B.; White, H.D.; Siegel, J.B.; Krebs, E.G. Cyclic AMP-dependent protein kinase I: cyclic nucleotide binding, structural changes, and release of the catalytic subunits. *Proc. Natl. Acad. Sci. USA* 78:1591–1595, 1981.

24. Alemany, S.; Pelech, S.; Brierley, C.H.; Cohen, P. The protein phosphatases involved in cellular regulation. Evidence that dephosphorylation of glycogen phosphorylase and glycogen synthase in the glycogen and microsomal fractions of rat liver are catalysed by the same enzyme: protein phosphatase-1. *Eur. J. Biochem.* 156:101–110, 1986.

Ingebritsen, T.S.; Cohen P. Protein phosphatases: properties and role in cellular regulation. *Science* 221:331–338, 1983.

25. Babu, Y.S.; et al. Three-dimensional structure of calmodulin. *Nature* 315:37–40, 1985.

Cheung, W.Y. Calmodulin. *Sci. Am.* 246(6):48–56, 1982.

Gerday, C.; Gilles, R.; Bolis, L., eds. Calcium and Calcium Binding Proteins. Berlin: Springer-Verlag, 1988.

Klee, C.B.; Crouch, T.H.; Richman, P.G. Calmodulin. *Annu. Rev. Biochem.* 49:489–515, 1980.

26. Cohen, P. Protein phosphorylation and hormone action. *Proc. R. Soc. Lond. (Biol.)* 234:115–144, 1988.

27. Goldberg, N.D.; Haddox, M.K. Cyclic GMP metabolism and involvement in biological regulation. *Annu. Rev. Biochem.* 46:823–896, 1977.

Nakamura, T.; Gold, G.H. A cyclic nucleotide-gated conductance in olfactory receptor cilia. *Nature* 325:442–444, 1987.

Schnapf, J.L.; Baylor, D.A. How photoreceptor cells respond to light. *Sci. Am.* 256(4):40–47, 1987.

Stryer, L. Cyclic GMP cascade of vision. *Annu. Rev. Neurosci.* 9:87–119, 1986.

28. Cohen, P. Protein phosphorylation and hormone action. *Proc. R. Soc. Lond. (Biol.)* 234:115–144, 1988.

29. Schimke, R.T. On the roles of synthesis and degradation in regulation of enzyme levels in mammalian tissues. *Curr. Top. Cell. Regul.* 1:77–124, 1969.

30. Lewis, J.; Slack, J.; Wolpert, L. Thresholds in development. *J. Theor. Biol.* 65:579–590, 1977.

Miller, S.G.; Kennedy, M.B. Regulation of brain type II Ca^{2+}/calmodulin-dependent protein kinase by autophosphorylation: a Ca^{2+}-triggered molecular switch. *Cell* 44:861–870, 1986.

Mulvihill, E.R.; Palmiter, R.D. Relationship of nuclear estrogen receptor levels to induction of ovalbumin and conalbumin mRNA in chick oviduct. *J. Biol. Chem.* 252:2060–2068, 1977.

31. Lefkowitz, R.J., ed. Receptor regulation. Receptors and Recognition, Series B, Vol. 13. London: Chapman & Hall, 1981.

Soderquist, A.M.; Carpenter, G. Biosynthesis and metabolic degradation of receptors for epidermal growth factor. *J. Memb. Biol.* 90:97–105, 1984.

32. Sibley, D.R.; Benovic, J.L.; Caron, M.G.; Lefkowitz, R.J. Regulation of transmembrane signaling by receptor phosphorylation. *Cell* 48:913–922, 1987.

33. Kassis, S.; Fishman, P.H. Different mechanisms of desensitization of adenylate cyclase by isoproterenol and prostaglandin E_1 in human fibroblasts: role of regulatory components in desensitization. *J. Biol. Chem.* 257:5312–5318, 1982.

Klee, W.A.; Sharma, S.K.; Nirenberg, M. Opiate receptors as regulators of adenylate cyclase. *Life Sci.* 16:1869–1874, 1975.

Snyder, S.H. Opiate receptors and internal opiates. *Sci. Am.* 236(3):44–56, 1977.

34. Adler, J. The sensing of chemicals by bacteria. *Sci. Am.* 234(4):40–47, 1976.

Berg, H. How bacteria swim. *Sci. Am.* 233(2):36–44, 1975.

35. Koshland, D.E., Jr. Biochemistry of sensing and adaptation in a simple bacterial system. *Annu. Rev. Biochem.* 50:765–782, 1981.

Russo, A.F.; Koshland, D.E. Receptor modification and absolute adaptation in bacterial sensing. In Sensing and Response in Microorganisms (M. Eisenbach, M. Balaban, eds.), pp. 27–41. Amsterdam: Elsevier, 1985.

Springer, M.S.; Goy, M.F.; Adler, J. Protein methylation in behavioral control mechanisms and in signal transduction. *Nature* 280:279–284, 1979.

36. Hess, J.F.; Oosawa, K.; Kaplan, N.; Simon, M.I. Phosphorylation of three proteins in the signaling pathway of bacterial chemotaxis. *Cell* 53:79–87, 1988.

Oosawa, K.; Hess, J.F.; Simon, M.I. Mutants defective in bacterial chemotaxis show modified protein phosphorylation. *Cell* 53:89–96, 1988.

Cell Growth and Division

13

Cells reproduce by duplicating their contents and then dividing in two. Complex sequences of cell divisions, punctuated periodically by sexual cell fusion, generate multicellular organisms. Even after a higher animal or plant has reached maturity, cell division is usually required in order to make up for losses due to wear and tear. Thus an adult human being must manufacture many millions of new cells each second, simply to maintain the status quo; and if all cell division is halted—for example, by a large dose of ionizing radiation—he or she will die within a few days.

For most of the constituents of a cell, duplication need not be controlled exactly. If there are many copies of a particular type of molecule or organelle, it is sufficient that the number of copies be approximately doubled in one cycle and that the dividing parent cell allocate approximately equal shares to each daughter. But there is at least one obvious exception: the DNA must always be duplicated exactly and divided precisely between the two daughter cells, and this requires special machinery. In discussing the cell cycle, therefore, it is sometimes convenient to distinguish between the *chromosome cycle* and the parallel *cytoplasmic cycle*. In the **chromosome cycle,** *DNA synthesis*, in which the nuclear DNA is duplicated, alternates with *mitosis*, in which the duplicate copies of the genome are separated. In the **cytoplasmic cycle,** *cell growth*, in which the many other components of the cell double in quantity, alternates with *cytokinesis*, in which the cell as a whole divides in two.

We begin this chapter by discussing the coordination and control of these interdependent cycles. We examine the mechanisms that ensure that all the nuclear DNA gets replicated once and only once between one cell division and the next, and we consider how the events of the chromosome cycle are coordinated with those of the cytoplasmic cycle. We then explore the regulation of cell division in multicellular animals by factors in the cell's environment—a topic that has been greatly illuminated by recent advances in cancer research. Finally, we discuss the molecular machinery responsible for mitosis and cytokinesis. These two processes require that the *centrosome* (see p. 763) be inherited reliably and duplicated precisely in order to form the two poles of the mitotic spindle; this *centrosome cycle* can be considered a third component of the cell cycle.

727

The Steps of the Cell Cycle and Their Causal Connections

The division of a eucaryotic cell presents a striking spectacle under the microscope. In **mitosis** the contents of the nucleus condense to form visible chromosomes, which through an elaborately orchestrated series of movements are pulled apart into two equal sets; then, in **cytokinesis,** the cell itself splits into two daughter cells, each receiving one of the two sets of chromosomes. Because they are so easily seen, mitosis and cytokinesis were the chief focus of interest to early investigators. These two events, however, together occupy only a brief period, known as the **M phase** (M = mitosis), in the cell's reproductive cycle. The much longer time that elapses between one M phase and the next is known as **interphase.** Under the microscope interphase appears, deceptively, as an uneventful interlude in which the cell simply grows slowly in size. More sophisticated techniques reveal that interphase is actually a period in which elaborate preparations for division are occurring in a carefully ordered sequence. In this section we discuss how the sequence of events in interphase can be investigated and how the steps of the cell cycle are causally connected.

Replication of the Nuclear DNA Occurs During a Specific Part of Interphase[1]

In most cells the DNA in the nucleus is replicated during only a limited portion of interphase; this period of DNA synthesis is called the **S phase** of the cell cycle. Between the end of the M phase and the beginning of DNA synthesis, there is usually an interval, known as the G_1 *phase* (G = gap); a second interval, known as the G_2 *phase*, separates the end of DNA synthesis from the beginning of the next M phase. Interphase is thus composed of successive G_1, S, and G_2 phases, and it normally comprises 90% or more of the total cell-cycle time. For example, in rapidly proliferating cells of higher eucaryotes, M phases generally occur only once every 16 to 24 hours, and each M phase itself lasts only 1 to 2 hours. A typical cell cycle with its four successive phases is illustrated in Figure 13–1, and some of the major events are outlined in the legend.

Figure 13–1 The four successive phases of a typical eucaryotic cell cycle. After the *M phase*, which consists of nuclear division (*mitosis*) and cytoplasmic division (*cytokinesis*), the daughter cells begin interphase of a new cycle. Interphase starts with the G_1 *phase*, in which the biosynthetic activities of the cells, which proceed very slowly during mitosis, resume at a high rate. The *S phase* begins when DNA synthesis starts, and ends when the DNA content of the nucleus has doubled and the chromosomes have replicated (each chromosome now consists of two identical "sister chromatids"). The cell then enters the G_2 *phase*, which continues until mitosis starts, initiating the M phase. During the M phase, the replicated chromosomes condense and are easily seen in the light microscope. The nuclear envelope breaks down (except in some unicellular eucaryotes such as yeasts, where it remains intact), the sister chromatids separate, two new nuclei form, and the cytoplasm divides to generate two daughter cells, each with a single nucleus. Cytokinesis terminates the M phase and marks the beginning of the interphase of the next cell cycle. A typical 24-hour cycle is illustrated here, although cell-cycle times in eucaryotic cells vary widely, from less than 8 hours to more than a year in adult animals, with most of the variability being in the length of the G_1 phase.

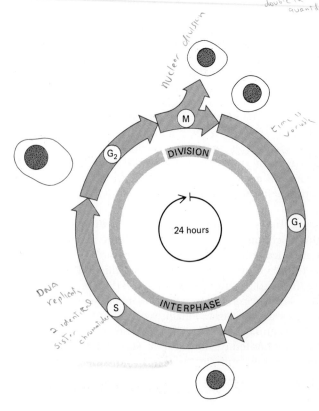

The time of DNA synthesis in the cell cycle was first demonstrated in the early 1950s by exploiting the technique of autoradiography to mark specifically those cells that are synthesizing DNA. The standard method employs ^{3}H-thymidine, a radioactive precursor of a compound that every cell uses exclusively for the synthesis of DNA. The ^{3}H-thymidine can either be injected into an animal to study the division cycles of cells in tissues or added to the culture medium of cells *in vitro* (Figure 13–2). In the former case, tissue is removed from the animal at a measured time after the injection of ^{3}H-thymidine, and autoradiographs are prepared from it. Those cells that have synthesized DNA at any time during the labeling period (and thus have been in S phase) can be identified by the silver grains over their nuclei. From the fraction of cells labeled in this way after different periods of exposure of an animal to ^{3}H-thymidine, and by scoring those cells that are in M phase, it is possible to show that the cell cycle has the four distinct phases described above and to measure the duration of each.

Suppose that a single injection of ^{3}H-thymidine is given and that the cells are fixed for autoradiography after a short time interval—say, half an hour. In a typical population of cells that are all proliferating rapidly but asynchronously, about 30% of the cells will be radioactively labeled. These are the cells that were synthesizing DNA during the brief exposure to ^{3}H-thymidine, and their frequency in the population reflects the fraction of the cell cycle that is occupied by S phase (Figure 13–3). Only around 5% of cells will be caught in mitosis at the moment of fixation (the small value of this *mitotic index* indicates that mitosis occupies only a small fraction of the cell cycle), and none of these will be radiolabeled, indicating that M phase and S phase are separate parts of the cycle. If, on the other hand, samples are fixed several hours after the injection of ^{3}H-thymidine, some of the cells in mitosis will be radiolabeled. These cells must have been synthesizing DNA at the time of injection. The minimum delay between injection and the time when radiolabeled mitotic cells appear will be equal to the duration of the G_2 phase. Proceeding along these lines, one can discover the durations of all four phases of the cycle. An example of the use of this method is explained in Figure 13–4.

Cells in different tissues, in different species, and at different stages of embryonic development have division cycles that vary enormously in duration, from less than an hour (for example, in the early frog embryo) to more than a year (for example, in the adult human liver). Although all phases of the cell cycle vary to some extent, by far the greatest variation occurs in the duration of G_1, which may be practically zero (as in the early frog embryo) or so long that the cell appears to have altogether ceased progressing through the division cycle and to have withdrawn into a quiescent state (as in the adult liver). Cells in such a quiescent G_1 state are often said to be in the G_0 state, as described below (see p. 749).

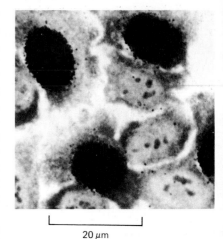

20 μm

Figure 13–2   An autoradiograph of cells that have been exposed for a short period to ^{3}H-thymidine. The technique is explained on page 176. The presence of silver grains in the photographic emulsion over a cell nucleus (*blackened area*) indicates that the cell incorporated ^{3}H-thymidine into its DNA, and thus was in S phase, sometime during the labeling period. (Courtesy of James Cleaver.)

Figure 13–3   The length of each phase of the cell cycle is approximately equal to the fraction of cells in that phase at any instant multiplied by the total cell-cycle time, assuming that the population of cells is growing steadily and that all the cells are proliferating at the same rate. A precise calculation of the length of each phase, however, involves a "correction factor" (ranging from approximately 0.7 for early G_1 cells to 1.4 for mitotic cells), which is needed because there are always more young (recently divided) cells than old cells in a steadily growing population.

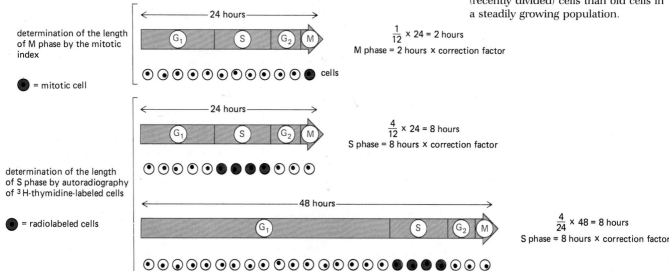

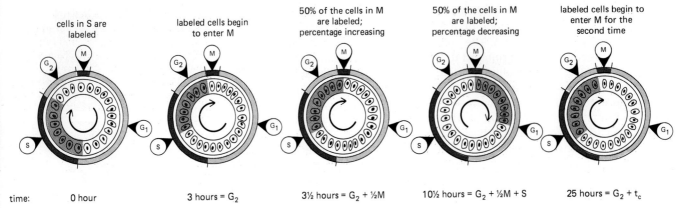

| time: | 0 hour | 3 hours = G_2 | 3½ hours = G_2 + ½M | 10½ hours = G_2 + ½M + S | 25 hours = G_2 + t_c |

At the top of the figure, the labels read:

cells in S are labeled

labeled cells begin to enter M

50% of the cells in M are labeled; percentage increasing

50% of the cells in M are labeled; percentage decreasing

labeled cells begin to enter M for the second time

Figure 13–4 A method commonly used to measure the lengths of the phases of the cell cycle. An asynchronously proliferating cell population is given a brief pulse of ^{3}H-thymidine and then washed, and the cells are left to continue their progress around the cycle. Autoradiographs are prepared from samples taken at various times after the pulse. To interpret the results, it is helpful to depict the cells as though they were distributed on a turntable, all moving around the cycle at the same rate and with the position of each cell on the turntable reflecting the stage it has reached in the cycle. Initially, cells in S phase are radiolabeled (*pale color*), while cells in G_2, M, and G_1 are not. After a time equal to the length of G_2, labeled cells begin to enter M phase; after a time G_2 + M, all the cells in M phase are labeled; and so on. By waiting and watching for the labeled cells to appear in M phase, to progress beyond it, and then to reappear in M phase, one can in principle discover the average durations of G_2, M, S, the total cycle time, and hence, by subtraction, G_1 also. In this example, these times are 3 hours, 1 hour, 7 hours, 22 hours, and 11 hours, respectively.

3-5 The Cell Cycle Is Most Easily Monitored in Culture[1,2]

It is difficult to analyze the mechanisms underlying the cell cycle in the complex and inaccessible tissues of an intact animal. The task is easier in cell culture. By time-lapse photography (see p. 142), for example, it is possible to watch an individual cell as it goes through mitosis, grows, and enters mitosis again; this allows direct measurement of the duration of M phase and of the total cell-cycle time. Cells undergoing DNA synthesis in culture can be detected, as in the intact animal, by ^{3}H-thymidine autoradiography. Alternatively, a cell's progress through the cycle can be followed by measuring its DNA content directly; this approach is greatly facilitated by the use of a *fluorescence-activated cell analyzer* (Figure 13–5).

Cell-cycle analysis is further simplified by using large populations of cultured cells that are all in the same phase of the cycle. Such *synchronous cell populations*

Figure 13–5 Typical results obtained for a growing cell population when the DNA content of its individual cells is determined with a fluorescence-activated cell analyzer, an electronic machine based on the same principles as the fluorescence-activated cell sorter (see p. 159). The cells are stained with a dye that becomes fluorescent when it binds to DNA, so that the amount of fluorescence is directly proportional to the amount of DNA in each cell. The cells fall into three categories: those that have an unreplicated complement of DNA (1 arbitrary unit) and are therefore in G_1 phase, those that have a fully replicated complement of DNA (2 arbitrary units) and are in G_2 or M phase, and those that have an intermediate amount of DNA and are in S phase. The distribution of cells in the case illustrated indicates that there are greater numbers of cells in G_1 than in G_2 + M, implying that G_1 is longer than G_2 + M in this population.

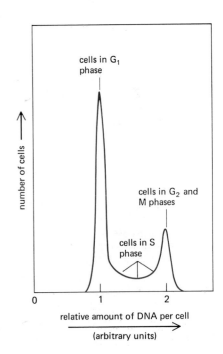

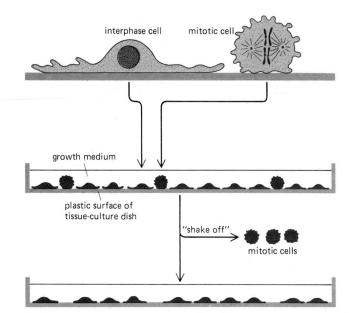

Figure 13–6 A commonly used method for obtaining a synchronously growing population of animal cells in culture. Mitotic cells are collected by shaking them off the dish on which they are growing. When transferred to a new dish, these collected cells continue through their cycles in synchrony. The synchrony is lost after a few cycles because of random variability in the rate of cycling of the individual cells.

can be obtained in several ways. One of the oldest techniques is to expose the cells to a drug that interferes with a specific step in the cycle and stops cells progressing beyond it: prolonged exposure causes all the cells to come to a halt at that same point, so that when the block is removed they resume cycling together and in step. During a cell cycle, however, many processes are occurring in parallel, and one cannot assume that all of them will be halted by the drug in a well-coordinated way. For this reason it is better, where possible, to use methods of preparing synchronized cell populations that do not interfere with the cell's normal progress through the cycle. For many mammalian cells in culture, the simplest and best procedure takes advantage of a cytoskeletal change that occurs in cells in M phase, causing them to "round up." The rounded M-phase cells adhere so weakly to the culture dish that they can be removed by gentle agitation (Figure 13–6). Mitotic cells collected by this technique of *mitotic shake-off* constitute a synchronous population that will almost immediately enter the G_1 phase of the cycle. Alternatively, because cells generally become progressively bigger as they proceed through the cycle, centrifugation can be used to select large numbers of cells at each stage of the cycle on the basis of their size.

Critical Cell-Cycle Events Occur Abruptly Against a Background of Continuous Growth[3]

Given a synchronous population of dividing cells, one can begin to examine in greater detail the chemical changes that occur in the course of the cell cycle. In conditions that favor growth, the total protein content of a typical cell increases more or less continuously throughout the cycle (Figure 13–7). Likewise, RNA synthesis continues at a steady rate, except during M phase, when the chromosomes are apparently too condensed to allow transcription, so that RNA synthesis is largely suppressed and the rate of protein synthesis is reduced. When the pattern of synthesis of individual proteins is analyzed (Figure 13–8), the vast majority are seen to be synthesized throughout the cycle. Thus, for most of the constituents of the cell, growth is a steady, continuous process, interrupted only briefly by the splitting of the cell into two.

Against this background of continuous growth, a series of abrupt changes occur, marking critical moments in the cell cycle. Some of these events, such as the onset of DNA synthesis, are easily detected. Others are more difficult to pinpoint. We shall see, for example, that for most cells there is a critical point in the G_1 phase of their cycle where they may pause if the environment is unfavorable to growth. In proceeding past this point, called the *restriction point*, or *Start*, a

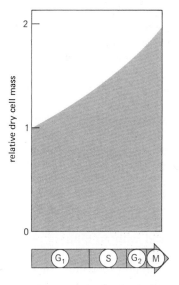

Figure 13–7 A plot of the increase in cell mass during the cell cycle. Most cell components are made more or less continuously throughout interphase, generally at an increasing rate as the cell (and its biosynthetic capacity) enlarges (with a brief lull during M phase). Note that each component must exactly double during the cycle if the average cell size in a proliferating population is to be maintained.

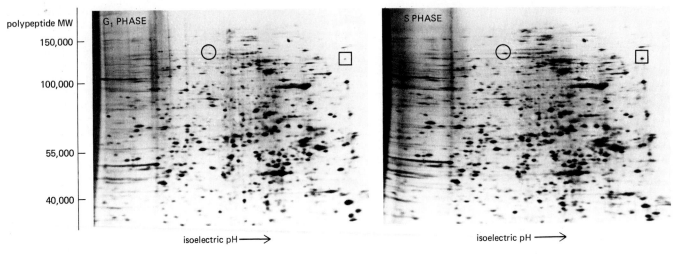

polypeptide MW

150,000

100,000

55,000

40,000

G₁ PHASE

S PHASE

isoelectric pH ⟶

isoelectric pH ⟶

cell undergoes an internal change that commits it to complete the subsequent steps of the cycle according to a rigid timetable.

As might be expected in a cycle marked by steplike transitions, it is possible to detect certain proteins—although only a very few—whose synthesis is switched on at a high rate at a specific stage of the cycle (see Figure 13–8). Thus histones, which are required for the formation of new chromatin, are made at a rapid rate only in S phase, and the same seems to be true for some of the proteins of the DNA replication apparatus.

But what controls the timing of these steplike events in the cell cycle, and how are they coordinated with one another and with the continuous process of cell growth? We shall first consider what initiates DNA synthesis so as to begin the S phase of the cycle.

DNA Synthesis Is Switched On by a Change in the Cytoplasm: The S-Phase Activator[4]

With the help of an appropriate agent in the culture medium (see p. 162), members of two populations of synchronized cells that are in different phases of the cell cycle can be made to fuse with each other. The results are remarkably informative.

When an S-phase cell is fused with a cell at an early stage of G_1, the G_1 nucleus promptly begins to replicate its DNA (Figure 13–9A). Evidently the G_1 nucleus is ready to replicate its DNA, but a signal (or set of signals) required to activate the machinery for DNA synthesis is not yet present in the normal G_1 cell. The S-phase cell apparently has an abundance of the activating signal(s) in its cytoplasm. The appearance of this *S-phase activator* presumably marks the boundary between G_1 phase and S phase in a normal cell.

Does the S-phase activator disappear after DNA synthesis is completed and the cell has entered G_2 phase? Cell-fusion experiments again provide a test. When a G_2-phase cell is fused with a G_1-phase cell (Figure 13–9B), the G_1 nucleus is *not* driven prematurely into DNA synthesis; but when the same type of G_2-phase cell is fused with an S-phase cell, the S nucleus continues replicating its DNA (see Figure 13–9A). Apparently the S-phase activator (or some crucial component of it) is eliminated soon after the end of S phase, and the G_2 cytoplasm contains neither a diffusible activator nor a diffusible inhibitor of DNA synthesis.

The Whole Genome Is Replicated Once and Only Once in Each Cycle[4,5]

As explained in Chapter 9 (p. 519), different parts of the genome are replicated at different times in S phase and are then prevented from undergoing a second round of replication by an unknown chemical change that occurs in each region of each chromosome as soon as the replication apparatus has passed over it. Because of this *DNA re-replication block*, when a G_2-phase cell is fused with an S-phase cell

Figure 13–8 An analysis of the proteins synthesized during the G_1 and S phases of the cell cycle by two-dimensional polyacrylamide-gel electrophoresis. Synchronized mouse lymphoma cells were briefly labeled in either the early G_1 or the late S phase with a mixture of radioactive amino acids. About 1000 different newly synthesized proteins can be detected in these autoradiographs, but only the two proteins enclosed by a circle or a square are regularly made at detectably different rates. Because they are so positively charged, the histones migrate off the gel and are not detected. (From P. Coffino and V.E. Groppi, *Adv. Cyclic Nucleotide Res.* 14:399–410, 1981.)

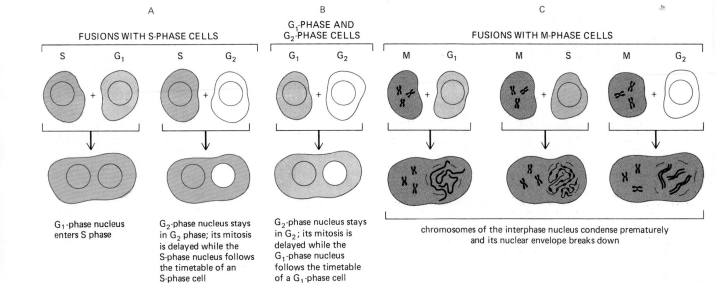

	A			B		C	
FUSIONS WITH S-PHASE CELLS				G_1-PHASE AND G_2-PHASE CELLS		FUSIONS WITH M-PHASE CELLS	

G_1-phase nucleus enters S phase

G_2-phase nucleus stays in G_2 phase; its mitosis is delayed while the S-phase nucleus follows the timetable of an S-phase cell

G_2-phase nucleus stays in G_2; its mitosis is delayed while the G_1-phase nucleus follows the timetable of a G_1-phase cell

chromosomes of the interphase nucleus condense prematurely and its nuclear envelope breaks down

Figure 13–9 Summary of the various results of fusing two mammalian cells that are in different phases of the cell cycle.

(Figure 13–9A), the G_2 nucleus is unresponsive to the S-phase activator and does not restart DNA synthesis. The block is removed when the cell passes through mitosis to begin a new G_1 phase.

An additional mechanism is needed to ensure that the S-phase activator remains present long enough for all of the DNA to be replicated once replication has begun. We have seen that cell-fusion experiments suggest that the cytoplasmic signal that activates the DNA-replication machinery at the beginning of S phase (the S-phase activator) disappears at the end of S phase. Yet when a cell is artificially blocked in the S phase by inhibitors of DNA synthesis, the DNA-replication machinery remains functional beyond the time at which S phase would normally have ended—so that DNA replication is completed when the inhibitor is removed. Apparently a chromosome that is partway through replication somehow keeps the replication machinery active.

The replication forks themselves might be responsible for this effect. As we have seen in Chapter 9 (p. 515), replication forks occur in pairs: the two forks of a pair move in opposite directions away from a common point of origin, and each replication fork terminates only when it reaches the end of the chromosome or collides with a fork moving in the opposite direction. Therefore, once a chromosome has begun replicating, at least one replication fork will persist until the whole chromosome has been fully duplicated. In some way not understood, this fork might ensure the production of more of the diffusible S-phase activator, causing the initiation of new replication forks to be catalyzed with high efficiency elsewhere on the DNA. Indeed, the initiation of the first pair of replication forks could be the all-or-none trigger for S phase to begin. Such a single initiation event might depend on a rare random collision between an origin sequence and an initiator molecule present at low concentration. In fact, the timing of the transition from G_1 to S phase shows a random variability that is consistent with this suggestion (see p. 746).

13-6 A Cytoplasmic Signal Delays Preparations for Mitosis Until DNA Replication Is Complete[4,6]

A nucleus that has just completed S phase and entered G_2 phase will normally condense its chromosomes and go through mitosis at a fixed time thereafter. When DNA synthesis is artificially blocked, however, mitosis is delayed until after the block has been removed and DNA synthesis completed. Similarly, when an S-phase cell is fused with a G_2 cell, the G_2 nucleus is delayed in G_2 phase until the other nucleus catches up, and eventually both nuclei go through mitosis together. In the simplest view, a cytoplasmic signal that is generated by incompletely replicated DNA delays the onset of mitosis. This signal might or might not be identical

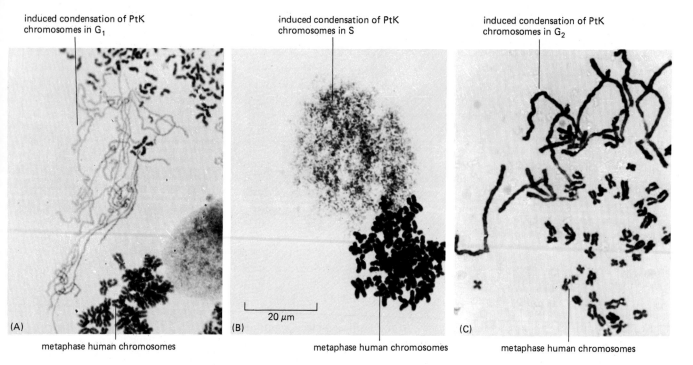

induced condensation of PtK chromosomes in G₁

induced condensation of PtK chromosomes in S

induced condensation of PtK chromosomes in G₂

(A)   metaphase human chromosomes

(B)   20 µm   metaphase human chromosomes

(C)   metaphase human chromosomes

with the S-phase activator; in either case a hint of a possible mechanism for its production comes from the observation that when the DNA of a cell in G₂ is damaged (by x-irradiation, for example), mitosis is delayed until the damage is repaired. Both DNA repair and DNA replication involve the presence of single-stranded DNA, and in bacteria an excess of such DNA is known to trigger the production of a cytoplasmic signal that delays cell division (the *SOS response*—see p. 225). It is possible that single-stranded DNA in the eucaryotic cell similarly generates an M-phase-delaying signal. There is as yet no further clue as to the nature of such a signal, apart from one intriguing observation: when mammalian cells in culture are artificially blocked in DNA synthesis—either by inhibitors or by DNA damage—they can be induced to enter mitosis prematurely, with their DNA still not fully replicated, by the addition of *caffeine* to the medium. The relevant target affected by the caffeine is not yet known.

An M-Phase-promoting Factor Triggers Mitosis[4,7]

The disappearance of M-phase-delaying signals is not by itself sufficient to trigger mitosis: another cytoplasmic factor is required. The normal G₂ phase can be regarded as a period of preparation for production of this crucial mitotic trigger following disappearance of the delaying factors. Evidence again comes from cell-fusion experiments.

When an M-phase cell is fused with a cell in any part of interphase—G₁, S, or G₂—the interphase nucleus is promptly driven into M phase, condensing its chromosomes in preparation for division, even though—in the case of the G₁- or S-phase nucleus—this spells disaster for further progress through the division cycle (Figures 13–9C and 13–10). Apparently the M-phase cytoplasm contains a powerful *M-phase-promoting factor* (abbreviated as *MPF*) to which a nucleus in any phase of the cycle is susceptible. Presumably the cytoplasmic factors just discussed that delay mitosis inhibit production of MPF but cannot block its effects once it has been produced.

The Events of the Chromosome Cycle Are Linked as a Dependent Sequence[8]

The experiments described above provide a functional classification for some of the molecules that must govern the events of the chromosome cycle. Three diffusible control factors have been discussed. For brevity, it is convenient to refer

Figure 13–10 Premature condensation of interphase chromosomes following fusion of interphase marsupial (PtK) cells with mitotic human cells. In (A) the PtK cell was in G₁ phase; consequently, its prematurely condensed chromosomes are still single chromatids. In (B) the PtK cell was in S phase, and its chromatin now adopts a "pulverized" appearance. In (C) the PtK cell was in G₂ phase, and now the chromatids, although very long compared to the normal (human) metaphase chromosome, are double. (From K. Sperling and P. Rao, *Humangenetik* 23:235–258, 1974.)

to these as though each were a single molecule, although they may in fact be more complex. They are (1) the *S-phase activator* that is normally present only in S-phase cytoplasm and switches on DNA synthesis, (2) the *M-phase-promoting factor (MPF)* that is present only in M-phase cytoplasm and causes chromosome condensation, and (3) the DNA-dependent *M-phase-delaying factor*—possibly identical with the S-phase activator—that is present in S-phase cytoplasm and inhibits the process leading to onset of MPF production. The abrupt appearances and disappearances of these diffusible factors in the cytoplasm are landmark events in the cell cycle, and the time between these landmarks determines the length of the total cycle.

The causal relationships among the factors we have listed (and presumably others not yet known) guarantee that the events of the chromosome cycle will always occur in a fixed order, prohibiting such fatal clashes as chromosome condensation occurring in the middle of DNA synthesis. Each successive step depends on a preceding one. Thus the cell cannot pass through mitosis until MPF has been produced; MPF cannot be produced until the M-phase-delaying factor has disappeared; the M-phase-delaying factor and the S-phase activator cannot disappear until DNA synthesis has ended; DNA synthesis cannot end until all of the DNA has replicated; the DNA cannot begin to replicate until the DNA re-replication block has been removed by passage through mitosis into G_1. We shall encounter yet another example later: a cell cannot progress from mitosis into G_1 until the chromosomes have separated on the mitotic spindle (see p. 773). All these observations, and others to be considered later (see p. 740), indicate that the many events and processes of the chromosome cycle are linked together as a *dependent sequence*.

The Cell Cycle Is Abridged During Early Cleavage Divisions in Which There Is No Cell Growth[9]

Studies of the egg and early embryo of the frog *Xenopus* have been especially useful for the analysis of molecules that control the chromosome cycle. The egg of *Xenopus*, like that of many other species, is an exceptionally large spherical cell. It measures just over a millimeter in diameter and contains stores of practically all the materials, other than DNA, that will be required for the construction of the early embryo. These stores accumulate during a long period of growth of the immature egg, known as the *oocyte*, which is meanwhile arrested in a state best described as the G_2 phase of its first meiotic division cycle. (Although this stage is usually referred to as prophase of the first meiotic division, it resembles an ordinary G_2 in many respects—see p. 854.) Hormones acting at the time of ovulation cause egg maturation, so that when the egg is laid, it has progressed further through meiosis and is arrested in M phase of its second meiotic division (see p. 857). Fertilization then triggers an astonishingly rapid sequence of cell divisions in which the single giant cell *cleaves* to generate an embryo consisting of thousands of smaller cells (Figure 13–11). Practically no growth occurs in this process: almost the only macromolecules synthesized are DNA, which is required to create the necessary number of nuclei, and a little protein. After a first division that takes about 90 minutes, the next 11 cleavage divisions occur, more or less synchronously, at 30-minute intervals, producing 4096 (2^{12}) cells within about 7 hours. The prior accumulation of materials in the egg makes these rapid cell cycles possible by eliminating the time normally required for cell growth in each cycle. The cycles

Figure 13–11 Within about 7 hours a *Xenopus* egg undergoes 12 exceptionally rapid synchronous cell divisions (cleavages), consisting of alternating S and M phases without perceptible G_1 or G_2 phases. These divisions subdivide the egg into 4096 (2^{12}) smaller cells. In each embryo a single cell is colored.

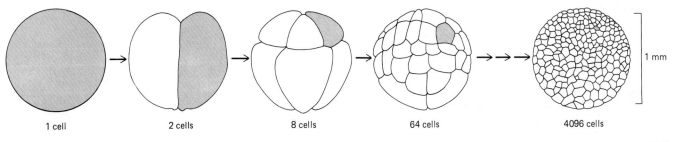

| 1 cell | 2 cells | 8 cells | 64 cells | 4096 cells |

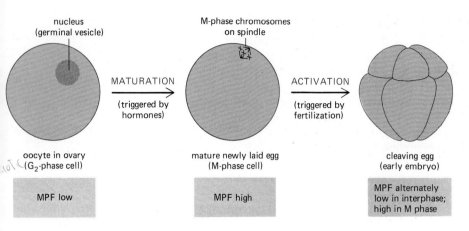

Figure 13–12 The levels of M-phase-promoting factor (MPF) in the *Xenopus* oocyte, egg, and early embryo. The oocyte is arrested in a meiotic G_2 phase, with a low level of MPF; the mature newly laid egg is arrested in a meiotic M phase, with a high level of MPF; following fertilization, the early embryo passes through alternate S and M phases, oscillating between low and high levels of MPF activity.

of DNA replication and division are abridged by speeding up both S phase and M phase and making the G_1 and G_2 phases so short as to be imperceptible.

MPF Induces Mitosis in a Wide Variety of Cells[10]

Because the *Xenopus* oocyte and egg are so big, it is easy to inject substances into their cytoplasm. Moreover, the oocyte, egg, and early embryo provide abundant sources of cytoplasm from defined stages of the cell cycle. This has been particularly important in the study of the **M-phase-promoting factor (MPF)** mentioned previously. MPF was first discovered in mature unfertilized *Xenopus* eggs (which are arrested in M phase). When cytoplasm from such an egg is injected into an oocyte, it releases the oocyte from its G_2 arrest and drives it into M phase. This initiates oocyte maturation, and the initials MPF originally stood for "maturation-promoting factor" (see p. 860). Active MPF also appears in the cleaving egg (embryo) during each M phase (Figure 13–12). Thus the *Xenopus* egg and oocyte provide both a good source of material for attempts to purify MPF and a convenient means to assay for it (Figure 13–13).

MPF is of universal importance to eucaryotic cells and has been highly conserved during evolution: extracts prepared from mitotic cells of the most diverse species, including mammals, sea urchins, clams, and yeasts, can be injected into *Xenopus* oocytes and will drive them into M phase. Material with MPF activity has been purified from mature *Xenopus* eggs. It behaves as a large protein that includes two types of subunits, one of which is a protein kinase and apparently can phosphorylate the other. Correspondingly, MPF seems to be able to activate itself: when a small amount of material with MPF activity is injected into a *Xenopus* oocyte, the cell responds by generating a very much larger amount of MPF from its own inactive reserves (see p. 860). This and other evidence suggest that the appearance and disappearance of MPF activity during the normal cell cycle depend on modification of the protein by phosphorylation and dephosphorylation, rather than on *de novo* synthesis and degradation. The normal triggering of MPF activation does, however, require the synthesis of another protein, identified as cyclin (see below); thus cells of all types are unable to progress from interphase to M phase when protein synthesis is blocked.

Many of the molecular changes that occur in mitosis seem to be brought about by phosphorylation. The MPF kinase directly phosphorylates several substrates, including, in particular, histone H1, thereby probably promoting chromosome condensation (see p. 503); and it may be through a cascade of phosphorylations that MPF triggers all the complex events of mitosis.

MPF Is Generated by a Cytoplasmic Oscillator[8,10,11]

The surge of MPF that occurs every 30 minutes in the cleaving *Xenopus* embryo is generated by a cytoplasmic oscillator that operates even in the absence of a nucleus. By constricting the activated egg before it has completed its first division, one can split it into two completely separate parts, one containing a nucleus, the

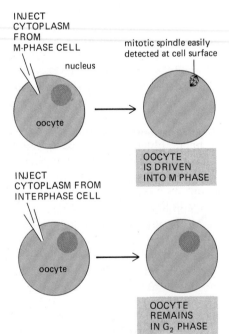

Figure 13–13 Assaying for MPF by injection into a *Xenopus* oocyte. MPF can be detected because it drives the oocyte into M phase. The large nucleus (or "germinal vesicle") of the oocyte breaks down as the mitotic spindle forms.

46 minutes　　　　　90 minutes　　　　　116 minutes　　　　⊢———⊣
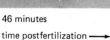　　　　　　　　　　　　　　　　　　　　　　　　　1 mm

Figure 13–14　Technique for demonstrating a cytoplasmic oscillation associated with the cell-division cycle in the cleaving *Xenopus* egg. A newly fertilized egg is split in two by constricting it with a loop of fine human hair; one half contains the nucleus and proceeds to divide, while the other half lacks a nucleus and does not divide. Time-lapse photography then shows that the nonnucleated half periodically changes its height through changes in the stiffness of the cell cortex, oscillating in close synchrony with the divisions of the nucleated half. (From K. Hara, P. Tydeman, and M. Kirschner, *Proc. Natl. Acad. Sci. USA* 77:462–466, 1980.)

other not (Figure 13–14). The nucleated part continues with the normal program of rapid cleavages. Remarkably, the nonnucleated part also goes through a series of oscillations, manifest in repeated cycles of slight periodic contraction and stiffening of its cortical cytoplasm. These recurrent spasms occur in almost perfect synchrony with the cleavage divisions of the nucleated half-egg. By taking samples of cytoplasm at intervals from the oscillating nonnucleated cell and assaying them by injection into oocytes, it can be shown that the visible oscillations are accompanied by, and perhaps caused by, oscillations in the concentration of active MPF.

These and other experiments suggest that the cleavage divisions in the early *Xenopus* embryo involve two parallel cyclic processes—a chromosome-replication cycle and a cytoplasmic MPF cycle—that normally stay in step with each other because each new chromosome cycle can be initiated only when the DNA re-replication block is lifted following the pulse of MPF during M phase. This interaction between the two cycles prevents the chromosome cycle from running faster than the MPF cycle and will keep them in step as long as there is no risk of the chromosome cycle running too slowly to complete DNA replication before MPF levels rise. In the *Xenopus* egg, with its exceptionally rapid S phases and regular division cycles, this risk is presumably slight, and the single interaction between the cycles seems to be sufficient. However, in the mammalian cells discussed earlier (and probably in most eucaryotic cells other than cleaving eggs) there is also another, complementary effect: as we have seen, unreplicated DNA generates an M-phase-delaying signal that prevents the cytoplasmic MPF cycle from running faster than the chromosome cycle. Experiments in which DNA replication is blocked with inhibitors show that in the early *Xenopus* cell cycle this additional control does not operate. Moreover, as suggested by the absence of a G_1 phase, an S-phase activator seems to be present at all times. Thus the early *Xenopus* cycle appears to be simplified and stripped down for speed.

The phenomena just described imply that a cytoplasmic oscillator may exist in all cells, but they do not tell us its mechanism. A possible clue is provided by another protein, called **cyclin,** that has been identified in cleaving eggs of *Xenopus*, sea urchins, and surf clams. Cyclin, like MPF, belongs to the small group of proteins whose activity depends markedly on the phase of the division cycle. Although cyclin is synthesized at a rapid, constant rate throughout the cycle, it is abruptly destroyed halfway through M phase. Thus in each cycle its concentration rises steadily from zero and then suddenly drops back to zero again. Cyclin genes have been cloned, allowing the preparation of pure cyclin mRNA. When this mRNA is injected into a *Xenopus* oocyte, it has the same effect as an injection of MPF, driving the oocyte out of G_2 and into M phase. From these and other observations, it has been suggested that the surge of MPF in M phase may be triggered by the increase of cyclin concentration above a certain threshold; cyclin destruction is caused by an event in M phase; and the subsequent disappearance of MPF could be a consequence of cyclin destruction (Figure 13–15). In this view the time from one mitosis to the next would be determined chiefly by the time required for the cyclin concentration to build up from zero to the threshold value; and, as is indeed observed, the cell cycle should be halted in interphase by inhibitors of protein synthesis.

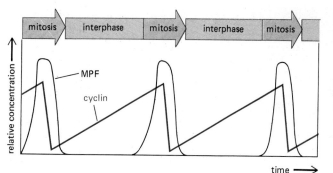

Figure 13–15 The correlated rise and fall in the levels of MPF and cyclin observed during the cell cycle. The cyclin measurements have been made chiefly in the eggs of marine invertebrates, where cyclin accounts for 5% of the protein synthesized during a brief pulse with radioactive amino acids.

Summary

The reproductive cycle of a typical eucaryotic cell can be divided into four phases, known as G_1 (gap 1), S (DNA synthesis), G_2 (gap 2), and M (mitosis). The beginning of S phase and the beginning of M phase are each signaled by soluble cytoplasmic factors (the S-phase activator and the M-phase-promoting factor [MPF], respectively). The S-phase activator continues to be generated throughout S phase, and it may act also as the control factor that delays preparations for M phase until after DNA replication has been completed. MPF can be detected in M-phase cells of many different species, from yeast to mammal, and its activity may be regulated by phosphorylation. In rapidly cleaving eggs, such as that of Xenopus, the cell cycle is speeded up and simplified. Here the cycle appears to be driven by a coupled oscillation in the activity of MPF and the concentration of cyclin.

Yeasts as a Model System[12]

Yeasts are unicellular fungi and constitute a large group of rather disparate organisms. Because they reproduce almost as rapidly as bacteria and have a genome size less than 1/100th that of a mammal, they are exceptionally useful for genetic studies of eucaryotic cell biology. While *Xenopus* eggs have been invaluable for cell biological and biochemical analysis of cell-cycle control, *Xenopus* is a poor organism for genetic analysis. Yeasts, on the other hand, have provided direct opportunities to identify, clone, and characterize the genes involved in controlling the cell cycle. The two species we shall discuss here are the **budding yeast** *Saccharomyces cerevisiae*, used by brewers and bakers, and the **fission yeast** *Schizosaccharomyces pombe*. The fission yeast divides symmetrically by splitting into two similar daughter cells; the budding yeast divides in a less conventional, asymmetric fashion in which the mother cell produces a small bud that grows throughout the cycle before finally separating from its parent (Figure 13–16).

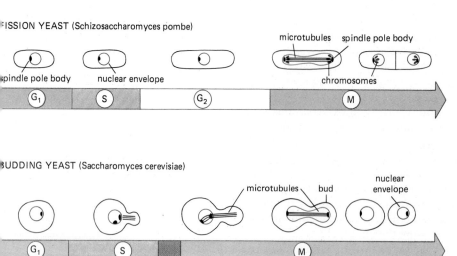

FISSION YEAST (Schizosaccharomyces pombe)

BUDDING YEAST (Saccharomyces cerevisiae)

Figure 13–16 A comparison of the cell cycles of fission yeasts and budding yeasts. The *fission yeast* shown in the upper panel has a typical eucaryotic cell cycle with G_1, S, G_2, and M phases. However, the nuclear envelope does not break down: the microtubules of the mitotic spindle form inside the nucleus and are attached to *spindle pole bodies* (SPBs) at its periphery. The cell divides by forming a partition (known as the cell plate) and splitting in two. The *budding yeast* has normal G_1 and S phases. However, a microtubule-based spindle begins to form very early in the cycle, during S phase, and thus there does not appear to be a normal G_2 phase. In contrast with fission yeasts, visible chromosome condensation during mitosis does not occur, and the cell divides by budding. As in fission yeasts but in contrast with higher eucaryotic cells, the nuclear envelope remains intact during mitosis.

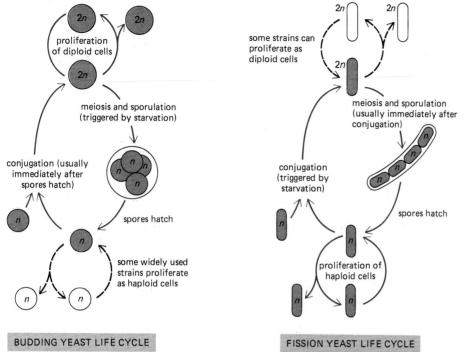

Figure 13–17 The life cycles of a budding yeast (*Saccharomyces cerevisiae*) and a fission yeast (*Schizosaccharomyces pombe*). The proportion of the life cycle spent in the diploid or haploid state varies with the species and according to the environment. When nutrients are plentiful, normal wild-type varieties of budding yeasts proliferate as diploid cells, with a cell-cycle time of about 2 hours. If starved, they go through meiosis to form haploid spores, which germinate when conditions improve to become haploid cells that can either proliferate or fuse sexually (conjugate) in G_1 phase to re-form diploid cells, depending on the environment and the genotype. Fission yeasts, by contrast, typically proliferate as haploid cells; these cells fuse in response to starvation to form diploid cells that promptly go through meiosis and sporulation to regenerate haploid cells. The most widely used laboratory strains of budding yeasts are mutants that proliferate, like fission yeasts, chiefly as haploids.

BUDDING YEAST LIFE CYCLE

FISSION YEAST LIFE CYCLE

The evolutionary lineages leading to budding and fission yeasts are thought to have diverged many hundreds of millions of years ago. Nevertheless, the two organisms have similar life cycles. Each can proliferate in either a diploid or a haploid state: the diploid cells, as an alternative to dividing in the ordinary way, can go through meiosis to form haploid cells (see Chapter 15); the haploid cells, as an alternative to dividing, can mate or *conjugate* with one another to form diploid cells (Figure 13–17 and see p. 571). The haploid phase makes genetic analysis easy and helps in the isolation of loss-of-function mutations whose phenotype would be recessive in a diploid organism (such as a cultured mammalian cell) and therefore less readily seen. In both species of yeasts, food and sex play important parts in controlling the cell-division cycle. Thus yeasts will bring us to the general question of how the division cycle is regulated by factors in a cell's environment.

13-10 ## Each Yeast Cell-Division-Cycle Mutant Halts or Misbehaves in a Specific Phase of the Cycle[12,13]

To identify the genes involved in cell-cycle control, one needs appropriate mutants and a way to breed from them. But a cell whose cell-cycle machinery is disrupted cannot reproduce. The way out of this dilemma is to search for *conditional mutants*, which display their mutant phenotype only under certain conditions. Usually one seeks a gene product whose molecular structure is slightly altered so that it fails to function in one temperature range—the *restrictive* temperature range—but is able to function normally in another—the *permissive* temperature range. For such *temperature-sensitive* mutants, low temperatures are usually permissive and high temperatures are usually restrictive. Thus one can generate a mutant at a low temperature and then, by raising the temperature, switch off the affected gene function and examine the mutant phenotype.

Mutations that specifically affect components of the cell-cycle machinery cannot be identified merely by the failure of the mutant cells to divide, since any lethal defect will have this consequence. The **cell-division-cycle (*cdc*) mutants** are defined more narrowly by the criterion that they become blocked, or misbehave, in a specific part of the cell cycle at the restrictive temperature (Figure 13–18). In the budding yeast, the presence and size of a bud provide a simple visible indication of the point in the cycle where a *cdc* mutant is blocked; in the fission yeast, more complex methods are required, using techniques of cell-cycle analysis such as those outlined earlier.

TEMPERATURE-SENSITIVE CELL-DIVISION-CYCLE (*cdc*) MUTANTS

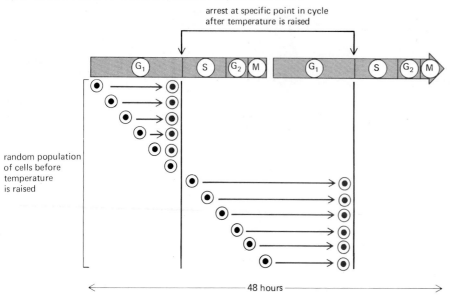

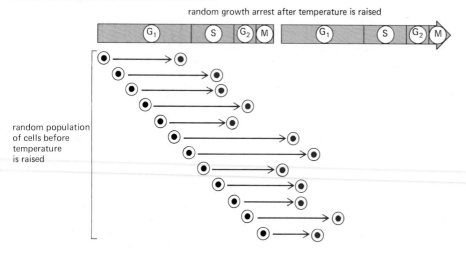

Between 40 and 50 *cdc* genes have been identified through mutations in each of the two yeast species. In several cases, biochemical studies have revealed the exact function of the gene product. For example, some *cdc* mutants that are blocked in their progress through S phase turn out to be defective in genes coding for DNA ligase or for enzymes required to synthesize DNA precursors. As will be described below, recombinant DNA techniques provide a general way to characterize all the proteins encoded by *cdc* genes. But some important insights can be gained even without this information.

Yeast *cdc* Mutants Can Be Used to Analyze the Coupling Between Cell-Cycle Events[12,13]

When the temperature is raised to the restrictive range, most *cdc* mutants are arrested at the stage in the cell cycle where the *cdc* gene product acts. Typically, the cell fails to proceed to subsequent steps in the cycle, indicating that the beginning of each process is conditional upon completion of the preceding process; thus, as for vertebrates, most steps of the yeast cell cycle seem to be organized in dependent sequences. This dependence has been more rigorously analyzed in experiments on cells that carry combinations of different *cdc* mutations. The results show that the events of the chromosome cycle form a dependent sequence that is not rigidly tied to the sequence of events of the cytoplasmic cycle (Figure

Figure 13–18   How a temperature-sensitive cell-division-cycle (*cdc*) mutant is distinguished from other temperature-sensitive mutants. Upon warming to the restrictive temperature, where the mutant gene product functions abnormally, the *cdc* mutant cell will continue to progress through the cycle until it reaches a specific step that it is unable to complete (the initiation of S phase in the example shown). Because cell growth continues despite the blocking of progress through the cycle, *cdc* mutant cells become abnormally large (not shown). Non-*cdc* mutant cells that are defective in processes (such as ATP production) necessary throughout the cycle for biosyntheses and growth will halt haphazardly at any stage of the cycle, as soon as their biochemical reserves run out.

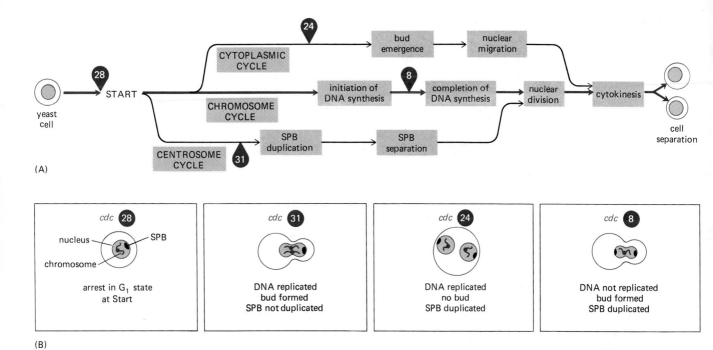

(A)

(B)

Figure 13–19 The causal connections between some of the events of the budding-yeast cell cycle and their relation to *cdc* genes. The spindle pole body (SPB) is the yeast equivalent of the centrosome. "Start" is where the cell becomes committed to complete a division cycle and loses the option to conjugate sexually (conjugation can only be performed in G₁ phase). (A) Outline of the cycle; an arrow pointing from an event *a* (or from events *a* and *b*) to an event *c* means that *c* cannot occur until *a* (or *both a* and *b*) has occurred. The numbers designate specific *cdc* mutants that will halt at that point in the cycle when shifted to their restrictive temperature. Thus, for example, the mutant *cdc8* stops during DNA synthesis. Note that the *chromosome cycle* (DNA cycle), the *cytoplasmic cycle* (bud cycle), and the *"centrosome cycle"* (SPB cycle) are partially independent. (B) The blocked states of the four *cdc* mutants in (A) at high temperature. (Based on L.H. Hartwell, *J. Cell Biol.* 77:627–637, 1978, and J.R. Pringle and L.H. Hartwell, in The Molecular Biology of the Yeast *Saccharomyces* [J.N. Stern *et al.*, eds.], pp. 97–142. Cold Spring Harbor Laboratory, 1981.)

13–19). Thus, whereas cytokinesis does not occur if nuclear division is prevented, those *cdc* mutants that fail to go through cytokinesis because they are defective in the machinery for bud formation will nevertheless proceed through repeated rounds of DNA synthesis and nuclear division. It seems to be a general rule, not only for yeasts but also for the cells of vertebrates, insects, and many other organisms, that the chromosome cycle can continue even when cytokinesis is prevented. In fact, multinucleate cells are often generated in this way during normal development (see p. 921).

The Regulation of Cell Size Depends on Cell-Cycle Controls That Operate at Start

The growth rates of simple free-living organisms such as yeasts depend chiefly on the supply of nutrients. Under conditions of poor nutrition, the daughter cells produced by rapid cell-division cycles would be disastrously small; so cells need a mechanism to control the rate of progress through the division cycle—and in particular through the chromosome cycle—according to the rate of cell growth (Figure 13–20). How is this control exerted?

DNA synthesis and mitosis are both complex dynamic processes that may be difficult to perform more slowly or to interrupt when nutrition is poor. In yeasts, as in most other eucaryotic organisms, the lengths of these phases of the cycle remain roughly constant despite wide variations in external conditions. Instead, it is generally the G₁ phase that is extended when cells are starved, although in fission yeasts there is also an important control, called the *mitotic control*, in G₂.

If the duration of a cell's G₁ phase can be adjusted in response to external factors and the duration of its S phase cannot, then there must be a critical point in G₁ where the S-phase chain of events is initiated and progress through the cycle is no longer influenced by external factors. This critical point is called **Start.** For most eucaryotic cells, Start (or its equivalent, the *restriction point* in mammalian cells) marks the moment of commitment to complete a cell-division cycle.

Cells Pass Through Start Only After Attaining a Critical Size[14]

For a budding yeast cell in a nutrient-poor medium, G₁ is a period of slow growth during which progress through the chromosome cycle appears to be arrested; exit from G₁ by passage through Start occurs when the cell reaches a certain standard

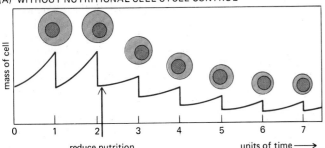

(A) WITHOUT NUTRITIONAL CELL-CYCLE CONTROL

mass of cell

0   1   2   3   4   5   6   7

reduce nutrition   units of time ⟶

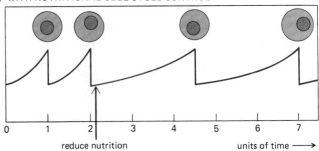

(B) WITH NUTRITIONAL CELL-CYCLE CONTROL

0   1   2   3   4   5   6   7

reduce nutrition   units of time ⟶

size (see Figure 13–20B). In a richer medium, G_1 is shorter but the size at passage through Start is practically the same; and if conditions of growth are manipulated so as to generate cells that are abnormally large or abnormally small at the time of their birth, these cells respectively curtail or prolong their stay in G_1 phase so as to pass through Start at approximately the standard size.

Very little is known about the mechanism by which cells sense their size, although there is evidence from many sources that they have some means of doing so. For example, if cytoplasm is repeatedly amputated from a growing giant amoeba (*Amoeba proteus*) so as to keep it unusually small, it will not divide, even over a period of many weeks, despite vigorous cell growth—whereas a control cell divides roughly once per day. One possible clue as to how cells sense their size is provided by the observation that the size of a eucaryotic cell is generally proportional to its ploidy: a diploid cell is twice as big as a haploid, and a tetraploid is twice as big as a diploid (see Figures 13–40 and 13–41). This suggests that the critical factor is the ratio of cell volume to the number of copies of a gene (or set of genes) or to the total quantity of DNA (and not, say, the ratio of cell volume to cell surface area). For example, some diffusible molecule M—a species of RNA, say—might be steadily synthesized in a DNA-dependent fashion; if M is unstable, with a constant half-life, the total quantity of M in each cell will always be constant and in a fixed, constant proportion to the quantity of DNA. As the cell enlarges in volume, the *concentration* of M will therefore fall; and decline of this concentration below a certain critical threshold value could be the signal to pass Start.

Whatever the mechanism, passage through Start must correspond to a flip in the state of some molecular switch. Four *cdc* genes in the budding yeast and two in the fission yeast act at or close to Start and may encode components of this control device. Cells with temperature-sensitive mutations of these genes fail to progress through the chromosome cycle and grow abnormally large if they are warmed to the restrictive temperature before the critical Start size has been attained. We shall have more to say below about one of the Start-control genes (*cdc28*) in the budding yeast and its counterpart (*cdc2*) in the fission yeast. These two genes are remarkable for an additional function, particularly prominent in the fission yeast: their products are required not only for passing Start, but also, at a second control point in the cycle, for the initiation of mitosis.

We shall see that higher eucaryotic cells have a G_1 control point analogous to Start, although the rules for passing this point are more complex than in yeasts. Cancer depends on violations of these rules. For that reason, if no other, the genes involved in the mechanism of Start have a special interest.

Passage Through Start Depends on a Protein Kinase Related to MPF[15]

Yeasts, with their rapid reproduction and simple unicellular life-style, are attractive organisms for genetic engineering, and they can be readily induced to take up and incorporate DNA that is added to the incubation medium. This makes it possible, in principle, to clone the normal (or "wild-type") form of any of the *cdc* genes. As explained in Figure 13–21, the desired clone can be picked out easily by its ability to rescue the corresponding *cdc* mutant from its abnormality.

This approach has been used to clone both the *cdc28* gene of the budding

Figure 13–20 The relationship between growth rate, cell size, and the cell-division cycle in a free-living organism such as a yeast. (A) If cell division continued at an unchanged rate when cells were starved, the daughter cells produced at each division would become progressively smaller until the size of each daughter was equal to the small amount of material synthesized in one cycle time. (B) Yeast cells actually respond to nutrition-poor conditions by retarding the rate of cell division: because a cell cannot proceed past a certain point in the division cycle until it has attained a certain standard size, the rate of cell division slows down and cell size remains more or less unchanged. (One unit of time is the cycle time observed when nutrients are in excess.)

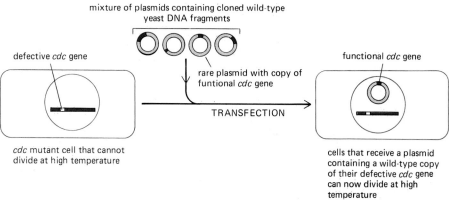

mixture of plasmids containing cloned wild-type
yeast DNA fragments

defective *cdc* gene

rare plasmid with copy of
funtional *cdc* gene

TRANSFECTION

functional *cdc* gene

cdc mutant cell that cannot
divide at high temperature

cells that receive a plasmid
containing a wild-type copy
of their defective *cdc* gene
can now divide at high
temperature

Figure 13–21 The procedure used to isolate *cdc* genes from a DNA library. The rare DNA clone that carries the normal, wild-type copy of a particular *cdc* gene is readily selected because the plasmid that carries it will enable the corresponding mutant cell to grow at high temperature. Both the *cdc*2 gene of fission yeasts and the *cdc*28 gene of budding yeasts were originally isolated in this way. By repeating this procedure with cDNA clones from a human DNA library in an appropriate plasmid, human genes capable of substituting for some of the yeast *cdc* genes have been isolated.

yeast and the *cdc*2 gene of the fission yeast, and some remarkable correspondences have thereby been discovered between the two types of yeasts and between yeasts and vertebrates with respect to the control of Start and of mitosis. The *cdc*2 and *cdc*28 genes of the two yeasts are homologous both in sequence and in function: mutant fission yeast cells defective in the *cdc*2 gene function can be rescued by introducing the *cdc*28 gene from budding yeasts. What is more, fission yeasts defective in *cdc*2 can be rescued by a cloned fragment of human DNA which turns out also to have sequence homology to *cdc*2/28, suggesting that this component of the cell cycle control machinery is common to yeasts, mammals, and presumably all eucaryotes.

Given the cloned *cdc*2/28 gene, it is relatively straightforward to isolate the protein that the gene encodes. It is a protein kinase, and by a whole array of structural and functional criteria it appears to be the yeast homologue of the kinase subunit of vertebrate MPF. Gene cloning reveals, moreover, that another fission-yeast *cdc* gene, *cdc*13, whose product interacts with that of *cdc*2, is closely homologous to the gene for cyclin (see p. 737). These findings indicate that MPF and cyclin are likely to have a universal importance in the eucaryotic cell cycle; and the dual role of *cdc*2/28 in yeasts, at the onset of M-phase and at Start, suggests that in vertebrates the corresponding subunit of MPF, or a related molecule, also might be somehow involved in controlling initiation of the division cycle during G_1. Indeed, from studies of fission yeast it seems that changes in the state of phosphorylation of this regulatory molecule may be the means by which cells adjust their readiness to initiate a division cycle according to environmental conditions.

Summary

Yeasts are single-celled eucaryotic organisms that are particularly amenable to genetic analyses. A large number of cell-division-cycle (cdc) mutants have been isolated in both budding and fission yeasts and the corresponding wild-type genes cloned. A standard cell size, in yeasts and in many other eucaryotic cells, is maintained in the face of variable nutrition by a control mechanism that prevents cells from passing a critical point called Start and embarking on a division cycle until they have reached a threshold size. Several of the key yeast cdc genes involved in this control have been identified and their nucleotide sequences determined. One of these (known as cdc2 in fission yeasts and cdc28 in budding yeasts) encodes a protein kinase that is homologous to MPF, another (cdc13) encodes the yeast homologue of cyclin.

Cell-Division Controls in Multicellular Organisms

In unicellular organisms such as yeasts, bacteria, or protozoa, there is strong selective pressure for each individual cell to grow and divide as rapidly as possible. For this reason the rate of cell division is generally limited only by the rate at which nutrients can be taken up from the medium and converted into cellular materials. The cells of a multicellular animal, by contrast, are specialized members of a complex cellular society, and it is the survival of the organism that is paramount, not the survival or proliferation of any of its individual cells. For the multicellular organism to survive, some cells must refrain from dividing even when nutrients are plentiful. Yet when a need arises for new cells, as in the repair of damage, previously nondividing cells must be rapidly triggered to reenter the division cycle; and where there is continual wear and tear, the cell birth rate and the cell death rate must be kept in balance. There must, therefore, be complex cell-division controls over and above those that operate in a simple organism such as a yeast. This section is concerned with these "social controls" at the level of the individual cell. In Chapters 17 and 21 we shall discuss how they function in a multicellular context to maintain and renew the tissues of the body, and how they malfunction in cancer; and in Chapter 16 we shall see how an even more complex system of controls governs cell division during development.

Cells Delay Division by Pausing After Mitosis for a Variable Length of Time[16]

The 10^{13} cells in the human body divide at very different rates. Some, such as neurons and skeletal muscle cells, do not divide at all; others, such as liver cells, normally divide only once every year or two; while certain epithelial cells in the gut divide more than twice a day so as to provide for constant renewal of the gut lining (Figure 13–22). Most cells in vertebrates fall somewhere between these extremes: they can divide, but normally do so infrequently. Almost all the variation lies in the time cells spend between mitosis and S phase, with slowly dividing cells remaining arrested after mitosis for weeks or even years. By contrast, the time taken for a cell to progress from the beginning of S phase through mitosis is brief (typically 12 to 24 hours in mammals) and remarkably constant, irrespective of the interval from one division to the next.

The time cells spend in a nonproliferative, so-called G_0 state varies not only according to cell type but also according to circumstances. Sex hormones stim-

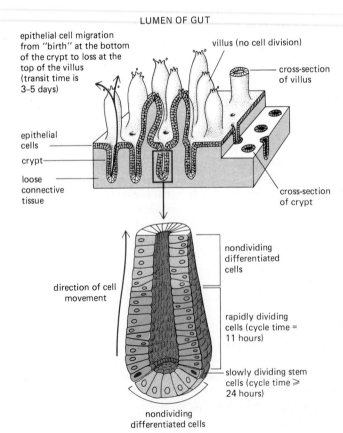

LUMEN OF GUT

epithelial cell migration from "birth" at the bottom of the crypt to loss at the top of the villus (transit time is 3–5 days)

villus (no cell division)

cross-section of villus

epithelial cells

crypt

loose connective tissue

cross-section of crypt

direction of cell movement

nondividing differentiated cells

rapidly dividing cells (cycle time = 11 hours)

slowly dividing stem cells (cycle time ⩾ 24 hours)

nondividing differentiated cells

Figure 13–22 Cell division and migration in the epithelium lining the small intestine of the mouse. All cell division is confined to the bottom portion of the tube-shaped epithelial infoldings known as *crypts*. Newly generated cells move upward to form the epithelium that covers the villi, where they function in the digestion and absorption of foodstuffs from the lumen of the gut. Most epithelial cells have a very short lifetime, being shed from the tip of a villus within 5 days after emerging from the crypt. However, a ring of about 20 slowly dividing "immortal" cells (*shown in dark color*) remain anchored near the base of each crypt. These "stem cells" will divide to give rise to two daughter cells: on average, one daughter remains in place as an undifferentiated stem cell, while the other usually migrates upward to differentiate and join the villus epithelium. (Adapted from C.S. Potten, R. Schofield, and L.G. Lajtha, *Biochim. Biophys. Acta* 560:281–299, 1979.)

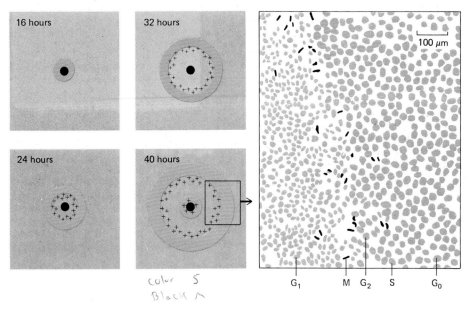

16 hours

32 hours

24 hours

40 hours

color S
Black M

100 μm

G₁   M   G₂   S   G₀

Figure 13–23 Cell proliferation in an epithelium in response to wounding. Lens epithelium was damaged with a needle; a specified number of hours later a pulse of ³H-thymidine was given to label cells in S phase (*color*), and the tissue was fixed and prepared for autoradiography. In the diagrams on the left, regions where cells are in S phase are shown in color, and cells in M phase are represented by black crosses; the central black spot marks the site of the wound. The stimulus for cell division propagates gradually outward from the wound, arousing cells from their quiescent G₀ state and causing an unusually extensive reaction to a relatively small wound. In the 40-hour specimen, cells far from the wound are entering S phase of their first division cycle, while cells close to the wound are entering S phase of their second division cycle. The drawing on the right, roughly corresponding to the boxed area in the diagram, is traced from a photograph of a 36-hour specimen stained to reveal the cell nuclei. (After C. Harding, J.R. Reddan, N.J. Unakar, and M. Bagchi, *Int. Rev. Cytol.* 31:215–300, 1971.)

ulate the cells in the wall of the human uterus to divide rapidly for a few days in each menstrual cycle to replace the tissues lost by menstruation; blood loss stimulates proliferation of blood cell precursors; acute liver damage provokes surviving liver cells to proliferate with a cycle time of only a day or two until the loss is made good. Similarly, epithelial cells in the neighborhood of a wound are stimulated to divide so as to repair the injured epithelium (Figure 13–23).

Delicately adjusted and highly specific controls exist to govern the proliferation of each class of cells in the body according to need. But while the importance of these controls is obvious, their mechanisms are very hard to analyze in the complex context of the intact body. Detailed studies of the regulation of cell division are therefore usually done in cell culture, where the environment can be easily manipulated and the cells can be continuously observed.

Like Yeast Cells, Animal Cells Halt at a Critical Point in G₁ When Conditions Are Unfavorable for Growth: The Restriction Point[17]

Most cell-cycle studies in culture use stably established cell lines (see p. 162), which will proliferate indefinitely. Even though such cell lines are variants specially selected for survival in culture, many of them—the so-called *nontransformed* cell lines—are widely used as models for the proliferative behavior of normal somatic cells.

Cells of fibroblast cell lines (such as the various types of mouse-derived 3T3 cells) typically proliferate most rapidly when settled on the surface of a culture dish, not too closely crowded together, and bathed in a medium that is rich in nutrients and includes *serum*—the fluid obtained by allowing blood to clot and discarding the insoluble clot and the blood cells. If the supply of some essential nutrient such as an amino acid is reduced, or if an inhibitor of protein synthesis is added to the medium, the cells behave in much the same way as the nutritionally deprived yeasts described earlier: the average duration of G₁ phase is prolonged, but the duration of the rest of the cycle is scarcely affected. Once a cell has passed out of G₁, it is committed to completing the S, G₂, and M phases regardless of environmental conditions. The point of no return occurs late in G₁ and is often called the **restriction point (R)** because it is here that cells pause when external conditions restrict their progress through the cycle. The restriction point corresponds to the Start point of the yeast cell cycle; as in yeast, it may in part represent the operation of a device for regulating cell size. However, it has a more complex significance for higher eucaryotic cells than for yeasts, and it is possible that there may be several slightly different restriction points in G₁ corresponding to the multiple controls of cell proliferation.

The Cycle Times of Proliferating Cells Seem to Depend on a Probabilistic Event[18]

Individual cells that are freely dividing in culture can be continuously observed by time-lapse cinematography. Such studies reveal that even cells that are genetically identical have widely variable cycle times (Figure 13–24). A quantitative analysis suggests that the time from one division to the next includes a randomly variable component, with the random variation occurring chiefly in the duration of the G_1 phase. It seems that, as cells approach the restriction point in G_1 (Figure 13–25), they must "wait" to embark on the rest of the cycle and that each waiting cell has a roughly constant probability per unit time of passing R. Thus, like a population of atoms undergoing radioactive decay, if half of the cells pass R within the first 3 hours, one-half of the remaining cells will pass R within the next 3 hours, one-half of the remainder after that will pass R within the next 3 hours, and so on. A possible mechanism that might explain such behavior was suggested previously when we discussed the production of the S-phase activator (see p. 733). But regardless of its cause, the random variability that is observed in the length of the cell-division cycle means that initially synchronous cell populations will drift out of synchrony within a few cycles. This is awkward for cell biologists but may be advantageous for the multicellular organism—otherwise, large clones of cells might pass through mitosis simultaneously; and since cells tend to round up and lose their attachments during mitosis, this could seriously disrupt the tissue they belong to.

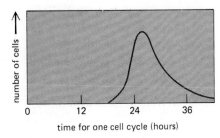

Figure 13–24 The variation in cell-cycle times typically observed for a homogeneous population of cells growing in tissue culture. Such data are obtained by observing individual cells in the microscope and directly measuring the time between their successive divisions.

Different Cell Types Require Different Growth Factors in Order to Divide[19,20]

The conditions that must be satisfied before a cell will grow and divide are considerably more complex for an animal cell than for a yeast. If vertebrate cells in a standard artificial culture medium are completely deprived of serum, they normally will not pass the restriction point, even though all the obvious nutrients are present; and they will halt their growth as well as their progress through the chromosome cycle. Painstaking analyses have revealed that the essential components of serum are highly specific proteins, mostly present in very low concentrations (on the order of 10^{-9} to 10^{-11} M). Different types of cells require different sets of these proteins. Some of the proteins in serum are directly and specifically involved in stimulating cell division and are called **growth factors.** One example is **platelet-derived growth factor,** or **PDGF.** The path to its isolation began with the observation that cultured fibroblasts proliferate when provided with serum but not when provided with plasma—the liquid component of blood, prepared by removing the blood cells without allowing clotting to occur. When blood clots, the platelets (see p. 974) are triggered to release the contents of their secretory vesicles; and among the products released (along with factors that promote clotting) is PDGF. It is mainly the PDGF in the serum that enables fibroblasts in culture to divide. PDGF presumably has the same effect in the body, where it is thought to stimulate connective tissue cells and smooth muscle cells to divide at the site of a wound to help repair the damage (Figure 13–26). The cells that respond to PDGF have specific receptors for PDGF—and for certain other growth factors—in their plasma membrane. Other cell types have other sets of receptors, mediating responses to other sets of growth factors (Table 13–1), several of which are also present in serum.

Figure 13–25 Cartoon indicating that rapidly dividing mammalian cells in culture are delayed, for a variable length of time, at a point late in G_1, which may be the same as the restriction point (R), at which they will halt indefinitely if protein synthesis is inhibited. From continuous observation of cells under a microscope, it seems that genetically identical cells, including the two daughter cells that result from a single division, are often delayed for very different amounts of time before passing R, suggesting that the delay is due to some random process (see text).

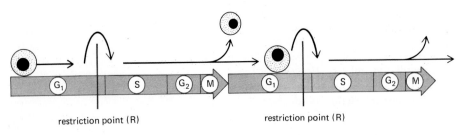

restriction point (R)   restriction point (R)

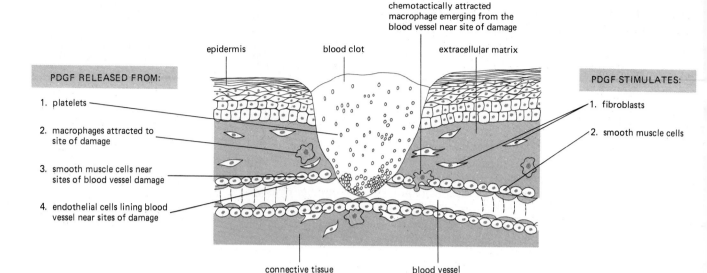

PDGF RELEASED FROM:

1. platelets

2. macrophages attracted to site of damage

3. smooth muscle cells near sites of blood vessel damage

4. endothelial cells lining blood vessel near sites of damage

chemotactically attracted macrophage emerging from the blood vessel near site of damage

epidermis   blood clot   extracellular matrix

PDGF STIMULATES:

1. fibroblasts

2. smooth muscle cells

connective tissue   blood vessel

Figure 13–26   The proposed role of platelet-derived growth factor (PDGF) in the healing of a wound. PDGF is secreted at a wound by platelets and macrophages and possibly by the endothelial cells and smooth muscle cells of injured blood vessels. It stimulates proliferation of both fibroblasts and smooth muscle cells, stimulates the fibroblasts to produce more extracellular matrix, and attracts fibroblasts and macrophages chemotactically. Healing is a complex process, and many other factors besides PDGF are also involved.

Table 13–1   Some Growth Factors and Their Actions

Factor	Composition	Representative Activities
Platelet-derived growth factor (PDGF)	AA, AB, or BB A chain = 125 aa B chain = 160 aa	Stimulates proliferation of connective tissue cells (see p. 746) and neuroglial cells (see p. 910)
Epidermal growth factor (EGF)	53 aa	Stimulates proliferation of many cell types (see p. 706)
Insulinlike growth factor I (IGF-I)	70aa ⎫	Collaborate with PDGF and EGF; stimulate proliferation of fat cells (see p. 986) and connective tissue cells
Insulinlike growth factor II (IGF-II)	73aa ⎭	
Transforming growth factor β (TGF-β)	Two chains, each 112 aa	Potentiates or inhibits response of most cells to other growth factors, depending on the cell type; regulates differentiation of some cell types (see pp. 987 and 894)
Fibroblast growth factor (FGF)	Acidic: 140 aa Basic: 146 aa	Stimulates proliferation of many cell types, including fibroblasts, endothelial cells (see p. 966), and myoblasts (see p. 984); induces mesoderm in *Xenopus* embryo (see p. 894)
Interleukin-2 (IL-2)	153 aa	Stimulates proliferation of T lymphocytes (see p. 1046)
Nerve growth factor (NGF)	Two chains, each 118 aa	Promotes axon growth and survival of sympathetic and some sensory and CNS neurons (see p. 1118)
Hemopoietic cell growth factors (IL-3, GM-CSF, M-CSF, G-CSF, erythropoietin)	See Table 17–2, p. 981	

Because they are secreted in small amounts, growth factors have been difficult to isolate. The difficulty is aggravated by the complexity of growth factor actions: most cell types probably depend on a specific *combination* of growth factors rather than a single specific growth factor. However, although relatively few distinct growth factors have so far been characterized—fewer than 30—many of these have turned up repeatedly in different contexts, receiving, at first, different names that were only later recognized to be aliases for a single molecule. This suggests that a fairly small number of growth factors may serve, in different combinations, to regulate selectively the proliferation of each of the many types of cells in a higher animal; and it is becoming clear that the same factors also act in certain contexts as regulators of other processes and especially of cell differentiation. Some growth factors are present in the circulation, but most act as local chemical mediators. The class of local chemical mediators probably includes a large number of as yet poorly understood factors that help to regulate cell division and differentiation during development (see p. 894). In addition to growth factors that stimulate cell division, there are counterbalancing factors that inhibit it, although these have for the most part been less well defined.

Neighboring Cells Compete for Growth Factors[20,21]

Through growth factors such as PDGF, one class of cells may control the proliferation of another class of cells. But it is also important that cells of the same class in a tissue interact with one another and adjust their rates of proliferation to maintain an appropriate population density. Social controls of this sort are seen clearly during responses to injury. When an epithelium is damaged, for example, the cells at the margin of the wound are stimulated to divide (see Figure 13–23) and to crawl over the denuded surface until it is once again fully covered; at this point the rapid proliferation and the movement stop. Similar phenomena are seen with dissociated cells in cell culture. Epithelial cells or fibroblasts plated on a dish in the presence of serum will adhere to the surface, spread out, and divide until a *confluent monolayer* is formed in which neighboring cells touch one another. At this point normal cells stop dividing—a phenomenon known as *density-dependent inhibition of cell division*. If such a monolayer is "wounded" with a needle so as to create a cell-free strip on the dish, the cells at the edges of the strip spread into the empty space and divide (Figure 13–27).

Such phenomena were originally described in terms of "contact inhibition" of cell division, but this may be misleading. The cell population density at which cell proliferation ceases in the confluent monolayer increases with increasing concentration of growth factors in the medium. Moreover, when culture medium is made to flow over the surface of a dish that is partly covered with cells and partly bare, it is found that cells bathed by medium that has just bathed other cells divide much more slowly than cells bathed by medium that has just passed over a cell-free region. It seems that the medium that has been bathing cells is depleted of essential nutrients or of growth factors. This could perhaps have been predicted. PDGF, for example, is typically present in the medium at concentrations of about 10^{-10} M (about one molecule in a sphere of 3 μm diameter). A fibroblast has about 10^5 PDGF receptors, each with a very high affinity for the growth factor. Each cell has enough receptors, therefore, to bind all the PDGF molecules within a sphere of diameter ~150 μm. Moreover, much of the PDGF that binds to cell-surface receptors is thought to be rapidly internalized by receptor-mediated endocytosis and degraded (see p. 333). It is clear, therefore, that neighboring cells are in competition for minute quantities of growth factors. This type of competition could be important for cells in tissues as well as in culture, preventing them from proliferating beyond a certain population density.

3-14 Normal Animal Cells in Culture Stop Dividing When They Lose Anchorage to the Substratum[22]

Competition for growth factors and nutrients is not the only influence affecting the rate of cell division observed in cell culture. The shape of a cell as it spreads and crawls out over a substratum to occupy vacant space also strongly affects its

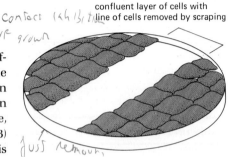

confluent layer of cells with line of cells removed by scraping

cells at margin spread and flatten, and increase their rate of protein synthesis

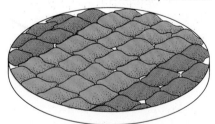

confluent layer of cells re-formed by cell division

Figure 13–27 Cells scattered on the surface of a culture dish normally proliferate until they touch one another, forming a *confluent monolayer*. The diagrams show the consequences of scraping away a strip of cells. The remaining cells at the margins of the vacant "wound" area flatten out and resume growth and division, which continue until the "wound" is "healed." Once the monolayer is again confluent, cell proliferation ceases almost entirely.

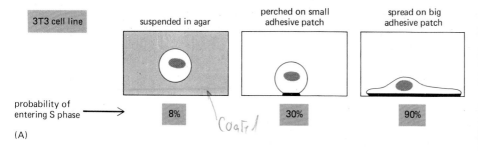

probability of entering S phase →

suspended in agar	perched on small adhesive patch	spread on big adhesive patch
8%	30%	90%

(coated)

(A)

(B)

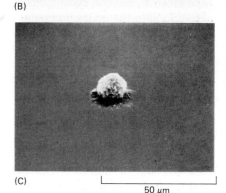

(C)

50 μm

ability to divide. When normal cells are cultured in suspension, unattached to any solid surface and therefore rounded up, they almost never divide—a phenomenon known as *anchorage dependence* of cell division. The relationship of cell spreading to proliferation can be demonstrated by culturing cells on substrata of varying stickiness or allowing them to settle on a nonsticky surface that is dotted with minute sticky patches, on which an individual cell may adhere but beyond which it cannot spread. The frequency with which a cell divides increases as the cell becomes more spread out. Perhaps well-spread cells can capture more molecules of growth factor and take up larger quantities of nutrients because of their larger surface area. But some types of cells (for example, 3T3 cells), although scarcely able to proliferate at all in suspension, will divide readily once they have touched down and formed a *focal contact*, even if the site of adhesion is a tiny patch on which there is no space for a cell to spread (Figure 13–28). Focal contacts are sites of anchorage of actin filaments inside the cell to extracellular matrix molecules on the outside (see p. 635), and these and other observations strongly hint that the control of cell division is somehow coupled to the organization of the cytoskeleton. Although the mechanism and function of this coupling are uncertain, anchorage dependence of cell division presumably helps to keep a tissue cohesive and to prevent the proliferation of cells that become detached from their proper surroundings.

Figure 13–28 The dependence of cell division on cell shape and anchorage. In the experiment shown here, cells are either held in suspension or allowed to settle on patches of an adhesive material (palladium) on a nonadhesive substratum; the patch diameter, which is variable, determines the extent to which an individual cell spreads and the probability that it will divide. ³H-thymidine is added to the culture medium, and after 1 or 2 days the culture is fixed and autoradiographed to discover the percentage of cells that have entered S phase. (A) Cells of the 3T3 cell line divide rarely when held rounded up in suspension, but adherence even to a very tiny patch— one that is too small to allow spreading—enables them to divide much more frequently. (B, C) Scanning electron micrographs showing a cell spread on a large patch as compared with a cell perched on a small patch. (Micrographs from C. O'Neill, P. Jordan, and G. Ireland, *Cell* 44:489–496, 1986. © Cell Press.)

Cell Division Involves Changes in Cell Contacts

The relationship between anchorage and cell division has several facets. On the one hand, most normal vertebrate cells require anchorage in order to pass the restriction point; on the other hand, cells that have passed this point no longer require anchorage to complete the division cycle, and they commonly loosen their attachments and round up as they pass through M phase. This cycle of attachment and detachment presumably allows rearrangement of both cell-cell and cell-matrix adhesive contacts, accommodating the daughter cells produced by cell division and then binding them securely into the tissue before they are permitted to begin the next division cycle.

The loosening of attachments seems to be an important part of the proliferative behavior of many types of cells. For example, PDGF causes fibroblasts to disassemble their focal contacts as an early part of the growth response (see p. 758). It is striking that the loss of growth control in cancer cells is almost always associated with a permanent reduction in cell adhesiveness, which is also detected as a loss of focal contacts in cancer cells grown in culture. As we shall discuss later, the relationship between cell division and attachment presents a complex puzzle, underlying an important gap in our understanding of the transformation that turns a normal cell into a cancer cell (see p. 759).

Cells That Are Not Licensed to Divide Come to Rest in a G_0 State[17,23]

When the preconditions for dividing are not met, a healthy cell will almost always be delayed in the G_1 phase of the cell cycle. When circumstances change so as to favor cell division, the cell resumes its progress through the cycle. A cell that has been deprived of the growth factors in serum, for example, will resume cycling when serum is put back in the medium. The lag from the time of addition of serum to the onset of S phase, however, is almost always long—generally several hours longer than the total time spent in G_1 by a normally proliferating cell. Growth-factor deprivation has driven the cell into a nonproliferative and profoundly altered state, in which it cannot pass the restriction point. Recovery from this state is a complex process that takes a long time and consists of a series of distinct stages, characterized by different sensitivities to growth factors.

The relationship between serum deprivation and the cell-division cycle has been clarified by studies on 3T3 cells in culture. A restriction point can be identified in the cell cycle 3.5 hours after the completion of mitosis. As little as an hour of serum deprivation (or of exposure to an inhibitor of protein synthesis) before this point halts the cell in G_1 and causes an all-or-none delay of 8 hours in the time taken to resume cycling after serum is restored to the medium. Similar deprivation after the restriction point does not cause this delay in the current division cycle, but it does cause a delay in progress through G_1 in the subsequent cycle. These observations suggest the following simple picture. In the proliferative state, a cell contains a set of molecules that license it to pass the restriction point; once the cell has passed this point, the molecules, although normally still present, are not necessary for completion of the S, G_2, and M phases, which proceed automatically. These "division-license" molecules are destroyed rapidly during a period of serum deprivation but take much longer to resynthesize when serum is restored. The destruction may occur in any phase of the cycle, but the consequences will not be manifest until the cell reaches the restriction point. In this view, the state of the cell is defined by two independent parameters: (1) the cell's position in the chromosome cycle and (2) the presence or absence of the division-license molecules, determining whether it is in the proliferative or the nonproliferative mode. A cell that is deprived of its license to divide will be unable to pass the restriction point and will come to a complete halt there; it is then said to be arrested in a **quiescent** or **G_0 state.**

Studies in culture emphasize the role of environmental factors in reversibly controlling the choice between proliferation and quiescence. In multicellular organisms, however, many types of cells enter G_0 as a result of terminal differentiation and lose the ability to divide, regardless of external stimuli. The production of such cells is commonly regulated by special control strategies involving *stem cells*, which will be discussed in Chapter 17.

The Chromosome Cycle Can Be Uncoupled from Cell Growth[24]

A G_0 cell, besides being quiescent with respect to the chromosome cycle, generally has an altered balance of protein synthesis and degradation compared with a proliferating cell: whereas a G_1 cell in the proliferative mode grows, a G_0 cell maintains a steady size. G_0 cells usually contain fewer ribosomes and less RNA than the corresponding cycling G_1 cells, and they typically synthesize protein at less than half the G_1 rate. When a G_0 cell is stimulated to proliferate by growth factors, changes in the rate of protein synthesis generally go hand in hand with effects on the chromosome cycle: as in yeasts, cell growth and division are coordinated so that a normal cell size is maintained.

The coupling between protein synthesis and the chromosome cycle is not rigid, however. With suitable combinations of protein synthesis inhibitors and growth factors, it is possible to depress protein synthesis in cultured cells without delaying progress through the cell cycle or, conversely, to stimulate protein synthesis without stimulating cell division. Moreover, differently specialized cells vary enormously in their ratio of cytoplasm to DNA, and some G_0 cells, such as neurons, can grow almost without limit without DNA replication (Figure 13–29).

13-17 The Probability of Entering G_0 Normally Increases with the Number of Times That a Cell Divides: Cell Senescence[25]

Most normal cells in the body of a mammal or a bird show a striking reluctance to continue proliferating forever. This distinguishes them from established cultured cell lines, such as 3T3 cells, which are thought to have undergone some genetic change that makes them "immortal." Fibroblasts taken from a normal

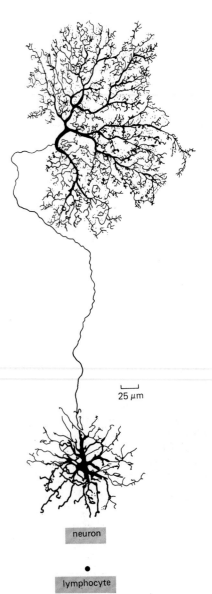

25 µm

neuron

lymphocyte

Figure 13–29 Comparison of the size of a mammalian neuron (from the retina) and a lymphocyte; both contain the same amount of DNA. A neuron grows progressively larger during its development while remaining in a G_0 state. During this time the ratio of cytoplasm to DNA increases enormously (by a factor of more than 10^5 for some neurons). (From B.B. Boycott in Essays on the Nervous System [R. Bellairs and E.G. Gray, eds.]. Oxford, U.K.: Clarendon Press, 1974.)

human fetus, for example, will go through only about 50 population doublings when cultured in a standard growth medium; toward the end of this time, proliferation slows down and finally halts, and the cells, after spending a period in a quiescent state, all die. Similar cells taken from a 40-year-old stop dividing after about 40 doublings, while cells from an 80-year-old stop after about 30 doublings. Fibroblasts from animals with a shorter life-span cease dividing after a smaller number of division cycles in culture. Because of the correspondence with aging of the body as a whole, this phenomenon has been called **cell senescence.**

Cell senescence is a puzzling phenomenon. Short programmed sequences of cell divisions terminating in differentiation are a familiar feature of embryonic development (see p. 907), but it is hard to imagine how a cell could keep a long-term count of its division cycles and halt after completing 50. According to one theory, cell senescence is the result of a catastrophic accumulation of self-propagating errors in a cell's biosynthetic machinery that is unimportant under the conditions of life in the wild, where most animals die from other causes long before a significant number of their cells become senescent. In this view, cell senescence merely reflects imperfections in cell physiology that are to be expected when there is little selection pressure to eliminate them. But further assumptions are then required to explain how germ-line cells and "immortal" cultured cell lines, and even ordinary somatic cells under special conditions (described below), are able to proliferate indefinitely. An alternative hypothesis is that cell senescence is the result of a mechanism that has evolved to protect us from cancer by limiting the growth of tumors. If this is the case, however, the protection seems inadequate, since 50 division cycles are enough to produce a very sizable tumor. A related suggestion is that cell senescence in the highly artificial conditions of cell culture reflects a tendency for cell proliferation in the body to slow down gradually with age and that this type of cell behavior has evolved as a means of stabilizing adult body size.

Whatever the function of cell senescence, there is good evidence that the process is strongly influenced by factors in the extracellular medium. Epidermal cells from the skin of a human baby, for example, will become senescent after roughly 50 division cycles in a medium that lacks epidermal growth factor (EGF), but only after 150 cycles when EGF is present. Conversely, "immortal" 3T3 cells have been shown to display senescent behavior when partially starved of growth factors. Moreover, cells from normal mouse embryos can be kept dividing indefinitely without any signs of senescence if they are placed in a chemically defined medium that contains a selection of purified growth factors instead of serum; addition of serum actually halts the proliferation. This suggests that senescence is due partly to factors in serum that inhibit cell proliferation, outweighing the effects of growth factors that promote it.

Some human mutants, such as sufferers from *Werner's syndrome*, grow old prematurely. Fibroblasts taken from patients with this disease, who typically die before the age of 50, stop proliferating after an abnormally small number of division cycles in culture. It is interesting that these fibroblasts are abnormally unresponsive to the growth factors PDGF and fibroblast growth factor (FGF), although they are able to proliferate vigorously in response to other growth factors.

Although senescence occurs at a predictable time for a cell population under specified conditions, it is not strictly programmed at the level of the individual cell. In a clone of apparently identical normal fibroblasts monitored under standard culture conditions, some cells divide many times, others only a few times. Individual cells seem to stop dividing as a result of a random transition that occurs with an increasing probability per unit time in each successive cell generation until there are no proliferating cells left in the population (Figure 13–30).

These studies on cell clones indicate that cells that seem otherwise identical are heterogeneous in their ability to divide. Perhaps the final transition of a senescent cell to a nondividing state is merely the culmination of a series of randomly timed downward steps in the expression of genes that regulate the cell's readiness to pass the restriction point. The molecular nature of these controls is beginning to emerge from studies of cancer cells, which are, among other things, "immortal" and not subject to senescence.

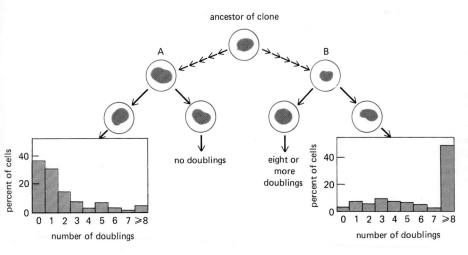

ancestor of clone

A

B

no doublings

eight or more doublings

percent of cells

40

20

0

0 1 2 3 4 5 6 7 ≥8

number of doublings

percent of cells

40

20

0

0 1 2 3 4 5 6 7 ≥8

number of doublings

Figure 13–30 Evidence for cell variation in a heritable ability to divide. Individual cells in a clone, even though genetically identical, vary in the number of division cycles they will undergo. Here, different pairs of sister cells from the same clone have been studied, and histograms have been drawn to show the numbers of cells that divide a given number of times. If one sister fails to divide at all, the other sister usually does likewise or divides only a few times (*left side*); and if one sister undergoes eight or more doublings, the other also usually undergoes eight or more doublings (*right side*). This shows that there are heritable differences between genetically identical cells in the numbers of division cycles of which they are capable. The different heritable states are not perfectly stable, however, so sister cells sometimes behave differently. Further studies show that, as the cell population ages, the cells undergo random transitions toward states of reduced division capability. (Data from J.R. Smith and R.G. Whitney, *Science* 207:82–84, 1980.)

Summary

Cell division in multicellular organisms depends on complex social controls, and the proliferation of different cell types is governed by different combinations of protein growth factors. These act at very low concentrations, and most serve as local chemical mediators to help regulate cell population densities. In addition, most normal cells are unable to divide unless they are anchored to the extracellular matrix. Cells that are starved of growth factors or deprived of matrix adhesion come to a halt after mitosis, entering a special quiescent state, G_0, from which they take several hours to recover when reexposed to growth factors. Once a cell has left G_0 and passed the restriction point in G_1, it will complete S, G_2, and M promptly, irrespective of growth factors or adhesion. In a proliferating cell population, passage through the restriction point is an all-or-none event that can be characterized by a transition probability, like a radioactive decay. In addition to immediate controls of cell proliferation, there are long-term controls that cause the normal somatic cells of mammals to become senescent and cease dividing after a limited number of division cycles in culture.

Genes for the Social Control of Cell Division[26]

As we have seen for yeasts, genetics provides a powerful approach to the problem of determining the molecular basis of the controls on cell division, provided that there are ways of selecting for mutations of the relevant genes. In multicellular animals, mutations in the genes that are involved specifically in the social controls on cell division, which we shall call **social control genes,** are selected all too easily. A cell that undergoes a mutation or set of mutations that disrupts the social restraints on division will divide without regard to the needs of the organism as a whole, and its progeny will become apparent as a tumor.

Cancers, by definition, are *malignant* tumors; that is, the tumor cells not only divide in an ill-controlled way, they also invade and colonize other tissues of the body to create widespread secondary tumors, or *metastases*. To generate a cancer, a cell must first undergo a number of mutations to escape the multiple controls on cell division and then accumulate further changes to become endowed with the capacity for invasion and metastasis. These aspects of cancer are discussed in Chapter 21. Here we shall not attempt to explain cancer but rather to see what can be learned from cancer cells about the genes that normally control cell division.

Cell Transformation in Culture Provides an Assay for Genes Involved in the Social Control of Cell Division[27]

Given a clone of tumor cells that are presumed to have originated by mutation, how does one identify the mutant gene or genes? Classical genetic mapping is not feasible because the cells do not reproduce sexually. A more direct approach

Table 13–2 Some Changes Commonly Observed When a Normal Tissue-Culture Cell Is Transformed by a Tumor Virus

1. Plasma-membrane-related abnormalities
   A. Enhanced transport of metabolites
   B. Excessive blebbing of plasma membrane
   C. Increased mobility of plasma membrane proteins

2. Adherence abnormalities
   A. Diminished adhesion to surfaces; therefore able to maintain a rounded morphology
   B. Failure of actin filaments to organize into stress fibers
   C. Reduced external coat of fibronectin
   D. High production of plasminogen activator, causing increased extracellular proteolysis

3. Growth and division abnormalities
   A. Growth to an unusually high cell density
   B. Lowered requirement for growth factors
   C. Less "anchorage dependence" (can grow even without attachment to solid surface)
   D. "Immortal" (can continue proliferating indefinitely)
   E. Can cause tumors when injected into susceptible animals

is to take genetic material from the tumor cells and search for fragments of it that will, when introduced into normal cells, cause these cells to behave like tumor cells. Techniques for achieving this feat were first devised in the late 1970s, but their development depended on earlier studies of a very similar process that occurs naturally.

Certain types of tumors are caused by viruses. The viruses shed from these tumors can infect normal cells and, as a result of the introduction of the RNA or DNA carried by the virus, *transform* them into tumor cells. The first **tumor virus** to be discovered causes connective tissue tumors, or *sarcomas*, in birds; the infectious agent—the *Rous sarcoma virus*—has been an important object of study ever since, along with a variety of more recently discovered tumor viruses.

The nature of tumorigenic **cell transformation** is most easily observed in culture. A few days after tumor viruses have been added to a culture of normal cells, small colonies of abnormally proliferating cells appear. Each such colony is a clone derived from a single cell that has been infected with the virus and has stably incorporated the viral genetic material. Released from the social controls on cell division, the transformed cells outgrow normal ones in the culture dish just as in the body and are therefore usually easy to select. The transformed cells commonly show a complex syndrome of abnormalities (summarized in Table 13–2): they tend not to be constrained by density-dependent inhibition of cell division (see p. 748) but pile up in layer upon layer as they proliferate (Figure 13–31); they often do not depend on anchorage for growth and are capable of dividing even when held in suspension; they have an altered shape and adhere poorly to the substratum and to other cells, maintaining a rounded appearance reminiscent of a normal cell in mitosis; they may be able to proliferate even in the absence of growth factors; they are immortal and do not undergo senescence in culture; and when they are injected back into a suitable host animal, they can give rise to tumors.

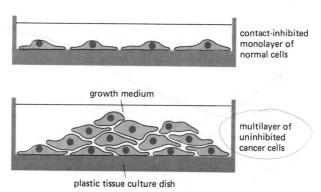

Figure 13–31 Cancer cells, unlike most normal cells, usually continue to grow and pile up on top of one another after they have formed a confluent monolayer.

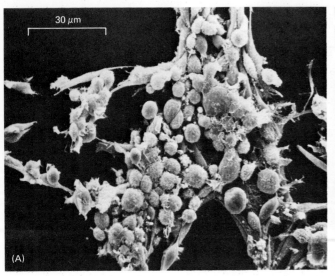

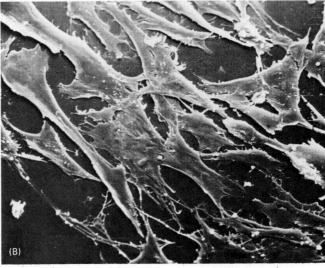

30 μm

(A)

(B)

₂₀ Tumor Viruses Provide a Source of Ready-cloned Oncogenes[28]

A tumor virus subverts the normal controls on cell division by causing a permanent change in the genetic constitution of its host cell such that the cell begins to make a protein that overrides the normal controls. Such viruses therefore provide ways to identify the mechanisms that are normally responsible for cell division controls. Thus far the most important revelations have come from studies of RNA tumor viruses, also called **retroviruses.** After a retrovirus infects a cell, its RNA is copied into DNA by reverse transcription and the DNA is then inserted into the host genome. Figure 5–75 (p. 255) outlines the life cycle of a retrovirus and shows how its genome undergoes reverse transcription, integration into host DNA, and exit from and entry into host cells.

When a retrovirus transforms a normal cell into a tumor cell, the misbehavior is often brought about by a gene that is carried by the virus but is not necessary for the virus's own survival or reproduction. This was first demonstrated by the discovery of mutant Rous sarcoma viruses that multiply normally but no longer transform their host cells. Some of these nontransforming mutants were found to have deleted all or part of a gene coding for a protein of molecular weight 60,000. Other mutations in this gene can make the transforming effect of the virus temperature-sensitive: infected cells show a transformed phenotype at 34°C, but when the temperature is raised to 39°C, they return promptly (within a matter of hours) to the normal phenotype (Figure 13–32). Apparently this specific gene in the tumor virus is responsible for cell transformation (thereby bringing the virus that contains it to our attention) but is superfluous baggage from the point of view of the virus's own propagation.

The transforming gene of the Rous sarcoma virus identified by these experiments is called the **v-src gene.** It is classified as an **oncogene** (from the Greek onkos, a mass or tumor) because when it is introduced into a normal cell, it can transform it into a tumor cell. What is the origin of this gene, and what is its normal function? When a radioactive DNA copy of the viral src gene sequence was used as a probe to search for related sequences by DNA-DNA hybridization (see p. 188), it was found that the genomes of normal vertebrate cells contain a sequence that is closely similar, but not identical, to the src gene of the Rous sarcoma virus. This normal cellular counterpart of the viral src gene is called **c-src,** and it is classified as a **proto-oncogene.** Evidently the viral oncogene has been picked up from the genome of a previous host cell but has undergone mutation in the process. One suspects that the proto-oncogene is a normal social control gene that the retrovirus has, in effect, cloned for us. A large number of other oncogenes have now been identified and analyzed in similar ways, and each has led to the discovery of a corresponding proto-oncogene.

Figure 13–32 Scanning electron micrographs of cells in culture infected with a form of the Rous sarcoma virus that carries a temperature-sensitive mutation in the gene responsible for transformation (the v-src oncogene). (A) The cells are transformed and have an abnormal rounded shape at low temperature (34°C), where the oncogene product is functional. (B) The same cells adhere strongly to the culture dish and thereby regain their normal flattened appearance when the oncogene product is inactivated by a shift to higher temperature (39°C). (Courtesy of G. Steven Martin.)

13-21 Tumors Caused in Different Ways Contain Mutations in the Same Proto-oncogenes[26,29]

Tumors often result from mutations occurring spontaneously or in response to chemical carcinogens or radiation rather than from viral infection. DNA can be purified from these tumor cells and assayed for the presence of oncogenes by introducing it into nontransformed cultured cells. 3T3 cells are often used for this assay because they divide indefinitely in culture and contain mutations that make them easy to transform with a single added oncogene. Using recombinant DNA techniques, the tumor-derived oncogene responsible for such a transformation can be identified, cloned, and sequenced (Figure 13–33). Remarkably, in most cases where this has been done, the oncogene has turned out to be a mutant form of one of the same proto-oncogenes identified by the retrovirus approach, although some new oncogenes have been discovered as well.

Related techniques reveal that transformation can also be caused by over-production of certain normal gene products. Tumors often contain an unchanged proto-oncogene that is overexpressed, either because it is present in an abnormally large number of copies or because a chromosomal rearrangement has brought it under the control of an inappropriate promoter, as discussed in Chapter 21.

More than 50 proto-oncogenes have thus far been identified (see pp. 1207 and 1209). These may represent a substantial sample of the proto-oncogenes in the normal cell. It is likely, however, that many social control genes have yet to be identified. The fibroblastlike 3T3 cells commonly used in the transformation assay may fail to be transformed by an oncogene that helped to transform some other differentiated cell type that gave rise to a tumor. Moreover, the cell-transformation assay is capable of detecting only *dominant* mutations of social control genes—those mutations that deregulate cell proliferation even when normal gene copies are present in the cell. Recessive mutations in social control genes—resulting from loss of function—may be the most common type in cancer cells but will not be detected by the assay. Genes whose products normally help stimulate cell division can therefore be identified more readily with current techniques than those whose products normally help inhibit it. Nevertheless, there is evidence that genes with inhibitory effects on cell division exist and that recessive mutations in them are a common cause of cell transformation and cancer. For example, when transformed cells are fused with nontransformed cells, the resulting hybrid cells very often appear nontransformed, implying that normal cell division control has been restored by restoring a protein that was missing from the transformed cell. Bearing in mind, therefore, that many important social control genes remain to be discovered, we shall now consider what the known ones do.

Figure 13–33 A procedure by which human oncogenes can be identified and cloned. Oncogenes present in a sample of DNA taken from a human tumor are detected by their ability to transform mouse 3T3 cells. Transformed 3T3 cells proliferate in an unrestrained way and can be recognized by the colonies they form on a culture dish. Repetitive DNA sequences in the *Alu* family (see p. 608) are scattered throughout the human genome. They therefore provide a convenient marker that can be used to identify the human DNA in a nonhuman cell. Used as a DNA probe, the *Alu* sequences allow the human oncogene that has transformed the 3T3 cell to be cloned. The cloned DNA that contains the oncogene will transform 3T3 cells with very high efficiency when retested in the same assay.

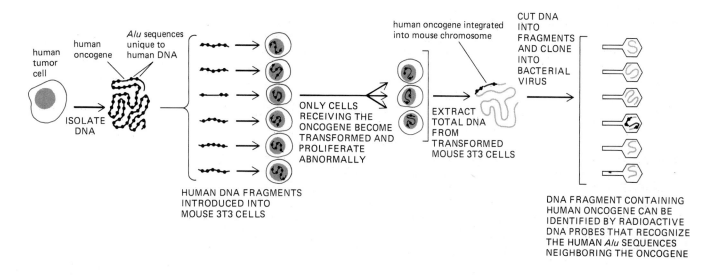

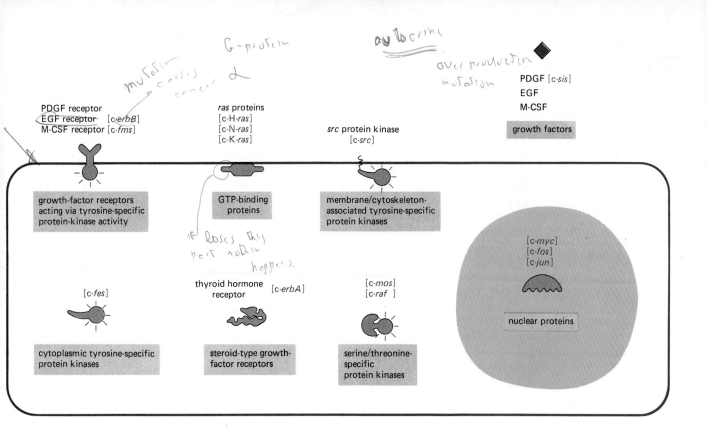

Figure 13–34 The activities and cellular locations of the main classes of known proto-oncogenes. The names of some representative proto-oncogenes in each class are shown in color. See also Figure 21–27, p. 1211.

Some Proto-oncogenes Code for Growth Factors or for Growth-Factor Receptors[26,30]

Once an oncogene has been cloned and sequenced, it is often possible to gain clues to the function of the corresponding social control gene (proto-oncogene) by looking for homologies between its sequence and the sequences of genes already known. In this way one proto-oncogene, known as c-sis, was discovered to be the gene that encodes a functionally active subunit of the growth factor PDGF. A cell that contains the corresponding v-sis oncogene manufactures the PDGF subunit constantly and inappropriately, and this protein, by binding to the cell's receptors for PDGF, constantly stimulates the cell to proliferate. At least three other proto-oncogenes, called c-erbB, c-fms, and c-erbA, code for receptors for growth factors or hormones: c-erbB encodes the receptor for epidermal growth factor (EGF, see p. 706); c-fms encodes the receptor for macrophage colony-stimulating factor (M-CSF)—a growth factor that stimulates proliferation of macrophage precursors (see p. 981); and c-erbA encodes the receptor for thyroid hormone (see p. 690). When mutated to oncogenes, these genes code for faulty receptors that behave as though a ligand were bound even when it is not, thereby stimulating the cell inappropriately (see p. 707).

Although the exact functions of most other proto-oncogenes are still unknown, one might expect many of them to code for proteins that are part of the signaling network inside the cell that enables growth factors to stimulate cell proliferation. We must now examine how far the known classes of proto-oncogenes, summarized in Figure 13–34, can account for the functions they might be expected to perform.

Some Proto-oncogenes Encode Intracellular Mediators Involved in Signaling Cells to Divide[26,31]

As discussed in Chapter 12 (p. 706), the receptors for many growth factors, including PDGF, are tyrosine-specific protein kinases that, when activated, phosphorylate themselves and various other proteins. One class of oncogenes encodes abnormal forms of such receptors; this class includes the altered EGF and M-CSF receptors just described (see Figure 13–34). Another small family of related proto-oncogenes, the c-*ras* genes (see Figure 13–34), codes for proteins that bind and hydrolyze GTP and may be distantly related to the G proteins involved in the transduction of many types of signals (see p. 705). Mutant *ras* genes that cause cell transformation are associated with raised concentrations or raised efficacy of the intracellular mediators inositol trisphosphate and diacylglycerol and make the cell hypersensitive to several growth factors that are believed to exert some of their effects by inducing production of inositol trisphosphate and diacylglycerol. Genes homologous to *ras* are present in budding yeasts, where they are involved in controlling the cell-division cycle according to the supply of nutrients in the medium.

In the normal response of an animal cell to growth-factor stimulation, many other intracellular changes occur—including changes in Ca^{2+}, pH, cyclic AMP, protein phosphorylation, gene transcription, mRNA processing and degradation, protein synthesis, and the cytoskeleton. While many of these occur within seconds, others take hours. Most of the proto-oncogene proteins that participate in this complex web of control systems can be classified only vaguely as yet. Some of them, like the growth-factor receptors mentioned above, are tyrosine-specific protein kinases that are associated with cell membranes. Others, like the *cdc*2/28 gene product in yeasts, are serine/threonine-specific protein kinases found in the cytoplasm. A third category comprises proteins that are located chiefly in the nucleus (see Figure 13–34); one of these, the c-*jun* protein, has been identified as the transcriptional regulator AP-1 (see p. 566) and combines with another member of the family, the c-*fos* protein, to form a DNA-binding complex.

Another protein in the nuclear category, corresponding to a proto-oncogene called c-*myc*, seems to be a marker of whether the cell is in a proliferative mode: in a cell that is rapidly cycling, the c-*myc* protein is present at a constant low level throughout the cycle, but it disappears when the cell enters a G_0 state and becomes quiescent. When growth factors are added to the medium bathing a quiescent cell, the concentration of c-*myc* protein rises steeply, reaching a peak within a few hours, and then falls to a lower nonzero level. In contrast to c-*myc*, the concentrations of the vast majority of other proteins in the cell change little between the proliferative and the quiescent state.

The Effects of Oncogenes on Cell-Division Control Are Closely Coupled with Effects on Cell Adhesion[32,33]

One of the most extensively studied proto-oncogenes is c-*src*, corresponding to the oncogene v-*src* of the Rous sarcoma virus. It belongs to a small family of homologous proto-oncogenes. It encodes the *src* protein, a tyrosine-specific protein kinase of molecular weight 60,000 (hence the alternative name p60[src]) that contains a covalently linked fatty acid that attaches it to the cytoplasmic side of the plasma membrane (see p. 417). The kinase is hyperactive in its oncogenic form (v-*src*), and the membrane attachment (Figure 13–35) is required for it to be able to cause cell transformation. Studies with antibodies indicate that the *src* protein becomes concentrated at *focal contacts*, where the cell is bound tightly to the substratum by a cell-matrix junction involving actin filaments on its intracellular side (see Figure 13–36). The *src* protein seems to be functionally involved with actin-membrane attachments, since activating a temperature-sensitive version of the v-*src* protein (by lowering the temperature) causes an immediate increase in membrane ruffling (an actin-mediated lamellipodial movement, see p. 638), as well as a general weakening of cell adhesions, including the disruption of focal contacts, causing the cell to round up (see Figure 13–30). At least two separate actions of the *src* protein have been implicated in the observed change in cell adhesion. First, the active v-*src* kinase phosphorylates a tyrosine residue on the cytoplasmic tail of the fibronectin receptor molecules in the cell (see p. 636). *In vitro* studies suggest that this phosphorylation reduces the affinity of the receptor for both talin (on the inside of the cell) and fibronectin (on the outside). In addition, cells trans-

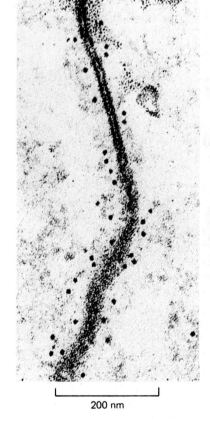

Figure 13–35 Electron micrograph demonstrating that the protein kinase synthesized by the v-*src* oncogene of Rous sarcoma virus is attached to the inner surface of the plasma membrane; the *src* protein synthesized by the c-*src* protein is thought to have a similar location but is harder to detect because it is normally present only in small quantities. The *src* protein has been localized in this preparation by reacting it with specific antibodies to which electron-dense ferritin particles are attached. (Courtesy of Ira Pastan; from M.C. Willingham, G. Jay, and I. Pastan, *Cell* 18:125–134, 1979. © Cell Press.)

200 nm

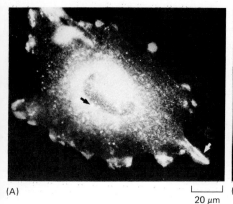
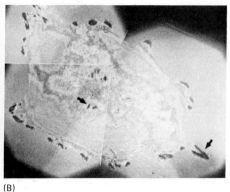

(A) (B)

20 μm

Figure 13–36 The *src* protein is present in many regions of the cell, but it appears especially concentrated at focal contacts and other sites of cell attachment to the extracellular matrix. (A) Immunofluorescence photograph showing the distribution of the *src* protein as revealed by the binding of *src*-specific antibodies. (B) A view of the same cell using an optical interference technique that makes sites of close adhesion of the cell to the substratum appear dark. The pattern of bright *src* protein spots in (A) matches the pattern of dark adhesion spots in (B) (see arrows). These pictures show the distribution of viral *src* protein in a cell transformed by Rous sarcoma virus; the *src* protein produced in a normal cell probably has a similar distribution but is harder to detect because it is present in smaller quantities. (From L.R. Rohrschneider, *Proc. Natl. Acad. Sci. USA* 77:3514–3518, 1980.)

formed by v-*src* secrete large amounts of a proteolytic enzyme called *plasminogen activator*. This enzyme is named for its ability to activate a second proteolytic enzyme, *plasmin*, by cleaving the precursor plasminogen; but plasminogen activator can also degrade other proteins directly, and, directly or indirectly, it evidently helps cells to dissolve their attachments (and to migrate through the extracellular matrix). When a monoclonal antibody against plasminogen activator is added to the culture medium, the cells become more adherent and tend to flatten down on the substratum. Thus the hyperactive v-*src* tyrosine-kinase activity appears to loosen cell adhesions in two separate ways: by phosphorylating the fibronectin receptor (and other transmembrane cell-matrix adhesion molecules in the integrin family—see p. 797), and by causing a protease to be secreted that destroys fibronectin (and other matrix molecules).

What might these findings mean for normal cell growth regulation? When a normal quiescent fibroblast is treated with PDGF, membrane ruffling is immediately induced and the cell's focal contacts change their structure within minutes: vinculin transiently disappears from focal contacts and the bundles of actin filaments that were anchored there are temporarily disrupted. Thus, in stimulating a quiescent cell to divide, PDGF induces many of the same types of changes as v-*src*. In fact, included among the immediate changes induced by PDGF is an increased phosphorylation of the c-*src* protein, which, by activating the *src* kinase activity, could explain the similarities directly (Figure 13–37). In this view the transformation of cells by v-*src* (and by many other oncogenes with similar effects) reflects an exaggeration of a normal mechanism of growth stimulation that involves loosening cell adhesions. The effects of oncogenes are dangerous for the organism because, unlike the PDGF stimulation, which is transitory and is soon turned off, proteins like v-*src* tend to drive cells out of G_0 permanently and thereby keep them in a proliferative state.

Figure 13–37 A speculative model of how rapid changes in cell adhesion might be mediated when a cell is stimulated to proliferate by PDGF. Binding of PDGF to its receptor leads (by an unspecified pathway) to phosphorylation of the c-*src* protein. This plasma-membrane-associated protein kinase is thereby activated and in turn phosphorylates tyrosine residues on neighboring transmembrane cell-adhesion proteins, including the fibronectin receptor. As a result, focal contacts and other sites of cell adhesion are partially disrupted and the associated actin filaments detach from the plasma membrane. The model is based partly on observations made on cells transformed by the Rous sarcoma virus, which carries a modified *src* protein (v-*src*) that is permanently active. The tyrosine protein kinases encoded by two other proto-oncogenes in the *src* family, c-*abl* and c-*yes*, may act in the same way as the c-*src* protein in the above pathway. However, many proteins are generally phosphorylated by such enzymes, and it is not certain which ones are crucial for the control of cell division. Some important targets may be present in only a few copies per cell—too few for detection by ordinary biochemical methods—and in different cell types the targets may be different. Above all, causal relationships are difficult to establish in a complex network of interacting components, where multiple factors may act in parallel and similar effects may be produced by different means.

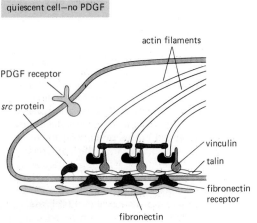

quiescent cell—no PDGF

actin filaments

PDGF receptor

src protein

vinculin

talin

fibronectin receptor

fibronectin

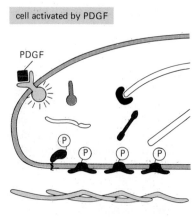

cell activated by PDGF

PDGF

P P P P

The Connection Between Cell Proliferation and Cell Adhesion Is Not Understood[33]

Our discussion of the control of normal vertebrate cell proliferation leads here to a paradox. On the one hand, it is clear that normal cells must form adhesions to the substratum (cell-matrix adhesions) in order to exit from G_0 and proliferate. This suggests that the transmembrane proteins that bind cells to the extracellular matrix (including the fibronectin receptor and other members of the integrin family) generate an intracellular signal that facilitates cell division when they are properly engaged. On the other hand, adherence by itself is not enough to trigger cell division: growth factors are also required. The paradox is that the growth factors seem to act, in part, by transiently weakening the adhesions on which normal cell proliferation depends (see Figure 13–37). This effect is reminiscent of a second observation: in many cases a brief exposure of normal growth-arrested cells to a protease such as trypsin, which causes the attached cells to lose their adherence to the culture dish and round up, has the side effect of triggering a single round of cell division. It seems that cell proliferation is transiently stimulated both by extracellular proteases that loosen cell-matrix adhesions directly, by digesting the external proteins responsible for adhesion, and by growth factors that loosen such adhesions indirectly, by acting on the focal contacts via intracellular mediators.

Studies on tumor cells reinforce the paradox. Most tumor cells, including those transformed by the well-studied oncogenes shown in Figure 13–34, differ from their normal counterparts in that they proliferate without requiring adhesion to a substratum. Since this *anchorage independence* enables the transformed cells to grow in new environments where normal cell-cell and cell-matrix attachments cannot be made (see p. 748), one might expect it to be an outcome of natural selection for cells that form tumors. But why do many tumor cells not merely divide without regard to anchorage but also fail to make firm attachments to the extracellular matrix even where the opportunity exists? A hint comes from observations of transformed cells that are forced to attach to the culture dish by artificial means. As noted above when fibroblasts from a chick embryo are transformed with v-*src*, they secrete large amounts of plasminogen activator, which loosens their attachments to the culture dish. If these cells are grown in the presence of an antibody that blocks the activity of this protease, they attach more firmly to the dish and, at the same time, become more obedient to the normal social controls on cell division: instead of piling up in multiple layers, they tend to stop dividing on reaching confluence. Thus, for these transformed cells, the formation of firm attachments to the extracellular matrix seems to inhibit growth.

The different growth requirements for normal and transformed cells and the seemingly contradictory effects of cell-matrix adhesion (Figure 13–38) must have a logical explanation. In some way the complex structure that forms at a focal contact between a cell and its substratum must play a central part in generating the intracellular signals that regulate cell division. The observations might be explained by supposing that normal triggering of cell division requires three steps: (1) anchoring of the cell to the matrix through cell-matrix attachments at which

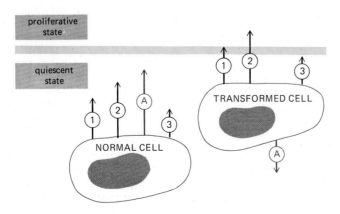

Figure 13–38 The general nature of the social control signals that act on normal and transformed cells. For both types of cells, different growth factors (here denoted by 1, 2, and 3) act in combination to "lift" the cell from G_0 to a proliferative state. Because the transformed cell is maintained at a position closer to the transition boundary (*colored line*), it can often be stimulated to proliferate by a single growth factor (or by an unusually low concentration of a mixture of growth factors). As discussed in the text, however, there is a marked difference in the effect of anchorage (denoted by A) on these two types of cells: whereas anchorage is required for normal cell proliferation, it tends to inhibit the proliferation of transformed cells.

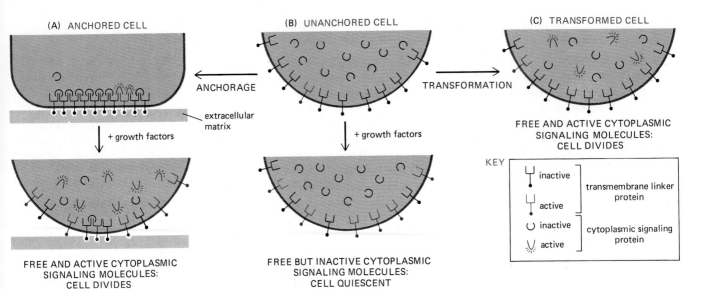

(A) ANCHORED CELL

(B) UNANCHORED CELL

(C) TRANSFORMED CELL

ANCHORAGE

TRANSFORMATION

extracellular matrix

FREE AND ACTIVE CYTOPLASMIC
SIGNALING MOLECULES:
CELL DIVIDES

+ growth factors

+ growth factors

KEY

inactive

transmembrane linker protein

active

inactive

cytoplasmic signaling protein

active

FREE AND ACTIVE CYTOPLASMIC
SIGNALING MOLECULES:
CELL DIVIDES

FREE BUT INACTIVE CYTOPLASMIC
SIGNALING MOLECULES:
CELL QUIESCENT

an ordered assembly of cytoskeletal proteins is created inside the cell (see p. 635); (2) activation of this assembly, typically by one or more growth factors, to release an intracellular signal for cell division; and (3) as a necessary part of the process of releasing the signal, partial disassembly of the cell-matrix attachments. In a transformed cell the requirement for step 1 would be bypassed, and disassembled intracellular components of the cell-matrix attachment structure would be necessary and sufficient to generate the signal to proliferate. A hypothetical model for such a growth control mechanism is presented in Figure 13–39.

Positional Signals and Cell-autonomous Programs Control Cell Division in the Growing Body[20,34]

Experiments conducted under the simplified artificial conditions of cell culture have told us most of what we know about the molecular mechanisms controlling cell growth and division in multicellular animals. Yet this work has so far revealed only some of the basic nuts and bolts of the much more complex social control systems that must operate in the intact body to regulate the proliferation of each group of cells according to its position and its developmental history.

As suggested by our discussion of cell senescence, cells often appear to be governed by long-term intracellular control programs, whereby the present proliferative behavior of a cell depends on its history of exposure to factors acting many cell generations before (see p. 751). While the relationship between the long-term and short-term controls is still mysterious, both seem to involve many of the same molecules—in particular, growth factors and products of proto-oncogenes. Programs of cell division can be remarkably complex and strictly defined during embryonic development. This is strikingly illustrated in the nematode worm *Caenorhabditis elegans*, whose fertilized egg divides so as to generate precisely 959 somatic cell nuclei in the adult body; some of the gene products that implement these programs are beginning to be characterized (see p. 905).

It should not be supposed, however, that the growth of embryos is regulated simply by counting cell divisions. This is made clear, for example, by comparing newts of differing ploidy. The cells of a pentaploid newt are roughly five times as big as those of a haploid, and yet, because there are roughly one-fifth as many cells in each tissue in the pentaploid, the dimensions of the body and of its organs are practically the same in the two types of animal (Figures 13–40 and 13–41). Evidently in vertebrates the mechanisms that control cell division to regulate bodily dimensions depend on measurements of distances rather than on simple counts of cell numbers or division cycles. Such mechanisms require complex positional controls, in which diffusible growth factors may well play an important part.

Positional control of cell division can operate with remarkable specificity. When a piece of epithelium from one cockroach leg is transplanted to a homologous site in another, it "heals in" without significant cell division. However, if it is

Figure 13–39 A speculative scheme to account for the observed relationships between cell-matrix adhesion and the proliferation of normal and transformed cells. The model focuses on two types of molecule: (1) a cytoplasmic protein that provides the intracellular signal for cell division and (2) a transmembrane linker protein that can bind both to the cytoplasmic signaling molecule on one side of the plasma membrane and to the extracellular matrix on the other side. The binding is cooperative, so that signaling molecules bound to the linker protein intracellularly stabilize the transmembrane array and promote its binding to the extracellular matrix, and, reciprocally, binding to the extracellular matrix promotes binding of signal molecules to the linker proteins inside the cell. For the cell to be signaled to divide, the signaling molecules must be released into the cytoplasm and must be in an activated conformation, in which they bind less strongly to the linker proteins. It is presumed that the signaling molecules are activated by phosphorylation catalyzed by a kinase activity associated with the linker protein.

When an anchored cell is stimulated by growth factors (A), the kinase activity is switched on: the signaling molecules become phosphorylated and dissociate from the linker proteins, signaling the cell to divide and weakening the extracellular adhesion. Normal cells in suspension (B) fail to divide in response to growth factors because very few of the intracellular signaling molecules are bound to the linker proteins that would enable them to become phosphorylated. Transformed cells (C) have reduced adherence and can divide even in suspension because there is a "short circuit" in the control system such that the signaling molecules are phosphorylated constitutively.

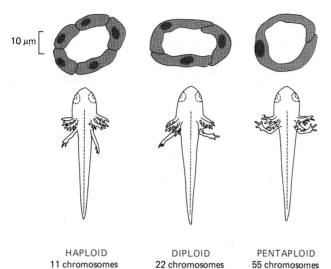

10 μm

HAPLOID	DIPLOID	PENTAPLOID
11 chromosomes	22 chromosomes	55 chromosomes

Figure 13–40 Drawings of representative sections of kidney tubules from salamander larvae of different ploidy. Pentaploid salamanders have cells that are bigger than those of haploid salamanders, but the animals and their individual organs are the same size because each tissue in the pentaploid animal contains fewer cells. This indicates that the number of cells is regulated by some mechanism based on size and distance rather than on the counting of cell divisions or of cell numbers. (After G. Fankhauser, in *Analysis of Development* [B.H. Willier, P.A. Weiss, and V. Hamburger, eds.], pp. 126–150. Philadelphia: Saunders, 1955.)

transplanted to a nonhomologous site, both the graft and the adjacent host cells proliferate and then differentiate to generate the cells that would normally lie between the region from which the graft was taken and the region to which it was transplanted (see p. 918). The molecular basis of this behavior is a complete mystery.

In general, cell division during embryonic development is governed by an interplay of cell-autonomous programs and cell-cell interactions, with the importance of each varying from species to species and from one part of the body to another. In adult tissues, too, cell division is regulated by a complex network of controls: when a deep skin wound heals in a vertebrate, about a dozen different cell types, ranging from fibroblasts to Schwann cells, must be regenerated in appropriate numbers to reconstruct the lost tissue. Moreover, there is redundancy in the social control system, with multiple restraints acting in parallel, so that the failure of a single control component in one cell (a common result of somatic cell mutation) will not endanger the whole organism by allowing the genesis of an enormous clone of cells that proliferates wildly. Studies of cancer incidence suggest that about four to six mutations must typically occur in a given cell lineage before it will give rise to a malignant tumor (see p. 1192).

Dissecting in molecular detail the elaborate social controls that enable an organ such as the kidney to develop and persist in an adult animal may well be an enterprise to occupy generations of cell biologists. But powerful tools are now available, such as antibodies to block specific growth factors or receptors, and transgenic animals engineered to produce such signaling molecules inappropriately in selected types of cells (see p. 268). With the help of these new approaches, the task, although formidable, does not seem impossible.

Summary

Abnormal cells that disobey the social constraints on cell division proliferate to form tumors in the body, and they also appear transformed in cell culture. Although often lethal to the organism as a whole, as individual cells they are favored by natural selection and therefore are easy to isolate. Cell transformation is often accompanied by mutation or overexpression of specific oncogenes, identified in many cases because they are carried by RNA tumor viruses (retroviruses). The normal counterparts of these viral oncogenes in the healthy cell are known as proto-oncogenes and are thought to encode key components of the normal system of social controls of cell division. Some proto-oncogenes code for growth factors, some for growth-factor receptors, some for intracellular regulatory proteins that are involved in cell adhesion, and some for proteins that help relay signals for cell division to the cell nucleus. Cells must contain alterations in multiple social control genes in order to become cancerous, reflecting a redundancy in the complex control systems that affect cell proliferation in tissues.

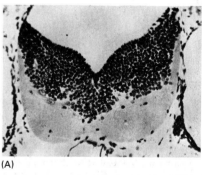

(A)

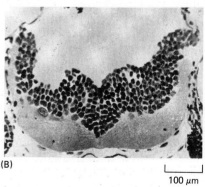

(B)

100 μm

Figure 13–41 Micrographs comparing cells in the brains of haploid and tetraploid salamanders (see also Figure 13–40). (A) Cross-section of the hindbrain of a haploid salamander. (B) Corresponding cross-section through the hindbrain of a tetraploid salamander showing how reduced cell numbers compensate for increased cell size. (From G. Fankhauser, *Int. Rev. Cytol.* 1:165–193, 1952.)

The Mechanics of Cell Division[35]

In this last section we discuss the events of M phase—the culmination of the cell cycle. In a comparatively brief period the chromosomes condense and the contents of the parental cell, which were doubled by the biosynthetic activities of the preceding interphase, are segregated into two daughter cells (Figure 13–42).

At the molecular level, M phase is thought to be initiated by a cascade of protein phosphorylations triggered by the appearance of MPF and terminated by dephosphorylations that restore the proteins to their interphase state (see p. 742). The protein phosphorylations present during M phase are in turn likely to be responsible for the many morphological changes that accompany mitosis, including chromosome condensation, nuclear envelope breakdown, and the cytoskeletal changes to be described below. The first readily visible manifestation of an impending M phase is a progressive compaction of the dispersed interphase chromatin into threadlike chromosomes. This condensation of the chromosomes is required for their subsequent organized segregation into daughter cells, and it is accompanied by phosphorylation of the many histone H1 molecules present in the cell (up to six phosphates per H1 molecule). Since histone H1 is present in a concentration of about one molecule per nucleosome and is known to be involved in packing nucleosomes together (see p. 499), its phosphorylation by the MPF kinase (see p. 742) at the onset of the M phase could be a major cause of chromosome condensation. While still quite tentative, this type of molecular explanation illustrates the level at which one must ultimately explain the entire cell cycle.

It has been said that the chromosomes in mitosis are like the corpse at a funeral: they provide the reason for the proceedings but play no active part in them. The active role is played by two distinct cytoskeletal structures that appear transiently in M phase. The first to form is a bipolar **mitotic spindle,** composed of microtubules and their associated proteins. The mitotic spindle first aligns the replicated chromosomes in a plane that bisects the cell; each chromosome then separates into two daughter chromosomes, which are moved by the spindle to opposite ends of the cell. The second cytoskeletal structure required in M phase in animal cells is a **contractile ring** of actin filaments and myosin that forms slightly later just beneath the plasma membrane. This ring pulls the membrane inward so as to divide the cell in two, thereby ensuring that each daughter cell receives not only one complete set of chromosomes but also half of the cytoplasmic constituents and organelles in the parental cell. The two cytoskeletal structures contain different sets of proteins and can be formed separately in some specialized cells. Their formation is usually closely coordinated, however, so that cytoplasmic division (*cytokinesis*) occurs immediately after the end of nuclear division (*mitosis*). The same is true for plant cells, even though, as we shall see, their rigid walls necessitate a different mechanism for cytokinesis.

The description of M phase just given applies only to eucaryotic cells. Bacterial cells do not contain either actin filaments or microtubules; they generally have only one chromosome, whose replicated copies are segregated to daughter cells by a mechanism that involves chromosome attachment to the bacterial plasma membrane (see p. 785). The need for complex mitotic machinery probably arose only with the evolution of cells that contained greatly increased amounts of DNA packaged in a number of discrete chromosomes. The primary function of this machinery is to divide the replicated chromosomes precisely between the two daughter cells. Its accuracy has been determined in yeast cells, where an error in chromosome segregation is made only about once every 10^5 cell divisions.

M Phase Is Traditionally Divided into Six Stages[35]

The basic strategy of cell division is remarkably constant among eucaryotic organisms. The first five stages of the M phase constitute mitosis; the sixth is cytokinesis. These six stages form a dynamic sequence, the complexity and beauty of which are hard to appreciate from written descriptions or from a set of static pictures. The description of cell division is based on observations from two sources:

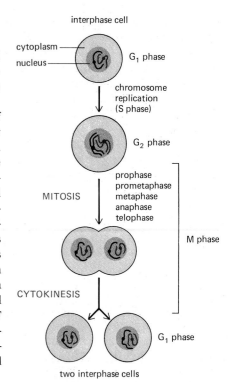

Figure 13–42 The M phase (or cell-division phase) of the cell cycle starts at the end of G_2 phase and ends at the start of the next G_1 phase. It includes the five stages of nuclear division (mitosis) and cytoplasmic division (cytokinesis).

light microscopy of living cells (often combined with microcinematography) and light and electron microscopy of fixed and stained cells. A brief summary of the various stages of cell division is given in Panel 13–1. As defined, the five stages of mitosis—*prophase, prometaphase, metaphase, anaphase,* and *telophase*—occur in strict sequential order, while cytokinesis begins during anaphase and continues through the end of the mitotic cycle (Figure 13–43). Light micrographs of cell division in a typical animal and a typical plant cell are shown in Figures 13–44 and 13–45, respectively.

Innumerable variations in all of the stages of cell division shown schematically in Panel 13–1 occur in the animal and plant kingdoms. We shall mention some of these when we take a closer look at the mechanisms of cell division, since they can help us understand the different parts of the mammalian mitotic apparatus.

Formation of the Mitotic Spindle in an M-Phase Cell Is Accompanied by Striking Changes in the Dynamic Properties of Microtubules[36]

We saw in Chapter 11 that the principal *microtubule organizing center (MTOC)* in most animal cells is the **centrosome,** a cloud of amorphous material surrounding a pair of centrioles (see p. 654). During interphase the centrosomal material nucleates the growth of microtubules, which project outward toward the cell perimeter with their minus ends attached to the centrosome. This *interphase microtubule array* radiating from the centrosome is a dynamic, constantly changing structure in which microtubules continually form and disassemble. New microtubules grow by adding tubulin molecules to their plus ends; sporadically, and apparently at random, individual microtubules become unstable and undergo rapid, "catastrophic" disassembly, yielding their subunit molecules to the pool of unpolymerized tubulin already present in the cytoplasm (see Panel 11–2, p. 655).

Changes occur in the centrosome throughout the cell cycle, as illustrated in Figure 13–46. Sometime during S phase, the centriole pair replicates while remaining embedded in a single cloud of centrosomal material. In **prophase** the centrosome splits and each daughter centrosome becomes the focal point of a separate starlike aster of microtubules, which have their ends embedded in the cloud of centrosomal material. The microtubules in each aster elongate where they contact each other, and the two centrosomes move apart. Then, at prometaphase, the nuclear envelope breaks down, allowing the microtubules from each centrosome to enter the nucleus and interact with the chromosomes. The two daughter centrosomes are now said to constitute the two **spindle poles.**

The events just described are thought to result from major changes that occur at prophase in both the inherent stability of microtubules and the properties of the centrosome. As mentioned previously, there is evidence that an M-phase-promoting factor (MPF) triggers entry into M phase by initiating a cascade of multiple protein phosphorylations (see p. 742). Certain molecules that interact with microtubules are presumably affected, since the half-life of an average microtubule decreases about twentyfold (from about 5 minutes to 15 seconds) as a cell enters prophase (see Figure 13–48). This change is thought to reflect a greatly increased probability that a typical growing microtubule will convert to a shrinking one by a change at its plus end (see p. 655), combined with an alteration in the centrosome whereby it acquires, at prophase, a greatly increased capacity for nucleating the growth of microtubules (this can be observed *in vitro*). These two changes are sufficient to explain why the onset of M phase is characterized by a rapid transition from relatively few long microtubules that extend from the centrosome to the cell periphery (the interphase microtubule array) to large numbers of short microtubules that surround each centrosome (see Prophase in Figure 13–46).

As mitosis proceeds, the elongating ends of the microtubules emanating from the spindle poles are believed to encounter structures that bind to them and stabilize them against catastrophic disassembly. Because each pole nucleates microtubules in random directions, it is thought that selective stabilization gives rise to the characteristic bipolar form of the mitotic spindle in which the majority of

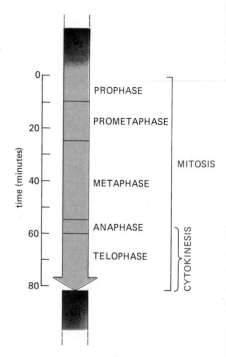

Figure 13–43 A typical time course for mitosis and cytokinesis (M phase) in a mammalian cell. The exact times vary for different cells. Note that cytokinesis begins before mitosis ends. The beginning of prophase (and therefore of M phase as a whole) is defined as the point in the cell cycle at which condensed chromosomes first become visible—a somewhat arbitrary criterion, since the extent of chromosome condensation appears to increase continuously during late G_2.

interphase early prophase late prophase prometaphase

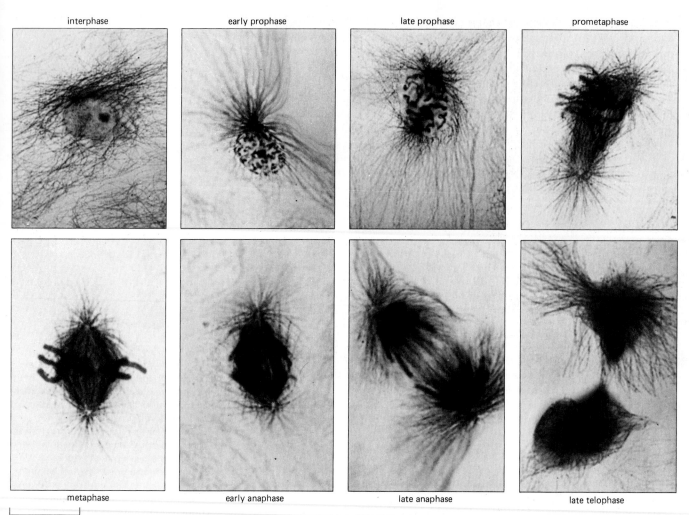

metaphase early anaphase late anaphase late telophase

20 μm

Figure 13–44 The course of mitosis in animal cells is illustrated here by selected micrographs of cultured marsupial (PtK) cells. The microtubules have been visualized using immunogold staining, while chromatin is stained with toluidine blue. At the light microscope level shown here, the major events of cell division have been known for more than 100 years. During *interphase* the centrosome, containing a centriole pair, forms the focus for the interphase microtubule array. By *early prophase* the single centrosome contains two centriole pairs (not visible); at *late prophase* the centrosome divides, and the resulting two asters move apart. The nuclear envelope breaks down at *prometaphase*, allowing the spindle microtubules to interact with the chromosomes. At *metaphase* the bipolar spindle structure is clear, and all the chromosomes are aligned across the middle of the spindle. The chromatids all separate synchronously at *early anaphase* and, under the influence of the spindle fibers, begin to move toward the poles. By *late anaphase* the spindle poles have moved farther apart, increasing the separation of the two groups of chromatids. At telophase the daughter nuclei re-form, and by *late telophase* cytokinesis is almost complete, with the midbody persisting between the daughter cells. (Photographs courtesy of M. deBrabander.)

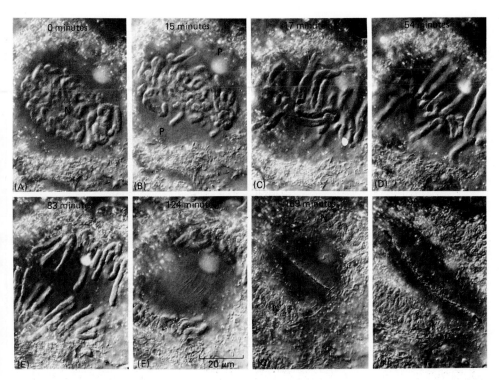

Figure 13–45 The course of mitosis in a typical plant cell. These micrographs were taken of a living *Haemanthus* (lily) cell at the times indicated using differential-interference-contrast microscopy (see p. 141). The cell has unusually large chromosomes, so they are easy to see. (A) *Prophase:* the chromosomes have condensed and are clearly visible in the cell nucleus (marked N). (B) and (C) *Prometaphase:* the nuclear envelope has broken down and the chromosomes are interacting with microtubules that emanate from the two spindle poles (marked P). Note that only 2 minutes have elapsed between the stages shown in (B) and (C). (D) *Metaphase:* the chromosomes have lined up at the metaphase plate with their kinetochores located halfway between the two spindle poles. (E) *Anaphase:* the chromosomes have separated into their two sister chromatids, which are moving to opposite poles. (F) *Telophase:* the chromosomes are decondensing to form the two nuclei that are seen later [marked N in (G)]. (G) and (H) *Cytokinesis:* two successive stages in the formation of the cell plate are shown; the cell plate appears as a line whose direction of outgrowth is indicated by arrows in (H). (Courtesy of Andrew Bajer.)

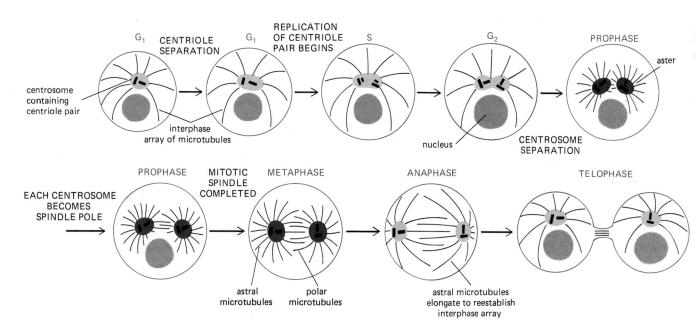

Figure 13–46 The centrosome cycle. The centrosome in an interphase cell duplicates to form the two poles of a mitotic spindle. In most animal cells (but not in plant cells) a centriole pair (shown here as a pair of black bars) is embedded in the centrosomal material (*color*) that nucleates microtubule outgrowth. At a discrete point in G₁ phase, the two centrioles separate by a few micrometers. During S phase a daughter centriole begins to grow near the base of each old centriole and at a right angle to it. The elongation of the daughter centriole is usually completed by G₂ phase. Initially the two centriole pairs remain embedded in a single mass of centrosomal material, forming a single centrosome. In early M phase each centriole pair becomes part of a separate microtubule-organizing center that nucleates a radial array of microtubules, the *aster* (aster = star). The two asters, which initially lie side by side and close to the nuclear envelope, move apart. By late prophase the bundles of *polar microtubules* that interact between the two asters preferentially elongate as the two centers move apart along the outside of the nucleus. In this way a mitotic spindle is rapidly formed.

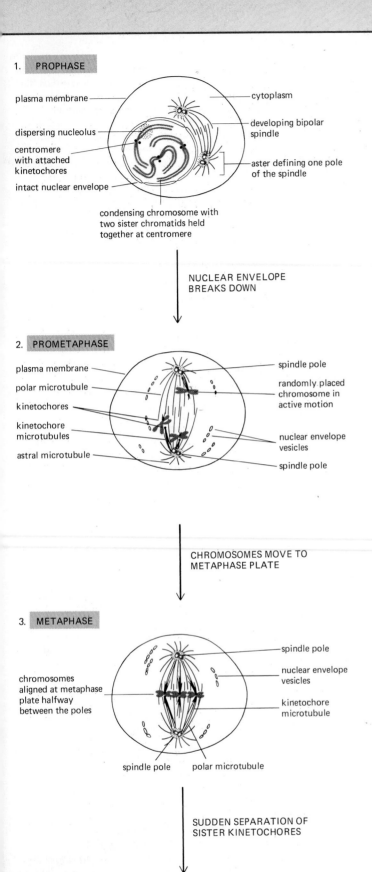

1. PROPHASE

plasma membrane

cytoplasm

dispersing nucleolus

developing bipolar spindle

centromere with attached kinetochores

intact nuclear envelope

aster defining one pole of the spindle

condensing chromosome with two sister chromatids held together at centromere

NUCLEAR ENVELOPE BREAKS DOWN

2. PROMETAPHASE

plasma membrane

spindle pole

polar microtubule

randomly placed chromosome in active motion

kinetochores

kinetochore microtubules

nuclear envelope vesicles

astral microtubule

spindle pole

CHROMOSOMES MOVE TO METAPHASE PLATE

3. METAPHASE

spindle pole

nuclear envelope vesicles

chromosomes aligned at metaphase plate halfway between the poles

kinetochore microtubule

spindle pole   polar microtubule

SUDDEN SEPARATION OF SISTER KINETOCHORES

1 PROPHASE

As viewed in the microscope, the transition from the G_2 phase to the M phase of the cell cycle is not a sharply defined event. The chromatin, which is diffuse in interphase, slowly condenses into well-defined chromosomes, the exact number of which is characteristic of the particular species. Each chromosome has duplicated during the preceding S phase and consists of two sister *chromatids;* each of these contains a specific DNA sequence known as a *centromere,* which is required for proper segregation. Toward the end of prophase, the cytoplasmic microtubules that are part of the interphase cytoskeleton disassemble and the main component of the mitotic apparatus, the *mitotic spindle,* begins to form. This is a bipolar structure composed of microtubules and associated proteins. The spindle assembles initially outside the nucleus.

2 PROMETAPHASE

Prometaphase starts abruptly with disruption of the nuclear envelope, which breaks into membrane vesicles that are indistinguishable from bits of endoplasmic reticulum. These vesicles remain visible around the spindle during mitosis. The spindle microtubules, which have been lying outside the nucleus, can now enter the nuclear region. Specialized protein complexes called *kinetochores* mature on each centromere and attach to some of the spindle microtubules, which are then called *kinetochore microtubules.* The remaining microtubules in the spindle are called *polar microtubules,* while those outside the spindle are called *astral microtubules.* The kinetochore microtubules extend in opposite directions from the two sister chromatids in each chromosome, exerting tension on the chromosomes, which are thereby thrown into agitated motion.

3 METAPHASE

The kinetochore microtubules eventually align the chromosomes in one plane halfway between the spindle poles. Each chromosome is held in tension at this *metaphase plate* by the paired kinetochores and their associated microtubules, which are attached to opposite poles of the spindle.

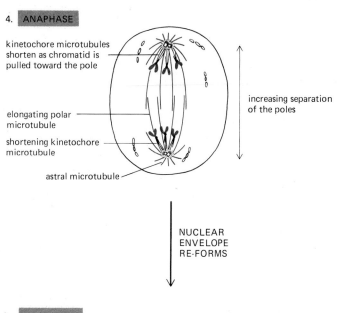

4. ANAPHASE

kinetochore microtubules
shorten as chromatid is
pulled toward the pole

increasing separation
of the poles

elongating polar
microtubule

shortening kinetochore
microtubule

astral microtubule

NUCLEAR
ENVELOPE
RE-FORMS

4 ANAPHASE

Triggered by a specific signal, anaphase begins abruptly as the paired kinetochores on each chromosome separate, allowing each chromatid to be pulled slowly toward the spindle pole it faces. All chromatids move at the same speed, typically about 1 μm per minute. Two categories of movement can be distinguished. During *anaphase A,* kinetochore microtubules shorten as the chromosomes approach the poles. During *anaphase B,* the polar microtubules elongate and the two poles of the spindle move farther apart. Anaphase typically lasts only a few minutes.

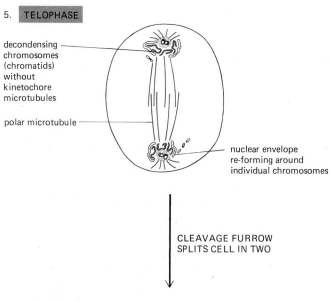

5. TELOPHASE

decondensing
chromosomes
(chromatids)
without
kinetochore
microtubules

polar microtubule

nuclear envelope
re-forming around
individual chromosomes

5 TELOPHASE

In telophase (*telos,* end) the separated daughter chromatids arrive at the poles and the kinetochore microtubules disappear. The polar microtubules elongate still more, and a new nuclear envelope re-forms around each group of daughter chromosomes. The condensed chromatin expands once more, the nucleoli—which had disappeared at prophase—begin to reappear, and mitosis is at an end.

CLEAVAGE FURROW
SPLITS CELL IN TWO

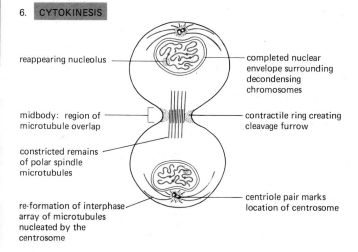

6. CYTOKINESIS

reappearing nucleolus

completed nuclear
envelope surrounding
decondensing
chromosomes

midbody: region of
microtubule overlap

contractile ring creating
cleavage furrow

constricted remains
of polar spindle
microtubules

re-formation of interphase
array of microtubules
nucleated by the
centrosome

centriole pair marks
location of centrosome

6 CYTOKINESIS

The cytoplasm divides by a process known as *cleavage,* which usually starts sometime during anaphase. The process is illustrated here as it occurs in animal cells. The membrane around the middle of the cell, perpendicular to the spindle axis and between the daughter nuclei, is drawn inward to form a *cleavage furrow,* which gradually deepens until it encounters the narrow remains of the mitotic spindle between the two nuclei. This thin bridge, or *midbody,* may persist for some time before it narrows and finally breaks at each end, leaving two separated daughter cells.

microtubules extend from the two poles of the spindle toward the equatorial plane halfway between them. Those microtubules that cross the equator may become selectively stabilized by microtubule-binding proteins that cross-link neighboring parallel microtubules of opposite polarity (Figure 13–47). At metaphase the spindles of higher animal and plant cells may contain up to several thousand microtubules, while the spindles of some fungi contain as few as 40.

Even though some of the microtubules in the spindle are partially stabilized against spontaneous disassembly, the majority continue to exchange their subunits with the pool of soluble tubulin molecules in the cytosol. The exchange can be measured directly by the method described in Figure 13–48. It can also be seen by subjecting mitotic cells to conditions that reversibly shift the equilibrium between tubulin polymerization and depolymerization and by observing the behavior of the birefringent spindle microtubules with polarized light (Figure 13–49). If mitotic cells are placed in heavy water (D_2O) or treated with taxol, either of which inhibits microtubule disassembly, the spindle fibers lengthen. Such stabilized spindles cannot move chromosomes, and mitosis is arrested. At the other extreme, mitosis is blocked when the spindle microtubules are reversibly disrupted by treatment with the drug colchicine, by low temperature, or by high hydrostatic pressure, all of which interfere with the assembly of tubulin molecules into microtubules. The observation that neither stabilized nor disassembled spindle microtubules can move chromosomes suggests that the spindle must be delicately poised at equilibrium between assembly and disassembly in order to perform mitotic movements. Before discussing the basis for these movements, we must describe in more detail how the spindle is organized and how the chromosomes are positioned.

13-25 Chromosomes Attach to Microtubules by Their Kinetochores During Mitosis[37]

Replicated chromosomes bind to the mitotic spindle via structures called **kinetochores.** At the start of M phase, each chromosome consists of two sister chromatids paired along their length but primarily joined near their **centromeres,** which consist of a specific DNA sequence required for chromosome segregation. During late prophase a mature kinetochore develops on each centromere, with the two kinetochores (one on each sister chromatid) facing in opposite directions. By metaphase, microtubules have become attached to each kinetochore (Figure 13–50). In most organisms the kinetochore is a large multiprotein complex that can be seen in the electron microscope as a platelike trilaminar structure (Figure 13–51). The number of microtubules linked to a kinetochore varies widely depending on the species. Human kinetochores, for example, have 20 to 40 microtubules, while yeasts and some other microorganisms have only one; a single microtubule must therefore be sufficient to move a chromosome.

The information that specifies the construction of a kinetochore at a specific site on a chromosome must come directly from the DNA sequence at the centromere. In yeasts, centromeric DNA can be identified genetically by its ability to

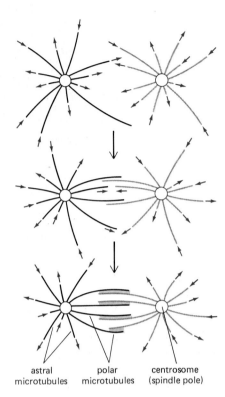

astral microtubules   polar microtubules   centrosome (spindle pole)

Figure 13–47 A model for how a bipolar mitotic spindle is thought to form by the selective stabilization of interacting microtubules. New microtubules grow out in random directions from two nearby centrosomes, to which they are anchored by their minus ends. Their plus ends are "dynamically unstable" and will switch suddenly from uniform growth to rapid shrinkage, during which the entire microtubule often depolymerizes (see p. 655). When two microtubules from opposite centrosomes interact in an overlap zone, microtubule-associated proteins are believed to cross-link the microtubules together (*gray shading*) in a way that caps their plus ends, stabilizing them by decreasing their probability of depolymerizing.

Figure 13–48 A study showing that the microtubules in an M-phase cell are much more dynamic, on average, than the microtubules at interphase. Mammalian cells in culture were injected with tubulin that had been covalently linked to a fluorescent dye. After the injected fluorescent tubulin had had time to become incorporated into a cell's microtubules, all of the fluorescence in a small region was bleached by an intense laser beam. The recovery of fluorescence in the bleached region of microtubules, caused by the exchange of their bleached tubulin subunits for unbleached fluorescent tubulin from the soluble pool, was then monitored as a function of time. The time $t_{1/2}$ for 50% recovery of fluorescence is thought to be equal to the time required for half of the microtubules in the region to depolymerize and re-form. (Data from W.M. Saxton *et al., J. Cell Biol.* 99:2175–2187, 1984, by copyright permission of the Rockefeller University Press.)

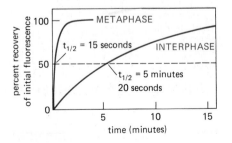

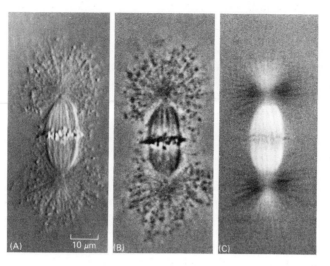

Figure 13–49 An isolated metaphase spindle viewed by three techniques of light microscopy: (A) differential-interference-contrast microscopy, (B) phase-contrast microscopy, and (C) polarized-light microscopy. (Courtesy of E.D. Salmon and R.R. Segall, from *J. Cell Biol.* 86:355–365, 1980. Reproduced by copyright permission of the Rockefeller University Press.)

allow plasmids to be stably inherited that otherwise segregate irregularly to daughter cells and are lost. Molecular genetic experiments show that each of the 17 chromosomes of the yeast *Saccharomyces cerevisiae* contains a different centromeric sequence about 110 base pairs long (Figure 13–52), although all the sequences contain substantial regions of homology and can be inverted or swapped from chromosome to chromosome without loss of function. The yeast centromere sequences bind specific proteins, which in turn are thought to initiate the formation of a multiprotein complex (the kinetochore) that binds the end of a single microtubule. Mammalian centromeres are thought to consist of different and much longer DNA sequences, and they form much larger kinetochores that bind multiple microtubules.

An unexpected opportunity to study the proteins of mammalian kinetochores came from the finding that human patients suffering from certain types of *scleroderma* (a disease of unknown cause that is associated with a progressive fibrosis of connective tissue in skin and other organs) produce autoantibodies that react specifically with kinetochores. When these antibodies are used to stain dividing cells by immunofluorescence, a pattern of fluorescent spots is obtained, each spot marking the position of a kinetochore. A similar speckled pattern is also obtained if nondividing cells are stained, and the number of spots per cell corresponds to the number of its chromosomes, suggesting that a kinetochore precursor is attached to each centromere even in interphase nuclei (Figure 13–53). Scleroderma antibodies have also made it possible to clone the genes that encode several of the many proteins associated with kinetochores, so that these normally rare proteins can now be produced in large quantities by recombinant DNA technologies; thus their interactions with one another, with DNA, and with microtubules can be characterized.

How are microtubules and kinetochores connected to each other? The linkage has some unique properties. If chemically marked tubulin is injected into mitotic

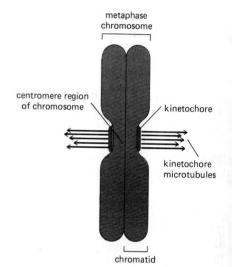

Figure 13–50 Schematic drawing of a metaphase chromosome showing its two sister chromatids attached to kinetochore microtubules.

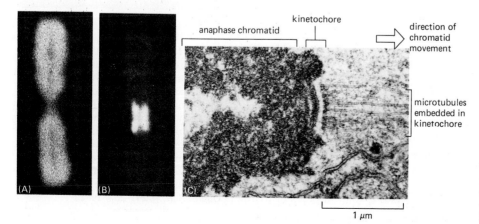

Figure 13–51 The kinetochore. A metaphase chromosome (A), when stained with human autoantibodies that react with specific kinetochore proteins, reveals two kinetochores, one associated with each chromatid (B). (C) Electron micrograph of an anaphase chromatid with microtubules attached to its kinetochore. While most kinetochores have a trilaminar structure, the one shown (from a green alga) has an unusually complex structure with additional layers. (A and B, courtesy of Bill Brinkley; C, from J.D. Pickett-Heaps and L.C. Fowke, *Aust. J. Biol. Sci.* 23:71–92, 1970. Reproduced by permission of CSIRO.)

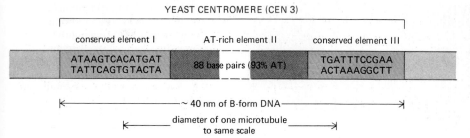

YEAST CENTROMERE (CEN 3)

conserved element I   AT-rich element II   conserved element III

ATAAGTCACATGAT
TATTCAGTGTACTA

88 base pairs (93% AT)

TGATTTCCGAA
ACTAAAGGCTT

|← ~ 40 nm of B-form DNA →|

|← diameter of one microtubule to same scale →|

Figure 13–52 The DNA sequence of a typical centromere in the yeast *Saccharomyces cerevisiae*. The DNA sequence shown is sufficient to cause faithful chromosome segregation; it serves to assemble the kinetochore proteins that attract a single microtubule to the kinetochore.

cells during metaphase, it is found to be continually incorporated into microtubules near their point of attachment to the kinetochore (Figure 13–54). Conversely, as we shall see below, the reverse reaction takes place during anaphase, when tubulin molecules are lost from these microtubules in the region close to the kinetochore as this structure moves toward a spindle pole. The puzzle is that both the addition and loss of tubulin molecules must take place while the kinetochore maintains a firm mechanical attachment to the microtubules, since it is through this point of attachment that chromosomes are pulled through the cytoplasm. The kinetochore thus seems to act like a sliding collar, maintaining a lateral association with polymerized tubulin subunits near the end of the microtubule while also allowing addition or loss of tubulin molecules to occur at that end (see Figure 13–61, below).

Kinetochores Are Thought to Capture the Plus Ends of Microtubules That Originate from a Spindle Pole[38]

The breakdown of the nuclear envelope, which signals the end of prophase and the beginning of **prometaphase,** enables the mitotic spindle to interact with the chromosomes. The eventual outcome of this interaction is that one chromatid of each chromosome will be faithfully segregated to each daughter nucleus. The kinetochore microtubules play a major role in this segregation process: they (1) orient each chromosome with respect to the spindle axis so that one kinetochore faces each pole and (2) move each chromosome into a plane (the *metaphase plate*) that bisects the mitotic spindle, a process that takes between 10 and 20 minutes in mammalian cells and reaches completion at the end of prometaphase.

Prometaphase is characterized by a period of frantic activity during which the spindle appears to be trying to contain and align the chromosomes at the metaphase plate. In actuality, the chromosomes are violently rotating and oscillating back and forth between the spindle poles because their kinetochores are capturing microtubules growing from one or the other spindle pole and are being pulled by the captured microtubules. The initial attachment of a chromosome usually takes place when it is close to a spindle pole, when microtubules attach to only one kinetochore; eventually the other kinetochore captures microtubules growing from the other pole. These random prometaphase movements, and the final chance orientation of the chromosome that results, guarantee the random segregation of chromatids to the daughter cells, which is important for gene mixing during the analogous nuclear division in meiosis (see p. 854).

Only the plus ends of microtubules extend from the poles, and it is these ends that attach to kinetochores. The kinetochore thereby acts as a "cap" that tends to protect the plus end from depolymerizing, just as the centrosome at the spindle pole tends to protect the minus end from depolymerizing. It is perhaps not surprising, therefore, that kinetochore microtubules, which are capped at both ends, are unusually stable. Other microtubules in the spindle (called *polar microtubules*) are less stable.

While the kinetochore microtubules tend to *pull* the chromosome toward their respective pole (see below), a separate force apparently repels any chromosomes that come too close to the pole. If the arms of a chromosome are cut free from the kinetochore by laser microsurgery, they tend to move away from the closest spindle pole, even though they seem not to be attached to microtubules or to any

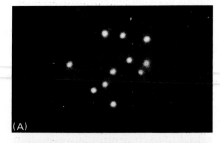

(A)

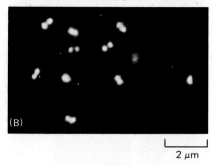

(B)

|— 2 μm —|

Figure 13–53 Immunofluorescent staining of kinetochores in interphase cells using an antibody that binds specifically to a kinetochore protein. These marsupial cells (PtK cells) contain relatively few chromosomes. (A) In cells in G_1 phase, one kinetochore is stained per chromosome. (B) In cells in G_2 phase, two kinetochores are stained per chromosome. (From S.L. Brenner and B.R. Brinkley, *Cold Spring Harbor Symp. Quant. Biol.* 46:241–254, 1982.)

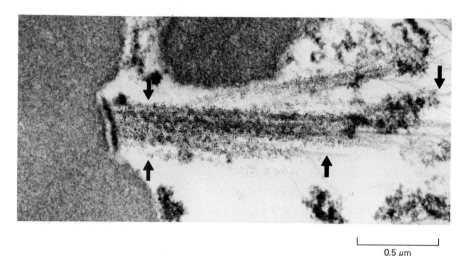

Figure 13–54 Experiment demonstrating that metaphase kinetochore microtubules grow at their kinetochore-attached (plus) end. Tubulin was covalently coupled to a small organic molecule (biotin) and then microinjected into a mammalian metaphase cell in culture. After 1 minute the cell was permeabilized, fixed, and stained with antibiotin antibodies bound to gold spheres, and then sections were prepared for electron microscopy. Regions of microtubules that incorporated the biotinylated tubulin during the minute following the injection are finely speckled with dark dots of gold (*colored arrows*), while regions of preexisting microtubule are unlabeled (*black arrows*). (Photograph courtesy of Louise Evans.)

0.5 μm

other obvious structure in the cell. One possibility is that the rapid polymerization of spindle microtubules outward from each pole might produce a general "wind" that carries any large unattached structures, such as the chromosome arms, away from the poles.

Sister Chromatids Attach by Their Kinetochores to Opposite Spindle Poles[39]

In early prometaphase the two kinetochores on a chromosome can become attached to the same spindle pole. However, this and other incorrect configurations, which would cause a failure of the chromosome to segregate properly if they persisted, are almost always corrected. It seems that a balanced arrangement, in which each sister kinetochore is attached to a different spindle pole and only to that spindle pole, has the greatest stability. Why this might be is suggested by experiments designed to test how chromosomes are attached to the mitotic spindle.

Delicate micromanipulation experiments, in which extremely fine glass needles are used to poke and pull at chromosomes inside a living mitotic cell, demonstrate that kinetochores are not committed to face one particular spindle pole as if polarized like a magnet, since a kinetochore of a manipulated chromosome will reengage to either pole if it is turned around. Moreover, by micromanipulation during prometaphase, it is possible to force the two kinetochores on a single chromosome to engage with the same spindle pole. If this configuration persists, the entire chromosome (with joined sister chromatids) is drawn toward the pole to which it is attached. Such an arrangement is generally unstable, however, and new kinetochore microtubules are usually captured from the other pole to produce the correct balanced configuration. On the other hand, if the movement of the incorrectly associated chromosome is restrained by a glass needle, the association of the chromosome with only one spindle pole becomes stabilized, suggesting that microtubules strengthen their linkage to the kinetochore by pulling against tension. Thus only chromosomes attached to both poles will normally retain their kinetochore microtubules and thereby interact stably with the spindle.

The tension generated by opposing kinetochore microtubules not only stabilizes the kinetochore-microtubule interaction, it also eventually aligns each chromosome on the metaphase plate, as will now be described.

Balanced Bipolar Forces Hold Chromosomes on the Metaphase Plate[40]

Why do all chromosomes line up at an equal distance from the two spindle poles at **metaphase,** thereby defining the metaphase plate? Experiments in which chromosomes are displaced with a glass needle suggest that the force exerted on a kinetochore is proportional to the length of the kinetochore fibers: that is, the pull

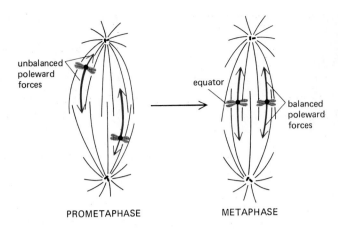

Figure 13–55 Chromosomes enter the spindle randomly during prometaphase and eventually line up at the equator (the metaphase plate) because the force on each kinetochore decreases as it gets closer to a pole. Thus the chromosomes are held under tension at the equator by balanced poleward forces. It is still unclear how the forces are generated.

on each kinetochore decreases the closer the kinetochore approaches to the attached pole (Figure 13–55). It is as though each chromosome is attached by a spring to each of the two spindle poles, so that any displacement toward one pole produces a restoring force in the opposite direction. The spindle that results at metaphase is illustrated in Figure 13–56.

The force on the chromosomes continues even after they have become aligned on the metaphase plate. Thus the chromosomes oscillate back and forth on the metaphase plate, continually adjusting their positions. Moreover, if one of a pair of metaphase kinetochore fibers is severed with a laser beam, the entire chromosome immediately moves toward the pole to which it is still attached. Similarly, if the attachment between the two chromatids is severed at metaphase, the two chromatids separate and move toward opposite poles, just as they do in anaphase. These experiments suggest that the same forces that move chromosomes to the metaphase plate also carry the chromatids to opposite poles as soon as the two kinetochores on each chromosome have separated.

Metaphase occupies a substantial portion of the mitotic period (see Figure 13–43), as if cells pause until all their chromosomes are lined up appropriately on the metaphase plate. Several experimental observations support this notion. Many cells arrest in mitosis for hours or days if treated with a drug, such as

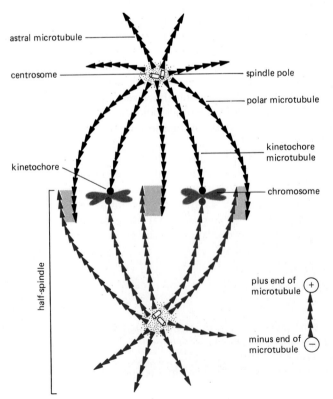

Figure 13–56 Simplified diagram of the mitotic spindle at metaphase. The spindle is constructed from two half-spindles (*black* and *color*), each composed of kinetochore, polar, and astral microtubules. The polarity of the microtubules is indicated by the arrowheads. The polar microtubules emanating from opposite spindle poles have a region of overlap (*shaded gray*), where microtubule-associated proteins may cross-link them; note that the microtubules are antiparallel in this overlap zone.

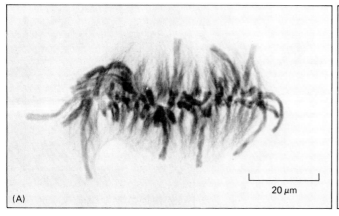
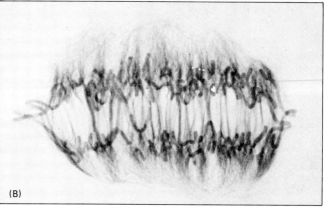

(A)   (B)

20 µm

Figure 13–57 Chromosomal movement at anaphase. In the transition from metaphase (A) to anaphase (B), chromosomes are pulled apart by spindle microtubules—as seen in these *Haemanthus* (lily) endosperm cells stained with gold-labeled antibodies to tubulin. (Courtesy of Andrew Bajer.)

colchicine or vinblastine, that depolymerizes their microtubules; in fact, this method of cell-cycle arrest is commonly used to collect large numbers of cells in mitosis so that their condensed chromosomes can be subjected to cytological analysis (see p. 505). The mitotic spindle rapidly regenerates if the drug is removed, and a normal mitosis often resumes once the chromosomes have been correctly positioned on the metaphase plate. It has been suggested that a chromosome with an unattached kinetochore may generate a diffusible signal that normally delays entry into anaphase, providing extra time for it to be attached correctly. If such a signal exists, disrupting the spindle with drugs would be expected to generate a strong signal and thereby prolong metaphase.

Sister Chromatids Separate Suddenly at Anaphase[41]

Metaphase, as we have just seen, is a relatively stable state, and in normal circumstances many cells remain for an hour or more with their chromosomes aligned and oscillating back and forth on the metaphase plate. **Anaphase** then begins abruptly with the synchronous splitting of each chromosome into its sister chromatids, each with one kinetochore (Figure 13–57). The signal to initiate anaphase is not a force exerted by the spindle itself, since chromosomes that are detached from the spindle will separate into chromatids at the same time as those that are still attached. Some experiments suggest that the signal could involve an increase in cytosolic Ca^{2+}: (1) continuous monitoring of living cells containing a fluorescent Ca^{2+} indicator dye (see p. 156) reveals a rapid, transient, tenfold rise in intracellular Ca^{2+} at anaphase in some cells; (2) microinjection of low levels of Ca^{2+} into cultured cells at metaphase can cause anaphase to begin prematurely; and (3) accumulations of membrane vesicles are usually seen at the spindle poles, and specialized electron microscopic techniques indicate that the vesicles are rich in Ca^{2+}. It is therefore possible that the spindle-associated vesicles release Ca^{2+} to initiate anaphase (Figure 13–58), much as the sarcoplasmic reticulum releases Ca^{2+} to initiate skeletal muscle contraction (see p. 624).

metaphase
chromosome   — kinetochore microtubule   membrane vesicles   pole
                                                                                  ↓

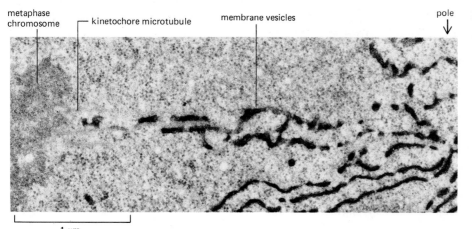

1 µm

Figure 13–58 Electron micrograph showing the collection of membrane vesicles (resembling cytoplasmic reticulum), specially stained for clarity, that are present at the spindle poles and seem to stretch out along the microtubules in the mitotic spindle. The cell, at metaphase, is from a barley leaf. (Courtesy of Peter Hepler, reproduced from *J. Cell Biol.* 86:490–499, 1980, by copyright permission of the Rockefeller University Press.)

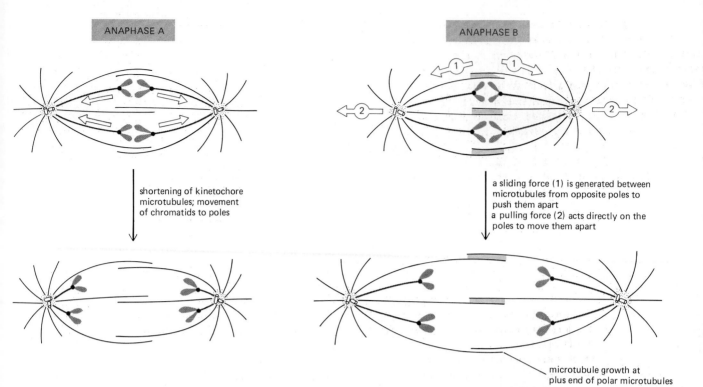

ANAPHASE A

shortening of kinetochore microtubules; movement of chromatids to poles

ANAPHASE B

a sliding force (1) is generated between microtubules from opposite poles to push them apart
a pulling force (2) acts directly on the poles to move them apart

microtubule growth at plus end of polar microtubules

Two Distinct Processes Separate Chromosomes at Anaphase[42]

Once each chromosome has split in response to the anaphase trigger, its two chromatids move to opposite spindle poles, where they will assemble into the nucleus of a new cell. Their movement appears to be the consequence of two independent processes within the spindle (Figure 13–59). The first is the poleward movement of chromatids accompanied by the shortening of the kinetochore microtubules, usually referred to as **anaphase A.** The second is the separation of the poles themselves accompanied by the elongation of the polar microtubules and is known as **anaphase B.** It is possible to distinguish the two processes by their differential sensitivities to certain drugs. A low concentration of chloral hydrate, for example, prevents the poles from separating and the polar microtubules from lengthening (anaphase B) but has no effect on the kinetochore microtubules or the poleward movement of the chromatids (anaphase A). The relative contribution of each of the two processes to the final separation of the chromosomes varies considerably depending on the organism. In mammalian cells, anaphase B begins shortly after the chromatids have begun their voyage to the poles and stops when the spindle is about 1.5–2 times its metaphase length. In some other cells, such as yeasts, anaphase B begins only after the chromatids reach their final destination; and in certain protozoa, anaphase B predominates and the spindles elongate to 15 times their metaphase length.

Kinetochore Microtubules Disassemble During Anaphase A[43]

A surprisingly large force acts on a chromosome as it moves from the metaphase plate to the spindle pole. Measurements using the deflection of fine glass needles give an estimate of 10^{-5} dynes per chromosome, which is more than 10,000 times greater than the force required simply to move chromosomes at their observed rate through the cytoplasm. Evidently there must be a powerful motor to move the chromosomes, but the speed of their movement must be limited by something other than viscous drag. As mentioned above, the same motor may be responsible for generating the tension on chromosomes on the metaphase plate.

As each chromosome moves poleward, its kinetochore microtubules disassemble, so that they have nearly disappeared at telophase. The site of subunit loss can be determined by injecting labeled tubulin into cells during metaphase. The

Figure 13–59 The several forces that act at anaphase to separate sister chromatids. (A) Chromatids are *pulled* toward opposite poles by forces associated with shortening of their kinetochore microtubules, a movement known as *anaphase A*. (B) Simultaneously the two spindle poles move apart, a movement known as *anaphase B*. It is likely that the forces that cause anaphase B are similar to those that cause the centrosome to split and separate into two spindle poles at prophase (see Figure 13–46). There is evidence that two separate forces are responsible for anaphase B: (1) the elongation and sliding of the polar microtubules *pushes* the two poles apart, while (2) outward forces act on the asters at each spindle pole to *pull* the poles away from each other.

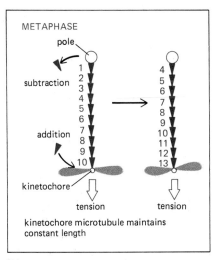

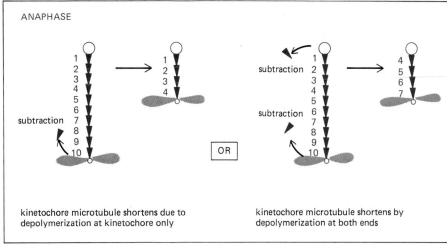

(A) (B)

Figure 13–60   The behavior of
kinetochore microtubules changes
during the transition from metaphase to
anaphase. (A) At metaphase, subunits
are added to the plus end of a
microtubule at the kinetochore and are
removed from the minus end at the
spindle pole. Thus a constant poleward
flux of tubulin subunits occurs, with the
microtubules remaining stationary and
under tension. (B) At anaphase the
tension is released, and the kinetochore
moves rapidly up the microtubule,
removing subunits from its plus end as
it goes (*left side*). Its attached chromatid
is thereby carried to a spindle pole. In
at least some organisms, part of the
chromatid movement is due to the
simultaneous shortening of the
microtubules at the pole (*right side*).

labeled subunits are found to be added to the kinetochore end of the kinetochore
microtubules and then lost as anaphase A proceeds, indicating that the kinetochore "eats" its way poleward along its microtubules at anaphase. This conclusion
is supported by the observation that anaphase kinetochores move toward a stationary mark generated on kinetochore microtubules by photobleaching. Microtubule disassembly at kinetochores, poles or both sites is probably necessary for
chromosome-to-pole movement (Figure 13–60), since this movement is inhibited
if microtubule depolymerization is blocked by the addition of taxol or D_2O.

The mechanism by which the kinetochore, and thus the chromosome, moves
up the spindle during anaphase A is still unknown. Two possible models are
presented schematically in Figure 13–61. In one the kinetochore hydrolyzes ATP
to move along its attached microtubule, with the plus end of the microtubule
depolymerizing as it becomes exposed. In the other, depolymerization of the microtubule itself causes the kinetochore to move passively to optimize its binding
energy on the microtubule. A third possibility, not illustrated in Figure 13–61, is
that microtubules are not directly responsible for the poleward force on kinetochores but serve merely to regulate movements generated by some other structure.
It has been suggested, for example, that a system of elastic protein filaments—
perhaps similar to the very long elastic filaments in striated muscle (see p. 624)—
might connect the kinetochore to the pole and pull the kinetochore steadily
poleward.

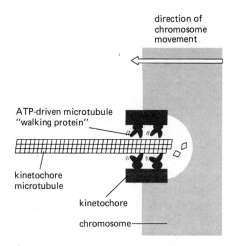

(A) ATP-driven chromosome movement drives
microtubule disassembly

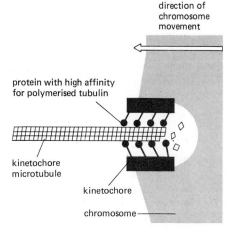

(B) Microtubule disassembly drives
chromosome movement

Figure 13–61   Two alternative models
of how the kinetochore might be able to
generate a poleward force on its
chromosome during anaphase.
(A) Microtubule-walking proteins that
resemble dynein or kinesin are part
of the kinetochore, and they use the
energy of ATP hydrolysis to pull
the chromosome along its bound
microtubules (see p. 660).
(B) Chromosome movement is driven by
microtubule disassembly: as tubulin
subunits dissociate, the kinetochore
tends to slide poleward in order to
restore its binding to the walls of the
microtubule. Similar mechanisms may
be used at the spindle pole, which
likewise seems to be able to hold onto
microtubules while permitting their
controlled depolymerization (see Figure
13–60).

Regardless of the force-generating mechanism, the dramatic change in microtubule polymerization at the kinetochore caused by the metaphase-to-anaphase transition (net polymerization in metaphase, net disassembly in anaphase—see Figure 13–60) remains to be explained. Perhaps it is simply attributable to the sudden release at anaphase of the tension exerted on the kinetochore; this could alter microtubule polymerization dynamics directly, or perhaps the release of tension could change the kinetochore chemically.

Two Separate Forces May Contribute to Anaphase B[44]

Anaphase B increases the distance between the two spindle poles and, in contrast to anaphase A, is accompanied by the *assembly* of microtubules. As the two poles move apart, the polar microtubules between them lengthen, apparently by assembly at their distal, plus ends.

Both the extent of spindle pole separation at anaphase and the degree of overlap of the polar microtubules in the midzone vary greatly from species to species. In many diatoms, which have the distinctive feature that mitosis occurs within the nuclear envelope (see p. 786), the overlapping arrays of spindle microtubules are especially prominent (Figure 13–62). Painstaking reconstruction of the three-dimensional architecture of complete diatom spindles from hundreds of serial thin sections examined in the electron microscope shows that the polar microtubules from each half-spindle overlap in a central region near the spindle equator. During anaphase these two sets of antiparallel polar microtubules appear to slide away from each other in the region of overlap.

Anaphase movements can also be studied in lysed diatom cells. In this model system the prominent mitotic spindle is freely accessible to macromolecules, so that the effects of various macromolecular probes (including specific antibodies) can be tested. Inhibitors that bind to either actin or myosin (such as anti-myosin antibodies) have no effect on anaphase chromosome movements, making it unlikely that a force-generating system involving actin and myosin like that in muscle is responsible for the chromosome movements. Alternative candidates for the force-generating system are proteins similar to *dynein*, a protein associated with microtubules in cilia and flagella (see p. 649), and *kinesin*, a protein implicated in fast axonal transport (see p. 660). Both of these proteins bind to microtubules and cause directed movement by hydrolyzing ATP, but it is still not known whether they play an important part in mitosis.

In higher cells the nuclear envelope disperses before the spindle is formed, and the astral microtubules (those directed away from the mitotic spindle, see p. 772) may play a more important part in anaphase B than they do in diatoms. The

Figure 13–62 Electron micrographs showing spindle elongation and the reduction in the degree of polar microtubule overlap during mitosis in a diatom. (A) Metaphase. (B) Late anaphase. (Courtesy of Jeremy D. Pickett-Heaps.)

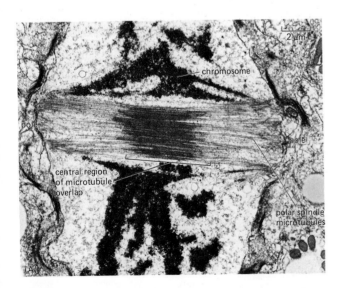

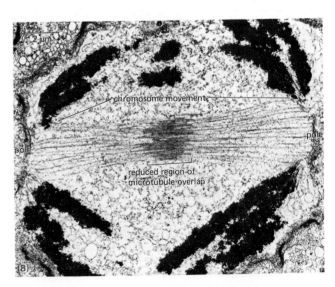

spindle microtubules of some marine invertebrate eggs, for example, can be cut without blocking anaphase B, suggesting that spindle poles are *pulled* apart—presumably through an attractive interaction between astral microtubules and the cell cortex. A similar type of interaction may help to guide asymmetric cell cleavages (see p. 779).

At Telophase the Nuclear Envelope Initially Re-forms Around Individual Chromosomes[45]

By the end of anaphase, the chromosomes have separated into two equal groups, one at each pole of the spindle. In **telophase,** the final stage of mitosis, a nuclear envelope reassembles around each group of chromosomes to form the two daughter interphase nuclei.

At least three parts of the nuclear-envelope complex must be considered during its breakdown and reassembly in mitosis: (1) the *outer and inner nuclear membranes* proper, which are continuous with the endoplasmic reticulum membrane (see p. 484); (2) the underlying *nuclear lamina* (a thin sheetlike network of intermediate filaments formed from the nuclear lamins), which interacts with the inner nuclear membrane, chromatin, and nuclear pores (see p. 665); and (3) the *nuclear pores*, which are formed by large complexes of incompletely characterized proteins (see p. 422). At prophase many proteins become phosphorylated. While phosphorylation of the histone H1 molecules is thought to help cause chromosome condensation (see p. 736), phosphorylation of the nuclear lamins helps regulate the disintegration and reconstruction of the nuclear envelope. The phosphorylation of the lamins occurs at many different sites in each polypeptide chain and causes them to disassemble, thereby disrupting the nuclear lamina. Subsequently, perhaps in response to a different signal, the nuclear envelope proper breaks up into small membrane vesicles.

The sudden transition from metaphase to anaphase is thought to initiate dephosphorylation of the many proteins, including histone H1 molecules and the lamins, that were phosphorylated at prophase. Shortly thereafter, at telophase, nuclear membrane vesicles associate with the surface of individual chromosomes and fuse to re-form the nuclear membranes, which partially enclose clusters of chromosomes before coalescing to re-form the complete nuclear envelope (Figure 13–63). During this process the nuclear pores reassemble and the dephosphorylated lamins reassociate to form the nuclear lamina; one of the lamina proteins

Figure 13–63 Schematic view of the nuclear-envelope cycle that occurs during mitosis. The nuclear membranes break down into small vesicles at prometaphase and reassemble at telophase. Between these two phases, while the nuclear envelope is disrupted and the nuclear pores and the nuclear lamina are dispersed into subunits, all of the mitotic movements take place that segregate the two sets of chromosomes to opposite poles. The new nuclear envelope in each daughter cell is generated by the fusion of the membrane vesicles on the surface of clustered individual chromosomes, as shown. In this way most of the cytoplasmic components are excluded from the new nucleus.

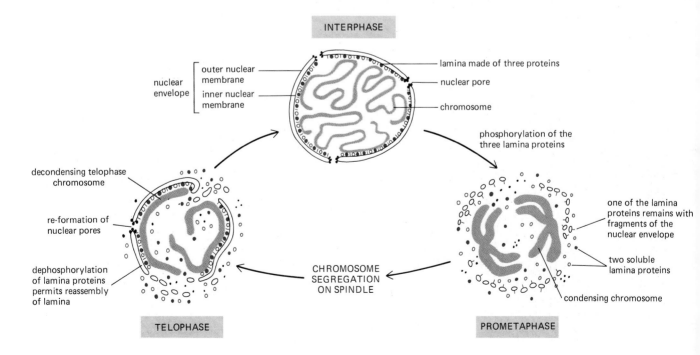

(lamin B) remains with the nuclear membrane fragments throughout mitosis and may help nucleate reassembly. After the nucleus re-forms, RNA synthesis resumes, causing the nucleolus to reappear as the chromatin uncoils and assumes its highly dispersed interphase conformation (see p. 543).

Both nuclear disassembly and nuclear assembly will occur in crude extracts of *Xenopus* eggs, provided that these extracts are prepared from cells of the correct cell-cycle stage (mitotic cells for disassembly and interphase cells for assembly). In these extracts the entire process involving the lamina, nuclear pores, and nuclear membranes appears to progress normally in response to phosphorylation and dephosphorylation cycles. *In vitro* systems of this type therefore provide an assay for the identification and purification of the proteins that catalyze nuclear envelope assembly and disassembly in the cell, including the proteins (such as MPF) that regulate these processes. While DNA must be added to these extracts to assemble a nucleus, a complete nuclear envelope will form around purified DNA molecules from virtually any organism, including a bacterial virus. Therefore, while DNA-binding proteins must be involved, it is unlikely that specific DNA sequences are recognized during nuclear reassembly.

Surprisingly, nuclear-envelope breakdown is not a requirement for mitosis. In fact, we shall see below that in lower eucaryotes the nuclear envelope does not disassemble during mitosis; these organisms are said to have a closed, rather than an open, spindle.

Metaphase and Interphase Can Be Viewed as Alternative "Stable" Cell States[46]

A current hypothesis concerning the mitotic cycle is sketched in Figure 13–64. It is, in a sense, a chemist's view of mitosis, in which interphase and metaphase are seen as two alternative "stable" states the cell can adopt, with the other stages of mitosis being merely necessary intermediates in the transition between these two states. It is postulated that a single "M-phase switch" is turned on at the end of interphase, causing the cell to pass through prophase and prometaphase to attain a more stable metaphase configuration. At the end of metaphase, the M-phase switch is suddenly turned off, and the cell passes through anaphase and telophase to return to interphase, which is the most stable state when the switch is off. This view of mitosis is supported by the nature of the changes that occur in the microtubule-based cytoskeleton (see Figure 13–48), as well as by the sudden changes that occur in the activities and phosphorylation states of several mitosis-related proteins at the interphase-prophase (switch on) and metaphase-anaphase (switch off) boundaries (see p. 736). The state of the switch might correspond to the level of MPF activity in the cell, as discussed earlier in this chapter when the MPF cycle was considered (see Figure 13–15).

Figure 13–64 A mitotic switch with only two states, "on" and "off," may be responsible for driving cells in and out of M phase. In this view the "on" state induces many protein phosphorylations that are unique to cells in mitosis. A succession of structural changes then occur during spindle assembly that do not require individual triggers. Rather they represent a series of energetically-favorable steps along a path that leads to the stable metaphase state. This state then persists until the anaphase trigger, when the switch is turned "off" and protein dephosphorylations reset the global parameters. This initiates a new series of structural changes (including chromosome separation) that leads back to the stable interphase state.

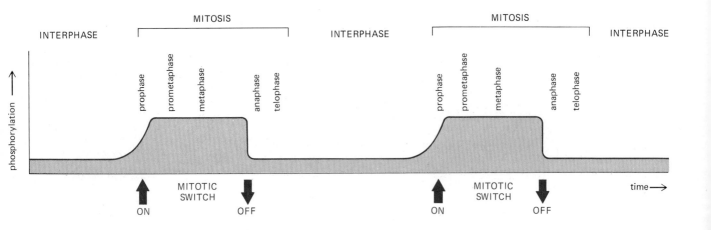

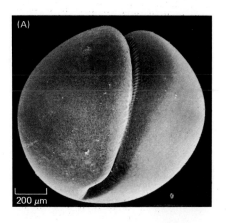

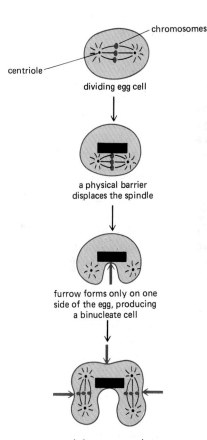

Figure 13–65 Scanning electron micrographs of early cleavage in a frog egg. The furrowing of the cell membrane is caused by the activity of the contractile ring underneath it. The cleavage furrow is unusually obvious and well defined in this giant spherical cell. (A) Low-magnification view of egg surface. (B) Surface of furrow at higher magnification. (From H.W. Beams and R.G. Kessel, *Am. Sci.* 64:279–290, 1976. Reprinted by permission of American Scientist, journal of Sigma Xi.)

13-30 The Mitotic Spindle Determines the Site of Cytoplasmic Cleavage During Cytokinesis[47]

During **cytokinesis,** the cytoplasm divides by a process called *cleavage.* Although nuclear and cytoplasmic division are generally linked, they are separable events. In some normal circumstances, nuclear division is not followed by cytokinesis. The early *Drosophila* embryo, for example, undergoes 13 rounds of nuclear division without cytoplasmic division, forming a single large cell containing 6000 nuclei arranged in a monolayer near the surface. Uninucleated cells are then generated by cytoplasmic cleavage around all of these nuclei (see p. 923).

Although cleavage does not necessarily accompany mitosis, the mitotic spindle plays an important part in determining where and when cleavage occurs. Cytokinesis usually begins in anaphase and continues through telophase and into the following interphase period. The first visible sign of cleavage in animal cells is a puckering and **furrowing** of the plasma membrane during anaphase (Figure 13–65). The furrowing invariably occurs in the plane of the metaphase plate, at right angles to the long axis of the mitotic spindle. If the spindle is moved by micromanipulation early enough in anaphase, the incipient furrow disappears and a new one develops in accord with the new spindle site. Ingenious experiments with fertilized sand dollar eggs show that a cleavage furrow will form midway between the asters originating from two centrosomes even when the centrosomes are not connected by a mitotic spindle (Figure 13–66). Later, once the furrowing process is well underway, cleavage proceeds even if the spindle and its asters are removed by suction or destroyed by colchicine.

Most cells divide symmetrically. The cleavage furrow forms around the equator of the parent cell so that the two daughter cells produced are of equal size and have similar properties. However, there are many instances during embryonic development in which cells divide asymmetrically: the furrow creates two cells that are different and destined to develop along divergent paths. Divisions of this kind are often precisely defined spatially. They may bear a precise relationship to the surface of an epithelial sheet, for example, or segregate regions of cytoplasm

chromosomes

centriole

dividing egg cell

a physical barrier displaces the spindle

furrow forms only on one side of the egg, producing a binucleate cell

second cleavage occurs between the centrioles linked by mitotic spindles and between the centrioles that are simply adjacent

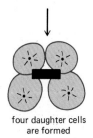

four daughter cells are formed

Figure 13–66 An experiment that shows the influence of spindle position on the subsequent plane of cleavage. If the spindle of a mitotic cell is mechanically pushed to one side, the membrane furrowing is incomplete, failing to occur on the opposite side of the cell. Subsequent cleavages occur not only in the conventional relation to each of the two mitotic spindles but also between the two adjacent asters that are not linked by a mitotic spindle but in this abnormal cell share the same cytoplasm. Apparently the contractile bundle of actin filaments that produces the cleavage furrow always forms in the region midway between two asters, which implies that the asters somehow alter the adjacent region of cell cortex.

that have a different complement of organelles. Whether the division is symmetrical or asymmetrical, the position of the furrow, and hence the plane of division, is always anticipated by the position of the mitotic spindle. To orient the plane of division, the spindle can rotate in a programmed manner to adopt a suitable position in the cell (Figure 13–67). It seems likely that these spindle movements are directed by programmed changes in local regions of the cell cortex and that the cortex then moves the spindle poles via their aster microtubules. A similar mechanism is thought to position the centrosome in a polarized cell (see p. 658). The structure of the actin-rich cortex is discussed in Chapter 11 (see p. 629).

The microtubules and any actin filament bundles present in the interphase cytoplasm are disassembled during mitosis. However, the cytoplasmic *intermediate filaments* remain intact in many cells. In such cells, the meshwork of cytoplasmic intermediate filaments that surrounds the interphase nucleus elongates during mitosis to enclose the two daughter nuclei and is finally divided in half by the cleavage furrow (Figure 13–68).

Actin and Myosin Generate the Forces for Cleavage[48]

Cleavage is accomplished by the contraction of a ring composed mainly of actin filaments. This bundle of filaments, known as the **contractile ring** (Figure 13–69), is bound to the cytoplasmic face of the plasma membrane by uncharacterized attachment proteins. The contractile ring assembles by an unknown mechanism in early anaphase; once assembled, it develops a force large enough to bend a fine glass needle inserted into the cell. There is compelling evidence that the muscle-like sliding of actin and myosin filaments in the contractile ring generates this force. In lysed mitotic cells, for example, the addition of an inactivated myosin subfragment blocks sites on actin that normally bind myosin, thereby stopping cleavage. Similarly, an injection of anti-myosin antibodies into sea urchin eggs causes the cleavage furrow to relax without affecting nuclear division. It is not known, however, exactly how the actin-myosin interaction pulls the plasma membrane down into a furrow.

During a normal cell division, the contractile ring does not get thicker as the furrow invaginates, suggesting that it continuously reduces its volume by losing filaments. The ring is finally dispensed with altogether when cleavage ends, as the plasma membrane of the cleavage furrow narrows to form the **midbody,** which remains as a tether between the two daughter cells. The midbody contains the remains of the two sets of polar microtubules, packed tightly together with dense matrix material (Figure 13–70).

Cytokinesis greatly increases the total cell-surface area as two cells form from one. Therefore the two daughter cells resulting from cytokinesis require more plasma membrane than the parent cell. In animal cells, net membrane biosynthesis increases just before cell division. This extra membrane seems to be stored as *blebs* on the surface of cells about to divide.

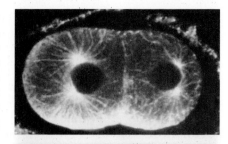

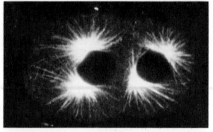

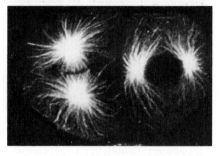

Figure 13–67 Light micrographs showing a precisely programmed rotation of a mitotic spindle in an embryo of the nematode worm *C. elegans* at the two-cell stage in preparation for cleavage to form four cells in a specific pattern. (Courtesy of Tony Hyman and John White.)

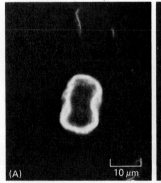
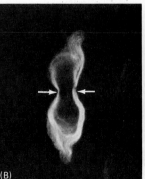
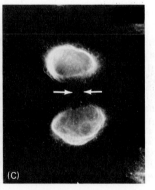

(A) 10 μm (B) (C)

Figure 13–68 A band of intermediate filaments surrounds the nuclear region during mitosis. The immunofluorescence micrographs shown here were obtained by staining permeabilized cells with fluorescent antibodies that bind to intermediate filaments: (A) anaphase, (B) early telophase (arrowheads indicate the position of the contractile ring), (C) late telophase. (From S.H. Blose, *Proc. Natl. Acad. Sci. USA* 76:3372–3376, 1979.)

Figure 13–69 Electron micrograph of the ingrowing edge of the cleavage furrow of a dividing animal cell. A schematic drawing of a cleavage furrow in a dividing cell is shown above the micrograph for orientation. (From H.W. Beams and R.G. Kessel, *Am. Sci.* 64:279–290, 1976. Reprinted by permission of American Scientist, journal of Sigma Xi.)

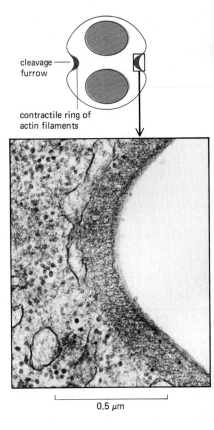

cleavage — furrow

contractile ring of actin filaments

0.5 μm

Cytokinesis Occurs by a Completely Different Mechanism in Higher Plant Cells[49]

Most higher plant cells are enclosed by a rigid *cell wall* and the mechanism of cytokinesis in these cells is very different from the one we have just described for animal cells. Rather than by pinching off the two daughter cells by means of a contractile ring at the cell surface, the cytoplasm of the plant cell is partitioned by the construction of a new cell wall inside the cell. This cross-wall precisely determines the relative positions that the two daughter cells will subsequently occupy in the plant. It follows that the planes of cell division, together with cell enlargement, determine plant form (see Chapter 20).

The new cross-wall, or **cell plate,** starts to assemble in a plane between the two daughter nuclei and in association with the residual polar spindle microtubules, which form a cylindrical structure, called the **phragmoplast.** This body, corresponding to the microtubules in the animal mid-body, contains two sets of oppositely oriented microtubules organized in a parallel array (see Figure 20–42, p. 1166). The microtubules are probably nucleated at the nuclear surface and have their plus ends embedded in an electron dense disc in the equatorial plane. As outlined in Figure 13–71, small membrane-bounded vesicles, largely derived from the Golgi apparatus and filled with cell-wall precursors, seem to contact the microtubules on each side of the phragmoplast and to be transported along them until they reach the equatorial region. Here they fuse to form a disclike membrane-

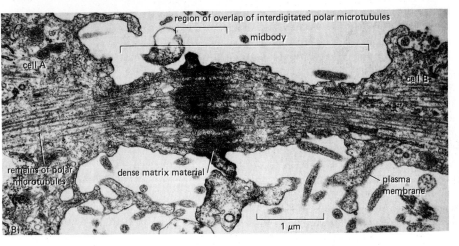

(A)   10 μm

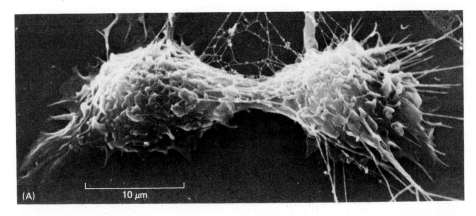

region of overlap of interdigitated polar microtubules

midbody

cell A

cell B

remains of polar microtubules

dense matrix material

plasma membrane

(B)   1 μm

Figure 13–70 (A) Scanning electron micrograph of an animal cell in culture in the process of dividing; the midbody still joins the two daughter cells. (B) Electron micrograph of the midbody of a dividing animal cell. Cleavage is virtually complete, but the daughter cells remain attached by this thin strand of cytoplasm. (A, courtesy of Guenter Albrecht-Buehler; B, courtesy of J.M. Mullins.)

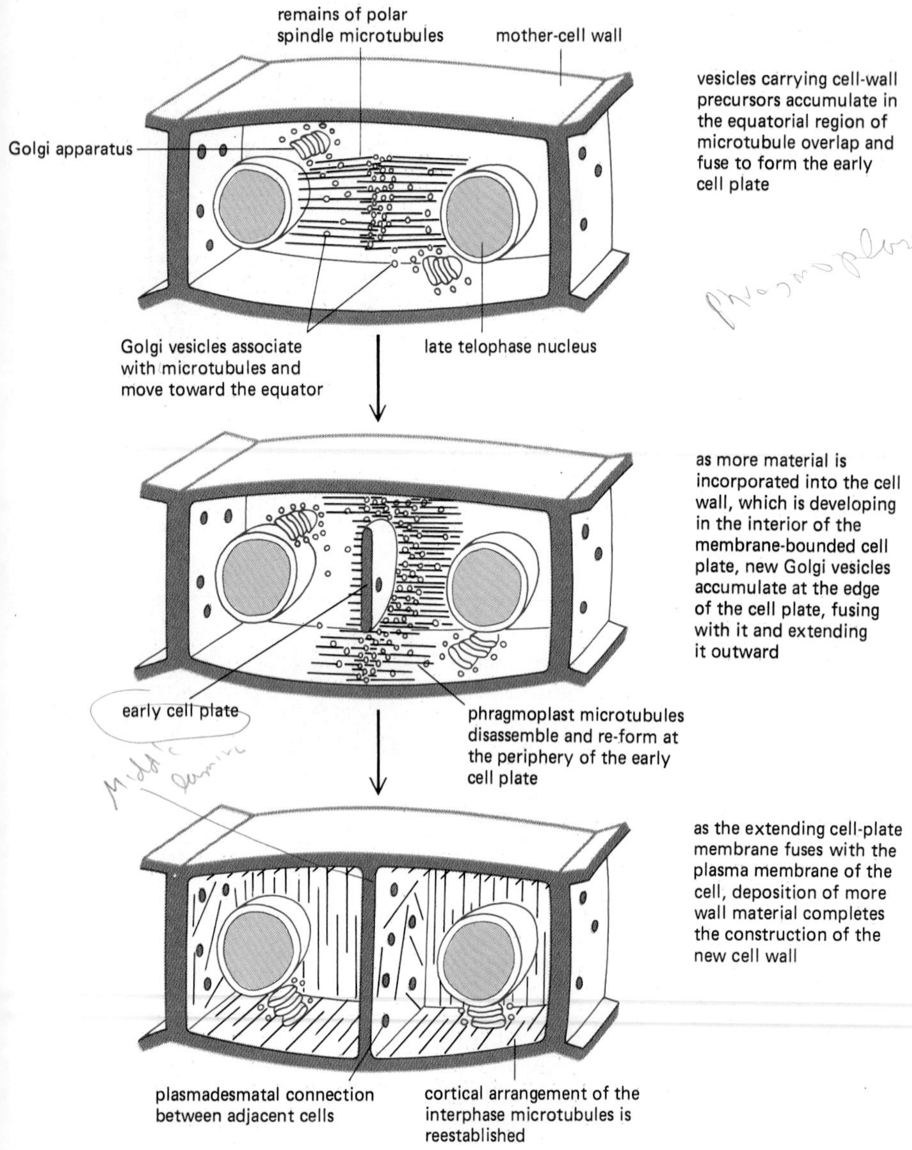

remains of polar
spindle microtubules

mother-cell wall

Golgi apparatus

vesicles carrying cell-wall
precursors accumulate in
the equatorial region of
microtubule overlap and
fuse to form the early
cell plate

Golgi vesicles associate
with microtubules and
move toward the equator

late telophase nucleus

as more material is
incorporated into the cell
wall, which is developing
in the interior of the
membrane-bounded cell
plate, new Golgi vesicles
accumulate at the edge
of the cell plate, fusing
with it and extending
it outward

early cell plate

phragmoplast microtubules
disassemble and re-form at
the periphery of the early
cell plate

as the extending cell-plate
membrane fuses with the
plasma membrane of the
cell, deposition of more
wall material completes
the construction of the
new cell wall

plasmadesmatal connection
between adjacent cells

cortical arrangement of the
interphase microtubules is
reestablished

Figure 13–71 The process of
cytokinesis in a higher plant cell
with a rigid cell wall.

bounded structure, the *early cell plate*. The polysaccharide molecules released from these vesicles assemble within the early cell plate to form pectin, hemicellulose, and other constituents of the primary cell wall. This disc now has to expand laterally to reach the original parent cell wall. To make this possible, the microtubules of the early phragmoplast are reorganized at the periphery of the early cell plate. There they attract more vesicles; these again fuse at the equator to extend the edge of the plate. This process is repeated until the growing cell plate reaches the plasma membrane of the parent cell and the membranes fuse, completely separating the two new daughter cells (see Figures 20–41 and 20–42). Cellulose microfibrils are laid down within the cell plate to complete the new cell wall (Figures 13–71 and 13–72).

Associated with the microtubules and vesicles of the forming cell plate are elements of the endoplasmic reticulum, which often remain trapped across the plate. These are subsequently transformed into *plasmodesmata*, the complex pores that traverse the mature cell wall and interconnect the cytoplasms of all the cells in the plant (see p. 1151 and Figure 20–20).

As in animal cells, mitosis and cytokinesis can be dissociated. In the endosperm tissue of seeds, for example, mitoses occur without cytokinesis, giving rise to a giant multinucleated cell. Much later, long after the mitotic spindle has disassembled, cross-walls are laid down between the separate nuclei to create new cells.

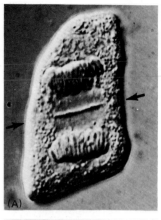

Figure 13–72 Light micrographs of cytokinesis in a plant cell. The cell plate (between the two arrows) is forming in a plane perpendicular to the micrograph. One cell (A) is viewed by differential-interference microscopy; the other (B) has been stained with gold-labeled antibodies to label the two arrays of microtubules that contribute to the phragmoplast. In both cases, arrows indicate the plane of the cell plate. (A, courtesy of Jeremy D. Pickett-Heaps; B, courtesy of Andrew Bajer.)

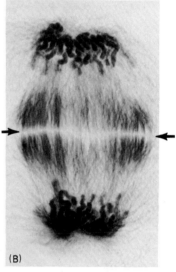

The mitotic spindle by itself is generally not sufficient to determine the exact position and shape of the cell plate. The plate's future site of junction with the mother cell wall seems to be defined very early—before mitosis has begun—by a narrow bundle of microtubules called the *preprophase band,* which is situated just beneath the plasma membrane (see p. 1177 and Figure 20–64). Although these microtubules disappear as the cell enters mitosis, they define a region of the cortex for the attachment of a radial network of actin filaments that persists throughout M phase and eventually guides the growing edge of the cell plate to the appropriate cortical site (Figure 13–73). Thus actin seems to have an important function in the division of walled cells, even though contraction plays no obvious part. Since actin is also associated with septum formation in fungal cells, it may help guide cytokinesis in all eucaryotes.

Cytokinesis Must Allow the Cytoplasmic Organelles to Be Faithfully Inherited[50]

The nucleus is only one of many organelles in a cell that cannot be duplicated without a preexisting copy. Ribosomes, for example, can assemble spontaneously from their component parts, but they cannot be made without preexisting ribosomes to synthesize the required proteins. Mitochondria and chloroplasts, on the other hand, cannot assemble spontaneously from their individual components. They can arise only from the growth and fission of the corresponding preexisting organelle (see p. 388). Likewise, the mechanisms by which other organelles, such as the Golgi apparatus and endoplasmic reticulum, grow suggest that it may not be possible to make new copies of them without the prior presence of at least part of the corresponding structure (see Chapter 8). Thus in some algae that have only one chloroplast or only one Golgi apparatus, the organelle that is present in a single copy splits in half prior to cytokinesis, and one of the two halves is segregated to each daughter cell when the cell divides (see Figure 7–67, p. 389).

radial actin network connecting phragmoplast to cell cortex

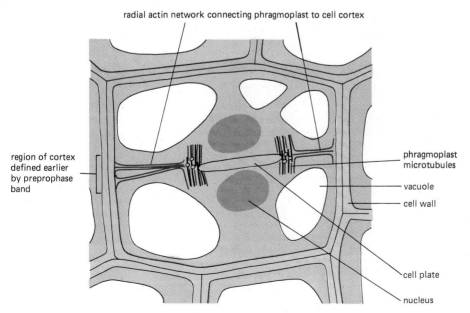

region of cortex defined earlier by preprophase band

phragmoplast microtubules

vacuole

cell wall

cell plate

nucleus

Figure 13–73 The organization of actin filaments in a plant cell undergoing cytokinesis. Actin filaments (*dark colored lines*) form a radial network that extends from the edge of the phragmoplast to the cell cortex, creating a ring around the cell. This network appears to determine the plane of cell-plate formation. A separate set of actin filaments runs parallel to the microtubules that are laying down the new cell plate in the phragmoplast. Further actin filaments (not shown here) extend to the cortex from the region of the two daughter nuclei through the large central vacuole common to plant cells (see p. 1172), helping to support thin transvacuolar strands of cytoplasm.

The orderly duplication and segregation of the centrosome in animal cells represents a similar phenomenon (see Figure 13–46).

How are the various membrane-bounded organelles besides the nucleus segregated when a higher eucaryotic cell divides? There are enough of most of them (see Table 8–1, p. 407) that a representative set of each would be expected to be partitioned into the two daughter cells when the cleavage furrow forms during cytokinesis so that each daughter cell receives a portion. Thus, while no mammalian cell can survive unless it inherits a mitochondrion, for example, it is possible that no special mechanism is required to ensure that each daughter cell ends up with one. Certainly organelles present in very large numbers will be safely inherited if, on average, their numbers simply double once each cell generation. Other organelles, such as the Golgi apparatus and the endoplasmic reticulum, break up into a set of smaller fragments and vesicles during mitosis. Their highly vesiculated forms may help guarantee their even distribution when cells divide.

In Special Cases, Selected Cell Components Can Be Segregated to One Daughter Cell Only[51]

Thus far we have tended to view cell division as a process with the sole purpose of producing two similar daughter cells. However, many of the cell divisions involved in developing and maintaining the tissues of complex multicellular organisms are manifestly unequal. This is especially true of the initial cleavages in which a large fertilized egg is subdivided into smaller cells destined to form different parts of the body (see p. 892). We have already discussed how the asymmetric positioning of the spindle during cell division can produce two cells of unequal size (p. 779), but the genesis of biochemically different daughter cells is a different problem.

A striking example of this process is provided by the behavior of a set of distinctive granules in the egg of the nematode worm *C. elegans*. These "P-granules" are spread evenly throughout the cytoplasm of the unfertilized egg, but they move to the posterior end of the cell just before the first cleavage and are therefore inherited by only one of the two daughter cells (Figure 13–74). The same type of segregation process is repeated in several subsequent cell divisions, so that the granules end up in only those cells that will form the nematode germ line (the

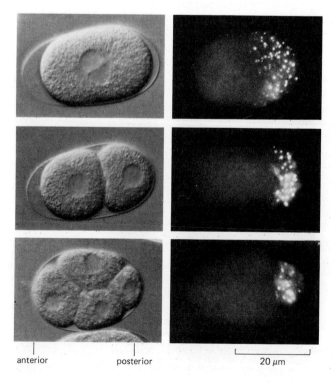

anterior   posterior   20 μm

Figure 13–74 The controlled asymmetric segregation of a cytoplasmic component to one daughter cell during each of the first two cell divisions of a fertilized egg of the nematode *C. elegans*. On the left are living cells viewed by differential-interference-contrast light microscopy; on the right are identical cells stained with an antibody against P-granules. These small granules of unknown function (0.5–1 μm in diameter) are distributed randomly throughout the cytoplasm in the unfertilized egg. (Courtesy of Susan Strome.)

eggs and sperm). It is possible that the granules play a part in controlling the distinctive fate of these cells.

The P-granules will still move to the posterior end of the cell even in mutants where the mitotic spindle is disoriented and turned at right angles to its normal position. Moreover, studies with cytoskeletal inhibitors suggest that the oriented movement of P-granules is independent of microtubules but dependent instead on actin filaments (the segregation is blocked by cytochalasin D). Although the unequal segregation of these components thus seems to be based on some asymmetric property of the actin-based cytoskeleton, the molecular mechanism for their oriented movement is unknown. It is tempting to speculate that similar mechanisms can move many normally invisible components in cells and that the unequal daughter cells that result are thereby programmed for different fates (see p. 912).

The Elaborate Mitotic Process of Higher Organisms Evolved Gradually from Procaryotic Fission Mechanisms[52]

In procaryotic cells, division of the DNA and of the cytoplasm are coupled in a very direct way. When DNA replicates, the two copies of the chromosome are attached to specialized regions of the cell membrane, which are gradually separated by the inward growth of the membrane between them. Fission takes place between the two attachment sites, so that each daughter cell captures one chromosome (Figure 13–75). With the evolution of the eucaryotes, the genome increased in complexity and the chromosomes increased in number and in size. For these organisms a more elaborate mechanism for dividing the chromosomes between daughter cells was apparently required.

Clearly, the mitotic apparatus could not have evolved all at once. In many primitive eucaryotes, such as the dinoflagellate *Crypthecodinium cohnii*, mitosis retains a membrane-attachment mechanism, with the nuclear membrane taking over the part played by the plasma membrane in procaryotes. The intermediate status of this large single-cell alga is also reflected in the biochemistry of its chromosomes, which, like those of procaryotes, have relatively little associated protein. The nuclear membrane in *C. cohnii* remains intact throughout mitosis, and the spindle microtubules remain entirely outside the nucleus. Where these spindle microtubules press on the outside of the nuclear envelope, the envelope becomes indented in a series of parallel channels (see Figure 13–75). The chromosomes become attached to the inner membrane of the nuclear envelope opposite these channels, and the separation of the chromosomes is entirely mediated on the inside of this channeled nuclear membrane. Thus the extranuclear "spindle" (which forms a rigid rod with no dynamic properties) is used merely to order the nuclear membrane and thereby define the plane of division.

A somewhat more advanced, though still extranuclear, spindle is seen in hypermastigotes, in which the nuclear envelope again remains intact throughout mitosis. These large protozoa from the guts of insects provide a particularly clear illustration of the independence of spindle elongation and the chromosome movements that separate the chromatids, since the sister kinetochores become separated by the growth of the nuclear membrane (to which they are attached) before becoming attached to the spindle. Only when the kinetochores are near the poles of the spindle do they acquire the kinetochore fibers needed to attach them to the spindle. Because the spindle fibers remain separated from the chromosomes by the nuclear envelope, the kinetochore fibers, which are formed outside the nucleus, must somehow attach to the chromosomes through the nuclear membranes. After this attachment has occurred, the kinetochores are drawn poleward in a conventional manner (see Figure 13–75).

A further stage in the evolution of mitotic mechanisms may be represented by organisms that have spindles inside an intact nucleus. In yeasts the spindle consists of a continuous intranuclear bundle of microtubules that extends from pole to pole and elongates during the course of mitosis (see Figures 13–75 and 13–16). In other species, such as diatoms, the continuous spindle is replaced by the familiar arrangement of two half-spindles, the microtubules of which inter-

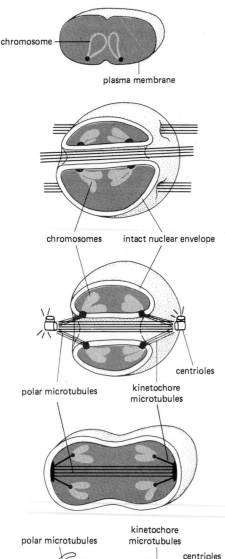

BACTERIA
daughter chromosomes attached to the plasma membrane are separated by the growth of membrane between them

chromosome

plasma membrane

TYPICAL DINOFLAGELLATES
several bundles of microtubules pass through tunnels in the intact nuclear envelope to establish the polarity of division; chromosomes move apart in association with the inner nuclear membrane without being attached to the microtubule bundles

chromosomes   intact nuclear envelope

HYPERMASTIGOTES AND SOME UNUSUAL DINOFLAGELLATES
a single central spindle between centrioles is formed in a tunnel through the intact nuclear envelope; chromosomes are attached by their kinetochores to the nuclear membrane and interact with the spindle poles via kinetochore microtubules

centrioles

polar microtubules   kinetochore microtubules

YEASTS AND DIATOMS
nuclear envelope remains intact; continuous polar spindle microtubules form inside the nucleus and run between the two spindle pole bodies associated with the nuclear envelope; a single kinetochore microtubule attaches each chromosome to a pole

polar microtubules   kinetochore microtubules

centrioles

HIGHER ORGANISMS
the spindle begins to form outside the nucleus; at prometaphase the nuclear envelope breaks down to allow chromosomes to capture spindle microtubules, which now become kinetochore microtubules

fragments of nuclear envelope

Figure 13–75 Different chromosome separation mechanisms are used by different organisms. Some of these may have been intermediary stages in the evolution of the mitotic spindle of higher organisms. For all the examples except bacteria, only the central nuclear region of the cell is shown.

digitate to form a zone of overlap. In both yeasts and diatoms the spindle is attached to chromosomes by their kinetochores, and the chromosomes are segregated in a way closely similar to that described for mammalian cells—except that the entire process generally occurs within the confines of the nuclear envelope. At present there is no convincing explanation for why higher plants and animals have instead evolved a mitotic apparatus that requires the controlled and reversible dissolution of the nuclear envelope.

Summary

The process of cell division consists of nuclear division (mitosis) followed by cytoplasmic division (cytokinesis). Mitosis begins with prophase, a transitional period during which the centrosome splits apart to form the two spindle poles that will organize the subsequent intracellular movements. At the same time, the beginning of M phase is accompanied by a marked increase in the phosphorylation of specific proteins. Perhaps as a result, a mitotic cell contains an unusually dynamic microtubule array. After the nuclear envelope breaks down in prometaphase, the kinetochores on condensed chromosomes can capture and stabilize subsets of microtubules from the large numbers that continually grow out from each spindle pole. Opposing pole-directed forces pull on these kinetochore microtubules, creating a tension that causes the chromosomes to line up at the spindle equator during metaphase. At anaphase this tension is suddenly released as sister chromatids detach from each other and are pulled to opposite poles. In addition, the two mitotic poles often move apart. In the final, telophase stage of mitosis, the nuclear envelope re-forms on the surface of each group of separated chromosomes as the proteins phosphorylated at the onset of M phase are dephosphorylated.

Cell division ends as the cytoplasmic contents are divided by the process of cytokinesis and the chromosomes decondense and resume RNA synthesis. Cytokinesis appears to be guided by organized bundles of actin filaments in eucaryotic cells as diverse as animals, plants, and fungi. Large membrane-bounded organelles such as the Golgi apparatus and the endoplasmic reticulum break up into smaller fragments and vesicles during M phase, which presumably ensures their even distribution into daughter cells. However, cytokinesis can also involve a programmed asymmetric distribution of materials. A particular cell can divide into one small cell and one large one, for example, or a specific cytoplasmic component can be moved to one side of a cell prior to cytokinesis so that it is inherited by only one of the two otherwise equal daughter cells.

References

General

Baserga, R. The Biology of Cell Reproduction. Cambridge, MA: Harvard University Press, 1985.

Beach, D.; Basilico, C.; Newport, J., eds. Cell Cycle Control in Eucaryotes. Cold Spring Harbor, NY: Cold Spring Harbor Laboratory, 1988.

John, P.C.L., ed. The Cell Cycle. Cambridge, U.K.: Cambridge University Press, 1981.

Mitchison, J.M. The Biology of the Cell Cycle. Cambridge, U.K.: Cambridge University Press, 1971.

Pollack, R. Readings in Mammalian Cell Culture, 2nd ed. Cold Spring Harbor, NY: Cold Spring Harbor Laboratory, 1981. (An anthology, including many important papers on cell growth and division.)

Prescott, D.M. Reproduction of Eucaryotic Cells. New York: Academic Press, 1976.

Cited

1. Baserga, R. The Biology of Cell Reproduction. Cambridge, MA: Harvard University Press, 1985.

   Howard, A.; Pelc, S.R. Nuclear incorporation of P^{32} as demonstrated by autoradiographs. Exp. Cell Res. 2:178–187, 1951.

   Lloyd, D.; Poole, R.K.; Edwards, S.W. The Cell Division Cycle. New York: Academic Press, 1982.

   Mitchison, J.M. The Biology of the Cell Cycle. Cambridge, U.K.: Cambridge University Press, 1971.

2. Van Dilla, M.A.; Trujillo, T.T.; Mullaney, P.F.; Coulter, J.R. Cell microfluorometry: a method for rapid fluorescence measurement. Science 163:1213–1214, 1969.

3. Creanor, J.; Mitchison, J.M. Patterns of protein synthesis during the cell cycle of the fission yeast Schizosaccharomyces pombe. J. Cell Sci. 58:263–285, 1982.

   Pardee, A.B.; Coppock, D.L.; Yang, H.C. Regulation of cell proliferation at the onset of DNA synthesis. J. Cell Sci., Suppl. 4:171–180, 1986.

4. Rao, P.N.; Johnson, R.T. Mammalian cell fusion: studies on the regulation of DNA synthesis and mitosis. Nature 225:159–164, 1970.

5. Xeros, N. Deoxyriboside control and synchronization of mitosis. Nature 194:682–683, 1962.

6. Mitchison, J.M. The Biology of the Cell Cycle, pp. 234–244. Cambridge, U.K.: Cambridge University Press, 1971.

   Schlegel, R.; Pardee, A.B. Caffeine-induced uncoupling of mitosis from the completion of DNA replication in mammalian cells. Science 232:1264–1266, 1986.

7. Johnson, R.T.; Rao, P.N. Mammalian cell fusion: induction of premature chromosome condensation in interphase nuclei. Nature 226:717–722, 1970.

8. Blow, J.J.; Laskey, R.A. A role for the nuclear envelope in controlling DNA replication within the cell cycle. Nature 332:546–548, 1988.

9. Kirschner, M.; Newport, J.; Gerhart, J. The timing of early developmental events in Xenopus. Trends Genet. 1:41–47, 1985.

10. Gerhart, J.; Wu, M.; Kirschner, M. Cell cycle dynamics of an M-phase-specific cytoplasmic factor in Xenopus laevis oocytes and eggs. J. Cell Biol. 98:1247–1255, 1984.

    Lohka, M.J.; Hayes, M.K.; Maller, J.L. Purification of maturation-promoting factor, an intracellular regulator of early mitotic events. Proc. Natl. Acad. Sci. USA 85:3009–3013, 1988.

    Newport, J.; Kirschner, M. Regulation of the cell cycle during early Xenopus development. Cell 37:731–742, 1984.

11. Hara, K.; Tydeman, P.; Kirschner, M. A cytoplasmic clock with the same period as the division cycle in Xenopus eggs. Proc. Natl. Acad. Sci. USA 77:462–466, 1980.

    Kimelman, D.; Kirschner, M.; Scherson, T. The events of the midblastula transition in Xenopus are regulated by changes in the cell cycle. Cell 48:399–407, 1987.

    Pines, J.; Hunt, T. Molecular cloning and characterization of the mRNA for cyclin from sea urchin eggs. EMBO J. 6:2987–2995, 1987.

    Swenson, K.I.; Farrell, K.M.; Ruderman, J.V. The clam embryo protein cyclin A induces entry into M phase and the resumption of meiosis in *Xenopus* oocytes. *Cell* 47:861–870, 1986.

12. Nurse, P. Cell cycle control genes in yeast. *Trends Genet.* 1:51–55, 1985.

Pringle, J.R.; Hartwell, L.H. The *Saccharomyces cerevisiae* cell cycle. In The Molecular Biology of the Yeast *Saccharomyces*, Life Cycle and Inheritance (J.N. Strathern, E.W. Jones, J.R. Broach, eds.), pp. 97–142. Cold Spring Harbor, NY: Cold Spring Harbor Laboratory, 1981.

Watson, J.D.; Hopkins, N.H.; Roberts, J.W.; Weiner, A.M. Molecular Biology of the Gene, 4th ed., pp. 550–592. Menlo Park, CA: Benjamin-Cummings, 1987.

13. Hartwell, L.H. Cell division from a genetic perspective. *J. Cell Biol.* 77:627–637, 1978.

Nurse, P. Genetic analysis of the cell cycle. *Symp. Soc. Gen. Microbiol.* 31:291–315, 1981.

14. Hayles, J.; Nurse, P. Cell cycle regulation in yeast. *J. Cell Sci.,* Suppl. 4:155–170, 1986.

Johnston, G.C.; Pringle, J.R.; Hartwell, L.H. Coordination of growth with cell division in the yeast *Saccharomyces cerevisiae*. *Exp. Cell Res.* 105:79–98, 1977.

Prescott, D.M. Changes in nuclear volume and growth rate and prevention of cell division in Amoeba proteus resulting from cytoplasmic amputations. *Exp. Cell Res.* 11:94–98, 1956.

15. Dunphy, W.G.; Brizuela, L.; Beach, D.; Newport, J. The *Xenopus* cdc2 protein is a component of MPF, a cytoplasmic regulator of mitosis. *Cell* 54:423–431, 1988.

Gautier, J.; Norbury, C.; Lohka, M.; Nurse, P.; Maller, J. Purified maturation-promoting factor contains the product of a *Xenopus* homolog of the fission yeast cell cycle control gene cdc2⁺. *Cell* 54:433–439, 1988.

Lee, M.G.; Nurse, P. Complementation used to clone a human homologue of the fission yeast cell cycle control gene cdc2. *Nature* 327:31–35, 1987.

Murray, A.W. A mitotic inducer matures. *Nature* 335:207–208, 1988.

Solomon, M.; et al. Cyclin in fission yeast. *Cell* 54:738–740, 1988.

16. Cheng, H.; LeBlond, C.P. Origin, differentiation and renewal of the four main epithelial cell types in the mouse small intestine. *Am. J. Anat.* 141:461–480, 1974.

Goss, R.J. The Physiology of Growth. New York: Academic Press, 1978.

17. Pardee, A.B. A restriction point for control of normal animal cell proliferation. *Proc. Natl. Acad. Sci. USA* 71:1286–1290, 1974.

18. Brooks, R.F. The transition probability model: successes, limitations and deficiencies. In Temporal Order (L. Rensing, N.I. Jaeger, eds.). Berlin: Springer, 1985.

Shields, R. Transition probability and the origin of variation in the cell cycle. *Nature* 267:704–707, 1977.

Smith, J.A.; Martin, L. Do cells cycle? *Proc. Natl. Acad. Sci. USA* 70:1263–1267, 1973.

19. Deuel, T.F. Polypeptide growth factors: roles in normal and abnormal growth. *Annu. Rev. Cell Biol.* 3:443–492, 1987.

Evered, D.; Nugent, J.; Whelan, J., eds. Growth Factors in Biology and Medicine. Ciba Foundation Symposium 116. London: Pitman, 1985.

Sato, G., ed. Hormones and Cell Culture. Cold Spring Harbor, NY: Cold Spring Harbor Laboratory, 1979.

Wang, J.L.; Hsu, Y.-M. Negative regulators of cell growth. *Trends Biochem. Sci.* 11:24–27, 1986.

20. Ross, R.; Raines, E.W.; Bowen-Pope, D.F. The biology of platelet-derived growth factor. *Cell* 46:155–169, 1986.

Stiles, C.D. The molecular biology of platelet-derived growth factor. *Cell* 33:653–655, 1983.

21. Dunn, G.A.; Ireland, G.W. New evidence that growth in 3T3 cell cultures is a diffusion-limited process. *Nature* 312:63–65, 1984.

Holley, R.W.; Kiernan, J.A. "Contact inhibition" of cell division in 3T3 cells. *Proc. Natl. Acad. Sci. USA* 60:300–304, 1968.

Stoker, M.G.P. Role of diffusion boundary layer in contact inhibition of growth. *Nature* 246:200–203, 1973.

22. Folkman, J.; Moscona, A. Role of cell shape in growth control. *Nature* 273:345–349, 1978.

O'Neill, C.; Jordan, P.; Ireland, G. Evidence for two distinct mechanisms of anchorage stimulation in freshly explanted and 3T3 mouse fibroblasts. *Cell* 44:489–496, 1986.

23. Brooks, R.F. Regulation of the fibroblast cell cycle by serum. *Nature* 260:248–250, 1976.

Larsson, O.; Zetterberg, A.; Engstrom, W. Consequences of parental exposure to serum-free medium for progeny cell division. *J. Cell Sci.* 75:259–268, 1985.

Zetterberg, A.; Larsson, O. Kinetic analysis of regulatory events in G1 leading to proliferation or quiescence of Swiss 3T3 cells. *Proc. Natl. Acad. Sci. USA* 82:5365–5369, 1985.

24. Baserga, R. The Biology of Cell Reproduction, pp. 103–113. Cambridge, MA: Harvard University Press, 1985.

Larsson, O.; Dafgard, E.; Engstrom, W.; Zetterberg, A. Immediate effects of serum depletion on dissociation between growth in size and cell division in proliferating 3T3 cells. *J. Cell. Physiol.* 127:267–273, 1986.

25. Bauer, E.; et al. Diminished response of Werner's syndrome fibroblasts to growth factors PDGF and FGF. *Science* 234:1240–1243, 1986.

Hayflick, L. The limited *in vitro* lifetime of human diploid cell strains. *Exp. Cell Res.* 37:614–636, 1965.

Holliday, R., ed. Genes, Proteins, and Cellular Aging. New York: Van Nostrand, 1986. (An anthology of papers on cell senescence.)

Loo, D.T.; Fuquay, J.I.; Rawson, C.L.; Barnes, D.W. Extended culture of mouse embryo cells without senescence: inhibition by serum. *Science* 236:200–202, 1987.

Rheinwald, J.G.; Green, H. Epidermal growth factor and the multiplication of cultured human epidermal keratinocytes. *Nature* 265:421–424, 1977.

Smith, J.R.; Whitney, R.G. Intraclonal variation in proliferative potential of human diploid fibroblasts: stochastic mechanism for cellular aging. *Science* 207:82–84, 1980.

26. Kahn, P.; Graf, T., eds. Oncogenes and Growth Control. Berlin: Springer, 1986.

Watson, J.D.; Hopkins, N.H.; Roberts, J.W.; Weiner, A.M. Molecular Biology of the Gene, 4th ed., pp. 961–1096. Menlo Park, CA: Benjamin-Cummings, 1987.

27. Feramisco, J.; Ozanne, B.; Stiles, C., eds. Cancer Cells 3: Growth Factors and Transformation. Cold Spring Harbor, NY: Cold Spring Harbor Laboratory, 1985.

Temin, H.M.; Rubin, H. Characteristics of an assay for Rous sarcoma virus and Rous sarcoma cells in tissue culture. *Virology* 6:669–688, 1958.

28. Bishop, J.M. Viral oncogenes. *Cell* 42:23–38, 1985.

Martin, G.S. Rous sarcoma virus: a function required for the maintenance of the transformed state. *Nature* 227:1021–1023, 1970.

Swanstrom, R.; Parker, R.C.; Varmus, H.E.; Bishop, J.M. Transduction of a cellular oncogene—the genesis of Rous sarcoma virus. *Proc. Natl. Acad. Sci. USA* 80:2519–2523, 1983.

Varmus, H. Cellular and viral oncogenes. In The Molecular Basis of Blood Diseases (G. Stamatoyannopoulos, A.W. Nienhuis, P. Leder, P.W. Majerus, eds.), pp. 271–345. Philadelphia: Saunders, 1987.

29. Harris, H. The genetic analysis of malignancy. *J. Cell Sci.,* Suppl. 4:431–444, 1986.

Klein, G. The approaching era of the tumor suppressor genes. *Science* 238:1539–1545, 1987.

Weinberg, R.A. A molecular basis of cancer. *Sci. Am.* 249(5):126–142, 1983.

30. Doolittle, R.F.; et al. Simian sarcoma virus oncogene, v-*sis*, is derived from the gene (or genes) encoding a platelet-derived growth factor. *Science*, 221:275–277, 1983.

Marshall, C.J. Oncogenes. *J. Cell Sci.*, Suppl. 4:417–430, 1986.

Waterfield, M.D.; et al. Platelet-derived growth factor is structurally related to the putative transforming protein p28^{sis} of simian sarcoma virus. *Nature* 304:35–39, 1983.

31. Almendral, J.M.; et al. Complexity of the early genetic response to growth factors in mouse fibroblasts. *Mol. Cell. Biol.* 8:2140–2148, 1988.

Hunter, T. The proteins of oncogenes. *Sci. Am.* 251(2):70–79, 1984.

Robertson, M. Molecular associations and conceptual connections. *Nature* 334:100–102, 1988.

Yu, C.L.; Tsai, M.H.; Stacey, D.W. Cellular *ras* activity and phospholipid metabolism. *Cell* 52:63–71, 1988.

32. Herman, B.; Pledger, W.J. Platelet-derived growth factor-induced alterations in vinculin and actin distributions in BALB/c-3T3 cells. *J. Cell Biol.* 100:1031–1040, 1985.

Hirst, R.; Horwitz, A.; Buck, C.; Rohrschneider, L. Phosphorylation of the fibronectin receptor complex in cells transformed by oncogenes that encode tyrosine kinases. *Proc. Natl. Acad. Sci. USA* 83:6470–6474, 1986.

Jove, R.; Hanafusa, H. Cell transformation by the viral *src* oncogene. *Annu. Rev. Cell Biol.* 3:31–56, 1987.

33. Burger, M.M. Proteolytic enzymes initiating cell division and escape from contact inhibition of growth. *Nature* 227:170–171, 1970.

Sullivan, L.M.; Quigley, J.P. An anticatalytic monoclonal antibody to avian plasminogen activator: its effect on behavior of RSV-transformed chick fibroblasts. *Cell* 45:905–915, 1986.

34. Adamson, E.D. Oncogenes in development. *Development* 99:449–471, 1987.

Bryant, P.J.; Bryant, S.V.; French, V. Biological regeneration and pattern formation. *Sci. Am.* 237(1):67–81, 1977.

Fankhauser, G. Nucleo-cytoplasmic relations in amphibian development. *Int. Rev. Cytol.* 1:165–193, 1952.

35. Bajer, A.S.; Mole-Bajer, J. Spindle Dynamics and Chromosome Movements. New York: Academic Press, 1972.

McIntosh, J.R. Mechanisms of mitosis. *Trends Biochem. Sci.* 9:195–198, 1984.

Mazia, C. Mitosis and the physiology of cell division. In The Cell (J. Brachet, A.E. Mirsky, eds.), Vol. 3, pp. 77–412. London: Academic Press, 1961.

Wilson, E.B. The Cell in Development and Heredity, 3rd ed. with corrections. New York: Macmillan Company, 1928. (Reprinted, New York: Garland, 1987.)

36. Inoué, S.; Sato, H. Cell motility by labile association of molecules: the nature of mitotic spindle fibers and their role in chromosome movement. *J. Gen. Physiol.* 50:259–292, 1967.

Karsenti, E.; Maro, B. Centrosomes and the spatial distribution of microtubules in animal cells. *Trends Biochem. Sci.* 11:460–463, 1986.

Kirschner, M.; Mitchison, T. Beyond self-assembly: from microtubules to morphogenesis. *Cell* 45:329–342, 1986.

Kuriyama, R.; Borisy, G.G. Microtubule-nucleating activity of centrosomes in Chinese hamster ovary cells is independent of the centriole cycle but coupled to the mitotic cycle. *J. Cell Biol.* 91:822–826, 1981.

Saxton, W.M.; et al. Tubulin dynamics in cultured mammalian cells. *J. Cell Biol.* 99:2175–2186, 1984.

37. Clarke, L.; Carbon, J. The structure and function of yeast centromeres. *Annu. Rev. Genet.* 19:29–56, 1985.

Earnshaw, W.C.; et al. Molecular cloning of cDNA for CENP-B, the major human centromere autoantigen. *J. Cell Biol.* 104:817–829, 1987.

Mitchison, T.J.; Evans, L.; Schulze, E.; Kirschner, M. Sites of microtubule assembly and disassembly in the mitotic spindle. *Cell* 45:515–527, 1986.

Peterson, J.B.; Ris, H. Electron microscopic study of the spindle and chromosome movement in the yeast *Saccharomyces cerevisiae*. *J. Cell Sci.* 22:219–242, 1976.

Rieder, C.L. The formation, structure and composition of the mammalian kinetochore fiber. *Int. Rev. Cytol.* 79:1–58, 1982.

38. Euteneuer, U.; McIntosh, J.R. Structural polarity of kinetochore microtubules in PtK$_1$ cells. *J. Cell Biol.* 89:338–345, 1981.

Mitchison, T.J.; Kirschner, M.W. Properties of the kinetochore *in vitro*. II: Microtubule capture and ATP-dependent translocation. *J. Cell Biol.* 101:766–777, 1985.

Rieder, C.L.; Davison, E.A.; Jensen, C.W.; Cassimeris, L.; Salmon, E.D. Oscillatory movements monooriented of chromosomes and their position relative to the spindle pole result from the ejection properties of the aster and half-spindle. *J. Cell Biol.* 103:581–591, 1986.

Roos, U.-P. Light and electron microscopy of rat kangaroo cells in mitosis. III. Patterns of chromosome behavior during prometaphase. *Chromosoma* 54:363–385, 1976.

39. Begg, D.A.; Ellis, G.W. Micromanipulation studies of chromosome movement. *J. Cell Biol.* 82:528–541, 1979.

Nicklas, R.B. The forces that move chromosomes in mitosis. *Annu. Rev. Biophys. Biophys. Chem.* 17:431–450, 1988.

Nicklas, R.B.; Kubai, D.F. Microtubules, chromosome movement, and reorientation after chromosomes are detached from the spindle by micromanipulation. *Chromosoma* 92:313–324, 1985.

40. Hays, T.S.; Wise, D.; Salmon, E.D. Traction force on a kinetochore at metaphase acts as a linear function of kinetochore fiber length. *J. Cell Biol.* 93:374–382, 1982.

McNeill, P.A.; Berns, M.W. Chromosome behavior after laser microirradiation of a single kinetochore in mitotic PtK2 cells. *J. Cell Biol.* 88:543–553, 1981.

Ostergren, G. The mechanism of coordination in bivalents and multivalents. The theory of orientation by pulling. *Hereditas* 37:85–156, 1951.

41. Hepler, P.K.; Callaham, D.A. Free calcium increases during anaphase in stamen hair cells of *Tradescantia*. *J. Cell Biol.* 105:2137–2143, 1987.

Murray, A.W.; Szostak, J.W. Chromosome segregation in mitosis and meiosis. *Annu. Rev. Cell Biol.* 1:289–315, 1985.

Wolniak, S.M. The regulation of mitotic spindle function. *Biochem. Cell Biol.* 66:490–514, 1988.

42. Ris, H. The anaphase movement of chromosomes in the spermatocytes of grasshoppers. *Biol. Bull. (Woods Hole)* 96:90–106, 1949.

43. Gorbsky, G.J.; Sammak, P.J.; Borisy, G.G. Microtubule dynamics and chromosome motion visualized in living anaphase cells. *J. Cell Biol.* 106:1185–1192, 1988.

Mitchison, T.J. Microtubule dynamics and kinetochore function in mitosis. *Annu. Rev. Cell Biol.* 4:527–549, 1988.

Nicklas, R.B. Measurements of the force produced by the mitotic spindle in anaphase. *J. Cell Biol.* 97:542–548, 1983.

Spurck, T.P.; Pickett-Heaps, J.D. On the mechanism of anaphase A: Evidence that ATP is needed for microtubule disassembly and not generation of polewards force. *J. Cell Biol.* 105:1691–1705, 1987.

44. Aist, J.R.; Berns, M.W. Mechanics of chromosome separation during mitosis in *Fusarium* (Fungi imperfecti): new evidence from ultrastructural and laser microbeam experiments. *J. Cell Biol.* 91:446–458, 1981.

Masuda, H.; Cande, W.Z. The role of tubulin polymerization during spindle elongation *in vitro*. *Cell* 49:193–202, 1987.

Pickett-Heaps, J.D. Mitotic mechanisms: an alternative view. *Trends Biochem. Sci.* 11:504–507, 1986.

45. Gerace, L.; Blobel, G. The nuclear envelope lamina is reversibly depolymerized during mitosis. *Cell* 19:277–287, 1980.

Gerace, L.; Burke, B. Functional organization of the nuclear envelope. *Annu. Rev. Cell Biol.* 4:335–374, 1988.

Swanson, J.A.; McNeil, P.L. Nuclear reassembly excludes large macromolecules. *Science* 238:548–550, 1987.

46. Kirschner, M.W.; Newport, J.; Gerhart, J. The timing of early developmental events in *Xenopus. Trends Genet.* 1:41–47, 1985.

47. Rappaport, R. Establishment of the mechanism of cytokinesis in animal cells. *Int. Rev. Cytol.* 105:245–281, 1986.

Tolle, H.G.; Weber, K.; Osborn, M. Keratin filament disruption in interphase and mitotic cells—how is it induced? *Eur. J. Cell Biol.* 43:35–47, 1987.

White, J.G.; Borisy, G.G. On the mechanism of cytokinesis in animal cells. *J. Theor. Biol.* 101:289–316, 1983.

48. DeLozanne, A.; Spudich, J.A. Disruption of the dictyostelium myosin heavy chain gene by homologous recombination. *Science* 236: 1086–1091, 1987.

Knecht, D.A.; Loomis, W.F. Antisense RNA inactivation of myosin heavy chain gene depression in Dictyostelium Discoideum. *Science* 236:1081–1086, 1987.

Mabuchi, I.; Okuno, M. The effect of myosin antibody on the division of starfish blastomeres. *J. Cell Biol.* 74:251–263, 1977.

Schroeder, T.E. Actin in dividing cells: contractile ring filaments bind heavy meromyosin. *Proc. Natl. Acad. Sci. USA* 70:1688–1692, 1973.

49. Gunning, B.E.S.; Wick, S.M. Preprophase bands, phragmoplasts and spatial control of cytokinesis. *J. Cell Sci. Suppl.* 2:157–179, 1985.

Huffaker, T.C.; Hoyt, M.A.; Botstein, D. Genetic analysis of the yeast cytoskeleton. *Annu. Rev. Genet.* 21:259–284, 1987.

Lloyd, C.W. The plant cytoskeleton: the impact of fluorescence microscopy. *Annu. Rev. Plant Physiol.* 38:119–139, 1987.

Pickett-Heaps, J.D.; Northcote, D.H. Organization of microtubules and endoplasmic reticulum during mitosis and cytokinesis in wheat meristems. *J. Cell Sci.* 1:109–120, 1966.

50. Lucocq, J.M.; Warren, G. Fragmentation and partitioning of the Golgi apparatus during mitosis in HeLa cells. *EMBO J.* 6:3239–3246, 1987.

51. Hill, D.P.; Strome, S. An analysis of the role of microfilaments in the establishment and maintenance of asymmetry in *Caenorhabditis elegans. Dev. Biol.* 125:75–84, 1988.

Kenyon, C. Cell lineage and the control of *Caenorhabditis elegans* development. *Phil. Trans. R. Soc. Lond. (Biol.)* 312:21–38, 1985.

52. Donachie, W.D.; Robinson, A.C. Cell division: parameter values and the process in *Escherichia coli* and *Salmonella typhimurium*. Cellular and Molecular Biology (F.C. Neidhardt, et al., eds.), pp. 1578–1593. Washington, DC: American Society for Microbiology, 1987.

Heath, I.B. Variant mitoses in lower eukaryotes: indicators of the evolution of mitosis? *Int. Rev. Cytol.* 64:1–80, 1980.

Kubai, D.F. The evolution of the mitotic spindle. *Int. Rev. Cytol.* 43:167–227, 1975.

Wise, D. The diversity of mitosis: the value of evolutionary experiments. *Biochem. Cell Biol.* 66:515–529, 1988.

Cell Adhesion, Cell Junctions, and the Extracellular Matrix

Most of the cells in multicellular animals are organized into cooperative assemblies called *tissues*, which are in turn associated in various combinations as larger functional units called *organs* (Figure 14–1). The cells in tissues are usually in contact with a complex network of secreted extracellular macromolecules referred to as the *extracellular matrix*. This matrix helps to hold cells and tissues together, and it provides an organized lattice within which cells can migrate and interact with one another. In some cases cells are attached to the matrix at specialized regions of their plasma membrane called *cell-matrix junctions;* cells in direct contact with neighboring cells are often linked to each other at specialized *intercellular junctions.*

It is helpful to distinguish two major categories of animal tissue, in which the parts played by the matrix and intercellular junctions are radically different. In *connective tissues* (see p. 802), extracellular matrix is plentiful and cells are sparsely distributed within it (see Figure 14–1). The matrix is rich in fibrous polymers, especially collagen, and it is the matrix—rather than the cells—that bears most of the stresses to which the tissue is subjected. The cells are attached to components of the matrix, on which they may exert force, but mechanical attachments between one cell and another are relatively unimportant. In *epithelial tissues*, by contrast, cells are tightly bound together into sheets (called *epithelia*); extracellular matrix is scanty, and most of the volume is occupied by cells. Here the cells themselves,

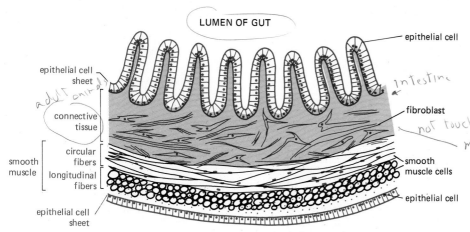

LUMEN OF GUT

epithelial cell

epithelial cell sheet

connective tissue

circular fibers

smooth muscle

longitudinal fibers

epithelial cell sheet

fibroblast

smooth muscle cells

epithelial cell

Figure 14–1 Simplified drawing of a cross section through the wall of the intestine, showing how this organ is constructed from epithelial, connective, and muscle tissues. Each tissue is an organized assembly of cells held together by intercellular junctions, extracellular matrix, or both.

rather than the matrix, bear most of the stresses, by means of strong *intracellular* protein filaments (components of the cytoskeleton) that criss-cross through the cytoplasm of each epithelial cell; the filaments are attached to the inside of the plasma membrane, where a specialized junction is formed with the surface of another cell or with the underlying extracellular matrix.

How are the epithelial tissues and the connective tissues in an organism related to each other? Epithelial cell sheets line all the cavities and free surfaces of the body, and the specialized junctions between the cells enable these sheets to form barriers to the movement of water, solutes, and cells from one body compartment to another. As illustrated in Figure 14–1, epithelial sheets almost always rest on a supporting bed of connective tissue, which may attach them to other tissues (such as muscle) that do not themselves have either strictly epithelial or strictly connective-tissue organization.

In this chapter we first discuss the structure and organization of cell-cell and cell-matrix junctions (collectively called *cell junctions*) and of the extracellular matrix, and then we consider how cells recognize one another in the process of assembling into tissues and organs.

Cell Junctions[1]

Specialized **cell junctions** are particularly important and plentiful in epithelia, but they occur at many points of cell-cell and cell-matrix contact in all tissues. Most of these junctions are too small to be resolved by light microscopy. They can be visualized, however, using either conventional or freeze-fracture electron microscopy, both of which show that the interacting plasma membranes (and often the underlying cytoplasm and the intervening intercellular space as well) are highly specialized in these regions. Cell junctions can be classified into three functional groups: (1) **occluding junctions,** which seal cells together in an epithelial cell sheet in a way that inhibits even small molecules from leaking from one side of the sheet to the other; (2) **anchoring junctions,** which mechanically attach cells (and their cytoskeletons) to their neighbors or to the extracellular matrix; and (3) **communicating junctions,** which mediate the passage of chemical or electrical signals from one interacting cell to its partner.

The major kinds of intercellular junctions within each class are listed in Table 14–1. *Tight junctions* are the main occluding junctions; *adherens junctions* and *desmosomes* are the main types of anchoring junctions; and *gap junctions, chemical synapses* formed by nerve cells, and *plasmodesmata* formed between plant cells are the main types of communicating junctions. Since chemical synapses and plasmodesmata will be considered in detail in Chapters 19 and 20, respectively, they will not be discussed further here.

Table 14–1   A Functional Classification of Cell Junctions

1. **Occluding junctions** (tight junctions)

2. **Anchoring junctions**
    a. actin filament attachment sites (adherens junctions)
        i. cell-cell (e.g., adhesion belts)
        ii. cell-matrix (e.g., focal contacts)
    b. intermediate filament attachment sites
        i. cell-cell (desmosomes)
        ii. cell-matrix (hemidesmosomes)

3. **Communicating junctions**
    a. gap junctions
    b. chemical synapses
    c. plasmodesmata (plants only)*

---

*These are the only junctions found in plants.

14-4 14-5 Tight Junctions Form a Permeability Barrier Across Epithelial Cell Sheets[2]

Despite the extensive structural and biochemical differences among various types of epithelia, all epithelia have at least one important function in common: they serve as selective permeability barriers, separating fluids on each side that have different chemical compositions. **Tight junctions** play a doubly important role in maintaining the selective-barrier function of cell sheets. This is well illustrated in the epithelium of the mammalian gut.

The epithelial cells lining the small intestine keep most of the gut contents in the inner cavity (the lumen). At the same time, however, they must pump selected nutrients across the cell sheet into the extracellular fluid permeating the connective tissue on the other side (see Figure 14–1), from which the nutrients diffuse into blood vessels. This *transcellular* (or *transepithelial*) *transport* depends on two sets of membrane-bound carrier proteins: one is confined to the *apical surface* (the surface facing the lumen) and actively pumps selected molecules into the epithelial cell from the lumen of the gut; the other, which is confined to the *basolateral* (basal and lateral) *surface,* allows the same molecules to leave the cell by facilitated diffusion into the extracellular fluid on the other side (Figure 14–2). If this directional pumping is to be maintained, the apical set of carrier proteins must not be allowed to migrate to the basolateral surface of the cell, and the basolateral set must not be allowed to migrate to the apical surface. Furthermore, the spaces between epithelial cells must be sealed so that the transported molecules cannot diffuse back into the lumen through the intercellular space down the concentration gradients created by the transepithelial transport process.

The tight junctions between the epithelial cells are thought to block both these kinds of diffusion. First, they function as barriers to the diffusion of membrane proteins between apical and basolateral domains of the plasma membrane (see Figure 14–2). This undesirable diffusion of membrane constituents occurs if tight junctions are disrupted, for example by removing the extracellular Ca^{2+}

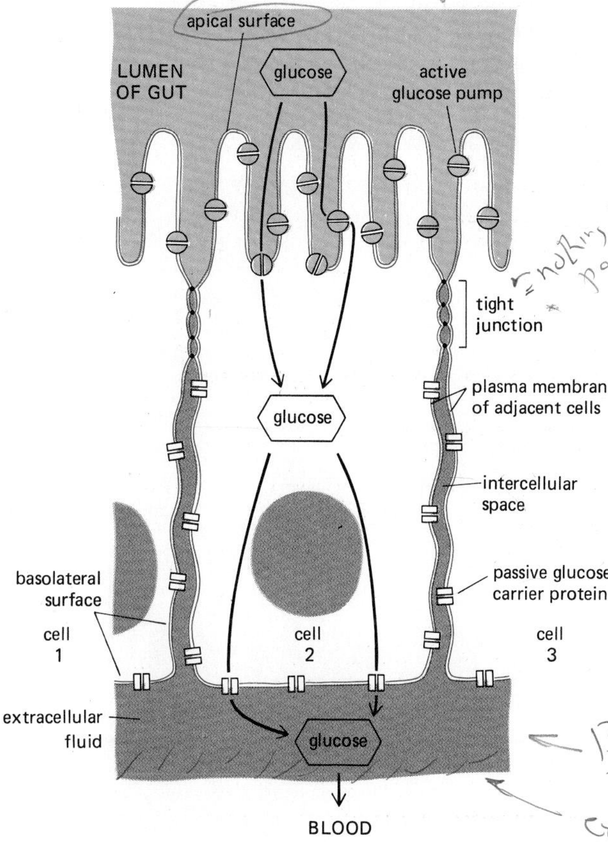

Figure 14–2 Transport proteins are confined to different regions of the plasma membrane in epithelial cells of the small intestine. This segregation permits nutrient transfer across the epithelial sheet from the gut lumen to the blood. In the example shown, glucose is actively transported into the cell by glucose pumps at the apical surface and diffuses out of the cell by facilitated diffusion mediated by passive glucose carrier proteins in the basolateral membrane. Tight junctions are thought to confine the transport proteins to their appropriate membrane domains by acting as diffusion barriers within the lipid bilayer of the plasma membrane; these junctions also block the diffusion of lipid molecules in the outer (but not the inner) leaflet of the lipid bilayer.

required for tight-junction integrity. Second, they seal neighboring cells together so that water-soluble molecules cannot leak between the cells: if a low-molecular-weight, electron-dense marker is added to one side of an epithelial cell sheet, it will usually not pass beyond the tight junction (Figure 14–3).

The molecular structure of tight junctions is still uncertain, but freeze-fracture electron microscopy shows them to be composed of an anastomosing network of strands that completely encircles the apical end of each cell in the epithelial sheet (Figure 14–4A and B). In conventional electron micrographs they are seen as a series of focal connections between the outer leaflets of the two interacting plasma membranes (Figure 14–4C). Although all tight junctions are impermeable to macromolecules, their permeability to small molecules varies greatly in different epithelia. Tight junctions in the epithelium lining the small intestine, for example, are 10,000 times more leaky to ions than those in the epithelium lining the urinary bladder. The ability of the junctions to restrict the passage of ions through the spaces between cells increases logarithmically with increasing numbers of strands in the network, as if each strand acts as an independent barrier. The strands are thought to be composed of long rows of specific transmembrane proteins in each of the two interacting plasma membranes, which join directly to each other to occlude the intercellular space (Figure 14–5).

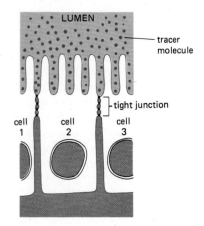

Figure 14–3 A soluble tracer molecule added on one side of an epithelial cell sheet cannot pass the tight junctions that seal adjacent cells together. The seal is not absolute and there is evidence that epithelial cells can alter their tight junctions in order to regulate the flow of solutes and water through the junctions.

Figure 14–4 Structure of a tight junction between epithelial cells of the small intestine shown schematically in (A) and in freeze-fracture (B) and conventional (C) electron micrographs. Note that the cells are oriented with their apical ends down. In (B), the plane of the micrograph is parallel to the plane of the membrane, and the tight junction consists of a beltlike band of anastomosing *sealing strands* that encircle each cell in the sheet. The sealing strands are seen as ridges of intramembrane particles on the cytoplasmic fracture face of the membrane (the P face) or as complementary grooves on the external face of the membrane (the E face). In (C) the junction is seen as a series of focal connections between the outer leaflets of the two interacting plasma membranes, each connection corresponding to a sealing strand in cross section. (B and C, from N.B. Gilula, in Cell Communication [R.P. Cox, ed.], pp. 1–29. New York: Wiley, 1974. Reprinted by permission of John Wiley & Sons, Inc.)

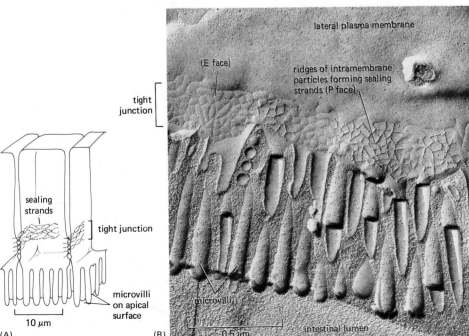

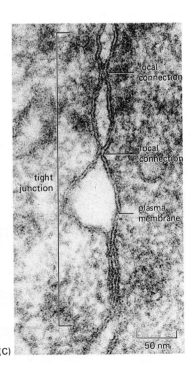

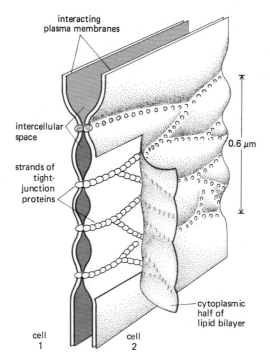

interacting
plasma membranes

intercellular
space

strands of
tight-
junction
proteins

0.6 μm

cytoplasmic
half of
lipid bilayer

cell
1

cell
2

Figure 14–5 A current model of a tight junction. It is postulated that adjacent plasma membranes are held together by continuous strands of transmembrane junctional proteins, which make contact across the intercellular space and create a complete seal. In this schematic, the cytoplasmic half of one membrane has been peeled back to expose the protein strands. In freeze-fracture electron microscopy the tight-junction proteins would remain with the cytoplasmic (P face) half of the lipid bilayer to give the pattern of intramembrane particles seen in Figure 14–4B, instead of staying in the other half as shown here.

Anchoring Junctions Connect the Cytoskeleton of a Cell to Those of Its Neighbors or to the Extracellular Matrix

Anchoring junctions are widely distributed in tissues. They enable groups of cells, such as those in an epithelium, to function as robust structural units by connecting the cytoskeletal elements of a cell either to those of another cell or to the extracellular matrix (Figure 14–6). They are most abundant in tissues that are subjected to severe mechanical stress, such as cardiac muscle, skin epithelium, and the neck of the uterus. They occur in two structurally and functionally different forms: (1) *adherens junctions* and (2) *desmosomes* and *hemidesmosomes*. Adherens junctions are connection sites for actin filaments; desmosomes and hemidesmosomes are connection sites for intermediate filaments.

Before we discuss the different classes of anchoring junctions, it is worth considering briefly the general principles of their construction. As illustrated in Figure 14–7, all of these junctions are composed of two classes of proteins:

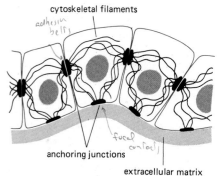

cytoskeletal filaments

adhesion belts

focal contact

anchoring junctions

extracellular matrix

Figure 14–6 Highly schematized drawing of how anchoring junctions join cytoskeletal filaments from cell to cell or from cell to extracellular matrix.

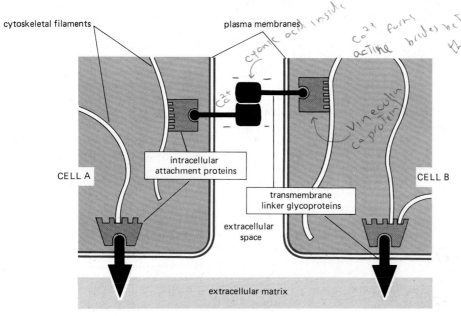

cytoskeletal filaments

plasma membranes

Ca²⁺ forms active bridges between the ① ions

Ca²⁺

cytonic acid inside

Vineculin co protein

intracellular
attachment proteins

transmembrane
linker glycoproteins

CELL A

CELL B

extracellular
space

extracellular matrix

Figure 14–7 Highly schematized drawing showing the two classes of proteins that constitute an anchoring junction: intracellular attachment proteins and transmembrane linker glycoproteins. In this example the extracellular domains of the transmembrane linker glycoproteins that join the cells together are shown interacting directly. In other cases they may be joined by additional proteins that are present in the extracellular space. Complexes of intracellular attachment proteins connect the linker glycoproteins to the cytoskeleton.

(1) *intracellular attachment proteins*, which connect the junctional complex to specific elements of the cytoskeleton (either actin filaments or intermediate filaments); and (2) *transmembrane linker glycoproteins*, whose intracellular domains bind to one or more intracellular attachment proteins, while their extracellular domains interact either with the extracellular matrix or with the extracellular domains of transmembrane linker glycoproteins on another cell.

Adherens Junctions Connect Bundles of Actin Filaments from Cell to Cell or from Cell to Extracellular Matrix[3]

Cell-cell adherens junctions occur in various forms. In many nonepithelial tissues they take the form of small punctate or streaklike attachments that connect actin filaments in the cortical cytoplasm of adjacent cells. In epithelial sheets they often form a continuous **adhesion belt** (or *zonula adherens*) around each of the interacting cells in the sheet, located near the apex of each cell just below the tight junction. The adhesion belts in adjacent epithelial cells are directly apposed, and the interacting membranes are held together by a Ca^{2+}-dependent mechanism. The transmembrane linker glycoproteins that mediate this adhesion are thought to be members of a family of Ca^{2+}-dependent cell-cell adhesion molecules called *cadherins* (see p. 830). An adhesion belt is sometimes called a belt desmosome, but this is misleading because the adhesion belt is chemically very different from a real desmosome (see below).

Within each cell a contractile bundle of actin filaments lies adjacent to the adhesion belt, running parallel to the plasma membrane, to which it is attached through a complex of intracellular attachment proteins containing *vinculin* (see p. 636). The actin bundles are thus linked, via transmembrane glycoproteins, in an extensive transcellular network (Figure 14–8), which is thought to help mediate a fundamental process in animal morphogenesis—the folding of epithelial cell sheets into tubes and other related structures (Figure 14–9). The oriented contraction of these bundles in the neural plate, for example, is thought to cause an apical narrowing of each epithelial cell, helping to initiate the rolling up of the plate to form the neural tube in early vertebrate development (see p. 887).

Cell-matrix adherens junctions connect cells and their actin filaments to the extracellular matrix. Fibroblasts cultured on an artificial substratum coated with extracellular matrix molecules, for example, adhere tightly at specialized regions of the plasma membrane called **focal contacts** or *adhesion plaques*, where

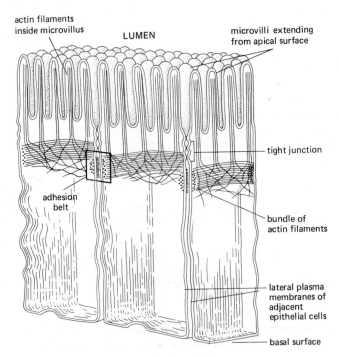

actin filaments inside microvillus

LUMEN

microvilli extending from apical surface

tight junction

adhesion belt

bundle of actin filaments

lateral plasma membranes of adjacent epithelial cells

basal surface

Figure 14–8 Adhesion belts between epithelial cells in the small intestine. This beltlike junction encircles each of the interacting cells and is characterized by a contractile bundle of actin filaments running along the cytoplasmic surface of the junctional plasma membrane.

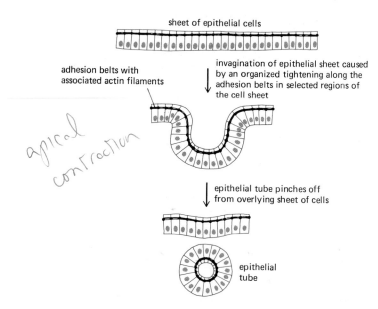

sheet of epithelial cells

adhesion belts with
associated actin filaments

invagination of epithelial sheet caused
by an organized tightening along the
adhesion belts in selected regions of
the cell sheet

apical contraction

epithelial tube pinches off
from overlying sheet of cells

epithelial tube

Figure 14–9 The folding of an epithelial sheet to form an epithelial tube (as in the formation of the neural tube). It is thought that the oriented contraction of the bundle of actin filaments running along adhesion belts causes the epithelial cells to narrow at their apex in selected regions of the cell sheet and that this plays an important part in the rolling up of the epithelial sheet into a tube, which then pinches off from the parent epithelial sheet.

bundles of actin filaments terminate (see p. 635). Many cells in tissues make analogous focal contacts with the surrounding extracellular matrix. A large transmembrane linker glycoprotein (which functions as a cell-surface receptor for the extracellular matrix glycoprotein called *fibronectin*—see p. 816) serves as one of the links between the matrix and the actin filament bundles in these plaques. The extracellular domain of this *fibronectin receptor* binds to fibronectin molecules attached to the culture dish, while its intracellular domain binds to an attachment protein called *talin*, which in turn binds to vinculin; and vinculin in turn binds to one or two other proteins, which bind actin (see Figure 11–38, p. 636).

The fibronectin receptor is only one member of a large family of related transmembrane linker glycoproteins, called *integrins* (see p. 821), that are thought to connect bundles of actin filaments to the extracellular matrix. Several integrins have been studied extensively (see p. 822), and they are believed to provide a model for the way that transmembrane linker glycoproteins involved in cell-cell adhesion, such as the cadherins, connect bundles of cortical actin filaments between adjacent epithelial cells; however, vinculin is present without talin in adhesion belts.

Desmosomes Connect Intermediate Filaments from Cell to Cell; Hemidesmosomes Connect Them to the Basal Lamina[4]

Desmosomes are buttonlike points of intercellular contact that rivet cells together in a variety of tissues, most of which are epithelial (Figure 14–10). They also serve as anchoring sites for intermediate filaments (see p. 661), which form a structural framework for the cytoplasm and provide tensile strength. Thus the intermediate filaments of adjacent cells are connected indirectly through these junctions to form a continuous network throughout the tissue. The particular type of intermediate filaments attached to the desmosomes depends on the cell type: they are *keratin filaments* in most epithelial cells, *desmin filaments* in heart muscle cells, and *vimentin filaments* in some of the cells that cover the surface of the brain (see Table 11–5, p. 662).

Electron microscopic and biochemical studies show that a desmosome consists of (1) a dense cytoplasmic plaque composed of a complex of intracellular attachment proteins, which are responsible for attachment to the cytoskeleton, and (2) transmembrane linker glycoproteins, which are connected to the plaque and interact through their extracellular domains to hold the adjacent plasma membranes together (Figure 14–11). The importance of desmosomes in holding cells together is demonstrated by some forms of the potentially fatal skin disease *pemphigus*, in which individuals make antibodies against one or more of their own desmosomal linker glycoproteins; these bind to and disrupt desmosomes between skin epithelial cells, causing severe blistering as a result of the leakage of body

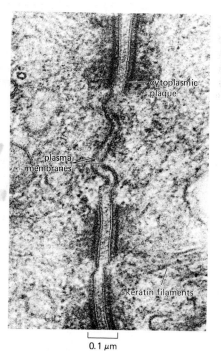

cytoplasmic plaque

plasma membranes

keratin filaments

0.1 μm

Figure 14–10 An electron micrograph of three desmosomes between two epithelial cells in the intestine of a rat. (From N.B. Gilula, in Cell Communication [R.P. Cox, ed.], pp. 1–29. New York: Wiley, 1974. Reprinted by permission of John Wiley & Sons, Inc.)

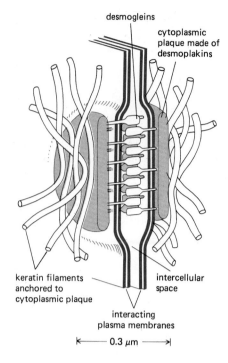

Figure 14–11 A highly schematized drawing of a desmosome. On the cytoplasmic surface of each interacting plasma membrane is a dense plaque composed of a mixture of intracellular attachment proteins called *desmoplakins*. Each plaque is associated with a thick network of keratin filaments, which pass along the surface of the plaque. Transmembrane linker glycoproteins called *desmogleins* bind to the plaques and interact through their extracellular domains to hold the adjacent membranes together by a Ca^{2+}-dependent mechanism. Although desmosomes and adhesion belts are morphologically and chemically distinct, they share at least one intracellular attachment protein (called plakoglobin).

fluids into the loosened epithelium. The antibodies disrupt desmosomes only in skin, suggesting that desmosomes in other tissues may be biochemically different.

Hemidesmosomes, or half-desmosomes, resemble desmosomes morphologically but are both functionally and chemically distinct. Instead of joining adjacent epithelial cell membranes, they connect the basal surface of epithelial cells to the underlying *basal lamina*—a specialized mat of extracellular matrix at the interface between the epithelium and connective tissue (see p. 818). Moreover, whereas the keratin filaments associated with desmosomes make lateral attachments to the desmosomal plaques (see Figure 14–11), many of those associated with hemidesmosomes terminate in the plaques (Figure 14–12).

Together desmosomes and hemidesmosomes act as rivets to distribute tensile or shearing forces through an epithelium and its underlying connective tissue.

14-6 Gap Junctions Allow Small Molecules to Pass Directly from Cell to Cell[5]

Perhaps the most intriguing cell junction is the **gap junction.** It is one of the most widespread, being found in large numbers in most tissues and in practically all animal species. It appears in conventional electron micrographs as a patch where the membranes of two adjacent cells are separated by a uniform narrow gap about 3 nm wide. Gap junctions mediate communication between cells by allowing inorganic ions and other small water-soluble molecules to pass directly from the cytoplasm of one cell to the cytoplasm of the other, thereby coupling the cells both electrically and metabolically. Such *cell coupling* has important functional implications, many of which are only beginning to be understood.

Cell-cell communication of this type was first demonstrated physiologically in 1958, but it took more than 10 years to show that this physiological coupling correlates with the presence of gap junctions seen in the electron microscope. The initial evidence for cell coupling came from electrophysiological studies of specific pairs of interacting nerve cells in the nerve cord of a crayfish. When a voltage gradient was applied across the junctional membrane after inserting an electrode into each of the two interacting cells, an unexpectedly large current flowed, indicating that inorganic ions (which carry current in living tissues) could pass freely from one cell interior to the other. Later experiments showed that small fluorescent dye molecules injected into one cell can likewise pass readily into adjacent cells without leaking into the extracellular space, provided that the molecules are no bigger than 1000 to 1500 daltons. This suggests a functional pore size for the connecting channels of about 1.5 nm (Figure 14–13), implying that coupled cells share their small molecules (such as inorganic ions, sugars, amino acids, nucleotides, and vitamins) but not their macromolecules (proteins, nucleic acids, and polysaccharides).

This sharing of small intracellular metabolites between cells is the basis of *metabolic cooperation*, which can be demonstrated in cells in culture. Mutant cell lines that lack the enzyme thymidine kinase, for example, can be cultured together with normal (wild-type) cells, which have thymidine kinase. The mutant cells on their own are unable to incorporate thymidine into their DNA because they cannot

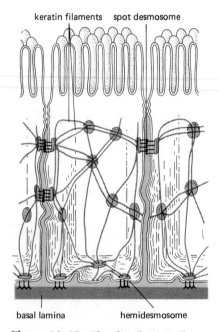

Figure 14–12 The distribution of desmosomes and hemidesmosomes in epithelial cells of the small intestine. The keratin filament networks of adjacent cells are indirectly connected to one another through desmosomes and to the basal lamina through hemidesmosomes. Whereas the keratin filaments make lateral attachments to the surface of the dense plaques associated with desmosomes, they tend to terminate in hemidesmosomes.

perform the initial step of converting the thymidine to thymidine triphosphate. But when these cells are co-cultured with wild-type cells and then exposed to radioactive thymidine, radioactivity is incorporated into the DNA of those mutant cells that are in direct contact with wild-type cells. This observation implies that a DNA precursor containing the radioactive thymidine—thymidine triphosphate, in fact—is passed directly from the wild-type cells into mutant cells in contact with them (Figure 14–14). Such metabolic cooperation does not occur when this type of experiment is performed with cells that cannot form gap junctions.

The evidence that gap junctions mediate electrical and chemical coupling between cells in contact with each other comes from several sources. Gap-junction structures can almost always be found where coupling can be demonstrated by electrical or chemical criteria. Conversely, coupling has not been demonstrated between vertebrate cells where there are no gap junctions. Moreover, dye and electrical coupling can be blocked if antibodies directed against the major gap-junction protein (see below) are microinjected into cells connected by gap junctions. Finally, when this gap-junction protein is reconstituted into synthetic lipid bilayers, or when mRNA encoding the protein is injected into frog oocytes, channels with many of the properties expected of gap-junction channels can be demonstrated electrophysiologically.

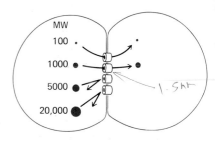

Figure 14–13 When fluorescent molecules of various sizes are injected into one of two cells coupled by gap junctions, molecules smaller than about 1000 to 1500 daltons (depending on the species and cell type) can pass into the other cell, but larger molecules cannot. This suggests that the functional diameter of the channel connecting the two cells is about 1.5 nm.

Gap-Junction Connexons Are Oligomers of a Multipass Transmembrane Protein[6]

Gap junctions are constructed from transmembrane proteins that form structures called *connexons*. When the connexons in the plasma membranes of two cells in contact are aligned, they form a continuous aqueous channel, which connects the two cell interiors (Figure 14–15). The connexons join in such a way that the interacting plasma membranes are separated by an interrupted gap—hence the term "gap junction"—emphasizing the contrast with a tight junction, where the membranes are more closely juxtaposed (compare Figures 14–5 and 14–15). Each connexon is seen as an intramembrane particle in freeze-fracture electron micrographs, and each gap junction can contain up to several hundred clustered connexons (Figure 14–16).

An unusual resistance to proteolytic enzymes and detergents has made it possible to isolate gap junctions from rodent liver (Figure 14–17). The junctions contain a single major protein of about 30,000 daltons. DNA sequencing studies suggest that the polypeptide chain (about 280 amino acid residues) crosses the lipid bilayer as four α helices. Six such protein molecules are thought to associate to form each connexon in a way similar to that postulated for the acetylcholine receptor channel, creating an aqueous pore lined by one α helix from each protein subunit (see Figure 6–64, p. 321).

Antibodies against the 30,000 dalton protein react with gap junctions in many tissues and species, suggesting that the connexon proteins in these tissues and organisms are similar (although biochemical and physiological analyses indicate that they are not identical). This accords with the finding that different cell types in culture will usually form gap junctions with each other, even across species.

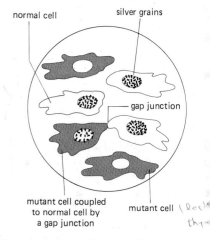

Figure 14–14 Schematic drawing of an autoradiograph demonstrating metabolic cooperation between cells in culture connected by gap junctions. The mutant cells lack the enzyme thymidine kinase and therefore cannot incorporate radioactive thymidine into DNA when thymidine is added to the medium. Normal cells can incorporate the thymidine into their DNA, and their nuclei are therefore stippled with black dots, representing developed silver grains in the autoradiograph. In mixed cultures of normal and mutant cells, if a mutant cell makes contact and forms gap junctions with a normal cell, its nucleus is also radiolabeled, as shown. This labeling occurs because the radioactive thymidine is phosphorylated by thymidine kinase to form thymidine triphosphate in the normal cell; the radioactive thymidine triphosphate then passes through the gap junctions into the mutant cell, where it is incorporated into DNA.

14-7 Most Cells in Early Embryos Are Coupled via Gap Junctions[7]

In some tissues, cell coupling via gap junctions serves an obvious function. For example, electrical coupling synchronizes the contractions of heart muscle cells and of smooth muscle cells responsible for the peristaltic movements of the intestine. Similarly, electrical coupling between nerve cells allows action potentials to spread rapidly from cell to cell without the delay that occurs at chemical synapses; this is advantageous where speed and reliability are crucial, as in certain escape responses in fish and insects. It is less obvious why gap junctions occur in tissues that are not electrically active. In principle, the sharing of small metabolites and ions provides a mechanism for coordinating the activities of individual cells in such tissues. For example, the activities of cells in an epithelial cell sheet, such as the beating of cilia, might be coordinated via gap junctions; and since intracellular mediators such as cyclic AMP can pass through gap junctions, responses of coupled cells to extracellular signaling molecules may be propagated and coordinated in this way.

Cell coupling via gap junctions appears to be important in embryogenesis. In early vertebrate embryos (beginning with the late eight-cell stage in mouse embryos), most cells are electrically coupled to one another. As specific groups of cells in the embryo develop their distinct identities and begin to differentiate, however, they commonly uncouple from surrounding tissue. As the neural tube closes, for instance, its cells uncouple from the overlying ectoderm (see Figure 14–9). Meanwhile the cells within each group remain coupled with one another and so tend to behave as a cooperative assembly, all following a similar developmental pathway in a coordinated fashion.

An attractive hypothesis is that the coupling of cells in embryos might provide a pathway for long-range cell signaling within a developing epithelium. For example, a small molecule could pass through gap junctions from a region of the tissue where its intracellular concentration is kept high to a region where it is kept low, thereby setting up a smooth concentration gradient. The local concentration could provide cells with "positional information" to control their differentiation according to their location in the embryo. Whether gap junctions actually function in this way is not known.

The importance of gap-junction-mediated communication in development is suggested by an experiment in which antibodies against the major gap-junction protein were microinjected into one cell of an eight-cell amphibian embryo; the injected antibodies not only selectively interrupted electrical coupling and dye transfer between progeny of the injected cell (assayed two cell cycles later in 32-cell embryos), they also grossly disrupted the development of the embryo (Figure 14–18). It is not clear how the inhibition of cell coupling at an early stage caused the later developmental defects seen in the injected embryos, but experiments of this type offer a promising beginning to an analysis of the role of gap junctions in embryonic development.

The Permeability of Gap Junctions Is Regulated[8]

The permeability of gap junctions is rapidly (within seconds) and reversibly decreased by experimental manipulations that decrease cytosolic pH or increase the cytosolic concentration of free Ca^{2+}, and in some tissues the permeability can be regulated by the voltage gradient across the junction or by extracellular chemical

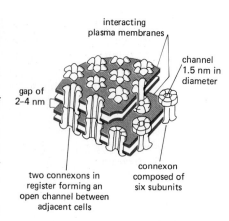

Figure 14–15 A model of a gap junction based on biochemical, electron microscopic, and x-ray diffraction observations. The drawing shows the interacting plasma membranes of two adjacent cells. The apposed lipid bilayers are penetrated by protein assemblies called *connexons*, each of which is thought to be formed by six identical protein subunits. Two connexons join across the intercellular gap to form a continuous aqueous channel connecting the two cells.

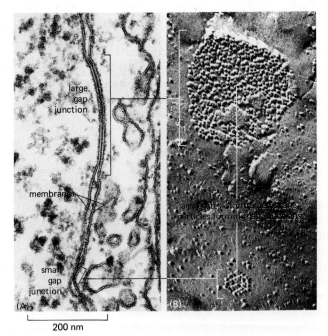

Figure 14–16 Thin-section (A) and freeze-fracture (B) electron micrographs of a large and a small gap junction between fibroblasts in culture. In (B) each gap junction is seen as a cluster of homogeneous intramembrane particles associated exclusively with the cytoplasmic fracture face (P face) of the plasma membrane. Each intramembrane particle corresponds to a connexon, illustrated in Figure 14–15. (From N.B. Gilula, in Cell Communication [R.P. Cox, ed.], pp. 1–29. New York: Wiley, 1974. Reprinted by permission of John Wiley & Sons, Inc.)

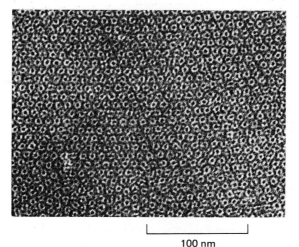

Figure 14–17 Electron micrograph of an area of an isolated gap junction from rat liver. The preparation has been negatively stained to show the connexons, which are organized as a hexagonal lattice. The densely stained central hole in each connexon has a diameter of about 2 nm. (From N.B. Gilula, in Intercellular Junctions and Synapses [Receptors and Recognition Series B, Vol. 2; J. Feldman, N.B. Gilula, and J.D. Pitts, eds.], pp. 3–22. London: Chapman & Hall, 1978.)

signals. These observations indicate that gap junctions are dynamic structures that can open and close in response to changes in the cell. Thus, like conventional ion channels, they are *gated* (see p. 313), although transitions between open and closed states occur much less frequently in a gap-junction channel than in most conventional ion channels.

The importance of voltage and pH regulation of gap-junction permeability to the normal function of cell assemblies is unknown. There is one case, however, where the reason for the Ca^{2+} control seems clear. When a cell dies or is damaged, its membrane becomes leaky. Ions such as Ca^{2+} and Na^+ move into the cell, and valuable metabolites leak out. If the cell were to remain coupled to its healthy neighbors, these too would suffer a dangerous disturbance of their internal chemistry. But the influx of Ca^{2+} into the sick cell, by closing the gap junction channels, effectively isolates it and prevents damage from spreading in this way.

Where gap-junction permeability is increased by extracellular chemical signals, the effect is to spread the response to neighboring cells that are not in direct contact with the signal. The hormone *glucagon*, for example, which stimulates liver cells to break down glycogen and release glucose into the blood, can also be shown to increase gap-junction permeability in rat liver cells. It does so by increasing the concentration of intracellular cyclic AMP, which activates cyclic AMP-dependent protein kinase (see p. 709), which in turn is thought to phosphorylate the major gap-junction protein. The breakdown of glycogen by liver cells is also mediated by an increase in cyclic AMP, so the simultaneous increase in gap-junction permeability, by facilitating the diffusion of cyclic AMP from cell to cell, tends to spread the glycogen-breakdown response through neighboring groups of liver cells.

Figure 14–19 summarizes the various types of junctions formed between cells in an epithelium. In the most apical portion of the cell, the relative positions of the junctions are the same in nearly all epithelia: the tight junction occupies the most apical portion of the cell, followed by the adhesion belt and then by a special parallel row of desmosomes; together these form a "junctional complex." Gap junctions and additional desmosomes are less regularly organized.

Summary

Many cells in tissues are linked to each other and to the extracellular matrix at specialized contact sites called cell junctions. Cell junctions fall into three functional classes: occluding junctions, anchoring junctions, and communicating junctions. Tight junctions are the main occluding junctions, and they play a critical part in maintaining the concentration differences of small hydrophilic molecules across epithelial cell sheets by (1) sealing the plasma membranes of adjacent cells together to create a continuous permeability barrier across the cell sheet and (2) acting as barriers in the lipid bilayer to restrict the diffusion of membrane transport proteins

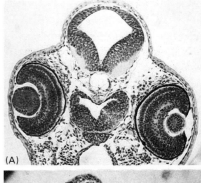

(A)

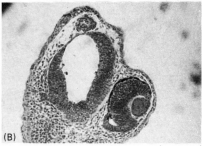

(B)

Figure 14–18 Effect of antibodies against the major gap-junction protein microinjected into one cell of an early *Xenopus* embryo. The micrographs are of transverse sections through a normal embryo (A) and through an embryo that was injected at the eight-cell stage (B). Note that the eye is missing and the brain is underdeveloped on the injected side of the embryo shown in (B). (From A. Warner, S. Guthrie, and N.B. Gilula, *Nature* 331:126–131, 1985. Copyright © 1985 Macmillan Journals Limited.)

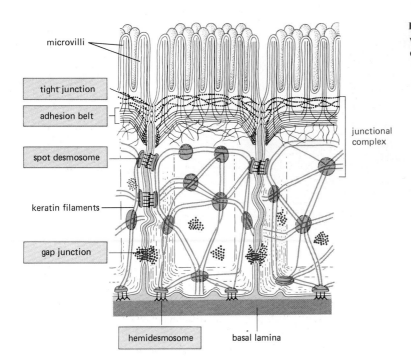

microvilli

tight junction

adhesion belt

spot desmosome

keratin filaments

gap junction

junctional complex

hemidesmosome

basal lamina

Figure 14–19 Distribution of the various cell junctions formed by epithelial cells of the small intestine.

between separate apical and basolateral domains of the plasma membrane in each epithelial cell.

There are two main types of anchoring junctions: adherens junctions *and* desmosomes. *Both join groups of cells together into strong structural units by connecting elements of their cytoskeletons;* adherens junctions *connect bundles of actin filaments, whereas* desmosomes *connect intermediate filaments.* Gap junctions *are communicating junctions composed of clusters of channel proteins that allow molecules of less than 1500 daltons to pass directly from the inside of one cell to the inside of the other. Cells connected by such junctions share many of their inorganic ions and other small molecules and are said to be chemically and electrically coupled.* Gap junctions *are important in coordinating the activities of electrically active cells, and they are thought to play a similar role in other groups of cells as well.*

The Extracellular Matrix[9]

Tissues are not composed solely of cells. A substantial part of their volume is *extracellular space*, which is largely filled by an intricate network of macromolecules constituting the **extracellular matrix** (Figure 14–20). This matrix comprises a variety of versatile polysaccharides and proteins that are secreted locally and assemble into an organized meshwork. Whereas we discussed cell junctions chiefly in the context of epithelial tissues, our account of extracellular matrix will focus chiefly on **connective tissues** (Figure 14–21). In these tissues the matrix is generally more plentiful than the cells and surrounds the cells on all sides, determining the tissue's physical properties. Connective tissues form the architectural framework of the vertebrate body, and the amounts found in different organs vary greatly: from skin and bone, in which they are the major component, to brain and spinal cord, in which they are only minor constituents.

Variations in the relative amounts of the different types of matrix macromolecules and the way they are organized in the extracellular matrix give rise to an amazing diversity of forms, each highly adapted to the functional requirements of the particular tissue. The matrix can become calcified to form the rock-hard structures of bone or teeth, or it can form the transparent matrix of the cornea, or it can adopt the ropelike organization that gives tendons their enormous tensile strength. At the interface between an epithelium and connective tissue, the matrix forms a *basal lamina*, an extremely thin but tough mat that plays an important

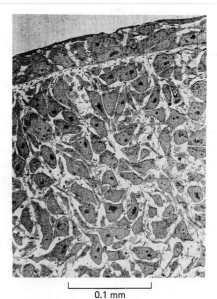

0.1 mm

Figure 14–20 Low-power electron micrograph showing cells surrounded by spaces filled with extracellular matrix. The particular cells shown are those in an early chick limb. The cells have not yet acquired their specialized characteristics. (Courtesy of Cheryll Tickle.)

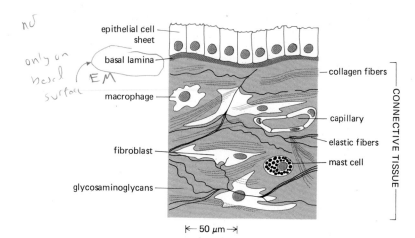

epithelial cell sheet

basal lamina

macrophage

fibroblast

glycosaminoglycans

collagen fibers

capillary

elastic fibers

mast cell

CONNECTIVE TISSUE

|← 50 μm →|

no

only on
basal
surface EM

Figure 14–21 The connective tissue underlying an epithelial cell sheet.

part in controlling cell behavior. We shall confine our account to the extracellular matrix of vertebrates, but unique and interesting related structures are seen in many other organisms, such as the cell walls of bacteria and plants, the cuticles of worms and insects, and the shells of mollusks. Plant cell walls are considered in detail in Chapter 20.

Until recently the vertebrate extracellular matrix was thought to serve mainly as a relatively inert scaffolding to stabilize the physical structure of tissues. But now it is clear that the matrix plays a far more active and complex role in regulating the behavior of the cells that contact it—influencing their development, migration, proliferation, shape, and metabolic functions. The extracellular matrix has a correspondingly complex molecular composition; although our understanding of its organization is still fragmentary, there has been rapid progress in characterizing some of its major components.

The Extracellular Matrix Consists Primarily of Fibrous Proteins Embedded in a Hydrated Polysaccharide Gel

The macromolecules that constitute the extracellular matrix are mainly secreted locally by cells in the matrix. In most connective tissues these macromolecules are secreted largely by *fibroblasts* (Figure 14–22). In some specialized connective tissues, however, such as cartilage and bone, they are secreted by cells of the fibroblast family that have more specific names: chondroblasts, for example, form cartilage, and osteoblasts form bone. The two main classes of extracellular macromolecules that make up the matrix are (1) polysaccharide *glycosaminoglycans (GAGs)*, which are usually found covalently linked to protein in the form of *proteoglycans*, and (2) fibrous proteins of two functional types: mainly structural (for example, *collagen* and *elastin*) and mainly adhesive (for example, *fibronectin* and *laminin*). The glycosaminoglycan and proteoglycan molecules form a highly hydrated, gel-like "ground substance" in which the fibrous proteins are embedded. The aqueous phase of the polysaccharide gel permits the diffusion of nutrients, metabolites, and hormones between the blood and the tissue cells; the collagen fibers strengthen and help to organize the matrix, and rubberlike elastin fibers give it resilience. The adhesive proteins help cells attach to the extracellular matrix: fibronectin promotes the attachment of fibroblasts and related cells to the matrix in connective tissues, while laminin promotes the attachment of epithelial cells to the basal lamina.

Glycosaminoglycan Chains Occupy Large Amounts of Space and Form Hydrated Gels[10]

Glycosaminoglycans (GAGs) are long, unbranched polysaccharide chains composed of repeating disaccharide units. They are called glycosaminoglycans because one of the two sugar residues in the repeating disaccharide is always an amino sugar (*N*-acetylglucosamine or *N*-acetylgalactosamine). In most cases this

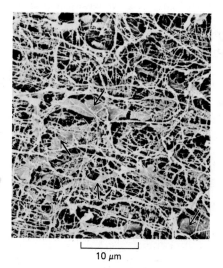

10 μm

Figure 14–22 Scanning electron micrograph of fibroblasts (*arrows*) in the connective tissue of the cornea in a chick embryo. The extracellular matrix surrounding the fibroblasts is composed largely of collagen fibers (there are no elastic fibers in the cornea). The glycosaminoglycans, which normally form a hydrated gel filling the interstices of the fibrous network, have collapsed onto the surface of the collagen fibers during the dehydration process necessary for specimen preparation. (Courtesy of Robert Trelstad.)

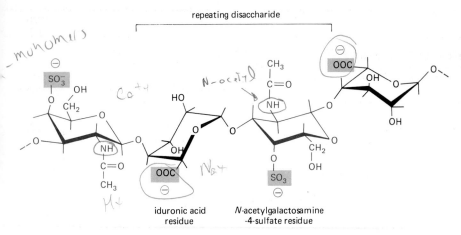

repeating disaccharide

iduronic acid
residue

N-acetylgalactosamine
-4-sulfate residue

Figure 14–23 Glycosaminoglycans are long linear polymers composed of a repeating disaccharide sequence. A small part of a dermatan sulfate chain is shown here; these chains are typically 70 to 200 sugar residues long. There is a high density of negative charges along the chain resulting from the presence of both carboxyl and sulfate groups.

amino sugar is sulfated and the second sugar is a uronic acid. Because of the sulfate or carboxyl groups on most of their sugar residues, glycosaminoglycans are highly negatively charged (Figure 14–23). Four main groups of glycosaminoglycans have been distinguished by their sugar residues, the type of linkage between these residues, and the number and location of sulfate groups: (1) *hyaluronic acid*, (2) *chondroitin sulfate* and *dermatan sulfate*, (3) *heparan sulfate* and *heparin*, and (4) *keratan sulfate* (Table 14–2).

Polysaccharide chains are too inflexible to fold up into the compact globular structures that polypeptide chains typically form. Moreover, they are strongly hydrophilic. Thus glycosaminoglycans tend to adopt highly extended, so-called random-coil conformations, which occupy a huge volume relative to their mass (Figure 14–24), and they form gels even at very low concentrations. Their high density of negative charges attracts a cloud of cations, such as Na^+, that are osmotically active, causing large amounts of water to be sucked into the matrix. This creates a swelling pressure, or turgor, that enables the matrix to withstand compressive forces (in contrast to collagen fibrils, which resist stretching forces). Cartilage matrix, for example, resists compression by this mechanism.

The amount of glycosaminoglycan in connective tissue is usually less than 10% by weight of the amount of the fibrous proteins. Because they form porous hydrated gels, however, the glycosaminoglycan chains fill most of the extracellular space, providing mechanical support to tissues while still allowing the rapid diffusion of water-soluble molecules and the migration of cells.

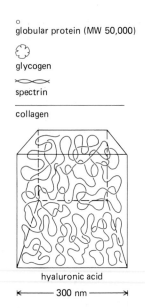

globular protein (MW 50,000)

glycogen

spectrin

collagen

hyaluronic acid

|← 300 nm →|

Figure 14–24 The relative volumes occupied by various proteins, a glycogen granule, and a single hydrated molecule of hyaluronic acid of about 8×10^6 daltons.

Hyaluronic Acid Is Thought to Facilitate Cell Migration During Tissue Morphogenesis and Repair[11]

Hyaluronic acid (also called hyaluronate or hyaluronan), which can contain up to several thousand sugar residues, is a relatively simple molecule consisting of a regular repeating sequence of nonsulfated disaccharide units (Figure 14–25). It is

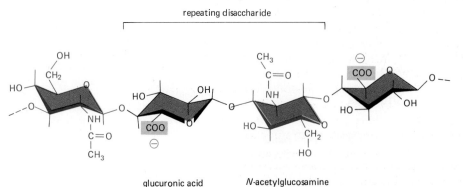

repeating disaccharide

glucuronic acid

N-acetylglucosamine

Figure 14–25 The repeating disaccharide sequence in hyaluronic acid, a relatively simple glycosaminoglycan that consists of a single long chain of up to several thousand sugar residues. Note the absence of sulfate groups.

Table 14-2 The Glycosaminoglycans

Group	Glycosamino-glycan	Molecular Weight	Repeating Disaccharide (A-B)$_n$		Sulfates per Disaccha-ride Unit	Linked to Protein	Other Sugar Compo-nents	Tissue Distribution
			Monosaccharide A	Monosaccharide B				
1	Hyaluronic acid	4,000 to 8 × 10⁶	D-glucuronic acid	N-acetyl-D-glucosamine	0	—	0	various connective tissues, skin, vitreous body, cartilage, synovial fluid
2	Chondroitin sulfate	5,000–50,000	D-glucuronic acid	N-acetyl-D-galactosamine	0.2–2.3	+	D-galactose D-xylose	cartilage, cornea, bone, skin, arteries
	Dermatan sulfate	15,000–40,000	D-glucuronic acid or *L-iduronic acid	N-acetyl-D-galactosamine	1.0–2.0	+	D-galactose D-xylose	skin, blood vessels, heart, heart valves
3	Heparan sulfate	5,000–12,000	D-glucuronic acid or *L-iduronic acid	N-acetyl-D-glucosamine	0.2–2.0	+	D-galactose D-xylose	lung, arteries, cell surfaces, basal laminae
	Heparin	6,000–25,000	D-glucuronic acid or *L-iduronic acid	N-acetyl-D-glucosamine	2.0–3.0	+	D-galactose D-xylose	lung, liver, skin, mast cells
4	Keratan sulfate	4,000–19,000	D-galactose	N-acetyl-D-glucosamine	0.9–1.8	+	D-galactos-amine D-mannose L-fucose, sialic acid	cartilage, cornea, intervertebral disc

*L-iduronic acid is produced by the epimerization of D-glucuronic acid at the position where the carboxyl group is located. Thus dermatan sulfate is a modified form of chondroitin sulfate, and the two types of repeating disaccharide sequences usually occur as alternating segments in the same glycosaminoglycan chain.

found in variable amounts in all tissues and fluids in adult animals and is especially abundant in early embryos. Because of its simplicity, hyaluronic acid is thought to represent the earliest evolutionary form of glycosaminoglycan, but it is not typical of the majority of glycosaminoglycans. All of the others (1) contain sulfated sugars, (2) tend to contain a number of different disaccharide units arranged in more complex sequences, (3) have much shorter chains, consisting of fewer than 300 sugar residues, and (4) are covalently linked to protein.

There is increasing evidence that hyaluronic acid has a special function in tissues where cells are migrating—such as during development or wound repair. It is produced in large amounts during periods of cell migration; and when cell migration ends, the excess hyaluronic acid is degraded by the enzyme *hyaluronidase.* This sequence of events has been demonstrated in a wide variety of tissues, suggesting that increased local production of hyaluronic acid, which attracts water and thereby swells the matrix, may be a general strategy for facilitating cell migration during morphogenesis and repair. Hyaluronic acid is also an important constituent of joint fluid, where it serves as a lubricant.

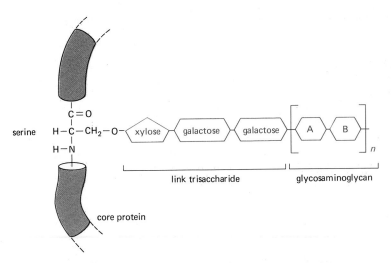

Figure 14–26 The linkage between a glycosaminoglycan chain and a serine residue of a core protein in a proteoglycan molecule. A specific "link trisaccharide" is first added to the serine, which is often located within the sequence Asp(or Glu)-Asp(or Glu)-X-Ser-Gly-X-Gly, where X is any amino acid. The rest of the glycosaminoglycan chain, consisting mainly of a repeating disaccharide unit (composed of the two monosaccharides A and B in Table 14–2), is then synthesized, one sugar residue being added at a time.

Proteoglycans Are Composed of Long Glycosaminoglycan Chains Covalently Linked to a Core Protein[12]

Except for hyaluronic acid, all glycosaminoglycans are found covalently attached to protein in the form of **proteoglycans.** As is the case for a glycoprotein (see p. 440), the polypeptide chain, or *core protein*, of a proteoglycan is made on membrane-bound ribosomes and threaded into the lumen of the endoplasmic reticulum. The polysaccharide chains are assembled on the core protein mainly in the Golgi apparatus: first a special *link trisaccharide* is attached to a serine residue on the core protein to serve as a primer for polysaccharide growth; then one sugar residue is added at a time by specific glycosyl transferases (Figure 14–26). While chain elongation is proceeding in the Golgi apparatus, many of the polymerized sugar residues are covalently modified by a sequential and coordinated series of sulfation reactions (see p. 456) and epimerization reactions that alter the configuration of the substituents around individual carbon atoms in the sugar molecule. The sulfation greatly increases the negative charge of proteoglycans.

Proteoglycans are usually easily distinguished from glycoproteins by the nature, quantity, and arrangement of their sugar side chains. Glycoproteins usually contain from 1% to 60% carbohydrate by weight in the form of numerous, relatively short, branched, *O*- and *N*-linked oligosaccharide chains, generally of fewer than 15 sugar residues and variable composition, which often terminate with sialic acid (see p. 453). Although the core protein in a proteoglycan can itself be a glycoprotein, proteoglycans can contain as much as 95% carbohydrate by weight, most of which takes the form of one to several hundred unbranched glycosaminoglycan chains, each typically about 80 sugar residues long and usually without sialic acid. Moreover, whereas glycoproteins are rarely larger than 3×10^5 daltons, proteoglycans can be much larger. For example, one of the best-characterized proteoglycan molecules is a major component of cartilage; it typically consists of about 100 chondroitin sulfate chains and about 50 keratan sulfate chains linked to a serine-rich core protein of more than 2000 amino acids. Thus its total mass is about 3×10^6 daltons, with approximately 1 glycosaminoglycan chain for every 20 amino acid residues (Figure 14–27). On the other hand, many proteoglycans are much smaller and have only 1 to 10 glycosaminoglycan chains.

In principle, proteoglycans have the potential for almost limitless heterogeneity. They can differ markedly in protein content, molecular size, and the number and types of glycosaminoglycan chains per molecule. Moreover, although there is always an underlying repeating pattern of disaccharides, the length and composition of the glycosaminoglycan chains can vary greatly, as can the spatial arrangement of hydroxyl, sulfate, and carboxyl side groups along the chains. It is thus an extremely complex problem to identify and classify proteoglycans in terms of their sugars. Many core proteins are now being sequenced with the aid of recombinant DNA techniques, and in the future it is likely that the classification of proteoglycans can be made more meaningful by characterizing them according to their core proteins rather than their glycosaminoglycan chains.

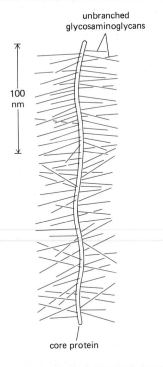

A PROTEOGLYCAN FROM CARTILAGE
(MW ~ 3×10^6)

unbranched glycosaminoglycans

100 nm

core protein

TYPICAL GLYCOPROTEIN
RIBONUCLEASE (MW ~ 15,000)

short, branched oligosaccharide side chain ——— polypeptide chain

Figure 14–27 The main proteoglycan molecule in cartilage. It consists of many glycosaminoglycan chains covalently linked to a protein core. The core protein contains a number of *N*-linked and *O*-linked oligosaccharide chains (not shown) in addition to the glycosaminoglycan chains. Most proteoglycans are smaller than the one shown, and their glycosaminoglycan chains are often restricted to particular regions of the core polypeptide chain. The lower drawing shows a typical glycoprotein molecule (pancreatic ribonuclease B), drawn to scale for comparison.

Glycosaminoglycan Chains May Be Highly Organized
in the Extracellular Matrix[13]

Given the structural heterogeneity of proteoglycan molecules, it seems highly unlikely that their function is limited to providing hydrated space around and between cells. Proteoglycans have been shown to bind various secreted signaling molecules in a test tube, and it seems likely that they do so in tissues, thereby localizing the action of the signaling ligand: fibroblast growth factor (FGF—see p. 747), for example, binds to heparan sulfate proteoglycans both *in vitro* and in tissues. Proteoglycans may form gels of varying pore size and charge density, thus functioning as sieves to regulate the traffic of molecules and cells according to their size and/or charge. There is evidence that proteoglycans function in this capacity in the basal lamina of the kidney glomerulus, which filters molecules passing into the urine from the bloodstream (see p. 820).

The organization of glycosaminoglycans and proteoglycans in the extracellular matrix is poorly understood. Biochemical studies indicate that these molecules bind to each other in specific ways, as well as to the fibrous proteins in the matrix. It would be surprising if such interactions are not important in organizing the matrix. The major keratan sulfate/chondroitin sulfate proteoglycan in cartilage discussed above has been shown to assemble in the extracellular space into large aggregates that are noncovalently bound through their core proteins to a large hyaluronic acid molecule. As many as 100 proteoglycan monomers are bound to a single hyaluronic acid chain, producing a giant complex with a molecular weight of 100 million or more and occupying a volume equivalent to that of a bacterium. When isolated from tissues, these complexes are readily seen in the electron microscope (Figure 14–28).

By contrast, attempts to determine the arrangement of proteoglycan molecules by electron microscopy while they are still in the tissues have been frustrating. Since they are highly water-soluble, they are readily washed out of the extracellular matrix when tissue sections are exposed to aqueous solutions during fixation. Recently proteoglycans have been seen in near-native state in cartilage that has been rapidly frozen at very low temperature ($-196°C$) under high pressure and then fixed and stained while still frozen (Figure 14–29). An alternative approach is to use a cationic dye with a relatively low charge density together with

Figure 14–28 (A) Electron micrograph of a proteoglycan aggregate from fetal bovine cartilage shadowed with platinum. Many free proteoglycan molecules are also seen. (B) Schematic drawing of the giant proteoglycan aggregate shown in (A). It consists of about 100 proteoglycan monomers (each like that shown in Figure 14–27) noncovalently bound to a single hyaluronic acid chain through two link proteins that bind to both the core protein of the proteoglycan and to the hyaluronic acid chain, thereby stabilizing the aggregate. The molecular weight of such a complex can be 10^8 or more, and it occupies a volume equivalent to that of a bacterium, which is about 2×10^{-12} cm^3. (A, courtesy of Lawrence Rosenberg.)

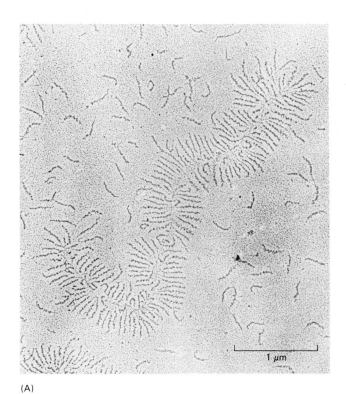

(A)

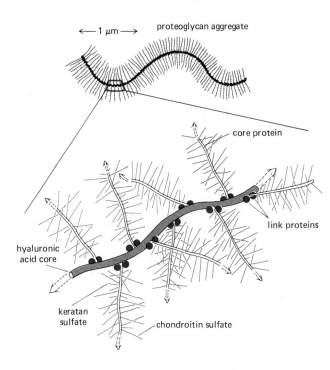

(B)

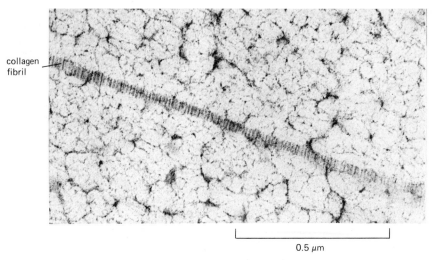

collagen fibril

0.5 μm

Figure 14–29 Electron micrograph of proteoglycans in the extracellular matrix of rat cartilage. The tissue was rapidly frozen at −196°C and fixed and stained while still frozen (a process called *freeze substitution*) to prevent the glycosaminoglycan chains from collapsing. The proteoglycan molecules are seen to form a fine filamentous network in which a single striated collagen fibril is embedded. The more darkly stained parts of the proteoglycan molecules are the core proteins; the faintly stained threads are the glycosaminoglycan chains. (Reproduced from E.B. Hunziker and R.K. Schenk, *J. Cell Biol.* 98:277–282, 1985 by copyright permission of the Rockefeller University Press.)

more conventional fixation. When the proteoglycans in rat tail tendon are stained with such a dye, they are seen as threadlike structures encircling collagen fibrils at regular intervals (Figure 14–30). They cross the collagen fibrils at intervals of about 65 nm, reflecting the staggered arrangement of collagen molecules in the collagen fibrils (see p. 812). Such ordered patterns of molecules are likely to be widespread in the extracellular matrix, and, given the diversity of both collagen molecules and proteoglycans, the patterns may be complex and varied.

Some polysaccharide chains are known to assemble into highly ordered helical or ribbonlike structures. In higher plants, for example, cellulose (polyglucose) chains are packed tightly together in ribbonlike crystalline arrays to form the microfibrillar component of the cell wall (see Figure 20–5, p. 1141). *In vitro*, two *different* polysaccharide chains can associate specifically with each other, producing regions with a regular helical structure (Figure 14–31); such polysaccharide-polysaccharide interactions may also occur in the extracellular matrix. If proteoglycan molecules can assume structural conformations as diverse as their chemistry, we have hardly begun to understand them.

Not all proteoglycans are secreted components of the extracellular matrix. Some are integral components of plasma membranes, and some of these have their core protein oriented across the lipid bilayer. The integral membrane proteoglycans usually contain only a small number of glycosaminoglycan chains, and they are thought to play a part in binding cells to the extracellular matrix and in organizing the matrix macromolecules that cells secrete.

Collagen Is the Major Protein of the Extracellular Matrix[14]

The **collagens** are a family of highly characteristic fibrous proteins found in all multicellular animals. They are secreted mainly by connective tissue cells and are the most abundant proteins in mammals, constituting 25% of their total protein. The characteristic feature of collagen molecules is their stiff, triple-stranded helical structure. Three collagen polypeptide chains, called *α chains* (each about 1000

collagen fibril

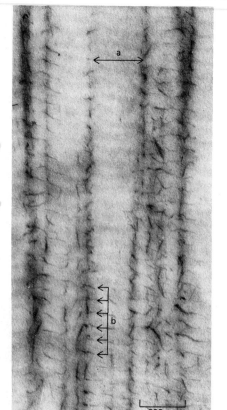

a

b

200 nm

Figure 14–30 Electron micrograph of a longitudinal section of rat tail tendon stained with a copper-containing dye to visualize proteoglycan molecules. A tendon consists of closely packed collagen fibrils, several of which are shown here. Proteoglycan molecules are seen as fine filaments encircling each of the collagen fibrils at regular intervals of about 65 nm (for example, those indicated by arrows b), revealing a specific interaction between proteoglycan and collagen molecules. In those regions where proteoglycan threads are not seen crossing a collagen fibril (such as the region of fibril indicated by the double arrow a), the plane of section presumably cuts through the interior of the fibril. (Reproduced with permission from J.E. Scott, *Biochem. J.* 187:887–891, 1980. Copyright 1980 American Chemical Society.)

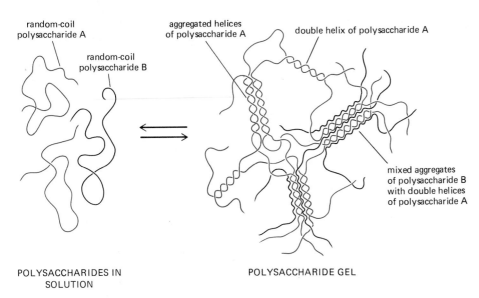

Figure 14–31 Some of the ordered conformations that two different polysaccharide chains, A and B, can assume in forming a gel *in vitro*. Since these interactions between molecules are confined to certain regions of the chains (so-called junctional regions) and are not propagated along the entire molecule, each chain can combine with more than one partner and thereby form a gel network. Examples of gel-forming polysaccharides are the agars (of algae) and the pectins (of higher plants).

random-coil polysaccharide A

random-coil polysaccharide B

aggregated helices of polysaccharide A

double helix of polysaccharide A

mixed aggregates of polysaccharide B with double helices of polysaccharide A

POLYSACCHARIDES IN SOLUTION

POLYSACCHARIDE GEL

amino acids long), are wound around one another in a regular superhelix to generate a ropelike collagen molecule about 300 nm long and 1.5 nm in diameter. Collagens are extremely rich in proline and glycine, both of which are important in the formation of the triple-stranded helix. Proline, because of its ring structure, stabilizes a left-handed helical conformation in each α chain, with three amino acid residues per turn. Glycine is the smallest amino acid (because it has only a hydrogen atom as a side chain); regularly spaced at every third residue throughout the central region of the α chain, it allows the three helical α chains to pack tightly together to form the final collagen superhelix (Figure 14–32).

So far, about 20 distinct collagen α chains have been identified, each encoded by a separate gene. Different combinations of these genes are expressed in different tissues. Although in principle more than 1000 types of triple-stranded collagen molecules could be assembled from various combinations of the 20 or so α chains, only about 10 types of collagen molecules have been found. The best defined are types I, II, III, and IV (Table 14–3). Types I, II, and III are the **fibrillar collagens.** They are the main types of collagen found in connective tissues, type I being by far the most common. After being secreted into the extracellular space, these three types of collagen molecules assemble into ordered polymers called **collagen fibrils,** which are thin (10–300 nm in diameter) cablelike structures, many micrometers long and clearly visible in electron micrographs (Figure 14–33). The collagen fibrils often aggregate into larger bundles, which can be seen in the light microscope as *collagen fibers* several micrometers in diameter. Type IV collagen molecules are found exclusively in basal laminae; instead of forming fibrils, they assemble into a sheetlike meshwork that constitutes a major part of all basal laminae (see p. 815). The arrangement of most of the other half-dozen or so types of collagen molecules in tissues is uncertain.

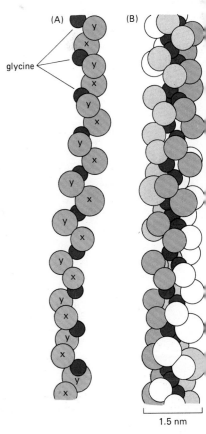

glycine

1.5 nm

Figure 14–32 (A) A model of a single collagen α chain in which each amino acid is represented by a sphere. The chain is arranged as a left-handed helix with three amino acid residues per turn and with glycine (*dark color*) as every third residue. Therefore an α chain is composed of a series of triplet Gly-X-Y sequences in which X and Y can be any amino acid (although one is commonly proline). (B) A model of a part of a collagen molecule in which three α chains are wrapped around one another to form a triple-stranded helical rod. One α chain is shown in light color, one in gray, and one in white. Glycine is the only amino acid small enough to occupy the crowded interior of the triple helix. Only a short length of the molecule is shown; the entire molecule is 300 nm long, with each α chain containing about 1000 amino acid residues. (Drawn from model by B.L. Trus.)

Table 14–3. Four Major Types of Collagen and Their Properties

Type	Molecular Formula	Polymerized Form	Distinctive Features	Tissue Distribution
I	$[\alpha 1(I)]_2\alpha 2(I)$	fibril	low hydroxylysine, low carbohydrate, broad fibrils	skin, tendon, bone, ligaments, cornea, internal organs (accounts for 90% of body collagen)
II	$[\alpha 1(II)]_3$	fibril	high hydroxylysine, high carbohydrate, usually thinner fibrils than type I	cartilage, intervertebral disc, notochord, vitreous body of eye
III	$[\alpha 1(III)]_3$	fibril	high hydroxyproline, low hydroxylysine, low carbohydrate	skin, blood vessels, internal organs
IV	$[\alpha 1(IV)]_2\alpha 2(IV)$	basal lamina	very high hydroxylysine, high carbohydrate, retains procollagen extension peptides	basal laminae

Note that types I and IV are each composed of two types of α chain, whereas types II and III are composed of only one type of α chain each. Only the four major types of collagen are shown, but, more than 10 types of collagen and about 20 types of α chain have been defined so far.

Many proteins that contain a repeated pattern of amino acids have evolved by duplications of DNA sequences (see p. 602). The fibrillar collagens apparently arose in this way. Thus the genes that encode the α chains of these collagens are very large (30–40 kilobases in length) and contain about 50 exons. Most of the exons are 54, or multiples of 54, nucleotides long, suggesting that these collagens arose by multiple duplications of a primordial gene containing 54 nucleotides; this is not the case for type IV collagen, which may therefore have evolved differently.

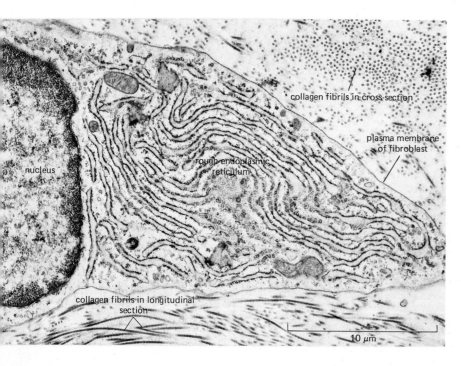

collagen fibrils in cross-section

plasma membrane of fibroblast

rough endoplasmic reticulum

nucleus

collagen fibrils in longitudinal section

10 μm

Figure 14–33 Electron micrograph showing part of a fibroblast surrounded by collagen fibrils in connective tissue. The extensive rough endoplasmic reticulum in the fibroblast cytoplasm reflects the cell's active synthesis and secretion of collagen and other extracellular matrix macromolecules. (Courtesy of Russell Ross.)

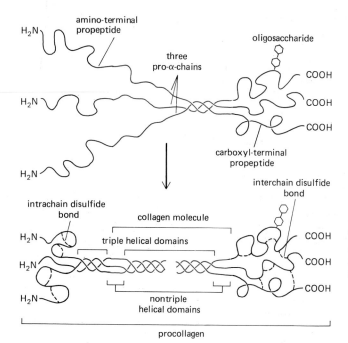

Figure 14–34 In fibrillar collagen molecules the α chains are initially synthesized in the form of pro-α chains, which contain extra propeptides at both ends (shown here in black) that will later be removed. The carboxyl-terminal propeptide is thought to help guide triple-helix formation during the assembly of the *procollagen* molecule. Note that the carboxyl-terminal propeptides in the procollagen molecule are covalently linked together by disulfide bonds and often contain an oligosaccharide chain. The amino-terminal propeptides form a short, triple-stranded "minicollagen" region. The final collagen molecule contains only the portion of the procollagen molecule that is shown in red; the rest is cleaved off.

14-13 Collagens Are Secreted with a Nonhelical Extension at Each End[14,15]

The individual collagen polypeptide chains are synthesized on membrane-bound ribosomes and injected into the lumen of the endoplasmic reticulum (ER) as larger precursors, called *pro-α chains*. These precursors not only have the short amino-terminal "signal peptide" required to thread secreted proteins through the membrane of the ER (see p. 438), they also have other extra amino acids, called *propeptides*, at both their amino- and carboxyl-terminal ends. In the lumen of the ER, selected proline and lysine residues are hydroxylated to form hydroxyproline and hydroxylysine, respectively. Each pro-α chain then combines with two others to form a hydrogen-bonded, triple-stranded helical molecule known as *procollagen* (Figure 14–34). The secreted forms of fibrillar collagens (but not of type IV collagen) are converted to *collagen molecules* in the extracellular space by the removal of the propeptides (see below).

Hydroxyproline and hydroxylysine residues (Figure 14–35) are rarely found in other proteins. Why are they present in collagen? There is indirect evidence that the hydroxyl groups of hydroxyproline residues form interchain hydrogen bonds that help stabilize the triple-stranded helix. For example, conditions that prevent proline hydroxylation (such as a deficiency of ascorbic acid [vitamin C]) inhibit procollagen helix formation. Normal collagens are continuously (albeit slowly) degraded by specific extracellular enzymes called *collagenases*. In scurvy, a human disease caused by a dietary deficiency of vitamin C, the defective pro-α chains that are synthesized fail to form a triple helix and are immediately degraded. Consequently, with the gradual loss of the preexisting normal collagen in the matrix, blood vessels become extremely fragile and teeth become loose in their sockets. This implies that in these particular tissues degradation and replacement of collagen is relatively rapid. In many other adult tissues, however, the "turnover" of collagen (and other extracellular matrix macromolecules) is thought to be normally very slow: in bone, to take an extreme example, collagen molecules persist for about 10 years before they are degraded and replaced. By contrast, most cellular proteins have half-lives of the order of hours or days.

The hydroxylation of lysine residues has a function different from that of the hydroxylation of proline residues. It is required for an unusual form of lysine-linked glycosylation found in collagen (the function of which is unknown) and is crucial for the extensive cross-linking of collagen molecules that occurs during collagen assembly in the extracellular space (see p. 813).

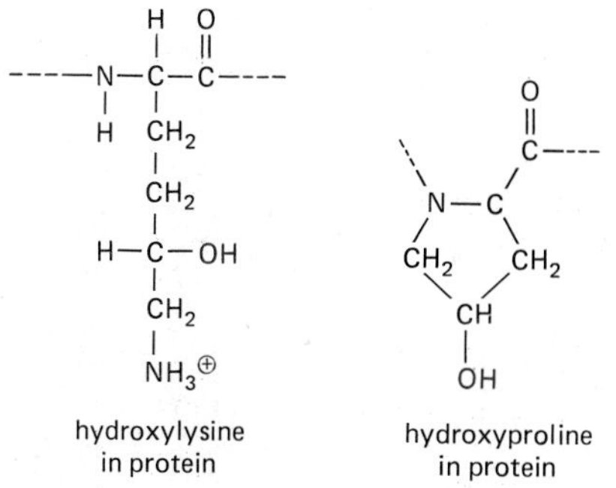

Figure 14–35 The structures of hydroxyproline and hydroxylysine residues, two modified amino acids that are common in collagen.

After Secretion, Types I, II, and III Procollagen Molecules Are Cleaved to Collagen Molecules, Which Assemble into Fibrils[16]

After secretion, the propeptides of types I, II, and III procollagen molecules are removed by specific proteolytic enzymes outside the cell. This converts the procollagen molecules to collagen (also called *tropocollagen*) molecules (1.5 nm in diameter), which then associate in the extracellular space to form the much larger collagen fibrils (10–300 nm in diameter). The process of fibril formation is driven, in part, by the tendency of the collagen molecules to self-assemble. However, the fibrils form close to the cell surface, often in deep recesses formed by the infolding of the plasma membrane, and the underlying cortical cytoskeleton can therefore influence the sites, rates, and orientation of fibril assembly (see p. 822).

The propeptides have at least two functions: (1) they guide the intracellular formation of the triple-stranded collagen molecules; and (2) because they are removed only after secretion, they prevent the intracellular formation of large collagen fibrils, which could be catastrophic for the cell. It is equally important, however, that the propeptides be removed from the fibrillar collagens once they have performed their functions. In some genetic diseases, such as Ehlers-Danlos syndrome, this process is defective and collagen fibril formation is impaired, resulting in fragile skin and hypermobile joints in affected individuals.

When isolated collagen fibrils are fixed, stained, and viewed in an electron microscope, they exhibit cross-striations every 67 nm. This pattern reflects the packing arrangement of the individual collagen molecules in the fibril: they are staggered, as shown in Figure 14–36, so that adjacent molecules are displaced longitudinally by almost one-quarter of their length (a distance of 67 nm). This arrangement presumably maximizes the tensile strength of the aggregate, and it gives rise to the striations seen in negatively stained fibrils (Figure 14–37). However, it is still not certain how these staggered molecules are packed in the three dimensions of a cylindrical fibril.

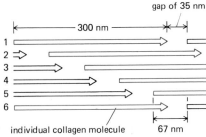

Figure 14–36 The staggered arrangement of collagen molecules in a collagen fibril. Adjacent molecules (shown as arrows) are displaced by 67 nm, with a 35-nm gap between successive molecules in a row. The gap size is such that the pattern repeats after five molecules have been lined up in this staggered fashion; thus the molecules in rows 1 and 6 are in register.

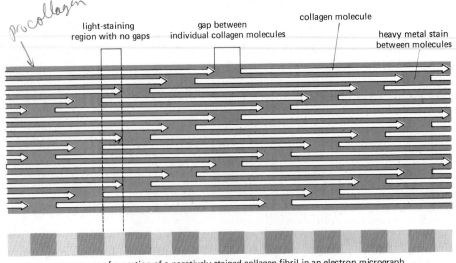

Procollagen

light-staining region with no gaps — gap between individual collagen molecules — collagen molecule — heavy metal stain between molecules

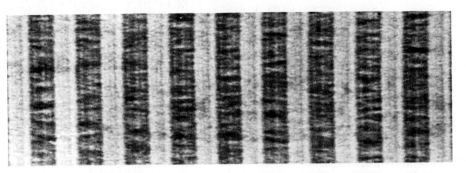

appearance of a portion of a negatively stained collagen fibril in an electron micrograph

67 nm

Figure 14–37 How the staggered arrangement of collagen molecules gives rise to the striated appearance of a negatively stained fibril. Since the negative stain fills only the space between the molecules, the stain in the gaps between the individual molecules in each row accounts for the dark staining bands. An electron micrograph of a negatively stained fibril is shown at the bottom of the figure. (Electron micrograph courtesy of Robert Horne.)

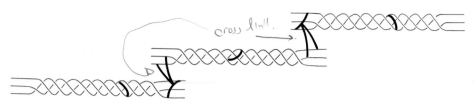

cross link

Figure 14–38 The covalent intramolecular and intermolecular cross-links formed between modified lysine side chains within a collagen fibril. The cross-links are formed in several steps. First, certain lysine and hydroxylysine residues are deaminated by the extracellular enzyme lysyl oxidase to yield highly reactive aldehyde groups. The aldehydes then react spontaneously to form covalent bonds with each other or with other lysine or hydroxylysine residues in which more than two amino acid side chains can be involved. Some of these bonds are relatively unstable and are ultimately modified to form a variety of more stable cross-links. Note that most of the cross-links form between the short nonhelical segments at each end of the collagen molecules (see Figure 14–35).

After collagen fibrils have formed in the extracellular space, they are greatly strengthened by the formation of covalent cross-links within and between lysine residues of the constituent collagen molecules (Figure 14–38). The types of covalent bonds involved are found only in collagen and elastin. If cross-linking is inhibited, collagenous tissues become fragile and structures such as skin, tendons, and blood vessels tend to tear. The extent and type of cross-linking varies from tissue to tissue. Collagen is especially highly cross-linked in the Achilles tendon, for example, where tensile strength is crucial.

The Organization of Collagen Fibrils in the Extracellular Matrix Is Adapted to the Needs of the Tissue[17]

Collagen fibrils come in a variety of diameters and are organized in different ways in different tissues. In mammalian skin, for example, they are woven in a wickerwork pattern so that they resist stress in multiple directions. In tendons they are organized in parallel bundles aligned along the major axis of stress on the tendon. And in mature bone and in the cornea, they are arranged like plywood in orderly layers, with the fibrils in each layer lying parallel to each other but nearly at right angles to the fibrils in the layers on either side. The same arrangement occurs in tadpole skin, which serves to illustrate this organization (Figure 14–39).

The connective tissue cells themselves determine the size and arrangement of the collagen fibrils. The cells can express one or more of the genes for the different types of fibrillar procollagen molecules (including minor types not listed in Table 14–3) and can regulate the disposition of the molecules after secretion. By controlling the order in which the amino- and carboxyl-terminal propeptides are sequentially cleaved, by secreting different kinds and amounts of noncollagen matrix macromolecules along with the collagen, and by guiding collagen fibril formation in close association with the plasma membrane, cells can determine the geometry and properties of the fibrils in their environment. Finally, the collagen is cross-linked to a greater or lesser degree depending on the tensile strength required. Figure 14–40 summarizes the steps in fibrillar collagen synthesis and assembly.

Cells Can Help Organize the Collagen Fibrils They Secrete by Exerting Tension on the Matrix[18]

There is yet another way that collagen-secreting cells determine the spatial organization of the matrix they produce. Fibroblasts work on the collagen they have secreted, crawling over it and tugging on it—helping to compact it into sheets and draw it out into cables. This mechanical role of fibroblasts in shaping collagen matrices has been demonstrated dramatically in culture. When fibroblasts are mixed with a meshwork of randomly oriented collagen fibrils that form a gel in a culture dish, the fibroblasts tug on the meshwork, drawing in collagen from the surroundings and causing the gel to contract to a small fraction of its initial volume; by similar activities, a cluster of fibroblasts will surround itself with a capsule of densely packed and circumferentially oriented collagen fibers.

If two small pieces of embryonic tissue containing fibroblasts are placed far apart on a collagen gel, the collagen becomes organized into a compact band of aligned fibers that connect the two explants (Figure 14–41). The fibroblasts subsequently migrate out from the explants along the aligned collagen fibers. Thus

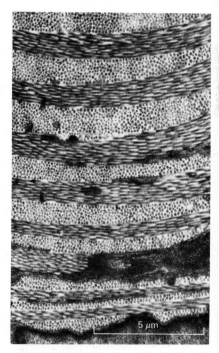

5 μm

Figure 14–39 Electron micrograph of a cross section of tadpole skin showing the plywoodlike arrangement of collagen fibrils, in which successive layers of fibrils are laid down at right angles to each other. This arrangement is also found in mature bone and in the cornea. (Courtesy of Jerome Gross.)

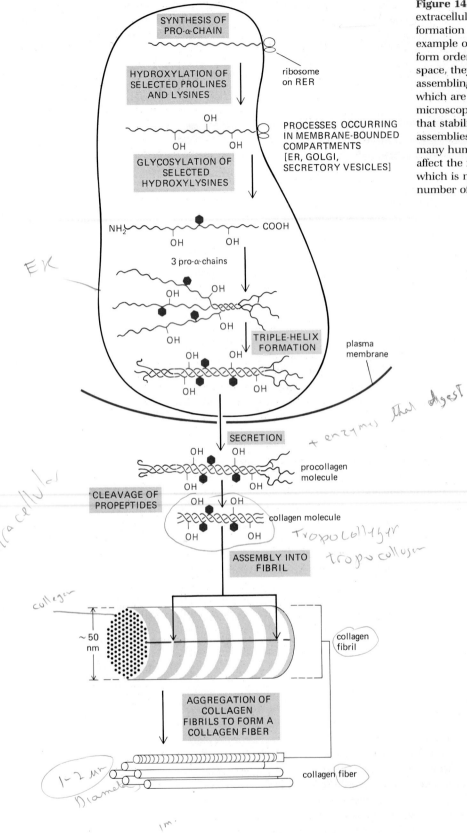

SYNTHESIS OF
PRO-α-CHAIN

ribosome
on RER

HYDROXYLATION OF
SELECTED PROLINES
AND LYSINES

OH

OH   OH

PROCESSES OCCURRING
IN MEMBRANE-BOUNDED
COMPARTMENTS
[ER, GOLGI,
SECRETORY VESICLES]

GLYCOSYLATION OF
SELECTED
HYDROXYLYSINES

NH₂   COOH

OH   OH

ER

3 pro-α-chains

OH

OH   OH

OH

OH

TRIPLE-HELIX
FORMATION

plasma
membrane

OH   OH

OH   OH

SECRETION

+ enzymes that digest

OH   OH

procollagen
molecule

OH   OH

extracellular

CLEAVAGE OF
PROPEPTIDES

OH   OH

collagen molecule

OH   OH

Tropocollagen
tropocollagen

ASSEMBLY INTO
FIBRIL

collagen

~ 50
nm

collagen
fibril

AGGREGATION OF
COLLAGEN
FIBRILS TO FORM A
COLLAGEN FIBER

1-2 μm
Diameter

collagen fiber

1m

Figure 14-40 The intracellular and extracellular events involved in the formation of a collagen fibril. As one example of how the collagen fibrils can form ordered arrays in the extracellular space, they are shown further assembling into large collagen fibers, which are visible in the light microscope. The covalent cross-links that stabilize the extracellular assemblies are not shown. There are many human genetic diseases that affect the formation of collagen fibrils, which is not surprising given the large number of enzymatic steps involved.

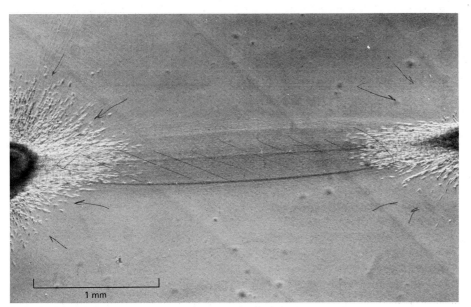

Figure 14–41 Photomicrograph of the region between two pieces of embryonic chick heart (rich in fibroblasts as well as muscle cells) growing in culture on a collagen gel for 4 days. Note that a dense tract of aligned collagen fibers has formed between the explants. (From D. Stopak and A.K. Harris, *Dev. Biol.* 90:383–398, 1982.)

the colagen stretched out cell's rear the ECM os so along.

the fibroblasts influence the alignment of the collagen fibers, and the collagen fibers in turn affect the distribution of the fibroblasts. Fibroblasts presumably play a similar role in generating long-range order in the extracellular matrix inside the body—in helping to create tendons and ligaments, for example, and the tough, dense layers of connective tissue that ensheathe and bind together most organs.

Type IV Collagen Molecules Assemble into a Laminar Meshwork[19]

Type IV collagen molecules differ from the fibrillar collagen molecules in several ways. First, the regular Gly-X-Y repeating amino acid sequence of the type IV α chain is interrupted in a number of regions, locally disrupting the triple-stranded helical structure of the collagen molecule. Second, type IV "procollagen" molecules are not cleaved after secretion and so retain their propeptides; the secreted molecules interact via their uncleaved propeptide domains to assemble into a sheetlike multilayered network rather than into fibrils. Electron microscopic studies of preparations of assembling type IV collagen molecules suggest that these molecules associate by their carboxyl-terminal propeptides to form head-to-head dimers, which then form an extended lattice by the further associations shown in Figure 14–42. Disulfide and other covalent cross-links between the collagen molecules stabilize these associations. Type IV collagen sheets are thought to form the core of all basal laminae; additional components of the basal lamina will be discussed below (see p. 818).

actually not in reading

Elastin Is a Cross-linked, Random-Coil Protein That Gives Tissues Their Elasticity[20]

Tissues such as skin, blood vessels, and lungs require elasticity in addition to tensile strength in order to function. A network of **elastic fibers** in the extracellular matrix of these tissues gives them the required ability to recoil after transient stretch. The main component of elastic fibers is **elastin,** a highly hydrophobic, nonglycosylated protein (about 830 amino acid residues long), which, like collagen, is unusually rich in proline and glycine but, unlike collagen, contains little hydroxyproline and no hydroxylysine. Elastin molecules are secreted into the extracellular space, where they form filaments and sheets in which the elastin molecules are highly cross-linked to one another to generate an extensive network (Figure 14–43). The cross-links are formed between lysine residues by the same mechanism that operates in cross-linking collagen molecules (see Figure 14–38). Elastin molecules are unlike most other proteins in that their function requires

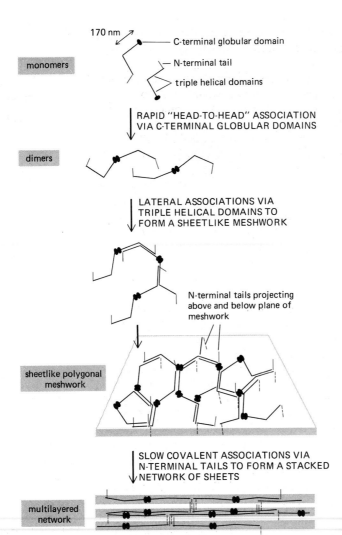

monomers

170 nm

— C-terminal globular domain

— N-terminal tail

triple helical domains

RAPID "HEAD-TO-HEAD" ASSOCIATION
VIA C-TERMINAL GLOBULAR DOMAINS

dimers

LATERAL ASSOCIATIONS VIA
TRIPLE HELICAL DOMAINS TO
FORM A SHEETLIKE MESHWORK

N-terminal tails projecting
above and below plane of
meshwork

sheetlike polygonal
meshwork

SLOW COVALENT ASSOCIATIONS VIA
N-TERMINAL TAILS TO FORM A STACKED
NETWORK OF SHEETS

multilayered
network

Figure 14–42 How type IV collagen molecules are thought to assemble into a multilayered network, which forms the core of all basal laminae. The model is based on electron micrographs of rotary-shadowed preparations of these molecules assembling *in vitro*. (Based on P.D. Yurchenco, E.C. Tsilibary, A.S. Charonis, and H. Furthmayr, *J. Histochem. Cytochem.* 34:93–102, 1986.)

their polypeptide backbones to remain unfolded as "random coils" (Figure 14–44). It is the cross-linked, random-coil structure of the elastic fiber network that allows the network to stretch and recoil like a rubber band (Figure 14–45). Elastic fibers are at least five times more extensible than a rubber band of the same cross-sectional area. Long, inelastic collagen fibrils are interwoven with the elastic fibers to limit the extent of stretching and thereby prevent the tissue from tearing.

Elastic fibers are not composed solely of elastin. They also contain a glycoprotein that is distributed mainly as microfibrils on the elastic fiber surface. Elastic fibers are assembled in close association with the plasma membrane of the cells that secrete elastin and the microfibrillar glycoprotein. The microfibrils appear before elastin and may help the cell organize the secreted elastin molecules into the elastic fibers and sheets that form in the extracellular matrix.

Fibronectin Is an Extracellular Adhesive Glycoprotein That Helps Mediate Cell-Matrix Adhesion[21]

The extracellular matrix contains a number of **adhesive glycoproteins** that bind to both cells and other matrix macromolecules and thereby help cells attach to the extracellular matrix. The best characterized of these is **fibronectin,** a large fibril-forming glycoprotein found throughout the animal kingdom. Fibronectin is a dimer composed of two similar subunits (each almost 2500 amino acid residues long); these subunits are joined by a pair of disulfide bonds near their carboxyl termini and are folded into a series of globular domains separated by regions of flexible polypeptide chain (Figure 14–46). Sequencing studies indicate that a

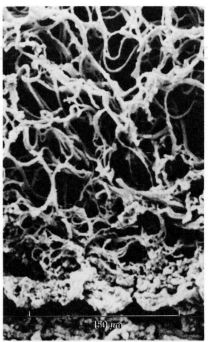

Figure 14–43 Scanning electron micrograph showing the extensive network of elastic fibers in a section of human skin connective tissue (dermis). The tissue has been heated under pressure to remove the collagen and glycosaminoglycans. (From T. Tsuji, R.M. Lavker, and A.M. Kligman, *J. Microscop.* 115:165–173, 1978.)

fibronectin molecule is composed mainly of three types of short amino acid sequences repeated many times, suggesting that the fibronectin gene evolved by multiple duplications of three small genes.

Fibronectin exists in three forms: (1) a soluble dimeric form, called *plasma fibronectin*, circulates in the blood and other body fluids, where it is thought to enhance blood clotting, wound healing, and phagocytosis; (2) oligomers of fibronectin can be found transiently attached to the surface of cells (*cell-surface fibronectin*); and (3) highly insoluble fibronectin fibrils form in the extracellular matrix (*matrix fibronectin*). In the cell-surface and matrix aggregates, fibronectin dimers are cross-linked to one another by additional disulfide bonds.

Fibronectin is a multifunctional molecule in which the various globular domains play different roles. For example, one domain binds to collagen, another to heparin, another to specific receptors on the surface of various types of cells, and so on (see Figure 14–46). In this way fibronectin contributes to the organization of the matrix and helps cells attach to it.

The parts played by the different domains, and in particular by the cell-binding domains, have been analyzed by cleaving the molecule into its separate domains with proteolytic enzymes or by synthesizing specific protein fragments either chemically or by recombinant DNA techniques. Thus a domain responsible for cell-binding activity has been isolated from proteolytic fragments and its amino acid sequence determined. Synthetic peptides corresponding to different segments of this domain were prepared and used to localize the cell-binding activity to a specific tripeptide sequence (Arg-Gly-Asp, or R-G-D). Peptides containing this **RGD sequence** compete for the binding site on cells and so inhibit the attachment of cells to fibronectin; and when these peptides are coupled to a solid surface, they cause cells to adhere to that surface. The RGD sequence is not confined to fibronectin. It is a common motif in a variety of extracellular adhesive proteins, and it is recognized by a family of homologous cell-surface receptors that bind these proteins (see p. 821). Despite the common tripeptide sequence found at the sites recognized by these receptors, each receptor specifically recognizes its own small set of adhesive molecules. Thus receptor binding must also depend on other parts of the adhesive protein sequence.

Fibronectin is important not only for cell adhesion but also for cell migration. In both invertebrate and vertebrate embryos, it seems to guide cell migration in many cases. For example, large amounts of fibronectin are found along the pathway followed by migrating prospective mesodermal cells during amphibian gastrulation (see p. 882). The migration of these cells can be inhibited either by injecting antibodies against fibronectin into the blastocoel cavity or by injecting peptides containing the cell-binding tripeptide but lacking the matrix-binding domains of fibronectin. Fibronectin presumably promotes cell migration by helping cells attach to the matrix. The effect must be delicately balanced so that the

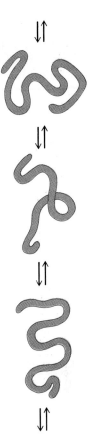

Figure 14–44 An elastin molecule in various "random-coil" conformations. Unlike most proteins, the elastin molecule does not adopt a unique structure but oscillates among a variety of partially extended, random conformations, as illustrated.

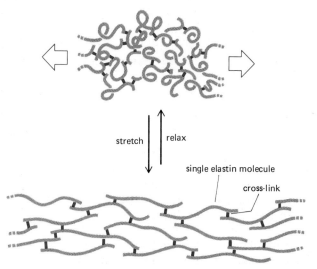

stretch | relax

single elastin molecule

cross-link

Figure 14–45 Elastin molecules are joined together by covalent bonds (indicated in color) to generate an extensive cross-linked network. Because each elastin molecule in the network can expand and contract as a random coil, the entire network can stretch and recoil like a rubber band.

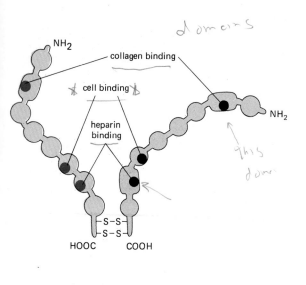

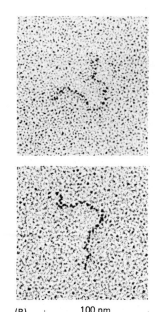

Figure 14–46 The structure of a fibronectin dimer shown schematically in (A) and in electron micrographs of individual molecules shadowed with platinum in (B). The two polypeptide chains are similar but not identical and are joined by two disulfide bonds near the carboxyl terminus. Each chain is folded into a series of globular domains connected by flexible polypeptide segments. Individual domains are specialized for binding to a particular molecule or to a cell, as indicated for three of the domains; for simplicity, not all of the known binding sites are shown. (B, from J. Engel et al., *J. Mol. Biol.* 150:97–120, 1981. © Academic Press Inc. [London] Ltd.)

(A)

(B) |———— 100 nm ————|

migrating cells get a grip on the matrix without becoming immobilized on it. We shall return later to the question of how this balance may be achieved by the many adhesive molecules believed to play a part in guiding morphogenetic movements.

Multiple Forms of Fibronectin Are Produced by Alternative RNA Splicing[22]

Aside from being a member of a large family of RGD-containing adhesion molecules, fibronectin itself, as mentioned earlier, can have a variety of forms; even the polypeptide chains in a single dimer may have minor differences. Yet all the different fibronectin polypeptide chains are encoded by a single large gene that, in the rat, is more than 70 kilobases long and contains about 50 exons, making it one of the largest genes characterized so far. Transcription produces a single large RNA molecule that is spliced in different ways to produce one or more of about 20 different messenger RNAs, depending on the cell type. It is not clear how these patterns of RNA splicing are determined or how the various polypeptide chains that result differ in function. There is some evidence that one function of alternative splicing of the human fibronectin RNA transcript is to add to selected fibronectin molecules an additional cell-binding domain that is distinct from the RGD-containing cell-binding site.

Fibronectin is not the only secreted glycoprotein involved in cell-matrix adhesion. For example, **tenascin** is also an extracellular adhesive glycoprotein, but it has a much more restricted distribution than fibronectin and is most abundant in embryonic tissues. In the nervous system it is secreted by glial cells, and some neurons are thought to adhere to it by means of a specific cell-surface proteoglycan. Tenascin is a large complex of six disulfide-linked polypeptide chains, which radiate from a center like the spokes of a wheel (see Figure 14–51).

Some cells, especially epithelial cells, secrete another type of extracellular adhesion glycoprotein, called *laminin*, which is a major protein in all basal laminae. It binds both to epithelial cells (as well as to some other cell types) and to type IV collagen, the main collagen type in the basal lamina.

The Basal Lamina Is a Specialized Extracellular Matrix Composed Mainly of Type IV Collagen, Proteoglycans, and Laminin[23]

Basal laminae are continuous thin mats of specialized extracellular matrix that underlie all epithelial cell sheets and tubes; they also surround individual muscle cells, fat cells, and Schwann cells (which wrap around peripheral nerve cell axons to form myelin). The basal lamina thus separates these cells and cell sheets from the underlying or surrounding connective tissue. In other locations, such as the kidney glomerulus and lung alveolus, a basal lamina lies between two different cell sheets, where it functions as a highly selective filter (Figure 14–47). However, basal laminae serve more than simple structural and filtering roles. They are able to determine cell polarity, influence cell metabolism, organize the proteins in adjacent plasma membranes, induce cell differentiation, and, like fibronectin, serve as specific "highways" for cell migration.

The basal lamina is largely synthesized by the cells that rest on it (Figure 14–48). In essence it is a tough mat of type IV collagen (see Figure 14–42) with specific additional molecules on each face that help bind it to the adjacent cells or matrix. Although the precise composition of basal laminae varies from tissue to tissue and even from region to region in the same lamina (see p. 821), all basal laminae contain type IV collagen together with proteoglycans (primarily heparan sulfates) and the glycoproteins *laminin* and *entactin*. **Laminin** is a large (~850,000 daltons) complex of three very long polypeptide chains arranged in the shape of a cross and held together by disulfide bonds (Figure 14–49). Like fibronectin, it consists of a number of functional domains: one binds to type IV collagen, one to heparan sulfate, and one or more to laminin receptor proteins on the surface of cells. A single dumbell-shaped entactin molecule is thought to be tightly bound to each laminin molecule where the short arms meet the long one.

As seen in the electron microscope after conventional fixation and staining, most basal laminae consist of two distinct layers: an electron-lucent layer (*lamina lucida* or *rara*) adjacent to the basal plasma membrane of the cells that rest on the lamina—typically epithelial cells—and an electron-dense layer (*lamina densa*) just below. In some cases a third layer containing collagen fibrils (*lamina reticularis*) connects the basal lamina to the underlying connective tissue. Some cell biologists use the term *basement membrane* to describe the composite of all three layers (Figure 14–50), which is usually thick enough to be seen in the light microscope. The detailed molecular organization of the basal lamina is still uncertain, although electron microscopic studies using antibody labeling suggest that the lamina densa is composed primarily of type IV collagen, with proteoglycan molecules located on either side; laminin is thought to be present mainly on the plasma-membrane side of the lamina densa, where it helps to bind epithelial cells to the lamina, while fibronectin helps to bind the matrix macromolecules and connective tissue cells on the opposite side.

The shapes and sizes of some of the major components of the basal lamina and other forms of extracellular matrix are compared in Figure 14–51.

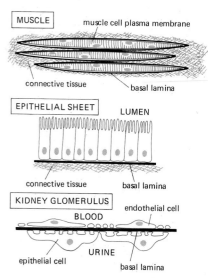

Figure 14–47 Three ways in which basal laminae (*black lines*) are organized: surrounding cells (such as muscle cells), underlying epithelial cell sheets, and interposed between two cell sheets (as in the kidney glomerulus). Note that in the kidney glomerulus, both cell sheets have gaps in them, so that the basal lamina serves as the permeability barrier determining which molecules will pass into the urine from the blood. Since the glomerular basal lamina develops as a result of the fusion of two basal laminae, one produced by the endothelial cells and the other by the epithelial cells, it is twice as thick as most basal laminae.

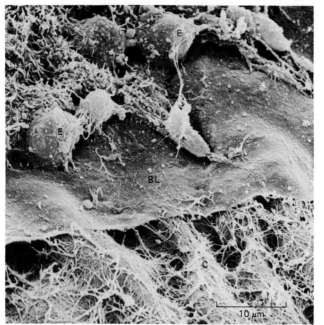

Figure 14–48 Scanning electron micrograph of a basal lamina in the cornea of a chick embryo. Some of the epithelial cells (E) have been removed to expose the upper surface of the matlike basal lamina (BL). Note the network of collagen fibrils (C) in the underlying connective tissue interacting with the lower face of the lamina. The macromolecules that comprise the basal lamina are synthesized by the epithelial cells that sit on it. (Courtesy of Robert Trelstad.)

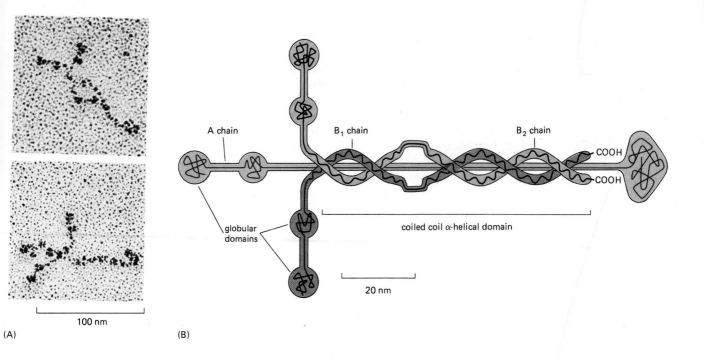

Figure 14–49 Electron micrographs of laminin molecules shadowed with platinum (A) and a schematic drawing of a model for the structure of laminin (B). The multidomain glycoprotein is composed of three polypeptides (A, B$_1$, and B$_2$) that are disulfide bonded into an asymmetric crosslike structure. Each of the polypeptide chains is more than 1500 amino acid residues long. (A, from J. Engel et al., *J. Mol. Biol.* 150:97–120, 1981. © Academic Press Inc. [London] Ltd.; B, based on B.L. Hogan *et al.*, in Basement Membranes [S. Shibata, ed.], pp. 147–154. Amsterdam: Elsevier, 1985.)

Basal Laminae Perform Diverse and Complex Functions[24]

The functions of basal laminae are surprisingly diverse. In the kidney glomerulus, an unusually thick basal lamina acts as a molecular filter, regulating the passage of macromolecules from the blood into the urine as urine is formed (see Figure 14–47). Proteoglycans seem to be important for this function: when they are removed by specific enzymes, the filtering properties of the lamina are destroyed. The basal lamina can also act as a selective cellular barrier. The lamina beneath epithelial cells, for example, usually prevents fibroblasts in the underlying connective tissue from making contact with the epithelial cells. It does not, however, stop macrophages, lymphocytes, or nerve processes from passing through it.

The basal lamina plays an important part in tissue regeneration after injury. When tissues such as muscles, nerves, and epithelia are damaged, the basal lamina survives and provides a scaffolding along which regenerating cells can migrate. In this way the original tissue architecture is readily reconstructed. A dramatic example of the importance of the basal lamina in regeneration comes from studies on the *neuromuscular junction*, where a nerve cell transmits its stimulus to a skeletal muscle cell.

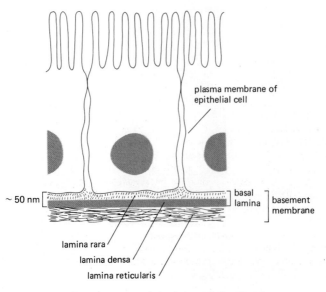

Figure 14–50 Schematic drawing of a basal lamina underlying an epithelial cell sheet as seen in cross section in an electron microscope.

At the site of neuromuscular contact (the synapse), the basal lamina has a chemically distinctive character, recognized, for example, by antibodies that bind to the lamina exclusively in this region. One of the functions of the junctional basal lamina, apparently, is to coordinate the spatial organization of the components on either side of the synapse. Evidence for the central role of the junctional basal lamina in reconstructing a synapse after nerve or muscle injury will be discussed in Chapter 19 (see p. 1124). These studies make it clear that we still have much to learn about the chemical and functional specializations of basal laminae. They also suggest that minor (but as yet undefined) components in the extracellular matrix may play a critical part in directing morphogenesis during embryonic development.

Integrins Help Bind Cells to the Extracellular Matrix[25]

To understand how the extracellular matrix interacts with cells, one has to define the cell-surface molecules that bind the matrix components as well as the extracellular matrix components themselves. As mentioned previously, some proteoglycans are integral components of the plasma membrane; their core protein may be either inserted across the lipid bilayer or covalently linked to it. By binding to most types of extracellular matrix components, these proteoglycans help link cells to the matrix. However, extracellular matrix components also bind to the cell surface via specific receptor glycoproteins. Because of the multiple interactions among matrix macromolecules in the extracellular space, it is largely a matter of semantics where the plasma membrane components end and the extracellular matrix begins. The glycocalyx of a cell, for example, often includes components of both (see p. 299).

The matrix receptors differ from cell-surface receptors for hormones and for other soluble signaling molecules in that they bind their ligand with relatively low affinity ($K_a = 10^6$–10^8 liters/mole) and are usually present at about 10- to 100-fold higher concentration on the cell surface. This suggests that the receptors might function cooperatively and that cells may respond to an organized group of ligands in the matrix rather than to individual molecules. In support of this suggestion, soluble cell-binding fragments of matrix components usually fail to elicit the cellular responses induced by the same components immobilized in a matrix.

A **fibronectin receptor** on mammalian fibroblasts is one of the best-characterized matrix receptors. It was initially identified as a plasma membrane glycoprotein that bound to a fibronectin affinity column and could be eluted with a small peptide containing the RGD cell-binding sequence (see p. 817). The receptor is a noncovalently associated complex of two distinct, high-molecular-weight polypeptide chains, called α and β. It functions as a transmembrane linker to mediate interactions between the actin cytoskeleton inside the cell and fibronectin in the extracellular matrix (Figure 14–52). We shall see later that these interactions across

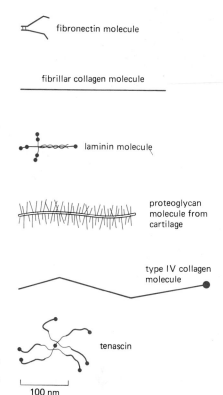

Figure 14–51 The comparative shapes and sizes of some of the major extracellular matrix macromolecules.

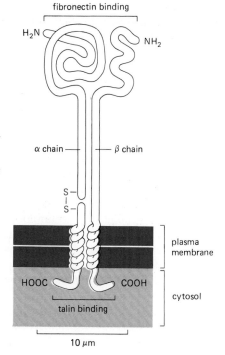

Figure 14–52 The subunit structure of a cell-surface fibronectin receptor. Electron micrographs of isolated receptors suggest that the molecule has approximately the shape shown, with the globular head projecting more than 20 nm from the lipid bilayer. By binding to fibronectin outside the cell and to the cytoskeleton (via the attachment protein talin) inside the cell, the protein serves as a transmembrane linker. The α and β chains are both glycosylated (not shown) and are held together by noncovalent bonds. The α chain is usually made initially as a single 140,000-dalton polypeptide chain, which is then cleaved into one small transmembrane chain and one large extracellular chain that remain held together by a disulfide bond. The extracellular part of the β chain contains a repeating cysteine-rich region, indicative of extensive intrachain disulfide bonding (not shown). The fibronectin receptor belongs to a large superfamily of homologous matrix receptors called *integrins*, most of which recognize RGD sequences in the extracellular proteins they bind.

the plasma membrane can orient both cells and matrix. Many other matrix receptors, including some that bind collagen and laminin, have been characterized and shown to be related to the fibroblast fibronectin receptor. Collectively called **integrins,** they are all heterodimers with α and β chains homologous to those of the fibronectin receptor. Most seem to recognize RGD sequences in the matrix components they bind.

There are at least three families within the large superfamily of integrins; the members of a family share a common β chain but differ in their α chains. One family includes a fibroblast fibronectin receptor and at least five other members. Another family includes a receptor found on blood platelets that binds several matrix components, including fibronectin and *fibrinogen*, which is a protein that interacts with platelets during blood clotting; humans with *Glanzmann's disease* are genetically deficient in these receptors and bleed excessively. A third family of integrins consists of receptors found mainly on the surface of white blood cells: one is called *LFA-1* (for lymphocyte function associated); another is called *Mac-1* because it is found mainly on macrophages. These receptors are involved in both cell-cell and cell-matrix interactions, and they are critically important in enabling these cells to fight infection. Humans with the disease called *leucocyte adhesion deficiency* are genetically unable to synthesize the β subunit. As a consequence, their white blood cells lack the entire family of receptors, and they suffer repeated bacterial infections. A number of cell-surface glycoproteins involved in position-specific cell adhesion in *Drosophila* larvae also belong to the integrin superfamily, but their relationship to the three families that are found in mammals is not certain.

Not all matrix receptors, however, belong to this superfamily. Some cells, for example, utilize an apparently unrelated transmembrane glycoprotein in binding to collagen; and many cells, as mentioned previously, have integral membrane proteoglycans that link cells to the extracellular matrix.

The Cytoskeleton and Extracellular Matrix Communicate Across the Plasma Membrane[26]

Extracellular matrix macromolecules have striking effects on the behavior of cells in culture, influencing not only their movement but also their shape, polarity, metabolism, and differentiation. Corneal epithelial cells, for example, make very little collagen when they are cultured on synthetic surfaces; but when they are cultured on laminin, collagen, or fibronectin, they accumulate and secrete large amounts of collagen. Other examples of extracellular matrix influences on cell metabolism and differentiation are discussed in Chapter 17 (see p. 987).

The matrix can also influence the organization of a cell's cytoskeleton. In general, the basal surfaces of epithelial cells cultured on plastic or glass are irregular, and the overlying cytoskeletons within the cells are disorganized. When the same cells are cultured on appropriate extracellular matrix macromolecules, the basal surfaces are smooth and the overlying cytoskeletons are highly organized, as they are in the intact tissue. Similar results have been obtained with neoplastically transformed fibroblasts in culture. Transformed cells often make less fibronectin than normal cultured cells and behave differently: for example, they adhere poorly to the substratum and fail to flatten out or develop the organized intracellular actin filament bundles known as *stress fibers* (see p. 627). In some of these cells, the fibronectin deficiency seems to be at least partly responsible for this abnormal behavior: if the cells are grown on a matrix of organized fibronectin fibrils, they will flatten and assemble intracellular stress fibers that are aligned with the extracellular fibronectin fibrils.

This interaction between the extracellular matrix and the cytoskeleton is reciprocal: intracellular actin filaments can influence the arrangement of secreted fibronectin molecules. In the neighborhood of cultured fibroblasts, for example, extracellular fibronectin fibrils assemble in alignment with adjacent intracellular stress fibers (Figure 14–53). If these cells are treated with the drug cytochalasin, which disrupts actin filaments, the fibronectin fibrils dissociate from the cell surface (just as they do during mitosis when a cell rounds up). Clearly there must be

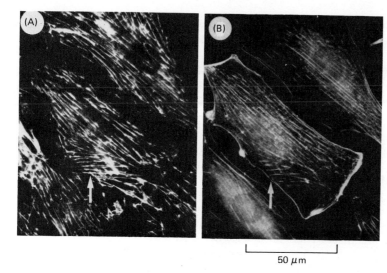

Figure 14–53 Immunofluorescence micrographs of extracellular fibronectin fibers (A) and intracellular actin filament bundles (B) in three rat fibroblasts in culture. The fibronectin is visualized by the binding of rhodamine-coupled anti-fibronectin antibodies and the actin by fluorescein-coupled anti-actin antibodies. Note that the orientation of the fibronectin fibers coincides with the orientation of the bundles of actin filaments. (From R.O. Hynes and A.T. Destree, *Cell* 15:875–886, 1978. © Cell Press.)

50 μm

a connection between extracellular fibronectin and intracellular actin filaments across the fibroblast plasma membrane. The connection is mediated by the fibronectin receptors discussed previously, which serve as transmembrane linkers between fibronectin and intracellular actin filaments via a set of intracellular attachment proteins, including talin (see p. 636 and Figure 14–52). The part of the receptor that binds talin contains a tyrosine residue that, when phosphorylated by tyrosine-specific protein kinases, seems to inactivate the talin binding site, thereby breaking the link between fibronectin and cortical actin filaments. It is thought that the attachment of cells to the matrix may be regulated in this way by specific growth factors that activate tyrosine-specific kinases (see Figure 13–37, p. 758).

Since the cytoskeletons of cells can order the matrix macromolecules they secrete, and the matrix macromolecules can in turn organize the cytoskeletons of cells that contact them, the extracellular matrix can in principle propagate order from cell to cell (Figure 14–54). Thus the matrix is thought to play a central part in generating and maintaining the orientations of cells in tissues and organs during development: the parallel alignment of fibroblasts and collagen fibrils in tendons, for example, may in part reflect this type of interaction between cells and matrix. The transmembrane matrix receptors serve as "adaptors" in this ordering process, mediating the interactions between cells and the matrix around them.

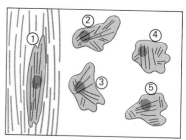

orientation of cytoskeleton in cell ① orients the assembly of secreted extracellular matrix molecules in the vicinity

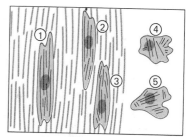

the oriented extracellular matrix reaches cells ② and ③ and orients the cytoskeleton of those cells

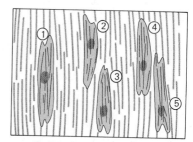

cells ② and ③ now secrete an oriented matrix in their vicinity; in this way the ordering of cytoskeletons is propagated to cells ④ and ⑤

Figure 14–54 A hypothetical scheme showing how the extracellular matrix could propagate order from cell to cell within a tissue. For simplicity, the figure shows one cell influencing the orientation of its neighboring cells, but by this scheme cells could mutually affect one another's orientation.

Summary

Cells in connective tissues are embedded in an intricate extracellular matrix that not only binds cells and tissues together but also influences the development, polarity, and behavior of the cells it contacts. The matrix contains various fiber-forming proteins interwoven in a hydrated gel composed of a network of glycosaminoglycan chains. The glycosaminoglycans are a heterogeneous group of long, negatively charged polysaccharide chains, which (except for hyaluronic acid) are covalently linked to protein to form proteoglycan molecules.

The fiber-forming proteins are of two functional types: mainly structural (collagens and elastin) and mainly adhesive (such as fibronectin and laminin). The fibrillar collagens (types I, II, and III) are ropelike, triple-stranded helical molecules that aggregate into long cablelike fibrils in the extracellular space; these in turn can assemble into a variety of highly ordered arrays. Type IV collagen molecules assemble into a sheetlike meshwork that forms the core of all basal laminae. Elastin molecules form an extensive cross-linked network of fibers and sheets that can stretch and recoil, imparting elasticity to the matrix. Fibronectin and laminin are examples of large adhesive glycoproteins in the matrix; fibronectin is widely distributed in connective tissues, whereas laminin is found mainly in basal laminae. By means of their multiple binding domains, such proteins help cells adhere to and become organized by the extracellular matrix. Many of these adhesive glycoproteins contain a common tripeptide sequence (RGD), which forms part of the structure recognized by a superfamily of homologous transmembrane matrix receptors, called integrins.

All of the matrix proteins and polysaccharides are secreted locally by cells in contact with the matrix, and they can be ordered by close associations with the exterior surface of the plasma membrane. Since the structure and orientation of the matrix in turn influence the orientation of the cells it contains, order is likely to be propagated from cell to cell through the matrix.

Cell-Cell Recognition and Adhesion[27]

We have so far considered how cell junctions and the extracellular matrix hold cells together in mature tissues and organs. But how do cells become associated with one another to form tissues in the first place? There are at least two distinctly different ways. Most commonly a tissue forms from "founder cells" whose progeny are prevented from wandering away by being attached to extracellular matrix macromolecules and/or to other cells (Figure 14–55). The precise pattern of adhesions determines the shape of the cell assembly. Epithelial cell sheets usually originate in this way, and much of animal development involves the formation, folding, and differentiation of such cell sheets to produce the tissues and organs of the mature organism. Typically, all of the cells in an early embryo are located in epithelia; only later do some cells change their adhesive properties and thereby escape to form other types of tissue (see pp. 882–888).

The other strategy for tissue formation seems more complex and involves cell migration: one population of cells invades another and assembles with them—and perhaps with other migrant cells—to form a tissue of mixed origin. In vertebrate embryos, for example, cells from the *neural crest* break away from the epithelial (neural) tube with which they are initially associated and migrate along specific paths to many other regions. There they assemble and differentiate into a variety of tissues, including those of the peripheral nervous system (Figure 14–56). Such a process requires some mechanism for directing the cells to their final destination, such as the secretion of a soluble chemical that attracts migrating cells (by *chemotaxis*) or the laying down of adhesive molecules, such as fibronectin (see p. 817), in the extracellular matrix to guide the migrating cells along the right paths (by *pathway guidance*).

Once a migrating cell reaches its destination, it must recognize other cells of the appropriate type in order to assemble into a tissue. Even in tissues that form without cell migration, there is evidence that the constituent cells specifically recognize one another: if the cells of such a developing tissue are dissociated into a suspension of single cells, they preferentially reassociate with one another rather than with cells of another tissue (see p. 827). Presumably this specific cell-cell recognition helps cells within a developing tissue stay together and remain segregated from the cells of neighboring tissues.

In an attempt to understand how cells in developing animal tissues recognize one another, many ingenious studies have been carried out on the social behavior displayed by some simple microorganisms that can alternate between unicellular and multicellular life-styles. Quite apart from their role as tentative models for cell-cell interactions in animals, these organisms are intriguing in their own right.

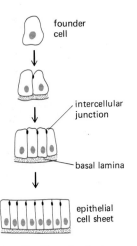

Figure 14–55 The simplest mechanism by which cells assemble to form a tissue. The progeny of the founder cells are retained in the epithelial sheet by the basal lamina and by cell-cell adhesion mechanisms, including the formation of intercellular junctions.

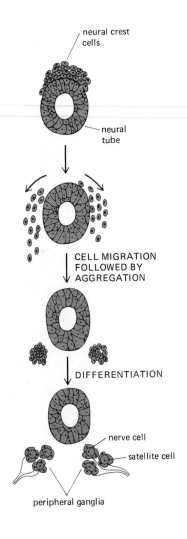

Figure 14–56 An example of a more complex mechanism by which cells assemble to form a tissue. Neural crest cells escape from the epithelium forming the upper surface of the neural tube and migrate away to form a variety of cell types and tissues throughout the embryo. Here they are shown assembling and differentiating to form two collections of nerve cells in the peripheral nervous system. Such a collection of nerve cells is called a *ganglion*. Other neural crest cells differentiate in the ganglion to become supporting (satellite) cells surrounding the neurons.

Slime Mold Amoebae Aggregate to Form Multicellular Fruiting Bodies When Starved[28]

The cellular slime mold, **Dictyostelium discoideum,** is a eucaryote whose genome is only 10 times larger than that of a bacterium and 100 times smaller than that of a human. These organisms live on the forest floor as independent motile cells called *amoebae*, which feed on bacteria and yeast and, under optimal conditions, divide every few hours. (In the laboratory they can be nourished with chemically defined liquid medium.) When their food supply is exhausted, the amoebae stop dividing and gather together to form tiny (1–2 mm), multicellular, wormlike structures, which crawl about as glistening slugs and leave trails of slime behind them (Figure 14–57).

Each slug is formed by the aggregation of up to 100,000 cells and shows a variety of behaviors that are not displayed by the free-living amoebae. The slug is extremely sensitive to light and heat, for example, and will migrate toward a light source as feeble as a luminous watch; presumably this behavior helps guide the slug to an advantageous environment. As the slug migrates, the cells begin to differentiate, initiating a process that ends with the production of a minute plant-like structure consisting of a stalk and a *fruiting body* some 30 hours after the beginning of aggregation (Figure 14–58). The fruiting body contains large numbers of *spores*, which can survive for long periods of time even in extremely hostile environments. The complex cell migrations that occur in stalk and fruiting-body formation are diagrammed in Figure 14–59. The cells in the front of the slug become the stalk region, those behind differentiate into spores, and those right at the rear form the foot plate. Both the stalk cells and the spore cells become covered with extracellular matrix (in the form of cellulose walls), and, in the end, all except the spore cells die. Only when conditions are favorable do the spores germinate to produce the free-living amoebae that start the cycle again (Figure 14–60).

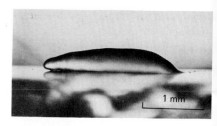

Figure 14–57 Light micrograph of a migrating slug of the cellular slime mold *Dictyostelium discoideum.* (Courtesy of David Francis.)

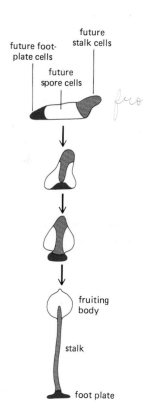

Figure 14–59 Cell migrations involved in the formation of a fruiting body in *Dictyostelium discoideum.* Cells in the front of the slug migrate down to become the stalk, while cells in the middle migrate up and differentiate into the collection of spores that forms the fruiting body.

Figure 14–58 Light micrographs of *Dictyostelium discoideum* showing various stages in fruiting-body formation. (Courtesy of John Bonner.)

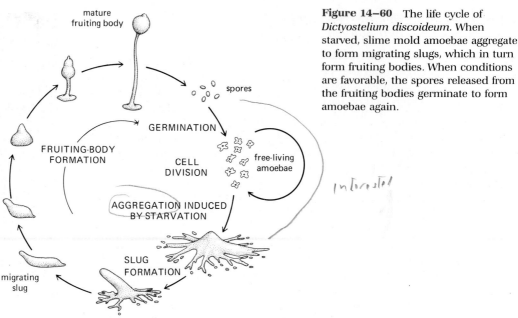

Figure 14–60 The life cycle of *Dictyostelium discoideum*. When starved, slime mold amoebae aggregate to form migrating slugs, which in turn form fruiting bodies. When conditions are favorable, the spores released from the fruiting bodies germinate to form amoebae again.

mature
fruiting body

spores

GERMINATION

FRUITING-BODY
FORMATION

CELL
DIVISION

free-living
amoebae

AGGREGATION INDUCED
BY STARVATION

SLUG
FORMATION

migrating
slug

Slime Mold Amoebae Aggregate by Chemotaxis[29]

In forming the slug, individual slime mold amoebae aggregate by *chemotaxis*, which we shall digress briefly to discuss before returning to the part played by cell-cell adhesion. One response of the amoebae to starvation is to start making and secreting cyclic AMP, which serves as a chemotactic signal that attracts other amoebae. (Cyclic AMP, as we have seen in Chapter 12, serves as an intracellular signal in procaryotic and animal cells; *Dictyostelium* is the only known organism in which cyclic AMP also acts as an extracellular signaling molecule.) Aggregation is apparently initiated at random: whichever cells first begin secreting cyclic AMP attract other cells and thereby become *aggregation centers*. The cyclic AMP that these "initiator" cells make is secreted in pulses and binds to specific receptors on the surface of neighboring starved amoebae, thereby orienting their normal locomotion in the direction of the source of cyclic AMP. This chemotactic response can be demonstrated by applying a tiny amount of cyclic AMP with a micropipette to any point on the surface of a starved amoeba cell. The result is the immediate formation of a pseudopod, which grows toward the micropipette (Figure 14–61); the pseudopod adheres to the surface on which the cell is placed and pulls the cell along in the same direction.

Once an aggregation center starts to form, its area of influence is rapidly enlarged because the aggregating cells not only respond to the cyclic AMP signal but relay it from cell to cell. Each pulse of cyclic AMP induces surrounding cells both to move toward the source of the pulse and to secrete their own pulse of cyclic AMP. In turn, this new pulse, released after a slight delay, orients the cells just beyond and induces a pulse of cyclic AMP from them, and so on. In this way regular pulsating waves of cyclic AMP flow from each aggregation center, causing more distant amoebae to move inward in surging concentric or spiraling waves that can be seen in time-lapse motion pictures (Figure 14–62). The advantage of such a relay system is that the signal is repeatedly renewed as it spreads from the center, so that it propagates without diminution over a large area. A signal that merely diffuses, by contrast, progressively fades out as it spreads. The difference can be appreciated by comparing the aggregation process in *Dictyostelium discoideum* with that in *Dictyostelium minutum*, a strain that lacks the relay system. In *D. minutum* the range of the signal released from each aggregation center is greatly reduced, and the slug and fruiting bodies that form are very small.

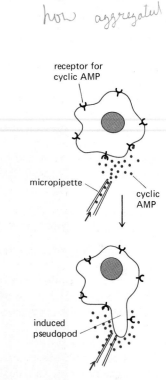

receptor for
cyclic AMP

micropipette

cyclic
AMP

induced
pseudopod

Figure 14–61 Application of a small amount of cyclic AMP to any point on the surface of a starved *Dictyostelium* amoeba (a single cell) results in the immediate formation of a pseudopod at that point. In this way the amoeba is induced to move toward a source of cyclic AMP. The cyclic AMP acts by binding to specific cell-surface receptors.

14-20 Cell-Cell Adhesion in Slime Molds Depends on Specific Cell-Surface Glycoproteins[30]

Besides activating the cyclic AMP signaling system, starvation of *Dictyostelium* amoebae induces the expression of hundreds of new genes, some of which encode the cell adhesion molecules involved in cell aggregation. For example, **discoidin-1,** which is a carbohydrate-binding protein (i.e., a lectin—see p. 299), is thought to be secreted by the starved cells to provide a primitive form of pathway guidance. By binding both to the surface of the amoebae and to the substratum on which they are migrating, it might help the amoebae move in streams toward aggregation centers, much as a trail of fibronectin guides cell migration during gastrulation in developing animals. In fact, cell binding to discoidin-1 depends on the same cell-binding tripeptide (RGD) sequence found in fibronectin and many other adhesive proteins (see p. 817).

Different newly synthesized proteins promote the cell-cell adhesion process that enables the migrating amoebae to adhere tightly to one another and assemble into a multicellular organism. During the first 8 hours of starvation, cells adhere by a Ca^{2+}-dependent mechanism involving a cell-cell adhesion molecule called **contact site B.** After 8 hours, a second adhesion system comes into play in which cells adhere by a Ca^{2+}-independent mechanism involving a cell-cell adhesion molecule called **contact site A.** Contact sites A and B were identified and characterized as integral plasma membrane glycoproteins by an ingenious immunological strategy, outlined in Figure 14–63. This method was later used to identify cell-cell adhesion molecules in vertebrates as well.

How do cell-surface glycoproteins such as contact sites A and B bind cells together? Three possibilities are illustrated in Figure 14–64: (1) molecules on one cell may bind to other molecules of the same kind on adjacent cells (so-called *homophilic* binding); (2) molecules on one cell may bind to molecules of a different kind on adjacent cells (so-called *heterophilic* binding); and (3) cell-surface receptors on adjacent cells may be linked to one another by secreted multivalent linker molecules. All of these mechanisms have been found to operate in animals.

Contact site A is thought to bind cells together by a homophilic mechanism: when the protein is coupled to synthetic beads, the beads bind only to cells that express contact site A, and the binding is blocked if the cells are pretreated with antibody against contact site A. DNA-sequencing studies show that contact site A is a single-pass transmembrane protein, apparently unrelated to any vertebrate cell-cell adhesion protein so far characterized (see below).

Figure 14–62 Light micrograph of waves of starved *Dictyostelium* amoebae moving toward an aggregation center. Individual amoebae cannot be distinguished at this low magnification. (Courtesy of Günter Gerisch.)

Dissociated Vertebrate Cells Can Reassemble into Organized Tissues Through Selective Cell-Cell Adhesion[31]

One of the attractions of using social microorganisms such as *Dictyostelium* to study cell aggregation is that the phenomenon proceeds normally in a culture dish, where it is accessible for investigation. Unfortunately, this is rarely the case with cell-cell recognition processes that occur during the development of multicellular animals. Usually the best one can do is to dissociate newly formed tissues into single cells, which can then be tested for their ability to reassemble *in vitro*. Unlike adult vertebrate tissues, which are difficult to dissociate, embryonic vertebrate tissues are easily dissociated by treatment with low concentrations of a proteolytic enzyme such as trypsin, sometimes combined with the removal of extracellular Ca^{2+} with a Ca^{2+} chelator (such as EDTA). These reagents disrupt the protein-protein interactions (many of which are Ca^{2+}-dependent—see p. 830) that hold cells together. Remarkably, such dissociated cells often reassemble *in vitro* into structures that resemble the original tissue. The tissue structure, therefore, is not just a product of history but is actively maintained and stabilized by the system of affinities that cells have for one another and for the extracellular matrix. Thus, by studying the reassembly of dissociated cells in culture, one can hope to illuminate the role of cell-cell and cell-matrix adhesion in creating and maintaining the organization of tissues in the body.

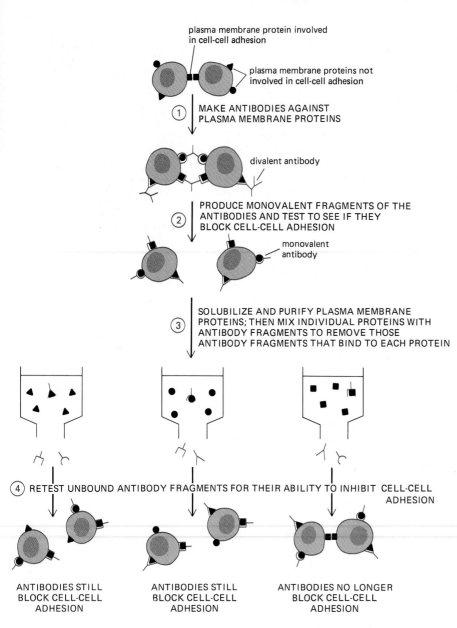

plasma membrane protein involved
in cell-cell adhesion

plasma membrane proteins not
involved in cell-cell adhesion

(1) MAKE ANTIBODIES AGAINST
PLASMA MEMBRANE PROTEINS

divalent antibody

(2) PRODUCE MONOVALENT FRAGMENTS OF THE
ANTIBODIES AND TEST TO SEE IF THEY
BLOCK CELL-CELL ADHESION

monovalent
antibody

(3) SOLUBILIZE AND PURIFY PLASMA MEMBRANE
PROTEINS; THEN MIX INDIVIDUAL PROTEINS WITH
ANTIBODY FRAGMENTS TO REMOVE THOSE
ANTIBODY FRAGMENTS THAT BIND TO EACH PROTEIN

(4) RETEST UNBOUND ANTIBODY FRAGMENTS FOR THEIR ABILITY TO INHIBIT CELL-CELL
ADHESION

ANTIBODIES STILL
BLOCK CELL-CELL
ADHESION

ANTIBODIES STILL
BLOCK CELL-CELL
ADHESION

ANTIBODIES NO LONGER
BLOCK CELL-CELL
ADHESION

CONCLUSION: PROTEIN ■ IS RESPONSIBLE FOR CELL–CELL ADHESION

Figure 14–63 An immunological strategy for identifying plasma membrane proteins involved in cell-cell adhesion. In step 1, antibodies are made (usually in rabbits) against the cell of interest or against plasma membranes isolated from such cells. In step 2, monovalent fragments of the antibodies are produced and tested to find an antibody preparation that inhibits cell-cell adhesion. (Monovalent fragments, made by protease digestion [see p. 1013], are used because they cannot cross-link cells and cause irrelevant adhesions.) To find the cell-surface molecules involved in cell-cell adhesion, the plasma membrane proteins are solubilized from the cells of interest, separated from one another, and individual fractions are tested for their ability to neutralize the aggregation-blocking effect of the antibody fragments (steps 3 and 4). Fractions that contain blocking activity are then further purified and retested until a pure protein is obtained (not shown). An alternative immunological strategy involves making monoclonal antibodies (see p. 178) against cell-surface antigens and screening large numbers to find those few that block cell-cell adhesion. Both of these immunological strategies depend on an important general finding: simply coating the surface of cells with antibodies does not in itself interfere with normal cell adhesion; adhesion is inhibited only when specific cell-surface molecules involved in the adhesion process are targets for antibody binding.

Experiments on cultured cells from the *epidermis* (the epithelium of the skin) provide an instructive example. In this tissue, Ca^{2+}-dependent adhesion systems play a crucial part in holding the cells, known as *keratinocytes*, together in a multilayered sheet resting on a basal lamina. The keratinocytes in the basal layer of the skin are relatively undifferentiated and proliferate steadily, releasing progeny into the upper layers, where cell division halts and terminal differentiation occurs (see p. 968). Given a suitable substratum, dissociated keratinocytes will likewise proliferate and differentiate in culture. If the concentration of Ca^{2+} in the culture medium is kept abnormally low, however, the Ca^{2+}-dependent cell-cell adhesion systems cannot operate, and the keratinocytes grow as a monolayer in which proliferating and differentiating cells are intermingled. If the Ca^{2+} concentration is then raised, the spatial organization of the cells is soon transformed: the monolayer is converted into a multilayered epithelium in which the proliferating cells form the basal layer adherent to the substratum and the differentiating cells are segregated into the upper layers, just as in normal skin. This result suggests that the normal stratified arrangement of keratinocytes, ordered according to their state of differentiation, is maintained by Ca^{2+}-dependent cell-cell adhesion mechanisms (see Figure 14–68, p. 832).

The Reassembly of Dissociated Vertebrate Cells Depends on Tissue-specific Recognition Systems[32]

The normal development of most tissues does not involve sorting out of randomly mixed cell types (see p. 911). Nevertheless, when dissociated embryonic cells from two vertebrate tissues such as liver and retina are mixed together, the mixed aggregates initially formed gradually sort out according to their tissue of origin. This assay presumably detects tissue-specific cell-cell recognition systems that keep the cells in a developing tissue together. Such recognition systems can also be demonstrated in another way. As described in Figure 14–65, disaggregated cells are found to adhere more readily to aggregates of their own tissue than to aggregates of other tissues. Thus similar results are obtained by two different assays, one determining the selective affinities of cells in aggregates during prolonged incubation (the sorting-out experiments) and the other measuring the rate at which cells bind to preformed aggregates.

What is the molecular basis of this selective cell-cell adhesion in vertebrates? As with slime molds, two distinct classes of cell-cell adhesion mechanisms, one Ca^{2+}-independent and the other Ca^{2+}-dependent, seem to be responsible, and each class is mediated by its own family of homologous cell-surface glycoproteins.

Plasma-Membrane Glycoproteins of the Immunoglobulin Superfamily Mediate Ca^{2+}-independent Cell-Cell Adhesion in Vertebrates: The Neural Cell Adhesion Molecule (N-CAM)[33]

The immunological strategy outlined in Figure 14–63 has been used to identify a number of cell-surface glycoproteins involved in cell-cell adhesion in vertebrates. In one of the best-studied examples, monovalent antibody fragments were made

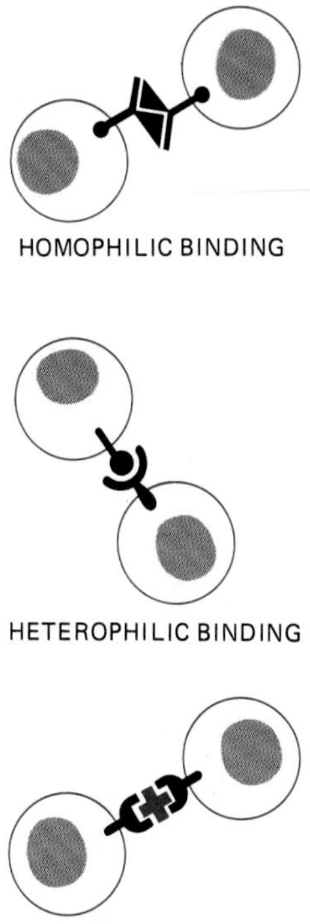

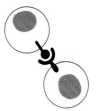

HOMOPHILIC BINDING

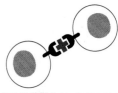

HETEROPHILIC BINDING

BINDING THROUGH AN EXTRACELLULAR LINKER MOLECULE

Figure 14–64 Three mechanisms by which cell-surface molecules can mediate cell-cell adhesion.

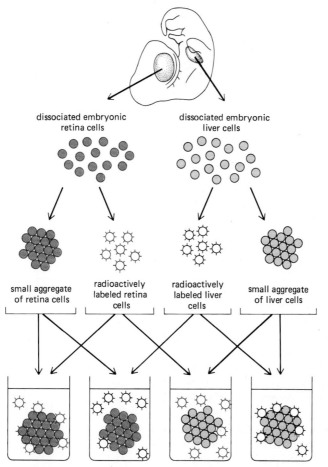

dissociated embryonic retina cells

dissociated embryonic liver cells

small aggregate of retina cells

radioactively labeled retina cells

radioactively labeled liver cells

small aggregate of liver cells

MIXING OF RADIOACTIVELY LABELED CELLS WITH CELL AGGREGATES

Figure 14–65 Tissue-specific adhesion of dissociated vertebrate embryo cells determined by a radioactive cell-binding assay. The rate of cell adhesion can be measured by determining the number of radioactively labeled cells bound to the cell aggregates after various periods of time. The rate of adhesion is greater between cells of the same kind. In a commonly used modification of this assay, cells labeled with a fluorescent or radioactive marker are allowed to bind to a monolayer of unlabeled cells in culture.

against cells from embryonic chick retina. Antibodies that inhibited the reaggregation of these cells *in vitro* were selected. Retinal cell membrane proteins were then fractionated and tested for their ability to neutralize the blocking activity of the antibodies. In this way a large, single-pass transmembrane glycoprotein (about 1000 amino acid residues long), called **neural cell adhesion molecule (N-CAM),** was identified. N-CAM is expressed on the surface of nerve cells and glial cells (see p. 1064) and causes them to stick together by a Ca^{2+}-independent mechanism. When these membrane proteins are purified and inserted into synthetic phospholipid vesicles, the vesicles bind to one another, as well as to cells that have N-CAM on their surface; the binding is blocked if the cells are pretreated with monovalent anti-N-CAM antibodies. These findings suggest that N-CAM binds cells together by a homophilic interaction that directly joins two N-CAM molecules (see Figure 14–64).

Anti-N-CAM antibodies disrupt the orderly pattern of retinal development in tissue culture and, when injected into the developing chick eye, disturb the normal growth pattern of retinal nerve cell axons. As explained in Chapter 19 (see p. 1116), these observations suggest that N-CAM plays an important part in the development of the central nervous system by promoting cell-cell adhesion. In addition, the neural crest cells that form the peripheral nervous system have large amounts of N-CAM on their surface when they are associated with the neural tube, lose it while they are migrating, and then reexpress it when they aggregate to form a ganglion (see Figure 14–56), suggesting that N-CAM may play a part in the assembly of the ganglion. N-CAM is also expressed transiently during critical stages in the development of many nonneural tissues, where it presumably helps specific cells stay together.

There are several forms of N-CAM, each encoded by a distinct mRNA. The different mRNAs are generated by alternative splicing of an RNA transcript produced from a single large gene. The large extracellular part of the polypeptide chain (~680 amino acid residues) is identical in most forms of N-CAM and is folded into five domains that are homologous to the immunoglobulin domains characteristic of antibody molecules (see p. 1021). Thus N-CAM belongs to the same ancient superfamily of recognition proteins to which antibodies belong (see p. 1053). The various forms of N-CAM differ mainly in their membrane-associated segments and cytoplasmic domains and therefore may interact with the cytoskeleton in different ways; in fact, one form does not cross the lipid bilayer and is attached to the plasma membrane only by a covalent linkage to phosphatidylinositol (see p. 448) (Figure 14–66), while another is secreted and incorporated into the extracellular matrix. It is not known how the various forms differ in function.

An increasing number of cell-surface glycoproteins that mediate Ca^{2+}-independent cell-cell adhesion in vertebrates are being discovered to belong to the immunoglobulin superfamily. But not all cell-surface proteins that mediate cell-cell adhesion are members of this superfamily: those that can function only in the presence of extracellular Ca^{2+} belong to another family.

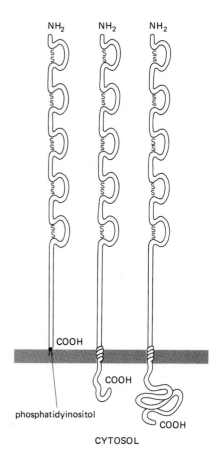

Figure 14–66 Schematic drawing of three forms of N-CAM. The extracellular part of the polypeptide chain is the same in each case and is folded into five immunoglobulinlike domains. Disulfide bonds connect the ends of each loop forming such a domain. (Based on data from B.A. Cunningham, J.J. Hemperly, B.A. Murray, E.A. Prediger, R. Brackenbury, and G.M. Edelman, *Science* 236:799–800, 1987. Copyright 1987 by the AAAS.)

The Cadherins, a Family of Homologous Cell-Surface Glycoproteins, Mediate Ca^{2+}-dependent Cell-Cell Adhesion in Vertebrates[34]

The immunological strategies explained in Figure 14–63 also played a crucial part in the discovery of three related cell-surface glycoproteins, collectively called **cadherins,** involved in Ca^{2+}-dependent cell-cell adhesion in vertebrate tissues. *E-cadherin* is found on many types of epithelial cells (and on cells in preimplantation mammalian embryos), *N-cadherin* on nerve, heart, and lens cells, and *P-cadherin* on cells in the placenta and epidermis; as for N-CAM, all are also found transiently on various other tissues during development. The three cadherins are homologous, single-pass transmembrane glycoproteins (each composed of about 700 amino acid residues). In these respects they resemble N-CAM. However, in the absence of Ca^{2+}, the cadherins undergo a large conformational change and, as a result, are rapidly degraded by proteolytic enzymes. Since some cells (such as endothelial cells) show Ca^{2+}-dependent adhesion but do not express any of the three known cadherins, it seems likely that additional members of the cadherin family remain to be discovered.

E-cadherin (also called liver cell adhesion molecule [*L-CAM*] or *uvomorulin*) is the best characterized. The large extracellular part of its polypeptide chain is folded into three homologous domains, which are apparently unrelated to immunoglobulin domains. It seems to play an important part in holding adjacent cells together in various epithelia. The Ca^{2+}-dependent reaggregation of dissociated epithelial cells from liver, for example, is blocked by antibodies to E-cadherin. In mature epithelial tissues, E-cadherin is usually concentrated in *adhesion belts*, where it is thought to function as a transmembrane linker, connecting the cortical actin cytoskeletons of the cells it holds together (see p. 796). It is also involved in the compaction of blastomeres in the early mouse embryo (see p. 894). During compaction, the blastomeres, at first loosely attached to one another, become tightly packed together and joined by intercellular junctions. Antibodies against E-cadherin block blastomere compaction, whereas antibodies that react with various other cell-surface molecules on these cells do not.

It seems likely that cadherins also play a crucial role in later stages of vertebrate development, since their appearance and disappearance correlate with major morphogenetic events in which tissues segregate from one another. For example, as the neural tube forms and pinches off from the overlying ectoderm (see p. 887), the cells of the developing neuroepithelium lose E-cadherin and acquire N-cadherin (as well as N-CAM) (Figure 14–67). As for N-CAM (see p. 830), when neural crest cells dissociate from the neural tube and migrate away, they lose N-cadherin, reexpressing it later when they aggregate to form a neural ganglion (see Figure 14–56).

The biological significance of the striking Ca^{2+}-dependence of the cadherin family of cell-cell adhesion proteins is unknown. There is no evidence as yet that the extracellular concentration of Ca^{2+} is regulated to control cell-cell adhesion during development, for example.

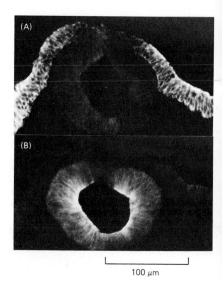

Figure 14–67 Immunofluorescence micrographs of a cross section of a chick embryo showing the developing neural tube labeled with antibodies against E-cadherin (A) and N-cadherin (B). Note that the overlying ectoderm cells express only E-cadherin, while the cells in the neural tube have lost E-cadherin and have acquired N-cadherin. (Courtesy of Kohei Hatta and Masatoshi Takeichi.)

Cell-Surface Molecules That Mediate Cell-Cell and Cell-Matrix Adhesion Can Be Viewed as Elements of a Morphogenetic Code[35]

Morphological, cell biological, and biochemical studies all indicate that even a single cell type utilizes multiple molecular mechanisms in adhering to other cells and to the extracellular matrix. Some of these mechanisms involve organized cell junctions; others do not (Figure 14–68). Because of the large number of adhesive systems used by an individual cell, nearly every cell type will share at least one cell-cell adhesion system with every other and therefore bind to it with some affinity. Usually cells from different tissues (and even from very different species) will also form desmosomes, gap junctions, and adherens junctions with each other, suggesting that the junctional proteins involved are highly conserved between tissues (and species). However, just as each cell in a multicellular animal contains an assortment of cell-surface receptors that enables the cell to respond specifically to a complementary assortment of soluble chemical signals (hormones and local chemical mediators), so each cell in a tissue is thought to have a particular combination (or concentration) of cell-surface receptors that enables it to bind in its own characteristic way to other cells and to the extracellular matrix.

Unlike receptors for soluble chemical signals, which bind their specific ligand with high affinity, the receptors that bind to molecules on cell surfaces or in the extracellular matrix usually do so with relatively low affinity. The latter receptors therefore rely on the enormous increase in binding strength gained through simultaneous binding of multiple receptors to multiple ligands on an opposing cell or in the adjacent matrix. The mixture of specific types of cell-cell adhesion molecules and matrix receptors present on any two cells, as well as their concentration and distribution on the cell surface, will therefore determine the total affinity with which the two cells bind to each other and to the matrix. It is this mixture that presumably constitutes a "morphogenetic code" to help determine how cells are organized in tissues. Since even closely related types of animal cells sort out correctly *in vitro*, cells must be able to detect relatively small differences in adhesion and to act on these differences so as to establish only the most adhesive of many possible cell-cell and cell-matrix contacts. Studies carried out on motile cells in culture suggest how this might occur.

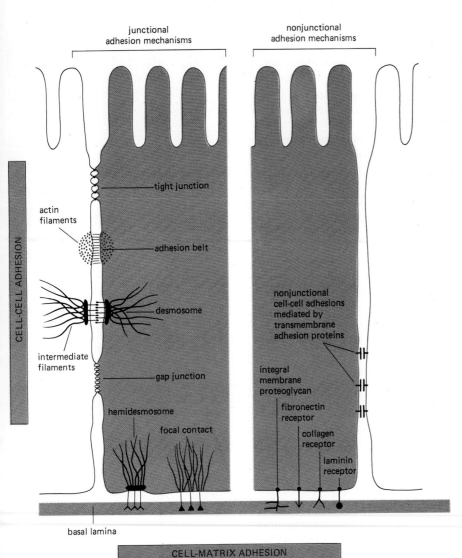

junctional adhesion mechanisms

nonjunctional adhesion mechanisms

CELL-CELL ADHESION

tight junction

actin filaments

adhesion belt

desmosome

intermediate filaments

gap junction

nonjunctional cell-cell adhesions mediated by transmembrane adhesion proteins

integral membrane proteoglycan

fibronectin receptor

hemidesmosome

collagen receptor

focal contact

laminin receptor

basal lamina

CELL-MATRIX ADHESION

Figure 14–68 A summary of the junctional and nonjunctional adhesive mechanisms used by typical epithelial cells in binding to one another and to the extracellular matrix (basal lamina). A junctional interaction is operationally defined as one that can be seen as a specialized region of contact by conventional and/or freeze-fracture electron microscopy. In some cases the same cell-surface glycoproteins are involved in binding cells to one another (or to the matrix) at both junctional and nonjunctional contacts. Except for gap junctions, all of the junctional adhesion mechanisms are Ca^{2+}-dependent; as discussed in the text, some nonjunctional adhesion mechanisms are Ca^{2+}-dependent and others are not.

Highly Motile Cells Are Sensitive Detectors of Small Adhesive Differences[36]

The cells involved in morphogenetic processes in embryos are often highly motile. If such cells are dissociated from embryos and observed in a culture dish, they are seen initially to extend microspikes and lamellipodia in all directions and then to crawl actively along the surface of the dish. The onset of motile behavior often coincides with the beginning of cell diversification and, therefore, with the period when specific cell recognition would be expected to become important. The cells in a *Xenopus* embryo, for example, suddenly become highly motile at the mid-blastula transition stage, when gene transcription begins (see p. 881).

Extensive studies on the mechanism of cell motility have been carried out in culture on fibroblasts, neutrophils, and regenerating neurons. The results, summarized in Chapter 11, indicate that a motile cell is a highly sensitive detector of small adhesive differences. The microspikes and lamellipodia that it extends in all directions appear to engage in a "tug-of-war," causing the cell to become polarized and move steadily in the direction of the most adhesive part of the substratum, even when the differences in adhesiveness are small (see p. 669). Fibroblasts, for example, will move steadily up a small adhesive gradient created on the surface of a culture dish. From studies of chemotaxis in neutrophils, it seems that a motile cell can detect differences in adhesiveness on either side of it as small as 1%. By similar means, cells in tissues may be able to decipher the "morphogenetic code" on cell surfaces with great sensitivity, moving steadily to establish intimate contacts with those neighboring cells to which they are most adherent.

Nonjunctional Contacts May Initiate Tissue-specific Cell-Cell Adhesion That Junctional Contacts Then Stabilize[37]

Which, if any, of the many types of intercellular junctions discussed at the beginning of this chapter are involved as cells migrate and recognize one another during the formation of tissues and organs? One way to find out is to use an electron microscope to examine the contacts between adjacent cells when they are moving over each other in developing embryos or in adult tissues undergoing repair after injury. Such studies show that these contacts generally do not involve the formation of organized intercellular junctions. Nevertheless, the interacting plasma membranes often come close together and run parallel, separated by a space of 10–20 nm. This is about the distance (~13 nm) that the influenza virus hemagglutinin glycoprotein (the first plasma membrane glycoprotein whose three-dimensional structure was determined—see p. 446) projects from the plasma membrane. Two cell-surface glycoproteins of this size could interact with each other across the 10–20-nm gap to mediate the adhesion. This type of nonjunctional contact may be optimal for cell locomotion—close enough to give traction but not tight enough to immobilize the cell.

Since junctional contacts are generally not seen between moving embryonic cells (except perhaps for small gap junctions), the formation of intercellular junctions may be an important mechanism for immobilizing cells within an organized tissue once it has formed. A reasonable hypothesis is that nonjunctional cell-surface adhesion proteins initiate tissue-specific cell-cell adhesion, which is then stabilized by the assembly of intercellular cell junctions. Since many of the transmembrane glycoproteins involved can diffuse in the plane of the plasma membrane, they can accumulate at sites of cell-cell contact and therefore be used for junctional as well as nonjunctional adhesions. Thus some cell-cell adhesion proteins, such as E cadherins (see p. 831), may help initiate cell-cell adhesion and then later become integral parts of intercellular junctions.

Ideally, one would like to be able to inactivate the various types of cell-cell adhesion proteins and matrix receptors individually and in various combinations in order to decipher the rules of recognition and binding used in the morphogenesis of complex tissues. As an increasing number of monoclonal antibodies and peptide fragments are characterized, each of which blocks a single type of cell-cell adhesion molecule or matrix receptor—and as the genes that encode these cell-surface proteins become available for manipulation in cells in culture and in transgenic animals—this dream of developmental biologists is becoming a reality.

Summary

Cells dissociated from various tissues of vertebrate embryos preferentially reassociate with cells from the same tissue when they are mixed together. The initial difficulties in studying the molecular mechanisms underlying the normal assembly of cells into complex tissues in higher animals, however, encouraged the study of simpler systems. The free-living amoebae of the cellular slime mold Dictyostelium discoideum aggregate to form multicellular fruiting bodies when they are starved. Their cell-cell adhesion is mediated by at least two cell-surface glycoproteins: one operates early in development and depends on extracellular Ca^{2+}, while the other operates later in development and does not require Ca^{2+}. The tissue-specific recognition process in vertebrates is likewise mediated by at least two families of cell-surface glycoproteins—one Ca^{2+}-dependent (the cadherins) and the other Ca^{2+}-independent (exemplified by N-CAM and other members of the immunoglobulin superfamily). Both families of cell-cell adhesion molecules seem to play important roles in guiding vertebrate morphogenesis. Since even a single cell type uses multiple molecular mechanisms in adhering to other cells (and to the extracellular matrix), the specificity of cell-cell adhesion seen in embryonic development must result from a mechanism that sums the affinity of a number of different adhesion systems. The demonstrated ability of motile cells to detect small adhesive differences suggests how the specific combination, concentration, and distribution of cell-cell adhesion molecules and matrix receptors present on each type of cell can function as the required "morphogenetic code."

References

Cited

1. Bock, G.; Clark, S., eds. Junctional Complexes of Epithelial Cells. Ciba Symposium 125. New York: Wiley, 1987.

   Farquhar, M.G.; Palade, G.E. Junctional complexes in various epithelia. *J. Cell. Biol.* 17:375–412, 1963.

   Gilula, N.B. Junctions between cells. In Cell Communication (R.P. Cox, ed.), pp. 1–29. New York: Wiley, 1974.

   Goodenough, D.A.; Revel, J.P. A fine structural analysis of intercellular junctions in the mouse liver. *J. Cell Biol.* 45:272–290, 1970.

   Staehelin, L.A.; Hull, B.E. Junctions between living cells. *Sci. Am.* 238(5):141–152, 1978.

2. Diamond, J.M. The epithelial junction: bridge, gate and fence. *Physiologist* 20:10–18, 1977.

   Madara, J.L. Tight junction dynamics: is paracellular transport regulated? *Cell* 53:497–498, 1988.

   Madara, J.L.; Dharmsathaphorn, K. Occluding junction structure-function relationships in cultured epithelial monolayer. *J. Cell Biol.* 101:2124–2133, 1985.

   Simons, K.; Fuller, S.D. Cell surface polarity in epithelia. *Annu. Rev. Cell Biol.* 1:243–288, 1985.

   van Meer, G.; Gumbiner, B.; Simons, K. The tight junction does not allow lipid molecules to diffuse from one epithelial cell to the next. *Nature* 322:639–641, 1986.

3. Burridge, K.; Fath, K.; Kelly, T.; Nuckolls, G.; Turner, C. Focal adhesions: transmembrane junctions between the extracellular matrix and the cytoskeleton. *Annu. Rev. Cell Biol.* 4:487–526, 1988.

   Geiger, B.; Volk, T.; Volberg, T. Molecular heterogeneity of adherens junctions. *J. Cell Biol.* 101:1523–1531, 1985.

4. Franke, W.W.; Cowin, P.; Schmelz, M.; Kapprell, H.-P. The desmosomal plaque and the cytoskeleton. In Junctional Complexes of Epithelial Cells, Ciba Foundation Symposium 125 (G. Bock, S. Clark, eds.), pp. 26–48. New York: Wiley, 1987.

   Garrod, D.R. Desmosomes, cell adhesion molecules and the adhesive properties of cells in tissues. *J. Cell Sci.* Suppl. 4:221–237, 1986.

   Jones, J.C.R.; Yokoo, K.M.; Goldman, R.D. Further analysis of pemphigus autoantibodies and their use in studies on the heterogeneity, structure, and function of desmosomes. *J. Cell Biol.* 102:1109–1117, 1986.

   Steinberg, M.S.; et al. On the molecular organization, diversity and functions of desmosomal proteins. In Junctional Complexes of Epithelial Cells, Ciba Foundation Symposium 125 (G. Bock, S. Clark, eds.), pp. 3–25. New York: Wiley, 1987.

5. Bennett, M.; Spray, D., eds. Gap Junctions. Cold Spring Harbor, NY: Cold Spring Harbor Laboratory, 1985.

   Furshpan, E.J.; Potter, D.D. Low-resistance junctions between cells in embryos and tissue culture. *Curr. Top. Dev. Biol.* 3:95–127, 1968.

   Gilula, N.B.; Reeves, O.R.; Steinbach, A. Metabolic coupling, ionic coupling and cell contacts. *Nature* 235:262–265, 1972.

   Hooper, M.L.; Subak-Sharpe, J.H. Metabolic cooperation between cells. *Int. Rev. Cytol.* 69:45–104, 1981.

   Loewenstein, W.R. The cell-to-cell channel of gap junctions. *Cell* 48:725–726, 1987.

   Neyton, J.; Trautmann, A. Single-channel currents of an intercellular junction. *Nature* 317:331–335, 1985.

   Pitts, J.D.; Finbow, M.E. The gap junction. *J. Cell Sci.* Suppl. 4:239–266, 1986.

   Young, J.D.-E.; Cohn, Z.A.; Gilula, N.B. Functional assembly of gap junction conductance in lipid bilayers: demonstration that the major 27 kd protein forms the junctional channel. *Cell* 48:733–743, 1987.

6. Caspar, D.L.D.; Goodenough, D.; Makowski, L.; Phillips, W.C. Gap junction structures. I. Correlated electron microscopy and x-ray diffraction. *J. Cell Biol.* 74:605–628, 1977.

   Gilula, N.B. Topology of gap junction protein and channel function. In Junctional Complexes of Epithelial Cells, Ciba Foundation Symposium 125 (G. Bock, S. Clark, eds.), pp. 128–139. New York: Wiley, 1987.

   Paul, D.L. Molecular cloning of cDNA for rat liver gap junction protein. *J. Cell Biol.* 103:123–134, 1986.

   Unwin, P.N.T.; Zampighi, G. Structure of the junction between communicating cells. *Nature* 283:545–549, 1980.

7. Caveney, S. The role of gap junctions in development. *Annu. Rev. Physiol.* 47:319–335, 1985.

   Warner, A.E. The role of gap junctions in amphibian development. *J. Embryol. Exp. Morphol.* Suppl. 89:365–380, 1985.

   Warner, A.E.; Guthrie, S.C.; Gilula, N.B. Antibodies to gap-junctional protein selectively disrupt junctional communication in the early amphibian embryo. *Nature* 311:127–131, 1984.

8. Rose, B.; Loewenstein, W.R. Permeability of cell junction depends on local cytoplasmic calcium activity. *Nature* 254:250–252, 1975.

   Saez, S.C.; et al. Cyclic AMP increases junctional conductance and stimulates phosphorylation of the 27-kDa principal gap junction polypeptide. *Proc. Natl. Acad. Sci. USA* 83:2473–2477, 1986.

   Spray, D.C.; Bennett, M.V.L. Physiology and pharmacology of gap junctions. *Annu. Rev. Physiol.* 47:218–303, 1985.

   Turin, L.; Warner, A.E. Intracellular pH in early *Xenopus* embryo: its effect on current flow between blastomeres. *J. Physiol. (Lond.)* 300:489–504, 1980.

9. Hay, E.D., ed. Cell Biology of Extracellular Matrix. New York: Plenum, 1981.

   McDonald, J.A. Extracellular matrix assembly. *Annu. Rev. Cell Biol.* 4:183–208, 1988.

   Piez, K.A.; Reddi, A.H., eds. Extracellular Matrix Biochemistry. New York: Elsevier, 1984.

10. Evered, D.; Whelan, J., eds. Functions of the Proteoglycans. Ciba Foundation Symposium 124. New York: Wiley, 1986.

    Hascall, V.C.; Hascall, G.K. Proteoglycans. In Cell Biology of Extracellular Matrix (E.D. Hay, ed.), pp. 39–63. New York: Plenum, 1981.

    Wight, T.N.; Meeham, R.P., eds. Biology of Proteoglycans. San Diego, CA: Academic Press, 1987.

11. Laurent, T.C.; Fraser, J.R.E. The properties and turnover of hyaluronan. In Functions of the Proteoglycans, Ciba Foundation Symposium 124 (D. Evered, J. Whelan, eds.), pp. 9–29. New York: Wiley, 1986.

    Toole, B.P. Glycosaminoglycans in Morphogenesis. In Cell Biology of Extracellular Matrix (E.D. Hay, ed.), pp. 259–294. New York: Plenum, 1981.

12. Dorfman, A. Proteoglycan biosynthesis. In Cell Biology of Extracellular Matrix (E.D. Hay, ed.), pp. 115–138. New York: Plenum, 1981.

    Hassell, J.R.; Kimura, J.H.; Hascall, V.C. Proteoglycan core protein families. Annu. Rev. Biochem. 55:539–567, 1986.

    Heinegård, D.; Paulsson, M. Structure and metabolism of proteoglycans. In Extracellular Matrix Biochemistry (K.A. Piez, A.H. Reddi, eds.), pp. 277–328. New York: Elsevier, 1984.

    Ruoslahti, E. Structure and biology of proteoglycans. Annu. Rev. Cell Biol. 4:229–255, 1988.

13. Fransson, L.-Å. Structure and function of cell-associated proteoglycans. Trends Biochem. Sci. 12:406–411, 1987.

    Höök, M.; Kjellén, L.; Johansson, S.; Robinson, J. Cell-surface glycosaminoglycans. Annu. Rev. Biochem. 53:847–869, 1984.

    Rees, D.A. Polysaccharide Shapes, Outline Studies in Biology, pp. 62–73. London: Chapman & Hall, 1977.

    Scott, J.E. Proteoglycan-collagen interactions. In Functions of the Proteoglycans, Ciba Foundation Symposium 124 (D. Evered, J. Whelan, eds.), pp. 104–124. New York: Wiley, 1986.

14. Burgeson, R.E. New collagens, new concepts. Annu. Rev. Cell Biol. 4:551–577, 1988.

    Linsenmayer, T.F. Collagen. In Cell Biology of Extracellular Matrix (E.D. Hay, ed.), pp. 5–37. New York: Plenum, 1981.

    Martin, G.R.; Timpl, R.; Muller, P.K.; Kühn, K. The genetically distinct collagens. Trends Biochem. Sci. 10:285–287, 1985.

15. Fleischmajer, R.; Olsen, B.R.; Kühn, K., eds. Biology, Chemistry, and Pathology of Collagen. Ann. N.Y. Acad. Sci., Vol. 460, 1985.

    Olsen, B.R. Collagen Biosynthesis. In Cell Biology of Extracellular Matrix (E.D. Hay, ed.), pp. 139–177. New York: Plenum, 1981.

    Woolley, D.E. Mammalian collagenases. In Extracellular Matrix Biochemistry (K.A. Piez, A.H. Reddi, eds.), pp. 119–157. New York: Elsevier, 1984.

16. Eyre, D.R.; Paz, M.A.; Gallop, P.M. Cross-linking in collagen and elastin. Annu. Rev. Biochem. 53:717–748, 1984.

    Piez, K.A. Molecular and aggregate structures of the collagens. In Extracellular Matrix Biochemistry (K.A. Piez, A.H. Reddi, eds.), pp. 1–39. New York: Elsevier, 1984.

17. Prockop, D.J.; Kivirikko, K.I. Heritable diseases of collagen. New Engl. J. Med. 311:376–386, 1984.

    Trelstad, R.L.; Silver, F.H. Matrix assembly. In Cell Biology of Extracellular Matrix (E.D. Hey, ed.), pp. 179–215. New York: Plenum, 1981.

18. Stopak, D.; Harris, A.K. Connective tissue morphogenesis by fibroblast traction I. Tissue culture observations. Dev. Biol. 90:383–398, 1982.

19. Yurchenco, P.D.; Furthmayr, H. Self-assembly of basement membrane collagen. Biochemistry 23:1839–1850, 1984.

    Yurchenco, P.D.; Ruben, G.C. Basement membrane structure in situ: evidence for lateral associations in the type IV collagen network. J. Cell Biol. 105:2559–2568, 1987.

20. Cleary, E.G.; Gibson, M.A. Elastin-associated microfibrils and microfibrillar proteins. Int. Rev. Connect. Tissue Res. 10:97–209, 1983.

    Gosline, J.M.; Rosenbloom, J. Elastin. In Extracellulr Matrix Biochemistry (K.A. Piez, A.H. Reddi, eds.), pp. 191–227. New York: Elsevier, 1984.

    Ross, R.; Bornstein, P. Elastic fibers in the body. Sci. Am. 224(6):44–52, 1971.

21. Dufour, S.; Duband, J.-L.; Kornblihtt, A.R.; Thiéry, J.P. The role of fibronectins in embryonic cell migrations. Trends Genet. 4:198–203, 1988.

    Hynes, R.O. Molecular biology of fibronectin. Annu. Rev. Cell Biol. 1:67–90, 1985.

    Hynes, R.O. Fibronectins. Sci. Am. 254(6):42–51, 1986.

    Hynes, R.O.; Yamada, K.M. Fibronectins: multifunctional modular proteins. J. Cell Biol. 95:369–377, 1982.

    Ruoslahti, E.; Pierschbacher, M.D. New perspectives in cell adhesion: RGD and integrins. Science 238:491–497, 1987.

22. Humphries, M.J.; Akiyama, S.K.; Komoriya, A.; Olden, K.; Yamada, K.M. Neurite extension of chicken peripheral nervous system neurons on fibronectin: relative importance of specific adhesion sites in the central cell-binding domain and the alternatively spliced type III connecting segment. J. Cell Biol. 106:1289–1297, 1988.

    Tamkun, J.W.; Schwarzbauer, J.E.; Hynes, R.O. A single rat fibronectin gene generates three different mRNAs by alternative splicing of a complex exon. Proc. Natl. Acad. Sci. USA 81:5140–5144, 1984.

23. Farquhar, M.G. The glomerular basement membrane: a selective macromolecular filter. In Cell Biology of Extracellular Matrix (E.D. Hay, ed.), pp. 335–378. New York: Plenum, 1981.

    Martin, G.R.; Timpl, R. Laminin and other basement membrane components. Annu. Rev. Cell Biol. 3:57–85, 1987.

    Sasaki, M.; Kato, S.; Kohno, K.; Martin, G.R.; Yamada, Y. Sequence of the cDNA encoding the laminin B1 chain reveals a multidomain protein containing cysteine-rich repeats. Proc. Natl. Acad. Sci. USA 84:935–939, 1987.

24. Reist, N.E.; Magill, C.; McMahan, U.J. Agrin-like molecules at synaptic sites in normal, denervated, and damaged skeletal muscles. J. Cell Biol. 105:2457–2469, 1987.

25. Anderson, D.C.; Springer, T.A. Leukocyte adhesion deficiency: an inherited defect in the Mac-1, LFA-1 and P150,95 glycoprotein. Annu. Rev. Med. 38:175–194, 1987.

    Buck, C.A.; Horwitz, A.F. Cell surface receptors for extracellular matrix molecules. Annu. Rev. Cell Biol. 3:179–205, 1987.

    Hynes, R.O. Integrins: a family of cell surface receptors. Cell 48:549–554, 1987.

    Ruoslahti, E. Fibronectin and its receptors. Annu. Rev. Biochem. 57:375–414, 1988.

26. Bornstein, P.; Duksin, D.; Balian, G.; Davidson, J.M.; Crouch, E. Organization of extracellular proteins on the connective tissue cell surface: relevance to cell-matrix interactions in vitro and in vivo. Ann. N.Y. Acad. Sci. 312:93–105, 1978.

    Burridge, K.; Fath, K.; Kelly, T.; Nuckolls, G.; Turner, C. Focal adhesions: transmembrane junctions between the extracellular matrix and the cytoskeleton. Annu. Rev. Cell Biol. 4:487–526, 1988.

    Horwitz, A.; Duggan, K; Buck, C.; Beckerle, M.C.; Burridge, K. Interaction of plasma membrane fibronectin receptor with talin—a transmembrane linkage. Nature 320:531–533, 1986.

    Hynes, R. Structural relationships between fibronectin and cytoplasmic cytoskeletal networks. In Cytoskeletal Elements and Plasma Membrane Organization (G. Poste, G.L. Nicolson, eds.), Vol. 7, pp. 100–137. Amsterdam: Elsevier, 1981.

    Watt, F.M. The extracellular matrix and cell shape. Trends Biochem. Sci. 11:482–485, 1986.

27. Le Douarin, N.; Smith, J. Development of the peripheral nervous system from the neural crest. Annu. Rev. Cell Biol. 4:375–404, 1988.

McClay, D.R.; Ettensohn, C.A. Cell adhesion and morphogenesis. *Annu. Rev. Cell Biol.* 3:319–346, 1987.

28. Loomis, W.F. *Dictyostelium discoideum.* A Developmental System. New York: Academic Press, 1975.

29. Bonner, J.T. Chemical signals of social amoebae. *Sci. Am.* 248(4):114–120, 1983.

Gerisch, G. Cyclic AMP and other signals controlling cell development and differentiation in *Dictyostelium. Annu. Rev. Biochem.* 56:853–879, 1987.

30. Gerisch, G. Interrelation of cell adhesion and differentiation in *Dictyostelium discoideum. J. Cell Sci.* Suppl. 4:201–219, 1986.

Gerisch, G. Univalent antibody fragments as tools for the analysis of cell interactions in *Dictyostelium. Curr. Top. Dev. Biol.* 14:243–270, 1980.

31. Hennings, H.; Holbrook, K.A. Calcium regulation of cell-cell contact and differentiation of epidermal cells in culture. An ultrastructural study. *Exp. Cell Res.* 143:127–142, 1983.

32. Moscona, A.A.; Hausman, R.E. Biological and biochemical studies on embryonic cell-cell recognition. In Cell and Tissue Interactions, Society of General Physiologists Series (J.W. Lash, M.M. Burger, eds.), Vol. 32, pp. 173–185. New York: Raven, 1977.

Roth, S.; Weston, J. The measurement of intercellular adhesion. *Proc. Natl. Acad. Sci. USA* 58:974–980, 1967.

33. Cunningham, B.A.; et al. Neural cell adhesion molecule: structure, immunoglobulin-like domains, cell surface modulation, and alternative RNA splicing. *Science* 236:799–806, 1987.

Edelman, G.M. Cell-adhesion molecules: a molecular basis for animal form. *Sci. Am.* 250(4):118–129, 1984.

Edelman, G.M. Cell adhesion molecules in the regulation of animal form and tissue pattern. *Annu. Rev. Cell Biol.* 2:81–116, 1986.

Rutishauser, U.; Goridis, C. N-CAM: the molecule and its genetics. *Trends Genet.* 2:72–76, 1986.

Williams, A.F.; Barclay, A.N. The immunoglobulin superfamily—domains for cell surface recognition. *Annu. Rev. Immunol.* 6:381–406, 1988.

34. Takeichi, M. The cadherins: cell-cell adhesion molecules controlling animal morphogenesis. *Development* 102:639–655, 1988.

35. Ekblom, P.; Vestweber, D.; Kemler, R. Cell-matrix interactions and cell adhesion during development. *Annu. Rev. Cell Biol.* 2:27–48, 1986.

Garrod, D.R. Desmosomes, cell adhesion molecules and the adhesive properties of cells in tissues. *J. Cell Sci.* Suppl. 4:221–237, 1986.

Jessel, T.M. Adhesion molecules and the hierarchy of neural development. *Neuron* 1:3–13, 1988.

Steinberg, M.S. The adhesive specification of tissue self-organization. In Morphogenesis and Pattern Formation (T.G. Connelly, et al, eds.), pp. 179–203. New York: Raven, 1981.

36. Devreotes, P.; Zigmond, S.H. Chemotaxis in eukaryotic cells. *Annu. Rev. Cell Biol.* 4:649–686, 1988.

37. Trinkaus, J.P. Cells into Organs, 2nd ed., pp. 69–178. Englewood Cliffs, NJ: Prentice-Hall, 1984.

A mouse embryo at 15 days of gestation.▶

From Cells to Multicellular Organisms

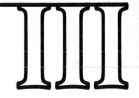

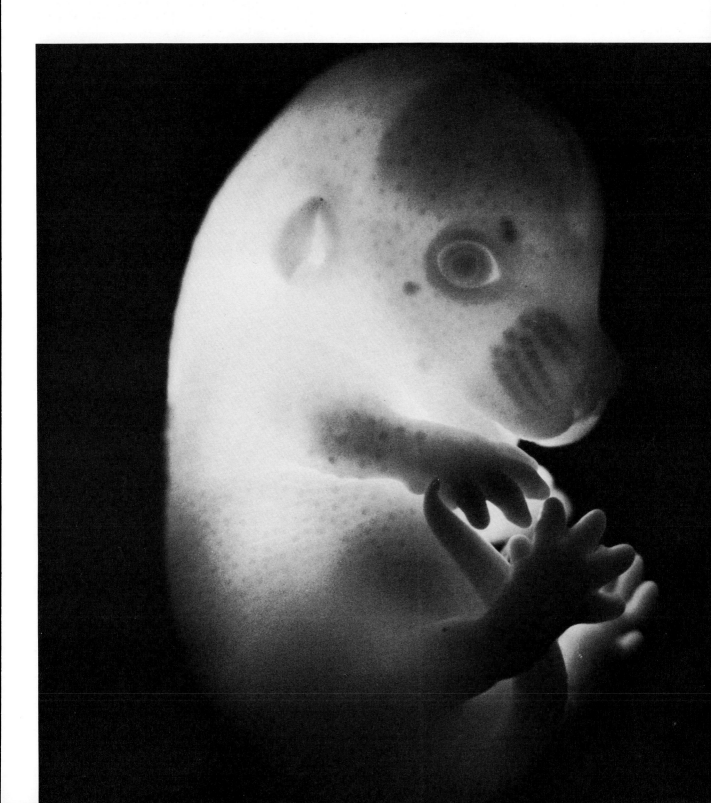

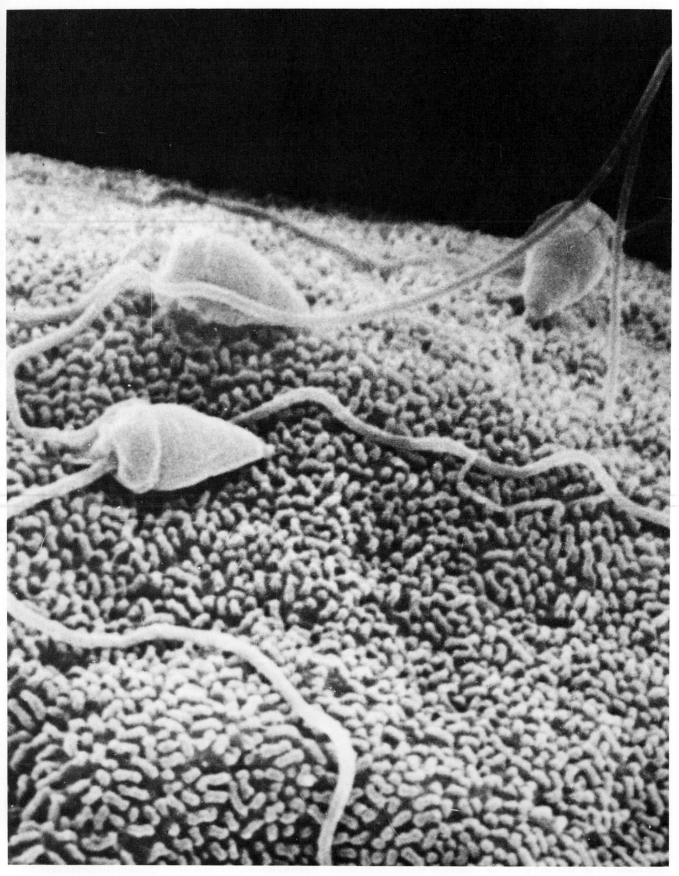

Scanning electron micrograph of sperm on the surface of a sea urchin egg. (Courtesy of Brian Dale.)

Germ Cells and Fertilization

15

Sex is not necessary for reproduction. Single-celled organisms can reproduce by simple mitotic division. Many plants propagate vegetatively by forming multicellular offshoots that later detach from the parent. Likewise, in the animal kingdom, a solitary multicellular *Hydra* can produce offspring by budding (Figure 15–1). Sea anemones and marine worms can split into two half-organisms, each of which then regenerates its missing half. There are even species of lizards that consist only of female individuals and reproduce without mating. While such **asexual reproduction** is simple and direct, it gives rise to offspring that are genetically identical to the parent organism. **Sexual reproduction,** on the other hand, involves the mixing of genomes from two different individuals to produce offspring that usually differ genetically from one another and from both their parents. This form of reproduction apparently has great advantages, since the vast majority of plants and animals have adopted it. Even many procaryotes and other organisms that normally reproduce asexually engage in occasional bouts of sexual reproduction, thereby creating new combinations of genes. This chapter is concerned with the cellular machinery of sexual reproduction. Before discussing in detail how the machinery works, we shall pause to consider why it exists and what benefits it brings.

The Benefits of Sex

The sexual reproductive cycle involves an alternation of **haploid** generations of cells, each carrying a single set of chromosomes, with **diploid** generations of cells, each carrying a double set of chromosomes (Figure 15–2). The mixing of genomes is achieved by fusion of two haploid cells to form a diploid cell. Later, new haploid cells are created when a descendant of this diploid cell divides by the process of *meiosis*. During meiosis the chromosomes of the double chromosome set exchange DNA by genetic recombination before being shared out, in fresh combinations, into single chromosome sets (see p. 847). In this way each cell of the new haploid generation receives a novel assortment of genes, with some genes on each chromosome originating from one ancestral cell of the previous haploid generation and some from the other. Thus, through cycles of haploidy, fusion, diploidy, and meiosis, old combinations of genes are broken up and new combinations are created.

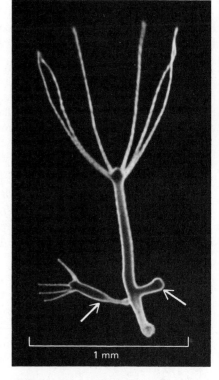

Figure 15–1 Photograph of a *Hydra* from which two new organisms are budding (*arrows*). The offspring, which are genetically identical to their parent, will eventually detach and live independently. (Courtesy of Amata Hornbruch.)

In Multicellular Animals the Diploid Phase Is Complex and Long, the Haploid Simple and Fleeting

Cells proliferate by mitotic division (see p. 762). In most organisms that reproduce sexually, this proliferation occurs during the diploid phase. Some primitive organisms, such as fission yeasts, are exceptional in that the haploid cells proliferate mitotically and the diploid cell, once formed, proceeds directly to meiosis. A less extreme exception occurs in plants, where cell divisions occur in both the haploid and the diploid phases. In all but the most primitive plants, however, the haploid phase is very brief and simple, while the diploid phase is extended into a long period of development and proliferation. For almost all multicellular animals, including vertebrates, practically the whole of the life cycle is spent in the diploid state: the haploid cells exist only briefly, do not divide at all, and are highly specialized for sexual fusion (Figure 15–3).

Haploid cells that are specialized for sexual fusion are called **gametes.** Typically, two types of gametes are formed: one is large and nonmotile and is referred to as the *egg* (or *ovum*); the other is small and motile and is referred to as the *sperm* (or *spermatozoon*) (Figure 15–4). During the diploid phase that follows fusion of gametes, the cells proliferate and diversify to form a complex multicellular organism. In most animals, although not so clearly in plants, a useful distinction can be drawn between the cells of the **germ line,** from which the next generation of gametes will be derived, and the **somatic cells,** which form the rest of the body and ultimately leave no progeny. In a sense, the somatic cells exist only to help the cells of the germ line (the **germ cells**) survive and propagate.

Sexual Reproduction Gives a Competitive Advantage to Organisms in an Unpredictably Variable Environment

The machinery of sexual reproduction is elaborate, and the resources spent on it are large. What benefits does it bring, and why did it evolve? Through genetic recombination, sexual individuals beget unpredictably dissimilar offspring, whose haphazard genotypes are at least as likely to represent a change for the worse as a change for the better. Why then should sexual individuals have a competitive

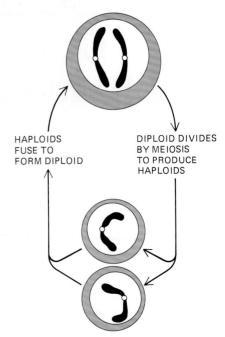

HAPLOIDS FUSE TO FORM DIPLOID   **DIPLOID DIVIDES BY MEIOSIS TO PRODUCE HAPLOIDS**

Figure 15–2 The sexual life cycle involves an alternation of haploid and diploid generations of cells.

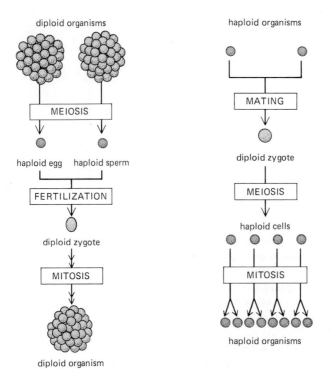

Figure 15–3 Drawing showing how cells in higher eucaryotic organisms proliferate in the diploid phase to form a multicellular organism; only the gametes are haploid. In some lower eucaryotes, by contrast, the haploid cells proliferate, and the only diploid cell is the *zygote*, which exists transiently following mating. The haploid cells are shown in color.

HIGHER EUCARYOTES   SOME LOWER EUCARYOTES

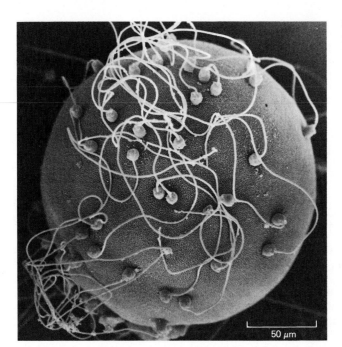

Figure 15-4 Scanning electron micrograph of a clam egg with many spermatozoa bound to its surface. (Courtesy of David Epel.)

advantage over individuals that breed true, by an asexual process? This problem continues to perplex population geneticists, but the general conclusion seems to be that the reshuffling of genes in sexual reproduction helps a species survive in an unpredictably variable environment. If a parent produces many offspring with a wide variety of gene combinations, there is a better chance that, whatever these circumstances may be, at least one of the offspring will have the assortment of features necessary for survival.

Many other ideas have been proposed to explain the competitive advantages of sexual reproduction. One of these suggests how one of the first steps in the evolution of sex might have occurred. Evolution depends to a large extent on competition among individuals carrying alternative *alleles*, or variants, created by mutation of particular genes. Suppose that two individuals in a population undergo beneficial mutations affecting a different *genetic locus* and therefore a different function in each case. In a strictly asexual species, each of these individuals will give rise to a clone of mutant progeny, and the two clones will compete until one or the other triumphs: one of the two beneficial mutations will spread through the population; the other will eventually be lost. But now suppose that one of the original mutants has an additional genetically determined peculiarity that enables it occasionally to incorporate genes from other cells. During the period of competition, acquisition of genes from a cell of the competing clone is likely to create a cell that carries both beneficial mutations. Such a cell will be the most successful of all, and its success will ensure the propagation of the trait that enabled it to incorporate genes from other cells. This rudimentary sexual capability will thus be favored by natural selection.

Whatever the origins of sex may be, it is striking that practically all complex present-day organisms have evolved largely through generations of sexual, rather than asexual, reproduction. Asexual organisms, although plentiful, seem mostly to have remained simple and primitive. Why should this be? A possible answer is that sexual reproduction creates special opportunities for genetic innovation leading to the evolution of complex organisms, as we shall argue below.

New Genes Evolve by Duplication and Divergence

The evolution of a complex organism involves more than simply introducing improved forms of existing genes: it means creating new genes to serve new functions. How does this occur?

Many of the proteins in a multicellular animal can be grouped into families: the collagens, the globins, the serine proteases, and so on. Proteins in the same family are related both in function and in amino acid sequence. There is little doubt that each family evolved from a single ancestral gene by a process of *duplication* and *divergence* (see p. 601). Different members of a protein family are often characteristic of different tissues of the body, where they perform analogous but distinctive tasks. The creation of new genes by diversification and specialization of existing genes has clearly been crucial for the evolution of complex multicellular organisms. In this respect, diploid species, with two copies of each gene, seem to enjoy an important advantage: one of two gene copies can be regarded as a spare and is available to mutate and serve as raw material for innovation. A haploid species does not have this easy means of taking the first step toward evolving a larger and more sophisticated genome. To make the mechanism clear, we must first explain more carefully the relationship between sex and diploidy.

Sexual Reproduction Helps to Keep a Diploid Species Diploid

A diploid organism contains two copies of each gene—one from each parent—yet in many cases a single copy is sufficient for good health and survival. A mutation that disrupts the function of an essential gene will be lethal in a haploid individual, but it may be relatively harmless in a diploid individual if only one of the two gene copies is affected—that is, if the individual is *heterozygous* for the mutation. In general, diploid organisms harbor many such *recessive lethal* mutant alleles in their genomes. Usually the heterozygous individuals will have slightly reduced fitness, so the spread and persistence of such alleles is limited. However, even when the heterozygotes have undiminished fitness, sexual reproduction sets a limit to the frequency of the recessive lethal alleles (Figure 15–5). Individuals whose parents both carry a recessive lethal mutation of the same gene may inherit two mutant copies of the gene and no good copies. Such *homozygous* individuals will die, and their mutant copies of the gene will die with them. The more common the mutant gene in the population, the higher the rate of its elimination by this mechanism. Thus a balance is struck between the rate of elimination of the mutant allele and its rate of production by new mutational events. At equilibrium the recessive lethal mutant allele, although far more common than it would be in a haploid population, is still rare: the vast majority of individuals will carry two similar functional copies of the gene. Essentially the same argument applies to recessive mutations that are merely deleterious (tending to reduce the number of offspring) rather than lethal. In general, when reproduction is sexual and involves genetic recombination, natural selection will tend to ensure that in most individuals, for the majority of gene loci, the two gene copies remain functionally interchangeable. Thus the genome is kept diploid.

Consider, by contrast, a population consisting initially of diploid individuals that reproduce asexually. Here, in the absence of genetic recombination, there will be nothing in the long term to prevent the two copies of each gene from evolving along divergent pathways, and there will often be only weak selection against mutations that inactivate one of the two copies. Without sexual recombination, therefore, the diploid species will not stay functionally diploid forever. One can get an idea of the time required for such evolutionary change from studies of a related phenomenon that has occurred in the evolution of the catostomid fish. These fish are descended from an ancestor that, some 50 million years ago, underwent a duplication of its entire genome, previously diploid, and so became tetraploid. It has been estimated that about 50% of the "spare" pairs of genes encoding enzymes have become nonfunctional since then.

A Diploid Species Has a Spare Copy of Each Gene That Can Mutate to Serve a New Function

Most mutations are deleterious, since they merely spoil the function of a gene that is already optimized by natural selection. At rare intervals, however, a mutation may occur that enables an existing gene to perform a valuable new function.

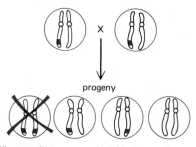

The recessive lethal mutant gene is created afresh whenever the original gene mutates. This creates a heterozygous individual, with one gene copy that is mutant and one that is functional in the original way.

SEXUAL POPULATION

progeny

When two heterozygous individuals sexually mate, some of their offspring inherit two copies of the recessive lethal gene and no copies of the functional gene and therefore die. This helps to eliminate the mutant gene from the population.

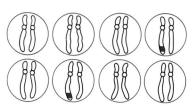

At equilibrium the recessive lethal gene is a rarity in the population: most individuals will have two functional copies of the gene.

Recessive lethal mutations arising at other loci are also kept at a low frequency in the same way. A typical member of the sexual population will carry such mutations at very few loci and will retain two functional copies of the majority of genes.

Figure 15–5 Diagram showing how sexual reproduction keeps diploid organisms diploid during the course of evolution. For simplicity, only recessive lethal mutations are considered. The argument for recessive deleterious mutations would be similar.

As a rule, such a mutation will at the same time spoil the gene for its original function. This change spells disaster for a haploid organism if the original function is vital. But in the diploid organism the occurrence of the mutation in one of the two gene copies is liable to be not merely tolerable, but actually beneficial. Even a small advantage provided by the new mutant gene will be sufficient to outweigh the small harm done by losing one of the two original gene copies: the heterozygous individual will have the benefit of both the old gene function and the new. Homozygotes, with two copies of the old allele or two copies of the new, will have inferior fitness. In these cases of *heterozygote advantage*, it can be shown that the novel mutant gene will spread rapidly through the sexual diploid population until an equilibrium is reached in which both the old and the new alleles are present at high frequency and the number of heterozygous individuals is large. This phenomenon is known as *balanced polymorphism*. There is a penalty to be paid, however: when two heterozygotes mate, a large proportion of their offspring, according to the usual Mendelian rules, will be homozygotes of inferior fitness. But this state of affairs will not persist forever: there is a way forward.

A Diploid Species Can Rapidly Enrich Its Genome by Adding New Genes

Spontaneous *gene duplications* occur in all organisms from time to time: a chromosome that contains one copy of a gene G gives rise, through an error of DNA replication, to a chromosome that contains two copies of the gene G in tandem. The duplication in itself brings no advantage and will, as a rule, be found only in a very few individuals. Suppose, however, that it occurs at a locus where a beneficial mutant allele G* is maintained by heterozygote advantage at high frequency in the population, in coexistence with the original allele G (Figure 15–6). There is a high probability, then, that in the diploid cell containing the GG chromosome (carrying the duplication) the homologous companion chromosome will carry the G* allele, giving a genotype GG/G*. By *genetic recombination* at meiosis (see below), gametes with a GG* chromosome can then be produced. In these, the original gene G and the mutant gene G* are arranged in tandem, no longer as two alleles competing to occupy the same locus but as two separate genes, each with a locus of its own. This is a winning combination, and it will be rapidly propagated until the whole population consists of GG*/GG* homozygotes (see Figure 15–6). With this genotype each individual will not only have the advantage of possessing both the old gene G and the new gene G*, it will also be able to pass on the advantage to all its progeny.

The suggestion, therefore, is that, in the sexual diploid species, new genes can be created by mutation of the spare copies of existing genes; the new genes can become common through heterozygote advantage without loss of the original genes; and finally, they can be inserted as additions to the genome through gene duplication and genetic recombination. This sequence of events is possible only for the diploid species. Innovation is more difficult for a haploid species. If it is not to lose the old gene in the process of acquiring the new, it must wait until the innovative mutation occurs in one of the few individuals that already carry a duplication at the appropriate locus. Under normal circumstances both the particular mutation and the particular duplication will occur only rarely, and the haploid species will have to wait a very long time before they occur in conjunction (Figure 15–7). Except for conditions where the rate of spontaneous gene duplication is very high, calculations show that the diploid organism will typically be able to enlarge and enrich its genome with new genes performing new functions at a rate that is hundreds or thousands of times faster than the haploid organism. The contrast is most marked in the case of mutations that occur with a very low frequency, and the mutations required for genetic innovation are likely to fall in this class.

In summary, sexual reproduction goes hand-in-hand with diploidy, and diploidy, in turn, may give a species special opportunities to evolve a larger, more complex, and more versatile genome. Evolution, admittedly, can occur by many routes, and the pathway for gene duplication and divergence that we have outlined

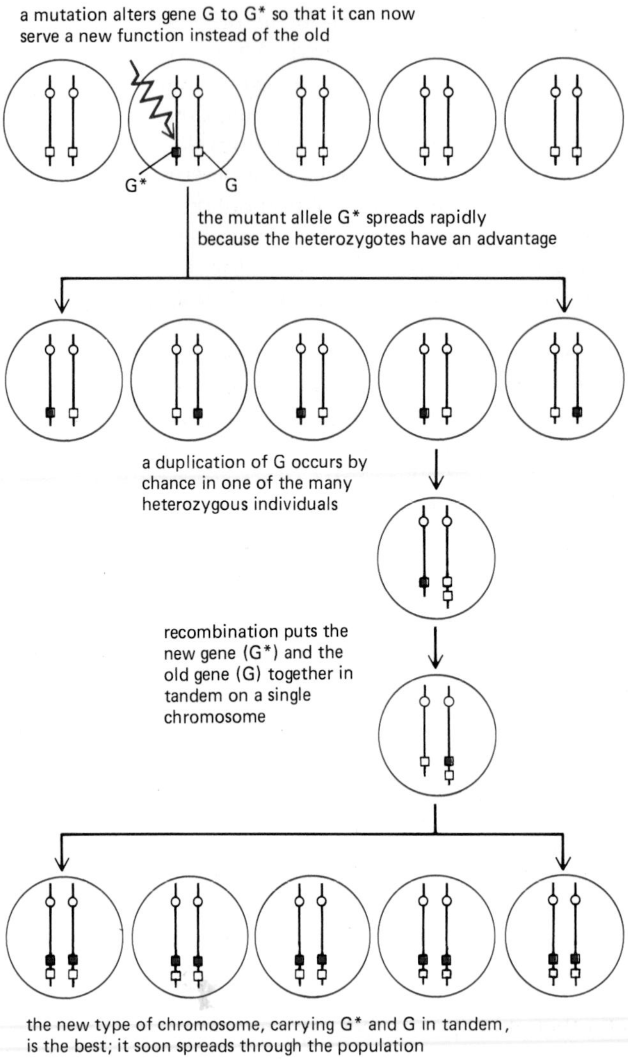

a mutation alters gene G to G* so that it can now
serve a new function instead of the old

Figure 15–6 The creation of a new gene (G*) in a sexually reproducing diploid organism via a "mutate, spread, and duplicate" pathway.

G*　　G

the mutant allele G* spreads rapidly
because the heterozygotes have an advantage

a duplication of G occurs by
chance in one of the many
heterozygous individuals

recombination puts the
new gene (G*) and the
old gene (G) together in
tandem on a single
chromosome

the new type of chromosome, carrying G* and G in tandem,
is the best; it soon spreads through the population

is certainly not the only possibility. Nevertheless, in one way or another, sex is likely to have profound consequences for the ways in which genetic innovations originate and spread through a population, making possible the evolution of such intricate organisms as ourselves.

We shall now examine the detailed cellular mechanisms of sex, beginning with the events of *meiosis*, in which genetic recombination occurs and diploid cells of the germ line divide to produce haploid *gametes*. Then we shall consider the gametes themselves and, finally, the process of *fertilization*, in which the gametes fuse to form a new diploid organism.

Summary

Sexual reproduction involves a cyclic alternation of diploid and haploid states: diploid cells divide by meiosis to form haploid cells, and the haploid cells from two individuals fuse in pairs at fertilization to form new diploid cells. In the process, genomes are mixed and recombined to produce individuals with novel assortments of genes. Most of the life cycle of higher plants and animals is spent in the diploid phase, and the haploid phase is very brief. Sexual reproduction has probably been favored by evolution because random genetic recombination improves the chances of producing at least some offspring that will survive in an unpredictably variable environment. Sex is also necessary for the long-term maintenance of diploidy, and diploidy may in turn have facilitated the rapid evolution of new genes in higher plants and animals.

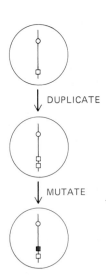

DUPLICATE

MUTATE

Figure 15–7 The creation of a new gene in a haploid organism. This sequence of events appears much simpler than that shown in Figure 15–6, but it is likely to proceed much more slowly.

Meiosis

The realization that germ cells are haploid, and must therefore be produced by a special type of cell division, came from an observation that was also among the first to suggest that chromosomes carry genetic information. In 1883 it was discovered that, whereas the fertilized egg of a particular worm contains four chromosomes, the nucleus of the egg and that of the sperm each contain only two chromosomes. The chromosome theory of heredity therefore explained the long-standing paradox that maternal and paternal contributions to the character of the progeny seem often to be equal, despite the enormous difference in size between the egg and sperm.

The finding also implied that germ cells must be formed by a special kind of nuclear division in which the chromosome complement is precisely halved. This type of division is called **meiosis,** from the Greek, meaning diminution. (There is no connection with the term mitosis, which is from the Greek *mitos*, meaning a thread, and refers to the threadlike appearance of the chromosomes as they condense during nuclear division—a process that occurs in both ordinary and meiotic divisions.) The behavior of the chromosomes during meiosis turned out to be considerably more complex than expected. Consequently, it was not until the early 1930s, as a result of painstaking cytological and genetic studies, that the important features of meiosis were established.

Meiosis Involves Two Nuclear Divisions Rather Than One

With the exception of the chromosomes that determine sex (the "sex chromosomes"), a diploid nucleus contains two closely similar versions of each chromosome, one from the male and one from the female parent. The two versions are called **homologues,** and in most cells they maintain a completely separate existence as independent chromosomes. When each chromosome is duplicated by DNA replication, the twin copies of the fully replicated chromosome at first remain closely associated and are called **sister chromatids.** In an ordinary cell division (described in Chapter 13) the sister chromatids line up on the spindle during mitosis with their kinetochore fibers pointing toward opposite poles. After the sister chromatids separate from each other at anaphase, they are called chromosomes. In this manner each daughter cell formed by ordinary cell division inherits one copy of each of the homologous versions of each chromosome (see p. 762). In contrast, a haploid gamete produced by the divisions of a diploid cell during meiosis must contain only one member of each homologous pair of chromosomes—either the maternal or the paternal homologue—and therefore only half the original number of chromosomes. This requirement makes an extra demand on the machinery for cell division, and the mechanism that has evolved to accomplish the additional sorting requires that homologues recognize each other and become physically paired before they line up on the spindle. This pairing of the maternal and the paternal copy of each chromosome is unique to meiosis (Figure 15–8) and will be discussed in detail later.

Given a mechanism for the pairing of the maternal and paternal homologues and their subsequent separation on the spindle, meiosis could, in principle, take place by modification of a single mitotic cell cycle in which chromosome duplication (S phase) is omitted: if the unduplicated homologues paired before M phase, the ensuing cell division could then produce two haploid cells directly. The actual meiotic process is, however, more complex. Before the homologues pair, each homologue replicates to produce two sister chromatids as in an ordinary cell division. It is only after DNA replication has been completed that the special features of meiosis become evident. Rather than separating, the sister chromatids behave as a unit, as if chromosome duplication had not occurred: each duplicated homologue pairs with its partner, forming a structure called a *bivalent*, which contains four chromatids. The bivalent lines up on the spindle, and at anaphase the two duplicated homologues (each consisting of two sister chromatids) separate and move to opposite poles. Because the joined sister chromatids behave as a unit, each daughter cell inherits two copies of one of the two homologues when

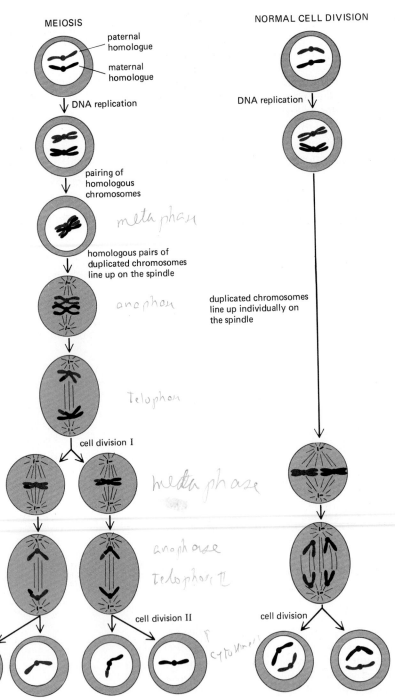

Figure 15–8 Comparison of meiosis and normal cell division. For clarity, only one set of homologous chromosomes is shown. The pairing of homologous chromosomes (homologues) is unique to meiosis. Each chromosome has been duplicated and exists as attached sister chromatids before the pairing occurs, so that two nuclear divisions are required to produce the haploid gametes. Each diploid cell that enters meiosis therefore produces four haploid cells. As shown, the chromosome pairing in meiosis involves crossing-over between homologous chromosomes, which will be explained below.

the meiotic cell divides. The two progeny of this division (**division I of meiosis**) therefore contain a diploid amount of DNA but differ from normal diploid cells in two ways: (1) both of the two DNA copies of each chromosome derive from only one of the two homologous chromosomes in the original cell (although, as we shall see, there has been some mixing of the maternal and paternal DNA caused by genetic recombination), and (2) these two copies are inherited as closely associated sister chromatids, as if they were a single chromosome (see Figure 15–8).

Formation of the actual gamete nuclei can now proceed simply through a second cell division, **division II of meiosis,** in which chromosomes align, without further replication, on a second spindle and the sister chromatids separate, as in normal mitosis, to produce cells with a haploid DNA content. Meiosis thus consists of two nuclear divisions following a single phase of DNA replication, so that four haploid cells are produced from each cell that enters meiosis (see Figure 15–8). Occasionally the meiotic process occurs abnormally and homologues fail to sepa-

Figure 15–9 Two major contributions to the reassortment of genetic material that occurs during meiosis. (A) The independent assortment of the maternal and paternal homologues during the first meiotic division produces 2^n different haploid gametes for an organism with n chromosomes. Here $n = 3$ and there are 8 different possible gametes, as indicated. (B) Crossing-over during meiotic prophase I exchanges segments of homologous chromosomes and thereby reassorts genes in individual chromosomes. Because of the many small differences in DNA sequence that always exist between any two homologues, both mechanisms increase the genetic variability of organisms that reproduce sexually.

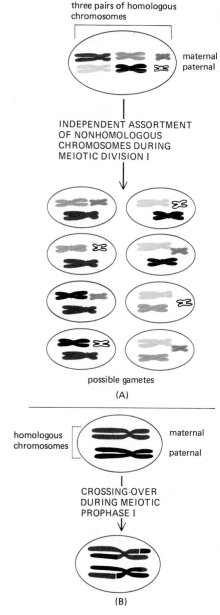

three pairs of homologous chromosomes

maternal
paternal

INDEPENDENT ASSORTMENT OF NONHOMOLOGOUS CHROMOSOMES DURING MEIOTIC DIVISION I

possible gametes

(A)

homologous chromosomes

maternal
paternal

CROSSING-OVER DURING MEIOTIC PROPHASE I

(B)

rate—a phenomenon known as **nondisjunction.** In this case some of the haploid cells that are produced lack a chromosome while others have more than one copy. Such gametes form abnormal embryos, most of which die.

Genetic Reassortment Is Enhanced by Crossing-over Between Homologous Nonsister Chromatids

We have seen that genes can be mixed by the fusion of gametes from two individuals. But genetic variation is not produced solely by this means. Unless they are identical twins, no two offspring of the same parents are genetically the same. This is because, long before the two gametes fuse, two kinds of genetic reassortment have already occurred during meiosis.

One kind of reassortment is a consequence of the random distribution of the maternal and paternal homologues between the daughter cells at meiotic division I, as a result of which each gamete acquires a different mixture of maternal and paternal chromosomes (Figure 15–9A). From this process alone, one individual could, in principle, produce 2^n genetically different gametes, where n is the haploid number of chromosomes. In humans, for example, each individual can produce at least $2^{23} = 8.4 \times 10^6$ genetically different gametes. But the actual number is very much greater than this because of **chromosomal crossing-over,** a process that takes place during the long prophase of meiotic division I, in which parts of homologous chromosomes are exchanged. On average, between two and three crossover events occur on each pair of human chromosomes. This process scrambles the genetic constitution of each of the chromosomes in gametes, as illustrated in Figure 15–9B.

Chromosomal crossing-over involves breaking the single maternal and paternal DNA double helices in each of the two chromatids and rejoining them to each other in a reciprocal fashion by a process known as **genetic recombination.** The molecular details of this process are outlined in Chapter 5 (see p. 239). Recombination takes place during prophase of meiotic division I at a time when the two sister chromatids are packed together so tightly that their individuality cannot be distinguished by light or electron microscopy (see below). Much later in this extended prophase, the two separate sister chromatids of each duplicated homologue become clearly visible, although they remain tightly apposed along their entire length and intimately connected to each other at their centromeres (Figure 15–10). The two duplicated homologues (maternal and paternal) remain attached to each other at those points where a crossover between a paternal and a maternal chromatid has occurred. At each attachment point, called a **chiasma** (plural **chiasmata**), two of the four chromatids are seen to have crossed over between

Figure 15–10 Paired homologous chromosomes during the transition to metaphase of meiotic division I. A single crossover event has occurred earlier in prophase to create one chiasma. Note that the four chromatids are arranged in two distinct pairs of sister chromatids and that the two chromatids in each pair are tightly aligned along their entire lengths as well as joined at their centromeres. The entire unit shown here is therefore frequently referred to as a *bivalent.*

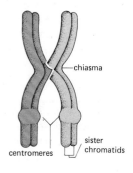

chiasma

sister chromatids

centromeres

the homologues (see Figure 15–10). Chiasmata are thus the morphological consequences of a prior, unobserved crossover event.

At this stage of meiosis, each pair of duplicated homologues, or *bivalent*, is held together by at least one chiasma. Many bivalents contain more than one chiasma, indicating that multiple crossovers can occur between homologues (Figures 15–11 and 15–12).

The Synaptonemal Complex Mediates Chromosome Pairing

Elaborate morphological changes occur in the chromosomes as they pair (*synapse*) and then begin to unpair (*desynapse*) during the first meiotic prophase. This prophase is traditionally divided into five sequential stages—*leptotene*, *zygotene*, *pachytene*, *diplotene*, and *diakinesis*—defined by these morphological changes (Figure 15–13). The most striking event is the initiation of intimate chromosome synapsis at **zygotene,** when a complex structure called the *synaptonemal complex* begins to develop between the two sets of sister chromatids in each bivalent. **Pachytene** is said to begin as soon as synapsis is complete, and it generally persists for days, until desynapsis begins the **diplotene** stage, in which the chiasmata are first seen.

Genetic recombination requires a close apposition between the recombining chromosomes. The **synaptonemal complex,** which forms just before pachytene and dissolves just afterward, keeps the homologous chromosomes in a bivalent together and closely aligned and is thought to be required for the crossover events to occur. It consists of a long ladderlike protein core, on opposite sides of which the two homologues are aligned to form a long linear chromosome pair (a bivalent, Figure 15–14). The sister chromatids in each homologue are kept tightly packed together, and their DNA extends from the same side of the protein ladder in a series of loops. Thus, while the homologous chromosomes are closely aligned along their length in the synaptonemal complex, the maternal and paternal chromatids that will recombine are kept separated by 100 nm on either side of the protein ladder.

From cytological studies, chromosome synapsis is seen to be preceded by the formation of a ropelike proteinaceous axis along each of the homologues. As pairing proceeds, the axes appear to become linked to each other to form the lateral elements of the synaptonemal complex—that is, the two sides of the protein ladder. Both the axes and the lateral elements contain a protein with unique silver-staining properties that make these structures visible by both light and electron microscopy (Figure 15–15).

It is not known what causes the homologous parts of chromosomes to become precisely aligned during zygotene. It is unlikely that homologous base-pairing all along the interacting chromosomes is required, since most of the chromatin of one homologue is positioned well away from the chromatin of its partner in the synaptonemal complex, and in special cases the synaptonemal complex is capable of joining regions of the two chromosomes that are not homologous. One possibility is that the initial pairing between chromosomes is mediated by complementary DNA base-pair interactions confined to specific regions on each chromosome; the synaptonemal complex then packs together the remainder of the prealigned chromosomes. An initial point-by-point matching mechanism of some type is required to explain the observation that the presence of an inverted section of chromosome in one of two pairing homologues usually (but not always) results in a temporary interruption of the normal zipperlike synapsis during zygotene, allowing homologous genes to synapse even within the inversion (Figures 15–16 and 15–17). The various stages of meiosis are outlined and described in detail in Figure 15–18.

Recombination Nodules Are Thought to Mediate the Chromatid Exchanges

Although the synaptonemal complex provides the structural framework for recombination events, it probably does not participate in them directly. The active recombination process is thought to be mediated instead by **recombination nodules,**

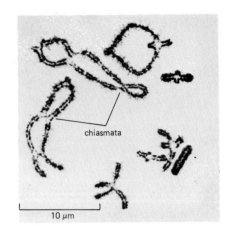

Figure 15–11 Light micrograph of several bivalents containing multiple chiasmata at diplotene, a late stage of meiotic prophase I (see Figure 15–18). These large grasshopper chromosomes provide particularly favorable material for cytological observations of this kind. (Courtesy of Bernard John.)

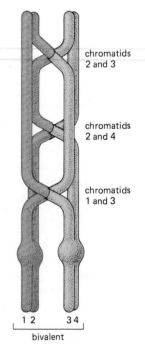

Figure 15–12 Three chiasmata resulting from three separate crossover events. Each of the two chromatids on each chromosome can cross over with either of the chromatids on the other chromosome in the bivalent. For example, here chromatid 3 has undergone an exchange with both chromatid 1 and chromatid 2.

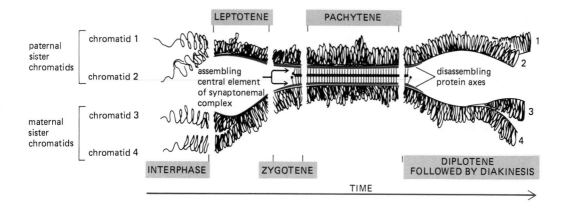

paternal sister chromatids — chromatid 1, chromatid 2

maternal sister chromatids — chromatid 3, chromatid 4

LEPTOTENE   PACHYTENE

assembling central element of synaptonemal complex

disassembling protein axes

INTERPHASE   ZYGOTENE   DIPLOTENE FOLLOWED BY DIAKINESIS

TIME

which are very large protein-containing assemblies with a diameter of about 90 nm. (For comparison, a large globular protein molecule of molecular weight 400,000 has a diameter of about 10 nm.) Recombination nodules sit at intervals on the synaptonemal complex, placed like basketballs on a ladder between the two homologous chromosomes (see Figure 15–14). They are thought to mark the site of a large multienzyme "recombination machine," which brings local regions of DNA on the maternal and paternal chromatids together across the 100-nm-wide synaptonemal complex.

The evidence that the recombination nodule serves this function is indirect: (1) The total number of nodules is about equal to the total number of chiasmata seen later in prophase. (2) The nodules are distributed along the synaptonemal complex in the same way that crossover events are distributed. For example, like the crossover events themselves, the nodules are absent from those regions of the synaptonemal complex that hold heterochromatin together. Moreover, both genetic and cytological measurements indicate that the occurrence of one crossover event prevents a second crossover event occurring at any nearby chromosomal site; similarly, the nodules tend not to occur very near one another. (3) Some *Drosophila* mutations cause an abnormal distribution of crossover events along the chromosomes, as well as a greatly diminished recombination frequency. In these mutants, correspondingly fewer recombination nodules are found, with a changed distribution that parallels the changed crossover distribution. This correlation strongly suggests that a recombination nodule determines the site of each crossover event. (4) Genetic recombination is thought to involve a limited amount of DNA synthesis at the site of each crossover event (see p. 244). Electron microscopic autoradiography shows that radioactive DNA precursors are preferentially incorporated into pachytene DNA at or near recombination nodules.

Because there are about as many recombination nodules as crossover events, their suggested role implies that recombination nodules are extremely efficient in causing the chromatids on opposite homologues to recombine. Nothing is yet known about their structure or mechanism of action.

Chiasmata Play an Important Part in Chromosome Segregation in Meiosis

In addition to reassorting genes, chromosomal crossing-over is crucial for the segregation of the two homologues to separate daughter nuclei. This is because each crossover event creates a chiasma, which plays a role analogous to that of the centromere in an ordinary mitotic division, holding the maternal and paternal homologues together on the spindle until anaphase I. In mutant organisms that have a reduced frequency of meiotic chromosome crossing-over, some of the chromosome pairs lack chiasmata. These pairs fail to segregate normally, and a high proportion of the resulting gametes contain too many or too few chromosomes—an example of nondisjunction.

There are at least two major differences in the way chromosomes separate in meiotic division I and in normal mitosis. (1) During normal mitosis the kineto-

Figure 15–13 Time course of chromosome synapsis and desynapsis during meiotic prophase I. A single bivalent is shown. The pachytene stage is defined as the period during which a fully formed synaptonemal complex exists.

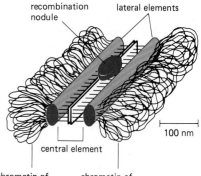

recombination nodule   lateral elements

100 nm

central element

chromatin of sister chromatids 1 and 2 (paternal)

chromatin of sister chromatids 3 and 4 (maternal)

Figure 15–14 A typical synaptonemal complex, showing the lateral and central elements of the complex. A recombination nodule is also shown. Only a short section of the long ladderlike complex is shown. Although a similar synaptonemal complex is present in organisms as diverse as yeast and human, very little is known about the protein molecules that form it.

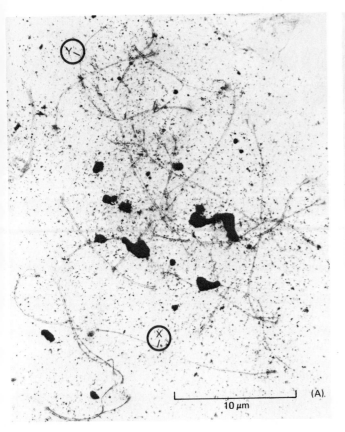

(A)

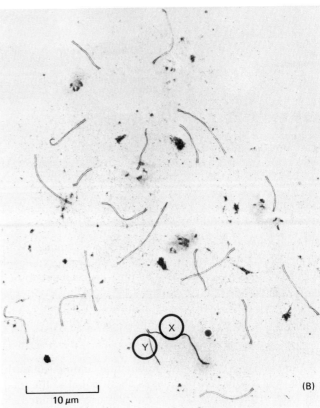

(B)

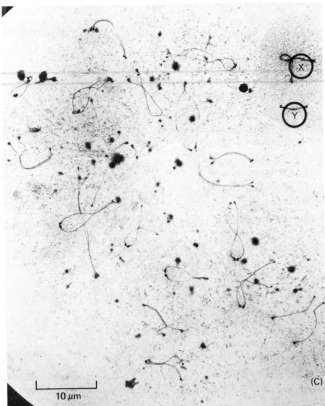

(C)

Figure 15–15 Electron micrographs of silver-stained, whole-mount spreads of chromosomes from mouse spermatocytes of (A) early (zygotene), (B) middle (pachytene), and (C) later (diplotene) stages of meiotic prophase I. (A) **Synapsing (zygotene):** The chromosome axes form before the chromosomes synapse, and they are initially seen as separate. The axes then move together and, when properly spaced at one or more synaptic initiation sites, the synaptonemal complex forms, often beginning at the end. The large separation between the sex chromosomes (X and Y) indicates that, in order to pair, chromosomes must often travel over substantial intranuclear distances, but it is not known how they do so. The dark bodies are nucleoli. (B) **Synapsed (pachytene):** Synapsis is complete when synaptonemal complexes have fully joined all homologous autosomes in pairs. The X and Y do not synapse fully. Crossing-over takes place between chromatids, which cannot be individually distinguished in these preparations. (C) **Desynapsing (diplotene):** Just before the axes disassemble, they separate, marking the end of synapsis. In places they are held together by persistent segments of synaptonemal complex thought to represent sites where crossing-over has occurred. Later, when the chromatin condenses and chromatids are distinguishable, chiasmata will indicate the crossovers. (Original micrographs courtesy of Montrose J. Moses.)

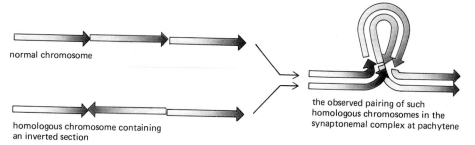

normal chromosome

homologous chromosome containing
an inverted section

the observed pairing of such
homologous chromosomes in the
synaptonemal complex at pachytene

Figure 15–16 Formation of the synaptonemal complex between one normal chromosome and its homologue carrying an inverted section. Such structures demonstrate that a local like-with-like pairing process brings homologous chromosomes together. Compare with Figure 15–17.

chores on each sister chromatid have attached kinetochore fibers pointing in opposite directions, whereas at metaphase I of meiosis the kinetochores on both sister chromatids appear to have fused so that their attached kinetochore fibers all point in the same direction (Figure 15–19). (2) During normal mitosis the movement of chromatids to the poles is triggered by a mechanism that detaches the two sister kinetochores from each other (thus beginning anaphase, see p. 773); in anaphase I of meiosis, however, this movement is initiated by the disruption of the poorly understood forces that have previously kept the arms of sister chromatids closely apposed, and this in turn dissolves the chiasmata that were holding the homologous maternal and paternal chromosomes together (see Figure 15–19). This explains not only why the chiasmata are necessary for the normal alignment of the chromosomes at metaphase I in many organisms, but also why the chromosomes produced at anaphase I tend to have nonadherent sister chromatid arms, giving each pair of sister chromatids an unusual "splayed-out" appearance compared with the arrangement of the chromatids in normal metaphase chromosomes (see Figure 15–19).

Pairing of the Sex Chromosomes Ensures That They Also Segregate

We have explained how homologous chromosomes pair so that they segregate between the daughter cells. But what about the sex chromosomes, which in male mammals are not homologous? Females have two X chromosomes, which pair and segregate like other homologues. But males have one X and one Y chromo-

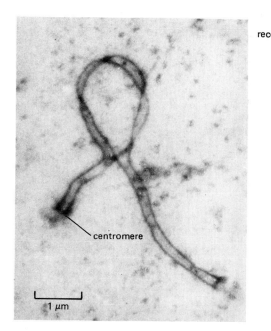

centromere

1 µm

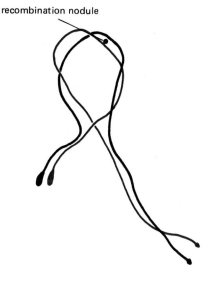

recombination nodule

Figure 15–17 Electron micrograph and drawing of two tightly synapsed mouse homologues at pachytene, one of which contains an inversion. A recombination nodule (see p. 848) is seen in the loop. (From P.A. Poorman, M.J. Moses, T.H. Roderick, and M.T. Davisson, *Chromosoma* 83:419, 1981.)

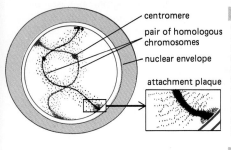

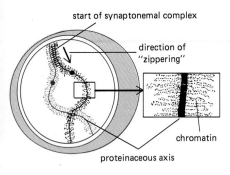

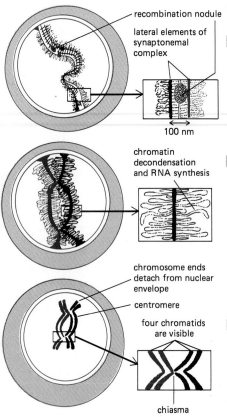

LEPTOTENE Prophase I begins at the leptotene stage, when each chromosome is first seen to have condensed from its interphase conformation to produce a long, thin thread with a proteinaceous central *axis*. Each chromosome is attached at both of its ends to the nuclear envelope via a specialized structure called an *attachment plaque*. Although each chromosome has replicated and consists of *two sister chromatids,* these chromatids are unusually closely apposed, and each chromosome therefore appears to be single (separate chromatids will not become visible until late in prophase, at either the diplotene stage or in diakinesis).

ZYGOTENE Leptotene is considered to end and the zygotene stage of prophase to begin as soon as synapsis, or intimate pairing, between the two homologues is initiated. Synapsis often starts when the homologous ends of the two chromosomes are brought together on the nuclear envelope, and it continues inward in a zipperlike manner from both ends, aligning the two homologous chromosomes side by side. In other cases, synapsis may begin in internal regions of the chromosomes and proceed toward the ends, producing the same type of alignment. Each gene is thus thought to be brought into juxtaposition with its homologous gene on the opposite chromosome. As the homologues pair, their ropelike proteinaceous axes are brought together to form the two *lateral elements*, or "sides," of the long ladderlike structure called the *synaptonemal complex*. Each resulting chromosome pair in meiotic prophase I is usually called a *bivalent*, but since each homologous chromosome in the pair consists of two closely apposed sister chromatids, it is better to think of each chromosome pair as a *tetrad*, another commonly used term.

PACHYTENE As soon as synapsis is complete all along the chromosomes, the cells are said to have entered the pachytene stage of prophase, where they may remain for days. At this stage, large *recombination nodules* appear at intervals on the synaptonemal complexes and are thought to mediate chromosomal exchanges. These exchanges result in crossovers between two nonsister chromatids—that is, one from each of the two paired homologous chromosomes. Although invisible at pachytene, each such crossover will appear later as a chiasma.

DIPLOTENE Desynapsis begins the diplotene stage of meiotic prophase I. The synaptonemal complex dissolves, allowing the two homologous chromosomes in a bivalent to pull away from each other to some extent. However, each bivalent remains joined by one or more chiasmata, representing the sites where crossing-over has occurred. In oocytes (developing eggs), diplotene can last for months or years, since it is at this stage that the chromosomes decondense and engage in RNA synthesis to provide storage materials for the egg. In the extreme, the diplotene chromosomes can become highly active in RNA synthesis and expand to an enormous extent, producing the lampbrush chromosomes found in amphibians and some other organisms.

DIAKINESIS Diplotene merges imperceptibly into diakinesis, the stage of transition to metaphase, as RNA synthesis ceases and the chromosomes condense, thicken, and become detached from the nuclear envelope. Each bivalent is clearly seen to contain four separate chromatids, with each pair of sister chromatids linked at their centromeres, while nonsister chromatids that have crossed over are linked by chiasmata.

Figure 15–18 Diagrams illustrating the appearance of two homologous chromosomes in a cell undergoing meiosis. Events are described and illustrated as they occur in mammals, although closely related chromosomal changes are seen in many other organisms. The five stages of meiotic prophase I (A) are illustrated on this page, while the remaining stages of meiosis (B) are shown on the facing page.

(B) THE REMAINING STAGES OF MEIOSIS

After the long prophase I has ended, two successive nuclear divisions without an intervening period of DNA synthesis bring meiosis to an end. These remaining meiotic stages typically occupy only 10% or less of the total time required for meiosis, and they are named in accord with the corresponding stages of mitosis of an ordinary division cycle. Thus the rest of meiotic division I is said to consist of metaphase I, anaphase I, and telophase I. At the end of division I, the chromosomal complement has been reduced from tetraploidy to diploidy, just as in an ordinary mitosis, and two cells have formed from one. The critical difference is that for each type of chromosome, *two sister chromatids* joined at their centromeres have been segregated to each cell rather than two separate chromatids as at mitosis. Division II quickly follows, consisting of a transient interphase II with no chromosome replication, followed by prophase II, metaphase II, anaphase II, and telophase II. In the end, four haploid nuclei are produced from each diploid cell that entered meiosis.

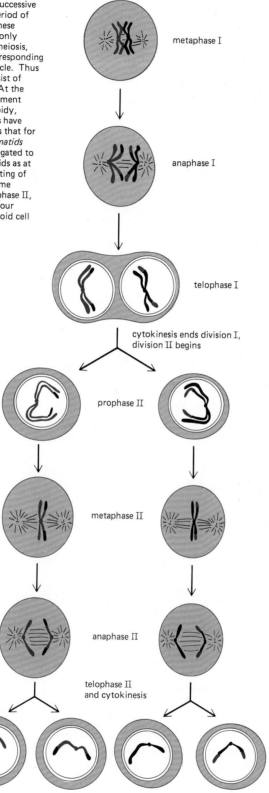

metaphase I

anaphase I

telophase I

cytokinesis ends division I, division II begins

prophase II

metaphase II

anaphase II

telophase II and cytokinesis

some, which must pair during the first metaphase if the sperm are to contain either one Y or one X chromosome and not both or neither. The necessary pairing is made possible by a small region of homology between the X and the Y at one end of these chromosomes. In this region they pair and cross over to form a chiasma during the first meiotic prophase (see Figure 15–15B). This small amount of genetic recombination is sufficient to keep X and Y chromosomes paired on the spindle so that only two types of sperm are normally produced: sperm containing one Y chromosome, which will give rise to male embryos, and sperm containing one X chromosome, which will give rise to female embryos.

Meiotic Division II Resembles a Normal Mitosis

Of the two successive cell divisions that constitute meiosis, division I occupies almost all of the time and is by far the more complex (Figure 15–20). It also has a number of unique features. The DNA replication during the preparatory S phase, for example, tends to take much longer than normal. Most strikingly, cells can remain in the first meiotic prophase for days, months, or even years, depending on the species and on the gamete being formed. (Although it is traditionally called prophase, this prolonged phase of meiotic division I resembles the G_2 phase of an ordinary cell division in that the nuclear envelope remains intact and disappears only when the spindle fibers begin to form, as prophase I gives way to metaphase I.)

After the end of meiotic division I, nuclear membranes re-form around the two daughter nuclei and a brief interphase begins. During this period, the chromosomes decondense somewhat, but they soon recondense and prophase II begins. As there is no DNA synthesis during this interval, in some organisms the chromosomes seem to pass almost directly from one division phase into the next. In all organisms, prophase II is brief: the nuclear envelope breaks down as the new spindle forms, after which metaphase II, anaphase II, and telophase II follow in quick succession. As in an ordinary mitosis, a separate set of kinetochore fibers forms on each sister chromatid, and these two sets of fibers extend in opposite directions from the centromere. Moreover, the two sister chromatids are kept together on the metaphase plate until they are released by the sudden separation of their kinetochores at anaphase (see Figure 15–19). Thus division II, unlike division I, closely resembles a normal mitosis. The difference is that one copy of each chromosome is present instead of two.

After nuclear envelopes have formed around the four haploid nuclei produced at telophase II, meiosis is complete (see Figure 15–18B). As we shall now see, by the end of meiosis a vertebrate egg is fully developed (and in some cases even fertilized), whereas a sperm that has completed meiosis has only just begun its development.

Summary

In meiosis, two successive cell divisions following one round of DNA replication give rise to four haploid cells from a single diploid cell. In animals the formation of both eggs and sperm begins in a similar way. In both cases meiosis is dominated by prophase of meiotic division I, which can occupy 90% or more of the total meiotic period. At this time each chromosome consists of two tightly joined sister chromatids. Chromosomal crossover events occur during the pachytene stage of prophase I, when each pair of homologous chromosomes is held in register by a synaptonemal complex. Each crossover event is thought to be mediated by a recombination nodule, and it results in the formation of a chiasma, which persists until anaphase I. In the first meiotic cell division, one member of each chromosome pair, still composed of linked sister chromatids, is distributed to each daughter cell. A second cell division, without DNA replication, then rapidly ensues in which each sister chromatid is segregated into a separate haploid cell.

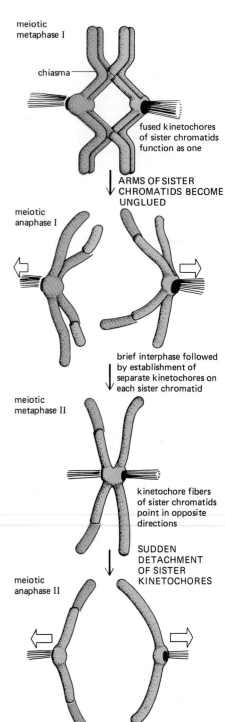

Figure 15–19 Comparison of the mechanisms of chromosome alignment (at metaphase) and separation (at anaphase) in meiotic division I and meiotic division II. The mechanisms used in meiotic division II are the same as those in normal mitosis (see Chapter 13).

Gametes[9]

In all vertebrate embryos, certain cells are singled out early in development as progenitors of the gametes. These **primordial germ cells** migrate to the developing gonads (ovaries in females, testes in males), where, after a period of mitotic proliferation, they undergo meiosis and differentiate into mature gametes—either eggs or sperm. Later the fusion of egg and sperm after mating initiates embryogenesis, with the subsequent production in the embryo of new primordial germ cells, which begin the cycle again.

For vertebrates it is unclear what determines which embryonic cells will become germ cells; but in at least some invertebrates the determining factor is known to be a component (or components) of the egg cytoplasm. In *Drosophila*, for example, a specialized region of cytoplasm, the *polar plasm*, at the posterior end of the egg contains small, RNA-rich granules (*polar granules*); cells that form at this end of the egg and contain polar granules become primordial germ cells and eventually migrate to the gonads to form gametes. If polar plasm is injected into the anterior pole of an egg, cells that would normally have developed into somatic cells develop instead into primordial germ cells.

An Egg Is the Only Cell in a Higher Animal That Is Able to Develop into a New Individual

In one respect at least, **eggs** are the most remarkable of animal cells: once activated, they can give rise to a complete new individual within a matter of days or weeks. No other cell in a higher animal has this capacity. Activation is usually the consequence of *fertilization*—fusion of a sperm with the egg—although, as we shall see, many eggs can be activated by other, surprisingly simple means (see p. 870). The activating stimulus initiates the program of development that produces a new individual.

In most nonmammalian eggs, the developmental program begins with a series of rapid cell divisions, or *cleavages*, in which the total mass of the embryo does not increase. Instead, the cytoplasm of the egg, which is very large to begin with, is partitioned between the cleaving cells, which become progressively smaller until they approach the size of a normal adult somatic cell. These early cleavage divisions require rapid DNA synthesis and some protein synthesis, yet almost no RNA synthesis (gene transcription) occurs. This is because before fertilization the egg has accumulated large reserves of mRNA, ribosomes, tRNA, and all the precursors needed for macromolecular synthesis, as well as the specialized mRNAs that encode the proteins required to guide the early stages of development.

In those eggs that undergo a long period of embryonic development outside the mother's body, especially large reserves of nutrients have to be stored to see the embryo through until it hatches. This is why, for example, amphibian and bird eggs are so much larger than mammalian eggs. Some aquatic invertebrates, such as sea urchins, develop from small eggs outside the mother's body, but in these cases the embryos develop very rapidly into feeding larvae.

Although an egg can give rise to every cell type in the adult organism, it is itself a highly specialized cell, uniquely equipped for the single function of generating a new individual. We shall now briefly consider some of its specialized features before discussing how an egg develops to the point at which it is ready for fertilization.

An Egg Is Highly Specialized for Independent Development, with Large Nutrient Reserves and an Elaborate Protective Coat[10]

The most obvious distinguishing feature of an egg is its large size. An egg is typically spherical or ovoid, and its diameter is about 100 μm in humans and sea urchins, 1 mm to 2 mm in frogs and fishes, and many centimeters in birds and reptiles (remember that a typical somatic cell has a diameter of only about 20 μm)

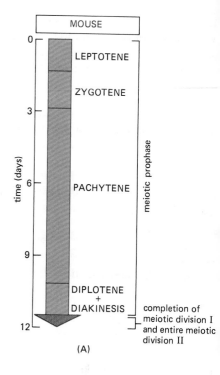

(A)

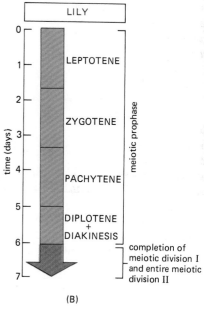

(B)

Figure 15–20 Comparison of times required for each of the stages of meiosis illustrated in Figure 15–18. Approximate times for both a male mammal (mouse) and the male tissue of a plant (lily) are shown. Times differ for male and female gametes (egg and sperm) of the same species, as well as for the same gametes of different species. For example, meiosis in a human male lasts for 24 days, compared with 12 days in the mouse. Meiotic prophase I, however, is always much longer than all the other meiotic stages combined.

(Figures 15–21 and 15–22). The size of the nucleus can be equally impressive: for example, in a frog egg with a diameter of 1500 μm, the diameter of the nucleus is about 400 μm; this nucleus stores nuclear proteins in preparation for the rapid nuclear divisions that occur following fertilization.

The egg cytoplasm contains nutritional reserves in the form of **yolk,** which is rich in lipid and protein and is usually contained within discrete structures called *yolk granules.* In eggs that develop into large animals outside the mother's body, yolk can account for more than 95% of the volume of the cell; whereas in mammals, whose embryos are largely nourished by their mothers, it constitutes less than 5% of the total egg volume.

The **egg coat** is another peculiarity of eggs. It is a specialized form of extracellular matrix consisting largely of glycoprotein molecules, some secreted by the egg and others by surrounding cells. In many species the major coat is a layer immediately surrounding the egg plasma membrane; in nonmammalian eggs, such as in sea urchins, it is called the **vitelline layer,** whereas in mammalian eggs it is called the **zona pellucida** (Figure 15–23). This layer protects the egg from mechanical damage; in some eggs it also acts as a species-specific barrier to sperm, admitting only those of the same or closely related species (see p. 869). Nonmammalian eggs often have additional layers overlying the vitelline layer that are secreted by surrounding cells. For example, as frog eggs pass from the ovary through the oviduct (the tube that conveys them to the outside), they acquire several layers of gelatinous coating secreted by epithelial cells lining the oviduct. Similarly, the "white" (albumin) and shell of chicken eggs are added (after fertilization) as the eggs pass along the oviduct. The vitelline layer of insect eggs is covered by a thick, tough layer called the *chorion,* which is secreted by the *follicle cells* that surround each egg throughout its development.

Many eggs (including those of mammals) contain specialized secretory vesicles just under the plasma membrane in the outer region, or *cortex,* of the egg cytoplasm (Figure 15–24). When the egg is activated by a sperm, these **cortical granules** release their contents by exocytosis; the contents of the granules act to alter the egg coat so as to prevent additional sperm from fusing with the egg (see p. 872).

Cortical granules are usually distributed evenly throughout the egg cortex, but other cytoplasmic components can have a strikingly asymmetrical distribution. In a frog egg, for example, most of the yolk is at one pole (the *vegetal pole*), while the nucleus is closer to the opposite pole (the *animal pole*). The polarity of the egg, which is generally determined by the environment in which it develops, often helps to determine the polarity of the embryo (see p. 880).

Eggs Develop in Stages[10,11]

A developing egg is called an **oocyte,** and its differentiation into a mature egg (or *ovum*) involves several highly specialized modifications of the normal cell cycle. We have already seen that instead of continuing to divide by ordinary mitosis, germ cells go through their last two divisions by the more elaborate process of meiosis. Moreover, oocytes have evolved special mechanisms for arresting the machinery for meiotic division: they remain suspended in prophase I for prolonged periods while the oocyte grows in size, and in many cases they later arrest in metaphase II while awaiting fertilization.

While the details of oocyte development (**oogenesis**) vary in different species, the general stages are similar, as outlined in Figure 15–25. Primordial germ cells migrate to the forming gonad to become **oogonia,** which proliferate by mitosis for a period before differentiating into **primary oocytes.** At this stage the first meiotic division begins: the DNA replicates so that each chromosome consists of two chromatids, the homologous chromosomes pair along their long axes, and crossing-over occurs between the chromatids of these paired chromosomes. After this the cell remains arrested in prophase of division I of meiosis for a period lasting from a few days to many years, depending on the species. During this prolonged prophase (or, in some cases, at the onset of sexual maturity), the pri-

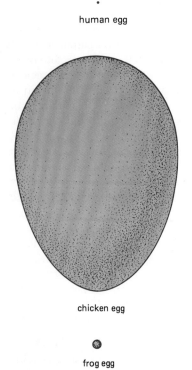

human egg

chicken egg

frog egg

Figure 15–21 The actual sizes of three eggs. The human egg is 0.1 mm in diameter.

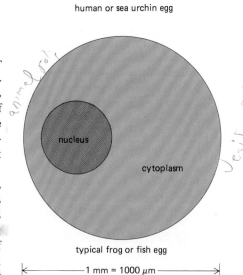

typical somatic cell

human or sea urchin egg

nucleus

cytoplasm

typical frog or fish egg

|← 1 mm = 1000 μm →|

Figure 15–22 The relative sizes of various eggs compared to a typical somatic cell.

mary oocytes synthesize a coat and cortical granules and accumulate ribosomes, yolk, glycogen, lipid, and the mRNA that will later direct the synthesis of proteins required for early embryonic growth and the unfolding of the developmental program. In many oocytes these activities are reflected in the appearance of the chromosomes, which decondense and form lateral loops, taking on the characteristic "lampbrush" appearance of a chromosome busily engaged in RNA synthesis (see p. 506).

The next phase of oocyte development is called *oocyte maturation* and usually does not occur until sexual maturity, when it is stimulated by hormones (see below). Under these hormonal influences, the cell resumes its progress through division I of meiosis: the chromosomes recondense, the nuclear envelope breaks down (this is generally taken to mark the beginning of maturation), and the replicated homologous chromosomes segregate at anaphase I into two daughter nuclei, each containing half the original number of chromosomes. To end division I, the cytoplasm divides asymmetrically to produce two cells that differ greatly in size: one is a small **polar body,** and the other is a large **secondary oocyte,** which contains all the developmental potential. At this stage each of the chromosomes is still composed of two sister chromatids. These chromatids do not separate until division II of meiosis, when they are partitioned into separate cells by a process that is identical to a normal mitosis, as previously described (see p. 854). After this final chromosome separation at anaphase II, the cytoplasm of the large secondary oocyte again divides asymmetrically to produce the mature **egg** (or **ovum**) and a second small polar body, each with a haploid number of single chromosomes (see Figure 15–25). Because of these two asymmetrical divisions of their cytoplasm, oocytes maintain their large size despite undergoing the two meiotic divisions. Both of the polar bodies are small, and they eventually degenerate.

In most vertebrates, oocyte maturation proceeds to metaphase of meiosis II and then arrests. At **ovulation** the arrested secondary oocyte is released from the ovary, and if fertilization occurs, the oocyte is stimulated to complete meiosis.

Oocytes Grow to Their Large Size Through Several Special Mechanisms[10,11]

A somatic cell with a diameter of 10 to 20 μm typically takes about 24 hours to double its mass in preparation for cell division. At this rate of biosynthesis, such a cell would take a very long time to reach the thousandfold greater mass of a mammalian egg with a diameter of 100 μm or the millionfold greater mass of an insect egg with a diameter of 1000 μm. Yet some insects live only a few days and manage to produce eggs with diameters even greater than 1000 μm. It is clear that eggs must have special mechanisms for achieving their large size.

One simple strategy for rapid growth is to have extra gene copies in the cell. Thus the oocyte delays completion of the first meiotic division so as to grow while it contains the diploid chromosome set in duplicate. In this way it has double the amount of DNA available for RNA synthesis compared to an average somatic cell in the G_1 phase of the cell cycle. Some oocytes go to even greater lengths to accumulate extra DNA: they produce many extra copies of certain genes. We have already seen in Chapter 9 that the somatic cells of most organisms require 100 to 500 copies of the ribosomal RNA genes in order to produce enough ribosomes for protein synthesis. Because eggs require even greater numbers of ribosomes to support protein synthesis during early embryogenesis, the rRNA genes are specifically amplified in the oocytes of many animals; some amphibian eggs, for example, contain 1 or 2 million copies of these genes (Figure 15–26).

Oocytes may also be partly dependent on the synthetic activities of other cells for their growth. For example, a major constituent of large eggs, the yolk, is usually synthesized outside the ovary and imported into the oocyte. In birds, amphibians, and insects, yolk proteins are made by liver cells (or their equivalents), which secrete these proteins into the blood. Within the ovaries, oocytes take up the yolk proteins from the extracellular fluid by receptor-mediated endocytosis (see Figure 6–72, p. 326). Nutritive help can also come from accessory cells in the ovary. Two

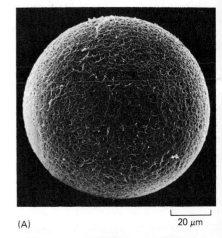

(A) ⊢————⊣ 20 μm

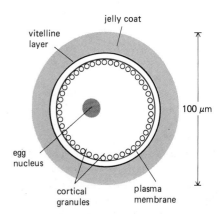

(B) ⊢————⊣ 20 μm

Figure 15–23 (A) Scanning electron micrograph of a hamster egg showing the zona pellucida. In (B) the zona (to which many spermatozoa are attached) has been peeled back to reveal the underlying plasma membrane of the egg, which contains numerous microvilli. (From D.M. Phillips, *J. Ultrastruct. Res.* 72:1–12, 1980.)

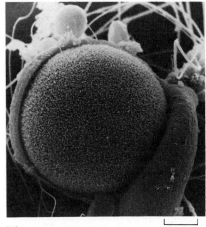

Figure 15–24 Schematic drawing of a sea urchin egg showing the location of cortical granules. The vitelline layer is covered by a jelly coat up to 30 μm thick.

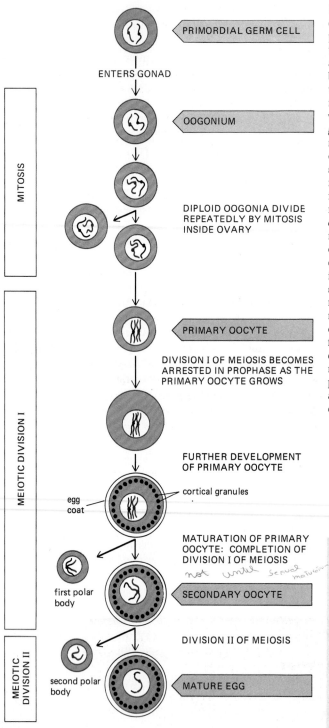

Figure 15-25 The various stages of oogenesis. Oogonia develop from primordial germ cells that migrate into the ovary early in embryogenesis. After a number of mitotic divisions, oogonia begin meiotic division I, after which they are called primary oocytes. In mammals, *primary oocytes* are formed very early (between 3 and 8 months of gestation in the human embryo) and remain arrested in prophase of meiotic division I until the female becomes sexually mature. At this point a small number periodically mature under the influence of hormones, completing meiotic division I to become *secondary oocytes*, which eventually undergo meiotic division II to become mature eggs (ova). The stage at which the egg or oocyte is released from the ovary and is fertilized varies from species to species. In most vertebrates, oocyte maturation is arrested at metaphase of meiosis II and the secondary oocyte completes meiosis II only after fertilization. All of the polar bodies eventually degenerate. However, in many animals, including mammals, the polar bodies remain inside the egg coat, and in some species the first polar body divides once before degenerating.

PRIMORDIAL GERM CELL

ENTERS GONAD

OOGONIUM

DIPLOID OOGONIA DIVIDE REPEATEDLY BY MITOSIS INSIDE OVARY

MITOSIS

PRIMARY OOCYTE

DIVISION I OF MEIOSIS BECOMES ARRESTED IN PROPHASE AS THE PRIMARY OOCYTE GROWS

MEIOTIC DIVISION I

FURTHER DEVELOPMENT OF PRIMARY OOCYTE

cortical granules

egg coat

MATURATION OF PRIMARY OOCYTE: COMPLETION OF DIVISION I OF MEIOSIS

not until sexual maturation

first polar body

SECONDARY OOCYTE

DIVISION II OF MEIOSIS

MEIOTIC DIVISION II

second polar body

MATURE EGG

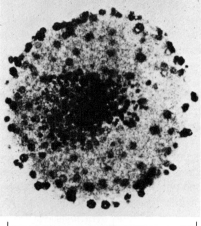

400 μm

Figure 15-26 Photomicrograph of an isolated nucleus from a frog egg stained with cresyl violet to show the large number of nucleoli, reflecting the enormous amplification of ribosomal RNA genes. (From D.D. Brown and I. Dawid, *Science* 160:273–275, 1968. Copyright 1968 by the American Association for the Advancement of Science.)

types of ovarian accessory cells function in this way in oogenesis, depending on the species. In some invertebrates, **nurse cells** surround the oocyte and are usually connected to it by cytoplasmic bridges through which macromolecules can pass directly into the oocyte cytoplasm. For the insect oocyte, the nurse cells manufacture many of the products—ribosomes, mRNA, protein, and so on—that a vertebrate oocyte has to manufacture for itself.

In some species, nurse cells are derived from the same oogonium that gives rise to the oocyte with which they are associated. In *Drosophila*, for example, an oogonium undergoes four mitotic divisions to form 16 cells. One of these cells becomes the oocyte, while the others become nurse cells and remain attached to

Figure 15–27 How a single *Drosophila* oogonium gives rise to 15 nurse cells and 1 oocyte, all connected by cytoplasmic bridges. At each mitosis, every cell divides once: in the first mitosis, cell 1 divides to produce cells 1 and 2; in the second mitosis, cell 1 divides to produce cells 1 and 3, while cell 2 divides to produce cells 2 and 4, and so on. Because cytoplasmic bridges form wherever remnants of the spindle had connected two daughter cells at telophase, these bridges join only cells that originate from a shared mitosis. Only cell 1 or cell 2 becomes the egg, perhaps because only these cells are connected by intercellular bridges to four others. An unusual feature of these cell divisions is that the cell volume does not double before mitosis, so the cells become smaller with each division. Later, as the oocyte grows, the nurse cells also become extremely large and make large amounts of macromolecules and organelles such as ribosomes and mitochondria, which are pumped into the oocyte through the cytoplasmic bridges.

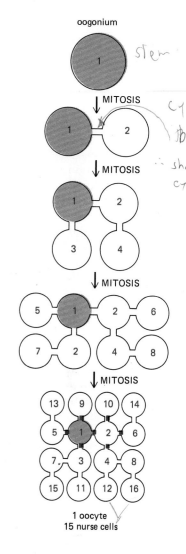

one another and to the oocyte by cytoplasmic bridges (Figure 15–27). In the nurse cells, DNA replication occurs repeatedly without cell division, so that eventually each cell reaches a very large size, with up to a thousand times the normal amount of DNA (arranged in *polytene chromosomes*, see p. 509). All 15 of the nurse cells, each with the equivalent of hundreds or thousands of genomes, are enlisted to synthesize the materials necessary for a single egg.

The other kind of accessory cells in the ovary that help nourish developing oocytes are **follicle cells,** which are found in both invertebrates and vertebrates. They are arranged as an epithelial layer around the oocyte (Figure 15–28), to which they are connected only by gap junctions, which permit the exchange of small molecules but not macromolecules (see p. 798). While these cells are unable to provide the oocyte with preformed macromolecules through these communicating junctions, they may help to supply the smaller precursor molecules from which macromolecules are made. In addition, follicle cells frequently secrete macromolecules that either contribute to the egg coat or are taken up by endocytosis into the growing oocyte.

The surrounding accessory cells make a further contribution to oocyte development: in both invertebrates and vertebrates it is the accessory cells that respond to the polypeptide hormones, called *gonadotropins*, that release the oocyte from prophase I arrest and enable it to mature further.

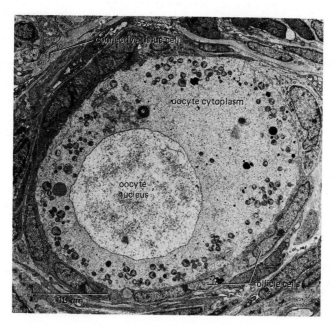

Figure 15–28 Electron micrograph of a rabbit primordial follicle, consisting of a central primary oocyte surrounded by a single layer of flattened follicle cells. The primordial follicle is enclosed by a basal lamina (not easily seen in this micrograph), which is in turn surrounded by the connective tissue of the ovary. Note that the primary oocyte does not have a zona pellucida or cortical granules at this stage of its development. (Copyright 1979. Urban & Schwarzenberg, Baltimore–Munich. Reproduced with permission from The Cellular Basis of Mammalian Reproduction, edited by Jonathan Van Blerkom and Pietro Motta. All rights reserved.)

Oocyte Maturation in Vertebrates Is Triggered by a Decrease in Intracellular Cyclic AMP[12]

In many animals, oocytes do not complete meiosis I until the onset of sexual maturity, when **gonadotropins** appear in the blood. These polypeptide hormones stimulate the accessory cells in the ovary to secrete a second molecule that acts on the oocyte to initiate its maturation. In most vertebrates, **oocyte maturation** is defined as the progression of the primary oocyte through meiosis I until it arrests again, as a *secondary oocyte,* at metaphase of meiosis II, where it remains until fertilization triggers the completion of meiosis (see Figure 15–25).

Oocyte maturation is best understood in amphibians, where gonadotropins released by the pituitary gland stimulate the follicle cells surrounding the oocytes to secrete the steroid hormone *progesterone.* When acting on most target cells, progesterone, like other steroid hormones, diffuses across the plasma membrane and binds to intracellular receptor proteins that regulate the transcription of specific genes (see p. 690). In oocyte maturation, however, progesterone is thought to act by binding to plasma membrane receptor proteins. This binding inactivates adenylate cyclase in the plasma membrane of the oocyte, decreasing the concentration of cyclic AMP in the cytosol, thereby reducing the activity of cyclic AMP-dependent protein kinase (A-kinase, see p. 709).

There is good evidence that this decrease in A-kinase activity is both necessary and sufficient to induce oocyte maturation. Microinjection of the catalytic subunit of the kinase, which is active in the absence of cyclic AMP (see p. 710), inhibits progesterone-induced maturation, whereas microinjection of specific inhibitors of the kinase induces maturation in the absence of progesterone. Similar microinjection experiments in mouse oocytes suggest that a similar intracellular signaling mechanism operates in mammals. The substrates for the A-kinase are unknown but have been suggested to be phosphoproteins that are part of a *meiotic division I arrest system* responsible for maintaining the oocyte in prophase I arrest. It is hypothesized that their activity depends on phosphorylations maintained by the A-kinase, so that a fall in cyclic AMP levels relieves the block, allowing meiosis I to proceed.

Oocyte Maturation Depends on the Activation of M-Phase Promoting Factor[13]

The events that follow the fall in cyclic AMP are poorly understood, but they eventually lead to the activation of a protein complex called **M-phase promoting factor (MPF),** which is thought to be required for the exit from meiotic prophase I. As discussed in Chapter 13, MPF has been highly conserved during eucaryotic evolution. It plays a crucial part in the ordinary cell division cycle by triggering cells to progress from G_2 into M phase (see p. 736). As mentioned previously, although traditionally referred to as prophase, the prophase of meiosis I closely resembles the G_2 phase of an ordinary division cycle: the DNA has replicated and is transcriptionally active, the nuclear envelope is intact, and the mitotic spindle has not yet formed. Moreover, like the transition from G_2 to M phase in ordinary dividing cells, the transition from meiotic prophase I to meiotic M phase is triggered by MPF. In fact, MPF was first discovered as a *maturation-promoting factor* for frog oocytes. Mature frog oocytes are blocked in metaphase II, where MPF levels are high (see p. 861). If a small aliquot of cytoplasm from such a mature oocyte is injected into an immature oocyte, the MPF it contains causes the breakdown of the nuclear membrane and the chromosomal condensation that occur at M phase and are diagnostic of oocyte maturation.

A remarkable property of MPF is that it can be transferred serially from one recipient oocyte to another indefinitely without any loss of activity, despite the enormous dilution involved in such serial transfers. This implies that the injection of MPF stimulates the production of still more MPF in the recipient oocyte. The amplification occurs even in enucleated oocytes and in the presence of protein synthesis inhibitors, showing that MPF activates preexisting MPF rather than inducing new MPF synthesis.

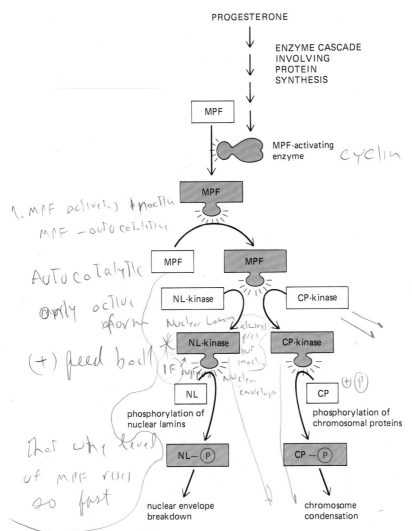

Figure 15–29 A model showing how M-phase promoting factor (MPF) may act to trigger a frog oocyte to progress out of prophase I into metaphase following stimulation by progesterone. It is hypothesized that progesterone indirectly causes production of an MPF-activating enzyme that activates a small amount of MPF. This activation of MPF leads to the activation both of more MPF (greatly amplifying the response) and of protein kinases that phosphorylate nuclear lamins and chromosomal proteins. This causes the nuclear envelope to break down and chromosomes to condense, thereby driving the cell to metaphase. A subsequent inactivation of MPF allows the nuclear envelope to re-form and chromosomes to decondense so that the cell can proceed into meiosis II (not shown). A similar mechanism may operate in mammalian oocyte maturation.

One component of MPF has been identified as a protein kinase. It is homologous to the protein kinase in yeasts that is encoded by the *cdc*2/28 gene, which plays a crucial part in controlling the cell cycle in yeast cells (see p. 743).

What is the mechanism underlying this self-amplifying autocatalytic reaction? Although MPF has not yet been completely characterized, there is evidence that it is phosphorylated, that phosphorylation is important in inducing its activity, and that at least one of its components is a protein kinase. Thus, MPF might phosphorylate, and thereby activate, itself. In frog oocytes it seems that the progesterone-induced *inactivation* of A-kinase initiates a complex set of events (requiring protein synthesis) that leads ultimately to the activation of a small amount of MPF, which in turn activates more MPF. Since extensive phosphorylations of structural proteins of the nuclear envelope and of the chromosomes are involved in the changes seen at M phase (see p. 777), it is tempting to propose that MPF also activates unknown protein kinases that cause these further changes (Figure 15–29).

Cyclic changes in the activation of MPF are thought to be crucial in the regulation of the ordinary mitotic cell cycle, driving cells in and out of mitosis (see p. 736). Oocyte maturation presumably involves similar cyclic changes in MPF. A rise in MPF first triggers the oocyte to progress through prophase into metaphase of meiotic division I. Then a subsequent inactivation of MPF allows oocytes to complete nuclear division, and its reactivation triggers the cell to progress to the M phase of meiotic division II. At this point a *meiotic division II arrest system* (sometimes attributed to a poorly characterized activity called *cytostatic factor*, or *CSF*) is thought to prevent MPF inactivation, thereby arresting the mature egg at metaphase II. Fertilization, possibly by increasing the concentration of Ca^{2+} in the cytosol (see p. 872), relieves the block and allows the completion of meiotic division II as well as the subsequent cycles of MPF activation and inactivation that drive the cells through repeated mitotic divisions to begin the process of embryonic development.

Figure 15–30 Photomicrograph of a developing follicle in the ovary of a rabbit. The primary oocyte is much larger than the younger oocyte shown in Figure 15–28, and it has acquired cortical granules (not readily seen in this micrograph) and a thick zona pellucida and is surrounded by several layers of follicle cells. Although not apparent in the micrograph, the innermost follicle cells extend processes through the zona and form gap junctions with the oocyte. The entire follicle is surrounded by a basal lamina.

In the Human Ovary Most Oocytes Die Without Maturing[14]

Although the molecular mechanisms underlying the main events in oocyte development and maturation are probably similar among vertebrates, the timing of these events and the extracellular signals that trigger them can vary greatly. In humans, for example, the signals are much more complex and less well understood than in amphibians. Moreover, the numbers of mature eggs produced in these two species are vastly different. Throughout its reproductive life, the female frog produces oocytes from self-renewing oogonia each mating season, when hundreds of oocytes mature and ovulate when stimulated by gonadotropins. In the human female, on the other hand, oocytes are produced from oogonia only during the first months of embryonic development, and all but a few hundred of these several million oocytes will eventually die in the ovary without maturing.

The primary oocytes in newborn girls are arrested in prophase of meiotic division I (like primary oocytes in amphibians), and most are surrounded by a single layer of follicle cells; such an oocyte with its surrounding follicle cells constitutes a **primordial follicle** (see Figure 15–28). Periodically, beginning sometime before birth, a small proportion of primordial follicles begin to grow to become **developing follicles:** the follicle cells enlarge and proliferate to form a multilayered envelope around the primary oocyte; the oocyte itself remains in prophase I while it enlarges and develops a zona pellucida and cortical granules (Figure 15–30). Some of the developing follicles eventually form a fluid-filled cavity, or *antrum*, to become **antral follicles** (Figure 15–31). It is not known what causes certain primordial follicles to begin growing to become developing follicles. However, the continuing development of such follicles seems to depend on gonadotropins (mainly *follicle-stimulating hormone (FSH)*) secreted by the pituitary gland and on estrogens secreted by the follicle cells themselves.

Each developing follicle continues to grow until it either degenerates or releases its oocyte by ovulation. However, oocyte maturation (progression beyond prophase I) and ovulation do not begin until puberty. Then, once each month (about halfway through the menstrual cycle), the pituitary secretes a surge of another gonadotropin, *luteinizing hormone (LH)*. LH acts to accelerate the growth of 15 to 20 of the thousands of developing follicles in the ovary. Of these, only one will complete its development with the maturation of the enclosed primary oocyte, which will progress to and arrest at metaphase of meiotic division II. The single stimulated follicle meanwhile rapidly enlarges and ruptures at the surface of the ovary, releasing the secondary oocyte still surrounded by a shell of follicle cells embedded in a gel-like matrix (Figure 15–32). The released oocyte is triggered to complete meiosis only if it is fertilized by a sperm within a day or so.

How does the midcycle LH surge initiate oocyte maturation? It seems that the follicle cells normally exert an inhibitory influence on the oocyte they surround, since oocytes in antral follicles resume meiosis spontaneously if they are removed from their follicles and cultured without stimulatory hormones. The LH surge may initiate oocyte maturation by relieving this inhibitory influence of the follicle cells.

One of the enigmatic features of oocyte maturation in humans is that only a small proportion of the many antral follicles present in the ovaries at the time of

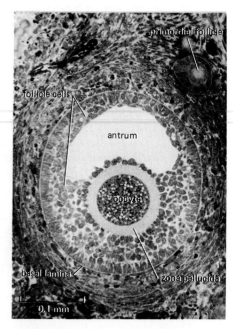

Figure 15–31 Photomicrograph of an antral follicle in the ovary of a rabbit. Many layers of follicle cells are now present. The nucleus of the oocyte is out of the plane of section and therefore is not seen. For comparison, note the much smaller primordial follicle in the upper right corner of the micrograph.

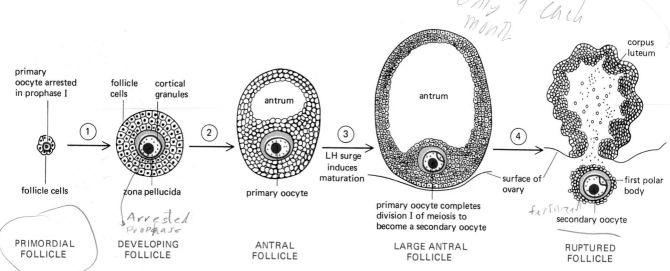

only 1 each monn

primary oocyte arrested in prophase I

follicle cells

follicle cells — cortical granules

zona pellucida

Arrested Prophase

antrum — primary oocyte

LH surge induces maturation

antrum

primary oocyte completes division I of meiosis to become a secondary oocyte

surface of ovary

fertilized

corpus luteum

first polar body

secondary oocyte

PRIMORDIAL FOLLICLE

DEVELOPING FOLLICLE

ANTRAL FOLLICLE

LARGE ANTRAL FOLLICLE

RUPTURED FOLLICLE

Figure 15–32 The stages in human oocyte development. (1) Starting before birth, a small proportion of primordial follicles sequentially begin to grow and are now called developing follicles. (2) Some developing follicles accumulate fluid to become antral follicles. (3) From the time of puberty, once each month a surge of luteinizing hormone (LH) activates about 20 antral follicles to accelerate their growth; only one of these, however, completes maturation and ovulates: the primary oocyte in this follicle completes meiosis I to form a polar body and a secondary oocyte (see Figure 15–25). (4) The secondary oocyte, arrested at metaphase of meiosis II, is released together with the polar body and a shell of surrounding follicle cells when the follicle ruptures at the surface of the ovary. The secondary oocyte will complete meiotic division II to become a mature egg only if it is fertilized.

the LH surge each month are stimulated to accelerate their growth, and, of these, only one completes maturation and releases its oocyte. The rest are destined to degenerate. Once the selected follicle has matured beyond a certain point, some feedback mechanism must operate to ensure that no other follicles complete maturation and ovulate during that cycle. Thus, during the 40 or so years of a woman's reproductive life, only 400 to 500 oocytes will be released. All the other 3 million primary oocytes present at birth degenerate. It is still a mystery why so many oocytes are formed only to die in the ovaries.

Toward the end of reproductive life, the secondary oocytes released from the human ovary at ovulation derive from primary oocytes that have been arrested in prophase I for up to 40 or 50 years. Damage to the oocytes during this long period is one possible explanation for the high incidence of certain genetic abnormalities among children born to older women. For example, 1% of children conceived by women over 40 years of age have *Down's syndrome*, a condition caused by an extra copy of chromosome 21 resulting from nondisjunction during meiotic division I or II (Figure 15–33).

After ovulation the emptied follicle transforms into an endocrine structure, the *corpus luteum*, which secretes progesterone to prepare the uterus for the fertilized oocyte. If fertilization does not occur, the corpus luteum regresses and the lining of the uterus is sloughed off during menstruation.

Sperm Are Highly Adapted for Delivering Their DNA to an Egg[15]

In most species there are just two types of gametes, and they are radically different. The egg is among the largest cells in an organism; the **sperm** (**spermatozoon**, plural **spermatozoa**) is often the smallest. The egg and the sperm are optimized in opposite ways for the propagation of the genes they carry. The egg aids the survival of the maternal genes by providing large stocks of raw materials for growth and development, as well as an effective protective wrapping. The sperm, by contrast, is usually optimized to propagate the paternal genes by being highly motile and streamlined for speed and efficiency in the task of fertilization. Competition between sperm is fierce, and the vast majority fail in their mission: of the billions of sperm released during the reproductive life of a human male, only a few ever manage to fertilize an egg.

Typical sperm (Figure 15–34) are "stripped-down" cells, equipped with a strong flagellum to propel them through an aqueous medium but unencumbered by cytoplasmic organelles such as ribosomes, endoplasmic reticulum, or Golgi apparatus, which are unnecessary for the task of delivering the DNA to the egg. On the other hand, sperm contain many mitochondria strategically placed where they can most efficiently power the flagellum. Sperm usually consist of two morpho-

Figure 15–33 The increase in the incidence of trisomy 21 (Down's syndrome) in newborn humans with increasing age of the mother. (Adapted from M.S. Golbus, *in* Fertilization and Embryonic Development *in Vitro*, edited by L. Mastroianni, Jr., and J.D. Biggers, p. 260. Plenum, 1981.)

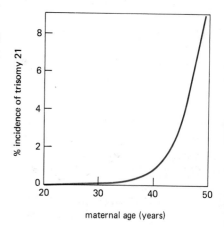

% incidence of trisomy 21

maternal age (years)

logically and functionally distinct regions enclosed by a single plasma membrane: the **head,** which contains an unusually highly condensed haploid nucleus, and the **tail,** which propels the sperm to the egg and helps it burrow through the egg coat. The DNA in the nucleus is inactive and extremely tightly packed, so that its volume is minimized for transport. The chromosomes of many sperm have dispensed with the histones of somatic cells and are packed instead with simple, highly positively charged proteins.

In the sperm head, closely apposed to the anterior end of the nuclear envelope, is a specialized secretory vesicle called the **acrosomal vesicle** (see Figure 15–34). This vesicle contains hydrolytic enzymes that help the sperm to penetrate the egg's outer coat. When a sperm contacts an egg, the contents of the vesicle are released by exocytosis in the so-called *acrosomal reaction*. In many invertebrate sperm this reaction also releases specific proteins that bind the sperm tightly to the egg coat (see p. 869).

The motile tail of a sperm is a long flagellum whose central axoneme emanates from a basal body situated just posterior to the nucleus. As described earlier (see p. 647), the axoneme consists of two central singlet microtubules surrounded by nine evenly spaced microtubule doublets. The flagellum of some sperm (including those of mammals) differs from other flagella in that the usual 9 + 2 pattern of the axoneme is further surrounded by nine *outer dense fibers* of unknown composition (Figures 15–35 and 15–36). The dense fibers are stiff and noncontractile, and it is not known what part they play in the active bending of the flagellum, which is caused by the sliding of the adjacent microtubule doublets past one another (see p. 648). Flagellar movement is powered by the hydrolysis of ATP generated by highly specialized mitochondria in the anterior part of the sperm tail (called the *midpiece*), where the ATP is needed (see Figures 15–34 and 15–35).

Sperm Are Produced Continuously in Many Mammals[16]

In mammals there are major differences in the way eggs are produced (oogenesis) and the way sperm are produced (**spermatogenesis**). In human females, for example, we have described how, early in embryogenesis, oogonia proliferate to generate a limited number of oocytes that ovulate at intervals, one at a time, beginning at puberty. In human males, on the other hand, spermatogenesis does not begin until puberty and then goes on continuously in the epithelial lining of very long, tightly coiled tubes, called *seminiferous tubules*, in the testes. Immature germ cells, called **spermatogonia,** are located along the outer edge of these tubes next to the basal lamina, where they divide continuously by mitosis. Some of the daughter cells stop proliferating and differentiate into **primary spermatocytes.** These cells enter the first meiotic prophase, in which their paired homologous chromosomes participate in crossing-over, and then proceed with division I of meiosis to produce two **secondary spermatocytes,** each containing 22 duplicated autosomal chromosomes and either a duplicated X or a duplicated Y chromosome. Each chromosome still consists of two sister chromatids until the two secondary spermatocytes undergo meiotic division II to produce four **spermatids,** each with a haploid number of single chromosomes. These haploid spermatids then undergo morphological differentiation into spermatozoa, which escape into the lumen of the seminiferous tubule (Figures 15–37 and 15–38). They subsequently pass into the *epididymis,* a coiled tube overlying the testis, where they are stored and undergo further maturation.

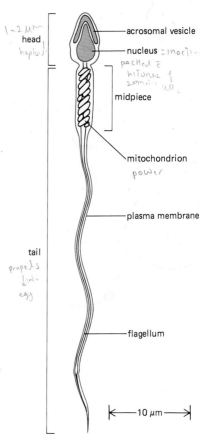

acrosomal vesicle

nucleus

midpiece

mitochondrion

plasma membrane

flagellum

tail

head

10 μm

Figure 15–34 A human sperm in longitudinal section.

Figure 15–35 Drawing of the midpiece of a mammalian sperm as seen in cross-section in an electron microscope. The core of the flagellum is composed of an axoneme surrounded by nine dense fibers. The axoneme consists of two singlet microtubules surrounded by nine microtubule doublets. The mitochondrion (shown in color) is well placed for providing the ATP required for flagellar movement; its unusual spiral structure results from the fusion of individual mitochondria during spermatid differentiation.

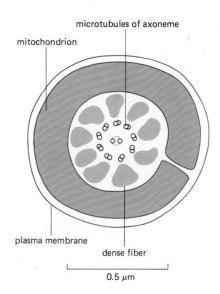

microtubules of axoneme

mitochondrion

plasma membrane

dense fiber

0.5 μm

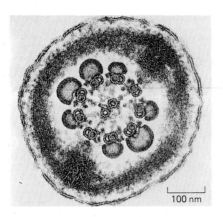

Figure 15–36 Electron micrograph of the flagellum of a guinea pig sperm seen in cross-section. Because this section is from a region of the sperm tail beyond the midpiece, no mitochondria are present. In the region shown, two of the nine outer dense fibers have terminated and have become continuous with an outer fibrous sheath. (Courtesy of Daniel S. Friend.)

Figure 15–37 The stages of spermatogenesis. Spermatogonia develop from primordial germ cells that migrate into the testis early in embryogenesis. When the animal becomes sexually mature, the spermatogonia begin to proliferate rapidly, generating some progeny that retain the capacity to continue dividing indefinitely (as stem-cell spermatogonia) and other progeny (maturing spermatogonia) that will, after a limited number of further normal division cycles, embark on meiosis to become primary spermatocytes. These continue through meiotic division I to become secondary spermatocytes. After they complete meiotic division II, the secondary spermatocytes produce haploid spermatids that differentiate into mature sperm. Spermatogenesis differs from oogenesis (Figure 15–25) in several ways: (1) new cells enter meiosis continually from the time of puberty, (2) each cell that begins meiosis gives rise to four mature gametes rather than one, and (3) mature sperm form by an elaborate process of cell differentiation that begins after meiosis is complete.

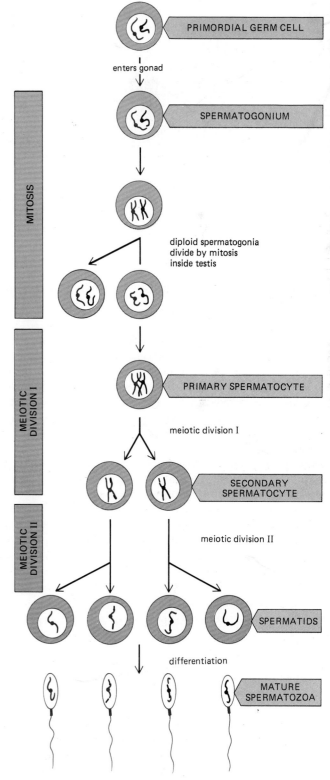

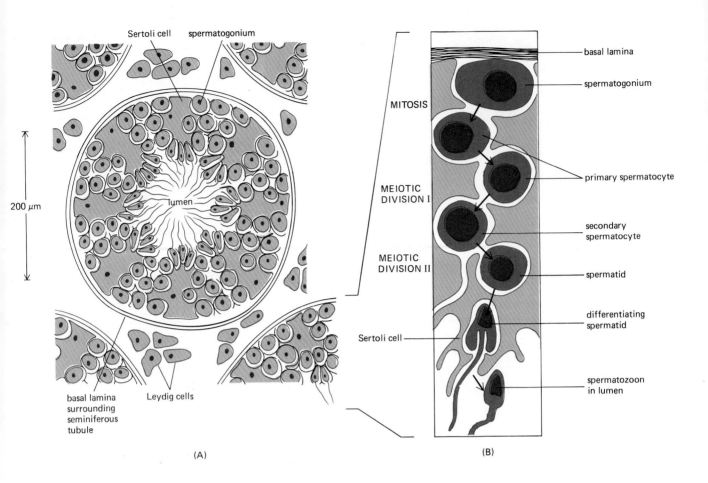

Labels in figure (A):
Sertoli cell   spermatogonium

200 μm

lumen

basal lamina
surrounding
seminiferous
tubule

Leydig cells

(A)

Labels in figure (B):
MITOSIS

MEIOTIC
DIVISION I

MEIOTIC
DIVISION II

Sertoli cell

basal lamina

spermatogonium

primary spermatocyte

secondary
spermatocyte

spermatid

differentiating
spermatid

spermatozoon
in lumen

(B)

As in the case of oogenesis, spermatogenesis is under hormonal control. Beginning at puberty, the pituitary gland in human males secretes the same gonadotropin that we encountered earlier in discussing mammalian oogenesis—luteinizing hormone (LH). LH stimulates *Leydig cells* located between the seminiferous tubules in the testis to secrete large amounts of the male sex hormone, *testosterone;* testosterone in turn stimulates spermatogenesis, probably through its action on *Sertoli cells,* which completely envelop the developing sperm, protecting and nourishing them (see Figure 15–38).

An intriguing feature of spermatogenesis is that the developing male germ cells fail to complete cytoplasmic division (cytokinesis) during mitosis and meiosis, so that all the differentiating daughter cells descended from one maturing spermatogonium remain connected by cytoplasmic bridges (Figure 15–39). These cytoplasmic bridges persist until the very end of sperm differentiation, when individual sperm are released into the tubule lumen. A group of cells joined in this way is known as a *syncytium.* This accounts for the observation that mature sperm arise synchronously in any given area of a seminiferous tubule. But what is the function of the syncytial arrangement?

Sperm Nuclei Are Haploid, but Sperm Cell Differentiation Is Directed by the Diploid Genome[17]

Unlike oocytes, sperm undergo most of their differentiation after their nuclei have completed meiosis to become haploid. In principle, the cytoplasmic bridges between them could allow each developing haploid sperm, by sharing a common cytoplasm with its neighbors, to be supplied with all the products of a complete diploid genome. There are two reasons why it is important that the diploid genome direct sperm differentiation, just as it directs egg differentiation. First, the diploid genome from which the sperm derives generally will include some defective gene copies, corresponding to recessive lethal mutations (see p. 842); a haploid cell receiving one of these defective gene copies is likely to die unless it is provided

Figure 15–38   Highly simplified drawing of a cross-section of a seminiferous tubule in a mammalian testis. (A) All of the stages of spermatogenesis shown take place while the developing gametes are in intimate association with *Sertoli cells,* which are large cells that extend from the basal lamina to the lumen of the seminiferous tubule. Spermatogenesis depends on testosterone secreted by *Leydig cells,* located between the seminiferous tubules. (B) Dividing spermatogonia are found along the basal lamina. Some of these cells stop dividing and enter meiosis to become primary spermatocytes. Eventually spermatozoa are released into the lumen. In man it takes about 24 days for a spermatocyte to complete meiosis to become a spermatid and another 5 weeks for a spermatid to develop into a spermatozoon. Spermatozoa undergo further maturation and become motile in the epididymis and are only then fully mature sperm.

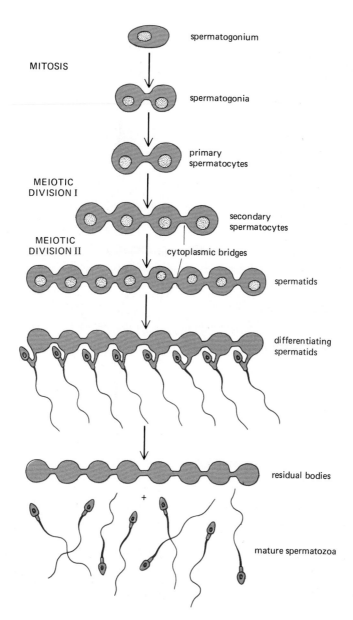

MITOSIS

spermatogonium

spermatogonia

primary
spermatocytes

MEIOTIC
DIVISION I

secondary
spermatocytes

MEIOTIC
DIVISION II

cytoplasmic bridges

spermatids

differentiating
spermatids

residual bodies

+

mature spermatozoa

Figure 15–39 Illustration showing how the progeny of a single maturing spermatogonium remain connected to one another by cytoplasmic bridges throughout their differentiation into mature sperm. For the sake of simplicity, only two connected maturing spermatogonia are shown entering meiosis to eventually form eight connected haploid spermatids. In fact, the number of connected cells that go through two meiotic divisions and differentiate together is very much larger than shown here.

with the functional gene products encoded by other nuclei that have the good gene copy. Second, in organisms such as humans, some sperm inherit an X chromosome at meiosis, while others inherit a Y chromosome. Since the X chromosome carries many essential genes that are lacking on the Y chromosome, it is unlikely that the Y-bearing sperm would be able to survive and mature without products of the X chromosome.

There is direct experimental evidence that sperm differentiation is governed by products of the diploid genome. In *disjunction-defective mutants* of *Drosophila*, for example, chromosomes are divided unequally between daughter cells during meiosis. As a result, some sperm contain too few chromosomes, some too many, and some contain none at all. Yet the major features of sperm differentiation occur in all such cells, even in those without chromosomes. An obvious explanation is that the products of missing chromosomes are supplied by diffusion through the cytoplasmic bridges that connect adjacent germ cells. However, it is also possible that the diploid spermatogonia or primary spermatocytes produce stable instructions for sperm differentiation (presumably in the form of long-lived mRNA) before meiosis, so that there is no need for the haploid sperm genome to function during the period of differentiation. In either case it is evident that the differentiation of a sperm, even though it occurs when the sperm nucleus is haploid, uses products specified by both sets of parental chromosomes.

Summary

Eggs develop in stages from primordial germ cells that migrate into the ovary very early in development to become oogonia. After mitotic proliferation, oogonia become primary oocytes that begin meiotic division I and then arrest at prophase for days or years, depending on the species. During this prophase-I arrest period, primary oocytes grow and accumulate ribosomes, mRNAs, and proteins, often enlisting the help of other cells, including surrounding accessory cells. Further development (oocyte maturation) depends on polypeptide hormones (gonadotropins) that act on the surrounding accessory cells, causing them to induce a proportion of the primary oocytes to mature. These oocytes complete meiotic division I to form a small polar body and a large secondary oocyte and proceed into metaphase of meiotic division II, where, in many species, the oocyte is arrested until stimulated by fertilization to complete meiosis and begin embryonic development.

A sperm is usually a small, compact cell, highly specialized for the task of delivering its DNA to the egg. Whereas in many female organisms the total pool of oocytes is produced early in embryogenesis, in males new germ cells enter meiosis continually from the time of sexual maturation, each primary spermatocyte giving rise to four mature sperm. Sperm differentiation occurs after meiosis, when the nuclei are haploid. However, because the maturing spermatogonia and spermatocytes fail to complete cytokinesis, the progeny of a single spermatogonium develop as a large syncytium. This may be why sperm differentiation is directed by the products of both parental chromosomes.

Fertilization[18]

Once released, egg and sperm alike are destined to die within minutes or hours unless they find each other and fuse in the process of **fertilization.** Through fertilization the egg and sperm are saved: the egg is activated to begin its developmental program, and the nuclei of the two gametes fuse to form the genome of a new organism. Much of what we know about the mechanism of fertilization has been learned from studies of marine invertebrates, especially sea urchins (Figure 15–40). In these organisms, fertilization occurs in sea water, into which huge numbers of both sperm and eggs are released. Such *external fertilization* is much more accessible to study than the *internal fertilization* of mammals, which occurs in the female reproductive tract following mating. For this reason most of our discussion of fertilization will deal with sea urchins. However, despite the great evolutionary distance between sea urchins and mammals, many of the cellular and molecular mechanisms involved in fertilization are similar in both.

Contact with the Jelly Coat of an Egg Stimulates a Sea Urchin Sperm to Undergo an Acrosomal Reaction[19]

A typical female sea urchin has 10^7 eggs and a typical male 10^{12} sperm, so sea urchin gametes can be obtained as pure populations in very large numbers, all at the same stage of development. When the two types of gametes are mixed together, the events of sperm-egg interaction begin synchronously within seconds. Fertilization begins when the head of a sperm contacts the jelly coat (see Figure 15–24) of an egg. The contact triggers the sperm to undergo an **acrosomal reaction,** in which the contents of the acrosomal vesicle are released to the exterior surroundings. In sea urchins and many other marine invertebrates, the discharge of the contents of the acrosomal vesicle is accompanied by the formation of a long, actin-containing **acrosomal process,** which projects from the anterior end of the sperm. As shown in Figure 15–41, the tip of this process becomes covered by the components of the old acrosomal vesicle membrane. It also becomes coated with the secreted contents of the acrosomal vesicle, including (1) hydrolytic enzymes that help the sperm penetrate the jelly coat and thus gain access to the vitelline layer, (2) specific binding proteins that mediate the attachment of the acrosomal process to the vitelline layer (see below), and (3) hydrolytic enzymes that then permit the acrosomal process to bore through this layer to the egg's plasma membrane. When

Figure 15–40 Photograph of two species of sea urchin that are commonly used in studies of fertilization. The one on top is *Strongylocentrotus purpuratus*, and the one below is *Strongylocentrotus franciscanus* (shown at approximate actual size). (Courtesy of Victor Vacquier.)

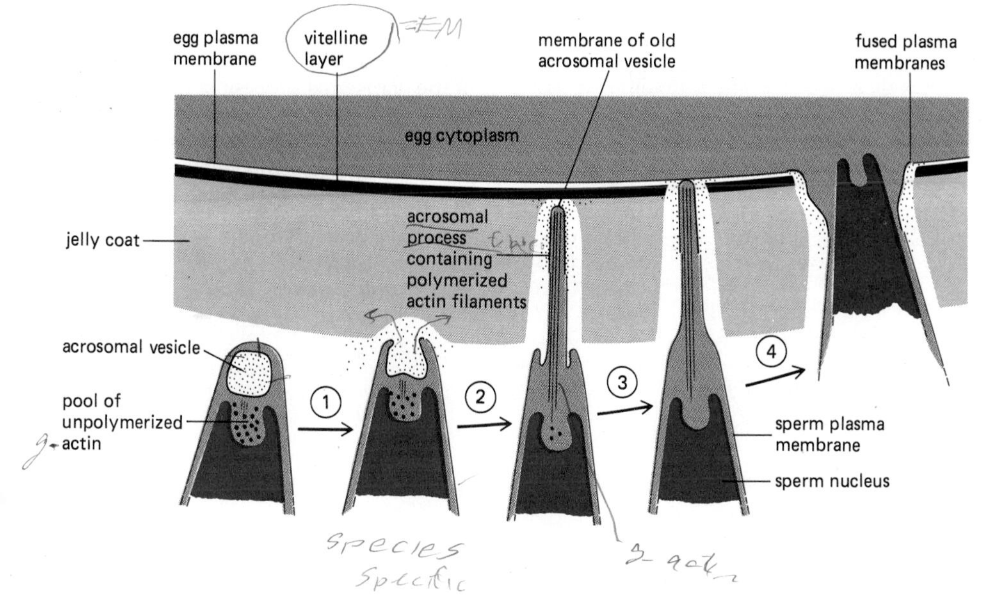

Figure 15–41 Details of the acrosomal reaction in sea urchins. When the sperm contacts the jelly coat, exocytosis of the acrosomal vesicle occurs (1), followed by the explosive polymerization of actin to form the long acrosomal process that penetrates the jelly coat (2). Proteins released from the acrosomal vesicle (*small black dots*) adhere to the surface of the acrosomal process and serve both to bind the sperm to the vitelline layer and to digest this layer (3). When the old acrosomal vesicle membrane (which forms the tip of the acrosomal process) contacts the egg plasma membrane (3), the two membranes fuse, the actin filaments disassemble, and the sperm enters the egg (4).

How do the sperm find an egg to fertilize after the gametes have been released into the seawater? Sea urchin eggs secrete a peptide called *resact,* which acts as a species-specific chemoattractant for sea urchin sperm. Resact binds to a transmembrane receptor on the sperm surface that has been shown to be an enzyme, guanylate cyclase, which catalyzes the production of cyclic GMP inside the sperm.

the membrane at the tip of the acrosomal process contacts the egg plasma membrane, the two membranes fuse, allowing the sperm nucleus to enter the egg (see Figure 15–41).

The trigger for the acrosomal reaction for sea urchin sperm is a large, fucose-sulfate-rich polysaccharide of the egg's jelly coat: when this polysaccharide is extracted from a sea urchin egg and added to sperm, it induces a normal acrosomal reaction within seconds. The jelly coat polysaccharide binds to a receptor glycoprotein in the sperm plasma membrane, causing the membrane to depolarize; the depolarization is thought to open voltage-gated Ca^{2+} channels in the membrane, allowing Ca^{2+} to enter the sperm. The jelly coat polysaccharide also activates a proton pump in the sperm plasma membrane that pumps H^+ out of the cell in exchange for Na^+. The resulting rise in pH inside the sperm head, together with the increase in cytosolic Ca^{2+}, initiates the acrosomal reaction. The rise in intracellular pH is thought to act, at least in part, by causing unpolymerized actin to dissociate from actin-binding proteins in the sperm cytoplasm that otherwise prevent actin polymerization (see p. 641); this initiates the explosive polymerization of actin, resulting in the formation of the acrosomal process.

Actin polymerization, however, is not the only mechanism driving the elongation of the acrosomal process. A net influx of ions (Ca^{2+}, Na^+, and Cl^-) increases the number of osmotically active molecules within the sperm head, thereby causing an influx of water. The sudden increase in hydrostatic pressure that results is thought to help extend the acrosomal process.

Sperm-Egg Adhesion Is Mediated by Species-specific Macromolecules[20]

The species-specificity of fertilization is especially important for aquatic animals that discharge their eggs and sperm into water, where they are liable to become mixed with eggs and sperm of other species. In sea urchins this specificity resides in the binding of the sperm to the vitelline layer beneath the jelly coat: sea urchin sperm will sometimes undergo an acrosomal reaction in response to eggs of other species, but they cannot bind to such eggs and therefore cannot fertilize them.

The molecule in sea urchin sperm that is thought to be responsible for the species-specific adhesion of the sperm to the egg's vitelline layer has been isolated. It is a protein called *bindin,* which is normally sequestered in the acrosomal vesicle. After its release in the acrosomal reaction, it coats the surface of the acrosomal process and mediates the attachment of the sperm to the egg. Each species of sea urchin makes a different type of bindin, which binds only to the vitelline layer of sea urchin eggs of the same species. The vitelline layer of one species of sea urchin egg has been found to contain a proteoglycan that acts as a *bindin receptor* in the adhesion process, and there is evidence that bindin acts as a lectin that rec-

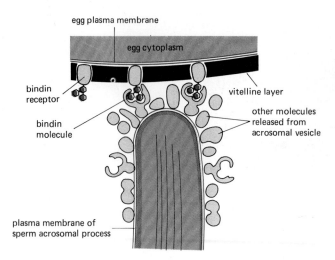

egg plasma membrane

egg cytoplasm

bindin receptor

bindin molecule

vitelline layer

other molecules released from acrosomal vesicle

plasma membrane of sperm acrosomal process

Figure 15–42 Schematic drawing of bindin molecules covering the surface of the acrosomal process of a sea urchin sperm. These proteins are thought to bind to a specific sugar sequence on a receptor molecule associated with the vitelline layer of the egg.

ognizes specific carbohydrate determinants on these proteoglycan molecules (Figure 15–42). Because bindin can induce the fusion of artificial lipid vesicles *in vitro*, it has been proposed that it may catalyze the fusion of the plasma membranes of the acrosomal process and the egg after bringing the two membranes together (see p. 336).

Egg Activation Involves Changes in Intracellular Ion Concentrations[21]

Once an activated sea urchin sperm attaches to an egg, the acrosomal process rapidly bores through the vitelline layer, and the membrane at the tip of the process fuses with the egg plasma membrane at the tip of a microvillus (Figure 15–43). Neighboring microvilli rapidly elongate and cluster around the sperm, which is then drawn head-first into the egg as the microvilli are resorbed.

The sperm activates the developmental program of the egg. Before fertilization an egg is metabolically dormant: it does not synthesize DNA, and it synthesizes RNA and protein at very low rates. Once released from the supportive environment of the ovary, an egg will die within hours unless rescued by fusion with a sperm. Metabolic activation by fusion with a sperm launches the dormant egg on a path leading to DNA synthesis and cleavage. The sperm, however, serves only to trigger a program that is already present in the egg. The sperm itself is not required. An egg can be activated artificially by a variety of nonspecific chemical or physical treatments; for example, a frog egg can be activated by pricking it with a needle. (The development of an egg that has been activated in the absence of a sperm is

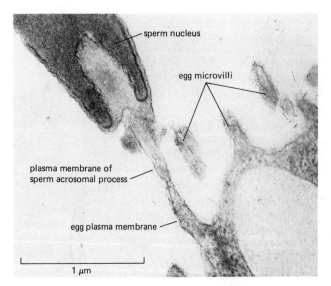

sperm nucleus

egg microvilli

plasma membrane of sperm acrosomal process

egg plasma membrane

1 μm

Figure 15–43 Electron micrograph showing a sea urchin sperm in the process of fertilizing an egg. The membrane at the tip of the acrosomal process of the sperm has fused with the egg plasma membrane at the tip of a microvillus on the egg surface. An unfertilized sea urchin egg is covered with more than 100,000 microvilli. (Courtesy of Frank Collins.)

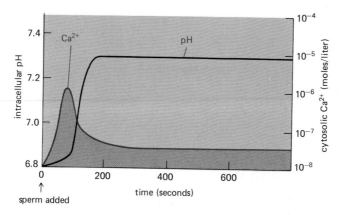

Figure 15-44 Two ionic changes involved in the activation of a sea urchin egg following fertilization. Beginning within 10 seconds after fertilization, Ca^{2+} is released into the cytosol from an intracellular Ca^{2+}-sequestering compartment, increasing the cytosolic concentration of free Ca^{2+} for about 2.5 minutes; afterwards the resting level of Ca^{2+} remains slightly higher than in the unfertilized egg. At about 60 seconds, a sustained efflux of H^+ coupled to an influx of Na^+ causes a permanent increase in intracellular pH.

called **parthenogenesis;** some organisms, including a few vertebrates, normally reproduce parthenogenically.) Moreover, the initial stages of egg activation cannot depend on the generation of any new proteins because they occur perfectly normally in the presence of drugs that inhibit protein synthesis.

In sea urchins many of the early steps in egg activation have been examined in detail and shown to be mediated by changes in ion concentrations within the egg. Three ionic changes occur within seconds or minutes of the addition of sperm to a suspension of eggs: (1) an increase in the permeability of the plasma membrane to Na^+ causes the membrane to depolarize within a few seconds; (2) a massive release of Ca^{2+} from an intracellular calcium-sequestering compartment (see p. 669) causes a marked increase in the concentration of Ca^{2+} in the cytosol within about 10 seconds; and (3) an efflux of H^+ coupled to an influx of Na^+ begins within 60 seconds and causes a large increase in intracellular pH (Figure 15-44). As we shall now describe, these three ionic changes have two consequences: first, they cause the egg to become impenetrable to further sperm; and second, they help trigger the initial steps in the developmental program of the egg.

The Rapid Depolarization of the Egg Plasma Membrane Prevents Further Sperm-Egg Fusions, Thereby Mediating the Fast Block to Polyspermy[22]

Although many sperm can attach to an egg, normally only one fuses with the egg plasma membrane and injects its nucleus into the cell. If more than one sperm fuses (a condition referred to as *polyspermy*), extra mitotic spindles are formed, resulting in the abnormal segregation of chromosomes during cleavage; non-diploid cells are produced, and development quickly stops. This means that eggs that are normally fertilized by the deposition of large numbers of sperm in their vicinity must somehow block the entry of extra sperm very soon after fertilization. Thus the eggs of many marine animals exhibit a **fast block to polyspermy,** brought about by different mechanisms, depending on the species.

Fish eggs have a small channel, called the *micropyle*, through which sperm must pass in single file. The passage of a single sperm through the channel stimulates the egg, causing the cortical granules to release their contents, which plug the hole so that no other sperm can enter. In most other organisms that reproduce using external fertilization, however, the eggs do not have a micropyle and can fuse with sperm over all or much of their surface. In some of these eggs (such as those of sea urchins and amphibians), the rapid depolarization of the plasma membrane caused by the fusion of the first sperm prevents further sperm from fusing. The membrane potential of a sea urchin egg is about -60 mV. Within a few seconds after sperm have been added, the membrane potential shifts precipitously to about $+20$ mV, where it remains for a minute or so before gradually returning to the original prefertilization level (Figure 15-45). If depolarization is prevented by fertilizing eggs in a solution containing an abnormally low concentration of Na^+, which reduces the sperm-triggered Na^+ influx that is largely responsible for depolarizing the membrane, there is an increased incidence of

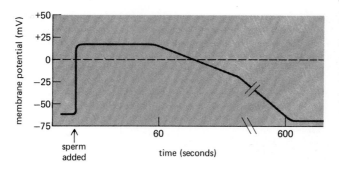

Figure 15–45 Changes in the sea urchin egg membrane potential after fertilization. The rapid depolarization somehow prevents further sperm from fusing with the egg plasma membrane, thereby creating the fast block to polyspermy.

polyspermy. Moreover, if an unfertilized egg is depolarized artificially by a current passed into it through a microelectrode, sperm can attach to the egg but cannot fuse; if the membrane is now repolarized with the microelectrode, the attached sperm fuse with and enter the egg. Although the molecular mechanism is unknown, it seems likely that the membrane depolarization that normally accompanies fertilization alters the conformation of a crucial protein in the egg plasma membrane so that a sperm membrane can no longer fuse with the egg membrane.

The egg membrane potential returns to normal within a few minutes after fertilization; therefore, a second mechanism must provide a longer-term barrier to polyspermy. In most eggs, including those of mammals, this barrier is provided by substances released from the cortical granules located just under the plasma membrane of the egg.

The Cortical Reaction Is Responsible for the Slow Block to Polyspermy[23]

The cortical granules in sea urchin eggs fuse with the plasma membrane and release their contents within 10 to 50 seconds of adding sperm. This **cortical reaction** is mediated by a large rise in the concentration of free Ca^{2+} in the cytosol. In an activated sea urchin egg, the Ca^{2+} concentration increases by about a hundredfold within less than a minute after the addition of sperm and then after a minute or so drops back toward normal (see Figure 15–44). The importance of Ca^{2+} in triggering the cortical reaction can be demonstrated directly in plasma membranes isolated from sea urchin eggs with cortical granules still attached to their cytoplasmic surfaces (Figure 15–46). When small amounts of Ca^{2+} are added to such preparations, exocytosis occurs within seconds.

In sea urchin eggs the cortical reaction has at least two separate effects: (1) proteolytic enzymes released from the cortical granules rapidly destroy the bindin receptors on the vitelline layer that are responsible for sperm attachment, and (2) components released from the cortical granules cause the vitelline layer to move away from the egg plasma membrane; at the same time, they enzymatically cross-link proteins in the vitelline layer, causing it to harden. In this way a *fertilization membrane* is formed that sperm cannot bind to or penetrate (Figure 15–47).

Egg Activation Is Mediated by the Inositol Phospholipid Cell-signaling Pathway[24]

Although membrane depolarization is the first detectable change following fertilization, it appears to serve only to prevent polyspermy. Artificially depolarizing the egg membrane does not activate the egg to begin biosynthesis; nor does blocking membrane depolarization at the time of fertilization inhibit this activation.

There is strong evidence that the transient increase in cytosolic Ca^{2+} concentration (which propagates as a wave across the egg from the site of sperm fusion—see Figure 4–35, p. 157) helps to initiate the program of egg development. If the cytosolic concentration of Ca^{2+} is increased artificially—either directly, by an injection of Ca^{2+}, or indirectly, by the use of Ca^{2+}-carrying ionophores, such as A23187 (see p. 322)—the eggs of all animals so far tested, including mammals, are activated. Moreover, preventing the increase in Ca^{2+} by injecting the Ca^{2+} chelator EGTA inhibits egg activation after fertilization. At least one way in which Ca^{2+} acts

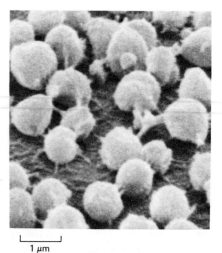

1 μm

Figure 15–46 Scanning electron micrograph of cortical granules attached to the cytoplasmic surface of the isolated plasma membrane of an unfertilized sea urchin egg. When Ca^{2+} is added to this preparation, the cortical granules fuse with the plasma membrane and release their contents by exocytosis. Since there are about 15,000 cortical granules in each cell, the cortical reaction causes the surface area of the egg to more than double in less than a minute; some of the extra membrane is accommodated by a lengthening of each microvillus on the egg surface, while the rest is endocytosed via coated pits and vesicles (see p. 328). (From V.D. Vacquier, *Dev. Biol.* 43:62–74, 1975.)

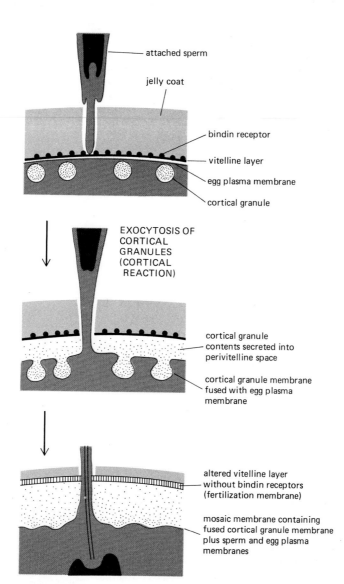

attached sperm

jelly coat

bindin receptor

vitelline layer

egg plasma membrane

cortical granule

EXOCYTOSIS OF
CORTICAL
GRANULES
(CORTICAL
REACTION)

cortical granule
contents secreted into
perivitelline space

cortical granule membrane
fused with egg plasma
membrane

altered vitelline layer
without bindin receptors
(fertilization membrane)

mosaic membrane containing
fused cortical granule membrane
plus sperm and egg plasma
membranes

Figure 15–47 The cortical reaction in a sea urchin egg prevents additional sperm from entering the egg. The released contents of the cortical granules raise the vitelline layer and alter it so that it no longer contains bindin receptors and is converted into a *fertilization membrane* that sperm cannot penetrate. This "hardening" of the vitelline layer is due mainly to the formation of covalent cross-links between protein tyrosine residues, which generates an extensive, insoluble protein network.

in cells is by binding to the Ca^{2+}-binding protein *calmodulin,* which in turn activates a variety of cellular proteins (see p. 711). Calmodulin has been found in large amounts in all eggs that have been studied.

How does fertilization lead to an increase in the concentration of Ca^{2+} in the cytosol of the egg? We discussed in Chapter 12 how extracellular ligands binding to cell-surface receptor proteins can lead to the hydrolysis of *phosphatidylinositol bisphosphate (PIP$_2$)* in the plasma membrane to produce *inositol trisphosphate (InsP$_3$)* and *diacylglycerol:* the InsP$_3$ in turn releases Ca^{2+} from an intracellular calcium-sequestering compartment (see p. 702) into the cytosol, while the diacylglycerol activates *protein kinase C* (see p. 703). Evidence obtained in sea urchin eggs suggests that fertilization increases the concentration of cytosolic Ca^{2+} via this pathway. The concentration of InsP$_3$ increases within seconds after fertilization, just before the concentration of Ca^{2+} increases in the cytosol, and if InsP$_3$ is injected into an unfertilized egg, it increases the concentration of Ca^{2+} in the cytosol and thereby activates the egg. As expected, sperm activation of this pathway appears to be mediated by a G protein that activates a specific phospholipase C to hydrolyze PIP$_2$ (see p. 702). It is not known, however, whether the sperm binds to a receptor in the egg plasma membrane that is functionally coupled to the phospholipase C via a G protein, or whether the sperm injects a G protein activator at the time of sperm-egg fusion.

Since the increase in Ca^{2+} concentration in the cytosol following fertilization is transient, lasting only a minute or so, it is clear that it cannot *directly* mediate the events observed during the later stages of egg activation, which in sea urchins include a gradual increase in protein synthesis beginning at 8 minutes and the initiation of DNA synthesis beginning at about 30 minutes. There is increasing evidence that the activation of protein kinase C plays an important part in these later events, mainly by increasing intracellular pH.

A Rise in the Intracellular pH in Some Organisms Induces the Late Synthetic Events of Egg Activation[25]

In sea urchins the activation of protein kinase C by diacylglycerol leads to the activation (presumably by phosphorylation) of a Na^+-H^+ exchanger in the egg plasma membrane. This membrane transport protein uses the energy stored in the Na^+ gradient across the membrane to pump H^+ out of the cell (see p. 309). The efflux of H^+ leads to an increase in the intracellular pH from 6.7 to 7.2, which is maintained throughout the rest of zygote development (see Figure 15–44). The unusually low intracellular pH in unfertilized sea urchin eggs is thought to be largely responsible for keeping the eggs metabolically inactive, and there is compelling evidence that the increase in pH following fertilization helps to induce the late synthetic events in these eggs: (1) When the intracellular pH is raised in unfertilized eggs by incubating them in a medium containing ammonia (Figure 15–48), protein synthesis is markedly increased and DNA replication is triggered, even in the absence of an increase in free intracellular Ca^{2+}. (2) When eggs are placed in Na^+-free sea water just after fertilization, so that there is no Na^+ gradient to drive the H^+ efflux, the intracellular pH does not rise and the late events do not occur. Such eggs can be rescued by adding ammonia to the medium: now the intracellular pH rises and the synthesis of protein and DNA is induced, even in the absence of extracellular Na^+.

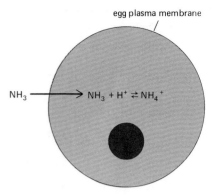

Figure 15–48 Increasing intracellular pH by incubating cells (such as eggs) in ammonia. The ammonia diffuses through the plasma membrane and combines with H^+ in the cytosol to form NH_4^+, thereby decreasing the intracellular concentration of H^+ and increasing the pH.

The marked increase in protein synthesis in fertilized sea urchin eggs does not require RNA synthesis, since it is unaffected by the drug actinomycin D, which inhibits RNA synthesis. It is thought to result from at least two separate changes: (1) preexisting mRNA molecules stored in the egg are made available for protein synthesis, and (2) the egg ribosomes are activated so that they translate the available mRNA molecules more rapidly. By contrast, the increase in protein synthesis in unfertilized eggs treated with ammonia results solely from increased recruitment of preexisting mRNA molecules. This suggests that, while the increase in intracellular pH is responsible for mRNA recruitment, some other factor normally increases the rate of ribosome movement along mRNAs. The detailed mechanisms involved in these two types of activation are unknown.

Although fertilization is a very special event, it relies on the same signaling pathways that regulate intracellular processes in somatic cells (see Chapter 12). The sequence of some of the events in sea urchin egg activation after fertilization is summarized in Table 15–1.

The Fusion of Sea Urchin Sperm and Egg Pronuclei Depends on Centrioles Donated by the Sperm[26]

Once fertilized, the egg is called a **zygote.** In most species, including some sea urchins, fertilization is not complete until the two haploid nuclei (called *pronuclei*) have fused. Because the egg is large, the sperm and egg pronuclei have to migrate substantial distances to find each other. Not surprisingly, this migration depends on the cytoskeleton.

The sea urchin sperm contributes more than DNA to the zygote: it donates two centrioles. The sperm centrioles are crucial because the egg loses its own centrioles during the last meiotic division. The sperm centrioles become the center of a radiating array of microtubules called the *sperm aster*, which seems to guide the male pronucleus toward the female pronucleus: if the microtubules are depolymerized by treatment with colchicine, the migration of the two pronuclei toward the center of the egg does not occur. Eventually the two pronuclei make contact and their membranes fuse to form the diploid nucleus of the zygote. The pair of sperm centrioles, with its associated aster, then divides to form the two poles of the mitotic spindle for the first cleavage division.

Table 15–1 Sequence of Events Following Fertilization of Sea Urchin Eggs

Event	Time After Fertilization	Mediator
1. Plasma membrane depolarization	<5 seconds	sperm-induced increase in plasma membrane permeability to Na^+ (and to some extent to Ca^{2+})
2. Hydrolysis of phosphatidylinositol bisphosphate	<10 seconds	activation of phospholipase C
3. Increased concentration of free cytosolic Ca^{2+}	10–40 seconds	$InsP_3$-induced release of Ca^{2+} from intracellular Ca^{2+}-sequestering compartment
4. Cortical granule exocytosis	10–50 seconds	increased intracellular Ca^{2+}
5. Increased intracellular pH	60 seconds	activation of Na^+-H^+ exchanger by protein kinase C
6. Increased protein synthesis	8 minutes	increased intracellular pH
7. Fusion of sperm and egg nuclei	30 minutes	
8. Initiation of DNA replication	30–45 minutes	increased intracellular pH

Mammalian Eggs Can Be Fertilized *in Vitro*[27]

Compared with sea urchin eggs, mammalian eggs are difficult to study. Whereas sea urchin eggs are readily available by the millions, investigators must be content to work with tens or hundreds of mammalian eggs. Nonetheless, it is now possible to fertilize mammalian eggs *in vitro*. (Although we shall continue to use the term egg, it should be recalled that in mammals it is a secondary oocyte that is fertilized—see p. 862.) This brings a medical benefit: mammalian eggs that have been fertilized *in vitro* can develop into normal individuals when transplanted into the uterus; in this way many previously infertile women have been able to produce normal children. *In vitro* fertilization of mammalian eggs also allows one to study the events of fertilization and activation. Such studies indicate that, whereas the sequence of events just described in sea urchin fertilization is followed in broad outline during mammalian fertilization, there are important differences in many of the individual steps.

Some of the differences are related to the sperm. Mammalian sperm are unable to fertilize an egg until they have undergone a process referred to as **capacitation,** induced by secretions in the female genital tract. The mechanism of capacitation is unclear; it seems to involve both an alteration in the lipid and glycoprotein composition of the sperm plasma membrane and an increase in sperm metabolism and motility. Capacitated mouse sperm penetrate the shell of follicle cells and bind specifically to a major glycoprotein in the *zona pellucida*—the protective coat equivalent to the vitelline layer of sea urchin eggs (see p. 856). At least in some species, this same egg glycoprotein is thought to activate sperm to undergo the acrosomal reaction. The zona pellucida of the mouse egg, for example, is composed of only three glycoproteins, which are secreted by growing oocytes and self-assemble into a three-dimensional network of interconnected filaments (see Figure 15–23). One of these, called ZP3, serves both to bind the sperm and then to induce the acrosomal reaction. The sperm is thought to recognize specific carbohydrate determinants on the ZP3 glycoprotein in binding to the zona. Unlike the case in sea urchins, the sperm molecules involved in the recognition process are located on the plasma membrane rather than on the acrosomal membrane. The zona pellucida, like the vitelline layer in sea urchin eggs, serves as a barrier to fertilization across species, and removing the zona often removes this barrier.

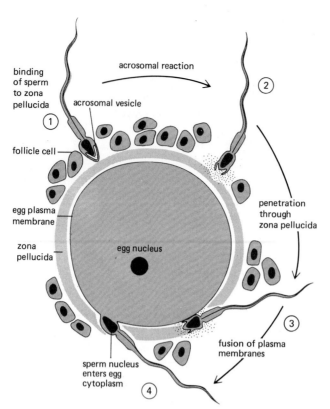

binding
of sperm
to zona
pellucida

acrosomal reaction

acrosomal vesicle

①

②

follicle cell

egg plasma
membrane

zona
pellucida

egg nucleus

penetration
through
zona pellucida

③

fusion of plasma
membranes

sperm nucleus
enters egg
cytoplasm

④

Figure 15–49 The acrosomal reaction that occurs when a mammalian sperm fertilizes an egg. In mice a single glycoprotein in the zona pellucida is thought to be responsible for both binding the sperm and inducing the acrosomal reaction. Note that a mammalian sperm interacts tangentially with the egg plasma membrane so that fusion occurs at the side rather than at the tip of the sperm head. In mice the zona pellucida is 7 μm in diameter and sperm cross it at a rate of about 1 μm/min.

For example, hamster eggs from which the zona pellucida has been removed with specific enzymes can be fertilized by human sperm. Not surprisingly, such hybrid "humsters" do not develop.

The acrosomal reaction of mammalian sperm releases proteases and hyaluronidase, which are essential for penetration of the zona pellucida. However, an acrosomal process similar to that formed by sea urchin sperm is not produced, and in most mammalian sperm it is the equatorial (postacrosomal) region of the plasma membrane that fuses with the egg, rather than the membrane of the acrosomal vesicle (Figure 15–49). Because relatively few sperm usually manage to reach an ovulated egg in mammals (fewer than 200 of the 3×10^8 human sperm released during coitus reach the site of fertilization), a fast block to polyspermy is apparently unnecessary: the rapid depolarization of the plasma membrane that accompanies fertilization and prevents polyspermy in sea urchin and amphibian eggs does not occur. On the other hand, the enzymes released by the cortical reaction of mammalian eggs change the structure of the egg coat to provide a slow block to polyspermy. They alter the ZP3 glycoprotein in the zona pellucida of the mouse egg, for example, so that it can no longer bind sperm or activate them to undergo an acrosomal reaction. In some other mammalian eggs, the cortical reaction alters the plasma membrane, rather than the zona, in a way that prevents further sperm from fusing with it.

Another distinctive feature of mammalian fertilization is that the egg contains centrioles while the sperm does not. Moreover, in fertilized mammalian eggs the two pronuclei do not fuse directly: they approach each other, but their chromosomes do not come together into one nucleus until after the membrane of each pronucleus breaks down in preparation for the first cleavage. Most of the other early events of fertilization in sea urchin eggs outlined in Table 15–1, such as the activation of the inositol phospholipid pathway and the resulting increase in cytosolic Ca^{2+} concentration, also occur in mammalian eggs. Subsequent events are part of the process of *embryogenesis*, in which the zygote develops into a new individual. It is perhaps the most remarkable phenomenon in all of biology and is the subject of the next chapter.

Summary

Fertilization begins when the head of a sperm makes contact with the protective coat surrounding the egg. This induces an acrosomal reaction in which the sperm releases the contents of its acrosomal vesicle, including proteins that help the sperm digest its way to the egg plasma membrane in order to fuse with it. Fertilization activates a cascade of changes in the egg that is initiated by the hydrolysis of phosphatidylinositol bisphosphate in the egg plasma membrane. Egg activation includes changes to the surface of the egg that prevent the fusion of additional sperm; one such block to polyspermy results from the cortical reaction in which cortical granules release their contents to the outside, altering the egg coat. Changes also take place in the interior of the egg in preparation for the subsequent development of the zygote after the sperm and egg pronuclei have come together.

References

General

Austin, C.R.; Short, R.V., eds. Reproduction in Mammals: I. Germ Cells and Fertilization, 2nd ed. Cambridge, U.K.: Cambridge University Press, 1982.

Browder, L. Developmental Biology, 2nd ed., Chapters 5, 6, and 8. Philadelphia: Saunders, 1980.

Epel, D. The program of fertilization. *Sci. Am.* 237(11):128–138, 1977.

Karp, G.; Berrill, N.J. Development, 2nd ed., Chapters 4 and 5. New York: McGraw-Hill, 1981.

Longo, F.J. Fertilization. London: Chapman & Hall, 1987.

Cited

1. Crow, J.F. The importance of recombination. In The Evolution of Sex: An Examination of Current Ideas (R.E. Michod, B.R. Levin, eds.), pp. 56–73. Sunderland, MA: Sinauer, 1988.

   Maynard Smith, J. Evolution of Sex. Cambridge, U.K.: Cambridge University Press, 1978.

   Williams, G.C. Sex and Evolution. Princeton, NJ: Princeton University Press, 1975.

2. Ayala, F.; Kiger, J. Modern Genetics, 2nd ed. Menlo Park, CA: Benjamin-Cummings, 1984.

   Ferris, S.D.; Whitt, G.S. Loss of duplicate gene expression after polyploidization. *Nature* 265:258–260, 1977.

   Fincham, J.R.S. Genetics. Boston: Jones and Bartlett, 1983.

3. Lewis, J.; Wolpert, L. Diploidy, evolution and sex. *J. Theor. Biol.* 78:425–438, 1979.

   Spofford, J.B. Heterosis and the evolution of duplications. *Am. Nat.* 103:407–432, 1969.

4. Evans, C.W.; Dickinson, H.G., eds. Controlling Events in Meiosis. *Symp. Soc. Exp. Biol.,* Vol. 38. Cambridge, U.K.: The Company of Biologists, 1984.

   Whitehouse, H.L. Towards an Understanding of the Mechanism of Heredity, 3rd ed. London: St. Martins. 1973. (Contains a lucid description of the development of our current understanding of chromosome behavior during meiosis.)

   Wolfe, S.L. Biology of the Cell, 2nd ed., pp. 432–470. Belmont, CA: Wadsworth, 1981.

5. John, B.; Lewis, K.R. The Meiotic Mechanism. Oxford Biology Readers (J.J. Head, ed.). Oxford, Eng.: Oxford University Press, 1976.

   Jones, G.H. The control of chiasma distribution. In Controlling Events in Meiosis (C.W. Evans; H.G. Dickinson, eds), *Symp. Soc. Exp. Biol.,* Vol. 38, pp. 293–320. Cambridge, U.K.: The Company of Biologists, 1984.

   Orr-Weaver, T.L.; Szostak, J.W. Fungal recombination. *Microbiol. Rev.* 49:33–58, 1985.

6. Heyting, C.; Dettmers, R.J.; Dietrich, A.J.; Redeker, E.J. Two major components of synaptonemal complexes are specific for meiotic prophase nuclei. *Chromosoma* 96:325–332, 1988.

   Moses, M.J. Synaptonemal complex. *Annu. Rev. Genet.* 2:363–412. 1968.

   Smithies, O.; Powers, P.A. Gene conversions and their relationship to homologous pairing. *Phil. Trans. R. Soc. Lond. (Biol.)* 312:291–302, 1986.

   von Wettstein, D.; Rasmussen, S.W.; Holm, P.B. The synaptonemal complex in genetic segregation. *Annu. Rev. Genet.* 18:331–413, 1984.

7. Carpenter, A.T.C. Gene conversion, recombination nodules, and the initiation of meiotic synapsis. *Bioessays* 6:232–236, 1987.

   Carpenter, A.T.C. Recombination nodules and synaptonemal complex in recombination-defective females of *Drosophila melanogaster. Chromosoma* 75:259–236, 1979.

8. Buckle, V.; Mondello, C.; Darling, S.; Craig, I.W.; Goodfellow, P.N. Homologous expressed genes in the human sex chromosome pairing region. *Nature* 317:739–741, 1985.

   Chandley, A. C. Meiosis in man. *Trends Genet.* 4:79–84, 1988.

   Solari, A.J. The behavior of the XY pair in mammals, *Int. Rev. Cytol.* 38:273–317, 1974.

9. Austin, C.R.; Short, R.V., eds. Reproduction in Mammals: I. Germ Cells and Fertilization. Cambridge, U.K.: Cambridge University Press, 1982.

10. Browder, L. Developmental Biology, pp. 173–231. Philadelphia: Saunders, 1980.

    Karp, G.; Berrill, N.J. Development, 2nd ed., pp. 116–138. New York: McGraw-Hill, 1981.

11. Browder, L.W., ed. Oogenesis. New York: Plenum, 1985.

    Davidson, E.H. Gene Activity in Early Development, 3rd ed., pp. 305–407. Orlando, FL: Academic, 1986.

    Metz, C.B.; Monroy, A., eds. Biology of Fertilization, Vol. 1: Model Systems and Oogenesis. Orlando, FL: Academic, 1985.

12. Bornslaeger, E.A.; Mattei, P.; Schultz, R.M. Involvement of cAMP-dependent protein kinase and protein phosphorylation in regulation of mouse oocyte maturation. *Dev. Biol.* 114:453–462, 1986.

    Maller, J.L. Regulation of amphibian oocyte maturation. *Cell Differ.* 16:211–221, 1985.

    Masui, Y.; Clarke, H.J. Oocyte maturation. *Int. Rev. Cytol.* 57:185–282, 1979.

Sadler, S.E.; Maller, J.L. Inhibition of *Xenopus* oocyte adenylate cyclase by progesterone: a novel mechanism of action. *Adv. Cyc. Nuc. Prot. Phosphor. Res.* 19:179–194, 1985.

13. Cyert, M.S.; Kirschner, M.W. Regulation of MPF activity *in vitro*. *Cell* 53:185–195, 1988.

Ford, C.C. Maturation promoting factor and cell cycle regulation. *J. Embryol. Exp. Morphol.* Suppl. 89:271–284, 1985.

Kirschner, M.; Newport, J.; Gerhart, J. The timing of early developmental events in *Xenopus*. *Trends Genet.* 1:41–47, 1985.

Lohka, M.J.; Hayes, M.K.; Maller, J.L. Purification of maturation-promoting factor, an intracellular regulator of early mitotic events. *Proc. Natl. Acad. Sci. USA* 85:3009–3013, 1988.

Maller, J.L. Regulation of amphibian oocyte maturation. *Cell Differ.* 16:211–221, 1985.

4. Peters, H.; McNatty, K.P. The Ovary: A Correlation of Structure and Function in Mammals, pp. 11–22, 60–84. Berkeley: University of California Press, 1980.

Richards, J.S. Hormonal control of ovarian follicular development. *Recent Prog. Horm. Res.* 35:343–373, 1979.

5. Bellvé, A.R.; O'Brien, D.A. The mammalian spermatozoon: structure and temporal assembly. In Mechanism and Control of Animal Fertilization (J.F. Hartmann, ed.), pp. 56–137. New York: Academic, 1983.

Fawcett, D.W. The mammalian spermatozoon. *Dev. Biol.* 44:394–436, 1975.

Fawcett, D.W.; Bedford, J.M., eds. The Spermatozoon. Baltimore: Urban & Schwarzenberg, 1979.

6. Browder, L. Developmental Biology, pp. 146–172. Philadelphia: Saunders, 1980.

Clermont, Y. Kinetics of spermatogenesis in mammals: seminiferous epithelium cycle and spermatogonial renewal. *Physiol. Rev.* 52:198–236, 1972.

Karp, G.; Berrill, N.J. Development, 2nd ed., pp. 100–116. New York: McGraw-Hill, 1981.

Metz, C.B.; Monroy, A., eds. Biology of Fertilization, Vol. 2: Biology of the Sperm. Orlando, FL: Academic, 1985.

7. Lindsley, D.L.; Tokuyasu, K.T. Spermatogenesis. In The Genetics and Biology of *Drosophila* (M. Ashburner, T.R.F. Wright, eds.), Vol. 2, pp. 225–294. New York: Academic, 1980.

Willison, K.; Ashworth, A. Mammalian spermatogenic gene expression. *Trends Genet.* 3:351–355, 1987.

8. Epel, D. Fertilization. *Endeavour* (New Series) 4:26–31, 1980.

Hendrick, J.L., ed. The Molecular and Cellular Biology of Fertilization. New York: Plenum, 1986.

Longo, F.J. Fertilization. London: Chapman & Hall, 1987.

Metz, C.B.; Monroy, A., eds. Biology of Fertilization, Vol. 3: The Fertilization Response of the Egg. Orlando, FL: Academic, 1985.

9. Shapiro, B.M. The existential decision of a sperm. *Cell* 49:293–294, 1987.

Shapiro, B.M.; Schackmann, R.W.; Tombes, R.M.; Kazazoglou, T. Coupled ionic and enzymatic regulation of sperm behavior. *Curr. Top. Cell Regul.* 26:97–113, 1985.

Tilney, L.G.; Inoué, S. Acrosomal reaction of the *Thyone* sperm. III. The relationship between actin assembly and water influx during the extension of the acrosomal process. *J. Cell Biol.* 100:1273–1283, 1985.

Trimmer, J.S.; Vacquier, V.D. Activation of sea urchin gametes. *Annu. Rev. Cell Biol.* 2:1–26, 1986.

20. Gao, B.; Klein, L.E.; Britten, R.J.; Davidson, E.H. Sequence of mRNA coding for bindin, a species-specific sea urchin sperm protein required for fertilization. *Proc. Natl. Acad. Sci. USA* 83:8634–8638, 1986.

Glabe, C.G. Interaction of the sperm adhesive protein bindin, with phospholipid vesicles. II. Bindin induces the fusion of mixed-phase vesicles that contain phosphatidylcholine and phosphatidylserine *in vitro*. *J. Cell Biol.* 100:800–806, 1985.

Rossignol, D.P.; Earles, B.J.; Decker, G.L.; Lennarz, W.J. Characterization of the sperm receptor on the surface of eggs of *Strongylocentrotus purpuratus*. *Dev. Biol.* 104:308–321, 1984.

Vacquier, V.D.; Moy, G.W. Isolation of bindin: the protein responsible for adhesion of sperm to sea urchin eggs. *Proc. Natl. Acad. Sci. USA* 74:2456–2460, 1977.

21. Shatten, G.; Hüsler, D. Timing the early events during sea urchin fertilization. *Dev. Biol.* 100:244–248, 1983.

Trimmer, J.S.; Vacquier, V.D. Activation of sea urchin gametes. *Annu. Rev. Cell Biol.* 2:1–26, 1986.

Whitaker, M.J.; Steinhardt, R.A. Ionic signalling in the sea urchin egg at fertilization. In Biology of Fertilization (C.B. Metz, A. Monroy, eds.), Vol. 3, pp. 168–222. Orlando, FL: Academic, 1985.

22. Jaffe, L.A.; Cross, N.L. Electrical regulation of sperm-egg fusion. *Annu. Rev. Physiol.* 48:191–200, 1986.

23. Kay, E.S.; Shapiro, B.M. The formation of the fertilization membrane of the sea urchin egg. In Biology of Fertilization (C.B. Metz, A. Monroy, eds.), Vol. 3, pp. 45–81. Orlando, FL: Academic, 1985.

Schuel, H. Functions of egg cortical granules. In Biology of Fertilization (C.B. Metz, A. Monroy, eds.), Vol. 3, pp. 1–44. Orlando, FL: Academic, 1985.

24. Eisen, A.; Reynolds, G.T. Source and sinks for the calcium released during fertilization of single sea urchin eggs. *J. Cell Biol.* 100:1522–1527, 1985.

Turner, P.R.; Jaffe, L.A.; Fein, A. Regulation of cortical vesicle exocytosis in sea urchin eggs in inositol 1,4,5-trisphosphate and GTP-binding protein. *J. Cell Biol.* 102:70–78, 1986.

Whitaker, M.; Irvine, R.F. Inositol 1,4,5-trisphosphate microinjection activates sea urchin eggs. *Nature* 312:636–639, 1984.

25. Dube, F.; Schmidt, T.; Johnson, C.H.; Epel, D. The hierarchy of requirements for an elevated intracellular pH during early development of sea urchin embryos. *Cell* 40:657–666, 1985.

Winkler, M. Translational regulation in sea urchin eggs: a complex interaction of biochemical and physiological regulatory mechanisms. *Bioessays* 8:157–161, 1988.

26. Schatten, H.; Schatten, G.; Mazia, D.; Balczon, R.; Simerly, C. Behavior of centrosomes during fertilization and cell division in mouse oocytes and in sea urchin eggs. *Proc. Natl. Acad. Sci. USA* 83:105–109, 1986.

27. Clegg, E.D. Mechanisms of mammalian sperm capacitation. In Mechanism and Control of Animal Fertilization (J.F. Hartmann, ed.), pp. 178–212. New York: Academic, 1983.

Grobstein, C. External human fertilization. *Sci. Am.* 240(6):57–67, 1979.

Wassarman, P.M. Early events in mammalian fertilization. *Annu. Rev. Cell Biol.* 3:109–142, 1987.

Wassarman, P.M. Zona pellucida glycoproteins. *Annu. Rev. Biochem.* 57:415–442, 1988.

Cellular Mechanisms
of Development

16

Almost every multicellular animal is a clone of cells descended from a single original cell, the fertilized egg. Thus the cells of the body, as a rule, are genetically alike. But phenotypically they are different: some are specialized as muscle, others as neurons, others as blood cells, and so on. The different cell types are arranged in a precisely organized pattern, and the whole structure has a well-defined shape. All these features are ultimately determined by the DNA sequence of the genome, which is reproduced in every cell. Each cell must act according to the same genetic instructions, but it must interpret them with due regard to time and circumstance, so as to play its proper part in the multicellular society.

Multicellular organisms are often very complex, but they are generated by a limited repertoire of cellular activities. Cells grow, divide, and die. They form mechanical attachments and exert forces for cell locomotion and deformation. They differentiate—that is, they switch on or off the production of specific proteins encoded by their genomes. They secrete, or display on their surface, molecules that influence the activities of their neighbors. These forms of cell behavior are the universal basis of animal development. In this chapter we shall present our current understanding of how such activities are brought into play at the right times and places to generate a whole organism.

Rather than follow any one organism in detail, we shall consider in turn the different aspects of cell behavior in development, illustrating the general principles by reference to the animals that display them best. We shall discuss how cell movements shape the embryo, how the genes inside cells and the interactions between cells control the development of the spatial pattern of differentiation, and how differentiated cells arising in different regions are assembled to form complex tissues and organs. To illustrate these topics we shall use amphibians, sea urchins, rodents, flies, birds, cockroaches, and a nematode worm. The development of plants will be discussed in Chapter 20.

Morphogenetic Movements and the Shaping
of the Body Plan[1]

In this section we consider how the geometrical structure of the early embryo is formed and what physical forces mold it. We shall take as our chief example the frog *Xenopus laevis* (Figure 16–1), whose early development has been particularly well studied. Like other amphibian embryos, it is robust and easy to manipulate experimentally.

To put the events that we are about to describe in context, it is helpful to regard the development of a vertebrate—and indeed of many other types of animal—as having three phases. In the first phase the fertilized egg *cleaves* to form many smaller cells, and these become organized into an epithelium and perform a complex series of *gastrulation* and *neurulation* movements, whose outcome is the creation of a rudimentary gut cavity and a *neural tube*. In the second phase the various organs, such as limbs, eyes, heart, and so on, are formed—a process called *organogenesis*. In the third phase the structures that have been generated in this way on a small scale proceed to grow to their adult size. These phases are not sharply distinct but overlap considerably in time. We shall use *Xenopus* to follow development from the fertilized egg to the beginning of organogenesis.

The Polarity of the Amphibian Embryo Depends on the Polarity of the Egg[2]

The amphibian egg is a large cell, about a millimeter in diameter, enclosed in a transparent extracellular capsule or jelly coat. Most of the cell's volume is occupied by yolk platelets, which are aggregates chiefly of lipid and protein. The yolk is concentrated toward the lower end of the egg, called the **vegetal pole;** the other end is called the **animal pole.** Fertilization initiates the conversion of this single egg cell into a multicellular tadpole, with a head and a tail, a back and a belly, and a median plane of symmetry dividing the body into a right side and a left. It is often useful to describe the animal by reference to three axes: *antero-posterior*, from head to tail; *dorso-ventral*, from back to belly; and *medio-lateral*, from the median plane outward to the left or to the right. In amphibians these polarities are established before cell division begins. Before fertilization the animal and vegetal poles of the egg contain different selections of mRNA molecules as well as different quantities of yolk and other cell components, but the egg is symmetrical about animal-vegetal axis. This initial animal-vegetal asymmetry is sufficient to generate a cylindrical embryo with an antero-posterior axis. Fertilization triggers a distortion of the amphibian egg contents that creates a dorsal-ventral difference. The outer, actin-rich cortex of the cytoplasm abruptly rotates as a coherent unit relative to the central core, so that the animal pole of the cortex is slightly shifted relative to the animal pole of the core, to the future ventral side (Figure 16–2). The direction of the rotation is determined by the point of sperm entry—perhaps through an effect of the centrosome that the sperm brings into the egg. Because pigment granules in the egg are displaced by the rotation, in many amphibians a band of pale pigmentation appears, called the *gray crescent*, opposite the sperm entry point. The sperm entry point corresponds, roughly speaking, to the belly; the opposite side will form the back and dorsal structures, such as the spinal cord (Figure 16–3). Treatments that block the rotation produce a radialized animal with a central gut and no dorso-ventral asymmetry.

Cleavage Produces Many Cells from One[3]

The cortical rotation is completed in about an hour after fertilization and sets the scene for *cleavage*, in which the single large egg cell subdivides by repeated mitosis into many smaller cells, or **blastomeres,** without any change in total mass. To survive, the embryo must quickly reach a stage where it can begin to feed, swim,

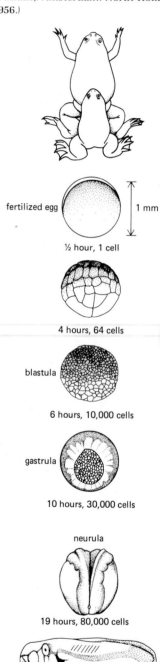

Figure 16–1 Synopsis of the development of *Xenopus laevis* from newly fertilized egg to feeding tadpole. The adult male and female are shown mating at the top. The developmental stages are viewed from the side, except for the 10-hour and 19-hour embryos, which are viewed from below and from above, respectively. All stages except the adults are drawn to the same scale. (After P.D. Nieuwkoop and J. Faber, Normal Table of *Xenopus laevis* [Daudin]. Amsterdam: North-Holland, 1956.)

fertilized egg — 1 mm

½ hour, 1 cell

4 hours, 64 cells

blastula — 6 hours, 10,000 cells

gastrula — 10 hours, 30,000 cells

neurula — 19 hours, 80,000 cells

32 hours, 170,000 cells

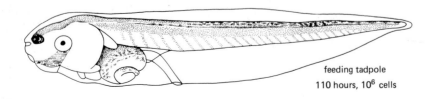

feeding tadpole
110 hours, 10^6 cells

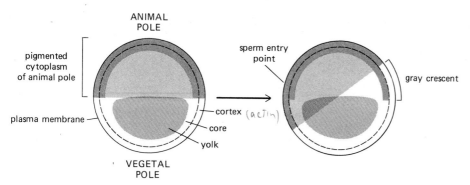

Figure 16–2 The first morphogenetic movement following fertilization of a frog's egg. The egg cortex (a layer a few micrometers deep) rotates through about 30° relative to the core of the egg in a direction determined by the site of sperm entry. In species where the cytoplasm capping the animal pole is appropriately pigmented, the rotation creates a visible gray crescent opposite the site of sperm entry.

and escape from predators, and these first cell divisions are extraordinarily rapid, with a cycle time of about 30 minutes (see p. 735). The very high rate of DNA replication and mitosis seems to preclude gene transcription, and the cleaving embryo is almost entirely dependent on reserves of RNA, protein, membrane, and other materials that accumulated in the egg while it developed as an oocyte in the mother. The only crucial biosynthesis obviously required is that of DNA, and unusually rapid DNA replication is made possible by an exceptionally large number of replication origins (see p. 521).

The first cleavage divides the egg vertically, or parallel to the animal-vegetal axis (Figure 16–4). The next cleavage is again vertical, but perpendicular to the first, giving four cells of similar size. The third cleavage passes horizontally through these four cells, slightly above the midline, to give four small cells stacked on top of four larger, more yolky cells. All the cells divide synchronously for the first 12 cleavages, but the divisions are asymmetric, so that the lower, vegetal cells, encumbered with yolk, are fewer and larger. After about 12 cycles of cleavage, the cell division rate slows down abruptly, synchrony is lost, and transcription of the embryo's genome begins. This change, known as the *mid-blastula transition*, seems to reflect the depletion of some maternally derived egg component that is used up by binding to newly synthesized DNA.

In other species the asymmetries of the egg and the detailed patterns of cleavage may be different. In very yolky eggs, such as those of birds, cleavage does not cut all the way through the yolk, and all the nuclei remain clustered at the animal pole; the embryo then develops from a cap of cells on top of the yolk.

The Blastula Consists of an Epithelium Surrounding a Cavity[4]

From the outset, the cells of the embryo are not only bound together mechanically, they are also coupled by gap junctions through which ions and other small molecules can pass, conveying messages that are thought to be important later (together with other types of signals) in coordinating the behavior of the cells (see p. 800). Meanwhile, in the outermost regions of the embryo, tight junctions (see p. 793) between the blastomeres create a seal, isolating the interior of the embryo from the external medium. At about the 16-cell stage, the intercellular crevices deep inside the embryo enlarge to form a single cavity, the **blastocoel,** by a mech-

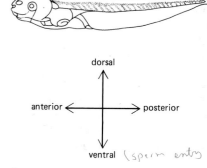

Figure 16–3 The asymmetries of the *Xenopus* egg are shown in (A). These determine the axes of the body of the tadpole, shown in (B). The exact relationship between the axes of the egg and the tadpole is complex and distorted by the movements of gastrulation. Very roughly, the animal pole corresponds to the anterior (head) end of the body, and the gray crescent corresponds to the dorsal side.

Figure 16–4 The stages of cleavage in *Xenopus*, as seen from the side.

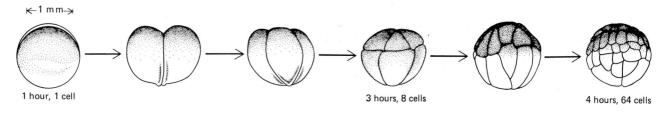

1 hour, 1 cell   3 hours, 8 cells   4 hours, 64 cells

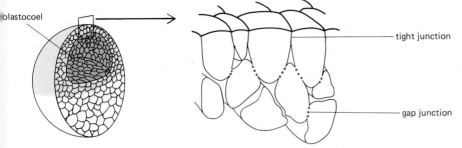

blastocoel

tight junction

gap junction

Figure 16–5 The blastula. At this stage the cells are arranged to form an epithelium surrounding a fluid-filled cavity, the blastocoel. The cells are electrically coupled via gap junctions, and tight junctions close to the outer surface create a seal that isolates the interior of the embryo from the external medium. Note that in *Xenopus* the wall of the blastocoel is several cells thick, and only the outermost cells are tightly bound together as an epithelium.

anism in which Na⁺ is pumped across the cell membranes into the spaces between cells in the interior of the embryo, and water follows as a result of the consequent osmotic pressure difference. The cells forming the exterior of the embryo have become organized into an epithelium, and the embryo is now termed a **blastula** (Figure 16–5). The organization of the cells as an epithelial sheet is crucial in coordinating their subsequent behavior.

Gastrulation Transforms a Hollow Ball of Cells into a Three-layered Structure with a Primitive Gut[5,6]

Once the cells of the blastula have become arranged into an epithelial sheet, the stage is set for the coordinated movements of **gastrulation.** This dramatic process transforms the simple hollow ball of cells into a multilayered structure with a central gut tube and bilateral symmetry: by a complicated invagination, a large area of cells on the outside of the embryo is brought to lie inside it. Subsequent development depends on the interactions of the inner, outer, and middle layers of cells thus formed. Gastrulation in one form or another is an almost universal episode of development throughout the animal kingdom. Since the geometry of gastrulation in amphibians is rather contorted, we shall first describe the process in the sea urchin, a close relative of the vertebrates (see Figure 15–40, p. 868).

Because the sea urchin embryo is transparent, both its internal and its external development can be followed in the living animal, and the activities of the individual cells can be analyzed. The starting point for gastrulation is a very simple blastula: a sheet of about 1000 cells, 1 cell thick, surrounding a spherical cavity. The embryo as a whole is enclosed in a thin layer of extracellular matrix, and vegetal and animal poles can be distinguished. Gastrulation begins with the detachment of a few dozen cells—the *primary mesenchyme cells*—from the epithelium at the vegetal pole (Figure 16–6A). These cells detach apparently because they lose their propensity to stick to other cells and to the matrix on the exterior of the embryo and develop instead an affinity for the fibronectin-rich matrix (see

Figure 16–6 Gastrulation in a sea urchin. (A) The primary mesenchyme cells break loose from the epithelium at the vegetal pole of the blastula. (B) These cells then crawl over the inner face of the wall of the blastula. (C) Meanwhile the epithelium at the vegetal pole is beginning to tuck inward. (D and E) The invaginating epithelium actively extends into a long gut tube through rearrangement of its cells. At the same time, filopodia project from the tip of the invaginating epithelial sheet and help to pull it in the right direction. (F) The end of the gut tube makes contact with the wall of the blastula; the site of contact coincides with the site of the future mouth opening. (After L. Wolpert and T. Gustafson, *Endeavour* 26:85–90, 1967.)

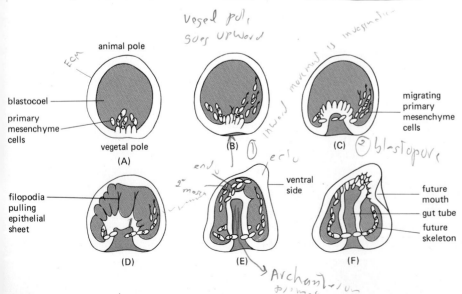

animal pole

blastocoel

primary mesenchyme cells

vegetal pole

(A)

migrating primary mesenchyme cells

(B)

(C)

filopodia pulling epithelial sheet

(D)

ventral side

(E)

future mouth

gut tube

future skeleton

(F)

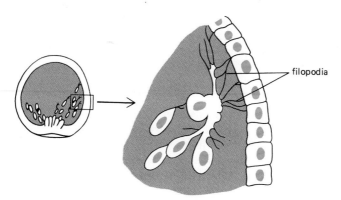

filopodia

Figure 16–7 The primary mesenchyme cells crawl over the inner face of the wall of the blastula by putting out contractile filopodia with sticky tips. (After L. Wolpert and T. Gustafson, *Endeavour* 26:85–90, 1967.)

p. 816) lining the interior of the blastocoel cavity. Having moved into the blastocoel cavity, they migrate over its lining, pulling themselves along by extending long thin processes, or *filopodia*, with sticky tips (Figure 16–7). When the tip of a filopodium contacts a surface to which it can attach firmly, it contracts, pulling the cell after it. Old filopodia seem to be withdrawn and new filopodia extended almost at random, pulling the cell hither and thither. The cells finally settle in well-defined positions that are probably specified by their specific adhesions to particular sites on the lining of the blastocoel cavity, inasmuch as studies with monoclonal antibodies show highly organized differences of surface chemistry among cells in different regions of the embryo (a topic that we shall return to later—see p. 940). Once the primary mesenchyme cells have settled, they begin to form the skeleton (Figure 16–6F).

As the primary mesenchyme cells are starting to migrate, the epithelium at the vegetal pole begins to invaginate, buckling inward into the blastocoel to form the gut (Figure 16–6C). To initiate the process, the epithelial cells in this region change their shape: the ends facing into the blastocoel become broader than the ends facing the external surface, causing the sheet to bend and bulge up into the blastocoel (Figure 16–8). The next stage in the invagination is driven by a different process: changes in cell packing. The invaginating cells actively rearrange themselves, without altering their average shape, so as to convert the initial squat dome-shaped invagination into a long narrow gut tube. At the same time, certain cells in the rounded tip of the invaginating sheet extend long filopodia into the blastocoel cavity; these contact the walls of the cavity, adhere there, and contract, thereby helping to steer the invagination movement (Figure 16–6D, E). The movement is complete when the blind end of the gut tube has been brought up against the epithelium near the opposite end of the embryo (Figure 16–6F). Where the two epithelia make contact, the mouth will develop—the epithelia fuse and a hole forms. Once their task is done, the cells whose filopodia helped steer the invagi-

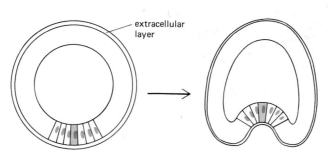

extracellular layer

change in the curvature of the epithelial sheet may be forced by change in the shape of the cells

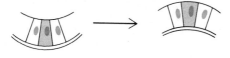

Figure 16–8 A possible mechanism for the initial intucking of the epithelium at the vegetal pole. For molecular details of a process of this type, see pp. 886 and 674.

nation of the gut break loose from the epithelium and move into the space between the gut tube and the body wall to become the so-called *secondary mesenchyme*, which will give rise to the coelom and the musculature.

Gastrulation transforms the hollow spherical blastula into a three-layered structure: the innermost layer, the tube of the primitive gut, is the **endoderm;** the outermost layer, the epithelium that has remained external, is the **ectoderm;** and between the two, the looser layer of tissue composed of primary and secondary mesenchyme cells is the **mesoderm.** These are the three primary **germ layers** common to higher animals. The organization of the embryo into the three layers corresponds roughly to the organization of the adult—gut on the inside, epidermis on the outside, and connective tissue in between. Very crudely, these three types of adult tissues may be said to derive from the endoderm, the ectoderm, and the mesoderm, respectively, although there are exceptions (see p. 887).

Gastrulation Movements Depend on Simple Cell Activities Coordinated in a Complex Pattern[1,6,7]

The movements of gastrulation depend on a relatively simple repertoire of basic cell activities. Cells can change their shape by extending or contracting; they can make and break adhesions to other cells or to extracellular matrix; they can secrete extracellular matrix materials that constrain or guide their movements. These activities, together with cell growth and cell division, underlie almost all morphogenetic movements, both of cells as individuals and of cells in groups, including, presumably, such processes as the changes in cell packing that drive the extension of the gut tube. Gastrulation, besides being a profoundly important event in its own right, exemplifies the varied ways in which these basic forms of cell behavior can be put to work.

The mechanisms by which an individual cell changes its shape or its adhesiveness have been discussed in earlier chapters. The special problem posed by embryonic development is to understand how these and other elementary cell activities are coordinated in space and time so that each part of the embryo behaves in the appropriate way. Although the beautifully choreographed movements of gastrulation are still largely a mystery from this point of view, experimental studies in amphibians have helped at least to clarify which parts of the gastrulating embryo move through their own exertions and which parts are pushed or pulled by forces generated elsewhere.

In a gastrulating amphibian the presence of very yolky cells impedes the movement of the invaginating cell sheets, making the geometry of the process more complex than in the sea urchin. The infolding of the endoderm begins not at the vegetal pole, but to one side of it: the site is marked at first by a short indentation, called the **blastopore,** in the exterior of the blastula. This indentation gradually extends (Figure 16–9), curving around to form a complete circle surrounding a plug of very yolky cells (destined to be enclosed in the gut and digested). Sheets of cells meanwhile turn in around the lip of the blastopore and move deep into the interior of the embryo. At the same time, the external epithelium in the region of the animal pole actively spreads to take the place of the cell sheets that have turned inward. Eventually the epithelium of the animal hemisphere extends in this way to cover the whole external surface of the embryo, and, as gastrulation reaches completion, the blastopore circle shrinks almost to a point.

The process of invagination in amphibian gastrulation seems to depend on the same basic mechanisms as in the sea urchin. It begins with changes in the shape of the cells at the site of the blastopore. In the amphibian these are called *bottle cells:* they have broad bodies and narrow necks that anchor them to the surface of the epithelium (see Figure 16–11), and they may help to force the epithelium to curve and so to tuck inward, producing the initial indentation seen from outside. Once this first tuck has formed, cells can continue to pass into the interior as a sheet to form the gut. As in the sea urchin, the movement seems to be driven mainly by active repacking of the epithelial cells, especially those in the

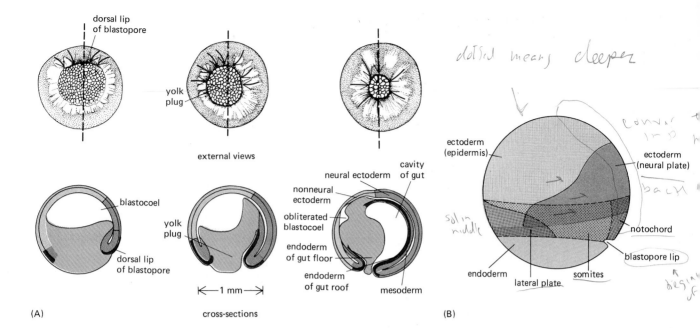

Handwritten annotations: "dotted means deeper", "convex in 9 h", "back", "sol in middle", "beginning of"

(A) external views

cross-sections

$\leftarrow$ 1 mm $\rightarrow$

Labels (A): dorsal lip of blastopore; yolk plug; blastocoel; dorsal lip of blastopore; yolk plug; neural ectoderm; nonneural ectoderm; obliterated blastocoel; endoderm of gut floor; endoderm of gut roof; cavity of gut; mesoderm

(B) ectoderm (epidermis); ectoderm (neural plate); notochord; blastopore lip; endoderm; lateral plate; somites

marginal zone neighboring the blastopore lip (Figure 16–11). Small square fragments of marginal-zone tissue isolated in culture will spontaneously narrow and elongate through a rearrangement of the cells (Figure 16–11B), just as they would in the embryo in the process of converging toward the blastopore and streaming inward around the blastopore lip. The mechanism of the cell rearrangements underlying this *convergent extension* is still a puzzle. A secondary driving or guiding force is provided by future mesoderm cells that crawl over the internal surface of the blastocoel roof, pulling the endoderm along with them, rather like the future secondary mesenchyme cells in the sea urchin. The blastocoel roof is lined with a fibronectin-rich matrix, and normal gastrulation is hindered, at least in newts and salamanders, when antibodies or peptides that block the interaction of fibronectin with cell-surface receptors (see p. 817) are injected into the blastocoel (Figure 16–10).

In the amphibian as in the sea urchin, gastrulation produces a three-layered structure: an outermost sheet of ectoderm, an innermost tube of endoderm forming the rudiment of the gut, and between them a layer of mesoderm. Again, the mouth develops as a hole formed at an anterior site where endoderm and ectoderm come into direct contact without intervening mesoderm.

Figure 16–9 (A) Gastrulation in *Xenopus.* The external views (*above*) show the embryo from the vegetal pole; the cross-sections (*below*) are cut in the plane indicated by the broken lines. The directions of cell movement are indicated by arrows. (B) A fate map for the early *Xenopus* embryo (viewed from the side) as it begins gastrulation, showing the origins of the cells that will come to form the three germ layers as a result of the movements of gastrulation. The various parts of the mesoderm (lateral plate, somites, and notochord) derive from deep-lying cells in the dotted region; the other cells, including the more superficial cells in the dotted region, will give rise to ectoderm (*gray, above*) or endoderm (*color, below*). (After R.E. Keller, *J. Exp. Zool.* 216:81–101, 1981.)

Figure 16–10 (A) A normal newt embryo toward the end of gastrulation; endoderm and mesoderm have invaginated into the interior, and the blastopore has closed down to a ring surrounding the yolk plug visible on the exterior. (B) A companion embryo in which gastrulation movements have been blocked by injecting antibodies against fibronectin into the cavity of the blastula. The cells of the vegetal hemisphere have failed to move into the interior and remain visible externally as the smooth round mass in the lower part of the picture; the epithelium of the animal pole, while remaining an empty bag, has undergone its usual spreading movements and so has been thrown into folds. (From J.C. Boucaut et al., *J. Embryol. Exp. Morphol.* 89[suppl]:211–227, 1985.)

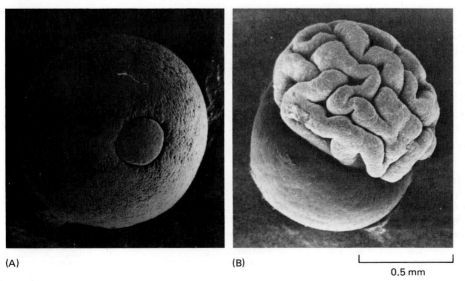

(A)   **(B)**

$\vdash$———$\dashv$
0.5 mm

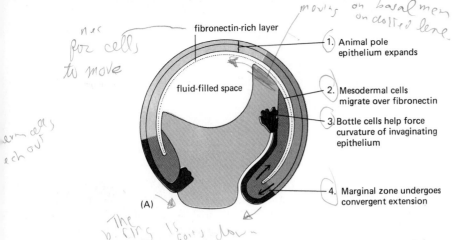

fibronectin-rich layer

fluid-filled space

1. Animal pole epithelium expands

2. Mesodermal cells migrate over fibronectin

3. Bottle cells help force curvature of invaginating epithelium

4. Marginal zone undergoes convergent extension

(A)

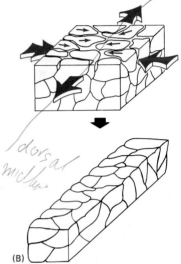

(B)

Figure 16–11 (A) A section through a gastrulating *Xenopus* embryo, cut in the same plane as in Figure 16–9, indicating the four main types of cell movement that gastrulation involves. (B) Cartoon of the cell repacking that brings about convergent extension, thought to be the chief driving force for gastrulation in Xenopus. (A, after R.E. Keller, *J. Exp. Zool.* 216:81–101, 1981; B, from J. Gerhart and R. Keller, *Annu. Rev. Cell Biol.* 2:201–229, 1986.)

Gastrulation Movements Are Organized Around the Blastopore[2,7,8]

The cell movements of gastrulation are complex but orderly, so that it is possible to plot on the surface of the pregastrulation embryo a *fate map* (Figure 16–9B) showing which cells of the very early stage will be carried into which parts of the mature animal. But how is the whole complex of gastrulation movements set in train and organized? In amphibians the rearrangement of the egg contents immediately after fertilization is an essential preliminary step. Invagination always begins at the site corresponding to the gray crescent (see p. 880): here, at the future dorsal lip of the blastopore, the fertilization-induced rotation of egg cortex relative to egg core (see Figure 16–2) evidently creates a combination of cell components with unique properties. If the dorsal lip of the blastopore is excised from a normal embryo at the beginning of gastrulation and grafted into another embryo but in a different position, the host embryo initiates gastrulation both at the site of its own dorsal lip and at the site of the graft (Figure 16–12). The movements of gastrulation at the second site entail the formation of a second whole set of body structures, and a double embryo (Siamese twins) results. By carrying out such grafts between species with differently pigmented cells, so that host tissue can be distinguished from implanted tissue, it has been shown that the grafted blastopore lip recruits host epithelium into its own system of invaginating endoderm and mesoderm. We shall see later that chemical, as well as physical, interactions between cells are crucial for the creation of the different germ layers at gastrulation. But first we must outline briefly the later development of the layers of endoderm, mesoderm, and ectoderm that constitute the vertebrate embryo just after gastrulation.

The Endoderm Gives Rise to the Gut and Associated Organs Such as the Lungs and Liver[9]

The endoderm forms a tube, the primordium of the digestive tract, from the mouth to the anus. It gives rise not only to the pharynx, esophagus, stomach, and intestines, but also to many associated glands. The salivary glands, the liver, the pancreas, the trachea, and the lungs, for example, all develop from extensions of the wall of the originally simple digestive tract and grow to become systems of branching tubes that open into the gut or pharynx. While the endoderm forms the epithelial components of these structures—the lining of the gut and the secretory cells of the pancreas, for example—the supporting muscular and fibrous elements arise from the mesoderm.

The Mesoderm Gives Rise to Connective Tissues, Muscles, and the Vascular and Urogenital Systems[9,10]

The mesodermal layer is divided in the post-gastrulation embryo into separate parts on the left and right of the body. Defining the central axis of the vertebrate body, and effecting this separation, is a very early specialization of the mesoderm known as the **notochord.** This is a slender rod of cells, about 80 μm in diameter, with ectoderm above it, endoderm below it, and mesoderm on either side (see Figure 16–15). The cells of the notochord become swollen with vacuoles, so that

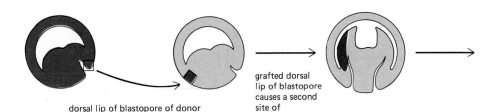

(A) dorsal lip of blastopore of donor is grafted to abnormal site in host

grafted dorsal lip of blastopore causes a second site of invagination

double embryo develops, with nearly all its tissues of host origin

the rod elongates and stretches out the embryo. In the most primitive chordates, which have no vertebrae, the notochord is retained as a primitive substitute for a vertebral column. In vertebrates it serves as a core around which mesodermal cells gather to form the vertebrae. Thus the notochord is the precursor of the vertebral column, both in an evolutionary and in a developmental sense.

In general, the mesoderm gives rise to the connective tissues of the body—at first to the loose, space-filling, three-dimensional mesh of cells known as *mesenchyme* (see Figure 14–20, p. 802), and ultimately to cartilage, bone, and fibrous tissue, including the dermis (the inner layer of the skin). Muscle cells also derive from mesoderm. In addition, most of the tubules of the urogenital system form from it and so does the vascular system, including the heart and the cells of the blood.

(B)

Figure 16–12 (A) Diagram of an experiment showing that the dorsal lip of the blastopore initiates and controls the movements of gastrulation and thereby, if transplanted, organizes the formation of a second set of body structures. (B) Photograph of a two-headed, two-tailed axolotl tadpole resulting from such an operation; the results are similar for *Xenopus*, though the extent of duplication is somewhat variable. (B, courtesy of Jonathan Slack.)

The Ectoderm Gives Rise to the Epidermis and the Nervous System[9,11]

At the end of gastrulation the sheet of ectoderm covers the embryo and thus eventually forms the epidermis (the outer layer of the skin). But that is not all: the entire nervous system also derives from it. In a process known as **neurulation,** a broad central region of the ectoderm thickens, rolls up into a tube, and pinches off from the rest of the cell sheet. This transformation is induced by an interaction with the underlying notochord and the mesoderm adjacent to it (see p. 941). The tube thus created from the ectoderm is called the **neural tube;** it will form the brain and the spinal cord. Along the line where the neural tube pinches off from the future epidermis, a number of ectodermal cells break loose from the epithelium and migrate as individuals out through the mesoderm. These are the cells of the **neural crest;** they will form almost all of the peripheral nervous system (including the sensory and sympathetic ganglia and the Schwann cells that make the myelin sheaths of peripheral nerves—see p. 1060) as well as the epinephrine-secreting cells of the adrenal gland and the pigment cells of the skin. In the head, many of the neural crest cells will differentiate into cartilage, bone, and other connective tissues, which elsewhere in the body arise from the mesoderm. This is one of several instances that run counter to the general scheme in which the three germ layers give rise to cells in three corresponding concentric layers of the adult body.

The sense organs, by which light, sound, smell, and so forth impinge on the nervous system, also have ectodermal origins: some derive from the neural tube, some from the neural crest, and some from the exterior layer of ectoderm (see Figure 19–55, p. 1109). The retina, for example, originates as an outgrowth of the brain and so is derived from cells of the neural tube, while the olfactory cells of the nose differentiate directly from the ectodermal epithelium lining the nasal cavity.

The Neural Tube Is Formed by Coordinated Changes in Cell Shape[6,12]

The formation of the neural tube (Figure 16–13) is a dramatic event to watch. At first the surface of the gastrula appears more or less uniform. But subtle changes are occurring: the ectoderm close to the midline begins to thicken, forming the *neural plate.* Then the lateral edges of the neural plate start to rear up in folds; these *neural folds* gradually roll together, while the midline of the plate sinks deeper. Eventually the folds meet and fuse to form the hollow neural tube, roofed over by a continuous sheet of ectoderm. As in gastrulation, the whole process

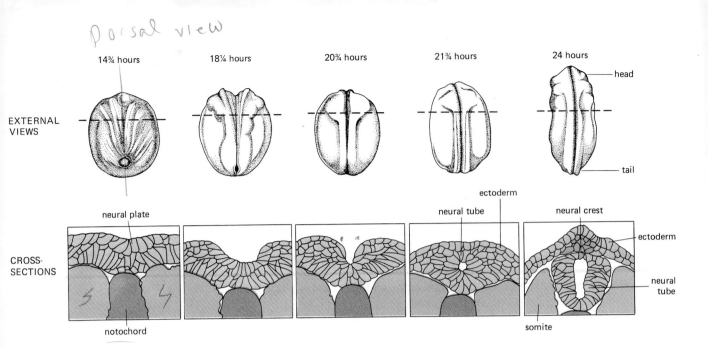

Dorsal view

EXTERNAL VIEWS

14¾ hours 18¼ hours 20¾ hours 21¾ hours 24 hours

head

tail

ectoderm

CROSS-SECTIONS

neural plate

notochord

neural tube

neural crest

ectoderm

neural tube

somite

Figure 16–13 Neural tube formation in *Xenopus*. The external views are from the dorsal aspect. The cross-sections are cut in a plane indicated by the broken lines. (After T.E. Schroeder, *J. Embryol. Exp. Morphol.* 23:427–462, 1970.)

depends on extension, adhesion, and contraction on the part of individual cells in an epithelial sheet.

The cells of the neural plate are bound together by strong lateral attachments. First the cells elongate in a direction perpendicular to the surface of the sheet. This elongation involves the deployment of microtubules and is essential for the continuation of neurulation, since drugs, such as colchicine, that disassemble microtubules prevent or reverse the elevation of the neural folds. The elongated cells then become wedge-shaped, with the narrower part of the wedges toward the upper, apical surface of the sheet. Since the cells are held tightly together at their lateral faces and their width does not change at the base, the cell sheet bends (Figure 16–14). The cells are made narrower at their apices by the contraction of the actin-filament bundles that run beneath the upper surfaces of the cells, where the cells are connected by adhesion belts (see p. 674).

Blocks of Mesoderm Cells Uncouple to Form Somites on Either Side of the Body Axis[13]

On either side of the newly formed neural tube lies a broad expanse of mesoderm (Figure 16–15). The thicker, more medial part of this mesoderm gives rise to the vertebrae, the ribs, practically all the skeletal musculature, and the dermis of the trunk. It consists at first of a single continuous slab of tissue on each side of the body. To form the repetitive segmented structures of the vertebrate body axis, this slab soon breaks up into separate "blocks," or **somites** (Figure 16–16). Each somite corresponds to one unit in the final sequence of articulated elements and will become differentiated into three regions that contribute three major tissues that recur in each segment. The part of the somite that faces the notochord is called the *sclerotome* and is the source of the cells that form the vertebrae and ribs; the part that lies outermost and closest to the dorsal surface of the embryo is called the *dermatome* and provides the cells that form the dermis of the trunk; and the remaining part, called the *myotome*, between the sclerotome and the dermatome, is the source of skeletal muscle cells—both for the segmental muscles of the trunk and for almost all the other muscles of the body (see p. 943).

The somites form one after another, starting near the head and ending at the tail. Segmentation is accompanied by changes in the connections between the mesoderm cells. The cells in the early unsegmented mesoderm of the amphibian are electrically and metabolically coupled via gap junctions, but this coupling between the cells disappears at or just before the time when the somites develop. Apparently the cells first alter their mutual attachments and then rearrange themselves in groups to form somites (see Figure 16–16). As we shall see later, the

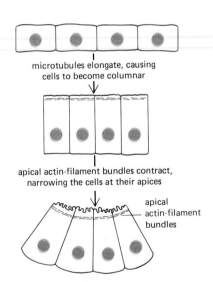

microtubules elongate, causing cells to become columnar

apical actin-filament bundles contract, narrowing the cells at their apices

apical actin-filament bundles

Figure 16–14 The bending of an epithelium through cell shape changes mediated by microtubules and actin filaments. As the apical ends of the cells become narrower, their upper surface membrane becomes puckered. This may be a symptom of rapid change in shape due to contraction, occurring in circumstances where strong lateral attachments prevent flow of excess membrane past the junctional complexes and down the sides of the cells.

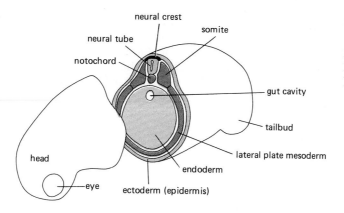

Figure 16–15 A cross-section (schematic) through the trunk of an amphibian embryo after the neural tube has closed. (After T. Mohun, R. Tilly, R. Mohun, and J.M.W. Slack, *Cell* 22:9–15, 1980.)

physical segmentation is correlated with a segmental zebra-stripe pattern of chemical differences between adjacent groups of somite cells (see p. 938), and selective cohesion based on differences of surface chemistry is likely to be the cause of the splitting into physically separate somites.

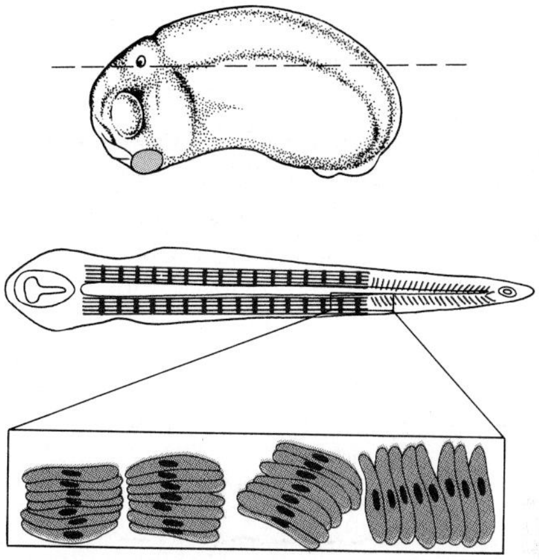

The Vertebrate Body Plan Is First Formed in Miniature and Then Maintained as the Embryo Grows[9]

The embryo at the stage when the somites are forming is typically a few millimeters long and consists of about 10^5 cells. While we have been speaking thus far of *Xenopus*, the scale and general form are much the same for a fish, a salamander, a chick, or a human (Figure 16–17). Later these species of embryo will grow to be very different in size and shape, but at this early stage they all share the basic vertebrate body plan. The details will be filled in later as the embryo grows. The

Figure 16–16 Somite formation in *Xenopus*. A side view of the embryo is shown at the top; the broken line indicates the plane of the horizontal section shown below. The bottom drawing is a schematic high-magnification view of the mesoderm cells in the process of rearranging to form somites. In *Xenopus* the future somite cells are initially all oriented at right angles to the body axis and then rotate in groups during somite formation.

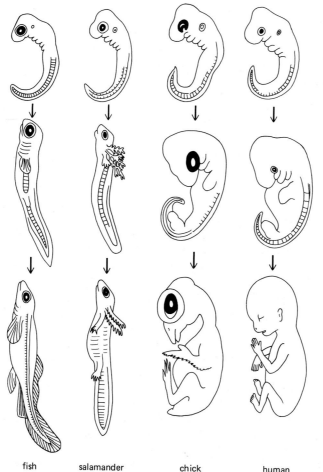

fish salamander chick human

Figure 16–17 Comparison of the embryonic development of a fish, an amphibian, a bird, and a mammal. The early stages (*above*) are closely similar; the later stages (*below*) are more divergent. The earliest stages are drawn roughly to scale; the later stages are not. (After E. Haeckel, Anthropogenie, oder Entwickelungsgeschichte des Menschen. Leipzig: Engelmann, 1874.)

central nervous system is represented by the neural tube, with an enlargement at one end for the brain; the gut and its derivatives by a tube of endoderm; the segments of the trunk by the somites; the other connective tissues, including the vascular system, by the more peripheral unsegmented mesoderm; and the epidermal layer of the skin by the ectoderm. During subsequent development, all of these components will enlarge, by a factor of as much as a hundred or more in length or a million or more in volume and cell number. But the same basic organization of the body will be preserved.

Summary

The eggs of most species are large cells, containing stores of nutrients and other cell components specified by the maternal genome. In amphibians the first major movement after fertilization is a rotation of the cortex of the egg relative to its core. The asymmetry created by this rotation, together with the original asymmetry in the distribution of the contents of the egg before fertilization, defines the future antero-posterior and dorso-ventral axes of the body. During the subsequent cleavage divisions, the egg subdivides into many smaller cells, but no growth occurs. Tight junctions develop between the cells in the outermost layers of the embryo, sealing off the interior from the external medium. Fluid is drawn into the interior of the embryo to form a cavity, the blastocoel, surrounded by an epithelium. The embryo at this stage is called a blastula.

In the process of gastrulation, part of the epithelium of the blastula invaginates as changes in the shapes of the cells of the blastula force the epithelium to bend inward. Convergent extension of the invaginating epithelium by repacking of its cells is thought to drive the later phases of gastrulation, which transform the embryo into a three-layered structure with an internal epithelial tube of endoderm, an external epithelial covering of ectoderm, and a middle layer of mesodermal cells that have broken loose from the original epithelial sheet. The endoderm will form the lining of the gut and its derivatives; the ectoderm will form chiefly the epidermis and the nervous system; and the mesoderm will form most of the muscles, connective tissues, vascular system, and urogenital tract.

Gastrulation brings together originally separate populations of cells, which then interact. The dorsal mesoderm, for example, induces the overlying ectoderm to thicken, roll up, and pinch off to form the neural tube and neural crest. This process, called neurulation, depends again on changes in the shape of the epithelial cells. In the middle of the dorsal mesoderm, a rod of specialized cells called the notochord constitutes the central axis of the embryo. The long slabs of mesoderm on either side of the notochord become segmented into somites, from which the vertebrae and skeletal muscles will be derived.

Cell Diversification and Cell Memory[14]

A fertilized egg may develop into a male or a female, a sea urchin, a frog, or a human being. The outcome is governed by the genome: the linear sequence of A, G, C, and T nucleotides in the DNA of the organism must direct the production of a variety of chemically different cell types, arranged in a precise pattern in space. Developmental biology aims to explain how. As an initial step toward answering this question, we consider how the cells in the early embryo become different even though they all inherit the same genetic information from the egg. To illustrate the principles, we focus first on amphibians, then on mammals.

During Development, the Genome Remains Constant
While the Pattern of Gene Expression Changes[15]

The many types of cells in the body differ essentially because, in addition to the many "housekeeping" proteins they all require, each makes a different set of specialized proteins. These include abundant proteins such as keratin in epidermal cells, hemoglobin in red blood cells, digestive enzymes in gut cells, crystallins in lens cells, and so on. Since the various cell types differ in their gene products, it might be thought that they contain different sets of genes. The lens cells, for example, might have lost the genes for hemoglobin, keratin, and so on, while retaining those for crystallins; or they might have selectively amplified the number of copies of the crystallin genes. There are, however, several lines of evidence that show that differentiation usually does not involve genetic changes and that almost every type of cell contains the same complete genome as is found in the fertilized egg. Genes can be switched on and off (see Chapter 10), and the cells of the body differ not because they contain different genes but because they *express* different genes.

Some of the most powerful evidence that the genome itself remains constant during cell differentiation comes from experiments on nuclear transplantation using amphibian eggs (Figure 16–18). A typical amphibian egg is so large that, using a fine glass pipette, one can readily inject into it a nucleus taken from another cell. The nucleus of the egg itself is destroyed beforehand by ultraviolet irradiation. The egg is activated to begin development by the prick of the fine pipette used to inject the transplanted nucleus (see p. 870). Thus one can test whether the nucleus from a differentiated somatic cell contains a complete genome, equivalent to that of a normal fertilized egg and equally serviceable for development. The answer is yes: a complete swimming tadpole can be produced, for example, from an egg whose own nucleus has been replaced by a nucleus derived from a keratinocyte cell from an adult frog's skin or by a nucleus from a frog red blood cell. These experiments admittedly have limitations: they have been successful only with nuclei from a limited range of differentiated cell types and in only a few species. But other studies point to the same conclusion: by and large it seems that the genome remains intact during development.

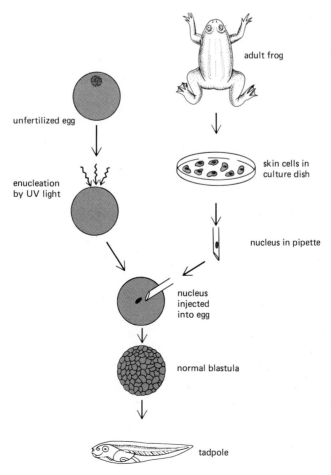

Figure 16–18 Diagram of an experiment showing that the nucleus of a differentiated cell from the skin of an adult frog contains all the genetic material necessary to control the formation of an entire tadpole. (Modified from J.B. Gurdon, Gene Expression During Cell Differentiation. Oxford, U.K.: Oxford University Press, 1973.)

adult frog

unfertilized egg

enucleation by UV light

skin cells in culture dish

nucleus in pipette

nucleus injected into egg

normal blastula

tadpole

A few exceptions to this rule are known. In certain invertebrates, for example, some of the chromosomes present in cells of the *germ line* (the lineage of the gametes) are eliminated early in embryogenesis from the *somatic* (non-germ-line) cells. In some other species, including *Xenopus laevis*, genes for ribosomal RNA are selectively amplified in the oocyte, and in some insect larvae the chromosomes become nonuniformly polytene, so that certain genes are amplified more than others (see p. 509). In vertebrates the production of antibodies and antigen-specific receptors by lymphocytes involves splicing together initially separate segments of DNA as these specialized cells differentiate (see Chapter 18).

Differences Among Nonmammalian Blastomeres Often Arise from Asymmetries in the Uncleaved Egg[16]

When the nucleus of a differentiated cell is transplanted into the enucleated *Xenopus* egg, it changes its pattern of gene expression to resemble that of a normal egg nucleus. Thus the behavior of the nucleus can be controlled by the cytoplasmic environment in which it is placed. In many species, as in *Xenopus*, the egg itself is chemically asymmetrical, with certain components concentrated in specific regions of the cytoplasm. As a result there are differences from the outset between the cells that form by cleavage, because their cytoplasms will necessarily contain different portions of the localized materials. The importance of such *localized determinants* in the egg varies from species to species. At one extreme are mammals, in which, as we shall see later, the egg seems to be symmetrical and all the early blastomeres are alike. At the other extreme are the so-called *mosaic eggs* of mollusks, ascidians (sea squirts), nematode worms (see p. 784), and some other phyla. These contain consistent patterns of localized determinants, and the initial cleavage planes are oriented in a strict relationship to the chemical pattern so that each blastomere inherits a specific, predictable set of molecules (Figure 16–19). If the early blastomeres are separated and allowed to develop in isolation from one another, most of them give rise to the same set of cell types that would have developed from that blastomere in the intact embryo. The controlling function of the materials in the egg can be demonstrated by artificially altering their distribution relative to the planes of cleavage. This has been done, for example, in the egg of the ascidian *Styela* (see Figure 16–19), where the different regions of cytoplasm can be easily distinguished because they are differently pigmented: in a disturbed embryo the mode of differentiation of a blastomere can be largely predicted from the type of cytoplasm it inherits.

Mosaic eggs raise an interesting cell-biological question: how do the cleavage pattern and the chemical pattern come to be correlated so precisely? It may be that both of them reflect an underlying pattern in the cytoskeleton. There is some evidence that the localized chemical determinants are attached to the cytoskeleton and that the cytoskeleton controls the planes of cell division (see p. 779).

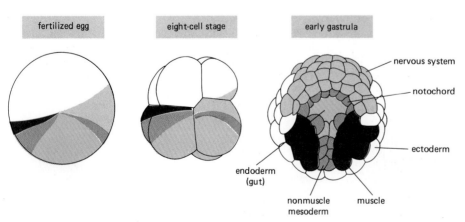

fertilized egg   eight-cell stage   early gastrula

nervous system
notochord
ectoderm
endoderm (gut)
nonmuscle mesoderm
muscle

Figure 16–19   Three successive stages in the early development of the ascidian *Styela*. The early gastrula is viewed from below, looking into the developing gut cavity. Different regions of the egg cytoplasm are differently pigmented in a regular and consistent pattern that is precisely related to the pattern of cleavage divisions. The fate of each blastomere can be predicted from the type of egg cytoplasm that it inherits, as indicated (see also Figure 16–29). (After P.P. Grassé, *Traité de Zoologie*. Paris: Masson, 1966.)

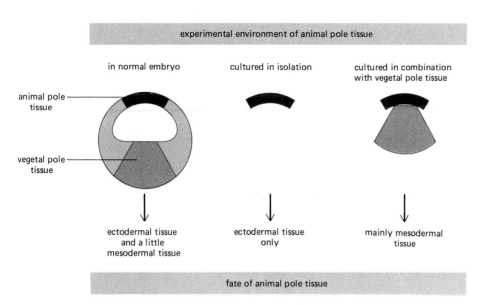

experimental environment of animal pole tissue

in normal embryo | cultured in isolation | cultured in combination with vegetal pole tissue

animal pole tissue

vegetal pole tissue

ectodermal tissue and a little mesodermal tissue

ectodermal tissue only

mainly mesodermal tissue

fate of animal pole tissue

Figure 16–20 Mesoderm induction in *Xenopus*. Cells from the animal pole of a blastula, normally destined to form only ectoderm, will form mesodermal tissues if they are cultured in conjunction with cells from the vegetal pole. In normal development, an inductive interaction of this sort presumably occurs at an earlier stage; the equatorial region of the blastula is already capable of forming mesodermal tissues when it is cultured in isolation.

Chemical Interactions Between Blastomeres Generate New Types of Cells in a More Fine-grained Pattern: Mesoderm Induction in *Xenopus*[17]

A *Xenopus* egg represents a typical intermediate case between the extremes exemplified by mosaic eggs and mammalian eggs. Asymmetry in the *Xenopus* egg cytoplasm creates a difference between blastomeres derived from its animal and vegetal poles, but this defines only the crude beginnings of the pattern of the embryo. To generate the full range of cell types, the blastomeres must interact with one another. If the early *Xenopus* embryo is placed in a medium devoid of Ca^{2+} and Mg^{2+}, the blastomeres lose their cohesiveness and can then be separated and allowed to develop on their own; some go on to develop features characteristic of ectoderm, while others develop features characteristic of endoderm, but none of them switches on expression of the muscle-specific actin genes characteristic of mesoderm. A converse experiment confirms that cells of mesodermal character are produced, in part at least, by cell-cell interactions: when cells from the animal pole of a blastula are placed next to vegetal cells, they are diverted from the ectodermal pathway of development into the mesodermal pathway (Figure 16–20). The switching of cells from one pathway into another by the influence of an adjacent group of cells is called **induction;** during normal development, inductive interactions may occur between cells that have been adjacent from the outset—as in mesoderm induction—or between cells that are brought together through morphogenetic movements such as gastrulation. By a series of successive inductions, it is possible to generate many different kinds of cells from interactions between a few kinds (Figure 16–21).

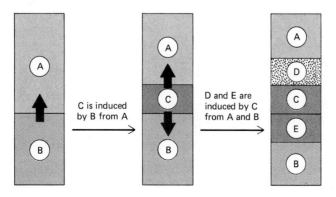

A | C is induced by B from A | A | D and E are induced by C from A and B | A
B | | C | | D
| | B | | C
| | | | E
| | | | B

Figure 16–21 A series of inductive interactions can generate many kinds of cells, starting from only a few.

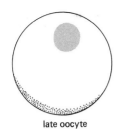
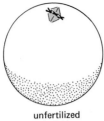

early oocyte   late oocyte   unfertilized
                                           newly laid egg   early blastula

Figure 16–22 The colored dots show the localization of messenger RNA coding for the Vg1 protein in the vegetal region of the *Xenopus* egg at different stages in its development, as visualized by *in situ* hybridization. The Vg1 protein is partly homologous to the growth factor TGF-β and may be a component of the mesoderm-inducing signal produced by vegetal cells in the early embryo. The mechanism that controls the changing localization of Vg1 mRNA is not understood.

For mesoderm induction in *Xenopus*, it is not necessary that the interacting cells be in direct contact, only that they be close. This suggests that the inducing agent is a diffusible molecule. It has been shown that *fibroblast growth factor (FGF)* (see p. 747) can act as a substitute for vegetal blastomeres, inducing cells from the animal pole to develop as mesoderm; and mesoderm can also be induced by *transforming growth factor-β2* (TGF-β2, one of two varieties of TGF-β—see p. 747). FGF tends to favor development of ventral mesodermal derivatives, such as blood cells, while TGF-β2 may favor more dorsal derivatives, such as muscle. Messenger RNA coding for FGF is present in the normal *Xenopus* embryo, and mRNA for a protein called *Vg1*, which is partly homologous to TGF-β, is not only present but has been shown to be localized at the vegetal end of the egg and early embryo (Figure 16–22). These observations suggest that FGF and TGF-β2, or related molecules, jointly mediate mesoderm induction. Indeed, from this and a rapidly growing number of other examples, it seems that the few dozen proteins conventionally called growth factors may play as large a part in controlling the pathways of development as they do in regulating cell division, differentiation, and repair in adult tissues (see Chapter 17); rather like neurotransmitters in the nervous system, the same few factors seem to be used in many different contexts to convey intercellular signals with many different meanings.

In Mammals the Protected Uterine Environment Permits an Unusual Style of Early Development[18]

The mammalian embryo does many things differently. Developing in the protected environment of the uterus, it does not have the same need as the embryos of most other species to complete the early stages of development rapidly. Moreover, the development of a placenta quickly provides for nutrition from the mother, so that the egg does not have to contain large stores of raw materials such as yolk. Thus the egg of a mouse has a diameter of only about 80 μm and therefore a volume about 2000 times smaller than that of a typical amphibian egg. Its cleavage divisions occur no more quickly than the divisions of many ordinary somatic cells, and gene transcription has already begun by the two-cell stage. Furthermore, while the later stages of mammalian development are fundamentally similar to those of other vertebrates such as *Xenopus*, mammals begin by taking a large developmental detour to generate a complicated set of structures—notably the amniotic sac and the placenta—that enclose and protect the embryo proper and provide for the exchange of metabolites with the mother. These structures, like the rest of the body, derive from the fertilized egg but are called *extraembryonic* because they are discarded at birth and form no part of the adult.

The early stages of mouse development are summarized in Figure 16–23. The egg is surrounded initially by a transparent cell coat, the *zona pellucida*. Upon fertilization the egg cleaves within this coat to form a mulberry-shaped cluster of cells called the **morula** (see Figure 16–23). Sometime between the 8-cell and 16-cell stages, the surface of the morula becomes smoother and more nearly spherical as the cells change their cohesiveness and become compacted together (Figure 16–24), with tight junctions forming between the outer cells and sealing off the interior of the morula from the external medium. Soon after, the internal intercellular spaces enlarge to create a central fluid-filled cavity—the blastocoel. At this stage the morula is said to have become a **blastocyst.** The cells of the blastocyst form a spherical shell enclosing the blastocoel, with one pole distinguished by a

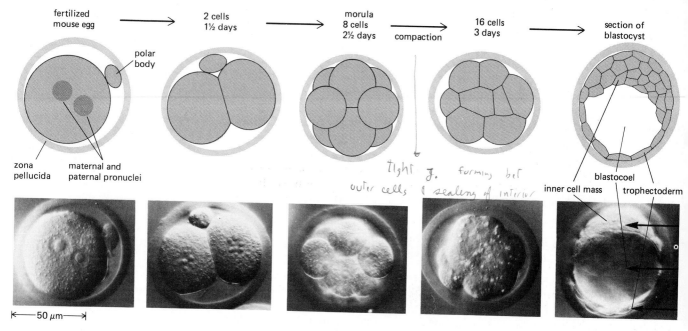

fertilized
mouse egg

2 cells
1½ days

morula
8 cells
2½ days

compaction

16 cells
3 days

section of
blastocyst

polar
body

zona
pellucida

maternal and
paternal pronuclei

tight J. forming bet
outer cells & sealing of interior

blastocoel

inner cell mass

trophectoderm

←—50 μm—→

Figure 16–23 The early stages of mouse development. (Photographs courtesy of Patricia Calarco, from G. Martin, *Science* 209:768–776, 1980. Copyright 1980 by the American Association for the Advancement of Science.)

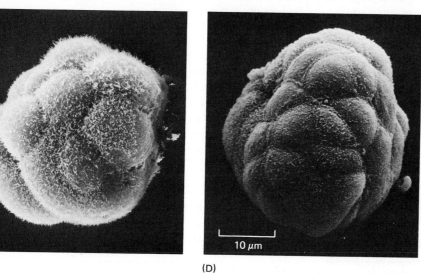

(A)

(B)

(C)

(D)

10 μm

Figure 16–24 Scanning electron micrographs of the early mouse embryo. The zona pellucida has been removed. (A) Two-cell stage. (B) Four-cell stage (a polar body is visible in addition to the four blastomeres—see p. 857). (C) Eight-to-sixteen-cell morula—compaction occurring. (D) Blastocyst. (Courtesy of Patricia Calarco; D, from P. Calarco and C.J. Epstein, *Dev. Biol.* 32:208–213, 1973.)

thicker accumulation of cells. As shown in Figure 16–23, the outer cell layer is the **trophectoderm;** the accumulation of cells inside the trophectoderm at one pole is the **inner cell mass.**

The whole of the embryo proper is derived from the inner cell mass. The trophectoderm is the precursor of the placenta and is the earliest component of the system of extraembryonic structures. Once the zona pellucida has been shed, the cells of the trophectoderm come into close contact with the wall of the uterus, in which the embryo becomes implanted. Meanwhile the inner cell mass grows and begins to differentiate: part of it gives rise to some further extraembryonic structures, such as the yolk sac, while the rest of it goes on to form the embryo proper by processes of gastrulation, neurulation, and so on, that are largely homologous to those seen in other vertebrates, although extreme distortions of the geometry sometimes make the homology hard to discern.

The Cells of the Very Early Mammalian Embryo Become Different Through Cell-Cell Interactions[19,20]

Up to the eight-cell stage, the early mammalian embryo is remarkably adaptable: each of its cells can form any part of the later embryo or adult. One example is seen in the formation of a pair of identical twins from a single fertilized egg—two complete normal individuals, each originating from only a part of the normal embryo. Similarly, if one of the cells in a two-cell mouse embryo is destroyed by pricking it with a needle and the resulting "half-embryo" is placed in the uterus of a foster mother to develop, in a fair proportion of cases a perfectly normal mouse will emerge.

Conversely, two eight-cell mouse embryos can be combined to form a single giant morula, which then develops into a mouse of normal size (Figure 16–25). Such creatures, formed from aggregates of genetically different groups of cells, are called **chimeras.** Chimeras can also be made by injecting cells from an early embryo of one genotype into a blastocyst of another genotype. The injected cells become incorporated into the inner cell mass of the host blastocyst, and a chimeric animal develops. It is even possible to make a chimera by injecting a single cell in this way; thus one can assay the developmental capabilities of the single cell. One of the major conclusions derived from these studies is that the cells of the very early mammalian embryo (up to the eight-cell stage) are initially identical and unrestricted in their capabilities: they are all *totipotent*.

The cells will become different subsequently as a result of their interactions with one another. In the mouse embryo the interaction that generates the first difference of cell character, that between inner cell mass and trophectoderm, depends on the arrangement of cell-cell contacts. At the eight-cell stage, all the blastomeres occupy roughly equivalent positions, with their inner faces making contact with other blastomeres and their outer faces exposed on the exterior of the embryo. Each blastomere has a polarity, marked by microvilli on the external face and corresponding to an asymmetrical distribution of components inside the cell. By manipulating blastomeres *in vitro*, it can be shown that this polarity is governed by the pattern of cell-cell contacts: the microvilli and the associated internal components assemble only where the cell surface makes no contacts with other cells. The plane of the next cleavage is apparently oriented so as to exploit this asymmetry to generate two different daughters, one (an inner-cell-mass cell) toward the interior and lacking the components associated with the microvilli, the other (a trophectoderm cell) facing the exterior and inheriting those components. In this way the pattern of cell-cell contacts seems to control the genesis of the first differences among the cells (Figure 16–26).

The Behavior of Teratocarcinoma Cells Demonstrates the Importance of Environmental Cues[21]

The mammalian style of early development, in which the fate of each cell is governed by interactions with its neighbors, corresponds to a system that is highly *regulative*. The mouse experiments described earlier illustrate this well: the cells

8-cell-stage mouse embryo whose parents are white mice

8-cell-stage mouse embryo whose parents are black mice

zona pellucida of each egg is removed by treatment with protease

embryos are pushed together and fuse when incubated at 37°C

development of fused embryos continues *in vitro* to blastocyst stage

blastocyst transferred to pseudopregnant mouse, which acts as a foster mother

the baby mouse has four parents (but its foster mother is not one of them)

Figure 16–25 A procedure for creating a chimeric mouse by combining two morulae of different genotypes.

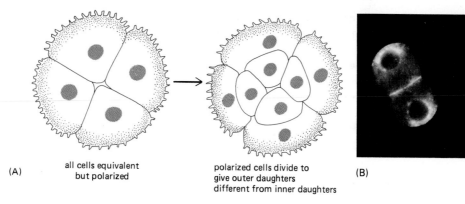

(A) all cells equivalent but polarized → polarized cells divide to give outer daughters different from inner daughters

(B) ⊢————⊣ 20 μm

Figure 16–26 (A) Polarization of mouse blastomeres at the 8-cell stage leads to chemically distinct inner and outer cells at the 16-cell stage. (B) A pair of blastomeres from a mouse embryo at the late 8-cell stage, labeled with fluorescent antibodies against clathrin (*on the left*) or with fluorescent concanavalin A (*on the right*), showing that the cells are polarized with respect both to internal components and to surface properties. (B, from B. Maro, M.H. Johnson, S.J. Pickering, and D. Louvard, *J. Embryol. Exp. Morphol.* 90: 287-309, 1985.)

in a half-embryo or in a chimeric double embryo adjust their behavior so as to generate an animal that is normal in both pattern and size. When the circumstances of development are more grossly abnormal, however, the embryonic cells can go wildly out of control. Some important lessons can be learned from these phenomena.

If a normal early embryo is grafted into the kidney or testis of an adult, it rapidly becomes disorganized, and the normal controls on cell proliferation break down. The result is a bizarre growth known as a *teratoma*: a disorganized mass of cells containing many varieties of differentiated tissue—skin, bone, glandular epithelium, and so on—mixed with undifferentiated stem cells that continue to divide and generate yet more of these differentiated tissues. Teratomas can also arise spontaneously as the result of developmental accidents. Cleavage and embryogenesis do not normally begin until the mammalian ovum has been activated through fertilization by a sperm. Occasionally, however, an oocyte is activated and begins development *parthenogenetically*, without fertilization (see p. 870). If this happens before the cell is released from the ovary, the oocyte develops there almost normally up to the blastocyst stage but then proceeds to form a teratoma. In a somewhat similar way, teratomas can originate in males, too, from the germ cells in the testis; or they can be provoked artificially (in rodents) by grafting a developing male gonad, which contains primitive germ cells, from a young embryo into an adult.

Teratomas that arise in any of these ways look similar, and from all of them it is possible to derive transplantable cancers known as *teratocarcinomas*. A teratocarcinoma will grow without limit until it kills its host. It can be maintained indefinitely by grafting samples of the tumor cells serially from one host to another, and it always includes some undifferentiated stem cells, together with a variety of differentiated cell types to which the stem cells give rise. Teratocarcinoma stem cells can be grown in culture as permanent cell lines, and in a suitable medium they will continue to proliferate indefinitely without differentiating. But if the medium is changed by adding an inducer of differentiation, such as retinoic acid, or if the cells are allowed to aggregate, the stem cells can be triggered to differentiate into a variety of apparently normal specialized cell types.

It might be thought that the curious properties of these teratocarcinoma stem cells originated, as in other cancers, from mutations in genes responsible for the normal controls of cell behavior (see p. 1190). The following observations, however, suggest that this is not the case. Stem cells with very similar properties can be derived by placing a normal inner cell mass in culture and dispersing the cells as soon as they proliferate. Once dispersed, some of the cells will continue dividing indefinitely without altering their character; these can then be used to establish permanent clonal lines with many properties of normal inner-cell-mass cells. These cell lines are almost indistinguishable from teratocarcinoma-derived cell lines, but they can be generated at such high frequency from normal embryos that it is most unlikely that they arise by mutation. Rather, it appears that separating the cells from their normal companions deprives them of cues that would ordinarily limit their proliferation and promote their progressive differentiation.

Moreover, the abnormal behavior of such *embryonal stem cells*—whether they are derived from teratocarcinomas or by direct culture of normal embryos—can be reversed if the cells are placed in a more normal environment. This can be achieved by injecting them into the blastocoel cavity of a normal blastocyst (Figure 16–27). The injected cells become incorporated in the inner cell mass of the blastocyst and can contribute to the formation of an apparently normal chimeric mouse: descendants of the injected embryonal stem cells can be found in practically any of the tissues of this mouse, where they differentiate in a well-behaved manner appropriate to their location and can even form viable germ cells. As this experiment directly demonstrates, the cues that a cell in an early mammalian embryo gets from its neighbors are critically important in determining its fate.

In Multicellular Animals the Behavior of a Cell Depends on Its History as Well as on Its Environment and Its Genome

Cells must not only become different, they must also *remain* different after the original cues responsible for cell diversification have disappeared. Despite the continual turnover and resynthesis of almost all cell components, most cell types in the adult body are stably and heritably different from one another and retain their specialized character even when the environment is changed. Thus, when a pigment cell divides, its daughters remain pigment cells; when a liver cell divides, its daughters remain liver cells; and so on. The differences between cell types are ultimately due to the different influences that the cells have been subjected to in the embryo, but the differences are maintained because the cells somehow remember the effects of those past influences and pass them on to their descendants. The humblest bacterium can rapidly adjust its chemical activities in response to changes in its environment. But the cells of a higher animal are more sophisticated than that. Their behavior is governed not only by their genome and their *present* environment, but also by their history.

Cells Often Become Determined for a Future Specialized Role Long Before They Differentiate Overtly[14,22]

The most familiar evidence of cell memory is seen in the persistence and stability of the differentiated states of cells in the adult body (see p. 752). But the final character of a cell has usually been decided by a complex sequence of cues delivered to its ancestors during development and may be fixed long before differentiation becomes manifest. Through a series of decisions taken before, during, and just after gastrulation, for example, certain cells in the somites of a vertebrate become specialized at a very early stage as precursors of skeletal muscle cells and migrate from the somites into various other regions including those where the limbs will form (see p. 943 for details). These muscle-cell precursors do not contain the large quantities of specialized contractile proteins found in mature muscle cells; indeed, they look superficially just like the other cells of the limb rudiment, which do not derive from the somites. But after several days they begin manufacturing large quantities of specialized muscle proteins, whereas the other limb cells with which they are mingled differentiate into various types of connective-tissue cells. Thus the developmental choice between muscle and connective tissue has been made long before it is expressed in overt differentiation, and it is presumably recorded in the cells as a chemical distinction of a much less obvious sort (probably through a change in the state of activation of a muscle-specific master regulatory gene—see p. 556).

A cell that has made a developmental choice is said to be **determined.** Determination is such an important and subtle concept in embryology, however, that we need a stricter definition. We should say: a cell is determined if it has undergone a self-perpetuating change of internal character that distinguishes it and its progeny from other cells in the embryo and commits them to a specialized course of development. The clauses of this definition are worth considering individually:

1. The change must distinguish the cell and its progeny from other cells and commit them to a specialized course of development: a cell is not determined simply because it is a step ahead of others in its maturation. Determination involves selecting a particular developmental pathway.

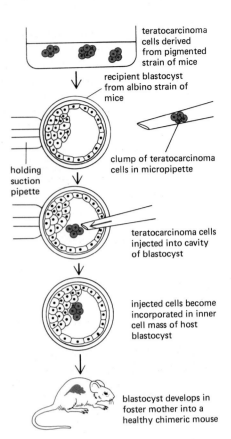

Figure 16–27 Experiment showing that teratocarcinoma cells can combine with the cells of a normal blastocyst to form a healthy chimeric mouse.

2. The change must be a self-perpetuating change of internal character: the element of cell memory is essential. Determination involves establishing differences that are *heritable* from one cell generation to the next. A cell is not considered to be determined simply because it has come to occupy a particular position in the body, from which its specialized prospective fate can normally be predicted, and it is not determined if it loses its distinctive character whenever the external influences that gave rise to the distinction disappear.

Determination is sometimes defined as an *irreversible* change. Because one cannot test a cell's behavior in all possible environments, however, reversibility cannot ever be entirely excluded; moreover, occasionally cell memory can be upset and the state of determination can be altered. For these reasons we prefer to be less absolute and to speak of determination as a *self-perpetuating* change in cell character.

The term **differentiation** is generally reserved for *overt* cell specialization, that is, for a specialization of cell character that is grossly apparent. Usually a cell becomes determined before it differentiates, although in some cases the two processes occur simultaneously. Indeed, it is possible in principle for differentiation to occur without determination, if the overt specialization of cell character is reversible.

The Time of Cell Determination Can Be Discovered by Transplantation Experiments[23]

To prove that a cell or group of cells is determined, one must show that it has a distinctive character that is maintained even when its circumstances are altered by experimental manipulation. The standard technique is to transplant the cell or cells to a test environment.

A simple example of such an experiment comes from studies on amphibian embryos. As noted earlier, one can plot a fate map for a blastula or an early gastrula, showing which of its parts will normally develop into what. The cells in one region, for example, are fated to become epidermis if development proceeds normally, while those in another region are fated to form brain. To establish when these two groups of cells become *determined* to follow their particular modes of differentiation, a block of cells is cut from the prospective epidermal region and put in the position of prospective brain, and vice versa. If the cells are already determined at the time of transplantation, they should develop autonomously according to their origins: the cells from the prospective epidermal region as misplaced epidermal tissue in the brain, and those from the prospective brain region as misplaced brain tissue in the epidermis. In fact, cells transplanted at the early gastrula stage show no memory of their origins and differentiate in the fashion appropriate to their new locations. If, however, the same experiment is done at a somewhat later stage, in the late gastrula, the prospective brain cells transplanted to an epidermal site will differentiate as misplaced neural tissue, while the prospective epidermal cells transplanted to a brain site differentiate there as misplaced epidermis. This shows that both groups of cells have become determined sometime between the early and late gastrula stages.

The State of Determination May Be Governed by the Cytoplasm or May Be Intrinsic to the Chromosomes[24,25]

Cell memory, as manifested in the phenomenon of determination, presents one of the most challenging problems in molecular biology: what makes certain patterns of gene expression self-sustaining? A detailed biochemical discussion of this question is given in Chapter 10. For our present purposes it is helpful to divide the possible molecular mechanisms into two broad classes, which may be called *cytoplasmic* and *nuclear*, respectively.

In **cytoplasmic memory,** components encoded by the set of active genes are present in the cytoplasm (or in the extracellular environment) and act back on the genome, directly or indirectly, to maintain the selective expression of that specific set of genes; different types of cells have cytoplasms containing different controlling factors. Thus, if a nucleus is taken from one type of differentiated cell and injected into the cytoplasm of another type, the pattern of gene expression should alter to match the character of the host cytoplasm. The nuclear transplantation experiments on amphibian eggs (see p. 891), as well as some studies involving artificial cell fusion, give evidence of this sort of behavior, indicating that cytoplasmic memory mechanisms can be important.

By contrast, **nuclear memory,** also known as *genomic imprinting,* depends on self-sustaining changes that are intrinsic to the chromosomes. Most of these changes leave the DNA sequence unaltered but define the selection of genes to be expressed. The best-understood example of a mechanism for nuclear memory depends on DNA *methylation:* as explained on page 583, through the operation of a maintenance methylase enzyme, an existing pattern of methylation of cytosines in the DNA can be preserved from one cell generation to the next.

Embryonic development provides a natural test for nuclear memory. The sperm and the unfertilized egg are cells in radically different states of differentiation, they contain almost identical sets of genes, and fertilization puts their chromosomes together in a single cell. Do sperm-derived and egg-derived chromosomes remain functionally different after they have come together in the zygote? Studies in the mouse show that the answer is yes.

The Sets of Chromosomes Derived from Egg and Sperm Retain the Imprint of Their Different Histories[25]

As mentioned earlier, an unfertilized egg can be triggered into cleavage without the help of sperm: whether the activation occurs spontaneously or is provoked artificially, the result is a *parthenogenetic* embryo (see p. 897). In many species, including some vertebrates (certain lizards and turkeys), such an embryo can develop into a normal healthy adult. In mammals, however, parthenogenetic embryos degenerate at an early stage: despite much interest and experimental effort, there are no convincingly authenticated cases of mammalian virgin birth. The reason for this failure has become apparent from experimental manipulations of the mouse egg.

A newly fertilized egg (a zygote) contains two pronuclei, one from the father and one from the mother. With a fine pipette, it is possible to suck out one of these pronuclei and replace it with a pronucleus from another egg (Figure 16–28). Thus one can construct a zygote that contains two maternal pronuclei or two paternal pronuclei. The genome in either case will be the same (assuming that the paternal pronuclei derive from sperm bearing an X chromosome rather than a Y chromosome). However, the zygote with two paternal pronuclei gives rise to an embryo that is defective in structures that would normally develop from the inner cell mass, while the zygote with two maternal pronuclei gives rise to an embryo that is defective in structures that would normally develop from the trophectoderm. Since the cytoplasm is the same in the two types of zygote, these

Figure 16–28 A technique for transplanting pronuclei between mouse eggs so as to make eggs whose two pronuclei are both of male (or female) origin. The membrane fusion required is promoted by injecting a little inactivated Sendai virus together with the transplanted pronucleus in the third step.

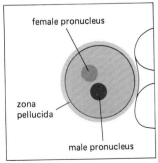

female pronucleus

zona
pellucida

male pronucleus

fertilized egg held by gentle
suction on a holding pipette

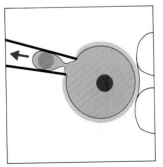

the female pronucleus is
removed by suction on a
micropipette

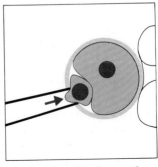

a male pronucleus from another
fertilized egg is injected inside
the zona pellucida

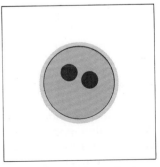

membrane fusion produces
an egg with two male
pronuclei

results strongly suggest that the maternal and paternal chromosomes express different genes as a consequence of nuclear memory; that is, some genes are expressed only when inherited from the father, while others are expressed only when inherited from the mother.

Observations on transgenic mice have shown that genes can behave in just this way and implicate DNA methylation as the mechanism of this **genomic imprinting.** Transgenic mice carrying a foreign DNA sequence, for example, can be bred to produce mice that have inherited the inserted sequence exclusively from the father or exclusively from the mother. In such mice the sequence is often differentially methylated in the zygote according to whether it originated from the sperm or the egg. The methylation pattern is preserved in non-germ-line tissues into adult life but can be reversed during the production of the next generation of gametes. For at least one gene that behaves in this way, the methylation pattern seems to determine gene expression; thus, if the gene is inherited from the mother, it is methylated and not expressed in the animal, while the same gene inherited from the father is nonmethylated and expressed.

Summary

In the course of development, many different types of cells are generated from the fertilized egg. With very rare exceptions, the genomes of the differentiated cells remain the same: it is the pattern of gene expression that changes. The differences between cells may originate from the unequal distribution of cytoplasmic determinants localized in the egg before cleavage or from subsequent differences in the environments of the cells in the embryo. In Xenopus, for example, the animal and vegetal blastomeres inherit different cytoplasmic determinants from the egg, and an influence from the vegetal blastomeres then induces some of the animal blastomeres to develop as mesoderm instead of ectoderm. The induction seems to be mediated by signaling molecules (FGF and TGF-β2, or their analogues) that also help regulate growth and differentiation in the mature organism.

Mammalian eggs are exceptional in that they are essentially symmetrical. Thus all the blastomeres in a mammalian embryo are initially alike and become different only through their interactions with one another. Through cell-cell interactions, cells from two different early mouse embryos can adjust their fates and collaborate to form a single chimeric mouse. Early mouse embryo cells removed from the normal influences of their neighbors can proliferate inappropriately to give rise to teratocarcinomas, from which embryonal stem cells can be obtained. But when they are implanted into a normal early embryo, such cells revert to normal behavior, and their progeny differentiate according to their environment and can contribute to the formation of a healthy chimeric animal.

Embryonic cells must not only become different, they must also remain different even after the influence that initiated cell diversification has disappeared. This requires cell memory, which enables cells to become determined for a particular specialized role long before they differentiate overtly. The mechanisms of cell memory may be either cytoplasmic, involving molecules in the cytoplasm that act back on the nucleus to maintain their own synthesis, or nuclear, involving processes of chromatin modification, such as DNA methylation.

Developmental Programs of Individual Cells: Lineage Analysis and the Nematode Worm[26]

For cells, as for computers, memory makes complex programs possible; and many cells together, each one stepping through its complex developmental control program, generate a complex adult body. Some of the steps that a cell takes in the course of development are autonomous, while others are affected by signals from surrounding cells. Thus the cells of the embryo can be likened to an array of computers operating in parallel and exchanging information with one another. Each cell contains the same genome and therefore the same built-in program, but it can exist in a variety of states; the program directs development along various alternative paths according to a combination of the past information the cell has remembered and the present environmental signals it receives. Computer modeling shows that even a very simple program can lead to the production of astonishingly complex patterns of cell states in such an array; one cannot deduce the program simply by observing the normal development of the pattern.

Because of the important role of cell memory, the behavior of a cell at any time will generally depend on choices made by its ancestors in previous division cycles. To describe and understand the program fully, therefore, one needs to know the pattern of cell divisions so as to be able to trace the ancestry of the individual cells in the embryo. This essential task of **lineage analysis** is often difficult, especially in large, complex animals such as vertebrates. Individual cells at early stages usually must be specially marked if they and their progeny are to be identified later. One approach is to microinject a tracer molecule—a fluorescent dye, for example, or the enzyme horseradish peroxidase (see p. 1064)—into the early embryonic cell (Figure 16–29). Alternatively, single cells in the embryo can be marked genetically—for example, by allowing them to become infected with a specially designed retrovirus (see p. 978) or by exposing the embryo to ionizing radiation, which induces occasional somatic mutations (see p. 932). These methods of lineage analysis are laborious, and each experiment yields only a small fragment of information about the pedigree of the cells of the embryo. To make matters worse, in vertebrates and many other organisms the details of cell lineage show random variations, even between genetically identical animals.

In some lower phyla, however, including mollusks, annelids, and nematode worms, the divisions and movements of individual cells are ordered with extreme precision and in an almost identical way in each member of the species. This strict reproducibility has been exploited in studies of a small transparent nematode, *Caenorhabditis elegans*. Because of its simple and invariant structure, its development can be followed cell by cell all the way from the egg to the adult animal, and one can draw up a complete account of the lineage of every cell in the body. Against this background, the effects of mutations and other experimental disturbances can be pinpointed very precisely. In this way one can begin to associate particular genes with particular steps in the cellular developmental control program. As we shall see, however, it is no easy task to discover the underlying logic of the program, especially where cell-cell interactions are involved. We shall end this section with an example showing how experiments in culture have provided a direct way to analyze small parts of the developmental control programs of individual mammalian cells.

Figure 16–29 The tracing of cell lineage in an ascidian, *Halocynthia roretzi*. (A) A 64-cell embryo. At this stage the tracer molecule horseradish peroxidase (HRP) is injected into a single blastomere. Individual blastomeres, such as the two marked in color on the drawing below the photograph, can be reproducibly identified in every specimen. (B) Larvae that have developed from a set of six such embryos, processed for HRP so as to reveal (*in black*) the cells descended from the chosen blastomere in each case. For each of the two larvae in the left-hand column, the ancestral labeled blastomere was the upper one marked in color in (A); for the two larvae in the middle column, it was the lower colored blastomere in (A); and for the two larvae in the right-hand column, it was a blastomere hidden from view in (A). In this species a given blastomere always gives rise to the same part of the larva (photos). (From H. Nishida, *Dev. Biol.* 121:526–541, 1987.)

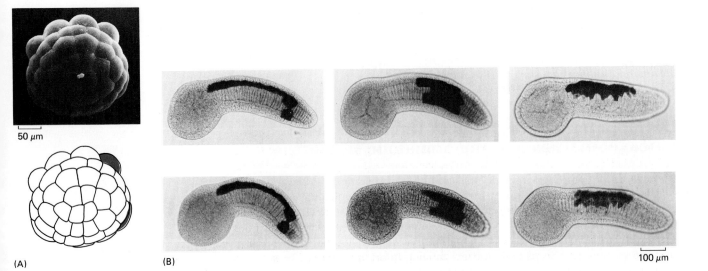

50 μm

100 μm

(A)   (B)

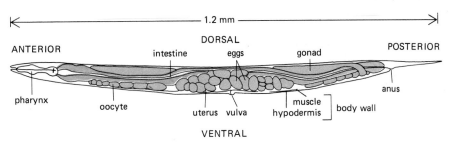

Figure 16–30 *Caenorhabditis elegans:* a side view of the adult hermaphrodite form. (From J.E. Sulston and H.R. Horvitz, *Dev. Biol.* 56:110–156, 1977.)

Caenorhabditis elegans Is Anatomically and Genetically Simple[27]

As an adult, *C. elegans* is about 1 mm long and consists of only some 1000 somatic cells and 1000–2000 germ cells (exactly 959 somatic cell nuclei plus about 2000 germ cells are counted in one sex, exactly 1031 somatic cell nuclei plus about 1000 germ cells in the other) (Figure 16–30). The anatomy has been reconstructed, cell by cell, by electron microscopy of serial sections. The body plan of this simple worm is fundamentally the same as that of most higher animals in that it has a bilaterally symmetrical, elongate body composed of the same basic tissues (nerve, muscle, gut, skin) organized in the same basic way (mouth and brain at the anterior end, anus at the posterior). The outer body wall is composed of two layers: the protective hypodermis, or "skin," and the underlying muscular layer. A simple tube of endodermal cells forms the intestine. A second tube, located between the intestine and the body wall, constitutes the gonad; the tube wall is composed of somatic cells, with the germ cells inside it. *C. elegans* has two sexes—a hermaphrodite and a male. The hermaphrodite can be viewed most simply as a female that produces a limited number of sperm: she can reproduce either by self-fertilization, using her own sperm, or by cross-fertilization after transfer of male sperm by mating. Self-fertilization tends to produce homozygous progeny, which helps to make *C. elegans* an exceptionally convenient organism for genetic studies.

The relative simplicity of *C. elegans* anatomy is reflected in a similar simplicity of its genome. The animal has an estimated total of 3000 essential genes on its six homologous pairs of chromosomes; it has 80×10^6 nucleotide pairs of DNA in its haploid genome, which is about 17 times more than *E. coli* and 38 times less than humans. Currently, about 800 genes have been identified by mutation. These include genes that influence visible features such as the shape or behavior of the worm, genes that code for known proteins such as myosin, and genes that control the course of development. In addition, nearly the entire genome has been mapped as a large set of overlapping DNA segments, represented by a library of ordered genomic clones (see p. 260).

Nematode Development Is Almost Perfectly Invariant[28]

C. elegans begins life as a single cell, the fertilized egg, which gives rise, through repeated cell divisions, to 558 cells that form a small worm inside the egg shell. After hatching, further divisions result in the growth and sexual maturation of the worm as it passes through four successive larval stages separated by molts. After the final molt to the adult stage, the hermaphrodite worm begins to produce its own eggs. The entire developmental sequence, from egg to egg, takes only about three days.

Since *C. elegans* is transparent, cells can be watched as they divide, migrate, and differentiate in living animals (Figure 16–31). By this simple technique of direct observation, the behavior and lineage of all of the cells from the single-cell egg to the adult animal have been described. These descriptive studies have revealed that the somatic structures of the animal develop by an invariant cell lineage and

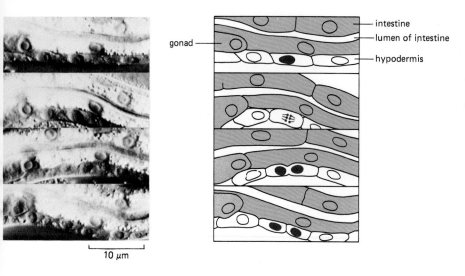

gonad

intestine
lumen of intestine
hypodermis

10 μm

Figure 16–31 Part of the midventral region of a living *C. elegans* larva photographed at four successive times, separated by intervals of a few minutes. A cell in the hypodermis can be seen dividing. (Courtesy of John Sulston and Judith Kimble.)

that each of the many cell divisions is precisely timed. This means that a given precursor cell follows the same pattern of cell divisions in every individual, and with very few exceptions, the fate of each descendant cell can be predicted from its position in the lineage tree (Figure 16–32). In contrast, the cells of the germ line do not develop by an invariant lineage: after hatching, the two germ-line precursor cells undergo a variable sequence of cell divisions, generating a large number of progeny in a partly random arrangement in the gonad.

The full description of cell lineage in *C. elegans* leads to an immediate answer to a fundamental question. The nematode, like most animals, is formed from a relatively large number of cells that can be classified into a much smaller number of differentiated cell types. Given the importance of cell ancestry, one might be tempted to guess that the differences of cell type correspond to different clonal

Figure 16–32 The lineage tree for the cells that form the intestine of *C. elegans*. The egg (*top*) is drawn to the same scale as the adult (*bottom*). Note that although the intestinal cells form a single clone (as do the germ-line cells), the cells of most other tissues do not.

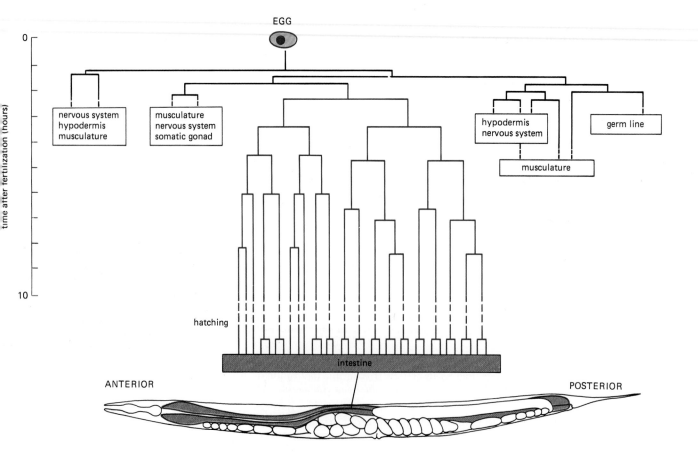

origins: all the cells of a given type might be descendants of a single "founder cell" committed exclusively to that developmental pathway. Lineage analysis shows, however, that this is not generally true, either for nematodes or for other animals, even though particular cells may become determined at a very early stage as progenitors of particular cell types. Thus in *C. elegans* (with a few exceptions such as the intestinal cells and the germ-line cells) each class of differentiated cells—hypodermal, neuronal, muscular, gonadal—is derived from *several* founder cells, originating in separate branches of the lineage tree (see Figure 16–32). Thus cells of similar character need not be close relatives. Conversely (but rarely), cells of very different character may be closely related by lineage; for example, some of the neurons in *C. elegans* are sisters of muscle cells.

The problem, then, is to understand the intricate rules that operate in each branch of the lineage tree to generate a specific array of cell types, each in appropriate numbers.

Developmental Control Genes Define the Program That Generates the Nematode Cell Lineage[29]

During development, a cell, like a digital computer, seems to make a series of discrete choices: divide or do not divide, become neuron or muscle cell or progenitor of this or that branch of the lineage tree. To explain the cellular developmental control program in genetic terms, one must be able to identify the genes that control these choices. Mutations in such genes will disturb development, but they are not the only mutations that do so. Some mutations, for example, will cut short all cell lineages and cause premature death of the embryo simply because they disrupt "housekeeping" genes that every cell needs in order to survive and proliferate. Other mutations will affect genes for proteins that particular types of differentiated cells require in order to carry out their specialized function; the body plan will then be essentially normal, but certain cell types, although still identifiable, will malfunction. Mutations in genes that are involved specifically in controlling developmental choices, by contrast, will disturb the body plan: they typically give rise to cells of the normal differentiated types arranged in an abnormal pattern or in abnormal numbers as a result of specific alterations in the lineage tree. Through such mutations one can identify **developmental control genes** and classify them in terms of cell lineage diagrams (Figure 16–33).

Mutations in some of these control genes simply cause replacement of one cell type by another in the final step of certain lineages. Others alter a choice made at an earlier stage, so that one whole branch of the tree is replaced by another. As illustrated in Figure 16–34, a particular lineage motif may often be repeated in several separate branches of the tree, and a single mutation may similarly alter all or many of the occurrences of this motif. Such phenomena suggest the existence of standard "subroutines" that are called into play in several locations. In general, each branch of the lineage tree appears to be controlled by a complex combination of genes, many of which are involved in the control of other branches as well.

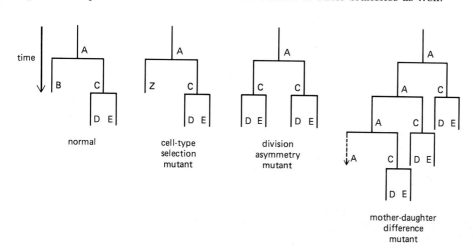

Figure 16–33 A simple fragment of a lineage diagram and some of the types of variation it can undergo as a result of mutations in developmental control genes. The mutant phenotypes suggest the normal functions of the mutated control genes, as listed below each lineage diagram. Horizontal lines represent cell divisions and vertical lines represent cells, with a letter beside each to indicate the cell's internal state.

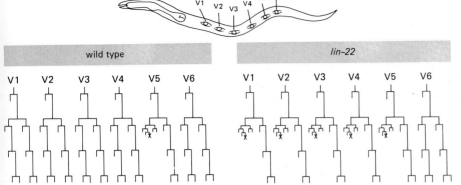

Figure 16–34 Lineage diagrams illustrating how a mutation in a developmental control gene (here a gene called *lin-22*) can cause the same substitution of one lineage motif for another in several separate branches of the lineage tree. Lines of descent are shown for the progeny of each of six precursor cells in the larval hypodermis of *C. elegans*; the locations of these cells are indicated above on a diagram of a young larva. The cross denotes a programmed cell death—a common feature of normal development, both in *C. elegans* and in other species. (After H.R. Horvitz et al., *Cold Spring Harbor Symp. Quant. Biol.* 48:453–463, 1983.)

As computer programmers know, small changes in a program can have drastic effects on the output produced when a program is executed. Likewise, a mutation in a single control gene can result in a grossly abnormal lineage tree. This is well illustrated by the *heterochronic mutations*, which cause certain sets of cells to behave in a way that would be appropriate for normal cells at a different stage in development. A daughter cell may behave like a parent or grandparent, for example, and the offspring of the daughter may behave again in the same way, and so on, with the result that a portion of the lineage pattern is reiterated several times and development is in effect retarded. Figure 16–35 shows the lineage diagrams for a set of mutations in a gene called *lin-14*, illustrating this phenomenon: instead of progressing through the normal series of cell divisions characteristic of the first, second, third, and fourth larval instars and then halting, many of the cells in certain *lin-14* mutants repeatedly go through the patterns of cell divisions characteristic of the first larval instar, continuing through as many as five or six molt cycles and persisting in the manufacture of an immature type of cuticle. Other mutations in the *lin-14* gene have the reverse effect, causing cells to adopt mature states precociously, skipping intermediate stages, so that the animal reaches its final state prematurely and with an abnormally small number of cells. This precocious development occurs in mutants deficient in normal *lin-14* activity, whereas retarded development occurs in mutants with abnormally high levels of *lin-14* activity. Thus the effect of the *lin-14* gene product is to keep the cells young, and normal development seems to depend on its progressive reduction as the animal matures.

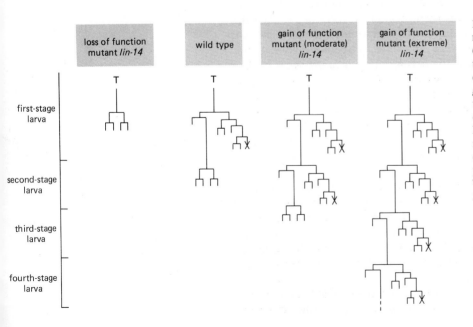

Figure 16–35 Heterochronic mutations in the gene *lin-14* of *C. elegans* and their effects on one of the many lineages they alter. The loss-of-function (recessive) mutation in this gene causes premature occurrence of the pattern of cell division and differentiation characteristic of a late larva; the gain-of-function (dominant) mutations in this gene have the opposite effect. The cross denotes a programmed cell death. (After V. Ambros and H.R. Horvitz, *Science* 226:409–416, 1984.)

The Program of Differentiation Is Coordinated with the Program of Cell Division[30]

Heterochronic mutations emphasize another important general point: the genome during development has to define a control program for cell division as well as cell differentiation, and the two aspects of cell behavior have to be synchronized. In the heterochronic mutant both division and differentiation are affected in a coordinated way, suggesting that both processes are regulated by one master timing mechanism that has been disturbed by the mutation.

One possibility is that the cell division cycle itself serves as the timer, but the evidence on the whole is against this: cells in developing embryos frequently differentiate on schedule even when cell division is artificially prevented by agents that block either cytokinesis or DNA synthesis. Thus for many species it seems that cell divisions usually are not in themselves the ticks of the clock that sets the tempo of biochemical development but rather are governed by this clock: the changing chemical state of the cell controls in parallel the decision to divide and the decision as to when and how to differentiate. The molecular mechanisms of cell-division control in embryos are almost completely unknown, and they pose one of the central unsolved problems of developmental biology. Cell-lineage mutants in the nematode may hold a key to the solution.

Development Depends on an Interplay Between Cell-autonomous Behavior and Cell-Cell Interactions[31]

One cannot begin to understand a cell's behavior during development until one knows whether it is autonomous or is governed by signals from the surroundings. The importance of signals from neighboring cells can be assessed in the nematode by observing the consequences when these cells are individually eliminated. This can be done using a focused laser beam with a diameter of about 0.5 μm (the average nucleus in *C. elegans* is about 2 μm in diameter). When a cell nucleus is exposed to repeated pulses of laser light, the cell dies with no apparent damage to other cells in the animal. Most often it is found that the surviving cells persist in their normal course of development even after their close neighbors have been destroyed. This suggests that most of the cells in the nematode, during much of their history, are following their developmental programs autonomously, without regard to signals from adjacent cells.

Nevertheless, intercellular signals play a crucial part in the development of *C. elegans*, as in that of other animals. A good example is provided by the development of the *vulva*, the egg-laying orifice in hermaphrodites. The vulva is a ventral opening in the hypodermis (skin) formed by 22 cells that arise by specific lineages from three precursor cells in the hypodermis. A single nondividing cell in the gonad, called the *anchor cell*, attaches or "anchors" the developing vulva to the overlying gonad (the uterus) to create a passageway through which the eggs can pass to the outside world. Laser destruction studies show that the anchor cell is responsible for inducing the three nearest hypodermal cells to form a vulva. If the anchor cell is killed, these cells, instead of following a vulval lineage, give rise to ordinary hypodermal cells. Thus the anchor cell induces vulval differentiation in *C. elegans* just as the vegetal blastomeres induce mesodermal differentiation in the early *Xenopus* embryo. Only the anchor cell is necessary for this induction: if all the gonadal cells except the anchor cell are killed, the vulva still develops normally (Figure 16–36A).

The inducing signal from the anchor cell ensures that the vulva develops in exactly the right place in relation to the gonad, and there is some potential for flexibility in the system: the three hypodermal cells that normally form the vulva are flanked by three others that are also capable of doing so if they are brought under the influence of the anchor cell. If the normal vulval precursor cells are destroyed with a laser beam, for example, these neighboring cells are diverted from their usual hypodermal fate (perhaps through a shift in their position) and form a vulva instead. Cells other than the three immediate neighbors, however, are unable to respond to the inducing signal and will not form a vulva under any

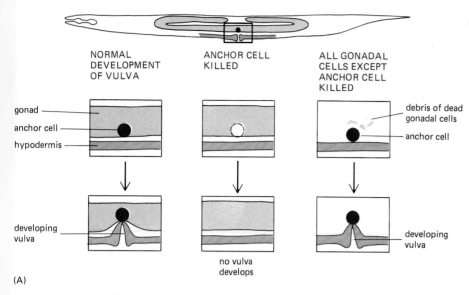

Figure 16–36 (A) Experiments showing that an inductive influence from the anchor cell is required for normal development of the vulva. (B) A magnified view of the six cells of the vulval equivalence group (*colored*) in the ventral hypodermis, with the normal lineage diagrams of their progeny sketched below. All six of these cells (and no others) are capable of responding to the vulva-inducing influence, but only three of them are normally exposed to it.

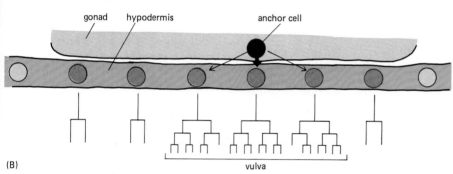

circumstances. Because they are functionally interchangeable in this way, the three normal vulval precursors together with their three similarly inducible neighbors are said to form an **equivalence group**—one of several such groups that have been identified in C. elegans (Figure 16–36B). The members of this equivalence group differ from surrounding cells because they have had a different developmental history, which has made them uniquely responsive to the inducing signal. Subsequently the cells in the group become different from one another depending on their exposure to the signal. Here we have a specific example of the way in which the cellular program of development depends on an interplay between cell memory—a cell-autonomous property—and cell signaling—an extrinsic control.

Experimental Manipulations Clarify the Role of Developmental Control Genes[32]

Experimental manipulations such as those just described help in the interpretation of data from genetic studies. Mutations in any of at least 5 different genes, for example, can cause a *multivulva* phenotype in which all 6 cells of the vulval equivalence group follow the vulval course of development, so that as many as 48 vulval cells are made rather than 22, and multiple vulvae are formed instead of only one. In these mutants all six cells behave as though they are locked into the anchor-cell-activated state, and destruction of the anchor cell no longer affects the vulval lineages. Evidently the mutations affect genes that normally act in the cells of the vulval equivalence group to determine their response to the anchor-cell signal. Mutations in another small set of genes cause an opposite, *vulvaless* phenotype—again, it seems, by altering the response properties of the cells in the vulval equivalence group.

Mutations in one particular multivulva gene, called *lin-12*, have a double effect on vulval development: they cause absence or duplication of the anchor cell, thereby altering the distribution of the inducing signal, and they change the rules by which the cells of the vulval equivalence group respond to that signal. Mutations in *lin-12* affect several other equivalence groups as well, suggesting that *lin-12* has some general role in the cell-cell interactions through which the members of an equivalence group become consigned to different fates. The sequence of the *lin-12* gene has been determined, and the protein it encodes turns out to be homologous to a family of proteins thought to be involved in cell-cell signaling in vertebrates and in insects (see p. 706). This family includes the mammalian epidermal growth factor (EGF), certain mammalian cell-surface receptors such as the low-density lipoprotein (LDL) receptor (see p. 328), and the product of a gene called *Notch*, which helps to control the developmental choice between neuronal and epidermal differentiation in *Drosophila*. Such findings suggest that the molecular lessons learned in the nematode will also be applicable to many other species.

The Developmental Programs of Individual Mammalian Cells Can Be Analyzed in Cell Culture: Glial Cell Differentiation in the Rat Optic Nerve[33]

The most direct way to distinguish cell-autonomous behavior from behavior controlled by cell-cell interactions is to test the effects of rearranging or isolating the cells and altering their environment artificially. This is generally difficult in *C. elegans:* cell destruction by a laser beam is a limited tool for the purpose, and the lack of information about cell-cell interactions in the developing nematode often makes it difficult to fathom the underlying cellular control program, even though the lineage of all the cells is known and many mutants in developmental control genes have been characterized. The environment of a cell can be manipulated much more freely in culture, and this has allowed a more direct analysis of the cellular control programs and cell-cell interactions that govern cell division and differentiation in the development of some components of the vertebrate body.

The optic nerve is one of the simplest parts of the mammalian central nervous system and provides a convenient model system for such analysis. It contains the long axons of retinal nerve cells that project from the eye to the brain. Three kinds of supporting, or *glial*, cells (see p. 1064) structurally and functionally support the axons in the nerve: oligodendrocytes and two types of astrocytes, known simply as type 1 and type 2 (Figure 16–37). Antibodies can be used to distinguish the three kinds of glial cells from one another and from their precursors. By monitoring cells taken from optic nerve at different stages of development and placed in culture, it has been shown that the three glial cell types (which normally divide infrequently, if at all) arise at different times from two distinct lineages of rapidly dividing precursor cells. *Type-1 astrocytes* develop before birth from one kind of precursor cell, and both *oligodendrocytes* and *type-2 astrocytes* develop after birth from a second, bipotential precursor, called an *O-2A progenitor cell*. Normally the oligodendrocytes start to appear at birth and the type-2 astrocytes about a week later.

Given this outline of the normal lineage and timetable, the problem is to discover the underlying rules of cell behavior. What, for example, makes O-2A

Figure 16–37 The rat optic nerve contains the axons of retinal ganglion cells (only one of which is shown) and three kinds of glial cells. *Type-1 astrocytes* provide the structural framework of the nerve, much like connective-tissue cells in a limb. *Oligodendrocytes* and *type-2 astrocytes* collaborate to ensheathe each axon: oligodendrocytes wrap around the axon to form an insulating myelin sheath (see p. 1073), while type-2 astrocytes extend fine processes that contact the axon where the myelin sheath is interrupted at nodes of Ranvier. A nerve impulse jumps from node to node as it passes along the axon (see p. 1073).

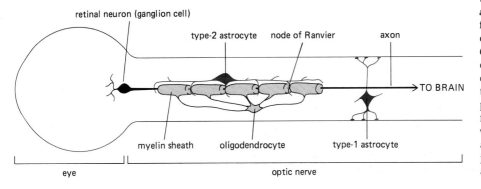

retinal neuron (ganglion cell)

type-2 astrocyte   node of Ranvier   axon

TO BRAIN

myelin sheath   oligodendrocyte   type-1 astrocyte

eye   optic nerve

progenitor cells differentiate when they do, and what causes those that differentiate at one time to become oligodendrocytes and those that differentiate at another time to become type-2 astrocytes? To what extent are these events governed by some autonomous timekeeping process inside each cell, to what extent by the timed delivery of external signals?

When O-2A progenitor cells are dissociated from the embryonic optic nerve and cultured in the absence of the type-1 astrocytes that are normally their neighbors, they stop dividing and differentiate prematurely; but if type-1 astrocytes are included in the culture, the O-2A progenitors carry on dividing and delay differentiation until the equivalent of the day of birth, just as *in vivo*. The effect of the type-1 astrocytes can be traced to a growth factor they secrete (believed to be *platelet-derived growth factor*, PDGF—see p. 746). This factor, however, is only half the story: when present, it prevents O-2A progenitors from differentiating prematurely; but even when its concentration is kept high, it does not delay their differentiation beyond the normal time. Apparently, O-2A progenitors in the embryonic optic nerve become committed to some program of internal change whereby, after a specific period of time or a specific number of division cycles, they lose sensitivity to the growth factor and differentiate whether it is present or not; individual O-2A progenitors become committed at different times, and the first of them normally reach the point of differentiation on the day of birth.

What then decides whether the cell will differentiate into an oligodendrocyte or a type-2 astrocyte? O-2A progenitors cultured alone in a serum-free medium differentiate only into oligodendrocytes: this seems to be the *constitutive* or, in computer terminology, the *default pathway*. In contrast, type-2 astrocyte differentiation seems to be an *induced pathway*: cells are directed into this mode of differentiation by a specific protein that can be extracted from the rat optic nerve—a protein that begins to be made there in appreciable quantities in the second week after birth, just when type-2 astrocyte differentiation begins. This protein appears to be identical with a factor known as ciliary neurotrophic factor, CNTF. It therefore seems likely that the time of production of CNTF determines the time of onset of type-2 astrocyte differentiation in the developing nerve. Although it is

Figure 16–38 Flowchart representing, in the style of a computer program, an oversimplified model of the type of intracellular control program that appears to govern cell division and differentiation in the oligodendrocyte–type-2 astrocyte (O-2A) lineage in the developing optic nerve of a mammal. Broad arrows represent "data inputs" and "data outputs"—that is, extracellular signals that the cell receives or emits. Incomplete or speculative parts of the flowchart are shown by dashed lines; the leftward-pointing broad arrow indicates the possibility that feedback from the O-2A lineage might help regulate the production of signals from the type-1 astrocyte lineage. The O-2A progenitor cells themselves are thought to be generated continuously *in vivo*, starting about one week before birth (not shown).

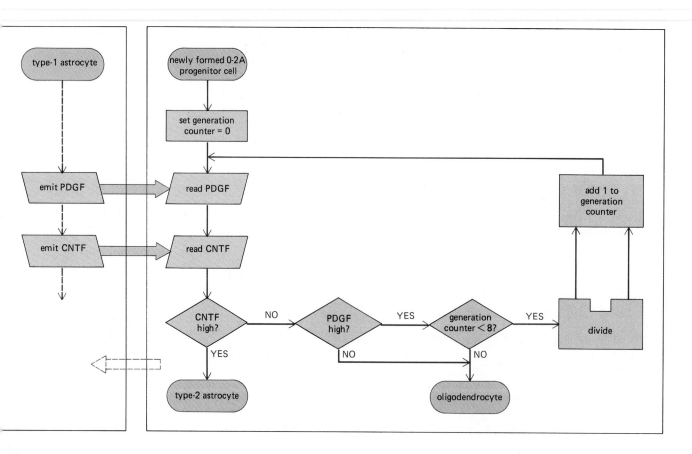

uncertain which cells make CNTF *in vivo* or what factors control the timing of its production, it is known that type-1 astrocytes can make it in culture.

These studies again reveal that developing cells are controlled through an interplay between cell-autonomous behavior and environmental signals, and they show the advantages of cell culture in analyzing the molecular basis of these processes. The findings can be tentatively summarized in a flowchart representing, in computer symbolism, the internal program of the O-2A progenitor cell—that is, the set of rules by which it reads signals from its environment and accordingly selects when and how it will differentiate (Figure 16–38). Such information about the behavior of the individual cell is essential if we are truly to understand embryonic development. By itself, however, it cannot answer the fundamental question of how the whole organism develops its multicellular spatial pattern. This is the problem to which we now turn.

Summary

To understand the developmental programs of individual cells in an embryo, one needs to be able to trace their ancestry—a procedure called lineage analysis. There is some random variability in cell lineage in vertebrates. But in certain nematodes and in some other invertebrate phyla, the pattern of cell divisions during development is controlled so precisely and predictably that a cell at a particular position in the body has the same lineage in every individual. The normal developmental lineage of all the cells of the nematode Caenorhabditis elegans *has been mapped in detail, and the consequences of experimental interference have been studied at the level of individual identified cells. Many of the genes that define the cellular developmental control program have been identified according to the alterations of cell lineage that they cause. In general, mutations in these developmental control genes affect cell differentiation and cell division in a coordinated way, suggesting that these two essential aspects of cell behavior in development are controlled by some common underlying mechanism. By destroying cells in the developing nematode with a laser microbeam, it can be shown that the invariance observed in* C. elegans *development is the result of an accurately meshed interplay between cell-cell interactions and cell-autonomous processes. Inductive interactions can be identified in the development of the vulva, for example, and can be correlated with the functions of specific developmental control genes.*

A different method for analyzing the developmental control programs of individual cells is to follow the cells' behavior in culture. This approach has been used to discover the rules governing cell differentiation and division in the glial cell lineages in a mammalian optic nerve, where again the cells are found to be controlled partly by intercellular signals and partly by cell-autonomous programs.

Principles of Spatial Patterning[34]

The different types of cells in the body are not generated in a random jumble and subsequently sorted out; rather, the differences are created in a spatially organized way. The spatial interactions that coordinate this process of **pattern formation** take many forms. In some cases—as in mosaic eggs—they occur inside a single cell, organizing an intracellular pattern of cytoplasmic determinants that is perpetuated in the pattern of progeny cells formed by cleavage. In other cases—as in the induction of *Xenopus* mesoderm or nematode vulva—the pattern is organized by interactions between separate cells. In any case, spatial signals—either intracellular or intercellular—are essential, and they may be said to provide cells with **positional information** to guide their specialization.

In this section we shall examine some of the ways in which positional information is generated and used. We shall see how, in many organisms, individual cells become endowed with increasingly precise molecular "address labels" or *positional values* as development proceeds. Thus a cell in the early embryo may know only how close it is to the head or the tail end, while its descendants in the adult many cell generations later may know that they are, for example, the bone cells at the tip of the third digit of the forelimb. This specification is built up gradually by mechanisms that depend on cell memory: early spatial signals specify crude global coordinates, and subsequent more local signals supply additional precise details of the address relative to local landmarks.

Asymmetries Imposed by a Patterned Environment Can Be Amplified by Positive Feedback[34,35]

Pattern formation begins with the generation of asymmetry: future head must be distinguished from tail, back from belly. We have seen that in most species the egg itself is asymmetrical (see p. 892). This asymmetry is often imposed on the egg during its development in the ovary. In insects, for example, the oocyte (the future egg) lies at one end of an ovarian follicle, where it is connected on one side to nurse cells that pass materials into it via cytoplasmic bridges (see p. 858). The asymmetrical flow of materials into the oocyte apparently creates chemical differences between regions of the egg cytoplasm that correspond to the future head and tail of the embryo (Figure 16–39).

The initial asymmetry does not have to be strong: a cell can use its own internal mechanisms to amplify a weak asymmetry so as to become highly polarized. The egg of the seaweed *Fucus*, for example, sends out a rhizoid or root process from one side only. The rhizoid normally grows downward and away from light, and gravity and light are both able to determine which side of the egg it will develop from. The establishment of this polarity depends on the flow of Ca^{2+} through the cell (see p. 1179): the incident light or the force of gravity is thought to initiate a slight asymmetry in the distribution of Ca^{2+} transport proteins in the plasma membrane that is then amplified by positive feedback (see Figure 20–66, p. 1180).

Positive feedback can function in a similar way at other stages in the development of multicellular organisms. When a polarized cell divides, for example, its daughters may inherit some degree of asymmetry and, by amplifying this, may themselves in turn become highly polarized, and so on through successive cell generations. Thus a complex pattern of different cell types can in principle be generated without the need for cell-cell signaling (Figure 16–40). It is possible that some of the apparently cell-autonomous pattern formation in the nematode might be explained by this type of mechanism.

More commonly, pattern formation in multicellular systems depends on signaling from cell to cell. Here again, positive feedback can play a crucial part by amplifying weak initial asymmetries so as to create localized groups of cells with highly distinctive properties. These cells may in turn emit signals that influence the development of cells in neighboring regions. The primitive multicellular organism *Hydra* provides a simple example. *Hydra* has a tubelike structure, with a "head" bearing tentacles at one end and a "foot" at the other (see Figure 1–33, p. 29), and it is well known for its ability to regenerate this structure after parts have been amputated. Various amputation and grafting experiments suggest that the pattern is created through a locally autocatalytic process that amplifies any weak tendency for cells to adopt a head character; cells that locally collaborate to develop this character generate a long-range inhibitory signal that diffuses along the body axis and prevents heads from developing elsewhere (Figure 16–41).

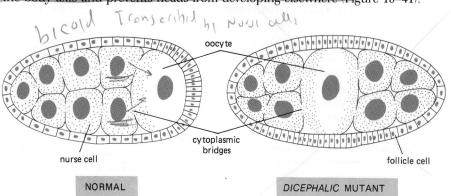

NORMAL   **DICEPHALIC MUTANT**

Figure 16–39 The control of *Drosophila* egg polarity by the environment of the oocyte in the ovary. Normally the nurse cells pass materials into one end of the oocyte, which corresponds to the future head. In the *Dicephalic* mutant the oocyte is placed centrally in the follicle and symmetrically in relation to the nurse cells. As a result, the egg is symmetrical and will give rise to an embryo with heads at both ends.

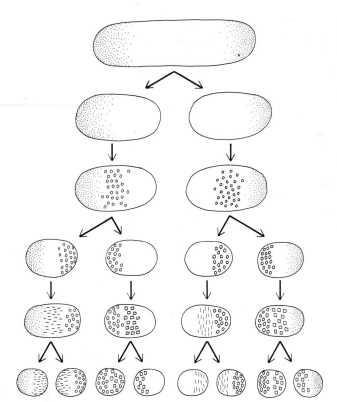

Figure 16–40 A complex multicellular pattern can in principle be generated by a series of asymmetrical cell divisions without any cell-cell interaction (other than that which keeps the cells bound together). An internal polarity inherited by each daughter cell at each cell division can suffice to guide a further round of intracellular pattern formation, so that the daughter cells produced at the next division will again be markedly different and internally polarized.

Localized Signaling Regions Often Act as Sources of a Morphogen Gradient[36]

The long-range influence of the head on the rest of the hydra illustrates a common phenomenon. In many developmental systems a small patch of tissue in a specific region acquires a specialized character and becomes the source of a signal that spreads into neighboring tissue and controls its behavior. The signal may, for example, take the form of a diffusible molecule secreted from the signaling region. Suppose that this substance is slowly degraded as it diffuses through the neighboring tissue. The concentration then will be high near the source and decrease gradually with increasing distance, so that a concentration gradient is established (Figure 16–42). Cells at different distances from the source will be exposed to different concentrations and may become different as a result. A hypothetical substance such as this, whose concentration is read by cells to discover their position relative to a certain landmark or beacon, is termed a **morphogen.** Through the morphogen, the signaling region can control the patterning of a broad field of adjacent tissue. Morphogen gradients are a simple and apparently a common way of providing cells with positional information. They will therefore be the focus of our discussion here; but it must be emphasized that cells in embryos can and do get positional information by many other strategies as well. Figure 16–21 illustrates one such alternative involving a sequence of short-range inductions.

Although there are many examples of signaling regions that are thought to exert a long-range influence by means of a morphogen gradient, there are remarkably few cases where a morphogen has been identified chemically. In most cases all that is known is that when the presumed signaling region is displaced, by grafting or otherwise, the pattern of the tissue nearby is correspondingly altered. It is not possible to tell from this sort of observation how far the pattern is due to direct responses of individual cells to a morphogen and how far it is dependent in addition on interactions between the responding cells. In most instances it is probable that the initial morphogen gradient imposes only the broad organization of the pattern and that local cell-cell interactions then generate the details. There is increasing evidence, for example, that the body plan becomes organized in this way in insects, as we shall discuss below. First, however, let us consider what kinds of pattern a morphogen gradient might engender in a field of cells reacting individually.

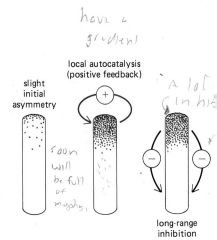

local autocatalysis (positive feedback)

slight initial asymmetry

long-range inhibition

Figure 16–41 A mechanism for producing a spatial pattern through an autocatalytic process. An "activator" substance accumulates autocatalytically to a high level in a certain small region, whose location is determined by a weak initial asymmetry in the system; the high concentration of activator in that region leads to production of an inhibitor substance that diffuses widely and inhibits other regions from making the activator. Thus, if a hydra is decapitated, the remaining tissue will normally regenerate a head; but if a head from another hydra is grafted onto the amputation surface or onto a site at some distance from that at which the regenerated head would normally appear, an additional new head will not form.

Threshold Reactions Can Convert a Smooth Morphogen Gradient into a Sharply Defined Spatial Pattern of Determination[37]

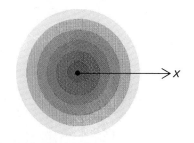

If the concentration gradient of a morphogen is smooth, one might expect that the consequent pattern of cell characters would also be smoothly graded. Smoothly graded patterns of cell character do indeed occur on a small scale in some tissues. But many of the differentiations of greatest interest in development are discrete. The ultimate cell types are sharply distinct; there is no graded series of mature kinds of cells intermediate between cartilage and muscle, for example. Sharp distinctions can arise in a population of initially uniform cells through a *threshold* in the response to a smoothly graded signal: positive feedback in each responding cell can amplify the effect of a small increment in the signal in such a way that cells exposed to only slightly different intensities of the signal are launched on radically different courses of development according to whether their exposure is above or below a certain threshold intensity. There may indeed be several thresholds of response to one signal, so that a single variable may control the pattern of several different choices. Once a cell is well launched on a given course, it will persist on that course even in the absence of the environmental influence that initially controlled the choice. In this way, transient, position-dependent influences can have effects that are "remembered" as discrete choices of cell state and thereby define the spatial pattern of determination. The choice of cell state represents the cell's *memory* of the positional information supplied. This record, registered as an intrinsic feature of the cell itself, may be called its **positional value.**

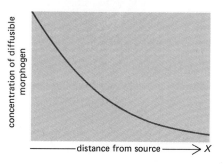

Figure 16–42 If a substance is produced at a point source and is degraded as it diffuses from that point, a concentration gradient results with a maximum at the source. The substance can serve as a morphogen, whose local concentration controls the behavior of cells according to their distance from the source.

Since Embryonic Fields Are Small, Gross Features of the Adult Must Be Determined Early Through Cell Memory[38]

Whatever their nature, the mechanisms for supplying positional information in an animal embryo generally act over only small regions, or *morphogenetic fields*, on the order of a millimeter long (or about 100 cell diameters) or less. There is clearly a limit to the amount of detail that can be defined in so small a space. For this fundamental reason, the final positional specification of a cell has to be built up as a composite of a sequence of items of positional information registered at different times. Cell memory, therefore, is crucial for the development of large complex animals. The distinction between head and tail has to be established when the rudiments of the head and the tail are no more than about a millimeter apart. The circumstances that gave rise to that distinction are ancient history by the time the animal is a centimeter or a meter long; if the distinction between head and tail is to be maintained, it must be through cell memory.

Thus the gross plan of the body is specified early, and successive levels of detail are filled in later as the rudiment of each part grows to a size at which

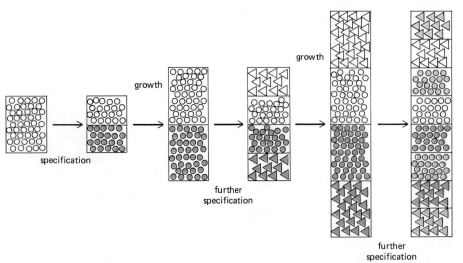

Figure 16–43 When an embryo is small, it becomes subdivided into a few distinct regions corresponding to the major subdivisions of the adult body. The cells in each of these coarse subdivisions are stamped with a crude positional value (represented here by the triangular or circular symbols). As the embryo grows, the subdivisions grow and themselves become subdivided, creating a progressively more fine-grained pattern of positional values.

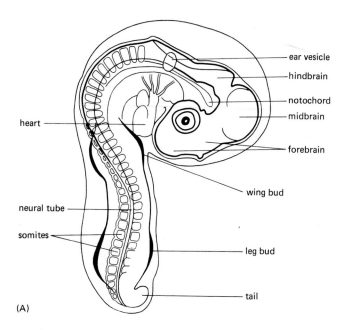

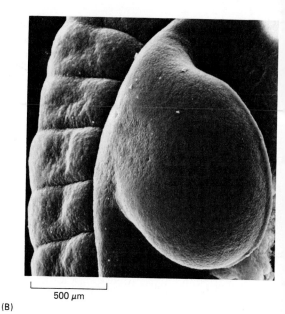

(A)

(B)

500 µm

Figure 16–44 (A) A chick embryo after 3 days of incubation, illustrating the positions of the early limb buds. (B) Scanning electron micrograph showing a dorsal view of the wing bud and adjacent somites one day later; the bud has grown to become a tongue-shaped projection about 1 mm long, 1 mm broad, and 0.5 mm thick. (A, after W.H. Freeman and B. Bracegirdle, An Atlas of Embryology. London: Heinemann, 1967; B, courtesy of Paul Martin.)

mechanisms for supplying additional positional information can conveniently act (Figure 16–43).

We shall now examine in more detail how this occurs, taking our examples from the development of limbs.

Positional Information in Limb Development Is Refined by Installments[39]

The development of the limbs—and of many other organs such as teeth or vertebrae or skin—involves relatively few modes of differentiation. In the case of the limbs, for example, the chief cell types are those of muscle, cartilage, bone, and loose connective tissue. But these few differentiated cell types are arranged in a complex spatial pattern. The forelimb differs from the hindlimb not because it contains different types of tissues but because it contains a different spatial arrangement of tissues. Transplantation experiments with the developing limbs of a chick show that intrinsic differences between the limb cells with respect to the patterns they will generate are determined long before differentiation begins.

In the chick embryo, the leg and the wing originate at about the same time in the form of small tongue-shaped buds projecting from the flank (Figure 16–44). The cells in the two pairs of limb buds appear similar and are undifferentiated at first, showing no hint of the subsequent skeletal pattern (see Figure 14–20, p. 802). A small block of undifferentiated tissue at the base of the leg bud, from the region that would normally give rise to part of the thigh, can be cut out and grafted into the tip of the wing bud. Developing there, the graft forms not the appropriate part of the wing tip, nor a misplaced piece of thigh tissue, but a toe (Figure 16–45). This experiment shows, first, that early leg cells are already determined as leg and, second, that although they are determined as leg, they have not yet been assigned their detailed positional values along the limb axis and can respond to cues in the wing so that they form structures appropriate to the tip of the limb rather than the base. We can thus conclude that the full specification of position in vertebrates is not supplied all at once but is built up from a series of items of positional information registered in the cell memory at different times. The final cell state is arrived at by a sequence of decisions.

Positional Values Make Apparently Similar Cells Nonequivalent[40]

The cells of the forelimb bud and the hindlimb bud, although they give rise to the same range of differentiated types of cells, are evidently *nonequivalent:* they are in intrinsically different states, corresponding to different positional values. The cells of the limb buds may retain the positional values that distinguish leg from wing

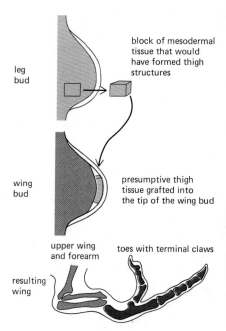

block of mesodermal tissue that would have formed thigh structures

leg bud

presumptive thigh tissue grafted into the tip of the wing bud

wing bud

upper wing and forearm

toes with terminal claws

resulting wing

Figure 16–45 Prospective thigh tissue grafted into the tip of a chick wing bud forms toes. (After J.W. Saunders et al., *Dev. Biol.* 1:281–301, 1959.)

PUT it in a new envir & became a distall
Tissue

even after they have differentiated; if so, it follows that mature cells of the same differentiated type, such as cartilage, can likewise be nonequivalent. This suggests that there may be a much greater variety among the cells of the body than the 200-odd categories into which the cells of a vertebrate are traditionally classified (see p. 995).

There are several pieces of evidence for the existence of such subtle differences between cells of the same differentiated type in different parts of the vertebrate body. Important examples of nonequivalence in the skin and in the nervous system will be discussed later in this chapter (see p. 942) and in Chapter 19 (see p. 1117). One of the most remarkable demonstrations comes from studies of limb regeneration in tailed amphibians (urodeles), which show that there is a persistent nonequivalence not only between forelimb and hindlimb cells, but also between cells at different levels along each limb axis.

If the leg of a newt or axolotl is amputated at any distance from its base, the missing parts are replaced. A *regeneration blastema*, consisting of a mound of apparently undifferentiated cells covered by a cap of epidermis, forms on the end of the amputation stump. The blastema cells proliferate and differentiate to generate precisely the parts that should lie distal to the cut: a hand, for example, if the hand was cut off, but a whole forearm and hand if the whole forearm and hand were cut off. It does not matter whether the blastema develops, as is usual, on the end of the proximal stump of the limb or, by experimental manipulation, on the cut face of the *distal* detached portion of the amputated limb (Figure 16–46). In either case the cells of the blastema give rise to just those limb parts that normally lie *distal* to the level of the cut, even though in the latter instance this means creating a mirror-image duplicate of what is already present. Evidently, cells at different levels along the axis of the mature limb have different remembered positional values, and these guide them in the execution of the program that generates the missing limb parts.

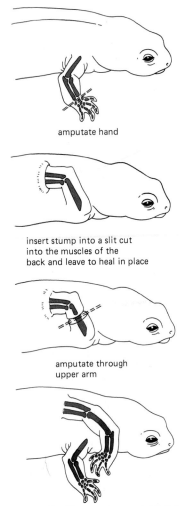

amputate hand

insert stump into a slit cut into the muscles of the back and leave to heal in place

amputate through upper arm

regeneration blastemas form on both cut faces; the cells of both blastemas give rise to just those limb parts that normally lie distal to the level of the cut

Figure 16–46 Scheme of an experiment on the salamander showing that the limb parts produced by a regeneration blastema depend on the level of the cut and not on the structures present in the limb stump. A forearm and a hand are generated from both the distal and the proximal portions of the original limb.

Cells in Separate Fields May Be Supplied with Positional Information in the Same Way but Interpret It Differently[36,41]

The experiment on chick limb development illustrated in Figure 16–45 serves to emphasize a further point about the way in which complex positional specifications are built up. It shows that the cues that provide cells with information about their position along the axis of the vertebrate limb are effectively the same in the leg and the wing. Leg cells grafted into the tip of the wing bud can correctly read the indications that their position is distal and that they should therefore make digits. But they interpret that positional information in their own way and make toes rather than fingers. Thus, with the help of cell memory, the same device for supplying positional information can be used repeatedly in different regions or fields of cells and yet produce a different pattern in each field.

These comments apply not only to the cues that control the pattern along the long axis of the limb—the proximo-distal axis—but also to those that control the pattern along the thumb-to-little-finger, or antero-posterior, axis. These two axes (Figure 16–47) correspond to different components of positional value, as in a Cartesian coordinate system, and they depend on different ways of specifying positional information. The antero-posterior control is of special interest because it provides one of the very few examples of a morphogen that has been chemically identified.

Retinoic Acid Is a Probable Morphogen in the Vertebrate Limb Bud[42]

At the posterior margin of each limb bud in a chick embryo, there lies a group of cells with a remarkable property that can be demonstrated by transplanting them to an anterior position in another limb bud (Figure 16–48). Within a day the host bud grows markedly wider under the influence of the graft, and eventually it develops into a limb with a drastically reorganized pattern: its skeleton and all the accompanying soft tissues are duplicated, forming a second set of elements arranged in mirror-image symmetry about the midline of the limb. The duplicate

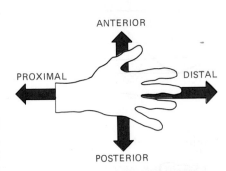

ANTERIOR

PROXIMAL   DISTAL

POSTERIOR

Figure 16–47 The proximo-distal and antero-posterior axes of the vertebrate limb.

elements are formed almost entirely from the host tissue and not from the graft itself. Evidently, the graft specifies a new pattern of structures in the host limb, ordered in a definite antero-posterior sequence in relation to itself. The region from which the graft is taken is called the **zone of polarizing activity** or **polarizing region.**

Both the leg bud and the wing bud possess a polarizing region, and the polarizing regions from the two pairs of limbs, and even from limbs of different species, are functionally interchangeable. Whether the polarizing tissue graft is from leg or wing, from chick or mouse, it will cause a chick wing bud to form additional chick wing digits and a chick leg bud to form additional chick toes. Thus the signal from the polarizing regions of different limbs and different species is the same, but the limb cells interpret the signal according to their own genome and history.

In a normal wing (whose three digits correspond to the three middle digits of a five-digit hand and so are designated by the numbers 2, 3, and 4), digit 4 forms next to the polarizing region, digit 3 farther away, and digit 2 farther still; and in the neighborhood of a grafted polarizing region the same rule applies, as though digit character were specified by the concentration of some morphogen secreted by polarizing tissue (Figure 16–49). The graded nature of the signal from the cells of the polarizing region is demonstrated directly by grafting these cells in small numbers. On average, a graft of 30 cells will cause formation of an extra digit 2. With about 80 polarizing cells, a digit 3 will be produced, and with about 130, a digit 4.

The action of a graft of cells from the polarizing region can be mimicked by implanting an inert carrier material impregnated with *retinoic acid*. The retinoic acid diffuses out of the carrier, setting up a concentration gradient across the limb bud, with a high point at least three times greater than the low point; extra digits begin to be evoked at concentrations on the order of 20 nM. Sensitive chromatographic methods reveal that in the normal limb bud there is a natural gradient of retinoic acid of just this magnitude, with its high point in the polarizing region. This makes it very likely that retinoic acid (Figure 16–50) is the natural morphogen.

The receptor for retinoic acid has recently been identified as a protein homologous to the receptors for steroid and thyroid hormones; it regulates gene transcription by binding to specific DNA sequences. A cDNA clone encoding the receptor has been isolated and sequenced, and so both DNA and antibody probes are now available for tracking the receptor in tissues such as the chick limb.

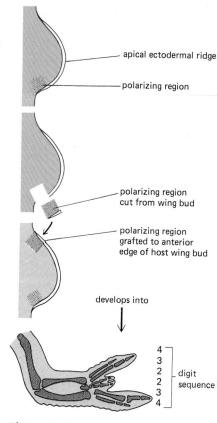

Figure 16–48 A graft of polarizing-region tissue causes a mirror-image duplication of the pattern of the host wing.

Figure 16–49 Distance from grafted and host polarizing regions governs the type of structure formed by wing-bud cells. Numbers 2, 3, and 4 show which digits develop from each part of the bud.

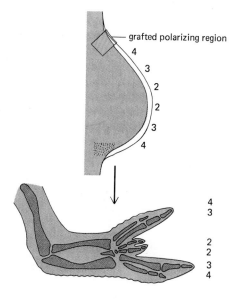
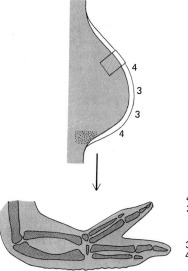
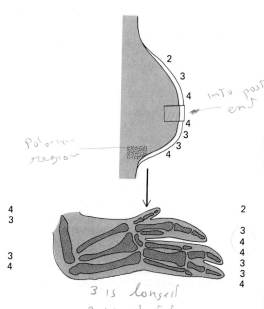

The Pattern of Positional Values Controls Growth and Is Regulated by Intercalation[43]

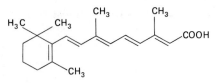

Figure 16–50 The structure of retinoic acid.

In the foregoing discussion of pattern formation and positional values, we have neglected one crucial aspect of the process: the regulation of growth, through which the parts of the pattern attain their appropriate sizes. In some cases this seems to depend on cell-autonomous programs initiated at an early stage in the creation of an organ rudiment. In many other cases, however, growth and the pattern of positional values both depend on continuing cell-cell interactions and depend on them in a closely coupled way. A simple and general rule has been deduced from studies of the regeneration that occurs in various organisms when fragments of tissue with different positional values are juxtaposed and allowed time to grow and adjust. The principles are perhaps most clearly illustrated by studies on the leg of the cockroach.

Cockroaches belong to the class of insects in which there is no radical metamorphosis from larva to adult but a gradual progression through a series of juvenile forms. The juvenile cockroach has well-differentiated limbs, but the differentiated cells—unlike those in human limbs—are still able to respond to the cues that governed the development of the limb pattern, and they can regenerate that pattern if it is disturbed. Thus the workings of the pattern-formation system can be tested by operations done long after the period of embryonic development.

The experiments to be described involve the epidermal sheet of cells and cuticle that covers the cockroach and forms the externally visible parts of the limbs. This outer covering grows by successive molts, in which the juvenile cockroach sheds its old cuticle and lays down a new and larger cuticle in its place. The cuticle is secreted by the epidermal cells, which are arranged underneath in a sheet one cell layer thick. Positional values in the epidermal sheet of cells are displayed in the pattern of the overlying cuticle that they lay down; the effect of experimental manipulation on the patterning of epidermal cells is detected in the cuticle after the animal has molted. Regeneration can be observed only in juveniles, since fully mature adults do not grow or molt.

The cockroach leg consists of several segments, called (in sequence from base to tip) coxa, trochanter, femur, tibia, and tarsus, the tarsus itself being a composite of several smaller segments and terminating in a pair of claws (Figure 16–51). If two legs are amputated through the tibia, say, but at different levels, the distal fragment of the one can be grafted onto the proximal stump of the other in such a way that the composite leg heals with the middle part of the tibia missing. Yet the leg that emerges after the animal has molted appears normal: the missing middle part has regenerated (Figure 16–52A). More surprising is the result of a variant of this operation. The tibia of one cockroach leg is cut through near the proximal end and that of another leg near the distal end. The large detached portion of the first leg is then stuck onto the large remaining stump of the second leg to give an excessively long leg with a middle part present in duplicate (Figure 16–52B). The animal is left to molt. The leg that results, far from being more nearly normal, is now even longer because a third middle part of a tibia has developed between the two already present. As shown in Figure 16–52B, the bristles on this freshly formed region point in the direction opposite to that of the bristles on the rest of the tibia.

Many different operations of this type can be performed. All of them point to the existence of a system of positional values that makes cells in different positions nonequivalent and that is intimately coupled to the control of cell proliferation. It is convenient to describe the positional value by a number that is graded from a maximum at one end of the limb segment to a minimum at the other. In the operations described above, epidermal cells with sharply different positional values are brought together. As a result, new cells are formed by proliferation of the epidermal cells in the neighborhood of the junction. These new cells acquire positional values smoothly interpolated between those of the two sets of cells that were brought into confrontation (Figure 16–52). This behavior is summed up in the **rule of intercalation:** *discontinuities of positional value provoke local cell proliferation, and the newly formed cells take on intermediate positional values so as to restore continuity in the pattern.* Cell proliferation ceases only when cells

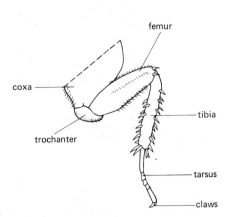

Figure 16–51 The cockroach leg. With each successive molt, the leg grows bigger but does not change its basic structure.

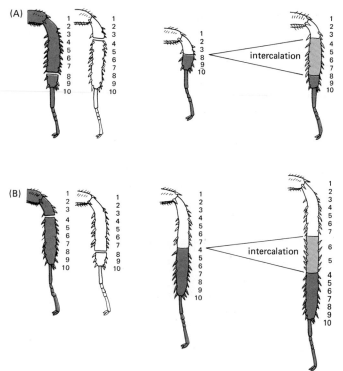

Figure 16-52 When mismatched portions of the cockroach tibia are grafted together, new tissue (*light color*) is intercalated to fill in the gap in the pattern of positional values (numbered from 1 to 10). In case (A), intercalation restores the missing part. In case (B), intercalation generates a third middle part of a tibia between the two middle parts already present. The bristles indicate the polarity of the intercalated tissue. In both cases, continuity is restored in the final pattern of positional values.

with all the missing positional values have been intercalated in the initial gap and have become spread out to the normal spatial separation from one another. This process as a whole is called **intercalary regeneration.**

The rule of intercalation, with the corollary that growth continues until a certain spacing of positional values has been attained, is a very powerful organizing principle in those systems to which it applies. Beginning with a pattern specified approximately and in miniature—for example, by a morphogen gradient—it can determine the construction of a complete accurate pattern of positional values and regulate the growth of each part of the pattern to a standard size: all that is necessary is that the initial pattern should be qualitatively—that is, topologically—correct. Applied to two- and three-dimensional patterns, the rule of intercalation accounts for a remarkably wide range of phenomena, including both normal regeneration of amputated parts and such bizarre effects as the genesis of supernumerary limbs after certain types of grafting procedures. It appears to govern many processes of organogenesis and regeneration not only in insects but also in crustaceans and amphibians. We shall see shortly that it can function in *Drosophila* to correct quite large early errors in pattern specification (see Figure 16-60). Even in creatures such as mammals, where lost structures generally do not regenerate in the adult, the rule of intercalation may help to regulate growth and pattern formation during embryonic development. The molecular mechanisms that underlie this crucial form of growth control are unknown.

Summary

The different kinds of cells in an embryo are produced in a regular spatial pattern. The formation of this pattern usually begins with asymmetries in the egg and continues by means of cell-cell interactions in the embryo. At each stage in the process, it is possible for a weak asymmetry to be amplified by positive feedback so as to create a well-defined pattern. The spatial signals that coordinate pattern formation may be said to supply cells with positional information. In the simplest case a graded concentration of a diffusible morphogen may serve to control the character of cells according to their distance from the source of the morphogen; retinoic acid appears to be a morphogen of this sort in the chick limb bud, controlling the pattern along the thumb-to-little-finger axis. Discrete differences of cell character would correspond to thresholds in the response to the morphogen.

The full positional specification of a cell may be built up from a combination of bits of positional information supplied at different times—a coarse specification first, when the embryo is small, and finer details later, as the parts grow. Cells in the early forelimb and hindlimb rudiments of a vertebrate embryo, for example, acquire different positional values, making forelimb and hindlimb cells nonequivalent in their intrinsic character, long before the detailed pattern of cell differentiation has been determined. To determine its fine-grained pattern, each organ rudiment then generates a more detailed internal framework of positional information. This system of detailed positional information seems to be the same in homologous organs, such as the forelimb and hindlimb. Because of cell memory, the cells in these separate fields interpret the same fine-grained positional information differently according to their different prior histories.

In many animals the pattern of positional values is closely coupled to the control of cell proliferation according to a simple rule of intercalation, derived mainly from studies of limb regeneration in insects and amphibians. According to the rule, discontinuities of positional value provoke local cell proliferation, and the newly formed cells take on intermediate positional values that restore continuity in the pattern. The same mechanism may also operate in normal embryonic development to correct inaccuracies in the initial specification of positional information.

Drosophila and the Molecular Genetics of Pattern Formation[44]

The structure of an organism is controlled by its genes: classical genetics is based on this proposition. Yet for almost a century, and even long after the role of DNA in inheritance had become clear, the mechanisms of the genetic control of body structure remained an intractable mystery. In recent years this chasm in our understanding has begun to be filled. The fly *Drosophila*, more than any other organism, has provided the new insights. Studies on *Drosophila* have revealed a class of developmental control genes whose specific function is to mark out the pattern of the body, and the combination of classical and molecular genetics has begun to show how these genes work. We shall see that not only the general strategies but also the specific genes controlling pattern in *Drosophila* may have close counterparts in vertebrates.

The first glimpses of this genetic system came with the discovery of mutations that cause bizarre disturbances of the adult *Drosophila* body plan. In the mutation *Antennapedia*, for example, legs sprout from the head in place of antennae (Figure 16–53); while in the mutation *bithorax*, portions of an extra pair of wings appear where normally there should be appendages called halteres. Such mutations, which transform parts of the body into structures appropriate to other positions, are called *homeotic*, and the normal genes that they disrupt are called *homeotic selector genes*. The discovery of homeotic mutants led to ingenious experiments showing that the body of the normal fly is formed as a patchwork of discrete regions, each expressing a different set of homeotic selector genes. The products of these genes act as molecular address labels, equipping the cells with a coarse-grained specification of their positional value. A homeotic mutation thus causes

Figure 16–53 The head of a normal adult *Drosophila* (A) compared with that of a fly carrying the homeotic mutation *Antennapedia* (B). The fly shown here displays the mutation in an extreme form; usually only parts of the antennae are converted into leg structures. (*Antennapedia* drawing based on a photograph supplied by Peter Lawrence.)

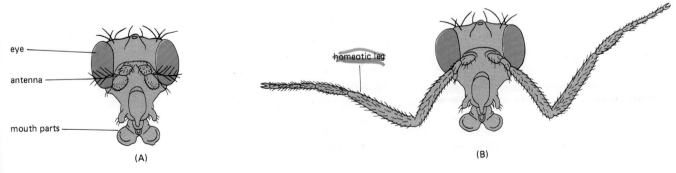

eye

antenna

mouth parts

(A)

homeotic leg

(B)

a whole patch of cells to be misinformed as to their location and consequently to make a structure appropriate to another region.

The homeotic selector genes are only one part of a larger system that creates the normal patchwork pattern of the insect body. In this section we shall describe the system as a whole and some of its molecular mechanisms, using the concepts of pattern formation discussed in the previous section. We shall see that the system consists of three classes of pattern-control genes: (1) The products of the *egg-polarity genes* act first to define the spatial coordinates of the embryo by setting up morphogen gradients in the egg. (2) The *segmentation genes* then serve to interpret the positional information provided by the initial morphogen gradients. These genes mark out the embryo into a series of *segments*—the basic modular units from which all insects are constructed. (3) The products of the segmentation genes influence the expression of the *homeotic selector genes*, which maintain the distinctions between one segment and another; and through the combined activities of the segmentation genes and the homeotic selector genes, the cells in each segment become imprinted with remembered positional values that guide their subsequent behavior. Lastly, within each segmental subdivision of the body, the cells communicate with one another to generate the finest details of the mature structure, apparently governed by the rule of intercalation.

The Insect Body Is Constructed by Modulation of a Fundamental Pattern of Repeating Units[45]

The timetable of *Drosophila* development, from egg to adult, is summarized in Figure 16–54. The period of embryonic development begins at fertilization and takes about a day, at the end of which the embryo hatches out of the egg shell to become a larva. The larva then passes through three stages, or *instars*, separated by molts in which it sheds its old coat of cuticle and lays down a larger one. At the end of the third instar it pupates. Inside the pupa a radical remodeling of the body takes place, and eventually, about nine days after fertilization, an adult fly, or *imago*, emerges.

The fly consists of a head, three thoracic segments (numbered T1 to T3), and nine abdominal segments (numbered A1 to A9). Each segment, although different from the others, is built according to a similar plan. Segment T1, for example, carries a pair of legs; T2 carries a pair of legs plus a pair of wings; and T3 carries a pair of legs plus a pair of halteres—small knob-shaped balancers important in flight, evolved from the second pair of wings that more primitive insects possess. The quasi-repetitive segmentation is more obvious in the larva, where the segments look more similar; and in the embryo it can be seen that the rudiments of the head, or at least the future adult mouth parts, are likewise segmental (Figure 16–55). It is partly a matter of convention where one draws the boundary between one segmental unit and the next; in discussing patterns of gene expression, we shall see that it is convenient to speak in terms of a total of 14 *parasegments* (numbered P1 to P14) that are half a segment out of register with traditionally defined segments (Figure 16–56). Finally, at the two ends of the animal are highly specialized structures that are not segmentally derived.

The groundplan of the whole structure, with its two specialized ends and between them a series of modulated repetitions of a basic segmental unit, is established by processes occurring in the egg and early embryo during the first few hours after fertilization.

Drosophila Begins Its Development as a Syncytium[46]

The egg of *Drosophila* is about 400 μm long and about 160 μm in diameter, with a clearly defined polarity. Like the eggs of other insects, it begins its development in an unusual way: a series of nuclear divisions without cell division creates a syncytium. The early nuclear divisions are synchronous and extremely rapid, occurring about every 8 minutes. The first nine divisions generate a cloud of nuclei, most of which migrate from the middle of the egg toward the surface, where they form a monolayer called the *syncytial blastoderm*. After another four rounds of nuclear division, plasma membranes grow inward from the egg surface to enclose

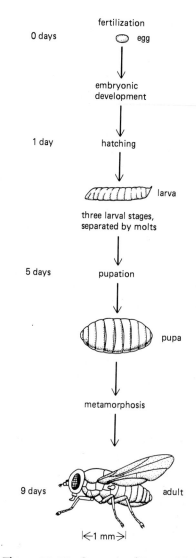

Figure 16–54 Synopsis of *Drosophila* development from egg to adult fly.

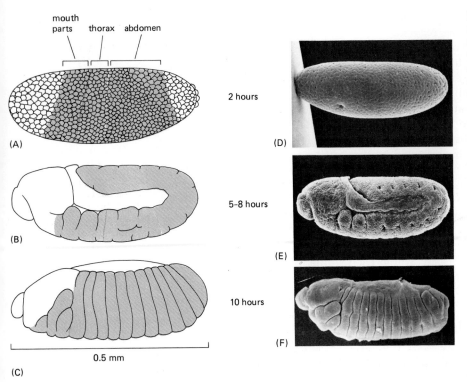

mouth
parts   thorax   abdomen

(A)

(B)

(C)

0.5 mm

2 hours

(D)

5-8 hours

(E)

10 hours

(F)

Figure 16–55 The origins of the *Drosophila* body segments during embryonic development, seen in side view in drawings (A–C) and corresponding scanning electron micrographs (D–F). (A, D) At 2 hours the embryo is at the syncytial blastoderm stage (see Figure 16–57) and no segmentation is visible, although a fate map can be drawn showing the future segmented regions (*pale color and gray shading*). (B, E) At 5–8 hours the embryo is at the "extended germ band" stage: gastrulation has occurred, segmentation has begun to be visible, and the segmented axis of the body has lengthened, curving back on itself at the tail end so as to fit into the egg shell. (C, F) At 10 hours the body axis has contracted and become straight again, and all the segments are clearly defined; the head structures, visible externally at this stage, will subsequently become tucked into the interior of the larva, to emerge again only when the larva goes through pupation to become an adult. (D and E, courtesy of Rudi Turner and Anthony Mahowald; F, courtesy of Jane Petschek.)

each nucleus, thereby converting the syncytial blastoderm into a *cellular blastoderm* consisting of some 5000 separate cells (Figure 16–57). A small subset of nuclei populating the extreme posterior end of the egg are segregated into cells a few cycles earlier; these *pole cells* are the primordial germ cells that will give rise to the next generation of eggs or sperm. As in a cleaving amphibian egg, the very rapid cycles of DNA replication seem to hinder transcription, so that up to the cellular blastoderm stage development depends largely—although not exclusively—on stocks of maternal mRNA and protein that accumulated in the egg before fertilization. After cellularization, cell division continues in a more conventional way, asynchronously and at a slower rate, and the rate of transcription increases dramatically.

The cellular blastoderm corresponds to the hollow blastula of an amphibian or a sea urchin, even though its interior is filled with yolk rather than being a fluid-filled cavity. Gastrulation follows as soon as cellularization is complete. Al-

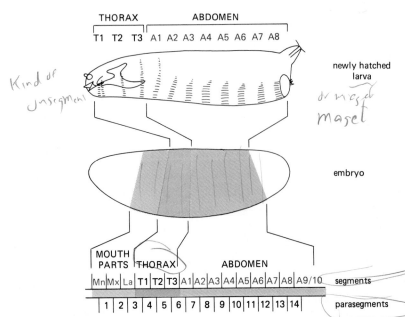

THORAX   ABDOMEN
T1 T2 T3 A1 A2 A3 A4 A5 A6 A7 A8

Kind of
Unsegment

newly hatched
larva

dr maged
maget

embryo

MOUTH
PARTS THORAX   ABDOMEN

| Mn | Mx | La | T1 | T2 | T3 | A1 | A2 | A3 | A4 | A5 | A6 | A7 | A8 | A9/10 | segments |

| 1 | 2 | 3 | 4 | 5 | 6 | 7 | 8 | 9 | 10 | 11 | 12 | 13 | 14 | parasegments |

Figure 16–56 The segments of the *Drosophila* larva and their correspondence with regions of the blastoderm. Note that the ends of the blastoderm correspond to nonsegmental structures that form largely internal parts of the larva, as do the segmental rudiments of the adult mouthparts. Segmentation in *Drosophila* can be described in terms of segments or parasegments: the relationship is shown in the bottom part of the figure. The exact number of abdominal segments is debatable: eight are clearly defined, and a ninth is probably present.

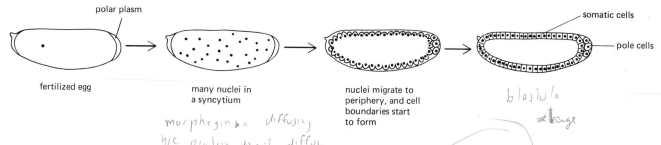

polar plasm

fertilized egg

many nuclei in
a syncytium

nuclei migrate to
periphery, and cell
boundaries start
to form

somatic cells

pole cells

[handwritten: morphogin·n diffusi·g h/c pr·t·in do n·t diffus·]

[handwritten: blastula stage]

though the geometry of this process is very different in the insect, the general outcome is similar: through coordinated cell movements, a gut tube consisting of endodermal cells is created along the axis of the embryo; mesoderm surrounds this tube and occupies the space between it and an enveloping layer of ectoderm on the exterior. During this process the long axis of the embryo becomes temporarily stretched out ("germ-band elongation") and then shortens again ("germ-band contraction"); while it is elongated, the axis becomes curved so as to fit into the egg shell (see Figure 16–55).

By following the cells through their complex gastrulation movements, one can draw a fate map for the monolayer of cells on the surface of the blastoderm (Figure 16–58). The fate map is especially simple for a cross-section through the middle of the embryo, with prospective mesoderm ventrally and ectoderm on each side above it. As in a vertebrate, the cords of nerve cells that run the length of the body derive from part of the ectoderm. Not only gastrulation and neurulation but also somite formation has its counterpart in the insect: as gastrulation nears completion, a series of indentations and bulges appear in the surface of the embryo, marking the subdivision of the body into parasegments along its antero-posterior axis (see Figure 16–55). More subtle tests show that the main features of this segmental pattern are already determined at the cellular blastoderm stage, before gastrulation begins.

Two Orthogonal Systems Control the Groundplan of the Embryo[47]

Two coordinates are needed to define each position in the blastoderm—a latitude and a longitude, so to speak. Correspondingly, genetic tests reveal that the organization of the body plan depends on two separate patterning mechanisms, one defining the dorso-ventral pattern, the other the antero-posterior pattern. A group of about 20 genes controls the dorso-ventral axis: mutations in any one of these give rise to embryos that are either "dorsalized" and lack ventral structures or "ventralized" and lack dorsal structures. Another group of about 50 genes (including the segmentation and homeotic selector genes already mentioned) controls the antero-posterior axis: mutations in these genes give rise to embryos that have defects or duplications in their antero-posterior pattern. Although many of the genes in each set have been cloned and their protein products identified, we shall discuss only the antero-posterior patterning system.

Figure 16–57 Development of the *Drosophila* egg from the time of fertilization to the cellular blastoderm stage. (After H.A. Schneiderman, in *Insect Development* [P.A. Lawrence, ed.], pp. 3–34. Oxford, U.K.: Blackwell, 1976.)

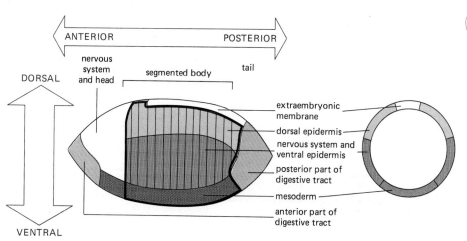

ANTERIOR
POSTERIOR

nervous
system
and head

segmented body

tail

DORSAL

extraembryonic
membrane

dorsal epidermis

nervous system and
ventral epidermis

posterior part of
digestive tract

mesoderm

anterior part of
digestive tract

VENTRAL

Figure 16–58 Fate map of a *Drosophila* embryo at the cellular blastoderm stage, in side view and in cross-section, showing the relationship between the dorso-ventral subdivision into future major tissue types and the antero-posterior pattern of future segments. A heavy line encloses the region that will form segmental structures. During gastrulation the cells along the ventral midline invaginate to form mesoderm, while the cells fated to form the digestive tract invaginate near each end of the embryo. (After V. Hartenstein, G.M. Technau, and J.A. Campos-Ortega, *Wilhelm Roux' Arch. Dev. Biol.* 194:213–216, 1985.)

Signals from the Two Ends of the *Drosophila* Egg Control the Antero-posterior Polarity of the Embryo[48]

The initial polarity of the egg depends on the distribution of materials produced before fertilization, when the egg or oocyte was still in the ovary (see Figure 16–39). If a *Drosophila* egg is carefully punctured at its anterior end, allowing a small amount of the most anterior cytoplasm to leak out, the embryo fails to develop head structures. Moreover, if cytoplasm from the posterior end of another egg is injected into the site from which the anterior cytoplasm has leaked, a second set of abdominal segments will develop, with reversed polarity, in the anterior half of the recipient egg (Figure 16–59).

Mutations have been identified that cause similar disturbances in the pattern in the anterior or the posterior half of the embryo. The **egg-polarity genes** defined by these mutations are the first elements in a hierarchical system that generates the antero-posterior pattern of the body. The egg-polarity genes are among those transcribed from the maternal genome during oogenesis, and the stored products begin to act very soon after fertilization. Thus the phenotype of the embryo is determined by the alleles present in the mother rather than by the combination of maternal and paternal genes possessed by the embryo itself. Genes expressed in this way are called **maternal-effect genes.**

Mothers that are homozygous for the egg-polarity mutation *bicoid* produce embryos that lack head and thoracic structures and have abdominal structures extended over an abnormally large fraction of the body length; conversely, the egg-polarity mutation *oskar* gives rise to embryos that lack all the abdominal segments. (The nonsegmental structures at the two extreme ends of the embryo have a special status: they do not disappear in either of these mutants but are eliminated by the mutation *torso* and certain others.) The *bicoid* and *oskar* mutants are deficient in the corresponding normal gene products, and they can be rescued by injections of normal cytoplasm. A *bicoid* mutant will develop more or less normally if cytoplasm from the anterior end of a normal egg is injected into its anterior end, while an *oskar* mutant will develop more or less normally if cytoplasm from the posterior end of a normal egg is injected into its prospective abdominal region. In each case the normal gene product seems to be localized at one end or the other of the egg and to act as the source of some long-range influence controlling the global pattern of antero-posterior positional values.

Molecular genetic experiments have illuminated the picture. Using *in situ* hybridization with a cloned *bicoid* cDNA probe, it can be shown that *bicoid* mRNA is concentrated at the anterior tip of the egg and that it is originally synthesized in the ovary by the nurse cells connected with the oocyte (see p. 912 and Figure 16–39). As the *bicoid* RNA passes through the cytoplasmic bridges into the oocyte, it becomes anchored to some component of the cytoplasm—perhaps to a part of

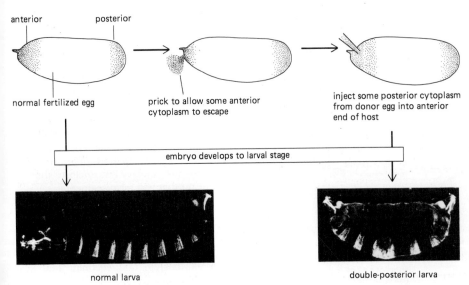

normal larva

double-posterior larva

Figure 16–59 Localized determinants at the ends of the *Drosophila* egg control its antero-posterior polarity. A little anterior cytoplasm is allowed to leak out of the anterior end of the egg and is replaced by an injection of posterior cytoplasm. The resulting double-posterior larva (*photograph on right*) is compared with a normal control (*photograph on left*); the substitution of cytoplasm at one end of the egg has had a long-range effect, converting all the more anterior segments into a mirror-image duplicate of the last three abdominal segments. The larvae are shown in dark-field illumination. (From H.G. Frohnhöfer, R. Lehmann, and C. Nüsslein-Volhard, *J. Embryol. Exp. Morphol.* 97[suppl]:169–179, 1986.)

graph

photograph

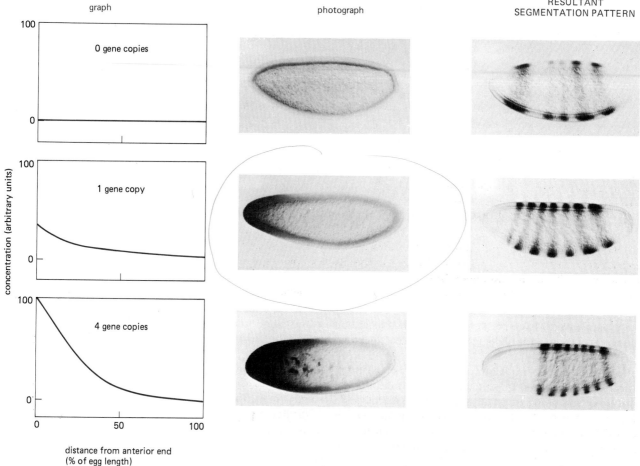

the cytoskeleton—at the oocyte's anterior end. Translation begins only when the egg is laid, giving rise to a concentration gradient of *bicoid* protein with its high point at the anterior end of the embryo. The concentration gradient can be altered genetically by constructing mutants that contain multiple copies of the normal *bicoid* gene: as the gene dosage increases in the mother, so does the protein concentration increase in the egg. The segments of the resultant embryo are correspondingly shifted toward the posterior pole, as though their locations were determined by positional information derived from the local concentration of *bicoid* protein (Figure 16–60). This protein therefore fits exactly the definition of a morphogen (see p. 913).

Three Classes of Segmentation Genes Subdivide the Embryo[49]

The graded global cues provided by the products of the egg-polarity genes have to guide the creation of a system of discrete segments. This process depends on a collection of about 20 **segmentation genes.** Mutations in these genes alter the number of segments or their basic internal organization without altering the global polarity of the egg. The segmentation genes act at later stages than the egg-polarity genes. Correspondingly, the phenotype of the embryo with regard to them is determined, in whole or in part, by the genotype of the embryo and not purely by the genotype of the mother; that is, they are **zygotic-effect** genes rather than maternal-effect genes.

Most mutations in segmentation genes are lethal: their effects are never seen in the adult fly because the mutant dies before it can develop that far. Lethal mutations of this sort can, however, be propagated if the lethal effects are recessive (as is commonly the case). Heterozygotes, with one mutant copy of the gene and one normal copy, are then viable. When a pair of heterozygous parents breed, one

Figure 16–60 The *bicoid* protein gradient in the *Drosophila* egg and its effects on the pattern of segments. The gradient is revealed by staining with an antibody against the *bicoid* protein; the segment pattern is revealed by an antibody against the product of a pair-rule gene, *even-skipped* (see p. 926). Three embryos are compared, containing zero, one, and four copies, respectively, of the normal *bicoid* gene. With zero dosage of *bicoid*, segments with an anterior character do not form; with increasing gene dosage they form progressively farther from the anterior end of the egg, as expected if their position is determined by the local concentration of the *bicoid* protein. Measurements of this concentration, as indicated by the intensity of staining, are shown in the graphs. Despite the considerable differences of position and spacing of the segment rudiments in the embryos with one and four doses of the gene, both embryos will develop into normally proportioned larvae and adults. The mechanism responsible for this regulation is discussed on page 918. (Slightly adapted from W. Driever and C. Nüsslein-Volhard, *Cell* 54:83–104, 1988.)

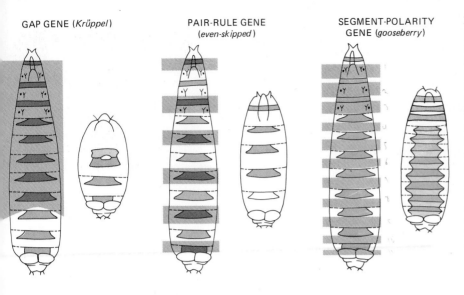

GAP GENE (*Krüppel*) PAIR-RULE GENE (*even-skipped*) SEGMENT-POLARITY GENE (*gooseberry*)

Figure 16–61 Examples of the phenotypes of mutations affecting the three types of segmentation genes. In each case the areas shaded in color on the normal larva (*left*) are deleted in the mutant or are replaced by mirror-image duplicates of the unaffected regions. By convention, dominant mutations are written with an initial capital letter and recessive mutations are written with a lowercase letter. Several of the patterning mutations of *Drosophila* are classed as dominant because they have a perceptible effect on the phenotype of the heterozygote, even though the characteristic major, lethal effects are recessive—that is, visible only in the homozygote. (Modified from C. Nüsslein-Volhard and E. Wieschaus, *Nature* 287:795–801, 1980.)

in four of the progeny are homozygotes, in which both copies of the gene are mutant. These progeny die prematurely, as late embryos or very young larvae, but survive long enough to reveal the mutant phenotype. In fact, almost all the segmentation genes were discovered by exposing flies to mutagens and laboriously screening tens of thousands of dying larvae produced by breeding from the mutant stocks.

The segmentation genes fall into three classes (Figure 16–61). The first to act are a set of at least three **gap genes,** whose products mark out the coarsest subdivisions of the embryo. Mutations in a gap gene eliminate a large block of contiguous segments, and mutations in different gap genes cause different but partially overlapping defects. In the mutant *Krüppel*, for example, the larva lacks eight segments, from T1 to A5 inclusive (approximately parasegments P3 to P10).

The next segmentation genes to act are a set of eight **pair-rule genes.** Mutations in these genes cause a series of deletions affecting alternate segments, leaving the embryo with only half as many segments as usual. While all the pair-rule mutants display this two-segment periodicity, they differ in the precise positioning of the deletions relative to the segmental or parasegmental borders. The pair-rule mutant *even-skipped*, for example, lacks the whole of each even-numbered parasegment, while the pair-rule mutant *fushi tarazu* (*ftz*) lacks the whole of each odd-numbered parasegment, and the pair-rule mutant *hairy* lacks a series of regions that are of similar width but out of register with the parasegmental units.

Finally, there are at least 10 **segment-polarity genes.** Mutations in these genes cause a part of each segment to be lost and replaced by a mirror-image duplicate of all or part of the rest of the segment. In *gooseberry* mutants, for example, the posterior half of each segment (that is, the anterior half of each parasegment) is replaced by an approximate mirror image of the adjacent anterior half-segment (see Figure 16–61).

The phenotypes of the various segmentation mutants suggest that the segmentation genes form a coordinated system that subdivides the embryo progressively into smaller and smaller domains distinguished by different patterns of gene expression. Again, molecular genetics provides the tools to investigate how this system works.

The Localized Expression of Segmentation Genes Is Regulated by a Hierarchy of Positional Signals[44,50]

Several representatives of each of the groups of segmentation genes have been cloned and used as probes to locate the gene transcripts in normal embryos by *in situ* hybridization (see p. 192). We have already seen how this technique has helped to show that the *bicoid* gene transcripts are the source of a positional

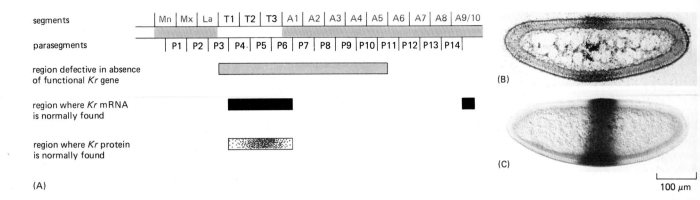

segments	Mn	Mx	La	T1	T2	T3	A1	A2	A3	A4	A5	A6	A7	A8	A9/10

parasegments | P1 | P2 | P3 | P4 | P5 | P6 | P7 | P8 | P9 | P10 | P11 | P12 | P13 | P14 |

region defective in absence
of functional *Kr* gene

region where *Kr* mRNA
is normally found

region where *Kr* protein
is normally found

(A)

(B)

(C)

100 µm

signal: the transcripts are localized at one end of the egg, even though the effects of a mutation in the gene are spread over a large part of the embryo. In a similar way it can be shown that certain segmentation genes—in particular the gap genes—in their turn generate (directly or indirectly) positional signals that help to control the pattern of development in more localized neighborhoods. Mutants that are defective in the gap gene *Krüppel*, for example, show abnormalities extending throughout the region where the gene transcripts are detected in a normal embryo and also for several segments beyond (Figure 16–62). The *Krüppel* gene has been sequenced and is homologous (as is another gap gene, *hunchback*) to a family of genes known to code in vertebrates for DNA-binding regulatory proteins, including the TFIIIA transcription factor of *Xenopus* (see p. 490). It is tempting to suggest, by analogy with *bicoid*, that the *Krüppel* protein spreads out as a diffusible morphogen from the site where *Krüppel* is transcribed, although the observed distribution of the protein seems less extensive than this hypothesis requires.

Some of the pair-rule genes may be involved in spatial signaling on a still finer scale, exerting effects on cells neighboring the regions where they are transcribed; others, by contrast, appear to affect the development only of those regions in which they are transcribed. For example, transcripts of the normal *ftz* gene at the blastoderm stage occur in seven circumferential "zebra stripes" (Figure 16–63), each of the stripes being roughly four cells wide, matching in width and location the rudiments of the even-numbered parasegments that would be missing in a *ftz* mutant.

Taken together, these observations suggest that the products of the egg-polarity genes provide global positional signals that cause particular gap genes to be expressed in particular regions, and the products of the gap genes then provide a second tier of positional signals that act more locally to regulate finer details of patterning by influencing the expression of yet other genes, including the pair-rule genes. In this way the global gradients produced by the egg-polarity genes organize the creation of a fine-grained pattern through a process of sequential subdivision, using a hierarchy of sequential positional controls. This is a reliable

Figure 16–62 The spatial domains of action of the gap gene *Krüppel*, mapped on the *Drosophila* blastoderm. (A) Diagram showing how the defect caused by an absence of functional *Krüppel* product extends far beyond the region where *Krüppel* transcripts are normally found. (B) The normal distribution of *Krüppel* transcripts, as seen by *in situ* hybridization at the blastoderm stage. (C) The normal distribution of *Krüppel* protein, as seen by antibody binding at the same stage. The protein may be distributed more widely at other stages. The phenotype of a mutant that lacks the functional *Krüppel* product is shown in Figure 16–61A. (B, from H. Jäckle, D. Tautz, R. Schuh, E. Siefert, and R. Lehmann, *Nature* 324:668–670, 1986; C, from U. Gaul, E. Seifert, R. Schuh, and H. Jäckle, *Cell* 50:639–647, 1987, copyright Cell Press.)

PAIR RULE

Figure 16–63 *In situ* hybridization of a radioactive *ftz* DNA probe to the *Drosophila* blastoderm, revealing that the gene is transcribed in a pattern of seven stripes corresponding to the pattern of defects in *ftz* mutants. The bands of *ftz* expression appear as black patches of autoradiographic silver grains in this longitudinal section. (Courtesy of Philip Ingham.)

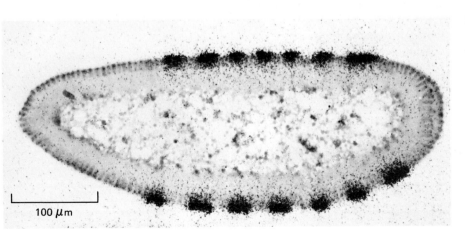

100 µm

strategy: because the global positional signals do not have to specify fine details, the individual nuclei that respond to them do not have to react with extreme precision to minute gradations in their magnitude.

Products of One Segmentation Gene Control the Expression of Another[44,50,51]

According to the scheme we have just outlined, the hierarchy of positional signals should correspond to a hierarchy of regulatory interactions among the genes governing the pattern. This can be confirmed by studying how a mutation in one gene affects the expression of another. The antero-posterior patterning genes are found to form a hierarchy with five major tiers, such that the products of the genes in each tier regulate the expression of the genes in the tiers below. Egg-polarity genes stand at the top, followed by gap genes, then pair-rule genes, then segment-polarity genes, and finally, homeotic selector genes.

For example, one can take a mutant embryo that lacks the normal *Krüppel* gene product and test the expression of the normal *ftz* gene by *in situ* hybridization with a cloned *ftz* gene probe. The usual *ftz* stripes fail to develop in just that region of the blastoderm corresponding to the defect in the *Krüppel* mutant. Thus the *Krüppel* product, directly or indirectly, regulates *ftz* gene expression. On the other hand, in a *ftz* mutant the distribution of the normal *Krüppel* product is not disturbed: the *ftz* product does not regulate *Krüppel* gene expression.

Interactions occur also between genes in the same tier of the hierarchy, and some of these interactions, unlike the interaction between *Krüppel* and *ftz*, are reciprocal. The gap genes *Krüppel* and *hunchback*, for example, have a relationship of mutual inhibition that prevents nuclei from expressing both of them at once. Thus normally they are expressed in adjacent regions of the blastoderm, with a sharp boundary between the *hunchback* territory anteriorly and the *Krüppel* territory posteriorly. But when the product of either one is absent, the territory of the other is extended beyond the usual demarcation line. The mutual inhibition experienced by these two genes, by forcing each cell nucleus to make a choice one way or the other, can explain how a set of crisply defined, nonoverlapping territories forms in response to a smooth gradient of a morphogen.

The regular periodic pattern of expression of the pair-rule genes is probably set up in a similar way, with graded positional signals that depend on gap gene expression providing rough guidance, and interactions between pair-rule genes (in particular, between the two known as *hairy* and *runt*) making the details precise and sharp. Thus an exactly reproducible arrangement of mutual exclusions and overlaps is formed by the territories of the different pair-rule genes, repeating itself reliably in every double-segment unit in the blastoderm of every normal embryo (Figure 16–64). Different bands of cells around the blastoderm are distinguished by different combinations of pair-rule gene expression, down to the finest possible level of detail—the width of a single cell, which corresponds to about a quarter of the width of a prospective segment or parasegment.

Figure 16–64 (A) The pattern of transcription of four of the eight known pair-rule genes in the *Drosophila* blastoderm and of one of the segment-polarity genes, *engrailed*. Although each pair-rule gene by itself defines only a simple alternation with a repeat distance of two segments, the whole set of pair-rule genes in combination, by their pattern of adjacency and overlap, potentially defines a much finer subdivision of the blastoderm into stripes only one cell wide, such as those in which the *engrailed* gene is expressed. (B) The pattern of expression of *engrailed* in a five-hour embryo (at the extended germ-band stage), a ten-hour embryo, and an adult (whose wings have been removed in this preparation). The pattern is revealed by an antibody against the *engrailed* protein (for the five-hour embryo), or (for the other two specimens) by constructing a strain of *Drosophila* containing the control sequences of the *engrailed* gene coupled to the coding sequence of a reporter enzyme, whose presence is easily detected histochemically through the colored product of a reaction that it catalyzes. Note that the *engrailed* pattern, once established, is preserved throughout the animal's life. (A, after M. Akam, *Development* 101:1–22, 1987, B, courtesy of Tom Kornberg.)

5 hour embryo

100 μm

10 hour embryo

100 μm

adult

500 μm

segments		Mn	Mx	La	T1	T2	T3	A1	A2	A3	A4	A5	A6	A7	A8	A9/10
parasegments			P1	P2	P3	P4	P5	P6	P7	P8	P9	P10	P11	P12	P13	P14

hairy

even-skipped

paired

fushi-tarazu

pair-rule genes

engrailed

segment-polarity gene

(A)

(B)

Egg-Polarity, Gap, and Pair-Rule Genes Create a Transient Pattern; Segment-Polarity and Homeotic Selector Genes Establish a Lasting Record[44,52]

The events just described occur within the first few hours after fertilization. The gap genes and pair-rule genes are activated one after another, and their mRNA products (as visualized by *in situ* hybridization) appear first in patterns that only approximate the final picture; the newly synthesized products themselves regulate gene expression, until—through a series of interactive adjustments—the initial distribution of gene products resolves itself into a regular, crisply defined system of stripes. But this system itself is unstable and transient. As the embryo proceeds through gastrulation and beyond, the regular segmental pattern of gap and pair-rule gene products disintegrates. Meanwhile, however, their actions have stamped a permanent set of positional values or labels on the cells of the blastoderm that maintain the segmental organization of the larva and adult. These positional labels are recorded in the persistent activation of the segment-polarity genes and the homeotic selector genes.

Segment-Polarity Genes Label the Basic Subdivisions of Every Parasegment[44,53]

Segment-polarity genes are expressed similarly in a portion of every parasegment. The gene *engrailed* provides a good example. Its RNA transcripts can be demonstrated by *in situ* hybridization in the cellular blastoderm, where they form a series of 14 bands, each approximately one cell wide, corresponding to the anteriormost portions of the future parasegments. These bands appear in a fixed relationship to the bands of expression of the pair-rule genes (see Figure 16–64); and from the altered pattern of *engrailed* bands in pair-rule mutants, one can deduce the rules for switching on *engrailed* transcription. One or other of two specific combinations of pair-rule gene products is apparently required: in even-numbered parasegments, *engrailed* is switched on in precisely those cells that express a combination including *ftz* with *odd-paired*, while in odd-numbered parasegments it is switched on in those cells that express a combination including *even-skipped* with *paired*. Other combinations of pair-rule gene products activate or repress other segment-polarity genes: transcription of the gene *wingless*, for example, is switched on in a series of bands one cell wide corresponding to the posteriormost portions of the parasegments. In this way each future parasegment is already subdivided at the cellular blastoderm stage into at least three distinct regions. The chemical distinctions will persist, maintained by continued transcription of at least some of the segment-polarity genes, after the pair-rule gene products have largely disappeared (Figure 16–64B). And some of the segment-polarity genes thus expressed—including, in particular, *wingless*—will give rise to products that act throughout subsequent development as spatial signals (local morphogens) within the parasegment to regulate the finest details of its internal patterning and growth.

The products of pair-rule genes not only regulate the segment-polarity genes, they also collaborate with the products of gap genes (and perhaps egg-polarity genes) to cause the precisely localized differential activation of the genes that form the fifth and final tier of the hierarchy—the homeotic selector genes, which permanently distinguish one parasegment from another. We shall next examine these selector genes in detail and consider more carefully their role in cell memory.

The Homeotic Selector Genes of the Bithorax Complex and the Antennapedia Complex Control the Differences Among Parasegments[54]

Many different homeotic mutations have been observed in *Drosophila*, each causing a substitution of one body part for another—wings for eyes, legs for antennae, and so forth, as illustrated earlier (see Figure 16–53). A particularly important subset of these mutations lie in one or the other of two tight gene clusters known as the **bithorax complex** and the **Antennapedia complex,** respectively. Each complex contains several genes with analogous functions: those in the bithorax complex control the differences among the abdominal and thoracic segments of the body, while those in the Antennapedia complex control the differences among thoracic and head segments. The genes in the two complexes are called **homeotic selector genes** because they define a choice between states of determination corresponding to different but homologous—or homeomorphic—structures. Each homeotic selector gene has a characteristic domain of action, defined as the region of the body that is transformed as a result of mutation in that gene. Typically, this domain has sharp boundaries that are roughly half a segment out of register with the conventional segment boundaries and correspond instead to the boundaries of parasegments (see Figure 16–56).

Although homeotic selector genes were first identified, as we saw on page 920, through the discovery of mutant adult flies, many of the mutations in this group have a recessive lethal phenotype: they allow the embryo to survive only as far as around the time of hatching. Thus, like the recessive lethal mutations in segmentation genes (see p. 925), their effects are never seen in the adult fly, but they can be observed in embryos or very early larvae. Observations of these immature stages therefore give the the clearest and in some respects most complete picture of the role of the homeotic selector genes.

Larvae that are deficient in all the genes of the bithorax complex have a particularly simple structure: the head and anterior thorax are normal as far as the P4 parasegment, but all of the remaining 10 parasegments are converted to the character of P4. Partial deletions of the bithorax complex cause transformations that are less extensive; for example, P5 and all the parasegments anterior to it may appear normal, while the parasegments posterior to P5 are converted to P5 (Figure 16–65). These observations, and analogous findings for the Antennapedia complex, illustrate the essential role of the homeotic selector genes in defining the differences among the parasegments: when the genes are missing, the distinctions are not made.

Homeotic Selector Genes Encode a System of Molecular Address Labels[44,55]

Like the segmentation genes, the homeotic selector genes are first activated in the blastoderm. Since all of the DNA in the Antennapedia and bithorax complexes has been cloned, DNA probes are available to map the spatial pattern of transcription of each of the homeotic selector genes by *in situ* hybridization. The conclusions from these studies are striking: to a first approximation, each homeotic selector gene is normally expressed in just those regions that develop abnormally, as though misplaced, when the gene is mutated or absent. The products of the selector genes can therefore be viewed as molecular address labels. Each parasegment has its own set, and if the address labels are changed, the parasegment behaves as though it were located somewhere else. Because activation of the homeotic selector genes is controlled by the segmentation genes, the pattern of homeotic selector gene expression is in exact register with the parasegmental boundaries defined by the pair-rule and segment-polarity gene products. In this way the combination of a particular homeotic selector gene product (or set of such products) with a particular segment-polarity gene product reliably defines a unique address, carried only by the cells in one subdivision of one segment.

Homeotic Selector Gene Products Help to Regulate Homeotic Selector Gene Expression[56]

As for the segmentation genes, the pattern of expression of the homeotic selector genes undergoes a series of complex adjustments following the first appearance of their transcripts in the blastoderm and becomes nonuniform within each parasegment; the definitive pattern is not established until after gastrulation is complete. Figure 16–66A shows the pattern of expression at the extended germ-band stage, about 5 hours after fertilization (see Figure 16–55).

The pattern is modulated partly by interactions between one homeotic selector gene and another. If the homeotic selector genes are arranged in sequence,

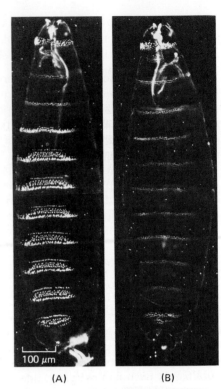

(A)   (B)

Figure 16–65 A normal *Drosophila* embryo (A), shown in dark-field illumination, compared with a mutant embryo (B) that lacks most of the genes of the bithorax complex. In the mutant the parasegments posterior to P5 all have the appearance of P5. (Courtesy of Gary Struhl; A, reprinted by permission from *Nature* 293:36–41. Copyright 1981 Macmillan Journals Limited.)

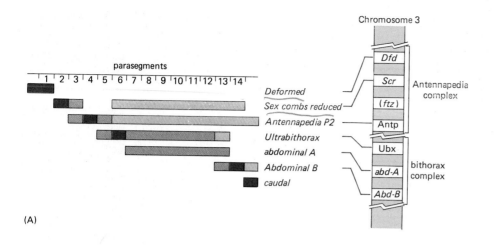

(A)

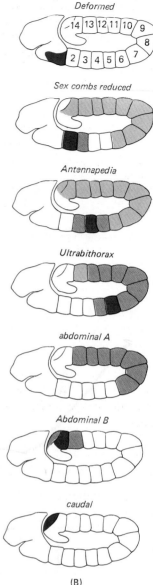

(B)

beginning with those expressed at the head end and ending with those expressed only at the tail end, it appears that each member of the series is subject to repression by the products of the subsequent members of the series. Thus, in the absence of the more "posterior" gene products, a given gene in the series will be strongly expressed both in its usual domain and in more posterior parasegments. This gene ordering according to their spatial pattern of expression and control relationships is mirrored in the sequence in which they are arranged along the chromosome in each of the two gene complexes (Figure 16–66B) as discussed on page 587.

The Adult Fly Develops from a Set of Imaginal Discs That Carry Remembered Positional Information[57]

The pattern of expression of the homeotic selector genes is established in the *Drosophila* embryo and determines the structure not only of the larva but also of the adult fly. To appreciate fully the role of these genes, it is necessary to have some idea of the curious way in which the adult, or *imago*, finally develops.

The adult fly is formed largely from groups of cells, called *imaginal cells*, that are set aside, apparently undifferentiated, in each segment of the larva. The imaginal cells originate from the embryonic epidermis—the epithelium that covers the body. They remain connected with the epidermis of the larva, and they will form mainly the epidermal structures of the adult fly. The imaginal cells for the head, thorax, and genitalia are organized into **imaginal discs;** the imaginal cells for the abdomen are grouped in clusters called *abdominal histoblast nests*. Detailed studies have focused chiefly on the discs. There are 19 of these, arranged as 9 pairs on either side of the larva plus 1 disc in the midline (Figure 16–67). The discs are pouches of epithelium, shaped like crumpled and flattened balloons, that evaginate and differentiate at metamorphosis. The eyes and antennae develop from one pair of discs, the wings and part of the thorax from another, the first pair of legs from another, and so on. The cells of one imaginal disc look just like those of another, and when they differentiate, they will give rise to generally similar sets of specialized cell types. But grafting experiments show that they are in fact already regionally determined and nonequivalent.

If one imaginal disc is transplanted into the position of another in the larva and the larva is then left to go through metamorphosis, the grafted disc is found to differentiate autonomously into the structure appropriate to its origin, regardless of its new site. This implies that the imaginal disc cells are governed by a memory of their original position. By an ingenious grafting procedure that lets the imaginal disc cells proliferate for an extended period before differentiating, it can be shown that this cell memory is stably heritable (with rare lapses) through an indefinitely large number of cell generations (Figure 16–68).

The homeotic selector genes are essential components of the memory mechanism. If they are eliminated from imaginal disc cells at any stage in the long period leading up to differentiation at metaphorphosis, the cells will differentiate

Figure 16–66 (A) The pattern of expression compared to the chromosomal locations of the genes in the Antennapedia and bithorax complexes. Note that the sequence of genes in each chromosomal complex corresponds to the spatial sequence in which the genes are expressed. (B) The pattern of transcription of the homeotic selector genes at the extended germband stage, about 5 hours after fertilization. All these genes except *caudal* lie in the Antennapedia complex or the bithorax complex, and all of them at this stage seem to be expressed in parasegmental domains. Most of them are expressed at a high level throughout one parasegment (*dark color*) and at a lower level in some adjacent parasegments (*medium color* where the presence of the transcripts is necessary for a normal phenotype, *light color* where it is not). (B, adapted from M. Akam, *Development* 101:1–22, 1987.)

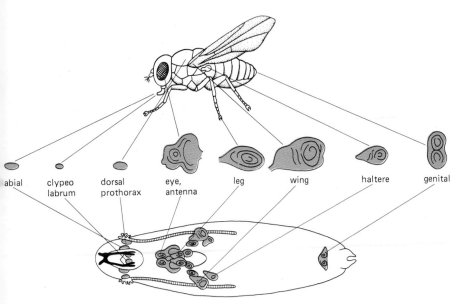

Figure 16–67 The imaginal discs in the *Drosophila* larva (schematic) and the adult structures they give rise to. Only one of the three pairs of leg discs is illustrated. (After J.W. Fristom et al., in Problems in Biology: RNA in Development [E.W. Hanley, ed.], p. 382. Salt Lake City: University of Utah Press, 1969.)

labial   clypeo labrum   dorsal prothorax   eye, antenna   leg   wing   haltere   genital

into incorrect structures, as though they belonged to a different segment of the body. This can be demonstrated by the very powerful technique of *x-ray-induced mitotic recombination*—in effect, a form of genetic surgery on individual cells by means of which mutant clones of cells of a specified genotype can be generated at any chosen time in development, as we shall now explain.

Homeotic Selector Genes Are Essential for the Memory of Positional Information in Imaginal Disc Cells[58]

A normal somatic cell contains two homologous sets of chromosomes: one homologue of each pair is derived from the father, and one is derived from the mother. Normally, exchange of DNA by crossing-over between the paternal and the maternal homologues happens only in germ cells, at meiosis. Occasionally, however, crossing-over between homologues occurs during the division cycle of an ordinary somatic cell. This process of **mitotic recombination**, although very rare under

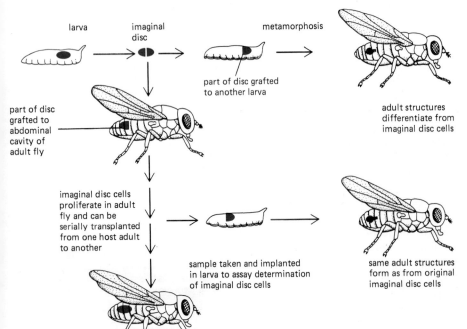

larva   imaginal disc   metamorphosis

part of disc grafted to another larva

part of disc grafted to abdominal cavity of adult fly

adult structures differentiate from imaginal disc cells

imaginal disc cells proliferate in adult fly and can be serially transplanted from one host adult to another

sample taken and implanted in larva to assay determination of imaginal disc cells

same adult structures form as from original imaginal disc cells

Figure 16–68 Experiments to test the state of determination of imaginal disc cells. The method of assay is to implant the cells in a larva that is about to undergo metamorphosis; the cells then differentiate to form recognizable adult structures, which lie, however, inside the body of the host fly after metamorphosis and are not integrated with it. The disc cells can either be assayed immediately or be implanted in the abdomen of adult flies, which serve as a natural culture chamber. Hormonal conditions in the adult allow the imaginal disc cells that have thus bypassed metamorphosis to continue to proliferate for an indefinite period, without differentiating, before the assay for cell determination is done. In both cases the cells generally differentiate to form the structures appropriate to the disc from which they derived originally.

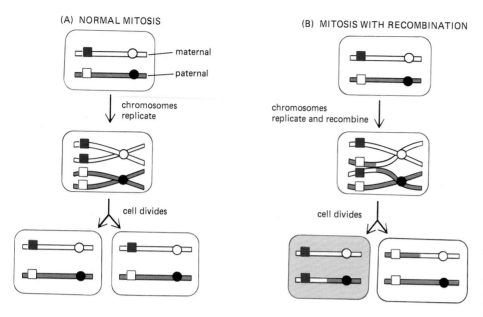

(A) NORMAL MITOSIS

maternal
paternal

chromosomes replicate

cell divides

(B) MITOSIS WITH RECOMBINATION

chromosomes replicate and recombine

cell divides

Figure 16–69 Mitotic recombination (B) compared with normal mitosis (A). The diagrams follow the fate of a single pair of homologous chromosomes, one from the father (*shaded*, with the centromere shown as a black disc), the other from the mother (*unshaded*, with the centromere shown as an open circle). These chromosomes contain a locus for a pigmentation gene (or other marker gene) with a wild-type allele *A* (small open square on paternal chromosome) and a recessive mutant allele *a* (small closed square on maternal chromosome) such that a homozygous *A/A* or heterozygous *A/a* cell has a normal appearance (*shown as white*) and a homozygous *a/a* cell has an altered appearance (*shown as colored*). Recombination by exchange of DNA between the maternal and paternal chromosomes can give rise to a pair of daughter cells, one homozygous *A/A* and therefore still normal in appearance, the other homozygous *a/a* and therefore visibly different. Mitotic recombination is a rare accidental event and occurs without the specialized apparatus that facilitates recombination during meiosis.

normal circumstances, can be provoked in insects by x-irradiation (presumably as a side effect of damage to the chromosomes). As explained in Figure 16–69, if the maternal and paternal chromosomes carry different alleles of a gene, so that the cell is heterozygous, a single mitotic recombination can lead to production of two homozygous daughter cells that are genetically different both from their parent cell and from each other: one inherits two copies of the paternal allele; the other, two copies of the maternal allele. Each daughter will then reproduce itself in the normal way to generate a pair of homozygous clones amid heterozygous tissue (Figure 16–70). If, for example, all the cells in the organism are originally heterozygous for a mutation that causes yellow pigmentation instead of the normal brown, but this mutation is recessive (so that the heterozygous cells have a brown

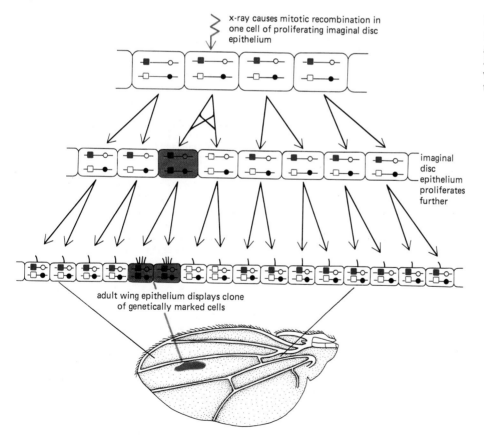

x-ray causes mitotic recombination in one cell of proliferating imaginal disc epithelium

imaginal disc epithelium proliferates further

adult wing epithelium displays clone of genetically marked cells

Figure 16–70 Mitotic recombination can be used to produce a clone of genetically marked mutant cells in the *Drosophila* wing. The earlier the stage at which recombination occurs, the larger the eventual clone will be.

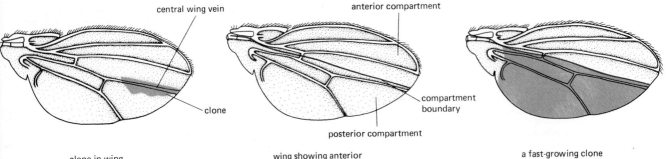

central wing vein

clone

(A) clone in wing

anterior compartment

compartment boundary

posterior compartment

wing showing anterior and posterior compartments

a fast-growing clone respects the boundary between the anterior and posterior compartments

phenotype), then mitotic recombination can create a clone of homozygous yellow cells against a brown background.

Of course, one cannot control which specific cells in the body will undergo mitotic recombination, or the exact chromosomal site of crossing-over. But one can control the timing of the event (through the timing of x-irradiation) and, most important, the initial genotype. In particular, one can breed heterozygotes in which a mutation in a gene of special interest—in the present case, a homeotic selector gene—is linked to a marker mutation—for example, one that causes abnormal pigmentation. The pigmentation marker will make homozygous clones produced by mitotic recombination stand out by virtue of their abnormal coloring, and these marked clones can then be safely assumed to be homozygous for the mutation of special interest as well.

The major effects of mutations in homeotic selector genes are generally recessive: only the homozygous mutant organism shows the homeotic transformation. By exploiting mitotic recombination as described, one can create a clonal patch of marked homozygous homeotic mutant cells in an imaginal disc and examine their behavior in a heterozygous, phenotypically normal background. The finding is that the marked cells, and only the marked cells, show the homeotic transformation (provided that they lie in the normal domain of action of the homeotic selector gene); and this applies whether the recombination event was provoked early in development or late. A 2-day larva heterozygous for a mutation that destroys the function of the *Ultrabithorax* (*Ubx*) gene (in the bithorax complex), for example, can be x-irradiated to produce isolated clones of homozygous *Ubx/Ubx* cells in its imaginal discs. These clones, if they lie in the haltere disc, will give rise to patches of wing-type tissue in the haltere. These and other observations indicate that each cell's memory of positional information depends on the continued activation of the normal homeotic selector gene and that this memory is expressed in a cell-autonomous fashion.

The Homeotic Selector Genes and Segment-Polarity Genes Define Compartments of the Body[59]

The remembered distinctions specified by the homeotic selector genes are discrete: there is an abrupt difference of gene expression between cells in adjacent parasegments. The same is true for at least one of the segment-polarity genes, *engrailed* (see Figure 16–64B), whose differential expression corresponds to an abrupt difference between cells in the posterior part of a parasegment and cells in its anterior part (or, equivalently, between cells in the anterior and posterior parts of a segment). Thus, through the differential expression of these two classes of genes, the body is subdivided into a series of discrete regions comprising cells in different states of determination.

At the frontier between one such region and the next, the cells appear to be prevented from mixing, as though selective cohesion between cells with the same molecular address label keeps them segregated from cells with a different label. Thus, for example, when a clone of genetically marked but otherwise normal cells is created in the wing by mitotic recombination, the clone is observed to be confined strictly to one side or the other of a precisely specified boundary marking

(B)

500 μm

Figure 16–71 (A) The shapes of marked clones in the *Drosophila* wing reveal the existence of a compartment boundary. The border of each marked clone is straight where it abuts the boundary. Even when a marked clone has been genetically altered so that it grows more rapidly than the rest of the wing, and is therefore very large, it respects the boundary in the same way (*last drawing*). Note that the compartment boundary does not coincide with the central wing vein. (B) The pattern of expression of the *engrailed* gene in the wing, revealed by the same technique as in Figure 16–64B. The compartment boundary coincides with the boundary of *engrailed* gene expression. (A, after F.H.C. Crick and P.A. Lawrence, *Science* 189:340–347, 1975; B, courtesy of Tom Kornberg.)

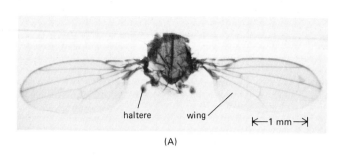

(A)

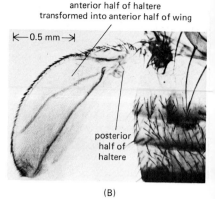

anterior half of haltere
transformed into anterior half of wing

←0.5 mm→

posterior
half of
haltere

(B)

the frontier between the two parasegments from which the wing is constructed. A subdivision of the body defined in this way—in the wing or any other organ—is called a **compartment** (Figure 16–71). Although the compartment boundary does not generally correspond to any visible structural feature of the normal organ, it does coincide with the boundary of a domain of homeotic selector gene action and of *engrailed* gene expression (Figure 16–71B): in appropriate mutants the tissue on one side of the compartment boundary is transformed, while the tissue on the other side is not (Figure 16–72).

A compartment boundary, by definition, is a frontier where two populations of cells in different states of determination are prohibited from mixing. Because the state of determination is not normally reversible, each compartment has to be a self-sufficient unit. It cannot recruit cells from the adjacent compartment or transfer surplus cells into it. It can, however, regulate its internal organization by adjustments that do not violate this constraint. Each compartment has its own characteristic detailed internal pattern, and this is generated through cell-cell interactions occurring in the imaginal disc at relatively late stages of development. The molecular basis of these late, local patterning mechanisms is not yet understood, although, as mentioned on page 929, the product of the segment-polarity gene *wingless* probably plays a part in them. It can be shown, however, that continuing cell-cell interactions govern not only the pattern but also the growth of the parts of the imaginal disc: the system obeys the rule of intercalation (see p. 918). In this regulation of pattern and growth, each compartment seems to behave as a more or less independent module. This is demonstrated by experiments in which mitotic recombination is used to create marked clones of cells with a genotype that makes them proliferate faster than their neighbors in the imaginal disc. Such clones grow to occupy a disproportionately large fraction of the compartment in which they lie, but they do not spill over into neighboring compartments; nor do they cause their own compartment to become abnormally large or distorted in its pattern (see Figure 16–71).

Figure 16–72 A normal wing and a normal haltere (A) compared with the haltere of a fly with the *bithorax* mutation (B). The anterior compartment of the mutant haltere is transformed into the anterior compartment of a wing. (A, courtesy of Peter Lawrence; B, from F.H.C. Crick and P.A. Lawrence, *Science* 189:340–347, 1975. Copyright 1975 by the American Association for the Advancement of Science.)

Self-Regulation of Homeotic Selector Genes May Contribute to Cell Memory[60]

From this excursion into the later stages of *Drosophila* development, we return now to examine the remarkable properties of the homeotic selector genes from a molecular point of view. First of all, how are they able to endow each cell with a memory of its segmental address? One way to achieve such an effect is by a positive feedback mechanism where the product of the gene stimulates its own transcription (see p. 899). There is evidence that homeotic selector genes may have this self-stimulating property. For example, cultured cells can be transfected with engineered plasmids containing either the protein-coding sequence of a homeotic selector gene coupled to some other strong promoter, or the homeotic selector gene's control sequence coupled to some other protein-coding sequence, such as a gene for an easily assayed enzyme. The first sort of plasmid lets one generate the protein encoded by the homeotic selector gene at will; the second serves to

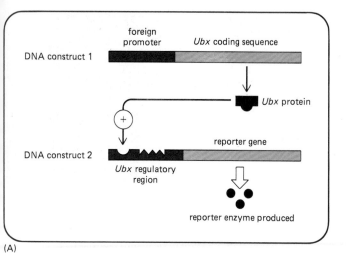

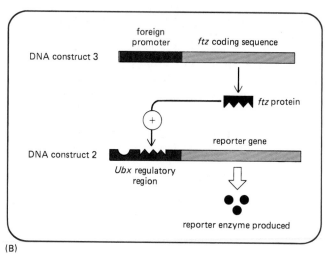

(A) (B)

test the controlling function of the protein, as indicated by the production of the "reporter" enzyme in a cell transfected with both types of plasmid together. Such experiments indicate that the *Ultrabithorax* protein can act back on an *Ultrabithorax* control sequence to stimulate *Ultrabithorax* expression (Figure 16-73). Moreover, it can act on an *Antennapedia* control sequence to inhibit *Antennapedia* expression, while *ftz* protein can stimulate both *Ultrabithorax* and *Antennapedia*. Although the culture system may be misleading as a model of conditions in the embryo, the observations demonstrate the capacity of homeotic selector genes and their relatives not only to stimulate or inhibit one another but also to stimulate themselves.

It is probable that such positive feedback is important in the initial establishment of the state of gene activation, but that other mechanisms serve to make the state permanent in the long term. Nevertheless, one can begin to see how the products of homeotic selector genes may serve simultaneously as address labels, memory records, and instruments to control expression of other genes.

Figure 16–73 Experiments using cultured cells transfected with artificially constructed plasmids carrying portions of the *Ultrabithorax* (*Ubx*) and *fushi tarazu* (*ftz*) genes show that both the *ftz* protein (B) and the *Ubx* protein (A) can act on the *Ubx* regulatory region to stimulate gene expression. The self-stimulation of *Ubx* (A) creates a positive feedback loop that would tend *in vivo* to make *Ubx* expression self-sustaining. Analogous effects are seen with the *Antennapedia* gene. The illustration is, however, drastically simplified. In a real embryo, there are probably many regulatory proteins that combine and/or compete to control expression of these genes.

The Expression of Homeotic Selector Genes Is Regulated Through Differential Splicing as Well as by Transcriptional Controls[56,61]

In broad outline the system of homeotic selector genes seems to have a simple, rational organization and function. There are, however, many complexities in the details. Interactions between the homeotic selector gene products and the genes themselves, for example, are not in fact sufficient on their own to stabilize the initial pattern of expression. An additional set of genes is required, and when these are inactivated by mutations, the homeotic selector genes become switched on indiscriminately all over the embryo (Figure 16–74). The permanent differences between parasegments may lie in heritable alterations of the control regions of the homeotic selector genes that become imprinted somehow in their chromatin structure (see p. 578). Moreover, although the bithorax complex defines the character of nine parasegments, it contains only three genes—*Ultrabithorax, abdominal A*, and *Abdominal B*. These genes—and likewise those of the Antennapedia complex—are unusually large, however, with very complicated control regions, and each produces a variety of differently spliced transcripts. The *Ultrabithorax* gene, for example, is about 75,000 nucleotide pairs long and gives rise to a family of transcripts that are spliced differently according to the stage of development and the location of the cell in the embryo. Mutations that map to different sites within the gene can affect different parasegments, or parts of parasegments. Thus mutations that map to one site within the *Ultrabithorax* gene convert the anterior half of the haltere into the anterior half of a wing (see Figure 16–72), while mutations at another site convert posterior haltere into posterior wing. Because these mutations lie in slightly different positions in the genetic map and affect distinct regions of the body, it was at first thought that they corresponded to different genes, and they were given different names: *bithorax* and *postbithorax*. It is now clear that many of the homeotic selector gene mutations that affect distinct parasegments or parts of parasegments in fact are alterations not of different genes but of different parts of the same large, complicated gene (see p. 587).

Many Pattern-Control Genes Share a Conserved Homeobox Sequence That Codes for Part of a DNA-binding Protein[62]

A surprising result of the cloning of homeotic selector genes was the discovery of a sequence of about 180 nucleotide pairs, called the **homeobox**, that is contained, with minor variations, in virtually all of them and in certain other genes as well. The pair-rule gene *ftz*, for example, contains a homeobox whose sequence is 77% identical to that of the homeobox in *Antennapedia*, and the match between the amino acid sequences encoded by these genes is even closer, with 83% identity. A total of at least 16 genes in *Drosophila* have been found to contain a homeobox (although the extent of the homology varies widely), and almost all of the homeobox-containing genes are key members of the set of about 50 control genes that define the antero-posterior pattern of the body. They include the egg-polarity gene *bicoid*; the pair-rule genes *ftz*, *even-skipped*, and *paired*; the segment-polarity genes *gooseberry* and *engrailed*; and essentially all the homeotic selector genes of the Antennapedia and bithorax complexes. (There is also at least one gene [*zerknüllt*] involved in dorsoventral patterning that contains a homeobox.) These findings suggest that the homeobox may be one of the characteristic signatures of the type of gene involved in the control of spatial pattern.

Genes that contain a homeobox generally code for proteins that are localized in the cell nucleus, suggesting direct involvement in the control of gene expression. Moreover, the homeobox amino acid sequence seems to enable the *Drosophila* proteins that contain it to bind to specific DNA sequences that act as enhancers and silencers (see p. 564) of gene expression, including the expression of other homeobox-containing genes.

The existence of specific regulatory DNA sequences that homeobox proteins recognize suggests that these too are a characteristic feature of the pattern-control genes that contain a homeobox. It is easy, then, to imagine how a set of genes of this type, with subtly varied protein-coding and regulatory sequences, might be knitted together into a complex control network in which the product of one gene controls the expression of another in just the way we have outlined in our account of the antero-posterior patterning of the *Drosophila* embryo. The property of cell memory and determination could have a simple explanation in such a system, through a self-regulation of gene expression similar to that described above for the *Ultrabithorax* gene (see Figure 16–73).

The Homeobox Is Highly Conserved in Evolution[62,63]

The striking homology of the homeobox-containing genes suggests that they have evolved by gene duplication and divergence. An origin by tandem duplication and divergence is also suggested by the close clustering of many of these genes in the *Drosophila* genome. The Antennapedia complex, for example, includes *bicoid*, *ftz*, and *zerknüllt* in addition to the set of homeotic selector genes. The clustering of homeobox-containing genes seems to be a reflection of their history and not of physiological necessity; flies with genetic rearrangements in which the *Ultrabithorax* gene is split away from its companions in the bithorax complex show no abnormality in their body plan.

It seems likely that, as an increasingly complex body plan evolved, a series of altered homeobox-containing genes were added to the genome. The phenotypes of mutants of *Drosophila* that lack particular sets of homeobox-containing genes give glimpses of the possible structure of the ancestral organism. Thus, stripping away the homeotic selector genes, one finds an animal with numerous identical

100 µm

Figure 16–74 A mutant embryo defective for the gene *extra sex combs* (*esc*) and derived from a mother also lacking this gene. Essentially all segments have been transformed to resemble the most posterior abdominal segment (compare with Figure 16–65). In this mutant the pattern of expression of the homeotic selector genes, which is roughly normal initially, is unstable in such a way that all these genes soon become switched on all along the body axis. (From G. Struhl, *Nature* 293:36–41, 1981.)

segments, like a millipede; and, speculatively retracing the course of evolution a step further by eliminating also the segment-polarity and pair-rule genes, one arrives at a hypothetical ancestor organized like a nematode, with no segmentation but still with a difference between its head and its tail. According to this picture of the sequence of events, the egg-polarity gene *bicoid* would be the primordial representative of the homeobox-containing gene family. Such an account is, of course, highly speculative, but it seems likely, in any case, that the archetypal homeobox-containing gene played a part in the spatial patterning of the body, since that is a function shared by almost all its descendants in *Drosophila*.

The discovery of the homeobox has illuminated some wider evolutionary questions. Homeobox-containing genes have been found not only in insects and other arthropods but also in the nematode *C. elegans*, in annelids (leech and earthworm), in sea urchins, in primitive chordates, and in vertebrates—including frogs, chickens, mice, and humans. The homeobox sequence is conserved with astonishing faithfulness at the protein level: one of the homeobox proteins of *Xenopus*, for example, has a sequence in which 59 of 60 amino acids are identical with the homeobox of the *Antennapedia* protein of *Drosophila*, despite more than 500 million years of independent evolution. Such conservation suggests that there may be fundamental similarities between insects and vertebrates in the mechanisms that control the development of the basic body plan.

Regional Cell Determination May Depend on Similar Machinery in Vertebrates and Insects[63,64]

In describing the early anatomical development of *Drosophila* (see p. 922), we emphasized parallels with vertebrates. Both classes of animals develop through a process of gastrulation, in which the different germ layers are established, and both have bodies that are in addition subdivided into segments, more or less at right angles to the subdivision into germ layers. Like the segments of a *Drosophila* embryo, the somites of a vertebrate embryo are similar in appearance but behave as though they carry distinct positional labels that will govern their subsequent differentiation. As for imaginal discs in *Drosophila*, this positional determination can be demonstrated by transplantation experiments. Thus, in the early chick embryo, one can graft newly formed somites from one position to another. At this stage, thoracic somites and neck somites are practically indistinguishable to the eye, yet thoracic somites grafted into the neck region will give rise to thoracic vertebrae bearing ribs, while neck somites grafted into the thorax will give rise to typical neck vertebrae. Indeed, the parallel goes further. Just as each *Drosophila* segment is subdivided into anterior and posterior compartments, distinguished by the expression of the *engrailed* gene, so also the anterior and posterior halves of each somite in the chick are chemically distinct (Figure 16–75), displaying different cell-surface properties that can be detected with certain lectins, as well as through their influence on the paths taken by growing nerve fibers.

With the discovery of the homeobox, it has begun to be possible to test whether these analogies reflect the deployment of similar molecules in similar ways. So far it has been shown that several vertebrate homeobox-containing genes have sequence homologies with *Drosophila* counterparts that extend beyond the homeobox region itself, and *in situ* hybridization has demonstrated that these vertebrate genes are expressed in strictly localized regions of the vertebrate embryo (Figure 16–76). Such evidence gives hope that the mechanisms of pattern formation in insects and vertebrates will prove to be fundamentally similar at the molecular level. The extent of the similarity is still far from clear, however. Vertebrates establish their pattern largely through cell-cell interactions following cleavage, for example, and not by processes that occur in a syncytial egg like that of the insect. Moreover, many experts on phylogeny would argue that the segmented body plan evolved independently in these two groups of animals, and it is disappointing to note that practically no clear-cut instances of homeotic mutations comparable to those seen in *Drosophila* have been observed in vertebrates. Be this as it may, there is little doubt that the studies in *Drosophila* have transformed our understanding of the general relationship between genes and body pattern in all organisms.

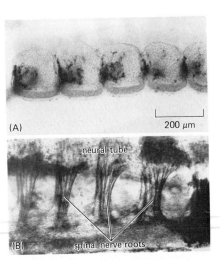

Figure 16–75 (A) A row of somites (anterior to the left, posterior to the right) from a two-day chick embryo, stained with an antibody that recognizes neural crest cells: these migrate only into the anterior half of each somite, responding to an intrinsic difference between anterior and posterior somite cells, the former being hospitable to migrating cells, the latter not. (B) A portion of a slightly older chick embryo, stained to reveal the spinal nerve roots as they pass through the somites; like the neural crest cells, the nerve roots travel only through the anterior halves of the somites. Grafting experiments show that this segmental patterning of the peripheral nervous system is preceded by, and imposed by, a prior pattern of differences in the state of determination of anterior and posterior somite cells, reminiscent of the pattern of expression of the *engrailed* gene in *Drosophila*. (A, from M. Rickmann, J.W. Fawcett, and R.J. Keynes, *J. Embryol. Exp. Morphol.* 90:437–455, 1985; B, courtesy of Claudio Stern.)

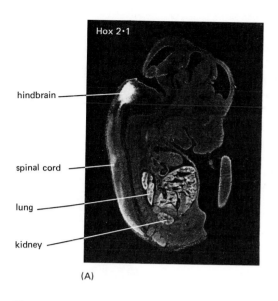

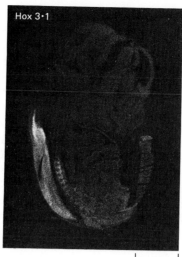

(A)

2 mm

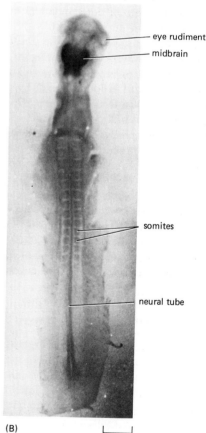

Figure 16–76 (A) The patterns of expression of two mouse homeobox-containing genes, *Hox 2.1* and *Hox 3.1*, revealed by *in situ* hybridization in adjacent longitudinal sections through a whole 13.5-day mouse embryo. High concentrations of transcripts correspond to bright patches of silver grains in these dark-field autoradiographs. The two genes are expressed in different but partly overlapping regions along the antero-posterior axis of the brain and spinal cord, as well as in some other tissues. (B) The pattern of expression of a chicken homeobox-containing gene, called chicken *en*, that has particularly close homology with the *Drosophila engrailed* gene. Here an antibody has been used to show the distribution of the protein encoded by the gene in a 2-day chick embryo. Unlike its *Drosophila* counterpart, the chicken *en* protein is localized in a single patch in the mid- and hindbrain, rather than in a segmental series of stripes. (A, from P.W.H. Holland and B.L.M Hogan, *Development* 102:159–174, 1988; B, courtesy of Tom Kornberg.)

Summary

Like other insects, Drosophila is constructed from a series of modulated repetitions of a basic unit—the segment—with specialized nonsegmental structures at each end of the body. Each major subdivision of each segment is distinguished by the expression of a particular selection of control genes that define its "address." The pattern originates with asymmetry in the egg: two groups of maternal-effect genes, called egg-polarity genes, can be distinguished, one responsible for the dorso-ventral pattern of germ layers and basic tissue types, the other responsible for the antero-posterior (head-to-tail) pattern of segments. The latter class is exemplified by the gene bicoid, whose mRNA is concentrated at the anterior end of the egg. Because the egg develops initially as a syncytium, the protein product of the bicoid gene is able to diffuse in the cytosol along the length of the embryo, providing a morphogen gradient to guide the global organization of the anterior half of the embryo. This gradient initiates the orderly expression of a hierarchy of gap genes, pair-rule genes, segment-polarity genes, and homeotic selector genes. These, through their interactions with bicoid and with one another, become expressed in some regions of the embryo and not others, progressively subdividing the body into a regular series of segmental and subsegmental units.

The pattern of expression of the gap and pair-rule genes is transitory but leaves its mark in the pattern of expression of segment-polarity genes and homeotic selector genes, which persists—with some adjustments—throughout subsequent development and provides the cells with remembered positional values. The memory may depend partly on a positive feedback mechanism whereby the protein products of homeotic selector genes act to stimulate their own gene transcription, and partly on heritable changes in chromatin structure. The necessity for some form of memory of positional values is highlighted by experiments showing that cells of the imaginal discs of the larva, from which the external structures of the adult will develop, retain a memory of their original address through an indefinite number of cell divisions. This behavior depends on continued presence of the homeotic selector genes in each individual imaginal disc cell. A compartment boundary, maintained apparently by selective cell-cell cohesion, separates cells in different states of determination with respect to the expression of these genes.

The homeotic selector genes are grouped in two clusters in the genome: the Antennapedia complex and the bithorax complex. They all contain a homeobox sequence, which codes for a highly conserved stretch of about 60 amino acids. This nucleotide sequence is also present in several egg-polarity, pair-rule, and segment-polarity genes, indicating a common evolutionary origin from a primordial gene involved in the control of spatial pattern. Homeobox-containing genes are found in many other types of animals, including nematodes and mammals, suggesting that the patterning of these animals may depend partly on molecular mechanisms similar to those that operate in Drosophila.

Organogenesis: The Coordinated Assembly of Complex Tissues

So far we have viewed pattern formation in terms of changes in the characters of cells rather than changes in their positions; and we have followed the progress of these changes in a single tissue—in *Drosophila* focusing on the epidermis, and in the vertebrate limb focusing on skeletal connective tissue. But cells in embryos can move about, and almost all organs consist of complex mixtures of tissues, composed of cells originating from different sources and obeying different rules. Cell movements bring these components together. In this final section we consider how positional information guides such cell movements and coordinates the construction of complex organs such as limbs.

Selective Cohesion Stabilizes the Pattern of Differently Determined Cells[65,66]

Cells in a developing animal often move about and jostle one another as they proliferate, so that their progeny tend to become mingled. In a chimeric mouse embryo, for example, because the cells of the two component morulae mingle, the tissues of the adult are a chaotic patchwork of the two genotypes (see p. 896). But haphazard displacements of cells after they have become determined will disrupt the organized spatial pattern of cells in different states. Thus, once cells have been assigned a character appropriate to their position, they must be prevented from randomly straying. The phenomenon of compartments in *Drosophila* (see p. 934) shows one way in which this may be achieved. Determined cells seem to be retained in their proper compartment through selective adhesion to cells in the same state of determination: cells that express the same molecular address label seem to stick more strongly to one another than to cells with a different address label. Apparently the positional information that is recorded in the activation of genes such as *engrailed* or *Ultrabithorax* is also displayed in the selection of adhesion molecules on the cell surface.

In the early amphibian embryo the stabilizing effects of selective cell-cell adhesion on the pattern of cell types are apparently so powerful that they can bring about an approximate reconstruction of the normal pattern even after the cells have been artificially dissociated into a random mixture. Mesoderm cells, neural plate cells, and epidermal cells, for example, will sort out from a random aggregate to form a structure with epidermis on the outside, mesoderm immediately beneath that, and an object resembling a neural tube buried in the interior (Figure 16–77). Studies on chick and mouse embryos suggest that this behavior depends, at least in part, on a family of homologous Ca^{2+}-dependent cell-cell adhesion glycoproteins: the *cadherins* (see p. 830). These molecules and other, Ca^{2+}-independent cell-cell adhesion molecules such as N-CAM (see p. 830), are differentially expressed in the various tissues of the early embryo, and antibodies against them interfere with the normal selective adhesion between cells of a similar type.

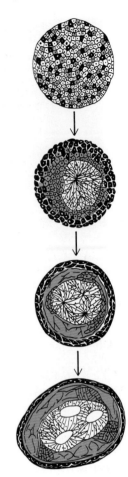

Figure 16–77 Cells from different parts of an early amphibian embryo will sort out according to their origins. In the classical experiment shown here, mesoderm cells, neural plate cells, and epidermal cells have been disaggregated and then reaggregated in a random mixture. They sort out into an arrangement reminiscent of a normal embryo, with a "neural tube" internally, epidermis externally, and mesoderm in between. (Modified from P.L. Townes and J. Holtfreter, *J. Exp. Zool.* 128:53–120, 1955.)

Spatial Patterns of Cell Adhesion Molecules Regulate Patterns of Morphogenetic Movement[66,67]

Changes in the patterns of expression of the various cadherins correlate closely with the changing patterns of association among cells during gastrulation, neurulation, and somite formation (Figure 16–78); these transformations of the early embryo may actually be regulated and driven in part by the cadherin pattern. In particular, cadherins appear to have a major role in controlling the formation and dissolution of epithelial sheets and clusters of cells. The movements that shape the early embryo thus seem to be governed by prior patterns of chemical differences between cells in different positions.

The same general point is illustrated by many other examples. The crumpled epithelium of a *Drosophila* imaginal disc, for instance, has to stretch, bend, and fold in precise ways in order to shape itself into an adult organ such as a wing. Although this process remains largely mysterious, it too seems to depend on

Figure 16–78 The changing patterns of expression of three cadherins at successive stages in the early chick or mouse embryo, as seen in cross-sections through the developing neural tube and somites. The pattern of cadherins may help to regulate the pattern of morphogenetic movements involved in formation of the neural tube, notochord, somites, neural crest, and sclerotomes. (After M. Takeichi, *Trends in Genetics* 8:213–217, 1987.)

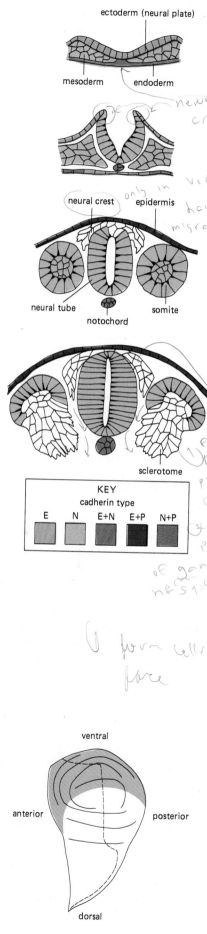

localized expression of specific cell-surface adhesion molecules, which are beginning to be identified by means of monoclonal antibodies. These *position-specific* cell-surface molecules have distributions that correlate closely with the pattern of folding of the disc at metamorphosis (Figure 16–79), and some of them have been shown to belong to the *integrin* family of cell-surface receptors (see p. 822). In contrast to the cadherins, the integrins mediate attachment of cells to components of the extracellular matrix. We have already seen that such cell-matrix interactions are important in gastrulation also (see p. 885). Cell-cell adhesion molecules and cell-matrix adhesion molecules can thus be viewed as two key types of device for translating patterns of positional information into patterns of morphogenetic movement.

In Vertebrates the Mesoderm Is the Primary Carrier of Positional Information[68]

Clearly, the positional information that cells carry can express itself in many ways. This is especially well illustrated in the development of complex organs, as we shall now see. The vertebrate limb, for example, is assembled from at least six distinct cell lineages, forming the connective tissues (skeleton, tendons, and so on), the epidermis, the muscles, the linings of blood vessels, the nerve axons and their glial sheaths, the pigmentation, and the hemopoietic tissue of the bone marrow. Although all of these components have their own well-defined patterns, they play very unequal parts in determining the pattern of the limb as a whole.

In flies, which are small, the surface of the body is large in relation to the volume, and the epidermis (derived from ectoderm) has a dominant role both in forming the exoskeleton, which provides mechanical support, and apparently in coordinating the process of pattern formation. In vertebrates, which are much bigger, the roles seem to be reversed. The supporting architectural framework of the body is provided internally by the connective tissues (derived mainly from mesoderm), and we shall see that these tissues also have the central coordinating role in pattern formation during organogenesis. In fact, experiments on amphibians in the early decades of this century indicated that the mesoderm has a master role in the control of pattern even at very early stages in the establishment of the body plan.

As mentioned earlier (see p. 887), neural structures in a vertebrate are formed from ectoderm through an inductive influence from the underlying mesoderm. If a piece of mesoderm is taken from the area just beneath the future neural tube of one gastrulating amphibian embryo and implanted directly beneath the ectoderm of another gastrulating embryo in, say, the belly region, the ectoderm in that region will thicken and roll up to form a piece of misplaced neural tube. The character of this piece of neural tube is found to depend on the origin of the

Figure 16–79 A *Drosophila* wing imaginal disc stained with a monoclonal antibody that detects the position-specific antigen PS2. The sharp boundary of the stained region (*color*) corresponds to the future wing margin, where dorsal and ventral wing epithelia meet and the epithelium must fold. The dashed line represents the antero-posterior compartment boundary. (After M. Wilcox, D.L. Bower, and R.J. Smith, *Cell* 25:159–164, 1981.)

grafted mesoderm. If the mesoderm has been taken from an anterior site, a piece of forebrain will be formed; if it is taken from a posterior site, a piece of spinal cord will form instead. This suggests that distinctive positional values are imprinted on the ectodermal cells according to the positional values of the underlying mesoderm cells.

The control of epithelial patterns by underlying connective tissue is a recurrent theme in the development of many organs in vertebrates. Thus the skin and the gut, with all their glands, appendages, and local specializations, derive their patterns largely from interactions in which the connective-tissue components provide the positional information. This is well exemplified by experiments on the skin that covers the limbs.

The Dermis Controls the Nature and Pattern of the Structures That Form from the Epidermis[66,69]

The skin consists of two layers: the epidermis, which is an epithelium derived from ectoderm, and the underlying dermis, which is a connective tissue formed by fibroblasts derived usually from mesoderm. Keratinized appendages, such as hairs, feathers, scales, and claws, develop from the epidermis, and so do many types of gland. The different types of appendages and their local arrangements are characteristic of particular regions of the body: the skin of a chick, for example, has tracts of feathers on the back, wings, and upper parts of the legs (Figure 16–80), but scales on the lower parts of the legs. Moreover, the feathers and scales vary in shape and color according to their exact location within each tract.

If embryonic epidermis from a region that normally forms scales is removed from its usual site and combined with embryonic dermis from a region that normally underlies feathers, then feathers, rather than scales, develop from the epidermis (Figure 16–81); a converse combination gives the converse result. In general the dermis controls both the detailed character of the appendages formed by the epidermis and their exact arrangement. Again we encounter the phenomenon of nonequivalence (see p. 915): morphologically similar dermal tissues from different regions have different positional values, which they manifest by inducing different behavior in overlying epidermis.

The molecular mechanisms by which connective tissue controls epithelial differentiation are not understood, but there has been some progress in identifying molecules involved in controlling the morphogenetic cell movements that give a hair, a feather, or a gland its shape. Again, cell-cell and cell-matrix adhesions are

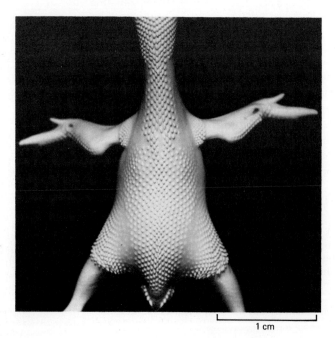

1 cm

Figure 16–80 The feather tracts on the back of a chick embryo at 9 days of incubation. Note the regular spacing of the feather rudiments in each tract. (Courtesy of A. Mauger and P. Sengel.)

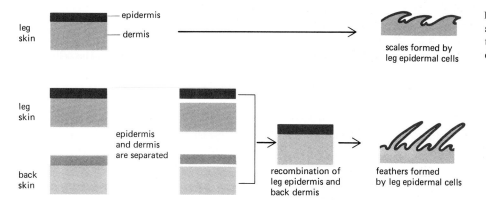

leg skin → epidermis / dermis → scales formed by leg epidermal cells

leg skin / back skin → epidermis and dermis are separated → recombination of leg epidermis and back dermis → feathers formed by leg epidermal cells

Figure 16–81 Scheme of experiments showing that the dermis controls the type of appendage formed by the epidermis.

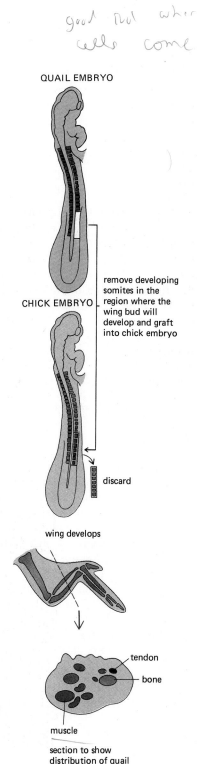

QUAIL EMBRYO

CHICK EMBRYO

remove developing somites in the region where the wing bud will develop and graft into chick embryo

discard

wing develops

section to show distribution of quail cells in forearm

tendon
bone
muscle

crucial. On the connective-tissue side of the structure, traction exerted by fibroblasts on the collagen they secrete (see p. 813) causes the fibroblasts to aggregate at the sites where appendages will form. Meanwhile, both in the connective tissue and in the overlying epithelium, changes are seen in the pattern of expression of cell-cell adhesion molecules such as N-CAM and E-cadherin, presumably regulating the packing and shaping of the cells (see p. 830). And at the interface between epithelium and connective tissue, basal lamina components (see p. 818), including laminin, proteoglycans, and collagen, are synthesized and degraded in a locally controlled fashion, helping to regulate the way in which the epithelium spreads and bends.

The Connective Tissue of a Vertebrate Limb Is Colonized by Many Types of Migrant Cells[11,70]

Connective tissue pervades the vertebrate body. In a limb it forms the bone and cartilage of the skeleton, the tendons and ligaments, the dermis, the sheathing for muscles, the outer coverings of blood vessels and nerves, and the intervening tissue that binds all these components together. All these forms of connective tissue are composed of fibroblasts or closely related cell types embedded in a collagen-rich extracellular matrix that they secrete, and all of them develop from the original undifferentiated stuffing, or *mesenchyme*, of the embryonic limb bud; their origin can be traced back to the *lateral plate mesoderm* adjacent to the somites of the very early embryo (see Figure 16–15). The other components of the limb, apart from its jacket of epidermis, consist of immigrant populations of cells originating from sources other than the lateral plate. These cells must travel through the embryonic connective tissue to reach their destinations and form their adult patterns.

Such migrations can be demonstrated by experiments in which cells are grafted from quail embryos into chick embryos. Although the quail is similar in most respects to the chick, its cells can be distinguished in histological sections by a large, strongly staining mass of heterochromatin associated with the nucleolus. This nucleolar marker makes it possible to identify grafted cells that have migrated from the site where they were implanted. For example, if quail somite tissue is substituted for the somite tissue of a very young chick embryo before the wing buds appear, all the muscle cells in the wings that subsequently develop have a quail origin (Figure 16–82). Evidently the future muscle cells migrate from the somites into the prospective wing region and remain there, inconspicuous but already determined until the time comes for them to differentiate.

Figure 16–82 If quail somite cells are substituted for the somite cells of a chick embryo at 2 days of incubation, and the wing of the chick is sectioned a week later, it is found that the muscle cells in the chick wing derive from the transplanted quail somites.

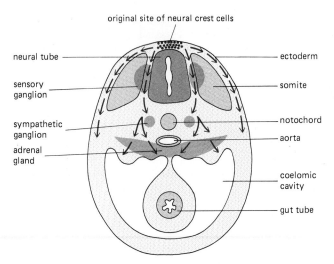

original site of neural crest cells

neural tube

sensory ganglion

sympathetic ganglion

adrenal gland

ectoderm

somite

notochord

aorta

coelomic cavity

gut tube

Figure 16–83 The main pathways of neural crest cell migration in a chick embryo, cross-sectioned through the middle part of the trunk. The cells that take the superficial pathway, just beneath the ectoderm, will form pigment cells of the skin; those that take the deep pathway via the somites will form sensory ganglia, sympathetic ganglia, and parts of the adrenal gland. (See also Figure 14–55, p. 824.)

Another important set of migratory cells originates from the **neural crest** at the site of closure of the neural tube (Figure 16–83) (see p. 887). This can be conveniently demonstrated in a similar way by substituting quail neural crest for chick neural crest in an early chick embryo and again using the quail nucleolar marker to identify the migrant cells. In the limbs, the pigment cells derive from the neural crest, as do the glial cells that envelop the nerve cell axons. The sensory and autonomic axons in a limb themselves are outgrowths from neurons that originated from the neural crest. (The voluntary motor axons, by contrast, are outgrowths from neurons in the spinal cord.)

The Patterning of the Limb's Connective Tissue Is Independent of the Cells That Colonize It[71]

By destroying the somite tissue or the neural crest and/or neural tube in the early embryo before the limb buds have begun to form, it is relatively easy to make limbs that lack particular classes of migratory cells or nerve axons. In general such limbs are remarkably normal in all other respects. One can make a limb that contains no muscle cells, for example, and yet has a normal skeleton, skin, and sensory nerves; even tendons develop in the normal places (although they subsequently degenerate in the absence of any muscles to pull on them). Likewise, a limb deprived of all nerves and neural-crest derivatives develops a normal skeleton, skin, and musculature (although the muscles subsequently degenerate for lack of neural stimulation). And although internal parts of the limb bud will not grow normally when deprived of their epidermal jacket, they can develop a largely normal pattern when a broad patch of epidermis has been destroyed, leaving the surface of the limb bud partially denuded.

Thus the patterning of the connective tissue is practically independent of the other limb components. Moreover, the other limb components have their patterns largely imposed on them by the connective tissue. We have already discussed the evidence for this in relation to the epidermis (see Figure 16–81), and by analogous tissue transplantation experiments it is possible to show that the same is true of the immigrant cell populations, such as muscle cells, nerve cell axons, and pigment cells. Cells of all these classes will by and large form the patterns appropriate to the host limb, whatever region of the embryo they come from. If the connective tissue is thus the central source and repository of positional information in the vertebrate limb, how does it exert its organizing influence on the immigrant cells?

Connective Tissue Defines Paths and Destinations for Migratory Cells[11,72]

In principle there are five main ways in which the connective tissue may control the immigrant cell populations that colonize it: it may determine (1) the paths along which they travel, (2) the sites at which they settle, (3) the extent to which they proliferate, (4) the manner in which they differentiate, and (5) the likelihood of their survival. The importance of each of these controls varies according to the migrant cell type, and none of them is well understood. Indeed, almost nothing is known about the molecular basis of the system of positional values in the connective tissue on which the whole pattern depends, except that monoclonal antibodies are beginning to reveal position-dependent differences in cell-surface molecules.

Some general principles of cell migration are, however, becoming clear. The behavior of all migratory cells depends on the mechanisms of cell adhesion and recognition discussed in Chapter 14, which are basic prerequisites for normal organogenesis. In particular, cells must be able to get a grip on the extracellular matrix or on the surfaces of other cells if they are to migrate. Fibronectin in the connective-tissue matrix seems to be an important component of the substratum (although not the only one) for many migrant cells; in some regions of the embryo, antibodies and peptides that block cell-surface fibronectin receptors will interfere with the migration of neural crest cells, for example, just as they block cell migration in amphibian gastrulation (see p. 885). The generality of some of the mechanisms involved in the homing of migrant cells to their target sites is evident from studies of mice bearing *Steel* or *dominant-spotting* mutations. In these mutants, pigment cells, hemopoietic cells, and germ cells all fail to colonize their normal sites in skin, bone marrow, and gonads, respectively. The *dominant-spotting* mutants have a defect in the migrant cells themselves; the *Steel* mutants have a defect in the connective tissues where the migrant cells should settle.

Although we are only beginning to understand the mechanisms by which the connective tissue of a limb guides the various migrant cells along specific paths to specific destinations, it seems almost certain that the guidance must depend on positional information displayed by the connective-tissue cells. Cells with different positional values may, for example, have different surface properties or may secrete slightly different extracellular matrix components. As a migrant cell moves through the connective tissue, it repeatedly extends projections that probe its immediate surroundings, testing for subtle cues to which it is particularly sensitive by virtue of its specific assortment of cell-surface receptor proteins. Inside the cell, these receptor proteins are connected to the cytoskeleton, which moves the cell along. An incessant tug-of-war between opposing tentative attachments made by the cell projections leads to a net movement in the most adherent direction (see p. 670), until the cell reaches a point where the forces of adhesion balance or are so strong that the cell cannot detach. Chemotaxis also may play an important part, as may interactions among the migratory cells (see p. 824), causing them, for example, to halt at the same site by aggregating with one another or to disperse widely by repelling one another.

The Development of the Nervous System Poses Special Problems[73]

Discussion of cell migration brings us to a topic that we have so far neglected: the development of the nervous system—the most intricate assembly process of all. The fundamental questions of developmental biology considered in this chapter can be summarized as follows: how do the different kinds of cells in the body arise and come to be arranged in their proper places? The development of the nervous system, however, poses an additional problem: how do the nerve cells come to be properly connected? Most other types of cells can be considered as pointlike objects, each one having a well-defined position and intrinsic character. But the essence of a neuron is that it is not pointlike but enormously extended, with a long axon and dendrites connecting it to other cells. If the connections are wrong, the nervous system malfunctions. The processes whereby the different kinds of neuron arise and their cell bodies come to lie in a regular arrangement can be explained in terms of the same principles that apply to other cells. But the orderly outgrowth of axons and dendrites and the formation of a regular system of synapses are phenomena of another class. The advancing tip of a growing axon or dendrite does indeed crawl along in the same manner as a migratory cell: it could be called a migratory organelle of a sessile cell, and to some extent the factors controlling its movement are the same as for the migratory cell, involving specific adhesions and so on. But its connection to its cell of origin, its involvement with other nerve fibers, and its capacity to form synapses raise new problems that require special treatment. We therefore leave the construction of the nervous system, the *tour de force* of development, for analysis in Chapter 19.

Summary

The positional values assigned to cells in the course of pattern formation are expressed in the adhesive properties of their surfaces as well as in their internal chemistry. Cells with the same character tend to adhere to one another and remain segregated from differently specified cells, thereby stabilizing the spatial pattern and enabling the cells to sort out spontaneously if they are mixed artificially. Changes in the pattern of adhesive properties underlie morphogenetic movements such as gastrulation, neurulation, and somite formation. Because the pattern of positional values in a given class of cells is expressed on the cell surface, it can guide the migrations of other embryonic cell populations during the assembly of complex tissues and organs. In vertebrates the connective-tissue cells appear to be the primary carriers of positional information. The connective-tissue cells in the dermal layer of the skin, for example, control the regional specialization of the epidermis to form feathers and scales. Similarly, connective-tissue cells in the limb control and coordinate the patterns formed by immigrant cell populations, such as muscle cells (derived from the somites), nerve cell axons (from the central nervous system and peripheral ganglia), and pigment cells (derived from the neural crest). Although many general-purpose cell-adhesion molecules have been identified, and some of them have been shown to be crucial for these phenomena, the molecular mechanisms by which migrant cells are guided along specific paths to precisely defined destinations in the limb are still not known.

References

General

Browder, L. Developmental Biology, 2nd ed. Philadelphia: Saunders, 1984.

Gilbert, S.F. Developmental Biology, 2nd ed. Sunderland, MA: Sinauer, 1988.

Molecular Biology of Development. *Cold Spring Harbor Symp.* 50, 1985.

Slack, J.M.W. From Egg to Embryo: Determinative Events in Early Development. Cambridge, U.K.: Cambridge University Press, 1983.

Spemann, H. Embryonic Development and Induction. New Haven: Yale University Press, 1938. (Reprinted, New York: Garland, 1988.)

Walbot, V.; Holder, N. Developmental Biology. New York: Random House, 1987.

Weiss, P.A. Principles of Development. New York: Holt, 1939.

Cited

1. Browder, L.W., ed. Developmental Biology: A Comprehensive Synthesis. Vol 2: The Cellular Basis of Morphogenesis. New York: Plenum, 1986.
   Gerhart, J.; et al. Amphibian early development. *Bioscience* 36: 541–549, 1986.
   Slack, J.M.W., ed. Early Amphibian Development. *J. Embryol. Exp. Morphol.* Suppl. 89, 1985.
   Trinkaus, J.P. Cells into Organs: The Forces that Shape the Embryo, 2nd ed. Englewood Cliffs, NJ: Prentice Hall, 1984.

2. Gerhart, J.C. Mechanisms regulating pattern formation in the amphibian egg and early embryo. In Biological Regulation and Development (R.F. Goldberger, ed.), Vol. 2, pp. 133–316. New York: Plenum, 1980.
   Gerhart, J.; Ubbels, G.; Black, S.; Hara, K.; Kirschner, M. A reinvestigation of the role of the grey crescent in axis formation in Xenopus laevis. *Nature* 292:511–516, 1981.
   Vincent, J.P.; Oster, G.F.; Gerhart, J.C. Kinematics of gray crescent formation in Xenopus eggs: the displacement of subcortical cytoplasm relative to the egg surface. *Dev. Biol.* 113:484–500, 1986.

3. Gilbert, S.F. Developmental Biology, 2nd ed., pp. 73–111. Sunderland, MA: Sinauer, 1988.
   Kirschner, M.; Newport, J.; Gerhart, J. The timing of early developmental events in Xenopus. *Trends Genet.* 1:41–47, 1985.
   Wilson, E.B. The Cell in Development and Heredity, 3rd ed. pp. 980–1034. New York: Macmillan, 1925. (Reprinted, New York: Garland, 1987.)

4. Furshpan, E.J.; Potter, D.D. Low-resistance junctions between cells in embryos and tissue culture. *Curr. Top. Dev. Biol.* 3:95–128, 1968.
   Kalt, M.R. The relationship between cleavage and blastocoel formation in Xenopus laevis. II. Electron microscopic observations. *J. Embryol. Exp. Morphol.* 26:51–66, 1971.
   Warner, A. The role of gap junctions in amphibian development. *J. Embryol. Exp. Morphol.*, Suppl. 81:365–380, 1985.

5. Fink, R.D.; McClay, D.R. Three cell recognition changes accompany the ingression of sea urchin primary mesenchyme cells. *Dev. Biol.* 107:66–74, 1985.

Gustafson, T.; Wolpert, L. Cellular movement and contact in sea urchin morphogenesis. *Biol. Rev.* 42:442–498, 1967.

Hardin, J.D.; Cheng, L.Y. The mechanisms and mechanics of archenteron elongation during sea urchin gastrulation. *Dev. Biol.* 115:490–501, 1986.

McClay, D.R.; Wesel, G.M. The surface of the sea urchin embryo at gastrulation: a molecular mosaic. *Trends Genet.* 1:12–00, 1985.

Wilt, F.H. Determination and morphogenesis in the sea urchin embryo. *Development* 100:559–575, 1987.

6. Ettensohn, C.A. Mechanisms of epithelial invagination. *Q. Rev. Biol.* 60:289–307, 1985.

McClay, D.R.; Ettensohn, C.A. Cell adhesion in morphogenesis. *Annu. Rev. Cell Biol.* 3:319–345, 1987.

Odell, G.M.; Oster, G.; Alberch, P.; Burnside B. The mechanical basis of morphogenesis. I. Epithelial folding and invagination. *Dev. Biol.* 85:446–462, 1981.

7. Gerhart, J.; Keller, R. Region-specific cell activities in amphibian gastrulation. *Annu. Rev. Cell Biol.* 2:201–229, 1986.

8. Spemann, H.; Mangold, H. Induction of embryonic primordia by implantation of organizers from a different species. *Roux's Archiv.* 100:599–638, 1924. (English translation in Foundations of Experimental Embryology, 2nd ed. [B.H. Willier, J.M. Oppenheimer, eds.] New York: Hafner, 1974.)

9. Balinsky, B.I. Introduction to Embryology, 5th ed. Philadelphia: Saunders, 1981.

Langman, J. Medical Embryology, 5th ed. Baltimore: Williams & Wilkins, 1985.

Romer, A.S.; Parsons, T. S. The Vertebrate Body, 6th ed. Philadelphia: Saunders, 1986.

10. Kitchin, I.C. The effects of notochordectomy in *Amblystoma mexicanum. J. Exp. Zool.* 112:393–411, 1949.

Smith, J.C.; Watt, F.W. Biochemical specificity of *Xenopus* notochord. *Differentiation* 29:109–115, 1985.

11. Le Douarin, N. The Neural Crest. Cambridge, U.K.: Cambridge University Press, 1982.

Newgreen, D.F.; Erickson, C.A. The migration of neural crest cells. *Int. Rev. Cytol.* 103:89–145, 1986.

12. Burnside, B. Microtubules and microfilaments in amphibian neurulation. *Am. Zool.* 13:989–1006, 1973.

Gordon, R. A review of the theories of vertebrate neurulation and their relationship to the mechanics of neural tube birth defects. *J. Embryol. Exp. Morphol.*, Suppl. 89:229–255, 1985.

Karfunkel, P. The mechanisms of neural tube formation. *Int. Rev. Cytol.* 38:245–271, 1974.

13. Blackshaw, S.E.; Warner, A.E. Low resistance junctions between mesoderm cells during development of trunk muscles. *J. Physiol.* 255:209–230, 1976.

Keynes, R.J.; Stern, C.D. Mechanisms of vertebrate segmentation. *Development* 103: 413–429, 1988.

14. Slack, J.M.W. From Egg to Embryo: Determinative Events in Early Development. Cambridge, U.K.: Cambridge University Press, 1983.

15. DiBerardino, M.A.; Orr, N.H.; McKinnell, R.G. Feeding tadpoles cloned from *Rana* erythrocyte nuclei. *Proc. Natl. Acad. Sci. USA* 83:8231–8234, 1986.

Gurdon, J.B. The Control of Gene Expression in Animal Development. Cambridge: Harvard University Press, 1974.

Gurdon, J.B. Transplanted nuclei and cell differentiation. *Sci. Am.* 219(6): 24–35, 1968.

McKinnell, R.G. Cloning—Nuclear Transplantation in Amphibia. Minneapolis: University of Minnesota Press, 1978.

16. Davidson, E.H. Gene Activity in Early Development, 3rd ed., pp. 411–524. Orlando, FL: Academic Press, 1986.

Jeffery, W.R. Spatial distribution of mRNA in the cytoskeletal framework of Ascidian eggs. *Dev. Biol.* 103:482–492, 1984.

Satoh, N. Towards a molecular understanding of differentiation mechanisms in Ascidian embryos. *Bioessays* 7:51–56, 1987.

Wilson, E.B. The Cell in Development and Heredity, 3rd ed. pp. 1035–1121. New York: Macmillan, 1928. (Reprinted New York: Garland, 1987.)

17. Gurdon, J.B. Embryonic induction—molecular prospects. *Development* 99:285–306, 1987.

Kimelman, D.; Kirschner, M. Synergistic induction of mesoderm by FGF and TGF-beta and the identification of an mRNA coding for FGF in the early *Xenopus* embryo. *Cell* 51:869–877, 1987.

Rosa, F.; et al. Mesoderm induction in amphibians: the role of TGF-β2-like factors. *Science* 239:783–785, 1988.

Slack, J.M.W.; Darlington, B.G.; Heath, J.K.; Godsave, S.F. Mesoderm induction in early *Xenopus* embryos by heparin-binding growth factors. *Nature* 326:197–200, 1987.

Weeks, D.L.; Melton, D.A. A maternal mRNA localized to the vegetal hemisphere in *Xenopus* eggs codes for a growth factor related to TGF-beta. *Cell* 51:861–867, 1987.

18. Austin, C.R.; Short, R.V., eds. Embryonic and Fetal Development, 2nd ed. Reproduction in Mammals, Ser., Book 2. Cambridge, U.K.: Cambridge University Press, 1982.

Hogan, B.; Costantini, F.; Lacy, E. Manipulating the Mouse Embryo: A Laboratory Manual. Cold Spring Harbor, NY: Cold Spring Harbor Laboratory, 1986.

Rugh, R. The Mouse: Its Reproduction and Development. Minneapolis: Burgess, 1968.

19. Gardner, R.L. Clonal analysis of early mammalian development. *Philos. Trans. R. Soc. Lond. (Biol.)* 312:163–178, 1985.

McLaren, A. Mammalian Chimeras. Cambridge, U.K.: Cambridge University Press, 1976.

20. Johnson, M.H.; Chisholm, J.C.; Fleming, T.P.; Houliston, E. A role for cytoplasmic determinants in the development of the mouse early embryo? *J. Embryol. Exp. Morphol.*, Suppl. 97:97–121, 1986.

Kelly, S.J. Studies of the developmental potential of 4- and 8-cell stage mouse blastomeres. *J. Exp. Zool.* 200:365–376, 1977.

Tarkowski, A.K. Experiments on the development of isolated blastomeres of mouse eggs. *Nature* 184:1286–1287, 1959.

21. Illmensee, K.; Stevens, L.C. Teratomas and chimeras. *Sci. Am.* 240(4):120–132, 1979.

Papaioannou, V.E.; Gardner, R.L.; McBurney, M.W.; Babinet, C.; Evans, M.J. Participation of cultured teratocarcinoma cells in mouse embryogenesis. *J. Embryol. Exp. Morphol.* 44:93–104, 1978.

Robertson, E.J. Pluripotential stem cell lines as a route into the mouse germ line. *Trends Genet.* 2:9–13, 1986.

22. Weiss, P.A. Principles of Development, pp. 289–437. New York: Holt, 1939.

23. Spemann, H. Über die Determination der ersten Organanlagen des Amphibienembryo I-VI *Arch. Entw. Mech. Org.* 43:448–555, 1918.

24. Hardeman, E.C.; Chiu, C.-P.; Minty, A; Blau, H.M. The pattern of actin expression in human fibroblast x mouse muscle heterokaryons suggests that human muscle regulatory factors are produced. *Cell* 47:123–130, 1986.

25. Barton, S.C.; Surani, M.A.H.; Norris, M.L. Role of paternal and maternal genomes in mouse development. *Nature* 311:374–376, 1984.

Monk, M. Memories of mother and father. *Nature* 328:203–204, 1987.

Reik, W.; Collick, A.; Norris, M.L.; Barton, S.C.; Surani, M.A. Genomic imprinting determines methylation of parental alleles in transgenic mice. *Nature* 328:248–251, 1987.

Sapienza, C.; Peterson, A.C.; Rossant, J.; Balling, R. Degree of methylation of transgenes is dependent on gamete of origin. *Nature* 328:251–254, 1987.

Swain, J.L.; Stewart, T.A.; Leder, P. Parental legacy determines methylation and expression of an autosomal transgene: a molecular mechanism for parental imprinting. *Cell* 50:719–727, 1987.

6. Kimmel, C.B.; Varga, R.M. Cell lineage and developmental potential of cells in the zebra fish embryo. *TIG* 4:68–74, 1988.

Gardner, R. L.; Lawrence, P., eds. Single Cell Marking and Cell Lineage. *Philos. Trans. R. Soc. Lond. (Biol.)* 312, 1985.

Price, J. Retroviruses and the study of cell lineage. *Development* 101:409–419, 1987.

Wolfram, S. Cellular automata as models of complexity. *Nature* 311:419–424, 1984.

7. Edgar, L.G.; McGhee, J.D. DNA synthesis and the control of embryonic gene expression in C. elegans. *Cell* 53:589–599, 1988.

Wilkins, A.S. Genetic Analysis of Animal Development. New York: Wiley, 1986.

Wood, W.B.; et al. The Nematode *Caenorhabditis elegans*. Cold Spring Harbor, NY: Cold Spring Harbor Laboratory, 1988.

8. Kenyon, C. Cell lineage and the control of *Caenorhabditis elegans* development. *Philos. Trans. R. Soc. Lond. (Biol.)* 312:21–38, 1985.

Sulston, J.E.; Horvitz, H.R. Post-embryonic cell lineage of the nematode, *Caenorhabditis elegans. Dev. Biol.* 56:110–156, 1977.

Sulston, J.E.; Schierenberg, E.; White J.G.; Thompson, J.N. The embryonic cell lineage of the nematode *Caenorhabditis elegans. Dev. Biol.* 100:64–119, 1983.

9. Ambros, V.; Horvitz, H.R. The *lin-14* locus of *Caenorhabditis elegans* controls the time of expression of specific post-embryonic developmental events. *Genes Dev.* 1:398–414, 1987.

Chalfie, M.; Horvitz, H.R.; Sulston, J.E. Mutations that lead to reiterations in the cell lineages of C. elegans. *Cell* 24:59–69, 1981.

Ellis, H.M.; Horvitz, H.R. Genetic control of programmed cell death in the nematode C. elegans. *Cell* 44:817–829, 1986.

Sternberg, P.W.; Horvitz, H.R. The genetic control of cell lineage during nematode development. *Annu. Rev. Genet.* 18:489–524, 1984.

0. Cooke, J. Properties of the primary organisation field in the embryo of *Xenopus laevis*. IV. Pattern formation and regulation following early inhibition of mitosis. *J. Embryol. Exp. Morphol.* 30:49–62, 1973.

Satoh, N. Towards a molecular understanding of differentiation mechanisms in Ascidian embryos. *Bioessays* 7:51–56, 1987.

Stephens, L.; Hardin, J.; Keller, R.; Wilt, F. The effects of aphidicolin on morphogenesis and differentiation in the sea urchin embryo. *Dev. Biol.* 118:64–69, 1986.

1. Kimble, J.E. Strategies for control of pattern formation in *Caenorhabditis elegans. Philos. Trans. R. Soc. Lond. (Biol.)* 295:539–551, 1981.

Priess, J.R.; Schnable, H.; Schnabel, R. The *glp-1* locus and cellular interactions in early C. elegans embryos. *Cell* 51:601–611, 1987.

Sternberg, P.W.; Horvitz, H.R. Pattern formation during vulval development in C. elegans. *Cell* 44:761–772, 1986.

Sulston, J.E., White, J.G. Regulation and cell autonomy during postembryonic development of Caenorhabditis elegans. *Dev. Biol.* 78:577–597, 1980.

2. Artavanis-Tsakonas, S. The molecular biology of the *Notch* locus and the fine-tuning of differentiation in *Drosophila. Trends Genet.* 4:95–100, 1988.

Ferguson, E.L.; Sternberg, P.W.; Horvitz, H.R. A genetic pathway for the specification of the vulval cell lineages of *Caenorhabditis elegans. Nature* 326:259–267, 1987.

Greenwald, I. The *lin-12* locus of *Caenorhabditis elegans. Bioessays* 6:70–73, 1987.

3. Hughes, S.M.; Lillien, L. E.; Raff, M.C.; Rohrer, H.; Sendtner, M. Ciliary neurotrophic factor induces type-2 astrocyte differentiation in culture. *Nature* 335:70–73, 1988.

Raff, M.C.; Abney, E.R.; Fok-Seang, J. Reconstitution of a developmental clock in vitro: a critical role for astrocytes in the timing of oligodendrocyte differentiation. *Cell* 42:61–69, 1985.

Raff, M.C.; Miller, R.; Noble, M. A glial progenitor cell that develops *in vitro* into an astrocyte or an oligodendrocyte depending on the culture medium. *Nature* 303:390–396, 1983.

Richardson, W.D.; Pringle, N.; Mosley, M.; Westermark, B.; Dubois-Dalcq, M. A role for platelet-derived growth factor in normal gliogenesis in the central nervous system. *Cell* 53:309–319, 1988.

Temple, S.; Raff, M.C. Clonal analysis of oligodendrocyte development in culture: evidence for a developmental clock that counts cell divisions. *Cell* 44:773–779, 1986.

34. Malacinski, G.M.; Bryant, S.V., eds. Pattern Formation: A Primer in Developmental Biology. New York: Macmillan, 1984.

Meinhardt, H. Models of Biological Pattern Formation. New York: Academic Press, 1982.

Theories of Biological Pattern Formation. *Philos. Trans R. Soc. Lond. (Biol.)* 295:425–617, 1981.

35. Bode, P.M.; Bode, H.R. Patterning in Hydra. In Pattern Formation: A Primer in Developmental Biology (G. Malacinski, S.V. Bryant, eds.), pp. 213–241. New York: Macmillan, 1984.

Jaffé, L.F. The role of ionic currents in establishing developmental pattern. *Philos. Trans. R. Soc. Lond. (Biol.)* 295:553–566, 1981.

Lenhoff, H.M.; Lenhoff, S.G. Trembley's polyps. *Sci. Am.* 258(4):86–91, 1988.

Lohs-Schardin, M. *Dicophalic*–a *Drosophila* mutant affecting polarity in follicle organization and embryonic patterning. *Wilhelm Roux's Arch.* 191:28–36, 1982.

Robinson, K.R.; Cone, R. Polarization of fucoid eggs by a calcium ionophore gradient. *Science* 207:77–78, 1980.

36. Wolpert, L. Positional information and pattern formation. *Curr. Top. Dev. Biol.* 6:183–224, 1971.

37. Lewis, J.; Slack, J.M.W.; Wolpert, L. Thresholds in development. *J. Theor. Biol.* 65:579–590, 1977.

38. Crick, F.H.C. Diffusion in embryogenesis. *Nature* 225:420–422, 1970.

39. Saunders, J.W., Jr.; Gasseling, M.T.; Cairns, J.M. The differentiation of prospective thigh mesoderm grafted beneath the apical ectodermal ridge of the wing bud in the chick embryo. *Dev. Biol.* 1:281–301, 1959.

40. Bryant, S.V.; Gardiner, D.M.; Muneoka, K. Limb development and regeneration. *Am. Zool.* 27:675–696, 1987.

Butler, E.G. Regeneration of the urodele forelimb after reversal of its proximodistal axis. *J. Morphol.* 96:265–282, 1955.

Goss, R.J. Principles of Regeneration. New York: Academic Press, 1969.

Wallace, H. Vertebrate Limb Regeneration. New York: Wiley, 1981.

41. Fallon, J.F.; et al., eds. Limb Development and Regeneration, Parts A and B. *Prog. Clin. Biol. Res.* 110, 1983.

Wolpert, L. Pattern formation in biological development. *Sci. Am.* 239(4):154–164, 1978.

42. Maden, M. Retinoids and the control of pattern limb development and regeneration. *Trends Genet.* 1:103–107, 1985.

Thaller, C.; Eichele, G. Identification and spatial distribution of retinoids in the developing chick limb bud. *Nature* 327:625–628, 1987.

Tickle, C. The number of polarizing region cells required to specify additional digits in the developing chick wing. *Nature* 289:295–298, 1981.

Tickle, C.; Summerbell, D.; Wolpert, L. Positional signalling and specification of digits in chick limb morphogenesis. *Nature* 254:199–202, 1975.

43. Bohn, H. Tissue interactions in the regenerating cockroach leg. In Insect Development (P.A. Lawrence, ed.), Royal Entomological Society of London Symposium No. 8, pp. 170–185. Oxford, U.K.: Blackwell, 1976.

Bryant, P.J.; Bryant, S.V.; French, V. Biological regeneration and pattern formation. *Sci. Am.* 237(1):66–81, 1977.

Bryant, P.J.; Simpson, P. Intrinsic and extrinsic control of growth in developing organs. *Q. Rev. Biol.* 59:387–415, 1984.

Lewis, J. Simpler rules for epimorphic regeneration: the polar-coordinate model without polar coordinates. *J. Theor. Biol.* 88:371–392, 1981.

44. Akam, M. The molecular basis for metameric pattern in the *Drosphila* embryo. *Development* 101:1–22, 1987.

Garcia-Bellido, A.; Lawrence, P.A.; Morata, G. Compartments in animal development. *Sci.Am.* 241(1):102–111, 1979.

Ingham, P.W. The molecular genetics of embryonic pattern formation in *Drosophila*. *Nature* 335:25–34, 1988.

Scott, M.P.; O'Farrell, P.H. Spatial programming of gene expression in early *Drosophila* embryogenesis. *Annu. Rev. Cell Biol.* 2:49–80, 1986.

45. Martinez-Arias, A.; Lawrence, P.A. Parasegments and compartments in the *Drosophila* embryo. *Nature* 313:639–642, 1985.

46. Foe, V.E.; Alberts, B.M. Studies of nuclear and cytoplasmic behavior during the five mitotic cycles that precede gastrulation in *Drosophila* embryogenesis. *J. Cell Sci.* 61:31–70, 1983.

Sander, K. Morphogenetic movements in insect embryogenesis. In Insect Development (P.A. Lawrence, ed.), Royal Entomological Society of London Symposium 8, pp. 35–52. Oxford, U.K.: Blackwell, 1976.

Technau, G.M. A single cell approach to problems of cell lineage and commitment during embryogenesis of *Drosophila melanogaster*. *Development* 100:1–12, 1987.

47. Anderson, K.V. Dorsal-ventral embryonic pattern genes of *Drosophila*. *Trends Genet.* 3:91–97, 1987.

48. Driever, W.; Nüsslein-Volhard, C. The *bicoid* protein determines position in the *Drosophila* embryo in a concentration-dependent manner. *Cell* 54:95–104, 1988.

Driever, W.; Nüsslein-Volhard, C. A gradient of *bicoid* protein in *Drosophila* embryos. *Cell* 54:83–93, 1988.

Lawrence, P.A. Background to *bicoid*. *Cell* 54:1–2, 1988.

Nüsslein-Volhard, C.; Frohnhöfer, H.G.; Lehmann, R. Determination of anteroposterior polarity in *Drosophila*. *Science* 238:1675–1681, 1987.

Sander, K. Embryonic pattern formation in insects: basic concepts and their experimental foundations. In Pattern Formation: A Primer in Developmental Biology (G. Malacinski, S.V. Bryant, eds.), pp. 245–268. New York: Macmillan, 1984.

49. Nüsslein-Volhard, C.; Wieschaus, E. Mutations affecting segment number and polarity in *Drosophila*. *Nature* 287:795–801, 1980.

50. Gaul, U.; Jäckle, H. How to fill a gap in the *Drosophila* embryo. *Trends Genet.* 3:127–131, 1987.

Gaul, U.; Jäckle, H. Pole region-dependent repression of the *Drosophila* gap gene. *Krüppel* by maternal gene products. *Cell* 51:549–555, 1987.

Hafen, E.; Kuroiwa, A.; Gehring, W.J. Spatial distribution of transcripts from the segmentation gene *fushi tarazu* during *Drosophila* embryonic development. *Cell* 37:833–841, 1984.

51. Jäckle, H.; Tautz, D.; Schuh, R.; Seifert, E.; Lehmann, R. Cross-regulatory interactions among the gap genes of *Drosophila*. *Nature* 324:668–670, 1986.

52. Ish-Horowicz, D.; Howard, K.R.; Pinchin, S.M.; Ingham, P.W. Molecular and genetic analysis of the hairy locus in *Drosophila*. *Cold Spring Harbor Symp.* 50:135–144, 1985.

Weir, M.P.; Kornberg, T. Patterns of *engrailed* and *fushi tarazu* transcripts reveal novel intermediate stages in *Drosophila* segmentation. *Nature* 318:433–439, 1985.

53. DiNardo, S.; Sher, E.; Heemskerk-Jongens, J.; Kastis, J.A.; O'Farrell, P.H. Two-tiered regulation of spatially patterned engrailed gene expression during *Drosophila* embryogenesis. *Nature* 332:604–609, 1988.

Ingham, P.W.; Baker, N.E.; Martinez-Arias, A. The products of the *ftz* and *eve* genes act as positive and negative regulators of *engrailed* and *wingless* expression in the *Drosophila* blastoderm. *Nature* 331:73–75, 1988.

Martinez-Arias, A.; Baker, N.E.; Ingham, P.W. Role of segment polarity genes in the definition and maintenance of cell states in the *Drosophila* embryo. *Development* 103:157–170, 1988.

Rijsewijk, F; et al. The *Drosophila* homolog of the mouse mammary oncogene *int-1* is identical to the segment polarity gene *wingless*. *Cell* 50:649–657, 1987.

54. Lewis, E.B. A gene complex controlling segmentation in *Drosophila*. *Nature* 276:565–570, 1978.

Morata, G.; Lawrence, P.A. Homoeotic genes, compartments and cell determination in *Drosophila*. *Nature* 265:211–216, 1977.

Struhl, G. A homoeotic mutation transforming leg to antenna in *Drosophila*. *Nature* 292:635–637, 1981.

Wakimoto, B.T.; Kaufman, T.C. Analysis of larval segmentation in lethal genotypes associated with the Antennapedia gene complex in *Drosophila melanogaster*. *Dev. Biol.* 81:51–64, 1981.

55. Akam, M.E. Segments, lineage boundaries and the domains of expression of homeotic genes. *Philos. Trans. R. Soc. Lond. (Biol.)* 312:179–187, 1985.

Harding, K.; Wedeen, C.; McGinnis, W.; Levine, M. Spatially regulated expression of homeotic genes in *Drosophila*. *Science* 229:1236–1242, 1985.

56. Akam, M. The molecular basis for metameric pattern in the *Drosophila* embryo. *Development* 101:1–22, 1987.

Lawrence, P.A.; Morata, G. The elements of the bithorax complex. *Cell* 35:595–601, 1983.

Struhl, G.; White, R.A. Regulation of the *Ultrabithorax* gene of *Drosophila* by other bithorax complex genes. *Cell* 43:507–519, 1985.

57. Ashburner, M.; Wright, T.F., eds. The Genetics and Biology of *Drosophila* Vol 2C. London, U.K.: Academic Press, 1978.

Gehring, W.; Nöthiger, R. The imaginal discs of *Drosophila*. In Developmental Systems: Insects (S. Counce, C.H. Waddington, eds.), Vol. 2, pp. 211–290. New York: Academic Press, 1973.

Hadorn, E. Transdetermination in cells. *Sci. Am.* 219(5):110–120, 1968.

58. Nöthiger, R. Clonal analysis in imaginal discs. In Insect Development (P.A. Lawrence, ed.), Royal Entomological Society of London Symposium No. 8, pp. 109–117. Oxford, U.K.: Blackwell, 1976.

Stern, C. Genetic Mosaics and Other Essays. Cambridge: Harvard University Press, 1968.

Struhl, G. Genes controlling segment specification in the *Drosophila* thorax. *Proc. Natl. Acad. Sci. USA*. 79:7380–7384, 1982.

59. Crick, F.H.C.; Lawrence, P.A. Compartments and polyclones in insect development. *Science* 189:340–347, 1975.

Garcia-Bellido, A.; Lawrence, P.A.; Morata, G. Compartments in animal development. *Sci. Am.* 241(1):102–111, 1979.

Kornberg, T.; Sidén, I.; O'Farrell, P.; Simon, M. The *engrailed* locus of *Drosophila*: *in situ* localization of transcripts reveals compartment-specific expression. *Cell* 40:45–53, 1985.

Simpson, P.; Morata, G. Differential mitotic rates and patterns of growth in compartments in the *Drosophila* wing. *Dev. Biol.* 85:299–308, 1981.

0. Desplan, C.; Theis, J.; O'Farrell, P.H. The sequence specificity of homeodomain–DNA interaction. *Cell* 54:1081–1090, 1988.

Hiromi, Y.; Gehring, W.J. Regulation and function of the *Drosophila* segmentation gene *fushi tarazu*. *Cell* 50:963–974, 1987.

Robertson, M. A genetic switch in *Drosophila* morphogenesis. *Nature* 327:556–557, 1987.

61. Bender, W.; et al. Molecular genetics of the bithorax complex in *Drosophila melanogaster*. *Science* 221:23–29, 1983.

Peifer, M.; Karch, F.; Bender, W. The bithorax complex: control of segmental identity. *Genes Dev.* 1:891–898, 1987.

Struhl, G.; Akam, M. Altered distributions of *Ultrabithorax* transcripts in *extra sex combs* mutant embryos of *Drosophila*. *EMBO J.* 4:3259–3264, 1985.

62. Desplan, C.; Theis, J.; O'Farrell, P.H. The sequence specificity of homeodomain-DNA interaction. *Cell* 54:1081–1090, 1988.

Gehring, W.J. Homeo boxes in the study of development. *Science* 236:1245–1252, 1987.

Scott, M.P.; Carroll, S.B. The segmentation and homeotic gene network in early *Drosophila* development. *Cell* 51:689–698, 1987.

63. Beeman, R.W. A homeotic gene cluster in the red flour beetle. *Nature* 327:247–249, 1987.

McGinnis, W.; Garber, R.L.; Wirz, J.; Kuroiwa, A.; Gehring, W.J. A homologous protein-coding sequence in *Drosophila* homeotic genes and its conservation in other metazoans. *Cell* 37:403–408, 1984.

Müller, M.M.; Carrasco, A.E.; De Robertis, E.M. A homeo-box-containing gene expressed during oogenesis in *Xenopus*. *Cell* 39:157–162, 1984.

64. Dressler, G.R.; Gruss, P. Do multigene families regulate vertebrate development? *Trends Genet.* 4:214–219, 1988.

Hogan, B.; Holland, P.; Schofield, P. How is the mouse segmented? *Trends Genet.* 1:67–74, 1985.

Holland, P.W.H.; Hogan, B.L.M. Expression of homeotic genes during mouse development; a review. *Genes Dev.* 2:773–782, 1988.

Kieny, M.; Mauger, A.; Sengel, P. Early regionalization of the somitic mesoderm as studied by the development of the axial skeleton of the chick embryo. *Dev. Biol.* 28:142–161, 1972.

Keynes, R.J.; Stern, C.D. Mechanisms of vertebrate segmentation. *Development* 103:413–429, 1988.

65. Gardner, R.L. Cell mingling during mammaliam embryogenesis. *J. Cell Sci.*, Suppl. 4:337–356, 1986.

Steinberg, M.S. Does differential adhesion govern self-assembly processes in histogenesis? Equilibrium configurations and the emergence of a hierarchy among populations of embryonic cells. *J. Exp. Zool.* 173:395–434, 1970.

Townes, P.L; Holtfreter, J. Directed movements and selective adhesion of embryonic amphibian cells. *J. Exp. Zool.* 128:53–120, 1955.

66. Edelman, G.M. Cell adhesion molecules in the regulation of animal form and tissue pattern. *Annu. Rev. Cell Biol.* 2:81–116, 1986.

McClay, D.R.; Ettensohn, C.A. Cell adhesion in morphogenesis. *Annu. Rev. Cell Biol.* 3:319–345, 1987.

Takeichi, M. The cadherins: cell-cell adhesion molecules controlling animal morphogenesis. *Development* 102:639–655, 1988.

67. Bogaert, T.; Brown, N.; Wilcox, M. The *Drosophila* PS2 antigen is an invertebrate integrin that, like the fibronectin receptor, becomes localized to muscle attachments. *Cell* 51:929–940, 1987.

Duband, J.L.; et al. Adhesion molecules during somitogenesis in the avian embryo. *J. Cell Biol.* 104:1361–1374, 1987.

Leptin, M.; Wilcox, M. The *Drosophila* position-specific antigens: clues to their morphogenetic role. *Bioessays* 5:204–207, 1986.

Nose, A.; Nagafuchi, A; Takeichi, M. Expressed recombinant cadherins mediate cell sorting in model systems. *Cell* 54:993–1001, 1988.

68. Spemann, H. Embryonic Development and Induction, pp. 260–296. New Haven: Yale University Press, 1938. (Reprinted, New York: Garland, 1988.)

Wessells, N.K. Tissue Interactions and Development. Menlo Park, CA: Benjamin-Cummings, 1977.

69. Bernfield, M.; Banerjee, S.D.; Koda, J.E.; Rapraeger, A.C. Remodelling of the basement membrane as a mechanism of morphogenetic tissue interaction. In The Role of the Extracellular Matrix in Development (R.L. Trelstad, ed.), pp. 545–572. *Symp. Soc. Dev. Biol.* 42. New York: Liss, 1984.

Ekblom, P.W.; Vestweber, D.; Kemler, R. Cell-matrix interactions and cell adhesion during development. *Annu. Rev. Cell Biol.* 2:27–47, 1986.

Harris, A.K.; Stopak, D.; Warner, P. Generation of spatially periodic patterns by a mechanical instability: a mechanical alternative to the Turing model. *J. Embryol. Exp. Morphol.* 80:1–20, 1984.

Sengel, P. Morphogenesis of Skin. Cambridge, U.K.: Cambridge University Press, 1976.

70. Chevallier, A.; Kieny, M.; Mauger, A. Limb-somite relationship: origin of the limb musculature. *J. Embryol. Exp. Morphol.* 41:245–258, 1977.

Christ, B.; Jacob, H.J.; Jacob, M. Experimental analysis of the origin of the wing musculature in avian embryos. *Anat. Embryol.* 150:171–186, 1977.

Stopak, D.; Harris, A.K. Connective tissue morphogenesis by fibroblast traction. I. Tissue culture observations. *Dev. Biol.* 90:383–398, 1982.

71. Bryant, S.V.; Gardiner, D.M.; Muneoka, K. Limb development and regeneration. *Am. Zool.* 27:675–696, 1987.

Chevallier, A. Role of the somitic mesoderm in the development of the thorax in bird embryos. II. Origin of thoracic and appendicular musculature. *J. Embryol. Exp. Morphol.* 49:73–88, 1979.

Kieny, M.; Chevallier, A. Autonomy of tendon development in the embryonic chick wing. *J. Embryol. Exp. Morphol.* 49:153–165, 1979.

Martin, P.; Lewis, J. Normal development of the skeleton in chick limb buds devoid of dorsal ectoderm. *Dev. Biol.* 118:233–246, 1986.

72. Bronner-Fraser, M. An antibody to a receptor for fibronectin and laminin perturbs cranial neural crest development *in vivo*. *Dev. Biol.* 117:528–536, 1986.

Chabot, B.; Stephenson, D.A.; Chapman, V.M.; Besmer, P.; Bernstein, A. The proto-oncogene *c-kit* encoding a transmembrane tyrosine kinase receptor maps to the mouse W locus. *Nature* 335:88–89, 1988.

Dufour, S.; Duband, J.L.; Konblihtt, A.R.; Thiery, J.P. The role of fibronectins in embryonic cell migrations. *Trends Genet.* 4:198–203, 1988.

Mackie, E.J.; et al. The distribution of tenascin coincides with pathways of neural crest cell migration. *Development* 102:237–250, 1988.

Weston, J. Phenotypic diversification in neural crest-derived cells: the time and stability of commitment during early development. *Curr. Top. Dev. Biol.* 20:195–210, 1986.

Wylie, C.C.; Stott, D.; Donovan, P.J. Primordial germ cell migration. In Developmental Biology: A Comprehensive Synthesis, Vol. 2: The Cellular Basis of Morphogenesis (L.W. Browder, ed.), pp. 433–448. New York: Plenum, 1986.

73. Purves, D.; Lichtman, J.W. Principles of Neural Development. Sunderland, MA: Sinauer, 1985.

Differentiated Cells and the Maintenance of Tissues

17

In the space of a few days or weeks, a single fertilized egg gives rise to a complex multicellular organism consisting of differentiated cells arranged in a precise pattern. As a rule, the pattern of the body is set up in this way on a small scale and then grows. During embryonic development, the different cell types become determined, each in its proper place. In the subsequent period of growth, the cells proliferate but—with certain exceptions—their specialized characters remain more or less fixed. The organism may continue to become bigger throughout life, as do most crustaceans and fish, or it may stop growing when it reaches a certain size, as do birds and mammals. In some types of animals with a fixed adult body size, such as flies and nematodes, proliferation of somatic cells ceases once the adult state has been attained. In many other such animals—including the higher vertebrates—cells continue to proliferate in the adult, replacing cells that die.

As the cells of vertebrate tissues such as skin, blood, and lung become worn out and are destroyed, new cells of the appropriate types take their places. Thus the adult body can be likened to a stable ecosystem in which one generation of individuals succeeds another but the organization of the system as a whole remains unchanged. This chapter will concentrate on the higher vertebrates, and in discussing the problems of tissue maintenance and renewal will try to convey something of the remarkable variety of structure, function, and life history to be found among their specialized cell types.

Maintenance of the Differentiated State[1]

Although the tissues of the body differ in many ways, they all have certain basic requirements, usually provided for by a mixture of cell types, as illustrated for the skin in Figure 17–1. They all need mechanical strength, which is often provided by a supporting framework of *extracellular matrix* (see p. 802) that is secreted by *fibroblasts*. In addition, almost all tissues need a blood supply to bring nutrients and remove waste products, and so they are pervaded by blood vessels lined with *endothelial cells*. Likewise, most tissues are innervated by *nerve cell* axons, which are ensheathed by *Schwann cells*. *Macrophages* are usually present, to dispose of debris from dying cells and to remove unwanted matrix, as are *lymphocytes* and other white blood cells, to combat infection. *Melanocytes* may be present to provide a protective or decorative pigmentation. Most of these cell types, ancillary to the specialized function of the tissue, originate outside it, and invade the tissue either early in the course of its development (endothelial cells, nerve cells, Schwann

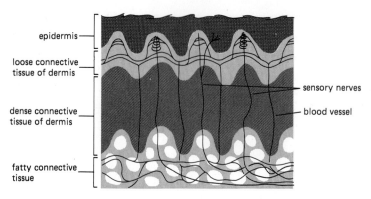

epidermis

loose connective
tissue of dermis

dense connective
tissue of dermis

fatty connective
tissue

sensory nerves

blood vessel

Figure 17–1 The composition of the skin, with epithelial tissue (the *epidermis*) outermost and, beneath it, connective tissue (the tough *dermis*, from which leather is made, and the fatty *hypodermis*). Each tissue is composed of a variety of cell types. The dermis and hypodermis are richly supplied with blood vessels and nerves. Some nerve fibers extend also into the epidermis.

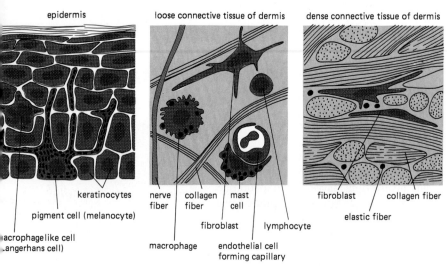

epidermis

loose connective tissue of dermis

dense connective tissue of dermis

keratinocytes

pigment cell (melanocyte)

macrophagelike cell
(Langerhans cell)

nerve
fiber

collagen
fiber

mast
cell

fibroblast

macrophage

endothelial cell
forming capillary

lymphocyte

fibroblast

collagen fiber

elastic fiber

cells, and melanocytes—see p. 943) or continually during life (macrophages and other white blood cells). This complex supporting apparatus is required to maintain the principal specialized cells of the tissue: the contractile cells of the muscle, the secretory cells of the gland, or the blood-forming cells of the bone marrow, for example.

Almost every tissue is therefore an intricate mixture of many cell types, which must remain different from one another while coexisting in the same environment. Moreover, the organization of the mixture must be preserved even though, in most adult tissues, cells are continually dying and being replaced. The retention of tissue form and function is made possible largely through two fundamental properties of cells. Because of *cell memory* (see p. 570), differentiated cells autonomously maintain their specialized character and pass it on to their progeny. At the same time, each type of differentiated cell continually senses its environment and adjusts its proliferation and properties to suit the circumstances. The intracellular mechanisms thought to be responsible for cell memory are discussed in Chapter 10; the ways in which cells respond to environmental signals are considered in Chapter 12. In this preliminary section on the behavior of cells in tissues, we briefly review some of the evidence for the stability and heritability of the differentiated state and consider to what extent this state can be modified by environmental influences.

Most Differentiated Cells Remember Their Essential Character Even in a Novel Environment[2]

Cell culture experiments demonstrate that even when cells are removed from their usual environment, they and their progeny generally remain true to their original instructions. Consider, for example, the epithelial cells that form the pigmented layer of the retina (Figure 17–2). Because they display their specialized character by manufacturing dark brown granules of melanin, it is easy to monitor their state of differentiation. When these cells are isolated from the retina of a chick embryo and grown in culture, they proliferate to form clones. Single cells taken from these clones breed true, giving subclones of similar pigment epithelial cells. The differentiated state can be maintained in this way through more than 50 cell generations.

The behavior of the cells is not, however, completely independent of their environment. In certain media, or in conditions of extreme crowding, they may survive but synthesize little or no pigment. But even when they fail to express their differentiated character, they remain *determined* as pigment cells: when they are returned to more favorable culture conditions, they synthesize pigment once again. There is one known exception to this rule: under certain conditions, they will "transdifferentiate" into lens cells; but no manipulation of the medium or of the culture conditions has been found to cause them to differentiate instead into blood cells, for example, or into liver cells or heart cells. Similarly, most types of differentiated cells, including blood cells, liver cells, and heart cells, maintain their essential character in culture.

In the body, just as in culture, most differentiated cells behave as though their basic character has been irreversibly determined by their developmental history. Epidermal cells, for example, remain epidermal cells even in the most alien surroundings: if a suspension of dissociated epidermal cells is prepared from the tail skin of a rat and injected beneath the capsule of the kidney, the cells grow there to form epidermal cysts, which contain hair follicles and sebaceous glands, like skin on the surface of the body.

The Differentiated State Can Be Modulated by a Cell's Environment[1,3]

Although radical transformations are largely forbidden, the character of most differentiated cells can be adjusted to some extent to suit the environment. The possible adjustments can mostly be classified as *modulations* of the differentiated state—that is, reversible changes between closely related cell phenotypes. For example, liver cells adjust their synthesis of specific enzymes (through changes in specific mRNA levels) according to the ambient concentrations of the steroid hormone hydrocortisone; and muscle cells modulate their pattern of gene expression according to the amount of stimulation they receive (see p. 985 and p. 1125). Fibroblasts and their relatives—the family of *connective-tissue cells*—represent a special case. These cells are exceptionally adaptable and can undergo various interconversions: fibroblasts, for example, can apparently change reversibly into cartilage cells. Such transformations are important in the healing of wounds and fractures and in other pathological processes, and they will be discussed later in this chapter (see p. 986), where we shall see how they may be controlled by cell shape, extracellular matrix, and diffusible signaling molecules. Even these conversions of one differentiated cell type into another, however, are narrowly restricted: the converted cell remains a member of the family of connective-tissue cells.

Some Structures Are Maintained by a Continuing Interaction Between Their Parts: Taste Buds and Their Nerve Supply[4]

Taste buds provide another unusually extreme example of a state of differentiation that depends on a continuing cell-cell interaction. These little structures, by which we perceive sweetness, sourness, saltiness, and bitterness, are formed chiefly in the epithelium on the upper surface of the tongue. Each consists of about 50 cells that are easily distinguished by their shape from the epithelial cells that surround them (Figure 17–3). The elongated cells of the taste bud are arranged like the staves of a barrel, extending through the full thickness of the epithelium and forming a small opening (the taste pore) to the exterior. Through this pore, presumably, must pass the molecules to be tasted. At least two types of cells can be distinguished in the taste bud, some appearing pale and others dark; and among these are the cells that act as the taste transducers. The sensory signal is conveyed to the brain by nerves whose fibers penetrate the taste bud and end in contact with

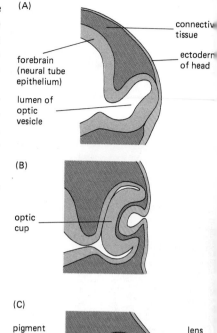

Figure 17–2 The development of the vertebrate eye. The retina develops from the *optic vesicle*, an epithelial outpocketing of the forebrain region of the neural tube. (A) The neural epithelium makes contact with the ectoderm covering the exterior of the head. (B) This contact induces the ectoderm to invaginate to form a lens. At the same time, the outer part of the optic vesicle invaginates, reducing the vesicle lumen to an interface between two layers that together form a cuplike structure. (C) The layer of the optic cup closest to the lens differentiates into the *neural retina*, which contains the photoreceptor cells and the neurons that relay visual stimuli to the brain (see Figure 17–6). The other layer differentiates into the retinal *pigment epithelium*. Its cells are heavily loaded with melanin granules and thus form a dark enclosure for the photoreceptive system (serving to reduce the amount of scattered light, much as a coat of black paint does inside a camera).

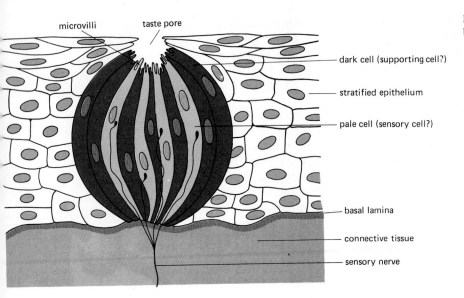

microvilli   taste pore

Figure 17–3   The structure of a taste bud.

— dark cell (supporting cell?)

— stratified epithelium

— pale cell (sensory cell?)

— basal lamina

— connective tissue

— sensory nerve

its cells. If the nerves are severed, the taste buds disappear entirely. When the nerves regenerate, they induce epithelial cells to change their state of differentiation to form new taste buds. Taste buds can be caused to form even in a region of epithelium that normally has none, such as that on the undersurface of the tongue, if it is cultured together with an appropriate sensory ganglion and thereby becomes innervated.

Despite such examples, however, the majority of adult tissues are composed of a number of distinct, irreversibly determined cell lineages. The numbers and spatial relationships of these components have to be maintained throughout life by mechanisms that do not require one type of cell to transform into another.

Summary

Most differentiated cells in adult tissues will maintain their specialized character even when placed in a novel environment. Although states of differentiation are generally stable and not interconvertible, even highly specialized cells can alter their properties to a limited extent in response to environmental cues. Especially striking cell transformations occur within the family of connective-tissue cells that includes fibroblasts and cartilage cells. Taste buds provide another unusual example of a state of differentiation that depends on a continuing cell-cell interaction, for their specialized cells disappear completely in the absence of nerves and reappear when innervation is restored.

Tissues with Permanent Cells[5]

Not all the populations of differentiated cells in the body are subject to cell turnover. Some cell types, having been generated in appropriate numbers in the embryo, are retained throughout adult life; they seem never to divide, and they cannot be replaced if they are lost. Almost all nerve cells are permanent in this sense. So are a few other types of cells, including—in mammals—the muscle cells of the heart and the lens cells of the eye.

While all these cells have extremely long life-spans and necessarily live in protected environments, they are dissimilar in other respects, and it is difficult to give a general reason why they should be permanent. For heart muscle cells it is difficult to give any reason at all. In the case of nerve cells (which will be discussed in detail in Chapter 19), it seems likely that extensive cell turnover in the adult would as a rule be disadvantageous, since it would be difficult to reestablish in the adult the precise and complex patterns of nerve connections that are set up under very different circumstances during development. Moreover, any memories recorded in the form of slight modifications of the structure or interconnections of individual nerve cells would presumably be obliterated. In the lens of the eye, on the other hand, the permanence of the cells appears to be simply an inevitable consequence of the way the tissue grows.

The Cells at the Center of the Lens of the Adult Eye Are Remnants of the Embryo[6]

Very little of the adult body consists of the same molecules that were laid down in the embryo. The **lens** of the eye is an exception: it is one of the few structures containing cells that are not only preserved but are preserved without turnover of their contents.

The lens is formed from the ectoderm at the site where the developing optic vesicle makes contact with it: the ectoderm here thickens, invaginates, and finally pinches off as a *lens vesicle* (see Figure 17–2). The lens thus originates as a spherical shell of cells formed from an epithelium, one cell layer thick, surrounding a central cavity. The cells at the rear of the lens vesicle (those facing the retina) soon undergo a striking transformation. They synthesize and become filled with *crystallins*, the characteristic proteins of the lens. In the process they elongate enormously, differentiating into *lens fibers* (Figure 17–4). Eventually their nuclei disintegrate and protein synthesis ceases. In this way the part of the lens vesicle epithelium facing the retina is expanded into a thick refractile body, consisting of many long, lifeless cells packed side by side (Figure 17–5). The central cavity of the vesicle is obliterated, and the front part of the epithelium of the lens vesicle— the part facing the external world—remains as a thin sheet of low cuboidal cells. Growth of the lens depends on the proliferation of these cells at the front, pushing some of the cells from this region around the rim of the lens and toward the back (see Figures 17–4 and 17–5A). As cells move to the rear, they stop dividing, step up their rate of synthesis of crystallins, and differentiate into lens fibers. Additional lens fibers continue to be recruited in this way throughout life, although at an ever-decreasing rate.

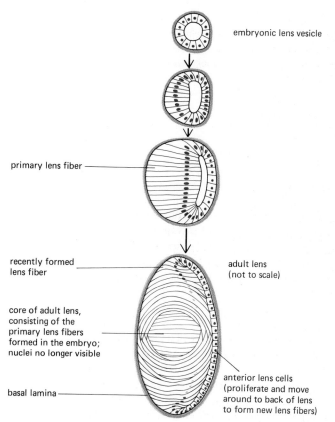

embryonic lens vesicle

Figure 17–4 The development of the lens of a human eye.

primary lens fiber

recently formed lens fiber

adult lens (not to scale)

core of adult lens, consisting of the primary lens fibers formed in the embryo; nuclei no longer visible

basal lamina

anterior lens cells (proliferate and move around to back of lens to form new lens fibers)

anterior lens
epithelium

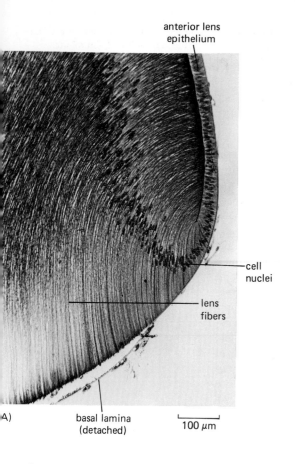

cell
nuclei

lens
fibers

(A)

basal lamina
(detached)

100 μm

(B)

20 μm

The types of crystallins filling the earliest generations of lens fibers are different from those of the later generations, just as the hemoglobins of fetal red blood cells are different from those of adult red blood cells. But whereas old red blood cells are discarded, old lens fibers are not. Thus at the core of the adult lens lie fibers that were laid down in the embryo and are still packed with the distinctive types of crystallins manufactured in that earlier period. Differences of refractive index between the early embryonic types of crystallins and those that are laid down later help to free the lens of the eye from the optical aberrations that bedevil simple lenses made out of homogeneous media such as glass.

Most Permanent Cells Renew Their Parts: The Photoreceptor Cells of the Retina[7]

There are few cells as immutable as lens fibers. As a rule, even those cells that persist throughout life without dividing undergo renewal of their component parts. Thus, while they do not divide, heart muscle cells and nerve cells are metabolically active and capable not only of synthesizing new RNA and protein, but also of altering their size and structure during adult life. Heart muscle cells, for example, replace the bulk of their protein molecules in the course of a week or two, and they will adjust the balance of protein synthesis and degradation so as to grow bigger if the load on the heart is increased—for example, by a sustained increase in blood pressure. Nerve cells also replace their protein molecules continuously; moreover, many nerve cells can regenerate axons and dendrites that have been cut off (see Chapter 19).

The turnover of cell components is dramatically illustrated in the highly specialized neural cells that form the **photoreceptors** of the retina. The neural retina (see Figure 17–2) consists of several cell layers organized in a way that seems perverse. The neurons that transmit signals from the eye to the brain (called *retinal ganglion cells*) lie closest to the external world, so that the light, focused by the lens, must pass through them to reach the photoreceptor cells. The photoreceptors, which are classified as **rods** or **cones,** according to their shape, lie with their

Figure 17–5 The structure of the mature lens. (A) Light micrograph of part of the lens, showing the junction between the thin sheet of anterior lens epithelium that covers the front of the lens and the differentiated lens fibers to the rear. (B) Scanning electron micrograph of part of the lens. The lens fibers are closely stacked, like planks in a lumberyard. Each one is a single, lifeless, elongated cell that can be up to 12 mm long. (A, courtesy of Peter Gould; B, from R.G. Kessel and R.H. Kardon, Tissues and Organs: A Text-Atlas of Scanning Electron Microscopy. San Francisco: Freeman, 1979. © 1979 W.H. Freeman and Company.)

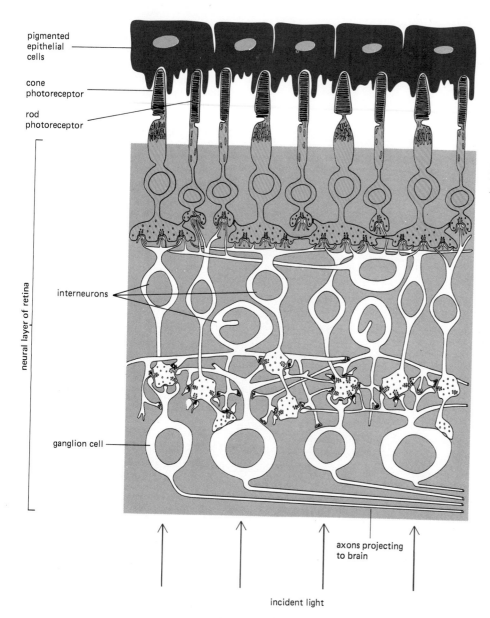

pigmented
epithelial
cells

cone
photoreceptor

rod
photoreceptor

neural layer of retina

interneurons

ganglion cell

axons projecting
to brain

incident light

Figure 17–6 The structure of the retina. The stimulation of the photoreceptors by light is relayed via the interneurons to the ganglion cells, which convey the signal to the brain. The spaces shaded light gray between neurons and between photoreceptors in the neural retina are occupied by a population of specialized supporting cells, whose individual outlines are not shown here. (Modified from J.E. Dowling and B.B. Boycott, *Proc. R. Soc. Lond. (Biol.)* 166:80–111, 1966.)

photoreceptive ends, or *outer segments,* partly buried in the pigment epithelium (Figure 17–6). Rods and cones contain different photosensitive complexes of protein with visual pigment: rods are especially sensitive at low light levels, while cones (of three types, each with different spectral responses) detect color and fine detail. The outer segment of a photoreceptor appears to be a modified cilium with a characteristic ciliumlike arrangement of microtubules in the region where the outer segment is connected to the rest of the cell (Figure 17–7). The remainder of the outer segment is almost entirely filled with a dense stack of membranes in which the photosensitive complexes are embedded. At their opposite ends the photoreceptors form synapses on interneurons.

The photoreceptors are permanent cells that do not divide. But the photosensitive protein molecules are not permanent. There is a steady turnover, which can be demonstrated by showing that injected radioactive amino acids are incorporated into these molecules. In rods (although not, curiously, in cones) this turnover is organized in an orderly production line, which can be analyzed by following the passage of a cohort of radiolabeled protein molecules through the cell after a short pulse of radioactive amino acid (Figure 17–8). The radiolabeled proteins can be followed from the Golgi apparatus in the inner segment of the cell to the base of the stack of membranes in the outer segment. From here they

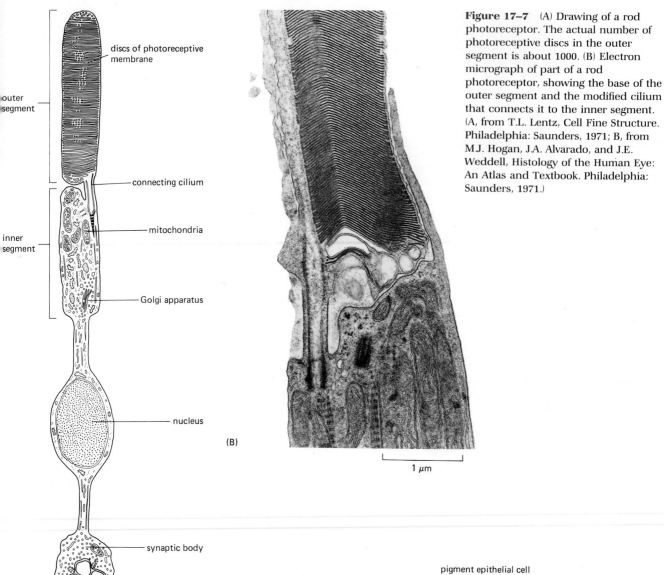

discs of photoreceptive
membrane

outer
segment

connecting cilium

inner
segment

mitochondria

Golgi apparatus

nucleus

(B)

synaptic body

(A)

Figure 17–7 (A) Drawing of a rod photoreceptor. The actual number of photoreceptive discs in the outer segment is about 1000. (B) Electron micrograph of part of a rod photoreceptor, showing the base of the outer segment and the modified cilium that connects it to the inner segment. (A, from T.L. Lentz, Cell Fine Structure. Philadelphia: Saunders, 1971; B, from M.J. Hogan, J.A. Alvarado, and J.E. Weddell, Histology of the Human Eye: An Atlas and Textbook. Philadelphia: Saunders, 1971.)

1 μm

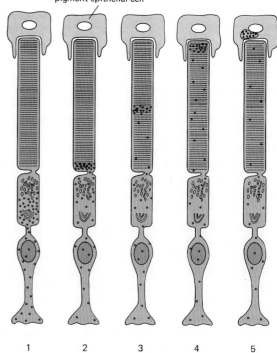

pigment epithelial cell

1   2   3   4   5

Figure 17–8 Turnover of membrane protein in a rod cell. Following a pulse of ³H-leucine, the passage of radiolabeled proteins through the cell is followed by autoradiography. Colored dots indicate sites of radioactivity. The method reveals only the ³H-leucine that has been incorporated into proteins; the rest is washed out during the preparation of the tissue. The incorporated leucine is first seen concentrated in the neighborhood of the Golgi apparatus (1), and from there it passes to the base of the outer segment into a newly synthesized disc of photoreceptive membrane (2). New discs are formed at a rate of three or four per hour (in a mammal), displacing the older discs toward the pigment epithelium (3–5).

are gradually displaced toward the tip as new material is fed into the base of the stack. Finally (after about 10 days in the rat), upon reaching the tip of the outer segment, the labeled proteins and the layers of membrane in which they are embedded are phagocytosed (chewed off and digested) by the cells of the pigment epithelium.

We shall have more to say about photoreceptors and their function in Chapter 19 (see p. 1105).

Summary

Nerve cells, heart muscle cells, and lens fibers persist throughout life without dividing and without being replaced. In mature lens fibers, the cell nuclei have degenerated and protein synthesis has stopped, so that the core of the adult lens consists of lens proteins laid down early in embryonic life. But in most other permanent cells, metabolic activity continues, and there is a steady turnover of cell components. In the rod cells of the retina, for example, new layers of photoreceptive membrane are synthesized close to the nucleus and are steadily displaced outward until they are eventually engulfed and digested by cells of the pigment epithelium.

Renewal by Simple Duplication[8]

Most of the differentiated cell populations in a vertebrate are not permanent: the cells are continually dying and being replaced. New differentiated cells can be produced during adult life in either of two ways: (1) they can form by the *simple duplication* of existing differentiated cells, which divide to give pairs of daughter cells of the same type; or (2) they can be generated from relatively undifferentiated *stem cells* by a process that involves a change of cell phenotype, as will be explained later in this chapter.

Rates of renewal vary from one tissue to another. The turnover time may be as short as a week or less, as in the epithelial lining of the small intestine (which is renewed by means of stem cells), or as long as a year or more, as in the pancreas (which is renewed by simple duplication). Many tissues whose normal rates of renewal are very slow can be stimulated to produce new cells at higher rates when the need arises.

In this section we shall discuss two examples of cell populations that are renewed by simple duplication—liver cells and endothelial cells.

The Liver Functions as an Interface Between the Digestive Tract and the Blood[9]

Digestion is a complex process. The cells that line the digestive tract secrete into the lumen of the gut a variety of substances, such as hydrochloric acid and digestive enzymes, to break down food molecules into simpler nutrients. The cells absorb these nutrients from the gut lumen, process them, and then release them into the blood for utilization by other cells of the body. All of these activities are adjusted according to the composition of the food consumed and the levels of metabolites in the circulation. The complex set of tasks is performed by a division of labor: some of the cells are specialized for the secretion of hydrochloric acid, others for the secretion of enzymes, others for absorption of nutrients, others for the production of peptide hormones, such as gastrin, that regulate digestive and metabolic activities, and so on (Figure 17–9). Some of these cell types lie closely intermingled in the wall of the gut; others are segregated in large glands that communicate with the gut and originate in the embryo as outgrowths of the gut epithelium.

The liver is the largest of these glands. It develops at a site where a major vein runs close to the wall of the primitive gut tube, and the adult organ retains a singularly close relationship with the blood. The cells in the liver that derive from the primitive gut epithelium—the **hepatocytes**—are arranged in folded sheets,

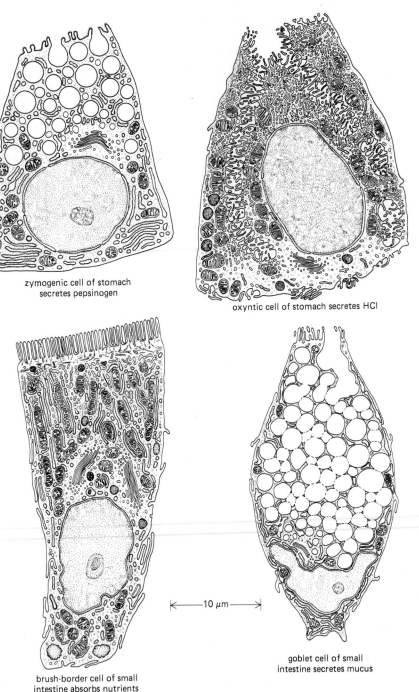

Figure 17–9 Some of the specialized cell types found in the epithelial lining of the gut. Neighboring positions in the epithelial sheet are often occupied by cells of dissimilar types (see Figure 17–17B). (After T.L. Lentz, Cell Fine Structure. Philadelphia: Saunders, 1971.)

zymogenic cell of stomach
secretes pepsinogen

oxyntic cell of stomach secretes HCl

brush-border cell of small
intestine absorbs nutrients

|← 10 µm →|

goblet cell of small
intestine secretes mucus

facing blood-filled spaces called *sinusoids* (Figure 17–10A). The blood is separated from the surface of the hepatocytes by a single layer of flattened endothelial cells that covers the sides of each hepatocyte sheet (Figure 17–10B). This structure facilitates the chief functions of the liver, which center on the exchange of metabolites between hepatocytes and the blood.

The liver is the main site at which nutrients that have been absorbed from the gut and then transferred to the blood are processed for use by other cells of the body; it receives a major part of its blood supply directly from the intestinal tract (via the portal vein). Hepatocytes are responsible for the synthesis, degradation, and storage of a vast number of substances; they play a central part in the carbohydrate and lipid metabolism of the body as a whole; and they secrete most of the protein found in blood plasma. At the same time, the hepatocytes remain connected with the lumen of the gut via a system of minute channels (or *canaliculi*)

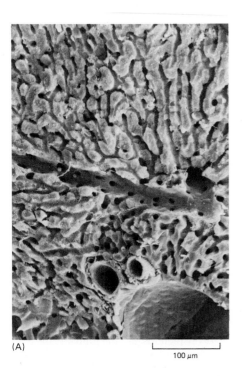

(A)

100 μm

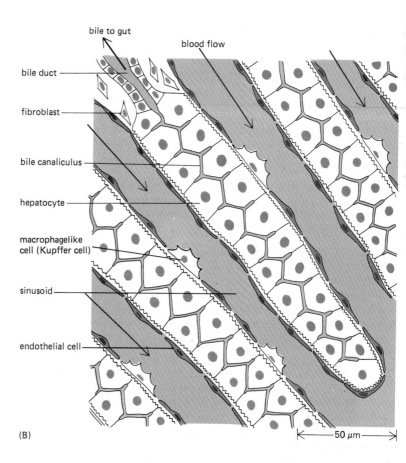

bile to gut

blood flow

bile duct

fibroblast

bile canaliculus

hepatocyte

macrophagelike
cell (Kupffer cell)

sinusoid

endothelial cell

(B)

50 μm

and larger ducts (see Figure 17–10B) and secrete into the gut by this route both waste products of their metabolism and an emulsifying agent, *bile*, which helps in the absorption of fats. In contrast to the rest of the digestive tract, there seems to be remarkably little division of labor within the population of hepatocytes: each hepatocyte appears to be able to perform the same broad range of metabolic and secretory tasks.

Hepatocytes have a life-style different from the cells that line the lumen of the gut. The latter, exposed to the abrasive and corrosive contents of the gut, cannot live for long and must be rapidly replaced by a continual supply of new cells (see Figure 17–17). Hepatocytes, removed from direct contact with the contents of the gut, live much longer and are renewed at a slow, precisely controlled rate.

Liver Cell Loss Stimulates Liver Cell Proliferation[10]

Even in a slowly renewing tissue, a small but persistent imbalance between the rate of cell production and the rate of cell loss will lead to disaster. If 2% of the heptatocytes in a human were to divide each week, but only 1% were lost, the liver would grow to exceed the weight of the rest of the body within 8 years. Some homeostatic mechanism must operate to adjust the rate of cell proliferation according to the mass of tissue present. The necessity for such control is especially great in an organ like the liver, whose cells may from time to time be destroyed by poisons (such as alcohol) in the diet.

Direct evidence for homeostatic control of liver cell proliferation comes from experiments in which large numbers of hepatocytes are removed surgically or are intentionally killed by poisoning with carbon tetrachloride. Within a day or so after either sort of damage, a surge of cell division occurs among the surviving hepatocytes, and the lost tissue is quickly replaced. If two-thirds of a rat's liver is removed, for example, a liver of nearly normal size can regenerate from the remainder within about 2 weeks. In cases of this kind, a signal for liver regeneration can be demonstrated in the circulation: if the circulations of two rats are con-

Figure 17–10 The structure of the liver. (A) Scanning electron micrograph of a portion of the liver, showing the irregular sheets of hepatocytes and the many small channels, or sinusoids, for the flow of blood. The larger channels are vessels that distribute and collect the blood that flows through the sinusoids. (B) The fine structure of the liver (highly schematized). The hepatocytes are separated from the bloodstream by a single thin sheet of endothelial cells with interspersed macrophagelike *Kupffer cells*. Small holes in the endothelial sheet allow exchange of molecules and small particles between the hepatocytes and the bloodstream while protecting the hepatocytes from buffeting by direct contact with the circulating blood cells. Besides exchanging materials with the blood, the hepatocytes form a system of minute bile canaliculi into which they secrete bile, which is ultimately discharged into the gut via bile ducts. The real structure is less regular than this diagram suggests. (A, from R.G. Kessel and R.H. Kardon, Tissues and Organs: A Text-Atlas of Scanning Electron Microscopy. San Francisco: Freeman, 1979 © 1979 W.H. Freeman and Company.)

nected surgically, and two-thirds of the liver of one of them is excised, cell division is stimulated in the unmutilated liver of the other. What the circulating signal is and how it acts remain debatable questions, despite much research. It may be that the signal consists of a complex combination of chemical conditions rather than a single growth factor. Somewhat similar regenerative phenomena are seen in the kidney, which seems to have its own separate but analogous system for controlling its growth.

Regeneration Requires Coordinated Growth of Tissue Components[11]

Like all organs, the liver comprises a mixture of cell types. Besides the hepatocytes and the endothelial cells that line its sinusoids, it contains both specialized macrophages (*Kupffer cells*), which engulf particulate matter in the bloodstream and dispose of worn-out red blood cells, and a small number of fibroblasts, which provide a tenuous supporting framework of connective tissue (see Figure 17–10B). All of these cell types are capable of division. For optimal regeneration, their proliferation must be properly coordinated.

The importance of balanced regeneration of cell types is demonstrated by what happens when an imbalance occurs. For example, if hepatocytes are poisoned repeatedly with carbon tetrachloride or with alcohol at such frequent intervals that they cannot recover fully between attacks, the fibroblasts take advantage of the situation and the liver becomes irreversibly clogged with connective tissue, leaving little space for the hepatocytes to grow even after the toxic agents are withdrawn. This condition, called *cirrhosis*, is common in chronic alcoholics. In a similar way, the regeneration of severely damaged skeletal muscle is often seriously hindered by the overgrowth of its connective tissue, so that scar tissue replaces the contractile muscle fibers. These imbalances, however, require unusual tissue damage; in ordinary circumstances of tissue renewal, poorly understood regulatory mechanisms ensure that the proper mixture of cell types is maintained.

Endothelial Cells Line All Blood Vessels[12]

By contrast with the above examples of ill-coordinated behavior of fibroblasts, the **endothelial cells** that form the lining of blood vessels have a remarkable capacity to adjust their numbers and arrangement to suit local requirements. Almost all tissues depend on a blood supply, and the blood supply depends on endothelial cells. They create an adaptable life-support system, ramifying into every region of the body. If it were not for endothelial cells extending and remodeling the network of blood vessels, tissue growth and repair would be impossible.

The largest blood vessels are arteries and veins, which have a thick, tough wall of connective tissue and smooth muscle (Figure 17–11A). The wall is lined by

Figure 17–11 (A) Schematic diagram of a part of the wall of a small artery. The endothelial cells, although inconspicuous, are the fundamental component. Compare with the capillary in Figure 17–12. (B) Scanning electron micrograph of a cross-section through an arteriole (very small artery), showing the inner lining of endothelial cells and the surrounding layer of smooth muscle and collagenous connective tissue. A slight contraction of the smooth muscle has thrown the endothelial lining of the vessel into folds. In fixation the endothelial lining has shrunk away from the muscular wall, leaving a small gap. (B, from R.G. Kessel and R.H. Kardon, Tissues and Organs: A Text-Atlas of Scanning Electron Microscopy. San Francisco: Freeman, 1979. © 1979 W.H. Freeman and Company.)

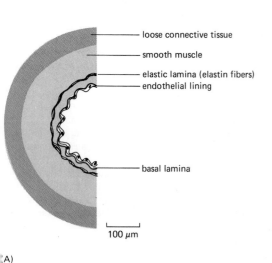

(A)

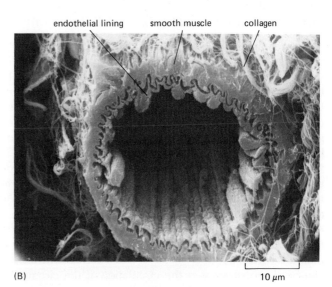

(B)

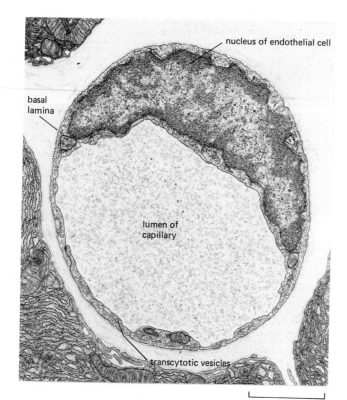

Figure 17–12 Electron micrograph of a small capillary in cross-section. The wall is formed by a single endothelial cell surrounded by a basal lamina. Note the small (80-nm) "transcytotic" vesicles, which, according to one theory, provide transport of large molecules in and out of this type of capillary: materials are taken up into the vesicles by endocytosis at the luminal surface of the cell and discharged by exocytosis at the external surface, or vice versa. (From R.P. Bolender, *J. Cell Biol.* 61:269–287, 1974. Reproduced by copyright permission of the Rockefeller University Press.)

Labels on figure: nucleus of endothelial cell; basal lamina; lumen of capillary; transcytotic vesicles; 2 μm

an exceedingly thin single layer of endothelial cells, separated from the surrounding outer layers by a basal lamina. The amounts of connective-tissue and smooth muscle in the vessel wall vary according to the vessel's diameter and function, but the endothelial lining is always present (Figure 17–11B). In the finest branches of the vascular tree—the capillaries and sinusoids—the walls consist of nothing but endothelial cells and a basal lamina (Figure 17–12). Thus endothelial cells line the entire vascular system, from the heart to the smallest capillary, and control the passage of materials—and the transit of white blood cells—into and out of the bloodstream. A study of the embryo reveals, moreover, that arteries and veins develop from small vessels constructed solely of endothelial cells and a basal lamina: connective tissue and smooth muscle are added later where required, under the influence of signals from the endothelial cells.

New Endothelial Cells Are Generated by Simple Duplication of Existing Endothelial Cells[13]

Throughout the vascular system of the adult, endothelial cells retain a capacity for cell division and movement. If, for example, a part of the wall of the aorta is damaged and denuded of endothelial cells, neighboring endothelial cells proliferate and migrate in to cover the exposed surface. Newly formed endothelial cells will even cover the inner surface of plastic tubing used by surgeons to replace parts of damaged blood vessels.

The proliferation of endothelial cells can be demonstrated by using ^{3}H-thymidine to label cells synthesizing DNA. In normal vessels the proportion of endothelial cells that become labeled is especially high at branch points in arteries, where turbulence and the resulting wear on the endothelial cells presumably stimulate cell turnover. On the whole, however, endothelial cells turn over very slowly, with a cell lifetime of months or even years.

Endothelial cells not only repair the lining of established blood vessels, they also create new blood vessels. They must do this in embryonic tissues to keep pace with growth, in normal adult tissues to support recurrent cycles of remodeling and reconstruction, and in damaged adult tissues to support repair.

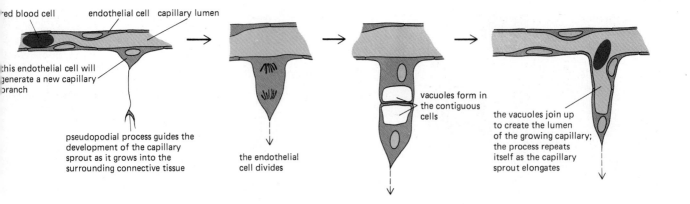

red blood cell   endothelial cell   capillary lumen

this endothelial cell will
generate a new capillary
branch

pseudopodial process guides the
development of the capillary
sprout as it grows into the
surrounding connective tissue

the endothelial
cell divides

vacuoles form in
the contiguous
cells

the vacuoles join up
to create the lumen
of the growing capillary;
the process repeats
itself as the capillary
sprout elongates

New Capillaries Form by Sprouting[14,15]

New vessels always originate as capillaries, which sprout from existing small vessels. This process of **angiogenesis** occurs in response to specific signals, and it can be readily observed in rabbits by punching a small hole in the ear and fixing glass cover slips on either side to create a thin transparent viewing chamber into which the cells that surround the wound can grow. Angiogenesis can also be conveniently observed in naturally transparent structures such as the cornea of the eye. Irritants applied to the cornea induce the growth of new blood vessels from the rim of the cornea, which has a rich blood supply, in toward the center, which normally has almost none. Thus the cornea becomes vascularized through an invasion of endothelial cells into the tough collagen-packed corneal tissue.

Observations such as these reveal that endothelial cells that will form a new capillary grow out from the side of a capillary or small venule by extending long processes or pseudopodia (Figure 17–13). The cells at first form a solid sprout, which then hollows out to form a tube. This process continues until the sprout encounters another capillary, with which it connects, allowing blood to circulate. Experiments in culture show that endothelial cells in a medium containing suitable growth factors will spontaneously form capillary tubes even if they are isolated from all other types of cells. The first sign of tube formation in culture is the appearance in a cell of an elongated vacuole that is at first completely encompassed by cytoplasm (Figure 17–14A). Contiguous cells develop similar vacuoles, and eventually the cells arrange their vacuoles end to end so that the vacuoles become continuous from cell to cell, forming a capillary channel (Figure 17–14B). The capillary tubes that develop in a pure culture of endothelial cells do not contain blood, and nothing travels through them, indicating that blood flow and pressure are not required for the formation of a capillary network.

Figure 17–13 A new blood capillary forms by the sprouting of an endothelial cell from the wall of an existing small vessel. This schematic diagram is based on observations of cells in the transparent tail of a living tadpole. (After C.C. Speidel, *Am. J. Anat.* 52:1–79, 1933.)

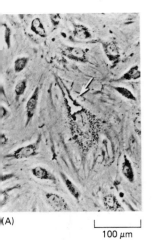

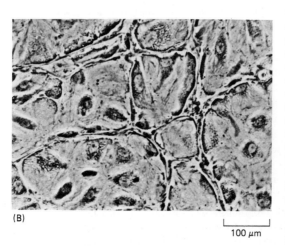

(A)   100 µm

(B)   100 µm

Figure 17–14 Endothelial cells in culture spontaneously develop internal vacuoles that join up, giving rise to a network of capillary tubes. Photographs (A) and (B) show successive stages in the process; the arrow in (A) indicates a vacuole forming initially in a single endothelial cell. The cultures are set up from small patches of two to four endothelial cells taken from short segments of capillary. These cells will settle on the surface of a collagen-coated culture dish and form a small flattened colony that enlarges gradually as the cells proliferate. The colony spreads across the dish, and eventually, after about 20 days, capillary tubes begin to form in the central regions. Once tube formation has started, branches soon appear, and after 5 to 10 more days, an extensive network of tubes is visible, as seen in (B). (From J. Folkman and C. Haudenschild, *Nature* 288:551–556, 1980. © Macmillan Journals Ltd.)

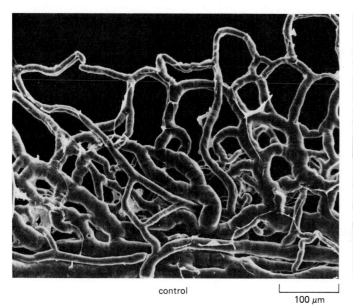

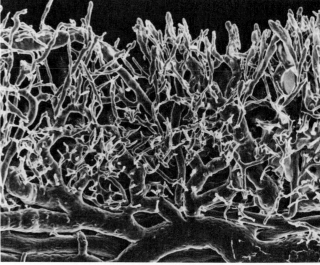

control   └─────┘
          100 µm

60 hours after wounding   └─
          100 µm

Figure 17–15 Scanning electron micrographs of casts of the system of blood vessels in the margin of the cornea, showing the reaction to wounding. The casts are made by injecting a resin into the vessels and letting the resin set; this reveals the shape of the lumen, as opposed to the shape of the cells. Sixty hours after wounding, many new capillaries have begun to sprout toward the site of injury, which is just above the top of the picture. Their oriented outgrowth reflects a chemotactic response of the endothelial cells to an angiogenic factor released at the wound. (Courtesy of Peter C. Burger.)

Growth of the Capillary Network Is Controlled by Factors Released by the Surrounding Tissues[15]

In living animals, endothelial cells form new capillaries wherever there is a need for them. It is thought that when cells in tissues are deprived of oxygen, they release angiogenic factors that induce new capillary growth. Probably for this reason, nearly all vertebrate cells are located within 50 µm of a capillary. Similarly, after wounding, a burst of capillary growth is stimulated in the neighborhood of the damaged tissue (Figure 17–15). Local irritants or infections also cause a proliferation of new capillaries, most of which regress and disappear when the inflammation subsides.

Angiogenesis is also important in tumor growth. The growth of a solid tumor is limited by its blood supply: if it were not invaded by capillaries, a tumor would be dependent on the diffusion of nutrients from its surroundings and could not enlarge beyond a diameter of a few millimeters. To grow further, a tumor must induce the formation of a capillary network that invades the tumor mass. A small sample of such a tumor implanted in the cornea will cause blood vessels to grow quickly toward the implant from the vascular margin of the cornea (Figure 17–16), and the growth rate of the tumor increases abruptly as soon as the vessels reach it.

In all of these cases the invading endothelial cells must respond to a signal produced by the tissue that requires a blood supply. The response of the endothelial cells includes at least three components. First, the cells must breach the basal lamina that surrounds an existing blood vessel; endothelial cells during angiogenesis have been shown to secrete *proteases*, such as *plasminogen activator*, which enable them to digest their way through the basal lamina of the parent capillary or venule. Second, the endothelial cells must move toward the source of the signal. Third, they must proliferate. In certain circumstances, one or two of

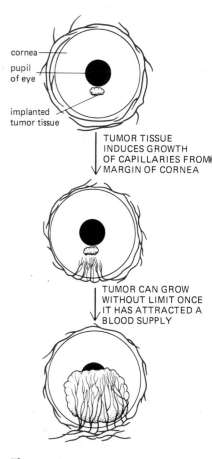

cornea
pupil of eye
implanted tumor tissue

TUMOR TISSUE INDUCES GROWTH OF CAPILLARIES FROM MARGIN OF CORNEA

TUMOR CAN GROW WITHOUT LIMIT ONCE IT HAS ATTRACTED A BLOOD SUPPLY

Figure 17–16 Tumor tissue implanted in the cornea releases a factor that causes the ingrowth of capillaries, supplying the tumor with blood-borne nutrients that allow it to grow.

he components of this tripartite response can be elicited in the absence of the others. Some new capillaries form, for example, even when endothelial cell proliferation has been blocked by irradiation, and a factor in fluid from wounds has been shown to attract endothelial cells and to stimulate them to secrete proteases without stimulating their proliferation.

Other factors can elicit on their own all three components of the endothelial cell response. Two examples are *acidic fibroblast growth factor (acidic FGF)* and *basic fibroblast growth factor (basic FGF)*. These two proteins, which have been purified independently from several different sources and therefore are known by various other names as well, have similar amino acid sequences (55% identity). In addition to their dramatic effect on endothelial cells, they stimulate the proliferation of fibroblasts and of several other cell types and are important regulators of early embryonic development (see p. 894). Their precise cellular origins are poorly understood. Other substances that can also act as angiogenic factors appear to be released during tissue repair, inflammation, and growth by a variety of cell types, including macrophages, mast cells, and fat cells. Angiogenesis, like the control of cell proliferation in general, seems to be regulated by complex—and perhaps redundant—combinations of signals, rather than by one signal alone.

Summary

Most populations of differentiated cells in vertebrates are subject to turnover through cell death and renewal. In some cases the fully differentiated cells simply divide to produce daughter cells of the same type. The rate of proliferation of hepatocytes, for example, is controlled to maintain appropriate total cell numbers, and if a large part of the liver is destroyed, the remaining hepatocytes increase their division rate to restore the loss. Under normal circumstances cell renewal keeps the numbers of cells of each type in a tissue in appropriate balance. In response to unusual damage, however, repair may be unbalanced, as when the fibroblasts in a repeatedly damaged liver grow too rapidly in relation to the hepatocytes and replace them with connective tissue.

Endothelial cells form a single cell layer that lines all blood vessels and regulates exchanges between the bloodstream and the surrounding tissues. New blood vessels develop from the walls of existing small vessels by the outgrowth of endothelial cells, which have the capacity to form hollow capillary tubes even when isolated in culture. In the living animal, anoxic or damaged tissues stimulate angiogenesis by releasing angiogenic factors, which attract nearby endothelial cells and stimulate them to proliferate and secrete proteases.

Renewal by Stem Cells: Epidermis[8,16]

We turn now from cell populations that are renewed by simple duplication to those that are renewed by means of **stem cells.** These populations vary widely, not only in cell character and rate of turnover, but also in the geometry of cell replacement. In the lining of the small intestine, for example, cells are arranged as a single-layered epithelium. This epithelium covers the surfaces of the *villi* that project into the lumen of the gut, and it lines the deep *crypts* that descend into the underlying connective tissue (Figure 17–17). The stem cells lie in a protected position in the depths of the crypts. The differentiated cells generated from them are carried upward by a sliding movement in the plane of the epithelial sheet until they reach the exposed surfaces of the villi, from whose tips they are finally shed. A contrasting example is found in the epithelium that forms the outer surface of the skin, called the *epidermis*. The epidermis is a many-layered epithelium, and the differentiating cells travel outward from their site of origin in a direction perpendicular to the plane of the cell sheet. In the case of blood cells, the spatial pattern of production is complex and appears chaotic. Before going into such details, however, we must pause to consider what a stem cell is.

LUMEN OF GUT

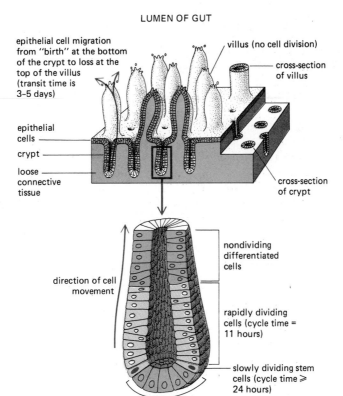

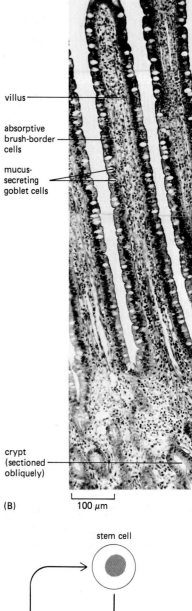

villus —

absorptive
brush-border —
cells

mucus-
secreting
goblet cells

crypt
(sectioned —
obliquely)

(B) ⊢ 100 μm ⊣

Figure 17–17 (A) The pattern of cell turnover and the proliferation of stem cells in the epithelium that forms the lining of the small intestine. (B) Photograph of a section of part of the lining of the small intestine, showing the villi and crypts. Note how mucus-secreting goblet cells (visible as pale ovals) are interspersed among the absorptive brush-border cells in the epithelium of the villi. See Figure 17–9 for the structure of these cells. (Courtesy of Peter Gould.)

Stem Cells Can Divide Without Limit and Give Rise to Differentiated Progeny[17]

The defining properties of a stem cell are as follows:

1. It is not itself terminally differentiated (that is, it is not at the end of a pathway of differentiation).
2. It can divide without limit (or at least for the lifetime of the animal).
3. When it divides, each daughter has a choice: it can either remain a stem cell, or it can embark on a course leading irreversibly to terminal differentiation (Figure 17–18).

Stem cells are required wherever there is a recurring need to replace differentiated cells that cannot themselves divide. In several tissues the terminal state of cell differentiation is obviously incompatible with cell division. For example, the cell nucleus may be digested, as in the outermost layers of the skin, or be extruded, as in the mammalian red blood cell. Alternatively, the cytoplasm may be heavily encumbered with structures, such as the myofibrils of striated muscle cells, that would hinder mitosis and cytokinesis. In other terminally differentiated cells the chemistry of differentiation may be incompatible with cell division in some more subtle way. In any such case, renewal must depend on stem cells.

The job of the stem cell is not to carry out the differentiated function but rather to produce cells that will. Consequently, stem cells often have a nondescript

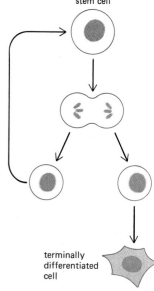

Figure 17–18 The definition of a stem cell. Each daughter produced when a stem cell divides can either remain a stem cell or go on to become terminally differentiated.

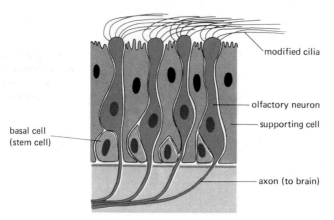

basal cell
(stem cell)

modified cilia

olfactory neuron

supporting cell

axon (to brain)

Figure 17–19 A schematic cross-section of olfactory epithelium (specialized for sensing smells). Three cell types can be distinguished: supporting cells, basal cells, and olfactory neurons. Autoradiographic experiments show that the basal cells are the stem cells for production of the olfactory neurons, which constitute one of the very few exceptions to the rule that neurons are permanent cells. Each olfactory neuron survives for about a month (in a mammal) before it is replaced. Six to eight modified cilia project from the globular head of the olfactory neuron and are believed to contain the smell receptors. The axon extending from the other end of the neuron conveys the message to the brain. A new axon must grow out and make appropriate connections in the brain whenever a basal cell differentiates into an olfactory neuron.

appearance, making them hard to identify. But that is not to say that stem cells are all alike. Although not terminally differentiated, they are nevertheless *determined* (see p. 898): the muscle satellite cell, as a source of skeletal muscle; the epidermal stem cell, as a source of keratinized epidermal cells; the spermatogonium, as a source of spermatozoa; the basal cell of olfactory epithelium, as a source of olfactory neurons (Figure 17–19); and so on. Those stem cells that give rise to only one type of differentiated cell are called *unipotent*, those that give rise to a small number of cell types are called *oligopotent*, and those that give rise to many cell types are called *pluripotent.*

Tissues that form from stem cells raise many important questions. We need to consider what factors determine whether a stem cell divides or stays quiescent, what decides whether a given daughter cell remains a stem cell or differentiates, and in what ways the differentiation of a daughter cell is regulated after it has become committed to differentiate. We begin our discussion with the epidermis, for its simple spatial organization makes it relatively easy to study the natural history of its stem cells and the fate of their progeny.

The Epidermis Is Organized into Proliferative Units[18,19]

The epidermal layer of the skin and the epithelial lining of the digestive tract are the two tissues that suffer the most direct and damaging encounters with the external world. In both, mature differentiated cells are rapidly lost from the most exposed positions and are replaced by the proliferation of less differentiated cells in more sheltered niches.

The **epidermis** is a multilayered epithelium composed largely of *keratinocytes* (so-called because their characteristic differentiated activity is keratin synthesis) (Figure 17–20). These cells change their appearance from one layer to the next.

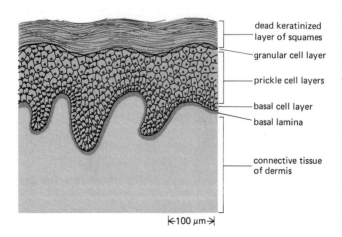

dead keratinized layer of squames

granular cell layer

prickle cell layers

basal cell layer

basal lamina

connective tissue of dermis

|←100 μm→|

Figure 17–20 Cross-section of moderately thick mammalian epidermis (see also Figure 17–1). The *granular cells* between the prickle cells and the flattened squames are in the penultimate stages of keratinization; they appear granular because they contain darkly staining aggregates of a material called *keratohyalin*, which is thought to be involved in the intracellular compaction and cross-linking of the keratin. Keratohyalin consists mainly of a protein known as *filaggrin*. In addition to the cells destined for keratinization, the deep layers of the epidermis include small numbers of cells of different character (not shown here)—including macrophagelike *Langerhans cells*, derived from bone marrow; *melanocytes*, derived from the neural crest; and *Merkel cells*, which are associated with nerve endings in the epidermis.

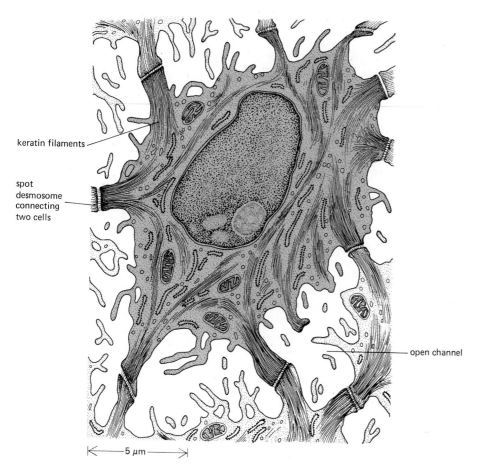

keratin filaments

spot
desmosome
connecting
two cells

open channel

←—5 μm—→

Figure 17–21 Drawing from an electron micrograph of a section of a prickle cell (*color*) from the epidermis, showing the bundles of keratin filaments that traverse the cytoplasm and are inserted at the desmosome junctions that bind the cell to its neighbors. Note that between adjacent cells there are open channels that allow nutrients to diffuse freely through the metabolically active layers of the epidermis. Further out, at the level of the granular cells, there is a waterproof barrier that is thought to be created by a sealant material that the granular cells secrete from vesicles called *membrane-coating granules*. (From R.V. Krstić, Ultrastructure of the Mammalian Cell: An Atlas. Berlin: Springer, 1979.)

Those in the innermost layer are termed *basal cells*, and it is chiefly these that undergo mitosis. Above the basal cells are several layers of larger *prickle cells* (Figure 17–21), whose numerous desmosomes—each a site of anchorage for thick tufts of keratin filaments—are just visible in the light microscope as tiny prickles around the cell surface (hence the name). Beyond the prickle cells lies the thin granular cell layer (see Figure 17–20). This marks the boundary between the inner, metabolically active strata and the outermost layer, consisting of dead cells whose intracellular organelles have disappeared. These outermost cells are reduced to flattened scales, or *squames*, filled with densely packed keratin. The plasma membranes of both the squames and the outer granular cells are reinforced on their cytoplasmic surface by a thin (12-nm), tough, cross-linked layer containing an intracellular protein called *involucrin*. The squames themselves are normally so compressed and thin that their boundaries are hard to make out in the light microscope, but soaking in sodium hydroxide makes them swell slightly, and with suitable staining a remarkably ordered geometric arrangement can be seen in regions where the skin is thin. The squames are found to be stacked in hexagonal columns that interlock neatly at their edges (Figure 17–22). The diameter of each column is such that about 10 basal cells form the foundation on which it rests. The basal cells can be classified as central or peripheral according to whether they lie beneath the center or the periphery of their column. The peripheral cells, but not the central cells (according to the account that we shall follow here), can at times be seen in the act of passing upward from the basal cell layer into the prickle cell layer. Each column is called an *epidermal proliferative unit*. Although the orderly columnar arrangement is found only in some regions of the skin, it will serve to illustrate the general principles of epidermal cell renewal.

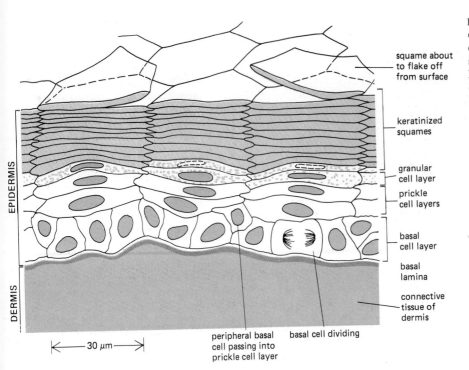

squame about to flake off from surface

keratinized squames

granular cell layer

prickle cell layers

basal cell layer

basal lamina

connective tissue of dermis

EPIDERMIS

DERMIS

←— 30 μm —→

peripheral basal cell passing into prickle cell layer

basal cell dividing

Figure 17–22 The columnar organization of squames in the epidermal layer of the skin. The structure is revealed by swelling the keratinized squames in a solution containing sodium hydroxide. This type of organization occurs only where the epidermis is thin.

Differentiating Epidermal Cells Synthesize a Sequence of Different Keratins as They Mature[19]

Having described the static picture, let us now set it in motion. The central basal cell divides, and some of its daughters, dividing in their turn, shift to peripheral basal positions. Peripheral basal cells slip out of the basal cell layer into the prickle cell layer, taking the first step on their outward journey. When they reach the granular layer, the cells start to lose their nuclei and cytoplasmic organelles and are transformed into the keratinized squames of the keratinized layer. These finally flake off from the surface of the skin (and become a main constituent of household dust). The period from the time a cell is born in the basal layer of the human skin to the time it is shed from the surface varies from 2 to 4 weeks, depending on the region of the body.

The accompanying molecular transformations can be studied by analyzing either thin slices of epidermis cut parallel to the surface or successive layers of cells stripped off by repeatedly applying and removing strips of adhesive tape. The keratin molecules, for example, which are plentiful in all layers of the epidermis, can be extracted and characterized. They are of many types (see p. 662), encoded by a large family of homologous genes, with the variety further increased through variable processing of the gene transcripts. As the new keratinocyte at the base of the column is transformed into the squame at the top (see Figure 17–22), it expresses a succession of different selections from its keratin gene repertoire. During this process, other characteristic proteins, such as involucrin, also begin to be synthesized as part of a coordinated program of terminal cell differentiation.

Stem Cell Potential Is Found Only in Cells That Contact the Basal Lamina[20]

If each epidermal proliferative unit is maintained indefinitely by proliferation of its basal cells, there must be among these basal cells at least one whose line of descendants will not die out in the lifetime of the animal. We shall call such a cell an *immortal stem cell* (Figure 17–23). In principle, the division of an immortal stem cell could generate two initially similar daughters whose different fates would be governed by subsequent circumstances. At the opposite extreme, the stem cell division could be always asymmetric: one and only one of the daughters would inherit a special character required for immortality, while the other would be somewhat altered already at the time of its birth in a way that forced it to differentiate and ultimately to die. In the latter case, there could never be any increase in the existing number of immortal stem cells; and this is contradicted by the facts. If a patch of epidermis is destroyed, the damage is repaired by surrounding healthy epidermal cells that migrate and proliferate to cover the denuded area. In this process, new epidermal proliferative units are formed whose central basal cells must have arisen from divisions that generated two immortal stem cells from one.

Thus the fate of the daughters of a stem cell must be governed at least partly by external circumstances. One possible determining factor might be contact with the basal lamina, with a loss of contact triggering the start of terminal differentiation. Cell-culture experiments lend some support to this suggestion: epidermal cells continue to proliferate if they are grown in contact with an appropriate substratum, such as a carpet of fibroblasts, but promptly differentiate if they are kept in suspension.

This sort of environmental control is probably not the whole story, however. Other evidence suggests an opposite chain of cause and effect, whereby the changes leading to terminal differentiation trigger detachment from the basal lamina, rather than vice versa. According to this hypothesis, only a small proportion of the basal cells have the capacity to serve as stem cells. These stem cells have special surface properties that enable them to adhere to the basal lamina, and they are programmed to generate a certain proportion of progeny that are committed to terminal differentiation, one aspect of which is a loss of this adhesiveness. In the special conditions of tissue repair, the ratio of differentiating to proliferating progeny would be altered by local growth factors so as to generate additional cells to cover the wound. Suggestive evidence in support of this view comes from experiments in which keratinocytes are grown in culture in a Ca^{2+}-deficient medium, which keeps them as a monolayer and therefore all in a basal position. Some cells in these conditions nevertheless embark on terminal differentiation, as indicated by the synthesis of involucrin; these differentiating cells emerge from the basal layer as soon as the Ca^{2+} concentration is raised (see p. 828).

Basal Cell Proliferation Is Regulated According to the Thickness of the Epidermis[21]

Whatever determines the choice between survival as a stem cell and death through terminal differentiation, additional controls must operate to regulate the rate of production of epidermal cells. If the outer layers of the epidermis are stripped away, for example, the division rate of the basal cells increases. After a transient overshoot, normal thickness is restored, and the division rate in the basal layer declines to normal. It is as though the removal of the outer differentiated layers releases the cells in the proliferative basal layer from an inhibitory influence, which is restored as soon as the outer layers regain their full thickness.

Although keratinocytes in culture are known to respond to a variety of hormones and growth factors, including epidermal growth factor (EGF), the molecular mechanisms that regulate their proliferation in the body remain an unsolved problem of great clinical importance. The consequences of faulty control of basal cell proliferation are seen in *psoriasis*. In this common skin disorder, the rate of basal cell proliferation is greatly increased—the epidermis thickens, and cells are shed from the surface of the skin within as little as a week after emerging from the basal layer, before they have had time to keratinize fully.

Secretory Cells in the Epidermis Are Secluded in Glands That Have Their Own Population Kinetics[22]

The skin has other functions besides providing a protective barrier, and in certain specialized regions other types of cells besides the keratinized cells described above develop from the epidermis. In particular, secretions are produced by skin cells segregated in deep-lying glands, which have patterns of renewal quite different from those of keratinizing regions.

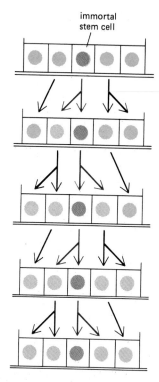

immortal stem cell

Figure 17–23 Each proliferative unit must contain, in each cell generation, at least one "immortal" stem cell, whose descendants will still be present in the unit in the distant future. The arrows indicate lines of descent. An immortal stem cell is shown here occupying the central basal position in each cell generation. Other basal cells might be born chemically different in a way that commits them to leave the basal layer and differentiate; or they too might be stem cells, equivalent to the immortal stem cell in character and mortal only in the sense that their progeny happen subsequently to be jostled out of the basal layer and shed from the skin.

A *sweat gland* is the simplest example of such a structure. It consists of a long tube with a blind end, and it develops as an ingrowth of the epidermis. Sweat is secreted by the cells in the bottom portion of the tube and is conveyed to the surface of the skin via an excretory duct (Figure 17–24). The secretory cells form an epithelium one layer thick, embraced by a small number of contractile *myoe-pithelial cells* (see Figures 17–25B and 17–38E), while the duct cells form an epithelium two layers thick with no myoepithelial component. The glands that produce tears, ear wax, saliva, and milk are similar. In the salivary glands and the mammary glands at least, the ducts contain the stem cells for the renewal of the secretory population.

The mammary gland is of special interest because of the hormonal control of its cell division and differentiation. Milk production must be switched on when a baby is born and switched off when the baby is weaned. A "resting" mammary

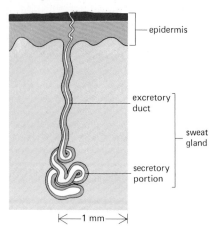

Figure 17–24 Schematic diagram of a sweat gland.

Figure 17–25 The mammary gland. (A) Schematic diagram of the growth of alveoli from the ducts of the mammary gland during pregnancy and lactation. Only a small part of the gland is shown. The "resting" gland contains a small amount of inactive glandular tissue embedded in a large amount of fatty connective tissue (*pale gray*). During pregnancy an enormous proliferation of the glandular tissue takes place at the expense of the fatty connective tissue, with the secretory portions of the gland developing preferentially to create alveoli. (B) One of the milk-secreting alveoli with a basket of myoepithelial cells (*color*) embracing it. The myoepithelial cells contract and expel milk from the alveolus in response to the hormone oxytocin, which is released as a reflex response to the stimulus of suckling. (C) A single type of secretory alveolar cell produces both the milk proteins and the milk fat. The proteins are secreted in the normal way by exocytosis, while the fat is released as droplets surrounded by plasma membrane detached from the cell. (B, after R. Krstić, Die Gewebe des Menschen und der Säugetiere. Berlin: Springer-Verlag, 1978; C, from D.W. Fawcett, A Textbook of Histology, 11th ed. Philadelphia: Saunders, 1986.)

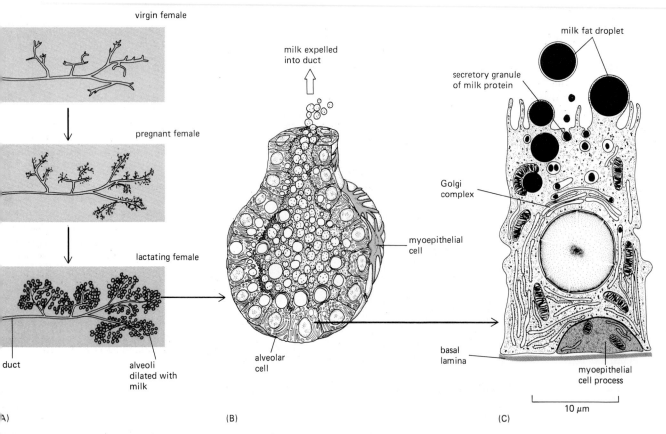

gland consists of branching systems of excretory ducts embedded in connective tissue; these ducts are lined, in their secretory portions, by a single layer of relatively inactive epithelial cells. As a first step toward large-scale milk production, the hormones that circulate during pregnancy cause the duct cells to proliferate and the terminal portions of the ducts to grow and branch, forming little dilated outpocketings, or *alveoli*, containing secretory cells (Figure 17–25). Milk secretion begins only when these cells are stimulated by the different combination of hormones circulating in the mother after the birth of the baby. Later, when suckling stops, the secretory cells degenerate and most of the alveoli disappear; macrophages clear away the debris, and the gland reverts to its resting state.

Summary

Many tissues, especially those with a rapid turnover—such as the lining of the gut, the epidermal layer of the skin, and the blood-forming tissues—are renewed by means of stem cells. Stem cells, by definition, are not terminally differentiated and have the ability to divide throughout the lifetime of the organism, yielding some progeny that differentiate and others that remain stem cells. In the skin the stem cells of the epidermis lie in the basal layer, in contact with the basal lamina. The progeny of the stem cells differentiate upon leaving this layer and, as they move outward, synthesize a succession of different types of keratin until eventually their nuclei degenerate, producing an outer layer of dead keratinized cells that are continually shed from the surface. In regions where the epidermis is thin, it is neatly organized into proliferative units or columns, with at least one "immortal" stem cell at the base of each. The fate of the daughters of a stem cell is controlled in part by poorly understood environmental factors. These factors allow two stem cells to be generated from one during repair processes, and they regulate the rate of basal cell proliferation according to the thickness of the epidermis. Glands connected to the epidermis, such as sweat glands and mammary glands, have their own stem cells and their own distinct patterns of cell renewal.

Renewal by Pluripotent Stem Cells: Blood Cell Formation[23,24]

The blood contains many types of cells with very different functions, ranging from the transport of oxygen to the production of antibodies. Some of these cells function entirely within the vascular system, while others use it only as a means of transport and perform their function elsewhere. All blood cells, however, have certain similarities in their life history. They all have limited life-spans and are produced throughout the life of the animal. Most remarkably, they are all generated ultimately from a common stem cell in the bone marrow. This *hemopoietic* (or blood-forming) *stem cell* is thus pluripotent, giving rise to all of the types of terminally differentiated blood cells.

Blood cells can be classified as red or white (Figure 17–26). The **red blood cells,** or **erythrocytes,** remain within the blood vessels and transport O_2 and CO_2

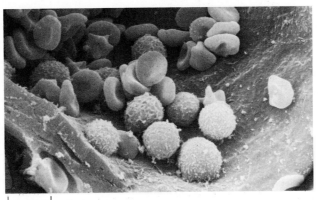

Figure 17–26 Scanning electron micrograph of mammalian blood cells in a small blood vessel. The larger, mo spherical cells with a rough surface ar white blood cells; the smaller, smoother, flattened cells are red bloo cells. (From R.G. Kessel and R.H. Kardon, Tissues and Organs: A Text-Atlas of Scanning Electron Microscopy San Francisco: Freeman, 1979. © 1979 W.H. Freeman and Company.)

10 μm

ound to hemoglobin. The **white blood cells,** or **leucocytes,** combat infection
nd, in some cases, phagocytose and digest debris. Leucocytes, unlike erythro-
ytes, must make their way across the walls of small blood vessels and migrate
to tissues to perform their tasks. In addition, the blood contains large numbers
platelets, which are not entire cells but small detached cell fragments or "mini-
lls" derived from the cortical cytoplasm of large cells called *megakaryocytes.*
atelets adhere specifically to the endothelial cell lining of damaged blood vessels,
here they help repair breaches and aid in the process of blood clotting.

here Are Three Main Categories of White Cells: Granulocytes, Ionocytes, and Lymphocytes[23,24]

l red blood cells are similar to one another, as are all platelets, but there are
any distinct types of white blood cells. They are traditionally grouped into three
ajor categories on the basis of their appearance in the light microscope: *gran-
ocytes, monocytes,* and *lymphocytes.*

The **granulocytes** all contain numerous lysosomes and secretory vesicles (or
anules) and are subdivided into three classes on the basis of the morphology
d staining properties of these organelles (Figure 17–27). The differences in stain-
g reflect major differences of chemistry and function: **Neutrophils** (also called
lymorphonuclear leucocytes because of their multilobed nucleus) are the most
mmon type of granulocyte; they phagocytose and destroy small organisms—
pecially bacteria. **Basophils** secrete histamine (and, in some species, serotonin)
help mediate inflammatory reactions. **Eosinophils** help destroy parasites and
odulate allergic inflammatory responses.

Once they leave the bloodstream, **monocytes** (see Figure 17–27D) mature into
acrophages, which together with neutrophils are the main "professional phag-
ytes" in the body (see p. 334). Both types of phagocytic cells contain specialized
ganelles that fuse with newly formed phagocytic vesicles (phagosomes), expos-
g phagocytosed microorganisms to a barrage of enzymatically produced, highly
active molecules of superoxide (O_2^-) and hypochlorite (HOCl, the active ingre-
ent in bleach), as well as to a concentrated mixture of lysosomal hydrolases.
acrophages, however, are much larger and longer-lived than neutrophils, and
ey are unique in being able to ingest large microorganisms such as protozoa.

There are two main classes of **lymphocytes,** both involved in immune re-
onses: *B lymphocytes* make antibodies, while *T lymphocytes* kill virus-infected
lls and regulate the activities of other white blood cells (see Chapter 18). In
dition, there are lymphocytelike cells called *natural killer (NK) cells,* which kill
me types of tumor cells and some virus-infected cells. The production of lym-
ocytes is a specialized topic that will be discussed in detail in Chapter 18. Here
shall concentrate mainly on the development of the other blood cells, often
erred to collectively as **myeloid cells.**

The various types of blood cells and their functions are summarized in
ble 17–1.

le Production of Each Type of Blood Cell in the Bone Marrow Individually Controlled[23,25]

st white blood cells function in tissues other than the blood, and the blood
nply transports them to where they are needed. A local infection or injury in
y tissue, for example, rapidly attracts white blood cells into the affected region
part of the **inflammatory response,** which helps fight the infection or heal
wound. The inflammatory response is complex and is mediated by a variety
signaling molecules produced locally by mast cells, nerve endings, platelets,
d white blood cells, as well as by the activation of complement (see p. 1031).
me of these signaling molecules act on nearby capillaries, causing the endo-
lial cells to adhere less tightly to one another but making their surfaces adhesive
passing white blood cells. The white blood cells are thus caught like flies on
paper and then can escape from the vessel by squeezing between the endo-

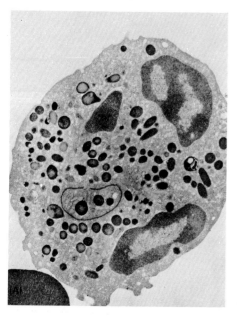

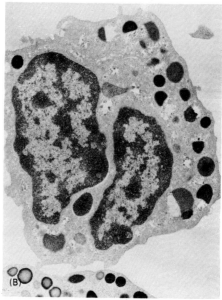

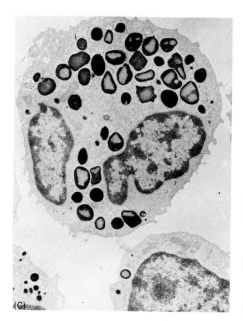

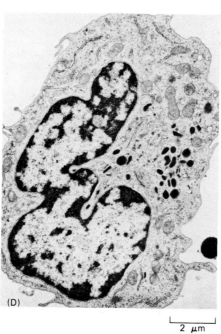

2 μm

Figure 17–27 Electron micrographs of four types of white blood cells (leucocytes): (A) neutrophil, (B) basophil, (C) eosinophil, (D) monocyte. Electron micrographs of lymphocytes are shown in Figure 18–4. Each of the cell types shown here has a different function, which is reflected in the distinctive types of secretory granules and lysosomes it contains. There is only one nucleus per cell, but it has an irregular lobed shape, and in (A), (B), and (C) the connections between the lobes are out of the plane of section. (Courtesy of Dorothy Bainton.)

thelial cells and crawling across the basal lamina with the aid of digestive enzymes. Other molecules act as chemoattractants for specific types of white blood cells, causing these cells to become polarized and crawl toward the source of the attractant (see p. 641). As a result, large numbers of white blood cells enter the affected tissue (Figure 17–28).

Other signaling molecules produced in the course of an inflammatory response escape into the blood and stimulate the bone marrow to produce more leucocytes and release them into the bloodstream. The bone marrow is the key target for such regulation because, with the exception of lymphocytes and some macrophages, most types of blood cells in adult mammals are generated only in the bone marrow. The regulation tends to be cell-type-specific: some bacterial infections, for example, cause a selective increase in neutrophils, while infections with some protozoa and other parasites cause a selective increase in eosinophils. (For this reason, physicians routinely use differential white blood cell counts to aid in the diagnosis of infectious and other inflammatory diseases.)

Table 17–1 Blood Cells

Type of Cell	Main Functions	Typical Concentration in Human Blood (cells/liter)
Red blood cells (erythrocytes)	transport O_2 and CO_2	5×10^{12}
White blood cells (leucocytes)		
Granulocytes		
Neutrophils (polymorphonuclear leucocytes)	phagocytose and destroy invading bacteria	5×10^{9}
Eosinophils	destroy larger parasites and modulate allergic inflammatory responses	2×10^{8}
Basophils	release histamine and serotonin in certain immune reactions	4×10^{7}
Monocytes	become tissue macrophages, which phagocytose and digest invading microorganisms, foreign bodies, and senescent cells	4×10^{8}
Lymphocytes		
B cells	make antibodies	2×10^{9}
T cells	kill virus-infected cells and regulate activities of other leucocytes	1×10^{9}
Natural killer (NK) cells	kill virus-infected cells and some tumor cells	1×10^{8}
Platelets (cell fragments, arising from megakaryocytes in bone marrow)	initiate blood clotting	3×10^{11}

Humans contain about 5 liters of blood, accounting for 7% of body weight. Red blood cells constitute about 45% of this volume and white cells about 1%, the rest being the liquid *blood plasma*.

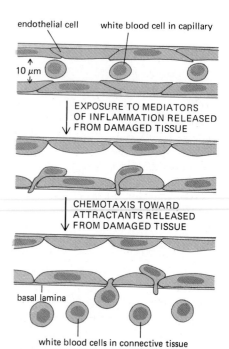

Figure 17–28 White blood cells migrate out of the bloodstream into an injured or infected tissue as part of the inflammatory response. The response is initiated by a variety of signaling molecules produced locally by cells (mainly in the connective tissue) or by complement activation. Some of these mediators act on capillary endothelial cells, causing them to loosen their attachments to their neighbors so that the capillaries become more permeable; the alteration in the endothelial cell surface also causes leucocytes in the blood to adhere to it. Other mediators act as chemoattractants, causing the bound leucocytes to crawl between the capillary endothelial cells into the tissue.

In other circumstances, erythrocyte production is selectively increased—for example, if one goes to live at high altitude, where oxygen is scarce. Thus blood cell formation (**hemopoiesis**) necessarily involves complex controls in which the production of each type of blood cell is regulated individually to meet changing needs. It is a problem of great medical importance to understand how these controls operate.

In intact animals, hemopoiesis is more difficult to analyze than is cell turnover in a tissue such as the epidermal layer of the skin. In epidermis there is a simple, regular spatial organization that makes it easy to follow the process of renewal and to locate the stem cells; this is not true of the hemopoietic tissues. On the other hand, the hemopoietic cells have a nomadic life-style that makes them more accessible to experimental study in other ways. Dispersed hemopoietic cells can be easily transferred, without damage, from one animal to another, and the proliferation and differentiation of individual cells and their progeny can be observed and analyzed in culture. For this reason, more is known about the molecules that control blood cell production than about those that control cell production in other mammalian tissues. Even so, there are still large gaps in our understanding.

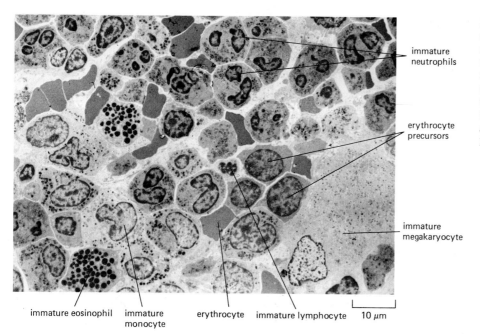

immature
neutrophils

erythrocyte
precursors

immature
megakaryocyte

immature eosinophil   immature   erythrocyte   immature lymphocyte   10 μm
                       monocyte

Figure 17–29 Low-magnification electron micrograph of a section of a region of bone marrow. This tissue is the main source of new blood cells (except for T lymphocytes). Note that the immature blood cells of a particular type tend to cluster in "family groups." (From J.A.G. Rhodin, Histology: A Text and Atlas. New York: Oxford University Press, 1974.)

Bone Marrow Contains Hemopoietic Stem Cells[23, 26]

The different types of blood cells and their immediate precursors can be recognized in the bone marrow by their distinctive appearances (Figure 17–29). They are intermingled with one another, as well as with fat cells and other *stromal cells* (connective-tissue cells) that form a delicate supporting meshwork of collagen fibers and other extracellular-matrix components. In addition, the whole tissue is richly supplied with thin-walled blood vessels (called *blood sinuses*) into which the new blood cells are discharged. **Megakaryocytes** are also present; these, unlike other blood cells, remain in the bone marrow when mature and are one of its most striking features, being extraordinarily large (diameter up to 60 μm), with a highly polyploid nucleus. They are normally plastered against blood sinuses, and they extend processes through holes in the endothelial lining of these vessels; platelets pinch off from the processes and are swept away into the blood (Figure 17–30).

Because of the complex arrangement of the cells in bone marrow, it is difficult to identify any but the immediate precursors of the mature blood cells. The corresponding cells at still earlier stages of development, before any overt differentiation has begun, are confusingly similar in appearance, and there is no visible feature by which the ultimate stem cells can be recognized. To identify and characterize the stem cells, one needs a functional test, which involves tracing the

Figure 17–30 (A) Schematic drawing of a megakaryocyte among other cells in the bone marrow. The enormous size of the megakaryocyte results from its having a highly polyploid nucleus. One megakaryocyte produces about 10,000 platelets, which split off from long processes that extend through holes in the walls of an adjacent blood sinus. (B) Scanning electron micrograph of the interior of a blood sinus in the bone marrow, showing the megakaryocyte processes. (B, from R.G. Kessel and R.H. Kardon, Tissues and Organs: A Text-Atlas of Scanning Electron Microscopy. San Francisco: Freeman, 1979. © 1979, W.H. Freeman and Company.)

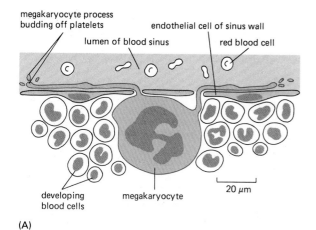

megakaryocyte process
budding off platelets

lumen of blood sinus

endothelial cell of sinus wall

red blood cell

developing
blood cells

megakaryocyte

20 μm

(A)

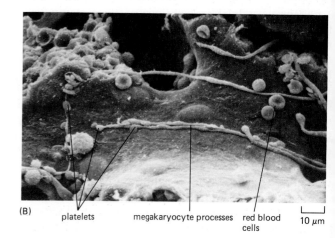

(B)

platelets   megakaryocyte processes   red blood
                                        cells

10 μm

rogeny of single cells. As we shall see, this can be done in culture simply by
xamining the colonies that artificially isolated cells produce. The hemopoietic
ystem, however, can also be manipulated so that such clones of cells can be
ecognized in the intact animal.

If an animal is exposed to a large dose of x-irradiation, the hemopoietic cells
e destroyed and the animal dies within a few days as a result of its inability to
anufacture new blood cells. The animal can be saved, however, by a transfusion
f cells taken from the bone marrow of a healthy, immunologically compatible
onor. Among these cells there are evidently some that can colonize the irradiated
ost and reequip it with hemopoietic tissue. One of the tissues where colonies
evelop is the spleen, which in a normal mouse is an important additional site of
emopoiesis. When the spleen of an irradiated mouse is examined a week or two
ter the transfusion of cells from a healthy donor, a number of distinct nodules
e seen in it, each of which is found to contain a colony of myeloid cells (Figure
7–31); after 2 weeks some colonies may contain more than a million cells. The
screteness of the nodules suggests that each might be a clone of cells descended
om a single founder cell, like a bacterial colony on a culture plate; and with the
elp of genetic markers, it can be established that this is indeed the case.

The founder of such a colony is called a **colony-forming cell,** or **CFC** (also
nown as a colony-forming unit, CFU). The colony-forming cells are heteroge-
eous. Some give rise to only one type of myeloid cell, while others give rise to
ixtures. Some go through many division cycles and form large colonies, while
hers divide less and form small colonies. Most of the colonies die out after
nerating a restricted number of terminally differentiated blood cells. A few of
e colonies, however, are capable of extensive self-renewal and produce new
lony-forming cells in addition to terminally differentiated blood cells. The found-
s of such self-renewing colonies are assumed to be the hemopoietic stem cells
the transfused bone marrow.

Pluripotent Stem Cell Gives Rise to All Classes of Blood Cells[27]

l the types of myeloid cells can often be found together in one spleen colony,
rived from a single stem cell. The hemopoietic stem cell, therefore, is *pluripotent:*
can give rise to many different cell types. Although the spleen colonies do not
em to contain lymphocytes, another approach shows that these also derive from
e same stem cell that gives rise to all of the myeloid cells. The demonstration
ploys genetic markers that make it possible to identify the members of a clone
en after they have been released into the bloodstream. Although several types
clonal markers have been used for this, a specially engineered retrovirus serves
e purpose particularly well. The marker virus, like other retroviruses, can insert
own genome into the chromosomes of the cell it infects, but the genes that
uld enable it to generate infectious virus particles have been removed. The
rker, therefore, is confined to the progeny of the cells that were originally
ected, and the progeny of one such cell can be distinguished from the progeny
another because the chromosomal sites of insertion of the virus are different.
analyze hemopoietic cell lineages, bone marrow cells are first infected with
e retrovirus (see p. 254) *in vitro* and then are transferred into a lethally irradiated
cipient; DNA probes (see p. 189) can then be used to trace the progeny of in-
vidual infected cells in the various hemopoietic and lymphoid tissues of the
st.

These experiments not only confirm that all classes of blood cells—both mye-
d and lymphoid—derive from a common stem cell (Figure 17–32), they also
ke it possible to follow the pedigrees of the blood cells over long periods of
e. A month or two after a transfusion of bone marrow cells, most of the blood
ls in an irradiated host mouse are found to be descendants of fewer than half
ozen pluripotent stem cells; this is still true several weeks later, but the blood
ls are now the progeny of a different handful of stem cells. These observations
gest that an individual stem cell has only a low probability in any given time
erval of initiating the formation of a clone of differentiated progeny, but that

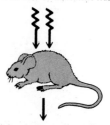

x-irradiation halts blood cell
production; mouse would die
if no further treatment was given

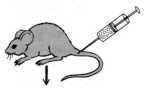

inject bone marrow cells
from healthy donor

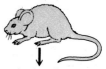

mouse survives; 2 weeks
after infection, many newly
formed blood cells are in circulation

examination of spleen reveals
large nodules on its surface

each spleen nodule contains a
clone of hemopoietic cells,
descended from one of the
injected bone marrow cells

Figure 17–31 The spleen colony
assay, in which the spleen of a heavily
irradiated animal is seeded with bone
marrow cells transfused from a healthy
donor. This assay revolutionized the
study of hemopoiesis by allowing
individual myeloid precursor cells to be
analyzed for the first time.

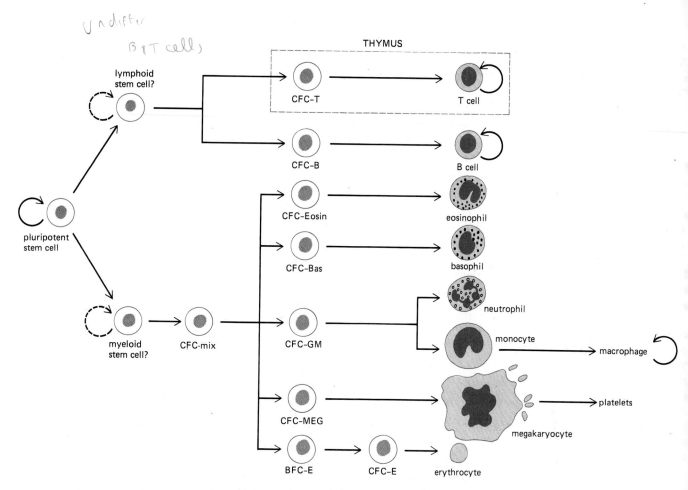

Undiffer
B & T cells)

lymphoid stem cell?

THYMUS

CFC-T → T cell

pluripotent stem cell

CFC-B → B cell

CFC-Eosin → eosinophil

CFC-Bas → basophil

myeloid stem cell? → CFC-mix → CFC-GM → neutrophil / monocyte → macrophage

CFC-MEG → megakaryocyte → platelets

BFC-E → CFC-E → erythrocyte

many cell divisions intervene between this initial step and the final differentiation so that the ultimate clone of progeny is very large—on the order of millions of cells. In spite of the seemingly rare and quantal nature of the initiating events, differentiated blood cells are produced at a smooth and steady rate; this reflects controls that operate at intermediate steps along the pathway to help regulate the final numbers of blood cells of each type.

The Number of Specialized Blood Cells Is Amplified by Divisions of Committed Progenitor Cells[23,28]

Once a cell has differentiated as an erythrocyte or a granulocyte or some other type of blood cell, there is no going back: the state of differentiation is not reversible. Therefore, at some stage in their development, the progeny of the pluripotent stem cell must become irreversibly committed or determined for a particular line of differentiation. It is clear from simple microscopic examination of the bone marrow that this commitment occurs well before the final division in which the mature differentiated cell is formed: one can recognize specialized precursor cells that are still proliferating but already show signs of having begun differentiation. It thus appears that commitment to a particular line of differentiation is followed by a series of cell divisions that amplify the number of cells of a given specialized type.

The hemopoietic system, therefore, can be viewed as a hierarchy of cells. **Pluripotent stem cells** give rise to **committed progenitor cells,** which are irreversibly determined as ancestors of only one or a few blood cell types. The committed progenitors are thought to divide rapidly but a limited number of times. At the end of this series of *amplification divisions*, they develop into **terminally differentiated cells,** which usually divide no further and die after several days or weeks. Studies in culture provide a way to find out how these cellular events are regulated.

Figure 17–32 A tentative scheme of hemopoiesis. The pluripotent stem cell normally divides infrequently to generate either more pluripotent stem cells (self-renewal) or *committed progenitor cells* (labeled CFC = colony-forming cells), which are irreversibly determined to produce only one or a few types of blood cells. The progenitor cells are stimulated to proliferate by specific growth factors but progressively lose their capacity for division and develop into terminally differentiated blood cells, which usually live for only a few days or weeks.

In adult mammals all of the cells shown develop mainly in the bone marrow—except for T lymphocytes, which develop in the thymus, and macrophages, which develop from monocytes in most tissues. The most controversial part of the scheme is the first branch point: the existence of stem cells committed exclusively to form T and B lymphocytes and of stem cells committed exclusively to form all the other classes of blood cells (myeloid cells), as shown, is uncertain. It is likely that the primary pluripotent stem cells also give rise to various types of tissue cells not shown in this scheme, such as NK cells, mast cells, osteoclasts, and a variety of classes of antigen-presenting cells (see p. 1045); but the pathways by which these cells develop are uncertain.

The Factors That Regulate Hemopoiesis Can Be Analyzed in Culture[29]

Hemopoietic cells will survive, proliferate, and differentiate in culture if, and only if, they are provided with specific growth factors or accompanied by cells that produce these factors. Long-term proliferation of pluripotent stem cells can be achieved, for example, by culturing dispersed bone-marrow hemopoietic cells on top of a layer of bone-marrow stromal cells, presumably mimicking the environment in intact bone marrow; such cultures can generate all the types of myeloid cells. Alternatively, dispersed bone-marrow hemopoietic cells can be cultured in a semisolid matrix of dilute agar or methylcellulose, and factors derived from other cells can be added artificially to the medium. In the semisolid matrix the progeny of each isolated precursor cell remain together as an easily distinguishable colony. A single committed neutrophil progenitor, for example, may be seen to give rise to a clone of thousands of neutrophils. Such culture systems provide a way to assay for the factors that support hemopoiesis and hence to purify them and explore their actions. These substances are found to be glycoproteins and are usually called **colony-stimulating factors,** or **CSFs.** Of the growing number of CSFs that have been defined and purified, some circulate in the blood and act as hormones, while others act as local chemical mediators (see p. 682). The best understood of the CSFs that act as hormones is the glycoprotein *erythropoietin*, which is produced in the kidney and regulates *erythropoiesis* (the formation of red blood cells).

Erythropoiesis Depends on the Hormone Erythropoietin[30]

The erythrocyte is by far the most common type of cell in the blood (see Table 17–1). When mature, it is packed full of hemoglobin and contains practically none of the usual cell organelles. In an erythrocyte of an adult mammal, even the nucleus, endoplasmic reticulum, mitochondria, and ribosomes are absent, having been extruded from the cell in the course of its development (Figure 17–33). The erythrocyte therefore cannot grow or divide; the only possible way of making more erythrocytes is by means of stem cells. Furthermore, erythrocytes have a limited life-span—about 120 days in humans or 55 days in mice. Worn-out erythrocytes are phagocytosed and digested by macrophages in the liver and spleen.

A lack of oxygen or a shortage of erythrocytes stimulates cells in the kidney to synthesize and secrete increased amounts of **erythropoietin** into the bloodstream. The erythropoietin in turn stimulates the production of more erythrocytes. Since a change in the rate of release of new erythrocytes into the bloodstream is observed as early as one or two days after an increase in erythropoietin levels in the bloodstream, the hormone must act on cells that are very close precursors of the mature erythrocytes.

The cells that respond to erythropoietin can be identified by culturing bone marrow cells in the presence of erythropoietin. In a few days, colonies of about 60 erythrocytes appear, each founded by a single committed erythroid progenitor cell. This cell is known as an **erythrocyte colony-forming cell,** or **CFC-E,** and it gives rise to mature erythrocytes after about six division cycles or less. The CFC-Es do not yet contain hemoglobin, and they are derived from an earlier type of progenitor cell whose proliferation does not depend on erythropoietin.

A second CSF, called **interleukin 3 (IL-3),** promotes the survival and proliferation of pluripotent stem cells and of most types of committed progenitor cells. In its presence, much larger erythroid colonies, each comprising up to 5000 erythrocytes, develop from cultured bone marrow cells in a process requiring a week or 10 days. These colonies derive from erythroid progenitor cells called **erythrocyte burst-forming cells,** or **BFC-Es.** The BFC-E is distinct from the pluripotent stem cell in that it has a limited capacity to proliferate and gives rise to colonies that contain erythrocytes only, even under culture conditions that enable other progenitor cells to give rise to other classes of differentiated blood cells. It is distinct from the CFC-E in that it is insensitive to erythropoietin, and its progeny must go through as many as 12 divisions before they become mature erythrocytes

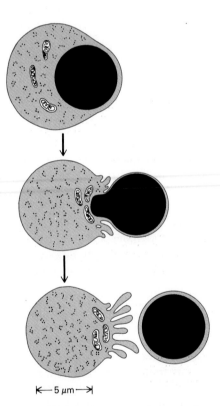

Figure 17–33 Schematic diagram of a developing red blood cell (erythroblast) extruding its nucleus to become an immature erythrocyte (*reticulocyte*) shortly before leaving the bone marrow and passing into circulation. The reticulocyte will lose its mitochondria and ribosomes within a day or two to become a mature erythrocyte. Erythrocyte clones develop in the bone marrow on the surface of a macrophage, which phagocytoses and digests the nuclei discarded by the erythroblasts.

(for which erythropoietin must be present). The cell also differs in size from the CFC-E and can be separated from it by sedimentation. Thus the BFC-E is thought to be a progenitor cell committed to erythrocyte differentiation and an early ancestor of the CFC-E (Figure 17–34).

Multiple CSFs Influence the Production of Neutrophils and Macrophages[29, 31]

The two professional phagocytic cells, neutrophils and macrophages, develop from a common progenitor cell called the **granulocyte/macrophage (or GM) progenitor cell.** Like the other granulocytes (eosinophils and basophils), neutrophils circulate in the blood for only a few hours before migrating out of capillaries into the connective tissues or other specific sites, where they survive for a few days and then die. Macrophages, by contrast, can persist for months or perhaps even years outside the bloodstream, where they can be activated by local signals to resume proliferation.

So far four distinct CSFs that stimulate neutrophil and macrophage colony formation in culture have been defined, and they are thought to act in different combinations to regulate the selective production of these cells *in vivo*. These CSFs are synthesized by various cell types—including endothelial cells, fibroblasts, macrophages, and lymphocytes—and their concentration in the blood increases rapidly in response to bacterial infection in a tissue, thereby increasing the number of phagocytic cells released from the bone marrow into the bloodstream. IL-3 is the least specific of the four, acting on pluripotent stem cells as well as on most classes of committed progenitor cells, including GM-progenitor cells. The other three factors act more selectively on committed GM-progenitor cells and their differentiated progeny (Table 17–2), although at high concentrations some of them affect other lineages as well.

All of these CSFs, like erythropoietin, are glycoproteins that act at low concentrations ($\sim 10^{-12}$ M) by binding to specific cell-surface receptors. They not only operate on the precursor cells to promote formation of differentiated colonies, they also activate the specialized functions (such as phagocytosis and target-cell killing) of the terminally differentiated cells. Proteins produced from the cloned genes for these factors are powerful stimulators of hemopoiesis in experimental animals, suggesting that they may prove useful clinically in stimulating the regeneration of hemopoietic tissue and in treating infections.

Factors that specifically promote the development of the other classes of myeloid cells, such as megakaryocytes, basophils, and eosinophils, have also been described, but they are less well characterized than the CSFs we have discussed.

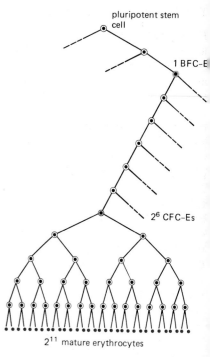

Figure 17–34 Pedigree showing the relationship of the pluripotent stem cell, the BFC-E, the CFC-E, and the mature erythrocyte. BFC-Es and CFC-Es are both committed erythroid progenitor cells. BFC-Es respond to the factor IL-3 but not to erythropoietin, whereas CFC-Es respond strongly to erythropoietin. The series of cell divisions that occur in this lineage under the influence of erythropoietin provides a powerful means of controlling the production of erythrocytes without upsetting the production of other types of blood cells.

Table 17–2 Some Colony-stimulating Factors (CSFs) That Influence Blood Cell Formation

Factor	Size (in mouse)	Target Cells	Producing Cells
Erythropoietin	51,000 daltons	CFC-E	kidney cells
Interleukin 3 (IL-3)	25,000 daltons	pluripotent stem cell, most progenitor cells, many terminally differentiated cells	T lymphocytes, epidermal cells
Granulocute/ macrophage CSF (GM-CSF)	23,000 daltons	GM progenitor cells	T lymphocytes, endothelial cells, fibroblasts
Granulocyte CSF (G-CSF)	25,000 daltons	GM progenitor cells and neutrophils	macrophages, fibroblasts
Macrophage CSF (M-CSF)	70,000 daltons (dimer)	GM progenitor cells and macrophages	fibroblasts, macrophages, endothelial cells

In addition to soluble CSFs, some of which are secreted locally by stromal cells in the bone marrow, there is evidence that cell-bound and extracellular-matrix-bound signals in the marrow also regulate hemopoiesis.

The Behavior of a Hemopoietic Cell Depends Partly on Chance[29,32]

The CSFs are defined as factors that promote the production of colonies of differentiated blood cells. It is not easy to discover precisely what influence a CSF has on an individual hemopoietic cell. The factor might increase the probability of cell survival; it might control the rate of cell division or the number of division cycles that the progenitor cell goes through before differentiating; it might act late in the hemopoietic lineage to facilitate differentiation; or it might act early to influence commitment (Figure 17–35). By monitoring the fate of isolated individual hemopoietic cells in culture, it has been possible to show that a single CSF, such as GM-CSF, can exert all these different effects. Nevertheless, it is still not clear which actions are important *in vivo*. The behavior of the pluripotent stem cells remains especially elusive: these crucial cells are few and far between—less than one in a thousand of the cells in the bone marrow—and are very difficult to identify unambiguously.

Studies *in vitro* indicate, moreover, that there is a large element of chance in the way a hemopoietic cell behaves. The CSFs act by regulating probabilities, not by dictating directly what the cell shall do. In hemopoietic cell cultures, even if the cells have been selected to be as homogeneous a population as possible, there is a remarkable variability in the sizes, and often in the characters, of the colonies that develop. And if two sister cells are taken immediately after a cell division and cultured apart under identical conditions, they will frequently give rise to colonies that contain different types of blood cells, or the same types of blood cells in different numbers. Thus both the programming of cell division and the process of commitment to a particular path of differentiation seem to involve random events at the level of the individual cell, even though the behavior of the multicellular system as a whole is regulated in a reliable way. The molecular mechanisms underlying these phenomena represent the most fundamental and challenging of the unsolved problems in hemopoiesis.

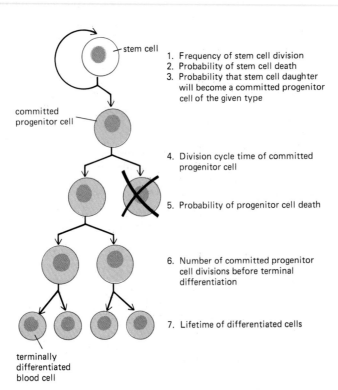

stem cell

1. Frequency of stem cell division
2. Probability of stem cell death
3. Probability that stem cell daughter will become a committed progenitor cell of the given type

committed progenitor cell

4. Division cycle time of committed progenitor cell

5. Probability of progenitor cell death

6. Number of committed progenitor cell divisions before terminal differentiation

7. Lifetime of differentiated cells

terminally differentiated blood cell

Figure 17–35 Some of the parameters through which the production of blood cells of a specific type might be regulated. Studies *in vitro* suggest that colony-stimulating factors (CSFs) can affect all these aspects of hemopoiesis.

Summary

The many types of blood cells all derive from a common pluripotent stem cell. In the adult the stem cells are found mainly in bone marrow, where they normally divide infrequently to produce more stem cells (self-renewal) and various committed progenitor cells, each able to give rise to only one or a few types of blood cells. The committed progenitor cells divide profusely under the influence of various glycoprotein signaling molecules (called colony-stimulating factors, or CSFs) and then differentiate into mature blood cells, which usually die after several days or weeks. Studies of hemopoiesis have been greatly aided by in vitro assays in which stem cells or committed progenitor cells form clonal colonies when cultured in a semisolid matrix. However, the pluripotent stem cells are rare and difficult to identify, and it is still not clear how they choose among alternative developmental pathways.

Genesis, Modulation, and Regeneration of Skeletal Muscle[33]

The term "muscle" covers a multitude of cell types, all specialized for contraction but in other respects dissimilar. As noted in Chapter 11, a contractile system involving actin and myosin is a basic feature of animal cells in general, but muscle cells have developed this apparatus to a high degree. Mammals possess four main categories of cells specialized for contraction: *skeletal muscle* cells, *heart* (or *cardiac*) *muscle* cells, *smooth muscle* cells, and *myoepithelial* cells (Figure 17–36). These differ in function, structure, and development. Although all of them appear to generate contractile forces by means of organized filament systems based on actin and myosin, the actin and myosin molecules employed are somewhat different in amino acid sequence, are differently arranged in the cell, and are associated with different sets of proteins to control contraction.

Skeletal muscle cells, whose contractile apparatus is discussed in detail in Chapter 11, are responsible for practically all movements that are under voluntary control. These cells can be huge (up to half a meter long and 100 μm in diameter in an adult human) and are often referred to as *muscle fibers* because of their highly elongated shape. Each one is a syncytium, containing many nuclei within a common cytoplasm. The other types of muscle cells are more conventional, having only a single nucleus. **Heart muscle cells** resemble skeletal muscle cells in that their actin and myosin filaments are aligned in very orderly arrays, so that the cells have a striated appearance. **Smooth muscle cells** (see p. 625) are so called because they, in contrast, do not appear striated. The functions of smooth muscle vary greatly, from propelling food along the digestive tract to erecting hairs in response to cold or fear. **Myoepithelial cells** also have no striations, but unlike all other muscle cells, they lie in epithelia and are derived from the ectoderm. They form the dilator muscle of the iris and serve to expel saliva, sweat, and milk from the corresponding glands (see Figure 17–36E).

The four main categories of muscle cells can be further divided into distinctive subtypes, each with its own characteristic features. We shall focus here on the skeletal muscle cell, which has a curious mode of development, a striking ability to modulate its differentiated character, and an unusual strategy for repair.

New Skeletal Muscle Cells Form by the Fusion of Myoblasts[2,34]

The previous chapter described how certain cells, originating from the somites of a vertebrate embryo at a very early stage, become determined as *myoblasts* (that is, as precursors of skeletal muscle cells) and migrate into the adjacent embryonic connective tissue, or mesenchyme (see p. 943). As discussed on page 556, the commitment to be a myoblast (rather than, say, a fibroblast) seems to correspond to the activation of a specific master regulatory gene. After a period of proliferation, the myoblasts fuse with one another to form multinucleate skeletal muscle cells

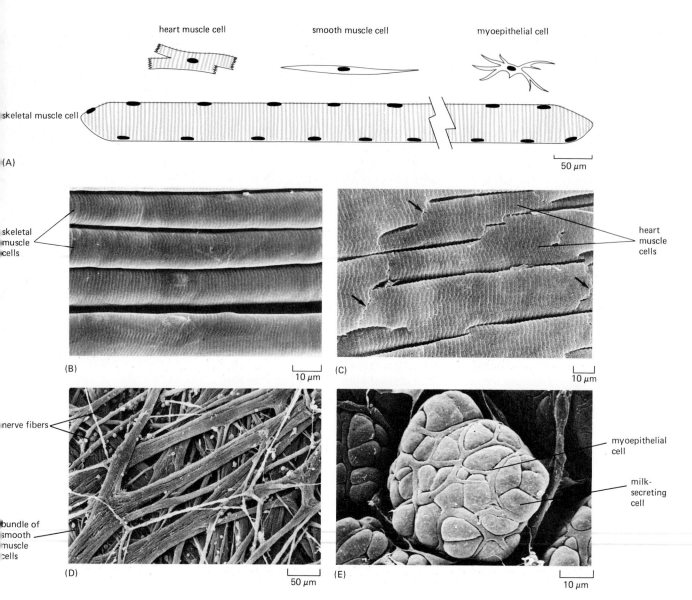

heart muscle cell   smooth muscle cell   myoepithelial cell

skeletal muscle cell

(A)   50 μm

skeletal
muscle
cells   heart
                                                         muscle
                                                         cells

(B)   10 μm   (C)   10 μm

nerve fibers   myoepithelial
                                                         cell

                                                         milk-
                                                         secreting
                                                         cell

bundle of
smooth
muscle
cells

(D)   50 μm   (E)   10 μm

(Figure 17–37). As they fuse, they undergo a dramatic switch of phenotype that depends on the coordinated activation of a whole battery of muscle-specific genes (see p. 556). Once fusion has occurred, the nuclei never again replicate their DNA. Fusion involves specific mutual recognition between myoblasts: they do not fuse with adjacent nonmuscle cells. The molecular basis of the recognition process is unknown.

Myoblasts that have been kept proliferating in culture for as long as 2 years still retain the ability to differentiate and will fuse to form muscle cells in response to a suitable change in culture conditions. *Fibroblast growth factor (FGF)* in the medium seems to be crucial for keeping the myoblasts proliferating and preventing their differentiation: if FGF is removed, the cells rapidly stop dividing, fuse, and differentiate. The system of controls is complex, however. In order to differentiate, myoblasts must, for example, attach to the extracellular matrix. Moreover, the process of fusion is cooperative: fusing myoblasts secrete unknown factors that encourage other myoblasts to fuse.

Muscle Cells Can Modulate Their Properties by a Change of Protein Isoforms[35]

A skeletal muscle cell, once formed, is generally retained for the entire lifetime of the animal. Over this period, it grows, matures, and modulates its character according to functional requirements. The genome contains multiple variant copies

Figure 17–36 The four classes of muscle cells of a mammal. (A) Schematic drawings (to scale). (B–E) Scanning electron micrographs, showing (B) skeletal muscle from the neck of a hamster, (C) heart muscle from a rat, (D) smooth muscle from the urinary bladder of a guinea pig, and (E) myoepithelial cells in a secretory alveolus from a lactating rat mammary gland. The arrows in (C) point to intercalated discs (see p. 624). Note that the smooth muscle is shown at a lower magnification than the others. (B, courtesy of Junzo Desaki; C, from T. Fujiwara, in Cardiac Muscle in Handbook of Microscopic Anatomy [E.D. Canal, ed.]. Berlin: Springer Verlag, 1986; D, courtesy of Satoshi Nakasiro; E, from T. Nagato, Y. Yoshida, A. Yoshida, and Y. Uehara, *Cell and Tissue Res.* 209:1–10, 1980.)

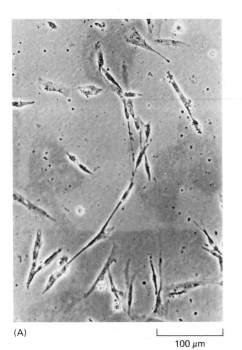
(A)

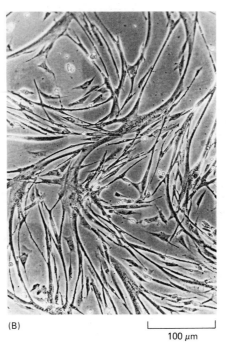
(B)

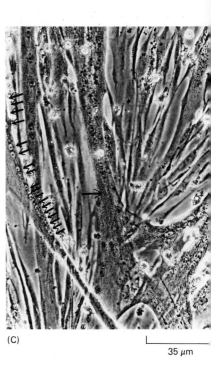
(C)

100 μm   100 μm   35 μm

of the genes encoding many of the characteristic proteins of the skeletal muscle cell, and the RNA transcripts of many of these genes can be spliced in several ways. As a result, a wealth of protein variants (traditionally called *isoforms*) can be produced for the components of the contractile apparatus. As the muscle cell matures, different selections of isoforms are produced, adapted to the changing demands for speed, strength, and endurance in the fetus, the newborn, and the adult.

Several distinct types of skeletal muscle cells, each with different sets of protein isoforms, can be found side by side in a single muscle. In the adult, two of these types are easily recognized even with the naked eye. *Red muscle cells*, such as those of the dark meat of a chicken, are rich in the oxygen-binding protein myoglobin. *White muscle cells*, such as those of the white meat of a chicken, have much less myoglobin. The different content of myoglobin—a molecule similar to hemoglobin—reflects different oxygen requirements: the red muscle cells are specialized for oxidative phosphorylation, the white for anaerobic glycolysis. The red muscle cells give a slow-twitch response to stimulation, are more resistant to fatigue, and are most efficient for generating sustained forces. The white muscle cells give a fast twitch, fatigue more easily, and are most efficient for quick, intermittent movements. A particular muscle, such as the biceps, usually contains a mixture of several muscle cell types in proportions suited to that muscle's function (Figure 17–38).

By surgically altering the innervation of muscles or by stimulating them artificially with implanted electrodes, it can be shown that the frequency of electrical excitation greatly influences a muscle cell's pattern of gene expression. A slow muscle cell stimulated at the frequencies appropriate to fast muscle becomes partially converted to a fast character, and vice versa, in part through alterations in the protein isoforms it makes. These changes are not unique to muscle cells: switches of gene expression that produce variant mRNAs often occur as a differentiated cell matures or reacts to its environment as we have already seen (p. 970).

Some Myoblasts Persist as Quiescent Stem Cells in the Adult[36]

A muscle can grow in three ways: its differentiated muscle cells can increase in number, in length, or in girth. Because skeletal muscle cells are unable to divide, more of them can be made only by the fusion of myoblasts. The adult number of multinucleated skeletal muscle cells is in fact attained early—before birth in hu-

Figure 17–37 Phase-contrast micrographs of myoblasts in culture, showing how the cells will proliferate, line up, and fuse to form multinucleate muscle cells. (C) is at higher magnification, showing the cross-striations that are just beginning to be visible as the contractile apparatus develops (*long arrow*), and the accumulations of many nuclei within a single cell (*short arrows*). (Courtesy of Rosalind Zalin.)

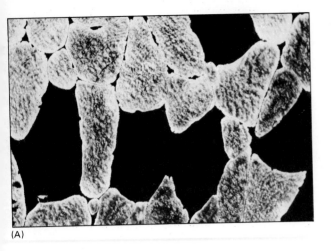

(A)

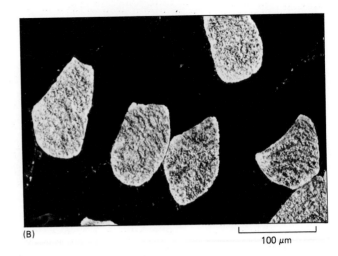

(B)

100 μm

Figure 17–38 Two consecutive sections of the same piece of adult chicken muscle, stained with two fluorescent antibodies, each specific for a different myosin isoform. In (A) the white (fast-twitch) cells are stained with antibodies against fast myosin; in (B) the red (slow-twitch) cells are stained with antibodies against slow myosin. (From G. Gauthier *et al., J. Cell Biol.* 92:471–484, 1982.)

mans. The subsequent enormous increase in muscle bulk is achieved by cell enlargement. Growth in length depends on recruitment of more myoblasts into the existing multinucleate cells, mainly by fusion at their ends, which increases the number of nuclei in each cell. In contrast, growth in girth, such as occurs in the muscles of weightlifters, depends on an increase in the size and numbers of the contractile myofibrils that each muscle cell contains (see p. 614), rather than on changes in the numbers of muscle cells or of their nuclei.

In the adult, nevertheless, a few myoblasts persist as small, flattened, and inactive cells lying in close contact with the mature muscle cell and contained within its sheath of basal lamina. If the muscle is damaged or if it is treated artificially with FGF, these so-called *satellite cells* are activated to proliferate (Figure 17–39), and their progeny can fuse to form new muscle cells. Satellite cells are thus the stem cells of adult skeletal muscle, normally held in reserve in a quiescent state but available when needed as a self-renewing source of terminally differentiated cells.

Summary

Skeletal muscle cells are one of the four main categories of vertebrate cells specialized for contraction, and they are responsible for voluntary movement. Each skeletal muscle cell is a syncytium and develops by the fusion of myoblasts. Myoblasts are stimulated to proliferate by growth factors such as FGF, but once they fuse, they can no longer divide. Myoblast fusion is generally coupled with the onset of muscle-cell differentiation, in which many genes encoding muscle-specific proteins are switched on coordinately. Subsequently the cells can modulate their differentiated character by altering the selection of protein isoforms they make. Some myoblasts persist (in a quiescent state) as satellite cells in adult muscle; when a muscle is damaged, these cells are reactivated to proliferate and to fuse to replace the muscle cells that have been lost.

muscle cell nuclei

20 μm

Figure 17–39 Autoradiograph of a single multinucleate muscle cell with associated satellite cells. The fiber has been isolated from an adult rat and transferred into culture medium containing ³H-thymidine plus an extract from damaged muscle that stimulates the satellite cells to divide. The dividing satellite cells (*arrows*) have become radioactively labeled (silver grains visible as black dots); the muscle cell nuclei are unable to proliferate and remain unlabeled. (From R. Bischoff, *Dev. Biol.* 115:140–147, 1986.)

Fibroblasts and Their Transformations: The Connective-Tissue Cell Family[37]

Many of the differentiated cells in the adult body can be grouped into families whose members are closely related by origin and by character. An important example is the family of **connective-tissue cells,** whose members are not only related but are also to an unusual extent interconvertible. The family includes *fibroblasts, cartilage cells,* and *bone cells,* all of which are specialized for secretion of collagenous extracellular matrix and are jointly responsible for the architectural framework of the body, as well as *fat cells* and *smooth muscle cells,* which appear to have a common origin with them. These cell types, and the interconversions that are thought to occur between them, are illustrated in Figure 17–40. Connec-

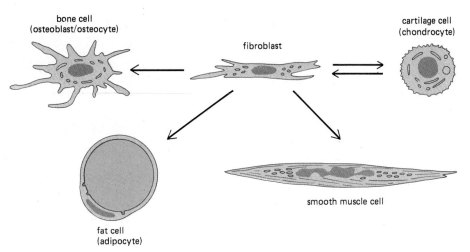

tive-tissue cells play a central part in the support and repair of almost every tissue and organ, and the adaptability of their differentiated character is an important feature of the responses to many types of damage.

Fibroblasts Change Their Character in Response to Signals in the Extracellular Matrix[37,38]

Fibroblasts appear to be the least specialized cells in the connective-tissue family. They are dispersed in connective tissue throughout the body, where they secrete a nonrigid extracellular matrix that is rich in type I and/or type III collagen, as discussed in Chapter 14 (see p. 809). When a tissue is injured, the fibroblasts nearby migrate into the wound, proliferate, and produce large amounts of collagenous matrix, which helps to isolate and repair the damaged tissue. Their ability to thrive in the face of injury, together with their solitary life-style, may explain why fibroblasts are the easiest of cells to grow in culture—a feature that has made them a favorite subject for cell biological studies (Figure 17–41).

As indicated in Figure 17–40, fibroblasts also seem to be the most versatile of connective-tissue cells, displaying a remarkable capacity to differentiate into other members of the family. Before going into details, we must state a caveat, however. There is good evidence that fibroblasts in different parts of the body are intrinsically different (see p. 995), and it is far from certain that all fibroblasts in a given region are equivalent. In the absence of firm evidence to the contrary, it is simplest to suppose that they are, but it is conceivable that connective tissue may contain a mixture of distinct fibroblast lineages, some capable of transformation into chondrocytes, others capable of transformation into fat cells, and so on, rather than just one type of fibroblast with multiple developmental capabilities. It is possible also that "mature" fibroblasts incapable of transformation may exist side-by-side with "immature" fibroblasts (often called *mesenchymal cells*) that can develop into a variety of mature cell types.

Despite these uncertainties, there is clear evidence, from studies both *in vivo* and *in vitro*, that connective-tissue cells can undergo radical changes of character. Thus, if a preparation of bone matrix, made by grinding bone into a fine powder and dissolving away the hard mineral component, is implanted in the dermal layer of the skin, some of the cells there—probably dermal fibroblasts—become transformed into cartilage cells and, a little later, others into bone cells, thereby creating a small lump of bone, complete with a marrow cavity. These experiments suggest that components in the extracellular matrix can dramatically influence connective-tissue cell differentiation. We shall see that similar cell transformations are important in the natural repair of broken bones. In fact, bone matrix has been found to contain, trapped within it, high concentrations of several growth factors that can affect the behavior of connective-tissue cells, including, for example, TGF-β (transforming growth factor β—see p. 747), which has been reported to induce cartilage differentiation *in vitro*.

Figure 17–41 (A) Phase-contrast micrograph of a mouse fibroblast in culture. (B) Drawings of a living fibroblastlike cell in the transparent tail of a tadpole, showing the changes in its shape and position on successive days. Note that while fibroblasts flatten out in culture, they can have more complex, process-bearing morphologies in tissues. (A, courtesy of Guenter Albrecht-Buehler; B, redrawn from E. Clark, *Am. J. Anat.* 13:351–379, 1912.)

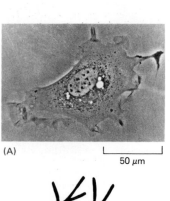

(A)

50 µm

day 1

day 2

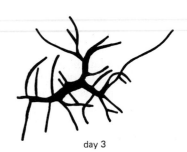

day 3

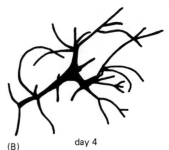

(B) day 4

The Extracellular Matrix May Influence Connective-Tissue Cell Differentiation by Affecting Cell Shape and Attachment[39]

The extracellular matrix may influence the differentiated state of connective-tissue cells through physical as well as chemical effects. This has been shown in studies on cultured cartilage cells, or *chondrocytes*. Under appropriate culture conditions, these cells will proliferate and maintain their differentiated character, continuing for many cell generations to synthesize large quantities of highly distinctive cartilage matrix, with which they surround themselves. However, under conditions where the cells are kept at relatively low density and remain as a monolayer on the culture dish, a transformation occurs: the cells lose the rounded shape that is typical of chondrocytes, flatten down on the substratum, and stop making cartilage matrix; in particular, they stop producing type II collagen—the type characteristic of cartilage—and instead start producing type I collagen—the type characteristic of fibroblasts. By the end of a month in culture, almost all the cartilage cells have switched their collagen gene expression and taken on the appearance of fibroblasts. The biochemical change must occur abruptly, since very few cells are observed to make both types of collagen simultaneously.

Several lines of evidence suggest that the biochemical change is induced at least in part by the change of cell shape and attachments. Cartilage cells that have made the transition to a fibroblastlike character, for example, can be gently detached from the culture dish and transferred to a dish of agarose. By forming a gel around them, the agarose holds the cells suspended without any attachment to a substratum, forcing them to adopt a rounded shape. In these circumstances the cells promptly revert to the character of chondrocytes and start making type II collagen again. How cell shape and anchorage might control gene expression is discussed on page 749.

For most types of cells, and especially for a connective-tissue cell, the opportunities for anchorage and attachment depend on the surrounding matrix, which is usually made by the cell itself. Thus a cell can create an environment that then acts back on the cell to reinforce its differentiated state. Furthermore, the extracellular matrix that a cell secretes forms part of the environment for its neighbors as well as for the cell itself, and thus tends to make neighboring cells differentiate in the same way. A group of chondrocytes forming a nodule of cartilage, for example, either in the developing body or in a culture dish, can be seen to enlarge by the conversion of neighboring fibroblasts into chondrocytes.

Different Signaling Molecules Act Sequentially to Regulate Production of Fat Cells[40]

Fat cells, or **adipocytes,** are also thought to develop from fibroblastlike cells, both during normal mammalian development and in various pathological circumstances—for example, in muscular dystrophy, where the muscle cells die and are gradually replaced by fatty connective tissue. Fat cell differentiation begins with the production of specific enzymes, followed by the accumulation of fat droplets, which then coalesce and enlarge until the cell is hugely distended, with only a thin rim of cytoplasm around the mass of lipid (Figure 17–42).

The process can be studied in culture (using cell lines such as certain strains of mouse 3T3 cells), so that the factors that influence it can be analyzed. It was initially found that the development of fat cells in culture required the presence of fetal calf serum, a common additive to culture media. The crucial factor in the

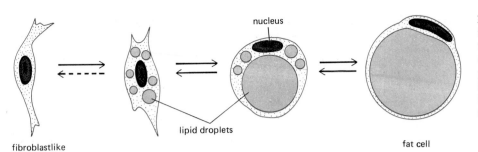

Figure 17–42 The conversion of a fibroblastlike precursor cell into a mature fat cell by the accumulation and coalescence of lipid droplets. The process is at least partially reversible, as indicated by the arrows. The cells in the early and intermediate stages can divide, but the mature fat cell cannot.

nucleus

lipid droplets

fibroblastlike precursor cell

fat cell

serum that triggers fat cell differentiation was later identified as *growth hormone—* a protein normally secreted into the bloodstream by the pituitary gland. There is evidence that growth hormone stimulates chondrocyte as well as fat cell differentiation and that it acts in this way *in vivo* as well as *in vitro*. But growth hormone is not the only secreted signaling molecule that regulates fat cell development. Fat cell precursors that have been stimulated by growth hormone become sensitive to *IGF-1 (insulinlike growth factor-1)*, which stimulates the proliferation of the differentiating fat cells.

The differentiation of fat cells, like that of chondrocytes, is also influenced by factors that affect cell shape and anchorage. The differentiation of 3T3 cells into fat cells, for example, is inhibited if the cells are allowed to flatten onto a culture dish coated with fibronectin, to which they adhere strongly. This inhibition is reversed, however, by treatment with the drug cytochalasin, which disrupts actin filaments and causes the cells to round up.

All these experiments on connective-tissue cells illustrate a recurrent theme: differentiation is regulated by a combination of soluble signals and contacts with the extracellular matrix. The effects of each of these factors depend on the character of the responding cell, and this in turn depends on the cell's developmental history.

Summary

The family of connective-tissue cells includes fibroblasts, cartilage cells, bone cells, fat cells, and smooth muscle cells. Fibroblasts seem to be able to transform into any of the other members of the family—in some cases reversibly—although it is not clear whether this is a property of a single type of fibroblast that is pluripotent or of a mixture of distinct types of fibroblasts with more restricted potentials. These transformations of connective-tissue cell type are regulated by the composition of the surrounding extracellular matrix, by cell shape, and by hormones and growth factors.

Soft Cells and Tough Matrix: Growth, Turnover, and Repair in Bone[41]

Bone is a very dense, specialized form of connective tissue. Like reinforced concrete, bone matrix is a mixture of tough fibers (type I collagen fibrils), which resist pulling forces, and solid particles (calcium phosphate as hydroxyapatite crystals), which resist compression. The volume occupied by the collagen is nearly equal to that occupied by the calcium phosphate. The collagen fibrils in adult bone are arranged in regular layers like plywood, with the fibrils in each layer lying parallel to one another but at right angles to the fibrils in the layers on either side.

For all its rigidity, bone is by no means a permanent and immutable tissue. Throughout its hard extracellular matrix are channels and cavities occupied by living cells, which account for about 15% of the weight of compact bone. These cells are engaged in an unceasing process of remodeling: one class of cells demolishes old bone matrix while another deposits new bone matrix. This mechanism provides for continuous turnover and replacement of the matrix in the interior of the bone.

Bone can grow only by *apposition*, that is, by the laying down of additional matrix and cells on the free surfaces of the hard tissue. This process must occur in the embryo in coordination with the growth of other tissues in such a way that the pattern of the body can be scaled up without its proportions being radically disturbed. For most of the skeleton, and in particular for the long bones of the limbs and trunk, the coordinated growth is achieved by a complex strategy. A set of minute "scale models" of the bones are first formed out of cartilage. Each scale model grows, and as new cartilage is formed, the older cartilage is replaced by bone. Cartilage growth and erosion and bone deposition are so ingeniously coordinated during development that the adult bone, though it may be half a meter long, is almost the same shape as the initial cartilaginous model, which was no more than a few millimeters long. Rather than go into the detailed geometry of this process, we shall concentrate on the cellular activities responsible for both the embryonic growth and the adult turnover of cartilage and bone, which illustrate especially well how the maintenance of body tissues depends on the interplay of multiple cell types.

Cartilage Can Grow by Swelling[39,42]

The collaboration of bone and cartilage depends on their contrasting properties. Both types of tissues are formed from mesenchymal cells that secrete large quantities of collagenous extracellular matrix. But whereas bone matrix is rigid, cartilage matrix is deformable, consisting largely of high concentrations of proteoglycans (see p. 806) and type II collagen. Consequently, **cartilage,** unlike bone, is able to grow by swelling, as cells already embedded in the matrix secrete more matrix around themselves.

Cartilage cells, or chondrocytes, are isolated from one another, each occupying a small cavity, or *lacuna*, in the matrix. Cartilage usually contains no blood vessels, and its cells are sustained by diffusion of nutrients and gases through the porous matrix to and from blood vessels that lie far away. A compact layer of collagenous connective tissue, the *perichondrium*, surrounds most of the cartilage mass (Figure 17–43). The cartilage expands from within as the chondrocytes secrete new matrix, while the fibrous perichondrium probably acts like a corset, constraining the expansion and thereby determining changes of shape. New cells are formed in this growth process: a chondrocyte, isolated in its lacuna in the matrix, will divide to give birth to two cells, each of which then proceeds to secrete more matrix, so that a layer of matrix is soon formed between them (Figure 17–44).

Figure 17–43 (A) Diagram of a section through a rod of developing cartilage, showing the surrounding fibrous perichondrium. Each chondrocyte fills a lacuna in the cartilage matrix. (B) Cross-section through a rod of cartilage from a chick embryo. This photomicrograph represents an early stage of development. As the tissue grows, the quantity of cartilage matrix per chondrocyte will become much greater and the boundary between cartilage and perichondrium will become more sharply demarcated. (Courtesy of Peter Gould.)

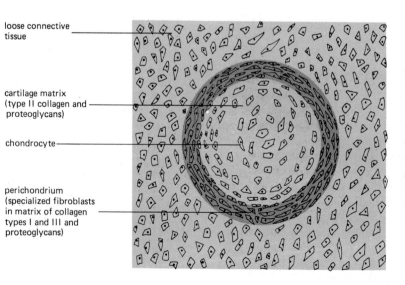

loose connective tissue

cartilage matrix (type II collagen and proteoglycans)

chondrocyte

perichondrium (specialized fibroblasts in matrix of collagen types I and III and proteoglycans)

(A)

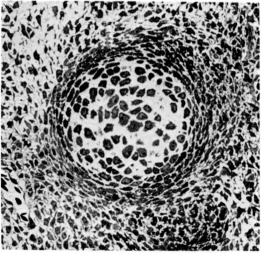

(B)

100 μm

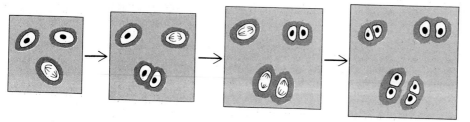

Figure 17–44 The growth of cartilage. The tissue expands as the chondrocytes divide and make more matrix. The freshly synthesized matrix with which each cell surrounds itself is shaded dark. Cartilage may also grow by recruiting fibroblasts from the perichondrium and converting them into chondrocytes (see Figure 17–43).

New cells may also be recruited into the cartilage from the perichondrium. Perichondrial cells resembling fibroblasts are thought to undergo a conversion in which they begin to secrete cartilage matrix around themselves and quickly become full-fledged chondrocytes. This appears to be an *in vivo* counterpart of the transformations described earlier (see p. 988) for connective-tissue cells in culture.

Osteoblasts Secrete Bone Matrix, While Osteoclasts Erode It[41,43]

Bone is a more complex tissue than cartilage. The bone matrix is secreted by **osteoblasts** that lie at the surface of the existing matrix and deposit fresh layers of bone onto it. Some of the osteoblasts remain free at the surface, while others gradually become embedded in their own secretion. This freshly formed material (consisting chiefly of type I collagen) is called *osteoid.* It is rapidly converted into hard bone matrix by the deposition of calcium phosphate crystals in it. Once imprisoned in hard matrix, the original bone-forming cell, now called an **osteocyte,** has no opportunity to divide, although it continues to secrete further matrix, in small quantities, around itself. The osteocyte, like the chondrocyte, occupies a small cavity or lacuna in the matrix, but unlike the chondrocyte, it is not isolated from its fellows. Tiny channels, or *canaliculi,* radiate from each lacuna and contain cell processes from the resident osteocyte, enabling it to form gap junctions with adjacent osteocytes (Figure 17–45). Although the networks of osteocytes do not themselves secrete or erode substantial quantities of matrix, they probably play a part in controlling the activities of the cells that do.

While bone matrix is deposited by osteoblasts, it is eroded by **osteoclasts** (Figure 17–46). These large multinucleated cells originate, like macrophages, from hemopoietic stem cells in the bone marrow. The precursor cells are released as monocytes into the bloodstream and collect at sites of bone resorption, where they fuse to form the multinucleated osteoclasts, which cling to surfaces of the bone matrix and eat it away. Osteoclasts are capable of tunneling deep into the substance of compact bone, forming cavities that are then invaded by other cells. A blood capillary grows down the center of such a tunnel, and the walls of the tunnel become lined with a layer of osteoblasts (Figure 17–47). To produce the

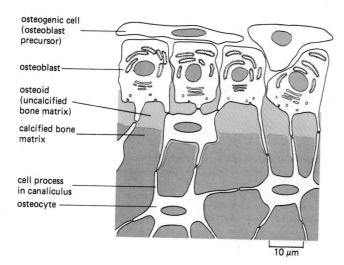

osteogenic cell (osteoblast precursor)

osteoblast

osteoid (uncalcified bone matrix)

calcified bone matrix

cell process in canaliculus

osteocyte

10 μm

Figure 17–45 How osteoblasts lining the surface of bone secrete the organic matrix of bone (osteoid) and are converted into osteocytes as they become embedded in this matrix. The matrix calcifies soon after it has been deposited. The osteoblasts themselves are thought to derive from osteogenic stem cells that are closely related to fibroblasts.

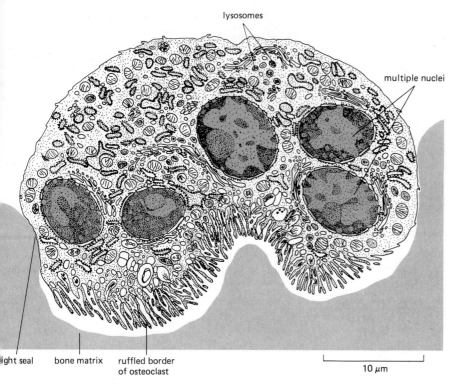

lysosomes

multiple nuclei

ght seal   bone matrix   ruffled border
of osteoclast

10 μm

Figure 17–46 An osteoclast is a giant cell that erodes bone matrix; the "ruffled border" is a site of secretion of acids (to dissolve the bone minerals) and hydrolases (to digest the organic components of the matrix). The cell is shown here in cross-section. Osteoclasts vary in shape, are motile, and often send out processes to resorb bone at multiple sites. They develop from monocytes and can be viewed as specialized macrophages. (From R.V. Krstić: Ultrastructure of the Mammalian Cell: An Atlas. Berlin: Springer, 1979.)

plywoodlike structure of compact bone, these osteoblasts lay down concentric layers of new bone, which gradually fill the cavity, leaving only a narrow canal surrounding the new blood vessel. Many of the osteoblasts become trapped in the bone matrix and survive as concentric rings of osteocytes. At the same time as some tunnels are filling up with bone, others are being bored by osteoclasts, cutting through older concentric systems. The consequences of this perpetual remodeling are beautifully displayed in the layered patterns of matrix observed in compact bone (Figure 17–48).

There are many unsolved problems about these processes. Bones, for example, have a remarkable ability to adapt to the load imposed on them by remodeling their structure, and this implies that the deposition and erosion of the matrix are somehow controlled by local mechanical stresses. We do not understand the

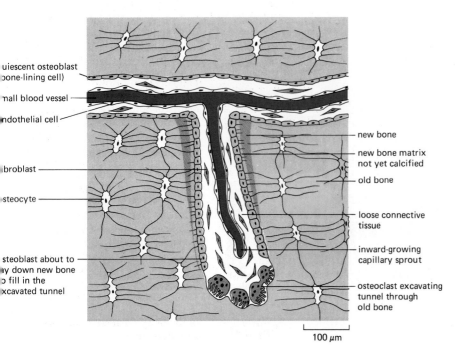

uiescent osteoblast
bone-lining cell)

nall blood vessel

ndothelial cell

new bone

new bone matrix
not yet calcified

old bone

broblast

steocyte

loose connective
tissue

inward-growing
capillary sprout

steoblast about to
ay down new bone
o fill in the
xcavated tunnel

osteoclast excavating
tunnel through
old bone

100 μm

Figure 17–47 The remodeling of compact bone. Osteoclasts acting together in a small group excavate a tunnel through the old bone, advancing at a rate of about 50 μm per day. Osteoblasts enter the tunnel behind them, line its walls, and begin to form new bone, depositing layers of matrix at a rate of 1 or 2 μm per day. At the same time, a capillary sprouts down the center of the tunnel. The tunnel will eventually become filled with concentric layers of new bone, with only a narrow central canal remaining. Each such canal, besides providing a route of access for osteoclasts and osteoblasts, contains one or more blood vessels bringing the nutrients the bone cells must have to survive. Typically, about 5–10% of the bone in a healthy adult mammal is replaced in this way each year. (After Z.F.G. Jaworski, B. Duck, and G. Sekaly, *J. Anat.* 133:397–405, 1981.)

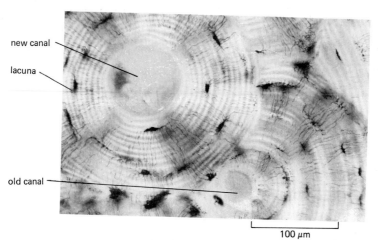

new canal

lacuna

old canal

100 µm

Figure 17–48 Photomicrograph of a transverse section through a compact outer portion of a long bone, showing the outlines of tunnels formed by osteoclasts and then filled in by osteoblasts during successive rounds of bone remodeling. The section has been prepared by grinding; the hard matrix has been preserved but not the cells. Lacunae and canaliculi that were occupied by osteocytes are clearly visible, however. The alternating bright and dark concentric rings correspond to an alternating orientation of the collagen fibers in the successive layers of bone matrix laid down by the osteoblasts that lined the wall of the canal during life. (This pattern is revealed here by viewing the specimen between partly crossed Polaroid filters.) Note how the older system of concentric layers of bone at the lower right, with the narrow central canal, has been partly cut through and replaced by the newer system, whose central canal is presumably still large because it is still in the process of being filled in.

mechanisms that determine whether matrix will be deposited by osteoblasts or eroded by osteoclasts at a given bone surface, but it seems likely that an important part is played by growth factors that are made by the bone cells, trapped in the matrix (see p. 987), and released, perhaps, when the matrix is degraded or suitably stressed.

During Development, Cartilage Is Eroded by Osteoclasts to Make Way for Bone[44]

The replacement of cartilage by bone in the course of development is also thought to depend on the activities of osteoclasts. As the cartilage matures, its cells in certain regions become greatly enlarged at the expense of the surrounding matrix, and the matrix itself becomes mineralized, like bone, by deposition of calcium phosphate crystals. The swollen chondrocytes die, leaving large empty cavities. Osteoclasts and blood vessels invade the cavities and erode the residual cartilage matrix, while osteoblasts following in their wake begin to deposit bone matrix. The only surviving remnant of cartilage in the adult long bone is a thin layer that forms a smooth covering on the bone surfaces at joints, where one bone articulates with another (Figure 17–49).

Some cells capable of forming new cartilage persist, however, in the connective tissue that surrounds a bone. If the bone is broken, the cells in the neighborhood of the fracture will carry out a repair by a rough-and-ready recapitulation of the original embryonic process, in which cartilage is first laid down to bridge the gap and is then replaced by bone.

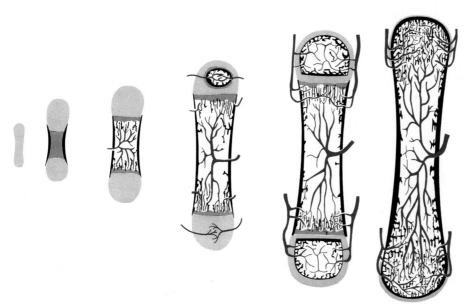

Figure 17–49 The development of a long bone (such as the femur or the humerus) from a miniature cartilage model. Uncalcified cartilage is shown in light gray, calcified cartilage in dark gray, bone in black, and blood vessels in color. The cartilage is not converted to bone but is gradually replaced by it through the action of osteoclasts and osteoblasts, which invade the cartilage in association with blood vessels. Osteoclasts erode cartilage and bone matrix, while osteoblasts secrete bone matrix. The process of ossification begins in the embryo and is not completed until the end of puberty. The resulting bone consists of a thick-walled hollow cylinder of compact bone enclosing a large central cavity occupied by the bone marrow. Note that not all bones develop in this way. The *membrane bones* of the skull, for example, are formed directly as bony plates, not from a prior cartilage model. (Adapted from D.W. Fawcett, A Textbook of Histology, 11th ed. Philadelphia: Saunders, 1986.)

The Structure of the Body Is Stabilized by Its Connective-Tissue Framework and by the Selective Cohesion of Cells[45]

A bone, like the body as a whole, is a dynamic system, maintaining its structure through a balance between the opposed activities of a variety of specialized cells. Any dynamic system poses a problem of stability, and this leads us to a general question about the maintenance of body structure. We have seen how cells in various types of tissues maintain their differentiated state, how new cells are produced in a controlled fashion to replace those that are lost, and how the extracellular matrix is remodeled and renewed. But why do the different types of cells not become progressively jumbled and misplaced? Why does the whole structure not sag, warp, or otherwise change its proportions as new parts are substituted for old?

To some extent, of course, the body does sag and warp with the passage of time—that is a part of aging. But it does so remarkably little. The skeleton, despite constant remodeling, provides a rigid framework whose dimensions scarcely change. This is partly because the parts of a bone are renewed not all at once but little by little, rather like a building whose bricks are replaced one at a time. Besides such conservatism in the mode of renewal, active homeostatic mechanisms are at work. Thus small departures of a bone from its normal shape set up altered patterns of stresses, which regulate bone remodeling in such a way as to restore the bone to its normal shape (Figure 17–50).

The growth and renewal of many of the soft parts of the body are also homeostatically controlled so that each component is adjusted to fit its niche. The epidermis spreads to cover the surface of the body, and the cells halt their migration by *contact inhibition* when that end is achieved (see p. 673); connective tissue grows to just the extent necessary to fill the gap created by a wound; and so on. In all this, something more than mere control of cell numbers is required. The various types of differentiated cells must be maintained not only in the correct relative quantities but also in the correct relative positions. Tissue turnover necessarily involves cell movements. Somehow those movements must be limited; the cells must be subject to territorial restraints.

These restraints are of various kinds. Glands and other masses of specialized cells are often contained, for example, within tough capsules of connective tissue. Some types of cells die if they find themselves outside their normal environment, deprived of specific growth factors on which their survival probably depends. Perhaps the most important strategy for keeping the different cells in their places, however, is the strategy of selective cell-cell adhesion: cells of the same type tend to stick together (see p. 829), either in solid masses, such as smooth muscle, or in epithelial sheets, such as the lining of the gut. As described on page 828, this mechanism enables dissociated epidermal cells, for example, to reassociate spontaneously to form a correctly structured epithelium. And, on a larger scale, stable epithelial sheets of cells serve to divide the body into compartments, thereby keeping other cells properly segregated and confined to their correct territories.

Clearly, the checks and balances that preserve the structure of the body and the organization of its cells in the face of continual turnover and renewal are intricate and subtle. The importance of these controls is illustrated with painful clarity when they go awry, as we shall see when we come to the topic of cancer in the final chapter of this book.

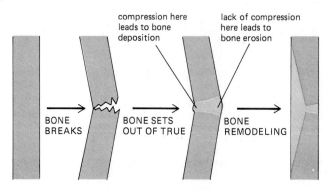

compression here leads to bone deposition

lack of compression here leads to bone erosion

BONE BREAKS

BONE SETS OUT OF TRUE

BONE REMODELING

Figure 17–50 Remodeling of a long bone in the leg after a fracture that has healed out of line. The deformity in the recently healed bone exposes it to abnormal stresses. Where the compressive forces are increased, the rate of bone deposition is increased relative to the rate of erosion; and where the forces are decreased, the rate of deposition is decreased relative to the rate of erosion. In this way the bone is gradually remodeled back to its normal form.

Summary

Cartilage and bone both consist of cells embedded in a solid matrix. Cartilage has a deformable matrix and can grow by swelling, whereas bone is rigid and can grow only by accretion at its surfaces. Bone is, nonetheless, subject to perpetual remodeling through the combined action of osteoclasts, which erode matrix, and osteoblasts, which secrete it. Some osteoblasts become trapped in the matrix as osteocytes and play a part in regulating the turnover of bone matrix. Most long bones develop from miniature cartilage "models," which, as they grow, serve as templates for the deposition of bone by the combined action of osteoblasts and osteoclasts. Similarly, in the repair of a bone fracture in the adult, the gap is first bridged by cartilage, which is later replaced by bone. Although bone, like most other tissues, is subject to continual turnover, this dynamic process is regulated so that the global structure is preserved. In this way, and through other mechanisms such as selective cell-cell adhesion, the organization of the body is stably maintained even though most of its components are continually being replaced.

Appendix

Cells of the Adult Human Body: A Catalog

How many distinct cell types are there in an adult human being? A large textbook of histology will mention about 200 that qualify for individual names. These traditional names are not, like the names of colors, labels for parts of a continuum that has been subdivided arbitrarily: they represent, for the most part, discrete and distinctly different categories. Within a given category there is often some variation—the skeletal muscle fibers that move the eyeball are small, while those that move the leg are big; auditory hair cells in different parts of the ear may be tuned to different frequencies of sound; and so on. But there is no continuum of adult cell types intermediate in character between, say, the muscle cell and the auditory hair cell.

The traditional histological classification is based on the shape and structure of the cell as seen in the microscope and on its chemical nature as assessed very crudely from its affinities for various stains. Subtler methods reveal new subdivisions within the traditional classification. Thus modern immunology has shown that the old category of "lymphocyte" includes more than 10 quite distinct cell types (see Chapter 18). Similarly, pharmacological and physiological tests reveal that there are many different varieties of smooth muscle cell—those in the wall of the uterus, for example, are highly sensitive to estrogen, and in the later stages of pregnancy to oxytocin, while those in the wall of the gut are not. Another major type of diversity is revealed by embryological experiments of the sort discussed in Chapter 16. These show that, in many cases, apparently similar cells from different regions of the body are nonequivalent, that is, they are inherently different in their developmental capacities and in their effects on other cells. For example, connective tissue cells from different regions of the dermis must be nonequivalent, since they provoke the overlying epidermal cells to behave differently (see p. 942). Thus, within categories such as "fibroblast" there are probably many distinct cell types, different chemically in ways that are not easy to perceive directly.

For these reasons, any classification of the cell types in the body must be somewhat arbitrary with respect to the fineness of its subdivisions. Here, we list only the adult human cell types that a histology text would recognize to be different, grouped into families roughly according to function. We have not attempted to subdivide the class of neurons of the central nervous system. Also, where a single cell type such as the keratinocyte is conventionally given a succession of different names as it matures, we give only two entries—one for the differentiating cell and one for the stem cell. With these serious provisos, the 210 varieties of cells in the catalog represent a more or less exhaustive list of the distinctive ways in which a given mammalian genome can be expressed in the phenotype of a normal cell of the adult body.

Keratinizing Epithelial Cells

keratinocyte of epidermis (= differentiating
    epidermal cell)
basal cell of epidermis (stem cell)
keratinocyte of fingernails and toenails
basal cell of nail bed (stem cell)
hair shaft cells
    medullary
    cortical
    cuticular
hair-root sheath cells
    cuticular
    of Huxley's layer
    of Henle's layer
    external
hair matrix cell (stem cell)

Cells of Wet Stratified Barrier Epithelia

surface epithelial cell of stratified squamous
    epithelium of cornea, tongue, oral cavity,
    esophagus, anal canal, distal urethra,
    vagina
basal cell of these epithelia (stem cell)
cell of urinary epithelium (lining bladder and
    urinary ducts)

Epithelial Cells Specialized for Exocrine Secretion

cells of salivary gland
    mucous cell (secretion rich in
        polysaccharide)
    serous cell (secretion rich in glycoprotein
        enzymes)
cell of von Ebner's gland in tongue (secretion
    to wash over taste buds)
cell of mammary gland, secreting milk
cell of lacrimal gland, secreting tears
cell of ceruminous gland of ear, secreting
    wax
cell of eccrine sweat gland, secreting
    glycoproteins (dark cell)
cell of eccrine sweat gland, secreting small
    molecules (clear cell)
cell of apocrine sweat gland (odoriferous
    secretion, sex-hormone sensitive)
cell of gland of Moll in eyelid (specialized
    sweat gland)
cell of sebaceous gland, secreting lipid-rich
    sebum
cell of Bowman's gland in nose (secretion to
    wash over olfactory epithelium)
cell of Brunner's gland in duodenum,
    secreting alkaline solution of mucus and
    enzymes
cell of seminal vesicle, secreting components
    of seminal fluid, including fructose (as
    fuel for swimming sperm)
cell of prostate gland, secreting other
    components of seminal fluid
cell of bulbourethral gland, secreting mucus
cell of Bartholin's gland, secreting vaginal
    lubricant
cell of gland of Littré, secreting mucus
cell of endometrium of uterus, secreting
    mainly carbohydrates
isolated goblet cell of respiratory and
    digestive tracts, secreting mucus
mucous cell of lining of stomach
zymogenic cell of gastric gland, secreting
    pepsinogen
oxyntic cell of gastric gland, secreting HCl
acinar cell of pancreas, secreting digestive
    enzymes and bicarbonate
Paneth cell of small intestine, secreting
    lysozyme

type II pneumocyte of lung, secreting
    surfactant
Clara cell of lung (function unknown)

Cells Specialized for Secretion of Hormones

cells of anterior pituitary, secreting
    growth hormone
    follicle-stimulating hormone
    luteinizing hormone
    prolactin
    adrenocorticotropic hormone
    thyroid-stimulating hormone
cell of intermediate pituitary, secreting
    melanocyte-stimulating hormone
cells of posterior pituitary, secreting
    oxytocin
    vasopressin
cells of gut and respiratory tract, secreting
    serotonin
    endorphin
    somatostatin
    gastrin
    secretin
    cholecystokinin
    insulin
    glucagon
    bombesin
cells of thyroid gland, secreting
    thyroid hormone
    calcitonin
cells of parathyroid gland, secreting
    parathyroid hormone
    oxyphil cell (function unknown)
cells of adrenal gland, secreting
    epinephrine
    norepinephrine
    steroid hormones
        mineralocorticoids
        glucocorticoids
cells of gonads, secreting
    testosterone (Leydig cell of testis)
    estrogen (theca interna cell of ovarian
        follicle)
    progesterone (corpus luteum cell of
        ruptured ovarian follicle)
cells of juxtaglomerular apparatus of kidney
    juxtaglomerular cell (secreting renin)
    macula densa
        cell
    peripolar cell
    mesangial cell

macula densa cell	(uncertain but probably related in function; possibly involved in secretion of erythropoietin)
peripolar cell	
mesangial cell	

Epithelial Absorptive Cells in Gut, Exocrine Glands, and Urogenital Tract

brush border cell of intestine (with microvilli)
striated duct cell of exocrine glands
gall bladder epithelial cell
brush border cell of proximal tubule of
    kidney
distal tubule cell of kidney
nonciliated cell of ductulus efferens
epididymal principal cell
epididymal basal cell

Cells Specialized for Metabolism and Storage

hepatocyte (liver cell)
fat cells
    white fat
    brown fat
    lipocyte of liver

Epithelial Cells Serving Primarily a Barrier Function, Lining the Lung, Gut, Exocrine Glands, and Urogenital Tract

type I pneumocyte (lining air space of lung)
pancreatic duct cell (centroacinar cell)
nonstriated duct cell of sweat gland, salivary
    gland, mammary gland, etc. (various)
parietal cell of kidney glomerulus
podocyte of kidney glomerulus
cell of thin segment of loop of Henle (in
    kidney)
collecting duct cell (in kidney)
duct cell of seminal vesicle, prostate gland,
    etc. (various)

Epithelial Cells Lining Closed Internal Body Cavities

vascular endothelial cells of blood vessels
    and lymphatics
    fenestrated
    continuous
    splenic
synovial cell (lining joint cavities, secreting
    largely hyaluronic acid)
serosal cell (lining peritoneal, pleural, and
    pericardial cavities)
squamous cell lining perilymphatic space of
    ear
cells lining endolymphatic space of ear
    squamous cell
    columnar cells of endolymphatic sac
        with microvilli
        without microvilli
    "dark" cell
    vestibular membrane cell
    stria vascularis basal cell
    stria vascularis marginal cell
    cell of Claudius
    cell of Boettcher
choroid plexus cell (secreting cerebrospinal
    fluid)
squamous cell of pia-arachnoid
cells of ciliary epithelium of eye
    pigmented
    nonpigmented
corneal "endothelial" cell

Ciliated Cells with Propulsive Function

of respiratory tract
of oviduct and of endometrium of uterus (in
    female)
of rete testis and ductulus efferens (in male)
of central nervous system (ependymal cell
    lining brain cavities)

Cells Specialized for Secretion of Extracellular Matrix

epithelial:
    ameloblast (secreting enamel of tooth)
    planum semilunatum cell of vestibular
        apparatus of ear (secreting proteoglycan)
    interdental cell of organ of Corti (secreting
        tectorial "membrane" covering hair cells
        of organ of Corti)
nonepithelial (connective tissue)
    fibroblasts (various—of loose connective
        tissue, of cornea, of tendon, of reticular
        tissue of bone marrow, etc.)
    pericyte of blood capillary
    nucleus pulposus cell of intervertebral disc
    cementoblast/cementocyte (secreting
        bonelike cementum of root of tooth)

odontoblast/odontocyte (secreting dentin of
   tooth)
chondrocytes
   of hyaline cartilage
   of fibrocartilage
   of elastic cartilage
osteoblast/osteocyte
osteoprogenitor cell (stem cell of
   osteoblasts)
hyalocyte of vitreous body of eye
stellate cell of perilymphatic space of ear

Contractile Cells

skeletal muscle cells
   red (slow)
   white (fast)
   intermediate
   muscle spindle—nuclear bag
   muscle spindle—nuclear chain
   satellite cell (stem cell)
heart muscle cells
   ordinary
   nodal
   Purkinje fiber
smooth muscle cells (various)
myoepithelial cells
   of iris
   of exocrine glands

Cells of Blood and Immune System

red blood cell
megakaryocyte
macrophages and related cells
   monocyte
   connective tissue macrophage (various)
   Langerhans cell (in epidermis)
   osteoclast (in bone)
   dendritic cell (in lymphoid tissues)
   microglial cell (in central nervous system)
neutrophil
eosinophil
basophil
mast cell
T lymphocyte
   helper T cell
   suppressor T cell
   killer T cell

B lymphocyte
   IgM
   IgG
   IgA
   IgE
killer cell
stem cells and committed progenitors for the
   blood and immune system (various)

Sensory Transducers

photoreceptors
   rod
   cones
     blue sensitive
     green sensitive
     red sensitive
hearing
   inner hair cell of organ of Corti
   outer hair cell of organ of Corti
acceleration and gravity
   type I hair cell of vestibular apparatus of ear
   type II hair cell of vestibular apparatus of ear
taste
   type II taste bud cell
smell
   olfactory neuron
   basal cell of olfactory epithelium (stem cell
     for olfactory neurons)
blood pH
   carotid body cell
     type I
     type II
touch
   Merkel cell of epidermis
   primary sensory neurons specialized for
     touch (various)
temperature
   primary sensory neurons specialized for
   temperature
     cold sensitive
     heat sensitive
pain
   primary sensory neurons specialized for
     pain (various)
configurations and forces in musculoskeletal
   system
   proprioceptive primary sensory neurons
     (various)

Autonomic Neurons

cholinergic (various)
adrenergic (various)
peptidergic (various)

Supporting Cells of Sense Organs and of Peripheral Neurons

supporting cells of organ of Corti
   inner pillar cell
   outer pillar cell
   inner phalangeal cell
   outer phalangeal cell
   border cell
   Hensen cell
supporting cell of vestibular apparatus
supporting cell of taste bud (type I taste bud
   cell)
supporting cell of olfactory epithelium
Schwann cell
satellite cell (encapsulating peripheral nerve
   cell bodies)
enteric glial cell

Neurons and Glial Cells of Central Nervous System

neurons (huge variety of types—still poorly
   classified)
glial cells
   astrocyte (various)
   oligodendrocyte

Lens Cells

anterior lens epithelial cell
lens fiber (crystallin-containing cell)

Pigment Cells

melanocyte
retinal pigmented epithelial cell

Germ Cells

oogonium/oocyte
spermatocyte
spermatogonium (stem cell for spermatocyte)

Nurse Cells

ovarian follicle cell
Sertoli cell (in testis)
thymus epithelial cell

References

General

Clark, W.E. Le Gros. The Tissues of the Body, 6th ed. Oxford, U.K.:
   Clarendon Press, 1971.
Cormack, D. Ham's Histology, 9th ed. Philadelphia: Lippincott,
   1987.
Fawcett, D.W. (Bloom and Fawcett) A Textbook of Histology, 11th
   ed. Philadelphia: Saunders, 1986.
Goss, R.J. The Physiology of Growth. New York: Academic Press,
   1978.
Weiss, L., ed. Histology: Cell and Tissue Biology, 5th ed. New York:
   Elsevier, 1983.
Wheater, P.R.; Burkitt, H.G.; Daniels, V.G. Functional Histology,
   2nd ed. Edinburgh: Churchill Livingstone, 1987.

Cited

1. Clark, W.E. Le Gros. The Tissues of the Body, 6th ed. Oxford,
   U.K.: Clarendon Press, 1971.

Montagna, W. The skin. Sci. Am. 212(2): 56–66, 1965.
Wessells, N.K. Tissue Interactions and Development. Menlo
   Park, CA: Benjamin-Cummings, 1977.
2. Cahn, R.D.; Cahn, M.B.. Heritability of cellular differentiation:
   clonal growth and expression of differentiation in retinal
   pigment cells in vitro. Proc. Natl. Acad. Sci. USA 55:106–114,
   1966.
Coon, H.G. Clonal stability and phenotypic expression of chick
   cartilage cells in vitro. Proc. Natl. Acad. Sci. USA 55:66–73,
   1966.
Itoh, Y.; Eguchi, G. In vitro analysis of cellular metaplasia from
   pigmented epithelial cells to lens phenotypes: a unique
   model for studying cellular and molecular mechanisms
   of "transdifferentiation." Dev. Biol. 115:353–362, 1986.
Yaffe, D. Retention of differentiation potentialities during prolonged cultivation of myogenic cells. Proc. Natl. Acad. Sci.
   USA 61:477–483, 1968.
3. Anderson, J.E. The effect of steroid hormones on gene transcription. In Biological Regulation and Development (R.F.

Goldberger, K. Yamamoto, eds.), Vol. 3B, pp. 169–212. New York: Plenum, 1983.

Okada, T.S.: Kondoh, H., eds. Commitment and Instability in Cell Differentiation. *Curr. Top. Dev. Biol.* 20, 1986.

4. Hosley, M.A.; Hughes, S.E.; Oakley, B. Neural induction of taste buds. *J. Comp. Neurol.* 260:224–232, 1987.

Kinnamon, S.C. Taste transduction: a diversity of mechanisms. *Trends Neurosci.* 11:491–496, 1988.

Zalewski, A.A. Neuronal and tissue specifications involved in taste bud formation. *Ann. N.Y. Acad Sci.* 228:344–349, 1974.

5. Goss, R.J. The Physiology of Growth. New York: Academic Press, 1978.

6. Clayton, R.M. Divergence and convergence in lens cell differentiation: regulation of the formation and specific content of lens fibre cells. In Stem Cells and Tissue Homeostasis (B. Lord, C. Potten, R. Cole, eds.), pp. 115–138. Cambridge, U.K.: Cambridge University Press, 1978.

Goss, R.J. The Physiology of Growth, pp. 210–225. New York: Academic Press, 1978.

Maisel, H., ed. The Ocular Lens: Structure, Function, and Pathology. New York: Dekker, 1985.

Wistow, G.J.; Piatigorsky, J. Lens crystallins: the evolution and expression of proteins for a highly specialized tissue. *Annu. Rev. Biochem.* 57:479–504, 1988.

7. Fawcett, D.W. (Bloom and Fawcett) A Textbook of Histology, 11th ed. Philadelphia: Saunders, 1986.

Gevers, W. Protein metabolism in the heart. *J. Mol. Cell. Cardiol.* 16:3–32, 1984.

Young, R.W. Visual cells. *Sci. Am.* 223(4):80–91, 1970.

8. Aherne, W.A.; Camplejohn, R.S.; Wright, N.A. An Introduction to Cell Population Kinetics. London: Edward Arnold, 1977.

Wright, N.A.; Alison, M.R. Biology of Epithelial Cell Populations, Vols. 1–3. Oxford, U.K.: Oxford University Press, 1984.

9. Fawcett, D.W. (Bloom and Fawcett) A Textbook of Histology, 11th ed., pp. 679–715. Philadelphia: Saunders, 1986.

Moog, F. The lining of the small intestine. *Sci. Am.* 245(5):154–176, 1981.

10. Fausto, N. New perspectives on liver regeneration. *Hepatol.* 6:326–327, 1986.

Gohda, E.; et al. Purification and partial characterization of hepatocyte growth factor from plasma of a patient with fulminant hepatic failure. *J. Clin. Invest.* 81:414–419, 1988.

Goss, R.J. The Physiology of Growth, pp. 251–266. New York: Academic Press, 1978.

Holder, N. Regeneration and compensatory growth. *Br. Med. Bull.* 37:227–232, 1981.

11. Anderson, J.R., ed. Muir's Textbook of Pathology, 12th ed. London: Edward Arnold, 1985.

Robbins, S.L.; Cotran, R.S.; Kumar, V. Pathologic Basis of Disease, 3rd ed. Philadelphia: Saunders, 1984.

12. Campbell, J.H.; Campbell, G.R. Endothelial cell influences on vascular smooth muscle phenotype. *Annu. Rev. Physiol.* 48:295–306, 1986.

Development of the Vascular System. *Ciba Symp.* 100. London: Pitman, 1983.

Fawcett, D.W. (Bloom and Fawcett) A Textbook of Histology, 11th ed., pp. 367–405. Philadelphia: Saunders, 1986.

Ryan, U.S., ed. Endothelial Cells, Vols. 1–3. Boca Raton, FL: CRC Press, 1988.

13. Goss, R.J. The Physiology of Growth, pp. 120–137. New York: Academic Press, 1978.

Hobson, B.; Denekamp, J. Endothelial proliferation in tumours and normal tissues: continuous labelling studies. *Br. J. Cancer* 49:405–413, 1984.

14. Folkman, J. The vascularization of tumors. *Sci. Am.* 234(5):58–73, 1976.

Folkman, J.; Haudenschild, C. Angiogenesis *in vitro. Nature* 288:551–556, 1980.

Madri, J.A.; Pratt, B.M. Endothelial cell-matrix interactions: *in vitro* models of angiogenesis. *J. Histochem. Cytochem.* 34:85–91, 1986.

15. Folkman, J.; Klagsbrun, M. Angiogenic factors. *Science* 235:442–447, 1987.

Kalebic, T.; Garbisa, S.; Glaser, B.; Liotta, L.A. Basement membrane collagen: degradation by migrating endothelial cells. *Science* 221:281–283, 1983.

Knighton, D.R.; et al. Oxygen tension regulates the expression of angiogenesis factor by macrophages. *Science* 221:1283–1285, 1983.

Leibovich, S.J.; et al. Macrophage-induced angiogenesis is mediated by tumour necrosis factor-alpha. *Nature* 329:630–632, 1987.

Schweigerer, L.; et al. Capillary endothelial cells express basic fibroblast growth factor, a mitogen that promotes their own growth. *Nature* 325:257–259, 1987.

16. Cairnie, A.B.; Lala, P.K.; Osmond, D.G., eds. Stem Cells of Renewing Cell Populations. New York: Academic Press, 1976.

Cheng, H.; Leblond, C.P. Origin, differentiation, and renewal of the four main epithelial cell types in the mouse small intestine. V. Unitarian theory of the origin of the four epithelial cell types. *Am. J. Anat.* 141:537–562, 1974.

17. Graziadei, P.P.C.; Monti Graziadei, G.A. Continuous nerve cell renewal in the olfactory system. In Handbook of Sensory Physiology, Vol. IX: Development of Sensory Systems (M. Jacobson, ed.), pp. 55–82. New York: Springer-Verlag, 1978.

18. Allen, T.D.; Potten, C.S. Fine-structural identification and organization of the epidermal proliferative unit. *J. Cell Sci.* 15:291–319, 1974.

Bereiter-Hahn, J.; Matoltsy, A.G.; Richards, K.S., eds. Biology of the Integument, Vol. 2: Vertebrates. New York: Springer, 1986.

MacKenzie, I.C. Ordered structure of the stratum corneum of mammalian skin. *Nature* 222:881–882, 1969.

Sengel, P. Morphogenesis of skin. Cambridge, U.K.: Cambridge University Press, 1976.

Stenn, K.S. The skin. In Histology: Cell and Tissue Biology (L. Weiss, ed.), 5th ed., pp. 569–606. New York: Elsevier, 1983.

19. Fuchs, E.; Green, H. Changes in keratin gene expression during terminal differentiation of the keratinocyte. *Cell* 19:1033–1042, 1980.

Green, H. The keratinocyte as differentiated cell type. *Harvey Lectures* 74:101–139, 1979.

Sawyer, R.H., ed. The Molecular and Developmental Biology of Keratins. *Curr. Top. Dev. Biol.* 22, 1987.

20. Barrandon, Y.; Green, H. Three clonal types of keratinocyte with different capacities for multiplication. *Proc. Natl. Acad. Sci. USA* 84:2302–2306, 1987.

Green, H. Terminal differentiation of cultured human epidermal cells. *Cell* 11:405–415, 1977.

Watt, F.M. Selective migration of terminally differentiating cells from the basal layer of cultured human epidermis. *J. Cell Biol.* 98:16–21, 1984.

21. Barrandon, Y.; Green, H. Cell migration is essential for sustained growth of keratinocyte colonies: the roles of transforming growth factor-alpha and epidermal growth factor. *Cell* 50:1131–1137, 1987.

Read, J.; Watt, F.M. A model for *in vitro* studies of epidermal homeostasis: proliferation and involucrin synthesis by cultured human keratinocytes during recovery after stripping off the suprabasal layers. *J. Invest. Dermatol.* 90:739–743, 1988.

22. Fawcett, D.W. (Bloom and Fawcett) A Textbook of Histology, 11th ed., pp. 568–576, 901–912. Philadelphia: Saunders, 1986.

Neville, M.C.: Neifert, M.R. Lactation: Physiology, Nutrition, and Breast-Feeding. New York: Plenum, 1983.

Patton, S. Milk. *Sci Am.* 221(1):58–68, 1969.

Richards, R.C.; Benson, G.K. Ultrastructural changes accompanying involution of the mammary gland in the albino rat. *J. Endocrinol.* 51:127–135, 1971.

Vonderhaar, B.K.; Topper, Y.J. A role of the cell cycle in hormone-dependent differentiation. *J. Cell. Biol.* 63:707–712, 1974.

23. Dexter, T.M.; Spooncer, E. Growth and differentiation in the hemopoietic system. *Annu. Rev. Cell Biol.* 3:423–441, 1987.

Weiss, L., ed. Histology: Cell and Tissue Biology, 5th ed., pp. 447–509. New York: Elsevier, 1983.

Wintrobe, M.M. Blood, Pure and Eloquent. New York: McGraw-Hill, 1980.

24. Stossel, T.P. The molecular biology of phagocytes and the molecular basis of nonneoplastic phagocytic disorders. In The Molecular Basis of Blood Diseases (G. Stamatoyannopoulos, A.W. Nienhuis, P. Leder, P.W. Majerus, eds.), pp. 499–533. Philadelphia: Saunders, 1987.

Zucker, M.B. The functioning of blood platelets. *Sci. Am.* 242(6):86–103, 1980.

25. Robbins, S.L.; Cotran, R.S.; Kumar, V. Pathologic Basis of Disease, 3rd ed. Philadelphia: Saunders, 1984.

Taussig, M.J. Processes in Pathology and Microbiology, 2nd. ed. Oxford, U.K.: Blackwell, 1984.

26. Magli, M.C.; Iscove, N.N.; Odartchenko, N. Transient nature of haematopoietic spleen colonies. *Nature* 295:527–529, 1982.

Till, J.E.; McCulloch, E.A. A direct measurement of the radiation sensitivity of normal mouse bone marrow cells. *Radiat. Res.* 14:213–222, 1961.

27. Lemischka, I.R.; Raulet, D.H.; Mulligan, R.C. Developmental potential and dynamic behavior of hematopoietic stem cells. *Cell* 45:917–927, 1986.

Wu, A.M.; Till, J.E.; Siminovitch, L.; McCulloch, E.A. Cytological evidence for a relationship between normal hematopoietic colony-forming cells and cells of the lymphoid system. *J. Exp. Med.* 127:455–462, 1968.

28. Till, J.E.; McCulloch, E.A. Hemopoietic stem cell differentiation. *Biochim. Biophys. Acta* 605:431–459, 1980.

29. Dexter, T.M.; Heyworth, C.; Whetton, A.D. The role of growth factors in haemopoiesis. *Bioessays* 2:154–158, 1985.

Metcalf, D. Clonal analysis of proliferation and differentiation of paired daughter cells: action of granulocyte-macrophage colony-stimulating factor on granulocyte-macrophage precursors. *Proc. Natl. Acad. Sci. USA* 77:5327–5330, 1980.

Metcalf, D. The Hemopoietic Colony-Stimulating Factors. Amsterdam: Elsevier, 1984.

30. Goldwasser, E. Erythropoietin and the differentiation of red blood cells. *Fed. Proc.* 34:2285–2292, 1975.

Heath, D.S.; Axelrad, A.A.; McLeod, D.L.; Shreeve, M.M. Separation of the erythropoietin-responsive progenitors BFU-E and CFU-E in mouse bone marrow by unit gravity sedimentation. *Blood* 47:777–792, 1976.

Ihle, J.N.; et al. Biologic properties of homogeneous interleukin 3. *J. Immunol.* 131:282–287, 1983.

Suda, J.; et al. Purified interleukin-3 and erythropoietin support the terminal differentiation of hemopoietic progenitors in serum-free culture. *Blood* 67:1002–1006, 1986.

31. Metcalf, D. The Wellcome Foundation Lecture, 1986. The molecular control of normal and leukaemic granulocytes and macrophages. *Proc. R. Soc. Lond. (Biol.)* 230:389–423, 1987.

Roberts, R.; et al. Heparan sulphate-bound growth factors: a mechanism for stromal cell-mediated haemopoiesis. *Nature* 332:376–378, 1988.

32. Spangrude, G.J.; Heimfeld, S.; Weissman, I.L. Purification and characterization of mouse hematopoietic stem cells. *Science* 241:58–62, 1988.

Suda, T.; Suda, J.; Ogawa, M. Disparate differentiation in mouse hemopoietic colonies derived from paired progenitors. *Proc. Natl. Acad. Sci. USA* 81:2520–2524, 1984.

33. Cormack, D. Ham's Histology, 9th ed., pp. 388–420. Philadelphia: Lippincott, 1987.

Pearson, M.L.; Epstein, H.F., eds. Muscle Development: Molecular and Cellular Control. Cold Spring Harbor, NY: Cold Spring Harbor Laboratory, 1982.

34. Clegg, C.H.; Linkhart, T.A.; Olwin, B.B.; Hauschka, S.D. Growth factor control of skeletal muscle differentiation: commitment to terminal differentiation occurs in G1 phase and is repressed by fibroblast growth factor. *J. Cell Biol.* 105:949–956, 1987.

Davis, R.L.; Weintraub, H.; Lassar, A.B. Expression of a single transfected cDNA converts fibroblasts to myoblasts. *Cell* 51:987–1000, 1987.

Devlin, R.B.; Emerson, C.P., Jr. Coordinate regulation of contractile protein synthesis during myoblast differentiation. *Cell* 13:599–611, 1978.

Konigsberg, I.R. Diffusion-mediated control of myoblast fusion. *Dev. Biol.* 26:133–152, 1971.

Menko, A.S.; Boettiger, D. Occupation of the extracellular matrix receptor, integrin, is a control point for myogenic differentiation. *Cell* 51:51–57, 1987.

Pinney, D.F.; Pearson-White, S.H.; Konieczny, S.F.; Latham, K.E.; Emerson, C.P. Myogenic lineage determination and differentiation: evidence for a regulatory gene pathway. *Cell* 53:781–793, 1988.

35. Buckingham, M.; et al. Actin and myosin multigene families: their expression during the formation and maturation of striated muscle. *Am. J. Med. Genet.* 25:623–634, 1986.

Lømo, T.; Westgaard, R.H.; Dahl, H.A. Contractile properties of muscle: control by pattern of muscle activity in the rat. *Proc. R. Soc. Lond. (Biol.)* 187:99–103, 1974.

Miller, J.B.; Stockdale, F.E. What muscle cells know that nerves don't tell them. *Trends Neurosci.* 10:325–329, 1987.

Sanes, J.R. Cell lineage and the origin of muscle fibre types. *Trends Neurosci.* 10:219–221, 1987.

36. Bischoff, R. Proliferation of muscle satellite cells on intact myofibers in culture. *Dev. Biol.* 115:129–139, 1986.

Goldspink, G. Development of muscle. In Differentiation and Growth of Cells in Vertebrate Tissues (G. Goldspink, ed.), pp. 69–99. London: Chapman and Hall, 1974.

Moss, F.P.; Leblond, C.P. Satellite cells as the source of nuclei in muscles of growing rats. *Anat. Rec.* 170:421–435, 1971.

37. Gabbiani, G.; Rungger-Brändle, E. The fibroblast. In Tissue Repair and Regeneration (L.E. Glynn, ed.), pp. 1–50. Handbook of inflammation, Vol. 3. Amsterdam: Elsevier, 1981.

Fawcett, D.W. (Bloom and Fawcett) A Textbook of Histology, 11th ed., pp. 136–187. Philadelphia: Saunders, 1986.

38. Conrad, G.W.; Hart, G.W.; Chen, Y. Differences *in vitro* between fibroblast-like cells from cornea, heart, and skin of embryonic chicks. *J. Cell Sci.* 26:119–137, 1977.

Hauschka, P.V.; Mavrakos, A.E.; Iafrati, M.D.; Doleman, S.E.; Klagsbrun, M. Growth factors in bone matrix: isolation of multiple types by affinity chromatography on heparin-Sepharose. *J. Biol. Chem.* 261:12665–12674, 1986.

Reddi, A.H.; Gay, R.; Gay, S.; Miller, E.J. Transitions in collagen types during matrix-induced cartilage, bone, and bone marrow formation. *Proc. Natl. Acad. Sci. USA* 74:5589–5592, 1977.

Schor, S.L.; Schor, A.M. Clonal heterogeneity in fibroblast phenotype: implications for the control of epithelial-mesenchymal interactions. *Bioessays* 7:200–204, 1987.

Seyedin, S.M.; et al. Cartilage-inducing factor-A: apparent identity to transforming growth factor-beta. *J. Biol. Chem.* 261:5693–5695, 1986.

9. Benya, P.D.; Schaffer, J.D. Dedifferentiated chondrocytes reexpress the differentiated collagen phenotype when cultured in agarose gels. *Cell* 30:215–224, 1982.

Caplan, A.I. Cartilage. *Sci. Am.* 251(4):84–94, 1984.

von der Mark, K.; Gauss, V.; von der Mark, H.; Müller, P. Relationship between cell shape and type of collagen synthesized as chondrocytes lose their cartilage phenotype in culture. *Nature* 267:531–532, 1977.

Zanetti, N.C.; Solursh, M. Induction of chondrogenesis in limb mesenchymal cultures by disruption of the actin cytoskeleton. *J. Cell Biol.* 99:115–123, 1984.

0. Spiegelman, B.M.; Ginty, C.A. Fibronectin modulation of cell shape and lipogenic gene expression in 3T3 adipocytes. *Cell* 35:657–666, 1983.

Sugihara, H.; Yonemitsu, N.; Miyabara, S.; Yun, K. Primary cultures of unilocular fat cells: characteristics of growth *in vitro* and changes in differentiation properties. *Differentiation* 31:42–49, 1986.

Zezulak, K.M.; Green, H. The generation of insulin-like growth factor-1-sensitive cells by growth hormone action. *Science* 233:551–553, 1986.

1. Cormack, D. Ham's Histology, 9th ed., pp. 264–323. Philadelphia: Lippincott, 1987.

Fawcett, D.W. (Bloom and Fawcett) A Textbook of Histology, 11th ed., pp. 188–238. Philadelphia: Saunders, 1986.

Jee, W.S.S. The skeletal system. In Histology: Cell and Tissue Biology (L. Weiss, ed.), 5th. ed., pp. 200–255. New York: Elsevier, 1983.

42. Hall, B.K. Cartilage, Vols. 1–3. New York: Academic Press, 1983.

43. Marcus, R. Normal and abnormal bone remodeling in man. *Annu. Rev. Med.* 38:129–141, 1987.

Osdoby, P.; Krukowski, M.; Oursler, M.J.; Salino-Hugg, T. The origin, development, and regulation of osteoclasts. *Bioessays* 7:30–34, 1987.

Vaughan, J. The Physiology of Bone, 3rd ed. Oxford, U.K.: Clarendon Press, 1981.

44. Cormack, D. Ham's Histology, 9th ed., pp. 312–320. Philadelphia: Lippincott, 1987.

45. Currey, J. The Mechanical Adaptations of Bone. Princeton, NJ: Princeton University Press, 1984.

Goss, R.J. The Physiology of Growth. New York: Academic Press, 1978.

Rogers, S.L. The Aging Skeleton. Springfield, IL: Thomas, 1982.

Sinclair, D. Human Growth After Birth, 4th ed. Oxford, U.K.: Oxford University Press, 1985.

The Immune System

18

Our immune system saves us from certain death by infection. Any child born with a severely defective immune system will soon die unless extraordinary measures are taken to isolate it from a host of infectious agents—bacterial, viral, fungal, and parasitic. Any vertebrate that is immunologically deficient runs the same deadly risk.

All vertebrates have an immune system. The defense systems found among invertebrates are more primitive, often relying chiefly on phagocytic cells. Such cells (mainly macrophages and neutrophils) also play an important role in defending vertebrates against infection (see p. 974), but they are only one part of a much more complex and sophisticated defense strategy.

Immunology, the study of the immune system, grew out of the common observation that people who recover from certain infections are thereafter "immune" to the disease; that is, they rarely develop the same disease again. Immunity is highly specific: an individual who recovers from measles is protected against the measles virus but not against other common viruses, such as mumps or chicken pox. Such specificity is a fundamental characteristic of immune responses.

Many of the responses of the immune system initiate the destruction and elimination of invading organisms and any toxic molecules produced by them. Because these immune reactions are destructive, it is essential that they be made in response only to molecules that are foreign to the host and not to those of the host itself. This ability to distinguish *foreign* molecules from *self* molecules is another fundamental feature of the immune system. Occasionally it fails to make this distinction and reacts destructively against the host's own molecules; such *autoimmune diseases* can be fatal.

Although the immune system evolved to protect vertebrates from infection by microorganisms and larger parasites, most of what we know about immunity has come from studies of the responses of laboratory animals to injections of non-infectious substances, such as foreign proteins and polysaccharides. Almost any macromolecule, as long as it is foreign to the recipient, can induce an immune response; any substance capable of eliciting an immune response is referred to as an **antigen** (*anti*body *gen*erator). Remarkably, the immune system can distinguish between antigens that are very similar—such as between two proteins that differ in only a single amino acid or between two optical isomers of the same molecule.

There are two broad classes of immune responses: (1) antibody responses and (2) cell-mediated immune responses. **Antibody responses** involve the production of antibodies, which are proteins called *immunoglobulins*. The antibodies circulate

1001

n the bloodstream and permeate the other body fluids, where they bind specifi-
ally to the foreign antigen that induced them. Binding by antibody inactivates
iruses and bacterial toxins (such as tetanus or botulinum toxin) by blocking their
bility to bind to receptors on target cells. Antibody binding also marks invading
microorganisms for destruction, either by making it easier for a phagocytic cell to
ngest them or by activating a system of blood proteins, collectively called *com-
lement*, that kills the invaders.

Cell-mediated immune responses, the second class of immune responses,
nvolve the production of specialized cells that react with foreign antigens on the
urface of other host cells. The reacting cell can kill a virus-infected host cell that
as viral proteins on its surface, thereby eliminating the infected cell before the
irus has replicated. In other cases the reacting cell secretes chemical signals that
ctivate macrophages to destroy invading microorganisms.

The main challenge in immunology has been to understand how the immune
ystem specifically recognizes and reacts aggressively to a virtually unlimited num-
er of different foreign macromolecules but avoids reacting against the tens of
housands of different self macromolecules made by host cells. We begin our
iscussion of how the immune system accomplishes this by considering the cells
hat are responsible for the two types of immunity. We shall then consider in turn
he function and structure of antibodies, the complement system, and the special
eatures of cell-mediated immunity.

The Cellular Basis of Immunity

The Human Immune System Is Composed of Trillions of Lymphocytes[1]

he cells responsible for immune specificity are a class of white blood cells known
s **lymphocytes.** They are found in large numbers in the blood and the lymph
he colorless fluid in the lymphatic vessels that connect the lymph nodes in the
ody) and in specialized **lymphoid organs,** such as the thymus, lymph nodes,
pleen, and appendix (Figure 18–1).

There are $\sim 2 \times 10^{12}$ lymphocytes in the human body, which makes the
mmune system comparable in cell mass to the liver or brain. Although lympho-
ytes have long been recognized as a major cellular component of the blood, their

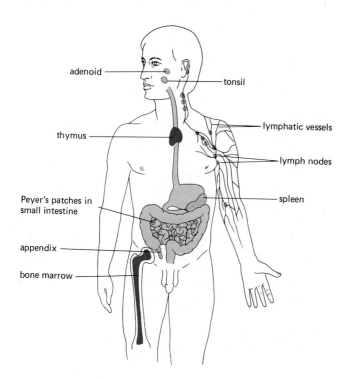

Figure 18–1 The human lymphoid organs. Lymphocytes develop in the thymus and bone marrow (*dark color*), which are therefore referred to as *primary lymphoid organs.* The newly formed lymphocytes migrate from these primary organs to *secondary lymphoid organs* (*light color*), where they can react with antigen. Only some of the secondary lymphoid organs are shown.

adenoid

tonsil

thymus

lymphatic vessels

lymph nodes

spleen

Peyer's patches in small intestine

appendix

bone marrow

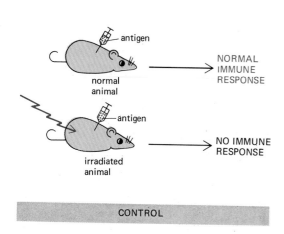

NORMAL
IMMUNE
RESPONSE

NO IMMUNE
RESPONSE

CONTROL

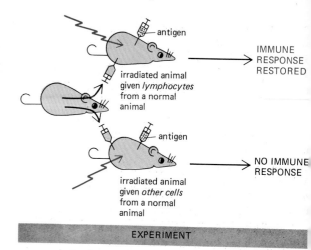

IMMUNE
RESPONSE
RESTORED

NO IMMUNE
RESPONSE

EXPERIMENT

central role in immunity was demonstrated only in the late 1950s. The proof came from experiments in which mice or rats were heavily irradiated in order to kill most of their white blood cells, including lymphocytes. Since such animals are unable to make immune responses, it is possible to transfer various types of cells into them and determine which ones reverse the deficiency. Only lymphocytes restored the immune responses of irradiated animals (Figure 18–2). Since both antibody and cell-mediated responses were restored, these experiments established that lymphocytes are responsible for both types of immune response. At the time these experiments were done, lymphocytes were among the least understood of vertebrate cells; now they are among the cells we understand best.

Figure 18–2 The classic experiment showing that lymphocytes are responsible for recognizing and responding to foreign antigens. An important feature of all such cell-transfer experiments is that cells are transferred between animals of the same *inbred strain*. Members of an inbred strain are genetically identical. If lymphocytes are transferred to a genetically different animal that has been irradiated, they react against the "foreign" antigens of the host and can kill the animal.

B Lymphocytes Make Humoral Antibody Responses; T Lymphocytes Make Cell-mediated Immune Responses[2]

During the 1960s it was discovered that the two major classes of immune responses are mediated by different classes of lymphocytes: **T cells,** which develop in the *thymus*, are responsible for cell-mediated immunity; **B cells,** which in mammals develop in the adult *bone marrow* or the fetal liver, produce antibodies. This dichotomy of the lymphoid system was initially demonstrated in animals with experimentally induced immunodeficiencies. It was found that removing the thymus from a newborn animal markedly impairs cell-mediated immune responses but has much less effect on antibody responses. In birds it was also possible to demonstrate the converse effect because B lymphocytes develop in a discrete gut-associated lymphoid organ, the *bursa of Fabricius*, that is unique to birds. Removing the bursa of Fabricius at the time of hatching impaired the bird's ability to make antibodies but had little effect on cell-mediated immunity. In addition, studies of children born with impaired immunity showed that some of these children could not make antibodies but had normal cell-mediated immunity, while others had the reverse deficiency; and those with selectively impaired cell-mediated responses almost always had thymus abnormalities.

One of the puzzling features of these studies on immunodeficient animals was that individuals deficient in T cells (because their thymus was removed at birth or was abnormal) were not only unable to make cell-mediated immune responses but also had somewhat impaired antibody responses. We now know that the antibody deficiency results because some of the T cells have a crucial regulatory role in immunity and help B cells make antibody responses.

In fact, the majority of T lymphocytes play a regulatory role in immunity, acting either to enhance or suppress the responses of other white blood cells. These cells, called *helper T cells* and *suppressor T cells*, respectively, are collectively referred to as **regulatory cells.** Other T lymphocytes, called *cytotoxic T cells*, kill virus-infected cells. Both cytotoxic T cells and B lymphocytes are involved directly in defense against infection and therefore are collectively referred to as **effector cells.**

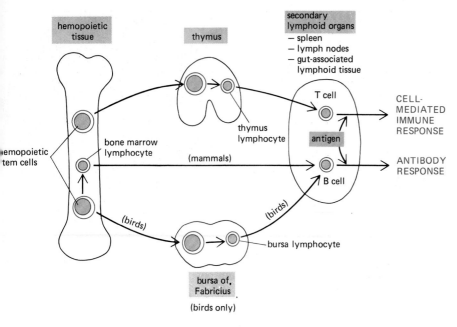

secondary
lymphoid organs
— spleen
— lymph nodes
— gut-associated
  lymphoid tissue

Figure 18–3 The development of T and B lymphocytes. In both mammals and birds, small numbers of precursor cells migrate via the blood to the thymus, where they differentiate into thymus lymphocytes. Although most of these lymphocytes die in the thymus, others migrate to secondary lymphoid organs to become thymus-derived lymphocytes (T cells). In birds, precursor cells migrate to the bursa of Fabricius, where they differentiate into bursa lymphocytes; many of these lymphocytes die, but others migrate to secondary lymphoid organs to become bursa-derived lymphocytes (B cells). In mammals the precursor cells destined to become B cells differentiate into lymphocytes in the hemopoietic tissue itself and then migrate to secondary lymphoid organs to become B cells. The terms T cell and B cell are also often used to include the lymphocytes in the thymus and bursa (or bone marrow), respectively. The stage at which precursor cells become committed to develop into T or B lymphocytes is uncertain. We will discuss why so many lymphocytes die in the primary lymphoid organs later in the chapter (see pp. 1028 and 1052).

Lymphocytes Develop in Primary Lymphoid Organs but React with Foreign Antigens in Secondary Lymphoid Organs[3]

Lymphocytes develop from *pluripotent hemopoietic stem cells*, which give rise to all of the blood cells, including red blood cells, white blood cells, and platelets (see p. 978). These stem cells are located primarily in *hemopoietic tissues*—the liver in fetuses and the bone marrow in adults. In mammals, B cells develop from stem cells in the hemopoietic tissues themselves; in birds, B cells develop in the bursa of Fabricius from precursor cells that migrate to the bursa from the hemopoietic tissues via the blood. In all vertebrates, T cells develop in the thymus from precursor cells that migrate in from the hemopoietic tissues via the blood. Because they are sites where lymphoctyes develop from precursor cells, the hemopoietic tissues, the bursa of Fabricius, and the thymus are referred to as **primary lymphoid organs** (see Figure 18–1).

Although most lymphocytes die soon after they develop in a primary lymphoid organ (see pp. 1028 and 1052), others migrate via the blood to the **secondary lymphoid organs**—mainly the lymph nodes, spleen, and gut-associated lymphoid tissues (appendix, tonsils, adenoids, and Peyer's patches in the small intestine) (see Figure 18–1). It is mainly in these secondary lymphoid organs that T cells and B cells react with foreign antigens (Figure 18–3).

Because most of the migration of lymphocytes from the thymus and bursa occurs early in development, removing these organs from *adult* animals has relatively little effect on immune responses, which is why their role in immunity remained undiscovered for so long. In contrast, the bone marrow in mammals continues to generate large numbers of new B cells ($\sim 5 \times 10^7$/day in a mouse) throughout life.

Cell-Surface Markers Make It Possible to Distinguish and Separate T and B Cells[4]

T and B cells become morphologically distinguishable only after they have been stimulated by antigen. Unstimulated ("resting") T and B cells look very similar, even in an electron microscope: both are small, only marginally bigger than red blood cells, and are largely filled by the nucleus (Figure 18–4A). Both are activated by antigen to proliferate and mature further. Activated B cells develop into antibody-secreting cells, the most mature of which are *plasma cells*, which are filled with an extensive rough endoplasmic reticulum (Figure 18–4B). In contrast, acti-

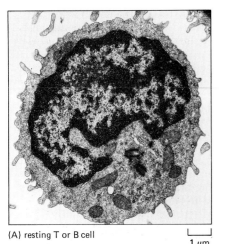

(A) resting T or B cell

1 μm

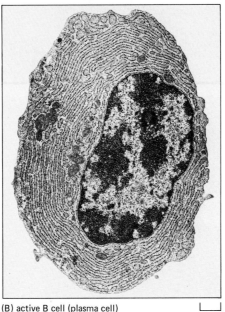

(B) active B cell (plasma cell)

1 μm

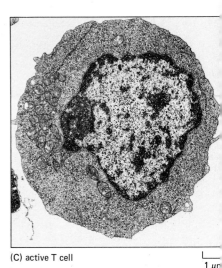

(C) active T cell

1 μm

vated T cells contain very little endoplasmic reticulum and do not secrete antibodies (Figure 18–4C).

Since both T and B lymphocytes occur in all secondary lymphoid organs, it has been necessary to find ways to distinguish and separate the two cell types and their various subtypes in order to study their individual properties. Fortunately, there are many differences in the plasma membrane glycoproteins of the different types of lymphocytes that can serve as distinguishing markers. For example, antibodies that react with the *Thy-1 glycoprotein* (see p. 1054), which is found on T but not B lymphocytes in mice, are widely used to remove or purify T cells from a mixed population of mouse lymphocytes. Similarly, antibodies against the *CD4* and *CD8 glycoproteins* (see p. 1050) are widely used to distinguish and separate helper T cells and cytotoxic T cells, respectively, in both mice and humans.

The Immune System Works by Clonal Selection[5]

The most remarkable feature of the immune system is that it can respond to millions of different foreign antigens in a highly specific way—for example, by making antibodies that react specifically with the antigen that induced their production. How can the immune system produce such a diversity of specific antibodies? One hypothesis, which dominated immunological thinking until the 1940s, was that antibodies are made as unfolded polypeptide chains whose final conformation is determined by the antigen around which they fold. At the time, this seemed the simplest explanation for the observation that animals can make antibodies specific to man-made molecules that do not exist in nature. This *instruction hypothesis*, however, had to be abandoned when protein chemists discovered that the three-dimensional folded structure of a protein molecule, such as an antibody, is determined solely by its amino acid sequence. In fact, a denatured (unfolded) antibody molecule sometimes can refold to form its original antigen-binding site even in the absence of antigen.

The instruction hypothesis was replaced in the 1950s by the **clonal selection theory,** which is based on the proposition that during development each lymphocyte becomes committed to react with a particular antigen, before ever being exposed to it. A cell expresses this commitment in the form of cell-surface receptor proteins that specifically fit the antigen. The binding of antigen to the receptors activates the cell, causing it both to proliferate and mature. Thus a foreign antigen

Figure 18–4 Electron micrographs of (A) a resting lymphocyte, (B) an active B cell, and (C) an active T cell. The resting lymphocyte could be a T cell or a B cell, for these cells are difficult to distinguish morphologically until they have been activated. The active B cell (a plasma cell) is filled with an extensive rough endoplasmic reticulum (ER) distended with antibody molecules, while the active T cell has relatively little rough ER but is filled with free ribosomes. The three cells are shown at the same magnification. (A, courtesy of Dorothy Zucker-Franklin; B, courtesy of Carlo Grossi; A and B, from D. Zucker-Franklin et al., Atlas of Blood Cells: Function and Pathology, 2nd ed. Milan, Italy: Edi. Ermes, 1988; C, courtesy of Stefanello de Petris.)

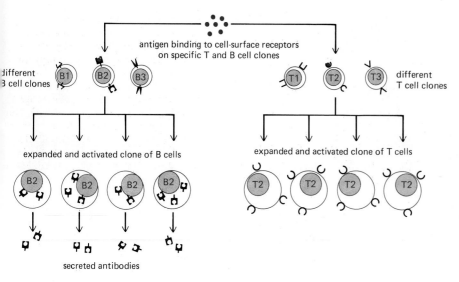

Figure 18–5 The clonal selection theory. An antigen activates only those T and B cell clones that are already committed to respond to it. The immune system is thought to consist of millions of different lymphocyte clones, hundreds of which may be activated by a particular antigen (see below). We shall see later that individual lymphocyte clones rarely respond autonomously to antigen, as shown here and as originally suggested by the clonal selection theory; their responses are usually regulated by interactions with other lymphocyte clones (see p. 1047). Moreover, T cells do not respond to free antigen as shown; they only respond to antigen that is bound to the surface of a host cell (see p. 1037).

selectively stimulates those cells that express complementary antigen-specific receptors and are thus already committed to respond to it. This is what makes immune responses antigen-specific (Figure 18–5).

The term "clonal" in clonal selection derives from the postulate that the immune system is composed of millions of different families, or *clones*, of cells, each consisting of T or B lymphocytes descended from a common ancestor. Since each ancestral cell is already committed to make one particular antigen-specific receptor protein, all cells in a clone have the same antigen specificity. Thus, according to the clonal selection theory, the immune system functions on the "ready-made" rather than the "made-to-measure" principle. The question of how an animal makes so many different antibodies therefore becomes a problem of genetics rather than one of protein chemistry.

There is compelling evidence to support the main tenets of the clonal selection theory. When lymphocytes from an animal that has not been immunized, for example, are incubated in a test tube with any of a number of radioactively labeled antigens—say, A, B, C, and D—only a very small proportion (<0.01%) bind each antigen, suggesting that only a few cells can respond to A, B, C, or D. This interpretation is confirmed by making antigen A so highly radioactive that any cell that binds it is lethally irradiated; the remaining population of lymphocytes is then no longer able to produce an immune response to A but can still respond normally to antigen B, C, or D. The same effect can be achieved by constructing an affinity column (see p. 168) of glass beads coated with antigen A and then passing the lymphocytes through the column. The cells with receptors for A stick to the beads, while other cells pass through; as a result, the cells that emerge from the column no longer respond to A but do respond normally to other antigens (Figure 18–6).

These two experiments indicate that (1) lymphocytes are committed to respond to a particular antigen before they have been exposed to it, and (2) the committed lymphocytes have receptors on their surface that specifically bind the antigen. Two major predictions of the clonal selection theory are therefore confirmed. Although almost all of the experiments of this kind have involved B cells and antibody responses, other experiments indicate that T-cell-mediated responses also operate by clonal selection.

Most Antigens Stimulate Many Different Lymphocyte Clones[6]

Most macromolecules, including virtually all proteins and most polysaccharides, can serve as antigens. Those parts of an antigen that combine with the antigen-binding site on an antibody molecule or on a lymphocyte receptor are called **antigenic determinants** (or *epitopes*). Molecules that bind specifically to such an antigen-binding site but cannot induce immune responses are called **haptens.** Haptens are usually small organic molecules; although they are too small to elicit

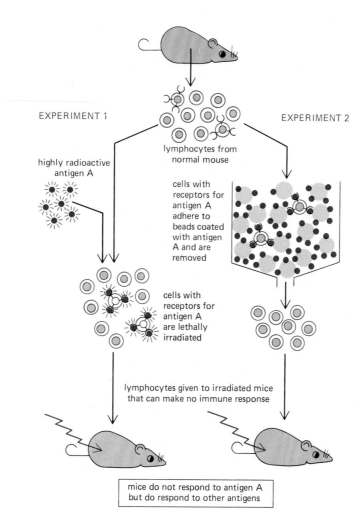

Figure 18–6 Two types of experiments that support the clonal selection theory. For simplicity, cell-surface receptors are shown only on those lymphocytes committed to respond to antigen A; in fact, however, all T and B lymphocytes have antigen-specific receptors on their surface. The experiments shown have been carried out mainly with B cells, since T cells recognize an antigen only when it is bound to the surface of a host cell (see p. 1037).

a response on their own, they become antigenic if they are coupled to a suitable macromolecule, called a *carrier*. Haptens have been important tools in experimental immunology. One of the most commonly used is the *dinitrophenyl* (*DNP*) group, which is usually coupled to a protein in order to make it antigenic (Figure 18–7).

Most antigens have a variety of antigenic determinants that stimulate the production of antibodies or T cell responses. Some determinants are more *immunogenic* (immunity-inducing) than others, so that the reaction to them may dominate the overall response; such determinants are said to be *immunodominant*.

As one might expect of a system that works by clonal selection, even a single antigenic determinant will, in general, activate many clones, each of which produces an antigen-binding site with its own characteristic affinity for the determinant. Even the relatively simple structure of the DNP group, for example, can be "looked at" in many ways. When it is coupled to a protein carrier, it usually stimulates the production of hundreds of species of anti-DNP antibodies, each made by a different B cell clone. Such responses are said to be *polyclonal*. When only a few clones respond, the response is said to be *oligoclonal*; and when the total response is made by a single B or T cell clone, it is said to be *monoclonal*. The responses to most antigens are polyclonal.

Even an antigen that activates many clones will stimulate only a tiny fraction of the total lymphocyte population. To ensure that these few lymphocytes are exposed to the antigen, antigens are collected in secondary lymphoid organs, through which T and B lymphocytes continuously recirculate. Antigens that enter through the gut are trapped by gut-associated lymphoid tissues, those that enter through the skin or respiratory tract are transported via the lymph to local lymph nodes, and those that enter the blood are filtered out in the spleen.

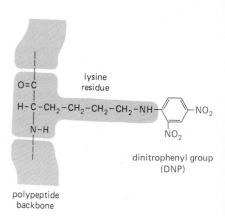

Figure 18–7 The simple hapten DNP, shown coupled covalently to a lysine side chain on a protein. Only when it is coupled to such a macromolecular carrier can a hapten induce an immune response.

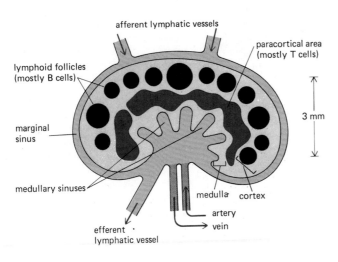

afferent lymphatic vessels

paracortical area
(mostly T cells)

lymphoid follicles
(mostly B cells)

3 mm

marginal
sinus

medullary sinuses

medulla   cortex

artery

vein

efferent
lymphatic vessel

Figure 18–8 A highly simplified drawing of a human lymph node. B lymphocytes are located primarily in the cortex, where they are clustered in structures called *lymphoid follicles*. T lymphocytes are found mainly in the *paracortical area*. Both types of lymphocytes enter the lymph node from the blood via specialized small veins in the paracortical area (not shown). T cells remain in this area, while B cells migrate to the lymphoid follicles. Eventually both T cells and B cells migrate to the medullary sinuses and leave the node via the efferent lymphatic vessel. This vessel ultimately empties into the bloodstream, allowing the lymphocytes to begin another cycle of circulation through a secondary lymphoid organ.

Foreign antigens that enter the lymph node are displayed on the surface of specialized *antigen-presenting cells*: one type presents antigen (in the form of antigen-antibody complexes) to B cells in the lymphoid follicles; another type presents antigen to T cells in the paracortical area (see p. 1045).

Most Lymphocytes Continuously Recirculate[7]

The majority of T and B lymphocytes continuously recirculate between the blood and the secondary lymphoid organs. In a lymph node, for example, lymphocytes leave the bloodstream, squeezing out between specialized endothelial cells; after percolating through the node, they accumulate in small lymphatic vessels that leave the node and connect with other lymphatic vessels, which then pass through other lymph nodes downstream (Figure 18–8). Passing into larger and larger vessels, the lymphocytes eventually enter the main lymphatic vessel (the *thoracic duct*), which carries them back into the blood. This continuous recirculation not only ensures that the appropriate lymphocytes will come into contact with antigen, it also ensures that appropriate lymphocytes encounter each other: we shall see that interactions between specific lymphocytes are a crucial part of most immune responses.

Lymphocyte recirculation depends on specific interactions between the lymphocyte cell surface and the surface of specialized endothelial cells lining small veins (called *postcapillary venules*) in the secondary lymphoid organs: of all the cell types in the blood that come into contact with these endothelial cells, only lymphocytes transiently adhere and then migrate through the postcapillary venules. Monoclonal antibodies (see p. 178) that bind to the surface of lymphocytes and inhibit their ability both to bind to the specialized endothelial cells in tissue sections of secondary lymphoid organs and to recirculate *in vivo* are helping to define the various "*homing receptors*" involved in lymphocyte migration. The majority of T and B cells have one type of glycoprotein on their surface that is required for them to recirculate through lymph nodes and another that is required for them to recirculate through Peyer's patches. Some lymphocytes have only the latter glycoprotein and recirculate selectively through Peyer's patches; they constitute, in effect, a gut-specific subsystem of lymphocytes specialized for responding to antigens that enter the body from the intestine. Other homing receptors on lymphocytes are presumably responsible for the segregation of T and B cells into distinct areas inside a lymphoid organ (see Figure 18–8). When they are activated by antigen, lymphocytes lose the homing receptors that mediate recirculation through lymphoid organs and acquire new ones that guide the activated cells to sites of inflammation.

Immunological Memory Is Due to Clonal Expansion and Lymphocyte Maturation[8]

The immune system, like the nervous system, can remember. This is why we develop lifelong immunity to many common viral diseases after our initial exposure to the virus. The same phenomenon can be demonstrated in experimental animals. If an animal is injected once with antigen A, its immune response (either antibody or cell-mediated) will appear after a lag period of several days, rise rapidly and exponentially, and then, more gradually, fall again. This is the characteristic

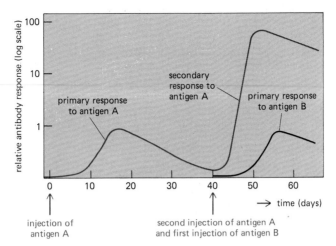

Figure 18–9 Primary and secondary antibody responses induced by a first and second exposure, respectively, to antigen A. Note that the secondary response is faster and greater than the primary response and is specific for A, indicating that the immune system has specifically "remembered" encountering antigen A before. Evidence for the same type of immunological memory is obtained if T-cell-mediated responses rather than B cell antibody responses are measured.

course of a **primary immune response,** occurring on an animal's first exposure to an antigen. If some weeks or months or even years are allowed to pass and the animal is reinjected with antigen A, it will produce a **secondary immune response** that is very different from the primary response: the lag period is shorter, the response is greater, and its duration is longer (Figure 18–9). These differences indicate that the animal has "remembered" its first exposure to antigen A. If the animal is given a different antigen (for example, antigen B) instead of a second injection of antigen A, the response is typical of a primary, and not a secondary, immune response; therefore the secondary response reflects antigen-specific **immunological memory** for antigen A.

The clonal selection theory provides a useful conceptual framework for understanding the cellular basis of immunological memory. In a mature animal the T and B cells in the secondary lymphoid organs are a mixture of cells in at least three stages of maturation, which can be designated *virgin cells, memory cells,* and *active cells.* When **virgin cells** encounter antigen for the first time, some of them are stimulated to multiply and become **active cells,** which we define as cells that are actively engaged in making a response (active T cells carry out cell-mediated responses, while active B cells secrete antibody). Some virgin cells, on the other hand, are stimulated to multiply and mature instead into **memory cells**—cells that do not themselves make a response but are readily induced to become active cells by a later encounter with the same antigen (Figure 18–10). Virgin lymphocytes are thought to survive in secondary lymphoid tissues for only a short time, probably dying within days unless they meet their specific antigen. Memory cells, on the other hand, are thought to live for many months or even years without dividing—continuously recirculating between the blood and secondary lymphoid organs. Moreover, memory cells respond more readily to antigen than do virgin cells. We shall see later (see p. 1026) that one reason for the increased responsiveness of memory B cells is that their receptors have a higher affinity for antigen.

According to this scheme, immunological memory is generated during the primary response because (1) the proliferation of antigen-triggered virgin cells

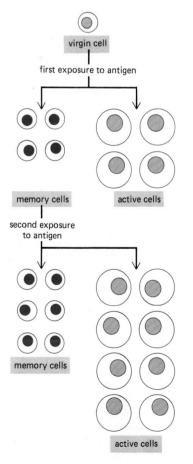

Figure 18–10 When virgin T or B cells are stimulated by their specific antigen, they proliferate and mature; some become activated to make a response, while others become memory cells. During a subsequent exposure to antigen, the memory cells respond more readily than did the virgin cells: they proliferate and give rise to active cells and to more memory cells. In the model shown, an individual virgin cell can give rise to either a memory cell or an activated cell, depending on the conditions. In an alternative model (not shown) the virgin cells that mature into memory cells are different from those that mature into activated cells. It is not known which of these models is correct.

creates many memory cells—a process known as *clonal expansion;* (2) the memory cells have a much longer life-span than virgin cells and recirculate between the blood and secondary lymphoid organs; and (3) each memory cell is able to respond more readily to antigen than does a virgin cell. Because of the changes induced during a primary response, most of the long-lived cells in the recirculating pool of lymphocytes are appropriate to the antigenic environment of the animal and are already primed and ready for action.

The Failure to Respond to Self Antigens Is Due to Acquired Immunological Tolerance[9]

How is the immune system able to distinguish foreign molecules from self molecules? One possibility might be that an animal inherits genes that encode receptors for foreign antigens but not self antigens, so that its immune system is genetically constituted to respond only to foreign antigens. Alternatively, the immune system could be inherently capable of responding to both foreign and self antigens but could "learn" early in development not to respond to self antigens. The latter explanation has been shown to be correct. The first evidence for this was an observation made in 1945. Normally, when tissues are transplanted from one individual to another, they are recognized as foreign by the immune system of the recipient and are destroyed. Dizygotic cattle twins, however, which develop from two fertilized ova and are therefore nonidentical, sometimes exchange blood cells *in utero* as a result of the spontaneous fusion of their placentas; such twins were shown to accept skin grafts from each other. These findings were later reproduced experimentally—in chicks, by allowing the blood vessels of two embryos to fuse, and in mice, by introducing cells from one strain of mouse into a neonatal mouse of another strain, where they survived for most of the recipient animal's life. In both cases, when the animals matured, grafts from the joined or donor animal were accepted (Figure 18–11), while "third-party" grafts from a different animal were rejected. Thus the continuous presence of nonself antigens from before the time the immune system matures leads to a long-lasting unresponsiveness to the specific nonself antigens. The resulting state of antigen-specific immunological unresponsiveness is known as **acquired immunological tolerance.**

There is strong evidence that the unresponsiveness of an animal's immune system to its own macromolecules (*natural immunological tolerance*) is acquired in the same way and is not inborn. Normal mice, for example, cannot make an immune response against their own blood complement protein C5 (see p. 1032), but mutant mice that lack the gene encoding C5 (but are otherwise genetically identical to the normal mice) can make an immune response to this protein. Thus it is clear that the immune system is genetically capable of responding to self but learns not to do so. In some cases at least, the learning process involves eliminating the self-reactive lymphocytes (see p. 1051), although it is not known how this is achieved. Many self-reactive lymphocytes are thought to be eliminated in the pri-

Figure 18–11 The skin graft seen here, transplanted from an adult brown mouse to an adult white mouse, has survived for many weeks only because the latter was made immunologically tolerant by injecting cells from the brown mouse into it at the time of birth. (Courtesy of Leslie Brent, from I. Roitt, Essential Immunology, 6th ed. Oxford, U.K.: Blackwell Scientific, 1988.)

mary lymphoid organs when they encounter their antigen. This negative response to antigen could be due to a special environment in these organs or to a unique responsiveness of a newly formed lymphocyte. Presumably because new self-reactive lymphocytes continue to be produced from stem cells throughout life, maintaining self-tolerance requires the constant presence of the self antigens. If an antigen such as C5 is removed, an animal regains the ability to respond to it within weeks or months.

Tolerance to self antigens sometimes breaks down, causing T or B cells (or both) to react against their own tissue antigens. *Myasthenia gravis* is an example of such an **autoimmune disease.** Affected individuals make antibodies against the acetylcholine receptors on their own skeletal muscle cells (see p. 319); the antibodies interfere with the normal functioning of the receptors so that such patients become weak and can die because they cannot breathe.

Immunological Tolerance to Foreign Antigens Can Be Induced in Adults[10]

It is generally much more difficult to induce immunological tolerance to foreign antigens in adult than in immature animals. But with some antigens it can be done experimentally by injecting the antigen (1) in very high doses, (2) in repeated very low doses, (3) together with an immunosuppressive drug, or (4) intravenously, after the antigen has been chemically coupled to the surface of B lymphocytes or ultracentrifuged to remove all aggregates, so that the normal mechanisms of antigen presentation (see p. 1046) are ineffective. Thus binding an antigen to its complementary receptors on a T or B lymphocyte can stimulate the lymphocyte to divide and mature to become an active cell or a memory cell, or it can eliminate or inactivate the lymphocyte, causing tolerance. The molecular mechanisms that determine the outcome are not well understood, but whether an antigen activates or induces tolerance depends largely on (1) the maturity of the lymphocyte, (2) the nature and concentration of the antigen, and (3) complex interactions between different classes of lymphocytes and between lymphocytes and specialized *antigen-presenting cells*, which will be discussed in a later section.

Summary

The immune system evolved to defend vertebrates against infection. It is composed of millions of lymphocyte clones. The lymphocytes in each clone share a unique cell-surface receptor that enables them to bind a particular "antigenic determinant" consisting of a specific arrangement of atoms on a part of a molecule. There are two classes of lymphocytes: B cells, which make antibodies, and T cells, which make cell-mediated immune responses.

Beginning early in lymphocyte development, many lymphocytes that would react against antigenic determinants on self macromolecules are eliminated or inactivated; as a result, the immune system normally reacts only to foreign antigens. Binding a foreign antigen to a lymphocyte initiates a response by the cell that helps to eliminate the antigen. As part of the response, some lymphocytes proliferate and mature into long-lived memory cells, so that the next time the same antigen is encountered, the immune response to it is faster and stronger.

The Functional Properties of Antibodies[11]

Vertebrates rapidly die of infection if they are unable to make antibodies. Antibodies defend us against infection by inactivating viruses and bacterial toxins and by recruiting the complement system and various types of white blood cells to kill invading microorganisms and larger parasites. Synthesized exclusively by B lymphocytes, antibodies are produced in millions of forms, each with a different amino acid sequence and a different binding site for antigen. Collectively called **immunoglobulins** (abbreviated as **Ig**), they are among the most abundant protein components in the blood, constituting about 20% of the total plasma protein by weight. In this section we shall describe the five classes of antibodies found in higher vertebrates, each of which mediates a characteristic biological response following antigen binding.

The Antigen-specific Receptors on B Cells Are Antibody Molecules[12]

As predicted by the clonal selection theory, all antibody molecules made by an individual B cell have the same antigen-binding site. The first antibodies made by a newly formed B cell are not secreted; instead, they are inserted into the plasma membrane, where they serve as receptors for antigen. Each B cell has approximately 10^5 such antibody molecules in its plasma membrane.

When antigen binds to the antibody molecules on the surface of a virgin or a memory B cell, it usually initiates a complicated series of events culminating in cell proliferation and maturation to produce either memory cells or active (antibody-secreting) cells (see p. 1047). The active cells make large amounts of soluble (rather than membrane-bound) antibody with the same antigen-binding site as the cell-surface antibody and secrete it into the blood. Active B cells can begin secreting antibody while they are still small lymphocytes, but the end stage of the maturation pathway is the large plasma cell (see Figure 18–4B), which secretes antibodies at the rate of about 2000 molecules per second. Plasma cells seem to have committed so much of their protein-synthesizing machinery to making antibody that they are incapable of further growth and division and die after several days.

B Cells Can Be Stimulated to Make Antibodies in a Culture Dish[13]

Two advances in the 1960s revolutionized research on B cells. The first was the development of the **hemolytic plaque assay,** which made it possible to identify and count individual active B cells secreting antibody against a specific antigen. In the simplest form of this assay, lymphocytes (commonly from the spleen) are taken from animals that have been immunized with sheep red blood cells (SRBCs). They are then embedded in agar together with an excess of SRBCs so that the dish contains a "lawn" of immobilized SRBCs with occasional lymphocytes in it. Under these conditions the cells are unable to move, but any anti-SRBC antibody secreted by a B cell will diffuse outward and coat all SRBCs in the vicinity of the secreting cell. Once the SRBCs are coated with antibody, they can be killed by adding complement (see p. 1031). In this way the presence of each antibody-secreting cell is indicated by the presence of a clear spot, or *plaque*, in the opaque layer of SRBCs. The same assay can be used to count cells making antibody to other antigens, such as proteins or polysaccharides, if these antigens are coupled to the surface of the SRBC.

The second important advance was the demonstration that B lymphocytes can be induced to make antibody by exposing them to antigen in culture, where cell interactions can be manipulated and the environment controlled. This led to the discovery that both T lymphocytes and specialized *antigen-presenting cells* (see p. 1045) are required for most antigens to stimulate B lymphocytes to secrete antibodies; the cell-cell interactions involved will be described later (see p. 1047).

Antibodies Have Two Identical Antigen-binding Sites[11]

The simplest antibodies are Y-shaped molecules with two identical **antigen-binding** sites, one at the tip of each arm of the Y (Figure 18–12). Because of their two antigen-binding sites, they are said to be *bivalent*. Such antibody molecules can cross-link antigen molecules into a large lattice as long as each antigen molecule has three or more antigenic determinants (Figure 18–13). Once it reaches a certain size, such a lattice precipitates out of solution. This tendency of large immune complexes to precipitate can be used to detect the presence of antibodies and antigens. The efficiency of antigen binding and cross-linking is greatly increased by a flexible *hinge region* in antibodies, which allows the distance between the two antigen-binding sites to vary (Figure 18–14).

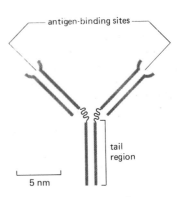

Figure 18–12 A simple representation of an antibody molecule.

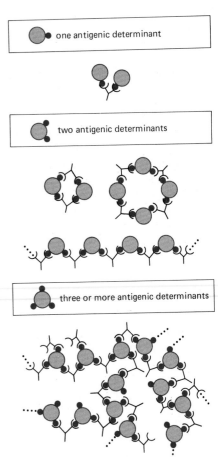

Figure 18–13 Because antibodies have two identical antigen-binding sites, they can cross-link antigens. The types of antibody-antigen complexes that form depend on the number of antigenic determinants on the antigen. Here a single species of antibody (a monoclonal antibody) is shown binding to antigens containing one, two, or three copies of a single type of antigenic determinant. Antigens with two antigenic determinants can form small cyclic complexes or linear chains with antibody, while antigens with three or more antigenic determinants can form large three-dimensional lattices that readily precipitate.

The protective effect of antibodies is not due simply to their ability to bind antigen. They engage in a variety of activities that are mediated by the tail of the Y-shaped molecule. This part of the molecule determines what will happen to the antigen once it is bound. Because of the way immunoglobulins are synthesized (see p. 1029), antibodies with the same antigen-binding sites can have any one of several different tail regions, each of which confers on the antibody different functional properties, such as the ability to activate complement (see p. 1032) or to bind to phagocytic cells (see p. 1015).

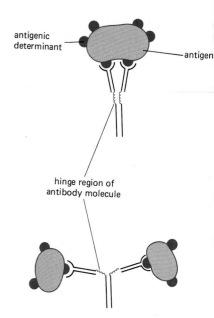

Figure 18–14   The hinge region of an antibody molecule improves the efficiency of antigen binding and cross-linking.

An Antibody Molecule Is Composed of Two Identical Light Chains and Two Identical Heavy Chains[14]

The basic structural unit of an antibody molecule consists of four polypeptide chains, two identical **light (L) chains** (each containing about 220 amino acids) and two identical **heavy (H) chains** (each usually containing about 440 amino acids). The four chains are held together by a combination of noncovalent and covalent (disulfide) bonds. The molecule is composed of two identical halves, each with the same antigen-binding site, and both L and H chains usually cooperate to form the antigen-binding surface (Figure 18–15).

The proteolytic enzymes papain and pepsin split antibody molecules into different characteristic fragments. *Papain* produces two separate and identical **Fab** (*fragment antigen binding*) **fragments,** each with one antigen-binding site, and one **Fc fragment** (so called because it readily crystallizes). *Pepsin,* on the other hand, produces one **F(ab')₂ fragment,** so called because it consists of two covalently linked F(ab') fragments (each slightly larger than a Fab fragment); the rest of the molecule is broken down into smaller fragments (Figure 18–16). Because F(ab')₂ fragments are bivalent, they can still cross-link antigens and form precipitates, unlike the univalent Fab fragments. Neither of these fragments has the other biological properties of intact antibody molecules because they lack the tail (Fc) region that is responsible for these properties.

There Are Five Classes of H Chains, Each with Different Biological Properties[11,15]

In higher vertebrates there are five *classes* of antibodies, IgA, IgD, IgE, IgG, and IgM, each with its own class of H chain—α, δ, ε, γ, and μ, respectively; IgA molecules have α chains, IgG molecules have γ chains, and so on (Table 18–1). In

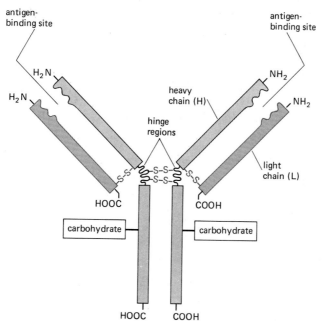

Figure 18–15   A typical antibody molecule is composed of two identical heavy (H) chains and two identical light (L) chains. Note that the antigen-binding sites are formed by a complex of the amino-terminal regions of both L and H chains, but the tail and hinge regions are formed by the H chains alone. Each H chain contains one or more oligosaccharide (carbohydrate) chains of unknown function.

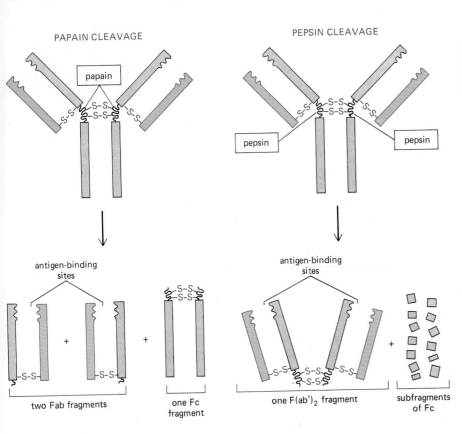

PAPAIN CLEAVAGE

papain

antigen-binding sites

two Fab fragments + one Fc fragment

PEPSIN CLEAVAGE

pepsin   pepsin

antigen-binding sites

one F(ab')₂ fragment + subfragments of Fc

Figure 18–16 The fragments produced when antibody molecules are cleaved with the proteolytic enzymes papain and pepsin.

addition, there are a number of subclasses of IgG and IgA immunoglobulins; for example, there are four human IgG subclasses (IgG1, IgG2, IgG3, and IgG4) having γ_1, γ_2, γ_3, and γ_4 heavy chains, respectively. The various H chains impart a distinctive conformation to the hinge and tail regions of antibodies and give each class (and subclass) characteristic properties of its own (see Table 18–1).

IgM is always the first class of antibody produced by a developing B cell, although many B cells eventually switch to making other classes of antibody (see p. 1029). The immediate precursor of a B cell, called a **pre-B cell,** initially makes only μ chains, which accumulate in the cell. Eventually the cell begins to synthesize light chains as well; these combine with the μ chains to form four-chain IgM molecules (each with two μ chains and two light chains), which become inserted into the plasma membrane. The cell now has cell-surface receptors with which it can bind antigen, and at this point it is called a *virgin B lymphocyte.* Many virgin B cells soon start to produce cell-surface **IgD** molecules as well, with the same antigen-binding site as the IgM molecules.

Table 18–1 Properties of the Major Classes of Antibody in Humans

Properties	Class of Antibody				
	IgM	**IgD**	**IgG**	**IgA**	**IgE**
Heavy chains	μ	δ	γ	α	ϵ
Light chains	κ or λ	κ or λ	κ or λ	κ or λ	κ or λ
Number of four-chain units	5	1	1	1 or 2	1
Percent of total Ig in blood	10	< 1	75	15	< 1
Activates complement	+ + + +	–	+ +	–	–
Crosses placenta	–	–	+	–	–
Binds to macrophages and neutrophils	–	–	+	–	–
Binds to mast cells and basophils	–	–	–	–	+

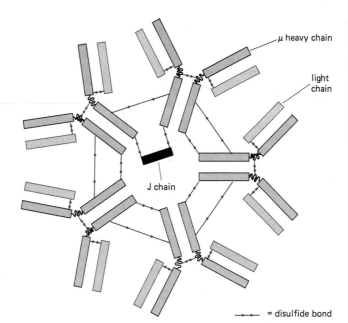

μ heavy chain

light chain

J chain

———•—— = disulfide bond

Figure 18–17 A pentameric IgM molecule. The five subunits are held together by disulfide bonds. A single J chain (~15,000 daltons), disulfide-bonded between two μ heavy chains, closes the ring structure and prevents further polymerization. The J chain is homologous to a single Ig domain (see p. 1020).

IgM is not only the first class of antibody to appear on the surface of a developing B cell, it is also the major class secreted into the blood in the early stages of a *primary* antibody response. In its secreted form, IgM is composed of five four-chain units and thus has a total of 10 antigen-binding sites. Each pentamer contains one copy of another polypeptide chain, called a *J (joining) chain,* which is produced by IgM-secreting cells and is covalently inserted between two adjacent Fc regions, where it closes the ring-shaped oligomer and prevents further polymerization (Figure 18–17).

The binding of antigen to the Fab regions of the secreted pentameric IgM molecule induces the Fc regions to bind to and thereby activate the first component of the *complement system.* When the antigen is on the surface of an invading microorganism, the resulting activation of the complement system unleashes a biochemical attack that kills the microorganism (see p. 1031). Unlike IgM, IgD molecules are rarely secreted by an active B cell, and their functions—other than as receptors for antigen—are unknown.

The major class of immunoglobulin in the blood is **IgG,** which is produced in large quantities during *secondary* immune responses. Besides activating the complement system, the Fc region of an IgG molecule binds to specific receptors on macrophages and neutrophils. Largely by means of such **Fc receptors,** these phagocytic cells bind, ingest, and destroy infecting microorganisms that have become coated with the IgG antibodies produced in response to the infection (Figure 18–18). Various types of white blood cells that express Fc receptors can also kill IgG-coated foreign eucaryotic cells without phagocytosing them. This process, called *antibody-dependent cell-mediated killing,* can be carried out by macrophages, neutrophils, and eosinophils (see below), as well as by killer cells (*K cells*), which are lymphocytelike cells apparently specialized for killing abnormal host cells (see p. 1039).

IgG molecules are the only antibodies that can pass from mother to fetus via the placenta. Cells of the placenta that are in contact with maternal blood have Fc receptors that bind IgG molecules and mediate their passage to the fetus. The antibodies are first ingested by receptor-mediated endocytosis and then transported across the cell in vesicles and released by exocytosis into the fetal blood (a process called *transcytosis,* see p. 332). Other classes of antibodies do not bind to these receptors and therefore cannot pass across the placenta.

IgA is the principal class of antibody in secretions (milk, saliva, tears, and respiratory and intestinal secretions). It exists mainly as either a four-chain monomer (like IgG) or a dimer of two such units, carrying a single J chain and a chain

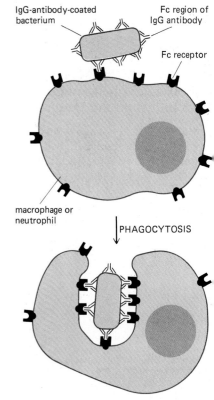

IgG-antibody-coated bacterium

Fc region of IgG antibody

Fc receptor

macrophage or neutrophil

PHAGOCYTOSIS

Figure 18–18 An IgG-antibody-coated bacterium is efficiently phagocytosed by a macrophage or neutrophil, which has cell-surface receptors able to bind the Fc region of IgG molecules. The binding of the antibody-covered bacterium to these Fc receptors activates the phagocytotic process (see p. 334).

called *secretory component* (Figure 18–19). In secretions, IgA is a dimer. It is transported from the extracellular fluid into the secreted fluid by the same kind of transepithelial transport process (transcytosis) that transfers IgG molecules from maternal blood to fetal blood. In this case the transport is mediated by a special class of Fc receptors that are present on the nonluminal surface of the epithelial cells lining the intestine, bronchi, or the milk, salivary, or tear ducts, where they bind IgA dimers present in the extracellular fluid (Figure 18–20).

The Fc region of **IgE** molecules binds with unusually high affinity ($K_a \sim 10^{10}$ liters/mole—see below) to yet another class of Fc receptors. These receptors are located on the surface of *mast cells* in tissues and on *basophils* (see p. 974) in the blood, and the IgE molecules bound to them in turn serve as receptors for antigen. Antigen binding triggers the cells to secrete a variety of biologically active amines (particularly *histamine* and, in some species, *serotonin*) (Figure 18–21). These amines cause dilation and increased permeability of blood vessels and are largely responsible for the clinical manifestations of such *allergic* reactions as hay fever, asthma, and hives. In normal circumstances the blood vessel changes are thought to help white blood cells, antibodies, and complement components to enter sites of inflammation. Mast cells also secrete factors that attract and activate a special class of white blood cells called *eosinophils* (see p. 974), which can kill various types of parasites, especially if the parasites are coated with IgG antibodies.

Antibodies Can Have Either κ or λ Light Chains but Not Both

In addition to the five classes of H chains, higher vertebrates have two types of L chains, κ and λ, either of which may be associated with any of the H chains. An individual antibody molecule always consists of identical L chains and identical H chains; therefore its antigen-binding sites are always identical. This symmetry is crucial for the cross-linking function of secreted antibodies. An Ig molecule, consequently, may have either κ or λ L chains, but never both. No difference in the biological function of these two types of L chain has yet been identified.

The Strength of an Antibody-Antigen Interaction Depends on Both the Affinity and the Number of Binding Sites[16]

The binding of an antigen to antibody, like the binding of a substrate to an enzyme, is reversible. It is mediated by the sum of many relatively weak noncovalent forces, including hydrophobic and hydrogen bonds, van der Waals forces, and ionic interactions. These weak forces are effective only when the antigen molecule is close

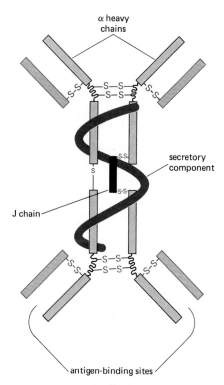

Figure 18–19 A highly schematized diagram of a dimeric IgA molecule found in secretions. In addition to the two IgA monomers that are disulfide-bonded through one of their α heavy chains, there is a single J chain and an additional polypeptide chain of 70,000 daltons called the *secretory component*, which is thought to protect the IgA molecules from being digested by proteolytic enzymes in the secretions.

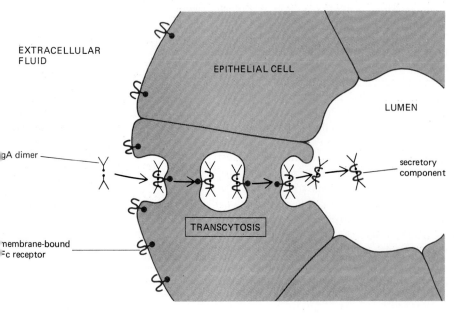

Figure 18–20 The mechanism of transport of a dimeric IgA molecule across an epithelial cell. The IgA binds to a specialized transmembrane Fc receptor protein on the nonluminal surface of the epithelial cell. The receptor-IgA complexes are ingested by receptor-mediated endocytosis, transferred across the epithelial cell cytoplasm in vesicles, and secreted into the lumen on the opposite side of the cell by exocytosis. When exposed to the lumen, the part of the Fc receptor protein that is bound to the IgA dimer (the *secretory component*) is cleaved from its transmembrane tail, thereby releasing the antibody as a complex into the lumen.

The IgA dimers enter the extracellular fluid in secretory organs from two sources. They are produced locally by IgA-secreting plasma cells in these organs, and they are produced in the spleen and lymph nodes, from where they travel in the bloodstream, leaking out of capillaries in various tissues.

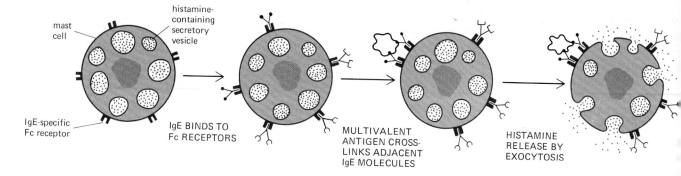

enough to allow some of its atoms to fit into complementary recesses on the surface of the antibody. The complementary regions of a four-chain antibody unit are its two identical antigen-binding sites; the corresponding region on the antigen is an *antigenic determinant* (Figure 18–22). Most antigenic macromolecules have many different antigenic determinants; if two or more of them are identical (as in a polymer with a repeating structure), the antigen is said to be *multivalent* (Figure 18–23).

The reversible binding reaction between an antigen with a single antigenic determinant (denoted Ag) and a single antigen-binding site (denoted Ab) can be expressed as

$$Ag + Ab \rightleftharpoons AgAb$$

The equilibrium point depends both on the concentrations of Ab and Ag and on the strength of their interaction. Clearly, a larger fraction of Ab will become associated with Ag as the concentration of Ag is increased. The strength of the interaction is generally expressed as the **affinity constant (K_a)** (see Figure 3–7, p. 94), where

$$K_a = [AgAb]/[Ag][Ab]$$

(the square brackets indicate the concentration of each component at equilibrium).

The affinity constant, sometimes called the *association constant*, can be determined by measuring the concentration of free Ag required to fill half of the antigen-binding sites on the antibody. When half the sites are filled, $[AgAb] = [Ab]$ and $K_a = 1/[Ag]$. Thus the reciprocal of the antigen concentration that produces half-maximal binding is equal to the affinity constant of the antibody for the antigen. Common values range from as low as 5×10^4 to as high as 10^{11} liters/mole. The affinity constant at which an immunoglobulin molecule ceases to be considered an antibody for a particular antigen is somewhat arbitrary, but it is unlikely that an antibody with a K_a below 10^4 would be biologically effective; moreover, B cells with receptors that have such a low affinity for an antigen are unlikely to be activated by the antigen.

Figure 18–21 Mast cells (and basophils) passively acquire cell-surface receptors that bind antigen. IgE antibodies secreted by active B lymphocytes enter the tissues and bind to Fc receptor proteins on the mast cell surface that specifically recognize the Fc region of these antibodies. Thus, unlike B cells, individual mast cells (and basophils) have cell-surface antibodies with a variety of antigen-binding sites. When an antigen molecule binds to these membrane-bound IgE antibodies so as to cross-link them to their neighbors, it activates the mast cell to release its histamine by exocytosis.

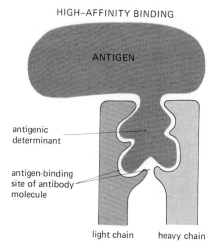

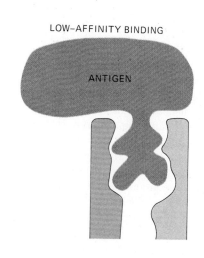

Figure 18–22 Highly schematized diagram of the binding of an antigenic determinant on a macromolecule to the antigen-binding site of two different antibody molecules, one of high and one of low affinity. The antigenic determinant is held in the binding site by various weak noncovalent forces. Note that both the light and heavy chains of the antibody molecule usually contribute to the antigen-binding site.

The **affinity** of an antibody reflects the strength of binding of an antigenic [de]terminant to a single antigen-binding site, and it is independent of the number [of] sites. However, the total **avidity** of an antibody for a multivalent antigen, such [as] a polymer with repeating subunits, is defined as the total binding strength of [all] of its binding sites together. When a multivalent antigen combines with more [th]an one antigen-binding site on an antibody, the binding strength is greatly in[cr]eased because all the antigen-antibody bonds must be broken simultaneously [be]fore the antigen and antibody can dissociate. Thus a typical IgG molecule will [bi]nd at least 1000 times more strongly to a multivalent antigen if both antigen-[bi]nding sites are engaged than if only one site is engaged.

For the same reason, if the affinity of the antigen-binding sites in an IgG and [an] IgM molecule is the same, the IgM molecule (with 10 binding sites) will have [a v]ery much greater avidity for a multivalent antigen than an IgG molecule (which [ha]s two sites). This difference in avidity, often 10^4-fold or more, is important be[ca]use antibodies produced early in an immune response usually have much lower [af]finities than those produced later. (The increase in the average affinity of anti[bo]dies produced with time after immunization is called *affinity maturation*—see [p.] 1026.) Because of its high total avidity, IgM—the major Ig class produced early [in] immune responses—can function effectively even when each of its binding sites [ha]s only a low affinity.

Summary

[A] typical antibody molecule is a Y-shaped protein with two identical antigen-binding [sit]es at the tips of the Y (the Fab regions) and binding sites for complement com[po]nents and/or various cell-surface receptors on the stem of the Y (the Fc region). [An]tibodies defend vertebrates against infection by inactivating viruses and bacterial [to]xins and by recruiting complement and various cells to kill and ingest invading [m]icroorganisms.

Each B cell clone makes antibody molecules with a unique antigen-binding site. [In]itially the molecules are inserted into the plasma membrane, where they serve [as] receptors for antigen. Antigen binding to these receptors activates the B cells [(u]sually with the help of T cells) to multiply and mature either into memory cells [or] into antibody-secreting cells, which secrete antibodies with the same antigen-[bi]nding site as the membrane-bound antibodies.

Each antibody molecule is composed of two identical heavy (H) chains and two [id]entical light (L) chains. Typically, parts of both the H and L chains form the antigen-[bi]nding sites. There are five classes of antibodies (IgA, IgD, IgE, IgG, and IgM), each [wi]th a distinctive H chain (α, δ, ε, γ, and μ, respectively). The H chains also form [th]e Fc region of the antibody, which determines what other proteins will bind to [th]e antibody and therefore what biological properties the antibody class has. Either [ty]pe of L chain (κ or λ) can be associated with any class of H chain, but the type of [L] chain does not seem to influence the properties of the antibody.

The Fine Structure of Antibodies

[Be]cause antibodies exist in so many forms, in an unimmunized individual any [on]e form will constitute less than one part in a million of the Ig molecules in the [bl]ood. This fact presented immunochemists with a uniquely difficult problem in [pr]otein chemistry: how to obtain enough of any one antibody molecule to deter[m]ine its amino acid sequence and three-dimensional structure.

The problem was solved by the discovery that the cells of a type of cancer [kn]own as **multiple myeloma** (because multiple tumors develop in the bone mar[ro]w, or "myelogenous" tissues) secrete large amounts of a single species of anti[bo]dy into the patient's blood. The antibody is homogeneous, or monoclonal, because [ca]ncer usually begins with the uncontrolled growth of a single cell (see p. 1190), [an]d in multiple myeloma the single cell is an antibody-secreting plasma cell. The [an]tibody, which accumulates in the blood, is known as a **myeloma protein.**

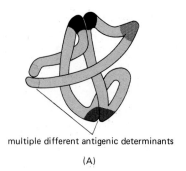

multiple different antigenic determinants

(A)

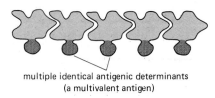

multiple identical antigenic determinants
(a multivalent antigen)

(B)

Figure 18–23 Molecules with multiple antigenic determinants. (A) A globular protein with a number of *different* antigenic determinants. Note that different regions of a polypeptide chain can come together in the folded structure to form a single antigenic determinant on the surface of the protein. (B) A polymeric structure with many *identical* antigenic determinants; such a molecule is called a *multivalent antigen*.

It had been known since the nineteenth century that the urine of patients with this disease often contains unusual proteins, called *Bence Jones proteins* after the English physician who first described them, but it was only in the 1950s that the proteins were recognized as free immunoglobulin L chains. The detailed structure of antibodies was initially determined by studying myeloma proteins from the urine or blood of patients, or from mice in which similar tumors had been purposely induced. More recently it has become possible to immortalize single antibody-secreting B cells by fusing them with non-antibody-secreting myeloma cells; the resultant *hybridomas* have provided a ready source of monoclonal antibodies, which can be produced in unlimited amounts against any desired antigen (see p. 178).

L and H Chains Consist of Constant and Variable Regions[11,14]

Comparison of the amino acid sequences of many myeloma proteins revealed a striking feature with important and surprising genetic implications. Both L and H chains have a variable sequence at their amino-terminal ends but a constant sequence at their carboxyl-terminal ends. When the amino acid sequences of many different myeloma κ chains (each about 220 amino acids long) are compared, for example, the carboxyl-terminal halves are the same or show only minor differences, whereas the amino-terminal halves are all different. Thus L chains have a **constant region** about 110 amino acids long and a **variable region** of the same size. The amino-terminal variable region of the H chains is also about 110 amino acids long, but the H-chain constant region is about 330 or 440 amino acids long, depending on the class (Figure 18–24).

It is the amino-terminal ends of the L and H chains that come together to form the antigen-binding site, and the variability of their amino acid sequences provides the structural basis for the diversity of antigen-binding sites. The existence of variable and constant regions raises important questions about the genetic mechanisms that produce antibody molecules, and we shall discuss these later. Before it became possible to investigate these genetic questions directly, other important features of antibody molecules emerged from structural studies on myeloma proteins.

The L and H Chains Each Contain Three Hypervariable Regions That Together Form the Antigen-binding Site[17]

Only part of the variable region participates directly in the binding of antigen. This conclusion was first deduced from estimates of the maximum size of an antigen-binding site. The measurements, which involved the use of oligomers of increasing size as "molecular rulers," were initially made on antibodies that react against dextran, a polymer of D-glucose. When disaccharides, trisaccharides, and higher oligosaccharides of glucose are used to inhibit the binding of dextran to anti-dextran antibodies, inhibition increases with chain length up to about six glucose units; larger oligosaccharides have no greater effect. This suggests that the largest antigen-binding sites can contact at most five or six sugar residues of the antigen. Thus it is most unlikely that all 220 amino acids of the variable regions of both L and H chains contribute directly to the antigen-binding site.

In fact, it is now clear that the binding site of antibodies is formed by only about 20 to 30 of the amino acid residues in the variable region of each chain.

Figure 18–24 Both light and heavy chains of an Ig molecule have distinct *constant* and *variable* regions. For light chains, the carboxyl-terminal halves of chains of the same type (either κ or λ) all have the same sequence (with occasional minor differences), while the amino-terminal halves are different. The amino-terminal variable region of the heavy chain, like that of the light chain, is about 110 amino acid residues long, but the constant region is three or four times the size of a light-chain constant region (depending on the antibody class).

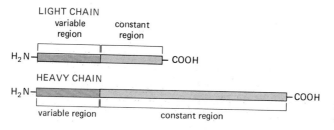

LIGHT CHAIN
variable region   constant region
H_2N—[   ]—COOH

HEAVY CHAIN
H_2N—[   ]—COOH
variable region   constant region

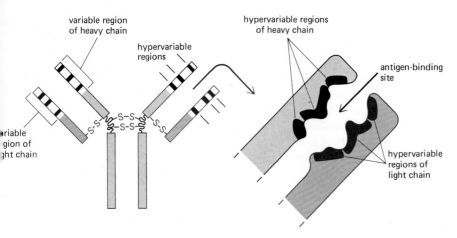

Figure 18–25 Highly schematized drawing of how the three hypervariable regions in each light and heavy chain together form the antigen-binding site of an antibody molecule. The hypervariable regions are sometimes called *complementarity-determining regions.*

This was first suggested by amino acid sequence data, which showed that the variability in the variable regions of both L and H chains is for the most part restricted to three small **hypervariable regions** in each chain. The remaining parts of the variable region, known as *framework regions*, are relatively constant. These findings led to the prediction that only the 5 to 10 amino acids in each hypervariable region form the antigen-binding site (Figure 18–25). This prediction has since been confirmed by x-ray diffraction studies of antibody molecules (see below).

The L and H Chains Are Folded into Repeating Homologous Domains[11,16,18]

With the sequencing of the first H chain, which was completed in the late 1960s, another important feature of Ig structure became apparent. The new insight came initially from the sequence of the constant region, which is about three times as long in most H chains as it is in L chains. It was found that the H-chain constant region consists of three homologous segments, each about 110 amino acids long and each containing one intrachain disulfide bond. The three segments are homologous not only to one another but also to the constant region of L chains. Similarly, the single variable domains in both the L and H chains are similar to each other and, to a lesser extent, to the constant domains.

It was correctly predicted from these findings that both L and H chains are made up of repeating segments, or **domains,** each of which folds independently to form a compact functional unit. Accordingly, as shown in Figure 18–26, an

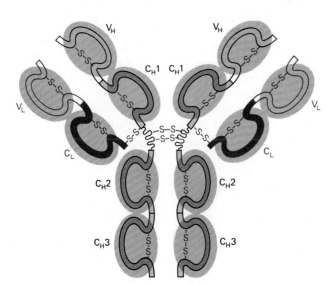

Figure 18–26 The light and heavy chains in an Ig molecule are each folded into repeating domains that are similar to one another. The variable domains of the light and heavy chains (V_L and V_H) make up the antigen-binding sites (see Figure 18–25), while the constant domains of the heavy chains (mainly C_H2 and C_H3) determine the other biological properties of the molecule. The heavy chains of IgM and IgE antibodies have an extra constant domain (C_H4).

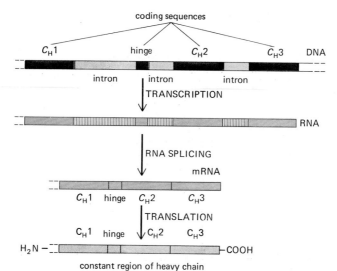

Figure 18–27 The organization of the DNA sequences that encode the constant region of an Ig heavy chain. The coding sequences (exons) for each domain and for the hinge region are separated by noncoding sequences (introns). The intron sequences are removed by splicing the primary RNA transcripts to form mRNA. The DNA that encodes the variable region of the heavy chain is not shown. The presence of introns in the DNA sequence may have facilitated the accidental duplications of DNA segments that gave rise to the antibody genes during evolution (see p. 536 and p. 602).

L chain consists of one variable (V_L) and one constant (C_L) domain, while most H chains consist of a variable domain (V_H) and three constant domains (C_H1, C_H2, and C_H3). (The μ and ε chains each have one variable and four constant domains.) The variable domains are responsible for antigen binding, while the constant domains of the H chains (excluding C_H1) form the Fc region that determines the other biological properties of the antibody.

The homology between their domains suggests that Ig chains probably arose during evolution by a series of gene duplications, beginning with a primordial gene coding for a single 110-amino-acid domain of unknown function (see p. 1053). This hypothesis is supported by the finding that each domain of the constant region of an H chain is encoded by a separate coding sequence (exon) (Figure 18–27).

X-ray Diffraction Studies Show the Structure of Ig Domains and Antigen-binding Sites in Three Dimensions[19]

Even when the complete amino acid sequence of a protein is known, it is not possible to deduce its three-dimensional structure; x-ray diffraction studies of protein crystals (see p. 151) are needed. Several myeloma protein fragments and one intact IgG molecule have been crystallized to date, and x-ray studies of their structures have confirmed the predictions of the immunochemists. More important, these studies have revealed the way in which millions of different antigen-binding sites are constructed on a common structural theme.

As illustrated in Figure 18–28, each Ig domain has a very similar three-dimensional structure based on what is now called the **immunoglobulin fold.** Each domain is roughly a cylinder (4 × 2.5 × 2.5 nm) composed of a "sandwich" of two extended protein layers: one layer contains three strands of polypeptide chain and the other contains four. In each layer the adjacent strands are antiparallel and form a β sheet (see p. 109). The two layers are roughly parallel and are connected by a single intrachain disulfide bond.

The variable domains are unique in that each has its particular set of three hypervariable regions, which are arranged in three *hypervariable loops* (see Figure 18–28). The hypervariable loops of both the L and H variable domains are clustered together to form the antigen-binding site, as had been predicted. An important principle to emerge from these studies is that the variable region of an antibody molecule consists of a highly conserved rigid framework, with hypervariable loops attached at one end. Therefore an enormous diversity of antigen-binding sites can be generated by changing only the length and amino acid sequence of the hypervariable loops, without disturbing the overall three-dimensional structure necessary for antibody function.

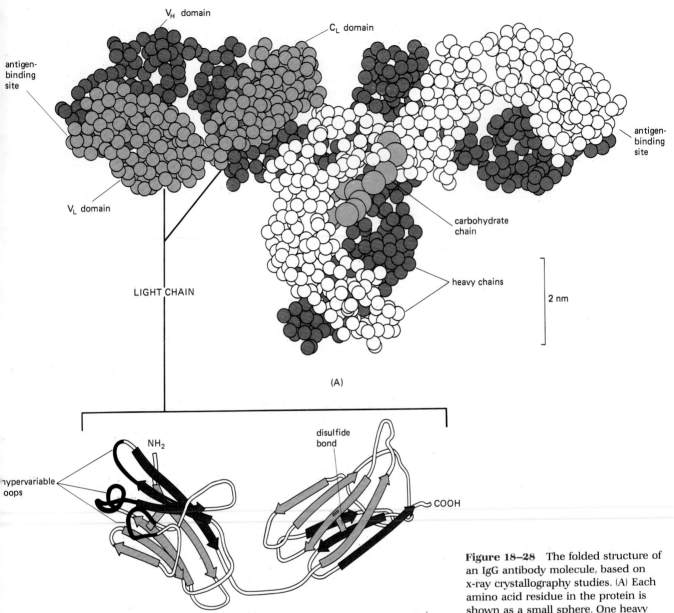

Figure 18–28 The folded structure of an IgG antibody molecule, based on x-ray crystallography studies. (A) Each amino acid residue in the protein is shown as a small sphere. One heavy chain is shown in white, the other in dark gray, with the light-chain domains in color. All antibody molecules are glycosylated: the oligosaccharide chain attached to a C_H2 domain is shown in light gray. (B) The path of the polypeptide chain for an entire light chain. Both the variable and constant domains consist of two β sheets (one composed of three strands and one composed of four strands). The chains in these two sheets are shown in different colors; the sheets are joined by a disulfide bond. Note that all the hypervariable regions form loops at the far end of the variable domain, where they come together to form part of the antigen-binding site. (A, after E.W. Silverton, M.A. Navia, and D.R. Davies, *Proc. Natl. Acad. Sci. USA* 74:5140, 1977; B, after M. Schiffer, R.L. Girling, K.R. Ely, and A.B. Edmundson, *Biochemistry* 12:4620, 1973. Copyright 1973 American Chemical Society.)

X-ray analysis of crystals of antibody fragments bound to an antigen or antigenic determinant (hapten) has revealed exactly how the hypervariable loops of the L and H variable domains cooperate to form an antigen-binding surface in particular cases (Figure 18–29). The dimensions and shape of each different site vary depending on the conformation of the polypeptide chain in the hypervariable loops, which in turn is determined by the sequence of the amino acid side chains in the loops. Thus the general principles of antibody structure are now clear.

Summary

Each immunoglobulin L and H chain consists of a variable region of about 110 amino acid residues at its amino-terminal end, followed by a constant region, which is the same size in the L chain and three or four times larger in the H chain. Each chain is composed of repeating, similarly folded domains: an L chain has one variable-region (V_L) and one constant-region (C_L) domain, while an H chain has one variable-region (V_H) and three or four constant-region (C_H) domains. The amino acid sequence variation in the variable regions of both L and H chains is for the most part

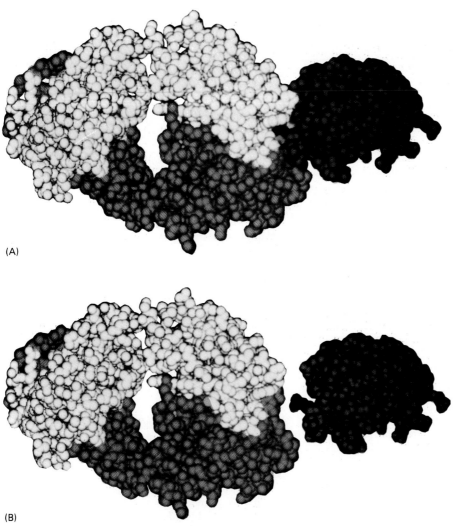

(A)

(B)

Figure 18–29   The three-dimensional structure of an antigen-antibody complex as determined by x-ray diffraction analysis. The antigen, which is the enzyme lysozyme, is shown in color. The antigen-binding site of the Fab fragment of the antibody is formed by both the light chain (*white*) and the heavy chain (*gray*). In (B) the antigen and antibody have been pulled apart to reveal their complementary contacting surfaces. The bit of the antigen that protrudes from the complementary surface is a glutamine residue. In several other antibody molecules that have been studied in this way, the antigen-binding site (for a small hapten) is formed by a much deeper cleft. (From A. Amit, R. Mariuzza, S. Phillips, and R. Poljak, *Science* 233:747–753, 1986. Copyright 1986 by the AAAS.)

confined to several small hypervariable regions, which come together to form the antigen-binding site. An antigen-binding site is generally only large enough to contact an antigenic determinant the size of five or six sugar residues.

The Generation of Antibody Diversity

It is estimated that even in the absence of antigen stimulation a mouse can make many millions of different antibody molecules—its *preimmune antibody repertoire.* Because the antigen-binding sites of many antibodies can cross-react with a variety of related but different antigenic determinants, the preimmune repertoire is apparently large enough to ensure that there will be an antigen-binding site to fit almost any potential antigenic determinant.

Antibodies are proteins, and proteins are encoded by genes. Antibody diversity therefore poses a special genetic problem: how can an animal make more antibodies than there are genes in its genome? (The human genome, for example, is thought to contain fewer than 10^5 genes.) This problem is not quite as formidable as it first appears. Because the variable regions of both the L and H chains contribute to an antigen-binding site, an animal with 1000 genes encoding L chains and 1000 genes encoding H chains could combine their products in 1000×1000 different ways to make 10^6 different antigen-binding sites (assuming that any L chain can combine with any H chain to make an antigen-binding site). Nonetheless, the immune system has evolved unique genetic mechanisms that enable it to generate an almost unlimited number of different L and H chains. The mechanisms can vary among species: chickens and mammals, for example, use very different strategies. We shall concentrate on the mechanisms used by mammals.

Antibody Genes Are Assembled from Separate Gene Segments During B Cell Development[20]

From the amino acid sequence studies of myeloma proteins, it was suspected that the variable (V) and constant (C) regions of each Ig chain might be encoded by two separate gene segments that were somehow joined together in the DNA before they were expressed. The first direct evidence that DNA is rearranged during B cell development came from experiments done in 1976 in which DNA from early mouse embryos, which do not make antibodies, was compared with the DNA of a mouse myeloma cell line, which does. The experiments showed that the specific V-region and C-region coding sequences used by the myeloma cells were present on the same DNA restriction fragment in the myeloma cells but on two different restriction fragments in the embryos, demonstrating that the DNA sequences encoding an antibody molecule are rearranged at some stage in the differentiation of B cells (Figure 18–30).

It is now known that for each type of Ig chain—κ light chains, λ light chains, and heavy chains—there is a separate pool of gene segments from which a single polypeptide chain is eventually synthesized. Each pool is on a different chromosome and usually contains a large number of gene segments encoding the V region of an Ig chain and a smaller number of gene segments encoding the C region. During B cell development, a complete coding sequence for each of the two Ig chains to be synthesized is assembled by site-specific recombination (see p. 246), bringing together the coding sequence for a V region and the coding sequence for a C region. In addition to bringing together the separate gene segments of the antibody gene, these rearrangements also activate transcription from the gene promotor, presumably through changes in the relative positions of the enhancers and silencers acting on the promotor (see p. 564). Thus an Ig chain can be synthesized only after DNA rearrangement has occurred. The process of joining gene segments contributes to the diversity of antigen-binding sites in several ways.

Each V Region Is Encoded by More Than One Gene Segment[21]

When genomic DNA sequences encoding V and C regions were analyzed, it was found that a single **C gene segment** encodes the C region of an Ig chain, but more than one gene segment encodes each V region. Each L-chain V region is encoded by a DNA sequence assembled from two gene segments—a long **V gene segment** and a short *joining*, or **J gene segment** (not to be confused with the protein *J chain*, which is encoded elsewhere in the genome—see p. 1015). Figure 18–31 illustrates the genetic mechanisms involved in producing an L chain.

Each H-chain V region is encoded by a DNA sequence assembled from three gene segments—a V segment, a J segment, and a *diversity*, or **D gene segment.** Figure 18–32 shows the organization of the gene segments used in making H chains.

The large number of inherited V, J, and D gene segments available for encoding Ig chains makes a substantial contribution on its own to antibody diversity, but the combinatorial joining of these segments (called *combinatorial diversification*) greatly increases this contribution. Any of the 300 or so V segments in the mouse light-chain gene-segment pool, for example, can be joined to any of the 4 J segments (see Figure 18–31), so that at least 1200 (300 × 4) different κ-chain V regions can be encoded by this pool. (The λ light-chain pool in mice contains only 2 V segments, but in humans it contains many more.) Similarly, any of the 500 or so V segments in the mouse H-chain pool can be joined to any of the 4 J segments and any of at least 12 D segments to encode at least 48,000 (500 × 4 × 12) different heavy-chain V regions. These are only rough estimates, since the exact numbers of V gene segments in these pools are not known.

The most important mechanism for diversifying antigen-binding sites of antibodies is combinatorial diversification resulting from the assembly of different combinations of inherited V, J, and D gene segments. As we have seen, by this mechanism alone it is estimated that a mouse could produce at least 1000 different V_L regions and on the order of 50,000 different V_H regions. These could then be

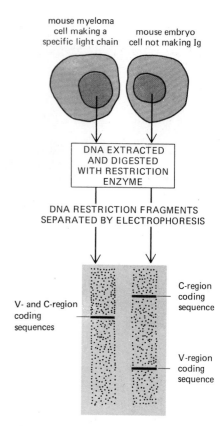

Figure 18–30 The experiment that directly demonstrated that DNA is rearranged during B cell development. DNA was extracted from a mouse plasma-cell tumor (a myeloma) synthesizing a specific Ig light chain and from a 13-day-old mouse embryo, which does not yet make antibodies. The two DNA preparations were digested with a restriction nuclease and electrophoresed through an agar gel. Those separated fragments containing the sequence encoding the C region of the L chain and those containing the sequence encoding the specific V region of the L chain were detected by Southern blot hybridization to two radioactive DNA probes—one complementary to the V-region coding sequence and one to the C-region coding sequence of the mRNA molecule specifying the specific myeloma L chain (see p. 1018). The sequences encoding the C and V regions were found on the same DNA fragment in the myeloma-cell DNA but on two separate fragments in the DNA extracted from the embryo (and in DNA extracted from a different myeloma tumor making a different L chain—not shown).

In the figure, labels read: mouse myeloma cell making a specific light chain; mouse embryo cell not making Ig; DNA EXTRACTED AND DIGESTED WITH RESTRICTION ENZYME; DNA RESTRICTION FRAGMENTS SEPARATED BY ELECTROPHORESIS; V- and C-region coding sequences; C-region coding sequence; V-region coding sequence.

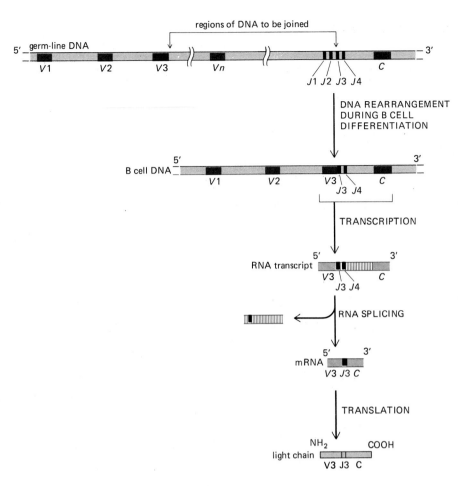

Figure 18–31 The *V-J* joining process involved in making a κ light chain in the mouse. In the "germ-line" DNA (where the immunoglobulin genes are not being expressed and are therefore not rearranged), the cluster of four *J* gene segments is separated from the *C* gene segment by a short intron and from the 300 or so *V* gene segments by thousands of nucleotide pairs. A *J* gene segment encodes the carboxyl-terminal 15 or so amino acids of the V region, and the *V-J* segment junction coincides with the third hypervariable region. Although not shown, the *V* gene segments are often clustered on the chromosome in groups of homologous families, which are thought to be spread over a region of more than 500,000 nucleotide pairs.

During B cell development, the chosen *V* gene segment (V3 in this case) is moved to lie precisely next to one of the *J* gene segments (*J3* in this case). The "extra" *J* gene (*J4*) and the intron sequence are transcribed (along with the joined V3, J3, and *C* gene segments) and then removed by RNA splicing to generate mRNA molecules in which the V3, J3, and *C* sequences are contiguous. These mRNAs are then translated into κ light chains.

combined to make 5×10^7 different antigen-binding sites. In addition, the joining mechanism itself, as we shall now discuss, greatly increases this number (probably more than 1000-fold), making it much greater than the total number of B lymphocytes in a mouse ($\sim 10^8$).

Imprecise Joining of Gene Segments Increases the Diversity of V Regions[21]

The mechanism by which gene segments that may be hundreds of thousands of nucleotide pairs apart are joined to form a functional V_L- or V_H-region coding sequence is not known in detail. Conserved DNA sequences flank each gene segment and presumably serve as recognition sites for a site-specific recombination enzyme (see p. 246), ensuring that only appropriate gene segments recombine. Thus, for example, a *V* segment will always join to a *J* or *D* segment and not to another *V* segment (Figure 18–33).

In most cases of site-specific recombination, DNA joining is precise but the joining of antibody gene segments is not. A variable number of nucleotides are often lost from the ends of the recombining gene segments, and, in the case of the heavy chain, one or more randomly chosen nucleotides may also be inserted. This random loss and gain of nucleotides at joining sites (called *junctional diversification*) enormously increases the diversity of V-region coding sequences cre-

Figure 18–32 The H-chain gene-segment pool in the mouse is thought to contain about 1000 *V* segments, at least 12 *D* segments, 4 *J* segments, and an ordered cluster of *C* segments, each encoding a different class of H chain. The *D* segment encodes from 1 to 15 of the amino acids in the third hypervariable region of the V region. The figure is not drawn to scale, and many details are omitted. For example (1) there are four C_γ gene segments (C_{γ_1}, $C_{\gamma_{2a}}$, $C_{\gamma_{2b}}$, and C_{γ_3}); (2) each *C* gene segment is composed of multiple exons (see Figure 18–27); and (3) about 200,000 nucleotide pairs separate the J_H1 and C_α gene segments, and the V_H gene segments are clustered on the chromosome in groups of homologous families, as in the case of the V_κ gene segments. The genetic mechanisms involved in producing an H chain are the same as those shown in Figure 18–31 for L chains except that two DNA rearrangement steps are required instead of one: first a *D* segment joins to a *J* segment, and then a *V* segment joins to the rearranged *DJ* segments.

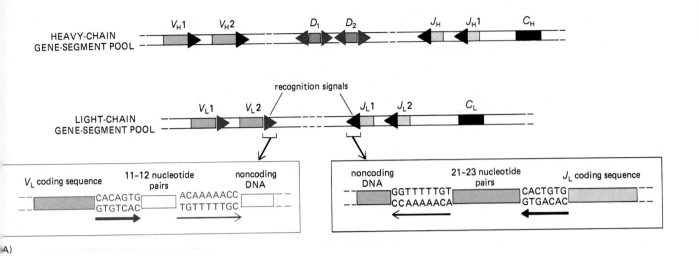

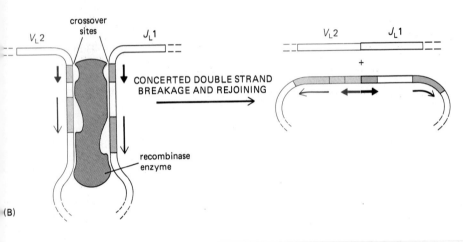

(A)

(B)

Figure 18–33 The two types of recognition signals used in joining V-region gene segments in L- and H-chain gene-segment pools. The location of the signals is shown in (A); for simplicity, only two of each type of gene segment are shown. Both types of signals consist of the conserved heptamer and nonamer sequences shown in (A) separated by a nonconserved spacer: in one case (◄) the spacer is 21 to 23 nucleotide pairs (about two turns of the DNA helix); in the other (►) it is 11 or 12 base pairs (about one turn of the double helix). Because joining occurs only when a "one-turn" spacer signal combines with a "two-turn" spacer signal (the so-called 12/23 joining rule), only appropriate gene segments recombine. As shown in (B), the recognition signals are thought to be recognized by a site-specific recombination enzyme (a recombinase) that catalyzes the joining.

The fate of the intervening DNA when two Ig gene segments join depends on the orientation of the segments. If the two segments have the same transcriptional orientation, the intervening DNA is released as a circle and discarded as shown (*deletional joining*); whereas if they have opposite orientations, the intervening DNA is retained in an inverted orientation (*inversional joining*). The joining processes are biochemically the same in both cases.

ated by recombination, specifically in the third hypervariable region. The increased diversification in this case comes at a price, since in many cases it will result in a shift in the reading frame so that a nonfunctional gene will be produced; such "nonproductive" joining occurs commonly in developing B cells.

Antigen-driven Somatic Hypermutation Fine-tunes Antibody Responses[22]

With the passage of time after immunization, there is usually a progressive increase in the affinity of the antibodies produced against the immunizing antigen. This phenomenon is known as **affinity maturation,** and it is due to the accumulation of somatic mutations in V-region coding sequences after antigen stimulation of B lymphocytes. This has been shown most convincingly by studies of oligoclonal antibody responses (see p. 1007) in inbred mice, in which the V regions of the L chains or H chains produced are encoded predominantly by a single combination of gene segments. Because of the initial uniformity of these V-region coding sequences, mutations that occur in them with time can be readily detected. For these studies, activated B cells are taken from individual mice at varying times after immunization and fused with non-antibody-secreting myeloma cells to generate hybridoma cells, each producing a single species of antibody (see p. 178). These immortal cells provide an unlimited source of RNA and DNA encoding the V regions of the antibodies; these nucleic acids can then be sequenced and examined for changes in the original V-region coding sequences. Such studies show that these sequences rapidly accumulate point mutations with time after repeated immunization. The rate of somatic mutation in these sequences has been estimated to be 10^{-3} per nucleotide pair per cell generation, which is about a million times greater than the spontaneous mutation rate in other genes. The process, therefore, is called **somatic hypermutation,** and it is thought to operate when B cells are activated by antigen to generate memory cells, rather than when they are activated to form antibody-secreting cells. The generation of memory B cells occurs principally in lymphoid follicles in secondary lymphoid organs (see Figure 18–8).

Because B cells are stimulated to proliferate by the binding of antigen, any mutation occurring during the course of an immune response that increases the affinity of a cell-surface antibody molecule will cause the preferential proliferation of the B cell making the antibody, especially when antigen concentration decreases with increasing time after immunization. Thus affinity maturation is the consequence of repeated cycles of somatic hypermutation followed by antigen-driven selection in the course of an antibody response.

Antibody Gene-Segment Joining Is Regulated to Ensure That B Cells Are Monospecific[23]

Abundant experimental evidence has shown that, as the clonal selection theory predicts, each individual B cell makes antibody with only a single type of antigen-binding site: that is, B cells are *monospecific.* The monospecificity of B cells may be important for at least two reasons. First, if each cell could produce more than one kind of antigen-binding site, some might produce self-reactive antibodies (autoantibodies) as well as useful ones, and this would make it more difficult to select for immunity to foreign antigens while maintaining tolerance to self antigens. Second, monospecificity ensures that all antibodies produced by one cell are composed of two identical halves and therefore contain two identical antigen-binding sites; this enables secreted antibodies to form large lattices of cross-linked antigens, thereby promoting antigen elimination (see Figure 18–13).

The requirement of monospecificity means that there must be some mechanism for ensuring that when Ig genes are activated during development, only one type of V_L region and one type of V_H region are made by each B cell. Since B cells, like other somatic cells, are diploid, each cell has six gene-segment pools encoding antibodies: two H-chain pools, one from each parent; and four L-chain pools, one κ and one λ from each parent. Thus, in principle, DNA rearrangements could occur independently in each H-chain and each L-chain pool; and by assembling different V-region coding sequences in each pool, a single cell could make up to eight different antibodies, each with a different antigen-binding site. In fact, however, each B cell uses only two of the six pools: one of the four L-chain gene pools and one of the two H-chain gene pools (Figure 18–34). Thus each B cell must choose not only between κ and λ L-chain pools but also between the maternal and paternal antibody gene pools. The expression of only the maternal or the paternal allele of a gene in any given cell is called **allelic exclusion** and seems to occur only in genes that encode antibodies (and in genes that encode the closely related T cell receptor proteins—see p. 1037). For other proteins that are encoded by autosomal genes, both maternal and paternal genes in a cell appear to be expressed about equally.

The mechanism of allelic exclusion and κ versus λ L-chain choice during B cell development is uncertain, but available evidence suggests that it occurs through *feedback regulation* of DNA rearrangement: a functional rearrangement in one gene-segment pool seems to suppress rearrangements in the remaining pools that encode the same type of polypeptide chain. Some of this evidence comes from experiments in which cloned genes that have already been assembled are injected into the nucleus of fertilized mouse eggs to generate *transgenic mice* (see p. 267), which carry the rearranged gene in all of their lymphocytes. In B cell clones isolated from transgenic mice carrying a rearranged L-chain gene, for example, the rearrangement of endogenous L-chain genes is usually found to be suppressed; similarly, introducing a rearranged μ gene usually suppresses the rearrangement of endogenous H-chain genes.

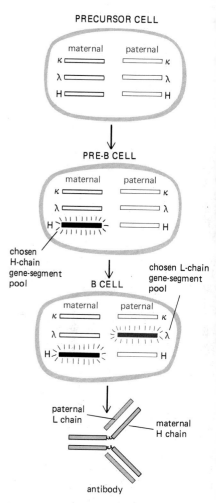

Figure 18–34 The sequential choices in Ig gene activation that a developing B cell must make in order to produce antibodies with only one type of antigen-binding site. The cell must choose one of four L-chain gene-segment pools and one of two H-chain gene-segment pools. During development, a precursor cell first activates one H-chain gene pool to become a *pre-B cell*, making only μ heavy chains. After a period of extensive proliferation, a pre-B cell activates one κ or λ light-chain pool to become a B cell that makes a unique IgM molecule.

The assembly of V-region coding sequences in a developing B cell seems to proceed in a strictly programmed order, starting with the H-chain pool, and occurs one segment at a time. First, D segments join to J_H segments on both parental chromosomes. Then V_H to DJ_H joining occurs on one of these chromosomes. If this rearrangement produces a functional gene, the resulting production of complete μ chains (always the first H chains made) shuts down further rearrangements of V_H-region-encoding gene segments and signals the onset of V_L rearrangements. The V_L-to-J_L joining occurs first in a κ gene-segment pool. If that fails, the other κ pool rearranges. If that too is unsuccessful, joining proceeds in one λ pool and then in the other. If, at any point, "in-phase" joining leads to the production of L chains, these combine with preexisting μ chains to form IgM antibody molecules, which shut down further assembly of V_L-region coding sequences.

Production of an intact μ chain seems to be sufficient to shut down further assembly of V_H-region coding sequences, but shutting down assembly of V_L-region coding sequences seems to require that a complete antibody molecule be made. If a developing B cell fails to assemble both a functional V_H- and a functional V_L-region coding sequence, it is unable to make antibody molecules and probably dies.

Although no biological differences between κ and λ light chains have been discovered, there is an obvious advantage in having two separate pools of gene segments that encode L chains: it increases the chances that a pre-B cell that has successfully assembled a V_H-region coding sequence will go on to successfully assemble a V_L-region coding sequence to become a B cell.

The Switch from a Membrane-bound to a Secreted Form of the Same Antibody Occurs Through a Change in the H-Chain RNA Transcripts[24]

We now turn from the genetic mechanisms that determine the antigen-binding site of an antibody to those that determine its biological properties—the mechanisms that determine what form of heavy-chain constant region is synthesized. The choice of the particular gene segments that encode the antigen-binding site is usually a commitment for the life of a B cell and its progeny, but the type of C_H region that is made changes during B cell development. The changes are of two types: changes from a membrane-bound form to a secreted form of the same C_H region, and changes in the class of the C_H region made.

Figure 18–35 A B cell switches from making a plasma-membrane-bound form to a secreted form of the same antibody molecule by altering the H-chain mRNAs it produces when it is activated by antigen. This is thought to result from a change in the way the initial H-chain RNA transcripts are cleaved and polyadenylated at their 3' ends. The two forms of H chain differ only in their carboxyl terminals: the membrane-bound form has a hydrophobic tail, which holds it in the membrane, whereas the secreted form has a hydrophilic tail, which enables it to escape from the cell. The long polyadenylated RNA transcript that specifies the membrane-bound form of H chain has a donor and an acceptor splice junction, which permits the RNA sequence that encodes the hydrophilic tail of the secreted form to be removed by RNA splicing; however, the short polyadenylated RNA transcript that specifies the secreted form has only a donor splice junction, and therefore the cross-hatched RNA sequence cannot be removed.

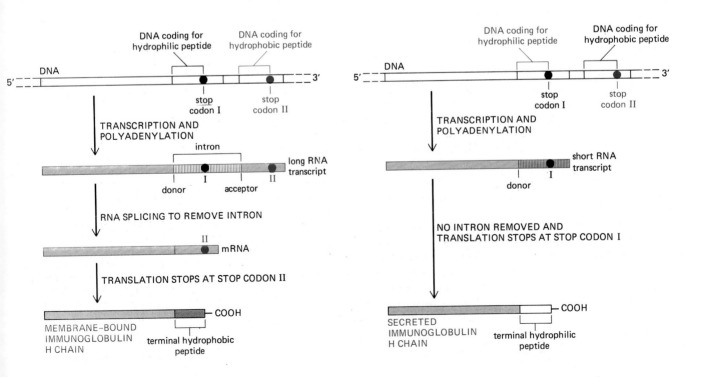

All classes of antibody can be made in a membrane-bound form as well as in a soluble, secreted form. The membrane-bound form serves as an antigen receptor on the B cell surface, while the soluble form is made only after the cell is stimulated by antigen to become an antibody-secreting cell. The sole difference between the two forms resides in the carboxyl terminus of the H chain: the H chains of membrane-bound IgM molecules, for example, have a hydrophobic carboxyl terminus, which anchors them in the lipid bilayer of the B cell plasma membrane, whereas those of secreted IgM molecules have instead a hydrophilic carboxyl terminus, which allows them to escape from the cell. Since a B cell contains only one copy of the C_μ gene segment per haploid genome and uses only one of its two H-chain gene pools to make antibodies, its ability to make μ chains with two types of constant regions at first seemed paradoxical. The paradox was resolved with the discovery that the activation of B cells by antigen induces a change in the way μ-chain RNA transcripts are processed in the nucleus, as explained in Figure 18–35 (see also p. 591). The switch from a membrane-bound to a secreted form of the other classes of antibodies involves a similar mechanism.

B Cells Can Switch the Class of Antibody They Make[25]

During B cell development, many B cells switch from making one class of antibody to making another—a process called **class switching.** All B cells begin their antibody-synthesizing lives by making IgM molecules and inserting them into the plasma membrane as receptors for antigen. Before they have interacted with antigen, most B cells then switch and make both IgM and IgD molecules as membrane-bound antigen receptors. Upon stimulation by antigen, some of these cells are activated to secrete IgM antibodies, which dominate the primary antibody response (see p. 1015). Other antigen-stimulated cells switch to making IgG, IgE, or IgA antibodies; memory cells express these molecules on their surface (often simultaneously with IgM molecules), while active B cells secrete them. The IgG, IgE, and IgA molecules are collectively referred to as *secondary* classes of antibodies because they are thought to be produced only after antigen stimulation and because they dominate secondary antibody responses.

Since the class of an antibody is determined by the constant region of its H chain (see p. 1021), the fact that B cells can switch the class of antibody they make without changing the antigen-binding site implies that the same assembled V_H-region coding sequence can sequentially associate with different C_H gene segments. This has important functional implications. It means that in an individual animal a particular antigen-binding site that has been selected by environmental antigens can be distributed among the various classes of immunoglobulin and thereby acquire the different biological properties characteristic of each class.

Class switching occurs by two distinct molecular mechanisms. When virgin B cells change from making membrane-bound IgM alone to the simultaneous production of membrane-bound IgM and IgD, the switch is thought to be due to a change in RNA processing. The cells produce large primary RNA transcripts that contain the assembled V_H-region coding sequence along with both the C_μ and C_δ sequences; IgM and IgD molecules are then produced by differential splicing of these transcripts (Figure 18–36). It is thought that the same mechanism underlies the switch to other classes of membrane-bound Ig when virgin B cells are stimulated by antigen to mature into memory cells that carry IgG, IgE, or IgA as antigen receptors on their surface.

By contrast, terminal maturation to an active B cell secreting one of the secondary classes of antibody is accompanied by an irreversible change at the DNA level—a process called *switch recombination*. It entails deletion of all the C_H gene segments upstream (that is, on the 5' side as measured on the coding strand) of the particular C_H segment the cell is destined to express (Figure 18–37). Evidence that this step in class switching involves DNA deletion comes from experiments on myeloma cells: myeloma cells that secrete IgG lack the DNA coding for C_μ and C_δ regions, and those that secrete IgA lack the DNA coding for all of the other classes of H-chain C regions.

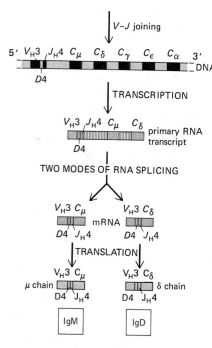

Figure 18–36 B cells that simultaneously make plasma-membrane-bound IgM and IgD molecules having the same antigen-binding sites produce long RNA transcripts that contain both C_μ and C_δ sequences. These transcripts are spliced in two ways to produce mRNA molecules that have the same V_H-region coding sequence joined to either a C_μ or a C_δ sequence. It is possible that the RNA transcripts produced by such cells are even longer than shown and contain all of the various C_H sequences.

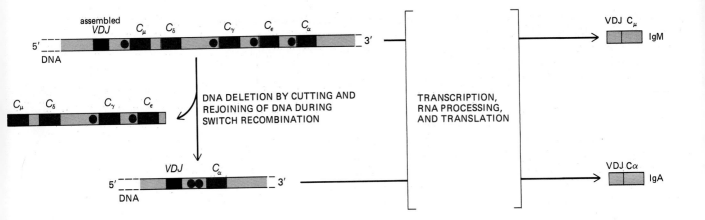

Idiotopes on Antibody Molecules Form the Basis of an Immunological Network[26]

Besides defending the body against infection, antibodies themselves play an important part in regulating immune responses. Termination of an antibody response to an antigen is brought about partly by the binding of secreted antibody to the antigen, which is thus prevented from binding to receptors on B cells; the B cells, consequently, cease to be stimulated. In addition to participating in this simple form of feedback inhibition, antibodies may play a more sophisticated role in immunoregulation as part of an intricate **immunological network.**

Antibodies are themselves antigenic, and it is possible to produce antibodies that will recognize antigenic determinants on both the constant and the variable regions of immunoglobulin chains. Antigenic determinants (epitopes) on the variable regions of L and H chains that are associated with the antigen-binding site of an antibody are called **idiotopes** (Figure 18–38). Each specific antigen-binding site has its own characteristic set of idiotopes, so that an animal with millions of different antigen-binding sites will have millions of different idiotopes. Since any individual idiotope is present in the body in only minute amounts, an animal is not tolerant to its own idiotopes and will make both T and B cell responses against them if immunized appropriately with any one of its own antibodies.

One might expect that an animal immunized with antigen A would first produce large amounts of anti-A antibodies and then produce antibodies against the idiotopes of these anti-A antibodies—and in turn antibodies against the anti-idiotope antibodies, and so on. In fact, this type of *network* reaction has been demonstrated in some immune responses, but its importance in immunoregulation is unknown.

As we shall see later, T cell receptors are much like membrane-bound antibodies: they exist in millions of forms, each with its own unique antigen-binding site and set of idiotopes. Presumably an animal contains both antibodies and T cell receptors that recognize most of its own idiotopes, with an average antigen-binding site probably recognizing at least one idiotope in its own immune system. Thus the antigen-binding sites of an immune system are potentially linked to-

Figure 18–37 An example of the DNA rearrangement that occurs in the "switch recombination" of C_H regions. When a B cell making IgM antibodies from an assembled *VDJ* sequence is stimulated by antigen to mature into an IgA-antibody-secreting cell, it deletes the DNA between the *VDJ* sequence and the C_α gene segment. Specific DNA sequences (*switch sequences*, shown as colored spheres) located upstream of each C_H gene segment (except C_δ) recombine with one another to delete the intervening DNA.

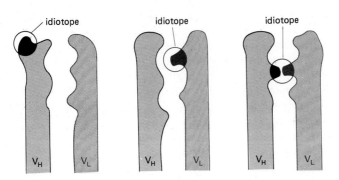

Figure 18–38 An *idiotope* is an antigenic determinant located in or close to the antigen-binding site of an antibody molecule. It can be formed by V_H, V_L, or both. Each different antigen-binding site has its own unique set of idiotopes that constitutes its *idiotype*. Idiotopes are also associated with the antigen-binding sites of T cell receptors.

gether in a complex network of anti-idiotope interactions (Figure 18–39). Like the nerve cells of the nervous system, many lymphocytes may interact with one another rather than with the outside world. An immune response might therefore be viewed as a reverberating perturbation of an immunological network rather than as a response of independent antigen-reactive lymphocytes.

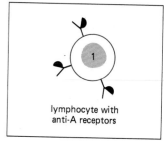

lymphocyte with anti-A receptors

Summary

Antibodies are produced from three pools of gene segments, encoding the κ, λ, and H chains, respectively. In each pool, separate gene segments that code for different parts of the variable regions of L and H chains are brought together by site-specific recombination during B cell differentiation. The L-chain pools contain one or more constant (C) gene segments and sets of variable (V) and joining (J) gene segments. The H-chain pool contains a set of C gene segments and sets of V, diversity (D), and J gene segments. To make an antibody molecule, a V_L gene segment is recombined with a J_L gene segment to produce a DNA sequence coding for the V region of the light chain, and a V_H gene segment is recombined with a D and a J_H gene segment to produce a DNA sequence coding for the V region of the heavy chain. Each of the assembled gene segments is then co-transcribed with the appropriate C-region sequence to produce an mRNA molecule that codes for the complete polypeptide chain. By variously combining inherited gene segments that code for V_L and V_H regions, vertebrates can make thousands of different L chains and thousands of different H chains that can associate to form millions of different antigen-binding sites. This number is further increased by the loss and gain of nucleotides during gene-segment joining and by somatic mutations that occur with very high frequency in these gene segments following antigen stimulation.

All B cells initially make IgM antibodies. Later some make antibodies of other classes but with the same antigen-binding site as the original IgM antibodies. Such class switching allows the same antigen-binding sites to be distributed among antibodies with varied biological properties.

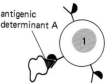

antigenic determinant A

lymphocyte 1 can bind antigenic determinant A

lymphocyte 1 can also interact with receptors on lymphocyte 2 that express an idiotope that resembles antigenic determinant A

The Complement System

Complement, so-called because it *complements* and amplifies the action of antibody, is the principal means by which antibodies defend vertebrates against most bacterial infections. It consists of a system of serum proteins that can be activated by antibody-antigen complexes or microorganisms to undergo a cascade of proteolytic reactions whose end result is the assembly of *membrane attack complexes.* These complexes form holes in a microorganism and thereby destroy it. At the same time, proteolytic fragments released during the activation process promote the defense response by dilating blood vessels and attracting phagocytic cells to sites of infection. Complement also enhances the ability of phagocytic cells to bind, ingest, and destroy the microorganisms being attacked.

Individuals with a deficiency in one of the central complement components (C3) are subject to repeated bacterial infections, just as are individuals deficient in antibodies themselves. Complement-deficient individuals may also suffer from *immune-complex diseases,* in which antibody-antigen complexes precipitate in small blood vessels in skin, joints, kidney, and brain, where they cause inflammation and destroy tissue; this suggests that complement normally helps to solubilize such complexes when they form during an immune response.

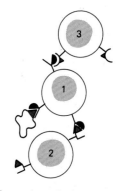

lymphocyte 1 can also interact with lymphocyte 3, which has a receptor that recognizes an idiotope on the anti-A receptor of lymphocyte 1

Figure 18–39 Any individual lymphocyte may be functionally connected to other lymphocytes through idiotope–anti-idiotope reactions. The extent of such an idiotope network is potentially great because each of the lymphocytes shown interacting with the anti-A lymphocyte can also interact with other lymphocytes in a similar way. It is still uncertain how important such anti-idiotope interactions are in regulating immune responses.

Complement Activation Involves an Amplifying Proteolytic Cascade Centered on the Cleavage of the C3 Protein[27]

Complement consists of about 20 interacting proteins, of which the reacting components are designated C1–C9, factor B, and factor D—the rest comprising a variety of regulatory proteins. The complement components are all soluble proteins. They are made mainly by the liver and circulate in the blood and extracellular fluid. Most are inactive unless they are triggered directly by an invading microorganism or indirectly by an immune response. The ultimate consequence of complement activation is the assembly of the **late complement components** (C5, C6, C7, C8, and C9) into a large protein complex (the membrane attack complex) that mediates microbial cell lysis.

Because its main function is to attack the membrane of microbial cells, the activation of complement is focused on the microbial cell membrane, where it is triggered either by antibody bound to the microorganism or by microbial envelope polysaccharides, both of which activate the **early complement components.** There are two sets of early components belonging to two distinct pathways of complement activation: C1, C2, and C4 belong to the *classical pathway*, which is triggered by antibody binding; factor B and factor D belong to the *alternative pathway*, which is triggered by microbial polysaccharides. The early components of both pathways ultimately act on C3, the most important complement component (Figure 18–40). The early components and C3 are proenzymes that are activated sequentially by limited proteolytic cleavage: as each proenzyme in the sequence is cleaved, it is activated to generate a serine protease, which cleaves the next proenzyme in the sequence, and so on. Many of these cleavages liberate a small peptide fragment and expose a membrane-binding site on the larger fragment. The larger fragment binds tightly to the target cell membrane by its newly exposed membrane-binding site and helps to carry out the next reaction in the sequence. In this way complement activation is confined largely to the cell surface where it began. The smaller fragment often acts independently as a diffusible signal that promotes an inflammatory response (see p. 1034).

The activation of C3 by cleavage is the central reaction in the complement-activation sequence, and it is here that the classical and alternative pathways converge (see Figure 18–40). In both pathways, C3 is cleaved by an enzyme complex called a **C3 convertase.** A different C3 convertase is produced by each pathway, formed by the spontaneous assembly of two of the complement components activated earlier in the cascade. Both types of C3 convertase cleave C3 into two fragments. The larger of these (C3b) binds covalently to the target-cell membrane and binds C5. Once bound, the C5 protein is cleaved by the C3 convertase (now acting as a C5 convertase) to initiate the spontaneous assembly of the late components—C5 through C9—that creates the membrane attack complex.

Since each activated enzyme cleaves many molecules of the next proenzyme in the chain, the activation of the early components consists of an amplifying *proteolytic cascade:* each molecule activated at the beginning of the sequence leads to the production of many membrane attack complexes.

The Classical Pathway Is Activated by Antibody-Antigen Complexes[27,28]

The **classical pathway** is usually activated by IgG or IgM antibodies bound to antigens on the surface of a microorganism. The binding of antigen by these antibodies enables their constant regions to bind in turn to the first component in the classical pathway, C1, which is a large complex composed of three subcomponents—C1q, C1r, and C1s (Figure 18–41). The binding of a globular head of C1q to an IgG or IgM antibody bound to antigen activates C1q to start the early proteolytic cascade of the classical pathway. Because more than one head must be bound before activation occurs, a cluster of foreign antigenic determinants is required to trigger the classical pathway (Figure 18–41B). This again serves to focus complement activation on the surface of microorganisms, where antigenic determinants are naturally clustered.

Activation of the C1q subcomponent of the C1 complex activates C1r to become proteolytic, and C1r in turn cleaves and activates C1s. Activated C1s then cleaves C4 into two fragments, C4a and C4b (by convention, the larger fragment in such cleavages is designated b and the smaller is designated a); C4b immediately binds covalently to the membrane and then binds C2. Once bound, C2 is also cleaved by activated C1s; the larger fragment C2a, remains bound to C4b, forming the complex C4b,2a, which is the C3 convertase produced by the classical pathway. As illustrated in Figure 18–42, C4b,2a cleaves C3 to produce two fragments,

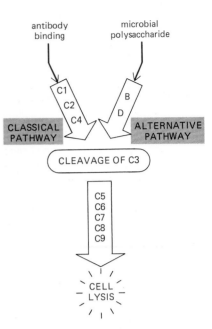

Figure 18–40 The principal stages in complement activation by the classical and alternative pathways. In both pathways the reactions of complement activation usually take place on the surface of an invading microbe, such as a bacterium.

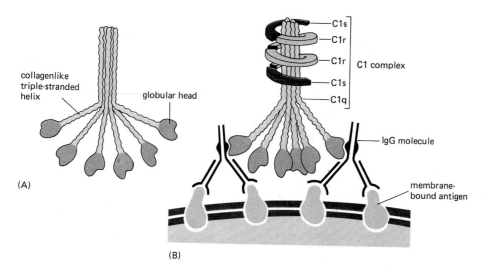

Figure 18–41 (A) The unusual structure of C1q. It is a large protein (~450,000 daltons) made up of six identical subunits, each composed of three different polypeptide chains. The carboxyl-terminal halves of each of the three polypeptide chains in a subunit are folded into a globular structure; the amino-terminal halves have a typical collagen amino acid sequence and are wound together to form a collagenlike triple-stranded helix (see p. 808). The six subunits are linked together by disulfide bonds (not shown) between their triple-helical stems, forming a structure that resembles a bunch of tulips. It is the heads of the tulips that bind to IgG or IgM antibody; thus, each C1q molecule has six antibody-binding sites. (B) C1 binding to a pair of IgG molecules bound to a cluster of antigenic determinants on the surface of a target cell. Each C1 complex consists of one molecule of C1q bound loosely to a tetramer (shown schematically) composed of two molecules of C1r and two molecules of C1s.

C3a and C3b; the latter rapidly binds to the target membrane and binds C5. Bound C5 is cleaved by C4b,2a (recall that C3 convertase is also the C5 convertase) into C5a and C5b, and C5b combines with C6 to initiate assembly of the late components to form the membrane attack complex.

Microorganisms Can Activate the Alternative Pathway Directly[27,29]

Polysaccharides in the cell envelopes of microorganisms can activate the **alternative pathway** even in the absence of antibody, and activation of the classical pathway also activates the alternative pathway through positive feedback. The alternative pathway therefore provides a first line of defense against infection before an immune response can be mounted and amplifies the effects of the classical pathway once an immune response has begun.

Although we have stated that microbial polysaccharides activate the alternative pathway directly, this is an oversimplification. In fact, it is C3b that activates the pathway, and the part played by the polysaccharides is to prevent C3b from being rapidly inactivated by a mechanism that protects normal cells in the vicinity from attack by activated complement (see below). The C3b that triggers the alternative pathway can be provided either by the classical pathway (hence the positive

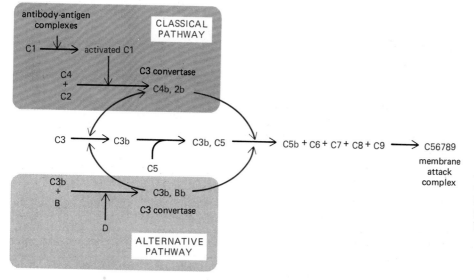

Figure 18–42 Generation of C3 convertase by the classical and alternative pathways. In both cases the C3 convertase is formed by the assembly of two complement components activated earlier in the cascade. C3 and C5 are homologous proteins; C5, by binding to C3b, comes to resemble C3 and is cleaved by the C3 convertase (acting as a C5 convertase) to produce C5b, which initiates assembly of the membrane attack complex. Activated complement components are often designated with a superscript bar—for example, activated C1 is C1̄. To simplify the text we have omitted these bars.

feedback loop) or by the spontaneous proteolysis of C3, which occurs at low rates even when the complement cascade has not been activated.

The next step in activating the alternative pathway is the binding of factor B to membrane-bound C3b. Factor D (which circulates in the blood in an active form) then cleaves the bound factor B to generate the active fragment Bb and thereby produce C3b,Bb, which is the C3 convertase for the alternative pathway, and therefore generates more C3b molecules, some of which bind C5. As in the classical pathway, the C3 convertase can also act as a C5 convertase and cleave the membrane-bound C5 molecules to initiate assembly of the membrane attack complex (see Figure 18–43 below).

Damage to normal host-cell membranes by complement activated via the classical pathway is prevented by the requirement for bound antibody to activate C1. The alternative pathway, however, is activated by C3b, which can bind to any membrane and, as mentioned previously, is produced spontaneously at low rates. Special mechanisms are therefore required to protect normal host cells from being attacked by this route. Unlike the large clusters of C3b produced by the classical pathway, most of the spontaneously formed C3b molecules bind as single molecules to membranes and are rapidly inactivated by a specific inhibitor protein called *factor H*, which competes with factor B for a binding site on C3b. C3b bound to microbial membranes is protected from this inhibitor protein (by an unknown mechanism) by the polysaccharides in the microbial cell envelope. Thus these C3b molecules can bind factor B and thereby activate the alternative pathway.

Complement Activation Promotes Phagocytosis and Inflammation[27]

C3b, produced by either the classical or the alternative pathways, has a number of important properties. As we have seen, it activates the alternative pathway to produce more C3b, and it binds C5, enabling the C5 to be cleaved by a C3 convertase. But complement does not act only to generate membrane attack complexes, and C3b also has a third important role. It binds to specific receptor proteins on macrophages and neutrophils, enhancing the ability of these cells to phagocytose a microbial cell to which C3b has attached; in this way C3b makes a crucial contribution to the defense against bacteria, independent of complement-mediated cell lysis.

In the course of the proteolytic complement cascade, several small, biologically active protein fragments are generated. The most important of these are C3a and C5a, both of which are highly basic peptides cleaved from the amino-terminal end of the parent proteins. Both cause smooth muscle to contract and stimulate mast cells and basophils to secrete histamine. The histamine increases the permeability of local blood vessels, allowing white blood cells and more antibodies and complement to enter the site of infection. C5a is also a potent chemoattractant for neutrophils and also enhances the mechanisms by which these cells kill ingested bacteria. The two peptides are mainly responsible for the local inflammatory response that is usually associated with the activation of complement.

Assembly of the Late Complement Components Generates a Giant Membrane Attack Complex[27,30]

Assembly of the late components begins when C5, already loosely bound to C3b on the target cell membrane, is split by the C3 convertase of either the classical or alternative pathway to give C5a and C5b. As just described, the C5a is released and promotes an inflammatory response. The C5b remains bound to the C3b and has the transient capacity to bind C6 to form C56 and then C7 to form C567. The C567 complex then binds firmly via C7 to the membrane. This complex adds one molecule of C8 to form C5678, which then binds 8 to 18 molecules of C9, which partially unfold and polymerize into a transmembrane channel (Figure 18–43). DNA sequencing studies suggest that the C6, C7, C8, and C9 proteins are all evolutionarily related.

The **membrane attack complexes** have a characteristic appearance in negatively stained electron micrographs, where they are seen to form aqueous pores

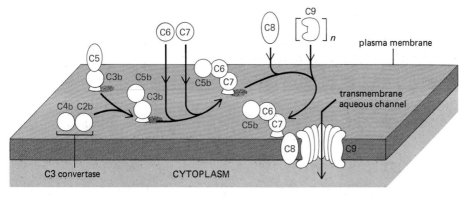

Figure 18–43 Assembly of the late complement components to form a membrane attack complex. The binding of a molecule of C9 to C5678 induces a conformational change in the C9 that exposes a hydrophobic region and causes the C9 to insert into the lipid bilayer of the target cell next to C8. This starts a chain reaction in which the altered C9 binds a second molecule of C9, which undergoes a conformational change and inserts into the bilayer, where it can bind another molecule of C9, and so on. In this way a large transmembrane channel is formed by a chain of C9 molecules. The final $C5678(9)_n$ complex can contain up to 18 C9 molecules and can have a molecular weight of almost 2 million (each C9 molecule contains 537 amino acid residues). The C5678 complexes themselves produce small transmembrane pores.

through the membrane (Figure 18–44). For this reason, and because they perturb the structure of the lipid bilayer in their vicinity, they make the membrane leaky. Since small molecules leak into and out of the cell around and through the complexes while macromolecules remain inside, the cell's normal mechanism for controlling water balance is disrupted (see Panel 6–1, p. 308). Water is therefore drawn into the cell by osmosis, causing it to swell and burst. The process is so efficient that a very small number of membrane attack complexes (perhaps even one) can kill a red blood cell. Even an enveloped virus that does not have a large osmotic pressure gradient across its membrane and is therefore not susceptible to such osmotic lysis can still be destroyed by these complexes, presumably because they disorganize the viral membrane.

The Complement Cascade Is Tightly Regulated[27,31]

The self-amplifying destructive properties of the complement cascade make it essential that key activated components be rapidly inactivated after they are generated to ensure that the attack does not spread to nearby host cells. Deactivation is achieved in at least two ways. First, specific inhibitor proteins in the blood terminate the cascade by either binding or cleaving certain components once they have been activated by proteolytic cleavage. Inhibitor proteins, for example, bind to the activated components of the C1 complex and block their further action, while other inhibitor proteins cleave C3b and thereby inactivate it. Without these inhibitors, all of the serum C3 might be depleted by the positive feedback loop created by the alternative pathway.

A second important mechanism of regulation is based on the instability of many of the activated components in the cascade; unless they bind immediately to an appropriate component in the chain or to a nearby membrane, they rapidly become inactive. C4b and C3b provide especially dramatic examples. When either of these components is formed by cleavage, it undergoes a series of rapid conformational changes that creates a short-lived active form. This active form has a hydrophobic site as well as a highly reactive glutamine side chain that is produced by the mechanical breakage of an unusual thioester bond in the protein (Figure 18–45). As a result, the glutamine forms a covalent bond with a protein or polysaccharide on a nearby membrane. Because of the very short half-life of their active forms (less than 0.1 millisecond), both C4b and C3b normally become bound only to membrane sites very close to the site where the complement components are activated. The complement attack is thereby confined to the surface membrane of a microorganism and is prevented from spreading to normal host cells in the vicinity.

How did such a complex system of complement components evolve? It presumably developed in gradual steps, with many of the more elaborate components, such as those of the large membrane attack complex, being added relatively late in the evolution of the system. It seems likely that the system originally evolved around C3, primarily to create the covalent complex between C3b and foreign cell membranes. This complex alone greatly enhances the ability of macrophages and neutrophils to ingest and destroy microorganisms. Indeed, humans who lack one

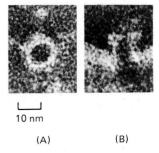

10 nm

(A)   (B)

Figure 18–44 Electron micrographs of negatively stained complement lesions in the plasma membrane of a red blood cell. The lesion in (A) is seen *en face*, while that in (B) is seen from the side as an apparent transmembrane channel. The negative stain fills the individual channels, which therefore look black. (From R. Dourmashkin, *Immunology* 35:205–212, 1978.)

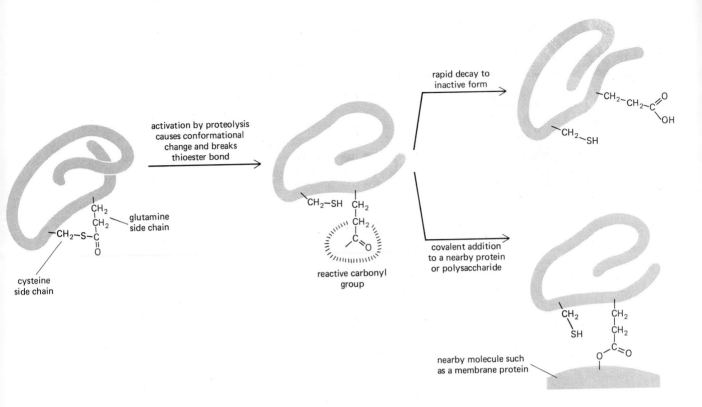

of the late components, and therefore are unable to assemble the membrane attack complex, are still protected against infection by all but a few types of bacteria—those that are able to survive inside phagocytic cells, so that complement-mediated lysis is especially important in their control. It is thought that both the alternative and classical pathways evolved from such a primitive complement system. The alternative pathway probably evolved first as a form of nonspecific innate protection against infection, the classical pathway evolving much later to link C3 activation to antibody binding and thereby to specific adaptive immune responses. Consistent with the view that the two pathways are evolutionarily related is the finding that many of their components are homologous, including the serine proteases C1r, C1s, C2, factor B, and factor D.

Summary

The complement system acts on its own and in cooperation with antibodies to defend vertebrates against infection. The early components are blood proenzymes that are sequentially activated in an amplifying series of limited proteolytic reactions, either by the classical pathway—which is triggered by IgG or IgM antibodies binding to antigen—or by the alternative pathway—which can be triggered directly by the cell envelopes of invading microorganisms. The most important complement component is the C3 protein, which is activated by proteolytic cleavage and then binds covalently to nearby membranes. Microorganisms with activated C3 (C3b) on their surface are readily ingested and destroyed by phagocytic cells. In addition, C3b helps to initiate assembly of the late complement components, which form a large membrane attack complex that can cause invading microorganisms to lyse. Complement activation also releases a variety of small, soluble peptide fragments that attract and activate neutrophils and stimulate mast cells to secrete histamine: this results in an inflammatory response at sites of complement activation. The complement proteolytic cascade remains focused on the membranes of invading microorganisms that activated the cascade, mainly because several complement components, including C3b, remain activated for less than 0.1 millisecond and therefore cannot spread the attack to nearby host cells.

Figure 18–45 The proteolytic activation of either C3 or C4 induces a conformational change in the protein, breaking the unusual intramolecular covalent bond shown. Breaking this thioester bond between protein side chains generates a very reactive carbonyl group that couples covalently to another macromolecule, forming an ester or an amide linkage. However, the ability of the protein to react this way decays with a half-life of 60 microseconds or so, confining the reaction to membranes that are very near the site where complement activation was initiated. Both C3 and C4 are composed of more than one polypeptide chain; the largest chain in each protein undergoes the reactions shown.

T Lymphocytes and Cell-mediated Immunity

The diverse responses of T cells are collectively called *cell-mediated immune reactions*. Like antibody responses, they are important in defending vertebrates against infection, particularly by certain viruses and fungi. Also like antibody responses, they are exquisitely antigen-specific.

T cells differ from B cells, however, in several important ways: (1) While some T cells fight infection directly by killing virus-infected cells, most regulate the activity of other effector cells, such as B cells and macrophages. (2) Both effector and regulatory T cells act mainly at short range, interacting directly with the cells they kill or regulate; B cells, on the other hand, secrete antibodies that can act far away. (3) Presumably for this reason, T cells bind foreign antigen only when it is on the surface of another cell in the body; the antigen is recognized in association with a special class of cell-surface glycoproteins, known as *MHC molecules* because they are encoded by a complex of genes called the *major histocompatibility complex (MHC)* (see p. 1040). This ensures that T cells are activated only when they contact another host cell and is the most important feature distinguishing antigen recognition by T and B cells.

T Cell Receptors Are Antibodylike Heterodimers[32]

Because T cells are activated only through close contact with other cells, their antigen receptors exist only in membrane-bound form. Thus, unlike antibodies, which are secreted (as well as being membrane-bound), T cell receptors for antigen were difficult to isolate, and it took much longer to identify these molecules and the genes that encode them. The receptor proteins were first identified in 1983, after it had become possible to grow pure clones of antigen-specific T cells in culture (see p. 1047), thus making available large numbers of T cells with identical receptors. Monoclonal antibodies could then be raised against the cloned cells, and those that recognized the T cell receptor were identified by their ability to block antigen-induced responses of the original cells but not those of other T cell clones. The antibodies were then used to purify the receptor molecules, which were found to be composed of two disulfide-linked polypeptide chains (called α and β). Each of these chains shares with antibodies the distinctive property of a variable amino-terminal region and a constant carboxyl-terminal region (Figure 18–46).

This antibodylike feature was an important element in an ingenious strategy used, a year or so later, to isolate the genes that encode T cell receptors. Because T and B lymphocytes are very closely related, most of the genes they transcribe are the same, and so they contain mostly the same mRNAs. Only the T cell, however, contains mRNA for the T cell receptor. By taking the total mRNAs from T cell clones and removing, by subtractive hybridization (see p. 262), the mRNAs also made by B cells, a small population of mRNAs unique to T cells was obtained. A cDNA library was then prepared from the mRNAs (see p. 260), and individual cDNA clones from the library were used to prepare radioactive DNA probes. On the assumption that the variability of the amino-terminal ends of the T cell receptor proteins, like that of antibody V regions, is generated by DNA rearrangements, each DNA probe was then used to examine the corresponding genomic DNA for evidence of rearrangement during T cell development. In this way the gene pools that encode the α and β chains were eventually localized on different chromosomes and shown to contain, like antibody gene pools, separate *V, D, J,* and *C* gene segments, which are brought together by site-specific recombination during T cell development in the thymus.

With one exception, all the mechanisms used by B cells to generate antibody diversity are also used by T cells to generate T cell receptor diversity. It is thought, however, that junctional diversification mechanisms play an especially important part in diversifying T cell receptors. The mechanism that seems not to operate in T cell receptor diversification is antigen-driven somatic hypermutation (see p. 1026). This is presumably because hypermutation would be likely to generate T

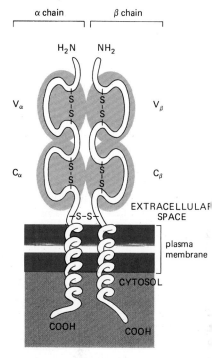

Figure 18–46 A T cell receptor heterodimer composed of an α and a β polypeptide chain, both of which are glycosylated (not shown). Each chain is about 280 amino acid residues long, and its large extracellular part is folded into two immunoglobulinlike domains—one variable (V) and one constant (C). From an analysis of amino acid sequences deduced from cDNA clones, it is thought that an antigen-binding site formed by a V_α and V_β domain is similar in its overall dimensions and geometry to the antigen-binding site of an antibody molecule. Unlike antibodies, however, which have two binding sites for antigen, T cell receptors have only one (probably because they are always bound to the plasma membrane, where they can act cooperatively). The α/β heterodimer shown is noncovalently associated with an invariant set of membrane proteins called the CD3 complex (not shown). A typical T cell has 20,000 to 40,000 α/β receptor proteins on its surface.

cells that react against self molecules. This is much less of a problem for B cells, since most self-reactive B cells could not be activated without the aid of self-reactive helper T cells (see p. 1047).

A second type of T cell receptor heterodimer, composed of γ and δ chains, has recently been discovered. These receptors are expressed on subpopulations of cells in the thymus, epidermis, and gut epithelium, whose functions are unknown.

Both α/β and γ/δ T cell receptors are physically associated on the cell surface with an invariant set of polypeptide chains called the *T3 (or CD3) complex*. This complex is present on the surface of all mature T cells and is thought to be involved in passing the signal from an antigen-activated T cell receptor to the cell interior.

Different T Cell Responses Are Mediated by Different Classes of T Cells[33]

T cells kill virus-infected cells, and they help or inhibit the responses of other white blood cells. These three functions are carried out by different classes of T cells—called *cytotoxic T cells, helper (or inducer) T cells,* and *suppressor T cells,* respectively. Cytotoxic T cells, together with B cells, are the main *effector cells* of the immune system; helper T cells and suppressor T cells are collectively referred to as *regulatory T cells.*

Of the three major T cell classes, we know least about suppressor cells. For instance, while both helper and cytotoxic T cells use the same receptors (α/β heterodimers) for recognizing antigen, the nature of the receptors used by suppressor T cells is still uncertain (although at least some seem to use α/β heterodimers). One reason we know so little about suppressor T cells is that, while it has been relatively easy to obtain antigen-specific cytotoxic and helper T cell clones in culture (see p. 1047), it has been exceedingly difficult to obtain suppressor T cell clones.

Although cytotoxic and helper T cells use antigen receptors encoded by the same gene-segment pools, they do not recognize the same MHC molecules on the surface of cells. This difference reflects the different functions of the two types of cells.

Cytotoxic T Cells Kill Virus-infected Cells[34]

Because viruses proliferate inside cells, where they are sheltered from attack by antibodies, the most efficient way to prevent them from spreading to other cells is to kill the infected cell before virus assembly has begun. This is the main function of **cytotoxic T cells.** Given their destructive potential, it is crucial that these lymphocytes confine their attack strictly to infected cells. Microcinematographic analysis has shown that a cytotoxic T cell can focus its attack on one target cell at a time, even when the cytotoxic cell has several target cells bound to it. How does it direct its attack with such accuracy?

The mechanism seems to depend on a cytoskeletal rearrangement in the cytotoxic cell that is triggered by specific contact with the target cell surface. When a cytotoxic T cell, caught in the act of killing its target, is labeled with anti-tubulin antibodies, its centrosome is seen to be oriented toward the point of contact with the target cell (Figure 18–47). Moreover, when labeled with antibodies against *talin,* a protein thought to help link cell-surface receptors to cortical actin filaments (see p. 636), the talin is found concentrated in the cortex of the cytotoxic cell at the contact site. There is evidence that the aggregation of T cell receptors at the contact site leads to a local, talin-dependent accumulation of actin filaments; a microtubule-dependent mechanism then orients the centrosome and associated Golgi apparatus toward the contact site, focusing the killing machinery on the target cell. A similar cytoskeletal polarization is seen when a helper T cell functionally interacts with the cell it helps.

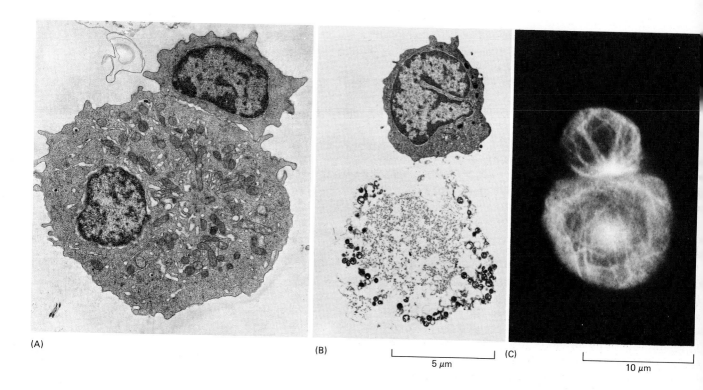

(A)

(B)

5 µm

(C)

10 µm

How Do Cytotoxic T Cells Kill Their Targets?[35]

Cytotoxic T cells defend us against virus-induced cancers, just as they do against conventional virus infections. Virus-induced tumors, however, are thought to account for less than 20% of human cancers, and there is little evidence that immune responses protect us against most other forms of cancer: whereas immunosuppressed patients and experimental animals are more susceptible to virus-induced tumors (and tumors suspected of being virus-induced), they are not more susceptible to spontaneous or chemically induced tumors. There is great interest, however, in the possibility that nonimmunological defense mechanisms may be important in defending us against cancer. Two possibilities are the killing of tumor cells by macrophages or by **natural killer (NK) cells.** NK cells are lymphocyte-like cells that are probably the same as *K cells,* which kill antibody-coated eucaryotic cells (see p. 1015). But NK cells can also spontaneously, and relatively nonspecifically, kill a variety of tumor and virus-infected cells in culture in the absence of antibody. How they distinguish abnormal cells from normal ones in these cases is unknown.

It is not known how cytotoxic T cells and NK cells kill their targets. Some NK cell lines and cytotoxic T cell lines, which can be maintained in culture indefinitely, seem to use a mechanism similar to that used by the complement system. Binding to the target stimulates these cytotoxic cells to release pore-forming proteins called **perforins,** which polymerize in the target cell plasma membrane to form transmembrane channels. By causing the membrane to become leaky, the channels are believed to help kill the cell. The perforins, which are homologous to the complement component C9, are stored in secretory vesicles and are released by local exocytosis at the point of contact with the target cell; the secretory vesicles also contain serine esterases, but it is not known if they play a part in target-cell killing. Under the electron microscope the perforin channels in target-cell membranes look very similar to the C9 channels, further supporting the notion that these cytotoxic cell lines and complement kill by related mechanisms. Normal cytotoxic T cells and NK cells, however, can kill target cells by a perforin-independent mechanism, whose molecular basis is not known. One possibility is that these cytotoxic cells activate an internal self-destruct mechanism in the target cell, which consequently commits suicide.

Figure 18–47 Cytotoxic T cells (the small cells) in the process of killing target cells in culture, visualized by electron microscopy in (A) and (B) and by immunofluorescence microscopy after staining with anti-tubulin antibodies in (C). The cytotoxic T cells were obtained from mice immunized with the target cells, which are foreign tumor cells. The T cells are shown binding to the target cell in (A) and (C) and having killed the target cell in (B). Note that the centrosome, and the microtubules radiating from it, are oriented toward the point of cell-cell contact in the T cell but not in the target cell (C). (A and B, from D. Zagury, J. Bernard, N. Thierness, M. Feldman, and G. Berke, *Eur. J. Immunol.* 5:818–822, 1975; C, reproduced from B. Geiger, D. Rosen, and G. Berke, *J. Cell Biol.* 95:137–143, 1982, by copyright permission of the Rockefeller University Press.)

While it is not known what molecules NK cells recognize on the cells they ~~k~~ill, cytotoxic T cells recognize viral molecules that are bound to MHC glycoproteins on the surface of virus-infected cells. But this crucial part played by MHC molecules in "presenting" antigen to T cells has been appreciated only recently.

MHC Molecules Induce Organ Graft Rejection[36]

MHC molecules were recognized long before their normal function was understood. They were initially defined as the main target antigens in **transplantation reactions.** When organ grafts are exchanged between adult individuals of either the same species (*allografts*) or of different species (*xenografts*), they are usually rejected. In the 1950s, experiments involving skin grafting between different strains of mice demonstrated that *graft rejection* is an immune response to the foreign antigens on the surface of the grafted cells. It was later shown that these reactions are mediated mainly by T cells and that they are directed against genetically "foreign" versions of cell-surface glycoproteins called *histocompatibility molecules* (from "histo," meaning tissue). By far the most important of these are the *major histocompatibility molecules*, a family of glycoproteins encoded by a complex of genes called the **major histocompatibility complex (MHC).** MHC molecules are expressed on the cells of all higher vertebrates. They were first demonstrated in mice and called **H-2 antigens** (histocompatibility-2 antigens). In humans they are called **HLA antigens** (human-leucocyte-associated antigens) because they were first demonstrated on leucocytes (white blood cells).

Three remarkable properties of MHC molecules baffled immunologists for a long time. First, MHC molecules are overwhelmingly the preferred target antigens for T-cell-mediated transplantation reactions. Second, an unusually large fraction of T cells is able to recognize foreign MHC molecules: whereas fewer than 0.001% of an individual's T cells respond to a typical viral antigen, more than 0.1% of them respond to a single foreign MHC antigen. Third, many of the loci that code for MHC molecules are the most *polymorphic* known in higher vertebrates; that is, within a species, there is an extraordinarily large number of *alleles* (alternative forms of the same gene) at each locus (often more than 100), each allele being present in a relatively high frequency in the population. For this reason, and because each individual has seven or more loci encoding MHC molecules (see below), it is very rare for two individuals to have an identical set of MHC glycoproteins, making it very difficult to match donor and recipient for organ transplantation in humans (except in the case of genetically identical twins).

But a vertebrate does not need to be protected against invasion by foreign vertebrate cells, so the apparent obsession of its T cells with foreign MHC molecules and the extreme polymorphism of these molecules were not only an obstacle to transplant surgeons, they were also a puzzle to immunologists. The puzzle was solved only after it was discovered that MHC molecules serve to focus T lympho-

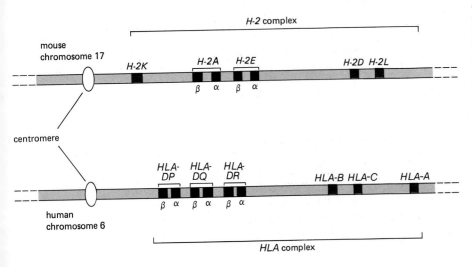

Figure 18–48 Schematic drawing of the *H-2* and *HLA* gene complexes, showing the location of loci that encode class I (*color*) and class II (*black*) MHC glycoproteins. There are three types of class I glycoproteins (H-2K, H-2D, and H-2L in mouse, and HLA-A, HLA-B, and HLA-C in human), each composed of an α chain, encoded by the loci shown, and a β₂-microglobulin chain, encoded on another chromosome. There are two types of class II MHC glycoproteins in the mouse—H-2A and H-2E, each composed of an α and a β chain. Three types of human class II molecules are shown—HLA-DP, HLA-DQ, and HLA-DR (each composed of an α and a β chain), but there are at least one or two others. Human DP and DR molecules are homologous to mouse H-2E, while DQ is homologous to mouse H-2A. All these loci are highly polymorphic except for H-2Eα and its homologues in humans, DPα and DRα, which are much less so. There are many other loci in the complex that encode class-I-MHC-like molecules, but their functions are unknown.

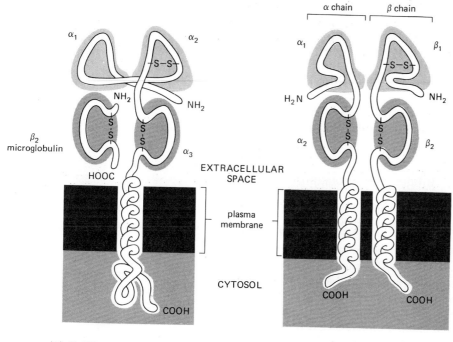

α chain   β chain

(A) CLASS I MHC GLYCOPROTEIN   (B) CLASS II MHC GLYCOPROTEIN

Figure 18–49 (A) Class I and (B) class II MHC glycoprotein molecules. The α chain of the class I molecule, which is about 345 amino acid residues long, has three extracellular domains, α_1, α_2, and α_3, encoded by separate exons. It is noncovalently associated with a smaller polypeptide chain, β_2-microglobulin (96 amino acids), which is not encoded within the MHC. The α_3 domain and β_2-microglobulin are homologous to immunoglobulin domains. While β_2-microglobulin is invariant, the α chain is extremely polymorphic, mainly in the α_1 and α_2 domains. By generating hybrid genes (using genetic engineering techniques) containing a mixture of α_1, α_2, and α_3 exons from different alleles at one locus and transfecting them into cultured fibroblasts, it has been shown that the antigenic determinants recognized by T cells are formed by the interaction of the α_1 and α_2 domains.

In class II MHC molecules, both chains are polymorphic (β more than α), mainly in the α_1 and β_1 domains; the α_2 and β_2 domains are homologous to immunoglobulin domains. Transfection experiments analogous to those described for class I molecules indicate that the antigenic determinants on class II molecules recognized by T cells are formed by the interaction of the α_1 and β_1 domains.

Thus there are striking similarities between class I and class II MHC glycoproteins: both have four extracellular domains, three of which have intrachain disulfide bonds. The two domains closest to the membrane are Ig-like. The other two domains interact to form a complex three-dimensional surface, which, as we shall see later, is thought to bind foreign antigen and present it to T cells. All of the chains are glycosylated except for β_2-microglobulin (not shown).

cytes on those host cells that have foreign antigen on their surface—for example, on virus-infected cells. How this discovery helped resolve much of the MHC puzzle will be discussed later (see p. 1043).

There Are Two Principal Classes of MHC Molecules[37]

There are two main classes of MHC molecules, *class I* and *class II*, each consisting of a set of cell-surface glycoproteins encoded in two linked clusters of genes that together comprise the major histocompatibility complex (Figure 18–48). Both classes of MHC glycoproteins are heterodimers with homologous overall structures; their amino-terminal domains are thought to be specialized for binding antigen for presentation to T cells.

Each **class I MHC gene** encodes a single transmembrane polypeptide chain (called α), most of which is folded into three extracellular globular domains (α_1, α_2, α_3). Each *α chain* is noncovalently associated with an extracellular, nonglycosylated small protein called *β_2-microglobulin*, which does not span the membrane and is separately encoded by a gene on a different chromosome (Figure 18–49A). β_2-microglobulin and the α_3 domain, which are closest to the membrane, are both homologous to an immunoglobulin domain. The two amino-terminal domains of the α chain, which are farthest from the membrane, contain the polymorphic (variable) residues that are recognized by T cells in transplantation reactions.

Like class I MHC molecules, **class II MHC molecules** are heterodimers with two conserved immunoglobulinlike domains close to the membrane and two polymorphic (variable) amino-terminal domains farthest from the membrane. In these molecules, however, both chains are encoded within the MHC, and both span the membrane (Figure 18–49B). The presence of immunoglobulinlike domains in class I and class II glycoproteins suggests that MHC molecules and antibodies have a common evolutionary history (see p. 1053).

There is strong evidence that the polymorphic regions of both classes of MHC molecules interact with foreign antigen and that it is the complex of MHC molecule and foreign antigen that is recognized by the T cell receptor. Before discussing this evidence, however, we shall consider the different parts played by class I and class II molecules in guiding cytotoxic and helper T lymphocytes, respectively, to their appropriate target cells.

The main functional difference between class I and class II MHC molecules reflected in their tissue distribution. Class I MHC molecules are expressed on virtually all nucleated cells, whereas class II molecules are confined largely to cells involved in immune responses. This presumably is because class I molecules are recognized by cytotoxic T cells, which must be able to focus on any cell in the body that happens to become infected with a virus, whereas class II molecules are recognized by helper T cells, which interact mainly with other cells involved in immune responses, such as B cells and *antigen-presenting cells* (Figure 18–50 and see p. 1045). The principal features of the two classes of MHC glycoproteins are summarized in Table 18–2.

Cytotoxic T Cells Recognize Foreign Antigens in Association with Class I MHC Molecules[38]

The first clear evidence that MHC molecules present foreign antigens to T cells came from an experiment on cytotoxic T cells performed in 1974. Mice of strain X were infected with virus A. Seven days later, the spleens of these mice contained active cytotoxic T cells that could kill virus-infected, strain-X fibroblasts within several hours in cell culture. As expected, they would kill the fibroblasts only if the fibroblasts were infected with virus A and not if they were infected with virus B; thus the cytotoxic T cells were virus-specific. Unexpectedly, however, the same T cells were unable to kill fibroblasts from strain-Y mice infected with the same virus A (Figure 18–51). The cytotoxic T cells were thus recognizing some difference between the two kinds of fibroblasts and not just the virus. By using special strains of mice (known as *congenic strains*) that were genetically identical except for their class I MHC loci, or genetically different except for their class I MHC loci, it was possible to show that infected target cells would be killed only if they expressed at least one of the same class I MHC molecules expressed by the original infected mouse. This indicated that class I MHC glycoproteins are necessary to present cell-surface-bound viral antigens to cytotoxic T cells. Because the T cells of any individual will recognize antigen only in association with the MHC molecules expressed by that individual, this MHC co-recognition is often known as *MHC restriction*. It was not until 10 years later, however, that a series of experiments on cytotoxic T cells responsive to influenza virus demonstrated the chemical nature of the viral antigens recognized by cytotoxic T cells.

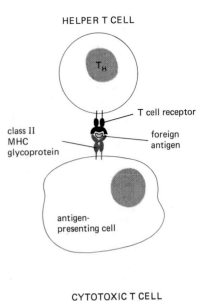

HELPER T CELL

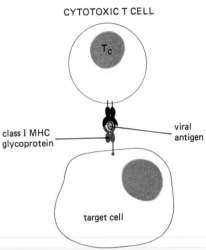

CYTOTOXIC T CELL

Figure 18–50 Cytotoxic T cells recognize foreign viral antigens in association with class I MHC glycoproteins on the surface of any host cell, whereas helper T cells recognize foreign antigens in association with class II MHC glycoproteins on the surface of an *antigen-presenting cell*. In transplantation reactions, too, helper cells react against foreign class II glycoproteins and cytotoxic cells react against foreign class I glycoproteins.

Table 18–2 Properties of Class I and Class II MHC Molecules

	Class I	Class II
Genetic loci	*H-2K, H-2D, H-2L* in mice *HLA-A, HLA-B, HLA-C* in humans	*I-A* and *I-E* clusters in mice; *DP, DQ, DR,* and one or two other clusters in humans
Chain structure	α chain (~45,000 daltons) + β₂-microglobulin (11,500 daltons)	α chain (29,000 to 34,000 daltons) + β chain (25,000 to 28,000 daltons)
Cell distribution	almost all nucleated cells	B cells, antigen-presenting cells, thymus epithelial cells, some others
Involved in presenting antigen mainly to	cytotoxic T cells	helper T cells
Polymorphic domains involved in T cell recognition and antigen binding	$\alpha_1 + \alpha_2$	$\alpha_1 + \beta_1$

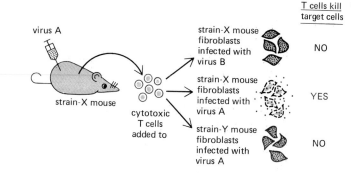

virus A

strain-X mouse

strain-X mouse
fibroblasts
infected with
virus B   NO

cytotoxic
T cells
added to

strain-X mouse
fibroblasts
infected with
virus A   YES

strain-Y mouse
fibroblasts
infected with
virus A   NO

T cells kill
target cells

Figure 18–51 The classic experiment showing that a cytotoxic T cell recognizes some aspect of the host target-cell surface in addition to a viral antigen. By repeating this experiment with target cells that differed from the cells of the infected mouse only in limited regions of their genome, the feature of the target-cell surface that the cytotoxic T cell recognizes was shown to be a class I MHC glycoprotein.

The most convenient way of measuring the ability of active cytotoxic T cells to kill uses the radioactive isotope ^{51}Cr, which is taken up by living cells and released only when the cells die. Thus the standard assay for cytotoxic T cells involves incubating them with ^{51}Cr-containing target cells for several hours and measuring the amount of ^{51}Cr radioactivity released from killed target cells.

Cytotoxic T Cells Recognize Fragments of Viral Proteins on the Surface of Virus-infected Cells[39]

It has been known since the 1960s that T cells, unlike B cells (and antibodies), do not usually recognize antigenic determinants on the folded structure of a protein (see Figure 18–23A, p. 1018); instead, they recognize determinants on the unfolded polypeptide chain. The reason for this became clear as evidence accumulated (initially from studies on how helper T cells recognize antigen—see p. 1044, below) that the antigens seen by T cells are normally degraded inside a host cell before they are presented on its surface. The first direct evidence for this mechanism of antigen presentation to cytotoxic T cells came from the observation that some of the cytotoxic T cells activated by influenza virus specifically recognize internal proteins of the virus that would not be accessible in the intact virus particle. Subsequent evidence suggested that the T cells were recognizing degraded fragments of the internal viral proteins. Since viruses are intracellular parasites, whose proteins are synthesized from viral genes inside the infected cell (see p. 250), it is thought that some fragments of viral proteins "leak" onto the surface of an infected cell after they are made, associating with MHC molecules either on the cell surface or somewhere inside the cell (Figure 18–52).

Two types of experiments support this view. First, if normal fibroblasts in culture are briefly exposed to fragments of an internal influenza virus protein (the nucleoprotein, NP—Figure 18–52), the cells are recognized and killed by cytotoxic T cells that originally were activated by influenza-virus-infected fibroblasts, but only if both fibroblasts express the same class I MHC glycoproteins. Second, if a DNA sequence encoding a fragment of the influenza nucleoprotein is introduced

Figure 18–52 A cytotoxic T cell will kill a virus-infected cell when it recognizes fragments of viral protein bound to class I MHC molecules on the surface of the infected cell. In the case shown, the peptide fragments are derived from the nucleoprotein (NP) of the influenza virus; for simplicity, this is the only internal viral protein shown. Only a very small proportion of the viral proteins synthesized in the target cell are degraded. It is not known how they are degraded, how the resulting peptide fragments get to the cell surface, or where the fragments first associate with the MHC glycoproteins.

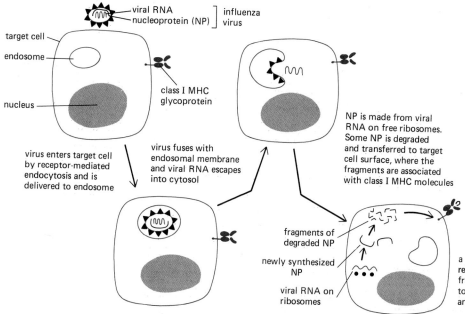

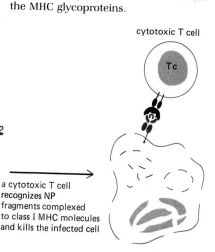

target cell

endosome

nucleus

viral RNA
nucleoprotein (NP)   influenza virus

class I MHC
glycoprotein

virus enters target cell by receptor-mediated endocytosis and is delivered to endosome

virus fuses with endosomal membrane and viral RNA escapes into cytosol

NP is made from viral RNA on free ribosomes. Some NP is degraded and transferred to target cell surface, where the fragments are associated with class I MHC molecules

fragments of degraded NP

newly synthesized NP

viral RNA on ribosomes

a cytotoxic T cell recognizes NP fragments complexed to class I MHC molecules and kills the infected cell

cytotoxic T cell

Tc

into normal fibroblasts, the transfected cells are killed by the cytotoxic T cells described in the first experiment. These and other experiments suggest that fragments of viral proteins can both find their way to the cell surface and associate with class I MHC molecules.

It is not difficult to understand how viral proteins can be degraded in infected cells, since almost all cellular proteins are known to be continually degraded (see p. 418). It is more difficult to understand how fragments of the influenza nucleoprotein get to the cell surface, since the protein is synthesized on cytoplasmic ribosomes and would not normally have access to the lumen of the endoplasmic reticulum, where proteins destined for the cell surface usually begin their journey (see p. 412). T cell recognition probably requires very small amounts of antigen, however, so misrouting only a small fraction of the nucleoprotein fragments to the cell surface may create a target cell that an appropriate cytotoxic T lymphocyte can recognize.

X-ray Diffraction Studies Show the Antigen-binding Site of a Class I MHC Glycoprotein[40]

A major advance in our understanding of how MHC molecules present antigen to T cells came in 1987, when the three-dimensional structure of a human class I MHC glycoprotein was obtained by x-ray crystallography. As shown in Figure 18–53A, the protein has a single putative antigen-binding site located at one end of the molecule. The site consists of a deep groove between two long α helices derived from the nearly identical α_1 and α_2 domains; the base of the groove is formed by eight β strands derived from the same two domains. The size of the groove is about 2.5 nm long, 10 nm wide, and 11 nm deep, which is large enough to accommodate a peptide of about 10 to 20 amino acid residues, depending on the extent to which the peptide is compressed by coiling or bending. Remarkably, the groove in the crystallized protein was not empty: it contained a small molecule of unknown origin, suspected to be a peptide, which co-purified and co-crystallized with the MHC glycoprotein (see Figure 18–53B). This finding strongly implicates the groove as the antigen-binding site and suggests that once a peptide binds to this site, it dissociates very slowly; this conclusion is supported by the observation that fibroblasts exposed for a short period to fragments of the influenza virus nucleoprotein remain targets for influenza-specific cytotoxic T cells for at least 3 days.

Most of the polymorphic amino acid residues in the MHC glycoprotein (those that vary between allelic forms of this type of molecule) are located inside the groove, where they would be expected to bind antigen, or on its edges, where they would be accessible for recognition by the T cell receptor. Presumably the variability in class I MHC molecules has been selected to allow them to bind and present many different virus-derived peptides. Nonetheless, it is still surprising that the small number of different antigen-binding sites associated with the class I MHC molecules in an individual (a maximum of six in humans) can bind the large number of virus-derived peptides that T cells can specifically recognize. Even more puzzling in this respect are the class II MHC glycoproteins, which are thought to have a three-dimensional structure very similar to that of class I molecules. Although an individual makes only about 10 to 20 types of class II molecules, each with its own unique antigen-binding site, these molecules seem to be able to bind and present an apparently unlimited variety of foreign peptides to *helper T cells*, which play a crucial part in almost all immune responses.

Helper T Cells Recognize Fragments of Foreign Antigens in Association with Class II MHC Glycoproteins on the Surface of Antigen-presenting Cells[41]

Helper T cells are required for most other types of lymphocytes to respond optimally to antigen. The crucial importance of helper T cells in immunity is dramatically demonstrated by the devastating epidemic of *acquired immunodeficiency syndrome (AIDS)*. The disease is caused by a retrovirus (human immuno-

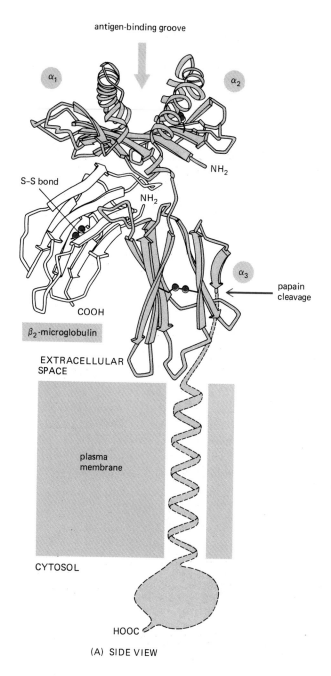

antigen-binding groove

α_1 α_2

NH$_2$

S–S bond

NH$_2$

α_3 → papain cleavage

COOH

β_2-microglobulin

EXTRACELLULAR SPACE

plasma membrane

CYTOSOL

HOOC

(A) SIDE VIEW

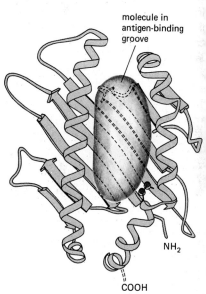

molecule in antigen-binding groove

NH$_2$

COOH

(B) TOP VIEW

Figure 18–53 (A) The structure of a human class I MHC glycoprotein as determined by x-ray diffraction analysis of crystals of the extracellular part of the molecule. The extracellular part was cleaved from the transmembrane segment by the proteolytic enzyme papain. Each of the two domains closest to the plasma membrane (α_3 and β_2-microglobulin) resembles a typical immunoglobulin domain (see Figure 18–28B), while the two domains farthest from the membrane (α_1 and α_2) are very similar to each other and together form a groove at the top of the molecule that is believed to be the antigen-binding site. Class II MHC molecules are thought to have a very similar structure. (B) The putative antigen-binding groove viewed from above, containing the small molecule (thought to be a peptide) that co-purified with the MHC protein. This is also the part of the molecule that interacts with the T cell receptor. (After P.J. Bjorkman, M.A. Saper, B. Samraoui, W.S. Bennett, J.L. Strominger, and D.C. Wiley, *Nature* 329:506–512, 1987.)

deficiency virus, HIV) that kills helper T cells, thereby crippling the immune system and rendering the patient susceptible to infection by microorganisms that rarely infect normal individuals. As a result, most AIDS patients die of infection within several years of the onset of symptoms.

Before they can help other lymphocytes respond to antigen, helper T cells must first be activated themselves. This activation occurs when a helper T cell recognizes a foreign antigen bound to a class II MHC glycoprotein on the surface of a specialized **antigen-presenting cell.** Antigen-presenting cells are found in most tissues. They are derived from bone marrow and comprise a heterogeneous set of cells, including *dendritic cells* in lymphoid organs, *Langerhans cells* in skin, and certain types of macrophages. Together with B cells, which can also present antigen to helper T cells (see below), and thymus epithelial cells (see p. 1052), these specialized antigen-presenting cells are the main cell types that normally express class II MHC molecules (see Table 18–2).

Many kinds of experiments demonstrate the central importance of class II MHC molecules in presenting foreign antigens to helper T cells. The binding of antibodies to class II molecules on antigen-presenting cells, for example, blocks

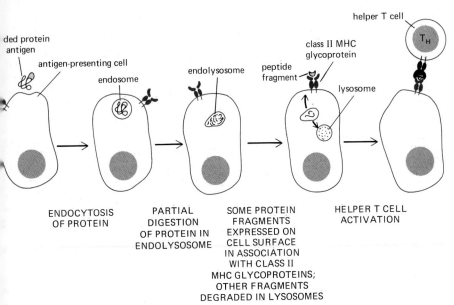

Figure 18–54 How protein antigens are thought to be "processed" and then displayed by an antigen-presenting cell. Since class II MHC glycoproteins have been found to recycle through the endosomal compartment, they may initially associate with peptide fragments in the endolysosomal compartment and then return to the cell surface with a bound peptide (not shown).

the ability of these cells to present foreign antigen to helper T cells. Moreover, fibroblasts, which do not make class II MHC molecules and cannot present foreign antigens to helper T cells, can be converted to effective antigen-presenting cells if they are transfected with a gene that codes for a class II MHC molecule.

Like the viral antigens presented to cytotoxic T cells, the antigens presented to helper T cells on antigen-presenting cells are usually degraded fragments of the foreign protein. These peptides are thought to be bound to class II MHC molecules in the same way that virus-derived peptides are bound to class I MHC molecules (see Figure 18–53). Unlike the virus-infected target of a cytotoxic T cell, however, the antigen-presenting cell does not synthesize the foreign protein. Instead, it is thought that the foreign protein is ingested by endocytosis and partially degraded in the acidic environment of endosomes or endolysosomes (see p. 331) before selected fragments are returned to the cell surface—a sequence of events collectively called **antigen processing** (Figure 18–54). Thus, if endocytosis is blocked by lightly fixing antigen-presenting cells with a chemical such as formaldehyde, or if proteolysis in endolysosomes and lysosomes is inhibited by a drug such as chloroquine, the cells are no longer able to process a foreign protein and present it to helper T cells. Cells treated in these ways, however, are still able to present the protein if it is cleaved into small peptides (10–15 amino acids long) before it is added to the cells.

A remarkable property of an antigen-presenting cell is that it can process and present virtually any antigen to an appropriate helper T cell. This lack of antigen specificity suggests that antigen-presenting cells take up antigen by fluid-phase rather than by receptor-mediated endocytosis (see p. 328). If this is so, then most of the proteins ingested and degraded will be host (self) proteins, whose peptide fragments will occupy the binding site of many of the class II MHC molecules. Presumably the binding of foreign peptides to only a small proportion of MHC molecules is sufficient to activate a helper T cell.

Helper T Cells Stimulate Activated T Lymphocytes to Proliferate by Secreting Interleukin-2[42]

Activation of a helper T cell is a complex process involving various secreted proteins called **interleukins,** which act as local chemical mediators. Activation is thought to begin when the T cell, by unknown means, stimulates the antigen-presenting cell to secrete one or more interleukins. The best characterized of these mediators is **interleukin-1 (IL-1).** The combined action of IL-1 (and probably other interleukins) and antigen binding, however, do not stimulate helper T cell proliferation directly. Instead, they cause the T cell to stimulate its own proliferation by inducing it to secrete a growth factor called **interleukin-2 (IL-2)** as well as to

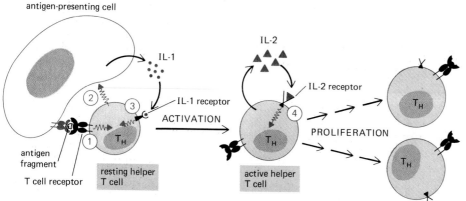

antigen-presenting cell

IL-1

IL-1 receptor

antigen fragment

T cell receptor

resting helper T cell

ACTIVATION

IL-2

IL-2 receptor

active helper T cell

PROLIFERATION

T_H

Figure 18–55 The sequence of signaling events believed to occur when antigen stimulates helper T cells to proliferate. Binding of the T cell to the antigen on the surface of an antigen-presenting cell signals the T cell receptor to trigger the inositol phospholipid cell-signaling pathway (see p. 702) (signal 1). This causes the T cell to stimulate the antigen-presenting cell by an unknown mechanism (signal 2). The antigen-presenting cell then secretes interleukins, such as interleukin-1 (IL-1) which help activate the T cell (signal 3). The activated T cell makes interleukin-2 (IL-2) receptors and secretes IL-2; the binding of IL-2 to its receptors (signal 4) stimulates the cell to grow and divide. When the antigen is eliminated, the T cells eventually stop producing IL-2 and IL-2 receptors, so cell proliferation stops.

synthesize cell-surface IL-2 receptors. It is the binding of IL-2 to these receptors that stimulates the T cell to proliferate. In this way the helper T cell can continue to proliferate, through an *autocrine mechanism* (see p. 690), after it has left the surface of the antigen-presenting cell (Figure 18–55). The helper T cell can also help stimulate the proliferation of any other T cells, including cytotoxic T cells, that have first been induced to express IL-2 receptors. Because the expression of IL-2 receptors is strictly dependent on antigen stimulation, however, this does not result in the indiscriminate proliferation of all T cells, but only those that have encountered antigen.

Once the requirements for T cell proliferation were discovered, it was possible to produce indefinitely proliferating, antigen-specific *T cell lines* in culture by continuously administering IL-2 and periodically stimulating the cells with antigen to maintain the expression of IL-2 receptors. Single cells from such lines could then be isolated to generate **T cell clones.** As we have seen, such clones have been critically important in T cell research. They made it possible, for example, to isolate T cell receptors and their genes; they have also been widely used to study the mechanisms of T cell activation and the role of helper T cells in stimulating the responses of other lymphocytes.

Helper T Cells Are Required for Most B Cells to Respond to Antigen[43]

Helper T cells are essential for B cell antibody responses to most antigens. This was first discovered in the mid-1960s through experiments in which either thymus cells or bone marrow cells were injected together with antigen into irradiated mice. Mice that had received only bone marrow or only thymus cells were unable to make antibody; but if a mixture of thymus and bone marrow cells was injected, large amounts of antibody were produced. It was later shown that the thymus provides T cells, while the bone marrow provides B cells (Figure 18–56). The use of a specific chromosome marker to distinguish between the injected T and B cells showed that the antibody-secreting cells are B cells, leading to the conclusion that T cells must help B cells respond to antigen.

There are some antigens, however, including many microbial polysaccharides, that can stimulate B lymphocytes to proliferate and mature without T cell help. Such *T-cell-independent antigens* are usually large polymers with repeating, identical antigenic determinants whose multipoint binding to the membrane-bound antibody molecules that serve as antigen receptors on B cells may generate a strong enough signal to activate B cells directly. There is evidence that the cells that respond to multimeric antigens in this way are mainly a separate subset of B cells that has evolved to react against microbial polysaccharides without T-cell help.

Helper T Cells Help Activate B Cells by Secreting Interleukins[44]

Once activated by foreign antigen on the surface of a specialized antigen-presenting cell, an appropriate helper T cell can help activate a B cell by binding to the same foreign antigen on the B cell surface. The antigen-presenting cell ingests and presents antigens nonspecifically (see p. 1046), but a B cell generally

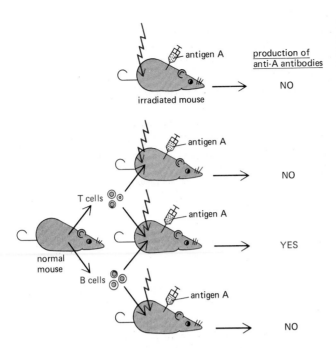

antigen A

production of
anti-A antibodies

NO

irradiated mouse

antigen A

NO

T cells

antigen A

YES

normal
mouse

B cells

antigen A

NO

Figure 18–56 The experiment that first suggested that both T cells and B cells are required if an animal is to make antibody responses. The dose of irradiation used kills the T cells and B cells of the irradiated mouse.

presents only an antigen that it specifically recognizes. The antigen is selected by its binding to the specific membrane-bound antibodies (antigen receptors) on the surface of the B cell; it is ingested by receptor-mediated endocytosis (see p. 328) and is then degraded and recycled to the cell surface in the form of peptides bound to class II MHC glycoproteins for recognition by the helper T cell. Thus the helper T cell recognizes the same antigen-MHC complexes on the B cell it helps as on the antigen-presenting cell that initially activated the T cell.

The specific contact between a helper T cell and a B cell initiates an internal rearrangement of the helper cell cytoplasm that orients the centrosome and Golgi apparatus toward the B cell, as described previously for a cytotoxic T cell contacting a target cell (see Figure 18–47). In this case, however, the orientation is thought to enable the helper T cell to direct the secretion of interleukins (and perhaps to focus membrane-bound signaling molecules) onto the B cell surface. These interleukins include IL-4, which helps initiate B cell activation, IL-5, which stimulates activated B cells to proliferate, and IL-6, which induces activated B cells to mature into antibody-secreting cells. Some of these and other interleukins can induce B cells to switch from making one class of antibody to making another (see p. 1029). Some of the signals thought to be involved in the initial activation of a B cell are illustrated in Figure 18–57.

How do signals pass from activated cell-surface receptors to the cell interior when B or T cells are stimulated by antigen and interleukins? The answer is not known for interleukin receptors, but there is strong evidence that receptors for antigen on both B and T cells signal the cell by activating the inositol phospholipid pathway discussed in Chapter 12 (see p. 702).

Some Helper T Cells Activate Macrophages by Secreting γ-Interferon[45]

Helper T cells do not confine their help to lymphocytes. Those helper T cells that secrete IL-2 when stimulated by antigen also secrete other interleukins, such as γ-interferon, that attract macrophages and activate them to become more efficient at phagocytosing and destroying invading microorganisms. The ability of T cells to attract and activate macrophages is especially important in defense against infections by microorganisms that can survive simple phagocytosis by nonactivated macrophages. Tuberculosis is one such infection.

The antigen-triggered secretion of γ-interferon and other macrophage-activating interleukins by helper T cells underlies the familiar tuberculin skin test. If tuberculin (an extract of the bacterium responsible for tuberculosis) is injected

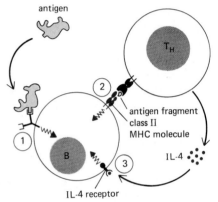

antigen

T_H

antigen fragment
class II
MHC molecule

IL-4

IL-4 receptor

Figure 18–57 At least three types of signals are likely to be involved in the initial stages of B cell activation. The relative importance of these signals is uncertain and may vary depending on the type of B cell and antigen. Signal 1 is caused by antigen binding and is thought to be mediated by the inositol phospholipid cell-signaling pathway (see p. 702); it helps activate the B cell and may induce the expression of receptors for some of the helper-T-cell-derived interleukins. The B cell then ingests and degrades the antigen (not shown) and presents small fragments of the antigen to the helper T cell in association with class II MHC molecules. It is not clear if T cell binding signals the B cell (shown here as signal 2) or only serves to focus the secretion of interleukin-4 (IL-4) and other interleukins (not shown) onto the B cell surface (signal 3). In addition to activating the B cell, signal 3 stimulates the cell to make more class II MHC glycoprotein, thereby increasing the ability of the B cell to receive T cell help. Once the B cell is activated, other helper-T-cell-derived interleukins (such as IL-5, IL-6, and γ-interferon) help induce the cell to proliferate and mature into an antibody-secreting cell (not shown).

into the skin of individuals who have had or have been immunized against tuberculosis, a characteristic immune response occurs in the skin. It is initiated at the site of injection by the secretion of interleukins by memory helper T cells that react to tuberculin. The interleukins attract macrophages and lymphocytes into the site, thereby causing the characteristic swelling of a positive reaction to tuberculin.

Another important effect of γ-interferon is to induce the expression of class II MHC glycoproteins on the surface of some cells (such as endothelial cells) that do not normally express them. This enables these cells to present antigen to helper T cells. In this way helper T cells can recruit extra antigen-presenting cells when the need arises.

There is evidence that there are at least two subclasses of helper T cells. One seems to be concerned mainly with helping B cells and secretes IL-4 and IL-5; the other seems to be concerned mainly with helping other T cells and macrophages and secretes IL-2 and γ-interferon. Some of the interleukins secreted by helper T cells (or antigen-presenting cells) are listed in Table 18–3.

Cell-Cell Adhesion Proteins Stabilize the Interactions Between T Cells and Their Targets[46]

The specific binding of antigen-MHC complexes on the surface of a target cell to α/β antigen receptors on the surface of a T cell is often not strong enough to mediate a functional interaction between the two cells. Various cell-cell adhesion proteins (see Chapter 14, p. 824) on T cells help stabilize such interactions by increasing the overall strength of cell-cell binding. We have already discussed in Chapter 14 (see p. 822) the role of the *lymphocyte-function-associated* protein *LFA-1*

Table 18–3   Properties of Some Interleukins*

Interleukin (IL)	Alternative Name	Approximate Molecular Weight	Source	Target	Action
IL-1	—	15,000	antigen-presenting cells	helper T cells	helps activate
IL-2	T cell growth factor	15,000	some helper T cells	all activated T cells	stimulates proliferation
IL-3	multi-CSF (see p. 980)	25,000	some helper T cells	various hemopoietic cells (see p. 981)	stimulates proliferation
IL-4	B cell stimulating factor-1 (BSF-1)	20,000	some helper T cells	B cells, T cells, mast cells	helps activate and promotes proliferation; increases class II MHC molecules on B cells
IL-5	B cell growth factor-2 (BCGF-2)	50,000 (dimer)	same helper T cells that make IL-4	B cells, eosinophils	promotes proliferation and maturation
IL-6	B cell stimulating factor-2 (BSF-2)	25,000	some helper T cells and macrophages	activated B cells, T cells	promotes B cell maturation to Ig-secreting cells; helps activate T cells
γ-Interferon	—	25,000 (dimer)	same helper T cells that make IL-2	B cells, macrophages, endothelial cells	induces class II MHC molecules and activates macrophages

*__Interleukins__ are secreted peptides and proteins that mediate local interactions between white blood cells (leucocytes) but do not bind antigen; those secreted by lymphocytes are also called **lymphokines**. The amino acid sequence is known for all the proteins listed. The sources, target cells, and actions listed are those most relevant to the immune system; most of the interleukins have many more sources, targets, and actions than are shown and are therefore more accurately called **cytokines**.

n helping T and B cells (and other white blood cells) to adhere to other cells and o the extracellular matrix. T cells also express a cell-surface protein called *CD2*, which helps them adhere to their target cells by binding to a complementary glycoprotein on the target cell surface called *LFA-3*.

Among the best characterized of the cell-cell adhesion proteins on T cells are he **CD4** and **CD8** glycoproteins, which are expressed on the surface of helper and cytotoxic T cells, respectively. Both glycoproteins have extracellular domains hat are homologous to immunoglobulin domains, and they are thought to bind o invariant parts of MHC molecules—CD4 to class II and CD8 to class I MHC glycoproteins (Figure 18–58).

Some of the accessory glycoproteins that are found on the surface of T lymphocytes are summarized in Table 18–4.

Suppressor T Cells Mainly Suppress Helper T Cells[47]

The discovery that T lymphocytes can *help* B cells make antibody responses was followed several years later by the discovery that they can also *suppress* the response of B cells or other T cells to antigens. Such T cell suppression was first demonstrated in mice that had been made specifically unresponsive (tolerant) to sheep red blood cells (SRBC) by repeated injections of large numbers of SRBCs. When T cells from tolerant mice were injected into normal mice, the latter also became specifically unresponsive to SRBC antigens. This implies that the tolerant state in this case is due to suppression of the response by T cells. Subsequent experiments using surface antigenic markers suggested that the cells responsible are a specialized class of T lymphocytes, called **suppressor T cells.** As we shall discuss below, however, not all forms of immunological tolerance are due to suppressor T cells.

Together, helper T cells and suppressor T cells are thought to control the activity of B cells and cytotoxic T cells, the major effector cells of the immune system. Helper T cells act directly on these effector cells, and suppressor T cells are thought to act indirectly by inhibiting the helper T cells on which the effector cells depend, although the mechanism of inhibition is unknown. How do suppressor T cells recognize the helper T cells they suppress? Because of the way helper cells recognize foreign antigens (see p. 1046), it seems unlikely that they would have sufficient foreign antigen (or fragments of foreign antigen) on their surface for suppressor T cells to recognize. Instead, it seems that suppressor T cells often interact with helper T cells by recognizing antigenic determinants associated with the antigen-binding sites of the helper T cell receptor—so-called *idiotopes* (see p. 1030), as illustrated in Figure 18–59.

The discovery of suppressor T cells raised the question of whether they play a part in self-tolerance by suppressing self-reactive lymphocytes. The available evidence is still controversial, but it suggests that self-tolerance is due mainly to the elimination of self-reactive lymphocytes (so-called *clonal deletion*) and does not depend on suppressor T cells. Since most B cells require helper T cells to respond to antigen, in principle it is necessary only to eliminate self-reactive helper T cells in order to avoid B cell responses to host macromolecules. This is the strategy used for many "self" antigens. Normal mice, for example, will not make antibodies against their own complement component C5. Their B cells, however, can be induced to make such antibodies if they are provided with helper T cells from mutant mice that lack C5 but are otherwise identical. Thus the only reason that normal mice do not make antibodies against this common serum protein is that the helper T cells that recognize C5 are absent or inactivated.

This mechanism, however, will not work for all self antigens. Those self macromolecules that can activate B cells without T cell help, for instance, appear to eliminate the B cells that recognize them, as do those self macromolecules that are present in high concentration. Similarly, cytotoxic T cells that could react against normal host cell-surface molecules must be eliminated, since cytotoxic T cells can be activated by antigen to some extent without helper T cells, although they respond much more vigorously with such help. It is likely that suppressor T cells play primarily a backup role in self-tolerance, being called into play only when the primary mechanism of clonal deletion fails.

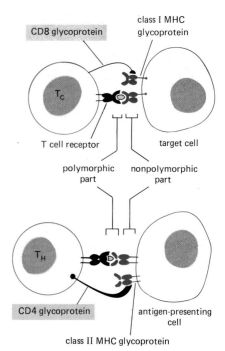

Figure 18–58 The role of two accessory receptor proteins on the surface of T cells. The CD8 glycoprotein on cytotoxic T cells is thought to bind to class I MHC molecules and the CD4 glycoprotein on helper T cells to class II MHC molecules. In both cases the binding is thought to be to nonvariable parts of the MHC molecules. These cell-cell adhesion proteins help stabilize the binding between T cell receptors and antigen-MHC complexes on the target cell, especially when the binding is weak; in these cases antibodies against these accessory receptor proteins inhibit T cell activation.

Antibodies against CD4 and CD8 are widely used to distinguish helper and cytotoxic T cells, respectively. The virus (HIV) that causes AIDS infects helper T cells by initially binding to CD4 molecules on the surface of these cells.

Table 18–4 Accessory Glycoproteins on the Surface of T Cells

Protein*	Alternative Name	Approximate Molecular Weight	Expressed on	Putative Function
CD2	T11	50,000	all T cells	promotes adhesion between T cells and their target cells by binding to LFA-3 on target cells
CD3	T3	γ chain = 25,000 δ chain = 20,000 ϵ chain = 20,000 ξ chain = 16,000	all T cells	helps transduce signal when antigen-MHC complex binds to T cell receptors
CD4	T4 in humans } L3T4 in mice	50,000	helper T cells	promotes adhesion to antigen-presenting cells and B cells, probably by binding to class II MHC molecules
CD8	T8 in humans Lyt2 } in mice Lyt3	60,000 (homodimer) 70,000 (heterodimer)	cytotoxic T cells	promotes adhesion to virus-infected target cells, probably by binding to class I MHC molecules
LFA-1	—	α chain = 190,000 β chain = 95,000	most white blood cells	promotes cell-cell and cell-matrix adhesion

*CD stands for *cluster of differentiation*, as each of the CD proteins was originally defined as a T cell "differentiation antigen" recognized by multiple monoclonal antibodies. Their identification depended on large-scale collaborative studies in which hundreds of such antibodies, generated in many laboratories, were compared and found to consist of relatively few groups (or "clusters"), each recognizing a single cell-surface protein.

Developing T Cells That React Strongly to Self MHC Molecules Are Eliminated in the Thymus[48]

As discussed previously, the early evidence that MHC glycoproteins are involved in T cell antigen recognition came from experiments showing that T cells can respond to antigen in association with self MHC molecules but not in association with foreign MHC molecules: that is, they showed *MHC restriction* (see p. 1042). Soon after, experiments with thymus transplants suggested that, during their development in the thymus, T cells *learn* to see antigen in association with self rather than foreign MHC molecules. For example, if a Y-strain thymus is transplanted into an X-strain mouse, which has been irradiated to eliminate all its mature T cells and then supplied with fresh bone marrow to provide new ones, new X-strain T cells will develop in the Y-strain thymus. In most cases the mature X-strain T cells produced recognize foreign antigen in association with Y-strain but not X-strain MHC glycoproteins. The simplest interpretation of this finding is that as T cells develop in the thymus, those with receptors that can recognize antigen in association with the types of MHC molecules expressed in the thymus are somehow selected to proliferate. A corollary of this hypothesis is that in the course of this *positive selection*, cytotoxic cells are selected for their recognition of class I MHC molecules while helper cells are selected for their recognition of class II MHC molecules. In support of this view, antibodies against class II MHC molecules specifically block helper T cell development, while antibodies against class I molecules specifically block cytotoxic T cell development.

This interpretation is not entirely satisfying, however, because it fails to explain how the selection is achieved in the absence of the foreign antigens that will later be recognized by the T cells. One possibility is that T cells need to bind weakly to self MHC molecules in order to survive and mature and are thus selected for a low degree of self MHC recognition that is insufficient on its own to activate mature T cells; activation would occur only when the addition of a foreign antigen to a self MHC molecule produces a structure to which the T cell receptor can bind strongly.

The evidence for *positive selection* for weak self-MHC recognition in the thymus is less convincing than the evidence for a *negative selection* process that

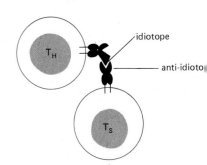

Figure 18–59 An interaction between a helper T cell (T_H) and a suppressor T cell (T_S) in which the receptor on one cell recognizes an idiotope (an antigenic determinant associated with the antigen-binding site) on the receptor of the other cell. An alternative possibility is that the suppressor cell recognizes an idiotope on a fragment of the helper cell receptor that is displayed on the helper cell surface in association with an MHC molecule (not shown). In either case the suppressor cell inhibits the function of the helper cell by an unknown mechanism.

eliminates those cells in the thymus that bind too strongly to self MHC molecules or to self MHC molecules in association with other self molecules. The most compelling evidence that strongly self-MHC-reactive T cells are eliminated in the thymus comes from genetic studies in mice. One such study depends on the fortuitous finding that one of the V gene segments that codes for the β chain of the T cell receptor confers on any T cell expressing it strong recognition of a specific class II MHC molecule (called H-2E), regardless of the β chain's associated D and regions, or of the V region of the α chain. (This finding suggests that T cell receptor V gene segments may have been selected in evolution for their ability to encode receptors that bind MHC molecules.) Not all strains of mice, however, express H-2E. In those that do not, T cells expressing the specific β-chain V segment are found among both immature and mature thymus lymphocytes. In those that do, such T cells are found only among the immature population of thymus lymphocytes; apparently such T cells are eliminated before they mature in the thymus.

Thymus transplantation experiments suggest that the positive selection proposed to explain MHC restriction and the negative selection conferring self MHC tolerance occur separately. The positive selection seems to occur on the surface of intrinsic thymus epithelial cells, while the negative selection seems to occur on the surface of cells that migrate into the thymus from the bone marrow; both types of cells express both class I and class II MHC molecules on their surface.

The mechanisms responsible for T cell selection in the thymus are unknown. Perhaps positive selection is mediated by growth or survival signals provided by the thymus epithelium to weakly binding T cells. In negative selection the bone-marrow-derived cell may act as a perverse antigen-presenting cell, killing rather than activating any T cell that recognizes it; cells with this property, called *veto cells*, have been demonstrated *in vitro*.

A remarkable feature of T cell development in the thymus is that more than 95% of the cells die without leaving the thymus. This waste is presumably due to the stringent selection that operates on developing T cells.

Some Allelic Forms of MHC Molecules Are Ineffective at Presenting Specific Antigens to T Cells: Immune Response (*Ir*) Genes[49]

Unlike class I MHC genes, which were first recognized by their effects on graft rejection, class II MHC genes were first recognized by their effects on T-cell-dependent immune responses to specific soluble antigens. When animals were immunized with a simple antigen, some made vigorous T-cell-dependent responses while others did not respond at all. Genetic studies indicated that the ability to respond to the antigen was controlled by a single gene, called an **immune response (*Ir*) gene**, and responses to different antigens were often controlled by different *Ir* genes. *Ir* genes that control the response of helper T cells to an antigen were the first to be mapped, and they defined the class II MHC loci; those that controlled the response of cytotoxic T cells to an antigen were later mapped to one or other of the class I MHC loci.

These observations were extremely puzzling until it was recognized that the MHC glycoproteins play a crucial role in presenting antigen to T cells. Now they can be explained by the simple proposal that a genetic nonresponder to a simple antigen (usually one with only a single antigenic determinant) lacks an MHC molecule that can bind and effectively present the antigenic determinant to an appropriate T cell. Strong support for this view has come from *in vitro* studies showing that purified class II MHC molecules from a *responder* animal can bind the relevant antigenic peptide, while those from a genetic *nonresponder* cannot. Further binding experiments have shown that class II molecules have a single antigen-binding site (as do class I MHC molecules—see Figure 18–57) that can bind a wide variety of peptides with an average affinity constant (K_a) of about 10^6 liters/mole (a ΔG of -8.5 kcal/mole, equivalent to the binding energy of about eight hydrogen bonds—see p. 89). Moreover, the binding rate is slow (about 10^5-fold slower than for a typical antibody-antigen reaction), and once bound, the peptide is released with a half-life of more than a day, suggesting that a slow conformational change may have to occur in the MHC molecule for the peptide to be released.

Another mechanism seems to be responsible for some cases of genetic non-responsiveness to specific antigens. Certain combinations of self MHC molecules and foreign peptides are likely to resemble other self MHC molecules. Because the helper T cells that react to such combinations are eliminated by negative selection during T cell development in the thymus (see p. 1052), an animal may be genetically unable to respond to these foreign peptides.

MHC Co-recognition Provides an Explanation for Transplantation Reactions and MHC Polymorphism

MHC co-recognition provides an explanation for why so many T cells respond to foreign MHC molecules and thereby reject foreign organ grafts. T cells may be obsessed with foreign MHC glycoproteins because these molecules (either alone or complexed with other molecules on the foreign cell surface) resemble various combinations of self MHC molecules complexed with foreign peptides. Thus, for example, some T cell clones that react to a viral antigen in association with a self class I MHC molecule have been shown to react to a foreign class I MHC molecule in the absence of the viral antigen.

MHC co-recognition can also explain the extensive polymorphism of MHC molecules. In the evolutionary war between pathogenic microorganisms and the immune system, microorganisms will tend to change their antigens to avoid associating with MHC molecules. When one succeeds, it will be able to sweep through a population as an epidemic. In such circumstances, the few individuals that produce a new MHC molecule that can associate with an antigen of the altered microorganism will have a large selective advantage. In addition, individuals with two different alleles for each MHC molecule (heterozygotes) will have a better chance of resisting infection than those with identical alleles at any given MHC locus. Thus selection will tend to promote and maintain a large diversity of MHC molecules in the population.

While MHC co-recognition has provided at least tentative answers to many of the questions raised initially by organ transplantation experiments, it has raised another in their place. How do fewer than two dozen different MHC molecules in an animal associate with enough different peptides to ensure that T cells can respond to virtually any protein antigen? The interactions of antigen with antibodies and with class I MHC glycoproteins have been clarified by x-ray diffraction studies of these molecules. Such analyses need to be extended to the interaction between the MHC-antigen complex and the T cell receptor. Recombinant DNA techniques should soon provide abundant amounts of T cell receptors in soluble form, making such projects feasible. Studies using recombinant DNA techniques have already shown that all of these proteins—MHC molecules, T cell receptors, and antibodies—have a common and ancient history.

Immune Recognition Molecules Belong to an Ancient Superfamily[50]

Most of the glycoproteins that mediate cell-cell recognition or antigen recognition in the immune system contain related structural elements, suggesting that the genes that encode them have a common evolutionary history. Included in this **Ig superfamily** are *antibodies, T cell receptors, MHC glycoproteins*, the *CD2, CD4* and *CD8* cell-cell adhesion proteins, some of the polypeptide chains of the *CD3 complex* associated with T cell receptors, and the various *Fc receptors* on lymphocytes and other white blood cells—all of which contain one or more immunoglobulin (Ig)-like domains (*Ig homology units*). Each of these domains is typically about 100 amino acids in length and is thought to be folded into the characteristic sandwichlike structure made of two antiparallel β sheets, usually stabilized by a conserved disulfide bond (see p. 1021). Many of these molecules are dimers or higher oligomers in which Ig homology units of one chain interact with those in another (Figure 18–60).

Each Ig homology unit is usually encoded by a separate exon, and it seems likely that the entire supergene family evolved from a gene coding for a single Ig

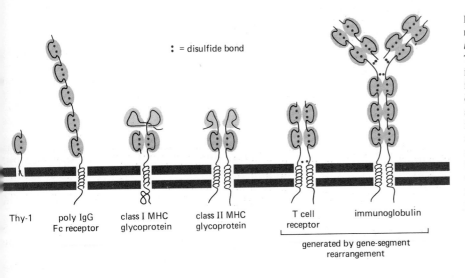

: = disulfide bond

Thy-1   poly IgG   class I MHC   class II MHC   T cell   immunoglobulin
      Fc receptor   glycoprotein   glycoprotein   receptor

generated by gene-segment
rearrangement

Figure 18–60 Some of the membrane proteins belonging to the *immunoglobulin superfamily.* The homologous immunoglobulin and immunoglobulinlike domains are shown in color; note that disulfide bonds connect the ends of each loop forming such a domain. Most of the domains interact with homologous domains of an associated polypeptide chain. It is also likely that some of the domains on the poly Ig Fc receptor chain interact with one another (not shown); this receptor binds both dimeric IgA (see p. 1016) and polymeric IgM, hence its name. The Thy-1 glycoprotein is linked to the membrane by covalent attachment to a glycosylated phospholipid molecule (see p. 448). Not included in the figure are the CD4 and CD8 accessory proteins on helper and cytotoxic T cells, respectively, the CD3 protein complex associated with T cell receptors, and CD2 (see Table 18–4), all of which also contain immunoglobulinlike domains. The immunoglobulin superfamily also includes cell-surface proteins involved in cell-cell interactions outside the immune system, such as the neural cell adhesion molecule (N-CAM, see p. 830).

homology unit—similar to that encoding Thy-1 or β_2-microglobulin (see Figure 18–54), which may have been involved in mediating cell-cell interactions. Since a Thy-1-like molecule has been isolated from the brain of squids, it is probable that such a primordial gene arose before vertebrates diverged from their invertebrate ancestors some 400 million years ago. New family members presumably arose by exon and gene duplications, and similar duplication events probably gave rise to the multiple gene segments that encode antibodies and T cell receptors.

Its remarkable powers of recognition make the immune system almost unique among cellular systems; only the nervous system is more complex. Both systems are composed of very large numbers of phenotypically distinct cells organized into intricate networks. Within the network, individual cells can interact either positively or negatively, and the response of one cell reverberates through the system by affecting many other cells. While the neural network is relatively fixed in space, the cells constituting the immunological network are constantly changing their locations and interact with one another only transiently. In the next chapter we shall consider the cells of the vertebrate nervous system, which is by far the most complex and sophisticated cellular system known.

Summary

There are at least three functionally distinct subclasses of T cells: (1) cytotoxic T cells, which can kill virus-infected cells directly; (2) helper T cells, which secrete a variety of local chemical mediators (interleukins) that help B cells make antibody responses, stimulate activated T cells to proliferate, and activate macrophages; and (3) suppressor T cells, which are thought mainly to inhibit the responses of helper T cells. Helper and suppressor T cells are the principal regulators of immune responses.

The T cell receptor is an antibodylike heterodimer encoded by genes assembled from multiple gene segments during T cell development in the thymus. T cells are activated when these receptors bind to fragments of foreign antigen that are bound to MHC glycoproteins on the surface of another host cell. This process of MHC co-recognition ensures that T cells recognize a foreign antigen only when it is bound to an appropriate target cell. There are two main classes of MHC molecules: (1) class I molecules are expressed on almost all nucleated somatic cells and present fragments of viral proteins to cytotoxic T cells; (2) class II molecules are expressed on B cells and specialized antigen-presenting cells and present fragments of foreign antigens to helper T cells. The finding that certain allelic forms of class I and class II MHC molecules are ineffective in presenting particular antigenic determinants to T cells probably explains why these molecules are so polymorphic.

References

General

Golub, E.S. Immunology: A Synthesis. Sunderland, MA: Sinauer, 1987.

Hood, L.E.; Weissman, I.L.; Wood, W.B.; Wilson, J.H. Immunology, 2nd ed. Menlo Park, CA: Benjamin-Cummings, 1984.

Male, D.; Champion, B.; Cooke, A. Advanced Immunology. London: Gower, 1987.

Paul, W.E., ed. Fundamental Immunology. New York: Raven, 1984.

Roitt, I.M.; Brostoff, J.; Male, D.K. Immunology. London: Gower, 1985.

Cited

1. Gowans, J.L.; McGregor, D.D. The immunological activities of lymphocytes. Prog. Allergy 9:1–78, 1965.
2. Greaves, M.F.; Owen, J.J.T.; Raff, M.C. T and B Lymphocytes: Origins, Properties and Roles in Immune Responses. Amsterdam: Excerpta Medica, 1973.
3. Cooper, M.; Lawton, A. The development of the immune system. Sci. Am. 231(5):59–72, 1974.
   Owen, J.J.T. Ontogenesis of lymphocytes. In B and T Cells in Immune Recognition (F. Loor, G.E. Roelants, eds.), pp. 21–34. New York: Wiley, 1977.
4. Möller, G., ed. Functional T Cell Subsets Defined by Monoclonal Antibodies. Immunol. Rev., Vol. 74, 1983.
   Raff, M.C. Cell-surface immunology. Sci. Am. 234(5):30–39, 1976.
   Reinherz, E.L.; Schlossman, S.F. The differentiation and function of human T lymphocytes. Cell 19:821–827, 1980.
5. Ada, G.L. Antigen binding cells in tolerance and immunity. Transplant. Rev. 5:105–129, 1970.
   Ada, G.L.; Nossal, G. The clonal selection theory. Sci. Am. 257(2):62–69, 1987.
   Burnet, F.M. The Clonal Selection Theory of Acquired Immunity. Nashville, TN: Vanderbilt University Press, 1959.
   Wigzell, H. Specific fractionation of immunocompetent cells. Transplant. Rev. 5:76–104, 1970.
6. Pink, J.R.L.; Askonas, B.A. Diversity of antibodies to cross-reacting nitrophenyl haptens in inbred mice. Eur. J. Immunol. 4:426–429, 1974.
7. Butcher, E.C.; Weissman, I.L. Lymphoid Tissues and Organs. In Fundamental Immunology (W.E. Paul, ed.), pp. 109–127. New York: Raven, 1984.
   Gallatin, M.; et al. Lymphocyte homing receptors. Cell 44:673–680, 1986.
   Gowans, J.L.; Knight E.J. The route of re-circulation of lymphocytes in the rat. Proc. R. Soc. Lond. (Biol.) 159:257–282, 1964.
   Sprent, J. Migration and lifespan of lymphocytes. In B and T Cells in Immune Recognition (F. Loor, G.E. Roelants, eds.), pp. 59–82. New York: Wiley, 1977.
   Woodruff, J.J.; Clarke, L.M.; Chin, Y.H. Specific cell-adhesion mechanisms determining migration pathways of recirculating lymphocytes. Annu. Rev. Immunol. 5:201–222, 1987.
8. Greaves, M.F.; Owen, J.J.T.; Raff, M.C. T and B Lymphocytes: Origins, Properties and Roles in Immune Responses, pp. 117–186. Amsterdam: Excerpta Medica, 1973.
   Rajewsky, K.; Forster, I.; Cumano, A. Evolutionary and somatic selection of the antibody repertoire in the mouse. Science 238:1088–1094, 1987.
9. Billingham, R.E.; Brent, L.; Medawar, P.B. Quantitative studies on tissue transplantation immunity. III. Activity acquired tolerance. Philos. Trans. R. Soc. Lond. (Biol.) 239:357–414, 1956.
   Harris, D.E.; Cairns, L.; Rosen, F.S.; Borel, Y. A natural model of immunologic tolerance. Tolerance to murine C5 is mediated by T cells and antigen is required to maintain unresponsiveness. J. Exp. Med. 156:567–584, 1982.
   Lindstrom, J. Immunobiology of myasthenia gravis, experimental autoimmune myasthenia gravis and Lambert Eaton syndrome. Annu. Rev. Immunol. 3:109–132, 1985.
   Nossal, G.J.V. Cellular mechanisms of immunologic tolerance. Annu. Rev. Immunol. 1:33–62, 1983.
   Owen, R.D. Immunogenetic consequence of vascular anastomoses between bovine twins. Science 102:400–401, 1945.
10. Howard, J.G.; Mitchison, N.A. Immunological tolerance. Prog. Allergy 18:43–96, 1975.
11. Davies, D.R.; Metzger, H. Structural basis of antibody function. Annu. Rev. Immunol. 1:87–118, 1983.
    Kabat, E.A. Structural Concepts in Immunology and Immunochemistry, 2nd ed. New York: Holt, Rinehart & Winston, 1976.
    Nisonoff, A.; Hopper, J.E.; Spring, S.B. The Antibody Molecule. New York: Academic Press, 1975.
12. Möller, G., ed. Lymphocyte Immunoglobulin: Synthesis and Surface Representation. Transplant. Rev., Vol. 14, 1973.
13. Dutton, R.W.; Mishell, R.I. Cellular events in the immune response. The in vitro response of normal spleen cells to erythrocyte antigens. Cold Spring Harbor Symp. Quant. Biol. 32:407–414, 1967.
    Jerne, N.K.; et al. Plaque forming cells: methodology and theory. Transplant. Rev. 18:130–191, 1974.
14. Edelman, G.M. The structure and function of antibodies. Sci. Am. 223(2):34–42, 1970.
    Porter, R.R. Structural studies of immunoglobulins. Science 180:713–716, 1973.
15. Ishizaka, T.; Ishizaka, K. Biology of immunoglobin E. Prog. Allergy 19:60–121, 1975.
    Koshland, M.E. The coming of age of the immunoglobulin J chain. Annu. Rev. Immunol. 3:425–454, 1985.
    Morgan, E.L.; Weigle, W.O. Biological activities residing in the Fc region of immunoglobulin. Adv. Immunol. 40:61–134, 1987.
    Solari, R.; Kraehenbuhl, J.-P. The biosynthesis of secretory component and its role in the transepithelial transport of IgA dimer. Immunol. Today 6:17–20, 1985.
    Underdown, B.J.; Schiff, J.M. Immunoglobulin A: strategic defense initiative at the mucosal surface. Annu. Rev. Immunol. 4:389–418, 1986.
    Unkeless, J.C.; Scigliano, E.; Freedman, V. Structure and function of human and murine receptors for IgG. Annu. Rev. Immunol. 6:251–282, 1988.
16. Berzofsky, J.A.; Berkover, I.J. Antigen-antibody interactions. In Fundamental Immunology (W.E. Paul, ed.), pp. 595–644. New York: Raven, 1984.
17. Capra, J.D.; Edmundson, A.B. The antibody combining site. Sci. Am. 236(1):50–59, 1977.
    Wu, T.T.; Kabat, E.A. An analysis of the sequences of the variable regions of Bence Jones proteins and myeloma light chains and their implications for antibody complementarity. J. Exp. Med. 132:211–250, 1970.
18. Sakano, H.; et al. Domains and the hinge region of an immunoglobulin heavy chain are encoded in separate DNA segments. Nature 277:627–633, 1979.
19. Alzari, P.M.; Lascombe, M.-B.; Poljak, R.J. Three-dimensional structure of antibodies. Annu. Rev. Immunol. 6:555–580, 1988.
    Capra, J.D.; Edmundson, A.B. The antibody combining site. Sci. Am. 236(1):50–59, 1977.
20. Dreyer, W.J.; Bennett, J.C. The molecular basis of antibody formation: a paradox. Proc. Natl. Acad. Sci. USA 54:864–869, 1965.

Hozumi, N.; Tonegawa, S. Evidence for somatic rearrangement of immunoglobulin genes coding for variable and constant regions. *Proc. Natl. Acad. Sci. USA* 73:3628–3632, 1976.

1. Alt, F.W.; Blackwell, T.K.; Yancopoulos, G.D. Development of the primary antibody repertoire. *Science* 238:1079–1087, 1987.

Leder, P. The genetics of antibody diversity. *Sci. Am.* 246(5):72–83, 1982.

Tonegawa, S. The molecules of the immune system. *Sci. Am.* 253(4):122–131, 1985.

Tonegawa, S. Somatic generation of antibody diversity. *Nature* 302:575–581, 1983.

2. Gearhart, P.J.; Johnson, N.D.; Douglas, R.; Hood, L. IgG antibodies to phosphorylcholine exhibit more diversity than their IgM counterparts. *Nature* 291:29–34, 1981.

Griffiths, G.M.; Berek, C.; Kaartinen, M.; Milstein, C. Somatic mutation and the maturation of immune response to 2-phenyl oxazolone. *Nature* 312:271–275, 1984.

Möller, G., ed. Role of Somatic Mutation in the Generation of Lymphocyte Diversity. *Immunol. Rev.*, Vol. 96, 1987.

Rajewsky, K.; Forster, I.; Cumano, A. Evolutionary and somatic selection of the antibody repertoire in the mouse. *Science* 238:1088–1094, 1987.

3. Alt, F.W.; et al. Ordered rearrangement of immunoglobulin heavy chain variable region segments. *EMBO J.* 3:1209–1219, 1984.

Yancopoulos, G.D.; Alt, F.W. Regulation of the assembly and expression of variable region genes. *Annu. Rev. Immunol.* 4:339–368, 1986.

4. Early, P.; et al. Two mRNAs can be produced from a single immunoglobulin μ gene by alternative RNA processing pathways. *Cell* 20:313–319, 1980.

5. Cebra, J.J.; Komisar, J.L.; Schweitzer, P.A. C$_H$ isotype switching during normal B-lymphocyte development. *Annu. Rev. Immunol.* 2:493–548, 1984.

Shimizu, A.; Honjo, T. Immunoglobulin class switching. *Cell* 36:801–803, 1984.

6. Jerne, N.K. The immune system. *Sci. Am.* 229(1):52–60, 1973.

Jerne, N.K. Toward a network theory of the immune system. *Ann. Immunol. Inst. Pasteur (Paris)* 125C:378–389, 1974.

Möller, G., ed. Idiotype Networks. *Immunol. Rev.*, Vol. 79, 1984.

Rajewsky, K.; Takemori, T. Genetics, expression and function of idiotypes. *Annu. Rev. Immunol.* 1:569–608, 1983.

7. Lachmann, P.J. Complement. In Clinical Aspects of Immunology, 4th ed. (P.J. Lachmann, D. Peters, eds.), pp. 18–49. Oxford, U.K.: Blackwell, 1982.

Müller-Eberhard, H.J. Molecular organization and function of the complement system. *Annu. Rev. Biochem.* 57:321–348, 1988.

Reid, K.B.M.; Porter, R.R. The proteolytic activation systems of complement. *Annu. Rev. Biochem.* 50:433–464, 1981.

8. Cooper, N.R. The classical complement pathway: activation and regulation of the first complement component. *Adv. Immunol.* 37:151–216, 1985.

9. Müller-Eberhard, H.J.; Schreiber, R.D. Molecular biology and chemistry of the alternative pathway of complement. *Adv. Immunol.* 29:1–53, 1980.

0. Müller-Eberhard, H.J. The membrane attack complex of complement. *Annu. Rev. Immunol.* 4:503–528, 1986.

1. Campbell, R.D.; Carroll, M.C.; Porter, R.R. The molecular genetics of components of complement. *Adv. Immunol.* 38:203–244, 1986.

Lachmann, P.J.; Hobart, M.J. Genetics of complement. *Trends Genet.* 1:145–149, 1985.

Reid, K.B. Activation and control of the complement system. *Essays Biochem.* 22:27–68, 1986.

2. Allison, J.P.; Lanier, L.L. The structure, function and serology of the T cell antigen receptor complex. *Annu. Rev. Immunol.* 5:503–540, 1987.

Clevers, H.; Alarcon, B.; Wileman, T.; Terhorst, C. The T cell receptor/CD3 complex: a dynamic protein ensemble. *Annu. Rev. Immunol.* 6:629–662, 1988.

Davis, M.M.; Bjorkman, P.J. T-cell antigen receptor genes and T-cell recognition. *Nature* 334:395–402, 1988.

Hedrick, S.M.; Cohen, D.; Nielsen, E.A.; Davis, M.M. Isolation of cDNA clones encoding T cell-specific membrane-associated proteins. *Nature* 308:149–153, 1984.

Kronenberg, M.; Siu, G.; Hood, L.E.; Shastri, N. The molecular genetics of the T-cell antigen receptor and T-cell antigen recognition. *Annu. Rev. Immunol.* 4:529–592, 1986.

Toyonaga, B; Mak, T. Genes of the T-cell antigen receptor in normal and malignant T cells. *Annu. Rev. Immunol.* 5:585–620, 1987.

Marrack, P.; Kappler, J. The T cell receptor. *Science* 238:1073–1079, 1987.

33. Cantor, H. T lymphocytes. In Fundamental Immunology (W.E. Paul, ed.), pp. 57–67. New York: Raven, 1984.

Fathman, C.G.; Frelinger, J.G. T-lymphocyte clones. *Annu. Rev. Immunol.* 1:633–656, 1983.

Parnes, J.R. Structure and function of T lymphocyte differentiation antigens. *Trends Genet.* 2:179–183, 1986.

34. Nabholz, M.; MacDonald, H.R. Cytolytic T lymphocytes. *Annu. Rev. Immunol.* 1:273–306, 1983.

35. Herberman, R.B.; Reynolds, C.W.; Ortaldo, J. Mechanisms of cytotoxicity by natural killer (NK) cells. *Annu. Rev. Immunol.* 4:651–680, 1986.

Möller, G., ed. Molecular Mechanisms of T-Cell Mediated Lysis. *Immunol. Rev.*, Vol. 103, 1988.

Young, J.D.-E.; Liu, C.-C. Multiple mechanisms of lymphocyte-mediated killing. *Immunol. Today* 9:140–145, 1988.

36. Bach, F.H.; Sachs, D.H. Transplantation immunology. *New Engl. J. Med.* 317:489–492, 1987.

Mellor, A. The class I MHC gene family in mice. *Immunol. Today* 7:19–24, 1986.

Widera, G. Molecular biology of the histocompatibility complex. *Science* 233:437–443, 1986.

37. Klein, J.; Figueroa, F.; Nagy, Z.A. Genetics of the major histocompatibility complex: the final act. *Annu. Rev. Immunol.* 1:119–142, 1983.

Möller, G., ed. Molecular genetics of class I and II MHC antigens. *Immunol. Rev.*, Vols. 84 and 85, 1985.

Sachs, D.H. The major histocompatibility complex. In Fundamental Immunology (W.E. Paul, ed.), pp. 303–346. New York: Raven, 1984.

38. Zinkernagel, R.M.; Doherty, P.C. MHC-restricted cytotoxic T cells: studies on the biological role of polymorphic major transplantation antigens determining T-cell restriction-specificity, function and responsiveness. *Adv. Immunol.* 27:51–177, 1979.

Zinkernagel, R.M.; Doherty, P.C. Restriction of *in vitro* T cell-mediated cytotoxicity in lymphocytic choriomeningitis within a syngeneic or semiallogeneic system. *Nature* 248:701–702, 1974.

39. Townsend, A.R. Recognition of influenza virus proteins by cytotoxic T lymphocytes. *Immunol. Res.* 6:80–100, 1987.

Wraitch, D.C. The recognition of influenza A virus-infected cells by cytotoxic T lymphocytes. *Immunol. Today* 8:239–245, 1987.

40. Bjorkman, P.J.; et al. Structure of the human class I histocompatibility antigen, HLA-A2. *Nature* 329:506–512, 1987.

Bjorkman, P.J.; et al. The foreign antigen binding site and T cell recognition regions of class I histocompatibility antigens. *Nature* 329:512–518, 1987.

41. Allen, P.M. Antigen processing at the molecular level. *Immunol. Today* 8:270–273, 1987.

Babbitt, B.P.; Allen, P.M.; Matsueda, G.; Haber, E.; Unanue, E.R. Binding of immunogenic peptides to Ia histocompatibility molecules. *Nature* 317:359–361, 1985.

Grey, H.M.; Chesnut, R. Antigen processing and presentation to T cells. *Immunol. Today* 6:101–106, 1985.

Shevach, E.M.; Paul, W.E.; Green, I. Histocompatibility-linked immune response gene function in guinea pigs. Specific inhibition of antigen-induced lymphocyte proliferation by alloantisera. *J. Exp. Med.* 136:1207–1221, 1972.

42. Durum, S.K.; Schmidt, J.A.; Oppenheim, J.J. Interleukin-1: an immunological perspective. *Annu. Rev. Immunol.* 3:263–288, 1985.

Fathman, C.G.; Frelinger, J.G. T-lymphocyte clones. *Annu. Rev. Immunol.* 1:633–656, 1983.

MacDonald, H.R.; Nabholz, M. T-cell activation. *Annu. Rev. Cell Biol.* 2:231–253, 1986.

Smith, K.A. Interleukin-2: inception, impact, and implications. *Science* 240:1169–1176, 1988.

43. Claman, H.N.; Chaperon, E.A. Immunologic complementation between thymus and marrow cells—a model for the two-cell theory of immunocompetence. *Transplant. Rev.* 1:92–113, 1969.

Davies, A.J.S. The thymus and the cellular basis of immunity. *Transplant. Rev.* 1:43–91, 1969.

44. Chesnut, R.W.; Grey, H.M. Antigen presentation by B cells and its significance in T-B interactions. *Adv. Immunol.* 39:51–94, 1986.

DeFranco, A.L. Molecular aspects of B-lymphocyte activation. *Annu. Rev. Cell Biol.* 3:143–178, 1987.

Kishimoto, T.; Hirano, T. Molecular regulation of B lymphocyte response. *Annu. Rev. Immunol.* 6:485–512, 1988.

Möller, G., ed. IL-4 and IL-5: Biology and Genetics. *Immunol. Rev.*, Vol. 102, 1988.

Paul, W.E.; Ohara, J. B-cell stimulating factor-1/Interleukin 4. *Annu. Rev. Immunol.* 5:429–460, 1987.

Cambier, J.C.; Ransom, J.T. Molecular mechanisms of transmembrane signaling in B lymphocytes. *Annu. Rev. Immunol.* 5:175–200, 1987.

45. Möller, G., ed. Gamma Interferon. *Immunol. Rev.*, Vol. 97, 1987.

Mosmann, T.R.; Coffman, R.L. Two types of mouse helper T-cell clone—implications for immune regulation. *Immunol. Today* 8:223–227, 1987.

Powrce, F.; Mason, D. Phenotypic and functional heterogeneity of CD4$^+$ T cells. *Immunol. Today* 9:274–277, 1988.

46. Littman, D.R. The structure of the CD4 and CD8 genes. *Annu. Rev. Immunol.* 5:561–584, 1987.

Springer, T.A.; Dustin, M.L.; Kishimoto, T.K.; Marlin, S.D. The lymphocyte-function associated LFA-1, CD2 and LFA-3 molecules: cell adhesion receptors of the immune system. *Annu. Rev. Immunol.* 5:223–252, 1987.

47. Asherson, G.L.; Colizzi, V.; Zembala, M. An overview of T suppressor cell circuits. *Annu. Rev. Immunol.* 4:37–68, 1986.

Dorf, M.E.; Benacerraf, B. Suppressor cells and immunoregulation. *Annu. Rev. Immunol.* 2:127–158, 1984.

Gershon, R.K. T-cell control of antibody production. *Contemp. Top. Immunobiol.* 3:1–40, 1974.

Harris, D.E.; Cairns, L.; Rosen, F.S.; Borel, Y. A natural model of immunologic tolerance. Tolerance to murine C5 is mediated by T cells and antigen is required to maintain unresponsiveness. *J. Exp. Med.* 156:567–584, 1982.

Mitchison, N.A.; Griffiths, J.A.; Oliveira, D.B.G. Immunogenetics of suppression. *Br. J. Clin. Pract.* 40:225–229, 1986.

48. Bevan, M. In a radiation chimera, host H-2 antigens determine the immune responsiveness of donor cytotoxic cells. *Nature* 269:417–419, 1977.

Kappler, J.W.; Roehm, N.; Marrack, P. T cell tolerance by clonal elimination in the thymus. *Cell* 49:273–280, 1987.

Sprent, J.; Lo, D.; Gao, E.-K.; Ron, Y. T cell selection in the thymus. *Immunol. Rev.* 101:5–19, 1988.

Sprent, J.; Webb, S. Function and specificity of T cell subsets in the mouse. *Adv. Immunol.* 41:39–133, 1987.

von Boehmer, H. The developmental biology of T lymphocytes. *Annu. Rev. Immunol.* 6:309–326, 1988.

49. Allen, P.M.; Babbitt, B.P.; Unanue, E.R. T cell recognition of lysozyme: the biochemical basis of presentation. *Immunol. Rev.* 98:171–187, 1987.

Buus, S.; Sette, A.; Grey, H.M. The interaction between protein-derived immunogenic peptides and Ia. *Immunol. Rev.* 98:115–142, 1987.

Mengle-Gaw, L.; McDevitt, H.O. Genetics and expression of murine I$_A$ antigens. *Annu. Rev. Immunol.* 3:367–396, 1985.

Schwartz, R.H. Immune response (Ir) genes in the murine major histocompatibility complex. *Adv. Immunol.* 39:31–201, 1986.

50. Williams, A.F.; Barclay, A.N. The immunoglobulin superfamily—domains for cell surface recognition. *Annu. Rev. Immunol.* 6:381–406, 1988.

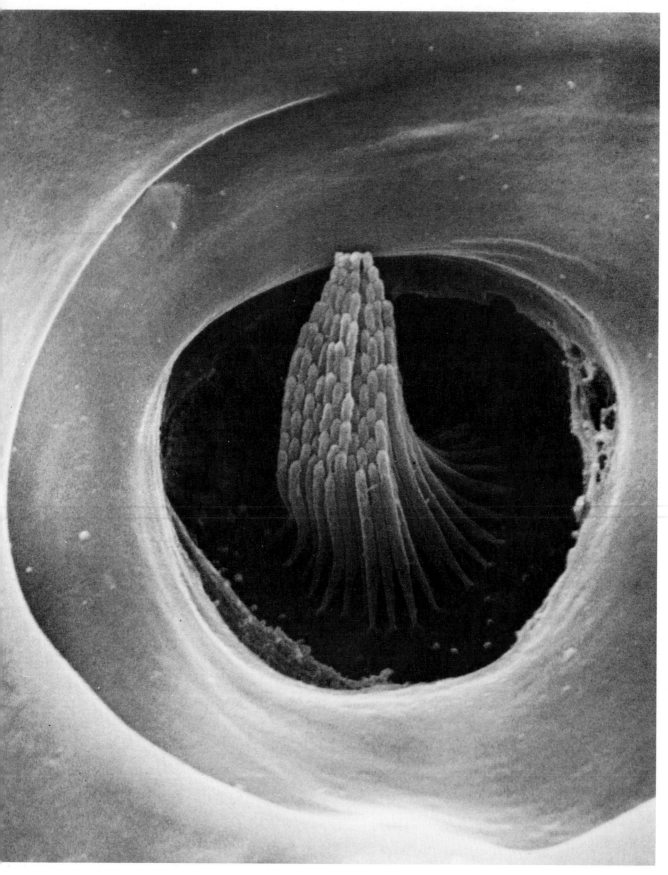

canning electron micrograph of a sensory hair cell from the
ner ear of a frog. (Courtesy of Richard Jacobs; from *Nature*
1(5733), 1979.)

The Nervous System

19

How can we hope to understand the workings of the human brain? This network of some 10^{11} nerve cells, with at least a thousand times that number of interconnections, is more complex, and seems in many ways more powerful, than even the largest of modern computers. Our present understanding of the nervous system is so rudimentary, however, that one can scarcely judge whether the comparison makes sense. We do not know, for example, how many functionally distinct categories of nerve cells the brain contains; nor can we give even an outline of the neural computations involved in hearing a word or reaching for an object, let alone proving a theorem or writing a poem.

And yet, paradoxically, while the brain as a whole remains the most baffling organ in the body, the properties of the individual nerve cells, or *neurons*, are understood better than those of almost any other cell type. At the cellular level at least, simple and general principles can be discerned. With their help, one can begin to see how small parts of the nervous system work. Important progress has been made, for example, in explaining the cellular machinery of simple reflex behavior, and even of visual perception. From a practical point of view, knowledge of the molecular biology of neurons provides a key to the biochemical control of brain function through drugs, and it holds out the promise of more effective treatment for many forms of mental illness.

In this chapter we shall focus on the nerve cell and try to illustrate how its properties give insight into neural organization at higher levels.

The Cells of the Nervous System: An Overview of Their Structure and Function[1]

The nervous system provides for rapid communication between widely separated parts of the body. Through its role as a communications network, it governs reactions to stimuli, processes information, and generates elaborate patterns of signals to control complex behaviors. The nervous system is also capable of learning: as it processes and records sensory information about the external world, it undergoes adjustments that result in altered future patterns of action.

The major neural pathways of communication were mapped out more than a hundred years ago, before the role of individual nerve cells was understood. Figure 19–1 shows the basic plan. Like a big computing facility, the vertebrate

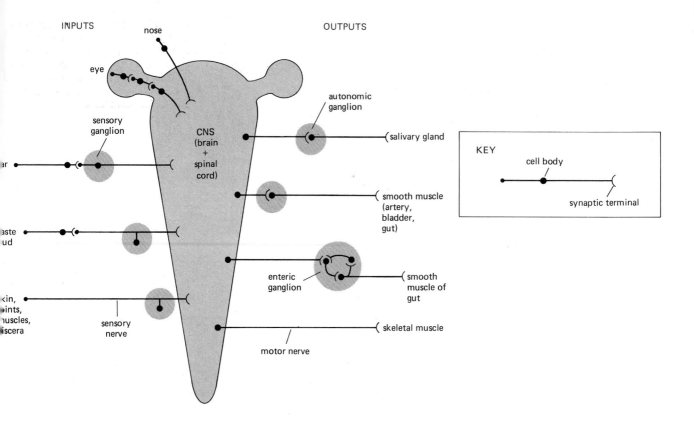

INPUTS   OUTPUTS

nose

eye

sensory ganglion

CNS (brain + spinal cord)

autonomic ganglion

salivary gland

ear

smooth muscle (artery, bladder, gut)

taste bud

enteric ganglion

smooth muscle of gut

skin, joints, muscles, viscera

sensory nerve

skeletal muscle

motor nerve

KEY

cell body

synaptic terminal

Figure 19–1 A highly schematized diagram of the nervous system of a vertebrate, showing how sensory inputs are relayed from the periphery to the central nervous system (CNS) and how motor outputs are relayed to the periphery from the CNS. Sensory and motor signals are conveyed from and to the peripheral organs by nerve cells whose cell bodies (*large black dots*) in many cases lie clustered in *ganglia* (*colored circles*) outside the CNS and whose axons are bundled together to form *nerves* (*black lines*). The nerves, ganglia, and sense organs together constitute the *peripheral nervous system*. Some ganglia are simple relay stations; others—notably the subclass of autonomic ganglia that control the peristaltic contractions of the gut (*enteric* ganglia)—are complex systems of interacting neurons capable of functioning even in isolation from the CNS. Within the CNS are masses of interconnected neurons that are not shown.

nervous system consists of a main processing unit, the **central nervous system,** comprising the brain and spinal cord, which is linked by cables, the *nerves,* to a large number of peripheral structures: sense organs for input, muscles (and to a lesser extent glands) for output. There are also connections to peripheral nerve cell clusters called *ganglia,* which serve in some cases simply for communication between periphery and center and in other cases as accessory minicomputers. The scheme is similar in invertebrates, although they usually do not have such conspicuous central nervous systems, and their ganglia have a larger role and more autonomy.

Within this broad scheme, the detailed patterns of neural connections in different species vary enormously. Yet the properties of the individual neurons are much the same, regardless of whether one looks at a mollusk, an insect, an amphibian, or a mammal.

The Function of a Nerve Cell Depends on Its Elongated Structure[2]

The fundamental task of the **neuron** is to receive, conduct, and transmit signals. To perform these functions, neurons in general are extremely elongated: a single nerve cell in a human being, extending, say, from the spinal cord to a muscle in the foot, may be a meter long. Every neuron consists of a *cell body* (containing the nucleus) with a number of long, thin processes radiating outward from it. Usually there is one long **axon,** to conduct signals away from the cell body toward distant targets, and several shorter branching **dendrites,** which extend from the cell body like antennae and provide an enlarged surface area to receive signals from the axons of other nerve cells (Figure 19–2). Signals are also received on the cell body itself. The axon commonly divides at its far end into many branches and so can pass on its message to many target cells simultaneously. Likewise, the extent of branching of the dendrites can be very great—in some cases sufficient to receive as many as 100,000 inputs on a single neuron. Neurons of different functional classes show an astonishing variety in the pattern of branching of their axons and dendrites (Figure 19–3).

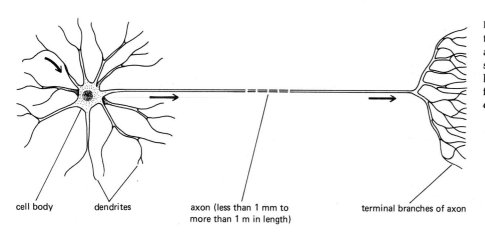

Figure 19–2 Schematic diagram of a typical neuron of a vertebrate. The arrows indicate the direction in which signals are conveyed. The longest and largest neurons in a human extend for about a meter and have an axon diameter of about 15 μm.

cell body   dendrites

axon (less than 1 mm to more than 1 m in length)

terminal branches of axon

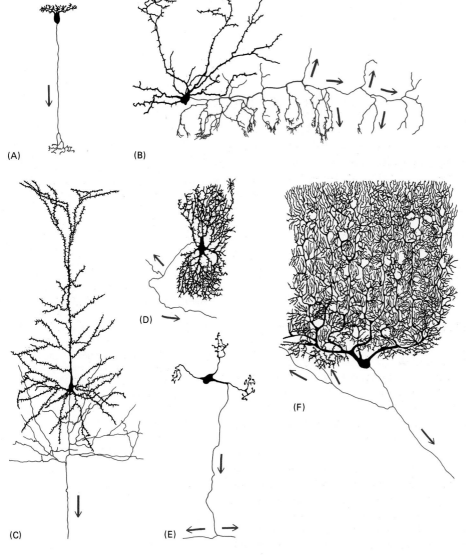

(A)

(B)

(C)

(D)

(E)

(F)

Figure 19–3 A few of the many types of neurons in the vertebrate nervous system as they appear when stained by the Golgi technique. This procedure, which involves immersing the tissue in a solution of metallic salts, picks out at random a small proportion of the cells, staining them an intense black so that all their delicate tracery of branching processes can be seen. Many dendrites, to receive signals, radiate from the cell body of each neuron, and a single thin, branching axon conducts signals away in the direction shown by the arrows. The axon is drawn in color, the cell body and dendrites in black. The cells in (A) and (B) have short axons, which are shown in their entirety. The cells in (C), (D), (E), and (F) have long axons, of which only the initial portion is shown. (A) shows a bipolar cell from the retina of a lizard, (B) is a basket cell from the cerebellum of a mouse, (C) is a pyramidal cell from the cerebral cortex of a rabbit, (D) is a cell from the brainstem of a human being, (E) is a granule cell from the cerebellum of a cat, and (F) is a Purkinje cell from the cerebellum of a human being. This last cell, with its vast array of dendrites, receives inputs from more than 100,000 other neurons. It forms part of the brain's machinery for controlling complex movements. The drawings are not to scale: cell (A) is about 100 μm long, whereas the part of cell (F) shown in the drawing is about 400 μm across (the length of its axon—not shown—is on the order of centimeters). (From S. Ramón y Cajal, Histologie du Système Nerveux de l'Homme et des Vertébrés. Paris: Maloine, 1909–1911; reprinted, Madrid: C.S.I.C., 1972.)

Nerve Cells Convey Electrical Signals[3]

The significance of the signals carried by a neuron depends on the part played by the individual cell in the functioning of the nervous system as a whole. In a *motor neuron* the signals represent commands for the contraction of a particular muscle. In a *sensory neuron* they represent the information that a specific type of stimulus, such as a light, a mechanical force, or a chemical substance, is present at a certain site in the body. In an *interneuron*, forming a connection between one neuron and another, the signals represent parts of elaborate computations that combine information from many different sources and regulate complex behavior.

Despite the varied significance of the signals, their *form* is the same, consisting of changes in the electrical potential across the neuron's plasma membrane. Communication occurs because an electrical disturbance produced in one part of the cell spreads to other parts. Such a disturbance becomes weaker with increasing distance from its source unless energy is expended to amplify it as it travels. Over short distances this attenuation is unimportant, and in fact many small neurons conduct their signals passively, without amplification. For long-distance communication, however, such passive spread is inadequate. Thus the larger neurons employ an active signaling mechanism, which is one of their most striking features: an electrical stimulus that exceeds a certain threshold strength triggers an explosion of electrical activity that is propagated rapidly along the neuron's plasma membrane and is sustained by automatic amplification all along the way. This traveling wave of electrical excitation, known as an *action potential* or *nerve impulse*, can carry a message without attenuation from one end of a neuron to the other at speeds as great as 100 m/sec or more.

Nerve Cells Communicate Chemically at Synapses[4]

Neuronal signals are transmitted from cell to cell at specialized sites of contact known as **synapses**. The usual mechanism of transmission appears surprisingly indirect. The cells are electrically isolated from one another, the *presynaptic cell* being separated from the *postsynaptic cell* by a *synaptic cleft*. A change of electrical potential in the presynaptic cell triggers it to release a chemical known as a *neurotransmitter*, which is stored in membrane-bounded *synaptic vesicles* and released by exocytosis. The neurotransmitter then diffuses across the synaptic cleft and provokes an electrical change in the postsynaptic cell (Figure 19–4). As we shall see, transmission via such *chemical synapses* is far more versatile and adaptable than direct electrical coupling via gap junctions (see p. 799), which is also used, but to a much lesser extent.

The chemical synapse is a site of intense biochemical activity, involving continual degradation, turnover, and secretion of proteins and other molecules. The biosynthetic center of the neuron, however, is in the cell body, where the ultimate instructions for protein synthesis lie. The neuron must therefore have an efficient intracellular transport system to convey molecules from the cell body to the outermost reaches of the axon and dendrites. How is this transport system organized, and what molecules are actually transported?

Slow and Fast Transport Mechanisms Carry Newly Synthesized Materials from the Nerve Cell Body into the Axon and Dendrites[5]

Electron microscopy reveals that the cell body of a typical large neuron contains vast numbers of ribosomes, some crowded together in the cytosol, some attached to rough endoplasmic reticulum (ER) (Figure 19–5A). Although dendrites often contain some ribosomes, there are no ribosomes in the axon, and its proteins must therefore be provided by the many ribosomes in the cell body (Figure 19–5B). The needs of the axon are considerable: a large motor neuron in a human being, for example, may have an axon 15 μm in diameter and a meter long, corresponding to a volume of about 0.2 mm³, which is about 10,000 times the volume of a liver

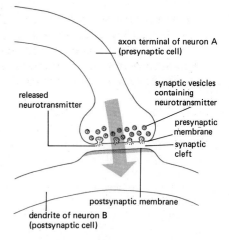

Figure 19–4 Schematic diagram of a typical synapse. An electrical signal arriving at the axon terminal of neuron A triggers the release of a chemical messenger (the neurotransmitter), which crosses the synaptic cleft and causes an electrical change in the membrane of a dendrite of neuron B. A broad arrow indicates the direction of signal transmission.

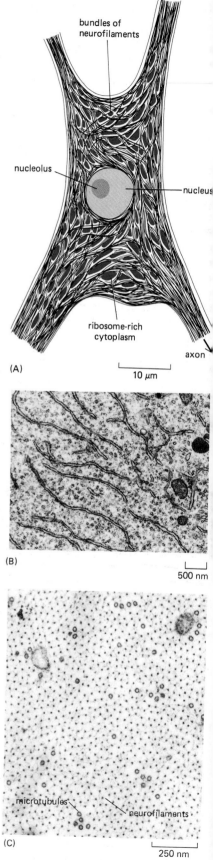

Figure 19–5 The organization of the cytoplasm in a typical large nerve cell (a motor neuron in the spinal cord). (A) A sketch of the cell body at low magnification, showing how regions of cytoplasm rich in ribosomes are packed in the gaps between bundles of neurofilaments and other cytoskeletal proteins. (B) An electron micrograph of one such ribosome-rich region; some of the ribosomes are free, others are attached to rough ER. (C) An electron micrograph showing part of a cross-section of the axon; large numbers of neurofilaments and microtubules can be seen, but no ribosomes are present. The membranous vesicles in the axon are probably traveling along the adjacent microtubules by fast axonal transport. (B, courtesy of Jennifer La Vail; C, courtesy of John Hopkins.)

cell. Because such a neuron contains only a single nucleus, its ratio of cytoplasm to DNA is far greater than that of any nonneuronal cell type in the human body.

The most plentiful proteins in the axon are those that form microtubules, neurofilaments (a class of intermediate filaments), and actin filaments (Figure 19–5C). These cytoskeletal proteins are exported from the cell body and move along the axon at speeds of 1 to 5 mm per day by the process of **slow axonal transport.** (A similar transport occurs in the dendrites, which contain a slightly different set of microtubule-associated proteins—see p. 659.) Other cytosolic proteins, including many enzymes, are also carried by slow axonal transport, whose mechanism is not understood.

Noncytosolic materials required at the synapse, such as secreted proteins and membrane-bound molecules, move outward from the cell body by a much faster mode of transport. These proteins and lipids pass from their sites of synthesis in the endoplasmic reticulum to the Golgi apparatus, which lies close to the nucleus, often facing the base of the axon. From here, packaged in membrane vesicles, they are carried by **fast axonal transport,** at speeds of up to 400 mm per day, along tracks formed by microtubules in the axon or the dendrites (see p. 660); mitochondria are conveyed by the same means. Since different populations of proteins are sent out in this way along axons and dendrites, the transported molecules are presumed to be sorted in the cell body into separate and distinctive types of transport vesicles (see p. 463).

Among the proteins rapidly transported along the axon are those to be secreted at the synapse, such as the *neuropeptides* that many neurons release as neurotransmitters, often in conjunction with nonprotein transmitters. From the point of view of their internal organization, neurons can thus be thought of as secretory cells in which the site of secretion has been removed to an enormous distance from the site where proteins and membranes originate (Figure 19–6).

Retrograde Transport Allows the Nerve Terminal to Communicate Chemically with the Cell Body[5,6]

Fast axonal transport is required during development for the growth of axons and dendrites, which elongate by adding new membrane to their tips. Fast axonal transport also occurs in a full-grown neuron, in which there is no net accumulation of membrane at the ends of the axon and dendrites. In this case the fast transport of membrane outward from the cell body, called *fast anterograde transport,* must be exactly balanced by *fast retrograde transport* of membrane back from the ends of the cell processes. The mechanisms of fast transport in the two directions are similar but not identical. The fast retrograde transport has a speed about half that of fast anterograde transport, is driven by a different motor protein (see p. 660), and carries somewhat larger vesicles on average. The structures returning to the cell body consist partly of aging cytoplasmic organelles, such as mitochondria, and partly of vesicles formed by the extensive endocytosis required for membrane retrieval at the axon terminal after neurotransmitter release (see

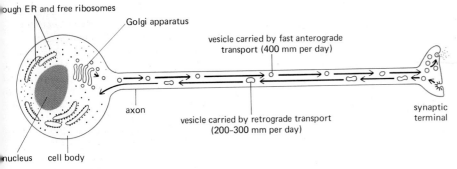

rough ER and free ribosomes

Golgi apparatus

vesicle carried by fast anterograde transport (400 mm per day)

axon

vesicle carried by retrograde transport (200-300 mm per day)

synaptic terminal

nucleus   cell body

Figure 19–6 A neuron viewed schematically as a secretory cell in which the site of secretion (the axon terminal) lies at a great distance from the site of macromolecular synthesis (the cell body). This mode of organization creates a need for a rapid axonal transport mechanism. The diagram is not meant to imply that all synaptic vesicles have to be transported from the cell body; in most neurons, synaptic vesicles are formed largely by local recycling of membrane in the axon terminal.

Figure 19–20). Molecules present in the extracellular medium surrounding the axon terminal are liable to be captured in these endocytic vesicles and thereby carried back from the axon terminal to the cell body. Thus the biosynthetic machinery in the cell body can sample conditions at the axon terminal and make an appropriate response, as we shall see later (see p. 1119).

Retrograde transport is extremely useful to neuroanatomists, who routinely exploit it to trace neural connections, as explained in Figure 19–7.

Neurons Are Surrounded by Various Types of Glial Cells[7]

All neural tissue, both peripheral and central, consists of two major classes of cells. Neurons play the star role, but they are outnumbered, by about 10 to 1 in the mammalian brain, by a supporting cast of **glial cells.** The glial cells surround neurons (both their cell bodies and their processes) and occupy the spaces between them. The best understood are the *Schwann cells* in vertebrate peripheral nerves and the *oligodendrocytes* in the vertebrate central nervous system, which both wrap themselves around axons to provide electrical insulation in the form of a *myelin sheath* (see p. 1073, below). The three other types of glial cell in the central nervous system are microglia, ependymal cells, and astrocytes (Figure 19–8). The *microglia* belong in a class apart: they are functionally akin to macrophages (see p. 974) and, like them, originate from hemopoietic tissue. With this exception, all the glial cells share a common embryonic origin with the neurons with which they are associated; unlike most neurons, however, they are not as a rule electrically excitable. Moreover, whereas neurons cannot divide after they have differentiated, most glial cells remain capable of dividing throughout life.

Ependymal cells line the internal cavities of the brain and spinal cord (see Figure 19–8), and their epithelial arrangement is a memento of the origin of the central nervous system from an epithelial tube (see p. 1109).

Astrocytes (see Figure 19–8) are the most plentiful and diverse of the glial cells but also the most enigmatic: their functions are still largely uncertain, although it seems clear that they play an important part in guiding the construction of the

1. INJECTION OF HRP

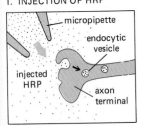

micropipette

endocytic vesicle

injected HRP

axon terminal

HRP is injected at the site of interest and is taken up by endocytosis into axon terminals in the vicinity

2. TRANSPORT OF HRP

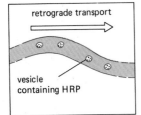

retrograde transport

vesicle containing HRP

Several hours or days are allowed to elapse to give time for the vesicles containing HRP to travel back along the axon by retrograde transport

3. LOCALIZATION OF HRP

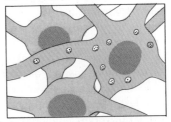

Tissue sections are treated with reagents to generate colored products wherever HRP is located. Nerve cell bodies that send axons to the remote site of HRP injection are thus identified

Figure 19–7 How fast retrograde transport is exploited to identify and locate remote nerve cell bodies whose axons project to a given site of interest. The enzyme horseradish peroxidase (HRP) is the most widely used tracer molecule for this purpose, since it can be detected in very small quantities by the colored products of the reaction that it catalyzes.

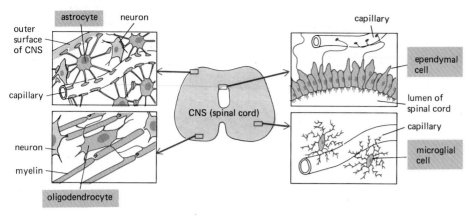

Figure 19–8 The four major classes of glial cells in the vertebrate central nervous system. The glial cells are shown in color. Astrocytes, which are the most plentiful, have many radiating processes. Some of these processes end on the surfaces of neurons; others, with expanded tips, go to form the external surface layer of the CNS and the wrapping that surrounds blood vessels in the CNS, collaborating with endothelial cells of the blood capillaries to create the blood–brain barrier. Ependymal cells form the ciliated epithelial lining of the central cavity of the CNS and, like astrocytes, often have processes ending on blood vessels. Oligodendrocytes form insulating sheaths of myelin around axons in the CNS. Microglial cells are related in function and origin to macrophages and are involved in reactions to tissue damage and infection; they tend to be found in the neighborhood of blood vessels.

nervous system (see p. 1110) and in controlling the chemical and ionic environment of the neurons. Thus one type of astrocyte extends processes that are expanded into "endfeet," which, linked by junctional complexes such as one finds in epithelia (see p. 792), form a sealed barrier at the external surface of the central nervous system. They also extend processes that form similar endfeet on blood vessels, where they induce the endothelial cells (in the case of capillaries and venules) to become sealed together by unusually well-developed tight junctions so as to form the *blood–brain barrier*. This barrier prevents water-soluble molecules from passing into brain tissue from the blood unless they are specifically carried by transport proteins in the plasma membranes of the endothelial cells that line the vessels. The neurons thus occupy a sheltered and controlled environment, on which the molecular machinery for electrical signaling is critically dependent.

Summary

Nerve cells, or neurons, are exceptionally elongated cells that convey electrical signals in the form of action potentials—traveling waves of electrical excitation. Typically, several branching dendrites and a single long axon project from the nerve cell body. Signals are usually received on the dendrites and cell body, sent out along the axon, and passed on to other cells at chemical synapses. Here the electrical signal in the presynaptic axon terminal triggers the secretion of neurotransmitter, which provokes an electrical change in the postsynaptic cell.

The neuron can be viewed as a secretory cell that releases its secretion—the neurotransmitter—at a very large distance from the cell body, where macromolecules are synthesized. Newly synthesized membrane and proteins for secretion are exported along the axon and dendrites by fast axonal transport, in which small membrane vesicles are propelled along tracks formed by microtubules. The microtubules and other non-membrane-bound components of the neuronal cytosol are exported from the cell body by a quite different, slow axonal transport mechanism. Fast axonal transport also operates in a retrograde direction, conveying membrane vesicles back from the axon terminals to the cell body.

Neurons are surrounded by glial cells, which help in various ways to control the chemical and electrical environment of the neurons.

Voltage-gated Ion Channels and the Action Potential[3,4,8]

As discussed in Chapter 6, the voltage difference across a cell's plasma membrane—the **membrane potential**—depends on the distribution of electric charge (see p. 314). Charge is carried back and forth across the nerve cell membrane by small inorganic ions—chiefly Na^+, K^+, Cl^-, and Ca^{2+}—which traverse the lipid bilayer by passing through selective ion channels formed by specific transmembrane proteins (see p. 312). When the ion channels open or close, the charge distribution shifts and the membrane potential changes. Neuronal signaling thus depends on channels whose permeability is regulated—so-called **gated ion channels.**

Two classes of gated channels are of crucial importance: (1) *voltage-gated channels*—especially voltage-gated Na^+ channels—play the key role in the explosions of electrical activity by which action potentials are propagated along an axon; and (2) *ligand-gated channels*, which convert extracellular chemical signals into electrical signals, play a central part in the operation of synapses. The account in Chapter 6 (pp. 312–319) of ion channels and of their role in electrical signaling forms the basis for the further discussion of neuronal signaling to be given here. Some principles of electrochemistry that are of special relevance to nerve cells are reviewed in Panel 19–1.

Voltage Changes Can Spread Passively Within a Neuron[3,4,8,9]

Action potentials are typically triggered at one end of an axon and propagate along its length. To understand the mechanism it is helpful to consider first how electrical disturbances spread along a nerve cell in the absence of action potentials. As mentioned earlier, such *passive spread* is common, especially in the many neurons that have very short axons or no axon at all; these cells often have few or no voltage-gated Na^+ channels and rely for their signaling entirely on passive spread, manifest as smoothly graded *local potentials*.

In an axon at rest, the membrane potential is uniformly negative, with the interior of the axon everywhere at the same negative potential relative to the external medium. As explained in Chapter 6 (see p. 314), the potential difference depends on the large concentration gradients of Na^+ and K^+, built up by the Na^+-K^+ pump. K^+ *leak channels* make the resting membrane permeable chiefly to K^+, so that the resting potential is close to the K^+ equilibrium potential— typically about -70 mV (see Panel 19–1). An electrical signal may take the form of a *depolarization,* in which the voltage drop across the membrane is reduced, or a *hyperpolarization,* in which it is increased. To illustrate the passive spread of an electrical signal, let us consider what happens when an axon is locally depolarized by injecting current through a microelectrode inserted into it. If the current is small, the depolarization will be *subthreshold:* practically no Na^+ channels open, and no action potentials are triggered. A steady state is quickly reached in which the inflow of current through the microelectrode is exactly balanced by the outflow of current (carried mainly by K^+ ions) across the axonal membrane. Some of this current flows out in the neighborhood of the microelectrode, while some travels down the interior of the axon for some distance in either direction before escaping. The consequence is that the membrane potential is disturbed by an amount that decreases exponentially with the distance from the source of the disturbance (Figure 19–9). This passive spread of an electrical signal along a nerve cell process is analogous to the spread of a signal along an undersea telegraph cable; as the current flows down the central conductor (the cytoplasm), some leaks out through the sheath of insulation (the membrane) into the external medium, so that the signal becomes progressively attenuated. For this reason the electrical characteristics involved in passive spread are often referred to as *cable properties* of the axon.

Axons, though, are much poorer conductors than electric cables, and passive spread is inadequate to transmit a signal over a distance of more than a few millimeters, especially if the signal is brief and transient. This is not only because of current leakage but also because the change in membrane potential that results from current flow is not instantaneous but takes a while to build up. The time required depends on the membrane *capacitance,* that is, on the quantity of charge that has to be accumulated on either side of the membrane to produce a given membrane potential (see Panel 19–1). The membrane capacitance has the effect both of slowing down the passive transmission of signals along the axon and of distorting them, so that a sharp, pulselike stimulus delivered at one point will be detected a few millimeters away as a slow, gradual rise and fall of the potential,

1. Separated layers of charge create a voltage gradient

The voltage gradient across the cell membrane, or *membrane potential*, is created by an excess of positive charge on one side and a matching excess of negative charge on the other. The charge is concentrated in a thin (< 1 nm) layer on each side of the membrane.

2. The membrane capacitance determines the charge required to create a given voltage difference

The amount of charge (in coulombs) required on each side of the membrane to create a voltage difference of 1 V is called the membrane *capacitance* (in farads).

Cell membranes typically have a capacitance of about 1 μF/cm^2, or 0.01 pF/μm^2. Therefore a movement of 0.001 pC of charge across 1 μm^2 of membrane will alter the membrane potential by 100 mV.

6. A flow of current builds up a charge

For a squid axon membrane at the peak of the action potential,

$$g_{Na} = 300 \text{ pS/}\mu m^2;$$

the corresponding Na$^+$ current is roughly

$$i_{Na} = 5 \text{ pA/}\mu m^2.$$

If no other ions crossed the membrane, the charge transferred by the Na$^+$ current if it were sustained for 0.2 ms at the peak value seen during the action potential would be 0.001 pC/μm^2. This charge would alter the membrane potential—see (1) and (2), above.

UNITS

Charge: couloumb (C) (6.2 $\times$ 10^{18} $\times$ charge on one electron)
Electric potential: volt (V)
Current: ampere (= coulombs per second) (A)
Capacitance: farad (= coulombs per volt) (F)
Conductance: siemens (= amperes per volt) (S)
mV: millivolt (10^{-3} V)
μF: microfarad (10^{-6} F)
nC: nanocoulomb (10^{-9} C)
pS: picosiemens (10^{-12} S)

3. The number of ions that go to form the layer of charge adjacent to the membrane is minute compared with the total number inside the cell

One coulomb is the charge carried by roughly 6 $\times$ 10^{18} univalent ions, so that 0.001 pC is equivalent to 6000 univalent ions. Therefore the movement of 6000 Na$^+$ ions across 1 μm^2 of membrane will carry sufficient charge to shift the membrane potential by about 100 mV. Because there are about 3 $\times$ 10^7 Na$^+$ ions in 1 μm^3 of bulk cytoplasm, such a movement of charge will generally have a negligible effect on the ion concentration gradients across the membrane.

4. The electrochemical "driving force" is the sum of an effect of the membrane potential and an effect of the concentration gradient

For a univalent positive ion, such as Na$^+$ or K$^+$, at room temperature, the net "driving force" across the membrane is proportional to

$$V - 58 \log_{10}\left(\frac{C_o}{C_i}\right),$$

where V is the membrane potential in millivolts, and C_o and C_i are, respectively, the extracellular and intracellular concentrations of the ion. The "driving force" for positive ions is zero when

$$V = 58 \log_{10}\left(\frac{C_o}{C_i}\right) \text{ mV.}$$

This is the *Nernst equation* in its simplest form (see p. 315). It defines the *equilibrium potential* for the given positive ion. For the squid axon, the equilibrium potentials V_{Na}, V_K, and V_{Cl} for Na$^+$, K$^+$, and Cl$^-$, are, respectively, about +55 mV, −75 mV, and −65 mV. The net driving forces for each ion are proportional to $V - V_{Na}$, $V - V_K$, and $V - V_{Cl}$.

5. Ion current is proportional to driving force multiplied by membrane conductance

The current of, say, Na$^+$ ions passing across the membrane (measured in amperes) is

$$i_{Na} = g_{Na} \times (V - V_{Na}),$$

where g_{Na} is the *Na$^+$ conductance* of the membrane. The Na$^+$ conductance is proportional to the number of Na$^+$ channels that are open at any instant. The conductance of a single open Na$^+$ channel is about 4 pS in the squid axon, and there are about 75 Na$^+$ channels/μm^2 of membrane.

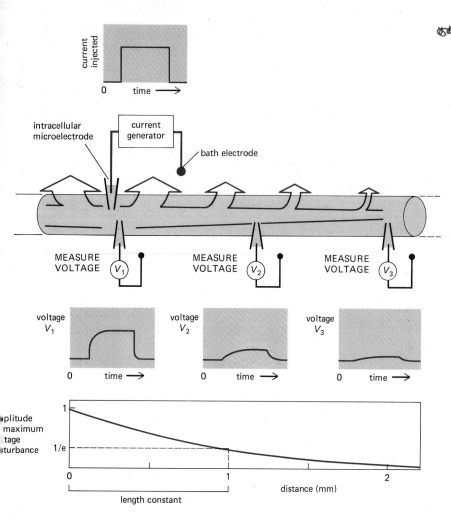

🐙 **Figure 19–9** Current injected into an axon through a microelectrode flows out again across the plasma membrane; the magnitude of the outflowing current falls off exponentially with distance from the microelectrode. The current flow is assumed to be small, causing a subthreshold depolarization of the membrane. The graphs show how the disturbance of membrane potential produced by injection of a pulse of current falls off with distance from the source of the disturbance. The length constant is the distance over which the amplitude of the disturbance of the membrane potential falls off by a factor of 1/e. The length constant ranges from about 0.1 mm (for a very small axon with a relatively leaky membrane) to about 5 mm (for a very large axon with a relatively nonleaky membrane). Here it is 1 mm.

with a greatly diminished amplitude (see Figure 19–9). To transmit faithfully over more than a few millimeters, therefore, an axon requires, in addition to its passive cable properties, an active mechanism to maintain the strength and waveform of the signal as it travels. This automatically amplified signal is the *action potential.*

Voltage-gated Na⁺ Channels Generate the Action Potential; Voltage-gated K⁺ Channels Keep It Brief[3,4,8,10]

The electrochemical mechanism of the **action potential** was first established in the 1940s and 1950s. Techniques for studying electrical events in small single cells had not yet been developed, and the experiments were made possible only by the use of a giant cell, or rather a part of a giant cell: a giant axon from a squid (Figure 19–10). Subsequent work has shown that the neurons of most animals conduct their action potentials in a similar way. Panel 19–2 outlines some of the key original experiments. Despite the many technical advances that have been made since then, the logic of the original analysis continues to serve as a model for present-day work. The crucial insight was that the permeability of the membrane to Na⁺ and K⁺ is changed by changes in the membrane potential: in other words, the membrane contains channels for Na⁺ and K⁺ that are voltage-gated. The voltage-

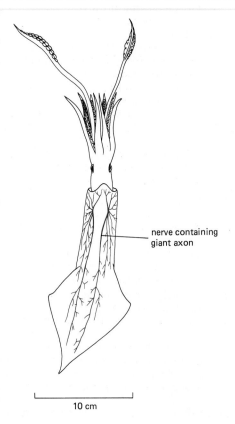

nerve containing giant axon

Figure 19–10 A squid, showing the location of the giant axons whose large size made possible the original analysis of the mechanism of the action potential. (From H. Curtis, Biology, 4th ed. New York: Worth, 1983; after Keynes, R.D. The nerve impulse and the squid. *Scientific American,* December 1958. Copyright © 1958 by Scientific American, Inc. All rights reserved.)

10 cm

1 Action potentials are recorded with an intracellular electrode

The squid giant axon is about 0.5–1 mm in diameter and several centimeters long (Figure 19–10). An electrode in the form of a glass capillary tube containing a conducting solution can be thrust down the axis of the cell so that its tip lies deep in the cytoplasm. With its help, one can measure the voltage difference between the inside and the outside of the cell—that is, the membrane potential—as an action potential sweeps past the electrode. The action potential is triggered by a brief electrical shock to one end of the axon. It does not matter which end, because the excitation can travel in either direction; and it does not matter how big the shock is, as long as it exceeds a certain threshold: the action potential is *all or none*.

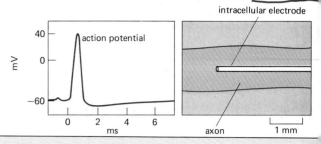

2 Action potentials depend only on the neuronal plasma membrane and on gradients of Na⁺ and K⁺ across it

The three most plentiful ions, both inside and outside the axon, are Na^+, K^+, and Cl^-. As in other cells, the Na^+-K^+ pump maintains a concentration gradient: the concentration of Na^+ is about 9 times lower inside the axon than outside, while the concentration of K^+ is about 20 times higher inside than outside. Which ions are important for the action potential?

The squid giant axon is so large and robust that it is possible to extrude the cytoplasm from it, like toothpaste from a tube, and then to perfuse it internally with pure artificial solutions of Na^+, K^+, and Cl^- or SO_4^{2-}. Remarkably, if (and only if) the concentrations of Na^+ and K^+ inside and outside approximate those found naturally, the axon will still propagate action potentials of the normal form as shown above. The important part of the cell for electrical signaling, therefore, must be the membrane; the important ions are Na^+ and K^+; and a sufficient source of free energy to power the action potential must be provided by their concentration gradients across the membrane, because all other sources of metabolic energy have presumably been removed by the perfusion.

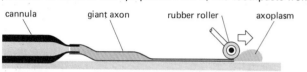

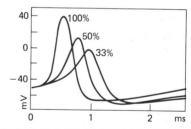

3 At rest, the membrane is chiefly permeable to K⁺; during the action potential, it becomes transiently permeable to Na⁺

At rest the membrane potential is close to the equilibrium potential for K^+. When the external concentration of K^+ is changed, the resting potential changes roughly in accordance with the Nernst equation for K^+ (see Panel 19–1 and p. 315). At rest, therefore, the membrane is chiefly permeable to K^+: K^+ leak channels provide the main ion pathway through the membrane.

If the external concentration of Na^+ is varied, there is no effect on the resting potential. However, the height of the peak of the action potential varies roughly in accordance with the Nernst equation for Na^+. During the action potential, therefore, the membrane appears to be chiefly permeable to Na^+: Na^+ channels have opened. In the aftermath of the action potential, the membrane potential reverts to a negative value that depends on the external concentration of K^+ and is even closer to the K^+ equilibrium potential than the resting potential is: the membrane has lost its permeability to Na^+ and has become even more permeable to K^+ than before—that is, Na^+ channels have closed, and additional K^+ channels have opened.

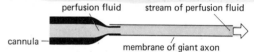

The form of the action potential when the external medium contains 100%, 50%, or 35% of the normal concentration of Na^+

4 Voltage clamping reveals how the membrane potential controls opening and closing of ion channels

The membrane potential can be held constant ("voltage clamped") throughout the axon by passing a suitable current through a bare metal wire inserted along the axis of the axon while monitoring the membrane potential with another intracellular electrode (see Figure 19–11). When the membrane is abruptly shifted from the resting potential and held in a depolarized state (A), Na^+ channels rapidly open until the Na^+ permeability of the membrane is much greater than the K^+ permeability; they then close again spontaneously, even though the membrane potential is clamped and unchanging. K^+ channels also open but with a delay, so that the K^+ permeability becomes large as the Na^+ permeability falls (B). If the experiment is now very promptly repeated, by returning the membrane briefly to the resting potential and then quickly depolarizing it again, the response is different: prolonged depolarization has caused the Na^+ channels to enter an *inactivated* state, so that the second depolarization fails to cause a rise and fall similar to the first. Recovery from this state requires a relatively long time—about 10 milliseconds—spent at the repolarized (resting) membrane potential.

In a normal unclamped axon, an inrush of Na^+ through the opened Na^+ channels produces the spike of the action potential; inactivation of Na^+ channels and opening of K^+ channels bring the membrane back down to the resting potential.

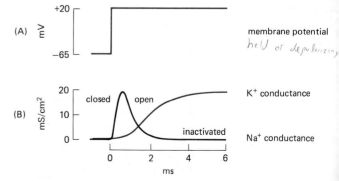

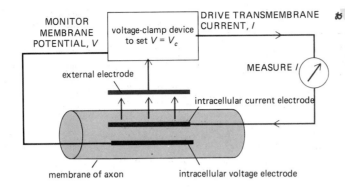

Figure 19-11 The voltage-clamp technique for studying the behavior of ion channels by measuring the current that flows across the plasma membrane when the membrane potential is held fixed at any chosen level. Two intracellular electrodes are used, one to monitor the membrane potential, the other to pass a measured current into the cell. The current passed in via the current electrode flows out again through ion channels in the plasma membrane; the circuit is shown in color. So long as the membrane potential is constant, the current, I, passed into the axon via the current electrode is exactly equal to the total current flowing out again across all regions of its membrane (otherwise the net charge inside the cell would change, causing a change in membrane potential). The membrane potential can be altered by forcing more or less current to flow across the membrane. The electronic apparatus marked "voltage-clamp device" monitors the membrane potential, V, and controls the current I so as to hold V constant: any slight departure from a chosen preset "command voltage," V_c, automatically causes a compensating adjustment of the current in such a way that the membrane potential is *clamped* at $V = V_c$. To study the time-dependent behavior of the membrane channels, one can switch abruptly from one fixed value of V_c to another and observe the corresponding current on an oscilloscope. By repeating the current measurements with different concentrations of Na^+ and K^+ in the medium, it is possible to deduce how much of the transmembrane current is carried by each of these ions and so distinguish the contributions due to Na^+-selective and K^+-selective channels. The voltage-clamp technique can be adapted to analyze the behavior of individual ion channel molecules in a tiny patch of membrane covering the mouth of a microelectrode; this is the method of *patch-clamp* recording.

clamp technique (Figure 19–11) made it possible to determine the detailed rules by which a change of membrane potential opens and closes the channels, and the action potential was then shown to be a direct consequence of the operation of these rules.

An action potential is triggered when the membrane is momentarily depolarized beyond a certain *threshold* value. As explained in Chapter 6, such a depolarization, applied to a given area of membrane, causes voltage-gated Na^+ channels here to open; this permits an influx of Na^+ down its electrochemical gradient, which causes still further depolarization, causing more Na^+ channels to open, and so on, in an explosive self-amplifying fashion, until the membrane in that region is driven almost all the way to the Na^+ equilibrium potential (see Panel 19–1). At this point, two factors conspire to bring the membrane back down toward its original negative potential: the Na^+ channels convert spontaneously to a closed, inactivated state, and voltage-gated K^+ channels open. These K^+ channels respond to changes of membrane potential in much the same way as the Na^+ channels do, but with slightly slower kinetics (for this reason they are sometimes called *delayed K^+ channels*). Once the K^+ channels are open, the transient influx of Na^+ is rapidly overwhelmed by an efflux of K^+, and the membrane is driven back toward the K^+ equilibrium potential, even before the inactivation of the Na^+ channels is complete. The repolarization causes the voltage-gated K^+ channels to close again and allows the inactivated Na^+ channels to regain their original closed but activatable state. In this way the patch of membrane can be made ready to fire another action potential in less than a millisecond.

Subsequent experiments have shown that not all neurons depend on voltage-gated K^+ channels to terminate the action potential. In particular, in mammalian myelinated axons (see p. 1073), voltage-gated K^+ channels are present only in very small numbers, and the return to rest is brought about simply by the inactivation of the Na^+ channels. Although voltage-gated K^+ channels are thus not essential for the propagation of action potentials, we shall see (p. 1089) that they play a crucial part in the mechanism for triggering action potentials in response to stimulation of the nerve cell body.

Propagated Action Potentials Provide for Rapid Long-distance Communication[3,4,8,11]

Because of the cable properties of the axon, the large local influx of Na^+ ions during an action potential causes some current to flow along its length, depolarizing neighboring regions of the membrane to threshold levels, so that they in their turn produce action potentials (Figure 19–12). This process continues down the axon, with one region "igniting" the next, at speeds that in a vertebrate range from less than 1 m/sec to more than 100 m/sec, depending on the type of axon. The speed of propagation depends mainly on the cable properties of the axon: the larger the membrane capacitance, the larger the charge required for depolarization to threshold; and the larger the internal resistance of the axonal cytoplasm, the smaller the currents that will flow along it and the longer it will take for the requisite charge to accumulate. Both the resistance and the capacitance of a unit length of axon depend on the axon's cross-sectional area, and a simple calculation

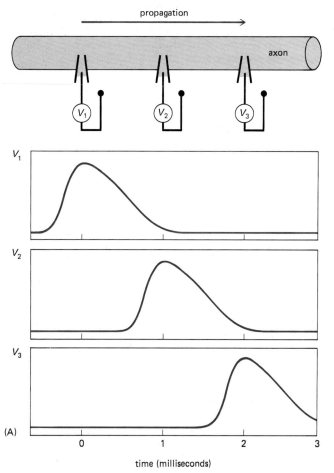

propagation

axon

V_1

V_2

V_3

(A)

time (milliseconds)

Figure 19–12 The propagation of an action potential. (A) shows the voltages that would be recorded from a set of intracellular electrodes placed at intervals along the axon. (B) shows the changes in the Na$^+$ channels and the current flows (*colored lines*) that give rise to the traveling disturbance of the membrane potential. The region of the axon with a depolarized membrane is shown in color.

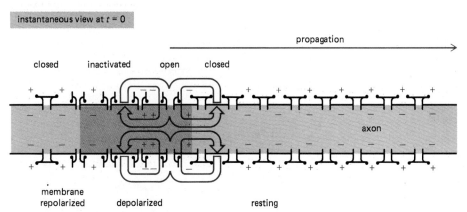

instantaneous view at $t = 0$

propagation

closed   inactivated   open   closed

axon

membrane repolarized   depolarized   resting

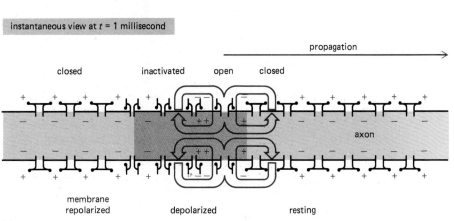

instantaneous view at $t = 1$ millisecond

propagation

closed   inactivated   open   closed

axon

membrane repolarized   depolarized   resting

(B)

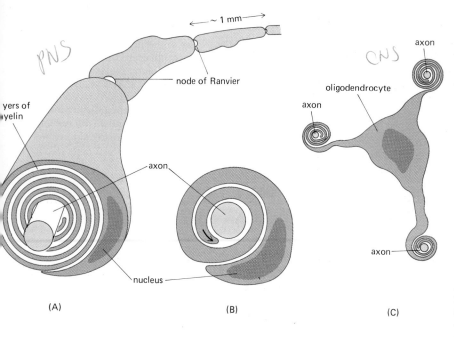

PNS

~ 1 mm

node of Ranvier

layers of myelin

axon

nucleus

(A)

(B)

CNS

axon

oligodendrocyte

axon

axon

(C)

Figure 19–13 (A) Schematic diagram of a myelinated axon from a peripheral nerve. Each Schwann cell wraps its plasma membrane concentrically around the axon to form a segment of myelin sheath about 1 mm long. For clarity, the layers of myelin are not shown so tightly compacted together as they are in reality (see part D). (B) Schematic diagram of a Schwann cell in the early stages of forming a spiral of myelin around an axon during development. Note that it is the inner tongue of the Schwann cell (marked with an arrow) that continues to extend around the axon, thereby adding turns of membrane to the myelin sheath. (C) Schematic diagram of an oligodendrocyte, which forms myelin sheaths in the central nervous system. A single oligodendrocyte myelinates many separate axons. (D) Electron micrograph of a section from a nerve in the leg of a young rat. Two Schwann cells can be seen: one is just beginning to myelinate its axon, the other has formed an almost mature myelin sheath. (E) Electron micrograph of an oligodendrocyte in the spinal cord of a kitten extending processes to myelinate at least two axons. (D and E, from C. Raine, in Myelin [P. Morell, ed.]. New York: Plenum, 1976.)

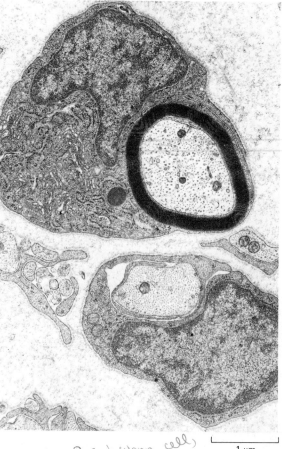

2 schwann cells can be seen

1 μm

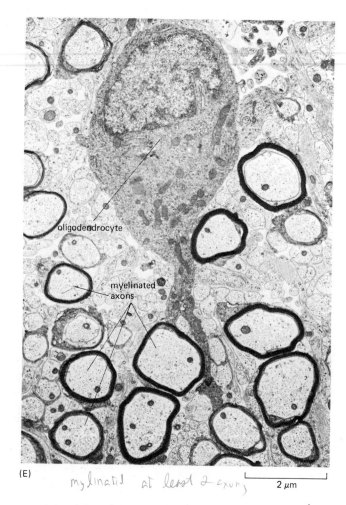

oligodendrocyte

myelinated axons

(E)

myelinated at least 2 axons

2 μm

shows that the net effect of enlarging the axon diameter is to increase the speed of propagation of action potentials. The squid and various other invertebrates have thus achieved rapid signaling by evolving axons of giant diameter. Vertebrates, however, have achieved equally high speeds of conduction in a much more compact manner by insulating many of their axons with a *myelin sheath*.

Myelination Increases the Speed and Efficiency of Propagation of Action Potentials in Vertebrates[8,12]

The **myelin sheath** is formed by specialized glial cells—*Schwann cells* in peripheral nerves and *oligodendrocytes* in the central nervous system. These cells wrap layer upon layer of their own plasma membrane in a tight spiral around the axon (Figure 19–13). Each myelinating Schwann cell devotes itself to a single axon, forming a segment of sheath that is about 1 mm long and may consist of up to 300 concentric layers of membrane; oligodendrocytes form similar segments of sheath, but they do so for many separate axons simultaneously.

The insulating layer formed by the myelin sheath drastically reduces the effective capacitance of the axon membrane and at the same time prevents almost all current leakage across it. Between one segment of sheath and the next, small regions of axon membrane remain bare (Figure 19–14). These so-called *nodes of Ranvier*, only about 0.5 μm long, are foci of electrical activity. Practically all the Na^+ channels of the axon are concentrated at the nodes, giving a density of several thousand channels per square micrometer there, with almost none in the membrane covered by myelin sheath. Thus the ensheathed portions of the axon membrane are not excitable but have excellent cable properties—a low capacitance and a high resistance to current leakage. Consequently, when an action potential is triggered at a node, the resulting currents are funneled efficiently by passive spread to the next node, depolarize it rapidly, and trigger it to fire another action potential. Thus conduction is *saltatory:* the signal propagates along the axon by leaping from node to node. Myelination brings two main advantages: action potentials travel faster, and metabolic energy is conserved because the active excitation is confined to the small nodal regions.

Summary

Electrical signaling in nerve cells depends on changes of membrane potential due to movements of small numbers of ions through gated ion channels. The Na^+-K^+ pump builds up a large store of energy to drive these movements by generating

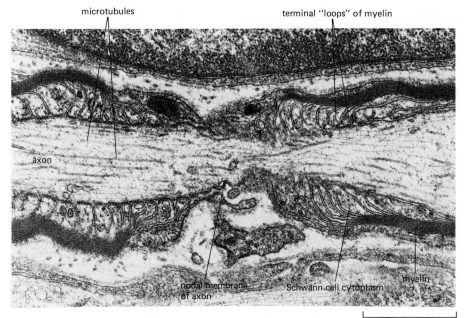

microtubules

terminal "loops" of myelin

axon

nodal membrane of axon

Schwann cell cytoplasm

myelin

1 μm

Figure 19–14 Electron micrograph of a longitudinal section of a myelinated axon from a peripheral nerve showing a node of Ranvier, where a small portion of the axon's plasma membrane is left exposed between the ends of two adjacent segments of myelin sheath. (Courtesy of Richard Bunge.)

arge concentration gradients of Na⁺ and K⁺ across the nerve cell membrane. In
he resting neuron the K⁺-selective leak channels in the membrane make it more
ermeable to K⁺ than to other ions, and the membrane potential is consequently
close to the K⁺ equilibrium potential of about −70 mV. An action potential is
riggered when a brief depolarizing stimulus causes voltage-gated Na⁺ channels to
open, making the membrane more permeable to Na⁺ and further depolarizing the
membrane potential toward the Na⁺ equilibrium potential. This positive feedback
causes still more Na⁺ channels to open, resulting in an all-or-none action potential.
n each region of membrane, the action potential is rapidly terminated by the
nactivation of the Na⁺ channels and, in many neurons, by the opening of voltage-
ated K⁺ channels.

The propagation of an action potential along a nerve fiber depends on the fiber's
assive cable properties: when the membrane is locally depolarized and fires an
ction potential, the current entering through open Na⁺ channels at that site spreads
assively to depolarize neighboring regions of the membrane, where action poten-
als are triggered in turn. In many vertebrate axons the speed and efficiency of
ropagation of action potentials are increased by insulating sheaths of myelin, which
hange the cable properties of the axon and leave only small regions of excitable
membrane exposed.

Ligand-gated Ion Channels and Fast Synaptic Transmission[13]

he simplest way for one neuron to pass its signal to another is by direct electrical
upling through gap junctions. Such **electrical synapses** have the virtue that
ansmission occurs without delay. But they are far less rich in possibilities for
djustment and control than are the **chemical synapses** that provide the majority
 nerve cell connections. Electrical communication through gap junctions was
nsidered in Chapter 14 (pp. 798–801). Here we shall confine our discussion to
hemical synapses.

The principles of chemical communication at a synapse are the same as those
 chemical communication by water-soluble hormones, as discussed in Chapter
. In both cases a cell releases a chemical messenger that acts on another cell,
 set of cells, by binding to membrane receptor proteins. Unlike a hormone,
owever, the chemical messenger at a synapse—the **neurotransmitter**—acts at
ry close quarters.

Electrical stimulation of the presynaptic cell causes the release of a neuro-
ansmitter by exocytosis (see Figure 19–4); once the neurotransmitter has crossed
e gap—typically a small fraction of a micrometer—between the pre- and post-
naptic cells, the chemical signal must be converted back into an electrical one.
his conversion is mediated by the receptors in the plasma membrane of the
stsynaptic cell, which fall into two distinct categories: *channel-linked receptors*
d *non-channel-linked receptors* (Figure 19-15). Channel-linked receptors can be
scribed equivalently as *ligand-gated channels*. A channel-linked receptor, upon
nding neurotransmitter, promptly changes its conformation so as to create an
en channel for specific ions to cross the membrane, thereby altering the mem-
ane permeability. This type of receptor underlies the most familiar and the best-
derstood mode of chemical synaptic signaling, where transmission is very rapid.

Non-channel-linked receptors work by the same mechanisms that mediate
sponses to water-soluble hormones and local chemical mediators throughout
e body (see p. 694). In such receptors the neurotransmitter-binding site is func-
nally coupled to an enzyme that, in the presence of neurotransmitter, usually
talyzes the production of an intracellular messenger such as cyclic AMP. The
racellular messenger in turn causes changes in the postsynaptic cell, including
difications of the ion channels in its membrane. Compared with channel-linked
ceptors, these receptors generally provide for neurotransmitter actions that are
atively slow in onset and long in duration. Some of them are believed to mediate
 long-lasting neuronal changes that underlie learning and memory (see p. 1094).

(A) CHANNEL-LINKED RECEPTOR

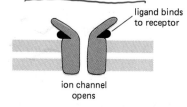

ligand binds
to receptor

ion channel
opens

(B) NON-CHANNEL-LINKED RECEPTOR

ligand binds
to receptor

ion channel
opens

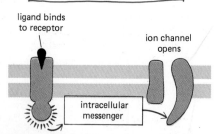

intracellular
messenger

Figure 19–15 A neurotransmitter can
exert its effect on a postsynaptic cell by
means of two fundamentally different
types of receptor proteins: channel-
linked receptors and non-channel-
linked receptors. Channel-linked
receptors are also known as ligand-
gated channels.

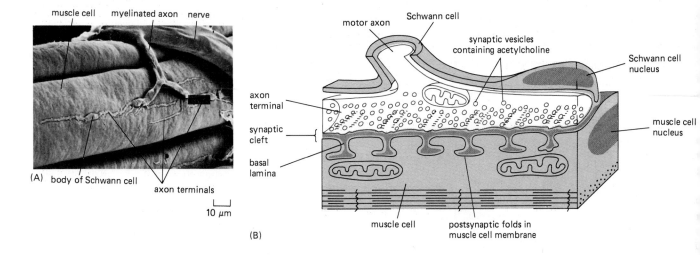

muscle cell   myelinated axon   nerve

axon terminals

10 μm

(A) body of Schwann cell

motor axon   Schwann cell

synaptic vesicles
containing acetylcholine

Schwann cell
nucleus

muscle cell
nucleus

axon
terminal

synaptic
cleft

basal
lamina

muscle cell   postsynaptic folds in
muscle cell membrane

(B)

In this section we shall discuss rapid synaptic transmission based on ligand-gated ion channels. The special features of synaptic signaling based on non-channel-linked receptors, and their role in long-term synaptic change, will be discussed in a later section (see p. 1091).

The Neuromuscular Junction Is the Best-understood Synapse[14]

The central nervous system is so densely packed with neurons that it is extremely difficult to perform experiments on single synapses within it. Detailed understanding of synaptic function has come instead chiefly from work on the junctions between nerve and skeletal muscle in the frog and, to a lesser extent, on synapses between giant neurons in the squid and other mollusks.

Skeletal muscle cells in vertebrates, like nerve cells, are electrically excitable, and the **neuromuscular junction** (Figure 19–16) has proved to be a valuable model for chemical synapses in general. A motor nerve and its muscle can be dissected free from the surrounding tissue and maintained in a bath of controlled composition. The nerve can be stimulated with extracellular electrodes, and the response of a single muscle cell can be monitored relatively easily with an intracellular microelectrode (Figure 19–17). Figure 19–18 compares the fine structure of a neuromuscular junction with that of a typical synapse between two neurons in the central nervous system.

The neuromuscular junction has been the focus of a long and fruitful series of investigations that began in the 1950s. The background to the early experiments was the discovery, in the early 1920s, that acetylcholine is released upon stimulation of the vagus nerve to the heart and acts on heart muscle to slow its beating. This was the first clear evidence of chemical neurotransmission, and it soon led to the demonstration, in the 1930s, that stimulation of a motor nerve innervating a skeletal muscle also causes the release of acetylcholine and that acetylcholine in turn stimulates skeletal muscle to contract. Acetylcholine was thereby identified as the neurotransmitter at the neuromuscular junction. But how is the release of acetylcholine brought about, and how does it exert its effect on the muscle?

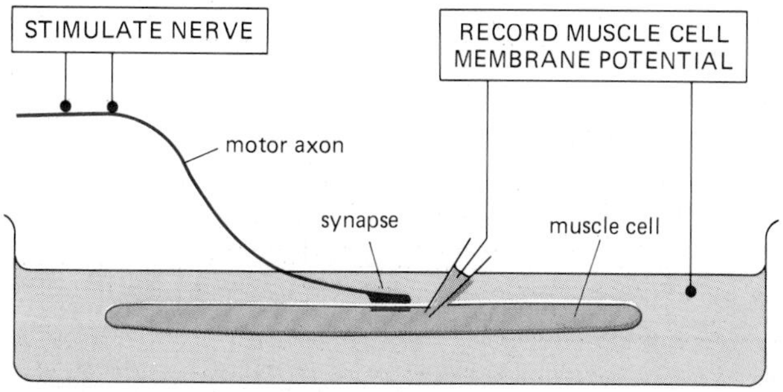

Figure 19–16 A neuromuscular junction in a frog. (A) Low-magnification scanning electron micrograph of the termination of a single axon on a skeletal muscle cell. (B) Schematic drawing of the part of the junction boxed in (A), showing the major features visible by transmission electron microscopy. The pattern of small terminal branches of the axon at the junction varies with the species and with the type of skeletal muscle cell. From its appearance in mammals, the neuromuscular junction is often called the *end plate*. (A, from J. Desaki and Y. Uehara, *J. Neurocytol.* 10:101–110, 1981, by permission of Chapman & Hall.)

STIMULATE NERVE

RECORD MUSCLE CELL
MEMBRANE POTENTIAL

motor axon

synapse   muscle cell

Figure 19–17 An experimental arrangement used to study synaptic transmission at the neuromuscular junction.

Neuromuscular junction (handwritten)

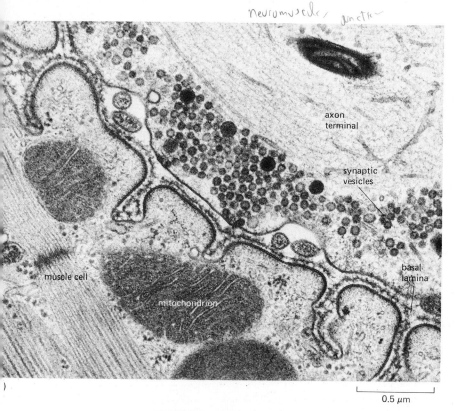

axon
terminal

synaptic
vesicles

basal
lamina

muscle cell

mitochondrion

0.5 μm

Figure 19–18 (A) Electron micrograph of part of a neuromuscular junction. (B) Electron micrograph of a small region from the brain of a rat. Two synapses are clearly visible in (B), each showing pre- and postsynaptic membranes, a synaptic cleft between them, and synaptic vesicles on the presynaptic side, as in (A). The two synapses labeled in (B) differ from each other in the size and shape of their vesicles: the vesicles at the *type I* synapse are round, whereas those at the *type II* synapse are flattened and are believed to contain a different neurotransmitter. Note the characteristic "thickened" appearance of the postsynaptic membrane and, to a lesser extent, of the presynaptic membrane in both (A) and (B). There is no basal lamina interposed between the pre- and postsynaptic membranes at synapses in the brain, although some extracellular material is faintly apparent in the cleft. The absence of a basal lamina represents the chief structural difference between a synapse in the central nervous system and a neuromuscular junction. (A, courtesy of John Heuser; B, courtesy of G. Campbell and A.R. Lieberman.)

Brain (handwritten)

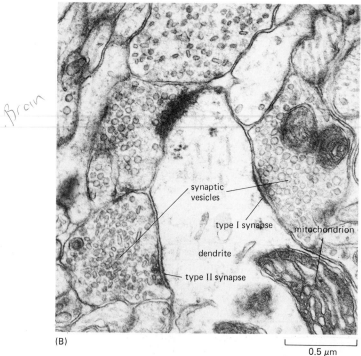

synaptic
vesicles

type I synapse

mitochondrion

dendrite

type II synapse

(B)

0.5 μm

No Basal lamina in CNS (brain) (handwritten)

Basal lamina present in neurojunction (handwritten)

Voltage-gated Ca²⁺ Channels Couple Action Potentials to Neurotransmitter Release[15]

The action potential is propagated along the axon by the opening and closing of Na⁺ channels until it reaches the neuromuscular junction. Here the action potential opens *voltage-gated Ca²⁺ channels* in the plasma membrane of the axon terminal, allowing Ca²⁺ to enter and trigger the release of acetylcholine (Figure 19–19).

Three simple observations showed that this influx of Ca²⁺ into the axon terminal is <u>essential</u> for synaptic transmission. <u>First</u>, if there is no Ca²⁺ in the extra-

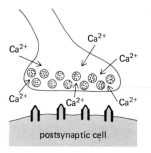

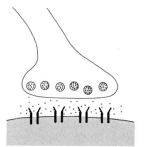
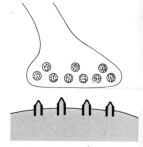

| action potential triggers entry of Ca²⁺ into presynaptic terminal | synaptic vesicles fuse with presynaptic membrane releasing transmitter | transmitter binds to proteins in the postsynaptic membrane changing their conformation | transmitter is removed from cleft and postsynaptic proteins revert to original conformation |

Figure 19–19 Summary of the essential events at a chemical synapse following the arrival of an action potential in the axon terminal.

cellular medium bathing the axon terminal when the action potential arrives, no neurotransmitter is released and transmission fails. Second, if Ca²⁺ is injected artificially into the cytoplasm at the axon terminal through a micropipette, transmitter is immediately released even without electrical stimulation of the axon. (This microinjection experiment is difficult to do at the neuromuscular junction because the axon terminal is so small, but it has been done at a synapse between giant neurons in the squid.) Third, artificial depolarization of the axon terminal (again at the squid giant synapse), in the absence of an action potential and with the Na⁺ and K⁺ channels blocked by specific toxins, leads to Ca²⁺ entry and transmitter release; furthermore, if the depolarization reverses the membrane potential so far as to reduce the electrochemical driving force for Ca²⁺ entry to zero, no transmitter release occurs.

The channel protein that lets the Ca²⁺ into the cell—the **voltage-gated Ca²⁺ channel**—has a uniquely important role. It provides the only known means of converting electrical signals—fleeting depolarizations of the membrane—into chemical changes inside nerve cells. As explained in Panel 19–1, voltage-gated channels for Na⁺, K⁺, or Cl⁻ are of no use for this purpose: the ion fluxes driven through them by a single action potential are so small that they do not significantly alter the ion concentrations in the cytosol. The ion flux through voltage-gated Ca²⁺ channels is no larger in absolute terms and generally makes only a small contribution to the electrical current across the membrane; but it is very much larger in relation to the free Ca²⁺ concentration inside the cell, which is normally kept at about 10^{-7} M, corresponding to less than 100 Ca²⁺ ions/μm³. In 1 millisecond, a single open Ca²⁺ channel would typically pass several hundred Ca²⁺ ions, driven by the membrane potential and the relatively high extracellular concentration of Ca²⁺ (usually ~1–2 mM). Thus a small number of voltage-gated Ca²⁺ channels in the presynaptic terminal, opening in response to an action potential, can easily raise the intracellular concentration of free Ca²⁺ by a factor of 10 to 100. The surge of free Ca²⁺ then acts as an intracellular messenger, triggering the release of neurotransmitter at a rate that increases very steeply with the free Ca²⁺ concentration.

The increase of free Ca²⁺ concentration is short-lived because Ca²⁺-binding proteins, Ca²⁺-sequestering vesicles, and mitochondria rapidly take up the Ca²⁺ that has entered the axon terminal, while Ca²⁺ pumps in the plasma membrane, driven either by ATP hydrolysis or by the Na⁺ electrochemical gradient, pump it out of the cell (see p. 307 and p. 700). In this way the terminal is ready to transmit another signal as promptly as the axon is ready to deliver one.

Neurotransmitter Is Released Rapidly by Exocytosis[16]

The axon terminal at the neuromuscular junction is crammed with thousands of uniform (~40 nm diameter) secretory vesicles, called *synaptic vesicles*, each containing acetylcholine (see Figure 19–18). The entry of Ca²⁺ induces a synchronized burst of exocytosis in which the vesicles fuse with the presynaptic membrane, discharging their contents into the synaptic cleft to act on the postsynaptic cell.

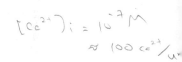

<image type="margin_note">$[Ca^{2+}]_i = 10^{-7} M$
$\approx 100\ Ca^{2+}/\mu^3$</image>

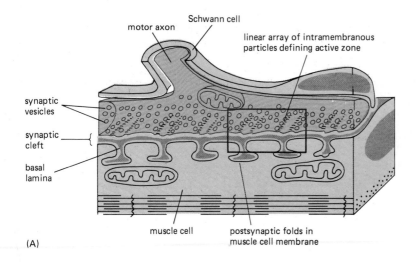

motor axon

Schwann cell

linear array of intramembranous particles defining active zone

synaptic vesicles

synaptic cleft

basal lamina

(A)

muscle cell

postsynaptic folds in muscle cell membrane

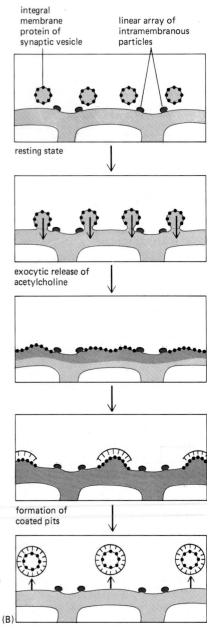

integral membrane protein of synaptic vesicle

linear array of intramembranous particles

resting state

exocytic release of acetylcholine

formation of coated pits

(B)

membrane retrieval

The exocytosis is restricted to specialized regions known as *active zones*, exactly opposite the receptors on the postsynaptic cell; in this way the delay associated with diffusion of neurotransmitter across the cleft is made negligibly short. The membrane of the discharged synaptic vesicles is subsequently retrieved from the presynaptic plasma membrane by endocytosis.

There is evidence that, as well as triggering exocytosis, the influx of Ca^{2+} into the axon terminal activates a Ca^{2+}/calmodulin-dependent protein kinase (Ca-kinase II—see p. 711), which phosphorylates a number of proteins in the terminal, including *synapsin I*, a protein attached to the surface of synaptic vesicles. Phosphorylation is thought to release the synapsin I and thereby allow the vesicles to dock at the active zone of the presynaptic membrane, where they are needed to replace the vesicles lost from that region by exocytosis. The whole cycle of events that is initiated by a single nerve impulse has been vividly demonstrated by very rapidly freezing the nerve and muscle tissue and then preparing it for electron microscopy. Some of the results are shown in Figure 19–20.

Neurotransmitter Release Is Quantal and Probabilistic[17]

An axon terminal at a neuromuscular junction typically releases a few hundred of its many thousands of synaptic vesicles in response to a single action potential. Each vesicle, by discharging its contents into the synaptic cleft, contributes to the production of a voltage change in the postsynaptic muscle cell, which can be recorded with an intracellular electrode (Figure 19–21). The muscle cell membrane is thus depolarized beyond its threshold and fires an action potential. This excitation sweeps over the cell (Figure 19–22), causing a contraction, as described on page 621.

Even when the axon terminal is electrically quiet, occasional brief depolarizations of the muscle membrane are observed in the neighborhood of the synapse. These **miniature synaptic potentials** typically have an amplitude of only about 1 mV—far below threshold—and they occur at random, with a certain low probability per unit time—typically about once per second (Figure 19–23). Each miniature potential results from a single synaptic vesicle fusing with the presynaptic membrane so as to discharge its contents. The amplitude as recorded in a given muscle cell is roughly uniform because each vesicle contains practically the same number of molecules of acetylcholine, on the order of 5000. This number represents the minimum packet or *quantum* of transmitter release. Larger signals are made up of integral multiples of this basic unit. The Ca^{2+} that enters the axon terminal during an action potential acts for a fraction of a millisecond to increase the rate of occurrence of the exocytic events more than 10,000-fold above the resting spontaneous frequency. Nonetheless, the process remains probabilistic, and identical stimulations of the nerve do not always produce exactly the same postsynaptic effect: if, for example, 300 quanta are released on average, more or less than this number may be released on any particular occasion.

Figure 19–20 The cycle of membrane events in the axon terminal at a neuromuscular synapse following stimulation. To follow the action, samples of tissue are prepared by sudden freezing at measured times after the stimulus. To make the task easier technically, the conditions of excitation are artificially adjusted so as to slow down the normal time course by a factor of 5 or 10 and increase the number of vesicles that undergo exocytosis. (A) Simplified drawing of a neuromuscular junction, showing the active zones where transmitter release occurs. (B) The boxed area in (A) is enlarged and shown schematically in cross-section at a series of different times after stimulation of the nerve.

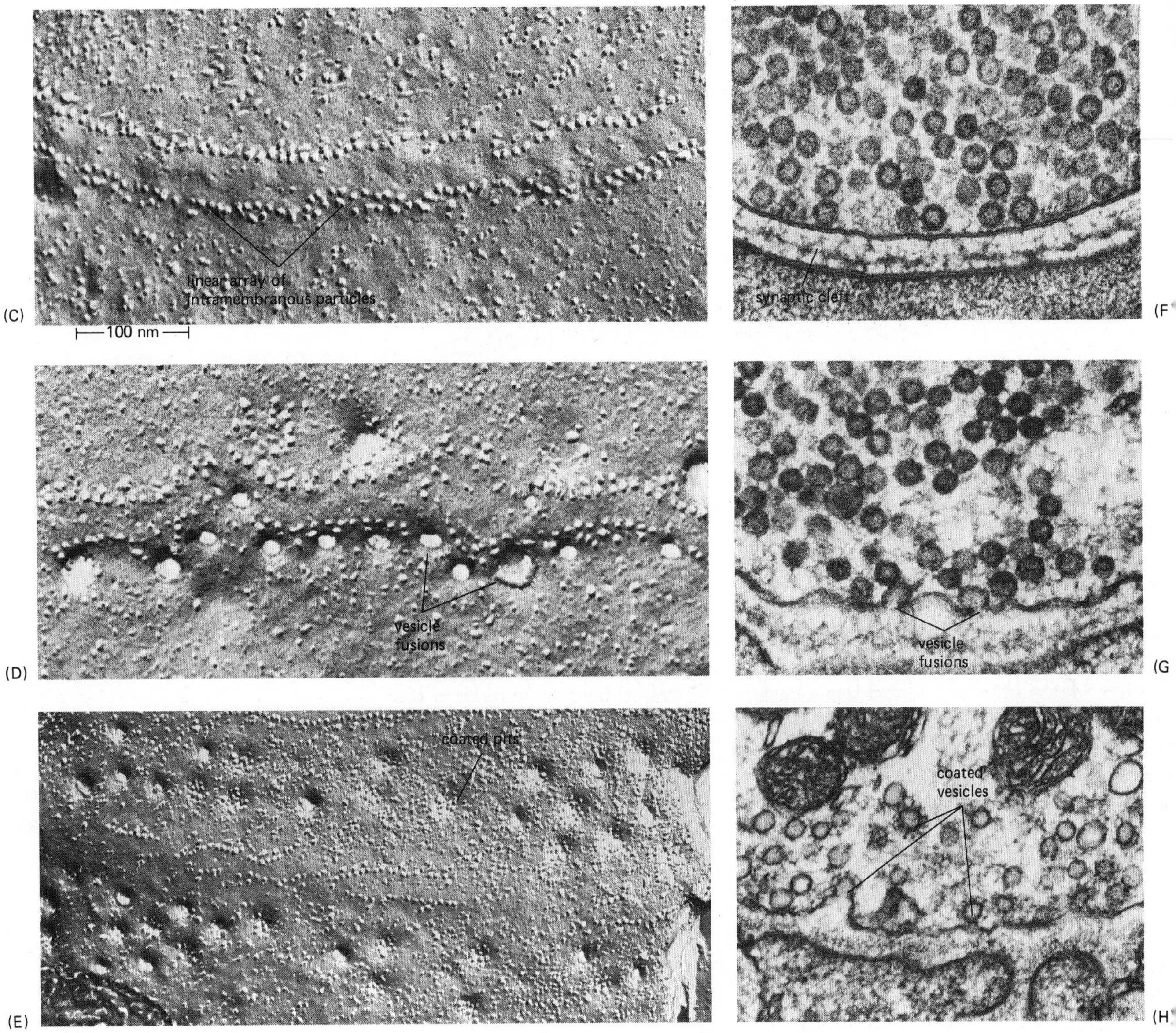

(C)

├──100 nm──┤

linear array of
intramembranous particles

(F)

synaptic cleft

(D)

vesicle
fusions

(G)

vesicle
fusions

(E)

coated pits

(H)

coated
vesicles

Figure 19–20 *(Cont'd)* (C-H) The actual appearance of the membrane as viewed by electron microscopy. Freeze-fracture electron micrographs of the cytoplasmic half of the presynaptic membrane are shown on the left; thin-section micrographs are shown on the right. (C, F) Resting state. (D, G) Fusion of synaptic vesicles with the plasma membrane at an active zone (marked by the linear arrays of intramembranous particles). (E, H) Retrieval of synaptic vesicle membrane via coated pits and coated vesicles.

Synaptic vesicles can be seen to have begun fusing with the plasma membrane within 5 milliseconds after the stimulus (D, G); each of the openings in the plasma membrane apparent in (D) represents the point of fusion of one synaptic vesicle. Fusion is complete within another 2 milliseconds. The first signs of membrane retrieval become apparent within about 10 seconds as coated pits (see p. 326) form and then, after a further 10 seconds, begin to pinch off by endocytosis to form coated

vesicles (E, H). These vesicles include the original membrane proteins of the synaptic vesicle and also contain molecules captured from the external medium. The cycle ends when the coat dissociates from the coated vesicle, which refills with acetylcholine to form a smooth-surfaced, regenerated synaptic vesicle. This scheme probably accounts for the strikingly uniform size of the synaptic vesicles, a size defined by the dimensions of the latticelike coat of clathrin (see p. 327).

Further evidence for this retrieval scheme can be obtained by stimulating the nerve in the presence of electron-dense extracellular markers such as ferritin. These markers quickly appear within coated vesicles and eventually show up in synaptic vesicles.

Note, however, that some experts have been skeptical about these experiments, interpreting some of the phenomena as artefacts. (C–H, courtesy of John Heuser.)

1079

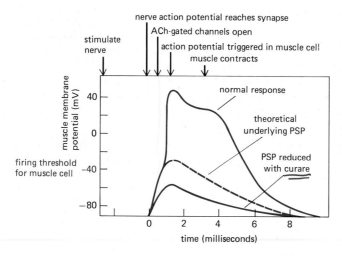

Figure 19–21 The postsynaptic response to a single nerve impulse at the neuromuscular junction: a graph of the voltage change in a frog muscle cell recorded, as in Figure 19–17, with an intracellular electrode close to the synapse. Normally the postsynaptic potential (PSP)—the depolarization directly produced by the neurotransmitter acting on the muscle cell membrane—is large enough to trigger an action potential, which complicates the analysis. A pure PSP, uncomplicated by an action potential, can be obtained by adding a moderate concentration of *curare* to the extracellular medium. This toxin, by binding to some of the receptors and blocking their response to the neurotransmitter, reduces the size of the PSP to the point where no action potential is triggered.

(A)

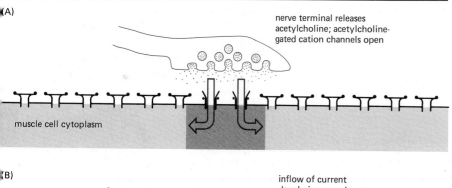

nerve terminal releases acetylcholine; acetylcholine-gated cation channels open

muscle cell cytoplasm

(B)

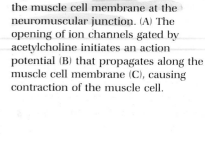

inflow of current depolarizes membrane, opening voltage-gated Na^+ channels

Figure 19–22 Electrical events in the muscle cell membrane at the neuromuscular junction. (A) The opening of ion channels gated by acetylcholine initiates an action potential (B) that propagates along the muscle cell membrane (C), causing contraction of the muscle cell.

(C)

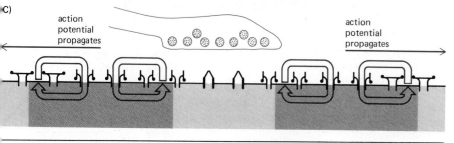

action potential propagates

action potential propagates

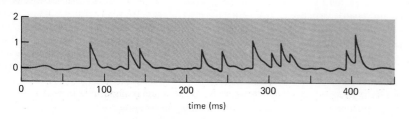

Figure 19–23 Miniature synaptic potentials (often called "miniature end-plate potentials") recorded from a frog muscle cell with an intracellular electrode inserted close to the neuromuscular junction. Each blip in the record is a miniature synaptic potential generated by release of the contents of a single synaptic vesicle from the axon terminal. (Redrawn from P. Fatt and B. Katz, *J. Physiol.* 117:109–128, 1952.)

→ insensitive to Vm

Ligand-gated Channels Convert the Chemical Signal Back into Electrical Form[18]

The muscle cell membrane at the synapse behaves as a (transducer) that converts a chemical signal in the form of a neurotransmitter into an electrical signal. The conversion is achieved by **ligand-gated ion channels** (that is, **channel-linked receptors**) in the postsynaptic membrane: when the neurotransmitter binds to these proteins, they change their conformation—opening to let ions cross the membrane—and thereby alter the membrane potential. The shift of membrane potential, if it is large enough, will in turn cause voltage-gated channels to open, thereby triggering an action potential (Figure 19–24). Unlike voltage-gated ion channels, the ligand-gated ion channels are relatively insensitive to the membrane potential. They cannot by themselves, therefore, produce an all-or-none, self-amplifying excitation. Instead they produce an electrical change that is graded according to the intensity and duration of the external chemical signal—that is, according to how much transmitter is released into the synaptic cleft and how long it stays there. This feature of ligand-gated ion channels is important in information processing at synapses, as will be discussed later.

Postsynaptic ligand-gated channels have two other important properties. First, in their role as receptors, they have an enzymelike specificity for particular ligands so that they respond only to one neurotransmitter—the one released from the presynaptic terminal; other transmitters are virtually without effect. Second, in their role as channels, they are characterized by different ion selectivities: some may be selectively permeable to K^+, others to Cl^-, and so on, whereas still others may, for example, be relatively nonselective among the cations but exclude anions. We shall see that the ion selectivity of the ligand-gated channels determines the nature of the postsynaptic response.

The Acetylcholine Receptor Is a Ligand-gated Cation Channel[19]

The channel in the skeletal muscle cell membrane gated by acetylcholine and known as the **acetylcholine receptor** is the best understood of all ligand-gated ion channels, and its molecular properties have already been discussed (see p. 319).

Like the voltage-gated Na^+ channel, the acetylcholine receptor has a number of discrete alternative conformations (Figure 19–25). Upon binding acetylcholine it jumps abruptly from a closed to an open state and then stays open, with the ligand bound, for a randomly variable length of time, averaging about 1 millisecond or even less, depending on the temperature and the species. In the open conformation the channel is indiscriminately permeable to small cations, including Na^+, K^+, and Ca^{2+}, but it is impermeable to anions (Figure 19–26).

Since there is little selectivity among these cations, their relative contributions to the current through the channel depend chiefly on their concentrations and on the electrochemical driving forces. If the muscle cell membrane is at its resting potential, the net driving force for K^+ is near zero because the voltage gradient nearly balances the K^+ concentration gradient across the membrane. For Na^+, on the other hand, the voltage gradient and the concentration gradient both act in the same direction to drive Na^+ into the cell. (The same is true for Ca^{2+}, but the extracellular concentration of Ca^{2+} is so much lower than that of Na^+ that Ca^{2+} makes only a small contribution to the total inward current.) Opening the acetylcholine receptor channel therefore leads chiefly to a large influx of Na^+, causing membrane depolarization.

Acetylcholine Is Removed from the Synaptic Cleft by Diffusion and by Hydrolysis[20]

If the postsynaptic cell is to be accurately controlled by the pattern of signals sent from the presynaptic cell, the postsynaptic excitation must be switched off promptly when the presynaptic cell falls quiet. At the neuromuscular junction this is achieved by rapidly removing the acetylcholine from the synaptic cleft through two mechanisms. First, the acetylcholine disperses by diffusion, a rapid process because

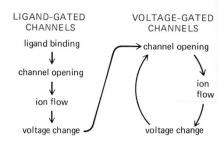

Figure 19–24 Diagram summarizing the function of ligand-gated and voltage-gated channels in the response to a neurotransmitter. The arrows indicate causal connections.

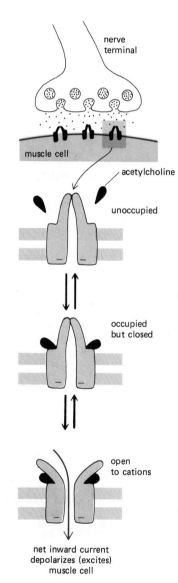

Figure 19–25 The response of the acetylcholine receptor to acetylcholine. Prolonged exposure to high concentrations of acetylcholine causes the receptor to enter yet another state (not shown), in which it is inactivated and will not open even though acetylcholine is present.

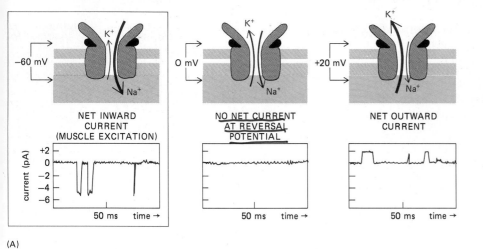

(A)

(B)

Figure 19–26 Measurements of the current through the open acetylcholine-receptor channel at different values of the membrane potential. Such measurements can be used to discover the channel's ion selectivity. A current carried through an open channel by a particular type of ion will vary with the membrane potential in a characteristic way that differs from one type of ion to another according to its concentration gradient across the membrane. Knowing the concentration gradients of the major ions present, one can thus get a useful clue to the channel's ion selectivity simply by measuring its current/voltage relationship; and fuller information can be obtained by repeating the measurements with altered ion concentrations.

(A) Patch-clamp recordings of the current through a single channel kept in a bath containing a fixed concentration of acetylcholine, with the membrane potential clamped at three different voltages. The channel flips randomly between open and closed states in a similar way in each case, but at a certain value of the membrane potential, called the *reversal potential*, the current is zero even when the channel is open. In this particular example the reversal potential happens to be approximately 0 mV. (B) The same phenomenon can be observed by monitoring the total current that flows through the large population of acetylcholine-receptor channels in the postsynaptic membrane at a neuromuscular synapse following a single stimulus to the nerve. The graphs show voltage-clamp recordings of this current made with intracellular electrodes. The channels open during the brief exposure to acetylcholine, but again the current is zero when the membrane potential is clamped at the reversal potential. Because the open channels are permeable to both Na^+ and K^+ and the electrochemical driving forces for these ions are different, zero net current corresponds to balanced nonzero currents of Na^+ and K^+ in opposite directions. (The channels are also permeable to Ca^{2+}, but the Ca^{2+} current is small because Ca^{2+} concentrations are low.) The value of the reversal potential and its sensitivity to ion concentrations in the external medium give a useful indication of the relative permeability of the channel to the different ions. For example, some other ligand-gated channels are selectively permeable to Cl^- (see p. 1083); they can be recognized because they have a reversal potential of about -60 mV, close to the Cl^- equilibrium potential, and the value of this potential depends on the extracellular concentration of Cl^- but not of Na^+ or K^+. (A, data from B. Sakmann, J. Bormann, and O.P. Hamill, *Cold Spring Harbor Symp. Quant. Biol.* 48:247–257, 1983; B, data based on K.L. Magleby and C.F. Stevens, *J. Physiol.* 223:173–197, 1972.)

the dimensions involved are small. Second, the acetylcholine is hydrolyzed to acetate and choline by *acetylcholinesterase*. This enzyme is secreted by the muscle cell and becomes anchored by a short collagenlike "tail" to the basal lamina that lies between the nerve terminal and the muscle cell membrane. Each acetylcholinesterase molecule can hydrolyze up to 10 molecules of acetylcholine per millisecond, so that all of the transmitter is eliminated from the synaptic cleft within a few hundred microseconds after its release from the nerve terminal. Consequently, acetylcholine is available only for a fleeting moment to bind to its receptors and drive them into the open conformation that produces the conductance change in the postsynaptic membrane (Figure 19–27). The sharply defined timing of presynaptic signals is thus preserved in sharply timed postsynaptic responses.

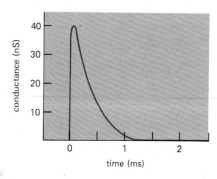

Figure 19–27 The conductance change produced in the postsynaptic membrane by a single quantum (one vesicle) of acetylcholine at the frog neuromuscular junction. About 1600 channels are open at the time of peak conductance, and each channel remains open for an average of 400 microseconds.

Fast Synaptic Transmission Is Mediated by a Small Number of Neurotransmitters[13,21]

Everything about the neuromuscular junction appears to be designed for speed: the large myelinated motor axon, the active zones in the axon terminal, with synaptic vesicles held ready to release their acetylcholine precisely opposite the postsynaptic receptors, the narrow synaptic cleft, the ligand-gated channels in the postsynaptic membrane, ready to open instantly when the transmitter binds, the acetylcholinesterase in the cleft to terminate transmission promptly. The synaptic delay, from the peak of the presynaptic action potential to the peak of the postsynaptic action potential, is on the order of a millisecond or less. Increasing evidence suggests that fast chemical synapses in the central nervous system also employ ligand-gated channels and are constructed on the same principles, with active zones, a narrow cleft, and receptors precisely localized opposite the sites of exocytosis. Moreover, it seems that there are only a handful of neurotransmitters that mediate such rapid signaling. These generalizations must be tentative, however: it is surprisingly difficult to identify conclusively the neurotransmitter acting at a given synapse.

It is probable that the rapid synaptic signaling systems evolved long ago, since the same neurotransmitters are used in the most disparate species of animals, from mollusks to mammals. The rapidly acting neurotransmitters include acetylcholine, γ-aminobutyrate (GABA), glycine, glutamate—and probably aspartate and ATP (Figure 19–28); and in general, a given neuron secretes only one (or occasionally two) of these transmitters—the same at all the synapses it makes. There is direct evidence from patch-clamp studies that the receptors for acetylcholine, GABA, glycine, and glutamate are channel-linked; for the others this is probable but not proved. DNA-sequencing studies indicate that the receptors for acetylcholine, GABA, and glycine are homologous, suggesting that all ligand-gated ion channels share a common evolutionary origin.

Acetylcholine and Glutamate Mediate Fast Excitation; GABA and Glycine Mediate Fast Inhibition[19,22]

Neurotransmitters can be classified according to their actions. We have seen that acetylcholine acting on its receptor in the skeletal muscle cell membrane opens a cation channel and so depolarizes the cell toward the threshold for firing an action potential. This transmitter receptor therefore mediates an *excitatory* effect. **Glutamate** appears to act on a similar type of receptor; it has been shown to be an excitatory transmitter at the neuromuscular junction of an insect and is thought to be the major excitatory transmitter in the central nervous system of vertebrates—a counterpart to acetylcholine, which is the main excitatory transmitter in the vertebrate peripheral nervous system (and also has important central actions). *Aspartate* may act on the same receptors as glutamate, with similar consequences. There is some evidence that *ATP* serves as a fast excitatory transmitter at synapses on certain types of *smooth* muscle.

Counterbalancing these excitatory effects, **GABA** and **glycine** mediate fast *inhibition*. The receptors to which they bind are linked to channels that, when open, admit small negative ions—chiefly Cl⁻—but are impermeable to positive ions. The

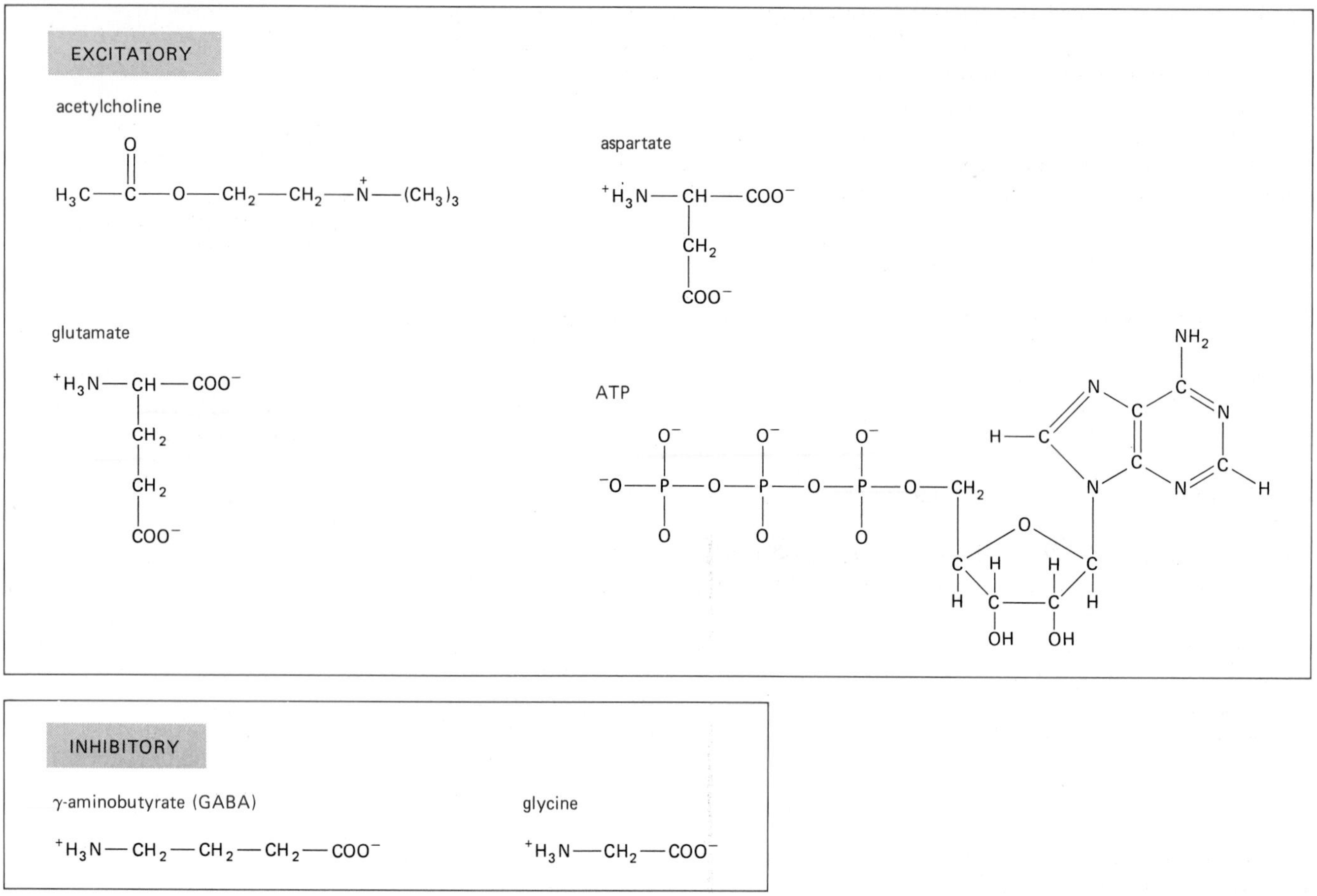

EXCITATORY

acetylcholine

$$H_3C-\overset{\overset{O}{\|}}{C}-O-CH_2-CH_2-\overset{+}{N}-(CH_3)_3$$

aspartate

$$^+H_3N-\underset{\underset{COO^-}{|}}{\underset{CH_2}{|}}\overset{|}{CH}-COO^-$$

glutamate

$$^+H_3N-\underset{\underset{\underset{COO^-}{|}}{\underset{CH_2}{|}}}{\underset{CH_2}{|}}\overset{|}{CH}-COO^-$$

ATP

$$^-O-\underset{O}{\overset{O^-}{P}}-O-\underset{O}{\overset{O^-}{P}}-O-\underset{O}{\overset{O^-}{P}}-O-CH_2$$

INHIBITORY

γ-aminobutyrate (GABA)

$$^+H_3N-CH_2-CH_2-CH_2-COO^-$$

glycine

$$^+H_3N-CH_2-COO^-$$

Figure 19–28 The chemical structures of the major neurotransmitters believed to act on channel-linked receptors so as to mediate fast synaptic transmission.

concentration of Cl^- is much higher outside the cell than inside, and the equilibrium potential for Cl^- is close to the normal resting potential or even more negative. The opening of these Cl^- channels, therefore, tends to hold the membrane potential at its resting value or even at a hyperpolarized value, making it more difficult to depolarize the membrane and hence to excite the cell (Figure 19–29). GABA and glycine are thought to be the major transmitters that mediate fast inhibition in the vertebrate central nervous system, and GABA is known also to perform the same function at neuromuscular junctions in insects and crustaceans. The importance of the inhibitory transmitters is demonstrated by the effects of toxins that block their action: strychnine for example, by binding to glycine receptors and blocking the action of glycine, causes muscle spasms, convulsions, and death.

Several Types of Receptors Often Exist for a Single Neurotransmitter[23]

The action of a neurotransmitter is defined not by its own chemistry but by the receptor to which it binds. In fact there are often several types of receptors for the same neurotransmitter. Acetylcholine in vertebrates, for example, acts in opposite ways on skeletal muscle cells and on heart muscle cells, exciting the former and inhibiting the latter. The acetylcholine receptors are different in the two cases; a non-channel-linked receptor is thought to mediate the inhibitory effect, which is much slower than the excitatory effect on skeletal muscle.

The channel-linked receptors that mediate rapid excitatory actions of acetylcholine are called *nicotinic* because they can be activated by nicotine; the non-channel-linked receptors that mediate the slow actions of acetylcholine, which can be either inhibitory or excitatory, are called *muscarinic* because they can be activated by muscarine (a toxin from a fungus). In addition to such receptor-specific activators (so-called *agonists*), there are also potent receptor-specific blockers (so-called *antagonists*) that distinguish the two types of acetylcholine receptors. For example, curare and α-bungarotoxin bind specifically to nicotinic acetylcholine receptors, blocking their activity, whereas atropine acts in a similar way on muscarinic receptors. Other agonists and antagonists distinguish among receptors for other neurotransmitters, and it is common to identify, localize, and assay the different receptors according to the agonists and antagonists that bind to them.

Synapses Are Major Targets for Drug Action[23,24]

The receptors for neurotransmitters are important targets for toxins and drugs. A snake paralyzes its prey by injecting α-bungarotoxin to block the nicotinic acetylcholine receptor. A surgeon can make muscles relax for the duration of an operation by blocking the same receptors with curare. The heart, meanwhile, continues to beat normally because the curare does not bind to muscarinic acetylcholine receptors. Thus the distinct ligand-binding properties of the two acetylcholine receptors allow drug action to be precisely targeted.

Most of the psychoactive drugs exert their effects at synapses, and a large proportion of them act by binding to specific receptors. GABA receptors provide an example. The best-studied type, known as $GABA_A$ receptors, are ligand-gated Cl^- channels, mediating rapid inhibition as described above. They are acted upon both by the benzodiazepine "tranquilizers," such as Valium and Librium, and by the barbiturate drugs used in the treatment of insomnia, anxiety, and epilepsy. GABA, benzodiazepines, and barbiturates bind cooperatively to three different sites on the same receptor protein: the drugs apparently alter behavior by allowing lower concentrations of GABA to open the Cl^- channel, thus *potentiating* the inhibitory action of GABA.

Synaptic transmission can also be disturbed in many other ways—for example, by interfering with the degradation or removal of the transmitter from the synaptic cleft. There are drugs that inhibit acetylcholinesterase activity at neuromuscular junctions so that acetylcholine lingers on the muscle cell for a longer time. This helps to relieve the weakness of patients suffering from myasthenia gravis, who have a shortage of functional acetylcholine receptors (see p. 1011). Other neurotransmitters, such as GABA, are not degraded enzymatically in the synaptic cleft but instead are retrieved by the presynaptic terminals that secreted them or by neighboring glial cells. Typically the nerve terminals and glial cells have specific transport proteins in their plasma membranes for active uptake of the neurotransmitter. Some psychoactive drugs block or potentiate the retrieval mechanism at specific classes of synapse, causing clinically useful effects.

Summary

Neural signals pass from cell to cell at synapses, which can be either electrical (gap junctions) or chemical. At a chemical synapse the depolarization of the presynaptic membrane by an action potential opens voltage-gated Ca^{2+} channels, allowing an influx of Ca^{2+} to trigger exocytic release of neurotransmitter from synaptic vesicles. The neurotransmitter diffuses across the synaptic cleft and binds to receptor proteins in the membrane of the postsynaptic cell; it is rapidly eliminated from the cleft by diffusion, by enzymatic degradation, or by reuptake into nerve terminals or glial cells. The receptors for neurotransmitters can be classified as either channel-linked or non-channel-linked. Channel-linked receptors, also known as ligand-gated ion channels, mediate rapid postsynaptic effects, occurring within a few milliseconds. Only a handful of neurotransmitters are known to act on such receptors. In particular, acetylcholine and glutamate (and probably aspartate and ATP) open ligand-gated channels that are permeable only to cations and thereby produce rapid excitatory postsynaptic potentials, whereas GABA and glycine open homologous channels that are permeable chiefly to Cl^- and thereby produce rapid inhibitory postsynaptic potentials. All these neurotransmitters, as well as many others, may also act on non-channel-linked receptors, with slower and more complex consequences.

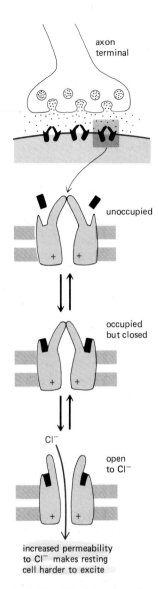

Figure 19–29 The behavior of channel-linked receptors for GABA. In response to binding of GABA, these form an open channel that is selectively permeable to Cl^-. In this way they mediate an inhibitory effect: the open Cl^- channels tend to hold the membrane close to the Cl^- equilibrium potential, which is itself close to the resting potential.

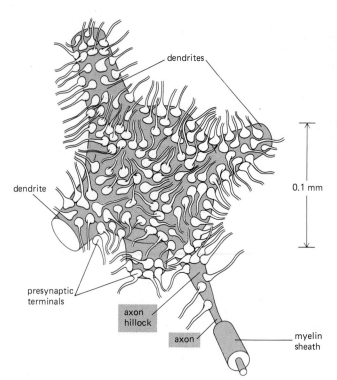

dendrites

dendrite

0.1 mm

presynaptic
terminals

axon
hillock

axon

myelin
sheath

Figure 19–30 A motor neuron cell body in the spinal cord, showing some of the many thousands of nerve terminals that synapse on the cell and deliver signals from other parts of the organism to control its firing. The regions of the motor-neuron plasma membrane that are not covered with synaptic endings are covered by glial cells (not shown).

The Role of Ion Channels in Neuronal Computation[25]

In the central nervous system, neurons typically receive inputs from many pre-synaptic cells—the number may be anything from one to many thousands. For example, several thousand nerve terminals, from hundreds or perhaps thousands of neurons, make synapses on a typical motor neuron in the spinal cord; its cell body and dendrites are almost completely covered with them (Figure 19–30). Some of these synapses transmit signals from the brain, others bring sensory information from muscles or from the skin, and still others supply the results of computations made by interneurons in the spinal cord. The motor neuron must combine the information received from these many sources and react either by firing signals along its axon or by remaining quiet.

The motor neuron provides a typical example of the way in which neurons play their individual parts in the fundamental task of computing an output from a complex set of inputs. Of the many synapses on the motor neuron, some will tend to excite it, others to inhibit it. Although the motor neuron secretes the same neurotransmitter at all its axon terminals, it makes many different types of receptor proteins, concentrating them at different postsynaptic sites on its surface. At each such site, firing of the presynaptic cell causes a specific set of channels to open or close, leading to a characteristic voltage change or **postsynaptic potential (PSP)** in the motor neuron. An *excitatory PSP* (produced, for example, by the opening of channels permeable to Na^+) is generally a depolarization; an *inhibitory PSP* (produced, for example, by the opening of Cl^- channels) is usually a hyperpolarization. The PSPs generated at the different synapses on a single neuron are highly variable in size and duration. At one synapse on the motor neuron, an incoming nerve impulse might produce a depolarization of less than 0.1 mV, whereas at another there might be a depolarization of 5 mV. However, as we shall now discuss, the properties of the system are such that even small PSPs can combine to produce a large effect.

The Shift of Membrane Potential in the Body of the Postsynaptic Cell Represents a Spatial and Temporal Summation of Many Postsynaptic Potentials[25,26]

The membrane of the dendrites and cell body of most neurons, although rich in receptor proteins, contains few voltage-gated Na^+ channels and so is relatively inexcitable. An individual PSP generally does not trigger the postsynaptic membrane to fire an action potential. Instead, each incoming signal is faithfully reflected in a PSP of graded magnitude, which falls off with distance from the site of the synapse. If signals arrive simultaneously at several synapses in the same region of the dendritic tree, the total PSP in that neighborhood will be roughly the sum of the individual PSPs, with inhibitory PSPs making a negative contribution to the total. At the same time, the net electrical disturbance produced in one postsynaptic region will spread to other regions through the passive cable properties of the dendritic membrane.

The cell body, where the effects of the PSPs converge, is relatively small (generally smaller than 100 μm in diameter) compared with the dendritic tree (whose branches may extend for millimeters). The membrane potential in the cell body and its immediate neighborhood will therefore be roughly uniform and will be a composite of the effects of all the signals impinging on the cell, weighted according to the distances of the synapses from the cell body. The **grand postsynaptic potential** of the cell body is thus said to represent a **spatial summation** of all the stimuli received. If excitatory inputs predominate, it will be a depolarization; if inhibitory inputs predominate, it will usually be a hyperpolarization.

While spatial summation combines the effects of signals received at different sites on the membrane, **temporal summation** combines the effects of signals received at different times. The neurotransmitter released when an action potential arrives at a synapse evokes a PSP in the postsynaptic membrane that rises rapidly to a peak (through the transient opening of ligand-gated ion channels) and then declines to baseline with a roughly exponential time course (which depends on the membrane capacitance). If a second action potential arrives before the first PSP has decayed completely, the second PSP adds to the remaining tail of the first. If, after a period of inactivity, a long train of action potentials is delivered in quick succession, each PSP adds to the tail of the preceding PSP, building up to a large sustained average PSP whose magnitude reflects the rate of firing of the presynaptic neuron (Figure 19–31). This is the essence of temporal summation: it translates the *frequency* of incoming signals into the *magnitude* of a net PSP.

The Grand PSP Is Translated into a Nerve Impulse Frequency for Long-distance Transmission[27]

Temporal and spatial summation together provide the means by which the rates of firing of many presynaptic neurons jointly control the membrane potential in the body of a single postsynaptic cell. The final step in the neuronal computation made by the postsynaptic cell is the generation of an output, usually in the form of action potentials, to relay a signal to other cells that are often far away. The output signal reflects the magnitude of the grand PSP in the cell body. However, while the grand PSP is a continuously graded variable, action potentials are all-or-none and uniform in size. The only variable in signaling by action potentials is the time interval between one action potential and the next. For long-distance transmission, the magnitude of the grand PSP is therefore translated, or *encoded*, into the *frequency* of firing of action potentials (Figure 19–32). This encoding is achieved by a special set of voltage-gated ion channels present at high density at the base of the axon, adjacent to the cell body, in a region known as the **axon hillock** (see Figure 19–30).

Before we discuss how these channels operate, a word of qualification is necessary. The firing of an action potential itself causes drastic changes of the membrane potential of the cell body, which therefore no longer directly reflects the net synaptic stimulation that the cell is receiving. It is therefore a complex

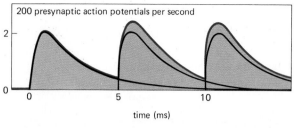

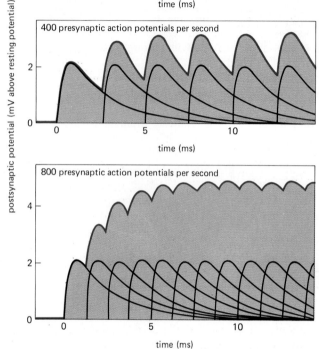

Figure 19–31 Temporal summation. The overlapping curves within the shaded region of each graph represent the individual contributions to the total postsynaptic potential evoked by the arrival of the successive presynaptic action potentials.

problem to give a rigorous analysis of the encoding mechanism. In the nonrigorous, qualitative account that follows, we shall loosely refer to "the strength of synaptic stimulation" or to "the grand PSP," meaning the grand PSP that would be observed if action potentials were somehow prevented from firing; and we shall suppose that this underlying grand PSP is the cause of the firing of action potentials.

Encoding Requires a Combination of Different Ion Channels[28]

The propagation of action potentials depends chiefly, and in many vertebrate axons almost entirely, on voltage-gated Na^+ channels. The membrane of the axon hillock is where action potentials are initiated, and Na^+ channels are plentiful there. But to perform its special function of encoding, the membrane in that neighborhood typically contains in addition at least four other classes of ion channels—three selective for K^+ and one selective for Ca^{2+}. The three varieties of K^+ channels have different properties; we shall refer to them as the *delayed*, the *early*, and the Ca^{2+}-*activated K^+ channels*. The functions of these channels in encoding have been studied most thoroughly in giant neurons of mollusks, but the principles appear to be similar for most other neurons.

To understand the necessity for multiple types of channels, consider first the behavior that would be observed if the only voltage-gated ion channels present in the nerve cell were the Na^+ channels. Below a certain threshold level of synaptic stimulation, the depolarization of the axon hillock membrane would be insufficient to trigger an action potential. With gradually increasing stimulation, the threshold would be crossed: the Na^+ channels would open, and an action potential would fire. The action potential would be terminated in the usual way by inactivation of the Na^+ channels. Before another action potential could fire, these channels would have to recover from their inactivation. But that would require a return of the membrane voltage to a very negative value, which would not occur as long as the

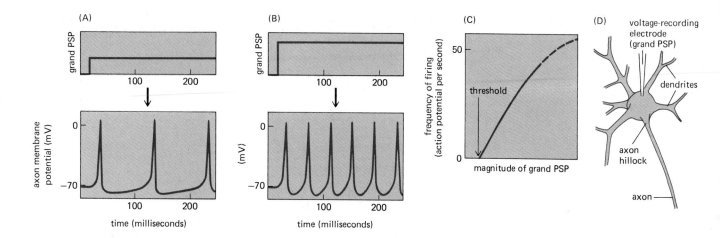

Figure 19–32 The encoding of the grand PSP in the form of the frequency of firing of action potentials by an axon. A comparison of (A) and (B) shows how the firing frequency of an axon increases with an increase in the grand PSP, while (C) summarizes the general relationship. In (D) the experimental setup for measuring the grand PSP is shown. In (A) and (B) the upper graphs (marked "grand PSP") show the net intensity of synaptic stimulation as received by the cell body, while the lower graphs show the resulting trains of action potentials that are transmitted along the axon. The upper graphs can be thought of as representations of the grand PSP that would be observed if the firing of action potentials were somehow blocked.

strong depolarizing stimulus (from PSPs) was maintained. An additional channel type is needed, therefore, to repolarize the membrane after each action potential to prepare the cell to fire another. This task is performed by the **delayed K$^+$ channels,** which we discussed previously in relation to the propagation of the action potential (see p. 1070). They are voltage-gated and respond to membrane depolarization in much the same way as the Na$^+$ channels, but with a longer time delay. By opening during the falling phase of the action potential, they permit an efflux of K$^+$, which short-circuits the effect of even a sustained depolarizing stimulus and drives the membrane back toward the K$^+$ equilibrium potential. This potential is so far negative that the Na$^+$ channels recover from their inactivated state. In addition, the K$^+$ conductance turns itself off: repolarization of the membrane causes the delayed K$^+$ channels themselves to close again (without ever entering an inactivated state). Once repolarization has occurred, the depolarizing stimulus from synaptic inputs becomes capable of raising the membrane voltage to threshold again so as to cause another action potential to fire. In this way, sustained stimulation of the dendrites and cell body leads to repetitive firing of the axon.

However, repetitive firing in itself is not enough: the frequency of the firing has to reflect the intensity of the stimulation. Detailed calculations show that a simple system of Na$^+$ channels and delayed K$^+$ channels is inadequate for this purpose. Below a certain threshold level of steady stimulation, the cell will not fire at all; above that threshold it will abruptly begin to fire at a relatively rapid rate. The **early K$^+$ channels** (also known as *A channels*) solve the problem. These too are voltage-gated and open when the membrane is depolarized, but their specific voltage sensitivity and kinetics of inactivation are such that they act to reduce the rate of firing at levels of stimulation that are only just above the threshold. Thus they help to remove the discontinuity in the relationship between the firing rate and the intensity of stimulation. The result is a firing rate that is proportional to the strength of the depolarizing stimulus over a very broad range (see Figure 19–32).

Adaptation Lessens the Response to an Unchanging Stimulus[29]

The process of encoding is usually further modulated by the two other types of ion channels in the axon hillock that were mentioned at the outset—*voltage-gated Ca^{2+} channels* and *Ca^{2+}-activated K$^+$ channels*. The former are similar to the Ca^{2+} channels that mediate release of neurotransmitter at axon terminals: those present in the neighborhood of the axon hillock open when an action potential fires, allowing Ca^{2+} into the axon. The **Ca^{2+}-activated K$^+$ channel** is different from any of the channel types described earlier. It opens in response to a raised concentration of Ca^{2+} at the cytoplasmic face of the nerve cell membrane.

Suppose that a strong depolarizing stimulus is applied for a long time, triggering a long train of action potentials. Each action potential permits a brief influx of Ca^{2+} through the voltage-gated Ca^{2+} channels, so that the intracellular Ca^{2+}

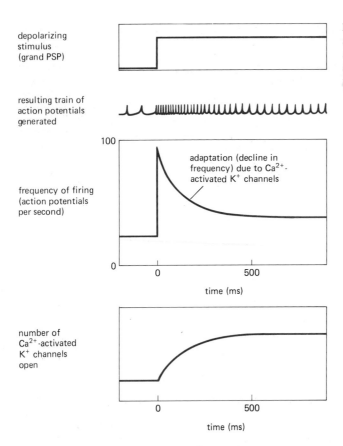

depolarizing
stimulus
(grand PSP)

resulting train of
action potentials
generated

frequency of firing
(action potentials
per second)

100

adaptation (decline in
frequency) due to Ca^{2+}-
activated K$^+$ channels

0

0   500

time (ms)

number of
Ca^{2+}-activated
K$^+$ channels
open

0   500

time (ms)

Figure 19–33 Adaptation. When steady stimulation is prolonged, the stimulated cell gradually reduces the strength of its response, as expressed in the rate of firing of action potentials.

concentration gradually builds up to a high level. This opens the Ca^{2+}-activated K$^+$ channels, and the resulting increased permeability of the membrane to K$^+$ makes the membrane harder to depolarize and increases the delay between one action potential and the next. In this way a neuron that is stimulated continuously for a prolonged period becomes gradually less responsive to the constant stimulus. The phenomenon, which can also occur by other mechanisms, is known as **adaptation** (Figure 19–33). It allows a neuron, and indeed the nervous system generally, to react sensitively to *change*, even against a high background level of steady stimulation (see p. 1107). It is one of the strategies that help us, for example, to feel a touch on the shoulder and yet ignore the constant pressure of our clothing.

Not All Signals Are Delivered via the Axon[30]

In the typical neuron that we have been describing, there is a clear distinction, in both structure and function, between dendrites and axon. Some neurons, however, do not conform to this model, although the molecular principles of their operation are the same. In most invertebrates, for example, the majority of neurons have a *unipolar* organization: the cell body is connected by a single stalk to a branching system of cell processes, among which it is not always easy to see a structural difference between dendrites and axon (Figure 19–34). The functional distinction can also be blurred, in both vertebrates and invertebrates: processes that are classified structurally as dendrites often form presynaptic as well as postsynaptic specializations and deliver signals to other cells as well as receive them. Conversely, synaptic inputs are sometimes received at strategic sites along the axon—for example, close to the axon terminal, where they can inhibit or facilitate the release of neurotransmitter from that particular terminal without affecting transmission at the terminals of other branches of the same axon (Figure 19–35). We shall discuss later (see p. 1097) an example of this important device of *presynaptic inhibition* or *presynaptic facilitation*.

Synapses at which a dendrite delivers a stimulus to another cell play a large part in communication between neurons that lie close together, within a few millimeters or less. Over such distances, electrical signals can be propagated pas-

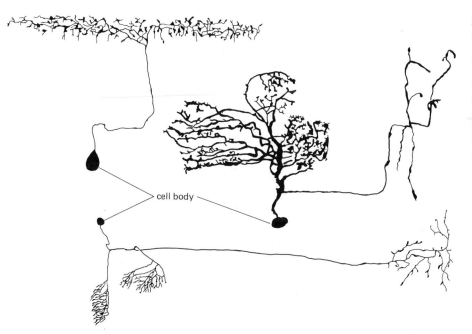

cell body

sively, spreading from postsynaptic sites on the dendritic membrane, where they are received, to presynaptic sites on the same dendritic membrane, where they then control transmitter release. Indeed, there are neurons that possess no axon, do not conduct action potentials, and perform all of their signaling via processes that are conventionally referred to as dendrites. Moreover, if the dendritic tree is large, separate parts of it can behave as more or less independent pathways for communication and for information processing. In some neurons the range of possibilities is still further complicated by the presence of voltage-gated ion channels in the dendritic membrane, which enable the dendrites to conduct action potentials. Thus even a single neuron can behave as a highly complex computational device.

Summary

A typical neuron receives on its dendrites and cell body many different excitatory and inhibitory synaptic inputs, which combine, by spatial and temporal summation, to produce a grand postsynaptic potential in the cell body. The magnitude of the grand postsynaptic potential is translated (encoded) for long-distance transmission into the rate of firing of action potentials, by a system of ion channels in the membrane of the axon hillock. The encoding mechanism often shows adaptation, so that the cell responds weakly to a constant stimulus but strongly to a change of stimulus. There are many variants of this basic scheme: for example, not all neurons produce an output in the form of action potentials, dendrites can be presynaptic as well as postsynaptic, and axons can be postsynaptic as well as presynaptic.

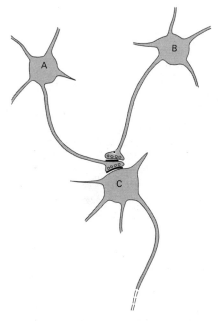

Figure 19–35 An axo-axonic synapse. The neurotransmitter released from the axon terminal of cell B acts on channels in the axon terminal of cell A, thereby altering the number of quanta of neurotransmitter released onto C when A fires. If firing of B causes a reduction in the stimulus delivered by A to C, then B is said to exert a *presynaptic inhibition;* the contrary effect is called *presynaptic facilitation.*

Non-Channel-linked Receptors and Synaptic Modulation[13,31]

At synapses that use channel-linked receptors, the effect of the neurotransmitter is immediate, simple, and brief, and the site of reception of the message is defined with pinpoint accuracy; the transmitter released from one axon terminal acts only on a single postsynaptic cell. By contrast, non-channel-linked receptors allow for effects that are slow, complex, long-lasting, and often spatially diffuse; the transmitter released from one terminal may act on many cells in the neighborhood of that terminal. These slow effects are often described as examples of *neuromodulation* because they modulate the rapid responses mediated by channel-linked receptors on the same cell. The non-channel-linked receptors act by the same molecular mechanisms as the receptors for hormones and local chemical mediators outside the nervous system—indeed, many of them probably are identical.

As discussed in Chapter 12 (see p. 695), non-channel-linked cell-surface receptors for signaling molecules fall into two large families: (1) **catalytic receptors,** most of which are tyrosine-specific protein kinases, which, when activated by ligand binding, directly phosphorylate tyrosine residues on proteins inside the cell; and (2) **G-protein-linked receptors,** which transmit signals into the cell interior by activating a GTP-binding regulatory protein, or *G protein,* which in turn activates or inactivates a membrane-bound enzyme or ion channel. Most of the non-channel-linked receptors for neurotransmitters studied so far seem to be G-protein-linked, employing their G protein in one of at least three ways:

1. The G proteins may activate or inactivate adenylate cyclase, thereby controlling the cyclic AMP level in the postsynaptic cell. The cyclic AMP then regulates the activity of the cyclic-AMP-dependent protein kinase (*A-kinase*—see p. 709), which, among other target proteins, can phosphorylate ion channels in the plasma membrane, altering their properties. Cyclic AMP may also regulate some ion channels by binding to them directly.
2. The G protein may activate the inositol phospholipid pathway (see p. 702), thereby activating protein kinase C (C-kinase) and releasing Ca^{2+} into the cytosol from a Ca^{2+}-sequestering compartment in the postsynaptic cell. The C-kinase may regulate the behavior of ion channels by phosphorylating them. The Ca^{2+} may alter ion channel behavior directly, or indirectly via a Ca^{2+}-dependent protein kinase that phosphorylates the channel (see p. 711).
3. The G protein may interact directly with ion channels, causing them to open or close.

In each case a set of molecules in the postsynaptic cell act as go-betweens or *intracellular messengers,* diffusing within the cell to relay the signal from the receptor to other cell components. The more steps there are in this cascade of intracellular messengers, the more opportunities there are for amplification and regulation of the signal (see p. 713).

More than 50 neurotransmitters have been identified that act on non-channel-linked receptors to produce these varied and complex effects. Some, such as acetylcholine, also bind to channel-linked receptors, whereas others, such as neuropeptides (see below), apparently do not.

Non-Channel-linked Receptors Mediate Slow and Diffuse Responses[32]

Whereas channel-linked receptors take only a few milliseconds or less to produce electrical changes in the postsynaptic cell, non-channel-linked receptors typically take hundreds of milliseconds or longer. This is to be expected, since a series of enzymatic reactions must intervene between the initial signal and the ultimate response. Moreover, the signal itself is often not only temporally but also spatially diffuse.

A clear example is seen in the innervation of smooth muscle by axons releasing *norepinephrine,* which activates adenylate cyclase via a G-protein-linked receptor. Here, the transmitter is released not from nerve terminals but from swellings or *varicosities* along the length of the axon (Figure 19–36). These varicosities contain synaptic vesicles but no active zones to define the exact sites of release. Moreover, the varicosities are not closely apposed to specialized receptive sites on a postsynaptic cell; instead, the transmitter diffuses widely to act on many smooth muscle cells in the neighborhood, in the manner of a local chemical mediator (see p. 682). It is likely that many of the signaling molecules that operate on catalytic and G-protein-linked receptors in the central nervous system also act in this *paracrine* mode. Indeed, many of these neurotransmitters also serve as hormones or as local chemical mediators outside the nervous system: for example, norepinephrine, together with its close relative *epinephrine,* is also released as a hormone from the adrenal gland.

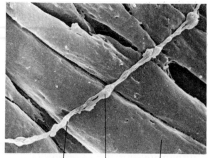

axon bundle   varicosity   smooth muscle cell

|———————————|
5 μm

Figure 19–36   Scanning electron micrograph of a small bundle of autonomic motor axons innervating smooth muscle cells in the wall of the ureter. The varicosities (swellings) contain synaptic vesicles loaded with the neurotransmitter norepinephrine. The synapses here are ill-defined structures, with a gap that may be as large as 0.2 μm between the site of release of the neurotransmitter and the nearest muscle cell membrane on which it must act. (From S. Tachibana, M. Takeuchi, and Y. Uehara, *J. Urol.* 134:582–586, 1985. © by Williams & Wilkins, 1985.)

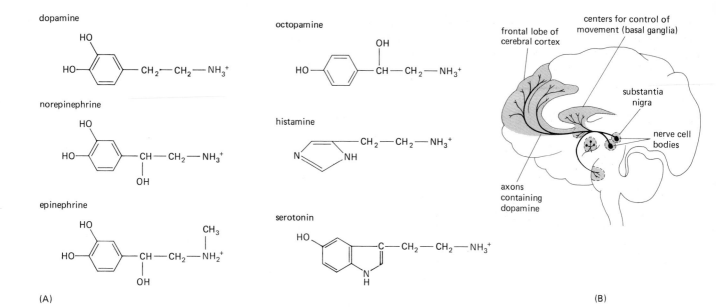

Figure 19–37 (A) The monoamine family of neurotransmitters. (B) Schematized diagram of the distribution of dopamine-containing neurons in the human brain. The movement disorders of Parkinson's disease are due to the death of many of the cells in a particular set of dopamine-containing neurons (those in the substantia nigra); the symptoms can be relieved by drug treatments that boost the synthesis of dopamine and inhibit its breakdown. The distribution of monoamine-containing neurons can be made visible by treating tissue sections with formaldehyde, which reacts with monoamines to give fluorescent products.

Epinephrine and norepinephrine are representatives of the family of **mono-amine** neurotransmitters, which have widespread functions both in vertebrates and in invertebrates and are of great medical importance (Figure 19–37A). It is possible to design drugs that interfere with the synthesis, uptake, or breakdown of particular monoamines, or that interact with particular subclasses of mono-amine receptors; and some of these drugs have proved to be valuable in the treatment of psychiatric and neurological diseases. Schizophrenia, for instance, can often be treated successfully with drugs that block certain classes of *dopamine* receptors, while drugs that increase the concentrations of dopamine in the brain give dramatic relief from the movement disorders of Parkinson's disease (Figure 19–37B). Drugs that raise synaptic concentrations of noradrenaline and/or serotonin are often effective in the treatment of severe depression.

The Neuropeptides Are by Far the Largest Family of Neurotransmitters[32,33]

Most of the signaling molecules used elsewhere in the body are also employed by neurons. This is true in particular of the array of small protein molecules or peptides that serve as hormones and local chemical mediators to control such bodily functions as the maintenance of blood pressure, the secretion of digestive enzymes, and the proliferation of cells.

Rapid advances in this area over the last ten years or so have largely depended on immunocytochemistry. Once a peptide has been identified in one tissue, it is possible to make antibodies against it and to use these to search elsewhere for that peptide and for others that are structurally related. In this way neurons have been found to contain peptides that were not previously suspected to have a neural function, including many newly discovered varieties. The evidence that these **neuropeptides** (Figure 19–38) serve as neurotransmitters is in most cases persuasive but incomplete. For example, an antipeptide antibody might be shown to label certain neurons and their axon terminals, while the peptide itself, when supplied locally, might mimic the effect of activity of these neurons. Most convincingly, the peptide might be shown to be secreted when the neurons are active, and the effects of activity of the neurons might be shown to be blocked by antibodies against the peptide. Neuropeptides appear to be particularly important in regulating feelings and drives, such as pain, pleasure, hunger, thirst, and sex.

The nonpeptide neurotransmitters are synthesized by enzymes that are usually present both in the cell body and in the axon terminals, so that even if the axon is long, the stores of neurotransmitter at the synapse can be rapidly replen-

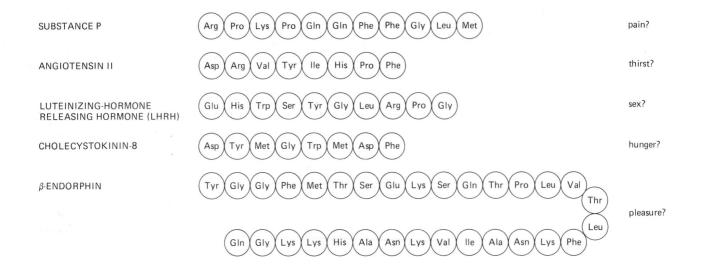

| SUBSTANCE P | Arg Pro Lys Pro Gln Gln Phe Phe Gly Leu Met | pain? |

| ANGIOTENSIN II | Asp Arg Val Tyr Ile His Pro Phe | thirst? |

| LUTEINIZING-HORMONE RELEASING HORMONE (LHRH) | Glu His Trp Ser Tyr Gly Leu Arg Pro Gly | sex? |

| CHOLECYSTOKININ-8 | Asp Tyr Met Gly Trp Met Asp Phe | hunger? |

| β-ENDORPHIN | Tyr Gly Gly Phe Met Thr Ser Glu Lys Ser Gln Thr Pro Leu Val Thr | pleasure? |

Leu
Gln Gly Lys Lys His Ala Asn Lys Val Ile Ala Asn Lys Phe

Figure 19–38 A small selection of neuropeptides, with a tentative indication of some of the sensations and drives in which they are thought to be involved.

ished. The neuropeptides, by contrast, are made on ribosomes on rough endoplasmic reticulum in the cell body and must be exported to the axon terminals by fast axonal transport—a journey that may take a day or more for a long axon. Neuropeptides are derived from larger precursor proteins, from which they are cleaved enzymatically; in many cases more than one functional peptide is cleaved from a single precursor molecule, which for this reason is called a *polyprotein*. Synaptic vesicles loaded with neuropeptides can usually be recognized by their large size compared with vesicles containing acetylcholine, amino acid transmitters, or monoamines.

At many of the synapses where neuropeptides are secreted, a nonpeptide neurotransmitter is also released, and the two transmitters act side by side but in different ways. The presynaptic axon terminals in certain autonomic ganglia of a bullfrog, for example, contain both acetylcholine and a peptide that closely resembles the reproductive hormone LHRH (luteinizing-hormone releasing hormone). The postsynaptic cell membrane contains at least three types of receptors: a nicotinic (channel-linked) acetylcholine receptor that mediates a fast response, a muscarinic (G-protein-linked) acetylcholine receptor that mediates a much slower response, and a receptor (probably G-protein-linked) for the LHRH-like peptide that mediates the slowest response of all (Figure 19–39A). The action of the LHRH-like peptide is not only slower than that of acetylcholine but also more diffuse, so that the peptide molecules released at a synapse on one postsynaptic cell also evoke a postsynaptic potential in other cells in the neighborhood (Figure 19–39B).

If, as seems likely, other neuropeptide transmitters have similar properties, one can see why the number of neuropeptides needs to be large. Since the peptides diffuse widely, their site of release does not define their site of action; consequently, if peptides released from different presynaptic terminals in the same neighborhood are to act on different postsynaptic targets, the peptides and their receptors must be chemically different.

Long-lasting Alterations of Behavior Reflect Changes in Specific Synapses[34]

The responses mediated by non-channel-linked receptors are long-lasting as well as slow in onset. Therein lies much of their special importance for the control of behavior: they bring about a persistent change in the rules that govern the immediate reaction of the nervous system to the inputs it receives and thus appear to be the basis for at least some forms of memory. This is most strikingly illustrated by studies on the sea snail *Aplysia*, a type of mollusk (Figure 19–40). In this animal, changes in behavior with experience can be traced to identified neural circuits and their molecular mechanisms can be deciphered.

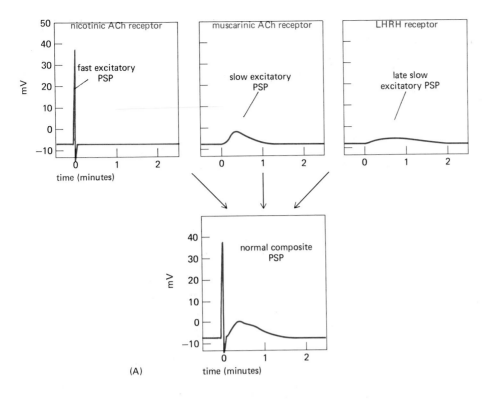

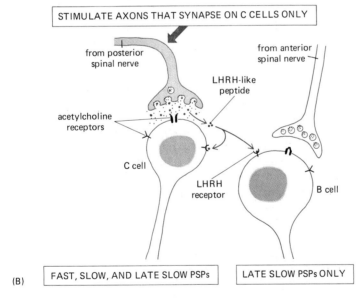

Figure 19–39 The pattern of responses to a peptide neurotransmitter.

(A) The three components of the postsynaptic potential observed in a ganglion cell of a frog following stimulation of the presynaptic nerve. The presynaptic axon terminal releases two neurotransmitters—acetylcholine and a peptide that closely resembles LHRH (luteinizing-hormone releasing hormone). The normal complex PSP is a composite of responses mediated by three kinds of receptors—two for acetylcholine and one for the LHRH-like peptide. Each of the three components can be observed independently by blocking the receptors responsible for the other two components with specific toxins. Only the fast excitatory PSP, mediated by a channel-linked acetylcholine receptor, is large enough to trigger an action potential. The two slow components, mediated presumably by non-channel-linked receptors, serve to modulate the excitability of the cell, making it more responsive to any stimuli that follow soon after the initial stimulus.

(B) Schematic diagram of an experiment on the same ganglion demonstrating the diffuse mode of action of the LHRH-like neuropeptide. This is released together with acetylcholine at synapses made on one population of cells (the C cells), but it diffuses over distances of several tens of micrometers to produce a late slow PSP in other, neighboring cells also (the B cells). (A, after Y.N. Jan *et al.*, *Cold Spring Harbor Symp. Quant. Biol.* 48:363–374, 1983.)

Aplysia withdraws its gill if its siphon is touched (see Figure 19–40). If the siphon is touched repeatedly, the animal becomes **habituated** and ceases to respond. Habituation is similar in function to adaptation, although it operates on a longer time scale and, as we shall see, at a different point in the neural pathway. An unpleasant experience, such as a hard bang or an electrical shock, removes the habituation and leaves the animal **sensitized** so that it responds vigorously again to being touched. The sensitization persists for many minutes or hours, according to the severity of the brief noxious stimulus that caused it, and represents a simple form of *short-term memory*. If the animal is struck or shocked repeatedly on successive days, the sensitization—that is, the memory—becomes

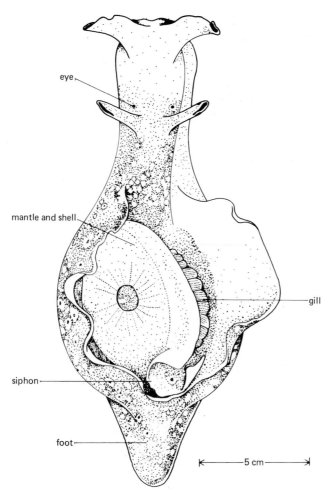

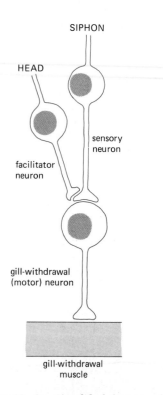

Figure 19–40 The sea snail *Aplysia punctata* viewed from above. An overlying flap of tissue has been drawn aside to reveal the gill under the protective mantle and shell. (After J. Guiart, *Mem. Soc. Zool. France* 14:219, 1901.)

long term, lasting for weeks. These modifications of behavior can be traced to changes occurring in a particular class of synapses in the neural circuit that controls the gill-withdrawal reflex.

The neurons of *Aplysia* are large (~100 μm), relatively few in number (~10^5), and stereotyped in their individual appearance. Touching the siphon stimulates a set of *sensory neurons* to fire. The sensory neurons make excitatory synapses on the *gill-withdrawal motor neurons*, which drive the muscles for gill withdrawal, and changes in these synapses underlie the behavioral phenomena. During habituation, the PSP evoked in the gill-withdrawal neurons is observed to become weaker with repeated firing of the sensory cells. Sensitization has the reverse effect, increasing the PSP. In both cases the changes are due to alterations in the amount of neurotransmitter released from the presynaptic axon terminals of the sensory neurons when they fire. The problem therefore reduces to the question of how transmitter release at these synapses is modulated.

G-Protein-linked Receptors Mediate Sensitization in *Aplysia*[35]

As mentioned on page 1077, the amount of neurotransmitter released at a synapse is controlled by the amount of Ca^{2+} that enters the terminal during the action potential. In habituation, repeated firing of the sensory cells leads to a modification of channel proteins in the terminals such that Ca^{2+} entry is reduced and the amount of neurotransmitter released decreases; in sensitization, by contrast, Ca^{2+} entry is increased, so that more neurotransmitter is released. The detailed molecular changes underlying these simple forms of memory are best understood in the case of sensitization.

In sensitization—provoked, for example, by shocks to the head—the alteration in transmitter release from the sensory neurons is brought about by the firing of

Figure 19–41 Simplified diagram of the neuronal pathways involved in habituation and sensitization of the gill-withdrawal reflex in *Aplysia*. Only one representative neuron of each class is shown.

another set of neurons that are responsive to the noxious stimulus. These *facilitator neurons* synapse on the presynaptic terminals of the sensory neurons (Figure 19–41), where they release serotonin (as well as certain neuropeptides). Their actions can be mimicked by applying serotonin directly to the membrane of the *sensory* neurons, whose presynaptic axon terminals contain serotonin receptors. These receptors operate via a G protein: binding of serotonin activates adenylate cyclase, thereby causing a rise in the intracellular concentration of cyclic AMP, which in turn activates A-kinase (see p. 709). It is this protein kinase that alters the electrical properties of the membrane of the sensory neuron by phosphorylating a special class of K^+ channels (Figure 19–42).

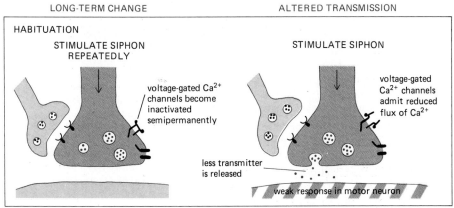

NORMAL TRANSMISSION

STIMULATE SIPHON

facilitator neuron

sensory neuron

gill-withdrawal motor neuron

① action potential in presynaptic terminal

② voltage-gated Ca^{2+} channels open in terminal

③ Ca^{2+}-mediated transmitter release

Ca^{2+}

K^+

④ influx of K^+ through K^+ channels helps end action potential and so terminates Ca^{2+} entry

⑤ transmitter activates motor neuron controlling gill withdrawal

goes out

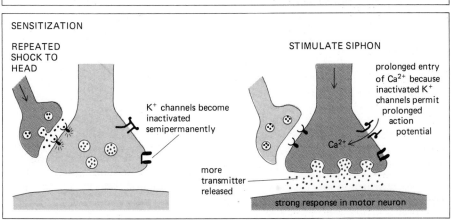

LONG-TERM CHANGE

ALTERED TRANSMISSION

HABITUATION

STIMULATE SIPHON REPEATEDLY

STIMULATE SIPHON

voltage-gated Ca^{2+} channels become inactivated semipermanently

voltage-gated Ca^{2+} channels admit reduced flux of Ca^{2+}

less transmitter is released

weak response in motor neuron

SENSITIZATION

REPEATED SHOCK TO HEAD

STIMULATE SIPHON

K^+ channels become inactivated semipermanently

more transmitter released

prolonged entry of Ca^{2+} because inactivated K^+ channels permit prolonged action potential

Ca^{2+}

strong response in motor neuron

Figure 19–42 The mechanisms that bring about habituation and sensitization of the gill-withdrawal reflex in *Aplysia*. In each diagram the neurons that are electrically active are shown in color. The upper diagram shows the normal mechanism of transmission from the sensory neuron to the gill-withdrawal motor neuron. In each of the lower diagrams, the left-hand drawing shows the nature of the persistent change in the sensory nerve terminal that underlies the memory phenomenon, while the right-hand drawing shows how this change affects synaptic transmission from the sensory neuron to the gill-withdrawal motor neuron. The indicated mechanisms are less certain for habituation than for sensitization.

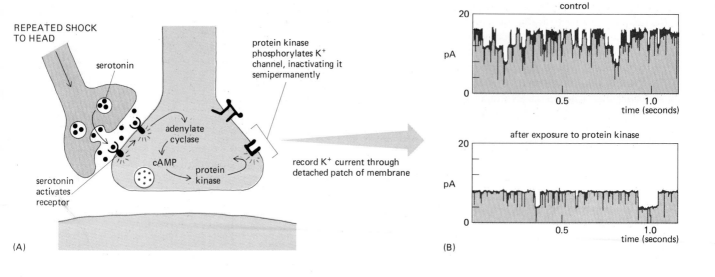

REPEATED SHOCK TO HEAD

serotonin

protein kinase phosphorylates K+ channel, inactivating it semipermanently

adenylate cyclase

cAMP

protein kinase

serotonin activates receptor

record K+ current through detached patch of membrane

(A)

control

20

pA

0

0.5   1.0
time (seconds)

after exposure to protein kinase

20

pA

0

0.5   1.0
time (seconds)

(B)

The behavior of these K+ channels, called *S channels*, can be analyzed by patch-clamp recording (see p. 318). They become locked shut when serotonin is applied to the exterior of the cell (Figure 19–43). Moreover, they close in the same way when the patch of membrane containing them is detached from the cell and transferred to a bath of artificial medium in which the channels are directly subjected to phosphorylation by the catalytic subunit of the A-kinase. This strongly suggests that phosphorylation of the S channels (or of proteins tightly associated with them) is the cause of their long-lasting closure. Because it is the flow of K+ ions that normally helps to restore the resting potential, closing the S channels prolongs action potentials invading the axon terminal. The prolonged action potentials hold the voltage-gated Ca^{2+} channels open for a longer time, permitting a greater influx of Ca^{2+}, which in turn triggers the release of a larger number of synaptic vesicles; and this produces a larger postsynaptic potential, causing a more vigorous withdrawal of the gill.

These experiments demonstrate how a G-protein-linked receptor can enable a transient signal to cause a persistent change in the electrical properties of a synapse, and hence in the behavior of the animal. The phosphorylation of the S channels represents a form of memory, but it is only a short-term memory, easily erased by the action of phosphoprotein phosphatases (which dephosphorylate the S channels) and limited by the finite lifetime of the S-channel proteins. The mechanism of the *long-term memory* that follows repeated noxious stimulation is not known, but unlike the short-term memory, it requires new RNA and protein synthesis and seems to involve changes in the structure as well as the chemistry of the presynaptic terminals (see p. 1132). Cyclic AMP and A-kinase seem to mediate these changes too, presumably by phosphorylating other proteins in the cell and thereby probably altering the pattern of gene expression. The details are not yet understood, but an intermediate step in the creation of the long-term memory trace appears to be a prolonged activation of the A-kinase itself as a result of a reduction in the concentration of the regulatory subunits that inhibit it (see p. 710). These regulatory subunits are thought to be degraded during the period when cyclic AMP levels are high because, on binding cyclic AMP, they dissociate from the catalytic subunits and thereby become exposed to proteolysis.

Ca²⁺ and Cyclic AMP Are Important Intracellular Messengers in Associative Learning in Invertebrates[36]

Habituation and sensitization as presented above are very simple kinds of learning. An essential feature of the more complex types of learning most widely studied by psychologists is that they are *associative*: thus, in Pavlov's famous experiments,

Figure 19–43 (A) The chain of events during sensitization of the gill-withdrawal reflex that leads to inactivation of a special class of K+ channels (so-called S channels) in the sensory nerve terminal (see Figure 19–42). (B) Patch-clamp recordings of the current through these channels as they flicker between open and closed states. The patch, which has been detached from the cell, contains four channels, and in the control condition these are open most of the time. When the catalytic subunit of the cyclic AMP-dependent protein kinase (A-kinase) is added to the medium bathing the cytosolic side of the patch, two of the four channels become phosphorylated and are thereby locked shut, while the other two continue to spend most of their time in the open state; this reduces the average current through the patch to half its control value. (Patch-clamp data reprinted by permission from M.J. Schuster, J.S. Camardo, S.A. Siegelbaum, and E.R. Kandel, *Nature* 313:392–395, 1985. Copyright © 1985 Macmillan Journals Limited.)

the dog learned to associate the sound of a bell with food. *Aplysia* is also capable of associative learning. For example, if a sensitizing stimulus (a severe electrical shock, as before) is repeatedly paired in time with a particular mild stimulus that normally excites only a weak withdrawal reflex, the animal behaves as though it has learned that the specific mild stimulus is associated with the shock and becomes strongly and specifically sensitized to the mild stimulus. The same classes of neurons are thought to be involved as in the simple sensitization described earlier. The paired stimuli to different parts of the body cause sensory neurons and facilitator neurons to fire at the same time. Thus, while an action potential is invading the sensory axon terminals, causing their voltage-gated Ca^{2+} channels to open, serotonin (or a neuropeptide) is being released onto their exterior from the facilitator neurons, causing an increase in the intra-axonal cyclic AMP concentration. The cyclic AMP by itself would cause simple sensitization; the simultaneous influx of Ca^{2+} is thought to intensify this effect, producing a much stronger sensitization than would result from firing of facilitator neurons while the sensory neuron was quiet.

It is not clear how far one can extrapolate from these findings in *Aplysia*. Whether memories in other animals are generally recorded in presynaptic changes or in postsynaptic changes, in synaptic chemistry or in synaptic structure, or indeed in synapses at all, are open questions. Experiments on mutants of the fruit fly *Drosophila*, however, suggest that molecular mechanisms like those described above in *Aplysia* may operate in many other forms of learning. In particular, normal *Drosophila* can be trained to avoid a specific odor if they repeatedly receive an electrical shock in association with the odor. Flies that rapidly forget or fail to learn the association can easily be picked out because they will stray into regions where the smell is strong. In this way it has been possible to isolate dim-witted and forgetful mutants. Two of them, *dunce* (*dnc*) and *rutabaga* (*rut*), are capable of learning but have a drastically reduced memory span—on the order of tens of seconds in the case of *dunce*. In *dunce* the mutation turns out to be in a phosphodiesterase that breaks down cyclic AMP; in *rutabaga* it is in a Ca^{2+}-dependent adenylate cyclase, which makes cyclic AMP. It seems that either too much or too little cyclic AMP can interfere with memory formation. Another mutant, called *Ddc*, seems to be unable to learn in the first place; here the deficiency is in a gene for the enzyme dopa decarboxylase, which catalyzes an essential step in the production of serotonin and dopamine. All these mutants with defects in associative learning also show defects in their susceptibility to sensitization. Evidently the two processes share some common mechanisms, and it seems that these mechanisms, like sensitization in *Aplysia*, involve a monoamine neurotransmitter at an initial step and protein phosphorylations—controlled by cyclic AMP and by Ca^{2+}— for the production of a lasting effect.

Learning in the Mammalian Hippocampus Depends on Ca^{2+} Entry Through a Doubly Gated Channel[37]

Practically all animals can learn, but mammals seem to learn exceptionally well (or so we like to think). This may reflect the operation of some unique molecular mechanisms. In a mammal's brain the *hippocampus*, a part of the cerebral cortex, seems to play a special role in learning: when it is destroyed on both sides of the brain, the ability to form new memories is largely lost, although previous long-established memories remain. Correspondingly, some synapses in the hippocampus show dramatic functional alterations with repeated use. Whereas occasional single action potentials in the presynaptic cells leave no lasting trace, a short burst of repetitive firing causes **long-term potentiation,** such that subsequent single action potentials in the presynaptic cells evoke a greatly enhanced response in the postsynaptic cells. The effect lasts hours, days, or weeks, according to the number and intensity of the bursts of repetitive firing. Only the synapses that were activated show the potentiation; synapses that have remained quiet on the same postsynaptic cell are not affected. But if, while the cell is receiving a burst of repetitive stimulation via one set of synapses, a single action potential is delivered at *another* synapse on its surface, that latter synapse also will undergo long-term potentiation, even though a single action potential delivered there at another time would leave no such lasting trace. Clearly this provides a basis for associative learning.

The underlying rule in the hippocampus seems to be that *long-term potentiation occurs on any occasion where a presynaptic cell fires (once or more) at a time when the postsynaptic membrane is strongly depolarized* (either through recent repetitive firing of the same presynaptic cell or by other means). There is good evidence that this rule reflects the behavior of a particular class of ion channels in the postsynaptic membrane. Most of the depolarizing current responsible for the excitatory PSP is carried in the ordinary way by ligand-gated ion channels that bind glutamate. But the current has in addition a second and more intriguing component, which is mediated by a distinct subclass of channel-linked glutamate receptors, known as **NMDA receptors** because they are selectively activated by the artificial glutamate analog N-methyl-D-aspartate. The NMDA-receptor channels are doubly gated, opening only when two conditions are satisfied simultaneously: the membrane must be strongly depolarized (the channels are subject to a peculiar form of voltage gating that depends on extracellular Mg^{2+} ions), and the neurotransmitter glutamate must be bound to the receptor. The NMDA receptors are critical for long-term potentiation. When they are selectively blocked with a specific inhibitor, long-term potentiation does not occur, even though ordinary synaptic transmission continues. An animal treated with this inhibitor fails in learning tasks of the type thought to depend on the hippocampus but behaves almost normally otherwise.

How do the NMDA receptors mediate such a remarkable effect? The answer seems to be that these channels, when open, are highly permeable to Ca^{2+}, which acts as an intracellular messenger close to its site of entry into the postsynaptic cell, triggering the local changes responsible for long-term potentiation. Long-term potentiation is prevented when Ca^{2+} levels are held artificially low in the postsynaptic cell by injecting the Ca^{2+} chelator EGTA into it and can be induced by transiently raising extracellular Ca^{2+} levels artificially high. The nature of the long-term changes triggered by Ca^{2+} is uncertain, but they are thought to involve structural alterations in the synapse.

Despite the differences between this example of a memory mechanism in mammals and the previous examples in invertebrates, there is a common theme. Neurotransmitters released at synapses, besides relaying transient electrical signals, can also alter concentrations of intracellular messenger molecules, which activate enzymatic cascades that bring about lasting changes in the efficacy of synaptic transmission. But several major mysteries remain, and we do not yet know how these changes endure for weeks, months, or a lifetime in the face of the normal turnover of cell constituents. We shall see later that the development of the nervous system raises some closely related problems.

Summary

Unlike channel-linked receptors, non-channel-linked neurotransmitter receptors respond to their ligand by initiating a cascade of enzymatic reactions in the postsynaptic cell. In most cases studied so far, the first step in this cascade is the activation of a G protein, which may either interact directly with ion channels or control the production of intracellular messengers such as cyclic AMP or Ca^{2+}. These in turn regulate ion channels directly or activate kinases that phosphorylate various proteins, including ion channels. At many synapses both channel-linked and non-channel-linked receptors are present, responding either to the same or to different neurotransmitters. Responses mediated by non-channel-linked receptors have a characteristically slow onset and long duration, and they may modulate the efficacy of subsequent synaptic transmission, thus providing the basis for at least some forms of memory. Channel-linked receptors that allow Ca^{2+} to enter the cell, such as the NMDA receptor, can also mediate long-term memory effects.

Sensory Input[38]

We have seen how nerve cells conduct electrical signals, compute with them, record them, and transmit them to muscles to bring about movement. But how do the signals originate? There are two types of source: spontaneous firing and sensory input. Examples of spontaneously active neurons, such as those in the brain that generate the rhythm of breathing, are common: quite complex patterns of spontaneous firing can be produced in a single cell by appropriate combinations of gated ion channels of the types we have already encountered in discussing neuronal computation. Sensory input likewise involves principles that are already familiar, but they are embodied in cells of the most diverse and remarkable types.

Sense organs have evolved to meet exceptionally stringent engineering specifications: they discriminate precisely between stimuli of different types, operate over phenomenally wide ranges of stimulus intensity, and approach the utmost sensitivity that the laws of physics will allow. An olfactory cell in the male gypsy moth can detect a single molecule of a specific sexual attractant (a so-called *pheromone*) released into the air by a female a mile away. The human eye can see both in bright sunlight and on a starlit night, where the illumination is 10^{12} times fainter; and five photons absorbed in the human retina are perceived as a flash.

We shall concentrate on just two examples where the cellular mechanisms of sensory input are beginning to be understood: the ears and the eyes of vertebrates. At each of these gateways into the nervous system, there stands a highly specialized type of **sensory cell,** very different in the two cases but in both cases remarkable for its selectivity, its operating range, and its sensitivity. Before going into details, however, it will be helpful to discuss some general principles.

Stimulus Magnitude Is Reflected in the Receptor Potential[38,39]

Any signal that is to be fed into the nervous system must first be converted to an electrical form. The conversion of one kind of signal into another is known as transduction, and all sensory cells are therefore **transducers.** Indeed, in a general sense almost every neuron is a transducer, receiving chemical signals at synapses and converting these into electrical signals. Thus, although some sensory cells respond to light, some to temperature, some to a particular chemical, some to a mechanical force or displacement, and so on, transduction in all of them involves many of the same basic principles that were discussed earlier for synaptic activation by neurotransmitters. In some sense organs the transducer is part of a *sensory neuron* that propagates action potentials. In others it is part of a *sensory cell* specialized for transduction but not for long-distance communication; such a cell then passes its signal to an adjacent neuron via a synapse (Figure 19–44).

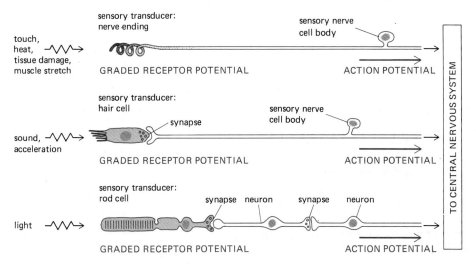

Figure 19–44 Different ways in which sensory stimuli are transmitted to the nervous system. In some cases the sensory transducer is part of a neuron (top drawing); in other cases it is a separate sensory cell (lower two drawings). In all three cases a graded receptor potential (indicated in color) is evoked in the sensory transducer and translated into the frequency of firing of action potentials, which carry the signal rapidly into the CNS.

But in every case the effect of the external stimulus is to cause a voltage change, called the **receptor potential,** in the transducer cell. This is analogous to a post-synaptic potential and likewise serves ultimately to control the release of neuro-transmitter from another part of the cell.

Moreover, just as at a synapse, the external stimulus can exert its electrical effect either directly, by acting on an ion channel, or indirectly, by acting on a receptor that generates an intracellular messenger that affects ion channels. Although there are still some uncertainties, it seems that the sensory cells in the ear use the direct mode, based on channel-linked receptors, whereas those in the eye use the indirect mode, based on G-protein-linked receptors.

Hair Cells in the Ear Respond to Tilting of Their Stereocilia[40]

The ear is not only for hearing. It also provides information on acceleration and on the direction of gravity and so is important for balance and coordination of movements. All these sensory functions of the ear depend on *mechanoreception*—that is, the detection of small movements produced by forces acting in the environment of the ear's sensory cells. The movements are rapid oscillations in the case of sound and slower, more sustained displacements in the case of gravity and acceleration. The cells responsible for the various types of mechanoreception in the ear all have a similar and characteristic form: each of them has a tuft of giant microvilli, confusingly called *stereocilia*, projecting from its upper surface (Figure 19–45 and see p. 675). They are consequently known as **hair cells.**

The hair cells in higher vertebrates all lie in the epithelium of the *membranous labyrinth* of the inner ear, where they are grouped in several separate sensory patches. The hair cells in each group are held in place by a framework of inter-posed *supporting cells*, while above them lies a sheet of gelatinous extracellular matrix attached to the tips of the tufts of stereocilia (Figure 19–46). The movement of this overlying sheet of matrix tilts the stereocilia and produces a mechanical deformation of the hair cells that gives rise to the receptor potential (Figure 19–47). The specific functions of the different groups of hair cells are determined mainly by the nature of the surrounding structures that transmit forces to them. In the case of those hair cells that respond to linear acceleration and to the force of gravity, the overlying matrix is weighted with dense crystals of calcium carbonate: when the head is accelerated or tilted, the weighted matrix shifts relative to the hair cells and the stereocilia are deflected. In contrast, the hair cells that sense rotational acceleration are arranged so that a sideways force is exerted on their overlying matrix by the swirling of fluid in the semicircular canals of the inner ear when the head is turned.

The most elaborate mechanical setting is provided for the hair cells that detect sound in the ears of mammals (see Figure 19–46). These *auditory hair cells* are arrayed on a thin, resilient sheet of tissue—the *basilar membrane*—that forms a long, narrow dividing partition between two fluid-filled spiral channels running in parallel within the portion of the inner ear known as the *cochlea*. Airborne sounds cause vibrations of the eardrum, which are conveyed via the tiny bones in the middle ear to the fluid-filled channels of the inner ear, where they result

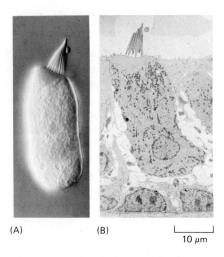

(A) (B) 10 µm

Figure 19–45 (A) Photograph of a sensory hair cell isolated from the inner ear of a bullfrog, showing the bundle of stereocilia on its apical surface. (B) Low-magnification transmission electron micrograph of the hair cell in its normal context, amid supporting cells. (From A.J. Hudspeth, *Science* 230:745–752, 1985. Copyright 1985 by the AAAS.)

Figure 19–46 Diagrammatic cross-section of the auditory apparatus (the organ of Corti) in the inner ear of a mammal, showing the auditory hair cells held in an elaborate structure of supporting cells and overlaid by the tectorial membrane (a mass of extracellular matrix). The inner hair cells are thought to be the receptors that are primarily responsible for hearing, through the transduction mechanism discussed in the text; they synapse with neurons that convey the auditory signals inward from the ear to the brain. The outer hair cells, by contrast, are richly innervated by an additional set of axons that convey signals outward from the brain, and their function is still a puzzle. There is some evidence to suggest that they are capable of acting (by an unknown mechanism) as transducers in a reverse direction—as loudspeakers rather than microphones—and that they serve as part of a feedback system to modulate the mechanical stimulus delivered to the inner hair cells.

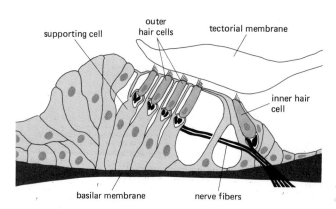

supporting cell

outer hair cells

tectorial membrane

inner hair cell

basilar membrane

nerve fibers

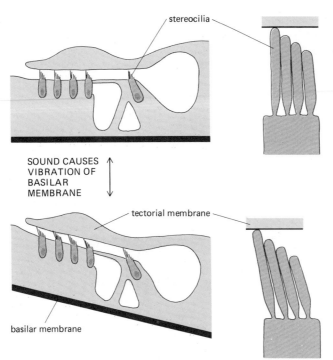

Figure 19–47 How a relative movement of the overlying extracellular matrix (the tectorial membrane) tilts the stereocilia of auditory hair cells in the inner ear of a mammal. The stereocilia behave as rigid rods hinged at the base. The tips of the bundles of stereocilia can be mechanically coupled to the overlying matrix by direct attachment, or indirectly through viscous drag via the intervening fluid.

stereocilia

SOUND CAUSES
VIBRATION OF
BASILAR
MEMBRANE

tectorial membrane

basilar membrane

in vibrations of the basilar membrane and hence of the stereocilia of the auditory hair cells. Hair cells in different positions report on sounds of different pitch because of the mechanics of the cochlea, which resonates most strongly at different positions along its length according to the pitch of the incident sound.

Mechanically Gated Cation Channels at the Tips of Stereocilia Open When the Stereociliary Bundles Tilt[40,41]

When the sheet of matrix overlying a patch of hair cells is abruptly shifted sideways so as to tilt the stereocilia by a few degrees, the hair cells respond by changing their membrane permeability so that a current, called the *receptor current*, flows into them (Figure 19–48). The response reaches a plateau within 100–500 microseconds, which is about the same as the speed of opening of the acetylcholine-activated cation channel at the neuromuscular junction and much faster than the electrical changes produced by any known non-channel-linked receptor. It seems very probable, therefore, that the mechanical stimulus directly opens an ion channel. Studies in which the extracellular ion concentrations are varied have shown that this *mechanically gated ion channel*, like the acetylcholine receptor, is rather

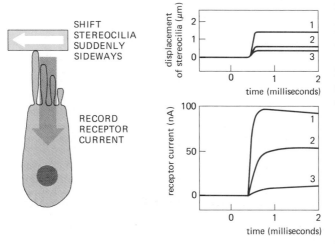

SHIFT
STEREOCILIA
SUDDENLY
SIDEWAYS

RECORD
RECEPTOR
CURRENT

displacement of stereocilia (μm)

time (milliseconds)

receptor current (nA)

time (milliseconds)

Figure 19–48 Recordings of the receptor current that enters hair cells in the inner ear of a bullfrog in response to a sudden deflection of the bundle of stereocilia. The amount of current is larger for larger deflections. (Data from D.P. Corey and A.J. Hudspeth, *J. Neurosci.* 3:962–976, 1983.)

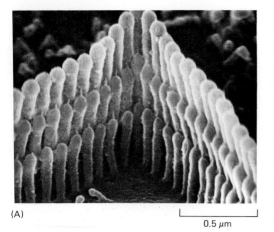

(A)

|——————| 0.5 μm

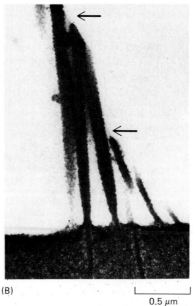

(B)

|——————| 0.5 μm

Figure 19–49 (A) Scanning electron micrograph of an auditory hair cell from a mammal, showing the fine filaments that run from the tips of the shorter stereocilia to attach higher up on the sides of stereocilia in the next, taller, row. (B) A cross-section of the same structures as seen by transmission electron microscopy. Arrows point to the filaments. (B, photograph by M.P. Osborne; A and B, reprinted with permission from J.D. Pickles, *Prog. Neurobiol.* 24:1–42, 1985. Copyright 1985, Pergamon Press plc.)

indiscriminately permeable to small cations and that the current through it is largely carried by K^+. (Ionic conditions in the ear are unusual, and there is a large electrochemical gradient for K^+ across the hair-cell membrane.) But where in the hair cell are such channels located, and how is the elaborate structure of the cell related to the process of transduction?

The stereocilia on each cell are arranged like organ pipes in close-packed ranks of graded height (see Figures 19–45 and 19–49). There is often a single true cilium, or *kinocilium*, behind the middle of the tallest row; this is always present during development, but it plays no part in transduction and is sometimes (as in mammalian auditory hair cells) lost during maturation. When deflected with a microprobe, the stereocilia behave as a coherent bundle of rigid rods, individually pivoted about their points of attachment to the apex of the hair cell and therefore sliding parallel to one another so that their tips undergo a relative displacement. Electron microscopy reveals, in addition to lateral attachments that hold the stereocilia together in a bundle, a fine filament running more or less vertically upward from the tip of each shorter stereocilium to attach at a higher point on its adjacent taller neighbor (Figure 19–49). When a microelectrode is used to record from hair cells, the maximal depolarizing response is seen when the stereocilia are deflected in the direction that would be expected to stretch the fine vertical filaments maximally. Moreover, the transmembrane current induced by the deflection (which gives rise to the receptor potential) appears to enter the hair cell near the tips of the stereocilia. Thus the whole structure seems to be designed so that ion channels near the tips of the stereocilia will be opened by a mechanical tug when the bundle of stereocilia is deflected (Figure 19–50).

The hair-cell mechanism is astonishingly sensitive: the faintest sounds that we can hear have been estimated to stretch the vertical filaments attached to the tops of the stereocilia by an average of about 0.04 nm, which is just under half the diameter of a hydrogen atom. Moreover, analyses of the receptor current in a hair cell indicate that there are probably only between one and five mechanically gated ion channels in each stereocilium; the hair cells responsible for human hearing each have about 100 stereocilia; and there are about 3500 such hair cells in each ear. Apparently, we owe our hearing to fewer than 4 million transducer molecules.

Photoreceptors Are Sensitive and Adaptable but Relatively Slow[42]

The sensitivity of the photoreceptors in the vertebrate eye approaches the ultimate limit set by the quantal nature of light. Moreover, the operating range, from the brightest light tolerable to the dimmest light perceptible, is extraordinarily wide.

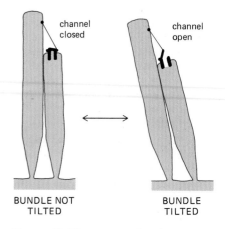

channel closed   channel open

BUNDLE NOT TILTED   BUNDLE TILTED

Figure 19–50 Cartoon drawing suggesting how the tilting of stereocilia on a hair cell may be coupled to the opening of ion channels. By extraordinarily delicate mechanical measurements, correlated with electrical recordings from a single hair cell as the bundle of stereocilia is deflected by pushing with a flexible glass probe, it is in fact possible to detect an extra "give" of the bundle as the mechanically gated channels yield to the applied force and are pulled open, just as the model would predict. In this way it can be shown that the force required to open a single one of the hypothesized channels is about 2×10^{-13} newtons and that its gate swings through a distance of about 4 nm as it opens.

But the speed of response, by comparison with the auditory transducers, is very slow. Under optimal conditions the fastest of the photoreceptors in the human eye take about 25 milliseconds to reach the peak of their electrical response to a flash of light—more than a hundred times longer than the response time of a typical hair cell. The relative slowness of the visual response probably reflects an essential limitation imposed by the nature of the visual transduction mechanism.

The Receptor Potential in a Rod Cell Is Due to Closure of Na$^+$ Channels[43]

As described in Chapter 17 (see p. 956), the photoreceptors in the vertebrate eye are of two classes. **Cone cells** (or **cones**) serve for color vision and perception of fine detail, and they require fairly bright light. **Rod cells** (or **rods**) provide for monochromatic vision in dim light, and they can produce a measurable electrical signal in response to a single photon (Figure 19–51). Rods and cones appear to operate on similar principles, but rods have been studied more intensively.

The rod cell (Figure 19–52) consists of an *outer segment*, containing the photoreceptive apparatus; an *inner segment*, containing many mitochondria; a *nuclear region*; and, at the base, a *synaptic region*, which makes synaptic contact with nerve cells of the retina (see Figure 17–6, p. 957). In the dark, paradoxically, the cell is strongly depolarized; the depolarization holds voltage-gated Ca^{2+} channels open in the synaptic region, and the resulting influx of Ca^{2+} produces a steady release of neurotransmitter. The depolarization is due to open Na$^+$ channels in the plasma membrane of the outer segment. Illumination causes these channels to close, so that the receptor potential takes the form of a *hyperpolarization*, which leads to a decrease of Ca^{2+} influx at the synapse and consequently a decrease in the rate of transmitter release (see Figure 19–52). Because the transmitter acts to inhibit many of the postsynaptic neurons, illumination frees these neurons from inhibition and thus, in effect, excites them. The rate of release of transmitter from the photoreceptors is graded according to the intensity of the light: the brighter the light, the greater the hyperpolarization and the greater the decrease in transmitter release. When background lighting is very dim, so that the cell is in its most

retinal rod cell held in tight-fitting suction pipette

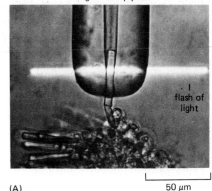

(A)   50 μm

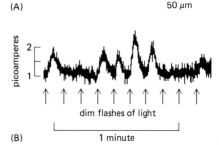

dim flashes of light

(B)   1 minute

Figure 19–51 The electrical responses of a rod cell to single photons. (A) Photomicrograph to show the recording technique. A fragment of the retina of a toad is dissected out, and the outer segment of a single rod cell is sucked into the mouth of a tight-fitting glass pipette, which then serves as an electrode to collect and record the current passing through the rod cell membrane. (B) Recording of the changes in the rod cell current in response to a series of 10 dim flashes; the number of photons absorbed by the rod from each flash is randomly variable but always a whole number; most of the large peaks in the record correspond to absorption of one or two photons, but many of the flashes fail to evoke a response because no photon is absorbed. (From D.A. Baylor, T.D. Lamb, and K.-W. Yau, *J. Physiol.* 288:589–611, 1979.)

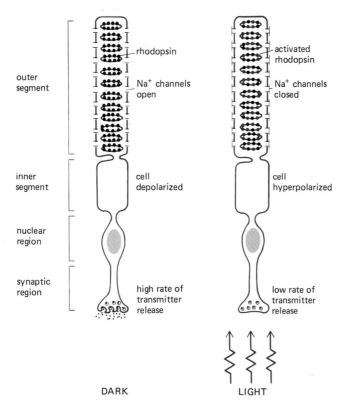

Figure 19–52 The response of a rod photoreceptor cell to illumination. Photons are absorbed, at random, by rhodopsin molecules in the outer segment; this leads to the closure of Na$^+$ channels in the plasma membrane and reduces the rate of transmitter release from the synaptic region.

sensitive, "dark-adapted" state, the absorption of a single photon blocks the influx of a million or more Na^+ ions that would otherwise have entered the cell and thereby generates a hyperpolarization of about 1 mV.

Photons Change the Conformation of Rhodopsin Molecules[43,44]

How is the light initially detected by the cell, and what chain of events then causes the Na^+ channels to close? These crucial steps in the transduction of the light signal occur in the outer rod segment, which is a cylindrical structure containing a stack of about a thousand *discs* (see Figure 17–7). Each disc is formed by a closed membrane in which photosensitive **rhodopsin** molecules are embedded, packed closely together at a density of about $10^5/\mu m^2$. Each rhodopsin molecule consists of a transmembrane glycoprotein called *opsin* (348 amino acid residues long), with a covalently attached prosthetic group, 11-*cis*-retinal, embedded in its interior. The 11-*cis*-retinal is light-sensitive; absorption of a photon causes it to isomerize almost instantaneously to all-*trans*-retinal, changing its own shape and thereby forcing a slower change in the conformation of the opsin protein, which takes about 1 millisecond. After a further delay of about 1 minute, the all-*trans*-retinal dissociates from the opsin by hydrolysis of the bond between them and is released into the cytosol, where it eventually reverts to the 11-*cis* form, which then reassociates with opsin to regenerate a photosensitive rhodopsin molecule. It is the early conformational change in the rhodopsin in response to light that causes Na^+ channels in the plasma membrane to close. But since the rhodopsin is in the discs, some distance from the Na^+ channels, a messenger system is required to couple the two events.

A Light-induced Fall in Cyclic GMP in the Cytosol of the Photoreceptor Cell Closes Na^+ Channels in the Plasma Membrane[43,44,45]

When light falls on a rod cell, changes occur in the intracellular concentrations of both Ca^{2+} and cyclic GMP, so that either of these molecules could, in principle, be the intracellular messenger. The technique of patch recording (see p. 156) has revealed that a fall in the cytosolic concentration of cyclic GMP is the crucial signal. In the key experiment a microelectrode was pressed against the side of the outer segment and then pulled away, taking with it a patch of membrane whose cytosolic side was thus exposed to the external medium (Figure 19–53). Cyclic GMP was then added to the medium and an electrochemical potential difference was applied across the membrane patch; this caused a Na^+ current to flow. When the cyclic GMP was removed, the current stopped, regardless of the concentration of Ca^{2+}. Thus cyclic GMP opens the Na^+ channels, and light, by causing a fall in cyclic GMP concentration, makes the channels close. Whereas cyclic nucleotides usually work by activating a protein kinase to phosphorylate specific proteins (see p. 708), cyclic GMP in rod cells acts directly on the Na^+ channels to keep them open. But how does the light-triggered change in the conformation of rhodopsin decrease the concentration of cyclic GMP in the rod cell cytosol?

The absorption of a single photon of light by a single molecule of rhodopsin leads to the hydrolysis of many molecules of cyclic GMP. This amplification is

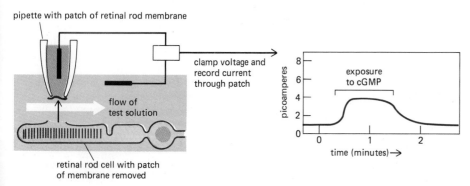

Figure 19–53 An experiment showing that cyclic GMP applied to the cytoplasmic face of the rod cell plasma membrane acts directly to control the opening and closing of ion channels. The relatively slow, smooth rise and fall of the current as cyclic GMP is added and then removed reflect the time taken to change the composition of the test solution; the rate of response of the channels to cyclic GMP is far too rapid to be resolved by this flow technique. (After E.E. Fesenko, S.S. Kolesnikov, and A.L. Lyubarsky, *Nature* 313:310–313, 1985.)

achieved through a catalytic cascade. The single activated rhodopsin molecule catalyzes the activation of a G protein, called *transducin,* at the very rapid rate of about 1000 transducin molecules per second. Transducin is homologous to the G_s protein (see p. 696) that functionally couples various receptors to adenylate cyclase (and rhodopsin itself is homologous to such receptors—see p. 706). Instead of interacting with adenylate cyclase, however, each activated transducin molecule activates a molecule of *cyclic-GMP phosphodiesterase,* which specifically hydrolyzes cyclic GMP at a rate of about 4000 molecules of cyclic GMP per second, causing cyclic GMP levels to fall rapidly. This catalytic cascade lasts for about 1 second and results in the hydrolysis of more than 10^5 cyclic GMP molecules for a single quantum of light absorbed, which transiently closes 250 Na^+ channels in the plasma membrane (Figure 19–54).

The Photoreceptor Adapts to the Brightness of the Light[46]

Each reaction in the light-activated catalytic cascade must be counterbalanced by a corresponding inactivation reaction so as to restore the photoreceptor to its resting state after it has been excited by light. It seems that light speeds up the inactivation reactions as well as the activation reactions, but with a slight delay between the two effects, so that when a light is switched on, there is momentarily a strong net response, which is then soon damped down. Besides helping to ensure a brief response to a brief flash of light, this delayed inactivation enables the photoreceptor to *adapt:* steady light, instead of simply driving the cell into a saturated state with a near-zero concentration of cyclic GMP, exerts two contrary effects that nearly cancel, leaving the cell still able to signal subsequent changes of illumination.

A light-induced fall in Ca^{2+} concentration appears to be crucial both in terminating the response to a flash of light and in mediating adaptation. When a Ca^{2+} buffer is artifically introduced into a photoreceptor so that changes in intracellular Ca^{2+} concentration are abnormally delayed, the electrical responses to flashes of light are greatly prolonged, and the cell becomes excessively slow to adapt to steady illumination; and if a photoreceptor is bathed in a solution that completely blocks movement of Ca^{2+} across the plasma membrane, the same effects are seen to an even greater degree, and adaptation is completely abolished. Normally, the channels through which Na^+ enters the outer segment of the photoreceptor are somewhat permeable to other cations, including Ca^{2+}. Light, by closing the channels, blocks the influx of Ca^{2+}, while Ca^{2+} efflux (mediated by a Ca^{2+}-Na^+ antiporter in the rod plasma membrane) continues. Consequently the intracellular concentration of Ca^{2+} falls. This is thought to accelerate those enzymatic reactions (including, in particular, the synthesis of cyclic GMP by guanylate cyclase) that counteract the light-induced fall in cyclic GMP concentration, thereby helping the cell to adapt.

Neuronal Computations Process the Raw Data from the Sensory Receptor Cells[47]

A flood of sensory information enters the nervous system via the sensory receptor cells. The brain must process this information to extract the significant features: it must pick out words from the hubbub of sound, recognize a face in the pattern of light and dark, and so on. This represents a second, neural stage of sensory processing, which is far more subtle and complex than that which occurs in the receptor cells. It involves computations performed by intricate networks of neurons, each one typically receiving a multitude of convergent inputs, some excitatory, some inhibitory. The individual neuron responds by generating an output that represents the presence or absence of some specific pattern in the data supplied by the receptor cells. Certain classes of cells in the visual centers of the brain, for example, fire action potentials when the eye sees a line with a specific orientation. The output signals from one set of neurons are received by yet other neurons, which take the whole process a step further, and so on to increasingly

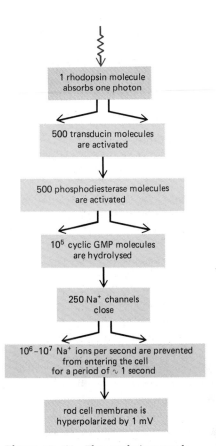

Figure 19–54 The catalytic cascade leading from absorption of a single photon in a dark-adapted rod cell to the production of a receptor potential. Divergent arrows indicate the steps at which amplification occurs.

higher levels of feature detection, culminating in the recognition of such subtle and complex entities as meaningful words and the expressions on faces.

Neuronal computations such as these depend on the bewilderingly complex arrangement of anatomical connections among the nerve cells. The details of the relationship between neuroanatomy and higher neural functions fall outside the scope of a cell biology book. But what are the basic mechanisms by which the complicated yet orderly anatomy develops? This will be the topic of the next section.

Summary

Sensory stimuli are translated into electrical signals by specialized transducers. Sensory hair cells in the vertebrate ear, for example, are mechanoreceptors: each has a bundle of stereocilia (giant microvilli) on its exposed surface, and when this bundle is tilted, the cell responds by opening or closing ion channels so as to change its membrane potential. Photoreceptor cells in the vertebrate eye change their membrane potential when light falls on the rhodopsin molecules that they contain. In both cases the electrical change in the sensory cell—the receptor potential—is signaled to adjacent neurons via chemical synapses. However, the two cases illustrate contrasting strategies for the production of receptor potentials, one based on channel-linked receptors, the other based on non-channel-linked receptors. In the hair cell it appears that physical linkages between the stereocilia bring mechanical forces directly to bear on ion channels in the plasma membrane, which very promptly open or close in response. In the rod cell, light-activated rhodopsin molecules trigger a cascade of enzymatic reactions that lead to the hydrolysis of cyclic GMP in the cytosol, causing Na^+ channels in the plasma membrane to close. The catalytic response mechanism, although intrinsically slow, allows detection of single photons.

Birth, Growth, and Death of Neurons[48]

The nervous system poses a unique developmental problem. How do the axons and dendrites from the billions of neurons find their right partners, so as to create a functional network?

Most of the components of the nervous system—the various classes of neurons, sensory cells, and muscles—originate in widely separate locations in the embryo and are initially unconnected. Thus, in the first phase of neural development, the different parts develop according to their own local programs, following principles of cell diversification common to other tissues of the body, as discussed in Chapter 16. The next phase involves a type of morphogenesis unique to the nervous system, in which a provisional but orderly set of connections is set up between the parts of the nervous system through the outgrowth of axons and dendrites along specific routes. Parts that were originally separate can now begin to interact. In the third and final phase, which continues into adult life, the connections are adjusted and refined through interactions among the far-flung components in a way that depends on the electrical signals that they transmit and receive.

Neurons Are Generated Through Finite Programs of Cell Division[48,49]

With a few exceptions, three principles apply to the production of neurons in almost all species, from nematode worms to vertebrates: (1) mature neurons do not divide; (2) once the adult complement of neurons has been generated, no stem cells persist to generate more; and (3) each small region of the developing nervous system generates neurons through its own program of cell divisions, independently of influences from the distant groups of cells with which neural connections will later be made.

In vertebrates the nervous system develops chiefly from two sets of cells—those of the *neural tube* (see p. 887) and those of the *neural crest* (see p. 944)—both originating from the ectoderm. The neural tube forms the central nervous system (the brain and spinal cord), while the neural crest gives rise to most of the neurons and supporting cells of the peripheral nervous system. In addition, thickenings or *placodes* in the ectoderm on the head give rise to some of the sensory cells and neurons in that region, including those of the ear and the nose (Figure 19–55).

The **neural tube,** with which we shall be mainly concerned here, consists initially of a single-layered epithelium that gives rise to both the neurons and the glial cells of the central nervous system (Figure 19–56). In the process the simple epithelium is transformed into a thicker and more complex structure, with many layers of cells of various types. The pattern of cell proliferation has been analyzed by labeling the cells that are in S phase of the division cycle with ³H-thymidine and fixing the tissue either immediately, to see which cells are currently dividing (see p. 729), or after a long delay, to see which of the mature cells are derived from ancestors that were dividing at the time the ³H-thymidine was given. Such studies on the developing nervous system are especially useful in revealing the "birthdays" of the various neurons: because differentiated neurons do not divide, each neuronal precursor cell must undergo its final division on a particular day before beginning to mature as a neuron. In both vertebrates and invertebrates, the birthdays of the neurons of a given type are generally all found to lie within a strictly limited period of development, after which no further neurons of that type are produced. Each region of the developing neural tube has its own program of cell divisions, and neurons with different birthdays are generally destined for different functions. Since stem cells usually do not persist once the production of nerve cells is complete and neural connections have begun to form, nerve cell numbers thereafter can only be regulated downward, through cell death (see p. 1120, below).

Radial Glial Cells Form a Temporary Scaffold to Guide the Migrations of Newborn Neurons[50]

Before sending out its axon and dendrites, the immature neuron commonly migrates from its birthplace and settles in some other location. It is possible to trace these migrations with the help of ³H-thymidine: precursor cells going through their final division in a certain location become labeled, and the labeled neurons

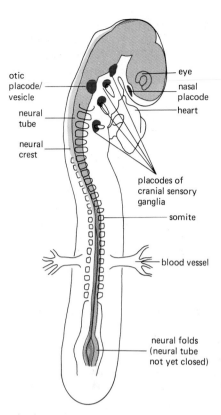

Figure 19–55 Diagram of an early (2½-day) chick embryo, showing the origins of the nervous system. The neural tube (*pale color*) has already closed, except at the tail end, and lies internally, beneath the ectoderm, of which it was originally a part (see Figure 16–13). The neural crest (*gray*) lies dorsally between the roof of the neural tube and the ectoderm. The placodes are thickenings of the surface ectoderm that will give rise to certain sensory cells and neurons. By this stage the otic placode has practically completed its invagination to form the otic vesicle—the rudiment of the inner ear and the source of the neurons in its associated ganglion; the nasal placode will invaginate to form the lining of the nose, including the olfactory neurons responsible for smell; the other cranial placodes will contribute cells to the cranial sensory ganglia, which provide most of the sensory innervation of the head and neck, apart from hearing, smell, and vision. Unlike other sensory cells, those of the eye originate as part of the neural tube.

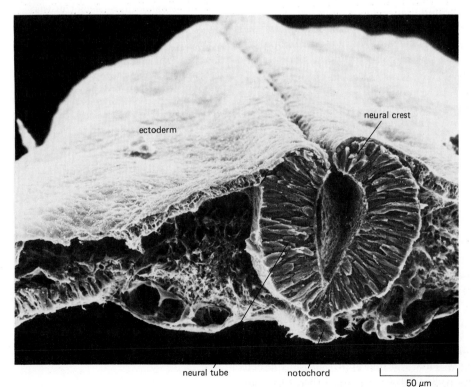

Figure 19–56 Scanning electron micrograph of a cross-section through the trunk of a 2-day-old chick embryo. The neural tube is about to close and pinch off from the ectoderm; at this stage it consists of an epithelium that is only one cell thick. (Courtesy of Jean-Paul Revel.)

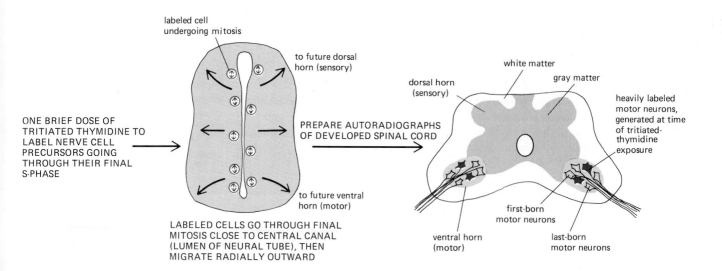

ONE BRIEF DOSE OF TRITIATED THYMIDINE TO LABEL NERVE CELL PRECURSORS GOING THROUGH THEIR FINAL S-PHASE

labeled cell undergoing mitosis

to future dorsal horn (sensory)

PREPARE AUTORADIOGRAPHS OF DEVELOPED SPINAL CORD

to future ventral horn (motor)

LABELED CELLS GO THROUGH FINAL MITOSIS CLOSE TO CENTRAL CANAL (LUMEN OF NEURAL TUBE), THEN MIGRATE RADIALLY OUTWARD

white matter

gray matter

dorsal horn (sensory)

heavily labeled motor neurons, generated at time of tritiated-thymidine exposure

first-born motor neurons

ventral horn (motor)

last-born motor neurons

descended from them can later be seen elsewhere. The motor neurons that will innervate the limbs, for example, undergo their final division close to the lumen of the neural tube and then move outward to settle in the *ventral horn* of the future spinal cord (Figure 19–57).

Nerve cell bodies are guided in their migrations by a specialized class of cells in the neural tube—the *radial glial cells* (Figure 19–58A). These can be considered as persisting cells of the original columnar epithelium of the neural tube that become extraordinarily stretched as the wall of the tube thickens: each cell extends from the inner to the outer surface of the tube, a distance that may be as much as 2 cm in the cerebral cortex of the developing brain of a primate. Three-dimensional reconstructions from serial electron microscope sections reveal that the immature migrating neurons cling closely to the radial glial cells and evidently crawl along them (Figure 19–58B and C).

The radial glial cells remain for many days—in some species for months—as a nondividing population, clearly distinct from the neurons and their precursors. Eventually, toward the end of development, they disappear from most regions of the brain and spinal cord; it has been suggested that many of them transform into astrocytes, but this has yet to be directly demonstrated. Thus the radial glial cells can be viewed as a developmental apparatus, necessary—like scaffolding—for the complex process of construction but not retained in most parts of the completed structure.

The Character and Future Connections of a Neuron Depend on Its Birthday[50,51]

There is a regular relationship between the birthday of a neuron in the vertebrate central nervous system and the site where it comes to rest (an echo, perhaps, of the rigid relationship between cell lineage and cell location that one sees in invertebrates such as nematodes—see p. 902). In the cerebral cortex, for example, the neurons are arranged in layers according to their birthdays through a migration in which the cells that are born later migrate outward past those born earlier. The cells in the successive layers of the cortex, as they mature, will come to differ in their shape, size, and patterns of connections with other cells. Thus small pyramidal cells, born late, lie in an outer layer and send their axons to other regions of the cerebral cortex, whereas large pyramidal and irregularly shaped cells, born earlier, lie in inner layers and send their axons to regions outside the cerebral cortex.

Is it the birthday or the final location that governs these differences? The *reeler* mouse provides an answer. In this mutant, named for its uncoordinated gait, there is a defect in the mechanism of nerve cell migration, so that the cells born late settle in an inner layer and the cells born early settle in an outer one.

Figure 19–57 The origins of motor neurons in the spinal cord, as revealed by autoradiography following a brief dose of tritiated thymidine given at an early stage. The diagrams represent cross-sections of the early neural tube (*on the left*) and of the relatively mature spinal cord that develops from it (*on the right*); radioactively labeled cells are shown in color. Cells that are heavily labeled at the late stage are those that were going through their final round of DNA synthesis in the early embryo when tritiated thymidine was given. For simplicity, only the motor neurons are indicated in the mature spinal cord, whose *gray matter* (*shaded*) also contains many other nerve cell bodies. The *white matter* (*unshaded*) consists chiefly of bundles of axons traveling along the length of the spinal cord and connecting one region of gray matter to another. (These regions appear white in the adult because they contain large amounts of myelin.) For an account of the production of glial cells during development, see Chapter 16, p. 909.

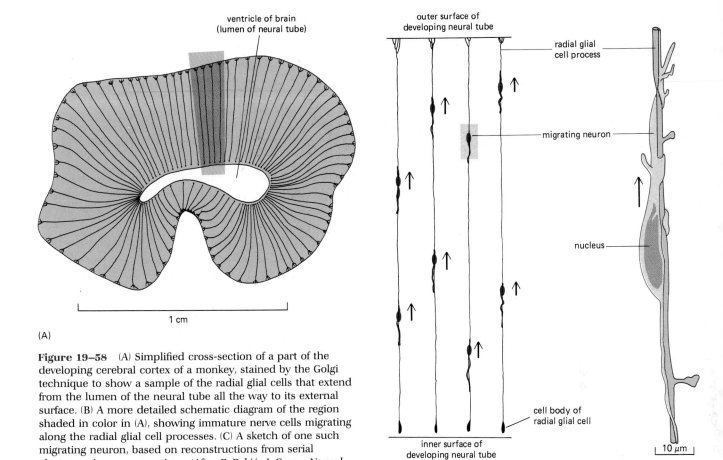

ventricle of brain
(lumen of neural tube)

outer surface of
developing neural tube

radial glial
cell process

migrating neuron

nucleus

1 cm

(A)

cell body of
radial glial cell

inner surface of
developing neural tube

(B)

10 μm

(C)

Figure 19–58 (A) Simplified cross-section of a part of the developing cerebral cortex of a monkey, stained by the Golgi technique to show a sample of the radial glial cells that extend from the lumen of the neural tube all the way to its external surface. (B) A more detailed schematic diagram of the region shaded in color in (A), showing immature nerve cells migrating along the radial glial cell processes. (C) A sketch of one such migrating neuron, based on reconstructions from serial electron microscope sections. (After P. Rakić, *J. Comp. Neurol.* 145:61–84, 1972.)

Despite this inversion of their normal positions, the cortical cells differentiate according to their birthdays: late-born cells become small pyramidal neurons, whereas early-born cells become large pyramidal or irregularly shaped neurons. In this system, therefore, it is the birth date rather than the final location that determines cell character (Figure 19–59). Indeed, it seems that in general the character of a neuron is dictated largely by its ancestry and the place and time of its birth.

The intrinsic character of the cell in turn governs the connections it will form—an important general principle about which more will be said later (see p. 1118). Thus the misplaced neurons in the *reeler* mouse, with relatively few errors, make the connections appropriate to their birthdays rather than to their

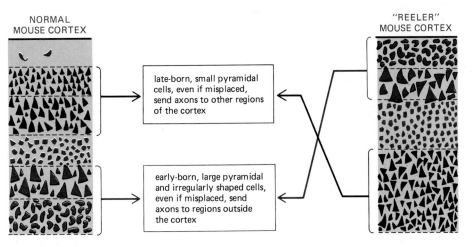

NORMAL
MOUSE CORTEX

"REELER"
MOUSE CORTEX

late-born, small pyramidal cells, even if misplaced, send axons to other regions of the cortex

early-born, large pyramidal and irregularly shaped cells, even if misplaced, send axons to regions outside the cortex

Figure 19–59 Comparison of the layering of neurons in the cortex of normal and *reeler* mice. In the *reeler* mutant an abnormality of cell migration causes an approximate inversion of the normal relationship between neuronal birthday and position. The misplaced neurons nevertheless differentiate according to their birthdays and make the connections appropriate to their birthdays.

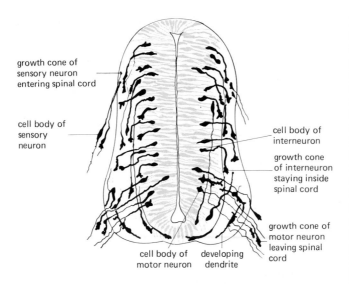

growth cone of
sensory neuron
entering spinal cord

cell body of
sensory
neuron

cell body of
interneuron

growth cone
of interneuron
staying inside
spinal cord

growth cone of
motor neuron
leaving spinal
cord

cell body of
motor neuron

developing
dendrite

Figure 19–60 Growth cones in the developing spinal cord of a 3-day chick embryo as seen in a cross-section stained by the Golgi technique. Most of the neurons, apparently, have as yet only one elongated process—the future axon. The growth cones of the interneurons remain inside the spinal cord, those of the motor neurons emerge from it (to make their way toward muscles), and those of the sensory neurons grow into it from outside (where their cell bodies lie). Many of the cells in the more central regions of the embryonic spinal cord are still proliferating and have not yet begun to differentiate as neurons or glial cells. (From S. Ramón y Cajal, Histologie du Système Nerveux de l'Homme et des Vertébrés. Paris: Maloine, 1909–1911; reprinted, Madrid: C.S.I.C., 1972.)

positions: the small pyramidal cells send axons to other regions of cortex, while the large pyramidal and irregularly shaped cells send their axons to regions outside the cortex. To understand how such selective connections are made, we must first examine the machinery by which axons and dendrites are produced.

Each Axon or Dendrite Extends by Means of a Growth Cone at Its Tip[52]

As a rule the axon and then the dendrites begin to grow out from the nerve cell body soon after it has reached its final location. The sequence of events was originally observed in intact embryonic tissue by the method of Golgi staining (Figure 19–60). This technique and other methods developed subsequently reveal an irregular, spiky enlargement at the tip of each developing nerve cell process. This structure, which is called the **growth cone,** appears to be crawling through the surrounding tissue. It comprises both the engine that produces the movement and the steering apparatus that directs the tip of each process along the proper path.

Most of what we know about the properties of growth cones has come from studies in tissue or cell culture. Embryonic nerve cells in culture send out processes that are often hard to identify as axon or dendrite and are therefore given the noncommittal name of *neurite.* The growth cone at the end of each neurite moves forward at a speed of about 1 mm per day. It consists of a broad, flat expansion of the neurite, like the palm of a hand, with many long *microspikes* or *filopodia* extending from it like fingers (Figure 19–61). These are continually active: some

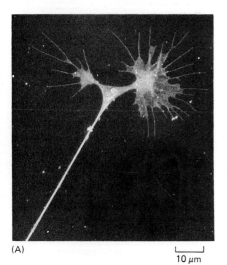

(A)

10 μm

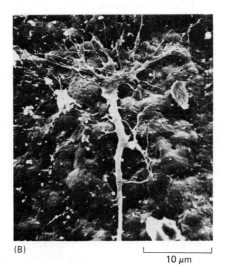

(B)

10 μm

Figure 19–61 (A) Scanning electron micrograph of growth cones at the end of a neurite put out by a chick sympathetic neuron in culture. Here a previously single growth cone has recently divided in two. Note the many filopodia and the taut appearance of the neurite, due to tension generated by the forward movement of the growth cones, which are often the only firm points of attachment to the substratum. (B) Scanning electron micrograph of the growth cone of a sensory neuron *in vivo,* crawling over the inner surface of the epidermis of a *Xenopus* tadpole. (A, from D. Bray, in Cell Behaviour [R. Bellairs, A. Curtis, and G. Dunn, eds.]. Cambridge, U.K.: Cambridge University Press, 1982; B, from A. Roberts, *Brain Res.* 118:526–530, 1976.)

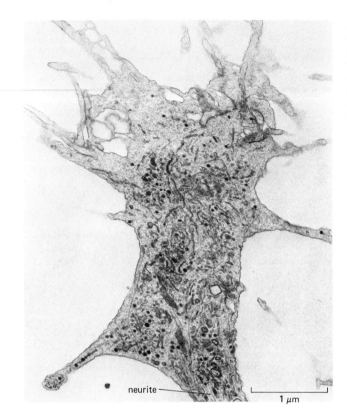

Figure 19–62 Electron micrograph of a section through a growth cone, showing the many membrane-bounded vesicles of varied shapes that it contains. These probably reflect a high rate of exocytosis and endocytosis at the growth cone. (Courtesy of Gerald Shaw.)

neurite

1 μm

are retracting back into the growth cone while others are elongating, waving about, and touching down and adhering to the substratum. The "webs" or "veils" between the filopodia have a ruffling membrane (see p. 638). Electron microscopy shows that the microtubules and neurofilaments present in the neurite terminate in the growth cone and that the broad "palm" is full of flattened membranous sacs and vesicles, together with some mitochondria (Figure 19–62). Immediately beneath the ruffling margins of the growth cone, and filling the filopodia, is a dense meshwork of actin filaments. All these microscopic observations suggest that the growth cone is crawling forward in much the same way as the leading edge of a cell such as a neutrophil or fibroblast (see p. 670).

The Growth Cone Is a Site of Assembly of Materials for Growth[53]

Besides acting as the locomotive for neurite elongation, the growth cone serves as a moving construction site where new components of the growing cell are inserted (Figure 19–63). Because the ribosomes of a developing neuron are largely confined to the cell body, the cell body must be the site of synthesis of the proteins used to extend the neurite. New membrane also is synthesized in the cell body and carried outward in the form of small vesicles by fast axonal transport (see p. 1063) toward the growth cone. Upon arriving there the vesicles are inserted into the plasma membrane by exocytosis. Although some of the membrane is retrieved by endocytosis and recycled, there is a net addition of new membrane at the growth cone. Evidence for this mode of growth has come from watching the movement of small particles of dust clinging to the external surface of the developing neuron: on the growth cone itself these move rapidly, but when bound to the more proximal part of the neurite, they remain stationary relative to the cell body even as the neurite elongates. Microtubules are the "railroad tracks" for fast axonal transport (see p. 660), and membrane vesicles traveling along them are evidently delivered to the sites where the tracks end. Various experiments suggest that microtubules determine where growth cones can occur, and they may do so by controlling the delivery of membrane in this way.

At the same time the microtubules themselves must grow, as must the rest of the cytoskeleton. Tubulin is shipped out from the cell body by slow axonal

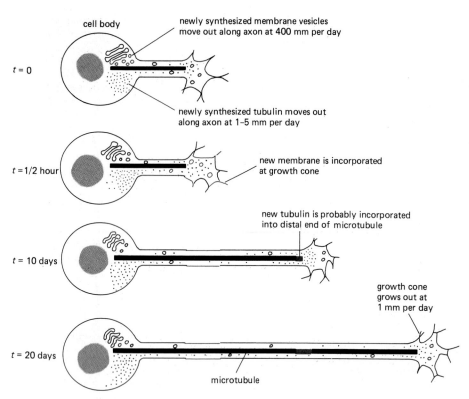

Figure 19–63 How materials for neurite growth are delivered from the nerve cell body to the growth cone and become incorporated there. For simplicity, only one microtubule is shown. Microtubules serve as the tracks for the fast axonal transport of membrane. Tubulin itself is carried outward from the cell body by slow axonal transport; the site of addition of the subunits for microtubule extension is still controversial.

transport (see p. 1063), but it is not certain where the subunits assemble into microtubules. It is known, however, that the microtubules are generally oriented with their "plus" or potentially fast-growing ends (see p. 653) at the growth cone and that the growth cone is uniquely sensitive to local application of drugs that interfere with microtubule assembly. This suggests that microtubules are extended by addition of subunits in the growth cone.

The Growth Cone Can Be Guided *in Vitro* by Selective Adhesion, Chemotaxis, and Electric Fields[52,54]

In the simplified conditions of cell culture, one can explore the mechanisms that might guide the movements of growth cones in the intact animal. Like a neutrophil or a fibroblast, a growth cone confronted with a choice of substrata shows a preference for the surface to which it can adhere most strongly (Figure 19–64). As it moves forward, it continually extends microspikes into the regions that lie ahead and on either side. Some of these projections may contact a less adhesive sub-

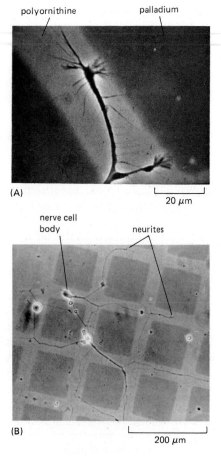

Figure 19–64 Growth cones can be offered a choice of substrata by culturing the cells in a dish whose surface has first been coated with polyornithine and then has had patches of palladium deposited on top of that. Because cell surfaces are largely negatively charged, they adhere strongly to polyornithine, which is positively charged. The growth cones advance along the lanes of polyornithine and stay off the palladium; given a choice between palladium and an even less adhesive substratum, they would travel on the palladium. (A) is a phase-contrast photograph at high magnification showing growth cones at the boundary between the two substrata. (B) is a phase-contrast photograph at lower magnification; the routes taken by the growth cones are recorded in the disposition of the neurites laid out behind them, which have remained adherent to the polyornithine. (From P. Letourneau, *Dev. Biol.* 44:92–101, 1975.)

stratum, in which case they are withdrawn relatively promptly; others contact a more adhesive substratum and persist longer. It appears as though the microspikes serve as feelers to test the surfaces in the neighborhood of the growth cone and to steer it along the most adhesive track.

The stickiness of the substratum is not the only guiding influence to which a growth cone responds, however. The shape of the terrain is also important: growth cones clinging to fibers, for example, will tend to follow their orientation—a phenomenon known as *contact guidance*. Substances dissolved in the extracellular fluid also seem to play a part. When the embryonic ganglion containing the sensory neurons that will innervate the jaw, for example, is placed in culture at a distance of about a millimeter from the embryonic jaw rudiment, neurites grow out predominantly in the direction of the jaw rudiment, suggesting that this target tissue secretes molecules that have a chemotactic effect. Electric fields also have a powerful orienting action, causing growth cones in nerve cell cultures to move toward the negative electrode; a field as small as 7 mV/mm is effective.

Although such culture experiments show what types of factors can guide growth cones, they do not indicate which influences are important in the developing animal. Are growth cones normally constrained to follow specific paths, or do they go exploring at random? Studies of the behavior of growth cones in their natural environments provide some answers.

The Growth Cone Pilots the Developing Neurite Along a Precisely Defined Path *in Vivo*: The Doctrine of Pathway Guidance[55]

In general, growth cones in living animals travel toward their targets along precisely specified routes. The detailed mechanism of such **pathway guidance** is difficult to study in most vertebrates; it is somewhat easier to examine in certain invertebrates, such as the grasshopper, where the innervation of the developing limb has been analyzed in detail (Figure 19–65).

The sensory neurons in a grasshopper limb originate from cells in the embryonic limb epithelium; the nerve cell bodies remain in the periphery and send axons back toward the central nervous system along precisely defined zigzag paths. These paths are initially followed in each limb bud by one or two "pioneer" axons, which can be selectively stained by antibodies or by injecting the fluorescent dye lucifer yellow into their cell bodies. At each turning in the path, the pioneer growth cone is seen to make contact with a specific "guidepost" or "stepping-stone" cell (Figure 19–66), with which it transiently forms gap junctions: if lucifer yellow has been injected into the axon, the stepping-stone cell will also become brightly stained. A pioneer growth cone extends microspikes that are as much as 50 or even 100 μm long—long enough to reach the next stepping-stone cell along the pathway. Microspikes that make contact with this particular cell are stabilized, while others are withdrawn. In this way the growth cone is guided step by step toward the central nervous system. If a specific stepping-stone cell is destroyed by a laser beam before the pioneer growth cone has reached it, the growth cone goes astray at that point in its pathway (see Figure 19–66). In those places where there are no stepping-stone cells along the normal pathway, the pioneer growth cone is guided instead by graded differences in the adhesivity of the basal lamina underlying the limb epithelium (see p. 819). In all its travels, a growth cone depends on specific molecules on its surface that enable it to adhere to the appropriate substratum. Some of these adhesion molecules are beginning to be identified.

Growth Cones Use Specific Adhesion Molecules to Adhere to Cell Surfaces and to Extracellular Matrix[56]

Once the first neurites have pioneered a given route, others follow by contact guidance: growth cones cling to the existing neurites and advance along them. This is a universal tendency, in both vertebrates and invertebrates. Since there is a strong cohesion between neurite and neurite as well as between neurite and growth cone, the consequence is that nerve fibers in a mature animal are usually

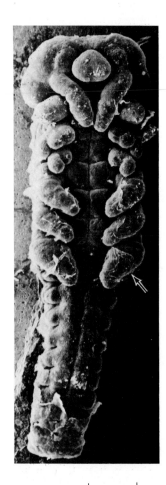

200 μm

Figure 19–65 Scanning electron micrograph of a grasshopper embryo seen from the ventral side, showing the leg buds (*arrow*). (From D. Bentley and H. Keshishian, *Science* 218:1082–1088, 1982. Copyright 1982 by the AAAS.)

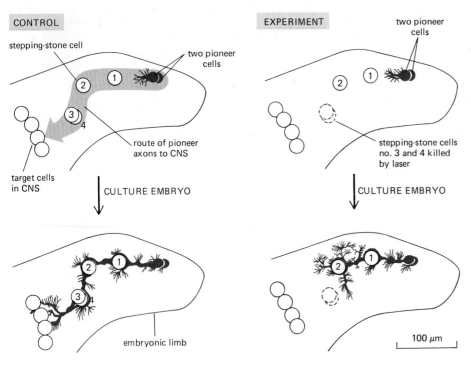

CONTROL

stepping-stone cell

two pioneer cells

2 1

3 4

route of pioneer axons to CNS

target cells in CNS

CULTURE EMBRYO

2 1

3 4

embryonic limb

TWO PIONEER AXONS REACH CNS AND ESTABLISH PATH OF MAJOR ADULT NERVE TRUNK

EXPERIMENT

two pioneer cells

2 1

stepping-stone cells no. 3 and 4 killed by laser

CULTURE EMBRYO

2 1

100 μm

TWO PIONEER AXONS FOLLOW ABERRANT PATHS BEYOND STEPPING-STONE CELL NO. 2 WHEN STEPPING-STONE CELLS NO. 3 AND 4 ARE MISSING

Figure 19–66 The guidance of pioneer axons by specific "stepping-stone" cells in the epithelium of an embryonic grasshopper limb. If stepping-stone cells are destroyed, the pioneer axons fail to advance along their normal path. The stepping-stone cells themselves are destined later to differentiate into neurons whose axons will travel to the CNS by following the routes taken by the pioneers. (After D. Bentley and M. Caudy, *Nature* 304:62–65, 1983.)

found grouped together in tight parallel bundles, or *fascicles*. The large peripheral nerves of a vertebrate, which are visible to the naked eye, originate in this way (although subsequently the axons become individually enveloped and insulated from one another by Schwann cells). Specific integral membrane glycoproteins that mediate this cohesion of neurites have been identified in vertebrates. Two widely studied examples are the so-called *neural cell-adhesion molecule*, or N-CAM (see p. 829), and the *L1 glycoprotein*, also known as the *neuron-glia cell-adhesion molecule*, or Ng-CAM. Antibodies directed against these glycoproteins, both of which are members of the immunoglobulin superfamily (see p. 830 and p. 822), inhibit the tendency of developing neurites to form fascicles and can be shown to disrupt the normal pattern of axon outgrowth (to varying degrees in different parts of the nervous system). N-CAM is found not only on neurons but also on glial cells and, during development and regeneration, on many nonneural cells, including muscle cells. In the latter case it may help to entice the growth cones of motor neurons into the regions where they are to form synapses.

Growth cones are guided not only by their adhesion to the surfaces of other cells but also by their adhesion to various components of the extracellular matrix. An important example is seen in nerve regeneration. When a peripheral nerve is severed, the axons will commonly regenerate by forming growth cones at their cut ends. These growth cones crawl down tunnels of basal lamina, secreted originally by the Schwann cells that ensheathed the axons in the degenerated distal portion of the nerve. There is evidence that *laminin* (see p. 819), or a complex of laminin and heparan sulfate proteoglycan, plays a crucial part in such guidance by binding to matrix receptors of the *integrin* family (see p. 822) in the growth cone membrane. Both materials promote neurite outgrowth *in vitro*, and antibodies that bind to the laminin/proteoglycan complex inhibit nerve regeneration in intact animals.

While proteins such as N-CAM, Ng-CAM, and laminin appear to play an important part in promoting cell adhesion and guiding growth cone migration, they do not provide an obvious answer to the crucial question of why some growth cones take one route while others take another.

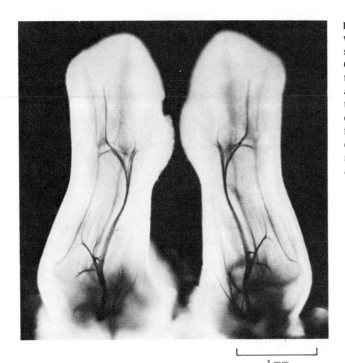

Figure 19–67 Light micrograph of the wings of an 8-day chick embryo, silver-stained to show the patterns of nerves. Compare the right wing with the left: the routes followed by the nerves are almost exactly symmetrical on the two sides of the body, implying the existence of a precise guidance system for nerve outgrowth. Individual growth cones face choices at points where the routes branch; the choices are made according to strict rules.

1 mm

The Pattern of Nerve Connections Is Governed by the Nonequivalent Characters of Neurons: The Doctrine of Neuronal Specificity[57]

In the developing limb of a chick embryo, as in other systems, the growing axons are confined to a precisely defined set of paths (Figure 19–67). These paths branch along their course, with different branches leading to different targets. Thus individual growth cones face a series of choices at successive branch points. The choices are made according to precise rules, with the consequence that a highly ordered system of connections is set up between neurons and their target cells. The connections can be traced by the horseradish peroxidase technique (see p. 1064), which reveals that the cell bodies of the motor neurons innervating a given muscle lie clustered at a site in the spinal cord that is the same in every animal but different for different muscles (Figure 19–68).

How is the choice of pathway made? Are the growth cones somehow channeled to their different destinations as a direct consequence of their different starting positions, like drivers on a multilane highway where it is forbidden to change lanes? This possibility was tested by cutting out a short portion of the neural tube of an embryo at an early stage, before any axons had begun to grow out, and replacing it with its anteroposterior axis reversed. Thus the precursors

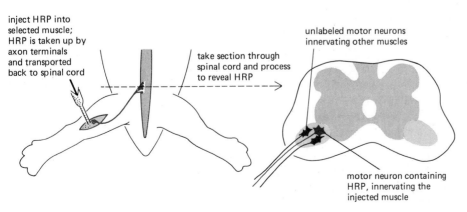

inject HRP into selected muscle; HRP is taken up by axon terminals and transported back to spinal cord

take section through spinal cord and process to reveal HRP

unlabeled motor neurons innervating other muscles

motor neuron containing HRP, innervating the injected muscle

Figure 19–68 The use of retrograde transport of horseradish peroxidase (HRP) to identify which motor neurons in the spinal cord innervate a particular muscle. For clarity, the sizes of the motor neurons are exaggerated in the cross-section, and only three are shown; in reality each muscle is supplied by a nerve that contains the processes of many individual nerve cells (typically a few hundred) whose cell bodies are clustered together in the spinal cord.

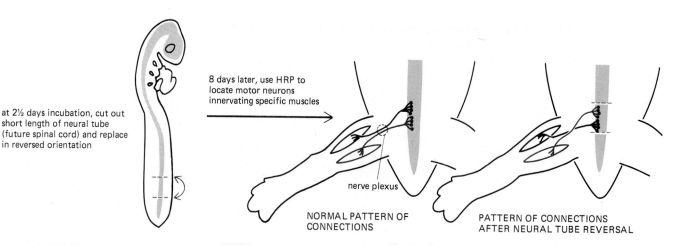

at 2½ days incubation, cut out short length of neural tube (future spinal cord) and replace in reversed orientation

8 days later, use HRP to locate motor neurons innervating specific muscles

nerve plexus

NORMAL PATTERN OF CONNECTIONS

PATTERN OF CONNECTIONS AFTER NEURAL TUBE REVERSAL

of the neurons originally destined to innervate muscle A were put in the place of those originally destined to innervate muscle B, and vice versa. Provided that the shift of position was not too extreme, the growth cones of the misplaced neurons traveled out by altered routes to connect with the muscle appropriate to their *original* position in the neural tube (Figure 19–69). This implies that the motor neurons destined to innervate different muscles are nonequivalent (see p. 915): like the neurons in the *reeler* mouse cerebral cortex, they are distinguished from one another not simply by their positions but by their intrinsic chemical characters. Such nonequivalence among neurons is commonly referred to as **neuronal specificity.** As discussed in Chapter 16, connective tissue cells in different regions of the limb bud are also nonequivalent and may provide the markers that enable a specific growth cone to select a specific branch of the highway system.

In the central nervous system there is also evidence, from both vertebrates and invertebrates, that particular subsets of neurons or glial cells display specific labels that are recognized by other neurons and so help to guide the formation of selective nerve connections. But so far little is known about the molecules involved in either the central or peripheral nervous system.

Figure 19–69 An experiment on a chick embryo demonstrating that motor neurons, even when misplaced, nevertheless send their axons to the muscles appropriate to their original positions in the embryonic spinal cord. Note that the axons from motor neurons at different levels along the spinal cord are funneled together into a *plexus* at the base of the limb and then separate again to innervate their separate targets. A growth cone passing through the region of the plexus has a large choice of targets available to it.

Target Tissues Release Neurotrophic Factors That Control Nerve Cell Growth and Survival[58]

During the initial part of its journey, the growth cone is generally guided by the tissues through which it is passing; as it nears its destination, it comes under the influence of the target itself, often even before cell-to-cell contact has been made, through the action of *neurotrophic factors* that emanate from the target cells. As we have seen in the example of the trigeminal ganglion innervating the embryonic jaw, such factors may serve as chemotactic attractants for growth cones. More fundamentally, however, they control the *survival* of growth cones, of axon branches, and of entire neurons.

The first neurotrophic factor to be identified, and by far the best characterized, is known simply as **nerve growth factor,** or **NGF.** It was discovered by accident in the course of experiments in which foreign tissues and tumors were transplanted into chick embryos. Transplants of one particular tumor became exceptionally densely innervated and caused a striking enlargement of certain groups of peripheral neurons in the vicinity of the graft. Just two classes of neurons were affected: *sensory neurons* and *sympathetic neurons* (a subclass of the peripheral autonomic neurons that control contractions of smooth muscle and secretion from exocrine glands). Soluble extracts from the tumor also stimulated neurite outgrowth from these neurons in culture. Further work showed that one particular tissue, the salivary gland of the male mouse, produced the same factor in enormous quantities. This quirk of nature is still puzzling, since bulk production of NGF by male mouse salivary gland cells bears no obvious relation to the major functions of the factor, but it made it possible to purify NGF in large enough quantities to discover its chemistry and explore its functions. The activity was

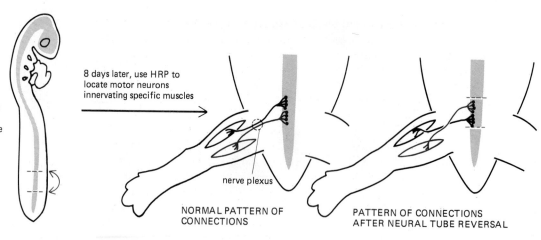

found to lie in a protein dimer composed of two identical polypeptide chains 118 amino acids long. Once NGF had been purified, it was possible to raise antibodies that would block its activity. If anti-NGF antibodies are administered to mice while the nervous system is still developing, most sympathetic neurons and some sensory neurons die.

Likewise in culture, sympathetic neurons and some sensory neurons die in the absence of NGF; if NGF is present, they survive and send out neurites (Figure 19–70). Neuronal survival and neurite production represent two distinct effects of NGF. This has been neatly shown by placing the cells in the central compartment of a three-chambered culture dish whose two side compartments are separated from the central one by barriers that prevent mixing of the media in the three compartments but allow neurites to pass (Figure 19–71). If NGF is present in all three compartments, neurites extend into all three. If NGF is absent from one of the side compartments, no neurites will extend into it; and if all NGF is removed from one of the side compartments when neurites are already there, they will wither and retract as far as the barrier. The cells in the central compartment will not survive or send out neurites unless NGF is present there initially; but if NGF is provided initially in all three compartments and then withdrawn from the central one after the neurites have extended into the side compartments, the cells survive and neurite outgrowth continues in the side compartments.

Thus NGF acts both locally at the periphery of the cell, maintaining and stimulating those neurites and growth cones that are exposed to it, and centrally as a survival factor for the cell as a whole. The local effect on growth cones is direct, rapid, and independent of communication with the cell body; when medium devoid of NGF is substituted for medium containing NGF, the deprived growth cones halt their movements within a minute or two. Besides responding directly to NGF, the growth cones of NGF-sensitive cells take up NGF by endocytosis into vesicles, which are carried by retrograde transport back to the cell body, where the NGF (or some intracellular messenger) presumably exerts its effect on cell survival.

Figure 19–70 (A) Dark-field photomicrographs of a sympathetic ganglion cultured for 48 hours with (*above*) or without (*below*) NGF. Neurites grow out from the sympathetic neurons only if NGF is present in the medium. Each culture also contains Schwann cells that have migrated out of the ganglion; these are not affected by NGF. (B) Phase-contrast photomicrographs showing the behavior of sensory neurons cultured for 24 hours either with NGF (*left*), or in control medium without NGF (*center*), or in medium containing an extract of skeletal muscle (*right*). The sensory neurons in the upper row of pictures are from a group that normally innervates chiefly the body surface (but also sends a few fibers to skeletal muscle and elsewhere); most of these cells respond to NGF like the sympathetic neurons in (A). The sensory neurons in the lower row are from a group that normally innervates skeletal muscle (to provide sensory feedback); these are not responsive to NGF but respond strongly to an extract prepared from skeletal muscle. (A, courtesy of Naomi Kleitman; B, from A.M. Davies, *Dev. Biol.* 115:56–67, 1986.)

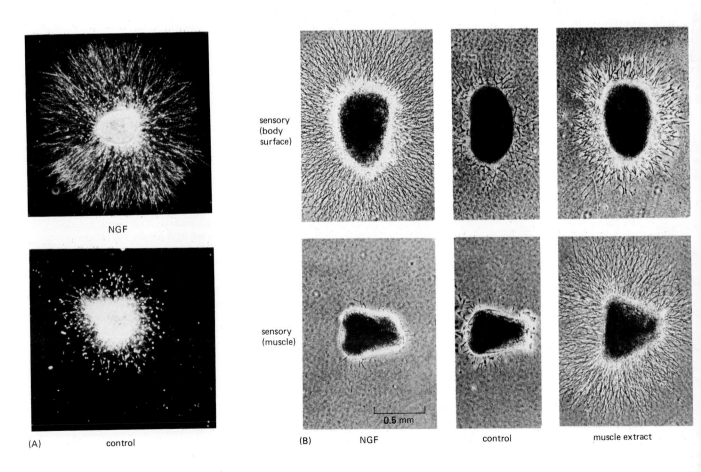

(A)   NGF   control   sensory (body surface)   sensory (muscle)   (B)   NGF   control   muscle extract   0.5 mm

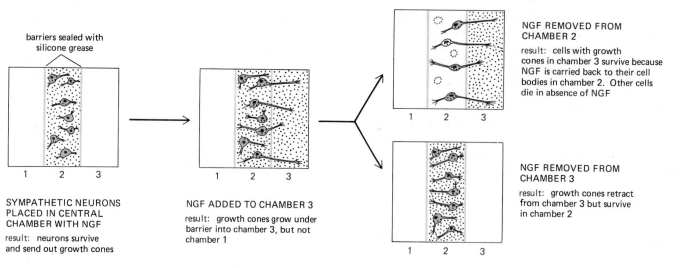

SYMPATHETIC NEURONS
PLACED IN CENTRAL
CHAMBER WITH NGF

result: neurons survive
and send out growth cones

NGF ADDED TO CHAMBER 3

result: growth cones grow under
barrier into chamber 3, but not
chamber 1

NGF REMOVED FROM
CHAMBER 2

result: cells with growth
cones in chamber 3 survive because
NGF is carried back to their cell
bodies in chamber 2. Other cells
die in absence of NGF

NGF REMOVED FROM
CHAMBER 3

result: growth cones retract
from chamber 3 but survive
in chamber 2

Cell Death Adjusts the Number of Surviving Neurons According to the Amount of Target Tissue[59]

In a vertebrate the spinal sensory ganglia are generated in a regular segmental pattern corresponding to the series of vertebrae. Each ganglion consists of a cluster of sensory neurons derived from the neural crest, each of which sends one neurite outward to the periphery of the body and one neurite inward to the spinal cord. The rudiments of the ganglia at first are all similar in size, but in the mature animal the ganglia that innervate the body segments that have limbs attached are much bigger and contain more neurons than the ganglia that innervate the thoracic segments, where there are no limbs (Figure 19–72). This disparity is brought about chiefly by cell death: a larger proportion of the ganglion neurons at the thoracic levels die. If a limb bud is cut off at an early stage, the adjacent ganglia are reduced to the size of thoracic ganglia; conversely, if an extra limb bud is grafted onto the embryonic thorax, it becomes innervated and an abnormally large number of ganglion neurons survive at that level. The control of the survival of ganglion neurons according to the quantity of target tissue is thought to be mediated in large part by NGF secreted by the target. If extra NGF is injected into the embryo during the appropriate period of development, a large proportion of the thoracic ganglion neurons that would ordinarily die are saved, as are ganglion neurons adjacent to an amputated limb bud.

It may seem wasteful to generate excess neurons and then adjust the numbers by cell death according to the amount of the target tissue. Yet this strategy is

Figure 19–71 Schematic diagram of tissue-culture experiments showing that sympathetic neurons, besides requiring NGF in order to survive, can send out growth cones and maintain neurites only in regions where NGF is present. Note that cell survival does not require NGF in the neighborhood of the cell body as long as the cell has a neurite extending into a region that contains NGF.

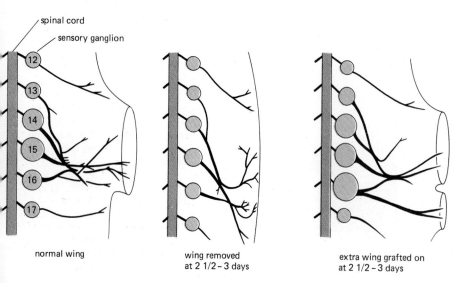

normal wing

wing removed
at 2 1/2 – 3 days

extra wing grafted on
at 2 1/2 – 3 days

Figure 19–72 The control of nerve cell survival in the spinal sensory ganglia of a chick embryo. The spinal ganglia (*colored*) and nerves (*black*) are sketched from embryos 8–9 days old. The size of each ganglion reflects the number of neurons that have survived, which in turn is governed by the quantity of tissue that is available for the ganglion to innervate. (After V. Hamburger, *J. Exp. Zool.* 68:449–494, 1934 and *J. Exp. Zool.* 80:347–389, 1939.)

commonplace throughout the nervous system in vertebrates—both in sensory and in motor cell groups, and in the central nervous system as well as in the periphery. About 50% of all the motor neurons that send axons to skeletal muscles, for example, die in the course of embryonic development within a few days after making contact with their target muscles. A variety of target-cell-derived trophic factors analogous to NGF appear to regulate neuronal survival in these systems. The strategy has several major advantages. First, it provides an automatic device to correct for variations in the relative sizes of different parts of the body. Second, it facilitates evolution: if mutations alter the size of one part of the body, the numbers of neurons connecting with it will be automatically adjusted without other mutations being needed to alter the programs that generate the neurons. Finally, a small number of neurotrophic factors such as NGF can regulate the individual quantitative matching of a large number of paired targets and sources of innervation, even if the system of connections is very intricate. Through axonal transport, the factor produced by a given target is delivered selectively to the neurons that innervate that target and not to other neurons that may have similarly located cell bodies and similar receptors but send their axons elsewhere. Thus cell death regulated by neurotrophic growth factors can help to set up precise and detailed correspondences between the numbers of cells in different parts of the nervous system.

Neural Connections Are Made and Broken Throughout Life[60]

Even in normal, undamaged nervous tissue, there is evidence that dendrites and axon terminals continually retract and regrow. In a mature autonomic ganglion of a mouse, for example, identified individual neurons can be seen to withdraw some dendrite branches and to sprout others over the course of a month (Figure 19–73). Such remodeling occurs slowly and on a limited scale in normal circumstances, but it is called into play in a striking way when a proportion of the target cells in a tissue are deprived of innervation. In the case of a skeletal muscle, this can be done by cutting some but not all of the axons that innervate it. The denervated muscle fibers then apparently secrete a diffusible "sprouting factor," which stimulates profuse sprouting of new growth cones from the surviving axon terminals on neighboring innervated muscle fibers (Figure 19–74). The sprouting factors produced by denervated skeletal muscle have not yet been identified, but for smooth muscle, NGF has been shown to play an exactly analogous role. Denervation leads to an increase in the amount of NGF available from the smooth muscle (at least in part because there are fewer nerve terminals transporting the NGF away), and the excess NGF stimulates growth of axons toward the muscle so as to restore its innervation.

Evidently NGF acts in the intact animal just as it does in a culture dish, both as a survival factor, to determine whether cells shall live or die, and as a local stimulus for growth cone activity, to control the sprouting of axon terminals. The first action is prominent during development; the second is important throughout life. But both actions contribute to the same end: they adjust the supply of innervation according to the requirements of the target. Evidence for the existence of other neurotrophic growth factors, performing similar functions in relation to

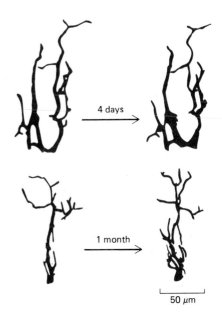

Figure 19–73 Remodeling of the dendrites of neurons in the superior cervical (autonomic) ganglion of a mouse. The ganglion is exposed by careful dissection in the anesthetized animal, and a fluorescent dye is microinjected into one of the nerve cell bodies so as to make its dendrites visible. The wound is then stitched up, and after an interval of some days or weeks, the dissection is repeated and the dye is again injected into the same neuron. The upper and lower pictures show two neurons left for different periods of time. The longer the interval between the first and second injections, the greater is the change seen in the pattern of the cell's dendrites. (Reprinted by permission from D. Purves and R.D. Hadley, *Nature* 315:404–406, 1985. Copyright © 1985 Macmillan Journals Limited.)

Figure 19–74 By cutting some of the axons that innervate a skeletal muscle, it is possible to deprive some of the muscle cells of innervation while leaving other, adjacent muscle cells with their innervation intact. The cut axons degenerate; the surviving axons, although they have suffered no direct disturbance, are provoked to sprout where they lie close to denervated muscle fibers. Within a month or two, those sprouts that have found their way to vacated sites on the denervated muscle fibers have formed stable synapses on them, restoring their innervation, and the other sprouts have been retracted. Such phenomena suggest that denervated muscle fibers release a diffusible "sprouting factor." (Reproduced with permission from M.C. Brown, R.L. Holland, and W.G. Hopkins, *Ann. Rev. Neurosci.* 4:17–42, 1981. © 1981 by Annual Reviews Inc.)

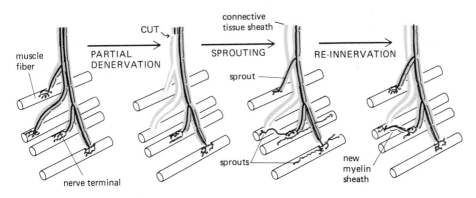

other classes of nerve cells, is rapidly accumulating (see Figure 19–70). In the next section we shall see that such factors may well mediate some of the important effects of electrical activity on the developing pattern of nerve connections.

Summary

The development of a nervous system can be conveniently divided into three phases, which partly overlap. In the first phase, the neurons are generated by finite programs of cell proliferation and the newborn cells migrate from their birthplaces to settle in an orderly fashion in other locations. In the second phase, axons and dendrites extend from the cell bodies by means of growth cones. The growth cones travel along precisely specified paths, guided for the most part by contact interactions with other cell surfaces or with components of the extracellular matrix. Neurons destined to connect with different targets behave as though they have intrinsically different characters (neuronal specificity), expressed in distinctive surface characteristics that enable their growth cones to select different paths. At the end of its path, a growth cone encounters the target cell with which it is to synapse and falls under the influence of target-derived neurotrophic factors. These govern the sprouting and movement of the growth cone in the neighborhood of the target and also control the survival of the neuron from which the growth cone originates. In these two ways neurotrophic factors, such as nerve growth factor (NGF), regulate the density of innervation of target tissues. In the third phase of neural development, to be discussed in the next section, synapses are formed and the pattern of connections is adjusted by mechanisms that depend on electrical activity.

Synapse Formation and Elimination[61]

The encounter of a growth cone with its target cell is a crucial moment in neuronal development: both the growth cone and the target cell undergo a transformation, and synaptic communication can begin. But the developmental process does not end there: many of the synapses formed initially are later eliminated, and new synapses form elsewhere on the same target cell. This local remodeling of the pattern of synaptic connections provides an opportunity for error correction and fine tuning: first the system is roughed out through pathway guidance as growth cones migrate along specific routes to the vicinity of their target cells; then tentative synaptic connections are made, allowing pre- and postsynaptic cells to communicate; and lastly, the initial connections are revised and adjusted by mechanisms that involve both neurotrophic factors and electrical signals in the form of action potentials and synaptic transmission. Thus external stimuli that excite electrical activity in the nervous system can influence the development of the pattern of nerve connections.

In this section we examine the molecular events of synapse formation, the rules that determine whether synapses are to be formed or eliminated, and the part played by electrical activity in controlling these processes. We begin with synapses between motor neurons and skeletal muscle cells because most is known about them.

Synaptic Contact Induces Specializations for Signaling in Both the Growing Axon and the Target Cell[62]

The early events of neuromuscular synapse formation can be observed best in culture. Here it can be seen that much of the molecular machinery for synaptic transmission is present even before a growth cone has made contact with a muscle cell. As the growth cone crawls forward, it releases tiny pulses of acetylcholine in response to electrical excitation of the nerve cell body (Figure 19–75). Its membrane already contains voltage-gated Ca^{2+} channels to couple excitation to secretion, and these channels also serve to propagate action potentials along the embryonic neurite (which at first lacks Na^{+} channels). Before the muscle cell is innervated,

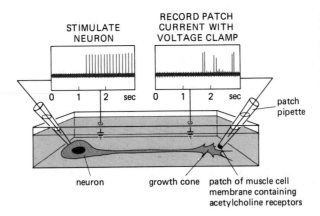

Figure 19–75 An experiment showing that a growing motor neuron in culture discharges pulses of acetylcholine from its growth cone in response to stimulation of the cell body. The minute quantities of acetylcholine released are detected by measuring their effect on the current through a detached patch of muscle cell membrane, rich in acetylcholine receptors, covering the mouth of a patch pipette. The release of acetylcholine from the growth cone is much less plentiful and less reliable than the release from a mature synaptic terminal.

it already has acetylcholine receptors (of an embryonic type) and responds to acetylcholine by depolarization and contraction.

A relatively inefficient form of synaptic transmission can be demonstrated within minutes after the first contact of growth cone with muscle cell. To form a mature synapse, however, both the growth cone and the target cell must develop structural and biochemical specializations—a process that typically takes several days. The growth cone halts its movements, accumulates synaptic vesicles in its interior, and constructs "active zones" for rapid and localized release of acetylcholine (see p. 1078). The muscle cell concentrates its acetylcholine receptors at the synapse and removes them from other regions of its plasma membrane. How is this rearrangement of neurotransmitter receptors achieved? The question is as relevant to neurons as it is to muscle cells, since neurons also, as we have seen, must be able to concentrate particular classes of receptors and ion channels in specific regions of their plasma membranes in order to function in signaling and computation.

Acetylcholine Receptors Diffuse in the Muscle Cell Membrane and Become Tethered at the Forming Synapse[63]

In an adult muscle cell the concentration of acetylcholine receptors at the synapse is more than a thousand times greater than elsewhere in the plasma membrane. Fluorescence bleaching experiments (see p. 295) show that the receptors at the synapse are tethered in place and not free to diffuse in the plane of the membrane. In the uninnervated embryonic muscle cell, by contrast, the receptors are initially spread out over the whole surface and diffuse more freely. When a motor axon makes contact with the muscle cell, these acetylcholine receptors begin to aggregate beneath the axon terminal; moreover, newly synthesized acetylcholine receptors are now preferentially inserted at the developing synapse (Figure 19–76). The receptors become locked into place—perhaps by adhering to one another, perhaps by anchoring to the underlying cytoskeleton or to the overlying extracellular matrix. Some important clues as to how the axon terminal marks out the site of the synapse come from studies of the regeneration of neuromuscular connections.

Figure 19–76 The aggregation of acetylcholine receptors in the membrane of a developing muscle cell at the site where a motor axon terminal makes contact to form a synapse. The aggregation depends partly on diffusion of receptors toward that site from neighboring regions of the muscle cell membrane and partly on the insertion of newly synthesized receptors into the membrane there. The aggregation at the synapse seems to be independent of neurotransmitter release from the nerve terminal, for it occurs even in the presence of agents that block action potentials in the nerve cells and even when the extracellular medium contains high concentrations of α-bungarotoxin, a poison from snake venom that binds to the acetylcholine receptors and blocks their interaction with acetylcholine. The receptors aggregated at the synapse are somehow trapped there; they have a much slower rate of turnover than receptors elsewhere in the membrane, surviving for 5 days or more before they are degraded and replaced.

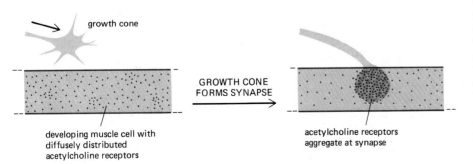

The Site of a Neuromuscular Synapse Is Marked by a Persistent Specialization of the Basal Lamina[64]

Each muscle cell in an adult muscle is enveloped in a basal lamina (see Figures 19–16 and 19–18A). If the muscle is badly damaged, it degenerates and dies, and macrophages move in to clear away the debris. The basal lamina, however, remains and provides a scaffolding within which new muscle fibers can be constructed from surviving stem cells (see p. 986). Moreover, even if a muscle fiber and its axon terminal have both been destroyed, the site of the old neuromuscular junction is still recognizable from the corrugated appearance of the basal lamina there. This *junctional basal lamina* has a specialized chemical character, and it is possible to make antibodies that bind selectively to it. Remarkably, it is the junctional basal lamina that controls the localization of the other components of the synapse.

The importance of the basal lamina at the neuromuscular junction has been demonstrated in a series of experiments on amphibians. By destroying both the nerve and the muscle cells, leaving only empty shells of basal lamina, it is easily shown that the acetylcholinesterase molecules that hydrolyze the acetylcholine released by the axon terminal are tethered in the junctional basal lamina. Moreover, the junctional basal lamina holds the nerve terminal in place: if the muscle cell but not the nerve is destroyed, the nerve terminal remains attached to the basal lamina for many days. On the other hand, removing the basal lamina with collagenase causes the nerve terminal to detach even if the muscle cell is still present.

Indeed, it appears that the basal lamina by itself can guide the regeneration of an axon terminal. This has been demonstrated by destroying both the muscle and the nerve and then allowing the nerve to regenerate while the basal lamina remains empty: a regenerating axon regularly seeks out the original synaptic site and differentiates there into a synaptic ending. The junctional basal lamina also controls the localization of the acetylcholine receptors at the junctional region. If the muscle and the nerve are both destroyed, but now the muscle is allowed to regenerate while the nerve is prevented from doing so, the acetylcholine receptors synthesized by the regenerated muscle localize predominantly in the region of the old junctions, even though the nerve is absent (Figure 19–77). As might be expected, extracts prepared from junctional basal lamina contain a protein, called *agrin*, that promotes receptor clustering in cultured muscle cells.

Evidently, where an axon terminal contacts a muscle cell, it deposits, or causes the muscle cell to deposit, specialized macromolecules, including agrin, that sta-

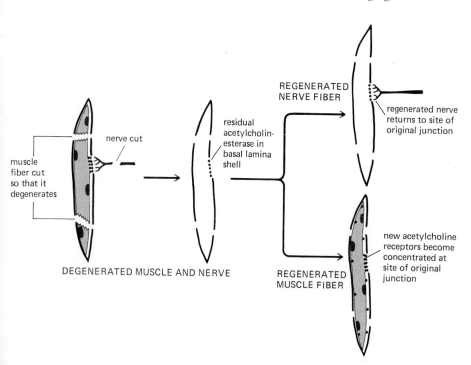

Figure 19–77 Experiment showing that the specialized character of the basal lamina at the neuromuscular junction controls the localization of the other components of the synapse.

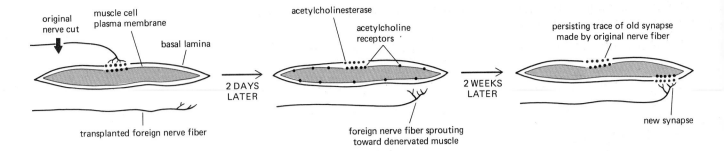

original nerve cut

muscle cell plasma membrane

basal lamina

2 DAYS LATER

transplanted foreign nerve fiber

acetylcholinesterase

acetylcholine receptors

foreign nerve fiber sprouting toward denervated muscle

2 WEEKS LATER

persisting trace of old synapse made by original nerve fiber

new synapse

bilize the synaptic connection. The role of the basal lamina, however, is only part of the story of neuromuscular synapse formation. For not all encounters between axon and muscle cell lead to formation of a synapse, nor are all newly formed synapses absolutely stable.

The Receptivity of a Muscle Cell Is Controlled by Its Electrical Activity[63,65]

If a nerve in a rat is cut and the cut end is deflected so that it lies over an adjacent normal healthy muscle, the severed axons will regenerate and grow over the surface of this muscle; but as long as the muscle's own nerve supply is intact, these foreign axons do not make contact with the individual muscle cells or form synapses on them. If now the normal nerve to the muscle is cut, striking changes occur. Within a few days the muscle cells alter their membrane properties and metabolism: in particular, large quantities of new acetylcholine receptors are synthesized and inserted in the membrane of each muscle cell over its whole surface, making it supersensitive to acetylcholine. At the same time the muscle cells become receptive to new synapse formation by the foreign axons that had grown over the surface of the muscle. Although the axons show a preference for sites where synapses existed before, they can also make synapses at new locations on the muscle cells. Once synapses have formed, the diffuse distribution of acetylcholine receptors disappears, as in embryonic development, leaving a high concentration of receptors only at the sites of the synapses (Figure 19-78).

A denervated muscle cell is deprived of stimulation from its nerve, and it is principally the lack of electrical activity in the muscle cell that brings about the changes described above, as well as evoking release of the "sprouting factor" mentioned previously (see p. 1121). All of these effects of denervation, which make the muscle more receptive to synapse formation, can be mimicked by applying a local anesthetic to the intact nerve, thereby blocking the stimulation of the muscle. Conversely, if a denervated muscle is stimulated artificially through implanted electrodes, the extrajunctional sensitivity to acetylcholine is suppressed and new synapses are prevented from forming. Normally the electrical activity triggered by a neuron that has already established a synapse prevents the muscle cell from receiving unwanted additional innervation.

In a related fashion, electrical activity regulates the elimination of synapses during development. In the vertebrate embryo, where many nerve terminals encounter an uninnervated muscle cell more or less simultaneously, many superfluous nerve connections are initially formed. The adult pattern, in which each muscle cell normally receives only one synapse, is achieved in two distinct steps separated in time. The first involves the death of surplus motor neurons (*neuronal death*); the second involves pruning of axon branches (*synapse elimination*).

Electrical Activity in Muscle Influences the Survival of Embryonic Motor Neurons[59,66]

As mentioned earlier, about 50% of embryonic motor neurons die shortly after making synaptic contact with muscle cells. This death of surplus neurons is prevented if neuromuscular transmission is blocked by means of a toxin such as

Figure 19-78 An experiment on the soleus muscle of a rat, showing how a muscle cell becomes receptive to synapse formation by a transplanted foreign nerve fiber when (and only when) the original nerve is cut. Note that the distribution of acetylcholine receptors in the muscle cell membrane changes as a result of denervation: new extrajunctional receptors become distributed over the cell's entire surface, although the concentration of receptors remains especially high at the site of the old neuromuscular junction. The electrical excitability of the membrane also changes following denervation, through the appearance in the membrane of a new class of voltage-gated channels that are relatively resistant to tetrodotoxin.

α-bungarotoxin and is increased if the muscle is given direct electrical stimulation. These findings suggest that electrical activity in the muscle controls production of a muscle-derived neurotrophic factor necessary for survival of embryonic motor neurons. By analogy with NGF, this might be identical with the "sprouting factor" thought to cause sprouting of axon terminals toward a muscle cell that is denervated. A muscle that is inactive, either because of a block of synaptic transmission or because it is not innervated, would produce the factor in large quantities as a signal of its need for innervation; electrical activation of the muscle, either by artificial stimulation or by the normal spontaneous firing of motor neurons that innervate it, would depress production of the factor, and in the embryo some of the young motor neurons would die in the competition for what little there was.

Electrical Activity Regulates the Competitive Elimination of Synapses According to a Temporal Firing Rule[61,67]

Even after half the embryonic motor neurons have died, the developing muscles are left with a large excess of synaptic inputs. Each motor neuron branches profusely, making synapses on many muscle cells; and a typical muscle cell becomes innervated by branches from several neurons. To attain the adult configuration, all but one of the synapses on each muscle cell must be eliminated. The process of **synapse elimination** during development has been well studied in the soleus muscle of the rat leg, where about three motor neurons on average innervate each muscle cell at birth. During the next 2 or 3 weeks, each neuron retracts a large proportion of its terminal branches until each muscle cell is innervated by a single branch of one motor axon (Figure 19–79).

If the surplus axon branches were eliminated at random, some muscle cells would be left with no synapse at all while others would retain several. The fact that each muscle cell retains exactly one synapse implies that the process of synapse elimination is competitive. Indeed, competitive synapse elimination throughout the nervous system is one of the most important processes governing the development of neural connections and, as we shall see later, their subsequent modification by environmental input. Although the molecular mechanisms of competitive synapse elimination are not understood, the competition in most cases seems to be governed by a simple and general set of principles, applicable both to neuromuscular synapses and to synapses of neuron on neuron.

First, when competition occurs, it involves an element of chance; but the final outcome is clear-cut, and each synapse either survives or is completely eliminated. Second, competition generally occurs only between synapses that are relatively close together and on the same target cell. Thus, in the normal development of a typical mammalian skeletal muscle cell, the incoming nerve terminals all initially synapse in the same small "end-plate" region, and they then compete until only

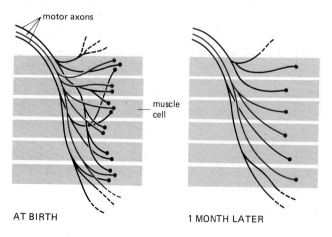

AT BIRTH   1 MONTH LATER

Figure 19–79 Elimination of surplus synapses in a mammalian skeletal muscle in the period after birth. In this schematic diagram the number of terminal branches of each motor axon is underrepresented for the sake of clarity; in reality a single motor axon in a mature muscle typically branches to innervate several hundred muscle cells. All the axon branches innervating one immature muscle cell normally make their synapses in the same small neighborhood on the muscle cell and compete at close range until only one synapse is left.

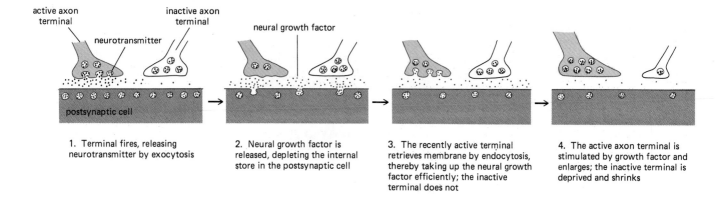

1. Terminal fires, releasing neurotransmitter by exocytosis

2. Neural growth factor is released, depleting the internal store in the postsynaptic cell

3. The recently active terminal retrieves membrane by endocytosis, thereby taking up the neural growth factor efficiently; the inactive terminal does not

4. The active axon terminal is stimulated by growth factor and enlarges; the inactive terminal is deprived and shrinks

one is left. When multiple synapses are artificially induced to form on the same cell at sites separated by 1 mm or more, the multiple innervation is retained.

Third, and most important to the theme of this section, the competitive elimination of synapses depends on electrical activity in both the axons and the target cells that they innervate. For example, synapse elimination is delayed if excitation of a developing muscle is blocked by applying local anesthetic to the nerve or α-bungarotoxin to the neuromuscular junction. And in most systems that have been studied, if some of the innervating axons are paralyzed while others remain active, the active axons gain control of more target cells. It seems as though in the neighborhood of an active synapse, the stimulated target cell either produces something that tends to destroy other synapses on that region of its surface or fails to produce something that is required for their maintenance.

But this presents a paradox. If synaptic stimulation of the target cell drives off synaptic contacts, how can the synapse through which the cell gets its stimulation remain and consolidate itself? The answer seems to lie in the following **firing rule:**

> Each excitation of the target cell tends to consolidate any synapse where the presynaptic axon terminal has just been active and to cause rejection of any synapse where the presynaptic axon terminal has just been quiet.

Thus the relative timing of activity is all-important, and where several independently active neurons make neighboring contacts with a single target cell, each of them tends to consolidate its own synapse while promoting elimination of the synapses made by the others.

The molecular mechanisms underlying the firing rule are unknown; Figure 19–80 outlines one speculative suggestion that has been put forward, and the legend mentions another. Nonetheless, there is evidence for the firing rule from many different systems, and we shall now examine some of its applications to synapses other than the neuromuscular junction.

Synchronously Firing Axon Terminals Make Mutually Supportive Synapses[68]

One of the corollaries of the firing rule is illustrated in the submandibular ganglion of the rat, where each neuron is innervated at birth by axons from about five presynaptic neurons located in the brainstem. By the end of the first month of postnatal life, through competitive synapse elimination, each neuron in the ganglion is innervated by only one such axon. But meanwhile that axon has formed many new terminal branches, synapsing on the same cell at many sites, so that the total number of synapses is larger finally than it was initially (Figure 19–81). The branches of a single axon have one obvious property in common that distinguishes them from branches of other axons of the same type: they all fire at the same time. In accordance with the firing rule, neighboring axon terminals that fire synchronously have collaborated in forming synapses, whereas terminals that fire asynchronously have competed.

Figure 19–80 One of several molecular mechanisms that have been tentatively suggested to underlie the firing rule. In this view, maintenance of a synapse depends on a neural growth factor released from the postsynaptic cell. Release is triggered locally by the electrical stimulation that follows delivery of an action potential at the synapse; at other times there is some spontaneous release, at a lower rate. The factor is taken up into recently active axon terminals during the endocytic retrieval of membrane that immediately follows the exocytosis of neurotransmitter (see p. 1079). Those axon terminals that endocytose the factor respond by enlarging and depositing materials that reinforce the synaptic bond. The more frequently the muscle is stimulated, the more its internal store of the factor is depleted and the lower is the rate of spontaneous release in the absence of stimulation. Thus inactive axon terminals competing with active ones fail to get adequate growth factor and consequently shrink and are eventually withdrawn. Two axon terminals that are active at different times will compete for the limited amount of growth factor that the postsynaptic cell contains. If large terminals take up more growth factor, which in turn makes them larger still, the outcome of such competition may depend on slight differences in initial size.

An alternative hypothesis proposes that stimulation of the postsynaptic cell causes local release or activation of a protease that tends to destroy inactive synapses but against which recently active synapses are somehow protected.

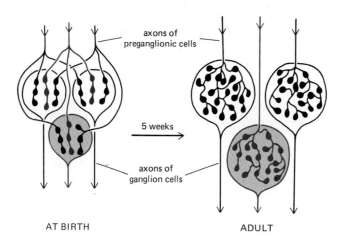

Figure 19–81 The changes that occur soon after birth in the pattern of synapses on neurons in the submandibular ganglion of a rat. Initially each cell is innervated by several axons. These compete until, by synapse elimination, one axon is left in sole command; this one axon, meanwhile, increases the number of its synapses on the cell, which show no sign of competing with one another. (After D. Purves and J.W. Lichtman, *Physiol. Rev.* 58:821–862, 1978.)

AT BIRTH

ADULT

The Number of Surviving Inputs Depends on the Number of Dendrites on the Postsynaptic Neuron[69]

Because the competition among synapses for survival depends in part on the distances between them, the final outcome depends on the structure of the postsynaptic cell. The submandibular ganglion neuron is a somewhat atypical neuron in that it has no dendrites and, following a synaptic competition fought out at close quarters on the cell body, retains input from only one axon. Most other neurons have multiple dendrites and continue in adult life to receive inputs from multiple sources, this being essential for their integrative function. The role of dendrites in regulating synapse elimination is illustrated in the ciliary ganglion of a rabbit, where some of the neurons have many dendrites while others have few or none (Figure 19–82). At birth all the neurons are similarly innervated by about four or five presynaptic axons. But in the adult the cells without dendrites receive input from just one axon, while the number of axons providing input to the other cells increases in direct proportion to the number of main dendrites. The synapses on a single dendrite, however, tend to be made by branches of a single axon. Thus

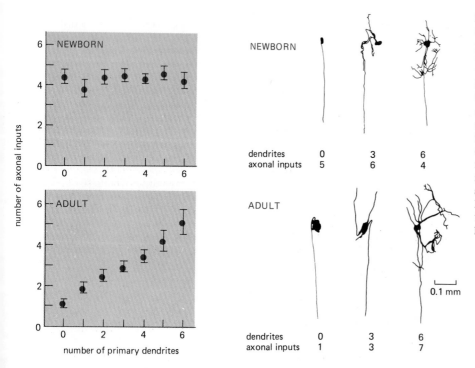

Figure 19–82 The developmental relationship between the number of primary dendrites and the number of axonal inputs to individual cells in the ciliary ganglion of a rabbit. At birth the average number of inputs is independent of the number of dendrites; in the adult the average number of inputs that have survived the period of competitive synapse elimination is proportional to the number of dendrites. At the right the relationship is illustrated by drawings of selected individual ganglion cells. (After D. Purves and R.I. Hume, *J. Neurosci.* 1:441–452, 1981; and R.I. Hume and D. Purves, *Nature* 293:469–471, 1981.)

it seems that each dendrite provides a separate and independent territory, such that synapses on one dendrite do not compete with those on another. As in skeletal muscle, the competition is local, and it obeys the predictions of the firing rule.

Perhaps the most profound implications of the firing rule, however, concern the ways in which stimuli from the external world control adjustments of the anatomical connections between neurons. This is especially clear from studies on the development of the vertebrate visual system. We shall concentrate here on the evidence from mammals.

Visual Connections in Young Mammals Are Adjustable and Sensitive to Visual Experience[70]

The visual system of a mammal is not mature at birth. The first few years of postnatal life (in humans), or the first few months (in cats or monkeys), are a *sensitive* (or *critical*) *period*, during which the pattern of neural connections is still adjustable, and abnormal visual experience can have drastic and irreversible consequences. A common example is the "lazy eye" that can result from a childhood squint. Children with a squint frequently fall into the habit of using one eye only and neglecting the input from the other eye, which is perpetually misdirected and rarely receives a sharply focused image on its retina. If the squint is corrected early and the child is taught to use both eyes, both eyes will continue to function normally. But if the squint goes uncorrected throughout childhood, the unused eye becomes almost completely blind in a permanent way that no lens can correct—a condition known as *amblyopia*. The eye itself remains normal: the defect lies in the brain. Before explaining the nature of the defect, we must outline some of the anatomy of the adult mammalian visual system.

Active Synapses Tend to Displace Inactive Synapses in the Mammalian Visual System[71]

Each eye in a mammal such as a human or a cat sees almost the same visual field, and the two views are combined in the brain to provide binocular stereoscopic vision. This is possible because axons relaying input from equivalent regions in the two retinas make synapses in the same region of the brain (Figure 19–83). Thus, on the *primary visual cortex* of the left side of the adult brain, there are two orderly maps of the right half of the visual field—one from the left eye, the other from the right eye. These two maps, however, are not precisely superimposed on

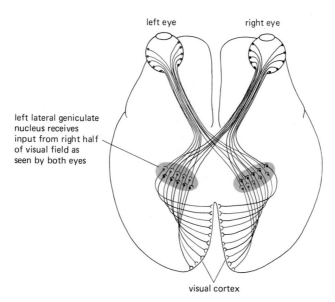

left eye   right eye

left lateral geniculate nucleus receives input from right half of visual field as seen by both eyes

visual cortex

Figure 19–83 The major human visual pathway, showing how the inputs from the right and left eyes are distributed so that related streams of information are brought together in the same region of the brain. Note that all the information obtained by the left side of each eye (relating to the right side of the visual field) is relayed to the left side of the brain, and vice versa.

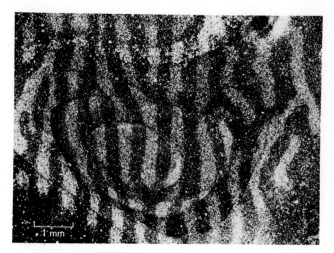

Figure 19–84 Ocular dominance columns in the visual cortex of a normal monkey. Radioactive proline is injected into one eye and the animal is then allowed to survive for 10 days, during which the radioactive label is transported to the parts of the cortex that receive their input from that eye. Sections of the cortex are cut tangentially to its surface, and autoradiographs are prepared. With dark-field illumination the silver grains covering the radioactive regions appear bright against a dark background. The picture is a montage composed of photographs of several successive sections cut at slightly different depths through the thickness of the cortex. The ocular dominance columns connected to the labeled eye (*bright bands*) are of the same width as those connected to the unlabeled eye (*dark bands*). (From D.H. Hubel, T.N. Wiesel, and S. Le Vay, *Philos. Trans. R. Soc.* [*Biol.*] 278:377–409, 1977.)

each other: instead the inputs arriving from the two eyes are segregated in a pattern of narrow, alternating stripes known as **ocular dominance columns.** This arrangement is represented schematically in Figure 19–83 and can be demonstrated directly by a labeling technique that involves injecting radioactive amino acids into one eye. The labeled molecules are taken up by the retinal neurons and incorporated into proteins, which are carried by axonal transport toward the visual cortex; the labeled proteins somehow pass from one neuron to the next at a synaptic relay station (the lateral geniculate nucleus) on the way. Autoradiographs of sections of the visual cortex of the adult monkey brain, for example, show labeled bands about half a millimeter wide, receiving their input from the labeled eye, alternating with unlabeled bands of equal width, receiving their input from the unlabeled eye (Figure 19–84).

During development, however, when the visual connections are first made, no ocular dominance columns can be seen: the projections from the two eyes overlap completely. It is only later (typically during the first few weeks after birth) that the projections segregate into the adult pattern of alternating stripes by means of the competitive elimination of overlapping axon terminals. The pattern apparently develops through the operation of the firing rule: axons relaying excitations from adjacent sites in one eye tend to fire in close synchrony with one another but often out of synchrony with axons relaying excitations from the other eye. The axons that fire in synchrony with one another collaborate in establishing their own set of synapses on a given cortical cell while driving off the synapses from other axons. The segregation into alternating stripes can be halted either by artificially stimulating both optic nerves so as to force the axons from the left eye to fire in strict synchrony with those from the right, or by suppressing electrical activity with injections of tetrodotoxin (which blocks voltage-gated Na^+ channels) into both eyes.

The most striking effects, from a functional point of view, are seen when one eye is simply kept covered and thereby deprived of visual stimulation during the sensitive period. When the cover is removed, the animal behaves as though blind or semiblind in the deprived eye. Autoradiographic tracing shows that the ocular dominance columns connected to the deprived eye have shrunk drastically, while those connected to the experienced eye have widened to occupy the vacated space (Figure 19–85). Again in accordance with the firing rule, synapses made by inactive axons have been eliminated, whereas active axons have consolidated their synapses and made more. In this way cortical territory is allocated to axons that carry information and is not wasted on those that are silent. The effect is irreversible once the sensitive period has ended. Thus the stimulation that a sensory pathway receives in early life determines the amount of cortical machinery—the number of neurons and synapses—that will be available to deal with that input in the adult.

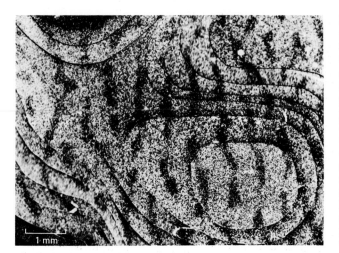

Figure 19–85 Ocular dominance columns in the visual cortex of a monkey that has had one eye covered during the sensitive period of development. The other eye has received an injection of radioactive proline, and autoradiographs have been prepared as described in the caption to Figure 19–84. The ocular dominance columns connected to the eye deprived of visual experience (*dark bands*) are abnormally narrow, whereas those connected to the other eye are abnormally wide. If the deprived eye is labeled, a converse picture is seen, with narrow bright bands alternating with broad dark bands. (From D.H. Hubel, T.N. Wiesel, and S. Le Vay, *Philos. Trans. R. Soc.* [*Biol.*] 278:377–409, 1977.)

Convergent Binocular Visual Connections Depend on Synchronized Binocular Stimulation[70,71,72]

Early visual experience is also important is subtler ways in establishing the nerve connections that enable us to see. For example, some children with an uncorrected squint, instead of neglecting one eye, will use both eyes but in alternation rather than together. Both eyes then remain functional, but the ability to see depth (stereopsis) is permanently lost. Studies of the behavior of single cells in the brain show that this phenomenon too can be explained by the firing rule.

Stereopsis depends on *binocularly driven* neurons—that is, neurons that receive and respond to convergent synaptic inputs from both eyes at once. Such neurons can be identified in experimental animals by inserting a recording microelectrode into the visual cortex of the brain and observing the firing of single cells in response to visual stimuli presented to the two eyes. Such cells are found in particular layers of the visual cortex, above and below the layer containing monocularly driven neurons arranged in clear-cut ocular dominance columns. In a normal animal, binocularly driven neurons are plentiful. But there are scarcely any to be found in an animal that has been specifically deprived of synchronous binocular stimulation during the sensitive period (either by covering different eyes on alternate days or as a result of a severe squint). Evidently, the inputs from each eye to a binocularly driven neuron are maintained only if the two inputs frequently fire in synchrony. When synchronous stimulation is prevented, the axons bringing input from one eye compete with those bringing input to the same neuron from the other eye, in accordance with the firing rule for synapse elimination and maintenance, until one eye gains sole control of that neuron and the possibility of stereopsis is lost.

The Firing Rule Guides the Organization of the Nervous System in the Light of Experience[73]

The development of binocular vision illustrates a general organizing principle: synchronous firing tends to establish convergent connections. This principle, which follows from the firing rule, may help to explain how the brain comes to contain neurons that are specifically responsive to particular complex combinations of sensations that are repeatedly evoked by common objects in the world around us. The brains of primates, for example, contain neurons that appear to fire specifically in response to the sight of a particular face. More generally, one can begin to understand how the brain may be adjusted in the light of experience so as to represent in its own structure and function the existence of relationships between one external phenomenon and another. In this sense the rules for synapse formation and elimination early in life provide a basis for early learning and memory.

It was suggested earlier in the chapter that memory in the adult involves modulation of synaptic transmission through long-lasting chemical changes initiated by neurotransmitters binding to certain types of receptors. Is there any connection between this and the developmental phenomena we have just described, involving relatively gross structural changes in the pattern of synaptic connections? In some cases at least, chemical and structural synaptic changes seem to be intimately related. For example, when *Aplysia* is subjected to long-term habituation or sensitization (see p. 1095) by repeated training on successive days, the chemical modulation of synaptic transmission becomes reinforced by alterations in the size of the presynaptic structures. It is noteworthy also that ocular dominance columns in a frog are altered by exposure to agonists or antagonists of the NMDA receptor, which is thought to play a part in memory in the adult hippocampus (see p. 1100).

Beyond such hints, however, one can only speculate; memory and the mechanisms of synapse formation and elimination are not yet adequately understood. But it is clear that they are among the problems that go to the heart of nerve cell biology and whose solutions promise to illuminate and unify our understanding of the brain at almost every level.

Summary

Synapses first form early in development, but the initial pattern of connections undergoes prolonged remodeling through the elimination of old synapses and the creation of new ones. The creation of a synapse between a motor neuron and a muscle cell involves changes in both cells and the deposition of specialized components in the basal lamina that lies between them. The specialized junctional basal lamina survives the destruction of both motor axons and muscle cells and controls the location of synaptic specializations in both muscle and axon terminal when they regenerate.

In a mammal at the time of birth, each muscle cell generally has several synapses on it, all but one of which are subsequently eliminated by a competitive mechanism. Activation of a muscle cell at a synapse tends to cause disappearance of other synapses in the neighborhood of the one that is active if these synapses are not themselves active at the same time; but where pre- and postsynaptic activity occur synchronously, the synapse tends to be consolidated. This "firing rule," whose molecular mechanism is unknown, appears to explain the phenomena of synapse formation and elimination in many different parts of the developing nervous system. In particular, it helps to explain how synaptic connections in the brain become adjusted according to an animal's experience of the world.

References

General

The Brain. *Sci. Am.* 241(3), 1979. (A whole issue on neurobiology.)
Cooke, I; Lipkin, M., eds. Cellular Neurophysiology: A Source Book. New York: Holt, Rinehart and Winston, 1972. (An anthology.)
Hille, B. Ionic Channels of Excitable Membranes. Sunderland, MA: Sinauer, 1984.
Kandel, E.R.; Schwartz, J.H. Principles of Neural Science, 2nd ed. New York: Elsevier, 1985.
Kuffler, S.W.; Nicholls, J.G.; Martin, A.R. From Neuron to Brain, 2nd ed. Sunderland, MA: Sinauer, 1984.
Molecular Neurobiology. *Cold Spring Harbor Symp. Quant. Biol.* 48, 1983.
Patterson, P.H.; Purves, D. Readings in Developmental Neurobiology. Cold Spring Harbor, NY: Cold Spring Harbor Laboratory, 1982. (An anthology.)
Purves, J.; Lichtman, J.W. Principles of Neural Development. Sunderland, MA: Sinauer, 1985.

Ramón y Cajal, S. Histologie du Systéme Nerveux de l'Homme et des Vértebrés. Paris: Maloine, 1909–1911. (Reprinted, Madrid: Consejo Superior de Investigaciones Científicas, Instituto Ramón y Cajal, 1972.)

Cited

1. Bullock, T.H.; Orkand, R.; Grinnell, A. Introduction to Nervous Systems, pp. 6 and 393–496. San Francisco: Freeman, 1977.
   Nauta, W.J.H.; Feirtag, M. Fundamental Neuroanatomy. New York: Freeman/Scientific American Library, 1986.
2. Ramón y Cajal, S. Recollections of My Life (E.H. Craigie, trans.). In Memoirs of the American Philosophical Society, Vol. 8. Philadelphia, 1937. Reprinted, New York: Garland, 1988. (The most entertaining of all introductions to cellular neurobiology, by the founding father of the subject.)
   Stevens, C.F. The neuron. *Sci. Am.* 241(3):48–59, 1979.

3. Hodgkin, A.L. The Conduction of the Nervous Impulse. Liverpool, U.K.: Liverpool University Press, 1964.

4. Katz, B. Nerve, Muscle, and Synapse. New York: McGraw-Hill, 1966.

5. Bunge, M.B. The axonal cytoskeleton: its role in generating and maintaining cell form. *Trends Neurosci.* 9:477–482, 1986.

    Grafstein, B.; Forman, D.S. Intracellular transport in neurons. *Physiol. Rev.* 60:1167–1283, 1980.

    Peters, A.; Palay, S.L.; Webster, H. deF. The Fine Structure of the Nervous System. Philadelphia: Saunders, 1976.

    Schwartz, J.H. The transport of substances in nerve cells. *Sci. Am.* 242(4):152–171, 1980.

6. Jones, E.G. Pathways to progress—the rise of modern neuroanatomical techniques. *Trends Neurosci.* 9:502–505, 1986.

    LaVail, J.H.; LaVail, M.M. Retrograde axonal transport in the central nervous system. *Science* 176:1416–1417, 1972.

    Schnapp, B.J.; Vale, R.D.; Sheetz, M.P.; Reese, T.S. Single microtubules from squid axoplasm support bidirectional movement of organelles. *Cell* 40:455–462, 1985.

    Vale, R.D. Intracellular transport using microtubule-based motors. *Annu. Rev. Cell Biol.* 3:347–378, 1987.

7. Goldstein, G.W.; Betz, A.L. The blood-brain barrier. *Sci. Am.* 255(3):74–83, 1986.

    Kuffler, S.W.; Nicholls, J.G.; Martin, A.R. From Neuron to Brain, 2nd ed., pp. 323–375. Sunderland, MA: Sinauer, 1984.

    Perry, V.H.; Gordon, S. Macrophages and microglia in the nervous system. *Trends Neurosci.* 11:273–277, 1988.

8. Hille, B. Ionic Channels of Excitable Membranes, pp. 21–75. Sunderland, MA: Sinauer, 1984.

    Kandel, E.R.; Schwartz, J.H. Principles of Neural Science, 2nd ed., pp. 49–86. New York: Elsevier, 1985.

    Kuffler, S.W.; Nicholls, J.G.; Martin, A.R. From Neuron to Brain, 2nd ed., pp. 97–206. Sunderland, MA: Sinauer, 1984.

9. Hodgkin, A.L.; Rushton, W.A.H. The electrical constants of a crustacean nerve fibre. *Proc. R. Soc. Lond. (Biol.)* 133:444–479, 1946.

    Roberts, A.; Bush, B.M.H., eds. Neurones Without Impulses. Cambridge, U.K.: Cambridge University Press, 1981.

10. Baker, P.F.; Hodgkin, A.L.; Shaw, T. The effects of changes in internal ionic concentrations on the electrical properties of perfused giant axons. *J. Physiol.* 164:355–374, 1962.

    Chiu, S.Y.; Ritchie, J.M.; Rogart, R.B.; Stagg, D. A quantitative description of membrane currents in rabbit myelinated nerve. *J. Physiol.* 292:149–166, 1979.

    Hodgkin, A.L. Chance and design in electrophysiology: an informal account of certain experiments on nerve carried out between 1934 and 1952.

    Hodgkin, A.L.; Huxley, A.F. Currents carried by sodium and potassium ions through the membrane of the giant axon of Loligo. *J. Physiol.* 116:449–472, 1952.

    Hodgkin, A.L.; Huxley, A.F.; Katz, B. Measurement of current-voltage relations in the membrane of the giant axon of Loligo. *J. Physiol.* 116:424–448, 1952.

    Hodgkin, A.L.; Katz, B. The effect of sodium ions on the electrical activity of the giant axon of the squid. *J. Physiol.* 108:37–77, 1949.

11. Hodgkin, A.L.; Huxley, A.F. A quantitative description of membrane current and its application to conduction and excitation in nerve. *J. Physiol.* 117:500–544, 1952.

12. Bray, G.M.; Rasminsky, M.; Aguayo, A.J. Interactions between axons and their sheath cells. *Annu. Rev. Neurosci.* 4:127–162, 1981.

    French-Constant, C.; Raff, M.C. The oligodendrocyte-type-2 astrocyte cell lineage is specialized for myelination. *Nature* 323:335–338, 1986.

    Morell, P.; Norton, W.T. Myelin. *Sci. Am.* 242(5):88–118, 1980.

13. Brown, D.A. Synaptic mechanisms. *Trends Neurosci.* 9:468–470, 1986.

    Katz, B. Nerve, Muscle and Synapse, pp. 97–158. New York: McGraw-Hill, 1966.

    Kuffler, S.W.; Nicholls, J.G.; Martin, A.R. From Neuron to Brain, 2nd ed., pp. 207–320. Sunderland, MA: Sinauer, 1984.

14. Dale, H.H.; Feldberg, W.; Vogt, M. Release of acetylcholine at voluntary motor nerve endings. *J. Physiol.* 86:353–380, 1936.

    Fatt, P.; Katz, B. An analysis of the end-plate potential recorded with an intracellular electrode. *J. Physiol.* 115:320–370, 1951.

    Feldberg, W. The early history of synaptic and neuromuscular transmission by acetylcholine: reminiscences of an eyewitness. In The Pursuit of Nature (A.L. Hodgkin, et al.), pp. 65–83. Cambridge, U.K.: Cambridge University Press, 1977.

15. Hille, B. Ionic Channels of Excitable Membranes, pp. 76–98. Sunderland, MA: Sinauer, 1984.

    Katz, B.; Miledi, R. The timing of calcium action during neuromuscular transmission. *J. Physiol.* 189:535–544, 1967.

    Llinas, R. Calcium in synaptic transmission. *Sci. Am.* 247(4):56–65, 1982.

    Miller, R.J. Calcium signalling in neurons. *Trends Neurosci.* 11:415–419, 1988. (Introducing collection of reviews.)

16. Couteaux, R.; Pécot-Dechavassine, M. Vésicules synaptiques et poches au niveau des zones actives de la jonction neuromusculaire. *Comptes Rendus Acad. Sci. (Paris)* D 271:2346–2349, 1970.

    Heuser, J.E.; et al. Synaptic vesicle exocytosis captured by quick freezing and correlated with quantal transmitter release. *J. Cell Biol.* 81:275–300, 1979.

    Heuser, J.E.; Reese, T.S. Structural changes after transmitter release at the frog neuromuscular junction. *J. Cell Biol.* 88:564–580, 1981.

    Smith, S.J.; Augustine, G.J. Calcium ions, active zones and synaptic transmitter release. *Trends Neurosci.* 11:458–464, 1988.

17. Del Castillo, J.; Katz, B. Quantal components of the end-plate potential. *J. Physiol.* 124:560–573, 1954.

    Fatt, P.; Katz, B. Spontaneous subthreshold activity at motor nerve endings. *J. Physiol.* 117:109–128, 1952.

18. Hille, B. Ionic Channels of Excitable Membranes, pp. 117–147. Sunderland, MA: Sinauer, 1984.

19. Lester, H.A. The response to acetylcholine. *Sci. Am.* 236(2):106–118, 1977.

    Sakmann, B.; Bormann, J.; Hamill, O.P. Ion transport by single receptor channels. *Cold Spring Harbor Symp. Quant. Biol.* 48:247–257, 1983.

20. Massoulié, J.; Bon, S. The molecular forms of cholinesterase and acetylcholinesterase in vertebrates. *Annu. Rev. Neurosci.* 5:57–106, 1982.

    Taylor, P.; Schumacher, M.; MacPhee-Quigley, K.; Friedmann, T.; Taylor, S. The structure of acetylcholinesterase: relationship to its function and cellular disposition. *Trends Neurosci.* 10:93–95, 1987.

21. Barnard, E.A.; Darlison, M.G.; Seeburg, P. Molecular biology of the GABA-A receptor: the receptor/channel superfamily. *Trends Neurosci.* 10:502–509, 1987.

    Grenningloh, G.; et al. The strychnine-binding subunit of the glycine receptor shows homology with nicotinic acetylcholine receptors. *Nature* 328:215–220, 1987.

    Hille, B. Ionic Channels of Excitable Membranes, pp. 371–383. Sunderland, MA: Sinauer, 1984. (Evolution of channels.)

22. Gottlieb, D.I. GABAergic neurons. *Sci. Am.* 258(2):38–45, 1988.

    Mayer, M.L.; Westbrook, G.L. The physiology of excitatory amino acids in the vertebrate central nervous system. *Prog. Neurobiol.* 28:197–276, 1987.

    Sneddon, P.; Westfall, D.P. Pharmacological evidence that adenosine triphosphate and noradrenaline are co-transmitters in the guinea pig vas deferens. *J. Physiol.* 347:561–580, 1984.

23. Snyder, S.H. Drug and neurotransmitter receptors in the brain. *Science* 224:22–31, 1984.

Nathanson, N.M. Molecular properties of the muscarinic acetylcholine receptor. *Annu. Rev. Neurosci.* 10:195–236, 1987.

24. Bormann, J. Electrophysiology of GABA-A and GABA-B receptor subtypes. *Trends Neurosci.* 11:112–116, 1988.

Snyder, S.H. Drugs and the Brain. New York: W.H. Freeman/Scientific American Books, 1987.

Tallman, J.F.; Gallager, D.W. The GABAergic system: a locus of benzodiazepine action. *Annu. Rev. Neurosci.* 8:21–44, 1985.

25. Kuffler, S.W.; Nicholls, J.G.; Martin, A.R. From Neuron to Brain, 2nd ed., pp. 407–430. Sunderland, MA: Sinauer, 1984.

26. Barrett, J.N. Motoneuron dendrites: role in synaptic integration. *Fed. Proc.* 34:1398–1407, 1975.

27. Coombs, J.S.; Curtis, D.R.; Eccles, J.C. The generation of impulses in motoneurones. *J. Physiol.* 139:232–249, 1957.

Fuortes, M.G.F.; Frank, K.; Becker, M.C. Steps in the production of motoneuron spikes. *J. Gen. Physiol.* 40:735–752, 1957.

28. Connor, J.A.; Stevens, C.F. Prediction of repetitive firing behaviour from voltage clamp data on an isolated neurone soma. *J. Physiol.* 213:31–53, 1971.

Hille, B. Ionic Channels of Excitable Membranes, pp. 99–116. Sunderland, MA: Sinauer, 1984.

Rogawski, M.A. The A-current: how ubiquitous a feature of excitable cells is it? *Trends Neurosci.* 8:214–219, 1985.

29. Meech, R.W. Calcium-dependent potassium activation in nervous tissues. *Annu. Rev. Biophys. Bioeng.* 7:1–18, 1978.

Tsien, R.W.; Lipscombe, D.; Madison, D.V.; Bley, K.R.; Fox, A.P. Multiple types of neuronal calcium channels and their selective modulation. *Trends Neurosci.* 11:431–438, 1988.

30. Bullock, T.H.; Horridge, G.A. Structure and Function in the Nervous System of Invertebrates, pp. 38–124. San Francisco: Freeman, 1965.

Llinas, R.; Sugimori, M. Electrophysiological properties of *in vitro* Purkinje cell dendrites in mammalian cerebellar slices. *J. Physiol.* 305:197–213, 1980.

Shepherd, G.M. Microcircuits in the nervous system. *Sci. Am.* 238(2):92–103, 1978.

31. Breitwieser, G.E.; Szabo, G. Uncoupling of cardiac muscarinic and beta-adrenergic receptors from ion channels by a guanine nucleotide analogue. *Nature* 317:538–540, 1985.

Levitan, I.B. Modulation of ion channels in neurons and other cells. *Annu. Rev. Neurosci.* 11:119–136, 1988.

Nairn, A.C.; Hemmings, H.C.; Greengard, P. Protein kinases in the brain. *Annu. Rev. Biochem.* 54:931–976, 1985.

Pfaffinger, P.J.; Martin, J.M.; Hunter, D.D.; Nathanson, N.M.; Hille, B. GTP-binding proteins couple cardiac muscarinic receptors to a K channel. *Nature* 317:536–538, 1985.

32. Iversen, L.L. The chemistry of the brain. *Sci. Am.* 241(3):118–129, 1979.

Snyder, S.H. Drugs and the Brain. New York: W.H. Freeman/Scientific American Library, 1988.

33. Bloom, F.E. Neuropeptides. *Sci. Am.* 245(4):148–168, 1981.

Hokfelt, T.; Johansson, O.; Goldstein, M. Chemical anatomy of the brain. *Science* 225:1326–1334, 1984.

Jan, Y.N.; Bowers, C.W.; Branton, D.; Evans, L.; Jan, L.Y. Peptides in neuronal function: studies using frog autonomic ganglia. *Cold Spring Harbor Symp. Quant. Biol.* 48:363–374, 1983.

Scheller, R.H.; Axel, R. How genes control an innate behavior. *Sci. Am.* 250(3):44–52, 1984. (Neuropeptides in Aplysia).

34. Farley, J.; Alkon, D. Cellular mechanisms of learning, memory, and information storage. *Annu. Rev. Psychol.* 36:419–494, 1985.

Kandel, E.R. Small systems of neurons. *Sci. Am.* 241(3):60–70, 1979.

Morris, R.G.M.; Kandel, E.R.; Squire, L.R., eds. Learning and Memory. *Trends Neurosci.* 11:125–181, 1988.

35. Greenberg, S.M.; Castellucci, V.F.; Bayley, H.; Schwartz, J.H. A molecular mechanism for long-term sensitization in Aplysia. *Nature* 329:62–65, 1987.

Montarolo, P.G.; et al. A critical period for macromolecular synthesis in long-term heterosynaptic facilitation in Aplysia. *Science* 234:1249–1254, 1986.

Schwartz, J.H.; Greenberg, S.M. Molecular mechanisms for memory: second-messenger induced modifications of protein kinases in nerve cells. *Annu. Rev. Neurosci.* 10:459–476, 1987.

Siegelbaum, S.A.; Camardo, J.S.; Kandel, E.R. Serotonin and cyclic AMP close single K^+ channels in Aplysia sensory neurons. *Nature* 299:413–417, 1982.

36. Dudai, Y. Neurogenetic dissection of learning and short-term memory in Drosophila. *Annu. Rev. Neurosci.* 11:537–563, 1988.

Kandel, E.R.; et al. Classical conditioning and sensitization share aspects of the same molecular cascade in Aplysia. *Cold Spring Harbor Symp. Quant. Biol.* 48:821–830, 1983.

37. Bliss, T.V.P.; Lomo, T. Long-lasting potentiation of synaptic transmission in the dentate area of the anaesthetized rabbit following stimulation of the perforant path. *J. Physiol.* 232:331–356, 1973.

Collingridge, G.L.; Bliss, T.V.P. NMDA receptors—their role in long-term potentiation. *Trends Neurosci.* 10:288–293, 1987. (The same issue of the journal contains an excellent collection of reviews of various aspects of NMDA receptors.)

Cotman, C.W.; Monaghan, D.T.; Ganong, A.H. Excitatory amino acid neurotransmission: NMDA receptors and Hebb-type synaptic plasticity. *Annu. Rev. Neurosci.* 11:61–80, 1988.

Lisman, J.E.; Goldring, M.A. Feasibility of long-term storage of graded information by the Ca^{2+}/calmodulin-dependent protein kinase molecules of the postsynaptic density. *Proc. Natl. Acad. Sci. USA* 85:5320–5324, 1988.

Mishkin, M.; Appenzeller, T. The anatomy of memory. *Sci. Am.* 256(6):80–89, 1987.

Morris, R.G.; Anderson, E.; Lynch, G.; Baudry, M. Selective impairment of learning and blockade of long-term potentiation by an N-methyl-D-aspartate receptor antagonist, AP5. *Nature* 319:774–776, 1986.

38. Barlow, H.B.; Mollon, J.D., eds. The Senses. Cambridge, U.K.: Cambridge University Press, 1982.

Schmidt, R.F., ed. Fundamentals of Sensory Physiology. New York: Springer, 1978.

Shepherd, G. Neurobiology, 2nd ed., pp. 205–353. New York: Oxford University Press, 1988.

39. Katz, B. Depolarization of sensory terminals and the initiation of impulses in the muscle spindle. *J. Physiol.* 111:261–282, 1950.

40. Hudspeth, A.J. The cellular basis of hearing: the biophysics of hair cells. *Science* 230:745–752, 1985.

Roberts, W.M.; Howard, J.; Hudspeth, A.J. Hair cells: transduction, tuning, and transmission in the inner ear. *Annu. Rev. Cell Biol.* 4:63–92, 1988.

von Bekesy, G. The ear. *Sci. Am.* 197(2):66 - 78, 1957.

41. Corey, D.P.; Hudspeth, A.J. Kinetics of the receptor current in bullfrog saccular hair cells. *J. Neurosci.* 3:962–976, 1983.

Howard, J.; Hudspeth, A.J. Compliance of the hair bundle associated with mechanoelectrical transduction channels in the bullfrog's saccular hair cell. *Neuron* 1:189–199, 1988.

Pickles, J.O. Recent advances in cochlear physiology. *Prog. Neurobiol.* 24:1–42, 1985.

42. Barlow, H.B.; Mollon, J.D., eds. The Senses, pp. 102–164. Cambridge, U.K.: Cambridge University Press, 1982.

43. Baylor, D.A.; Lamb, T.D.; Yau, K.-W. Responses of retinal rods to single photons. *J. Physiol.* 288:613–634, 1979.

Schnapf, J.L.; Baylor, D.A. How photoreceptor cells respond to light. *Sci. Am.* 256(4):40–47, 1987.

44. Stryer, L. The molecules of visual excitation. *Sci. Am.* 257(1): 32–40, 1987.

45. Fesenko, E.E.; Kolesnikov, S.S.; Lyubarsky, A.L. Induction by cyclic GMP of cationic conductance in plasma membrane of retinal rod outer segment. *Nature* 313:310–313, 1985.

    Stryer, L. The cyclic GMP cascade of vision. *Annu. Rev. Neurosci.* 9:87–119, 1986.

46. Koch, K.-W.; Stryer, L. Highly cooperative feedback control of retinal rod guanylate cyclase by calcium ions. *Nature* 334: 64–66, 1988.

    Matthews, H.R.; Murphy, R.L.W.; Fain, G.L.; Lamb, T.D. Photoreceptor light adaptation is mediated by cytoplasmic calcium concentration. *Nature* 334:67–69, 1988.

    Nakatani, K.; Yau, K.-W. Calcium and light adaptation in retinal rods and cones. *Nature* 334:69–71, 1988.

47. Hubel, D.H. Eye, Brain, and Vision. New York: W.H. Freeman/ Scientific American Library, 1988.

    Kuffler, S.W.; Nicholls, J.G.; Martin, A.R. From Neuron to Brain, 2nd ed., pp. 19–96. Sunderland, MA: Sinauer, 1984.

    Masland, R.H. The functional architecture of the retina. *Sci. Am.* 255(6):102–111, 1986.

48. Cowan, W.M. The development of the brain. *Sci. Am.* 241(3):106–117, 1979.

    Hopkins, W.G.; Brown, M.C. Development of Nerve Cells and their Connections. Cambridge, U.K.: Cambridge University Press, 1984.

    Parnavelas, J.G.; Stern, C.D.; Stirling, R.V., eds. The Making of the Nervous System. Oxford, U.K.: Oxford University Press, 1988.

    Purves, D.; Lichtman, J.W. Principles of Neural Development. Sunderland, MA: Sinauer, 1985.

49. Alvarez-Buylla, A.; Nottebohm, F. Migration of young neurons in adult avian brain. *Nature* 335:353–354, 1988. (An example of neural stem cells persisting in adult life.)

    Le Douarin, N.M.; Smith, J. Development of the peripheral nervous system from the neural crest. *Annu. Rev. Cell Biol.* 4:375–404, 1988.

    Stent, G.S.; Weisblat, D.A. Cell lineage in the development of invertebrate nervous systems. *Annu. Rev. Neurosci.* 8:45–70, 1985.

    Williams, R.W.; Herrup, K. The control of neuron number. *Annu. Rev. Neurosci.* 11:423–453, 1988.

50. Hollyday, M.; Hamburger, V. An autoradiographic study of the formation of the lateral motor column in the chick embryo. *Brain Res.* 132:197–208, 1977.

    Rakic, P. Mode of cell migration to the superficial layers of the fetal monkey neocortex. *J. Comp. Neurol.* 145:61–84, 1972.

51. Caviness, V.S. Neocortical histogenesis in normal and reeler mice: a developmental study based on [3H]thymidine autoradiography. *Dev. Brain Res.* 4:293–302, 1982.

    McConnell, S.K. Development and decision-making in the mammalian cerebral cortex. *Brain Res. Rev.* 13:1–23, 1988.

    Rakic, P. Specification of cerebral cortical areas. *Science* 241: 170–176, 1988.

52. Bray, D.; Hollenbeck, P.J. Growth cone motility and guidance. *Annu. Rev. Cell Biol.* 4:43–61, 1988.

    Harrison, R.G. The outgrowth of the nerve fiber as a mode of protoplasmic movement. *J. Exp. Zool.* 9:787–846, 1910.

    Yamada, K.M.; Spooner, B.S.; Wessells, N.K. Ultrastructure and function of growth cones and axons of cultured nerve cells. *J. Cell Biol.* 49:614–635, 1971.

53. Bamburg, J.R. The axonal cytoskeleton: stationary or moving matrix? *Trends Neurosci.* 11:248–249, 1988.

    Bamburg, J.R.; Bray, D.; Chapman, K. Assembly of microtubules at the tip of growing axons. *Nature* 321:788–790, 1986.

    Bray, D. Surface movements during the growth of single ex-

planted neurons. *Proc. Natl. Acad. Sci. USA* 65:905–910, 1970.

    Letourneau, P.C.; Ressler, A.H. Inhibition of neurite initiation and growth by taxol. *J. Cell Biol.* 98:1355–1362, 1984.

54. Davies, A.M. Molecular and cellular aspects of patterning sensory neurone connections in the vertebrate nervous system. *Development* 101:185–208, 1987.

    Kater, S.B.; Mattson, M.P.; Cohan, C.; Connor, J. Calcium regulation of the neuronal growth cone. *Trends Neurosci.* 11:315–321, 1988.

    Letourneau, P.C. Cell-to-substratum adhesion and guidance of axonal elongation. *Dev. Biol.* 44:92–101, 1975.

    Lumsden, A.G.S.; Davies, A.M. Earliest sensory nerve fibres are guided to peripheral targets by attractants other than Nerve Growth Factor. *Nature* 306: 786–788, 1983.

    Patel, N.; Poo, M.-M. Orientation of neurite growth by extracellular electric fields. *J. Neurosci.* 2:483–496, 1982.

55. Bentley, D.; Caudy, M. Navigational substrates for peripheral pioneer growth cones: limb-axis polarity cues, limb-segment boundaries, and guidepost neurons. *Cold Spring Harbor Symp. Quant. Biol.* 48:573–585, 1983.

    Berlot, J.; Goodman, C.S. Guidance of peripheral pioneer neurons in the grasshopper: adhesive hierarchy of epithelial and neuronal surfaces. *Science* 223:493–496, 1984.

    Goodman, C.S.; Bastiani, M.J. How embryonic nerve cells recognize one another. *Sci. Am.* 251(6):58–66, 1984.

56. Bixby, J.L.; Pratt, R.S.; Lilien, J.; Reichardt, L.F. Neurite outgrowth on muscle cell surfaces involves extracellular matrix receptors as well as Ca^{2+}-dependent and -independent cell adhesion molecules. *Proc. Natl. Acad. Sci. USA* 84:2555–2559, 1987.

    Chang, S.; Rathjen, F.G.; Raper, J.A. Extension of neurites on axons is impaired by antibodies against specific neural cell adhesion molecules. *J. Cell Biol.* 104:355–362, 1987.

    Jessell, T.M. Adhesion molecules and the hierarchy of neural development. *Neuron* 1:3-13, 1988.

    Sanes, J.R.; Schachner, M.; Covault, J. Expression of several adhesive macromolecules (N-CAM, L1, J1, NILE, uvomorulin, laminin, fibronectin, and a heparan sulfate proteoglycan) in embryonic, adult, and denervated adult skeletal muscle. *J. Cell Biol.* 102:420–431, 1986.

    Tomaselli, K.J.; et al. N-cadherin and integrins: two receptor systems that mediate neuronal process outgrowth on astrocyte surfaces. *Neuron* 1:33–43, 1988.

57. Lance-Jones, C.; Landmesser, L. Motoneurone projection patterns in the chick hindlimb following early partial reversals of the spinal cord. *J. Physiol.* 302:581–602, 1980.

    Landmesser, L. The development of specific motor pathways in the chick embryo. *Trends Neurosci.* 7:336–339, 1984.

    Sperry, R.W. Chemoaffinity in the orderly growth of nerve fiber patterns and connections. *Proc. Natl. Acad. Sci. USA* 50:703–710, 1963.

    Udin, S.B.; Fawcett, J.W. Formation of topographic maps. *Annu. Rev. Neurosci.* 11:289–327, 1988.

58. Campenot, R.B. Local control of neurite development by nerve growth factor. *Proc. Natl. Acad. Sci. USA* 74:4516–4519, 1977.

    Davies, A.; Lumsden, A. Relation of target encounter and neuronal death to nerve growth factor responsiveness in the developing mouse trigeminal ganglion. *J. Comp. Neurol.* 223:124–137, 1984.

    Greene, L.A. The importance of both early and delayed responses in the biological actions of nerve growth factor. *Trends Neurosci.* 7:91–94, 1984.

    Levi-Montalcini, R.; Calissano, P. The nerve growth factor. *Sci. Am.* 240(6):68–77, 1979.

59. Cowan, W.M.; Fawcett, J.W.; O'Leary, D.D.M.; Stanfield, B.B. Regressive events in neurogenesis. *Science* 225:1258–1265, 1984.

Davies, A.M. Role of neurotrophic factors in development. *Trends Genet.* 4:139–144, 1988.

Hamburger, V.; Levi-Montalcini, R. Proliferation, differentiation and degeneration in the spinal ganglia of the chick embryo under normal and experimental conditions. *J. Exp. Zool.* 162:133–160, 1949.

Hamburger, V.; Yip, J.W. Reduction of experimentally induced neuronal death in spinal ganglia of the chick embryo by nerve growth factor. *J. Neurosci.* 4:767–774, 1984.

Williams, R.W.; Herrup, K. The control of neuron number. *Annu. Rev. Neurosci.* 11:423–453, 1988.

60. Brown, M.C.; Holland, R.L.; Hopkins, W.G. Motor nerve sprouting. *Annu. Rev. Neurosci.* 4:17–42, 1981.

Davies, A.M. The survival and growth of embryonic proprioceptive neurons is promoted by a factor present in skeletal muscle. *Dev. Biol.* 115:56–67, 1986.

Ebendal, T.; Olson, L.; Seiger, A.; Hedlund, K.-O. Nerve growth factors in the rat iris. *Nature* 286:25–28, 1980.

Purves, D.; Voyvodic, J.T. Imaging mammalian nerve cells and their connections over time in living animals. *Trends Neurosci.* 10:398–404, 1987.

61. Purves, D.; Lichtman, J.W. Principles of Neural Development. Sunderland, MA: Sinauer, 1985.

62. Hume, R.I.; Role, L.W.; Fischbach, G.D. Acetylcholine release from growth cones detected with patches of acetylcholine receptor-rich membranes. *Nature* 305:632–634, 1983.

Jaramillo, F.; Vicini, S.; Schuetze, S.M. Embryonic acetylcholine receptors guarantee spontaneous contractions in rat developing muscle. *Nature* 335:66–68, 1988.

Spitzer, N.C. Ion channels in development. *Annu. Rev. Neurosci.* 2:363–397, 1979.

Young, S.H.; Poo, M.-M. Spontaneous release of transmitter from growth cones of embryonic neurones. *Nature* 305:634–637, 1983.

63. Poo, M.-M. Mobility and localization of proteins in excitable membranes. *Annu. Rev. Neurosci.* 8:369–406, 1985.

Schuetze, S.M.; Role, L.W. Developmental regulation of nicotinic acetylcholine receptors. *Annu. Rev. Neurosci.* 10:403–457, 1987.

64. Burden, S.J.; Sargent, P.B.; McMahan, U.J. Acetylcholine receptors in regenerating muscle accumulate at original synaptic sites in the absence of the nerve. *J. Cell Biol.* 82:412–425, 1979.

Nitkin, R.M.; et al. Identification of agrin, a synaptic organizing protein from Torpedo electric organ. *J. Cell Biol.* 105:2471–2478, 1987.

Sanes, J.R.; Hall, Z.W. Antibodies that bind specifically to synaptic sites on muscle fiber basal lamina. *J. Cell Biol.* 83:357–370, 1979.

65. Frank, E.; Jansen, J.K.S.; Lømo, T.; Westgaard, R.H. The interaction between foreign and original motor nerves innervating the soleus muscle of rats. *J. Physiol.* 247:725–743, 1975.

Jones, R.; Vrbová, G. Two factors responsible for denervation hypersensitivity. *J. Physiol.* 236:517–538, 1974.

Lømo, T.; Jansen, J.K.S. Requirements for the formation and maintenance of neuromuscular connections. *Curr. Top. Dev. Biol.* 16:253–281, 1980.

Lømo, T.; Rosenthal, J. Control of ACh sensitivity by muscle activity in the rat. *J. Physiol.* 221:493–513, 1972.

66. Oppenheim, R.W. Cell death during neural development. In Handbook of Physiology, Vol. 1: Neuronal Development, (W.M. Cowan, ed.). Washington, DC: American Physiological Society, 1988.

Pittman, R.; Oppenheim, R.W. Cell death of motoneurons in the chick embryo spinal cord. IV. Evidence that a functional neuromuscular interaction is involved in the regulation of naturally occurring cell death and the stabilization of synapses. *J. Comp. Neurol.* 187:425–446, 1979.

67. Brown, M.C.; Jansen, J.K.S.; Van Essen, D. Polyneuronal innervation of skeletal muscle in new-born rats and its elimination during maturation. *J. Physiol.* 261:387–422, 1976.

Callaway, E.M.; Soha, J.M.; Van Essen, D.C. Competition favouring inactive over active motor neurons during synapse elimination. *Nature* 328:422–426, 1987. (Describes an apparent exception to the general rule.)

O'Brien, R.A.D.; Ostberg, A.J.C.; Vrbová, G. Protease inhibitors reduce the loss of nerve terminals induced by activity and calcium in developing rat soleus muscles in vitro. *Neurosci.* 12:637–646, 1984.

Ribchester, R.R.; Taxt, T. Motor unit size and synaptic competition in rat lumbrical muscles reinnervated by active and inactive motor axons. *J. Physiol.* 344:89–111, 1983.

68. Lichtman, J.W. The reorganization of synaptic connexions in the rat submandibular ganglion during post-natal development. *J. Physiol.* 273:155–177, 1977.

Purves, D.; Lichtman, J.W. Principles of Neural Development, pp. 271–328. Sunderland, MA: Sinauer, 1985.

69. Purves, D. Modulation of neuronal competition by postsynaptic geometry in autonomic ganglia. *Trends Neurosci.* 6:10–16, 1983.

70. Barlow, H. Visual experience and cortical development. *Nature* 258:199–204, 1975.

Wiesel, T.N. Postnatal development of the visual cortex and the influence of environment. *Nature* 299:583–591, 1982.

71. Fawcett, J.W. Retinotopic maps, cell death, and electrical activity in the retinotectal and retinocollicular projections. In The Making of the Nervous System (J.G. Parnavelas; C.D. Stern; R.V. Stirling, eds.), pp. 395–416. Oxford, U.K.: Oxford University Press, 1988.

Hubel, D.H.; Wiesel, T.N. Ferrier lecture: functional architecture of macaque monkey visual cortex. *Proc. R. Soc. Lond. Biol.* 198:1–59, 1977.

Hubel, D.H.; Wiesel, T.N.; Le Vay, S. Plasticity of ocular dominance columns in monkey striate cortex. *Philos. Trans. R. Soc. Lond. Biol.* 278:377–409, 1977.

Rakic, P. Prenatal genesis of connections subserving ocular dominance in the rhesus monkey. *Nature* 261:467–471, 1976.

Stryker, M.P.; Harris, W.A. Binocular impulse blockade prevents the formation of ocular dominance columns in cat visual cortex. *J. Neurosci.* 6:2117–2133, 1986.

72. Hubel, D.H.; Wiesel, T.N. Binocular interaction in striate cortex of kittens reared with artificial squint. *J. Neurophysiol.* 28:1041–1059, 1965.

Le Vay, S.; Wiesel, T.N.; Hubel, D.H. The development of ocular dominance columns in normal and visually deprived monkeys. *J. Comp. Neurol.* 191:1–51, 1980.

73. Baylis, G.C.; Rolls, E.T.; Leonard, C.M. Selectivity between faces in the responses of a population of neurons in the cortex in the superior temporal sulcus of the monkey. *Brain Res.* 342:91–102, 1985.

Cline, H.T.; Debski, E.A.; Constantine-Paton, M. N-methyl-D-aspartate receptor antagonist desegregates eye-specific stripes. *Proc. Natl. Acad. Sci. USA* 84:4342–4345, 1987.

Greenough, W.T.; Bailey, C.H. The anatomy of a memory: convergence of results across a diversity of tests. *Trends Neurosci.* 11:142–147, 1988.

Perrett, D.I.; Mistlin, A.J.; Chitty, A.J. Visual neurones responsive to faces. *Trends Neurosci.* 10:358–364, 1987.

Special Features of Plant Cells

20

Anyone can distinguish a flowering plant from a mammal. Even deciding whether an individual cell is of plant or animal origin is usually a simple matter, although there are problematic cases. But as the level of analysis extends into the cell—to cytoplasm, organelle, and molecule—the similarities between the two kingdoms begin to outweigh the differences. It requires sophisticated procedures to distinguish higher-plant mitochondria, nuclei, ribosomes, or cytoskeletal components from their animal counterparts. Indeed, the disparities between plants and animals tend not to be in fundamental molecular features such as DNA replication, protein synthesis, mitochondrial ATP production, or the basic design of cell membranes. Rather they relate to higher-order functions of cells and tissues.

Comparisons of ribosomal RNA sequences suggest that plants, yeasts, invertebrates, and vertebrates diverged from one another relatively late in the evolution of eucaryotic organisms (see Figure 1–16, p. 13). The differences between plants and animals apparently arose over a period of about 600 million years, and most of them can be traced to two fundamental acquisitions by the progenitors of plants: the ability to fix carbon dioxide by photosynthesis (discussed in Chapter 7), and the production of a rigid *cell wall*. The first acquisition gave plants an internal source of carbon compounds to use in growth and metabolism; the second placed severe constraints on how plants behave. In this chapter we shall consider the special features of plant cells that result from these two acquisitions.

The Importance of the Cell Wall

The plant cell wall is an elaborate extracellular matrix that encloses each cell in a plant. While animal cells also have extracellular matrix components on their surface (see p. 802), the plant cell wall is generally much thicker, stronger, and—most important—more rigid. In fact, most of the differences between plants and animals—in nutrition, digestion, osmoregulation, growth, reproduction, intercellular communication, defense mechanisms, as well as in morphology—can be traced to the plant cell wall. In evolving relatively rigid cell walls, which vary from 0.1 μm to many micrometers in thickness, plants cells forfeited the ability to crawl about; this nonmotile life-style has persisted in multicellular plants. It was the thick cell walls of cork, visible under a microscope, that in 1663 enabled Robert Hooke to distinguish cells clearly and to name them as such.

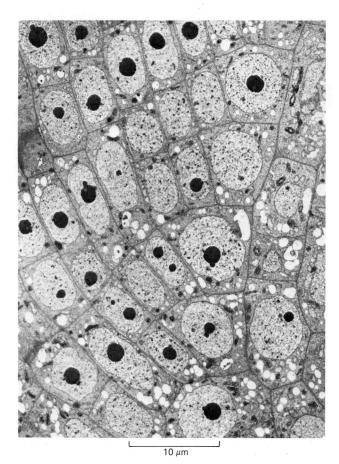

Figure 20–1 Electron micrograph of the root tip of a rush, showing the organized pattern of cells that results from an ordered sequence of cell divisions in cells with rigid cell walls. (Courtesy of Brian Gunning.)

10 μm

The cell wall provides a protective home for the plant cell within. Each cell wall binds to that of its neighbors, cementing the cells together to form the intact plant (Figure 20–1). Even though each plant cell is encased within its own "wooden box," direct cell-cell communication is still possible through the *plasmodesmata*. Thousands of these plasma-membrane-lined channels of cytoplasm cross the cell wall, connecting adjacent cells and allowing small molecules to move from cell to cell. In addition, fluids percolate along and through cell walls. The plant cell wall thus has transport functions as well as protective and skeletal functions.

When plant cells become specialized, they generally produce walls that are particularly well adapted for one or another of these functions. Thus the different types of cells in a plant can be recognized and classified by the shape and nature of their cell walls. In this section we shall examine the ways in which plant cells have evolved to exploit their walled environment. The starting point is a description of the nature of the wall itself.

The Cell Wall Is Composed of Cellulose Fibers Embedded in a Matrix of Polysaccharide and Protein[1]

Most newly formed cells in a multicellular plant are produced in special regions of the plant called *meristems*, as will be explained later (see p. 1173). These new cells are generally small in comparison to their final size. To accommodate subsequent cell growth, the walls of such cells, called **primary cell walls** (Figure 20–2), are thin and only semirigid. Once growth stops, the wall no longer needs to be able to expand. Although the mature nongrowing cell may simply retain its primary cell wall, far more commonly it augments it by producing a **secondary cell wall,** either by thickening the primary wall or by depositing inside it new, tough wall layers with a different composition (see p. 1146).

Although the primary cell walls of higher plants vary greatly in both composition and detailed organization, like all extracellular matrices they are constructed

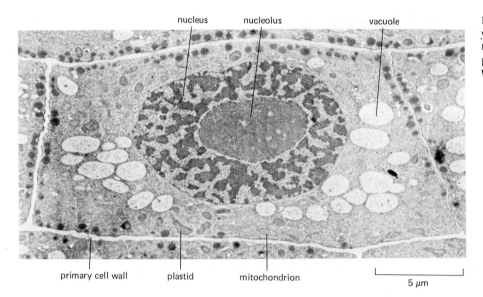

nucleus   nucelolus   vacuole

primary cell wall   plastid   mitochondrion   5 μm

Figure 20–2 Electron micrograph of a young root tip cell of onion, showing the main organelles and the thin primary cell wall. (Courtesy of Brian Wells.)

according to a common principle: they derive their tensile strength from long fibers held together by a matrix of protein and polysaccharide, which is highly resistant to compression. The same architectural principle (strong fibers resistant to tension embedded in an amorphous matrix resistant to compression) is used in the construction of animal bones (see p. 989) and in such common building materials as fiber glass and reinforced concrete. In the cell walls of higher plants, the fibers are generally made from the polysaccharide *cellulose*, the most abundant organic macromolecule on earth. The matrix is composed predominantly of two other sorts of polysaccharide—*hemicellulose* and *pectin*—together with structural glycoproteins (Figure 20–3). The fibers and matrix molecules are cross-linked by a combination of covalent bonds and noncovalent forces into a highly complex structure whose composition is generally cell-specific (Figure 20–4). The major molecules that form the cell wall are known for many cell types, but it is not known what all the component molecules are in any cell type, nor how they are cross-linked and organized in three dimensions.

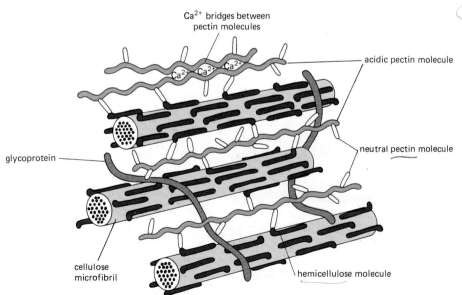

Ca²⁺ bridges between pectin molecules

acidic pectin molecule

neutral pectin molecule

glycoprotein

cellulose microfibril

hemicellulose molecule

Figure 20–3 Interconnections between the two major components of the primary cell wall, the cellulose microfibrils and the matrix. Hemicellulose molecules (for example, xyloglucans) are linked by hydrogen bonds to the surface of the cellulose microfibrils. Some of these hemicellulose molecules are cross-linked in turn to acidic pectin molecules (for example, rhamnogalacturonans) through short neutral pectin molecules (for example, arabinogalactans). Cell-wall glycoproteins are tightly woven into the texture of the wall to complete the matrix.

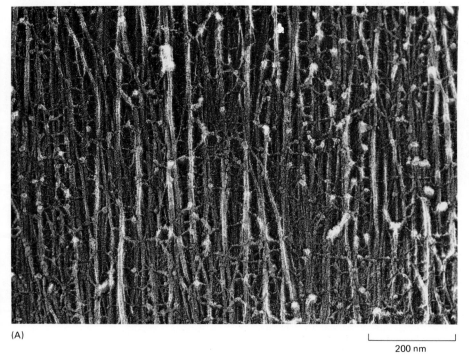

(A)

200 nm

Cellulose Microfibrils Are Cross-linked into a Complex Network with Hemicellulose, Pectin, and Glycoprotein Molecules[1,2]

A **cellulose** molecule consists of a linear chain of at least 500 glucose residues covalently linked to one another by $\beta1{\rightarrow}4$ glycosidic bonds, giving each molecule a ribbonlike structure that is stabilized by intramolecular hydrogen bonds. Intermolecular hydrogen bonds between adjacent cellulose molecules cause them to adhere strongly to one another in overlapping parallel arrays, forming a bundle of 60 to 70 cellulose chains that all have the same polarity. These highly ordered crystalline aggregates, many micrometers long, are called **cellulose microfibrils** (Figure 20–5).

Hemicellulose is the name given to a heterogeneous group of branched matrix polysaccharides that bind tightly but noncovalently to the surface of each cellulose microfibril and to one another, coating the microfibrils and helping to cross-link them via hydrogen bonds into a complex network (see Figure 20–3). There are many classes of hemicelluloses, but they all have a long, $\beta1{\rightarrow}4$-linked, linear backbone composed of one type of sugar, from which short side chains of other sugars protrude (Figure 20–6) It is the sugar molecules in the backbone that form hydrogen bonds along the outside of the cellulose microfibrils. Both the backbone sugar and the side-chain sugars are characteristic of the plant species and its stage of development.

The third major type of cell-wall polysaccharide, the **pectins,** are heterogeneous branched molecules that contain many negatively charged galacturonic acid residues (Figure 20–7). Because of their negative charge, pectins are highly hydrated and avidly bind cations. When Ca^{2+} is added to a solution of pectin molecules, it cross-links them to produce a semirigid gel. (It is pectin that is added to fruit juice to make jelly.) Such Ca^{2+} cross-links are thought to help hold cell-wall components together. Certain pectins are particularly abundant in the *middle lamella*, the specialized central region that cements together the cell walls of adjacent cells (see Figure 20–17). This layer ruptures in some places, forming the intercellular air gaps found in many tissues (Figure 20–8).

In addition to these three classes of polysaccharides, the primary cell wall also contains glycoproteins, which may account for up to 10% of the wall mass. The DNA that encodes the major glycoproteins has been cloned and sequenced, revealing that these proteins are unusual in containing many repeating sequences.

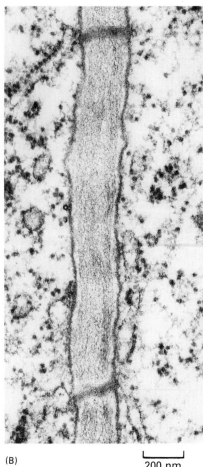

(B)

200 nm

Figure 20–4 (A) Electron micrograph of the primary cell wall of a carrot cell prepared by the fast-freeze, deep-etch technique (see p. 147). The cellulose microfibrils are cross-linked by a complex web of matrix molecules. Compare this micrograph with the model shown in Figure 20–3. (B) Thin section of a typical primary cell wall separating two adjacent plant cells. (A, courtesy of Brian Wells and Keith Roberts; B, courtesy of Jeremy Burgess.)

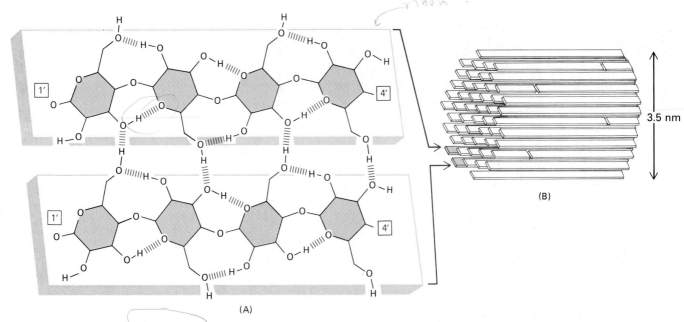

(A)

Figure 20–5 The structure of cellulose. (A) A small portion of two cellulose molecules, each consisting of a long planar chain of β1→4-linked glucose units, which may be many micrometers in length. Intramolecular hydrogen bonds stabilize each chain; intermolecular hydrogen bonds tightly cross-link adjacent chains. Other hydrogen bonds (not shown) cross-link each chain with other chains above and below it in the microfibril. (B) A cellulose microfibril in which many parallel cellulose molecules are hydrogen bonded together. In most higher plants a microfibril is about 3.5 nm in diameter, but in some algae it may be 10 times wider. Each cellulose molecule has a polarity (with a 1' and a 4' end), and all the molecules in each microfibril have the same polarity.

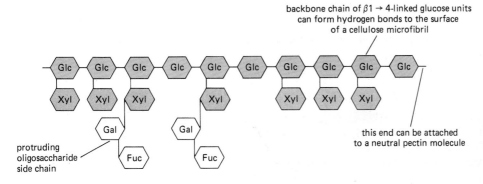

backbone chain of β1 → 4-linked glucose units can form hydrogen bonds to the surface of a cellulose microfibril

protruding oligosaccharide side chain

this end can be attached to a neutral pectin molecule

Figure 20–6 A hemicellulose molecule from the cell wall of a typical flowering plant. The backbone in this example (a xyloglucan) consists of a chain of glucose residues, which become hydrogen bonded to the surface of cellulose microfibrils. Hemicelluloses differ from cellulose, however, in having sugar side branches such as xylose. As indicated, other sugars, such as galactose and fucose, may also be present. In monocotyledenous plants the main hemicellulose has a backbone chain of β1→4-linked xylose residues instead of glucose residues.

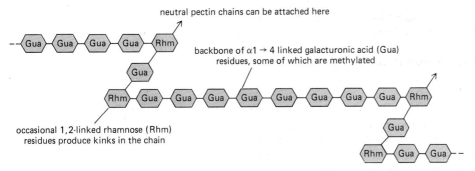

neutral pectin chains can be attached here

backbone of α1 → 4 linked galacturonic acid (Gua) residues, some of which are methylated

occasional 1,2-linked rhamnose (Rhm) residues produce kinks in the chain

Figure 20–7 An acidic pectin molecule (a rhamnogalacturonan) from the cell wall of a higher plant. Kinks in the straight backbone chain of negatively charged galacturonic acid residues are introduced by the occasional rhamnose residues, which also serve as attachment sites for neutral pectins, which cross-link the pectin molecule to hemicellulose (see Figure 20–3).

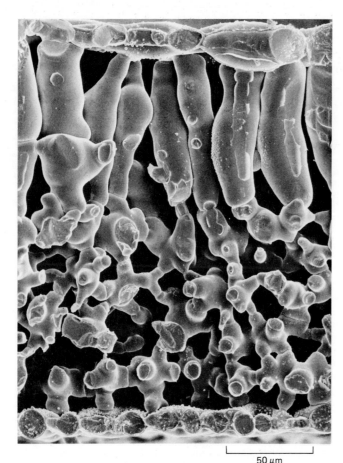

Figure 20–8 Scanning electron micrograph of the inside of a bean leaf that has been rapidly frozen and snapped open to expose the cells within. At the top is the single sheet of cells that forms the upper epidermis, and at the bottom is the lower epidermis. Between them lie large numbers of mesophyl cells. As these photosynthetic cells of the leaf grow and expand, their walls separate at defined regions along the middle lamella, resulting in an open network of cells, all of which have ready access to the carbon dioxide in the large air spaces that surround them. (From C.E. Jeffree, N.D. Read, V.A.C. Smith, and J.E. Dale, *Planta* 172:20–37, 1987.)

50 μm

Up to 30% of their residues are hydroxyproline, an amino acid formed by the post-translational hydroxylation of proline residues (as in collagen, see p. 811). Many short oligosaccharide side chains are attached to hydroxyproline and serine side chains, so that more than half the weight of each glycoprotein is carbohydrate. It is difficult to extract these glycoproteins without destroying the structure of the cell wall, suggesting that they are tightly woven into the complex three-dimensional polysaccharide network of the wall. They are thought to act like glue to increase the strength of the wall, and, significantly, the levels of some of the mRNAS that encode these glycoproteins rise dramatically as part of the host response to infection or wounding.

In order for a plant cell to grow or change its shape, the cell wall has to stretch or deform. Because of their crystalline structure, individual cellulose microfibrils are unable to stretch, and therefore such changes must involve either the sliding of microfibrils past one another and/or the separation of adjacent microfibrils. As we shall discuss later, the direction in which the growing cell enlarges depends on the way that strain-resisting cellulose microfibrils are deposited in the primary wall (see p. 1168).

The Limited Porosity of Cell Walls Restricts the Exchange of Macromolecules Between Plant Cells and Their Environment[3]

All cells take in nutrients and expel waste products across their plasma membrane. They also respond to chemical signals in their environment. In the case of plant cells, such molecules and signals must penetrate the cell wall. Since the matrix of the wall is a highly hydrated polysaccharide gel (the primary cell wall being 60% water by weight), water, gases, and small water-soluble molecules diffuse rapidly through it. The cross-linked structure of the cell wall only slightly impedes the diffusion of small molecules such as water, sucrose, or K^+. (Even in the case of a cell with a wall 15 μm thick, only 10% of the resistance to water flow between the cytoplasm and the external medium is contributed by the wall; the remaining 90% is contributed by the plasma membrane.) The average diameter of the spaces between the cross-linked macromolecules in most cell walls, however, is about 5 nm; this is small enough to make the movement of any globular macromolecule with a molecular weight much above 20,000 extremely slow. Therefore plants must subsist on molecules of low molecular weight, and any intercellular signaling molecules that have to pass through the cell wall must also be small and water-soluble. In fact, most of the known plant signaling molecules, such as the growth-regulating substances—auxins, cytokinins, and gibberellins—have molecular weights of less than 500 (see p. 1181).

The Tensile Strength of Cell Walls Allows Plant Cells to Generate an Internal Hydrostatic Pressure Called Turgor[4]

The mechanical strength of the cell wall allows plant cells to survive in an extracellular environment that is hypotonic with respect to the cell interior. The extracellular fluid in higher plants consists of the aqueous phase in all of the cell walls plus the fluid in the long tubes formed by the cell walls of dead *xylem vessel elements* (see p. 1153). These tubes carry water (the *transpiration stream*) from the roots to sites of evaporation, mainly in the leaves. Although the extracellular fluid contains more solutes than does the dilute solution in the plant's external milieu (for example, soil), it is still hypotonic in comparison to the intracellular fluid. This can be readily demonstrated by removing the cell wall with cellulases and other wall-degrading enzymes and then observing the behavior of the wall-less cell, called a *protoplast* (see Figure 20–71). If such a spherical protoplast is exposed to the hypotonic fluid that normally bathes the plant cell, it takes up water by osmosis, swells, and eventually bursts (Figure 20–9). A walled cell placed in the same environment, in contrast, takes up water but can swell to only a limited extent. The cell develops an internal hydrostatic pressure that pushes outward on the cell wall, rather like an inner tube pressing against a bicycle tire. This hydrostatic pressure brings the cell to osmotic equilibrium and prevents any further net influx of water. (For a detailed discussion of osmosis refer to Panel 6–1, p. 308.)

The outward **turgor pressure** (or **turgor**) in all plant cells, caused by the osmotic imbalance between its intracellular and extracellular fluids, is vital to plants. It is the main driving force for cell expansion during growth and provides much of the mechanical rigidity of living plant tissues. Compare the wilted leaf of a dehydrated plant with the turgid leaf of a well-watered one.

Plant Cell Growth Depends on both Turgor Pressure and a Controlled Yielding of the Wall[5]

Turgor pressure does far more than keep plant tissues distended. Whenever the cell wall yields to the internally generated turgor pressure, an irreversible increase in cell volume results; in other words, the cell grows. Cell growth can occur only if the turgor pressure in the cell exceeds the local tensile strength of the wall. In principle, then, the plant cell could use two strategies to grow: it could increase its turgor, or it could weaken the cell wall in local areas. There is good evidence that plant cells adopt the second strategy and weaken their walls by a variety of mechanisms, including the local secretion of H^+ into the wall by a H^+-pumping ATPase in the plasma membrane. Although the molecular details are not clear, it is thought that the locally lowered pH reduces the number of weak bonds holding the wall together, so that its component macromolecules slip past each other under the influence of turgor pressure. To further facilitate wall growth, other, more complex changes also occur, including the activation of enzymes that hydrolyze glycosidic and other covalent bonds.

In most cases the increase in cell volume is nonuniform with respect to the cell's contents, the major increase being accounted for by enlargement of the *vacuole* rather than of the cytoplasm. The retention of a nearly constant amount of nitrogen-rich cytoplasm reduces the metabolic "cost" of making large cells (Figure 20–10). This is a particular advantage for plants, for which nitrogen is often

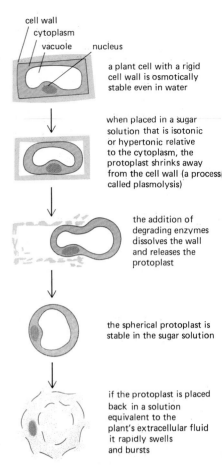

cell wall
cytoplasm
vacuole   nucleus

a plant cell with a rigid cell wall is osmotically stable even in water

when placed in a sugar solution that is isotonic or hypertonic relative to the cytoplasm, the protoplast shrinks away from the cell wall (a process called plasmolysis)

the addition of degrading enzymes dissolves the wall and releases the protoplast

the spherical protoplast is stable in the sugar solution

if the protoplast is placed back in a solution equivalent to the plant's extracellular fluid it rapidly swells and bursts

Figure 20–9 A plant cell without its cell wall, called a protoplast, is osmotically unstable and will swell and burst if it is placed in water or the hypotonic extracellular fluid that bathes plant cells. Inside its rigid cell wall, however, it can swell only as far as the boundaries of the wall. The pressure developed by the cell pressing against the wall keeps the cell turgid and in osmotic equilibrium, so that no more water enters the cell (see Panel 6–1, p. 308).

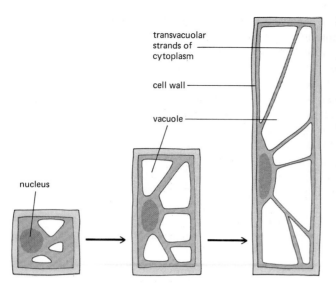

transvacuolar
strands of
cytoplasm

cell wall

vacuole

nucleus

Figure 20–10 A large increase in cell volume can be obtained without increasing the volume of the cytosol. Wall relaxation orients a turgor-driven cell expansion that accompanies the uptake of water into an expanding vacuole. The cytoplasm is eventually confined to a thin peripheral layer, which is connected to the nuclear region by transvacuolar strands of cytoplasm that contain bundles of actin filaments (see Figure 20–53).

a limiting nutrient. In order to maintain the turgor pressure required for continuing cell expansion, solutes must be actively accumulated in the growing vacuole to maintain its osmolarity.

Plant cell growth, however, is more subtle than a simple balloonlike inflation of the cell. Cells also need to be appropriately shaped, and in practice cell growth is a blend of two complex processes. First there is a cyclical process, with appropriate feedback controls, of progressive cell-wall yielding followed by the reestablishment of turgor. Second there is local control of wall yielding, which allows some regions of the wall to remain rigid while others expand, thereby determining the shape of the cell. The basis of this control of cell shape resides in cytoplasmic events to be described later (see p. 1169).

Turgor Is Regulated by Feedback Mechanisms That Control the Concentrations of Intracellular Solutes[6]

In view of the importance of turgor pressure to the plant, it is not surprising that plant cells have evolved sensitive mechanisms for regulating their size. The turgor pressure varies greatly from plant to plant and from cell to cell, from the equivalent of half an atmosphere in some large-celled algae to nearly 50 atmospheres in the stomatal guard cells of some higher plants (Figure 20–11). Cells can increase their turgor pressure by increasing the concentration of osmotically active solutes in the cytosol—either by pumping them in from the extracellular fluid across the plasma membrane or by generating them from osmotically inactive polymeric stores, usually in the vacuole. In both cases, feedback loops monitor the level of turgor and regulate it.

How do such feedback control systems work? Experiments indicate that a "turgor-pressure detector" in the plasma membrane induces inward ion transport,

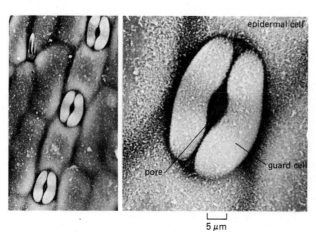

epidermal cell

pore

guard cell

5 μm

Figure 20–11 Scanning electron micrographs of open stomata in the leaf epidermis of a tropical grass, shown at two magnifications. **Stomata** are pores formed in the leaf surface by two guard cells, whose turgor-regulated movements control the size of the pore and thereby regulate the gas exchange between the leaf and the environment. In most plants the stomata are open during the day to admit carbon dioxide and to allow the products of photorespiration to escape; at night they are usually closed. The epidermal cells are coated by a waterproof cuticle with an outer waxy layer (see also Figure 20–18). (Courtesy of H.W. Woolhouse and G.J. Hills.)

most commonly the active pumping of K^+ into the cell, in response to a sudden fall in turgor pressure, while a sudden increase in turgor brings about a K^+ efflux. These responses are very rapid and reflect changes in the activity of specific transport proteins in the plasma membrane.

Membrane-bound turgor-pressure detectors presumably also induce alterations in the rate at which osmotically active solutes are synthesized in the cytoplasm and vacuole. These changes occur much more slowly but are critical to turgor regulation in plants exposed to environments with extreme or fluctuating osmotic properties. Plants that live in a high-salt habitat, for example, must accumulate very high internal solute concentrations in order to maintain turgor. Since the accumulation of ions such as K^+ to such high levels would probably alter the activities of vital enzymes, plant cells in these environments accumulate specific organic solutes instead: polyhydroxylic compounds such as glycerol or mannitol, amino acids such as proline, or N-methylated derivatives of amino acids such as glycinebetaine. These solutes can reach very high concentrations (0.5 M) in the cytosol without affecting the cell's metabolism. The vacuole and its contents are closely involved in the regulation of turgor pressure in response to environmental fluctuations (see p. 1163).

Regulated changes in turgor also generate the limited movements of plants. Stomatal guard cells, for example, control the rate of exchange of gases between leaves and the surrounding air by movements that open or close the pore that they form (see Figure 20–11). During the day, when the stomata are open, light activates K^+-selective pumps in the plasma membrane of the guard cells, causing an influx of K^+ that increases the turgor pressure so that the guard cells swell outward to open the pores. Very rapid changes in turgor in strategically located cells account for more spectacular movements such as trap closure in some carnivorous plants and the rapid movements of flower parts in some pollination mechanisms. The changes in turgor responsible for these phenomena are brought about by drastic increases in the permeability of membranes in critically positioned cells that act as a turgor-regulated "hinge." How a slight touch on a receptive surface induces a turgor collapse in these cells is not understood; it has been suggested that voltage-gated ion channels may be involved.

The Cell Wall Is Modified During the Creation of Specialized Types of Cell[7]

There are large differences between groups of land plants in their structures and reproductive strategies (Figure 20–12). Nevertheless, they are all constructed on similar principles from a very small repertoire of tissues and cell types. Almost all plants are constructed along modular lines, a typical module consisting of a stem,

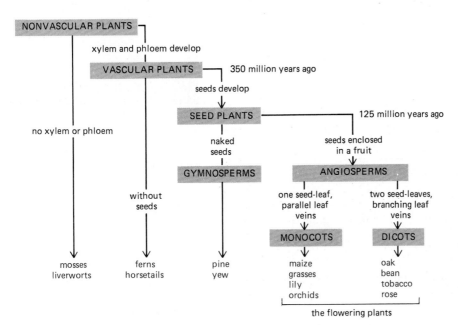

Figure 20–12 The evolution of land plants. A major division in the plant kingdom, excluding the algae, is between the *vascular plants* (which have phloem and xylem tissues for transport) and their progenitors, the *nonvascular plants* (such as mosses), which are small and relatively primitive. As indicated, vascular plants took a further step about 350 million years ago when seeds evolved. The seed provides a protective environment for the developing embryo, which can remain dormant until it encounters the right conditions for further development. Because seed-bearing vascular plants have become the dominant plants on land, they form the main focus of this chapter.

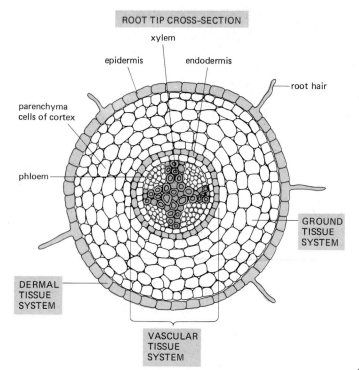

ROOT TIP CROSS-SECTION

xylem
epidermis — endodermis
parenchyma
cells of cortex — root hair
phloem
GROUND
TISSUE
SYSTEM
DERMAL
TISSUE
SYSTEM
VASCULAR
TISSUE
SYSTEM

Figure 20–13 The various organs of a higher plant (for example, leaf, stem, and root) are each composed of three easily recognizable *tissue systems*—vascular, ground, and dermal. In this cross-section of a root tip, the *vascular tissue system* is embedded in the *ground tissue system*, which is in turn enclosed by the *dermal tissue system*. The same three tissue systems, in different arrangements, make up all the parts of a higher plant. Each is composed of a relatively small number of common cell types, some of which are illustrated in Panel 20–1.

a leaf, and a bud (see Figure 20–58). Moreover, all contain the same specialized cell types organized into the same three main tissue systems: *dermal* (providing covering), *ground* (providing support and nutrition), and *vascular* (providing fluid transport) (Figure 20–13).

The major cell types are shown in Panel 20–1. They arise from cells with a *primary cell wall* by a process of cell expansion followed by cell differentiation. As cells differentiate, the primary cell wall is elaborated to form the *secondary cell wall*. In some cases this involves simply adding more layers of cellulose, while in other cases new layers of different composition are laid down. The cellulose molecules laid down in secondary walls are in general much longer (~15,000 glucose residues) than those found in primary walls (500 to 5000 glucose residues). Moreover, the highly hydrated pectin components characteristic of the primary cell wall are largely replaced by other polymers, with the result that the secondary wall is considerably more dense and less hydrated than the primary wall.

The secondary cell walls provide most of the plant's mechanical support. They also provide essential components of many animals' diet and are the basis for such products as wood and paper. The form and composition of the final cell wall are closely related to the function of the particular specialized cell type: each cell type is readily distinguishable by its morphology, as exemplified by the highly species-specific secondary cell wall pattern laid down on the surface of mature pollen grains (Figure 20–14).

All major changes in the composition and architecture of both primary and secondary cell walls reflect cytoplasmic events, a principle that is most clearly illustrated by xylem vessel development. During their initial differentiation, xylem vessel elements in young, growing tissues lay down cellulose-rich wall thickenings in patterns defined by bundles of cytoplasmic microtubules that lie in the cortex just under the plasma membrane. These microtubule bundles are arranged generally in helical or hooplike patterns (Figure 20–15) and arise from a regrouping of the more evenly distributed microtubules that constitute the normal cortical array (see p. 1169). This is a specific example of the general rule, discussed later (p. 1169), that extracellular cellulose microfibrils are deposited parallel to cortical cytoplasmic microtubules.

The regions of thickened cellulose wall in developing xylem cells later become strengthened by the deposition of *lignin*, a highly insoluble polymer of aromatic phenolic units that forms an extensive cross-linked network within the cell wall

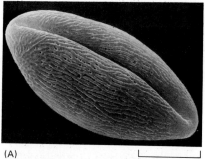

(A)   10 μm

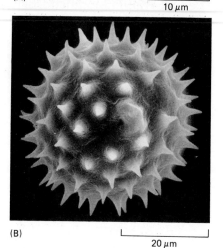

(B)   20 μm

Figure 20–14 Scanning electron micrographs of pollen grains: (A) petunia, (B) sunflower. The sculptured wall is made from sporopollenin, a complex and very tough polymer of hydrocarbons that forms a characteristic and species-specific pattern. Pores exist in the wall, through which the pollen tube will emerge when the grain germinates. (Courtesy of Colin MacFarlane and Chris Jeffree.)

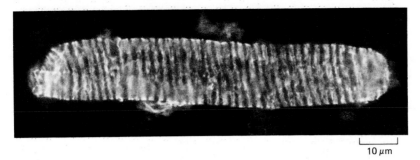

Figure 20–15 Microtubules in the cortex of a cell from the developing xylem of a pea shoot. Immunofluorescence staining shows the grouping of cortical microtubules into a helical array, which defines the area of the cell wall that will be thickened initially with cellulose and then with lignin (see Panel 20–1). The cell is large, and the depth of focus allows only one side of the helical array to be seen. (Courtesy of Ian Roberts.)

10 μm

and is the main constituent of wood. The localized removal of material from the end walls finally creates the low-resistance, rigid vessels used to transport water in the xylem (Figure 20–16). A similarly dramatic remodeling of the primary cell walls takes place during the development of phloem sieve tubes in the vascular tissues of plants (see Figure 20–16 and Panel 20–1).

The secondary cell wall is usually deposited between the plasma membrane and the primary cell wall, sometimes in successive layers (Figure 20–17). In other cases, however, special macromolecules are also deposited either *within* the primary wall (as is lignin in xylem cells) or on its outer surface. Epidermal cells, for example, which cover the external surface of the plant, usually have thickened primary cell walls whose outer face is coated with a thick, waterproof *cuticle*, which helps to protect the plant from infection, mechanical damage, water loss, and harmful ultraviolet light (see Panel 20–1, pp. 1148–1149). The cuticle is secreted as the epidermal cells differentiate. It is made primarily of *cutin* (or the related compound suberin in bark), which is a polymer of long-chain fatty acids that forms an extensive cross-linked network on the plant surface. The cutin layer is frequently impregnated and overlaid by a complex mixture of *waxes*, which are esters between long-chain alcohols and fatty acids (Figure 20–18). The plant cell cuticle has a very different composition from insect and crustacean cell cuticles, which are composed of proteins and polysaccharides.

Even the Mature Cell Wall Is a Dynamic Structure[8]

The composition and architecture of the cell wall in a mature plant are not fixed: components can be added or removed, and the cross-links between components can change. The localized removal of wall material during the development of the end walls of both xylem vessels and phloem sieve-tube elements provides a particularly striking example of such changes (see Figure 20–16).

Figure 20–16 Two examples of the ways in which the cell wall is elaborated and modified during the formation of specialized cell types. (A) A schematic longitudinal section of a small developing vessel element of the xylem. This cell forms annular wall thickenings, but many other patterns are also found. Ultimately the protoplast and end walls disappear to create an open-ended tube. The mature element is dead, having lost its protoplast. (B) A schematic longitudinal section of the development of a sieve-tube element in phloem. The primary cell wall becomes thickened, and the end walls become perforated to form the sieve plates that connect adjacent elements of the sieve tube. The mature cell retains its plasma membrane, but the nucleus and much of the cytoplasm are lost.

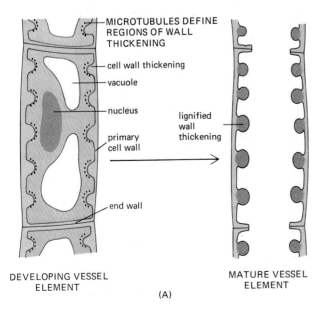

MICROTUBULES DEFINE REGIONS OF WALL THICKENING

cell wall thickening

vacuole

nucleus

lignified wall thickening

primary cell wall

end wall

DEVELOPING VESSEL ELEMENT

MATURE VESSEL ELEMENT

(A)

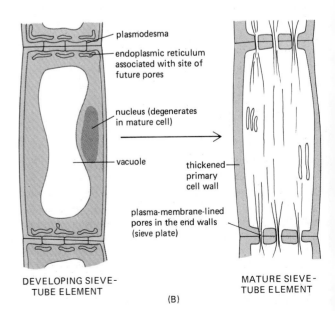

plasmodesma

endoplasmic reticulum associated with site of future pores

nucleus (degenerates in mature cell)

vacuole

thickened primary cell wall

plasma-membrane-lined pores in the end walls (sieve plate)

DEVELOPING SIEVE-TUBE ELEMENT

MATURE SIEVE-TUBE ELEMENT

(B)

THE PLANT

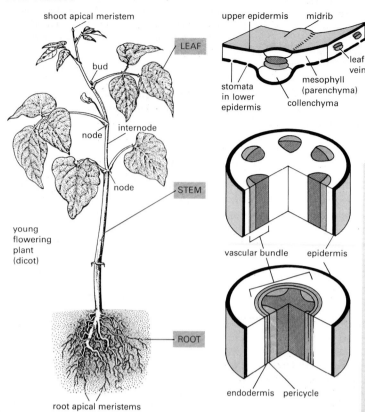

shoot apical meristem

LEAF

bud

node — internode

node

STEM

young
flowering
plant
(dicot)

ROOT

root apical meristems

upper epidermis — midrib

leaf
vein

stomata
in lower
epidermis

mesophyll
(parenchyma)

collenchyma

vascular bundle — epidermis

endodermis — pericycle

The young flowering plant shown on the left is constructed from three main types of organs: leaves, stems, and roots. Each plant organ is in turn made from three tissue systems: ground, dermal, and vascular (see Figure 20-13).

All three tissue systems derive ultimately from the cell proliferative activity of the shoot or root apical meristems, and each contains a relatively small number of specialized cell types. These three common tissue systems, and the cells that comprise them, are described in this panel.

THE THREE TISSUE SYSTEMS

Cell division, growth, and differentiation give rise to tissue systems with specialized functions.

DERMAL TISSUE (▬▬) This is the plant's protective outer covering in contact with the environment. It facilitates water and ion uptake in roots and regulates gas exchange in leaves and stems.

VASCULAR TISSUE Together the phloem (▭) and the xylem (▬) form a continuous vascular system throughout the plant. This tissue conducts water and solutes between organs and also provides mechanical support.

GROUND TISSUE (▭) This packing and supportive tissue accounts for much of the bulk of the young plant. It functions also in food manufacture and storage.

GROUND TISSUE

The ground tissue system contains three main cell types called parenchyma, collenchyma, and sclerenchyma.

Parenchyma cells are found in all tissue systems. They are living cells, generally capable of further division, and have a thin primary cell wall. These cells have a variety of functions. The apical and lateral meristematic cells of shoots and roots provide the new cells required for growth. Food production and storage occur in the photosynthetic cells of the leaf and stem (called mesophyll cells); storage parenchyma forms the bulk of most fruit and vegetables. Because of their proliferative capacity, parenchyma cells also serve as stem cells for wound healing and regeneration.

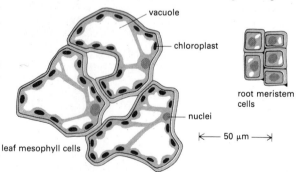

vacuole

chloroplast

root meristem
cells

nuclei

|← 50 μm →|

leaf mesophyll cells

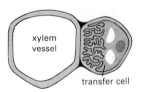

xylem
vessel

transfer cell

Transfer cell, a specialized form of the parenchyma cell, is readily identified by elaborate ingrowths of the primary cell wall. The increase in the area of the plasma membrane beneath these walls facilitates the rapid transport of solutes to and from cells of the vascular system.

Collenchyma are living cells similar to parenchyma cells except that they have much thicker cell walls and are usually elongated and packed into long ropelike fibers. They are capable of stretching and provide mechanical support in the ground tissue system of the elongating regions of the plant. Collenchyma cells are especially common in subepidermal regions of stems.

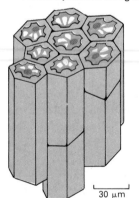

30 μm

typical locations
of supporting groups
of cells in a stem

sclerenchyma fibers

vascular bundle

collenchyma

Sclerenchyma, like collenchyma, have strengthening and supporting functions. However, they are usually dead cells with thick, lignified secondary cell walls that prevent them from stretching as the plant grows. Two common types are *fibers* (see Figure 20-17), which often form long bundles, and *sclereids*, which are shorter branched cells found in seed coats and fruit.

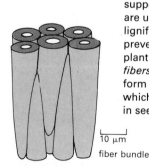

10 μm

fiber bundle

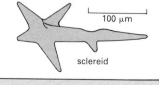

100 μm

sclereid

DERMAL TISSUE

The epidermis is the primary outer protective covering of the plant body. Cells of the epidermis are also modified to form stomata and hairs of various kinds.

Epidermis

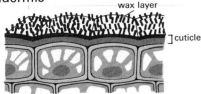

wax layer

cuticle

The epidermis (usually one layer of cells deep) covers the entire stem, leaf and root of the young plant. The cells are living, have thick primary cell walls, and are covered on their outer surface by a special cuticle with an outer waxy layer (see Figure 20-18). The cells are tightly interlocked in different patterns.

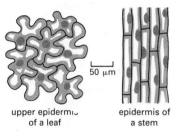

50 μm

upper epidermis
of a leaf

epidermis of
a stem

Stomata

guard cells

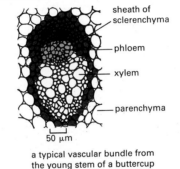

air space

5 μm

Stomata are openings in the epidermis, mainly on the lower surface of the leaf, that regulate gas exchange in the plant. They are formed by two specialized epidermal cells called guard cells (see Figure 20-11), which regulate the diameter of the pore. Stomata are distributed in a distinct species-specific pattern within each epidermis.

Vascular bundles

In roots there is usually a single vascular bundle, but in stems there are several. These are arranged with strict radial symmetry in dicots, but they are more irregularly dispersed in monocots.

sheath of
sclerenchyma

phloem

xylem

parenchyma

50 μm

a typical vascular bundle from
the young stem of a buttercup

Hairs (or trichomes) are appendages derived from epidermal cells. They exist in a variety of forms and are commonly found in all plant parts. Hairs function in protection, absorption, and secretion.
examples:

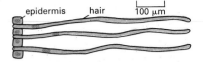

epidermis   hair   100 μm

young, single-celled hairs in the epidermis of the cotton seed (see Figure 20-45). When these grow, the walls will be secondarily thickened with cellulose to form cotton fibers.

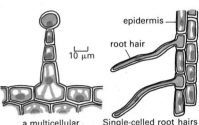

epidermis

root hair

10 μm

a multicellular
secretory hair from
a geranium leaf

Single-celled root hairs
have an important function
in water and ion uptake.

VASCULAR TISSUE

The phloem and the xylem together form a continuous vascular system throughout the plant. In young plants they are usually associated with a variety of other cell types in *vascular bundles*. Both phloem and xylem are complex tissues. Their conducting elements are associated with parenchyma cells that maintain and exchange materials with the elements. In addition, groups of collenchyma and sclerenchyma cells provide mechanical support.

Phloem

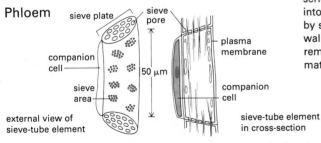

sieve plate

sieve
pore

companion
cell

plasma
membrane

sieve
area

50 μm

companion
cell

external view of
sieve-tube element

sieve-tube element
in cross-section

Phloem is involved in the transport of organic solutes in the plant. The main conducting cells (elements) are aligned to form tubes called *sieve tubes*. The sieve-tube elements at maturity are living cells, interconnected by perforations in their end walls formed from enlarged and modified plasmodesmata (sieve plates). These cells retain their plasma membrane, but they have lost their nuclei and much of their cytoplasm; they therefore rely on associated *companion cells* for their maintenance. These companion cells have the additional function of actively transporting soluble food molecules in and out of sieve-tube elements through porous sieve areas in the wall.

Xylem carries water and dissolved ions in the plant. The main conducting cells are the vessel elements shown here, which are dead cells at maturity that lack a plasma membrane. The cell wall has been secondarily thickened and heavily lignified. As shown below, its end wall is largely removed, enabling very long, continuous tubes to be formed.

Xylem

The wall that connects the serially aligned vessel elements into a tube may be perforated by several small pores. This wall, however, is often removed completely in more mature vessels.

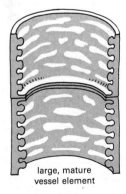

small vessel
element in root tip

large, mature
vessel element

The vessel elements are closely associated with xylem parenchyma cells, which actively transport selected solutes in and out of the elements across their plasma membrane.

xylem parenchyma cells

vessel element

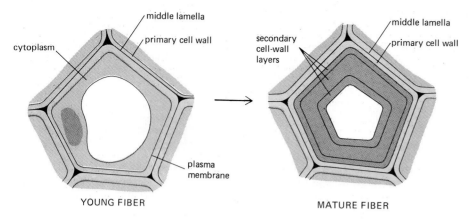

YOUNG FIBER   MATURE FIBER

Figure 20–17 Secondary cell-wall deposition shown schematically in a cross-section of a fiber cell. Here, three new cell-wall layers have been laid down underneath the primary cell wall. Because the net orientation of the cellulose microfibrils is different in each layer, a strong plywoodlike effect is created. In many mature fibers, as shown here, the cell inside the wall dies.

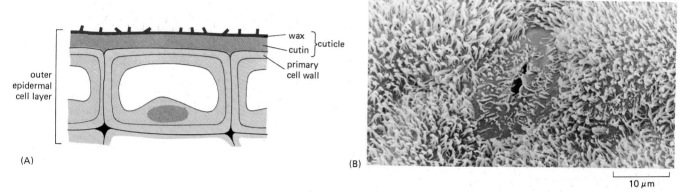

(A)   (B)

Figure 20–18 (A) Drawing of a section of a typical mature epidermal cell of a leaf. On the outer face of the thick primary cell wall, a waterproof cutin layer has been deposited together with a layer of wax to produce a cuticle. The wax layer often gives the cuticle an elaborate sculptured shape. (B) Scanning electron micrograph of the lower epidermis of a pea leaf, showing the pattern of wax deposition around a stomatal pore. (Courtesy of Paul Linstead.)

The dynamic behavior of mature cell walls is also illustrated by the changes that occur during *abscission*—the controlled loss of a part of the plant, such as a dead leaf. When a leaf dies, cellular polymers are degraded and sugars, amino acids, and ions are recycled to the plant. In addition, *ethylene* gas is produced in small quantities by the senescing leaf. A particular layer of cells in a zone located between the base of the leaf stalk and the stem (the *abscission zone*) responds to a complex and ill-understood combination of ethylene and other endogenous plant growth regulators (see Figure 20–67) by making and secreting wall-degrading enzymes, such as pectinase and cellulase, which act locally to partially dissolve the cell walls in a separation layer (Figure 20–19). At the same time, the layer of cells on the stem side of the abscission zone deposits water-resistant suberin to protect the "wound" that will be left when the leaf finally drops off as a result of enzymatic digestion.

A similar localized cell response occurs during fruit ripening. Here low levels of ethylene (one part per million) stimulate target cells in fruit (oranges or bananas, for example) to secrete pectin-degrading enzymes that weaken the adhesions between adjacent cells. This results in a softening, or "ripening," of the fruit.

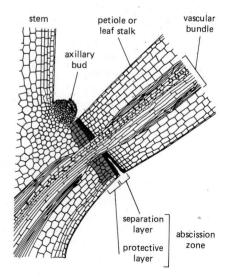

Figure 20–19 Specialized layers of small cells at the base of the leaf stalk are involved in leaf abscission. Two or three layers of small cells secrete wall-degrading enzymes that lead to cell separation. The petiole parts from the stem in this region, where cell walls are thinner, and few lignified cells cross the abscission zone. Cells that remain exposed at abscission deposit suberin and form a protective layer over the wound.

Summary

Higher plants are composed of large numbers of cells cemented together in fixed positions by the rigid cell walls that surround them. Many of the unique features of plants are related directly or indirectly to the presence of this cell wall, whose composition and appearance reflect the different plant cell types and their functions. The underlying structure of all cell walls, however, is remarkably consistent. Tough fibers of cellulose are embedded in a highly cross-linked matrix of polysaccharides, such as pectins and hemicelluloses, and glycoproteins. The result is a primary cell wall that possesses great tensile strength and is permeable only to relatively small molecules. When placed in water, a plant cell without a wall (a protoplast) will take up water by osmosis, swell, and burst. Inside its cell wall, however, it swells and presses against the wall, creating a pressure known as turgor. Turgor is closely regulated and is vital for both cell expansion and the mechanical rigidity of the young plant.

A relatively small number of specialized cell types is used in the construction of the great variety of higher plants. During the formation of these specialized cell types—for example, cells of the two vascular tissues, xylem and phloem—the cell wall is extensively modified. Certain regions of the wall may be strengthened, often by the addition of one or more new layers to form a secondary cell wall. Other regions of the wall may be selectively removed, as are the end walls in the formation of a conducting tube from a long row of cylindrical cells. These changes in the cell wall are controlled by temporal and spatial changes in the cytoplasm of the developing cell. The cell wall is a dynamic structure whose composition and form can change markedly, not only during cell growth and differentiation but also after cells have matured.

Transport Between Cells

In the preceding section we described how the rigid wall of a plant cell limits the exchange of molecules between the cell's cytoplasm and its environment. The cell wall also restricts the ways in which a plant cell can interact and communicate with its neighbors. Plant cells have evolved ingenious ways to overcome these limitations, however. Direct cell-cell communication is just as important in multicellular plants as it is in multicellular animals, and special channels have evolved to connect the cytoplasm of one plant cell to that of its neighbors—thereby allowing the controlled passage of ions and small molecules. In vascular plants, moreover, long columns of cylindrical cells become connected end to end by perforations to generate long tubes through which water and nutrients flow.

Plant Cells Are Connected to Their Neighbors by Special Cytoplasmic Channels Called Plasmodesmata[9]

Except for a very few specialized cell types, every living cell in a higher plant is connected to its living neighbors by fine cytoplasmic channels, each of which is called a **plasmodesma** (plural, **plasmodesmata**), which pass through the intervening cell walls. As shown in Figure 20–20, the plasma membrane of one cell is continuous with that of its neighbor at each plasmodesma. A plasmodesma is a roughly cylindrical, membrane-lined channel with a diameter of 20 to 40 nm. Running from cell to cell through the center of most plasmodesmata is a narrower cylindrical structure, the *desmotubule*, which electron micrographs show to be continuous with elements of the endoplasmic reticulum membrane of each of the connected cells (Figure 20–21). Between the outside of the desmotubule and the inner face of the cylindrical plasma membrane is an annulus of cytosol (see Figure 20–20), which often appears to be constricted at each end of the plasmodesma. These constrictions may be of great significance, for they are located at sites where each cell could, in principle, regulate the flux of molecules through the annulus that joins the two cytosols.

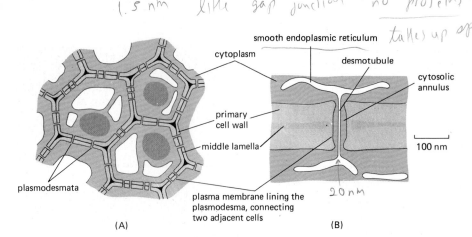

Handwritten annotations at top: 1.5 nm · like gap junction · no proteins · only small molecules + hormones. · takes up space · 20 nm

smooth endoplasmic reticulum

cytoplasm

desmotubule

cytosolic annulus

primary cell wall

middle lamella

plasmodesmata

plasma membrane lining the plasmodesma, connecting two adjacent cells

100 nm

(A)   (B)

Figure 20–20 (A) Cytoplasmic channels called plasmodesmata pierce the cell wall and connect all cells in the plant together. (B) The plasmodesma is lined with plasma membrane common to two connected cells and usually contains a fine tubular structure, the desmotubule, derived from smooth endoplasmic reticulum.

Plasmodesmata are normally formed around elements of endoplasmic reticulum that become trapped during cytokinesis within the new cell wall that will bisect the parental cell (see Figure 13–71). Plasmodesmata are also found in nondivision walls separating nonsister cells, however, and they can increase in number during cell growth, demonstrating that they can arise de novo.

Plasmodesmata Allow Molecules to Pass Directly from Cell to Cell[9,10]

The measured rates of solute movement from plant cell to plant cell are much faster than can be accounted for by plasma membrane permeability alone, and it seems likely that the cytoplasmic continuity provided by plasmodesmata is responsible. Besides their suggestive structure, there is good evidence that plasmodesmata function in intercellular communication. They are especially common in the walls of columns of cells that lead toward sites of intense secretion, such as in nectar-secreting glands; in such cells there may be 15 or more plasmodesmata per square micrometer of wall surface, whereas there is often less than 1 per square micrometer in other cell walls.

The most direct evidence for intercellular transport via plasmodesmata, however, comes from experiments involving the intracellular injection of dyes or pulses of electric current. If fluorescent dyes that do not easily cross the plasma membrane are microinjected into one cell of a leaf, they readily move into neighboring cells. Similarly, if pulses of electrical current are injected through an electrode into one cell, a measuring electrode in an adjacent cell will detect the same pulses, albeit in attenuated form. The degree of attenuation is found to vary both with the density of plasmodesmata in the intervening walls and with the number of cells between the two electrodes; since a measuring electrode placed outside the cell membrane does not detect these pulses, they must follow an intracellular pathway.

It seems likely, therefore, that plasmodesmata mediate transport between adjacent plant cells, much as gap junctions mediate transport between adjacent animal cells (see p. 798). Evidence obtained by microinjecting fluorescent dyes coupled to peptides of different sizes suggests that plasmodesmata allow the passage of molecules with molecular weights of less than about 800 (Figure 20–22), which is about the same as the molecular-weight cutoff for gap junctions. Several lines of evidence suggest that plasmodesmatal transport is subject to complex regulation. Thus dye-injection experiments show that there can be barriers to the movement of even low-molecular-weight materials between certain tissue systems despite the presence of apparently normal plasmodesmatal connections. No dye movement is observed, for example, between cells of the root cap and the root apex, or between epidermal and cortical cells in either roots or shoots. The regulatory mechanisms that restrict communication in these cases are not understood, although there is some evidence that both Ca^{2+} and protein phosphorylation may be involved. Conversely, certain plant viruses can enlarge plasmodesmata in order to use this route to pass from cell to cell (Figure 20–23). It has been

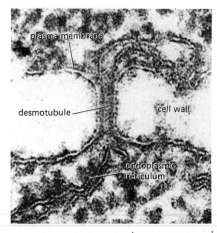

plasma membrane

desmotubule

cell wall

endoplasmic reticulum

(A)   0.1 μm

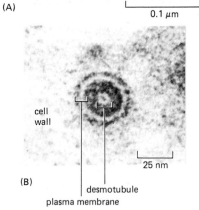

cell wall

25 nm

(B)   desmotubule

plasma membrane

Figure 20–21 Plasmodesmata as seen in the electron microscope. (A) Longitudinal section of a plasmodesma from a water fern. The plasma membrane lines the pore and is continuous from one cell to the next. Endoplasmic reticulum and its association with the central desmotubule can be seen. (B) A similar plasmodesma in cross-section. (Courtesy of R. Overall.)

demonstrated for tobacco mosaic virus that this ability to override the normal controls on plasmodesmatal transport resides in a single 30,000-dalton viral protein (P30): a defective virus that cannot spread from cell to cell in normal tobacco plants will do so in transgenic tobacco plants that contain the P30 gene and express low levels of the P30 protein.

The Fluids in a Vascular Plant Are Segregated into One Large Intracellular Compartment and One Large Extracellular Compartment

Plasmodesmata connect both the plasma membranes of adjacent cells and their cytoplasm and thereby convert a plant into a interconnected commune of living protoplasts. It follows that the whole plant body can be viewed as consisting of two compartments separated by a continuous plasma membrane: (1) an intracellular compartment, known as the **symplast,** which is made up of the total protoplast community—including the sieve tubes of the phloem—bounded by the combined plasma membranes of all these living cells; and (2) an extracellular compartment, or **apoplast,** composed of all the cell walls, the empty dead cells of the xylem tubes, and the water contained in them (Figure 20–24). Both apoplast and symplast have their own internal transport processes, can be locally sealed off, and are subject to local modifications that regulate the flow of fluids and materials between them.

Photosynthetic and Absorptive Cells Are Linked by Vascular Tissues That Contain Xylem and Phloem[11]

Multicellularity permits division of labor, with different cell types supporting one another by virtue of the specialized functions that emerge during their differentiation. Two examples of plant cells with specialized functions are **photosynthetic cells,** which contain chloroplasts and are sources of carbon-containing compounds, such as sucrose, and **absorptive cells,** which take up mineral nutrients from the environment. In the majority of higher plants, these two functions cannot be carried out by the same cells because one process requires light and the other occurs below the soil surface in darkness. Each process also has other special requirements. Photosynthesis requires a special microenvironment in which the relative humidity and carbon dioxide supply are closely controlled. This is achieved by the *stomatal pores* located in the cuticle-covered epidermis of the leaf, which are opened and closed by the turgor-regulated movements of their guard cells (see Figure 20–11). Absorption, in contrast, requires a very large surface area, which is provided by the roots, as well as membrane transport systems that are often augmented by symbiotic associations with microorganisms. The photosynthetic and absorptive cells must cross-feed each other, in addition to supplying other regions of the plant with both the organic compounds and minerals for biosynthesis. To facilitate this long-distance transport, photosynthetic and absorptive tissues are linked by a network of vascular tissue containing two types of conducting elements—the *xylem* and the *phloem*—each consisting of chains of cylindrical cells that become joined end to end to form tubes (see Panel 20–1, pp. 1148–1149, and Figure 20–16).

Water and Dissolved Salts Move Through the Xylem[12]

The **xylem** vessels consist of long tubes in the vascular tissue. Although they arise from thin-walled living cells, the mature functional xylem elements are dead cells consisting only of the strengthened wall with all the cytoplasm removed. These tubes have unusually thick secondary cell walls that are strengthened by high local concentrations of lignin, which accounts for 20% to 30% of the weight of the wall (see Panel 20–1, pp. 1148–1149). They are responsible for conducting water and dissolved inorganic ions from the roots to the rest of the plant body (Figure 20–25). (The xylem also provides a vital support, particularly in woody plants.)

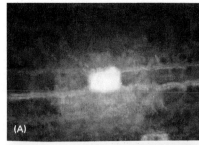

Figure 20–22 The use of fluorescein-coupled peptides of various sizes to determine the functional size of the channels in plasmodesmata. In (A) the fluorescent peptide was >850 daltons, while in (B) it was <850 daltons. In both cases the peptide was injected into a single epidermal cell of a small pondweed. Note that the larger probe has hardly moved at all, while the smaller probe has passed to many neighboring cells. (Courtesy of P.B. Goodwin.)

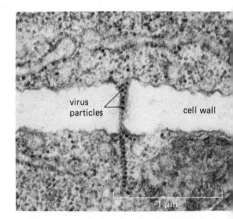

Figure 20–23 An electron micrograph showing small spherical plant viruses passing from one cell to another through a plasmodesma. These viruses have a diameter of 25 nm, which is much larger than that of peptides that do not pass through these openings. (Courtesy of Kitty Plaskitt.)

THE FLOWER

Flowers, which contain the reproductive cells of higher plants, arise from vegetative shoot apical meristems (see Figures 20–60 and 20–62). They terminate further vegetative growth from that meristem. Environmental factors, often the rhythms of day length and temperature, trigger the switch from vegetative to floral development. The germ cells thus arise late in plant development from somatic cells rather than from a germ cell line as in animals.

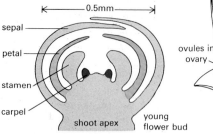

0.5mm

sepal
petal
stamen
carpel
shoot apex
young flower bud

Flower structure is both varied and species specific but generally comprises four concentrically arranged sets of structures that may each be regarded as modified leaves.

stigma   style

ovules in ovary

mature flower

Petal: Distinctive leaflike structures, usually brightly colored, they facilitate pollination by, for example, attracting insects.

Stamen: The stamens contain cells which undergo meiosis and form haploid pollen grains (see Figure 20–14) each of which contains two male sperm cells. Pollen transferred to a stigma germinates and the pollen tube delivers the two nonmotile sperm to the ovary.

pollen grain
sperm cells
pollen tube nucleus

Carpel: The carpel contains one or more ovaries that each contain ovules. Each ovule houses cells that undergo meiosis and form an embryo sac containing the female egg cell. At fertilization, one sperm cell fuses with the egg cell and will form the future diploid embryo, while the other fuses with two cells in the embryo sac to form the triploid endosperm tissue (see p. 1173).

Sepals are leaflike structures that form a protective covering during early flower development.

THE SEED

A seed contains a dormant embryo, a food store, and a seed coat. At the end of its development a seed's water content can drop from 90% to 5%. The seed is usually protected in a *fruit* whose tissues are of maternal origin.

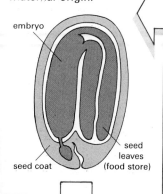

embryo
seed coat
seed leaves (food store)

THE EMBRYO

fertilized egg

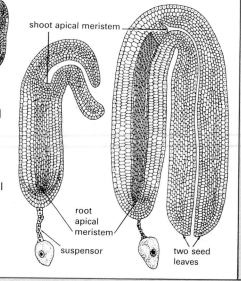

shoot apical meristem

root apical meristem
suspensor
two seed leaves

The fertilized egg within the ovule will grow to form an embryo using nutrients transported from the endosperm by the suspensor (see p. 1173). A complex series of cell divisions, illustrated here for the common weed called shepherds purse, produces an embryo with a root apical meristem, a shoot apical meristem and either one (monocots) or two (dicots) seed leaves, or cotyledens.

Development is arrested at this stage and the ovule, containing the embryo, now becomes a seed, adapted for dispersal and survival.

GERMINATION

For the embryo to resume its growth the seed must germinate, a process dependent upon both internal factors (dormancy) and environmental factors including water, temperature, and oxygen. The food reserves for the early phase of germination may either be the endosperm (maize) or the cotyledens (pea and bean).

The primary root usually emerges first from the seed to ensure an early water supply for the seedling. The cotyleden(s) may appear above the ground, as in the garden bean shown here, or they may remain in the soil, as in peas. In both cases the cotyledens eventually wither away.

The apical meristem can now show its capacity for continuous growth, producing a typical pattern of nodes, internodes, and buds (see p. 1174).

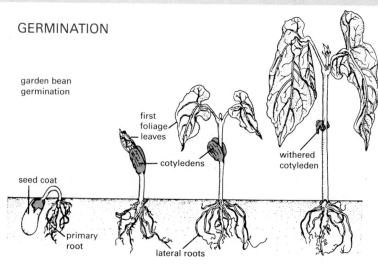

garden bean germination

first foliage leaves
cotyledens
withered cotyledon
seed coat
primary root
lateral roots

Transport in the xylem is unidirectional, toward surfaces of evaporation; the water is pulled up through the capillarylike tubes by the suction generated by evaporation. The pattern of lignin deposition around the xylem vessel creates a structure that is highly resistant to compression, a required property for a tube that carries fluid under negative pressure. Without such reinforcement, the long tubes would collapse like a weakened drinking straw.

The solutes carried in the xylem fluid, mainly mineral salts and nitrogen-containing compounds, are actively pumped into it by specialized cells in the roots. Back-leakage from the tubes into the apoplast near the site of xylem loading is inhibited by apoplastic seals (*Casparian bands*), which are analogous in function to the tight junctions between animal epithelial cells (Figure 20–26). Higher up in the plant, cells equipped with specific membrane transport proteins pump solutes from the xylem fluid into receiver cells in the photosynthetic tissues. Most of the solvent (water) that percolates through the xylem eventually evaporates, primarily from the surface of the photosynthetic tissue in the leaf.

As a fail-safe mechanism, the pumping of solutes into the xylem of roots can cause an osmotic flow capable of delivering solutes to leaves and shoots even when atmospheric conditions are unfavorable for evaporation. The beads of fluid that may be seen at the tips of leaves on a lawn in early morning have been extruded from special pores by this phenomenon of osmotically induced *root pressure*. (For a description of some of the cell types involved, see Panel 20–1, pp. 1148–1149.)

Sugars Move Under Pressure in the Phloem[13]

Phloem is a complex set of cells in the vascular tissue that is responsible for transporting the main product of photosynthesis, sucrose, from the photosynthetic cells in the leaf to the rest of the plant (see Figure 20–25). The principal conducting component of the phloem is the *sieve tube*, a long column of living cylindrical cells (called sieve-tube elements) that are interconnected by the perforations in their end walls (sieve plates). Sucrose is pumped into the sieve-tube cells in the leaves, creating a concentrated sucrose solution (typically 10–25%).

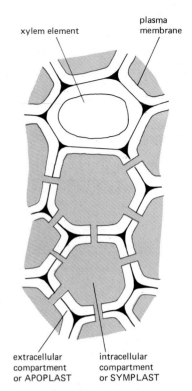

Figure 20–24 Highly schematic diagram of a group of plant cells interconnected by plasmodesmata lined by plasma membrane. This membrane divides the plant into an extracellular space (the apoplast, shown in white) and an intracellular space (the symplast, shown in color). For clarity, no organelles are shown. Note that the xylem element lacks a plasma membrane and is therefore part of the apoplast.

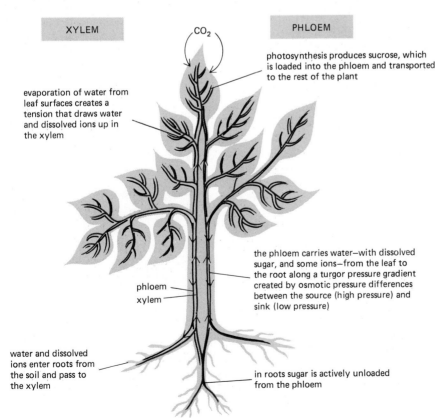

Figure 20–25 The two major routes, the xylem and the phloem, by which water and solutes are transported throughout the plant. These routes have been highly simplified in the drawing; for example, an extensive lateral water exchange that occurs between the xylem and the phloem is not shown.

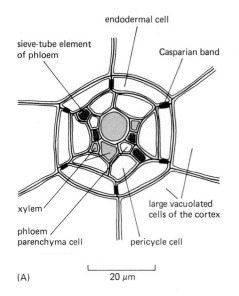

endodermal cell

sieve-tube element
of phloem

Casparian band

xylem

phloem
parenchyma cell

large vacuolated
cells of the cortex

pericycle cell

(A)

20 μm

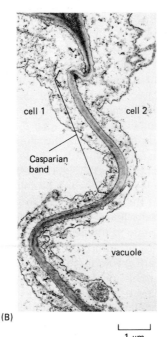

cell 1

cell 2

Casparian
band

vacuole

(B)

1 μm

Figure 20–26 (A) The vascular network in the root of a simple plant—the water fern *Azolla*—depicted in cross-section. This plant has an extremely simple arrangement of conducting cells, with four xylem vessels and four phloem sieve-tube elements. Most higher plants have much more complex arrangements of cells in their vascular tissue. Note that the xylem and phloem are surrounded by endodermal cells and that the strategically placed Casparian bands in the endodermal cell walls prevent water leakage outward through the apoplast. (B) The Casparian band between two root endodermal cells similar to those shown in (A). Compared with a normal primary cell wall, the band has a smooth texture, the result of heavy impregnation with water-resistant cutin, which makes this specialized wall impermeable to water. (B, from B. Gunning and M. Steer, Ultrastructure and the Biology of Plant Cells. London: Arnold, 1975.)

Osmotic forces drive water into the sieve tube, creating a high local turgor pressure (up to 30 atmospheres) that drives the sucrose solution from cell to cell down the length of the tube. The thick-walled cells form a low-resistance tube for transporting fluid under high pressure. Although the mature sieve-tube elements are living cells, with functioning plasma membranes (and thus are part of the symplast), they have lost their nuclei and much of their cytoplasm. For sustenance they rely on associated *companion cells*, which transport soluble food molecules and other components in and out of the sieve-tube elements through groups of clustered plasmodesmata in shared side walls (see Panel 20–1, pp. 1148–1149).

Transport in the phloem is more complex than transport in the xylem because it is not limited to one direction: it takes solutes, mainly sucrose, from various sites of production (sources) to sites of consumption and storage (sinks), irrespective of location. Sucrose is actively transported into and out of sieve-tube elements by specialized donor and receiver *transfer cells* at sources and sinks, respectively (Figure 20–27). The increased sugar concentration at sources causes more water to enter the phloem at these points, creating the pressure needed for a mass flow of fluid through the sieve tubes to the sinks. Here the sugar is actively unloaded and the water is then removed by osmosis, largely to the xylem. The fluid moves through the phloem at rates of about a meter per hour, far in excess of the rate of diffusion.

The transport of fluid from one part of a plant to another differs in two important respects from fluid transport in animals. First, animals have only one main transport system for all nutrients—the blood system—while plants have two distinct systems—the phloem and the xylem—carrying different solutions. Second, fluid does not circulate continuously in plants, the way blood does in an animal; instead, there are two open pipelines with net movement of water and solutes in each (see Figure 20–25).

Figure 20–27 Specialized **transfer cells** in a small leaf vein. A phloem sieve-tube element (S) is surrounded by three transfer cells (T), whose convoluted infoldings of cell-wall material, lined by plasma membrane, increase the surface area of the cell by as much as twentyfold. Transfer cells are found at many sites in the plant body where the rate of transport of solutes across the plasma membrane is especially high—for example, in those regions where inorganic ions are pumped from the xylem into the tissues or, as in this case, where sucrose is pumped into the phloem.

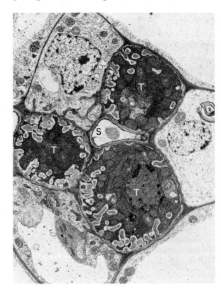

Summary

The presence of a tough, relatively impermeable cell wall profoundly influences the ways that plant cells communicate with one another. All of the living cells in a plant are connected to one another by plasmodesmata—small cytoplasmic channels, lined by plasma membrane, that cross the cell wall and through which the controlled movement of solutes occurs. All of the living protoplasts of the plant body are thereby linked to form a large, interconnected intracellular space called the symplast. The extracellular space, through which most water transport takes place, is also interconnected and is called the apoplast. The photosynthetic cells of the plants produce sugars that are actively transported into phloem tubes. These tubes, composed of living cells, are part of the symplast, and they carry the sugars to the rest of the plant by exploiting osmotic pressure gradients. The cells of the root absorb water and actively take up dissolved minerals from the soil and transport them to the leaves through tubes formed from the dead cells of the xylem, part of the apoplast; again physical forces, in this case evaporation, pull a continuous column of water from root to leaf.

Interactions Between Plants and Other Organisms

Both the root and shoot systems of all plants come into intimate contact with myriads of organisms, including bacteria, fungi, and a variety of invertebrates such as nematodes and insects. Plants have evolved complex ways—genetic, chemical, and anatomical—of interacting with these organisms, ranging from defense responses against deadly pathogens to symbiotic relationships with organisms that then become essential to the survival of the plant. We shall briefly consider four examples of such interactions that illustrate the complexity of the signaling events and responses involved.

Most Vascular Plants Have a Symbiotic Relationship with Soil Fungi[14]

If seedlings of a forest tree grown under sterile conditions are planted on grassland soil, they run a high risk of dying from malnutrition, even though there are adequate nutrients in the new soil. If the seedling is exposed to a small quantity of forest soil, however, normal growth is ensured (Figure 20–28). The crucial ingredients in the forest soil are fungi that form a close symbiotic association with the roots of the plant and produce a *mycorrhiza*, or "fungus-root," found in more than 90% of vascular plants. The hyphal filaments of the fungus have a very large surface area and form a feltlike covering around the root, penetrating between, and in some cases into, the cortical cells of the root. The fungus secretes growth factors that stimulate the cortical cells to enlarge and proliferate, thereby causing the roots to swell and branch. By altering soil pH, the fungus also increases the supply of inorganic nutrients from the soil, in particular phosphate, making them available for absorption by the plant; in return the plant supplies the fungus with sugar and amino acids. Each plant forms mycorrhiza with only a limited number of fungal species, and in many cases the fungus is essential for the plant's survival. Successful growth following germination of orchid seeds, for example, depends absolutely on the presence of the correct mycorrhizal fungi. It is probable that the geographic limits of many plants are dictated by the requirements of their particular fungi. The molecular details of the fungus–plant interaction in mycorrhizae are still obscure. Considerably more is known about the symbiotic association of some plant roots with bacteria.

Figure 20–28 The effect of mycorrhizae on plant growth. Both sets of pine seedlings spent their first 2 months in a sterile nutrient solution and are now 9 months old. One set (A) was transferred directly to soil from the open prairie, while another (B) spent 2 weeks in forest soil before being moved to the prairie soil. Only the latter plants (B) have benefited from the appropriate mycorrhizal fungi found in the forest soil. (Courtesy of Dr. Jaya G. Iyer, Department of Soil Science, University of Wisconsin, for the late Dr. S.A. Wilde, University of Wisconsin.)

Symbiotic Bacteria Allow Some Plants to Use Atmospheric Nitrogen[15]

The raw materials required for plant growth are inorganic compounds—mainly O_2, CO_2, H_2O, and minerals. Most of the minerals taken up from the soil by the roots come from weathered rocks. An exception is nitrogen. Much of the nitrogen in living organisms ultimately derives from atmospheric nitrogen that has been incorporated (fixed) into organic compounds by processes that require a good deal of energy (hence the high cost of artificially produced nitrogen-based fertilizers). The only organisms that can fix atmospheric nitrogen are procaryotic (cyanobacteria and certain eubacteria). While they are all free-living bacteria, some (for example, *Rhizobium*) can also form close symbiotic associations with the roots of certain plants, such as the *legumes* (peas, beans, and clovers), where they make a major contribution to soil nitrogen resources.

The first step in establishing these important symbiotic associations is specific recognition by the bacteria of the thin root hairs that project from specialized

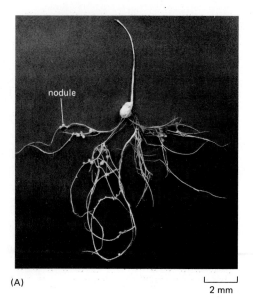

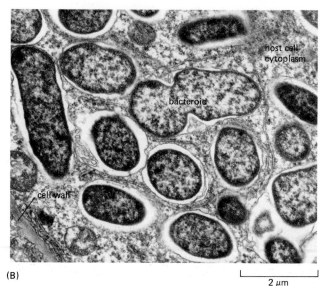

(A)

2 mm

(B)

2 µm

epidermal cells of the host plant. After binding to the root epidermal cells, growing bacteria enter the plant via an *infection thread* and stimulate underlying cortical cells to divide and form a large *root nodule* (Figure 20–29A). The bacteria invade the new cortical cells, colonizing the cytoplasm. Each mature nodule contains about half its weight in intracellular bacteria, which have lost most of their own cell wall. The plasma membrane of each bacterium is surrounded by another membrane derived from the host-cell plasma membrane. It is these altered bacteria, called *bacteroids*, that fix the nitrogen eventually used by the plant (Figure 20–29B).

Nitrogenase, the bacterial enzyme that catalyzes the fixation of nitrogen, is a complex of three polypeptide chains. In symbiotic *Rhizobia* this protein complex catalyzes the conversion of atmospheric nitrogen to ammonia, which is then rapidly released into the cytoplasm of the host cell, where it is converted to glutamine. Eventually the fixed nitrogen is incorporated into all the other amino acids.

Genetic analyses show that establishing and maintaining this symbiotic interaction requires the coordinated expression of many genes in both the bacterium and the host plant cell. The binding and invasion of the host root hair by the bacterium is the first step in a dialogue between host cell and bacterium that eventually activates a set of host genes encoding plant cell proteins called *nodulins*, which are essential for the growth and function of the nodule. The dialogue begins when a flavanoid produced by the host plant cell binds to and activates a protein encoded by a bacterial *nod* gene (Figure 20–30). This activated protein, called *nod*D, then switches on the synthesis of other bacterial *nod* gene products, whose functions include inducing production of host plant nodulins. A large plasmid carried by the bacterium contains most of the bacterial *nod* genes, as well as the *nif* genes that encode the enzymes involved in nitrogen fixation, including the nitrogenase. A *Rhizobium* that normally nodulates only beans can be converted into one that nodulates only peas if the appropriate plasmid genes are replaced with those from the plasmid carried by a pea-specific *Rhizobium*.

The nodulins of the host cell include proteins involved in stimulating root cortical cell division, structural components of the nodule, enzymes that enable the plant to assimilate fixed nitrogen compounds, and specialized proteins that support bacteroid function. One of the most important of these is *leghemoglobin*, a cytoplasmic oxygen-binding protein similar to mammalian myoglobin. The bacterial nitrogenase complex is irreversibly inactivated by free oxygen, and elaborate mechanisms are required to ensure that the bacteroids remain in an environment in the root whose oxygen tension is buffered so as to support the respiration required for the energetically demanding nitrogen-fixing process without inactivating nitrogenase. As part of this mechanism, the *Rhizobium* induce the host cells

Figure 20–29 (A) A young pea seedling that has already formed a symbiotic association with the nitrogen-fixing bacterium *Rhizobium*. The root nodules containing the bacteria are clearly visible. (B) Electron micrograph of a thin section through a pea root nodule like those shown in (A). The nitrogen-fixing *Rhizobium* bacteroids, surrounded by host-cell-derived membrane, fill the host-cell cytoplasm. (A, courtesy of Andy Johnston; B, courtesy of B. Huang and Q.S. Ma.)

Figure 20–30 Various flavones and related compounds released by the host root bind to and activate specific nodulating bacteria. Luteolin, shown here, is a flavone from alfalfa that induces *nod* genes in *Rhizobium meliloti*. This signaling molecule probably binds to the nodD protein, which, in its activated form, then switches on the early bacterial nodulation genes. The flavones are closely related to the anthocyanin pigments of flowers and fruit. Different plants release different combinations of flavanoid molecules, which selectively activate specific *Rhizobium* species.

to produce large amounts of leghemoglobin to act as an oxygen buffer. The globin part of the molecule is encoded by a host gene, while the heme prosthetic group may be supplied by the bacterial partner—a remarkable example of evolutionary co-adaptation.

Nitrogen fixation consumes a great deal of energy, supplied ultimately by the sun through photosynthesis. It is estimated that the fixation of one molecule of nitrogen (N_2) by *Rhizobium* requires between 25 and 35 ATP molecules. These procaryotes contribute vastly more fixed nitrogen for plant growth ($\sim 2 \times 10^8$ tons/year), largely in unmanaged ecosystems, than is provided by nitrate fertilizers.

Agrobacterium Is a Plant Pathogen That Transfers Genes to Its Host Plant[16]

Another soil bacterium that is closely related to *Rhizobium* is called *Agrobacterium tumefaciens*. This bacterium causes *crown-gall* disease in plants: when they are exposed to *Agrobacterium*, normal plant cells become transformed into gall-forming tumor cells by a process that involves the transfer of genes from the bacterium to the plant (Figure 20–31).

The tumor cells induced by *Agrobacterium* are remarkable in several ways. First, unlike most normal plant cells, they can be removed from the gall and grown indefinitely in culture in the absence of added growth factors and without the continued presence of *Agrobacterium* cells. Second, they synthesize a variety of unusual chemicals called *opines*; these amino acid derivatives can be catabolized and used preferentially by the strain of bacteria that induced the plant cells to make them—an arrangement with an obvious survival advantage for the *Agrobacterium*.

Explanations for many of the properties of *Agrobacterium*-induced tumor cells have come from molecular genetic studies. The ability of *Agrobacterium* to induce tumors is associated with a large DNA plasmid called Ti (Tumor-inducing), a small part of which, the **T-DNA** (*transferred DNA*), integrates into the nuclear genome of the plant cell. The bacteria can infect a susceptible plant only at the site of a wound, where plant cells secrete unusual phenolic compounds that include *acetosyringone* (Figure 20–32). This compound, produced in response to wounding, identifies susceptible sites of entry for the pathogen and activates a series of reactions in the bacterium that lead to the excision of T-DNA from the Ti plasmid and its transfer to the host cell genome (Figure 20–33). Once integrated into a host cell chromosome, the T-DNA is transcribed and translated by the host cell to produce three classes of proteins. One is the enzyme that causes the plant to produce a specific opine, while the other two are enzymes that catalyze synthesis of the plant growth regulators *indole-acetic acid* and *cytokinin* (see Figure 20–67). The abnormal balance and increased levels of these two growth regulators, produced as a result of the activity of the integrated T-DNA genes, is responsible for the continued growth and division of the transformed plant cells and explains why these cells can continue to grow in the absence of both added growth regulators and the original bacteria.

The ability of *Agrobacterium* T-DNA to integrate stably into its host's genome has lead to its widespread exploitation as a vector for the genetic transformation of higher plant cells with recombinant DNA molecules (see p. 1183).

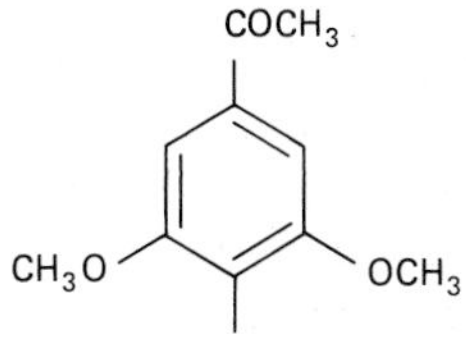

Cell-Wall Breakdown Products Often Act as Signals in Plant–Pathogen Interactions[17]

A large number of highly specialized plant metabolites serve as protective agents for plants. Some compounds, such as the glycosides in mustard oil and the wide variety of alkaloids (for example, caffeine, morphine, strychnine, and colchicine), probably act as deterrents to herbivores, although there are always some predators that can tolerate or detoxify such chemical weapons. In addition to these constitutive defense mechanisms, plants have evolved more complex, adaptive defense mechanisms that are activated only when host and pathogen interact.

Figure 20–31 Tumors induced by *Agrobacterium tumefaciens* on a succulent pot plant. (Courtesy of Paul J.J. Hooykaas, from *Genetic Eng.* 1:155, 1979.)

Figure 20–32 Acetosyringone, a signaling molecule found in the exudates of wounded but metabolically active plant cells, specifically activates the virulence genes on the Ti plasmid of *Agrobacterium* that promote the formation of the T-DNA strand that is transferred to the host plant (see Figure 20–33).

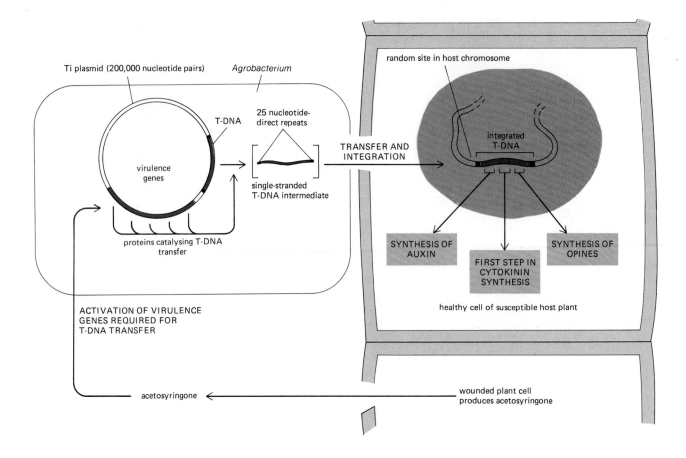

Figure 20–33 diagram labels:

Ti plasmid (200,000 nucleotide pairs) *Agrobacterium*

T-DNA

25 nucleotide-direct repeats

virulence genes

single-stranded T-DNA intermediate

TRANSFER AND INTEGRATION

random site in host chromosome

integrated T-DNA

proteins catalysing T-DNA transfer

SYNTHESIS OF AUXIN

FIRST STEP IN CYTOKININ SYNTHESIS

SYNTHESIS OF OPINES

healthy cell of susceptible host plant

ACTIVATION OF VIRULENCE GENES REQUIRED FOR T-DNA TRANSFER

acetosyringone

wounded plant cell produces acetosyringone

Many of these responses involve cell-wall components of the host cell or pathogen. As a common reaction to infection by a bacterial or fungal pathogen, for example, plant cells will strengthen their walls by increased deposition of cell-wall polymers—including lignin and hydroxyproline-rich glycoproteins. But pathogens have their own weapons in this war between plant and pathogen. Consider, for instance, what happens when a pathogenic fungus (*Fusarium solani*) attacks a pea plant. When the fungal spores land on the plant, they secrete very low levels of the enzyme cutinase, which break down a small amount of the cutin in the cuticle overlying the leaf. The released cutin monomers stimulate the fungal spores in a positive feedback loop to increase cutinase production. As a result, the cuticle is dissolved and the germinating fungal hyphae can penetrate the leaf.

The products of the cell-wall breakdown caused by an invading pathogen can, however, act as useful early-warning signals to the host plant cell. Plant cells exposed to a pathogen commonly synthesize low-molecular-weight molecules called *phytoalexins*, which are antibiotics that are toxic to some pathogenic bacteria and fungi. Several of the molecules responsible for stimulating phytoalexin biosynthesis by the plant (called *elicitors*) have now been identified as short oligosaccharides, derived from cell-wall polysaccharides, that are active at a very low concentration (10^{-9}–10^{-10} M). The first such elicitor to be extensively characterized was a hepta-β-glucoside derived from the cell wall of a fungus that attacks soybeans (Figure 20–34). Oligosaccharide elicitors of phytoalexin synthesis can also be derived from plant cell walls. In this case they appear to be fragments of the galacturonic acid backbone of pectin molecules, which can be released from the plant cell wall by enzymes secreted by an invading pathogen or, in some cases, by enzymes in the plant cell itself that are activated by injury.

Simple mechanical injury to plants can induce the synthesis of proteins that inhibit both insect and microbial proteases (protease inhibitors). This normal protective response is thought to be locally induced by small fragments of pectic polysaccharides released following damage. Interestingly, an as yet unidentified long-range signaling mechanism must also exist, since leaves some distance from the wounded one also eventually produce protease inhibitors.

Figure 20–33 Some of the early events in the infection of a susceptible plant by the bacterium *Agrobacterium tumifaciens*. The T-DNA, transferred from the bacterial tumor-inducing (Ti) plasmid to the plant cell, is integrated at a random site in a host chromosome. Among its gene products are enzymes involved in the synthesis of plant growth regulators, which are responsible for tumor formation.

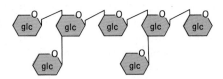

Figure 20–34 This heptaglucoside (an oligosaccharide containing seven glucose [glc] residues) is released from the cell wall of *Phytophthora megasperma*, a fungal pathogen of soybeans. It has a β1→6 glucose backbone with two β1→3 glucose side chains, and at very low concentrations it specifically activates several soybean genes, including those involved in phytoalexin biosynthesis. Altering the linkages or positions of any of the glucose residues results in loss of activity.

The Cell Wall May Be an Important Source of Normal Plant Cell Signals[18]

Recent experiments suggest that cell signaling that depends on the enzymatic release of oligosaccharides from plant cell walls also operates during normal plant growth and development. Cell elongation, for example, is promoted by the plant growth regulator auxin (see p. 1181), but this effect can be overcome by very low concentrations of a complex nine-sugar fragment derived from hemicellulose. Production of this wall fragment is stimulated by auxin, suggesting that it is part of a regulatory feedback system for controlling normal cell expansion (see p. 1168). More generally, it has been suggested that the extracellular matrix of plant cells—the plant cell wall—may serve as an immobilized store of many signaling molecules that are released locally to influence cell development. This hypothesis is currently being tested in many different laboratories.

Summary

Plants inevitably encounter a wide range of bacteria and fungi, and they have evolved complex ways of interacting with them. Many of these interactions are beneficial to the plant. Thus most plant root systems coexist with an elaborate fungal network that increases the mutual availability of nutrients. In addition, much of the nitrogen found in the molecules of living organisms derives ultimately from atmospheric nitrogen fixed by procaryotes such as Rhizobium, *which can form complex symbiotic associations with the roots of certain plants. Other interactions seem to benefit only the microorganism.* Agrobacterium, *a plant pathogen, causes plant tumors by inserting its T-DNA into the host cell genome; expression of the newly introduced genes results in an imbalance of the plant's normal growth-regulating molecules. Specific low-molecular-weight molecules released by the plant activate genes in both* Rhizobium *and* Agrobacterium *to begin their invasive processes. Other small molecules serve as signals that trigger the host's defensive response to pathogens; these signaling molecules include specific oligosaccharide breakdown products of cell-wall polysaccharides. Similar cell-wall-derived products may serve as signals in normal plant development.*

The Distinctive Internal Organization of Plant Cells

The cells of higher plants contain all the intracellular compartments previously described for animal cells, including the cytosol, Golgi apparatus, endoplasmic reticulum, nucleus, mitochondria, peroxisomes, and lysosomes. Similarly, plant cells have cytoskeletons composed of actin filaments, microtubules, and intermediate filaments comparable to those found in animal cells. Plant cells are readily distinguished from animal cells, however, by the presence of two types of membrane-bounded compartments—*vacuoles* and *plastids*. Both organelles are related to the immobile life-style of plant cells. These and other special features of the interior of plant cells are the subject of this section.

The Chloroplast Is One Member of a Family of Organelles, the Plastids, That Is Unique to Plants[19]

In general, plants produce all of the organic nutrients that they need themselves by means of photosynthesis, which takes place in **chloroplasts** (see p. 366). The products of photosynthesis can be used directly by the photosynthetic cells for biosyntheses, stored as an osmotically inert polysaccharide (usually starch), or converted to a low-molecular-weight sugar (usually sucrose) that is exported to meet the metabolic needs of the other tissues in the plant.

Chloroplasts are just one member of a family of organelles, the **plastids.** Plastids are present in all living plant cells, each cell type having its own characteristic complement. All plastids share certain features. Most notably, all plastids in a particular plant species contain multiple copies of the same relatively small genome (see Figure 7–69) and are enclosed by an envelope composed of two concentric membranes. Since the structure and function of the chloroplast is discussed in detail in Chapter 7, the focus here will be on other members of the family.

All plastids, including chloroplasts, develop from **proplastids,** which are relatively small organelles present in meristematic cells (Figure 20–35). Proplastids develop according to the requirements of each differentiated cell. If a leaf is grown in darkness, its proplastids enlarge and develop into *etioplasts.* These have a semicrystalline array of internal membranes that contain *protochlorophyll* (a yellow chlorophyll precursor) instead of chlorophyll (Figure 20–36A). When exposed to light, the etioplasts rapidly develop into chloroplasts by converting protochlorophyll to chlorophyll and by synthesizing new membrane, pigments, photosynthetic enzymes, and components of the electron-transport chain (Figure 20–36B).

Not surprisingly, light regulates many processes in plants. In ways that are still unclear, photoreceptors including *phytochrome* (see p. 1180) control the transcription of many light-activated genes involved in chloroplast development, not only in the chloroplast itself but also in the nucleus. Approximately one-fifth of the 120 chloroplast genes (see p. 391) are regulated in a light-dependent manner.

Chromoplasts are another form of plastid (Figure 20–37A). They accumulate carotenoid pigments and in many species are responsible for the yellow-orange-red coloration of petals (for example, daffodils) and fruits (for example, tomatoes). **Leucoplasts** are plastids that occur in many epidermal and internal tissues that do not become green and photosynthetic. They are little more than enlarged proplastids. A common form of leucoplast is the *amyloplast* (Figure 20–37B), which accumulates starch in storage tissues. In some plants, such as potatoes, the amyloplasts can grow to be as large as an average animal cell.

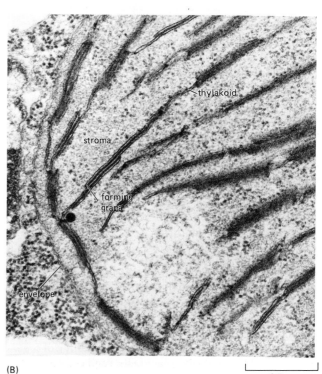

Figure 20–35 Electron micrograph of a typical proplastid from a root tip cell of a bean plant. It is surrounded by the plastid envelope, which consists of two membranes; the inner membrane gives rise to the sparse internal membrane system. (From B. Gunning and M. Steer, Ultrastructure and the Biology of Plant Cells. London: Arnold, 1975.)

Figure 20–36 Electron micrographs of two forms of plastids found in oat seedlings. (A) An etioplast from a seedling grown in the dark. The semicrystalline array of internal membranes contains protochlorophyll. (B) Upon exposure to light, the etioplast develops into a chloroplast. A portion of a young greening chloroplast is shown. The rearrangement of the etioplast membranes is underway: chlorophyll is now present, and small grana are beginning to form. (Courtesy of Brian Gunning.)

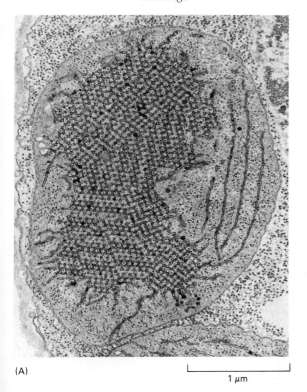

(A)

1 μm

(B)

0.5 μm

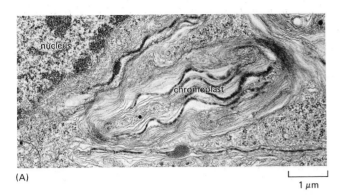

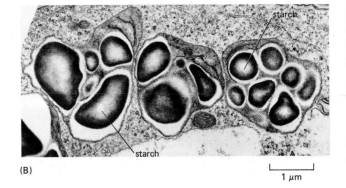

(A)

1 µm

(B)

1 µm

It is not known how the choice of developmental pathway from the proplastid to one of the more differentiated forms of plastid is regulated, but it is clear that the nuclear genome plays a dominant role. Nuclear mutations can switch development between chromoplast and chloroplast pathways, or they can block development, giving rise either to the various forms of leucoplast or to the immature chloroplasts with pale pigmentation that are found in many variegated ornamental plants.

Plastids can proliferate in the cytoplasm by growing and dividing (see p. 389). The only type of living cell in higher plants that can lose its population of plastids is the male sperm cell in certain species. As a consequence, some plants (for example, maize) acquire their plastids solely from the egg cell: plastids in these plants, like mitochondria in animals, are maternally inherited (see p. 395).

It is important to realize that plastids are not just sites for photosynthesis and the deposition of storage materials. Plants have exploited their plastids in the cellular compartmentalization of intermediary metabolism. Plastids produce more than the energy and reducing power (as ATP and NADPH) that is used for the plant's biosynthetic reactions. Purine and pyrimidine, most amino acid, and all of the fatty acid synthesis of plants takes place in the plastids, whereas they are made in the cytosol of animal cells.

The Plant Cell Vacuole Is a Remarkably Versatile Organelle[20]

The most conspicuous compartment in most plant cells is a very large, fluid-filled vesicle called a **vacuole** (Figure 20–38). There may be several vacuoles in a single cell, each separated from the cytoplasm by a single membrane called the **tonoplast.** Vacuoles generally occupy more than 30% of the cell volume, but this may vary from 5% to 90%, depending on the cell type. By convention, most plant cell biologists do not consider this organelle to be part of the cytoplasm; thus the interior of the plant cell is said to be composed of nucleus, vacuole, and cytoplasm—the latter containing all the other membrane-bounded organelles, including the plastids.

Vacuoles arise initially in young dividing cells, probably by the progressive fusion of vesicles derived from the Golgi apparatus. They are structurally and functionally related to lysosomes in animal cells and may contain a wide range of hydrolytic enzymes. The vacuole has a variety of functions. It can act as a storage organelle for both nutrients and waste products, as a lysosomal compartment, as an economical way of increasing the size of cells, and as a controller of turgor pressure. Different vacuoles with distinct functions (for example, lysosomal and storage) are often present in the same cell. We considered the space-filling role of vacuoles earlier in the context of cell growth (see Figure 20–10); here we shall examine their storage functions.

Vacuoles Can Function as Storage Organelles[20]

Vacuoles can store many types of molecules. In particular, they can sequester substances that are potentially harmful if present in bulk in the cytoplasm. The vacuoles of certain specialized cells, for example, contain such interesting prod-

Figure 20–37 Plastid diversity. (A) Pigment-containing chromoplasts from a cell in the yellow-orange petals of a daffodil. The outlines of these plastids are convoluted, and dissolved in their chaotic internal membranes is the pigment β-carotene, which gives the petals their color. (B) Three *amyloplasts* (a form of leucoplast), or starch-storing plastids, in a root tip cell of soybean. (From B. Gunning and M. Steer, Ultrastructure and the Biology of Plant Cells. London: Arnold, 1975.)

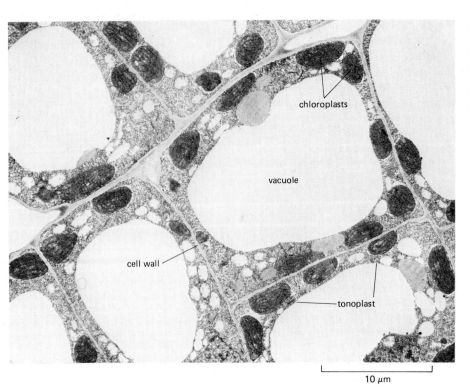

Figure 20–38 Electron micrograph of cells in a young tobacco leaf. The cytoplasm in these highly vacuolated cells is confined to a thin layer, containing numerous chloroplasts, pressed against the cell wall. (Courtesy of J. Burgess.)

chloroplasts

vacuole

cell wall

tonoplast

10 μm

ucts as rubber (in *Hevea brasiliensis*) or opium (in *Papaver somniferum*). Even ubiquitous molecules such as Na$^+$ are stored in these organelles, where their osmotic activity contributes to turgor pressure. Analysis of the giant cells of the alga *Nitella* indicates that Na$^+$ pumps located in the tonoplast maintain low concentrations of Na$^+$ in the cytosol and four- to fivefold higher concentrations in the vacuole; and since the vacuole occupies a much greater volume than the cytoplasm in *Nitella*, the great bulk of cellular Na$^+$ is in the vacuole.

The vacuole has an important homeostatic function in plant cells that are subjected to wide variations in their environment. When the pH in the environment drops, for example, the flux of H$^+$ into the cytoplasm is buffered—at least in part—by increased transport of H$^+$ into the vacuole. Similarly, many plant cells maintain turgor pressure at remarkably constant levels in the face of large changes in the tonicity of the fluid in their immediate environment. They do so by changing the osmotic pressure of the cytoplasm and vacuole—in part by the controlled breakdown and resynthesis of polymers such as polyphosphate in the vacuole, and in part by altering transport rates across the plasma membrane and the tonoplast. The permeability of these two membranes is partly regulated by turgor pressure and is determined by the distinct set of membrane transport proteins that transfer specific sugars, amino acids, and other metabolites across each lipid bilayer (see Chapter 6). The substances in the vacuole differ qualitatively and quantitatively from those in the cytoplasm. Because the tonoplast has little mechanical strength, however, the hydrostatic pressure must remain roughly equal in cytoplasm and vacuole, and the two compartments must act together in osmotic balance to maintain turgor.

Some of the products stored by vacuoles have a metabolic function. Succulent plants, for example, open their stomata and take up carbon dioxide at night (when transpiration losses are less than in the day) and convert it to malic acid. This acid is stored in vacuoles until the following day, when light energy can be used to convert it to sugar while the stomata are kept shut. As a second example, proteins can be stored for years as reserves for future growth in the vacuoles of the storage cells of many seeds, such as those of peas and beans. When the seeds germinate, the proteins are hydrolyzed and the amino acids mobilized to form a food supply for the developing embryo (Figure 20–39).

Other molecules stored in vacuoles are involved in the interactions of the plant with animals or with other plants. The anthocyanin pigments, for instance,

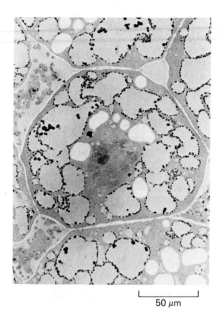

Figure 20–39 A light micrograph of a cell from a forming pea seed. The early stages of protein deposition can be seen at the edges of the extensive system of vacuoles. Legumes such as peas and beans store large amounts of protein in such vacuoles in order to sustain embryo growth when the seed germinates. (Courtesy of S. Craig.)

50 μm

color the petals of some flowers so that they attract pollinating insects. Other molecules defend the plant against predators. Noxious metabolites released from vacuoles when the cells are eaten or otherwise damaged range from poisonous alkaloids to unpalatable inhibitors of digestion. The trypsin inhibitors commonly found in seeds and the wound-induced protease inhibitors of leaf cells mentioned earlier (see p. 1160) both accumulate in the vacuole and are presumably designed to interfere with the digestive processes of herbivores. Both plants and animals have used such weapons in a subtle chemical warfare during evolution, the balance shifting when a potent new deterrent to herbivores develops or, conversely, when an insect evolves a means of breaking down a toxic plant metabolite so that it can feed upon, and even be attracted to, plants that possess the substance.

Golgi Vesicles Deliver Cell-Wall Material to Specific Regions of the Plasma Membrane[21]

Most cell-wall matrix components are transported via vesicles derived from the Golgi apparatus and secreted by exocytosis at the plasma membrane—a pathway described in detail in Chapter 8. The Golgi apparatus in plant cells, however, is engaged mainly in producing and secreting a very wide range of extracellular polysaccharides, rather than the glycoproteins typical of animal cells (Figure 20–40). The precise pathways involved in the synthesis of these polysaccharides are not known. This is perhaps not surprising, since (1) each cell-wall matrix polysaccharide is made of two or more sugars, (2) at least 12 types of monosaccharides are used, (3) most of the polysaccharides are branched, and (4) many covalent modifications are introduced in the polysaccharides after they are synthesized. It has been estimated that several hundred different enzymes are involved in the assembly of the polysaccharide components needed to construct a typical primary cell wall. Most of these enzymes are found in the endoplasmic reticulum and Golgi apparatus, although some concerned with the later covalent modifications of the polysaccharides are found in, and are sometimes covalently bound to, the cell wall itself.

Because the wall varies in composition and morphology at different locations around the cell, Golgi-derived vesicles must be directed to specific regions of the plasma membrane. This is accomplished, at least in part, by the cytoskeleton.

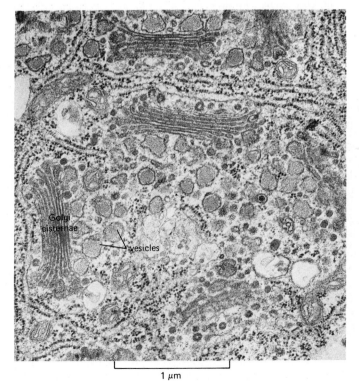

Figure 20–40 Electron micrograph of a hair cell from the green alga *Bulbochaete*. Numerous discrete stacks of Golgi cisternae can be seen, together with associated transport vesicles containing polysaccharide to be secreted. A clear polar organization can be seen from the ER through *cis*, *medial*, and *trans* Golgi cisternae to transport vesicles. Unlike most animal cells, the Golgi cisternae in different stacks are not interconnected. (Courtesy of T.W. Frazer, from B. Gunning and M. Steer, Ultrastructure and the Biology of Plant Cells. London: Arnold, 1975.)

Golgi cisternae

vesicles

1 µm

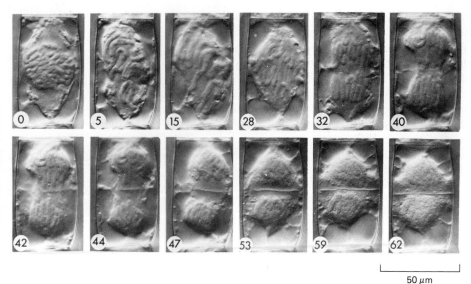

Figure 20–41 Sequential light micrographs of a dividing stamen hair cell. The elapsed time in minutes is shown at the bottom left corner of each photograph. The vesicles that align to form the cell plate can be seen after 42 minutes. The plate then extends sideways until it reaches and fuses with the mother cell wall. (Courtesy of Peter Hepler.)

50 μm

Consider, for example, the formation of the new primary cell wall that separates two daughter cells after mitosis (Figure 20–41). At the end of telophase, a disclike region—the **phragmoplast**—forms in the plane of the former spindle equator. The phragmoplast contains a double ring of short microtubules on either side of, and terminating at, the division plane (Figure 20–42) and a set of actin filaments coaligned with the microtubules. Golgi-derived vesicles containing cell-wall precursors, especially pectin, are guided inward along these oriented microtubules until they reach the phragmoplast, where they fuse with one another to form the **cell plate.** The microtubular ring moves centrifugally outward as Golgi vesicles continue to add precursors to the growing cell plate. Finally, the plate fuses with the mother cell wall to create two separate daughter cells (see Figures 20–41 and 13–71). It is not clear which components of the phragmoplast, the microtubules or the actin filaments (or both), are responsible for the movement and guidance of the Golgi-derived vesicles.

Not all of the polysaccharides made and secreted by the Golgi apparatus are destined for the cell wall: a fucose-rich slime secreted by root tip cells, for example, moves through the wall and serves to lubricate the outer surface of the root on its passage through the soil (Figure 20–43).

Figure 20–42 (A) Electron micrograph of the phragmoplast in a dividing plant cell. The microtubules guide vesicles containing cell-wall precursors toward the growing cell plate. For details see Figure 13–71, page 782. (B) Fluorescence micrographs of an onion root tip cell at cytokinesis. The developing cell plate and the two sets of phragmoplast microtubules on either side of it are revealed by immunofluorescence (*left*), while a fluorescent stain for DNA (*right*) shows the positions of the two daughter nuclei that will be separated by the new cell wall. (A, courtesy of Jeremy Pickett-Heaps.)

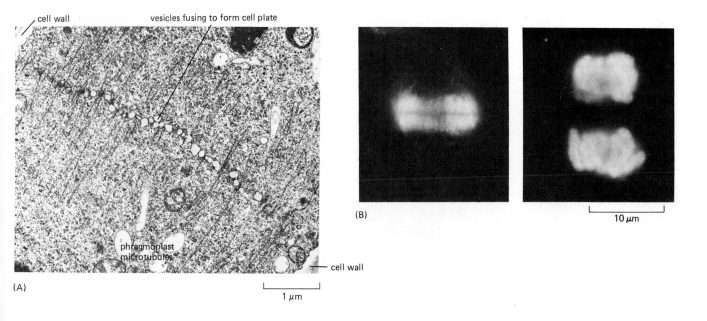

cell wall

vesicles fusing to form cell plate

phragmoplast microtubules

cell wall

(A)

1 μm

(B)

10 μm

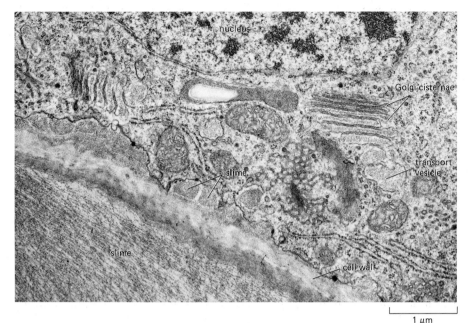

Figure 20–43 Electron micrograph of a cell from the root cap of Timothy grass. In such cells the Golgi apparatus functions mainly in manufacturing, packaging, and delivering mucilage (slime) to the surface of the root tip to lubricate its passage through the soil. Material identical to that seen in the Golgi cisternae can also be seen in vesicles and just outside the plasma membrane. (From B. Gunning and M. Steer, Ultrastructure and the Biology of Plant Cells. London: Arnold, 1975.)

Membrane Is Rapidly Recycled by the Process of Fluid-Phase Endocytosis[22]

What happens to the large amount of new membrane that is added to the existing plasma membrane by repeated vesicle fusions? In some actively secreting plant cells, the number of Golgi vesicles involved in exocytosis would lead to a doubling of the plasma membrane area every 20 minutes. Clearly, if the plasma membrane is to maintain a constant surface area, some form of membrane retrieval must operate. Numerous coated pits occur in the plasma membrane of plant cells (Figure 20–44A), and they are believed to function in membrane recycling, just as they do in animal cells (see p. 326). This pathway of fluid-phase endocytosis has recently been demonstrated in plant cells by monitoring the uptake of electron-dense markers such as ferritin or colloidal gold into plant cell protoplasts. The electron-dense marker taken up by protoplasts is rapidly delivered to a complex set of membranous tubes that has been called the *partially coated reticulum* (Figure 20–44B and C), an organelle that is thought to be functionally equivalent to the endosomal compartment in animal cells (see p. 460). From here the marker is delivered to large vesicles and finally ends up in the vacuole.

Thus the major intracellular pathways of synthesis, sorting, packaging, targeting, and secretion, as well as the pathway of endocytosis, are very similar in plant and animal cells.

Figure 20–44 Endocytosis in a soybean protoplast, revealed by electron microscopy. (A) A view of the cytoplasmic face of the plasma membrane. The protoplast was attached to an electron microscope grid and then osmotically lysed, washed, and negatively stained. Cortical microtubules can be seen in addition to numerous coated pits that are involved in endocytosis. (B) The "partially coated reticulum" seen in a thin section. It consists of a complex membrane-bounded tubular compartment with numerous clathrin-coated vesicles forming from it, and it is thought to be equivalent to the endosomal compartment of animal cells. (C) Endocytosed material—here the electron-dense marker ferritin—first appears in the partially coated reticulum. (Courtesy of Larry Fowke.)

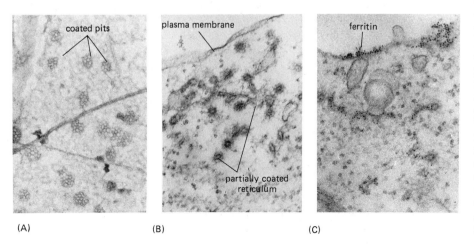

Cellulose Synthesis Occurs at the Surface of Plant Cells[23]

There is one important exception to the rule that cell-wall polysaccharides are produced in the Golgi apparatus and secreted by exocytosis. In most plants, cellulose is synthesized on the exterior surface of the cell by a plasma-membrane-bound enzyme complex, **cellulose synthase,** which uses a sugar nucleotide precursor supplied from the cytosol, probably UDP-glucose. As they are being synthesized, the nascent cellulose chains spontaneously assemble into microfibrils that form a layer on the surface of the plasma membrane (a lamella) in which all the microfibrils have more or less the same alignment. Because cellulose is synthesized at the plasma membrane, each new wall lamella forms internally to the last. The wall therefore consists of concentrically arranged lamellae, with the oldest on the outside.

Despite much effort, it has not been possible to reconstitute the reactions of cellulose synthesis *in vitro*, at least partly because the reaction appears to require a functionally intact plasma membrane. Electron micrographs of freeze-fractured cells of higher plants have shown suggestive "rosettes" of intramembrane particles on the protoplasmic fracture face of the plasma membrane. These rosettes may be cellulose synthase complexes, since they are concentrated in areas where new cellulose microfibrils are deposited—for example, in the membrane overlying the ordered wall thickenings of young xylem vessel elements (Figure 20–45 and see p. 1147).

The Organization of Cellulose Microfibrils Determines the Shape of a Growing Plant Cell[24]

The final shape of each growing plant cell, and hence the final form of the plant, is determined by controlled cell expansion. As described earlier (see p. 1143), the general mechanism for expansion relies on a combination of turgor pressure and wall plasticity. But we have not yet explained how this nonpolarized pressure is harnessed to produce a *directional* cell elongation. The most recently laid down microfibrils in the side walls of elongating cells commonly lie perpendicular to the axis of elongation, surrounding the cell with lamellae whose constituent microfibrils are arranged helically. Although the orientation of the microfibrils in the outer lamellae that were laid down earlier may be different, it is the orientation of these inner lamellae that have a dominant influence on the direction of cell expansion.

If the cellulose microfibrils in newly formed lamellae are transversely oriented, they will restrain lateral expansion, but because the neighboring cellulose microfibrils within a lamella can be separated by turgor pressure, cell elongation will be encouraged (Figure 20–46). As the cell expands, new matrix and microfibrillar components need to be added to the existing cell wall to maintain its thickness and strength. The newly secreted matrix components are probably intercalated throughout the thickness of the wall, while additional microfibrillar elements are deposited in new lamellae at the outer face of the plasma membrane.

Plant cells, therefore, anticipate their future morphology when they produce their primary cell walls. Exactly how the precise pattern of all the wall components is determined is not clear, but cellulose deposition is apparently guided by the pattern of microtubules in the cell cortex.

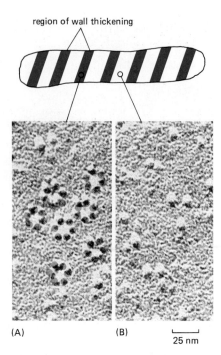

Figure 20–45 "Rosettes," seen by freeze-fracture electron microscopy (see p. 147), in the plasma membrane of a developing xylem vessel element from a cress root (compare Figure 20–15). These hexagonal groups of membrane proteins are concentrated over the area where the cell wall is being thickened (A) and are scarce in the areas between (B). It has been suggested that the "rosettes" are cellulose synthase complexes. Each of these complexes presumably synthesizes the 60 to 70 cellulose chains that form each microfibril. (Courtesy of Werner Herth.)

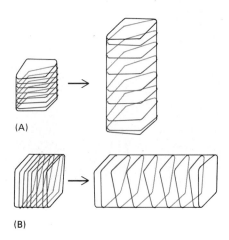

Figure 20–46 Schematic illustration of how the net orientation of cellulose microfibrils within the cell wall influences the direction in which the cell will elongate. (A) and (B) start as identically shaped cells (shown here as cubes) with different cellulose microfibril orientations. The combination of cell-wall weakening and turgor pressure causes each cell to elongate in a direction perpendicular to its orientated microfibrils. In turn, the final shape of an organ, such as a shoot, is profoundly influenced by the direction in which all of its constituent cells expand. Although the fibrils appear hooplike, numerous studies have suggested that they are helically wound as shown here.

Extracellular Cellulose Microfibrils Are Deposited Parallel to Cortical Microtubules[25]

As we have just seen, the orientation of cellulose microfibrils is crucial in determining the final shape of plant cells. An important clue to understanding how this orientation is brought about was provided by the discovery that most cytoplasmic microtubules are arranged in the cortex of the plant cell with the same orientation as the cellulose microfibrils that are currently being deposited in that region.

The cortical microtubules lie close to the cytoplasmic face of the plasma membrane (Figure 20–47) and are oriented mainly perpendicular to the axis of elongation of the cell. When examined by immunofluorescence microscopy, this *cortical array* is seen to consist of microtubules of indeterminate number and length wound around the cell in compressed helices (Figure 20–48). The microtubules are anchored to the plasma membrane by poorly characterized proteins.

A congruent orientation of the cortical array of microtubules (lying just inside the plasma membrane) and cellulose microfibrils (lying just outside the plasma membrane) is seen in many types and shapes of cells and is present during both primary and secondary cell-wall deposition (Figure 20–49). The cortical microtubules are especially concentrated in regions of intensive local wall deposition—as when xylem cell-wall thickenings are formed at specific sites on the cell surface (see Figure 20–15).

If the entire system of cortical microtubules is disassembled by treating a plant tissue with a microtubule-depolymerizing drug, the consequences for subsequent cellulose deposition are not as straightforward as might be expected. The drug treatment has no effect on the production of new cellulose microfibrils, and in some cases cells can continue to deposit new microfibrils in the preexisting orientation. Any developmental change in the microfibril pattern that would normally occur between successive lamellae, however, is invariably blocked. In addition, a cell that would normally start to develop regular wall thickenings to form a xylem vessel deposits instead a disorganized smear of wall material in the presence of the drug (Figure 20–50). The general conclusion is that a preexisting orientation of microfibrils can be propagated, even in the absence of microtubules,

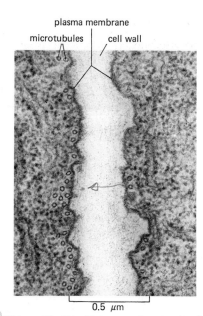

plasma membrane
microtubules | cell wall

0.5 μm

Figure 20–47 Electron micrograph of two adjacent wheat root tip cells, showing in cross-section the numerous cortical microtubules typical of interphase cells.

Figure 20–48 The cortical microtubule array. (A) A grazing section of a root tip cell from Timothy grass, showing cortical microtubules lying just below the plasma membrane. These microtubules are oriented perpendicular to the long axis of the cell. (B) An isolated onion root tip cell. (C) The same cell stained by immunofluorescence to show the transverse cortical microtubule array. (A, courtesy of Brian Gunning; B and C, courtesy of Kim Goodbody.)

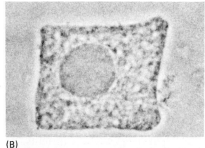

(B)

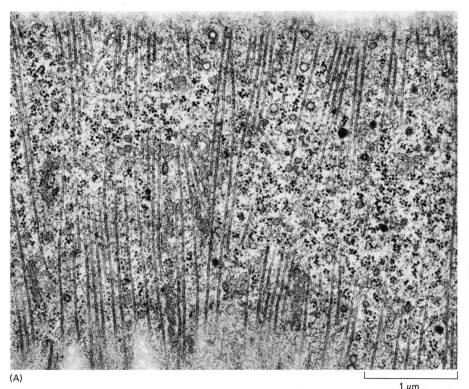

(A)

1 μm

(C)

10 μm

(A)

(B)

|_____|
20 μm

Figure 20–49 Microtubules and cellulose microfibrils in developing cotton fibers. The microtubules (A) are stained by immunofluorescence and are helically arranged. The cell is flattened, and both the front and back of the helices can be seen. Newly deposited cellulose microfibrils in a similar cell (B) are stained with the dye calcofluor white, which fluoresces when bound to nascent cellulose molecules. A similar helical arrangement of microfibrils is seen. This arrangement of cellulose, like the reinforcing threads in a garden hose, imparts great strength to the wall of the cotton fiber. (Courtesy of Robert Seagull.)

but that any change in the deposition of cellulose microfibrils requires that intact microtubules be present to determine the new orientation.

The cellulose-synthesizing complexes embedded in the plasma membrane are believed to spin out long cellulose molecules. As their coupled biosynthesis and self-assembly proceeds, the distal end of the microfibril presumably forms indirect cross-links to the previous layer of wall material. At the growing (proximal) end the synthesizing complexes would therefore need to move along the membrane in the direction of synthesis. In principle there are two ways in which microtubules could influence the direction of this movement and hence the orientation of the microfibrils. The cytoplasmic domain of the cellulose synthase complex could be linked, directly or indirectly, to the cortical microtubules. Alternatively, the direction in which the synthase complexes move could be restricted by microtubule-determined boundaries in the plasma membrane that act like the banks of a canal to constrain movement of the synthase complexes to a parallel axis (Figure 20–51). This second model has the advantage that cellulose synthesis is viewed as an activity that is independent of microtubules, being constrained spatially only when cortical microtubules are present to define membrane domains within which the enzyme complex can move. It is not known how the organization of the cortical microtubules is controlled. It is worth recalling that the cytoskeleton of an animal cell likewise determines the orientation of extracellular matrix components that the cell secretes and that elongated proteins such as collagen and fibronectin are laid down by these cells in close association with their plasma membrane (see p. 822).

The Movement of Materials in Large Plant Cells Is Driven by Cytoplasmic Streaming[26]

Cell metabolism demands that substrates, intermediates, cofactors, intracellular signaling molecules, and enzymes be able to move from one part of the cytoplasm to another. In small cells, such as bacteria or even small animal cells, small solutes move by diffusion over distances comparable to the size of the cell in fractions of a second. Plant cells, however, because of their cell walls, vacuoles, and turgor, are able to become very large. They are commonly more than 100 μm long, and some are a few millimeters or even centimeters long. Diffusion is relatively ineffective over such distances, as the time taken for a molecule to reach its destination by diffusion alone increases with the square of the distance involved (see p. 92). Furthermore, cells in the adult plant are often many cells distant from their vital sources of oxygen and nutrients. It is not surprising, therefore, that large plant cells display an active directional **cytoplasmic streaming,** which stirs their cytoplasm and rapidly moves material around from one end of the cell to the other.

Examination of living plant cells shows that the larger the cell, the more extensive are the movements of its cytoplasm. As in animal cells, small plant cells exhibit discontinuous movements of their organelles known as *saltations*, in which the organelles stop and go, being suddenly propelled at a high speed in a particular direction. In larger plant cells there is more directionality to the cytoplasmic movements; and in cells with a thin rim of cytoplasm surrounding a large central vacuole, it is common for the cytoplasm to circulate almost continuously, at a rate of several micrometers per second. In the cells of the giant algae *Nitella*, this rapid movement is driven by myosin moving along oriented bundles of cortical actin

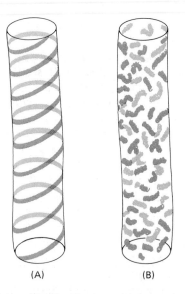
(A)   (B)

Figure 20–50 The organized pattern of wall thickenings formed during normal xylem cell differentiation (A) depends on the presence of organized arrays of cortical microtubules (see Figure 20–15). In the presence of colchicine the cortical microtubules are depolymerized and a disorganized pattern of wall thickenings results (B).

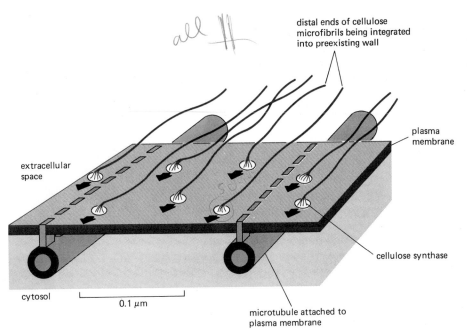

distal ends of cellulose
microfibrils being integrated
into preexisting wall

plasma
membrane

extracellular
space

cellulose synthase

cytosol

0.1 μm

microtubule attached to
plasma membrane

Figure 20–51 A simple model that could explain how the orientation of newly deposited cellulose microfibrils can be determined by the orientation of cortical microtubules. The large cellulose synthase complexes are integral membrane proteins that continuously assemble cellulose microfibrils on the outer face of the plasma membrane. Because the distal ends of the stiff microfibrils become integrated into the texture of the wall, their elongation at the proximal end pushes the synthase complex along in the plane of the membrane. Because microtubules are attached to the plasma membrane in a way that confines this complex to defined membrane channels, the microtubule orientation can determine the axis along which the microfibrils are laid down.

filaments (see p. 632). Such cytoplasmic movements promote not only intracellular traffic but also intercellular transport by delivering solutes to the openings of plasmodesmata connecting adjacent cells.

The thin cells that form the hairs on the surfaces of plants are transparent; therefore, their cytoplasmic movements can be readily observed in the living cell. These cells contain large vacuoles through which run thin transvacuolar strands of cytoplasm about 1 μm in diameter (Figure 20–52). Individual particles, such as mitochondria, can be seen moving rapidly through the cytoplasmic strands. The strands, which contain bundles of actin filaments but do not appear to contain microtubules, seem to emanate from a region near the cell nucleus (Figure 20–53). They can be observed to change their shape and position continuously, merging, branching, collapsing, and forming anew.

The Plant Cell Cytoskeleton Responds to Extracellular Signals[27]

In plant cells, constrained by their walls, subtle responses to changes in the environment, particularly to light, are of great importance. As we have seen, changes in the direction of plant cell growth are often mediated by the cytoskeleton, and so it is not surprising that both the actin and microtubule components of the cytoskeleton can respond in complex ways to external stimuli.

Many plant cells can respond to changes in the intensity and direction of light by altering the position of their chloroplasts. At low levels of illumination, the chloroplasts tend to become aligned in a monolayer perpendicular to the light source, thereby maximizing their exposure to the light. High levels of illumination, however, induce a protective response in which the chloroplasts become aligned against the cell walls, parallel to the incident light, thereby minimizing their exposure (Figure 20–54). The molecular basis of such movements has been explored in an unusual alga.

Mougeotia is an alga in which each cylindrical cell contains a single platelike chloroplast. In response to light, the chloroplast rotates until it is either edge-on or perpendicular to the incident light, depending on the intensity. The photoreceptor molecules that initiate the response are *phytochrome* (see p. 1180) and a blue-light receptor located on or very near the plasma membrane. Illumination of these receptors with a microbeam causes an influx of Ca^{2+} by an unknown mechanism; the Ca^{2+} binds to calmodulin, which in turn activates a network of actin filaments, attached both to the outer membrane of the chloroplast envelope and to the plasma membrane, which mediates the rotation. If a localized microbeam is used to illuminate only a small part of the cell, only the illuminated part of the

20 μm

Figure 20–52 Stamens from the flower of *Tradescantia* are covered in long thin hairs, each made from a single row of large cells. Light micrographs taken 5 seconds apart show rapid streaming along the cytoplasmic strands that cross the vacuole, with organelles moving at speeds of up to 5 μm/second.

chloroplast bends, indicating that different regions of the actin cytoskeleton in a single cell can respond independently (Figure 20–55).

The cortical microtubule array can also respond rapidly to external stimuli. As discussed earlier, the shape of a plant cell, and hence the shape of the plant, depends on the ordered deposition of oriented layers of cellulose, with the innermost layer being the most important (see p. 1168). Outer layers of cellulose in the wall are commonly found to have different orientations from the more recently laid-down inner layer. There are at least two mechanisms by which the new and old layers could end up with different orientations, and both seem to operate in plant cells: (1) the cellulose microfibrils in older layers of the wall could become rearranged with respect to one another as they move outward by breaking and re-forming bonds that cross-link wall polysaccharides, and (2) new wall layers could be deposited on the plasma membrane with orientations different from those outside them.

Since the orientation of new cell wall deposition is related to the orientation of the cortical microtubule array, any change in one implies a corresponding change in the other. Therefore, the second mechanism requires that microtubules change their orientation. Such dynamic changes in the helical array can be observed directly. The plant growth regulators ethylene and gibberellic acid, for example, have opposite effects on the orientation of microtubule arrays (and hence on the direction of cell expansion) in epidermal cells of young pea shoots. Gibberellic acid promotes a net orientation of the cortical microtubule array that is perpendicular to the long axis of the cell; this causes a corresponding cellulose deposition that allows the cells only to elongate, producing thin, elongated shoots. But if these shoots are treated with ethylene, the microtubule arrays reorient within an hour to a net longitudinal direction. Cellulose deposition in this direction ensues, and now the cells expand only laterally, producing fat, stunted shoots (Figure 20–56). The molecular mechanisms underlying these dramatic cytoskeletal rearrangements are still unknown.

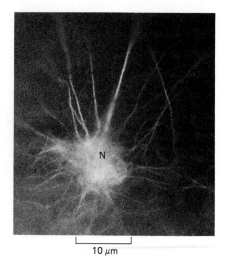

Figure 20–53 Fluorescence micrograph of a small portion of a large vacuolate stem cell showing actin filament bundles in the transvacuolar strands. The cell has been stained with rhodamine-labeled phalloidin, a drug that binds tightly and specifically to actin filaments (see p. 642). The actin bundles, radiating from the region around the nucleus (N), are believed to mediate the rapid streaming found in most vacuolate higher plant cells, presumably by a myosin-based process. (Courtesy of Clive Lloyd.)

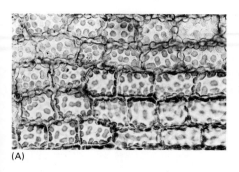
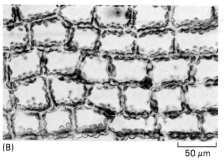

Figure 20–54 Light micrographs of leaf cells from a moss, showing how the chloroplasts move in response to light. The direction of illumination is perpendicular to the plane of the page. (A) In low light the disclike chloroplasts orient to maximize light absorption. (B) When the same area is examined after 30 minutes of bright illumination, the chloroplasts appear to have migrated, so that they are now lined up against the cell walls parallel to the incident light.

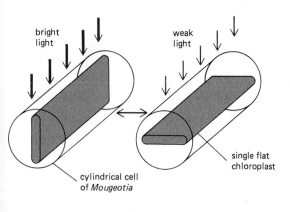
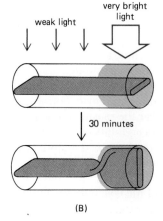

Figure 20–55 Chloroplast movements in the green alga *Mougeotia*. This cylindrical cell contains a single, large, flat chloroplast, which orients itself to regulate the amount of the light absorbed (A). A portion of the chloroplast can be induced to orient independently of the rest if it is exposed to a very bright light, showing that the response is locally mediated (B).

Summary

Two conspicuous organelles are unique to plant cells, the plastids and vacuoles. The plastids are a varied group of organelles that contain a common genome. The most familiar example is the photosynthetic chloroplast found in all green tissue. The vacuole is a large water-filled space surrounded by a membrane called the tonoplast. Vacuoles are used by plant cells in a variety of ways—as an economical means of space filling during cell enlargement, for example, and for the storage of food reserves or toxic metabolites. Although plant cells cannot move, their cytoplasm, particularly in cells with very large vacuoles, is kept stirred by means of active cytoplasmic streaming. In some cells this streaming has been shown to depend on actin filaments.

The internal organization of the plant cell and its cytoskeleton is important in orienting the deposition of the cell wall, which in turn determines the direction of cell expansion and therefore cell shape. The matrix components of the cell wall are made and exported by the Golgi apparatus, but the cellulose microfibrils are synthesized in situ at the cell surface. Both the site of deposition of the wall components and the specific orientation of the cellulose fibers are controlled by arrays of cortical microtubules. The components of the plant cytoskeleton, including both the microtubules and actin filaments, can reorganize rapidly in response to extracellular stimuli such as growth factors and light.

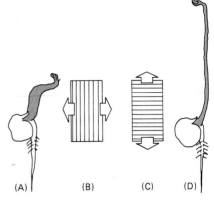

Figure 20–56 The plant growth regulators ethylene and gibberellic acid exert rapid and opposing effects on the orientation of the cortical microtubule array in cells of young pea shoots. Most ethylene-treated cells (B) show a net longitudinal orientation of microtubules, while most gibberellic acid-treated cells (C) show a net transverse orientation. New cellulose microfibrils are deposited parallel to the microtubules. Since this influences the direction of cell expansion, gibberellic acid and ethylene encourage growth in opposing directions: ethylene-treated seedings will develop short, fat shoots (A), while gibberellic acid-treated seedlings will develop long, thin shoots (D).

Cellular Aspects of Plant Development

Despite the staggering variety of flowering plants in the world, certain features of their form and development are remarkably constant. The presence of a cell wall constrains plants to strategies for reproduction, growth, and development that are very different from those adopted by animals. In this section we shall examine some of these general features and discuss their cellular basis. The special strategies that plants use to reproduce are summarized in Panel 20–2, p. 1154. We begin with the fertilized zygote and follow some of the early events in its development. Plants, like animals, make extensive use of spatially regulated cell differentiation. In place of the migrations and reassortments of cells that play such a major part in the development of animal embryos (see Chapter 16), however, oriented cell divisions and strictly regulated cell expansions are crucial to the morphogenesis of plants. These processes are influenced by environmental factors such as light, nutrients, and gravity, all acting synergistically with low-molecular-weight regulators of plant cell growth. The availability of these plant "growth factors" has made it possible to grow plant cells and tissues in culture, providing powerful ways to study plant growth and development and to manipulate plants genetically.

Embryonic Development Starts by Establishing a Root-Shoot Axis and Then Halts Inside the Seed[28]

In the earliest stage of development, the zygote of a higher plant divides asymmetrically to establish the polarity of the future embryo. One product of this division is a small cell with dense cytoplasm, which will become the embryo proper. The other is a large vacuolated cell that divides further and forms a structure called the *suspensor*, which in some ways is comparable to the placenta in mammals. The suspensor attaches the embryo to the maternal tissue and also provides a pathway for the flow of nutrients to the developing embryo from the surrounding *endosperm* tissue (the endosperm is a triploid nutritive tissue that is formed separately during the fertilization process—see Panel 20–2, p. 1154).

During the next step in development, the diploid embryo cell proliferates to form a ball of cells in which the first *meristems* of the plant are established. Cell division in plants is confined almost entirely to specialized regions called **meristems,** which are composed of self-renewing stem cells. Meristems are usually of two types: (1) *apical meristems*, at the tips of growing roots and shoots, are involved primarily in adding new cells that will elongate and differentiate into root tissues on the one hand and stem tissues plus leaf primordia on the other; and (2) *lateral meristems*, circumferentially arranged inside the plant, are involved primarily in producing cells that increase the girth of the plant (Figure 20–57). An important lateral meristem in all plants is called the *cambium*—the layers of cells that give rise to the vascular tissues.

The ball of embryonic cells quickly becomes polarized into two groups of meristematic cells—one at the suspensor end of the embryo, which will become a root apical meristem, and one at the opposite pole, which will become a shoot apical meristem. These two meristems define the main root-shoot axis of the future plant. Subsequently, as the embryo elongates, the vascular cambium is formed to provide the future transport tissue between root and shoot. Slightly later in development, the apical meristem of the shoot produces the embryonic seed leaves, or cotyledons—one in the case of monocots and two in the case of dicots (see Figure 20–12). At this stage, further development usually ceases and the embryo becomes packaged in a *seed*, which is specialized for dispersal and for survival in harsh conditions.

The embryo in a seed is stabilized by dehydration, and it can remain dormant for a very long time—even thousands of years, as documented in the case of cereal grains from Egyptian tombs. When rehydrated, the seeds germinate and embryonic development continues. Embryos depend for their development on the large food reserves accumulated in the triploid tissue of the endosperm (see Panel 20–2). During development of the seed, this food material may be transferred to a varying extent to the cotyledons. Thus the seeds of monocots (such as wheat and corn) and some dicots keep large endosperm reserves, whereas the seeds of the majority of dicots (including peas and beans) rely on the nutrients stored in the fleshy cotyledons.

Meristems Continually Produce New Organs and New Meristems, Creating a Repeating Series of Similar Modules[29]

A vertebrate embryo is a small version of the adult, already possessing most of its characteristic organs. The plant embryo inside the seed, in contrast, looks nothing like the adult plant. Each of its two meristems, however, has the capacity to produce a full complement of new organs at the appropriate time. As soon as the seed coat ruptures during germination, rapid cell expansion occurs, with the root emerging first to establish an immediate foothold in the soil. This is followed by rapid and continual cell divisions in the apical meristems: in the apical meristem of a maize root, for example, cells divide every 12 hours, producing 5×10^5 cells per day. The rapidly growing roots and shoots probe the environment—the roots increasing their capacity for taking up water and minerals from the soil, the shoots increasing their capacity for photosynthesis (see Panel 20–2, p. 1154).

Two sorts of plant structures are formed by apical meristems: (1) the main root and shoot system, whose growth perpetuates the meristems themselves, and (2) structures such as leaves and flowers, whose growth is limited. The former are responsible for continuous development and persist for the life of the plant (and in the case of grafts, like the McIntosh or Cox's Orange Pippin apples, for considerably longer). The latter have a short lifetime, invariably undergoing programmed senescence. Contrast the seasonal turnover of leaves with the virtual absence of turnover among the somatic cells of the rest of the plant.

As the shoot elongates, the apical meristem produces an orderly sequence of *nodes* and *internodes*. In this way the continuous activity of the meristem produces an ever-increasing number of similar **modules,** each consisting of a stem, a leaf, and a bud (Figure 20–58). The modules are connected to one another by supportive and transport tissue, and successive modules are precisely located relative to each other, giving rise to the patterned structures that are characteristic of plants (Figure 20–59).

Each individual module has a degree of autonomy, so that in many respects one can regard mature plants as having some of the characteristics of colonial organisms such as sponges and corals. Thus each module begins as an identically sized swelling on the apical meristem, but the local environment it encounters

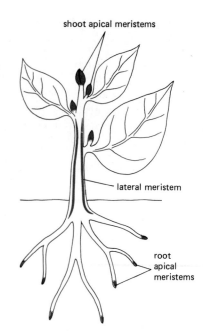

Figure 20–57 Highly schematic diagram of a higher plant, showing the locations of the main meristems, the areas of most rapid and continuous cell division. The activity of the shoot and root apical meristems is responsible for increasing the length, while the activity of a lateral meristem, such as the cambium, is responsible for increasing the girth of the various parts of the plant.

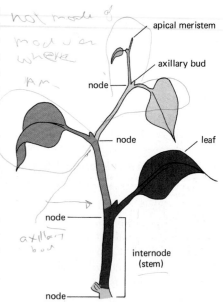

Figure 20–58 A simple example of the modular construction of plants. Each module (shaded in different colors) consists of a stem, a leaf, and a bud containing a potential meristem. Modules arise sequentially from the continuous activity of the apical meristem.

Figure 20–59 Accurate placing of successive modules from a single apical meristem produces these elaborate but regular patterns in leaves (A), flowers (B), and fruits (C). (Courtesy of Andrew Davies.)

(A)　　　　　(B)　　　　　(C)

during growth can modify its final form. Unusually large, dark green leaves will usually develop in a shady spot, for example, while short, thick leaf stems will develop in a windy one. Experiments show that these local adaptations are properties of each module, not of the entire plant. Moreover, in the acute angle between the stem and leaf branch (the axil), a bud is formed in each module that contains a new apical meristem, which can potentially produce another branch. Many of these new meristems, however, will become active only if the original apical meristems are damaged or removed—a fact well known to gardeners, who pinch off branch tips to stimulate side growth.

The modular nature of plants is emphasized by the experimental practice of meristem culture. Using suitable growth media, one can induce excised apical meristems to grow and regenerate complete normal plants, and this method is now used in horticulture to propagate valuable specimens, roses, orchids, and some crop plants such as strawberries and sugar cane.

The Form of the Plant Is Determined by Pattern-Formation Mechanisms in the Apical Meristems[30]

The repeating pattern of internode shoot growth followed by node differentiation requires that the cells in the apical meristem be able to sense their distance and orientation from the previous node, so as to differentiate appropriately. As in animal development, the pattern-formation mechanisms are required to operate over distances of less than a millimeter and therefore involve only relatively small groups of cells (see p. 914). Thus the precursors of each new node (an incipient branch point) can be detected as successive swellings that form near the very tip of the meristem, each forming from perhaps 100 cells (Figure 20–60). In the sim-

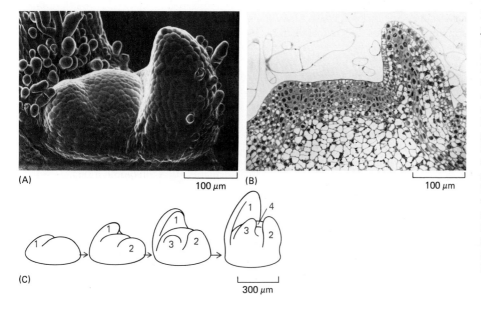

(A)　　　　　100 μm　　　　(B)　　　　　100 μm

(C)

300 μm

Figure 20–60 A shoot apex from a young tobacco plant. (A) A scanning electron micrograph shows the shoot apex with two sequentially emerging leaf primordia, seen here as lateral swellings on either side of the domed apical meristem. (B) A thin section of a similar apex shows that the youngest leaf primordium arises from a small group of cells (about 100) in the outer four or five layers of cells. (C) A very schematic drawing showing that the sequential appearance of leaf primordia takes place over a small distance and very early in shoot development. Growth of the apex will eventually form internodes that will separate the leaves in order along the stem (see Figure 20–58). (A and B, from R.S. Poethig and I.M. Sussex, *Planta* 165:158–169, 1985.)

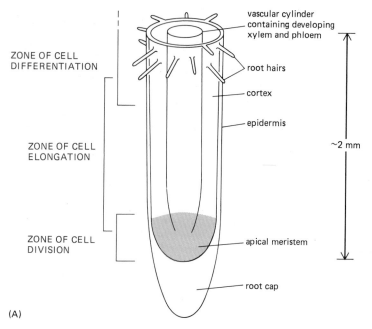

(A)

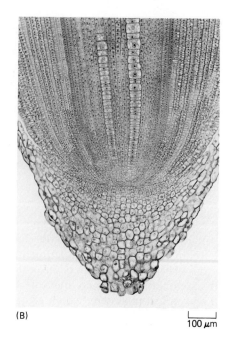

(B)

100 μm

Figure 20–61 (A) The organization of the final 2 mm of a growing root tip. The approximate zones in which cells can be found dividing, elongating, and differentiating are indicated. (B) The apical meristem and root cap of a corn root tip, showing the orderly files of cells produced. (B, from P.H. Raven, R.F. Evert, and S.E. Eichhorn, Biology of Plants, 4th ed. New York: Worth, 1986.)

plest models proposed to explain this process, each new node produces a diffusible *morphogen* that inhibits the formation of the next node until the morphogen is diluted out by a fixed distance of internode growth. Almost nothing is known, however, about the molecular mechanisms that mediate this central patterning mechanism in the plant kingdom.

The Generation of New Structures Depends on the Coordinated Division, Expansion, and Differentiation of Cells[31]

Our discussion of higher plant development has so far been largely at the descriptive level of tissues and organs. What is the cellular basis of these complex events? Since plant cells are immobilized by their cell walls, plant morphogenesis must depend on regulated cell division coupled to strictly oriented cell expansion. Most cells produced in the root tip meristem, for example, go through three distinct phases of development: division, growth (elongation), and differentiation. These three steps, which overlap in both space and time, give rise to the characteristic architecture of a root tip. Although the process of cell differentiation often begins while a cell is still enlarging, it is comparatively easy to distinguish in a root tip a zone of cell division, a zone of cell elongation (which accounts for the growth in length of the root), and a zone of cell differentiation (Figure 20–61). When differentiation is complete, some of the differentiated cell types remain alive (for example, phloem cells) while others die (for example, xylem vessel elements).

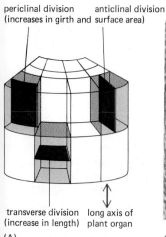

(A)

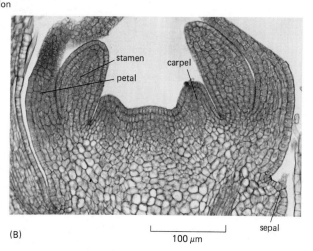

(B)

100 μm

Figure 20–62 The relationship among division plane, cell expansion, and morphogenesis. (A) Three planes of cell division found in a typical plant organ. Variations in the relative proportion of each, combined with oriented cell expansion, can account for the morphogenetic patterns found in plants. (B) A longitudinal section of a young flower bud of a periwinkle. The small domes of cells destined to become the different floral parts have arisen by a combination of new planes of cell division and directional cell expansion determined by the reinforcing hoops of cellulose in the cell wall. (From N.H. Boke, *Am. J. Bot.* 36:535–547, 1949.)

The exact plane in which cells divide is crucial to plant morphogenesis, and many developmental processes, such as the development of stomata or leaf primordia, begin with a change in the plane of cell division. The continual cell divisions in a meristem often give rise to columns of daughter cells oriented parallel to the growth axis (see Figure 20–61), and changes in the plane of division are often associated with morphogenetic events. *Anticlinal* divisions occur frequently in the epidermis, and provide additional cells in the surface layers. These can accommodate the increase in girth that will result from *periclinal* divisions (Figure 20–62A). A combination of local changes in the frequency and orientation of cell divisions is responsible for the formation of organs such as flowers, leaves, and lateral roots (Figure 20–62B).

We have already seen that the turgor-driven expansion of a plant cell, which can often increase its volume fiftyfold or more, is oriented primarily by the orientation of cellulose microfibrils in the cell wall, which in turn is determined by the orientation of the cortical microtubule array. The cytoskeleton also plays a role in defining the plane of cell division.

A Cytoskeletal Framework Determines the Plane of Cell Division[32]

The organized growth of a plant requires that cells in selected sites divide in a particular plane and at a particular time so as to set up correctly oriented cell lineages. Microtubules play an early part in determining the plane of cell division. Thus the first visible sign that a higher plant cell has become committed to divide in a particular plane is seen just after the interphase cortical array of microtubules disappears in preparation for mitosis. At this time a narrow circumferential band of microtubules appears and forms a ring around the entire cell just beneath the plasma membrane (Figure 20–63). Because this array of microtubules appears in G_2 before prophase begins, it is called the **preprophase band.** The band disap-

Figure 20–63 The preprophase band of microtubules in a higher plant cell. (A) Simplified diagram of the changing distribution of microtubules during the cell cycle in higher plant cells, emphasizing four types of arrays. As cells leave interphase and enter prophase, the cortical array of microtubules (1) becomes bunched up into a dense *preprophase band* (2), which is 1–3 μm wide and often contains more than a hundred microtubules. This preprophase band predicts the subsequent plane of cell division. The mitotic spindle (3) then forms and aligns the chromosomes on the metaphase plate (4). Finally, microtubules are organized during cytokinesis into the phragmoplast (5), which lays down the new cell wall (6) (see Figure 13–71, p. 782). The interphase array is reestablished (8) in the daughter cells from microtubules polymerized from amorphous centrosomal material at the surface of the nucleus (7). (B) Immunofluorescence staining of microtubules in the preprophase band of an onion root tip cell. At the left is shown the isolated cell, while on the right two planes of focus clarify the relationship of the band to early spindle formation outside the nucleus. (C) Even though the two cells shown here are of similar shape, the preprophase band predicts that the top cell will divide transversely while the bottom cell will divide longitudinally. (B and C, courtesy of Kim Goodbody and Clive Lloyd.)

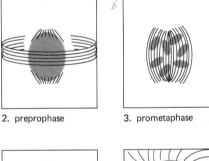

1. interphase 2. preprophase 3. prometaphase 4. metaphase

5. telophase 6. cytokinesis 7. early interphase 8. interphase

(A)

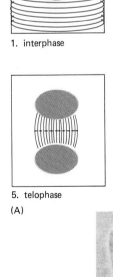

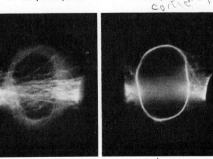

(B)

(C)

10 μm

10 μm

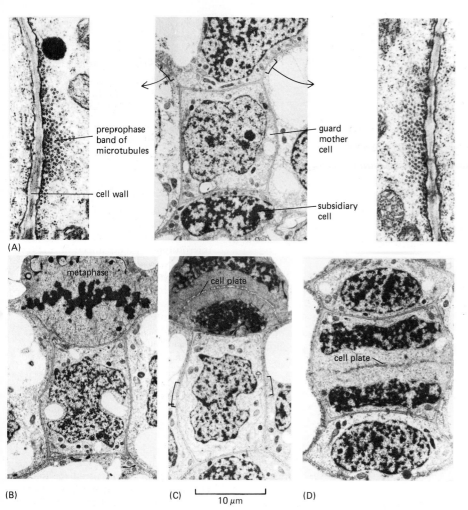

preprophase
band of
microtubules

cell wall

(A)

guard
mother
cell

subsidiary
cell

(B) metaphase

(C) cell plate

(D) cell plate

10 μm

Figure 20–64 Sequence of electron micrographs of cells in the epidermis of a sugar cane leaf, showing how the formation of stomata involves a precise sequence of symmetric and asymmetric cell divisions. The plane of each division is accurately predicted by a preprophase band. (A) The uppermost cell is about to divide asymmetrically to form a subsidiary cell (the one at the bottom has already formed). The position of the preprophase band is indicated by brackets, and the band is shown in greater detail on either side. (B) Somewhat later, the upper cell is in metaphase, and the preprophase band has gone. (C) The cell plate is now forming in the upper cell during cytokinesis and is curving down toward the position of the former preprophase band. Meanwhile, the large central cell, the guard mother cell, is about to divide symmetrically, and the position of its preprophase band is indicated by brackets. (D) Cytokinesis is complete in the upper cell and almost complete in the guard mother cell. The latter division will produce the two guard cells that flank the stomatal pore (see Figure 20–11). (Courtesy of C. Busby.)

pears before metaphase is reached, yet the boundary of the division plane has somehow been imprinted: when the new cell plate forms later during cytokinesis, it grows outward to fuse with the parental wall precisely at the zone that was formerly occupied by the preprophase band (Figures 20–64). Even if the cell contents are displaced by centrifugation after the preprophase band has disappeared, the growing cell plate will tend to find its way back to the plane defined by the former preprophase band.

It is now known that the preprophase band contains numerous actin filaments in addition to microtubules. The actin filaments are not confined to the cell cortex but also form a radial, disclike array of strands, which crosses the vacuole and connects to and supports the central dividing nucleus. After the microtubules in the preprophase band depolymerize, these radial actin strands remain and provide a "memory" of the predetermined division plane. During cytokinesis, as the phragmoplast grows out centrifugally like a circular ripple in a pond, the edges of the growing cell plate are connected to the site of the preprophase band by actin filaments (see Figure 13–73, p. 783).

Regardless of whether the division is symmetric or asymmetric, or whether it is transverse, periclinal, or anticlinal, the preprophase band in a plant cell always specifies where the cell is going to divide *before* it enters mitosis (see Figure 20–63C). Such spatial controls are especially important in the asymmetric cell divisions that create two daughter cells with different developmental fates: stomatal cells, root hair cells, and the generative cells of pollen grains, for example, all develop from the smaller of two daughter cells. In such divisions the nucleus moves to the appropriate position in the cell before mitosis (Figure 20–65). Although the mechanism of nuclear migration is uncertain, there is evidence that both microtubules and actin filaments are involved.

The precise control over cell division planes that we have been describing is possible only if the plant tissue and its constituent cells have a structural polarity that can subsequently be either consolidated or modified. Although the structural basis for the establishment of cell polarity in higher plants is not known, there are well-studied examples among lower plants that may provide relevant models.

The Polarity of Plant Cells Depends on the Asymmetric Distribution of Membrane-bound Ion Channels and Carrier Proteins[33]

The polarity of a plant and its constituent cells is remarkably stable: small lengths of stem (even upside down) always regenerate roots and shoots at their original basal and apical ends, respectively. It is still unclear how such cell polarity is established, maintained, and transmitted to daughter cells, but there is evidence that asymmetrically distributed ion channels and carrier proteins in the plasma membrane may help to establish cell polarity by generating intracellular ion currents.

A particularly well-studied example is the common brown seaweed *Fucus*. This large alga releases thousands of free-living sperm and eggs into the sea water around it. The large spherical fertilized eggs of *Fucus* are initially homogeneous, with no intrinsic polarity. Within 18 hours, however, the zygotes become polarized and undergo an asymmetric division in which the smaller, basal cell will develop into a structure for anchoring the seaweed to a rock and the larger, apical cell will develop into the thallus (the photosynthetic part). A clue to the mechanism of polarization of the early zygote is a small current that flows across it. This current is generated and maintained in part by a passive Ca^{2+} influx at the basal pole and an active efflux of Ca^{2+} that occurs elsewhere, which presumably reflect an asymmetrical distribution of Ca^{2+} channels and Ca^{2+} pumps, respectively, in the plasma membrane. This current flow through the egg is capable of moving highly charged intracellular components by electrophoresis, and there is evidence that one result of the polarization is an accumulation of secretory vesicles at the basal pole, followed by the localized deposition of specific sulfated polysaccharides in the cell wall there.

The unpolarized *Fucus* zygote becomes polarized in response to external stimuli, usually in the form of a continuous gradient—for example, of light or gravity. The information in the gradient is thought to lead to an initial inhomogeneity in the distribution of Ca^{2+} in the egg (Figure 20–66). Even small inhomogeneities in Ca^{2+} could have large effects, since they could be rapidly amplified and propagated by feedback loops to produce the sustained, self-driven Ca^{2+} flux that is observed. Although it is unclear how the zygote axis becomes stabilized, the process requires both actin filaments and a cell wall. Perhaps the cytoskeletal elements help connect the Ca^{2+} channels to wall fibrils by transmembrane bridges at the future basal pole.

Intracellular Ca^{2+} gradients are thought to exist in many other types of plant cells that exhibit polarized growth. Electrical currents carried in part by Ca^{2+}, for example, have been shown to enter the growing apices of a variety of polarized structures—including root hairs, pollen tubes, and whole roots—and to leave at nongrowing regions. Thus the observations on *Fucus* eggs may have general relevance for the generation of plant cell polarity.

Plant Growth and Development Are Modulated in Response to Environmental Cues[34]

Environmental conditions often have a much more profound influence on the development of plants than they do on the development of animals. Plants have evolved intricate systems for monitoring gravity, nutrient status, temperature, and the intensity and duration of light. The responses to these stimuli are complex and can be either rapid and brief (as in the light-induced chloroplast movements discussed on p. 1171) or slow and long-lasting (as in the long periods of daylight required to induce some plants to flower, or the long periods of cold required before many seeds will germinate).

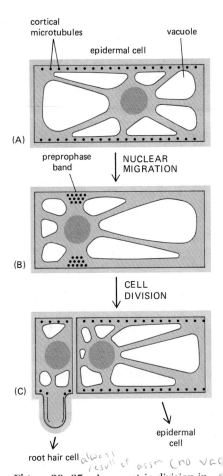

Figure 20–65 Asymmetric division in an elongated epidermal cell from a root. In interphase (A), cortical microtubules are distributed along the length of the cell wall. During preprophase (B), however, they congregate in a discrete band circling the cell. This preprophase band of microtubules accurately predicts where the new cell wall will join the old one when the cell divides (C). This epidermal cell divides asymmetrically to form a large daughter cell that will continue as an epidermal cell and a small daughter cell that will become a root hair cell.

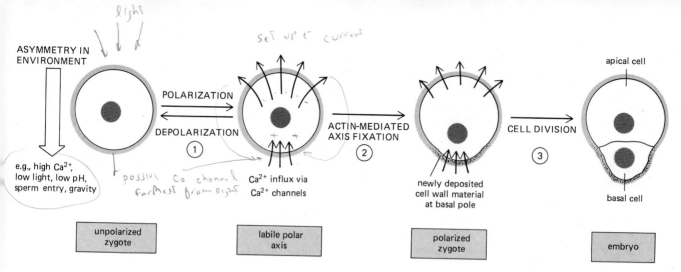

The handwritten annotations on the figure read: "light", "set up e⁻ current", "possible Ca channel farthest from light"

Figure 20–66 Schematic diagram illustrating how the fertilized egg of the brown seaweed *Fucus* might become polarized (1). In response to one of the indicated asymmetries in the environment, Ca^{2+} channels and pumps in the plasma membrane become localized (or activated) and establish a Ca^{2+} flux across the cell (2). This initially labile axis becomes fixed in an actin-dependent step, and negatively charged cell-wall polysaccharides are then deposited at the future basal end (3). This initial structural asymmetry is perpetuated by means of the first cell division, which produces a large apical cell and a small basal cell. The polarity of higher plant cells may reside in a similar asymmetric distribution of ion channels and pumps in the plasma membrane.

The responses to light are among the most elaborate, and they fall into two classes. Long-term responses generally involve changes in gene expression, while short-term responses usually do not. Although there are several types of photoreceptor molecules in plants, the best characterized is the large (124,000-dalton) protein **phytochrome.** This protein contains a chromophore that is responsive to light and can exist in two interconvertible forms: an *inactive* form produced by exposure to far red light (between visible red and infrared), and an *active* form produced by red light. Phytochrome is known to mediate many light-activated responses, including plastid differentiation, seed germination, stem elongation, leaf initiation, and flowering.

Some clues to how activated phytochrome triggers competent cells to respond have been provided by experiments with the fern *Onoclea*. Spores of this fern require red light to germinate, but germination is inhibited by far red light. The initial signal generated by an activated phytochrome molecule appears to be a transient influx of Ca^{2+} from the cell exterior, which increases cytosolic Ca^{2+} levels. Ca^{2+} ionophores promote germination in the absence of light. It is not clear how phytochrome mediates the longer-term, light-induced responses that involve alterations in gene transcription; perhaps protein kinases triggered by a Ca^{2+} influx phosphorylate, and thereby activate, transcription factors that are specific for light-regulated genes.

Phytochrome can act over a remarkably wide range of light levels, from single photons to full sunlight, with each process being saturated at a different light intensity. In almost all such processes, however, light acts synergistically with other environmental cues, as well as with one or more plant growth regulatory molecules.

Plant Growth and Development Are Regulated by Chemical Mediators[35]

Many compounds, including inorganic nutrients, have profound effects on normal plant growth. But one group of signaling molecules in particular, the **plant growth regulators** (sometimes called *plant hormones*), acts specifically in this capacity and plays a crucial part in plant development. There are five known classes of plant growth regulators: **auxins, gibberellins, cytokinins, abscisic acid,** and the gas **ethylene.** As shown in Figure 20–67, all are small molecules that readily penetrate cell walls. They are all synthesized by most plant cells and can either act locally or be transported to influence target cells at a distance. For example, auxin (which, among its other properties, induces vascular bundle formation) is transported from cell to cell at a rate of about 1 cm per hour from the tip of a shoot to its base. Each growth regulator has multiple effects, and these are modulated by the other growth regulators, as well as by environmental cues and nutritional status. Thus auxin alone can promote root formation; but in conjunction with gibberellin it can promote stem elongation, with cytokinin it can suppress lateral shoot outgrowth, and with ethylene it can stimulate lateral root growth.

An important general principle in chemical signaling, in both plants and animals, is that each cell type is programmed to respond to particular signals in specific ways (see p. 684). Moreover, a given cell type often responds differently to the same signal depending on its stage of maturation. Thus the ripening of fruit is triggered by ethylene (see p. 1150), but cells acquire their sensitivity to ethylene only late in fruit development. Similarly, in barley seeds, gibberellic acid released by the embryo triggers the release of enzymes that mobilize starch reserves in the endosperm, but the responsive cells in the seed coat become sensitive to the gibberellic acid only after they have been through the dehydration process required for seed maturation.

It is not known with certainty how plant growth regulators signal their target cells. The corresponding receptor proteins have not yet been purified or well characterized. As in animal cells, however, there seem to be at least two classes of receptors: (1) those in the plasma membrane that activate ion channels or the inositol phospholipid signaling pathway (and possibly other pathways still to be defined), and (2) those inside the cell that act on the nucleus to regulate gene expression.

The availability of pure plant growth regulators has made it possible to grow plant cells and tissues in culture, providing a powerful way to study certain aspects of plant growth and development.

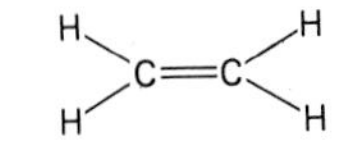

Cell and Tissue Culture Are Powerful Tools for Studying Plant Development[36]

Under certain conditions the somatic cells in a mature plant can resume cell division or differentiate into another cell type. This plasticity enables plants to repair themselves after damage. If a stem is cut, for example, some nearby cells are stimulated to divide and repair the wound, while others redifferentiate into xylem cells to re-form the vascular continuity around the wound (Figure 20–68).

Plant cells also display their plasticity in culture. When a piece of plant tissue is cultured in a sterile medium containing nutrients and appropriate growth regulators, many of the cells are stimulated to divide. Such cultured cells often proliferate indefinitely in a disorganized manner, producing a mass of relatively undifferentiated cells called a *callus*. By carefully manipulating nutrients and growth regulators, shoot and then root apical meristems can be induced to form within the callus, and in many species a whole new plant can be regenerated (Figure 20–69).

Callus cultures can be mechanically dissociated into single cells that will grow and divide as a suspension culture. The cells in such cultures all look much alike, having thin primary cell walls and large vacuoles crossed by thin cytoplasmic strands (Figure 20–70). In a number of plants—including tobacco, petunia, carrot, and potato—a single cell from a suspension culture can be grown into a small clump (a clone) from which a whole plant can be regenerated. This ability of a single somatic cell to give rise to a whole new plant dramatically illustrates the *totipotency* of many plant cells.

Plants regenerated from cell culture display an unexpectedly high degree of genetic variability, a phenomenon referred to as *somaclonal variation*. The same phenomenon is seen when intact plants are stressed. Under stressful conditions (and this includes culture conditions), plant cells undergo a variety of chromosomal rearrangements that rapidly generate a wide variety of genetic changes (see p. 607). Plants appear to be much more tolerant than animals of major genomic changes, and when new gametes form from the altered somatic cells, some of the mutant plants produced may have an advantage in adapting to a changing and hostile environment. In fact, somaclonal variation is being exploited by modern agricultural scientists to generate improved strains of crop plants.

Molecular Genetics Is Helping to Solve Problems in Plant Development

The ability to regenerate a plant from cloned somatic cells in culture has made it possibile to manipulate cells genetically to produce transgenic plants. One way to facilitate the introduction of foreign DNA into these cells is to use plant cells that have had their cell walls removed with cell-wall-degrading enzymes to produce

ethylene

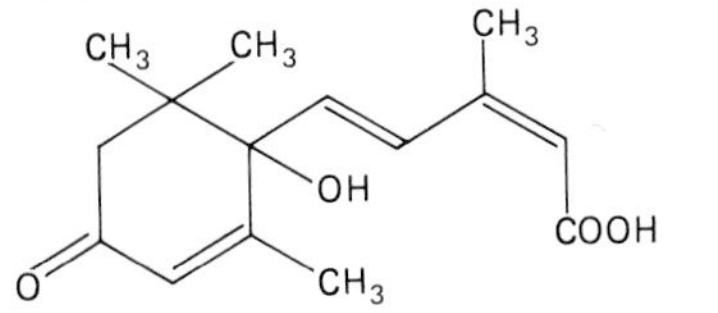

abscisic acid (ABA)

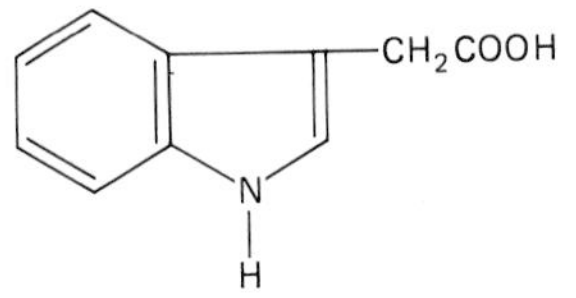

indole-3-acetic acid (IAA) [an auxin]

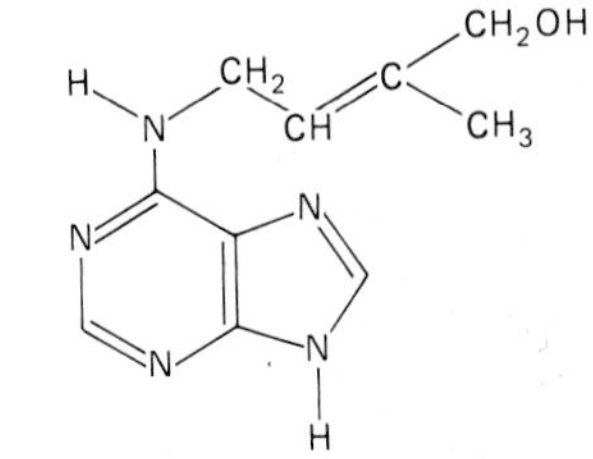

zeatin [a cytokinin]

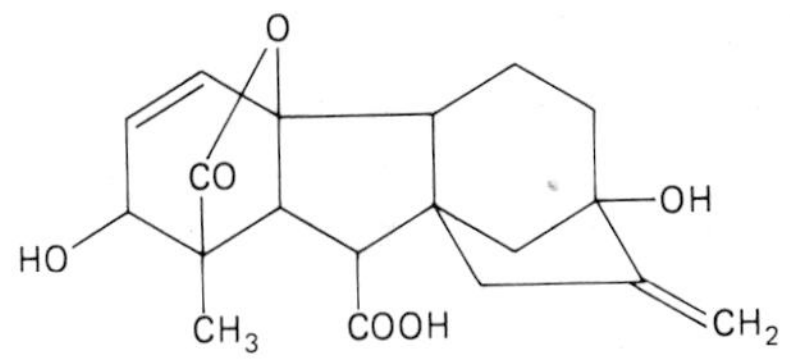

gibberellic acid (GA3) [a gibberellin]

Figure 20–67 The formula of one naturally occurring representative molecule from each of the five groups of plant growth regulatory molecules.

protoplasts (Figure 20–71). After foreign DNA has been introduced into these naked spherical plant cells, they can often be encouraged to re-form their walls, divide, and regenerate into a new plant.

A common vector used to introduce foreign DNA directly into normal plant cells is the T-DNA of *Agrobacterium* described earlier (see p. 1159). For this purpose the genes in the T-DNA that cause plant tumors are removed by recombinant DNA methods and replaced by a desired new gene. After the modified T-DNA is transferred into an appropriate Ti plasmid in *Agrobacterium*, the bacteria are cocultured with a piece of a leaf. This system allows the integration of the T-DNA into a plant chromosome, thereby adding the new gene permanently to the entire plant (Figure 20–72). At present this method works efficiently only on certain families of dicots.

Recent developments such as these have provided plant cell biologists with the same range of recombinant DNA techniques used by animal cell biologists, greatly facilitating progress in such important research areas as isolating receptors for plant growth regulators, unraveling morphogenetic pathways, and analyzing mechanisms of gene expression. In addition, plant recombinant DNA technology has opened up many new possibilities in agriculture that could benefit both the farmer and the consumer. These possibilities include the ability to modify the lipid, starch, and protein storage reserves in seeds, to impart pest and virus resistance to plants, and to create modified plants that tolerate extreme habitats such as salt marshes or waterlogged soil.

Many of the major advances in understanding animal development have come from studies on invertebrates that are amenable to extensive genetic analysis as well as to experimental manipulation, such as the fruit fly *Drosophila* (see p. 920) and the nematode worm *Caenorhabditis* (see p. 901). Progress in plant developmental biology has been relatively slow in comparison. Many of the organisms that have proved most amenable to genetic analysis, such as maize and tomato, have long life cycles and very large genomes, which have made both classical and molecular genetic analysis time-consuming. Increasing attention is consequently being paid to a small weed, the common wall cress (*Arabidopsis thaliana*), which has several major advantages as a "model plant." It is so small (Figure 20–73) that it can be grown indoors in test tubes in large numbers. With a minimum generation time of only 5 weeks, it can produce thousands of offspring per plant after 8 to 10 weeks. *Arabidopsis* also has the smallest plant genome known (7×10^7

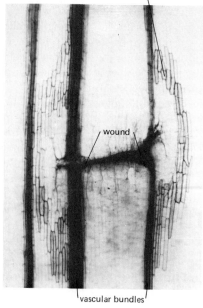

Figure 20–68 Light micrograph showing a wound response by the parenchymal cells that surround the vascular bundle in the common houseplant *Coleus*, seen in a longitudinal section. The vascular bundles have been severed at the wound. Seven days later, after stimulation of division and redifferentiation in the nearby cortical cells, a series of new xylem and phloem cells have been generated that restore vascular continuity around the wound. A supply of auxin is normally carried from the shoot apex to the base of the plant, and it is thought that the wounding provides a newly directed flux of auxin that in turn induces the vascular regeneration shown here. (Courtesy of N.P. Thomson.)

[handwritten annotations: callus; anxin cytokinis on relation conc.; High conc. of cytokin; High Auxin low cytokines → growth roots]

Figure 20–69 The production of self-supporting plants from callus cultures. The *Freesia* callus culture in (A) was induced to produce shoots (B) and then roots (C) by altering the available ratios of an auxin and a cytokinin. A high cytokinin ratio appears to be required for shoot and leaf induction. (Courtesy of Graham Hussey.)

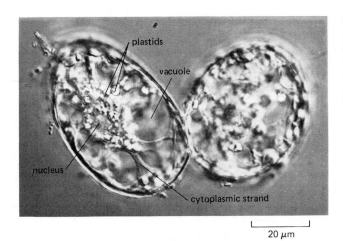

Figure 20–70 Callus cells can be grown as free-living, single cells suspended in liquid media. Two such cells derived from a sycamore callus are shown here. They are highly vacuolated cells with cytoplasmic strands radiating from the region of the nucleus, and they have a primary cell wall.

20 μm

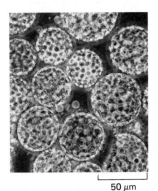

Figure 20–71 Light micrograph of protoplasts prepared from the green leaf cells of a tobacco plant. Without their walls the cells round up and need to be stabilized in a sugar solution that matches the osmotic pressure of their cytoplasm. Numerous chloroplasts can be seen in each protoplast. (Courtesy of J. Burgess.)

50 μm

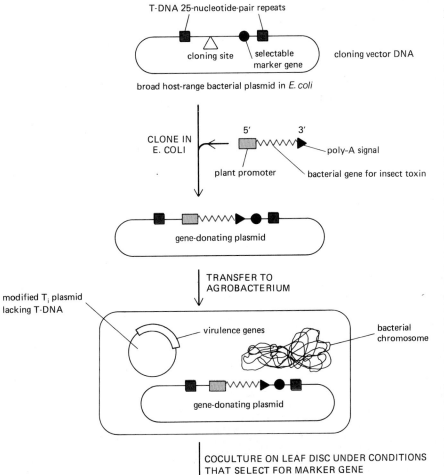

Figure 20–72 A strategy used to obtain transgenic plants. In this example the gene being transferred codes for a bacterial protein that is toxic to insects. To allow the gene to be expressed in a plant cell, the 5′ end of the gene is attached to a plant promoter and the 3′ end to a poly-A addition site. The modified toxin gene is inserted into a plasmid that also contains a marker gene (such as a kanamycin resistance gene) that can be selected for in the plant (see p. 259). The plasmid is designed so that both the marker gene and the toxin gene are flanked by the special 25-nucleotide-pair repeats that normally flank T-DNA. The plasmid is grown in *E. coli* and then transferred to an *Agrobacterium* that contains the virulence genes on a separate plasmid. When the *Agrobacterium* is cocultivated with a leaf disc, the virulence gene products recognize the T-DNA repeats and transfer the DNA that contains the marker and toxin genes to a plant chromosome. All of the cells in the leaf disc are stimulated to proliferate by the growth medium, but only those plant cells that contain the selectable marker gene are able to divide and form a callus. The callus is then used to produce transgenic plants, which express the bacterial gene for the insect toxin and are therefore unusually resistant to insect attack.

nucleotide pairs), comparable to yeast (2×10^7 nucleotide pairs), *C. elegans* (8×10^7 nucleotide pairs), and *Drosophila* (10×10^7 nucleotide pairs). Cell culture and transformation methods have been established, large numbers of interesting mutants have been isolated, and work is underway to create a complete ordered collection of genomic DNA clones (see p. 260). Taken together, the attractive features of *Arabidopsis* suggest that it may soon become the *Drosophila* of plant developmental biology.

Summary

In the earliest stages of embryonic development in a higher plant, cell division takes place throughout the body of the embryo. As the embryo grows, however, addition of new cells becomes restricted to certain regions of the plant body known as meristems. The adult plant can be viewed as a series of repeating modules constructed by the pattern-formation mechanisms that occur in these meristems.

Plant morphogenesis depends on the coordinated division, expansion, and differentiation of nonmotile cells. Control over both the plane of cell division and the oriented expansion of plant cells is exerted in part by microtubules associated with the inner face of the plasma membrane. The growth and division of plant cells is influenced by light, gravity, temperature, and other environmental factors, as well as by specific low-molecular-weight growth regulators, such as auxins and cytokinins.

Many aspects of plant cell growth and development can be studied in tissue and cell culture. Plant somatic cells show a plasticity of developmental potential not shown by the somatic cells of animals. A dramatic demonstration of the totipotency of many plant cells is the routine regeneration of a complete plant from a single cultured somatic cell.

Protoplasts are plant cells without their cell walls. They can be manipulated in culture in much the same way that animal cells can, but they have the added property that whole plants can be regenerated from them.

Both protoplasts and intact cells can be manipulated genetically by recombinant DNA methods to introduce any desired new gene into their chromosomes; in many cases these cells can subsequently be used to produce transgenic plants.

Figure 20–73 *Arabidopsis thaliana*, a small member of the mustard—or crucifer—family, is a weed of no economic value but of potentially great value for experiments in the molecular genetics of plants. Its small genome (7×10^7 nucleotide pairs) is comparable in size to the *Drosophila* genome (10×10^7 nucleotide pairs). (Courtesy of Chris Sommerville.)

References

General

Cutter, E.G. Plant Anatomy, 2nd ed., Part 1, Cells and Tissues; Part 2, Organs. London: Arnold, 1978.

Esau, K. Anatomy of Seed Plants, 2nd ed. New York: Wiley, 1977.

Grierson, D.; Covey, S.N. Plant Molecular Biology, 2nd ed. New York: Methuen, 1988.

Gunning, B.E.S.; Steer, M.W. Ultrastructure and the Biology of Plant Cells. London: Arnold, 1975.

Raven, P.H.; Evert, R.F.; Eichhorn, S.E. Biology of Plants, 4th ed. New York: Worth, 1986.

Roberts, K.; Johnston, A.W.B.; Lloyd, C.W.; Shaw, P.J.; Woolhouse, H.W., eds. The Cell Surface in Plant Growth and Development. *J. Cell. Sci.*, Suppl. 2. Cambridge, U.K.: Company of Biologists Ltd, 1985.

Salisbury, F.B.; Ross, C.W. Plant Physiology, 3rd ed. Belmont, CA: Wadsworth, 1985.

Cited

1. Bacic, A.; Harris, P.J.; Stone, B.A. Structure and Function of Plant Cell Walls. In Biochemistry of Plants: A Comprehensive Treatise (J. Preiss, ed.), Vol. 14: Carbohydrates, pp. 298–371. San Diego, CA: Academic Press, 1988.

Brett, C.T.; Hillman J.R., eds. Biochemistry of Plant Cell Walls. SEB Seminar Series, 28. New York: Cambridge University Press, 1985.

McNeil, M.; Darvill, A.G.; Fry, S.C.; Albersheim, P. Structure and function of the primary cell walls of plants. *Annu. Rev. Biochem.* 53:625–663, 1984.

Rees, D.A. Polysaccharide Shapes. London: Chapman and Hall, 1977.

2. Cassab, G.I.; Varner, J.E. Cell wall proteins. *Annu. Rev. Plant Physiol.* 39:321–353, 1988.

Fry, S. Cross-linking of matrix polymers in the growing cell walls of angiosperms. *Annu. Rev. Plant Physiol.* 37:165–186, 1986.

Selvendran, R.R. Developments in the chemistry and biochemistry of pectic and hemicellulosic polymers. *J. Cell Sci.* (Suppl. 2):51–88, 1985.

3. Carpita, N.; Sabularse, D.; Montezinos, D.; Delmer, D.P. Determination of the pore size of cell walls of living plant cells. *Science* 205:1144–1147, 1979.

Milburn, J.A. Water Flow in Plants. London: Longman, 1979.

4. Street, H.E.; Öpik, H. The Physiology of Flowering Plants, 3rd ed. London, U.K.: Edward Arnold, 1984.

5. Fry, S.C. The Growing Plant Cell Wall: Chemical and Metabolic Analysis. New York: Wiley, 1988.

Masuda, Y.; Yamamoto, R. Cell-wall changes during auxin-induced cell extension. Mechanical properties and constituent polysaccharides of the cell wall. In Biochemistry of Plant Cell Walls; (C.T. Brett, J.R. Hillman, eds.), pp. 269–300. SEB Seminar Series 28. New York: Cambridge University Press, 1985.

6. Baker, D.A.; Hall, J.L. Ion Transport in Plant Cells and Tissues. Amsterdam: North-Holland, 1975.

Cheeseman, J.M. Mechanisms of salinity tolerance in plants. *Plant Physiol.* 87:547–550, 1988.

Morgan, J.M. Osmoregulation and water stress in higher plants. *Annu. Rev. Plant Physiol.* 35:299–319, 1984.

Zimmerman, U. Cell turgor pressure regulation and turgor pressure-mediated transport processes. *Symp. Soc. Exp. Biol.* 31:117–154, 1977.

7. Esau, K. Anatomy of Seed Plants, 2nd ed. New York: Wiley, 1977.

Tanner, W.; Loewus, F.A., eds. Encyclopedia of Plant Physiology, New Series. Vol. 13B: Plant Carbohydrates II, Extracellular Carbohydrates. Heidelberg: Springer-Verlag, 1982.

8. Dugger, W.M.; Bartnicki-Garcia, S., eds. Structure, Function, and Biosynthesis of Plant Cell Walls. *Proc. 7th Annu. Symp. Botany* Rockville, MD: American Society of Plant Physiologists, 1984. (Several relevant papers appear in this collection.)

Fry, S. The Growing Plant Cell Wall: Chemical and Metabolic Analysis. New York: Wiley, 1988.

9. Gunning, B.E.S.; Overall, R.L. Plasmodesmata and cell-to-cell transport in plants. *Bioscience* 33:260–265, 1983.

Gunning, B.E.S.; Robards, A.W., eds. Intercellular Communication in Plants: Studies on Plasmodesmata. New York: Springer-Verlag, 1976.

10. Baron-Epel, O.; Hernandez, D.; Jiang, L-W.; Meiners, S.; Schindler, M. Dynamic continuity of cytoplasmic and membrane compartments between plant cells. *J. Cell Biol.* 106:715–721, 1988.

Gunning, B.E.S.; Hughes, J.E. Quantitative assessment of symplastic transport of pre-nectar into the trichomes of *Abutilon* nectaries. *Aust. J. Plant Physiol.* 3:619–637, 1976.

Terry, B.R.; Robards, A.W. Hydrodynamic radius alone governs the mobility of molecules through plasmodesmata. Planta 171:145–157, 1987.

Zaitlin, M.; Hull, R. Plant virus-host interactions. *Annu. Rev. Plant Physiol.* 38:291–315, 1987.

11. Aloni, R. Differentiation of vascular tissue. *Annu. Rev. Plant Physiol.* 38:179–204, 1987.

Moorby, J. Transport Systems in Plants. New York: Longman, 1981.

12. Baker, D.A. Transport Phenomena in Plants. London, U.K.: Chapman and Hall, 1978.

Clarkson, D.T. Factors affecting mineral nutrient acquisition by plants. *Annu. Rev. Plant Physiol.* 36:77–115, 1985.

Milburn, J.A. Water Flow in Plants. London: Longman, 1979.

Passioura, J.B. Water transport in and to roots. *Annu. Rev. Plant Physiol. Plant Mol. Biol.* 39:245–265, 1988.

13. Cronshaw, J. Phloem structure and function. *Annu. Rev. Plant Physiol.* 32:465–484, 1981.

Cronshaw, J.; Lucas, W.J.; Giaquinta, R.T., eds. Phloem Transport. New York: Liss, 1986.

Gunning, B.E.S. Transfer cells and their roles in transport of solutes in plants. *Sci. Prog. (Oxford)* 64:539–568, 1977.

Ho, L.C. Metabolism and compartmentation of imported sugars in sink organs in relation to sink strength. *Annu. Rev. Plant Physiol. Plant Mol. Biol.* 39:355–378, 1988.

14. Gianinazzi-Pearson, V.; Gianinazzi, S., eds. Physiological and Genetical Aspects of Mycorrhizae. Paris: INRA, 1986.

Smith, S.E.; Gianinazzi-Pearson V. Physiological interactions between symbionts in vesicular-arbuscular mycorrhizal

plants. *Annu. Rev. Plant Physiol. Plant Mol. Biol.* 39:221–244, 1988.

15. Downie, J.A.; Johnston, A.W.B. Nodulation of legumes by *Rhizobium*: the recognized root? *Cell* 47:153–154, 1986.

Peters, N.K.; Frost, J.W.; Long, S.R. A plant flavone, luteolin induces expression of *Rhizobium meliloti* nodulation genes. *Science* 233:977–980, 1986.

Rolfe, B.G.; Gresshoff, P.M. Genetic analysis of legume nodule initiation. *Annu. Rev. Plant Physiol. Plant Mol. Biol.* 39:297–319, 1988.

16. Buchanan-Wollaston, V.; Passiatore, J.E.; Cannon, F. The *mob* and *oriT* mobilization functions of a bacterial plasmid promote its transfer to plants. *Nature* 328:172–175, 1987.

Klee, H.; Horsch, R.; Rogers, S. *Agrobacterium*-mediated plant transformation and its further applications to plant biology. *Annu. Rev. Plant Physiol.* 38:467–486, 1987.

Stachel, S.E.; Messens, E.; Van Montagu, M.; Zambryski, P. Identification of the signal molecules produced by wounded plant cells that activate T-DNA transfer in *Agrobacterium tumefaciens. Nature* 318:624–629, 1985.

17. Ayers, A.R.; Ebel, J.; Finelli, F.; Berger, N.; Albersheim, P. Host pathogen interactions. *Plant Physiol.* 57:751–759, 1976. (This should be read in conjunction with the three related papers that follow it.)

Darvill, A.G.; Albersheim, P. Phytoalexins and their elicitors—a defence against microbial infection in plants. *Annu. Rev. Plant. Physiol.* 35:243–275, 1984.

Ralton, V.E.; Smart, M.G.; Clarke, A.E. Recognition and infection process in plant pathogen interactions. In Plant-Microbe Interactions. (T. Kosuge, E.W. Nestor, eds.), Vol. 2, pp. 217–252. New York: Macmillan, 1987.

Ryan, C.A. Oligosaccharide signalling in plants. *Annu. Rev. Cell Biol.* 3:295–317, 1987.

18. McDougall, G.J.; Fry, S.C. Inhibition of auxin-stimulated growth of pea stem segments by a specific monosaccharide of xyloglucan. *Planta* 175:412–416, 1988.

Thanh Van, K.T.; et al. Manipulation of the morphogenetic pathways of tobacco explants by oligosaccharins. *Nature* 314:615–617, 1985.

19. Anderson, J.M. Photoregulation of the composition, function, and structure of thylakoid membranes. *Annu. Rev. Plant Physiol.* 37:93–136, 1986.

Mullet, J.E. Chloroplast development and gene expression. *Annu. Rev. Plant Physiol. Plant Mol. Biol.* 39:475–502, 1988.

Thomson, W.W.; Watley, J.M. Development of nongreen plastids. *Annu. Rev. Plant Physiol.* 31:375–394, 1980.

20. Boller, T.; Kende, H. Hydrolytic enzymes in the central vacuole of plant cells. *Plant Physiol.* 63:1123–1132, 1979.

Boller, T.; Wiemken, A. Dynamics of vacuolar compartmentation. *Annu. Rev. Plant Physiol.* 37:137–164, 1986.

Marin, B., ed. Plant Vacuoles: Their Importance in Solute Compartmentation in Cells and Their Applications in Plant Biotechnology. NATO ASI Series, Vol. 134. New York: Plenum, 1987.

Matile, P. Biochemistry and function of vacuoles. *Annu. Rev. Plant Physiol.* 29:193–213, 1978.

21. Mollenhauer, H.H.; Morré, D.J. The Golgi apparatus. In The Biochemistry of Plants—A Comprehensive Treatise. (N.E. Tolbert, ed.), Vol. 1, pp. 437–488. New York: Academic Press, 1980.

Moore, P.J.; Staehelin, L.A. Immunogold localization of the cell-wall-matrix polysaccharides rhamnogalacturonan I and xyloglucan during cell expansion and cytokinesis in *Trifolium pratense* L.; implications for secretory pathways. *Planta* 174:433–445, 1988.

Northcote, D.H. Macromolecular aspects of cell wall differentiation. In Encyclopedia of Plant Physiology, New Series. Vol. 14A: Nucleic Acids and Proteins in Plants I (D. Boulter, B. Parthier, eds.), pp. 637–655. Berlin: Springer-Verlag, 1982.

22. Coleman, J.; Evans, D.; Hawes, C.; Horsley, D.; Cole, L. Structure and molecular organization of higher plant coated vesicles. *J. Cell Sci.* 88:35–45, 1987.

Robinson, D.G.; Depta, H. Coated vesicles. *Annu. Rev. Plant Physiol. Plant Mol. Biol.* 39:53–99, 1988.

Tanchak, M.A.; Griffing, L.R.; Mersey, B.G.; Fowke, L.C. Endocytosis of cationized ferritin by coated vesicles of soybean protoplasts. *Planta* 162:481–486, 1984.

23. Brown, R.M. Cellulose microfibril assembly and orientation: recent developments. *J. Cell Sci.*, (Suppl. 2):13–32, 1985.

Delmer, D.P. Cellulose biosynthesis. *Annu. Rev. Plant Physiol.* 38:259–290, 1987.

Schneider, B.; Herth, W. Distribution of plasma membrane rosettes and kinetics of cellulose formation in xylem development of higher plants. *Protoplasma* 131:142–152, 1986.

24. Green, P.B. Organogenesis—a biophysical view. *Annu. Rev. Plant Physiol.* 31:51–82, 1980.

Herth, W. Plant cell wall formation. In Botanical Microscopy (A.W. Robards, ed.), pp. 285–310. New York: Oxford University Press, 1985.

25. Lloyd, C.W., ed. The Cytoskeleton in Plant Growth and Development. New York: Academic Press, 1982. (Chapters 5–8 are particularly relevant.)

26. Kersey, Y.M.; Hepler, P.K.; Palevitz, B.A.; Wessels, N.K. Polarity of actin filaments in characean algae. *Proc. Natl. Acad. Sci. USA* 73:165–167, 1976.

Parthasarathy, M.V. F-actin architecture in coleoptile epidermal cells. *Eur. J. Cell Biol.* 39:1–12, 1985.

Sheetz, M.P.; Spudich, J.A. Movement of myosin-coated fluorescent beads on actin cables *in vitro. Nature* 303:31–35, 1983.

Williamson, R.E. Organelle movements along actin filaments and microtubules. *Plant Physiol.* 82:631–634, 1986.

27. Haupt, W. Light-mediated movement of chloroplasts. *Annu. Rev. Plant Physiol.* 33:205–233, 1982.

Roberts, I.N.; Lloyd, C.W.; Roberts, K. Ethylene-induced microtubule reorientations: mediation by helical arrays. *Planta* 164:439–447, 1985.

Virgin, H.I. Light and chloroplast movements. *Symp. Soc. Exp. Biol.* 22:329–352, 1968.

Wagner, G.; Klein, K. Mechanism of chloroplast movement in *Mougeotia. Protoplasma* 109:169–185, 1981.

28. Cutter, E.G. Plant Anatomy, 2nd ed., Part 2, Organs. London: Arnold, 1978.

Johri, B.M. Embryology of Angiosperms. Berlin: Springer-Verlag, 1984.

Raven, P.H.; Evert, R.F.; Eichhorn, S.E. Biology of Plants, 4th ed. New York: Worth, 1986. (Chapters 19–22 provide a good general outline of plant development.)

29. Harper, J.L., ed. Growth and Form of Modular Organisms. London: Royal Society, 1986. (There are several relevant papers in this collection.)

Walbot, V. On the life strategies of plants and animals. *Trends Genet.* 1:165–169, 1985.

30. Green, P.B. A theory for influorescence development and flower formation based on morphological and biophysical analysis in *Echeveria. Planta* 175:153–169, 1988.

McDaniel, C.N.; Poethig, R.S. Cell lineage patterns in the shoot apical meristem of the germinating corn embryo. *Planta* 175:13–22, 1988.

Sachs, T. Controls of cell patterns in plants. In Pattern Formation (G.M. Malacinski, S.V. Bryant, eds.). New York: Macmillan, 1984.

Steeves, T.A.; Sussex, I.M. Patterns in Plant Development, 2nd ed. New York: Cambridge University Press, 1988.

31. Gunning, B.E.S. Microtubules and cytomorphogenesis in a developing organ: the root primordium of *Azolla pinnata*. In Cytomorphogenesis in Plants (O. Kiermayer, ed.), pp. 301–325. New York: Springer, 1981.

Poethig, R.S. Clonal analysis of cell lineage patterns in plant development. *Am. J. Bot.* 74:581–594, 1987.

32. Gunning, B.E.S.; Wick, S.M. Preprophase bands, phragmoplasts and spatial control of cytokinesis. *J. Cell Sci.*, Suppl. 2:157–179, 1985.

Lloyd, C. Actin in plants. *J. Cell Sci.* 90:185–188, 1988.

Lloyd, C.W. The plant cytoskeleton: the impact of fluorescence microscopy. *Annu. Rev. Plant Physiol.* 38:119–139, 1987.

Pickett-Heaps, J.D.; Northcote, D.H. Organization of microtubules and endoplasmic reticulum during mitosis and cytokinesis in wheat meristems. *J. Cell Sci.* 1:109–120, 1966.

Wick, S.M.; Seagull, R.W.; Osborn, M.; Weber, K.; Gunning, B.E.S. Immunofluorescence microscopy of organized microtubule arrays in structurally stabilized meristematic plant cells. *J. Cell Biol.* 89:685–690, 1981.

33. Hepler, P.K.; Wayne, R.O. Calcium and plant development. *Annu. Rev. Plant Physiol.* 36:397–439, 1985.

Kropf, D.L.; Kloareg, B.; Quatrano, R.S. Cell wall is required for fixation of the embryonic axis in *Fucus* zygotes. *Science* 239:187–190, 1988.

Quatrano, R.S.; Griffing, L.R.; Huber-Walchli, V.; Doubet, R.S. Cytological and biochemical requirements for the establishment of a polar cell. *J. Cell Sci.*, Suppl. 2:129–141, 1985.

Schnepf, E. Cellular polarity. *Annu. Rev. Plant Physiol.* 37:23–47, 1986.

34. Jordan, B.R.; Partis, M.D.; Thomas, B. The biology and molecular biology of plytochrome. *Ox. Sur. Plant Mol. Cell Biol.* 3:315–362, 1986.

Kuhlemeier, C.; Green, P.J.; Chua, N. Regulation of gene expression in higher plants. *Annu. Rev. Plant Physiol.* 38:221–257, 1987.

Tobin, E.M.; Silverthorne, J. Light regulation of gene expression in higher plants. *Annu. Rev. Plant Physiol.* 36:569–593, 1985.

Verma, D.P.S.; Goldberg, R.B. Temporal and Spatial Regulation of Plant Genes. New York: Springer-Verlag, 1988.

von Wettstein, D.; Chua, N-H., eds. Plant Molecular Biology. NATO ASI Series A, Vol. 140. New York: Plenum, 1987. (Several relevant papers in this collection.)

35. Hoad, G.V.; Lenton, J.R.; Jackson, M.B.; Atkin, R.K. Hormone Action in Plant Development: A Critical Appraisal. London: Butterworth, 1987.

Salisbury, F.B.; Ross, C.W. Plant Physiology, 3rd ed. Belmont, CA: Wadsworth, 1985. (Chapters 16 and 17 are relevant.)

Theologis, A. Rapid gene regulation by auxin. *Annu. Rev. Plant Physiol.* 37:407–438, 1986.

Trewavas, A.J. Growth substance sensitivity: the limiting factor in plant development. *Physiol. Plant.* 35:60–72, 1982.

36. Applications of Plant Cell and Tissue Culture. CIBA Foundation Symposium 137. New York: Wiley, 1988.

Lee, M.; Phillips, R.L. The chromosomal basis of somaclonal variation. *Annu. Rev. Plant Physiol. Plant Mol. Biol.* 39:413–437, 1988.

Vasil, I.K., ed. Perspectives in Plant Cell and Tissue Culture. *Int. Rev. Cytol.*, Suppl. 11A and B, 1980.

37. Bevan, M. Binary *Agrobacterium* vectors for plant transformation. *Nuc. Acids Res.* 12:8711–8721, 1984.

De Block, M.; Herrera-Estrella, L.; Van Montagu, M.; Schell, J.; Zambryski, P. Expression of foreign genes in regenerated plants and in their progeny. *EMBO J.* 3:1681–1689, 1986.

Finkelstein, R.; Estelle, M.; Martinez-Zapater, J.; Somerville, C. *Arabidopsis* as a tool for the identification of genes involved in plant development. In Temporal and Spatial Regulation of Plant Genes (D.P.S. Verma, and R.B. Goldberg, eds.), pp. 1–25. New York: Springer-Verlag, 1988.

Myerowitz, E.M. *Arabidopsis thaliana. Annu. Rev. Genet.* 21:93–111, 1987.

Vaeck, M.; et al. Transgenic plants protected from insect attack. *Nature* 328:33–37, 1987.

Cancer

Roughly one person in five, in the prosperous countries of the world, will die of cancer; but that is not the reason for devoting a chapter of this book to the subject. Heart disease causes more deaths, and many other illnesses result in just as much distress; in the world as a whole, other health problems, such as malnutrition and parasitic infections, are more serious. In the context of cell biology, however, cancer has a unique importance, for the family of diseases grouped under this heading reflect disturbances of the most fundamental rules of behavior of the cells in a multicellular organism. To understand cancer and to devise rational ways to treat it, we have to understand both the inner workings of cells and their social interactions in the tissues of the body. Thus basic cancer research has been the source of many advances in our knowledge of normal cells. By-products of cancer research have become crucial tools in the current revolution of cell biology—tools such as reverse transcriptase from RNA tumor viruses, used to make cDNA (see p. 183), and myeloma cell lines derived from cancerous B lymphocytes, used to make monoclonal antibodies. One may debate how much the massive resources devoted to laboratory research on cancer have so far contributed directly toward improvements in cancer treatment; but there can be no doubt that, by contributing to progress in cell biology, the cancer research effort has profoundly benefited a much wider area of medical knowledge than that of cancer alone.

We have already discussed how cancer research has begun to reveal the molecular mechanisms underlying the normal controls of cell growth and division (see pp. 752–761). In this concluding chapter we examine the disease itself. In the first section we shall consider the nature of cancer and the natural history of the disease from a cellular standpoint; in the second section we focus on its molecular basis.

Cancer as a Microevolutionary Process[1]

The body of an animal can be viewed as a society or ecosystem whose individual members are cells, reproducing by cell division and organized into collaborative assemblies or tissues. In our earlier discussion of the maintenance of tissues (in Chapter 17), our concerns were similar to those of the ecologist: cell births, deaths, habitats, territorial limitations, the maintenance of population sizes, and the like. The one ecological topic conspicuously absent was that of natural selection: we said nothing of competition or mutation among somatic cells. The reason is that a healthy body is in this respect a very peculiar society, where self-sacrifice, rather than competition, is the rule for every class of cells except one: all somatic cell lineages are committed to die, leaving no progeny but dedicating their existence to support of the germ cells, which alone have a chance of survival. There is no mystery in this, for the body is a clone, and the genome of the somatic cells is the same as the genome of the germ cells; thus, by their self-sacrifice for the sake of the germ cells, the somatic cells help to propagate copies of their own genes.

In contrast, therefore, with free-living cells such as bacteria, which compete to survive, the cells of a multicellular organism are committed to collaboration. Any mutation that gives rise to nonaltruistic behavior by individual members of the cooperative will jeopardize the future of the whole enterprise. Thus mutation, competition, and natural selection operating *within* the population of somatic cells are ingredients for a disaster. It is, in essence, just this type of disaster that occurs in cancer: cancer is a disease in which individual cells begin by prospering selfishly at the expense of their neighbors but in the end destroy the whole cellular society and die.

In this section we discuss the development of cancer as a microevolutionary process, occurring on a time scale of months or years in a population of cells in the body, but dependent on the same principles of mutation and natural selection that govern the long-term evolution of all living organisms.

Cancers Differ According to the Cell Type from Which They Derive[2]

Cancer cells are defined by two heritable properties: they and their progeny (1) reproduce in defiance of the normal restraints and (2) invade and colonize territories normally reserved for other cells. It is the combination of these features that makes cancers peculiarly dangerous. An isolated abnormal cell that does not proliferate more than its normal neighbors does no significant damage, no matter what other disagreeable properties it may have; but if its proliferation is out of control, it will give rise to a tumor or **neoplasm**—a relentlessly growing mass of abnormal cells. So long as the neoplastic cells remain clustered together in a single mass, however, the tumor is said to be **benign,** and a complete cure can usually be achieved by removing the mass surgically. A tumor is counted as a cancer only if it is **malignant,** that is, only if its cells have the ability to invade surrounding tissue. Invasiveness usually implies an ability to break loose, enter the bloodstream or lymphatic vessels, and form secondary tumors or **metastases** at other sites in the body (Figure 21–1). The more widely a cancer metastasizes, the harder it becomes to eradicate.

Cancers are classified according to the tissue and cell type from which they arise. Cancers arising from epithelial cells are termed **carcinomas;** those arising from connective tissue or muscle cells are termed **sarcomas.** Cancers that do not fit in either of these two broad categories include the various **leukemias,** derived from hemopoietic cells, and cancers derived from cells of the nervous system. Table 21–1 lists the types of cancers that are common in the United States, together with their incidence and the death rate from them. Each of the broad categories has many subdivisions according to the specific cell type, the location in the body, and the structure of the tumor; many of the names used are fixed by tradition and have no modern rational basis. In parallel with the set of names for malignant tumors, there is a related set of names for benign tumors: an *adenoma*, for example, is a benign epithelial tumor with a glandular organization, the corresponding type of malignant tumor being an *adenocarcinoma* (Figure 21–2); a *chondroma* and a *chondrosarcoma* are, respectively, benign and malignant tumors of cartilage. About 90% of human cancers are carcinomas, perhaps because most of the cell proliferation in the body occurs in epithelia, or perhaps because epithelial tissues are most frequently exposed to the various forms of physical and chemical damage that favor the development of cancer.

Each cancer has characteristics that reflect its origin. Thus, for example, the cells of an epidermal *basal-cell carcinoma*, derived from a keratinocyte stem cell in the skin, will generally continue to synthesize cytokeratin intermediate filaments, whereas the cells of a *melanoma*, derived from a pigment cell in the skin, will often (but not always) continue to make pigment granules. Cancers originating

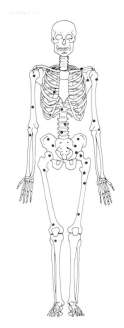

Figure 21–1 Malignant tumors typically give rise to metastases, making the cancer hard to eradicate. The drawing shows common sites in the bone marrow for metastases from carcinoma of the prostate gland. (From Union Internationale Contre le Cancer, TNM Atlas: Illustrated Guide to the Classification of Malignant Tumors, 2nd ed. Berlin: Springer, 1986.)

Table 21-1 Cancer Incidence and Cancer Mortality in the United States, 1986

Type of Cancer	New Cases per Year		Deaths per Year	
Total cancers	930,000		472,000	
Cancers of epithelia: carcinomas	789,000	(85%)	381,400	(81%)
Oral cavity and pharynx	29,500	(3%)	9,400	(2%)
Digestive organs (total)	217,800	(23%)	119,700	(25%)
Colon and rectum	130,000	(14%)	60,000	(13%)
Pancreas	25,500	(3%)	24,000	(5%)
Stomach	24,700	(3%)	14,300	(3%)
Liver and biliary system	13,600	(1%)	10,600	(2%)
Respiratory system (total)	164,500	(18%)	135,400	(29%)
Lung	149,000	(16%)	130,100	(28%)
Breast	123,900	(13%)	40,200	(9%)
Skin (total)	(>400,000)*		7,500	(2%)
Malignant melanoma	23,000	(2%)	5,600	(1%)
Reproductive tract (total)	169,800	(18%)	49,400	(10%)
Prostate gland	90,000	(10%)	26,100	(6%)
Ovary	19,000	(2%)	11,600	(2%)
Uterine cervix	14,000	(2%)	6,800	(1%)
Uterus (endometrium)	36,000	(4%)	2,900	(1%)
Urinary organs (total)	60,500	(7%)	19,800	(4%)
Bladder	40,500	(4%)	10,600	(2%)
Cancers of hemopoietic and immune system: leukemias and lymphomas	70,100	(8%)	41,100	(9%)
Cancers of central nervous system and eye: gliomas, retinoblastoma, etc.	15,600	(2%)	10,600	(2%)
Cancers of connective tissues, muscles, and vasculature: sarcomas	7,100	(1%)	4,200	(1%)
All other cancers + unspecified sites	48,200	(5%)	34,800	(7%)

*Nonmelanoma skin cancers are not included in total of all cancers, since almost all are cured easily and many go unrecorded.

In the world as a whole, the five most common cancers are those of lung, stomach, breast, colon/rectum, and uterine cervix, and the total number of new cancer cases per year is just over 6 million. Note that only about half the number of people who develop cancer die of it. (Data for USA from American Cancer Society, Cancer Facts and Figures, 1986.)

Figure 21-2 An adenoma (a benign glandular tumor) and an adenocarcinoma (a malignant glandular tumor) contrasted. There are many forms that such tumors may take; this diagram illustrates schematically types that might be found in the breast.

ADENOMA (BENIGN)

ADENOCARCINOMA (MALIGNANT)

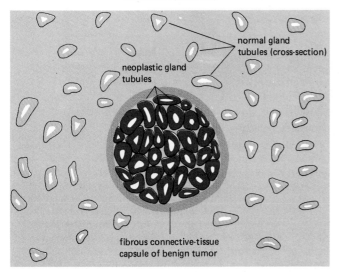
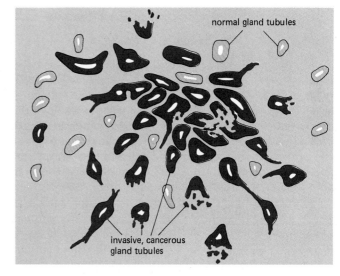

from different cell types are, in general, very different diseases. The basal-cell carcinoma, for example, is only locally invasive and rarely forms metastases, whereas the melanoma is much more malignant and rapidly gives rise to many metastases (behavior that recalls the migratory tendencies of the normal pigment-cell precursors during development—see p. 944). The basal-cell carcinoma is usually easy to remove by surgery, leading to complete cure; but the malignant melanoma, once it has metastasized, is often impossible to extirpate and consequently fatal.

Most Cancers Derive from a Single Abnormal Cell[3]

The origins of most cancers can be traced to a single isolated **primary tumor;** this suggests that they are derived by cell division from a single cell that has undergone some heritable change that enables it to outgrow its neighbors. By the time it is first detected, however, a typical tumor already contains about a billion cells or more (Figure 21–3), often including many normal cells—fibroblasts, for example, in the supporting connective tissue that is associated with a carcinoma. It is not easy to prove that the cancer cells are a clone descended from a single abnormal cell; but where evidence is available, it usually confirms that the cancer has a monoclonal origin. In almost all patients with *chronic myelogenous leukemia,* for example, the leukemic white blood cells are distinguished from the normal cells by a specific chromosomal abnormality (the so-called Philadelphia chromosome, created by a translocation between the long arms of chromosomes 22 and 9, as shown in Figure 21–4). It is unlikely that the genetic accident responsible for this abnormality would have occurred in several cells at once in the same individual; it is much more likely that all the leukemic cells are descendants of the same single mutant cell. Indeed, when the DNA at the site of translocation is cloned and sequenced, it is found that the site of breakage and rejoining of the translocated fragments is identical in all the leukemic cells in any given patient, but differs slightly (by a few hundred or thousand base pairs) from one patient to another, as expected if each case of the leukemia arises from a unique accident occurring in a single cell.

Another way to show that a cancer has a monoclonal origin is by exploiting the phenomenon of X-chromosome inactivation (see p. 577). A normal woman is a random mixture, or mosaic, of two classes of cells—those in which the paternal X chromosome is inactivated and those in which the maternal X chromosome is inactivated. The inactivation of one X chromosome in each cell occurs early in embryonic development, and thereafter the daughters of a dividing somatic cell always have the same X chromosome inactivated as the parent cell. Consequently, the state of X-chromosome inactivation—maternal or paternal—can be used as a heritable marker to trace the lineage of cells in the body. In the great majority of tumors that have been analyzed—both benign and malignant—all the tumor cells have been found to have the same X chromosome inactivated, strongly suggesting that they are derived from a single deranged cell (Figure 21–5).

Most Cancers Are Probably Initiated by a Change in the Cell's DNA Sequence[4]

If a single abnormal cell is to give rise to a tumor, it must pass on its abnormality to its progeny: the aberration has to be heritable. A first problem in understanding a cancer is to discover whether the heritable aberration is due to a genetic change— that is, an alteration in the the cell's DNA sequence—or to an *epigenetic* change— that is, a change in the pattern of gene expression without a change in the DNA sequence. Heritable epigenetic changes, reflecting cell memory (see p. 570 and p. 898), are a familiar feature of normal development, as manifest in the stability of the differentiated state (see p. 952) and in such phenomena as X chromosome inactivation (see p. 577); and there is no obvious a priori reason why they should not be involved in cancer. For one rare and extraordinary type of cancer—the teratocarcinoma (see p. 897)—there is indeed evidence in favor of an epigenetic origin. There are, however, good reasons to think that most cancers are initiated by genetic change (although epigenetic changes may play a part in the subsequent

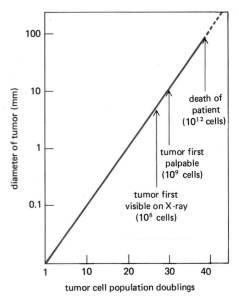

Figure 21–3 The growth of a typical human tumor, with the diameter of the tumor plotted on a logarithmic scale. Years may elapse before the tumor becomes noticeable.

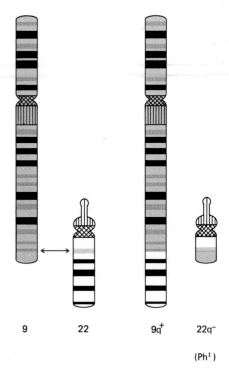

Figure 21–4 The translocation between chromosomes 9 and 22 responsible for chronic myelogenous leukemia. The smaller of the two resulting abnormal chromosomes is called the Philadelphia chromosome, after the city where the abnormality was first recorded.

development of the disease). Thus cells of a given cancer can often be shown to have a shared abnormality in their DNA sequence, as we have just seen for chronic myelogenous leukemia; many other examples will be discussed in the second half of this chapter. But this does not prove that genetic change is an essential *first* step in the causation of cancer. A more cogent argument is that most of the agents known to cause cancer cause genetic change; and, conversely, agents that cause genetic change cause cancer. This correlation between **carcinogenesis** (the generation of cancer) and *mutagenesis* is clear for three classes of agents: chemical carcinogens (which typically cause simple local changes in the nucleotide sequence), ionizing radiation such as x-rays (which typically cause chromosome breaks and translocations), and viruses (which introduce foreign DNA into the cell). The role of viruses in cancer will be discussed later; we pause here to discuss **chemical carcinogens.**

In general, a given cancer cannot be blamed entirely on a single event or a single cause: as we shall see, cancers as a rule result from the chance occurrence in one cell of several independent accidents, with cumulative effects. The cell's environment influences the frequency of these accidents in a variety of ways, and most cancers should be viewed as the outcome of a random process that is made more probable by a mixture of contributory environmental factors (see p. 1195). There are, however, some unusually carcinogenic agents that increase the likelihood of the critical events to the point where it becomes virtually certain, given a high enough dosage, that at least one cell in the body will turn cancerous. The compound 2-naphthylamine, used in the chemical industry in the early part of this century, is one notorious example: in one British factory, all of the men who had been employed in distilling it (and were thereby subjected to prolonged exposure) eventually developed bladder cancer.

Many quite disparate chemicals have been shown to be likewise carcinogenic when they are fed to experimental animals or painted repeatedly on their skin. Some of these carcinogens act directly on the target cells; many others take effect only after they have been changed to a more reactive form by metabolic processes—notably by a set of intracellular enzymes known as the cytochrome P-450 oxidases, which normally help to convert ingested toxins and foreign lipid-soluble materials into harmless and easily excreted compounds but which fail in this task with certain substances, converting them instead into direct carcinogens (Figure 21–6). Although the known chemical carcinogens are very diverse, most of them have at least one property in common: they cause mutations. The mutagenicity can be demonstrated by various methods, one of the most convenient being the *Ames test*, in which the carcinogen is mixed with an activating extract prepared from rat liver cells and added to a culture of specially designed test bacteria; the resulting mutation rate of the bacteria is then measured (Figure 21–7). Most of the compounds scored as mutagenic by this bacterial assay also cause mutations and/ or chromosome aberrations when tested on mammalian cells, and they have chemical structures that can be seen to imply an ability to react with DNA. When mutagenicity data from these various sources are combined and compared with carcinogenicity data from studies of cancer induction *in vivo*, it is found that the majority of known carcinogens are mutagenic and, conversely, that the majority of mutagens are carcinogenic.

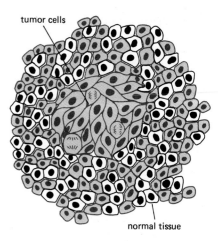

Figure 21–5 Evidence from the analysis of X-inactivation mosaics demonstrating the monoclonal origin of cancers. As a result of a random process that occurs in the early embryo, practically every normal tissue in a woman's body is a mixture of cells with different X chromosomes heritably inactivated (indicated here by the mixture of colored cells and black cells in the normal tissue). When the cells of a cancer are tested for their expression of an X-linked marker gene, however, they are usually all found to have the same X chromosome inactivated. This implies that they are all derived from a single cancerous founder cell.

Figure 21–6 Many chemical carcinogens have to be activated by a metabolic transformation before they will cause mutations by reacting with DNA. The compound illustrated here is *aflatoxin B1*, a toxin from a mold (*Aspergillus flavus oryzae*) that grows on grain and peanuts when they are stored under humid tropical conditions. It is thought to be a contributory cause of liver cancer in the tropics.

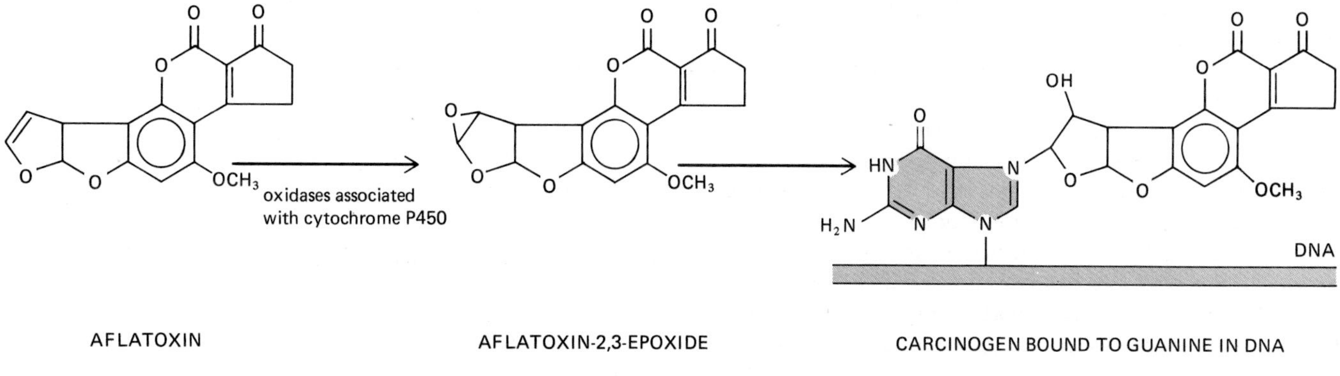

AFLATOXIN   AFLATOXIN-2,3-EPOXIDE   CARCINOGEN BOUND TO GUANINE IN DNA

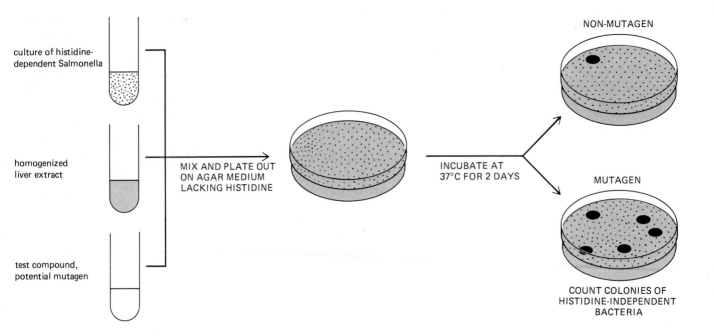

culture of histidine-
dependent Salmonella

homogenized
liver extract

MIX AND PLATE OUT
ON AGAR MEDIUM
LACKING HISTIDINE

INCUBATE AT
37°C FOR 2 DAYS

MUTAGEN

test compound,
potential mutagen

COUNT COLONIES OF
HISTIDINE-INDEPENDENT
BACTERIA

There is, nevertheless, a significant minority of carcinogens that do not appear to be mutagenic. We shall discuss below (p. 1196) how non-mutagenic substances may promote the development of cancer by affecting the behavior of pre-existing mutant cells. But first we must consider how frequently such mutant cells are likely to arise in the normal course of events.

A Single Mutation Is Not Enough to Cause Cancer[1,5]

Something on the order of 10^{16} cell divisions take place in a human body in the course of a lifetime; in a mouse, with its smaller number of cells and its shorter lifespan, the number is about 10^{12}. Even in an environment that is free of mutagens, mutations will occur spontaneously, at an estimated rate of about 10^{-6} mutations per gene per cell division—a value set by fundamental limitations on the accuracy of DNA replication and repair (see p. 229). Thus, in a lifetime, every single gene is likely to have undergone mutation on about 10^{10} separate occasions in any individual human being, or about 10^6 occasions in a mouse. Among the resulting mutant cells one might expect that there would be many that have disturbances in genes involved in the regulation of cell division and that consequently disobey the normal restrictions on cell proliferation. From this point of view, the problem of cancer seems to be not why it occurs, but why it occurs so infrequently.

Evidently, the survival of mammals must depend on some form of double— or more than double—insurance in the mechanisms that protect us from being overrun by mutant clones of cells that have a selective advantage over our healthy normal cells: if a single mutation in some particular gene were enough to convert a typical healthy cell into a cancer cell, we would not be viable organisms. Many lines of evidence indicate that the genesis of a cancer does indeed require that several independent rare accidents occur together in one cell. One such indication comes from epidemiological studies of the incidence of cancer as a function of age. If a single mutation were responsible, occurring with a fixed probability per year, the chance of developing cancer in any given year should be independent of age. In fact, for most types of cancer the chance goes up very steeply with age— typically as the third, fourth, or fifth power (Figure 21–8). From such statistics it has been estimated that somewhere between three and seven independent random events, each of low probability, are typically required to turn a normal cell into a cancer cell; the smaller numbers apply to leukemias, the larger to carcinomas.

Now that specific mutations responsible for the development of cancer have been identified, it has become possible to test the effects of the mutant genes in transgenic mice (see p. 267); as we shall see later (p. 1212), the results give addi-

Figure 21–7 The Ames test for mutagenicity. The test uses a strain of *Salmonella* bacteria that require histidine in the medium because of a defect in a gene necessary for histidine synthesis. Mutagens can cause a further change in this gene that reverses the defect, creating revertant bacteria that do not require histidine. To increase the sensitivity of the test, the bacteria also have a defect in their DNA repair machinery that makes them especially susceptible to agents that damage DNA. A majority of compounds that are mutagenic by tests such as this are also carcinogenic, and vice versa.

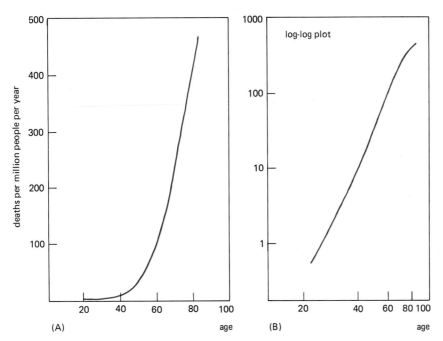

Figure 21–8 Number of deaths from cancer of the large intestine in the United States in one year, plotted as a function of age at death; the same data are shown on an ordinary linear scale in (A) and on a logarithmic scale in (B). The incidence of cancer rises steeply as a function of age—roughly as the fifth power, in this example [that is, the slope of the log-log plot in (B) is about 5]. This suggests that a cell must accumulate the disruptive effects of several independent accidents before it will give rise to a cancer. (From U.S. Department of Health, Education and Welfare: Vital Statistics of the United States, Vol. II: Mortality. Washington, D.C.: U.S. Government Printing Office, 1968.)

tional and more direct evidence for the hypothesis that a single mutation is insufficient to cause cancer. The hypothesis is also supported by many older studies of the phenomenon of **tumor progression,** whereby an initial mild disorder of cell behavior evolves gradually into a full-blown cancer. These observations of how tumors develop, moreover, provide insight into the nature of the multiple changes that must occur for a normal cell to become a cancer cell and into the factors that control their occurrence.

Cancers Develop in Slow Stages from Mildly Aberrant Cells[1,5,6]

For those cancers that have a discernable external cause, there is almost always a long delay between the causal event(s) and the onset of the disease: the incidence of lung cancer does not begin to rise steeply until after 10 or 20 years of heavy smoking; the incidence of leukemias in Hiroshima and Nagasaki did not show a marked rise until about 5 years after the explosion of the atomic bombs, and it did not reach its peak until 8 years had elapsed; industrial workers exposed for a limited period to chemical carcinogens do not usually develop the cancers characteristic of their occupation until 10, 20, or even more years after the exposure (Figure 21–9); and so on. During this long incubation period, the prospective cancer cells undergo a succession of changes. Chronic myelogenous leukemia, mentioned earlier, provides a clear and simple example. This disease begins as a disorder characterized by a nonlethal overproduction of white blood cells and

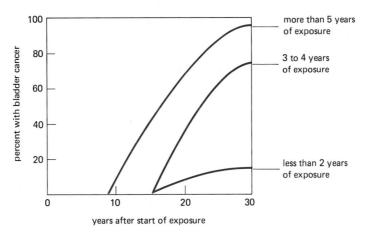

Figure 21–9 The length of the delay before onset of bladder cancer in a set of 78 men who had been exposed to the carcinogen 2-naphthylamine, grouped according to the duration of their exposure. (Modified from J. Cairns, Cancer: Science and Society. San Francisco: Freeman, 1978. After M.H.C. Williams, in Cancer, Vol. III (R.W. Raven, ed.). London: Butterfield, 1958.)

continues as such for several years before changing into a much more rapidly progressing illness that usually ends in death within a few months. In the chronic early phase the leukemic cells in the body are distinguished simply by their possession of the chromosomal translocation mentioned previously (see p. 1190). In the subsequent acute phase of the illness, the hemopoietic system is overrun by cells that show not only this chromosomal abnormality but also several others. It appears as though members of the initial mutant clone have undergone further mutations that make them proliferate more rapidly (or divide more times before they terminally differentiate), so that they come to outnumber both the normal hemopoietic cells and their cousins that have only the primary disorder.

Carcinomas and other solid tumors are thought to evolve in a similar way. Although most such cancers in humans are not diagnosed until a relatively late stage, in a few cases it is possible to observe the early steps in the development of the disease. Cancers of the *uterine cervix* (the neck of the womb) provide a typical example. These cancers derive from the multilayered cervical epithelium, which has an organization similar to that of the epidermis of the skin (see p. 968). Normally, proliferation occurs only in the basal layer, generating cells that then move outward toward the surface, differentiating into flattened, keratin-rich, nondividing cells as they go, and finally being sloughed off from the surface (Figure 21–10A). When many specimens of this epithelium from different women are examined, however, it is not unusual to find patches of **dysplasia,** where dividing cells are no longer confined to the basal layer and there is some disorder in the process of differentiation (Figure 21–10B). Cells are sloughed from the surface in abnormally early stages of differentiation, and the presence of the dysplasia can be detected by scraping a sample of cells from the surface and viewing it under the microscope (the "Pap smear" technique—Figure 21–11). Left alone, the dysplastic patches will often remain harmless or even regress spontaneously; more rarely, however, they may progress, over a period of several years, to give rise to patches of so-called **carcinoma *in situ*** (Figure 21–10C). In these more serious lesions (somewhat misleadingly named, since they are not yet fully malignant), the usual pattern of cell division and differentiation is much more severely disrupted, and all the layers of the epithelium consist of undifferentiated proliferating cells, which are often highly variable in size and karyotype; the abnormal cells are still confined, however, to the epithelial side of the basal lamina. At this stage it is still easy to achieve a complete cure by destroying or removing the abnormal tissue surgically. Without such treatment the abnormal patch may still remain harmless or regress; but in an estimated 20–30% of cases it will develop, again over a period of several years, to give rise to a truly malignant cervical carcinoma (Figure 21–10D), whose cells break out of the epithelium by crossing the basal lamina and begin to invade the underlying connective tissue. Surgical cure becomes progressively more difficult as the invasive growth spreads.

Figure 21–10 The stages of progression in the development of cancer of the epithelium of the uterine cervix. In dysplasia, the most superficial cells still show some signs of differentiation; but this is incomplete, and proliferating cells are seen abnormally far above the basal layer. In carcinoma *in situ*, the cells in all the layers are proliferating and apparently undifferentiated. True malignancy begins when the cells cross the basal lamina and begin to invade the underlying connective tissue. Several years may elapse from the first signs of dysplasia to the onset of full-blown malignant cancer.

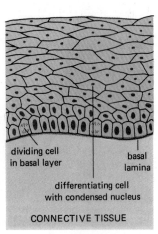

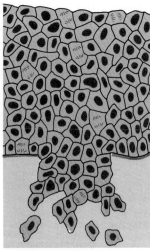

dividing cell in basal layer

basal lamina

differentiating cell with condensed nucleus

CONNECTIVE TISSUE

(A) normal   (B) dysplasia   (C) carcinoma *in situ*   (D) malignant carcinoma

(A)

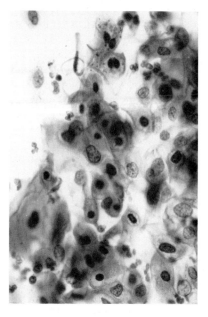

(B)

(C)

20 mm

Tumor Progression Involves Successive Rounds of Mutation and Natural Selection[6,7]

As illustrated by the two very different examples just discussed, cancers in general seem to arise by a process in which an initial population of slightly abnormal cells, descendants of a single mutant ancestor, evolves from bad to worse through successive cycles of mutation and natural selection. This evolution involves a large element of chance and usually takes many years; most of us die of other ailments before cancer has had time to develop. To understand the causation of cancer it is essential to understand the factors that may speed up the process.

In general, the rate of evolution, whether in a population of cells exploiting the opportunities for cancerous behavior in the body or in a population of organisms adapting to a new environment on the surface of the Earth, would be expected to depend on four main parameters: (1) the *mutation rate*, that is, the probability per gene per unit time that any given member of the population will undergo genetic change; (2) the *number of individuals in the population*; (3) the *rate of reproduction*, that is, the average number of generations of progeny produced per unit time; and (4) the *selective advantage* enjoyed by successful mutant individuals, that is, the ratio of the number of surviving fertile progeny they produce per unit time to the number of surviving fertile progeny produced by non-mutant individuals. The selective advantage depends both on the nature of the mutation and on environmental conditions, and further complications arise if heritable epigenetic changes occur, either randomly or in reaction to specific cues.

Experimental studies on the induction of cancer in animals illustrate these evolutionary principles. In the light of such studies, one can begin to make sense of the confusing variety of factors that affect the incidence of human cancers—factors ranging from cigarette smoke (for cancer of the lung) to the age at which a woman has her first baby (for cancer of the breast). The mutation rate per cell is not the only significant variable in the development of cancer.

The Development of a Cancer Can Be Promoted by Factors That Do Not Alter the Cells' DNA Sequence[6,8]

The stages by which an initial mild lesion progresses to become a cancer can be most easily observed in the skin. Skin cancers can be elicited in mice, for example, by repeatedly painting the skin with a mutagenic chemical carcinogen such as benzo[a]pyrene (a constituent of coal tar and tobacco smoke) or the related com-

Figure 21–11 Photographs of cells collected by scraping the surface of the uterine cervix (the Papanicolaou or "Pap smear" technique). (A) Normal; the cells are large and well differentiated, with highly condensed nuclei. (B) Dysplasia; the cells are in a variety of stages of differentiation, some quite immature. (C) Invasive carcinoma; the cells all appear undifferentiated, with scanty cytoplasm and a relatively large nucleus; debris in the background includes blood cells that have leaked out at the site of the ulcer created by the carcinoma. (Courtesy of Edward Miller.)

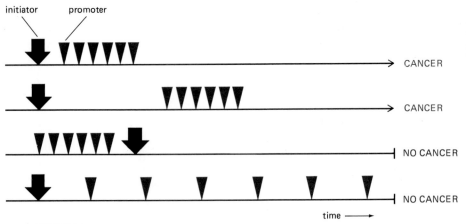

initiator promoter

CANCER

CANCER

NO CANCER

NO CANCER

time ⟶

Figure 21–12 Some possible schedules of exposure to a tumor initiator (mutagenic) and a tumor promoter (nonmutagenic), and their outcomes. Cancer ensues only if the exposure to the promoter follows exposure to the initiator, and only if the intensity of exposure to the promoter exceeds a certain threshold. Cancer can also occur as a result of repeated exposure to the initiator alone.

pound dimethylbenz[a]anthracene (DMBA). A single application of the carcinogen, however, usually does not by itself give rise to a tumor or any other obvious lasting abnormality. Yet it does cause latent genetic damage, and this can be detected through a greatly increased incidence of cancer when the cells are exposed either to further treatments with the same substance or to certain other, quite different, insults. A carcinogen that sows the seeds of cancer in this way is said to act as a **tumor initiator.** Simply wounding skin that has been exposed once to such an initiator can cause cancers to develop from some of the cells at the edge of the wound. Alternatively, repeated exposure over a period of months to certain substances known as **tumor promoters,** which are not themselves mutagenic, can cause cancer selectively in skin previously exposed to a tumor initiator. The most widely studied tumor promoters are *phorbol esters*, such as tetradecanoylphorbol acetate (TPA), which we have already encountered in another context as artificial activators of protein kinase C (and hence as agents that activate part of the phosphatidylinositol intracellular signaling pathway—see p. 704). These substances cause cancers at high frequency only if they are applied *after* a treatment with a mutagenic initiator (Figure 21–12).

As one might expect for genetic damage, the hidden changes caused by a tumor initiator are irreversible: thus they can be uncovered by treatment with a tumor promoter even after a long delay. The immediate effect of the promoter is apparently to stimulate cell division (or to cause cells that would normally undergo terminal differentiation to continue dividing instead); and in the region that had previously been exposed to the initiator this results in the growth of many small, benign, wartlike tumors, called *papillomas*. The greater the prior dose of initiator, the larger the number of papillomas induced; it is thought that each papilloma (at least for low doses of the initiator) consists of a single clone of cells descended from a mutant cell that the initiator has engendered. Both wounding and the application of the promoter probably act by inducing the expression of some of the "social control" genes that directly or indirectly affect cell proliferation (see p. 752). Such genes may remain quiescent in the resting epithelium, so that any mutations they have undergone in response to the initiator go undetected; by

Figure 21–13 One hypothesis proposed to explain the observed effect of tumor promoters on the development of tumors. Alternatively, the mutant gene might be constitutively expressed but have no effect until the promoter activates other genes required for cell proliferation.

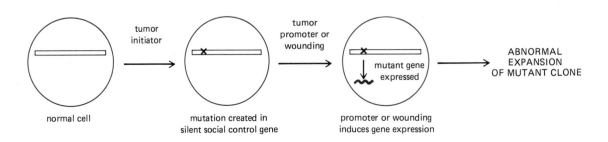

normal cell | tumor initiator → mutation created in silent social control gene | tumor promoter or wounding → promoter or wounding induces gene expression | mutant gene expressed → ABNORMAL EXPANSION OF MUTANT CLONE

inducing gene expression, the promoter or the stimulus of wounding may uncover the mutations and enable them to begin influencing cell proliferation (Figure 21–13).

A typical papilloma might contain about 10^5 cells—more than a thousand times as many as there are in a normal epidermal proliferative unit (see p. 969). If exposure to the tumor promoter is stopped, almost all the papillomas regress, and the skin regains a largely normal appearance—as expected from the hypothesis illustrated in Figure 21–13. In a few of the papillomas, however, further changes occur that enable growth to continue in an uncontrolled way, even after the promoter has been withdrawn. These changes seem to originate in occasional single papilloma cells, at about the frequency expected for spontaneous mutations. In this way a small proportion of the papillomas progress to become cancers. Thus the tumor promoter apparently favors the development of cancer, in this system at least, by expanding the population of cells that carry an initial mutation: the more such cells there are, the greater the chance that at least one of them will undergo another mutation carrying it one more step along the road to malignancy. Although naturally occurring cancers do not necessarily arise through the specific sequence of distinct initiation and promotion steps just described, their evolution must be governed by similar principles. They too will evolve at a rate that depends both on the frequency of mutations and on influences affecting the survival, proliferation, and spread of certain types of mutant cells once they have been created.

Most Cancers Result from Avoidable Combinations of Environmental Causes[9]

The development of a cancer generally involves many steps, each governed by multiple factors, some dependent on the genetic constitution of the individual, others dependent on his or her environment and way of life. By changing our surroundings or our habits, therefore, we should, in principle, be able to reduce drastically our chance of developing almost any given type of cancer. This is demonstrated most clearly by a comparison of cancer incidence in different countries: for almost every cancer that is common in one country, there is another country where the incidence is several times lower (Table 21–2); and migrant populations tend to take on the pattern of cancer incidence typical of the host country, implying that the differences are due to environmental, not genetic, factors. From such data it is estimated that 80–90% of cancers should be avoidable. Unfortunately, different cancers have different environmental risk factors, and a country that happens to escape one such danger is no more likely than other countries to escape the rest; thus the incidence of all cancers combined (among individuals of a given age) is similar from country to country. There are, however, some subgroups whose abstinent way of life does seem to reduce the total cancer death rate: the incidence of cancer among strict Mormons in Utah, for example, is only about half that among Americans in general.

While such epidemiological observations indicate that cancer can be avoided, it remains difficult to identify the specific environmental risk factors or to establish how they act. Some certainly operate as mutagenic tumor initiators, directly provoking genetic change; others presumably serve as tumor promoters that help to enlarge the population of cells liable to progress, through further mutation, to full-blown cancer. The carcinogens in tobacco smoke, like the aflatoxin on tropical peanuts (see Figure 21–6), probably belong mostly in the first category, while the reproductive hormones that circulate in a woman's body at different stages of her life may belong in the second category (Figure 21–14). It is possible that some factors act in still other ways—for example, by causing heritable epigenetic changes. Of course, it is not necessary to understand how cancer-causing agents act in order to identify them and show how to avoid them. In this task, cancer epidemiology has had some notable successes and promises more to come; simply by revealing the role of smoking, it has shown a way to reduce the total cancer death rate in North America and Europe by as much as 30%. The prevention of cancer is not only better than cure but seems also, given our present state of knowledge, to be much more readily attainable.

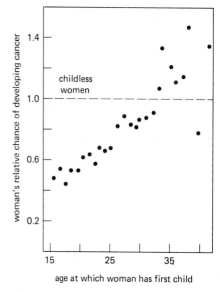

Figure 21–14 The relative probability of breast cancer developing at some time in a woman's life plotted as a function of the age at which she gives birth to her first child. The graph shows the value of the probability relative to that for a childless woman. The longer the delay before bearing the first child, the higher the probability of breast cancer, suggesting that exposure to certain combinations of reproductive hormones may promote development of the cancer. There is some evidence from laboratory studies that the first full-term pregnancy may result in a permanent epigenetic change in the cells of the breast, altering their subsequent responses to hormones. Many other factors—for example, the amount of fat in the diet—are also highly correlated with breast cancer. (From J. Cairns, Cancer: Science and Society. San Francisco: Freeman, 1978. After B. MacMahon, P. Cole, and J. Brown, J. Natl. Cancer Inst. 50:21–42, 1973.)

Table 21–2 Variation Between Countries in the Incidence of Some Common Cancers

Site of Origin of Cancer	High–Incidence Area	Cumulative Incidence (%) in High–Incidence Area	Low–Incidence Area	Ratio of Rates in High– and Low–Incidence Areas
Skin	Australia (Queensland)	20	India (Bombay)	>200
Esophagus	Iran	20	Nigeria	300
Lung	England	11	Nigeria	35
Stomach	Japan	11	Uganda	25
Uterine cervix	Columbia	10	Israel (Jewish)	15
Prostate	United States (blacks)	9	Japan	40
Liver	Mozambique	8	England	100
Breast	Canada	7	Israel (non-Jewish)	7
Colon	United States (Connecticut)	3	Nigeria	10
Uterus	United States (California)	3	Japan	30
Oral cavity	India (Bombay)	2	Denmark	25
Rectum	Denmark	2	Nigeria	20
Bladder	United States (Connecticut)	2	Japan	6
Ovary	Denmark	2	Japan	6
Nasopharynx	Singapore (Chinese)	2	England	40
Pancreas	New Zealand (Maori)	2	India (Bombay)	8
Larynx	Brazil (São Paulo)	2	Japan	10
Pharynx	India (Bombay)	2	Denmark	20
Penis	Parts of Uganda	1	Israel (Jewish)	300

Data for uterine cervix, breast, uterus, and ovary are for women; others are for men. The cumulative incidence is defined as the percentage of the population that would develop the specified cancer by the age of 75, in the absence of other causes of death; the ratio of rates is calculated for the 35- to 64-year age group. (Slightly modified from R. Doll and R. Peto, The Causes of Cancer. New York: Oxford University Press, 1981.)

The Search for Cancer Cures Is Hard but Not Hopeless[10]

The difficulty of curing a cancer is like the difficulty of getting rid of weeds. Cancer cells can be removed surgically or destroyed with toxic chemicals or radiation; but it is hard to eradicate every single one of them. Surgery can rarely ferret out every metastasis, and treatments that kill cancer cells are generally toxic to normal cells as well. If even a few cancerous cells remain, they can proliferate to produce a resurgence of the disease; and unlike the normal cells, they may evolve resistance to the poisons used against them. Yet the outlook is not hopeless. In spite of the difficulties, effective cures using anticancer drugs (alone or in combination with other treatments) have been devised for some formerly highly lethal cancers (notably Hodgkin's lymphoma, testicular cancer, choriocarcinoma, and some leukemias and other cancers of childhood). For several of the more common cancers, moreover, appropriate surgery or local radiotherapy enables a large proportion of patients to recover if the illness is diagnosed at a reasonably early stage; and even where a cure at present seems beyond our reach, there are treatments that will prolong life or at least relieve distress.

A great deal of clinical cancer research centers on the problem of how to kill cancer cells selectively. For the most part, current methods exploit relatively subtle differences between normal and neoplastic cells with respect to proliferation rate, metabolism, and radiosensitivity, and they have unpleasant toxic side effects. A few types of cancer cells are especially vulnerable to selective attack because they depend on specific hormones or because their surfaces have unusual chemical features that can be recognized by antibodies. In general, however, progress with the vexing problem of anticancer selectivity has been slow—a matter of trial and error and guesswork as much as rational calculation.

In the search for better ways of curbing the survival, proliferation, and spread of cancer cells, it is important to examine more closely the strategies by which they thrive and multiply.

Cancerous Growth Often Depends on Derangements of Cell Differentiation[11]

We have so far emphasized that cancer cells defy the normal controls on cell division: this is their central property. But many tissues are organized in such a way that even an uncontrolled increase in the frequency of cell division will not by itself produce a steadily growing tumor. The example of the uterine cervix, discussed above on page 1194, illustrates this point. Like the epidermis of the skin and many other epithelia, the epithelium of the uterine cervix normally renews itself continually by shedding terminally differentiated cells from its outer surface and generating replacements from stem cells in the basal layer (see p. 970). On average, each normal stem cell division generates one daughter stem cell and one cell that is condemned to terminal differentiation and a cessation of cell division. If the stem cell simply divides more rapidly, terminally differentiated cells will be produced and shed more rapidly, and a balance of genesis and destruction will still be maintained. Thus if a transformed stem cell is to generate a steadily growing clone of progeny, the basic rules must be upset: either more than 50% of the daughter cells must remain as stem cells, or the process of differentiation must be deranged so that daughter cells embarked on this route retain an ability to carry on dividing indefinitely and avoid being discarded at the end of the production line (Figure 21–15).

Presumably, the development of such properties underlies the progression from a mild dysplasia of the uterine cervix to carcinoma *in situ* and malignant cancer (see Figure 21–10). Similar considerations apply to the development of cancer in other tissues that rely on stem cells, such as the skin, the lining of the gut, and the hemopoietic system. Several forms of leukemia, for example, seem to arise from a disruption of the normal program of differentiation, such that a committed progenitor of a particular type of blood cell continues to divide indefinitely, instead of differentiating terminally in the normal way after a strictly limited num-

Figure 21–15 The stem-cell strategy for producing new differentiated cells, and two types of derangement that can give rise to the unbridled proliferation characteristic of cancer. Note that an excessive cell-division rate for the stem cells will not by itself have this effect.

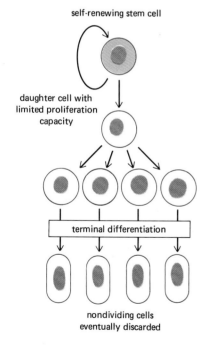

self-renewing stem cell

daughter cell with limited proliferation capacity

terminal differentiation

nondividing cells eventually discarded

(A) NORMAL PATHWAY

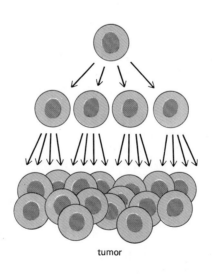

tumor

(B) STEM CELL FAILS TO PRODUCE ONE NON-STEM-CELL DAUGHTER IN EACH DIVISION AND THEREBY PROLIFERATES TO FORM A TUMOR

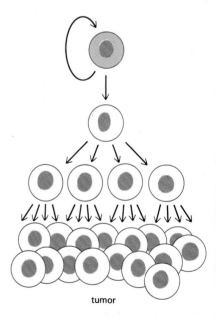

tumor

(C) DAUGHTER CELLS FAIL TO DIFFERENTIATE NORMALLY AND THEREBY PROLIFERATE TO FORM A TUMOR

ber of division cycles (see p. 979). In general, mutations or epigenetic changes that block the normal maturation of cells toward a nondividing, terminally differentiated state must play an essential part in many cancers. In the treatment of cancer, therefore, there is some prospect that drugs that promote cell differentiation may turn out to be a useful alternative or supplement to drugs that simply kill dividing cells.

To Metastasize, Cancer Cells Must Be Able to Cross Basal Laminae[12]

It is the ability to metastasize that makes cancers hard to eradicate surgically or by localized irradiation. To disseminate widely in the body, the cells of a typical solid tumor must be able to loosen their adhesion to their original neighbors, escape from the tissue of origin, burrow through other tissues until they reach a blood vessel or a lymphatic vessel, cross the basal lamina and endothelial lining of the vessel so as to enter the circulation, make an exit from the circulation elsewhere in the body, and survive and proliferate in the new environment in which they find themselves (Figure 21–16). The final steps are probably the most difficult: many tumors release quite large numbers of cells into the circulation, but only a minute proportion of these cells succeed in founding metastatic colonies.

A few types of normal cells—notably white blood cells—already have some or all of the properties needed to disseminate through the body, but for most cancers the ability to metastasize probably requires additional mutations or epigenetic changes. Such transformations, like the others involved in the development of cancer, are thought to occur at random in the initial tumor population: only

Figure 21–16 Steps in the process of metastasis. This example illustrates the spread of a tumor from the lung to the liver. Tumor cells may enter the bloodstream directly by crossing the wall of a blood vessel, as diagrammed here, or, more commonly perhaps, by crossing the wall of a lymphatic vessel. The lymphatic vessels ultimately discharge their contents (lymph) into the bloodstream, but tumor cells that have entered a lymphatic vessel often become trapped in lymph nodes along the way, giving rise to lymph-node metastases. Studies in animals show that typically less than one in every thousand malignant tumor cells that enter the bloodstream will survive to produce a tumor at a new site. The likelihood of metastasis depends on the characteristics of the host tissue as well as on those of the cancer cell.

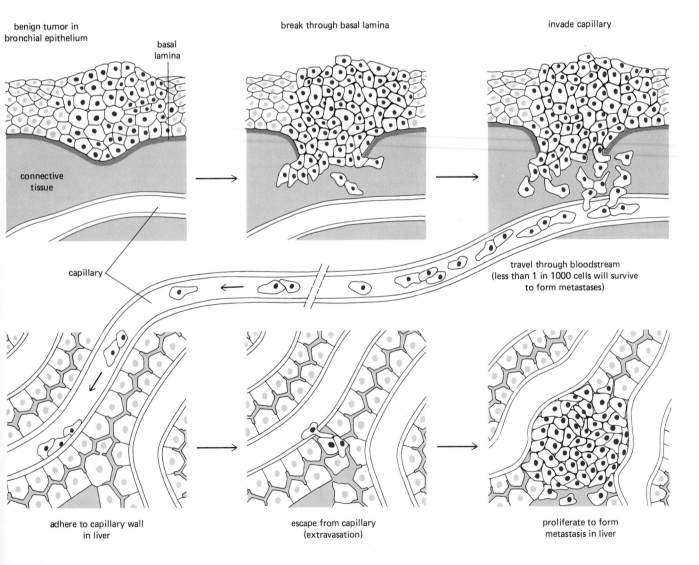

benign tumor in bronchial epithelium

basal lamina

break through basal lamina

invade capillary

connective tissue

capillary

travel through bloodstream (less than 1 in 1000 cells will survive to form metastases)

adhere to capillary wall in liver

escape from capillary (extravasation)

proliferate to form metastasis in liver

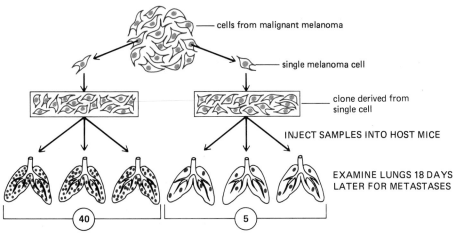

cells from malignant melanoma

single melanoma cell

clone derived from single cell

INJECT SAMPLES INTO HOST MICE

EXAMINE LUNGS 18 DAYS LATER FOR METASTASES

40

5

median number of metastatic tumors per mouse

Figure 21–17 Experiment showing that there are clonally heritable differences between the cells of a single tumor with respect to the ability to metastasize. Cells derived from a single cancer cell line are subcloned, and standard aliquots of each subclone are tested by injection into the bloodstream of host mice. The subclones differ markedly in the number of resulting metastases per mouse.

those few cells that acquire the properties needed for metastasis and that happen to land in a suitable environment will be able to found secondary tumors. In accordance with this concept of evolution through random variation and natural selection, it is found that the cells of a single tumor are heterogeneous in metastatic capacity (Figure 21–17).

An understanding of the molecular mechanisms of metastasis should eventually allow the design of treatments to block it. Some progress is being made along these lines. It has been shown, for example, that for tumor cells to cross a basal lamina they must have laminin receptors (see p. 822), which enable the cells to adhere to the lamina, and they must secrete type IV collagenase, which helps them digest the lamina (Figure 21–18). Antibodies or other reagents that block either laminin attachment or the activity of type IV collagenase have been found to block metastasis in experimental animals. It remains to be seen whether human cancer patients can be helped by treatments that halt metastatic spread in this way.

Defects of DNA Repair, Replication, and Recombination Accelerate the Development of Cancer[1,13]

As we have emphasized, the incidence of tumors and their rate of progression from benign to malignant depends on the frequency of mutations. The mutation rate may be high because of mutagens in the environment or because of intracellular defects in the machinery governing replication, recombination, and repair of DNA. People with the rare genetic disorder *xeroderma pigmentosum*, for example, have a defect in the system of enzymes required to repair the type of damage done to DNA by ultraviolet irradiation (see p. 225); as a result, the slightest exposure of the skin to sunlight is liable to provoke skin cancers. A more general predisposition to cancer is seen in *Bloom's syndrome*, where there is a defect in the enzyme *DNA ligase I*, required for DNA replication and repair, and in *Fanconi's anemia* and *ataxia-telangiectasia*, where there are less well characterized defects in the same functions. In these rare genetic disorders, the abnormality is inherited through the germ line and is therefore present in all the cells of the body. Similar genetic defects in DNA metabolism can also arise, however, through mutations originating in somatic cells, and there is evidence to suggest that such aberrations are a common and important factor in the development of many cancers.

Cancer cells often display an abnormal variability in the size and shape of their nuclei (Figure 21–19) and in the number and structure of their chromosomes; indeed, abnormal nuclear morphology is one of the key features used by pathologists to diagnose cancer. When cancer cells are grown in culture, they are often found to have an extraordinarily unstable karyotype: genes become amplified or deleted and chromosomes become lost, duplicated, or translocated with a far higher frequency than in normal cells in culture. Such chromosomal variability might simply be the consequence of hasty cell-division cycles occurring in a dif-

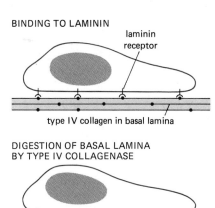

BINDING TO LAMININ

laminin receptor

type IV collagen in basal lamina

DIGESTION OF BASAL LAMINA BY TYPE IV COLLAGENASE

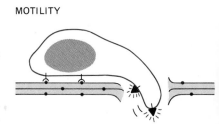

MOTILITY

Figure 21–18 Three steps in crossing a basal lamina—a task that invasive tumor cells must be able to perform.

ferentiated cell that is poorly adapted for rapid proliferation. Alternatively, it might reflect a heritable deficiency in the machinery or control of DNA replication, repair, or recombination, arising by somatic mutation in any one of the many genes involved in these complex processes. Such a mutation would be liable to increase the likelihood of subsequent mutations in other classes of genes. For this reason one might expect it to be a common feature of cells that undergo the multiple mutations required to become cancerous. Suppose, for example, that three mutations in genes governing the social behavior of cells are required to convert a normal cell into a cancer cell and that each of these mutations normally occurs at a rate of 10^{-4} per cell per human lifetime. Then the probability for a single cell with the normal level of mutability to accumulate these three mutations in the course of a lifetime would be $10^{-4} \times 10^{-4} \times 10^{-4} = 10^{-12}$ per cell. But now let us suppose that each of these mutations occurs at a rate of 10^{-2} per cell per lifetime if there has been a prior mutation in some particular enzyme involved in DNA replication or repair. If this latter mutation itself has a probability of 10^{-4}, cancer cells will arise most frequently by the route that begins with the mutation that increases mutability: this route (given the simplest possible assumptions) involves a combination of four events whose joint probability is on the order of $10^{-4} \times 10^{-2} \times 10^{-2} \times 10^{-2} = 10^{-10}$ per cell per lifetime; it is thus 100 times more likely than the route involving only the minimal three mutations.

The Enhanced Mutability of Cancer Cells Helps Them Evade Destruction by Anticancer Drugs[10,14]

Whatever the origins of the abnormally high mutability of cancer cells, most malignant tumor cell populations are heterogeneous in many respects and capable of evolving at an alarming rate when subjected to new selection pressures. This aggravates the difficulties of cancer therapy. Repeated treatments with drugs that are selectively toxic to dividing cells can be used to kill the majority of neoplastic cells in a cancer patient, but it is rarely possible to kill them all: usually some small proportion are drug-resistant, and the effect of the treatment is to favor the spread and evolution of cells with this trait. To make matters worse, cells that are exposed to one drug often develop a resistance not only to that drug, but also to other drugs to which they have never been exposed.

This phenomenon of **multidrug resistance** is frequently correlated with a curious change in the karyotype: the cell is seen to contain additional pairs of miniature chromosomes—so-called *double minute chromosomes*—or to have a *homogeneously staining region* interpolated in the normal banding pattern of one of its regular chromosomes. Both these aberrations consist of massively amplified numbers of copies of a small segment of the genome (see Figures 21–26 and 21–31, below). Cloning of this amplified DNA has revealed that it often contains a specific gene, known as the *multidrug resistance (mdr1)* gene, which codes for a plasma-membrane-bound transport ATPase that is thought to prevent the intracellular accumulation of certain classes of lipophilic drugs by pumping them out of the cell. The amplification of other types of genes can also give the cancer cell a selective advantage: thus the gene for the enzyme dihydrofolate reductase (DHFR) often becomes amplified in response to cancer chemotherapy with the folic-acid antagonist methotrexate; and we shall see that some proto-oncogenes involved in cell-division control are similarly amplified in some cancers (see p. 1214).

While defects in DNA replication, recombination, or repair may help cancer cells to evolve by increasing their mutability, they may also make the cells more vulnerable to certain types of attack. This may explain the observation—exploited in therapy—that the cells of many tumors are killed more easily than normal cells by irradiation or by exposure to specific drugs that interfere with DNA metabolism. As we learn more about the molecular mechanisms of DNA replication, recombination, and repair, it should become possible to devise tests to pinpoint defects in these functions in individual cases of cancer. Using such information we may be better able to kill the delinquent cells by designing drugs that exploit their particular weaknesses.

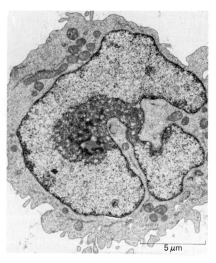

Figure 21–19 Typical abnormalities in the appearance of the nucleus of a cancer cell—in this example, an erythroleukemia cell. The cancer cell nucleus is large in relation to the amount of cytoplasm, with an irregularly indented envelope and a nucleolus that is also abnormally large and complex in its structure. (Courtesy of Daniel Friend.)

Summary

Cancer cells, by definition, proliferate in defiance of normal controls (that is, they are neoplastic) and are able to invade and colonize surrounding tissues (that is, they are malignant). By giving rise to secondary tumors, or metastases, they become hard to eradicate surgically. Cancer cells usually retain many features of the specific cell type from which they are derived. Most cancers are thought to originate from a single cell that has undergone a somatic mutation, but the progeny of this cell must undergo further changes, probably requiring several additional mutations, before they become cancerous. This phenomenon of tumor progression, which usually takes many years, reflects the operation of evolution by mutation and natural selection among somatic cells; the rate of the process is accelerated both by mutagenic agents (tumor initiators) and by certain nonmutagenic agents (tumor promoters) that affect gene expression, stimulate cell proliferation, and alter the ecological balance of mutant and nonmutant cells. Thus many factors contribute to the development of a given cancer, and since some of these factors are avoidable features of the environment, a large proportion of cancers are in principle preventable.

Much effort in cancer research has been devoted to the search for ways to cure the disease by exterminating cancer cells while sparing their normal neighbors. A rational approach to this problem requires an understanding of the special properties of cancer cells that enable them to evolve, multiply, and spread. Thus neoplastic cell proliferation often seems to be associated with a block in differentiation, whereby the progeny of a stem cell are enabled to continue dividing instead of entering a terminal nondividing state; in principle, the proliferation could be curbed by promoting cell differentiation. To become malignant, tumor cells must be able to cross basal laminae; antibodies can be designed that interfere with this ability, thereby hindering metastasis. Cancer cells are often found to be abnormally mutable; this hastens evolution of the complex set of properties required for neoplasia and malignancy and helps the cancer cells develop resistance to anticancer drugs. At the same time, however, defects of DNA metabolism underlying such mutability may make the cancer cells uniquely vulnerable to a suitably designed therapeutic attack.

The Molecular Genetics of Cancer[15]

Because cancer is the outcome of a series of random genetic accidents subject to natural selection, no two cases even of the same variety of the disease are likely to be genetically identical. Nevertheless, all cancers can be expected to involve a disruption of the normal restraints on cell proliferation, and for each cell type there is a finite number of ways in which such disruption can occur. Moreover, some parts of the machinery for regulating cell proliferation are likely to be the same in many or all cell types, and similarly vulnerable. In fact, changes in a relatively small set of genes appear to be responsible for much of the deregulation of cell division in cancer. The identification and characterization of many of these genes has been one of the great triumphs of molecular biology in the past decade.

Cell proliferation can be regulated directly, through the mechanism that determines whether a cell passes the restriction point or "Start" of the cell-division cycle (see p. 745), or indirectly—for example, through regulation of the commitment to terminal differentiation (see p. 967). In either case the normal regulatory genes can be loosely classified into those whose products help stimulate cell proliferation and those whose products help inhibit it. Correspondingly, there are two mutational routes toward the uncontrolled cell proliferation that is characteristic of cancer. The first is to make a stimulatory gene hyperactive: this type of mutation has a dominant effect—only one of the cell's two gene copies need undergo the change—and the altered gene is called an **oncogene** (the normal allele being a **proto-oncogene**). The second is to make an inhibitory gene inactive: this type of mutation has a recessive effect—both the cell's gene copies must be inactivated or deleted to free the cell of the inhibition—and the lost gene is sometimes called, for want of a better term, a **tumor suppressor gene.**

In addition to ordinary mutations, there is another type of genetic change that can lead to cancer: the cell-division control system can be subverted by foreign DNA introduced into the cell by a virus. In fact, insight into the molecular genetics of cancer came first from the study of such **tumor viruses,** which paved the way for an explosion of discoveries of oncogenes and proto-oncogenes. More recently, progress has been made in the more difficult task of identifying and cloning tumor suppressor genes, as we shall see at the end of this section.

Both DNA Viruses and RNA Viruses Can Cause Tumors[15,16]

Both DNA viruses and RNA viruses—retroviruses in particular—can play a part in transforming healthy cells into cancer cells. This can be demonstrated experimentally both in laboratory animals, where certain viruses cause cancers, and in cell culture, where the same viruses transform the behavior of infected cells, enabling them to divide in circumstances where untransformed cells will not (see p. 753 for an account of the properties of neoplastically transformed cells in culture). Two intensively analyzed examples are the *SV40 virus*, a DNA virus isolated from monkey cells, and the *Rous sarcoma virus*, a retrovirus isolated from chickens. It is harder, however, to incriminate viruses in human cancer: they do not stand out among the multiple causative factors involved in almost every variety of the disease, and it is thought likely that in most of the common human cancers they play no part. This may be because the immune system, by destroying virus-infected cells, protects us from many virus-induced cancers that might otherwise arise. Nonetheless, there is now good evidence that viruses do play a causative role in several types of human cancers (Table 21–3)—in some cases probably through

Table 21–3   Viruses Associated with Human Cancers

Virus	Associated Tumors	Areas of High Incidence	Other Suspected Risk Factors
DNA viruses			
Papovavirus family			
Papillomavirus (many distinct strains)	warts (benign) carcinoma of uterine cervix	worldwide worldwide	— smoking
Hepadnavirus family			
Hepatitis-B virus	liver cancer (hepatocellular carcinoma)	Southeast Asia; tropical Africa	aflatoxin from fungal contamination of food; alcoholism; smoking; other viruses
Herpesvirus family			
Epstein-Barr virus	Burkitt's lymphoma (cancer of B lymphocytes)	West Africa; Papua New Guinea	malaria
	nasopharyngeal carcinoma	southern China; Greenland (Inuit)	histocompatibility genotype (?); salted fish in infancy (?)
RNA viruses			
Retrovirus family			
Human T-cell leukemia virus type I (HTLV-I)	adult T-cell leukemia/ lymphoma	Japan (Kyushu); West Indies	—
Human immunodeficiency virus (HIV-1, the AIDS virus)	Kaposi's sarcoma [cancer of endothelial cells of blood vessels (?)]	Central Africa	immune deficiency or suppression; infection with other virus (?)

For all the above viruses, the number of people infected is much larger than the number who develop cancer: the viruses must act in conjunction with other factors. Moreover, some of the viruses probably contribute to cancer only indirectly; for example, HIV-1, by obliterating cell-mediated immune defenses, may allow endothelial cells transformed by some other agent to thrive as a tumor instead of being destroyed by the immune system.

indirect promoting actions, in other cases by helping directly to cause neoplastic transformation of the cells they infect.

If a cell is to be stably transformed by a virus, a stable parasitic association must be established: the virus must not kill the cell, and the cell must retain the viral genes from one cell generation to the next—usually by integrating those genes into one or more of its own chromosomes, occasionally by retaining them as an extrachromosomal plasmid that replicates in step with the chromosomes. SV40 and the retroviruses, for example, are integrated into the chromosomes of the cells they transform; the *papillomaviruses*—a class of DNA viruses responsible for human warts and implicated in carcinomas of the uterine cervix (see Table 21–3)—are maintained in some conditions as plasmids, in other conditions as integrated elements. In either case the effect is to endow the cell with a heritably altered genome. DNA viruses and retroviruses differ fundamentally, however, in the nature of the viral genes that cause neoplastic transformation and in the relationship between cell transformation and the virus's usual life cycle.

DNA Tumor Viruses Subvert Cell Division Controls as Part of Their Strategy for Survival[15,17]

As explained in Chapter 5 (p. 250), a **DNA tumor virus** such as SV40 normally propagates in the wild by a process that does not depend on the production of cancer. An SV40 virus, having entered a host cell, typically does not become stably incorporated into the host cell genome. Instead, a protein (or set of proteins) encoded by a viral gene rapidly activates the host cell's machinery for DNA replication, and the virus then uses this subverted equipment to replicate its own genome, which in turn directs the synthesis of other viral components at the expense of the host cell until the host is killed, releasing a horde of new infectious virus particles. Much more rarely, the virus may enter a host cell of a type that is *nonpermissive* for viral proliferation and may persist there as a result of a genetic accident through which the virus becomes stably incorporated into one or more of the host cell's chromosomes. In these circumstances the viral gene that activates the host's machinery for DNA replication may still be transcribed, thereby driving the host cell into S phase and forcing it to undergo repeated division cycles: the viral gene acts as an oncogene, causing a cancerous transformation. It is, however, unlike the other classes of oncogenes to be discussed below in that it has no counterpart in the normal host cell genome; it has evolved as an essential part of the virus's equipment for its usual mode of propagation, and not by mutation from a cellular proto-oncogene.

DNA viruses are a diverse group, but the general principles just described apply, with some variations, to most of those that are implicated in cancer. One important variant is illustrated by the papillomaviruses, for which persistent association with the host cell is a normal part of the life cycle. These viruses belong to the *papovavirus* family that includes SV40, but they can apparently switch between a nonproductive mode of infection, in which they replicate in step with the host cell without harming it, and a productive mode, in which they reproduce rapidly and kill the host cell, so that large numbers of new virus particles are released to infect other cells. Like SV40, these viruses have to be able to subvert the host cell's DNA synthesis machinery, and the viral genes that have this function can act as oncogenes. Studies of cancer of the uterine cervix have suggested that carcinogenesis by papillomaviruses in humans requires the integration of a specific viral replication gene into a host chromosome, as illustrated in Figure 21–20.

Retroviruses Pick Up Oncogenes by Accident[15,18]

By contrast with DNA viruses, most **retroviruses** (see p. 254 and p. 754) are more or less harmless to the host cell, from which they are released continuously by budding from the plasma membrane, without causing neoplastic transformation. Occasionally, however, a retrovirus accidentally acquires from its host a cellular control gene—or a corrupted copy or fragment of such a gene—that has no useful

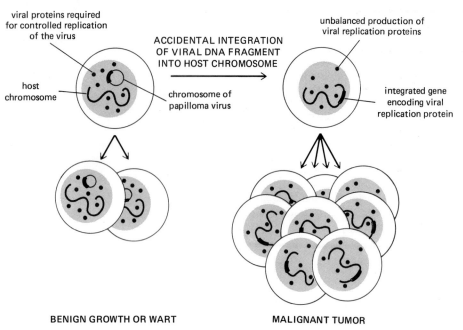

BENIGN GROWTH OR WART

MALIGNANT TUMOR

Figure 21–20 How certain papillomaviruses are thought to give rise to cancer of the uterine cervix. Papillomaviruses have double-stranded circular DNA chromosomes of about 8000 nucleotide pairs. In a wart or other benign infection, these chromosomes are stably maintained in the basal cells of the epithelium as plasmids whose replication is regulated so as to keep step with the chromosomes of the host (*left*). Rare accidents can cause the integration of a fragment of such a plasmid into a chromosome of the host, altering the environment of the viral genes and disrupting the control of their expression. The consequent unregulated production of a viral replication protein tends to drive the host cell into S phase, thereby helping to generate a cancer (*right*).

function in the viral life cycle but may drastically influence subsequent host cells. In particular, as we saw in Chapter 13 (see p. 754), a retrovirus such as the Rous sarcoma virus that has picked up a host-derived oncogene (Figure 21–21) will be easily detected because of its dominant transforming effect on infected host cells, which are driven to proliferate excessively. The oncogene can be identified by genetic dissection of the retrovirus; DNA probes can then be prepared from it to fish for homologous genes in normal cells. More than 20 oncogenes have been discovered by this approach (Table 21–4), each corresponding to its own closely similar counterpart—a proto-oncogene—present in the normal vertebrate cell genome (see p. 754). As discussed on page 756, these oncogenes fall into several

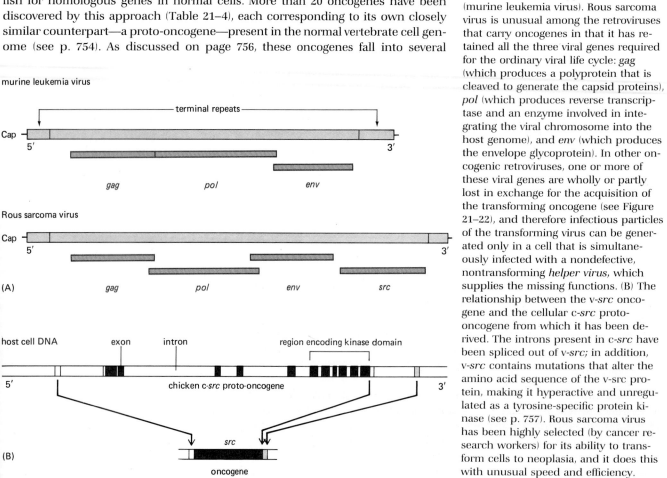

Figure 21–21 The structure of the Rous sarcoma virus. (A) The organization of the viral genome as compared with that of a more typical retrovirus (murine leukemia virus). Rous sarcoma virus is unusual among the retroviruses that carry oncogenes in that it has retained all the three viral genes required for the ordinary viral life cycle: *gag* (which produces a polyprotein that is cleaved to generate the capsid proteins), *pol* (which produces reverse transcriptase and an enzyme involved in integrating the viral chromosome into the host genome), and *env* (which produces the envelope glycoprotein). In other oncogenic retroviruses, one or more of these viral genes are wholly or partly lost in exchange for the acquisition of the transforming oncogene (see Figure 21–22), and therefore infectious particles of the transforming virus can be generated only in a cell that is simultaneously infected with a nondefective, nontransforming *helper virus*, which supplies the missing functions. (B) The relationship between the v-*src* oncogene and the cellular c-*src* proto-oncogene from which it has been derived. The introns present in c-*src* have been spliced out of v-*src*; in addition, v-*src* contains mutations that alter the amino acid sequence of the v-src protein, making it hyperactive and unregulated as a tyrosine-specific protein kinase (see p. 757). Rous sarcoma virus has been highly selected (by cancer research workers) for its ability to transform cells to neoplasia, and it does this with unusual speed and efficiency.

Table 21–4 Oncogenes Originally Identified Through Their Presence in Transforming Retroviruses

Oncogene	Proto-oncogene Function (where known)	Source of Virus	Virus-induced Tumor
abl	protein kinase (tyrosine)	mouse / cat	pre-B-cell leukemia / sarcoma
akt	?	mouse	T-cell lymphoma
crk	activator of tyrosine-specific protein kinase(s)	chicken	sarcoma
erb-A	thyroid hormone receptor	chicken	(supplements action of v-erb-B)
erb-B	protein kinase (tyrosine): epidermal growth factor (EGF) receptor	chicken	erythroleukemia; fibrosarcoma
ets	nuclear protein	chicken	(supplements action of v-myb)
fes/fps	protein kinase (tyrosine)	cat/chicken	sarcoma
fgr	protein kinase (tyrosine)	cat	sarcoma
fms	protein kinase (tyrosine): macrophage colony-stimulating factor (M-CSF) receptor	cat	sarcoma
fos	nuclear transcription factor	mouse	osteosarcoma
jun	nuclear protein: AP-1 transcription factor	chicken	fibrosarcoma
kit	protein kinase (tyrosine)	cat	sarcoma
mil/raf	protein kinase (serine/threonine)	chicken/mouse	sarcoma
mos	protein kinase (serine/threonine)	mouse	sarcoma
myb	nuclear protein	chicken	myeloblastosis
myc	nuclear protein	chicken	sarcoma; myelocytoma; carcinoma
H-ras	G protein	rat	sarcoma; erythroleukemia
K-ras	G protein	rat	sarcoma; erythroleukemia
rel	nuclear protein	turkey	reticuloendotheliosis
ros	protein kinase (tyrosine)	chicken	sarcoma
sea	protein kinase (tyrosine)	chicken	sarcoma; leukemia
sis	platelet-derived growth factor, B chain	monkey	sarcoma
ski	nuclear protein	chicken	carcinoma
src	protein kinase (tyrosine)	chicken	sarcoma
yes	protein kinase (tyrosine)	chicken	sarcoma

distinct families, the largest group being the protein kinase gene family to which the oncogene of the Rous sarcoma virus (the src oncogene) belongs (Figure 21–22).

There are two ways in which a proto-oncogene can be converted into an oncogene upon incorporation into a retrovirus: the gene sequence may be altered or truncated so that it codes for a protein with abnormal activity, or the gene may be brought under the control of powerful promoters and enhancers in the viral genome, which cause its product to be made in excess or in inappropriate circumstances; often both effects occur together. Retroviruses can also exert similar oncogenic effects in another way that does not involve picking up host genes and carrying them from cell to cell: DNA copies of the viral RNA may simply insert themselves into the host cell genome at sites close to, or even within, proto-oncogenes. The resulting genetic disruption is called **insertional mutagenesis,** and the altered genome is inherited by all the progeny of the original host cell. Random insertion of DNA copies of the viral RNA into the host DNA is a part of the normal retroviral life cycle, and insertion anywhere within about 10,000 nucleotide pairs from a proto-oncogene is liable to cause abnormal activation of that

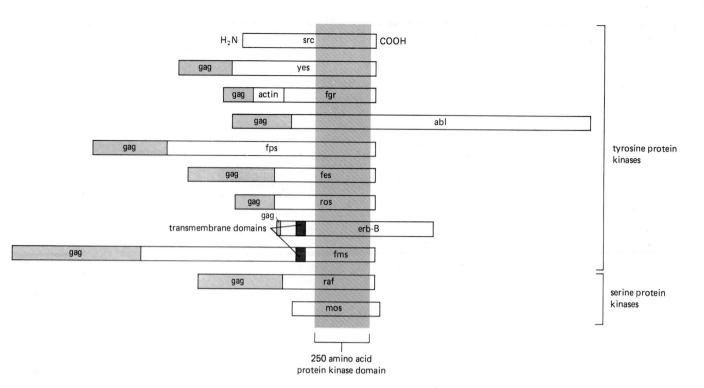

tyrosine protein kinases

serine protein kinases

250 amino acid
protein kinase domain

gene. Insertional mutagenesis provides an important means of identifying proto-oncogenes, which can be tracked down by their proximity to the inserted retrovirus. Proto-oncogenes identified in this way often turn out to be the same as those discovered in the other way, as counterparts to oncogenes that retroviruses carry from cell to cell, but some new ones have been discovered as well (Table 21–5). An example is the *int*-1 gene, activated by insertional mutagenesis in breast cancers in mice infected with the mouse mammary tumor virus (Figure 21–23). When cloned and sequenced, this gene turned out to be closely homologous to the *Drosophila* gene *wingless* (see p. 929), which is apparently involved in cell-cell communications that regulate details of the segmental pattern of the fly.

Figure 21–22 Many of the proteins produced by the oncogenes carried by transforming retroviruses are protein kinases; the diagram compares their primary structures schematically. Note that most of the oncogenes have disrupted normal viral genes and thereby produce fusion proteins that contain the amino terminus of the viral gag protein. (After T. Hunter, *Sci. Am.* 251(2):70–79, 1984.)

Different Searches for the Genetic Basis of Cancer Converge on Disturbances in the Same Small Set of Proto-oncogenes[15,19]

While some researchers pursued the line of investigation leading from retroviruses to oncogenes, others took a more direct approach and searched for DNA sequences in human cancer cells that would provoke uncontrolled proliferation when introduced into noncancerous cells. The assay was done in cell culture, using a particular line of mouse-derived 3T3 cells—NIH 3T3 cells—as the noncancerous hosts and transfecting them with DNA taken from human tumor cells (see p. 755 and Figure 13–33). The findings were dramatic: oncogenes were detected in many lines of human cancer cells, and in several cases these oncogenes turned out to be mutant alleles of some of the same proto-oncogenes that had been identified by the retroviral approach, or of genes very closely related to them. For example, about one in four human tumors was found to contain a mutated mem-

Figure 21–23 Insertional mutagenesis that activates a gene called *int*-1 and produces breast cancer in mice infected with the mouse mammary tumor virus (MMTV). The sites of MMTV integration observed in 19 different tumor isolates are indicated by arrows. Note that the insertions can activate transcription of the *int*-1 gene from distances of more than 10,000 nucleotide pairs away and from either side of the gene. This effect is attributed to a powerful enhancer DNA sequence present in the terminal repeats of the MMTV genome.

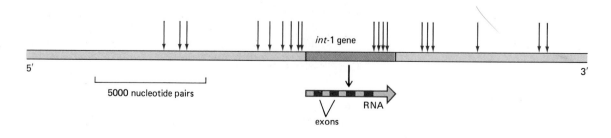

Table 21–5 Some Oncogenes Originally Identified by Means Other Than Their Presence in Transforming Retroviruses

Means of Detection	Oncogenes
Amplification:	L-*myc*, N-*myc*, Gli
Transfection:	*mas*, *met*, *neu*, N-*ras*, *trk*, *onc*-F, "*thy*"
Translocation:	*bcl*-1, *bcl*-2, *tcl*-1, *tcl*-2, *tcl*-3a, *tcl*-3b
Insertional mutation:	*evi*-1, *int*-1, *int*-2, *int*-3, *int*-4, *Mlvi*-2, *Mlvi*-3, *Pim*-1, *Flvi*-1, *Gin*-1, *fis*-1, *lck*, *dsi*-1, *fim*-1, *ahi*-1, *mis*-1, *mis*-2, *mis*-3, *mis*-4, *spi*

ber of the *ras* gene family (see p. 705), first discovered as oncogenes carried by retroviruses that cause sarcomas in rats. Thus two independent lines of enquiry converged on the same genes.

Yet another approach that led to some of the same proto-oncogenes was based on the karyotyping of tumor cells. As mentioned earlier (see p. 1190), in almost all patients with chronic myelogenous leukemia, the leukemic cells show the same chromosomal translocation, between chromosomes 9 and 22; likewise, in Burkitt's lymphoma there is regularly a translocation between chromosome 8 and one of the three chromosomes containing the genes that encode antibody molecules. In both these types of cancer the translocation breakpoint, where part of one chromosome is joined to another, was found to coincide exactly with the location of a proto-oncogene already known from retroviral studies—c-*abl* in chronic myelogenous leukemia, c-*myc* in Burkitt's lymphoma. Analogous chromosome translocations are similarly associated with some other types of cancer. From DNA sequencing studies it seems that in some cases the translocation turns a proto-oncogene into an oncogene by fusing the proto-oncogene to another gene in such a way that an altered protein is produced (Figure 21–24); in other cases the translocation moves a proto-oncogene into an inappropriate chromosomal environment that activates its transcription so that the normal protein is produced in excess.

A Proto-oncogene Can Be Made Oncogenic in Many Ways[14,15,20]

So far, about 60 proto-oncogenes have been discovered (see Tables 21–4 and 21–5); each of these can be perverted into an oncogene that plays a dominant part in cancers of one sort or another. Most such genes have been encountered repeat-

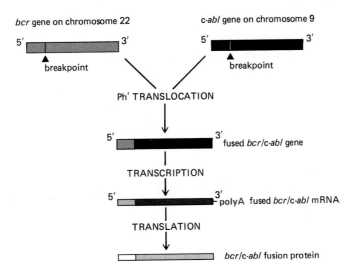

Figure 21–24 The conversion of the c-*abl* proto-oncogene into an oncogene in patients with chronic myelogenous leukemia. The chromosome translocation responsible joins the *bcr* gene on chromosome 22 to the c-*abl* gene from chromosome 9, thereby generating a Philadelphia chromosome (see Figure 21–4). The resulting fusion protein has the amino terminus of the bcr protein joined to the carboxyl terminus of the abl tyrosine protein kinase; in consequence, the abl kinase domain presumably becomes inappropriately active, driving excessive proliferation of a clone of hemopoietic cells in the bone marrow.

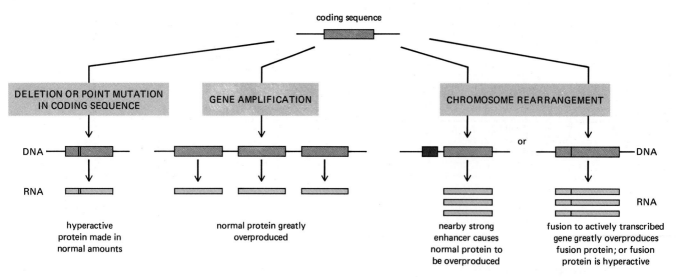

Figure 21–25 Three ways in which a proto-oncogene can be converted into an oncogene. A fourth mechanism (not shown) involves recombination between retroviral DNA and a proto-oncogene: this has effects similar to those of chromosome rearrangement, bringing the proto-oncogene under the control of a viral enhancer and/or fusing it to a viral gene that is actively transcribed.

edly, in a variety of mutant guises and in several different forms of cancer, suggesting that the majority of mammalian proto-oncogenes may already have been identified. There are many conceivable modes of genetic disruption that could convert a proto-oncogene into an oncogene, and examples of almost all of them are known (Figure 21–25). The gene may be altered by a point mutation, or through a chromosomal translocation, or by insertion of a mobile genetic element such as a retrovirus. The change can occur in the protein-coding region so as to yield a hyperactive product, or it can occur in adjacent control regions so that the gene is simply overexpressed. Alternatively, the gene may be overexpressed because it has been amplified to a high copy number through a process that probably begins with anomalous chromosome replication (Figure 21–26). Specific types of abnormality are characteristic of particular genes and of the responses to particular carcinogens. Members of the *myc* gene family, for example, are frequently overexpressed or amplified, while 90% of the skin tumors evoked in mice by the tumor initiator dimethylbenz[a]anthracene (DMBA) have an A-to-T alteration at exactly the same site in a mutant H-*ras* gene.

The molecular functions of the proto-oncogenes mutated in these various ways are beginning to be elucidated, and an increasing number of them can be assigned to particular cell-signaling pathways, as discussed in Chapters 12 and 13 (see p. 705 and p. 756). Many of the gene products appear to interact with one another as components of an elaborate control network (Figure 21–27). Thus some proto-oncogenes code for growth factors, some for growth-factor receptors and other protein kinases; still others code for G proteins of the *ras* family (see p. 705) or for gene regulatory proteins located in the nucleus. The molecular mechanisms of carcinogenesis are correspondingly beginning to make sense: through mutation of a gene (c-*sis*) for a growth factor (PDGF), leading to its inappropriate expression, cells become able to stimulate themselves continually to divide; through mutation of a gene (c-*erbB*) for a growth-factor receptor (the EGF receptor), they behave as though the corresponding growth factor were constantly present; through mutations that exaggerate expression of a gene (c-*myc*) whose product is thought to mediate nuclear changes required for cell proliferation, cells become committed to proliferate indefinitely. A great deal still remains unclear, however. In spite of a wealth of information about the DNA sequences and molecular structures of proto-oncogenes and oncogenes and their products, we still have only a vague notion of the parts that most of them play in the physiology of the cell.

The Actions of Oncogenes Can Be Assayed in Transgenic Mice[21]

The concept of an oncogene is paradoxical. As we have argued at length in the first part of this chapter, a single mutation is not enough to cause a cancer; yet an oncogene is defined as a dominantly acting gene and is typically assayed by its ability to cause neoplastic transformation of cultured cells on its own. This

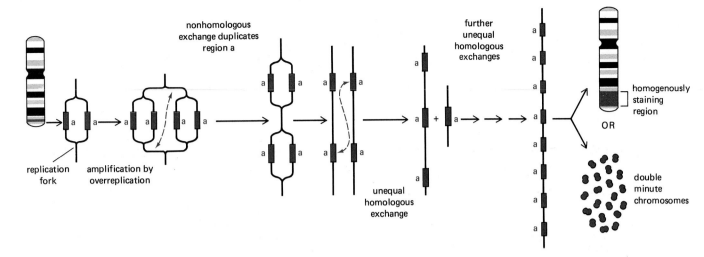

Figure 21–26 A possible mechanism for gene amplification leading to overproduction of a protein. The process begins with a gene-duplication event whose mechanism is not well understood but must involve illegitimate recombination. The mechanism illustrated here incorporates the suggestion that the illegitimate recombination may be a consequence of a destabilizing effect of illegitimate over-replication of the DNA. Once a gene duplication has occurred, unequal sister-chromatid exchanges due to recombination between identical gene copies in the course of DNA replication can further amplify the number of gene copies (see p. 600), until the chromosome contains tens or hundreds of them. When many DNA repeats are present, the segment containing them may become visible as a *homogeneously staining region* in the chromosome, or it may—probably through another recombination event—become excised from its original locus and give rise to independent *double minute chromosomes* (see p. 1202). The DNA sequence that becomes repeated by such an amplification mechanism is generally more than 100,000 nucleotide pairs long.

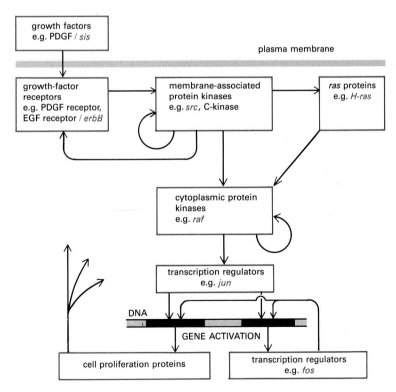

Figure 21–27 A tentative outline of the relationships of the major classes of proto-oncogenes in the intracellular control network through which external signals stimulate cell proliferation. A single representative proto-oncogene is indicated for each class. An arrow from A to B indicates that in a normal cell the activation of A exerts its effect by activating B (directly or indirectly). The diagram is based partly on studies of the molecules in cell-free systems and partly on studies of cells in which particular components are activated by introducing oncogenes or inactivated by microinjecting appropriate antibodies. Note that each of the classes of regulatory molecules has many members, so that each arrow in the diagram potentially stands for many parallel arrows linking the individual members of one class to those of another. Moreover, the members of a given class in many instances can interact with one another—for example, by mutual phosphorylation—as well as with the members of the classes upstream and downstream from them. The existence of multiple parallel interconnected pathways presumably helps to make the cell fail-safe, so that a single oncogenic mutation will not normally suffice to turn it into a cancer cell; but this complexity, and the presence of feedback loops, makes the system hard to analyze experimentally. For further details, see pp. 566, 705–708, and 756–758.

apparent contradiction reflects the gulf between the simplified models of cancer most widely studied by molecular biologists and the complexity of the actual human disease. The standard assay for identification of oncogenes does not test their effects on normal human somatic cells but on a mouse-derived 3T3 cell line; and the 3T3 cells, through establishment in culture, have already undergone mutations that make them abnormally easy to transform by a single further genetic change. Moreover, as explained on page 1192, mice, with their shorter life-spans and smaller cell numbers, run an intrinsically smaller risk of cancer than do human beings, so that their cells may be less securely protected against the consequences of carcinogenic mutations than are human cells.

Even in a mouse, however, a single oncogene is not usually sufficient to turn a normal cell into a cancer cell. This can be strikingly demonstrated by studies of transgenic mice. An oncogene in the form of a DNA fragment, derived from either a virus or a tumor cell, can be linked to a suitable promoter DNA sequence and then injected into a mouse egg nucleus. Often this recombinant DNA molecule will become integrated into a mouse chromosome, leading to the generation of a strain of transgenic mice that carry the oncogene in all their cells. The oncogene introduced in this way may be expressed in many tissues or only in a select few, according to the tissue specificity of the associated promoter. Typically, in mice that are thus endowed with a *myc* or H-*ras* oncogene, some of the tissues that express the oncogene grow to an exaggerated size; and occasional cells, with the passage of time, undergo further changes and give rise to cancers. The vast majority of the cells in the transgenic mouse that express the *myc* or H-*ras* oncogene, however, do not give rise to cancers, showing that the single oncogene is not enough to cause neoplastic transformation.

This approach has been taken a step further by breeding from a pair of transgenic mice, one carrying a *myc* oncogene, the other carrying an H-*ras* oncogene, so as to obtain progeny that carry both oncogenes together. These offspring develop cancers at a much higher rate than either parental strain (Figure 21–28), but again the cancers originate as scattered isolated tumors among noncancerous cells. Thus, even with two expressed oncogenes, the cells must undergo further, randomly generated changes to become cancerous.

Although this is a typical finding, some exceptions have been reported where a type of transgenic mouse carrying a single oncogene has all the cells of a particular tissue transformed simultaneously into cancer cells. But in these cases the oncogene apparently has to be doubly activated, through a mutation in its coding sequence as well as through coupling to an abnormal control sequence that drives it to be overexpressed, or inappropriately expressed, in the affected tissue. Thus, even though a single oncogene may sometimes be enough to cause transformation of a normal cell into a cancer cell, a single genetic lesion probably is not.

Cancer Involves the Loss of Tumor Suppressor Genes as Well as the Acquisition of Oncogenes[22]

From all the evidence, it seems that the control system that regulates cell proliferation must be fail-safe, in the sense that no single genetic accident is disastrous for its proper operation: proliferation is usually kept within safe bounds even if any single component goes awry. As a rule, a single oncogene can exert its dominant effect in provoking cancerous behavior only if the control system is already severely disturbed; the discoverers of oncogenes were fortunate in the choice of suitable cell lines for their assays.

It has proved considerably harder to identify and isolate tumor suppressor genes—the genes that normally act to inhibit excessive cell proliferation. There are, however, several lines of evidence that the loss of such genes plays a part in cancer. As we saw on page 755, for example, when a transformed cell in culture is fused with a nontransformed cell, the resulting hybrid cell often behaves as though nontransformed; only if a specific chromosome derived from the nontransformed partner is lost from the hybrid does the cell revert to cancerous behavior. In such a case the specific chromosome presumably contains the cell's only copy of a particular tumor suppressor gene.

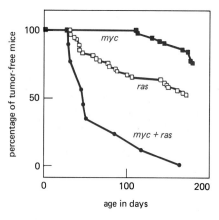

Figure 21–28 The occurrence of tumors in three types of transgenic mice, one carrying a *myc* oncogene, one carrying a *ras* oncogene, and one carrying both oncogenes. For these experiments, two lines of transgenic mice were constructed. One carries an inserted copy of an oncogene created by fusing the proto-oncogene c-*myc* with the mouse mammary tumor virus promoter/enhancer (which then drives c-*myc* overexpression in specific tissues such as the mammary gland). The other line carries an inserted copy of the oncogene v-H-*ras* under control of the same promoter/enhancer. Both strains of mice develop tumors much more frequently than normal, most often in the mammary or salivary glands. Mice that carry both oncogenes together were obtained by crossing the two strains. These hybrids develop tumors at a far higher rate still, much greater than the sum of the rates for the two oncogenes separately. Nevertheless, the tumors arise only after a delay and only from a small proportion of the cells in the tissues where the two genes are expressed: some further accidental change, in addition to the two oncogenes, is apparently required for the development of cancer. (After E. Sinn, et al., *Cell* 49:465–475, 1987.)

NORMAL HEALTHY INDIVIDUAL

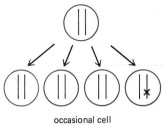

occasional cell
inactivates one of its two
good *Rb*1 genes

RESULT: NO TUMOR

HEREDITARY RETINOBLASTOMA

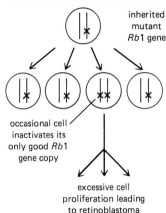

inherited
mutant
*Rb*1 gene

occasional cell
inactivates its
only good *Rb*1
gene copy

excessive cell
proliferation leading
to retinoblastoma

RESULT: MOST PEOPLE WITH
INHERITED GENE DEVELOP TUMOR

NONHEREDITARY RETINOBLASTOMA

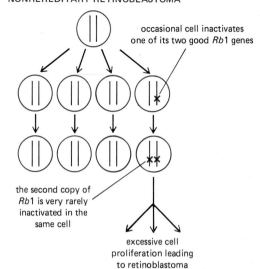

occasional cell inactivates
one of its two good *Rb*1 genes

the second copy of
*Rb*1 is very rarely
inactivated in the
same cell

excessive cell
proliferation leading
to retinoblastoma

RESULT: ONLY ABOUT 1 IN 30,000
NORMAL PEOPLE DEVELOP TUMOR

In an ordinary diploid cell, both copies of a tumor suppressor gene will generally have to be lost or inactivated before there is a loss of growth control: a mutation that contributes to cancer by deleting or inactivating the gene will be recessive. A mutation that creates an oncogene, by contrast, can have an effect on cell behavior even in the presence of a normal copy of the corresponding proto-oncogene. Thus, in principle, defective tumor suppressor genes can be recognized and distinguished from oncogenes by simple genetic criteria. An illustration is provided by a rare childhood cancer known as *retinoblastoma*, in which tumors develop from neural precursor cells in the immature retina. There are two forms of the disease, one hereditary, the other not. In the hereditary form, multiple tumors usually arise independently, affecting both eyes and sometimes occurring also in bone; in the nonhereditary form, only one eye is affected, and only by one tumor. Moreover, some hereditary sufferers from retinoblastoma are found to have a visibly abnormal karyotype, with a deletion of a specific band on chromosome 13; and deletions of this same locus are also seen in tumor cells from patients with the nonhereditary disease. This points to an explanation of the cancer in terms of loss of a tumor suppressor gene rather than acquisition of an oncogene. Specifically, the observations suggest that the cells of patients with the hereditary disease have a predisposition to become cancerous because they all lack one of the two normal copies of a certain tumor suppressor gene, called the retinoblastoma or *Rb*1 gene; a single somatic mutation that spoils the remaining good copy of the gene in one of the million or more cells in the growing retina will then suffice to initiate a cancer. In children without the hereditary predisposition, retinoblastomas would be very rare because they require the coincidence of two somatic mutations in a single retinal cell to destroy both copies of the *Rb*1 gene (Figure 21–29). The *Rb*1 gene has now been cloned and shown to encode a protein that is normally expressed in the retina (and in many other tissues), contains "zinc fingers" similar to those found in many DNA binding proteins (see p. 490), and also binds to the proteins encoded by the oncogenes of some DNA viruses. DNA probes prepared from the cloned *Rb*1 gene have been used to confirm that the tumor cells differ from their nontransformed neighbors in having deletions or inactivating lesions of *Rb*1 in both their maternal and their paternal chromosome sets; the loss can occur in a variety of ways (Figure 21–30). Similar mechanisms (involving other genes) have been shown to be responsible for some other cancers of early childhood, such as Wilms' tumor of the kidney, that have a hereditary form. Although these cancers are rare and somewhat exceptional, there is increasing evidence to suggest that loss or inactivation of tumor suppressor genes also plays a part in many common cancers occurring in adult life.

Figure 21–29 The genetic mechanisms underlying retinoblastoma. In the hereditary form, all cells in the body lack one of the normal two functional copies of a tumor suppressor gene, and tumors occur where the remaining copy is lost or inactivated by a somatic mutation. In the nonhereditary form, all cells initially contain two functional copies of the gene, and the tumor arises because both copies are lost or inactivated through the coincidence of two somatic mutations in one cell.

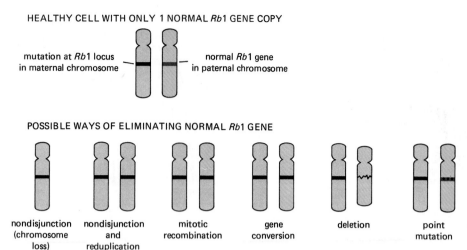

HEALTHY CELL WITH ONLY 1 NORMAL *Rb*1 GENE COPY

mutation at *Rb*1 locus in maternal chromosome — — normal *Rb*1 gene in paternal chromosome

POSSIBLE WAYS OF ELIMINATING NORMAL *Rb*1 GENE

nondisjunction (chromosome loss) — nondisjunction and reduplication — mitotic recombination — gene conversion — deletion — point mutation

Figure 21–30 Possible mechanisms by which a cell that is defective in one of its two copies of a tumor suppressor gene may come to lose the function of the other copy as well, and thereby progress toward cancer. Cloned DNA probes can be used in conjunction with restriction-fragment length polymorphisms (see p. 270) to analyze the tumor DNA and thereby discover which type of event has occurred in a given patient. Note that most of the mechanisms result in a cell that totally lacks either the maternal or the paternal copy of the tumor suppressor gene. The remaining version is itself defective and may be present in duplicate or in a single copy. Such *reduction to homozygosity or hemizygosity* is a characteristic hallmark of a cancer dependent on loss of the function of a tumor suppressor gene. (After W.K. Cavenee et al., *Nature* 305:779–784, 1983.)

Molecular Analysis of Lung Tumors Highlights the Complexity and Variability of Human Cancer[23]

Each year, about 140,000 new cases of lung cancer are diagnosed in the United States. At present, very little can be done to save these patients, and more than 90% of them die within a year of diagnosis. To get some insight into the molecular biology of the disease, detailed studies have been made of one of its forms, *small-cell carcinoma of the lung*, which accounts for about 20% of all lung cancers. The small-cell carcinomas are thought to be derived from pulmonary neuroendocrine cells, which secrete *gastrin-releasing peptide* (*GRP*, also known as *bombesin* from its similarity to an amphibian neuropeptide of that name) and other local chemical mediators. More than a hundred cultured cell lines have been established from the tumors of different patients with this cancer and subjected to molecular analysis. They display a rich array of anomalies, differing from one cell line to another. There are instances of disturbances in at least ten known oncogenes and tumor suppressor genes, numerous chromosome deletions and translocations, and disorders that cause secretion of signaling molecules that affect cell proliferation.

Thus, for example, many of the cell lines show overexpression of the *myc* proto-oncogene or of the related genes N-*myc* and L-*myc*, often as a result of gene amplification, manifest in the karyotype as double minute chromosomes or homogeneously staining chromosome regions (Figure 21–31; see also p. 1202). In the same cells, genes of the *ras* family are often found to have undergone specific point mutations. Amplification of *myc* is regularly correlated with especially rapid growth of the tumor cells in culture and with reduced expression of differentiated properties; and patients who have tumors with this gene amplification have a particularly short life expectancy.

Together with other chromosomal abnormalities, several of the tumor cell lines have gross chromosomal defects in the region of the *Rb*1 gene, and more than half of the lines make no detectable *Rb*1 messenger RNA, even though this is made in normal lung cells and in most other types of cancers. In addition, many of the tumor cell lines have grossly visible deletions that include a specific band on chromosome 3, hinting that loss of a gene in this region may play a part in development of the cancer. Cloned DNA probes complementary to chromosome 3 in this region provide a precise and reliable way to test for deletions of the region and to identify any remaining copies of it according to whether they are versions inherited from the patient's mother or the patient's father. (The technique depends on exploiting restriction-fragment length polymorphisms—see p. 270.) This approach reveals that in almost every case (of small-cell carcinoma and of the other major forms of lung cancer) one of the two parental copies of the identified region of chromosome 3 has been lost. Noncancerous cells from the same patients do not show such a loss, and it is not generally observed in other types of cancers in other patients. There is a strong suggestion, therefore, that a crucial step in the development of lung cancer is the elimination of a specific tumor suppressor gene

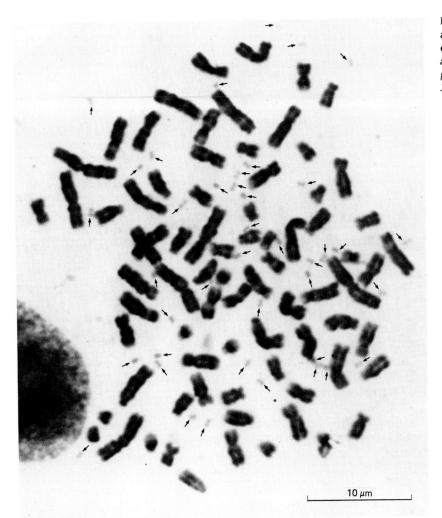

Figure 21–31 Karyotype of a cell from a small-cell lung cancer, showing double minute chromosomes (*see arrows*) reflecting amplification of a *myc* gene. (Courtesy of J. Whang-Peng and J.D. Minna.)

10 μm

on chromosome 3, by one of the same types of mechanisms outlined for retinoblastoma in Figure 21–30: presumably, in a typical lung cancer cell, one of the two original good copies of the lung cancer suppressor gene is lost by deletion of the chromosome segment that contains it, while the other is inactivated by a less conspicuous, more pointlike mutation. The analogy with retinoblastoma suggests that some people may inherit a recessive defect at this locus, predisposing them to lung cancer; because lung cancer probably requires changes in other genes as well and is strongly dependent on environmental factors, a hereditary predisposition of this sort might be less marked than in retinoblastoma and could easily have gone unnoticed in the past (although there are in fact some reports of genetic variation in the susceptibility to lung cancer from smoking). Similar circumstantial evidence has recently been obtained for loss of tumor suppressor genes at a different locus in another very common cancer, *colorectal carcinoma* (carcinoma of the large intestine); in this case there is known to be a rare hereditary form of the disease, although its molecular relationship to the nonhereditary form is unclear. Molecular studies such as those just described should help in identifying individuals who are at high risk for certain specific types of cancer and who may benefit from appropriate preventive measures.

Another clinically important finding from the molecular analysis of small-cell lung cancers centers on the GRP/bombesin peptide, which acts as a growth factor, stimulating proliferation of the tumor cell lines. Some, but not all, of the cell lines both respond to this factor and secrete it, thereby creating an autocrine loop through which they stimulate themselves to divide. In culture the proliferation of such cells can be inhibited by an antibody that binds to the peptide, blocking the stimulation. This holds out the hope that it may be possible to inhibit tumor growth by a similar growth-factor blockade in patients who have this particular variant of small-cell lung cancer.

Each Case of Cancer Represents a Separate Experiment in Cell Evolution

All medical progress depends on accurate diagnosis. If one cannot identify a disease correctly and distinguish it from others, one cannot discover its causes, predict its outcome, select the appropriate treatment for a given patient, or make trials on a population of patients to judge whether a proposed treatment is effective. As we have just seen for small-cell carcinoma of the lung, the traditional classification of cancers is inadequate: a single one of the conventional categories turns out on close scrutiny to be a heterogeneous collection of disorders, each characterized by its own array of genetic lesions. Molecular biology is beginning to provide tools to find out precisely which genes are amplified, which deleted, and which mutated in the tumor cells of any given patient. Such information may prove to be as important for the management and prevention of cancer as is the identification of microorganisms in patients with infectious diseases.

The discovery of oncogenes and, more recently, of tumor suppressor genes has marked the end of an era of groping in the dark for clues to the biochemical basis of cancer. But it is still sobering to turn from the simplified laboratory models of neoplasia that have made these triumphs possible to the complexity of the common human cancers. We are far from understanding these diseases fully, and there has been painfully little progress toward devising effective rational treatments for them. We know the DNA sequences of many oncogenes and proto-oncogenes but the precise physiological functions of only a few. We need better understanding of how these and other molecules interact to govern the behavior of the individual cell, better understanding of the sociology of cells in tissues, and better understanding of the cell population genetics that govern the genesis of cancer cells through mutation and natural selection.

Each patient with cancer represents an independent, unfortunate natural experiment in cell evolution. The complexity of the evolutionary possibilities accounts for the complexity of the phenomenon of cancer, and at the same time the principles of evolution provide a unifying perspective from which that complexity can be surveyed and understood. Thus cancer returns us to the theme with which this book began: as it has been said, nothing in biology makes sense except in the light of evolution.

Summary

The proliferation of normal cells is regulated by both inhibitory and stimulatory control molecules, corresponding to tumor suppressor genes and proto-oncogenes, respectively. A tendency toward cancerous behavior can result from loss or inactivation of both copies of a tumor suppressor gene or from amplification or hyperactivation of one of the two copies of a proto-oncogene. The control of cell proliferation may also be heritably upset by foreign, viral genetic material. Retroviruses can become oncogenic by picking up a copy of a proto-oncogene from a host cell and, in the process, converting it into an oncogene; they can also create an oncogene by acting as an insertional mutagen when they integrate next to a proto-oncogene in the host genome. Although most human cancers are thought not to be caused by viruses, they frequently involve mutations that affect the same proto-oncogenes identified through studies of retroviruses. In human cancers, proto-oncogenes have been found to give rise to oncogenes through point mutations, through amplification of the number of gene copies, or through chromosome translocations that either upset the control of their expression or join them to another gene to produce a fusion protein. Likewise, tumor suppressor genes may be functionally lost through a variety of types of mutations; a person who inherits a deletion or a defective copy of one of these genes may show a strong predisposition to a specific type of cancer, a notable example being retinoblastoma. Molecular analysis of tumor cells from patients sharing one of the common forms of cancer reveals a complex and heterogeneous collection of genetic lesions, including both oncogene activation and tumor suppressor gene loss. These observations reflect the randomness of the evolutionary process by which cancers arise, and suggest that each case of cancer is likely to represent a molecularly unique illness.

References

General

Cairns, J. Cancer: Science and Society. San Francisco: Freeman, 1978.

Cancer Biology: Readings from Scientific American. New York: W.H. Freeman, 1986.

Darnell, J.E.; Lodish, H.F.; Baltimore, D. Molecular Cell Biology, pp. 1035–1080. New York: W.H. Freeman/Scientific American Books, 1986.

De Vita, V.T.; Hellman, S.; Rosenberg, S.A., eds. Cancer: Principles and Practice of Oncology, 2nd ed. Philadelphia: Lippincott, 1985.

Farmer, P.B.; Walker, J.M., eds. The Molecular Basis of Cancer. New York: Wiley, 1985.

Franks, L.M.; Teich, N., eds. Introduction to the Cellular and Molecular Biology of Cancer. Oxford, U.K.: Oxford University Press, 1986.

Prescott, D.M.; Flexer, A.S. Cancer: The Misguided Cell, 2nd ed. Sunderland, MA: Sinauer, 1986.

Watson, J.D.; Hopkins, N.H.; Roberts, J.W.; Steitz, J.A.; Weiner, A.M. Molecular Biology of the Gene, 4th ed., pp. 962-1096. Menlo Park, CA: Benjamin-Cummings, 1987.

Cited

1. Cairns, J. Mutation selection, and the natural history of cancer. Nature 255:197–200, 1975.
   Nowell, P.C. The clonal evolution of tumor cell populations. Science 194:23–28, 1976.

2. Doll, R.; Peto, R. Epidemiology of cancer. In Oxford Textbook of Medicine (D.J. Weatherall, J.G.G. Ledingham, D.A. Warrell, eds.), 2nd ed., pp. 4.95–4.123. Oxford, U.K.: Oxford University Press, 1987.
   Parkin, D.M.; Laara, E.; Muir, C.S. Estimates of the worldwide frequency of sixteen major cancers in 1980. Int. J. Cancer 41:184–197, 1988.
   Ritchie, A.C. The classification, morphology, and behaviour of tumours. In General Pathology (H. Florey, ed.), 4th ed., pp. 668–719. London: Lloyd-Luke, 1970.
   Robbins, S.L.; Cotran, R.S.; Kumar, V. Pathologic Basis of Disease, 3rd ed., pp. 214–272. Philadelphia: Saunders, 1984.

3. Fearon, E.R.; Hamilton, S.R.; Vogelstein, B. Clonal analysis of human colorectal tumors. Science 238:193–197, 1987.
   Fialkow, P.J. Clonal origin of human tumors. Biochim. Biophys. Acta 458:283–321, 1976.
   Groffen, J.J.; et al. Philadelphia chromosomal breakpoints are clustered within a limited region, bcr, on chromosome 22. Cell 36:93–99, 1984.
   Nowell, P.C.; Hungerford, D.A. A minute chromosome in human granulocytic leukemia. Science 132:1497, 1960.

4. Ames, B.; Durston, W.E.; Yamasaki, E.; Lee, F.D. Carcinogens are mutagens: a simple test system combining liver homogenates for activation and bacteria for detection. Proc. Natl. Acad. Sci. USA 70:2281–2285, 1973.
   Ashby, J.; Tennant, R.W. Chemical structure, Salmonella mutagenicity and extent of carcinogenicity as indicators of genotoxic carcinogenesis among 222 chemicals tested in rodents by the U.S. NCI/NTP. Mutation Res. 204:17–115, 1988.
   Case, R.M.P. Tumours of the urinary tract as an occupational disease in several industries. Ann. R. Coll. Surg. Engl. 39: 213–235, 1966.
   Doll, R. Strategy for detection of cancer hazards to man. Nature 265:589–597, 1977.
   Hay, A. How to identify a carcinogen. Nature 332:782–783, 1988.
   Walker, J.M. Testing for carcinogens. In The Molecular Basis of Cancer (P.B. Farmer, J.M. Walker, eds.), pp. 181–200. New York: Wiley, 1985.

5. Armitage, P.; Doll, R. The age distribution of cancer and a multi-stage theory of carcinogenesis. Br. J. Cancer 8:1–12, 1954.
   Peto, R. Cancer epidemiology, multistage models, and short-term mutagenicity tests. In Origins of Human Cancer (H.H. Hiatt, J.D. Watson, J.A. Winsten, eds.), pp. 1403–1428. Cold Spring Harbor, NY: Cold Spring Harbor Laboratory, 1977.

6. Cairns, J. Cancer: Science and Society, pp. 144–156. San Francisco: Freeman, 1978.
   Campion, M.J.; McCance, D.J.; Cuzick, J.; Singer, A. Progressive potential of mild cervical atypia: prospective, cytological, colposcopic, and virological study. Lancet 2:237–240, 1986.
   Farber, E.; Cameron, C. The sequential analysis of cancer development. Adv. Cancer Res. 31:125–225, 1980.
   Foulds, L. The natural history of cancer. J. Chronic Dis. 8:2–37, 1958.
   McIndoe, W.A.; McLean, M.R.; Jones, R.W.; Mullins, P.R. The invasive potential of carcinoma in situ of the cervix. Obstet. Gynecol. 64:451–464, 1984.

7. Albanes, D.; Winick, M. Are cell number and cell proliferation risk factors for cancer? J. Natl. Cancer Inst. 80:772–775, 1988.
   Schnipper, L.E. Clinical implications of tumor-cell heterogeneity. N. Engl. J. Med. 314:1423–1431, 1986.

8. Berenblum, I. A speculative review: the probable nature of promoting action and its significance in the understanding of the mechanism of carcinogenesis. Cancer Res. 14: 471–477, 1954.
   Berenblum, I.; Shubik, P. The role of croton oil applications associated with a single painting of a carcinogen in tumour induction in the mouse's skin. Br. J. Cancer 1:379–383, 1947.
   Dolberg, D.S.; Hollingsworth, R.; Hertle, M.; Bissell, M.J. Wounding and its role in RSV-mediated tumor formation. Science 230:676–678, 1985.
   Lamph, W.W.; Wamsley, P.; Sassone-Corsi, P.; Verma, I.M. Induction of proto-oncogene JUN/AP-1 by serum and TPA. Nature 334:629–631, 1988.
   Reddy, A.L.; Fialkow, P.J. Influence of dose of initiator on two-stage skin carcinogenesis in BALB/c mice with cellular mosaicism. Carcinogenesis 9:751–754, 1988.

9. Cairns, J. The treatment of diseases and the war against cancer. Sci. Am. 253(5):51–59, 1985.
   Cohen, L.A. Diet and cancer. Sci. Am. 257(5):42–48, 1987.
   Doll, R.; Peto, R. The Causes of Cancer. New York: Oxford University Press, 1981.
   Enstrom, J.E. Cancer and mortality among active Mormons. Cancer 42:1943–1951, 1978.
   Pike, M.C.; Ross, R.K. Breast cancer. Br. Med. Bull. 40:351–354, 1984.

10. Fentiman, I. The local treatment of cancer. In Introduction to the Cellular and Molecular Biology of Cancer (L.M. Franks, N. Teich, eds.), pp. 350–362. Oxford, U.K.: Oxford University Press, 1986.
    Malpas, J. Chemotherapy. In Introduction to the Cellular and Molecular Biology of Cancer (L.M. Franks, N. Teich, eds.), pp. 363–377. Oxford, U.K.: Oxford University Press, 1986.

11. Sachs, L. Growth, differentiation, and the reversal of malignancy. Sci. Am. 254(1):40–47, 1986.

12. Fidler, I.; Hart, I. Biological diversity in metastatic neoplasms: origins and implications. Science 217:998–1003, 1982.
    Liotta, L.A. Tumor invasion and metastases—role of the extracellular matrix: Rhoads Memorial Award lecture. Cancer Res. 46:1–7, 1986.
    McCarthy, J.B.; Skubitz, A.P.N.; Palm, S.L.; Furcht, L.T. Metas-

tasis inhibition of different tumor types by purified laminin fragments and a heparin-binding fragment of fibronectin. *J. Natl. Cancer Inst.* 80:108–116, 1988.

Nicholson, G. Cancer metastasis. *Sci. Am.* 240(3):66–76, 1979.

Schirrmacher, V. Cancer metastasis: experimental approaches, theoretical concepts, and impacts for treatment strategies. *Adv. Cancer Res.* 43:1–73, 1985.

13. German, J., ed. Chromosome Mutation and Neoplasia. New
York: Liss, 1983.

Hanawalt, P.; Sarasin, A. Cancer-prone hereditary diseases with
DNA processing abnormalities. *Trends Genet.* 2:124–129,
1986.

Lindahl, T. Regulation and deficiencies in DNA repair. *Br. J.
Cancer* 56:91–95, 1987.

Nowell, P.C. Mechanisms of tumor progression. *Cancer Res.*
46:2203–2207, 1986.

14. Alt, F.W.; Kellems, R.E.; Bertino, J.R.; Schimke, R.T. Selective
multiplication of dihydrofolate reductase genes in
methotrexate-resistant variants of cultured murine cells.
J. Biol. Chem. 253:1357–1370, 1978.

Gerlach, J.H.; et al. Homology between P-glycoprotein and a
bacterial haemolysin transport protein suggests a model
for multidrug resistance. *Nature* 324:485–489, 1986.

Roninson, I.B.; Abelson, H.T.; Housman, D.E.; Howell, N.; Varshavsky, A. Amplification of specific DNA sequences correlates with multi-drug resistance in Chinese hamster
cells. *Nature* 309:626–628, 1984.

Schimke, R.T. Gene amplification in cultured cells. *J. Biol.
Chem.* 263:5989–5992, 1988.

15. Klein, G. The approaching era of the tumor suppressor genes.
Science 238:1539–1545, 1987.

Watson, J.D.; Hopkins, N.H.; Roberts, J.W.; Steitz, J.A.; Weiner,
A.M. Molecular Biology of the Gene, 4th ed., Chapters 26–
27. Menlo Park, CA: Benjamin-Cummings, 1987.

Wyke, J.A. Viruses and Cancer. In Introduction to the Cellular
and Molecular Biology of Cancer (L.M. Franks, N. Teich,
eds.), pp. 176–199. Oxford, U.K.: Oxford University Press,
1986.

16. Kripke, M.L. Immunoregulation of carcinogenesis: past, present,
and future. *J. Natl. Cancer Inst.* 80:722–727, 1988.

17. Lazo, P.A. Human papillomaviruses in oncogenesis. *Bioessays*
9:158–162, 1988.

Steinberg, B.M.; Brandsma, J.L.; Taichman, L.B. Cancer Cells
5 / Papillomaviruses. Cold Spring Harbor, NY: Cold Spring
Harbor Laboratory, 1987.

18. Bishop, J.M. Oncogenes. *Sci. Am.* 246(3):80–92, 1982.

Rijsewijk, F.; et al. The *Drosophila* homolog of the mouse
mammary oncogene int-1 is identical to the segment polarity gene wingless. *Cell* 50:649–657, 1987.

Varmus, H.E. The molecular genetics of cellular oncogenes.
Annu. Rev. Genet. 18:553–612, 1984.

19. Bishop, J.M. The molecular genetics of cancer. *Science* 235:
305–311, 1987.

Croce, C.M.; Klein, G. Chromosome translocations and human cancer. *Sci. Am.* 252(3):44–50, 1985.

Weinberg, R.A. A molecular basis of cancer. *Sci. Am.* 249(5):
126–142, 1983.

20. Bradshaw, R.A.; Prentis, S. Oncogenes and Growth Factors.
Amsterdam: Elsevier, 1987. (An anthology of articles from
Trends Biochem. Sci., Trends Genet., and *Immunol. To-
day.*)

Hunter, T. The proteins of oncogenes. *Sci. Am.* 251(2):70–79,
1984.

Kahn, P.; Graf, T., eds. Oncogenes and Growth Control. Berlin:
Springer, 1986. (An excellent collection of short reviews.)

Quintanilla, M.; Brown, K.; Ramsden, M.; Balmain, A. Carcin-
ogen-specific mutation and amplification of Ha-ras during mouse skin carcinogenesis. *Nature* 322:78–80, 1986.

Smith, M.R.; DeGudicibus, S.J.; Stacey, D.W. Requirement for
c-ras proteins during viral oncogene transformation. *Nature* 320:540–543, 1986. (Illustrates how antibodies can be
used to analyze the organization of the proto-oncogene
control network.)

21. Hanahan, D. Dissecting multistep tumorigenesis in mice. *Annu.
Rev. Genet.* 22:[in press], 1988.

Land, H.; Parada, L.F.; Weinberg, R.A. Tumorigenic conversion
of primary embryo fibroblasts requires at least two cooperating oncogenes. *Nature* 304:596–602, 1983.

Quaife, C.J.; Pinkert, C.A.; Ornitz, D.M.; Palmiter, R.D.; Brinster,
R.L. Pancreatic neoplasia induced by *ras* expression in
acinar cells of transgenic mice. *Cell* 48:1023–1034, 1987.

Sinn, E.; et al. Coexpression of MMTV/v-Ha-*ras* and MMTV/c-
myc genes in transgenic mice: synergistic action of oncogenes *in vivo. Cell* 49:465–475, 1987.

22. Hansen, M.F.; Cavenee, W.K. Retinoblastoma and the progression of tumor genetics. *Trends Genet.* 4:125–129, 1988.

Harris, H. The genetic analysis of malignancy. *J. Cell Sci.* Suppl.
4:431–444, 1986.

Knudson, A.G. Hereditary cancer, oncogenes, and antionco-
genes. *Cancer Res.* 45:1437–1443, 1985.

Lee, E.Y.-H.P.; et al. Inactivation of the retinoblastoma susceptibility gene in human breast cancers. *Science* 241:218–221,
1988.

Weinberg, R.A. Finding the anti-oncogene. *Sci. Am.* 259(3):34–41,
1988.

23. Bodmer, W.F.; et al. Localization of the gene for familial ade-
nomatous polyposis on chromosome 5. *Nature* 328:614–
627, 1987.

Harbour, J.W.; et al. Abnormalities in structure and expression
of the human retinoblastoma gene in SCLC. *Science* 241:
353–357, 1988. (SCLC = small-cell lung cancer.)

Kok, K.; et al. Deletion of a DNA sequence at the chromosomal
region 3p21 in all major types of lung cancer. *Nature* 330:
561–578, 1987.

Leppert, M.; et al. The gene for familial polyposis coli maps
to the long arm of chromosome 5. *Science* 258:1411–1413,
1987.

Minna, J.D.; et al. Molecular genetic analysis reveals chromosomal deletion, gene amplification, and autocrine
growth factor production in the pathogenesis of human
lung cancer. *Cold Spring Harbor Symp. Quant. Biol.* 51:
843–853, 1986.

List of Tables

Index

Page numbers in **bold face** refer to a major text discussion of the entry; page numbers with an F refer to a figure, with an FF to figures that follow consecutively; page numbers with a T refer to a table; cf. means compare.

A

A band, of sarcomere, 614F, 615
A-form of DNA, 492–494, 493F
A-kinase, 707F, 709–710, 709FF, 1092
  allosteric regulation of, 710, 710F
  effect on Ca²⁺ levels, 712
  in memory
    long-term, 1098
    short-term, **1097–1098**, 1097FF
  oocyte maturation, inactivation in, 860
  substrates, different target cells, 710
A-site, *see* Aminoacyl-tRNA, binding site
A subfiber, of axoneme and centriole, 645, 645F, 647, 647F, 650F
A23187 ionophore, 322, 872
abl genes, 1207T, 1209
Abscissic acid (ABA), 1180–1181, 1181F
Abscission, of plant parts, 1150, 1150F
Absorptive cells, of plants, 1153
Absorptive epithelia, catalog of, 996
Accelerating positive feedback, 716–717, 717F
Acceptor splice junction of intron, 533, 535, 534FF
Accessory glycoprotein, on T lymphocytes, 1049–1050, 1050F, 1051T
Acetosyringone, 1159, 1159FF
Acetyl CoA, 45
  cholesterol synthesis from, 81F
  in citric acid cycle, oxidation in, 64–65, 64F, 67–69, 68F, 348, 348F, 350F
  as coenzyme, 76
  model, 345F
  polymerization in lipid synthesis, 88T
  structural formula, 65F
  synthesis, 345–346, 346F
Acetyl group, transfer by coenzyme A, 76T
Acetylation, of proteins, in cytosol, 417, 417F
Acetylcholine, 319–320, 321F, 685T, 703T
  see also Acetylcholine receptor
  ligand-gated membrane channel, stimulation of, **1081**, 1081FF, **1083**
  neuromuscular junction, exocytosis at, **1075–1078**, 1077FF
  neuropeptide synapses, release at, 1094, 1095F
  removal from synaptic cleft, **1081**, **1083**
  skeletal muscle contraction, stimulation of, 684, 688F, 1075
  storage in synaptic vesicles, 1077, 1077F, 1079F
Acetylcholine receptor
  see also Ligand-gated ion channels; Neurotransmitter receptors; Transmitter-gated ion channels
  acetylcholine binding sites, 321F
  aggregation on innervation, 1123–1125, 1123FF
  conformational states of, 320, 320F, 1081, 1081F
  differences, in heart and skeletal muscle, 684, 688F, 1084–1085
  extrajunctional, 1125, 1125F
  as gated cation channel, **319–322**, 319FF, **1081**, 1081FF, **1083**
  genes, 320
  ion currents
    in open, 320
    patch-clamp recordings, 1082F
  isolation, purification, reconstitution, 319–320
  nicotinic vs. muscarinic, 1084–1085
  structure, 320, 321F

synthesis in denervated muscle, 1125
in uninnervated muscle, 1123, 1123F
Acetylcholinesterase, 1083, 1124
Acid, defined, 47F
Acid anhydride, 49F
Acid hydrolases, of lysosomes, 459, 459F
Acid phosphatase
  as marker for lysosomes, 459F
  as TGN marker, 458F
Acidic fibroblast growth factor (acidic FGF),
    endothelial cell, effects on, 966
Acinar cells
  enzymes, 411
  secretory pathway in, 409–414, 410FF
Acquired immunodeficiency syndrome (AIDS),
    1044–1045
  see also Human immunodeficiency virus
Acrosomal process, 641, 641F, 868–869, 869F
Acrosomal reaction, 864, 868–869, 869F, 876, 876F
  in mouse fertilization, 875–876, 876F
Acrosomal vesicle, 864, 864F, 868, 869F, 876
Actin, 642T, 643
  see also Actin filaments
  amounts in cells, 629–630
  antibodies, for visualization, 635F
  G (globular), 616, 617F
  gel, **630–631**, 630F
  isoforms, 628
  isolation, 616
  location in metaphase cells, 140T
  monomer, percentage in cells, 630
  myosin ATPase, activation of, 619–620, 621F
  in phagocytosis, 335
  polymerization, 616, **637–639**, 637FF
    acrosomal process, **641**, 641F, 869, 869F
    control via nucleation sites, **641**
    drugs affecting, **642–643**, 642F
    profilin control of, 641
    in stereocilia development, 676, 677F
  treadmilling
    in leading edge, **640**, 640F
    of subunits in polymer, **638–640**, 638F, 640F
  tubulin, comparison to, 652–654, 654T
Actin-binding proteins
  see also individual proteins
  bundling, **634**, 634F, 642T, 643
  cooperative and competitive interactions
      between, **643**, 643F
  cross-linking, **630–631**, 630F, 633, 633F, 642T, 643, 643F
  severing, **631**, 642T, 643
  summary of major classes in vertebrates, 642T, 643
Actin filaments, 21, 21F, 613, 1038
  see also Actin-binding proteins; Thin filament of skeletal muscle; and individual proteins
  in acrosomal reaction, 641, 641F
  in adherens junctions, 796–797, 796F
  in adhesion belts, 674–675, 675F
  attachment
    sites in smooth muscle, 625F
    to plasma membrane at focal contacts,
      635–636, 635FF, 642T, 643
  bundling proteins, 634, 634F, 642T, 643
  capping protein and 636, 636F, 642T
  capping, role in 334, 334F
  of cardiac muscle, 624, 625F
  of cell cortex
    arrangement in stretched, 669–670, 669F
    nonmuscle **629–643**, 629FF
  in cell division, asymmetric molecule
      segregation, 785

in cell locomotion mechanism, **671–672**, 672F
in cell migration, directed, **670–671**, 671F
cross-linking of, **630–631**, 630F, **633**, 642T, **643**, 643F
in cytoplasmic streaming
  in algae, 632, 632F
  plants, 1170–1171
as dynamic assembly, **637–643**, 637FF
in erythrocyte membrane organization, 289, 290F
extracellular fibronectin, organization of, 822–823, 823F
of focal contacts, 749, 757, 758F
  pull against substratum, **635–636**, 635FF
growth asymmetry of, **637–639**, 637FF
in growth cone and fibroblast, arrangement, 670–671, 671F
inhibition of depolymerization, 642–643
in lamellipodia, 638, 640, 640F
in microspikes, 638, 640, 640F
microtubules, comparison, 652–654, 654T
microvilli, stabilizing of, **633–635**, 634F
minimyosin binding in microvilli, 634F, 635
of muscle, **616–617**, 616FF, **619–623**, 620FF, 623T
in musclelike assemblies, transient, **626–627**, 627F
myosin heads
  binding to, **619–622**, 620F, 626
  decoration with, 619, 620F
  walking movement of, **620**, 621F
in neural fold formation, 888, 888F
in nonmuscle cell cortex, **629–643**, 629FF
organelle movement and, 632, 632F
of plant cell
  in cell-plate formation, 783, 783F, 1166, 1166F
  chloroplast alignment, **1171–1172**, 1172F
  nuclear migration, 1178, 1179F
  preprophase band, **1178**
plasma membrane, attachments to, **632–636**, 633FF
polarity, 617, 631, 631F
  in microvillus, **634–635**, 634F
  polymerization at plus and minus ends, **637–639**, 637FF
sarcomere, **619–620**, 620F
in sarcomere, anchoring, 624, 624F
severing, by gelsolin, 631, 642T
sliding
  in cytoplasmic gel, 631, 631F
  in muscle, *see* Sliding filament model
in smooth muscle, organization, 625, 625F
steady state, 639
in stereocilia of cochlea, **676**, 676FF
structure, 92, 119, 120F, **616–617**, 616F
in terminal web, 634, 634F
tropomyosin and troponin with, **622–623**, 622FF
Actin genes, 616, 628
Actin-myosin assemblies
  in nonmuscle cells, adhesion belts, 627, 627F
  in nonmuscle cells, transient, **626–627**, 627F
Actin-myosin interaction
  in adhesion belts, 627, 627F
  in contractile ring, 626–627, 627F
  in muscle contraction, *see* Muscle contraction
  in nonmuscle cell locomotion, **672**
  troponin effects on, 622–623, 623F
α-actinin, 623T, 624, 624F, 642T, 643, 643F
  in dense bodies, 628F
  in focal contacts, 636, 636F

I-1

Anaphase A, 774–776, 774FF
Anaphase B, 774, 774F, 776–777, 776F
Anchorage dependence of cell division, 748–749, 749F
Anchorage independence, of transformed cells, 759, 759F
Anchoring junctions, **795–798**, 795FF, 802F
  defined, 792
  kinds, 792, 792T
Angiogenesis, **964–966**, 964FF
Angiogenic factors, 966
Angiosperms, 1145F
Angiotensin II, 1093, 1094F
Animal pole, of egg, 856, 880, 881F
Anionic detergent, 286, 286F
Anisomycin, 218T
Ankyrin, 289, 289FF, 633, 633F, 634
Antenna complex, 373–374, 374F, 378F
Antennapedia complex, 587, 587F, 929–930, 931F
  genes, 937
  control of, 936, 936F
Antennapedia mutation, 555, 920, 920F
Anthocyanin pigments, 1164–1165
Antibiotic resistance, in DNA cloning, 259, 259F
Antibiotics
  as inhibitors
    of protein synthesis, 218, 218T, 388, 388F
    of RNA synthesis, 218, 218T
  ionophores as, 322
  phytoalexins, 1160
Antibody, 33, **1011–1031**, 1012FF
  see also Immunoglobulin; and inidividual antibodies under their antigens
  as antigen, 1030–1031, 1030FF
  antigen, detection of, **177–180**, 177FF
  antigen-binding sites, 1012–1013, 1012F, **1016–1018**, 1017FF, 1030–1031, 1030FF
    somatic hypermutation in genes for, 1026–1027
  avidity, 1018, 1018F
  bivalent, 1012, 1012F
  classes and their properties, **1013–1016**, 1014T, 1015FF
  effector functions, 1011–1013
    see also individual antibodies
    cell lysis by complement, **1031–1036**, 1032FF
    of different classes, 1013–1016, 1019T, 1015FF
  in humoral response, **1003–1004**, 1004F
    see also B lymphocyte; Immune response
  idiotypes, **1030–1031**, 1030FF
  membrane bound, 1012
    transition to secreted, **1028–1029**, 1028F
  microinjection into cells, 180
  monoclonal, 1018–1019
    from hybridomas, 178–179, 179F
  for newborns, transcytosis of, 332
  patching and capping, 333–334, 334F
  as phagocytosis triggers, 334–335
  proteolysis by papain and pepsin, 1013, 1014F
  as receptor for antigen on B cell, 1012, 1029
  shape, 1012, 1012FF
  synthesis
    class switching, 1029, 1029FF
    in pre-B and virgin B cell, 1014
    membrane-bound to secreted switch, mechanism, **591**, 592F
  as tools in cell biology, **177–180**, 177FF
  use
    adhesion molecules, identifying by, 828F
    in cell separation, 159
    in FRAP, 296, 296F
    to identify cloned gene products, 265–266
    in indirect immunocytochemistry, 177, 178F
    in mRNA selection for cDNA synthesis, 262
    in visualizing macromolecules, 139, 140F
Antibody diversity
  combinational diversification, 1024–1025, 1025F
  genetic basis of, **1023–1029**, 1024FF
  junctional diversification, 1025–1026, 1026F
  somatic hypermutation, 1026–1027
Antibody genes, 1021, 1021F, **1024–1029**, 1024FF

allelic exclusion and expression of, 1027–1028, 1027F
C gene segment, 1024, 1025FF, 1028–1029, 1028FF
C region genes, 1021, 1021F
D and *J* gene segments, 1024–1026, 1025FF, 1028, 1029F
  somatic hypermutation in, 1026–1027
  switch recombination, 1029, 1030F
V gene segment, 1024–1028, 1025FF
V-DJ joining, 1025F, 1028
V-J joining, 1025–1026, 1025FF, 1028
Antibody response, 1001–1003, 1003F
  complement, cell lysis by, 1031–1036, 1032FF
  effector functions of antibody classes, 1011–1016, 1014T, 1011FF
  primary vs. secondary, 1009, 1009F
  regulation, 1030–1031, 1030FF
Antibody structure
  antigen-binding sites, **1019–1022**, 1020FF
  constant (C) region, 1019–1021, 1019F
  domains, 1020–1022, 1020F, 1022F
  fine, **1018–1022**, 1019FF
  fragments, 1013, 1014F
  hypervariable region, 1020–1022, 1020F, 1022F, 1026
  molecular, **1013–1016**, 1012FF, 1014T
  variable (V) region, 1019–1021, 1019FF
Antibody-dependent cell-mediated killing, **1015–1016**
Anticlinal divisions, 1176F, 1177
Anticodon, 206, 206F, 208FF, 217, 217F
  of tRNA, 103–104, 103FF
Antigen, 177, 703T
  see also MHC glycoproteins
  binding to cell receptor, 1005–1007, 1006FF
  defined, 1001
  processing
    in antigen-presenting cell, 1046, 1046F
    in cytotoxic T target cell, 1043–1044, 1043F
  purification by affinity chromatography, 168, 168F
  recognition
    by B lymphocytes, **1005–1007**, 1006FF, 1009–1011, 1009F, 1012, **1047–1048**, 1048F
    by T cells, **1040–1046**, 1041FF, 1042T, 1053
    in association with MHC glycoproteins, **1042–1046**, 1042FF, 1053
  self vs. nonself, 1001–1002, 1010–1011, 1010F, 1050–1052
  specificity of immune response, 1005–1007, 1006FF
  tissue-specific, 1010–1011
  trapping in secondary lymphoid organs, 1007
  valency, 1017–1018, 1018F
  viral, 1038–1040, 1042–1044, 1042FF
Antigen-antibody complex
  complement classical pathway activation, 1032–1033, 1033F
  3-D structure, 1023F
Antigen-antibody interaction, 1012–1013
  affinity maturation, 1026–1027
  binding affinity and number of sites, 1016–1018, 1017FF
Antigen-binding site
  and antibody structure, **1019–1022**, 1020FF
  on MHC glycoproteins, 1044, 1045F
Antigen-presenting cell, 1008F, 1011, 1042F, 1042T, **1045–1047**, 1046FF
  B cell, stimulation of, 1012
  foreign antigen processing by, 1046, 1046F
  helper T cell activation, 1042F, 1045–1047, 1046FF
  interleukin secretion, 1046–1047, 1047F
  macrophages of, 334
Antigen-specific receptors
  on B cells, 1012, 1029
  on T cells, 1037–1038, 1037F
Antigenic determinant, 1017, 1017F, 1030, 1030FF
  defined, 1006
Antigenic stimulation
  of B lymphocytes, 1004–1007, 1005FF, 1012
  of multiple lymphocyte clones, 1006–1007
  of T lymphocytes, 1004–1007, 1005FF, 1012, **1041–1046**, 1042FF, 1042T
Antimitotic drugs, 653, 653F, 653T
Antiport, 303–304, 303F

Antiport systems
  in bacteria, 365, 365F
  in mitochondria, 353, 353F
  to regulate intracellular pH, 309–310
Antiporters
  see also individual antiporters
  sodium-calcium ion, 307, 700, 700F, 1107
Antisense RNA, 195–196, 195F, 598
Antiserum, 178
Antogonists, 1085
Antral follicles, 862–863, 862FF
AP endonuclease, in DNA repair, 223, 223F
AP1, 566T
AP2, 566T
Apes, hemoglobin gene map, 183F
Apical meristems, **1148–1149**, 1154, **1173–1176**, 1174FF
Aplysia punctata
  associative learning, 1098–1099
  short-term memory in, 1095–1098, 1096FF
Apolipoprotein-B, novel RNA editing, 597
Apoplast, 1153
Arabidopsis thaliana, for plant developmental studies, 1182–1184, 1184F
Arachidonic acid, 704
Archaebacteria, 11, 11F
Aromatic ring, 48F
Arrestin, 719
ARS element, 484, 485F, 516
  as replication origin in yeast, 516, 517F
Artery structure, 962, 962F
Ascidian, cell lineage, tracing of, 902, 902F
Ascorbic acid, deficiency, collagen synthesis and, 811
Asexual reproduction, 32, 839
  in evolution of genes, 841–843
Aspartate, as neurotransmitter, 1083, 1084F
Aspartate transcarbamoylase, allosteric regulation of, 129F
Aspirin, 689F, 690
Association constant, 94F
Association rate, 94F
Associative learning, 1098–1099
Asters, 652, 656, 656F, 763, 764FF
  cleavage furrow location and, 779–780, 779FF
  sperm, 874
Astral microtubules, 766F, 767F, 770, 772F, 776–777
Astrocytes, 1064–1065, 1065F
  type-1 and type-2, optic nerve, 909, 909F
Asymmetric cell division, 912, 913F
  in plant growth and embryo, 1173, **1177–1179**, 1177FF
Ataxia-telangiectasia, 1201
ATF, 566T
Atherosclerosis, 329
ATP, 44
  ADP ratio, feedback control and, 364
  antiport with ADP in mitochondria, 353–354, 353F
  chemical structure, 58F
  as coenzyme, 75, 76T
  collision rates, 94
  as energy currency in cells, 62–63, 63F
  function, 45
  high-energy phosphate bond, 62
  hydrolysis, *see* ATP hydrolysis
  measurement in cells, by NMR, 154, 154F
  as neurotransmitter, 1083, 1084F
  in nitrogen fixation, 1159
  radioactive forms, 176F
  synthesis, *see* ATP synthesis
  transport vesicle budding, 473, 474F
  turnover rate in cells, 126
ATP hydrolysis, 74–75, 74FF
  in actin
    polymerization, 616
    treadmilling, 638–639
  in active transport, 303
    Ca^{2+} pump, 307
    membrane pumps, 131, 132F
    Na^+-K^+ pump, 287F, **304–305**, 305FF
    periplasmic substrate-binding protein-mediated, 312, 313F
  allosteric conformational changes and, **129–131**, 130FF
  in amino acid activation, 207, 207F
  ATP synthetase, function in, 357–358, 357FF
  and biological order, 62–63, 63F

Bacteriorhodopsin, 292–293, 293F, 298, 298F, 357, 357F
　hydropathy plot, 444F
Bacteroids, 1158–1159, 1158F
Band 3 protein, of erythroctye, 289, 289FF, **291–293**
Band 4.1 protein, 289, 289FF
Banding pattern of chromosomes
　mitotic metaphase, **503–506**, 505F
　polytene, **507–510**, 508FF
Barbiturates, 1085
Barr body, 577–578, 578F
Barrier epithelia, catalog of, 996
Basal body, of ciliary axoneme, 650–651
Basal cell carcinoma, 1188, 1190
Basal lamina, 802, 803F, **818–821**, 819FF
　in blood vessels, 962FF, 963
　cancer cells, migration through, 1194, 1194F, **1200–1201**, 1200FF
　components, 819, 820FF
　distribution, 818–819, 819F
　of follicle, 862F
　functions, 819, **820–821**, 819F, 824, 824F
　at hemidesmosomes, 798, 798F
　lack in neuronal synapses, brain, 1076F
　of muscle, 1124, 1124FF
　at neuromuscular junction, 1075FF, **1124**, 1124FF
　in regeneration
　　of muscle cells, 1124, 1124F
　　of nerve, 1116, 1124, 1124FF
　stem cell potential, maintenance of, **970–971**, 971F
　structure, 819, 819FF
　type IV collagen in, 809, 810T, 815, 816F
Base, defined, 47F
Base of nucleotide, 95–96, 96F, 98FF, 101
Base-pairing of nuceotides, 5–6, 5F, 6F, 58
　catalytic RNA, within, 106, 106F
　codon with anticodon, 103–104, 104F, 206, 208FF, 209, 212, 212F, 217, 217F
　in DNA, 96, 96F, 99F
　　cohesive ends of restriction fragments, 182, 183F, 259, 259F
　in gene conversion, mismatch, 244, 246
　in general recombination, **239–243**, 239FF
　repair, **223**, 223FF, **225–227**
　replication, **227–231**, 228FF
　in RNA, 98F, 101
　in RNA synthesis, 202FF, 203
　of rRNA with mRNA, 219
　in rRNAs, 211F
　by snRNPs, 533
　tRNA, within, 103, 103F
　in transcriptional attentuation, 588, 589F
Base triples of nucleotides, in tRNA, 103F
Basement membrane, 819
　see also Basal lamina
Basic fibroblast growth factor (basic FGF)
　endothelial cell, effects on, 966
Basic pancreatic trypsin inhibitor, see BPTI
Basilar membrane, of cochlea, 1102–1103, 1102FF
Basket cell, of cerebellum, 1061F
Basophil, 974, 975F, 976T, 979F
　allergic reactions, mechanism, 1017, 1017F
Behavior, animal, neural bases for, 35–37, 36F
Bence Jones proteins, 1019
Benign tumor, 1188, 1189F
Benzo(a)pyrene, as carcinogen, 1195–1196, 1197
Benzodiazepines, 1085
β (beta) particle, 175
beta sheets
　in immunoglobulin molecule, 1021, 1022F
　in protein, 109, 109F, 111FF, 112
β-links, in sugars, 51F
Bicoid gene, homeobox in, 937
Bicoid mutation and protein, 924–925, 925F
Bilayer adherence and joining
　in exocytosis and endocytosis, 323, 324F
　in membrane fusion, 336–337, 336F
Bile, 961, 961F
Bindin, 869–870, 870F
Bindin receptor, 869–870, 870F, 872, 873F
Binding protein (BiP), for protein folding, 445
Binding site, on protein for ligand, 122, 123F
Biological clock, 34, 34F
　from oscillations in metabolic intermediates, 84

Biosphere, 60
Biosynthesis
　see also Metabolic reactions
　ATP hydrolysis and coupled reactions, 62–63, 71, 73–75, 74FF, 78, 79F
　and biological order, **71–78**, 74FF, 76T
　of chloroplasts, 387–388, 388F
　of cholesterol, 77, 77F, 81F
　coenzymes, **75–76**, 76T, 77F
　of glutamine, 74F
　of mitochondria, 387–388, 388F
　of nucleotides, 76F
　of polymers by repeated dehydrations, 78, 79F
　reducing power requirement, 77, 77FF
　thermodynamics, **71–75**, 74FF
Biotin, 124
　as coenzyme, 76, 76T, 77F
　in *in situ* hybridization, 192, 193F
　in indirect immunocytochemistry, 177, 178F
　tubulin indicator, 771F
Bipolar cell, of retina, 1061F
Bithorax complex, **929–930**, 931F
　deletion, effects, 930, 930F
　differential splicing of transcripts, 936–937
　organization of, **587**, 587F
Bithorax mutation, 935, 935F, 936
Bivalent
　antibody, 1012, 1012FF
　chromosome, see Meiosis, bivalent
Black membrane, 278, 278F
Blastema, 916, 916F
Blastocoel
　in amphibians, **881–882**, 882F
　of mammal, 894, 895F
　in sea urchin, 882–883, 882F
Blastocyst, 894, 895F, 896
　in creation of chimeras, 896, 896F, 898F
Blastoderm
　cellular, of *Drosophila*, 922, 923F
　homeotic selector gene activation, 930
Blastomere, 880–881, 881F, 895F
　compaction, E-cadherin and, 831
　differences in nonmammalian, 892, 892F
　interaction, for full development, 893–894
　polarization of mouse, 896, 897F
Blastopore, dorsal lip, 884–886, 885F, 887F
Blastula, **881–882**, 880F, 882F
　fate map, 886, 885FF
Blebs, of membrane, predivision, 780
Blood brain barrier, 1065
Blood cells, 25F
　see also individual blood cells
　catalog of, 997
　formation, see Hemopoiesis
Blood sinuses, of bone marrow, 977
Bloom's syndrome, 224F, 1201
Blue-green algae, see Cyanobacteria
Body, axes, defined, 880
Bombesin, 1214
Bone, **989–994**, 990FF
　cell renewal, **989–994**, 990FF
　collagen in, 810T, 813, 813F
　growth
　　by apposition, 990
　　in embryo, 990, 993, 993F
　matrix, 989, 991, 991F
　osteoblasts, bone formation, 991–993, 991FF
　osteoclasts, bone erosion, 991–993, 992FF
　remodeling, 991–992, 992F, 994, 994F
　skeleton, growth from cartilage, 993, 993F
Bone marrow
　hemopoiesis, site of, 975, **977–982**, 977FF
　lymphocytes, 1003–1004, 1004F
　transplantation, 978, 978F
Bottle cells, of amphibian gastrula, 884, 886F
BPTI, molecular models, 112, 113F
Brain
　neurons, number of, 1059
　synapse, fine structure, 1076F
Branch migration, in DNA, 242–243, 243FF
Breast cancer, probability of developing, 1197F
Bright-field microscopy, 141F
Bromoconduritol, 454T
5–bromodeoxyuridine, 519, 519F
Brown fat cells, respiration, 365
Brush border cell, 960F
Budding
　from endosome, 328, 329F
　of enveloped viruses, 250, 250F

of vesicles
　intracellular protein transport, 411, 412FF
　non-clathrin-coated, 473–474, 473FF
　transport, 411, 412FF, 460, 462FF
　in yeast, 573, 573F, 738, 738F
Bulky lesion repair, 225
Bundle-sheath cells, of plants, 371, 372F
α-bungarotoxin, 320, 1085
　synapse formation and, 1126–1127
Bunyavirus, 471
Buoyant density, of cell components, 165
Burkitt's lymphoma, 1209
Bursa of Fabricius, 1003, 1004F

C

C gene segment, 1024
　see also Antibody genes
C-kinase, in neurotransmitter action, 1092
C protein, of skeletal muscle, 623T
C region, of antibody molecule, 1019–1021, 1019FF
C subfiber, of centriole, 650, 650F
C. elegans, see Caenorhabditis elegans
C1 complex, of complement, 1032, 1033F
C1q subcomponent, complement, 1032, 1033F
C3 convertase, of complement system, 1032–1036, 1033F
C$_3$ and C$_4$ plants, 371, 372F
Ca-kinase, 711–712
　Ca^{2+} independence after autophosphorylation, 717
Ca-kinase II, 711
　in active presynaptic cell, 1078
Ca^{2+}/calmodulin-dependent protein kinases, see Ca-kinase
Ca^{2+}/calmodulin-regulated multifunctional kinase, see Ca-kinase II
Cadherins, **830–831**, 831F
　in adhesion belts, 796
　morphogenic movements in embryo and, 940, 941F
Caenorhabditis elegans
　see also Nematode worm
　early development, 784–785, 784F
　mapping of genomic clones for, 264
Caffeine, 734
Calcium ion
　acetylcholine release, as trigger, 321–322, 321F
　actin-myosin interaction, control of, **621–623**, 622FF, **626**, 626F
　anaphase initiation by, 773, 773F
　in associative learning, 1098–1099
　calmodulin complex formation, 711, 712F
　in cell junctions
　　gap, closing by, 801
　　integrity, 793–794, 796, 798F, 801
　cell-cell adhesion, epidermis, 828
　chelation by EDTA, 159
　chloroplast alignment in plants, **1171–1172**, 1172F
　concentration
　　extracellular, 301T, 1077
　　gradient across plasma membrane, 321, 699, 701
　　gradient, intracellular, and cellular polarity, plants, **1179**, 1180F
　　intracellular, 301T, 699, 701, 710, 1077
　　maintenance of intracellular, 307
　cyclic AMP, interaction in cellular regulation, 711–712, 712F
　in egg polarity, seaweed, 912
　exocytosis, role in, 321–322, 321F, 325
　in fertilized egg
　　development, initiation by, 872–874, 875T, 876
　　increase, 871, 871F, 875T, 876
　gel-to-sol transition and gelsolin binding, 631
　glycogen synthesis and, 128
　indicators
　　light-emitting, 156–157, 157F
　　to permeabilize cells, 157–158, 158F
　inositol-phospholipid signaling pathway, release in, 701F, 702–703, 703FF
　as intracellular messenger, 694F, 695, **701–704**, 701FF, 708, 710–717, 711FF
　in acetylcholine release, **1076–1078**, 1077FF

Cell wall, plant, Continued
  functions, overview, 1138
  growth, Golgi vesicles and, **1165–1166**,
    1165FF
  localized deposition by electrophoresis, 1179,
    1180F
  matrix, **1139–1142**, 1139FF, 1168
  modifications in specialized cells, **1145–1147**,
    1146FF
  permeability, 1142–1143
  in plant cell growth
    signals from, 1160
    weakening of, 1143–1144
  primary, **1138–1142**, 1139FF, 1146
  secondary, 1138, **1146–1150**, 1146FF
  structure and compositon, **1138–1142**,
    1139FF
  synthesis
    in Golgi, 1165
    at plasma membrane, **1168–1170**, 1168FF
  tensile strength and turgor, 1143, 1143F
Cellulose, 808, 1139–1140, 1139FF
  of secondary cell wall, **1146–1147**
  synthesis, **1168–1170**, 1168FF
Cellulose microfibrils, 1139FF, 1140
  organization
    cell shape and, 1168, 1168F, 1172, 1173F
    cortical microtubules and, 1146,
      1169–1170, 1169FF, **1172**, 1173F
Cellulose synthase, **1168**, 1168F, **1170**, 1171F
CEN sequence, 484, 484FF, 497
Central nervous system, 1060, 1060F
  glial cells of, **1064–1065**, 1065F
  neural tube, origin, 1109, 1109F
Centriole, **650–652**, 650F
  as basal body, 650–651
  as cell center, 651
  formation
    by duplication in cell cycle, **651–652**, 651F,
      763, 765F
    de novo, 651–652
  lack in some MTOCs, 656
  pericentriolar regions as MTOCs, 656, 657F
  segregation, 763, 765F
  structure, 650, 650F
  in zygote formation, 874, 876
Centromere, 484, 484FF, 503, 503F, 513, 852F
  DNA sequences at, 768–769, 770F
  heterochromatin of, 513, 513F
    constitutive, 577, 577F
  kinetochore formation, 768–769, 769FF
  late replication, 520
  satellite DNA of, 513, 577, 605
Centrosome, 651, 763, 764FF
  in early prophase, 763, 765FF
  ER location and, 661
  microtubule polarity and, 653, 654F
  as microtubule-organizing center, **656**, 656FF
  in mitotic spindle, 772, 772F
  perinuclear location, 481F
  protein from pericentriolar region, 656, 657F
Centrosome cycle, 727
  partial independence from other cycles, yeast,
    740–741, 741F
  steps in, 765F
Ceramide, 282, 282F, 449
Cerebral cortex, embryogenesis, 1110–1112,
  1111F
Cervical cancer
  development stages, 1194, 1194FF, 1199
  papillomavirus, cause of, 1205, 1206F
Cesium chloride, density gradient and, 165,
  166F
CFC, *see* Colony-forming cells
CG islands, 585–586, 586F
Channel-former ionophore, 322–323, 322F
Channel-linked receptor, **1074**, 1074FF,
  1076–1100
  in auditory perception, 1103–1104, 1104F
Channel proteins, 302, 302FF, **312–323**, 313FF
  see also Ion channels; and individual gated
    channels
Charge, defined, 1067
CheA, in bacterial chemotaxis, 723, 724F
Chemical coupling between cells, 798–800,
  799F
Chemical pathways, radioisotopes in study of,
  175–176, 175F
Chemical signaling, *see* Hormone; Local
  chemical mediators; Neurotrasmitters

Chemical synapses, *see* Neuromuscular
  junction; Synapse
Chemiosmotic coupling, 341, 341F, **349**, 349FF,
  349T
  in bacteria, 365, 365F
  sodium ion-based, 365
Chemoattractant, 720, 720F, 722–723, 722FF
  cyclic AMP for slime mold, 826, 826FF
  in inflammatory response, 974–975, 976F
  for sea urchin sperm, 869F
Chemorepellent, 720, 720F, 724F
Chemotaxis, 824
  bacterial, **720–724**, 720FF
    adaptation, 721–724, 723F
    molecular mechanism, 721–724, 722FF
    receptors, 721–722, 722F
  in cell migration in organogenesis, 945
  growth cone of neurite, guidance of, 1115
  by neutrophil, 641, **670**, 671F
  and periplasmic substrate-binding proteins,
    312
  slime mold, 826, 826FF
CheW, CheY, CheZ, in bacterial chemotaxis,
  723, 724F
Chiasma, defined, 847, 847F
Chick embryo, 889F, 915F
  feather and scales, induction by dermis,
    942–943, 942FF
  neural crest and muscle cell migration,
    943–945, 943F
  pattern formation, analogy to insects, 938,
    938FF
  quail cell transplants, 943–944, 943F
Chimera, 896, 896F, 898F
Chlamydomonas, 27, 28F, 451F
  centriole and basal body, interconversion,
    651
  ciliary mutants, 649–650, 650F
Chloral hydrate, 774
Chloramphenicol, 218, 218T
Chloramphenicol acetyltransferase (CAT), 565F
  as marker in gene fusion, 266–267
Chloride ion
  concentration, intracellular and extracellular,
    301T
  equilibrium potential (V_{Cl}), 1067
  fluxes
    at inhibitory synapses, 1083–1084, 1085F
    across membranes, 314, 316
Chlorophyll, in antenna complex, 373–374,
  374F
  exited, decay pathways, 373, 373F
  light energy capture, 368, 373, 373F, 375,
    375FF, 377–379, 378F
  in photosynthesis, 14
  protochlorophyll, conversion from, 1162,
    1162F
  structure, 373F
Chloroplast, **366–380**, 367F, 406, 406F, 1161
  alignment in cortical cytoplasm, 1171–1172,
    1172F
  ATP synthesis
    cyclic photophosphorylation, 378–379,
      379F
    noncyclic photophosphorylation, 377,
      378F, 379
  ATP synthetase, 356, 380, 380F
  biosynthesis in, 380
  division, 388–389, 389F
  DNA
    importance of, 396
    sizes, amounts, packaging, 389–390, 389TT
  electrochemical proton gradient, 341–342,
    341F, 344F
    cf. mitochondria, 379–380, 380F
  electron-transport chain, outline, 387F
  etioplast, development from, 1162, 1162F
  evolution from bacteria, 18, 366, **385–386**,
    386F, 391, **398–399**, 399F, **408–409**,
    409F
  genes
    biparental non-Mendelian inheritance, 396
    introns, 394
    maternal inheritance, 396
    regulation by light, 1162
    and their products, 391, 391F
  genome, **387–391**, **393–401**, 388FF, 389T,
    390T, 392T
  organization, liverwort, 391, 391F
  reasons for, 400–401

  sequencing, 186
  similarity to bacteria, 391
  grana, 367F
  in green algae, 1171–1172, 1172F
  inner membrane, 367FF, 426, 426F
    carrier proteins, 380
  outer membrane, 367FF, 426, 426F
  proplastid, development from, 1162
  protein synthesis, resemblance to bacterial,
    390
  proteins
    import from cytosol, **430**, 430F
    intracellular transport, ribosomes to
      mitochondria, 412, 413F
    synthesis overview, 388, 388F
  ribosomes, 390
  RNA, self-splicing of some, 537–538, 538F
  size, 368F
  stroma, 367F, 368, 379, 380F, 430
  structure, 15F, 17F, 366, 367F, 368
  thylakoid, 367FF, **368**, 377, 378F, 379, 380F,
    426, 426F, 430, 430F
    isolation, 380
  tissue-specific proteins, 397
Chloroplast biogenesis
  of genetic system, overview, 400F
  lipid synthesis, 398
  nuclear genes, control by, 396
  overview, 387–388, 388F
  protein import, 397–398, 398F
Chloroquine, 462
Cholate, detergent, 361F
Cholera toxin
  bindings to ganglioside G_{M1}, 283
  G_s inhibitor, 697
  mechanism of action, 283
Cholesterol
  biosynthesis, 77, 77F, 81F
  modification in smooth ER, 436
  in plasma membrane, 278
    effect on, 279–280, 279F
    flip-flop of, 279
  regulation of synthesis and uptake, 329, 329F
  structural formula, 53F, 279F
  synthesis in ER, 449
  uptake by receptor-mediated endocytosis,
    328–331, 329FF
Choline, in phospholipids, 45
Chondrocyte, **990–991**, 990FF, 993
  switch to fibrobast collagen, 988
Chondroitin sulfate, 804, 805T
Chorion, 856
Chromatin, 23, 26, 26F, 236, **496–513**, 496FF
  see also Chromosome; Euchromatin;
    Heterochromatin
  active
    early replication, 520–521, 521F
    structure and biochemistry, **506–513**,
      506FF
  condensation
    and DNA replication in S phase, 520
    in gene inactivation, 577–579, 578FF
  condensed location of in nucleus, 545, 545F
  decondensation, DNA transcription
    independent, 583F
  decondensation in gene activation, 499, 500F,
    506–513, 506FF
    bithorax complex activation and, 587,
      587F
  DNase I sensitivity, 511, 512F
    domain control region and, 580–582,
      580FF
  lampbrush chromosomes, 506–507, 506FF
  models, 582, 583F
  polytene chromosomes, 509, 510FF
  proteins in, 512–513
  defined, 496
  electron microscopy, 527, 527FF
  fiber, 30-nm, 497FF, **498–499**, 502F
  heterochromatin, 513, 513FF
  histones, **496–500**, 496FF
  maturation after replication, 518
  nonhistone proteins, 496
    sequence-specific DNA-binding proteins,
      488–496, 488FF
  nucleosomes, **496–500**, 497FF
    lack at hypersensitive sites, 498, 498F
    preferential positioning, 497–498, 498F
    transcribed regions, 500, 501F
  polarity, H1–based, 499, 500F

DNA, Continued
one molecule per, 483
sequences needed for functional, 483–484, 484F
circular, of organelles, **389**, 390F
cleavage by restriction nucleases, **181–183**, 182FF
cloning, *see* DNA cloning
coding, **485–488**, 486T, 487F, 506–507, 507F, 523
distinguishing regions of, 487–488
deamination, 222F, 223, 225–226, 225FF
deletion
immunoglobulin gene joining, 1026F
in class switching, **1029**, 1030F
denaturation, 188
deoxyribose, 56FF, 58, 59F, 95–96, 96F, 99F
depurination, 222F, 223
evolution of hereditary function, 10, 10F
flexibility, 563F
fluorescence labeling, in cell analyzer, 730, 730F
genetic engineering, 181
genetic information in, 95–97
as hereditary material, 10
histones, associated with, 496, 499, 500F
interspersed repeated, 604
inversion, immunoglobulin gene joining, 1026F
junk, 393
length in human chromosome, 483, 503
localization by autoradiography, 176
location in metaphase cells, 140F
methylation, *see* DNA methylation
mitochondrial, *see* Mitochondrial DNA
noncoding, **485–488**, 486T, 487F, 507, 507F, 523, **604–609**, 604FF, 605T
determining regions of, 487–488
in nucleosome, **496–498**, 497FF
nucleotide sequences, *see* DNA nucleotide sequences
packaging, in organelles, 390
percent of cell weight, 88T
phosphodiester bonds, 95–96, 96F, 99F
plasmids, 259, 259F
probes in recombinant DNA technology, 189–193, 189FF
to protein, overview, 101–104, 101FF
radioisotope labeling of, 185, 185F
rearrangements, *see* DNA rearrangements
recombinant technology, **180–196**, 181T, 182FF
repair, *see* DNA repair
replication, *see* DNA replication
restriction fragments, 182–183, 182FF, **258–266**, 259F
radioisotope labeling, 185, 185F
separation by gel electrophoresis, 184–185, 184F
sequence analysis, **185–187**, 186FF
restriction maps, 182, 182FF
satellite, 513, **604–605**, 604F
selfish, 605
separation
gel electrophoresis, 184–185, 184F
single-stranded from double-stranded, 262F
single-stranded
as probe for complementary nucleotide sequences, 189–193, 189FF
mitosis delay, 734
spacer, between rRNA genes, 539, 540F
spontaneous damage, 222F, 222–223
steroid-receptor complex, binding of, 691, 691F
structure, *see* DNA structure
synthetic oligonucleotides, 191
tandem repeats, 600–601, 600F
transposable elements, 23
cDNA
advantages for use, 261
comparison with genomic DNA clones, 261
enrichment of specific mRNAs for, 261–262, 262F
hybridization, to RNA, in subtractive hybridization, 262, 262F
preparation of clones and libraries, 260–261, 260FF
use in overproduction of specific proteins, 265
cDNA clones, 183

T-DNA, **1159**, 1159FF, **1182**, 1183F
DNA-affinity chromatography, 489
DNA-binding proteins
binding-site sequences for same, 566T
demethylation by, in germ cells, 586
detection, gel retardation assay, 489, 489F
DNA helix distortion by, 494, 494F
DNA sequences, determination of, 186–187, 188F
eucaryote transcription factors (TF), **525–527**, 526FF
functions of sequence-specific, 488
as gene regulatory proteins, 553–554
helix-turn-helix family, 491, 491FF
histones, **496–500**, 496FF
homeobox-containing gene products, 937
interaction
competitive binding between, 492, 493F, 495, 495F
cooperative binding between, **492**, 493F, **495**, 495F
of separated, DNA-bound, 562, 563F, 569, 569F
with DNA, hydrogen bond, 488, 488F
isolation
by reduced stringency hybridization, 192
of sequence-specific, 489–490
at nuclease hypersensitive sites, 498, 498FF
nucleoprotein particle formation by, 495, 495F
proto-oncogenes encoded, 756F, 757
*Rb*1 gene product as, 1213
recA protein, 242–243, 243F
retinoic acid receptor, 917
sequence-specific, **488–496**, 488FF
symmetric DNA sequences, recognition, **490–491**, 491FF
zinc finger domains, 490, 490F
DNA clones
see also DNA cloning
cDNA, selection of rare mRNAs for, 261–262, 262F
genomic and complementary, 183, 260–261, 261F
identification
by chromosome walking, **262–265**, 264F
by hybrid selection, 265, 265F
in situ hybridization, 262, 263F
of protein products, 265–266
DNA cloning, 181, 183, **258–266**, 259FF
see also DNA clones; DNA library
in vitro, PCR, **271–273**, 271F
minor proteins, obtaining by, 193
of proto-oncogenes from human tumors, 755, 755F
of yeast *cdc* genes, 742–743, 743F
DNA footprinting, 187, 188F
to map steroid-receptor binding sites, 691, 691F
of RNA polymerase binding site, 204, 204FF
DNA gyrase, 581, 582F
DNA helicases, **231–233**, 232FF, 235, 235F
directional movement from allosteric change, 130, 131FF
recBCD protein as, 240, 241F
DNA hybridization, 181, **188–193**, 189FF
gene copy number and, 189, 604
gene expression, study of, 189–191, 189FF
in situ, 192–193, 193F
process *in vitro*, 240–241, 242F
to RNA
A-form of DNA, 492, 493F
in hybrid selection, 265, 265F
Northern blotting, 190–191, 190F
nucleotide sequences, detecting specific, **189–191**, 189FF, 553, 553FF
in reverse transcription, 254–255, 255FF
Southern blotting, 191, 190F
stringent vs. reduced-stringency conditions, 191–192, 192F
DNA library
cDNA, 260–261, 261F
subtractive hybridization, 262, 262F
in expression vectors, 265–266
genomic, 260–261
and chromosome walking, 263–265, 264F
human, YAC vector for propagation, 484, 485F
identification of clones, **262–265**, 263FF
preparation of, 259–260, 259F

DNA ligase
in DNA cloning, 259, 259F
in DNA repair, 223, 223FF
in DNA replication, 229, 231
DNA methylation
of C in CG sequence, 583, 583F
establishment in egg, 584
inheritance in vertebrate cells, 583–584, 584F
lack in invertebrate gene repression, 583
maintenance methylase for, 584, 584F
in mismatch proofreading, DNA replication, 234, 234F
in nuclear memory, 900–901
nucleases, protection from, 182, 250
DNA nucleotide sequences
centromere, conserved elements, 770F
colinearity with protein amino acid sequence, **100–102**, 101F
consensus, RNA polymerase promoter, 204, 205F
detecting specific, **188–193**, 189FF
for DNA-binding proteins, 186–187, 188F
evolution in related organisms, 13, 13F
excision by restriction nucleases, 182–183, 182FF
and genetic information, 97, 97F
of β-globin gene, 97F
helix bending, induced by, 494, 494F
identification, 182, 182F
inversion of, 570–571, 571F
localization of specific, **192–193**, 193F
loss of CG, in evolution, 585, 586F
maintenance of, 220, 221F
mutation rate, **220–221**, 221F, 599
organization and evolution, *see* Genome; Genome evolution
preferential for nucleosomes, 497–498, 498F
rearrangements, 117–118, 118F, **570–574**, 571F, 606, 606F
recognition, for steroid-receptor complexes, 691, 691F
replication origins, yeast, 516, 517F
symmetric, protein binding by, **490–491**, 491FF
of telomeres, 519, 519F
DNA oligonucleotides
oligonucleotides, synthesis, 191
oligonucleotides, synthetic, in PCR, 271, 271F
in prenatal diagnosis, 191
DNA polymerase, 97, 100F, 225
in cDNA synthesis, 260, 260F
in DNA repair, 223, 223FF
in DNA replication, **227–229**, 228FF, **231–233**, 231FF, **235**, 235F
exonuclease activity and proofreading, 230, 233–234, 230F
and gene conversion, 247F
in PCR, 271, 271F
primer, need for, 230–232, 230F
reverse transcriptase as, 254, 255F
in site-specific recombination, 247, 248F
DNA polymerase, alpha and delta, 517
DNA labeling and, 185, 185F
DNA primers
in cDNA synthesis, 260F
in polymerase chain reaction (PCR), 271, 271F
DNA rearrangements, 117–118, 118F, **570–574**, 571F, 606, 606F
in B cell development, **1024–1029**, 1024FF
in carcinogenesis, 1209–1210, 1209FF
gene regulation by, bacterial, 570–571, 571F
in T cell receptor genes, 1037
by transposable elements, 606, 606F
transpositions, during, 256
DNA renaturation, *see* DNA hybridization
DNA repair
bulky lesion repair, 225
double-helix, importance of, 226–227
in gene conversion, 244, 246, 247F
genes coding for, 225
inducible, SOS response, 225–226
mechanisms, 220, **223–227**, 221FF, 221T
nucleases, in DNA repair, 223, 225, 223F, 225F
stability of genes and, 223, 225
systems in DNA replication, 231, 232
DNA replication, **514–522**, 515FF
accuracy, 97, 100
active chromatin, early replication, 520–522, 521F

Egg, Continued
  localized determinants in nonmammalian,
    892, 892F
  MPF in, 736, 736F
  nurse cells, and growth, 858–859, 859F
  parthenogenic activation, 870–871
  polarity, 856, 880, 881F, 921, 923D, 924
  protein synthesis, translational control of, 594
  size, 855, 856F, 857–859
  structural components, 856, 857F
  symmetry in mammals, 892
  yolk, 856
Egg-polarity genes, 921, **924–925**, 925F,
    927–928, 937
EGTA, 872
Ehidium bromide, 184, 184F
Eicosanoids, 689F
eIF-2, in protein synthesis, 213, 214F, 215
Elastase, 115, 115FF
Elastic fibers, 815–816, 816FF
Elastin, 110–112, 111F, 803, **815–816**, 817F
Electric charge
  isoelectric focusing, 171–172, 171FF
  of protein, variation with pH, 171, 171F
Electric field, growth cone of neurite, guidance
    of, 1115
Electric shock, to permeabilize cells, 157–158,
    158F
Electrical coupling of cells, **798–799**
  plant cells, 1152
Electrical synapse, gap junctions, 798–799
Electrical units, 1067
Electrically excitable cells, voltage-gated
    channels and action potentials,
    316–319, 317FF
Electricity, and membranes, summary of
    principles, 1067
Electrochemical gradient
  ATP synthetase, driven by, 356–357, 357F
  defined, 302
  for permeable ions and membrane potential,
    314–316, 316F
Electrochemical proton gradient, 69, 69F,
    341–342, 341F
  ATP synthetase, driving, **356–358**, 357F
  bacterial flagella rotation, 720F
  calcium pump, driving, 353
  in chloroplast, cf. mitochondria, 379–380,
    380F
  control of electron flow and, 364
  evolution, 383, 383F
  free-energy change of proton transfer,
    357–358, 357F
  ΔG for ATP hydrolysis and, **73**
  generation
    by respiratory chain, **351**, 352F, **362**, 363F
    in Z scheme, **377**, 378F, 379
  pH and voltage components, 352, 352F
  protein import, requirement, 427, 428F
  purple bacteria, ATP and NADH synthesis,
    375–377, 376F
  transport of molecules, driving, 352–353,
    353F
Electrochemistry, principles, outline, 1067
Electron acceptor, in photosynthesis, 373,
    374–375, 375F
Electron carriers, of respiratory chain, 359–360,
    359F
Electron donors, in photosynthesis, 373,
    374–375, 375F
Electron microscope
  high-voltage, 147
  scanning (SEM), 144F, **146–147**, 146F
  transmission (TEM), **142–145**, 143T, 144FF,
    147–149, **142–150**, 142FF, 143T
Electron microscopy, **142–150**, 142FF, 143T
  antibody use in, 177, 177FF
  cryoelectron microscopy, **149**, 150F
  developments, important, 143T
  freeze-etch, **147–148**, 148FF
  freeze-fracture, **147–148**, 149F
  image-processing techniques, 149–150
  negative staining, **148–149**, 149F
  preparation of specimens, fixing, embedding,
    sectioning, staining, 144–145, 144F
  replica methods, 147–148, 147FF
  resolution, 142, 142F
    image reconstruction, 150
  shadowing, **147**, 147F, 148
  three-dimensional images, 146, 146F

Electron-transport chain, 69–70, 69F, 341
  see also Electrochemical proton gradient;
    Respiratory chain; and individual
    components
  in chloroplast, **377–379**, 378FF
  evolution, **381–387**, 382FF
  summary of three types, 387F
  using inorganic molecules, 383–384, 384F
Electrons, transfer in electron-transport chain,
    69–70, 69F
Electrophoresis, 169–172, 170FF, 173T
  development, historical, 173T
  of DNA, 184–185, 184F
Elicitors, of plants, 1160, 1160F
Elongation factor (EF), in kinetic proofreading,
    217, 217F
Embedding of specimens
  for light microscopy, 138, 138F
  for TEM, 144
Embryo
  see also individual species
  axes, 880, 881F
  cleavage divisions, early Xenopus, 735–736,
    735F
  development patterns, evolution, 31–32, 31F
  fate map, 886, 885F
  MPF oscillations in, early, Xenopus, 736–737,
    737F
  parthenogenetic, 900
  of plant, 1154, **1173–1174**
  totipotency of early cells, **896**, 896F
Embryogenesis
  see also individual class or species
  body plan, shaping of in early, **879–890**,
    880FF
  cell adhesions, effect on cell fate, 896–898,
    897FF
  cell diversification and cell memory,
    890–901, 891FF
  cell-division control, 907
  cell memory and, **898–901**
  cell migration in, **943–945**, 943FF
  cleavage and blastula formation, **880–882**,
    880FF
  coupling via gap junctions, 800, 801F
  early mouse, **894–898**, 895FF
  environmental effect on cell fate, 897–898,
    898F
  gastrulation, **882–886**, 882FF, 896
  growth factors in early, 894, 894F
  induction, **941–943**, 942FF
  lens formation, **955–956**, 955F
  localized determinants in eggs and blastom-
    eres, 892, 892F
  mammalian, comparison to others, 894
  mesoderm induction in Xenopus, 893–894,
    893FF
  of nervous system, **1108–1132**, 1109FF
    see also Neuromuscular junction;
      Synapse elimination; Synapse
      formation; Visual map
  neurulation, **887–888**, 888FF, 890, 896
  nuclear memory in, **900–901**, 900F
  somite formation, 888–889, 889F
  vertebrate, comparison among, 889–890,
    889F
Embryonic vertebrate cells, aggregation, 829,
    829F
Endocrine cells, 682–683, 682FF
Endocrine signaling, 682, 682F
  synaptic signaling, compared with, 682–683,
    683F
Endocrine system, 34, 35F
  regulation by hypothalamus, 683–684, 684F
Endocytic cycle, cell migration and, **672–673**,
    673F
Endocytosis, 323–324, 324F, **326–337**, 326FF
  see also Phagocytosis; Pinocytosis; Receptor-
    mediated endocytosis
  antigen uptake, antigen-presenting cells,
    1046, 1046F
  cell locomotion and capping, role in,
    333–334, 333F
  coated pits, **326–327**, 326FF
  defined, 20
  fate of endocytic vesicles, see Endol-
    ysosomes; Endosomes
  fluid-phase, see Fluid-phase endocytosis
  lysosome formation, 460, 461F
  in plant cells, 1167, 1167F

  in presynaptic cell, 1063–1064, 1064F, 1078,
    1078FF
  receptor-mediated, see Receptor-mediated
    endocytosis
  transcytosis, 332, 332F
  virus uptake by, 336–337, 336F
Endoderm, 29–30, 29F
  development from single blastomeres, 893
  formation in gastrulation, 884, 885F
  gut and associated structures, embryonic
    origin, 886, 890
Endodermal cells, of plant, 1146F, 1156F
Endoglycosidase, 454F
Endolysosome, 331, 331F, 1046, 1046F
  formation and fate, **460**, 461F
  M6P receptor-enzyme dissociation, 461–462,
    462F
  membrane recycling, between Golgi and,
    461–462, 462F
Endoplasmic reticulum (ER), 16F, 20, 20F, 406,
    406FF, 407TT, **433–450**, 433FF
  see also Microsomes
  autophagosome formation, 460, 461F
  evolution, 409, 409F
  Golgi apparatus, relationship with, 434, 434F
  in intracellular sorting, see Protein, sorting;
    Sorting signals
  location in cell, microtubule-dependent,
    660–661, 661F
  lumen
    BiP in, 445
    protein disulfide isomerase, 445, 445F
    protein folding and disulfide bond
      formation, 445, 445F
  membrane
    continuity with nuclear, 421, 422F, 434F,
      481, 481F
    lateral diffusion in, 437–438
    lipid biosynthesis, **448–450**, 449FF
    lipid composition, 281F
    percent of total cell membrane, 407, 408T
    phospholipid translocator, 449, 450F
    protein translocators, 440–441, 440FF
    signal peptidase, 441, 442F
    SRP receptor, 440, 440F
  microsome formation, 436
  oligosaccharide
    processing, 448, 453, 454F
    synthesis, 446, 447F
  of plant cell, 1151, 1152F, 1161
  plasma membrane synthesis, phospholipid
    asymmetry, 281
  protein
    glycosylation, **446–448**, 446FF
    linkage to glycolipid, 448, 448F
    synthesis, 434–435, 435F, **438–448**, 439FF
  protein pathway
    ribosomes to ER, 412, 413F
    signal sequences, 414, 415T
  ribosomes, 434–435, 435F
  rough, **434–435**, 435F
    ribosome-binding proteins, 438
  sarcoplasmic reticulum, 621, 622F
  in secretory pathway, 411–414, 410FF
  separation of rough and smooth, 436–437, 437F
  smooth, **436**, 435F,
  transitional elements, 411, 412F, 436
  vesicles from, non-clathrin-coated, 328
Endorphins, 719F, 1093, 1093F
Endosomal compartment
  in endocytosis pathway, 330–331, 330FF
  in plant cells, 1167, 1167F
Endosomes, 406, 406FF, 407TT, 460, 461F
  in endocytosis, 326
  fusogenic protein activation in, 336, 336F
  perinuclear, 330–331, 331F
  peripheral, 330, 331F
  pH changes in, 330, 461
  sorting of ligand-receptor complexes, and
    acidity, 329F, **331–332**, 331FF
Endosperm, of plant, 1154, 1173
Endosymbiont hypothesis, 398–399, 399F
Endothelial cells, 951–952, 952F
  basal lamina, associated, 962FF, 963
  blood vessel, fundamental unit of, **962–966**,
    962FF
  in inflammatory response, 974–975, 976F
  of liver, 961F, 962
  renewal in repair or angiogenesis, **963–966**,
    964FF

F

Fibroblasts, Continued
 differentiation, into other cell types, 556,
  556F, **987–989**, 987F, 989F
 extracellular matrix
  organizing by, 813, 815, 815F
  production of, 803, 803F, 951, 952F
 fat cell, conversion to, 988–989, 989F
 locomotion *in vitro*, 638, 638F, 640, 640F
 major properties, **987**, 988F
 migration
  contact inhibition, 673–674, 674F
  directed microtubule-dependent, 670, 671F
  muscle cell, differentiation into, 556, 556F
 nonequivalence of, 987
 as osteogenic stem cells, 991F
 PDGF, effects on, 746, 749
 PDGF receptors, 748
 SEM of *in vitro*, 161F
 stress fibers, **627**, 627F
 ultrastructure, 810F
Fibroin, 102, 109
Fibronectin, 636, 636F, 689, 797, 803, **816–818**, 818F
 adhesiveness of PDGF-stimulated and
  transformed cells, 757, 758F
 in cell adhesion, 636, 636F, 757, **816–817**
 forms, 817
  synthesis of multiple, 818
 functions, 817
 gastrulation, effect on, 885, 885FF
 gene, 818
 for neural crest cell migration, 945
 organization, by actin filaments, 822–823, 823F
 structure, 816–817, 818F
 and transformed cells, 757, 822
Fibronectin receptor, **821–822**, 821F
 in focal contacts, 636, 636F, 797
 phosphorylation
  adhesion, *src* protein and, 757–758, 758F
  cell-matrix uncoupling, 823
Fibrous protein, 108, 119, 120F
Filamin, **630–631**, 630F, 642T, 643, 643F
Filopodia, 638
 in mesenchyme cell migration, 883, 882FF
Fimbrin, 634, 634F, 642T, 643
Fingerprint of protein, 172, 173F
Firing rule, for synapse elimination, 1127–1132, 1127FF
First law of thermodynamics, 72
Fish, embryogenesis, 889F
Fixation of specimens
 for light microscopy, 138, 138F
 for TEM, 144, 144F
Fixed anions, and membrane potential, 316
Flagellin
 bacterial, 720, 720F
 H1 and H2, switching between, 570–571, 571F
Flagellum
 axoneme, *see* Axoneme
 bacterial, **720**, 720F, **722**, 722F, 724F
 formation and regeneration, 650–651
 movements of, 645, 645F
 mutants, 649–650
 of sperm, 863, 864F
 structure, **645–647**, 645FF, 648T
Flavivirus, 471
Flavones, 1158, 1158F
Flip-flop motion in lipid bilayer, 278–279, 278F
Flower, 1154
Fluid-phase endocytosis, 328, 333
 in plant cell, 1167, 1167F
Fluidity of lipid bilayer, **278–280**, 278FF
Fluorescein, 139, 140F, 178F, 295F
Fluorescence-activated cell analyzer, 730, 730F
Fluorescence-activated cell sorter, **159**, 160F
Fluorescence microscopy, **139–140**, 141
 antibody techniques with, 177, 178F
 in DNA sequence analysis, 187F
 light-emitting indicators for, 157, 157F
Fluorescence recovery after photobleaching
 (FRAP), 295–296, 296F
Focal contacts, 627, **635–636**, 635FF, 749, 749F, 797
 src protein in, 757, 758F
Follicle, 862–863, 862FF
Follicle cells, 856, 859, 859F
Follicle-stimulating hormone (FSH), 686T, 862
Forensic medicine, PCR in, 272

Formaldehyde, as fixative in microscopy, 138
Formic acid oxidation, 383, 383F
fos genes, 1207F, 1210, 1211F
Founder cell
 of nematode structure, 904–905, 904F
 in tissue formation, 824, 824F
FRAP, *see* Fluorescence recovery after
 photobleaching
Free energy, *see* Energy
Freeze-drying, in electron microscopy, 147, 148F
Freeze-etch electron microscopy, **147–148**, 148FF
Freeze-fracture electron microscopy, 147–148, 148F
 of erythrocyte membrane, 291–292, 291FF
Fructose diphosphatase, 83
Fructose 1,6–diphosphate, 66F, 83–84, 83F
Fructose 6–phosphate, 66F, 83F
Fruit fly, *see* Drosophila
Fruiting body, slime mold, 825, 825FF
ftz (*fushi tarazu*) gene and mutants
 actions and control, 193FF, 926–928, 927F, 936, 936F
 homeobox in, 937
Fucus, cell polarity, basis for, **1179**, 1180F
Fumarate, in citric acid cycle, 68F, 348F
Fumaric acid, formic acid oxidation in electron-
 transfer chain, 383, 383F
Fungus, symbiosis with plants, 1157, 1157F
Fura-2, 157, 157F
Fusion protein, 266
Fusogenic proteins, viral, 336–337, 336F

G

G (globular) actin, 616, 617F
ΔG, *see* Energy
G bands, in metaphase chromosomes, 505, 505F, 521
G-kinase, *see* Cyclic GMP-dependent kinase
G nucleotide, in catalysis by RNA, 105–106, 105FF
G proteins
 see also individual proteins
 adaptation, role in, 719, 719F
 calcium ion pathway and, 694F
 cAMP pathway, 694F
 discovery, 696
 gated channels, 314
 in inositol-phospholipid pathway, 702
 and morphine addiction, 719, 719F
 in neurotransmitter release, 1092, 1094, 1095F
 ras oncogenes, encoded by, 705
 receptors, similarities, 706, 706F
 regulation of hormone action at, **696–699**, 698F
 in short-term memory, **1097–1098**, 1097FF
 structural homologies, 706
 transducin, 705–706, 706F
 types and functions, 699T
G_0 cells, properties, 750, 750F
G_0 phase, 729
G_0 state
 length, variations, 744–745
 triggers to leave, 744–745, 745F
G_1 phase, 728, 728F
 arrest in, 731
 length, variability
  higher eucaryotes, **746**, 746F
  in starvation, 741–742, 742F
  measurement of, **729–730**, 729FF
 restriction point (R) in, 745, 746F
 Start in, 741
 transition to S, diffusible initiator, 522
G_1-phase cells
 chromosome condensation in, MPF effect, 734, 734F
 S-phase activator, effect on, 732, 733F
G_2 phase, 728, 728F
 DNA replication, block to, 522, 523F
 length, measurement of, 729–730, 729FF
 meiosis prophase I as, 860
 oocyte, arrest in, 735, 736F
G_2-phase cells
 chromosome condensation in, MPF effect, 734, 734F

DNA re-replication block in, 522, 523F, 732–733, 733F
 membrane biosynthesis, 780
 S-phase activator and, 732–733, 733F
GABA, *see* γ-aminobutyrate
gag gene, of RSV, 1206F
GAG, *see* Glycosaminoglycans
Gag proteins, 595, 595F
Gal4 protein, 567, 568F
Galactocerebroside, 282F, 283
Galactose, 452FF, 453
Galactosyl transferase, as marker for Golgi
 apparatus, 453
Galacturonic acid, 1140, 1141F
Gametes
 see also Egg; Sperm
 defined, 840
 haploidy, 845
γ (gamma) class of antibody heavy chain,
 1013–1015, 1014T, 1015F, 1029, 1029FF
Ganglia, 1060, 1060F
Gangliosides, 283, 283F
Gap genes, **926–929**, 926FF
Gap junctions, 681, 681F, 792, 792T, **798–801**, 799FF
 connexons, 799, 780FF
 of electrical synapses, 1074
 embryonic cells, coupling of, 800, 801F, 881, 882F, 888
 follicle cell and egg, between, 859, 859F, 862F
 functions in cell coupling, **798–800**, 799FF
 growth cone and guidepost cell, 1115, 1116F
 in heart muscle, 624, 625F
 of mesoderm, lack in somite formation, 888
 osteocytes, between, 991, 991F
 permeability, **798–799**, 799F
 regulation, 800–801
Gastrin-releasing peptide (GRP), 1214
Gastrula, 880F
 ascidian, 899
 late, cell determination in, 899
Gastrulation, **882–886**, 882FF, 896
 in amphibians, 884–886, 885FF
 blastopore, dorsal lip, 884–886, 885F, 887F
 cadherin changes during, 940, 941F
 cell movements in, types, 884–885, 885FF
 convergent extension, 885, 886F
 defined, 880
 in *Drosophila*, 922–923, 923F
 epithelial cells, shape change of, 883–884, 883F, 886F
 in sea urchin, 882–884, 882FF
  computer model, 675, 675F
Gated ion channels, *see* Ligand-gated ion
 channels; Voltage-gated ion channels;
 and individual channels
Geiger counter, 175
Gel
 actin, 630–631, 630F
 glycosaminoglycan, 804, 804F
 polysaccharide, *in vitro*, 808, 809F
 solation, 631
Gel electrophoresis
 blotting techniques and, 189–191, 190F
 in DNA molecule separation, 184–185, 184F
 in protein separation, **169–172**, 170FF, 173T
Gel-filtration chromatography, 168, 168FF
Gel-retardation assay, 489, 489F
Gelsolin, **631**, 642T, 643
Gene, 95
 see also DNA; Gene expression; Gene
  expression, control of
 addition, 195, 195F
  to study protein function, 195–196, 195F
 altered
  methods and uses, 266–267, 267F
  reinsertion into organism, 267–269, 268F
 amplification, *see* Gene amplification
 analysis
  by blotting techniques, 190–191, 190F
  by hybridization, 189, 189F
 chloroplast, 391, 391F
 chromosomal location, bands and interbands, 510
 cloning, *see* DNA cloning; cDNA; Genetic
  engineering; DNA clones; DNA
  libraries
 constancy, in multicellular organism, 552, 552F

General recombination, Continued
  initiation at single-strand nick, 240, 241F
  isomerization, 244, 245F
  recA protein, 242–243, 243F
  recBCD protein and, 240, 241F
  synapsis, 242, 243F
  synaptonemal complex, 848–849, 849FF,
    852F
Genetic code, 102, 103F, 209, 210F
  defined, 8
  degeneracy, 102, 103F, 209, 210F
  mitochondrial, 392, 392T
Genetic determination of protein sequence,
  100–101
Genetic diseases
  cancer and, 1201
  detection with PCR, 272
  Ehlers-Danlos, 812
  gene mapping with RFLP, 272–273, 272F
  identifying molecular defects for, 264
  of integrins, 822
  LDL receptor defect, 329, 329F
  lysosomal storage diseases, 464–465
  prenatal diagnosis, 191
  thalassemias, 536–537, 537F
  Werner's syndrome, 751
  xeroderma pigmentosum, 225
Genetic engineering, 181, 265–269, 267FF
Genetic information in DNA, 97, 97F, 100F
Genetic locus, 841
Genetic map, chloroplast, liverwort, 391, 391F
Genetic processes, outline of basic, 201
Genetic recombination, 839
  in evolution
    of genome, 599–602, 600FF
    of new genes, 843, 844F
  of exons, 602
  gene replacement or addition by, 195, 195F
  insertion of foreign genes into genome by,
    194–196, 194FF
  introns, facilitation by, 602
  in mitotic cells, 932–934, 933F
  in sexual reproduction, 844–854, 844FF
Genetic transformations, 194–196, 194FF
Genome
  see also DNA; Gene; Haploid genome
  of chloroplasts, 387–391, 393–401, 388FF,
    389T, 390T, 392T
  constancy in multicellular organism, 552,
    552F
  defined, 483
  as dictionary, 551
  differentiated cell, unchanged in, 891
  mapping and analysis, with recombinant
    DNA methods, 272–273, 272F
  mapping of human, C. elegans, Drosophila,
    264
  of mitochondria, 387–401, 388FF, 389T,
    390T, 392T
  mitochondrial organization, special features,
    392–393, 392T
  size
    E. coli, 483, 486F
    human, 483, 486F
Genome evolution, 599–609, 600FF
  DNA rearrangements, by transposable
    elements, 605–606, 606F
  exon shuffling by transposable elements, 606,
    606F
  gene conversion and, 600–601, 600F
  by gene duplication, 600–602, 600FF
  genetic recombination and, 599–602, 600F
  point mutation, 599
  recombination of exons, 602
  transposable elements and, 605–609, 605T,
    606FF
  by transposition bursts, 607–608, 607F
  by transposition of regulatory elements,
    606–607, 607F
Genomic DNA, see DNA clones; DNA library
Genomic DNA clones, 183, 260–261, 261F
Genomic imprinting
  see also Nuclear memory, 900–901
Genotype, defined, 6
Germ cells, 25F, 32, 840
  catalog of, 997
  differentiation in mammalian embryo, 855
  primary, migration to gonad, 856, 858F
  primordial, 855
    in spermatogenesis, 865F

stability, 221–222
survival, at expense of somatic cells, 1188
tissue-specific genes, inactivation and
  methylation, 586
Germ layers, defined, 884
Germ line, cells, 840
Germination
  of fern spores, Ca2+ and, 1180
  of seed, 1154, 1174
Ghosts, erythrocyte, 288–289, 288F
G protein, 699, 699T
Gibberellic acid, microtubule array, orientation
  and, 1172, 1173F
Gibberellins, 1180–1181, 1181F
  cell-wall permeability to, 1143
Gibbs free energy, see Energy
Glands
  see also individual glands
  cell renewal, 971–973, 972F
GlcNAc phosphotransferase, lysosomal
  hydrolases and, 463–465, 464F
Glial cells, 1064–1065, 1065F
  catalog of, 997
  developmental programs of individual,
    909–911, 909FF
  in embryogenesis of nervous system,
    1109–1110, 1111F
  glial fibrillary acidic protein of, 662–663,
    662T, 663F, 667F
  neural tube, origin, 1109, 1109F
  O-2A progenitor cell, induced pathway,
    910–911, 910F
  PDGF, effects on development, 910, 910F
  reuptake of neurotransmitter, 1085
  types in optic nerve, 909, 909F
Glial fibrillary acidic protein, 662–663, 662T, 663F
Glial filaments, 662, 667F
Globin
  see also Hemoglobin; Hemoglobin gene
  oxygen binding, 601
  structure and components, 601, 601F
β-globin
  hnRNA, abnormal processing of, 536–537,
    537F
  mRNA, half-life, 595, 596F
β-globin gene, 486T, 497
  DNA sequence
    exons and introns, 532F
    of human, 97F
  enhancer, identification and regulation, 565,
    565F
  evolution, 601–602, 601F
  expression
    in development, controls, 565–566, 567FF,
      580–581, 580FF
    domain control locus, 580–581, 580FF
    intron, demonstration, 531, 531F
Globin gene family, evolution of, 601–602, 601F
Globular protein, 108–110, 109FF
  α helices in, 110, 110F
  domains in, 116–117, 116FF
  in protein assemblies, 118–121, 118FF
Glucagon, on gap-junction permeability, 801
Glucocorticoid receptor, 567–568, 568F
Gluconeogenesis
  bypass reactions, 82–84, 83F
  glycolysis
    comparison with, 82–84, 83F
    switch from, 80, 82, 83F
Glucose
  ATP yield from complete catabolism, 354
  catabolism, compartmentalization, 85, 85F
  as food compound of cells, 44
  formula, 43
  glycolysis of, 65–67, 66FF
  insulin, effect on metabolism of, 684
  lac operon control and, 560, 560F
  structures, forms, 43F, 50F
  synthesis, see Gluconeogenesis
  transport
    active and passive, 304
    sodium ion gradient-dependent, 309, 310FF
    transepithelial, 310–311, 311F, 793–794,
      793FF
Glucose 6–phosphate, 66F, 83F, 709
Glutamate, as excitatory neurotransmitter, 1083,
  1084F
Glutamate receptor, long-term potentiation, 1100
Glutamic acid, 45F, 55F, 58F
Glutamine, biosynthesis, 74F

Glutamine synthetase, 561
Glutamine synthetase gene, regulation of
  bacterial, 561–562, 561FF
Glutaraldehyde, as fixative
  for electron microscopy, 144, 144F
  in light microscopy, 138
Glutathione, 445, 445F
Glyceraldehyde 3–phosphate, 65–66, 66FF, 83F,
  380
  in carbon-fixation cycle, 369–371, 370F
Glycerol
  in glycolipid synthesis, 282
  in phospholipids, 45, 52F
  in triglycerides, 52F
  turgor maintenance in plants, 1145
Glycine, 685T
  in collagen, 809, 809F
  as inhibitory neurotransmitter, 1083–1084,
    1084F
Glycinebetaine, 1145
Glycocalyx, 299–300, 299FF, 821
Glycogen, 44, 51F, 345, 346, 346F
  breakdown, 346
  catabolism, 709–710, 709F, 712, 712F
  metabolism
    cyclic AMP dependence, 709–710, 709FF
    epinephrine effect, 709
    synthesis, allosteric regulation by calcium
      ion, 128
Glycogen phosphorylase, 347F, 695
  crystals, 152F
  phosphorylation-dephosphorylation,
    709–710, 709F
Glycogen synthase, 347F
  phosphorylation-dephosphorylation,
    709–710, 709FF
Glycolate, 371
Glycolipids, 44, 53F, 282–283, 282FF
  membranes
    amounts of different in, 281T
    asymmetric orientation in, 454, 455F
  of plasma membrane, 278, 298–300, 299FF
    asymmetric distribution, 281F, 282
  as receptors in intercellular signaling, 283
Glycolysis, 65–67, 66FF, 341
  ATP, net synthesis, 66–67, 66FF
  defined, 12
  gluconeogenesis, comparison with, 82–84, 83F
  irreversible steps, 66F, 81, 82
    bypass in gluconeogenesis, 82–84, 83F
  localization in cytosol, 66, 85, 85F
  NADH, production, 66–67, 66FF
  rate, control, 364
Glycophorin, 289F, 291, 292F
  hydropathy plot, 444F
Glycoproteins, 44
  adhesive, RGD sequence in all, 817
  in cell adhesions, 827
  distribution, 446
  envelope proteins, 469–470, 469F
  major histocompatibility complex, see MHC
    glycoproteins
  microfibrillar, with elastic fibers, 816
  molecular structure, 446F
  oligosaccharide side chains, 452F
  of plant cell wall, 1139F, 1140–1142
  of plasma membrane, 298–300, 299FF
  proteoglycans, compared to, 806, 806F
  synthesis in ER, 446–448, 446FF
Glycosaminoglycans (GAGs), 803–807, 804FF,
  805T
  composition, 803–804, 804F
  groups of, 804, 805T
  hydrated gel formation by, 803–804, 804F
  organization in extracellular matrix, 807–808,
    808FF
  proteoglycan, formation from, 806, 806F
  sulfation in Golgi, 456
Glycosphingolipids, 282, 282F
  synthesis in Golgi, 449
Glycosyl transferase
  in Golgi apparatus, 453, 453F, 456
  protein glycosylation in ER, 447F
Glycosylases, in DNA repair, 223, 225–226,
  223FF
Glycosylation of proteins
  in cytosol, 417, 417F
  in ER, 446–448, 446FF
  in Golgi, O-linked, 456
  inhibitors, 454T

Glyoxylate cycle, 432
Glyoxysomes, 432, 432F
Goblet cell of intestinal epithelium, 452, 452F, 960F
Golgi apparatus, 16F, 20, 20F, 406, 406FF, 407TT, **451–458**, 451FF
  cis, medial, and *trans* compartments, 451F, 452, 457–458, 457FF
  coated vesicles from, 328, 462, 462F
  exocytosis, role in, 324, 324F
  galactosyl transferase as marker, 453
  and glycolipid, asymmetric distribution, 282
  glycosylation of proteins, *O*-linked, 456
  location in cell, microtubule-dependent, **661**, 661F
  lysosomal hydrolases, M6P addition, 463, 464F
  M6P receptor protein, **461–462**, 462F
  membrane recycling, Golgi and endolysosomes, **461–462**, 462FF
  oligosaccharides, modification of, **452–458**, 452FF, 454T, 464
  of plant cell, 1161, **1165–1168**, 1165FF
    in cell-wall formation, 781–782, 782F, **1165–1166**, 1165FF
    mucilage synthesis and packaging, 1166, 1167F
  polarization, 451F, 452
    biochemical and functional, **457–458**, 457FF
  proteins of, intracellular transport, ribosomes to Golgi, **411–415**, 410FF
  proteoglycan assembly, 456
  in secretory pathway, 410FF, **411–414**
  structure, **451–452**, 451FF
  sulfations, 456
  trans Golgi network (TGN), 452, 457–458, 458F
  see also Trans Golgi network
Gonadotropins, oocyte maturation and, 859–863, 861F
Gooseberry mutants and gene, 926, 926F, 937
Graft
  immunological tolerance, 1010
  rejection, 1010, 1040
Gramicidin A, 322–323, 322F
Grana, of chloroplast, 367F
Grand postsynaptic potential, action potential, translation to, **1087–1090**, 1089FF
Granule cell, of cerebellum, 1061F
Granulocyte, 974, 975F, 976T
Granulocyte CSF (G-CSF), 981, 981T
Granulocyte/macrophage CSF (GM-CSF), 981, 981T
Granulocyte/macrophage progenitor cell, 981, 981T
Gray crescent, 880, 881F
  of amphibian egg, 886
Gray matter, of spinal cord, 1110F
Green algae, 27, 28F
Green sulfur bacteria, photosynthesis by, 384, 384F
Ground substance, of extracellular matrix, 803
Ground tissue system, of plants, 1146, 1146F, **1148–1149**
Group translocation, 311–312, 312F
Group-transfer reactions, 75–76, 76T, 77F
Growth cone of developing neurite, **1112–1119**, 1112FF
  actin dynamics in, 638–640, 638F, 640F
  contact inhibition, 674
  cytoskeleton, 1113–1114, 1113FF
  guidance
    contact guidance, **1115–1116**, 1116F
    extracellular matrix, 1116
    in vitro, 1114–1115, 1114F
    neurotrophic factors, **1118–1119**, 1120F
    pioneer cone and guidepost cell, **1115**, 1116F
    specific adhesion molecules, **1115–1116**
  membrane synthesis, **1113–1114**, 1114F
  microspikes, movement, 1112–1113, 1112FF
  migration, cytoskeletal basis, **670–671**, 671F
  pathway to target muscles, **1117–1118**, 1117F
  precontact specialization, 1122, 1123F
  target cell contact, effects on, **1122–1123**, 1123F
  target contact or cell death, **1119–1122**, 1120F

Growth factor receptors
  carcinogenesis and, 1207T, 1211F
  ras protein intermediate for, 705
Growth factors, 746–748, 747F, 747T
  see also specific growth factors
  activation of Na$^+$-H$^+$ exchange carrier, 310
  all-or-none, effects of some, 715
  carcinogenesis and, 1211F, 1215
  cell senescence and, 751
  concentrations, 748
  in culture media, 161–162
  discovery and isolation by gene cloning, 266
  intracellular changes, induced by, 757
  mRNA, stability of, 595, 596F
  of plants, *see* Plant growth regulators; and individual regulators
Growth hormone, fat cell development with, 988–989
Growth signals, inositol-phospholipid signaling pathway and, 705
G$_s$ protein, 713
  activation, 697, 698F
  amplification of signal by, 697, 713–714, 714F
  inactivation by cholera toxin, 697
  prostaglandin PGE$_1$, desensitization to, 719
  role in signal transduction, 697, 698F, 699T
  structure, 697
GTP
  hydrolysis
    by G protein, 697, 698F
    in gluconeogenesis, 82–84, 83F
    in microtubule dynamics, **654–655**
    proofreading in protein synthesis, 217, 217F
    in protein synthesis, 212
  synthesis, citric acid cycle, 67–69, 68F, 348, 348F
GTP-binding regulatory protein, *see* G protein
Guanine, 45, 56FF
  in DNA, 95–96, 99F
  in RNA, 5, 5F, 98F, 101
Guanylate cyclase, 713
  as receptor on sperm for resact, 869F
Guard cells of stoma, **1149**
  formation, 1178F
  movement, 1144F, 1145
Guidepost cell, in nerve trunk development, 1115, 1116F
Gut
  cancers, and cell differentiation, 1199–1200, 1199F
  cilia of, 646, 646F
  embryonic formation from endoderm, 883–886, 882FF, 890
  epithelia, catalog of, 996
  specialized cells of, 959, 960F
Gylceraldehyde 3–phosphate dehydrogenase, molecular structure, 116, 117F

H

H and ΔH, *see* Enthalpy
H chain, *see* Heavy (H) chain of antibody
H-2 antigens, 1040, 1040F, 1042T
Habituation, **1095–1096**
Hair cells of cochlea
  see also Auditory hair cell
  development, **675–676**, 676FF
Hair cells of plants, 1171, 1171F, 1178, 1179F
Hairpin, RNA, transcription stop signal, 205F
Hairs, of plant cells, **1149**
Half-life
  of intracellular molecules, **714–715**, 715F
  of radioisotopes, 175T
Halobacterium halobium, 293, 293F, 408
Haploid cells, 839, 840F
  meiosis, production in, 845–847, 846F
  newt, 760, 761F
  in yeasts, 739, 739F
Haploid genome
  new gene, creation of, 843, 844F
  rRNA genes, amounts, 539
  variation in DNA among species, 485, 486F
Haploid species, evolutionary disadvantage, 842–843, 842F
Hapten, 1006, 1007F, 1022, 1023F
Hapten-carrier complex, 1007

Head polymerization, 78, 79F, 208, 228
Heart muscle, *see* Cardiac muscle
Heat, energy equivalent of, 59–60
Heat-shock proteins, **420–421**
Heat-shock response, 225
Heavy (H) chain of antibody, 1013, 1013F
  classes and properties, **1013–1016**, 1014T, 1015FF
  constant region, **1019–1021**, 1019FF
  domains, **1020–1022**, 1020F, 1022FF
  gene pool, **1024**, 1025FF, **1027–1029**, 1028F
  hypervariable region, **1020–1022**, 1020F, 1022F, 1026
  light chain, combination with, 1019, 1020F, 1021, 1023F
  production, genetic mechanism, **1024**, 1025FF
  synthesis, different C regions, **1029**, 1029FF
  variable region, **1019–1022**, 1019FF
Heavy chain of myosin, 617, 617F
HeLa cell line, 162T
Helix
  see also α-helix; DNA structure
  of DNA, 96, 96F, 99F
  handedness, 92, 92F
  in macromolecules, types, 89, 92, 92F
  in supramolecular assemblies, **119**, 119FF
Helix-destabilizing protein, 232, 232FF
  cooperative binding, 492, 493F
  and general recombination, 241–242
Helix nucleation, in DNA renaturation, 240, 242F
Helix-turn-helix protein
  family of DNA-binding proteins, 491, 492F
  tryptophan repressor protein, 558F
Helper T cells, 1003, 1038, 1042, 1042T, **1044–1050**, 1045FF, 1049T, 1051T
  antigen, recognition on, antigen-presenting cells, **1045–1046**, 1046F
  B cell activation by, **1047–1048**, 1048F
  control, *Ir* genes, 1052–1053
  cytoskeletal rearrangments in, 1038, 1039F, 1048
  lack in AIDS, 1044–1045
  macrophage activation by, 1048–1049, 1049T
  MHC class II proteins and, 1042F, **1045–1046**, 1046F
  proliferation, interleukin-2 and, 1046–1047, 1047F
  subclasses, 1049, 1049T
  suppression, by suppressor T cells, 1050, 1051F
Hemagglutinin, 833
  3–D structure, 446F
Hematoxylin, 139
Heme, 124, 125F, 373F, 433, 1159
  in cytochromes, 359, 359F
Hemicellulose, 782, 1139–1140, 1139FF
Hemidesmosomes, 792T, 795, **798**, 798
  location in cell, 798F, 802F
Hemoglobin
  evolution and, 221–222, 221F, 221T
  heme coenzyme, 124
Hemoglobin gene
  conservation, 182, 183F
  restriction maps, 183F
Hemolytic plaque assay, 1012
Hemopoiesis, **973–982**, 973FF, 976T, 981T
  blood cell types, 973F, 974, 975F, 976T
  committed progenitor cells, **979–982**, 979F
  erythrocyte production, **980–981**, 980FF, 981T
  pluripotent stem cell, **978–979**, 979F
  regulation
    cell-type-specific, 974–976
    CSFs, **980–982**, 980FF, 981T
    possible sites for, 982, 982F
  tentative scheme of, 978–979, 979F
  terminally differentiated cells, **979**, 979F
  white blood cell precursors, formation, 981–982, 981F
Hemopoietic cells, 1188, 1189T
  cell culture of, 980
  differentiation pathway, randomness of choice, 982
Hemopoietic stem cells, 973
  clones
    identification, *in vitro*, 977–978, 978F
    retrovirus marker for, 978
  colony-forming cell, 978, 979F

M

Membrane proteins, Continued
  in cell-cell adhesion, *see* Cell-adhesion molecules
  channel proteins, *see* Ion channels; and individual gated channels
  and cytoskeleton, association with, 289, 290F, 291
  electric field, effect on conformation, 318
  erythrocyte, **288–292**, 288FF
  fluorescent labeling, 296
  glycoprotein mobility, 296
  integral, 284FF, 285–287
  lateral diffusion, **295–296**, 295FF
    in heterocaryon, 162
    restrictions on, 297–298, 297FF
  linkage to glycolipid, 448, 448F
  lipid bilayer
    asymmetry and, 281–282
    types of association with, 284–285, 284F
  and lipid composition of bilayers, 280–281
  multipass, 291, 292F
  peripheral, 285, 284F
  reconstitution of purified, 286–287, 287F
  single-pass, 291, 292F
  solubilization, 286–287, 286FF
  synthesis in rough ER, 285, **438–443**, 439FF
  transmembrane, 284–289, 284FF, 302–304, 302FF
  transport
    see also Membrane transport
    acetylcholine receptor, **319–322**, 320FF
    bacteriorhodopsin, 292–293, 293F
    band 3, **291–293**, 289FF
    carriers, **302–305**, 302FF
    channel, 302, 302FF, **312–323**, 313FF
    of chloroplast inner membrane, 380
    freeze-fracture electron microscopy, 291–293, 291FF
    gated channels, **316–322**, 317FF
    segregation in epithelial cell, 793–794, 793F
    turnover in rod cell, 957, 958F, 959
Membrane resistance
  defined, 315
  propagation of action potential, speed and, 1070, 1073
Membrane transport
  see also Active transport; Facilitated diffusion; Membrane proteins, transport
  ATP synthesis driven by, 307, 309
  ATP synthetases, 309
  calcium ion pump, 307
  channel proteins, **312–323**, 313FF
  by ionophores, 322–323, 322F
  of macromolecules and particles, **323–337**, 324FF
  of proteins, 300, **302–314**, 302FF, **316–323**, 317FF, 380
    see also Protein translocation
  of small molecules, **300–323**, 301T, 301FF
  sodium ion gradient driven, 309–311, 310FF
  sodium-potassium ion pump, **304–307**, 305FF
Membrane transport proteins, *see* Ion channels; Membrane proteins
Membrane vesicles
  to introduce impermeant molecules into cells, 158, 158F
  in plant cell plate, 781–782, 782F
Membrane-zippering mechanism, in phagocytosis, 335, 335F
Memory
  immunological, 1008–1010, 1009F
  long-term, 1098
    in hippocampus, **1099–1100**
  non-channel-linked receptors and, **1094–1098**, 1096FF
  short-term, mechanism, **1094–1098**, 1096FF
  synapse formation and, 1132
Memory cells of immune system, 1009–1010, 1009F
  somatic hypermutation in, 1026–1027
Menstrual cycle, ovulation during, 862–863
Mercaptoethanol, 170, 170F
Meristems, 1138, **1154**, **1173–1176**, 1174FF
Mesenchymal cells, 987
  bone and cartilage, formation from, 990
  vimentin and, 662–663, 662T, 663F, 665, 666
Mesenchyme
  connective tissue development from, 943

mesoderm, formation from, 887
  primary, of sea urchin embryo, 882–884, 882FF
Mesoderm
  in *Drosophila*, 923, 923F
  formation in gastrulation, 884, 885F
  induction of ectoderm, 941–942
  induction in *Xenopus*, 893–894, 893FF
  lateral plate, 943
  pattern formation, control in vertebrates, 941–942
  segmentation, in somite formation, 888–890, 889F
  tissues, arising from, 886–887, 888, 890
Mesophyll cells, 1142F, **1148–1149**
  carbon dioxide pump of, 371, 372F
Mesosome, 409F
Messenger RNA, *see* mRNA
Metabolic cooperation between cells, gap junctions, 798–799, 799F
Metabolic pathways
  defined, 11
  direction of, and enzymes, 62, 63F
  evolution, 11–12, 12F
  feedback regulation, **80–84**, 82FF
  NMR in study of, 154, 154F
  organization in cytosol, 668
  radioisotopes in study of, 175–176, 175F
  regulation
    by reversible covalent modification of proteins, 712
    simple, **80–85**, 81FF
Metabolic reactions
  compartmentalization in cells and tissues, 84–85, 85F
  evolution, 12–13, 13F
  organization, 80, 81
  rate, effect of multienzyme complexes on, 126–127
Metabolic regulation
  allosteric proteins in, 128, 128FF
  enzyme activity changes, **80–84**, 82FF
Metabolism and storage, catalog of specialized cells, 996
Metabolites, transport across mitochondrial membrane, 352–353, 353F
Metamorphosis, of frog, hormonal control, 688, 688F
Metaphase
  location of actin, DNA, tubulin, 140F
  meiotic, 853F, 849, 851, 854F
  mitotic, 763, 763FF, **766**, **771–773**
Metaphase plate, 1177FF
Metastasis, 752, 1188, 1188F, **1200–1201**, 1200FF
Methyl-accepting chemotaxis proteins (MCPs), 721, 722F
Methyl esterase, 723F
Methyl group, 42, 48F
  transfer by S-adenosylmethionine, 76T
Methyl transferase, 723, 723F
N-methyl deoxynojirimycin, 454T
Methylation
  in bacterial chemotaxis, 721–724, 722FF
  of DNA, *see* DNA methylation
  of proteins, in cytosol, 417, 417F
Methylation-demethylation cycle, in bacterial chemotaxis, 722–723, 723F
5-methylcytosine, in vertebrate DNA, 583, 583F
MHC (major histocompatibility complex), **1040–1041**, 1040F, 1042T
  Ir gene locus, 1052
MHC glycoprotein, **1040–1046**, 1041FF, 1042T, **1051–1054**, 1050FF
  allelic, inactive forms, 1052–1053
  as Ig superfamily member, 1053–1054, 1054F
  polymorphism, 1053
  self-tolerance, development, 1051–1052
  in transplantation reactions, 1053
MHC glycoprotein, class I, **1040–1044**, 1040FF, 1042T, 1051–1054, 1054F
  antigen-binding site, 1044, 1045F
  CD8 protein and, 1050, 1050F, 1051T
  genetic loci, 1040F
  polymorphism, 1040, 1040F, 1044
  structure, 1041, 1041F, 1042T, 1044, 1045F
  tissue distribution, 1042, 1042T
  in viral antigen presentation to cytotoxic T cells, **1042–1044**, 1042FF

MHC glycoprotein, class II, 1040FF, **1041–1042**, 1040FF, 1042T, **1044–1046**, 1046FF, 1051–1054, 1054F
  antigen-binding site, 1052
  on antigen-presenting cells, helper T cell activation, 1045–1046, 1046F
  B cell activation and, 1048, 1048F
  CD4 glycoprotein and, 1050, 1050F, 1051T
  genetic loci, 1040F
  interferon effect on, 1049
  polymorphism, 1040,1040F, 1044
  structure, 1041, 1041F, 1042T, 1044, 1045FF
  tissue distribution, 1042, 1042T
MHC restriction, 1042
  development of, 1051–1052
Micelle, 53F, 277, 277F
  detergent, 286, 286F
Microbody, *see* Peroxisome
Micrococcal nuclease, 496, 497F
Microelectrodes, intracellular, **155–157**, 155FF, 318, 1152, 1153F
Microglia, 1064, 1065F
β₂-microglobulin, 1041, 1041F, 1044, 1045F
Microinjection
  of cells, 180
  of impermeant molecules, 157, 157FF
Micropipette, patch-clamp recording, 318F
Micropyle, of fish eggs, 871
Microscopy
  see also Electron microscopy; Light microscopy
  units of length, 136F
Microsomes, 436–438, 437F
  sedimentation in ultracentrifuge, 164–165, 165F
Microspikes, **638**, 638F, 640, 640F, **669–671**, 669FF, 674
  actin treadmilling in, 640, 640F
  cell cortex in locomotion and, 669–670, 670F
  inhibition by cytochalasins, 671
  of neurite growth cone, 1112–1113, 1112FF, 1115
Microsporidia, stages in evolution and, 18
Microtome, 138, 138F
Microtubule-associated proteins (MAPs), 646, **659–661**, 659FF
Microtubule-binding proteins, 768, 768F
Microtubule-organizing center (MTOC), **656**, 657FF, 763
Microtubule-walking proteins, in kinetochore, 775F
Microtubules, 21, 613
  A and B tubules, 645, 645F, 647, 647F
  actin filament location and, 673
  actin filaments, comparison, 652–654, 654T
  aster formation, 656, 656F
  in axon, 1061F
  in axoneme, **645–647**, 645FF, 648T
  C tubule, 650F
  in cell migration, **672–673**, 673F
    directed, **670–671**, 671F
  cell shape and, 673
  of centrioles, 650
  colchicine, effect of, 653, 653T, 768
  dynamic instability, **654–655**, 656F
  in early prophase, 763, 765FF
  fibroblast, arrangement in, 670, 671F
  growth cone, arrangement in, 670, 671F
  in interphase array, 763, 764FF
  kinetochore, *see* Kinetochore, microtubules
  metaphase cells, location in, 140F
  mitochondria, associated, 342, 342F
  in mitotic spindle, 772, 772F
  in neural-fold formation, 888, 888F
  neurite growth, requirement for, 1113–1114, 1114F
  polarity, 649
    plus and minus ends, **653–656**, 653F
  pole, *see* Polar microtubules
  polymerization-depolymerization in mitotic spindle formation, 763, 768, 768F, 775F
  protofilaments, 646–647, 646F
  sliding mechanism in ciliary movement, **648–650**, 648FF
  in sperm flagellum, 864FF
  structure, **646**, 646F
  subunit, slow axonal transport in neurite, 1113–1114, 1114F
  in T cells, in cellular interaction, 1038, 1039F

Morula
blastomere polarity in, 896, 897F
formation, 894, 895F
Mosaic eggs, 892, 892F
Moss, 1172F
Motor neuron, 1062
action potential, *see* Action potential
death in development, 1121, **1125–1126**
embryogenesis, 1110, 1110F, 1112, 1112F, 1117–1118, 1117FF
identification by retrograde tracing, 1117, 1117FF
innervation of muscle
elimination of excess junctions, 1126–1127, 1126FF
specificity, **1117–1118**, 1117FF
sprouting for reinnervation, 1121, 1121F, 1125, 1125F
synaptic inputs, 1086, 1086F
Mougeotia, chloroplast alignment in light, 1171–1172, 1172F
Mouse
embryogenesis, **894–898**, 895FF
fertilization, **875–876**, 876F
gene replacement, direct, 269
meiosis timing, 855F
transgenic 267–269, 268F
MPF, *see* M phase promoting factor
μ (mu) class of antibody heavy chain, 1013–1015, 1014T, 1015F, 1020F, 1021, 1028F, 1029
Mucopolysaccharide, *see* Glycosaminoglycans
Mucus, 452F, 456, 960F
Multicellular organisms
see also individual organisms
behavior, bases for, 35–37, 36F
cell communication in development of, 30–31
cell division
controls, cellular level, **743–752**, 744FF, 747T
random variability, 746, 746F
rates of, 744
cell memory in development, 31–32
cell specialization in, 85, 85F
cohesion between cells, **28–29**
colony-forming cells, **27–28**, 28F
epithelial sheets and internal environment, 29–31, 29F
evolution from single cells, 22–23, **27–37**, 28FF
evolutionary relationships, 38F
hormonal and neuronal signaling, 34–35, 35F
Multi-CSF, *see* Interleukins
Multidomain, structure of protein, 116–117, 116FF
Multidrug resistance (*mdr* 1) gene, 1202
Multienzyme complex, 126–127
Multipass transmembrane proteins, transport proteins as, 302, 304–305
Multiple myeloma, 1018
Murine leukemia virus, 1206F
Muscarinic acetylcholine receptors, 706
Muscle, 1060, 1060F, 1188, 1189T
see also Cardiac muscle; Insect, flight muscle; Skeletal muscle; Smooth muscle
embryonic origin, 887, 888, 943, 943F
phosphorous compounds and activity, 154, 154F
proteins
alternative RNA splicing, 628–629
multigene families for, 628
satellite cell, 968
types in vertebrates, 624
Muscle cell, 25F
see also Actin filaments; Myosin filaments; Neuromuscular junction
action potentials, *see* Action potentials
activity
effect on axon, 1125–1126
effect on membrane and metabolism, 1125–1126
elimination of excess synapses, 1126–1127, 1126FF
categories in mammals, 983, 984F
characteristics, cell-type-specific, 556
deep-etching, rapid-freeze EM, 149F
denervated, 1124–1125, 1125F
reinnervation of, 1121, 1121F, 1125, 1125F
desmin of, 662–663, 662T, 666
differentiation, **983–984**, 985F

gene expression, changes in pattern of, 985
intracellular membrane systems, 621, 622F
migration from somites, 943, 943F
neurotrophic factor production, 1126
overview, 32F, 33
preinnervated, 1123, 1123F
protein isoforms, production, 984–985, 986F
regeneration, 1124–1125, 1124F
Muscle contraction, **613–629**, 614FF, 623T
actin filaments, **616–617**, 616FF
myosin heads, interaction with, **619–620**, 620FF
troponin and tropomyosin association, 621–623, 622FF
ATP hydrolysis, **618–620**, 621F, 626
creatine kinase activity in, 619F
heart muscle, **624**, 625FF
initiation by Ca^{2+}, **621–623**, 622FF
myofibrils, **614–615**, 614FF
myosin, **619–620**, 620FF
phosphocreatine, 618, 619F
regulation, by troponin and tropomyosin, **621–623**, 622FF
sequence of events, diagram, **620**, 621F
skeletal muscle, **613–624**, 614FF, 623T
sliding filament model, **615–616**, 615F
smooth muscle, **625–626**, 625FF
Muscle fiber, 614, 614F, 983
Muscular dystrophy, 988
Mutagenesis, 1191–1192, 1192F
Mutagens, 194
Ames test for, 1191, 1192F
Mutant(s)
albumin gene, analysis, 190–191, 190F
bicoid and *oskar*, 924–925, 925F
conditional, 739
disjunction defective, 867
in embryonic migratory cells, 945
gene insertion into foreign genome, 194–196, 194FF
genetic diseases, 191
of β-globin gene enhancer, 565, 565F
heterochromatin, *Drosophila*, 579, 579F
memory-impaired, *Drosophila*, 1099
protein function, determination from, **194–196**, 194FF
reeler mice, 1110–1112, 1111F
of RNA splice sites, 536–537, 537F
for secretory pathway, 473
Mutant genes, detection by hybridization, 190F, 191
Mutation(s)
alleles, generation of favorable, 841–843, 842FF
in carcinogenesis, **1191–1197**, 1191FF, **1201–1202**
in cell-type-specific protein genes, 905
in coding and noncoding DNA, 485–486
deleterious, recessive, 842–843, 842F
in developmental control genes, nematode worm, **905–906**, 906F, 908–909
dominant
creation in higher eucaryotes, 195–196, 195F
in proto-oncogenes, 1203
elimination by natural selection, 221, 221T
from error in DNA replication, 97, 100
heterochromic, 906, 906F
in housekeeping genes, 905
lethal recessive, 842, 842F
mapping of cloned genes, use of, 263–264
meaning to organism, 1188
in membrane transport proteins, 302
mispairing of tautomeric forms of bases, 229–230, 230F
point, genome evolution and, 599
recessive, in tumor suppressor gene, 1213, 1213F
in segmentation genes, **925–926**, 926FF
silent, 100
somatic hypermutation antibody genes, 1026–1027
by transposable elements, 606–608, 607F
at transposition sites, 256
Mutation rate
and cancer, 221–222
determination of, **220–221**, 221F
by Ames test, 1191, 1192
and evolution of related species, 221
and human evolution, 222

necessity for low, 221–222
spontaneous, 1192
in ssDNA and RNA viruses, 227
Myasthenia gravis, 1011, 1085
myc genes, 1207T, **1209–1210**, 1212, 1212F, 1214
c-*myc* gene product, as marker of proliferative mode, 757, 756F
Mycoplasma, **9–10**, 10F
Mycorrhizal fungus, 1157, 1157F
Myelin
see also Myelin sheath
galactocerebroside, 283, 282F
lipid composition, 281T
Myelin sheath, 1064, 1072F, **1073**
Myeloid cells, 974, 975F, 979F
Myeloma protein, 1018
Myoblasts
determination, 983
differentiation, control by myoD1, 556, 556F
fusion, and differentiation, **983–984**, 985F, 986
proliferation, 983–984, 985F
as satellite cells, **986**, 986F
MyoD1, 556, 556F
Myoepithelial cells, 624, 972, 972F, 983, 984F
Myofibrils, 614–615, 614FF
see also Muscle contraction
major proteins of, 623–624, 623T
Myoglobin, 985, 110FF
Myomesin, 623T, 624
Myosin, 147F, **619–621**, 620FF, 623T, **624–628**, 624FF, 631
absence in some migrating cells, 672
anchoring in sarcomere, 624, 624F
antibody, cytokinesis inhibition, 780
cell division in egg and, 180
in contractile ring, 626–627, 627F
in cytoplasmic streaming, 632, 632F, 1170
directional movement from allosteric change, 130, 131FF
extraction from skeletal muscle, 617, 617F
filament formation
in vitro, 617–618, 619F
phosphorylation and, 626, 627F
heads, **617–620**, 617FF
binding to actin filaments, **619–622**, 620F, 626
decoration of actin filaments, 619, 620F
number in thick filaments, 620
walking along actin filaments, **620**, 621F
heavy chain, 617–618, 617F
isoforms, fast and slow, 986F
light chain, 617, 617F, 625–626, 626F
phosphorylation of, 626, 626FF
minimyosins, 632, 635
nonmuscle, **626–627**, 627F, 631, 642T, 643, 643F
bipolar assembly and sliding mechanism, 631, 631F
cytoplasmic movement and, **631**
in cytoplasmic streaming, 632, 632F, 1170
papain digestion, 617, 618F
phosphorylation, calcium-dependent in smooth muscle, **626**, 626FF
polarity, 617–620, 617FF
rod, 617–618, 618F
S1 fragments, 617, 618F, 619, 620F
of skeletal muscle, **617–620**, 617FF, 623T
in smooth muscle, **625–626**, 625FF
structure, **617–618**, 617FF
in thick filaments, **617–618**, 619F
in transient musclelike assemblies, **626–627**, 627F
Myosin ATPase, 619
actin-activated
skeletal muscle, **619–622**, 620FF
smooth muscle, 626
Myosin heavy-chain genes, 628
Myosin light-chain kinase, **626**, 626FF, 711
Myotome, 888
Myristic acid, 417–418, 418F
Myxobacteria, 27, 28F

———————————————————— N

O

Origin of life
  cell age, 7
  pairing of nucleotides and, 5–6, 5F
Oscillator, biological, 84, 84F
Oskar mutants, 924
Osmium tetroxide, as fixative for electron
    microscopy, 144, 144F
Osmolarity, intracellular, sources and control,
    306, **308**
Osmosis, defined, 47F
Osteoblast, 991, 991F
Osteoclasts, 991–993, 992FF
Osteocyte, formation from osteoblast, 991, 991F
Osteoid, 991, 991F
OTF1, 566T
Ouabain, 304–306, 305F
Outer membrane
  of chloroplasts, 367FF, 368, 426F
  of mitochondria, **343**, 344F, 426, 426F
Ovalbumin gene
  intron, demonstration, 531, 531F
  processing of hnRNA of, 535F
Ovary, 859–863
Overlapping genes, viral, 257
Ovulation, 857
  hormonal induction, 862, 863F
Ovule, of plants, 1154
Ovum, 840, 840FF
Oxaloacetate, 67, 68F, 69, 83F, 348, 348F
Oxidation, defined, 61
Oxidation of biological molecules
  see also Citric acid cycle; Oxidative
      phosphorylation
  electron movements, 61, 62F
Oxidative catabolism, evolution and efficiency,
    67
Oxidative metabolism, evolution, 14–15
Oxidative phosphorylation, **69–70**, 69F, 341,
    343
  ATP formation, 69–70, 69F
  chemiosmotic process, **349**, 349FF, 349T
  compartmentalization requirement, 358, 406
  efficiency, 354–355
  electron-transport chain, *see* Electron-
      transport chain; Respiratory chain
  evolution, 382–383, 383F
  mechanism, general, 352, 352F
  mitochondrial location, 85, 85F
Oxygen
  electron sink in respiratory chain, 350,
      350FF, 362
  in evolution, first production of, 384–385,
      385FF
  in photorespiration, 371, 372F, 377, 378F
  photosynthesis, production in, 369, 369F
Oxyntic cell of stomach, 960F

P

P element, and gene replacement in *Drosophila*,
    268–269, 268F
P face of lipid bilayer, 291FF, 292
P-granules, germ-cell development and,
    784–785, 784F
P-site, *see* Peptidyl-tRNA binding site
p60^{src}, *see* src protein
Pachytene, 848, 849FF, 852F, 855F
PAGE, *see* Polyacrylamide-gel electrophoresis
    (PAGE)
Pair-rule genes, **926–929**, 926FF, 937
  control of segment-polarity genes by, 928F,
    929
Palmitic acid, 417–418, 418F
  ATP yield from, 354
  structure, formula, 44, 44F
Pap smear, 1194, 1195FF
Papain, 617, 618F, 1013, 1014F
Papillomas, 1196–1197
Papillomaviruses, as tumor virus, 1204T, 1205,
    1206F
Papovavirus family, 1204T
Paracrine signaling, 682, 682F
  in nervous system, 1092
Parathormone, 686T, 696T
Parenchyma, **1148–1149**
Parkinson's disease, 1093, 1093F
Parthenogenesis, 652, 871

Partially coated reticulum, in plant cells, 1167,
    1167F
Partition chromatography, 167, 167F
Parvoviruses, 250, 251F
Passive transport, *see* Facilitated diffusion
Patch-clamp recording, 156, 156F, 318–319,
    318FF
Patching
  of multivalent ligands, 333–334, 334F
  of plasma membrane proteins, 295
Pathway guidance, 1115, 1116F
Pattern formation
  see also Positional information
  asymmetrical cell divisions, amplification,
    912, 913F
  via cell-cell interactions
    mesoderm induction, 893–894, 893FF
    nematode worm, 907–909, 908F
  in chick limb, **915–917**, 915FF
  in *Drosophila*, **920–938**, 920FF
  in early embryogenesis, **879–890**, 880FF
  in epidermal cells of cockroach, 918–919,
      918FF
  genes for
    evolution of, 937–938
    homeobox in, **937–938**
    mutual inhibition among, 928
  hierarchy of controls for, 927–928
  localized determinants in egg, 892, 892F
  localized signaling regions, 913, 914F
  molecular genetics of, **920–938**, 920FF
  in plant development, **1175–1176**, 1175F
  principles of spatial, **911–919**, 912FF
  rule of intercalation in, **918–919**, 919F
  spatial signal gradient, **912–914**, 913FF, **917**,
      917FF
  threshold response, 914
  stabilization by cell-cell adhesions, 940, 940F
  transient vs. permanent patterns, 929
  in vertebrates, comparison to insects, 938
PCR, *see* Polymerase chain reaction
Pectin, 782, 1139–1140, 1139F, 1146, 1160
Pelomyxa palustris, stages in evolution and, 18
Pemphigus, 797
Pentaploid cells, 760, 761F
Pentoses, structures, 50F
Pepsin, 1013, 1014F
Pepsinogen, 960F
Peptide
  hormones, 683, 686T
  local chemical mediators, 685T
  neurotransmitters, 685T
Peptide bond, 45, 45F, 54F, 100
  cleavage, 172, 173T
  formation, 54F, 208, 209F
Peptide map, 172, 173F
Peptidoglycan, in periplasmic space, 312F
Peptidyl transferase, 212
Peptidyl-tRNA binding site, 210, 212, 212F
  initiator tRNA binding, 213, 214F
Perforins, 1039
Pericentriolar regions, 656, 657F
Perichondrium, 990–991, 990F
Periclinal divisions, 1176F, 1177
Perinuclear space, 421, 422F
Peripheral membrane proteins, 284F, 285
  spectrin, 289, 289FF
Peripheral nerves
  contact guidance to development, 1116
  myelination by Schwann cells, 1072F
Peripheral nervous system, 1060, 1060F
  development, and N-CAM changes during,
    830
  neural crest origin, 1109
Peripheral neurons, support cells, catalog, 997
Periplasmic space, 312F
Periplasmic substrate-binding proteins, 312,
    313F
  in chemotaxis, 722, 722F
Permanent cells, **954–959**, 955F
  see also individual cells
Permeability
  of cell sheets, 793–794, 793FF
  of gap junctions, 798–799, 799F
  of lipid bilayer, 301, 301F
  and membrane potentials, 316
Permeability coefficients
  defined, 301
  lipid bilayer, synthetic, 301F
Permissive cells, 254

Peroxidase, as indicator in TEM, 145
Peroxidative reaction, 431–432
Peroxisome, 16F, 20, 406, 406FF, 407TT,
    431–433, 431FF
  assembly from cytosol molecules, 433, 433F
  lipid import from ER, 450, 450F
  oxidative reactions, 431–432
  photorespiration, 371, 372F
  of plant cells, 432, 432F, 1161
  proteins, intracellular transport, ribosomes to
      mitochondria, 412, 413F
  sedimentation in ultracentrifuge, 164–165,
      165F
Pertussis toxin, inhibition of G$_i$ protein, 699
Petite mutants
  cytoplasmic, 396, 396F
  of *Euglena*, 396
  nuclear, 397
  in yeasts, 396–397, 396F
pH
  defined, 47F
  in fertilized egg, increase, 871, 871F, 874,
      874F, 875T
  fluorescent indicators, 157
  fusogenic protein activation and, 336–337,
      336F
  measurement in cells, by NMR, 154, 154F
  regulation of intracellular by antiport
      systems, 309–310
  sorting in endocytic pathway, 331–332,
      331FF
Phagocytic cells, 1015, 1015F, 1016
  see also Macrophages; Neutrophil
Phagocytosis, 33, 33F, 326, **334–335**, 334FF
  antibody dependent, 1015, 1015F
  enhancement by complement, 1034
  inhibition by cytochalasin, 642, 642F
  interleukins, stimulation by, 1048–1049, 1049T
  local actin filament disassembly in, 631
  phagolysosome formation, 460, 461F
  process and mechanism, **334–335**, 334FF
Phagolysosomes, 326, **334**, **460**, 461F
Phagosome, 326, 334, 334FF, **460**, 461F, 974
Phalloidin, **642–643**
Phase-contrast microscope, **141–142**, 141F
Phase transition and separations, in lipid
      bilayers, 279
Phase variation, of bacterial flagellins, 570–571,
      571F
Phenobarbitol, detoxification in smooth ER, 436
Phenotype, defined, 6
Phenylalanine, 45F, 55F
Philadelphia chromosome, 1190, 1190F
Phloem, 1147, 1147F, **1149**
  companion cells, 1156
  development, 1176–1177, 1176F
  regeneration in wound response, 1181, 1182F
  sieve tube, *see* Sieve tube
  sucrose transport, 1155–1156, 1155FF
Phorbol esters
  to mimic diacylglycerol effects, 704, 704F
  as tumor promoters, 705, 1196
Phosphoenolypyruvate, 372F
Phosphatase inhibitor protein, cyclic AMP-
      dependent, 710, 711F
Phosphate group transfer
  in allosteric control proteins, 129–131, 130FF
  by ATP, 75, 76T
Phosphates, 49F, 56F
Phosphatidylcholine, 280, 280F, 281T, 398
  synthesis in ER, 448–449, 449F
Phosphatidylethanolamine, 280, 280F, 281T,
    398
Phosphatidylinositol (PI), 448, 448F
  in signal transduction, 702, 702FF
Phosphatidylserine, 80, 280, 280F, 281T, 398,
    704
Phosphocreatine, 618, 619F
  changes with muscle activity, by NMR, 154F
Phosphocreatine kinase, 618
Phosphodiester bond
  cleavage in DNA footprinting, 187, 188F
  in DNA, 95–96, 96F, 99F
  DNA ligase and, 224F
  DNA topoisomerases and, 236, 237F
  in RNA, 98F
Phosphodiester linkage, in nucleic acids, 57F,
    58, 59F
Phosphoenolpyruvate, 66F, 83F
  in group translocation, 312, 312F

Proton-motive force
across bacterial membranes, 365, 365F
in chloroplast, 379–380, 380F
of electrochemical gradient, 352
Proton pump
ATP synthetase, **357–358**, 357F, **382–383**, 383F
of bacteria, 365, 365F, 382–383, 383F
bacteriorhodopsin, 293, 293F, 357, 357F
in chloroplast, cf. mitochondra, **379–380**, 380F
in endosomal membranes, 330
evolution, 382–383, 383F
of lysosomes, 459, 459F
in photophosphorylation, 377, 378F, 379
of plant cell plasma membrane, 1143
in purple photosynthetic bacteria, 375–376, 376F
respiratory chain, 349T, 351–352, 352F, 358–359, **363–364**
in sperm, acrosomal reaction, 869
Protoplast, 1143, 1143F, 1167, 1167F, 1181–1182, 1183F
Protozoa, 21–22, 22F
osmoregulation, 308
phagocytosis in, 334
Provirus, 253, 253F
integration in host genome, 257
Pseudogenes, 602
Pseudopod, of epithelial cells, in angiogenesis, 964, 964F
Psoriasis, 971
Puffing, *see* Polytene chromosomes
Pulse-chase labeling, 175, 175F
Pulsed-field gel electrophoresis, 184, 184F
Purine, 45, 56FF
Purkinje neuron, 1061F
Puromycin, 218, 218T
Purple bacteria
electrochemical proton gradient, ATP and NADPH synthesis, 375–377, 376F
electron transfers in photochemical reaction, 375, 375F
nonsulfur, electron-transport chain, outline, 387F
Purple membrane, of *Halobacterium halobium*, 293, 293F, 408
Pyramidal cell, cerebral cortex, 1061F
Pyramidal neurons, differentiation into, 1111–1112, 1111F
Pyrimidine, 45, 56F-57F
Pyrimidine dimers, repair, 225
Pyrophosphate
hydrolysis, in DNA replication, 97, 100F
product of ATP hydrolysis, 74–75, 75F
Pyruvate dehydrogenase, structure, 84F
Pyruvate dehydrogenase complex, 347, 347F
Pyruvate metabolism
to acetyl CoA in mitochondria, 345–348, 346FF
in fermentation, 66–67, 381–382, 382F
to glucose, gluconeogenesis, 82–84, 83FF
from glucose in glycolysis, 65–67, 66FF
mitochondrial location, 85, 85F
transport into mitochondria, 345, 353, 353F

Q

Quaternary structure, of protein, 114F
Quiescent state, *see* G$_0$ cells; G$_0$ state
Quiescent stem cell, renewal from, 986, 986F
Quin-2, 157, 157F
Quinones, 360, 360F
in NADPH synthesis, 376, 376F, 378F, 379
in photosynthesis, purple bacteria, 374–377, 374FF

R

R bands, in metaphase chromosomes, 505, 505F, 521
R point of cell cycle, *see* Restriction point of cell cycle
Rabl orientation, of chromosomes, 544–545, 544F
Radial glial cells, 1109–1110, 111FF

Radial spokes, of axoneme, 647–650, 647FF, 648T
Radioactive tracing of molecules, **174–176**, 175FF, 175T
Radioisotopes, **175–176**, 175T
DNA, labeling with, 185, 185F
raf genes, 1207F, 1210, 1211F
ras oncogenes, G protein, encoding by, 705
ras proteins, 699T
as intermediates for growth factor receptors, 705
intracellular mediators, effects on, 757, 756F
membrane, attachment to for transformation, 417–418, 418F
*Rb*1 gene, in retinoblastoma, 1213, 1213FF
Reading frame, in protein synthesis, 102, 102F, 213
recA protein, 225
cooperative binding, 492, 493F
in genetic recombination, 242–243, 243F
recBCD protein, in general recombination, 240, 241F
Receptor current, in hair cells of cochlea, 1103, 1103F
Receptor down-regulation, 332, 332F, 718, 719F
Receptor-mediated endocytosis, **322–332**, 328FF
antigen uptake, B cell, 1048, 1048F
of catalytic receptors, 707
of IgA, transcytosis of, **332**, 1016, 1016F
insulin uptake, 694
LDL uptake and processing, **328–331**, 329FF
of lysosomal enzymes, 461
receptor concentration, regulation by, 718, 719F
receptor-mediated transport, analogy to, 461–462, 462F
recycling of receptors, **328**, 329FF, **331**
viral uptake and processing, 469, 1043, 1043F
of yolk proteins in oocyte, 857
Receptor-mediated transport, of lysosomal hydrolases, 461–462, 462FF
Receptor potential
transducer cell, 1102
vertebrate rods, 1105, 1105F
Receptor sequestration, 718, 719F
Receptors, **682–684**, 683F, 688F, **688**
see also individual receptors
for plant growth regulators, **1181**
Receptors, cell-surface, **693–699**, 694FF, **702**, 704FF, **705–707**, 709
actin nucleation sites and cell locomotion, **641**
antigen-specific, 1005–1006, 1006F, 1012
bacterial chemotaxis, 721–724, 722F
catalytic, 695, **706–708**, 707F
in nervous system, 1092
to cell-adhesion molecules, 831, 832F
in cell migration in organogenesis, 945
channel-linked, 694–695, **1074**, 1074FF, **1081–1091**
classes, **694–695**
in coated pits, 329–330
diffusion in plasma membrane, 696, 696F
extracellular matrix, 821–822, 821F
G-protein-linked, 694F, **695–706**, 976FF
in nervous system, **1092–1099**, 1092FF
in short-term memory, **1096–1098**, 1096FF
G$_i$ protein, interaction with, 699
G$_s$ protein, interaction with, 697, 698F, 713–714, 714F
IL-2 on T cells, 1046–1047, 1047F
intracellular messengers, need for, 694
intracellular signal generation, 694F, 695
ligand bindings, 693–694
mechanically gated, in hair cells, 1103–1104, 1104F
of methylation, 723, 723F
non-channel-linked, **1074**, 1074F, **1091–1100**, 1092FF
phagocytosis-promoting, 334–335
for progesterone, 860
recycling, in endocytosis, **328**, 329F, **331**, 331FF
in regulated exocytosis, 325, 325F
regulation
down-regulation via lysosomes, **331–332**, 332F
by endocytosis, 718, 719F
by reversible phosphorylation, 718–719, 719F

seven-pass transmembrane receptor proteins, 706, 706F
transcytosis of, 331–332, 332F, 1016, 1016F
Receptors, intracellular
for nuclear protein uptake, 423
recycling, **461–462**, 462F
for steroid hormones, **690–693**, 691FF
Recessive mutant alleles, 842–843, 842F
Recombinant DNA technology, restriction nucleases, **181–183**, 181T, 182FF, **258–259**, 259FF
Recognition
intercellular, **824–833**, 824FF
molecular, **87–95**, 88TT, 89FF
diffusion and, 92–95, 92FF
errors and repair, 94–95
Recognition sequences, DNA, for steroid receptor, 691, 691F
Recombinant DNA molecule, 259
Recombinant DNA technology, **180–196**, 181T, 182FF, **258–269**, 259FF
see also DNA; DNA cloning; Cloning vectors; Expression vectors
amino acid sequences in proteins using, 174
development, historical, 181T
gene regulatory regions, identifying by, 564–565, 565F
for plants, 1182–1184, 1183FF
restriction nucleases, **181–183**, 181T, 182FF, **258–259**, 259FF
specific nucleotide sequence identification with DNA probes, **189–193**, 189FF
transgenic organisms, 194–196, 194FF, 267–269, 268F, 1181–1184, 1183FF
Recombinase, 1026F
Recombination nodule, 848–849, 849F
Reconstitution, of functional sodium-potassium ion pump, 287F, 305
Red blood cells, *see* Erythrocyte
Redox potentials, of respiratory chain, 362, 363F
Reduced-stringency hybridization, 192F
Reducing agents, disulfide bonds and, 108, 108F
Reducing power, in biosynthetic reactions, 77, 77FF, 383–384, 384F
Reduction, defined, 61
Reflection-interference microscopy, 635, 635F
Regeneration
in *Hydra*, mechanism, 912, 913F
of liver, **961–962**
Regulated secretory pathway, 324–325, 324F, 409, **465–467**, 466FF, 468
see also Constitutive secretory pathway; Exocytosis
Regulatory T cell, 1038
Release factors, in protein synthesis, 212–213, 213F
Renaturation of DNA, *see* DNA hybridization
Reovirus, 250, 251F
Repetitive DNA, **600–601**, 600F
Replica methods of electron microscopy, 147–148, 147FF
Replicas, for *in situ* hybridization, 262, 263F
Replicases, 251, 252F
Replication of DNA, *see* DNA replication
Repressor protein, in bacterial gene expression, control, **557–559**, 558FF
Reproduction, 32
see also Asexual reproduction; Sexual reproduction
Resact, 869F
Residual bodies, 334
from spermatid differentiation, 867F
Resonance energy transfer, in photosynthesis, 373, 373F
Resonance frequency, 153, 154F
Respiration
defined, 15
efficiency of cellular, **354–355**
Respiratory chain, 350–352, 350FF, 356, **358–365**, 359FF, 387F
see also individual enzyme complexes
in bacteria, 365, 365F
comparison of mitochondrial, chloroplast, bacterial 387F
cytochrome oxidase complex, 360–362, 361FF
electrochemical proton gradient, *see* Electrochemical proton gradient

Thin filament of skeletal muscle, **614–618**, 614FF
  actin filaments, *see* Actin filaments
  anchoring, 624, 624F
  polarity, 619–620, 620F
  in sliding filament model, **619–620**, 621F
  tropomyosin and troponin, **621–623**, 622FF
Thin-layer chromatography, 167, 167F
Thioester bond, 418F, 1035, 1036F
Thioredoxin, 445, 445F
Threonine
  phosphorylation and enzyme activity, 84
  phosphorylation, 129
Thrombin, 703T
Thromboxanes, 689F
Thy-1 glycoprotein, 1005, 1053–1054, 1054F
Thylakoid membranes, freeze-fracture EM of, 148F
Thylakoid space, 430, 430F
Thylakoids
  chloroplast, 367FF, 368
  isolation and study, 380
Thymidine, 57F, 176
Thymine, 45, 56FF
  dimer, 222F, 223
    repair, 225
  in DNA, 95–96, 99F
Thymus
  lymphocytes, 1003–1004, 1004F
  removal, effect of, 1003
  T cell development and selection in, 1051–1052
Thyroglobulin gene, 486T
Thyroid hormone (TH), 687T
  carrier protein, 688, 689F
  mechanism of action, **690–693**, 691FF
  in metamorphosis of frog, 688, 688F
  regulation by pituitary hormones, 684, 684F
Thyroid hormone (TH) receptor, 567–568, 568F
  c-*erbA* gene for, 756, 756F
Thyroid-stimulating hormone (TSH), 684F, 686T
  cyclic AMP role, 696T
Thyroid-stimulating hormone (TSH) receptors, stimulation by antibodies, 694
Ti plasmid, **1159**, 1159FF
Tight junctions, **793–794**, 793FF, 802F
  Casprian band analogy of plant, 1155, 1156F
  embryonic cells, between, 881, 882F, 894
  function, 793–794, 793F
  lateral diffusion, block by, 297, 297F
  location in cell, 793F, 802F
  structure, 794, 794FF
  and transcellular tranport, intestinal, 311, 311F
Tissue culture
  see also Cell culture
  of plants, 1181, 1182FF
Tissue regeneration, basal lamina, role in, 820–821
Tissue-specific genes, methylation and loss of CG, 585–586, 586F
Tissue systems, of higher plants, 1146, 1146F, **1148–1149**
Tissues
  see also individual tissues
  blood supply, 951, 952F
  cytoskeleton organization in, **674**
  defined, 791
  disruption, techniques, 159
  formation, two strategies, 824, 824F
  maintenance and renewal, **951–954**, 952FF, 1187–1188
  primary cell cultures, 161
  reassembly *in vitro*, 827–829, 829F
  supporting apparatus for all, **951–952**, 952F
Titin, 623T, 624, 624F
Tobacco mosaic virus (TMV), 250, 251F
  self-assembly, 119, 121, 121F
  transport through plasmodesmata, 1152–1153, 1153F
Tobacco smoke, carcinogens in, 1195, 1197
Tomato bushy stunt virus (TBSV), 120F
Tonicity, of solutions, 307F
Tonoplast, **1163–1164**, 1164F
Totipotency
  of differentiated cell nucleus, 891, 891F
  of plant cells, 1181, 1183F
  of teratocarcinoma cells, 897–898, 898F
TPP, 124, 124F

Tradescantia, cytoplasmic streaming in stamens, 1171, 1171F
Trans Golgi network (TGN), 452, 457–458, 458F
  see also Golgi apparatus
  constitutive secretion from, **467–468**, 467F
  lysosomal hydrolases, sorting, packaging of, **460–466**, 462F
  recycling of receptors, 461–462, 462F
  secretory vesicle, sorting and packaging, **465–466**, 466FF
Transcellular transport, *see* Transcytosis
Transcription complex, for RNA polymerase II, 563
Transcription of DNA, *see* DNA transcription
Transcription factor IID, *see* TATA factor
Transcription factor III A (TFIIIA), 489, **490**, 490F, 525–526
Transcription factor Sp1, 490, 490F
Transcription factor(s) (TF), for RNA synthesis, eucaryotes, **525–527**, 526FF
Transcriptional attenuation, 588, 589F
Transcriptional control
  see also Gene expression, control of
  defined, 552F, 553
  evidence, nucleic acid hybridization, 553, 553FF
Transcriptional enhancers, 692
Transcytosis, **310–311**, 311F, 328, **331–332**, 332F
  of IgA, 1016, 1016F
  of IgG, placenta, 1015
Transcytotic vesicles, of capillary, 963F
Transducer cell, 1101–1102, 1101F
  receptor potential, 1102
Transducin, 699T, **705–706**, 706F, **713**, **1107**, 1107F
Transduction
  see also individual transducers
  sensory, **1101–1108**, 1101FF
Transfection
  in DNA cloning, 259, 259F
  in genetic engineering, 266–267, 269
  of yeast, 743F
Transfer cells, of plant, 1148, 1156, 1156F
Transfer RNA, *see* tRNA
Transferrin, in iron uptake by endocytosis, 331
Transferrin receptor, synthesis, control of, 595, 597F
Transformants, permanent, 266–267
Transformation, neoplastic
  by DNA tumor viruses, 254
  in vitro, 162
Transformed cell
  anchorage independence, 759, 759FF
  difference from normal, 753, 753T
  fibronectin levels and adhesion, 822
  growth requirements, general, 759–760, 759FF
  lines, 162
Transforming growth factor-β (TGF-β), 747T
Transforming growth factor-β2 (TGF-β2), mesoderm induction by, 894
Transgene, 579–580, 580T
Transgenic animals, **267–269**, 268F
  gene expression, position effects, 579–580, 580T
Transgenic flies, enhancer analysis in, 608F
Transgenic mice
  genomic imprinting in, 901
  oncogene-carrying, 1212, 121F
Transgenic organisms, **194–196**, 194FF
Transgenic plants, **1181–1184**, 1183FF
Transhydrogenase, in purple bacteria, 376, 376F
Transitional elements
  association with *cis* Golgi, 452, 451F
  of endoplasmic reticulum, 411, 412F, 436
Translation activator, 594, 594F
Translation enhancer sequence, 593–594, 594F
Translation repressor proteins, 593, 593F
Translation, *see* Protein synthesis
Translational control of protein synthesis, **593–595**, 593FF
  defined, 552F, 553
  extent, 594
  negative control, 593, 593F
  positive control, translation enhancers, 593–594, 594F
  translational frameshifting, 595, 595F
Translational frameshifting, 595, 595F
Translocation, in carcinogenesis, 1209–1210, 1209FF, 1214

Transmembrane proteins, **284–289**, 284FF
  see also Membrane proteins
  acetylcholine receptor, 320, 321F
  asymmetrical orientation
    maintenance, 454, 455F
    maintenance, in vesicular transport, 411, 412F, 443, 454, 455F
  bacteriorhodopsin, 292–293, 293F
  band 3, 291–293, 289F
  in cell junctions
    adherens, 795–797, 795FF
    desmosomes, 797–798, 798F
    gap connexon, 799, 780FF
    tight, 794, 795F
  fatty acid attachment, 418, 418F
  freeze-fracture electron microscopy, 291–292, 291FF
  gated channels, **316–322**, 317FF
  glycophorin, 290F, 291, 292F
  hydropathy plot, conformation predictor, 443, 444F
  linker, 636, 636F
    desmosome, **797**
    E-cadherin, 831
    fibronectin receptor, 821F
  multipass
    calcium ion pump, 304
    synthesis in ER, **441–443**, 443FF
  photosynthetic reaction center, 293–294, 294F
  rhodopsin, 1106–1107
  seven-pass, family of receptors, 706, 706F
  single-pass
    catalytic receptors, **706–708**, 707F
    synthesis in ER, **441**, 442F, 444F
  synthesis in ER, 434–435, 435F, **438–443**, 439FF
  transport of small molecules, 300
  vectorial labeling, 288–289
Transmission electron microscopy (TEM), **142–145**, 143T, 144FF, **147–149**
Transmitter-gated ion channels, 314, **319–322**, 319FF
  acetylcholine receptor, **319–322**, 320FF
  structural similarities and evolution of, 321
Transpiration stream, of plants, 1143
Transplantation, of pronucleus, 900, 900F
Transplantation reactions, immunological, 1040–1041, 1042F, 1053
Transport
  see also Membrane transport
  axonal, *see* Axonal transport
  in bacteria, 365, 365F
  intracellular, *see* Protein transport, intracellular; Sorting signals; Transport vesicles
  passive vs. active, 302–303, 302F
  in plant, **1151–1156**, 1152FF, 1171
  of proteins across membranes, *see* Protein translocation
  transcellular, **310–311**, 311F, **322**, 793–794, 793FF, 1016, 1016F
Transport vesicles, 324, 324F, 409, 411, 412F, 413–414
  see also Secretory vesicles
  clathrin vs. non-clathrin-coated, 472, 472FF
  ER to Golgi, 434, 434F
  in plant cells, **1165–1166**, 1165F
  recycling of intracellular receptors, 462, 463F
  ten distinct, need for, 471
Transposable elements, 23, 26F, 248–249, **255–257**, 256FF, **605–609**, 605T, 606F
  DNA sequences, effects on, 605–606, 606F
  exon shuffling by, 606, 606F
  gene regulation, effects on, 606–607, 607FF
  inverted repeat sequences, 256F
  L1 and *Alu* sequence of humans, 608–609, 605T, 609F
  major families, characteristics, 605, 605T
  movement within chromosomes, 256, 256FF
  as mutagens, unusual features, 606–608, 607F
  retrotransposons, **255–256**, 255F
  site-specific recombination, **255–256**, 256FF
  transposition bursts and biological diversity, 607–608, 607FF
Transposases, 255–257, 256FF, 269, 606, 606F
Transposition
  in gene replacement in *Drosophila*, 268F, 269
  of transposable elements, mechanism, **255–256**, 255FF

Visual map, development, **1129–1131**, 1129FF
  critical period, 1129–1131, 1131F
  ocular dominance columns, development, 1130, 1130F
  synapse elimination in, 1130–1131, 1130FF
  visual experience requirement, 1130–1131, 1131F
Vitamin C, *see* Ascorbic acid
Vitamins
  as coenzyme precursors, 76
  in culture media, 161T
Vitelline layer of egg, 856, 857F
  bindin receptor, 869–870, 870F, 872, 873F
Voltage clamp, of axons, 1069, 1070F
Voltage-gated calcium ion channel, **318–319**, 321–322, 321F, 701–702, 701F
  acrosomal reaction and, 869, 869F
  action potential coupling to presynaptic acetylcholine release, 321–322, 321F, **1076–1078**, 1077FF
  in adaptation of postsynaptic cell, 1089–1090
  in axon hillock, 1088–1089, 1089F
  in developing neurite, 1122
  in rod cell, 1103–1107
  in short-term memory, 1097F, 1098
Voltage-gated ion channels, 313
  see also individual voltage-gated channels
  action potential, generation of, **316–319**, 317FF, **1065–1073**, 1068FF
  in axon hillock, 1087–1090, 1090F
  in dendrite membrane, 1091
  ligand-gated channels, comparison with, 1081, 1081F
  in postsynaptic cell, 321F, **322**
  in presynaptic cell, **1076–1078**, 1077FF
Voltage-gated potassium ion channels
  delayed and early in axon hillock, **1088–1089**, 1089F
  in repetitive firing, 1089, 1089F
  repolarization, after action potential, 1070, 1070F
  voltage clamp of, 1069, 1070F
Voltage-gated sodium ion channels
  in action potential
    generation, **317–318**, 317FF, **1068–1070**, 1070F
    propagation, 1070, 1071F, 1073
  in axon hillock, 1088–1089
  conductance, 1067
  conformational changes, 317, 317F
  on dendrites and cell body, scarcity, 1087
  inhibitors, 318
  in neuromuscular transmission, 321F, 322, 716
  at nodes of Ranvier, 1073
  patch recording, 318–319, 318FF
  structure, 318
  voltage clamp of, 1069, 1070F
Voltage gating, and electrical field, 318
Voltage gradient across membranes, *see* Membrane potential
Volvox, 28, 28F
Vulva, of nematode worm, development, 907–909, 908F

W

Water
  cell mass, percent of, 87,88T
  chemical properties, 46FF
  macromolecular structure, effect on, 90F-91F, 123
  permeability of lipid bilayers, 301, 301F
  physical properties, 42
  structure, 46F
  transport in plants, 1153–1155, 1155F
Wavelength of electron, resolution in electron microscopy, 142
Wavelength of light, resolution in light microscopy, 137, 138F
Waxes, of plants, 1147, 1149, 1150F
Weak chemical bonds, **88–91**, 89F
  see also Noncovalent bond of macromolecule
Weak noncovalent forces, antigen-antibody interactions, 1016–1018, 1017F
Werner's syndrome, 751
Western blotting, defined, 170
White blood cells
  see also individual cell types

White blood cells, Continued
  categories, **974**, 976T
  in inflammatory response, 974–975, 976F
  turnover rates, 975
White matter, of spinal cord, 1110F
Wilms' tumor, 1213
Wobble
  in codon-anticodon base-pairing, 209, 210F
  in mitochondria, relaxing of, 392
Wound healing
  angiogenesis in, 965–966, 965FF
  contact inhibition of movement, 674
  epidermal proliferative units in, 971
  fibroblasts and, 987
  hyaluronic acid gel in, 805
  in plants, 1181, 1182F

X

X chromosome, 850F, 851, 854, 867
  inactivated
    inheritance of, **577–578**, 578F
    late replication, 520
    to show monoclonal origin of cancer, 1190, 1191F
  methylation of DNA in, 585
X-irradiation, destruction of hemopoietic cells by, 978, 978F
X-ray diffraction analysis, **150–152**, 150FF
  of cAMP and CAP binding site, 123F
  of bacteriorhodopsin, 293, 293F
  of crystalline structure, 151–152, 151FF
  developments, historical, 152T
  hemagglutinin, 446F
  of immunoglobulin molecule, 1021–1022, 1022FF
  of MHC glycoprotein, 1044, 1045F
  of photosynthetic reaction center, 294, 294F
  protein conformation and, 109
  resolution, 152
  viral capsid formation, 120F
Xenografts, 1040
Xenopus laevis
  cleavage and blastula formation, **879–882**, 880FF
  early embryo, cleavage divisions, 735–736, 735FF
  eggs, *in vitro* nuclear assembly in, 778
  gastrulation, **884–886**, 885FF
  homeobox protein in, 938
  induction of mesoderm, 893–894, 893FF
  neurulation, 887–888, 888FF, 890
  oocyte, assaying MPF in, 736, 736F
  rRNA gene amplification, 892
  somite formation, 888–889, 889F
Xeroderma pigmentosum, 225, 1201
Xylem, 1143, 1146–1147, 1147F, **1149**, 1176
  in apoplast, 1153, 1155F
  development, **1146–1147**, 1147F, 1176–1177, 1176F
  regeneration in wound response, 1181, 1182F
  thickening, **1168–1170**, 1168F, 1170FF
  transport, of water and solutes, 1153, 1155, 1155F
Xyloglucan, 1139F, 1141F
Xylose, 1141F

Y

Y chromosome, 850F, 851, 854, 867
Yeast, 2F
  ARS elements, as replication origins, 516, 517F
  artificial chromosome vector, 484, 485F
  asymmetric inheritance in, 573–574, 573F
  cdc mutants, genes and gene products, **739–743**, 740FF
  *cdc*2/28 Start-control gene, and product, 742–743, 743F
  cell cycle, 738, 738F
    dependent sequences, 740–741, 741F
    as model for study of, **738–743**, 738FF
    in starvation, 741–742, 742F
    Start for, **741–743**, 741FF
  chromosome
    separation, 184, 184F

TEL, CEN, ARS sequences needed for function, 483–484, 484FF
gene replacement, 195–196, 195F
haploid cells, 394, 395F
haploid and diploid stages, 571, 572FF
heat-shock genes, 421
histone H1 lack, 496
introns, loss of, 603–604
life cycle, 739, 739F
mating, 571, 739, 739F, 741F
mating-type interconversion, **571–574**, 572FF
Mendelian inheritance, 395F
mitochondrial DNA
  junk, 393
  introns, 394
mitochondrial genes, cytoplasmic inheritance, 394–395, 395F
mitosis, haploid cells, 840
mitotic spindle, inside nucleus, 785–786, 786F
petite mutants
  cytoplasmic, 396, 396F
  nuclear, 397
protein synthesis controls
  gal4 protein, 567, 568F
  lack of some higher eucaryote, 583
  positive translational, 593–594, 594F
reproduction, sexual and asexual, 394–395, 395F
Ty1 element of, 255–256, 255F
Yolk, 856, 880, 881F
  gastrulation, amphibian, effect on, 884, 885F
Yolk proteins, synthesis in liver, 857

Z

Z disc of sarcomere, 614, 614F
  actin filament
    anchoring at, 624, 624F
    polarity of, 619–620, 620F
  desmin filaments in, 666
Z-form of DNA, 493F, 494
Z scheme, of photosynthesis, 377, 378F, 379
Zinc fingers, of gene regulatory proteins, 490, 490F
Zona pellucida
  see also Jelly coat, of egg
  of egg, 856, 857F, 862FF, 894, 895F
  in mouse, fertilization, 875–876, 876F
Zone of cell differentiation, in root tip, 1176, 1176F
Zone of cell division, in root tip, 1176, 1176F
Zone of cell elongation, in root tip, 1176, 1176F
Zone of polarizing activity, in chick limb bud, morphogen from, 917, 917F
ZP3 glycoprotein, in fertilization in mouse, 875–876, 876F
Zygote, 840F, 874, 876
  fusion of sperm and egg nuclei, 874, 876
  polarization mechanism, algae, **1179**, 1180F
Zygotene, 848, 849FF, 852F, 855F
Zygotic-effect genes, defined, 925
Zymogen granules, 411, 411F
Zymogenic cell of stomach, 960F
Zymogens, 411

The Genetic Code

1st position (5′ end) ↓	U	C	A	G	3rd position (3′ end) ↓
U	Phe	Ser	Tyr	Cys	U
	Phe	Ser	Tyr	Cys	C
	Leu	Ser	STOP	STOP	A
	Leu	Ser	STOP	Trp	G
C	Leu	Pro	His	Arg	U
	Leu	Pro	His	Arg	C
	Leu	Pro	Gln	Arg	A
	Leu	Pro	Gln	Arg	G
A	Ile	Thr	Asn	Ser	U
	Ile	Thr	Asn	Ser	C
	Ile	Thr	Lys	Arg	A
	Met	Thr	Lys	Arg	G
G	Val	Ala	Asp	Gly	U
	Val	Ala	Asp	Gly	C
	Val	Ala	Glu	Gly	A
	Val	Ala	Glu	Gly	G

Amino Acids and Their Symbols

Codons

			Codons					
A	Ala	Alanine	GCA	GCC	GCG	GCU		
C	Cys	Cysteine	UGC	UGU				
D	Asp	Aspartic acid	GAC	GAU				
E	Glu	Glutamic acid	GAA	GAG				
F	Phe	Phenylalanine	UUC	UUU				
G	Gly	Glycine	GGA	GGC	GGG	GGU		
H	His	Histidine	CAC	CAU				
I	Ile	Isoleucine	AUA	AUC	AUU			
K	Lys	Lysine	AAA	AAG				
L	Leu	Leucine	UUA	UUG	CUA	CUC	CUG	CUU
M	Met	Methionine	AUG					
N	Asn	Asparagine	AAC	AAU				
P	Pro	Proline	CCA	CCC	CCG	CCU		
Q	Gln	Glutamine	CAA	CAG				
R	Arg	Arginine	AGA	AGG	CGA	CGC	CGG	CGU
S	Ser	Serine	AGC	AGU	UCA	UCC	UCG	UCU
T	Thr	Threonine	ACA	ACC	ACG	ACU		
V	Val	Valine	GUA	GUC	GUG	GUU		
W	Trp	Tryptophan	UGG					
Y	Tyr	Tyrosine	UAC	UAU				